UNITS/CONVERSIONS

	SI	"Chemical"		
Force	N/newton (kg m s^{-2})	d/dyne (g cm s^{-2})		
	1 N =	10^5		
Pressure	(N m^{-2})/Pascal	(atm)	(torr)	
	1 unit =	9.869 × 10^{-3}	7.501	
Work	J/joule (N m)	erg (d cm)	cal	eV
	1 J =	10^7	0.239	6.24 × 10^{18}
Power	W/watt (J s^{-1})			
Electric Charge	C/coulomb	esu (cm$^{3/2}$ g$^{1/2}$ s^{-1})	electron charge	
	1 C =	2.998 × 10^9	6.242 × 10^{18}	
Electric Current	A/ampere (C s^{-1})			
Electric Potential	V/volt (W A^{-1})			
Frequency	Hz/hertz (s^{-1})	ω/angular velocity (rad s^{-1})		
	1 Hz =	2π		
Magnetic Flux	Wb/weber (V s)			
Magnetic Flux Density	T/tesla (Wb m^{-2})	G/gauss (sometimes incorrectly called the magnetic field strength)		
	1 T =	10^4		
Magnetic Field Strength	(A m^{-1})	Oe/oersted		
	1 unit =	1.2566 × 10^{-2}		
Electron Magnetic Moment	m_B/Bohr magneton (A m^2 or J T^{-1}) 9.273 × 10^{-24}	β or μ_B/Bohr magneton (cm$^{5/2}$ g$^{1/2}$ s^{-1} or erg gauss^{-1}) 9.273 × 10^{-21}		
Molar Magnetic Susceptibility	χ_M (m^3 mole^{-1} = 10^7 JT^{-2} mole^{-1}) 1 unit =	(cm^3 mole^{-1} = erg G^{-2} mole^{-1} or cgs units) 7.958 × 10^4		

UNIT PREFIXES

		power of ten
d	deci	−1
c	centi	−2
m	milli	−3
μ	micro	−6
n	nano	−9
p	pico	−12
da	deka	1
h	hecto	2
k	kilo	3
M	mega	6
G	giga	9
T	tera	12

INORGANIC CHEMISTRY

KEITH F. PURCELL
Kansas State University,
Manhattan, Kansas

JOHN C. KOTZ
State University of New York
College at Oneonta

SAUNDERS GOLDEN SUNBURST SERIES

W. B. SAUNDERS COMPANY
PHILADELPHIA · LONDON · TORONTO

W. B. Saunders Company: West Washington Square
Philadelphia, PA 19105

1 St. Anne's Road
Eastbourne, East Sussex BN21 3UN, England

1 Goldthorne Avenue
Toronto, Ontario M8Z 5T9, Canada

Listed here is the latest translated edition of this book together with the language of the translation and the publisher.

Spanish — Editorial Reverte S.A., Barcelona, Spain

Library of Congress Cataloging in Publication Data

Purcell, Keith F

Inorganic chemistry.

(Saunders golden sunburst series)
Includes index.

1. Chemistry, Inorganic.   I. Kotz, J. C., joint author.
II. Title.

QD151.2.P87   546   76-8585

ISBN   0-7216-7407-0

Front cover illustration is *Simultaneous Contrasts; Sun and Moon (Soleil, Lune, Simultané 2)* by Robert Delaunay (1913. Dated 1912 on Painting). Oil on canvas, 53″ diameter. Collection, The Museum of Modern Art, New York. Mrs. Simon Guggenheim Fund.

Inorganic Chemistry   ISBN   0-7216-7407-0

© 1977 by W. B. Saunders Company. Copyright under the International Copyright Union. All rights reserved. This book is protected by copyright. No part of it may be reproduced, stored in a retrieval system, or transmitted in any form or by any means, electronic, mechanical, photocopying, recording, or otherwise, without written permission from the publisher. Made in the United States of America. Press of W. B. Saunders Company. Library of Congress catalog card number 76-8585.

Last digit is the print number: 9   8   7   6   5   4   3   2

PREFACE

One of the problems facing the instructor of inorganic chemistry—and the authors of inorganic textbooks—is the antiquated view that a single semester of three lectures per week will suffice to cover the subject with any degree of thoroughness. This is an outrageously difficult situation, even if it were not for the awesome growth of the field in recent years. Because of this situation, inorganic texts of the past have tended to follow one or the other of two philosophies: to enumerate chemical facts without developing for the student the beauty of their systematics; or, to select a few "topics" for a more detailed examination. Our experience has been that both philosophies fail to give a student a sense of the integrity and continuity of inorganic chemistry. Our effort to write about inorganic chemistry at the senior/graduate interface arose from the frustrations of our students in dealing with such organizational schemes. Too often we found that our students simply developed an attitude of survival. A more serious consequence of this was that the student left the encounter with inorganic chemistry feeling that the field was not systematized; in his view, inorganic chemists must be a disorganized group of specialists largely ignoring each other. Does one have to be a transition metal chemist, *or* a rare gas chemist, *or* a boron chemist? Without an integrated view, the student often missed the excitement and challenge of inorganic chemistry and the almost limitless potential for new discoveries in this field.

Another view that we have developed over the past decade of teaching inorganic chemistry is that the usual senior course uniquely challenges the student to integrate concepts from thermodynamics, kinetics, and bonding theory for their application to chemistry at large. We have written this text in the hope of presenting inorganic chemistry as a focus of many of the concepts our students have previously learned.

The degree to which each of us can realize such goals depends, of course, on our skill as instructors but is limited by the fact that most of us labor under the yoke of approximately forty-five lectures to make our case. *It is not possible to cover all of the material in a text of this size in a semester of the usual length;* we certainly do not do so in our own courses, nor do we advocate that others try. Rather, our attempt to resolve the dilemma is to recognize that each of us will skimp on some aspects and delve more deeply into others. What we have attempted to do, therefore, is to write a text that allows the instructor the flexibility to pursue his convictions on the topics to be stressed and, at the same time, aids the student in integrating those topics. It is our hope that the student is spared the frustrations of an otherwise dislocated treatment of chemistry. Specifically, when a principle or "fact" is encountered we make an effort to note its previous and later use, if these are not obvious.

In other words, we intend to remind the student of what he has seen before and where he will see it again. With the material integrated in this manner, the systematics should develop with little effort on the student's part; equally important, the tentatively non-systematic facts, which are harbingers of new principles, stand out in relief.

At the crudest level of organization, this text is structured into theoretical, non-metal, and metal topics and, in this regard, it is more or less standard. Students often have difficulty with the first of these (bonding treatments) because, while of intrinsic interest, the concepts are soon forgotten by authors in succeeding descriptive sections. We have avoided this because theory without application to observed phenomena is as lifeless as phenomena without a basis for their interpretation. The first four chapters develop theoretical concepts *and their chemical interpretations*. Ultimately, our desire for a unified electronic theory has led us to emphasize *the molecular orbital model,* rather than the myriad approximations to it.

The descriptive non-metal chemistry begins with a chapter on the Lewis donor/acceptor concept and solvent properties for they, along with the structure and bonding concepts of the first four chapters, provide the foundation for the rest of the text. This is followed by two chapters which examine main group structures and reactivities in terms of thermodynamic and mechanistic principles. The second of these (Chapter 7) is especially important since a considerable amount of inorganic chemistry can be organized around mechanistic types, and this book reflects throughout our feeling that mechanisms should be emphasized more strongly.

The descriptive chemistry encountered under the aegis of the more physical aspects of inorganic chemistry is brought to practical focus through the systematic synthesis of non-metal compounds in Chapter 8. Again, this reflects our belief that theory is given special meaning through understanding of, and interest in, the purposeful synthesis of molecules. While Chapter 8 overtly addresses this point, in fact we have attempted to emphasize the synthetic aspects of inorganic chemistry throughout.

The bonding concepts used for non-metal compounds are extended to interpret the special magnetic and electronic properties of *transition metal complexes* in Chapter 9, and Chapters 10 and 11 are devoted to the topologies and stereochemistries of this class of compounds. It is here that the angular overlap model (AOM), a molecular orbital approach, is first brought into use. Although the AOM has been used in the chemical literature, this text is its first use in such a manner that the average undergraduate can cope with the theory. Although the model is as useful as the ligand field model in interpreting spectral and magnetic properties of coordination complexes, the AOM approach finds its greatest use in predicting and interpreting the structural problems in coordination chemistry. It is the only approach of which we are aware that allows a student to assess, in a simple and straightforward manner, the electronic energy bias toward a favored structure of a transition metal coordination complex.

Following these chapters on bonding and structural theory for coordination complexes, there are chapters on the reactions of such compounds. Our approach has been to start "within" the complex to look at electron transfer between metal ions and work outward to reactions of coordinated ligands. Each of these reactivity chapters continues the thermodynamic, kinetic, synthetic sequence of the non-metal chapters and each culminates in an application of the principles developed therein to the synthesis of coordination complexes.

Two additional chapters on metal chemistry—organometallic chemistry—follow. The average undergraduate gains little exposure in his other courses to this very important and rapidly expanding field. It is our belief that most students need some knowledge of organometallic chemistry, and most are intrinsically interested in such material. Organometallic chemistry is developed here along the same lines as the previous chapters on metal chemistry, again with an emphasis on reaction mechanisms.

The final two chapters describe two other expanding areas of inorganic chemistry: metal and non-metal cluster chemistry and inorganic aspects of biochemistry. Both are areas of considerable importance, and we have treated both within the framework of the structural and bonding principles laid down in the earlier chapters. (The topic of polyhedral boron hydrides received recent recognition by the announcement that Professor W. N. Lipscomb was awarded the 1976 Nobel Prize in Chemistry for his pioneering work on the bonding in such compounds.) Finally, the chapter on inorganic biochemistry is perhaps different from other published approaches in that, rather than simply recounting inorganic prosthetic groups, a synopsis is given of the bio-processes in which such reagents are necessary.

We are greatly indebted to our reviewers (Mike Bellema, Bob Fay, Ron Gillespie, Bill Hatfield, Galen Stucky, and Jerry Zuckerman) for critically examining the manuscript and insisting on clarity of presentation and accuracy of facts. Some of our students and faculty colleagues (especially Tay Tahk and Bruce Knauer) have generously given time and effort to the improvement of this text; to all of them we are sincerely grateful. The editors and artists of W. B. Saunders Company and York Graphic Services have done an outstanding job in bringing this project to its conclusion. John Vondeling gave support beyond what is customary from editors in such endeavors and Jay Freedman time and again proved an invaluable partner. Finally, to our wives, Susan and Katie, and to our children (Kristan and Karen, David and Peter) go our deepest appreciation and thanks, as well as a promise; there will be more time now for sailing/backpacking and evenings at home.

K. F. PURCELL

J. C. KOTZ

SUGGESTED OUTLINE FOR A SEMESTER COURSE

This text is deliberately large for reasons important to our students and to all of us as instructors.

BREADTH OF COVERAGE

Each chemistry major should have in his personal library a text which will serve him in subsequent years.

A text that is broad in scope does not constrain course content.

PEDAGOGICAL ASPECTS

The format is "open" for readability and to provide margins for jotting down notes.

Each concept is supported by artwork, explanations, and applications to firmly establish in the student's mind the value of the concept.

Careful text explanation of concepts and their applications frees lecture time for emphasis on the concepts themselves and eliminates the need for supplementary materials.

As noted in the Preface, however, time limitations preclude complete coverage of this text in a semester course. Below are identified those sections of the book which constitute the equivalent of a 400-page text suitable for 40 to 45 lectures at the senior level. These sections cover all those topics rated "important to essential" by the recent survey of inorganic chemistry instructors by the ACS Examinations Committee.

Chapter	Pages
1	17–60
2	75–95
3	107–114; 117–124
4	130–145; 147–149; 155–169
5	188–218; 225–235
6	276–291
7	366–373; 386–430
8	452–465; 468–470; 475–479; 483–503
9	517–520; 543–560; 575–583
10	590–603
11	610–636
12	654–660; 684–693
13	694–700; 710–719; 725–755
15	792–804
16	843–855; 861–863; 871–876
17	906–910; 938–943; 962–979

CONTENTS

A NOTE ON UNITS xix

PROLOGUE
THE COMING OF AGE: PERSPECTIVE 1

1

USEFUL ATOMIC CONCEPTS 12
   Probability Density Functions 13
      The Electron as a Matter Wave 13
      Probability Density Functions 17
      Radial Density Functions and Orbital Energies 18
      Angular Functions and Orbital Shapes 25
      Total Density Functions 30
   Polyelectronic Atoms 33
      Atom Electron Configurations and the Long Form Periodic Table . 33
      Slater Orbitals and Their Uses 44
      Atomic Configurations and Atomic Terms 47
      Atom and Orbital Electronegativities 54
   Epilog 60
   Appendices to Chapter 1 61
      A. Hydrogen Type Wavefunctions
      B. First Ionization Potentials of the Main Group and
         Transition Elements
      C. Electron Affinity Values of the Elements

2

BASIC CONCEPTS OF MOLECULAR TOPOLOGIES 64
   Shared and Lone Pairs and Lewis Structures 65
   Electron Pair Repulsion Model 75
   Symmetry Concepts 82
      Point Groups 82
      Character Tables 89

Epilog 95
Appendix to Chapter 2 96
    Point Group Character Tables

3

THE DIRECTED ATOMIC ORBITAL VIEW OF CHEMICAL BONDS 98

Directional Atomic Orbitals 99
Molecular Properties Conveniently Interpreted with the Directed
    Atomic Orbital Concept 107
    Bond Distances 108
    Force Constants 110
    Dipole Moments 114
    Nuclear Spin Coupling 117
    Bond Energies 119
Epilog 124

4

THE UNDIRECTED ORBITAL VIEW OF CHEMICAL BONDS 126

Introductory Comments 127
Molecular Orbital Probability Functions 130
Principles of MO Construction and Interpretation 134
    The Hydrogen Molecule 134
    Beryllium Hydride and Hydrogen Chloride 135
    Summary 140
Main Group Diatomic Molecules 141
    Homonuclear Molecules 141
    The $ps\sigma$, $p\pi$ Order Controversy 145
    Heteronuclear Molecules 147
Structure, VSEPR, and LCAO-MO 149
    The EF_3 Series 151
    The EF_4 Series 153
"Linear" (One-Dimensional) Molecules 155
Cyclic (Two-Dimensional) Molecules 158
Polyhedral (Three-Dimensional) Molecules 163
    Stratagem 163
    EX_3 (D_{3h}) 164
    EX_4 (T_d) 169
    EX_5 (D_{3h}) 171
    EX_4 (D_{4h}) 174
    EX_6 (O_h) 177
The Equivalence of the Localized and Delocalized Models 183
Epilog 185

5

THE DONOR/ACCEPTOR CONCEPT 186

Introduction 186
Survey of Adduct Types 188

Acidic and Basic Hydrogen ... 188
Non-Metal Acids and Bases ... 194
    Boron and Aluminum ... 194
    Carbon and Silicon ... 196
    Nitrogen and Phosphorus ... 203
    Oxygen and Sulfur ... 205
    Halogens ... 209
    Xenon ... 212
Metals ... 213
Acid-Base Strengths ... 216
  The Thermodynamic Definition ... 216
  Quantitative Prediction of Relative Adduct Stabilities ... 218
  Illustrative Interpretations ... 225
    Proton Affinities ... 226
    $(CH_3)_3N$, $(SiH_3)_3N$ ... 228
    Picolines ... 229
    Et_3N, $HC(C_2H_4)_3N$ ... 229
    $Me_3N \cdot MMe_3$, M = B, Al, Ga, In ... 229
    π Resonance Effects ... 230
    BF_xMe_{3-x} ... 232
    Retrodative Bonding ... 232
An Acid-Base View of Solvation Phenomena ... 235
  Liquid Ammonia ... 241
  Hydrofluoric and Sulfuric Acids ... 249
  Sulfur Dioxide ... 255
  HSO_3F and Superacids ... 259
Epilog ... 263

6

ENERGETICS AND STRUCTURES AS GUIDES TO MAIN GROUP CHEMISTRY ... 264

Free Energy, Reaction Potential, and Equilibrium ... 265
  Review ... 265
  Estimation of Reaction Spontaneity ... 266
    The Non Sequitur "Stability" ... 267
    Heats of Reactions from Bond Energies ... 268
    Higher Oxidation State Stabilization ... 271
    A Caveat Concerning Condensed Phase Reactions ... 274
Metal-Containing Solids ... 276
  Lattice Energies ... 276
  Structure and Stoichiometry ... 279
  Structure and the Concept of Ion Radii ... 283
  Layer Lattices and Incipient Covalency ... 286
  Synthesis Principles ... 290
    Cation Oxidation States ... 291
    Ionic Fluorinating Agents ... 292
    Ion Size and Compound Isolation from Solution ... 293
  Metals ... 296
    Structures and Bonding ... 296
    Oxidative Stabilities ... 299

x ☐ CONTENTS

    Non-Metal Compounds 300
        Boron and Aluminum 300
        Carbon and Silicon 316
        Nitrogen and Phosphorus 329
        Oxygen and Sulfur 340
        Halogens 349
        Noble Gases 354
    Epilog 357

7

REACTION PATHWAYS 358

    Basic Concepts 359
    Rate Expressions and Interpretations 364
        General Formulation 364
        Second Order Rate Law 366
        Pseudo-First Order Rate Law 368
        First Order Rate Law 369
    Activation Parameters 371
    A Caveat Concerning Solvent Effects 372
    The Constraint of Orbital Following 373
    The Principle of Microscopic Reversibility 378
    Survey of Reactions 382
        One Valence Pair 383
            Hydrogen 383
        Three Valence Pairs 386
            Boron 386
        Four Valence Pairs 393
            Boron 393
            Silicon 401
            Nitrogen and Phosphorus 410
            Oxygen and Sulfur 422
            Halogens 429
        Five Valence Pairs 430
            Phosphorus and Sulfur 430
    Epilog 444

8

SYNTHESIS OF IMPORTANT CLASSES OF NON-METAL COMPOUNDS 446

    Special Techniques 447
        The Chemical Vacuum Line 447
        Plasmas 449
        Photochemical Apparatus 450
        Electrolysis 450
    An Overview of Strategy 452
    Synthesis of Fluorides and Chlorides 453
        Boron and Aluminum 454
        Silicon 457
        Nitrogen and Phosphorus 457

Oxygen and Sulfur	459
Fluorine and Chlorine	462
Xenon	463
Fluorinating Agents	465
SbF_3 and SbF_5	466
ClF, ClF_3, BrF_3	467
Sulfur Tetrafluoride	468
Dinitrogen Tetrafluoride	470
Chloropentafluorosulfur	472
Chloride Substitution by Hydrogen and Organics	474
Organometal Reagents	475
Boron and Aluminum	476
Silicon	477
Nitrogen and Phosphorus	477
Oxygen and Sulfur	479
Halogens	481
Hydrometalation and Others	483
Catenation by Coupling	486
Solvolysis Reactions	488
Boron and Aluminum	488
Silicon	494
Nitrogen and Phosphorus	496
Oxygen and Sulfur	503
Halogens and Xenon	511
Epilog	513

9

FUNDAMENTAL CONCEPTS FOR TRANSITION METAL COMPLEXES ... 514

A Sampler	515
The Language of Coordination Chemistry	517
Spectral/Magnetic Characteristics of Transition Ion Complexes	520
Structure Puzzles	525
The Transition Metals and Periodic Properties	527
The Molecular Orbital Model	531
General View of MD_6 and MD_4 Structures	533
MD_6 (O_h)	533
MD_4 (D_{4h})	537
MD_4 (T_d)	541
The Angular Overlap Model	543
Tenets	543
d^* Orbital Energies and Occupation Numbers	545
Complex Structures and Preferences	549
Experimental Evidence for a Structural Preference Energy and the Ligand Field Stabilization Energy	551
Jahn-Teller Distortions from O_h Geometry	553
Electronic States and Spectra	559
Ground and Excited States	560
The Energies of Electronic Transitions	567
The Spectrochemical Series	575
Paramagnetism of Complexes	577

Epilog 583
Appendix to Chapter 9 584
    State Energy Diagrams for MD_6 (O_h)

10

COORDINATION CHEMISTRY: STRUCTURAL ASPECTS 586

General Considerations 587
Low Coordination Numbers 590
    Two-Coordinate Complexes 590
    Three-Coordinate Complexes 591
    Four-Coordinate Complexes 593
        Tetrahedral Complexes 593
        Square Planar Complexes 594
    Five-Coordinate Complexes 596
    Six-Coordinate Complexes 600
Polyhedra of High Coordination Number 603
    Seven-Coordinate Complexes 604
    Eight-Coordinate Complexes 606
    Complexes Having Coordination Numbers of Nine or Higher 609
Epilog 609

11

COORDINATION CHEMISTRY: ISOMERISM 610

Constitutional Isomerism 613
    Hydrate Isomerism 613
    Ionization Isomerism 614
    Coordination Isomerism 614
    Polymerization Isomerism 614
    Linkage Isomerism 615
Stereoisomerism 619
    General Aspects of Stereochemical Notation in Coordination
        Chemistry 622
    Four-Coordinate Complexes 625
    Six-Coordinate Complexes 628
        Isomerism from Ligand Distribution and Unsymmetrical
            Ligands; Isomer Enumeration 629
        Isomerism from Ligand Conformation and Chirality 636
    Chirality and the Special Nomenclature of Chiral
        Coordination Complexes 638
    Optical Activity, ORD, and CD 644
    Absolute Configurations of Chiral Coordination Complexes 647
    Ligand Conformation 650
Epilog 653

12

COORDINATION CHEMISTRY: REACTION MECHANISMS AND METHODS OF SYNTHESIS; ELECTRON TRANSFER REACTIONS 654

Mechanisms of Electron Transfer Reactions 659

Key Ideas Concerning Electron Transfer Between
    Transition Metals 660
Outer Sphere Electron Transfer Reactions 660
    Chemical Activation and Electron Transfer 662
    Cross Reactions and Thermodynamics 667
Inner Sphere Electron Transfer Reactions 669
    Formation of Precursor Complexes 671
    Rearrangement of the Precursor Complex and Electron
        Transfer 673
    Electronic Structure of the Oxidant and Reductant 673
    The Nature of the Bridge Ligand 675
    Fission of the Successor Complex 679
Two-Electron Transfers 680
Non-Complementary Reactions 681
Synthesis of Coordination Compounds Using Electron
    Transfer Reactions 684
Epilog 693

13

COORDINATION CHEMISTRY: REACTION MECHANISMS AND METHODS OF SYNTHESIS; SUBSTITUTION REACTIONS 694

Replacement Reactions at Four-Coordinate Planar Reaction
    Centers 696
    The General Mechanism of Square Planar Substitution
        Reactions 697
    Factors Affecting the Reactivity of Square Planar
        Complexes of Pt(II) and Other d^8 Metal Ions 700
        Influence of the Entering Group 700
        Influence of Other Groups in the Complex—Ligands
            trans to the Entering Group 702
        Influence of Other Groups in the Complex—Ligands
            cis to the Entering Group 707
        The Nature of the Leaving Group 707
        Effect of the Central Metal 708
    The Intimate Mechanism for Replacement at Four-Coordinate
        Planar Reaction Centers 708
Substitution Reactions of Octahedral Complexes 710
    Replacement of Coordinated Water 713
        The Mechanism of Water Replacement 713
        Rates of Water Replacement 716
        Orbital Occupation Effects on Substitution Reactions
            of Octahedral Complexes 719
    Solvolysis or Hydrolysis 721
        Hydrolysis under Acidic Conditions 721
        Base-Catalyzed Hydrolysis: The Conjugate Base or
            CB Mechanism 725
Synthesis of Coordination Compounds by Substitution
    Reactions 731
    Thermodynamic Stability of Coordination Compounds 733
    The Synthesis and Chemistry of Some Cobalt Compounds ... 742
    The Synthesis and Chemistry of Some Platinum Compounds . 750
Epilog 755

14

COORDINATION CHEMISTRY: REACTION MECHANISMS AND METHODS OF SYNTHESIS; MOLECULAR REARRANGEMENTS AND REACTIONS OF COORDINATED LIGANDS 756

    Molecular Rearrangements 757
        Four-Coordinate Complexes 757
        Six-Coordinate Octahedral Complexes 764
    Reactions at Coordinated Ligands 773
        Reactions Due to Metal Ion Polarization of Coordinated Ligands 774
            Hydrolysis of Amino Acid Esters and Amides and of Peptides 774
            Aldol Condensation 781
            Imine Formation, Hydrolysis, and Substituent Exchange 782
        The Template Effect and Macrocyclic Ligands 783
    Epilog 790

15

FROM CLASSICAL TO ORGANOMETALLIC TRANSITION METAL COMPLEXES AND THE SIXTEEN AND EIGHTEEN ELECTRON RULE 792

    The Sixteen and Eighteen Electron Rule 793
    Theoretical Aspects of the 16 and 18 Electron Rule 804
    Epilog 807

16

ORGANOMETALLIC CHEMISTRY: SYNTHESIS, STRUCTURE, AND BONDING 810

    Introduction 811
    A Note on the Organization of Organometallic Chemistry 815
    The Literature of Organometallic Chemistry 816
    Carbon σ Donors 817
        The Synthesis of Metal Alkyls and Aryls 817
            Direct Reaction of a Metal with an Organic Halide 818
                Thermodynamic Considerations 818
                Experimental Considerations 820
            Reactions of Anionic Alkylating Agents with Metal Halides or Oxides 824
            Reaction of a Metal with a Mercury Alkyl or Aryl 825
            Metalation Reactions: Metal-Hydrogen Exchange 826
            Oxidative Addition Reactions 829
            Reactions of Metal-Containing Anions with Organic Halides 830
            1,2-Addition of Metal Complexes to Unsaturated Substrates 832
                Hydrometalations 832
                Oxymetalations 837
                Halometalations 838

Organometalations	839
Structure and Bonding in Metal Alkyls and Aryls	843
Metal Carbonyls	855
The Synthesis of Metal Carbonyls	856
Metal Carbonyls: Properties and Structures	858
Bonding in Metal Carbonyls	861
Metal-Carbene and -Carbyne Complexes	863
Carbon π Donors	866
Chain π Donor Ligands (Olefins, Acetylenes, and π-Allyl)	867
Synthesis of Olefin, Acetylene, and π-Allyl Complexes	868
Olefin Complexes	868
Acetylene Complexes	870
π-Allyl Complexes	871
Structure and Bonding in Olefin, Acetylene, and π-Allyl Complexes	871
Complexes with Cyclic π Donors	876
Synthesis and Properties	877
Structure and Bonding	883
Epilog	898
Appendix to Chapter 16	899
Metal Carbonyls and Infrared Spectroscopy	

17

ORGANOMETALLIC COMPOUNDS: REACTION PATHWAYS 906

The 16 and 18 Electron Rule and Reactions of Transition Metal Organometallic Compounds	909
Association Reactions	910
The Lewis Acidity and Basicity of Organometallic Compounds	910
Ligand Protonation	916
Substitution Reactions	918
Nucleophilic Ligand Substitutions	918
Electrophilic and Nucleophilic Attack on Coordinated Ligands	923
Addition and Elimination Reactions	927
1,2-Additions to Double Bonds	927
1,1-Addition to CO: Carbonylation and Decarbonylation	933
Oxidative Addition Reactions	938
General Considerations	939
Stereochemistry of Oxidative Additions	941
Influence of Central Metal, Ligands, and Addend on Oxidative Addition	943
Mechanism of Oxidative Addition	947
Elimination Reactions and the Stability of Metal-Carbon σ Bonds	948
Rearrangement Reactions	953
Redistribution Reactions	953
Fluxional Isomerism or Stereochemical Non-Rigidity	957
Catalysis Involving Organometallic Compounds	962
Olefin Hydrogenation	963
The Oxo Reaction	966

	The Wacker Process (Smidt Reaction)	967
	Polymerization	970
	Cyclooligomerization, Olefin Isomerization and Metathesis, and Polymer-Bound Catalysts	976
Epilog		979

18

MOLECULAR POLYHEDRA: BORON HYDRIDES AND METAL CLUSTERS ... 980

The Boron Hydrides	983
The Neutral Boron Hydrides, $(BH)_p H_q$	988
Structure and Bonding	988
The Topological Approach to Boron Hydride Structure: the *styx* Numbers	990
Molecular Orbital Concepts	994
Synthesis and Reactivity of the Neutral Boron Hydrides	997
A General Organizational Scheme for the Neutral Boron Hydrides, the *Closo* Polyhedral Hydroborate Ions, and the Carboranes	1005
A Molecular Orbital View of *Closo*-Hydroborate Anions and Carboranes	1009
Closo-Hydroborate Ions	1011
The Carboranes	1015
Metallocarboranes	1020
Metal-Metal Bonds and Metal Clusters	1026
Binuclear Compounds	1027
Three-Atom Clusters	1033
Four-Atom, Tetrahedral Clusters	1035
Five- and Six-Atom Clusters	1040
Epilog	1045
Appendix to Chapter 18	1047
Bonding Model for M_6 Clusters	

19

BIOCHEMICAL APPLICATIONS ... 1050

The Cell	1051
Processes Coupled to Phosphate Hydrolysis	1053
Nucleotide Transfer—DNA Polymerase	1055
Phosphate Transfer	1057
General Comments	1058
Pyruvate Kinase	1058
Glucose Storage—Phosphoglucomutase	1060
Phosphate Storage in Muscle—Creatine Kinase	1061
Na^+/K^+ Ion Pump—ATPase	1063
Oxygen Carriers—Hemoglobin and Myoglobin	1064
Cobalamins; Vitamin B_{12} Coenzyme	1073

    Electron Transfer Agents 1078
        Cytochromes 1078
        Iron-Sulfur Proteins 1082
            1-Fe—S and 2-Fe—S 1082
            2-Fe—S* 1083
            8-Fe—S* 1084
            N_2 Fixation 1086
        Chlorophyll 1090
    Epilog 1095

COMPOUND INDEX 1097

SUBJECT INDEX 1105

A NOTE ON UNITS

After much soul searching, and in view of the primarily American usage of this text, your authors have opted to use the existing hodge-podge of units to which you are accustomed. Inside the front cover you will find a table of units and constants in the SI (Système International d'Unités)[1] and the conversion relations between these and the traditional units used in this text. You would be well advised to learn to use the SI units (they derive from the rationalized MKS units used in physics texts). Current research journals in Britain and on the Continent are using the SI exclusively, and it is beginning to appear in the American research journals as well. The advantages of a standardized system of units are obvious; the reluctance of research scientists to universally adopt them stems from the adoption of units convenient for the phenomena, and instruments, of their studies. For example, the convenient calorimetry unit is the calorie, ionization potential and electrochemical measurements are traditionally reported in electron volts, while the spectroscopist, using Einstein's photoelectric equation, prefers the erg unit.

---

[1] Probably the most readily accessible material, with applications, is to be found in articles by A. C. Norris, *J. Chem. Educ.*, **48**, 797 (1971); (for magnetic units) T. I. Quickenden and R. C. Marshall, *ibid.*, **49**, 114 (1972) and J. I. Hoppeé, *ibid.*, **49**, 505 (1972); and N. H. Davies, *Chem. Britain*, **6**, 344 (1970); *ibid.*, **7**, 331 (1971). Corrections to the latter papers are to be found in *ibid.*, **8**, 36 (1972).

PROLOGUE
THE COMING OF AGE: PERSPECTIVE

What is modern inorganic chemistry and what direction is it likely to take in the near future? What does one teach undergraduate and graduate students in general, and prospective inorganic chemists in particular, that they gain perspective on the past and are prepared for the future? It is these questions that inorganic chemists ask one another at American Chemical Society meetings, at Gordon Research Conferences, and at other "gatherings of the clan." Not surprisingly, there are nearly as many answers as there are inorganic chemists, and the answers change with time.

From an historical view inorganic chemistry is synonymous with "general" chemistry; in addition to his area of specialization the "inorganic" chemist was expected to be conversant, if not knowledgeable, with the chemistries of all the elements. At various times the inorganic chemist has been closely allied with analytical chemistry (both qualitative and quantitative chemical methods), with physical chemistry, and even with organic chemistry. These strong ties are still existent today, but the rapid development of our field since the late fifties has created great depth in research specialties so that the "generalist" posture has waned slightly. Concomitant with this evolution, the field gained formal recognition in this country with the establishment of the journal *Inorganic Chemistry* by the American Chemical Society in 1962; since then the "Division of Inorganic Chemistry of the American Chemical Society" has produced offspring in the form of the "Organometallic" and "Solid State" subdivisions.

To illustrate what modern inorganic chemistry has become, we surveyed the papers published on topics in the field in *Accounts of Chemical Research* since the inception of the journal in 1968. Such a survey at least partially measures the current status of inorganic chemistry and the probable direction of future work because *Accounts* . . . "publishes concise, critical reviews of research areas currently under active investigation . . . written by scientists personally contributing to the area reviewed." Although our division of papers was perhaps arbitrary at times, the results summarized in Table 1 are quite enlightening. Metals make up about 80% of the Periodic Table, and it is interesting that the fraction of papers on metal chemistry is approximately the same, although all but six of the papers are concerned with transition metal chemistry. Nonetheless, the results of this crude survey give a fairly clear indication of the current emphasis—the bulk of the papers is on organometallic chemistry, with a bias toward catalysis, while a large number are concerned with coordination chemistry and the biochemical role of metals.

The chemistry of organometallic compounds, especially of the transition metals, has become increasingly important in the past twenty years, as attested to by the fact that the Nobel Prizes in Chemistry in 1963 and 1973 were awarded for work falling in

TABLE 1
REVIEWS ON INORGANIC CHEMISTRY IN *ACCOUNTS OF CHEMICAL RESEARCH*, 1968–1974

General Area and Sub-area	Number of Reviews
Metal chemistry	Total = 61
Transition metals	Total = 55
Organometallic chemistry (incl. 10 on catalysis)	27
Coordination chemistry	7
Bioinorganic chemistry	9
Theoretical and physical papers	12
Non-transition metals: Organometallic chemistry	6
Non-metal chemistry	Total = 22
Boron chemistry (4 with significant metal chemistry)	8
Silicon chemistry	4
Theoretical papers	4
Other (chiefly halogen compounds)	6
TOTAL REVIEWS	83

this general area.[1,2] Karl Ziegler of Germany and Giulio Natta of Italy shared the Prize in 1963 for their finding that the stereospecific polymerization of olefins was catalyzed by alkyl aluminum–transition metal halide mixtures.[3-5] Although the mechanism of the process is not completely understood, the aluminum compound is thought to act as an initiator by alkylating the metal halide (step 1); the olefin to be polymerized forms a π complex with the alkylated metal (step 2) so that the olefin and alkyl group have the proper stereochemical relationship, and the metal–carbon bond can then in effect add to the olefin (step 3). More olefin arriving at the newly vacated coordination site (step 4) forms the next link in the continuing polymerization process. The fact that this is carried out on the ordered surface of an insoluble metal halide is the reason for the stereospecificity of the polymer.

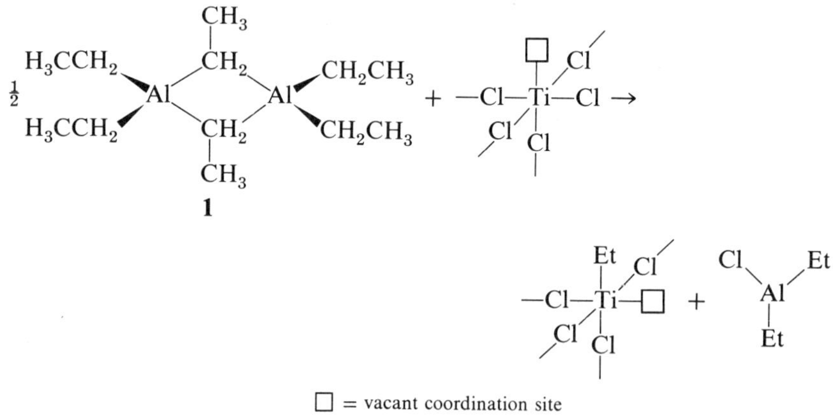

$$\tfrac{1}{2}\,\mathbf{1} + \text{TiCl}_4 \longrightarrow \text{Cl}_3\text{Ti}(\square) + \text{AlEt}_2\text{Cl} \tag{1}$$

$\square$ = vacant coordination site

---
[1] C. G. Overberger, *Science*, **142**, 938 (1963).
[2] D. Seyferth and A. Davison, *Science*, **182**, 669 (1973).
[3] G. Natta and I. Pasquon, *Advan. Catalysis*, **11**, 1 (1959).
[4] P. Cossee, *in* "The Stereochemistry of Macromolecules," A. D. Ketley (ed.), Vol. 1, Marcel Dekker, New York, 1967, p. 145.
[5] T. Mole and E. A. Jeffrey, "Organoaluminum Compounds," Elsevier, New York, 1972.

$$\text{(diagram for reaction 2)} \qquad (2)$$

$$\text{(diagram for reaction 3)} \qquad (3)$$

$$\text{(diagram for reaction 4)} \qquad (4)$$

This example serves to illustrate much of the fascination of inorganic chemistry. For example, why is the alkylating agent, triethylaluminum (1), a dimeric species under normal conditions, whereas triethylborane, $B(CH_2CH_3)_3$, is monomeric? How is one AlR_3 unit bound to the other? In what way is an olefin bound to a metal, and why do metals form kinetically stable olefin complexes only when the metal is in a low oxidation state? What are the details of the alkyl group–olefin interaction—that is, does the olefin insert into the metal alkyl bond or does the alkyl group migrate to the olefin?

Professor Wilkinson at Imperial College, London, and Professor Fischer at Munich earned the Nobel Prize in 1973 for the impetus they provided to the study of metallocenes, or "metal sandwiches" as they are often called. Ferrocene (2) in

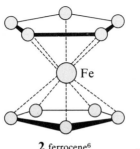

2 ferrocene[6]
[bis(η^5-cyclopentadienyl)iron
orange, sublimable solid (mp, 173°)]

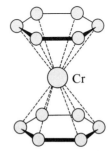

3 dibenzenechromium[7]
[bis(η^6-benzene)chromium
black-brown, sublimable solid (mp, 284°)]

---

[6] G. Wilkinson, *Org. Syn.*, **36**, 31 (1956).
[7] E. O. Fischer, *Inorg. Syn.*, **6**, 132 (1960).

particular has been widely studied, as it is readily synthesized or can be purchased inexpensively. The compound is electron-rich and has an extensive chemistry based on electrophilic substitution reactions.[8] As prototypes of many other similar complexes, **2** and **3** have been thoroughly studied to uncover the reasons for the tremendous stability achieved when a low-valence metal interacts with the π electrons of an unsaturated or aromatic molecule.

Perhaps the most active area of inorganic chemistry in the 1950's and 1960's was coordination chemistry: the study of complexes of higher valence metals. The most important development during that period was in our understanding of the electronic spectra of these complexes, and many excellent books have been published on "ligand field theory" as a result.[9-11] Another area of special activity in coordination chemistry has been the study of the stereochemistry of chiral complexes (**4**),[12] or those with high coordination numbers (**5**)[13] or with unusual coordination geometries (**6**).[14]

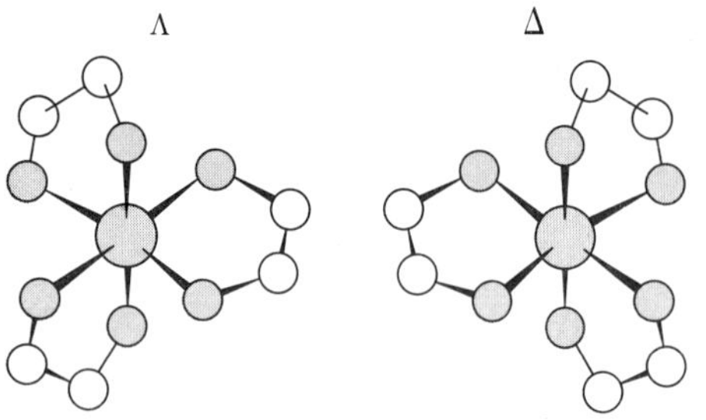

4
$M\{C_2H_4(NH_2)_2\}_3^{n+}$

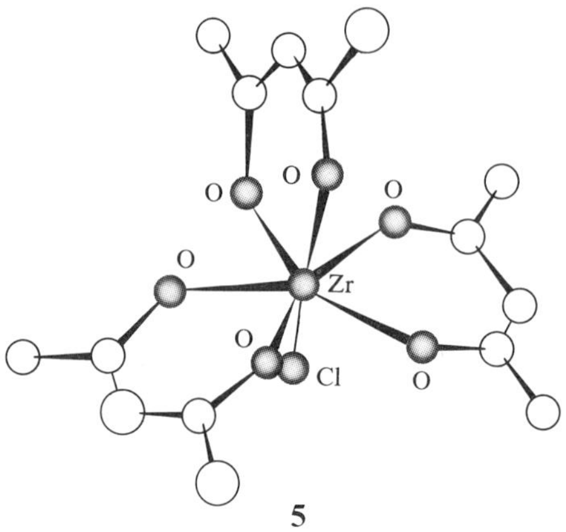

5
$Zr(acac)_3Cl$[15]

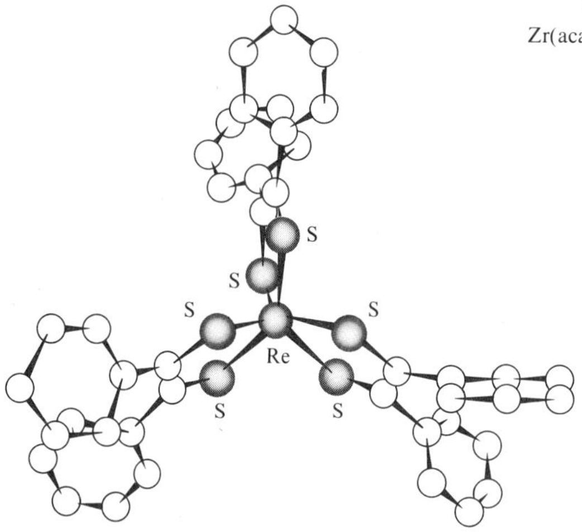

6
$Re[S_2C_2(C_6H_5)_2]_3$[16]

---

[8] M. Rosenblum, "Chemistry of the Iron Group Metallocenes," Interscience Publishers, New York, 1965.

[9] C. J. Ballhausen, "Introduction to Ligand Field Theory," McGraw-Hill, New York, 1962.

[10] B. N. Figgis, "Introduction to Ligand Fields," Interscience, New York, 1966.

[11] A. B. P. Lever, "Inorganic Electronic Spectroscopy," Elsevier, New York, 1968.

[12] C. J. Hawkins, "Absolute Configuration of Metal Complexes," Interscience, New York, 1971.

[13] S. J. Lippard, "Eight Coordination Chemistry," *Prog. Inorg. Chem.*, **8**, 109 (1967); E. L. Muetterties and C. M. Wright, *Quart. Revs.*, **21**, 109 (1967).

[14] E. Larsen, G. N. LaMar, B. E. Wagner, J. E. Parks, and R. H. Holm, *Inorg. Chem.*, **11**, 2652 (1972).

[15] R. B. VorDreele, J. J. Stezowski, and R. C. Fay, *J. Amer. Chem. Soc.*, **93**, 2887 (1971).

[16] R. Eisenberg and J. A. Ibers, *Inorg. Chem.*, **5**, 411 (1966).

Complexes of unsaturated dithiolates (**6**) were also widely studied in the 1960's, not only because such ligands led to unusual geometries, but also because their complexes could often exist in several stable oxidation states.[17] More recently, the discovery of metal complexes of molecular nitrogen (**7**) has had considerable impact on inorganic chemistry[18] and the search for the mysteries of natural nitrogen fixation. New light has been shed on metal-O_2 complexes (**8**)[19] and on metalloporphyrins (**9**).[20]

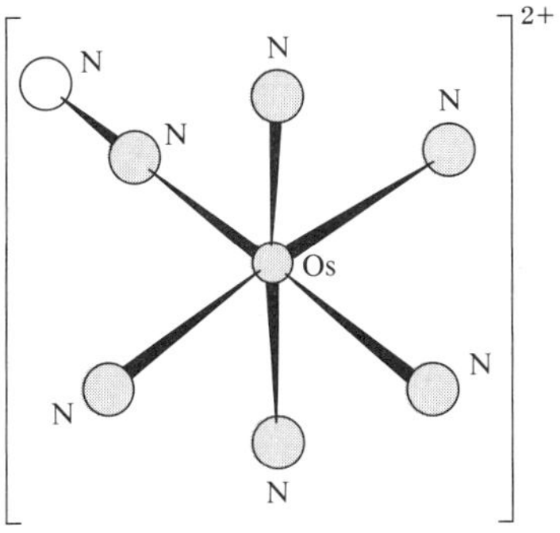

7 $[Os(NH_3)_5N_2]Cl_2$[21]

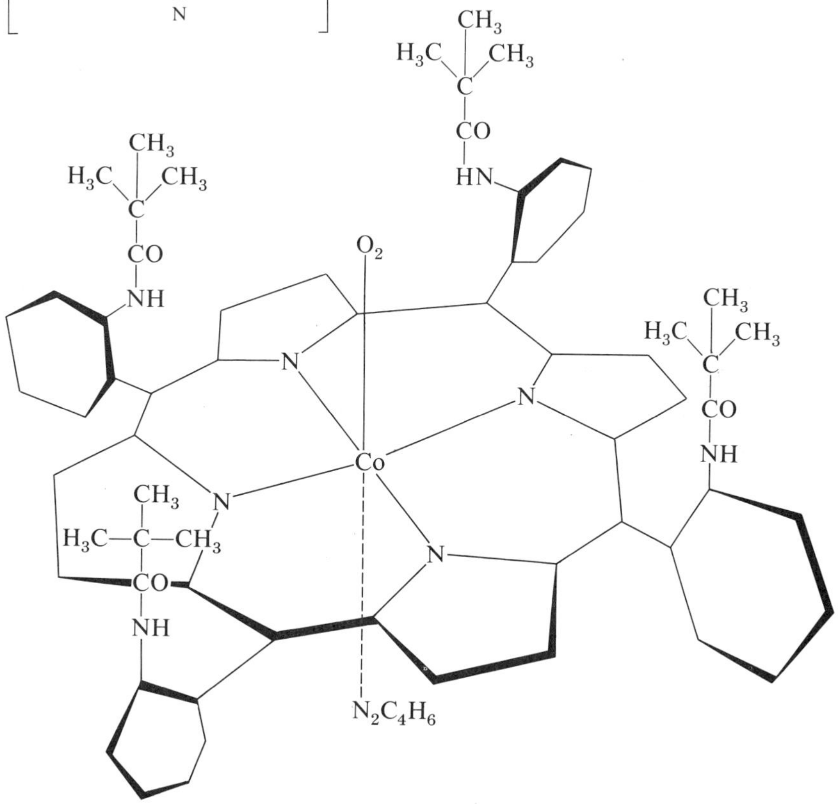

8   An O_2 adduct of a "picket fence porphyrin"[22]

---

[17] J. A. McCleverty, "Metal 1,2-Dithiolene and Related Complexes," *Prog. Inorg. Chem.,* **10,** 49–221 (1969); G. N. Schrauzer, *Acc. Chem. Res.,* **2,** 72 (1969).

[18] A. D. Allen and F. Bottomley, *Acc. Chem. Res.,* **1,** 360 (1968); see also E. E. van Tamelen. J. A. Gladysz, and C. R. Brulet, *J. Amer. Chem. Soc.,* **96,** 3020 (1974) for a recent publication dealing with the fixation of N_2 by an inorganic "model" system.

[19] J. P. Collman, *et al., J. Amer. Chem. Soc.,* **97,** 1427 (1975); J. Valentine, *Chem. Rev.,* **73,** 235 (1973).

[20] E. B. Fleischer, *Acc. Chem. Res.,* **3,** 105 (1970).

[21] J. E. Fergusson, J. L. Love, and W. T. Robinson, *Inorg. Chem.,* **11,** 1662 (1972).

[22] J. P. Collman, R. G. Gagne, J. Kouba, and H. Ljusberg-Wahren, *J. Amer. Chem. Soc.,* **96,** 6800 (1974).

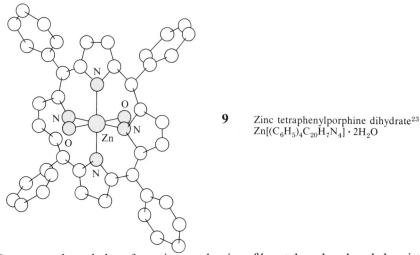

9 Zinc tetraphenylporphine dihydrate[23]
$Zn[(C_6H_5)_4C_{20}H_7N_4] \cdot 2H_2O$

By now our knowledge of reaction mechanisms,[24] metal–carbon bond chemistry, coordination complex stereochemistry, metal-sulfur complexes, and complexes with O_2 and N_2 has increased to the point where a systematic attack on the vast area of bioinorganic chemistry seems possible and potentially fruitful.[25,26] Indeed, examination of model complexes (**10**)[27] has led to a better understanding of the Co—C bond in vitamin B_{12} (**11**) and the B_{12} coenzymes,[28] and models have very recently been generated that duplicate some of the properties of O_2-hemoproteins.[29,30] Further, models of the iron-sulfur proteins (e.g., **12**) have been discovered,[31] and simple platinum complexes such as cis-$Pt(NH_3)_2Cl_2$ (**13**) have been found to be active against certain tumors.[32]

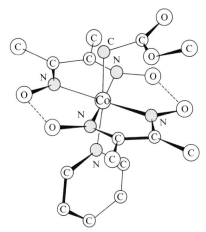

10 O-methyl-(Co-C)-carboxymethyl-[bis(dimethylglyoximato)pyridine]cobalt

---

[23] E. B. Fleischer, C. K. Miller, and L. E. Webb, *J. Amer. Chem. Soc.*, **86**, 2342 (1964).
[24] F. Basolo and R. Pearson, "Mechanisms of Inorganic Reactions," Wiley, New York, 1967.
[25] M. N. Hughes, "The Inorganic Chemistry of Biological Processes," John Wiley and Sons, New York, 1972.
[26] G. L. Eichhorn (ed.), "Inorganic Biochemistry," Elsevier, New York, 1973.
[27] G. N. Schrauzer, *Acc. Chem. Res.*, **1**, 97 (1968).
[28] J. M. Pratt, "Inorganic Chemistry of Vitamin B_{12}," Academic Press, New York, 1972.
[29] See ref. 22.
[30] *Newsweek,* March 24, 1975, page 91; J. Almog, J. E. Baldwin, and J. Huff, *J. Amer. Chem. Soc.*, **97**, 227 (1975).
[31] S. J. Lippard, *Acc. Chem. Res.*, **6**, 282 (1973); J. J. Mayerle, S. E. Denmark, B. V. DePamphilis, J. A. Ibers, and R. H. Holm, *J. Amer. Chem. Soc.*, **97**, 1032 (1975).
[32] A. J. Thompson, R. J. P. Williams, and S. Reslova, *Structure and Bonding*, **11**, 1 (1972).

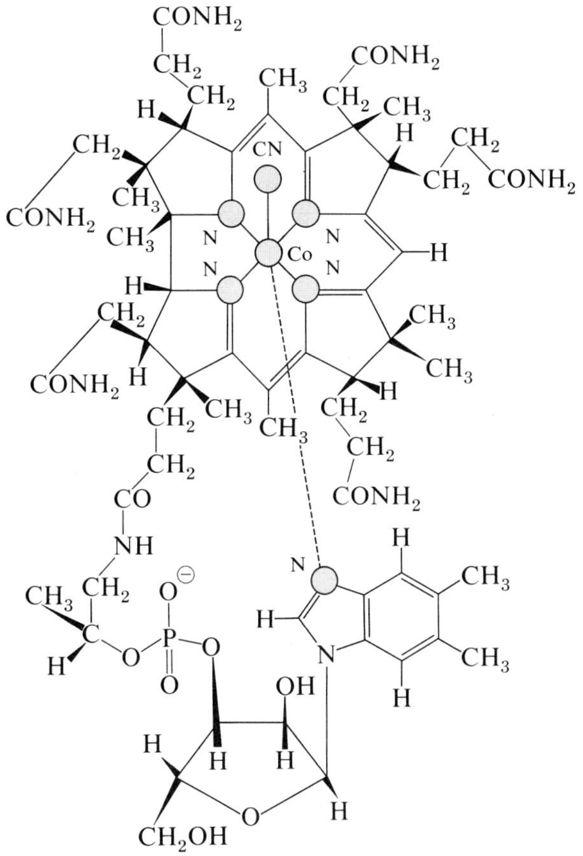

11 Vitamin B$_{12}$

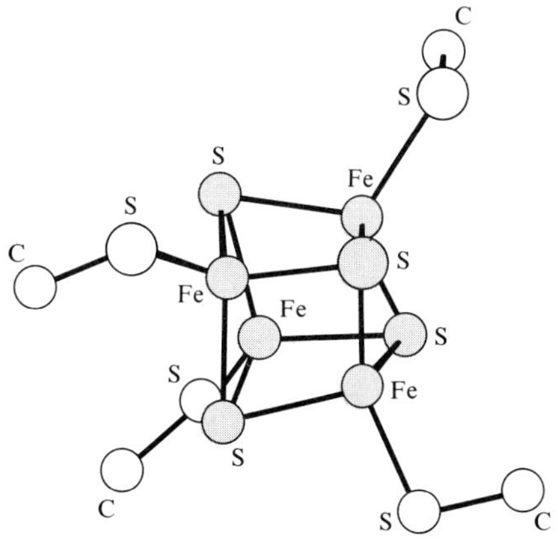

12 [Fe$_4$S$_4$(SC$_6$H$_5$)$_4$]$^{2-}$ [33]

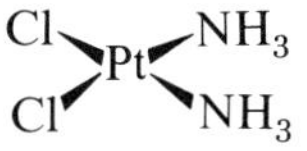

13 cis-PtCl$_2$(NH$_3$)$_2$

The recent emphasis on metal chemistry does not mean that non-metal chemistry has been neglected. When your authors were at your stage of their chemical education (~1960), we were taught that the compounds HOF (hypofluorous acid, **14**), BrO$_4^-$ (perbromate ion, **15**), and XeF$_n$ (the xenon fluorides, **16**) were not likely to exist or would have only the most fleeting existence at room temperature, and would

---

[33] L. Que, Jr., M. A. Bobrik, J. A. Ibers, and R. H. Holm, *J. Amer. Chem. Soc.*, **96**, 4168 (1974).

8 □ THE COMING OF AGE: PERSPECTIVE

be of little interest to the chemist. The intervening 15 years have seen the syntheses of all these compounds![34,35,36] These impressive achievements accentuate the need by all chemists for a good understanding of theoretical, thermodynamic, and mechanistic concepts in application to chemical synthesis.

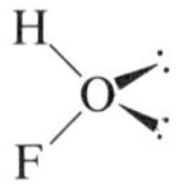

14, 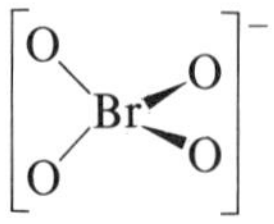 **15**

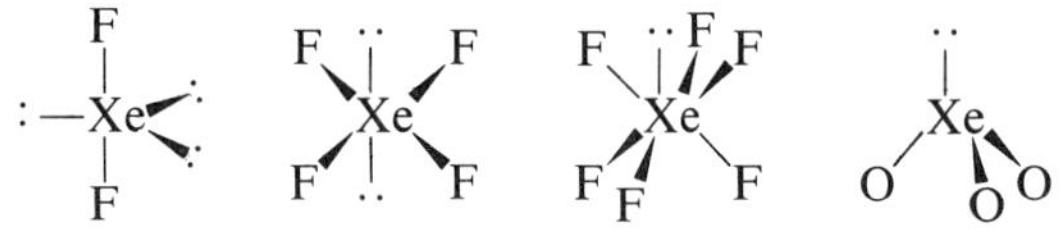

16

The need of industries for new stable polymers, particularly those that could withstand extremes in thermal and physical stress, has spurred great interest in inorganic polymers with skeletons consisting of boron, aluminum, silicon, and phosphorus atoms **(17-21)**.[37]

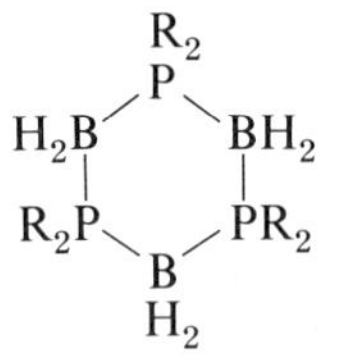

17, 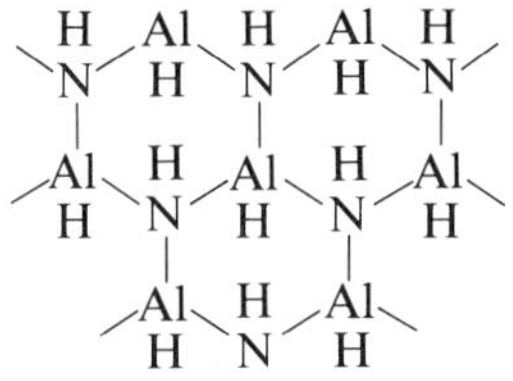 **18**

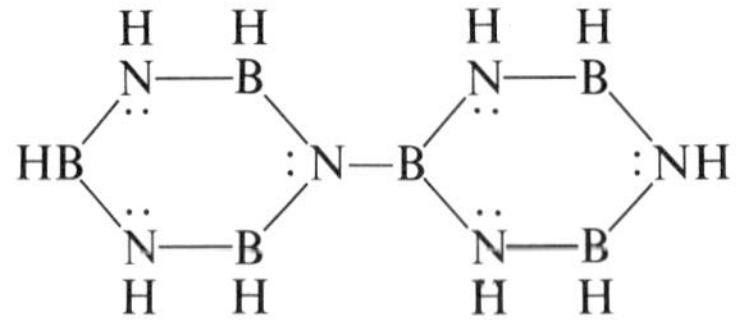

19

---

[34] M. H. Studier and E. H. Appleman, *J. Amer. Chem. Soc.*, **93**, 2349 (1971).

[35] G. K. Johnson, *et al.*, *Inorg. Chem.*, **9**, 119 (1970); J. R. Brand and S. A. Bunck, *J. Amer. Chem. Soc.*, **91**, 6500 (1969).

[36] H. H. Hyman (ed.), "Noble Gas Compounds," University of Chicago Press, Chicago (1963) lets you feel the pulse of these history making events. For synthesis methods see E. H. Appleman and J. G. Malm, *Prep. Inorg. Reactions*, **2**, 341 (1965); J. H. Holloway, "Noble-Gas Chemistry," Methuen, London (1968).

[37] D. A. Armitage, "Inorganic Rings and Cages," Edward Arnold, London (1972).

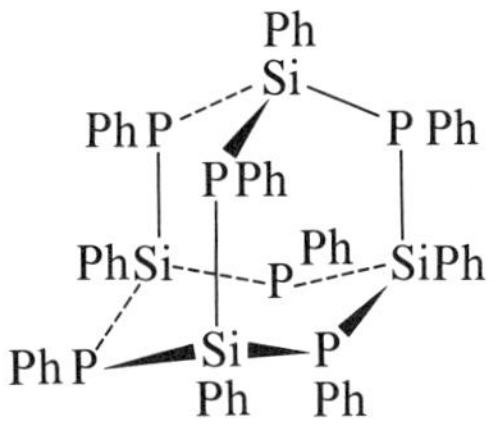

20

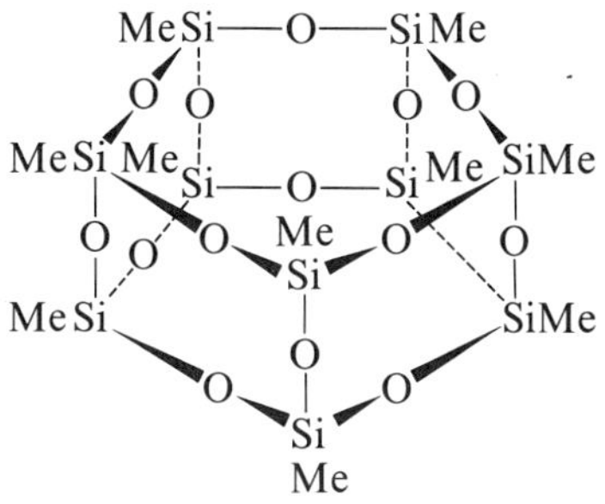

21

The agricultural problems of the world have spurred the development of many phosphorus, sulfur, and halogen compounds useful as toxins for control of insects, fungi, and weeds. The space program in particular, and industrial and military concerns in general, have generated continued interest in the development of techniques for synthesizing and handling high energy oxidizers **(22-24)**,[38] while the pharmaceutical promise of fluorine-containing compounds has sustained interest in

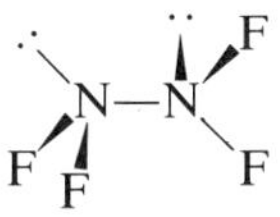

22 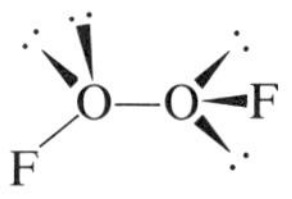   **23** 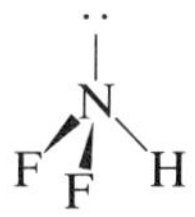   **24**

stable organo compounds containing fluorine. Although a considerable number of papers have been published concerning the boron hydrides, additional developments can be expected in this area as new order was recently imposed on the field by the discovery of simple rules for interrelating structure and bonding.[39,40] Indeed, a most promising area for future development in boron hydride chemistry is in the metal complexes of the boranes and carboranes **(25, 26)**.[41]

Some of the reasons for the rather spectacular recent progress in inorganic chemistry have been the continuing development of a theoretical framework and the application of crystallographic and spectroscopic techniques[42] as well as other instrumental techniques (e.g., electrochemistry[43]). With regard to the former, the recognition of symmetry constraints on chemical problems has had a great impact on

---

[38] E. W. Lawless and I. C. Smith, "Inorganic High Energy Oxidizers," Edward Arnold, London (1968).
[39] K. Wade, *Adv. Inorg. Chem. Radiochem.,* **18**, 1 (1976).
[40] R. W. Rudolph and W. R. Pretzer, *Inorg. Chem.,* **11**, 1974 (1972).
[41] G. B. Dunks and M. F. Hawthorne, *Acc. Chem. Res.,* **6**, 124 (1973).
[42] R. S. Drago, "Physical Methods in Inorganic Chemistry," W. B. Saunders Company, Philadelphia, 1976.
[43] R. E. Dessy and L. A. Bares, *Acc. Chem. Res.,* **5**, 415 (1972); J. B. Headridge, "Electrochemical Techniques for Inorganic Chemists," Academic Press, New York, 1969.

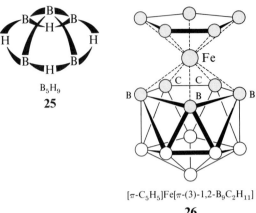

B_5H_9
25

$[\pi\text{-}C_5H_5]Fe[\pi\text{-}(3)\text{-}1,2\text{-}B_9C_2H_{11}]$
26

inorganic chemistry[44] and has encouraged the wider application of molecular orbital concepts to inorganic structures and reactions. Among the more important advances in the area of structures of molecules are models[45] for predicting and correlating the molecular topologies of non-metal compounds and the recognition that many of these molecules are not "rigid" but rather "floppy." The more sophisticated expression is "pseudorotation" of terminal groups about the central atom. For example, NH_3 has long been known to undergo inversion at a rate of 10^{10} sec^{-1} at room temperature. Many other non-metal and metal compounds are now known to experience such rapid intramolecular rearrangement, and it requires very low temperatures and/or very fast detection methods[46] to observe the "static" structures of such compounds. These

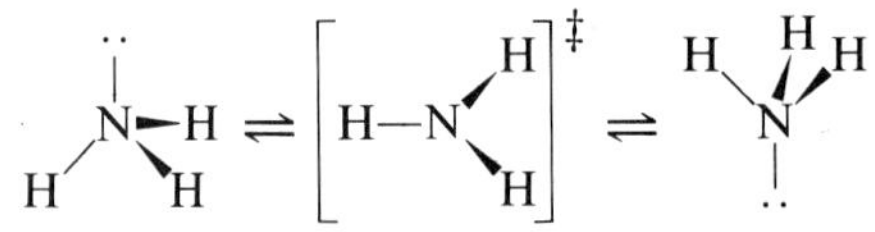

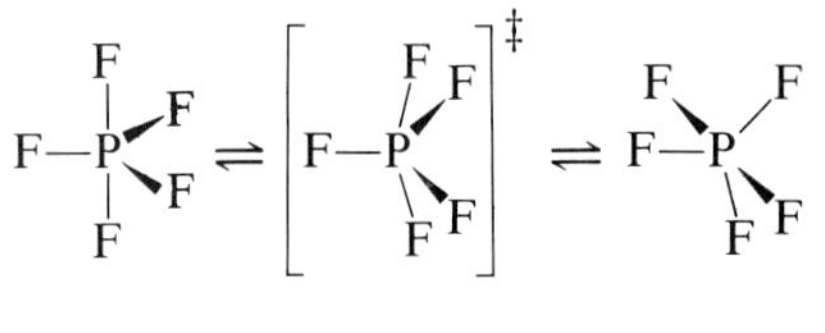

advances have been very important in understanding the reaction modes of compounds of four-coordinate phosphorus and silicon.[47] For example, how is it possible for a nucleophile to attack Si opposite the group to be displaced and *not* lead to inversion (change in molecular chirality) of the silicon stereochemistry?

---

[44] F. A. Cotton, "Chemical Applications of Group Theory," 2nd edition, John Wiley and Sons, New York, 1971.

[45] R. J. Gillespie, "Molecular Geometry," Van Nostrand Reinhold Co., London (1972); *J. Chem. Educ.*, **51**, 367 (1974); **47**, 18 (1970); L. S. Bartell, *J. Chem. Educ.*, **45**, 457 (1968); *Inorg. Chem.*, **5**, 1635 (1966); R. M. Gavin, *J. Chem. Educ.*, **46**, 413 (1969).

[46] See ref. 42.

[47] F. H. Westheimer, *Accts. Chem. Res.*, **1**, 70 (1968); K. Mislow, *ibid.*, **3**, 321 (1970); *J. Amer. Chem. Soc.*, **91**, 7031 (1969).

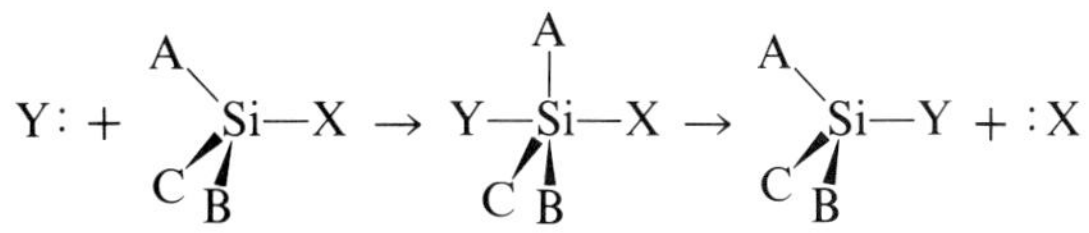

$$\text{Y:} + \underset{\underset{B}{C}}{\overset{A}{\diagdown}}\text{Si}-\text{X} \rightarrow \text{Y}-\underset{\underset{B}{C}}{\overset{A}{|}}\text{Si}-\text{X} \rightarrow \underset{\underset{B}{C}}{\overset{A}{\diagdown}}\text{Si}-\text{Y} + \text{:X}$$

The scope of inorganic chemistry is immense, and the texture of the field is changing rapidly. In the chapters that follow we cannot hope to be comprehensive. Rather, we have attempted first to present you with an invaluable theoretical framework upon which modern concepts of structure, bonding, and reactivity in inorganic chemistry are built. In later chapters, you will sample—in as un-superficial a manner as possible—both "established" areas and those of greatest current interest, areas sharing in common the promise to dominate the field in the coming years.

PROBABILITY DENSITY FUNCTIONS

   The Electron as a Matter Wave
   Radial Density Functions and Orbital Energies
   Angular Functions and Orbital Shapes
   Total Density Functions

POLYELECTRONIC ATOMS

   Atom Electron Configurations and the Long Form Periodic Table
   Slater Orbitals and Their Uses
   Atomic Configurations and Atomic Terms
   Atom and Orbital Electronegativities

EPILOG

APPENDICES

1. USEFUL ATOMIC CONCEPTS

This chapter broaches what is usually a difficult and mysterious subject for the undergraduate student of chemistry. Yet the need for the inorganic chemist to have a rudimentary understanding of the presently accepted theory of atomic structure cannot be disputed. In fact, many of the results of this theory have greatly spurred the development of synthetic and physical inorganic chemistry. You should realize at the outset that neither this text nor any other written at a comparable level will explain quantum mechanics to you. Rather, the point is to present the principal ideas (many of which are quite radical) and identify their utility to you in interpreting chemistry. The proof of any theory is its success in helping the understanding of experimental observations and, in this capacity, the wave model of the atom has been eminently successful.

PROBABILITY DENSITY FUNCTIONS

THE ELECTRON AS A MATTER WAVE

The Bohr model[1] of the atom was the first successful model of the hydrogen atom, in that it quantitatively accounted for the emission spectrum and ionization potential of atomic hydrogen. To achieve this quantitative agreement, Niels Bohr followed the lead of Planck in adopting the *quantum* concept and applied it to the hydrogen atom. Bohr had to make assumptions that ran counter to the accepted dogma of classical mechanics. Nevertheless, his postulates yielded some results that agreed quantitatively with experiment and were not to be taken lightly.

Table 1-1 summarizes various properties derived from Bohr's theory for the hydrogen atom. The values for $n = 1$ define a system of units, called atomic units, widely used in theoretical chemistry.

One point on which Bohr was in error concerns the magnetic properties of the $n = 1$ level. The ground state velocity of the electron of 2.2×10^8 cm sec^{-1} (about 1% of the speed of light) results in a magnetic field at the nucleus that is quite large ($\sim$125,000 oersted). Experimentally, this result is incorrect, as there is no magnetic field at the hydrogen nucleus from the orbital motion of the electron in the $n = 1$ level. However, there is orbital magnetism for some of the higher states. It is one of the triumphs of the quantum wave model that it correctly accounts for the variation in orbital magnetism as a function of atomic state.

At the turn of the century and shortly thereafter, physicists performed many new experiments, the results of which could not be explained by classical mechanics. Most

---

[1] N. Bohr, *Phil. Mag.*, **26**, 1, 476, 875 (1913).

14 □ USEFUL ATOMIC CONCEPTS

TABLE 1–1
HYDROGEN ATOM PROPERTIES FROM BOHR MODEL

Property	Formula*	Value for $n = 1$
r (radius)	$n^2 \left(\dfrac{\hbar^2}{me^2} \right)$	0.529 Å $= a_0$
v (velocity)	$\dfrac{1}{n}(e^2/\hbar)$	2.19×10^8 cm sec^{-1} $= v_0$
t (period)	$n^3 \left(\dfrac{2\pi \hbar^3}{me^4} \right)$	1.52×10^{-16} sec $= \tau_0$
E (energy)	$-\dfrac{1}{n^2}\left(\dfrac{e^4 m}{2\hbar^2} \right) = -\dfrac{1}{n^2}\left(\dfrac{e^2}{2a_0} \right)$	-13.6 eV $= E_0$
μ (magnetic moment)	$\dfrac{evr}{2} = n\left(\dfrac{e\hbar}{2m} \right)$	0.927×10^{-20} erg/gauss $= \beta_e$
H (magnetic field at nucleus)	$\dfrac{e/t}{2r} = \dfrac{1}{n^5}\left(\dfrac{m^2 e^7}{4\pi^2 \hbar^5} \right)$	1.25×10^5 oersted

*$\hbar = h/2\pi = 1.055 \times 10^{-27}$ erg sec
$m = 9.109 \times 10^{-28}$ gm
$e = 4.803 \times 10^{-10}$ e.s.u.
$n =$ orbital quantum number
a_0 is called the "Bohr radius"; β_e is called the "Bohr magneton".

of the experiments had to do with energy in the form of radiation, such as the frequencies of light radiated by a "black body," Einstein's photoelectric effect,[2] and the line spectrum of excited hydrogen atoms. Prior to these experiments, a theory of electromagnetic radiation had been developed in terms of wave forms to represent the behavior of energy radiation. The new experiments modified existing thought on the meaning of these wave functions by showing that the energy of the radiation could not be thought of as being distributed over the entire wave form; rather, the energy is concentrated in a packet, for which the term *photon* was coined. Compton[3] convincingly demonstrated that when light interacts with matter, as in a collision between X-rays and electrons, it behaves as though it consists of particles with momentum, a property normally associated with particles having mass and velocity (Figure 1-1).

---

[2] A. Einstein, *Ann. Phys.*, **17**, 132 (1905).
[3] A. H. Compton, *Phys. Rev.*, **21**, 483 (1923).

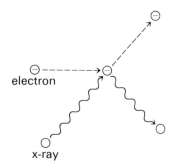

Figure 1–1. The Compton effect (a change in electron momentum and photon wavelength upon an e$^-$/photon "collision").

The wave theory was not invalidated by these findings; instead, the scientists' view of radiation had to be modified. As things now stand, the only view that is consistent with both the fact that light is *diffracted* from slits, pin holes, and other openings (a wave form property) and the fact that radiative energy interacts with matter as though that energy were particulate in nature (a classical concept) is the *dualistic nature of light.* The radiation is believed to be particulate in nature, consisting of photons, and the motion of these photons is best described by wave equations and wave functions.

Light had long been believed to be a form of wave motion, with energy distributed continuously over the wave; the new experiments demonstrated the duality of light by showing that a particulate nature is also called for. The development of a theory of *electrons* had just the inverse course. Early experiments were successfully interpreted by application of the laws of classical mechanics. In the mid and late 1920's experiments were performed by Davisson and Germer[4] that showed that electrons are *diffracted* in the same manner as are photons. Therefore, *electrons also exhibit a duality of character;* their particle nature can be convincingly demonstrated, but so can the fact that their motion is governed by the laws of wave motion just as is the motion of the photons of radiative energy. Thus, we have *matter waves*.

In a tremendously significant theoretical advance, de Broglie[5] proposed his relation for the connection between the momentum of a classical particle (momentum $\equiv p = mv$) and the length of the wave that describes its motion, λ:

$$\lambda = h/p$$

Davisson and Germer were able to verify de Broglie's relation experimentally by showing that electrons are diffracted with a wavelength inversely related to their momentum. As a result of these developments, the previously derived theory of wave motion for light was then borrowed to describe electrons and their motion in the form of Schrödinger's wave model[6] of the hydrogen atom.

Aside from the philosophical aspects of a duality principle for matter, two practical features of this theory are difficult. First is the fact that it is a wave theory—a theory requiring a philosophical view not fully developed in most chemists by prior experience. Second, the mathematics involved is not conventional algebra but differential calculus.

An inorganic chemistry text is not the proper place to develop the philosophy of quantum mechanics; rather, our interest lies primarily in the utilization of the results of that theory to the extent that our limited understanding will permit. Consequently, we outline in what follows the major points[7] and proceed to chemically useful interpretations.

    A. *For wave forms in general:*
        1. A physical wave form is a function of space and time coordinates: $\Psi(x,y,z,t)$
        2. It is important that in the differential equations describing wave motion there be no cross-terms between space and time coordinates: $\Psi(x,y,z,t)$ can be written $\psi(x,y,z) \cdot \psi'(t)$.
        3. A characteristic of wavefunctions is that they can be complex; the imaginary part generally appears as Ke^{im}, where m is a "catch-all" term containing the independent variable(s) and other constants determined by the

---

[4] C. J. Davisson and L. H. Germer, *Nature,* **119,** 558 (1927).

[5] L. de Broglie, *Ann. de. Phys.,* **3,** 22 (1925). de Broglie argued that, for a photon, $E = mc^2$ and $E = h\nu$ implies $mc = h\nu/c = h/\lambda$, where $mc =$ the momentum of the photon. He felt that a similar relationship was appropriate for matter.

[6] E. Schrödinger, *Ann. Phys.,* **81,** 109 (1926).

[7] A good discussion is to be found in a paper by R. S. Berry, *J. Chem. Educ.,* **43,** 283 (1966).

physical phenomenon being described. In general, because the function contains imaginary terms, *no physical significance can be given to* Ψ.
4. The wave intensity is an entirely real (measurable) phenomenon and must be given by

$$\Psi^*\Psi = \psi^*(x,y,z)\psi(x,y,z)\psi'^*(t)\psi'(t)$$

The symbol Ψ^* means that we replace $i\,(=\sqrt{-1})$ wherever it appears in Ψ with $-i$. This formalism is appropriate because the complex component of Ψ makes a constant contribution to $\Psi^*\Psi$, by virtue of the fact that $Ke^{-im} \cdot Ke^{im} = K^2$ for each complex part of Ψ.
5. $\psi'(t)$ has the form $e^{i\nu t}$, where ν is the wave frequency, and thus $\psi'^*\psi' = 1$ makes no contribution to $\Psi^*\Psi$.
6. You see that while Ψ usually has no physical interpretation, $\Psi^*\Psi$ does because the latter is real at all points in space at all times.
7. Accordingly, $\psi(x,y,z)$ is called the *amplitude function* and $\psi'(t)$ is called the *phase function*.
8. As long as we do not perform a time-dependent experiment, we need not concern ourselves with $\psi'(t)$, but only with $\psi(x,y,z)$.

B. *For the electron in the atom:*
1. We are to deal with a *matter wave* for which "wave intensity" means "particle probability." Viewing an electromagnetic wave as particulate (a photon), the same semantics and conceptualization apply to electrons.
2. For a single electron, the sum probability at all points in space must equal unity. Mathematically,

$$\int^{\text{all space}} \psi^*\psi(x,y,z)\, d\tau = 1$$

This is the *normalization condition* whereby, for consistency with the notion that the electron must be somewhere in space, the *normalization is to unity*.
3. Schrödinger proposed a matter wave analog of the classical wave-mechanical wave equation, which, upon solution, leads to the electron space wavefunction in terms of polar rather than cartesian coordinates (Figure 1-2). The conversion to polar coordinates is for convenience in solving the second-order differential matter wave equation. The result Schrödinger obtained has the form

$$\psi(r,\theta,\phi) = R_{n,\ell}(r)\Theta_\ell^{|m|}(\theta)\Phi_m(\phi) = R(r)Y_\ell^m(\theta,\phi)$$

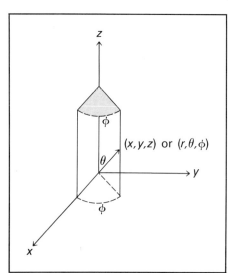

Figure 1-2. Spherical polar coordinates.

4. The function $R(r)$ is called the *radial function;* it is real and makes a contribution to the total probability function of $R^2(r)$. The primary and secondary quantum numbers n and l arise in this function. The *angular functions* $\Theta(\theta)$ and $\Phi(\phi)$ are known collectively as *spherical harmonics* $Y(\theta,\phi)$. The quantum numbers l and m_l arise in these functions. While $\Theta(\theta)$ is entirely real, $\Phi(\phi)$ is not—it has the form $e^{im\phi}/\sqrt{2\pi}$. The angular contribution to the total probability function, consequently, is simply $Y^*Y = \Theta^2(\theta)/2\pi$.

With this brief summary behind us, we may proceed with the important business of analyzing the total density function $R^2(r)Y^*Y(\theta,\phi)$ and finding the physical meaning of the quantum numbers n, l, and m_l.

PROBABILITY DENSITY FUNCTIONS

As suggested in the preceding outline, by calculating the value of $\psi^*\psi$ at some point (r,θ,ϕ) in space we determine the *probability density* for the electron at that point.

$$\psi^*\psi(r,\theta,\phi) = \text{probability density for the point } (r,\theta,\phi)$$

The unit is 1/volume, as you will see in the next section, and this is why we think of $\psi^*\psi$ as a probability *density* function. Multiplication of this density function by some small volume of space yields the *occupation number* for the electron in that small volume element, $d\tau$:

$$\psi^*\psi(r,\theta,\phi)\,d\tau = \text{occupation number for the infinitesimal volume } d\tau$$
$$\text{about the point } (r,\theta,\phi).$$

This function is dimensionless. Finally, by integrating the occupation number for each infinitesimal volume element in all of the space over all space, you will obtain the total occupation of the electron in all space, which physically must equal unity.

$$\int \psi^*\psi(r,\theta,\phi)\,d\tau = 1$$

The volume element $d\tau$ represents $(dx\,dy\,dz)$ if the coordinates are the conventional cartesian coordinates. On conversion to spherical coordinates, $(dx\,dy\,dz)$ becomes $(r^2 \sin\theta\,dr\,d\theta\,d\phi)$. Because of the separability of the wavefunction into $R(r)Y(\theta,\phi)$, the density integral takes the form

$$\int_0^\infty R^2(r)r^2\,dr \int_0^\pi \int_0^{2\pi} Y^*(\theta,\phi)Y(\theta,\phi)\sin\theta\,d\theta\,d\phi$$

Thus, we think of the total probability density function as being a product of a radial function, $R^2(r)$, and an angular function, $Y^*(\theta,\phi)Y(\theta,\phi)$. For convenience, each of these is separately normalized.

$$\int_0^\infty R^2(r)r^2\,dr = 1$$

$$\int_0^\pi \int_0^{2\pi} Y^*(\theta,\phi)Y(\theta,\phi)\sin\theta\,d\theta\,d\phi = 1$$

In the next section we will examine these density functions in some detail for certain values of n, l, and m. To do this will give you some feeling for the *shapes and symmetries of atomic orbitals* through analysis of the *angular density functions*. The *radial density functions* are important in developing a link between theory and experiment on the subjects of *atom sizes, ionization potentials,* and *electronegativity.*

RADIAL DENSITY FUNCTIONS AND ORBITAL ENERGIES

Judging from the comments of the preceding section, radial density distribution can be given either as $R^2(r)$ or as $R^2(r)r^2$. The latter is more useful, and we need to note carefully the difference between the two. The function $R^2(r)$ is a *density function* for the distance of the electron from the nucleus, *exclusive of any angular variation*. It has units of 1/volume. $R^2(r)r^2$, on the other hand, is a *1/distance function* that is derived from $\psi^*\psi\,d\tau$ by integrating over the angular coordinates θ and ϕ only. Now, $\psi^*\psi\,d\tau$ takes explicit account of angular variation but, because $Y^*(\theta,\phi)Y(\theta,\phi)$ is normalized, integration over θ and ϕ actually reduces $\psi^*\psi\,d\tau$ to $R^2(r)r^2\,dr$. This is the *electron occupation function for infinitesimally thin spherical shells of radius r and thickness dr*. The function $R^2(r)\,r^2$ represents a *spherical* surface density function for a sphere of radius r and center at the nucleus. *We will call this the surface density function, $S(r)$*.

Table 1–2 lists the radial functions for (n,ℓ) pairs through $n=4$ (note that all have dimensions of $a_0^{-3/2}$), and Figure 1–3 is a plot of three radial function types for the hydrogen atom with electron quantum numbers $n=1$, $\ell=0$, and $m=0$. (Actually, only n and ℓ need to be specified, since the radial wave function depends only on these two quantum numbers and not on m.) The solid curve is for $R(r)$ as a function of r, and distance is measured in atomic units, a_0. The dashed curve is the radial function $R^2(r)$. An important characteristic of "s" radial functions ($\ell=0$) is their non-zero amplitude at $r=0$. For example, $R_{1,0}^2(r=0) = 4/a_0^3 = 27.0/\text{Å}^3$. Since the spherical harmonic $Y_0^0 = 1/\sqrt{4\pi}$, the probability density for the 1s electron *at the nucleus* is $1/\pi a_0^3 = 2.15/\text{Å}^3$. You are alerted to the fact that $R^2(r)$ does not by itself give the *total* point density of the electron but only a part of it.

Also shown in Figure 1–3 is the surface density function (dotted curve) $R^2(r)r^2 = S(r)$. As mentioned above, this function is special in that it is the surface density function for a sphere of radius r. Consequently, $S(r)$ has a value of zero at $r=0$ (the surface area of a sphere of zero radius is nil).

TABLE 1–2
NORMALIZED RADIAL FUNCTIONS FOR HYDROGEN ($\rho = Zr/a_0$)

$R_{1,0}$: $N_{1,0}[e^{-\rho}]$	$N_{1,0} = 2\left(\dfrac{Z}{a_0}\right)^{3/2}$
$R_{2,0}$: $N_{2,0}[(2-\rho)e^{-\rho/2}]$	$N_{2,0} = \dfrac{1}{2\sqrt{2}}\left(\dfrac{Z}{a_0}\right)^{3/2}$
$R_{2,1}$: $N_{2,1}[\rho e^{-\rho/2}]$	$N_{2,1} = \dfrac{1}{2\sqrt{6}}\left(\dfrac{Z}{a_0}\right)^{3/2}$
$R_{3,0}$: $N_{3,0}[(27 - 18\rho + 2\rho^2)e^{-\rho/3}]$	$N_{3,0} = \dfrac{2}{81\sqrt{3}}\left(\dfrac{Z}{a_0}\right)^{3/2}$
$R_{3,1}$: $N_{3,1}[(6\rho - \rho^2)e^{-\rho/3}]$	$N_{3,1} = \dfrac{4}{81\sqrt{6}}\left(\dfrac{Z}{a_0}\right)^{3/2}$
$R_{3,2}$: $N_{3,2}[\rho^2 e^{-\rho/3}]$	$N_{3,2} = \dfrac{4}{81\sqrt{30}}\left(\dfrac{Z}{a_0}\right)^{3/2}$
$R_{4,0}$: $N_{4,0}[192 - 144\rho + 24\rho^2 - \rho^3]e^{-\rho/4}$	$N_{4,0} = \dfrac{1}{768}\left(\dfrac{Z}{a_0}\right)^{3/2}$
$R_{4,1}$: $N_{4,1}[80 - 20\rho + \rho^2]\rho e^{-\rho/4}$	$N_{4,1} = \dfrac{3}{768\sqrt{15}}\left(\dfrac{Z}{a_0}\right)^{3/2}$
$R_{4,2}$: $N_{4,2}[12 - \rho]\rho^2 e^{-\rho/4}$	$N_{4,2} = \dfrac{1}{768\sqrt{5}}\left(\dfrac{Z}{a_0}\right)^{3/2}$
$R_{4,3}$: $N_{4,3}\rho^3 e^{-\rho/4}$	$N_{4,3} = \dfrac{1}{768\sqrt{35}}\left(\dfrac{Z}{a_0}\right)^{3/2}$

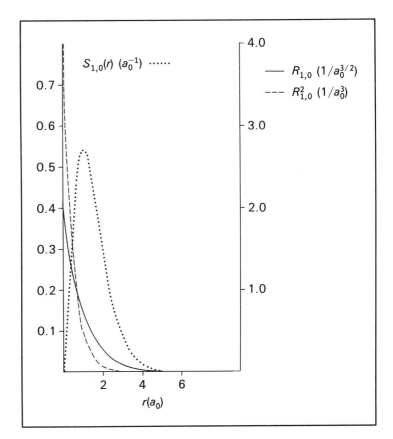

Figure 1-3. The 1s radial function, radial density function, and surface density function.

You should be careful to notice how the functions $R^2(r)$ and $S(r)$ differ. The $R^2(r)$ function is an exponential function decreasing from a maximum value at $r = 0$ to zero at $r = \infty$. As far as density is concerned, the electron with quantum numbers $n = 1$, $\ell = 0$, and $m = 0$ is most highly concentrated at the nucleus. If you ask the question, "What is the radius of the spherical surface over which the electron is most likely to be found?" the answer comes from the maximum in the $S(r)$ curve and is 1 a_0, or 0.5292 Å. Don't miss the further point that *the area under the $S(r)$ curve is unity*, since $\int_0^\infty S(r)\, dr = 1$.

You may be surprised that the Schrödinger and Bohr models for the hydrogen atom agree perfectly on the value for this distance, since Bohr has the electron on a fixed circular orbit while Schrödinger allows it full radial freedom. One of the important differences between the Bohr and Schrödinger models is seen to be the whole idea of whether the electron (with a given total kinetic and potential energy) must be thought of as traveling in a fixed orbit or whether it is, in fact, free to roam over all of space with greater probability of being in some regions than in others.

> An interesting sidelight to this conceptual difference regarding a subatomic particle has been advanced[8] by Heisenberg in the form of his Uncertainty Principle. The principle, if we are to accept it, warns us that if we determine the momentum of such a small particle, we cannot be sure where the particle is with that momentum. Conversely, we cannot determine the momentum of the particle at any precisely defined point. The same comments pertain to the pair (energy, time). This is a difficult philosophical concept and you may be relieved to know that Einstein believed, and Bohr disagreed, that one could design an experiment *not* subject to the Uncertainty Principle.[9] Though many experiments were devised by Einstein, in the end Bohr succeeded in arguing that all were subject to Heisenberg's Principle.

---

[8] W. Heisenberg, *Z. Phys.*, **43**, 172 (1927).
[9] N. Bohr, "Atomic Physics and Human Knowledge," John Wiley and Sons, New York (1958), p. 38ff.

Interestingly, this *most probable* spherical distance is not the same as the *average* value of the electron's distance from the proton. The average value is computed to be $\frac{3}{2}a_0$. That the average value is larger than the most probable value stems directly from consideration of the equation defining the average value. The average value of any quantity is calculated from its distribution function by weighting each possible value of the variable by its distribution function, summing these adjusted values, and dividing by the total distribution. In the present instance,

$$\langle r \rangle = \frac{\sum_r S(r) \cdot r}{\sum_r S(r)} = \int_0^\infty S(r) \cdot r \, dr$$

The value of this integral could be obtained analytically or by integration of the area under the curve $S(r) \cdot r$ as a function of r. Comparison of the $S(r)$ and $S(r) \cdot r$ curves in Figure 1-4 reveals the effect of asymmetry about the maximum in $S(r)$; values of $r > a_0$ have greater probability than their counterparts at $r < a_0$. The area under the $S(r)$ curve is unity, while that under $S(r) \cdot r$ is $\frac{3}{2}a_0$.

An alternate view of the comparison is that of the scaling of $S(r)$ by the factor r, which increases steadily with r. The scaling down of $S(r)$ at $r < a_0$ and scaling up of $S(r)$ at large r are the characteristic differences in these two curves.

Figure 1-5 shows the radial density functions for the hydrogen atom with its electron in the $n = 1$ and $n = 2$ states. For $n = 2$, it is possible for the electron to have an ℓ value of 0 or 1. Thus, the figure gives the radial density curves for the 1s, 2s, and 2p orbitals. (For $n = 2$ and $\ell = 1$, m could equal -1, 0, or $+1$ but we need not consider these different possibilities for m, since $R(r)$ is the same for each of these m values.) *The important feature of this figure is that the 1s and 2s point densities are greatest for $r = 0$; that is, an electron whose motion is described by the 1s or 2s wave forms has its greatest concentration at the nucleus. This is a striking characteristic of*

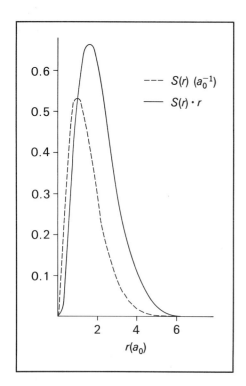

Figure 1-4. The surface density functions $S(r)$ and $S(r) \cdot r$ for the hydrogen 1s orbital graphically illustrate the calculation of $\langle r \rangle$.

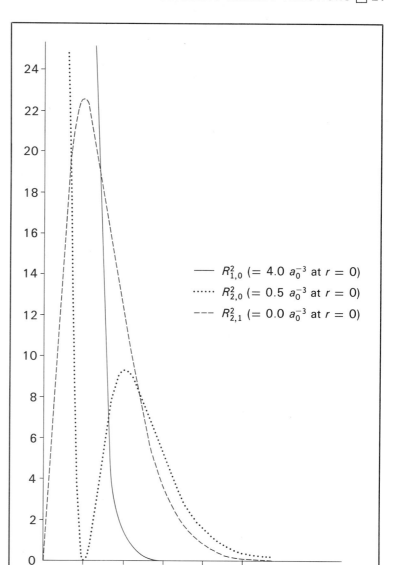

Figure 1-5. Comparison of the 1s, 2s, and 2p radial density functions. Units along ordinate = $10^{-3} a_0^{-3}$.

atomic s functions and is probably their most significant feature when contrasted with the other radial functions—p, d, and f—all of which have nodes at the nucleus. A node is a point or surface at which a function is zero-valued. (Later we will encounter nodal surfaces for the angular functions, and at that time the relationship between quantum numbers, energy, and nodes will be clarified.)

Figure 1-6 shows the 1s, 2s, and 2p surface density functions for the hydrogen atom. Several points are to be noted from this figure. First, the $n = 2$ functions are *radially more diffuse* than the 1s. Accordingly, the $n = 2$ curve maxima occur at greater r, and their average values of r are greater than that of the 1s. It stands to reason (but see the later discussion of potential energies) that an electron with these wavefunctions should have a higher energy than one with the 1s wavefunction. (The Schrödinger values are $-e^2/8a_0$ and $-e^2/2a_0$, respectively.) Other points that are very significant are that the 2s centroid (the average value of r for the 2s function) is greater than that of the 2p function and that the 2s distribution has a node at $r = 2a_0$

22 ☐ **USEFUL ATOMIC CONCEPTS**

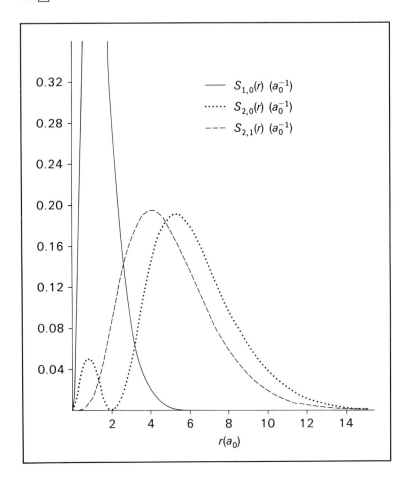

Figure 1-6. Comparison of the 1s, 2s, and 2p surface density functions.

with a second maximum at approximately $0.7a_0$. The occurrence of the node is not a surprising result, since standing wave forms generally have nodes.[10] Also worth noting are the dramatic differences between the $R^2(r)$ and $S(r)$ functions of the 2s wave at small and large r.

Of considerable significance, as you shall see later, is the greater concentration of the 2s function than of the 2p function in the range $r = 0$ to $1.3\ a_0$, in contrast to the smaller distribution values over the range $r = 1.3$ to $4.8\ a_0$. *With reference to Figure 1-6, the 2s and 2p both penetrate the region of 1s distribution, but the 2p does so to a much greater extent than the 2s (56% of the 2p and 35% of the 2s densities occur within the sphere of radius $5a_0$, which contains essentially all of the 1s density).* This *penetration effect* will be of great importance later when we discuss the order of filling of atomic orbitals in an attempt to account for the electronic structures of atoms.

---

[10] When we think of the electron in terms of its particle (classical) properties, it is difficult to imagine surfaces of zero probability (nodal surfaces). The fallacy of this artificial dilemma lies in the duality concept—if we must describe the electron in terms of its wave property, we must automatically abandon the classical particle concept. When conceptualizing experiments requiring a classical view, we must abandon the wave concept. The classical view of the electron as a particle is inappropriate for the Schrödinger atom and for the diffraction part of the Davison-Germer experiment. An interesting point, of little *chemical* consequence, is made by A. Szabo in *J. Chem. Educ.*, **46,** 678 (1969). Here it is pointed out that the Schrödinger treatment ignores relativity (incurring a negligible error for chemical purposes); were relativity to be incorporated, the nodal surfaces would become surfaces of low, but not zero, probability. In this case, the "dilemma" of mixing classical and wave concepts doesn't even arise, though it still is improper to invoke both simultaneously.

The point was made earlier that the 2s density function gives a larger average value of r than does the 2p function. From this you might erroneously conclude that the 2s function implies a higher (less negative) electron energy than the 2p. In fact, the energy of an electron whose motion is described by the 2s function is identical to that of one whose motion is described by the 2p function. As you have already seen for the Bohr theory, the energy of the electron in the hydrogen atom is given by half the potential energy,

$$E = -\frac{1}{2}\frac{Ze^2}{r}$$

This same result can be shown to hold for the wave model, in which case we must use the *average* potential energy,

$$V = -e^2 \left\langle \frac{Z}{r} \right\rangle = -e^2 \int_0^\infty \frac{Z}{r} S(r)\, dr.$$

That is, the energy is half the area under a plot of $(Z/r)S(r)$ versus r. These plots are given in Figure 1-7 for the 2s and 2p wave forms. The areas under these two curves are equal and correspond to the value 6.39 eV. As a special note, Z is a constant, independent of r, for the hydrogen atom; the energy may be given, in this case, as

$$2E = -Ze^2 \left\langle \frac{1}{r} \right\rangle.$$

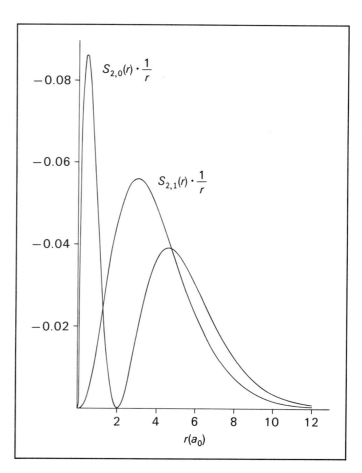

Figure 1-7. Potential energy, $V(r)$, plots for 2s and 2p wave forms.

24 □ USEFUL ATOMIC CONCEPTS

The paradox of $\langle r \rangle_{2s}$ greater than $\langle r \rangle_{2p}$ but $E_{2s} = E_{2p}$ is resolved by realizing that $1/\langle r \rangle$ is not equal to $\langle 1/r \rangle$. The $\langle r \rangle$, as given above, is determined by the area under a curve of $S(r) \cdot r$, while $\langle 1/r \rangle$ has a value given by the area under a curve of $S(r) \cdot 1/r$. That is,

$$\langle r \rangle = \int_0^\infty S(r) r \, dr$$

and

$$\left\langle \frac{1}{r} \right\rangle = \int_0^\infty \frac{S(r)}{r} \, dr$$

The functions under the integral signs are the $S(r)$ functions weighted by r in one case and by $1/r$ in the other. Thus, $\langle r \rangle$ is more sensitive to differences in $S(r)$ plots at large r (where the weighting factor r is largest) than at small r, while $\langle 1/r \rangle$ is more sensitive to differences in $S(r)$ at small r (where the weighting factor $1/r$ is the largest) than at large r. The plot of Z/r given in Figure 1–8 shows just how strong the differentiation is. In comparing the 2s and 2p functions, *you are aware that the 2s function has greater amplitude at both large and small r.* Thus, $\langle r \rangle$ depends mainly on differences in $S(r)$ at large r and $\langle 1/r \rangle$ depends mainly on differences at small r. The differences in 2s and 2p $S(r)$ functions are such that $\langle 1/r \rangle$ is the same for both. It is instructive to compare Figures 1-6 and 1-7 to note visually the effect of $1/r$ on $S(r)$. Table 1–3 lists $\langle r \rangle$ and $\langle 1/r \rangle$ for the first ten orbitals of the hydrogen atom.

An additional point to be noticed about the entries in the $\langle 1/r \rangle$ column of the table is that the difference in electron energies decreases as the principal quantum number n increases. This is an important feature of the wave model and is one that we will have occasion to use when the bonding between atoms in molecules is discussed.

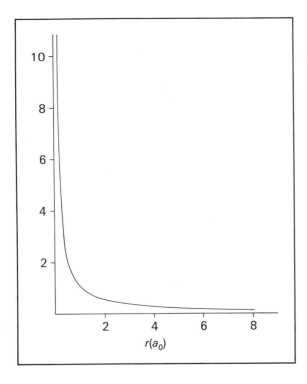

Figure 1–8. Plot showing that $\frac{1}{r}$ has greatest magnitude in the region $0 < r < 2$.

PROBABILITY DENSITY FUNCTIONS 25

TABLE 1-3

$\langle r \rangle$ AND $\langle \frac{1}{r} \rangle$ FOR ORBITALS OF THE HYDROGEN ATOM

	$\langle r \rangle$, Å	$\langle \frac{1}{r} \rangle$, Å^{-1}	Difference
1s	0.8	1.89	
			1.42
2s	3.2	0.47	
2p	2.6		
			0.26
3s	7.1	0.21	
3p	6.6		
3d	5.6		
			0.09
4s	12.7	0.12	
4p	12.2		
4d	11.1		
4f	9.5		

STUDY QUESTION

1. Construct plots of $S(r) \cdot r$ and $S(r) \cdot 1/r$ for the hydrogen atom 3s, 3p, and 3d functions. Comment on the relative values of $\langle r \rangle$ and $\langle r^{-1} \rangle$ for these.

ANGULAR FUNCTIONS AND ORBITAL SHAPES

Before inspecting the angular probability functions, it is important to discuss the symmetries of the orbital functions themselves. We could pursue the inspection of the angular functions $\Theta(\theta)$ and $\Phi(\phi)$ individually or as the composite $\Theta(\theta) \cdot \Phi(\phi)$. *The composite functions belong to a class of functions widely known as the spherical harmonics,* symbolized Y_l^m, where l and m are quantum numbers. It will suit our purposes better to examine the spherical harmonics than the individual components. Furthermore, we could begin the inspection by retaining the angles θ and ϕ or by converting to the cartesian coordinates x, y, and z. As you will soon see, the cartesian forms of the spherical harmonics are more easily interpreted, especially since the cartesian coordinates are more familiar to us.

To effect the conversion to cartesian coordinates we make use of the relationships

$$x/r = \sin \theta \cos \phi$$
$$y/r = \sin \theta \sin \phi$$
$$z/r = \cos \theta$$
$$r = [x^2 + y^2 + z^2]^{1/2}$$

Table 1-4 summarizes the cartesian forms of the spherical harmonics of greatest

26 □ USEFUL ATOMIC CONCEPTS

TABLE 1-4
THE SCHRÖDINGER AND REAL SPHERICAL HARMONICS

	Symbol	Schrödinger	Symbol	Real
$Y_{0,0}$	s	$1/\sqrt{4\pi}$	s	$1/\sqrt{4\pi}$
$Y_{1,0}$	p_0	$\sqrt{\dfrac{3}{4\pi}}[z/r]$	p_z	$\sqrt{\dfrac{3}{4\pi}}[z/r]$
$Y_{1,\pm 1}$	$p_{\pm 1}$	$\sqrt{\dfrac{3}{8\pi}}[(x \pm iy)/r]$	p_x	$\sqrt{\dfrac{3}{4\pi}}[x/r]$
			p_y	$\sqrt{\dfrac{3}{4\pi}}[y/r]$
$Y_{2,0}$	d_0	$\sqrt{\dfrac{5}{16\pi}}[(2z^2 - (x^2 + y^2))/r^2]$	d_{z^2}	$\sqrt{\dfrac{5}{16\pi}}[(2z^2 - (x^2 + y^2))/r^2]$
$Y_{2,\pm 1}$	$d_{\pm 1}$	$\sqrt{\dfrac{30}{16\pi}}[(xz \pm iyz)/r^2]$	d_{xz}	$\sqrt{\dfrac{60}{16\pi}}[xz/r^2]$
			d_{yz}	$\sqrt{\dfrac{60}{16\pi}}[yz/r^2]$
$Y_{2,\pm 2}$	$d_{\pm 2}$	$\dfrac{1}{2}\sqrt{\dfrac{30}{16\pi}}[(x \pm iy)^2/r^2]$	$d_{x^2-y^2}$	$\sqrt{\dfrac{15}{16\pi}}[(x^2 - y^2)/r^2]$
			d_{xy}	$\sqrt{\dfrac{60}{16\pi}}[xy/r^2]$

interest to us.[11,12] The striking feature of these functions is that when $m \neq 0$, the spherical harmonics are complex. This, of course, creates problems for us when we attempt to present the angular functions graphically. To skirt this difficulty, the theoretician makes use of an important mathematical property of the Y_l^m functions; namely, that any linear combination of members of a Y_l^m set of common l is also an acceptable function. *Thus, it cannot make any difference whether we use Schrödinger's original Y_l^m or linear combinations of them in calculations of atomic properties.* For the purpose of developing an intuitive grasp of these functions, it would make quite a difference if we could find linear combinations of the complex functions that produce (guaranteed equivalent) real functions. The linear combinations required are simply[13]

---

[11] As an aside, you may have noticed that in converting trigonometric functions of (θ,ϕ) to their cartesian forms we appear to have increased the number of coordinates by one. In fact, this means that two cartesian coordinates are interdependent (required to define one of the angles). Clearly this pair is x and y, since $x/y = \tan \phi$. A further interesting point is to be made concerning the values of the x, y, and z coordinates. Let us evaluate one of the spherical harmonics for a particular choice of x, y, and z. All other coordinates that stand in the same ratio to each other (mx, my, and mz) yield the same value of the harmonic. [For any of the harmonics in Table 1-4, make a substitution $x \to mx$, $y \to my$, $z \to mz$ and satisfy yourself that $Y(x,y,z) = Y(mx,my,mz)$.] In fact, the choice of a particular set of x, y, and z serves simply to define a direction in space (just as the two coordinates θ and ϕ do) and the values of x, y, and z are simply direction cosines for a unit vector in that direction.

[12] Complete orbital functions are given in cartesian form in Appendix A.

[13] Any two normalized, orthogonal functions f_1 and f_2

$$\int f_1^* f_1 = \int f_2^* f_2 = 1 \quad \text{normalized}$$
$$\int f_1^* f_2 = 0 \quad \text{orthogonal}$$

may be combined in the following manner, to maintain the orthogonality:

$$f_1' = af_1 + bf_2$$
$$f_2' = bf_2 - af_1$$

Normality requires $a^2 + b^2 = 1$.

$$Y_l^{\text{real}} = \frac{1}{\sqrt{2}} [Y_l^m + Y_l^{-m}]$$

$$Y_l^{\text{real}'} = \frac{i}{\sqrt{2}} [Y_l^{-m} - Y_l^m]$$

The final column of Table 1-4 presents the real equivalents of Schrödinger's angular functions. *An important point, not to be overlooked, is that there is no one-to-one correspondence between one of the Y_l^m and any of the Y_l^{real} counterparts, except when $m = 0$.* That is, p_1 cannot be claimed to be equivalent to either p_x or p_y alone. *Only the pair (p_1, p_{-1}) is equivalent to the pair (p_x, p_y).*

The cartesian forms of the spherical harmonics make very easy our visualization of the angular functions. *For each harmonic, those coordinates x, y, and z that cause the harmonic to have zero value define nodal surfaces.* For the p_x harmonic, the wave value is zero whenever the x coordinate is zero (regardless of the values of y and z). Consequently, the yz plane is a nodal plane; the function is positive in the $+x$ direction and negative in the $-x$ direction. The d_{xy} harmonic has two nodal surfaces (as do all the d harmonics); the xz and yz planes define surfaces over which d_{xy} has zero value. When x and y are both positive or both negative, the function has positive phase, and when x and y are of opposite sign, the harmonic lobes are negatively phased. Figure 1-9 summarizes these features for all the s, p, and d (real) harmonics. You should compare the real functions in Table 1-4 with their sketches in Figure 1-9 to verify the location of the nodal surfaces and the signs of the orbital lobes. The harmonic plots of Figure 1-9 are simply surfaces in which each point on the surface defines a direction in space and the distance of the surface from the origin is the magnitude of Y in that direction (these are polar plots).

To obtain the spherical harmonic probability functions for the Y's of Table 1-4 we need simply square the Y_l^{real}. For our purposes, there is little difference between the form of the real harmonic and that of its probability function. That is, the nodal surfaces do not change while the probability functions become positive everywhere (there are no negative lobes). The probability functions for the Schrödinger harmonics are real, however, in contrast to the functions themselves. *All the Schrödinger angular probability functions are rotationally symmetric about the z axis, and those of equal but oppositely signed m are indistinguishable* (Figure 1-10). This is of great importance to the physical interpretation of the m quantum number.

The relation between angular momentum about the z axis and the m quantum number can be simply demonstrated by consideration of the $\Phi_m(\phi)$ function, with the assistance of the de Broglie relation. The $\Phi_m(\phi)$ function has the form $e^{im\phi}$, which implies a standing wave about the z axis proportional to $[\cos(m\phi) \pm i \sin(m\phi)]$. Such a wave repeats itself in $2\pi/m$ radians, which means that the wave length λ is $2\pi/m$. From the de Broglie relationship

$$p_\phi = h/\lambda = m\hbar$$

Thus, the angular momentum about the z axis is given by m, in units of $\hbar$. A positive m value implies a clockwise motion for the electron, while a negative m value implies a counterclockwise motion. This has an obvious connection with the fact that Y_l^m and Y_l^{-m} have indistinguishable probability functions. For $m = 0$, the electron motion (remember that we really have no information as the details of the motion) is such that no net angular momentum is developed about z.

On the other hand, the real Y functions arise from *equal* mixtures of Y_l^m and Y_l^{-m}. The result of such equal clockwise and counterclockwise motions is net zero angular momentum about z for each such real function. This appears in Figure 1-9 in the form of probability functions that are *not* rotationally uniform about the z axis.

You will recall the statement in the discussion of the Bohr atom concerning the error of that model in predicting an orbital magnetism for the electron in the ground

28 □ USEFUL ATOMIC CONCEPTS

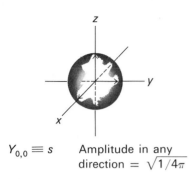

$Y_{0,0} \equiv s$ Amplitude in any direction $= \sqrt{1/4\pi}$

Amplitude along $z = \pm\sqrt{3/4\pi}$
$Y_{1,0} \equiv p_z$ along z axis
$Y_{1,x} \equiv p_x$ along x axis
$Y_{1,y} \equiv p_y$ along y axis

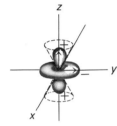

$Y_{2,0} \equiv d_{z^2}$

Amplitude along $z \equiv \sqrt{\dfrac{5}{4\pi}}$

Amplitude in xy plane $\equiv -\sqrt{\dfrac{5}{16\pi}}$

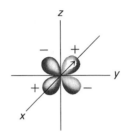

Amplitude along $(z = \pm y) = \pm\sqrt{\dfrac{15}{16\pi}}$
$Y_{2,yz} \equiv d_{yz}$
$Y_{2,xz} \equiv d_{xz}$
$Y_{2,xy} \equiv d_{xy}$

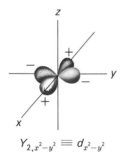

$Y_{2,x^2-y^2} \equiv d_{x^2-y^2}$

Amplitude along x and y axes $= \pm\sqrt{\dfrac{15}{16\pi}}$

Figure 1-9. Sketches of the Y_l^{real} in Table 1-4.

state hydrogen atom (Table 1-1). Experimentally, the orbital magnetism is null, and the wave model is in perfect agreement with this finding. In fact, all states of the hydrogen atom that place the electron in an s orbital ($l = 0$) have zero orbital magnetism. For states that place the electron in p or d orbitals ($l > 0$), the electron exhibits orbital magnetism, and again the wave model agrees with the experimental results perfectly. Later we will discuss atomic magnetism and will return to examine the nature of the orbital angular momentum in a little more detail.

A frequent question is whether the quantum number n has a physical meaning (the orbital magnetism relation to l and m was mentioned above). The answer is yes, in that the number of nodal surfaces for each orbital function is determined by n and l.

$n - \ell - 1$: derives from the radial function and equals the number of spherical nodal surfaces (each *radial* nodal point distant from the nucleus becomes a spherical surface in three dimensions).

ℓ: derives from the angular functions and equals the number of nodal surfaces through the nucleus, for real orbitals.

$n - 1$ = the total number of nodal surfaces.

A little reflection also reveals that the more nodal surfaces there are in the electron's wave, the higher is the energy of the electron. In fact, the energy order $1s < 2s = 2p < 3s = 3p = 3d\ldots$ correlates exactly with increasing number of nodes: 0, 1, 2, ... A more explicit physical consequence of n derives from the energy expression

$$E = -\frac{1}{n^2}(e^2/2a_0)Z^2$$

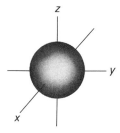

$Y_{0,0}^2$
Probability in any direction = $1/4\pi$

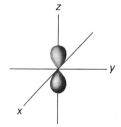

$Y_{1,0}^2$
Probability along $z = 3/4\pi$

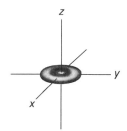

$Y_{1,1}^* Y_{1,1} = Y_{1,-1}^* Y_{1,-1}$
Probability along any direction
in xy plane = $3/8\pi$

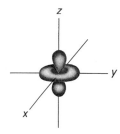

$Y_{2,0}^2$
Probability along $z = 5/4\pi$
Probability along any direction
in xy plane = $5/16\pi$

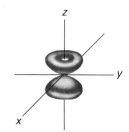

$Y_{2,1}^* Y_{2,1} = Y_{2,-1}^* Y_{2,-1}$
Probability along any line
45° to $z = 15/32\pi$

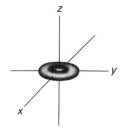

$Y_{2,2}^* Y_{2,2} = Y_{2,-2}^* Y_{2,-2}$
Probability along any line
in xy plane = $15/32\pi$

Figure 1-10. Sketches in the Y_ℓ^m probability functions.

30 □ USEFUL ATOMIC CONCEPTS

STUDY QUESTIONS

2. Show that $[Y_1^{1*}Y_1^1 + Y_1^{-1*}Y_1^{-1}] = (Y_1^x)^2 + (Y_1^y)^2$.

3. Which has the greater probability:

   a) along the x axis, an s orbital or a p_y orbital?
   b) along the z axis, a p_0 orbital or a p_1 orbital?
   c) along the line $x = z$, a p_z orbital or a d_{yz} orbital?

TOTAL DENSITY FUNCTIONS

As a conclusion to this section on probability functions, we wish to put the radial and angular probability functions together to arrive at a *total* density distribution function for the electron. The graphical presentation of the total density function presents a special problem; one must give the density as a function of three space variables, and such a situation requires a four-dimensional graph. Many techniques have been used, but the most precise technique—and the one that most clearly emphasizes the orbital nodal characteristics—is the construction of contour maps. For any arbitrary plane through the atom, one gets the corresponding cross section of the three-dimensional density function. *All points in this plane for which the electron density is constant are connected by a smooth curve or curves.* These curves are called contour lines, and the entire figure is referred to as a contour map. Such maps are illustrated in Figure 1–11 for $2p_z$, $3p_z$ and some $3d$ orbitals. The point near the center of each closed set of contour curves locates the point of greatest probability. A clear

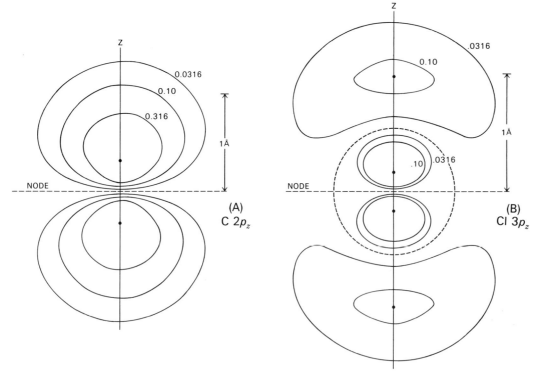

Figure 1–11. Contour maps for various a.o.'s For (a) through (d) the cross-sectional plane could be any plane containing the z axis. For (e) the cross-sectional plane is the xy plane. (From E. A. Ogryzlo and G. B. Porter, *J. Chem. Educ.*, **40**, 258 (1963).

Illustration continued on opposite page.

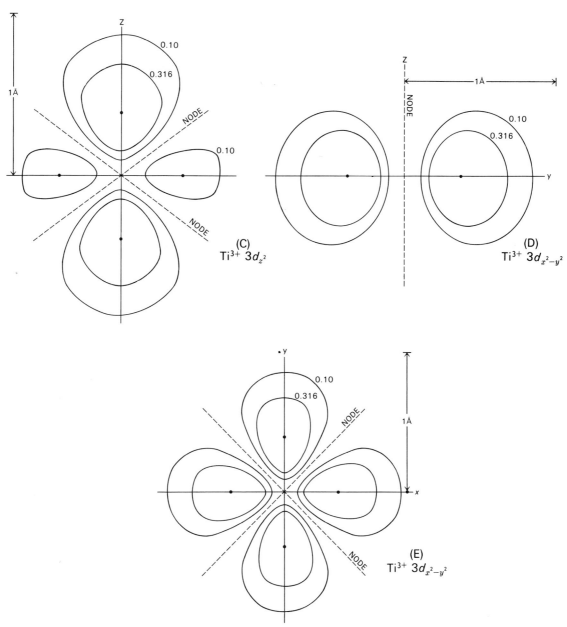

Figure 1–11. *Continued.*

advantage of such contour cross sections is that the location of radial nodal surfaces is very simple. The dashed circle in the map for the $3p_z$ function is the nodal line that arises from the node in the $3p$ radial function. *In three dimensions, all radial nodal surfaces are spheres.* The $3p_z$ function also has an angular nodal plane, the xy plane.[14]

---

[14] It is our experience that students do best by first visualizing the nodal surfaces in three dimensions. Then introduce the intersecting contour plane and deduce the forms of the lines of intersection. These can be sketched as dashed line curves. For those desiring a more mathematical approach (such as would be used in writing a computer program to construct the intersection lines), we include the following. The contour maps are derived from the square of the product of the radial functions of Table 1-2 and the spherical harmonics of Table 1-4. The key to construction of contour maps is the location of the radial and harmonic nodal surfaces. For the radial part of the orbital function, nodes occur for those values of r that render the function zero-valued. The number of radial nodes is equal to $n - l - 1$ (Table 1-2), and each is defined mathematically as

$$r = [x^2 + y^2 + z^2]^{1/2} = \text{a constant (different for each node)}$$

This equation is that of a sphere, so the radial nodes are spheres in three dimensions. In the contour plane, radial nodes appear as circles. (For the xy plane, the equation is $[x^2 + y^2]^{1/2} = $ constant.)

The same technique applied to the spherical harmonics also leads to a cartesian equation defining the nodal surface. As we have seen, these are planes or conical surfaces such as that appropriate to the d_{z^2} harmonic,

$$2z^2 = x^2 + y^2$$

The choice of a contour plane always reduces such an equation to a two-dimensional expression. For example, choice of the xy plane means $z = 0$. For the d_{z^2} orbital, this means that the nodal line is given by the equation

$$x^2 = -y^2$$

which is possible only for the point $x = y = 0$. Choice of the xz plane yields ($y = 0$)

$$z^2 = \frac{1}{2}x^2$$

A plot of this equation gives two lines (for $\pm x$) making a tetrahedral angle to each other, bisected by the z axis.

STUDY QUESTIONS

4. Calculate the probability density for a H $2p_0$ electron at the point $r = a_0$, $\theta = 60°$, $\phi = 45°$. What are the units of the probability? Does the probability change with the angle ϕ?

5. Sketch a contour map for a $4p_1$ orbital, indicate all nodal lines as dashed lines, and identify their origin. Be sure to identify the cross-sectional plane in your figure.

6. Sketch a contour diagram for $4d_{x^2-y^2}$ in the xy plane. Show nodal lines as dashed lines.

7. Carefully sketch the xy plane contour maps for the $5d_0$ and $5d_{xy}$ orbitals. Indicate nodal lines as dashed lines and identify their source (radial or angular).

8. Consider the following diagram of an atom in the electric field of a "spy satellite" electron located on the positive z axis. Because of e^-, e^- repulsions, the d orbital energies will not be equal when the "spy" is present. Sketch the $3d$ orbitals in relation to this "spy" and indicate the order of their energies.

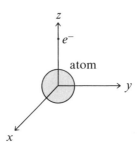

POLYELECTRONIC ATOMS
ATOM ELECTRON CONFIGURATIONS AND THE LONG FORM PERIODIC TABLE

A most unfortunate situation exists when it comes to describing atoms with more than one electron by Schrödinger's wave theory. While the second-order differential equation can be written down, a closed form, or analytical, solution to the equation is not possible as was the case for the hydrogen atom. This is not a difficulty inherent in the wave formulation but is characteristic of any system of more than two interacting particles. The problem is one of separation of variables, and it has not been solved in classical mechanics either. Nevertheless, just as with systems that behave classically and that can be handled by approximate methods to any desired accuracy, approximate methods have been developed for wave mechanics that will allow the theoretician to achieve an answer to any desired level of accuracy—providing he is willing to spend the time and effort required.

For most of our purposes in inorganic chemistry, we find little reward in going to great lengths to get a quantitatively correct answer, for often all that is needed is a basic understanding of the qualitative aspects of the electronic structures of many-electron atoms. In this section we will not pursue the wave equation in as much detail as previously, but we will use our understanding of the hydrogen wave functions to develop an appreciation for the electronic structures of the elements and their energy states.

A very simplistic approach to the energies of electrons in polyelectronic atoms is to view each electron as moving in an effective nuclear potential of $+1$ charge unit. *This idea is a very approximate way of dealing with electron-electron repulsions that arise when more than one electron is present in an atom.* If such a model were to be

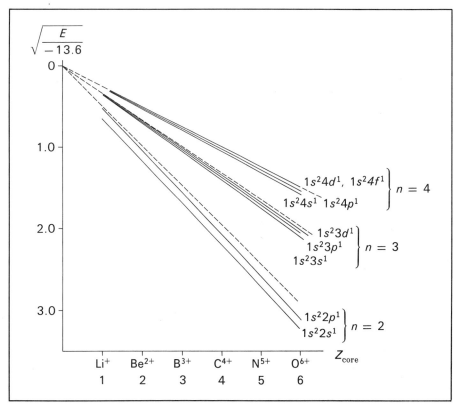

Figure 1–12. Energies of one electron beyond a $1s^2$ core versus the core charge.

even qualitatively correct, the absorption and emission spectra of the elements would be little different from that of the hydrogen atom. You can readily appreciate that nature presents us with a greater challenge than that.

Figure 1-12 is a plot of the energies of the outer electrons of three-electron ions (two electrons in the $1s$ orbital) as a function of the atomic number of the ion less two. The dashed lines indicate the variation in the ion energy according to the assumption that the two $1s$ electrons completely neutralize or *screen* two units of positive nuclear charge from the remaining electron. According to the equation derived for the hydrogen atom,

$$E = -13.6 \frac{(Z^*)^2}{n^2}$$

or

$$\sqrt{\frac{E}{-13.6}} = \frac{Z^*}{n}$$

with $Z^* = Z - 2$. The interesting feature is that the solid lines drawn for the experimental ion energies are displaced from the "theoretical" lines by an amount that depends on the principal quantum number of the unpaired electron, the deviation being largest for the $1s^2 2s^1$ configuration and decreasing for higher ion energy; the $1s^2 nd^1$ and $1s^2 nf^1$ configurations essentially agree exactly with the perfect screening concept.

As evidenced by displacement of energy of the $1s^2 2s^1$ configuration to a lower or more stable energy, *the $2s$ electron experiences a nuclear charge that is greater than $Z - 2$*. In other words, the two $1s$ electrons do an incomplete job of screening nuclear charge from the $2s$ electron. A similar comment can be made for the $1s^2 2p^1$ configuration, but there the $1s$ electrons obviously do a better job of screening. For the $1s^2 3s^1$ configuration the deviation from the $n = 3$ line is greater than that for $1s^2 3p^1$, which in turn is greater than that for $1s^2 3d^1$, which nearly coincides with the ideal line. Therefore, the screening property of the $1s$ pair is least for $3s$ and greatest for $3d$. A similar order is found for the $n = 4$ configurations.

The variation in ability of inner electrons to screen nuclear charge from the outer electrons lies at the heart of an understanding of the electron configurations and their relation to the periodic table. In building up the periodic table by starting with the hydrogen atom and successively adding a proton to the nucleus and an electron to the extranuclear shell (this process is called the *Aufbau principle*), we must allow for the screening effects evidenced above. In other words, we must account for the fact that orbitals of the same n quantum number but different ℓ quantum number may have different energies in a given atom. The only *ad hoc* assumption we need introduce at this point is that of the Pauli exclusion principle, which permits only two electrons to occupy the same spatial orbital or to have the same spatial wave function. Taking a clue from the Moseley diagram (Figure 1-12), we may deduce the following *empirical rule concerning the order of filling of the hydrogen atom orbitals:*

(a) *Orbital energies increase as $(n + \ell)$ increases.*
(b) *For two orbitals of the same $(n + \ell)$, the one with the smaller n lies lowest in energy.*

Thus, the order of filling becomes

$$1s, 2s, \underbrace{2p, 3s}, \underbrace{3p, 4s}, \underbrace{3d, 4p, 5s}, \ldots$$
$$n + \ell = 1 2 3 4 5$$

The radial wave functions, which depend on n and ℓ, may be used to formulate a basis for this rule. In what follows, your attention will be directed to the penetration of inner electron density by outer radial functions, thereby defining an *effective nuclear charge* for the outer electron. A qualitative estimate of the relative energies of outer electrons can be obtained from estimates of their potential energies as given (classically) by

$$V = -e^2 \frac{Z}{r}$$

According to the wave formulation of the problem, we need to consider the average potential energy given by

$$\langle V \rangle = -e^2 \left\langle \frac{Z}{r} \right\rangle = -e^2 \int_0^\infty \frac{Z}{r} \cdot S(r)\, dr \int_0^\pi \Theta(\theta)^2 \sin\theta\, d\theta \int_0^{2\pi} \Phi(\phi)^* \Phi(\phi)\, d\phi$$

$$= -e^2 \int_0^\infty \frac{Z}{r} \cdot S(r)\, dr$$

A graphical evaluation of the potential energy then requires a determination of the area under a plot of $Z/r \cdot S(r)$ as a function of r. To compare the potential energies of an electron in one outer orbital to those of another, we need to compare the areas under the appropriate potential energy curves. Note that each point on such a curve gives the potential energy of the electron for the given distance from the nucleus.

According to the wave model expression for the point-by-point potential energy of an electron, we need to evaluate the potential energy of the electron at each point $(= -e^2 Z/r)$ and weight this by the probability that the electron will be at that point. In a polyelectronic atom, the nuclear positive charge experienced by an electron will be Z when $r = 0$ and will decrease toward unity as r approaches infinity. The decrease in Z with increasing r is a consequence of the shielding of nuclear charge by the other electrons closer to the nucleus.[15] For example, consider the Li$^+$ ion with configuration $1s^2$; calculate the effective nuclear charge (Z^*) as a function of distance from the nucleus by integrating the probability function for the two $1s$ electrons from zero to r, and subtract this from the atomic number Z. Repeating this for every r value from $r = 0$ on, you get the effective nuclear charge Z^* as a function of r. Figure 1-13(a) shows the variation in Z^* as a function of r for such a situation. Also shown in this figure is Z^*/r as a function of r. Z^*/r is infinitely large at $r = 0$ and fairly rapidly decreases to a very small value. Figure 1-13(b) shows the $2s$ and $2p$ surface probability functions in relation to the probability function for the $1s$ pair of electrons, from which it is readily apparent that both $2s$ and $2p$ penetrate the $1s$ pair charge distribution, the $2p$ to a seemingly greater extent than the $2s$. (Refer to page 22.) *However, because of the form of the function Z^*/r, the difference in $2s$ and $2p$ radial functions is greatly magnified at small r while the difference is greatly diminished at large r, so that the function $Z^*/r \cdot S(r)$ takes on the appearance of the curves for $2s$ and $2p$ in Figure 1-13(c). A $2s$ electron penetrates best where (0 to 1.5 a_0) it counts most and realizes a significantly lower energy than a $2p$ electron. It therefore is no surprise that, for Li, the lowest energy configuration for three electrons is $1s^2 2s^1$ rather than $1s^2 2p^1$. In this example we find a physical basis for the empirical rule that the lower the $(n + \ell)$ value for an orbital, the more stable is that orbital.*

The second part of the rule deals with the eventuality that $(n + \ell)$ is the same for two orbitals. Such a situation is first encountered with $2p$ and $3s$. That is, will the proper configuration for boron be $1s^2 2s^2 2p^1$ or $1s^2 2s^2 3s^1$? Figure 1-14(a,b,c) is a series of plots like those of Figure 1-13 and provides a physical basis for the answer that the lowest energy configuration is $1s^2 2s^2 2p^1$.

---

[15] In adopting this view, we are implicitly ignoring repulsion from electron density further out from the nucleus than (that is, "behind") the electron whose energy we are calculating.

36 ☐ USEFUL ATOMIC CONCEPTS

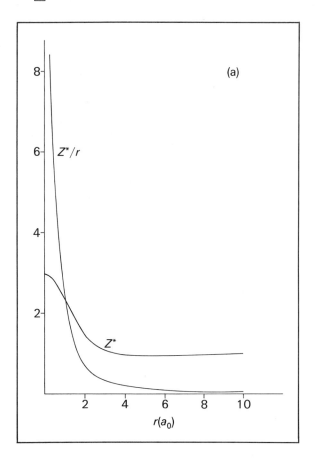

Figure 1-13. (a) Z^* and Z^*/r versus r for the third electron of Li.
Illustration continued on opposite page.

The penetration of the $1s^2 2s^2$ core by a $2p$ electron is far more extensive than that of the $3s$ electron. Therefore [Figure 1-14(c)], the $2p$ orbital provides a lower energy motion for the boron electron than does the $3s$ orbital.

Up to this point, we have pretty much glossed over a very important property of an electron. That is, the electron possesses a spin angular momentum; (relativistic quantum) theory, backed by experiment, tells us that this angular momentum is quantized, with the quantum number m_s taking only two possible values, $+1/2$ or $-1/2$. A more sophisticated statement of the Pauli exclusion principle is that no two electrons in an atom may have all four quantum numbers the same. Two electrons can have the three space quantum numbers n, ℓ, and m the same, but they must differ in the value of the fourth or spin quantum number. This leads directly, as we apply the Aufbau principle, to the statement that only two electrons are permitted in the same orbital. The physical phenomenon on which the Pauli principle is based is that *two particles with the same spin do a better job of avoiding each other in space than do two electrons of opposite spin quantum number.* In fact, this phenomenon is a spin-quantum effect, has no classical counterpart, and is observed for any particles that possess spin angular momentum. Even neutrons, which are uncharged and so do not experience a Coulomb or electrostatic repulsion, still avoid each other better in their motions if they have the same spin (both $+1/2$ or both $-1/2$) than if they have opposite spins.

The chemical significance of this *spin correlation of electron motions* is that, since electrons are charged and repel one another, *the repulsion is less, by an amount called the "exchange energy," for electrons of like spin than for electrons of different spin.* We have an opportunity to see this effect in operation on passing from boron to carbon, which has two electrons in the $2p$ orbitals. *Charge correlation* (e^2/r) *dictates that a*

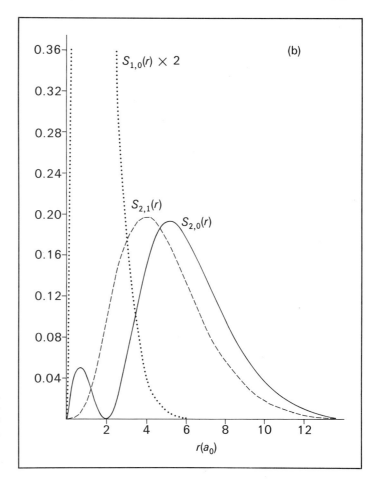
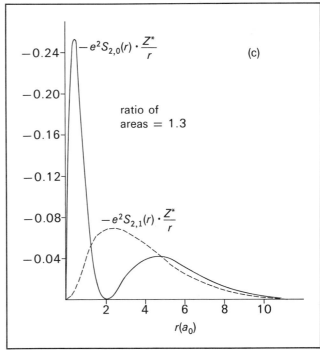

Figure 1-13. *Continued.*
(b) Surface probability plots for 2s and 2p functions in relation to the same plot for two 1s electrons. (c) The 2s and 2p potential energy curves for $Z = 3$ with a pair of 1s electrons present.

38 □ USEFUL ATOMIC CONCEPTS

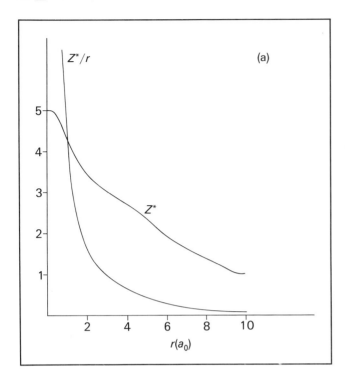

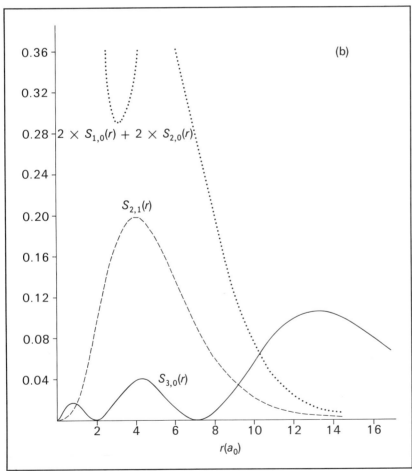

Figure 1–14. (a) Z^* and Z^*/r versus r for the third valence electron of boron. (b) Surface probability plots for $2p$ and $3s$ functions in relation to the same plot for two $1s$ and $2s$ electrons.

Illustration continued on following page.

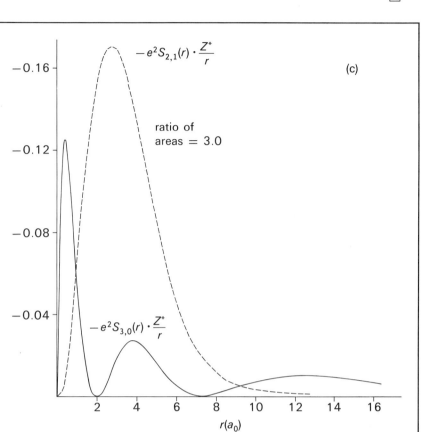

Figure 1-14. *Continued.*
(c) The 2p and 3s potential energy curves for $Z = 5$ with 1s and 2s electron pairs present.

lower energy situation exists if the two p electrons have different angular functions. Spin correlation further affords a lower (more negative) energy for like spins.[16] In the conventional box diagram notation, the lowest energy configuration of the carbon atom will be $1s^2 2s^2 2p^2$, with the requirement that the two p electrons reside in different orbitals with the same m_s value.

$$\boxed{\uparrow\downarrow}\quad\boxed{\uparrow\downarrow}\quad\boxed{\uparrow\,|\,\uparrow\,|\,}$$
$\quad$ 1s $\qquad$ 2s $\qquad\quad$ 2p

Similarly, in the lowest energy configuration for nitrogen, all three p electrons have the same spin; therefore, the configuration is one of filled 1s and 2s orbitals and a half-filled 2p shell.

$$\boxed{\uparrow\downarrow}\quad\boxed{\uparrow\downarrow}\quad\boxed{\uparrow\,|\,\uparrow\,|\,\uparrow}$$
$\quad$ 1s $\qquad$ 2s $\qquad\quad$ 2p

---

[16] While it is true that "like spins" afford lower energy than "opposite spins," the origin for this energy difference is being closely examined. R. L. Snow and J. L. Bills, *J. Chem. Educ.*, **51**, 585 (1974) have argued that, in some atoms, the "like spin" case may have a lower energy because of a lower *kinetic energy*, not lower electron repulsion energy.

Spin correlation and its "exchange energy" are the physical basis for the often cited stability of the half-filled shell. It is simply a matter of choosing the electron arrangement that leads to the lowest electron-electron repulsions. Obviously, placing two electrons in the same orbital will lead to higher repulsions than if they move in different spatial orbitals (the electrons have the same charge). In addition to the lower energy made possible by allowing separate orbital motions for the two charged particles, the spin correlation allows for a further reduction in repulsion if both have the same spin. To summarize, we would rank the stability of p^2 electron arrangements for carbon as follows:

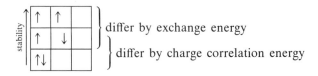

Very often, then, the electron configuration and energy of an atom are dictated by fine details of electron repulsions that are not apparent in the averaged repulsion (screening) concept of orbital energies. A good illustration can be found in the trends in ionization potential and electron affinity[17] (Appendices B and C) for the main group elements from groups III through VIII. (*Electron affinity is defined as the energy for the process* atom$^-$ $\to$ atom $+$ e^-.) The valence configurations of these elements are ns^2np^x, where x progresses from 1 through 6. The trend of increasing I and E across this series is broken at the configuration s^2p^4. That is, the configurations with $x = 1, 2, 3$ follow a smoothly increasing trend in I, as do the configurations with $x = 4, 5, 6$. The latter series is simply shifted to lower I from the values that would be extrapolated from $x = 1, 2, 3$. The origin of the dislocation at s^2p^4 is easily traced to a discontinuity of electron repulsions, as the nuclear charge smoothly increases by 1 across the entire series. The following box diagrams should clarify the anomaly.

$$B^+ \to B : \boxed{\uparrow\downarrow}\ \boxed{\uparrow\downarrow}\ \boxed{\uparrow\ \ }\ \boxed{\ \ }\ \boxed{\ \ }$$

$$C^+ \to C : \boxed{\uparrow\downarrow}\ \boxed{\uparrow\downarrow}\ \boxed{\uparrow\ \ }\ \boxed{\uparrow\ \ }\ \boxed{\ \ }$$

$$N^+ \to N : \boxed{\uparrow\downarrow}\ \boxed{\uparrow\downarrow}\ \boxed{\uparrow\ \ }\ \boxed{\uparrow\ \ }\ \boxed{\uparrow\ \ }$$

$$O^+ \to O : \boxed{\uparrow\downarrow}\ \boxed{\uparrow\downarrow}\ \boxed{\uparrow\downarrow}\ \boxed{\uparrow\ \ }\ \boxed{\uparrow\ \ }$$

$$F^+ \to F : \boxed{\uparrow\downarrow}\ \boxed{\uparrow\downarrow}\ \boxed{\uparrow\downarrow}\ \boxed{\uparrow\downarrow}\ \boxed{\uparrow\ \ }$$

$$Ne^+ \to Ne : \boxed{\uparrow\downarrow}\ \boxed{\uparrow\downarrow}\ \boxed{\uparrow\downarrow}\ \boxed{\uparrow\downarrow}\ \boxed{\uparrow\downarrow}$$

The trend in I from $x = 1, 2, 3$ is determined by a balance between regularly increasing nuclear charge and regularly increasing electron repulsion for the added electron; in combination, these factors establish a trend of smoothly increasing effective nuclear charge and I. In each case, the added electron has the same spin as those already present in the p sub-shell. On adding a p electron to the s^2p^3 configuration to produce the s^2p^4 configuration, the effective nuclear charge for the added electron will appear to drop because this electron must be added with spin opposite to those already present and must be placed in an orbital already occupied by an opposite-spin electron. This electron does not benefit from the "exchange energy" stabilization possible for the preceding three members of the series, and further suffers from larger coulombic repulsion. This configuration establishes a new "origin" for a smooth trend, shifted from the original, in effective nuclear charge and I for the remaining series members.

---

[17] E. C. M. Chen and W. E. Wentworth, *J. Chem. Educ.*, **52**, 486 (1975).

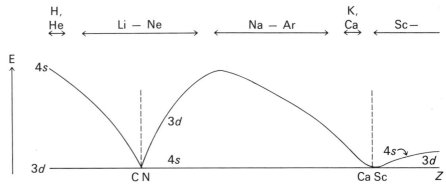

Figure 1-15. Relative 3*d* and 4*s* orbital energies as a function of Z. Reversals occur at the dotted lines.

Another illustration of the breakdown of the screening concept arises for the transition elements.[18] The orbital energies estimated from shielding or penetration effects adequately account for the fact that the 4*s* orbital is filled prior to the 3*d* on passing from Ar to K to Ca. However, proceeding across the 3*d* series of metals (Sc through Zn), there are so-called anomalies at Cr and Cu, which have [Ar] $4s^1 3d^5$ and [Ar] $4s^1 3d^{10}$ configurations. As Figure 1-15 shows, the relative 3*d* and 4*s* orbital energies are nearly the same for the first transition elements, and under such circumstances the orbital occupancies are as much influenced by the finer points of electron repulsions as by the "orbital energies." Reference to Figure 1-16 shows just how complex the situation becomes; of the ten transition metal groups, five show irregularities in electron configuration. Thus, while "the stability of half-filled shells" seems a good phrase to invoke for Cr, the "principle" fails for W. It is certainly better to realize that the electron configurations of these elements depend on fine points of electron interactions (the balance between d/d, d/s, and s/s repulsions) and accept them for what they are than to propose a rule that must be violated 50% of the time.

In the preceding paragraph it was mentioned that the shielding concept was capable of accounting for the slightly more stable 4*s* than 3*d* orbitals at potassium. Actually, the energy of the 4*s* orbital falls below that of the 3*d* long before the nuclear charge increases to 19. Observation of the high energy configurations of the nitrogen atom shows that even at atomic number 7 the configuration $1s^2 2s^2 2p^2 4s^1$ lies at lower energy than the configuration $1s^2 2s^2 2p^2 3d^1$. Clearly, in the hydrogen atom the 4*s* orbital energy is greater than that of the 3*d*; yet, by adding six protons to the nucleus and six electrons to the 1*s*, 2*s*, and 2*p* orbitals, the 4*s* moves to lower energy than the 3*d*. This is easily traced to the effect on potential energy of the relative 4*s* and 3*d* penetrations of the $1s^2 2s^2 2p^2$ core of electrons. This penetration is illustrated in Figure 1-17(a). The stabilization of the 4*s* relative to the 3*d* continues throughout the second period. Across the third period, where the $n = 3$ shell begins to fill, the penetration of

---

[18] An excellent discussion of the transition metal electron configurations is presented by R. M. Hochstrasser, *J. Chem. Educ.*, **42**, 154 (1965).

Sc	Ti	V	Cr	Mn	Fe	Co	Ni	Cu	Zn
$d^1 s^2$	$d^2 s^2$	$d^3 s^2$	$d^5 s^1$	$d^5 s^2$	$d^6 s^2$	$d^7 s^2$	$d^8 s^2$	$d^{10} s^1$	$d^{10} s^2$
Y	Zr	Nb	Mo	Tc	Ru	Rh	Pd	Ag	Cd
$d^1 s^2$	$d^2 s^2$	$d^4 s^1$	$d^5 s^1$	$d^5 s^2$	$d^7 s^1$	$d^8 s^1$	d^{10}	$d^{10} s^1$	$d^{10} s^2$
La	Hf	Ta	W	Re	Os	Ir	Pt	Au	Hg
$d^1 s^2$	$d^2 s^2$	$d^3 s^2$	$d^4 s^2$	$d^5 s^2$	$d^6 s^2$	$d^7 s^2$	$d^9 s^1$	$d^{10} s^1$	$d^{10} s^2$

Figure 1-16. The *nd* and $(n+1)s$ electron configurations for transition elements.

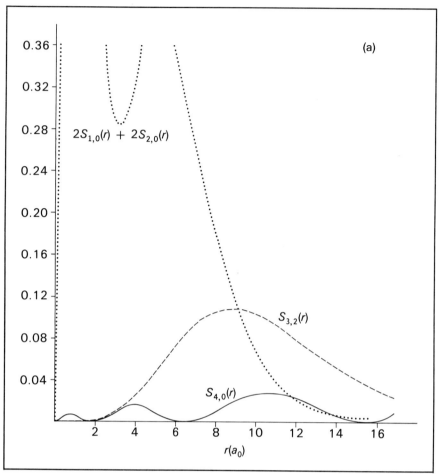

Figure 1-17. (a) The 3d and 4s surface probabilities in relation to the $1s^22s^2$ core.
Illustration continued on opposite page.

the $3s^x3p^y$ configurations of these elements by the 3d function is much greater than by the 4s [Figure 1–17(b)], with the result that the 3d begins to drop sharply in energy, approaching the energy of the 4s at potassium. Quite obviously, the 3d thoroughly penetrates the 4s function, so as electrons are added to form potassium and calcium, the 3d continues its sharp drop to become as stable as the 4s, or slightly more so at Sc.

While the electron configurations of the transition elements in the zero oxidation states are difficult to interpret, the configurations of the divalent and trivalent ions are much simpler. *All di- and trivalent ions of the transition elements have d^n configurations. For the ions,* the energies of the d orbitals are considerably lower than the energies of the next higher s orbital. There is a simple interpretation possible here. After ionization of two or three electrons, we reexamine the relative stabilities of the d and s orbitals in the ion. The ionization process that forms a 2+ cation produces an ion that could also have arisen by the addition of two protons to the nucleus of an atom of the preceding element with the same number of d and s electrons. *Regardless of how the cation is formed, it has but a single most stable electron configuration.* The effective nuclear charge experienced by the 3d electrons is greatly enhanced over that of any 4s electrons as a direct consequence of the greater penetration of the $3s^23p^6$ core by a 3d electron than by a 4s electron. Consequently, the 3d orbitals are expected to drop significantly below the 4s (the orbital energy concept is reinstated), and any 4s electrons will move to the 3d orbitals. Thus, ionization of two or more electrons from

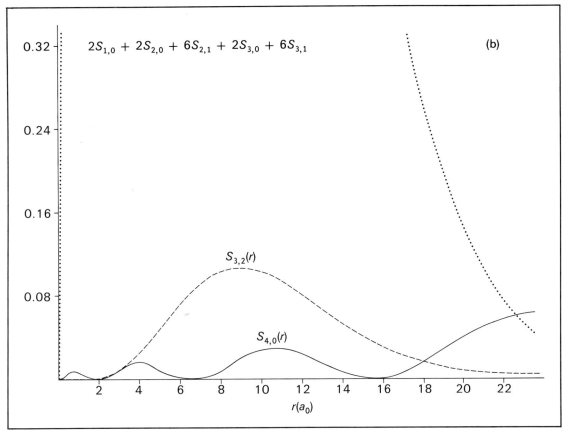

Figure 1–17. *Continued.*
(b) In relation to the $1s^22s^22p^63s^23p^6$ core.

an atom of a transition element will make it appear as though the s electrons have been preferentially removed.

Summarizing the relation between the positions of the elements in the periodic table and their electron configurations, we can make the following generalization. As far as the main-group elements are concerned, all elements in a group have the same valence orbital configuration. This is not always true for the transition elements. Ionization of the representative elements results in ions that appear to have simply lost the least stable electrons in the valence shell. For the transition ions, the electron configurations are those to be expected because the *d* orbitals are clearly more stable than the *s* orbitals of one greater quantum number, *n*. Finally, you should be aware that elements, like potassium and copper, that have the same number of unfilled valence orbitals of the same type would not necessarily be expected to show strong similarities in their chemistry. Using the potassium-copper pair as an example, both have $4s^1$ valence configurations, but this electron in potassium is outside a $1s^22s^22p^63s^23p^6$ core that the 4s electron does not penetrate as well as the 4s electron of copper penetrates the $1s^22s^22p^63s^23p^63d^{10}$ core of that atom. The 10 extra *d* electrons of copper do not completely screen the additional 10 units of nuclear charge. Clearly, the loss of an electron (the 4s) from potassium should require much less energy than from copper. Thus, the correlation between valence configuration and group in the periodic table is not perfect (hence the use of A and B groups in the long form of the periodic table), but at least it is understandable in terms of the hydrogen radial functions.

One more consequence of orbital penetration is orbital contraction. Understandably, the greater the orbital penetration of the core electrons, the greater will be the contraction of the orbital function—that is, the maximum in the probability curve will occur at a smaller distance from the nucleus, and the wave amplitude (electron probability) will increase at small r and decrease at large r.[19] For example, the radial functions for the hydrogen 2s and 2p orbitals will be altered in the lithium atom so that the penetration differences discussed above will be reinforced, and it is found that the 2s function is not as radially diffuse as the 2p. This phenomenon is of some importance in interpretations of atom hybridization, to be discussed in Chapter 3.

STUDY QUESTIONS

9. In the discussion of why the 2s orbital of Li$^+$ fills with an electron before the 2p orbitals, we encountered the idea that the 2s penetration of the 1s^2 cloud is greater than that of the 2p. Figures 1-13, 1-14, and 1-17, illustrating the consequences of this concept, were constructed from H atom orbitals. Would the nuclear charge of Li, being greater than that of H, cause a change in the Z^* versus r plot of Figure 1-13? Would this further enhance the difference in 2s and 2p probabilities at low r? Which orbital, 2s or 2p, would be more greatly contracted toward the nucleus? Can we still argue that the $1s^22s^1$ is a lower-energy electron configuration than $1s^22p^1$?

10. What are the *spdf* configurations for Tl$^+$, Ca, Sc, O$^+$, Cr, Ni, Cu$^+$, Mn^{2+}, and Fe^{3+}?

11. What is the valence shell configuration for Tl? Pt^{2+}? Group III main-group elements? Group V main-group elements?

12. Why is it more difficult to remove the valence electron from K than to remove the valence electron from Na after the latter has been excited to the 4s level?

13. Why has Zn a smaller atomic volume than Ca?

14. Which atom in each pair has the higher experimental I.P. (see Appendix) and why?

    S or P   Al or Mg
    K or Cu   Cs or At
    Ca or Rb   Mo or W

SLATER ORBITALS AND THEIR USES

While we have made only qualitative use of the concept of orbital penetration and electron screening, various attempts have been made to put these ideas on a more quantitative basis. In particular, we will focus on the method proposed by Slater, and the rules bearing his name, for calculating screening effects.[20] Slater developed a method for approximating the radial functions for real atoms by assuming that only the highest-order term of each Schrödinger radial function is important (cf. Table 1-2). His radial function, in general form, is (distance in a_0 units)

$$R(r) = N_r r^{n^*-1} e^{-(Z^*/n^*)r}$$

The constant N_r is a normalizing constant that is different for different orbitals (it depends on Z^* and n^*). These functions are quite a bit different from the hydrogen atom solutions, in that they do not depend explicitly on the ℓ quantum number (2s

---

[19] In fact, the contraction is sufficient to yield a total radial distribution for all electrons that decreases monotonically with r. That is, the *total* distribution function has no maxima or minima, as suggested by the "core" density plots from hydrogen a.o.'s in Figures 1-14(b) and 1-17(a). This is discussed by H. Weinstein, *et al.*, in *Theoret. Chim. Acta,* **38,** 159 (1975).

[20] J. C. Slater, *Phys. Rev.,* **36,** 57 (1930).

and 2p have the same radial function) and they all have only one node, which occurs at $r = 0$ (i.e., the nucleus). Because of this, Slater's approximate functions poorly mimic an electron's behavior in the inner regions of the atom. However, it is found that his functions are very similar to the correct functions at distances from the nucleus in the range of covalent bond distances. Thus, they have been found useful for calculations of bonding between atoms in molecules.

Slater's rules pertain to prescription of values of Z^* and n^* to be used for the various atomic orbitals. To develop these rules, Slater used his functions to calculate the energies of a large number of atoms and found for each atom the values of Z^* and n^* that gave the best agreement with experimental values. Then, by inspection, he determined a simplified set of rules that would yield Z^* and n^* values near these optimum values.

RULE 1   Orbitals are grouped in the following manner—(1s) (2s,2p) (3s,3p) (3d) (4s,4p), . . .

RULE 2   The effective atomic number for each orbital group is determined by the actual atomic number decreased by a certain amount (the screening constant, s_i) for each electron in the *same or preceding groups*.

$$Z^* = Z - \sum_i s_i$$

Electrons in following groups have zero screening ability according to this approximate scheme.

RULE 3   Each electron in a group screens 0.35 nuclear charge unit ($s_i = 0.35$) from all other electrons in the same group—except for a 1s electron screening another 1s, in which case the screening is reduced to 0.30.

RULE 4   When calculating Z^* for a d or f group electron, each electron in all preceding groups screens 1.00 charge unit. For (ns,p) group electrons, each electron in the group $n' = n - 1$ screens only 0.85 charge unit; all electrons in further preceding groups screen 1.00 unit.

RULE 5   Values of n^* are not the same as n for larger n:
$n = 1, 2, 3, \ 4, \ \ 5, \ \ 6$
$n^* = 1, 2, 3, \ 3.7, 4.0, 4.2$

Electron Group	s_i for Electrons in			
	All Higher Groups	Same Group	Groups with $n' = n - 1$	Groups with $n' < n - 1$
(1s)	0	0.30		
(ns,p)	0	0.35	0.85	1.00
(nd), (nf)	0	0.35	1.00	1.00

A useful consequence of Slater's "average repulsion" model is that for a series of elements with the same (ns,p) or (nd) group electrons, Z^* regularly increases by 0.65. The model suggests, then, that atomic properties should smoothly vary from one element to the next.

Slater's functions can be used to calculate the energies of atoms and ions and their radii with some success. The energy of the atom or ion is proportional to the sum of the $(Z^*/n^*)^2$ for each electron in the atom or ion (cf. the Bohr and Schrödinger results). The radii can be estimated from the maximum radial charge density for the electrons in the last group. This maximum occurs at (n^{*2}/Z^*) atomic units.

To illustrate the application of Slater's rules, we will calculate the ionization potentials of Li and F. The lithium atom has the configuration $1s^2 2s^1$, and the ion has $1s^2$. The first ionization potential is just the difference in energy between the $+1$ ion and the atom:

$$I = E_{\text{Li}}^+ - E_{\text{Li}}^0$$

To calculate the energies of the ion and atom, we sum the energies of their electrons as given by $-13.6(Z^*/n^*)^2$ (units of electron volts). Since the $2s$ electron of the atom is presumed not to screen nuclear charge from the $1s$ electrons, the energies of the $1s$ electrons will be the same in the atom and in the ion. In this case the ionization potential is simply the energy of the $2s$ electron in the atom.

$$\begin{aligned} I &= -13.6[\{2(Z^*/n^*)_{1s}^2\}_{\text{ion}} - \{2(Z^*/n^*)_{1s}^2 + (Z^*/n^*)_{2sp}^2\}_{\text{atom}}] \\ &= +13.6(Z^*/n^*)_{2sp}^2 \end{aligned}$$

Since each of the two $1s$ electrons of the atom screens 0.85 charge unit from the $2s$ electron, $Z^* = Z - 2(0.85) = 1.30$. The calculated ionization potential for Li is therefore 5.8 eV to be compared with an experimental value of 5.4 eV.

For the fluorine atom, the calculation is similar. The ionization potential is the difference between the ion and atom energies. The atom has the configuration $1s^2 2s^2 2p^5$ or $(1s)^2(2sp)^7$, while the ion has the configuration $(1s)^2(2sp)^6$. Again, since the $1s$ electrons "see" the same nuclear charge before and after ionization, their contributions to the energies of the atom and ion are the same and will cancel when the difference is taken. The ionization potential is

$$I = -13.6[\{6(Z^*/n^*)_{2sp}^2\}_{\text{ion}} - \{7(Z^*/n^*)_{2sp}^2\}_{\text{atom}}]$$

Since Z^* for the $2sp$ group before ionization is different from that after ionization (there is a change in the number of electrons in the group, so the energies of the remaining electrons change), we must compute Z^* for the $2sp$ group both in the ion and in the atom. In the ion there are two $(1s)$ group electrons and five $(2sp)$ group electrons screening each $(2sp)$ electron, and the effective nuclear charge for any one of the $(2sp)$ electrons is $Z^* = Z - 5(0.35) - 2(0.85) = 5.55$. For the atom, there are two $(1s)$ and six $(2sp)$ electrons screening each $(2sp)$ electron, so the effective nuclear charge of a $(2sp)$ electron is $Z^* = Z - 6(0.35) - 2(0.85) = 5.20$. The calculated ionization potential for fluorine is 15 eV, to be compared with the experimental value of 17 eV.

The agreement of these calculated values with experiment is typical; the error is on the order of 10%. This prevents the approximate method from being quantitative, but it is nevertheless useful for estimates. More accurate results[21] can be obtained by sacrificing the simplicity and the minimum number of rules in Slater's prescription of Z^*.

---

[21] E. Clementi and D. L. Raimondi, *J. Chem. Phys.*, **38**, 2686 (1963).

STUDY QUESTIONS

15. Which, according to Slater's rules, has the greater first ionization potential, N or O? Experiment reveals the order to be N > O. Analyze any discrepancy by comparing the "averaged" e^-/e^- repulsions for atoms inherent in the Slater scheme with the details of e^-/e^- repulsion (charge and spin correlation effects) in the ground configurations of the atoms N and O.

16. Calculate from Slater's rules the *second* ionization potential for carbon ($C^+ \to C^{2+}$, expt. = 24.4 eV).

17. Calculate the electron affinities of F and Cl by Slater's rules and compare with the experimental order, Cl(+3.7) > F(+3.5).

ATOMIC CONFIGURATIONS AND ATOMIC TERMS

In the discussion of the Aufbau principle, on several occasions it was necessary to take a closer look at the influence of electron repulsions on the energies of possible electron configurations. We were able to improve our understanding of atomic structure by observing that an electron's spin has a great deal to do with its motion when other electrons are present. Thus, we considered both the coulombic and spin correlations of the electrons' motions. In this section we will consider these interactions in a more formal way by introducing the atomic spectroscopist's system for cataloging the lowest energy and higher energy states of atoms. The problem is that the experimentalist observes far more electronic states for polyatomic atoms than can be accounted for with concepts so far discussed. An example that will be discussed at some length later is the carbon atom. It is found that no less than three different energy *terms* arise from the single *configuration* $2p^2$. These three terms arise because of the multitude (15, as you will see) of different ways of arranging two electrons in a set of three p atomic orbitals of equal energy. Some of these arrangements have the same electron-electron repulsion energies as others, and some are different. You will see that the 15 arrangements possible are *resolved by e^-/e^- repulsions* into a group of nine arrangements of one energy, five of another, and one of yet another. These three groups account for the three experimentally observed terms for the $2p^2$ configuration.

While theoretical methods have been developed that nicely account for the energies of these atomic terms or states, the expositions of the wave mechanical and mathematical principles are far removed from the intent of a text such as this (they are generally covered in a graduate level course on the theory of atomic structure). However, a great deal of insight can be obtained by presenting the spectroscopists' *vector model* of the atom, a model that was originally developed to catalog the origin of atomic terms as observed experimentally and without reference, *per se*, to any wave mechanical description of electron motions.

In any assemblage of classical systems (like a room full of spinning bicycle wheels), each of which can be characterized by its angular momentum, a characteristic entity of the entire assemblage is the total angular momentum. *If the angular momentum of one of the component systems changes, the angular momentum of the assemblage changes. Furthermore, the angular momentum and energy of the assemblage are related. This is the basic principle behind the vector model of the atom. State energies are distinguished on the basis of their angular momentum.* Each electron in a polyelectronic atom has two characteristic angular momenta—an orbital angular momentum (represented by the l quantum number) and the intrinsic spin angular momentum (represented by s). In classical mechanics, the angular momentum of a collection of particles, each with angular momentum, is determined from the *vector sum* of the individual angular momenta. A new twist in the atomic problem is that these angular momenta are quantized and the problem is not, strictly speaking, a classical one. Nevertheless, as far as the vector model goes, there is no disagreement with wave formulations nor, of course, of its predictions with experiment.

In general, there are two limiting situations that can arise for the coupling of these individual angular momenta. (i) One extreme is that the individual orbital angular momenta ($\vec{l}$) couple strongly to give a resultant, total orbital angular momentum ($\vec{L}$), and the spin angular momenta ($\vec{s}$) strongly couple to give a total spin angular momentum ($\vec{S}$). The resultant total orbital and total spin angular momenta couple to a weaker extent to produce a total angular momentum ($\vec{J}$) for the atom. (ii) The other extreme situation is that the individual electron orbital ($\vec{l}$) and spin angular momenta ($\vec{s}$) strongly couple to give a resultant spin + orbital angular momentum ($\vec{j}$) for each electron, followed by weaker coupling of the individual electron angular momenta to give rise to a total angular momentum for the atom ($\vec{J}$). *The former of these two extremes is called Russell-Saunders or L-S coupling. The latter is called j-j coupling.* It is observed experimentally that the interaction of two electrons' orbital (and spin) angular momenta decreases as the radial functions become more diffuse, with the consequence that the L-S scheme becomes less appropriate and the j-j scheme more appropriate. Thus, the lighter elements are generally treated by the L-S scheme, while the heavier elements (say beyond bromine) are usually treated by the j-j scheme. It is to be understood, however, that the relative importance of spin-orbit coupling increases in a fairly smooth manner with atomic number, so there is a gradual transition from the L-S to the j-j scheme. The atoms and ions of most interest in this text will be those that follow the L-S scheme, so we will fully discuss the application of this method.

To identify the terms of a polyelectronic atom, we really need only to determine the appropriate quantum numbers for each term. Let us begin the discussion with a review of the single-electron quantum numbers l and m_l.

1. l, defined as the orbital quantum number, has integral values ≥ 0. The familiar alphabetic symbolism is:

$$
\begin{aligned}
l = 0 &\Rightarrow s \text{ orbital} \\
1 &\quad p \\
2 &\quad d \\
3 &\quad f
\end{aligned}
$$

2. m_l, the azimuthal quantum number, may have the $2l + 1$ integral values from $-l$ to $+l$.

3. Strictly analogous comments apply to the s and m_s quantum numbers.

For polyelectronic atoms we expect to use entirely analogous concepts, with L representing a total orbital quantum number and M_L representing a total azimuthal quantum number, and an analogous alphabetic code is used for L:

$$
\begin{aligned}
L = 0 &\Rightarrow S \text{ term} \\
1 &\quad P \\
2 &\quad D \\
3 &\quad F \\
4 &\quad G
\end{aligned}
$$

A most important connection to make is that $L = $ maximum M_L just as $l = $ maximum m_l; $S = $ maximum M_S just as $s = $ maximum m_s. Thus, if we can determine the maximum M_L and M_S quantum numbers for a term, we can immediately write down the L and S quantum numbers.

The reason that we attempt to identify terms through their M_L and M_S quantum numbers, rather than directly through the L and S quantum numbers, is intimately related to the physical meaning of these quantum numbers. The individual electron m_l values represent angular momentum vectors oriented in the same direction (the z axis), so that the *vector* addition reduces to simple *algebraic* addition. Addition of the general angular momentum vectors related to l and s is much more difficult because those vectors, in general, are not parallel.

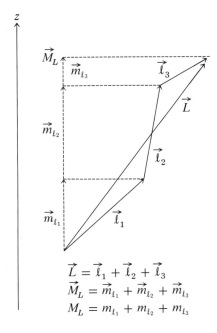

$$\vec{L} = \vec{l}_1 + \vec{l}_2 + \vec{l}_3$$
$$\vec{M}_L = \vec{m}_{l_1} + \vec{m}_{l_2} + \vec{m}_{l_3}$$
$$M_L = m_{l_1} + m_{l_2} + m_{l_3}$$

For each way of assigning four quantum numbers to each electron in the atom, we arrive at an M_L value ($= \Sigma m_l$) and M_S value ($= \Sigma m_s$). Each of these (M_L, M_S) pairs defines a *microstate* of the atom. It will be important later to remember that each term with quantum numbers L and S consists of $(2L + 1) \cdot (2S + 1)$ such (M_L, M_S) pairs or microstates. The strategy is pretty obvious now: *find out all possible microstates, identify them by (M_L, M_S) values, and deduce the allowed L and S quantum numbers.* Finally, write the atomic term symbol in the form $^{(2S+1)}L$, and use the alphabetic code for L (that is, $S, P, D, \ldots$). To illustrate how one implements this strategy, we will work through the example of the carbon atom with the lowest energy configuration $1s^2 2s^2 2p^2$.

To begin, note that all filled shells and sub-shells make zero contribution to the total M_L. This is because for every electron in such shells with a $+m_l$ quantum number there is another with $-m_l$ quantum number. The sum of all these will equal zero. The same is true for the M_S quantum number. All orbitals containing a pair of electrons have one with $m_s = +1/2$ and another with $m_s = -1/2$. Their contribution to M_S totals to zero. *So the problem immediately simplifies to consideration of only partially filled sub-shells.* In the carbon atom configuration given above, you need consider only the two p electrons.

Since there are three p atomic orbitals, and since an electron that has any of these three spatial motions can be given either of two spin quantum numbers, in effect we are faced with how many different ways there are to arrange two particles in six *spin orbitals* (each spin orbital is characterized by an m_l and an m_s quantum number).

$$m_l = \underbrace{\quad 1 \quad}\; \underbrace{\quad 0 \quad}\; \underbrace{\quad -1 \quad}$$
$$m_s = \tfrac{1}{2}\; -\tfrac{1}{2}\quad \tfrac{1}{2}\; -\tfrac{1}{2}\quad \tfrac{1}{2}\; -\tfrac{1}{2}$$

The number of different combinations possible is determined from the theory of permutations and combinations. There are six choices for the first of the two electrons, and five choices remain for the second. Since the electrons are indistinguishable, the number of different arrangements is

$$\frac{6 \cdot 5}{1 \cdot 2} = 15 \text{ choices}$$

50 □ USEFUL ATOMIC CONCEPTS

Each of these fifteen choices will be called a microstate for the atom. Each such microstate will be characterized by an M_L and an M_S value given by the sums of the individual m quantum numbers. These microstates and the resultant M_L and M_S values are summarized in the following diagram.

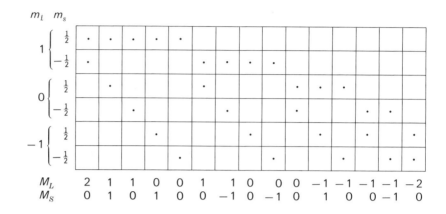

Scanning this list of (M_L, M_S) pairs, you see that the *maximum* M_L quantum number is 2. This immediately signals the occurrence of a term with $L = 2$. Note, however, that the possible (M_L, M_S) pairs of the form $(2, M_S)$ are limited to $M_S = 0$. Therefore, *maximum* M_S for $L = 2$ is 0. The term quantum numbers are therefore $L = 2$, $S = 0$; the term symbol is ^1D, where the pre-superscript 1 is read "singlet" and is the value of $2S + 1$ (the spin multiplicity).

Since the ^1D term accounts for $(2L + 1) \cdot (2S + 1) = 5(M_L, M_S)$ pairs (microstates), eliminate from the original list of (M_L, M_S) pairs the following pairs: $(2,0)$, $(1,0)$, $(0,0)$, $(-1,0)$, $(-2,0)$. This leaves 10 microstates:

$$\begin{array}{lcccccccccc} M_L & 1 & 0 & 1 & 1 & 0 & 0 & 0 & -1 & -1 & -1 \\ M_S & 1 & 1 & 0 & -1 & 0 & -1 & 0 & 1 & 0 & -1 \end{array}$$

Scanning this list for *maximum* M_L, we settle on the value $+1$. Of the pairs $(1, M_S)$, the largest M_S is $+1$. Consequently, we have just determined the presence of a term $L = 1$, $S = 1$: a ^3P term (read triplet P). The associated 9 microstates are: $(1,1)$, $(1,0)$, $(1,-1)$, $(0,1)$, $(0,0)$, $(0,-1)$, $(-1,1)$, $(-1,0)$, $(-1,-1)$.

This leaves the single pair

$$\begin{array}{ll} M_L & 0 \\ M_S & 0 \end{array}$$

signaling the presence of ^1S ($L = 0$, $S = 0$).

In summary, then, the $2p^2$ configuration gives rise to three atomic terms: ^1D, ^3P, and ^1S in the order in which they were determined. The existence of three terms is confirmed by experiment.

If there were no coupling of the spin and orbital angular momentum vectors L and S, the problem would be complete at this point. Such coupling is sometimes of importance, and we must continue onward to find the (generally several) total angular momenta J for each of these terms. *The mnemonic for finding the values of J for each term is to compute $L + S$ and $L - S$, and to assign to J all values from $L + S$ through $|L - S|$ in integral steps.* To see how this rule is achieved, proceed as before by arranging a list of M_J values for each microstate of a term. We will illustrate this for the ^3P term. The possible M_J values are

M_L	M_S	M_J
1	1	2
1	0	1
1	−1	0
0	1	1
0	0	0
0	−1	−1
−1	1	0
−1	0	−1
−1	−1	−2

The *largest* M_J possible for the ^3P term is seen to be $+2$, which implies that the largest J vector has magnitude 2. Since a J of this magnitude can take any of five orientations relative to the z axis [that is, there are $(2J+1)$ values of M_J: $+2, +1, 0, -1, -2$], remove from the list five microstates for $J = 2$. That leaves four microstates with M_J values of 1, 0, 0, −1. The presence of $M_J = 1$ requires a $J = 1$. Removing three microstates to account for $M_J = 1, 0$, and −1, you are left with one microstate for which M_J and $J = 0$. Thus, allowing for spin-orbit coupling, the ^3P term should consist of three terms of different total angular momenta: $^3P_2$, $^3P_1$, and $^3P_0$. This agrees with the rule on $L + S$ $(= 2)$ through $|L - S|$ $(= 0)$.

Applying the same procedure to the ^1S and ^1D terms, we find the following: $^1S_0$, $^1D_2$. You see that the general form of the complete term symbol is $^{(2S+1)}L_J$.

As a convenient check on these accounting procedures, apply the following logic. We know that each M_J requires a microstate of that M_J value. Since there are $(2J + 1)$ values of M_J for each J, the number of microstates at the start (15 in this example) must equal the sum of $(2J + 1)$ possible values.

	$(2J + 1)$
$^1S_0$	1
$^1D_2$	5
$^3P_2$	5
$^3P_1$	3
$^3P_0$	1
Total	15

To summarize, we started with 15 microstates of equal energy and, by accounting for electron-electron repulsions, found that the 15 microstates grouped themselves into three groups of different energy: one term of one microstate (^1S), another of five microstates (^1D), and another of nine microstates (^3P). The number of microstates in any one group corresponds to the orbital (spatial) degeneracy $(2L + 1)$ times the spin degeneracy $(2S + 1)$. By further allowing the spin and orbital angular momenta to interact weakly, the microstates of each term further split into different energy sets of 1 ($^1S_0$), 5 ($^1D_2$), 5 ($^3P_2$), 3 ($^3P_1$), and 1 ($^3P_0$). The number of microstates in any one spin-orbit term is $(2J + 1)$. These progressive changes are illustrated in Figure 1-18. Electronic transitions between these atomic terms are the bases for atomic emission and absorption spectra.

A final useful generalization that was empirically discovered in *Hund's rule*, a principle supported by the wave theory. This rule identifies the lowest-energy term from all those for a given configuration of an atom or ion. The rule consists of three parts. *First, of all the terms that arise from a configuration, that which lies lowest in energy is the term with the largest value of S*. The rule has its basis in electron charge and spin correlation energies, since it requires the maximum number of unpaired electrons. Applied to the carbon atom, the ^3P term lies lowest. *Second, if more than one term is obtained with the same largest value of S, that term with the largest L lies lowest. Finally,* considering the spin-orbit interaction, *the term with lowest J value lies lowest*

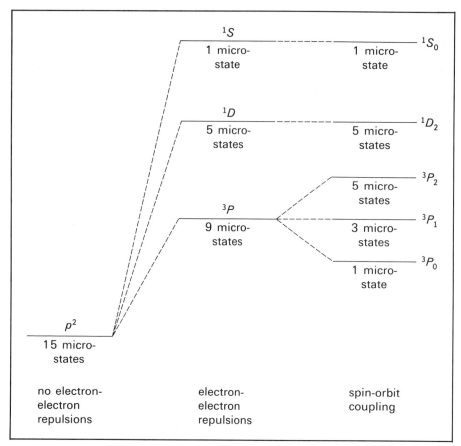

Figure 1-18. A flow chart of electron-electron interactions for the p^2 case, showing the creation of atomic terms and spin-orbit states.

in energy if the partially filled sub-shell is less than half-filled. If the sub-shell is more than half-filled, the term with largest J lies lowest in energy. For exactly half-filled sub-shells, $L = 0$ and only one value of $J (= S)$ is possible. For the carbon atom, the lowest term is $^3P_0$.

Unfortunately, no rules can be given for the relative energies of the other terms that arise from a configuration. The order of other terms has to be determined by experiment or theory.

To further illustrate Hund's rules, let us consider nitrogen. The nitrogen atom has a ground configuration of $1s^2 2s^2 2p^3$, from which arise ^4S, ^2D, and ^2P terms. The $^4S_{3/2}$ term lies lowest in energy.

While this full procedure is necessary if one wants to determine all the terms that can arise from a given electronic configuration, *a much simpler procedure, based on Hund's rules, may be used to determine the lowest energy atomic term.* A box diagram for the atomic orbitals can be used; electrons are assigned to boxes so as to yield the maximum number of unpaired electrons and the maximum value of the sum of m_l values. For example, for the configuration $3d^2$ (Ti^{2+}), diagram the following:

$$m_l \quad +2 \; +1 \;\; 0 \;\; -1 \; -2$$

↑	↑			

With two unpaired electrons, $S = 1$; arranging these electrons so as to yield the maximum $M_L = 3$ (which implies an L of 3), the ground term for Ti^{2+} is found to be ^3F. Allowing for spin-orbit coupling, the lowest term is $^3F_2$.

An interesting illustration of the importance of spin-orbit coupling is found in the trend of ionization potential for the series Pb, Bi, and Po. The trend is unusual in that the value for Bi (7.3 eV) is lower than that of both Pb (7.4 eV) and Po (8.4 eV), in contrast to the normal trend of a Group V element having a larger I than the elements to either side. With reference to the discussion of atomic terms and the effects of spin-orbit coupling, we expect such effects to be largest for Pb, Bi, and Po of the Group IV, V, and VI elements, and herein lies the reason[22] for the anomalous behavior. By definition, the value of the I for any atom is the difference in energy between the atom ground term and the ion ground term. These are diagrammed below to show the effects of spin-orbit coupling on the relative term energies.

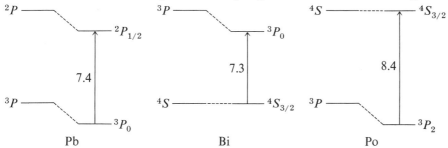

The striking feature of these diagrams is that there is no spin-orbit stabilization of the ground term of Bi and the ground term of Po$^+$, while the spin-orbit stabilization of the ground term of Bi$^+$ and the ground term of Po should be quite large. Thus, the spin-orbit effects, which should be largest for this series of atoms relative to their group congeners, dominate the usual trend in electron effective nuclear charges.

Having learned the procedure by which one determines the different terms that arise from a given configuration, you may be left with little feeling for atomic terms. It is possible, by pursuit of the wave formulation of the problem, to determine wavefunctions for these terms and to examine the shapes (probability functions) of atoms whose energies correspond to the various terms. Figure 1–19 shows the shapes of a carbon atom[23] when it is at its lowest energy and when it has been excited to the ^1D and ^1S terms. The probability functions, in terms of real $2p$ probability functions, are also given so that you can establish for yourself the relationship of each shape to its probability function.

The electron cloud sketches are simply related to the p orbital densities or probability formulas given below each sketch (these formulas are not easy to derive and require advanced techniques). For example, below the sketch for ^1S you find the formula $\frac{2}{3}\{p_x^2 + p_y^2 + p_z^2\}$. The orbital probability function within the brackets is that which would arise for one electron in each of p_x, p_y, and p_z. Such a distribution is spherical in shape. The factor 2/3 is required because the carbon atom has only two p type electrons. This term, accordingly, exhibits a spherical probability distribution because the electrons are evenly distributed among the three p ao's. Because the term is a singlet, they have paired spins. The term wavefunction is actually a hybrid of p_x^2, p_y^2, and p_z^2 microstates (the superscripts 2 designate two electrons). The $^3P(M_L = 0)$ sketch derives from the probability function $p_x^2 + p_y^2$, which indicates one electron in p_x and one electron in p_y with like spins (a triplet term). The work required to excite the carbon atom from the ^3P term to the ^1S term amounts to forcing the electrons into the same p orbital, with pairing of their spins. The ^1D term (examine the $M_L = 0$ sketch) does not require as much work to attain (is at lower energy). The electron distribution here also arises from spin-pairing (it is a singlet term), but the electron orbital occupancies derive from a different hybrid of the p_x^2, p_y^2, and p_z^2 microstates; though the wavefunction is more complex than that for the ^1S and ^3P terms, its probability function appears equivalent to a simple assignment of one p_z electron and another distributed evenly (spherically) among all p ao's (including p_z).

---

[22] D. W. Smith, *J. Chem. Educ.*, **52**, 576 (1975).

[23] Recall that each term is represented by $(2L + 1)(2S + 1)$ two-electron wavefunctions. Only the orbital parts are given here. As with the hydrogen atomic orbitals, the $\pm M_L$ functions are different imaginary functions, but their probability functions are indistinguishable.

54 USEFUL ATOMIC CONCEPTS

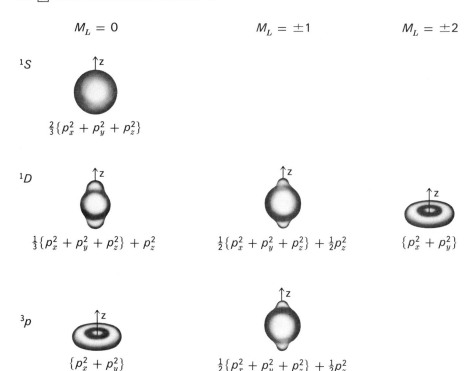

Figure 1-19. Sketches of the atomic probability functions for terms of the ground configuration carbon atom.

STUDY QUESTIONS

18. What term symbol denotes the lowest energy assemblage of microstates for the beryllium atom valence state: $1s^2 2s^1 2p^1$? What are the term symbols of the other states arising from this configuration?

19. Will the $^2S_{1/2}$ and $^2P_{1/2}$ terms of $1s^2 2s^1$ and $1s^2 2p^1$ have the same energy for Li? For Li^{2+}?

20. Given below are the ground configuration term symbols of the representative elements. From which group(s) do these terms arise?

    a) ^1D, ^3P, ^1S d) ^2D, ^2P, ^4S
    b) ^2S e) ^2P
    c) ^1S

21. A transition metal ion has the ground configuration term symbols ^1S, ^3P, ^1D, ^1G, ^3F. Considering only 2+ and 3+ ions, which elements are possible?

ATOM AND ORBITAL ELECTRONEGATIVITIES

One of the more important uses to which the energies of atomic terms have been put is the definition of electronegativities. The original ideas were advanced by Mulliken[24] and more recently amplified by Hinze and Jaffe.[25]

---

[24] R. S. Mulliken, *J. Chem. Phys.*, **2**, 782 (1934); **3**, 573 (1935).
[25] J. Hinze and H. H. Jaffe, *J. Am. Chem. Soc.*, **84**, 540 (1962); J. Hinze, M. A. Whitehead, and H. H. Jaffee, *J. Am. Chem. Soc.*, **85**, 148 (1963); J. Hinze and H. H. Jaffe, *J. Phys. Chem.*, **67**, 1501 (1963).

Mulliken began his argument for the definition of electronegativity by considering electron transfer between two atoms, A and B,

$$A^-B^+ \leftrightarrow A^+B^-$$

He argued that if the electronegativities of A and B were identical, the two ionic pairs would be of equal energies. The energy of the first form depends on the ionization potential of B and the electron affinity of A, while the energy of the second form depends on the ionization potential of A and the electron affinity of B. The ionization potential and electron affinity of A are defined by the energies of the following:

$$I_A: A \rightarrow A^+ + e^-$$
$$E_A: A^- \rightarrow A + e^-$$

If the energies of the two ionic structures are equal, then $I_B - E_A = I_A - E_B$ or $I_A + E_A = I_B + E_B$ is the condition that specifies equal A and B electronegativities. If A is more electronegative than B, then $I_A + E_A > I_B + E_B$. Therefore, Mulliken defined the electronegativity of an atom to be the sum of its ionization potential and electron affinity.

One of the more interesting ramifications of this definition of atomic electronegativity is that the latter depends on the configuration of electrons in the valence shell and, furthermore, on the specific term arising from that configuration. For example, the ionization potential of a carbon atom is different for the lowest terms ($^3P_0$ and $^5S_2$) of the configurations $2s^22p^2$ and $2s^12p^3$. As ionization potential and electron affinity data are generally available for ions, Mulliken's definition also allows for the determination of ion electronegativity.

As an example of the calculation of atom electronegativity by this definition, we will work through an example of the carbon atom. Figure 1-20 should be consulted throughout the example.

Since a given valence configuration of an atom generally results in several atomic term levels, the energy of the configuration is taken to be the average energy of all terms, each weighted by the ratio of its number of microstates to the total for the configuration.[26] Thus, for $C(s^2p^2)$ the ground configuration energy $= \frac{1}{15}\{E(^1S) + 5E(^1D) + 9E(^3P)\}$. The energies of the $C^-(s^2ppp)$, $C(s^2pp)$, and $C^+(s^2p)$ configurations have been determined in this way and are represented by horizontal solid lines in Figure 1-20. The difference in energy between C^+ and C is simply the first ionization potential for a carbon atom (I_C) with its lowest energy configuration. Similarly, the energy difference between C^- and C is the electron affinity for the ground configuration carbon atom (E_C). The carbon atom electronegativity (actually the p orbital electronegativity) calculated for the configuration s^2p^2 is 12.4 eV.

In many carbon compounds the four electrons of a tetrahedral carbon atom are often described as evenly distributed among the four valence atomic orbitals, giving the configuration s^1p^3. Again, the energy of this carbon atom valence configuration can be calculated from the average energy of the term levels that arise from it. This, an excited configuration of the atom, is still not very close to describing the state of the atom called "sp^3 hybridized," since in the hybridized state the orbital electronegativities should all be equal and the electrostatic repulsions between each electron and the other three should be the same (the electrons are supposed to move in entirely equivalent orbitals). For the $sppp$ configuration of the atom, Hinze and Jaffee find the s orbital to have a Mulliken electronegativity of 29.9 eV and the p orbitals to have an electronegativity of only 11.6 eV. In consequence, the configuration $sppp$ for the carbon does not very accurately describe the electron interactions that we expect to obtain in a carbon atom in a compound such as methane; we would have to assign

---

[26] This average is simply the average energy of the 15 microstates possible for the s^2p^2 configuration. It is interesting to note that like-spin microstates number only two of the total 15, so that a complete averaging of the electron spin orientations is *not* achieved.

56 ☐ **USEFUL ATOMIC CONCEPTS**

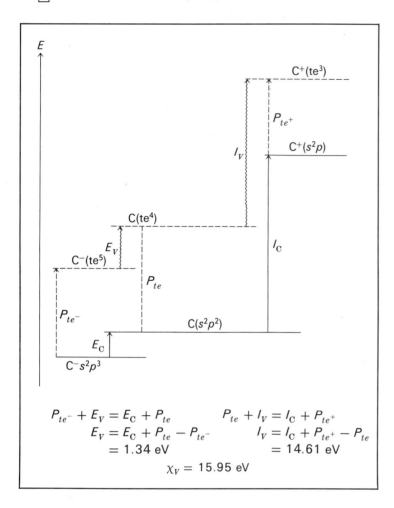

Figure 1-20. Diagram of the relationships between average configuration energy, valence state energy, and promotion, ionization, and electron affinity energies.

two different electronegativity values simultaneously to the same atom. Thus far we have said nothing specific about spin correlation.[27] *In a molecule* such as methane, where the four electrons about carbon have averaged spins, *an atomic valence state, defined to mimic the condition of the carbon atom electrons in a tetrahedral molecule, must allow for this averaging.* This averaging, a somewhat complicated procedure, was carried out by Hinze and Jaffe with the aid of a computer. The energies of the hybridized configurations C (tetetete), C^+ (tetete), and C^- (te²tetete) calculated by them are also indicated in Figure 1-20. For each ion and its neutral atom, *the difference between the term-averaged $s^m p^n$ configuration energy and the corresponding random-spin te^{m+n} configuration energy is called the valence state promotion energy* (P_{te}) *for that ion or atom.* Knowledge of values for the promotion energies and for the ionization potential and electron affinity of the carbon atom allows us to calculate an electronegativity for the te valence state of carbon. The valence state ionization potential and electron affinity are computed in Figure 1-20 and lead to a Mulliken electronegativity for tetrahedral carbon of 15.95 eV.

For the sake of convenience in working with smaller numbers, it is customary to use the simple average of I and E rather than $(I + E)$. Values of electronegativities for the representative elements, in various valence states corresponding to known compounds, are reported in Table 1-5 for future reference.

---

[27] The averaging of the 40 microstate energies for $s^1 p^3$ involves an incomplete averaging of electron orientations.

TABLE 1-5
SOME MULLIKEN ELECTRONEGATIVITIES (eV)

	H													
s	7.2													

	Li		Be		B		C		N		O		F	
s	3.1	di^2	4.8	tr^3	6.4	$di^2\pi^2$	10.4, 5.7	$di^3\pi^2$	15.7, 7.9	$tr^5\pi$	17.1, 20.2	s	31.3
p	1.8	te^2	3.9	te^3	6.0	$tr^3\pi$	8.8, 5.6	$tr^4\pi$	12.9, 8.0	$di^2\pi^4$	19.1	p	12.2
						te^4	8.0	te^5	11.6	te^6	15.3		

	Na		Mg		Al		Si		P		S		Cl	
s	2.9	di^2	4.1	tr^3	5.5	$di^2\pi^2$	9.0, 5.7	$di^3\pi^2$	11.3, 6.7	$tr^4\pi^2$	10.9	s	19.3
p	1.6	te^2	3.3	te^3	5.4	$tr^3\pi$	7.9, 5.6	$tr^4\pi$	9.7, 6.7	te^6	10.2	p	9.4
						te^4	7.3	te^5	8.9				

	K		Ca		Ga		Ge		As		Se		Br	
s	2.9	di^2	3.4	tr^3	6.0	$di^2\pi^2$	9.8, 6.5	$di^3\pi^2$	9.0, 6.5	$tr^4\pi^2$	10.6	s	18.3
p	1.8	te^2	2.5	te^3	6.6	$tr^3\pi$	8.7, 6.4	$tr^4\pi$	8.6, 7.0	te^6	9.8	p	8.4
						te^4	8.0	te^5	8.3				

	Rb		Sr		In		Sn		Sb		Te		I	
s	2.1	di^2	3.2	tr^3	5.3	$di^2\pi^2$	9.4, 6.5	$di^3\pi^2$	9.8, 6.3	$tr^4\pi^2$	10.5	s	15.7
p	2.2	te^2	2.2	te^3	5.1	$tr^3\pi$	8.4, 6.5	$tr^4\pi$	9.0, 6.7	te^6	9.7	p	8.1
								te^5	8.5				

Values can be computed only for orbitals holding 1 electron. For the carbon and nitrogen families it is possible to have both hybrid and π atomic orbitals half-filled. *di*gonal $\equiv sp$ hybrid, *tri*gonal $\equiv sp^2$ hybrid, *tetra*hedral $\equiv sp^3$ hybrid.

Just as orbital electronegativities were found to be different for the 2s and 2p orbitals of the carbon atom, *orbital electronegativities vary with the type of hybrid—that is, with the relative s and p character of the hybrid*. To illustrate this, the results for nitrogen are given in Table 1-6. For the two unhybridized configurations, two important observations arise. The *p* orbital electronegativity increases upon promotion of an *s* electron to the *p* sub-shell, and this increase in electronegativity can be traced to a decrease in shielding for the *p* electrons resulting from a reduction in the number of *s* electrons. Even more important is the large difference between the *s* and *p* orbital electronegativities (a factor of ~2). *This is traced to the better penetration of the electron core by 2s than by 2p.* For the *sp* (or *di*gonally) hybridized nitrogen atom, as found in nitriles, the high *s* character of the hybrid yields a high electronegativity for the orbital. For sp^2 (or *tri*gonally) hybridized nitrogen, as found in pyridine, the

TABLE 1-6
ORBITAL ELECTRONEGATIVITIES OF NITROGEN

Configuration	Orbital	χ_M (eV)
s^2ppp	p	7.4
sp^2pp	p	8.5
	s	20.5
di^2dipp	di	15.7
	p	7.9
tr^2trtrp	tr	12.9
	p	8.0
$te^2tetete$	te	11.5

di $\equiv sp$ hybrid; tr $\equiv sp^2$ hybrid; te $\equiv sp^3$ hybrid

58 ☐ USEFUL ATOMIC CONCEPTS

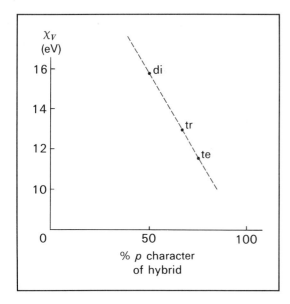

Figure 1-21. The variation in hybrid orbital electronegativity with % p character (for nitrogen).

hybrid is reduced in electronegativity because of the decrease in s character. The sp^3 (or *te*trahedrally) hybridized nitrogen atom shows the lowest electronegativity of all hybrids, since the s character is lowest. This relation between hybrid s character and Mulliken electronegativity is shown graphically in Figure 1-21. In later chapters we will find this variation in orbital electronegativity with s content to be of great value.

One of the outstanding features of the Mulliken definition of electronegativity is that allowance can be made for atom hybridization and atom charge. Other electronegativity definitions, which suffer from allowing less insight into the variation of electronegativity with hybridization and charge, are more widely known. Historically, Pauling[28] was the first to introduce the concept of electronegativity; he defined the electronegativity difference for two atoms A and B in terms of the deviation of the AB bond energy from the mean of AA and BB bond energies. Since it is generally found that the AB bond energy is greater than the average of AA and BB bond energies, Pauling attributed this difference to contributions of ionic resonance forms to the AB bond energy.

$$A\!-\!B \leftrightarrow A^+B^-$$

The greater the deviation, the greater the importance of the ionic structure and the greater the difference between A and B electronegativities. Allred[29] has revised Pauling's original table of atom electronegativities, using more recent and more extensive thermochemical data. In an unexpected development, the Mulliken electronegativities are generally (but there are exceptions) linearly related to the newer Pauling values, in spite of the considerably different definitions used by the two workers.

$$\chi_P = 0.34\chi_M - 0.2$$

A final definition of electronegativities (there have been many others worthy of mention, but restrictions of time and space do not permit a review of all methods), which is related to the previous discussion of Slater's rules, is one by Allred and

---

[28] L. Pauling, "The Nature of the Chemical Bond," Cornell University Press, Ithaca, N.Y. (1960, 3rd edition).

[29] A. L. Allred, *J. Inorg. Nucl. Chem.*, **17**, 215 (1961).

TABLE 1-7
NOMINAL ELECTRONEGATIVITY VALUES FOR THE ELEMENTS

H 2.2																	
Li 1.0	Be 1.5											B 2.0	C 2.5	N 3.1	O 3.5	F 4.1	
Na 1.0	Mg 1.2											Al 1.5	Si 1.7	P 2.1	S 2.4	Cl 2.8	
K 0.9	Ca 1.0	Sc 1.2	Ti 1.3	V 1.5	Cr 1.6	Mn 1.6	Fe 1.6	Co 1.7	Ni 1.8	Cu 1.8	Zn 1.7	Ga 1.8	Ge 2.0	As 2.2	Se 2.5	Br 2.7	
Rb 0.9	Sr 1.0	Y 1.1	Zr 1.2	Nb 1.2	Mo 1.3	Tc 1.4	Ru 1.4	Rh 1.4	Pd 1.4	Ag 1.4	Cd 1.4	In 1.5	Sn 1.7	Sb 1.8	Te 2.0	I 2.2	
Cs 0.9	Ba 1.0	La 1.0	Hf 1.1	Ta 1.2	W 1.3	Re 1.4	Os 1.5	Ir 1.6	Pt 1.4	Au 1.4	Hg 1.4	Tl 1.4	Pb 1.6	Bi 1.7	Po 1.8	At 2.0	
Fr 0.9	Ra 1.0																

Lanthanides ~ 1.1

Rochow.[30] They considered the force acting on a valence electron of an atom to be a good measure of the ability of the atom to attract electrons. This force is given by

$$|F| = e^2 \frac{Z^*}{r^2}$$

Instead of computing an average of this force—i.e., $\langle Z^*/r^2 \rangle$—from radial functions, the force is taken to be that exerted by the effective nuclear charge at a distance corresponding to that of the valence electrons at the outer periphery of the atom; hence, $r = r_{cov}$, the covalent radius of the atom. Again, these forces are generally linearly related to the Pauling electronegativities of the atoms, χ_P.

$$\chi_P = 0.36 \frac{Z^*}{r^2} + 0.7$$

Values of the electronegativities computed in this way are given in Table 1-7. *These are nominal values and give only a rough estimate of each atom's ability to attract electrons.* As such, they are useful guides but are not to be interpreted too quantitatively.

Interestingly, the large values for elements from Ga through Se in the fourth period reflect the enhanced effective nuclear charges for the *p* shell electrons in elements immediately following the completion of the 3*d* shell. The effect is also noticed in the fifth and sixth row *p* block elements. The 5*d* elements are similarly affected by the poor shielding of nuclear charge by their $4f^{14}$ shells. This moderately effective shielding of nuclear charge is evident in other atomic properties, such as atomic radii, and has been given the special name "Lanthanide Contraction" when such effects are found in elements following the lanthanide elements.

STUDY QUESTIONS

22. Arrange the following list of atomic orbitals in ascending order of electronegativity.

    a) F(2*s*), O(2*p*), Na(3*s*), F(di)
    b) N(*p*), O(tr), N(te), C(*p*)

23. For the *nd* elements, the order of electronegativities in a given group is 3*d* > 4*d* < 5*d* (example: Fe > Ru < Os). Why are the electronegativities of the 5*d* elements so high?

24. On the basis of electronegativities, state which member of each pair has the more polar bond, and indicate the direction of polarity.

    C—N or N—O
    P—S or S—Cl
    Sn—I or C—I

EPILOG

In this chapter we have reviewed selected aspects of the currently accepted *matter wave* model of electrons in atoms. The concepts focused on are those to be of greatest value to us in succeeding chapters on molecular structure, bonding, and reactivity. It is critical that you have developed an intuition for the relationship between an *electron radial function,* its *effective nuclear charge,* and the *orbital electronegativity.* In addition, the concept of *orbital radial diffuseness* will be highly

---

[30] A. L. Allred and E. Rochow, *J. Inorg. Nucl. Chem.,* **5,** 264 (1958).

useful in interpreting trends in bond strengths. Similarly, you must be fully acquainted with *orbital angular functions,* for these will be closely tied into the concept of directional bonds and bond angles in molecules. We will be particularly interested in what orbitals on adjacent atoms have lobe orientations proper for effective interpenetration.

The *vector model* of the atom will be important to us in the later chapters on transition metals and their compounds. Now is a good time to review the idea that a given *electron configuration* gives rise to many different *atomic terms* and that the origin of these terms from a given configuration can be traced to the details of electron-electron repulsion. The concepts of orbital degeneracy $(2L + 1)$ and spin multiplicity (or degeneracy) $(2S + 1)$ will be put to good use in Chapter 9. Finally, make certain you have clearly in mind the difference in repulsion energy for two electrons of the same spin and for two of opposite spin.

Appendices to Chapter 1

Appendix A: Hydrogen Type Wavefunctions ($\rho = Zr/a_0$, $\xi = Z/a_0$)

$1s$	$(\xi^{3/2}/\sqrt{\pi})\, e^{-\rho}$	
$2s$	$(2 - \rho)$	
$2p_x$	ξx	
$2p_y$	ξy	$\times\, (\xi^{3/2}/\sqrt{32\pi})\, e^{-\rho/2}$
$2p_z$	ξz	
$3s$	$(27 - 18\rho + 2\rho^2)/\sqrt{3}$	
$3p_x$	$\xi\sqrt{2}(6-\rho)x$	
$3p_y$	$\xi\sqrt{2}(6-\rho)y$	
$3p_z$	$\xi\sqrt{2}(6-\rho)z$	
$3d_{z^2}$	$\xi^2(3z^2 - r^2)/\sqrt{6}$	$\times\, (\xi^{3/2}/81\sqrt{\pi})\, e^{-\rho/3}$
$3d_{xz}$	$\xi^2\sqrt{2}(xz)$	
$3d_{yz}$	$\xi^2\sqrt{2}(yz)$	
$3d_{xy}$	$\xi^2\sqrt{2}(xy)$	
$3d_{x^2-y^2}$	$\xi^2(x^2 - y^2)/\sqrt{2}$	
$4s$	$(1 - 3\rho/4 + \rho^2/8 - \rho^3/192)/8$	
$4p_x$	ξx	
$4p_y$	ξy $\quad \times\, (\sqrt{5}/32)(1 - \rho/4 + \rho^2/80)$	
$4p_z$	ξz	
$4d_{z^2}$	$\xi^2(3z^2 - r^2)$	
$4d_{xz}$	$\xi^2\sqrt{12}(xz)$	
$4d_{yz}$	$\xi^2\sqrt{12}(yz)$ $\quad \times\, (1 - \rho/12)/256$	$\times\, (\xi^{3/2}/\sqrt{\pi})\, e^{-\rho/4}$
$4d_{xy}$	$\xi^2\sqrt{12}(xy)$	
$4d_{x^2-y^2}$	$\xi^2\sqrt{3}(x^2 - y^2)$	
$4f_{z^3}$	$(1/\sqrt{5})(5z^2 - 3r^2)z$	
$4f_{xz^2}$	$(3/\sqrt{30})(5z^2 - r^2)x$	
$4f_{yz^2}$	$(3/\sqrt{30})(5z^2 - r^2)y$	
$4f_{xyz}$	$(6/\sqrt{3})xyz$ $\quad \times\, (\xi^3/3072)$	
$4f_{z(x^2-y^2)}$	$\sqrt{3}(x^2 - y^2)z$	
$4f_{x(x^2-3y^2)}$	$(1/\sqrt{2})(x^2 - 3y^2)x$	
$4f_{y(3x^2-y^2)}$	$(1/\sqrt{2})(3x^2 - y^2)y$	

Appendix B: First Ionization Potentials (eV) of the Main Group and Transition Elements

H 13.6																	He 24.6
Li 5.4	Be 9.3											B 8.3	C 11.3	N 14.5	O 13.6	F 17.4	Ne 21.6
Na 5.1	Mg 7.6											Al 6.0	Si 8.1	P 10.5	S 10.4	Cl 13.0	Ar 15.8
K 4.3	Ca 6.1	Sc 6.5	Ti 6.8	V 6.7	Cr 6.8	Mn 7.4	Fe 7.9	Co 7.9	Ni 7.6	Cu 7.7	Zn 9.4	Ga 6.0	Ge 7.9	As 9.8	Se 9.8	Br 11.8	Kr 14.0
Rb 4.2	Sr 5.7	Y 6.4	Zr 6.8	Nb 6.9	Mo 7.1	Tc 7.3	Ru 7.4	Rh 7.5	Pd 8.3	Ag 7.6	Cd 9.0	In 5.8	Sn 7.3	Sb 8.6	Te 9.0	I 10.5	Xe 12.1
Cs 3.9	Ba 5.2	La 5.6	Hf 7.0	Ta 7.9	W 8.0	Re 7.9	Os 8.7	Ir 9.0	Pt 9.0	Au 9.2	Hg 10.4	Tl 6.1	Pb 7.4	Bi 7.3	Po 8.4	At 9.5	Rn 10.7

Appendix C: Electron Affinity Values (eV) of the Elements (estimated values in parentheses)

H 0.75									He (−0.22)
Li 0.62	Be (−2.5)			B 0.24	C 1.27	N 0.0	O 1.47	F 3.34	Ne (−0.30)
Na 0.55	Mg (−2.4)		Cu 1.28	Al 0.46	Si 1.24	P 0.77	S 2.08	Cl 3.61	Ar (−0.36)
K 0.50	Ca (−1.62)		Ag 1.30	Ga (0.37)	Ge 1.20	As 0.80	Se 2.02	Br 3.36	Kr (−0.40)
Rb 0.49	Sr (−1.24)		Au 2.31	In (0.35)	Sn 1.25	Sb 1.05	Te 1.97	I 3.06	Xe (−0.42)
Cs 0.47	Ba (−0.54)			Tl (0.5)	Pb 1.05	Bi 1.05	Po (1.8)	At (2.8)	Rn (−0.42)
Fr (0.46)									

SHARED AND LONE PAIRS AND
LEWIS STRUCTURES

ELECTRON PAIR REPULSION MODEL

SYMMETRY CONCEPTS

   Point Groups

   Character Tables

EPILOG

APPENDIX: POINT GROUP CHARACTER TABLES

2. BASIC CONCEPTS OF MOLECULAR TOPOLOGIES

It is the goal of this and the following two chapters to develop increasingly sophisticated models for the electronic and geometric structures of molecules. We begin here with a review of basic electron accounting procedures familiar to you from a general chemistry course, and proceed to a model for predicting three-dimensional molecular structures and to the use of symmetry to classify molecular shapes. The succeeding two chapters will explore the application of the wave model to these structures. It is our specific intent to develop a hierarchy of working models of molecular electronic structure that the practicing chemist uses daily. As experimental probes of molecular properties become more sophisticated, so must the chemist appeal to higher levels of conceptualization of molecular electronic structure. As the concepts are developed, we will pause to forecast, with examples, their use in later chapters.

SHARED AND LONE PAIRS AND LEWIS STRUCTURES

One of the very first concepts one encounters in describing the electronic structures of molecules is that of the shared pair of electrons defining a bond between two atoms. This "perfect pairing of electrons" concept of G. N. Lewis[1] is where we start in developing the hierarchy of models for molecular structure and models for chemical bonding.

Drawing Lewis structures for electron pairs in molecules is simply a matter of deciding how to partition the valence electrons of a molecule into pairs—either bond pairs to be shared by adjacent atoms or lone pairs to be localized about one atom. Lever[2] has given a useful logic for determining this partitioning in the cases where all atoms satisfy the "inert gas rule."[3] In algebraic terms, the number of valence electron pairs (V) in the molecule is equal to the number of σ pairs (n_σ), plus the number of π pairs (n_π), plus the number of lone pairs (n_ℓ).

$$V = n_\sigma + n_\pi + n_\ell$$

V is simply half of the sum of valence electrons for all atoms in the molecule minus the molecular charge. A useful point of view to assume initially is that enough electrons are taken from some "pool" to give each atom alone four pairs (an octet) of

---

[1] G. N. Lewis, *J. Amer. Chem. Soc.*, **38**, 762 (1916).
[2] A. B. P. Lever, *J. Chem. Educ.*, **49**, 819 (1972).
[3] The "inert gas rule" is based on the tendency for the number of electron pairs about an atom to equal the number of valence orbitals of the atom: H = 1, s/p block elements = 4 (or 5 or 6 for those having nd valence orbitals), transition elements $\leqq$ 9.

electrons (one pair is required for each hydrogen). If the molecule contains q heavy atoms and h hydrogen atoms, this view means that we start with $4q + h$ electron pairs for the atomic anions. To permit formation of one bond between two anions, we must give back a pair to the "pool."

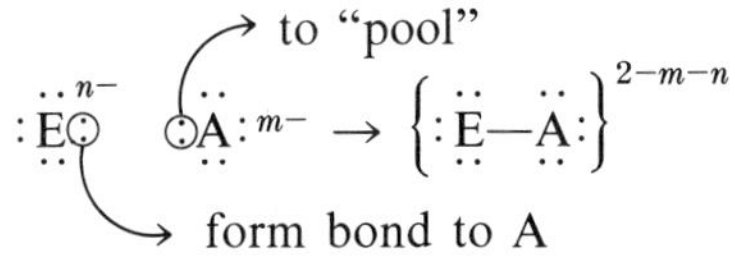

$$\underset{\underset{\text{form bond to A}}{\longrightarrow}}{:\overset{..}{\underset{..}{E}}\overset{n-}{\odot}\overset{\overset{\text{to ``pool''}}{\longrightarrow}}{\odot}\overset{..}{\underset{..}{A}}:^{m-}} \rightarrow \left\{:\overset{..}{\underset{..}{E}}-\overset{..}{\underset{..}{A}}:\right\}^{2-m-n}$$

To account for any double bonds between atoms we must return to the "pool" additional pairs (n_π). For each pair returned, a remaining pair is utilized to form a bond of the σ or π type. The unused pairs are lone pairs (n_l). Consequently we may quantify this accounting by writing

$$(4q + h) - (n_\sigma + n_\pi) = (n_\sigma + n_\pi) + n_l = V$$
$$n_\pi = (4q + h) - n_\sigma - V$$

The last equation has two unknowns, n_σ and n_π. If the heavy atoms are linked in a cyclic structure, then $n_\sigma = h + q$.

If the heavy atoms do not form a cyclic topography, then $n_\sigma = h + (q - 1)$.

Hence,

$$n_\pi = 3q - V + 1 \quad \text{branched case}$$
$$= 3q - V \quad \text{cyclic case}$$

All that remains is to connect atoms by sigma bonds, and to follow with second bonds and lone pairs distributed so as to satisfy the octet rule for each atom.

Let us illustrate the application of this procedure with four examples: HCN, OCN⁻, NO, and S_2Cl_2.

HCN: linear triatomic

$$V = 5 \; (N = 2\tfrac{1}{2}, C = 2, H = \tfrac{1}{2})$$
$$q = 2$$
$$h = 1$$
$$n_\pi = 3 \cdot 2 - 5 + 1 = 2 \qquad H—C\equiv N:$$

OCN⁻: linear triatomic

$$V = 8 \; (O = 3, C = 2, N = 2\tfrac{1}{2}, \text{charge} = \tfrac{1}{2})$$
$$q = 3$$
$$h = 0$$
$$n_\pi = 9 - 8 + 1 = 2 \qquad :\overset{..}{\underset{..}{O}}{=}C{=}\overset{..}{N}:^- \quad \text{or} \quad :\overset{..}{\underset{..}{O}}{-}C{\equiv}N:^- \quad \text{or} \quad :O{\equiv}C{-}\overset{..}{\underset{..}{N}}:^-$$

**TABLE 2–1
TYPICAL LEWIS STRUCTURES
FOR p BLOCK ELEMENTS**

	Number of Valence Pairs			
Group	Total	Shared	Lone	Examples
III	3	3	0	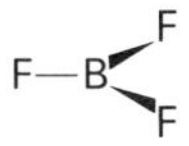 F–B(F)(F)
IV	4	4	0	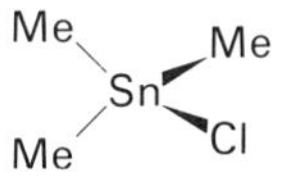 Me₃Sn–Cl
	3	2	1	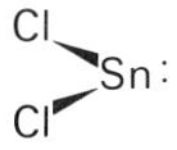 Cl₂Sn:
V	4	3	1	 P₄
	5	5	0	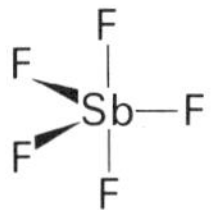 SbF₅
VI	4	2	2	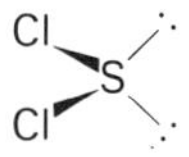 Cl₂S:
	5	4	1	 SF₄
	6	6	0	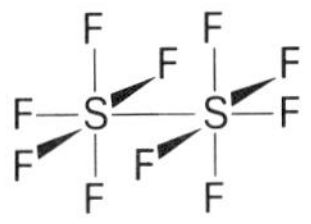 S₂F₁₀
VII	4	1	3	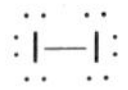 :I–I:
	5	3	2	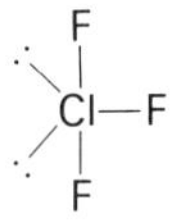 ClF₃
	6	5	1	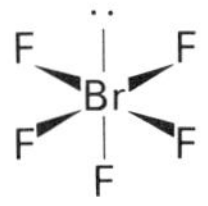 BrF₅
	7	7	0	 IF₇
VIII	4	3	1	 XeO₃
	5	2	3	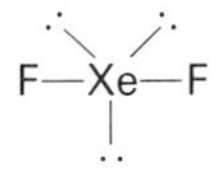 F–Xe–F
	6	4	2	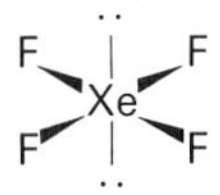 XeF₄

70 ▢ BASIC CONCEPTS OF MOLECULAR TOPOLOGIES

TABLE 2-2
EXAMPLES OF DATIVE STRUCTURES

Group	Behavior	Shared Pairs	Example
IV	acceptor	4 → 5	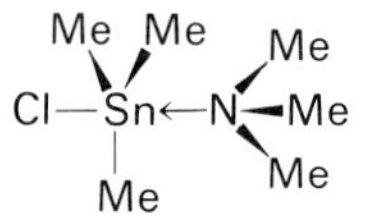 Cl—Sn←N(Me)₃ with Me substituents on Sn
V	donor	3 → 4	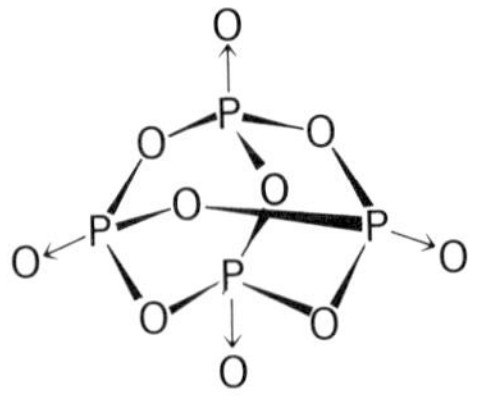 P₄O₁₀ cage structure
	acceptor	5 → 6	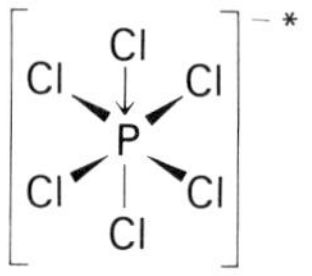 [PCl₆]⁻ *
VI	donor	2 → 3	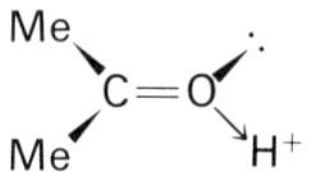 Me₂C=O→H⁺
VII	donor	1 → 2	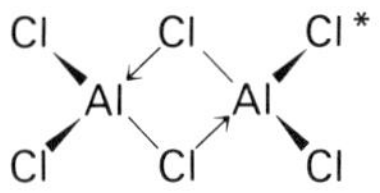 Al₂Cl₆ *
	acceptor	1 → 2	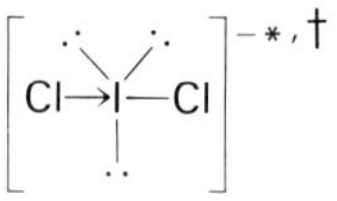 [Cl→I—Cl]⁻ *,†
		3 → 4	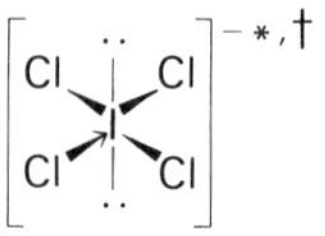 [ICl₄]⁻ *,†
		5 → 6	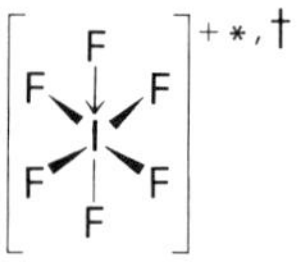 [IF₆]⁺ *,†

*In these examples all shared pairs are equivalent.
†Cf. XeF_2, XeF_4, XeF_6 (of the "typical" category).

bonds are indistinguishable. Similarly indistinguishable are the two pairs for each bridging chloride of Al_2Cl_6, the two chlorides of ICl_2^-, the four chlorides of ICl_4^-, and the six fluorides of IF_6^+.

Implicit to the working through of each of these examples was the knowledge or prediction of how the atoms were to be arranged—which atoms were to be bonded together. Some guidelines to molecular topography are:

1. Draw upon analogies with known, similar compounds.

SHARED AND LONE PAIRS AND LEWIS STRUCTURES

2. The lower electronegativity atoms tend to be central, rather than terminal.
3. Heavier, larger elements tend to exhibit expanded valence shells (more than 4 pairs).
4. Check atom formal charges for reasonableness (zero formal charges preferred, otherwise more electronegative atoms should have negative formal charge).

Let us illustrate the application of these guidelines with some examples. A tetra-atomic molecule like S_2Cl_2 could have one of three topographies

$$O-O-O-O \qquad \overset{O}{\underset{O}{\diagdown}}O-O \qquad \begin{array}{c} O-O \\ | \quad | \\ O-O \end{array}$$

(chain)   (branched)   (cyclic)

With deference to guide 2, we expect sulfur to be central and chlorine to be terminal, where possible:

$$Cl-S-S-Cl \qquad \overset{Cl}{\underset{Cl}{\diagdown}}S-S \qquad \begin{array}{c} S-Cl \\ | \quad | \\ Cl-S \end{array} \quad \text{or} \quad \begin{array}{c} S-S \\ | \quad | \\ Cl-Cl \end{array}$$

Better known analogs for the first two topographies would be H_2O_2 and R_2SO. Following the octet rule, we determine the Lewis structures:

$$:\!\ddot{C}l\!-\!\ddot{S}\!-\!\ddot{S}\!-\!\ddot{C}l\!: \qquad \overset{:\ddot{C}l\!:\oplus \ominus}{\underset{:\ddot{C}l\!:}{\diagdown}}\!S\!-\!\ddot{S}\!: \qquad \begin{array}{c} :\ddot{S}\!-\!\ddot{C}l\!:\oplus \\ | \quad | \\ \oplus:\ddot{C}l\!-\!\ddot{S}\!: \end{array} \qquad \begin{array}{c} :\ddot{S}\!-\!\ddot{S}\!: \\ | \quad | \\ \oplus:\ddot{C}l\!-\!\ddot{C}l\!:\oplus \end{array}$$

$n_\sigma = 3$ $n_\sigma = 3$ $n_\sigma = 4$
$n_\pi = 0$ $n_\pi = 0$ $n_\pi = -1$
$n_\ell = 10$ $n_\ell = 10$ $n_\ell = 10$

Clearly, the first two possibilities make more sense; the cyclic structures become, after inclusion of the extra pair:

$$\begin{array}{c} :\ddot{S}\!-\!\ddot{C}l\!:\!\ominus \\ | \quad | \\ \oplus:\ddot{C}l\!-\!\ddot{S}\!: \end{array} \qquad \begin{array}{c} :\ddot{S}\!-\!\ddot{S}\!: \\ | \quad | \\ \oplus:\ddot{C}l\!-\!\ddot{C}l\!:\!\ominus \end{array}$$

and would likely be so reactive that they would have no finite existence. Of the two non-cyclic structures, only the first is found in the laboratory (see the later discussions in Chapters 6 and 8).

As another example, consider the compound of molecular formula NOH_3. The hydrogen atoms may be assumed to be terminal atoms, so we have the two possible topographies:

$$\underset{H}{\overset{H}{\diagdown}}\!\!\underset{|}{N}\!-\!O \qquad \underset{H}{\overset{H}{\diagdown}}N\!-\!O\!-\!H$$

(You may wish to consider the topography H_2ONH on your own.) Following the octet rule, we get

$$\underset{H}{\overset{H}{\underset{|}{H-N^{\oplus}\!-\!\ddot{O}\!:\!\ominus}}} \qquad \begin{array}{l} n_\sigma = 4 \\ n_\pi = 0 \\ n_\ell = 3 \end{array} \qquad H\!-\!\underset{H}{\overset{|}{\ddot{N}}}\!-\!\ddot{O}\!-\!H$$

72 □ BASIC CONCEPTS OF MOLECULAR TOPOLOGIES

Nature prefers the second of these (hydroxyl amine).

Finally, we examine thiosulfate, $S_2O_3^{2-}$. There are many topographical possibilities for a five-atom system. Perhaps the most obvious choice is the one with a four-branch point (the tetrahedron). Analogy with SO_4^{2-} (hence the name thiosulfate) strongly favors this choice (note that all atoms come from Group VI). Following the octet rule, we develop

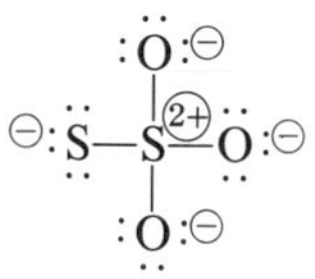

$$\overset{\displaystyle :\overset{..}{O}:^{\ominus}}{\underset{\displaystyle :\overset{..}{O}:^{\ominus}}{^{\ominus}:\overset{..}{\underset{..}{S}}-\overset{(2+)}{\underset{|}{S}}-\overset{..}{O}:^{\ominus}}}$$

Further examples are given in Table 2-3, while Tables 2-1 and 2-2 may be consulted for yet others (since no formal charges are given in the first two tables, you might wish to try your hand at those). Perhaps needless to say, the "derivation" of possible molecular topographies by this procedure is cumbersome for large molecules; after a little experience one learns to rely heavily on analogy with simpler topographies and to build up complex ones from these.

TABLE 2-3
FORMAL CHARGES

Species	Structure		Formal Charge
SF_4 (Sulfur tetrafluoride)	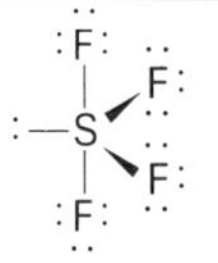 [structure]	S F	0 0
XeF_2 (Xenon difluoride)	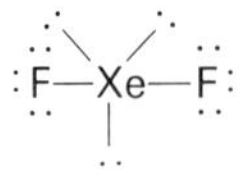 [structure]	Xe F	0 0
NO (Nitrogen oxide)	$:N=O:$	N O	0 0
H_3NO (Hydroxyl amine)	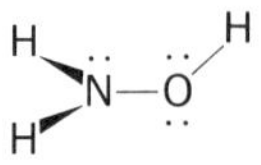 [structure]	N O	0 0
$POCl_3$ (Phosphoryl chloride)	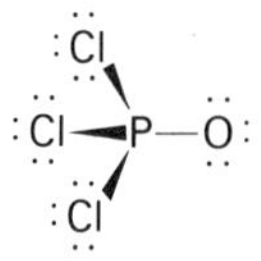 [structure]	P Cl O	+1 0 −1
NO_2^+ (Nitronium ion)	$[:O=N=O:]^+$	N O	+1 0
$S_2O_8^{2-}$ (Peroxydisulfate ion)	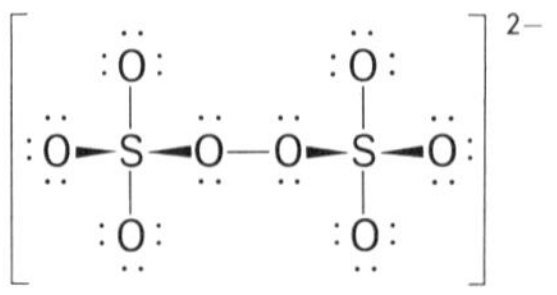 [structure]	S O_c* O_t*	+2 0 −1

*The subscripts c and t stand for central and terminal.

In drawing electron dot structures for molecules with $n_\pi > 0$, one frequently encounters the possibility of writing more than one such structure. Often the possibilities are equivalent, as in the case of NO_2^-. (Note that the net oxygen formal charges = -0.5.)

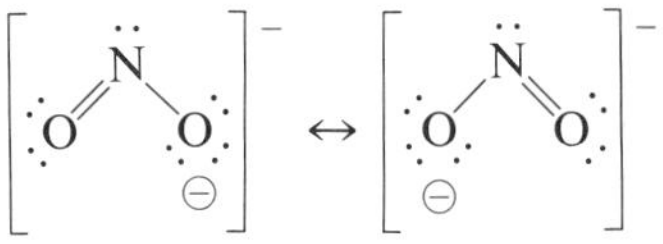

In connection with such structures with more than one shared pair per pair of atoms, the chemist finds it useful to define a quantity called the *bond order* for two atoms. The bond order for two connected atoms is defined as *the average number of shared pairs*. In the NO_2^- example, each NO bond order is 1.5.

Another example arises for the boron trihalides; in BF_3, for example, the fluorine atoms are equivalent. This is an interesting case; satisfying the octet rule for boron requires net F formal charges of $+0.33$. In this sense, BF_3 is electron deficient. Three equivalent structures need to be given, and each BF bond order is between 1.0 and 1.33.

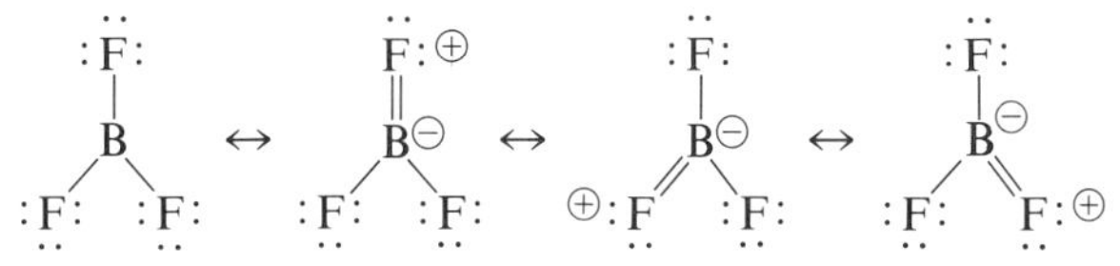

A more complex example of resonance structures is found in $S_2O_3^{2-}$. The thiosulfate anion with the "all octet" resonance structure yields a rather excessive formal charge for sulfur of $+2$. Many more structures may be written to reduce the sulfur charge by allowing some *bonding character for lone pairs;* this is done by designating multiple bonds between the central sulfur and the terminal sulfur and oxygen atoms.

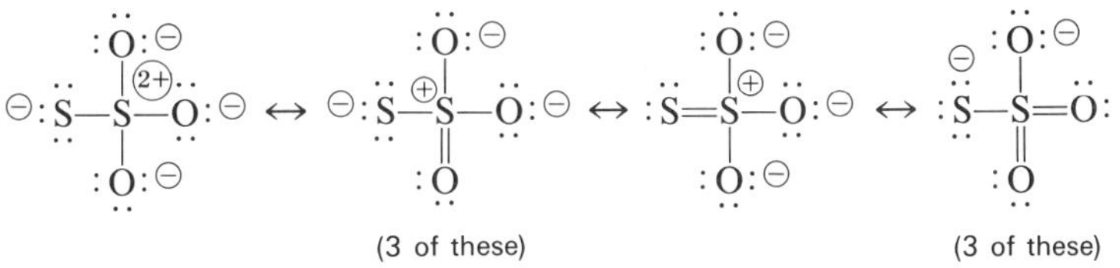

From the last example given, it can be appreciated that some guidelines must be followed to estimate the relative importance of the structures and thus limit the often unbelievably large number of structures that can be written for some molecules. Four such guidelines are listed below and illustrated with one final example, that of thiocyanate ion, SCN^- (similar to cyanate, OCN^-, discussed earlier).

1. *Do not change the relative positions of the nuclei in writing resonance structures.* To do so involves writing structures for different isomers of the same molecular formula (refer to H_3NO, H_2NOH; ClS_2Cl, Cl_2SS).

2. *All structures must have the same number of unpaired electrons.* That is, do not draw one structure with no unpaired electrons and another with two unpaired electrons. Only one possibility exists for the lowest energy electron configuration of the molecule.

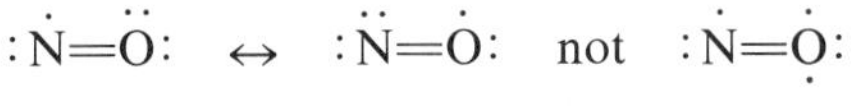

3. Generally, *adhere to the octet or inert gas rule,* but remember that elements from the third or later periods may have expanded octets.

4. Generally, *zero formal charge at each atom is preferred.* In deference to the other rules above, and certainly for ions, it may be reasonable to write structures with non-zero atom formal charges. *Structures that locate negative formal charge at the more electronegative atoms are to be preferred.* In such cases, *avoid a large separation of opposite charge and close proximity of like charges.*

Now let's consider several structures that one might draw for SCN$^-$, and estimate with the aid of the preceding guides the importance of each. By analogy with cyanate (see the earlier discussion) we have (I), (II), and (III).

(I) $^\ominus$:S̈—C≡N: (II) :S̈=C=N̈:$^\ominus$ (III) $^\oplus$:S≡C—N̈:$^{2\ominus}$

Structures (I) and (II) are sound choices, as they adhere to all the rules. (II) is preferred to (I) by item 4. That at least one atom must carry a formal negative charge is dictated by the fact that the species is an anion. *A useful check of any structure drawn is that the sum of the atom formal charges must equal the overall charge on the molecule.* This is simply a consequence of the fact that the algebraic sum of core charges and electron charges must equal the molecule charge. (III) is judged by item 4 not to be very important.

With structures (I) and (II) as the only important ones, and with (II) somewhat more important than (I), we may estimate the SC bond order to be between $1\frac{1}{2}$ and 2. The sulfur formal charge is taken to be less negative than -0.5, while the nitrogen atom formal charge is estimated to be a little more negative than -0.5.

Before continuing to the next section, it is well to point out here that formal charges are often somewhat unrealistic and that use of them in chemical arguments must be made judiciously. The difficulty lies in the convention of considering each shared pair to be *equally* shared by the two atoms in question. For example, reconsidering the molecule BF_3, we estimate from the three double-bond and one single-bond resonance structures that the boron atom carries a formal negative charge (between zero and -1), while each fluorine bears a positive formal charge (between zero and $+0.33$). The assumption that the BF σ pair is equally shared is not a fair representation of the electron distribution in the BF_3 molecule. It is for this reason that the phrase "formal charge" has been used, and such charges should rarely be taken to give an accurate representation of atom charges.

STUDY QUESTIONS

1. Two of the three possible isomers of the anion NCO$^-$ are known (cyanate and fulminate). Develop Lewis structures for these, and compare atom formal charges and bond orders.

2. Draw a Lewis dot structure and give the atom formal charges for $XeOF_4$. Repeat for XeO_3 and ICl_2^+.

3. Draw the important resonance structures for

   (a) NNO (nitrous oxide)
   (b) allyl radical (H_2CCHCH_2)
   (c) F_2SO (thionyl fluoride)
   (d) *N,N*-dimethylacetamide

$$CH_3-C\begin{smallmatrix}\nearrow O \\ \searrow N-Me \\ | \\ Me\end{smallmatrix}$$

   (e) HOCN and OCN$^-$; how do you expect the OC and CN bond strengths to differ in these two molecules?

ELECTRON PAIR REPULSION MODEL

The next step in the construction of a working model of molecular structure concerns the three-dimensional aspects of structure. Here we will use the qualitative but highly useful idea that geometries of molecules can be predicted from consideration of repulsions between pairs of electrons about an atom.[6] Simply put, *we assume (i) that each valence electron pair of an atom is stereochemically significant, and (ii) that repulsions between these pairs determine molecular shape.* The orientations of bond pair electrons will serve to define the spatial arrangement of nuclei in a molecule.

The angular distribution of valence electron pair clouds about a nucleus follows the dictates of some repulsive force law which must reflect the coulombic correlation forces between electrons.[7] The exact form of the force law ($F = 1/r^2$ is appropriate for *point* charges) seems to be of little importance for two to six such electron pairs; for seven pairs, however, a unique result is not possible. Table 2-4 summarizes the results of this model. The first six basic structural types are common; you should become facile in recognizing into which basic structural pattern a molecule will fall, once you have written an electron dot structure for the molecule.

---

[6] R. J. Gillespie, "Molecular Geometry," Van Nostrand Reinhold Co., London (1972); *J. Chem. Educ.,* **51**, 367 (1974); *ibid.,* **47**, 18 (1970).

[7] It has been assumed by many that "Pauli forces" between the spin-bearing electrons also determine the stereochemistry of valence electron pairs. This idea is under attack: J. L. Bills and R. L. Snow, *J. Amer. Chem. Soc.,* **97**, 6340 (1975); R. L. Snow and J. L. Bills, *J. Chem. Educ.,* **51**, 585 (1974). See also footnote 16 on p. 39.

**TABLE 2-4
IDEALIZED ORIENTATIONS OF e⁻ PAIRS ABOUT AN ATOM**

No. of e⁻ pairs	Idealized Structure of e⁻ pairs		Angles
2	linear	:—E—:	180°
3	trigonal plane		120°
4	tetrahedron		109.5°
5	trigonal bipyramid	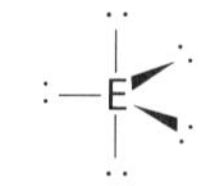	90°, 120°
6	octahedron		90°
7	pentagonal bipyramid		90°, 72°
	capped octahedron "1:3:3"		
	capped trigonal prism "1:4:2"		

76 BASIC CONCEPTS OF MOLECULAR TOPOLOGIES

Considerable refinement of these basic structural units can be made by distinguishing between lone pairs of electrons and bond pairs of electrons. With good justification, *we assume that the spatial requirements of lone pairs are greater than those of bond pairs.* Such a premise is not unreasonable when it is realized that lone pairs move under the effective nuclear charge of only one atom, while bond pair electrons experience the effective nuclear charges of two atoms. We expect bond pairs to be less angularly diffuse and more radially distended than lone pairs about an atom.

$$E\,\substack{:\\ } \qquad E\,\substack{\cdot\cdot\,} X$$

lone pair bond pair

An equivalent view of this difference between lone and bond pairs is that the amount of electron density near the valence core is different. Lone pairs are assumed to be localized at the atom in question and thus create two negative charges near the valence core. Bond pairs are shared, with the result that the net electron density near a core is less than two electrons. Consequently, *we expect the order of repulsion between pairs to decrease as*

$$lp/lp \gg lp/bp > bp/bp$$

lp/lp lp/bp bp/bp

A final aspect of the model concerns the interpair angles in Table 2-4. The repulsive forces should markedly decrease with increasing angle. Thus, while repulsions may be significant at all angles between 90° and 180°, they are much more severe at 90° than at 120° and are lowest at 180°. As you will see in the following examples, repulsions at 90° provide a key to molecular geometry. All the foregoing conceptualizations and assumptions define what is referred to as the *valence shell electron pair repulsion model (VSEPR model)*.

A consequence of these hypotheses, considering the stereochemistry of four pairs for example, is that four lone pairs or four bond pairs will be distributed with interpair angles of 109° 28′. If one of the pairs is a lone pair while the other three are bond pairs, the forces acting on the bond pairs will not be balanced in the regular tetrahedron configuration; the bond pairs will be forced together slightly until the forces on all pairs do balance. In this case we expect the *interbond pair angles* to be less than 109.5°.

To give two examples of the utility of the electron repulsion concept in distinguishing structural isomers, we will consider ICl_4^- and ClF_3. In drawing an electron dot structure for ICl_4^-, one quickly sees that there are six electron pairs about the iodine atom.

$$\begin{array}{c} Cl \quad \substack{\cdot\cdot\\|} \quad Cl \\ \diagdown I \diagup \\ Cl \diagup \, | \, \diagdown Cl \\ \substack{\cdot\cdot} \end{array}$$

Consequently, the basic stereochemical arrangement for the six pairs is the octahedron. Interestingly, two possibilities arise when a stereochemical view of the ion is sketched. One possibility is that all four chlorines are coplanar with the iodine (see sketch A below). The other possibility has one chlorine perpendicular to the plane defined by the iodine and the remaining three chlorine atoms. The observed structure is the square planar arrangement of atoms; this is nicely rationalized by the electron repulsion model, which predicts a higher molecular energy for the juxtaposition of two lone pairs at 90° to each other (as in sketch B) than for their separation by 180°.

In making the valence shell electron pair repulsion (VSEPR) analysis of the two structures, we list below each structure the number of pair interactions of each type (all at 90° in the ICl_4^- example). On comparison of (A) with (B), we see that two lp/bp repulsions in (A) are replaced by one lp/lp and one bp/bp interaction. The lp/lp repulsion at 90° is expected to be quite large and disfavors (B).

(A) Cl–I(Cl)(Cl)(Cl)Cl (B) Cl–I(Cl)(Cl)(Cl)Cl

lp/lp = 0 → 1 = lp/lp
lp/bp = 8 → 6 = lp/bp
bp/bp = 4 → 5 = bp/bp

The second example, that of ClF_3, leads to the prediction of the basic trigonal bipyramidal orientation of electron pairs of chlorine. Three isomeric structures are encountered: (A), the two lone pairs equatorial; (B), one lone pair equatorial and the other axial; and (C), both lone pairs axial. The observed structure is that of a T-shaped molecule, i.e., the first possibility. The summary of interpair repulsion types below each sketch should make it clear that the order of stability is (A) > (B) or (C). In making the VSEPR analysis of ClF_3, we focus our attention on pair interactions at 90°, recognizing that the interactions at 120° will be much less important. The (A) → (B) comparison reveals conversion of a lp/bp repulsion into a lp/lp repulsion. For (A) → (C) we find conversion of two bp/bp interactions into two of the lp/bp type. Hence, (A) is preferred.

(A) :–Cl(F)(F)(F): (B) :–Cl(F)(F)(F) (C) F–Cl(F)(F)

(A)	(B)	(C)
lp/lp = 0	lp/lp = 1	lp/lp = 0
lp/bp = 4	lp/bp = 3	lp/bp = 6
bp/bp = 2	bp/bp = 2	bp/bp = 0

As a finer point of the geometry of (A), we expect the axial Cl—F bonds to be somewhat longer than the equatorial bond because the axial pairs are acted upon by two lp and one bp (at $<120°$) whereas the equatorial pair is acted upon by only two bp (at $<120°$). Similarly, the FClF angles should be less than 90°. Both expectations are borne out by the experimental structure (Cl—F_{eq} = 1.598 Å, Cl—F_{ax} = 1.698 Å; FClF = 87.5°).

Be sure to note that in the preceding examples we have carefully distinguished between *molecular structure* and *electron pair orientations*. The former describes the relative nuclear positions and is subject to experimental determination; electron pair orientations cannot be experimentally determined.

Thus far we have dealt only with examples for which we conceive of a single bond pair. The occurrence of double and triple pairs in a given internuclear region certainly enhances repulsions between that region's bond pair(s) and other pairs. Note that second and third pairs do not define new stereochemical directions, for their orientation mimics that of the first pair. An interesting example of this is found in the series NO_2^+, NO_2, and NO_2^-, which differ progressively by one electron and by a large reduction in NO bond order from NO_2^+ to NO_2^-. The most important resonance structures, combined with the electron repulsion model, lead us to expect—in agreement with experiment—a progressively decreasing ONO angle over the series.

$$:\ddot{O}=\overset{\oplus}{N}=\ddot{O}: \rightarrow \text{linear} \quad O—N—O \quad 180°$$

$$\overset{\ominus}{:}\ddot{O}—\overset{\oplus}{N}=\ddot{O}: \rightarrow \text{bent} \quad O \overset{N}{\underset{\leftrightarrow}{}} O \quad 132°$$

$$\overset{\ominus}{:}\ddot{O}—\ddot{N}=\ddot{O}: \rightarrow \text{bent} \quad O \overset{N}{\underset{\leftrightarrow}{}} O \quad 115°$$

As an aside, you might note that NO_2 is an enigma, like NO, when one tries to satisfy the octet rule for all atoms. To avoid exceeding the octet for oxygen the odd electron is assigned to nitrogen.

In addition to being useful for isomer prediction, the VSEPR model can be used to account for some otherwise perplexing trends in bond angles about an atom. The Group V hydrides present an interesting case, and one that will help us further generalize the model.

$$\underset{107°}{NH_3} > \underset{94°}{PH_3} \gtrsim \underset{92°}{AsH_3} \sim \underset{92°}{SbH_3}$$

We note first that all HMH angles are less than the 109° expected for the idealized tetrahedral stereo-arrangement of four electron pairs about an atom. This is presumed to be due to the greater lp/bp repulsions than bp/bp repulsions in the idealized tetrahedral configuration. The decrease in angle with increasing atomic number in Group V can be related to the *decreasing difference in central and terminal atom electronegativities.*[8] In suggesting this, we are attempting a refinement of the basic electron repulsion model in order to increase its utility. The effect of decreasing Group V electronegativities is that the bond pairs are drawn further away from the central atom core, thereby decreasing the bp/bp repulsions more than the lp/bp (*both* pairs of a bp/bp repulsion defining a bond angle are affected). This argument clearly accounts for the trend of decreasing bond angle with increasing atomic number of the central atom. To expect more quantitative information from a strictly qualitative and simplified model of molecular electronic structure would be an exercise in frustration. Thus, an attempt to rationalize the small differences in angle after PH_3 solely in terms of electron repulsions is probably not really worth the effort.

Another interesting comparison that emphasizes the importance of the relative atom electronegativities is that of NH_3 and NF_3, for which the interbond angles are 107° and 102°, respectively. This order is the opposite of that expected on the basis of steric interactions between substituents. On the other hand, the electron repulsion model is consistent with this order of bond angles if the reasonable proposition is made that the bond pair repulsions are less when the terminal atoms are fluorines. This follows from the expected overall displacement of the bonding electrons away from the nitrogen when the substituents are extremely electronegative. Additionally, we expect the withdrawal of charge from the valence core to increase the effective

---

[8] Be careful here. We expect an increase in radial diffuseness of both lone and bond pairs as the central atom becomes heavier. This alone would tend to *reduce* the tensions between *all* valence pairs and would lead us to expect less distortion from the ideal tetrahedral angles as the central atom becomes heavier.

nuclear charge for the lone pair, leading to a contraction of that density and greater lp/bp repulsion.

Two comparisons that show the necessity of considering the effects of multiple bonding are found in the pair PH_3 and PF_3, and in the series OF_2, OH_2, and OCl_2. For the former the bond angles are 94° and 98°, respectively—just the opposite of the order for NH_3 and NF_3. The facts that fluorine has three "unshared" pairs while hydrogen has none and that phosphorus need not follow the octet rule (while nitrogen must) mean that resonance forms such as

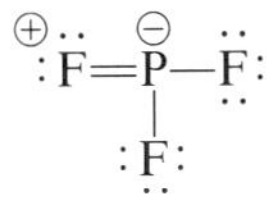

$$:\overset{\oplus}{\ddot{F}}=\overset{\ominus}{\underset{\underset{:\ddot{F}:}{|}}{\ddot{P}}}-\ddot{F}:$$

are possible for PF_3 but not for NF_3 and PH_3. Even though the expected bond order is certainly less than two, it will nevertheless be greater than one, and the extra electron density in the PF internuclear region will tend to increase electron repulsions between bond pairs more than between lone and bond pairs (because both bonds defining a bond angle are affected).

A striking example is to be found in the series OF_2, OCl_2, and OH_2, for which the order of increasing angle should be

$$OF_2 < OCl_2 < OH_2$$

but is found to be

$$OF_2 < OH_2 < OCl_2.$$

The resonances structures

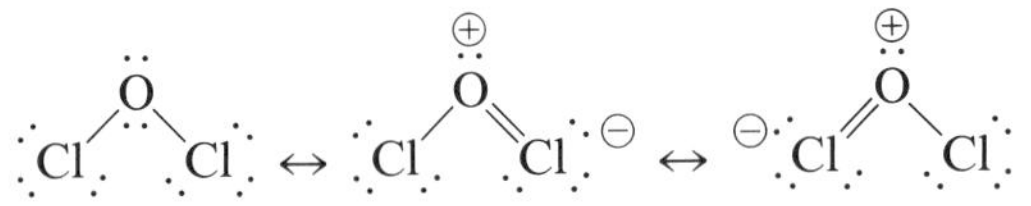

are not possible for X = F or H, and thus could account for the misplaced position of OCl_2 relative to OH_2. This example is important, for not only do we anticipate enhanced bp/bp repulsions, but the "double bond" structures also imply only three stereochemically active pairs about oxygen and therefore a larger angle.

With regard to the effects of multiple bonding on molecular geometry, the OCl_2 example warns that the occurrence of resonance structures involving lone pair electrons at a central atom usually can influence structure more drastically than is indicated by the PF_3 example. For example, nitrous acid exists in *cis* and *trans* isomeric forms, which attests to the importance of the central oxygen-nitrogen double bond resonance structure.[9]

---

[9] The fact that *cis* and *trans* forms have discrete existence is associated with a significant barrier to rotation about the (HO)—(NO) bond. A single bond would permit essentially "free" rotation, whereas double bond character would tend to favor a planar structure (see the discussion of double bonding on p. 80).

This leads us to expect an HON bond angle of something greater than 109° but less than 120°. A similar situation is expected for the HOB angles of boric acid:

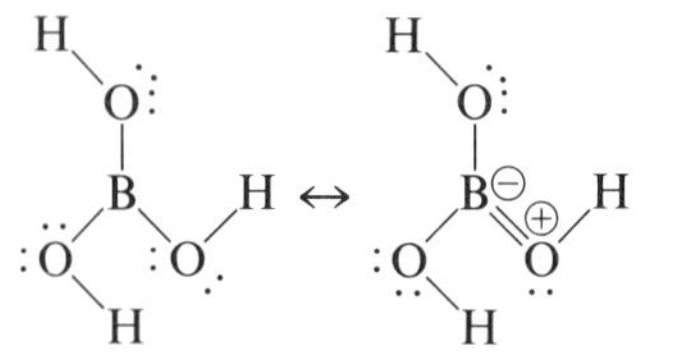

$$\text{(resonance structures of boric acid)}$$

The near planarity of formamide is also rationalized by consideration of the second resonance structure below.

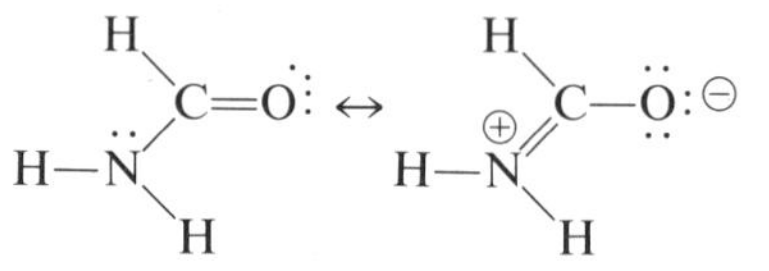

$$\text{(resonance structures of formamide)}$$

In spite of the notable successes of the electron pair repulsion model, there are some failings (it is not unexpected that such a qualitative model should have some failings). Some have been listed and discussed in terms of non-bonded (steric) repulsions by Bartell.[10] Two examples are given here. First, the XCX angles of $H_2C=CX_2$ (X = H, Cl, F) are all observed to decrease when the CH_2 moiety is replaced by oxygen. The second example centers on the ambiguity in structure for seven electron species. The Xe of XeF_6 has 6 bp + 1 lp. If the one lone pair is presumed to be stereochemically significant, one might expect (from the precedent of IF_7) a pentagonal bipyramidal arrangement of electron pairs. In fact, the structure of XeF_6 depends on the sample phase. In the solid state[11] one finds tetrameric and trimeric "rings" of $(XeF_5^+F^-)$ ion pairs, such that each XeF_5^+ cation has the distorted octahedral structure expected from VSEPR considerations. Notice that the trimeric "chair" rings are fused in pairs so the molecular unit is actually a hexamer. Each F_5Xe^+ bridges $3F^-$ and conversely.

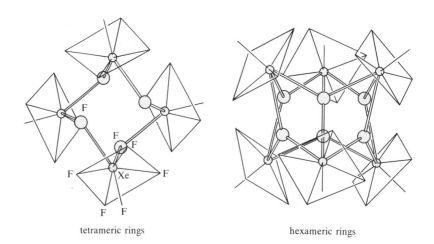

tetrameric rings   hexameric rings

---

[10] L. S. Bartell, *J. Chem. Educ.*, **45**, 755 (1968).
[11] R. D. Burbank and G. R. Jones, *J. Amer. Chem. Soc.*, **96**, 43 (1974).

The structure of monomeric (vapor phase) XeF_6 also reveals a stereochemically active role for the Xe lone pair, in that the structure is based on a *distorted* octahedron of F about Xe.[12] Whether the lowest energy geometry is that of a "1:4:2 capped trigonal prism" **(1)**, a pentagonal bipyramid **(2)**, or a "1:3:3 capped octahedron" **(3)** has been unsettled (at present the 1:3:3 structure is favored) because the geometry is not static—an intramolecular rearrangement process rapidly interconverts fluorine positions. An intriguing possibility is that of "1:3:3" interconversion via the pentagonal bipyramid. The mystery is compounded by the report[13] that the ions $TeCl_6^{2-}$, $TeBr_6^{2-}$, and $SbCl_6^{3-}$ all have octahedral structures. These latter structures are, of course, for the ions in an ionic lattice, where the adjacent cations may exert a structure-determining influence on anion structure.

In comparison with IF_7, all this points up the difficulty in accounting for any observed structure with seven valence pairs, particularly when one of those pairs is unshared.

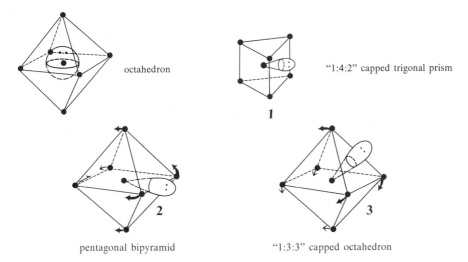

[12] L. S. Bartell and R. M. Gavin, Jr., *J. Chem. Phys.,* **48**, 2460 (1968); S. Y. Wang and L. L. Lohr, *ibid.,* **61**, 4110 (1974); U. N. Nielsen *et al., ibid.,* **61**, 3581 (1974); K. S. Pitzer and L. S. Bernstein, *ibid.,* **63**, 3849 (1975).

[13] S. Lawton and R. Jacobsen, *Inorg. Chem.,* **5**, 743 (1966).

STUDY QUESTIONS

4. Sketch the possible isomers of SF_4, predict the stable isomer, and describe the FSF angles as $>$, $=$, or $<$ the "ideal" angles.

5. Which in each pair will have the larger bond angle?
   (a) CH_4 or NH_3
   (b) HOCl or HOH
   (c) NO_2^- or O_3

6. Give an explanation for the fact that $\angle$ HCH is greater than $\angle$ FCF (C_2H_4 *vs.* C_2F_4), while $\angle$ HPH is less than $\angle$ FPF (PH_3 *vs.* PF_3).

7. Describe the bond angles of $XeOF_4$ in relation to the "ideal" angles. Repeat for $SnCl_3^-$.

SYMMETRY CONCEPTS[14]

POINT GROUPS

Now that we have discussed a working hierarchy of models involved in reaching a prediction of a molecule's structure, it is convenient to introduce a formal system for classifying molecules according to their structures. To be specific, we will characterize a molecular structure in terms of the elements of symmetry that it possesses and introduce a system of special symbols, called *point group symbols,* that convey all the symmetry elements possessed by a molecule. *Molecular symmetry elements are planes, axes, and points that relate equivalent nuclear positions.* Every molecule belongs to a *point group,* which is defined by the numbers and kinds of symmetry elements (called a group of elements) having the property of a common intersection, usually a point (hence the name point group).

Fortunately, there are only two basic symmetry elements used to classify the structure of a molecule—the *proper rotation axis* (C_n) and the *improper rotation axis*

---

[14] M. Orchin and H. H. Jaffe, *J. Chem. Educ.,* **47,** 246, 372, 510 (1970) is an excellent supplementary source for the topics of this section. Also, G. Davidson, "Introductory Group Theory for Chemists," Applied Science Publishers, London, 1971, is highly recommended at this level. Also, F. A. Cotton, "Chemical Applications of Group Theory," 2nd Ed., Wiley-Interscience, N.Y., 1971.

Figure 2–1. Two examples (*Circle Limit IV* and *Whirlpools*) of symmetry in art from M. C. Escher, famous for his graphic (and sometimes paradoxical) designs. (With permission of the Escher Foundation—Haags Gemeentemuseum—The Hague.)

Illustration continued on opposite page.

Figure 2-1. *Continued.*

TABLE 2-5
MOLECULAR SYMMETRY ELEMENTS, OPERATIONS, AND SCHOENFLIES SYMBOLS

Element	Operation	Symbol
Axis	Rotation by 2π	E^*
	Rotation by $\frac{2\pi}{n}$	C_n
Axis $+ \perp$ plane	(Improper rotation)	S_n
	Rotation by $\frac{2\pi}{n}$ followed by reflection through a plane perpendicular to rotation axis	
Plane	Reflection	σ^* (σ, σ_v, σ_d, σ_h)
Point	Inversion	i^*

*Note that E is a special case of C_n, and that i and σ are special cases of S_n: $S_1 = \sigma$ and $S_2 = i$.

(S_n), as defined in Table 2-5. The special cases of C_1, S_1, and S_2 are commonly identified as the *identity* (E), the *reflection* (σ), and the *inversion* (i), respectively. Notice that subscripts (to be explained later) are given for the reflection element.

We will now proceed to examine some specific examples. The molecule PF_2H is predicted by the electron repulsion scheme to be a pyramidal molecule, as shown in Figure 2-2. Only one symmetry element, other than the identity, can be identified here—a reflection plane containing the phosphorus and hydrogen atoms and bisecting the F—P—F angle. A molecule possessing these two elements, and only these two, is said to belong to the C_s point group. Conversely, a molecule with C_s symmetry possesses only these two symmetry elements. A molecule with no symmetry beyond the identity is classified as C_1; an example is $NDHCH_3$, shown in Figure 2-2.

The water molecule is the next example. As shown in Figure 2-2, H_2O possesses three distinct symmetry elements beyond the identity. First, there is the two-fold rotation axis (after rotation by 180° about this axis one could not tell that the operation had been effected); secondly, there is a reflection plane coincident with the molecular plane; and thirdly, there is a reflection plane perpendicular to the molecular plane and bisecting the HOH angle (the space to one side of a reflection plane is the mirror image of the space to the other side). As the two reflection planes are *not* equivalent (related by any of the other symmetry elements of the group), they are given separate designations. The first plane we identify as σ_v', and the second plane is given the symbol σ_v' to distinguish it from the first; the axis is called C_2. The collection of symmetry elements E, C_2, σ_v, and σ_v' is referred to as the C_{2v} point group. The Schoenflies point group symbol itself carries the complete information as to which elements are present, that is, the C_2 rotation axis and two σ_v planes. *The meaning of*

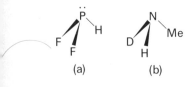
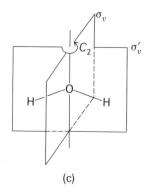

Figure 2-2. Examples of (a) C_s and (b) C_1 symmetry, and (c) the elements constituting C_{2v} symmetry.

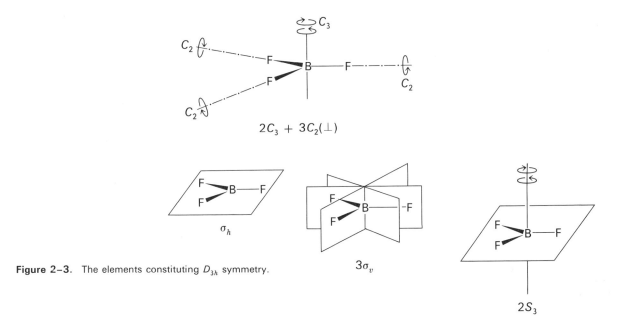

Figure 2-3. The elements constituting D_{3h} symmetry.

the subscript v (for vertical) for the reflection planes is that the planes contain (their intersection defines) the C_2 rotation axis. Note that the "v" subscript was not used to describe the plane in the C_s point group, which has no rotation axis. The molecule ClF_3 is another example of C_{2v} symmetry (see structure A, page 77).

The next example is BF_3. On the basis of the VSEPR model, the molecule is predicted to be precisely trigonal planar. As shown in Figure 2-3, there are several symmetry elements associated with the trigonal planar structure. First, there is the three-fold rotation axis, C_3. Actually there are two coincident three-fold axes, one corresponding to the operation of clockwise rotation and the other to the operation of counterclockwise rotation of the molecule by 120°. These are related by each of the C_2 elements to be discussed next (any of the C_2 operations converts a clockwise C_3 arrow into a counterclockwise arrow), and are therefore not distinguished from each other but are identified as the class $2C_3$. *Elements interconverted by other operations define a class, designated by* n *(Schoenflies symbol).* Unlike the σ_v and σ'_v elements of C_{2v}, such elements share a common designation (Schoenflies symbol).

Second, there are the two-fold rotation axes defined by the BF internuclear axes. As these are related by the C_3 axes, they are not separately designated but are identified as the class $3C_2$. There are no further proper rotation axes.

Examining the predicted structure for reflection planes, we find several. Perhaps the most obvious is the molecular plane. *As this plane is perpendicular to the highest-fold rotation axis,* which is the C_3 axis, *we designate it as σ_h (for horizontal) to avoid confusion with reflection planes,* identified with the v subscript, *which contain the highest-fold rotation axis.* In the vertical category there are three reflection planes, all related by the C_3 element, and these are identified as the class $3\sigma_v$. There is no inversion element for this molecule, but there are further improper rotation elements. The C_3 axis combined with the σ_h reflection plane serves to define S_3. As there are two of these (clockwise and counterclockwise C_3 rotation), and they are related by the C_2 axes, they are designated collectively as the class $2S_3$. The identity (E) completes the specification of symmetry elements for BF_3. This collection of elements is identified by the symbol D_{3h}. *The symbol D_n is meant to convey the presence of an n-fold rotation axis, C_n, and n 2-fold axes perpendicular to C_n*. The presence of a horizontal reflection plane is identified by adding the subscript h to the point group symbol. The combination of C_n, n 2-fold axes perpendicular to it, and σ_h guarantees the existence of n σ_v planes (see Study Questions). The presence of σ_h combined with C_n also guarantees the presence of several S_n elements. Hence, the basic information of a C_n, n 2-folds perpendicular to C_n, and the σ_h is all that is needed to characterize the collection of elements for BF_3, and the point group symbol becomes D_{3h}. Compare this with the pyramidal molecule NF_3, which lacks σ_h and n C_2 axes perpendicular to C_3. Only E, $2C_3$, and $3\sigma_v$ are left; therefore, the point group symbol is C_{3v}.

Our final detailed example is for the ICl_4^- molecule (Figure 2-4). The considerations of the VSEPR model led us to predict that this molecule is planar with equivalent I—Cl bonds. The highest-fold rotation axis is C_4. There are two of these (clockwise and counterclockwise rotations by 90°) related by any of four perpendicular rotation axes, so they are grouped together as the class $2C_4$. Coincident with C_4 elements, there is a C_2 axis. In the plane of the molecule we find four C_2 axes, which may be grouped into two pairs (members of different pairs are not symmetry-related). To avoid confusion, one pair (those along the I—Cl axes) are designated as $2C_2'$, while the other pair (those bisecting the Cl—I—Cl axes) are represented by $2C_2''$.

For reflection planes, we find first the molecular plane, called σ_h in keeping with the convention encountered with BF_3. *Recognition of these elements is sufficient to identify the point group as D_{4h}.* The four reflection planes defined by the C_4 axis and the perpendicular C_2 axes (and required by C_4, $4 \perp C_2$, and σ_h) may be grouped into pairs corresponding to the two pairs of C_2 axes. One pair is identified as $2\sigma_v$ and the other pair is identified as $2\sigma_d$. Out of deference to the fact that the second pair of reflection planes has the special property of *bisecting the angles between equivalent C_2 axes* (the pair $2C_2'$), they are given the special d (for diagonal) subscript rather than a prime superscript as in the H_2O example. There is a center of inversion (at the iodine atom) for this molecule (a result of $C_2 + \sigma_h$), and we identify the element i. Finally, there are two S_4 axes, defined by the two C_4 axes and σ_h, and we group them as $2S_4$. This collection, including the identity, is given the point group symbol D_{4h}. The analogy with the D_{3h} example just given is appropriate. With an even value for n of D_{nh}, there must be a C_2 coincident with C_n, and this requires the presence of i (or S_2).

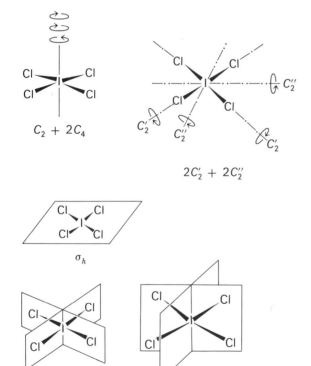

Figure 2-4. The planes and axes of D_{4h} symmetry.

**TABLE 2-6
COMMON POINT GROUPS
AND THEIR ELEMENTS**

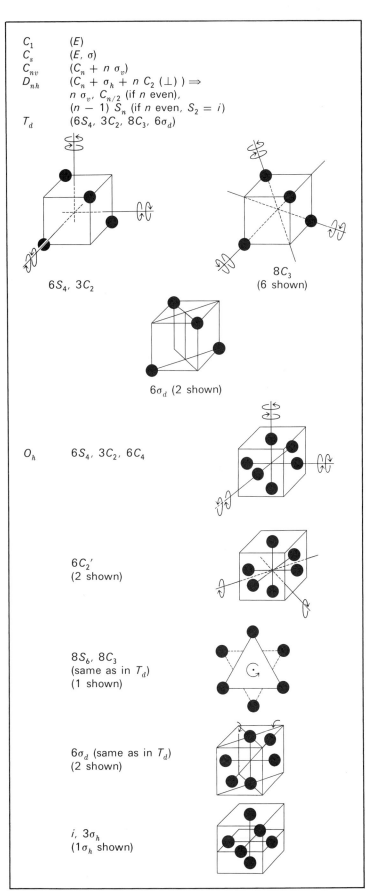

BASIC CONCEPTS OF MOLECULAR TOPOLOGIES

Number of Stereochemically Active e⁻ Pairs	Example	Point Group
2	O=C=O	$D_{\infty h}$
	H—C≡N	$C_{\infty v}$
3	[NO₃]⁻	D_{3h}
	[CSO₂]²⁻	C_{2v}
4	BFClH	C_s
	[PCl₄]⁺	T_d
	XeO₃ (with lone pair)	C_{3v}
	SnH₂(CH₃)₂	C_{2v}
5	PF₂Cl₃ (axial F)	D_{3h}
	PF Cl₃ (with additional Cl)	C_{3v}
	ClF₃	C_{2v}
6	SF₆	O_h
	[ICl₄]⁻	D_{4h}
	BrF₅	C_{4v}

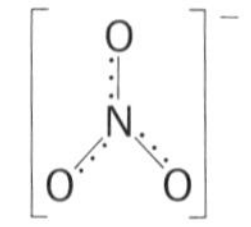

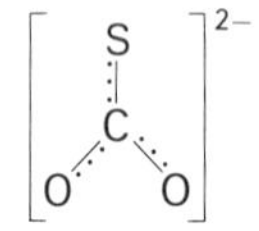

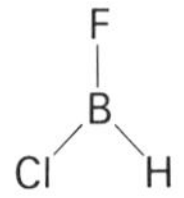

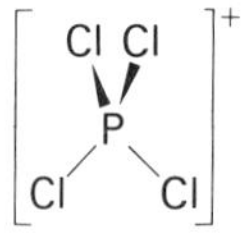

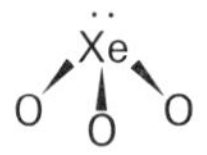

TABLE 2-7
EXAMPLES OF STRUCTURE–POINT GROUP RELATIONS

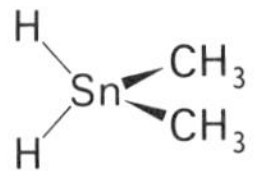

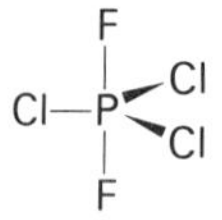

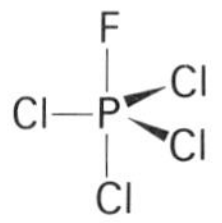

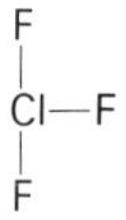

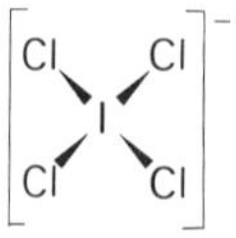

To summarize our discussion, thus far we have identified the special point groups C_1 (which consists of only the identity, E) and C_s (which consists of the identity, E, and a single reflection plane σ). The two general point groups C_{nv} and D_{nh} were discussed because they are appropriate for a great number of molecular structures. Two further special cases need to be mentioned: the tetrahedron, with point group symbol T_d, and the octahedron, with symbol O_h. A summary of the elements of these point groups is given in Table 2-6. Figures that will help in location of the symmetry elements are included for the tetrahedron and octahedron.

For practice in identifying the symmetries of commonly occurring molecular structures, Table 2-7 gives examples of molecules identified by the number of stereochemically active electron pairs and by point group. You should familiarize yourself with the content of this table, paying particular attention to the relation of the stereochemical structural types to their associated point groups.

STUDY QUESTIONS

8. State the point groups of the following species.

    (a) Al_2Cl_6
    (b) HOCN
    (c) an atomic d_{z^2} orbital
    (d) $SnCl_3^-$
    (e) BrF_5
    (f) FNH_2
    (g) Cl_3PO
    (h) the stable isomer of ICl_4^-
    (i) the stable isomer of ClF_3
    (j) $XeOF_4$
    (k) XeO_3
    (l) ICl_2^+
    (m) XeO_3F_2

9. How many chlorines of CCl_4 would need to be replaced with hydrogens to give molecules of C_{3v} and C_{2v} symmetries? Repeat for BF_3, to give C_{2v} symmetry. Are any of these molecules without S_n axes (remember $S_1 = \sigma$, $S_2 = i$) and therefore optically active?

10. Using the adjacent figure, demonstrate by successive application of C_3, $C_2^\perp$, and σ_h that the result is identical to that achieved by a σ_v alone (proving that C_3, $C_2^\perp$, σ_h imply the presence of σ_v).

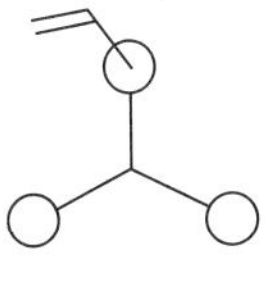

CHARACTER TABLES

In discussing molecular electronic structures at levels more advanced and quantitative than those of the Lewis and electron repulsion concepts, we deal with three-dimensional functions (atomic and molecular orbitals). It is most useful to be able to categorize the symmetries of atomic and molecular orbitals according to their behavior under the symmetry operations of the molecular point group. One advantage of determining the symmetries of orbital functions with respect to the elements of the molecular point group is that, by so doing, we systematize our thoughts on electronic structure. Orbital functions can be grouped according to their likenesses and differences under the operations of the point group.

For each point group there are certain inherent, basic symmetry patterns (for functions of space coordinates such as x, y, z or r, θ, ϕ) in terms of which we may attempt to describe complex patterns (functions). These complex patterns may or may not match the basic patterns; those that do not can be identified as combinations of the simpler, basic patterns. In this sense, the complex patterns can be decomposed or reduced to combinations of the basic types and are called *reducible*. The basic symmetry patterns are called irreducible.[15]

---

[15] A simple analogy is the resolution of a vector into its cartesian components. Unit vectors along the x, y, and z axes are irreducible, while the general vector is reducible into its components.

By way of example, we can imagine in the case of the C_{2v} point group any number of functions that are symmetric to all the operations of that point group. *By "symmetric to an operation" we mean that the function is transformed into itself by that operation.* An example would be the oxygen $2s$ orbital in the water molecule (see Figure 2–2). Similarly, the oxygen $2p_z$ orbital, if the z axis is defined to coincide with the C_2 axis, is also totally symmetric in the C_{2v} point group. On the other hand, the oxygen $2p_y$ orbital (if defined as perpendicular to σ_v) is symmetric with respect to the identity and σ_v' elements, but is antisymmetric (is transformed into the negative of itself) with respect to the C_2 and σ_v elements.

Continuing with the H_2O example, it is easy to see that one of the hydrogen $1s$ orbitals *alone* has neither symmetric nor antisymmetric behavior with respect to the C_{2v} symmetry elements. The difference between the simplicity of the oxygen orbital symmetries and the complexity of the hydrogen $1s$ orbital in this example arises from the fact that the center of the oxygen atom orbitals lies along the intersection of all the C_{2v} elements, whereas the hydrogen orbital center lies elsewhere. For the time being we will concern ourselves only with the symmetry properties of orbitals whose centers are located along the intersection (remember that for most point groups, the intersection is a point). Later we will return to examine more complex circumstances such as that incurred by the hydrogen atoms of H_2O.

Table 2–8 presents the *character table* for the C_{2v} point group. Our approach here will not be to present a lengthy discussion of the mathematical derivation of this table, but will be to familiarize you with the terminology of the symbols appearing in the table and to show you how the table conveys valuable information about the basic (irreducible) symmetry patterns for the point groups. Later we will see how to reduce the symmetries of more complex three-dimensional functions into irreducible components.

In the upper left-hand corner of the character table is given the (Schoenflies) point group symbol for that table. Across the top of the table are listed the symmetry elements by classes. These classes are used to identify columns of the table. Along the left-hand side of the table, arranged vertically so as to identify rows of the table, are given the *Mulliken irreducible representation symbols.* These are equal in number to the number of distinct classes of elements for the group, ensuring that the table is a square array. The fact that the number of irreducible representation symbols is equal to the number of distinct classes can be proved in general, but the proof will be deferred to a course that gives a rigorous mathematical development of character tables. *The numbers that appear within the body of the table are called characters,* and these are unambiguously derived by an elaborate procedure for each point group. Each row of characters defines an *irreducible representation* or basic symmetry pattern for the point group. It is these characters that, in effect, *represent* the effects of the individual symmetry operations on the basic (irreducible) symmetry functions. The irreducible representation symbols are shorthand notations for the string of characters in each row and thus identify the basic patterns, just as C_{2v} is a shorthand notation for the E, C_2, σ_v, and σ_v' elements. Characters of $+1$ and -1 identify the symmetric and antisymmetric properties of the basic patterns with respect to each operation of the point group. A function is said to have A_1 symmetry if it is symmetric with respect to all four operations of the C_{2v} point group. A function that is symmetric

C_{2v}	E	C_2	σ_v (xz)	σ_v' (yz)	
A_1	1	1	1	1	z, x^2, y^2, z^2
A_2	1	1	-1	-1	xy
B_1	1	-1	1	-1	x, xz
B_2	1	-1	-1	1	y, yz

TABLE 2–8

THE C_{2v} CHARACTER TABLE

with respect to the E and σ_v elements but antisymmetric with respect to the C_2 and σ'_v elements is said to have B_1 symmetry under the C_{2v} point group.[16]

To the right side of the character table are listed the cartesian forms of atomic orbitals. An orbital function possesses the basic symmetry pattern with which it is listed when the coordinate system origin falls at the "point" of intersection of the group symmetry elements.[17] Note that these functions are found in the orbital spherical harmonics of Table 1–4.

Before illustrating how one uses the character table to represent the symmetries of cartesian functions $f(x,y,z)$, we must locate the C_{2v} symmetry elements on the cartesian axis system. Since the reflection planes of the C_{2v} point group are perpendicular to each other, we may identify the C_{2v} symmetry elements with respect to a cartesian coordinate system as follows (see Figure 2–5). The highest-fold rotation axis is (by established convention) taken to be coincident with the z axis. The choice of σ_v and σ'_v to coincide with the xz and yz planes, respectively, is arbitrary but agreed to by convention. These choices greatly simplify the characterization of the symmetry properties of functions with respect to the coordinate system.

---

[16] Just as rules have been agreed upon for deriving a point group symbol from the group of symmetry elements, there is a code for the representation symbols. A and B are reserved for symmetry patterns with 1 in the identity column. A (or B) distinguishes symmetry ($+1$) (or antisymmetry (-1)) under the highest-fold rotation operation. Subscripts 1 and 2 designate symmetry ($+1$) or antisymmetry (-1) under the σ_v operations. Verify this code for the C_{2v} table.

[17] With reference to the C_{2v} table, x^2, y^2, and z^2 each have the A_1 symmetry pattern; consequently, $(x^2 - y^2)$, $(x^2 + y^2 + z^2)$, and $2z^2 - (x^2 + y^2)$ all have A_1 symmetry.

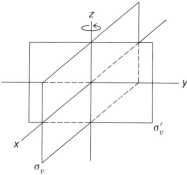

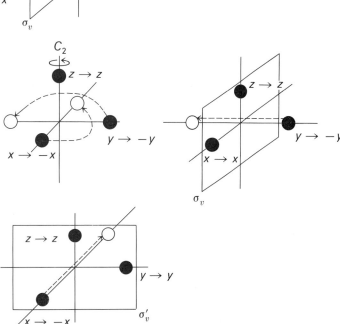

Figure 2–5. The C_{2v} elements placed on the cartesian axes and the effects of the C_{2v} operations on the coordinates x, y, and z.

Now we are ready to examine some cartesian functions for their symmetry properties. We need to determine the change in a given function when we perform a symmetry operation upon it. Should the function simply change sign $[f(x,y,z) \to -f(x,y,z)]$, we signify this by a character of -1. If the function is not affected by the operation $[f(x,y,z) \to f(x,y,z)]$, we say that the function is symmetric to the operation, and we represent this symmetry by a character of $+1$. Should the function be completely altered into an entirely different function $[f(x,y,z) \to f'(x,y,z)]$, we represent this property with a character of 0. As you will see later, an intermediate case is sometimes encountered when the symmetry operation is only partially effective in transforming $f(x,y,z) \to f'(x,y,z)$, so that some of the original function is retained. In such cases the character is the fraction of the original function retained. To summarize, *we recognize that the symmetry operation can have the following general effect on a cartesian function:*

$$\widehat{R}f(x,y,z) = c \cdot f(x,y,z) + c' \cdot f'(x,y,z) + \cdots$$

where $\widehat{R}$ is the operation, c represents the fraction of the original function retained, and f' represents some other cartesian function. Following the early comments of this paragraph, c can have a value from -1 to $+1$, including zero.

Since we want to determine the symmetry properties of *general* functions of the x, y, and z coordinates, *a general approach would be first to determine the symmetries of the simple functions* $f(x) = x$, $f(y) = y$, and $f(z) = z$. Their individual symmetry properties can then be used to find the symmetry of a general $f(x,y,z)$. For the C_{2v} operations, points on the x, y, and z axes are transformed according to the following scheme (cf. Figure 2-5).

		E	C_2	$\sigma_v(xz)$	$\sigma'_v(yz)$
x	$\to$	x	$-x$	x	$-x$
y	$\to$	y	$-y$	$-y$	y
z	$\to$	z	z	z	z

The characters are

x	1	-1	1	-1
y	1	-1	-1	1
z	1	1	1	1

The characters for the x coordinate match those of the B_1 representation of the C_{2v} character table, the characters of the coordinate y match those of B_2, and the coordinate z has the same characters as the representation A_1.

Now let us analyze the function $f(x,y,z) = x \cdot y \cdot z$. Using the transformation properties of x, y, and z from above, we conclude that

		E	C_2	σ_v	σ'_v
$x \cdot y \cdot z$	$\to$	$x \cdot y \cdot z$	$x \cdot y \cdot z$	$-x \cdot y \cdot z$	$-x \cdot y \cdot z$

The characters are

1	1	-1	-1

and we conclude that the symmetry of $f(x,y,z)$ under the C_{2v} point group is A_2.

The preceding discussion has centered on examples of cartesian functions with symmetry properties matching those of the basic symmetry patterns for the C_{2v} group of operations. It is now time to examine some functions that do not possess irreducible symmetries. Consider the function $f(x,y,z) = x \cdot y + z$. Under the C_{2v} set of operations, the function has the following behavior:

$$
\begin{array}{cccc}
E & C_2 & \sigma_v & \sigma_v' \\
x \cdot y + z \quad \rightarrow \quad x \cdot y + z & x \cdot y + z & -x \cdot y + z & -x \cdot y + z
\end{array}
$$

with characters

$$
\begin{array}{cccc}
1 & 1 & 0 & 0
\end{array}
$$

These characters tell that this function does not possess a symmetry matching that of one of the irreducible representations, nor does its symmetry arise as a superposition of those of any of the irreducible representations. Clearly, the σ_v and σ_v' operations are significant in that they transform the original function into the completely different function $-x \cdot y + z$. Thus, we are forced to consider the *pair* of functions

$$
\begin{aligned}
f(x,y,z) &= x \cdot y + z \\
f'(x,y,z) &= -x \cdot y + z
\end{aligned}
$$

together. You may quickly verify the following behaviors of f and f':

$$
\begin{array}{ccccc}
 & E & C_2 & \sigma_v & \sigma_v' \\
f \rightarrow & f & f & f' & f' \\
f' \rightarrow & f' & f' & f & f
\end{array}
$$

Counting a character of 1, -1, 0 as $f \rightarrow f, -f, f'$ and as $f' \rightarrow f', -f', f$, the *pair* of functions f and f' generate the characters

$$
\begin{array}{cccc}
E & C_2 & \sigma_v & \sigma_v' \\
2 & 2 & 0 & 0
\end{array}
$$

These two functions must be considered simultaneously under the C_{2v} elements just as the two hydrogen s orbitals of H_2O must be considered simultaneously. Inspection of the C_{2v} character table reveals that the sum of the characters for the *pair* of representations $A_1 + A_2$ matches those of the pair f and f'. Accordingly, the pair of functions $x \cdot y + z$ and $-x \cdot y + z$ have a symmetry that is reducible into A_1 and A_2 components. Clearly, neither function alone has either symmetry. It should not escape your attention that the blend of A_1 and A_2 symmetries is carried into each of f and f' by the z and $x \cdot y$ components, respectively (cf. the right side of Table 2-8). As an exercise, show that the *pair* of hydrogen $1s$ orbitals have a symmetry designated as $A_1 + B_2$ when the molecular plane is taken to be the yz plane, σ_v'.

It is not generally necessary that a pair of functions possess a reducible symmetry, as in the preceding example. *For point groups containing a three-fold or higher axis, it is possible to have so-called degenerate irreducible representations.* An example is the C_{3v} point group, the character table for which is given in Table 2-9. For C_{3v} (a molecule with this symmetry is NH_3) there are three classes of symmetry elements and three irreducible representations. The characters for the last representation, E, are quite different from those discussed above for C_{2v}. Here we find the occurrence of the characters 2 and 0 describing one of the basic symmetry patterns. *Because of the*

TABLE 2-9
THE C_{3v} CHARACTER TABLE

C_{3v}	E	$2C_3$	$3\sigma_v$	
A_1	1	1	1	z, z^2
A_2	1	1	-1	
E	2	-1	0	(x, y) $(x^2 - y^2, xy)$ (xz, yz)

character of 2 for the identity operation (meaning that a pair of functions are necessary for this symmetry pattern) we say that the E symmetry pattern is a two-fold or doubly degenerate representation.

Before examining the symmetries of cartesian functions under the C_{3v} operations, we again need to locate the symmetry elements with respect to cartesian coordinate axes. This is done in Figure 2-6, where the C_3 and z axes are taken to be coincident and σ_v and the xz plane are coincident. As the angle between σ_v and the other two σ_v's is 120°, the yz plane is not a symmetry plane for this point group.

You may easily confirm that the function $f(x,y,z) = z$ goes into itself upon all operations of the C_{3v} point group and therefore possesses the totally symmetric symmetry pattern called A_1 (as noted to the right of the character table). Let us now examine the symmetry of the pair $f(x,y,z) = x$, $f'(x,y,z) = y$.

The identity operation (see Figure 2-6), of course, leaves the x and y coordinates unshifted and produces a character of 2 for the pair. The C_3 rotation operation gives a more complex result. As shown in Figure 2-6, a rotation about the z axis by 120° moves a point originally on the x axis to a position on neither the x nor y axis. The x component of the new position is $-(1/2)x$ and its y component is $+(\sqrt{3}/2)y$. The original function (a vector along the x axis) becomes a new function (vector) with components consisting of the original function, f, and its partner, f'. Thus we may write

$$\widehat{C}_3 f(x,y,z) = -\frac{1}{2}f + \frac{\sqrt{3}}{2}f'$$

Similarly, for the original y vector we find x and y components of $-\sqrt{3}/2\,x$ and $-(1/2)y$ respectively. Accordingly,

$$\widehat{C}_3 f'(x,y,z) = -\frac{\sqrt{3}}{2}f - \frac{1}{2}f'$$

For the pair (x,y), we find then that x goes into $-1/2$ times itself and y goes into $-1/2$ times itself, for a total character for the (f,f') pair of -1. The result for each reflection plane will be the same, since they all belong to the same class; it will be easiest, however, to examine the character for the (x,y) pair for the σ_v coincident with

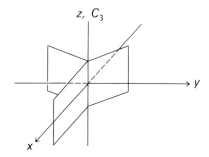

Figure 2-6. The C_{3v} elements placed on a cartesian axis system and the effects of C_3 and $\sigma_v(xz)$ on the x and y coordinates.

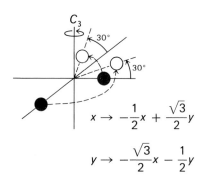
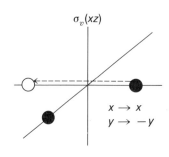

the cartesian plane xz. Referring again to Figure 2-6, reflection in σ_v carries x into itself (for a contribution to the pair character of $+1$), while σ_v carries y into minus itself (for a character contribution of -1). The character under σ_v for the (x,y) pair therefore is 0. Because of mixing by C_3 and σ_v, f and f' are inseparable, just as $(x \cdot y + z)$ and $(-x \cdot y + z)$ were inseparable under the C_{2v} symmetry elements. *But in this case, f and f' under the C_{3v} symmetry elements are examples of a degenerate pair of functions with a basic symmetry called E.*

The same procedure can be followed through the cubic point groups T_d and O_h to show that there the triad $f(x) = x, f(y) = y, f(z) = z$ is inseparable, with symmetry properties matching that of the triply degenerate representation called T_2 in the case of T_d, and called T_{1u} in the case of O_h. (See the tables at the end of this chapter.)

This concludes our "exposition" on character tables. Of necessity, you have been asked to accept several properties of character tables until such time as you undertake serious study of group theory and the derivation of character tables. As you will see in the following chapters, a great deal of use can be made of character tables without first formally studying their origin. Here, the character tables will be used in conjunction with pictorial representations of the functions whose symmetry properties are important, so there is little chance that you will feel lost in an abstract development of a chemical problem. Later in your career, when you have a fuller appreciation of symmetry principles, you will find the abstract use of group theory makes it an even more powerful technique than can be illustrated here.

STUDY QUESTIONS

11. What is the symmetry of the function $f(x) = x(1 + e^{x^2})$ in the C_{2v} point group?

12. Classify an equilateral triangle according to its point group and then determine the irreducible representation symbols for the symmetry pattern of the three vertices. Repeat for an isosceles triangle.

13. Using the information

    $\hat{E}x \to x$ $\hat{C}_3 x \to -(1/2)x + (\sqrt{3}/2)y$ $\hat{\sigma}_v x \to x$ (where σ_v is the
    $\hat{E}y \to y$ $\hat{C}_3 y \to -(\sqrt{3}/2)x - (1/2)y$ $\hat{\sigma}_v y \to -y$ xz plane)

    for the C_{3v} group, show that $d_{z^2} \sim 2z^2 - x^2 - y^2$ has A_1 symmetry, while $d_{xz} \sim xz$ and $d_{yz} \sim yz$ are symmetry-degenerate, E.

14. Determine the transformation coefficients for $f(x) = x$ and $f'(y) = y$ subject to a C_5 operation, where the C_5 axis coincides with the z axis:

    $$\hat{C}_5 f = af + bf'$$
    $$\hat{C}_5 f' = -bf + af$$

15. Recall the text discussion of the pair of functions $f = xy + z$ and $f' = -xy + z$, which are reducible under C_{2v} operations. Are the new functions $F = f + f'$ and $F' = f' - f$ reducible or irreducible under C_{2v} operations?

EPILOG

In this, the first of three chapters to acquaint you with molecular architecture, we began with the rather crude representation of electron pairs by topological diagrams (Lewis structures) and refined the premise to develop some systematics of molecular stereochemistry. By the elementary technique of counting the average number of bonding electron pairs for two atoms in reasonable resonance structures, you find it possible to make crude estimates of relative bond strengths. At this point you also should find yourself capable of translating a molecular formula into a three-dimensional representation of relative nuclear positions, including the description of those positions according to the symmetry elements inherent in their spatial disposition. Further development of symmetry concepts allows you to represent concisely the symmetry of three-dimensional functions in terms of their behavior under point group symmetry operations. The stage is set now for translating the Lewis and VSEPR concepts into the more sophisticated matter-wave model of electrons.

Appendix to Chapter 2

Point Group Character Tables

C_s	E	$\sigma(xy)$		
A'	1	1	x, y	$x^2 - y^2, 2z^2 - x^2 - y^2, xy$
A''	1	-1	z	yz, xz

C_1	E
A	1

C_2	E	C_2		
A	1	1	z	$x^2 - y^2, 2z^2 - x^2 - y^2, xy$
B	1	-1	x, y	yz, xz

C_{2v}	E	C_2	$\sigma_v(xz)$	$\sigma_v'(yz)$		
A_1	1	1	1	1	z	$x^2 - y^2, 2z^2 - x^2 - y^2$
A_2	1	1	-1	-1		xy
B_1	1	-1	1	-1	x	xz
B_2	1	-1	-1	1	y	yz

C_{3v}	E	$2C_3$	$3\sigma_v$		
A_1	1	1	1	z	$2z^2 - x^2 - y^2$
A_2	1	1	-1		
E	2	-1	0	(x, y)	$(x^2 - y^2, xy), (xz, yz)$

C_{4v}	E	$2C_4$	C_2	$2\sigma_v$	$2\sigma_d$		
A_1	1	1	1	1	1	z	$2z^2 - x^2 - y^2$
A_2	1	1	1	-1	-1		
B_1	1	-1	1	1	-1		$x^2 - y^2$
B_2	1	-1	1	-1	1		xy
E	2	0	-2	0	0	(x, y)	(xz, yz)

$C_{\infty v}$	E	$2C_\infty^\Phi$	$\cdots$	$\infty\sigma_v$		
$A_1 = \Sigma^+$	1	1	$\cdots$	1	z	$2z^2 - x^2 - y^2$
$A_2 = \Sigma^-$	1	1	$\cdots$	-1		
$E_1 = \Pi$	2	$2\cos\Phi$	$\cdots$	0	(x, y)	(xz, yz)
$E_2 = \Delta$	2	$2\cos 2\Phi$	$\cdots$	0		$(x^2 - y^2, xy)$
$E_3 = \Phi$	2	$2\cos 3\Phi$	$\cdots$	0		
$\cdots$						

D_{2h}	E	$C_2(z)$	$C_2(y)$	$C_2(x)$	i	$\sigma(xy)$	$\sigma(xz)$	$\sigma(yz)$		
A_g	1	1	1	1	1	1	1	1		$x^2 - y^2, 2z^2 - x^2 - y^2$
B_{1g}	1	1	-1	-1	1	1	-1	-1		xy
B_{2g}	1	-1	1	-1	1	-1	1	-1		xz
B_{3g}	1	-1	-1	1	1	-1	-1	1		yz
A_u	1	1	1	1	-1	-1	-1	-1		
B_{1u}	1	1	-1	-1	-1	-1	1	1	z	
B_{2u}	1	-1	1	-1	-1	1	-1	1	y	
B_{3u}	1	-1	-1	1	-1	1	1	-1	x	

D_{3h}	E	$2C_3$	$3C_2$	σ_h	$2S_3$	$3\sigma_v$		
A'_1	1	1	1	1	1	1		$2z^2 - x^2 - y^2$
A'_2	1	1	−1	1	1	−1		
E'	2	−1	0	2	−1	0	(x, y)	$(x^2 - y^2, xy)$
A''_1	1	1	1	−1	−1	−1		
A''_2	1	1	−1	−1	−1	1	z	
E''	2	−1	0	−2	1	0		(xz, yz)

D_{4h}	E	$2C_4$	C_2	$2C'_2$	$2C''_2$	i	$2S_4$	σ_h	$2\sigma_v$	$2\sigma_d$		
A_{1g}	1	1	1	1	1	1	1	1	1	1		$2z^2 - x^2 - y^2$
A_{2g}	1	1	1	−1	−1	1	1	1	−1	−1		
B_{1g}	1	−1	1	1	−1	1	−1	1	1	−1		$x^2 - y^2$
B_{2g}	1	−1	1	−1	1	1	−1	1	−1	1		xy
E_g	2	0	−2	0	0	2	0	−2	0	0		(xz, yz)
A_{1u}	1	1	1	1	1	−1	−1	−1	−1	−1		
A_{2u}	1	1	1	−1	−1	−1	−1	−1	1	1	z	
B_{1u}	1	−1	1	1	−1	−1	1	−1	−1	1		
B_{2u}	1	−1	1	−1	1	−1	1	−1	1	−1		
E_u	2	0	−2	0	0	−2	0	2	0	0	(x, y)	

T_d	E	$8C_3$	$3C_2$	$6S_4$	$6\sigma_d$			
A_1	1	1	1	1	1			xyz
A_2	1	1	1	−1	−1			
E	2	−1	2	0	0		$(2z^2 - x^2 - y^2, x^2 - y^2)$	
T_1	3	0	−1	1	−1			see ** in O_h
T_2	3	0	−1	−1	1	(x, y, z)	(xy, xz, yz)	see * in O_h

O_h	E	$8C_3$	$6C'_2$	$6C_4$	$3C_2(=C_4^2)$	i	$6S_4$	$8S_6$	$3\sigma_h$	$6\sigma_d$			
A_{1g}	1	1	1	1	1	1	1	1	1	1			
A_{2g}	1	1	−1	−1	1	1	−1	1	1	−1			
E_g	2	−1	0	0	2	2	0	−1	2	0		$(2z^2 - x^2 - y^2, x^2 - y^2)$	
T_{1g}	3	0	−1	1	−1	3	1	0	−1	−1			
T_{2g}	3	0	1	−1	−1	3	−1	0	−1	1		(xz, yz, xy)	
A_{1u}	1	1	1	1	1	−1	−1	−1	−1	−1			
A_{2u}	1	1	−1	−1	1	−1	1	−1	−1	1			xyz
E_u	2	−1	0	0	2	−2	0	1	−2	0			
T_{1u}	3	0	−1	1	−1	−3	−1	0	1	1	(x, y, z)		*
T_{2u}	3	0	1	−1	−1	−3	1	0	1	−1			**

* $([2z^2 - 3(x^2 + y^2)] \cdot (x, y, \text{ and } z))$
** $((x^2 - y^2)z, (x^2 - 3y^2)x, (3x^2 - y^2)y)$ } see Figure 16–20, pp. 894–895

$D_{\infty h}$	E	$2C_\infty^\Phi$	$\cdots$	$\infty \sigma$	i	$2S_\infty^\Phi$	$\cdots$	∞C_2		
Σ_g^+	1	1	$\cdots$	1	1	1	$\cdots$	1		$2z^2 - x^2 - y^2$
Σ_g^-	1	1	$\cdots$	−1	1	1	$\cdots$	−1		
Π_g	2	$2\cos\Phi$	$\cdots$	0	2	$-2\cos\Phi$	$\cdots$	0		(xz, yz)
Δ_g	2	$2\cos 2\Phi$	$\cdots$	0	2	$2\cos 2\Phi$	$\cdots$	0		$(x^2 - y^2, xy)$
$\cdots$	$\cdots$	$\cdots$	$\cdots$	$\cdots$	$\cdots$	$\cdots$	$\cdots$	$\cdots$		
Σ_u^+	1	1	$\cdots$	1	−1	−1	$\cdots$	−1	z	
Σ_u^-	1	1	$\cdots$	−1	−1	−1	$\cdots$	1		
Π_u	2	$2\cos\Phi$	$\cdots$	0	−2	$2\cos\Phi$	$\cdots$	0	(x, y)	
Δ_u	2	$2\cos 2\Phi$	$\cdots$	0	−2	$-2\cos 2\Phi$	$\cdots$	0		
$\cdots$	$\cdots$	$\cdots$	$\cdots$	$\cdots$	$\cdots$	$\cdots$	$\cdots$	$\cdots$		

DIRECTIONAL ATOMIC ORBITALS

MOLECULAR PROPERTIES CONVENIENTLY INTERPRETED WITH THE DIRECTED ATOMIC ORBITAL CONCEPT

  Bond Distances

Force Constants
Dipole Moments
Nuclear Spin Coupling
Bond Energies

EPILOG

3. THE DIRECTED ATOMIC ORBITAL VIEW OF CHEMICAL BONDS

In this chapter you will learn of one matter-wave conceptualization of electron pairs in molecules, the familiar hybrid atomic orbital model. Not only will you see that such orbitals have neither more nor less intrinsic significance than the Schrödinger atomic orbitals but, more importantly, you will learn of their application to the interpretation of some chemical experiments. Herein lies the raison d'être *for the hybrid concept: the interpretation of some molecular properties is more simply achieved from a directed atomic orbital view than from the equivalent Schrödinger, non-directed orbital view* (the topic of the next chapter).

DIRECTIONAL ATOMIC ORBITALS

To review our progress to this point in developing a physical description of electrons in molecules, we have utilized the following concepts: (1) perfect pairing of electrons between different atoms to form electron pair bonds, (2) π resonance electronic structures to allow for alternate designations of electron pairs as lone or π bond pairs, (3) repulsive forces between electron pairs as a guide to the basic structures of molecules, with refinements to allow more detailed descriptions of molecular structure and trends in structures of related molecules, and (4) a formal symbolism for the characterization of molecular structures according to the inherent symmetry of molecules.

The concept of molecular symmetry was broadened into the use of character tables to identify and classify the basic symmetry of three-dimensional functions under the operations of point groups. Therefore, it is now time to examine the relationship between atomic orbital symmetry and molecular point group in order to attempt quantification of the previous classical ideas of electron distribution in molecules. The question of great importance then is whether the wave model can be extended from atoms to molecules. Fortunately the answer is yes, and we will begin the transition from classical to wave models by introducing the concept of *hybrid atomic orbitals* (hao).

The principal feature of molecular structure with which the wave model must be consistent is the angular distribution of electron pairs about given atoms. Consequently, we will be primarily interested in the use of the angular parts of atomic orbital functions (*as opposed to the radial parts*).

We begin the examination of the wave model with the example of the linear triatomic molecule BeH_2. We will concentrate our attention on the Be atom. The VSEPR model has us thinking of the Be—H bond electron pair distributions as radially distributed from, and to each side of, the Be atom. The question, then, is whether we can use the (centrosymmetric) Schrödinger valence orbital functions for the isolated Be atom ($2s$ and $2p$ orbitals) to develop alternate Be atomic orbitals,

100 ☐ THE DIRECTED ATOMIC ORBITAL VIEW OF CHEMICAL BONDS

which should have the characteristic that one is directed to one side, and the other to the other side, of the Be atom. Note that the centrosymmetry of the 2s and 2p orbitals means that each is equally directed to both sides of the Be atom. *You should clearly understand that such an "alternate orbital" model has no physical significance; the significance lies in a convenient conceptualization of bonding electron pair distributions.*

An alternate statement of what we are about to do and why we're doing it goes as follows. Nature provides, in the BeH_2 molecule, a valence electron density distribution in the Be—H internuclear regions that we hope to mimic successfully by the use of atomic matter waves. To account for electron density between Be and each H, we are free to

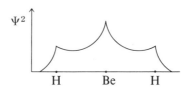

use the Schrödinger form atomic orbitals (there are six of these, an *s* and three *p* orbitals from Be and one *s* orbital from each H) *or* some equivalent orbitals. To account for density along the internuclear axes, we should use the four Schrödinger waves with non-zero amplitude along the axis: an *s* and *p* orbital from Be and the two hydrogen orbitals.

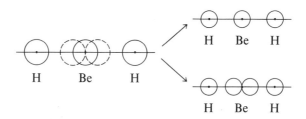

Wave superposition of the Be 2s and H 1s ao's will produce one molecular valence electron wave, and superposition of the Be 2p and the H 1s waves produces another wave. These two waves contribute equally to both Be—H "bonds," so the Schrödinger ao approach means that we attempt to account for electron density in *each* Be—H region by *adding the contributions from two separate molecular waves.* That is, *no one molecular wave is concentrated predominately to one side or the other of the Be atom.* This approach to electrons in molecules is highly important and convenient at times and will be explored in Chapter 4. For the present we hope to find a set of Be ao's, *equivalent* to Schrödinger's 2s and 2p, that will allow us to account for the electron density between Be and each H by the superposition of the H 1s and a *single* Be valence orbital (*directed* toward the H).

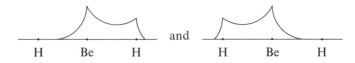

You may well ask at this point, "Why bother to create from the Schrödinger orbitals an equivalent set of hybrid atomic orbitals?" The answer is simple and direct. First, we achieve a unification of the concept of *localized* electron pairs (lone and bond pairs) with the wave model of electrons in atoms and molecules. Secondly, chemical interpretations of molecular properties are often particularly convenient when based on the ideas of localized lone and bond pairs of electrons.

In order to develop the alternate orbitals, we will make use of the $D_{\infty h}$ character

DIRECTIONAL ATOMIC ORBITALS 101

TABLE 3-1
THE SYMMETRY OF Be-DIRECTED AO'S IN BeH$_2$

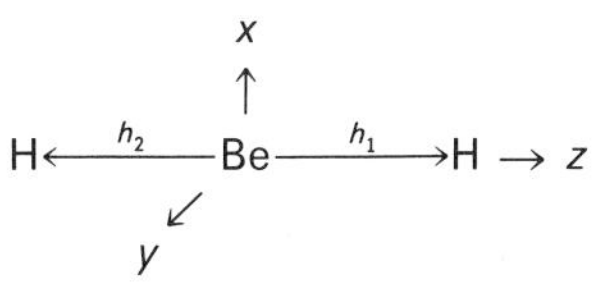

$$H \xleftarrow{h_2} Be \xrightarrow{h_1} H \to z$$
(with x up, y into page)

$D_{\infty h}$	E	$\cdots 2C_\phi$	C_2	$\infty \sigma_v$	i	$\cdots 2S_\phi$	σ_h	∞C_2	
Σ_g^+	1	1	1	1	1	1	1	1	s
Σ_u^+	1	1	1	1	-1	-1	-1	-1	z
Σ_g^-	1	1	1	-1	1	1	1	-1	
Σ_u^-	1	1	1	-1	-1	-1	-1	1	
$\vdots$	.	.	.						
h_1	h_1	h_1	h_1	h_1	h_2	h_2	h_2	h_2	
h_2	h_2	h_2	h_2	h_2	h_1	h_1	h_1	h_1	
(h_1, h_2)	2	2	2	2	0	0	0	0	

table appropriate to the BeH$_2$ molecule. To represent the alternate (directed) atomic orbitals on the Be atom, we will use vectors originating at the Be atom and directed at the hydrogen nuclei, as shown in Table 3-1. To help us keep track of the symmetry properties inherent to these vectors, we will identify them as h_1 and h_2. *We may now ask, "What is the symmetry pattern or representation appropriate to these two vectors?"* Applying the $D_{\infty h}$ symmetry operations to these vectors, we may keep track of the results as shown below the character table in Table 3-1. An operation that takes h_1 into itself contributes a character of $+1$ to the total character for the pair (h_1, h_2). In this way we generate a *reducible* representation for the two vectors, the characters for which are given in the final line of the character table in Table 3-1. Inspection of the table then shows that the reducible representation generated by the vectors is a composite of Σ_g^+ and Σ_u^+ representations of $D_{\infty h}$. The only *valence* orbitals on Be with these symmetries are the $2s$ (Σ_g^+) and $2p_z$ (Σ_u^+). Therefore, we must describe the directed atomic orbital functions of Be in terms of these atomic orbitals.

To accomplish this, we will form from the pair $(2s, 2p_z)$ two new atomic orbitals, to be called hybrid atomic orbitals (hao), by sum and difference. Thus, let us represent h_1 by the combination

$$h_1 = a\phi_s + b\phi_z$$

and h_2 by the orthogonal combination

$$h_2 = b\phi_s - a\phi_z$$

The equivalence of these new orbitals by symmetry (the inversion operation interconverts them) implies the following relation between h_1 and h_2:

$$\hat{i} \text{ on } h_1 = h_2$$
$$\hat{i} \text{ on } (a\phi_s + b\phi_z) = b\phi_s - a\phi_z$$
$$a\phi_s - b\phi_z = b\phi_s - a\phi_z$$

This last relation requires that

$$a = b$$

102 ☐ THE DIRECTED ATOMIC ORBITAL VIEW OF CHEMICAL BONDS

If h_1 is to be normalized, then:

$$\int h_1^2 \, d\tau = a^2 \int \phi_s^2 \, d\tau + a^2 \int \phi_z^2 \, d\tau + 2a^2 \int \phi_s \phi_z \, d\tau = 1$$
$$= 2a^2 = 1$$
$$a = \sqrt{1/2}$$

This means that the analytical forms of the hao's are

and
$$h_1 = \sqrt{1/2} \, (\phi_s + \phi_z)$$
$$h_2 = \sqrt{1/2} \, (\phi_s - \phi_z)$$

By making sketches of the angular parts of the hao wave functions, we readily see that h_1 and h_2 satisfy our imposed requirement that the hao represent directed (toward the hydrogen nuclei) electron functions for the Be atom. In making such sketches we use the fact that the superposition (algebraic addition) of two wave forms of opposite sign at some points in space leads to *destructive interference* of the waves and a reduction of the wave amplitude at those points. Similarly, superposition of two wave forms of the same sign at certain points in space leads to *constructive interference* at those points. The sketches in Figure 3–1 illustrate the results for h_1 and h_2 of the Be atom in BeH$_2$.

To emphasize for you that there is no *physical* distinction to be made between the *pair* of functions (h_1, h_2) and the *pair* of functions (ϕ_s, ϕ_z), the angular probability function $h_1^2 + h_2^2$ is seen to be identical to that of $\phi_s^2 + \phi_z^2$:

$$h_1^2 = \tfrac{1}{2}\phi_s^2 + \phi_s\phi_z + \tfrac{1}{2}\phi_z^2$$
$$h_2^2 = \tfrac{1}{2}\phi_s^2 - \phi_s\phi_z + \tfrac{1}{2}\phi_z^2$$
$$h_1^2 + h_2^2 = \phi_s^2 + \phi_z^2$$

The "whole" of the electron distribution is the same from both points of view; only the "pieces" [(h_1, h_2) and (ϕ_s, ϕ_z)] are different. The middle terms in these expressions are quite important. For h_1, at those points in space where ϕ_s and ϕ_z have the same sign (the region of positive z coordinate), the product $\phi_s \cdot \phi_z$ is positive and the probability is increased by $\phi_s \cdot \phi_z$; whereas at points where ϕ_s and ϕ_z are opposite in sign (negative z), $\phi_s \cdot \phi_z$ is negative and the probability is diminished. One envisions displacement of wave amplitude from points of $z < 0$ to points of $z > 0$. Equal displacement in the opposite sense occurs for h_2. In relation to $h_1^2 + h_2^2$ vs. $\phi_s^2 + \phi_z^2$, the displacements offset each other so that there is *no* net change in *total* point density, i.e.,

$$h_1^2 + h_2^2 = \phi_s^2 + \phi_z^2$$

Finally, it is because of the equal contributions of $2s$ and $2p_z$ functions to the total probability that we call these particular hao's "*sp*" (or di, for *digonal*) hybrids (the $s:p$ ratio is $1:1$).

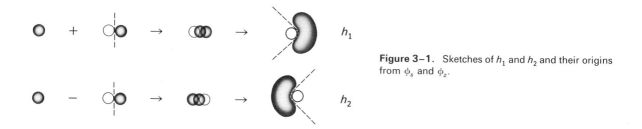

Figure 3–1. Sketches of h_1 and h_2 and their origins from ϕ_s and ϕ_z.

The hao schemes for other commonly encountered VSEPR structures are given in Table 3–2. These can be worked out by the same basic procedure as was used for the *sp* hao of BeH_2 but, understandably, the complexity of the derivation increases as the number of directional orbitals increases. In the upper left-hand corner you find the "*sp*" notation for the hao along with the Schoenflies symbol for the *set* of hybrids. Each hybrid is represented by an appropriately directed vector and, using the procedure as in Table 3–1, the symmetry patterns (Mulliken symbols) generated by the set of vectors is given next. This is followed by a listing of the *specific* Schrödinger ao's of the *identical* symmetry pattern (these are read directly from the character tables at the end of Chapter 2). The linear combinations of these Schrödinger ao's that, by wave superposition, are directed along the hybrid "vectors" are given in the center; to the right of these you find the square of the *s* ao coefficient and the sum of the squares of the *p* and/or *d* ao coefficients. It is, of course, these numbers which determine the "sp^xd^y" designation of the hybrids. The fact that the sum of these numbers across a row equals 1 derives from the hao normalization. The fact that the sum of these numbers down a column equals 1 for *s*, "*x*" for *p*, and "*y*" for *d* derives from the complete distribution of 1*s*, "*x*"*p* and "*y*"*d* Schrödinger ao's among the hybrids.

While the linear combinations of Schrödinger ao's look complicated in each case, the relative signs of the *p/d* ao coefficients are simply determined by the hybrid "vector" orientations. Realizing that the p_x and p_y ao's themselves can be represented as unit vectors along the *x* and *y* axes, the unit vector tr_1 of "sp^2" is seen to have only a $+x$ component and no *y* component; thus, ϕ_y does not appear and the coefficient of ϕ_x is positive. The vector tr_2, on the other hand, has a $-x$ component and a $+y$ component. Turning to te_3, you find *x*, *y*, and *z* components with $-$, $+$, and $-$ signs. The *d* ao's cannot be represented by simple vectors; the coefficient signs are, however, not difficult to explain. Taking oc_3 as an example, the phasing of all *s*, *p*, and *d* ao's in the direction of the $+y$ axis must lead to wave "build-up" in that direction:

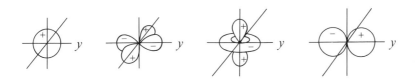

Thus, ϕ_x and ϕ_z do not contribute; $\phi_{x^2-y^2}$ and ϕ_{z^2} both contribute to oc_3 with negative coefficients.

Finally, we call your attention to the fact that the axial hybrids of the trigonal bipyramid geometry are not symmetry-equivalent to the equatorial hao. The equatorial hybrids must involve ϕ_x and ϕ_y but not ϕ_z, while the axial hao's involve ϕ_z but not ϕ_x or ϕ_y. Each of these two sets requires, in addition, an A_1' ao—*either or both* of ϕ_s and ϕ_{z^2}. The ambiguity is not resolved by symmetry constraints. As you predicted in Chapter 2 (see the VSEPR discussion of ClF_3), the axial bonds are usually longer than the equatorial bonds. Later you will see that *for the hao model to be consistent with this topological feature*, the equatorial hybrids must have greater ϕ_s contribution, while the axial hao's have greater ϕ_{z^2} character. In Table 3–2 we have presented the limiting case of ϕ_{z^2} in di_4 and di_5, and ϕ_s in tr_1, tr_2, and tr_3.

In addition to the so-called sigma orbital framework formed by the hao, we need an orbital description for the double bond encountered with some resonance structures. In contrast to the sigma hao's, which have the property of a maximum in electron probability *along the internuclear axis*, the so-called pi electron pairs are characterized by electron probability maxima at points *off the internuclear axis*. The most common instances arise when the two atoms for which we accept a double bond resonance structure have either atomic *p* or *d* orbitals, as shown in Figure 3–2. The various possibilities are designated as *p—p* π bonding, *d—p* π bonding, and *d—d* π bonding. Examples of the first two types are given.

TABLE 3-2
BASIC HYBRID ATOMIC ORBITALS

---

sp^2 (D_{3h})
(tr_1, tr_2, tr_3) $A_1' + E'$
Schrödinger: $s + p_x, p_y$

$$tr_1 = \frac{1}{\sqrt{3}}(\phi_s + \sqrt{2}\phi_x)$$

$$tr_2 = \frac{1}{\sqrt{3}}\left(\phi_s - \sqrt{\frac{1}{2}}\phi_x + \sqrt{\frac{3}{2}}\phi_y\right)$$

$$tr_3 = \frac{1}{\sqrt{3}}\left(\phi_s - \sqrt{\frac{1}{2}}\phi_x - \sqrt{\frac{3}{2}}\phi_y\right)$$

ao character

	s	p
tr_1	1/3	2/3
tr_2	1/3	2/3
tr_3	1/3	2/3

---

sp^3 (T_d)
$(te_1, \ldots, te_4)$ $A_1 + T_2$
Schrödinger: $s + p_x, p_y, p_z$

$$te_1 = \frac{1}{2}(\phi_s + \phi_x + \phi_y + \phi_z)$$

$$te_2 = \frac{1}{2}(\phi_s + \phi_x - \phi_y - \phi_z)$$

$$te_3 = \frac{1}{2}(\phi_s - \phi_x + \phi_y - \phi_z)$$

$$te_4 = \frac{1}{2}(\phi_s - \phi_x - \phi_y + \phi_z)$$

ao character

	s	p
te_1	1/4	3/4
te_2	1/4	3/4
te_3	1/4	3/4
te_4	1/4	3/4

---

dsp^3 (D_{3h})
(tr_1, tr_2, tr_3) $A_1' + E'$ (equatorial)
(di_4, di_5) $A_1' + A_2''$ (axial)
Schrödinger: $s, d_{z^2} + p_z + (p_x, p_y)$

$$tr_1 = \sqrt{\frac{1}{3}}(\phi_s + \sqrt{2}\phi_x)$$

$$tr_2 = \sqrt{\frac{1}{3}}\left(\phi_s - \sqrt{\frac{1}{2}}\phi_x + \sqrt{\frac{3}{2}}\phi_y\right)$$

$$tr_3 = \sqrt{\frac{1}{3}}\left(\phi_s - \sqrt{\frac{1}{2}}\phi_x - \sqrt{\frac{3}{2}}\phi_y\right)$$

$$di_4 = \sqrt{\frac{1}{2}}(\phi_z + \phi_{z^2})$$

$$di_5 = \sqrt{\frac{1}{2}}(\phi_z - \phi_{z^2})$$

ao character

	s	p	d
tr_1	1/3	2/3	—
tr_2	1/3	2/3	
tr_3	1/3	2/3	
di_4		1/2	1/2
di_5		1/2	1/2

Table continued on the following page.

TABLE 3-2
BASIC HYBRID ATOMIC ORBITALS (Continued)

d^2sp^3
$(oc_1, \ldots, oc_6)$ $A_{1g} + E_g + T_{1u}$
Schrödinger: $s + (d_{z^2}, d_{x^2-y^2}) + (p_x, p_y, p_z)$

orbital	expression	ao character s	d	p
$oc_1 = \frac{1}{\sqrt{6}}\left(\phi_s + \sqrt{2}\,\phi_{z^2} + \sqrt{3}\,\phi_z\right)$		1/6	1/3	1/2
$oc_2 = \frac{1}{\sqrt{6}}\left(\phi_s - \sqrt{\frac{1}{2}}\,\phi_{z^2} + \sqrt{\frac{3}{2}}\,\phi_{x^2-y^2} + \sqrt{3}\,\phi_x\right)$		1/6	1/3	1/2
$oc_3 = \frac{1}{\sqrt{6}}\left(\phi_s - \sqrt{\frac{1}{2}}\,\phi_{z^2} - \sqrt{\frac{3}{2}}\,\phi_{x^2-y^2} + \sqrt{3}\,\phi_y\right)$		1/6	1/3	1/2
$oc_4 = \frac{1}{\sqrt{6}}\left(\phi_s - \sqrt{\frac{1}{2}}\,\phi_{z^2} + \sqrt{\frac{3}{2}}\,\phi_{x^2-y^2} - \sqrt{3}\,\phi_x\right)$		1/6	1/3	1/2
$oc_5 = \frac{1}{\sqrt{6}}\left(\phi_s - \sqrt{\frac{1}{2}}\,\phi_{z^2} - \sqrt{\frac{3}{2}}\,\phi_{x^2-y^2} - \sqrt{3}\,\phi_y\right)$		1/6	1/3	1/2
$oc_6 = \frac{1}{\sqrt{6}}\left(\phi_s + \sqrt{2}\,\phi_{z^2} - \sqrt{3}\,\phi_z\right)$		1/6	1/3	1/2

To summarize this description of hao's, you will find in Table 3-3 examples of molecules with two through six electron pair clouds about some atom and the *idealized* hybridization scheme for each. Two points are important in examining this table. First, in cases where there is more than one pair shared between two atoms, one of these pairs is treated as arising from a combination of sigma hao's from each atom. Additional pairs are accounted for in terms of combinations of π type atomic orbitals from each atom. The second observation to be made is that the idealized hao's are only approximately correct descriptions. A case in point is the oxygen atom of H_2COH^+. Here, in an approximate sense, the oxygen atom is described as sp^2 hybridized even though the local symmetry about the oxygen is *not* D_{3h} (a similar comment applies to the carbon atom; the correct molecular symmetry is C_s).

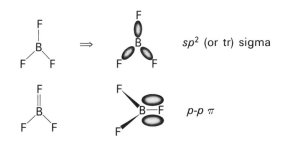

Figure 3-2. Sketches of σ, p-p π, and d-p π orbitals.

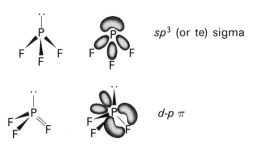

Stereochemically Distinct Electron Pairs	Example	HAO
2	Me—Be*—Me	sp
	O=C*=O	$sp + 2p\text{-}p\,\pi$
3	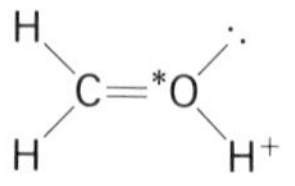 H₂C=*O—H⁺ (formaldehyde protonated)	$sp^2 + p\text{-}p\,\pi$
	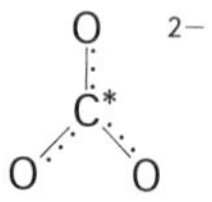 CO₃²⁻ (C*)	$sp^2 + p\text{-}p\,\pi$
4	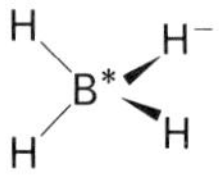 BH₄⁻ (B*)	sp^3
	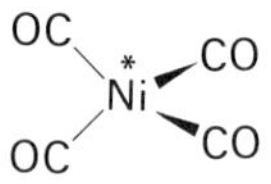 Ni*(CO)₄	sp^3
5	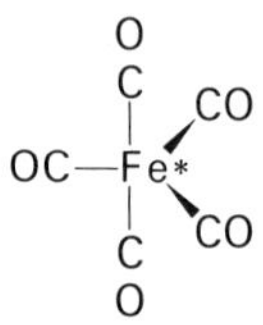 Fe*(CO)₅	dsp^3
6	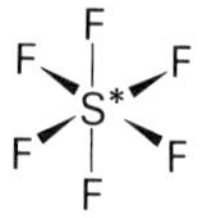 SF₆ (S*)	sp^3d^2
	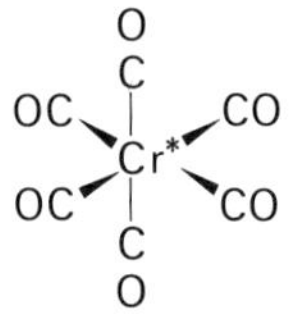 Cr*(CO)₆	d^2sp^3

TABLE 3-3
IDENTIFICATION OF STRUCTURAL AND HAO PATTERNS

STUDY QUESTIONS

1. Derive the irreducible symmetry components of the set of three boron hybrid orbitals of BF_3.

2. Sketch the hybrids given by
   a) $\sqrt{\tfrac{1}{2}}(s + p_z)$
   b) $\sqrt{\tfrac{1}{2}}(d_{z^2} + p_z)$
   c) $\sqrt{\tfrac{1}{2}}(s + d_{x^2-y^2})$

   Which of these gives rise to a non-zero orbital dipole (non-centrosymmetric wave)?

3. For a general hybrid ao given by

$$h = a\phi_s + b\phi_p \quad \text{(where } a^2 + b^2 = 1\text{)}$$

how does the contribution of the wave interference term ($2ab\phi_s\phi_p$) vary as the *ratio* of s/p character (a^2/b^2) varies? A good way to demonstrate this is to plot $2ab$ as a function of a^2.

4. List the idealized central atom hybrid atomic orbitals of

a) $XeOF_4$
b) SO_3
c) $SnCl_3^-$
d) P in P_4
e) ClF_3
f) Al in Al_2Cl_6

MOLECULAR PROPERTIES CONVENIENTLY INTERPRETED WITH THE DIRECTED ATOMIC ORBITAL CONCEPT

In the preceding chapter we have developed a qualitative model for molecular structure, which has now led us to the use of directed or hybrid atomic orbitals. In the next few sections we will examine the utility of such an approach to conceptualization of bond formation, and several of the interpretative schemes for molecular properties that have been developed from the directed orbital point of view.

One of the properties of *sp* hybrid atomic orbitals discussed in the preceding section was that the *centroid of electron density for each hybrid is shifted away from the nucleus relative to that of either the valence s or the valence p orbital.* In wave mechanical terms, this is important to the concept of localized electron pair bonds because constructive interference of the central atom hybrid wave and its neighbor atom hybrid wave is greater in the internuclear region than would be possible for a combination of two Schrödinger ao's alone. The basic idea behind bond formation is much the same as that behind the hybridization concept. Just as hybridization is the mixing of two atomic orbitals (on the same center) *to enhance the orbital amplitude along some internuclear axis,* the combination or mixing of two orbitals from different atoms leads to constructive interference in the internuclear region. In terms of probability functions, this means that the greater the reinforcement, the greater is the probability of the electron pair in the internuclear region (where the probability must be large if the electron pair is to do a good job of holding the two nuclei together).

In analogy to the formation of hybrid atomic orbitals by superposition of pure atomic orbitals, we allow for the constructive interference of the two hybrid atomic orbitals on different atoms by writing the wavefunction for the bond orbital as

$$\psi_b = ah_1 + bh_2$$

where h_1 and h_2 are hybrids, on adjacent atoms, directed at each other.

As is necessary, the bond function must be normalized to unity, so

$$\int \psi_b^2 \, d\tau = a^2 \int h_1^2 \, d\tau + b^2 \int h_2^2 \, d\tau + 2ab \int h_1 h_2 \, d\tau = 1$$

As h_1 and h_2 are already normalized, the first two integrals on the right side of this equation have values of 1. The third integral is commonly called the *overlap integral* (symbol $= S$) and has the important physical significance that the greater it is, the greater is the probability that an electron in the bond orbital ψ_b will be found in the internuclear region. In other words, if *both* h_1 and h_2 have large amplitudes or probabilities in the internuclear region, then their product $h_1 \cdot h_2$ will also be large, and a large overlap integral results when the two hybrid orbitals are strongly

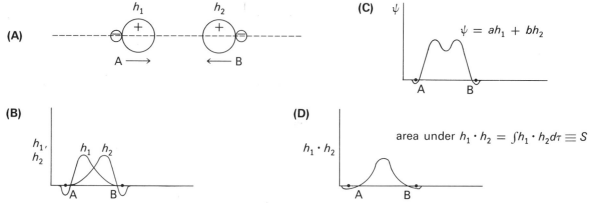

Figure 3-3. Bond formation as viewed from interfering hybrid waves: (A) hybrid angular sketches, (B) the hybrid radial functions, (C) the bonding orbital $ah_1 + bh_2$, (D) the product $h_1 \cdot h_2$ illustrating overlap of h_1 and h_2.

reinforcing. *Theoretically, then, a useful criterion for the strength of a bond is the magnitude of the overlap integral between the hybrid orbitals that define the bond.* The sketches of Figure 3-3 should help in giving a pictorial presentation of the reinforcement of overlapping atomic orbitals.

BOND DISTANCES

At this point it is interesting to comment on the results of calculations by Maccoll[1] that show the variation of overlap with fractional valence orbital s character

---

[1] A. Maccoll, *Trans. Faraday Soc.*, **46**, 369 (1950).

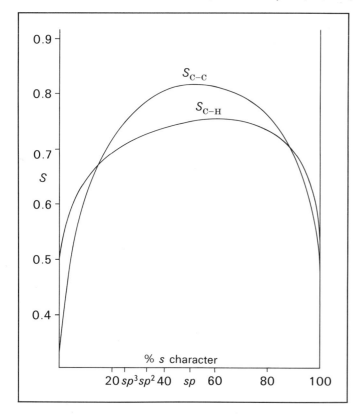

Figure 3-4. The overlap integral for C—C and C—H bonds as a function of % s character. [From A. Maccoll, *Trans. Faraday Soc.*, **46**, 369 (1950).]

in overlapping *sp* hybrid orbitals. Figure 3-4 is a summary of the results, from which it may readily be seen that overlap of two atomic *s* orbitals is larger than that of two atomic *p* orbitals, and both are smaller than the overlaps of the customary sp^3, sp^2, and *sp* hybrids. It is important to note that the overlap (and presumably the bond strength) increases as the *s/p* character of the hybrids becomes better balanced (did you work study question 3?).

As two valence *s* orbitals are brought from large separation to the range of chemical bond distances, the overlap product $\phi_s \phi_s'$ is everywhere positive and continuously increases, whereas the $\phi_p \phi_p'$ product increases and then decreases because of the negative sign of

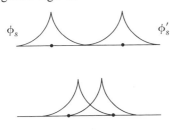
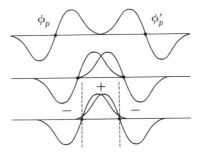

$\phi_p \phi_p'$ outside the internuclear region. Thus, at shorter bond distances we expect to find greater ϕ_s character in the hao's that best describe the bond pair density. That is, the shorter the chemical bond distance, the more *s* character there is in the best hao's. These developments markedly demonstrate one conceptual advantage to viewing chemical bond formation from an overlapping hao vantage point. The C—H distances cited below illustrate this simple correlation, as does the fact that the C—C bond of methylacetylene (sp^3/sp overlap) is shorter and stronger than that of ethane (sp^3/sp^3 overlap).

	r_{CH}	Csp^n
H—C≡C—H	1.057 Å	*sp*
H₂C=CH₂	1.079 Å	sp^2
CH₄	1.094 Å	sp^3

We are now nearly in a position to illustrate the utility of the hybrid orbital model in the interpretation of other molecular properties. One last, very useful concept must be introduced. Up to this point in the discussion, we have idealized atom hybridization in the sense that all hybrids about a central atom are considered equivalent. In terms of *s* and *p* character, this means that all have been considered to have equal *s*:*p* ratios. This is, of course, true when symmetry requires it. For example, the carbon hybrids in methane are all of the same *s*:*p* character. Similarly, all four carbon hybrids have identical *s*:*p* character in CF_4. What about a compound like CH_3F? Here symmetry does not require equivalence of all four carbon hybrids; rather, three (the carbon hybrids directed toward the hydrogens) must have identical *s*:*p* ratios, but the hybrid toward fluorine is expected to be different. In what way? *The concept of isovalent hybridization* as promoted by Bent[2] clarifies the situation. The following statements are the basis for the isovalent hybridization idea:

a. The more electronegative an atom bound to the central atom, the greater will be its demand for electron density from the central atom.

---

[2] H. A. Bent, *Chem. Rev.*, **61**, 275 (1961); also, C. A. Coulson, "Valence," Oxford University Press, New York (1952), pp. 198–201.

b. The greater the p character of a hybrid orbital, the less its electronegativity (refer to the hybrid orbital electronegativities of Chapter 1 and the relationship between orbital electronegativity and effective nuclear charge for s and p atomic orbitals).
c. Accordingly, a central atom will direct greater p character, and less s character, into its hybrid directed toward the more electronegative substituent.
d. Conversely, a central atom will direct less p character, and greater s character, into the hybrids directed toward the less electronegative substituents.

With respect to H_3CF, the $C \rightarrow H$ hao's should have $>25\%$ s character, while the $C \rightarrow F$ hybrid should have $<25\%$ s character.

Several important aspects of molecular electronic structures and geometries can be interpreted in terms of this concept of hybridization. The inductive transmission of charge withdrawal takes on greater meaning with this model. For example, replacing a hydrogen atom of CH_4 with a fluorine leads to loss of charge from carbon to fluorine, via increased p character in the carbon hybrid toward fluorine. The increased s character in the remaining hybrids toward hydrogen makes the carbon appear to have increased electronegativity from the point of view of the hydrogens. In this way, the hydrogens participate in the intramolecular charge flow toward fluorine. Additionally, since greater hybrid s character tends to lead to shorter bonds, the model is consistent with the observed shortening of the CH bonds.

Atoms containing lone pairs as well as bond pairs tend to direct greater s character into the hybrid identified with the lone pair. This also has a reasonable physical basis, since the lone pair (unlike the bond pair) moves under the influence of only a single nuclear charge and "needs" to experience greater *central* atom nuclear charge than the bond pairs. Within the hybrid orbital model, then, it is reasonable that the lone pair hybrid possesses greater s character. Given the greater angular diffuseness of s than p atomic orbitals, this greater s character of lone pair hybrids correlates well with the VSEPR idea that lone pair electrons subtend greater solid angles about an atom than do bond pairs.

A final structural example involving pentavalent phosphorus is appropriate. You will recall from the discussion surrounding the sp^3d hybridization scheme in Table 3–2 that the axial and equatorial hybrids are not required by symmetry to be equivalent. The bond lengths of axial P—Cl bonds and equatorial P—Cl bonds of PCl_5 are different, the axial (2.19 Å) being greater than the equatorial (2.04 Å). This result is expected from application of the VSEPR model, for the axial pairs are acted upon by three equatorial pairs, whereas the equatorial pairs are acted upon by only the two axial pairs. In hybrid orbital terms, one describes the shorter equatorial bonds as greater in phosphorus s character than the axial bonds. This is an argument in favor of viewing the phosphorus hybrids as *principally sp²* (equatorial) and *dp* (axial). Similarly, PCl_3F_2 exhibits a structure in which both fluorines are axial. This is successfully interpreted with the VSEPR model; and again the isovalent hybridization model of (ideal) $sp^2 + dp$ central atom hybridization is used to describe the phosphorus hybrids. Comparing PCl_5 and PCl_3F_2, the axial hybrids of the latter require less s character than do those of the former.

FORCE CONSTANTS

While the problem of the vibrational motion of atoms in molecules is best handled in general terms by quantum theory, it has been possible to use a classical theory of these motions to determine the vibrational wavefunctions in terms of stretching of bonds and bending of bond angles. The classical model, in crude terms, may be viewed as assuming that Hooke's law of restoring forces is operative between atoms defining bonds and between atoms defining bond angles. This approach has great intuitive value for the chemist, and we will briefly discuss some aspects of this model for their value in characterizing bonds and changes in bonds incurred by chemical reaction.

While it is true that a molecular vibration is just that, i.e., motion involving all atoms simultaneously, it has been a great boon to the chemist that for qualitative purposes many such molecular modes seem to have the greatest part of the motion concentrated in a small region of the molecule, that is, the stretching of one bond or the bending of a certain angle. It is therefore worthwhile to inquire into the treatment of a diatomic oscillator for some guiding principles in dealing with bond stretching motions in larger molecules. The problem is usually treated in some detail in the junior level physical chemistry course and so will only be outlined here.

Two oscillating nuclei whose motion is such that the center of gravity of the molecule remains fixed (no molecular translational motion) can be treated by the expedient of an equivalent system of a single atom of effective or reduced mass given by

$$\frac{1}{\mu} = \frac{1}{m_a} + \frac{1}{m_b}$$

oscillating about some equilibrium point; m_a and m_b are the masses of the two atoms defining the diatomic molecule. According to Hooke's law, the restoring force for a displacement, x, of this effective mass from its equilibrium value is given by $F = -kx$, where k is the force constant. Using Newton's first law ($F = ma$), the equation of motion for the oscillating effective mass is

$$\mu \frac{d^2x}{dt^2} = -kx$$

The solution to this simple second-order differential equation for the displacement as a function of time is

$$x = x_0 \cos 2\pi\nu(t - t_0)$$

In this equation x is the displacement at time t, x_0 is the displacement at time t_0, and $2\pi\nu$, with units of sec^{-1}, is the oscillation frequency characteristic of the oscillator. Substitution of the solution for x into the equation of motion yields

$$\nu = \frac{1}{2\pi}\sqrt{\frac{k}{\mu}}$$

Often the experimentalist expresses the frequency of the oscillation in units of reciprocal distance by using the relation $\lambda\nu = c$. That is,

$$\tilde{\nu} = \frac{1}{\lambda} = \frac{1}{2\pi c}\sqrt{\frac{k}{\mu}}$$

Two important generalizations emerge from this equation. First, the larger the force constant, the higher is the vibrational frequency. Second, the smaller the reduced mass of the oscillator (the lighter the atoms defining the chemical bond), the higher is the frequency. These trends are nicely illustrated by the following comparisons:

	k	μ	$\sqrt{k/\mu}$	$\tilde{\nu}$
C—N	5.3 md/Å	6.46	0.91	~1000 cm^{-1}
C=N	10.5 md/Å	6.46	1.27	~1500 cm^{-1}
C≡N	17.2 md/Å	6.46	1.63	~2000 cm^{-1}
C—C	~5 md/Å	6.0	.91	~1000 cm^{-1}
O—H	~5 md/Å	0.941	2.30	~3600 cm^{-1}

From comparison of these nominal values we find a regular, but of course non-linear, increase in vibrational frequency with increasing force constant and constant reduced mass, and in the last two instances, increased frequency with decreased reduced mass and constant force constant. Also interesting is the rough proportionality of CN force constant to bond order. An association such as this has led chemists to expect a rough correlation of bond energy with force constant.

In instances where the localization of a particular molecular vibrational mode is not appreciably changed upon chemical reaction, *the change in vibrational frequency for a bond may be diagnostic of the change in force constant for the bond and thus of the nature of the reaction.* That is, the change in frequency of a particular vibration may be qualitatively indicative of a change in bond strength accompanying a reaction. For example, the coordination of the carbonyl oxygen of ethyl acetate by a Lewis acid such as BBr_3 should decrease the double bond character of the carbonyl link and thereby reduce the C=O vibrational frequency because of a reduction in C=O bond order.

$$CH_3-C\underset{O-CH_3}{\overset{O:}{\diagup}} + BBr_3 \rightarrow \underset{\underset{CH_3}{|}}{\overset{CH_3}{\diagdown}}\underset{O}{\overset{C=O}{\diagup}}BBr_3$$

The reduction in C=O bond order is predicted by consideration of the change in relative importance of the three important π resonance structures. That is, the single bond structures for the carbonyl oxygen are more important in coordinated than in "free" ester.

$$R-C\underset{O-R'}{\overset{O:}{\diagup}} \leftrightarrow R-\overset{\oplus}{C}\underset{O-R'}{\overset{\ominus O:}{\diagup}} \leftrightarrow R-C\underset{\overset{\oplus}{O}-R'}{\overset{\ominus O:}{\diagup}}$$

In fact, the C=O stretching frequency of ethyl acetate, at ~ 1600 cm^{-1}, is observed to decrease by about 190 cm^{-1} on adduct formation with BBr_3. It is useful to note, too, that the low energy shift strongly suggests that it is the *carbonyl oxygen* that acts as the donor atom in this reaction and not the *ester oxygen*. (This has been confirmed by x-ray structural analysis of the adduct.) The resonance structures given above suggest that *ester oxygen* coordination would incur an *increase in the carbonyl frequency*.

In using the arguments advanced above, one must be aware that the sigma orbitals have been ignored. Often this is done with good justification, since the sigma bonds are viewed as overlapping valence sp hybrids. Because of the s character (high effective nuclear charge) of the σ hybrids, σ electron pairs are less polarizable than $p\pi$ electron pairs. In short, the carbon $p\pi$ ao is less electronegative than the carbon "sp^2" sigma hao, so the polarization of the C=O electron density occurs mainly through the π bond pair as opposed to the σ bond pair.

At this point it is interesting to recall Bent's isovalent hybridization concept. Applying it to the carbonyl oxygen atom, we think of the oxygen atom as being in a trigonally hybridized valence state; two of the hybrids contain unshared pairs of electrons, while the third hybrid overlaps a hybrid from the carbon to form the C=O σ bond. As the electronegativity of the carbon atom is certainly greater than that of free space (the approximate environment of the lone pairs), the oxygen hybrid toward the carbon atom is expected to possess higher p orbital character than the lone pair hybrids. This situation will change when the oxygen coordinates with the boron atom of BBr_3. The increase in electronegativity of the lone pair neighbor (free space → boron atom) will cause the oxygen to increase the p character of the lone pair hybrid.[3] The increase in p character of this hybrid will come at least partially at the expense of the hybrid directed toward carbon, thereby strengthening the C=O σ bond somewhat (see Figure 3-4). This oxygen "rehybridization" effect tends to oppose the weakening effect of polarization of the C—O σ orbital; the net effect of coordination on the C—O σ bond is believed to be of less importance than the polarization of the C—O π bond.

A particularly interesting application[4] of these ideas can be made for the CN stretching frequencies of CN^-, HCN, and SCN^-. We will examine the changes in σ and π bonding between C and N as a result of addition of H^+ and S to the cyanide carbon. We expect stiffening of the C≡N σ bond in both cases, with a lesser change for S than H^+ (sulfur should be less electronegative than the *bare* proton). In both cases, the increase in carbon hybrid s character *and* polarization of the σ pair toward carbon act in concert to strengthen the bond. Considering the π resonance structures

$$:\overset{\ominus}{C}{\equiv}N: \leftrightarrow :C{=}\overset{..}{\underset{..}{N}}:^{\ominus}$$

$$H{-}C{\equiv}N: \leftrightarrow H{-}\overset{\oplus}{C}{=}\overset{..}{\underset{..}{N}}:^{\ominus}$$

$$^{\ominus}:\overset{..}{\underset{..}{S}}{-}C{\equiv}N: \leftrightarrow :\overset{..}{\underset{..}{S}}{=}C{=}\overset{..}{\underset{..}{N}}:^{\ominus}$$

we readily predict an increase in C≡N π bond order for HCN but a decrease in π bond order for SCN^-. Combining the σ and π bonding effects, it is straightforward that HCN should have a greater CN force constant than CN^-, but a prediction is not so easy for SCN^- because the σ and π effects are oppositely directed. The experimental CN stretching frequencies are 2080 cm^{-1} (CN^-), 2097 cm^{-1} (HCN), and 2006 cm^{-1} (SCN^-). Clearly, the predicted increase is realized for HCN, and the π effect is seen to dominate the σ effects for SCN^-.

One further example of the use of vibrational frequency–bond strength correlations concerns the mode of coordination of dimethylsulfoxide (DMSO), $(CH_3)_2SO$, which possesses two donor atoms. The two important π resonance structures for DMSO are

$$\underset{CH_3 \; CH_3}{\overset{\overset{\oplus}{S}{-}\overset{\ominus}{O}:}{\diagup\hspace{-0.5em}\diagdown}} \leftrightarrow \underset{CH_3 \; CH_3}{\overset{\overset{..}{S}{=}\overset{..}{O}:}{\diagup\hspace{-0.5em}\diagdown}}$$

Coordination of the oxygen atom enhances the contribution of the first structure relative to the second, while sulfur coordination favors the second relative to the first. By analogy with RCO_2R', oxygen coordination should have a slight effect on the S—O σ bond, while sulfur coordination should strengthen the S—O σ bond (cf. CN^-). Consequently, the combined σ/π effects suggest a decreased SO frequency for oxygen coordination and increased $\tilde{\nu}_{SO}$ for sulfur coordination. Examples of ambidentate behavior are found[5] in $Cr(DMSO)_6^{2+}$, which exhibits oxygen coordination and an SO stretching frequency 27 cm^{-1} lower than that of the free DMSO, and in $PtCl_2(DMSO)_2$, which exhibits sulfur coordination and an SO absorption band 61 cm^{-1} above that of the free ligand.

---

[3] K. F. Purcell and T. G. M. Dolph, *J. Amer. Chem. Soc.*, **94**, 2693 (1972).
[4] K. F. Purcell, *J. Amer. Chem. Soc.*, **89**, 247, 6139 (1967); **91**, 3487 (1969).
[5] J. H. Price, A. N. Williamson, R. F. Schramm, and B. B. Wayland, *Inorg. Chem.*, **11**, 1280 (1972).

DIPOLE MOMENTS

Dipole moments are often helpful to the chemist in deducing molecular structures, but they are not easy to interpret. As an example of the utility of dipole moment measurements, the moments of *cis* and *trans* square planar complexes of Pt(II) can be used to distinguish the isomers. An interesting specific case is the complex $PtCl_2(Et_2S)_2$. The *cis* complex has a moment of 9.5 Debye, while the *trans* moment is much reduced to 2.4 Debye. That the *trans* structure has a non-zero moment is due mainly to the fact that the sulfur atoms have lone pair electrons, which possess considerable dipolar character. This will be discussed and amplified shortly.

In order to determine the dipole moment of a molecule, we must locate the centroid of nuclear charges and the centroid of electron charges. Calling the distance between the centroids r, the dipole moment is given by $n \cdot r$, where n is the number of electrons. Only if the two centroids are not coincident will a dipole moment result. The natural viewpoint to take is to break down the total electron density into pairs of electrons, bond pairs and lone pairs, and to recognize that associated with each pair there are two proton charges required to maintain electroneutrality for the molecule. We must estimate the centroid of each $+2$ nuclear charge and each -2 electron charge. In other words, each electron pair–proton pair is considered to develop a dipole, and all of these must be added up to produce the total.

From the hybrid orbital view, *lone pair dipoles* are quite large and vary slightly with the $s{:}p$ character of the hybrid (di $>$ tr $>$ te).

$$\mu_{l_p} = 2e\langle r \rangle$$

$2+\quad 2-$

Remembering that a hybrid atomic orbital is a superposition of pure atomic orbitals, we can derive the expression for the hybrid centroid by writing (refer to Chapter 1 for a discussion of computing the average value of r)

$$\phi_h = C_s\phi_s + C_p\phi_p$$
$$\int \phi_h r \phi_h \equiv \langle r \rangle_h = C_s^2 \langle r \rangle_s + C_p^2 \langle r \rangle_p + 2C_sC_p\langle r \rangle_{sp}$$
$$= (C_s^2 + C_p^2)\langle r \rangle_s + 2C_sC_p\langle r \rangle_{sp}$$
$$= \langle r \rangle_s + 2C_sC_p\langle r \rangle_{sp}$$

The last step is possible because the hybrid is a normalized atomic orbital and because the s and p pure ao centroids are coincident. By taking the nucleus as the origin for measuring the nuclear and electron centroids, the hybrid ao dipole moment for an electron pair becomes just

$$\mu = 2e\langle r \rangle_h = e \cdot 4C_sC_p\langle r \rangle_{sp}$$

and arises solely from the "interference" part of the hybrid wavefunction (cf. p. 102). For two electrons in a carbon hybrid, the dipole moment varies[6] from about 4.4 Debye (di hybrid) to 3.7 Debye (te hybrid); obviously, the hybrid moments make large contributions to a molecular dipole.

Bond pair moments are much more difficult to estimate, both in magnitude and in direction. Location of the $2+$ centroid for a bond pair is not difficult (it is simply the

---

[6] C. A. Coulson, "Valence," Oxford University Press, New York (1952), p. 208.

mid-point of the bond), but complications set in when we attempt to estimate qualitatively the centroid for the electron pair. The problem can be illustrated with a simple bond involving a hydrogen atom and a heavier atom, using a hybrid atomic orbital. The bond pair wavefunction is given by

$$\psi = a\phi_H + b\phi_h \qquad \begin{pmatrix} \phi_H = \text{hydrogen 1s ao} \\ \phi_h = \text{other atom's hybrid ao} \end{pmatrix}$$

and we wish to know the average position of the bond pair along the internuclear axis.

$$\int \psi r \psi = a^2 \int \phi_H r \phi_H + b^2 \int \phi_h r \phi_h + 2ab \int \phi_H r \phi_h$$
$$\langle r \rangle = a^2 \langle r \rangle_H + b^2 \langle r \rangle_h + 2ab \langle r \rangle_{Hh}$$

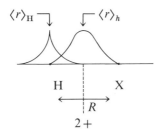

The first term in the expression gives the contribution of electron density (of magnitude a^2) in the hydrogen 1s ao, the second term gives the contribution from density (of magnitude b^2) in the other atom's hybrid ao, and the third term gives the contribution of overlap density (magnitude $2ab$) to the centroid of negative charge. Clearly, the density centroid for the hydrogen ao is to the "left" of the bond midpoint, at $\langle r \rangle_H = -R/2$. It is generally difficult to estimate where the hybrid orbital density centroid will be located relative to the bond mid-point. Using the preceding result for $\langle r \rangle_h$,

$$\langle r \rangle_h = \langle r \rangle_s + 2C_s C_p \langle r \rangle_{sp}$$
$$= \frac{R}{2} + 2C_s C_p \langle r \rangle_{sp}$$

For the bond pair, then,

$$\langle r \rangle = a^2 \left(-\frac{R}{2}\right) + b^2 \left(\frac{R}{2}\right) + 2b^2 C_s C_p \langle r \rangle_{sp} + 2ab \langle r \rangle_{Hh}$$
$$= i \left(\frac{R}{2}\right) + (2b^2 C_s C_p) \langle r \rangle_{sp} + (2ab) \langle r \rangle_{Hh}$$

where $i = b^2 - a^2$ is the "ionic" character of the bond [i varies from zero for a truly covalent bond ($a^2 = b^2$) to ± 1 for an ionic bond (a^2 or $b^2 = 1$)]. The first term allows for an unequal distribution of charge between the two atoms, and the second term is the hybrid orbital "interference" contribution (which is lacking for the hydrogen orbital, of course). This term could tend to shift the centroid to either side of the bond mid-point and depends on the radial diffuseness of the hybrid. The last term, the contribution from the overlap density, is also an "interference" contribution, this time between orbitals centered on different nuclei. The following sketch of the overlap product $\phi_H \phi_h$ makes it apparent that the centroid is in the direction of the more radially compact atomic orbital (the hydrogen 1s in this case).

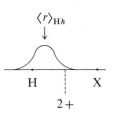

To summarize:

$i\left(\dfrac{R}{2}\right)$: tends to orient the bond dipole toward the more electronegative atom

$2ab\langle r\rangle_{Hh}$: tends to orient the bond dipole toward the atom with the more radially compact orbital

$2b^2 C_s C_p \langle r\rangle_{sp}$: tends to orient the bond dipole to one side or the other, depending on the radial diffuseness of the hybrid ao

If we were to consider a bond pair between two atoms, both of which use hybrid ao's, we would need to add another term like the last to allow for the hybrid orbital moments of both atoms. It is helpful to note, however, that the hybrid moments will tend to cancel one another.

From these considerations we are drawn to conclude that often the bond dipole moment will be smaller than a lone pair moment, unless the two atoms are quite different in electronegativity, and could even be directed from the other atom toward hydrogen.

To give an example of the application of these concepts, we can compare the dipole moments of NH_3 and NF_3, which are 1.5 Debye and 0.2 Debye, respectively. In NH_3 the NH bond dipoles may be small ($\chi_N > \chi_H$, but hydrogen is more compact and does *not* have a hybrid moment to cancel that of nitrogen), so we find the lone pair moment to be marginally canceled or reinforced by the bond pair moments.

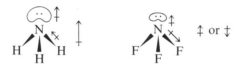

For NF_3, the bond moments should be directed toward fluorine ($\chi_F > \chi_N$, the fluorine ao's are more compact than those of nitrogen). Furthermore, the fluorines have lone pair moments aligned with those of the bond pairs. On the other hand, the nitrogen lone pair of NF_3 has greater s character than that of NH_3. With the fluorine lone pair moments and the bond pair moments directed oppositely to the nitrogen lone pair moment, we may even wonder whether the NF_3 moment is not oppositely directed to that of NH_3!

Carbon monoxide is an interesting example because the moment is small (0.1 Debye) and oriented toward carbon (in the sense C^-O^+). Bent's hybridization model implies that both the carbon and oxygen lone pairs have *greater* than 50% s character, with carbon greater than oxygen. This tends to make the oxygen moment greater than that of carbon. On the other hand, the carbon hybrid is radially more diffuse than that of oxygen, so that the oxygen hybrid moment is less than that of carbon. The net lone pair moment is of uncertain magnitude and direction. Both the sigma and pi bond pair centroids are clearly oriented toward oxygen because the oxygen ao's are more compact and because the electronegativity difference reinforces the compactness term.

In locating the centroid of positive charge for the bond pairs, note that the six proton charges are unsymmetrically distributed, $2+$ on carbon and $4+$ on oxygen, with a mean position one-third of the CO distance from oxygen. The bond moment will accordingly be small, but *probably* oriented C^+O^-. The small net moment suggests near cancellation of lone pair and bond pair moments, with probable relative orientations

To anticipate some points of chemical significance about these last two examples, you will discover in Chapter 5 that NF_3 is very weakly basic in the Lewis base sense and that CO exhibits a normal tendency to act as a Lewis base through the carbon, rather than the oxygen, lone pair. Both behaviors are conveniently linked to the nature of the lone pair orbitals as described above.

To return to the structure/dipole moment relationship for $PtCl_2(Et_2S)_2$, the large moment for the *cis* complex is not surprising. For *trans*-$PtCl_2(Et_2S)_2$, the Pt—S and Pt—Cl bond moments cancel, by symmetry, as do the Cl lone pair moments. For free rotation of the Et_2S molecules about the Pt—S axes, we would measure a net dipole of zero. The large observed value implies the absence of a center of symmetry, as in the structures below, where the origin of $\mu \neq 0$ is seen to arise from the difference in sulfur lone pair and S—Et moments.

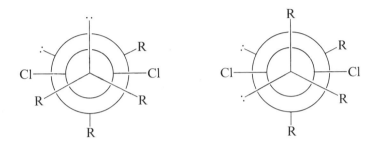

NUCLEAR SPIN COUPLING

Another experimental result that is easily interpreted from the hao view is that of the coupling of a hydrogen nuclear spin to the nuclear spin of an atom to which the hydrogen is bound. You are already aware that the proton spin property makes possible the nmr absorption experiment. In the presence of a laboratory magnetic field, the nuclear magnetic moments (intrinsic to the spin angular momentum) of a sample of protons will exhibit nearly equal numbers of aligned and anti-aligned nuclear magnets; the nmr experiment detects the aligned magnets by measuring the work required to "flip" them to an anti-aligned condition. The position of the absorption line on an energy scale represents the energy absorbed by the proton upon flipping. If the hydrogen is bound to an atom that also possesses a nuclear magnet, the pair of hydrogen/other-atom magnets gives rise to four spin microstates:

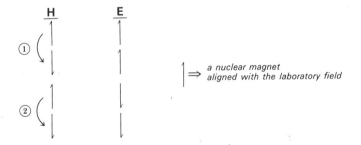

These four possibilities represent different energy states for the system of four magnets, and a real sample will have molecules nearly equally distributed (by Boltzmann statistics) over the four possibilities.

If there were no communication between the hydrogen and other-atom magnets, flipping the hydrogen spin as in ① would require the same energy as in ② (because the proton is acted upon by only the *external* magnetic field). If by some mechanism the *internal* field of the other-atom magnet augments the *external* field, as in ① say, and opposes the *external* field, as in ② for example, "flipping" the hydrogen spin in

① would require more energy than in ②. This will give rise to two different energy absorption phenomena for the protons.

It is the coupling mechanism in which we are interested here. The electron pair defining the H—E bond may be treated as two particles also possessing opposite magnetic moments. If neither H nor the other atom possessed a nuclear magnet, the electrons would have equal probability of occurring at the hydrogen atom and at the other atom, say carbon:

If carbon, as in the ^{13}C isotope, has a nuclear spin, this equal distribution of electrons will be upset with the result

$$\text{(H)}\,\phi\,\phi\,C\!\uparrow \;>\; \text{(H)}\,\phi\,\phi\,\uparrow C$$

because of the preference for alignment of electron/nuclear magnets. The ↑ and ↓ moment electrons no longer have the same probability functions (the orbitals are slightly different). Accordingly, the hydrogen experiences a net *internal* magnetic field opposed to the external field. Molecules with the *carbon magnet anti-aligned with the external field* exhibit an *internal field at hydrogen aligned with the external field*.

Carbon nuclear alignment	Internal field at hydrogen
C↑	↓
C↓	↑

Including the electron moments, the net direction of the internal field at hydrogen is determined from

```
           H    e      e    C
      ↑-------↓   ↑-------↑
  ①  ⎨
      ↓-------↓   ↑-------↑

      ↑-------↑   ↓-------↓
  ②  ⎨
      ↓-------↑   ↓-------↓
```

Now the "flip" ② requires more energy than the "flip" ① and the H absorption occurs at two photon frequencies instead of one. Clearly, the *internal* field at hydrogen depends on the electron probability at the proton. For hydrogen, the $1s$ ao has non-zero amplitude at the nucleus, and this enhances the *internal* field felt by the proton nuclear magnet. Similarly, the *spin polarization* of the C—H bond pair will increase as the $2s$ character of the carbon hybrid increases, because the carbon nuclear field felt by the electrons increases. This phenomenon is called *Fermi contact coupling* (of electron/nuclear magnets).

To illustrate the sensitivity of the difference in absorption processes ① and ② to the s character of the carbon hybrid, we list the following CH coupling constants, J (defined as the energy difference between the two proton absorption events): CH_4 (sp^3, 125 Hz), C_2H_4 (sp^2, 156 Hz), C_2H_2 (sp, 249 Hz). Hz is the Herz frequency unit, sec^{-1}, and is proportional to energy through $E = h\nu$. Even more subtle changes in

TABLE 3–4
SOME REPRESENTATIVE J_{EH} DATA

Structure	J (Hz)	Structure	J (Hz)
CH_4	125	$HC(F)CH_2$	200
C_2H_4	156	H_2CO	172
C_2H_2	249		
HCF_3	239	$HC(O)(Me)$	172
$HC(OMe)_3$	186	(NMe_2)	191
		(OH)	222
$H_3C(SiH_3)$	122	$HC\equiv(P)$	211
(PH_2)	128	(CPh)	251
(NH_2)	133	(N)	269
(CN)	136	(NH^+)	320
(NH_3^+)	145		
(OH)	141	H_2P^-	139
(NO_2)	146	H_2PH	182
(F)	149	$H_2PH_2^+$	547
		H_2PCH_3	186
$H_3Si(H)$	203	H_2PCF_3	200
(SiH_3)	196		
H_3N	40		
H_3NH^+	54		

carbon *s* character can be detected by CH coupling, as illustrated by CH_4 (sp^3, 125 Hz) compared to HCF_3 (sp^3, 239 Hz). Here again we find application of Bent's isovalent hybridization scheme. Table 3–4 summarizes typical data that are interpretable in this fashion.[7] Before leaving this topic we should pause to note that correlating *J* with hybrid %*s* character alone ignores what may be another important consideration. Electron withdrawal by a substituent descreens the central atom nucleus and enhances the valence *s* orbital amplitude at the nucleus (a contraction effect). This also may enhance *J*.[7]

BOND ENERGIES

From the thermodynamics section of your physical chemistry course, you will recall the utility of the concept of additivity of bond energies. That such a concept works in practical applications reaffirms the idea that it is possible to think of localized bond pairs of electrons as characteristic of the atoms defining the bond. Here we will pursue the concept of an experimental bond energy and its relation to atom hybridization.

The definition of a bond energy is not ambiguous at all for a diatomic molecule. Experimentally the bond energy is simply the internal energy change required to separate the molecule into its two constituent atoms (the molecule and atoms in the gas phase):

$$A\text{—}B_{(g)} \rightarrow A_{(g)} + B_{(g)} \qquad D_{AB} = \Delta E$$

This definition of a bond energy, applied to a polyatomic molecule, must be carefully considered in relation to the valence state concepts discussed in Chapter 1. For the simpler case of a binary compound AB_n we must consider the following:

(1) Removal of one terminal atom to leave the fragment AB_{n-1} will alter the bonding between A and the remaining B atoms because of changes in AB multiple bonding, changes in A hybridization, or both.

---

[7] G. E. Maciel, *et al.*, *J. Amer. Chem. Soc.*, **92**, 1 (1970); A. H. Cowley and W. D. White, *ibid.*, **91**, 1917 (1969).

(2) The ground state electron configuration of the B atom dissociated may be quite different from its valence state when bound to A. The valence state promotion energy for B may make a significant contribution to the D_{AB} for dissociation of that atom.

Several interesting examples of the operation of these effects may be cited. For example, the stepwise dissociation of CO_2 shows the following:

$$O=C=O \rightarrow O=C + O \rightarrow C + 2O$$

$$D_{CO} = 127 \text{ kcal/mole} \qquad D_{CO} = 256 \text{ kcal/mole}$$

The dissociation of the first oxygen is the net result of breaking the CO double bond, the relaxation of the departing oxygen from the averaged spin valence state $(di^1 di^2 \pi^1 \pi^2 = s^{1.5} p^{4.5})$ to the ground atomic electronic term $(s^2 p^4 \rightarrow {}^3P)$, the relaxation of carbon from (ideally) $didi\pi\pi(s^1 p^3)$ to $didi^2\pi(s^{1.5} p^{2.5})$ and, to some extent, the increased bonding between the carbon and the remaining oxygen. Dissociation of the remaining oxygen entails oxygen atom relaxation again, the breaking of a CO diatomic bond with bond order ~ 2.5, and the carbon atom relaxation. The fact that the second dissociation requires more energy than the first is probably a reflection of the lower CO bond order and greater[8] electronic relaxation for carbon during the first dissociation.

A distinctly different ordering of bond dissociation energies is found with mercuric chloride:

$$Cl-Hg-Cl \rightarrow HgCl + Cl \rightarrow Hg + 2Cl$$

$$D_{HgCl} = 81 \text{ kcal/mole} \qquad D_{HgCl} = 25 \text{ kcal/mole}$$

Here the same general factors as mentioned for the CO_2 case are at play, but with the interesting result that dissociation of the second HgCl bond requires less energy than dissociation of the first. This is apparently due to a large relaxation energy for the Hg atom in the second dissociation, when the Hg falls from a valence sp state of $(di^1 di^1 = s^1 p^1)$ to the ground state configuration of $6s^2({}^1S)$.

In both of these examples, one is faced with unambiguously defining what is meant by the "bond energy." For general usage, the *bond energies* of AB_n molecules are taken to be $1/n$ times the internal energy change for the process

$$AB_{n(g)} \rightarrow A_{(g)} + nB_{(g)}$$

$$E_{AB} = \frac{1}{n} \Delta E$$

and are to be distinguished from the concept of individual *bond dissociation energies*, D_{AB}.

Consideration of the CO_2 and $HgCl_2$ examples should make clear the fallacy of attempting to distinguish the energies of the axial and equatorial bonds of a molecule like PCl_5. In the first place, there seems to be no way, experimentally, to dissociate selectively either an axial or an equatorial bond. Second, the reactant and product for the dissociation of one chlorine atom must be the same, whether an axial or equatorial bond is actually broken, since only one structure for PCl_4 is possible. As the internal energy change of the dissociation process is a so-called state function, the

---

[8] Remember that the lone pair (di^2) of CO has greater s character than does the di hybrid of CO_2. Thus, the change toward the free atom configuration $(s^2 p^2)$ is greater on loss of the first oxygen than on loss of the second.

energy measured is the same for breaking of either type of bond. If we presume, with theoretical justification, that breaking of an axial PCl bond requires less energy than breaking of an equatorial PCl bond, then the relaxation of the PCl_4 fragment in the former case is less exothermic than in the latter. The sum of the energies of bond breaking and PCl_4 fragment reorganization is the same for both processes.

These examples show the care with which one must approach the interpretation of experimental bond energies. The theoretical interpretation of bond energies requires the factorization of atom relaxation energies to get at the bond energies in which the products represent atoms not in their ground electronic states but with the valence state configurations appropriate to the molecule under consideration. In this way we arrive at more meaningful bond energies for interpretation in terms of the effects of multiple bond formation, sigma orbital hybridization, and the degree of overlap of atomic hybrids.

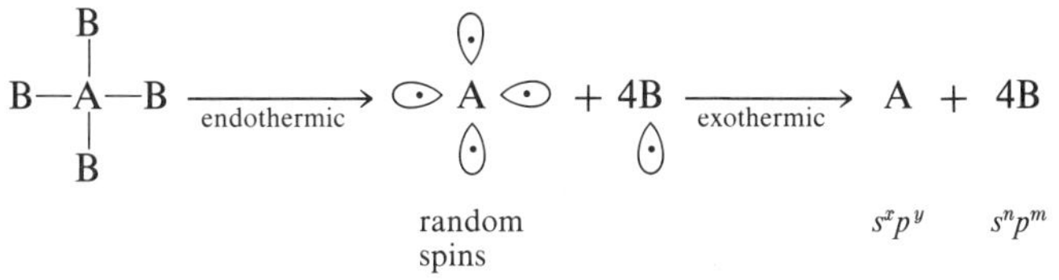

$$B-\underset{\underset{B}{|}}{\overset{\overset{B}{|}}{A}}-B \xrightarrow{\text{endothermic}} A + 4B \xrightarrow{\text{exothermic}} A + 4B$$

random spins $\qquad s^x p^y \quad s^n p^m$

As an interesting application of the ideas thus far discussed, the bond energies (kcal/mole) for the hydrides of some of the second and third row elements are given in the following table (taken from Table 6–3):

C	99	N	93	O	111
Si	76	P	78	S	88

Given that the heavy atom promotion energies to the tetrahedral valence states are

C	150	N	228	O	194	F	120
Si	115	P	133	S	109	Cl	62

it is apparent that the bonds formed by the lighter elements in each group are truly stronger than those of the heavier elements. To illustrate the comparison of CH_4 with SiH_4, for example, we write

$$4E_{CH} = 4(99) = 4E_{CH}^V - P_C - 4P_H$$
$$4E_{SiH} = 4(76) = 4E_{SiH}^V - P_{Si} - 4P_H$$
$$396 = 4E_{CH}^V - 150 - 0$$
$$304 = 4E_{SiH}^V - 115 - 0$$
$$E_{CH}^V = \tfrac{1}{4}(546) = 137$$
$$E_{SiH}^V = \tfrac{1}{4}(419) = 105$$

and the CH bond is 32 kcal/mole stronger than the SiH bond, exclusive of atom relaxation energies. This is a ramification of the greater diffuseness of the atomic orbitals of the heavier elements. The diffuseness has its origin in the radial parts of the hybrid orbital functions, and it results in smaller overlap integrals, and consequently weaker bonds, for the orbitals of the heavier elements with hydrogen. For methyl compounds the same trend is found:

C	83	N	79	O	86
Si	73	P	63	S	62

For fluorides and oxides, however, the trend is often reversed, presumably because fluorine and oxygen have unshared electron pairs for multiple bond formation with the heavier elements. The bond energies are:

C	117	N	67	O	51	Fluorides
Si	143	P	119	S	68	
C	86	N	39	O	34	Oxides
Si	111	P	88	S	—	

Correction for the promotion energies shows, for example, that the P—F bond is 20 kcal/mole stronger than the N—F bond. This reversal is probably due not only to P—F multiple bond character enhancing the P—F energy:

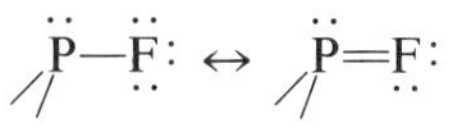

$$\ddot{P}-\ddot{F}: \leftrightarrow \ddot{P}=F:$$

but also to greater lp/lp repulsions for N—F, acting to weaken that bond more:

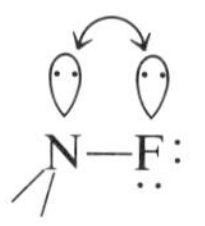

$$N-\ddot{F}:$$

Another interesting example of the application of these ideas is to be found in the instability of the heavy elements of Groups III and IV with oxidation states of III and IV. Boron and carbon form stable compounds of formulae BX_3 and CX_4 with X = halide, whereas the elements Tl and Pb form as the most stable halides the compounds TlX and PbX_2. Drago[9] has discussed this problem thoroughly and finds that the energy gained by forming two more bonds in each case does not offset the promotion energy involved in achieving trigonal and tetrahedral valence states (tr^3 from s^2p^1 and te^4 from s^2p^2, respectively). It does not appear that the promotion energies are inordinately large for Tl and Pb, but rather that the bonds formed by trigonal Tl and tetrahedral Pb are not very strong. Again, the weakness of the bonds of Tl and Pb is to be attributed to the highly diffuse character of the atomic orbitals of these elements.

To help in visualizing the importance of orbital compactness in producing large overlap integrals, Figure 3-5 schematically illustrates the effect.

---

[9] R. S. Drago, *J. Phys. Chem.*, **62**, 353 (1958).

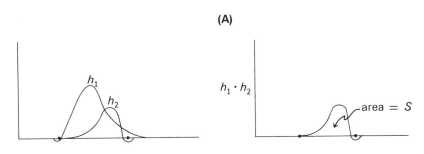
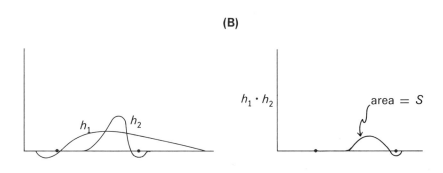

Figure 3-5. The effect of increasing radial diffuseness of an atomic hybrid on the overlap of that hybrid with another. (A) h_1 "compact," (B) h_1 "diffuse."

STUDY QUESTIONS

5. Use the VSEPR model to predict the stable structures for PCl_2F_3, $PCl_2(CF_3)_3$, and SF_4; then describe the "ideal" central atom hybrids for each electron pair. For each of these molecules, tell how the hybrids differ in s character in actuality ("non-ideal"). Which molecules have non-zero dipole moments? For those, identify the symmetry axis along which the molecular dipoles lie.

6. Compare, on the basis of the hao model, the predicted bond strengths of the P—F bonds in PF_3 and PF_5. Rank the bonds in order of decreasing Fermi contact coupling (J_{PF}).

7. Predict a VSEPR structure for HCP. The nmr spectra of this compound reveal $J_{HC} = 211$ Hz, $J_{CP} = 54$ Hz, and $J_{HCP} = 44$ Hz. How is it that these data (and the facts that the J_{CP} of PR_3 compounds are much smaller than 54 Hz and that J_{HCP} is typically 0 to 25 Hz) support the predicted structure?

8. Predict an order of vibrational frequencies for:

   a) Xe—O in XeO_2, $XeOF_4$
   b) I—Cl in ICl_3
   c) M—H in CH_4, SiH_4

9. Consider the compound $Me_3N \cdot SnMe_3Cl$, derived from the Lewis acid-base addition of Me_3N: to $SnMe_3Cl$.

   a) Sketch the structure of the compound, showing the geometry of the N, C, and Cl atoms about the Sn atom. (Be sure to tell *briefly* why you pick one isomer or another.)
   b) Predict a direction of change for the Sn—Cl stretching frequency upon formation of the adduct and justify it, briefly.
   c) Do you expect $^2J_{SnCH}$ to increase or decrease upon adduct formation?

10. HF has a large dipole moment (1.98 D). Analyze the contributions to the total dipole. Are there any offsetting contributions?

11. Which should have the stiffer C=O bond, CO_2 or cyanate?

12. Which group should show the greater σ orbital electronegativity toward X? Explain.

   a) $\text{\textbackslash B—X}$ or $\text{\textbackslash C—X}$ b) C_6H_5—X or $\text{H}_3\text{C—X}$ c) $\ddot{\text{O}}{=}\text{X}$ or $\text{H}\ddot{\text{O}}{—}\text{X}$

13. Consider the molecule SOF_4.

   a) Draw Lewis structures for the two principal resonance forms.
   b) Estimate limits for the S—O bond order.
   c) Predict the stable isomer.
   d) Are all S—F bonds of equal length? If not, tell which are longer.
   e) Tell how the bond angles differ from ideal.
   f) Give the correct molecular point group.
   g) Rank the sulfur hybrids in order of decreasing s character.
   h) Which should have the stronger, shorter S—O bond, SOF_4 or SOF_2?
   i) The answer to (f) tells you the axis of the molecular dipole. What is it? Now that you know where the dipole is, determine its direction by comparing the various bond and lone pair moments.

14. For each of the following stoichiometries, give the "main group" group number(s) from which the central atom may come.

   a) EF_3 b) EF_4 c) EF_5 d) EF_6

15. a) For set EF_3 of question 14, answer each of the following questions for an example from *each* group:

   i) What is the *spd* notation for the central atom hybrids of your example, assuming an "ideal" VSEPR structure?
   ii) Does the example possess a permanent molecular dipole moment? If so, identify the symmetry axis along which the dipole lies.

   b) Comparing bond pairs in the different EF_3 molecules, rank the bonds in order of predicted strength on the basis of E hybridization alone.
   c) Rank the EF_3 bonds in order of predicted magnitude of E—F Fermi contact coupling (assume the F hybridization to be constant).

16. Consider the molecular fragments (i) $\overset{\ominus}{M}$—N≡C—R and (ii) $\overset{\ominus}{M}$—C≡N—R, and determine for which of these the lone pair on M is more likely to become a bonding pair (M=N··· or M=C···). Which species, on the basis of this answer alone, will have higher ν_{CN}?

EPILOG

In the last two chapters we have attempted to outline and exemplify to some extent a generally successful hierarchy of molecular structural models that form the basis for chemists' thoughts on the geometrical and electronic structures of molecules. The progression of models is sufficient for many purposes, while for other purposes it provides only a working system for "starting point" ideas of molecular structure; it is well to note, as we have at various points, that there are instances in which the hierarchy is insufficient and refinements have to be made in those special cases that demand them. Nevertheless, the value of the basic framework of ideas lies in the systematization of experimental molecular geometries and properties that are results of the electronic structures of molecules. This, after all, is a prime goal of chemical studies: to identify some generally applicable principles of chemical phenomena and to refine them when new experimental results call for more sophistication in our understanding.

We now leave, temporarily, the ideas developed so far in terms of *localized* bond electron pairs, in order to consider a theory of molecular electronic structure that has made possible tremendous advances in our understanding, primarily through the ease with which the alternate model lends itself to quantitative calculations. This model will at first seem diametrically opposed to the concept of localized sigma electron pairs, but we will ultimately return to this point to establish a strong link between the new model and the concepts of localized bond pairs and hybrid orbitals.

INTRODUCTORY COMMENTS

MOLECULAR ORBITAL PROBABILITY FUNCTIONS

PRINCIPLES OF MO CONSTRUCTION AND INTERPRETATION

   The Hydrogen Molecule
   Beryllium Hydride and Hydrogen Chloride
   Summary

MAIN GROUP DIATOMIC MOLECULES

   Homonuclear Molecules
   The $ps\sigma$, $p\pi$ Order Controversy
   Heteronuclear Molecules

STRUCTURE, VSEPR, AND LCAO-MO

   EF_3 Series

   EF_4 Series

"LINEAR" (ONE-DIMENSIONAL) MOLECULES

CYCLIC (TWO-DIMENSIONAL) MOLECULES

POLYHEDRAL (THREE-DIMENSIONAL) MOLECULES

   Stratagem
   EX_3 (D_{3h})
   EX_4 (T_d)
   EX_5 (D_{3h})
   EX_4 (D_{4h})
   EX_6 (O_h)

THE EQUIVALENCE OF THE LOCALIZED AND DELOCALIZED ORBITAL MODELS

EPILOG

4. THE UNDIRECTED ORBITAL VIEW OF CHEMICAL BONDS

In the preceding chapters you have learned of working models for molecular structure derived from the localized electron pair concept. In this chapter we will continue the process of developing progressively more sophisticated models for molecular electronic structure. Although the LCAO technique was initially developed shortly after Schrödinger published his model of the atom, the chemical community is just now beginning to probe the power of this theoretical technique and is finding it to be tremendously useful in chemical interpretations of molecular spectra, structure, and reaction mechanisms. The LCAO-MO model is less restrictive at the outset than the localized pair approach by refusing to be predisposed on the ideas of valence s and p ao mixing (hybridization) and localization of electron pairs. Orbital symmetries play a very important role in this model, as do the ideas of matter wave interference and orbital nodal structure.

INTRODUCTORY COMMENTS

According to the wave model for the electrons in an atom, all available information about the electrons can be obtained from solution of the time-dependent Schrödinger equation (refer to Chapter 1). For discussions of the electron probability and quantities related to the probability, you found that the time dependency could be factored out; considerable chemical information is held in the time-independent portion of the wavefunction, which is obtained from the time-independent Schrödinger equation:

$$H\Psi = E\Psi$$

Now, there is no reason to expect that the wave character of the electron should be altered by the requirement that the electron be attracted to more than one nucleus, and so it is reasonable to assume the time-independent Schrödinger equation should also apply to molecules. In fact, the equation has been solved for the hydrogen molecule ion, H_2^+, with great success.

The fact that more than one nuclear charge is present in a molecule means that the molecular Hamiltonian now contains some additional terms involving the several nuclei. In fact, the molecular Hamiltonian for a molecule with more than one electron is a fairly simple extension of that for a polyelectronic atom:

$$\text{K.E.}(1) + \text{K.E.}(2) + \cdots + \text{P.E.}(1) + \text{P.E.}(2) + \cdots$$
$$+ \text{E.R.}(1,2) + \text{E.R.}(1,3) + \cdots + \text{N.R.}(A,B) + \text{N.R.}(A,C) + \cdots$$

In this expression, K.E.(1) represents the operator for the kinetic energy of electron 1; P.E.(1) represents the operator giving the attractive potential between electron 1 and the several nuclei A, B, . . .; the term E.R.(1,2) stands for the operator representing the coulombic repulsion between electrons 1 and 2; and, finally, the term N.R.(A,B) is the operator for the coulombic repulsion between nuclei A and B. In mathematical terms:

$$\text{K.E.}(1) = -\frac{\hbar^2}{2m}\left[\frac{\partial^2}{\partial x_1^2} + \frac{\partial^2}{\partial y_1^2} + \frac{\partial^2}{\partial z_1^2}\right]$$

$$\text{P.E.}(1) = \sum_X \frac{Z_X e}{r_{1,X}}$$
where Z_X is the nuclear charge of atom X and $r_{1,X}$ is the distance between electron 1 and the nucleus of atom X

$$\text{E.R.}(1,2) = \frac{e^2}{r_{12}}$$
where r_{12} is the distance between electrons 1 and 2

$$\text{N.R.}(A,B) = \frac{Z_A Z_B}{R_{AB}}$$
where R_{AB} is the distance between the nuclei of atoms A and B

In the time-independent Schrödinger equation, $H\Psi = E\Psi$, the quantity E now represents the molecular *electronic* energy (that is, the energy of the entire assemblage of electrons) and Ψ is the molecular electronic wavefunction (that is, Ψ is a function of the space coordinates of *all* electrons of the molecule):

$$\Psi = f(x_1, y_1, z_1, x_2, y_2, z_2, \cdots, x_m, y_m, z_m)$$

As in the polyelectronic atom, a predicament arises in attempting to solve the wave equation in closed analytic form for so many interacting particles—in short, the mathematical machinery is not available to reach such a solution. (This is true even for a system that obeys the "laws" of classical mechanics.) Consequently, approximations must be made, as is done for systems that follow classical mechanics. (A particularly relevant case is the problem of a spaceship interacting with the earth, sun, and moon, for which it is understandable that very good approximate methods are used!) In the problem of electrons in a molecule, one can also choose among approximations of varying degrees of accuracy, depending on the amount of effort (nowadays meaning the amount of computer time) he is willing to put into the calculation. It is an unfortunate fact of life, however, that as the approximations are increasingly refined, the complexity of the wavefunction increases, and it soon loses value to the chemist who hopes to use simpler models for predictive and systematization purposes. In deciding to retain conceptual simplicity, it is apparently inescapable that some accuracy must be forgone.

This brings us to the approximation known as the linear combination of atomic orbitals (LCAO) method,[1] *which has been important to the chemist by yielding chemically useful wavefunctions for molecules.* The basic assumption in this method is as follows: molecules can be thought of as aggregates of atoms close enough that they perturb one another; however, the atoms are not so perturbed that we have to forgo the useful idea that the electrons, when in the vicinity of each nucleus, behave much like they do in the isolated atom. Thus, it is assumed that the molecular wavefunction for each electron can be written as a linear combination of atomic orbitals for the various atoms. In making this assumption, the theoretician implicitly extends the concept (used for polyelectronic atoms) that each electron moves in the average electrostatic field of all other electrons; he can therefore write a separate wavefunction for each electron, which depends only on the spatial coordinates of that electron and does not

---

[1] R. S. Mulliken, *Phys. Rev.*, **32**, 186, 761 (1928); **41**, 49 (1932); F. Hund, *Z. Physik*, **51**, 759 (1928); **73**, 1 (1931); S. R. La Paglia, "Introductory Quantum Chemistry," Harper & Row, New York (1971), Chapter 5.

include the spatial coordinates of any other electrons. *Such an approximate "one-electron" wavefunction for a molecule is called a molecular orbital.* This is a greatly simplifying assumption, because one can then look for a solution to the Fock equation, which involves a one-electron energy operator called the Fock operator:

$$F_i(1)\psi_i(1) = \varepsilon_i \psi_i(1)$$

The Fock operator F_i contains the K.E. operator for an electron in orbital ψ_i, the P.E. operator for the electron in ψ_i, and the *average* repulsion energy (also a potential energy) for the electron in ψ_i interacting with all other electrons in the molecule. It is important to understand that the last named quantity depends on the probability distributions for all other electrons in the molecule. This puts us in the interesting situation of having to know the orbital functions of all other electrons before we can calculate the orbital function for electron 1! In short, we must use an iterative procedure; that is, we start out *assuming* orbital functions for all other electrons and use these to obtain the average repulsion energy for electron 1. Using the orbital function that solves the Fock equation for electron 1, and the assumed functions for electrons 3, 4, . . ., we go on to find a new orbital function for electron 2. The process is continued until a new function has been obtained for each electron, and the process is started all over again. It is continued until all orbital functions have converged to our satisfaction, meaning that successive cycles of the calculations yield negligibly small changes in the orbital functions (the theoretician is left to decide what he will accept as a negligible change). You can see the value of having a computer available for these calculations!

To get back to the concept of LCAO, the molecular orbital functions are assumed to be of the form

$$\psi_i = c_1 \phi_1 + c_2 \phi_2 + \cdots$$

where the ϕ's are the atomic orbitals of the atoms making up the molecule. The linear combination could include an infinity of atomic orbitals (the more terms carried, the greater the accuracy of the calculation) but, for our purpose, it suffices to include only the valence atomic orbitals of each atom.

The goal of the iterative search for the orbital functions amounts to determination of the coefficients c, which are different for each molecular orbital. To do this, the theoretician makes use of a theorem called the Variation Theorem, which states that the energy calculated from an approximate molecular wavefunction, Ψ, will always be greater than the true molecular energy, and that as the approximate orbital wavefunctions get better the molecular energy calculated from them will more closely approach the true energy. Mathematically, the condition to be met is that the molecular energy is minimized with respect to each of the coefficients, c, in each molecular orbital. This means that the first derivative of the energy with respect to each unknown coefficient is set equal to zero; this yields as many equations for the unknowns as there are unknowns.

Finally, if the molecular energy calculation is repeated for all possible nuclear configurations (molecular structures), the lowest energy value obtained corresponds to the theoretically predicted geometry for the molecule.

You should be impressed by now with the immensity of the problem of calculating even approximate orbital functions for electrons in molecules. In this text we will treat the results of such calculations as empirical findings with the intent of systematizing orbital patterns associated with a few basic nuclear geometries. There are quite a few systematic principles to be found in the results of the "number crunching" process of determining molecular orbitals. *Most important to us will be the concepts of symmetry, orbital probabilities given by ψ_i^2, and the phase relationships between the atomic orbitals as combined into LCAO molecular orbitals.*

MOLECULAR ORBITAL PROBABILITY FUNCTIONS

As was thoroughly discussed in the preceding chapter and in the atomic structure chapter, if the orbital functions are to have physical meaning, each function must be normalized to unity, corresponding to unit probability that the electron will be found in all of space. The mathematical condition is

$$\int \psi_i^2 \, d\tau = 1$$

where $\psi_i^2(x,y,z)$ is the point probability for an electron in ψ_i at the point (x,y,z).
Since the molecular orbital function ψ_i is given by

$$\psi_i = c_1 \phi_1 + c_2 \phi_2 + \cdots$$

the probability function is given by (if we use real ϕ's, then ψ_i will be real):

$$\psi_i^2 = c_1^2 \phi_1^2 + c_2^2 \phi_2^2 + 2c_1 c_2 \phi_1 \phi_2 + \cdots$$

and the integrated probability is

$$\int \psi_i^2 \, d\tau = c_1^2 + c_2^2 + 2c_1 c_2 S_{12} + \cdots$$

The last equation follows from the preceding one if the individual atomic orbitals are normalized ($\int \phi_m^2 \, d\tau = 1$), as usual, and S_{12} is the *overlap integral between atomic orbitals* ϕ_1 and ϕ_2, as discussed in the earlier section on overlapping hybrid ao's.

Mulliken[2] has given a very useful interpretation of this probability function that will form the basis for our later interpretation of molecular orbital functions.

---

[2] R. S. Mulliken, *J. Chem. Phys.*, **23**, 1833, 1841, 2338, 2343 (1955).

Type	Examples	
Sigma	⬭⬭	s,s
	 ⬭⬤⬭	s,p
	 ⬭⬤⬭	p,p
	 ✾	d,d
Pi	 8	p,p
	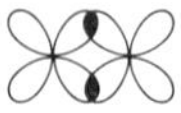 ✾	d,d
	 ✾	p,d
Delta	 8	d,d

TABLE 4–1
THE COMMON TYPES OF ATOMIC ORBITAL OVERLAP

Since the sum of terms on the right side of the integrated probability equation constitutes the total probability of the electron in space, it is logical to view the term c_1^2 as the probability that the electron behaves like an electron does in atomic orbital ϕ_1, and thus we say that c_1^2 is the probability that the electron is to be found in atomic orbital ϕ_1. A similar interpretation is given to the term c_2^2. The cross term $2c_1c_2S_{12}$ represents the *overlap probability* between atomic orbitals ϕ_1 and ϕ_2, as already discussed for hybrid atomic orbitals. The greater the magnitude of this term, the greater is the probability that the electron is in the overlap region of the atomic orbitals ϕ_1 and ϕ_2, and this quantity should be related to the strength of the chemical bond between the atoms for which ϕ_1 and ϕ_2 are atomic orbitals.

Before proceeding to a more in-depth examination of the significance of these terms in the probability function, it will be necessary to examine the types of orbital overlap one commonly encounters in molecular orbital functions, the significance of the relative phases of the overlapping atomic orbitals, and the signs of overlap integrals. In Table 4–1 the usual overlap situations are identified; the identification of each type is based on whether the overlap density arises from interference of just one lobe of each atomic orbital (σ overlap), two lobes (π overlap), or four lobes (δ overlap).

Missing from the figures in Table 4–1 are two things. First, there are many more overlapping orbital situations that could be conceived of, *viz.*,

$s,p\pi$ $s,d\pi$

The omission of these situations is directly related to the fact that also missing from the figures in Table 4–1 are the $+$ and $-$ signs that indicate the signs of the angular functions (phases) on opposite sides of the orbital nodes. These signs are very important in characterizing the bonding or anti-bonding nature of the interaction of two atomic orbitals that appear in the linear combination of atomic orbitals for a molecular orbital.

To be specific, the overlap integral between two atomic orbitals could lie anywhere between -1 and $+1$; that is, $-1 < S < +1$. First consider the special case of $S = 0$, which could arise if the two atomic orbitals are at an infinite distance from one another or, more importantly and more to the point, because symmetry ensures a null value. Using the specific case (see Figure 4–1a) of overlapping atomic s and atomic p orbitals in which the nucleus of the atom with the s orbital lies in the nodal plane of the p atomic orbital of the second atom, the overlap product $\phi_s\phi_p$ has positive values at all points above the nodal plane and negative values at all points below this plane. Furthermore, because the nodal plane is a reflection plane, for each point above the plane (with a positive value for $\phi_s\phi_p$) there is a corresponding point below the plane with the same magnitude of the $\phi_s\phi_p$ product but with a negative sign. Consequently, when the integration is performed (the $\phi_s\phi_p$ products are summed over all points), a value for S identically equal to zero results. The symmetry of a three-dimensional plot of the overlapping s and $p\pi$ angular functions is characterized by only three elements (the point group is C_{2v}): C_2, σ_v, and σ_v'. The s orbital function is symmetric with respect to all of these, whereas the p orbital function is symmetric with respect to only σ_v' and is antisymmetric with respect to C_2 and σ_v. *This is a specific example of the general rule that two orbital functions of different symmetry with respect to common symmetry elements have an overlap integral identically equal to zero.*

We shall now consider a situation for which S is non-zero. The second example of sigma overlap in Table 4–1 involves the "head-on" interference of an atomic s orbital with an atomic p orbital. Depending upon whether the interfering lobes have the same or different signs, the overlap integral would have a positive value or a negative value (see Figure 4–1b). Notice that in both instances, ϕ_s and ϕ_p have the same symmetry with respect to the C_ϕ's and σ_v's (the point group in both cases is $C_{\infty v}$). *This is a specific example of the general rule that two orbitals of the same symmetry with*

(A)

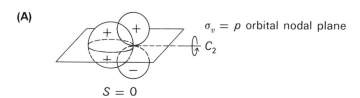

$S = 0$

(B)

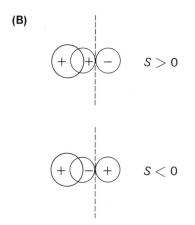

Figure 4-1. (a) $S = 0$, by symmetry, for s, p overlap when the s orbital center lies in the p orbital nodal plane; (b) non-zero overlap for sigma interaction of an s with a p orbital.

respect to common symmetry elements will have a non-zero overlap integral, but the sign of the integral depends on the relative phasing of the two orbitals.

It is logical to ask at this point, "What is the physical significance of the sign of an overlap integral?" To answer this question with an illustration, presume we have an LCAO molecular orbital function at hand, that this function involves a linear combination of an atomic s and an atomic p_z orbital, and that the nucleus of the atom with the s orbital lies on the z axis and in the negative z direction with respect to the atom with the p_z orbital. Two situations can arise from this linear combination:

$$\Psi = c_1\phi_s + c_2\phi_z$$

or else

$$\Psi = c_1\phi_s - c_2\phi_z$$

where c_1 and c_2 have positive values. In **1** the ϕ_s and ϕ_z waves *destructively* interfere in the internuclear region, with the creation of a *node* in that region. This is characterized by sketch **1'** of the molecular orbital.

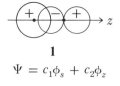

1

$$\Psi = c_1\phi_s + c_2\phi_z$$

In **2** the ϕ_s and ϕ_z orbitals *constructively* interfere (no node) in the internuclear region. This is illustrated by sketch **2'**.

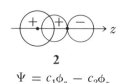

2

$$\Psi = c_1\phi_s - c_2\phi_z$$

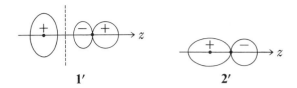

Orbital **2** and its equivalent (**2′**) characterize a bonding interaction between the two orbitals, while **1** and **1′** illustrate what is called an anti-bonding situation. Now, it is useful to identify and characterize these two cases in terms of the electron probability functions for the molecular orbital. The integrated probability functions for the two molecular orbitals differ only by the sign of the overlap term:

$$\int \psi^2 = c_1^2 + c_2^2 \pm 2c_1 c_2 S_{sz} = 1$$

S_{sz} is the same for both mo's and is inherently negative in sign because the s orbital overlaps the negative p lobe.[3] Now, for the anti-bonding mo the overlap term gives a negative contribution to the probability, whereas for the bonding mo the overlap contribution is positive. Because the sum of c_1^2, c_2^2, and $2c_1 c_2 S_{sz}$ must equal unity in both cases, we find that the sum $c_1^2 + c_2^2$ in the anti-bonding situation must be greater than 1, and in the bonding situation the sum must be less than 1.

$$c_1^2 + c_2^2 > 1 > c_1^2 + c_2^2$$
(anti-bonding case)   (bonding case)

For the bonding situation the electron probability in the atomic orbitals is reduced below the value of unity that is appropriate to the case in which the orbitals do not interact at all (the non-bonding case, characterized by $S_{sz} = 0$), whereas the anti-bonding situation results in greater electron probability in the atomic orbitals than for the non-bonding case. In summary:

1) Two non-interacting orbitals have $S = 0$.
2) Two bonding orbitals have $c_1 c_2 S > 0$ and $c_1^2 + c_2^2 < 1$ by the amount $2c_1 c_2 S$.
3) Two anti-bonding orbitals have $c_1 c_2 S < 0$ and $c_1^2 + c_2^2 > 1$ by the amount $2c_1 c_2 S$.

Relative to the non-interacting situation, electron density flows out of the atomic orbitals into the internuclear region for a bonding interaction. For the anti-bonding interaction, electron density flows in opposite directions out of the internuclear region. In simple terms, the sign of $c_1 c_2 S$ identifies the interaction of two atomic orbitals in a molecular orbital as bonding or anti-bonding.

Based on these simple physical arguments of electron probability, you should expect the electron to be *stabilized* (have a lower energy) in the *bonding case* and *destabilized* in the *anti-bonding case*. The effect of in-phase and out-of-phase combinations of atomic orbitals on the energy of electrons "occupying" them is graphically characterized in terms of an mo energy level diagram (**3**) for a general two-orbital case. The energies of the interacting atomic orbitals are indicated by horizontal lines to the left and right of **3**, and the energies of electrons in bonding and anti-bonding molecular orbitals are given in the middle; note that the diagram conveys the information that the electron energy for the *bonding* orbital is *lower* than that of either atomic orbital, whereas the electron energy for the *anti-bonding* orbital is *higher* than that of either atomic orbital.

---

[3] Were we to have the z axis pointing to the left, S_{sz} would be inherently positive, and the labels "bonding" and "anti-bonding" assigned to the difference and sum of ϕ_s and ϕ_z would be reversed.

134 □ THE UNDIRECTED ORBITAL VIEW OF CHEMICAL BONDS

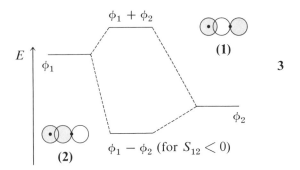

With these basic ideas of orbital probability and energy in mind, we will now proceed to examine and discuss the qualitative orbital patterns and energy level diagrams for some basic structural arrangements of nuclei.

PRINCIPLES OF MO CONSTRUCTION AND INTERPRETATION

THE HYDROGEN MOLECULE

This problem is treated in the junior level physical chemistry course in a more formally correct way than we will do here. Our goal is to establish the qualitative forms of the orbital wavefunctions and energy level diagram on the basis of the physical concepts discussed in the preceding section and with the help of symmetry and the character table appropriate to the molecular point group.

The point group for H_2 is $D_{\infty h}$, the character table for which is reproduced in part in Figure 4-2. Also shown in Figure 4-2 are sketches of the hydrogen atom $1s$ angular functions (the only valence atomic orbital for the atom).

You now face for the first time a situation in which the orbitals whose symmetry we need to know do not lie at the "point" of intersection of all molecular symmetry operations. To be specific, certain operations of the $D_{\infty h}$ point group do not send s_1 into itself, or some fraction of itself, but rather completely into s_2. The functions s_1

$D_{\infty h}$	E	$2C_\phi$	C_2	σ_v	i	$2S_\phi$	σ_h	C_2'
Σ_g^+	1	1	1	1	1	1	1	1
Σ_u^+	1	1	1	1	−1	−1	−1	−1
Σ_g^-	1	1	1	−1	1	1	1	−1
Σ_u^-	1	1	1	−1	−1	−1	−1	1
Π_g	2	$2\cos\phi$	−2	0	2	$-2\cos\phi$	−2	0
Π_u	2	$2\cos\phi$	−2	0	−2	$2\cos\phi$	2	0
s_1, s_2	2	2	2	2	0	0	0	0

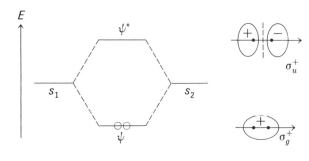

Figure 4-2. The $D_{\infty h}$ character table (in part), sketches of the valence orbital angular functions for H_2, and the LCAO molecular orbitals and the orbital energy diagram.

and s_2 are symmetry-related. Treating s_1 and s_2 as a *pair* of symmetry-related functions, you must deduce the characters of the irreducible representation for the *pair* in the way learned before (refer to the BeH$_2$ hybrid ao discussion of the last chapter). An operation that takes s_1 into itself contributes $+1$ to the character for that operation, and similarly for s_2. Operations that carry s_1 completely into s_2 (not at all into itself) count zero toward the character for that operation, and similarly for s_2. In this way, the reducible representation for the pair of symmetry-related functions (s_1,s_2) is generated and given in Figure 4–2. Also shown are the characters for several of the irreducible representations of $D_{\infty h}$. It is easy to see that the representation for the pair (s_1,s_2) is made up of the Σ_g^+ and Σ_u^+ representations. This means that, in finding the linear combinations of s_1 and s_2, you should create *two* LCAO molecular orbitals whose symmetries are the same as those of s_1 and s_2. *Now, it is shown in a standard course on the chemical applications of group theory*[4] *that molecular orbitals with symmetries corresponding to the irreducible representations of the molecular point group automatically satisfy the Fock equation on page 129.* We will not prove this theorem here, but will make use of the fact that the molecular orbitals have symmetries that generate the irreducible representations of the point group. In the present case, the theorem means that we must look for two linear combinations of s_1 and s_2 such that one combination has Σ_g^+ symmetry while the other has Σ_u^+ symmetry. The problem is really trivial in this case, since only two linear combinations of s_1 and s_2 are possible:

$$\psi \cong (s_1 + s_2) \qquad \sigma_g^+$$

and

$$\psi^* \cong (s_1 - s_2) \qquad \sigma_u^+$$

You can easily verify that ψ has σ_g^+ symmetry while ψ^* has σ_u^+ symmetry.‡ As written, the functions are not normalized; but the sketches of ψ and ψ^* in Figure 4–2 make it clear that the first is a bonding orbital and the second is an anti-bonding orbital. Also shown is the orbital energy diagram for the linear combinations of s_1 and s_2 to form ψ and ψ^*.

If we now assume that the Aufbau principle for atoms (Chapter 1) applies equally well for molecules, the electronic configuration for the H$_2$ molecule is $(\psi)^2(\psi^*)^0$ or just $(\psi)^2$. This ground configuration is indicated on the orbital energy diagram by drawing two circles on the ψ level. With two electrons occupying the bonding molecular orbital and none occupying the anti-bonding orbital, we would "predict" H$_2$ to be a stable molecule with a "bond order" of one.

BERYLLIUM HYDRIDE AND HYDROGEN CHLORIDE

The example of H$_2$ has given us a simple application of the ideas that (i) two overlapping orbitals are to be linearly combined by sum and difference, and (ii) the resulting combinations are constrained to have irreducible representation symmetries. To escalate to a slightly more complex case, we shall consider the linear BeH$_2$ molecule discussed in the last chapter; this time we will use the centrosymmetric Schrödinger ao's. A convenient view of BeH$_2$ is that of inserting a Be atom between the hydrogen atoms. The molecular point group is still $D_{\infty h}$ as it is for H$_2$; the ψ and ψ^* wavefunctions for the H atoms are still good but, because the overlap of s_1 with s_2 is now so small, the ψ and ψ^* orbital energies are little different from that of an isolated H 1s ao. These mo's (ψ,ψ^*) are symmetrized atomic orbitals for terminal atoms; *we will refer to all such terminal atom symmetry orbitals in this chapter with the acronym TASO.*

---

[4] F. A. Cotton, "Chemical Applications of Group Theory," Wiley Interscience, New York (1971).

‡ Here we are utilizing a long established convention for writing orbital irreducible representation symbols: a lower case letter is used for orbitals (*e.g.*, σ_g^+) while the upper case letter (*e.g.*, Σ_g^+) is to be used when referring to the representation itself or to a molecular electronic term. This convention parallels that of referring to ao's by *s, p, d, f,* . . . and to atomic terms as *S, P, D, F,* . . . .

The problem at hand is to *consider the valence ao's of the central atom* (Be) *for possible overlap* (symmetry likeness) *with the* H_2 *TASO's*. This can be done by inspection and checked by consulting the $D_{\infty h}$ character table for the symmetries of the s, x, y, and z ao's of Be. To illustrate the inspection procedure, we will locate the molecule on a coordinate axis system, sketch the TASO's, sketch the central atom ao's, and compare for non-zero overlaps.

From these sketches, it should be apparent that the Be s ao has $S > 0$ with ψ, that the Be p_z ao has $S < 0$ with ψ^*, and that all other comparisons yield $S = 0$. Construction of the mo energy diagram is straightforward. Recognizing first that the Be 2s and 2p ao's are less electronegative than the H 1s ao, the starting ao levels are "skewed" to reflect that difference in Figure 4–3. The Be 2s is to be taken in sum and difference combinations with ψ, and the same is true for the Be p_z with ψ^*. The combination $\phi_s + \psi$ accounts for a bonding mo because the Be—H overlaps are constructive; $\phi_s - \psi$, on the other hand, causes destructive Be—H overlaps and this corresponds to anti-bonding. We can distinguish these by calling the first $s\sigma$ and the second $s\sigma^*$. For the (ϕ_z, ψ^*) combinations it is $\phi_z - \psi^*$ that leads to bonding (call this mo $p\sigma$) and $\phi_z + \psi^*$ that leads to anti-bonding (call this mo $p\sigma^*$). Finally, note that the Be ϕ_x and ϕ_y ao's are unshifted and non-bonding (call them $p\pi_n$).

Chemical interpretations follow after you decide which mo's are occupied with

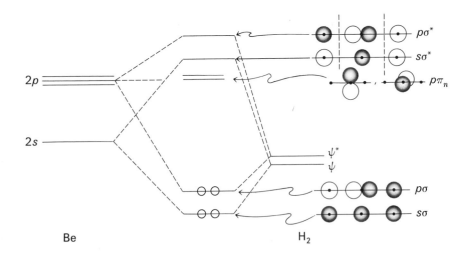

Figure 4–3. The energy diagram and orbital sketches for BeH$_2$.

electrons. Since BeH$_2$ has four valence electrons (two from Be and one each from the H atoms), the molecular electron configuration is

$$(s\sigma)^2(p\sigma)^2(p\pi_n)^0(s\sigma^*)^0(p\sigma^*)^0$$

There are two bonding electron pairs (a reassuring result!). *Both* contribute to defining *each* Be—H "bond," so that we might say, in rough terms, that each Be—H bond has a bond order of 1 ($\frac{1}{2}$ from each of $s\sigma$ and $p\sigma$). In short, we have arrived at the same qualitative point as was reached by the directed ao approach of the last chapter. Finally, we must introduce the important concepts of *Highest Occupied Molecular Orbital* (*HOMO*) and *Lowest Unoccupied Molecular Orbital* (*LUMO*). In BeH$_2$, the $p\sigma$ mo is the HOMO and the $p\pi_n$ pair of ao's is the LUMO. In later chapters you will see that the energies and shapes of HOMO and LUMO have an important bearing on the understanding of molecular reactions. Another important interpretation to be recognized is that, because H is more electronegative than Be, the electron pair probabilities for $s\sigma$ and $p\sigma$ should be greater at H than at Be. That is, writing

$$s\sigma = a\phi_s + b\psi$$
$$p\sigma = c\phi_z - d\psi^*$$

we expect $a < b$ and $c < d$. The $s\sigma$ and $p\sigma$ mo's are *polarized* toward the hydrogen atoms. Another way of saying this is that $s\sigma$ has greater ψ character than ϕ_s character, and similarly for $p\sigma$. A convenient idea here, to be used in the upcoming discussion of HCl, is that the lower-energy TASO's are contaminated with the Be ao's and move to lower energy, while the higher-energy Be ao's are contaminated by the TASO's and shift to higher energy. In short, *lower levels shift lower while higher levels shift higher.*‡

As a final point, to which we shall refer later, note that the Be s ao has a symmetry entirely different from that of the Be p ao's; consequently, there is no s,p mixing in this case. The next example will show such mixing.

Hydrogen chloride is a simple heteronuclear diatomic molecule from which we may learn several useful simplifying concepts for later application to more complex molecules. The valence ao's for this molecule are simply the H $1s$ and the $3s$, $3p_x$, $3p_y$, and $3p_z$ ao's of chlorine. Making ao sketches on a coordinate axis system produces the following (note that the irreducible representation symbols from the $C_{\infty v}$ point group are also given):

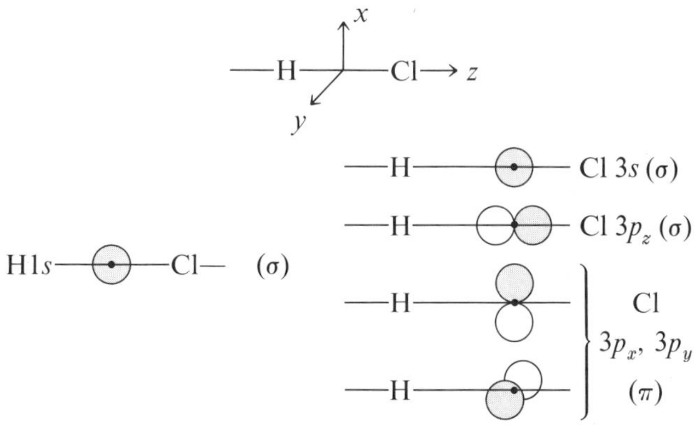

Matching H and Cl ao's for overlap, you find that the H $1s$ ao overlaps both ϕ_s and ϕ_z of Cl, and the ϕ_x and ϕ_y ao's are again a degenerate non-bonding pair. Unlike the BeH$_2$ problem, however, two heavy atom (Cl) valence ao's (s and p_z) have a common symmetry.

---

‡R. W. Jotham, *J. Chem. Educ.*, **52**, 377 (1975).

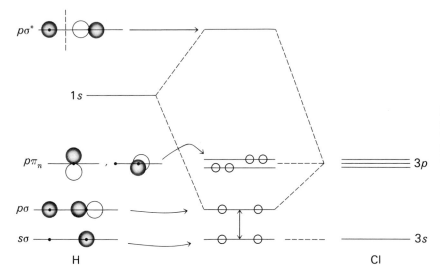

Figure 4-4. The "first approximation" energy diagram for HCl, with the second approximation mixing anticipated by the arrow between $s\sigma$ and $p\sigma$.

How do we handle a problem with *three overlapping orbitals*? Three linear combinations must be taken, but how is it to be done? While a digital computer with the proper program could handle this situation directly, we will take a more controlled, stepwise approach by the brazen technique of simply ignoring one of these ao's *at the start*, and later we will dispense with such convenient shoddiness to arrive at the final answer.

The energy level diagram of Figure 4-4 is developed as follows. The chlorine ao's are placed at lower energy than that of hydrogen. The ϕ_x and ϕ_y ao's of chlorine are unshifted and non-bonding ($p\pi_n$). The three-orbital situation for bonding is simplified initially by ignoring the low energy ϕ_s ao of chlorine. The choice of the ϕ_s ao rather than the ϕ_z ao is dictated by the fact that *mixing of two ao's depends not only on their overlap integral* (greater overlap, greater mixing) *but also on their likeness in energy* (better energy match leads to greater mixing). This is a result we take without apology from the mathematics of molecular orbital calculations. Completion of the energy level diagram is now straightforward. The ϕ_s level is shown unshifted and non-bonding, while the H 1s ao is combined in-phase (to produce $p\sigma$) and out-of-phase (to produce $p\sigma^*$) with ϕ_z.

HCl has four pairs of valence electrons, yields the electronic configuration

$$(3s)^2(p\sigma)^2(p\pi_n)^4(p\sigma^*)^0$$

and shows three lone pairs (the 3s and $p\pi_n$ pairs) and one bond pair (from $p\sigma$). The bond order is qualitatively one, and HOMO = $p\pi_n$ while LUMO = $p\sigma^*$. The $p\sigma$ mo is polarized toward chlorine, while $p\sigma^*$ is polarized toward hydrogen.

What do we now do about ignoring the fact that ϕ_s and the H ao do overlap and should be mixed? Two guides we have used for mixing of ao's to form mo's are also applicable for mixing simple mo's to form better mo's. (i) The mo's ϕ_s, $p\sigma$, and $p\sigma^*$ all have the same symmetry and so overlap, the fact we have been ignoring. *The overlap arises because ϕ_s overlaps the hydrogen ao appearing in $p\sigma$ and $p\sigma^*$.* (ii) The mixing of $p\sigma$ with ϕ_s will be greater than the $\phi_s/p\sigma^*$ overlap because of the better energy match. Now, what are the phasings of $p\sigma$ and $p\sigma^*$ when they are to be mixed into ϕ_s?

We know that the 3s ao must be stabilized by the mixing with $p\sigma$ and $p\sigma^*$ (lowest goes lower); to achieve this with $p\sigma$ and $p\sigma^*$ *as drawn* in Figure 4-4, both $p\sigma$ and $p\sigma^*$ must contaminate ϕ_s with positive phases ($3s \rightarrow 3s + p\sigma + p\sigma^*$) so as to introduce H 1s/Cl 3s bonding. Interestingly, the ϕ_z contributions are out-of-phase in $p\sigma$ and $p\sigma^*$, and the new 3s mo has less ϕ_z character than might otherwise have been the case; the new ϕ_s mo is primarily H 1s/Cl 3s in character, and the new mo is somewhat bonding between H and Cl. What about the new $p\sigma$ mo? Its contamination by ϕ_s must shift it to higher energy, a requirement met by $p\sigma \rightarrow p\sigma - \phi_s$ so as to introduce H $1s$/Cl $3s$ anti-bonding character into the $p\sigma$ mo. (Note that $p\sigma$ contaminates ϕ_s with a positive phase, and the converse is necessary for ϕ_s contamination of $p\sigma$.) Similarly, ϕ_s contaminates $p\sigma^*$ with a negative phase.

To illustrate these improvements in ϕ_s, $p\sigma$, and $p\sigma^*$ in an exaggerated way, we would sketch **4**.

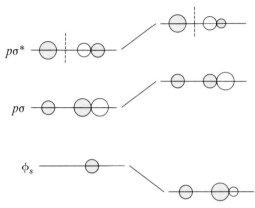

4, the improved σ mo's of HCl

As the sketches imply, ϕ_s is stabilized by new bonding character, while $p\sigma$ is destabilized by some anti-bonding character. In qualitative terms, bonding character has been transferred from $p\sigma$ to ϕ_s. Perhaps the most important point to come out of this second-level approximation is that *when a valence s and a valence p ao (ϕ_z in this case) have the same symmetry under a molecular point group, they are mixed.* Thus, s/p mixing (hybridization) does arise when the molecular symmetry is low enough not to distinguish the s from the p ao's (this was *not* the case for BeH_2).

Before leaving this discussion of the mo's of HCl, we should mention the photoelectron spectrum of HCl, an experimental result that is straightforwardly interpreted by means of the mo level diagram of Figure 4-4. The photoelectron spectroscopy experiment is based on the idea that an electron may be ejected from a molecule by "striking" the molecule with a photon of energy in excess of that required for ionization.[5] After measuring the kinetic energy of the ejected electron, one calculates the work required for the ionization as the difference ($h\nu - $ K.E.).

$$HCl + h\nu \rightarrow HCl^+ + e^-$$

In fact, the experiment should reveal several such ionization processes, for not all electrons in HCl feel the same effective nuclear charges.

$$HCl + h\nu \rightarrow HCl^+ + e^-$$
$$HCl + h\nu \rightarrow HCl^{+*} + e^-$$
etc.

$\left.\right\}$ ejected e^- have different kinetic energies

Figure 4-5 shows that in the region between 12 and 18 eV there are in fact two ionization processes possible for HCl. The lower-energy ionization reveals two bands of equal intensity, telling us that there are two states for HCl^+ at very similar energies. According to the mo level diagram, ejection of a $p\pi_n$ electron should correspond to this lowest-energy photo-ionization. How are we to account for the doublet (two *molecular terms,* to use the language of Chapter 1 for atoms)? The ion, of course, has an unpaired electron with a spin quantum number of $\frac{1}{2}$. Can there also be an orbital angular momentum? The answer is yes, because the odd electron occurs in a degenerate pair of p ao's at chlorine and so may freely circulate about the C_∞ axis, developing angular momentum as in a free chlorine atom. In both the atomic and

---

[5] An excellent little book on this subject has been written by J. H. Eland, "Photoelectron Spectroscopy," Wiley, New York (1974).

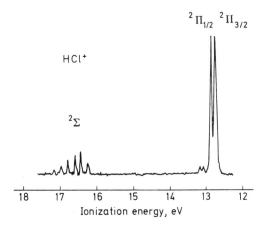

Figure 4-5. The photoelectron spectrum of HCl in the region $12 < h\nu < 18$ eV. [From J. H. Eland, "Photoelectron Spectroscopy," Butterworths, London (1974).]

molecular cases, the orbital angular momentum quantum number is 1 ($L = 1$, $S = \frac{1}{2}$). Spin-orbit coupling in the chlorine *atom* resolves the ^2P ground term into the spin-orbit terms $^2P_{1/2}$ and $^2P_{3/2}$. The notation $^2\Pi_{1/2}$, $^2\Pi_{3/2}$ is different for molecules (the $C_{\infty v}$ symmetry quantum number symbol is Π, not P) but the physical principle is the same. Thus, the fact that the lowest ionization band is actually a doublet confirms the ionization of an electron from a $p\pi_n$ orbital of HCl. The molecular term symbols for the electronic configuration $(3s)^2(p\sigma)^2(p\pi_n)^3$ are $^2\Pi_{1/2}$ and $^2\Pi_{3/2}$.

The second ionization band at ~16.5 eV arises from ejection of an electron from the $p\sigma$ mo. The appearance of a progression of bands is due to the possibility of ionization leaving the ion in any one of several vibrationally excited states (all of the same ... $(p\sigma)^1(p\pi_n)^4$ electronic configuration, however). The vibrational excitation arises because a bonding electron is removed and this ion has a reduced H—Cl bond strength; i.e., the ion has a longer bond distance than the HCl molecule. In contrast, removal of a $p\pi_n$ electron has little effect on the H—Cl bond and no pronounced vibrational progression is observed. The Cl $3s$ and core electrons of HCl can also be ionized in this fashion; but, as they experience greater Z^*, it requires photon energies much greater than 18 eV to eject them.

SUMMARY

Now is a good time to stop and reflect on what has been learned from the mo formalisms of H_2, BeH_2, and HCl.

1. Two orbitals (ao's *or* mo's) that overlap (have $S \neq 0$) and that are of equal energy are taken in equal weights by sum and difference to form bonding and anti-bonding orbitals.

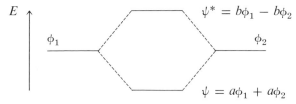

Electrons in ψ and ψ^* will be equally distributed between ϕ_1 and ϕ_2; ψ and ψ^* are *not* "polarized."

2. Two overlapping orbitals of unequal energy will *not* be combined with equal weights in sum and differences. The bonding combination (lower in energy) retains greater character of the orbital most like it in energy. (See diagram at top of page 141.) The condition $a < b$ means that ψ is "polarized" toward the region of ϕ_2; electrons with orbital wave ψ have greater probability at ϕ_2 than at ϕ_1. The converse applies to ψ^*.

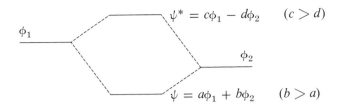

3. In both (1) and (2), ψ is lower in energy than ψ^* because ϕ_1 and ϕ_2 constructively interfere in the internuclear region; they destructively interfere in ψ^*.

4. When faced with three overlapping orbitals, at least one of which is of different energy (in a later section you will learn to handle the case in which all three are of the same energy), it is convenient *initially* to ignore one of them so as to reduce the problem to consideration of *two* overlapping orbitals.

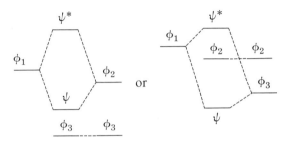

As a second step, we may remove this artificial convenience by allowing the ignored orbital to overlap ψ and ψ^*. Always, the lowest level moves lower, the highest moves higher, and the intermediate level shifts but remains intermediate.

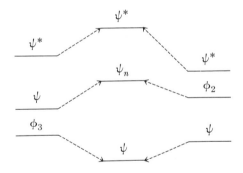

If worked to completion, the final answer is the same. Remember that *higher levels are mixed into lower levels in a bonding way, while lower levels are mixed into higher levels in an anti-bonding way.* The extent of mixing, as always, depends on the energy match and overlap.

MAIN GROUP DIATOMIC MOLECULES

HOMONUCLEAR MOLECULES

Staying with the basic structural unit of the diatomic molecule, we will now examine the slightly more complex case of two atoms of the same element, each of which has s and p atomic orbitals in the valence shell. *The first step will be (this is always the first step) to identify the symmetry-equivalent atomic orbitals.* Proceeding in analogy with the H_2 example, you are to form sum and difference combinations of those orbitals to get a *first approximation* to the molecular orbitals. Figure 4-6 gives sketches of the atomic orbital angular functions for the orbitals to be combined. It is quite easy to see in this case that the s orbitals are symmetry-equivalent, the p_z orbitals are symmetry-equivalent to each other but not to the s orbitals, and the p_x

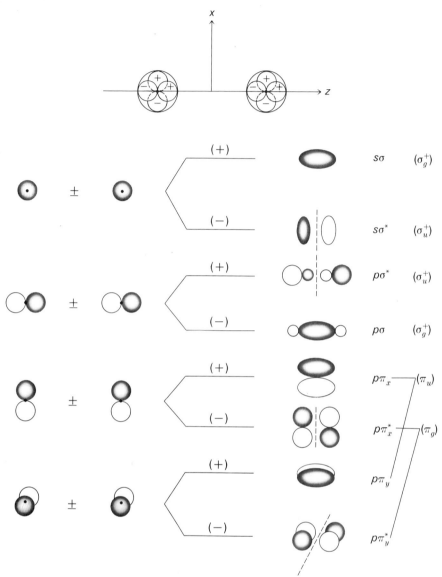

Figure 4-6. Sketches of the angular parts of the atomic and LCAO orbitals for the homonuclear diatomic molecule with valence s and p orbitals.

and p_y orbitals are also equivalent (not only does the relation in this case include p_x on one atom with p_x on the other and the same for the p_y orbitals; also, p_x and p_y on each atom are an inseparable pair and generate a doubly degenerate representation for the $D_{\infty h}$ point group). The reducible representations for the (s_1,s_2) and the (p_{z1},p_{z2}) pairs and for the $(p_{x1},p_{y1},p_{x2},p_{y2})$ quartet are given in the following table.

	E	$2C_\phi$	C_2	σ_v	i	$2S_\phi$	σ_h	C_2'
(s_1,s_2)	2	2	2	2	0	0	0	0
(p_{z1},p_{z2})	2	2	2	2	0	0	0	0
$(p_{x1},p_{x2},p_{y1},p_{y2})$	4	$4\cos\phi$	-4	0	0	0	0	0

MAIN GROUP DIATOMIC MOLECULES 143

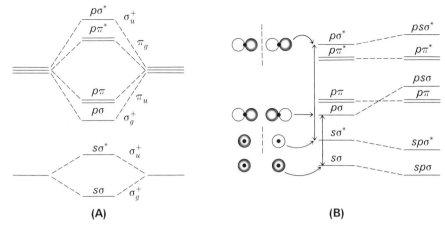

Figure 4–7. The orbital energy diagrams for the homonuclear diatomic case. (A) The first approximation with no s, p mixing permitted; (B) the improved theory with s, p mixing.

From these reducible representations it is not difficult to see (refer to Figure 4–2) that the (s_1, s_2) pair is taken to form σ_g^+ and σ_u^+ LCAO's, the (p_{z1}, p_{z2}) pair is also taken to form σ_g^+ and σ_u^+ LCAO's, while the $(p_{x1}, p_{y1}, p_{x2}, p_{y2})$ quartet is combined to form two degenerate pairs π_g and π_u. All these combinations are sketched in Figure 4–6; the "first approximation" orbital energy diagram is given in Figure 4–7a. In sketching these diagrams we have made use of our knowledge that the atomic s orbital is more stable than the atomic p orbitals. The levels are labeled according to their irreducible representation symbols and also according to a common notation that identifies the molecular orbital according to type (σ or π) and according to origin in terms of atomic orbitals (s or p).

Another point necessary to bring up now is that the "first approximation" orbital treatment has completely ignored the fact that an s atomic orbital on one atom has a non-zero overlap with the p_z atomic orbital on the other atom. This deficiency is disconcerting, and its removal will establish a closer link between LCAO and hybrid orbital models.

To be specific, we have not allowed for interactions between the s_1 and p_{z2} atomic orbitals and between the s_2 and p_{z1} atomic orbitals, which are permitted by symmetry to have non-zero overlap integrals. To handle this by merely refining the simplistic orbital energy level diagram, we need only note that the $s\sigma$ and $p\sigma$ molecular orbitals have the same symmetry and will have a non-zero overlap integral (the "first approximation" molecular orbitals are not orthogonal). Figure 4–8 illustrates this overlap with sketches (the crossed dashed lines draw your eye to the s/p overlap responsible for the mixing). The phases of $p\sigma$ and $s\sigma$ as drawn indicate that $s\sigma$ is stabilized by $+p\sigma$, while $p\sigma$ is destabilized by contamination with $-s\sigma$.

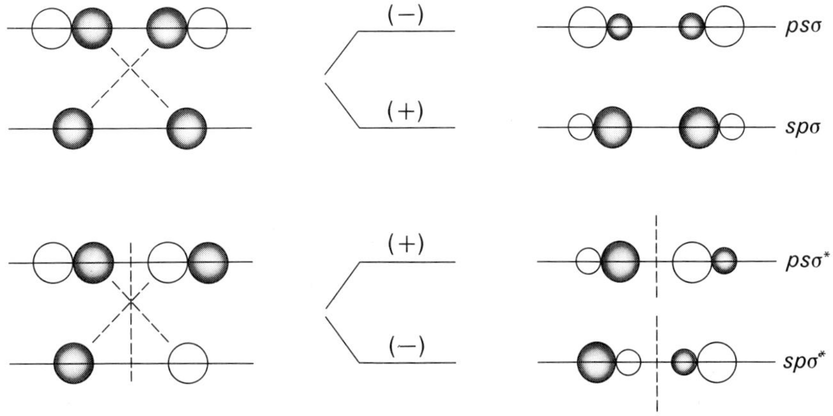

Figure 4–8. The non-zero overlap for $s\sigma$, $p\sigma$, $s\sigma^*$, and $p\sigma^*$, and the new mixed LCAO molecular orbitals.

144 ☐ THE UNDIRECTED ORBITAL VIEW OF CHEMICAL BONDS

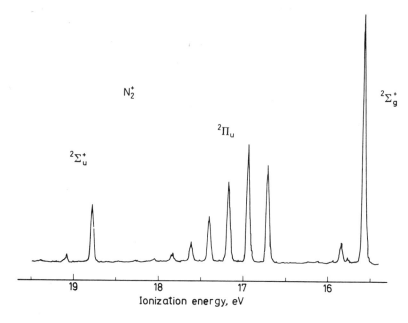

Figure 4-9. The photoelectron spectrum of N_2 in the region $15 < h\nu < 20$ eV. [With the kind permission of Professor W. C. Price.]

The effect of this mixing on the $s\sigma$ and $p\sigma$ orbital energies is shown in Figure 4-7b. The $sp\sigma$ orbital takes on a more stable energy (becomes more bonding), while the $ps\sigma$ orbital is destabilized (loses bonding character). The renaming of $s\sigma$ and $p\sigma$ is meant to emphasize the mixing. A similar interaction between $s\sigma^*$ and $p\sigma^*$ is also possible, and this is indicated in Figures 4-7 and 4-8.

You will notice in Figure 4-7 that $ps\sigma$ is shown at higher energy than $p\pi$. Both accurate calculations[6] and photoelectron spectra[7] indicate that $ps\sigma$ can lie higher in energy than $p\pi$. Figure 4-9 is the photoelectron spectrum of N_2, which has a valence electron configuration

$$(sp\sigma)^2(sp\sigma^*)^2(p\pi)^4(ps\sigma)^2$$

The lowest-energy ionization is characteristic of electron removal from an mo that is not strongly bonding between the nitrogens (there is only a weak vibrational progression), while the ionization process at ~17 eV indicates that a strongly bonding electron has been removed. The third ionization band arises from ejection of an electron from $sp\sigma^*$, an orbital that is weakly anti-bonding between nitrogens.

With the orbital diagram thus confirmed for N_2, we can make some chemical interpretations. The LUMO is the degenerate pair of $p\pi^*$ mo's, and the HOMO is the weakly bonding $ps\sigma$ mo. There are two π bonding electron pairs ($p\pi$), and the σ "bond" is best described as a superposition of the $ps\sigma$ and $sp\sigma$ mo's (where the $sp\sigma^*$ electron pair contributes in a somewhat antibonding way).

Continuing on to the other important second row homonuclear diatomics, O_2 and F_2, a summary is as follows:

O_2: 1. Configuration $= (sp\sigma)^2 \cdots (p\pi^*)^2$
2. Hund's rule predicts the ground term to be a spin triplet (two unpaired electrons in $p\pi^*$), and O_2 is paramagnetic.
3. The bond order is ~2 (net $\sigma = 1$, net $\pi = 1$)
4. HOMO $= p\pi^*$, LUMO $= ps\sigma^*$

---

[6] C. W. Scherr, *J. Chem. Phys.*, **23**, 569 (1955).
[7] See footnote 5.

F_2: 1. Configuration = $(sp\sigma)^2 \cdots (p\pi^*)^4$
2. All electrons paired.
3. The net bond order is 1 (no net π bonding)
4. HOMO = $p\pi^*$, LUMO = $ps\sigma^*$

THE $ps\sigma$, $p\pi$ ORDER CONTROVERSY

In the discussion of N_2 we noted that the $ps\sigma$ mo appears to lie above the $p\pi$ mo's. It was suggested to you that this ordering is a result of s/p mixing. Such mixing is greatest (on an energy match basis) for light elements, in which the core penetration difference of $2s$ and $2p$ ao's is smallest. As protons are added to the nuclei, passing from Li to Ne, this s/p ao energy gap widens so the $s\sigma/p\sigma$ mixing diminishes and $p\sigma$ should be less destabilized. This factor tends to order $ps\sigma$ below $p\pi$. Another factor may also be important to the relative ordering of $ps\sigma$ and $p\pi$ across the series $Li_2 \rightarrow F_2$. At *short bond distances* the p ao/p ao overlap giving rise to the $p\sigma$ mo will be diminished because the ao's interpenetrate too greatly:

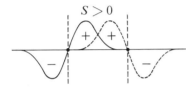 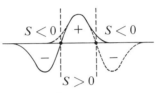

while the p ao/s ao overlap is less seriously diminished:

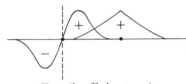 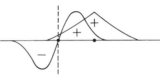

The former effect (inefficient $p\sigma/p\sigma$ overlap) tends to keep $p\sigma$ from being strongly stabilized, while the s/p overlap increases in a relative sense, both favoring the energy order $\varepsilon_{ps\sigma} > \varepsilon_{p\pi}$. To summarize the trends in factors affecting the $ps\sigma/p\pi$ energy ordering across the second row diatomics:

a. The extent of $s\sigma/p\sigma$ mixing (judged by energy match) decreases, favoring a trend from $\varepsilon_{ps\sigma} > \varepsilon_{p\pi}$ to $\varepsilon_{p\pi} > \varepsilon_{ps\sigma}$.

b. Bond shortening, and then lengthening, as the mo's are occupied from B_2 to Ne tends to interchange the order $\varepsilon_{p\pi} > \varepsilon_{ps\sigma}$ to $\varepsilon_{p\pi} < \varepsilon_{ps\sigma}$ and back to $\varepsilon_{p\pi} > \varepsilon_{ps\sigma}$, with the second inequality most likely for triply bonded N_2 (note that *both* $p\pi/p\pi$ and $p\sigma/p\sigma$ ao overlaps are sensitive to bond distance).

c. The extent of $s\sigma/p\sigma$ mixing (judged by overlap) as the bond distance shortens and then lengthens favors a trend from $\varepsilon_{ps\sigma} > \varepsilon_{p\pi}$ to $\varepsilon_{p\pi} \sim \varepsilon_{ps\sigma}$ and back to $\varepsilon_{ps\sigma} > \varepsilon_{p\pi}$.

These relative trends are summarized as follows:

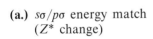

(a.) $s\sigma/p\sigma$ energy match (Z^* change)

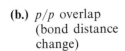

(b.) p/p overlap (bond distance change)

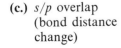

(c.) s/p overlap (bond distance change)

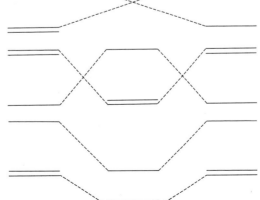

That B_2 is paramagnetic (spin triplet) while C_2 is diamagnetic tells us that trends **a** and **c** dominate early in the series ($\varepsilon_{ps\sigma} > \varepsilon_{p\pi}$):

$$B_2: (sp\sigma)^2(sp\sigma^*)^2(p\pi)^2(ps\sigma)^0$$
$$C_2: (sp\sigma)^2(sp\sigma^*)^2(p\pi)^4(ps\sigma)^0$$

We know that this is reinforced by **b** at N_2 ($\varepsilon_{ps\sigma} > \varepsilon_{p\pi}$). For O_2 and I_2 (to take an example from beyond the second row), the photoelectron spectra suggest that **a** and **b** now reinforce to determine the orbital ordering ($\varepsilon_{p\pi} > \varepsilon_{ps\sigma}$):

$$O_2: (sp\sigma)^2(sp\sigma^*)^2(ps\sigma)^2(p\pi)^4(p\pi^*)^2$$
$$I_2: (sp\sigma)^2(sp\sigma^*)^2(ps\sigma)^2(p\pi)^4(p\pi^*)^4$$

The spectra are shown in Figure 4–10. The multitude of bands in the O_2 spectrum is a little confusing at first, but their occurrence actually confirms the mo level ordering

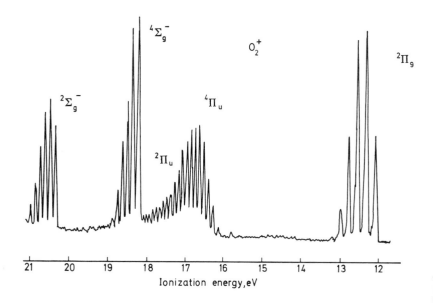

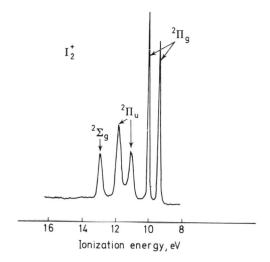

Figure 4–10. Photoelectron spectra of O_2 and I_2. [O_2 spectrum from J. H. Eland, "Photoelectron Spectroscopy," Butterworths, London (1974). I_2 spectrum from S. Evans and A. F. Orchard, *Inorg. Chim. Acta*, **5**, 81 (1975).]

for O_2. The first three ionization processes for O_2 should leave an ion (O_2^+) with the following configurations:

1. $\ldots (ps\sigma)^2(p\pi)^4(p\pi^*)^1$ (1 odd e^-)
2. $\ldots (ps\sigma)^2(p\pi)^3(p\pi^*)^2$ (3 odd e^-)
3. $\ldots (ps\sigma)^1(p\pi)^4(p\pi^*)^2$ (3 odd e^-)

The first configuration yields only a single spin doublet term ($^2\Pi$) corresponding to the band at $\sim$12 eV; the second and third ionizations could produce doublet and quartet ion terms, and these are observed (the two terms are close in energy for the second configuration; that there are two terms in Figure 4-10 is not apparent to the untrained eye). Two important conclusions follow: (i) a $ps\sigma$ electron is more tightly held than a $p\pi$ electron; and (ii) all bands show vibrational activation of the ions. Ionization of a $p\pi^*$ electron should cause shortening of the O—O distance, while $p\pi$ electron ejection should show O—O lengthening. Ionization from $ps\sigma$ in O_2 reveals a significant O—O distance change, unlike the analogous ionization of N_2. This finding substantiates the arguments given for $\varepsilon_{p\pi} > \varepsilon_{ps\sigma}$ because the longer O—O distance than N—N distance and greater O Z^* than N Z^* confer greater bonding character on $ps\sigma$ in O_2.

HETERONUCLEAR MOLECULES

In inorganic chemistry perhaps the most important heteronuclear diatomic molecules (aside from the hydrogen halides) are CO, CN$^-$, and NO. All appear as important ligands in transition metal chemistry, and we need to pause here to discuss their mo's. The *first approximation* will be invoked initially, and the development closely follows that for the homonuclear diatomic molecule. The primary difference now is the polarization of the mo's toward oxygen and toward carbon or nitrogen. Because the oxygen atom ao's are lower in energy than the carbon atom ao's, the "skewing" results in the $p\sigma$ mo being lower than the $s\sigma^*$:

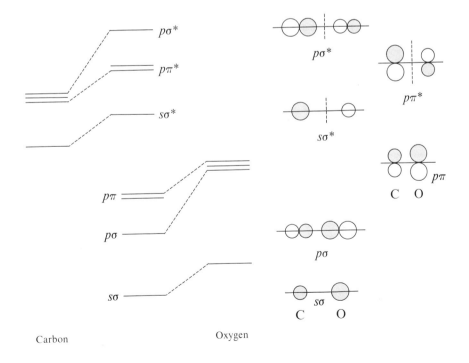

9, first approximation mo's for CO

The orbital sketches to the right in **9** are meant to reflect the mo polarizations through the relative ao sizes drawn. To proceed to the second level of approximation, we need only mix $s\sigma/p\sigma$ and $s\sigma^*/p\sigma^*$.[8] The refined mo's now take on the appearances (exaggerated for clarity) shown in **10**. The phasing relationships are:

$s\sigma^*$ into $p\sigma^*$: + $ps\sigma^*$

$p\sigma^*$ into $s\sigma^*$: − $sp\sigma^*$

$s\sigma$ into $p\sigma$: − $ps\sigma$

$p\sigma$ into $s\sigma$: + $sp\sigma$

10, improved σ mo's for CO

Carbon monoxide possesses 10 valence electrons and therefore has a valence configuration

$$(sp\sigma)^2(ps\sigma)^2(p\pi)^4(sp\sigma^*)^2$$

The HOMO is the "carbon lone pair orbital" $sp\sigma^*$, while the $p\pi^*$ mo's are the LUMO. You will later see that the polarizations of these mo's are important to the Lewis acid/base chemistry of CO (and NO^+ and CN^-, which are isoelectronic with CO). There are roughly three net bonding pairs (two π and one σ), describing the usual formulation C≡O. Similarly, NO has an electron configuration

$$(sp\sigma)^2(ps\sigma)^2(p\pi)^4(sp\sigma^*)^2(p\pi^*)^1$$

and so should be paramagnetic with a single unpaired electron. A familiar aspect of NO is its ready oxidation to NO^+. This corresponds to loss of the $p\pi^*$ electron, an associated increase in NO bond order (from $2\frac{1}{2}$ to 3), and a decreased NO distance. As you will see in Chapter 6, the HOMO of NO ($p\pi^*$) plays an important role in the structures of its compounds.

---

[8] While the $C_{\infty v}$ symmetry here makes it correct also to mix $s\sigma^*$ with $p\sigma$ and $p\sigma^*$ with $s\sigma$, the poor energy match in the latter case means that the $p\sigma^*/s\sigma$ mixing will be minimal, while we shall simply ignore for convenience sake the $s\sigma^*/p\sigma$ mixing in this discussion. See study questions at the end of this section.

STUDY QUESTIONS

1. The mixing of $s\sigma^*$ into $p\sigma$ for CO was ignored in the text. Comment on the effect of such mixing on both the $ps\sigma$ and $sp\sigma^*$ mo energies, and make mo sketches to reflect any changes in mo shape and polarization.

2. Compare the HOMO's of CO and N_2 and draw a conclusion about the relative Lewis basicities of the C and N atoms of these molecules (consider not only the HOMO amplitudes at C and N, but also the Z^* experienced by the HOMO pairs).

3. From the polarization and shape of the HOMO of CO, which "end" (C or O) should be more basic in the Lewis sense? Is this mo predominately C_{2s} or C_{2p} in character?

4. From the form of the HOMO of NO, deduce the more likely point of attack (N or O) by an F atom upon NO to form nitrosyl fluoride. Does this agree with the topology you would predict from the guides given in Chapter 2? Is formation of the other isomer possible?

5. Sketch a photoelectron spectrum of BeH_2.

6. Describe the LUMO of I_2. Describe, by giving the ground and excited electronic configurations, the lowest-energy electronic transition responsible for the color of I_2 (if you need help, see Chapter 5).

7. Using the mo scheme (both level ordering and orbital sketches) of the text, account for the three ionization bands and their vibrational structure in the photoelectron spectrum of CO.

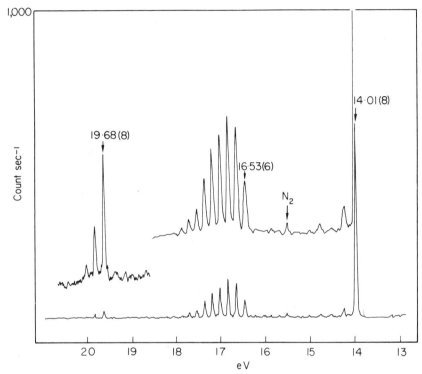

[From D. W. Turner, et al., "Molecular Photoelectron Spectroscopy," John Wiley and Sons, New York (1970).]

STRUCTURE, VSEPR, AND LCAO-MO[9]

For a theory that purports to be generally applicable to the electronic and three-dimensional structures of molecules, it would be a catastrophic inadequacy were it to fail to account for molecular structures at least as well as the VSEPR model. The ability to account for molecular geometries is one of the basic requirements for any electronic theory. In fact, the mo model does account for the geometry changes $BF_3 \rightarrow NF_3 \rightarrow ClF_3$, $SiF_4 \rightarrow SF_4 \rightarrow XeF_4$, and $PF_5 \rightarrow ClF_5$. *Of even greater significance is the ability of the model to account for the assumption of the VSEPR model*

---

[9] A similar approach to this problem has been taken by L. S. Bartell, *J. Chem. Educ.*, **45**, 754 (1968). We recommend this paper for its perceptive analysis, even though the viewpoint involves "second order Jahn-Teller" concepts not brought up in this text. Particularly helpful in comparing the VSEPR and LCAO-MO tenets is his Table 3.

that all valence pairs about an atom are stereochemically active. The LCAO-MO scheme in a simplified form, wherein only central atom–terminal atom σ bonding is considered, does provide an explanation of MF_x geometries. Before demonstrating this, we need to identify the concepts, developed in discussing the BeH_2, HCl, and CO structures, that will be of use.

1. (From BeH_2) When the central atom s ao has a symmetry different from those of any of the p ao's, the s ao gives rise to bonding and anti-bonding mo's of its own.

2. (From CO and HCl) When the central s ao and p_z ao have a common symmetry they will, in the end, be mixed. A simplification is possible by anticipating this with replacement of the s and p_z Schrödinger ao's by their hybrid counterparts:

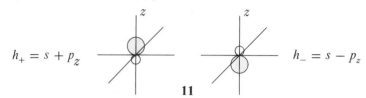

$$h_+ = s + p_z \qquad h_- = s - p_z$$

11

3. (From BeH_2) For each central atom ao or hao overlapped by a terminal atom $p\sigma$ ao (*defined to lie along the internuclear axis*) there is, in fact, a p TASO of the same symmetry as the central ao. This TASO is combined in sum and difference with the central ao. In the BeH_2 case, both the Be p_z and s ao's are overlapped by a hydrogen s ao. The respective TASO's have the forms

for s ⊖—⊖ → z

for p_z ⊖ ⊖ → z

Later in this chapter you will learn more about the shapes of such TASO's in general. For now, you need only appreciate that they exist.

4. Because the central atom ao's are less electronegative than the F ao's, the ao levels appear strongly skewed and the *general* form of the mo diagram will appear as

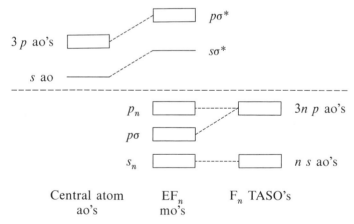

5. By noting that EF_n can be regarded as arising from the union of $E^{n+} + nF^-$, the mo occupation count is especially simple. The n s ao's and the $3n$ p ao's of the fluorides are all filled with electron pairs, so that all the non-bonding TASO's and σ bonding mo's below the dashed line in the mo diagram of item 4 are filled. Excess electron pairs on E^{n+} (zero for B^{3+}, one pair for N^{3+}, two pairs for Cl^{3+}) must populate anti-bonding and/or non-bonding mo's above the dashed line in item 4.

6. The key to the mo prediction of a stable structure, as you will see, derives from the number of such excess electron pairs at E^{n+}. Generally, if a highly symmetric structure forces these excess pairs to occupy σ anti-bonding mo's, a lower-symmetry structure (usually a simple bond angle deformation) will give these electrons a non-bonding role to play.

THE EF₃ SERIES

We will begin with the series EF₃ to account for the structure changes

BF₃   NF₃   ClF₃
(D_{3h}) (C_{3v}) (C_{2v})

F—B(F)(F)   N(F)(F)(F)   F—Cl—F
                              |
                              F

The mo level diagrams for these geometries are given in Figure 4–11. It is the construction of these that we want to discuss first.

For D_{3h} symmetry, the boron s and p ao's are all of different symmetry [from the D_{3h} character table you may look up $s = a_1'$, $p_z = a_2''$, $(p_x, p_y) = e'$]. Separate $p\sigma^*$ and $s\sigma^*$ mo's result; by ignoring B—F π bonding, the boron p_z ao is non-bonding.[10] As B^{3+} has no excess electron pairs, none of the anti-bonding mo's are occupied. For NF₃ and ClF₃ with one and two excess pairs, respectively, we would encounter first p_z and then p_z and $s\sigma^*$ electron pairs for D_{3h} symmetry.

For C_{3v} symmetry, the central atom s and p_z ao's have a_1 symmetry and should be mixed to form hao's h_+ and h_-:

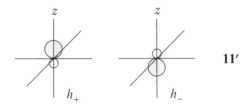

As h_+ is directed away from the F's, it cannot strongly overlap the F $p\sigma$ ao's and will be treated as non-bonding. The nitrogen p_x, p_y, and h_- ao's are overlapped by F $p\sigma$ ao's and so give rise to bonding and anti-bonding combinations with TASO's. As NF₃ has one excess pair of electrons, h_+ is occupied and an "sp" hybrid, non-bonding role is assigned to that pair (which would be more stable than the pure p ao non-bonding role in D_{3h} symmetry). ClF₃ has two excess pairs, one of which must serve an anti-bonding role in both D_{3h} and C_{3v} symmetries.

For C_{2v} geometry, the central atom ao symmetries are $s = a_1$, $p_z = a_1$, $p_y = b_2$, and $p_x = b_1$. Clearly, p_x is non-bonding. Since s and p_z have the same symmetry, we again form h_+ and h_-, as in **11'**, where an h_+ electron pair will play a non-bonding role. The sigma bonding involves the p_y and h_- ao's of the central atom. Clearly, the two excess pairs of Cl^{3+} are *both* given lone pair roles (p_x and h_+) in ClF₃, and this is the stable molecular geometry.

As a brief digression, you might notice that the bonding of the *two* fluorines along the y axis is achieved only by the electron pair (bond order $\sim \frac{1}{2}$) in the p mo (**12**),

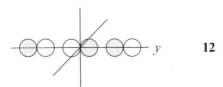

---

[10] In the later section (Figure 4–15) where a more complete development of mo's is given by including E—F π bonding, you will find the p_z ao replaced by a π^* mo.

152 ☐ THE UNDIRECTED ORBITAL VIEW OF CHEMICAL BONDS

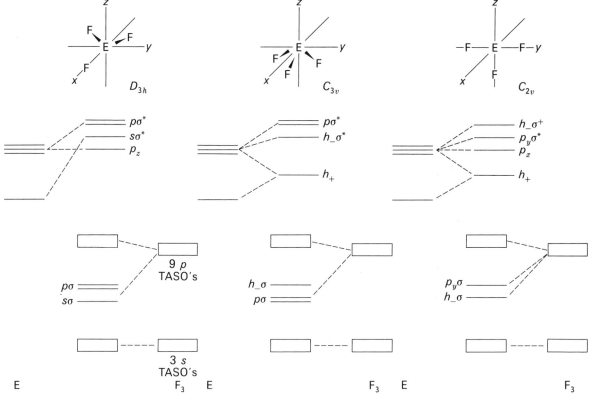

Figure 4–11. Simplified, σ-only mo diagrams for EX$_3$ structures.

whereas the bonding (bond order 1) along the z axis arises from **13**.

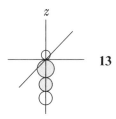

It is not surprising that the symmetry-equivalent fluorine bond distances are longer than that of the unique fluorine.

At this point you may be asking, "Why doesn't NF$_3$ adopt a C_{2v} geometry, which also provides an h_+ non-bonding role for the excess pair?" The answer comes from the higher bond orders for the C_{3v} structure and bp-bp and F—F repulsions, which favor greater inter-fluorine angles. Finally, you should recognize the simple bond angle deformation relationship between these structures.

THE EF$_4$ SERIES

The important structures here are those exemplified by SiF$_4$, SF$_4$, and XeF$_4$.

$$\begin{array}{ccc} \text{F}\diagdown\,\diagup\text{F} & & \text{F} \\ \text{Si} & \rightarrow \text{F}-\text{S}-\text{F} \rightarrow & \text{F}-\text{Xe}-\text{F} \\ \text{F}\diagup\,\diagdown\text{F} & \text{F}\diagup\,\diagdown\text{F} & \text{F} \\ (T_d) & (C_{2v}) & (D_{4h}) \end{array}$$

Figure 4-12 summarizes the salient features of the mo diagrams for these three geometries. Only in the C_{2v} case do we find s/p_z mixing. In SiF$_4$ the HOMO is one of the F$_4$ TASO's (lone pairs on fluorine atoms), as there are no excess Si^{4+} pairs to populate the σ^* mo's. SF$_4$, in which S^{4+} possesses an excess pair, avoids population of a σ^* mo in the T_d geometry by deforming the bond angles to the C_{2v} geometry; the HOMO in this case is the h_+ orbital directed away from the fluorines. XeF$_4$, interestingly, with two excess pairs, finds it impossible to avoid at least one σ^* pair. The D_{4h} geometry permits one of the excess pairs to occupy a Xe p_z ao, and the second pair must occupy a σ^* mo. While it is clear that the T_d geometry causes both pairs to play σ^* roles and is thus strongly disfavored, it is not immediately clear why D_{4h}, and not C_{2v}, is favored. Both geometries are characterized by four 90° inter-fluorine interactions, so these are not likely to be the determining factor. To account for the preference of XeF$_4$ for the D_{4h} rather than the C_{2v} geometry, we will have to compare the $s\sigma^*$ mo of D_{4h} (14) with the counterpart σ^* mo of C_{2v} (15).

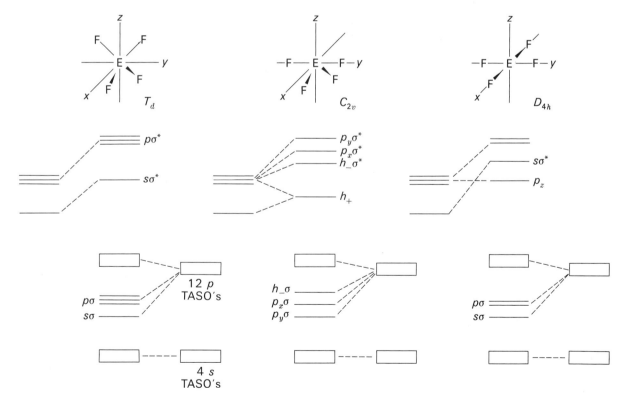

Figure 4-12. Simplified, σ-only mo diagrams for EF$_4$ structures.

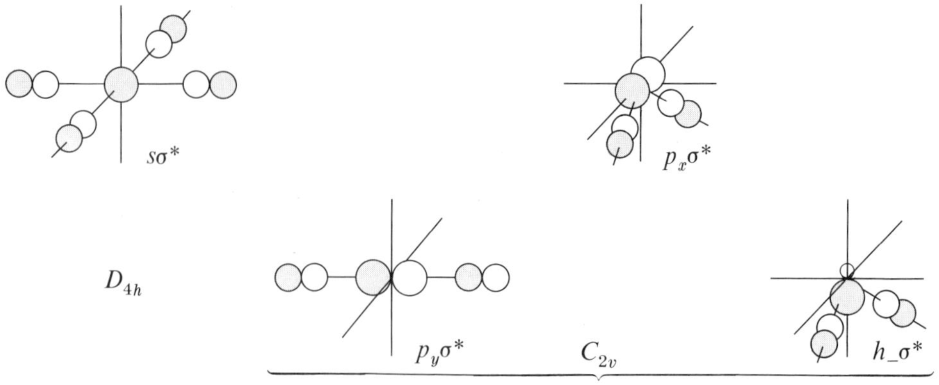

14 D_{4h}

15 C_{2v}

It should be clear from the overlaps in $p_y\sigma^*$ and $p_x\sigma^*$ of **15** that the former lies higher in energy. Does it make much difference whether $p_x\sigma^*$ or $h_-\sigma^*$ lies lowest? The answer is no. In D_{4h} geometry the $s\sigma^*$ occupation cancels the bonding action of the electron pair in the $s\sigma$ mo, leaving the Xe—F bonding to be described by the "head-on" overlap situation of the $p\sigma$ mo's **(16)**.

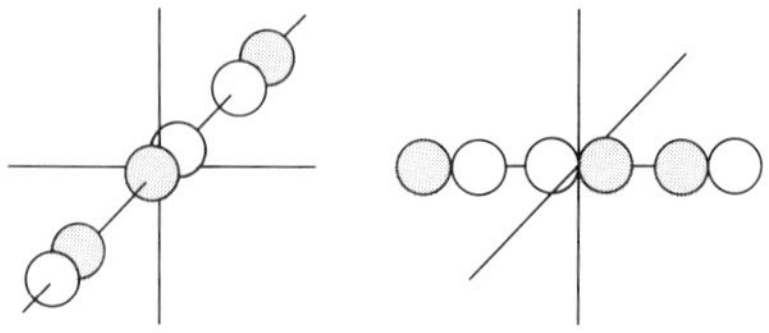

16a **16b**

Each Xe—F bond is characterized by a bond order of $\frac{1}{2}$. (The same description applies to the F atoms along the y axis of the C_{2v} structure.) The bonding of the atoms in the xz plane of the C_{2v} geometry is described by both the $p_x\sigma$ and $h_-\sigma$ mo's; one of

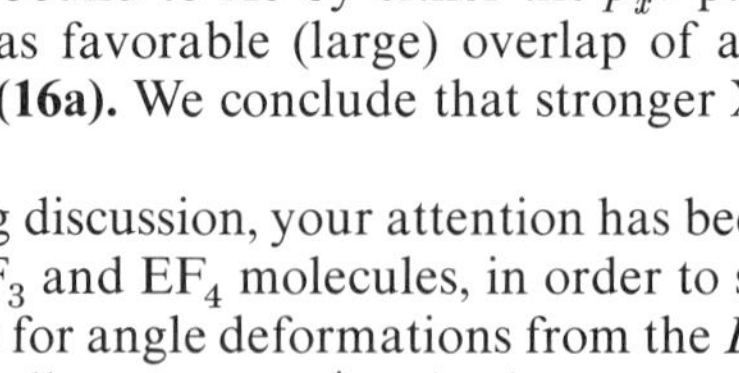

these is canceled by the occupation of either $p_x\sigma^*$ or $h_-\sigma^*$. The net result is that the xz plane fluorines would be bound to Xe by either the $p_x\sigma$ pair or the $h_-\sigma$ pair *alone*. Neither of these affords as favorable (large) overlap of ao's as does the head-on overlap in D_{4h} symmetry **(16a)**. We conclude that stronger Xe—F bonds arise in the planar topology.

In all of the preceding discussion, your attention has been restricted to structures known to arise for real EF_3 and EF_4 molecules, in order to see how $\sigma^* \rightarrow$ lp conversion acts as a driving force for angle deformations from the D_{3h} and T_d geometries. In the process, stereochemically *inactive* σ^* pairs become stereochemically active sp hybrid lone pairs. We have not, you notice, fully explored in any one case *all* orientations of the F's about E. This is a job for the quantitative theoretician. Such studies do show, however, that when the number of bond pairs plus excess pairs becomes too large, very subtle factors determine the preferred geometry. Operationally, "too large" is at hand when the process of avoiding σ^* roles for electron pairs incurs large inter-fluorine repulsions. XeF_4 provides a good example. Alternatives to the D_{4h} geometry for XeF_4 are the C_{4v} and C_{2v} geometries, wherein a lone pair "protrudes" along the C_n axis. In both cases, the inter-fluorine angles drop below 90°, signaling the onset of severe terminal atom repulsions (bp/bp or atom-atom or both). In the C_{4v} case, there is no compensating electron pair stabilization to offset the increased terminal atom repulsions; the deformation $D_{4h} \rightarrow C_{4v}$ incurs transformation of the p_z lone pair to an h_+ lone pair (stabilizing) at the expense of converting an $s\sigma^*$ pair to a σ^* pair derived from h_- (destabilizing).

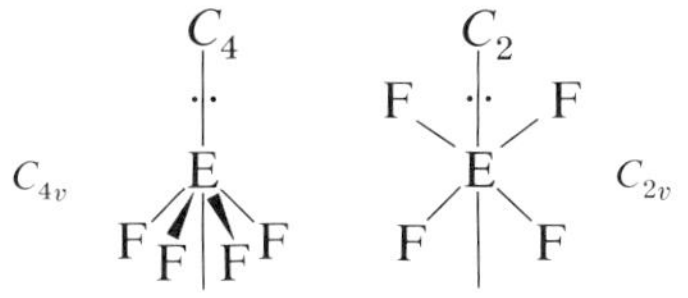

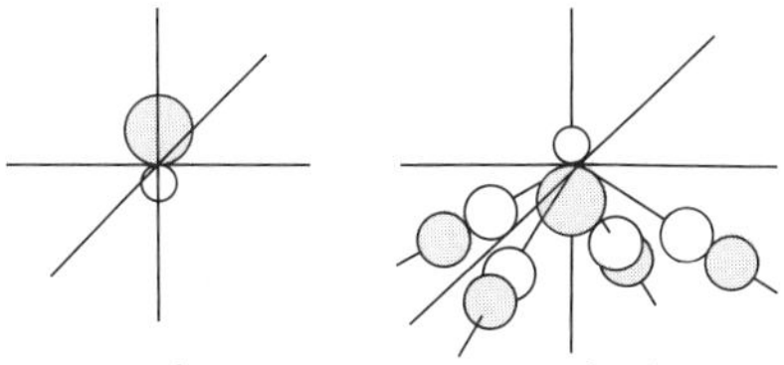

Deformation of $D_{4h} \rightarrow C_{2v}$ to allow transformation of the $s\sigma^*$ pair to an h_+ lone pair must come at the expense of even greater terminal atom repulsions than in the $D_{4h} \rightarrow C_{4v}$ distortion (note that the p ao lone pair role is unchanged by $D_{4h} \rightarrow C_{2v}$). It is *not clear at this simplified level* which geometry is to be preferred. Only experiment and very precise calculations can reveal the preferred geometry.

This section concludes a preliminary treatment of the principles of mo diagram construction, orbital sketching, and chemical interpretations. The next few sections introduce you to a *systematic* treatment of chain, cyclic, and three-dimensional structures. Along the way you will learn to develop the TASO's taken for granted in this section. With the overview gained to this point, you should be impressed in what follows with the power of symmetry concepts in unifying what would otherwise seem to be a bewildering variety of orbital types.

STUDY QUESTIONS

8. Why does BF_3 adopt D_{3h} symmetry, rather than C_{3v} or C_{2v}? Why is the T_d symmetry favored for SiF_4, instead of C_{2v} or D_{4h}?

9. Analyze the D_{3h} and C_{4v} structures for EF_5 to account for the structures of PF_5 and BrF_5.

10. Why does BrF_4^+ not adopt a D_{4h} geometry?

11. Develop an mo diagram like those in Figures 4-11 and 4-12 for XeF_6 with an O_h geometry. Describe the HOMO. Contrast the change in role for this electron pair on the one hand, and terminal atom repulsions on the other, when O_h deforms to the 1:3:3 geometry. (Don't bother to work out the 1:3:3 mo diagram; a simple comparison of the two factors involved will suffice.) Comment on how these changes tend to reinforce or cancel each other.

"LINEAR" (ONE-DIMENSIONAL) MOLECULES

Thinking back to Chapter 1, one of the more important distinctions between ao's of different energy is the number of nodal surfaces. This is no less true for molecules.

Realizing that each mo is characterized by a specific nodal pattern (number and placement) and accepting the statement on page 135 that every mo possesses an irreducible representation symmetry, we may make the following key generalization:

All molecular orbital nodes must be symmetrically disposed.

For linear arrangements of nuclei, the consequences of this statement are far-reaching in the development of an "intuition" of orbital shapes. The orbital nodal patterns calculated for the general cases of two through six atoms are collected in Figure 4–13 for future reference. In all cases, *successively higher energy orbitals have mo nodes symmetrically placed and increasing in number by one, in order from most stable to least stable.* An mo node through a nucleus means that the ao from that atom does not contribute. The mo energy increases with increasing numbers of nodes because the net number of bonding interactions (between neighboring atoms) decreases by two as the number of mo nodes increases by one. Furthermore, there is a unifying relationship between the placement of mo nodes in one orbital and that in the next higher orbital. A central mo node "splits" and shifts outward by $\sim d$ (d = bond distance), and this new pair of nodes continues to shift outward by $\sim d$ until placed just inside the terminal atoms (for the cases in which n is odd, the last shift, of necessity, is by $\sim d/2$). This pattern is continued for all successive central mo nodes (check the $n = 5$ and $n = 6$ cases).

These mo nodal patterns are basic; they apply equally well to something as esoteric as a linear string of two to six hydrogen atoms and to the more practical example of π molecular orbitals of the unsaturated hydrocarbons C_2H_4 through C_6H_8. To be specific about the application of these general orbital patterns, we shall focus on the $p\pi$ ao's of a triatomic molecule (NO_2^+, CO_2, O_3, NO_2, C_3H_5, etc.). In all these cases the $p\pi$ mo's have the form **17**.

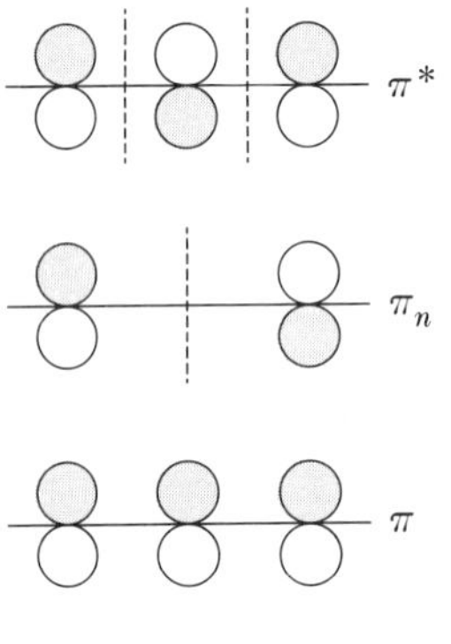

17

Electrons populating π and π^* are distributed over all three atoms, while electrons in π_n are constrained to the terminal atoms. The case of the allyl radical is interesting, for the π electron configuration is $(\pi)^2(\pi_n)^1$ and the odd electron is concentrated at the terminal carbon atoms; the same description is achieved by π resonance structures:

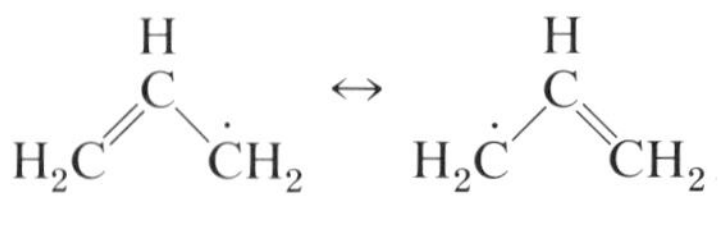

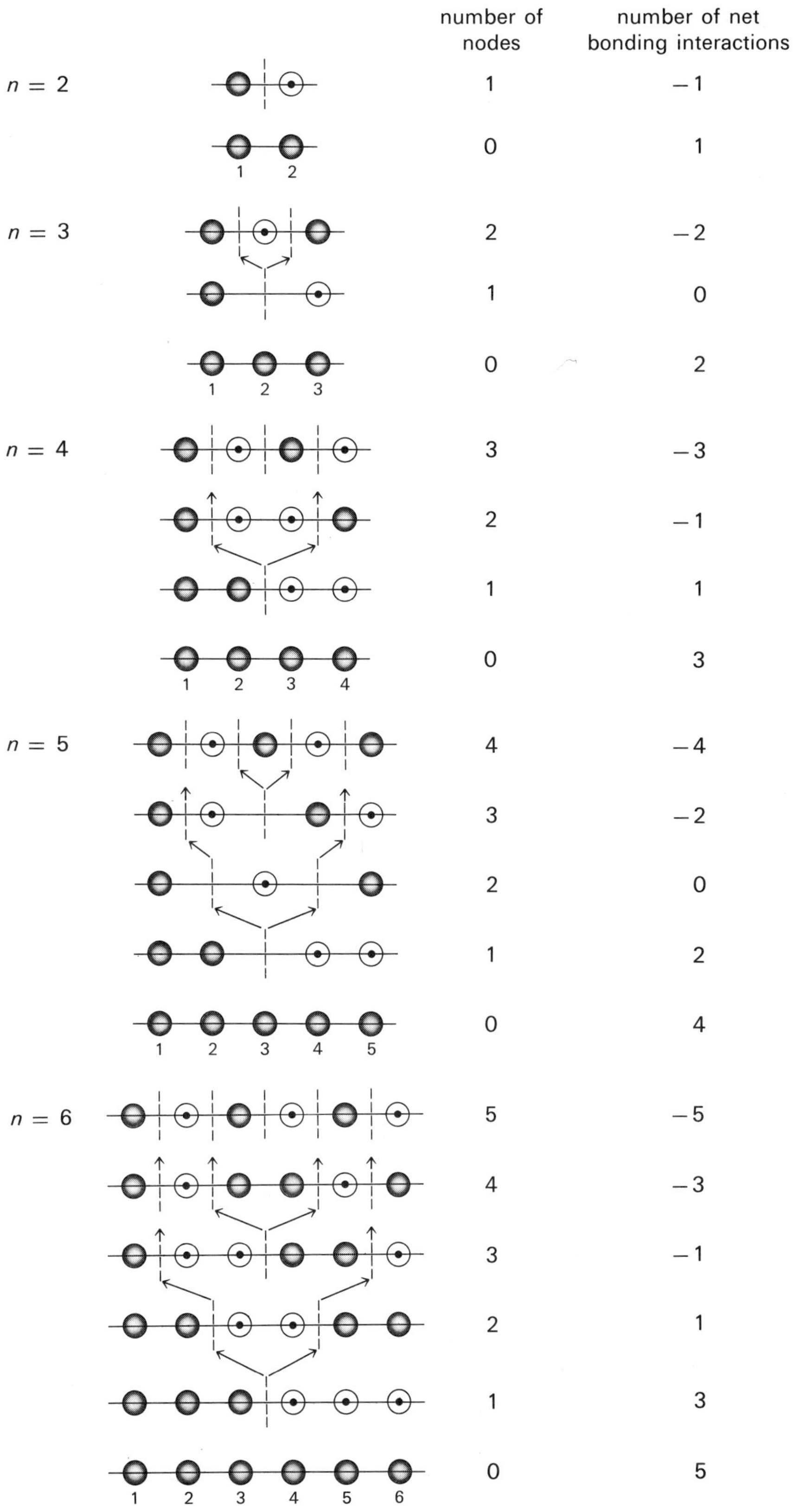

LINEAR (ONE-DIMENSIONAL) MOLECULES ☐ 157

		number of nodes	number of net bonding interactions
$n = 2$		1	−1
		0	1
$n = 3$		2	−2
		1	0
		0	2
$n = 4$		3	−3
		2	−1
		1	1
		0	3
$n = 5$		4	−4
		3	−2
		2	0
		1	2
		0	4
$n = 6$		5	−5
		4	−3
		3	−1
		2	1
		1	3
		0	5

Figure 4–13. Molecular orbital nodal patterns for two to six atoms in a "linear" chain, where each atom contributes one atomic orbital to the LCAO mo's.

The same pattern is equally applicable to the $p\sigma$ **(18)**† and $s\sigma$ **(19)** mo's of molecules like XeF_2 and CO_2.

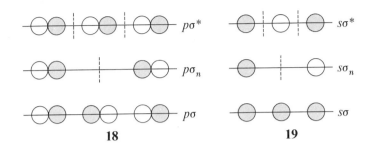

CYCLIC (TWO-DIMENSIONAL) MOLECULES

The results given in Figure 4-13 for "linear" molecules provide a nice basis for introducing the forms of molecular orbitals of cyclic compounds. A convenient mnemonic device to use for the cyclic structures is to start with the orbital sketches of the linear molecule with the same number of atoms, and to "wrap" the linear arrangement into the cyclic arrangement:

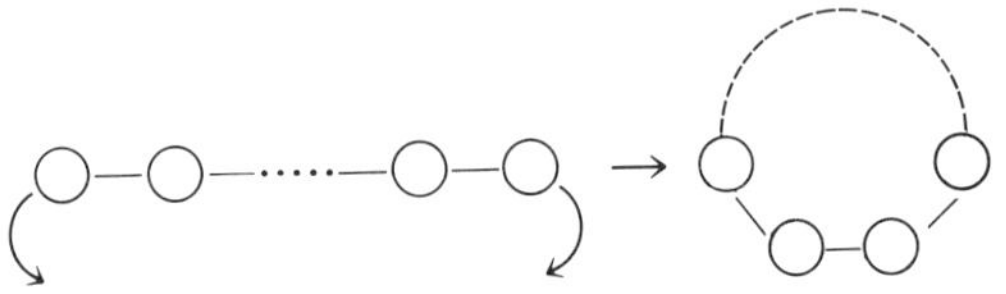

This may cause separate nodal planes in certain of the linear mo's to coincide or shift. The key to the cyclic mo structures lies in the nodal structure. *The all-important symmetry requirement is that the nodal lines pass through the center of the polygonal arrangement of nuclei.* It is straightforward now to sketch the cyclic orbital patterns. This is done in Figure 4-14 for the cyclic cases of three through six atoms.

Let us now consider each case in detail. For three atoms, the three linear-case mo's with 0 through 2 nodes transform into a single mo with no nodes and two mo's with one node each; these nodes are mutually perpendicular. As perhaps expected, the latter two cyclic mo's have the same energy (they have the same number of nodes) and are called degenerate orbitals—even though the sketches appear very different. (Trust your intuition!)

For the four-atom case, the non-degenerate mo's of the linear configuration transform into a most stable orbital with no nodes, a degenerate pair of orbitals with a single node each (which are mutually perpendicular), and a higher-lying orbital with two nodes. Pay particular attention to the fact that symmetry dictates that the outer nodes of the third and fourth linear mo's must coalesce into a single node (drawn horizontally) in the third and fourth cyclic mo's.

The five-atom case presents us with our first instance of nodes that, while passing through atoms in the linear case, shift off those atoms to adhere to the symmetry requirement that the nodes must pass through the center of the polygon. The most stable linear mo transforms into the most stable cyclic mo with no nodes. The next two linear mo's, with one and two nodes, transform into a degenerate pair with one node each; these nodes are mutually perpendicular. Note that the nodes at the second and fourth atoms in the third linear mo must shift off these atoms and coalesce in the

---

†The "node" in $p\sigma_n$ does not have its usual meaning ($\psi = 0$) here. In context, the "node" means no contribution from the central atom p ao.

third cyclic structure. The two highest linear mo's transform into a degenerate pair of cyclic mo's with two nodes each. Again note the node coalescence.

For the six-atom case, the no-node linear mo becomes the most stable no-node orbital of the hexagon. The one- and two-node linear orbitals become the degenerate, single node cyclic orbitals. The three- and four-node linear orbitals become a second degenerate pair at higher energy than the previous pair, owing to the presence of two nodes each. Finally, the five-node linear orbital becomes a three-node high-energy orbital of the hexagon. (The arrows to pairs of nodes in the fifth and sixth linear mo's identify those that coalesce.)

As the number of nodes is directly related to the orbital energy, it is quite straightforward to rank the orbitals in order of increasing energy and thus to determine the electron configuration of the molecule. To illustrate, the molecule benzene, with six pi electrons, is predicted to have the π electron configuration $(\pi_1)^2(\pi_2,\pi_3)^4$. The π_1 orbital is most stable, with six bonding interactions between nearest neighbors; while π_2 and π_3 (the HOMO), each with a *net* of two bonding interactions, are

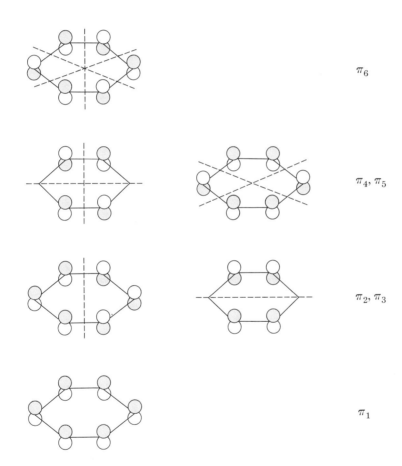

next most stable, degenerate, and bonding. The orbitals π_4 and π_5 (LUMO) each have a *net* of two anti-bonding interactions and so are properly described as anti-bonding orbitals, while π_6 is most anti-bonding with six anti-bonding interactions. An interesting aside here is that $C_6H_6^-$ should be a powerful one-electron reducing agent while $C_6H_6^+$ should be a strong one-electron oxidant.

The cyclopentadiene molecule is predicted to be a radical with a $(\pi_1)^2(\pi_2,\pi_3)^3$ configuration; it should be an oxidant forming the stable cyclopentadienide anion, $C_5H_5^-$, and this is a known species. In fact, the anion will occur repeatedly throughout

160 ☐ THE UNDIRECTED ORBITAL VIEW OF CHEMICAL BONDS

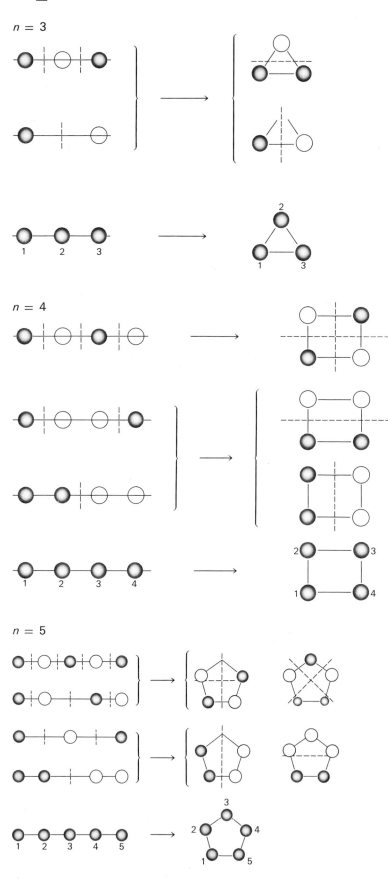

Figure 4-14. Cyclic mo nodal structures from linear counterparts.

Illustration continued on opposite page.

CYCLIC (TWO-DIMENSIONAL) MOLECULES □ 161

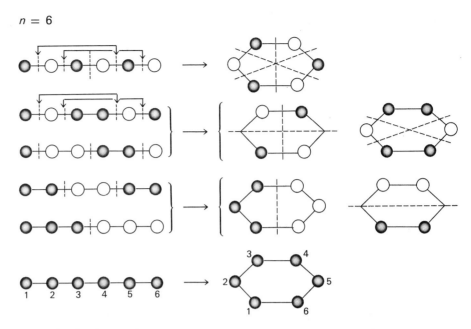

Figure 4–14. *Continued*

later sections of the text in connection with transition metal complexes, at which time the orbital sketches presented here will be used.

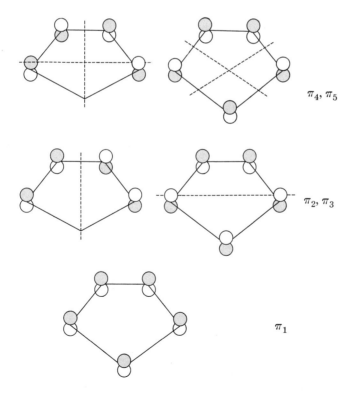

The symmetries of the cyclic molecular orbitals in Figure 4–14 can be summarized by writing down the phase (+ or −) of each atomic orbital in LCAO form. (We are interested only in the sign of the mo at each atom, not its amplitude there.) This is done in Table 4–2 for the triangle, square, and pentagon. *This is an important table to which we will frequently refer; familiarize yourself with its contents.*

Triangle	$1 + 2 + 3$	
	$\left.\begin{array}{l} 1 - 3 \\ 1 - 2 + 3 \end{array}\right\}$	degenerate
Square	$1 + 2 + 3 + 4$	
	$\left.\begin{array}{l} 1 + 2 - 3 - 4 \\ 1 - 2 - 3 + 4 \end{array}\right\}$	degenerate
	$1 - 2 + 3 - 4$	
Pentagon	$1 + 2 + 3 + 4 + 5$	
	$\left.\begin{array}{l} 1 + 2 - 4 - 5 \\ 1 - 2 - 3 - 4 + 5 \end{array}\right\}$	degenerate
	$\left.\begin{array}{l} 1 - 2 + 4 - 5 \\ 1 - 2 + 3 - 4 + 5 \end{array}\right\}$	degenerate

TABLE 4–2

FORMS (RELATIVE AO PHASING) OF THE CYCLIC LCAO MOLECULAR ORBITALS

STUDY QUESTIONS

12. From π resonance structures of NO_2^- you are aware that there are two π electron pairs:

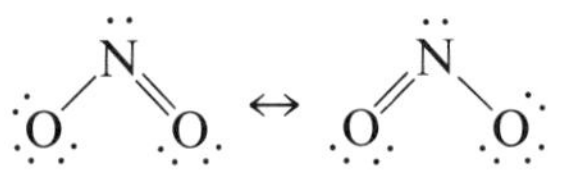

    $$\ddot{\underset{\ddot{O}:}{\overset{\ddot{N}}{\diagdown}}}\underset{:\ddot{O}:}{\diagup} \leftrightarrow \underset{:\ddot{O}:}{\overset{\ddot{N}}{\diagdown}}\underset{\ddot{O}:}{\diagup}$$

    What does the mo model predict for the bond order of each N—O bond? What would be the effect on the NO distance of photoionization of an electron from the π HOMO?

13. Sketch the $s\sigma$, $p\sigma$, and $p\pi$ mo's of I_3^-.

14. With which π mo's of cyclopentadienide do the s and p ao's of a silver ion overlap when the cation is positioned along the C_5 rotation axis?

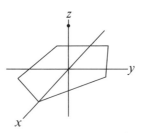

(Hint: Look for coincidence of Ag^+ p ao nodes with those of the cyclopentadienide mo's.)

> *NOTE TO INSTRUCTORS*
>
> The section that follows may be approached in one of three ways, depending on course goals:
>   i) complete, in-depth coverage of the polyhedral structures with the intent of solidifying for the student his practice of the concepts and techniques of mo construction;
>   ii) elimination of the details of TASO and mo construction altogether, with emphasis given to chemical interpretation of the mo diagrams and the occupied orbital forms (in particular the HOMO and LUMO) as given "facts" to be interpreted;
>   iii) a hybrid of these approaches whereby the EX_3 and/or EX_4 (D_{4h}) examples are followed through in their entirety, with the other examples treated as in approach ii.
>
> The last technique has been used by us for several years with considerable success. Our students have shown great interest in exploring the details of the EX_4 (T_d), EX_5, and EX_6 cases on their own. (It is most rewarding to watch their confidence and fluency with orbital concepts develop through their independent efforts.)

POLYHEDRAL (THREE-DIMENSIONAL) MOLECULES

With the basic orbital patterns of the linear and cyclic cases in mind, you may now enlarge your horizons by using those simpler results as a guide to the qualitative description of the mo's of more complicated molecules. Of particular importance will be the symmetry basis for TASO's and their qualitative use in developing more complete mo diagrams. We will not be comprehensive, but we will consider a few of the more common structural types and continue building toward slightly more complex molecules.

Before proceeding to an examination of those structural types already considered in Chapter 2 and earlier in this chapter, it will be well to lay down some organizational rules for the development of molecular orbitals for these three-dimensional molecules. The rules are simply an extension and generalization of the useful concepts already encountered.

STRATAGEM

1. Group the terminal atoms into symmetry-equivalent sets.
2. For each equivalent atom set, group the ao's according to their symmetry equivalence. This is particularly easy if you adopt a convention for the terminal atom p ao's of taking one p orbital (to be called $p\sigma$) to lie along the bond axis to the central atom and taking the other two p orbitals (to be called $p\pi$) to lie perpendicular to this axis. It is then generally possible to classify the terminal ao's into three symmetry sets: (a) the s ao's, (b) the $p\sigma$ ao's, and (c) the $p\pi$ ao's.

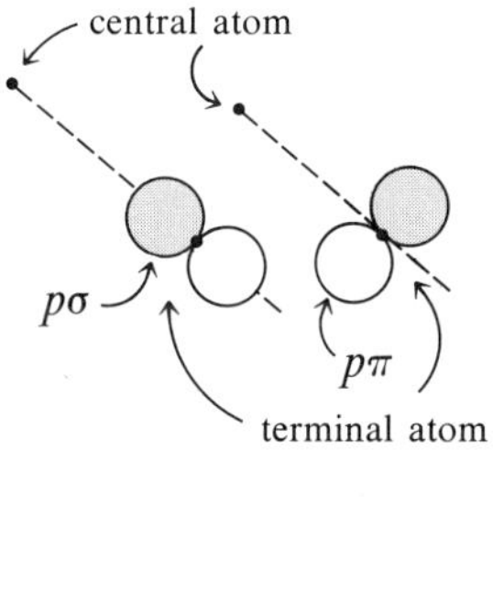

3. For each of these ao sets, using Table 4–2, sketch the TASO's.

4. Make sketches of the central atom ao's and decide which have non-zero overlaps (common nodal structure) with any of the TASO's. This allows you to group ao's into symmetry-distinct sets.

5. Consider each of these central ao/TASO sets separately.
   a. Recognize that the s and $p\sigma$ TASO sets will have identical symmetry patterns. As a "first approximation," ignore the s TASO sets, in effect taking the s TASO's to be stable, lone orbitals.
   b. If any $p\pi$ TASO's have the same symmetry patterns as the $p\sigma$ TASO's, assume that the central atom ao's interact (overlap) more strongly with the $p\sigma$ sets than with the $p\pi$.

The emphasis, based on the "first approximation" as in a and b, is on viewing the central atom–terminal atom bonding as arising primarily via the terminal $p\sigma$ TASO's.

   c. The need might arise on occasion to go beyond the "first approximation" by including the interactions of the s TASO's with the mo's formed from adoption of the "first approximation" (in some instances you will even encounter a $p\pi$ TASO with a symmetry matching that of the σ mo's). You are assured that the most stable orbital of such a set will be stabilized (lowered in energy owing to amplification of the central/terminal bonding for that orbital) and the least stable orbital of the set will be destabilized (amplification of the anti-bonding interaction for that orbital). The remaining orbitals of the set are more or less non-bonding and may experience an energy shift. To say in advance how big and in what direction the shift will be for a particular orbital will require detailed theoretical analysis and often amounts to more work than the answer justifies (certainly this is true for the level of this text). You need to remember that the s TASO/$p\sigma$ TASO mixing introduces hybridization (s, p mixing) for the terminal atoms. In the event of $p\sigma$ TASO/$p\pi$ TASO mixing, you find further hybridization (p, p mixing) and accordingly lose the ability to characterize mo's as exclusively σ or π.

Now you are ready to apply this stratagem to the important class of planar EX_3 molecules. BF_3 is an example of this class, and we will refer back to this section when we discuss the chemistry of BF_3 as a Lewis acid. The EX_3 problem will also be used shortly as a stepping stone to the EX_5 class of D_{3h} symmetry.

In what follows, we will use irreducible representation symbols to label the central ao's and TASO's. It is not intended, and certainly not necessary, that you be able to confirm these labels. Their use is to be preferred to some arbitrary labeling scheme, and it has the added advantage that you are already familiar with the meaning a or $b \Rightarrow$ non-degenerate, $e \Rightarrow$ doubly degenerate pair, $t \Rightarrow$ triply degenerate trio.

EX_3 (D_{3h})

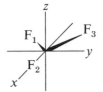

POLYHEDRAL (THREE-DIMENSIONAL) MOLECULES 165

1. All three fluorines belong to a symmetry-equivalent class.
2. a) The three fluorine $2s$ ao's will produce a set of TASO's.
   b) The three fluorine $2p\sigma$ ao's will produce a set of TASO's.
   c) The six fluorine $2p\pi$ ao's will produce a set of TASO's.

Notice that here we may distinguish in-plane and out-of-plane $p\pi$ orbitals.

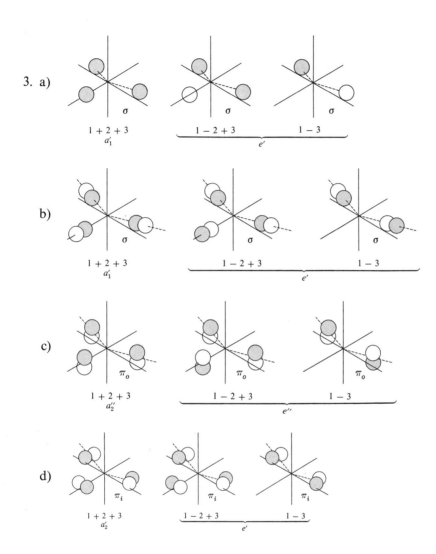

In applying the formulae of Table 4–2 to the $p\pi$ ao's, note that the standard orientation of the out-of-plane ao is parallel to the C_3 axis and in the $+z$ direction. The standard orientation of the in-plane ao is perpendicular to the C_3 axis and pointing in a counterclockwise direction.

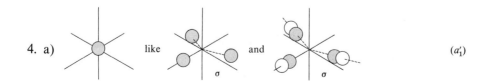

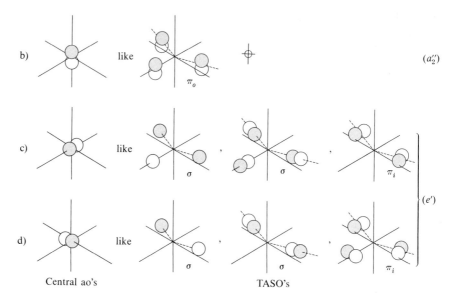

Note that all three s and $p\sigma$ TASO's match symmetries with a boron ao. Only one of the out-of-plane $p\pi$ TASO's matches symmetry (boron $2p_z$). The second and third $p\pi_o$ TASO's (e'') are non-bonding. Only the second and third in-plane $p\pi$ TASO's (e') match symmetry with boron p_y and p_x, respectively. The first in-plane $p\pi$ TASO (a_2') has a unique symmetry and will be non-bonding.

5. Make the "first approximation" that s and in-plane $p\pi$ TASO's are not involved.

As a final important comment, you should note that the boron $2p_x$ and $2p_y$ orbitals comprise a degenerate pair in D_{3h} symmetry (the character table shows (x,y) to be a pair of e' symmetry). Consequently, *all* mo's involving them occur in degenerate pairs.

To summarize the preceding conclusions regarding the qualitative molecular orbital scheme for planar BF_3, we present in Figure 4–15 the energy level diagram. To the left of the diagram you find the boron $2s$ and $2p$ ao energies, labeled according to their orbital forms in the sketches under items 4 and 5 above. To the right of the diagram you find the energies of the fluorine TASO's with the $2s$ set distinguished from the $2p$. At the center of the diagram you find the qualitative ordering of molecular orbitals for EX_3. All levels are labeled with their representation symbols to make it easier to see how the diagram was constructed.

While the details of relative ordering of the bonding and "non-bonding" orbitals are uncertain without an elaborate quantitative calculation, the difference between the B and F ao energies assures a sizeable energy gap between the 12 lowest-energy mo's and the four higher-energy (anti-bonding) mo's. An EX_3 species such as BF_3, NO_3^-, or CO_3^{2-} has 24 valence electrons, so you can conclude that the lowest 12 mo's are occupied with electron pairs and the highest four are empty. It is uncertain from this relatively unsophisticated treatment which of the "lone pairs" represents the highest energy occupied orbital(s) (HOMO), but you can be reasonably certain that the lowest unoccupied orbital (LUMO) is the out-of-plane π^* orbital. Because the bonding partner (a_2'') to the LUMO ($a_2''^*$) will have greater fluorine than boron character (because the fluorine $p\pi$ ao energies are lower in energy than those of boron), the LUMO is more heavily concentrated on boron and may be viewed as a boron $2p\pi$ ao somewhat delocalized onto the fluorines. Thus, the LUMO is the orbital

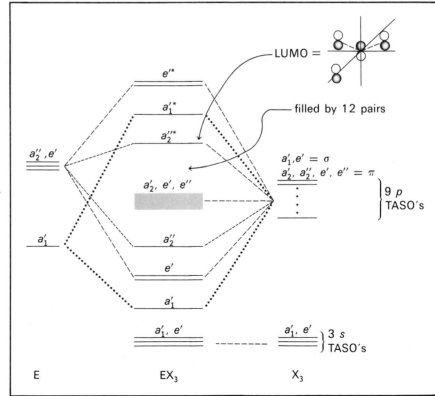

Figure 4–15. The mo energy level diagram for EX_3.

primarily responsible for the Lewis acid character of BF_3. We will return to this important point in the chapter on Lewis acids and bases.

It should be apparent from the diagram and orbital sketches that there are three[11] strongly bonding electron pairs from in-plane σ orbitals (labeled a_1' and e') and one bonding pair from out-of-plane orbitals (labeled a_2''). The net number of bonding pairs per B—F link is $1\frac{1}{3}$, a value that agrees with the prediction of the resonance structure model. The remaining eight electron pairs are of essentially non-bonding character and of either of two types: six are in-plane (σ and π) pairs and two are out-of-plane pairs.

In Figure 4–15, three of the in-plane lone pairs are treated as arising from fluorine $2s$ ao's but this is, of course, the "first approximation" simplification. Mixing of these s TASO's with bonding molecular orbitals of like symmetry produces the now-familiar s/p mixing of terminal ao's, so that the s "lone pairs" take on $2p$ character and the bonding $p\sigma$ mo's take on fluorine $2s$ character. It is an interesting point that *there is no mixing of central atom s and p ao's*. This is precluded by the symmetries of the central atom orbitals. As you will see later in this chapter, this does not mean that the directed orbital model of an sp^2 hybridized boron is incorrect, but only that the hybrid orbital model is a different but equivalent model of atomic bonding in molecules.

---

[11] Note that these mo's involve the central atom s, p_x, and p_y ao's, and note the connection with the use of these same ao's by the hybrid model for trigonal planar hybrids.

STUDY QUESTIONS

15. Construct an mo level diagram for FHF^- (bifluoride ion, point group = $D_{\infty h}$) by adopting a first approximation approach (ignore the s TASO's). Assign a bond order value to each F—H "bond." Make sketches of the σ, σ_n, and σ^* mo's. Identify for each of these the s TASO's of the same symmetry. Make a sketch of the new σ_n mo *after* mixing of the old σ_n with the like-symmetry s TASO.

16. Derive the mo diagram for XeF_2 (by adopting the "first approximation" in the sense of the text).

17. Given below are some of the molecular orbital wavefunctions for CO_2.

    $$\begin{array}{c} y \\ \uparrow \\ O_1\text{—}C\text{—}O_2 \rightarrow z \end{array}$$

    $\Psi = 0.7(p_{O1y}) - 0.7(p_{O2y})$
    $\Psi' = 0.6(s_{O1}) + 0.1(p_{O1z}) - 0.5(p_{Cz}) - 0.6(s_{O2}) + 0.1(p_{O2z})$
    $\Psi'' = 0.4(s_{O1}) - 0.5(p_{O1z}) - 0.4(s_C) + 0.4(s_{O2}) + 0.5(p_{O2z})$

    (a) Sketch each mo.
    (b) Identify each mo as sigma or pi and give the proper labels to the sigma orbitals $(\sigma_g^+, \sigma_g^-, \sigma_u^+, \sigma_u^-)$.
    (c) Tell whether each mo is bonding, non-bonding, or anti-bonding, between O_1 and C and between C and O_2.
    (d) Which mo exhibits significant hybridization of O_1 and O_2?

18. Derive the mo diagram for H_2O (the answer appears at the top of Figure 4–20).

19. In Figure 4–15 for BF_3, which non-bonding orbitals are strictly localized on the F atoms, even after the "first approximation" is dropped?

POLYHEDRAL (THREE-DIMENSIONAL) MOLECULES

EX_4 (T_d)

The next structural type in the hierarchy of molecular structures is the tetrahedron of formula EX_4. The mo patterns for this structure can be derived from the basic patterns for the cyclic four-member structure, and this is the approach to be

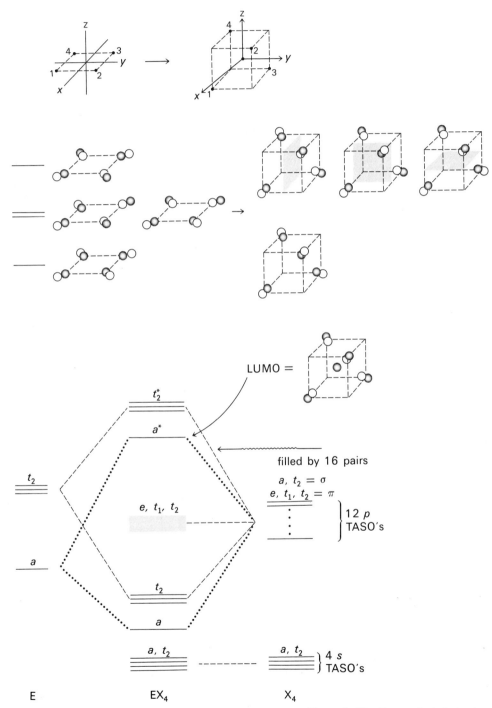

Figure 4-16. The tetrahedral structure.

taken. In the interest of simplicity, we will restrict your attention to the s and $p\sigma$ TASO's. The $p\pi$ TASO's should be included, but the "first approximation" would have you treat them as non-bonding orbitals anyway. Another good reason for deleting them is that the orbital sketches are very difficult to view in two dimensions, and an attempt to present them would more than likely only obscure the essentials of the E—X σ interaction. Suffice it to say that the eight $p\pi$ TASO's consist of three sets: a degenerate pair (symmetry label e) and two triply degenerate sets with different nodal structures (symmetry labels t_1 and t_2; the latter will be seen to have the same symmetry as one of the s and one of the $p\sigma$ TASO's and warns you that separate σ and π mo's are not possible for tetrahedral EX_4).

The nodal structures of the $p\sigma$ TASO's for the cyclic four-member ring atom mo's are depicted in Figure 4–16. To generate the tetrahedron orbital patterns from these, it is necessary only to realize that by displacing opposite terminal atoms in the same direction (one pair above the square plane and the other pair below that plane) you ultimately arrive at a tetrahedral disposition of terminal atoms. (This takes you through item 3 of the stratagem.)

With these orbital sketches at hand (see Figure 4–16), it is now a simple matter to reason that the central atom valence s atomic orbital has a non-zero overlap with only the nodeless TASO, so these two are taken in sum and difference combinations. The remaining three TASO's for the tetrahedron constitute a three-fold or triply degenerate set of orbitals with non-zero overlaps with, respectively, the central atom valence p_y, p_x, and p_z atomic orbitals.[12] (The key is found in the mutually perpendicular nodal planes.) Consequently, the valence p atomic orbitals of the central atom form triply degenerate bonding and anti-bonding mo's with the triply degenerate TASO's. It is interesting to note that, in a manner similar to what occurs in the linear $\rightarrow$ cyclic transformation, the two-node mo for the square planar configuration becomes degenerate with the pair of already degenerate cyclic mo's. It is easy to satisfy yourself of the degeneracy of the three TASO's, since they differ only in their orientation in space (just like the three atomic p orbitals for the central atom). Any one of the triply degenerate TASO's may be rotated into either of the other two by rotations about the $x, y,$ or z axes. The final mo sketches and energy diagram are given in Figure 4–16, where it is seen that any molecule that satisfies the octet rule has completely filled bonding mo's and empty anti-bonding mo's.

Molecules in the class of the C_{3v} EX_3 (pyramidal) structure may be thought of as being derived from the EX_4 (T_d) case by removal of an X atom and its one (X=H) or four (X = p block element) ao's. In this event, the three-fold degeneracy (insured by symmetry) of the central atom p orbitals is lost. The C_{3v} mo scheme for EX_3 stands in even greater similarity to the D_{3h} EX_3 case after loss of the σ_h plane, but is easiest to develop by applying the stratagem to the pyramidal structure from the start. This constitutes an exercise for you (the simple C_{3v} diagram in Figure 4–11 may be consulted for a check).

---

[12] The hybrid ao model utilizes the $s, p_x, p_y,$ and p_z central atom ao's for tetrahedral hybrids. The LCAO model depicts these same ao's as responsible for the four bond pairs.

STUDY QUESTIONS

20. Sketch the LUMO and HOMO of BH_4^-.

21. Would ionization of a valence electron from CH_4 weaken or strengthen the CH bonds? Repeat for electron capture by NH_4^+ to form NH_4.

22. Construct the mo diagrams for planar BH_3 and for pyramidal BH_3 (go beyond the first approximation). Describe each LUMO in terms of its amplitude at boron and the constitution of boron $2s$ and $2p_z$ ao's, and point out their different bonding and non-bonding characters.

EX_5 (D_{3h})

Because (stratagem item 1) the axial and equatorial X's of a molecule such as PX_5 belong to different symmetry sets, we need only build on the description of the orbital patterns already developed for the trigonal planar case. To implement the combination of EX_3 mo's with the axial X_2 orbitals, you need to characterize the TASO's for the X_2 set.

20 like a_1' of EX_3 like a_2'' of EX_3

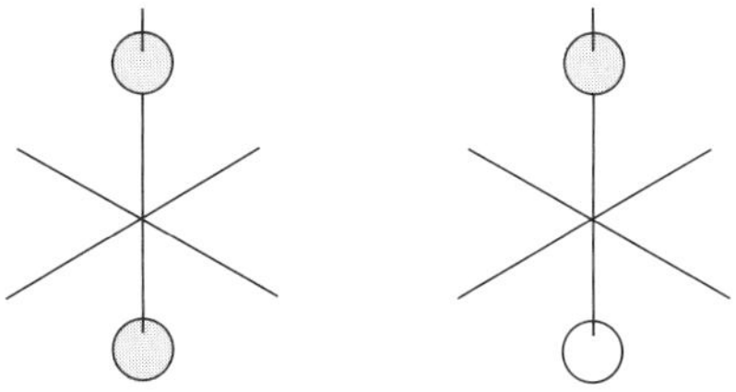

21 like a_1' of EX_3 **22** like a_2'' of EX_3

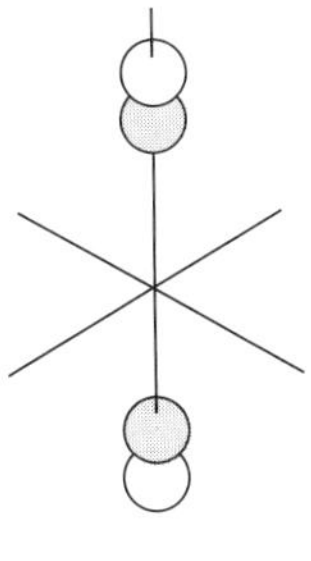

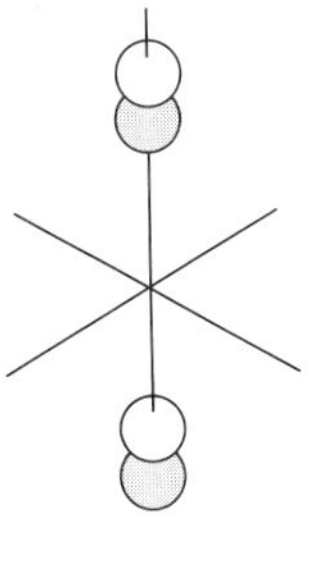

23
24 like e' of EX_3 like e'' of EX_3

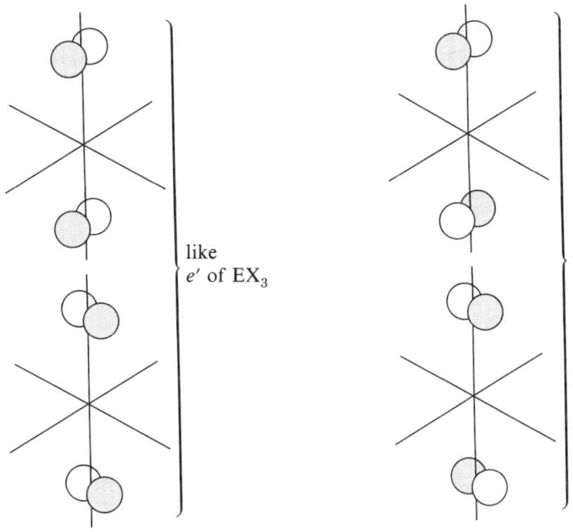

Having sketched the axial TASO's, you see that each axial TASO can interact ($S \neq 0$) with only certain of the EX_3 (planar) mo's (p. 166). These possible interactions have been noted above for each axial TASO. By making the simplifying "first approximation," only the axial TASO's **21** and **22** and the first of each of **23** and **24** need be considered.

This is done as follows: The mixing of the TASO **21'** serves to lower the energy of a_1', raise the energy of $a_1'^*$, and create a nominally non-bonding "axial" mo from **21'**. In fact, all three new σ orbitals are delocalized over all nuclei.

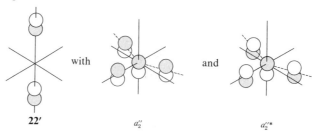

The mixing of the TASO

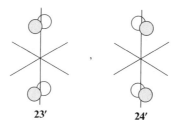

has the effect of stabilizing a_2'' (formerly a π bonding mo for EX_3 that now is dominated by axial σ bonding character), of destabilizing $a_2''^*$, and creating another nominally non-bonding axial mo from **22'**.

Analogous results arise from combination of the doubly degenerate pair

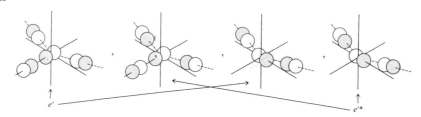

These interactions are summarized in Figure 4-17. In this figure, the bonding mo labeled a_2'' was, in the EX_3 case, an out-of-plane π mo; but with the axial X_2 present, the axial σ interaction should dominate the equatorial π interaction with the p_z orbital of E. Consequently, the electron pair in this mo must serve as the sigma pair for two axial E—X links, and also serves as a π bonding pair for three equatorial E—X links. Thus you find a reasonable bond-order explanation for the relative weakness of the axial bonds in EX_5. The same conclusion arises, however, by ignoring π interactions altogether: a_1' is E—X bonding with respect to both axial and equatorial X's, e' is E—X bonding for equatorial X's only, and a_2'' is E—X bonding for axial X's only. As a first estimate of E—X bond orders, you could total the individual mo contributions as follows:

	a_1'	e'	a_2''	
E—X_{eq}	1/5	2/3	0	= 13/15
E—X_{ax}	1/5	0	1/2	= 7/10

POLYHEDRAL (THREE-DIMENSIONAL) MOLECULES □ 173

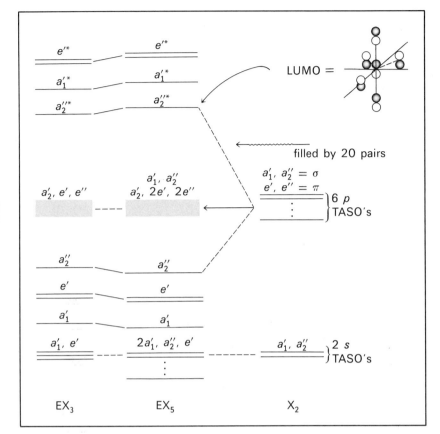

Figure 4-17. The mo energy diagram for EX$_5$ (D_{3h}) derived from planar EX$_3$ and axial X$_2$.

Including E—X π bonding via central atom d ao's further discriminates in favor of stronger equatorial bonds.[13]

---

[13] The VSEPR model had a complementary explanation (p. 110, Chapter 3). The hybrid ao model for the central atom in a trigonal bipyramid invokes dsp^3 hybridization. An interesting point of departure has arisen for the LCAO model, for here we have ignored the central atom d ao's. Given the shorter EX$_{eq}$ than EX$_{ax}$ distances, the equatorial σ bonding arises from s, p_x, and p_y (tr or sp^2 hao, in directed orbital terms) while the axial σ bonding depends primarily on p_z alone. Now you see that the inclusion of central atom d ao's is not necessary. Their inclusion would have the following result (remember that the d ao's will be at higher energy than the s and p ao's): d_{z^2} has a_1' symmetry; $(x^2 - y^2, xy)$ have e' symmetry; and (xz, yz) have e'' symmetry. The a_1', e', and e'' bonding and non-bonding mo's and TASO's will be stabilized by the inclusion of the high-energy d ao's. As the low-energy a_1' bonding mo involves principally the equatorial atoms, you see that d_{z^2} is involved in equatorial bonding. However, the d_{z^2} effect on the a_1' "non-bonding" TASO will be more important (greater overlap and better energy match), so you expect the d_{z^2} ao to be most effective in binding the axial atoms. This is a conclusion reached in the last chapter in the discussion of the hybrid ao's, and ties in with the hybrid ao model of dsp^3 for sigma bonds.

The e' ao's $(x^2 - y^2, xy)$ stabilize not only the e' bonding mo (primarily σ for equatorial groups) but, to a greater extent, the e' "non-bonding" TASO representing the in-plane π TASO's. Though permitted by symmetry, there is in fact no stabilization by $(x^2 - y^2, xy)$ of the axial π TASO of e' symmetry. Reference to the sketches of this TASO on p. 171 shows the overlap integrals to be zero. Thus, $(x^2 - y^2, xy)$ are devoted entirely to equatorial bonding. Axial π bonding is handled by (xz, yz), of e'' symmetry, through stabilization of the e'' TASO representing the axial π ao's (cf. p. 171). In addition, the (xz, yz) ao's stabilize the other e'' TASO, representing the equatorial out-of-plane ao's, so as to effect better equatorial π bonding. In short, equatorial groups may form π bonds by all of $(x^2 - y^2, xy, xz,$ and $yz)$ while axial groups may utilize only (xz, yz).

174 ☐ THE UNDIRECTED ORBITAL VIEW OF CHEMICAL BONDS

STUDY QUESTIONS

23. Make sketches of the bonding, non-bonding, and anti-bonding a_1' mo's of the EX_5 structure by analyzing the phasing of the a_1' and $a_1'^*$ mo's of the EX_3 structure (Figure 4-15) with the a_1' TASO (21) of the axial X_2 atoms. Is it a good approximation to view the non-bonding mo as E—X non-bonding?

24. Considering the a_1', e', and a_2'' *bonding* mo's of EX_5, characterize each as to its content of one or more of the following ao interactions: E—$X_{ax}\sigma$, E—$X_{ax}\pi$, E—$X_{eq}\sigma$, and E—$X_{eq}\pi$.

25. Referring to footnote 13 on page 173, decide whether equatorial or axial π bonding to the central atom is greater for the trigonal bipyramid. Now predict a structure for $POCl_3 \cdot py$ (where py is pyridine). Does this agree with your prediction based on the VSEPR model?

26. Construct an mo diagram for PH_3F_2.

EX_4 (D_{4h})

The last case to be considered for non-metal compound structures is that for six pairs about a central atom. In this class we find molecules of D_{4h} symmetry (square planar) such as ICl_4^- and XeF_4, and those of O_h symmetry such as PCl_6^- and SF_6. These geometries are also common for transition metal complexes in which the central atom valence orbitals are, of course, primarily the d orbitals rather than the s and p orbitals of the non-metals. A common feature of both the non-metal and metal structures is, obviously, the TASO's. We will begin with the non-metals and deal first with the square planar case, and then build on it to get the O_h structure.

The TASO's of the square planar case are easily sketched[14] as (refer to Table 4-2):

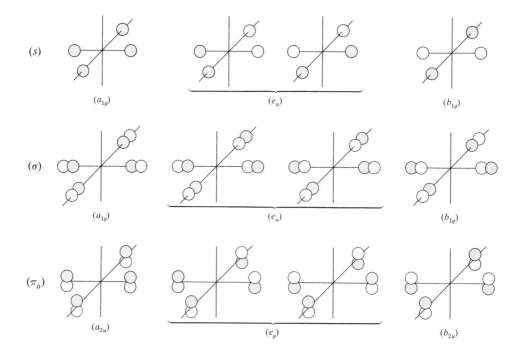

---

[14] Note the standard, counterclockwise orientation of the in-plane $p\pi$ ao's.

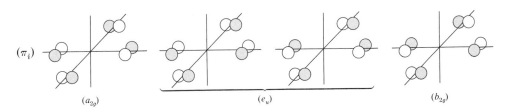

The *s* TASO's are sketched in the group (*s*) and, by adopting the "first approximation," are taken to be non-bonding. Group (σ) is the set of *p*σ TASO's. Group (π_o) is the set of out-of-plane *p*π TASO's, and group (π_i) is the set of in-plane *p*π TASO's. The central atom *s* and *p* ao's and their symmetry likenesses with the TASO's are now determined:

Central ao's   TASO's

Ignoring, in the "first approximation," the interaction of central ao's with the π_i TASO's labeled e_u, you conclude that eight of the 12 *p* TASO's are "non-bonding." Each central atom ao is to be taken in simple sum and difference combinations with the appropriate *p* TASO's to form four bonding and four anti-bonding mo's (three of σ character and one of π_o nature). These combinations are summarized in Figure 4–18.

Species such as ICl_4^- and XeF_4 have 18 valence pairs of electrons, so that all orbitals but the highest two σ* mo's are filled with electrons. This creates the interesting situation of a net of only two pairs of electrons binding the four terminal atoms to the central atom. In fact, inspection of the orbital sketches for the degenerate e_u mo's (the only bonding mo's not offset by electron occupation of the partner

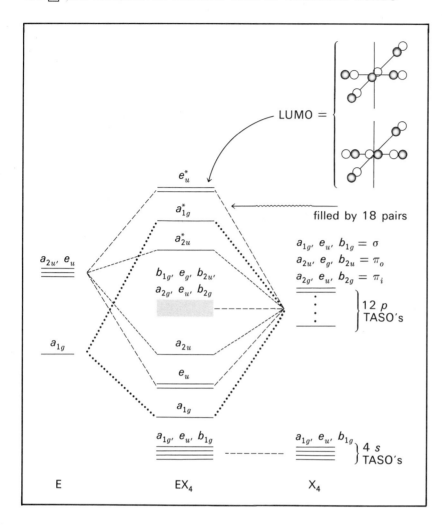

Figure 4-18. The "first approximation" mo's for square planar EX_4, E = non-metal atom.

anti-bonding mo's) reveals that the central-terminal bonding may be viewed as two mutually perpendicular three-center mo systems.[15] Weak bonding seems to be indicated, and this is borne out to some extent by somewhat long E—X bonds and low E—X force constants for ICl_4^- and XeF_4.

---

[15] The hybrid ao model depicts the sigma bonding in the square planar structure as involving dsp^2 hybrids for the central atom. The LCAO model does not require the inclusion of central atom d ao's to account for the bonding, but we can note that the $d_{x^2-y^2}$ ao would strengthen the σ bonding between E and X by stabilizing the b_{1g} TASO (non-bonding in Figure 4–18). Similarly, inclusion of d_{z^2} would somewhat stabilize the σ bonding via the a_{1g} mo. The remaining d ao's act through π bonding, d_{xy} leading to in-plane π bonding by stabilizing the b_{2g} TASO, and d_{xz} and d_{yz} causing out-of-plane π bonding through interaction with the e_g TASO.

STUDY QUESTIONS

27. You are given the following $p\sigma$ TASO definitions:

$$\chi_a = \frac{1}{2}(\sigma_1 + \sigma_2 - \sigma_3 - \sigma_4) \qquad \chi_a' = \frac{1}{\sqrt{2}}(\sigma_1 - \sigma_3)$$

$$\chi_b = \frac{1}{2}(\sigma_1 - \sigma_2 - \sigma_3 + \sigma_4) \qquad \chi_b' = \frac{1}{\sqrt{2}}(\sigma_2 - \sigma_4)$$

(a) Prove the identities

$$\frac{1}{\sqrt{2}}[\chi_a + \chi_b] = \chi'_a$$

$$\frac{1}{\sqrt{2}}[\chi_a - \chi_b] = \chi'_b$$

(b) Make sketches of χ_a, χ_b, χ'_a, and χ'_b, and argue on the bases of nodal structure and number of *cis* and *trans* ao interactions that the set (χ_a,χ_b) is equivalent to the set (χ'_a,χ'_b).

(c) Generate the characters for (χ'_a,χ'_b) under the D_{4h} operations. Give the irreducible representation symbol(s) for the TASO pair.

28. Predict the appearance of the first three photoionization bands of XeF$_4$ (base your answer on the relative, predicted, changes in Xe—F bond distance).

EX$_6$ (O_h)

We are now ready to expand the EX$_4$ case into the EX$_6$ (O_h) case by a procedure analogous to that used to derive the EX$_5$ case from the EX$_3$ case. One very important difference between these two systems needs to be noted. Whereas in the EX$_5$ case the axial and terminal sets of X's were symmetry-distinct, the axial and terminal X's of the O_h case belong to the same symmetry set. Following stratagem item 1, we must construct TASO's for all six X atoms as a group. Some simplification is possible (stratagem item 2), as there will be an *s* TASO set and a *pσ* TASO set with similar symmetry properties. The *pπ* TASO set will be more complex because you must deal with 12 ao's to get 12 *pπ* TASO's (no in-plane *vs.* out-of-plane distinction is possible). Unlike the EX$_3 \to$ EX$_5$ conversion, you must begin by generating the complete set of TASO's by combining the equatorial orbital pattern with the axial pattern.

Beginning with the *pσ* TASO's, you discover that the symmetric combination of axial TASO's has the same symmetry as the nodeless equatorial TASO.

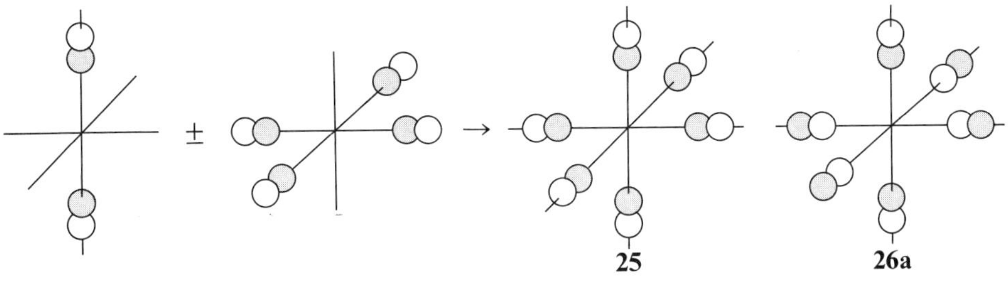

Note there are no nodes in the sum combination (**25**) and there are two nodal planes (above and below the equatorial plane) in the difference combination (**26a**). The anti-symmetric axial combination, since its nodal plane coincides with the equatorial plane, is not to be combined with any of the remaining three equatorial TASO's. The following sketches, however, reveal that the anti-symmetric axial TASO is symmetry-equivalent to the degenerate equatorial TASO's (labeled e_u on page 174).

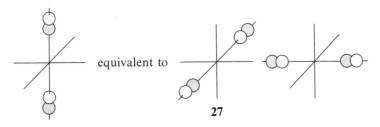

The last two sketches look different from those given on page 174, but remember that functions defining a symmetry-degenerate pair are not unique and that sum and difference combinations of the members of one pair give an equivalent pair (this fact was used in Chapter 1 to derive real atomic p orbitals from the imaginary ones). Study question 27 (preceding this section) takes you through the equivalency transformation. From these sketches, you see again the relation between symmetry, degeneracy, and nodal properties; it is no surprise that the three TASO's **(27)** constitute a *triply* degenerate set (t_{1u}) of TASO's for O_h symmetry.

There remain the two-node equatorial TASO **(26b)**

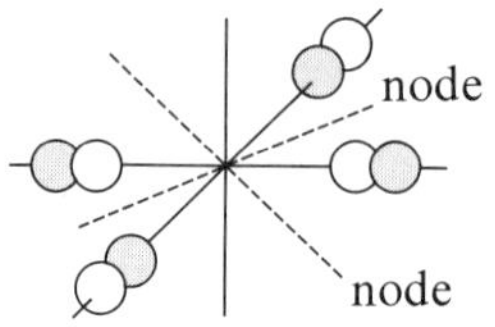

26b

and the 2-node difference combination **(26a′)**

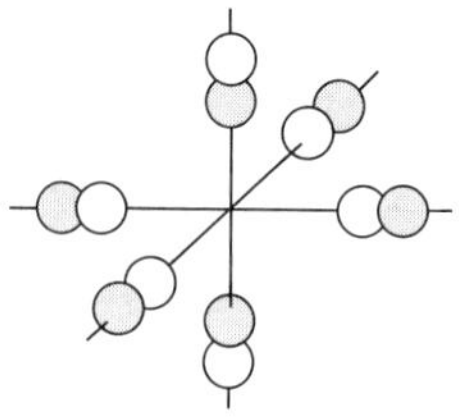

26a′

As you have probably anticipated by now, these TASO's constitute a *doubly* degenerate pair (e_g) for O_h symmetry. They have the same nodal structure as do central atom $d_{x^2-y^2}$ and d_{z^2} ao's, which possess e_g symmetry under the O_h operations (see the Appendix for Chapter 2).

To summarize the s and $p\sigma$ TASO's for the O_h case:

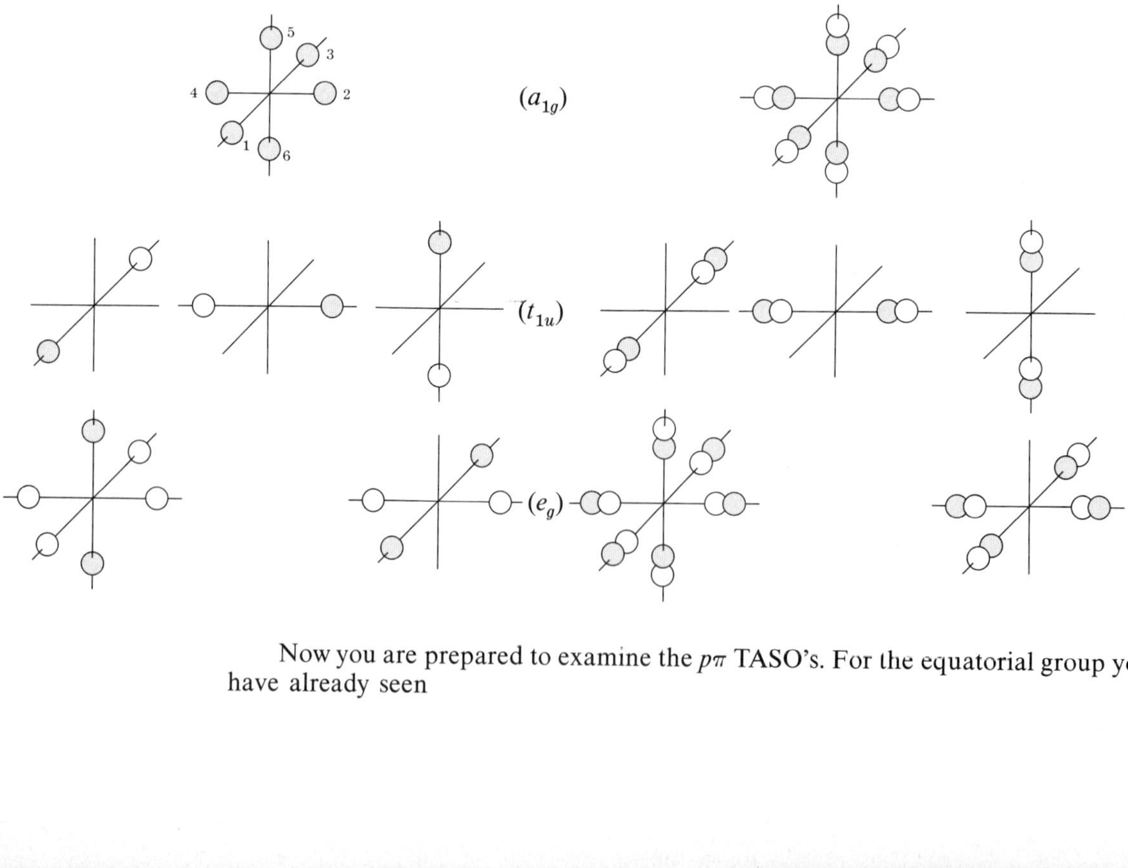

(a_{1g}) 25′

(t_{1u}) 27′

(e_g) 26′

Now you are prepared to examine the $p\pi$ TASO's. For the equatorial group you have already seen

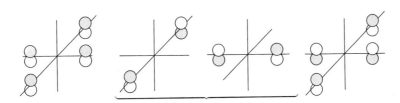

and

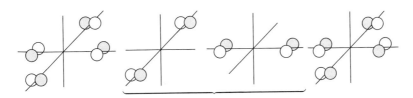

For the axial group there are the simple sum and difference combinations

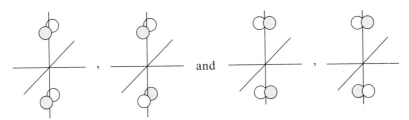

like those of the EX$_5$ case.

You must now determine, as for the s and $p\sigma$ TASO's, which of the four axial combinations have non-zero overlaps with the eight equatorial TASO's. Of 32 possible combinations (!), only four survive the test.

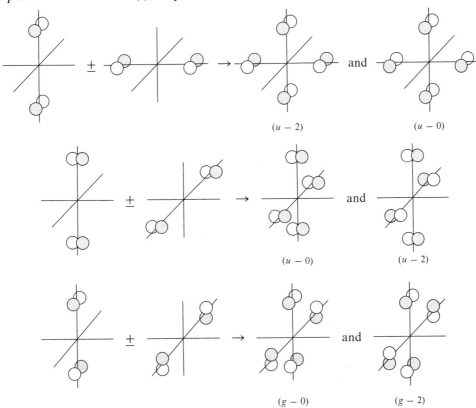

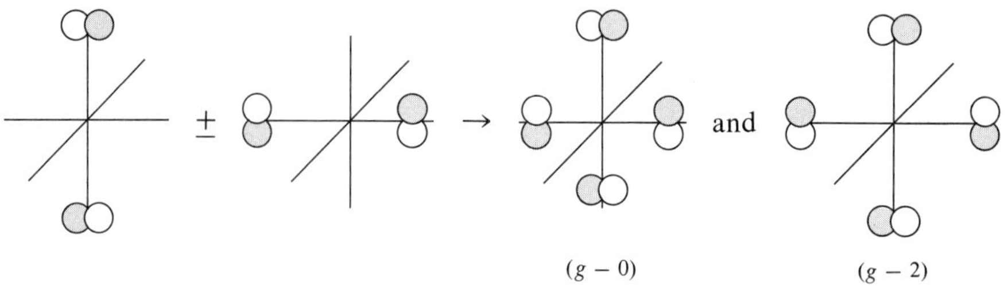

180 ☐ THE UNDIRECTED ORBITAL VIEW OF CHEMICAL BONDS

Left over are the equatorial TASO's

Some way is desperately needed to organize these 12 TASO's. You should suspect that they may be grouped into degenerate sets. Careful inspection reveals that the 12 TASO's can be grouped first according to their symmetry (u or g) with respect to the inversion operation. This produces two groups of six; each group can be further divided into two groups of three according to the useful criterion of number of nodes. The label such as (g-0) for each TASO conveys the even-odd inversion symmetry and the number of nodes between ao's on adjacent atoms. Thus, it is expected (and a detailed group theoretical analysis confirms) that the 12 TASO's may be divided into four sets of *triply degenerate* TASO's.

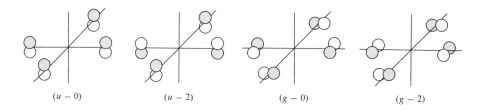

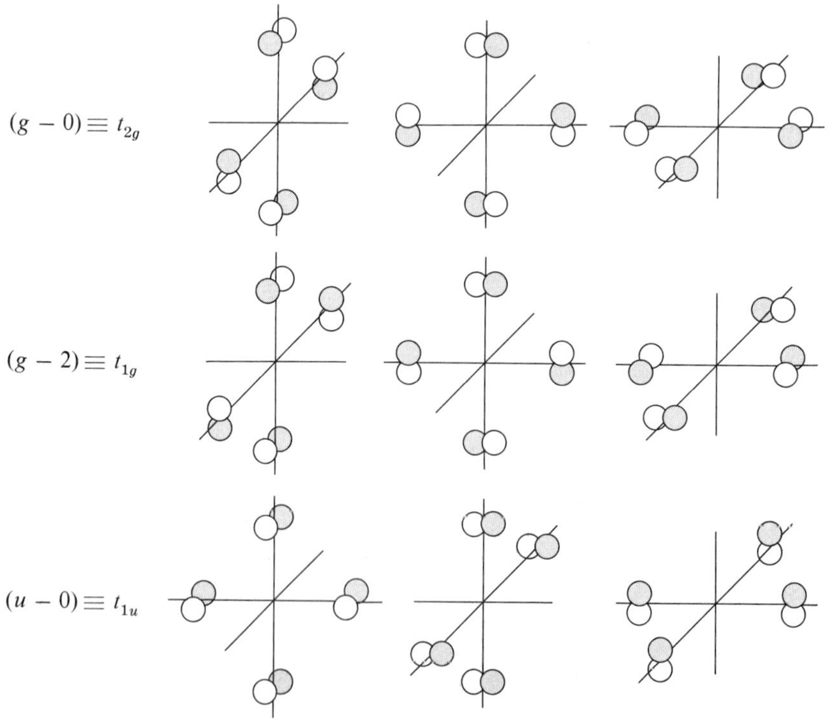

$(g - 0) \equiv t_{2g}$

$(g - 2) \equiv t_{1g}$

$(u - 0) \equiv t_{1u}$

$(u-2) \equiv t_{2u}$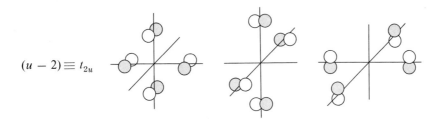

It may be helpful in viewing these TASO's to observe that in each the four p ao's either lie in (g-symmetry) or out of (u-symmetry) the plane of their four nuclei; furthermore, don't miss the fact that those planes are at right angles to each other in each set (the planes are the cartesian planes). It will help now to drop the rather crude inversion-node symbolism and use the proper irreducible representation symbols for these TASO's. They are

$$(g\text{-}0) = t_{2g}$$
$$(g\text{-}2) = t_{1g}$$
$$(u\text{-}0) = t_{1u}$$
$$(u\text{-}2) = t_{2u}$$

Before constructing a qualitative orbital energy diagram for the O_h structure, recall that the s and $p\sigma$ TASO's each formed three sets:

Relative ao phasing

$(1 + 2 + 3 + 4 + 5 + 6) = a_{1g}$	See **25'**
$\begin{matrix}(1-3)\\(2-4)\\(5-6)\end{matrix} = t_{1u}$	See **27'**
$\begin{matrix}(1+2+3+4-5-6)\\(1-2+3-4)\end{matrix} = e_g$	See **26'**

The next step is to examine the central atom valence orbitals for non-zero overlaps with the TASO's. The central s ao matches symmetry with the a_{1g} TASO's; the central p_x, p_y, and p_z ao's match symmetry with the t_{1u} TASO's (three sets, one from s ao's, one from $p\sigma$ ao's, and one from $p\pi$ ao's). The s and $p\sigma$ TASO's of e_g symmetry are non-bonding, as are the $p\pi$ TASO's of t_{2g}, t_{1g}, and t_{2u} symmetry.

Making the "first approximation" and (as in the EX$_3$ and EX$_5$ (D_{3h}) cases) regarding the central atom t_{1u} orbital interactions with the $p\sigma$ TASO's of t_{1u} symmetry to be dominant over the interaction with the t_{1u} $p\pi$ TASO's, you should arrive at the diagram in Figure 4–19. The basic features of Figure 4–19 are not much altered by recognizing (i) that all the s TASO's find higher-lying orbitals with which to interact, and (ii) that the t_{1u} π TASO's generated by the terminal $p\pi$ ao's generally interact with the bonding and anti-bonding t_{1u} σ mo's. The orbital sketches, on the other hand, can become quite complex; all a_{1g}, t_{1u}, and e_g mo's display terminal atom s/p mixing while the t_{1u} mo's reveal mixed $p\sigma/p\pi$ character.

A particularly simple view of the relation between the EX$_4$ and EX$_6$ cases is that the introduction of two X atoms along the z axis stabilizes the out-of-plane π mo of the EX$_4$ case (a_{2u}) to an extent to make it degenerate with the formerly e_u σ bonding pair, the set of three mo's now being called t_{1u}.

For molecules like SF$_6$ (24 valence pairs), you can confirm from the O_h energy level diagram that there remain four net bonding pairs to describe all six S—F bonds, in apparent contradiction with the *assumptions* of the Lewis formulation that there

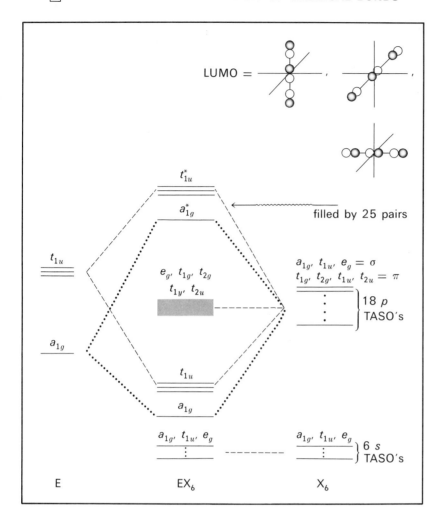

Figure 4-19. The simplified orbital energy diagram for EX_6 (O_h), where E = non-metal atom.

are six S—F bond pairs. Inclusion of sulfur $3d$ ao's brings the mo description closer to that assumed in Chapter 2.[16]

We might pause here to note that studies[17] of the valence shell isoelectronic series SF_6^-, SeF_6^-, TeF_6^-, ClF_6, BrF_6, and IF_6 reveal the odd electron to reside in an mo of considerable central atom s character (this is deduced from the large Fermi contact interaction between the odd electron and the nuclear magnet, as discussed in Chapter 3). This finding is consistent with Figure 4-19, in which the odd electron is to reside in the s^* (a_{1g}^*) mo. *The σ^* electron* apparently is not stereochemically active, whereas the σ^* *pair of electrons* of O_h XeF_6 becomes stereochemically active in a fluxional sense (as discussed in Chapter 2 and in problem 11 of this chapter).

---

[16] The hybrid ao scheme depicts the central atom as d^2sp^3 hybridized. Inclusion of central atom d ao's in the LCAO scheme leads to stabilization of the "non-bonding" e_g and t_{2g} TASO's of Figure 4-19; the d_{z^2} and $d_{x^2-y^2}$ have e_g symmetry and lead to enhanced σ bonding (there are now six σ bonding mo's involving the central atom s, p_x, p_y, p_z, $d_{x^2-y^2}$ and d_{z^2} ao's), while d_{xz}, d_{yz}, and d_{xy} (of t_{2g} symmetry) strengthen E—X π bonding.

[17] A. R. Boate *et al.*, *Inorg. Chem.*, **14**, 3127 (1975).

29. Based on bond order interpretations from Figures 4-18 and 4-19, compare bond strengths of XeF_4, SF_6, and $SbCl_6^{3-}$.

30. In anticipation of a discussion of transition metal complexes in Chapter 9, consider the central atom in EX_4 (planar) and EX_6 (octahedral) to use valence d orbitals rather than s and p. Apply the stratagem used in the LCAO section to determine the qualitative mo description of the EX_4 and EX_6 structures. To check that you have the d ao's and TASO's properly grouped, compare the d orbital symmetry labels from the D_{4h} and O_h character tables with those given in this chapter for the TASO's.

31. Consult the appropriate character tables and confirm that the central atom valence s and p ao's have the representation symbols assigned to them in the discussions of EX_3 (planar), EX_4 (tetrahedral and planar), EX_5 (trigonal bipyramidal), and EX_6 (octahedral). (Hint: what are the symmetry symbols for $x^2 + y^2 + z^2, x, y, z$?)

STUDY QUESTIONS

THE EQUIVALENCE OF THE LOCALIZED AND DELOCALIZED MODELS

At the end of the last chapter on the localized pair model for molecular electronic structure, we remarked that the LCAO-MO method would appear to be a radically different model but that the similarity would be explained later. Actually, at several points along the way in developing the mo method, you discovered that the hybrid (s/p mixing) character of the TASO's arose naturally after dropping the "first approximation." Only when the molecular symmetry was sufficiently low that the central atom s and p_z ao's were of common symmetry did central atom s/p mixing arise.

In fact, it has been known for some time that the LCAO orbitals that are based on molecular symmetry properties are not the only ones that can be used to characterize a molecular electronic wavefunction. It is possible to derive an alternate set of molecular orbitals from the symmetry-based orbitals of this chapter by quantifying the concept of the VSEPR model. That is, one takes linear combinations of the occupied, symmetry-based molecular orbitals in such a way as to minimize the coulombic repulsion between electrons in different orbitals.[18] This method produces *localized orbitals,* that is, lone pair mo's that are highly localized (of high probability) on an atom and bond pair mo's that are localized between two atoms. By way of illustration, we will consider the molecules H_2O and C_2H_4.

For H_2O, symmetry-adapted molecular orbitals can be derived from the linear triatomic orbital pattern as illustrated in Figure 4-20. It is quite apparent from the mo sketches (after dropping the "first approximation") at the center of the figure that both OH bond pair orbitals are fully delocalized over all three atoms, and one oxygen lone pair, while identifiable, is delocalized onto the hydrogen atoms. The one truly lone pair, the pair assigned to the oxygen p orbital perpendicular to the molecular plane in Figure 4-20, does not contain oxygen s character. The VSEPR model, on the other hand, has us envision the oxygen atom as being approximately tetrahedrally hybridized, with two oxygen lone pairs containing some 25% s character.

By taking sum and difference combinations, for example, of the two *bonding* orbitals (a_1 and b_2) as shown near the bottom of Figure 4-20, you can see how localized OH bond orbitals can be created from the symmetry-adapted molecular orbitals. By using a computer to actually find[19] the proper linear combinations of all

---

[18] C. E. Edmiston and K. Ruedenberg, *J. Chem. Phys.,* **43,** S97 (1965).
[19] K. F. Purcell *et al.,* Program 198.1, Quantum Chemistry Program Exchange, University of Indiana, Bloomington, Ind.

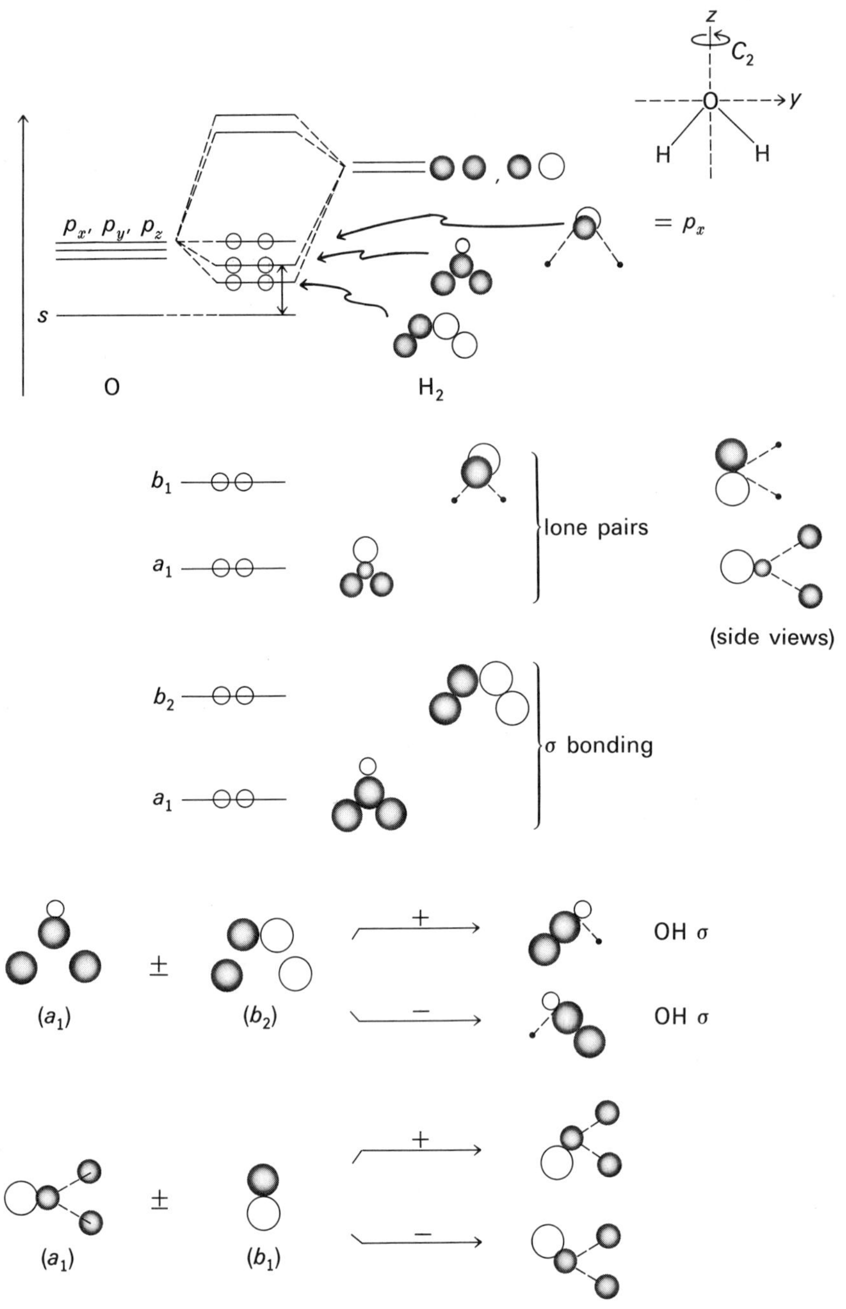

Figure 4-20. The symmetry-based and approximate localized bond and lone pair orbitals of H_2O.

four symmetry-based molecular orbitals that exhibit minimum repulsions between electrons in different orbitals, one obtains two localized OH bond orbitals and two localized equi-oxygen lone pair orbitals. Approximate shapes for the "rabbit-ear" lone pair mo's can be derived by taking sum and difference combinations of the two symmetry-based "lone pairs" (bottom, Figure 4-20).

The molecule C_2H_4 presents the interesting case of a molecule with a pi bond. Here the localized orbitals are four CH orbitals and two CC orbitals of the bent or "banana bond" variety; this is an orbital model, taught many years ago and now regaining legitimacy, based on two carbon tetrahedra sharing an edge! The "banana" bonds arise from sum and difference of the pi mo and a localized CC σ orbital.

$$\text{CC}_\sigma \quad \pm \quad \text{CC}_\pi$$

STUDY QUESTIONS

32. What are the irreducible representation labels for the two delocalized mo's of H_2O described as O—H bonding? Are they symmetry-equivalent? Sketch them. From a localized orbital view, there are also two O—H bond mo's. Sketch them. Are they symmetry-equivalent? Can you label each alone with a C_{2v} irreducible representation symbol? Can you form + and − combinations of them so as to produce mo's that have irreducible symmetries? Sketch these combinations and compare them with those in the first part of this question.

33. Refer to the text discussion of BeH_2, and combine the two occupied symmetry mo's ($s\sigma$ and $p\sigma$) so as to form symmetry-equivalent localized mo's. Based on sketches of these mo's, which mo ($s\sigma$ or one of the localized mo's) incurs greater repulsion between electrons within the mo? Do you expect the repulsion between an electron in $s\sigma$ and an electron in $p\sigma$ to be greater or less than that between one electron in one localized mo and one electron in the other localized mo?

34. Refer to Table 3-2 and the entries for the sp^3 (T_d) structure. By writing four localized EH bond pair waves as

$$b_1 = a\text{te}_1 + b\phi_{H1}$$
$$b_2 = a\text{te}_2 + b\phi_{H2}$$
$$b_3 = a\text{te}_3 - b\phi_{H3}$$
$$b_4 = a\text{te}_4 + b\phi_{H4}$$

write out the linear combination $b_1 + b_2 + b_3 + b_4$ as a function of $\phi_s, \phi_x, \phi_y, \phi_z$, and ϕ_H. Make a sketch of this mo. Now compare this function with those of Figure 4–16 for EH_4 so as to identify the ($b_1 + \cdots$) combination function with one of the symmetry mo's.

EPILOG

A primary goal of this chapter has been to draw your attention to the fact that the practicing chemist has at his disposal a second approximate view of electron orbitals in molecules. It is important that you recognize that the "difference" between localized and delocalized orbital models is superficial when no approximations are made. Both models are widely used because some experiments are more directly interpreted in a qualitative way in the language of one model than in terms of the other. Mathematically, any experimental observation can be approached as accurately by one model as the other. Experimental phenomena such as molecular structure, Fermi contact coupling, dipole moments, force constants, and bond energies are most often couched in the language of localized pair orbitals. Electronic spectra, ionization phenomena, and reaction mechanisms are usually discussed more readily in the language of delocalized mo's. Nevertheless, earlier in this chapter you saw how the delocalized mo model could be used for basic molecular structure analysis, and that model's corroboration of the VSEPR assumption that central atom electron pairs are stereochemically active. Proceeding through the remainder of this text, you can use with confidence whichever orbital model most directly helps you understand the reactions, structures, and other properties of molecules (whether inorganic or organic).

INTRODUCTION

SURVEY OF ADDUCT TYPES

   Acidic and Basic Hydrogen
   Non-Metal Acids and Bases
      Boron and Aluminum
      Carbon and Silicon
      Nitrogen and Phosphorus
      Oxygen and Sulfur
      Halogens
      Xenon
   Metals

ACID-BASE STRENGTHS

   The Thermodynamic Definition
   Quantitative Prediction of Relative Adduct Stabilities

   Illustrative Interpretations
      Proton Affinities
      $(CH_3)_3N$ and $(SiH_3)_3N$
      Picolines
      Et_3N and $HC(C_2H_4)_3N$
      $Me_3N \cdot MMe_3$, $M = B, Al, Ga, In$
      π Resonance Effects
      BF_xMe_{3-x}
      Retrodative Bonding

AN ACID-BASE VIEW OF SOLVATION PHENOMENA

   Liquid Ammonia
   Hydrofluoric and Sulfuric Acids
   Sulfur Dioxide
   HSO_3F and Superacids

EPILOG

5. THE DONOR/ACCEPTOR CONCEPT

The impact of Lewis's definition of the acid and base and the concept of the electron pair bond has been felt throughout all of chemistry. Often an understanding of nucleophilic and electrophilic reaction steps depends on knowledge of "adduct-like" formation of intermediates or activated complexes, and in favorable instances the intermediates are even sufficiently stable to be isolated. In fact, proposed mechanisms are often judged reasonable or unreasonable against a background of donor/ acceptor chemistry. Another area of great practical impact by the donor/acceptor concept is compound solubility and the choice of a solvent for chemical synthesis. Later in this chapter we will explore these solvation ideas, and a later chapter deals with mechanisms of reactions. The fact that this topic has been chosen to precede all but the atomic and molecular structure discussions stems from its unifying value to the study of "inorganic" reactions. Now that you have a background in molecular structure and HOMO/LUMO concepts, we begin a rewarding journey into the reactions of molecules.

INTRODUCTION

Few concepts have been so enduring as the concept of acids and bases. The development of the concept of intermolecular interactions from Arrhenius' restrictive original idea [acids are sources of protons (H^+) and bases are sources of hydroxide ion (OH^-)] through Brönsted's generalization[1] [acids are proton sources (HA) and bases are proton consumers (A^-, the conjugate base of HA)] and Lewis' idea[2] of the electron pair bond (acids are electron pair acceptors, while bases are electron pair donors) to the modern concept (acids are molecules with a relatively low-energy LUMO and bases are molecules with a fairly high-lying HOMO) illustrates most dramatically how chemical concepts evolve over a period of years. Nowadays we recognize that Brönsted's acid is itself an adduct (for example, HCN) and generally can act as a Lewis acid via hydrogen bonding or by proton transfer to the conjugate base of a weaker acid:

$$H-C\equiv N: + Et_2O: \rightleftarrows Et_2O \cdots HCN$$

HCN also happens to be an example of a Lewis adduct that can act as a Lewis base:

$$H-C\equiv N: + SbF_5 \rightarrow HCN \cdots SbF_5$$

---

[1] J. N. Brönsted, *Recl. Trav. Chim. Pays-Bas*, **42**, 718 (1923).
[2] G. N. Lewis, "Valence and the Structure of Molecules," The Chemical Catalogue Co., New York (1923).

188 ☐ THE DONOR/ACCEPTOR CONCEPT

To help introduce you to the scope of the acid-base concept and how the detection of new chemical species by modern instrumental techniques has added even greater depth to its basis, we will begin this chapter with a sampling of examples of acids, bases, and adducts from the various groups of the periodic table. We forecast what is to be found with the generalizations contained in the following chart.

Type	Acid	Base
hydrogen	protonic (characteristically bound to electronegative atoms)	hydridic (characteristically bound to electropositive atoms)
non-metals	coordinatively unsaturated (may exhibit valence shell expansion)	generally non-bonding electron pairs
metals	coordinatively unsaturated (may exhibit valence shell expansion)	generally non-bonding electron pairs (low oxidation states, organometals)

As you examine each of the samples to follow, ask yourself the following questions:

Which atom(s) is the acceptor and which is the donor?
In which category in the chart does each example fit?
Are the structures of the acid molecular fragment and the donor fragment in the adduct much different from their structures as separate entities?
Is the presence of an acceptor orbital (LUMO) in the acid (or the donor orbital, HOMO, in the base) readily apparent for the separated fragment or is that orbital more apparent when the fragment assumes the structure it is to have in the adduct?

SURVEY OF ADDUCT TYPES

ACIDIC AND BASIC HYDROGEN[3]

In addition to "complete" proton transfer reactions such as

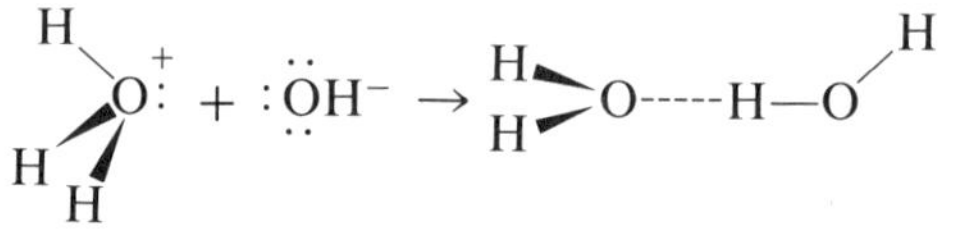

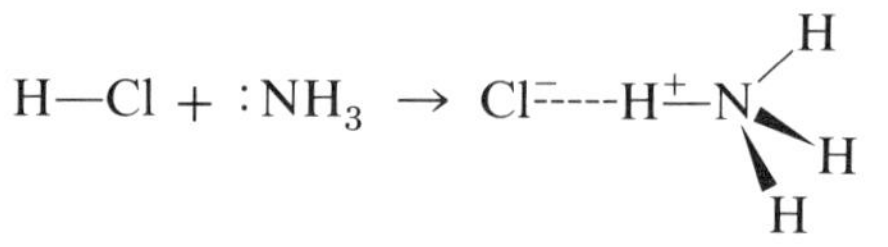

$$H_3O^+ + :OH^- \rightarrow H_2O\text{-----}H\text{---}OH$$

$$H\text{---}Cl + :NH_3 \rightarrow Cl^-\text{-----}H^{\pm}\text{---}NH_3$$

we must also include the hydrogen bonding interaction further exemplified by

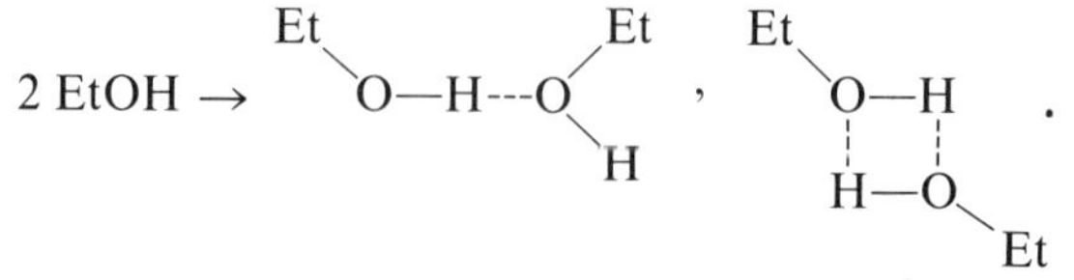

$$2\ EtOH \rightarrow \underset{Et}{O}\text{---}H\text{---}\underset{Et}{O}\text{---}H \quad , \quad \underset{Et}{O}\text{---}H \cdots H\text{---}\underset{Et}{O}$$

---

[3] Good general references on the subject of hydrogen bonds are: G. C. Pimentel and A. L. McClellan, "The Hydrogen Bond," Freeman, San Francisco (1960). W. C. Hamilton and J. A. Ibers, "Hydrogen Bonding in Solids," Benjamin, New York (1968).

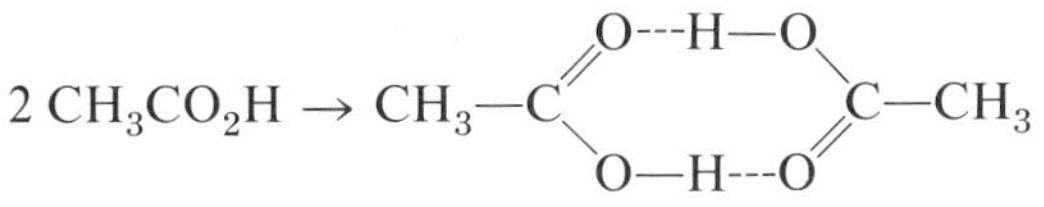

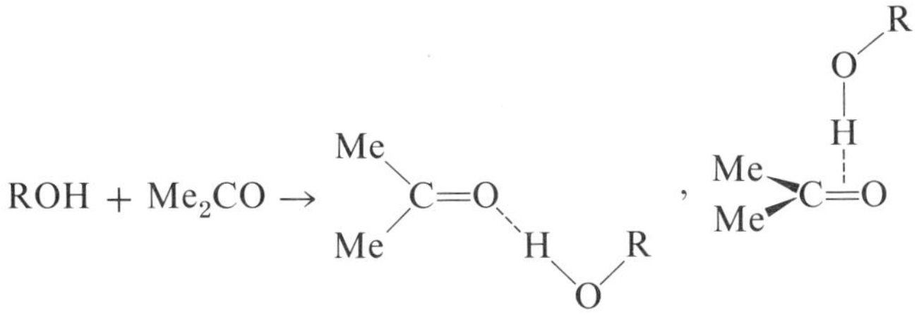

$$2\ CH_3CO_2H \rightarrow CH_3-C\begin{matrix}O---H-O\\ \diagup\quad\quad\quad\diagdown\\ O-H---O\end{matrix}C-CH_3$$

$$ROH + Me_2CO \rightarrow \begin{matrix}Me\\ \diagdown\\ Me\end{matrix}C=O\cdots H-O-R \quad , \quad \begin{matrix}Me\\ \diagdown\\ Me\end{matrix}C=O\cdots H-O-R$$

The HOMO/LUMO interaction in the adducts from "complete" proton transfer is usually easy to identify. For the proton the LUMO, of course, is the H $1s$ atomic orbital. The HOMO depends on the base, but most often is an sp^n lone pair. The adduct bond is a strong, essentially two-center/two-electron bond. The more novel bonding description arises with the so-called hydrogen bond. Again the HOMO of the donor is often an sp^n hybridized lone pair (however, see the later discussion of π bonding pairs in a donor role). The LUMO of the hydrogen-bonding acid must be the AH σ^* molecular orbital. We must not neglect the AH bonding orbital, however, and the hydrogen bond (see study question 15 in Chapter 4 on FHF^-) entails a three-center/four-electron situation. An important characteristic of most hydrogen bonds is that they are asymmetric (unlike FHF^-); that is, the proton is not symmetrically placed between donor atoms.

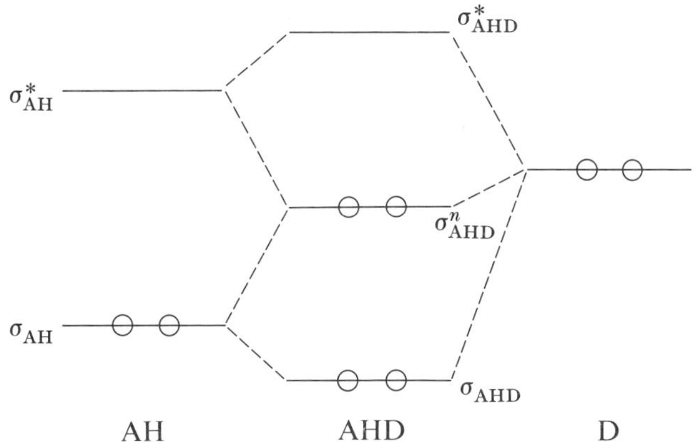

There are several important chemical features embodied in the mo diagram for the asymmetric hydrogen bond adduct. First let's look at the sketches for the AH and donor orbitals.

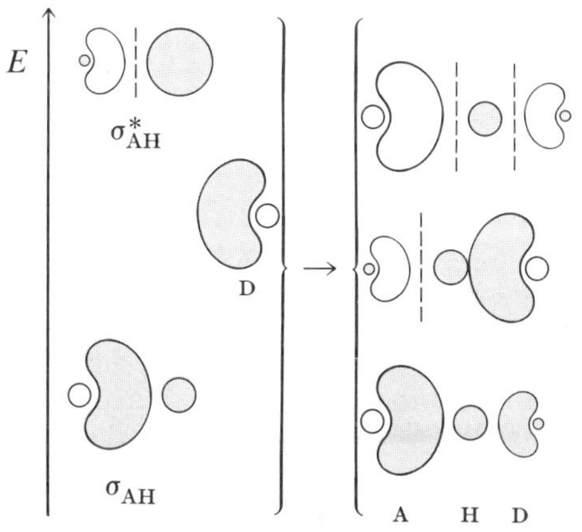

It is obvious that the donor orbital overlaps both the σ_{AH} and σ^*_{AH} mo's; the lowest energy orbital is stabilized by the overlap, while the highest energy orbital is destabilized, as you fully expect from the last chapter. Were it not for the σ^*_{AH} mo, the hydrogen bond interaction would be nonexistent, for the stabilization of the AH pair would be canceled by the destabilization of the donor pair. The important role of the σ^*_{AH} mo is thus seen to be the stabilization of the donor electron pair. We would predict that the characteristic of A, in AH, that maximizes the interaction of σ^*_{AH} with the donor orbital would be high electronegativity, and this arises in two ways. The greater the electronegativity of A, the more polarized will be σ_{AH} toward A and the smaller will be the overlap of that mo with the donor orbital; conversely, the greater will be the polarization of σ^*_{AH} toward the H atom and the greater will be the overlap between σ^*_{AH} and the donor orbital. The net result will be less destabilization of the donor pair by interaction with σ_{AH} and greater stabilization by interaction with σ^*_{AH}. The second influence A has on the interaction of σ_{AH} and σ^*_{AH} with the donor orbital is through the energies of those orbitals relative to that of the donor orbital. Higher A electronegativity means that both σ_{AH} and σ^*_{AH} will appear at lower energy, with poorer energy match between the donor and σ_{AH} orbitals but with better energy match between σ^*_{AH} and the donor orbital. The experimental observation that more electronegative substituents in HA favor its action as a hydrogen-bonding acid finds a logical interpretation in terms of the HOMO/LUMO concept.

These considerations also lead us to explanations of other important properties of the hydrogen bond. First of all, the bonds are weak as bonds go; typical values for the enthalpy of hydrogen bond formation fall in the range from 1 to 15 kcal/mole. Because of the limited covalent energy of many hydrogen bonds, and the fact that the AH bond and donor lone pair dipoles are large, it is generally important to recognize that dipole-dipole attractions make a relatively large contribution to the hydrogen bond energy. A second important property of hydrogen bonded adducts is the greatly decreased AH stretching frequency (and longer bond length) in the adduct AHD than in free AH. This is expected because the electron pair in σ_{AH} becomes delocalized over A, H, and D, and the donor lone pair orbital takes on A-H anti-bonding character (see mo sketches above). Finally, you might expect to find that the hydrogen bond is not highly directional. While the donor orbital overlap with σ_{AH} and σ^*_{AH} would be maximum with a linear AHD arrangement, "bending" the hydrogen bond with an angular AHD arrangement might not be too detrimental. In fact, both structural types are found. In the ethanol dimer example,

$$\begin{array}{c} \text{Et} \\ \backslash \\ \text{O}-\text{H} \\ | | \\ \text{H}-\text{O} \\ \backslash \\ \text{Et} \end{array}$$

we find a bent hydrogen bond resulting from the structural constraint imposed on the formation of two hydrogen bonds.

The occurrence of such dimers (and even higher oligomers) of alcohols is made interesting by the fact that alcohol viscosities and boiling points are related to the extent of such "self-association." In seeming contradiction to the preceding discussion of the effect of electronegativity on the strength of the hydrogen bond, greater electronegativity of the alkyl group causes *less* self-association. With *heterosystems*, greater electronegativity of the alkyl group causes greater association between the alcohol and the donor; but for *self-association*, where the donor atom is the oxygen atom of another alcohol molecule, the donor character of the oxygen atom is reduced as the substituent increases in electronegativity, causing a decrease in self-association.[4] This follows from the fact that the oxygen is directly bound to the substituent.

---

[4] A. D. Sherry and K. F. Purcell, *J. Amer. Chem. Soc.*, **94**, 1853 (1972).

The withdrawal of charge from the oxygen by the substituent serves to expose the oxygen lone pairs to greater effective nuclear charge; this leads to a contraction of the lone pair orbitals and to a lowering of the energy of the donor electrons (the former effect results in a smaller overlap, and the latter causes a poorer energy match between the donor electron pair and σ_{AH}^*).

In the acetic acid dimer

$$CH_3-C\underset{O-H---O}{\overset{O---H-O}{\diagdown\hspace{-0.5em}\diagup}}C-CH_3$$

we find two linear, asymmetric hydrogen bonds between carboxylic acid functional groups. The linearity of the O—H---O structure is constrained by the structure of the carboxyl groups. In a series of RCO_2H, there is again an inverse correlation between acidity and self-association because of the greater sensitivity of the oxygen basicity than the hydrogen acidity to the alkyl group electronegativity. A noteworthy feature of the carboxylic acid dimer hydrogen bond is that its stability is greater than that of the corresponding alcohol dimer (for example, ethanol vs. acetic acid). The acetyl substituent on the OH group of acetic acid is certainly more electronegative than the ethyl group of ethanol, and this causes the OH group of the acid to function as a stronger acid. That the OH oxygen atom of the acid is a poorer donor than its alcohol counterpart is of little consequence, since in the acid it is the carbonyl oxygen group that functions as the donor atom.

The third example of hydrogen bonding ($ROH + Me_2CO$) is interesting for several reasons. First, there are no structural constraints on the angle of the oxygens about the acid proton; second, using a ketone as an example of a Lewis base, we have an opportunity to note that evidence has been cited[5] for the Lewis base character of the carbonyl π bonding pair of electrons. In this sense, and at least toward hydrogen-bonding acids, the carbonyl group can behave as a π donor as well as a σ donor (by using one of the lone pairs about the carbonyl oxygen). It was not too many years ago that such an idea would have been viewed with skepticism, but now such behavior is more readily accepted (we will see more examples of π donors when we consider aromatic organic molecules) and points up the necessity of dropping Lewis' original concept of Lewis bases as acting through lone pair electrons only. We see here the first instance of what are normally considered to be *bond pair electrons acting in a Lewis base capacity* (more examples will follow).

Hydridic hydrogen, meaning hydrogen bound to an electropositive or low electronegativity element, quite frequently acts as a donor atom in intermolecular interactions. In this capacity hydrides may be added to that class of molecules for which bond pair electrons may be involved in adduct formation. A very important example arises in the dimerization of borane,[6] wherein we see the BH_3 molecule acting as both an acid (through the boron atom) and as a base (through the hydrogen atom).

$$H_3B + BH_3 \rightarrow B_2H_6$$

[diborane(6)
colorless gas
spontaneously flammable
b.p. = −93°C]

---

[5] P. V. R. Schleyer, as cited by M. S. Nozari and R. S. Drago, *J. Amer. Chem. Soc.*, **94**, 6877 (1972).
[6] The enthalpy for the dimerization is ∼ −35 kcal/mole; K. Wade, "Electron Deficient Compounds," Nelson, London (1971).

B_2H_6 could therefore be considered an adduct. As a simple extension of these ideas, we may view BH_4^- as an adduct with the capacity to act as a donor, as the following examples[7] illustrate.

$Al(BH_4)_3$ [structure shown] $\begin{bmatrix} \text{liquid} \\ \text{m.p.} = -65° \\ \text{b.p.} = 45° \end{bmatrix}$

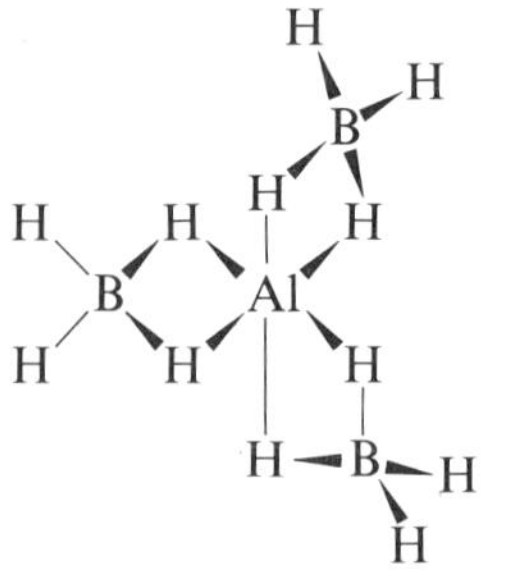

$BH_4^- + \tfrac{1}{2} B_2H_6 \rightarrow$ [structure shown]

$BH_4^- + Cl-Ag(PPh_3)_3 \rightarrow (Ph_3P)_2Ag(H_2BH_2) + Cl^- + PPh_3$

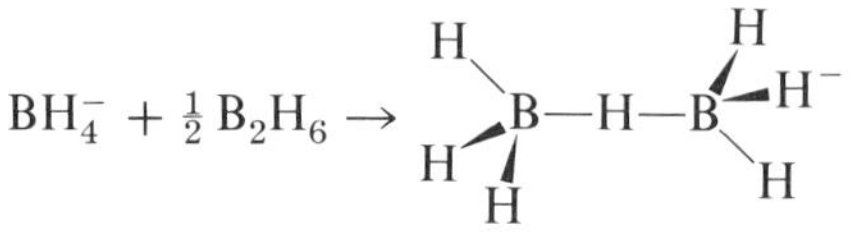

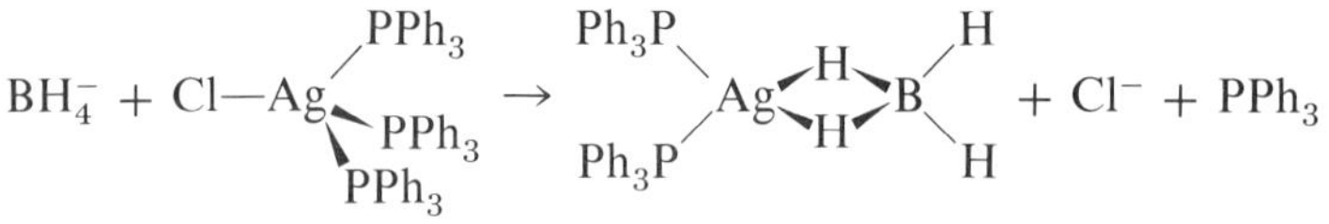

$Al(BH_4)_3$ and $(Ph_3P)_2AgBH_4$ are especially interesting and worth remembering because they illustrate the dibasic or bidentate ability of BH_4^-.

In these examples we encounter the so-called three-center bridge bond, a recurring unit in stable inorganic structures. In a molecule like $B_2H_7^-$ the bridging hydrogen atom is symmetrically positioned and the molecular point group is D_{3d}; that is, the boron atoms are symmetry-equivalent. To describe the $B-H_b-B$ bonding we may proceed as follows. Each $B(H_t)_3$ unit has mo's much like those of NH_3. The H_b 1s ao principally overlaps the BH_3 LUMO's, which individually look a great deal like the lp (HOMO) of NH_3.

$H_3B \circ)$ (H) $(\circ BH_3$

At this point, the overlapping ao situation bears a strong resemblance to the BeH_2 structure discussed in Chapter 4, with the qualification that the central atom here (H_b) possesses only the single s valence ao. Alternately, you should recognize this problem as one of the "three orbital linear chain" type. Accordingly, the $B-H_b-B$ mo's appear as

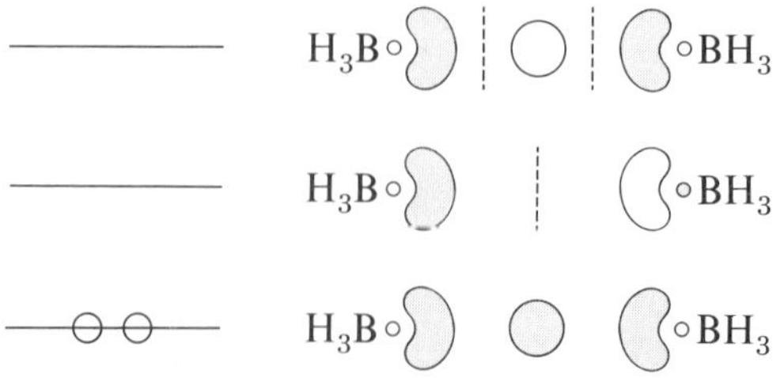

---

[7] B. D. James and M. G. H. Wallbridge, *Progr. Inorg. Chem.*, **11**, 99 (1970); S. J. Lippard and K. M. Melmed, *Inorg. Chem.*, **6**, 2223 (1967); E. B. Baker, et al., *J. Inorg. Nucl. Chem.*, **23**, 41 (1961).

After accounting for the six B—H_t bond electron pairs, there remains a single pair to occupy the bonding mo of the B—H_b—B unit. Such a situation is called "three-center/two-electron" bonding. Note that the "bridge bond" corresponds to a bond order of $\frac{1}{2}$ for each boron. You should not miss the obvious parallel of this mo pattern to that of the hydrogen bond on page 189, a "three-center/four-electron" case.

To describe the double bridge, let us take the most symmetric example to develop a qualitative model. B_2H_6, following the same tack taken for $B_2H_7^-$, may be considered as two BH_2 units doubly bridged by hydrogen atoms. These BH_2 units are structurally analogous to H_2O. The two lp orbitals of H_2O are vacant in $B(H_t)_2$ and are oriented so as to overlap well with the two bridging hydrogen ao's.

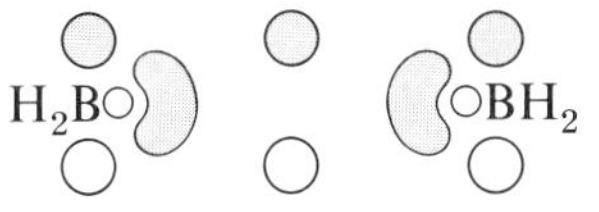

By forming sum and difference combinations constrained to symmetry-equivalent ao's, we arrive at

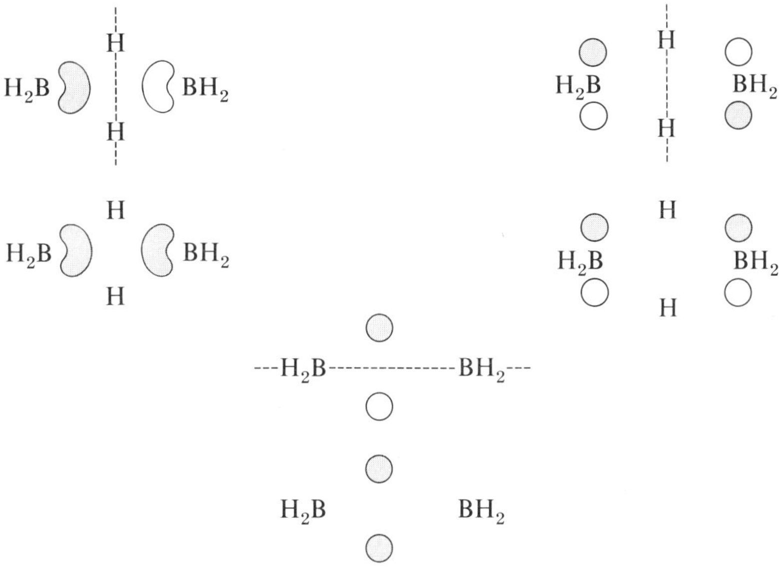

Comparing the $(H_b)_2$ mo's with those of the $(BH_2)_2$ unit, you should readily identify the following combinations as having $S \neq 0$:

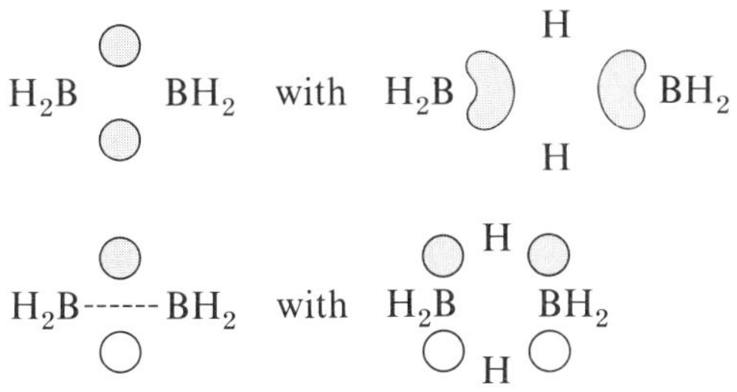

The orbitals

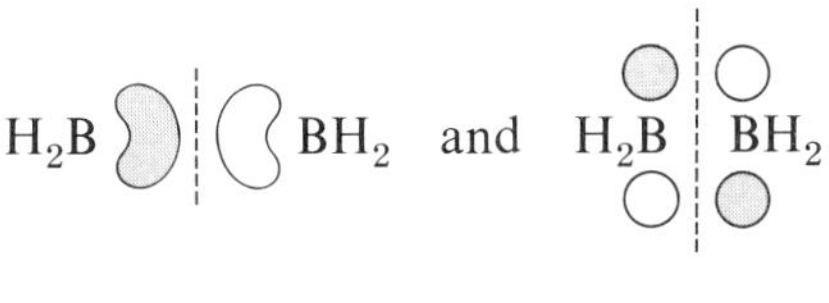

194 ☐ **THE DONOR/ACCEPTOR CONCEPT**

are non-bonding as far as the $(H_b)_2$ ao's are concerned. The orbital level diagram and mo sketches therefore take the form

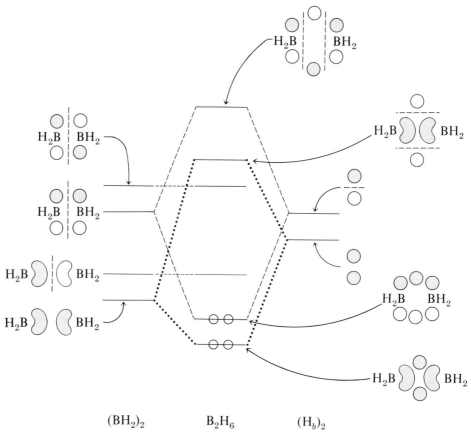

$\quad\quad\quad\quad (BH_2)_2 \quad\quad\quad B_2H_6 \quad\quad\quad (H_b)_2$

Allowing for four valence electron pairs to describe the four B—H_t bonds, there remain two pairs to occupy bridge mo's. That is, there are two "four-center/two-electron" bonds in B_2H_6: one derives from boron $p\pi$ type ao's and the other from boron "sp" type ao's. A localized mo view of these two pairs is derived (see the end of Chapter 4) from their sum and difference (to minimize repulsions between electron pairs in the two symmetry mo's).

The obvious similarity of these three-center/two-electron bond mo's to that of $B_2H_7^-$ should not go unnoticed.

NON-METAL ACIDS AND BASES

Boron and Aluminum

Probably the most familiar examples of Lewis acids come from the compounds formed by these elements. Normally trivalent, these elements do not thereby satisfy the "octet rule," which in modern terminology means that a valence shell orbital is unoccupied and of fairly low energy (a stable LUMO). This orbital plays the role of the acceptor mo or ao. The acid-base concept nicely accommodates many of the chemical and structural features of this class of molecules. In the next chapter we will examine more closely the tendency of organo- and hydrido-boranes and the lighter organo- and hydrido-alanes to dimerize with the characteristic of bridging H and C atoms.[8] Similarly, the heavier halides of aluminum (all solids) dimerize with bridging halogens[9] but, interestingly, AlF_3 is a polymeric solid (m.p. = 1290°) with six fluorides about each aluminum (see also the next chapter).

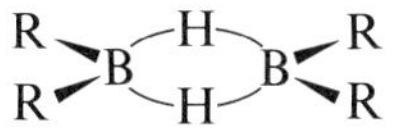

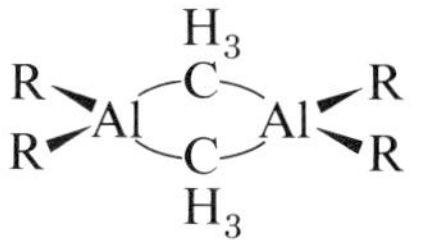

As the boron halides do not dimerize like those of aluminum, an intramolecular acid/base interaction, described as π bonding between boron and halogen, seems to inhibit formation of the B_2X_6 dimer.

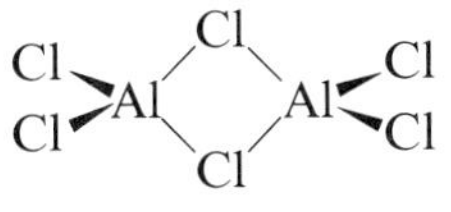

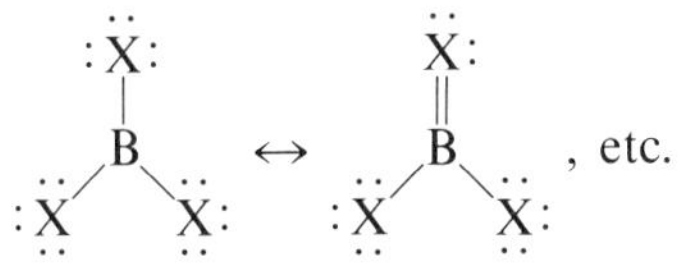

This intramolecular "adduct" formation is of great importance to the ability of the boron halides to function as Lewis acids in general. In terms of the molecular orbital diagram for the π ao interaction, we see that an otherwise stable, vacant boron p ao is transformed into a higher lying vacant π^* mo (the LUMO of BX_3).

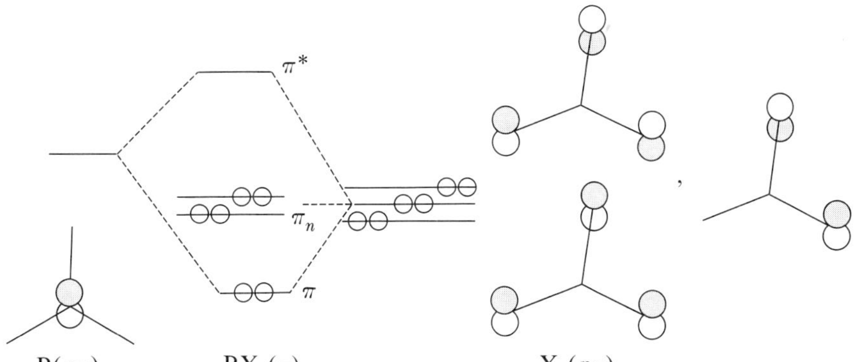

---

[8] The dimerization enthalpy of $AlMe_3$ has been determined to be −20 kcal/mole; C. H. Henrickson and D. P. Eyman, *Inorg. Chem.*, **6**, 1461 (1967). With CH_3 bridging groups, one still may envision three-center/two-electron bonding; comparing CH_3 with hydrogen, an ∼sp^3 carbon hybrid takes the place of the $1s$ ao of hydrogen. See also J. C. Huffman and W. E. Streib, *Chem. Comm.*, 911 (1971).

[9] Unlike H and CH_3, the halides possess additional lone pair electrons. Thus, a three-center mo description is not *required* here, but is still possible. Accordingly, the halogen may be viewed, like CH_3, to provide an sp^n ao with an electron for an analogous "no-node" three-center mo; the halogen p ao lying in the bridging plane possesses a node coincident with the "one-node," "non-bonding" combination of aluminum hybrids. Consequently, the "one-node" mo becomes bonding (hybrid$_1$ + halogen p − hybrid$_2$) and is occupied by a pair of electrons. Two bonding electron pairs are delocalized over three centers, and this is equivalent to the localized orbital view of two separate halogen-aluminum bond pairs.

196 ☐ THE DONOR/ACCEPTOR CONCEPT

Nevertheless, it appears safe to say that these dimerization and intramolecular donor/acceptor interactions are not generally effective in achieving "saturation" of the chemical valency of boron and aluminum; the trivalent compounds of both elements are known to be readily involved in adduct formation with other Lewis base molecules.

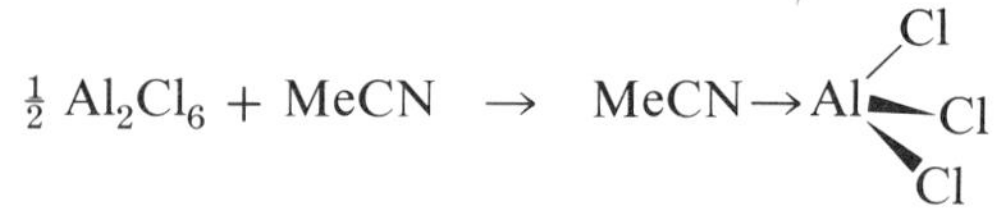

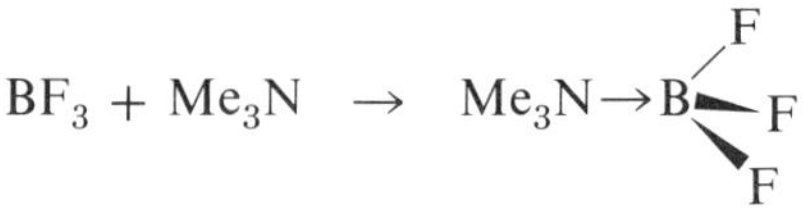

$$\tfrac{1}{2}\,Al_2Cl_6 + MeCN \rightarrow MeCN \rightarrow AlCl_3$$

$$BF_3 + Me_3N \rightarrow Me_3N \rightarrow BF_3$$

Carbon and Silicon

Among the strongest known Lewis acids are the carbonium ions[10] so important in organic chemistry. You will recognize a resemblance of many of these compounds to isoelectronic boron acids, but of course the greater nuclear charge of carbon renders the carbonium ions particularly effective as acids (but very poor as donors; to wit, CH_3BH_2 dimerizes but $CH_3CH_2^+$ apparently does not). The high acidity of carbonium ions is illustrated by the reaction

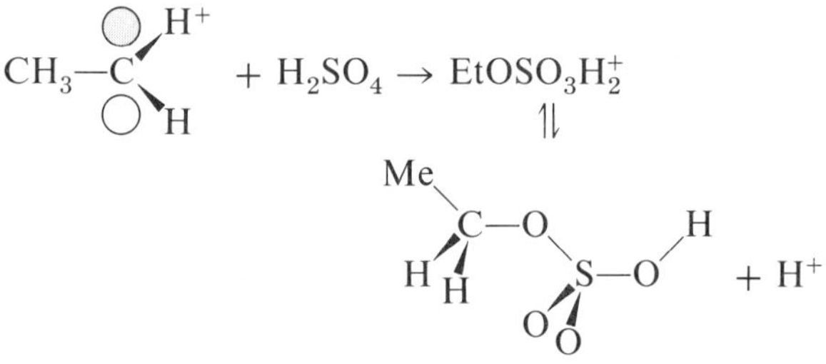

$$CH_3-C(OH^+)(OH) + H_2SO_4 \rightarrow EtOSO_3H_2^+$$

$$\updownarrow$$

(Me)(H)(H)C-O-S(O)(O)-O-H + H^+

Normally, the oxygens of H_2SO_4 are weakly basic, but the carbonium ion is such an avid acid that complexation occurs.

An important class of carbonium ions, similar to the boroxines, $(RBO)_3$, except that they do not tend to polymerize, are the acylium ions. Formed from carbonyl halides and a strong halide ion acceptor, these species are very reactive and can serve as precursors for substitution at the carbonyl carbon.

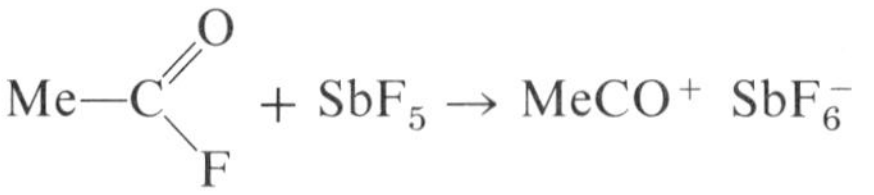

$$Me-C(=O)F + SbF_5 \rightarrow MeCO^+\ SbF_6^-$$

$$Me-\overset{\oplus}{C}\equiv O: \leftrightarrow Me-\overset{\oplus}{C}=\ddot{O}:$$

If your imagination is working for you, you have no doubt noticed that CH_3CO^+ could itself be considered an adduct between CO and CH_3^+ (not so long ago, before the acceptance of carbonium ions, such a view would have seemed preposterous; but in this enlightened age the viewpoint has value in predicting the structure and reactivity of CH_3CO^+).

---

[10] G. A. Olah and P. V. R. Schleyer, Eds., "Carbonium Ions," Vols. I–IV, Interscience, New York (1968).

Perhaps more startling than carbonium ions as acids are the acids and bases of conjugated carbon compounds. These are the π acids and bases referred to earlier in the outline of hydrogen bond interactions. For example, the aromatic phenyl ring of anisole is known[11] to perturb the OH function by acting as a π donor.

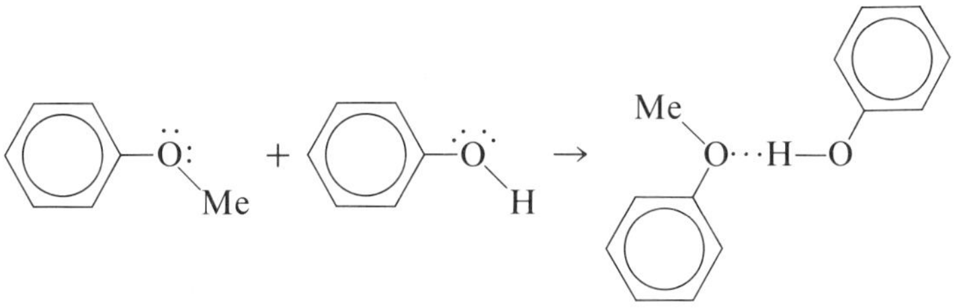

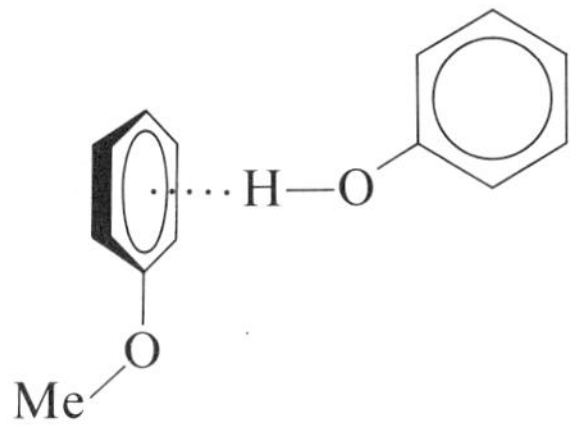

Benzene itself is known to form weak adducts with acceptors like tetracyanoethylene. The structure of the adduct is believed to be that of parallel benzene and TCNE planes with C_{2v} symmetry.

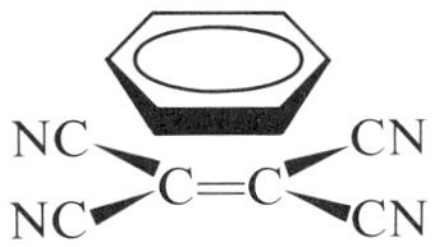

Study of the electronic structures of such organic π-acid/π-base adducts has aroused considerable interest.[12] A characteristic feature of the stronger adducts is that they can be isolated as solids, with the acid-base molecules alternately layered upon one another in chains. These solids have the intriguing property of being anisotropic conductors of electricity, current flowing easily normal to the molecular planes but negligibly in directions parallel to the aromatic planes (in this sense, they are one-dimensional organic "metals"). That the compounds are often highly colored means that a low-energy electronic transition is possible (in the visible region) and suggests their use in photoelectric devices.

In Chapter 4, the π HOMO of benzene was depicted as a pair of degenerate mo's possessing a single node each. LCAO calculations for TCNE reveal a reasonably stable LUMO with similar nodal properties. The interaction between these as donor-acceptor mo's is held responsible for the formation of the adduct.

top view of C_6H_6 HOMO's

---

[11] B. B. Wayland and R. S. Drago, *J. Amer. Chem. Soc.,* **86,** 5240 (1964).
[12] R. Foster, "Organic Charge-Transfer Complexes," Academic Press, New York (1969).

198 ☐ THE DONOR/ACCEPTOR CONCEPT

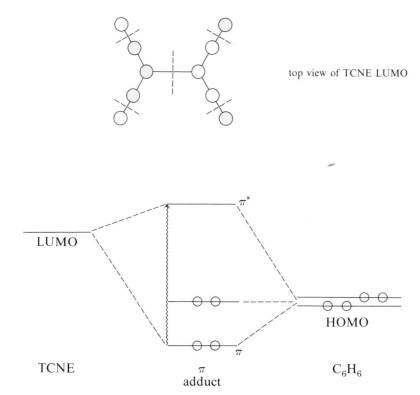

top view of TCNE LUMO

As reflected by the diagram, the delocalization of HOMO electrons into LUMO is weak by virtue of the fact that LUMO energy is higher than HOMO energy.

A photon-induced electronic excitation is indicated by the vertical wavy line. Because the absorption process causes *transfer of an electron from an mo with greater amplitude on the donor to an mo with greater amplitude on the acceptor, the transition is called a charge transfer absorption—conveying the idea of photoinduced transfer of charge from one fragment to another.* When the energy gap between LUMO and HOMO is not too large, the absorption band appears in the visible region, in which case the adduct will be colored. With deference to this occurrence of a "low" energy electronic transition, such adducts are often referred to as "charge transfer" complexes.

Finally, we should mention that Cu^+ and Ag^+ ions (among others) are known[13] to form adducts with olefins and acetylenes. Many stable solids have been isolated and found to have structures in which the metal ion is symmetrically positioned with respect to the carbons defining the olefin or alkyne link.‡ The LUMO in such cases would be

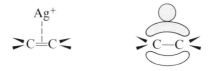

the metal valence s atomic orbital, while the HOMO is the π orbital of the hydrocarbon. This orbital overlap situation is somewhat unusual (but see the many examples discussed in the later chapter on bonding in organometallic compounds)

---

[13] C. D. M. Beverwijk, *et al., Organometallic Chem. Revs., A*, **5**, 215 (1970).
‡ More on this subject in Chapter 16.

and is sometimes referred to as a μ-bond. An interesting application of these adducts is their formation in aqueous solutions of silver salts by extraction of olefins and alkynes from mixtures with saturated hydrocarbons.

Before leaving the survey of carbon donors, we must recognize that carbanions in general should be very effective donors. Probably the most stable carbanion is cyanide, and it is known to function as a donor in many reactions[14] such as

$$BH_3 \cdot THF + CN^- \rightarrow BH_3CN^- \xrightarrow{BH_3 \cdot THF} BH_3CNBH_3^-$$

$$2Me_3SiBr + 2CN^- \rightarrow Me_3SiCN + Me_3SiNC + 2Br^-$$

and particularly with metal ions (you will see such complexes throughout Chapters 10 to 14 on transition metals).

Although much weaker in its donor ability than CN^- (because oxygen has greater nuclear charge than nitrogen), CO is known to form a wide variety of adducts, in particular with organometallic species; some examples beyond the oxocarbonium ions mentioned before are

$$\begin{bmatrix} \text{colorless liquid} \\ \text{b.p.} = 43° \end{bmatrix} \quad \begin{bmatrix} \text{yellow liquid} \\ \text{b.p.} = 103° \end{bmatrix} \quad \begin{bmatrix} \text{colorless solid} \\ \text{sublimes } in \text{ } vacuo \end{bmatrix}$$

For many years, it was puzzling to inorganic chemists that metal atoms (which did not seem likely candidates for good Lewis acids) should form adducts with CO (recognized to be a rather poor donor toward most other Lewis acids; *e.g.*, BF_3 and SbF_5 do not form adducts with CO). The LCAO model helps to explain the stability of these adducts and focuses our attention on a two-way acid-base interaction between the metal and CO. You will see that the metal synergically interacts with CO, with the metal acting as an acid *and* a base while CO simultaneously displays donor *and* acceptor character.

In Chapters 9 and 15 we will give a fuller account of the orbital diagram for these compounds; for the present purposes a simplified version will suffice. Counting valence electrons and orbitals for $Cr(CO)_6$ as an example, we find nine atomic orbitals (the 3d, 4s, and 4p aos) and nine electron pairs (three from the Cr atom and, with a carbon lone pair from each CO, six from the ligands). Reference to the EX_6 diagram of Chapter 4 will convince you that the six metal ao's directed toward the six CO's (to be specific, the $d_{x^2-y^2}$ and d_{z^2} ao's of e_g symmetry, the s ao of a_{1g} symmetry, and the p_x, p_y, and p_z ao's of t_{1u} symmetry) all find σ TASO's of the same symmetry with which to form bonding and anti-bonding combinations. The six bonding mo's of the σ type so formed accommodate six of the nine valence electron pairs. The remaining three metal ao's, d_{xz}, d_{yz}, and d_{xy}, are at this point non-bonding ao's of t_{2g} symmetry and contain the remaining three electron pairs. In this sense it should come as no surprise that a compound such as $Cr(CO)_6$ could exist.

---

[14] J. R. Berschied, Jr., and K. F. Purcell, *Inorg. Chem.*, **9**, 624 (1970); R. C. Wade, *et al.*, *ibid*, **9**, 2146 (1970); E. C. Evers, *et al.*, *J. Amer. Chem. Soc.*, **81**, 4493 (1959).

200 □ THE DONOR/ACCEPTOR CONCEPT

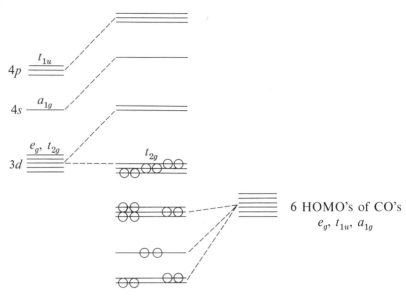

If we were to stop here, we would miss one of the more important aspects of the metal-carbonyl bonding. We need to ask whether there exist any CO TASO's with non-zero overlap with the d_{xz}, d_{yz}, and d_{xy} metal ao's. If there are, we need to form bonding and anti-bonding combinations from them, fill in the electrons, and ask whether there is net stabilization or destabilization of the molecule from those interactions. In fact (refer to Chapter 4), there are possible $p\pi$ TASO's of t_{2g} symmetry. In the examples of Chapter 4, the terminal atoms were just that, atoms with $p\pi$ ao's. Here we have a slightly more complicated set of $p\pi$ TASO's because the carbon $p\pi$ ao's have already been combined with the oxygen $p\pi$ ao's to form π_{CO} and π^*_{CO} mo's for the diatomic molecule. Nevertheless, the CO π and π^* mo's retain the symmetry, as viewed from the metal, that the carbon p ao's have. Thus, we conclude that there exists a set of π TASO's for the CO molecules of t_{2g} symmetry and a set of π^* CO TASO's of the same symmetry. These are sketched below.

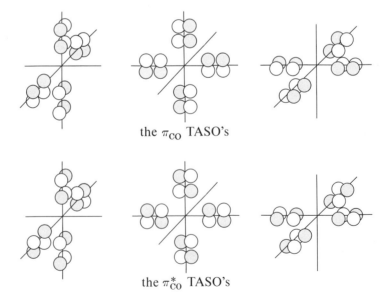

By focusing your attention upon the carbon $p\pi$ ao's in these TASO's, you readily see the relation between these sketches and the simpler ones in Chapter 4 (p. 180). Now, form the linear combinations of these TASO's with the metal t_{2g} ao's (in the order of the TASO sketches above, these are d_{xz}, d_{yz}, and d_{xy}). Aside from the fact that we are dealing with triply degenerate sets of mo's, you should notice a similarity to the energy diagram for hydrogen-bonding acids (here it is the "donor" CO that has both bonding and anti-bonding mo's to be considered, whereas with AH · D it was the acid that had the low-energy HOMO and higher-energy LUMO).

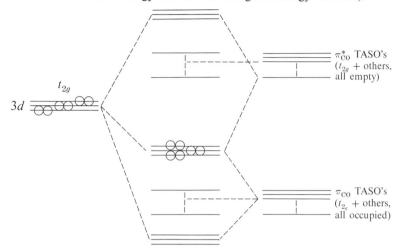

This situation is characterized by the interesting fact that the metal $d\pi$ ao's play the role of HOMO's while the CO π^* TASO's are cast in a role of LUMO's. The interaction between the metal ao's and CO π TASO's has the effect of destabilizing the metal ao's, whereas the interaction between the metal ao's and the CO π^* TASO's has the effect of stabilizing the metal ao's. In analogy with the hydrogen bond analysis, which effect is more important depends on the relative extents of interaction, metal $\leftrightarrow \pi_{CO}$ or metal $\leftrightarrow \pi^*_{CO}$, and this depends on the relative energy matches and overlaps. Because oxygen is more electronegative than carbon, the occupied π_{CO} orbitals of the ligands will be polarized toward oxygen (i.e., the π_{CO} electrons have smaller probability on carbon than on oxygen). This means that the overlap of the metal ao's with the π_{CO} TASO's at the carbon atoms will be somewhat small. On the other hand, the π^*_{CO} mo's are polarized toward the carbon and should experience greater overlap with the metal ao's. Generally, it is found that the stable π_{CO} mo's will be further in energy from the metal ao's than are the π^*_{CO} mo's, although this is difficult to justify in qualitative terms. The upshot of all this is that we expect, and find, greater interaction of the metal ao's with the π^*_{CO} mo's; and the metal ligand $d\pi$-$p\pi$ bonding has a stabilizing effect on the molecule.

To recap the interaction between the metal and the CO's, we find the bonding to be of the normal variety with each ligand supplying a pair of electrons in forming six σ bonds to the metal. The interesting twist is that the metal $\leftrightarrow$ ligand π interaction is of a "retro" sense relative to the sigma interaction—the metal appears as a π donor and the CO ligands appear as π acids. This *back-bonding* or *retro-bonding* increases the metal-ligand bond order and confers greater stability upon the metal-carbon bonds than would be attainable from the σ interaction alone. Another important example of CO as a donor occurs in its reaction with diborane, and you should not miss the analogy with the isoelectronic acylium cation mentioned above.[15]

$$:CO: + \tfrac{1}{2} B_2H_6 \rightarrow \begin{array}{c} H \\ H \nearrow B-CO \\ H \end{array} \begin{bmatrix} \text{gas} \\ \text{m.p.} = -137° \\ \text{b.p.} = -64° \end{bmatrix}$$

---

[15] Of course, the analogy cannot be taken too far because the greater nuclear charge of carbon than boron confers greater strength to the C—C bond, renders the hydrogens much less hydridic, and means that CH_3CO^+ is a cation whereas BH_3CO is a gas at room temperature and is considerably dissociated.

The other elements of the fourth group also exhibit acid and base properties. As with third and later row elements of the main groups, we find for silicon the ability to expand the "octet" to form five and six coordinate adducts. This, of course, is a primary feature of the reactions of silicon with nucleophiles; in Chapter 7 we will examine in some detail the nature of the five-coordinate intermediate occurring in nucleophilic displacements at silicon. Significant in this regard has been the synthesis and isolation of a 1:1 adduct of SiF_4 (b.p. = $-86°$) and F^-. The anion SiF_5^- was unknown until it was stabilized[16] in crystalline form with a quaternary ammonium cation of like size and charge (this highly useful principle will be developed in Chapter 6).

$$SiF_4 + [R_4N]F \rightarrow [R_4N][SiF_5]$$

An important, stable adduct of silicon, occurring as an undesirable product from use of excess HF when synthesizing SiF_4 from SiO_2, is hexafluorosilicic acid and its Brönsted conjugate base, SiF_6^{2-}:

$$SiF_4 + 2HF \rightarrow H_2SiF_6$$

which also appears in the hydrolysis of SiF_4

$$SiF_4 \xrightarrow{H_2O} \tfrac{1}{2} SiO_2 + \tfrac{1}{2} SiF_6^{2-} + HF + H^+$$

Another element from this group that has been studied extensively is tin. A large variety of adducts in which tetravalent Sn becomes five-coordinate are known; the trigonal bipyramidal structure[17] of Me_3SnCl adducts is believed to feature the Cl and donor atoms in apical positions (recall the electronegativity/pair repulsion arguments of Chapter 3 for the preferred trigonal bipyramidal isomer).

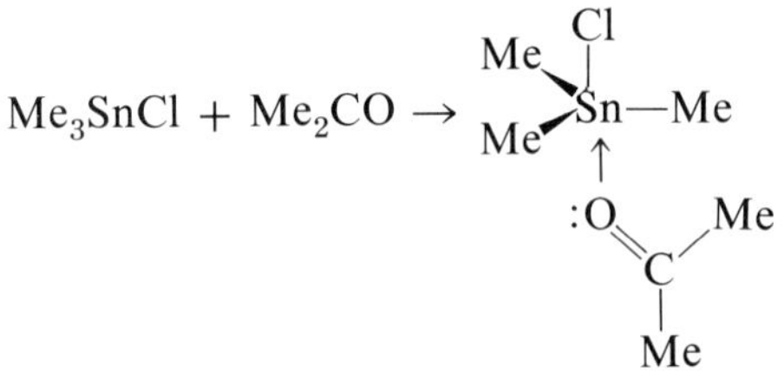

As examples of Si and Sn species with basic properties, we can cite the anions R_3M^-. In Chapter 8 we will examine the utility of the former as a nucleophile for synthesis of unsymmetrically substituted organo-silanes. There it will be noted that R_3Si^- is capable of displacing Cl^- from alkyl chlorides. The trichlorostannate ion is also a useful nucleophilic base[18] and the reaction, featuring formation of a metal-metal adduct bond,

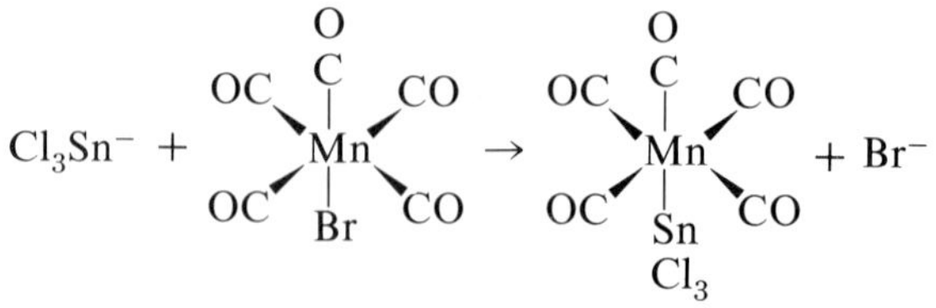

---

[16] H. C. Clark, et al., *Inorg. Chem.*, **8**, 450 (1969); F. Klanberg and E. L. Muetterties, *Inorg. Chem.*, **7**, 155 (1968); J. J. Harris and B. Rudnor, *J. Amer. Chem. Soc.*, **90**, 575 (1968).
[17] T. F. Bolles and R. S. Drago, *J. Amer. Chem. Soc.*, **88**, 3921, 5730 (1968).
[18] J. F. Young, *Adv. Inorg. Chem. Radiochem.*, **11**, 91 (1968).

may be simply viewed as a displacement of one donor (Br⁻) from an adduct ($Mn(CO)_5Br$) by another. This interesting reaction is one of many leading to metal-metal bonds and will be mentioned again in Chapter 16 on organometallic compounds and in Chapter 18.

The trichlorostannate ion may itself be regarded as an acid-base adduct between Cl^- and $SnCl_2$. Diamagnetic stannous chloride is a "violator" of the "octet rule" and could act as an acid or base (draw an electron dot structure for $SnCl_2$). $SnCl_3^-$ is one example of the many adducts in which the Sn atom acts as an acid. With organic solvents, 1:1 adducts of the general formula $SnCl_2 \cdot S$ are common.[19] Unfortunately, no examples can be cited at this time of $SnCl_2$ acting solely as a Lewis base. Perhaps it is not surprising that such adducts are difficult to form, if you recall the orbital contraction effects for elements immediately following the transition element series (the effective nuclear charge concept as discussed in Chapter 1) and the valence promotion ideas related to increased stability of lower valent compounds of Tl and Sn (as developed in Chapter 3). At any rate, we can cite[20] one example of $SnCl_2$ acting as a donor and acceptor in the same compound, via an "insertion" reaction:

$$(C_5H_5)(CO)_2Fe\text{—}Fe(CO)_2(C_5H_5) + SnCl_2 \rightarrow \underset{Cl}{\overset{Cl}{>}}Sn\underset{Fe(CO)_2(C_5H_5)}{\overset{Fe(CO)_2(C_5H_5)}{<}}$$

Here the $SnCl_2$ "inserts" into the Fe-Fe linkage, acting, in effect, as an acid toward an anion, $(C_5H_5)Fe(CO_2)^-$, and as a base toward a cation, $(C_5H_5)Fe(CO_2)^+$. This behavior is reminiscent, with good reason, of that of the isoelectronic carbenes (R_2C) of organic chemistry. Many more examples of this reaction type are explored in Chapter 17.

Nitrogen and Phosphorus

This family affords many common examples of Lewis donors (the trivalent compounds of nitrogen and phosphorus) and a few less widely appreciated examples of Lewis acids (the pentavalent compounds of phosphorus). The amines and ammonia are among the most commonly recognized and generally the strongest of the nitrogen donors. It was noted in the introduction that HCN, like the organic nitriles, is known also to act as a weak donor. As a class, the pyridines are of intermediate basicity and complete the correlation of lone pair hybridization with donor strength, a correlation to be more fully explored in the next section. As a striking example of a substituent effect on basicity, we might note that NF_3 (b.p. = −129°) is a notoriously weak amine donor but that it does, like other amines, participate in an oxidation reaction with oxygen atoms that is reminiscent of the Lewis interaction[21]:

$$F_3N + \frac{1}{2}O_2 \xrightarrow[\text{dischg.}]{\text{elec.}} F_3N\text{—}O$$

Similarly, trivalent phosphines undergo this reaction to form the highly stable (in both the thermodynamic and kinetic senses) phosphine oxide link (see Chapter 6).

A particularly interesting phosphine donor, to be encountered again in the study of organometallic compounds in Chapters 16 and 17, is PF_3. While the reaction with

---

[19] R. J. H. Clark, *et al.*, *J. Chem. Soc. (A)*, 2687 (1970); J. D. Donaldson, *et al.*, *J. Chem. Soc. (A)*, 2928 (1968).
[20] D. E. Fenton and J. J. Zuckerman, *Inorg. Chem.*, **8**, 1771 (1969); F. Bonati and G. Wilkinson, *J. Chem. Soc.*, 179, 1964; R. D. Garisch, *J. Amer. Chem. Soc.*, **84**, 2486 (1962); J. F. Young, *Adv. Inorg. Chem. Radiochem.*, **11**, 91 (1968).
[21] W. B. Fox, *et al.*, *J. Amer. Chem. Soc.*, **92**, 9240 (1970).

diborane is more or less conventional, the ability of gaseous PF_3 to convert solid nickel metal to the gaseous tetrakis(trifluorophosphine) nickel complex

$$PF_{3(g)} + \frac{1}{2} B_2H_{6(g)} \rightarrow F_3P-BH_{3(g)}$$

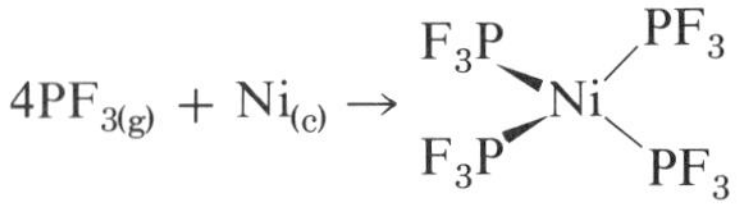

$$4PF_{3(g)} + Ni_{(c)} \rightarrow \begin{array}{c} F_3P \diagdown \diagup PF_3 \\ Ni \\ F_3P \diagup \diagdown PF_3 \end{array}$$

is quite remarkable. Normally one might not think of nickel atoms as being very good Lewis acids. In fact, CO (as noted in the previous section) and PF_3 share this unusual ability to form MD_n complexes.[22]

An interesting aspect of PCl_5 chemistry is that, while the vapor is monomeric, the solid is a crystalline salt (sublimes at 162°) and thereby illustrates an important class of acid-base behavior—the halide transfer reaction. It is interesting to speculate that lattice energies have a great deal to do with this

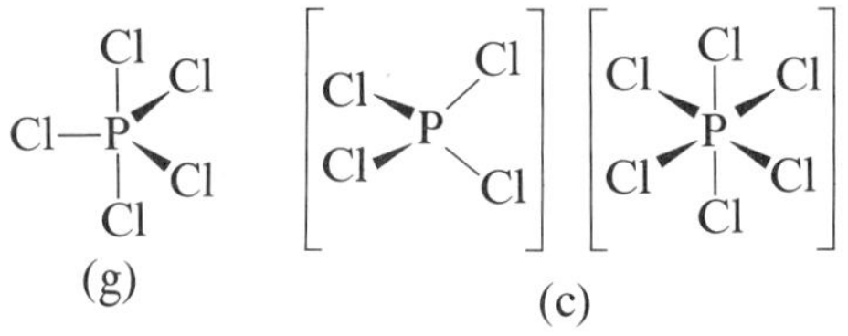

property of PCl_5, because PF_5 (b.p. = −75°) does not form an ionic solid (but PF_6^- salts are known). Complementary to this aspect of PCl_5 as a Lewis acid is the occurrence[23] of the simple 1:1 adduct $PCl_5 \cdot py$. Important examples of halide ion transfer arise particularly with pentavalent antimony (recall the reaction of CH_3COF).

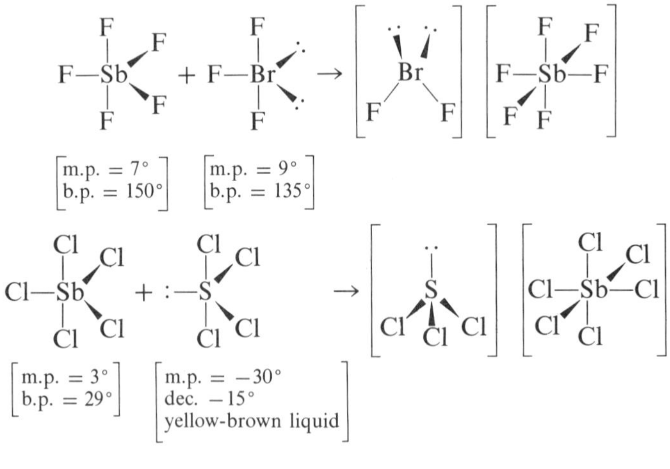

With other reagents, such as arsenic trifluoride, the fluoride transfer is incomplete and bridging fluoride structures often result. AsF_2^+ should be, particularly as an

---

[22] H. J. Plastas, et al., *J. Amer. Chem. Soc.*, **91**, 4326 (1969); J. C. Green, et al., *Chem. Comm.*, 1121 (1970); I. H. Hillier, et al., *Chem. Comm.*, 1316 (1970).

[23] I. R. Beattie and M. Webster, *J. Chem. Soc.*, 1730 (1961); I. R. Beattie, et al., *J. Chem. Soc. (A)*, 2772 (1968).

electron "deficient" cation, more highly acidic than SbF_5 (compare AsF_2^+ with BrF_2^+ and SCl_3^+ with regard to the "octet" rule).

$$:-\underset{\underset{F}{|}}{\overset{\overset{F}{|}}{As}}\overset{F}{\underset{F}{\diagdown\diagup}}\underset{\underset{F}{|}}{\overset{\overset{F}{|}}{Sb}}\overset{F}{\diagup}\overset{F}{\diagdown} \quad \text{or} \quad :\overset{\overset{F}{|}}{\underset{\underset{F}{|}}{As}}-F-\overset{\overset{F}{|}}{\underset{\underset{F}{|}}{Sb}}-F$$

Both complete halide transfer and association via bridging halide are also encountered with the halogen fluorides. It is well to be alert to the fact that this *halide transfer/association behavior is characteristic of halides of heavier elements and plays an important role in the chemistries of such compounds.*

Our final example of pentavalent phosphorus concerns its acid character. The formation of PCl_6^- with a properly sized cation and of $PCl_5 \cdot py$ have already been mentioned. A less conventional example[24] is the reaction of phosphoryl trichloride with an organic base like Et_3N:

$$\underset{Cl}{\overset{Cl}{\underset{|}{Cl}}}P=O + Et_3N \rightarrow \underset{\underset{Et}{\overset{|}{N}}\underset{Et}{\diagdown}Et}{Cl-\overset{\overset{Cl}{|}}{\underset{|}{P}}\diagdown O}\overset{Cl}{\diagup} \rightleftharpoons \underset{Cl}{\overset{Cl}{\diagdown}}\underset{NEt_3}{\overset{\overset{O}{||}}{P}}\overset{\oplus}{\diagup} + Cl-\underset{\underset{Cl}{|}}{\overset{\overset{Cl}{|}}{P}}\underset{O}{\overset{Cl}{\diagdown}}\overset{\ominus}{}$$

$$\begin{bmatrix} \text{m.p.} = 2° \\ \text{b.p.} = 105° \end{bmatrix}$$

leading to conducting solutions with $POCl_3$ as the solvent (see the later discussion of non-aqueous solvents and their ability to support solute ionization).

Oxygen and Sulfur

While many examples of oxygen as a donor can be cited [water, ethers, ketones, amine and phosphine oxides, and thionyl ($>$SO) and sulfuryl ($>SO_2$) compounds are a few of the important classes], we are hard pressed to find examples in which oxygen acts as an acid. The amine and phosphine oxides and sulfur oxides mentioned above could be considered, in fact, to be examples of adducts with the oxygen atom playing the role of the acid:

$$\left.\begin{matrix} R_3N: \\ X_3P: \\ R_2\ddot{S}: \\ R_2SO \end{matrix}\right\} \xrightarrow{\text{oxygen atom transfer}} \begin{cases} R_3N\ddot{O}: \\ X_3PO \\ R_2SO \\ R_2SO_2 \end{cases}$$

As you will see in the next three chapters, such oxygen atom transfer reactions play an important role in phosphorus and sulfur chemistry.

A contemporary example of dioxygen adduct formation arises with the so-called "oxygen carrier" transition metal complexes. Complexes with the ability to reversibly add O_2 are of great interest as analogs to the biological carriers such as hemoglobin and myoglobin. Recent studies in inorganic laboratories[25] have confirmed that such carrier complexes are characterized by an angular M—O—O structure.

$$\underset{D\diagup\underset{D}{|}\diagdown D}{\overset{O\diagdown O}{\underset{M}{}}}\underset{D}{\diagup} \quad D = \text{other ligand(s)}$$

---

[24] M. Baaz and V. Gutmann, *Monatsch. Chem.*, **90**, 276 (1959).
[25] F. Basolo, B. M. Hoffman, and J. A. Ibers, *Acc. Chem. Res.*, **8**, 384 (1975); L. D. Brown and K. N. Raymond, *Inorg. Chem.*, **14**, 2595 (1975).

In Co(II) systems it appears that adduct formation actually incurs oxidation of the metal [to Co(III)] and reduction of dioxygen to superoxide, O_2^-. The evidence for this comes

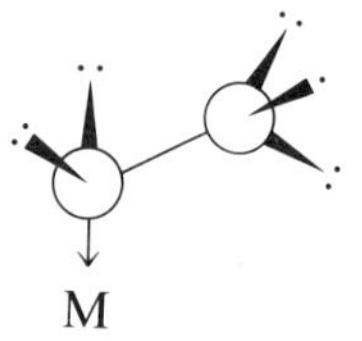

M

from electron spin resonance that shows weak Fermi coupling of the unpaired electron with the cobalt nuclear magnet; the O—O distance of ~1.25 Å is intermediate between those of O_2 and O_2^- (O_2, 1.21 Å; O_2^-, 1.28 Å; O_2^{2-}, 1.49 Å); the O—O stretching frequency of ~1400 cm^{-1} is considerably above that for O_2^{2-} (~700 to 800 cm^{-1}), below that for O_2 (~1500 cm^{-1}), and above that for O_2^- (~1150 cm^{-1}). The homonuclear diatomic mo view of O_2^- helps greatly in understanding these properties of coordinated O_2^-. The HOMO of O_2^- is the degenerate pair of π^* mo's [the configuration is $(sp\sigma)^2(sp\sigma^*)^2(p\pi)^4(ps\sigma)^2(p\pi^*)^3$].

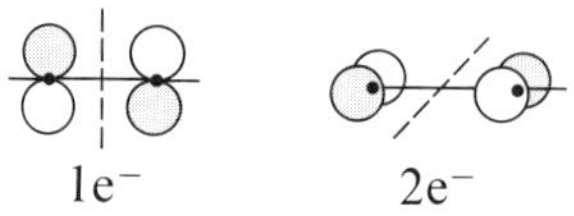

1e$^-$ 2e$^-$

In the presence of a Lewis acid, the doubly occupied π^* mo becomes the donor orbital. Bonding with the metal delocalizes these electrons out of the O—O region and so reduces their π^* role.

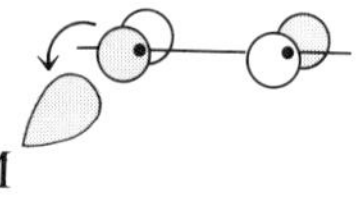

M

In resonance structure terminology, the equivalent description arises from

$$\overset{\ominus}{M} \leftarrow :\!\ddot{\underset{..}{O}}\!-\!\dot{\underset{..}{O}}\!: \text{ more important than } \overset{\ominus}{M} \leftarrow :\!\overset{\oplus}{\underset{..}{O}}\!-\!\ddot{\underset{..}{O}}\!:$$

and coordination of the oxygen pair reduces the pair repulsion found in O_2^-.

Yet another reaction of dioxygen with certain square planar metal complexes is known to give a symmetrically bonded O_2 group.[25a] Called "oxidative addition," the reaction produces an adduct that *could* be viewed as an adduct of O_2^{2-},

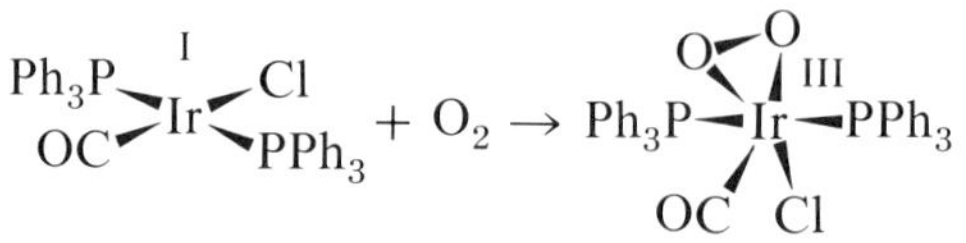

$$\begin{array}{c}Ph_3P\diagdown\overset{I}{Ir}\diagup Cl \\ OC\diagup\quad\diagdown PPh_3\end{array} + O_2 \rightarrow \begin{array}{c} O\diagdown\overset{O}{\overset{|}{\overset{III}{Ir}}}\diagup \\ Ph_3P-Ir-PPh_3 \\ OC\quad Cl \end{array}$$

[yellow]   [orange]

From an mo view of O_2^{2-}, the O_2^{2-} donor mo (HOMO) is a π^* mo having non-zero overlap with an Ir d ao (d_{xy}, for example).

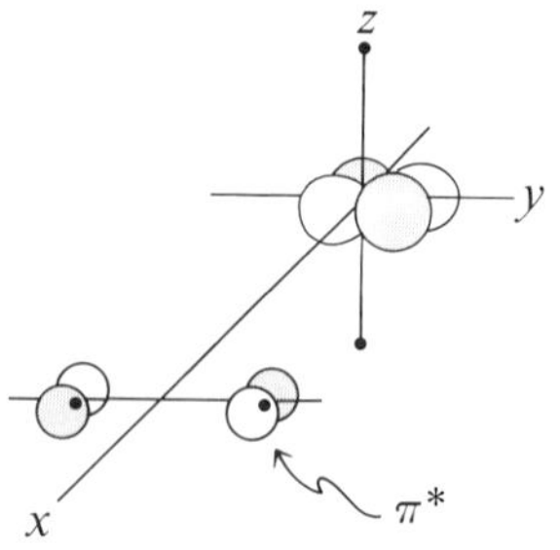

[25a] A current brief review of all O_2 complexes is given by L. Vaska, *Acc. Chem. Res.*, **9**, 175 (1976).

The delocalization of these O—O anti-bonding electrons onto the metal incurs an O—O linkage strengthening, as evidenced by an O—O distance of 1.30 Å. This bond distance is very close to that in superoxide, where there are only three π^* electrons. Finally, there is an interesting substituent effect operating in such complexes. Replacement of Cl with the less electronegative I results in O—O lengthening to 1.51 Å, which is at least partly due to decreased $\pi^* \to d$ electron transfer from O_2^{2-} to Ir as a result of less Z_{Ir}^* as seen by O_2^{2-}.‡

Sulfur has a richer acid-base chemistry than does oxygen. Not only do we find examples of sulfides[26] and phosphine sulfides as donors

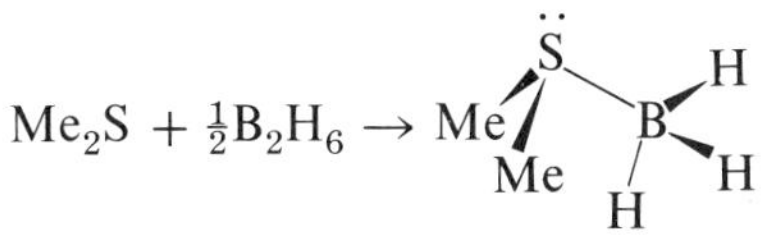

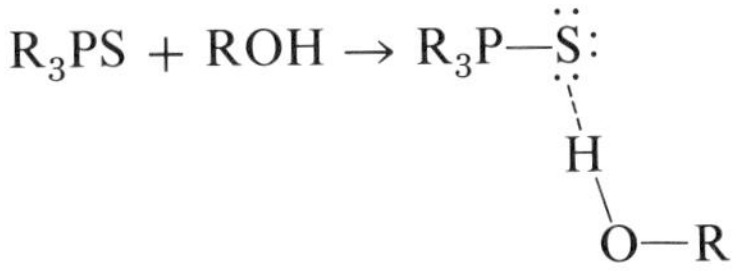

$$Me_2S + \tfrac{1}{2}B_2H_6 \to Me_2S\!\cdot\!BH_3 \qquad R_3PS + ROH \to R_3P\!-\!S\!:\!\cdots H\!-\!O\!-\!R$$

but the sulfoxides can also act as Lewis bases (at either sulfur or oxygen).[27]

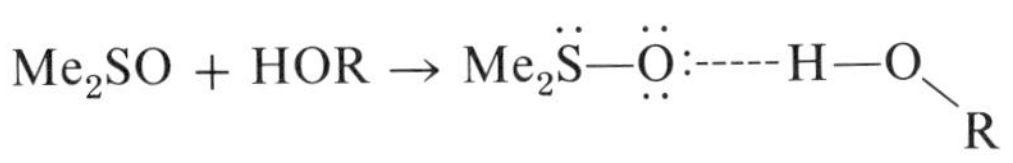

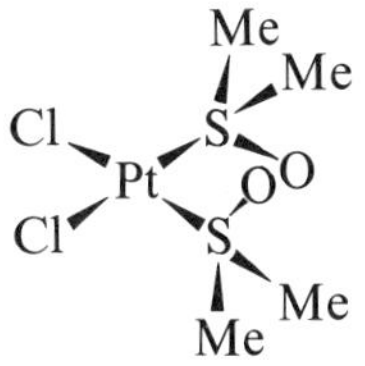

$$Me_2SO + HOR \to Me_2\ddot{S}\!-\!\ddot{O}\!:\!\cdots\!-\!H\!-\!O\!-\!R$$

At this point you might wish to review the discussion in Chapter 3 (p. 113) on the use of the S—O stretching frequency to diagnose the mode of coordination.

Two of the more interesting sulfur compounds exhibiting acid-base properties are the binary oxides, SO_3 and SO_2. The trioxide is a voracious acid but a notoriously weak (oxygen) donor. Accordingly, SO_3 is often encountered as a vapor at room temperature (b.p. = 45°) but, interestingly, has polymeric structures as a solid. The trimeric ring and chain structures of two different crystalline forms suggest that an acid-base self-interaction takes place in the solid.

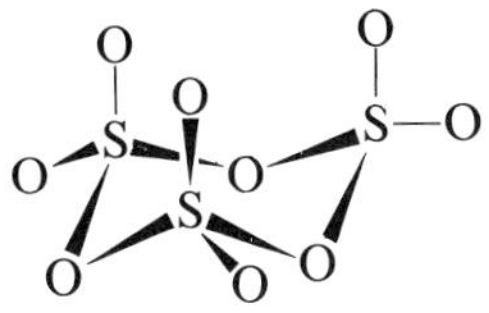

[m.p. = 17°] 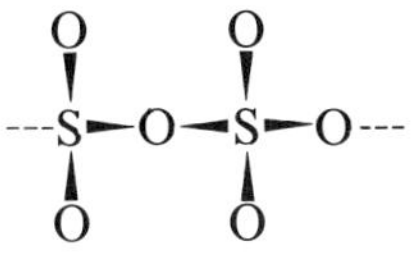   [m.p. = 33°]

More conventional examples involve SO_3 acting as an acid toward donors like trimethylamine, alcohols, and thioalcohols:

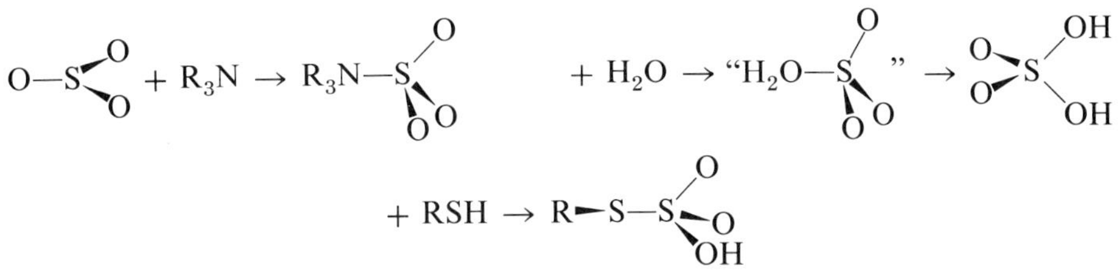

$$O\!=\!SO_2 + R_3N \to R_3N\!-\!SO_3$$
$$+ H_2O \to \text{``}H_2O\!-\!SO_3\text{''} \to (HO)_2SO_2$$
$$+ RSH \to R\!-\!S\!-\!SO_2OH$$

---

‡ A more complete mo treatment is given in Chapter 17.

[26] $Me_2S\cdot BH_3$ (m.p. = $-38°C$, b.p. = $97°C$) is thought to present a particularly convenient source of BH_3 for use in the laboratory because it is 18% by weight BH_3, is not a gas (like B_2H_6), and avoids the hazards associated with ethereal BH_3 (THF/BH_3 solutions, only 1.5 per cent by weight BH_3, are commercially available). J. Beres, et al., Inorg. Chem., **10**, 2072 (1971).

[27] J. H. Price, et al., Inorg. Chem., **11**, 1280 (1972).

Simple amine and dioxane adducts are particularly useful as sulfonating reagents, for they are solids and afford less vigorous, more easily controlled reactions than does SO_3 itself.[28]

Even more interesting than SO_3 is sulfur dioxide. Its acid-base character is weak but unusual; not only can the sulfur act as an acid site, but both the sulfur and oxygen atoms can act as donors.[29]

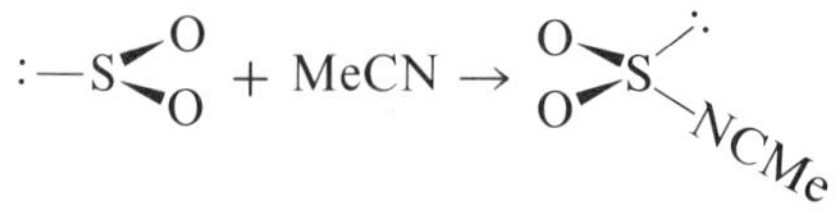

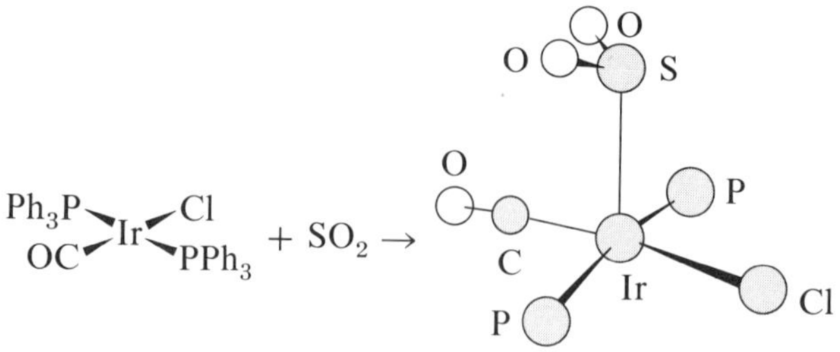

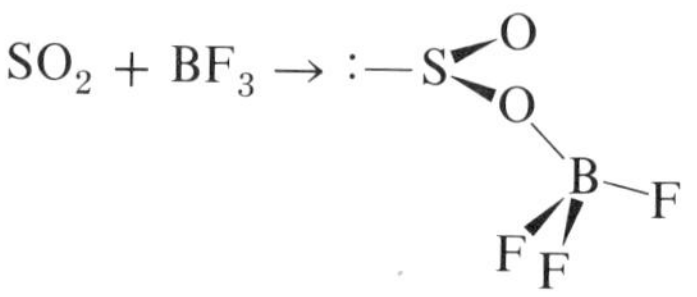

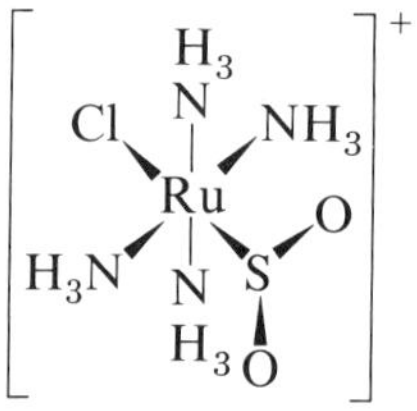

To help you tie things together, notice that the SO_2 valence shell is isoelectronic with $SnCl_2$ and that they form structurally similar adducts. The analogy extends to the insertion reaction also, for SO_2 is known to insert into metal-carbon and metal-metal bonds, sometimes with formation of isomeric products.[30]

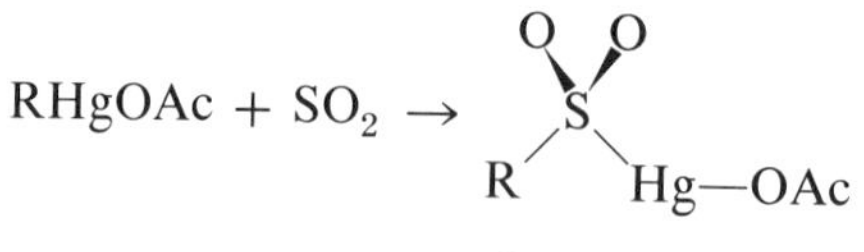

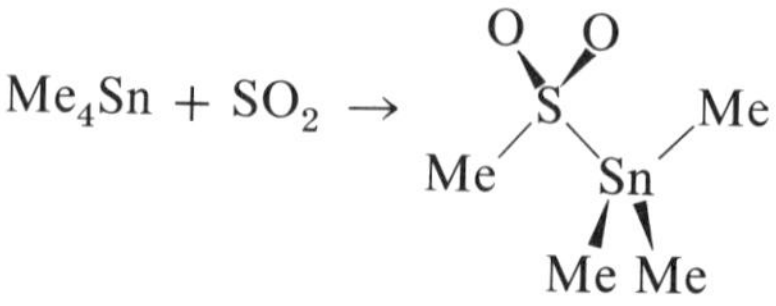

---

[28] H. H. Sisler and L. F. Audrieth, "Inorganic Syntheses," Vol. II, 173 (1946).

[29] S. J. LaPlaca and J. A. Ibers, *Inorg. Chem.*, **5,** 405 (1966); L. H. Voight, J. L. Katz, and S. E. Wiberley, *ibid*, **4,** 1157 (1965); J. W. Moore, *et al., J. Amer. Chem. Soc.*, **90,** 1358 (1968).

[30] C. W. Fong and W. Kitching, *J. Amer. Chem. Soc.*, **93,** 3791 (1971); M. R. Churchill and J. Warmald, *Inorg. Chem.*, **10,** 572 (1971); A. Wojcicki, *et al., ibid*, **10,** 2130 (1971); *J. Amer. Chem. Soc.*, **93,** 2535 (1971).

$$(C_5H_5)(CO)_2Fe-CH_2CH=CHPh + SO_2 \rightarrow$$
$$(C_5H_5)(CO)_2Fe-\overset{O_2}{S}-CH_2CH=CHPh$$

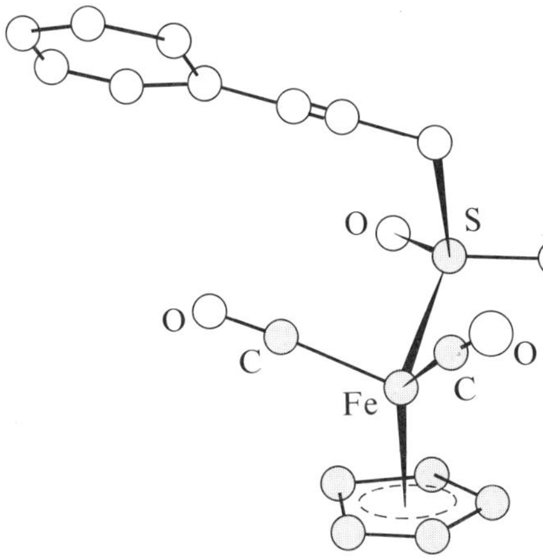

$$(C_5H_5)(CO)_2Fe-CH_2CH=CMe_2 + SO_2 \rightarrow$$
$$(C_5H_5)(CO)_2Fe-\overset{O_2}{S}-CH_2CH=CMe_2$$
$$+$$
$$(C_5H_5)(CO)_2Fe-\overset{O_2}{S}-CMe_2CH=CH_2$$

Because SO_2 is highly polar and weakly acidic-basic, and has a usefully high boiling point of $-10°C$, it is an important non-aqueous solvent, either alone or in combination with SbF_5 and HF as a "super acid" solvent. We will explore the uses of this solvent system later in this chapter.

Finally, we should mention that SF_4 has been reported[31] to act as a fluoride ion acceptor in the reaction with quaternary fluoride salts

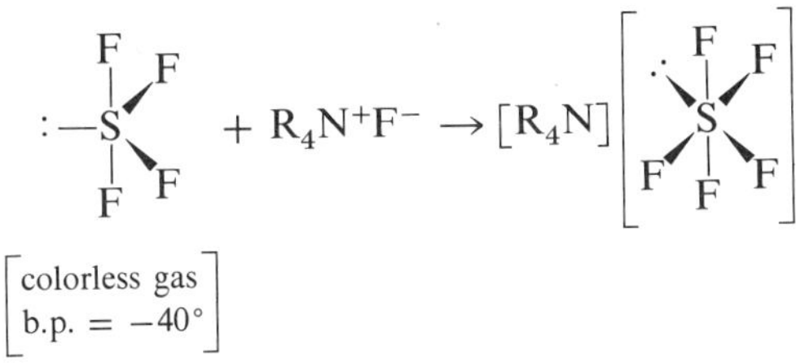

$$:-S\begin{matrix}F\\|\\F\end{matrix}\diagup\begin{matrix}F\\F\end{matrix} + R_4N^+F^- \rightarrow [R_4N]\begin{bmatrix}\ddot{}\\F\diagdown|\diagup F\\S\\F\diagup|\diagdown F\\F\end{bmatrix}$$

$$\begin{bmatrix}\text{colorless gas}\\\text{b.p.} = -40°\end{bmatrix}$$

where again (refer to the SiF_5^- and PCl_6^-, PCl_4^+ examples given earlier) lattice stabilization of the adduct is of importance.

Halogens

The halide ions are well known as Lewis bases, particularly toward metal ions but also in the ion transfer and bridging capacities noted above with the Group V elements and the halides of aluminum. The elemental forms of the halogens are known to have primarily acidic properties, but one example of Cl_2 acting as a base has been reported[32] in the reaction

$$ClF + Cl_2 + SbF_5 \rightarrow [Cl_3][SbF_6]$$

---

[31] K. O. Christie, et al., *Inorg. Chem.*, **11**, 1679 (1972).
[32] R. J. Gillespie and M. J. Morton, *Quart. Rev.*, **25**, 553 (1971); O. Glemser and A. Sinali, *Angew. Chem. Internat. Edn.*, **8**, 517 (1969).

which may be viewed as F⁻ transfer from one acid (Cl⁺) to another (SbF$_5$) with stabilization of Cl⁺ by adduct formation with Cl$_2$.

The acid functions of the halogens are largely restricted to iodine for the simple reason that the other elements, particularly fluorine and chlorine, are such strong oxidants that their reactions with donors carry on beyond adduct formation to oxidation products. One example of Cl$_2$ acting as an acid can be cited[33]:

$$[Ph_4As][Cl] + Cl_2 \xrightarrow{-78°} [Ph_4As][Cl_3]$$

The product anion is linear, as expected from the VSEPR model.

One of the more colorful halogens, iodine (purple in the gas phase), gives solutions of varying shades of color when dissolved in organic solvents. In the simplest terms, such solutions contain both "free" I$_2$ and complexed iodine in equilibrium with each other, and the color of the solution is simply a superposition of the different colors of I$_2$ and I$_2$ · D. A well-known example of the Lewis acid character of I$_2$ is its reaction with iodide ion to form the yellow triiodide ion, I$_3^-$. (This reaction is made use of when aqueous solutions of I$_2$, normally susceptible to air oxidation, are stabilized by addition of I⁻.)

To account for the role of I$_2$ as an acid, and the colors of I$_2$ and I$_2$ · D, we will undertake an analysis much like that for hydrogen-bonding acids. As before, we need to identify the LUMO of I$_2$ and examine its overlap with the donor orbital of a general base. I$_2$ is a homonuclear diatomic molecule, and you will recognize the mo pattern as developed in Chapter 4.

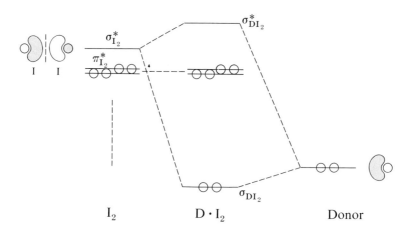

Adduct bond formation between the donor and the I$_2$ molecule stabilizes the lone pair of electrons on the donor and at the same time weakens the I—I bond (because the donor orbital is mixed with the σ* orbital of I$_2$, the donor electron pair takes on to some extent the I—I anti-bonding character of the I$_2$ σ* mo). Because the σ* mo of I$_2$ is not particularly low in energy, nor the D/I$_2$ overlap very large, the degree of covalency or strength of the adduct bond is expected to be low (1 to 10 kcal/mole).

A very important consequence of the adduct bond formation is that while the occupied donor orbital is stabilized by the constructive mixing-in of a little of the I$_2$ σ* mo, the σ* mo of I$_2$ is destabilized by the destructive mixing-in of a little of the donor orbital (remember that the destabilization is associated with anti-bonding character in the *adduct bond* region). Another concept from Chapter 4 of use here is that, since the acid-base interaction is weak, the adduct bonding pair of electrons is still pri-

---

[33] E. F. Riedel and R. D. Willett, *J. Amer. Chem. Soc.*, **97**, 701 (1975).

marily associated with the donor (the wave amplitude is greater at the donor atom) and the adduct anti-bonding mo is primarily associated with the I_2 molecule. (Compare the description of hydrogen bonding and organic π acid/π base adducts given earlier).

These ideas, which follow from the LCAO model of adduct formation, have important consequences for the spectral properties of I_2 complexes. To begin with, I_2 in the gas phase has a purple color owing to absorption of photons in the middle of the visible range of radiation (~500 nm). The purple color arises from the transmission of the red and blue wavelengths at opposite extremes of the visible region. The electronic transition responsible for this absorption of light involves excitation of an electron from the I_2 π^* mo into the I_2 σ^* mo.

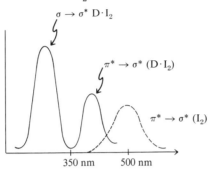

Adduct formation, as you have seen, results in a destabilization of the I_2 σ^* mo, with the consequence of making the $\pi^* \to \sigma^*$ absorption process more energetic. That is, radiation of shorter wavelength is now required for the electronic transition. We expect, and find, that the stronger the adduct bond formed with iodine, the more shifted to the blue is the absorbed photon. As far as transmitted light is concerned, blue is filtered out and replaced by hues from the middle of the visible region.

Another interesting feature of the spectrum of I_2 adducts is the occurrence of a strong absorption in the near ultraviolet region, which is not present in the spectrum of I_2 alone. This new absorption band is assigned to excitation of an electron from the adduct bond orbital (mainly the donor lone pair orbital) into the adduct anti-bonding mo (mainly the I_2 σ^* mo). Because charge density within the adduct shifts from the donor toward the I_2 region, this transition is referred to as a charge-transfer (sometimes abbreviated C-T) transition. As we might expect, there is also a correlation[34] between the energy of this electronic transition and the energy of the adduct bond between I_2 and donors.

Still other examples of acid behavior are to be found for the interhalogen compounds. ICl, for example, is normally a stronger acid than I_2, with the acid site being the iodine atom.

$$:\ddot{I}-\ddot{Cl}: \quad + \text{ MeCN}: \to \text{MeCN}-I-Cl:$$
$$:\ddot{I}-\ddot{Cl}: \quad + \text{ Cl}^- \to [\text{Cl}-I-\text{Cl}]^-$$

$\begin{bmatrix}\text{red solid/liquid} \\ \text{m.p.} = 27°\end{bmatrix}$ [yellow]

Similarly, ICl_3 is a good chloride ion acceptor,

$$\begin{array}{c}\text{Cl}\\|\\ \ddot{\ddot{}}\text{I}-\text{Cl}\\|\\\text{Cl}\end{array} \quad + \text{ Cl}^- \to \begin{bmatrix}\text{Cl} & \ddot{|} & \text{Cl}\\ & \text{I} & \\ \text{Cl} & | & \text{Cl}\end{bmatrix}^-$$

$\begin{bmatrix}\text{yellow-brown solid}\\ \text{red liquid}\\ \text{dec. } 77°\end{bmatrix}$ [yellow]

---

[34] R. S. Mulliken and W. B. Person, "Molecular Complexes," Wiley, New York (1969); ref. 12.

a property also revealed by its dimerization (I_2Cl_6, D_{2h}) in the solid. Further, BrF_3 self-associates and appears to auto-ionize (the liquid is conducting)

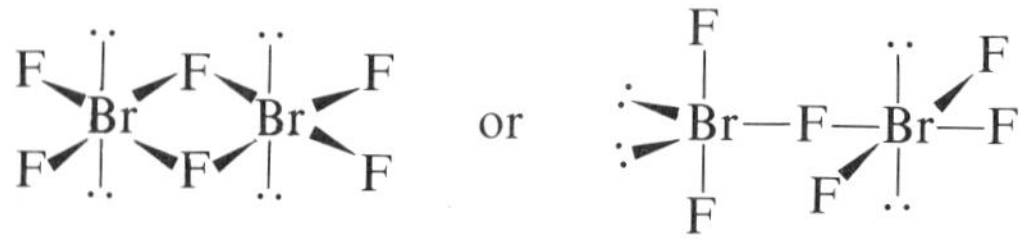

$$\text{[Br}_2\text{F}_6\text{ bridged structures]}$$

[yellow-green liquid
m.p. = 9°
b.p. = 126°]

and acts as a fluoride ion acceptor and fluoride ion donor (p. 204).

$$ClF_3 + BrF_3 \rightarrow [ClF_2][BrF_4]$$

Xenon

Of the three fluorides of Xe (XeF_2, XeF_4, and XeF_6 are all colorless solids of m.p. 129°, 117°, and 50°), all are known to act as fluoride donors, and the hexafluoride is also known to function in an acid capacity.[35] From the difluoride, two different salts have been obtained, depending on whether an excess of XeF_2 or of fluoride acceptor is used:

$$XeF_2 + 2SbF_5 \rightarrow XeF_2 \cdot 2SbF_5$$
$$2XeF_2 + SbF_5 \rightarrow 2XeF_2 \cdot SbF_5$$

In the first (1:2) adduct we find a dinuclear anion with F^- bridging two SbF_5 groups, and in the (2:1) adduct there is a dinuclear cation (of C_{2v} symmetry; cf. the VSEPR prediction) with F^- bridging, in effect, two XeF^+ moieties.

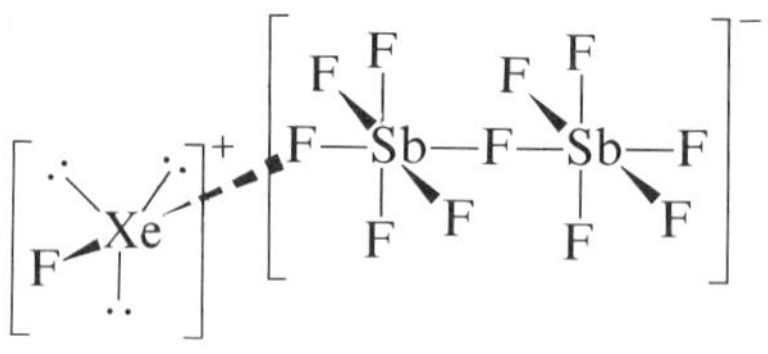

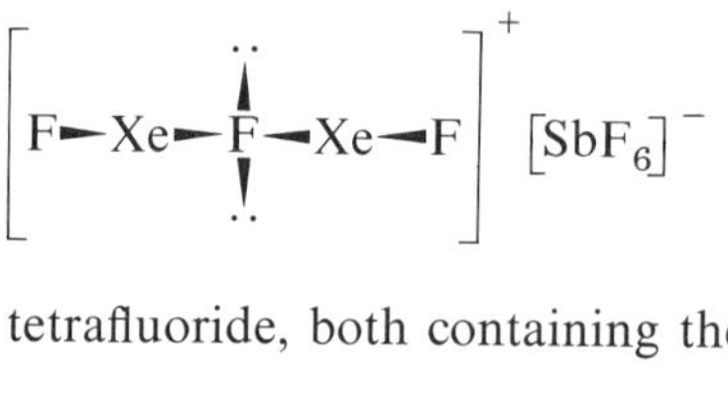

$$[XeF]^+ [F_5Sb-F-SbF_5]^-$$

$$[F-Xe-F-Xe-F]^+ [SbF_6]^-$$

Two adducts of the tetrafluoride, both containing the XeF_3^+ cation, have been reported:

$$XeF_4 + SbF_5 \rightarrow [XeF_3][SbF_6]$$
$$XeF_4 + 2SbF_5 \rightarrow [XeF_3][Sb_2F_{11}]$$

Other examples[36] of fluoride transfer from Xe to SbF_5 involve formation (from $XeOF_4$) of the cation $[XeOF_3]^+$ with $[SbF_6]^-$ and $[Sb_2F_{11}]^-$, and formation (from XeO_2F_2) of $[XeO_2F]^+$ with $[Sb_2F_{11}]^-$. In a similar way, XeF_6 and antimony pentafluoride form a compound with the composition $2XeF_6 \cdot SbF_5$, which presumably also

---

[35] R. J. Gillespie, *et al.*, *Chem. Comm.*, 1543 (1971); G. R. Jones, *et al.*, *Inorg. Chem.*, **9**, 2264 (1970); J. H. Holloway and J. G. Knowles, *J. Chem. Soc.* (A), 756 (1969); F. O. Sladky, *et al.*, *ibid*, 2179, 2188 (1969); V. M. McRae, *et al.*, *Chem. Comm.*, 62 (1969).
[36] R. J. Gillespie and G. J. Schrobilgen, *Inorg. Chem.*, **13**, 2370 (1974).

contains a dinuclear cation $[Xe_2F_{11}]^+$. That the hexafluoride can also act as an acid is evidenced by the reactions

$$XeF_6 + F^- \rightarrow [XeF_7]^- \underset{\Delta}{\rightarrow} \frac{1}{2}XeF_6 + \frac{1}{2}[XeF_8]^{2-}$$

The crystalline form of XeF_6 itself reflects the dual acid-base nature of the compound, for in the solid phase one finds fluorines bridging xenon atoms in tetrameric ring and hexameric cage structures (refer to Chapter 2, p. 80).

At least one oxide of xenon has been shown to exhibit Lewis acid behavior, in that XeO_3 will react with alkali fluorides to form the complex anion XeO_3F^-:

$$XeO_3 + KF \rightarrow K[XeO_3F]$$

As you might well expect by now, the anions are linked, by sharing F atoms, into chains.[37]

METALS

To complete this brief survey of the utility of the acid-base concept, we will emphasize a few more unusual properties of metals. Metal cations are widely known to act as acids—this is the basis for a great deal of transition metal ion chemistry, and we will give this topic considerable attention in subsequent chapters. You have seen that with the right donors (*e.g.*, CO and PF_3) even metal *atoms* appear to act synergically as Lewis acids and bases (σ acceptor, π donor). A closely related aspect of metal acid-base chemistry is their function as σ donor Lewis bases. The low-oxidation-state organometallic complexes have received considerable study recently with a view to understanding their chemistry in terms of Lewis base properties.[38] A few examples will suffice at this point. The manganese pentacarbonyl anion exhibits base character in the reactions

$$Mn(CO)_5^- + H^+ \rightleftharpoons H-Mn(CO)_5$$
$$+ Me_3SiCl \rightarrow Me_3Si-Mn(CO)_5 + Cl^-$$
$$+ Cr(CO)_6 \rightarrow [(OC)_5Cr-Mn(CO)_5]^- + CO$$

The last example might be viewed as another example of a low-oxidation-state metal atom (Cr) acting as an acid; the last two examples further illustrate (see p. 202) the concept of metal-metal bonds.

While we will examine the relationship between the electronic structures of $Mn(CO)_5^-$ and $HMn(CO)_5$ later, at this point we will adopt a *very* simplified view of the electronic structure of $Mn(CO)_5^-$. The Mn metal atom carries with it seven valence electrons. Together with the pair of electrons donated by each carbon monoxide and the electron making an anion of $Mn(CO)_5$, we find that the Mn atom has an equal number of electron pairs and valence orbitals (nine of each). With a trigonal bipyramid structure [cf. $Fe(CO)_5$], it would appear that only the five pairs constituting the Mn—C bonds are stereochemically active, with the other four pairs constituting a "valence shell core" of electrons. A better, simpler orbital model for this structure amounts to use of Mn d_{z^2}, s, p_x, p_y, and p_z as acceptor orbitals, with the

---

[37] D. J. Hodgson and J. A. Ibers, *Inorg. Chem.*, **8**, 326 (1969).
[38] More attention will be given to this important property of organometal compounds in Chapters 16 and 17. J. C. Kotz and D. G. Pedrotty, *Organometal. Chem. Rev. A*, **4**, 479 (1969); D. F. Shriver, *Accts. Chem. Res.*, **3**, 231 (1970); R. B. King, *ibid*, **3**, 417 (1970).

(d_{xz}, d_{yz}) and $(d_{x^2-y^2}, d_{xy})$ orbital pairs (occupied) remaining for retro π bonding with the CO's, in analogy with the earlier discussion of the electronic structure of $Cr(CO)_6$. It is obvious that these "lone pairs" are chemically (if not stereochemically) active, for the structural rearrangement on protonation to form the octahedral (actually C_{4v}) symmetry is fairly easy [the pK_a in H_2O of $HMn(CO)_5$ is about 7]. In a sense, a net of one π bonding pair of metal electrons is diverted into forming a σ bond between the metal and a proton. Were it not for the experimental fact that manganese pentacarbonyl hydride is an acid (in the classic Brönsted sense), one might have assumed that the compound would exhibit chemical properties normally associated with the presence of highly hydridic hydrogen. The reversion of the Mn—H σ bonding pair to a Mn—CO π bonding pair in the conjugate base is fundamental to the acidity of $HMn(CO)_5$.

To illustrate that this base behavior of organometal compounds is not just a peculiarity of $Mn(CO)_5^-$, we note that tetracarbonylcobaltate(1−) [isoelectronic with $Ni(CO)_4$] is capable of reacting in a conventional Lewis manner with indium tribromide.[39]

$$Co(CO)_4^- + InBr_3 \rightarrow \begin{bmatrix} & & O & & \\ & & \overset{\|}{C} & & Br \\ OC & \diagdown & | & \diagup & \\ & & Co\!-\!In & & \\ OC & \diagup & | & \diagdown & Br \\ & & \overset{\|}{C} & & Br \\ & & O & & \end{bmatrix}$$

Again we find that the metal atom has a filled valence shell with four pairs defining the Co—C σ bonds and five non-stereochemically active pairs involved in Co—C π bonding. Structural rearrangement to accommodate the presence of the indium atom about the cobalt again indicates that a net of one π pair is easily diverted into a sigma valence role.

Having called to your attention the structural similarities between $Mn(CO)_5^-$ and $Fe(CO)_5$ and between $Co(CO)_4^-$ and $Ni(CO)_4$, we should quickly point out that the neutral carbonyls seem to be much less basic than the anionic analogs. While $Fe(CO)_5$, for example, exhibits a proton affinity in the gas phase[40] of 204 kcal/mole (similar to the NH_3 affinity of 207 kcal/mole), only in strongly acidic solvents does *solution* protonation of $Fe(CO)_5$ occur.[41]

$$Fe(CO)_5 + H^+ \rightarrow HFe(CO)_5^+$$

In solution, of course, the free H^+ does not exist (see the later section in this chapter on solvation energies). Protonation of $Fe(CO)_5$ in solution actually involves an equilibrium of the sort

$$Fe(CO)_5 + HA \rightleftharpoons Fe(CO)_5H^+ + A^-$$

where $Fe(CO)_5$ finds itself in competition with A^-, and solvation energies may be important to the position of the equilibrium. Nevertheless, a reduced basicity of the neutral compounds is expected from their higher metal nuclear charge. Replacement of one or more CO ligands by PR_3 ligands appears to enhance the metal basicity. In this vein, cationic complexes $HNiL_4^+$ [where $L = P(OR)_3$] and $HNiL_2^+$ (where $L = Ph_2PC_2H_4PPh_2$) have been reported recently;[42] the phosphorus ligands in these examples are probably better σ donors to Ni and poorer π acceptors than is CO, both of these properties enhancing metal basicity.

---

[39] J. K. Ruff, *Inorg. Chem.*, **7**, 1499 (1968).
[40] M. S. Foster and J. L. Beauchamp, *J. Amer. Chem. Soc.*, **97**, 4808, 4814 (1975).
[41] A. Davison, *et al.*, *J. Chem. Soc.*, 3653 (1962).
[42] R. A. Schunn, *Inorg. Chem.*, **9**, 394 (1970); G. K. McEwen, *et al.*, *ibid*, **13**, 2800 (1974).

STUDY QUESTIONS

1. Compare RCH_2^+ and RCO^+ with the isoelectronic CH_3BH_2 and RBO with regard to self-association, and explain why the latter do, but the former don't, dimerize and trimerize, respectively.

2. Propose a rationale for the good solubility of BF_3 in benzene. Can you identify a HOMO and LUMO for the interaction?

3. What two changes in the ao make-up of the LUMO of BF_3 occur to lower the LUMO energy as BF_3 is distorted to a pyramidal geometry?

4. Describe the SO_2 LUMO which acts as the acceptor orbital on approach of a donor.

5. Interpret the finding that the boiling points of the following compounds are in the order shown.
   a) $H_2O > NH_3$
   b) $CH_3OH > CF_3CH_2OH > (CF_3)_2CHOH > (CF_3)_3COH$
   c) $CH_3CO_2H > ClCH_2CO_2H > CF_3CO_2H$

6. Which acid should show the largest shift to lower energy in AH stretching frequency upon hydrogen bond formation with a Lewis base?

   a) H_2O
   b) EtOH
   c) NH_3
   d) CF_3CH_2OH

7. What is the predicted effect of metal-CO retro-bonding on the CO bond strength? Would ν_{CO} be larger, the same, or smaller in free CO? Refer to the discussion of diatomics in Chapter 4 and predict the effect on ν_{CO} from σ bonding alone upon coordination of CO.

8. Consider the atom formal charges in the valence structures below and suggest a preferred site for attack by a nucleophile (M has zero formal charge in the third structure).

$$\overset{..}{M}-C=\overset{..}{\overset{..}{O}}: \leftrightarrow \overset{..}{M}-C\equiv O: \leftrightarrow M=C=\overset{..}{\overset{..}{O}}:$$

9. Given that $[AlMe_2Ph]_2$ has a structure with bridging phenyl groups and the phenyl groups are perpendicular to the four-membered ring of Al and bridging carbons, comment on the similar roles of the phenyl groups and the Cl atoms in Al_2Cl_6. (For help, see Chapter 6, p. 311.)

10. Predict products for the reactions

$$2ClF + AsF_5 \rightarrow$$
$$ClF_3 + AsF_5 \rightarrow$$
$$IF_5 + CsF \rightarrow$$
$$ClF_3 + CsF \rightarrow$$
$$ClF + CsF \rightarrow$$

11. Considering the nodal patterns for the occupied π mo's of benzene, would you expect Ag^+ to complex C_6H_6 with C_{6v} or lower symmetry? Does your answer depend on whether you consider only the $5s$ ao of Ag^+ as the acceptor orbital (LUMO) or the $5s$ and $5p$ ao's (the observed structure is asymmetric)?

12. Predict structures for

    a) B_2H_5Me
    b) $Al_2Me_2Cl_4$
    c) $(R_2AlH)_3$
    d) $(RBO)_3$
    e) $Al_2Cl_7^-$
    f) $AlH_3(NMe_3)_2$
    g) $Sn_2F_5^-$

13. Dissect each of the adducts below into two fragments: an "easily recognized" acid and base. Within each fragment, which are the acceptor and donor atoms? As the criterion for

"easily recognized," at least one of the fragments must have a reasonable chemical identity.

a) C₆H₅N·HOEt (phenyl ring with N·HOEt)
b) $B_2H_7^-$
c) B_2H_6
d) $GaCl_4^-$
e) CH_3COF
f) $C_6H_6 \cdot C_2(CN)_4$
g) $Mn(CO)_5SnCl_3$
h) $SnCl_3^-$
i) F_3NO
j) $POCl_3 \cdot N$(pyridine ring)
k) $SbF_5 \cdot AsF_3$
l) $SCl_4 \cdot SbCl_5$
m) $(H_3N)_5RuSO_2^{2+}$
n) $BF_3 \cdot SO_2$
o) $ClF_3 \cdot BrF_3$
p) XeO_3F^-
q) $(XeF_2)_2SbF_5$
r) $HCo(CO)_4$
s) HSO_4^- (not H^+, SO_4^{2-})
t) XeF_8^{2-}

14. In the presence of ketones, aldehydes, and aryl ethers, alcohols often exhibit three OH stretching vibrational absorption bonds. (One of these, that at highest energy, corresponds to the vibration in the free molecule.) Discuss the origins of the two lower energy bands and assign structures to the species responsible for them.

15. Suggest a structure for the chains of $[XeO_3F^-]_n$ anions in $K[XeO_3F]$. Will the Xe—F—Xe angles deviate appreciably from 180°?

ACID-BASE STRENGTHS

THE THERMODYNAMIC DEFINITION[43]

The preceding survey of acid-base compounds should have convinced you that a large number of compounds are characterized by conveniently high-energy occupied orbitals that can serve as the source of Lewis basicity, and conveniently low-energy unoccupied mo's that can serve as the basis for Lewis acid character. As a qualitative guide to intermolecular interactions, the concept of acids and bases is certainly useful. Nevertheless, we soon reach the point of asking quantitative questions such as, "If I place two donors in competition for a given acid, which will preferentially form the adduct?" or "Just how strong is the adduct formed by a given acid-base pair?"

These questions are not easy to answer in general, and research continues as experimental and theoretical answers are sought. One of the first problems we face is how to measure the "strength" of the adduct bond. A thermodynamic approach seems most reasonable, but even after agreeing on this we are faced with the question, "Which thermodynamic property of the reaction

$$A \cdot D \rightarrow A + D$$

are we to use—the internal energy change or the free energy change?" The latter seems to have more direct bearing on practical experience but is more difficult to interpret than the former, because of the inclusion of entropy changes [the $\Delta(PV)$ term can easily be allowed for]. To give one example in which different conclusions are reached regarding the strengths of adducts, we give below some data[44] for dissociation of the $Me_3N \cdot BMe_3$ and $Me_3P \cdot BMe_3$ adducts, all species in the gas phase.

	$Me_3N \cdot BMe_3$	$Me_3P \cdot BMe_3$
$K_d^{100°C}$	0.47	0.13
$\Delta G_d^{100°C}$	+0.6	+1.5
$\Delta H_d^{100°C}$	17.6	16.5

---

[43] See, for example, T. D. Coyle and F. G. A. Stone, Some Aspects of the Coordination Chemistry of Boron, in "Progress in Boron Chemistry," H. Steinberg and A. L. McCloskey, Eds., Macmillan, New York (1964).

[44] H. C. Brown, *J. Chem. Soc.*, 1248 (1956).

According to the gas phase equilibrium constants (free energies) for adduct dissociation, the phosphine adduct is more than three times more stable—yet the bond energies of the adducts (more properly, the enthalpies) suggest that the N—B bond is 7 per cent stronger than the P—B bond. In both of these examples, the entropy change upon adduct dissociation (at 100°C) is nearly as important to ΔG as is ΔH.

In addition, there remains the difficult question of the phase of the reaction. An entirely gas phase reaction will be simplest to interpret because the complication of condensation energies (heats or free energies of solution, vaporization, and/or crystallization) is avoided.

These problems are summarized by the Hess' Law cycle

$$\begin{array}{ccc} AD_{(g)} & \xrightarrow{\Delta H_g} & A_{(g)} + D_{(g)} \\ \downarrow \Delta H_c^{AD} & & \downarrow \Delta H_c^A \quad \downarrow \Delta H_c^D \\ AD_{(c)} & \xrightarrow{\Delta H_c} & A_{(c)} + D_{(c)} \end{array}$$

$$\Delta H_c = \Delta H_g + [(\Delta H_c^A + \Delta H_c^D) - \Delta H_c^{AD}]$$

We see that the gas phase enthalpy and the condensed phase enthalpy are the same *only* when the differential of condensation energies is null. A particularly striking example of the importance of condensation energies to adduct formation comes from calorimetric studies[45] of the reactions

$$Me_3As_{(g)} + BF_{3(g)} \rightarrow Me_3As \cdot BF_{3(c)}$$
$$Me_3Sb_{(g)} + BCl_{3(g)} \rightarrow Me_3Sb \cdot BCl_{3(c)}$$

The enthalpy changes for these reactions are -20 and -27 kcal/mole, respectively. With adduct sublimation energies in excess of 25 kcal/mole, the first reactant pair, and probably the second also, *do not produce stable adducts in the gas phase.* Adduct formation actually depends on the exothermicity of the crystallization of the product!

When one wishes to compare the strengths of two different adducts, the order of condensed phase enthalpies will be the same as the order of gas phase enthalpies only when the differential condensation energies are sufficiently small to be negligible in determining the ordering of adduct strengths or, if large, simply reinforce the ordering of gas phase energy changes;

$$\Delta H_c(A \cdot D) - \Delta H_c(A \cdot D') = \Delta H_g(A \cdot D) - \Delta H_g(A \cdot D') + [(\Delta H_c^D - \Delta H_c^{D'}) - (\Delta H_c^{AD} - \Delta H_c^{AD'})]$$

in the latter instances, however, one is misled as to the origin of the difference in adduct stabilities.

These difficulties are mentioned here not in a disparaging sense, for it is often true that the entropy changes and/or condensation energies do fulfill the necessary requirement. The point is that one must be alert to the possibility that in a given comparison such may not be the case. It is foolish to attempt an erudite explanation of why one donor forms a stronger bond to an acid than does another when the difference is actually to be found in entropy effects or solvation energies.

---

[45] D. C. Mente, *et al., Inorg. Chem.,* **14,** 123 (1975).

STUDY QUESTIONS

16. Do you expect any difference in the spontaneity of the following reactions?

$$HOF_{(g)} \rightarrow HF_{(g)} + \frac{1}{2}O_{2(g)}$$

$$HOF_{(aq)} \rightarrow HF_{(aq)} + \frac{1}{2}O_{2(g)}$$

17. For the reaction

$$Ni(OH_2)_6^{2+} + L(H_2O)_y \xrightarrow{H_2O} NiL(OH_2)_2^{2+} + (4+y)H_2O$$

greater exothermicity is found for L = [cyclic tetraamine (cyclam-like) with four NH groups connected by ethylene bridges] than for

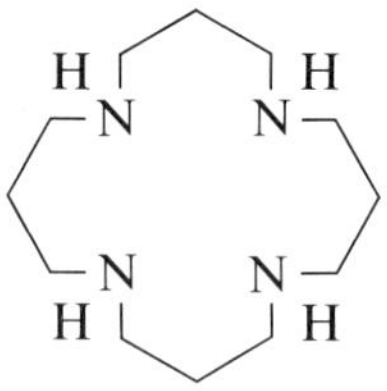

L = $H_2N(C_3H_6)NH(C_3H_6)NH(C_3H_6)NH_2$. How could solvation differences of the amines account for the difference in reaction enthalpies?

QUANTITATIVE PREDICTION OF RELATIVE ADDUCT STABILITIES[46]

Now that we understand the variety of adduct bonds possible and have seen the factors that may contribute to an experimental definition of an adduct's stability, a fair conclusion is that the *a priori* prediction of the relative strengths of two adduct bonds is very difficult. As practicing chemists, we would like to be able to rank acids and bases according to their ability to coordinate each other; that is, to assign a "strength factor" to each acid or base, such that a stronger base can always be relied upon to displace a weaker base from its adducts. Such a ranking of acids and bases is referred to as a single scale concept (the electromotive series is an example with which you are familiar), in that each molecule is endowed with a single property (in this case, donor or acceptor strength) that determines its affinity for acids or bases.

Let's pause before going further to look at some data that bear on this question of a single scale of acidities and basicities.

	ΔH_d
$Me_3Ga \cdot Me_3N$	21
$\cdot Et_2O$	9.5
$\cdot Me_2S$	8
$I_2 \cdot Me_3N$	12
$\cdot Me_2S$	8
$\cdot Et_2O$	4

---

[46] An excellent review of the concepts introduced in this section is to be found in R. S. Drago, *Structure and Bonding*, **15**, 73 (1973). Also, R. S. Drago, *J. Chem. Educ.*, **51**, 300 (1974). This review is rather critical of the natural desire to oversimplify the concepts of Lewis acidity and basicity, and will suggest to you that a controversy is being waged between two philosophies: A similar review of the other side's point of view is to be found in R. G. Pearson, Ed., "Hard and Soft Acids and Bases," Dowden, Hutchinson and Ross, Stroudsburg (1973). However, be sure to read R. T. Meyers, *Inorg. Chem.*, **13**, 2040 (1974).

We see that Me₃N forms stronger bonds to both the Ga and I acid atoms than do either the sulfide or the ether oxygen. Thus, at least with these acids, a single scale concept appears to work for the comparison of an amine with an ether or sulfide. On the other hand, the answer to the question "Is an ether a better donor than a sulfide?" clearly depends on the acid used to answer the question. Toward Ga the oxygen is *slightly* better than sulfur, while toward I the sulfide is *significantly* better. This is an example of a *reversal* in donor-acceptor interaction; *such reversals preclude the applicability of a single scale concept.* We are forced to expand our ideas of acid-base interaction to allow for *at least* two properties of each acid and base that are important in determining the affinity of each for a species of opposite type. Whatever these two properties are, the combination of acid and base that achieves the best mutual matching of these properties will be the combination yielding the strongest adduct.

As an attempt to quantify the application of a double scale relationship for acidities and basicities (and, incidentally, to determine whether a two-parameter model would work), Drago has successfully determined that the *dissociation enthalpies* of a large number of adduct bonds can be fit by an equation of the form

$$\Delta H = E_a E_b + C_a C_b \tag{5-1}$$

We find in this equation two parameters (E and C) for the acid (subscript a) and two for the base (subscript b). The enthalpies used in this equation are restricted to those determined under conditions that minimize or eliminate the condensation energies referred to earlier in this chapter—that is, enthalpies from gas phase reactions or from solution phases with solvents of poor solvating ability.

The determination of unique parameters for each acid and each base according to the dictates of equation (5-1) is impossible. Although the combination of four acids and four bases to yield 16 equations in 16 unknowns would appear to lead to a unique solution, the form of the equation prevents the unique determination of all of the E and C parameters. [For those interested, the simple equation $f = a \cdot b$ cannot be solved with $f' = a \cdot b'$, $f'' = a' \cdot b$, and $f''' = a' \cdot b'$ for unique values of a, a', b, and b' because the four equations are not independent ($f = f' \cdot f''/f'''$). The best that can be done is to compute the ratios b/b' and a/a'. That is, two independent variables (any of a, a', b, and b') must be assigned values, called the reference values, and the others determined from them.] For equation (5-1), one must either assign as reference values four of the E and C parameters or else introduce some model of adduct formation in the form of additional independent equations so as to constrain the solution (reduce the number of unknowns). To free the E and C values from a bias toward a particular theoretical description and thereby preserve their empirical objectivity, the following four reference values have been selected: for acids, $E = C = 1$ for I_2; for bases, $E = 1.32$ for N,N-dimethylacetamide and $C = 7.40$ for diethylsulfide.[47] Use of a computer to obtain the best fit of the data for a large number of adducts to equation (5-1) produces the E and C values in Tables 5-1 and 5-2.

Because of the arbitrariness surrounding the choice of four reference E and C values, it is *not* possible to draw any inferences from the relative magnitudes of E and C for a given molecule. That is, the *magnitude of C/E for one molecule actually contains no useful information.* Rather, one is limited to comparing the C value for one acid with that of another, the E value of one acid to that of another, and similarly for donors. Consequently, it is also possible to compare the C/E ratio of one acid to the C/E ratio of another, and similarly for donors. In short, as the pursuit of a physical basis for the E and C parameters continues, it is only the *relative* E_a, E_b, C_a, and C_b values, and the *relative* (C_a/E_a) and (C_b/E_b) values, that may be physically significant.

---

[47] These values have only historical significance; see ref. 46.

Acid	C/E	C	E
I_2	1.0	1.00	1.00
IBr	0.647	1.56	2.41
ICl	0.163	0.83	5.10
SO_2	0.878	0.81	0.92
$SbCl_5$	0.695	5.13	7.38
$Cu(hfac)_2$*	0.393	1.36	3.46
BMe_3	0.277	1.70	6.14
BF_3	0.164	1.62	9.88
$Zn[N(SiMe_3)_2]_2$	0.221	1.09	4.94
C_6H_5SH	0.201	0.20	0.99
C_4H_9OH	0.147	0.30	2.04
$m\text{-}CF_3C_6H_4OH$	0.118	0.53	4.48
CF_3CH_2OH	0.116	0.45	3.88
C_4H_4NH	0.116	0.30	2.54
$m\text{-}FC_6H_4OH$	0.114	0.51	4.42
$p\text{-}ClC_6H_4OH$	0.110	0.48	4.34
$p\text{-}FC_6H_4OH$	0.107	0.45	4.17
C_3F_6HOH	0.105	0.62	5.93
C_6H_5OH	0.102	0.44	4.33
C_4H_9OH	0.100	0.73	7.34
$p\text{-}CH_3C_6H_4OH$	0.097	0.40	4.18
$p\text{-}C_4H_9C_6H_4OH$	0.095	0.39	4.06
$C_7F_{15}H$	0.092	0.23	2.45
NCOH	0.080	0.26	3.22
CCl_3H	0.053	0.16	3.02
NCSH	0.043	0.23	5.30
$MeCo(oxime)_2$*	0.167	1.53	9.14
$AlEt_3$	0.163	2.04	12.5
$AlMe_3$	0.085	1.43	16.9
$GaMe_3$	0.066	0.88	13.3
$GaEt_3$	0.047	0.59	12.6
$InMe_3$	0.043	0.65	15.3
Me_3SnCl	0.005	0.03	5.76

*hfac = $HC(C(O)CF_3)_2^-$; oxime = $[(CH_3\overset{|}{C}\!\!=\!\!NO)_2H]^-$

TABLE 5–1
ACID E AND C PARAMETERS GROUPED BY ACCEPTOR ATOMS AND IN DECREASING ORDER OF C/E

Even with the non-uniqueness of the C and E parameters, there are several quantitative and qualitative uses to which they may be put. A new donor or acid may be included in the correlation by measuring its enthalpy of interaction with at least two of the acids or donors (preferably of greatly different C/E ratios) for which C and E values have been determined; enthalpies with the other acids or donors can then be estimated, usually within a few tenths of a kcal/mole.

Another important application is in the identification of steric repulsion between acid and base, as for example with Me_3NBMe_3. Using the E and C parameters for Me_3N and BMe_3 (determined, of course, from adducts with no steric congestion), the adduct dissociation enthalpy is calculated to be 24.5 kcal/mole. The experimental value is 17.6 kcal/mole. The calculated value is about 7 kcal/mole too large, a discrepancy close to that found in the heat of formation of the highly analogous Me_3CCMe_3.

Returning to an idea expressed at the beginning of this section, we might ask

TABLE 5-2
DONOR E AND C PARAMETERS GROUPED BY DONOR ATOM AND IN DECREASING ORDER OF C/E

Donor	C/E	C	E
Me$_2$Se	39.39	8.33	0.22
(CH$_2$)$_4$S	23.17	7.90	0.34
Et$_2$S	21.83	7.40	0.34
Me$_2$S	21.75	7.46	0.34
(CH$_2$)$_5$S	19.73	7.40	0.38
(CH$_2$)$_3$S	19.43	6.84	0.35
HC(C$_2$H$_4$)$_3$N	18.75	13.2	0.70
Me$_3$N	14.28	11.5	0.81
Et$_3$N	11.19	11.1	0.99
Et$_2$NH	10.20	8.83	0.87
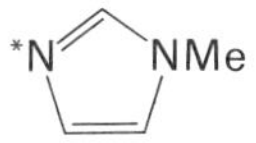 *N⟨⟩NMe (N-methylimidazole)	9.59	8.96	0.93
(CH$_2$)$_5$NH	9.23	9.32	1.01
Me$_2$NH	8.00	8.73	1.09
p-MeC$_5$H$_4$N	6.88	7.71	1.12
C$_5$H$_5$N	5.47	6.40	1.17
MeNH$_2$	4.52	5.88	1.30
EtNH$_2$	4.39	6.02	1.37
NH$_3$	2.54	3.46	1.36
Me$_2$NCN*	1.64	1.81	1.10
MeCN	1.55	1.34	0.89
ClCH$_2$CN	0.56	0.53	0.94
EtC(CH$_2$O)$_3$P*	11.70	6.41	0.55
Me$_3$P	7.82	6.55	0.84
(MeO)$_3$P*	5.82	5.99	1.03
2,2,6,6-Me$_4$C$_5$H$_6$NO*	6.79	6.21	0.92
Me$_3$PO	5.82	5.99	1.03
(CH$_2$)$_4$O	4.37	4.27	0.98
p-CH$_3$OC$_5$H$_4$NO*	4.21	5.77	1.37
(CH$_2$)$_5$O	4.12	3.91	0.95
p-CH$_3$C$_5$H$_4$NO*	3.67	4.99	1.36
HC(C$_2$H$_4$)$_2$CHO	3.48	3.76	1.08
Et$_2$O	3.38	3.25	0.96
C$_5$H$_5$NO*	3.37	4.52	1.34
(C$_4$H$_9$)$_2$O	3.11	3.30	1.06
(C$_3$H$_7$)$_2$O	2.87	3.19	1.11
(Me$_2$N)$_2$CO	2.58	3.10	1.20
Me$_2$CO	2.36	2.33	0.99
(Me$_2$N)$_3$PO	2.34	3.55	1.52
(CH$_2$)$_4$SO	2.29	3.16	1.38
O(C$_2$H$_4$)$_2$O	2.18	2.38	1.09
Me$_2$SO	2.13	2.85	1.34
(Me$_2$N)(H)CO*	2.02	2.48	1.23
(Me$_2$N)(CH$_3$)CO*	1.96	2.58	1.32
MeCO$_2$Et	1.79	1.74	0.98
MeCO$_2$Me	1.78	1.61	0.90
p-Me$_2$C$_6$H$_4$	4.28	1.78	0.42
sym-Me$_3$C$_6$H$_3$	3.82	2.19	0.57
C$_6$H$_6$	1.30	0.68	0.53

*signifies the donor atom

how the E and C values could be used to tell us whether one donor (or acid) always forms stronger adducts than another or whether a reversal is possible. The answer to this question is straightforward and can be illustrated for the case of an acid reacting with two different donors. The E and C equations for the two adducts are

$$\Delta H_b = E_a E_b + C_a C_b$$
$$\Delta H_{b'} = E_a E_{b'} + C_a C_{b'}$$

so that the difference in adduct enthalpies is

(5-2)
$$\delta H = E_a \Delta E_b + C_a \Delta C_b \qquad \gtreqless 0 \Rightarrow A \cdot B \gtreqless A \cdot B'$$

where
$$\begin{cases} \Delta E_b = E_b - E_{b'} \\ \Delta C_b = C_b - C_{b'} \end{cases}$$

We now clearly see that the difference in enthalpies is accounted for by differences in E parameters and differences in C parameters of the donors. If both the E and C parameters of one donor are greater than the E and C parameters of the other (ΔE_b and ΔC_b have the same sign), the former will form stronger adducts with all acids (remembering that steric congestion might have to be allowed for). On the other hand, if one donor has a larger E but smaller C than the other donor (ΔE_b and ΔC_b have opposite signs), then a reversal is possible. Which donor interacts more strongly then *depends critically on the acid* through E_a and C_a. For an acid with a large C term and a small E term, the sign of ΔC_b will tend to determine whether δH is positive or negative and which donor is the stronger toward that acid, *subject to the additional requirement that ΔC_b is not too small*. For an acid with a large E term and small C term, the difference in E_b's will determine the more stable adduct, as long as ΔC_b is not too large.

We can illustrate these ideas with some specific examples. For Et_3N, $E_b = 0.99$ and $C_b = 11.1$; for Et_2O, $E_b = 0.96$ and $C_b = 3.25$. Each parameter of Et_3N is greater than its counterpart for Et_2O, and we would conclude that Et_3N is always a stronger donor than Et_2O (barring appreciably greater steric congestion in some amine adducts).

For Et_2S, $E_b = 0.34$ while $C_b = 7.40$. Again, Et_3N is the stronger donor. Comparing the ether and sulfide, however, ΔE_b and ΔC_b are of opposite signs, so it is possible for some acids to interact more strongly with the sulfide than with the ether and for other acids to prefer the ether. Now, it is possible to determine the acid E_a and C_a parameters that will allow no preference for the sulfide or ether. Rearranging equation (5-2),

(5-3)
$$\delta H = E_a \left(\Delta E_b + \frac{C_a}{E_a} \Delta C_b \right)$$

The condition that the acid shows no preference for the sulfide or the ether means that

$$0 = E_a \left[\Delta E_b + \left(\frac{C_a}{E_a}\right)_0 \Delta C_b \right] \quad \text{and} \quad \left(\frac{C_a}{E_a}\right)_0 = -\Delta E_b / \Delta C_b$$

where the subscript zero on (C_a/E_a) signifies the acid (C/E) ratio of null preference. Applying this concept to the ether/sulfide case, we see there will be no difference in the enthalpies when the acid $C/E = 0.15$.

$$\left(\frac{C_a}{E_a}\right)_0 = \frac{0.62}{4.15} = 0.15$$

Acids with $C/E > 0.15$ will prefer the sulfide (ΔC_b dominates ΔE_b), while acids with $C/E < 0.15$ favor the ether (ΔE_b dominates ΔC_b). A case in point is I_2 with $C/E = 1.0$ and phenol with $C/E = 0.10$. Note carefully, however, that we cannot say whether the preference of I_2 for Et_2S is greater than the preference of ROH for Et_2O, until we compare magnitudes of δH's.

From the preceding example you might be tempted, as a simplification, to generalize that the greater the C/E ratio of the acid, the greater will be its preference for Et_2S. In fact, this is a single scale concept that will not generally apply. Consider I_2 ($C/E = 1.0$) and IBr ($C/E = 0.65$) with Et_2S and Et_2O. You can see that the reversal possible for the donors cannot be realized by this pair of acids (Et_2S is the stronger donor for both). The question is whether I_2 (with a larger C/E) shows a greater preference for Et_2S than does IBr. An affirmative prediction would be in error, as shown by the enthalpies

$$\text{IBr} \cdot Et_2S \quad 12.4 \qquad I_2 \cdot Et_2S \quad 7.7$$
$$\text{IBr} \cdot Et_2O \quad 7.4 \qquad I_2 \cdot Et_2O \quad 4.2$$

The source of difficulty is that we cannot tell from C/E ratios alone which acid forms the stronger adduct with the preferred donor; in this case both E_a and C_a of IBr are greater than E_a and C_a of I_2, and IBr is simply a stronger acid than I_2.

In terms of equation (5–3), we summarize this situation by writing

$$\delta H_{I_2} = E_a \left[\Delta E_b + \frac{C_a}{E_a} \Delta C_b \right] \qquad C_a/E_a = 1.0 \Rightarrow I_2 \cdot Et_2S > I_2 \cdot Et_2O$$

$$\delta H'_{\text{IBr}} = E_{a'} \left[\Delta E_b + \frac{C_{a'}}{E_{a'}} \Delta C_b \right] \qquad C_{a'}/E_{a'} = 0.65 \Rightarrow \text{IBr} \cdot Et_2S > \text{IBr} \cdot Et_2O$$

It is the quantity in square brackets in each case that tells us that Et_2S is preferred over Et_2O by both acids; the degree of preference (δH vs. $\delta H'$) is determined not only by the magnitude of the terms in square brackets (greater for I_2) *but also by the relative magnitudes of E_a and $E_{a'}$* (greater for IBr). In the present example, the ratio of $\delta H_{I_2}:\delta H'_{\text{IBr}}$ is less than one because the ratio of E_a's is 0.41, in spite of the fact that the ratio of square bracket terms is 1.7. In conclusion, we find that when a large C_a/E_a favors that acid relative to another, we can be *sure* that "the larger the C/E the greater the preference" *only* when E_a is also larger, that is, when *both* C_a and E_a are larger (one acid is inherently stronger than the other). Otherwise, the relative magnitudes of the E_a's may or may not allow us to think in single scale terms.

A good question to ask at this point is, "Just where does all this leave us?" (1) The E-C concept is quantitative for all adducts studied thus far under conditions free of solvation effects and large steric congestion. (2) Since it works so well, we can be encouraged (and likely surprised) that only two factors are required to account for adduct stabilities, and not surprised that we cannot confidently predict, *a priori,* the relative importance of these parameters for two acids or two bases. (3) The range of C/E values shown in Tables 5-1 and 5-2 warns that substituents about a donor or acceptor atom can have a pronounced influence on its acid or base properties and can prevent us from making very general statements about the relative acid/base properties of different *atoms*. (4) We are reminded that single scale concepts are not dependable; we can find examples of relative adduct energies that are correctly rationalized by a single scale variable like C/E and other examples in which the single scale concept of acid or base strength (defined below) will successfully predict adduct stabilities, but unfortunately there are many others in which each concept alone fails to make the correct prediction. In fact, a statistical survey[48] of the acids

---

[48] K. F. Purcell, previously unpublished. The idea behind combining the E and C parameters for a molecule into a strength is as follows: The E and C values are viewed as orthogonal components of the strength *vector* such that $\vec{S} = \vec{E} + \vec{C}$ or $|S| = (E^2 + C^2)^{1/2}$. According to vector algebra, $\Delta H = \vec{S}_a \cdot \vec{S}_b = \vec{E}_a \cdot \vec{E}_b + \vec{C}_a \cdot \vec{C}_b$. Alternately, one could write $\Delta H = |S_a||S_b| \cdot \cos\theta$, where $\theta = \tan^{-1}(C_a/E_a) - \tan^{-1}(C_b/E_b)$. This last formulation emphasizes to us that ΔH depends not only on the acid and base strengths (through $|S_a||S_b|$) but also on how well matched are the C/E ratios of the acid and base [θ is smallest, and $\cos\theta$ largest, when (C_a/E_a) is closest to (C_b/E_b)].

and bases for which E and C parameters are known reveals the following: Taking, as Drago has suggested, the root mean square of E and C ($= [E^2 + C^2]^{1/2}$) as a measure of the inherent strength of an acid or base, we ask, "What percentage of the time can we successfully predict the more stable adduct of an acid and either of two donors by simply saying that the stronger donor forms the more stable adduct?" *Of the 37,000 adduct combinations, we do no better than an out-and-out guess:* 50%. Similarly, if we assume that the acid favors the donor with a C/E ratio more like that of the acid, we make the correct prediction only 40% of the time, meaning that we do slightly better (about 60%) if we assume that the stronger adduct will be formed by the donor with a C/E ratio more different from that of the acid. In either case, we don't do well at all.

As a summary, we have constructed the following array by collecting together those acids of common acceptor atom (the number of members in each set is given in parentheses) and determining from inspection of Table 5-1 whether reversals are possible between classes. To use this array to compare one acid class with another, note the positions of the two classes in the list at the left (third and eighth, for example). Locate the entry where the position of the first class is the column number (3) and the position of the second class is the row number (8). In this case the entry is R, which stands for reversal. An entry of S or W means that the first class is stronger or weaker than the second class. Because some groups are represented by only a single molecule and because substituents bound to the acid atom can have a pronounced effect on the acidity of the atom, an entry of S or W is subject to change as more members are added to the set. The groups are well balanced, since they show reversals half the time. Noteworthy is SO_2 (only two class reversals) because the low C and E parameters of SO_2 reveal it to be a quite weak acid. This array is a major simplification of the preceding discussion, in that it does not reflect the facts that some pairs *within the same acid class* show reversals in acidities and that, for two classes showing a reversal, some members of one can be stronger or weaker than some or all members of the other.

(3)	IX	↓										
(1)	SO_2	W	↙									
(1)	$SbCl_5$	S	S	↙								
(1)	$Cu(hfac)_2$	R	S	W	↙							
(2)	BX_3	S	S	R	S	↙						
(1)	$Zn[N(SiMe_3)_2]$	R	S	W	R	W	↙					
(17)	HX	R	R	W	R	R	R	↙				
(1)	$MeCo(oxime)_2$	S	S	R	S	R	S	S	↙			
(2)	AlR_3	S	S	R	S	R	S	S	S	↙		
(2)	GaR_3	R	S	R	R	R	R	R	R	R	↙	
(2)	InR_3	R	S	R	R	R	R	S	R	R	R	↙
(1)	SnR_3Cl	R	R	W	R	W	R	R	W	W	W	W	←
# reversals for each class		6	2	5	6	6	6	7	4	4	9	8	5

18. Given that $SO_2 \cdot BF_3$ is oxygen bonded, make a prediction regarding the donor atom of SO_2 in the adduct $SO_2 \cdot SbF_5$. Qualitatively compare the C/E expected for S vs. O coordination, and comment on whether the acid characteristics of BF_3 and SbF_5 are such as to favor S or O coordination.

19. Predict the favored direction of the equilibria

   a) 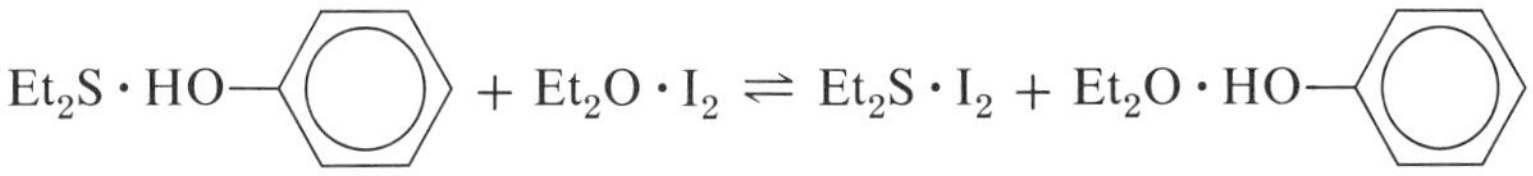 $Et_2S \cdot HO$—⟨○⟩ + $Et_2O \cdot I_2$ ⇌ $Et_2S \cdot I_2$ + $Et_2O \cdot HO$—⟨○⟩

   b) $Me_2S \cdot I_2 + Et_2O \cdot GaMe_3$ ⇌ $Me_2S \cdot GaMe_3 + Et_2O \cdot I_2$

20. For which of the following pairs are reversals in apparent donor or acceptor strength possible? Might steric effects alter any of your predictions?

   I_2, ICl $EtOAc$, Me_2CO
   BMe_3, BF_3 Me_3P, Me_3N
   SO_2, $SbCl_5$ $AlMe_3$, $GaMe_3$
   py, MeCN

STUDY QUESTIONS

ILLUSTRATIVE INTERPRETATIONS

The preceding section has alerted you to the existence of a successful empirical but quantitative model for breaking down an adduct dissociation energy into two components, $E_a E_b$ and $C_a C_b$. The *a priori* prediction of the magnitudes of these components in terms of the concepts of Chapters 2 through 4 is unfortunately far beyond our grasp. The situation is really no different in principle from that in Chapter 3, where you learned of *post facto* rationalizations of homolytic bond energies. Adduct dissociation involves heterolytic bond cleavage but, like the usual bond energy concept, depends on promotion energies (of molecules this time) and the formation of a "dative chemical bond." In this section you will see a similar Hess' Law rationale for adduct formation and its qualitative application to selected adducts. The rationale is to be developed according to the following outline for a reaction

$$AX_m + :DY_n \rightarrow X_m A\!-\!DY_n:$$

1. Structural change in the acid and base generally
   a. occurs at the expense of bonding between the acid or base atom and its substituents (A and X, and D and Y)
   b. gives rise to increased VSEPR (A—X ↔ A—X′) and steric interaction between substituents (X ↔ X′)

2. Donor-acceptor interaction (bond formation)
   a. depends on the relative energies of donor (HOMO) and acceptor (LUMO) orbitals and their overlap (often substituent dependent)
   b. is inhibited by non-bonded (steric) effects (X ↔ Y) and VSEPR effects between the adduct bond and acid-substituent bond pairs (A—D ↔ A—X)

The importance of these points can be best visualized in terms of an energy diagram.

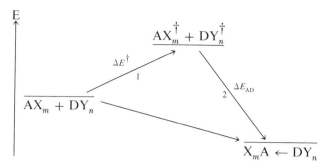

In preparing the acid and base for interaction with each other, we may visualize the distortion of each molecule into the structural configuration that it will adopt in the adduct. This will require the expenditure of energy and usually is greater for the acid because one of its atoms is to experience valence shell expansion. Generally, this means greater bond pair/bond pair and non-bonded repulsions between the substituent atoms directly bound to the donor or the acceptor atoms (such a steric interaction is often referred to as *back-strain* or B-strain). Even when the back-strain is negligible, the distorted acid or base molecular structure cannot be of the lowest energy because of weakened bonding between the donor or acceptor atoms and their substituents. The second step of the energy cycle corresponds to the release of energy as the fragments are brought together. To annalyze this we must, as a minimum, estimate *how well the donor and acceptor orbitals overlap one another and approach each other in energy*. Additionally, we may have to decide whether there will be steric interaction (*front-strain* or F-strain) between the substituents on the acid and base sites. There is always the added VSEPR about the acid as the donor pair enters the acid atom valence shell. Finally, the interaction may be facilitated or hindered by any special bonding interactions between the substituents and the acid or base atom.

Proton Affinities

The simplest acid-base reaction possible is that of donor protonation in the gas phase,

$$D + H^+ \rightarrow DH^+$$

We begin with these reactions because they allow us to analyze the donor-acceptor interaction unencumbered by acid reorganization and front-strain. Furthermore, data have recently been collected[49] that clearly reveal the effects of donor lone pair hybridization and structural reorganization on basicity. For example, the *proton affinities* of pyridine and piperidine are 225 and 230 kcal/mole, respectively. The sp^3 nitrogen of piperidine is the stronger donor. Recalling the discussions of Chapter 3 on the relationship between bond energy and hybrid percentage *s* character, you may be perplexed that the greater lone pair *s* character occurs with the weaker D—H$^+$ bond. In fact, there is a fundamental difference between these two bond cleavage processes: the process of Chapter 3 requires homolytic bond scission, while the proton affinity definition requires heterolytic cleavage.

$$D-H^+ \rightarrow D\cdot^+ + H \quad \text{HA (hydrogen affinity)}$$
$$D-H^+ \rightarrow D: \; + H^+ \quad \text{PA (proton affinity)}$$

In the present context, the energy of the homolytic scission is called the *hydrogen affinity*, HA. The two energies are related by the ionization potential difference between H and D.

$$PA = HA - (IP_D - IP_H)$$

---

[49] D. H. Aue, *et al.*, *J. Amer. Chem. Soc.*, **97**, 4136, 4137 (1975); **98**, 311, 318 (1976).

Accordingly, the HA's of pyridine and piperidine are 135 and 116 kcal/mole, respectively, and thus are in the order expected on the basis of N hybrid s character ($sp^2 > sp^3$). While the IP_H correction to HA is the same for all donors, the IP_D factor varies with lone pair Z^*; the expected order pyridine > piperidine ($sp^2 > sp^3$) is found (224 vs. 200 kcal/mole). In short, the difference between pyridine and piperidine PA's depends not only on the difference of HA's (primarily reflecting HOMO/LUMO overlap differences) but also on the difference in IP's (strictly reflecting the difference in HOMO/LUMO energy match):

$$(PA_{py} - PA_{pip}) = (HA_{py} - HA_{pip}) - (IP_{py} - IP_{pip})$$
$$-5 \quad = \quad 19 \quad - \quad 24$$

A brief physical interpretation of the difference between PA and HA is simply that with PA there is greater transfer of lone pair density from D to H^+ than is the case for HA, for which there is much less net electron transfer between $D\cdot^+$ and H. Both δHA and δIP depend on the difference in lone pair s character, and δIP is more important.

A change in donor atom from nitrogen to oxygen should also incur a change in PA through a change in lone pair Z^* (HOMO/LUMO energy match). Accordingly, Et_2O has a smaller PA than Et_2NH (198 vs. 230 kcal/mole), where both donors are "sp^3."

Given such Z^* effects on donor basicity, the lower PA of manxine than $(i\text{-}Pr)_3N$ (235 vs. 238 kcal/mole) is a good illustration of the donor reorganization contribution to basicity. Manxine is a "strained" molecule for an "sp^3" nitrogen because of transannular CH_2 repulsions; consequently, the bridgehead CCC and CNC angles are ~115°. The nitrogen lone pair should be predominately p ao in character and thus

manxine

exhibit greater basicity than that of $(i\text{-}Pr)_3N$. On the other hand, protonation of manxine should incur significant reorganization about the nitrogen as the p lone pair is polarized to form the $N-H^+$ bond.

That is, VSEPR forces should strongly favor a pyramidal nitrogen, in opposition to CH_2 repulsions constraining the nitrogen to be planar. This, then, is an (extreme) example of how donor reorganization is an influential factor in donor basicity. For comparison, quinuclidine and Et_3N exhibit the same PA of 236 kcal/mole, and space-filling models show that no strain energy is to be expected for quinuclidine.

quinuclidine

$(CH_3)_3N$, $(SiH_3)_3N$

As the next example of analyzing donor abilities of related donors, we will consider the adducts formed by trimethylamine and trisilylamine with boron trimethyl. With the same acid in both adducts, your attention should focus on the differences in the donors. As a first guess you might figure that the lower electronegativity of silicon than of carbon would lead to a stronger adduct by trisilylamine, because the greater shielding of nitrogen nuclear charge should yield a higher energy, less tightly held HOMO pair (point 2a). A check of the adduct dissociation enthalpies reveals[50] this expectation to be wrong—the silylamine adduct has a bond enthalpy of less than 1 kcal/mole, while that of the normal amine is on the order of 18 kcal/mole. A clue to the flaw in the prediction based on point 2a is that trisilylamine has a planar structure[51] while the normal amine is pyramidal—the structural reorganization of the silylamine will be greater than that of the normal amine, and this will inhibit the base character of the silylamine (points 1a and b).

The origin of the planar Si_3N grouping is a matter of some debate; most agree that the silicon valence d ao's act in a Lewis acid capacity so as to involve the nitrogen lone pair in π bonding, and this would be maximized by a planar structure. An

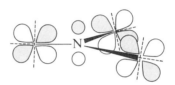

additional factor might be that the electron releasing character of the silicon, with regard to the Si—N σ bond, tends to flatten the molecule (this idea derives from increased bond-bond pair repulsion energy). In any event, the distortion of the planar molecule to the pyramidal shape needed for the adduct represents an energy cost apparently not made up for by adduct bond formation. The effect of N—Si π delocalization of the lone pair is double-edged. Once the molecule is in the pyramidal configuration, there would still be competition for the nitrogen lone pair by the silicon d orbitals; that is, the lone pair orbital would still be delocalized to some extent over the three silicon atoms and thus not highly localized on the nitrogen for good overlap with the boron acceptor orbital (point 2a). In a sense, the silicon atoms compete with

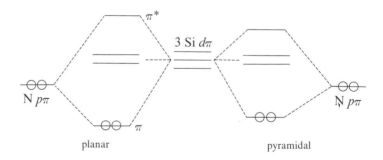

the Lewis acid for the amine lone pair. Furthermore, the N lone pair is stabilized by the $p\pi$-$d\pi$ interaction, worsening the HOMO/LUMO energy match. In summary, we expect all points from 1a through 2b to favor Me_3N as a donor.

---

[50] S. Sujiski and S. Witz, *J. Amer. Chem. Soc.*, **76**, 4631 (1954); A. B. Burg and E. S. Kuljian, *ibid*, **72**, 3103 (1950).
[51] B. Beagley and A. Conrad, *Trans. Faraday Soc.*, **66**, 2740 (1967).

Picolines

Many examples[52] exist that illustrate rather clearly the operation of steric effects. A fairly straightforward one is the comparison of pyridine, 2-picoline, and 4-picoline as donors toward the common acid BMe_3. The adduct "bond energies" are 17, 10,

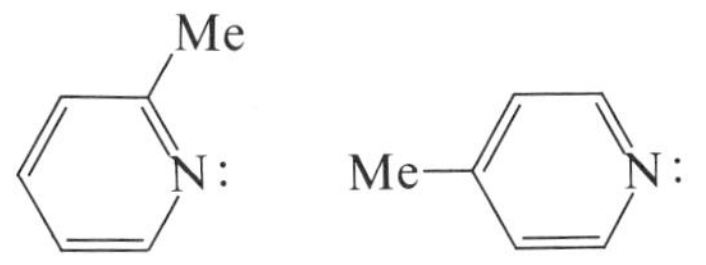

2-picoline   4-picoline

and 18 kcal/mole, respectively. Clearly, the reorganization energies (points 1a,b) of the donors are small and should be very similar. The inductive effect of the methyl group should, by increased screening of the N nuclear charge, make the *ortho* isomer the stronger donor (2a, energy match). This example, then, is one showing the effects of F-strain (2b) with 2-picoline.

Et_3N, $HC(C_2H_4)_3N$

An example involving both F- and B-strain is to be found in the comparison of triethylamine and quinuclidine (1-aza[2.2.2]bicyclooctane). You have already seen that these donors have the same PA. Construction of a molecular model of triethylamine reveals considerable steric congestion (B-strain) if all three ethyl groups

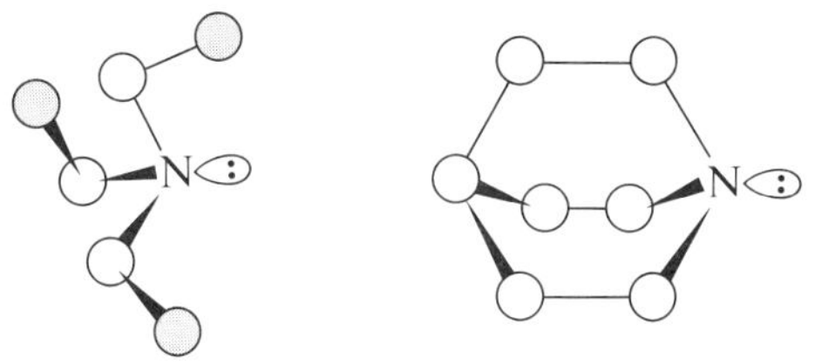

are to be "folded back" behind the nitrogen atom. In quinuclidine, the ethyl groups are constrained in the back position, so the idea of B-strain is not applicable. Toward boron trimethyl as the acid, quinuclidine is the stronger donor[53] (20 kcal/mole compared with 10 kcal/mole). The difference is ascribed to the steric congestion encountered by the third terminal methyl group of triethylamine as it attempts to avoid the methyls of BMe_3 (F-strain) and the terminal methyls of the other two ethyl groups (B-strain).

$Me_3N \cdot MMe_3$, M = B, Al, Ga, In

	ΔH (kcal/mole)[54]
$Me_3N \cdot BMe_3$	18
$\cdot AlMe_3$	30
$\cdot GaMe_3$	21
$\cdot InMe_3$	20

In interpreting the relative acidities of this series of acids, primary attention should be directed to the fact that it is the acid atom that varies; the interpretation should be based on the expected variation in acceptor properties of these atoms. The prediction of the variation of acceptor properties down this series may be complicated by the "transition series" contraction effects noted at the end of Chapter 1. Therefore, a smoothly varying set of acceptor properties may not be realized. With regard to

---

[52] A good compilation of amine $\cdot BR_3$ adduct stabilities is ref. 44.
[53] See ref. 44.
[54] G. E. Coates and R. A. Whitcombe, *J. Chem. Soc.*, 3351, 1956; G. E. Coates, *ibid*, 2003, 1951; C. H. Henrickson, *et al.*, *Inorg. Chem.* **7**, 1047 (1968); ref. 44.

point 1a, one would forecast a general favoring of the heavier acceptor because the M—C bonds tend to become weaker in MMe$_3$ (the M radial functions become more diffuse and the M—C energy match grows poorer), so the change in M—C bond energy on reorganization should become less important down the group. Similarly, point 1b should increasingly favor the heavier acids because the increasing size of the central atom should reduce CH$_3$ congestion and the increasing radial diffuseness of its hybrids toward carbon should correspond to diminishing VSEPR. Similarly (point 2b), F-strain should diminish down the series and thus increasingly favor the heavier acid. The single factor favoring the lighter acids involves point 2a. The increasing radial diffuseness of the acid acceptor hybrid, in overlap with the nitrogen lone pair hybrid, should lead to weaker adduct bonds. Similarly, the energy match between the acid LUMO and the nitrogen HOMO should grow poorer as the heavier acid atom orbital becomes less stable. It would appear that factor 2a is dominant; for, excepting boron, an overall decrease in adduct stability is observed. The discrepancy of BMe$_3$ has a partial explanation in point 2b. It is known[55] that Me$_3$CCMe$_3$ experiences about 7 kcal/mole steric strain; and the E-C model suggests a similar degree of steric congestion for Me$_3$NBMe$_3$. Allowance for F- and B-strain still leaves BMe$_3$ as the second most stable adduct in the series. (Note, however, that applying a steric congestion correction to the borane adduct and not to the others has value only if steric interactions are much smaller in the other adducts.) You should ask whether AlMe$_3$ is always a stronger acid than BMe$_3$; the E and C parameters for these acids suggest that the answer is no—reversals are possible. However, the donor $(C/E)_0$ value of no distinction is so large (40) that it is not likely that, in practice, BMe$_3$ will ever appear to form stronger adducts than AlMe$_3$.

The failure of BMe$_3$ to follow the smoothly varying trend in MMe$_3$ acidities, established by M = Al, Ga, and In, must be due to nonuniform variation of points 1a through 2b, exclusive of excessive F- and B-strain on passing from B to Al. A potential source of nonuniformity from points 1a and 2a is B—CH$_3$ π bonding. In resonance structure language, we would write

$$\ce{>B-CH_3} \leftrightarrow \ce{>B^{-}=CH_2} \; H^{+} \leftrightarrow \ce{>B^{-}=CH^{+}} \; H \; H$$

Called *hyperconjugation* when it is σ bond pairs (rather than lone pairs or π bond pairs) that are utilized to describe π bonding, such intramolecular dative bond formation competes with intermolecular bonding. A fuller exposition of the effect of such intramolecular dative π bonding on the LUMO is given in the next section and in the later section on "retrobonding."

In summary, the order AlMe$_3$ > BMe$_3$ can be attributed to greater hyperconjugation for BMe$_3$ and, for bulky donors, to greater F-strain. For (BMe$_3$, AlMe$_3$) > GaMe$_3$ > InMe$_3$ with Me$_3$N, it appears that acid Z^* and ao radial diffuseness are the most important discriminators.

π Resonance Effects

You should recall from Chapter 3 that π resonance effects can be important in determining a molecule's donor or acceptor ability. To repeat an example given before, it is generally observed that N,N-dimethylacetamide is a better donor than is acetone (the E and C parameters tell us that this should always be true) and that it is the oxygen atom, rather than the nitrogen atom, of the amide that acts as the site of basicity. Both observations naturally arise from the involvement of the single lone pair at nitrogen in three-center π bonding. (Oxygen, of course, has two in-plane lone pairs, in addition to the π pair of electrons involved in the conjugation).

---

[55] H. C. Brown and R. B. Johannesen, *J. Amer. Chem. Soc.*, **75**, 16 (1953).

$$\text{Me}_2\text{C}=\ddot{\text{O}}: \qquad \underset{\underset{\text{Me}}{|}}{\overset{\text{Me}}{\underset{\ddot{\text{N}}}{\text{C}}=\text{O}}} \leftrightarrow \underset{\underset{\text{Me}}{|}}{\overset{\text{Me}}{\underset{\overset{\oplus}{\text{N}}}{\text{C}}-\ddot{\text{O}}:^{\ominus}}}$$

A very similar explanation is proposed for the observation that tris(dimethylamino)phosphine is a stronger donor than is trimethyl phosphine[56] and that $(Me_2N)_x PMe_{3-x}$ ($x = 1$ to 3) all coordinate BH_3 through phosphorus, whereas $(Et_2NCH_2)_3P$ will easily add up to three BH_3 ligands, all at nitrogen.[57] In all these examples, the NR_2 function should be more strongly σ electron withdrawing than is the methyl group (point 2a, favoring Me_2CO) but more strongly π donating (point 2a, favoring the amide). Point 1a (stretching the C=O bond) also favors the amide. There should be little distinction in the other points. It is apparently the case that the π donation or conjugation is more important than the σ induction.

On reflecting upon the nature of these substituent effects, it is worth noticing that the π conjugation effects just mentioned do not primarily involve the donor orbitals but indirectly affect their Z^* and energies. This is to be contrasted with the direct effects of π bonding in the example of trisilyamine given earlier.

Similar π factors are known to operate in Lewis acids. Direct conjugation of the

boron acceptor orbital into the phenyl aromatic π system in triphenyl boron lowers the acidity of BPh_3 relative to BMe_3 (where only hyperconjugation is operative). This π interaction between acid site and substituent will persist even in the adduct pyramidal configuration [cf. $(H_3Si)_3N$], with several (2a, 1a) important consequences. First, the π interaction destabilizes the acceptor orbital (it is a partially B—C π* mo rather than a simply hybrid ao on boron). Second, the LUMO is delocalized over the carbon atoms, causing the LUMO overlap with donor orbitals to be reduced. In Chapter 7 you will see this conjugation has important rate and mechanism implications for bromination of the phenyl group. Furthermore, the change from planar to pyramidal geometry is opposed by the B—C interaction. Were it not for these π factors (as well as F- and B-strain), you might have predicted that BPh_3 would be a stronger acid than BMe_3 on the basis of σ effects (the sp^2 hybrid of C_6H_5 is more

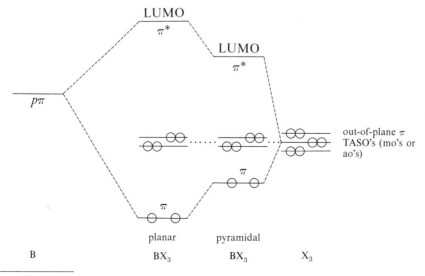

[56] See E. Fluck, *Topics in Phosphorus Chemistry*, **4**, 291 (1967).
[57] C. Jouany, et al., *J. Chem. Soc. (Dalton)*, 1510 (1974).

electronegative than the sp^3 of CH_3 and so more effectively descreens the boron nucleus).

Similar effects are found in the rather surprising trend[58] of acidities $BF_3 < BCl_3 < BBr_3 \sim BI_3$. Simple substituent inductive effects (descreening of the boron nuclear charge via withdrawal of σ electron density) would lead you to expect that BF_3 would be the strongest acid of the series and the iodide the weakest. Rather, the boron-halogen π interaction is greatest for fluorine and decreases toward iodine. This translates into a "stiffer" molecule for BF_3 (point 1a and perhaps the VSEPR part of 1b; more energy would be consumed in folding BF_3 to the extent of BI_3 or, alternately, BF_3 folds less). The degree of folding is important because, as before, the greater the folding the less anti-bonding, more localized, and more electronegative the boron acceptor mo becomes (point 2a, both energy match and overlap). A general statement implicating all these factors would be that iodine less strongly competes against the donor for the boron acceptor orbital than does fluorine.

BF_xMe_{3-x}

This series provides us with an example of the importance of σ orbital effects. Replacing F with CH_3 leads to a series of decreasing adduct stabilities[59]: BF_3 (large)[60] > BF_2Me (23) > $BFMe_2$ (18) $\sim$ BMe_3 (18) when Me_3N is the donor (the numbers in parentheses are ΔH of dissociation, in kcal/mole). A prediction of relative acidities based only on the importance of boron-substituent π bonding would suggest exactly the opposite order. Rather, increased screening of boron nuclear charge (a σ effect) by replacing F with CH_3, as well as the increased importance of F-strain and B-strain, helps us understand the observed substituent effect.

Retrodative Bonding

One of the more interesting series of donors to compare consists of the trimethyl and trifluoro amines and phosphines. Toward BH_3, the order of donor strengths is found to be $Me_3P > Me_3N > F_3P > F_3N$. (Recall from the previous discussions of NF_3 that it exhibits negligible basicity, F_3NO being the only stable compound of NF_3 that could be considered an adduct.) For each of phosphorus and nitrogen, it is not surprising that the trimethyl compounds are more basic than the trifluoro compounds. What is perplexing is that the phosphine is a stronger donor than the correspondingly substituted amine. You have already seen that Me_3N forms a stronger bond with BMe_3, even with the steric interaction, than does Me_3P. Similarly, with boron trifluoride Me_3N forms a more stable adduct (>38 kcal/mole) than does Me_3P (19 kcal/mole), and F_3P shows negligible basicity.[61] Consequently, there seems to be some special property of BH_3 as an acid that makes Me_3P and F_3P appear to be better donors than expected.

This interesting reversal in donor strengths is also apparent in the adducts of BH_3 with organo sulfides and ethers (the sulfides forming the stronger adducts with BH_3, but weaker adducts with BF_3). Even carbon monoxide (CO and NF_3 have similar PA's of $\sim$140 kcal/mole,[62] not very different from that of CH_4!) is known to form a moderately stable adduct with BH_3 and shares with PF_3 an ability to coordinate metal atoms, as we mentioned earlier in the survey of acid-base interactions.

Earlier in this chapter you learned that borane hydrogens exhibit some tendency to act in a Lewis base capacity. Coupled with the idea that phosphorus (but not nitrogen) and sulfur (but not oxygen) have vacant d valence ao's that could act in an acceptor (LUMO) capacity, it could be that back-bonding or retro-bonding occurs

---

[58] D. G. Brown, et al., J. Amer. Chem. Soc., **90**, 5706 (1968); D. F. Shriver and B. Swanson, Inorg. Chem., **10**, 1354 (1971).

[59] A. B. Burg and Sr. A. A. Green, J. Amer. Chem. Soc., **65**, 1838 (1943). While the BF_3 and BMe_3 E and C values suggest that reversals are possible, it is unlikely they can be realized in practice.

[60] $\Delta H > \sim 25$ kcal/mole cannot be reliably determined by study of $\ln K$ as a function of T^{-1}.

[61] See ref. 43. Also W. A. G. Graham and F. G. A. Stone, J. Inorg. Nucl. Chem., **3**, 164 (1956).

[62] M. A. Haney and J. L. Franklin, Trans. Faraday Soc., **65**, 1794 (1969); W. L. Jolly and D. N. Hendrickson, J. Amer. Chem. Soc., **92**, 1863 (1970); D. Holtz, et al., Inorg. Chem., **10**, 201 (1971).

between the B—H bond density and the heavy atoms; such an effect would help stabilize an adduct directly and further stimulate greater σ donor character in the base. In resonance structure terminology, we would write

$$\overset{\oplus}{>}P-\overset{\ominus}{B}\diagup\!\!\!\!\overset{H}{\underset{H}{\diagdown}}H \leftrightarrow >P=B\diagup\!\!\!\!\overset{H}{\underset{H}{\diagdown}}\overset{H}{H}, \quad >P=B\diagup\!\!\!\!\overset{H}{\underset{\ominus}{\diagdown}}\overset{H}{H^{\oplus}}, \text{ etc.}$$

in analogy with the B—CH$_3$ hyperconjugative structure.

Now, a little thought about the fact that Me$_3$P forms a more stable adduct than F$_3$P with BH$_3$ should alert you not to overestimate the importance of this retro-bonding to the energy of the adduct bond.[63] The normal σ inductive effects seem to dominate a comparison of Me$_3$P with F$_3$P. In addition, there is evidence that OR and NR$_2$ substituents enhance the phosphorus donor ability by $p\pi$-$d\pi$ interaction of their lone pairs with phosphorus [refer to (Me$_2$N)$_3$P, mentioned earlier].

Just as phosphorus has valence d ao's available for acceptor behavior, carbon monoxide has vacant π^* mo's suitable for acceptor behavior. To be sure, these mo's (just as the phosphorus ao's) are higher in energy than the bonding B—H mo's, so the

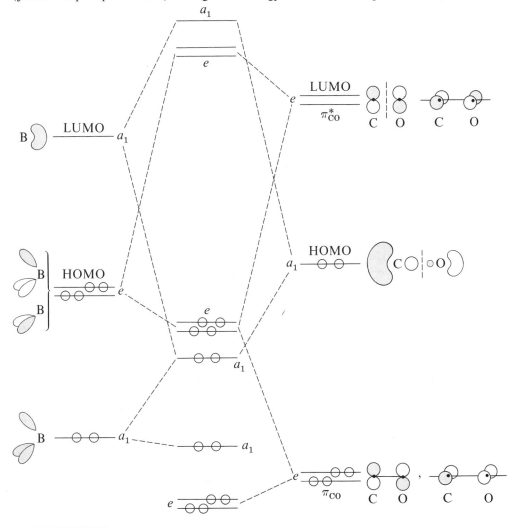

---

[63] E. L. Lines and L. F. Centofanti, *Inorg. Chem.*, **13**, 2796 (1974); A. H. Cowley and M. C. Damasco, *J. Amer. Chem. Soc.*, **93**, 6815 (1971); R. W. Rudolph and C. W. Schultz, *ibid.*, **93**, 6821, (1971); R. W. Parry, *et al.*, *Inorg. Chem.*, **11**, 1, 1237 (1972).

importance of such retro-bonding may be due as much to stimulated σ donor behavior as to the primary effect of B—C multiple bond character. This synergism appears to be quite important but is difficult to estimate and probably varies greatly, depending on the specific acid-base combination under examination.[64] The idea of retro-bonding has a convenient interpretation in terms of LCAO theory, and we can briefly outline the principal features of such a description with the diagram on p. 233. You may wish to refer back to Chapter 4 to reacquaint yourself with the origin of the BH_3 and CO orbitals and energies, and to the earlier discussion in this chapter of metal carbonyls.

In this diagram are shown the acid and base orbital sketches appropriate to the folded-back BH_3 fragment (problem 22 of Chapter 4). The borane σ acceptor mo is the LUMO (of a_1 symmetry under C_{3v}) directed along the three-fold axis and nearly localized on the boron as an "sp^3" hybrid. The primary σ donor orbital on the CO molecule is the HOMO of a_1 symmetry (called psσ in Chapter 4) and directed along the three-fold axis toward the boron (since it is considerably localized on the carbon, we might loosely refer to this mo as the "carbon lone pair" orbital). Overlap between these fragment orbitals accounts for the formation of the adduct σ bond.

Now for the retro-bonding description: The HOMO of BH_3 is the degenerate pair of mo's of e symmetry, and the LUMO of CO is the degenerate pair of π^* mo's, also of e symmetry. Each of these e symmetry mo's has a single node (within a pair, the nodal planes are at right angles to each other). The overlap between these π donor-acceptor mo's accounts for the retro-bonding and stabilization of the e symmetry electron pairs of BH_3. In an analogous fashion, PF_3 exhibits retrobonding by using d_{xz} and d_{yz} ao's in place of the π^*_{CO} mo's.

---

[64] K. F. Purcell and R. L. Martin, *Theoret. Chim. Acta* (Berl.), **35**, 141 (1974).

STUDY QUESTIONS

21. SO_3 and BF_3 are (valence shell) isoelectronic and have analogous mo structures, but are emphatically different in acid strengths. Comment on the important difference in the number of S and B valence atomic orbitals and the role of π bonding, O to S and F to B, as a factor in the magnitude of $\Delta E^{\dagger}$ for adduct formation.

22. Rationalize the linear SiNCS structure of H_3SiNCS, knowing that CH_3NCS is angular about N. Similarly, draw a conclusion about the difference in Si—P and Si—N bonding in $(H_3Si)_3P$ and $(H_3Si)_3N$ from the fact that (ignoring the hydrogens) the former has C_{3v} symmetry while the latter has D_{3h} symmetry.

23. Predict relative donor characters of Me_2O and $MeOSiH_3$.

24. "π bonding" is thought to be of prime importance to the order of adduct stabilities: $BI_3 \sim BBr_3 > BCl_3 > BF_3$. For each of the four contributions to adduct dissociation energy, tell what trend is expected from the trend in π bonding: $BF_3 > BCl_3 > BBr_3 > BI_3$.

25. For boranes and alanes (= AlR_3) that dimerize

$$2A \rightarrow A_2$$

the enthalpy of dimerization is in the range from 20 to 30 kcal/mole of dimer. Compute the range of bridge adduct bond energies; would you say that the (B, Al)–(H, R) bond pairs are weakly, moderately, or strongly basic? How does this "self-association" affect adduct stabilities in terms of ΔH for the reaction

$$D \cdot A \rightarrow D + \frac{1}{2}A_2?$$

26. The dissociation enthalpies for py·BMe$_3$ and py·BCl$_3$ are 17 and 38 kcal/mole. Rationalize the order.

27. Interpret the order of adduct stabilities (based on ΔH_d) of methyl amines with BMe$_3$, which is

$$NH_3 < MeNH_2 < Me_3N < Me_2NH$$

28. Rationalize the following order of adduct stabilities (ΔH_d):

$$Me_3N \cdot BF_3 \gg Me_3P \cdot BF_3 > Me_3N \cdot BMe_3 > Me_3P \cdot BMe_3 \gg H_3P \cdot BF_3 \gg H_3P \cdot BMe_3$$

29. Knowing that BH$_3$ and BF$_3$ bond to different donor atoms in Me$_2$NPF$_2$, predict which acid bonds to which donor atom.

30. The reaction

$$M(CO)_n + PR_3 \rightarrow M(CO)_{n-1}PR_3 + CO$$

tends to be disfavored by R groups of good π donor character (like NR$_2$). Discuss this trend in terms of P σ donor and π acceptor (with respect to the metal atom) ability.

31. Account for the basicity order

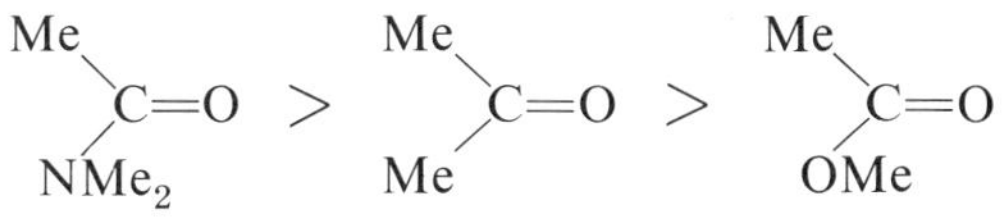

$$\underset{NMe_2}{\overset{Me}{>}}C=O \;>\; \underset{Me}{\overset{Me}{>}}C=O \;>\; \underset{OMe}{\overset{Me}{>}}C=O$$

32. Predict whether NH$_2^-$ or PH$_2^-$ has the greater proton affinity.

AN ACID-BASE VIEW OF SOLVATION PHENOMENA[65]

By far the overwhelming majority of chemical studies are performed in the solution phase, so it behooves us to understand the role of the solvent in determining the species present in the solutions we study. It is difficult to predict quantitatively the effects of various solvent properties on solution behavior of compounds, but some generalizations—derived from experience—are possible that provide useful guidelines. In a general way, we appreciate that solubilization of a solute will occur when the attraction forces between solvent molecules and solute molecules are greater than those between the solvent molecules themselves and between the solute molecules themselves. These solvent-solute attractions may be *non-specific* (van der Waals and dipolar) or *specific* (with reference to an identifiable acid-base interaction between solvent and solute) in origin. The terms specific or non-specific thereby suggest solvation by a specific donor orbital-acceptor orbital interaction or by more general intermolecular attractions, respectively. Heterolytic cleavage of solute bonds to form ionic species in solution is another important aspect of solute-solvent interactions and

---

[65] R. S. Drago and K. F. Purcell, *Progr. Inorg. Chem.*, **6**, 217, Wiley, New York (1964); R. S. Drago and K. F. Purcell, "Non-Aqueous Solvent Systems," T. C. Waddington, Ed., p. 211, Academic Press, New York (1965); D. W. Meek, "The Chemistry of Non-Aqueous Solvents," Vol. I, J. J. Lagowski, Ed., Academic Press, New York (1966).

requires efficient ion separation and solvation. These ideas are implied by equations such as

$$MX_{n(c)} \stackrel{S}{\to} [MX_n]_{(s)} \stackrel{S}{\to} [MX_{n-1}^+]_{(s)} + [X^-]_{(s)}$$

where $[\ ]_{(s)}$ symbolizes solvation of the species in brackets.

To give a preliminary illustration of these ideas, it is known that $FeCl_3$ (an ionic solid that easily sublimes to form Fe_2Cl_6 dimers in the gas phase) gives solutions of Fe_2Cl_6 in some solvents, of $FeCl_3 \cdot S$ adducts in others, of $[FeCl_{3-x}S_{3+x}]^{x+} + [FeCl_4]^-$ in yet others, or $[FeCl_{3-x}S_{3+x}]^{x+} + [Cl]^-$ in still others. In the series of equilibria

$$FeCl_{3(c)} \rightleftharpoons [Fe_2Cl_6] \rightleftharpoons [FeCl_3 \cdot S] \rightleftharpoons [FeCl_{3-x}S_{3+x}]^{x+} + [FeCl_4]^-$$

$$\rightleftharpoons [FeCl_{3-x}S_{3+x}]^{x+} + [Cl]^-$$

solvent properties play an understandable role in determining what species actually arise in solution.

From a thermodynamic viewpoint, an important aspect of the solution process is the entropy change associated with the randomization of the solute/solvent phases. The net entropy change for dissolution (and ionization) of a solute need not, however, always be positive, for on a molecular level the association between solute and solvent might be more constrained than that between solute molecules on the one hand and that between solvent molecules on the other. Whether the enthalpy change accompanying a solution process is net exo- or endothermic similarly depends on the balance between solute interaction energies and solvent interaction energies relative to the solvent-solute interaction energies. This is where the acid-base concept makes an important contribution. If we recognize the solute to be a Lewis acid (or base) and consider its solution in a basic (or acidic) solvent, it stands to reason that a sufficiently strong solute-solvent interaction (adduct formation) might be possible to effect the solution process and may even inhibit further solute reactions. Examples of this will be examined later.

Another important consideration with many potential solutes is whether solute ionization can occur in a solvent. In more general terms, what species do we expect to find in a solution? The heterolytic cleavage of solute bonds to form cationic and anionic species will depend on the ability of the solvent to stabilize the incipient cation and anion, and ion solvation becomes an important consideration. Cations are often good Lewis acids and anions are often good Lewis bases so that, again, it is convenient to consider solute ionization and ion solvation in terms of the acid-base properties of the solvent. Specific examples will follow.

One solvent characteristic of great importance in causing ion formation by a solute is the dielectric constant. The dielectric constant of a continuous medium (and it *is* inherently *inconsistent* to view a liquid phase as a uniform continuum if we subscribe to the atomic and molecular concept of matter) is defined in terms of the medium's ability to attenuate the force between two ions:

$$F = (q_1 q_2 / r^2)(1/\varepsilon)$$

where ε is the dielectric constant of the medium. The generalization we need to recognize is that a high solvent dielectric constant ($\gtrsim 15$ to 20) tends to support a high degree of solute ionization. As an empirical observation, high dielectric solvents often possess high dipole moments and/or the combination of acidic hydrogens and donor atoms in the same molecule; in either case, considerable self-association of solvent molecules is possible. Such solvent attributes should be important to ion separation, since the first few layers of polar solvent molecules about the ion would rapidly attenuate its electric field.

Table 5-3 illustrates the basis for the expectation that a high solvent dielectric supports solute ionization. The solvents in this table are listed in order of decreasing

AN ACID-BASE VIEW OF SOLVATION PHENOMENA 237

**TABLE 5-3
DATA ILLUSTRATING THE RELATIONSHIP BETWEEN DIELECTRIC AND SOLUTE IONIZATION**

Solvent	Salt	ε	K_{as}
HC(O)N(H)CH$_3$	Et$_4$NBr	182	0
CH$_3$C(O)N(H)CH$_3$	Et$_4$NBr	166	0
(CH$_3$)$_2$SO	Bu$_4$NI	47	0
(CH$_2$)$_4$SO	Ph$_4$AsCl	44	0
CH$_3$C(O)N(CH$_3$)$_2$	Et$_4$NBr	38	20
HC(O)N(CH$_3$)$_2$	Me$_4$NBr	37	37
	Me$_4$NI		15
CH$_3$CN	Me$_4$NCl	36	78
	Me$_4$NBr		41
	Me$_4$NI		28
CH$_3$OH	Bu$_4$NBr	33	26
(CH$_3$)$_2$CO	Bu$_4$NI	21	164
NOCl	[NO][FeCl$_4$]	20	268
POCl$_3$	Et$_4$NCl	14	1390
	Et$_4$NBr		530
	Et$_4$NI		260
(C$_2$H$_5$O)$_3$PO	Et$_4$N Picrate	13	1600

dielectric constant; the ion association constants for the simple salts listed are seen generally to increase with decreasing dielectric constant of the solvent. The ion association constant is the equilibrium constant for the reaction

$$[M^+]_{(s)} + [X^-]_{(s)} \rightleftharpoons [M^+X^-]_{(s)}$$

and the quaternary ammonium salts are used in order to eliminate specific cation-solvent interactions as a contribution to the K_{as}.

To give a quick illustration of the application of these ideas, we will compare the solvents H$_2$O, NH$_3$, HF, and Me$_2$SO. Of these, water is often the best solvent, for it has a high dielectric constant (the dielectric constants of the four solvents are 78, 23, 175, and 47 in the order above), possesses a moderately high dipole moment (the dipole moments are 1.85, 1.47, 1.82, and 3.96 Debyes), and has both acidic protons and lone pairs on oxygen so that it solvates bases and acids well. While ammonia is probably a better donor than water, it is also a poorer acid (less acidic hydrogens). Solute ionization is often not as great in ammonia as in water because the dielectric constant is lower and the anion solvation (via hydrogen bonding) is poorer. In situations where solvent basicity is the important property, NH$_3$ may be the better solvent. HF should be an exceptionally acidic solvent and should facilitate solution of donor molecules. Its high dielectric constant would favor appreciable solute ionization, but the fluorine atom is a very weak donor, making cation solvation not as favorable as with water or ammonia. Me$_2$SO possesses a sufficiently high dielectric constant to lead to solute ionization, and is a fairly good Lewis base. It should solvate cations and acidic solutes well and the high dipole moment helps in solvating anions (nevertheless, it lacks the acidic hydrogens of the other solvents).

To show how specific and non-specific solvation properties combine to determine the species in solution, we return to the FeCl$_3$ case brought up earlier. Of the solvents (CH$_3$)$_2$SO, pyridine, and acetonitrile, we expect the sulfoxide to show good donor ability toward the metal ion (a specific interaction), to show good anion solvation (a non-specific interaction of the ion-dipole type), and to be generally supportive of ionization. Accordingly, FeCl$_4^-$ and FeCl$_3$ solutes are dissociated into cationic chloro complexes [FeCl$_{3-x}$S$_{3+x}$]$^{x+}$ and chloride ion. Pyridine is generally a better donor than the sulfoxide (see Table 5-2) but is much poorer at non-specific ion solvation. Accordingly, FeCl$_3$ forms the simple adduct FeCl$_3 \cdot$ pyridine, and FeCl$_4^-$ is stable (does not dissociate to FeCl$_3 \cdot$ py + Cl$^-$). Acetonitrile is a generally weak donor of moderate ion solvating ability, with the result that dissolution of $FeCl_3$ in acetonitrile effects ionization only to the extent of $FeCl_2S_4^+$ and $FeCl_4^-$ (the $FeCl_4^-$ anion does not appear to dissociate appreciably into cationic chloro complexes and Cl^- ion).

Thus, the choice of a particular solvent for a chemical reaction is dictated by its chemical properties, including, but sometimes going beyond, the simple acid-base properties of the individual solvent molecules. A few examples will show how important the acid-base properties are in determining the feasibility of a chemical reaction. A good illustration is the well-known reaction in water

$$BaCl_2 + 2AgNO_3 \rightarrow 2AgCl_{(c)} + Ba(NO_3)_2$$

which occurs in the *reverse* direction in liquid ammonia

$$2AgCl + Ba(NO_3)_2 \rightarrow BaCl_{2(c)} + 2AgNO_3$$

In a qualitative sense, $Ba(NO_3)_2$ is "soluble" in both solvents, as is $AgNO_3$. The dramatic solvent effect is due to the difference that $BaCl_2$, but not $AgCl$, is soluble in water, whereas $AgCl$ is moderately soluble in ammonia while $BaCl_2$ is not. In short, the relative solubilities of $AgCl$ and $BaCl_2$ reverse on changing from H_2O to NH_3. Water is expected to solvate Cl^- more energetically than ammonia, but ammonia should solvate Ag^+ better than water. Obviously, the Ag^+ and Cl^- ion *solvation energies* are less than the $AgCl$ *lattice energy* in water but not in ammonia. This would seem to be due to a greater difference between H_2O and NH_3 cation solvations than anion solvations. For $BaCl_2$ and $AgCl$ the anion solvation order is, of course, the same ($H_2O > NH_3$); but the cation solvation appears to reverse, as the $Ba^{2+} + 2Cl^-$ ion solvation energy in water is greater than the lattice energy, whereas it is not in ammonia.

If one wished to synthesize an ether adduct with boron trimethyl, he would be foolish to use a basic solvent like pyridine, which is a stronger donor than ether. The pyridine solvent would complex the boron trimethyl, and the ether molecules would be unable to displace pyridine. Conversely, an ether solvent would be a good choice for synthesis of the pyridine adduct. Acetonitrile (a less basic donor than ether) would be a logical choice for solvent in synthesis of the ether adduct.

Similarly, experience has shown[66] that HCl is not a good choice for synthesis of $GeCl_6^{2-}$ from $GeCl_4$ and Cl^- because the highly acidic HCl solvent competes with $GeCl_4$ for Cl^- through formation of the hydrogen-bonded complex HCl_2^-. $MeNO_2$, which is sufficiently polar to dissolve many quaternary ammonium chlorides but not basic enough to strongly complex the $GeCl_4$, does not solvate Cl^- well enough to inhibit formation of the hexachlorogermanate anion.

In an entirely analogous fashion, water is too basic to permit a reaction between phosphine and hydrogen ion to form PH_4^+ salts, whereas the very weakly basic solvent HCl makes synthesis of $PH_4^+BCl_4^-$ straightforward[66]:

$$PH_3 \xrightarrow{HCl} PH_4^+ + HCl_2^- \xrightarrow{BCl_3} [PH_4][BCl_4]_{(c)}$$

A reaction unknown in mixed aqueous/organic solutions is

$$KF + Cl-C_6H_4-X \rightarrow F-C_6H_4-X + KCl$$

---

[66] T. C. Waddington, "Non-Aqueous Solvents," Appleton-Century-Crofts, New York (1969).

However, in Me$_2$SO we find the reaction possible,[67] probably because of poorer solvation of F$^-$ by Me$_2$SO than by water.

As a final interesting example of the solvent effect on the apparent relative basicities of donors (earlier in this chapter you saw that solvation or condensation energies could play an important role in such equilibria), the pK_b's (defined by the following reaction) of amines fall in the unexpected order NH$_3$ > Me$_3$N > MeNH$_2$ > Me$_2$NH

$$\text{amine} + \text{H}_2\text{O} \xrightarrow{\text{H}_2\text{O}} \text{amine} \cdot \text{H}^+ + \text{OH}^-$$

(the smaller pK_b, the greater the basicity), whereas the PA's fall in the order[68] NH$_3$ < MeNH$_2$ < Me$_2$NH < Me$_3$N for the protonation reaction

$$\text{amine} + \text{H}^+ \rightarrow \text{amine} \cdot \text{H}^+$$

The key to understanding the aqueous system ordering lies first in realizing that the proton basicity of the alkyl amines as a group is greater than that of ammonia; this dominates the comparison of R$_x$NH$_{3-x}$ with NH$_3$. Within the alkyl amine series, alkyl substitution does lead to enhanced basicity in the gas phase, so we look to the difference in amine/ammonium ion solvation energies to account for the relatively large pK_b of Me$_3$N. The solvation (via N $\cdots$ H$_2$O hydrogen bonding) of the amines should *inhibit* ammonium ion formation most for Me$_3$N and least for MeNH$_2$. Solvation of the ammonium cations (via N—H $\cdots$ OH$_2$) should *favor* ammonium ion formation most for MeNH$_3^+$ and least for Me$_3$NH$^+$. Thus, the change in solvation energies in the reaction amine $\rightarrow$ ammonium ion would tend to make Me$_3$N appear the weakest base and MeNH$_2$ the strongest base. Increasing methyl substitution increases the nitrogen atom gas phase basicity but decreases the aqueous basicity through solvation energies. The anomalous position of Me$_3$N in the aqueous series arises from the solvation inhibition of its basicity; this is an excellent example of the point on page 217 that solvation energies may confuse interpretation of adduct bond energies.

For protonic and halide solvent systems, the ideas developed above for solvent-solute interaction can be expressed in terms of the concept of acid and base *leveling* by the solvent.[69] With water, any acid HA that extensively reacts with H$_2$O to form the hydronium ion is said to be too strong and is leveled to the acidity of H$_3$O$^+$. This is nothing more than a statement that, in water as solvent, H$_2$O is a stronger Brönsted base than is the anion A$^-$. Similarly, all bases stronger than OH$^-$ are leveled in water to OH$^-$.

$$\text{HA} + \text{H}_2\text{O} \rightarrow \text{H}_3\text{O}^+ + \text{A}^-$$
$$\text{B} + \text{H}_2\text{O} \rightarrow \text{BH}^+ + \text{OH}^-$$

For ammonia, NH$_4^+$ is the strongest acid and NH$_2^-$ is the strongest base. In liquid HF, HF$_2^+$ and HF$_2^-$ are the strongest acid and base permitted.

$$2\text{NH}_3 \rightarrow \text{NH}_4^+ + \text{NH}_2^-$$
$$2\text{HF} \rightarrow \text{H}_2\text{F}^+ + \text{HF}_2^-$$

This concept leads to the idea that acetic acid is a strong acid in liquid ammonia (much as HCl is a strong acid in water) and NH$_4$OAc is a strong acid in ammonia.

---

[67] G. C. Finger and C. W. Kruse, *J. Amer. Chem. Soc.*, **78**, 6034 (1956).
[68] M. S. B. Munson, *J. Amer. Chem. Soc.*, **87**, 2332 (1965); J. I. Brauman and L. K. Blair, *ibid*, **90**, 6561 (1968); J. I. Brauman, *et al., ibid*, **93**, 3914 (1971).
[69] H. H. Sisler, "Chemistry in Non-Aqueous Solvents," Reinhold, New York (1961).

These ideas were, of course, implied in the examples of the preceding paragraphs, where we found PH_3 to act like a strong Brönsted donor in liquid HCl but a weak Brönsted donor in H_2O. Cl^- is a strong base in HCl and accordingly is leveled to the basicity of HCl_2^-, making difficult the synthesis of $GeCl_6^{2-}$ (which must accordingly be considered a strong base, too) in liquid HCl.

The application of these ideas to solvent choice for synthesis work can be further illustrated by selecting a solvent for preparation of a salt of the urea anion. The choice could be approached in two equivalent ways: find a solvent that makes urea appear to be a strong acid,

$$H_2NC(O)NH_2 + S \rightarrow SH^+ + HNC(O)NH_2^-$$

or find a solvent that will tolerate the urea anion as not a very strong base.

$$HNC(O)NH_2^- + S \not\rightarrow H_2NC(O)NH_2 + (S-H)^-$$

Liquid ammonia is a good choice, for it is highly basic; acids that are weak in water appear much stronger in ammonia. Similarly, we could predict that $(H_2NCO)NH^-$ would be a weaker base than $(H)NH^-$. Reaction of an amide salt with urea in ammonia would effect the desired synthesis of H_2NCONH^-.

$$NH_2^- + H_2NC(O)NH_2 \xrightarrow{NH_3} NH_3 + HNC(O)NH_2^-$$

This same concept can be applied to the choice of solvents for working with oxidizing and reducing agents.[70] While there is sometimes difficulty in defining a set of oxidized and reduced forms for a particular solvent, it stands to reason that some solutes will be too oxidizing or reducing for use in a particular solvent. We can, of course, set rough upper limits for each solvent, based on its most highly oxidized and reduced forms. In aqueous chemistry we are accustomed to the concept that oxidants more oxidizing than O_2 will oxidize water solvent, and that those more reducing than H_2 will reduce water (often, however, we can seemingly violate such dictates of thermodynamics because the gas overvoltages, a kinetic feature, are often appreciable). By analogy, one can work in HCl solvent with oxidizing agents up to the power of Cl_2 but only with reducing agents less powerful than H_2. In Chapter 6 you will find a good application of this idea in the choice of HSO_3F as a medium for synthesis of halogen and sulfur cationic species.

Solvent	"Reduced Form"	"Oxidized Form"
H_2O	H_2	O_2
HCl	H_2	Cl_2
HSO_3F	H_2	$S_2O_6F_2$
$AsCl_3$	As	Cl_2

To conclude this section on solvents, we will examine a few inorganic solvents that are commonly used for synthesis purposes and that have distinctive properties quite unlike those of the more customary aqueous and organic solvents you are familiar with. These solvents are NH_3, HF, H_2SO_4, SO_2, and HSO_3F.[71]

---

[70] See ref. 66.

[71] In addition to those references already given, excellent in-depth summaries by authorities in the field are to be found in: J. J. Lagowski, Ed., "The Chemistry of Non-Aqueous Solvents," Vols. I and II, Academic Press, New York (1966); T. C. Waddington, Ed., "Non-Aqueous Solvent Systems," Academic Press, New York (1965).

LIQUID AMMONIA[72]

M.P. (°C)	−78
B.P. (°C)	−33
η (centipoise)	0.25 (−33°C)
ρ (g cm^{-3})	0.69 (−40°C)
ε	23
Λ (ohm^{-1} cm^{-1})	10^{-11}
K_{ion}	10^{-27}

Liquid ammonia is a colorless fluid only 70% as dense as water and only $\frac{1}{4}$ as viscous (an obvious advantage in filtration). In spite of the low boiling point of liquid ammonia, the liquid has a sufficiently high heat of vaporization (~5 kcal/mole) that it can be handled in unsilvered vacuum dewars at room temperature with tolerable loss to evaporation. Further comparing ammonia with water, the auto-ionization constant is considerably smaller (that for water is $\sim 5 \times 10^{-14}$)

$$2NH_3 \rightleftharpoons NH_4^+ + NH_2^-$$

and the specific conductance is much smaller (that for water is $\sim 10^{-7}$). The Grotthuss (or proton "switching") mechanism for the high conductance of acidic

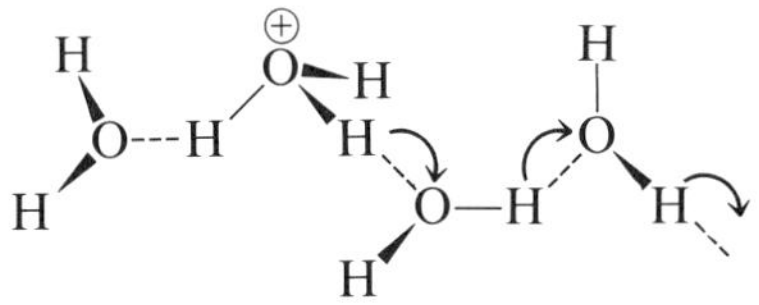

or basic water solutions is clearly not applicable to liquid ammonia solutions of NH_4^+ and NH_2^-. The data below reveal the anomalously high mobilities of H^+ and OH^- in

Ion	H_2O λ_0	NH_3 λ_0
H_3O^+	350	
NH_4^+	73	142
Na^+	50	158
OH^-	198	
NH_2^-		166
NO_3^-	71	177

aqueous solvent and the perfectly normal mobilities of NH_4^+ and NH_2^- in liquid ammonia. This marked difference in acid and base mobilities is, of course, expected from the point of view that NH_4^+ is a weaker hydrogen-bonding acid than is H_3O^+, in spite of the fact that NH_3 is a stronger donor than H_2O. Further evidence (in addition to the difference in H_2O and NH_3 boiling points) for the weaker hydrogen-bonding interactions and more difficult proton transfer in NH_3 than in H_2O comes from nmr studies of the exchange rates

$$NH_2^- + NH_3 \rightleftharpoons NH_3 + NH_2^- \quad k \sim 10^8 \text{ M}^{-1} \text{ sec}^{-1}$$
$$NH_4^+ + NH_3 \rightleftharpoons NH_3 + NH_4^+$$
$$OH^- + H_2O \rightleftharpoons H_2O + OH^- \quad k \sim 10^{10} \text{ M}^{-1} \text{ sec}^{-1}$$
$$H_3O^+ + H_2O \rightleftharpoons H_2O + H_3O^+$$

where those for H_2O fall in the range of diffusion-controlled reaction velocities at 25°C.

---

[72] J. J. Lagowski and G. A. Moczygemba, in "The Chemistry of Non-Aqueous Solvents," Vol. II, J. J. Lagowski, Ed., Academic Press, New York (1967); W. L. Jolly and C. J. Hallada, in "Non-Aqueous Solvent Systems," T. C. Waddington, Ed., Academic Press, New York (1965).

TABLE 5-4

SOLUBILITIES OF METAL HALIDES IN LIQUID AMMONIA AT 0°C (g/100 g soln.)

Cation	Cl$^-$	Br$^-$	I$^-$	NO$_3^-$
Li$^+$	1.43	—	—	—
Na$^+$	11.37	29.00	56.88	56.05
K$^+$	0.132	21.18	64.81	9.52
Rb$^+$	0.289	18.23	68.15	—
Cs$^+$	0.381	4.38	60.28	—
Ag$^+$	0.280	2.35	84.15	—
NH$_4^+$	39.91	57.96	76.99	—
Mg^{++}	—	0.004	0.156	—
Ca^{++}	—	0.009	3.85	45.13
Sr^{++}	—	0.008	0.308	28.77
Ba^{++}	—	0.017	0.231	17.88

Solubilities in liquid ammonia are interesting to contrast with those in water. It has been estimated that the non-specific solvation possible with H$_2$O is dominated by the high polarity (dipole moment) of the molecule, with the polarizability and van der Waals contributions being some five times less effective. With ammonia there is a more even balance between dipole and van der Waals forces, with the dipolar contribution being some two times smaller than that for water but the van der Waals contribution being two times greater. Consequently, it is not surprising to learn that liquid ammonia is the better solvent for solutes that have a large number of electrons or that are highly polarizable. This is illustrated by the fact that organic compounds such as hydrocarbons tend to have higher solubilities in ammonia than in water. Data that reflect this tendency of ammonia to best solvate species with high polarizabilities are given in Tables 5-4 and 5-5.

As the solubility data suggest, salts of divalent ions are often much less soluble in ammonia than in water, but exceptions are known. Particularly for heavy transition metal ion salts (note the silver halides above), appreciable solubility can be traced to the energetic coordination of the metal ion by ammonia. This is an example of the

TABLE 5-5

SOLUBILITIES IN LIQUID AMMONIA AT 25°C (g/100 g NH$_3$)

NH$_4$NO$_3$	390.0	NaCl	3.02
NH$_4$I	368.4	KClO$_3$	2.52
NH$_4$SCN	312.0	H$_3$BO$_3$	1.92
(NH$_4$)(CH$_3$CO$_2$)	253.2	KCNO	1.70
LiNO$_3$	243.7	AgCl	0.83
NH$_4$Br	237.9	NaF	0.35
AgI	206.8	Na$_2$S$_2$O$_3$	0.17
NaSCN	205.5	ZnI$_2$	0.1
KI	182.0	KCl	0.04
NaI	161.9	MnI$_2$	0.02
NaBr	138.0	NaNH$_2$	0.004
NH$_4$ClO$_4$	137.9	KBrO$_3$	0.002
(NH$_4$)$_2$S	120.0	(NH$_4$)$_2$SO$_3$	0.0
NH$_4$Cl	102.5	(NH$_4$)$_2$HPO$_4$	0.0
NaNO$_3$	97.6	(NH$_4$)HCO$_3$	0.0
Ba(NO$_3$)$_2$	97.2	(NH$_4$)$_2$CO$_3$	0.0
Sr(NO$_3$)$_2$	87.1	Li$_2$SO$_4$	0.0
AgNO$_3$	86.0	Na$_2$SO$_4$	0.0
Ca(NO$_3$)$_2$	80.2	KI$_3$	0.0
KBr	13.5	K$_2$SO$_4$	0.0
KNO$_3$	10.4	K$_2$CO$_3$	0.0
AgBr	5.9	BaCl$_2$	0.0
KNH$_2$	3.6	ZnO	0.0

importance of the donor strength and polarizability of ammonia. Some interesting data to ponder at this point are the heats of solution of salts in water and in ammonia. At 25°C and for formation of 0.002 mole ratio (solute:solvent) solutions (~0.1 M) the following results have been obtained (in kcal/mole):

	$\Delta H_{solution}$	
	H_2O	NH_3
KI	+5	−9
CsI	+8	−5
NH_4I	+3	−16
NaCl	+1	−7
NH_4Cl	+4	−8
$Ba(NO_3)_2$	+10	−15

While these data make ammonia appear to be the better solvent, we must be careful not to overinterpret the data. The dielectric constant of ammonia is only a third of that of water (23 vs. 78). Consequently, the extent of endothermic ion separation achieved in liquid ammonia is far less than that for water, perhaps accounting for the different signs of the heats. Thus, the ion pairing is not so completely disrupted by solvent ammonia as by water. Heats of solution at infinite dilution would allow more direct inferences regarding relative ion solvation. (See the later discussion of electrode potentials for estimates of ion solvation energies.) Some data that illustrate the degree of ion association in ammonia are the following (at −34°C):

	K_{as}
KCl	10^3
KBr	5×10^2
KI	10^2
NaBr	3×10^2
KNH_2	10^4

In many respects the most intriguing solutes for liquid ammonia are the *alkali metals*. Since they are strong reducing agents, we might expect the alkali metals to cause H_2 evolution, as happens very rapidly with H_2O. Quite often, however, there is a kinetic inhibition of the reduction reaction in ammonia, so that alkali metal solutions are kinetically (but not thermodynamically) stable. On the other hand, insoluble metals do not generally exhibit an overvoltage effect. Before continuing with a summary of the properties of such solutions, an examination of the thermodynamics of redox behavior in ammonia from a general point of view is in order.

The H_2/NH_3 and NH_3/N_2 half-cell potentials for liquid ammonia show very little difference under standard state acid (1 M NH_4^+) and base (1 M NH_2^-) conditions.

$$NH_3 + \frac{1}{2} H_2 \to NH_4^+ + e^- \qquad E° = 0$$

$$4NH_3 \to 3NH_4^+ + \frac{1}{2} N_2 + 3e^- \qquad E° = -0.04$$

$$NH_2^- + \frac{1}{2} H_2 \to NH_3 + e^- \qquad E° = 1.59$$

$$3NH_2^- \to 2NH_3 + \frac{1}{2} N_2 + 3e^- \qquad E° = 1.55$$

These pH-dependent $E°$'s are diagrammed as solid lines in Figure 5-1; the narrow span in $E°$'s of only 0.04 V indicates that it should be very difficult to find solutes that would *not* be capable of oxidizing or reducing ammonia. For a chemical species to be

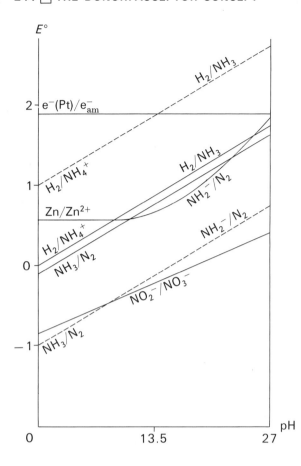

Figure 5-1. The oxidation potential diagram for liquid ammonia, illustrating the "overvoltage" range and pH dependence. e^-_{am} represents the solvated electron.

stable in liquid ammonia, its redox half-cell potential would have to fall between the parallel solid lines at the center of Figure 5-1. Reducing agents stronger than H_2 should reduce NH_4^+ and ammonia, and oxidizing agents stronger than N_2 should oxidize NH_2^- and ammonia. Fortunately, the H_2 and N_2 couples commonly, but not always, exhibit overvoltages of about 1 volt so that, as the dotted lines in Figure 5-1 indicate, the practical working range for liquid ammonia is considerably extended beyond that thermodynamically possible. Species that we would normally characterize as strong oxidizing agents (MnO_4^-, O_3, O_2^-) and reducing agents (Na, Cs) can accordingly be handled in liquid ammonia.

Some reagents exhibit a very different potential dependence on pH than do the H_2/NH_3 and N_2/NH_3 couples and can be utilized only in basic or acidic solutions. For example, the solvated electron is stable in solutions at least a little more basic than neutral (pH = 13.5), nitrate ion is stable to the acid side of neutral solutions, and Zn metal (insoluble) reduces ammonia at all but two narrow pH ranges.

Table 5-6 presents some electrode potential data[73] for metals in liquid ammonia and, for comparison, similar data for aqueous solution. The data for liquid ammonia, of course, constitute a self-consistent set of data for making comparisons within that solvent. While it is not possible to make a precise thermodynamic measurement of the relative H_2 electrode potentials in the two solvents, useful estimates place the reference potential in ammonia approximately 1 volt more negative than that in water. The data as presented allow us to conclude that the degree of solvation of the alkali metal ions is nearly the same for water and ammonia. For the transition metal ions,

---

[73] H. Strehlow, in "The Chemistry of Non-Aqueous Solvents," Vol. I, J. J. Lagowski, Ed., Academic Press, New York (1966).

TABLE 5-6
STANDARD (1 M ACID)
ELECTRODE POTENTIALS

	H_2O (25°C)	NH_3 (−35°C)
Cs/Cs^+	2.92	1.95
Rb/Rb^+	2.93	1.93
K/K^+	2.93	1.98
Na/Na^+	2.71	1.85
Li/Li^+	3.05	2.24
Zn/Zn^{2+}	0.76	0.53
Cd/Cd^{2+}	0.40	0.20
Pb/Pb^{2+}	0.13	−0.32
H_2/H^+	0	0
Cu/Cu^{2+}	−0.34	−0.43
Cu/Cu^+	−0.52	−0.41
Ag/Ag^+	−0.80	−0.83
I_2/I^-	1.06	
Br_2/Br^-	1.60	
Cl_2/Cl^-	1.97	

however, ammonia solvation is greater than water solvation (recall the previous discussion of solubilities), and this is to be expected from the recognized greater basicity of ammonia. Ion solvation energy values derived from thermodynamic data and shown in Table 5-7 are consistent with these conclusions.

Now we can return to the topic of metal solutions. The metal properties that favor their dissolution are those of low metal lattice energy, high cation solvation energy, and—interestingly enough—low metal ionization potential. Consequently, it is the alkali metals, the heavy alkaline earth metals, and divalent rare earth metals that show greatest solubility. Solubilities of the alkali metals fall in the range from 10 to 20 moles of metal/1000 grams of NH_3, so that quite concentrated solutions can be prepared (Table 5-8).

The three most striking physical properties of these solutions are their color (dilute solutions are blue, concentrated solutions are bronze), their conductivity (dilute solutions have conductivities an order of magnitude greater than those of salts in water, while the concentrated solutions have conductivities approaching those of pure metals), and their magnetic properties (at infinite dilution the magnetic susceptibilities approach one equivalent of unpaired electrons, and the solutions become progressively less paramagnetic as the concentration of metal increases). The conductivity/concentration curves for Na and K are shown in Figure 5-2. While the full

TABLE 5-7
FREE ENERGIES OF
SOLVATION (KCAL/MOLE)

	H_2O	NH_3
Cl^-	−75	−68
Br^-	−69	−66
I^-	−61	−60
H^+	−257	−278
Li^+	−117	−121
Na^+	−94	−97
K^+	−77	−77
Rb^+	−73	−71
Cs^+	−63	−63
Ag^+	−111	−132
Zn^{2+}	−493	−530
Cd^{2+}	−428	−457

246 ☐ THE DONOR/ACCEPTOR CONCEPT

	gram-atoms metal/kg NH$_3$	ΔH (kcal/mole)	
Li	15.7	−9.7	(0.07 M)
Na	10.9	+1.4	(0.4 M)
K	11.9	0.0	(0.07 M)
Rb	—	0.0	(0.13 M)
Cs	25.1 (−50°C)	0.0	(0.19 M)
Ca	—	−19.7	(0.02 M)
Sr	—	−20.7	
Ba	—	−19.0	

TABLE 5–8
SOLUBILITIES AND HEATS OF SOLUTION OF METALS IN LIQUID AMMONIA (−33°C)

nature of the species present in these solutions is not understood at present, the conductivity and magnetic data seem best interpreted in terms of the presence of five species (M, M$_2$, M$^+$, M$^-$, and e$^-_{am}$) related according to the equilibria

$$M \rightleftharpoons M^+ + e^-_{am} \qquad K \sim 10^{-2}$$
$$2M \rightleftharpoons M_2 \qquad K \sim 10^{-4}$$
$$e^-_{am} + M \rightleftharpoons M^- \qquad K \sim 10^3$$

Particularly intriguing is the concept of a solvated electron, e$^-_{am}$. While the identity of this species is still subject to considerable research, agreement has centered on the qualitative correctness of a "cavity" (on the order of 3 Å in diameter) of NH$_3$ molecules in which the electron is trapped. The quantitative aspects of the conductance and magnetic data seem to indicate that the solvated electron behaves in a normal fashion for an anionic species in liquid ammonia, in that it is extensively ion-paired with a cationic species.

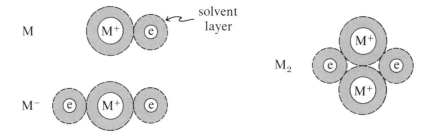

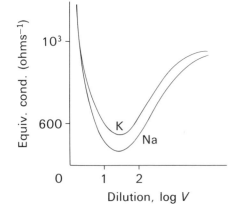

Figure 5–2. Equivalent conductance of metal-ammonia solutions at −33.5°. V = liters of ammonia (of density 0.674) in which one gram-atom of metal is dissolved.

Liquid ammonia is a very versatile solvent for effecting many reactions that are impossible in aqueous media. In particular, it is a very useful solvent for working with materials that are too strongly basic for aqueous media—examples are alkoxides, sulfides, and organic amide anions. Just as in aqueous systems, many reactions can be effected in NH_3 by simple metathesis, if the solubility relationships are right. Owing to the general high solubility of NO_3^- in ammonia, we find reactions like

$$Ba(NO_3)_2 + 2AgCl \rightarrow BaCl_2\downarrow + 2AgNO_3$$
$$2KOR + M(NO_3)_2 \rightarrow M(OR)_2\downarrow + 2KNO_3$$
$$(NH_4)_2S + Ba(NO_3)_2 \rightarrow BaS\downarrow + 2(NH_4)_2NO_3$$

In further analogy with aqueous chemistry, wherein metals or basic materials react with hydronium ion (H_3O^+) to evolve H_2 or form "hydrides," we find reactions in ammonia like

$$Co + 2NH_4NO_3 \rightarrow Co(NH_3)_6^{2+} + H_2 + 2NO_3^-$$
$$\text{"}Mg_xSi_y\text{"} + NH_4NO_3 \rightarrow SiH_4$$

The counterpart of OH^- in water chemistry is played, in ammonia chemistry, by NH_2^-, often with novel results (generally KNH_2 has sufficient solubility to be useful—$LiNH_2$ and $NaNH_2$ do not):

$$C_2H_2 + KNH_2 \rightarrow KC_2H + NH_3$$
$$NaNH_2 + 2KNH_2 \rightarrow K_2[Na(NH_2)_3]$$
$$AgNH_2 + KNH_2 \rightarrow K[Ag(NH_2)_2]$$
$$Zn + 2KNH_2 + 2NH_3 \rightarrow K_2[Zn(NH_2)_4] + H_2$$

Similarly, many non-metal halides undergo so-called solvolytic reactions in water to eliminate HX and form a hydroxy derivative of the non-metal halide; in liquid ammonia we find many analogous reactions leading to amide derivatives in place of the hydroxide derivatives:

$$BX_3 + 6NH_3 \rightarrow B(NH_2)_3 + 3NH_4X$$
$$SOCl_2 + 4NH_3 \rightarrow SO_2(NH_2)_2 + 2NH_4X$$
$$SO_3 + 2NH_3 \rightarrow [NH_4][SO_3NH_2]$$

In this last reaction the initial adduct $SO_3 \cdot NH_3$ is too strong an acid for NH_3 solvent (compare the solvation of SO_3 by H_2O) and is "leveled" to the acidity of NH_4^+.

Among the more interesting reactions possible in liquid ammonia are those of alkali metal solutions. These solutions are strongly reducing and make possible the formation of interesting reduced species. Among the simple electron transfer reactions we find the reaction with O_2

$$e_{am}^- + O_2 \rightarrow O_2^-$$

to form superoxides that are stable in NH_3, you recall, because the overvoltage effect is large. The production of complexes of metal *atoms* from complexes of cationic metals is particularly intriguing:

$$2e_{am}^- + Pt(NH_3)_4^{2+} \rightarrow Pt(NH_3)_4$$
$$+ Ni(CN)_4^{2-} \rightarrow Ni(CN)_4^{4-}$$

The large majority of reduction reactions by metals in ammonia are not simple electron transfer reactions but entail further reaction of the initially produced reduction product. Here we find reduction of non-metal hydrogen bonds,

$$e^-_{am} + AH \rightarrow A^- + \frac{1}{2} H_2 \quad A = OR, NH_2, PH_2$$

of carbon-sulfur bonds,

$$e^-_{am} + R_2S \rightarrow \frac{1}{2} R_2 + RS^-$$

and, more rarely, the production of stable radicals by control of the stoichiometry

$$e^-_{am} + Et_3SnBr \rightarrow Et_3Sn\cdot + Br^-$$

More generally, the electron transfer reactions lead to heterolytic bond cleavage by net two-electron processes. A few examples are

$$2e^-_{am} + Ge_2H_6 \rightarrow 2GeH_3^-$$

$$2e^-_{am} + RX \rightarrow R^- + X^- \xrightarrow{NH_3} RH + X^- + NH_2^-$$

(the occurrence of the last step depends on R^- being more basic than NH_2^-).

$$2e^-_{am} + PhHN-NH_2 \rightarrow PhNH^- + NH_2^-$$

$$2e^-_{am} + NNO \rightarrow N_2 + O^{2-} \xrightarrow{NH_3} OH^- + NH_2^-$$

(O^{2-} is too strong a base for ammonia and is leveled to OH^- and NH_2^-.)

$$2e^-_{am} + RCH=CH_2 \rightarrow [R\bar{C}H-\bar{C}H_2] \xrightarrow{2NH_3} RCH_2CH_3 + 2NH_2^-$$

Of some interest to those developing techniques for forming catenated species are the reactions of Li, Na, and K with elemental sulfur to open and cleave the eight-membered sulfur ring, thereby forming various anionic polysulfides

$$\frac{16}{n} e^-_{am} + S_8 \rightarrow \frac{8}{n} S_n^{2-} \quad n = 1 \text{ to } 7$$

Even more interesting is the reduction of diorganotin dihalides to form the diorganostannate anion. The possibility of R_2Sn^{2-} being a good nucleophile accounts for the succeeding reactions with RX and R_3SnX:

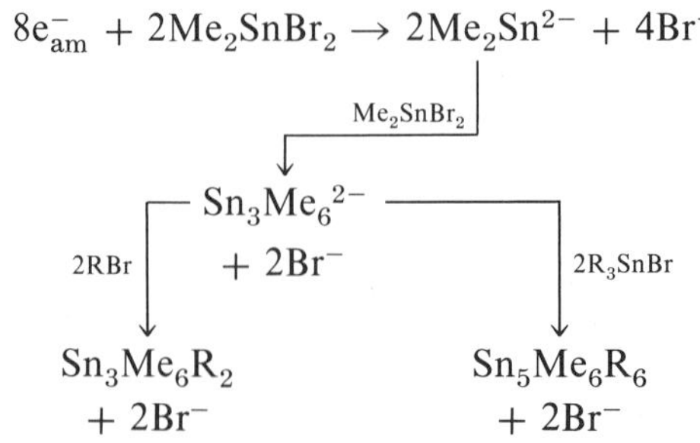

$$8e^-_{am} + 2Me_2SnBr_2 \rightarrow 2Me_2Sn^{2-} + 4Br^-$$

$$\downarrow Me_2SnBr_2$$

$$Sn_3Me_6^{2-} + 2Br^-$$

$$\swarrow 2RBr \qquad \searrow 2R_3SnBr$$

$$Sn_3Me_6R_2 + 2Br^- \qquad Sn_5Me_6R_6 + 2Br^-$$

HYDROFLUORIC AND SULFURIC ACIDS[74]

	HF	H_2SO_4
M.P. (°C)	−89	10
B.P. (°C)	20	290
η (centipoise)	0.26 (0°C)	24.5 (25°C)
ρ (g cm^{-3})	1.00 (0°C)	1.83 (25°C)
ε	84 (0°C)	100 (25°C)
Λ (ohm^{-1} cm^{-1})	10^{-6} (0°C)	10^{-2} (25°C)
K_{ion}	10^{-12}	10^{-4}

These solvents are interesting from several points of view. With regard to the oxidizing properties of solutes, both can tolerate the presence of highly oxidizing ions. At the same time, we must realize that such solutes are often powerful Lewis acids, too strong for *simple* existence in either solvent, so that coordination of F^- or HSO_4^- achieves a leveling of solute acidity. Of the two, H_2SO_4 is the stronger oxidant in the usual sense and provides somewhat more limited opportunities for synthetic application (Br^- and I^- are readily oxidized). However, a limitation for HF is its propensity for degradative fluorination of many materials.

Of the two solvents, H_2SO_4 is the more highly associated (refer to the viscosities above). Both solvents are associated; HF exists as chains and rings of various sizes (with an average of ~3.5 HF per structure),

while H_2SO_4 has a three-dimensional network. The high association is a result of hydrogen bond interactions between molecules. For HF, the one-dimensional nature of these interactions and the weakness of the van der Waals attractions between the rings and/or chains result in the fairly low heat of vaporization of 1.8 kcal/mole. This can be an advantage in synthesis work with HF, for the solvent can be easily vacuum distilled at low temperature from the reaction mixture. Such a technique is very difficult to implement with H_2SO_4.

Tables 5-9 and 5-10 present some solubility data for the two acidic solvents. Given the high self-association of H_2SO_4, an important aspect of the solution process for H_2SO_4 is the disruption of the strong bonding between solvent molecules. High solvation energy of the solute is required for appreciable solubility. Electrolytes normally satisfy this requirement because of high ion solvation energies, and non-electrolytes usually undergo protonation if they are soluble.

Because of the high dielectric constants of both acids, they facilitate ionization of a solute that is soluble in them. The role of the dielectric constant in this process is supported by the specific (hydrogen bonding) solvation of the anions formed. In some cases noted later, the solvation carries on to proton transfer when the anion is a stronger base than HF_2^- or HSO_4^-.

Another ramification of the extent of H_2SO_4 association and the basicity of the solvent molecule is seen in the following comparison of ion mobilities in H_2SO_4.

---

[74] M. Kilpatrick and J. G. Jones, in "The Chemistry of Non-Aqueous Solvents," Vol. II, J. J. Lagowski, Ed., Academic Press, New York (1967); W. H. Lee, H. H. Hyman, and J. J. Katz, in "Non-Aqueous Solvent Systems," T. C. Waddington, Ed., Academic Press, New York (1965); R. J. Gillespie and E. A. Robinson, *ibid.*

TABLE 5-9
SOLUBILITY OF SOME FLUORIDES IN
HYDROGEN FLUORIDE

Fluoride	Solubility in HF (g/100 g)	Temperature °C	Fluoride	Solubility in HF (g/100 g)	Temperature °C
LiF	10.3	12	PbF_2	2.62	12
NaF	30.1	11	FeF_2	0.006	12
KF	36.5	8	CoF_2	0.036	14
RbF	110.0	20	NiF_2	0.037	12
CsF	199.0	10	AlF_3	<0.002	11
NH_4F	32.6	17	CeF_3	0.043	12
AgF	83.2	19	TlF_3	0.081	12
Hg_2F_2	0.87	12	SbF_3	0.536	12
TlF	580.0	12	ClF_3	0.10	12
CuF_2	0.010	12	BiF_3	0.010	12
AgF_2	0.048	12	MnF_3	0.164	12
CaF_2	0.817	12	FeF_3	0.008	12
SrF_2	14.83	12	CoF_3	0.257	12
BaF_2	5.60	12	ZrF_4	0.009	12
BeF_2	0.015	11	CeF_4	0.10	12
MgF_2	0.025	12	ThF_4	<0.006	18
ZnF_2	0.024	14	NbF_5	6.8	25
CdF_2	0.201	14	TaF_5	15.2	25
HgF_2	0.54	12	SbF_5	∞	25

Clearly, the bisulfate ion and $H_3SO_4^+$ are strongly hydrogen bonded to the solvent, and a Grotthuss mechanism, based on proton transfer between molecules, is called for to explain the high mobilities of HSO_4^- and $H_3SO_4^+$.

Ba^{2+}	2
Na^+	3
K^+	5
$H_3SO_4^+$	242
HSO_4^-	171

TABLE 5-10
SOLUBILITIES OF METAL SULFATES
IN SULFURIC ACID AT 25°C

Sulfate	Solubility (mole %)	Sulfate	Solubility (mole %)
Li_2SO_4	14.28	$PbSO_4$	0.12
Na_2SO_4	5.28	$CuSO_4$	0.08
K_2SO_4	9.24	$FeSO_4$	0.17
Ag_2SO_4	9.11	$NiSO_4$	very small
$MgSO_4$	0.18	$HgSO_4$	0.78
$CaSO_4$	5.16	Hg_2SO_4	0.02
$BaSO_4$	8.85	$Al_2(SO_4)_3$	<0.01
$ZnSO_4$	0.17	$Tl_2(SO_4)_3$	<0.01

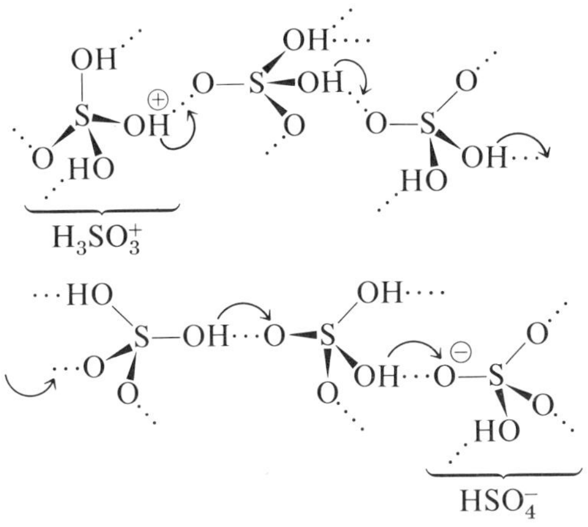

[Structural diagrams showing H₃SO₃⁺ and HSO₄⁻ formation via proton transfers among H₂SO₄ molecules]

The property of making normally weak bases appear strong in solution is not unexpected for such strong acids and is dramatically illustrated, for H_2SO_4 in particular, by the solvent auto-ionization constants. For HF the ionization is usually represented as

$$3HF \rightleftarrows H_2F^+ + HF_2^-$$

where HF exhibits base character. Here you should reflect for a moment on the role of ion solvation as a factor in self-ionization.

H_2SO_4 presents a more complicated case. The ionization constant cited earlier is for the reaction

$$2H_2SO_4 \rightleftarrows H_3SO_4^+ + HSO_4^- \quad K_1 = 10^{-4}$$

This is the largest known solvent auto-ionization constant and makes H_2SO_4 appear to be a reasonably good Lewis base. Surely, however, this apparent basicity is largely due to particularly effective solvation of $H_3SO_4^+$ and HSO_4^- by H_2SO_4. Other important equilibria for pure H_2SO_4 are

$$2H_2SO_4 \rightleftarrows H_3O^+ + HS_2O_7^- \quad K_2 \sim 10^{-5}$$
$$H_2S_2O_7 + H_2SO_4 \rightleftarrows H_3SO_4^+ + HS_2O_7^- \quad K_3 \sim 10^{-2}$$

Consequently H_2SO_4, $H_3SO_4^+$, HSO_4^-, H_3O^+, $H_2S_2O_7$, and $HS_2O_7^-$ are the species present in pure H_2SO_4 (other than H_2SO_4, to the extent of 0.04 M). In the second of these equilibria we see an example of the well-known tendency of H_2SO_4 to act as a strong desiccant—in this instance dehydrating itself, so to speak. The reaction could be viewed as arising from dissociation of H_2SO_4 into its Lewis acid-base components

$$H_2SO_4 \rightarrow H_2O + SO_3$$

followed by protonation of the "strong" base H_2O and leveling of the strong acid SO_3 by H_2SO_4. The third equilibrium again illustrates the idea that H_2SO_4 exhibits some basic character in H_2SO_4 solvent. Furthermore, the fact that $K_3 \sim 100\, K_1$ implies that (at least in H_2SO_4) pyrosulfuric acid is stronger than H_2SO_4. $H_2S_2O_7$ is one of the few acids strong enough to protonate H_2SO_4. In this regard, HCl and HF react to form the halosulfuric acids, HSO_3X. Like $HClO_4$, neither of these act as Brönsted acids or bases, while HNO_3 and H_3PO_4 are protonated by H_2SO_4 solvent.

As mentioned earlier, a characteristic chemical process for solute/solvent interaction in these highly acidic solvents is solute protonation. When the solute possesses an OH grouping, the protonation of the OH oxygen can be followed by elimination of

H_2O in the form of H_3O^+ (reflect that H_2O would be a good base in these solvents). The "onium" ion left behind can exist in a nonspecifically solvated condition or, as is more often the case, it has a high acidity that is leveled by coordination of F^- or HSO_4^-. In this way we account for the formation of some rather unusual chemical species.

$$XOH + HA \rightarrow XOH_2^+ + A^- \xrightarrow{HA} X^+ + H_3O^+ + 2A^-$$

or

$$\xrightarrow{HA} XA + H_3O^+ + A^-$$

where X is some substituent.

The solution of metal chlorides in either acid results in evolution of HCl. Similarly, NO_3^- and ClO_4^- salts yield, at least initially, HNO_3 and $HClO_4$.

$$NO_3^- + H_2SO_4 \rightarrow HNO_3 + HSO_4^-$$
$$NO_3^- + 2HF \rightarrow HNO_3 + HF_2^-$$

These reactions have been proposed to account for the fact that one formula weight of a perchlorate salt in HF, for example, yields a solution with an average of 2.7 equivalents of dissolved species (as determined from the boiling point elevation of HF). The nitronium ion, NO_2^+, is a well-known species, and you should not be particularly surprised to learn that NO_3^- solutions in HF or H_2SO_4 ultimately yield this ion through the dehydration phenomenon (the process is quite familiar to the organic chemist, who recognizes the HNO_3/H_2SO_4 mixed solvent as a powerful nitrating agent).

$$HNO_3 + HA \rightarrow H_2NO_3^+ + A^- \xrightarrow{HA} H_3O^+ + NO_2^+ + 2A^-$$

Boric acid presents another example of acid dehydration in H_2SO_4 solvent. It is an interesting reaction because $HB(SO_4H)_4$ (in addition to $H_2S_2O_7$) appears to be one of the few strong acids in H_2SO_4:

$$B(OH)_3 + 6H_2SO_4 \rightarrow B(OSO_3H)_4^- + 3H_3O^+ + 2HSO_4^-$$

The same type of solvolysis reaction follows for borate esters in HF.

$$B(OR)_3 + 7HF \rightarrow 2ROH_2^+ + ROH \cdot BF_3 + 2HF_2^-$$

The Lewis acid BF_3 competes successfully with HF for ROH as a base, since

$$ROH_2^+ + BF_4^- \rightarrow ROH \cdot BF_3 + HF$$

Additional examples worth mentioning are iodic acid (HIO_3) and phosphoric acid.

$$HIO_3 + 4HF \rightarrow [IO_2]^+ + H_3O^+ + 2HF_2^- \rightarrow IO_2F + H_3O^+ + HF_2^- + HF$$

Similarly, in H_2SO_4,

$$HIO_3 + 2H_2SO_4 \rightarrow IO_2OSO_3H + H_3O^+ + HSO_4^-$$

Notice the feature in both solvents of the high acidity of IO_2^+.

A point of difference in the behaviors of HF and H_2SO_4 arises with phosphoric acid:

$$H_3PO_4 + H_2SO_4 \rightarrow H_4PO_4^+ + HSO_4^-$$
$$H_3PO_4 + 3xHF \rightarrow POF_x(OH)_{3-x} + xH_3O^+ + xHF_2^-$$

Finally, we might note the slow reaction between H_2SO_4 and HF in liquid HF:

$$H_2SO_4 + HF \xrightarrow{HF} H_3O^+ + SO_3F^-$$

The protonation of alcohols by these solvents is another reaction they have in common. However, in H_2SO_4 a dehydration reaction follows; if the substituents about the carbonium center sufficiently stabilize the cation, coordination of HSO_4^- does not follow. Primary and secondary alcohols generally yield insufficiently stable carbonium ions to elude the acid leveling effect of HSO_4^-. The presence of two or three phenyl groups, however, seems sufficient to stabilize the carbonium ion.

$$EtOH + H_2SO_4 \rightarrow EtOH_2^+ + HSO_4^- \xrightarrow{H_2SO_4} EtSO_4H + H_3O^+ + HSO_4^-$$

$$Ph_3COH \xrightarrow{2H_2SO_4} Ph_3C^+ + H_3O^+ + 2HSO_4^-$$

An interesting stereochemical study of sulfonation of secondary alcohols using optically active butane-2-ol shows retention of configuration as OH is replaced by SO_4H. This certainly seems to militate against the unimolecular loss of H_2O from ROH_2^+, since the planar R^+ intermediate formed in such a step would be susceptible to attack by HSO_4^- from "either side" of the plane to yield a racemic mixture of products. Similarly, a Walden inversion mechanism is inappropriate. Rather, it may be speculated that HSO_4^- displacement of H_2O may occur through a "front" attack on ROH_2^+ or on a "loose" complex between R^+ and H_2O,

$$ROH + 2H_2SO_4 \nrightleftharpoons \left[\overset{|}{\underset{|}{C^+}} \right] + H_3O^+ + 2HSO_4^-$$

[structure: C+ with OSO$_3$H$^-$ and OH$_2$ attached] $\rightarrow$ C—OSO$_3$H + H_3O^+ + HSO_4^-

or by nucleophilic attack by ROH upon $H_3SO_4^+$ or one of the other important species present in pure H_2SO_4.

[structure showing H$_2$O$^+$ and R—OH attacking sulfur center]

Carboxylic acids behave as fairly strong bases in HF and H_2SO_4, as evidenced by the conductivity and cryoscopic properties of their solutions. In H_2SO_4 we again encounter the dehydration process, if the acylium ion to be formed is sufficiently stable and/or if the carboxylate group is sterically crowded.[75]

$$CH_3CO_2H + HA \rightleftharpoons CH_3C(OH)_2^+ + A^-$$

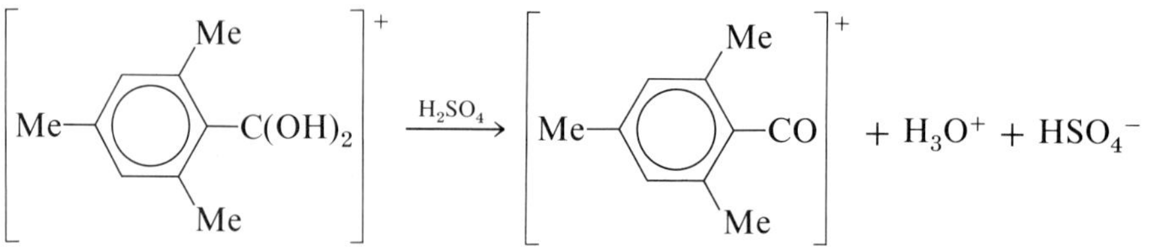

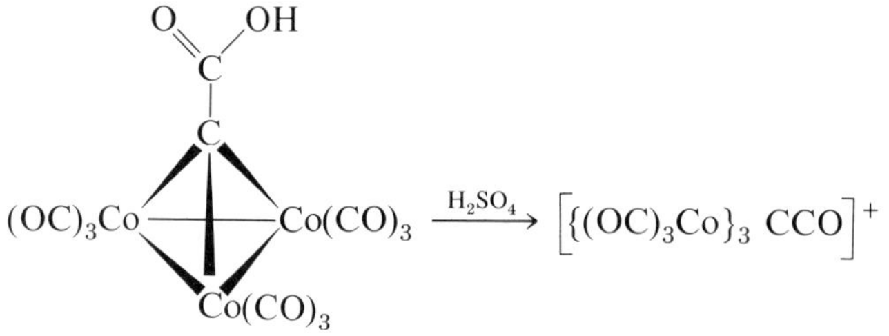

$$\left[Me-\underset{Me}{\underset{|}{\overset{Me}{\overset{|}{C_6H_2}}}}-C(OH)_2\right]^+ \xrightarrow{H_2SO_4} \left[Me-\underset{Me}{\underset{|}{\overset{Me}{\overset{|}{C_6H_2}}}}-CO\right]^+ + H_3O^+ + HSO_4^-$$

$$(OC)_3Co\underset{\underset{Co(CO)_3}{|}}{\overset{\overset{O\diagup\diagdown OH}{C}}{\underset{|}{\overset{|}{C}}}}Co(CO)_3 \xrightarrow{H_2SO_4} \left[\{(OC)_3Co\}_3\ CCO\right]^+$$

An interesting variant in this general behavior is shown by formic and oxalic acids, which evolve CO and CO/CO_2, respectively, on treatment with H_2SO_4.

$$HCO_2H \xrightarrow{H_2SO_4} [HCO^+] + H_3O^+ + HSO_4^-$$
$$\xrightarrow{H_2SO_4} CO + H_3SO_4^+$$

$$HO_2C-CO_2H \xrightarrow{H_2SO_4} [HO_2C-CO^+] + H_3O^+ + HSO_4^-$$
$$\hookrightarrow CO + [HO_2C^+]$$
$$\xrightarrow{H_2SO_4} CO_2 + H_3SO_4^+$$

It appears that HCO^+ and CO_2H^+ are strong acids in H_2SO_4 (*i.e.*, CO and CO_2 are weak donors toward the proton, as noted earlier by the PA of CO; here they are seen to be weaker than H_2SO_4).

Earlier in this chapter we surveyed the class of donors known as π donors. While normally considered very weak bases (refer to the *E* and *C* values), olefins and polynuclear aromatic hydrocarbons dissolve in H_2SO_4 with protonation. In the case of olefins, if the substituents about the carbonium center do not sufficiently stabilize the center, sulfonation occurs. Mono-aryl olefins are a case in point.

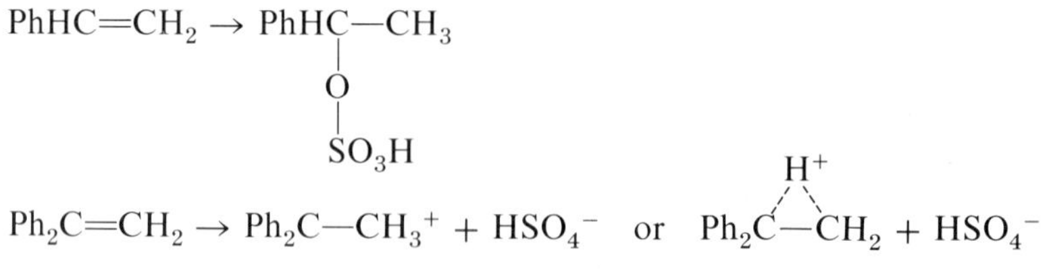

$$PhHC=CH_2 \rightarrow PhHC-CH_3$$
$$\underset{SO_3H}{\underset{|}{\overset{|}{O}}}$$

$$Ph_2C=CH_2 \rightarrow Ph_2C-CH_3^+ + HSO_4^- \quad \text{or} \quad Ph_2\overset{H^+}{\overset{\diagup\diagdown}{C-CH_2}} + HSO_4^-$$

---

[75] D. Seyferth, *et al.*, *J. Amer. Chem. Soc.*, **96**, 1730 (1974).

1,1-Diaryl olefins appear to fall in the class of "stable" carbonium ions. A final interesting point is that if the olefin is not sufficiently basic to undergo *complete* conversion to the carbonium ion, the unprotonated olefin present may react with the carbonium ion to effect polymerization of the olefin.

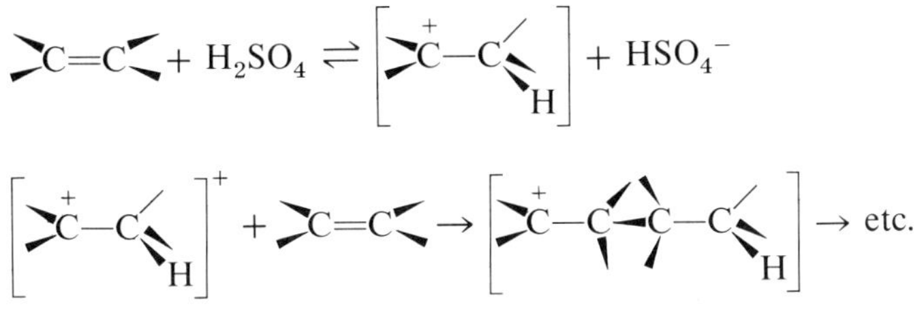

$$\text{>C=C<} + H_2SO_4 \rightleftharpoons \left[\text{>}\overset{+}{C}\text{-C<}_H\right] + HSO_4^-$$

$$\left[\text{>}\overset{+}{C}\text{-C<}_H\right]^+ + \text{>C=C<} \rightarrow \left[\text{>}\overset{+}{C}\text{-C-C-C<}_H\right] \rightarrow \text{etc.}$$

Quite analogous phenomena are encountered in liquid HF, *i.e.*, polymerization, addition of HF, and so forth.

Before leaving this discussion of acidic solvents, we need to identify two other aspects of HF chemistry. First is the behavior of non-metal fluorides in HF. The two general classes of behavior are summarized in the equations

$$EF_n + HF \rightarrow [EF_{n-1}]^+ + HF_2^- \quad \text{(fluoride donors)}$$
$$EF_n + 2HF \rightarrow [EF_{n+1}]^- + H_2F^+ \quad \text{(fluoride acceptors)}$$

The pentavalent fluorides of P, As, and Sb are the most often encountered examples of the F^- acceptor behavior, while the halogen fluorides are the principal examples of the donor class. Quite obviously, the EF_{n-1}^+ cations are strongly oxidizing and require a non-reducing solvent to stabilize them. Additionally, a high solvent ε value assists these ion formation reactions.

The final useful property of HF as a solvent to be mentioned is also related to its non-reducing character. Anodic oxidations in HF have made possible the synthesis of fully fluorinated derivatives of hydrogen compounds. While the mechanisms of most of these reactions have not been delineated, it is highly plausible that atomic fluorine is the reactive species:

$$F^- \rightarrow F\cdot + e^-$$
$$NH_4F \rightarrow NH_xF_{3-x}$$
$$H_2O \rightarrow OF_2$$
$$Et_2O \rightarrow (C_2F_5)_2O$$
$$CH_3CN \rightarrow CF_3CN, C_2F_5NF_2$$

SULFUR DIOXIDE[76]

M.P. (°C)	-75
B.P. (°C)	-10
η (centipoise)	0.43 ($-10°C$)
ρ (g cm^{-3})	1.46 ($-10°C$)
ε	15.6 (0°C)
Λ (ohm^{-1} cm^{-1})	10^{-8} ($-10°C$)

Molecular solids and liquids tend to be much more soluble in liquid SO_2 than do ionic materials. Among the ionic materials, iodides seem to be most soluble; often

---

[76] T. C. Waddington, Ed., "Non-Aqueous Solvent Systems," Academic Press, New York (1965).

TABLE 5-11

SOLUBILITIES OF IONIC COMPOUNDS IN LIQUID SULFUR DIOXIDE AT 0°C (MILLIMOLES SOLUTE/1000 g SOLVENT)

	I^-	Br^-	Cl^-	F^-	SCN^-	CN^-	ClO_4^-	CH_3COO^-	SO_4^{2-}	SO_3^{2-}	CO_3^{2-}
Li^+	1490.0	6.0	2.82	23.0				3.48	1.55		
Na^+	1000.0	1.36	insol.	6.9	80.5	3.67		8.90	insol.	1.37	
K^+	2490.0	40.0	5.5	3.1	502.0	2.62		0.61	insol.	1.58	
Rb^+			27.2							1.27	
NH_4^+	580.0	6.0	1.67		6160.0		2.14	141.0	5.07	2.67	
Tl^+	1.81	0.60	0.292	insol.	0.915	0.522	0.43	285.0	0.417	4.96	0.214
Ag^+	0.68	0.159	<0.07	insol.	0.845	1.42		1.02	insol.	insol.	
Be^{2+}			5.8								
Mg^{2+}	0.50	1.3	1.47							insol.	
Ba^{2+}	18.15	insol.	insol.		insol.						
Zn^{2+}	3.45		11.75		40.4			insol.			
Cd^{2+}	1.17		insol.								
Hg^{2+}	0.265	2.06	3.80		0.632	0.556		2.98	0.338		
Pb^{2+}	0.195	0.328	0.69	2.16	0.371	0.386		2.46	insol.		
Co^{2+}	12.2		1.00		insol.						
Ni^{2+}	insol.		insol.					0.08	insol.		insol.
Al^{3+}	5.64	0.60	v. sol.								
Sb^{3+}	0.26	21.8	575.0	0.56							
Bi^{3+}		3.44	0.60								

cyanide, thiocyanate, and acetate salts have appreciable solubility (Table 5-11). At the beginning of this chapter some adducts of SO_2 were mentioned that illustrated the ability of SO_2 to function as a weak base (by either O or S coordination) and as a weak acid (by S coordination). This is thought to be a major factor in the solvation of solutes by SO_2. However, the weakness with which SO_2 specifically solvates ions is probably a major factor in the low solubilities of salts. Nevertheless, there is evidence for the occurrence of such specific solvation effects in the isolation of solvates. Only a few examples of oxygen-coordinated solvates are known, *viz.*, $BF_3 \cdot SO_2$ and $SbF_5 \cdot SO_2$, while many more are believed to be of the sulfur *acid* variety. Such solvate formation is common for halide salts, particularly the iodides, thiocyanates, and sulfates.

The formation of conducting solutions by many normally covalent halogen compounds is thought to be due to the effective solvation of the halide, an effect supporting solute ionization in spite of a rather low dielectric constant for liquid SO_2. We must be careful to note, however, that such solutions are only weakly conducting, presumably because of the low ε of SO_2. Similarly, even ionic materials are highly ion paired and therefore weakly conducting in SO_2.

$$EX \rightleftharpoons (E^+, X^-) \rightleftharpoons E^+_{solv} + X^-_{solv}$$

Interesting examples of solute ionization arise with triarylchloromethanes and HBr. The presence of three aryl substituents about the carbon facilitates the formation of a stable carbonium ion, as you saw in the previous discussion of acidic solvents. That SO_2 effects partial ionization of such chlorocarbons, while benzene with a higher dielectric constant (25) does not, is thought to be due to a specific solvation of the chloride ion. While HBr is a non-electrolyte (weak acid) in liquid SO_2, an equimolar mixture of HBr/H_2O in SO_2 is conducting. Electrolysis of such a solution liberates H_2O and H_2 at the cathode and bromine at the anode. The cathodic products strongly suggest the presence of H_3O^+ in liquid SO_2.

Before continuing on to a discussion of the chemical properties of SO_2, you should know that liquid SO_2 has a sufficiently high boiling point and a low enough viscosity to be useful as a preparative solvent. With a heat of vaporization like that of NH_3 (6 kcal/mole for SO_2), the solvent can often be used at room temperature in unsilvered vacuum dewars but can be easily removed at low temperatures by vacuum distillation. These factors, together with its property of being neither strongly acidic nor strongly basic, make liquid SO_2 an especially useful solvent for reactions between polar and/or highly polarizable solutes.

One of the most interesting ideas concerning the species present in liquid SO_2 is that of Jander, who suggested[77] that SO_2 auto-ionizes according to

$$2SO_2 \rightleftharpoons SO^{2+} + SO_3^{2-} \quad \text{(actually } S_2O_3^{2+}(?) + S_2O_5^{2-}\text{)}$$

The sulfite ion in this proposed ionization reaction is, of course, a well-known chemical species. Controversy settles in the existence of SO^{2+} (or $S_2O_3^{2+}$). Unfortunately, no *direct* evidence has been presented to refute the existence of SO^{2+} in liquid SO_2. On the other hand, no compounds are known containing SO^{2+}, but this does not in itself preclude the formation of small amounts of the cation in solution.

Radio-exchange experiments[78] with isotopically labeled *SO_3 and S^*O_3 reveal the occurrence of rapid oxygen exchange between SO_3 and SO_2 but a negligibly slow sulfur exchange. Both experimental results are consistent with (1) a weak ionic equilibrium

$$SO_2 + SO_3 \rightleftharpoons SO^{2+} + SO_4^{2-}$$

as a means to effect *O exchange but no *S exchange, and (2) a bimolecular exchange process through an intermediate or transition state species that is sufficiently asymmetric to maintain the identities of the sulfur atoms. Given the strongly acidic nature

of SO_3 and acidic (S atom) and basic (O atom) properties of SO_2, the bimolecular mechanism makes good sense. On the other hand, the equivalent conductance of 0.1 M SO_3 in SO_2 was reported long ago[79] to be approximately 10^{-1} ohm^{-1} cm^{-1}, suggesting the presence of ionic species. Whether the conductivity is due to SO^{2+} and SO_4^{2-} or to the presence of impurities appears open to question; confirmation of the conductivity study is needed.

Most damaging to the hypothesis of the occurrence of SO^{2+} in liquid SO_2 is the observation[80] that there is no sulfur exchange and no oxygen exchange between SO_2 and the thionyl halides $SOCl_2$ and $SOBr_2$. Reactions such as

$$SO^{2+} + SOX_2 \rightleftharpoons SOX^+ + SOX^+$$

would seem to be a distinct possibility in these solutions if SO^{2+} were present to any significant extent. Accordingly, one would expect both sulfur and oxygen exchange to occur; thus, the exchange studies militate against the postulate of SO^{2+}. Finally, the observation[81] that SO_2 exchanges oxygen with itself very rapidly in *both liquid and*

---

[77] G. Jander, "Die Chemie in Wasserahnlichen Losungsmitteln," p. 209, Springer, Berlin (1949).
[78] S. Nakata, *Nippon Kagaku Zasshi*, **64**, 635 (1943); J. L. Huston, *J. Amer. Chem. Soc.*, **73**, 3049 (1951); *J. Phys. Chem.*, **63**, 389 (1959).
[79] See ref. 77.
[80] B. J. Masters and T. H. Norris, *J. Amer. Chem. Soc.*, **77**, 1346 (1955); E. C. M. Grigg and I. Lauder, *Trans. Faraday Soc.*, **46**, 1039 (1950); R. E. Johnson, *et al., J. Amer. Chem. Soc.*, **73**, 3052 (1951).
[81] N. N. Lichtin, *et al., Inorg. Chem.*, **3**, 537 (1964).

gas phases further damages the auto-ionization concept of Jander and reinforces models of oxygen atom transfer between neutral molecules like the SO_3/SO_2 exchange above.

We can now consider a few reactions to illustrate the uses to which SO_2 may be put as a solvent. The simplest kinds of reactions are metatheses based on solubility relationships—for example, the preparations of thionylthiocyanate (thionyl is the functional name for the $=\!SO$ group) and thionylacetates:

$$2NH_4SCN + SOCl_2 \rightarrow 2NH_4Cl\downarrow + (NCS)_2SO$$
$$2Ag[XCH_2CO_2] + SOCl_2 \rightarrow 2AgCl\downarrow + (XCH_2CO_2)_2SO \qquad X = Cl, C_6H_5$$

With regard to the latter, it is known that when $X = H$, the thionyl acetate decomposes into acetic anhydride and SO_2. The rationale for the stability of the thionylacetates when X is Cl or phenyl is not clear.

Two interesting solvolysis reactions may be cited, in addition to the direct reaction of acetates with SO_2. Organometallic compounds, acting in effect as sources of carbanions, tend to form the metal sulfite and the organosulfoxide:

$$ZnEt_2 + 2SO_2 \rightarrow ZnSO_3 + Et_2SO$$

This reaction perhaps proceeds *via* an oxide transfer from RSO_2^- or $R_2SO_3^{2-}$ to SO_2. An oxygen atom transfer reaction is known for the reaction of PCl_5 with SO_2:

$$PCl_5 + SO_2 \rightarrow POCl_3 + SOCl_2$$

but not for the chlorides of elements surrounding phosphorus in the periodic table ($SiCl_4$, SCl_2, SCl_4, $SbCl_3$, $SbCl_5$). The known dissociation of PCl_5 to PCl_3 and Cl_2 may well explain this curiosity:

$$PCl_5 \rightleftharpoons Cl_2 + PCl_3$$
$$Cl_2 + SO_2 \rightarrow Cl_2SO_2$$
$$PCl_3 + SO_2Cl_2 \rightarrow POCl_3 + SOCl_2$$

Perhaps even more startling, but highly useful, is the formation of diphosphorus trioxide from PCl_3 and sulfite:

$$2PCl_3 + 3SO_3^{2-} \xrightarrow{SO_2} P_2O_3 + 6Cl^- + 3SO_2$$

At present we can only speculate about the intermediates formed in this reaction. To illustrate further this characteristic oxygen transfer feature of the S—O bond, SO_2 exchanges O for NR_2 in the reactions[82]

$$2R_2'BNR_2 + SO_2 \rightarrow R_2'B\!-\!O\!-\!BR_2' + OS(NR_2)_2$$
$$3R'B(NR_2)_2 + 3SO_2 \rightarrow (R'BO)_3 + 3OS(NR_2)_2$$

$(R'BO)_3 =$ [cyclic structure with three B atoms and three O atoms alternating in a six-membered ring, each B bearing an R' substituent]

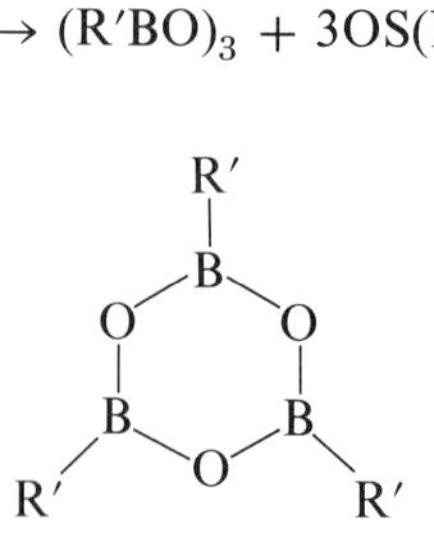

[82] M. Nöth and P. Schweizer, *Chem. Ber.*, **97**, 1464 (1964).

Because of the low acidity and basicity of SO$_2$ as a solvent, many complexation reactions are easily performed that might be difficult with other solvents. For example, there are the convenient syntheses of triiodide salts

$$MI + I_2 \xrightarrow{SO_2} M[I_3]$$

and the unusual hexachloro complex of Sb(III)

$$SbCl_3 + 3Cl^- \xrightarrow{SO_2} SbCl_6^{3-} \quad \text{(see p. 81, Chapter 2)}$$

In a more anticipated fashion, SbCl$_5$ readily forms the SbCl$_6^-$ anion in SO$_2$:

$$SbCl_5 + Cl^- \xrightarrow{SO_2} SbCl_6^-$$

and this has been used as the basis for the synthesis of acylium ion salts:

$$\overset{O}{\underset{\|}{RC}}-Cl + SbCl_5 \xrightarrow{SO_2} RCO^+ SbCl_6^-$$

Friedel-Crafts reactions are similarly possible in liquid SO$_2$.

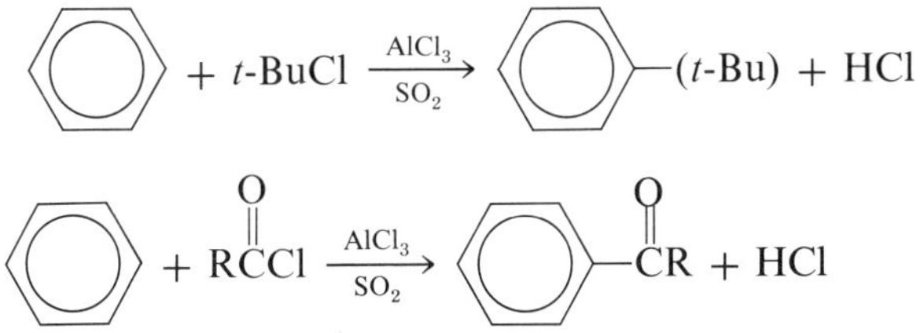

$$C_6H_6 + t\text{-BuCl} \xrightarrow[SO_2]{AlCl_3} C_6H_5(t\text{-Bu}) + HCl$$

$$C_6H_6 + R\overset{O}{\underset{\|}{C}}Cl \xrightarrow[SO_2]{AlCl_3} C_6H_5\overset{O}{\underset{\|}{C}}R + HCl$$

Other examples of organic reactions possible in SO$_2$ are the 1,2-bromination of aryl substituted ethylenes and the sulfonation of aryl compounds.

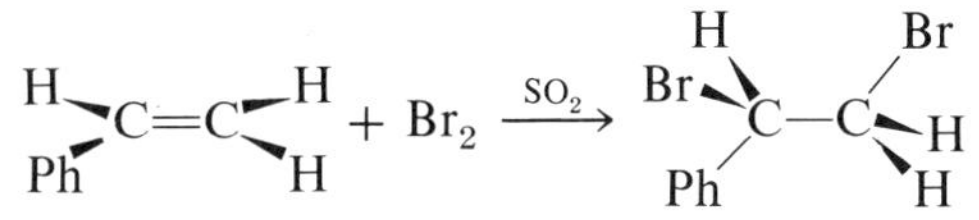

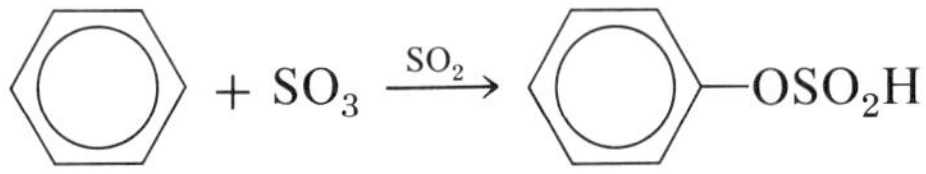

$$\underset{Ph}{\overset{H}{\diagdown}}C=C\underset{H}{\overset{H}{\diagup}} + Br_2 \xrightarrow{SO_2} \underset{Ph}{\overset{Br}{\diagdown}}\overset{H}{\underset{}{C}}-\overset{Br}{\underset{H}{C}}$$

$$C_6H_6 + SO_3 \xrightarrow{SO_2} C_6H_5-OSO_2H$$

HSO$_3$F AND SUPERACIDS[83]

M.P. (°C)	−89
B.P. (°C)	163
η (centipoise)	1.56 (25°C)

---

[83] R. J. Gillespie, *Accts. Chem. Res.*, **1**, 202 (1968); A. W. Jache, *Adv. Inorg. Chem. Radiochem.*, **16**, 177 (1974); R. J. Gillespie and J. Passmore, *Accts. Chem. Res.*, **4**, 413 (1971).

ρ (g cm^{-3})	1.73 (25°C)
Λ (ohm^{-1} cm^{-1})	10^{-4} (25°C)
K_{ion}	10^{-8}

This is a truly important acid solvent with distinct advantages over HF and H_2SO_4. When free of HF, the liquid is easily handled in glass vessels. Its liquid range is more useful, for its solutions can be studied at very low temperatures as well as ambient ones, and its boiling point is low enough to allow its ready removal after use as a synthesis medium. Its viscosity also makes it much more desirable as a solvent. The high boiling point indicates a significant degree of self-association, while the K_{ion} value denotes less auto-ionization than H_2SO_4. There are essentially none of the solvent-originated species present that make difficult the interpretation of chemistry in H_2SO_4.

The low K_{ion} could be taken to mean that HSO_3F is a weaker acid or base toward itself than is H_2SO_4.

$$2HSO_3F \rightleftharpoons H_2SO_3F^+ + SO_3F^-$$

That the latter is the correct view is shown by the fact that p-nitrochlorobenzene has $K_b = 0.76$ in HSO_3F, but $K_b = 0.004$ in H_2SO_4. Even $C_6H_3(NO_2)_3$ has a K_b of 4×10^{-5} in HSO_3F, but is an immeasurably weak π base in H_2SO_4. Furthermore, K_2SO_4 reacts with HSO_3F to form H_2SO_4 and small amounts of $H_3SO_4^+$ ($K_b \sim 10^{-4}$).

$$K_2SO_4 + 2HSO_3F \rightarrow 2K^+ + 2SO_3F^- + H_2SO_4$$
$$\Updownarrow HSO_3F$$
$$H_3SO_4^+ + SO_3F^-$$

These properties of HSO_3F have opened doors to several areas of organic and inorganic chemistry that were previously inaccessible. One of the most important reactions is that between SbF_5 and HSO_3F. The significance of this reaction is that the product

$$SbF_5 + HSO_3F \rightarrow H[SbF_5(OSO_2F)] \xrightleftharpoons{HSO_3F} (H_2SO_3F)^+ + [SbF_5(OSO_2F)]^-$$

is an even stronger acid than HSO_3F itself! Even more remarkable is the effect of adding SO_3 up to a 3:1 ratio of SO_3 to SbF_5. Successive equilibria lead to $H[SbF_2(OSO_2F)_4]$.

$$SO_3 + SbF_5 \rightleftharpoons SbF_4(OSO_2F)$$
$$2SO_3 + SbF_5 \rightleftharpoons SbF_3(OSO_2F)_2$$
$$3SO_3 + SbF_5 \rightleftharpoons SbF_2(OSO_2F)_3$$
$$2HSO_3F + SbF_2(OSO_2F)_3 \rightleftharpoons [H_2SO_3F]^+ + [SbF_2(OSO_2F)_4]^-$$

which has *strong* acid properties in HSO_3F solvent. Both SbF_5 and 3:1 SO_3/SbF_5 solutions in HSO_3F are justifiably called "superacids." In fact, $C_6H_3(NO_2)_3$ appears as a strong base in the $SO_3/SbF_5/HSO_3F$ solvent (cf. the K_b of 10^{-5} in HSO_3F). (As an aside, the SbF_5/HSO_3F solutions are more viscous than HSO_3F alone. To achieve a more fluid solvent, SO_2 is sometimes added as a diluent, for it seems to be "non-interacting" with the other components.)

Before getting into some of the new compound syntheses possible in such media, some important fluorination reactions should be mentioned. Most of these could be viewed as occurring through "onium" formation after dehydration of an oxide (in analogy with such characteristic reactions of HF and H_2SO_4). A few such reactant/product pairs are listed here:

$$B(OH)_3/BF_3$$
$$SiO_2/SiF_4$$
$$KMnO_4/MnO_3F$$
$$CrO_3/CrO_2F_2$$
$$P_4O_{10}/POF_3$$
$$KClO_4/ClO_3F$$

The major new developments in organic chemistry have come from carbonium ion studies. For example, dehydration of protonated secondary and tertiary alcohols in particular is facile:

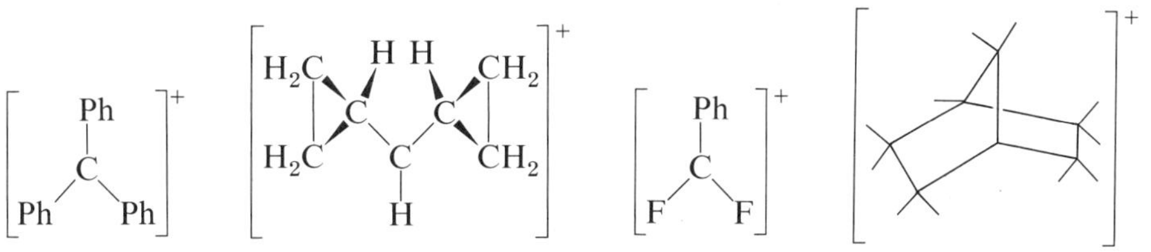

The occurrence of such cations is made possible by the weak basicity of SO_3F^-.

From an inorganic chemist's point of view, the most exciting new systems to be studied in HSO_3F and superacid media are the catenated halogen and sulfur *cations*. Clearly, such species should be strongly oxidizing and/or acidic, and the requirement for a non-basic/non-reducing solvent medium is strict. Using peroxydisulfuryldifluoride ($S_2O_6F_2$; $FO_2SO-OSO_2F$) as the oxidant, I_2 is converted into one of three products by stoichiometry control:

$$I_2 + 3S_2O_6F_2 \rightarrow 2I(SO_3F)_3$$
$$3I_2 + S_2O_6F_2 \rightarrow 2I_3^+ + 2SO_3F^-$$
$$2I_2 + S_2O_6F_2 \rightarrow 2I_2^+ + 2SO_3F^-$$

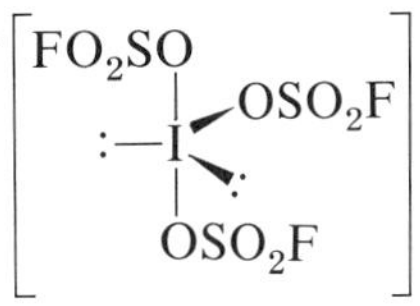

The latter cation is paramagnetic (the ground term symbol is $^2\Pi$, and I_2^+ is valence isoelectronic with superoxide) and has a characteristic blue color. Such solutions are unstable with respect to disproportionation:

$$I_2^+ + \frac{1}{2}SO_3F^- \rightleftharpoons \frac{1}{8}[I(SO_3F)_4]^- + \frac{5}{8}I_3^+$$

which accounts for the loss of paramagnetism as I_2^+ solutions are made more concentrated. The fact that I_2^+ solutions turn red and diamagnetic at $-90°C$ has caused speculation on the formation of the I_2^+ dimer, I_4^{2+}, with the following presumed structure.

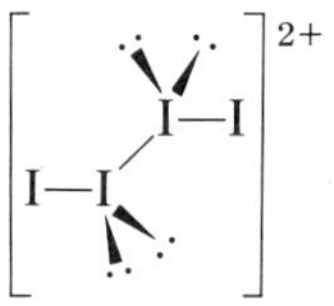

Oxidation of S_8 by $S_2O_6F_2$ has similarly been studied in "superacid" media, where S_8^+, S_8^{2+}, and S_4^{2+} are formed. S_8^{2+} and S_4^{2+} are diamagnetic; the former has a bicyclo structure and the latter is a simple planar ring.

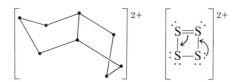

The paramagnetic species S_8^+ apparently tends to dimerize slowly to diamagnetic S_{16}^{2+}, the structure of which is not known (the simplest formulation would be two S_8 rings connected by an S—S bond).

As a final interesting use of $SbF_5/HSO_3F/SO_2$ media, Verkade has found it possible[84] to form and study trihalophosphonium cations in "superacid." In the series of weak σ donors beginning with PF_3 and following through the successive replacement of F by Cl and then Cl by Br, the $^1J_{PH}$ of $[HPX_{3-n}X'_n]^+$ decreases from 1191 Hz for $[HPF_3]^+$ to 810 Hz for $[HPBr_3]^+$. Note, too, that Bent's isovalency concept and its relation to Fermi contact coupling holds up well in this series (see Chapter 3 if you need a review of these concepts).

---

[84] L. J. VandeGriend and J. G. Verkade, *J. Amer. Chem. Soc.*, **97**, 5958 (1975).

STUDY QUESTIONS

33. Describe the solvent properties that

    a) would give ClO_4^- strong apparent Brönsted base character;
    b) would give $B(OH)_3$ strong apparent Brönsted acid character;
    c) would give HNO_3 and HI different apparent (Brönsted) acid strengths;
    d) would distinguish apparent basicities of NH_2^- and PH_2^-;
    e) would permit amphoteric behavior of AsF_5.

34. $FeCl_4^-$ ion is yellow in color, whereas Fe_2Cl_6 has a reddish color; 0.1 M solutions of $FeCl_3$ in $POCl_3$ *and* in $PO(OR)_3$ are red and *both* turn yellow on dilution. Titration of red $FeCl_3/POCl_3$ solutions with colorless $Et_4NCl/POCl_3$ solutions causes the color to change from red to yellow with a sharp end-point at 1:1 $FeCl_3$ to Et_4NCl. Furthermore, as far as is known, oxychloride solvents form adducts with Lewis acids wherein the oxygen, rather than chlorine, is the donor atom.

    a) Are these observations best represented by the equilibria
        a) $Fe_2Cl_6 + 2POCl_3 \rightleftharpoons 2FeCl_4^- + 2POCl_2^+$
        b) (a) shifted to right on dilution
        c) $FeCl_4^- + POCl_2^+ + Et_4N^+Cl^- \rightleftharpoons Et_4N^+ + FeCl_4^- + POCl_3$
        or by
        a) $Fe_2Cl_6 + 4POCl_3 \rightleftharpoons FeCl_2(OPCl_3)_4^{2+} + FeCl_4^-$
        b) (a) shifted to right on dilution
        c) $FeCl_2S_4^{2+} + 2Cl^- \rightleftharpoons FeCl_4^-$
    b) Does your "intuition" suggest $FeCl_3$ or $POCl_2^+$ to be the stronger acid? In other words, do you believe that $FeCl_3$ is a strong acid in $POCl_3$, so as to be leveled to $POCl_2^+$?
    c) Vibrational spectra of $AlCl_3 \cdot OPCl_3$ and $GaCl_3 \cdot OPCl_3$ as solids and as melts give no evidence for $POCl_2^+$. How could one distinguish $POCl_2^+$ from O-coordinated $POCl_3 \cdot$ acid on the basis of vibrational spectra?

35. Do you think it would be possible to determine the end-point of a titration such as

$$BCl_3 + Et_3N \rightarrow Et_3N \cdot BCl_3$$

in CCl_4, using a "normal" organic indicator like phenolphthalein?

36. Account for the facts that 1:1 and 2:1 ratios of $FeCl_3$ to $ZnCl_2$ in $POCl_3$ lead to quantitative formation of $FeCl_4^-$.

37. Account for the high conductivity of $AlCl_3$ in CH_3CN.

38. The exchange of radioactive sulfur between SO_2 and $SOCl_2$ is catalyzed by Cl^- and by $SbCl_5$. Suggest mechanisms for these reactions, with the first step in each case being formation of an appropriate adduct.

39. Propose mechanisms for the exchange of labeled sulfur between SO_2 and $S_2O_5^{2-}$ and the exchange of oxygen between labeled and unlabeled SO_2.

40. Consider the π mo's of S_4^{2+} and sketch the HOMO and LUMO. Estimate the S—S π bond order from the occupied mo's and compare it with the bond order estimated from π resonance structures.

EPILOG

The goal of this chapter has been to present you with a feeling for the applicability of the donor/acceptor concept as a strongly unifying point of view of molecular interactions. You have seen a broad survey of typical, and not so typical, chemical facts regarding self- and hetero-association of molecules and molecular fragments. The application of the molecular structure and bonding ideas of the first four chapters is important to the interpretation of these facts; while many of these applications have been only qualitative, they are powerful tools for lending insight on the strengths of intermolecular association and even for predicting the mode of intermolecular interactions. The exceedingly important topic of solvent influence (even control) of solute reactions is one area of extension of the ideas presented earlier.

Very often, but not always, new compounds may be quickly recognized and incorporated into your existing knowledge of chemical properties by viewing them as adducts of known molecules and molecular fragments. This view often suggests synthetic approaches to new compounds. In later chapters of this book you will experience greater understanding of the chemical facts and their interpretation by having mastered the ideas in these first five chapters. Your proficiency in predicting (or remembering, as the case may be) the reactions of molecules with each other often reduces to the simple expedient of recognizing more or less stable fragments within a molecule. This often suggests molecular weak links, with respect to molecular fragmentation by heterolytic bond cleavage, and the points of attachment of two reacting molecules. To be sure, you will encounter observations that are not readily compartmentalized, but by having developed an appreciation for the "normal" compounds and their potential for reaction you will have a solid background against which to view the anomalies.

> FREE ENERGY, REACTION POTENTIAL, AND EQUILIBRIUM
>
>   Review
>   Estimation of Reaction Spontaneity
>   The Non Sequitur "Stability"
>   Heats of Reaction from Bond Energies
>   Higher Oxidation State Stabilization
>   A Caveat Concerning Condensed Phase Reactions
>
> METAL-CONTAINING SOLIDS
>
>   Lattice Energies
>   Structure and Stoichiometry
>   Structure and the Concept of Ion Radii
>   Layer Lattices and Incipient Covalency
>
>   Synthesis Principles
>   Cation Oxidation States
>   Ionic Fluorinating Agents
>   Ion Size and Compound Isolation from Solution
>   Metals
>   Structures and Bonding
>   Oxidative Stabilities
>
> NON-METAL COMPOUNDS
>
>   Boron and Aluminum
>   Carbon and Silicon
>   Nitrogen and Phosphorus
>   Oxygen and Sulfur
>   Halogens
>   Noble Gases
>
> EPILOG

6. ENERGETICS AND STRUCTURES AS GUIDES TO MAIN GROUP CHEMISTRY

In this chapter we will look into the more (and some less) highly developed aspects of main-group element compounds. Against a background of bond energies, we will summarize and compare the known classes of compounds formed by the elements primarily from Groups III through VIII and from the second and third periods of Mendeleev's Periodic Table in its long form. Our interest here is in surveying briefly the structural units on which the chemistries of these elements are based and in focusing your attention on the similarities and dissimilarities between these units from element to element. Continuing in the spirit of preceding chapters, our intention is to inspect the compounds in terms of bond energy concepts to see "why" they are stable or unstable and to determine which bonds within them are most susceptible to attack by other reagents or to thermal cleavage. This overview will give perspective to the succeeding chapters on mechanisms and synthetic methods. Additionally, we hope to establish a background against which less conventional chemistry may be viewed and contrasted.

FREE ENERGY, REACTION POTENTIAL, AND EQUILIBRIUM

REVIEW

The thermodynamic quantity that has the most direct bearing on the course of a chemical reaction is the Gibbs free energy change for the reaction, ΔG. But you are familiar with other quantities that describe the stability of a chemical system and that are related to ΔG. Of great practical use are the reaction potential,[1] E, and the equilibrium constant, K;

$$\Delta G = -nFE$$

and

$$\Delta G = -RT \ln(K/Q)$$

In the first of these, n is the number of Faradays of electricity involved in the reaction [$F = 96{,}500$ coul $= 23.06$ kcal volt^{-1} (gram equivalent)$^{-1}$] and E is the

---

[1]The standard reaction potential, $E°$, is derived from the weighted difference of the standard reduction potentials.

$$A + ne^- \rightarrow A^{n-} \quad :E_A°$$
$$B + me^- \rightarrow B^{m-} \quad :E_B°$$
$$mA + nB^{m-} \rightarrow mA^{n-} + nB \quad :E° = (mE_A° - nE_B°)/(m+n)$$

reaction potential. In the relationship connecting ΔG and K we find also the reaction quotient Q, which is evaluated for specific non-equilibrium concentrations (technically activities) of reactants and products from an expression analogous to that for K. Thus, Q in relation to K is a measure of the deviation of the system from equilibrium. In the event that K is known, we can quantify this deviation and predict the spontaneity with which chemical systems approach equilibrium in the direction of products or reactants: $Q > K$ means that the concentrations of the products are too high ($\Delta G > 0$) so that formation of reactants from products will proceed spontaneously, while $Q < K$ means that the concentrations of the products are too low ($\Delta G < 0$) and products will be spontaneously formed from reactants. After such an unbalanced system reaches equilibrium ($Q = K$), the equation reduces to the condition $\Delta G = 0$. The other special condition of use to chemists, for reference purposes, is that in which reactants and products are all in their standard states (activities = 1) and $Q = 1$; in this event,

$$\Delta G \equiv \Delta G° = -RT \ln K = -nFE°$$

Thus, the sign and magnitude of $\Delta G°$ depend upon to which side and to what extent the standard state system deviates from equilibrium.

ESTIMATION OF REACTION SPONTANEITY

In the estimation of the relative stabilities of reactants and products (these stabilities always being measured in terms of ΔG), we find it most important to consider the relative enthalpies and entropies of reactants and products:

$$\Delta G = \Delta H - T\Delta S$$
$$= \Delta E + \Delta(PV) - T\Delta S$$

If we are ever to develop an intuition for the spontaneity of reactions we must, explicitly or implicitly, develop a "feel" for the relative importance of each of these terms and how each favors or disfavors the reaction as written.

Perhaps the quantity most difficult for which to develop an intuition is the entropy change, ΔS. Because ΔS is a measure of the relative "randomness" of products and reactants, we can predict that a reaction that increases the number of gaseous molecules is favored by $T\Delta S$. Similarly, in condensed phases, entropy generally favors an increase in the number of molecules.

The $\Delta(PV)$ term is negligible for condensed phase reactions and usually for reactions involving gases; in the cases of reactions involving gases, $\Delta(PV) \cong \Delta(nRT)$, which amounts to a contribution (at 300°K) of less than 0.6 kcal for each mole of gas produced or consumed. Because of this, the practicing chemist often interprets ΔH and ΔE as equivalent quantities.

With regard to ΔE, it is obvious that the arrangement of atoms to form molecules with the greatest binding energies will be favored. It is also likely that the arrangement with the greatest binding energies will be the one most constrained, so that often

Compound	$\Delta H_f°$	$\Delta G_f°$
$SO_{2(g)}$	−71	−72
$H_2O_{(l)}$	−58	−55
$H_2S_{(g)}$	−5	−8
$NH_{3(g)}$	−11	−4
$PH_{3(g)}$	+2	+4
$KNO_{3(c)}$	−118	−94

TABLE 6-1

STANDARD ENTHALPIES AND FREE ENERGIES OF FORMATION (kcal/mole)

TABLE 6-2

SOME REACTION FREE ENERGIES AND ENTHALPIES AT 300°K (kcal/mole)

	$\Delta H°$ (kcal/mole)	$\Delta G°$ (kcal/mole)
$SF_{6(g)} + 3H_2O_{(g)} \rightarrow SO_{3(g)} + 6HF_{(g)}$	-45	-76
$Xe_{(g)} + 3F_{2(g)} \rightarrow XeF_{6(g)}$	-96	-67
$PCl_{3(g)} + AlBr_{3(c)} \rightarrow PBr_{3(g)} + AlCl_{3(c)}$	-3	-4
$CH_{4(g)} + 2O_{2(g)} \rightarrow CO_{2(g)} + 2H_2O_{(g)}$	-192	-191
$CaO_{(c)} + CO_{2(g)} \rightarrow CaCO_{3(c)}$	-43	-31
$AsCl_{3(l)} + 3NaF_{(c)} \rightarrow AsF_{3(g)} + 3NaCl_{(c)}$	-25	-32
$CH_3OH_{(l)} + NH_{3(g)} \rightarrow CH_3NH_{2(g)} + H_2O_{(g)}$	$+4$	-4

ΔE and $T\Delta S$ are of opposite sign. The important caveat from Chapter 5 that intermolecular interaction energies in condensed phase reactions may be a significant portion of ΔE will be repeated and illustrated in a later section.

On occasion, the spontaneity of a reaction is controlled by $T\Delta S$ (whenever $|T\Delta S| > |\Delta H|$); but most often it is found that, at temperatures near 300°K, the $T\Delta S$ term amounts to only a few kcal/mole whereas ΔH is one to two orders of magnitude larger. Some data[2] are given in Table 6-1 to illustrate the point that ΔH is often dominant over $T\Delta S$.

Table 6-2 presents some data (derivable from free energies and enthalpies of formation) with which we can make some additional, useful points. First of all, the first two reactions involve an appreciable change in the number of molecules and exhibit reasonably large entropy changes (nevertheless, ΔH dominates $T\Delta S$). For the next two reactions, with no change in numbers of gas molecules, the entropy change at 300°K is quite small. The final three reactions exhibit entropy contributions ($T\Delta S$) on the order of 10 kcal/mole. The last of these reactions makes the interesting point that when the enthalpy change for a reaction is less than 10 kcal/mole, the entropy contribution to the free energy may determine the sign of ΔG. Finally, only the third, fifth, and sixth reactions proceed, unassisted, at a significant rate at 300°K. This, of course, warns you that *thermodynamic spontaneity is a necessary but not sufficient condition for the observation of a chemical reaction*. The requirement that a low energy pathway must exist for the reaction to occur will be discussed in the next chapter. Moreover, the data in Table 6-2 serve to illustrate the commonly made assumption that spontaneity of a reaction can be judged by the enthalpy change and that entropy contributions must be allowed for when ΔH is small.

The Non Sequitur "Stability"

Before proceeding to the use of thermochemical data for defining the strengths of chemical bonds, we need to mention briefly that chemists are often sloppy in their thinking when they refer to a compound as "stable" or "unstable." Unfortunately, the practice is widespread of using the word "stable" to connote either thermodynamic or kinetic stability. Even when the context in which the word is used is obvious, it is strictly meaningful to use the word only in reference to a particular kind of reaction. Thus, compounds can be thermodynamically unstable but kinetically stable (the interconversion of the carbon allotropes diamond and graphite is the textbook

---

[2] There are many collections of such data, some newer than others and having more reliability for difficult-to-handle compounds. To be internally consistent, we have relied heavily on D. A. Johnson's "Some Thermodynamic Aspects of Inorganic Chemistry," Cambridge University Press, London (1968). This little book is highly recommended. In addition to the sources used by Johnson, supplementary data may be found in: L. Brewer, *et al.*, *Chem. Rev.*, **61**, 425 (1961); *ibid.*, **63**, 111 (1963); R. C. Feber, Los Alamos Report LA-3164 (1965); B. Darwent, *Nat. Stand. Ref. Data Ser., Nat. Bur. Stand.*, 1970. NSRDS-NBS31.

example); and compounds can be stable with respect to their elements but reactive (unstable) with other compounds and *vice versa*. A good example is CuH, which is "unstable" with respect to the elements and "stable" with respect to reaction with aqueous acid:

$$CuH_{(c)} \rightarrow Cu_{(c)} + \tfrac{1}{2}H_{2(g)} \qquad \Delta H° = -8 \text{ kcal/mole}$$
$$CuH_{(c)} + H^+_{(aq)} \rightarrow Cu^+_{(aq)} + H_{2(g)} \qquad \Delta H° = +5 \text{ kcal/mole}$$

For comparison, NaH is "stable" with respect to its elements but thermodynamically "unstable" in aqueous acid:

$$NaH_{(c)} \rightarrow Na_{(c)} + \tfrac{1}{2}H_{2(g)} \qquad \Delta H° = +14 \text{ kcal/mole}$$
$$NaH_{(c)} + H^+_{(aq)} \rightarrow Na^+_{(aq)} + H_{2(g)} \qquad \Delta H° = -45 \text{ kcal/mole}$$

(Understanding this difference in reactivity with H^+ is a good example of the point to be made in a later section on Haber cycles, because the cation solvation energies and to some extent the lattice energies are important to the relative reactivities of CuH and NaH.) In defense of current usage, however, compounds that do not readily decompose at room temperature are referred to as "stable," and as long as this meaning is understood we can live with the usage.

Heats of Reactions from Bond Energies

In Chapter 3 we first broached the subject of bond dissociation energies and bond energies. The semantic/conceptual distinction to be drawn here derives from the usage that *bond dissociation energy* (D) refers to the energy involved in homolytically breaking one mole of a particular bond in a molecule, whereas *bond energies* (E) are defined so that their sum (one term for each bond in the molecule) equals the atomization energy of a mole of the molecules. For example, each step of the reaction

$$CH_4 \rightarrow CH_3 + H \rightarrow CH_2 + 2H \rightarrow CH + 3H \rightarrow C + 4H$$

involves a CH bond dissociation energy (D_{CH} values for this example are 104, 106, 106, and 81 kcal/mole, respectively), whereas the C—H bond energy for methane is the average of these, $E_{CH} = 99$ kcal/mole. A bond dissociation energy value depends on the valence states of both atoms defining the bond and on the substituents about them. It is the bond energies with which you are most familiar from your organic and physical (thermochemistry) chemistry courses; for organic molecules in particular the bond energies are often additive, making possible an effective estimate of the heat of formation of a molecule.

Table 6-3 contains a listing of bond energies,[3] which will be of use to you in the later discussion of the stabilities of various non-metal compounds.

An often convenient view of the tendency of a system of chemicals to undergo a net reaction (not to be at equilibrium) is the following. Consider the likelihood of separated atoms forming reactants as opposed to forming products. If the products are favored, then $\Delta G < 0$; conversely, $\Delta G > 0$ if the reactants are favored. It is through the development of an ability to estimate whether products or reactants are favored that the chemist creates his intuition for synthesis of new materials. (This, of course, is an understatement, for the importance of a low-energy pathway to get from reactants to products is of immense practical concern; herein lies the impetus for the study of reaction mechanisms and development of an intuition for understanding reaction mechanisms.)

---

[3] See ref. 2, p. 267 of Johnson's book.

While the atomization energy view is philosophically convenient, at the present time it is exceedingly difficult to predict *a priori* that ΔH_f for NH_3, for example, is negative. To do so requires that one be able to estimate the bond energy of N_2 (you can appreciate that it should be quite large), the bond energy of H_2, and the atomization energy of NH_3:

$$\Delta H_f(NH_3) \cong \frac{1}{2}D_{NN} + \frac{3}{2}D_{H_2} - 3E_{NH}$$

However, our inability to make good *a priori* estimates of bond energies does not mean that we cannot make intelligent use of bond energies for a few smaller molecules to estimate the atomization energies of larger molecules. By way of example, we can use $D_{N_2} = 226$ kcal/mole, $D_{H_2} = 104$ kcal/mole, and $E_{NH} = 93$ kcal/mole with $E_{N-N} = 38$ kcal/mole to find for the reactions

$$\left(\frac{n}{2}\right)N_{2(g)} + \left(\frac{n+2}{2}\right)H_{2(g)} \rightarrow H-(NH)_n-H_{(g)}$$

an estimated $\Delta H_{(g)}$ of $(34n - 44)$ kcal/mole. In this case you predict that catenation of N to form hydronitrogens for which $n > 2$ is highly unfavored. Even hydrazine (N_2H_4) should be, and is, highly unstable with respect to decomposition into the elements (note that, because of the $(n + 1):1$ mole ratio of reactants to products, the $T\Delta S$ factor further reduces the stability of the higher homologs of ammonia and hydrazine). In a similar vein, using the bond energies for N—H, N—O (from hydroxylamine, NH_2OH), and O—H, we could estimate the energy of formation of the unknown compound $N(OH)_3$ to be only -2 kcal/mole.

Implicit in the foregoing discussions is an assumed accuracy in the transferability of bond energies. We need to note carefully that the number of bonds formed by an atom (its valency) can appreciably affect the average energies of those bonds. A case in point concerns the compounds ClF and ClF_3. The bond energy of Cl—F is 60 kcal/mole, while that of ClF_3 is but 42 kcal/mole.[4] Thus, ClF_3 should be a better oxidant than would be predicted on the basis of three times the atomization energy of ClF (*viz.,* atomization of ClF_3 is 54 kcal/mole less endothermic than would have been predicted).

To make the use of bond energies easier for the prediction of the spontaneity of reactions in which pairs of atoms in different molecules exchange partners, some of the data in Table 6-3 are presented in a useful form in Figure 6-1. A couple of examples will show how easy it is to use the bond energy terms presented in this way.[5]

You may be interested in whether HF and BrF might react to produce HBr and F_2.

$$F-H + Br-F \rightarrow Br-H + F-F$$

Locate the F—H and Br—H terms in column 1 (the H column) and the Br—F and F—F terms in the second column. The conversion F—H → Br—H incurs a loss of energy, as does the conversion Br—F → F—F. Consequently, you predict—correctly—that H—F and Br—F will not react to form significant amounts of HBr and F_2.

---

[4] In directed ao language, Cl—F is *sp* hybridized while ClF_3 is dsp^3 (one ClF is $\sim sp^2$ while the other two are *dp*). In addition, the relaxation energy of chlorine in ClF_3 will be larger than in ClF. The weakness of the "axial" F—Cl—F "bonds" in three-center mo language is discussed in Chapter 4 (p. 151ff and p. 172).

[5] Be alert to the fact that the bond energies are defined to be positive. If you use the rule "final state energy minus initial state energy" to estimate the reaction energy, you must change the sign of the numerical result.

TABLE 6-3
SOME BOND ENERGIES

Bond	E (kcal/mole)	Compounds
O=O	119	O_2
H—H	104	H_2
O—H	111	H_2O
O—O	34	H_2O_2
F—F	38	F_2
F—O	51	F_2O
F—H	136	HF
Cl—Cl	58	Cl_2
Cl—O	49	Cl_2O
Cl—H	103	HCl
Cl—F	60	ClF
Br—Br	46	Br_2
Br—H	88	HBr
Br—F	60	BrF
I—I	36	I_2
I—H	71	HI
I—F	58	IF
I—Cl	50	ICl
I—Br	42	IBr
S=S	103	S_2
S=O	125	SO
S—H	88	H_2S
S—S	63	S_8, H_2S_n
S—Cl	65	SCl_2
Se—Se	38	Se_2Cl_2, Se_2Br_2
Se—H	73	H_2Se
Se—Cl	58	$SeCl_2$
N≡N	226	N_2
N≡O	151	NO
N≡O$^+$	252	NO^+
N—H	93	NH_3
N—O	39	NH_2OH
N—F	67	NF_3
N—Cl	45	NCl_3
N—N	38	N_2H_4, N_2F_4
N=O	142	NOF, NOCl
P≡P	125	P_2
P—H	78	PH_3
P—P	50	P_4, P_2H_4
P—F	119	PF_3
P—Cl	79	PCl_3
P—Br	64	PBr_3
P—O	88	P_4O_6
As—As	43	As_4
As—O	79	As_4O_6
As—F	116	AsF_3
As—Cl	74	$AsCl_3$
As—Br	61	$AsBr_3$
As—H	71	AsH_3
Sb—Sb	34	Sb_4
Sb—Cl	75	$SbCl_3$
Sb—Br	63	$SbBr_3$
Sb—H	61	SbH_3
C—C	83	Organic
C=C	146	Organic
C≡C	200	Organic
C—H	99	Organic
C—N	73	Organic
C—O	86	Organic

TABLE 6-3 (Continued)

Bond	E (kcal/mole)	Compounds
C=O	178	Organic
C=O	192	CO_2
C≡O	257	CO
C—F	117	CF_4
C—Cl	78	CCl_4
C—Br	65	CBr_4
Si—H	76	Si_nH_{2n+2}
Si—Si	54	$Si_{(c)}$, Si_nH_{2n+2}
Si—O	111	$SiO_{2(c)}$
Si=O	153	$SiO_{2(g)}$
Si≡O	192	$SiO_{(g)}$
Si—F	143	SiF_4
Si—Cl	96	$SiCl_4$
Si—Br	79	$SiBr_4$
Si—C	73	$Si(CH_3)_4$, $SiC_{(c)}$
Si—N	80	$Si_3N_{4(c)}$
Ge—Ge	45	$Ge_{(c)}$, Ge_nH_{2n+2}
Ge—H	68	Ge_nH_{2n+2}
Ge—O	86	$GeO_{2(c)}$
Ge—F	113	GeF_4
Ge—Cl	81	$GeCl_4$
Ge—Br	67	$GeBr_4$
Ge—I	51	GeI_4
Ge—N	61	$Ge_3N_{4(c)}$
Sn—Sn	36	$Sn_{(c)}$
Sn—H	60	SnH_4
Sn—C	50	$Sn(CH_3)_4$, $Sn(CH_3)_3H$
Sn—Cl	75	$SnCl_4$
Sn—Br	64	$SnBr_4$
B—F	154	BF_3
B—Cl	106	BCl_3
B—Br	88	BBr_3
B—I	65	BI_3
B—B	72	B_2F_4
B—O	125	Boron esters, ester chlorides and hydroxides

Another example involves the H and OR exchange between an amine and a phosphate ester:

$$N-H + P-OR \rightarrow P-H + N-OR$$

This sort of reaction is quickly seen not to be favorable. On the other hand, the conversion (reduction) of a fluorophosphine to a phosphine with R_3SiH is a possibility[6]:

$$Si-H + P-F \rightarrow P-H + Si-F$$

Higher Oxidation State Stabilization

Were you to be asked to name a few compounds of P, S, Cl, and Xe exhibiting the maximum oxidation state of the central atom (and therefore the largest number of terminal atoms) you would, probably inadvertently, name all oxides and/or fluorides. This is because you have not encountered any other examples in your training. For example, PH_5 and $S(Me)_6$ are not known. Two of the characteristic physico-chemical

---

[6] This exchange has actually been used to prepare some R_xPH_y compounds: J. W. Gilje, *et al.*, *Chem. Comm.*, 813 (1973).

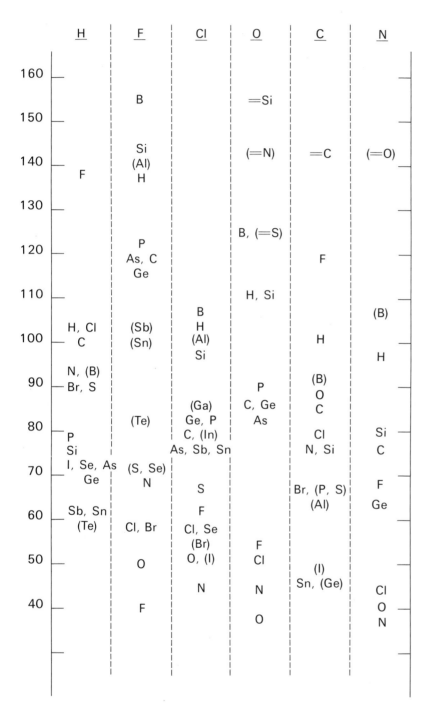

Figure 6-1. A graphical presentation of "nominal bond energies" from Table 6-3. Values in parentheses are either dissociation energies or estimated values.

properties of F and O are their high electronegativity and small size. The importance of small size to maximizing n in EX_n is rather obvious. Simply put, if EX_n is to avoid loss of X, as in

$$EX_n \rightarrow EX_{n-2} + X_2$$

mechanistic and/or thermodynamic resistance to the reaction is required. Here we want to explore the thermodynamic basis for such a reaction, the ΔE for which is given by

$$\Delta E = \{nE_{EX}(EX_n) - (n-2)E_{EX}(EX_{n-2})\} - E_{X_2}$$

When $nE_{EX}(EX_n)$ is greater than $(n-2)E_{EX}(EX_{n-2})$ by more than E_{X_2}, thermodynamic stability of EX_n is indicated. VSEPR/steric congestion in EX_n lowers the bond energy of that structure. What about the role of electronegativity? A rather clear answer is provided by the mo concepts of Chapter 4. Take the SX_6/SX_4 case and consider only the orbital energy and occupancy diagrams appropriate to the O_h and C_{2v} geometries. To keep things as simple as they need be, we will consider only the mo's arising from the overlaps of the sulfur valence s and p ao's with just the σ ao's of the terminal atoms.

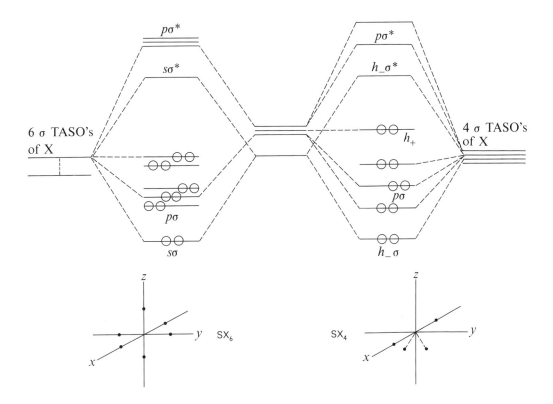

The key question now is how lowering the electronegativity of X, as in $SF_6 \rightarrow SH_6$, reduces the stability of SX_6. In SX_6 you will notice that there are two occupied non-bonding TASO's; these electron pairs must be localized on the X atoms. In SX_4, there is one non-bonding mo localized on sulfur (the h_+ ao, as in Figure 4–12). Thus, an important consideration in the SX_6 structure is the stability of the non-bonding TASO electron pairs. If X is highly electronegative (high Z^* for X electrons), these pairs are more stable than they would be if X had a low electronegativity:

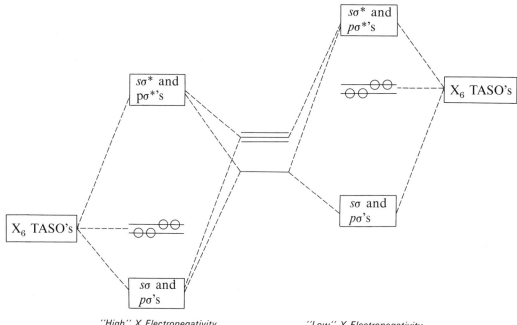

"High" X Electronegativity   "Low" X Electronegativity

Quite clearly, then, low X electronegativity, acting through the inability of X to stabilize the SX_6 non-bonding pairs, reduces the stability of SX_6 (*i.e.*, makes the average S—X bond energy lower). You may also note that the mo model suggests that the average E—X bond energy in EX_n is lower than that in EX_{n-2} (as noted earlier for ClF_3/ClF) regardless of X electronegativity. This is suggested in the SX_6/SX_4 comparison by the average σ bond orders of 4/6 = 2/3 for SX_6 and 3/4 for SX_4. In fact, rather detailed calculations[7] on molecules such as SH_6, SH_4, and PH_5 reveal that these molecules are capable of existence, but that they are unstable with respect to loss of H_2.

A Caveat Concerning Condensed Phase Reactions

In a preceding section we implicitly introduced the idea of using a Haber cycle at the molecular level to interpret gas phase heats of reaction in terms of a two-step process (bond breaking *vs.* bond making). More generally, we deal with reactions in condensed phases, and this complicates our use of bond energies to predict the feasibility of new reactions or to understand known chemical systems. Haber cycles, which are practical applications of Hess' Law, help us to understand such reactions. As exemplified by the following general reaction, the Haber cycle shows us in what way we must allow for the "condensation" energies of reactants and products.

$$A_{(g)} + B_{(g)} \xrightarrow{\Delta G_{(g)}} C_{(g)} + D_{(g)}$$
$$\Delta G^A_{(cond)} \downarrow \quad \Delta G^B_{(cond)} \downarrow \qquad \Delta G^C_{(cond)} \downarrow \quad \Delta G^D_{(cond)} \downarrow$$
$$A_{(cond)} + B_{(cond)} \xrightarrow{\Delta G_{(cond)}} C_{(cond)} + D_{(cond)}$$

---

[7] G. M. Schwenzer and H. F. Schaefer, III, *J. Amer. Chem. Soc.*, **97**, 1393 (1975); A. Rauk, *et al., ibid.*, **94**, 3035 (1972).

Using Hess' Law,

$$\Delta G_{(cond)} = \Delta G_{(g)} + \left\{ \sum^{products} \Delta G_{(cond)} - \sum^{reactants} \Delta G_{(cond)} \right\}$$

and an entirely analogous equation can be written for enthalpies. In this formalism the condensed phase for any one compound could be any of the following: solid, pure liquid, or solution. Correspondingly, the $\Delta H_{(cond)}$ are the lattice energy (ignoring ΔPV), the heat of condensation (or the negative of the heat of vaporization), and the heat of solution. The *corrective term* in brackets (so called because the term corrects the gas phase free energy or enthalpy to that for the condensed phase) is often small because it is the difference between two quantities that frequently are of similar magnitude. Nevertheless, the corrective term must be considered whenever the $\Delta G_{(g)}$ we predict is small. Thus the corrective term, typically up to 10 kcal/mole, could have a critical effect on the existence of compounds with $|\Delta G_{(g)}|$ estimated at ~10 kcal/mole. Such a circumstance is most likely when molecules self-associate by hydrogen bonding or other acid-base interaction or by non-specific interactions (dipole-dipole or van der Waals attractions), as discussed in Chapter 5.

While we will encounter many examples later in this chapter, it will help to illustrate how the corrective term can be of importance. The ΔH_f° of the alkali metal fluorides is an interesting case. The Haber cycle can be written as:

$$\begin{array}{ccc} M_{(g)} + \tfrac{1}{2}F_{2(g)} & \xrightarrow{\Delta H_{(g)}^\circ} & M_{(g)}^+ + F_{(g)}^- \\ {\scriptstyle -S}\downarrow & & \downarrow{\scriptstyle E_0} \\ M_{(c)} + \tfrac{1}{2}F_{2(g)} & \xrightarrow{\Delta H_f^\circ} & MF_{(c)} \end{array}$$

Here S is the sublimation energy of the pure metal, while E_0 is the lattice energy of the salt MF. Hess' Law tells us that

$$\Delta H_f^\circ = \Delta H_{(g)}^\circ + (E_0 + S)$$

The $\Delta H_{(g)}^\circ$ can itself be broken down, using a Haber cycle, to give

$$\Delta H_{(g)}^\circ = I_M + \frac{1}{2}D_{F_2} - A_F$$

where I_M is the ionization potential of the metal and A_F is the electron affinity of F. Thus,

$$\Delta H_{(g)}^\circ = I_M + 19 - 80 = I_M - 61 \text{ kcal/mole}$$

Consequently we have

$$\Delta H_f^\circ = (I_M - 61) + (E_0 + S) \text{ kcal/mole}$$

The following table shows these two terms for Li through Cs:

	$I_M - 61$	$(E_0 + S)$	ΔH_f°
Li	+63	−208	−145
Na	+57	−193	−136
K	+39	−173	−134
Rb	+35	−166	−133
Cs	+29	−160	−131

Here is a good example of the importance of the "corrective" condensation energy term. The formation of $M^+ + F^-$ in the gas phase is thermodynamically unfavored (assuming that the sign of $\Delta G°_{(g)}$ is determined by that of $\Delta E°_{(g)}$) and it falls to the lattice energy of the product to alter this state of affairs.

STUDY QUESTIONS

1. How large must be K for the reaction $A^- + B^+ \rightarrow C^+ + D^-$ if the products are to be contaminated with no more than 0.1% of each reactant? What value of $\Delta G°$ does this condition require? How large a change in log K is incurred by a change in $\Delta G°$ of 1.38 kcal/mole? Reflecting on the influence of "condensation energies," would you say that choice of solvent could be critical to reaction efficiency?

2. Referring to the reaction in problem 1, what is the minimum difference in half-cell $E°$'s required to meet the product purity criterion?

3. For a reaction $A + B \rightarrow C$, for which $K = 10^2 \, M^{-1}$, what ratio of A_0 to B_0 is required to produce a product contaminated by no more than 0.1% of B? (A_0 and B_0 are the initial concentrations of A and B). If $B_0 = 1 \, M$, what must A_0 be?

4. Is the reaction
$$R_3P{=}O + Si_2Cl_6 \rightarrow R_3P + (Cl_3Si_2)O$$
exothermic?

5. Do you suppose that the thermochemical bond energies in Table 6–3 are larger or smaller than simple homolytic dissociation energies? Why? (You may find it helpful to re-read the bond energy section of Chapter 3.)

6. With reference to Figure 6–1, predict whether

$$>\!\!B{-}F + HOR \rightarrow >\!\!BOR + HF$$

or

$$>\!\!B{-}Cl + HOR \rightarrow >\!\!BOR + HCl$$

is more exothermic. To what do you ascribe the greater driving force in your choice?

7. Would
$$S_2Cl_2 + 2H_2S_n \rightarrow H_2S_{2n+2} + 2HCl$$
be a reasonable (from a thermodynamic view) scheme for synthesizing sulfanes?

8. Is the reaction
$$R_3SiOR + BX_3 \rightarrow R_3SiX + BX_2(OR)$$
a potentially useful way to convert a silyl ether to a silyl fluoride or chloride?

9. How would entropy factors and solvent polarity affect scrambling reactions like
$$BX_3 + BY_3 \rightleftharpoons BX_2Y + BY_2X \quad ?$$

METAL-CONTAINING SOLIDS

LATTICE ENERGIES

In these sections of the chapter our attention will focus on compounds that are stable as crystalline materials at ordinary conditions, with specific interest in crystalline forms of ionic compounds rather than those of uncharged compounds. The latter usually have relatively low condensation energies, which are determined by weak van der Waals and dipole-dipole forces between molecules; consequently, they have lattice energies that are not particularly decisive in the course of a chemical reaction.

The electrostatic forces acting to hold ions together in an ordered array are quite large, as you saw in the preceding section on condensation energies for the alkali metal fluorides; in fact, on a formula weight basis the electrostatic energy of interacting ions is comparable to the "covalent" bond energies in Table 6–3. In that previous section you also were left with the impression that, were it not for the condensation of a mole of each metal ion and fluoride ion to form the crystalline lattice, the reactions between the alkali metals and fluorine would not be spontaneous. This impression is misleading and is simply a result of our decision to represent the products in the gas phase as separated cations and anions. In fact, the alkali metals would react with fluorine at temperatures above the boiling point of the ionic salt formed. The apparent discrepancy is due to the fact that even at such temperatures, ion pairs are expected to form:

$$M_{(g)} + \frac{1}{2}F_{2(g)} \to M^+_{(g)} + F^-_{(g)} \to MF_{(g)}$$

Thus, the enthalpy change for the first step of this reaction is given by the expression $I_M + \frac{1}{2}D_{F_2} - A_F = (I_M - 61)$ kcal/mole^{-1}, and this was found to be net endothermic for all the alkali metals. However, the electrostatic energy of attraction of the M^+ and F^- ions is quite large, and it is the chemist's working model for this attraction that we wish to inspect.

The electrostatic attraction between two *point charge* ions separated by a distance d is given by Coulomb's Law:‡

$$E = q_+ q_- e^2/d$$

For a mole of such ion pairs, the energy is simply Avogadro's number times the energy for a single pair. The Coulomb's Law approach has to be incomplete because no allowance is made for the repulsion between the electron clouds of the ions, and this repulsion becomes increasingly important as the two ions draw closer together. This repulsion energy has been treated by Born in terms of the function B/r^n, where n is characteristic of each cation/anion pair and can be determined by compressibility measurements. Incorporation of this repulsion term into the expression for the energy of attraction of two ions leads to the equation:

$$E = Ne^2 q_+ q_-/d + B/d^n$$

for a mole of ion pairs. The equilibrium value of d is determined by the balance of electrostatic attraction and repulsion between the ions; minimization of E with respect to d, by setting $dE/dd = 0$, gives a relation between the coefficient B and the other parameters in the equation for E. Substitution of this expression for B back into the original equation for E yields the final form of the Born equation for the ion-pair energy:

$$E_{\min} = \frac{Ne^2 q_+ q_-}{d_0}\left(1 - \frac{1}{n}\right)$$

By signifying the equilibrium value of d as d_0, we emphasize that this equation is valid only at the equilibrium ion-pair distance. To get a feeling for how large this energy is, you may insert the nominal values of $q_+ = +1$, $q_- = -1$, $d = 3$ Å, and

---

‡This form is appropriate for the cgs system; for a discussion pertaining to SI and engineering units, see N. H. Davies, *Chem. Britain*, **7**, 331 (1971). Corrections appear in *ibid.*, **8**, 36 (1972).

$n = 9$ into the Born equation to calculate a value for E_{min} on the order of -100 kcal/mole. This is clearly of sufficient magnitude to make the formation of MF ion-pairs in the gas phase thermodynamically feasible.

To consider further the problem of formation of the ionic lattice, we need only extend the equation for ion pairs to accommodate the formation of an ordered crystal of such pairs. To do this, however, requires that we know the structure of the crystal lattice. By way of example, we will consider in some detail the formation of a compound with the face-centered cubic structure of the NaCl lattice depicted below. (As an aside, you should realize that terms such as "face-centered" and "body-centered" are given with reference to ions of the *same type*. The NaCl structure is that of two interpenetrating face-centered lattices, one for Na$^+$ and the other for Cl$^-$.)

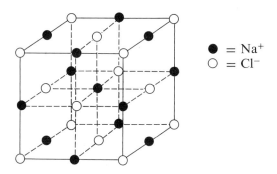

The geometric characteristics of this lattice are the following: the lattice consists of Na$^+$ and Cl$^-$ ions, each of which is surrounded by six counter-ions at a radius of d_0; furthermore, beyond this first coordination sphere there are found twelve ions of the same charge as that at the center and at a distance of $\sqrt{2}\, d_0$ from it; the next sphere of ions has eight counter-ions at a distance of $\sqrt{3}\, d_0$; the next sphere consists of six ions of the same type as that at the origin and at a distance of $2d_0$; and so on. Expressing each of these interactions in the same form as that used for the ion pairs, we arrive at a series of terms of the form

$$E = \frac{Ne^2 q_+ q_-}{d_0} \cdot \left(1 - \frac{1}{n}\right) \cdot \left[6 - \frac{12}{\sqrt{2}} + \frac{8}{\sqrt{3}} - \frac{6}{2} + \frac{24}{\sqrt{5}} + \cdots\right]$$

The series of fractions in this expression is convergent on the value 1.748, so that the expression for the lattice energy of a compound with the NaCl structure reduces to the simple expression

$$E = 1.748 N \frac{e^2 q_+ q_-}{d_0}\left(1 - \frac{1}{n}\right)$$

The quantity 1.748 for the NaCl structure is, of course, dependent solely on the geometry of the crystal and is known as the *Madelung constant* for the NaCl structure. It is interesting to note that the condensation of the ion-pairs to form a three-dimensional NaCl type lattice

$$N\,\text{NaCl}_{(g)} \rightarrow (\text{NaCl})_{N(c)}$$

results in a further electrostatic stabilization of the ions of ~75 per cent. Thus, more than 50 per cent of the lattice energy is accounted for simply in terms of the basic ion-pair attraction.

Other crystal geometries have derivable Madelung constants; these have been calculated for the common lattice types and are presented in Table 6-4 for future reference.

TABLE 6-4
MADELUNG CONSTANTS FOR COMMON LATTICES

Structure	M
NaCl	1.748
CsCl	1.763
ZnS	1.638
CaF$_2$	2.519
TiO$_2$	2.408

STUDY QUESTIONS

10. Using the expression

    $$E = e^2 q_+ q_-/d + B/d^n$$

    for a single ion pair, determine how d_0 depends on q_+, q_-, and B. What is the effect of ion charge on the value of d_0? Assuming n to be the same for NaF and MgO (both cations and anions have the same inert gas configuration), what is the ratio $d_0(\text{NaF})/d_0(\text{MgO})$?

11. Assuming that the Na$^+$ and F$^-$ ions are just in contact in NaF and knowing that $d_0 = 2.31$ Å, calculate the radii of Na$^+$ and F$^-$ from Slater's effective nuclear charges (Chapter 1), where $r = \text{const.}/Z^*$.

12. Using the Born equation as a basis for your answer, to what can you attribute the facts that MgO is much harder, higher melting, and less soluble in H$_2$O than NaF (they both exhibit the same crystal structure)?

13. What plane through a face-centered cubic unit cell contains the largest number of ion nuclei? How many nuclei are in that plane?

STRUCTURE AND STOICHIOMETRY[8]

As we begin to survey the common lattice structures of ionic solids, we need to realize that these materials have highly ordered structures in which each ion (cation or anion) appears in a regular array or lattice in three dimensions. It is on the basis of the lattice of each ion that we may systematize our thoughts on structures of solids. By far the simplest close packed arrangement of ions of one type is that of the cube; it is interesting, however, that two different types of cubic lattices are commonly found in nature—one of these is called *simple cubic* (sc) and the other is called *face centered cubic* (fcc), as diagrams **1** and **2** illustrate. The most common structures of salts with formula MX, MX$_2$, and M$_2$X are best visualized as interpenetrating sub-lattices of these sc and/or fcc types. The solid lines connecting the ions are not meant to represent bonds in the sense of the Lewis system of structure drawing, but rather help to outline a region in the lattice known as the unit cell. A proper definition of a unit cell is that it be of the smallest possible dimensions, have opposite sides parallel, and exhibit the full symmetry of the cation and anion lattice networks. A unit cell completely defines the full lattice structure *by simple translation of the cell along its edges*. (The figures given on p. 280 exhibit the sc (**1**) and fcc (**2**) structures for only one type of ion. For specific compounds, more complete sketches showing both cations and anions are required.)

---

[8] The most comprehensive source book for the non-specialist on the subject of solid state structures is A. F. Wells, "Structural Inorganic Chemistry," 4th Edition, Oxford, London (1975).

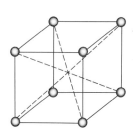

The simple cubic lattice showing the cubic "hole"

1

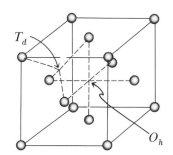

The face centered cubic lattice showing the complete octahedral hole and one of the tetrahedral holes

2

One important property of the unit cell is the number of ions contained within the boundaries of the cell, and discussion of a few general rules concerning the method of counting the ions within the boundaries of the cell is in order. Note that any ion located at a corner of the unit cell contributes equally to a total of eight adjacent cells, so each corner ion is counted as one-eighth of an ion for its contribution to one cell. An ion situated in a face of the cell contributes to only two adjacent cells and so is counted as one-half of an ion for any one cell; finally, an ion located along a cell edge contributes symmetrically to four adjacent cells, and thus is counted as one-fourth of an ion for any one cell. In the examples given above, each unit cell of the sc structure contains a total of one ion of the type defining the simple cubic lattice, while the face centered lattice contains four ions of the type defining the cell.

Another geometrical property of the ions defining the unit cell is the number and symmetries of the holes in the lattice. For the *simple cubic lattice* (**1**) there is only one hole of any reasonable size (at the cell center) and we say the hole is cubic, meaning that the arrangement of ions about the hole is cubic. With one ion/unit cell and one cubic hole/unit cell, the ratio of ions to holes for the sc lattice is 1:1. Therefore, only salts of formula MX can exhibit an sc structure of anions with *all* cubic holes filled by cations. Such a salt is CsCl, and the structure **3** is referred to as the CsCl structure.

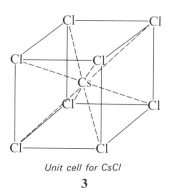

Unit cell for CsCl

3

For the *face centered lattice* (**2**) there are two distinct types of holes. A complete hole of octahedral symmetry is seen at the center of the cell, and 12 one-quarter holes (one at the mid-point of each edge) define three more octahedral holes/unit cell; furthermore, there are eight holes of tetrahedral symmetry, one each at the centers of the cell octants. For the fcc lattice we conclude that there are two T_d holes and one O_h hole *per ion defining the fcc structure*. Consequently, only salts of formula MX can be expected in which *all* octahedral holes are filled with cations. As you have seen (in the last section), this structure is appropriate for common salt and is referred to as the NaCl structure (**4**). If all of the tetrahedral holes created by the fcc anion lattice are filled with cations, the compound formula must be M_2X. As will be clarified later, this structure is known as the anti-fluorite structure (5). Finally, if only half of the tetrahedral holes in the fcc anion lattice are filled, the formula must be MX. This structure is identified as the ZnS structure (6).

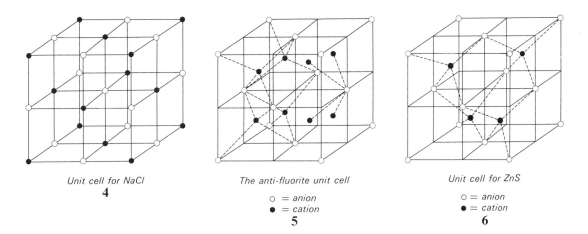

Unit cell for NaCl
4

The anti-fluorite unit cell
○ = anion
● = cation
5

Unit cell for ZnS
○ = anion
● = cation
6

A somewhat different way to express these same ideas is to identify the coordination number of an ion, were one to be located at a lattice hole: the cubic hole of the simple cubic lattice (**1** and **3**) has a coordination number of 8; each octahedral hole in the face centered cubic lattice has a C.N. of 6 (**2** and **4**), and each of the tetrahedral holes has a C.N. of 4 (**2, 5,** and **6**).

In view of the above considerations, a compound MX could be expected to exhibit the CsCl, NaCl, or ZnS structures. The simplest structure possible for an ionic compound of formula MX is that exhibited by CsCl, CsBr, and CsI (CsF has, interestingly, a different structure). This structure (**3**) is simply that of *two interpenetrating simple cubic lattices* of Cs and Cl ions. Both Cs and Cl ions sit in cubic holes formed by the sc lattice of the other, each ion has a C.N. of 8, and the unit cell contains one pair of ions. Salts of complex ions of high symmetry such as $[Be(H_2O)_4]SO_4$, $K[SbF_6]$, and $[Ni(H_2O)_6][SnCl_6]$ have been found to exhibit this type of lattice structure, as have compounds of ions of considerably less symmetry such as CsCN and CsSH, where the non-spherical anions in these examples achieve an effective spherical shape at temperatures near ambient by rapid tumbling. Notice in Table 6-4 that the Madelung constant for the CsCl structure is only slightly different from (greater than) that of the NaCl structure.

The next most common structure for MX is that formed by *two interpenetrating face centered cubic lattices,* as exemplified by NaCl (**4**). Notice in the sketch of the unit cell for this lattice that only the octahedral holes of one ion's lattice are occupied by the counter-ions. Each ion, therefore, has a C.N. of 6 and the unit cell contains four NaCl ion pairs. Examples of salts with the NaCl structure are all the alkali metal halides (with the exception of the heavier cesium halides), the alkaline earth chalcogens, the silver halides (except AgI), and more complex materials such as $[Co(NH_3)_6]\{TiCl_6\}$ and NH_4I. Even salts such as KOH, KCN, and KSH exhibit the NaCl structure by rapid tumbling of the otherwise non-spherical anions.

Finally, there are numerous examples of MX salts with the *fcc lattice for anions* and with only *half of the tetrahedral holes* in the face centered lattice *occupied by counter-ions* (**6**). The S, Se, and Te salts of Zn, Cd, and Hg form this type of lattice (known as the *zinc blende* structure after the geologist's name for ZnS). Note that this structure is appropriate to compounds of the formula MX, as the ratio of face centered ions to those in the tetrahedral holes is 1:1. Each ion in the zinc blende structure has a C.N. of 4, and each is located in a tetrahedral environment of the counter-ions. Refer to Table 6-4 and notice that the Madelung constant for zinc blende is not too different from (smaller than) those of NaCl and CsCl.

We now turn to the common structures for salts of formula MX_2 and M_2X. The simplest such structure is based on interpenetrating simple cubic and face centered cubic lattices of ions. These structures are called the *fluorite* (the common name for CaF_2) structure or the *anti-fluorite* structure, depending on the anion and cation sub-lattice structure:

| sc = anion | fcc = anion |
fcc = cation	sc = cation
fluorite	anti-fluorite

The unit cell for the fluorite structure may be represented in two ways, each having a particular visual advantage. Cell 7 emphasizes the sc lattice of anions (○) with cations

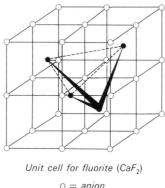

Unit cell for fluorite (CaF_2)

○ = anion
● = cation

7

(●) in half of the cubic holes. This structure bears a strong resemblance to the ZnS cell (6) in that the cation arrangement is the same; the difference is that ZnS has a fcc anion lattice while CaF_2 has a sc anion lattice. The alternate cell for representing the fluorite structure is **5**, also used for the anti-fluorite structure, except that for fluorite the open circles are to represent fcc cations, and the solid circles represent sc anions. The advantage to **5** is that it emphasizes the cation/anion interchange relation between fluorite and anti-fluorite. The primary differences in these two structures are that: (1) as you shall see, the lattice holes are bigger in relation to the anion size for the sc anion lattice, and (2) the stoichiometry changes from MX_2 to M_2X.

At this point you should be wondering whether there are any examples of salts of fcc anions in which all O_h and T_d holes are filled with cations. Such salts would of necessity have a formula M_3X, or, with the roles of cation and anion reversed, MX_3. In fact, several compounds are known to have the cryolite structure (Na_3AlF_6) and the anti-cryolite structure (an example is $Co(NH_3)_6I_3$).

Finally, we should note that the most common lattice structure for MX_2 compounds that permits "octahedral" sites for the cations (fluorite provides for cubic sites) is the *rutile* structure (**8**). Rutile (the common name for TiO_2) presents a low

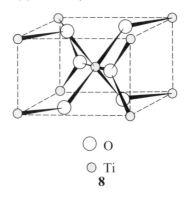

○ O
○ Ti

8

symmetry sub-lattice of anions that does not fit well into the systematics of sc and fcc lattices discussed before; on the other hand the cation sub-lattice of **8** is *body centered*. The fluorite structure is common to difluorides and is found for *some* dioxides of metals, while the chalcogenides of Li, Na, and K(M_2X) are known to have the anti-fluorite structure.

Before considering what factors might be important in determining which of the three structures (NaCl, CsCl, or ZnS) is most reasonable for a salt of formula MX and which (fluorite/anti-fluorite or rutile/anti-rutile) is reasonable for compounds of formula MX_2 or M_2X, we should note that many compounds exhibit slightly distorted versions of these basic structural types. For example, CaC_2, $CaCO_3$, and $NaNO_3$ all exhibit the basic NaCl structure, but each is distorted slightly to accommodate the non-spherical shapes of the anions. (It is perhaps of some interest that some anions such as CN^- and SH^- appear to tumble rapidly in their lattice positions at ambient temperatures while others do not.) The CaC_2 lattice is elongated in the direction of the Ca---CC---Ca axis, while the $NaNO_3$ lattice is compressed along the direction of the C_3 axis of the NO_3^- anions, so that the unit cell structure reflects the ion structure.

STRUCTURE AND THE CONCEPT OF ION RADII

In the preceding discussion of the most basic, simplest structural types it was convenient to make use of the assumption that the ions of a lattice could be thought of as hard spherical objects. If such a view is reasonable, and there will be more about this later, then we might attempt to define the holes (cubic, octahedral, and tetrahedral) in terms of the size of an ion that will fit into them. Simple geometrical considerations then allow us to determine the radius of the hole in relation to the radius of the ion defining the simple cubic and face centered cubic lattices.

For the ideal sc structure (**9**) in which the hard spheres constituting the lattice are just touching, the length of the edge of the cube is $2r_l$, where r_l is the radius of the lattice ion. The body diagonal of the cube is $2(r_l + r_h)$, where r_h stands for the radius of the hole. The Pythagorean Theorem gives an equation that can be solved for the ratio r_h/r_l, with the result that the ratio in the ideal case of just-touching lattice spheres is 0.73—the radius of the hole is about 3/4 that of the ions defining it.

For the face centered lattice (**10**) we must consider both the octahedral and tetrahedral holes. Considering the octahedral holes first, we observe that the distance between two opposite fcc ions is $2(r_l + r_h)$ and that the cube edge measures $2\sqrt{2}r_l$.

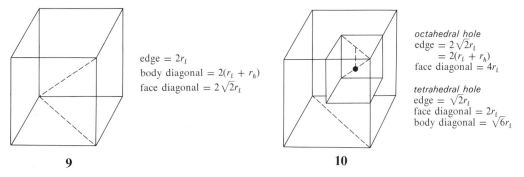

edge = $2r_l$
body diagonal = $2(r_l + r_h)$
face diagonal = $2\sqrt{2}r_l$

octahedral hole
edge = $2\sqrt{2}r_l$
      = $2(r_l + r_h)$
face diagonal = $4r_l$

tetrahedral hole
edge = $\sqrt{2}r_l$
face diagonal = $2r_l$
body diagonal = $\sqrt{6}r_l$

9   10

(The latter dimension derives from the Pythagorean Theorem and the fact that a face diagonal has a length of $4r_l$.) Thus, the ratio r_h/r_l is 0.41. Notice that the size of the octahedral hole of the face centered lattice is *very* roughly about half as large as both the cubic hole of the simple cubic lattice *and* the ion defining the fcc lattice. Finally, for the tetrahedral hole of the face centered lattice, we are dealing with a sub-cube of edge-length $\sqrt{2}r_l$, face diagonal = $2r_l$, and body diagonal = $\sqrt{6}r_l$. Since the body diagonal also equals $2(r_l + r_h)$, we find the ratio r_h/r_l to be 0.22, so the hole radius is about half the size of the octahedral hole and a quarter the size of the lattice ion.

In summary, we expect that when the cation has a radius not much bigger than one-fourth that of the cubic lattice anion, the cation will take up a tetrahedral site in an anion face centered lattice. When the cation has a radius in the range from one-fourth to approximately one-half that of the anion, it will prefer the larger octahedral site in an anion fcc lattice. Finally, as the cation size approaches that of the cubic lattice ion, we expect the fcc anion lattice to open up to the simple cubic structure so that the cation will occupy a cubic hole.

While these observations of relative hole sizes to ion sizes for the different lattice structures are interesting, we have engaged in a rather impractical exercise if we cannot do something about defining the radii of various cations and anions. A potential shortcoming of the idea of structures based on ion sizes (the hard sphere model) is that the radial wavefunctions of electrons in atoms and ions do not abruptly fall to zero at particular distances from the nuclei. The "hard sphere" has a rather fuzzy boundary. On the other hand, we are seeking an operational definition of ion radius from which we need only qualitative and semi-quantitative information.

Now, there are probably as many different ways of estimating ion radii as there are ways of estimating atom and ion electronegativities, another less-than-uniquely-defined concept. One of the ways to arrive at a set of ion radii is the following. In comparing the internuclear distances of salts of common structure, such as the alkali metal halides, one is struck by the fact that the difference in alkali metal bromide and chloride internuclear distances is essentially constant for each cation at about 0.16 Å ($\pm$0.02 Å)(Br^- > Cl^-). Similarly, for the chlorides and bromides of K and Na, the differences in internuclear radii are constant for each anion at about 0.30 Å (K^+ > Na^+). Many more such comparisons could be made, and all such results are supportive of attempts to define operationally unique, constant radii for ions. Furthermore, many compounds are known for which variation of the cation has essentially no effect on the cation-anion internuclear distance. This phenomenon is interpreted to mean that for such compounds the cations just fit into or are somewhat smaller than the holes they occupy in the anion lattices. In other words, the distance between the anion nucleus and the hole center is fixed by the fact that the anions are just in contact with one another. According to this interpretation, the anion radius is simply half of the anion-anion internuclear distance. In this vein, "absolute" radii can be established for anions, and these in turn can be used to determine the cation radii in salts of those same anions where the cation-anion distance is *not* independent of the cation. By systematically considering various salts, one expands a list of cation and anion radii. The procedure inevitably produces some discrepancies, which have been studied from several points of view. Probably the most widely accepted ion radii

TABLE 6–5
SOME IONIC RADII (Å)

								H^- 2.08
Li^+ 0.60	Be^{2+} 0.31			B^{3+} 0.20	C^{4+} 0.15	NH_4^+ 1.48	O^{2-} 1.40	F^- 1.36
Na^+ 0.95	Mg^{2+} 0.65			Al^{3+} 0.50	Si^{4+} 0.41		S^{2-} 1.84	Cl^- 1.81
K^+ 1.33	Ca^{2+} 0.99	Cu^+ 0.96 *	Zn^{2+} 0.74	Ga^{3+} 0.62	Ge^{4+} 0.53		Se^{2-} 1.98	Br^- 1.95
Rb^+ 1.48	Sr^{2+} 1.13	Ag^+ 1.26	Cd^{2+} 0.97	In^{3+} 0.86	Sn^{4+} 0.71		Te^{2-} 2.21	I^- 2.16
Cs^+ 1.69	Ba^{2+} 1.35	Au^+ 1.37	Hg^{2+} 1.10	Tl^{3+} 0.95	Pb^{4+} 0.84			
				Tl^+ 1.40	Pb^{2+} 1.21			

*Mn^{2+} (0.80), Fe^{2+} (0.76), Co^{2+} (0.74), and Ni^{2+} (0.69).

TABLE 6-6

SOME OBSERVED CRYSTAL STRUCTURES AND RADIUS RATIOS

Formula	MX			MX_2		M_2X
Lattice Type	ZnS	NaCl	CsCl	CaF_2	TiO_2	anti-fluorite
Theoretical Radius Ratio	(0.23)	(0.41)	(0.75)	(0.75)		(0.23)
	BeO (0.22)			$SrCl_2$ (0.62)		Li_2O (0.43)
	MgTe (0.29)			CaF_2 (0.73)		Na_2O (0.68)
		MgSe (0.33)		CdF_2 (0.71)		K_2O (0.95)
		MgS (0.35)		HgF_2 (0.81)		Li_2S (0.33)
		CaTe (0.45)			MgF_2 (0.48)	Na_2S (0.52)
		MgO (0.46)			SnO_2 (0.51)	
			CsCl (0.93)		ZnF_2 (0.54)	
			CsBr (0.87)		PbO_2 (0.60)	

are those of Pauling,[9] who introduced corrections for the greater internal compression forces in crystals with ions of high charge. These corrected radii are termed *crystal radii*, and some of them are presented in Table 6-5. A word of caution is in order regarding these values. Their absolute magnitudes are of questionable meaning; they have their greatest use in the operational sense of *comparing* relative ion sizes and internuclear distances in crystals.

We are now in a position to test our ideas on ion size as a factor in lattice structure. Some exemplary salts are given in Table 6-6. The data support the basic ideas of using radius ratios (r_h/r_l) to predict structure once the stoichiometry has dictated a set of possible structures. (Some discrepancies are noted when the radius ratio is intermediate between ideal values.) Considering the MX compounds first, the radius ratio concept works pretty well, but the MgTe and MgSe examples illustrate the weakness of the concept for r_h/r_l in an intermediate range. For the M_2X compounds we do find many examples of the anti-fluorite structure suggested by the stoichiometry, but the radius ratio conditions are not met too well. For the MX_2 compounds, those with larger ratios are in keeping with the theoretical values for the fluorite structure, and it appears that the smaller ratios favor the rutile structure.

---

[9] L. Pauling, "The Nature of the Chemical Bond," 3rd Edition, Cornell University Press, Ithaca (1960).

STUDY QUESTIONS

14. How many formula weights of MX_2 occur in the unit cell of the rutile structure? What is the coordination number of each cation and anion?

15. What is the ratio r_h/r_l for the triangular holes in a layer of close packed ions? Were such a structure (with all trigonal holes filled) found for a compound, what would be its

empirical formula? Taking the lattice ion to be O^{2-}, are there any cations (except H^+) small enough to fit in the trigonal holes?

16. Calculate the volume of the unit cell of KCl and then the density of the salt (experimental value = 1.98 g/cm^3).

17. Which of the ionic structure types (CsCl or NaCl) is favored for a given compound MX if severe pressure is exerted on the crystal?

18. When the cation in the anion lattice holes is smaller than the hole, will the lattice energy be different from that predicted by the Born equation? How and why?

19. A simple binary compound, M_mX_n, is known to have a structure of ccp anions with all T_d holes filled with M cations. What is the empirical formula of the compound?

20. Use the following data (in kcal/mole) to calculate the electron affinity of Cl.

	ΔH_f°	ΔH_{sub}^M	I_M	D_{Cl_2}	shortest inter-ion distance
NaCl	98.3	25.9	5.14	57.8	2.81 Å
KCl	104.4	19.8	4.34		3.14 Å

21. Which would you predict would have higher *solid state* ionic conductivity, a salt MX with the NaCl structure or a salt MX with the zinc blende structure?

22. The solid Ag_2HgI_4 (yellow) consists of cubic close packed I^- with Ag^+ and Hg^{2+} ions in tetrahedral holes. Are there sufficient cations to occupy all the T_d holes? In the range from 40° to 50°C, the color changes to red and the conductivity increases to a value ~1000 times greater than normal for an ionic solid at 50°, yet the anion lattice remains cubic close packed. What do you suppose happens within the crystal in the 40°–50° range?

LAYER LATTICES AND INCIPIENT COVALENCY

Now that we have considered the most common structures for compounds that may be termed ionic, we must be careful to note that there are many common salts that have lattice energies considerably in excess of those expected on the basis of the ionic model of crystalline materials. Some examples of compounds and the deviations between the theoretical and experimental lattice energies are cited in Table 6-7. A characteristic feature of those compounds for which the deviations are large is that they have either a small, highly charged cation or a large, polarizable cation combined with a relatively large, polarizable anion. These characteristics are consonant in a vague way with the idea of emerging covalency in the cation-anion attraction.

In some instances these deviations become particularly marked so that layer lattices are formed. For example, $CdCl_2$ possesses a lattice of face centered cubic Cl^- ions with half of the octahedral holes occupied, as required by the stoichiometry. The interesting feature, however, is that only *alternate layers* of holes are occupied by the Cd^{2+} ions.

It is exceedingly difficult to make and view a two-dimensional sketch of the lattice structure of such a material when we want to show clearly both the fcc structure of the Cl^- lattice *and* the layer structure. Careful inspection of the fcc lattice of anions shows that this structure can also be drawn as parallel sheets of anions.[10] Figure 6-2 shows two different perspectives of the fcc structure; the fcc structure is defined by anions from four different parallel sheets. Figure 6-3 gives presentations of one and two such sheets. Notice, in Figure 6-3(b), how the ions of the upper layer

---

[10] S-M. Ho and B. E. Douglas, *J. Chem. Educ.*, **46**, 207 (1969) gives further good photos and drawings of ion packing and the lattice holes.

TABLE 6–7
SOME EXPERIMENTAL AND CALCULATED LATTICE ENERGIES
(kcal/mole)

Compound	Calculated	Experimental	Difference	Structure
NaF	−216	−216	0	NaCl
NaCl	−181	−185	−4	NaCl
KCl	−165	−168	−3	NaCl
KI	−149	−155	−6	NaCl
CsF	−173	−175	−2	NaCl
CsCl	−149	−157	−8	CsCl
CaF_2	−620	−619	+1	CaF_2
$SrCl_2$	−494	−508	−12	CaF_2
BaF_2	−559	−553	+6	CaF_2
AgF		−228		NaCl
AgCl	−167	−216	−49	NaCl
$MgBr_2$	−511	−572	−61	Layer
MgI_2	−474	−553	−79	Layer
CaI_2	−453	−494	−41	Layer
MnF_2	−660	−663	−3	Rutile
$MnBr_2$	−520	−583	−63	Layer
MnI_2	−480	−569	−89	Layer
CdF_2	−632	−663	−31	CaF_2
CdI_2	−475	−582	−107	Layer
HgF_2	−624	−660	−36	CaF_2

(full circles) rest in the cusps of the lower layer (broken circles). Such a "packing" arrangement is designated *cubic close packing* to distinguish it from simple *cubic packing*, in which the ions of the upper layer are situated directly above those of the lower layer.

In the previous discussion of the fcc lattice we were able to identify one octahedral hole per anion and two tetrahedral holes per anion. This, of course, can also be done for the partial fcc structure in Figure 6–3(b), where several such holes are identified by the labels *O* and *T*. Picking any ion in the upper layer, we can readily identify between the upper and lower layers four tetrahedral holes to which the chosen ion contributes. If a third layer of anions were to be laid on top of the second, the pattern would repeat itself so that, with any one ion contributing 1/4 to each of its tetrahedral holes, we conclude, as before, that there are 8/4 = 2 tetrahedral holes/anion. Similarly, any one ion contributes to the creation of three octahedral

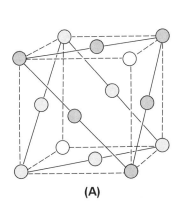
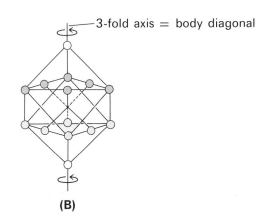

Figure 6–2. Two different perspectives of the fcc lattice to emphasize the parallel sheets of anions. Note O_h hole at unit cell center.

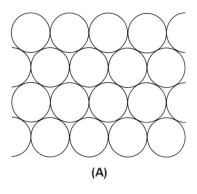 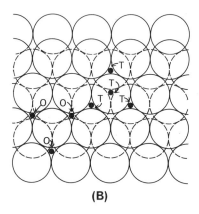

Figure 6-3. (a), One of the sheets of Figure 6-2 extended in two directions. (b), Two close packed layers with typical octahedral and tetrahedral holes identified.

holes below it and to three above it, for a net of one octahedral hole/anion. In this way, you see that the octahedral holes themselves may be thought of as occurring in parallel sheets mid-way between the anion sheets. If all the octahedral holes are filled, we have the NaCl structure discussed earlier.

Returning to the $CdCl_2$ structure, note that the MX_2 stoichiometry requires that only half of the octahedral holes be filled by Cd^{2+} ions. This could be done by Cd^{2+} ions occupying half of the octahedral holes in each sheet of holes; but, in fact, $CdCl_2$ has the layer structure in which *all* the holes in alternate hole-sheets are occupied. This, of course, means that the structure may be thought of as a sheet of cations sandwiched between two sheets of anions (Figure 6-4), and this three-layer arrangement repeats itself in the third direction. Consequently, we find adjacent layers of anions with no cations between them to hold them together. Such a structure is easily sheared, as the coulombic forces holding the sandwiches or layers together are those between the anions of one sandwich and the cations of another.

Related to the $CdCl_2$ structure is that of $CrCl_3$, which, because of the stoichiometry, has only one-third of the octahedral holes occupied with Cr^{3+} ions. Again we encounter the layer structure, with only two-thirds of the holes in alternate sheets occupied.

The mercuric halides (HgX_2, where X = F, Cl, Br, I) all have different structures ranging from ionic (HgF_2) to molecular ($HgCl_2$), with $HgBr_2$ and HgI_2 "intermediate." The radius ratio values of 0.81 (F), 0.61 (Cl), 0.56 (Br), and 0.51 (I) provide at least a partial explanation of the irregular trend. HgF_2 exhibits the fluorite structure, which is not unexpected from the radius ratio, stoichiometry, and ionic nature of metal fluorides. $HgCl_2$ presents somewhat of a problem because Hg^{2+} is too small to accommodate the cubic holes of a fluorite anion sub-lattice of Cl^- and yet is too big to adapt to the smaller O_h holes of a layer structure like $CdCl_2$ (for which $r_h/r_l = 0.41$).

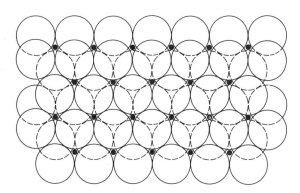

• = cations in O_h holes between anion sheets

Figure 6-4. The $CdCl_2$ layer structure.

METAL-CONTAINING SOLIDS 289

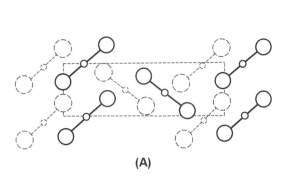
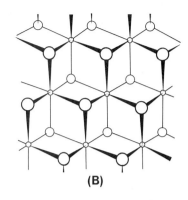

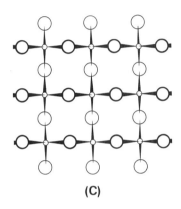

Figure 6–5. The lattice structures of (a) HgCl$_2$, (b) HgBr$_2$, and (c) HgI$_2$. [From A. F. Wells, "Structural Inorganic Chemistry," Oxford University Press, New York (1975).]

Consequently, the fact that the HgCl$_2$ structure is based on packing of linear Cl-Hg-Cl molecules is not all that surprising [Figure 6-5(a)].

The Hg^{2+}/Br$^-$ combination has a radius ratio more accommodating to a CdCl$_2$ type of lattice, but Hg^{2+} is still a little too large. Thus, HgBr$_2$ has a structure intermediate between those of HgCl$_2$ and CdCl$_2$. That is, there are two *trans*, short Hg-Br distances and four longer Hg-Br distances defining an axially compressed octahedron of I about Hg [Figure 6-5(b)]. Finally, the Hg^{2+}/I$^-$ combination adopts an unusual layer lattice *not* based on cubic close packed layers of anions (the incipient covalency in HgI$_2$ should be very large, yet Hg^{2+} is still too large to fit an O_h or T_d hole of a cubic anion lattice). This structure is conveniently viewed as HgI$_4$ tetrahedra sharing all corners (every I$^-$ bridges two Hg^{2+}) [Figure 6-5(c)].

As an alternative to this layer lattice type of HgI$_2$, there are a few examples in which the role of bridging halides is carried to the extreme of "one-dimensional" chains. Most notable among these are BeCl$_2$, CuCl$_2$, and PdCl$_2$. In Be^{2+} we find a small cation of high charge, and in Pd^{2+} a polarizable cation of high charge. The BeCl$_2$ chains may be viewed as edge-sharing *tetrahedra* of BeCl$_4$ sub-units

$$\cdots \mathrm{Be} \begin{smallmatrix} \mathrm{Cl} \\ \mathrm{Cl} \end{smallmatrix} \mathrm{Be} \begin{smallmatrix} \mathrm{Cl} \\ \mathrm{Cl} \end{smallmatrix} \mathrm{Be} \begin{smallmatrix} \mathrm{Cl} \\ \mathrm{Cl} \end{smallmatrix} \mathrm{Be} \cdots$$

The PdCl$_2$ and CuCl$_2$ chains are similarly viewed as edge-sharing *squares* of MCl$_4$ sub-units

$$\cdots \mathrm{Pd} \begin{smallmatrix} \mathrm{Cl} \\ \mathrm{Cl} \end{smallmatrix} \mathrm{Pd} \begin{smallmatrix} \mathrm{Cl} \\ \mathrm{Cl} \end{smallmatrix} \mathrm{Pd} \cdots$$

$$\cdots \mathrm{Cu} \begin{smallmatrix} \mathrm{Cl} \\ \mathrm{Cl} \end{smallmatrix} \mathrm{Cu} \begin{smallmatrix} \mathrm{Cl} \\ \mathrm{Cl} \end{smallmatrix} \mathrm{Cu} \cdots$$

This structure of edge-sharing squares for PdCl$_2$ is called the α-form. Of greater thermodynamic stability, however, is a hexameric β-form of corner-shared squares,[11] as follows:

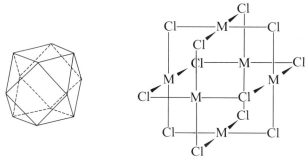

In this cuboctahedral structure, Cl appears at the polyhedral vertices and Pd at the face centers.

[11] K. Brodersen et al., Z. Anorg. Allgem. Chem., **337**, 120 (1965); R. Mattes, Z. Anorg. Chem., **364**, 290 (1969); U. Wiese et al., Angew. Chem. Internat. Edn., **9**, 158 (1970).

STUDY QUESTIONS

23. Describe predicted structural types for AlP, $FeCl_3$, NaH, MnF_2, CuI, $NaBH_4$, and TlCl.

24. Both $FeCl_2$ and $FeCl_3$ have layer structures based on close-packed anions. How do their structures differ? Aside from the anions, in what sense are their structures alike? Make sketches like those of Figure 6-4 to illustrate your points.

25. Which salt in each pair do you expect to show the larger deviation from the ideal ionic behavior?

    $CaCl_2/ZnCl_2$ MgF_2/TiO_2 $MgCl_2/BeCl_2$
    $CdCl_2/CdI_2$ Al_2O_3/Ga_2O_3
    ZnO/ZnS $CaCl_2/CdCl_2$

26. Make a sketch, like those of Figure 6-4, for $CrCl_3$ (each Cr^{3+} has a C.N. = 6, each Cl^- has a C.N. = 2).

SYNTHESIS PRINCIPLES

The use of the Born equation for the computation of lattice energies has the undesirable feature that we must know the lattice structure before we can calculate the lattice energy of a salt. Kapustinskii[12] noticed that this problem could be effectively circumvented by dividing and multiplying the Born equation by the number of ions in the salt formula. This is effective because the ratio of the Madelung constant to the number of ions per formula is essentially constant, ranging from 0.80 for rutile to 0.88 for cesium chloride. From a survey of more lattice types than we have covered here, it has been decided that the most effective value of M/v, where v is the number of ions per unit formula, is 0.874. Substituting this value into the Born equation, and also noting that the parameter n in the Born repulsion term has a nominal value of 9, Kapustinskii advocated the use of the simpler formula

$$E = \frac{256 \, v q_+ q_-}{r_+ + r_-}$$

for the computation of the lattice energy of ionic solids of unknown lattice structure.

[12] A. F. Kapsutinskii, Quart. Revs. Chem. Soc., 283 (1956), and Z. Phys. Chem., **22B**, 257 (1933).

In this equation, q_+ is the cation charge, q_- is the anion charge, r_+ is the cation radius, and r_- is the anion radius. The substitution of $r_+ + r_-$ for d_0 is also an approximation, as discussed the preceding sections.

An additional benefit to be gained by the use of the Kapustinskii equation is that it is of simple enough form to use in attempts to generalize the effects of ion size and charge on lattice energies—both of which are important considerations to the synthetic chemist dealing with solid ionic materials. The next few sections of this chapter will deal with these topics, using the Kapustinskii equation as a means for carrying out the analyses.

Cation Oxidation States

One of the more useful applications of the Kapustinskii equation is in consideration of decomposition reactions such as

$$MX_n \rightarrow MX_{n-1} + \frac{1}{2}X_2$$

where X is an anion from the halogen group. The internal energy change for such a reaction can be analyzed so as to introduce the lattice energies of MX_n and MX_{n-1} by using a Haber cycle, with the result that

$$\Delta E = \{E_{n-1} - E_n\} - I_n - \left\{\frac{1}{2}D_{X_2} - A_X\right\}$$

In this equation E_n is the lattice energy of MX_n, I_n is the nth ionization potential of M, D_{X_2} is the bond dissociation energy of X_2, and A_X is the electron affinity of X. $-I_n$ always favors the decomposition, so the first two terms and the last two terms must be decisive in determining the stability of MX_n relative to MX_{n-1}. Let us focus our attention for the moment on the difference in lattice energies.

Using the Kapustinskii equation for the case where X is a halogen, the lattice energy difference is given by

$$\{E_{n-1} - E_n\} = 256(n)\left\{\frac{n+1}{r_n + r_-} - \frac{n-1}{r_{n-1} + r_-}\right\}$$

where r_n and r_{n-1} are the radii of M^{n+} and $M^{(n-1)+}$ cations. As r_n will always be slightly less than r_{n-1}, we are assured that the lattice energy difference will be positive. In other words, the relative lattice energies of MX_n and MX_{n-1} tend to favor the higher oxidation state (do not forget that this conclusion depends on the accuracy of treating both materials as ionic compounds). Furthermore, the last equation suggests that the smaller the anion, the more will the higher oxidation state of the cation be favored. This is certainly one factor in the well-known "stabilization" of higher oxidation states of cations by F^-. For example, Ag^{2+}, Co^{3+}, Mn^{4+}, and the tetravalent ions of cerium, terbium, and praseodymium form only "stable" fluorides, among the possibilities of halogens. Interestingly, while the other halides of Cu^{2+} and Fe^{3+} are known, the iodides are unstable at room temperature and decompose to CuI and FeI_2.

Because of the change in stoichiometry and change in anion charge, the Kapustinskii expression for the lattice energy difference with X = chalcogenide is slightly different from that for the halides; but the general conclusion is the same. Of the chalcogens, oxygen should best stabilize the higher oxidation states of metals. In keeping with this, oxides of Ag^{2+}, Co^{3+}, Pb^{4+}, and Ce^{4+} are "stable" at room temperature.

To complete the discussion of the $MX_n \rightarrow MX_{n-1}$ decomposition, we should also consider the effects of the final two terms in the expression for the internal energy change. For the halogens, the terms $\{\frac{1}{2}D_{X_2} - A_X\}$ vary as -63, -57, -58, and -53 kcal for F through I, respectively. Thus, we see that these terms also oppose the decomposition and effect an ordering $F > Cl \sim Br > I$ with respect to the halogen's ability to stabilize high metal oxidation states.

In comparing oxides and fluorides on the basis of energy per mole of the metal in the higher state, you might wish to confirm on your own that the lattice energy difference for higher and lower oxidation state salts tends to make oxide better at stabilizing the higher metal oxidation state than fluoride (it is helpful in quantifying this comparison that the oxide and fluoride radii are nearly equal). However, the much greater O_2 bond energy and the electron affinity for $O \rightarrow O^{2-}$ tend to disfavor the oxides very strongly relative to the fluoride. Generally, this is what is observed experimentally. $PtO_2 \cdot H_2O$ cannot be dehydrated above 200°C without decomposition to the metal, whereas PtF_4 is, in fact, prepared by fluorination of a red-hot Pt wire. Similarly, AgO decomposes at 200°C, while AgF melts without decomposition at 435°C. HgF_2 is reported to boil at 650°C, whereas you learned early in your career that HgO decomposes to produce O_2 upon fairly mild heating.

Another interesting application of the Kapustinskii equation is to the reaction

$$3MX_2 \rightarrow M + 2MX_3$$

This disproportionation reaction is the one appropriate to the question of "stability" of low metal ion oxidation states. The anion effect on this reaction is restricted to its role in determining the relative lattice energies of MX_2 and MX_3. Using the Kapustinskii equation again, we find that the lattice energy contribution to the internal energy change of the metal disproportionation reaction is

$$2E_3 - 3E_2 = \frac{256}{d}(-6)$$

where again we have simplified the expression by ignoring the difference in radii of divalent and trivalent cations. Clearly, the lattice energy difference between MX_2 and MX_3 tends to favor the disproportionation (assuming that the ionic model is applicable), but the tendency is diminished with the larger cation/anion combinations. Evidence of this can be found in the fact that the only Au^+ and Cu^+ halides unknown at room temperature are the fluorides. Here, the disproportionation reactions must be of the type

$$3AuX \rightarrow 2Au + AuX_3$$

or

$$2CuX \rightarrow Cu + CuX_2$$

for which lattice energies also favor disproportionation. (Note that Au^{2+} is unknown.) Interestingly, Ga^+, otherwise difficult to obtain, is known in the form of the salt $Ga(GaCl_4)$, where stabilization appears to occur with the large anion $GaCl_4^-$.

Ionic Fluorinating Agents

A final application of the idea of lattice energies as a significant factor in the thermodynamic ease of reactions is found in halogen exchange reactions, of which fluorination is an important example.

$$RCl + MF \rightarrow RF + MCl$$

RCl and RF represent polar covalent bonds. The dependency on cation size of the internal energy change for such a reaction arises solely from the difference in MF and MCl lattice energies. From the Kapustinskii equation, this energy difference

$$E_{MCl} - E_{MF} = 512 \left\{ \frac{1}{r_{M^+} + r_{F^-}} - \frac{1}{r_{M^+} + r_{Cl^-}} \right\} > 0$$

diminishes as the cation size increases. Because the lattice energy difference actually opposes the reaction, the halogen exchange reaction is best attempted with fluoride salts of larger cations. In keeping with this prediction, cesium fluoride is the most

TABLE 6-8

DIFFERENCES IN HEATS OF FORMATION OF FLUORIDES AND CHLORIDES

Salt	$\frac{1}{n}\{\Delta H_f^\circ (MCl_n) - \Delta H_f^\circ (MF_n)\}$ (kcal/mole^{-1})
LiF	49.7
NaF	38.1
KF	30.3
CsF	23.4
AgF	18.1
MgF_2	55.9
CaF_2	50.1
SrF_2	46.2
BaF_2	40.7
HgF_2	23.

powerful fluorinating agent of all the alkali metal fluorides and has found extensive application in such reactions. As you might expect by now, the effect of increased cation charge with no change in cation radius is to disfavor the halogen exchange reaction. This, too, can be predicted from comparison of Kapustinskii equation lattice energies for

$$RCl + MF \rightarrow RF + MCl$$

and

$$RCl + \frac{1}{2}MF_2 \rightarrow RF + \frac{1}{2}MCl_2$$

In practice, LiF is a much better fluorinating agent than is MgF_2, and KF is better than BaF_2.

It is in reactions such as the halogen exchange reactions that we find a significant synthetic application for salts that show deviation from the ionic model. Because the extra stability of the lattice attributed to covalency in the cation-anion interaction increases as the size and polarizability of the anion increases, it should come as no surprise that AgF is a much better fluorinating agent for chlorides than is NaF (the sizes of Na^+ and Ag^+ cations are nearly the same) and that HgF_2 is much better for Cl/F exchange than is SrF_2. In fact, HgF_2 rivals AgF in its ability to exchange F^- for Cl^-, in spite of the predicted (from the purely ionic model) detrimental effect of the higher cation charge for Hg^{2+}. Some data on the differences in heats of formation of chlorides and fluorides are given in Table 6-8.

Ion Size and Compound Isolation from Solution

Some general examples of the application of the ideas presented above have been summarized by Basolo.[13] In reactions with oxygen, Li produces Li_2O whereas Cs forms the superoxide, CsO_2, because the Li^+ radius is more compatible with the smaller O^{2-} ion while the larger Cs^+ ion packs better with the larger O_2^- ion. For carbonate decomposition reactions into the alkaline earth oxide and CO_2, the smaller cations increasingly favor the MO lattice with the opposite trend for the CO_3^{2-} lattice, with the result that the thermal stability of the MCO_3 increases as the cation size increases. Similarly, isolation of the anions ICl_4^- and BCl_4^- is facilitated by use of large cations; whereas NaCl will not react with BCl_3, CsCl does so readily. The mercurous ion (Hg_2^{2+}) is a well-known species, but the isolation and identification of the cation Cd_2^{2+} (cadmous ion) was finally achieved by lattice formation with $AlCl_4^-$. A particularly interesting reaction is that of halogen exchange between LiI and CsF. The experimental heats of formation of these salts readily show that the exchange reaction is exothermic as a result of the more favorable LiF combination than the LiI combination (but be careful to note that the CsF packing is also more favorable than CsI).

---

[13] F. Basolo, *Coord. Chem. Revs.*, **3**, 213 (1968).

$$\text{LiI}_{(c)} + \text{CsF}_{(c)} \rightarrow \text{LiF}_{(c)} + \text{CsI}_{(c)}$$
$$-65 \quad\quad -127 \quad\quad -146 \quad\quad -80 \quad \Delta H° = -34$$

In each of these examples we find it possible to rationalize observed reaction products in terms of the qualitative concept of ion size and packing efficiency. Considering that salt solubilities depend in a direct way on the lattice energy, we may cite the following as exemplary: AgCl is less soluble than either Ag_2SO_4 or Ag_3PO_4; similarly, $BaSO_4$ is less soluble than $BaCl_2$ or $Ba_3(PO_4)_2$. A particularly interesting example of the use of the lattice energy concept in rational synthesis is the preparation of the elusive $CuCl_5^{3-}$ complex ion by precipitation with $Cr(NH_3)_6^{3+}$ from aqueous solution, in which $CuCl_3^-$ and $CuCl_4^{2-}$, but *not* $CuCl_5^{3-}$, have been detected. Apparently only trace amounts of the pentachloro complex are present in solution, and the equilibrium

$$2Cl^- + CuCl_3^- \rightleftharpoons Cl^- + CuCl_4^{2-} \rightleftharpoons CuCl_5^{3-}$$

is shifted in favor of the pentachloro complex by its removal as $\{Co(NH_3)_6\}\{CuCl_5\}$. Similarly, the hexachloro complex ions MCl_6^{3-}, which are not stable in aqueous systems, can be isolated from the same systems with the cation $\{Co(pn)_3\}^{3+}$. Here, pn = $H_2N(C_3H_6)NH_2$, and examples of M are Cr, Mn, and Fe.

Examples of large non-metal ions that have been found to be useful for isolation of large counter-ions are: R_4N^+ (tetraalkylammonium ions), R_4P^+ (tetraalkyl- and tetraarylphosphonium ions), Ph_4B^- (tetraphenyl borate), PF_6^- (hexafluorophosphate), BF_4^- (tetrafluoroborate), and SiF_6^{2-} (hexafluorosilicate), to name but a few.

Lest we over-commit ourselves to the importance of ion size in determining (via lattice energy) solubilities, we must be careful to note that ion solvation energies, as well as lattice energies, can be important in determining the solubilities of ionic materials. Fajans,[14] and more recently Morris,[15] have noted a definite relation between salt solubility in water and the difference in solvation energies of cation and

---

[14] K. Fajans, *Verhandl. Deut. Physik. Ges.*, **21**, 549, 709 (1919), and *Naturwiss.*, **9**, 729 (1921).
[15] D. F. C. Morris, *Struct. Bond.*, **6**, 157 (1969).

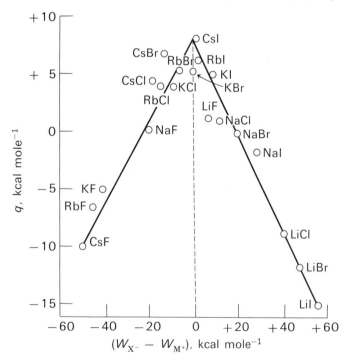

Figure 6-6. Relationship between the standard heat of solution of a crystalline alkali halide and the difference between the absolute heats of hydration of the corresponding gaseous anion and cation. [From D. F. C. Morris, *Structure and Bonding*, **6**, 157 (1969).]

TABLE 6-9

ABSOLUTE ENTHALPIES OF HYDRATION AT 298°K (kcal/mole)

Ion	H_{hyd}
H^+	−264
Li^+	−128
Na^+	−100
K^+	−80
Rb^+	−73
Cs^+	−69
F^-	−120
Cl^-	−88
Br^-	−80
I^-	−70

anion. Their findings are summarized in Figure 6-6, where we see, at least for the alkali metal halides, that salt solubility is greatest when cation and anion solvation energies differ most. It is particularly striking that CsI, with a lattice energy of 80 kcal/mole, is much less soluble than CsF (lattice energy of 175 kcal/mole) and NaF (lattice energy of 216 kcal/mole). While Figure 6-6 is a relation between standard heat of solution and difference in ion heats of solvation, a consideration of free energies in place of enthalpies yields the same general correlation. A possible interpretation of this relationship is that solubility is higher when the polar water molecules exhibit much stronger binding with one ion than the other, in effect more efficiently separating the ions in solution than when both ions strongly compete for the solvent molecules.

For reference purposes, Table 6-9 is a listing of enthalpies of hydration of alkali metal and halide ions, all based on the best estimate of the hydration enthalpy of the proton of −264 kcal/mole. Note the decrease in solvation enthalpy with increase in ion size, and also note that for cations and anions of the same size (Na^+, F^-; Rb^+, Cl^-; Cs^+, Br^-) *it is the anion that is more strongly solvated by water.*

As one further example of the influence of solvation energies on salt solubilities, the solubilities in water of the silver halides follow the order $AgCl > AgBr > AgI$. This ordering is not consistent with the lattice energies of 216, 212, and 212 kcal/mole, respectively. Reference to Table 6-9 shows us that the observed solubility order follows that of the anion solvation enthalpies, *i.e.*, $Cl^- > Br^- > I^-$. Furthermore, the successive differences in anion solvation enthalpies are greater than the successive differences in lattice energies. (The almost negligible differences in lattice energies are important, therefore, and serve as a reminder that the covalent character of the cation/anion interaction increases along the series.) As a final point, you might recall (Chapter 5) that in liquid ammonia as solvent the order of silver halide solubilities is just the reverse of that for water. While the heat of solution data for the anions in liquid ammonia appear not to be available, it is not unreasonable to suppose that the lower hydrogen bond acidity of the ammonia molecule would make it less discriminating than water as far as anion solvation is concerned. It is interesting to note, however, that, as a group, the silver halide solubilities in liquid ammonia are greater than in water, presumably because of the greater cation solvation by the more basic ammonia molecule.

STUDY QUESTIONS

27. Account for the observed trends in ΔH_f° of the following series (use a Haber cycle to identify critically important influences on ΔH_f°). Remember that $\Delta H_f^\circ < 0$.

   a) $NaF < NaCl < NaBr < NaI$
   b) $LiF < NaF < KF < RbF$
   c) $LiI > NaI > KI > RbI > CsI$

28. Why can BPh_4^- be used to determine K^+ gravimetrically in aqueous solution, while ClO_4^- cannot (the anions are very nearly the same size)?

29. Suggest reasons why $MClO_4$ salts (M = alkali metal) exhibit a minimum in aqueous solubility at M = Rb^+.

METALS

Structures and Bonding

It is appropriate at this point to consider the structures of the metallic elements because, generally, these are closely related to the lattice structures already discussed for ionic solids. Briefly, the three principal lattice types for metals are the *body centered* (*cf.* the CsCl structure), *face centered cubic* (*cf.* the anion or cation sublattice of NaCl), and *hexagonal close packed;* sketches of these structures are given in Figure 6-7. Notice that the C.N. of each atom in the bcc structure is 8, whereas the C.N. is 12 for both fcc and hcp structures.[16] With such high coordination numbers for each atom, we immediately encounter great difficulty in proposing a stereo-related bonding model for metals according to the covalent, electron-pair bond concept.

In many ways the most successful model of the electronic structures of metals (but this model is not particularly helpful with regard to lattice structure) is an extended form of molecular orbital theory that deals with physical properties of metals such as electrical conductivity. The theory is known as the *band theory* of metals, for reasons that will become obvious from the following brief outline.

With reference to the concepts introduced in Chapter 4 concerning the LCAO-MO theory, the overlapping of orbitals on adjacent atoms leads to bonding, non-bonding, and anti-bonding orbitals. Within limits, the greater the overlap between orbitals, the greater the energy gap between bonding and anti-bonding orbitals. Applying these ideas to a crystal of a metallic element (a "molecule" of an astronomical number of atoms), we are confronted with so many lattice orbitals that the band of orbital energies for the valence electrons of the metal atoms approaches a continuum. Figure 6-8 depicts the creation of such a band from atomic orbitals (*s* ao's, say) for a linear chain of metal atoms (Chapter 4, p. 157). A key point is that as the number of atoms in the chain increases, an energy band develops. Because the band width for any particular type of atomic orbital depends on orbital overlap at the normal metal-metal distances of metallic lattices, the inner or core orbitals form fairly narrow bands separated by appreciable energy gaps from each other and from the valence orbital bands (Figure 6-9). Furthermore, since these atomic orbitals contain pairs of electrons, all lattice orbitals arising from core atomic orbitals are filled with electrons.

---

[16] There is a close similarity between the fcc and bcc structures, which appears on close inspection. Both structures present the central atom with 8 neighbors at the corners of a cube. In the bcc structure these are shown in Figure 6-7(c) at unit cell corners. In the fcc structure, the 8 neighbors are at unit cell face mid-points. The additional 4 neighbors in the fcc cell are at the unit cell corners and are somewhat closer to the central atom.

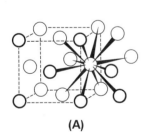
(A)
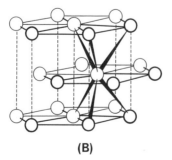
(B)
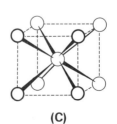
(C)

Figure 6-7. The three common metal lattices: (a) face centered cubic, (b) hexagonal close packed, and (c) body centered cubic. In (a) and (b), atoms in the adjoining unit cell are shown to help in visualizing the coordination number.

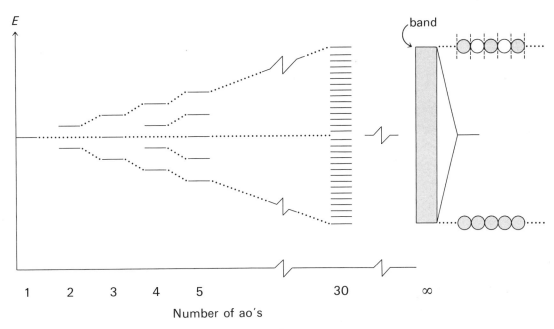

Figure 6-8. Correlation of one atom/one s ao with the energy band for a linear chain of N atoms/N s ao's.

Unoccupied lattice orbitals must be available for the transport of valence electrons throughout the crystal. An *insulator* is represented (Figure 6-10) as a case in which completely filled bands are separated by a considerable energy gap from unoccupied bands. *Conductors* are identified as cases in which a particular valence band is only partially filled or, if enough electrons are available to fill a particular band, two adjacent bands are contiguous so as to provide a composite partially filled band. In either case, application of an electric field (a voltage drop across the crystal) will induce conduction. Either description could apply to the alkali metals with only half-filled s orbital bands, while contiguous valence s and p bands are required to

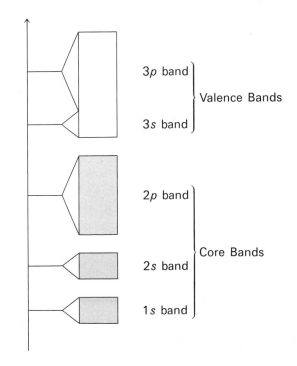

Figure 6-9. Illustration of core and valence bands for a third period metal.

298 ENERGETICS AND STRUCTURES AS GUIDES TO MAIN GROUP CHEMISTRY

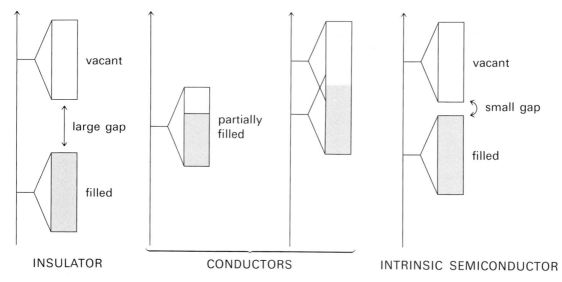

Figure 6-10. Schematics of the band concept for insulators, conductors, and intrinsic semiconductors.

account for the conductance observed with alkaline earth metals. According to this model, *intrinsic semiconductors* are identified as elements for which sufficient electrons are available to occupy one band completely, but in which another band is to be found across an energy gap that is small in comparison with thermal energies. Silicon and germanium are familiar to you as belonging to this class. In this way we can account for the increase in conductivity with temperature, which is a characteristic property of semiconductors but not of metals. *Photoelectric* materials fit into this model as cases for which the gap between a filled valence band and an empty valence band is comparable to visible/ultraviolet photon energies. Here, electronic absorption causes both higher and lower bands to be partially filled. Another optical property of metals that fits into this model is their high reflectivity or luster. With a near continuum of unoccupied orbital energies in the range of visible radiation, a metal could absorb and subsequently emit radiation of nearly all wavelengths in the visible range, so as to appear "white" and highly reflective.

The semiconductor properties of many materials can be controlled by "doping" semiconductors with elements having fewer or greater numbers of valence electrons than possessed by the semiconductor atoms (doped semiconductors are distinguished by the term "extrinsic"). In these ways, normally filled valence bands can be made less than filled (such a semiconductor is called a *p*-type, for positive or "hole" charge carrier) or normally unoccupied valence bands can become partially occupied (called an *n*-type, for negative charge carrier), and the conductivity is a direct function of crystal content of the doping element.

STUDY QUESTIONS

30. Consider the body centered unit cell for metals (all atoms of like size, in contact along the body diagonal). Find the unit cell volume as a function of r, the atom radius. What is the packing efficiency (the fraction of the unit cell volume actually occupied by atoms)? Is the body centered structure more or less efficient than the face centered geometry?

31. Do you expect GaAs to be an *ionic* solid? What do you expect would happen to the electrical conductivity of GaAs if Se were added? If Ga were added? If a crystal of GaAs is doped with Se at one end and with Ga at the other, in which direction would electric current (electrons) flow more readily? Would you anticipate greater or less discrimination in direction of electron flow from such an extrinsic semiconductor, compared to one doped at one end only?

Oxidative Stabilities

One of the most interesting aspects of the chemistry of the representative element metals, and at the same time most difficult to understand, is their oxidation potentials. In an attempt to identify factors that influence the activities of these elements, we may use a Born-Haber cycle for the reaction of the elemental metal with aqueous hydrogen ion.

$$\begin{array}{ccc}
M_{(c)} + nH^+_{(aq)} & \xrightarrow{\Delta H^\circ} & M^{n+}_{(aq)} + \tfrac{1}{2}nH_{2(g)} \\
S_M \downarrow \quad \downarrow -n\Delta H^\circ_{H^+} & \downarrow -H^\circ_{M^{n+}} & \downarrow \tfrac{1}{2}nD_{H_2} \\
M_{(g)} + nH^+_{(g)} & \xrightarrow{I_M - nI_H} & M^{n+}_{(g)} + nH_{(g)}
\end{array}$$

In terms of enthalpies for the various steps (we still ignore the differences between enthalpies and internal energy changes) we may write as an approximation for the enthalpy of the reaction of the metal with hydrogen ion:

$$\Delta H^\circ = (S_M - \tfrac{n}{2}D_{H_2}) + (I_M - nI_H) + (\Delta H^\circ_{M^{n+}} - n\Delta H^\circ_{H^+})$$

In this equation, S_M is the sublimation (lattice) energy of the metal, I_M is the sum of the first n ionization potentials of the metal, and $\Delta H^\circ_{M^{n+}}$ is the hydration energy (heat of solution) of the metal ion. Since oxidation potentials of metals are expressed in volts/electron, we must divide the enthalpy by the charge of the ion, n; in comparing enthalpies with E°'s we must also assume that entropy changes will parallel or be dominated by enthalpy differences. Table 6-10 summarizes values for each of the three terms in this equation, divided by the valence of the metal, for Groups I, II, and III of the representative metals. From comparison of the values in the column headed by the summation symbol with those in the E°(obs) column, it is apparent that in some cases the entropy contributions are significant, so that detailed comparisons of enthalpies are unwarranted. In general, however, the trends in enthalpies within the three groups follow the observed E°'s, particularly in showing that Li is slightly more reactive than the other alkali metals, that Be is the least reactive of the alkaline earths (which tend to increase in reactivity with

TABLE 6-10
ANALYSIS OF THE ENTHALPY TERMS AFFECTING THE E°'s OF METALS
(UNITS ARE eV)

	$\tfrac{1}{n}(S_M - \tfrac{n}{2}D_{H_2})$	$\tfrac{1}{n}(I_M - nI_H)$	$\tfrac{1}{n}(\Delta H^\circ_{M^{n+}} - n\Delta H^\circ_{H^+})$	$\sum$	E°(obs)
Li	−0.63	−8.21	+5.84	−3.0	3.0
Na	−1.11	−8.46	+6.97	−2.6	2.7
K	−1.31	−9.26	+7.84	−2.7	2.9
Rb	−1.31	−9.42	+8.14	−2.6	2.9
Cs	−1.42	−9.71	+8.44	−2.7	2.9
Be	−0.58	0.15	+1.22	−1.6	1.9
Mg	−1.46	−2.3	+1.08	−2.7	2.4
Ca	−1.24	−4.6	+2.88	−3.0	2.9
Sr	−1.39	−5.25	+3.58	−3.0	2.9
Ba	−1.33	−6.0	+4.33	−3.0	2.9
Al	−1.14	4.17	−5.05	−2.0	1.7
Ga	−1.27	5.5	−5.09	−0.9	0.5
In	−1.41	4.0	−3.22	−0.7	0.3
Tl	−1.64	5.2	−3.05	+0.5	−0.7

increased atomic number), and that the reactivities of the Group III metals tend to decrease drastically with atomic number. (In making comparisons among the $E°$'s, don't forget the relation $\Delta G = -nFE°$, which means that the reaction with the most positive $E°$ is the most thermodynamically spontaneous one.) Among the alkali metals, the reactivity of Li in comparison with Na and the other members of the group can be traced to the *solvation* energy of the Li$^+$ ion. For the remainder of the group, however, the trends in S_M and the other terms are largely self-concealing. For the alkaline earths, the low reactivity of Be seems to be due to a combination of high *sublimation* and *ionization* energies, as solvation energy seems to play a less important role than it does for Li. Again, for heavier Group II elements, the variations of S_M and the other terms tend to cancel. Interpretation of the trend in $E°$'s within the Group III metals is no easier than for Groups I and II. This group shows the most pronounced regular trend in $E°$'s of all three groups, but the regularity is not a reflection of analogous regularity in just one or two of the ionization, sublimation or hydration energies. For example, Al appears more reactive than Ga because of a significantly lower *ionization potential* sum, while Al is more reactive than In because of its much higher *solvation* energy.

NON-METAL COMPOUNDS

In these remaining sections, our emphasis will be on covalently bonded elements. Of special interest are the tendencies of the elements to catenate via both saturated and unsaturated linkages, and properties of their bonds to hydrogen, carbon, oxygen, and the halogens.

BORON AND ALUMINUM

Bond Energies for Reference

B—B	72	Al—H	(65)
B—H	(93)	Al—C	(60)
B—C	(89)	Al—F	(139)
B—F	154	Al—Cl	(100)
B—Cl	106		
B—O	125		

Boron is a most interesting element, in that it tends to form trivalent compounds typical of most non-metallic elements but, in doing so, is left in a state of valency unsaturation (more valence ao's than electron pairs). This is a "mild" form of the extreme valency unsaturation found in metals. Thus, we are not surprised to find some boron compounds with rather bizarre structures, at least for a non-metallic element. In many ways, boron exhibits compounds that are in the ill-defined region between metallic and "normal" covalent compounds. This is no more strikingly shown than by the element and its binary compounds with metals, the so-called *borides*.

Elemental boron exists in three different crystallographic forms, but all are characterized by icosohedral B_{12} clusters linked together in a three-dimensional network.[17] The icosohedron is a regular polyhedron with 12 vertices and 20 faces. There are two high-symmetry polyhedra with this same number of vertices; the cuboctahedron has six square faces and eight triangular faces, while the icosohedron has only equilateral triangular faces.[18]

---

[17] E. L. Muetterties, Ed., "The Chemistry of Boron and Its Compounds," Wiley, New York (1967); R. M. Adams, Ed., "Boron, Metallo-Boron Compounds, and Boranes," Interscience, New York (1964); K. Wade, "Electron Deficient Compounds," Nelson, London (1971).

[18] The cuboctahedron has 24 edges, while the icosohedron has 30. The difference of 6 edges affords a convenient view of the relation between the two structures. Notice the 6 square faces of the cuboctahedron. By connecting a pair of diagonally opposite vertices in each square face, each square is converted to two triangles. "Relaxation," so that all edges are of equal length, produces the icosohedron.

icosohedron *cuboctahedron*

The boron atoms are located at the vertices of the icosohedron, and this leads us to note that each atom has *five* nearest neighbor atoms. Quite obviously we cannot identify an electron pair bond with each edge of the icosohedron (there are 30 edges), so it is difficult (but not impossible) to rationalize the bonding of the boron atoms in such a structure in terms of a "localized" orbital model. As you will see in Chapter 18, the bonding in such polyhedra is more conveniently described in terms of the delocalized molecular orbital model.

It would appear that formation of such a polyhedral structure with extensive delocalization of valence electrons is highly effective in stabilizing the element. Even though the more common, impure, amorphous boron is highly reactive, truly crystalline boron is extremely hard and resistant to chemical attack, even by hot, concentrated solutions of most oxidizing agents. In short, the HOMO must be of quite low energy and not significantly localized on any particular group of boron atoms (which would facilitate electrophilic attack by other reagents).

Boron's proclivity for catenation is also shown in its compounds with highly electropositive elements. A bewildering number of MB_x structures are now known.[19] One class that bears a close resemblance to elemental boron is that of stoichiometry MB_{12}. The metal atoms are dispersed in a fcc sub-lattice among the boron atoms. Two views of the boron network are diagrammed below. In **11** you see emphasized a grouping of boron atoms into fcc cuboctahedra so as to emphasize the relationship to elemental boron and to the NaCl lattice. In **12** the boron atoms have been interconnected in a different way so as to depict each metal at the center of a cage of 24 boron atoms.

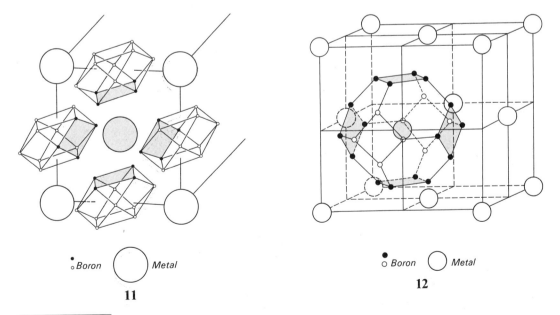

•/○ Boron   ○ Metal
11

•/○ Boron   ○ Metal
12

---

[19] See ref. 17; R. Thompson, in "Progress in Boron Chemistry," R. J. Botherton and H. Steinberg, Eds., Vol. 2, p. 173, Pergamon Press, New York (1970); B. Aronson, *et al.*, "Borides, Silicides, and Phosphides," Methuen, London (1965).

Another class of borides (MB$_6$) have structures based on a three-dimensional network of B$_6$ octahedra (connected by B—B bonds), with metal atoms located in holes between the packed octahedra. As shown by **13**, this lattice bears a strong resemblance to the CsCl structure (simple cubic packing of both metal atoms and B$_6$ octahedra). Structure **14** is simply a redrawing of **13** so as to show the metal caged in a 24-atom cell of boron atoms. Unlike **12**, which has hexagonal and square cage faces, **14** has octagonal and triangular faces. Still other examples of boron catenation in its compounds with metals could be cited in which we would find B$_2$ units, one- and two-dimensional chains of boron atoms, and even a class in which the boron atoms occur in layers of hexagonal (fused) rings between layers of metal atoms.

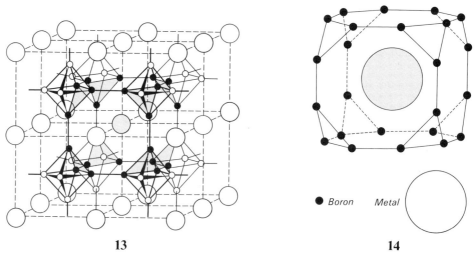

13 **14**

With hydrogen,[20] boron could ostensibly form a simple borane, BH$_3$. This compound, as noted early in Chapter 5, demonstrates dramatically the effect of valency unsaturation in trivalent boron compounds. BH$_3$ is a quite unstable species and readily dimerizes into B$_2$H$_6$, known as diborane [or more frequently as diborane(6), where the (6) signifies the number of hydrogens in the molecule]. This simplest member of the borane series exhibits the general feature of all *neutral boranes,* the bridging hydrogen atom.

$$\begin{matrix} H & H & H \\ & B & B & \\ H & H & H \end{matrix} \quad \begin{bmatrix} \text{m.p. } -165° \\ \text{b.p. } -93° \\ \text{pyrophoric} \end{bmatrix}$$

The enthalpy change for the dimerization reaction

$$2BH_3 \rightarrow B_2H_6$$

is about 35 kcal/mole exothermic.[21] As cleavage of a terminal B—H bond is endothermic by about 90 kcal/mole (estimated), the B—H—B bridge has a somewhat higher energy of about 110 kcal/mole.[22]

Hydrogen in a bridging role is not really all that unique in chemistry; if you recall, we previewed several other examples of three-center orbitals in Chapters 4 and 5 (examples were FHF$^-$, the π system of the allyl group, and the axial atom bonding

---

[20] See ref. 17; W. N. Lipscomb, "Boron Hydrides," W. A. Benjamin, New York (1963); R. J. Brotherton and H. Steinberg, "Progress in Boron Chemistry," Vols. 2 and 3, Pergamon Press, New York (1970); H. C. Brown, "Boranes in Organic Chemistry," Cornell University Press, Ithaca (1972).

[21] L. H. Long, *Prog. Inorg. Chem.,* **15**, 1 (1972).

[22] This value is derived from the energy cycle:

$$2BH_3 \rightarrow 2BH_2 + 2H \rightarrow B_2H_6$$

in trigonal bipyramidal structures such as PF_5 and the closely related XeF_2). The unique feature here, of course, is the $2e^-$ (as opposed to $4e^-$) nature of this structural unit (Chapter 5, p. 193).

While we will save until a later chapter on polyhedra in chemistry a complete discussion of these interesting boron hydrides, a few examples are given now to impress on you that boron catenation, hydrogen bridge bonding, and polyhedral structures are the characteristic features of the hydrides of boron.

After diborane, the next most complex neutral borane is tetraborane(10):

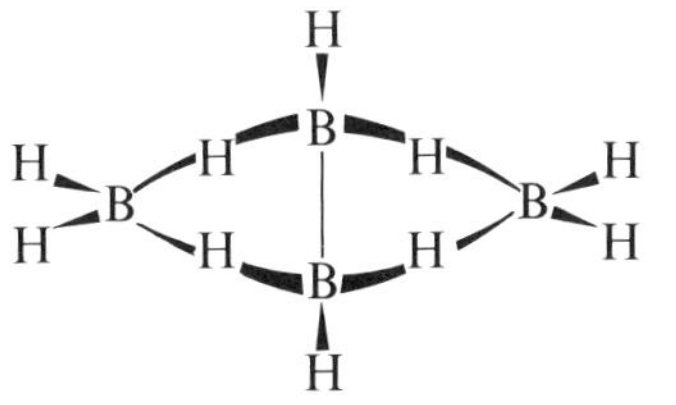

[m.p. −120°
b.p. 16°
pyrophoric]

Note the bicyclo (C_{2v}) structure of four borons and four bridging hydrogens, the B—B bond, and the presence of two borons with two terminal hydrogens.

Pentaborane(9), the most thermally stable[23] of the lower boranes, has the regular (C_{4v}) structure of a square-based pyramid of boron atoms, each with a single terminal hydrogen, and four bridging hydrogens between basal borons. The apical boron seems to be bound to the basal borons through five-center, six-electron bonding (the $6e^-$ character of the apical/basal boron binding is determined from the fact that 24 valence electrons are present, with two required for each terminal and bridging H link).

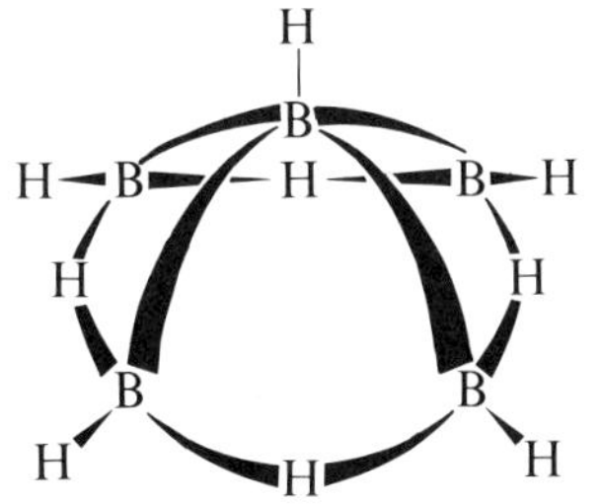

[m.p. −47°
b.p. 60°
detonates in air]

In Chapter 18 we will discuss the bonding in these polyhedral structures in a general, integrated way, using the delocalized mo model. The localized mo counterpart, derived from the delocalized scheme as at the end of Chapter 4, reveals[24] the occurrence of a three-center BBB electron pair within the triangular faces (because the molecule has C_{4v} symmetry and not the C_s symmetry appropriate to the structure **15**, we must actually write down four equivalent resonance structures).

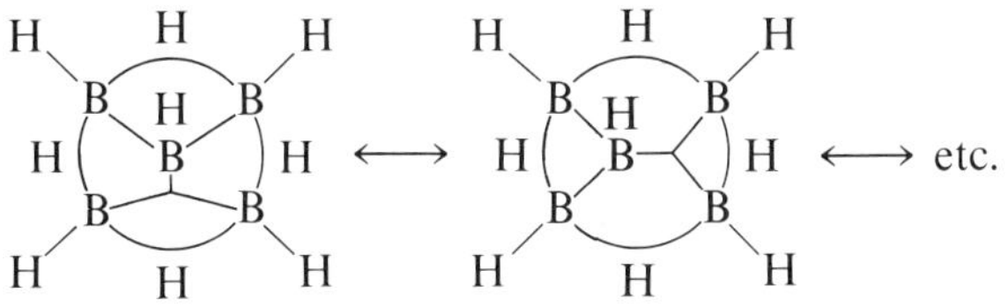

15

An interesting example of the polyhedral boron structure arises for decaborane(14). This structure (where each vertex represents a B—H unit) is seen to be an incomplete icosohedron of boron atoms, each with a terminal hydrogen. The missing

---

[23] All the lower B_nH_m are thermodynamically unstable with respect to elemental B and H_2. Their "stability" derives from mechanistic factors. See K. Wade (ref. 17). For a review of B_5H_9, see D. F. Gaines, *Accts. Chem. Res.*, **6**, 416 (1973).

[24] E. Switkes, *et al.*, *J. Amer. Chem. Soc.*, **92**, 3847 (1970).

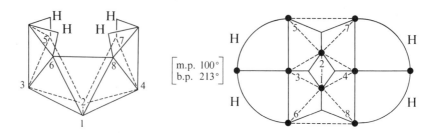

eleventh and twelfth B—H units are effectively replaced with four hydrogens of the bridging type. In the localized electron pair scheme at the right, we find two kinds of three-center BBB "bonds." Four of the so-called "closed" type are located in faces outlined by dashed lines. Two "open" three-center BBB "bonds" (reminiscent of the BHB bridge) are defined by the 5,3,6 and 7,4,8 edges.

The neutral boranes, as a class of compounds, have no simple stability characteristics. Using Wade's[25] suggested bond energies (BH = 89, BB = 74, BHB = 108, BBB = 91), you may readily verify that the B_nH_m are thermodynamically unstable with respect to the elements, to air oxidation (products = B_2O_3, H_2O) and hydrolysis (products = $B(OH)_3$, H_2). However, the rates of such reactions are highly variable, owing to the feasibility (or lack of it) of mechanistic pathways for such reactions. The only significant generalization to be made is that the neutral boranes owe their existence to kinetic factors. Were it not for the fact that B_4H_{10} is much less susceptible to hydrolysis than is B_2H_6, Stock's hydrolysis[26] of MgB_2 would not have led to his isolation of B_4H_{10}

$$MgB_2 \xrightarrow{H_3O^+} \text{primarily } B_4H_{10}$$

and the discovery of neutral boranes.

Boron also forms binary, anionic compounds with hydrogen. Some examples would be BH_4^-, $B_2H_7^-$ (again with bridging hydrogen, as the structure [C_{3v} more likely than D_{3h}] is BH_3—H—BH_3^-), and the *closo*[27] *boranes* of generic formula $(BH)_n^{2-}$. These latter compounds all display polyhedral structures, with known examples having $n = 6$ to 12. The BH units of these anions are arranged in as regular a fashion as possible so that the polyhedra have triangular faces. The octahedral B_6 and icosahedral B_{12} groupings are reminiscent of the borides mentioned earlier.

The simplest hydride anion of boron, BH_4^-, is very important in chemical synthesis, where it finds use as a reducing agent and has the striking property of being "stable" in water, as well as soluble in ether-type solvents. (As will be discussed in the next chapter, this stability is kinetic in origin. The facts that the speed of hydrolysis depends on [H$^+$] and that solutions of BH_4^- become basic on hydrolysis,

$$BH_4^- + 4H_2O \rightarrow 4H_2 + B(OH)_3 + OH^-$$

translate into a sharply decreasing rate of hydrolysis after only a slight degree of decomposition.) Readily available commercially, salts of BH_4^- are convenient sources of B_2H_6 through treatment with acids such as 85% H_3PO_4 and BF_3.

$$NaBH_4 + H_3PO_4 \rightarrow \frac{1}{2}B_2H_6 + H_2 + NaH_2PO_4$$

$$3NaBH_4 + 4BF_3 \xrightarrow{ether} 3NaBF_4 + 2B_2H_6$$

---

[25] See K. Wade (ref. 17).
[26] A. Stock, "Hydrides of Boron and Silicon," Cornell University Press, Ithaca, New York (1933).
[27] The word "closo" designates a closed, polyhedral boron skeleton; see Chapter 18.

Given that BH_3 dimerizes to relieve the unsaturation at boron and that CH_3 is known to exhibit a tendency to bridge like hydrogen in many organometallic compounds (see Chapters 5 and 16), it is somewhat surprising that $B(CH_3)_3$, and *triorganoboranes* in general, do not dimerize. It has been suggested that steric compression (particularly for branched hydrocarbons) is at least partly responsible for this, but it also may be correct that hyperconjugation of the alkyl groups with the boron may also be responsible for partially relieving the unsaturation.

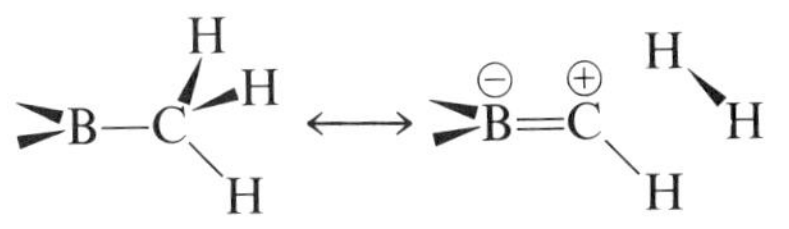

This seems to be particularly true for the triaryl boranes:

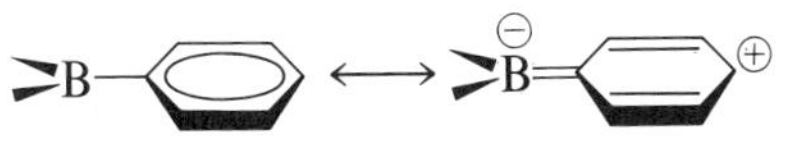

Consistent with such a view of internal valency saturation, the triarylboranes are less highly susceptible to nucleophilic attack[28] and are generally non-flammable compounds, whereas the lighter alkylboranes are pyrophoric. More will be said about the reactivities of the organoboranes in the chapters on organometallic compounds.

The terminal groups that seem to relieve best the unsaturation inherent to trivalent boron are those with lone pairs (OR, NR_2, and halogens). The halogens are fairly effective in this regard, and the $p\pi-p\pi$ bonding between boron and fluorine is an important factor in the high B—F bond strength and its relative unreactivity compared with the other boron halides (recall the discussion of this point in Chapter 5). These molecules are stable as monomers, in spite of the fact that the halogens possess three lone pairs, one of which could be used to form bridge bonds between borons. Quite obviously, nature finds halogen bridging an unattractive alternative (at least partly for entropic and steric reasons).

A further consequence of the differences in B—X bond strengths appears on comparison of hydrolytic stabilities. BF_3 is stable in aqueous solution and renders the solutions acidic by the reaction

$$BF_3 \cdot OH_2 \rightleftharpoons BF_3OH^- + H^+$$

and it is known to form solids with the composition $BF_3 \cdot OH_2$ and $BF_3 \cdot 2OH_2$; the other trihalides of boron rapidly hydrolyze to form HX and boric acid ($B(OH)_3$). These hydrolysis reactions have a negative enthalpy change for all halogens except fluorine, and appear to be general in that alcohols and amines effect similar solvolyses. Such reactions constitute a basis for polymerization reactions.

To preview some polymerization reactions (their mechanisms are to be discussed in the next chapter, and a more comprehensive view will be taken in Chapter 8), boron trihalides react with many secondary amines with loss of HX to give species of the type R_2N-BX_2. These *aminoboranes* may then oligomerize to give species such as

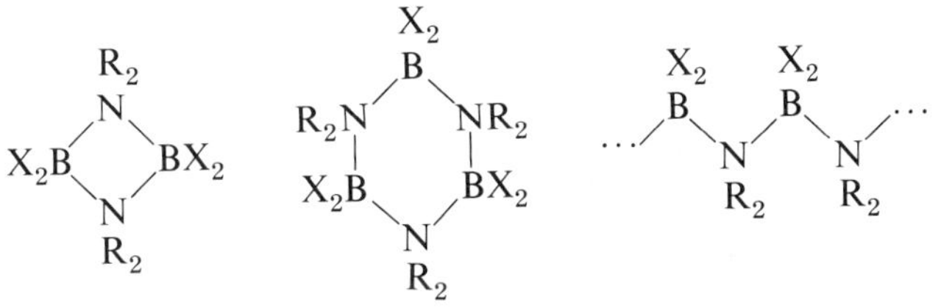

---

[28] The consequence of such π conjugation on the BX_3 LUMO nature was discussed on p. 231, Chapter 5.

depending on steric and kinetic factors. Again, R—X combinations having incompatable steric requirements disfavor the acid/base condensation of the R_2NBX_2 monomers, which can be stabilized by intramolecular dative bonding. Given[29] that the dimerization enthalpies are typically no larger than 20 kcal/mole dimer (exothermic), while $T\Delta S \approx +10$ kcal/mole dimer, it is not surprising that steric congestion (recall Me_3NBMe_3 from Chapter 5) has an observable effect on the magnitude of ΔG for dimerization.

An important reaction of this type is also observed to occur on pyrolysis of $NH_3 \cdot BH_3$ (where H_2 instead of HX is evolved).

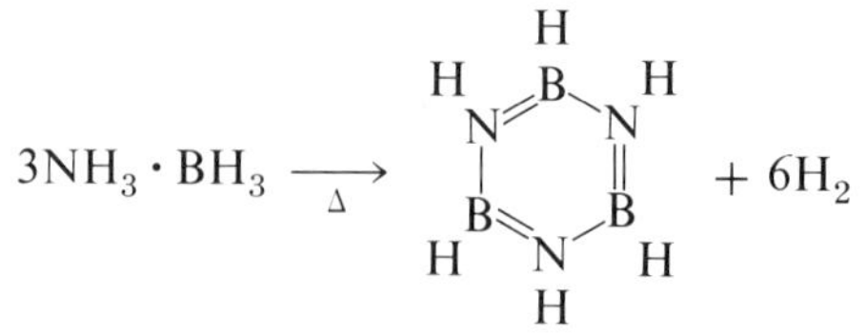

$$3NH_3 \cdot BH_3 \xrightarrow{\Delta} (HBNH)_3 + 6H_2$$

This interesting reaction produces the inorganic analog of benzene, $(HBNH)_3$, a compound commonly called *borazine*. In contrast to benzene, borazine is much more reactive; this is generally attributed to the more localized nature of borazine π electrons:

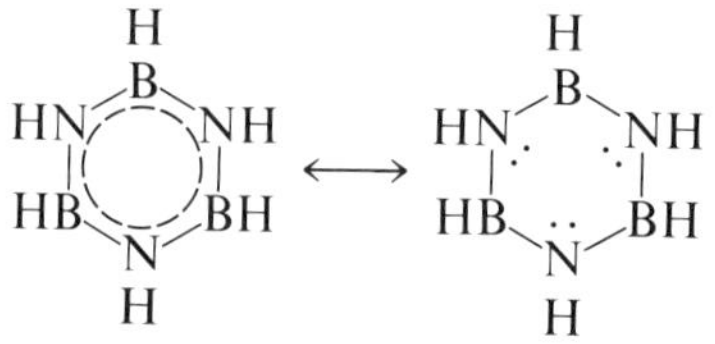

The delocalized mo model of the π orbitals of such systems (you will encounter others later) is worth a comment. In the borazine system you recognize interpenetrating cyclic three-atom sets, the mo's of which are (compare the π TASO's of BF_3 in Chapter 4):

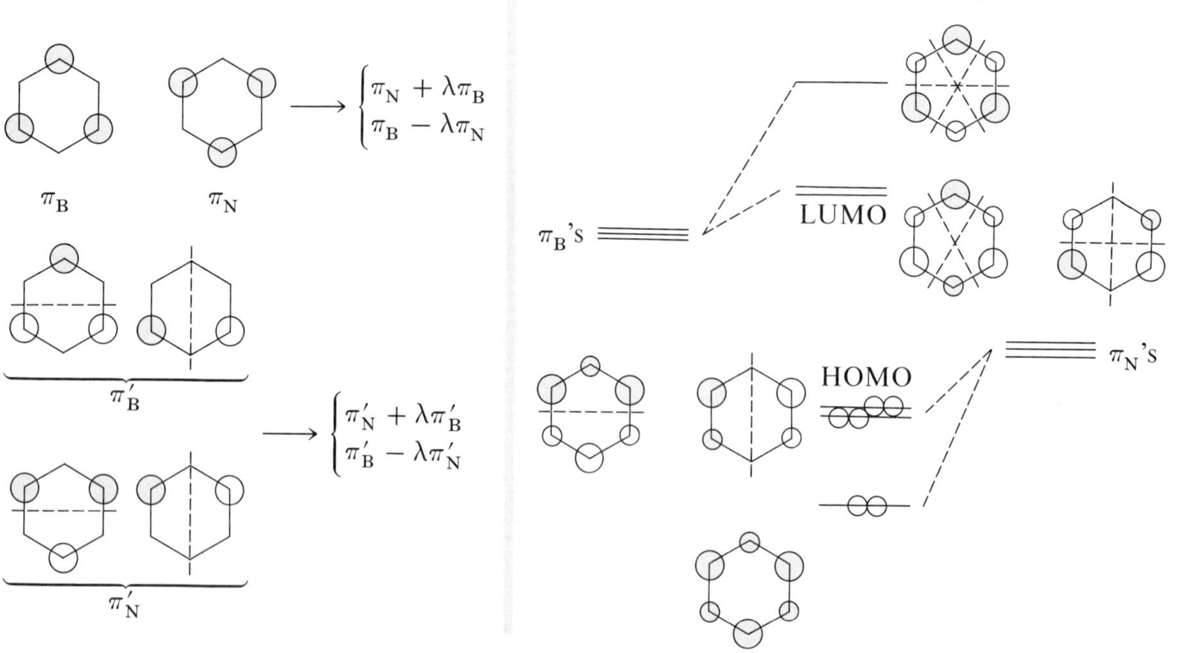

___

[29] H. Nöth and H. Vahrenkamp, *Chem. Ber.*, **100**, 3353 (1967).

There is an obvious similarity between these mo's and those of benzene (see Chapter 4, p. 159). The principal distinction is the polarization of each net bonding mo toward the nitrogen atoms, with inverse polarization toward boron atoms in the anti-bonding mo's. Nucleophilic attack upon $B_3N_3H_6$ occurs via the LUMO and thus at a boron atom; electrophiles attack at a nitrogen atom through the HOMO.

Reaction of NH_4Cl with BCl_3 at elevated temperatures

$$3NH_4Cl + 3BCl_3 \xrightarrow{\Delta} (HNBCl)_3 + 9HCl$$

leads to the borazine analog with Cl replacing H at boron. This product is useful for introducing three organo groups at boron by Grignard reactions (see Chapter 8).

The other reasonably well known halides of boron are the (pyrophoric) *diboron tetrahalides* B_2X_4, which exhibit boron catenation.[30] As an interesting example of how the structure of a molecule can change when packed into a crystal lattice, both the fluoride (m.p. $-56°$) and the chloride (m.p. $-98°$) are known to have mutually perpendicular BX_2 groups (D_{2d} symmetry) in the gas phase but are planar molecules (D_{2h}) in the solid.

gas (D_{2d}) solid (D_{2h})

While it is not surprising that the B—B bond suffers cleavage in air (B_2F_4 is pyrophoric) so as to form B_2O_3 (see below), it is interesting that the B—B bond of B_2Cl_4 survives solvolytic reaction with liquid H_2O and alcohols to yield $B_2(OH)_4$ and its esters $B_2(OR)_4$, in analogy with the trihalides.

A dominant feature of boron chemistry is the strength of its bonds to oxygen.[31] Given that the dissociation energy of BO is only about 190 kcal/mole for an estimated (from molecular orbital theory) bond order of 2.5, it is not to be expected that boron-oxygen compounds will exhibit —B=O as a structural feature. This is because the B—O "single" bond has an energy of about 125 kcal/mole. Thus, the —B=O link would be expected to undergo dimerization. The high "single" bond energy is an important consequence of the dative p-$p\pi$ interaction between boron and oxygen alluded to earlier. Consequently, the stable (solid) *boric oxide*, B_2O_3, has a structure based on chains of BO_3 units (triangular), rather than the simpler molecular arrangement of O=B—O—B=O.[32] Boric oxide is a major constituent (14%) of borosilicate glass, a type of glass sold under the trade names Pyrex (Corning Glass Works) and Kimax (Owens-Illinois). In fact, the triangular BO_3 unit is a structural feature common to almost all boron-oxygen compounds, including a large number of anionic borate salts that have chain, ring, and three-dimensional structures. The structures **16** through **20** illustrate the chain and ring structures possible with 3- and 4-coordinate boron.

as in $Mg_2B_2O_5$
16

$B_2O(OH)_6^{2-}$
17

---

[30] Less well characterized are $(BX)_4$ (T_d), $(BCl)_8$ (D_{2d}, dodecahedron), $(BCl)_9$, and $(BCl)_{12}$; A. G. Massey, *et al.*, *J. Inorg. Nuclear Chem.*, **33**, 1195, 1569 (1971).

[31] See ref. 17.

[32] The chains may be most easily viewed as oligomerization of terminal B=O units of the hypothetical O=B—O—B=O monomers.

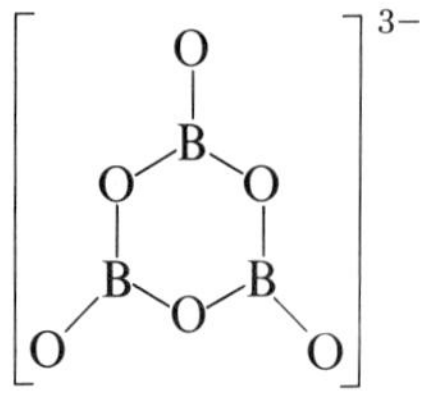

308 ◻ ENERGETICS AND STRUCTURES AS GUIDES TO MAIN GROUP CHEMISTRY

$$\left[\begin{array}{c} O \\ \parallel \\ B \\ O \diagup \quad \diagdown O \\ | \qquad | \\ B \qquad B \\ O \diagdown \diagup O \\ O \end{array}\right]^{3-}$$

as in
$Na_3B_3O_6$
18

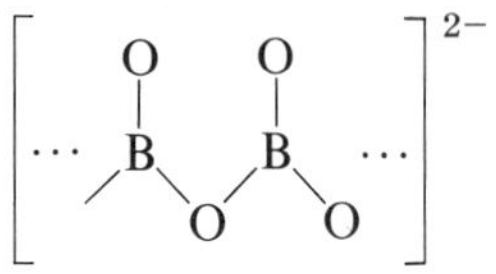

$$\left[\cdots \overset{O}{\underset{O}{B}} - \overset{O}{\underset{O}{B}} \cdots \right]^{2-}$$

as in
CaB_2O_4
19

$$\left[\text{structure} \right]^{2-}$$

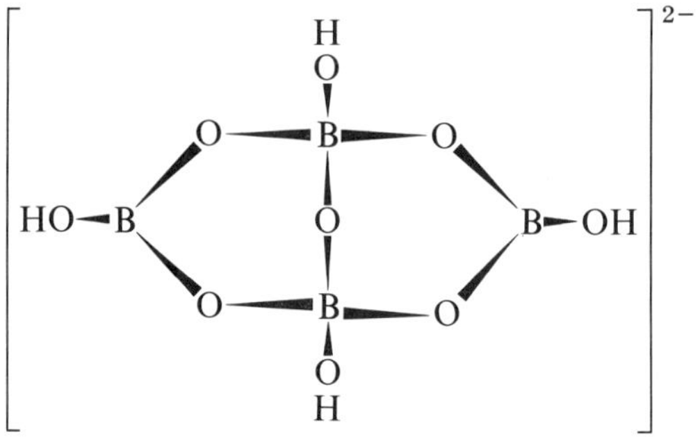

as in
"$Na_2B_4O_7 \cdot 10H_2O$"
20

Compound **20** is *borax,* easily the most important polyborate in both dollar sales and tonnage. Perhaps you are already familiar with its use as a household cleaner; a few years ago, when environmentalists objected to the use of phosphates in laundry detergent, some consumers returned to their earlier use of borax cleaners. Borax is found in arid regions of the world, especially the desert Southwest of the United States. These were apparently regions of volcanic activity in the past, and boron was transported to the surface by hot springs or steam in the form of boric acid; there the acid combined with surface rocks such as carbonates or salines and was converted into alkali or alkaline earth metal borates. If this activity occurred in an arid region, the water soon evaporated after borate formation (or after the solution accumulated in a lake), leaving immense beds of borate salts.

Two oxyacids of boron are known. The "simpler" one, HBO_2, called *metaboric acid,* is the intermediate anhydride of boric acid and exhibits various crystalline forms, either with layers of trimeric hexagonal rings $((HOBO)_3)$ connected by hydrogen bonds between rings, or a related form in which each boron atom is

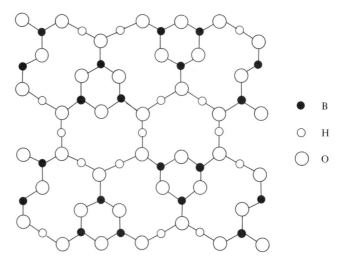

coordinated by four oxygens. These two forms, therefore, exhibit the two common structural configurations of boron: trigonal planar and tetrahedral. Again, notice the lack of a B=O structural sub-unit.

Boric acid is soluble in water, where it behaves as a weak acid. Interestingly, the acidic behavior is not due to the simple ionization

$$B(OH)_3 \rightleftarrows (HO)_2BO^- + H^+$$

but rather to ionization of the acid-base adduct $(HO)_3B \cdot OH_2$

$$B(OH)_3 + H_2O \rightleftarrows B(OH)_4^- + H_3O^+ \qquad pK_a = 9.2$$

in analogy with $BF_3 \cdot OH_2$. In many other respects boric acid behaves like most compounds with acidic hydrogens. Consistent with such a formulation is the fact that ΔS is much more negative (-31 e.u.) than usual for simple proton ionization.[33] An important medicinal application of this mild acid is in eye-washes.

Reactions of boric acid with alcohols in the presence of a strong desiccant such as H_2SO_4 lead to *borate esters*,

$$B(OH)_3 + 3ROH \xrightarrow{H_2SO_4} B(OR)_3 + H_2O \text{ (solvated)}$$

and reactions with carboxylic anhydrides lead to boron *tricarboxylates*, $B(O_2CR)_3$.

$$B(OH)_3 + 3(RCO)_2O \rightarrow B(O_2CR)_3 + 3RCO_2H$$

Finally, we might mention that $B(OH)_3$ has a layer structure with fairly strong hydrogen bonding between $B(OH)_3$ units.

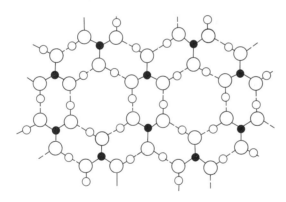

The relationship between the $B(OH)_3$ structure and that given above for $BO(OH)$, obtained by thermally dehydrating $B(OH)_3$ (below 130°C), is not surprising.

The final examples of boron oxyacids are those with one or two OH groups of boric acid replaced by organo groups.[34] The former, $RB(OH)_2$, are referred to as *boronic acids* while the latter, R_2BOH, are called *borinic acids*. They can be prepared by hydrolysis of the corresponding diorgano boron halide, while the boronic acids have the interesting property of self-dehydrating, often quite easily.

$$3RB(OH)_2 \rightarrow (RBO)_3 + 3H_2O$$

---

[33] R. P. Bell, *et al.*, in "The Chemistry of Boron and Its Compounds," E. L. Muetterties, Ed., Wiley, New York (1967).
[34] K. Torssell, in "Progress in Boron Chemistry," Vol. 1, H. Steinberg and A. L. McCloskey, Ed., Pergamon Press, New York (1964).

The molecular species RBO is not expected to be stable, however, so that trimerization to form the alkyl *boroxine* (RBO)$_3$ occurs (note the analogy with bor*azine*).

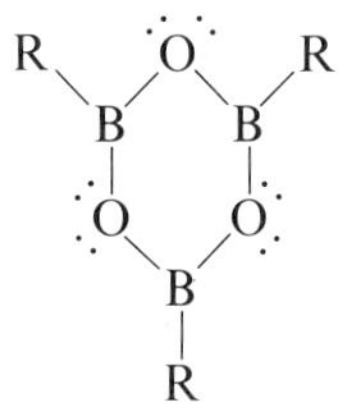

As an element of the third row of the Periodic Table, aluminum has *d* orbitals in its valence shell, and it is a common practice to expect valence shell expansion for these elements. Thus, the structural chemistry of aluminum is based not only on tetrahedral structures but also on octahedral geometries. Whether such Al *d* ao involvement is critical to the formation of 6-coordinate Al^{3+} structures is far from established. In Chapter 4, the discussions of EX$_5$ and EX$_6$ structures showed that it is not necessary to bring *d* ao's on E into the discussion. Rather, nothing more esoteric than the difference between B^{3+} and Al^{3+} sizes and electronegativity (VSEPR) may account for the 6-coordinate structures of many Al compounds.

As a point of contrast with boron, aluminum is not known[35] to form catenated polyhedral hydride species. Instead we find the simple *alane,* AlH$_3$, to be a highly polymerized solid with bridging hydrogen atoms[36] (its heat of formation[37] is only ~ -3 kcal/mole). Thus, we find here a good example of aluminum "going one better" than boron on valency saturation. The structure of the solid is indicated by the following figure, where you find each Al in an octahedral environment with each hydrogen serving in a bridging capacity.

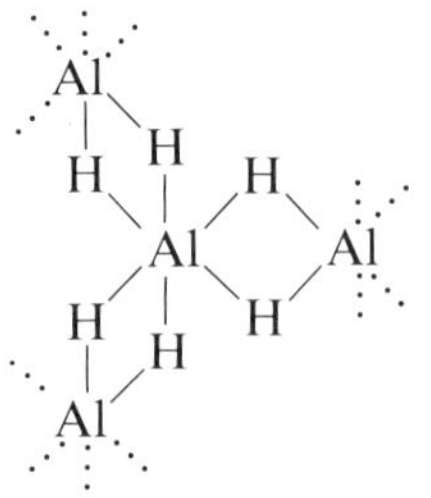

It is an interesting fact that treatment[38] of LiAlH$_4$ in THF (tetrahydrofuran) with anhydrous H$_2$SO$_4$ can be controlled so as to produce the solid AlH$_3$. With water and alcohols, as you

$$H_2SO_4 + 2LiAlH_4 \xrightarrow{THF} 2AlH_{3(c)} + Li_2SO_4 + 2H_{2(g)}$$

could expect for a metal hydride, AlH$_3$ solvolyzes to form "hydrous oxides" or aluminum alkoxides, Al(OR)$_x$H$_{3-x}$.

The simple hydridic anion AlH$_4^-$ is known in combination with several cations, and, like BH$_4^-$, it is a *very* useful reducing agent for inorganic and organic materials. The structures of tetrahydridoaluminate salts of second row cations (Li$^+$, Be^{2+}, and B^{3+}) all feature bridging hydrogen structures, and this has some important conse-

---

[35] S. Cucinella, *et al., Inorg. Chem. Acta Revs.,* **2,** 51 (1970); E. C. Ashby, *Adv. Inorg. Chem. Radiochem.,* **8,** 283 (1966).
[36] J. W. Turley and H. W. Rinn, *Inorg. Chem.,* **8,** 18 (1969).
[37] G. C. Sinke, *et al., J. Chem. Phys.,* **47,** 2759 (1967).
[38] N. M. Yoon and H. C. Brown, *J. Amer. Chem. Soc.,* **90,** 2926 (1968).

quences for the particularly useful Li⁺ salt. The association of Li⁺ with AlH_4^- is thought to yield "molecular" species in ether and aromatic hydrocarbon solvents, and this is the structural feature that accounts for the solubility. It is an interesting observation that *tertiary* amines react with $LiAlH_4$ to produce amine adducts of alane, R_3NAlH_3, and Li_3AlH_6. While both Na_3AlH_6 and Li_3AlH_6 are known[39] (made simply by reaction of MH with $MAlH_4$), only $LiAlH_4$ reacts with tertiary amines to form a 1:1 adduct. Thus, it must be that the high degree of association of Li⁺ with the hydrogens of AlH_4^- is important kinetically and thermodynamically for the displacement of H⁻ from AlH_4^-.

$$3MAlH_4 + 2R_3N \rightarrow 2R_3N \cdot AlH_3 + M_3AlH_6 \quad (M = Li, \neq Na)$$

Finally, we should note that there is a marked difference in the hydrolytic stabilities of BH_4^- and AlH_4^-. Whereas BH_4^- is kinetically stable in aqueous solution as explained earlier, AlH_4^- violently hydrolyzes. Clearly, a difference in mechanisms for the hydrolyses of tetrahydridoborate and tetrahydridoaluminate is indicated, and this will be discussed in the next chapter.

Another useful class of alanes consists of the *organoalanes*. These are most often considered to be organometallic compounds, and this view of their chemical properties will be covered in some detail in a later chapter of this text. Again calling to mind the size difference of B and Al, the organoalanes are dimeric in condensed phases but somewhat dissociated in the gas phase ($\Delta H \sim 20$ kcal/mole).[40] Even $Al(C_6H_5)_3$ possesses the dimeric structure, and the mixed compound $Al_2(CH_3)_4(C_6H_5)_2$ exhibits bridging phenyl groups rather than bridging methyls (there was a discussion of this in Chapter 5).

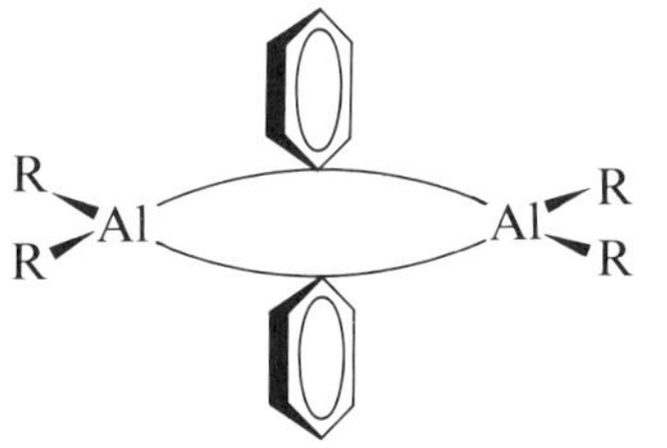

An interesting idea to ponder at this point is the role of $C_6H_5^-$ as a π donor. In BPh_3, you saw this action in the form

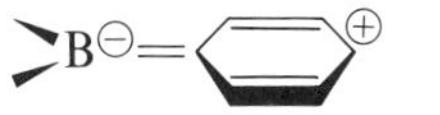

In the $Al_2R_4Ph_2$ structure above it seems that Ph, in contrast with CH_3, may act much like a bridging *halogen*, involving both the carbon sp^2 ao and a π HOMO in bridge bonding:

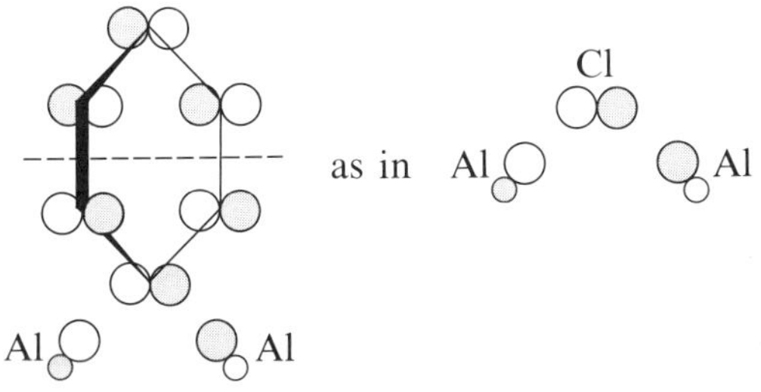

---

[39] Note that M_3BH_6 is unknown, and considered not likely to exist.
[40] C. H. Henrickson and D. P. Eyman, *Inorg. Chem.,* **6,** 1461 (1967).

as well as

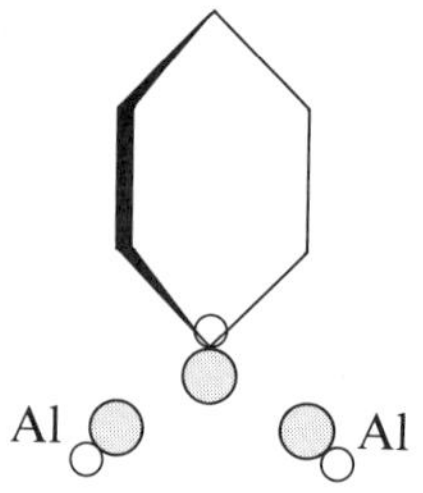

as in

With monofunctional bases, the organoalanes form simple 1:1 adducts like the borane analogs; but a point of demarcation, illustrating valence shell expansion for Al, is reached with alane adducts.[41] For example, trimethyl amine forms a 2:1 adduct with AlH_3; as you could predict, the H atoms take up equatorial sites and the amine nitrogens take up axial sites in the trigonal bipyramid.

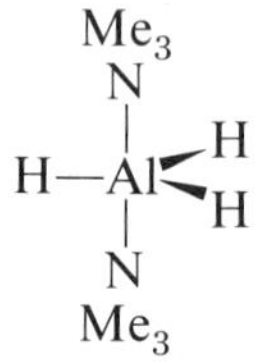

With 1,4-diazobicyclo(2.2.2)octane (a common acronym is DABCO), on the other hand, the 1:1 stoichiometry $AlH_3 \cdot N(C_2H_4)_3N$ actually entails a polymeric chain structure.

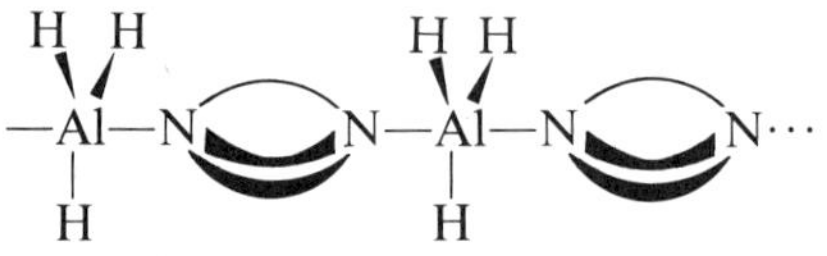

This behavior is not found with primary and secondary amines because "solvolytic" reactions eliminate alkane and form *amidoalanes*, R'_2NAlR_2, and

$$R'_2HN{-}AlR_3 \rightarrow R'_2NAlR_2 + RH$$

imidoalanes, R'NAlR. As we found for the boron analogs, these monomeric structures (with, formally, $\dot{N}{-}Al$ and $\dot{N}{=}Al$ linkages) are unstable with respect to condensation, where the amide and imide groups function to bridge Al atoms.[42] With primary amines, from which solvolysis would ostensibly lead to monomers with the empirical formula $R'N{=}AlR$, nature prefers condensation into three-dimensional tetrameric "cubane-like" imido structures of D_{2d} symmetry.

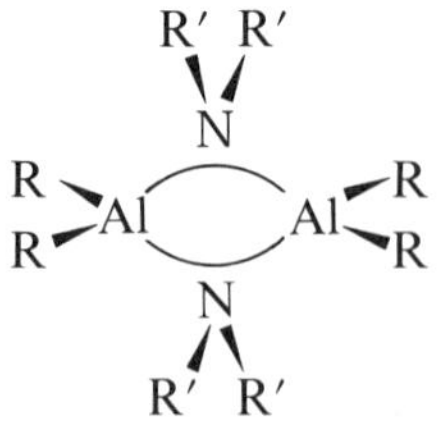

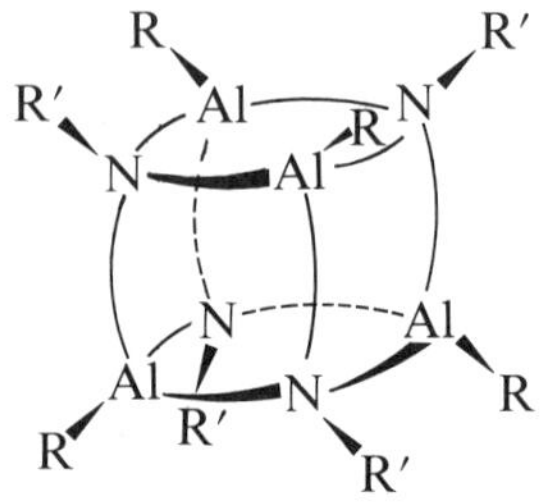

---

[41] K. M. MacKay, "Hydrogen Compounds of the Metallic Elements," Spon, London (1966).
[42] J. K. Ruff, in "Developments in Inorganic Nitrogen Chemistry," Vol. 1, Elsevier, New York (1968).

Again you are reminded of less severe steric problems in Al compounds than in their boron analogs.

Having seen the theme of bridging roles played by hydrogen, carbon, and nitrogen, we fully expect to find the halides serving a similar function. All four halides of empirical formula AlX_3 are known. The fluoride is thermally the most stable member of the series and exhibits a bridged structure like AlH_3. The chloride possesses a layer structure similar to $CrCl_3$ (p. 288), but it is easily vaporized (below 200°C) into gaseous dimers. The bromide and iodide exclusively exhibit the D_{2h} dimeric structure and are easily vaporized.[43] The dimers are reasonably stable but have dissociation enthalpies sufficiently small (~30 kcal/mole)[44] that D_{3h} symmetry monomers can be observed at high temperatures.

Solution of the aluminum halides in hydroxylic solvents leads to displacement of halide ions in a manner typical of metal halide salts. Partly owing to factors related to the size of the aluminum cation and the fact that the cation achieves 6-coordination with formation of $Al(OH_2)_6^{3+}$, these solutions are not as acidic ($pK_a \approx 5$) as those of boron halides (which, except for BF_3, completely hydrolyze to $B(OH)_3$ and HX). Even hydrated forms of the aluminum halides are known, such as $AlF_3(OH_2)_3$, $Al(OH_2)_6F_3$, and $Al(OH_2)_6Cl_3$. It is suggestive of a low polarizing power of the 6-coordinate hydrated aluminum cation that the trihydrate of aluminum fluoride is dehydrated, rather than solvolyzed, by the application of heat, even though the last water is driven out only at red heat. In solution,[45] the hexaquo cation and its derivatives from successive replacement of H_2O by fluoride are present, as well as the anion AlF_4^-. In solutions with high concentrations of added F^-, the hexafluoroaluminate anion[46] makes an appearance and completes the structural analogy with the three hydride species (AlH_3, AlH_4^-, and AlH_6^{3-}).

It is the Lewis acid properties of the aluminum halides, like those of the boron halides, that make them such useful synthesis reagents. They are best known as catalysts in Friedel-Crafts reactions of organic acid chlorides. The role of $AlCl_3$ and $AlBr_3$ in these reactions is to associate strongly with or abstract the halides from the acid chlorides, forming *in situ* the tetrahaloaluminate anion and the highly reactive carbonium ion. While the carbonium ion salts are not usually isolated, the salt $CH_3CO^+Al_2Cl_7^-$ has been isolated.[47] With one bridging chloride, the $Al_2Cl_7^-$ anion has a C_{3v} geometry.

While only one oxide of aluminum is known, it exists in two important forms. The so-called γ-form of Al_2O_3 is man-made and has a somewhat disordered arrangement of aluminum ions surrounded tetrahedrally and octahedrally by oxide ions. Prepared by thermally dehydrating "hydrous oxides" of aluminum, *alumina* is characterized by small particle size and a fairly reactive nature. It readily absorbs water and is soluble in acids and bases; doping γ-Al_2O_3 with iron and titanium oxides produces synthetic blue sapphire. Its most important use is as chromatographic alumina, and as such it is an important reagent in most laboratories.

Heating the γ-form above 1000°C leads to the more orderly structure of the α-form. Here all aluminum ions are octahedrally surrounded by oxide ions, and each oxide bridges four aluminum ions. That there are too many cations per anion for a

---

[43] The r_+/r_- ratios are (0.37, Al/F), (0.28. Al/Cl), (0.26, Al/Br) and (0.23, Al/I). The ideal r_h/r_i for cubic close packed anions is 0.41 at O_h sites. This ideal is met well by Al^{3+} in a F^- sub-lattice, not at all for Br^- and I^-, and not too well for Cl^-. Thus, it is much easier for $AlCl_3$ than AlF_3 to pass into the gas phase as Al_2Cl_6. Again, in the trend from 6- to 4-coordination of Al^{3+}, we see a feature of main group chemistry determined by VSEPR/steric forces.

[44] K. Wade, *J. Chem. Educ.*, **49,** 502 (1972).

[45] Review, if necessary, the introductory section on solvents in Chapter 5.

[46] This anion occurs naturally as cryolite (Na_3AlF_6), so important to the large scale electrolytic manufacture of Al (Hall process).

[47] M. E. Peach, *et al., J. Chem. Soc. A*, 366 (1969).

CdCl$_2$ type layer lattice is consistent with the hardness of Al$_2$O$_3$. Rather, two out of every three octahedral holes between oxide layers contain Al^{3+} in the fashion

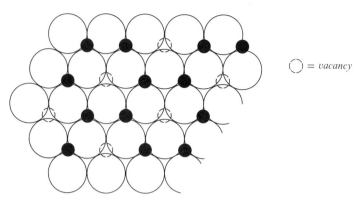

The α-form, as expected, is much less reactive than the γ-form and is very hard (you are probably familiar with the use of *corundum* as an abrasive material for grinding and polishing; corundum is the α-form of Al$_2$O$_3$).[48]

Solid solutions of Cr$_2$O$_3$/corundum constitute the gem ruby, which has the well-known red color at less than 8 per cent Cr$_2$O$_3$ but a green color at higher concentrations (Cr$_2$O$_3$ is green, while Al$_2$O$_3$ is white). The red color at low Cr^{3+} concentration seems to arise from the size difference between Al^{3+} and Cr^{3+}, the latter having a radius about 0.15 Å greater. In essence, Cr^{3+} ions are forced to occupy the smaller Al^{3+} holes in the Al$_2$O$_3$ lattice (Cr$_2$O$_3$ has an expanded corundum-like lattice). This causes the electronic transitions within the Cr^{3+} ion that are responsible for its green color to shift to higher energy and allow red light to be transmitted by the crystal (Chapter 9 explores the basis for the colors of transition metal ions).

There are two "oxyacids" of aluminum—the compound Al(OH)$_3$, more often called aluminum hydroxide, and its anhydride HAlO$_2$. As the empirical formulae of these compounds would suggest, and in analogy with the boron acids, these "oxy-acids" are amphoteric. They differ considerably in structure from the boron analogs, however. The OH groups in both aluminum compounds, and the "terminal" oxide in the "meta" acid,[49] bridge two aluminum ions so as to make each aluminum 6-coordinate. The simpler of the two structures is that for Al(OH)$_3$, which is similar to the CrCl$_3$ layer lattice (p. 288). An interesting detail of the structure of Al(OH)$_3$ is that the oxygen atoms within each layer are connected by hydrogen bonds, and oxygen atoms in two adjacent double layers are also connected by hydrogen bonds.

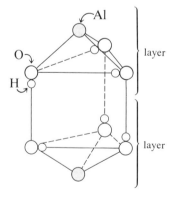

---

[48] The surface oxide responsible for rendering metallic aluminum unreactive is yet another crystal form, similar to that of NaCl, but with (as required by the 2:3 stoichiometry) every third cation missing.

[49] As with B and as with N═Al, the Al═O linkage is thermodynamically unstable with respect to the Al—O "single" bond.

The aqueous chemistry of Al^{3+} is quite complicated because of the amphoteric nature of the hydroxy species present and the propensity of oxygen to bridge different Al^{3+} ions. Thus, solutions of $Al(OH_2)_6^{3+}$ salts are acidic, although the acidity does not simply arise from the equilibrium

$$Al(OH_2)_6^{3+} \rightleftharpoons Al(OH_2)_5OH^{2+} + H^+_{(aq)}$$

but is complicated by the dimerization[50]

$$2Al(OH_2)_5OH^{2+} \rightleftharpoons (H_2O)_4Al(OH_2)_2Al(OH_2)_4^{4+} + 2H_2O$$

and further polymerization[51] to form species of general formula $Al\{Al_3(OH)_8\}_x^{3+x}$ (it is not known whether each aluminum is 6-coordinate or 4-coordinate, so additional coordinated water molecules are not written). The presence of these polymeric species can apparently be avoided[52] only in solutions that are less than 1.5 M in Al^{3+} and more basic than pH = 13. Under these conditions the tetrahydroxoaluminate anion is formed ($Al(OH)_4^-$). At this pH, but with an Al^{3+} concentration greater than 1.5 M, a dimerization reaction becomes important and produces an analog of $Al_2Cl_7^-$, i.e., $\{(HO)_3Al\}_2O^{2-}$ (which, interestingly, is the anhydride of $Al(OH)_4^-$).

---

[50] E. Grunwald and F. W. Fong, *J. Phys. Chem.*, **73**, 650 (1969); J. W. Akitt, et al., *J. Chem. Soc. A.*, 803 (1969) and *Chem. Comm.*, 988 (1969).

[51] N. Dezelie, et al., *J. Inorg. Nuclear Chem.*, **33**, 791 (1971).

[52] M. Yoshio, et al., *J. Inorg. Nuclear Chem.*, **32**, 1365 (1970); R. J. Moolenaar, et al., *J. Phys. Chem.*, **74**, 3629 (1970); L. A. Cameron, et al., *J. Chem. Phys.*, **45**, 2216 (1966).

STUDY QUESTIONS

32. The "closo" borane anion $(BH)_{12}^{2-}$ has an icosohedral structure. From a valence electron point of view, does it seem reasonable to you that the carborane $B_{10}C_2H_{12}$ exists? Of the three geometric isomers [(1,2), (1,7), and (1,12)] possible for the carborane, two may be readily interconverted at 450°C. Starting with the icosohedron below, discover which pair [(1,2 $\rightleftharpoons$ 1,7), (1,2 $\rightleftharpoons$ 1,12) or (1,7 $\rightleftharpoons$ 1,12)] are interconverted simply by expansion of the icosohedron into a cuboctahedron and then relaxation back to an icosohedron.

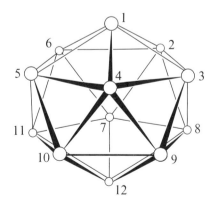

33. Knowing that B_2H_6 and BH_4^- hydrolyze in aqueous acid, why do you suppose 85% H_3PO_4 (1:1 mole ratio H_2O/H_3PO_4) can be used in the synthesis of diborane? What is the likely experimental arrangement and order of mixing of the reagents in this synthesis?

34. Knowing that $Al_2Cl_7^-$ has a C_{3v} geometry, what do you conclude about the Al—Cl—Al angle? Is this expected for the central Cl^- sharing only two electron pairs with the two Al ions? Suggest a feature of the Al—Cl—Al bonding to account for that angle.

35. Propose single step syntheses of $B(OR)_3$ starting with a boron hydride, a boron halide, and a boron oxide.

36. By analogy with carbon allotropes, suggest two structures for boron nitride, $(BN)_x$.

37. Compare the thermal yields of

    and
    $$CH_{4(g)} + O_{2(g)} \rightarrow CO_{2(g)} + H_2O_{(g)}$$
    $$B_2H_{6(g)} + O_{2(g)} \rightarrow B_2O_{3(g)} + H_2O_{(g)}$$

    on the basis of kcal/(g hydride). Would use of B_2H_6 as a substitute for natural gas be practical? (What are the physical properties of the products?)

38. Predict products for the following reactions.

    a) $2NaBH_4\ +\ 2I_2\ \rightarrow 2NaI\ +\ ?$
       (reducing   (oxidizing
       Lewis base) Lewis acid)

    b) $B_2Cl_4 + C_2H_4 \rightarrow$? (intermediate species from π acid/base concept)

    c) $3Me_2PH + \frac{3}{2}B_2H_6 \xrightarrow{\Delta} H_2 + ?$

    d) $R_2BH + C_2H_4 \xrightarrow{\Delta}$? (consider the HOMO and LUMO of both reagents)

    e) the product from (d) $+ H_2O \xrightarrow{H^+} C_2H_6 + ?$

39. Why do you think the order and rate of mixing of H_2SO_4 and $LiAlH_4$/THF is important to the preparation of AlH_3?

40. Why do you suppose HCl is an effective choice of solvent for the reaction

    $$CH_3COCl + Al_2Cl_6 \rightarrow [CH_3CO][Al_2Cl_7]\ ?$$

41. Predict products for the following reactions.

    a) $LiAlH_4 + 4RCHCH_2 \rightarrow$? (a Li^+ salt of 4-coordinate Al; related to question 38(d))
    b) $LiAlH_4 + Me_3NH^+Cl^- \rightarrow H_2 + ?$
    c) $LiAlH_4 + 4ROH \rightarrow 2H_2 + ?$
    d) $4LiH + AlCl_3 \xrightarrow{Et_2O} 3LiCl + ?$
    e) $Al_2X_6 + Al_2R_6 \rightarrow$? (product has bridging X)

CARBON AND SILICON

Bond Energies for Reference

Si=Si	76	C≡C	200
Si—Si	54	C=C	146
Si—H	76	C—C	83
Si—C	73	C—H	99
Si—O	111	C—O	86
Si—F	143	C—F	117
Si—Cl	96	C—Cl	78

Having studied organic chemistry in some depth, you are already well aware of carbon chemistry. Our emphasis in this section will be on silicon, but at the end of this section you will find study questions on some less familiar "inorganic" carbon compounds. These questions will introduce those compounds by asking you to make predictions, comparisons, and interpretations based on structural, bonding, and thermodynamic ideas.

One of the important features of carbon chemistry is the propensity of the element to undergo catenation; therefore, an important question is why silicon does not seem to exhibit the same property, or at least why catenated silicon compounds are so much more reactive. In large measure, the reactivity of catenated silicon compounds with other compounds is due to the fact that the silicon valence shell includes d orbitals that are believed to have an important effect on the rates of reactions involving silicon compounds. This idea will be more fully developed in the next chapter; right now, we want to examine the problem from the thermodynamic point of view.

The list of catenated silicon compounds is not particularly extensive at the moment. *Silanes* of formula Si_nH_{2n+2} are known[53] for $n \leq 8$, and the larger values of n correspond to greater branching of the Si backbone. For lower values of n, hydrosilanes, chlorosilanes, and organosilanes have been well characterized. Among the non-cyclic chlorosilanes[54] are the compounds $SiCl_4$, Si_2Cl_6, Si_5Cl_{12}, and Si_6Cl_{14}. Pyrolysis (at 1150°) of SiF_4 in the presence of Si leads to a polymeric solid $(SiF_2)_n$ via condensation of the "carbene-like" SiF_2 at lower temperature.[55] Controlled pyrolysis of this chain polymer provides a route to chain perfluorosilanes, Si_nF_{2n+2}, up to $n = 14$.

In addition to the binary hydrosilanes and chlorosilanes, an extensive series of organosilanes, such as the cyclic Si_5R_{10} and Si_4R_8 (where $R = C_6H_5$), have been prepared.[56] From evidence thus far available, it appears that the phenyl group is a stabilizing substituent for organosilanes, presumably because of the possibility of partial double bond formation between the 1-carbon of the phenyl group and the silicon atom via the d orbitals of the latter (compare the comments on pp. 305 and 311 for B and Al). This, of course, is potentially important in compounds formed by chlorine, an element having lone pairs for use in partial double bond formation with silicon. More will be said subsequently about silicon-oxygen bonding, in which such double bond formation is highly important.

The thermal stability of the hydrosilanes, as a series, can be analyzed in terms of a Haber cycle for the reaction

$$Si_{n-1}H_{2n(g)} + Si_{(c)} + H_{2(g)} \rightarrow Si_nH_{2n+2(g)}$$

Elemental silicon possesses the diamond structure, in which each silicon atom is tetrahedrally bound to four other silicon atoms. The rupture of four Si—Si bonds to liberate one silicon atom has a net effect of breaking only two Si—Si bonds. One such bond is to be formed when $Si_{n-1}H_{2n}$ expands to Si_nH_{2n+2}. Therefore, to a first

---

[53] A classical writing from many points of view, you should read A. Stock, "Hydrides of Boron and Silicon," Cornell University Press, Ithaca (1933); K. Borer and C. S. G. Phillips, *Proc. Chem. Soc.*, 189 (1959); B. J. Aylett, *Adv. Inorg. Chem. Radiochem.*, **11**, 249 (1968).

[54] G. Urry, *J. Inorg. Nucl. Chem.*, **26**, 409 (1964).

[55] P. L. Timms, *et al.*, *J. Amer. Chem. Soc.*, **87**, 2824 (1965).

[56] H. Gilman, *et al.*, *Adv. Organomet. Chem.*, **4**, 1 (1966); K. M. McKay and R. Watt, *Organometal Chem. Rev.*, **4**, 137 (1969); R. K. Ingham and H. Gilman, in "Inorganic Polymers," F. G. A. Stone and W. A. G. Graham, Eds., Academic Press, New York (1962); E. Carberry and R. West, *J. Amer. Chem. Soc.*, **91**, 5440 (1969).

approximation there is a net endothermicity to the above reaction of 54 kcal/mole from the breaking and making of Si—Si bonds. The other bond energy changes associated with this reaction are the endothermic breaking of the H_2 bond and the exothermic making of two Si—H bonds (a net change of -48 kcal/mole). For silane, SiH_4, this balance of bond breaking and bond making amounts to an endothermicity of 2 kcal/mole. The endothermicity increases by some 6 kcal/mole for each SiH_2 unit added. Consequently, all silanes, like neutral boranes, are unstable with respect to the elements; but SiH_4 and Si_2H_6 seem to be capable of prolonged stability at room temperature (a kinetic effect). By comparison, CH_4 has a heat of formation of -22 kcal/mole and a net *exothermicity* of 11 kcal/mole for each CH_2 unit of expansion. The low heats of formation of hydrosilanes are easily traced to the relative weakness of the Si—H bonds ($3sp$/H overlap not as good as $2sp$/H overlap) and not to the weakness of the Si—Si bond. Even though the Si—Si bond energy is some 30 kcal/mole less than that of carbon, relative to the hydrocarbons this actually favors the formation of the hydrosilanes. Nevertheless, an important conclusion is that the atomization of each Si_nH_{2n+2} requires much less energy than atomization of each C_nH_{2n+2} so that conversion of Si—H and Si—Si bonds to Si—X (X = halogen) bonds tends to be easier than the corresponding conversions in C_nH_{2n+2}.

Owing to the much smaller Cl—Cl bond energy (58 kcal/mole) than H—H bond energy (104 kcal/mole), and to the much higher Si—Cl bond energy (96 kcal/mole) than Si—H bond energy (76 kcal/mole), it is understandable that the chlorosilanes appear to have greater thermodynamic stability than the hydrosilanes. The greater Cl—Si bond energy in this comparison could be attributed to Si—Cl partial multiple bond formation.

Thus far we have mentioned only silicon compounds of the saturated variety. An important question arises concerning the possibility of Si=Si bonding to form a series of unsaturated silanes. Here again a comparison with carbon chemistry is useful because unsaturated carbon compounds are quite prevalent. For carbon, the bond energy associated with C=C from a variety of organic materials is about the same as that for the diatomic molecule C_2 (:C=C:), that is, 146 kcal/mole; by making the assumption that the bond energy of Si_2 (76 kcal/mole) is indicative of that for unsaturated silanes, you see that the formation of Si_nH_{n+2} compounds will be, on a relative basis, considerably less favorable than those for carbon. In fact, the heat of formation of Si_2H_4 is estimated to be $+46$ kcal/mole. The reason for the small increase (40%) in Si—Si bond energy for doubly bound silicons relative to singly bound silicons (the increase is only about 22 kcal/mole) is not well understood[57] and has been widely disputed (the corresponding increase for carbon is 80%, or 63 kcal/mole). Nevertheless, if we accept the value of 76 kcal/mole for Si=Si, then hydrogenation reactions such as

$$Si_2H_{4(g)} + H_{2(g)} \rightarrow Si_2H_{6(g)}$$

have a net exothermicity of 26 kcal/mole of hydrogen consumed. Even more importantly, because the Si=Si double bond falls short of the stability of two Si—Si single bonds, the cyclic dimer Si_4H_8 should be stable with respect to $2Si_2H_4$. Another consequence of the low Si=Si energy is that Si is not known to form a graphite-like allotrope; only the diamond structure is found. This is not to say that species with Si=Si have not been postulated; more and more frequently, species with this functional feature are being postulated as short-lived reaction intermediates.

An important aspect of silicon chemistry is the behavior of the hydrosilanes in oxidizing media. For example, combustion of Si_nH_{2n+2} homologs is exceedingly

---

[57] W. E. Dasent, "Nonexistent Compounds," Dekker, New York (1965) gives some background on this problem. Fun to read; this author dares to suggest reasons why certain compounds and structural features have not been unearthed yet. Since 1965, the "nonexistent" compounds HOF and BrO_4^-, for example, have been synthesized. A more recent book by this author is "Inorganic Energetics," Penguin Books, Baltimore (1970).

exothermic (and kinetically rapid, so that the hydrosilanes are pyrophoric); this is attributable to the high Si—O bond energy on one hand, and low Si—H and Si—Si energies on the other. Furthermore, using bond energies from Table 6-3 (p. 270), the hydrolysis reactions

$$MH_{4(g)} + 2H_2O_{(g)} \rightarrow MO_{2(g)} + 4H_{2(g)}$$

for carbon and silicon are, respectively, *endothermic* by 40 kcal/mole and 26 kcal/mole. As $SiO_{2(g)}$ is much less stable than quartz ($SiO_{2(g)} \rightarrow SiO_{2(c)}$ is *exothermic* by 138 kcal/mole), the hydrolysis reaction to produce quartz is thermodynamically spontaneous with an estimated $\Delta E°$ of -112 kcal/mole. As was noted for Si=Si bonding, the Si=O double bond energy (153 kcal/mole) is much less than twice the Si—O bond energy (111 kcal/mole), presumably in this case because the Si—O "single" bond in quartz has appreciable double bond character.

Judging from the bond energies cited at the beginning of this section, you might surmise that halosilanes (except, as usual, for fluorides) and organosilanes tend to be thermodynamically quite unstable in combination with O_2 and hydroxylic solvents. While this is true in principle, in practice the reactions of halo- and organosilanes with O_2 are kinetically slow so that both can often be handled in air; other than the fluorosilanes, the halosilanes vigorously solvolyze in hydroxylic solvents, e.g.,

$$SiCl_4 + 4HOR \rightarrow Si(OR)_4 + 4HCl$$

The high stability of the Si—O bond is responsible for the utility of *organosiloxane* polymers (silicones), which are prepared by hydrolysis of the organohalosilanes:

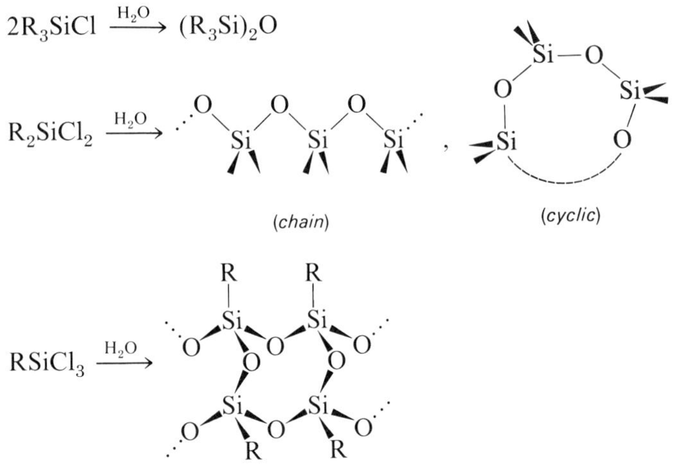

$$2R_3SiCl \xrightarrow{H_2O} (R_3Si)_2O$$

$R_2SiCl_2 \xrightarrow{H_2O}$ (chain), (cyclic)

$RSiCl_3 \xrightarrow{H_2O}$

The dichlorosilanes are useful for forming linear chains of Si—O backbones, the trichlorosilanes make possible the cross-linking of linear chains, and the monochlorosilanes are chain terminators. Consequently, hydrolysis of mixtures of all three produce polymers with properties (degree of cross-linking) dependent on the starting mixture.[58]

The polymerization reactions, of course, owe their occurrence to the instability of *silanols* (*cf.* boronic acids) presumably initially formed from the hydrolysis.

$$2R_3SiOH \rightarrow (R_3Si)_2O + H_2O$$

Note that this reaction is approximately thermally neutral and largely entropy-driven.

---

[58] An interesting, short exposé of the utility and preparation of "silicone liquid rubbers" is by J. A. C. Watt, *Chem. Britain*, **6**, 519 (1970).

As a final note on silicon halide chemistry, we might mention that SiF_4 may also be hydrolyzed to "hydrous oxides," but the decomposition is incomplete owing to formation of the hexafluorosilicate ion,

$$2SiF_4 + 2H_2O \rightarrow SiO_{2(hyd)} + SiF_6^{2-} + 2HF + 2H^+$$

a species also formed by the etching action of HF on glass. As mentioned in Chapter 5, SiF_4 (and $RSiF_3$ and R_2SiF_2 as well) also acts as a fluoride ion acceptor in 1:1 ratio with M^IF salts (with the proper size of cation) to form the SiF_5^- anion (M^ISiF_5). In solution the anion appears to have a D_{3h} structure[59]; its existence establishes the possibility of base (F^-) attack upon 4-coordinate silicon (SiF_4) to form a stable trigonal bipyramidal unit for silicon.

Nature achieves unsurpassed variety in its structural variations of binary combinations of silicon and oxygen. Such natural polymer chemistry achieves heights at which the synthetic chemist can only marvel. In the interest of correlating the structures of these polymers, we will limit your exposure to a few of the more important structural identities. The simplest natural binary combination of silicon and oxygen is, of course, SiO_2; in the gas phase, it has a structure like that of its congener, CO_2. However, owing to the low energy of two Si=O bonds (153 kcal/mole each) relative to four Si—O "single" bonds (111 kcal/mole each), SiO_2 undergoes self-association to form two important structural arrangements. *Cristobalite* (**21a**) is an orderly structure derived from the diamond structure of elemental silicon.[60] In ordinary hydrocarbon chemistry, adamantane (**21b**) exhibits the same 10-atom skeletal cage. Insertion of an oxygen atom between each pair of silicon atoms considerably expands (~50 percent) the Si frame, creating a cage with a large hole at the center. This openness is a characteristic feature of silicate materials and, as you shall see in a moment, it is important to the highly useful ion-exchange and molecular sieve (chromatographic) properties of these materials. A second structural feature you will see again and again is the six-membered ring of silicon atoms (where oxygen atoms are imposed between each Si—Si pair).

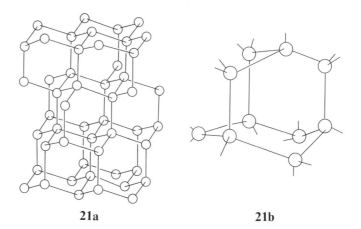

21a   **21b**

The more stable form of silica, *quartz,* has the intriguingly useful property of being an optically active polymer. This physico-chemical characteristic is a direct consequence of its structure. The very complicated arrangement of silicon and oxygen atoms (though each silicon is surrounded by a tetrahedral arrangement of oxygen atoms) is difficult to show without the use of three-dimensional models. The most

---

[59] F. Klanberg and E. L. Muetterties, *Inorg. Chem.*, **7**, 155 (1968).
[60] In fact, you first encountered this structure in the earlier section on ionic lattices. See the unit cell perspective of zinc blende, structure **6** on p. 281.

direct way to describe the structure here is to imagine stacking springs so as to form the hexagonal arrangement shown in the center of **22**:

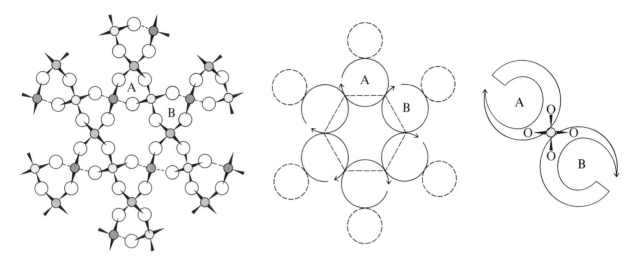

22 Views down the helical axis of quartz. Shaded circles represent Si atoms at three depths, with the darkest ones closest to you.

The points of tangency of the springs define, in projection, a hexagon and locate the silicon atoms along the coils.[61] These are separated by oxygen atoms. Notice that three silicon and three oxygen atoms complete a 360° turn along a coil. These coils *must* all be of the same "handedness" in order for the silicon atoms to find themselves in a tetrahedral environment of oxygen atoms, and this imposes the optical activity feature upon the structure. Notice that the SiO_2 units manage to create rings of six silicons bridged by oxygen atoms, in spite of an obviously more intricate scheme of self-association than found in cristobalite.

Fusion of silica with alkaline earth oxides or carbonates (in the latter case CO_2 is evolved, emphasizing the greater effective acidity of SiO_2 than CO_2) will yield materials, called "metasilicates," which are the stoichiometric analogs of CO_3^{2-} and SO_3.

$$MO + SiO_2 \rightarrow \text{``}MSiO_3\text{''}$$
$$MCO_3 + SiO_2 \rightarrow \text{``}MSiO_3\text{''} + CO_2$$

If a 2:1 ratio of oxide to silica is used, the silicate anion (SiO_4^{4-}) results, for which no carbon analog is known.

$$2MO + SiO_2 \rightarrow M_2SiO_4$$

The silicon in SiO_4^{4-} is coordinatively saturated and so does not undergo self-polymerization. Were a 1.5:1 ratio of oxide to SiO_2 to be fused, the product anion produced would be $Si_2O_7^{6-}$. This "pyrosilicate" ion is conveniently thought of as an adduct of SiO_4^{4-} and SiO_3^{2-}. The bridging oxygen of this anion shares with its counterpart in the silica structures the characteristic of a large Si—O—Si angle, approaching 180° in many cases.

The remaining important silicates derive their structures from the self-association of metasilicate anions (SiO_3^{2-}) into chain and ring polymers. The simplest of these is the linear chain characteristic of the minerals known as *pyroxenes*. The

---

[61] Be careful to realize that the silicon atoms at these points are not, in fact, coplanar.

strands of these chains are held together by the metal ions coordinated to the terminal oxygen atoms of the chains. In the sketch below, the repeating SiO_3^{2-} unit is identified by the dotted lines.

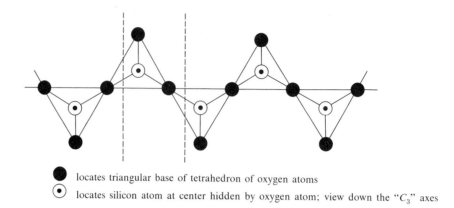

● locates triangular base of tetrahedron of oxygen atoms
◉ locates silicon atom at center hidden by oxygen atom; view down the "C_3" axes

Nature achieves cross-linking of these polymeric chains in two ways; one way, characteristic of the *amphibole* minerals, is to form ribbons of double strands of the chain and the other way, characteristic of *micas*, is to fully cross-link the chains so as to form sheets. The cross-linking of two chains is most readily visualized in terms of diagram **23**.

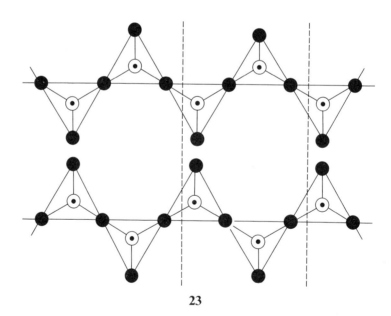

23

The chains are not interlocking in this diagram; replacement of one of each pair of face-to-face terminal oxygen atoms by the other achieves cross-linking of two chains into a ribbon. The repetitive structural unit is the $Si_4O_{11}^{6-}$ fragment, as indicated by the dotted lines. Complete cross-linking in this manner achieves the sheet structure characteristic of the micas. As the cross-linking is carried further and further, the ratio of Si to O increases from the value $1:3$ for pyroxenes to a limit of $1:2.5$, characteristic of the micas. Diagram **24** shows the "ultimate" structure of mica and identifies the $Si_2O_5^{2-}$ repetitive unit.

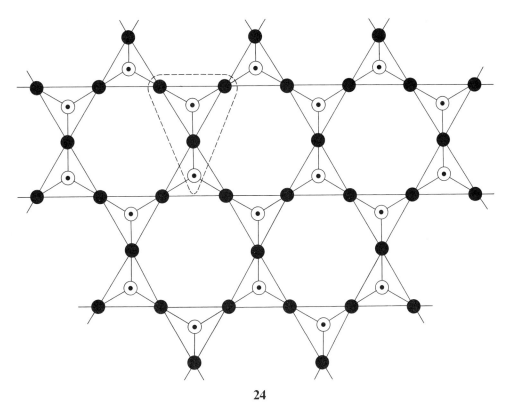

24

The pyroxenes and amphiboles are the basic *chain* polymers derived from the SiO_3^{2-} unit, and you have now seen their relation to each other and to the "mica" sheet; these same mica sheets can also be arrived at by condensation of *cyclic* silicates, and this establishes a point of convergence of the chain and cyclic structures, as you will now see. The simplest cyclic silicate is the trimer **25**; notice that each silicon has two terminal oxygens and two bridging oxygens. Recall that you have seen this structural unit (but puckered because all oxygens played a bridging role) as responsible for the optical activity of quartz structure. Expansion of the three-silicon ring into a six-silicon ring produces the hexamer (**26**) of SiO_3^{2-}, and again you recognize a unit (but with all oxygen atoms bridging) found in quartz and also in cristobalite. The hexameric SiO_3^{2-} structure and that of the trimer are, however, more regular than those found in the two forms of silica, in that these rings are not puckered—all

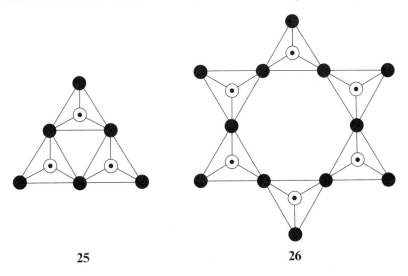

25   **26**

324 □ ENERGETICS AND STRUCTURES AS GUIDES TO MAIN GROUP CHEMISTRY

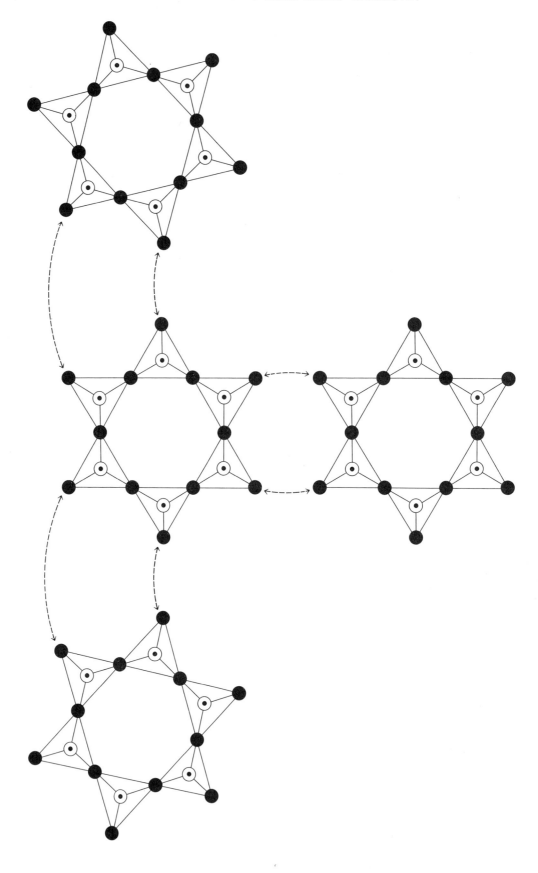

silicons, all bridging oxygens, and all terminal oxygens lie in respective planes. You should also recall having seen this planar six-silicon ring in the mica structure, and this establishes the point of convergence of the chain and cyclic polymers of SiO_3^{2-}. The relationship between the hexamer and mica structures is most easily visualized by removing from the hexamer the basal terminal oxygens and then linking the six-pointed stars together (see opposite page) so that a terminal oxygen of one star occupies the vacant position of another. The net effect is fusion of $Si_2O_5^{2-}$ units, as we previously deduced by approaching the mica structure **24** from the point of view of fusion of amphiboles.

An interesting situation arises for materials in which the silicon atoms of silica are replaced by aluminum atoms. We may think of this either as replacement of Si^{4+} ions by Al^{3+} ions or as replacement of Si atoms by Al^- ions—the views are equivalent, for they maintain the number of valence electrons fixed. Because of the excess negative charges so generated, counter-ions must also be present to maintain electrical neutrality. Presumably any number of Si atoms may be replaced by Al, but only in certain materials containing equal numbers of Si and Al atoms are the structures related[62] to those of silica. The quartz structure is so compact that the channels between helices can accommodate only counter-ions as small as Li^+, Be^{2+}, and B^{3+}. $LiAlSiO_4$ is an example. To accommodate larger cations, a more open structure is needed; this can be provided by the cristobalite structure, an example of which is $NaAlSiO_4$. Note in both of these examples, and in the ones to follow, that the ratio of (Al + Si) atoms to oxygen is 1:2. This is dictated by the fact that each derives from replacement of Si in an SiO_2 polymer by Al.

Of much more practical significance are minerals with more open structures. There are a great variety of these materials presenting a number of complicated structures. We will mention here one of the more symmetric structures found in *chazabite*, $(Ca, Na_2)(Al_2Si_4)O_{12} \cdot 6H_2O$. Remembering that an oxygen atom bridges each Al and Si atom, we can most simply describe the structure by drawing lines to connect the Al and Si atoms, with the basket-like result shown below.

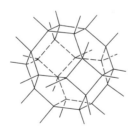

By counting the vertices and edges of this polyhedron, you can determine that the unit contains 24 (Al + Si) atoms and 48 oxygen atoms. You should also feel that you have seen this polyhedron before, because you first encountered it as **12** for MB_{12}; it is simply an expanded form of the cuboctahedron.

The linking together of these cuboctahedra to form a three-dimensional lattice is achieved by the 24 bridging oxygens *exo* to the surface of the polyhedron. Considering the two most symmetric ways in which these "inter-cuboctahedra bridging" oxygens can be grouped leads to two well-known *zeolite* structures. Grouping the 24 bridging oxygens into four mutually tetrahedrally disposed hexagons leads to the *faujasite* structure $\{NaCa_{0.5}(Al_2Si_5O_{14}) \cdot 10H_2O\}$, where the cuboctahedra take up a diamond-like structure (p. 326; and see **21** for a different perspective).

---

[62] Note that some distortion of the O—Si—O skeleton is inevitable because of the differing sizes of Al^{3+} and Si^{4+}.

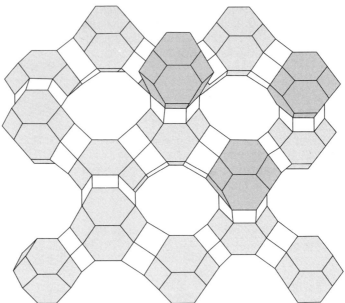

[From A. F. Wells, "Structural Inorganic Chemistry," Oxford University Press, New York, 1975.]

By considering the 24 bridging oxygens to be grouped into six octahedrally disposed squares, we are led to the structure of the synthetic zeolite called *Zeolite A* {$Na_{12}(Al_{12}Si_{12}O_{48}) \cdot 27H_2O$}.

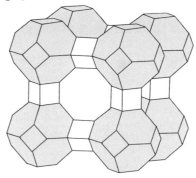

The baskets, as well as the prominent channels in these networks, provide for the chromatographic and ion exchange properties characteristic of this class of minerals. By now you should understand more clearly the idea behind calling such materials "molecular sieves."

Before leaving this discussion of silicates, you should know that such materials are continuing to find application in organic chemistry, in particular as substrates for heterogeneous catalysis of hydrocarbon reactions. Pt deposited onto silica has an important application as a cracking catalyst in the petroleum industry. In Chapters 16 and 17 you will learn about the important use of transition metal compounds as homogeneous catalysts in organic synthesis. Here we will pause briefly to comment on a recent use[63] of mica-like structures to incorporate such complexes for use as heterogeneous catalysts. Referring to Figure 6-4 and the octahedral holes between close packed anion layers, you should be able to identify the pattern of anions **27** in the vicinity of any such hole.

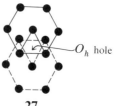

27

---

[63] T. J. Pinnavia and P. K. Welty, *J. Amer. Chem. Soc.*, **97**, 3819 (1975).

Reference to **24** shows the same hexagonal arrangement of terminal oxygens in a mica layer, but *lacking* the central anion of **27**. Diagram **28** is a sketch of the hexagonal Si—O_t groups corresponding to the arrangement in **27** and in which the upper mica sheet has the terminal oxygens suspended below the Si plane.

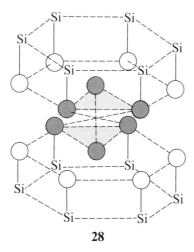

28

With an additional anion (like F^- or OH^-) incorporated at the center of each of the $(O_t)_6$ hexagons, the shaded circles define an octahedral hole. This "double-layer" mica is quite common in naturally occurring clays. *Talc,* $Mg_3(OH)_2Si_4O_{10}$, is a well-known example, in which the "extra" anions are OH^-, and Mg^{2+} ions fill octahedral holes in **28**. The role of the Mg^{2+} is to hold the mica sheets in the double layer configuration. A material like *hectorite* ($Na_{2/3}[Li_{2/3}Mg_{5 1/3}](OH)_4Si_8O_{20}$) is derived from talc by the replacement of 2/3 out of 6 (or 2 out of 18) of the Mg^{2+} ions by Li^+ ions. To maintain electroneutrality, additional 2/3 + charge is provided by Na^+ ions. As all the O_h holes between mica sheets are occupied by Mg^{2+} and Li^+, these additional Na^+ ions take up positions *between* the mica double-layers. Schematically, the situation is as in **29**.

These Na^+ ions are not tightly incorporated into the structure and may be readily exchanged with H^+ and other cations.

As a dramatic illustration of how the presence of the silicate substrate can affect the properties of the interlamellar ion, Rh_2^{4+} (exchanged for Na^+) reacts with H_2 at 25°C (methanol can be used to suspend the silicate). H_2 does *not* react with Rh_2^{4+} under these conditions in the absence of the silicate. The rhodium hydride complex can then be used to reduce olefins to alkanes, with regeneration of the rhodium species for re-entry into the catalytic cycle.

STUDY QUESTIONS

42. Recall or look up the structure of graphite and predict the C—C bond order. Does your result fit with the fact that the C—C distance in graphite is a little greater than that in benzene?

43. Is the fluorination of graphite to produce poly(carbon fluoride) exothermic?

$$(C)_n + \frac{n}{2}F_2 \rightarrow (CF)_n$$

Which compound, graphite or $(CF)_n$, should be more stable to air oxidation to CO_2? To be of advantage as a substitute for graphite lubricant, $(CF)_n$ must have what structural characteristic? Knowing that the hexagons of $(CF)_n$ have a "boat" structure, make a sketch of several fused "boats" (show the CF bonds, too) so as to compare the lamellar structure directly with that of graphite.

44. The *simpler* metal carbides are of three types with distinctly different hydrolytic behavior:

$$Be_2C \text{ or } Al_4C_3 \xrightarrow{H_2O} CH_4 + \text{metal hydroxide}$$

$$Na_2C_2 \text{ or } CaC_2 \xrightarrow{H_2O} C_2H_2 + \text{metal hydroxide}$$

$$Mg_2C_3 \xrightarrow{H_2O} C_3H_4 + Mg(OH)_2$$

$$SiC \xrightarrow{H_2O} \text{inert}$$

Characterize each of these as covalent or ionic; in the latter cases, infer a structure for the carbon anion present. Carborundum (SiC) is an extremely hard material used as an abrasive. Identify a sketch in this chapter that describes its structure.

45. Construct π mo diagrams for the linear molecules C_nO_2 for $n = 1,3,5$ and for $n = 2,4$. Which set yields free radicals? What is their spin multiplicity? Only one of these sets (n even or odd) is known; which is it?

46. Cyanogen, $(CN)_2$, is a flammable gas. Although it is generally stable, impurities can cause its polymerization to solid paracyanogen, $(CN)_x$, at high temperature. Estimate the exothermicity of this reaction, using $E_{C \equiv N} = 212$ kcal/mole, $E_{C=N} = 147$ kcal/mole, and $E_{C-N} = 73$ kcal/mole.

$$(CN)_{2(g)} \rightarrow \frac{1}{x}\left(\begin{array}{c}\text{ring structure with C, N}\end{array}\right)_{x(g)}$$

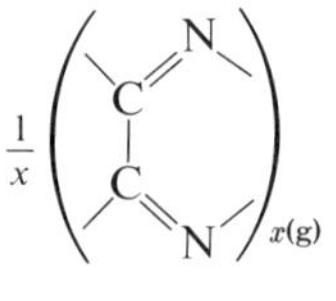

Is your estimate in the range where the energy of $(CN)_{x(g)} \rightarrow (CN)_{x(c)}$ and the overall entropy change might seriously affect $\Delta G°$?

47. Liquid HCN is dangerously unstable with respect to trimer formation. Propose a structure for the trimer and estimate the energy of the trimerization reaction. (See the preceding problem for bond energy values.) What all-carbon and boron/nitrogen compounds exhibit related structures?

48. Account for the large Si—O—Si angle of $\sim 150°$ in $(H_3Si)_2O$.

49. Is it surprising to you to learn that Me_3SiOH is as good a hydrogen-bonding acid as t-BuOH? If so, why and how can you rationalize this finding? If not, why not? Do you expect a large Si—O—H angle?

50. Ph_3SiX reacts with bipyridine (structure shown) to form an ionic solid of formula $Ph_3SiX \cdot bipy$. Propose a geometry for the silicon cation; note and comment on any unusual aspect regarding the positions of the carbon and nitrogen atoms about silicon.

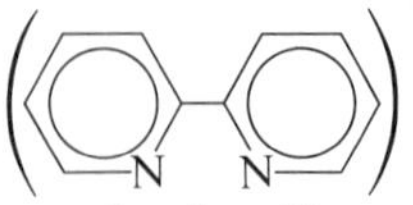

51. Propose a structure for dimeric *bis*(trimethylsilyl)amide, $(Me_3Si)_2N^-$, and comment on any "unusual" aspect of the disposition of C and N about Si.

52. Predict products for

   a) $X_3SiH + C_2H_4 \xrightarrow{\text{(metal catalyst)}}$ (see problems 6-38, 6-41)

   b) $2SiF_{4(g)} + H_2O_{(g)} \rightarrow$

53. Which silicate anion structure most resembles those of B_2O_3 and $(CF)_n$? Does it have the expected shear properties?

54. What are the point groups for $(SiO_3^{2-})_3$ and $(SiO_3^{2-})_6$?

55. Make a "bed-spring" sketch of hypothetical racemic quartz (*i.e.*, similar to **22**). With silicon ions at the points of tangency, satisfy yourself that the silicon is in a *planar* environment of four oxygens in racemic quartz, whereas coils of like symmetry produce a tetrahedral environment for silicon.

56. Using an $Si-O_b$ distance of 1.7 Å and an $Si-O_b-Si$ angle of 140° in $(SiO_3^{2-})_6$, calculate the distance between nearest silicon atoms. How big (diameter) is the "bowl" of the hexamer? Does this seem large enough to pass an O_2 molecule ($\sim 1.2 \times 1.4$ Å)?

NITROGEN AND PHOSPHORUS

Bond Energies for Reference

N≡N	226	P≡P	125
N=N	100	P—P	50
N—N	38	P—H	78
N—H	93	P—C	64
N—C	73	P—O	88
N—O	39	P—F	119
N—F	67	P—Cl	79
N—Cl	45		

Whereas the order of single bond energies for the Group IV elements is C—C > Si—Si, the converse appears to apply for the Group V elements, with $E_{N-N} < E_{P-P}$. A nominal value for the N—N single bond is 38 kcal/mole, while that for P—P is 50 kcal/mole.[64] Interestingly, this ordering of light and heavy atom single bond energies is maintained for the remaining groups of the representative elements. It has been generally associated with the first appearance at Group V of lone pairs on the neighboring atoms and may be attributed to lone pair-lone pair repulsions on the neighboring atoms. Whatever the reason, the ordering of single bond energies has an important connection with the number and types of catenated N—N and P—P compounds known to date. As noted on p. 269, hydrazine, the simplest catenated N—N compound, is unstable with respect to its elements and the higher homologs become increasingly unstable. This, of course, derives primarily from the fact that the N—N single bond energy is only about one-sixth of the N≡N triple bond energy (226 kcal/mole).

---

[64] More striking is the comparison

$$B-B\ (72) < C-C\ (83) \gg N-N\ (38) \sim O-O\ (34) \sim F-F\ (38)$$
$$Si-Si\ (54) \sim P-P\ (50) < S-S\ (63) \sim Cl-Cl\ (58).$$

In addition to the hydride N_2H_4, nitrogen is known to form the analogous fluoride N_2F_4, a gas with the interesting property of dissociating ($\Delta H \approx 20$ kcal/mole) into NF_2 radicals. The remaining examples of catenated nitrogen involve oxygen as a substituent: N_2O_4 is a dimer of NO_2 (N—N bond dissociation energy of ~ 15 kcal/mole); N_2O_2 is a dimer found in solid NO; and N_2O_3 may be thought of as the association product of $NO + NO_2$ (N—N bond dissociation energy of ~ 10 kcal/mole). All have weak N—N bonds and are highly dissociated at room temperature (an unstable solid form of N_2O_3 is believed to have a structure with a weak bond between the N of NO and the O of NO_2). A little later, we shall consider these oxides further.

These characteristics of catenated N—N compounds contrast sharply with the apparently greater stability of analogous P—P compounds. A good case in point is that elemental nitrogen has only one allotropic form, the N_2 molecule. Elemental phosphorus, on the other hand, exists in no fewer than three allotropic forms—all of which are solids at room temperature. All three forms are characterized by P—P single bonds. *White phosphorus* (**29**) consists of P_4 tetrahedra such that each P forms single bonds to the other three P atoms of the tetrahedron.

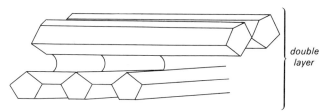

Red phosphorus has an intriguing double layer structure[65] consisting of mutually perpendicular, connected, pentagonal tubes.

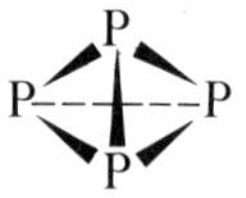

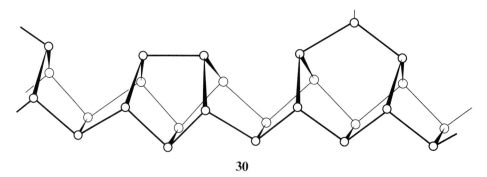

The pentagonal shape arises from the repeating pattern of fused bicyclo(3.3)P_8 units, as in **30**. Notice that the fused three-membered rings of P_4 have opened to a fused five-membered ring system. In **30** you notice also the insertion of a P atom at alternate bridgehead positions; these provide the "connectors" to the tubes of the partner layer.

30

The most stable allotrope of the element, a flaky *black* solid, has a layer lattice, **31** (notice the occurrence of the "chair" six-membered ring, as in carbon and silicon), where each P is connected to three other P atoms via single bonds (the internal strain of the white and red allotropes is fully removed in this structure).

---

[65] H. Thurn and H. Krebs, *Acta Cryst.*, **B25**, 125 (1969).

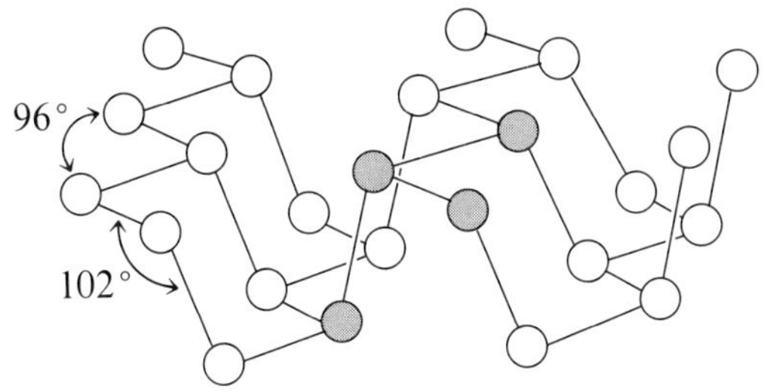

31

With a bond energy of only 125 kcal/mole for P_2 (P≡P), the $\Delta E°$ for the reaction

$$2P_{2(g)} \rightarrow P_{4(g)}$$

is nominally -50 kcal/mole while that for

$$2N_{2(g)} \rightarrow N_{4(g)}$$

is estimated to be $+224$ kcal/mole (with no allowance for internal strain), because E≡E > 3E—E for nitrogen but E≡E < 3E—E for phosphorus.

With regard to the relative stabilities of N=N and P=P bonds, we find a trend like that for C=C vs. Si=Si. You may have learned in organic chemistry that some (often explosively dangerous) $(-N=N-)_x$ chain and cyclic compounds are known. There is one inorganic analog of these, N_2F_2, called difluorodiazene. The stable azide ion (N_3^-) and unstable hydrazoic acid (HN_3) are more familiar to you; phosphorus analogs of all these materials are unknown.

Owing in part to the reversal (relative to the carbon/silicon trend) in order of single bond energies, phosphorus has yielded more catenated compounds than has nitrogen.[66] Thus, the chain molecules P_2H_4, P_3H_6, and P_4H_6 have been reported, as has P_2F_4. Several organopolyphosphines with chain [P_2R_4, $P_3(C_6H_5)_5$, $HP(CF_3)-P(CF_3)-PH(CF_3)$] and cyclic [$P_4R_4$, $P_5(CF_3)_5$, and $(PCF_3)_x$ for $x > 5$] structures are known. A few diphosphines with oxygen substituents,[67] such as $HP_2O_5^{3-}$ (**32**) and $P_2O_6^{4-}$, add to the list of examples of the P—P link.

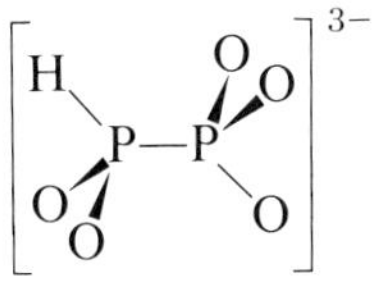

32

One further striking difference between nitrogen and phosphorus concerns their compounds with oxygen. Phosphorus is known to form two stable binary oxygen compounds: P_4O_6 and P_4O_{10} (more commonly written as P_2O_3 and P_2O_5), both of which have structures based on insertion of oxygen atoms into the P—P bonds of the tetrahedral P_4 unit **29** (and reminiscent of the Si ↔ cristobalite relation). As both structures below suggest, each phosphorus atom is singly bound to three (P_4O_6) or four (P_4O_{10}) oxygen atoms in these structures.

[garlic odor
m.p. 24°
b.p. 174°]

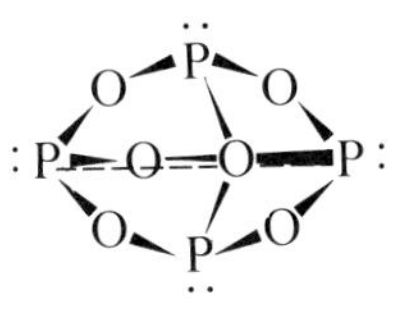

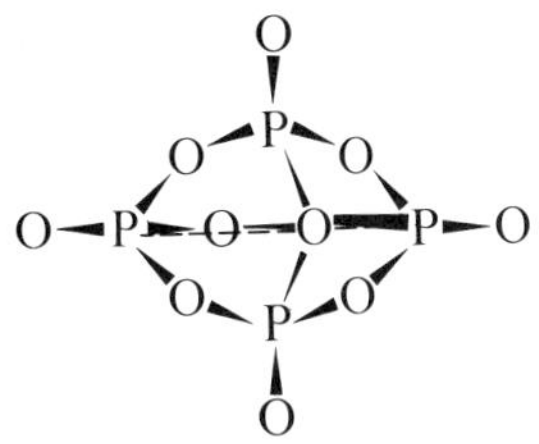

[sublimes 360°]

---

[66] A. H. Cowley, "Compounds Containing Phosphorus-Phosphorus Bonds," Dowden, Hutchinson and Ross, Stroudsberg, Pa. (1973).

[67] J. E. Huheey, *J. Chem. Educ.*, **40**, 153 (1963).

As was mentioned in Chapter 3 in the discussion of bond energies, and also in the preceding discussions of Si—O bonds, these P—O linkages are most likely only formally "single" bonds. At this stage in your development as a chemist you know that no such oxides of nitrogen have ever been prepared; rather unusual reagents or conditions (and precautions!) will be necessary to do so, as the bond energy data allow you to estimate the heat of formation of N_4O_6 (with a structure analogous to that of P_4O_6) to be greater than +350 kcal/mole.

Related to these binary oxide structures for phosphorus are the structures of the phosphorus oxyacids and their esters. H_3PO_4 and its ester, R_3PO_4, present the phosphorus atom in a C_{3v} environment of oxygen atoms, one of which is the terminal type found in P_4O_{10}. In one sense, P_4O_6 is somewhat unique because phosphorous acid, H_3PO_3, and most of its esters do not have the simple C_{3v} structures $P(OR)_3$; rather, they also contain a terminal PO linkage so that their formulae are better written as $RPO(OR)_2$ (33). This tendency extends to the lower phosphorus acid

O   O
‖   ‖
P   P
/|\   /|\
RO OR   H   OH
 R   H
33   34

H_3PO_2 (34) (hypophosphorus acid), which is only a monobasic acid. (The P—H hydrogens in H_3PO_3 and H_3PO_2 are not acidic.) In this connection it is interesting to note that the hypothetical compound $P(OH)_3$ has an estimated heat of formation (from Table 6-3) of −182 kcal/mole. Consequently, with a net endothermicity of 33 kcal/mole estimated for the breaking of an OH bond and formation of a PH bond, we can judge that the energy gained by the P=O dative bonding for the terminal oxygen must be in excess of 33 kcal/mole, a not unreasonable lower limit. In fact, one estimate of the PO_t bond energy is 130 kcal/mole, about 40 kcal/mole in excess of the P—O— bond energy. This tendency for terminal P=O formation is further evident in the condensed (poly) phosphoric acids, $H\left(O—\overset{\overset{O}{\|}}{\underset{OH}{P}}—\right)_n OH$ and phosphoryl halides, $X_3P{=}O$.

Further interest in the terminal P=O bond is stimulated by recalling that the heat of formation of the (unknown) molecule $N(OH)_3$ was estimated earlier (p. 269) to be only −2 kcal/mole. Were such a compound ever to be prepared, rearrangement to $HNO(OH)_2$ (35) would not be predicted because $\Delta H°$ for

$$N(OH)_3 \rightarrow HNO(OH)_2$$

O
|
N—OH   35
/ \
H   OH

should be much greater than +18 kcal/mole.

We will complete this preliminary discussion of basic phosphorus compounds by noting that all the phosphorus trihalides, pentahalides, and oxyhalides vigorously solvolyze in hydroxylic solvents to form phosphoric and phosphorous esters, except for the fluorides. Refer to the bond energy terms at the beginning of this section and you will see that this is not unexpected. The unique role of fluorine among the halogens probably stems from a tendency toward P=F character, as with oxygen. Given this character for oxygen and fluorine, you might wonder whether PN bonds tend to show the same characteristic. In fact, the phosphorus halides and oxyhalides do solvolyze in secondary ammines to form PN bonds,

$$X_3P=O + 6R_2NH \rightarrow (R_2N)_3PO + 3(R_2NH_2)Cl$$

and an interesting structural feature of these compounds is the planarity of the NR_2 groups. This strongly implicates P=N bond character.[68]

Unlike phosphorus, which forms only two binary oxides, nitrogen forms many. The simplest, of course, is the colorless diatomic π^* radical NO, *nitric oxide*, which tends to dimerize weakly ($\Delta H \approx 4$ kcal/mole) in the liquid (b.p. $-152°$) and solid (m.p. $-164°$) states to $(NO)_2$ and is readily oxidized to the nitrosyl cation $(NO)^+$. The structure of the colorless[69] dimer is planar and cyclic (C_{2v}) with long (2.2 Å), weak bonds between the nitrogens. In the dimerization, the weak "N—N bond" energy and the planar structure are easily understood to arise from the half-occupied HOMO of NO. The constructive overlap of both the nitrogen p and oxygen p ao components of the π^* HOMO introduces both N—N and (to a lesser extent) O···O bonding and an ONN angle of $\sim 100°$.

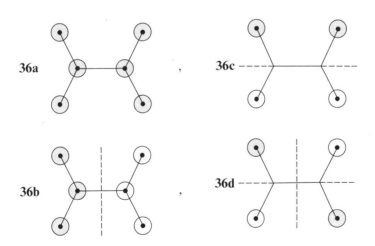

NO_2 (m.p. $-11°$, b.p. $21°$) is another familiar oxide of nitrogen. Like NO, *nitrogen dioxide* is a free radical (but a σ, not π, radical) with a slight tendency to dimerize to N_2O_4 and thereby lose its brown color. The N—N bond in this compound is weak (distance = 1.75 Å and bond energy ≈ 15 kcal/mole). The most stable structure of the dimer is planar. An interesting bonding question arises for the N_2O_4 and B_2F_4 structures (p. 307) with regard to the role of π bonding as a factor favoring a D_{2h} geometry. Using the principles of Chapter 4, you can quickly settle on the following out-of-plane π mo's for a planar E_2X_4 molecule.

36a, 36b, 36c, 36d

---

[68] A. H. Cowley and J. R. Schweiger, *J. Amer. Chem. Soc.*, **95**, 4179 (1973) reviews evidence (preferred donor atoms and rotational barriers for OP—N compounds) indicating P=N character; I. G. Csizmadia, et al., *J. Chem. Soc., Chem. Comm.*, 1147 (1972), however, make a case for the importance of σ release (P $\rightarrow$ N) as a factor in the planarity of N in H_2PNH_2.

[69] In one of the "Textbook Errors" features of *J. Chem. Educ.*, J. Mason [**52**, 445 (1975)] notes that $(NO)_2$ is *not* blue and does *not* have a C_{2h} structure, as is often written. Her article also contains structural and thermodynamic data references for $(NO)_2$.

These mo's increase in energy in the order **36a** → **f** (simply consider the number of bonding versus anti-bonding interferences). With four π valence pairs to occupy these mo's, **36a** through **d** are occupied. Orbitals **36a** and **36b** account for net N=O character but no net N=N character. Similarly, the bonding (long range) between *syn* X atoms in **36c** is cancelled by the anti-bonding between these atoms in **36d**. Thus, the out-of-plane π electrons cannot discriminate in favor of a D_{2h} structure. These same conclusions arise from the π resonance structures

$$\text{structure} \longleftrightarrow \text{etc.}$$

What is not apparent from π resonance structures, but is seen from a full mo study,[70]

is that the in-plane HOMO of NO_2 has appreciable O character, affording some long range bonding between *syn*-oxygens in the N_2O_4 mo otherwise characterizing the N—N bond. In B_2F_4, the BF_2 HOMO should be more highly localized on the boron; and the *syn*-F, F bonding in the B—B mo will not be nearly as effective as is the case for N_2O_4. In both molecules, of course, the D_{2d} geometry is favored by *syn*-VSEPR/steric repulsions. Thus, the difference between BF_2 and NO_2 HOMO's accounts for a stronger B—B than N—N bond and the difference in preferred structures.

Dinitrogen trioxide, which may be thought of as an association product of NO and NO_2, exists in a pure form only as a blue solid (m.p. −102°). In solution and as a liquid, extensive reversible dissociation into NO and NO_2 occurs (dissociation energy ≈10 kcal/mole). The stable form[71] of this molecule has a weak N—N bond (C_s symmetry, distance ≈1.86 Å), but the interesting fact that nitrogen isotope studies[72] reveal rapid scrambling of nitrogen atoms may indicate the presence

$$^*NO + NO_2 \rightleftarrows NO + {^*NO_2}$$

in fluid phases of a species (perhaps an intermediate) with an ONO—NO structure.

In addition to NO and NO_2 as oxides with an identifiable existence at room temperature, *nitrous oxide* (or dinitrogen oxide) is a well-known gas (b.p. −89°) with a $C_{\infty v}$ structure. Until recently, this and N_3^- were the only clear examples of stable compounds of N_2. N_2O seems to be the least reactive of the nitrogen oxides and has

---

[70] R. Alrichs and F. Keil, *J. Amer. Chem. Soc.*, **96**, 7615 (1974); J. M. Howell and J. R. Van Wazer, *ibid.*, **96**, 7902 (1974).

[71] A. H. Brattain, *et al.*, *Trans. Faraday Soc.*, **65**, 1963 (1969).

[72] E. Leifer, *J. Chem. Phys.*, **8**, 301 (1940); W. Spindel in "Inorganic Isotope Synthesis," R. H. Herber, Ed., W. A. Benjamin, New York (1962).

uses as an anesthetic (laughing gas) and as the propellant in aerosol cans of dessert toppings.

The least stable (tends to explode) of the nitrogen oxides is the white solid (sublimes 33°), *dinitrogen pentoxide*. An interesting property of this material is that its structure varies with phase. The solid is actually ionic, consisting of *nitronium* (NO_2^+) cations and *nitrate* anions, whereas the gas phase structure is that of a "dinitro ether" [$O(NO_2)_2$].

Before leaving this review of the binary oxides of nitrogen, we need to bring up some intriguing observations relating to possible isomeric forms of N_2O_4. Many reactions of N_2O_4 produce products containing or appearing to derive from reactions with NO and NO_3 radicals. These observations are often rationalized by proposing the presence (at admittedly undetectable levels[73]) of $ONONO_2$ and its possible dissociation products:

$$O_2NNO_2 \rightleftharpoons ONONO_2 \rightleftharpoons NO + NO_3$$

In ionizing solvents (refer to the discussion in Chapter 5 of solvent properties conducive to solute ionization) such as ethyl acetate, the reaction products appear to stem from the presence of NO^+ and NO_3^-. Both viewpoints provide "chemical" support for the idea of isomeric forms of N_2O_4 (*cf.* the N_2O_3 discussion above).

$$ONONO_2 \underset{\text{solvent}}{\overset{\text{polar}}{\rightleftharpoons}} NO^+ + NO_3^-$$

Related to this behavior of N_2O_4, N_2O_5 produces solutions of NO_2^+ in the anhydrous acids H_3PO_4, HNO_3, and H_2SO_4. Knowing that solid N_2O_5 consists of NO_2^+ and NO_3^-, and that these solvents should be highly ionizing, you could correctly suspect that the following reaction can occur:

$$N_2O_5 \underset{\text{acid}}{\rightleftharpoons} NO_2^+ + NO_3^-$$

The desiccating power of H_2SO_4 (Chapter 5) will transform NO_3^- into NO_2^+:

$$NO_3^- + 2H_2SO_4 \rightarrow NO_2^+ + 2HSO_4^- + H_2O_{\text{(solvated)}}$$

Similarly, in H_2SO_4 and HNO_3 (both anhydrous), we find

$$N_2O_4 \rightarrow NO^+ + NO_3^- \rightleftharpoons NO^+ + NO_2^+ + H_2O_{\text{(solvated)}}$$

A few oxyhalides of nitrogen are known; they have structures derived from the radical binary nitrogen oxides, NO and NO_2, or (alternately) from the cations NO^+ and NO_2^+. For example, the gases XNO (the *nitrosyl halides,* where X = F, Cl, Br) have structures of C_s symmetry as expected from overlap of an ao from X (or X^-) with the π^* HOMO of NO (or the π^* LUMO of NO^+).[74] The nitrosyl halides are very reactive oxidizing agents and, as the acid halides of HNO_2, vigorously solvolyze in water to produce *nitrous acid* (HONO). Similarly, the gases *nitryl fluoride* and *chloride* are conveniently viewed as combinations of X and NO_2 or of X^- and NO_2^+. Considering them as acid halides of HNO_3, the idea of heterolytic cleavage of the X—N bond most directly accounts for the solvolysis reaction:

$$XNO_2 + H_2O \rightarrow HX + HNO_3$$

---

[73] By diffusing N_2O_4 vapor onto a surface at 4°K, a species possibly with a structure $ONONO_2$ has been detected; W. G. Fateley, *et al., J. Chem. Phys.,* **31,** 204 (1959); H. A. Bent, *Inorg. Chem.,* **2,** 747 (1963).

[74] Both views are supported by chemical evidence: neat ClNO and BrNO are slightly dissociated into X_2 + NO, while in solution they often undergo reaction via X^- + NO^+.

Halogen nitrates, XNO_3, are known for X = F and Cl. Both are *highly* reactive and tend to react explosively with other materials, particularly organic compounds. While the common name for these compounds makes us think of them as resulting from formation of a bond between the halogen atom and the oxygen of a nitrate radical, you will find it useful to think of them as resulting from an association of NO_2 with XO (a hypohalite radical) or as an adduct of hypohalite anion (XO^-) with nitronium ion (NO_2^+). No matter which way one chooses to relate these structures to those of the simpler nitrogen oxides, a weak and reactive X—O bond is ultimately expected.

One nitrogen oxyhalide seems to stand out as being related to the phosphorus oxyhalides: *trifluoramine oxide,* F_3NO, has a C_{3v} structure. It happens to be a kinetically stable, though toxic, gas (b.p. $-88°$) that is not easily hydrolyzed and is probably the most unreactive member of the class of amine oxide compounds (R_3NO). It shares with hydroxylamine the unusual characteristic of being an oxide of a tetrahedral nitrogen. (Note that while the terminal oxygen of F_3NO possesses lone pairs, the nitrogen atom does not.)

A consistent structural thread runs through all the N/O compounds discussed thus far: with the exception of R_3NO and R_2NOH compounds, at least one oxygen is bound to nitrogen via multiple bonds. This feature is no surprise when we compare the strength of the N=O bond to that of the N—O single bond (~ 140 *vs.* ~ 40 kcal/mole). In comparison with C—O (~ 85 kcal/mole) and C=O (~ 180 kcal/mole) bonds, however, it appears that both N=O and N—O bonds are ~ 40 kcal/mole weaker.

To understand the "low" N/O bond energies, we must inspect more closely the compounds from which these bond energies are derived. As is suspected to be the case with hydrazine, the N—O lp/lp repulsions of *hydroxyl amine* could appreciably lower the bond energy in comparison to the CO bond of, say, methanol.

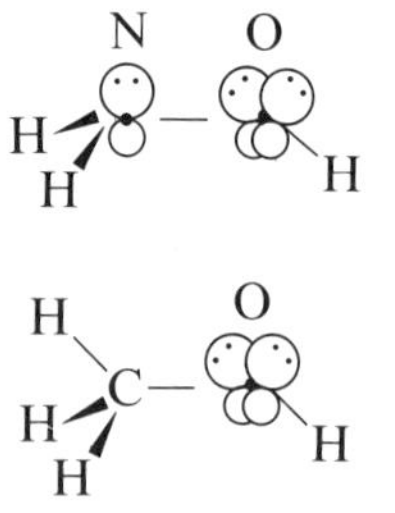

An analogous lp/lp versus bp/lp relation arises for XNO and the organo carbonyl function:

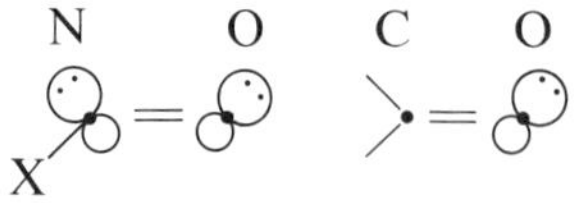

Exceptions to this generalized (relative) weakness of the NO bonds arise for cases in which the nitrogen lacks a lone pair. The unusual stability of F_3NO, for example, is partially attributed to an important contribution from the N=O resonance structure[75]:

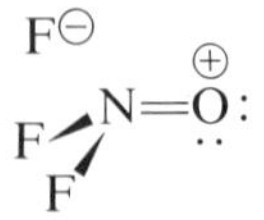

---

[75] The mo description of this resonance structure corresponds to delocalization of the degenerate oxygen $p\pi$ ao lone pairs (of e symmetry under C_{3v}) into the N—F σ^* mo's of e symmetry. (Refer to Chapter 4 and the analogous BH_3CO problems in Chapter 5.) This retro-bonding of O → N is possibly significant because the high electronegativity of F translates into reasonably low-lying σ^* LUMO's with fairly large amplitude at nitrogen.

The remaining cases exhibiting no nitrogen lone pair arise for derivatives of NO_2, *viz.*, HNO_3.

In another instance deserving of comment, exemplified by HNO_2, we find a nitrogen "singly" bound to a non-terminal oxygen. The simultaneous presence of a bond to a terminal oxygen does insure the presence of a nitrogen lone pair (N lp/O lp repulsions) but also suggests a greater strength of sigma bonds (expected for sp^2 nitrogen) and the possibility of N=O character for the formally "single" N—O bond:

$$\mathrm{H{-}O{-}N{=}O \longleftrightarrow H{-}O{=}N{-}O}$$

Quite obviously, the 39 kcal/mole figure for N—O underestimates the energy of such "single N—O" bonds (in Chapter 2 we noted that *cis-* and *trans-* forms are known, a fact confirming the partial double bond nature of the "single" bond).

The characteristic ⟩N=O and ⟩N=O structural features for the oxides and oxyhalides are, of course, also observed in the oxyacids formed by nitrogen. The two most commonly known oxyacids of nitrogen are *nitrous* (HNO_2) and *nitric* (HNO_3). The admittedly unusual view of HNO_3 that it is an adduct of OH^- and NO_2^+ is of some value in systematizing the nitrating power of HNO_3 as a liquid or in strongly acidic and/or ionizing media. Sulfuric acid (as noted in Chapter 5) apparently dehydrates HNO_3 to form NO_2^+ according to the reaction

$$HNO_3 \xrightarrow{H_2SO_4} NO_2^+ + HSO_4^- + H_2O_{(solvated)}$$

At this point it is well to recall that N_2O_5 is the anhydride of HNO_3 and is known to exist in the ionic form $NO_2^+ NO_3^-$.

Nitrous acid is considerably less stable than nitric acid, in the sense that acid solutions tend to disproportionate into NO and HNO_3 (note that there is no change in the number of OH bonds, but an increase in N=O π bonding—and an associated loss of effective N lp/O lp repulsions—is achieved in this reaction)

$$3\ \mathrm{H{-}O{-}N{=}O} \rightarrow 2\ :N{\equiv}O: + HO{-}N({=}O){=}O + H_2O$$

$3\pi_{NO} \rightarrow 4\pi_{NO}$

3 lp/lp → 0 lp/lp

Consistent with this behavior is the fact that HNO_2 is both a good oxidizing agent and a good reducing agent, forming NO in the first instance and NO_3^- in the second.

The structure of HNO_2 shown in the preceding equation is easily related to that of the nitrosyl halides if we take the point of view that it is an adduct between OH^- and NO^+. This means, of course, that the acidic proton is bound to oxygen, not to nitrogen, and the molecular symmetry is C_s. (While an N—H structure would eliminate N lp/O lp repulsion, such stabilization occurs at the expense of converting O—H to N—H.) In contrast with this structure, it is an interesting property of NO_2^- that it can act as a Lewis base toward metal ions via either N-coordination or O-coordination. Analogy is found, then, with organic nitro and nitrite compounds.[76]

---

[76] Were there any evidence for a C_{2v} structure for HNO_2, we could point out an analogy with phosphorous acid.

Just as phosphorus has a "hypo-ous" acid, so does nitrogen ($H_2N_2O_2$). Although little of the chemistry of this acid has been developed, it has been used as a reducing agent. Salts of the $N_2O_2^{2-}$ anion are known. In fact, syntheses of *hyponitrous acid* solutions usually entail *in situ* generation by acidification of a previously isolated salt. The hyponitrite anion has C_{2h} symmetry with the dominant resonance structure **37** (note the analogy with N_2F_2, since F and O^- are isoelectronic).

$$\left[\begin{array}{c} \ddot{\text{O}}: \\ \text{N}=\text{N} \\ :\ddot{\text{O}} \end{array} \right]^{2-} \quad 37$$

Presumably the acid has a similar *trans-* structure that is distinct from that of the isomer, *nitramide* (H_2N—NO_2), to be discussed in the next chapter.

It is instructive to pause for a moment and consider the relationship between $(NO)_2$ and $(NO)_2^{2-}$. The neutral dimer has a Lewis dot structure (**38**) *similar* to that of *cis*-butadiene, to which are added two electrons

$$\begin{array}{c} \dot{\text{N}}-\dot{\text{N}} \\ :\ddot{\text{O}} \quad \ddot{\text{O}}: \end{array} \quad 38$$

giving the dianion. The out-of-plane π mo's of $(NO)_2$ are simply

with the two π electron pairs of **38** occupying **39a** and **b**. (Note from **39a** and **b** that there is net NO but no net NN π bonding.) Orbital **39c** is the LUMO of $(NO)_2$, to which are added the two electrons to form the hyponitrite anion. In balance, you find net NN π bonding and loss of NO π bonding, as suggested by **37**.

Our survey of comparative nitrogen and phosphorus structures is not complete until we note that both elements form *trihalides,* while only phosphorus forms *pentahalides* (except with iodine). The nitrogen trihalides are rather unimportant as *common* laboratory materials because of their unpredictable nature. The least dangerous to handle (with regard to explosion hazards) is the gas NF_3. The infamous solid, "nitrogen triiodide," is actually formulated as $NI_3 \cdot NH_3$; it is safe to handle in aqueous solution but is shock sensitive as a solid. The black color of the crystals forecasts a rather unusual structure[77]: chains of NI_4 tetrahedra built by bridging iodine atoms. The presence of NH_3 molecules serves to link (and stabilize?) the chains via hydrogen bonding.

---

[77] Ammonia is present because the synthesis of NI_3 is by I_2 oxidation of aqueous ammonia. The structure is reported by H. Hartl, *Z. Anorg. Chem.*, **375**, 225 (1968).

STUDY QUESTIONS

57. Using resonance structures and bond order arguments, comment on the less hazardous nature of NaN_3 than of HN_3, with respect to decomposition to produce N_2. Does knowing that r_{NN} = 1.16 Å in N_3^- and 1.24 Å, 1.13 Å in HN_3 relate to your theoretical arguments?

    Which in each pair is more likely to detonate?

    RbN_3, AgN_3
    $Hg(N_3)_2$, $Ba(N_3)_2$
    KN_3, CuN_3

58. The problem of chemical conversion (nitrogen "fixation") of N_2 to NH_3 is a significant one that illustrates the importance of thermodynamic concepts. Consider the relative ease of the steps

$$N_2 \xrightarrow{H_2} N_2H_2 \xrightarrow{H_2} N_2H_4 \xrightarrow{H_2} 2NH_3$$

Which step is most costly (in terms of energy), and which of the species N_2, N_2H_2, or N_2H_4 therefore requires the greatest chemical activation (either by extreme temperature conditions or by catalytic combination) so as to weaken the NN link? The first three species exhibit a lone pair on each nitrogen. How does the lp/lp repulsion increase for $N_2 \rightarrow N_2H_2$ compare with that of $N_2H_2 \rightarrow N_2H_4$?

59. Which compound, NF_3 or NCl_3, is more unstable with respect to reversion to the elements? Are both unstable in this sense? Which presents the greater hazard in handling?

60. Were you to attempt synthesis of $N(OH)_3$, what decomposition pathways might you have to guard against? Consider only known, "stable" oxides and oxyacids of nitrogen as possible products, *viz.*,

$$N(OH)_3 \rightarrow HNO_2 + H_2O$$

61. How might you explain why the fumes generated by HNO_3 oxidation of Cu are colorless for a few millimeters above the solution surface, but brown thereafter?

62. With regard to the structure of $N_2O_{5(g)}$, would a linear N—O—N geometry favor a D_{2h} or D_{2d} (twisted NO_2 groups) structure?

63. Is nitric oxide thermodynamically stable with respect to nitrous oxide and nitrogen dioxide?

	NO	N_2O	NO_2
ΔH_f°	+22	+19	+8

64. Using the approach of Chapter 4, develop an mo diagram applicable to NO_2 and NF_2. Identify the HOMO in each case so as to predict whether the radicals are σ or π in nature.

65. Following the idea of association of NO_2 and NO radicals by HOMO overlap, extract from the text the N—N bond *dissociation* energies of N_2O_4, N_2O_3, and N_2O_2 and correlate their order with the variation in HOMO—HOMO overlaps.

66. Sketch the LUMO of NO_2^+ and account for the structure of XNO_2 in terms of attack by X^- upon NO_2^+ (what is there about the LUMO that renders its amplitude at N larger than at O?).

67. With reference to the parenthetical comment on p. 337 regarding the O—H, rather than N—H, bond in HNO_2, argue that an F—N, rather than F—O, bond is preferred in FNO_2.

68. Since there is no difference in the number of P—P bonds per phosphorus atom in the white and black allotropes, why is the former so much more (a) volatile and (b) reactive than the latter?

69. With regard to the NO σ bond strength as a function of nitrogen hybrid *s* character, how do the fluorines of F_3NO affect the NO bond strength relative to organic amine oxides? (Consult the section on proton affinities in Chapter 5 before answering.)

70. Estimate the enthalpy of

$$2P_2R_2 \rightarrow (PR)_4$$

in order to identify whether the cyclophosphine may be produced during attempts to synthesize compounds containing —P=P—.

71. While PF_5 is a colorless gas, PCl_5 is a white solid, often described as having a pale green-yellow tinge to it. PCl_3 is a colorless liquid (b.p. 76°). Account for the "green-yellow" tinge to PCl_5 and infer from this some ideas about the strengths of bonds in PCl_5.

72. Diphosphine, a pyrophoric volatile liquid, is photosensitive, yielding P_4 and phosphine

$$3P_2H_4 \xrightarrow{h\nu} 4PH_3 + \frac{1}{2}P_4$$

Argue that this reaction is approximately thermally "neutral" ($\Delta E \simeq 0$) but free-energy spontaneous.

73. Draw resonance structures for HPO_3 (obtained from thermal dehydration of H_3PO_4 at $\gtrsim 300°$) and the conjugate base PO_3^-. Comment on the anion structure of $Na_3P_3O_9$ in relation to that of monomeric BO_3^{3-} salts.

OXYGEN AND SULFUR

Bond Energies for Reference

O=O	119	S=S	103
O—O	34	S—S	63
O—H	111	S—H	88
O—C	86	S—C	(65)
O—F	51	S—F	(68)
O—Cl	49	S—Cl	65
		S—O	?
		S=O	125

The trend in relative second and third row element single bond energies is carried on by the members of Group VI: O—O, 34 kcal/mole, and S—S, 63 kcal/mole. On the other hand, the (formal) double bond energies for O=O and S=S are 119 kcal/mole and 103 kcal/mole, respectively, an order you should expect by now from the properties of the preceding groups. These data in themselves reflect the general instability of catenated oxygen compounds and the much wider occurrence of catenated sulfur compounds. For example, even the elemental forms mirror this different propensity for multiple bond formation by oxygen and catenation by sulfur. Oxygen exists in two allotropic forms, O_2 (dioxygen) and O_3 (ozone), but not as cyclic O_n. Both allotropes are characterized by O—O multiple bonding. Sulfur, on the other hand, is found naturally in the form of S_8 rings of a D_{4d} crown structure **(40)** but S_2 is found only at high temperatures in the gas phase.

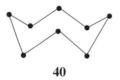

40

The simple *halides* of oxygen include the fluoride, OF_2, and chloride, OCl_2. The former is a highly toxic compound of marginal stability and high oxidizing power,

$$\frac{1}{2}O_{2(g)} + F_{2(g)} \rightarrow OF_{2(g)} \qquad \Delta E° = -4 \text{ kcal/mole}$$

$$OF_2 + 2H^+ + 4e^- = H_2O + 2F^- \qquad E° = 2.1 \text{ v}$$

while the chloride violently decomposes on heating:

$$\frac{1}{2}O_{2(g)} + Cl_{2(g)} \rightarrow OCl_{2(g)} \qquad \Delta E° = +20 \text{ kcal/mole}$$

The high reactivities of both compounds could be attributed to oxygen-halogen lone pair repulsions, to the stability of O_2 (where it is a product), and to the stability of oxygen and fluorine or chlorine bonds to other elements.

Consider the general reaction[78]

$$\begin{array}{c}X \qquad\quad X\\ \diagdown\quad\diagup\\ O{-}O\end{array} \rightarrow O_2 + X_2$$

for which the energy change is estimated to be $2E_{OX} - (E_{O=O} - E_{O-O}) - E_{XX} = (2E_{OX} - E_{XX}) - 85$ kcal/mole. The condition for thermodynamic stability is met for H_2O_2 ($\sim +30$ kcal/mole), not met for O_2F_2 (~ -20 kcal/mole),[79] and nearly met for ROOR (~ -4 kcal/mole). The difficulty with O_2F_2 clearly lies in the very low O—F bond energy (~ 50 kcal/mole), which in turn (refer to the earlier discussion of N—N) could arise from O lp/F lp repulsions. In reactions with other compounds, O_2F_2 is expected to be quite reactive as a fluorinating agent because of this weakness of the O—F bonds. For analogous reasons, the O—O bond in all peroxides is quite weak and may suffer cleavage. In fact, products from reactions of O_2F_2 contain or derive from F, O_2F, and OF. Mechanistic considerations are often of prime importance, for some reactions are believed to be radical-initiated and propagated by F, O_2F, and OF.

Based on the examples given thus far, it is uncertain which terminal group properties are desirable in stabilizing oxygen catenation. Certainly, the terminal groups must form strong bonds to oxygen, as do H and R, and as F probably would were it not for lone pairs. Based on Bent's idea of bond strength and s orbital distribution among hybrids (Chapter 3, p. 109), high terminal group electronegativity should help strengthen the O—O bond. You should also expect that oxygen lone pair involvement in π bonding with the substituent would be of primary value to strengthening the O—O bond.

Organic peroxides are known to be generally dangerous compounds. Particularly hazardous are the hydroperoxides (HO_2R), commonly found in the laboratory in bottles of ethers possessing α H atoms (Et_2O, C_4H_8O). It is an interesting observation, however, that fluorocarbon peroxides tend to be more stable than their hydrocarbon analogs. This observation has led to the preparation of a few trioxides [$(CF_3)_2O_3$, $(C_2F_5)_2O_3$, and $CF_3O_3C_2F_5$] and even a mixed dioxide-trioxide ($CF_3{-}O_3{-}CF_2{-}O_2{-}CF_3$). One must appreciate, however, that these still are not routine reagents; they most often are highly unstable (thermodynamically and kinetically) with respect to reactions with oxidizible compounds and sometimes to thermal and shock-initiated decomposition. They are best handled in small quantities at low temperatures.

In addition to "stabilization" by perfluoro groups as termini, terminal groups such as SO_3, FSO_2, and SF_5 seem to make possible the preparation of peroxy compounds. Specific examples are

$CF_3O_2{-}SF_5$

$HO_2{-}SF_5$

$HO_2{-}SO_3H$ (commonly called Caro's acid)

$FSO_2{-}O_2{-}SO_2F$ (commonly called peroxydisulfuryldifluoride)

$CF_3{-}O_2{-}SOF_2$

$SO_3{-}O_2{-}SO_3^{2-}$ (commonly called peroxydisulfate)

While the relation between CF_3, for example, and sulfur in the $+6$ oxidation state may seem obscure, it is reasonable to view the "stabilization" of the O—O link by

---

[78] Probably of greater importance to the stability of peroxides is the reaction $O_2X_2 \rightarrow OX_2 + \tfrac{1}{2}O_2$. Satisfy yourself that this reaction does not depend on E_{OX} and, therefore, only in a secondary way on the identity of X.

[79] The orange solid melts at $-164°C$ and reverts to the elements within hours at $-50°C$.

these groups as arising from their conjugation[80] with the oxygen lone pairs so as to confer some bonding character to the lone pair. Perhaps more important may be the fact that, with reduced lp/lp repulsion, better advantage may be taken of the O—O σ orbital overlap, which improves with electronegative termini.

$$-\ddot{\text{O}}-\ddot{\text{O}}-\text{C}\overset{F}{\underset{F}{\diagdown}}F \longleftrightarrow -\ddot{\text{O}}-\overset{\oplus}{\ddot{\text{O}}}=\text{C}\overset{F^{\ominus}}{\underset{F}{\diagdown}}F$$

$$-\ddot{\text{O}}-\ddot{\text{O}}-\text{SF}_5 \longleftrightarrow -\ddot{\text{O}}-\overset{\oplus}{\ddot{\text{O}}}=\overset{\ominus}{\text{SF}}_5$$

In contrast to the exciting (!) chemistry associated with oxygen catenation, sulfur has a rich, mild chemistry of catenation. As there is no change in the number of H—S and S—S bonds for the reaction

$$H_2S_{(g)} + \frac{n-1}{8} S_{8(g)} \rightarrow H_2S_{n(g)}$$

it should not surprise you that the hydrogen persulfides (*sulfanes*) have been well characterized.[81] Similarly, the *polysulfide anions* have been studied in aqueous solutions and the compounds from Na_2S through Na_2S_5 have been identified. Again, because there is no net change in the number of S—S bonds, reactions such as the following are nearly thermally neutral:

$$S_4^{2-}{}_{(aq)} + S_{(c)} \rightarrow S_5^{2-}{}_{(aq)}$$

In addition to the anionic polysulfides, recent studies have shown the existence of *cationic polysulfides*. Working with highly acidic, non-reducing solvents (this use of HF and HSO_3F as solvents was forecast in Chapter 5), Gillespie[82] has succeeded in preparing and characterizing the species S_8^{2+}, S_{16}^{2+}, and S_4^{2+}.

Two-electron oxidation of S_8 by the peroxide $S_2O_6F_2$ apparently leads to some degree of bond formation between S atoms opposite each other in the crown ring (recall the allotropes of P, with which S^+ is isoelectronic). Because of ring strain, the formation of the "bicyclo" structure necessitates the conversion of the crown structure (**40**) to the *endo-exo* form (**41**) observed for S_8^{2+}.

41

Incomplete oxidation of S_8 apparently leads to the formation of paramagnetic S_8^+, which tends to dimerize to S_{16}^{2+}. The structure of this dication may be simply that of a dimer of two S_8 crowns connected by an S—S bond, $(S_8)_2^{2+}$. Strong oxidation of

---

[80] Compare the CF_3—O— unit with F_3NO and see ref. 75. Similar arguments can be made for SF_5, by considering the SF σ* mo's *cis*- to the oxygen and perhaps by virtue of the presence of S valence *d* ao's.

[81] In fact, while this reaction is thermally neutral, entropy factors favor $H_2S + S_8$. The real reason the sulfanes have been well characterized lies in the fact that no suitable *low* energy pathway connecting H_2S_n to H_2S and S_8 is available: T. K. Wiewioroski, *Endeavor*, **29**, 9 (1970); E. Muller and J. B. Hyne, *Can. J. Chem.*, **46**, 2341 (1968).

[82] R. J. Gillespie and J. Passmore, *Accts. Chem. Res.*, **4**, 413 (1971).

S_8^{2+} can lead to isolation of the S_4^{2+} cation, which has a planar structure and is diamagnetic. Interestingly, this molecule can be described in terms of S=S bonding. The π out-of-plane mo's of this species are, of course, those characteristic of the cyclic four-atom case (Chapter 4), with a net, single π-bonding electron pair.

$$\begin{bmatrix} \ddot{\text{S}}=\ddot{\text{S}} \\ | \quad | \\ :\ddot{\text{S}}-\ddot{\text{S}}: \end{bmatrix}^{2+} \longleftrightarrow \begin{bmatrix} :\ddot{\text{S}}-\ddot{\text{S}} \\ | \quad || \\ :\ddot{\text{S}}-\ddot{\text{S}}: \end{bmatrix}^{2+} , \text{etc.}$$

In summary, one of the striking features of the chemistry of elemental sulfur is its propensity for forming cyclic species.[83] Whereas the S_8 ring can be opened on reduction of sulfur to form a variety of chain-like polyanions, oxidation of S_8 causes oligomerization of the rings in the formation of S_{16}^{2+}, introduces cross-ring bonding in the formation of S_8^{2+}, and results in symmetrical cleavage of the ring to form the S_4^{2+} cation.

The examples of sulfur catenation go far beyond those mentioned thus far. Fluorine is known to form only the simpler catenated species such as the gas S_2F_2 and the liquid *disulfurdecafluoride,* the latter featuring sulfur in the +5 oxidation state. The structure[84] of S_2F_{10} is, as expected, based on the octahedron, with the sixth position about each sulfur being occupied by the other sulfur atom. The overall

$$\text{F}_5\text{S}-\text{SF}_5 \quad \begin{bmatrix} \text{m.p.} & -55° \\ \text{b.p.} & 29° \end{bmatrix}$$

symmetry is C_{4v} with "gauche" S—F bonds. While the mean S—F distance is shorter by ~0.15 Å than that of SF_6 (1.58 Å), the S—S distance of 2.21 Å is unexpectedly *longer* than that of lower valent species such as S_8 (2.04 Å) or ClSSCl (1.93 Å) and more like those of S_3^{2-} (2.20 Å) and $S_2O_5^{2-}$ (2.21 Å) (this latter compound may be viewed as an adduct of O_3S^{2-} and SO_2 and so may be considered as having only a moderately strong S—S bond). The SF_5 dimer is not, however, reactive in the general sense (much like SF_6); but at elevated temperatures it is a vigorous oxidizing agent, even causing degradative fluorination of substrates—presumably by radical reactions of SF_5 and/or F.

$$\tfrac{1}{2} S_2F_{10} \xrightarrow{\Delta} SF_5 \rightarrow SF_4 + F$$

One product of this reaction, *sulfur tetrafluoride,* is a synthetically useful fluorinating agent that will be discussed at some length in Chapter 8.

An intriguing situation arises for *disulfurdifluoride.* The more stable isomer of this compound has the C_s structure[85] **42** while the less stable isomer, and probably the one most of us would have "intuitively" expected to be the more stable, has the normal (C_2) peroxide structure **43**

$$\text{F}_2\text{S}=\ddot{\text{S}} \quad \textbf{42}$$

$$\text{F}-\ddot{\text{S}}-\ddot{\text{S}}-\text{F} \quad \textbf{43}$$

---

[83] Other all-sulfur ring structures are known (S_{6-10} and S_{12}); all are unstable at room temperature (the decomposition reactions are favored by entropy and internal ring strain energy) and can be retained only at low temperatures.
[84] R. B. Harvey and S. H. Bauer, *J. Amer. Chem. Soc.,* **75**, 2840 (1953).
[85] R. D. Brown, *et al., Australian J. Chem.,* **18**, 627 (1965).

The thermodynamics of the isomerization reaction

$$XSSX \rightarrow X_2SS$$

do not depend on the magnitude of E_{XS} (if we assume E_{XS} to be the same in both structures). That is,[86]

$$\Delta E = (E_{SS}^{XSSX} - E_{SS}^{X_2SS}) = -40$$

Thus, it seems reasonably consistent for nature to prefer the "sulfoxide" structure **42**. Why, then, H_2S_2 and Cl_2S_2 have never been reported to exhibit the "sulfoxide" structure is unexplained. Perhaps the mechanisms involved in the synthesis of these two materials dictate the formation of the higher energy isomer only.

$$Na_2S_n + 2HCl \xrightarrow{H_2O} 2NaCl + H_2S_{n(l)} \xrightarrow{\Delta} H_2S_2$$

$$S_{8(l)} + Cl_2 \xrightarrow{120°} S_2Cl_2$$

For H_2S_2, Cl_2S_2, and FSSF we are left with the interesting question: Why, at room temperature, is the rate of isomerization to the preferred isomer so slow?[87]

In the previous discussions of peroxide compounds we learned that complex oxysulfur(VI) groups apparently act as stabilizers of the peroxide linkage. A similar characteristic is found for polysulfur linkages. Thus, the *sulfanemonosulfuric* acids and their salts have been prepared; however, HS_nSO_3H and salts of $S_nSO_3^{2-}$ are not very stable at room temperature. The simplest member of this series is well known as *thiosulfate;* while its conjugate acid is not particularly stable, you are well aware that sodium thiosulfate is available at any store dealing in photographic supplies.

A simple extension of the ideas in the previous paragraph should lead you to expect the existence of *polythionate* anions, and indeed many ions of the type $O_3S-S_n-SO_3^{2-}$ are known in alkaline solutions (the acids are unstable). The best characterized members of this series are those with $n = 0$ through 4 (dithionate, trithionate, etc.); the simplest, dithionate, is best known and least reactive of the series.

At least formally related to the dithionate ion is the *dithionite* ion, $S_2O_4^{2-}$ (the conjugate base of dithionous acid, $H_2S_2O_4$). While aqueous solutions of the anion are not very stable for extended periods,[88] when fresh they make good scavengers for O_2 impurities in gases (Fieser's reagent is a solution of dithionite and 2-anthraquinone). The ease with which $S_2O_4^{2-}$ can be oxidized is probably related to the fact that the ion dissociates into SO_2^- radicals (solutions of $S_2O_4^{2-}$ are paramagnetic). The C_{2v} structure[89] **(44)** of the dithionite ion is intriguing, especially since the O—S—O planes deviate from parallelity by only 30°, compared to 70° expected for tetrahedral sulfur atoms. (Study question 81 leads you through an accounting of this structure.)

$$\left[\begin{array}{c} S\text{---}S \\ /\backslash\ /\backslash \\ O\ O\ O\ O \end{array}\right]^{2-} \quad \mathbf{44}$$

---

[86] The use of 40 kcal/mole from $(E_{S=S} - E_{S-S})$ is an approximation, since the S=S in X_2SS is somewhat different from that in S_2

$$X\diagdown\overset{..}{\underset{X}{S}}=S: \longleftrightarrow X\diagdown\overset{\oplus}{\underset{X}{S}}-\overset{\ominus}{\underset{..}{S}}:$$

Both the S—S σ and π bonds should be weaker than the nominal values used to arrive at 40 kcal/mole, making it difficult to estimate by how much $(E_{S=S} - E_{S-S})$ may be in error.

[87] That mechanistic factors may control the reaction product is further suggested by the facts that fluorination of S_8 with AgF produces a mixture of both isomers of S_2F_2 and that alkali fluorides catalyze the isomerization of FSSF to SSF_2: F. Seel and R. Budenz, *Chimia*, **17**, 355 (1963); *ibid.*, **22**, 79 (1968).

[88] The products are $S_2O_3^{2-}$ and HSO_3^-.

[89] J. D. Dunitz, *Acta Cryst.*, **9**, 579 (1950).

The final example of catenated sulfur is a hybrid of the dithionite and dithionate ions. *Disulfite* ion, $S_2O_5^{2-}$, plays an important role in the aqueous chemistry of sulfurous acid, arising as it does in concentrated solutions of bisulfite by elimination of H_2O from two molecules of HSO_3^-. Similarly, thermal dehydration of bisulfite salts produces disulfite salts. A convenient view of this derives from the known donor/acceptor reaction

$$SO_2 + SO_3^{2-} \rightarrow S_2O_5^{2-}$$

Easily the most important acid of sulfur is H_2SO_4, the solvent properties of which were explored in Chapter 5. The existence of sulfurous acid, on the other hand, has never been proved; rather, SO_2, which is highly soluble in water, simply forms a hydrated species much as CO_2 does. The bisulfite ion, however, is well known and exhibits an interesting tautomerism like H_3PO_3. Two species are believed to be in equilibrium in dilute aqueous media:

$$\left[\overset{..}{\underset{O}{\overset{O}{S}}} \diagdown_{O}\diagup^H \right]^- \rightleftharpoons \left[\overset{H}{\underset{O}{\overset{|}{\overset{S}{\diagup\diagdown}}}}_{O}{}_{O} \right]^-$$

As mentioned in the preceding paragraph, the formation of detectable amounts of disulfite ion results from an equilibrium ($pK \sim 1$) requiring $2HSO_3^-$, but the mechanism of the reaction seems not to have been determined.

Judging from examples given already, the S—O bond is an important feature of sulfur chemistry. The two best known simple oxides are SO_2 and SO_3. Both molecules are believed to involve strong $p\pi$—$p\pi$ multiple bonding between sulfur and oxygen, which is necessary to avoid valency unsaturation at sulfur; the persistence of strong, short S—O bonds in *thionyl* ($>\overset{..}{S}=O$) and *sulfuryl* ($>SO_2$) compounds is generally taken to implicate some degree of $d\pi$—$p\pi$ character in these bonds, too. In fact, the remarkably small variation in S—O distance for a wide variety of SO compounds seems to imply a characteristic constant multiple bond nature for those bonds.

	r_{SO} (Å)
S_2O	1.47
SO	1.49
X_2SO	1.45
SO_2	1.43
SO_3	1.43
$S_2O_4^{2-}$	1.50
$S_2O_5^{2-}$	1.50 (SO_2); 1.44 (SO_3)
$S_2O_3^{2-}$	1.47
$S_2O_6^{2-}$	1.43
SO_4^{2-}	1.44

While the oxyanions in this list have been mentioned previously, a few comments are in order for the simple binary compounds. *Sulfur monoxide* is known only as an intermediate in reactions and is highly reactive. *Disulfurmonoxide* (of C_s symmetry, compare to O_3) is also quite unstable to disproportionation into the dioxide and polymeric materials. *Sulfur dioxide* is, of course, well known as a public nuisance, but SO_2 has some redeeming chemical properties; some of these were surveyed in the last chapter (particularly its use as a solvent).

Sulfur trioxide is an extremely strong Lewis acid, and this aspect of its chemistry was discussed in Chapter 5. There you learned that SO_3 rarely acts as a Lewis base to any significant degree; an exception to this statement is its self-association in the solid state. Three crystalline forms of SO_3 are known. Rapid condensation of the gas at low temperatures ($< -80°C$) leads to cyclic trimers (compare **25**) with a melting point of about 16°C (γ-form). The more stable β-form consists of helical chains of —O—S—O—S backbones. The form with the highest melting point (α,60°C) has an unknown structure, but, like the β-form, has an asbestos-like appearance and so may consist of cross-linked chains. (The γ-form melts to a liquid of monomers and trimers and boils [45°C] at temperatures below the melting point of the α-form.)

For the simple *halides* formed by sulfur, we find that the red liquids SCl_2 (b.p. 59°) and SCl_4 (m.p. $-31°$) and the gases SF_4 (b.p. $-40°$) and SF_6 (sublimes $-64°$) are known (S_2F_{10} and S_nCl_2 were mentioned earlier). The initially surprising non-existence of SF_2 is often rationalized in thermodynamic terms by a highly favorable energy for the disproportionation reaction,[90]

$$3SF_{2(g)} \rightarrow \frac{1}{4}S_{8(g)} + SF_{6(g)} \qquad \Delta E \approx -130 \text{ kcal/mole}$$

even though SF_2 should be quite stable with respect to the elements:

$$\frac{1}{8}S_{8(g)} + F_{2(g)} \rightarrow SF_{2(g)} \qquad \Delta E \approx -35 \text{ kcal/mole}$$

It is interesting to note, after this, that SF_4 does not disproportionate into S and SF_6, though the reaction should be thermodynamically spontaneous. Similarly, SCl_4 does not disproportionate[91] into S and SCl_6, but then neither does SCl_2. The failure of both SF_4 and SCl_4 to disproportionate must be kinetic in origin, although for chlorine there could be a general thermodynamic barrier to formation of SCl_6. It is supposed that steric congestion about S in SCl_6 may be too severe. All of this should still leave some questions in your mind. How does SF_2 disproportionate to SF_6 without also producing SF_4? Why does SCl_2 not decompose to S_8 and SCl_4? An interesting complication arises from the observation that SCl_2 does slowly decompose to S_2Cl_2:

$$2SCl_2 \rightarrow S_2Cl_2 + Cl_2$$
[red liquid]   [yellow liquid; b.p. 137°]

At this point we can only conclude that some interesting mechanistic constraints are operating.

The remarkable property of SF_6 and S_2F_{10} is their chemical inertness. Neither compound experiences acid or base hydrolysis near room temperature, even though the reactions are expected to be strongly spontaneous from the thermodynamic point of view. While S_2F_{10} appears to be highly toxic, SF_6 gas is so safe to handle and so stable that it is used as an insulator in high-voltage electrical equipment. One facile reaction of SF_6 is electron transfer; metallic sodium, in glycolic solvents, effects the reduction of SF_6 to sodium fluoride and sodium sulfide. Similarly, S_2F_{10} oxidizes I^- to I_2. In the case of SF_6^-, occupation of the $s\sigma^*$ mo (see Figure 4-20) leads to labilization of the S—F bonds.

The *chlorosulfanes* have found use as "solvents" for sulfur in the rubber industry:

$$\frac{y-x}{8}S_8 + S_xCl_2 \rightarrow S_yCl_2$$

---

[90] Note that this reaction is exothermic, no matter what the value of E_{S-F}, unless that value should be greatly different for SF_2 and SF_6.

[91] In this regard, SCl_4 is quite a bit less stable than its analog, SF_4. The tetrachloride is synthesized by the low temperature ($-80°C$) chlorination of SCl_2 but decomposes to the lower chlorides at $\sim -31°C$.

If you wish to prepare a specific chlorosulfane, a good method to use is

$$H_2S_x + 2SCl_2 \rightarrow 2HCl + S_{x+2}Cl_2$$

Generally speaking, the chlorosulfanes are readily hydrolyzed and must be kept free from moisture.

Of greater chemical utility is SF_4. While the hydrolytic stability of SF_4 is low, as expected, it is a very useful fluorinating agent. Generally speaking, the compound can be used to convert metal and non-metal oxide compounds into the corresponding fluorides without destroying other groups or oxidizing the metals and non-metals. Particularly useful is the fluorination of keto ($>C=O$) and phosphoryl ($>P=O$) groups. Special attention is given to these properties in Chapter 8.

As we found to be the case for phosphorus, sulfur is known to form mixed oxyhalide compounds. The simple molecular compounds are the *thionyl halides* [(F, Cl, or Br)$_2$SO, with C_s symmetry] and the *sulfuryl halides* [(F or Cl)$_2$SO$_2$, with C_{2v} symmetry]. In striking analogy with other non-metal halides and oxyhalides, the thionyl and sulfuryl fluorides are relatively stable to hydrolysis, whereas thionyl chloride and bromide, and sulfuryl chloride, are vigorously solvolyzed by water and other hydroxylic solvents. As mentioned earlier, the terminal SO bonds in these compounds are remarkably similar to SO bonds in other compounds and are believed to possess considerable multiple bond character. All are stable to decomposition at normal temperatures.

A structural unit common to many more-or-less stable oxyhalide compounds is that of the fluorosulfate group (FSO_3^-). We have noted the occurrence of this unit in peroxydisulfuryldifluoride, but it is also found in $S_2O_5F_2$ (**45**), SO_3F_2 (**46**), and $S_3O_8F_2$ (**47**).

You saw in Chapter 5, and in the previous discussion of S_n^{x+}, that *fluorosulfuric acid* is very useful as a solvent medium. When one needs a very weakly basic anion in a salt, FSO_3^- is a good choice. Interestingly, *fluorosulfurous acid* does not have a discrete existence, but alkali salts are known; the fluorosulfite salts represent an example of the Lewis acid property of SO_2 because they are prepared by reaction of a fluoride salt with SO_2.

A few chlorine analogs of the fluorosulfates have been characterized. Examples are $S_2O_5Cl_2$, $S_3O_8Cl_2$, and HSO_3Cl. Their structures are analogous to those of the fluorides but, as expected from the properties of Cl_2SO and Cl_2SO_2, they are generally more reactive (hydrolysis is quite vigorous).

The SF_5 group also is a common feature of sulfur-fluorine chemistry. In addition to the compounds SF_6 and S_2F_{10}, we find SF_5Cl, SF_5NF_2, $(SF_5)_2O$, $(SF_5O)_2$, SF_5OF, and SF_5R (where R = halogenated alkanes, olefins, and imines).

A derivative chemistry of SF_4 is beginning to develop, with the existence of SOF_4 (C_{2v} symmetry) and $SF_4(OSF_5)_2$ having been established. It has been confirmed, by the formation of $R_4N^+SF_5^-$ and $SF_3^+BF_4^-$, that SF_4 has acid and base properties. In this connection, note that SOF_4 may be thought of as an adduct of SF_4 and O, much like the amine and phosphine oxides from Group V.

To close this section on Group VI compounds, we might point out a further analogy with the chemistries of silicon and, particularly, phosphorus. *Esters* of sulfurous and sulfuric acids are readily achieved by solvolysis of thionyl chloride and sulfuryl chloride:

$$X_2SO_n + 2ROH \rightarrow (RO)_2SO_n + 2HCl \qquad \begin{array}{l} n = 1 \text{ thionyl} \\ n = 2 \text{ sulfuryl} \end{array}$$

In the case of sulfurous esters, the tautomerization reaction (seen before for HSO_3^-) to produce a second terminal SO bond is found

$$(RO)_2SO \rightarrow (RO)SO_2(R)$$

(Trifluoromethyl)sulfuric acid, with properties much like those of HSO_3F, possesses such a structure.

An unfortunate situation in the nomenclature of sulfur-oxygen compounds has arisen over the years, and it seems best to pause for a moment to identify the different usages. The $>\!\ddot{S}\!-\!O$ group is variously referred to as *thionyl, sulfoxide,* and *sulfinyl,* while $>\!SO_2$ goes by the names *sulfuryl, sulfone,* and *sulfonyl.* All are so commonly used that they need to be committed to memory.

STUDY QUESTIONS

74. Compare the approximate energies for

$$O_2F_2 \rightarrow O_2 + F_2$$

$$S_2F_2 \rightarrow \frac{1}{4}S_8 + F_2$$

Is a distinction to be made in the care with which the difluorides must be handled? Which of the two difluorides is potentially more useful for *selective* fluorination reactions?

75. Generalize in equation form the energy of

$$XO_nX \rightarrow XO_2X + \frac{n-2}{2}O_2$$

What is the prognosis for preparation of polyoxide materials?

76. Would decomposition of RO_2H more likely produce $(RH + O_2)$ or $(ROH + \frac{1}{2}O_2)$?

77. Comment on the order of thermochemical bond energies $E_{O-O} < E_{O-F}$ with respect to lp/lp repulsions, σ bond strengths, and atom relaxation energies (for the latter you may wish to consult the section on bond energies in Chapter 3).

78. Why should S_8^{2+} prefer the bicyclo(3.3) rather than bicyclo(1.5) or bicyclo(2.4) structures?

79. a) Use Pauling's defined relation between bond energy and atom electronegativity to estimate E_{S-Br} and E_{S-I}.

$$\Delta \chi = \frac{1}{23}[E_{S-X} - (E_{S-S} \cdot E_{X-X})^{1/2}]$$

From these, estimate the enthalpy change for

$$\frac{1}{8}S_{8(g)} + X_{2(g)} \rightarrow SX_{2(g)}$$

Finally, might condensation energies and entropy changes be important enough to affect your decision to attempt synthesis of SBr_2 or SI_2?

b) Were you to attempt synthesis of SBr_2 or SI_2, would there be cause to anticipate product disproportionation to S_8 and SX_6?

80. Can the stabilities of SF_6 and S_2F_{10} toward hydrolysis be thermodynamic in origin? In answering, propose as hydrolysis products reasonable oxyacids, exclusive of redox reactions with H_2O.

81. Should SO_2^- be a σ or π radical (what does the HOMO, Chapter 5, look like)? Does your prediction help to explain the structural features of dithionite [long S—S bond of 2.4 Å, nearly parallel SO_2 planes, and eclipsed (C_{2v}) geometry]? Is there an analogy here with $(NO)_2$ and N_2O_4?

82. From donor/acceptor concepts, predict possible structures for $S_2O_5^{2-}$. Identify the donor and acceptor atoms in each case.

83. By analogy with the hydrolysis of SO_3, propose a synthesis method for HS_nSO_3H.

84. Propose a mechanism for the incorporation

$$S_2O_4^{2-} + {}^*SO_2 \rightleftarrows O_2S^*SO_2^{2-} + SO_2$$

of labeled sulfur dioxide into dithionite.

HALOGENS

Bond Energies for Reference

F—F	38	Cl—Cl	58	Br—Br	46	I—I	36
F—H	136	Cl—H	103	Br—H	88	I—H	71
F—C	117	Cl—C	78	Br—C	65	I—C	51
F—O	51	Cl—O	44	Br—O	(48)	I—O	(48)
F—Cl	60	Cl—F	60	Br—F	60	I—F	58
				Br—Cl	52	I—Cl	50
						I—Br	42

The question of the X=X double bond as a primary functional unit does not arise for the elements themselves, since single bond formation completes the octet of each atom in the pair. (From the molecular orbital description of the diatomics, recall that there are predicted to be two fully occupied π bonding orbitals and two fully occupied π^* anti-bonding orbitals—a situation giving no net double bonding; the refinement of including d orbitals could conceivably somewhat stabilize the π and π^* orbitals and presumably raise the bond energy slightly.) The order of single bond energies of F_2 and Cl_2 is as expected from the trend started by the earlier groups: F—F, 38 kcal/mole and Cl—Cl, 58 kcal/mole. This factor in itself should tend to make compounds of the type MF_x more stable with respect to reversion to the elements than those of the type MCl_x with respect to reversion to their elements. While this is what is generally found, the relative bond energies of the halogens are only partly responsible for the observed greater stabilities of the covalent fluorides.

A second factor of considerable importance is that, quite generally (see Figure 6-1), the M—F bond energies are greater than the M—Cl bond energies. This observation is believed to be associated with a variety of factors such as (1) smaller size for F when steric congestion may be a factor, (2) the higher F electronegativity (Z^*) when high oxidation states are considered (see p. 273), (3) better σ ao overlap when E of EX_n is a second period element, (4) the occurrence of E=F bonding for third period elements, and (5) the advantage of high Z^* when "electron rich" species like XeX_n are considered. This is the thermodynamic analysis of the existence of compounds such as NF_3, OF_2, and XeF_2 on the one hand, but the "non-existence" or instability of NCl_3 (easily detonated), OCl_2 (explodes easily at room temperature), and $XeCl_2$ on the other.

Since fluorine, and sometimes chlorine, promotes formation of compounds with the central element in several oxidation states, it is pertinent to inquire as to the relative stabilities of the higher and lower halides in such cases:

$$MX_{x(g)} \rightarrow MX_{y(g)} + \frac{x-y}{2} X_{2(g)}$$

The energy change for such a reaction can be analyzed in terms of the different M—X bond energies and the X—X bond energy:

$$\Delta E = \{xE_{MX}(MX_x) - yE_{MX}(MX_y)\} - \frac{x-y}{2} E_{XX}$$

As mentioned earlier in the discussions (p. 269) on variation in bond energies with central atom hybridization or valency, E_{MX} in the higher-valent compound will often be less than E_{MX} in the lower-valent compound; but the term in brackets above will still be positive (exceptions to this generalization were mentioned at the end of Chapter 1 in the discussion of the "inert pair" effect). That this first term in the expression for ΔE is positive and larger than the second term is the thermodynamic requirement for the stability of the higher-valent compound. Furthermore, in comparing F with Cl, it is generally true that E_{MX} for either valency of M is greater for fluorine than for chlorine, which means that the first term in brackets will be a larger positive number when X = F than when X = Cl. This tends to make covalent fluorides more stable toward loss of halogen than chlorides. Finally, the second term is smaller for fluorine than for chlorine, a situation that amplifies the difference between the thermodynamic stabilities of fluorides and chlorides toward the decomposition reaction.

The heavier halogens are known to form several well-defined *catenated cations*[92] and *anions*. The element iodine, as perhaps expected, has yielded the greatest number of cationic species (I_2^+, I_3^+, and I_5^+); but even bromine is known to form Br_2^+ and Br_3^+, while the cation Cl_3^+ has been identified for chlorine. The iodine species I_2^+ (blue) and I_3^+ (red-brown) have been observed in fluorosulfuric acid (HSO_3F) upon treatment of I_2 with the oxidant peroxydisulfuryldifluoride:

$$2I_2 + S_2O_6F_2 \rightarrow 2I_2^+ + 2SO_3F^-$$
$$3I_2 + S_2O_6F_2 \rightarrow 2I_3^+ + 2SO_3F^-$$

The cation I_3^+ has a bent triatomic structure, as expected from the VSEPR concept. The bromine cations Br_2^+ and Br_3^+ have also been prepared by oxidation with $F_2S_2O_6$ in the so-called "superacid" $HSO_3F/SbF_5/3SO_2$ as solvent. The structure of the triatomic ion is presumed to be angular like that of I_3^+.

Thus far, attempts to prepare Cl_3^+ by oxidation of Cl_2 in fluorosulfuric acid have been unsuccessful; rather unusual is its synthesis by the displacement of F^- from ClF by Cl_2 (at $-78°C$) in the presence of the strong fluoride ion acceptor AsF_5:

$$Cl_2 + ClF + AsF_5 \rightarrow [Cl_3][AsF_6]^{93}$$

---

[92] R. J. Gillespie, and J. Passmore, *Adv. Inorg. Chem. Radiochem.*, **17**, 49 (1975).
[93] The salt reverts to reactants at room temperature; R. J. Gillespie and M. J. Morton, *Inorg. Chem.*, **9**, 811 (1970).

With respect to catenated anionic halides, fluorine (the light element of the group) is not known to form any, chlorine appears to form small amounts of Cl_3^- in saturated aqueous solutions of Cl_2/Cl^-, and iodine is known[94] to form several species: I_3^- is the best known (recall from analytical chemistry that I_2 is more highly soluble in KI solutions than in H_2O alone, and that such solutions are useful as a way to stabilize I_2 solutions against air oxidation), and I_n^- anions with $n = 5, 7, 8,$ and 9 have also been reported. Whereas I_3^- has $D_{\infty h}$ symmetry, the higher *polyiodides* are best thought of as an iodide or triiodide anion weakly bound to one or more I_2 molecules.

2.91 Å

I_3^- ($D_{\infty h}$)

2.81 Å
3.17 Å

I_5^- (C_{2v})

I_7^-

2.83 Å
2.86 Å
3.42 Å 3.00 Å

I_8^{2-} (C_{2h})

Of all these species, only I_5^- seems *not* to involve I_3^- adducts with I_2. Furthermore, $(Et_4N)(I_7)$ is probably best viewed as $(Et_4N)(I_3) \cdot 2I_2$, where both cation and I_3^- sub-lattices are fcc (the NaCl structure) with $8I_2$ per unit cell packed between the ions.[95] With the I—I distance in I_2 of 2.66 Å as a reference, the complexation of I_2 in these species reflects the expected lengthening of the I_2 bond from overlap of the I_2 $p\sigma^*$ LUMO with a donor lone pair (review, if necessary, the discussion of I_2 as an acid in Chapter 5). Another interesting point about I_5^- in particular is that its structure is not the D_{4h} geometry found for ICl_4^-. To what extent this is due to lattice forces and to what extent to anticipated intra-anion steric congestion in a D_{4h} structure is uncertain.

The heavier halides are known to form highly reactive compounds with fluorine. Chlorine and bromine exhibit the first three positive oxidation states with fluorine by forming EF, EF_3, and EF_5, whereas the more reactive element iodine apparently does not form a stable monofluoride but readily forms IF_3, IF_5, and IF_7. It is most interesting that the thermal stabilities of the *halogen fluorides* increase with the number of fluorines, a fact that may have both kinetic and thermodynamic roots (refer to the generalizations on page 349). While these substances are certainly reactive enough to be used as fluorinating agents, they are also such strong oxidizers that they tend to degradatively fluorinate most other compounds, in an often violent way. Their use as fluorinating agents for inorganic materials is often possible but, unless diluted with an inert carrier or solvent, they are dangerous to handle in the presence of organic compounds.

---

[94] L. E. Topol, *Inorg. Chem.*, **10**, 736 (1971); A. E. Wells, "Structural Inorganic Chemistry," 4th Edition, Oxford, London (1975).
[95] E. E. Havinga and E. H. Wiebenga, *Acta Cryst.*, **11**, 733 (1958).

An important, characteristic feature of these interhalogen fluorides, brought out in the survey section of Chapter 5, is their simple Lewis acid action toward the fluoride ion, which is difficult to oxidize; examples are ClF_2^-, ClF_4^-, BrF_4^-, BrF_6^-, and IF_6^- (these ions are all isoelectronic with the inert gas compounds XeF_n). Alternately, when in contact with compounds like BF_3, AsF_5, and SbF_5 (good F^- acceptors), the halogen fluorides readily form cationic species by loss of F^-. Examples of this behavior are ClF_2^+, ClF_4^+, BrF_2^+, BrF_4^+, IF_4^+, and IF_6^+. An interesting example is ClF, which, much like the preparation of Cl_3^+ given above, loses F^-; but the Cl^+ so generated is readily coordinated by a ClF molecule to form Cl_2F^+. Knowing that such fluoride transfer reactions are a characteristic feature of the chemistries of these materials, it comes as no surprise that most of them exhibit conductivity to some degree by virtue of equilibria such as

$$IF_5 \rightleftarrows \frac{1}{2} IF_4^+ + \frac{1}{2} IF_6^-$$

Even though we implied above that F^- transfer leads to ion formation, at least in the solid state the situation is far from ideal. With the "salt" $ClF_2^+SbF_6^-$, for example, the cations and anions pack in such a way as to leave the Cl atom in a distorted square planar environment of fluorides. That is, two of the fluorines from a nearby SbF_6^- appear[96] to be shared with the Cl.

As with the catenated halogen ions and the interhalogen compounds, the *oxides* of the halogens are highly reactive, often violently temperamental materials of limited use to anyone other than the specialist skilled in handling them. The known oxides of fluorine, OF_2 and O_2F_2, have been mentioned in the section on oxygen compounds. Of all the chlorine oxides (OCl_2, ClO_2, Cl_2O_3, Cl_2O_4, Cl_2O_6, and Cl_2O_7), perhaps it is not unexpected that the heptoxide, with the chlorines of highest oxidation state and least vulnerable to nucleophilic attack, is the most stable (note the similarity with SF_6 and F_3NO "stabilities"). The structure[97] of this oily compound (b.p. 82°) is similar to that of an ether [$(ClO_3)_2O$, where the Cl—O—Cl unit is not linear] and is reminiscent of the isoelectronic species $S_2O_7^{2-}$. An interesting relationship exists between the structure of ClO_2 (a yellow gas, b.p. 10°) and Cl_2O_4 (a liquid, b.p. 45°). Reasoning by analogy, we might suppose that *dichlorine tetroxide* is simply the dimer of chlorine dioxide (recall nitrogen dioxide and, in particular, SO_2^-), but no such dimer is known; rather, the structure of the tetroxide can be described as that of *chlorine perchlorate* (Cl—$OClO_3$). This appears to be a not uncommon arrangement, for both $FClO_4$ and $BrClO_4$ are known to have similar structures.

The halides form quite a few useful oxyacids. With the recent preparation[98] of HOF, all halogens can be included in the list of *hypohalous* acids. All but fluorine form *halic* acids, HXO_3, and, with the recent synthesis[99] of perbromic acid, all but fluorine form *perhalic* acids, HXO_4.

Typical of the hypohalous acids, HOF is quite unstable and decomposes, at room temperature, into O_2 and HF with a half-life of something less than an hour. The acid rapidly decomposes in aqueous media by oxidation of H_2O to H_2O_2 with generation of HF (refer to the next chapter for notes on the mechanism of this reaction). Similarly, the other hyphalous acids are not best prepared by oxidation of H_2O with X_2. The standard potentials for the reactions

---

[96] K. O. Christie and C. J. Schack, *Inorg. Chem.*, **9**, 2296 (1970); A. J. Edwards and R. J. C. Sills, *J. Chem. Soc. A.*, 2697 (1970).

[97] J. D. Witt and R. M. Hammaker, *J. Chem. Phys.*, **58**, 303 (1973).

[98] M. H. Studier and E. H. Appleman, *J. Amer. Chem. Soc.*, **93**, 2349 (1971).

[99] G. K. Johnson, *et al.*, *Inorg. Chem.*, **9**, 119 (1970); J. R. Brand and S. A. Bunck, *J. Amer. Chem. Soc.*, **91**, 6500 (1969).

$$\tfrac{1}{2}X_2 + H_2O \rightarrow H^+ + HOX + e^-$$

$$X = Cl \quad E^\circ = -1.63 \text{ v}$$
$$Br \quad\quad\quad -1.59 \text{ v}$$
$$I \quad\quad\quad\quad -1.45 \text{ v}$$

$$\tfrac{1}{2}X_2 + e^- \rightarrow X^-$$

$$X = F \quad E^\circ = 1.44 \text{ v}$$
$$Cl \quad\quad\quad 1.36 \text{ v}$$
$$Br \quad\quad\quad 1.07 \text{ v}$$
$$I \quad\quad\quad\quad 0.54 \text{ v}$$

$$H_2O \rightarrow \tfrac{1}{2}O_2 + 2H^+ + 2e^- \quad\quad E^\circ = -1.23 \text{ v}$$

are pH dependent and can be used to show that the equilibrium constants for formation of HOX from X_2 and H_2O are all less than 10^{-3}, while those for oxidation of water to O_2 range from 10^{28} for F_2 down to 10^{-3} for I_2. A better method of synthesis seems to be to use the strong, insoluble base HgO to remove the halide ion (formed as insoluble HgX_2) and to avoid the presence of H^+ that naturally arises in the reaction

$$X_2 + H_2O \rightarrow HOX + X^- + H^+$$

Of all the hypohalous acids, only HOF actually effects the oxidation of water to H_2O_2 or O_2 at an appreciable rate and accounts for the difficulties in past synthesis attempts. An additional interesting point is that the pH dependence of the E°'s for the formation of HOX suggests that it should be possible to stabilize the hypohalite ions in basic solution. This, in fact, is the basis for preparation of chlorine laundry bleaches by electrolysis of brine (this produces OCl^- in a basic solution). With Br and I the chemistry is more interesting and complex. Hypobromite and hypoiodite are both unstable with respect to disproportionation into halide and halate, and the disproportionation reaction at room temperature is fast for Br and I. Thus, OCl^- is relatively long-lived at room temperature, OBr^- solutions need to be cooled, and IO^- simply decomposes rapidly at temperatures down to the freezing point of its solutions. The instability of the OX^- anion is not unexpected in view of the reactivity of peroxides and halogens with which the hypohalites are (valence shell) isoelectronic.

Disproportionation of the halate ions

$$4XO_3^- \rightarrow X^- + 3XO_4^-$$

into halides and perhalate ions is thermodynamically favored to a high degree for chlorine, but the reaction is very slow. For bromate and iodate, the disproportionation is highly unfavorable in a thermodynamic sense. Thus, the halate ions are quite stable in aqueous solutions, but for different reasons. In acid solution the perhalates are all strong oxidizing agents, but it is interesting that the reduction potentials for the XO_4^-/XO_3^- half-cells are in the order Br(1.76) > Cl(1.23), again reflecting a relatively high thermodynamic stability for ClO_4^-. Nevertheless, all the halate anions and acids in concentrated form can be dangerous in contact with organic materials, so that such contact should be effected, if necessary, under highly controlled and protective conditions.

The aqueous chemistry of periodic acid is easily the most intriguing of the series. HIO_4 rapidly aquates to form the so-called *orthoperiodic* acid, H_5IO_6, by stepwise addition (accompanied by proton transfer between oxygens) of two water molecules to the iodine atom to leave the structure $(HO)_5IO$. Again, we note a tendency for iodine to exhibit high oxidation states with highly electronegative substituents.

In closing this section on the oxyhalides, we should mention that it is generally believed that multiple bonding between the halogen and oxygen is an important feature of the electronic structures of the higher oxidation states. Supposed failure of fluorine to support multiple bonds with oxygen in fluorate and perfluorate is often mentioned as a reason for believing that they cannot be synthesized. This argument is based on the assumption that halogen valence d orbitals are necessary for multiple bonding to oxygen, but you should remember that molecules such as PF_5 and F_3NO are calculated[100] to be stable without invoking participation of central atom d orbitals. A more reasonable hypothesis for postulating the "non-existence" of FO_4^- and FO_3^- would be the thermodynamic VSEPR/steric congestion about the small F atom.

STUDY QUESTIONS

85. Compare the series of $S_xO_y^{n-}$ and Cl_xO_y compounds for correspondence. As the structures of Cl_2O_3 and Cl_2O_6 are not known, do you feel confident in assuming, by analogy, they have the structures of $S_2O_3^{2-}$ and $S_2O_6^{2-}$? What other reasonable structures for Cl_2O_3 and Cl_2O_6 can you suggest?

86. Why do you suppose that oxygen, but not fluorine, is found combined with chlorine in the +7 oxidation state? Is the reason for the chemical inertness of SF_6 related to the basis for your answer?

87. Contrast the structures of Cl_2O_7, $B_2H_7^-$, $Al_2Cl_7^-$, $S_2O_7^{2-}$, and $(Cl_3Si)_2O$.

NOBLE GASES[101]

One of the more exciting developments in inorganic chemistry in recent years has been the delineation of xenon chemistry. All compounds that are currently well characterized contain the highly electronegative fluorine and/or oxygen atoms. Less well characterized and presently less useful compounds are KrF_2[102] ($E_{Kr-F} \approx 12$ kcal/mole) and $XeCl_2$.[103] Very recently there has been a report[104] of $FXeN(SO_2F)_2$, which is claimed to possess a Xe—N bond (no structural or thermodynamic data are available, however) and which therefore extends to nitrogen the elements known to form bonds to a rare gas atom.

The Xe—F bond energy[105] is estimated to be about 30 kcal/mole and that[106] of Xe—O is lower yet, about 20 kcal/mole. Thus, the simple fluorides XeF_2, XeF_4, and XeF_6 are all thermodynamically stable with respect to the elements. On the other hand, the simple oxides XeO_3 (solid) and XeO_4 (gas) explosively revert to the elements.

$$XeF_n \to Xe + \frac{n}{2} F_2 \qquad \Delta E \cong +11n$$

$$XeO_n \to Xe + \frac{n}{2} O_2 \qquad \Delta E \cong -40n$$

$$XeCl_n \to Xe + \frac{n}{2} Cl_2 \qquad \Delta E < n$$

---

[100] F. Keil and W. Kutzelnigg, *J. Amer. Chem. Soc.*, **97**, 3623 (1975).
[101] J. H. Holloway, "Noble Gas Chemistry," Methuen, London (1968).
[102] S. R. Gunn, *J. Phys. Chem.*, **71**, 2934 (1967); D. R. MacKenzie and I. Fajer, *Inorg. Chem.*, **5**, 669 (1966); J. J. Turner and G. C. Pimentel, *Science*, **140**, 974 (1963).
[103] L. Y. Nelson and G. C. Pimentel, *Inorg. Chem.*, **6**, 1758 (1967).
[104] R. D. LeBlond and D. D. DesMarteau, *Chem. Comm.*, 555 (1974).
[105] B. Weinstock, *et al.*, *Inorg. Chem.*, **5**, 2189 (1966); M. Karplus, *et al.*, *J. Chem. Phys.*, **40**, 3738 (1964).
[106] S. R. Gunn, *J. Amer. Chem. Soc.*, **87**, 2290 (1965).

This difference in XeO_n/XeF_n stabilities is due not so much to the slightly stronger XeF bonds as to the much lower bond energy of F_2 than O_2. At least the oxygen compounds are seen to benefit from some degree of kinetic stability. The fact that E—Cl bonds are normally weaker than E—F bonds suggests that $XeCl_n$ compounds should also be less thermodynamically stable than XeF_n, but more stable than XeO_n. Given this background on XeF_n and XeO_n, you should not be surprised to learn that $XeOF_4$ (liquid, m.p. $-46°$), $XeOF_2$ (solid, m.p. $31°$), XeO_2F_2, and XeO_3F_2 have been synthesized, in spite of their expected instability[107] toward loss of O_2.

The xenon fluorides all act as F^- donors in the presence of strong F^- acceptors, a behavior much like that of the halogen fluorides. Also as for the halogen fluorides, the transfer of the fluoride ion to the acceptor is at best only partially complete. The XeF_n^+ cations all interact more or less strongly with a fluoride in the anion or with another XeF_{n+1} molecule. Examples are $XeF_2 \cdot 2SbF_5(XeF^+Sb_2F_{11}^-)$ and $2XeF_2 \cdot AsF_5(Xe_2F_3^+AsF_6^-)$, where the cation has C_{2v} symmetry similar to that of I_5^-.

XeF_6 is an interesting species since, as mentioned in Chapter 2 (p. 80), its structure in the solid state is that of tetrameric and hexameric rings formed by square pyramidal XeF_5^+ units bridged by F^- ions. This tendency for XeF_6 to expand its coordination shell and to act as a F^- acceptor is also evident in the fact that it forms salts such as $M^+XeF_7^-$ (M must be a large cation like Rb or Cs). Even more interesting is the fact that these salts decompose on mild heating to evolve XeF_6 and form the *eight*-coordinate species M_2XeF_8.

While $XeOF_4$ has not been shown to act as a fluoride ion acceptor, the other Xe(VI) compound, XeO_3, does accept F^- from alkali fluorides (KF and CsF) to form salts $MXeO_3F$. The crystal structures[108] of these compounds reveal that the anions themselves tend to expand the Xe coordination sphere further by associating, via fluoride bridges, into anionic chains.

A similar consequence of the Lewis acid nature of XeO_3 is that aqueous OH^- coordinates XeO_3 ($K \approx 10^{-3}$) to form solutions of the anion, $HXeO_4^-$. It is the base-induced disproportionation of this ion that leads to the *perxenate* anion, XeO_6^{4-}:

$$OH^- + HXeO_4^- \rightarrow \frac{1}{2}O_2 + \frac{1}{2}Xe + \frac{1}{2}XeO_6^{4-} + H_2O$$

In aqueous solution this reaction equation actually misrepresents the fact that perxenate ion is a fairly strong base, with $HXeO_6^{3-}$ being the principal species, even when the pH is as high as 13. Sodium and barium salts of XeO_6^{4-} have been isolated, and all exhibit O_h symmetry for the anion. Protonation and dehydration of the barium salt with concentrated sulfuric acid is the technique used to synthesize the dangerously explosive gas XeO_4 (T_d geometry). The use of concentrated H_2SO_4 for this reaction may be contrasted with the fact that dilute acid solutions effect an alternate rapid decomposition mode of XeO_6^{4-}, leading to $HXeO_4^-$, O_2, and H_2O as products.

$$HXeO_6^{3-} + 2H^+ \rightarrow HXeO_4^- + \frac{1}{2}O_2 + H_2O$$

---

[107] Stability here is used in an operational sense; it would appear that kinetic factors may be important for these compounds. On the other hand, the Xe—F bonds might enhance the Xe—O bond strength as you have seen for F_3NO [the Xe—O bond stretching force constants are 7.1 ($XeOF_4$) and 5.8 (XeO_3,XeO_4) md/Å].

[108] D. J. Hodgson and J. A. Ibers, *Inorg. Chem.*, **8**, 326 (1969).

Such reactions, as opposed to complete decomposition, should reinforce in your thinking the idea that Xe—O compounds seem to enjoy a certain degree of mechanistic stability.

Mechanistic considerations must also be important in the hydrolytic behavior of the higher oxidation state fluorides. XeF_4 and XeF_6 are rapidly hydrolyzed in water to form relatively stable solutions of XeO_3 (HF is the only other product in XeF_6 hydrolysis, but the hydrolysis of XeF_4 is more complex in that it involves disproportionation with Xe and O_2 as products along with XeO_3 and HF):

$$3XeF_4 + 6H_2O \rightarrow XeO_3 + 2Xe + \frac{3}{2}O_2 + 12HF$$

Controlled hydrolysis of XeF_6 with the stoichiometric quantity of H_2O leads to $XeOF_4$ and 2HF.

To close this section on noble gas compounds, let us note that reactions of KrF_2 are under development. In analogy with XeF_2, ionic materials of stoichiometry $(KrF_2)_2SbF_5$ (actually $[Kr_2F_3][SbF_6]$) and $KrF_2(SbF_5)_2$ (actually $[KrF][Sb_2F_{11}]$) have been prepared.[109] The latter has been studied as an oxidative fluorinating agent in the reactions[110]:

$$IF_5 + [KrF][Sb_2F_{11}] \rightarrow [IF_6][Sb_2F_{11}] + Kr$$

$$2Au + 7KrF_2 \rightarrow 2[KrF][AuF_6] + 5Kr$$

$$[KrF][AuF_6] + 2XeF_2 \rightarrow [Xe_2F_3][AuF_6] + F_2 + Kr$$

$$+ O_2 \rightarrow [O_2][AuF_6] + \frac{1}{2}F_2 + Kr$$

$$+ Xe \rightarrow [XeF_5][AuF_6] + \text{other products}$$

$$+ 2XeOF_4 \rightarrow [XeOF_4 \cdot XeF_5][AuF_6] + \frac{1}{2}O_2 + Kr$$

In the last reaction, O appears to be replaced by the isoelectronic F^+ cation; while the structure of the cation has not been fully delineated, one would expect bridging fluorides

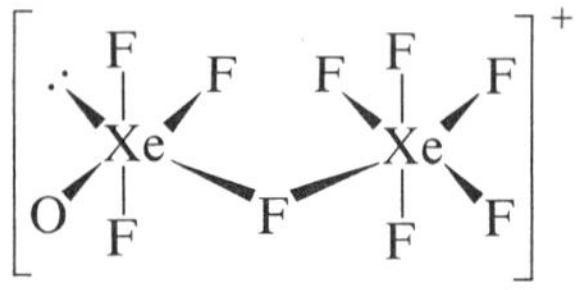

where the "XeF_6" unit may exhibit a distorted fluxional structure.

---

[109] H. Selig and R. D. Peacock, *J. Amer. Chem. Soc.*, **86**, 3895 (1964); R. J. Gillespie and G. J. Schrobilgen, *J. Chem. Soc., Chem. Comm.*, 90 (1974).

[110] D. E. McKee, *et al.*, *J. Chem. Soc., Chem. Comm.*, 26 (1973); J. H. Holloway and G. J. Schrobilgen, *ibid.*, 622 (1975).

STUDY QUESTIONS

88. Estimate the energy for the reactions

$$FXeN(SO_3F)_2 \rightarrow Xe + FN(SO_3F)_2$$
$$\rightarrow Xe + \frac{1}{2}F_2 + \frac{1}{2}((FSO_3)_2N_2)$$

in order to estimate the thermodynamically more favored decomposition products (the XeN bond energy is unknown). Assuming the favored reaction to be exothermic, place an upper limit on E_{XeN}.

89. Predict structures and point group symbols for $XeOF_2$, $XeOF_4$, XeO_2F_2, and XeO_3F_2.

90. Would the synthesis of $XeOF_4$ by XeF_6 hydrolysis in the gas phase probably be exothermic?

91. a) Using Pauling's thermochemical method (see problem 79, p. 348) for estimating atom electronegativities, insert the XeF and XeO thermochemical bond energies to estimate χ_{Xe}. Xe is like what halogens in its χ values?
    b) After considering the Lewis dot structure for XeO_4 and/or after reading the discussion supporting Figure 4–16, criticize the view that the Xe—O bond is a "normal" σ bond.

92. To which neutral and ionic halogen compounds are the known XeF_n and XeO_n compounds similar?

EPILOG

The purpose of this chapter has been two-fold. First, with this material we extended your exposure, begun in the last chapter, to the descriptive chemistry of the main-group elements. Second, this was done from a structure/thermodynamic (bond energy) point of view for the very good reason that the practicing chemist invariably gives this point of view considerable weight in approaching the synthesis of a compound. Whether explicitly or implicitly, he chooses his reagents carefully from a knowledge of their structural and stability properties. Basically, one is faced with selecting reagents that are unstable with respect to a desired reaction, and this entails understanding the "weak links" in those molecular units. The second aspect of this important rational approach to chemistry concerns the mechanistic aspects of desired reactions. It obviously does no good to have conceived of a synthesis scheme involving a step that is expected to be impossibly slow. The need for such information is the basis for the next chapter.

BASIC CONCEPTS

RATE EXPRESSIONS AND INTER-
PRETATIONS

   General Formulation
   Second Order Rate Law
   Pseudo-First Order Rate Law
   First Order Rate Law

ACTIVATION PARAMETERS

A CAVEAT CONCERNING SOLVENT
EFFECTS

THE CONSTRAINT OF ORBITAL FOL-
LOWING

THE PRINCIPLE OF MICROSCOPIC
REVERSIBILITY

SURVEY OF REACTIONS

   One Valence Pair
      Hydrogen
   Three Valence Pairs
      Boron
   Four Valence Pairs
      Boron
      Silicon
      Nitrogen and Phosphorus
      Oxygen and Sulfur
      Halogens
   Five Valence Pairs
      Phosphorus and Sulfur

EPILOG

7. REACTION PATHWAYS

This chapter is dedicated to the question of what happens along the route from reactants to products in a chemical transformation. Of all chemical endeavors this subject is presently the most philosophical, for the simple reason that most often there are no direct *experimental observations of the molecular entities that arise along the reaction path. Without hard data to guide us, we are left in the often uncomfortable (for a scientist) position of making semi-objective guesses based on our knowledge of reactant and product topologies. The best hard data we can obtain come from the rate law expression, and this can tell us only what initial molecular species are involved prior to and during the slowest step of the reaction. The ideas about "how" these pieces fit together in one or more steps, and about what their subsequent fates are, define the concept of a mechanism and are often highly speculative. In this chapter your understanding of molecular ground state structures and bonding and of the concepts of intermolecular interaction (HOMO/LUMO, donor/ acceptor) will be given free rein. An important aspect of all this will be to learn the currently accepted semantics and concepts surrounding the ideas of chemical activation and reaction step types.*

BASIC CONCEPTS

Chemical reactions are often grouped for convenience into the following types: substitution, exchange (both intermolecular and intramolecular), addition or elimination (perhaps involving free radical species), solvolysis, and redox.

$$EX + D \rightarrow ED + X \quad \text{(substitution)}$$
$$EX + DY \rightarrow EY + DX \quad \text{(exchange)}$$
$$EX + DY \rightleftharpoons DEXY \quad \text{(addition/elimination)}$$
$$EX + HA \rightarrow EA + HX \quad \text{(solvolysis)}$$
$$Ox_1 + Red_1 \rightarrow Red_2 + Ox_2 \quad \text{(redox)}$$

Although such labels are meant to characterize the relationship between the compositions of products and reactants, they usually reflect little of the mechanisms of such reactions. It is often found, as you shall see in this chapter, that mechanistically these ostensibly different reaction types have a great deal in common. Consequently, only a few mechanistic concepts carry us a long way in understanding reaction types. It is hoped that, by approaching mechanism studies in a systematic way, sufficient understanding of how molecules react can be developed to make generalizations possible for rational syntheses of new materials. Many of the concepts treated in the preceding chapters on structure, bonding, and thermodynamic concepts will be of great use to us in the study of the highly reactive species that arise during the course of a reaction and that are so important in determining its speed and stereochemical course.

In this preliminary section we will examine in some detail the mechanisms commonly invoked to explain reactions in which one group replaces or substitutes for another. In the survey section to follow, you will find the concepts detailed here to be applicable also to the so-called exchange, addition, elimination, and solvolysis reactions, and even to some redox reactions.

The scientist who studies mechanisms of reactions has three important types of experiments to perform in unraveling the steps of a reaction: *determination of the rate law for the reaction* (this gives information about the composition of the species at the transition state of the rate controlling step), *determination of the stereochemistry of the reaction* (that is, the geometrical relationship between groups bound to the reactive atom in the reactants and products), and *determination of the effects of substituents* (groups bound to the reactive atom affect the speed of a reaction by their ability to facilitate or retard electronically the breaking and/or making of bonds at the reactive atom, by their nuclear mass, and by their steric bulk and stereochemical idiosyncracies).

As the majority of chemical reactions are carried out in solutions, we will confine our attention to the concepts developed for such reaction conditions. In a general reaction between R and R' to form P and P', we recognize that many steps may be required to complete the transformation from reactants to products and that some of these steps will be much faster than others. It appears that often, but not always, a single step is so much slower than all others that the speed of that step effectively determines the form of the rate law expression. *The rate law will then contain information as to the nature of all steps preceding and including the slowest one.* In the simplest terms we can envision the following alternatives:

(a) R and R' diffuse together to be trapped momentarily in a solvent cavity or "cage" as an *encounter* or *cage complex*, R:R'.

(b) R and R' smoothly transform into P and P' (in the form of a cage complex P:P').

(c) R and R' transform first into a new species, R—R', which subsequently evolves into P and P'.

Situation (b) as the slowest step characterizes what we will call an *interchange* (or *I*) *mechanism,* and (c) as the slowest step characterizes what we will refer to as an *associative* (or *A*) *mechanism,* distinguished by the formation of a bond between R and R' to produce an *intermediate,* R—R'. Situation (a) as the slow step characterizes a *diffusion controlled reaction.*

Yet another possibility exists. Perhaps the cage complex R:R' does not lead to products. Rather, R unimolecularly dissociates into species P' and X. The *intermediate* X can then recombine with P' (to form R or some stereochemical isomer of R) or be long-lived enough to diffuse into a solvent cavity with R' and undergo reaction to form P. This situation characterizes what we mean by a *dissociative* (or *D*) *mechanism.*

These possibilities for a *stoichiometric mechanism*[1] can be summarized by the equations (examples you have seen in organic chemistry are also given):

(*I*) $\quad R + R' \to \boxed{R:R' \to P:P'} \to P + P' \qquad\qquad H_3CCl + OH^- \to CH_3OH + Cl^-$

$$R_2CO + H_2O^* \to R_2C\begin{smallmatrix}OH\\ *OH\end{smallmatrix}$$

(*A*) $\quad R + R' \to \boxed{R:R' \to R-R'} \to P:P' \to P + P'$

$$\downarrow$$

$$R_2CO^* + H_2O$$

---

[1] Note that the stoichiometric mechanism notation tells us the composition (stoichiometry) of the transition state species. Good references with examples from transition metal reactions are: C. H. Langford and H. B. Gray, "Ligand Substitution Processes," W. A. Benjamin, New York, 1965; C. H. Langford and V. S. Sastri in "Inorganic Chemistry, Series One," Vol. 9, M. L. Tobe (Ed.), University Park Press, Baltimore, 1972; M. L. Tobe, "Inorganic Reaction Mechanisms," Nelson, London, 1972.

(D) $\boxed{R \rightarrow X:P'} \rightarrow X + P'$ $(t\text{-Bu})_3\text{CBr} \rightarrow (t\text{-Bu})_3\text{C}^+ + \text{Br}^-$

$X + R' \rightarrow X:R' \rightarrow P$ $(t\text{-Bu})_3\text{C}^+ + \text{OH}^- \rightarrow (t\text{-Bu})_3\text{COH}$

The *transition state structure* for the *I* mechanism is envisioned to be

$$\left[\begin{array}{c} \text{H} \; \text{H} \\ \text{HO}--\text{C}--\text{Cl} \\ \text{H} \end{array} \right]^-$$

and you will note there is no net change in the number of bond pairs in the carbon valence shell (the HO—C—Cl structure is a typical three-center/four-electron mo problem). Both bond-making and bond-breaking characterize this transition state. Net bond-making is often, but not always, the dominant characteristic of the *A* mechanism intermediate; in the example above, the CO π bond

$$\left[\begin{array}{c} R \\ R \diagdown \\ C-O \\ | \\ O \\ H_2 \end{array} \right]$$

is supplanted by a CO σ bond in the intermediate. The intermediate for the *D* mechanism is characterized by net bond breaking:

$$\left[\begin{array}{c} t\text{-Bu} \;\; t\text{-Bu} \\ \diagdown \; \diagup \\ C \\ | \\ t\text{-Bu} \end{array} \right]^+$$

These concepts are probably best understood in terms of what is called an *energy profile diagram* of the change in energy of the chemical system as the nuclei of the reacting parts execute the motion leading to the substitution (this motion, even though often not quantitatively known, is referred to as the *reaction coordinate*).

Figure 7-1 summarizes the differences between *D*, *A*, and *I* processes. The dashed curve illustrates that *I* process in which reactants are smoothly transformed into products by the formation of a single, unstable "high" energy chemical species

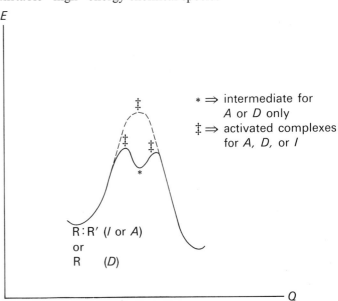

Figure 7-1. A reaction energy profile diagram illustrating the formation of an intermediate, characteristic of *A* and *D* processes (——), and the formation of an activated complex, characteristic of an *I* process (— — —).

* ⇒ intermediate for *A* or *D* only
‡ ⇒ activated complexes for *A*, *D*, or *I*

referred to as the *transition state* or *activated complex*. The A and D processes (solid line) involve two steps showing the appearance of two activated complexes, separated along the reaction coordinate by a minimum that is characteristic of the *intermediate*. The A mechanism implies the formation of the first activated complex by an *associative activation mode, a*, in which the reactant R:R' forms a bond creating the intermediate R—R'. The second step requires a much smaller activation energy for the conversion of R—R' into product P:P'. The D mechanism, on the other hand, has an energy profile similar to that of the A process, but the initial high activation energy step involves a *dissociative activation mode, d*, leading to complete fragmentation of R into one of the products, say P', and a species (intermediate X) of kinetically significant stability. The intermediate then rapidly reacts with R' to form the final product P. In contrast to the A mechanism, R' does *not* play an energetically significant role in the D mechanism. With regard to the intimate aspects of an I mechanism, one can envision a transition state complex involving considerable fragmentation of R with little progress being made in the formation of the new bond between R and R', or the converse. Such activation processes are then distinguished on the intimate level as I_d and I_a, respectively.

In summary, the symbols D and A convey not only the concepts of unimolecular and bimolecular reaction steps, but also the idea that these steps are the *slow* ones and proceed by d (bond-breaking) or a (bond-making or "adduct" formation) so as to produce reactive intermediates. The symbol I conveys the concept of a slow bimolecular reaction step that proceeds with predominantly bond-breaking (I_d) or bond-making (I_a) character so that no reactive intermediate is formed. With reference to the D and A reaction sequences above, the "boxed" steps are implied to be the slow ones. Were these steps not the slow ones, the A mechanism would become D in R—R' ($\rightarrow$ P:P') and the D mechanism would become A in X + R' ($\rightarrow$ P).

The important questions to us as chemists are: "How do we categorize a given reaction into one of the I, A, and D types?" and "If we settle on an I mechanism, how do we tell whether bond-breaking comes 'early' (I_d) or 'late' (I_a) along the reaction coordinate (that is, is bond-breaking or bond-making the important part of the activation process)?" Of particular help are studies of the effect on the rate constant of varying the steric bulk about the reactive centers in R and R', the strength of the bond of the leaving group to the reactive site in R, and the anticipated strength of the bond between R' and R. Such effects are often the only basis available on which to distinguish I from A mechanisms, whereas *the only really conclusive evidence for D or A is to observe the intermediate directly*.

Chemical Valency Ideas

If R is a structure representing valency unsaturation for the central atom, an I or A mode is anticipated. Compounds of elements for which valence shell expansion is established (Chapter 5) by the existence of "stable" adducts are prime candidates for the A mechanism. If the reagent R is known to be valency saturated, the most reasonable pathways are D and I.

Rate Law Expressions

The details of rate law interpretation are saved for the next section; to anticipate those conclusions, the charts at the top of page 363 express the correlation of stoichiometric mechanism with rate laws.

Very seldom are we in a position of knowing the mechanism and wanting to predict a rate law. It is important to realize the mechanism $\rightarrow$ rate law correlation is unique but that the rate law $\rightarrow$ mechanism translation is rarely so. Ambiguities in interpretation arise for the second order rate law in particular, but also for the first order law when [R'] is essentially constant throughout the reaction. Other evidences for a D mechanism as opposed to the I and A pathways are (i) marked *suppression* of the *forward* rate by the addition of P' (we will return to this point in study

Known Mechanism	Predicted Rate Law
I for R + R'	Rate = k[R][R']
A for R + R'	Same as I
D for R	Rate = k[R]

Known Rate Law	Implied Mechanism
Rate = k[R][R']	(i) I or A in R + R' (ii) D in R—R'
Rate = k[R]	(i) D in R (ii) Any of the second order mechanisms where [R'] is so large as to be approximately time independent

question 3) and (ii) optical or geometric isomerization of R (that is, X rotates in the cage before P' can recombine to form R) at a rate faster than the formation of P + P'. In the chart above you will notice that the rate law never distinguishes between I and A mechanisms. Analogs to (i) and (ii) above are usually of no help. For example, generally the addition of P' has the same effect (none) on the forward rates of *both* I and A mechanisms. (In the favorable cases where the steps R—R' $\rightleftharpoons$ P:P' $\rightleftharpoons$ P + P' of the A mechanism are very rapid equilibria, the addition of P (or P') will increase the rate of the *reverse* reaction P + P' until it becomes first order in P' (or P) only.) Neither does a clue arise from observed isomerization of R on forming P; such behavior could occur through an intermediate (R—R') sufficiently long-lived to effect a rearrangement (A mechanism) or could occur in an I mechanism through the stereochemistry of R' attack upon R.

Substituent Effects

Because of anticipated effects on bond-making and bond-breaking energies, variation of substituents can help complete the description of a mechanism. By implication, I_d and, particularly, D mechanisms should have rate constants sensitive to substituent changes that affect the ease of bond-breaking. On the other hand, I_a and, particularly, A mechanisms are sensitive to bond-making influences.

	Rate Constant Leaving Group Dependence	Rate Constant Entering Group Dependence
D	large	none
A	small	large
I_d	significant	less significant
I_a	less significant	significant

STUDY QUESTIONS

1. Sketch a reaction coordinate diagram for a reaction exhibiting an induction period (*i.e.,* one that *initially* shows a faster disappearance of reactants than formation of products).

2. Compare the following energy profile with that in Figure 7–1. Do the D and A mechanistic concepts still have meaning in this case? If so, with respect to what species as reactants? How are the rate laws for this figure and Figure 7–1 different for each of the A (with respect to R, R') and D (with respect to R) mechanisms? (The answers are to be found in the next section.)

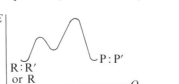

RATE EXPRESSIONS AND INTERPRETATIONS

GENERAL FORMULATION

We will now proceed to the inspection of commonly encountered mechanisms and the derivation of their corresponding rate laws. First to be considered is the elementary bimolecular reaction

$$R + R' \underset{k_r}{\overset{k_f}{\rightleftharpoons}} P + P'$$

for which the *overall* rate expression is

(7–1)
$$-\frac{d[R]}{dt} = \frac{d[P]}{dt} = k_f[R][R'] - k_r[P][P']$$

By envisioning the reaction to consist of:
(a) diffusion of R and R' together to become trapped in a solvent cage where R and R' collide with one another 10 to 100 times before one of them escapes the cage,
(b) either diffusion of R and R' apart without reacting or reaction to form P and P' in the solvent cage, and
(c) diffusion apart of P and P' or back reaction of P + P' in the cage to form R and R',

we write (k_2 and k_{-2} represent chemical conversions[2])

$$R + R' \underset{k_{-1}}{\overset{k_1}{\rightleftharpoons}} R:R' \underset{k_{-2}}{\overset{k_2}{\rightleftharpoons}} P:P' \underset{k_{-3}}{\overset{k_3}{\rightleftharpoons}} P + P'$$

If R:R' and P:P' have low concentrations but are sufficiently long lived that they can be thought of as intermediates, we may treat their concentrations as constant during most of the reaction (*this is called the steady-state approximation*). By writing out full expressions for $d[R:R']/dt$ and $d[P:P']/dt$ and setting these to zero,

$$\frac{d}{dt}[R:R'] = k_1[R][R'] - (k_2 + k_{-1})[R:R'] + k_{-2}[P:P'] = 0$$

$$\frac{d}{dt}[P:P'] = k_{-3}[P][P'] - (k_3 + k_{-2})[P:P'] + k_2[R:R'] = 0$$

we find, by elimination of [P:P'], a relation between [R:R'] and [R], [R'], [P], and [P']. This is substituted into the rate expression

$$-\frac{d[R]}{dt} = k_1[R][R'] - k_{-1}[R:R']$$

Rearrangement and comparison of the final expression with equation (7–1) yields the following results for k_f and k_r:

$$k_f = k_1/[1 + (k_{-1}/k_2) + (k_{-1}k_{-2}/k_2k_3)]$$
$$k_r = k_{-3}/[1 + (k_3/k_{-2}) + (k_2k_3/k_{-2}k_{-1})]$$

---

[2] The k_2 and k_{-2} represent overall rate constants for either one-step or multiple-step interconversions of R:R' and P:P'.

(A)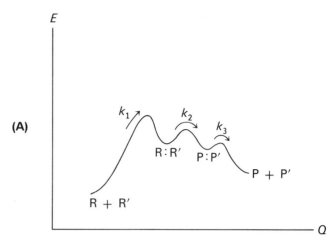

Figure 7–2. (A) An energy profile of a diffusion controlled reaction; (B) an energy profile of a chemical conversion controlled reaction.

(B)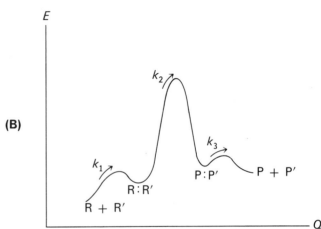

In this way you see that the forward rate constant for the overall second order reaction depends on the forward and reverse *diffusion* rate constants for the $R + R' \to R:R'$ step, on the forward and reverse rate constants for the *chemical conversion* step $R:R' \to P:P'$, and on the reverse diffusion rate constant for $P + P'$. From this rather complex result for k_f we may identify two limiting cases of general interest and utility. If k_{-1}/k_2 and k_{-2}/k_3 are quite small [refer to Figure 7-2(A)], the forward rate constant becomes simply

$$k_f = k_1$$

In this event, the reaction of R and R' is *diffusion controlled* and is generally quite fast (diffusion rate constants in aqueous systems are about 10^{10} M^{-1} sec^{-1}). Some examples of such reactions are proton transfer reactions, *e.g.*,

$$H_3O^+ + OH^- \xrightarrow{H_2O} 2H_2O$$
$$NH_4^+ + OH^- \xrightarrow{H_2O} NH_3 + H_2O$$

and some redox reactions, *e.g.*,

$$Cu^{2+} + e_{(aq)} \to Cu^+$$
$$Fe(CN)_6^{3-} + e_{(aq)} \to Fe(CN)_6^{4-}$$

The other important limiting case occurs when k_{-1}/k_2 is large and k_{-2}/k_3 is very small [Figure 7-2(B)]:

$$(1 + k_{-1}k_{-2}/k_2 k_3) \ll k_{-1}/k_2$$

In this event, the forward rate constant is given by

(7-2)
$$k_f = (k_1/k_{-1})k_2 = K_{\text{dif}} k_2$$

where K_{dif} is the *equilibrium constant*[3] for the diffusion together of R and R' to form the caged species R:R'. Interpretively, this means that a fast equilibrium establishes a relatively steady concentration of the species R:R' but that relatively few of these caged complexes have sufficient energy to chemically convert to P:P' so that the latter step becomes rate controlling. *This limiting case is the one most frequently encountered in solution reactions.*

Before passing on to discussion of other cases, you should be aware that neither of the two limiting conditions outlined above is met by some chemical systems, and cases have been documented[4] in which the more complete expression for k_f must be invoked. Finally, it is important to observe that the determination of the rate expression

$$\text{Rate}_f = k_f[\text{R}][\text{R}']$$

cannot distinguish between the two limiting cases. Recourse to other mechanistic information is necessary, such as comparison of the size of k_f (experimental) with expected diffusion rate constants.

While it is likely that all solution reactions could be broken down into a detailed series of diffusion and chemical conversion steps, it is useful for many reactions to disregard implicitly the diffusion steps (because they are such fast equilibria) and to concentrate on the chemical conversion steps. This amounts to invoking the pre-equilibrium condition for formation of cage complexes, as discussed in the second paragraph back. This view is appropriate when the chemical conversion rate is a few orders of magnitude smaller than the diffusion rate (that is, the chemical conversion rate is 10^8 sec^{-1} or slower).

SECOND ORDER RATE LAW

From equation (7-2), the expression for the forward rate of a second order reaction following an *I* mechanism is

(7-2a)
$$\text{Rate}_f = K_{\text{dif}} k_2[\text{R}][\text{R}'] = k_2[\text{R}:\text{R}']$$

It is logical to ask at this point what the effect will be on the rate expression *if the chemical transformation step involves formation of an intermediate* (*the A mechanism*). In this case the *single* chemical transformation step appropriate to the *I* mechanism is replaced by *two* steps:

$$\text{R}:\text{R}' \underset{k_{-2a}}{\overset{k_{2a}}{\rightleftarrows}} \text{R}-\text{R}' \underset{k_{-2b}}{\overset{k_{2b}}{\rightleftarrows}} \text{P} + \text{P}'$$

In those cases where R—R' may be regarded as an undetectable intermediate, an interesting situation develops concerning interpretations of the rate law expression. Making the steady-state approximation for [R—R']:

$$\frac{d[\text{R}-\text{R}']}{dt} = k_{2a}[\text{R}:\text{R}'] - (k_{-2a} + k_{2b})[\text{R}-\text{R}'] + k_{-2b}[\text{P}][\text{P}'] = 0$$

---

[3] When R and R' are ions, K_{dif} is often thought of as an ion-pair formation constant, K_{ip}.
[4] J. Halpern, *J. Chem. Educ.,* **45**, 372 (1968).

yields an expression for [R—R′] that on substitution into the expression

$$-\frac{d}{dt}[\mathrm{R:R'}] = k_{2a}[\mathrm{R:R'}] - k_{-2a}[\mathrm{R-R'}]$$

yields

$$\mathrm{Rate}_f = K_{\mathrm{dif}} \frac{k_{2a}k_{2b}}{k_{-2a} + k_{2b}}[\mathrm{R}][\mathrm{R'}] = K_{\mathrm{dif}} k_2[\mathrm{R}][\mathrm{R'}] \qquad (7\text{-}3)$$

Two limiting cases can be distinguished (see Figure 7-3):

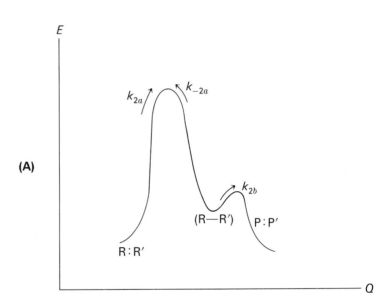

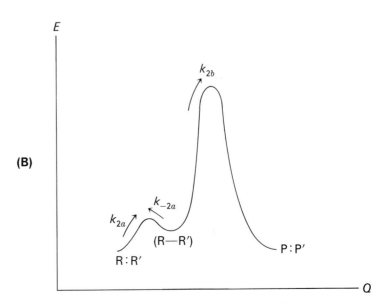

Figure 7–3. (A) An energy profile for an *A* mechanism in R + R′ ($k_{2b} \gg k_{-2a}$); (B) an energy profile for a *D* mechanism involving the species R–R′. Note that these two diagrams actually differ only in the interchange of the symbols R and R′ with P and P′!

(a) $k_{2b} \gg k_{-2a}$

[which means that there can be no pre-equilibrium formation of the intermediate, since k_{-2a} is so small; refer to Figure 7-3(A)] so that the forward rate constant reduces to

(7-3a)
$$k_f = K_{dif} k_{2a}$$

and the rate controlling step is formation of R—R' from R:R' via the *a* mode and *A* mechanism.

(b) $k_{2b} \ll k_{-2a}$

[amounts to pre-equilibrium conditions for the intermediate with $K_{eq} = k_{2a}/k_{-2a}$; see Figure 7-3(B)] so that the forward rate constant becomes

(7-3b)
$$k_f = K_{dif} K_{eq} k_{2b}$$

and the rate controlling step is conversion of the intermediate R—R' to P and P'. *Note that this limiting case,* in which the intermediate is formed *before* the activation step, *takes the mechanism out of the A class for R + R' and into the D class for R—R'* (i.e., Rate = k_{2b}[R—R']).[5]

Interestingly, for mechanistic purposes, the rate expression alone does not distinguish these two limiting cases from each other, nor does it distinguish either of them from the case in which the mechanism is of the "simple" *I* variety.

PSEUDO-FIRST ORDER RATE LAW

Even in the event of an *I* or an *A* mechanism with a second order rate law, if the reactant R' is present in sufficiently high concentration the rate law will become effectively first order, since the concentration of R' is essentially constant and is absorbed into the value for k_f. Hence,

$$\text{Rate}_f = \{K_{dif} k_2[R']\}[R] \qquad \text{for the I mechanism}$$

or

$$\text{Rate}_f = \left\{K_{dif} \frac{k_{2a} k_{2b}}{k_{-2a} + k_{2b}}[R']\right\}[R] \qquad \text{for all mechanisms with an R—R' intermediate}$$

In either case, the pseudo-first order nature of the rate constant can be checked[6] by a plot of k against [R'].

---

[5] An often used alternate form of equation (7-3b) is

$$\text{Rate} = \frac{Kk_{2b}}{1 + K[R']}[R]_t[R']$$

where $[R]_t = [R] + [R\text{—}R']$. Substitution of

$$[R] = [R]_t - [R\text{—}R'] = \frac{[R]_t}{1 + K[R']}$$

into (7-3b) yields the alternate form. Note that it is $[R]_t$, not [R], that appears in this alternate rate law. When K is very small, however, $[R]_t \cong [R]$ and $K[R'] \ll 1$; then equation (7-3b) is recovered. An application of this rate law form is to be found in Chapter 13 for the reaction $Co(NH_3)_5(OH_2)^{3+} + X^- \rightarrow Co(NH_3)_5X^{2+} + H_2O$.

[6] Of course, when R' = solvent, then it is not possible to vary [R'].

FIRST ORDER RATE LAW

First order rate expressions can arise for the D mechanism, where the chemical conversion involves the following steps (requiring the second step to proceed through a cage complex $X:R'$ has the effect of changing k_{2b} to $k_{2b}K_{\text{dif}}$ if the diffusion rates are fast):

$$R \underset{k_{-2a}}{\overset{k_{2a}}{\rightleftarrows}} X + P'$$

$$X + R' \underset{k_{-2b}}{\overset{k_{2b}}{\rightleftarrows}} P$$

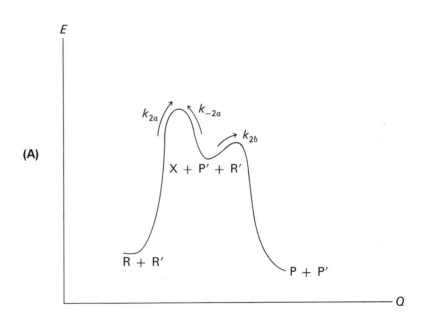

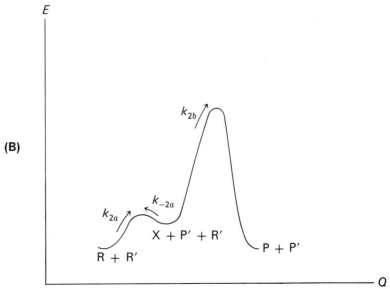

Figure 7-4. (A) An energy profile for a D mechanism with $k_{2b} \gg k_{-2a}$; (B) an energy profile for a "D mechanism" with $k_{2b} \ll k_{-2a}$ (really I or A in $X + R'$).

Application of the steady-state approximation for the intermediate X leads to the expression[7] for k_f:

(7-4) $$k_f = \frac{k_{2a}k_{2b}}{k_{-2a}[P'] + k_{2b}[R']}$$

Again two limiting cases can be identified (see Figure 7-4):

(a) $k_{2b} \gg k_{-2a}$

or $k_{2b}[R']$ can be made much larger than $k_{-2a}[P']$ by adjusting $[R']$, so that

(7-4a) $$\text{Rate}_f = \frac{k_{2a}k_{2b}}{k_{2b}[R']}[R][R'] = k_{2a}[R]$$

and the rate controlling step is formation of the intermediate X via the *d* mode and D mechanism for R;

(b) $k_{2b} \ll k_{-2a}$

or $k_{-2a}[P']$ is made much larger than $k_{2b}[R']$, so that

(7-4b) $$\text{Rate}_f = \frac{k_{2a}k_{2b}}{k_{-2a}}[R][R'][P']^{-1} = k_{2b}[X][R']$$

---

[7] In this development of k_2 of (7-2), the prefactor K_{dif} is omitted since the cage complex R:R' is *not* postulated to precede step 2a.

TABLE 7-1
SUMMARY OF RATE LAW INTERPRETATIONS

	Rate Law	Mechanism
(a)	$\text{Rate}_f = K_{dif}k_2[R][R']$	I_a or I_d in R + R'
(b)	$\text{Rate}_f = K_{dif}\dfrac{k_{2a}k_{2b}}{k_{-2a}+k_{2b}}[R][R']$	Any mechanism with R — R' as intermediate
	(i) $K_{dif}k_{2a}[R][R']$	A in R + R'
	(ii) $K_{dif}K_{eq}k_{2b}[R][R']$	D in R — R'
(c)	$\text{Rate}_f = \dfrac{k_{2a}k_{2b}}{k_{-2a}[P'] + k_{2b}[R']}[R][R']$	Any mechanism with X as intermediate
	(i) $k_{2a}[R]$	D in R
	(ii) $\dfrac{k_{2a}k_{2b}}{k_{-2a}}[R][R'][P']^{-1}$ $= k_{2b}[X][R']$	*I* or A in X + R'
	$\text{Rate}_f = C[R][R']$	$\Rightarrow$ (i) *I* or A in R + R' (ii) D in R — R'
	$\text{Rate}_f = C[R]$	$\Rightarrow$ (i) D in R (ii) pseudo-first order form of any of the second order possibilities
	$\text{Rate}_f = C[R][R'][P']^{-1}$	$\Rightarrow$ *I* or A in X + R'

and the rate controlling step is, in effect, the reaction step X + R' → P. Here again we encounter the intermediate as a *precursor* to the activation step, and this causes us to change our description of the reaction mechanism from *D* for species R to *I* or *A* for X + R'.

Table 7-1 summarizes the relationship between rate law and mechanism.

STUDY QUESTIONS

3. Refer to the legend of Figure 7-3 and justify the parenthetical comment on p. 363 regarding the use of [P'] to distinguish the *I* from the *A* mechanism. Refer to (7-4) to account for statement (i) on p. 362.

ACTIVATION PARAMETERS

When they are interpreted carefully, the enthalpy and entropy of activation (derived from the temperature dependence of the rate constant) can be suggestive of the *d* or *a* nature of the activation step. According to the *transition state* theory of chemical reactions (derived for gas phase reactions but assumed to be appropriate for solution reactions as well), the ascent of the activation barrier by reactant molecules is treated as a thermodynamic equilibrium between the pool of reactant molecules and that of the transition state complexes, so that a temperature dependent "equilibrium constant" follows:

$$RT \ln K^\ddagger = -\Delta G^\ddagger = -\Delta H^\ddagger + T\Delta S^\ddagger$$

The rate constant for the passage of reactant molecules over the activation barrier, according to the transition state theory, is given by

$$\ln k_{\text{rate}} = \ln\left(\frac{kT}{h}\right) - (\Delta H^\ddagger/RT) + (\Delta S^\ddagger/R)$$

so that a plot of $\ln k_{\text{rate}}$ versus T^{-1} is linear with a slope of $-\Delta H^\ddagger/R$ and an intercept of $\Delta S^\ddagger/R + \ln(kT/h)$ [the variation of the term[8] $\ln(kT/h)$ with T^{-1} is so small as usually to be well within the experimental error of the $\Delta H^\ddagger/R$ values]. In the gas phase, and in those solution reactions where the change in solvation of the reactants to form the transition state complex is negligible, *large positive enthalpies and entropies of activation are highly suggestive of d activation modes,* whereas *small positive enthalpies and negative entropies of activation are usually interpreted to reflect a activation (A or I_a mechanisms).*

In practice, one has to be careful in making such interpretations, since the rate constant extracted from the first and second order rate law expressions, as you have already seen, may contain equilibrium constants for steps preceding the activation step. In these cases, the enthalpies and entropies extracted from the temperature dependence of the rate constant are a *composite of those for all steps up to and including the activation step.* For example, for the *D* mechanism in R—R':

$$\text{Rate}_f = (K_{\text{dif}} K_{\text{eq}} k_{2b})[R][R']$$

$$\ln k_f = \ln K_{\text{dif}} + \ln K_{\text{eq}} + \ln k_{2b}$$

$$= -(\Delta H_{\text{dif}} + \Delta H_{\text{eq}} + \Delta H^\ddagger)/RT + (\Delta S_{\text{dif}} + \Delta S_{\text{eq}} + \Delta S^\ddagger)/R + \ln\left(\frac{kT}{h}\right)$$

---

[8] In this term *k* is the Boltzmann constant; $k/h \approx 2 \times 10^{10}$ (sec·deg)$^{-1}$ is thought of as the theoretically maximum reaction rate at 1°K.

Several applications of this situation will be discussed in detail in the survey section. We should note, however, that since precursor equilibria are relatively fast, ΔH_{dif} and ΔH_{eq} in some cases will tend to be much smaller than $\Delta H^{\ddagger}$. Similar expectations hold for the $\Delta S^{\ddagger}$'s.

A CAVEAT CONCERNING SOLVENT EFFECTS[9]

Since most reactions of interest to synthetic and reaction mechanism chemists are carried out in the liquid phase, we need to be aware of the effect the solvent might have in determining the pathway for a reaction; even for the same pathway for a series of related molecules, the way in which the rate constants vary may depend on the nature of the solvent. In many reactions, of course, *the solvent can play an important role by participating intimately in the rate controlling step of the reaction,* but it is often overlooked that *the activation free energy depends on the relative degree to which the initial reactants and the transition state species are solvated.*

Considering the A and D mechanisms, we may construct Haber cycles to analyze the free energies of the activation steps.

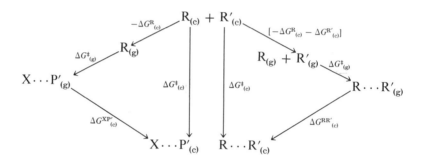

The subscripts (c) and (g) identify *c*ondensed phase species and *g*aseous species. For both the uni- and bimolecular mechanisms, $\Delta G^{\ddagger}$ for the gas phase reaction differs from that in solution by the difference in solvation energies of the reactant and transition state species. Let us suppose we have studied the reactions of a series of R molecules and all are found to undergo reaction with R' by the same mechanism. We might be interested in interpreting the differences in rate constants for the different R's in terms of steric and electronic properties of X and P', in the case of a D mechanism, and in terms of R and R', in the case of an A mechanism. Such interpretations would then be based on concepts appropriate to the gas phase reaction and would not include variation in solvation energy change upon generating the transition state species from the initial reactants. Even in a given solvent such variation might amount to as much as 1 to 2 kcal/mole, and this alone would cause a change in rate constant by a factor of 10 to 100. Consequently, *the presence of the solvent makes it difficult to interpret rate constant differences of this magnitude.*

In addition, we might also note that whether a particular reaction follows a D, an A, or an I mechanism might be determined by the differences in transition state solvation energies for these mechanisms. That is, solvation effects on reactants or transition state complexes might favor a particular mechanism. The solvent may also play an active molecular role in a reaction mechanism. This could occur in an overt way in which separate reaction channels exist for S (the solvent) and R'; that is, there are separate products formed. In such a case the interpretation of rate laws is as before with the substitution of S for R' in the solvent reaction channel. The other

---

[9] R. H. Prince, *Nucleophilic Displacement at Some Main Group Elements,* in "Inorganic Chemistry, Series One," Vol. 9, pp. 353ff, M. L. Tobe (Ed.), University Park Press, Baltimore, 1972.

active role for the solvent is as a catalyst. This role can be particularly difficult to detect, for there are no solvent-derived products and, though the solvent is intimately involved in the transition state structure and its concentration appears in the rate law expression, its concentration is time independent; thus, the catalytic role escapes direct detection. Typical solvent catalytic roles are as follows. In a mechanism that is judged to be of the D type from the rate expression, the solvent may merely "trap" X to await reaction of X·S with R′:

$$R \xrightarrow{\text{slow}} X + P' \xrightarrow[S]{\text{fast}} X \cdot S + P' \xrightarrow[R']{\text{fast}} P + P' + S$$

or it may actually assist the displacement of P′ in a mechanism that would properly be described as an I or A mechanism in R + S:

$$R + S \xrightarrow{\text{slow}} X \cdot S + P' \xrightarrow[R']{\text{fast}} P + P' + S$$

Even in a reaction following a second order rate law, the solvent might assist attack of R′ upon R:

$$R + S \xrightleftharpoons{\text{fast}} R \cdot S \xrightarrow[R']{\text{slow}} R'{-}R \cdot S \xrightarrow{\text{fast}} P + P' + S$$

or dissociation of R—R′ into products:

$$R + R' \xrightleftharpoons{\text{fast}} R{-}R' \xrightarrow[S]{\text{slow}} P' + P \cdot S \xrightleftharpoons{\text{fast}} P + P' + S$$

The first of these you recognize as an A process in R·S + R′, while the second, otherwise termed D in R—R′, is actually I in R—R′ + S. All of these possibilities will be encountered in the survey section to follow.

THE CONSTRAINT OF ORBITAL FOLLOWING

Before we go on to consideration of important examples of mechanisms of reactions at non-metal centers, we should pause briefly to consider the implications of the orbital concepts developed in Chapters 1 through 4. Many of the reactions you will see in the survey section involve nucleophilic or electrophilic attack by one molecule upon another. At least in the initial stages, such reactions begin by donor-acceptor interaction between these fragments. That means that we must look for LUMO's and HOMO's and ask how they can best interact to get the reaction started. What happens beyond this initial interaction is, of course, not easy to answer in a general way, for the answer depends heavily upon the ability of the atom under attack to expand its valence shell to accommodate the entering group[10] and upon how other groups already bound to this central atom respond to the presence of an intruder. This, in turn, requires a detailed understanding of the orbital characteristics of the new species, in terms of the orbitals of the central atom and its substituents.

As electron pairs are not destroyed in a chemical reaction, the entering group demands of the substrate that a bond be formed with it, and this generally means that a bond is to be broken somewhere else. In the simplest terms, a nucleophile enters with an electron pair (HOMO), populating with electrons a substrate mo (LUMO)

---

[10] That is, how well can the atom under attack "have its cake and eat it too" by forming a new bond without unduly weakening its other bonds?

with anti-bonding characteristics between the central atom and another substituent. This leads to weakening of this other bond. Attack by an electrophile (LUMO) demands that a formerly bonding mo (HOMO) in the substrate be depopulated so as to meet the electron demand of the electrophile. These ideas are summarized by the following sketches:

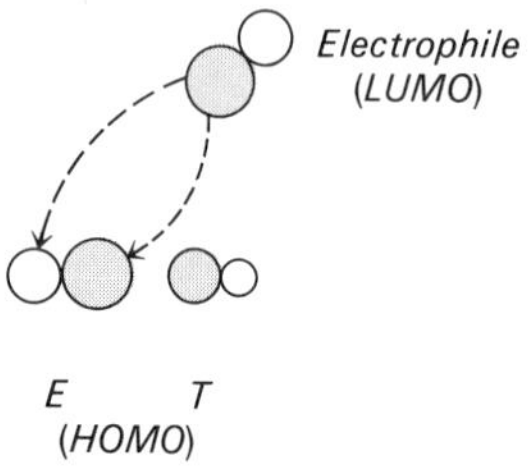

(a) Electrophile (LUMO)
E (HOMO)   T

In this case, HOMO has greater amplitude at the central atom E than at the terminal atom T. Two attack approaches shown.

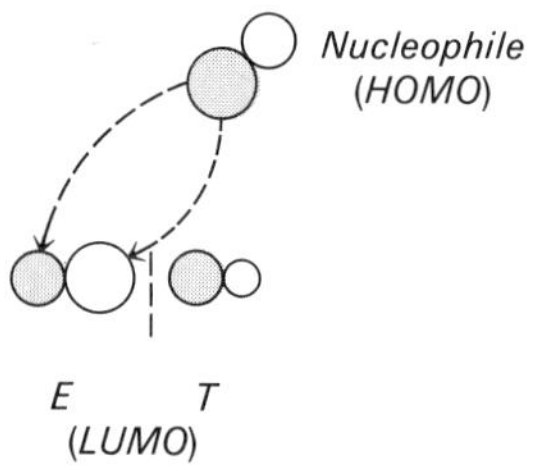

(b) Nucleophile (HOMO)
E (LUMO)   T

In this case, LUMO has greater amplitude at the central atom E than at the terminal atom T. Two attack approaches shown.

In (a) we find an electrophile (LUMO) entering the region of a substrate bond (HOMO) to create a three-center/two-electron structure. The electrophile will tend to associate itself most strongly with the central atom E or the terminal atom T, depending upon the effectiveness of its overlap with the bond hybrids of these atoms. Note that should E or T have a lone pair of electrons, it is more likely that the electrophile will attack that pair as HOMO, rather than a bond pair. Such attachment might or might not yield a species capable of further reaction. In sketch (b) we find a nucleophile seeking good overlap and energy match with a substrate LUMO. This is usually the most stable non-bonding or anti-bonding mo of the substrate. Generally the region (E or T) of attack will be that affording the greatest overlap with the attacking HOMO.[11] By populating the E-T anti-bonding mo the nucleophile succeeds in weakening the bond between these two atoms and establishing a bond between itself and either E or T, depending upon the polarization of anti-bonding mo. As you should be well aware by now, the bonding mo for E-T will be polarized toward the more electronegative of E and T, while the anti-bonding mo is more highly concentrated on the less electronegative of these two atoms. This effect determines the HOMO-LUMO overlap and, therefore, the initial entry of the electrophile or nucleophile. What happens beyond this point requires greater information about the nature of E, T, and the other groups bound to E. Possibilities are departure of T or E with the electrophile or nucleophile or, if the entry of the electrophile or nucleophile does not materially weaken the E-T bond, formation of an intermediate. In this last case, a second step may involve separation of E and T (as in the first possibility); or yet other orbital interactions in the intermediate may lead to reaction (electron flow accompanied by nuclear motion) elsewhere in the molecule, prior to or following an intramolecular rearrangement of the intermediate. Examples of these situations will be found in the mechanisms to be discussed in the following pages.

As we do not have the space to examine each example to be surveyed later in terms of a detailed description of the electron flow from orbital to orbital during the course of reaction, we will leave it to you to examine the cases of greatest interest to you. We will, however, illustrate the necessity of orbital following in a reaction by covering three cases. You will see these reaction steps several times in the survey section to follow.

The simplest example, bearing much in common with a normal *nucleophilic*

---

[11] Realize that steric interactions may prevent the preferred approach geometry of the attacking molecule and cause the entering group to find an entry point that is less well protected. This aspect of the stereochemistry is applicable to both nucleophilic and electrophilic attack.

attack at a terminal atom, is to be found in the reaction step called a proton shift. In our specific example this step involves two electron pairs: a lone pair is to become a bond pair and bond pair is to become a lone pair.

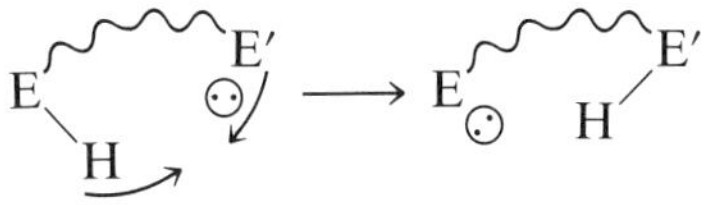

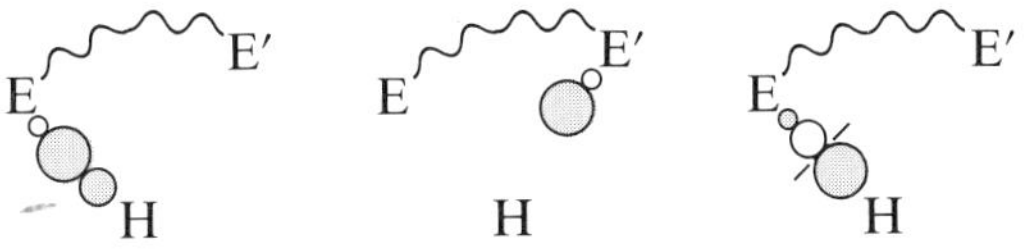

$$\sigma_{EH} \quad\quad lp_{E'} \quad\quad \sigma_{EH}^*$$

The critical mo's, then, are the E—H σ bond orbital, the E' lone pair orbital (HOMO), and the E—H σ* orbital (LUMO). A reasonable reaction coordinate for the proton transfer is simply the swinging of the proton from E toward E'. The required breaking of the E—H bond is facilitated by the overlap between the E' lone pair orbital and the E—H σ* orbital, so that not only does a bond begin to form between H and E', but the partial transfer of the E' lone pair into the E—H anti-bonding orbital weakens the E—H bond. With the phasing of the sketches above and the mo energy order $\sigma_{EH} < lp_{E'} < \sigma_{EH}^*$, the transition state mo's are described by the phasings (see Chapter 4, p. 138):

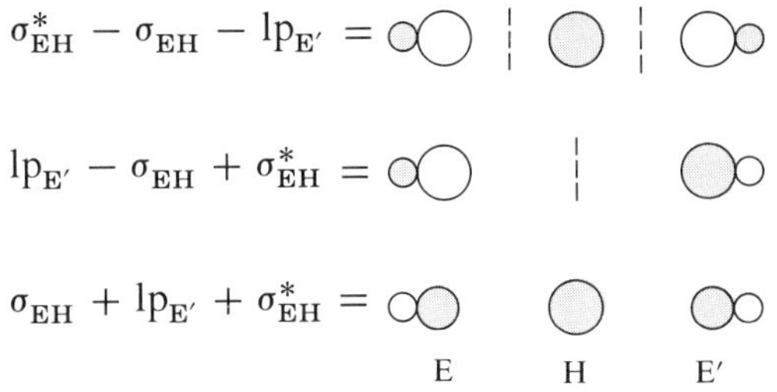

$$\sigma_{EH}^* - \sigma_{EH} - lp_{E'} =$$

$$lp_{E'} - \sigma_{EH} + \sigma_{EH}^* =$$

$$\sigma_{EH} + lp_{E'} + \sigma_{EH}^* =$$

You should recognize these as the three-center/four-electron mo's of hydrogen-bonded adducts from Chapter 5. As the reaction is completed by continued motion of H toward E', you can readily identify the correlation

$$\sigma_{EH} \rightarrow \sigma_{E'H}$$
$$lp_{E'} \rightarrow lp_E$$
$$\sigma_{EH}^* \rightarrow \sigma_{E'H}^*$$

Our conclusion is that *it is possible for the electron pairs to follow and support the proposed nuclear motion so as to convert reactant electron pairs smoothly into product electron pairs.*

Less trivial, and naturally more important, is the step called *1,2 elimination* or *vicinal elimination*. This step also involves two electron pairs: in the reactants these pairs serve to define the vicinal bonds, and in the products we find one pair defining an additional bond between the 1,2 atoms and the second pair defining a bond between the two vicinal substituents. This is illustrated by a unimolecular elimination of H_2 from diimide to form N_2 as the other product.

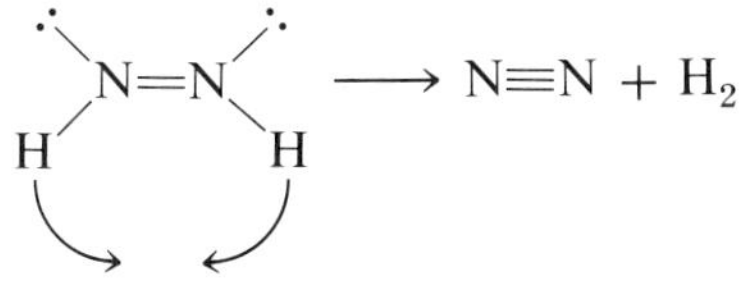

$$HN=NH \longrightarrow N\equiv N + H_2$$

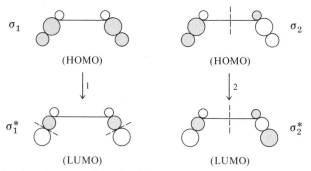

The critical orbitals to consider in this case are the two characterizing the N—H bonds and, since we wish to have these bonds weakened as the elimination proceeds, the two N—H σ^* orbitals. These orbitals are sketched above, with observance of the fact that nitrogen is more electronegative that hydrogen and with consideration of the symmetry of the *cis*-diimide. The reaction coordinate we wish to examine is the simplest possible; the reaction is to proceed by the hydrogens *symmetrically* departing the nitrogens and approaching each other. Under the symmetry of this reaction coordinate, σ_1 and σ_1^* have the same symmetry and so overlap or interact. The same is true of σ_2 and σ_2^*. Thus, the question boils down to whether the mixing of σ_1^* into σ_1 *and* the mixing of σ_2^* into σ_2 lead smoothly to formation of a bond between the hydrogen atoms and an additional bond between the nitrogens. The electron flow brought on by the HOMO/LUMO interactions will lead to bonding between some atoms and increased anti-bonding between others; we can keep track of these flows by constructing a table in which each atom-pair labels a column and the two HOMO/LUMO interactions define rows. Increased bonding will be counted as $+$, and decreased bonding will be signified by $-$. For the case at hand we find

	NH	N'H'	NN	HH
1	−	−	−	+
2	−	−	+	−
Net	−	−	∼0	∼0

Thus, both HOMO/LUMO interactions successfully weaken the NH bonds, but they have offsetting effects on the formation of the new N—N bond and the H—H bond. Our conclusion is that *symmetric withdrawal of the hydrogens in this 1,2 elimination case does not lead to successful orbital following; the electron flow does not fully support the proposed reaction coordinate.* In practical terms this means that the activation energy for the proposed pathway will be excessive; either the reactants must find a lower energy channel to the products, or the reaction will not occur under reasonable thermal conditions.

In a moment we will return to consider what does happen to diimide, but before doing that we consider a more general vicinal situation in which there is less inherent symmetry to the problem. A good, useful case would be that for which one of the N—H bonds is replaced by a B—H bond. The question now is whether this less symmetric case can lead to better orbital following of the desired elimination. The answer is yes, as we can show simply by the following argument. To the left side of the diagram below we show the energies of the mo's of a symmetric case like diimide.

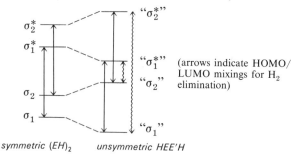

(arrows indicate HOMO/LUMO mixings for H_2 elimination)

symmetric $(EH)_2$ unsymmetric HEE'H

The lowering of symmetry that accompanies the replacement of one of the nitrogens with boron will bring about a mixing of σ_1 with σ_2 and of σ_1^* with σ_2^* to produce new occupied and unoccupied mo's, alternately somewhat *more localized* in the N—H and B—H regions.[12] The effect of this mixing on the energy level diagram will be to bring about an energy separation of σ_1 and σ_2 on the one hand, and of σ_1^* and σ_2^* on the other. An important consequence of this change is that the energy gap between σ_2 and σ_1^* *decreases;* this is important because the lowering of symmetry also makes possible a HOMO/LUMO interaction between these new "σ_2" and "σ_1^*" mo's that was not possible in the symmetric case. It is true that a HOMO/LUMO interaction between the new "σ_1" and "σ_2^*" orbitals is also possible, but the better energy match between "σ_2" and "σ_1^*" makes it more important in the orbital following effect. The consequence of the mixing of σ_1^* into σ_2 is an additional net decrease of N—H and B—H bonding density, a new net increase in N—B bonding density, and a new net increase in H—H bonding density. Our conclusion is that *the orbital following of the "symmetrical" elimination of H_2 from an unsymmetrical reactant is more successful,* and should therefore yield a *lower activation energy* for this pathway, than was encountered for the symmetrical case. In practical application we cannot, without detailed calculations, tell just how much lowering of the $\Delta E^{\ddagger}$ will be realized, but at least we are not surprised to learn that direct elimination is found in some unsymmetrical cases.

Now let's return to the question of what does happen to diimide. The answer is that it disproportionates into hydrazine and nitrogen by transfer of two hydrogens from one diimide to another.

$$2\text{HNNH} \rightarrow \text{N}_2 + \text{H}_2\text{NNH}_2$$

Orbital following arguments show how this can happen. The orbitals important to tracing possible changes in N—N bonding and N—H bonding are as follows: the σ_1, σ_2, σ_1^*, and σ_2^* orbitals of the preceding section will be necessary to account for electron flow germane to the loss of two hydrogens from one diimide and the formation of an N—N π bond within that diimide. For the second molecule our attention naturally focuses on the π and π^* mo's, for these must account for the breaking of an N—N π bond and acceptance of the two hydrogens. The interaction between these HOMO's and LUMO's will be best achieved by an approach geometry in which the molecular plane (yz) of one diimide coincides with the π "orbital plane" of the other diimide.

With this orientation, sketch the HOMO's and LUMO's and ask what HOMO/LUMO interactions are permitted (*i.e.*, have non-zero overlaps) by the symmetry of the reaction coordinate.

---

[12] Note that mixing of both σ_1 and σ_2 with both σ_1^* and σ_2^* is also permitted, but because of the poorer energy match it will be less important than mixing of σ_1 with σ_2 and σ_1^* with σ_2^*.

378 ☐ **REACTION PATHWAYS**

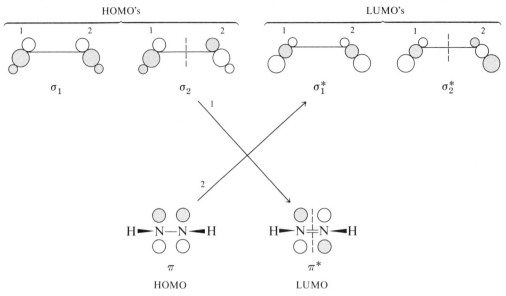

The HOMO and LUMO interactions with non-zero overlaps are indicated with arrows. The σ_2 to π^* charge flow supports strengthening of the N—N bond in the upper diimide, weakening of the original N—H bonds, weakening of the N—N bond of the second diimide, and formation of hydrogen bonds to the latter. The π to σ_1^* electron flow is synergic in all respects. Thus, the electron flow supports the transfer of the protons from one diimide to another in a single smooth step.

These examples should give you a feeling for the great utility of orbital concepts in application to reaction mechanisms. More examples, particularly from organic chemistry, can be found should you wish to pursue the topic.[13]

THE PRINCIPLE OF MICROSCOPIC REVERSIBILITY

The principle of microscopic reversibility[14] is such an important concept that we need to pause to outline its consequences to reaction pathways. To avoid writing out this rather cumbersome name, we shall abbreviate it as PMR. *This principle is actually nothing more than recognition of the common sense idea that a reacting system must be able to retrace its pathway from reactants to products by simply reversing, exactly, its stereochemical path.* (If the city fathers of your town observe this principle, the shortest path between your house and the chem lab involves only two-way streets; if you practice the principle, this is the route you follow.) This in itself is of considerable consequence for reactions in general, since any mechanistic path proposed for a "forward" reaction must also be consistent with rate and mechanism data for the reverse reaction.[15] The PMR contains great implications when the reactants and products are the same chemical species, as in an isotope exchange reaction

---

[13] A rather comprehensive, but at times indirect, treatment of organic reactions can be found in R. B. Woodward and R. Hoffmann, "The Conservation of Orbital Symmetry," Academic Press, New York, 1970. An equivalent approach, but somewhat different in appearance, may be found in R. G. Pearson, *Accts. Chem. Res.,* **4,** 152 (1971); *Chem. Eng. News,* **48,** 66 (Sept. 28, 1970); *J. Amer. Chem. Soc.,* **94,** 8287 (1967).

[14] R. C. Tolman, *Phys. Rev.,* **23,** 699 (1924); see also his "Principles of Statistical Mechanics," Oxford, London, 1938. The formal statement is that any molecular process at equilibrium occurs with equal forward and reverse velocities.

[15] A. A. Frost and R. G. Pearson, "Kinetics and Mechanism," Wiley, New York, 1961; D. S. Matteson, "Organo Metallic Reaction Mechanisms," Academic Press, New York, 1974; R. L. Burwell, Jr., and R. G. Pearson, *J. Phys. Chem.,* **70,** 300 (1966); D. S. Matteson and R. W. H. Mah, *J. Amer. Chem. Soc.,* **85,** 2599 (1963).

$$S-X + X^* \rightleftharpoons S-X^* + X$$

In this special case the forward and reverse reactions are identical (unlike your morning and nocturnal jaunts) and so must proceed with identical rate laws, activation energies, and stereochemistry. It is this last feature which is of great importance to us. The symmetry inherent to this special kind of reaction must be reflected in the reaction coordinate and the occurrence of transition states and/or intermediates along the path. With asymmetric substitution reactions such as

$$S-Y + X \rightleftharpoons S-X + Y$$

the symmetry inherent to the exchange reaction is lost and the forward and reverse reactions are not required to proceed with identical rate expressions, activation energies, and stereochemistry (it may take longer to get to the lab in the morning than to get home in the evening). It is important that you keep this intrinsic difference in mind. Understanding the symmetry constraint on the stereochemistry of the *exchange* reaction is of greatest value for those substitution reactions in which you would have good reason to expect the pathways for the exchange and substitution reactions to be little different from one another, although in the strictest sense they must be different. That is, in some reactions, X and Y will be so little different that the important features of the substitution and exchange pathways are essentially the same.

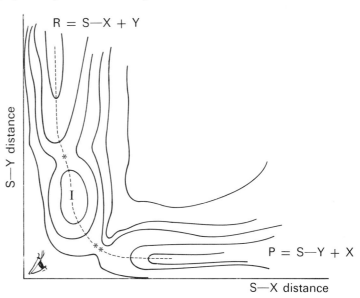

Figure 7-5. A hypothetical reaction surface and associated reaction coordinate diagram for the reaction $S-X + Y \rightleftharpoons S-Y + X$.

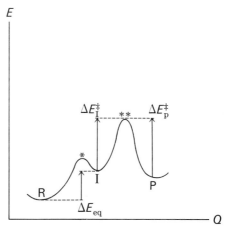

It will help us in exploring the implications of the PMR to think of the reaction in terms of the energy surface leading from reactants to products. Figure 7-5 is a hypothetical energy contour map for some reaction pathway involving an intermediate and two transition states between reactants and products. By following the least-energy path from reactants to products, we may trace the conventional energy profile diagram also shown in Figure 7-5. There are several features of this contour map to call to your attention. In this example there is but a single "least energy" channel connecting reactants and products, although there might be more than one path in a more general case. For the forward reaction you see a rapid pre-equilibrium to form the intermediate (I), followed by a second activation step to produce the products, whereas the reverse reaction proceeds through a major activation step followed by rapid formation of reactants from the intermediate I. The reactants and products are not identical, so the forward activation energy $(\Delta E_{eq} + \Delta E_I^\ddagger)$ is not the same as the reverse activation energy $(\Delta E_P^\ddagger)$.

If we now consider a less general reaction, specifically the intrinsically symmetric isotope exchange reaction, we must modify the contour map in two important ways. As the reactants and products are now identical, their energies must be identical and $(\Delta E_{eq} + \Delta E_I^\ddagger) = \Delta E_P^\ddagger$; otherwise, we would be faced with the non-sense that forward and reverse reactions proceed with different velocities! Even more important to us in visualizing structures for the transition states or intermediates is the following corollary. *For this instance of a single reaction channel, the intermediate I must occur mid-way along the channel and its structure must contain an element of symmetry relating X to X* so that it is not possible to distinguish entering and leaving X's.* The corresponding reaction coordinate diagram appears in Figure 7-6. This description and requirement for the stereochemical changes during the reaction can often be met, but there are times when it is impossible to write a *reasonable* intermediate molecular structure with the necessary symmetry element present. What do we do then? In this case in which only unsymmetrical structures can be written for species along the reaction coordinate, we must satisfy the *necessary* symmetry requirement by *allowing the existence of two reaction channels, containing only unsymmetrical molecular species, which exhibit an energy contour map with an inversion center of symmetry*. With two channels, *the PMR obviously requires that "forward" and "reverse" reactions use both channels with equal frequency*. When we say that it is impossible to write a reasonable symmetric structure along the reaction path, what we are usually saying is that such a structure is possible, but its energy is too high (we estimate) to fall reasonably along the least-energy path. The occurrence of this symmetric, high-energy species on the potential surface is between the least energy channels at the inversion center. Such a situation is diagrammed in Figure 7-7.

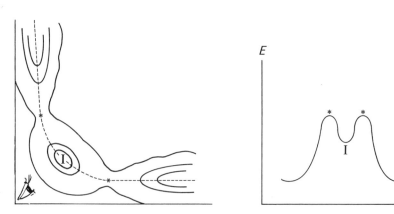

Figure 7-6. The energy contour map and reaction coordinate diagram for a single channel isotope exchange reaction with an intermediate.

THE PRINCIPLE OF MICROSCOPIC REVERSIBILITY 381

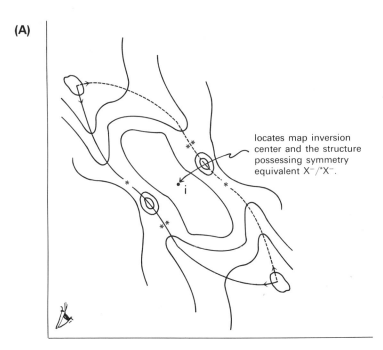

locates map inversion center and the structure possessing symmetry equivalent $X^-/{}^*X^-$.

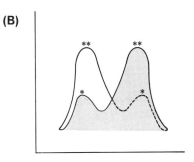

Figure 7-7. (A) An energy contour plot for a two-channel mechanism of unsymmetrical structures and (B) the symmetry-related reaction coordinate diagrams. The eye in (A) locates the view for (B), where the shaded curve is closest to the eye.

We can now summarize by the energy profile diagrams of Figure 7-8 what we may expect to find in surveying known reactions. Diagram (A) is appropriate to an I mechanism like that found for the Walden inversion reaction

$$CH_3Br + {}^*Br^- \rightleftharpoons CH_3Br^* + Br^-$$

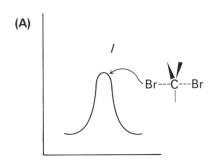
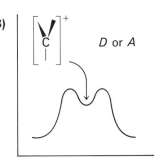

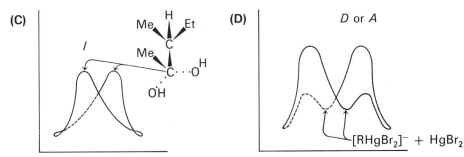

Figure 7-8. Some commonly encountered energy profile diagrams for reactions constrained by identical reactants and products.

Diagram (B) is appropriate to a *D* mechanism such as that exhibited by the reaction

$$(t\text{-Bu})_3\text{CBr} + {}^*\text{Br}^- \rightleftharpoons (t\text{-Bu})_3\text{CBr}^* + \text{Br}^-$$

Diagram (C) is the two-channel counterpart of (A) necessary to explain the oxygen exchange between OH⁻ and protonated methyl (2-butyl) ketone, where the chirality of the butyl 2-carbon atom (if nothing else) prevents us from writing a fully symmetric transition state structure. Diagram (D) is required for the mercury exchange mechanism proposed[16] for

$$\text{RHgBr} + {}^*\text{HgBr}_3^- \rightleftharpoons \text{R}^*\text{HgBr} + \text{HgBr}_3^-$$

i.e., $\text{RHgBr} + {}^*\text{HgBr}_3^- \rightleftharpoons \text{R}^*\text{HgBr}_2 + \text{HgBr}_2 \rightleftharpoons \text{R}^*\text{HgBr} + \text{HgBr}_3^-$

$$\left[\begin{array}{c} \text{Br} \diagdown \diagup \text{R} \diagdown \\ {}^*\text{Hg} \text{Hg}\text{—Br} \\ \text{Br} \diagup \diagdown \text{Br} \diagup \end{array} \right]^{\ddagger} \quad \left[\begin{array}{c} \text{R} \text{Br} \\ \diagdown \diagup \\ {}^*\text{Hg}\text{--Br--Hg} \\ \diagup \diagdown \\ \text{Br} \text{Br} \end{array} \right]^{\ddagger}$$

via (left) ; via (right)

These ideas will be of considerable help to us later in this chapter when we discuss some reactions of silicon and phosphorus.

SURVEY OF REACTIONS

In the remainder of this chapter we will present some proposed "inorganic" reaction mechanisms, supported by kinetic studies where possible, broadly organized on the basis of the number of electron pairs in the valence shell of the reactive atom. This is done to establish, so far as possible, a relationship between reactant structure and proposed mechanism.

As we pursue the ideas of reaction mechanisms, it will become apparent that any one of the overall reaction processes cited earlier in this chapter (substitution, solvolysis, addition, exchange, redox, and free radical) can occur by any of several mechanisms determined by the natures of the reagents involved. This seems particularly true of reactions in the substitution, solvolysis, certain redox, and exchange categories. In addition, while detailed kinetic studies of such things as substituent effects and activation parameters, can often distinguish *a* from *d* modes of activation (the intimate mechanisms), *the formulation of a stoichiometric mechanism*[17] *often involves a good deal of speculation*.

As you examine the reactions of selected compounds, look for the utility of the following concepts:
  (a) the estimated strengths of bonds to entering and leaving groups,
  (b) the "stability" of leaving groups,
  (c) the degree of "unsaturation" at the reactive centers, *e.g.*, the availability of reasonably low lying unoccupied orbitals (atomic or molecular) in the valence shells, and
  (d) the steric bulk of entering and leaving groups.

By way of generalization, we may begin by anticipating the following. For those molecules with only *three bonding pairs* of electrons in the valence shell (those that are electron deficient in the sense of the octet rule) you may expect many reactions certainly to involve an associative step somewhere along the reaction coordinate, but this does not always mean the rate controlling step or activation mode is necessarily *a*.

---

[16] For a good discussion see Matteson's book in Ref. 15.
[17] That is, it is difficult at times to distinguish I_a from A mechanisms and, when there may be a kinetic role for the solvent, to distinguish I_d from D mechanisms.

Those atoms with *four electron pairs* in the valence shell are by far the most numerous and can be divided into sub-classes according to whether they have none, one, two, three, or four lone pairs of electrons (corresponding to Groups IV through VIII). You will see examples of attack (via the lone pairs present) by these atoms upon others, and, conversely, attack (via valence shell expansion) upon the heavier atoms from these groups. The ideas developed for the four-electron-pair cases can often be carried over with simple modification to those molecules whose reactive centers have *five* and *six pairs* of valence electrons (the trigonal bypyramidal and octahedral structures).

ONE VALENCE PAIR

Hydrogen

Perhaps the most commonly encountered reactions involving attack upon a hydrogen atom are those belonging to the *solvolysis class*. Nucleophilic attack on hydrogen is encountered when solvents with a more or less acidic hydrogen atom (protonic solvents) are brought into contact with materials that are strongly basic toward the proton. You know from Chapter 5 that such solvents form hydrogen-bonded adducts with weaker Lewis base materials; with stronger bases, often the proton transfer is complete and leads to solvolysis. Many of these reactions are exceedingly fast—often diffusion controlled—and presumably involve simple, rapid transfer of the proton. Typical reactions are:

$$NaH + NH_3 \rightleftharpoons H_2 + NH_2^- + Na^+$$

$$\left.\begin{array}{l} CaS \\ CaO \\ CaC_2 \\ Ca_3N_2 \\ Ca_3P_2 \end{array}\right\} + H_2O \rightleftharpoons \left\{\begin{array}{l} HS^- \\ OH^- \\ C_2H_2 \\ NH_3 \\ PH_3 \end{array}\right\} + OH^-$$

Such reactions provide for convenient *preparations of deuterated hydrides* if D_2O is used. The reaction of CaC_2 is typical of most metal carbides, and is particularly interesting in that the reaction product suggests that the carbide is actually an acetylide. Be_2C and Al_4C_3 are unusual in that they produce methane on hydrolysis and could be called methides. Other examples of fast ($k \geq 10^8 \text{ M}^{-1} \text{ sec}^{-1}$, diffusion controlled) proton transfer reactions are those involving simple exchange such as

$$NH_3 + NH_4^+ \rightleftharpoons NH_4^+ + NH_3$$

$$NH_4^+ + H_2O \rightleftharpoons NH_3 + H_3O^+$$

Exchange reactions for non-acidic hydrogen seem to feature H in a bridging capacity following electrophilic attack:

$$BCl_3 + MePhNpSiH \rightleftharpoons MePhNpSiCl + BCl_2H$$

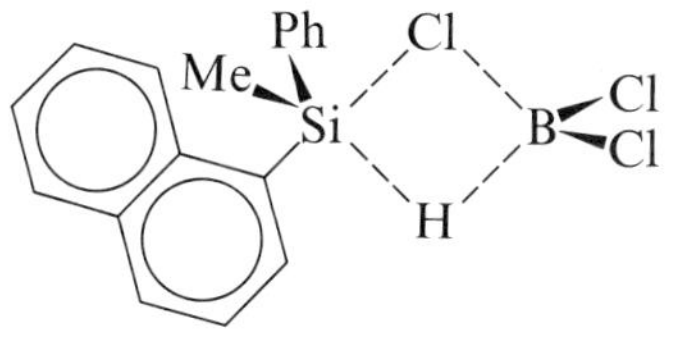

Because of the observed retention by the chiral silane of its configuration, it is believed[18] that the four-center structure shown for the exchange of H and Cl is correct. The concept of such a structure is not foreign to you if you reflect that many ground state structures are known in which H, and also Cl, serve as bridging functions. It is most likely that such a species is an intermediate.

Particularly interesting are the *catalytic hydrogenation reactions* (see Chapter 17 for more examples). Some of these reactions are thought[19] to proceed through a four-center species such as

$$\begin{array}{c} H \rightarrow M \\ | \\ B: \rightarrow H \end{array} \longrightarrow B\!-\!H^+ + H\!-\!M^-$$

for activation of molecular hydrogen (where base = pyridine, a carboxylate anion, or water; and metal = the monovalent silver, mercury, and copper ions, or Pt^{2+}). All of these metals are classified as Lewis acids, and the "push-pull" base-acid action favors heterolytic cleavage of the H_2 molecule to form a metal hydride and a protonated base. It is the metal hydride that then acts as the catalyst (reactive intermediate) in the hydrogenation reaction. A strictly inorganic reaction of this type is the oxidation of H_2 by aqueous Cr(VI), Fe(III), etc.,

$$H_2 + \frac{2}{3}Cr(VI) \xrightarrow{Cu^{2+}} 2H^+ + \frac{2}{3}Cr(III)$$

where the reaction steps are believed to be

$$Cu^{2+} + H_2 \underset{k_{-1}}{\overset{k_1}{\rightleftarrows}} CuH^+ + H^+$$

$$CuH^+ + Cu^{2+} \xrightarrow{k_2} 2Cu^+ + H^+$$

$$2Cu^+ + \frac{2}{3}Cr(VI) \longrightarrow 2Cu^{2+} + \frac{2}{3}Cr(III)$$

The rate law[20] for this reaction is analogous to equation (7-4):

$$-d[H_2]/dt = k_1 k_2 [H_2][Cu^{2+}]^2 / \{k_{-1}[H^+] + k_2[Cu^{2+}]\}$$

which expresses the importance of the catalytic role of Cu^{2+} in electrophilic cleavage of the H_2 bond (the possible role of the solvent is masked). The relative importance of both $k_{-1}[H^+]$ and $k_2[Cu^{2+}]$ means that the first two steps are rate controlling (slower than the third).

Another reaction[21] quite similar to this activation of molecular H_2 involves the deprotonation of olefins coordinated to Pd(II) to produce an allylic moiety stabilized by coordination to Pd, much as H^- was stabilized by coordination to Cu^{2+} in the last example.

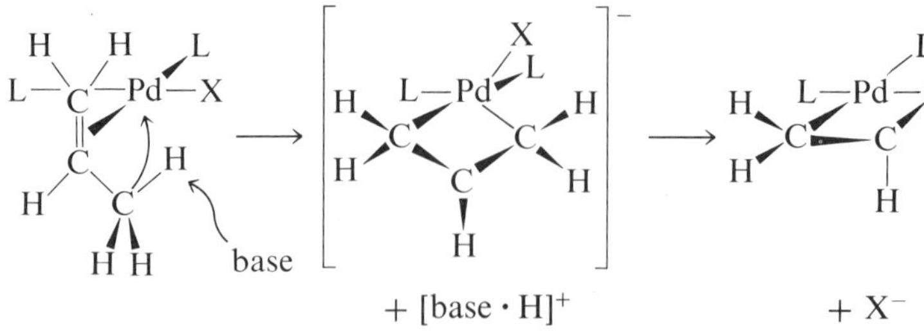

---

[18] This and other examples are discussed in Ref. 9.
[19] J. Halpern, *Quart. Rev.* (Chem. Soc. London), **10**, 463 (1965); *Ann. Rev. Phys. Chem.*, **16**, 103 (1965); see also Ref. 4.
[20] J. Halpern, *et al., J. Phys. Chem.*, **60**, 1455 (1956).
[21] R. Huttel, *et al., Chem. Ber.*, **94**, 766 (1961); G. W. Parshall and G. Wilkinson, *Inorg. Chem.*, **1**, 896 (1962).

The hydrides, in which hydrogen is bound to fairly electropositive metals, tend to be extremely reactive compounds. In addition to the simple binary hydrides (such as NaH noted above), which make excellent desiccants, you are aware of the class of complex anion species such as the hydrides of Group III elements (BH_4^-, AlH_4^-, $Al(BH_4)_3$, etc.), which are very important reagents as sources of H^- and for reduction reactions. The ions AlH_4^- and GaH_4^- are far more reactive (even explosively) with H_2O than is BH_4^-. The reasons for this have not been demonstrated[22] at this time, since mechanism studies appear to have been conducted only for BH_4^-.

The tetrahydroborate ion, for which the hydrolysis is drastically affected by solution pH, has been studied[23] in considerable detail. It is interesting that both hydrogen *exchange* and *hydrolysis* reactions can be observed.

$$BH_4^- + {}^*HOH \rightleftharpoons BH_3^*H^- + HOH$$

$$BH_4^- + 4H_2O \rightarrow B(OH)_4^- + 4H_2$$

Both can be acid catalyzed, with the acid catalyzed hydrolysis reaction being about 15 times faster than acid catalyzed exchange. A study of the *exchange* reaction at pD 14, where the D^+ catalyzed exchange and hydrolysis reactions are sufficiently slow to allow examination of the uncatalyzed exchange reaction, shows the exchange to be first order in $[BH_4^-]$ but zero order in $[OD^-]$. This suggests either a truly D mechanism to form H^- and BH_3 or else a pseudo-first order activation involving both BH_4^- and D_2O. At pH 7, the exchange reaction is dominated by the hydrogen ion dependent pathway for which the exchange rate law is first order in both $[BH_4^-]$ and $[H^+]$. The rate constant for this pathway is $\sim 10^5$ M^{-1} sec^{-1}.

The BH_4^- *hydrolysis* rate law expression also indicates both acid independent and acid catalyzed pathways. The activation enthalpy for the acid catalyzed path is only ~ 8 kcal/mole, while that of the uncatalyzed path is three times larger, ~ 22 kcal/mole. The rate constant for the acid dependent path is $\sim 2 \times 10^6$ M^{-1} sec^{-1}, so that elimination of H_2 is an order of magnitude (actually 15 times) faster than hydrogen exchange. A very important result, namely that less than 4% H_2 and less than 1% D_2 are evolved from reactions of BH_4^- in D_2O and BD_4^- in H_2O, strongly militates against formation of reactive *intermediates* with the structures **1** and **2**.

A mechanism based on loss of HD from equatorial positions in **1** would also permit loss of H_2 with equal facility. Axial/equatorial HD loss would also require equal loss of H_2. Analogous comments apply to **2**. Furthermore, assuming that the exchange and hydrolysis reactions proceed from the same species, this species must have a structure (such as **3**) in which two hydrogens are equivalent and different from the other three bound to boron:

---

[22] One could certainly argue greater nucleophilicity of the hydrogens in AlH_4^- and GaH_4^- and/or more ready nucleophilic attack upon Al and Ga because of their greater ease of valence shell expansion.

[23] W. L. Jolly and R. E. Mesmer, *J. Amer. Chem. Soc.*, **83**, 4470 (1961); *Inorg. Chem.*, **1**, 608 (1962); R. E. Davis, *et al.*, *J. Amer. Chem. Soc.*, **84**, 885 (1962); J. A. Gardiner, *ibid.*, **86**, 3165 (1964).

386 REACTION PATHWAYS

Such a formulation is consistent with the high specificity of HD elimination in the hydrolysis of BH_4^- (in D_2O) and BD_4^- (in H_2O). While the second order form of the rate law for the acid catalyzed path is consistent with the view that a BH_5 intermediate is rapidly formed in a pre-equilibrium step [a D mechanism in BH_5, as in Figure 7-3(B)], it is generally felt that the major activation step is the formation of the BH_5 species (an A mechanism in $BH_4^- + H^+$), with the subsequent exchange and H_2 elimination steps having much smaller activation energies (but comparable to each other). Recent studies[24] of BH_4^- hydrolysis in $MeOH/H_2O$ mixtures at $-78\,°C$ have permitted the identification of the following intermediates after the initial hydrolysis step: H_2OBH_3, $(H_2O)BH_2OH$, and $(H_2O)BH(OH)_2$. These neutral boranes, in which H^- has been replaced by less basic H_2O and/or OH^- donors, hydrolyze at rates slower than that of BH_4^-, presumably because of the solvent and temperature conditions and the less hydridic nature of the B—H hydrogens in the intermediates.

STUDY QUESTIONS

4. Sketch an energy profile for

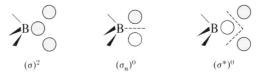

showing both reactions to proceed from a common intermediate. If formation of this intermediate were overwhelmingly the dominant activation process, would there be any detectable difference in the rates of formation of H_2 and H^+ exchange? Let your energy profile sketch roughly reflect the relative rates given on p. 385.

5. Do you find the C_s intermediate **3** for the BH_4^-/H^+ exchange reaction to be consistent with the PMR? Is this proposed mechanism of the single- or dual-channel variety?

6. Viewing the C_s intermediate **3** for the BH_4^-/H^+ reactions as an adduct between H_2 as donor and pyramidal BH_3 as acceptor, we can view the $B-H_2$ interaction as a three-center/two-electron case.

$(\sigma)^2 \qquad (\sigma_n)^0 \qquad (\sigma^*)^0$

Develop the orbital following argument to identify the HOMO/LUMO interaction supporting the departure of H_2 from the BH_5 intermediate.

THREE VALENCE PAIRS

Boron

The most important examples of three valence shell pairs come from Group III of the Periodic Table, and boron has been the most widely studied. A very general reaction for the boron halides is *solvolysis*. With protonic solvents the reaction products are hydrogen halide and a substituted borane. The reaction of the boron halides with water is rapid and vigorous when the halide is Cl, Br, or I, and it probably involves coordination of H_2O as a rapid first step, to be followed by elimination of HX to leave BX_2OH. The latter continues to react, forming $BX(OH)_2$ and finally $B(OH)_3$. The fluoride is unusual in that it can form (below 20°C) a

---

[24] F. T. Wang and W. L. Jolly, *Inorg. Chem.*, **11**, 1933 (1972).

relatively stable adduct with H_2O. This adduct melts at 10°C and, as could be expected, is a strong acid[25]:

$$BF_3 \cdot OH_2 + H_2O \rightleftharpoons H_3O^+ + BF_3OH^-$$

Studies[26] of the *elimination* of HX from the *solvolysis* reactions of amines with BX_3 and $PhBX_2$ in weakly basic aromatic solvents have revealed some interesting aspects of the solvolysis of B—X bonds. First we must recall (from Chapter 6) some generalizations about the relationship between reactants and products of these reactions. With tertiary amines, no solvolyses are observed so that only simple 1:1 adduct formation occurs:

$$R_3N + BX_3 \rightarrow R_3N \cdot BX_3$$

With secondary amines, one equivalent of HX is evolved per amine

$$R_2HN + BX_3 \rightarrow R_2NBX_2 + HX$$

and experimental control of the stoichiometry of reactants leads to the aminoboranes: BX_2NR_2, $BX(NR_2)_2$, or $B(NR_2)_3$. With less sterically hindered *primary* amines, much more interesting reactions occur, which are highly dependent on the nature of the amine substituents. When R of RNH_2 is an aryl group such as phenyl or *o*-tolyl, the first step of the reaction is the formation of a 1:1 adduct as an insoluble intermediate, which subsequently undergoes loss of HX upon heating.[27] In both of these cases, as well as with most primary alkyl amines, the final reaction product is the borazine **4** resulting from loss of two equivalents of HX per adduct.

$$RH_2N + BX_3 \rightarrow \frac{1}{3}[RNBX]_3 + 2HX$$

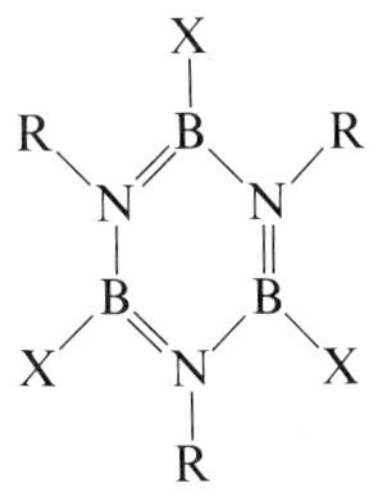

4

The following chart summarizes the findings for aniline and *o*-toluidine:

	o-toluidine	aniline
$RH_2N + BX_3 \rightleftharpoons RH_2N \cdot BX_3$	very fast	very fast
$RH_2N \cdot BX_3 \rightarrow RHNBX_2 + HX$	slow	slow
$2 RHNBX_2 \rightarrow RHN(BXNR)BX_2 + HX$	slower	fast
$RHN(BXNR)BX_2 + RHNBX_2 \rightarrow [RNBX]_3 + 2HX$	fast	fast

---

[25] Compare the ionization of $B(OH)_3$ in water, as discussed in Chapter 5.
[26] H. S. Turner and R. J. Warne, *J. Chem. Soc.*, 6421 (1965); R. K. Bartlett, *et al., J. Chem. Soc. (A)*, 479 (1966); J. C. Lockhart, *ibid.*, 809 (1966); J. R. Blackborow, *et al., J. Chem. Soc. (A)*, 49 (1971).
[27] Very often the elimination of HX can be accelerated (see the discussion of orbital following) by the presence of a tertiary amine, whose role is likely to be either creation of a reaction channel through $\supset$N—BX_3 by deprotonation of the amine adduct or a base assist, via hydrogen bonding, in the transition state.

With o-toluidine as the amine, the elimination of HX proceeds in two kinetically distinct phases. The first of these is the unimolecular evolution of HX from the adduct to form the aminoborane. The second phase more slowly evolves HX and ultimately leads to the borazine. If the reaction is quenched after the evolution of the first equivalent of HX, only the aminoborane (the product of the first step) is found. This means that the first step is essentially complete before the second phase begins.

Without the o-methyl group of the toluidine, the second phase of HX elimination with aniline is considerably faster. The rate expression smoothly progresses in time from first order in initial adduct to second order in that adduct. Here aminoborane so rapidly disappears from the reaction that quenching after the appearance of the first equivalent of HX shows only the presence of *unreacted* initial adduct and the borazine adduct. As the second HX elimination step in this scheme requires the coordination of two amine groups to the same boron atom, it is reasonable that steric congestion (in toluidinoborane) would inhibit the formation of the adduct precursor to HX elimination. In this way we account for the contrast in rate behavior between aniline and o-toluidine. In experiments with 2,6-dimethyl- and 2,6-diethylaniline, the subsequent steps are so severely inhibited that the final products are not the borazines but other (BN) chain polymers. Highly hindered primary alkyl amines behave similarly.

Of some interest is the finding that o-NO$_2$ and o,p-dinitro anilines react with BX$_3$ and PhBX$_2$ to give aminoboranes according to second order rate laws (first order in each of boron acid and amine) without the formation of isolable amounts of amine-borane adducts. The NO$_2$ substitution appears to have a double effect: the strongly electron withdrawing nitro groups lower the basicity of the amine lone pairs (so that the 1:1 adduct is no longer detected prior to HX evolution) and raise the acidity of the amine hydrogen (so that the activation energy for HX elimination from the 1:1 adduct to form aminoborane is reduced), and adduct formation and HX elimination become comparable in speed. Furthermore, the reactions stop at the aminoborane stage—*i.e.*, they do not proceed to formation of the corresponding borazines. This difference from the other anilines might be attributed to both steric congestion about the B—N bond of the aminoborane and a markedly reduced basicity of the nitroaniline nitrogen lone pairs. While the rate laws for these reactions appear to be the same for the solvents MeCN, Et$_2$O, and benzene, the rate constants are quite a bit larger in the hydrocarbon solvent than in the more strongly coordinating nitrile and ether solvents. Activation parameters were not studied for the reactions in benzene but were obtained for the reactions in acetonitrile. Since these reactions are between amine and BX$_3 \cdot$ S, we will postpone a discussion of their mechanism until the next section, where we deal with reactions at four-coordinate boron.

In Chapter 6 we commented on the thermodynamic spontaneity of the thermal decomposition reaction of NH$_3$BH$_3$ (actually performed by heating LiBH$_4$ and NH$_4$Cl above 250°C). For possible mechanistic aspects of this reaction you should note the analogy with the HX elimination reactions of aniline- and toluidine-BX$_3$ adducts to form the N- and B-substituted borazines. Given that

5a ↔ 5b

resonance structure **5a** should be quite important, you should not be surprised to learn that borazine has been found[28] to react with the hydrogen halides to form the

---

[28] A. W. Laubengayer, *et al., Inorg. Chem.*, **4**, 578 (1965).

analog of 1,3,5-trihalocyclohexane (the borons are the 1,3,5 atoms in borazine). This is a cleaner route than synthesis of $H_3N \cdot BHCl_2$, followed by thermal evolution of one equivalent of Cl. More interesting to contemplate are the mechanistic details of the reduction of (B-halo) borazines by sodium borohydride to produce the fully hydrogenated cyclohexane analogs.

In closely related thermal decomposition reactions, secondary amine adducts of borane (at temperatures above 100°C) lose H_2 and apparently dimerize[29] to generate cyclobutane analogs **6** (you should find it interesting to contrast this *dimer* formation of R_2NBX_2 when X = H with the aminoborane *monomers* formed when X = halogen).

$$R_2NHBH_3 \rightarrow H_2 + \frac{1}{2}[R_2NBH_2]_2$$

6

Solvolysis of alkyl boranes, BR_3, interestingly, is accelerated by the presence of aqueous carboxylic acids[30] (mineral acids are not nearly so effective); and hydroperoxide ion, OOH^-, is an excellent catalyst[31] for decomposition of organoboranes. The facility of both reactions in relation to the ineffectiveness of the mineral acids is believed to be a consequence of initial adduct formation by the carbonyl (or peroxo) oxygen atom followed by transfer of an alkyl group to the hydroxy groups (you will later see a mechanism for the disproportionation of peroxides bearing a good likeness to the second of these):

$$BR_3 + R'CO_2H \longrightarrow \quad \longrightarrow R_2BO_2CR' + RH$$
$$\xrightarrow{H_2O} R_2BOH + R'CO_2H$$

$$+ OOH^- \longrightarrow \quad \xrightarrow{H_2O} R_2BOH + ROH + OH^-$$

Both mechanisms maintain the chirality of the carbon bound to boron, in agreement with experiment. The relative strengths of C—O and O—O bonds appear to be as important to the nature of the final products as are the relative strengths (acidities) of the O—H bonds. Both mechanisms (and we must be frank in saying that they are not the only ones consistent with the nature of the products) bear a certain similarity to the "push-pull" concept mentioned earlier for the activation of H_2. More discussion of the organoboranes as alkylating agents will be encountered both in the next chapter and again in later chapters on organometallic compounds, at which times you will see greater application of these mechanistic ideas.

Analogously, it has been discovered[32] that bromination of aryl boronic acids is

---

[29] Trimerization is apparently inhibited by steric interactions between R groups in axial positions of what would be a cyclohexane structure.
[30] See Ref. 15 (Matteson's book); H. C. Brown and K. Murry, *J. Amer. Chem. Soc.*, **81**, 4108 (1959).
[31] H. C. Brown, *et al.*, *J. Amer. Chem. Soc.*, **78**, 5694 (1956); *ibid.*, **83**, 1001 (1961); H. G. Kuivila, *ibid.*, **76**, 870 (1954).
[32] H. G. Kuivila and E. K. Easterbrook, *J. Amer. Chem. Soc.*, **73**, 4629 (1951).

facilitated by the presence in organic solvents of oxygen donors (D = H_2O, OAc^-, etc.):

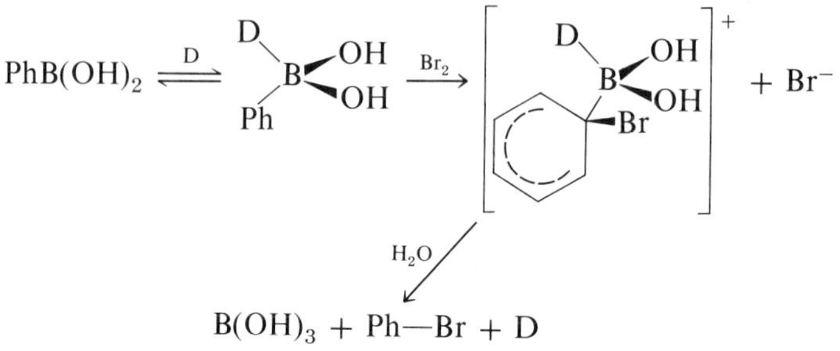

$$PhB(OH)_2 \underset{}{\overset{D}{\rightleftharpoons}} \underset{Ph}{\overset{D}{B}}\begin{matrix}OH\\OH\end{matrix} \overset{Br_2}{\longrightarrow} \left[\underset{Br}{\overset{D}{B}}\begin{matrix}OH\\OH\end{matrix}\right]^+ + Br^-$$

$$\overset{H_2O}{\longrightarrow} B(OH)_3 + Ph-Br + D$$

In fact, the rate law for the bromination is first order in the boronic acid, the oxygen donor, and bromine, which does indeed suggest the pre-equilibrium formation of the oxygen donor/borane adduct. Coordination of boron by D reduces the Ph-B π conjugation so as to enhance the nucleophilicity of Ph. Don't miss the opportunity to relate and contrast this reaction with the catalyzed reactions of H_2 in the last section.

Characteristic *substitution* reactions of Group III halides of some importance are the following:

$$3H^- + BF_3 \overset{R_2O}{\longrightarrow} BH_3 + 3F^-$$

$$4H^- + AlCl_3 \overset{R_2O}{\longrightarrow} AlH_4^- + 3Cl^-$$

$$MeO^- + BCl_3 \longrightarrow MeOBCl_2 + Cl^-$$

The first two are very important reactions, as they constitute methods for synthesis of the important hydrides B_2H_6 and lithium aluminumhydride (refer to the preceding chapter). The detailed mechanisms of these reactions have not been clarified, but the idea that the strong base H^- initially coordinates the boron and aluminum is very likely correct.[33] Subsequent reactions (*I* mechanism) with H^- to displace the less basic bases F^- and Cl^- from $[HEX_3]^-$ appear appropriate.

A final reaction type for boron that has been of some interest to kineticists studying inorganic systems has been the *exchange* reaction. You have already seen in the last section one example of this type in the exchange of hydrogen for chlorine at boron via a doubly bridged species.

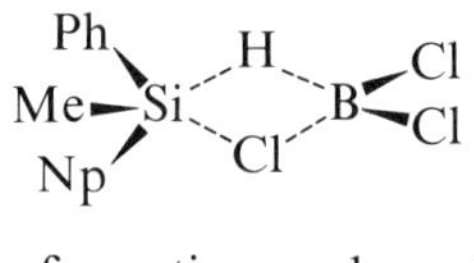

$$\underset{Np}{\overset{Ph}{\underset{Me}{Si}}}\begin{matrix}H\\Cl\end{matrix}B\begin{matrix}Cl\\Cl\end{matrix}$$

Another example of this type of reaction and mechanism is that of exchange of alkoxy groups between borate esters:

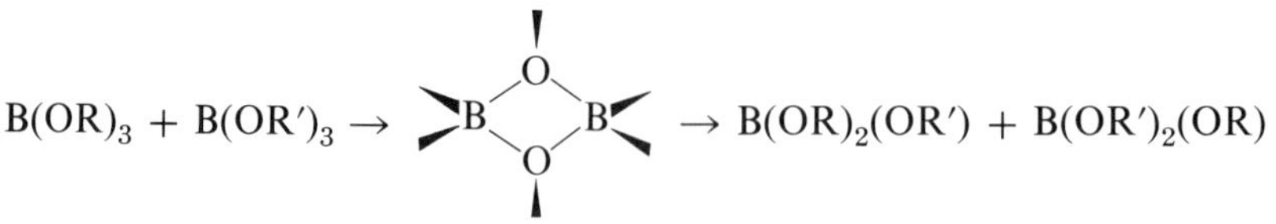

$$B(OR)_3 + B(OR')_3 \rightarrow B\overset{O}{\underset{O}{}}B \rightarrow B(OR)_2(OR') + B(OR')_2(OR)$$

where the reaction rates (which are quite fast) are accelerated for those borate esters that have greater Lewis acidity and are retarded in donor solvents. Similar reactions, with presumably the same doubly bridged type of intermediate or transition state, are known[34] for halogen exchange between the haloboranes, BX_3 and BX'_3.

---

[33] In fact, reactions of X^- with EX_3 will lead to EX_4^- as products.

[34] J. C. Lockhart surveys this general redistribution reaction type in her book, "Redistribution Reactions," Academic Press, New York, 1970.

Certainly one of the most important reactions of three-coordinate boron is the *addition* of BH_3 to olefins (alternately, an insertion of C=C into the BH bond). The so-called "hydroboration" reaction[35] is used to introduce one, two, or three alkyl groups at boron.

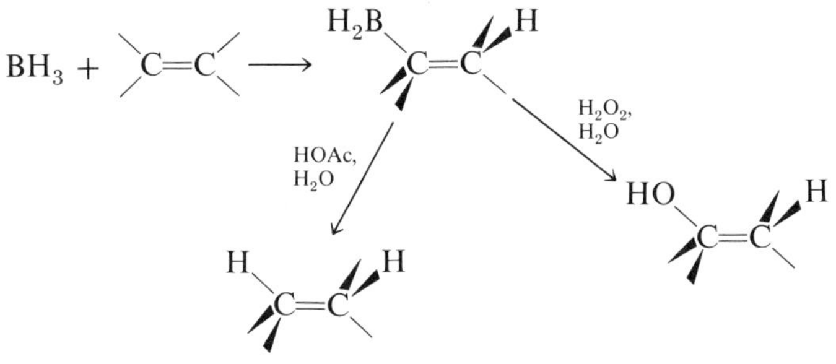

Several aspects of this highly important hydroboration reaction have been carefully determined. First of all, the reaction is strongly anti-Markownikoff, perhaps because the more highly substituted carbon better supports development of positive charge during the later stages of the reaction (see comments later). An example is 2-methyl-1-butene (7), for which 99% of the product has boron attached to the terminal carbon. With 4-methyl-2-pentene (8) there is no significant preference for boron attachment at one or the other carbon.

Given the quite different steric requirements of the methyl and isopropyl groups, it appears that steric factors are not of great importance to the stereochemistry of the hydroboration reaction. A steric effect can be observed, however, on the *extent* of the hydroboration reaction. With 2-methyl-2-butene (9) the reaction stops after formation of *bis*(3-methyl-2-butyl)borane, while 2,4,4-trimethyl-2-pentene (10) yields only 2,2,4-trimethyl-3-pentylborane.

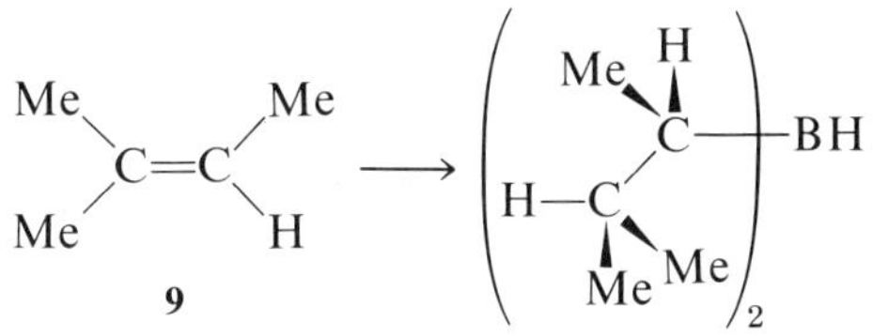

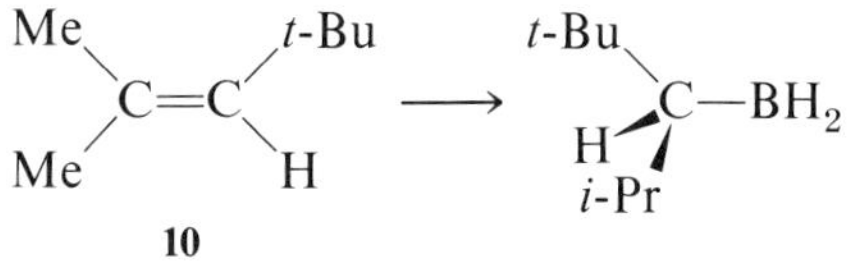

Another important aspect of the reaction is that it proceeds smoothly and quickly in ether solvents but is very slow in the gas phase or when the reagents are combined neat. Clearly, B_2H_6 is ineffective for addition to the double bond, whereas BH_3 is

---

[35] H. C. Brown's recounting ("Boranes in Organic Chemistry," Cornell University Press, Ithaca, 1972) of his and his students' experiences in this area is not only informative but highly entertaining reading. See also K. Wade, "Electron Deficient Compounds," Nelson, London, 1971.

much more reactive. A third feature concerns the *cis*-stereochemistry of the addition. This is deduced from the production of *trans*-2-methylcyclopentanol and *trans*-2-methylcyclohexanol from 1-methylcyclopentene and 1-methylcyclohexene, respectively.

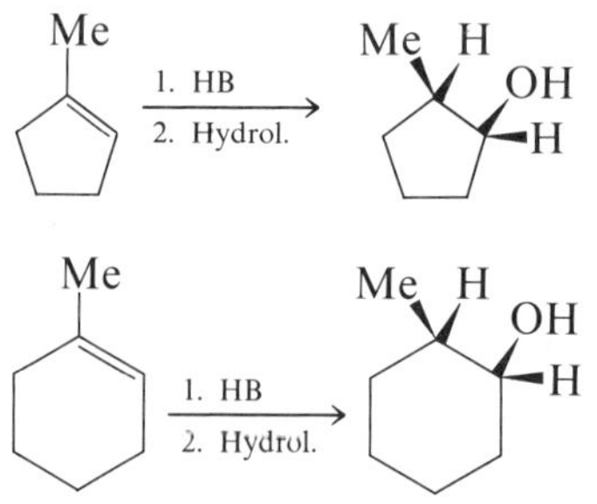

Finally, a few paragraphs back we discussed the mechanism of the carboxylic acid catalyzed decomposition of boranes to produce alkanes and the hydroperoxide catalyzed decomposition of such boranes to yield the alcohol. That both reactions preserve the 2-carbon chirality is critical to the proof of *cis*-hydroboration.

The details of the hydroboration reaction mechanism are not known, but an educated guess would have a weak olefin π-donor complex formed early in the reaction (review the discussion of such adducts in Chapter 5). While such an adduct probably would be symmetrical for symmetrical olefins, it would not be so for unsymmetrical olefins. The primary activation step should then be the shifting of the boron toward one carbon while the hydrogen is transferred to the other carbon. If you wish to pursue the orbital following of this two-step process, first consider the olefin π mo and the boron p ao for formation of the π adduct. Then trace the addition step as a greater HOMO/LUMO interaction between the olefin π mo and the boron acceptor orbital, and a HOMO/LUMO interaction between the B—H bond pair and the olefin π^* mo.

STUDY QUESTIONS

7. Analyze the fact that $BF_3 \cdot OH_2$ eliminates HF much more slowly than $BCl_3 \cdot OH_2$ eliminates HCl, in terms of bond-breaking and bond-making influences on the activation energy.

8. What feature of a boron bond to an aryl carbon is altered by boron adduct formation so as to facilitate bromination of the aryl group?

9. Is it true that the rate law (Rate = k [Donor] [Br_2] [Boronic acid]) *must* imply formation of an adduct of the boronic acid as *the precursor* in the bromination of boronic acids? What other "donor/acceptor" interaction, consistent with this rate law, could give rise to a different precursor? Explain the role of the Donor in this event.

10. In a synthesis of B_2H_6 by hydride reduction of BF_3, the reaction products are actually not consistent with

$$3MH + BF_3 \rightarrow \frac{1}{2} B_2H_6 + 3MF$$

but with

$$3MH + 4BF_3 \rightarrow \frac{1}{2} B_2H_6 + 3MBF_4$$

Viewing the reaction as proceeding through steps of H^- attack upon BF_xH_{3-x} (x = 3,2,1), what is a possible role for the excess BF_3 required?

FOUR VALENCE PAIRS

The examples that could be cited in this category are numerous indeed; in fact, examples can be drawn from all the elements in Groups IV through VII. In addition, the adducts of Group III elements fall into this class. We will consider some representative reactions from all these groups.

Boron

Several four-coordinate boron compounds have been carefully studied for nucleophilic displacement at the boron as the reactive center. The *exchange* reaction of excess Me_3N for that in Me_3NBMe_3 has been shown[36] to proceed with a rate law that is first order in adduct concentration. The enthalpy of activation (+17 kcal/mole) is indistinguishable from the adduct dissociation enthalpy, and the entropy of activation is positive and fairly large (+15 eu). These findings strongly support the view that the activation step in the exchange reaction is unimolecular dissociation of the adduct to give free BMe_3 (a D mechanism), which subsequently rapidly combines with free Me_3N to form the adduct.

It is particularly interesting to contrast this mechanism for donor exchange of Me_3NBMe_3 with those for $Me_3E \cdot NMe_xH_{3-x}$ (E = Ga, In), for which the exchange reactions are so rapid that attempts[37] to determine the rate laws have been unsuccessful. Certainly such rapid reactions proceed without adduct dissociation (contrast the BMe_3 case) since the adduct dissociation enthalpies are all about 20 kcal/mole. The somewhat larger sizes of Ga and In atoms presumably are responsible for the differences in displacement mechanisms of these Group III adducts (it may be that the availability of $4d$ and $5d$ atomic orbitals in the valence shells of Ga and In also facilitate the formation of a five-coordinate transition state for Ga and In). Arguments could be made for either I or A mechanisms.

Even more interesting are the results of studies[38] into the mechanism for the electrophilic acid-exchange reactions:

$$Me_3Ga + Me_3Ga \cdot NMe_xH_{3-x}$$

$$Me_3In + Me_3In \cdot NMe_xH_{3-x}$$

For both Lewis acids, the reactions are of the *D type when the amine is tertiary or secondary* [the activation energies are quite close to the adduct bond energies (10 to 20 kcal/mole) and the entropies of activation are positive and large (10 to 30 eu)]. With the primary amine and ammonia, however, the reactions are either A or I according to the rate laws, have activation energies of only half the adduct bond energies, and have negative entropies of activation (−10 eu). The dependence of this mechanism for exchange on the number of methyl groups about nitrogen is seen as a steric phenomenon (refer to Figure 7–9).

Rather thorough studies of the *substitution* of coordinated amine from borane adducts by phosphine and amine nucleophiles have been reported.[39] Some pertinent

---

[36] A. H. Cowley and J. H. Mills, *J. Amer. Chem. Soc.*, **91**, 2910 (1969).
[37] J. B. DeRoos and J. P. Oliver, *Inorg. Chem.*, **4**, 1741 (1965); *J. Amer. Chem. Soc.*, **89**, 3970 (1967).
[38] See Ref. 37.
[39] W. L. Buddle and M. F. Hawthorne, *J. Amer. Chem. Soc.*, **93**, 3147 (1971); D. E. Walmsley, *et al.*, *ibid.*, **93**, 3150 (1971).

Figure 7–9. (A) $(CH_3)_3GaN(CH_3)H_2$, assuming tetrahedral symmetry of the gallium and nitrogen atoms, shown with van der Waals radii for all protons. (B) $(CH_3)_3GaN(CH_3)_2H$, assuming tetrahedral symmetry of the gallium and nitrogen atoms, shown with van der Waals radii for all protons.

(A)   (B)

TABLE 7-2
NUCLEOPHILIC DISPLACEMENT AT BORON

Substrate	Nuc	k_1	$\Delta H_1^{\ddagger}$	$\Delta S_1^{\ddagger}$	k_2	$\Delta H_2^{\ddagger}$	$\Delta S_2^{\ddagger}$
amine · borane + Nuc → Nuc · borane + amine							
$Et_3N \cdot BH_3$ (50°C)	$(n\text{-Bu})_3P$	—	—	—	26	—	—
$Me_3N \cdot BH_3$ (50°C)	"	—	—	—	13	23	−5
(40°C)	"	—	—	—	4.4	—	—
$Me_3N \cdot BH_2R$							
R = n-Bu (40°C)	"	0.24	38	40	5.9	22	−9
i-Bu (40°C)	"	0.43	32	18	3.2	26	5
sec-Bu (40°C)	"	1.3	31	19	11.4	19	−7
t-Bu (40°C)	"	11.5	32	25	—		
phenyl (30°C)	"	—			11.6		
p-anisyl (30°C)	"	0.12			13.9		
p-Br phenyl (30°C)	"	—			12.6		
p-tolyl (30°C)	"	—			11.1		
o-tolyl (30°C)	"	—			51.5		
mesityl (30°C)	"	6.4			546		
$MeNH_2 \cdot BHR_2$	$NH_2CH_2CHPh_2$						
R = p-anisyl	"	22.9					
p-tolyl	"	3.2		26	11		
phenyl	"	.2		to	to		
p-F phenyl	"	.1		37	38		
p-Cl phenyl	"	.01					
p-Br phenyl	"	.009					

Values reported are: $k_1 \times 10^5$ sec^{-1}, $k_2 \times 10^5$ M^{-1} sec^{-1}, $\Delta H^{\ddagger}$ in kcal/mole, $\Delta S^{\ddagger}$ in eu.

data are presented in Table 7-2. It is to be noted that the rate law progresses *from second order to competitive first and second orders* when BH_3 is compared with BH_2R. In the case of the Me_3N adduct with BH_3, the activation parameters and the rate law point to attack by the phosphine as the rate controlling step (probably an I process). When one of the hydrogens of BH_3 is replaced by one of the butyl isomers or by an aryl group, several interesting effects are to be noted. Most important is the appearance of a competitive first order (in adduct only) pathway for the displacement reaction. This would seem to have its origin (barring solvation effects) in the steric bulk of the butyl group when the borane substituent is one of the butyl isomers (note the correlation between k_1 and the steric congestion about the B—N bond) and in electronic stabilization (π conjugation) of the transition state and intermediate when R is p-anisyl and mesityl (in this case steric congestion about the B—N bond due to *two* o-methyl groups is probably also important). The operation of a steric effect (interestingly, by facilitating rupture of the original N—B bond) in the second order pathway is one way to interpret (as an I_d process) the trend of k_2 values for the butyl boranes and the large value of k_2 for R = mesityl.

To support the hypothesis that electronic stabilization of the transition state for the unimolecular pathways is realistic, data have been collected for the reaction of methylamine-diarylborane adducts with 2,2-diphenylethylamine (bottom of Table 7-2). The activation parameters and rate law support the idea of unimolecular dissociation of the adduct as the activation step in these reactions. (Steric congestion in the transition state for an A or I mechanism can be cited as a factor in the lack of a competitive second order rate term.) Most important, however, is the smooth variation in k_1 with the ability of the substituted phenyl groups on boron to stabilize the incipient formation of BHR_2 via the resonance structures:

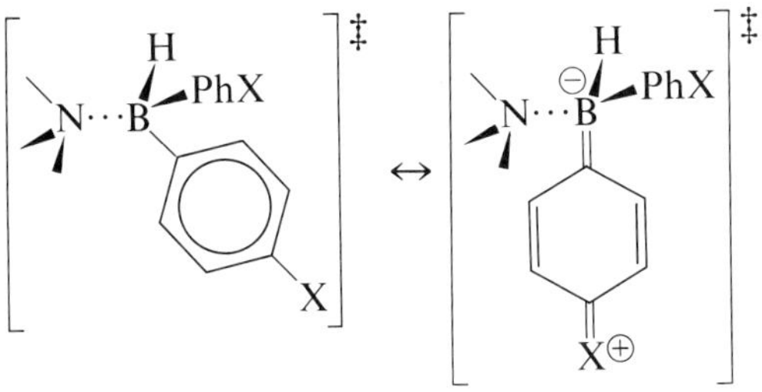

$$\left[\begin{array}{c} \diagdown \\ \diagup N \cdots B \diagdown \\ \diagup \end{array} \begin{array}{c} H \\ \ominus \\ \end{array} \begin{array}{c} PhX \\ \end{array} \right]^{\ddagger} \leftrightarrow \left[\begin{array}{c} \diagdown \\ \diagup N \cdots B \\ \diagup \end{array} \begin{array}{c} H \\ \ominus \\ \end{array} \begin{array}{c} PhX \\ \end{array} \right]^{\ddagger}$$

Finally, you should recall (p. 393) that the fully substituted borane adducts, $BMe_3 \cdot$ (amine base), exchange by a first order D mechanism only.

Related to the mechanisms for nucleophilic substitution at boron are the mechanisms[40] for HX *elimination-solvolysis* of BCl_3 and BBr_3 by amines in acetonitrile or ether as solvent (*in which the BX_3 are coordinated by the nitrile or ether*). On page 388 we considered the solvolysis reactions of the heavier boron halides with primary nitro-aryl amines in very weak-donor aromatic solvents:

$$RNH_2 + S \cdot BX_3 \rightarrow RNHBX_2 + HX + S$$

with R = (structures shown: 2,4-dinitrophenyl; 1,3-dinitronaphthyl (methyl-substituted); 2-nitrotolyl)

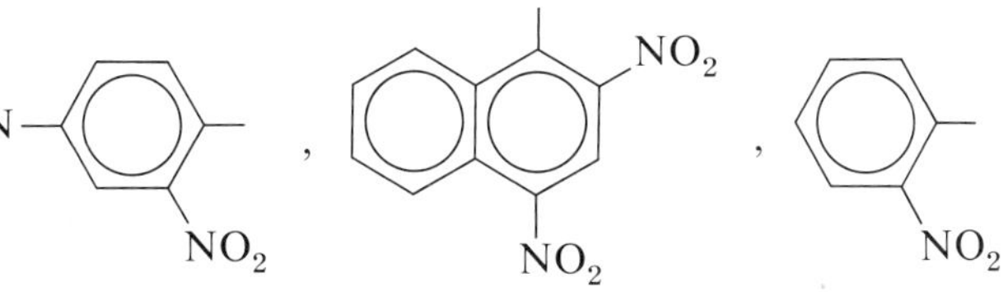

With the somewhat more basic nitrile and ether solvents (and the boron trihalide in large excess relative to the amine), a rate law of the form

$$\text{Rate} = k_{\text{obs}} [RNH_2]$$

was observed, and k_{obs} was shown to be proportional to $[S \cdot BX_3]$. Such a rate law could arise in any of three ways: (i) an I mechanism in amine + $S \cdot BX_3$ according to equation (7-3a); (ii) an *apparent D* process in $S \cdot BX_3$ according to equation (7-4b); or (iii) a D elimination of HX from $RH_2N \cdot BX_3$, following equation (7-3b). For (7-4b), R = $S \cdot BX_3$, P′ = S, and R′ = RNH_2,

$$S \cdot BX_3 \underset{k_{-2a}}{\overset{k_{2a}}{\rightleftarrows}} S + BX_3$$

$$BX_3 + RNH_2 \xrightarrow{k_{2b}} RNH_2BX_2$$

and the general (not limiting) rate expression is

$$\text{Rate} = \frac{k_{2a}k_{2b}[S \cdot BX_3][RNH_2]}{k_{-2a}[S] + k_{2b}[RNH_2]}$$

It is possible for this rate law to reduce to a second order rate law, if $k_{-2a}[S] \gg k_{2b}[RNH_2]$. In the actual experiments, $[S]/[RNH_2] \approx 10^5$, so the constraint for a second order rate law from a D mechanism becomes $(k_{-2a}/k_{2b}) > 10^{-3}$ (k_{-2a} must be no more than three orders of magnitude less than k_{2b}). The mechanism choices reduce to:

$$I: \quad S \cdot BX_3 + RNH_2 \rightarrow RNH_2 \cdot BX_3 + S \qquad \text{[see eqn. (7-3)]}$$

---

[40] See Ref. 26; J. R. Blackborow and J. C. Lockhart, *J. Chem. Soc., (A)*, 3015 (1968).

"D": $S \cdot BX_3 + RNH_2 \rightarrow S + BX_3 + RNH_2$ [see eqn. (7-4b)]

Elim: $RNH_2 \cdot BX_3 \rightarrow RNHBX_2 + HX$ [see eqn. (7-3b)]

for which

I: $\Delta H^{\ddagger}_{obs} = \Delta H^{\ddagger}_I$ $\Delta S^{\ddagger}_{obs} = \Delta S^{\ddagger}_I$ (<0)

"D": $\Delta H^{\ddagger}_{obs} = \Delta H_d + \Delta H^{\ddagger}_{2b}$ $\Delta S^{\ddagger}_{obs} = \Delta S_d + \Delta S^{\ddagger}_{2b}$ (>0)

$\cong \Delta H_d$ $\cong \Delta S_d$

Elim: $\Delta H^{\ddagger}_{obs} = \Delta H_{exch} + \Delta H^{\ddagger}_{elim}$ $\Delta S^{\ddagger}_{obs} = \Delta S_{exch} + \Delta S^{\ddagger}_{elim}$ (>0)

Now, in solution $\Delta H^{\ddagger}_{2b}$ (for BX_3 + amine → amine · BX_3) should be a small positive number, perhaps very close to zero, as should be $\Delta S^{\ddagger}_{2b}$. Also, ΔH_{exch} should be negative (amine a stronger donor than solvent), but smaller than $\Delta H^{\ddagger}_{elim}$, and $\Delta S^{\ddagger}_{elim}$ should be large and positive. When $X = Cl$, $\Delta H^{\ddagger}_{obs}$ is found to be $\sim \frac{1}{2} \Delta H_{diss}$ (10 vs. 18 kcal/mole), so the "D" mechanism is eliminated. That $\Delta S^{\ddagger}_{obs} \sim -20$ eu further eliminates the Elim mechanism. When $X = Br$, $\Delta H^{\ddagger}_{obs} = \Delta H_{diss}$ (20 kcal/mole), which doesn't necessarily eliminate any possibility but is strongly suggestive of "D". That $\Delta S^{\ddagger}_{obs}$ is small and positive ($\sim +10$ eu) does eliminate the Elim path. When $X = Cl$, we suspect that the I path is operative; when $X = Br$, the D path is suggested. A possible explanation for the mechanism change may lie in the steric bulk of Br as a suppressant of the I channel. It is also worth noting that in spite of a difference in mechanism, $S \cdot BCl_3$ and $S \cdot BBr_3$ react with very similar $\Delta G^{\ddagger}$ (very similar k_f).

Not surprising is the finding that replacement of one chlorine by a phenyl group causes no change in the rate law expression for amine solvolysis and only a modest increase in the activation enthalpy (from 10 to 12 kcal/mole). Consequently, it is believed that $CH_3CN \cdot BPhCl_2$ reacts with 2,4-dinitronapthylamine according to the same mechanism as the $CH_3CN \cdot BCl_3$ adduct.

A study[41] of the *hydrolysis* of substituted pyridine adducts of BCl_3 resulted in the conclusion that the hydrolysis reactions proceed by a dissociative mechanism

$$py \cdot BCl_3 + 3H_2O \rightarrow py + 3HCl + B(OH)_3$$

in which the activation step seems to involve heterolytic cleavage of a B—Cl bond. In comparison with $MeCN \cdot BCl_3$, the pyridine adducts have much stronger adduct bonds and should show more resistance to N—B cleavage. At least for the first hydrolysis step (the slow one) it is a B—Cl bond rather than the B—N bond that is cleaved. The activation energies and rate constants for the first order rate expression suggest the reactivity order

$$Me\text{-}py \cdot BCl_3 > py \cdot BCl_3 > 2\text{-}Me\text{-}py \cdot BCl_3$$

The relative rate constants are $3.5 : 1 : 0.5$ and the small differences do make it difficult (remember that H_2O is a basic, acidic, polar solvent) to interpret the ordering. This order is the exact reverse of that expected for rate determining cleavage of the pyridine bond to boron (there should be a steric enhancement of the rate for the 2-methylpyridine adduct if the process is D in $Am \cdot BCl_3$). A truly bimolecular rate step (I path) in which H_2O displaces pyridine or Cl^- is also inconsistent with this order, as the electrophilic character of the boron is expected to follow the order pyridine adduct > 4-methylpyridine adduct. In the event of an I mechanism, depending on which factor is dominant, the 2-methylpyridine adduct would be less

---

[41] G. S. Heaton and P. N. K. Riley, *J. Chem. Soc., (A)*, 952 (1966).

reactive than the other two because of intrinsic nitrogen donor strength or more reactive because of steric weakening of the N—B and B—Cl bonds. Unless some difficult-to-identify solvent (aqueous ethanol) effects are responsible for the observed rate constants, the order is that expected for cleavage of a B—Cl bond. The low rate constant for the 2-methylpyridine adduct is interpreted to mean that the transition state species for this adduct does not experience the expected degree of resonance stabilization (11) for a planar boron species

$$\left[\begin{array}{c}\text{Me}\\ \oplus\\ \text{N}=\text{B}\\ \text{Cl}\\ \text{Cl}\end{array}\right]^+ \leftrightarrow \left[\begin{array}{c}\text{Me}\\ \oplus\\ \text{N}-\text{B}\\ \text{Cl}\\ \text{Cl}\end{array}\right]^+$$

11

because the steric interaction between the 2-methyl group and the remaining chlorines prevents formation of the planar structure **11** by the incipient cation:

$$\left[\begin{array}{c}\text{Me}\\ \oplus\\ \text{N}-\text{B}\begin{array}{c}\text{Cl}\\ \text{Cl}\end{array}\end{array}\right]^+$$

In what *appear* to be mechanistically related reactions, the *hydrolyses* of trimethyl amine adducts of mono-haloboranes are believed[42] to proceed by d activation with B—X cleavage. In aqueous dioxane as a solvent, the rate law expression for

$$\text{Me}_3\text{N} \cdot \text{BH}_2\text{X} + 3\text{H}_2\text{O} \rightarrow \text{Me}_3\text{NH}^+ + \text{H}_3\text{BO}_3 + 2\text{H}_2 + \text{X}^-$$

is independent of H^+ and OH^- concentrations. Furthermore, the first order rate constants follow the order Cl (10^{-6} sec^{-1}), Br (5.1×10^{-4} sec^{-1}), I (4.0×10^{-3} sec^{-1}) and strongly argue against the possibility that d activation by B—N cleavage is operative here (see Chapter 5 for a discussion of the effect of X on the B—N bond strengths). The order observed is that expected for cleavage of the B—X bond. The role of H_2O in the activation step is uncertain, but it would be difficult to explain an activation step in which the H_2O attack at the boron plays a significant role when OH^- is essentially unreactive (OH^- is generally a much better nucleophile than H_2O). The activation parameters are quite interesting for these reactions. The $\Delta H^\ddagger$ values increase in order of decreasing rate constant (from 17.2 to 19.8 kcal/mole) while the entropies are negative: I (-12 eu), Br (-15 eu), and Cl (-19 eu). At first glance the negative values would seem to be in disagreement with the proposed mechanism; however, solvation of the X^- as it is formed in the transition state would be expected to be in the order $Cl^- > Br^- > I^-$. Thus, by assuming that anion solvation is more important to the entropy of activation than is the B—X bond-breaking, a perplexing situation is avoided.

Comparing hydrolysis (in H_2O) of $D \cdot BX_3$ with aminolysis (in MeCN) of $MeCN \cdot BX_3$, you are struck by a drastic change in reaction channel: B—X vs. D—B cleavage. The greater solvation of X^- by H_2O is likely as the source of this change, as you were advised in the section "Caveat Concerning Solvent Effects."

In contrast to these results for amine adducts of BH_2X, *hydrolysis* of amine adducts of BH_2(aryl) and BH_3 seem always to involve a process that is dependent on H_3O^+. When pyridine adducts of BH_2(aryl) were studied,[43] it was found that the rate

---

[42] J. R. Lowe, et al., *Inorg. Chem.*, **9**, 1423 (1970). C. Weidig et al. (*Inorg. Chem.*, **15**, 1783 (1976)) have reported that $Me_3N \cdot BH_2F$ experiences both uncatalyzed and H^+-catalyzed hydrolysis.

[43] R. E. Davis and R. E. Kenson, *J. Amer. Chem. Soc.*, **89**, 1384 (1967).

constant increases with the electron releasing ability of the substituent at the 4 position of the phenyl ring. Thus, the relative rate constants were X = Cl (0.5), H (1.0), CH$_3$ (4.3), and CH$_3$O (22.3).

[Structures 12 and 13: protonated transition states for pyridine–BH$_2$(C$_6$H$_4$X) + H$_3$O$^+$]

This order fits nicely with a concept (*A* path) of the transition state for these reactions as attack of H$^+$ upon the hydridic hydrogens of the borane unit (**12**) (compare the earlier discussion of BH$_4^-$ hydrolysis in acid media) *or upon the pyridine nitrogen* (**13**).

An extensive series of studies[44] into the mechanisms of *hydrolysis* of amine-boranes has led to the conclusion that these acid catalyzed reactions proceed by *attack at the amine nitrogen*. The BH$_3$ adducts of the tertiary amines Me$_3$N, Et$_3$N, and quinuclidine ([bicyclic N structure]) all hydrolyze according to a second order rate law: first order in each of amineborane and H$_3$O$^+$. The activation energies are all closely the same (24 kcal/mole) and the activation entropies are quite small (within a few eu of zero). The similarity in behavior of Et$_3$N and quinuclidine adducts seems to strongly militate against rearward attack by H$_3$O$^+$ at nitrogen (*cf.* Me$_3$Ga attack at nitrogen of Me$_3$N·GaMe$_3$ on page 393). Substituted aniline adducts of BH$_3$ also exhibit[45] a second order hydrolysis pathway, but a term that is first order in adduct concentration is also observed. Given the following information, the first order mechanism appears to involve adduct dissociation (a *D* path) while the second order term involves frontal H$_3$O$^+$ attack at nitrogen (an *I* or *A* path):

(a) electron-releasing phenyl substituents (of aniline) reduce k_1 and increase k_2, while withdrawing groups have the opposite effects;

(b) replacement of aniline hydrogens by alkyl groups increases k_1 but reduces k_2;

(c) the BD$_3$ adducts exhibit rates of hydrolysis no different from those of the isotopic BH$_3$ adducts (this seems to eliminate a BH$_5$-like intermediate);

(d) increasing alkyl substitution in the series RNH$_2$, R$_2$NH, R$_3$N results in decreasing k_2 (k_2 should increase if a BH$_5$-like intermediate were formed, but would decrease from steric congestion about nitrogen for H$^+$ attack at the N—B bond).

Presumably, an explanation for the appearance of a first order pathway for the anilines relative to the tertiary amine donors is their much reduced base strength. In contrasting the greater hydrolytic stability of the amine boranes with that of BH$_4^-$, we might cite the greater electronegativity of the amine nitrogen than of the hydride ion as rendering the borane hydrogens less hydridic [recall that with BH$_4^-$, rapid ($\Delta H^\ddagger \approx 8$ kcal/mole) reaction of H$_3$O$^+$ with the borane hydrogens occurs]. Most interesting is the general observation that phosphine borane adducts are stable toward hydrolysis in *hot* mineral acids. At present there appears to be no real understanding of this property of the phosphine boranes (but see study question 14).

---

[44] H. C. Kelly and J. A. Underwood, III, *Inorg. Chem.*, **8**, 1202 (1969).
[45] H. C. Kelly, *et al., Inorg. Chem.*, **3**, 431 (1964).

Finally, we might note that the *alcoholysis* by propanol of substituted pyridine boranes

$$X\text{-py} \cdot BH_3 \xrightarrow{ROH} X\text{-py} + 3H_2 + B(OR)_3$$

is first order in amine adduct and zero order in alcohol.[46] Substitution of the *ortho* position of the phenyl ring results in an increase in the first order rate constant as expected for a steric influence on the rate determining dissociation of the B—N bond; the activation enthalpies were in the range between 20 and 25 kcal/mole, consistent with rate determining dissociation of the B—N bond.

Particularly intriguing are the *solvolyses* by ammonia of simple borane adducts. For example, the simple ethane analog NH_3BH_3 can be formed by the reaction

$$R_2OBH_3 + NH_3 \xrightarrow{ether} NH_3BH_3 + R_2O$$

or by

$$LiBH_4 + NH_4Cl \xrightarrow{ether} LiCl + H_2 + NH_3BH_3$$

In the former reaction it is not known whether the borane substrate is the dimethyl ether adduct or free (monomeric) borane, but it seems unlikely that the reaction could proceed through attack on B_2H_6 by NH_3 (see the next paragraph). The stoichiometric mechanism for the second reaction is most likely to involve attack upon BH_4^- by NH_4^+ or H^+ to liberate BH_3 or the ether adduct (recall the acid catalyzed hydrolysis reaction of BH_4^-), either of which could be attacked by NH_3 as in the first reaction.

As an example of kinetic (as opposed to thermodynamic) control of reaction products, the reaction of B_2H_6 with liquid ammonia at temperatures lower than $-78°C$ produces[47] a mono-hydrogen bridged adduct by partial cleavage of the borane dimer (note the analogy with $B_2H_7^-$, Chapter 5):

$$\begin{array}{c} H_3N \\\\ \diagdown \\\\ B\text{—}H\text{—}B \\\\ \diagup\diagdown \\\\ HHH \end{array}$$

Warming the reaction mixture above $-78°C$ to remove the solvent ammonia stimulates further reaction to form the "diammoniate" of diborane with a salt structure

$$\left[\begin{array}{c} H_3N \\\\ \diagdown \\\\ B\overset{H}{\underset{H}{\diagup\!\!\!\diagdown}} \\\\ \diagup \\\\ H_3N \end{array}\right] [BH_4]$$

Note the *unsymmetrical* cleavage of diborane in the absence of ether solvent (whose role seems to be to cleave B_2H_6 *symmetrically*).

Whether a reaction of some donor with B_2H_6 will lead to symmetrical or unsymmetrical cleavage of the diborane cannot be predicted with great certainty yet. Review[48] of the known examples of both kinds of cleavage does bring out a pattern

---

[46] G. E. Ryschkewitsch and E. R. Birnbaum, *Inorg. Chem.*, **4**, 575 (1965).
[47] S. G. Shore and C. L. Hall, *J. Amer. Chem. Soc.*, **88**, 5346 (1966); D. R. Schultz and R. W. Parry, *ibid.*, **80**, 4 (1958); S. G. Shore and R. W. Parry, *ibid.*, **80**, 8 (1958).
[48] D. E. Young and S. G. Shore, *J. Amer. Chem. Soc.*, **91**, 3497 (1960).

and suggests a few important concepts. All such reactions are presumed to proceed by a two-step mechanism:

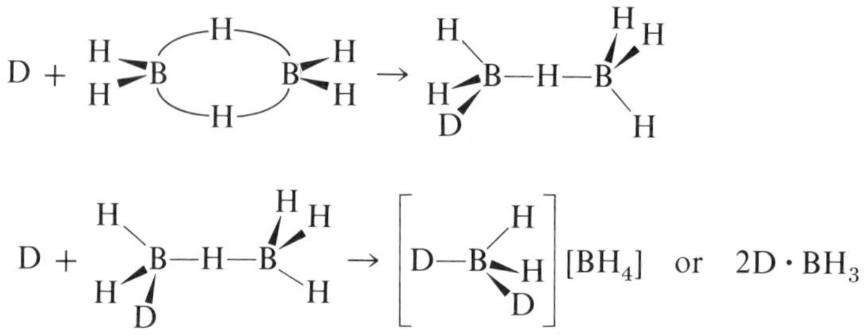

$$D + H_2B(\mu-H)_2BH_2 \rightarrow H_2(D)B-H-BH_3$$

$$D + H_2(D)B-H-BH_3 \rightarrow [D-BH_2(D)]^+[BH_4]^- \text{ or } 2D \cdot BH_3$$

Selection of one or the other of the two boron atoms as the point of attack by the second donor atom seems to determine the course of the reaction. By virtue of the greater electronegativity of most donor atoms than of H⁻, the relative partial charges of the two borons would tend to favor unsymmetrical cleavage. If the donor molecules are large, however, steric congestion during attempted attack at the same boron would tend to direct the second donor to the other boron. This presumably explains the results in Table 7–3 for $BH_{3-x}Me_x$. A similar pattern has been established for 1,2-tetramethylene diborane and 1,2-*bis*(tetramethylene) diborane:

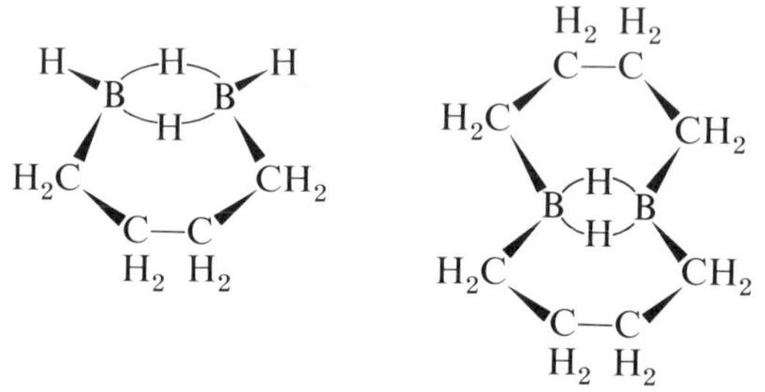

Since the latter borane is more sterically hindered than the first, which, in turn, is more hindered than diborane, the tendencies noted in Table 7–3 for the $NH_{3-x}Me_x$ series of donors make good sense. Steric factors may also play a role in the formation of symmetrical cleavage products for phosphines and sulfides, although the low electronegativities of the donor atoms in these Lewis bases may be an additional

Borane	Donor	Cleavage
$B_2C_8H_{18}$	NH_3	unsymmetrical
	$NMeH_2$	symmetrical
	NMe_2H	symmetrical
	NMe_3	symmetrical
$B_2C_4H_{12}$	NH_3	unsymmetrical
	$NMeH_2$	unsymmetrical
	NMe_2H	symmetrical
	NMe_3	symmetrical
B_2H_6	NH_3	unsymmetrical
	$NMeH_2$	unsymmetrical $\geqslant$ symmetrical
	NMe_2H	symmetrical $>$ unsymmetrical
	NMe_3	symmetrical
	R_3P	symmetrical
	R_2S	symmetrical
	R_2O	symmetrical
	H_2O	(unsymmetrical)

TABLE 7–3
SUMMARY OF BORANE CLEAVAGE PRODUCTS

factor. As noted earlier, ethers seem to effect symmetrical cleavage of diborane; this might be explained by steric congestion (recall the discussion of this in the "*E, C*" section of Chapter 5) but the relatively high electronegativity of the oxygen atom in these donors would seem to favor unsymmetrical cleavage. Quite possibly the weaknesses of the ether oxygen-boron bonds in the $BH_2(OR_2)_2^+$, if formed, make possible a rapid hydride transfer from BH_4^- to the cation with the equilibrium favoring the simple ether · borane adduct. Support for these ideas comes from the recent synthesis[49] of $BH_2(OH_2)_2^+ BH_4^-$ at $-130°C$, where the donor is the unhindered water molecule. The possibility of rapid hydride displacement of one of the H_2O molecules from the cation could not be examined, of course, owing to rapid H_2 evolution at the higher temperatures apparently necessary for such a displacement reaction.

STUDY QUESTIONS

11. A more complete analysis of the RNH_2 solvolysis of $S \cdot BX_3$ can be achieved by considering all three steps of the reaction:

    $$S \cdot BX_3 \underset{k_{-2a}}{\overset{k_{2a}}{\rightleftharpoons}} S + BX_3; \quad BX_3 + RNH_2 \underset{k_{-2b}}{\overset{k_{2b}}{\rightleftharpoons}} RNH_2 \cdot BX_3; \quad RNH_2 \cdot BX_3 \underset{k_{-2c}}{\overset{k_{2c}}{\rightleftharpoons}} RNHBX_2 + HX$$

    Such an analysis proceeds like that following equation (7-1), where in the final results for k_f, k_{-1}/k_2 is replaced by $k_{-2a}[S]/k_{2b}[Am]$, k_{-2}/k_3 by k_{-2b}/k_{2c}, and k_1 by k_{2a}. Show this and analyze the four limiting cases, to convince yourself that the text analysis is correct.

12. Can you draw a structural and chemical analogy between frontal attack of H^+ upon the N—B bonds of amine boranes and upon the B—H bond of BH_4^-?

13. Why could attack of H^+ upon the nitrogen atom of $py \cdot BH_2(aryl)$ and $py \cdot BH_3$ be reasonable when it is known that such attack does not occur for $py \cdot BX_3$?

14. Could hyperconjugation in $H_3P \cdot BH_3$ account for the stability of this adduct in hot mineral acids?

Silicon[50]

Not surprisingly, one of the most widely studied classes of reagents is the silanes, in particular the organosilanes. A very important aspect of the reactions of silanes is the absence of firm evidence for the formation of a siliconium ion, the silicon counterpart of the carbonium ion known to be important in many reactions of carbon. On the contrary, the evidence for a wide range of silanes is that substitution and solvolysis reactions involve an associative step (often the activation step) between a nucleophile and the silane substrate. Presumably, the ease with which silicon can expand its valence shell is of great importance to this general aspect of silane reactions. In fact, you saw several examples of stable five-coordinate silicon in the last chapter. *The ability of third row atoms to expand their valence shells is widely held to be responsible for many of the reactivity differences between second and third row elements.* Comparing carbon and silicon, for example, R_3CCl is known to hydrolyze slowly by either an *I* or *D* mechanism. For R_3SiCl, however, the solvolysis reactions are generally fast and are found to obey second order rate expressions; considerable data exist that support the *a* mode of activation for many of these reactions. For example, the lack[51] of a deuterium isotope effect on the hydrolysis of the cyclic silane

---

[49] P. Finn and W. L. Jolly, *Inorg. Chem.*, **11**, 1941 (1972).
[50] See Ref. 9, for a concise review of substitution reactions at silicon.
[51] D. R. Weyenberg, *Dissertation Abs.*, **19**, 2776 (1959).

is interpreted to mean that Si—H bond breakage is not very important in the activation step (the greater mass of D than of H should measurably raise the activation energy of Si—H bond breaking, were Si—H bond rupture the important feature of reaching the solvolysis transition state). However, the deuterium mass does have an effect on the relative rates of Si—H and Si—C cleavage in *subsequent* (fast) steps because the yield of silanol relative to silyl ether is decreased when one starts with the deuterated silane. Further evidence[52] for an *A* mechanism comes from studies of the class of silanes symbolized by the formula Ar_3SiO_2CR (Ar is a substituted aryl group). In the alcoholysis reactions (formation of Ar_3SiOR')

$$Ar_3Si(O_2CR) + R'OH \rightarrow Ar_3SiOR' + HO_2CR$$

of these compounds, it has been determined that electron-withdrawing substituents on the aryl ring and R groups enhance the rate of reaction, whereas electron-releasing substituents on the aryl moieties lower the rate constants. The effect of aryl substituents is, of course, opposite to that expected for an activation step of dissociation of RCO_2^-. Providing further characterization of the transition state for the solvolysis reactions, studies[53] of relative rate constants for the hydrolyses of the compounds Et_3SiH and the bicyclosilanes **14** and **15** have yielded the ratios $1:10:10^3$.

For the bicyclo compounds, front side attack is required and little difference is found between the triethyl and bicyclooctyl structures. The norbornane analog **15** has a structure of a tetrahedron highly distorted toward a geometry appropriate to a five-coordinate transition state (the hydrogen and ring carbons "equatorial," the bridgehead carbon "axial").‡ Hence the larger rate constant.

In tetrahedral carbon chemistry, where five-coordinate intermediates are uncommon, retention of configuration usually implies a *D* mechanism while inversion implies an *I* mechanism. The possibility of an *A* mechanism for silicon makes extension of such a correlation dangerous; studies of reactions of optically active silanes have shown that both retention and loss of the initial stereochemistry about Si can be observed.

In generalizing[54] the conditions for a change in silicon chirality, it appears that *good leaving groups, X, in combination with nucleophiles more basic than X,* favor inversion (a good leaving group is defined in this context as one with a pK_b greater than 8, such as Cl^-), whereas poor leaving groups usually lead to retention of configuration but sometimes exhibit inversion (a poor leaving group is one with a pK_b less than 4, such as H^- and OR^-). Thus, R_3SiCl and R_3SiO_2CR' (usually) change configuration in substitution reactions, whereas R_3SiOR' and $R_3SiOSiR_3'$ frequently

---

[52] G. Schott, *et al.*, *Chem. Ber.*, **99**, 291, 301, (1966); **100**, 1773, (1967).
[53] L. H. Sommer, *et al.*, *J. Amer. Chem. Soc.*, **79**, 3295 (1957); **81**, 251 (1959).
[54] L. H. Sommer, "Stereochemistry, Mechanism, and Silicon," McGraw-Hill, New York, 1965.
‡ R. L. Hilderbrandt, *et al.*, *J. Amer. Chem. Soc.*, **98**, 7476 (1976).

retain their configurations. By way of examples we can take the following as characteristic (Si* implies an optically active Si):

$$R_3Si^*OCH_3 + AlH_4^- \rightarrow R_3Si^* \underset{H}{\overset{Me}{\underset{O}{\diagdown}}} AlH_3 \xrightarrow[Et_2O]{retention} R_3Si^*H + AlH_3(OMe)^-$$

$$R_3Si^*Cl + H_2O \xrightarrow[Et_2O]{retention} R_3Si^*OH + HCl$$

$$R_3Si^*OAc + MeOH \xrightarrow[xylene]{inversion} R_3Si^*OMe + HOAc$$

The racemization reaction[55]

$$R_3Si^*F \xrightarrow{MeOH} racemization$$

is a striking example that not only makes us entertain the idea of a five-coordinate intermediate but also leads us to the important concept of *pseudorotation* as a pathway for intramolecular rearrangements.[56] For silicon compounds this concept provides (i) a mechanism for racemization *without* the necessity of siliconium ion formation and (ii) a mechanistic route for a rearward attack *with retention of configuration*. Before discussing how this last statement is possible, we must undertake a brief discussion into a topological method[57] for enumerating and relating the many isomers of a trigonal bypyramid.

To begin with, we note that for five different substituents about a central atom, 20 isomers are possible. This number arises from the choice of one of five groups for one apical position and one of the remaining four for the other apical position. Of these 20 possibilities, so far only 10 are unique. Two-fold isomerism remains for the equatorial groups. Adopting a convention of viewing the equatorial plane from the position of the lower-numbered apical group, the equatorial groups could be arranged in order of increasing number either clockwise or counterclockwise. We designate one of the 20 possible isomers by writing the numbers of its apical groups; we indicate the counterclockwise arrangement of equatorial groups by a bar over the numbers of the apical groups (no bar is used for the clockwise arrangement of equatorial groups). Illustrated below are the isomers (45) and ($\overline{45}$).

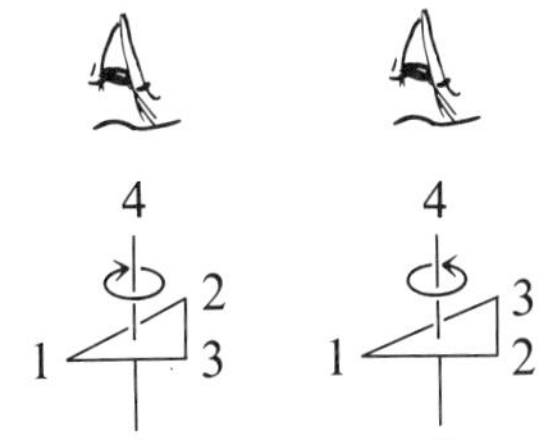

An important question to ask at this point is, "How may one isomer be converted into another?" While several mechanisms have been proposed, we will confine our attention to just one that is most often believed to be correct—the so-called *Berry*

---

[55] L. H. Sommer and P. G. Rodewald, *J. Amer. Chem. Soc.*, **85**, 3898 (1963).

[56] By this is meant any process internal to a molecule for permuting the positions of groups about a central atom. A recent discussion of the more general aspects is given by J. I. Musher, *J. Chem. Educ.*, **51**, 94 (1974).

[57] K. Mislow, *Accts. Chem. Res.*, **3**, 321 (1970); *J. Amer. Chem. Soc.*, **91**, 7031 (1969).

mechanism. The Berry mechanism[58] is a *pseudorotation process for simultaneously interchanging two equatorial groups with the two apical groups, while the third equatorial group* (called the *pivot group*) *remains an equatorial group*. The whole process is represented by the symbol ψ_i, where i is the pivot group. The transition state for the pseudorotation is (ideally) a square-based pyramidal structure, as illustrated for the conversion of the (45) isomer into the $(\overline{23})$ isomer:

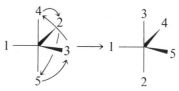

or the permutational equivalent

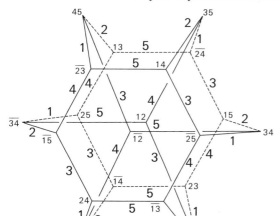

If we were to consider all 20 isomers and their interconversions, we could symbolize that information in graphical form by a map of points, labeled according to isomers those points represent, and connect each point with three straight lines to the points for isomers produced by the three Berry rotations.[59] *Each line is identified by the number of the pivotal (equatorial) group*. This diagram is a topological map of the relationship between different isomers and is called a *Desargues-Levi map*. In its most compact representation the map has the appearance shown in Figure 7–10.

Figure 7–10. A Desargues-Levi map of the Berry conversions between isomers of a trigonal bipyramid.

This map provides us with all information required to determine how one isomer can be converted into another *according to the Berry mechanism*. It should be emphasized that *this topological representation contains no information as to the ease (activation energy) of conversion of one isomer into another*. Such information must be deduced for a specific chemical system from other concepts; often it is possible to view certain interconversion pathways as high-energy and thus not possible or at least not likely for that system.

---

[58] R. S. Berry, *J. Chem. Phys.*, **32**, 933, (1960).
[59] The three Berry rotations derive from the occurrence of each of the three equatorial groups as a pivot.

We now can return to our original question concerning the formation of a five-coordinate intermediate from an optically active silane. In considering the addition of a fifth group to the enantiomers shown below, we recognize that there are 10 ways in which the fifth group could add to each tetrahedron,

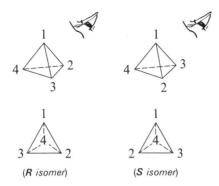

(*R* isomer)   (*S* isomer)

either rearward to one of the original four groups (face attack) or along one of the six edges of the tetrahedron (edge attack). Notice that all eight intermediates labeled (*n*5) or ($\overline{n5}$) necessarily arise by attack of 5 upon the face opposite group *n*. When 5 does not appear in the (*nm*) code, you know that 5 attacked along the *n*,*m* edge.

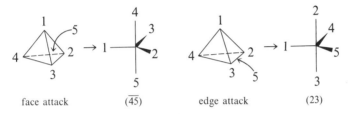

face attack   ($\overline{45}$)   edge attack   (23)

The Desargues-Levi map has a symmetry characteristic that makes it possible to distinguish the origins of the five-coordinate isomers in terms of the isomeric tetrahedra. All 10 isomers to the right of a vertical line through the Desargues-Levi graph arise from addition of the fifth group to the *R* tetrahedral isomer, whereas addition of the fifth group to the *S* tetrahedral isomer gives rise to all 10 isomers at the left of such a line. This symmetry is identified in Figure 7-11 with a vertical dashed line (labeled σ_5 because the dividing line bisects the "5" edges). The positions of (12) and ($\overline{12}$) in the map are deliberately made asymmetric to facilitate this division by σ_5.

Now we can return to consider how the five-coordinate intermediate makes possible the replacement of, say, group 4 by *trans*-attack *and* with configuration retention.

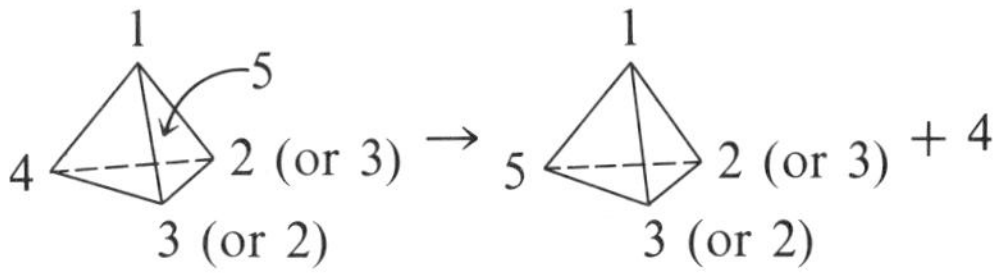

The product isomers contain groups 1, 2, 3, and 5; if we consider the reverse reaction of group 4 attacking these isomers, we find that we can again divide the Desargues-Levi graph by a second line as represented by σ_4 in Figure 7-11. [Note that this time (12) derives from *S*(1235) while ($\overline{12}$) derives from *R*(1235).] In order to identify a mechanism by which 5 attacks the silane to produce a five-coordinate intermediate, which can then undergo pseudorotations to give isomers from which 4 could be lost *and* produce a product with the same chirality as the original silane, we proceed as follows. Inspection of Figure 7-11 shows which trigonal bipyramidal isomers are *common to tetrahedral reactants and products of the same configuration*. Structures

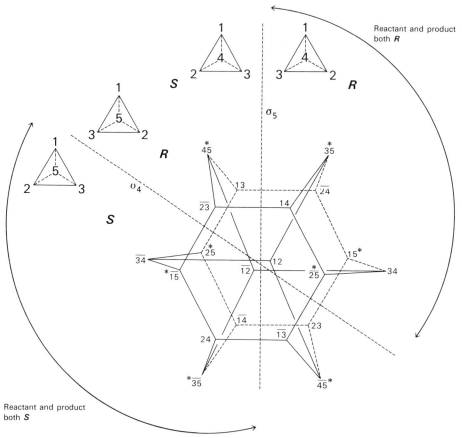

Figure 7-11. Origin of trigonal bipyramidal structures from tetrahedral optical isomers. The eight * isomers arise directly from rearward attack by group 5 on the original (1234) tetrahedra and by rearward attack by group 4 on the (1235) tetrahedra.

between 12 and 4 o'clock belong to the R configurations, and those between 6 and 10 o'clock arise from the S configurations. Between 4 and 6 o'clock and between 10 and 12 o'clock are the intermediates correlating with opposite reactant and product isomers. A mechanism in which 5 attacks the R silane opposite group 4 initially results in intermediate $\overline{45}$. Loss of 4 from its axial position (before pseudorotation) produces an inverted product (this is an I path). To produce a pseudorotamer from which 4 could be lost to yield an R product, we see that three paths, each with a minimum of two pseudorotations, are possible:

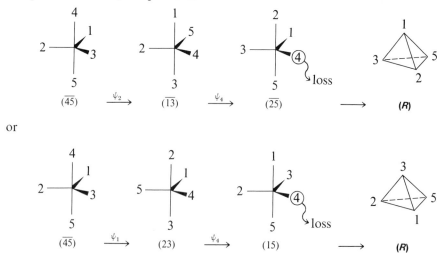

or

or

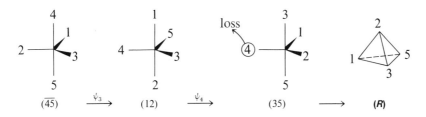

Were 4 to be lost from isomers ($\overline{13}$), (23), or (12), the product would be *S;* i.e., inversion would accompany substitution.

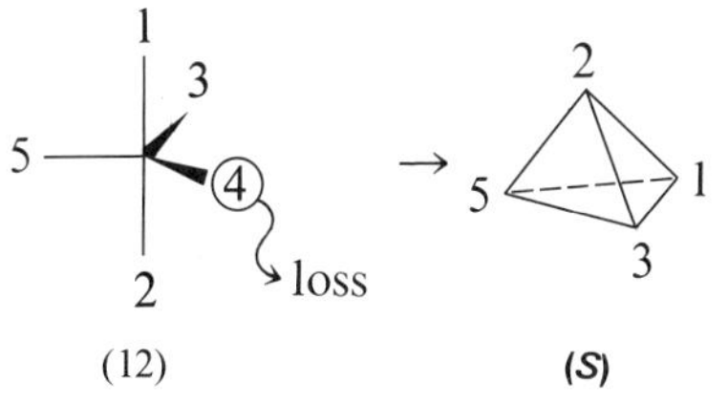

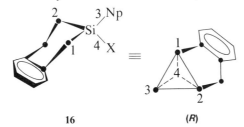

Thus, we see that it is possible for rearward attack to occur upon a silane to ultimately give a substitution product of unchanged chirality, and we have identified the three shortest pathways by which such a reaction could occur.[60] *In addition, you now see why retention of configuration does not demonstrate the occurrence of a D mechanism.*

The Desargues-Levi graph has been particularly valuable in understanding substitutions at asymmetric cyclic silanes such as **16** (where Np = α-naphthyl).

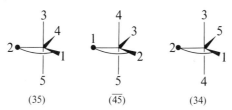

The tendency of this compound to *retain* configuration on substitution of X (=Cl) (a good leaving group) is much greater than that of PhMe(α-Np)SiCl, a fact attributable to the rigidity of the six-membered ring. Here the intra-ring angle about Si of ~120° makes unlikely the formation of intermediates in which the ring carbons are not both equatorial, because of large strain energy associated with compression to the 90° apical/equatorial angle in the trigonal bipyramid. Thus, labeling the adjacent ring carbons 1 and 2, all trigonal bipyramidal isomers labeled with 1 and/or 2 can be ignored [*i.e.,* (12), ($\overline{12}$), and all hexagonal vertices of the Desargues-Levi graph]. Upon attack on the *R* isomer shown above by a new group 5, you can see that the only intermediates possible are (35), ($\overline{45}$), and (34); (35) requires attack rearward of 3 (*i.e.,* α-Np), ($\overline{45}$) requires attack rearward of 4 (Cl), and (34) requires edge attack between 3 and 4 (between α-Np and Cl). (Note that any Berry pseudorotation of these isomers is highly unlikely because of ring strain.)

---

[60] To achieve loss of 4 from an axial (rather than equatorial) position requires at least one more rotation to produce (34), ($\overline{24}$), or (14).

As before, loss of Cl (group 4) directly from $(\overline{45})$ (at 5 o'clock in the Desargues-Levi map) leads to inversion and cannot account for an observed retention of configuration. The retention reactions must proceed either by attack of 5 opposite to 3 [to give (35)], with direct loss of Cl from an equatorial position, *or* by attack of 5 between Cl and α-naphthyl [to give (34)], with loss of Cl from an apical position. Consequently, either the nucleophile (5) that displaces Cl (4) enters apically and Cl departs from an equatorial position, or the nucleophile enters equatorially and Cl departs from an apical position. In either instance, *the positions of entry and departure are different*. In either case, the intermediate is the one that arises from "front" attack, normally diagrammed as

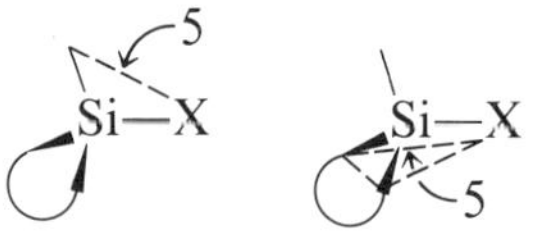

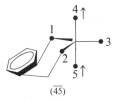

gives (34)   gives (35)

A most interesting and useful constraint on the reaction pathway arises when we attempt to decide which, (35) *or* (34), lies along the path from reactants $\left(\bigcirc\text{Si}\!<^{R}_{X} + N \right)$ to products $\left(\bigcirc\text{Si}\!<^{R}_{N} + X \right)$. In the case of X exchange (where, experimentally, N is a radioactive isotope of X) *the reactants and products are chemically identical and we must consider the constraints of the principle of microscopic reversibility*. Applied here, $(\overline{45})$ would *not* be a case covered by Figure 7-8(B) because of the asymmetry of the hexane ring.

$(\overline{45})$

Were *inversion* to accompany exchange, $(\overline{45})$ as the intermediate in Figure 7-8(D) would be appropriate. With *retention* during exchange, we must deal with intermediates (35) and (34), neither of which provides symmetry equivalence

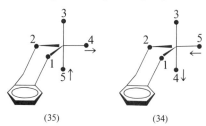

(35)   (34)

for positions 4 and 5. With a postulated inability of (35) or (34) to pseudorotate, we are locked into a mechanism featuring axial entry (face attack opposite 3) of 5 and equatorial departure of 4 in proposing (35) as the intermediate. The PMR requires that the exact reverse reaction (attack by 4 at the 3,5 edge) occur with equal velocity. Since 5 and 4 are the same species, 5 must also be able to enter by attack at the 3,4 edge. This reverse reaction entails intermediate (34), which is, of course, chemically identical to (35). Accordingly, the reaction coordinate diagram should appear as [a case of Figure 7-8(D)]:

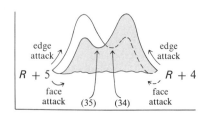

In summary, because of the chirality of the silane substrate, both retention and inversion require two-channel mechanisms. More significant is the effect of not permitting a pseudorotation of (35): the entry and departure points are different. *If* there were evidence to show that the equatorial entry [formation of (34) by edge attack] is impossible, the PMR would require us to discard the proposed mechanism. Later, when we discuss substitution at phosphorus, we will encounter just such a predicament. There we will find equatorial departure and entry to be "forbidden" in certain cases, but the likelihood of pseudorotation provides a "single-channel" mechanism with a symmetric intermediate.

You are well aware that, in addition to ring size restrictions on certain pseudorotation pathways, relative electronegativities and π-bonding properties of the groups bound to the central atom may influence the relative energies of the isomers. You have seen in earlier chapters that, for known ground state trigonal bipyramidal molecules, more electronegative and weakly π-bonding groups are directed to an apical position; pseudorotamers with the least electronegative and most strongly π-bonding groups in equatorial positions should be lower in energy than those that place such groups axially. Thus, we can often identify (via the preferred pivot atom) the preferred pseudorotations on the basis of electronegativity and π-bonding arguments.

Similarly, we find it possible to interpret the "rule" that good leaving groups tend to lead to inversion by *trans*-attack because the intermediates are not sufficiently long-lived to undergo pseudorotations. For example, rearward attack by 5 to displace 4 from an *R* substrate‡ produces the (45) isomer which, on losing 4 before pseudorotations → (23) → (15) (for example) can occur, produces the *S* product.

---

‡See study question 19 for a hint as to why attack *trans* to 4 is likely.

STUDY QUESTIONS

15. Propose a mechanism for substitution at tetrahedral silicon that could lead to a mixture of products, some exhibiting retention and others exhibiting inversion of the original chirality.

16. Predict whether retention or inversion of chirality about silicon is expected for

    a) $R_3SiOR' + KOH \rightarrow R_3SiOH + KOR'$

    b) $R_3SiO_2CR' + LiAlH_4 \xrightarrow{Et_2O} R_3SiH$

    c) $R_3SiCl + MeNO_2 \rightarrow R_3SiCl + MeNO_2$

    d) $R_3SiCl + KOH \xrightarrow{xylene} R_3SiOH + KCl$

17. Is it possible for a nucleophile to displace X, with inversion of silicon chirality,

    $$R_3SiX + Nuc \rightarrow R_3SiNuc + X$$

    by attacking along the Si—X bond?

18. What is the relationship between an *I* displacement of X from R_3SiX, where inversion accompanies displacement, and the *A* mechanism for the same reaction?

19. Why do you suppose that good leaving groups favor formation of the (45) isomers (4 = leaving group, 5 = entering group) relative to formation of the other isomers? (Hint: focus on the polarization of the Si—4 σ^* LUMO.)

Nitrogen and Phosphorus

Trivalent nitrogen provides for many interesting reaction pathways because of the great variety of compounds and structures known. Simple *electrophilic substitution* of coordinated BMe_3 by free BMe_3 in the adduct Me_3NBMe_3 has been shown[61] to proceed by the same d activation you discovered (two sections back) for exchange of coordinated and free Me_3N in the adduct. On the other hand, exchange of free and coordinated $GaMe_3$ and $InMe_3$ in adducts with primary amines and ammonia has been shown[62] to occur by an I mechanism, presumably through the transition state structure **17**.

$$Me\underset{Me}{\overset{Me}{\diagdown}}Ga\text{-----}\underset{H\ R}{\overset{H}{N}}\text{-----}Ga\underset{Me}{\overset{Me}{\diagup}}Me$$

17

With secondary and tertiary amine adducts, exchange of free and coordinated $GaMe_3$ and $InMe_3$ proceeds by a D mechanism like that of BMe_3.

Displacement reactions by *nucleophilic attack* upon substituted amines, NH_2X, have received considerable attention because they are involved in synthesis of hydrazine from chloramine and of hydroxylamine from nitrous acid, to name just two examples. Quite generally the rate laws are second order—that is, first order in both nitrogen species and nucleophile. Even so, as you learned in the introductory section of this chapter, different mechanistic pathways are consistent with such a rate law:

a) $NH_2X + Nuc \rightarrow NucNH_2^+ + X^-$
b) $NH_2X + Nuc \rightarrow NucX^+ + NH_2^-$
c) $NH_2X + Base \rightarrow NHX^- + BaseH^+$

or (when Base = Nuc)

$$NHX^- + Nuc \rightarrow NucNH + X^-$$
$$NHX^- \rightarrow products$$

Mechanism (a) is meant to imply attack by Nuc upon the nitrogen atom, (b) is meant to imply attack upon X, and (c) is called the *conjugate base* (CB) *mechanism,* for the first step is rapid generation of the conjugate base of the substrate. Which of these mechanisms is followed by a given system depends on the detailed nature of X and Nuc; generally speaking, pathways (a) and (b) are most commonly found and are of the I variety—that is, the rates depend on both the bond-making (Nuc attacking the substrate) and bond-breaking (X departing from the substrate) features of attaining the transition state.

A substituted amine that exhibits many of the characteristic features of the displacement reaction (a) for trivalent nitrogen is the hydroxylamine-O-sulfonate anion[63] ($H_2NOSO_3^-$). At low acidities the *anion* reacts with various nucleophiles such as triphenylphosphine (to form $Ph_3PNH_2^+$), triethylamine (to form $Et_3NNH_2^+$), and HI (to form NH_3 and I_2). The reaction with I^- is ostensibly an oxidation-reduction reaction, but all three reactions embody the steps characteristic of displacement reactions. That is, all have in common a second order rate expression that is first order in the nucleophile and first order in hydroxylamine-O-sulfonate. The second order rate constant is independent of H^+ concentration until the pH becomes low enough to cause appreciable concentration of the conjugate acid to build up. That the second

---

[61] See Ref. 36.
[62] See Ref. 37.
[63] P. A. S. Smith, *et al., J. Amer. Chem. Soc.,* **86,** 1139 (1964); J. H. Krueger, *et al., Inorg. Chem.,* **12,** 2714 (1973); *ibid.,* **13,** 1736 (1974).

order rate constant is sensitive to the nature of the nucleophile and exhibits a large negative activation entropy (~ -30 eu) emphasizes the importance of bond-making

$$Ph_3P(2.0 \text{ M}^{-1} \text{ sec}^{-1}) > I^-(0.07) > Et_3N(0.02) \gg Br^-(<10^{-4}) > Cl^-$$

in the transition state. That Ph_3P and I^- react fastest with the anion is believed to reflect an important contribution from retro-bonding between the nitrogen lone pair and P and I orbitals so as to lower the activation energy

$$H_3P\text{------}NH_2^+\text{------}OSO_3^-$$

of attack upon nitrogen. A further indication that the activation step is of the I type stems from the observation that N-alkyl substitution reduces the rate constants. The general mechanism then depends heavily on attack at the nitrogen with simultaneous displacement of the SO_4^{2-} group.

The reaction with I^- proceeds beyond the formation of NH_2I by subsequent fast reaction of NH_2I with HI according to the overall stoichiometry

$$NH_2I + 2HI \to NH_4^+ + I_3^-$$

Here we apparently find a reaction step of the type (b) above, although this has not been confirmed as yet because the $NH_2I + I^-$ reaction is too fast to study by conventional means.

$$NH_2I + I^- \to NH_2^- + I_2$$

A consideration of the relative electronegativities of nitrogen and iodine provides the basis for faster attack by I^- upon I than upon N.

The fact that the rate constants for these reactions *decrease* as the pH is lowered to a point where the anion is appreciably converted to its conjugate acid is most interesting. A decrease in rate is what is to be expected if the conjugate acid

$$H_2NOSO_3^- + H^+ \rightleftharpoons H_2NOSO_3H$$

is less reactive than the anion. Independent experiments, under such acid conditions that the anion is fully protonated, reveal no change in the form of the rate law (where the acid simply replaces the anion) and so no drastic change in mechanism is expected. The effect of anion protonation is confined to a reduction (by nearly an order of magnitude) in velocity of the reaction—the activation energy is simply slightly higher for attack on the acid than on the anion. Of considerable importance is the site of anion protonation.[64] Were an oxygen of the sulfato group to be protonated (H_2NOSO_3H), you are likely to be thinking that the rate constant should *increase;* for not only should HSO_4^- be a better leaving group (weaker base) than SO_4^{2-}, but also the less negative formal charge on the nitrogen should render it more susceptible to nucleophilic attack. Were protonation to occur at the nitrogen (to form the zwitterion form $H_3N^+OSO_3^-$), you should still predict that the increased nitrogen formal charge would facilitate the attack but *hinder* the N—O bond-breaking as SO_4^{2-} is displaced from NH_3^{2+}. Clearly, this view is consistent with the experimental observation of a lower rate constant for the acid than for the anion. Consequently, it appears that protonation occurs (in solution) at the nitrogen, and this incurs a slightly greater hindrance to bond-breaking than enhancement of bond-making.

---

[64] In the solid state it is the nitrogen that bears the proton: R. E. Richards and R. W. Yorke, *J. Chem. Soc.*, 2821 (1959).

The *hydrolysis* reaction of hydroxylamine-O-sulfonic acid is slower than those just discussed but is interesting in one important respect. Whereas at high pH the *anion* cleaves at the N—O bond so that attack by OH^- at the nitrogen is assured (and OH^- behaves as a nucleophile as do Ph_3P, etc.), under acid conditions ("no" OH^- present) it is the S—O bond of the *acid* that is cleaved. This, of course, is consistent with H_2O attack at the sulfur (either an I process or formation of a five-coordinate sulfur precursor) but also would arise from a D mechanism for the zwitterion

$$H_3N^+\text{—}O\text{—}SO_3^- \rightarrow H_3NO + SO_3$$

The key to understanding this change in mechanism seems to lie in recognizing that H_2O is a much poorer nucleophile than Ph_3P, I^-, and Et_3N. Thus, the stronger nucleophiles are able to make a clear distinction between nitrogen and sulfur in H_3NOSO_3 as electrophilic centers for attack. H_2O is such a poor nucleophile that it has "difficulty" attacking either position. Another important piece of information is that the rate law for the acid hydrolysis is (pseudo?) second order—first order in H_3NOSO_3 and first order in acidity. Thus, the transition state species consists of H_3NOSO_3 and H^+ (if the reaction is of the D variety) or H_3NOSO_3, H^+, and H_2O (if the mechanism belongs to the A or I types). Now, the site of protonation of H_3NOSO_3 is very important. It is not likely that the bridging oxygen (O_b) could compete effectively with the terminal oxygens as donors toward the proton, so we should postulate the species $H_3NOSO_2OH^+$ as the kinetically important one. The effect of terminal O protonation on the reaction velocity can be judged in the following way: the sulfur atom should become a stronger electrophile and the O_b—S bond should become more resistant to heterolytic cleavage. Clearly, the D mechanism is disfavored while the A and/or I mechanisms are favored. To sum up, H_3NOSO_3/H_2O is too weak an electrophile/nucleophile combination for hydrolysis to occur with much speed. Protonation at a terminal oxygen enhances the electrophilicity of both sulfur and nitrogen, but that of sulfur more than that of nitrogen because of its proximity to the site of protonation. The greater electrophilicity of $H_3NOSO_2OH^+$ than of H_3NOSO_3 now leads to a pseudo-second order rate constant of about $0.06\ M^{-1}\ hr^{-1}$, and a change of reaction type from (a) to (b). In a related reaction, this time with H_2NOH, organophosphines undergo a general acid catalyzed (H^+, BF_3) oxidation to phosphine oxides.

$$R_3P + NH_2OH \xrightarrow{acid} R_3PO + NH_3 \cdot Acid$$

In contrasting this reaction with those given above in which the phosphine attacks the *nitrogen* of $H_2NOSO_3^-$ and the *nitrogen* of H_3NOSO_3, note the difference in apparent electrophilicities of the bridging oxygens of H_3NOH^+ and H_3NOSO_3. Thus, Ph_3P appears to favor attack at oxygen of H_3NOH^+ but at the nitrogen of H_3NOSO_3. The dilemma of the fickle behavior of phosphorus is resolved by recognizing the considerable steric congestion in the transition state for attack at the bridging oxygen of hydroxylamine-O-sulfonate.

Related to these reactions in which phosphorus attacks oxygen is the pyrophoric character of dialkyl phosphines. A proposed[65] mechanism for these *oxidation* reactions involves addition of a P—H group to O_2, leading first to a peroxy derivative (alternately described as insertion of O_2 into the bond between the phosphorus and hydrogen):

$$R_2PH + O_2 \rightarrow R_2PO_2H \rightarrow products$$

As an interesting aside, it was long thought that PH_3 was pyrophoric but, in fact, it is not PH_3 that spontaneously combusts. Rather, impurity P_2H_4 is responsible for the

---

[65] B. Saville, *Angew. Chem. Int. Ed.* (Eng.), **6**, 928 (1967).

flammability of PH_3. In this connection it is worth reflecting upon the generally low basicity of PH_3 and the effect of replacing a hydrogen in PH_3 with a phosphino group (PH_2) or two hydrogens with R groups.

The *hydrolysis* of nitramide, H_2NNO_2, is most interesting in that the strongly conjugating and electron withdrawing NO_2 group alters the mechanism for hydrolysis from those observed [(a) and (b)] for the other H_2NX compounds so far discussed to the CB pathway[66] [type (c)]. The hydrolysis of nitramide is general-base catalyzed (not limited to OH^-), and the product is N_2O. [With OH^-, (a) or (b) would have led to $H_2NOH + NO_2^-$ or $NH_3 + HNO_3$ as products.]

$$H_2NNO_2 + \text{base} \rightleftharpoons HNNO_2^- + \text{base} \cdot H^+$$

$$\begin{bmatrix} H \\ N{-}N \\ \ddot{O}: \\ \ddot{O}: \end{bmatrix}^- \xrightarrow[\text{H}^+ \text{ shift}]{1,3} \begin{bmatrix} :\ddot{N}{=}N \\ \ddot{O}{-}H \\ \ddot{O}: \end{bmatrix}^- \rightarrow \; :N{=}N{=}\ddot{O}: + OH^-$$

The general rate expression for the hydrolysis is first order in nitramide and first order in each base present:

$$\text{Rate} = [H_2NNO_2]\left\{\sum k_B [B]\right\}$$

It is expected that the activation step is the D loss of OH^- from $N_2O_2H^-$, as both proton transfer steps are expected to be relatively fast. (Note the analogy of $NN(O)OH^-$ to $ON(O)OH$ and of NNO to ONO^+.)

The chloramines (NH_2Cl, $NHCl_2$, and NCl_3) provide us with some interesting, and often very complex, reactions.[67] The detailed kinetic information available for their reactions is somewhat limited by the difficulties in handling them. A few courageous chemists, with the requisite confidence in their laboratory technique, have provided what data is available. All these compounds are best handled in solution, where their detonation hazard is somewhat reduced. Their synthesis is achieved by chlorine *oxidation* of ammonia in aqueous media, where the active oxidizing species is actually hypochlorite or its conjugate acid (discussion of the formation of OCl^- from $H_2O + Cl_2$ will be postponed until the next section).

$$NH_3 + OCl^- \rightarrow NH_2Cl + OH^-$$

$$\underset{(8 < \text{pH} < 12)}{2NH_2Cl} + H^+ \rightleftharpoons \underset{(3 < \text{pH} < 5)}{NHCl_2} + NH_4^+$$

$$\underset{(3 < \text{pH} < 5)}{2NHCl_2} + H^+ \rightleftharpoons \underset{(\text{pH} < 3)}{NCl_3} + NH_4^+$$

All three compounds are related by strongly pH dependent equilibria, so pH control is critical to successful synthesis of one or the other. Of the three chloramines, only NCl_3 is sufficiently insoluble in water that it soon separates as a yellowish oil phase.[68] This is dangerous, for the "pure" liquid is shock- and light-sensitive. Fortunately, dilute CCl_4 solutions are not too unstable, so extraction[69] of NCl_3 into the CCl_4 layer of a two-phase H_2O/CCl_4 synthesis medium provides a way to avoid many of the dangers of NCl_3 synthesis.

---

[66] K. J. Laidler, "Chemical Kinetics," McGraw-Hill, New York, 1965; A. A. Frost and R. G. Pearson, "Kinetics and Mechanism," Wiley, New York, 1961; I. R. Wilson in "Comprehensive Chemical Kinetics," Vol. 6, C. H. Bramford and C. F. H. Tipper (Eds.), Elsevier, New York, 1972.

[67] A thorough review of the preparation and reactions of these compounds is given by J. Jander, *Nitrogen Compounds of Chlorine, Bromine, and Iodine*, in "Developments in Inorganic Nitrogen Chemistry," Vol. 2, C. B. Colburn (Ed.), Elsevier, New York, 1973.

[68] Note that the low solubility is expected for a very weakly basic amine.

[69] R. M. Chapin, *J. Amer. Chem. Soc.*, **51**, 2112 (1929).

Of the three chloramines, NH_2Cl is the easiest to work with and undergoes an important *reductive solvolysis*[70] by *liquid ammonia*. Attack by NH_3 at the chlorine would lead to no net reaction (although an exchange of N atoms would be effected, this appears not to have been studied), whereas attack by NH_3 on the nitrogen of NH_2Cl produces hydrazinium chloride (with the excess NH_3 that is present, the hydrazinium ion is converted to hydrazine). This reaction constitutes the basis for the *Raschig*[71] *synthesis of hydrazine*.

$$NH_3 + NH_2Cl \rightarrow N_2H_5^+ + Cl^- \xrightarrow{NH_3} N_2H_4 + NH_4^+ + Cl^-$$

Since the preparation of chloramine itself is conveniently carried out by the reaction of NH_3 with Cl_2, a useful synthesis of hydrazine entails the chlorination of liquid ammonia.[72]

In strongly basic aqueous solutions, second order ($k \approx 10^{-4}$ M^{-1} sec^{-1}) ammonia attack upon chloramine has been confirmed,[73] but there is also an important CB mechanism operative at $[OH^-] \geq 0.1$. The evidence from very thorough studies indicates that this is best treated as NH_3 attack upon the chloramide ion[74]:

$$NH_2Cl + OH^- \rightleftharpoons NHCl^- + H_2O$$
$$NHCl^- + NH_3 \rightarrow N_2H_4 + Cl^-$$

Under such strongly basic conditions, however, the reaction suffers from hydrolysis of NH_2Cl (see below) to produce hydroxylamine, which experiences a rapid decomposition reaction with chloramine so as to reduce the yield of hydrazine. An interesting complication to this reaction is that of a decomposition reaction between hydrazine and chloramine. That the aqueous ammonia/chloramine reaction to form hydrazine works at all requires a fine tuning of the reaction conditions to minimize the rates of all three competitive reactions. This is a fine example of the synthetic chemist's need to understand the mechanistic as well as the thermodynamic aspects of his chemistry.

The type (a) reaction with NH_3 appears to be general for chloramine, for a broad spectrum of nucleophiles appears to attack the nitrogen atom by way of a second order *I* mechanism. Examples are (the products shown are not necessarily the final ones, depending on other reagents present):

$$NH_2Cl + R_3P \rightarrow R_3PNH_2^+ + Cl^- \quad (R = aryl, alkyl)$$
$$NH_2Cl + R_3N \rightarrow R_3NNH_2^+ + Cl^- \quad (R = alkyl)$$
$$NH_2Cl + OH^- \rightarrow NH_2OH + Cl^-$$
$$[Ir(NH_3)_5(NH_2Cl)]^{3+} + OH^- \rightarrow [Ir(NH_3)_5(NH_2OH)]^{3+} + Cl^-$$
$$+ I^- \rightarrow [Ir(NH_3)_5(NH_2I)]^{3+} + Cl^-$$
$$NH_2Cl + N_2H_4 \rightarrow N_3H_6^+ + Cl^-$$

The first two reactions lead to stable products, and the phosphines appear to be better nucleophiles toward NH_2Cl than are amines, as is often the case.[75]

---

[70] F. N. Collier, *et al.*, *J. Amer. Chem. Soc.*, **81**, 6177 (1959).

[71] F. Raschig, "Schwefel und Stickstoffstudien," Verlag Chemie G.m.b.H., Berlin, 1924.

[72] Note that the formation of HOCl in water is analogous to the chlorination of NH_3 in ammonia. Also, the fact that neither $NHCl_2$ nor NCl_3 is found in liquid NH_3 means either that NH_3 attack upon them to form NH_2Cl is very rapid or that NH_2Cl is not sufficiently nucleophilic toward Cl_2 to compete with NH_3 attack on Cl_2.

[73] G. Yagil and M. Anbar, *J. Amer. Chem. Soc.*, **84**, 1797 (1962).

[74] The bimolecular rate constant for $NHCl^- + NH_3$ is greater than 10^{-2} sec^{-1} M^{-1}. From comparison to that for $NH_2Cl + NH_3$, it seems that the greater negative charges on N and Cl facilitate N—Cl cleavage more than they inhibit NH_3 attack. This might be taken to imply an I_d mechanism.

[75] R. M. Kren and H. H. Sisler, *Inorg. Chem.*, **9**, 836 (1970).

As you saw earlier for free NH_2I, subsequent reaction between I^- and coordinated NH_2I produces[76] coordinated NH_3 and I_2. Owing to the polarization of the N—I bond by the iridium complex, I^- attack on the iodine of coordinated NH_2I should be even more facile than attack upon free iodoamine.

The hydroxylamine formed from hydrolysis of chloramine suffers further "rapid" reaction with unreacted chloramine[77] ultimately to produce N_2O, N_2, and NH_3. This is the reaction alluded to earlier, which complicates the use of aqueous base as a medium for hydrazine synthesis from chloramine. The other deleterious side reaction in this synthesis is that of hydrazine and chloramine. As indicated above, this reaction is thought[78] initially to yield the triazane ($N_3H_6^+$) cation as an intermediate along the pathway to N_2 and NH_3 as final products. In fact, these reactions are just two of several *redox* reactions that involve the interaction of substituted amines to produce either (N_2 and NH_3) or (N_2O and NH_3) as products. The great similarity between reactants (and products) in these reactions suggests a common mechanism, but quantitative studies to confirm this are lacking. Accordingly, you will find an exercise at the end of this chapter that suggests a mechanistic scheme to correlate the reactions.

$$2NH_2X + NH_2Y \rightarrow N_2 + NH_3 + HY + 2HX$$
$$2NH_2Cl + NH_2NH_2 \rightarrow N_2 + 2NH_3 + 2HCl$$
$$2NH_2Cl + NH_2OH \rightarrow N_2 + NH_3 + H_2O + 2HCl$$
$$2NH_2OH + NH_2OH \rightarrow N_2 + NH_3 + 3H_2O$$
$$2NH_2NH_2 + NH_2NH_2 \rightarrow N_2 + 4NH_3$$
$$2NH_2Cl + NH_2Cl \rightarrow N_2 + NH_3 + 3HCl$$
$$2NH_2OH + NH_2Cl \rightarrow N_2 + NH_3 + 2H_2O + HCl$$

$$2NH_2X + 2NH_2OH \rightarrow N_2O + 2NH_3 + H_2O + 2HX$$
$$2NH_2Cl + 2NH_2OH \rightarrow N_2O + 2NH_3 + H_2O + 2HCl$$
$$2NH_2OH + 2NH_2OH \rightarrow N_2O + 2NH_3 + 3H_2O$$

In addition to the *disproportionation* reaction for decomposition of NH_2Cl as given above, acid influences a decomposition into NH_3 and $NHCl_2$. As mentioned earlier, both NH_2Cl and $NHCl_2$ coexist in the pH range from 5 to 8; for a solution pH in the 4 to 5 range, $NHCl_2$ may be obtained essentially free of NH_2Cl and NCl_3. The mechanism of this acid dependent conversion of chloramine to dichloramine has not been established and may occur through fast pre-equilibrium to form NH_3Cl^+, followed either by attack by NH_2Cl upon *chlorine* (**18**), or by attack upon the NH_3Cl^+ cation by H_2O (**19**).

$$NH_2Cl + H^+ \rightleftharpoons NH_3Cl^+$$

$$H_2O \searrow \qquad \swarrow NH_2Cl$$

$$NH_3 + H^+ + HOCl \qquad \rightarrow NH_3 + NHCl_2 + H^+$$

$$H_2O + NHCl_2 \swarrow \qquad \nwarrow NH_2Cl \qquad \mathbf{18}$$

$$\mathbf{19}$$

In fact, both paths may be important when H_2O is the solvent. These pathways imply the involvement of H^+ in the transition state for the reaction, so they implicate developing + charge (enhanced electrophilic character) at the chlorine and nitrogen of NH_2Cl.

---

[76] B. C. Lane, *et al.*, *J. Amer. Chem. Soc.*, **94**, 3786 (1972).
[77] See Ref. 73.
[78] K. Utvary and H. H. Sisler, *Inorg. Chem.*, **1**, 698 (1962); *ibid.*, **5**, 1835 (1966).

Of the three chloramines, dichloramine is the least stable in the sense of rapid reactions; thus, little detailed information about its reactions is available. One reaction you will recognize is the oxidation of HI by $NHCl_2$:

$$NHCl_2 + 4HI \rightarrow 2I_2 + NH_3 + 2HCl$$

Of greater interest is the basic *hydrolysis* (reduction) of $NHCl_2$. The reaction[79] proceeds with generation of N_2, Cl^-, and most importantly, OCl^-.

$$2NHCl_2 + OH^- \rightarrow N_2 + 3Cl^- + OCl^- + 3H^+$$

Again, mechanism studies are not available; but a consideration of the fact that hypochlorite is a product, but only in a 1:3 ratio with Cl^-, suggests the following pathway featuring attack at chlorine (by OH^-), attack at nitrogen of $NHCl_2$ (by NH_2Cl), and successive 1,2 eliminations of HCl from 1,2-dichlorohydrazine. Here we have made use of reaction steps that are of known or expected importance in the chemistry of nitrogen as an electrophile.

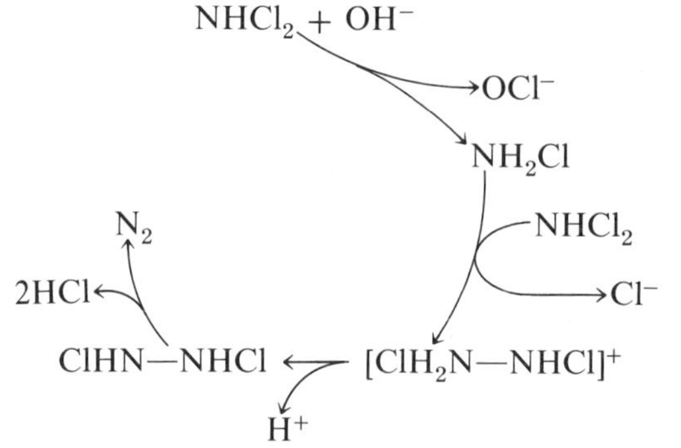

$$NHCl_2 + OH^- \rightarrow OCl^- + NH_2Cl \xrightarrow{NHCl_2} [ClH_2N-NHCl]^+ \xrightarrow{-H^+} ClHN-NHCl \rightarrow 2HCl + N_2$$ (+ Cl^-)

Simple extension of either of the pathways **(18** or **19)** envisioned for conversion of NH_2Cl to $NHCl_2$ accounts for the formation of NCl_3 from $NHCl_2$ at still lower pH. The decomposition[80] of NCl_3 in CCl_4 as a solvent is known to be first order in trichloramine with a high activation energy of +32 kcal/mole. Other evidences for the free radical nature of the decomposition

$$2NCl_3 \rightarrow N_2 + 3Cl_2$$

are the facts that it is photocatalyzed and that hexachlorobenzene is formed when benzene solutions are irradiated. The reaction of NCl_3 with liquid HCl or highly concentrated aqueous solutions of HCl effects a thermodynamic reversal of the NCl_3 synthesis reaction

$$NCl_3 + 4HCl \rightarrow NH_4Cl + 3Cl_2$$

and extends the concept of "chloronium" character (Cl^- attack upon Cl of N—Cl) of the chlorines in the chloramines and/or their respective ammonium cations. Thus, the trend for chloronium character, noted in the comparison of the basic hydrolyses of NH_2Cl and $NHCl_2$, appears to be amplified for NCl_3.[81]

$$2NCl_3 + 3OH^- \rightarrow N_2 + 3Cl^- + 3OCl^- + 3H^+$$

---

[79] R. M. Chapin, *J. Amer. Chem. Soc.*, **53**, 912 (1931); also Ref. 69.
[80] See Ref. 67.
[81] See Ref. 79.

This behavior of NCl_3 is to be contrasted with those of third row chlorides ($SiCl_4$, PCl_3, SCl_4), which do *not* hydrolyze to yield any hypochlorite. With NCl_3 several factors favor nucleophilic attack at chlorine rather than at nitrogen. The steric shielding of the nitrogen atom in the perchloro nitrogen compound should certainly render nitrogen inaccessible to nucleophilic attack, and the high electronegativity of the NCl_2 fragment to which each chlorine is bound should render the chlorine more electrophilic than normal. With the third row chlorides, the possibility for valence shell expansion by the central atom in the transition state and its "high" electrophilicity strongly favor nucleophilic attack there.

Yet another comparison you should make is that of NCl_3 and NF_3. As you have seen in the last two chapters, trifluoroamine is notoriously unreactive—the amine lone pair shows little basic character, the N—F bonds are stronger than N—Cl bonds, the nitrogen atom is sterically shielded from nucleophilic attack, and the fluorines should not develop nearly as much "onium" character as does chlorine. Finally, chlorine is a third row element and as such would show greater facility in valence shell expansion to accommodate nucleophilic attack.

Some preliminary results have appeared for the reactions of *difluoramine* with various nucleophiles.[82] All reactions studied proceed according to second order rate laws, first order in difluoramine and first order in nucleophile. The overall reactions (that is, steps beyond the rate controlling step) are not well understood because the products of the reactions depend so dramatically on the nucleophile. The reactions with readily oxidizable anions such as HSO_3^-, I^-, CN^-, SCN^-, and Br^- are similar to those between other NH_2X and NHX_2 compounds and I^-, in that the principal products are NH_3 and Nu_2 (or expected hydrolysis products of Nu_2). With the difficultly oxidized ions OAc^-, $H_2PO_4^-$, and HPO_4^-, the principal nitrogen-containing products are N_2O and (with the phosphates) N_2F_2. Finally, the catalytic action of OH^- results in a non-redox decomposition of HNF_2 to N_2F_2.

$$2HNF_2 \xrightarrow{OH^-} N_2F_2 + 2HF$$

You should recognize, moreover, that substitution of H for F produces a much more reactive amine than NF_3. Accordingly, we might speculate that the CB mechanism believed to be correct for nitramide should receive serious consideration for this reaction of HNF_2.

Some further interesting reactions of trivalent nitrogen can be found for the formally analogous nitrous acid and nitrosyl chloride. The latter compound undergoes an *hydrolysis* reaction to form the former:

$$H_2O + NOCl \rightarrow HNO_2 + H^+ + Cl^-$$

The mechanism could result either from a *D* mechanism ($ONCl \rightarrow ON^+ + Cl^-$) or from an *I* or *A* process involving attack at the nitrogen by oxygen of H_2O.

Many of the redox reactions of nitrous acid are analogous. For example, the nitrosation reactions of amines at low concentrations of HNO_2 follow a third order rate law[83]:

$$HNO_2 + R_2NH \rightarrow R_2NN{=}O + H_2O$$

$$\text{Rate} = k \, [\text{amine}][H^+][HNO_2]$$

Judging from the rate expressions, one mechanistic possibility is that a rapid protonation of HNO_2 establishes a steady-state concentration of $H_2NO_2^+$ (although it is not

---

[82] A. D. Craig and G. A. Ward, *J. Amer. Chem. Soc.*, **88**, 4526 (1966); W. T. Yap, *et al., ibid.*, **89**, 3442 (1967).

[83] A good, critical review of the kinetics and controversial mechanisms proposed for nitrosation reactions is given by J. H. Ridd, *Quart. Rev. (London)*, **15**, 418 (1961).

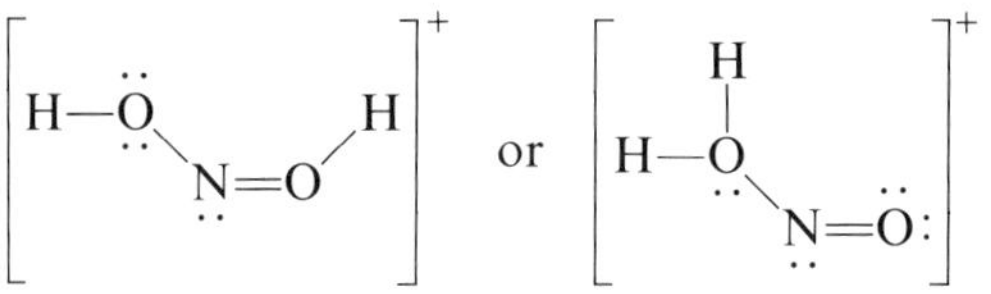

$$\left[\mathrm{H-\overset{..}{O}}\underset{..}{\overset{..}{N}}\!\!=\!\!\overset{H}{O}\right]^+ \quad \text{or} \quad \left[\mathrm{H-\overset{H}{\underset{..}{O}}}\underset{..}{N}\!\!=\!\!\overset{..}{\underset{..}{O}}:\right]^+$$

20

known which oxygen is protonated, **20** is more likely the kinetically important one), which experiences a rate controlling attack by the amine to form, after rapid loss of a proton, the product nitrosamine:

$$\mathrm{H_2NO_2^+ + Amine \xrightarrow{slow} Amine-NO^+ + H_2O}$$

$$\mathrm{R_2HN-NO^+ \xrightarrow{fast} R_2N-N=O + H^+}$$

Interestingly, the rate law expression changes drastically at high concentrations of HNO_2. The third order expression

$$\text{Rate} = k\,[\text{amine}][HNO_2]^2$$

most likely arises from disproportionation of HNO_2 into N_2O_3 (NO_2^- attack upon $H_2NO_2^+$ or HNO_2 upon HNO_2),

$$2HNO_2 \rightleftharpoons H_2O + \underset{O}{\overset{O}{\underset{\diagdown}{\diagup}}}N\!-\!N\overset{\overset{..}{O}:}{\underset{..}{}}$$

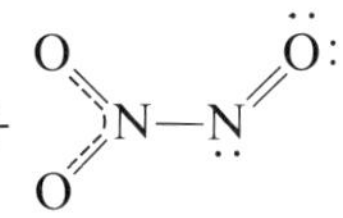

followed by rate controlling attack at the nitrosyl nitrogen of N_2O_3 to form, after elimination of H^+, the product nitrosamine:

$$\mathrm{Amine + N_2O_3 \xrightarrow{slow} Amine-NO^+ + NO_2^-}$$

$$\mathrm{R_2HNNO^+ \xrightarrow{fast} R_2N-N=O + H^+}$$

With secondary amines the reactions cease at production of the *N,N*-dialkyl nitrosamine. With primary amines, however, the initial product (*N*-alkyl nitrosamine) is unstable and decomposes, after a 1,3-tautomerization forming a hydroxyimide **(21)**, into N_2 and ROH.

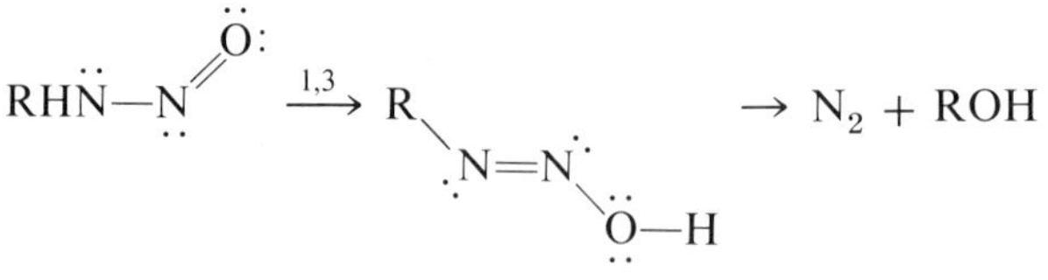

$$\mathrm{RHN\!-\!N}\!\!\overset{\overset{..}{O}:}{\diagup}\;\xrightarrow{1,3}\;\mathrm{R}\diagdown_{N=N\diagdown O-H}\;\rightarrow N_2 + ROH$$

21

These reactions have some current significance owing to the toxicity of nitrosamine to man and the facts that nitrites have been used to preserve meats and are found in abundance in watersheds in areas where cattle are fattened for slaughter.

In view of what has been learned about oxidation reactions of H_2NX and HNO_2, you are in a good position to develop a mechanistic scheme for the important *synthesis of hydroxyl amine* by reduction of nitrous acid with bisulfite (SO_3H^-) under weakly acid conditions. An intermediate hydroxylaminedisulfonate acid **(22)** can be isolated in this reaction

$$H\diagdown \overset{..}{\underset{..}{O}}-\overset{..}{N}\diagup \overset{\overset{..}{O}:}{} + H^+ \rightarrow \left[H\diagdown \overset{\overset{H}{|}}{\underset{..}{O}}-\overset{..}{N}\diagup \overset{\overset{..}{O}:}{} \right]^+ \xrightarrow{SO_3H^-} HO_3S-NO + H_2O$$

$$HO\diagdown \overset{..}{N}\diagup \overset{SO_3H}{\underset{SO_3H}{}} \xleftarrow{H^+} [ON(SO_3H)_2]^- \quad SO_3H^-$$
$$\phantom{HO\diagdown \overset{..}{N}\diagup \overset{SO_3H}{\underset{SO_3H}{}} \xleftarrow{H^+} } \mathbf{22}$$

so the scheme seems entirely reasonable. (Compare the nitrosation reactions of amines.) As in the hydrolysis of amine—O—sulfonates,[84] formation of the zwitterionic form of the acid (23), or an assist by the acidic conditions, could lead to solvolysis in successive steps. This time a monosulfonic acid intermediate (24) can be isolated.

$$(HO)N(SO_3H)_2 \rightleftharpoons \underset{\mathbf{23}}{} \quad HO\diagdown \overset{\overset{H}{|}}{N^+}\diagup \overset{SO_3H}{\underset{SO_3^-}{}} \xrightarrow{H_2O} [(HO)NH(SO_3H)] + H_2SO_4$$

$$HONH_2 + H_2SO_4 \xleftarrow{H_2O} \quad HO\diagdown \overset{\overset{H}{|}}{N^+}-H \diagup \overset{}{\underset{SO_3^-}{}}$$
$$\phantom{HONH_2 + H_2SO_4 \xleftarrow{H_2O} } \mathbf{24}$$

In application, this method of preparation of hydroxylamine suffers if one is careless and uses an excess of bisulfite, for hydroxylamine itself will oxidize bisulfite and form ammonia.

$$NH_2OH + SO_3H^- \rightarrow NH_3 + H_2SO_4$$

In view of the preceding mechanisms for H_2NX as an oxidizing agent, we might expect bisulfite to attack at nitrogen, first to displace OH^-; this could be followed by H_2O attack upon the H_2N-SO_3H form of H_3NSO_3. On the other hand, recalling the hydroxyl amine oxidation of Ph_3P, attack by sulfur could also occur directly at the OH oxygen with displacement of NH_2^-.

Throughout this chapter you can find numerous examples of trivalent phosphorus in the role of a nucleophile. One of the less conventional but more important of these reactions might be termed *oxidative addition*. A particularly simple example involves the chlorination of PCl_3:

$$PCl_3 + Cl_2 \rightarrow PCl_5$$

As was suggested in study question 6-71, this reaction is reversible to a slight extent ($\Delta G°$ only ~ 9 kcal/mole for the *gas phase* chlorination reaction). Though no mechanistic studies of this reaction have been performed, in the gas phase, at least, an asymmetric bimolecular *addition* seems plausible (and possible in orbital following terms):[‡]

$$\diagup \!\!\! P \odot + Cl_2 \rightarrow \left[\diagup \!\!\! \overset{P}{\underset{Cl}{}} \cdots Cl \right] \rightarrow \diagup \!\!\! \overset{|}{\underset{Cl}{P}}-Cl$$

---

[84]Note, however, the occurrence of an N—S, not N—O, bond in these compounds.

[‡]The HOMO/LUMO pair at phosphorus is lone pair/a d ao and the HOMO/LUMO pair at Cl_2 is σ/σ^*.

Just such a mechanism was proposed by Arbusov[85] in his studies of the reactions of alkyl phosphites with alkyl chlorides:

$$(RO)_3P + R'Cl \rightarrow (RO)_2P(O)R' + RCl$$

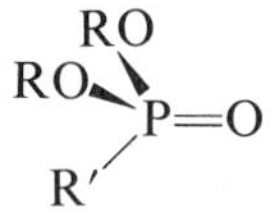

$$\begin{array}{c} RO \\ RO{-}P{=}O \\ R' \end{array}$$

Presumably the first step in this reaction is addition of R'X, followed by loss of X⁻, followed by X⁻ attack upon R.

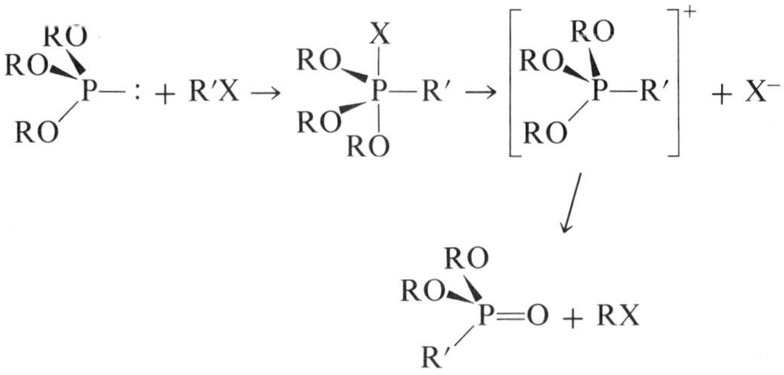

$$\begin{array}{c} RO \\ RO{-}P{:} \\ RO \end{array} + R'X \rightarrow \begin{array}{c} X \\ RO{-}P{-}R' \\ RO \;\; RO \end{array} \rightarrow \left[\begin{array}{c} RO \\ RO{-}P{-}R' \\ RO \end{array}\right]^+ + X^-$$

$$\downarrow$$

$$\begin{array}{c} RO \\ RO{-}P{=}O \\ R' \end{array} + RX$$

In support of just this mechanism, recent studies[86] of the chlorination of a cyclic catechol phosphite proved the formation of the five-coordinate intermediate

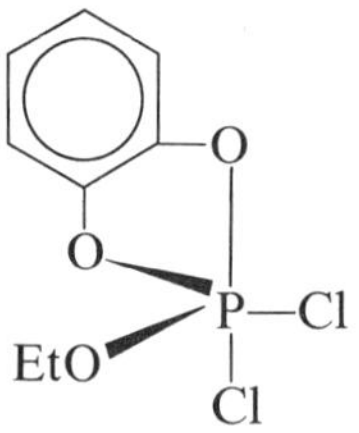

at −85°C. On warming to −40°C, the Arbusov product

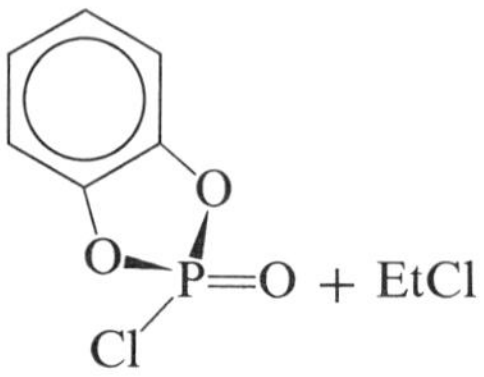

formed.

As a less conventional example[87] of the oxidative addition step, n-BuLi exchanges organo groups with the optically active phosphine MePhP(CH₂Ph) and changes the phosphine chirality:

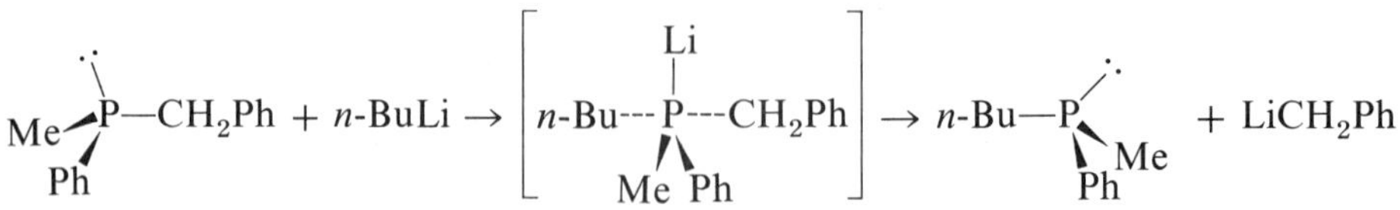

$$\text{Me}{-}\overset{..}{P}{-}CH_2Ph + n\text{-}BuLi \rightarrow \left[\begin{array}{c} Li \\ | \\ n\text{-}Bu{-}{-}{-}P{-}{-}{-}CH_2Ph \\ / \; \backslash \\ Me \;\; Ph \end{array}\right] \rightarrow n\text{-}Bu{-}\overset{..}{P}\underset{Ph}{\overset{Me}{\diagdown}} + LiCH_2Ph$$

Ph

---

[85] A. E. Arbusov, *J. Russ. Phys. Chem. Soc.*, **42**, 395 (1910); R. G. Harvey and E. R. Somber in Vol. 1 and B. Miller in Vol. 2 of "Topics in Phosphorus Chemistry," M. Grayson and E. J. Griffith (Eds.), Wiley, New York, 1964.
[86] A. Skowronska, *et al., J. Chem. Soc., Chem. Comm.*, 791 (1975).
[87] E. P. Kyba, *J. Amer. Chem. Soc.*, **97**, 2554 (1975). See also J. Omelanczuk and M. Mikolajczyk, *Chem. Comm.*, 1025 (1976).

The special features of this reaction are the need for the lone pair to take an equatorial position in the transition state and/or intermediate, the stabilization of the lone pair by Li$^+$, and the lack of the two successive pseudorotations (ψ_{Li}, then ψ_{benzyl}) required for retention. Though ψ_{Li} might be facile, pseudorotation about benzyl would cause Li to take an axial position prior to loss of benzyl.

STUDY QUESTIONS

20. Propose a mechanism for the decomposition of hyponitrous acid, via $HN_2O_2^-$, and compare it with the base catalyzed decomposition of the isomeric nitramide.

21. What analogy can you find between the dehydration reaction of HNO_2 at low pH and the nitrosation of amines by HNO_2?

22. Propose an experimental test of whether the acid hydrolysis of H_3NOSO_3 involves H_2O in the transition state. Would this resolve the uncertainty between D and A or I mechanisms? Assume that your experiment proves H_2O is *not* involved prior to and including the transition state; can you still account for the acid catalysis? (Hint: review the assumption made in the text about the protonation site of H_3NOSO_3.)

23. How might an activation energy analysis of the hydrolysis of $H_3NOSO_3H^+$ help clarify its D and A/I nature?

24. With reference to competing reactions, in what way would the use of liquid ammonia, rather than aqueous OH$^-$, be of advantage in hydrazine synthesis? In what way is it a disadvantage?

25. Why do you suppose chlorination of liquid ammonia fails to yield $NHCl_2$ and NCl_3 as products?

26. With regard to the two pathways suggested in the text for acid conversion of $NH_2Cl \rightarrow NHCl_2$, what steps have to be rate controlling for a distinction to be made on the basis of rate laws? What conditions would favor a common rate law, in practice, for both mechanisms?

27. Write a reaction scheme for the non-redox conversion of HNF_2 into N_2F_2 by OH$^-$.

28. In the discussion of the reactions of NH_2X and NH_2Y molecules, it was pointed out that a large number of these reactions produce one or all of the following sets of nitrogen-containing products: (N_2, NH_3), (N_2, N_2H_4), (N_2O, NH_3).

    On page 422 is a partially completed scheme that can account for the alternate pathways. Fill in the missing intermediates and products.

    a) With reference to the general scheme, identify each step as one of the following:
    nucleophilic displacement
    elimination
    proton shift
    acid ionization
    b) After reading the examples given in the section on "orbital following," explain how the presence of an olefin might open a reaction channel competitive with the disproportionation step. What would the products be? Must N_2H_2 have a *cis*-geometry? (The *trans-* ↔ *cis-* barrier for N_2H_2 is very large.[88])
    c) In this scheme the reactants and final products are circled. Find the paths followed by the general reactions

$$2NH_2X + NH_2Y \rightarrow N_2 + NH_3 + HY + 2HX$$
$$2NH_2X + 2NH_2OH \rightarrow N_2O + 2NH_3 + H_2O + 2HX$$
$$NH_2X + NH_2Y \rightarrow \frac{1}{2}N_2 + \frac{1}{2}N_2H_4 + HX + HY$$

---

[88] N. W. Winter and R. M. Pitzer, *J. Chem. Phys.*, **62**, 1269 (1975).

d) Refer to the text for examples of the reactions in (c) and note that examples with X = OH and/or Y = OH can be found for the first and second types. Taking X = Cl and Y = OH, what effect would increasing NH_2Y and decreasing NH_2X concentration have on the product distribution?

e) Though not shown as a "reactant" in the scheme, H^+ often assists the second general reaction in (c). At what point in the scheme could an assist from H^+ accelerate the formation of N_2O while retarding N_2 formation?

f) The first general reaction in (c) experiences base catalysis. At what steps would OH^- help this reaction and hinder the second general reaction? Note particularly that NH_2OH disproportionates differently under acid and base conditions.

g) What conditions of concentration and pH favor N_2H_4 as a product? Might you ever (under any conditions) expect to find N_2H_4 and N_2 as the exclusive products?

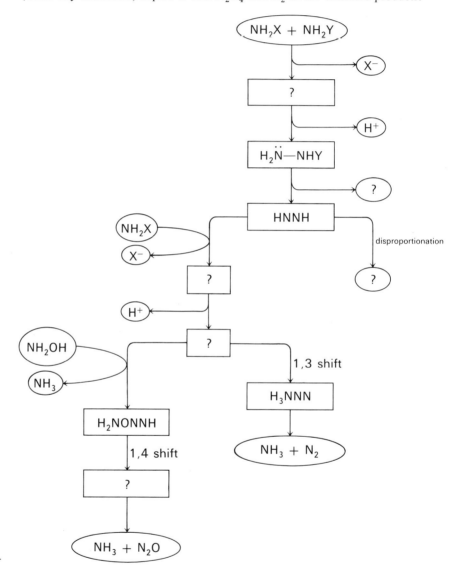

Scheme for study question 28.

Oxygen and Sulfur

For examples of reactions involving central atoms with four electron pairs, two of which are lone pairs, we look to the reactions of two-coordinate oxygen and sulfur compounds. In addition to the many examples encountered earlier in which oxygen acts as a nucleophile, there are other interesting examples in which the oxygen may appear electrophilic (refer to the last section for reactions of phosphines with NH_2OH and O_2). Reactions in which hydrogen peroxide acts as an *atom transfer oxidant* fit this category. With oxygen acting as an electrophile, we find a ready explanation for the oxidation of CN^- to cyanate, OCN^-, of sulfides to sulfoxides, of iodide to iodine, and of carbanions to ethers:

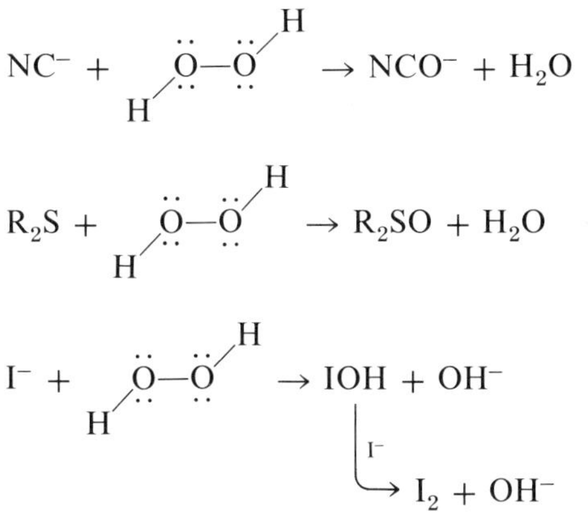

$$NC^- + H_2O_2 \rightarrow NCO^- + H_2O$$

$$R_2S + H_2O_2 \rightarrow R_2SO + H_2O$$

$$I^- + H_2O_2 \rightarrow IOH + OH^- \xrightarrow{I^-} I_2 + OH^-$$

$$R_3C^- + R_2'O_2 \rightarrow R_3COR' + R'O^-$$

By way of generalization,[89] it has been determined that many peroxides react with nucleophiles according to a second order rate law:

$$\text{Rate} = k[\text{peroxide}][\text{nucleophile}]$$

In the case of hydrogen peroxide it has been shown that the second order rate constant exhibits a pH dependence given by

$$k = k_2 + k_H[H^+]$$

with the implication that $H_3O_2^+$ can act as an electrophile with attack occurring at the OH oxygen to displace H_2O (compare this with the acid catalyzed reactions of HNO_2, H_2NOH, and H_3NOSO_3). From studies of the activation parameters for nucleophilic displacements from peroxide oxygen [*viz.*, reactions of H_2O_2, $(t\text{-butyl})O_2H$, HSO_5^-, and CH_3CO_3H] it appears that the $\Delta H^\ddagger$ for these reactions falls in the range from 10 to 15 kcal/mole and $\Delta S^\ddagger$ in the range from -16 to -39 eu; both are consistent with an *a* activation step, most reasonably I_a. The fairly large discrimination[90] between nucleophiles is also a good indication that bond formation is an important part of the activation step. Similarly, the reaction rates of various peroxy acids show a good correlation of rate constant with pK_b of the displaced group; for example, the order

$$H_3O_2^+(H_2O) > H_3PO_5(H_2PO_4^-) > HSO_5^-(SO_4^{2-}) > CH_3CO_3H(CH_3CO_2^-) > H_2O_2(OH^-)$$

is found for the oxidation of Br^- (the species in parentheses are the leaving groups). Whether this correlation should be taken to imply that bond-breaking is also important to the activation step or simply implies that the basicity of the leaving group determines the ease with which the new bond is formed is unclear.

By means of oxygen tracer studies it has been determined[91] that, in the oxidation

---

[89] An excellent analysis of peroxide reaction kinetics and mechanisms is given by J. O. Edwards, "Inorganic Reaction Mechanisms," W. A. Benjamin, New York, 1964.
[90] Remember that a variation in $\Delta G^\ddagger$ of ~ 5 kcal/mole causes a change in k of $\sim 10^4$.
[91] M. Anbar and H. Taube, *J. Amer. Chem. Soc.*, **76**, 6243 (1954).

of NO_2^- by peroxyacids, the NO_3^- product contains one oxygen atom derived from the peroxy acid. This, of course, lends credence to the hypothesis that the stoichiometric mechanism involves direct attack of the nucleophile upon a peroxide oxygen. Interestingly, H_2O and OH^- (the latter is generally a reasonably strong nucleophile) have been shown by isotope tracer studies to exchange oxygen atoms with peroxides at immeasurably slow rates. Comparison of rate constants for second order nucleophilic attacks at saturated carbon and peroxide oxygen reveals that oxygen is much more discriminating than carbon in these reactions. This has been taken to reflect the greater importance of lone-pair/lone-pair repulsions (between oxygen and nucleophile, **25**) in the peroxide reactions.

25

This is consistent with slow reactions for second row nucleophiles (H_2O, OH^-) and faster reactions with the more "polarizable" nucleophiles from the third and later periods. (See p. 411.) This concept is the kinetic equivalent (for *a* mechanisms) of the vicinal lone pair repulsion as a factor in bond energies.

Another aspect of peroxide chemistry that is important from a kinetic view is *disproportionation*. At constant pH, the rate laws appear to be generally valid far to either side of $pH = pK_a$ of the peroxide.

$$2RO_2H \rightarrow 2ROH + O_2$$
$$\text{Rate} = k_a[RO_2H]^2 \quad \text{and} \quad \text{Rate} = k_b[RO_2^-]^2$$

The rate constants k_a and k_b show a pH dependence such that at pH values below the pK_a of the peroxy acid, k_a is *inverse* first order in $[H^+]$, while at pH values above the pK_a of the peroxy acid, k_b is the first order in $[H^+]$. *Both* rate expressions derive from the ionization equilibrium of H_2O and

$$\text{Rate} = k_2[RO_2H][RO_2^-]$$

and fit the general concept of nucleophilic attack (RO_2^-) at peroxide oxygen (RO_2H). Furthermore, taking Caro's acid (H_2SO_5) as an example, the $\Delta H^\ddagger$ for the auto-redox reaction is 12 kcal/mole and $\Delta S^\ddagger$ is -20 eu, values in the "normal" range for nucleophilic attack at peroxide oxygen. You might pause to reflect on the relative nucleophilicities of H_2O and OH^- on the one hand and RO_2^- on the other. An intriguing question arises regarding the nature of the transition state species in such auto-redox reactions. In general, it is expected that the attacking nucleophile would select the OH oxygen of the peroxy acid so as to displace an RO^- group in the activation step.

By isotope (^{18}O) labeling both peroxide oxygen atoms of Caro's acid (O_3SOOH) and carrying out the reaction with unlabeled acid, it has been determined[92] that the oxygen atom isotopes in the product O_2 are highly "scrambled." The really interesting result[93] of *unscrambled* oxygen arises for peroxyacetic acid. Owing to the trigonal

---

[92] See Ref. 89; J. F. Goodman and P. Robson, *J. Chem. Soc.*, 2871 (1963).
[93] E. Koubek, *et al.*, *J. Amer. Chem. Soc.*, **85**, 2263 (1963); J. F. Goodman, *et al.*, *Trans. Faraday Soc.*, **58**, 1846 (1962).

planar structure about the carboxyl carbon and its expected greater electrophilicity than the peroxy oxygen, it seems reasonable that the entering nucleophile prefers to attack at carbon. With an assist from proton transfer, the following mechanism accounts for the fact that the evolved O_2 consists of oxygen atoms originally bound to each other.

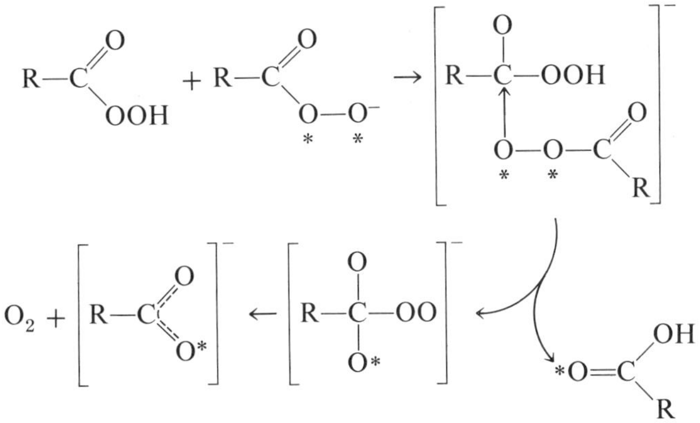

$$R-C(=O)OOH + R-C(=O)O-O^{*}-O^{*-} \rightarrow \left[R-C(=O)-OOH \cdots O^{*}-O^{*}-C(=O)R \right]^{-}$$

$$O_2 + \left[R-C(=O)O^{*} \right]^{-} \leftarrow \left[R-C(=O)-OO^{*} \right]^{-} \leftarrow {}^{*}O=C(OH)R$$

Be careful to note that this mechanism does *not* suggest that *all* peroxy linkages remain intact throughout the reaction. You see that half of them are ruptured, with those oxygen atoms appearing only in carboxylic acid groups.

In some respects you might expect the hypohalous acids (HOX) to react in the fashion of peroxy acids and hydroxylamine; the analogy is clear if the X group is viewed as a substitutent like OR of the peroxy acid or NH_2 of hydroxylamine. In this case, the oxygen of HOX could act as an electrophile or the oxygen of OX^- could act as a nucleophile. There are times, however, when it is more appropriate to view OX^- as a "pseudo" halogen. The peroxide analogy should be valid for HOF and OF^-, but when X = Cl, Br, or I, the electrophilicity of X may exceed that of oxygen and reactions occur by nucleophilic attack at the *halogen* (compare the syntheses of the chloramines and most reactions of hydroxylamine). In these cases, the site of attack chosen by the nucleophile may depend strongly on the relative "leaving" group stabilities and on the relative electrophilicities of O and X.

$$Nuc + HOX \rightarrow Nuc\text{---}O(H)(X) \rightarrow Nuc\text{---}OH^+ + X^-$$

or

$$Nuc\text{---}X\text{---}OH \rightarrow Nuc\text{---}X^+ + OH^-$$

The decomposition of HOF and HOCl in aqueous media appear to be good cases in point. The product O_2 could arise by a mechanism analogous to that discussed above for peroxide disproportionation,

$$OX^- + HOX \rightarrow O_2 + HX + X^-$$

or by the disproportionation of H_2O_2 formed as an intermediate

$$HOX + H_2O \rightarrow H_3O_2^+ + X^- \rightarrow \text{etc.}$$

HOF has been so recently prepared that its reactions with other materials are not well documented. For the better known hypochlorous acid, more reactions have been studied. In addition to the decomposition reaction just mentioned, oxygen *atom transfer oxidations* of SO_3^{2-}, IO_3^-, and NO_2^- proceed[94] according to second order rate

---

[94] M. W. Lister and P. Rosenblum, *Can. J. Chem.*, **39**, 1645 (1961).

laws, and in the case of nitrite the use of tracer oxygen confirms[95] that the third oxygen of the product nitrate derives from the hypochlorous acid.

$$HOCl + :SO_3^{2-} \rightarrow HSO_4^- + Cl^-$$

$$HOCl + :IO_3^- \rightarrow HIO_4 + Cl^-$$

$$HOCl + :NO_2^- \rightarrow H\overset{*}{O}NO_2 + Cl^-$$

In the next section on halogen reactions we will consider the evidence for nucleophilic attack upon the halogen of HOX. Before leaving the subject of nucleophilic attack at peroxy and hypohalous oxygen, we should note that not all such reactions are of the polar, displacement type. In the presence of trace metal ions these oxidants achieve reaction through free radical intermediates, where the role of the metal ion is to catalyze the free radical decomposition of R_2O_2 and HOX.[96]

Reactions involving *sulfenyl* sulfur (—S—) seem not to be much different in mechanism from those at oxygen, except that they are often much faster.[97] Generally speaking, reactions with nucleophiles follow a second order rate law

$$\text{Rate} = k[\text{RSX}][\text{Nuc}]$$

where Nuc is some nucleophile. It is often the case that such reactions are acid catalyzed, and a particularly interesting example is the acid catalyzed *exchange* of disulfides:

$$2\text{RSSR}' \rightarrow \text{RSSR} + \text{R}'\text{SSR}'$$

The mechanism of these exchange reactions has been clarified by studies of the related exchange of sulfur between Ph*SH and PhSSPh. This reaction is also catalyzed by acid, and the remarkable finding that the *rate constant* depends on the acid *anion* greatly aids formulation of a pathway.[98] Specifically, the relative rate constants are $HClO_4$ (0), HCl (1.0), HBr (10^2), and HI (10^4). Thus, the mechanism must account for *anion* catalysis:

$$\text{PhSSPh} + \text{HX} \rightleftharpoons \text{PhS}\overset{\overset{H}{|}}{\text{S}}\text{Ph}^+ + \text{X}^-$$

$$\text{PhS}\overset{\overset{H}{|}}{\text{S}}\text{Ph}^+ + \text{X}^- \underset{}{\overset{\text{slow}}{\rightleftharpoons}} \text{PhSX} + \text{HSPh}$$

Here we have written all steps so you can see that this mechanism satisfies the PMR as a single-channel reaction, with the chemical identity of PhSH and Ph*SH being lost at the pool of intermediates (PhSX + PhSH).[99]

$$\text{PhSX} + \text{Ph*SH} \overset{\text{slow}}{\rightleftharpoons} \text{PhS}(*\text{S})\overset{\overset{H}{|}}{}\text{Ph}^+ + \text{X}^-$$

$$\text{PhS}(*\text{S})\overset{\overset{H}{|}}{}\text{Ph}^+ + \text{X}^- \rightleftharpoons \text{PhS*SPh} + \text{HX}$$

---

[95] M. Anbar and H. Taube, *J. Amer. Chem. Soc.*, **80**, 1073 (1958). See Chapter 13 for the orbital following scheme.

[96] See Ref. 89.

[97] A good summary of nucleophilic substitution at sulfur is given by J. L. Kice in "Inorganic Reaction Mechanisms," Part II, J. O. Edwards (Ed.), Interscience, New York, 1972.

[98] A. Fava and G. Reichenbach, cited by E. Cuiffarin and A. Fava in "Progress in Physical Organic Chemistry," Vol. 6, A. Streitweiser and R. W. Taft (Eds.), Interscience, New York, 1968.

[99] In the second step we presume that X^- attacks the unprotonated sulfur because HSPh is a better leaving group (less basic) than PhS$^-$.

The HCl catalyzed exchange[100] of the organo groups of RSSR' is now seen to be actually an exchange of RS groups proceeding through steps like

$$Cl^- + \left[-S\overset{H}{-}S- \right]^+ \rightleftarrows -SCl + -SH$$

and

$$-SCl + -SS- \rightleftarrows \left[-S\overset{|}{S}S- \right]^+ + Cl^-$$

Because of the multiplicity of species present in the details of this scheme, we have avoided writing down all possible steps where the terminal organo groups (R, R') are specified. In fact, the shortened version above should satisfy you that scrambling is achieved by proper choices of R and R' in the second step (for example, R—SCl + R'—SS—R').

An interesting example of the effect of ring size on reactivities of cyclic disulfides is found when comparing the second order rate constants for the reactions

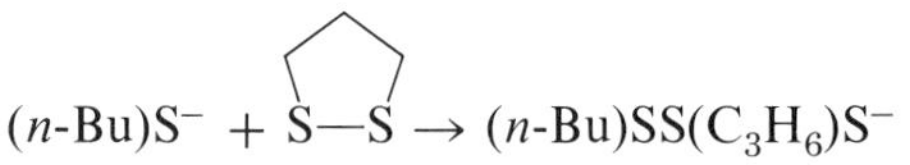

$$(n\text{-Bu})S^- + \overset{\frown}{S-S} \rightarrow (n\text{-Bu})SS(C_3H_6)S^-$$

and

$$(n\text{-Bu})^*S^- + (n\text{-Bu})SS(n\text{-Bu}) \rightarrow (n\text{-Bu})^*SS(n\text{-Bu}) + (n\text{-Bu})S^-$$

1,2-Dithiacyclopentane has a rate constant about 10^4 times greater ($\Delta G^{\ddagger} \approx 6$ kcal/mole less) than dibutyldisulfide.[101] Like peroxides, disulfides prefer a structural arrangement with the dihedral angle defined by R—S—S—R $\cong 90°$.

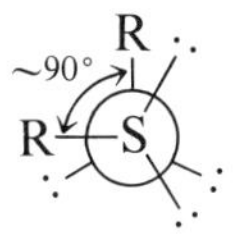

In a five-membered ring this preference cannot be met (1,2-dithiacyclopentane has a dihedral angle of only 27°),[102] so there is expected to be significant weakening of the S—S linkage. Consistent with this expectation, the enthalpy change[103] for hydrogenation of the 1,2-dithiacyclopentane

$$\overset{\frown}{S-S} \xrightarrow{H_2} HSC_3H_6SH$$

$$\overset{\frown}{S-S} \xrightarrow{H_2} HSC_4H_8SH$$

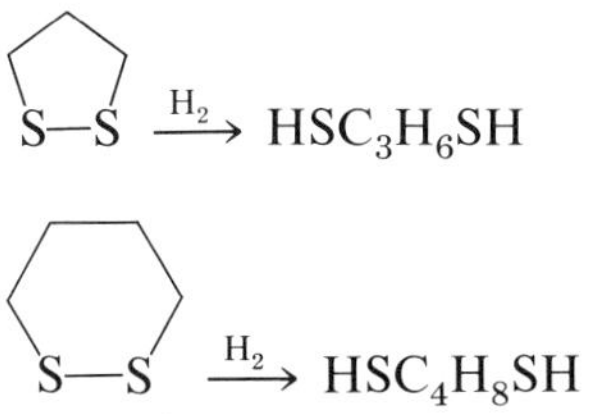

[100] R. E. Benesch and R. Benesch, *J. Amer. Chem. Soc.*, **80**, 1666 (1958). Note that the mechanism proposed in this reference is in error in that it does not recognize the acid anion effect treated in Ref. 101.

[101] A. Fava, et al., *J. Amer. Chem. Soc.*, **79**, 833 (1957).

[102] A. Hordvik, *Acta Chem. Scand.*, **20**, 1885 (1966).

[103] S. Sunner, *Nature*, **176**, 217 (1955).

is ~5 kcal/mole more negative than that of the 1,2-dithiacyclohexane.

At the present time there is insufficient information to prove whether nucleophilic substitutions at sulfenyl sulfur occur by an *I* mechanism or by an *A* mechanism. As mentioned in the introduction to this chapter, a proof is hard to come by. The steric inhibition by R observed[104] for the exchange reactions

$$*SO_3^{2-} + RSSO_3^- \rightarrow RS*SO_3^- + SO_3^{2-}$$

parallels almost exactly that for Br$^-$ exchange with RCH$_2$Br, a known *I* process. It would seem that bond formation is of some importance in the activation step, but the data cannot distinguish between *I* and *A* mechanisms. Relative rates for R = (Me, Et, *i*-Pr, *t*-Bu) are (2, 1, 10^{-2}, 10^{-5}) and are characteristic of displacement reactions at sulfenyl sulfur. The fact that the reaction

$$(n\text{-Bu})NH_2 + Ph_3C-S-Cl \xrightarrow{B} Ph_3C-S-NH(n\text{-Bu}) + BH^+ + Cl^-$$

is catalyzed by bases in general[105] is interesting but not very helpful in settling the intermediate question. One possibility for the role of the base is the pre-activation step CB mechanism noted earlier for amine reactions. A second possibility could be attack upon Ph$_3$CSCl by a hydrogen-bonded species B $\cdots$ HNH(*n*-Bu) where, in the transition state, the N—H bond is cleaved as the B—H and N—S bonds are formed. Neither of these pathways gives us information as to whether the S—Cl bond is broken at the time the N—S bond is formed or later. The idea that the base becomes involved *after* formation of a (*n*-Bu)H$_2$NSRCl intermediate **(26)** requires (from the rate law) that we imagine the formation of the intermediate to be faster than proton

$$\underset{\textbf{26}}{\overset{\displaystyle R \diagdown \quad \ddot{} \diagup}{\underset{H\diagup\,H}{N}-\underset{R}{\overset{\ }{S}}-Cl}}$$

abstraction by the base (as the rate controlling step). It is difficult to imagine the proton transfer from such an intermediate as being the "slow" step. In this regard, for what it is worth, there are no known adducts of the type D $\cdot$ SX$_2$ or SX$_3^-$. If you have not already done so, notice that this reaction belongs to the non-metal halide *solvolysis* category.

---

[104] A. Fava and A. Iliceto, *J. Amer. Chem. Soc.*, **80**, 3478 (1958).
[105] E. Cuiffarin and G. Guaraldi, *J. Amer. Chem. Soc.*, **91**, 1745 (1969).

STUDY QUESTIONS

29. Refer to the mechanism suggested for the auto-decomposition of Caro's acid and predict the ratios $*O_2 : *OO : O_2$, assuming that there are no isotope effects on the rates of any steps. Do assume rapid H$^+$ exchange between ROOH and ROO$^-$, and base your answer on the relative equilibrated amounts of the electrophile/nucleophile pairs: (labeled/unlabeled), (labeled/labeled), and (unlabeled/unlabeled).

30. Discuss the similarity between the acid catalyzed action of H$_2$O$_2$ as an electrophile and acid catalyzed oxidations by HNO$_2$ by identifying a common leaving group.

31. How might the enhanced nucleophilicities of third period, as contrasted with second period, elements toward peroxide oxygen be accounted for? Do you see a relation to their nucleophilicities toward NH$_2$SO$_4^-$ (p. 410)?

32. Verify that

$$\text{Rate} = k_2[\text{RO}_2\text{H}][\text{RO}_2^-]$$

may be converted into

$$\text{Rate} = k_a[\text{RO}_2\text{H}]^2 \text{ at low pH,}$$

$$\text{Rate} = k_b[\text{RO}_2^-] \text{ at high pH.}$$

Halogens

The reactions of the halogens bring us to the case of four electron pairs, three of which are lone pairs. Perhaps the most interesting reactions here are the *solvolysis* reactions (which could also be considered as redox reactions). We have already mentioned the solvolysis of Cl_2 by NH_3 (in liquid ammonia) leading to the formation of hydrazine and chloride ion. In aqueous systems the halogens initially yield halide ion and hypohalous acid.

$$X_2 + H_2O \rightarrow X^- + HOX + H^+$$

The kinetics of these reactions have been studied[106] in detail, with the finding that the major pathway for hydrolysis involves attack of H_2O and/or OH^- at the X_2 molecule to displace X^- and form HOX (note that these reactions could also be considered as simple substitution or displacement reactions). For the pathway involving attack of H_2O upon the halogen, the rate controlling step is the formation of the intermediate X_2OH^-; whereas for attack on X_2 by OH^-, the rate controlling step appears to be the dissociation of X^- from X_2OH^-.

$$\begin{array}{c} X_2 + H_2O \\ \searrow^{1} \\   X_2OH^- \xrightarrow{2} XOH + H^+ + X^- \\ \nearrow^{3} \\ X_2 + OH^- \end{array}$$

Rate constants for the three steps for Cl, Br, and I are reported to be:

	Cl	Br	I
k_1 (sec^{-1})	11.0	110	2.1
k_2 (sec^{-1})	—	5×10^9	3×10^7
k_3 (M^{-1} sec^{-1})	10^{10}	10^{10}	10^{10}

With H_2O solvent, k_1 is of necessity pseudo-first order; k_2 is truly first order, while k_3 is second order. Path 3 is obviously diffusion controlled.

An interesting feature of these reactions is that the HOX acids exhibit different kinetic stabilities in aqueous solution. As mentioned in the preceding section, HOF is apparently rapidly attacked by H_2O (or OH^-), and the final product O_2 results from decomposition of H_2O_2.

$$HOF + H_2O \rightarrow F^- + H^+ + H_2O_2$$

With X = Cl, the reaction of HOCl with H_2O is less rapid and the production of O_2 is much slower. HOBr solutions seem to be more unstable with respect to *dispropor-*

---

[106] M. Eigen and K. Kustin, *J. Amer. Chem. Soc.*, **84**, 1355 (1962).

tionation (oxygen transfer) to BrO_3^- and Br^- than to *solvolysis*. It appears that one HOBr reacts with another (Br attack at oxygen) to form Br^- and $HBrO_2$, while the latter reacts further with HOBr to yield BrO_3^- and more Br^-.

$$HOBr + {^*BrO^-} \rightarrow (HO)^*BrO + Br^-$$

or

$$^*BrOH \rightarrow HO^*BrO + Br^- + H^+$$

$$HOBr + {^*BrO(OH)} \rightarrow {^*BrO_2(OH)} + Br^- + H^+$$

Tracer studies[107] have revealed that, in addition to H_2O attack upon HOX to form H_2O_2, oxygen exchange between H_2O and HOCl and between H_2O and HOBr occurs in hypohalite solutions, illustrating the point made in the last section that attack on the heavier hypohalites can occur at either O or X. Recall, too, the NH_3/HOCl reaction as the basis for synthesis of the chloramines.

FIVE VALENCE PAIRS

Phosphorus and Sulfur

For examples involving reaction of molecules with five valence electron pairs, pentavalent phosphorus and tetravalent sulfur should come to your mind. As is well illustrated by silicon, a characteristic feature of such higher valency third row atoms is their ability to expand their valence shells to form six-valence-pair structures; therefore, it is not surprising that most chemists think of solvolysis and substitution reactions as proceeding through the formation of intermediates with six valence pairs about P and S. You have seen that SF_4 forms, in its reaction with $Me_4N^+F^-$, an adduct salt $Me_4N^+SF_5^-$ and also may be oxidized to the five-coordinate OSF_4. This behavior for sulfur(IV) contrasts with the uncertainty regarding a sulfenyl intermediate (two sections back). As examples of the tendency of phosphoranes to form six-coordinate species, recall from Chapters 5 and 6 the tendency of PCl_5 to form adducts with amines, ether, and Cl^-.

That the *trivalent halides* of phosphorus undergo *solvolysis* reactions with acidic amines, alcohols, and water to produce HX and the trivalent substituted phosphines was discussed in the last chapter. These reactions involve nucleophilic attack by nitrogen or oxygen at the phosphorus atom. With *pentavalent phosphorus halides* the reactions are much more exothermic and proceed with greater speed, so that care is sometimes required in bringing the PX_5 compounds into contact with acidic solvents. The phosphoryl halides (X_3PO) are similarly reactive; an important reaction for them is alcoholysis to form esters of phosphoric acid, $(RO)_3PO$. The only qualification to make concerning these comments is with reference to fluorides: in spite of the fact that solvolysis of fluorides can be thermodynamically spontaneous, such reactions are usually much slower than is the case with the heavier halides.

In highly related reactions, SF_4 and thionyl chloride ($SOCl_2$) hydrolyze to form SO_2 and HX (recall that sulfurous acid is actually the SO_2 solvate with formula $SO_2 \cdot H_2O$), and sulfuryl chloride (SO_2Cl_2) vigorously hydrolyzes to form sulfuric acid and HCl. Following the established pattern for fluorides, sulfuryl and thionyl fluorides are kinetically resistive to hydrolysis, in spite of a high thermodynamic spontaneity for formation of H_2SO_4 and SO_2, respectively. This resistivity to nucleophilic attack, as noted in the last chapter, is most pronounced for SF_6. While the electrophilicity of the central atom in these fluorides should be at its highest, the short E—F bonds and high charge density of the fluorines seem to interfere seriously with nucleophiles (O, N, etc.) crowding into the coordination sphere of the central atom.

A few reflections on the relation of these reactions to those of the chloramines are valuable at this point. You remember that the chlorines of NCl_3 can play the role of an electrophile, while with PX_3 it is the phosphorus atom that experiences nucleo-

---

[107] See Ref. 95.

philic attack, to the exclusion of the halogens. This is attributed to (i) the much greater electronegativity of the Cl_2N group than of the Cl_2P group, (ii) the more "open" PX_3 structure, and (iii) the ability of phosphorus to expand its valence shell. It might seem to you that in the pentavalent halides, the PX_4 group could render the "fifth" halogen sufficiently electrophilic to experience attack by nucleophiles. In fact, there are five halogens present to support the (+5) oxidation state of phosphorus in PX_5 and three to support the (+3) oxidation state in PX_3. Viewed in this way, it is difficult to see how the PX_5 halogens could be significantly more electrophilic than their PX_3 analogs. Rather, it is the central atom that should have a markedly increased electrophilicity, as is found.

In the next chapter we will consider many of the useful reactions of SF_4, so we will just interject here a note that SF_4 will not react with chlorine in the absence of CsF. A low nucleophilicity expected for SF_4 should be markedly enhanced by formation of SF_5^- as an intermediate. Attack upon Cl_2 to displace Cl^- now appears feasible.

$$SF_4 + Cl_2 + CsF \rightarrow SF_5Cl + CsCl$$

Many pentavalent phosphorus compounds have been found to be stereochemically non-rigid,[108] as might be anticipated from our earlier inspection of five-coordinate silicon intermediates. It is becoming clear that axial fluorines of PF_5, for example, rapidly interconvert with equatorial fluorines via the Berry mechanism of *pseudorotation*.[109] The first order rate constant for this intramolecular isomerization is so large that it cannot be measured, even at $-100°C$. The theoretical estimate[110] of 5 kcal/mole for the activation enthalpy of the pseudorotation appears quite reasonable in that, assuming a negligible activation entropy, it leads to a predicted rate constant at $173°K$ of $\sim 10^6$ sec^{-1}.

Detailed studies[111] of the fluorophosphorane $(CH_3)_2NPF_4$ have shown that this compound definitely undergoes a more restricted Berry pseudorotation ($\Delta H^\ddagger = 10$ kcal/mole) than PF_5. The pseudorotation of **27** is quite interesting because the preference[112] of the amine nitrogen to remain planar with N—C bonds eclipsing the axial P—F bonds requires that rotation of the NMe_2 group about the P—N bond accompany the pseudorotation about phosphorus (NMe_2 as pivot).

The simpler compounds CH_3PF_4 and HPF_4 have also been studied,[113,114] with the finding that rapid pseudorotation about phosphorus takes place. In all three of these RPF_4 compounds ($R = Me_2N$, Me, H), the R groups are equatorially positioned in the most stable isomeric forms and pseudorotation presumably occurs about

---

[108] This general topic is reviewed by E. L. Muetterties in Ref. 9.
[109] P. Russegger and J. Brickmann, *Chem. Phys. Letters*, **30**, 276 (1975); *J. Chem. Phys.*, **62**, 1086 (1975).
[110] A. Strick and A. Veillard, *J. Amer. Chem. Soc.*, **95**, 5574 (1973).
[111] G. M. Whitesides and H. L. Mitchell, *J. Amer. Chem. Soc.*, **91**, 5374 (1969).
[112] Owing to the equatorial site for NR_2 and the greater in-plane than out-of-plane $P=N$ π bonding (Ref. 13, Chapter 4); E. L. Muetterties, *et al.*, *J. Amer. Chem. Soc.*, **94**, 5674 (1972).
[113] L. S. Bartell and K. W. Hansen, *Inorg. Chem.*, **4**, 1777 (1965); P. Russegger and J. Brickmann, *J. Chem. Phys.*, **62**, 1086 (1975); R. R. Holmes, *Accts. Chem. Res.*, **5**, 296 (1972).
[114] P. M. Treichel, *et al.*, *J. Amer. Chem. Soc.*, **89**, 2017 (1967).

these groups as pivots. Interestingly, the compounds $(CH_3)_3PF_2$, $(CH_3)_2PF_3$, and H_2PF_3 do not undergo a rapid pseudorotation (presumably because the presence of two or more R groups raises the barrier to pseudorotation by forcing at least one of the R's into an axial position in the pseudorotamer), whereas PCl_2F_3 and PBr_2F_3 exhibit[115] quite low intramolecular exchange barriers (~ 6 kcal/mole). Rather than *intra*molecular exchange, the R_3PF_2 and R_2PF_3 compounds undergo (in sufficiently concentrated solutions) rapid exchanges of fluorine atoms between phosphorus atoms by dimeric, six-coordinate phosphorus intermediates or transition states.

$$2 \; Me-\!\!\!\overset{\overset{\displaystyle F}{|}}{\underset{\underset{\displaystyle F}{|}}{P}}\!\!\overset{\displaystyle Me}{\underset{\displaystyle Me}{\diagdown}} \rightleftharpoons \text{(dimeric six-coordinate structure)}$$

The activation parameters of these *inter*molecular exchange reactions[116] are quite similar and consistent with an *a* activation step: $\Delta H^\ddagger \approx 4$ kcal/mole and $\Delta S^\ddagger \approx -40$ eu. Other compounds believed to undergo such bimolecular intermolecular fluorine exchange are ClF_3, BrF_3. Note the striking contrast in exchange behavior for molecules with two or more electron pairs strongly preferring equatorial positions ($ClF_3, BrF_3, R_2PF_3, R_3PF_2$) and those with less than two such pairs.

On the basis of this limited series of phosphorus compounds, it would appear that we could generalize that facile pseudorotation usually occurs if there is present only one group that strongly prefers (because of π-bonding with P or low electronegativity) an equatorial position. When two or more such groups occur, there tends to be no exchange or halogen exchange by the intermolecular pathway. Note that the phrase "strongly prefers" is relative to the other groups present. For example, Cl_2PF_3 has a much lower barrier to pseudorotation than does Me_2PF_3. Such general guidelines are just that, for the scientific community has yet to establish the role of steric effects and the effect of synergic interaction between rotating and pivot groups on the barrier to pseudorotation. Nor is it fully understood how the central atom may affect the barrier magnitudes of inter- and intramolecular exchange processes. In this regard,[117] SF_4 is less labile than PF_5; $(C_2F_5)_2PF_3$ and $(CF_3)_2PF_3$, but *not* $(i-C_3F_7)SF_3$ and $(CF_3)SF_3$, are labile; $(i-C_3F_7)_2SF_2$ and $(i-C_3F_7)(CF_3)SF_2$ exhibit pseudorotation at room temperature. The first two comparisons suggest that Berry rotation about sulfur may be more hindered than that about phosphorus; the difference between $(i-C_3F_7)SF_3$ and $(i-C_3F_7)_2SF_2$ may signal an important steric distortion of the latter toward the square pyramidal transition state, thereby effectively lowering the activation barrier to $\psi_{\text{lone pair}}$.

Theoretical estimates[118] of the pseudorotation for the compound PF_3H_2 (to be taken as a model for R_2PF_3 compounds) places the activation energy at about 13 kcal/mole, far in excess of that determined for the intermolecular exchange reaction of $(CH_3)_2PF_3$. However, an interesting point to come out of the theoretical studies of pseudorotation in such compounds concerns the effects of replacing F by R on the ψ energy profile. The "square pyramidal" transition state energy for R_2PF_3 pseudorotation (one of the R groups is the pivot) is nearly the same as the energy of the rotamer with one R group equatorial and the other axial. In the case of R_4PF compounds, moreover, the rotamer having a trigonal bipyramidal structure with two R groups axial has an energy much higher than that of the square pyramidal transition state, so that the square pyramid no longer represents a transition state structure. Clearly, the effect of increasing the number of R groups on the transition

---

[115] E. L. Muetterties, *et al.*, *Inorg. Chem.*, **2**, 613 (1963); **3**, 1298 (1964); **4**, 1520 (1965).
[116] T. A. Furtsch, *et al.*, *J. Amer. Chem. Soc.*, **92**, 5759 (1970).
[117] W. G. Klemperer, *et al.*, *J. Amer. Chem. Soc.*, **97**, 7023 (1975); E. L. Muetterties, *et al.*, *Inorg. Chem.*, **3**, 1298 (1964).
[118] See Ref. 110.

state energy is less important than the effect on the energy of the pseudorotamer. This is so severe an effect for R_4PF that no *activation barrier* is encountered for conversion of $(R_2)_{ax}P(R_2F)_{eq}$ to $(RF)_{ax}P(R_3)_{eq}$!

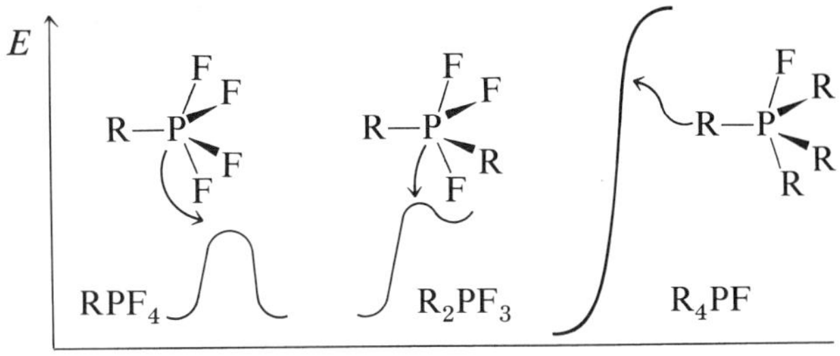

The reactions of *esters* and *acids* of phosphorus and sulfur make some interesting points that we need to review. While these compounds are properly thought of as examples with five valence electron pairs, one of the pairs participates in double bonding to a terminal oxygen; thus, the stereochemistry is approximately tetrahedral and much like that of the silicon compounds discussed previously.

The phosphate esters, and particularly the fluorophosphate esters, have recently attracted much attention because of their biochemical activity. Fluorophosphate esters of general formula

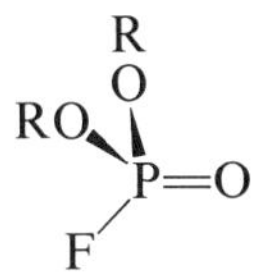

are exceedingly toxic to insects and animals because of a nucleophilic displacement reaction, *in vivo*, of the fluoride by the acetylcholinesterase enzyme. When phosphorylated in this way, the enzyme is rendered inactive and can no longer serve its normal function to assist the solvolysis of acetylcholine (**28**) in muscle tissue (the

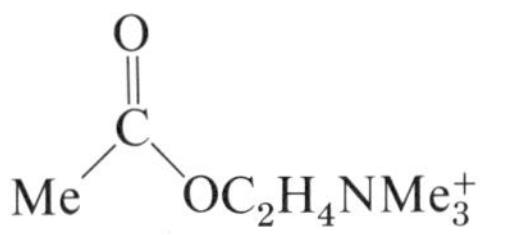

enzyme reacts with acetylcholine by nucleophilic attack at the acetyl group, which is subsequently hydrolyzed and the enzyme regenerated).

Studies of the reactions of the toxin Sarin (**29**) with various nucleophiles show[119]

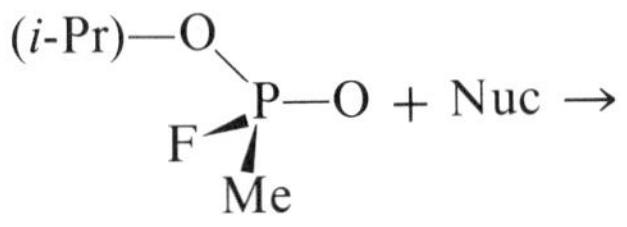

that the second order rate constants are sensitive to the nature of the nucleophile; this implies either that a rapid pre-equilibrium forms a five-coordinate phosphorus intermediate or that bond-making is important in reaching the transition state in the activation step. Some illustrative data are (k_2 values given in parentheses): F^- (highly reactive) > O_2H^- (10^5) > OH^- (10^3) > $S_2O_3^=$, Cl^-, Br^-, I^- (unreactive). Furthermore,

---

[119] A. L. Green, *et al., J. Chem. Soc.*, 1583 (1958); also Ref. 89.

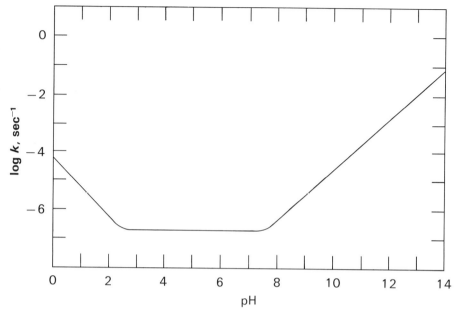

Figure 7-12. The rate of hydrolysis of diisopropylfluorophosphate as a function of pH at 25°C. [From J. O. Edwards, "Inorganic Reaction Mechanisms," W. A. Benjamin, New York, 1965.]

hydrolysis of organophosphates exhibits a pH dependence indicative of alternative mechanisms in different pH ranges. For example diisopropylfluorophosphate exhibits a dependence of the pseudo-first order rate constant on pH, as illustrated in Figure 7-12. In the low pH region the rate constant decreases as the H^+ concentration decreases, a behavior that suggests that an acid catalyzed reaction occurs at low pH:

$$\begin{array}{c} RO \\ \diagdown \\ RO \diagup P{=}O \\ | \\ F \end{array} + H^+ \rightleftharpoons (RO)_2POF \cdot H^+ \xrightarrow[\text{slow}]{H_2O} (RO)_2PO_2H + H^+ + F^-$$

Since protonation of the OR oxygen would actually facilitate ROH displacement, it appears that the most likely donor atom is the terminal oxygen, as you expect from the behavior of phosphate esters as Lewis bases. (Note the analogy with the $H_3NOSO_3H^+$ hydrolysis, p. 412.) In the high pH region we find the rate constant increasing with increasing OH^- concentration, which suggests direct attack on the fluorophosphate by OH^-.

$$(RO)_2POF + OH^- \rightarrow (RO)_2PO_2H + F^-$$

At moderate pH's, where neither the H^+ nor the OH^- concentration is sufficiently high to make the previous two pathways kinetically significant, we find k to be independent of pH; this is consistent with a relatively slower, uncatalyzed attack by the H_2O molecule.

Edwards has wisely pointed out[120] that some care must be exercised in reaching the conclusions given above. For example, the behavior noted for the dependence on OH^- at high pH of the pseudo-first order rate constant might only mean a base catalysis by OH^- of the attack by H_2O on the fluorophosphate:

$$\begin{array}{c} R \\ HO^-{-}{-}{-}H \quad O \diagup OR \\ \diagdown \diagup \\ O{-}{-}{-}P{-}{-}{-}F \\ | \quad \| \\ H \quad O \end{array}$$

30

---

[120] See Ref. 89.

In this mechanism the transition state structure **30** (or the formation of a precursor) involves OH^-, not as the attacking nucleophile, but in the role of assisting H_2O attack upon phosphorus by a hydrogen-bonded $H_3O_2^-$ species and subsequent (or concomitant) removal of a H^+ from H_2O, as part of the activation process. This pathway is like one of those possible as an explanation of general base catalysis for RNH_2 + Ph_3CSCl (p. 428).

In addition to the sensitivity of second order rate constants to the nature of the entering group (as in the reactions of Sarin), studies of the nucleophilic displacement reactions

$$CH_3\text{-}\underset{O(i\text{-}Pr)}{\overset{O}{\underset{\|}{P}}}\text{-}X + Nuc \rightarrow CH_3\text{-}\underset{O(i\text{-}Pr)}{\overset{O}{\underset{\|}{P}}}\text{-}Nuc + X^-$$

$X = O_2N\text{-}C_6H_4\text{-}O\text{-},\ Cl\text{-}C_6H_4\text{-}O\text{-},\ C_6H_5\text{-}O\text{-},\ CH_3O\text{-}C_6H_4\text{-}O\text{-}$

have shown that the rate constants decrease in the order of this listing of leaving groups. Whether this is best interpreted to mean that significant bond-breaking occurs in the transition states of the reactions (the *I* mechanism) or simply that the electron withdrawing character of the leaving group influences the ease of the bond-making process (either in the transition state or via a pre-equilibrium formation of five-coordinate phosphorus) is uncertain at this time.

Studies of trimethylphosphate using labeled (with ^{18}O) OH^- and H_2O have yielded the interesting result[121] that when OH^- is the nucleophile, the tagged oxygen appears primarily in the dimethyl phosphate product and not in the methanol product:

$$(MeO)_3PO + {}^*OH^- \rightarrow (MeO)_2PO{}^*O^- + MeOH$$

With $H_2{}^{18}O$ as a nucleophile, however, the tagged oxygen appears in the methanol product and not in the phosphate product:

$$(MeO)_3PO + H_2{}^*O \rightarrow (MeO)_2PO(OH) + Me{}^*OH$$

OH^- thus attacks the phosphorus atom much more readily than the carbon, while the converse holds for water. For aryl phosphates, which are not susceptible to nucleophilic attack by OH^- or H_2O at carbon, the ratio of relative second order rate constants for attack at phosphorus[122] is $k_{OH^-}/k_{H_2O} \approx 10^4$, a result further demonstrating the general distinction you have seen with other substrates between OH^- and H_2O as nucleophiles. Similarly, F^- attacks the phosphorus of alkyl phosphates, while the heavier halides tend to attack the carbon to produce methyl halides.

In the reaction of triphenylphosphate with OH^- it has been ascertained[123] that the activation enthalpy is only 9.6 kcal/mole and that the activation entropy is large and negative (-38 eu). The small activation energy distinctly implies con-

---

[121] P. W. C. Barnard, *et al.*, *J. Chem. Soc.*, 2670 (1961).
[122] C. A. Bunton, *et al.*, *J. Org. Chem.*, **33**, 29 (1968).
[123] See Ref. 121, and C. A. Bunton, *Accts. Chem. Res.*, **3**, 257 (1970).

siderable bond formation in the transition state and the negative $\Delta S^\ddagger$ adds strength to this argument: an A activation step to form a highly organized five-coordinate intermediate is indicated, as opposed to an I mechanism, because the latter mechanism is generally associated with a larger $\Delta H^\ddagger$ and, particularly, a smaller $|\Delta S^\ddagger|$.

In addition to the results given above, two findings are particularly germane to the question of intermediate formation and the detailed pathway of phosphate ester hydrolysis.[124] First of all, five-membered ring esters (31)

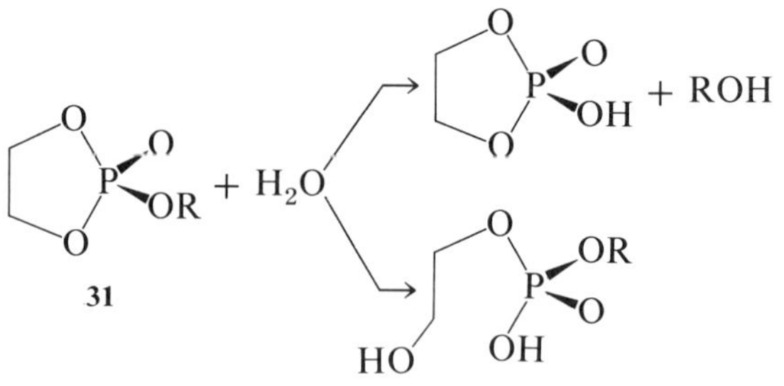

hydrolyze by loss of OR *and* by ring opening millions of times faster than do their acyclic analogs. Secondly, replacement of a cyclic phosphate ring oxygen with CH_2 gives a phostonate (32),

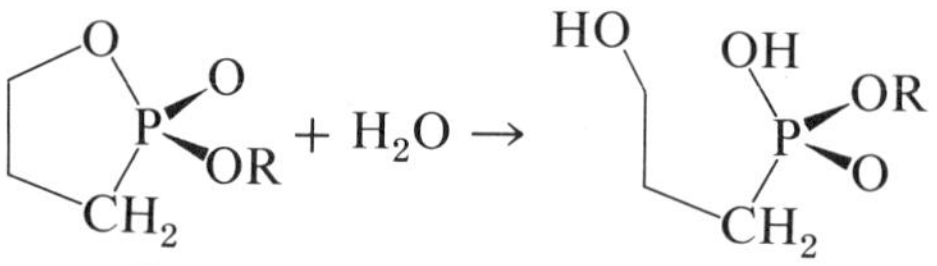

which hydrolyzes almost *exclusively* by ring opening, and the ring opening still proceeds at velocities many orders of magnitude faster than the analogous reaction of the acyclic analog. You might be thinking at this point that ring strain would account for the rate enhancement of the ring opening, but the important question is, "How does the ring feature so markedly enhance OR loss in the phosphate and so markedly inhibit OR cleavage in the phostonate?" The answer must lie in the structure of the intermediate and the position of the departing group within that structure.

Before passing on to consideration of the intermediate, let us briefly note that the ring O—P—O angle of 90° to 100° would make very unlikely a Walden inversion (I mechanism) path for OR loss. In such a transition state the O—P—O angle would become extremely strained (this translates into a high activation energy) with both ring oxygens in equatorial positions of a trigonal bipyramidal structure.

The analysis and elimination of mechanistic possibilities for the hydrolysis reactions of phosphates and phostonates are somewhat lengthy and cerebral. As the exercise is revealing and provides an excellent opportunity for you to see both structural and mechanism principles "in action," we will present it in the following paragraphs; to make it easy to by-pass the analysis, however, and get to the important conclusions, the following scheme summarizes the pathway for phosphate (E = oxygen) and phostonate (E = carbon) hydrolysis.

---

[124] The information to be covered in the next several pages is reviewed by F. H. Westheimer, *Accts. Chem., Res.,* **1**, 70 (1968).

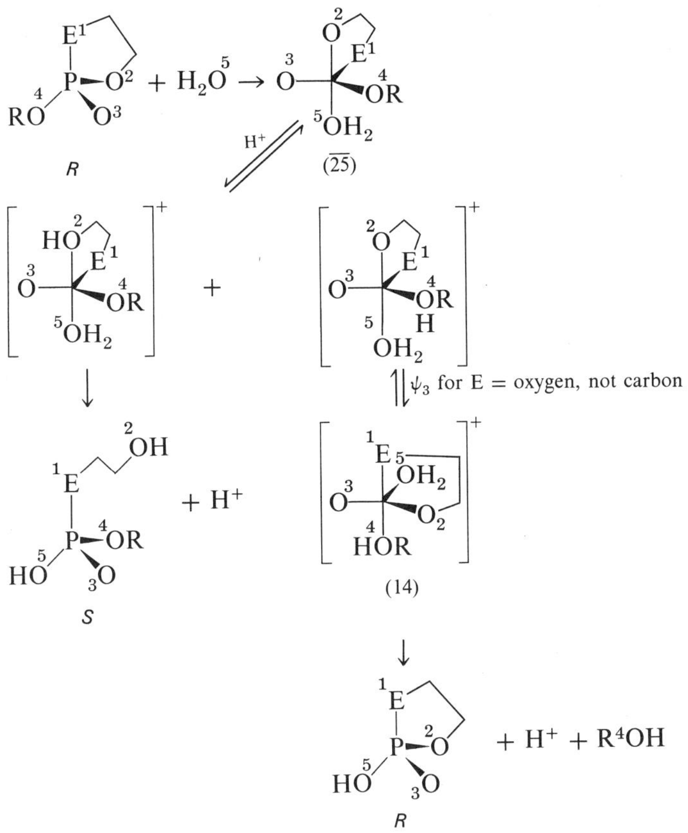

[Scheme showing hydrolysis mechanism with structures labeled (25), (14), R and S configurations, involving trigonal bipyramidal intermediates with H₂O attack, protonation, and pseudorotation steps]

Special features of this path are summarized as follows:

a) The initial attack by H_2O is upon the face opposite the ring oxygen.
b) There is rapid protonation of the ring- or OR-oxygen.
c) Dissociation of the axial P—O bond in the ring leads to ring opening of both phosphate and phostonate and a change in their chirality.
d) Pseudorotation about the terminal P—O group places the HOR group axially for rapid loss and net solvolysis of that group.
e) Pseudorotation is not likely for E = C, so phostonates experience only ring-opening.
f) Distortion by the ring of the tetrahedral substrate (as in **15**, p. 402) facilitates attack upon the (RO,E,O) face,

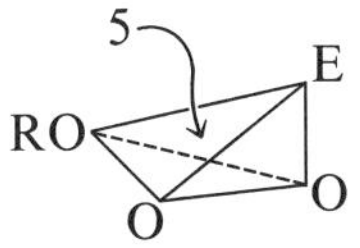

[Diagram showing tetrahedral distortion with RO, E, O, and position 5 labels]

making the *cyclic* substrate reactions faster than those of the *acyclic* analogs [either by faster formation of $\overline{(25)}$ in an *A* mechanism or by a more favorable equilibrium for formation of $\overline{(25)}$ in a mechanism that is *D* in protonated $\overline{(14)}$ and/or $\overline{(25)}$].

Let us now consider the structure of the trigonal bipyramidal intermediate for these reactions in more detail. (We will make use of the Desargues-Levi map of Figure 7-11.)

Without loss of generality, we will begin the analysis with H₂O attack upon the *R* isomer **33**.

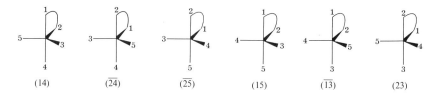

The ring O—P—O angle dictates that the lowest-energy intermediates will be only those with an axial/equatorial disposition of the ring. From Figure 7-11 we find these isomers to be (14), ($\overline{24}$), ($\overline{25}$), (15), ($\overline{13}$), and (23):

$$\text{(14)} \quad \text{($\overline{24}$)} \quad \text{($\overline{25}$)} \quad \text{(15)} \quad \text{($\overline{13}$)} \quad \text{(23)}$$

These isomers appear as a set of {(14), ($\overline{25}$), ($\overline{13}$)} on the "upper track" and a set of {($\overline{24}$), (15), (23)} on the "lower track" of Figure 7-11 and they are separated from each other by inaccessible pseudorotamers.

$$(14) \underset{}{\overset{\psi_3}{\rightleftharpoons}} (\overline{25}) \underset{}{\overset{\psi_4}{\rightleftharpoons}} (\overline{13})$$

$$(\overline{24}) \underset{}{\overset{\psi_3}{\rightleftharpoons}} (15) \underset{}{\overset{\psi_4}{\rightleftharpoons}} (23)$$

For the cyclic phosphates, where both atoms 1 and 2 are oxygens, these chiral triples are enantiomorphic sets. With a phostonate, however, this equivalence is lost. Again, without loss of generality, assign atom 3 to be the P=O oxygen. Doing so eliminates from our serious consideration the isomers ($\overline{13}$) and (23) as species lying along the reaction coordinate:

$$(\overline{13}) \qquad (23)$$

a) The P=O retro-π-bonding is least effectively realized when the terminal oxygen is axial [this concept was discussed in Chapter 4 (ref. 13, p. 173) and has been used several times since].
b) The short P=O bond and its steric size render H₂O attack along the 1,3 edge [to give ($\overline{13}$)] and the 2,3 edge [to give (23)] less likely than attack along the 1,4 edge [to give (14)] and the 2,4 edge [to give ($\overline{24}$)].
c) Postulated dissociation of the P—O₄ and ring P—O bonds in their equatorial positions in ($\overline{13}$), (23), ($\overline{25}$), and (15) leads to several difficulties.
  1) Ring opening by equatorial P—O dissociation from ($\overline{13}$), (23), ($\overline{25}$), and (15) should be several orders of magnitude faster than P—O₄ dissociation because of ring strain; in fact, H₂*O/(C₂H₄O₂)PO(OH) exchange is competitive with ring opening. This is an even more serious problem with (23) and ($\overline{25}$) when E = carbon.
  2) For P—O dissociation to *require* an equatorial site for the P—O bond, ring opening for the phostonate would require an *axial* ring carbon in ($\overline{13}$) *and* it would be impossible to explain why P—O₄ does *not* dissociate from such an intermediate. Thus, we may discard the notion that the P—O bond to be cleaved *must* lie in an equatorial position.

What, then, of the possibility of OR departing an *equatorial position* while the *ring opens at the axial position* as in ($\overline{25}$) and (15)? Assuming weaker P axial than P equatorial bonds, ring opening (P—$O_{2,1}$ breaking) should be much faster than OR loss (P—O_4 cleavage). This would be consistent with the phostonate hydrolysis but *not* with the ^{18}O exchange result with $(C_2H_4O_2)PO(OH)$.

Let us now turn our attention to the (14) and ($\overline{25}$) isomers. To explain why the phosphates and phostonates hydrolyze so much faster than their acyclic analogs, we need simply note that the ring O—P—(O,C) angles of the substrates produce "open" tetrahedral structures that are highly distorted toward a trigonal bipyramidal geometry, in much the same way as does **15** (p. 402). The formation of the intermediate also relieves some of the ring strain. With the phostonate it is possible to make a distinction between the (14), ($\overline{24}$), ($\overline{25}$), and (15) isomers after adopting an assignment of the CH_2 group to position #1. With this choice, isomers (14) and (15) are eliminated because they place the CH_2 group axially (pseudorotation is eliminated). Take a close look at ($\overline{24}$) and you will notice that the ring oxygen and the OR groups are *both axial*. How

are we to explain the *failure* of OR to depart axially when the ring opens from an axial position so rapidly? We can't. *Only isomer ($\overline{25}$) can provide a reasonable intermediate,* for here we find that OR is equatorial while the ring oxygen is axial. We must now make the reasonable presumption that axial loss is much faster than equatorial loss. *Our conclusion then is that phostonates hydrolyze via the intermediate ($\overline{25}$),* which arises from H_2O attack at the 1,3,4 face.

The remaining details of the reaction pathways for phosphates and phostonates are settled by the PMR, and we look to the acid catalyzed $(C_2H_4O_2)PO(OH)$ + $H_2{}^*O$ exchange to provide the details. *If* we were to postulate (14) as *the only*

intermediate formed for phosphate, we would have H_2O entering equatorially and H_2O departing axially. This is acceptable *as long as* the particles could *exactly* reverse their path with the same velocity—*i.e.*, with H_2O entering axially and H_2O departing equatorially. But the phostonate hydrolysis led us to conclude that equatorial departure is in fact so much slower than axial departure that it cannot be observed. This means that the reaction cannot proceed in forward and reverse directions with equal velocity. In practical terms, the reaction of $(C_2H_4O_2)PO(OH)$ with H_2O^* by this channel would produce a product containing something other than a random balance of $(C_2H_4O_2)PO(^*OH)$ and $(C_2H_4O_2)PO(OH)$! Here, then, we find an example of the PMR helping us to eliminate a mechanistic hypothesis. The pseudorotation ψ_3 (which would provide a single-channel path with chemical equivalence of 4 and 5 in the square pyramid) does not provide a "way out of the woods" because we are still trapped by having the leaving group (H_2O^4) depart equatorially (*i.e.*, at a rate too slow to compete with ring opening).

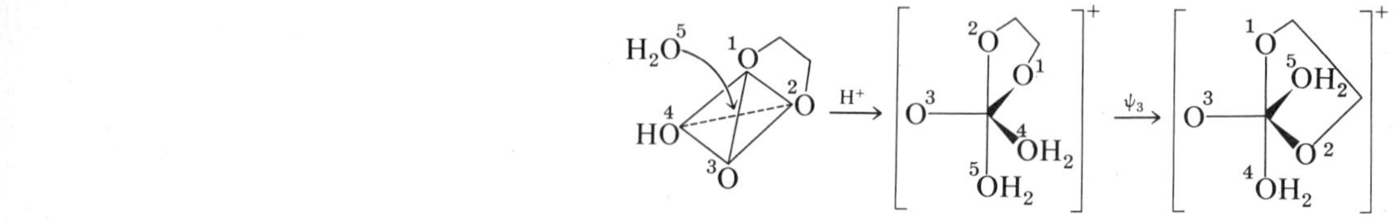

On the other hand, isomer $(\overline{25})$ and ψ_3 satisfy all the experimental observations and the principle of microscopic reversibility. The initial formation of $(\overline{25})$ for phosphates establishes the anticipated link with the phostonates by having initial attack at the axial position. H_2O^4 cannot leave via its initial equatorial position [for the same reasons given for (14)] but the pseudorotation ψ_3 moves the leaving group (H_2O^4) to the proper axial position for departure *and* satisfies the principle of microscopic reversibility by causing the two H_2O groups of the intermediate to become equivalent in the ψ_3 transition state:

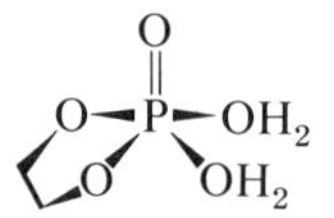

The reaction coordinate diagram for this *exchange* mechanism looks like this[125]:

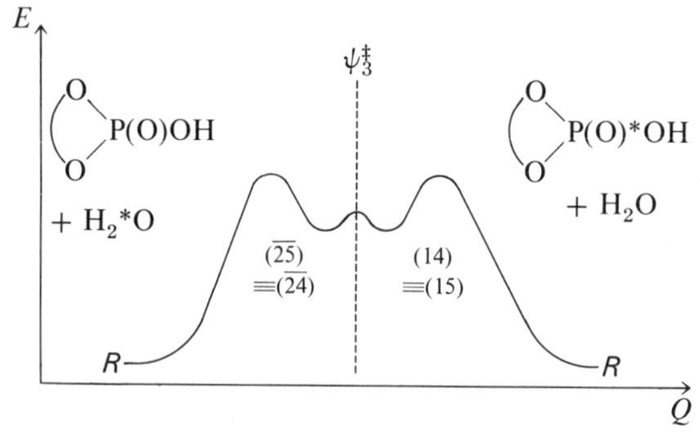

where the symmetry about $\psi_3^\ddagger$ is apparent, the forward and reverse reactions are entirely equivalent, and the labeled oxygen is randomly distributed between $(C_2H_4O_2)PO(OH)$ and H_2O. To account for the competitive ring opening and exchange reactions requires (i) that the ψ_3 process be *very fast* relative to the lifetime of $(\overline{25})$ and (ii) that axial ring opening proceed at a rate as fast as that of loss of H_2O. Both requirements are chemically acceptable. To conclude, we should note that solvolysis of $(C_2H_4O_2)PO(OR)$, instead of the acid, no longer must follow an entirely symmetrical pathway; in the strictest sense it can't, because the reactants and products are not identical and $\psi_3^\ddagger$ cannot be truly symmetrical. However, it is unlikely that the alkyl group will cause a major change in the structures of the intermediate and transition states, and we expect the reaction coordinate diagram to appear very much like that given above.

To summarize this rather lengthy and intricate analysis, we can say that all the hydrolysis data are handled by supposing that the initial intermediate is $(\overline{25})$, that axial cleavage is much faster than equatorial cleavage, and that pseudorotation is forbidden for the phostonate but necessarily fast for the phosphates:

---

[125] As an aside, you might consider the following:
(a) $(\overline{25})$ and $(\overline{24})$ are identical, as are (14) and (15),
(b) $(\overline{25})/(\overline{24})$ is enantiomorphic with (14)/(15),
(c) $(\overline{25})$ and (15) are *formed* by face attack by 5,
(d) $(\overline{24})$ and (14) are *formed* by edge attack by 5.
Which do you suppose is favored for representing "forward" and "reverse" exchange: formation of $(\overline{25})/(15)$ or $(\overline{24})/(14)$? In sketching a contour plot as in Figures 7-6 and 7-7, you should find a "high road and a low road to Scotland" (the intermediates).

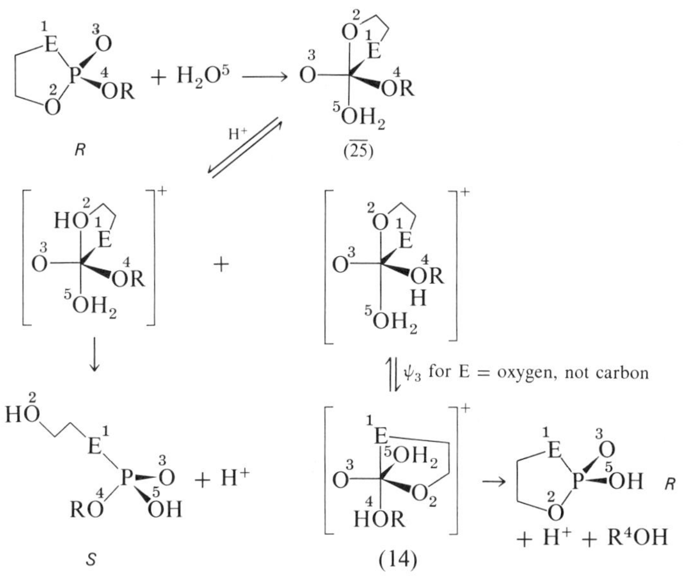

[Scheme showing mechanism with structures R, S, and intermediates (25), (14)]

The same ideas developed to understand the substitution reactions at phosphates have found application in the mechanism of reactions of sulfinyl ($-\overset{O}{\underset{}{\overset{\|}{S}}}-$) and sulfonyl ($-\overset{O}{\underset{O}{\overset{\|}{\underset{\|}{S}}}}-$) compounds. There are some intriguing differences, however. For example, the cyclic catechol *sulfate* (34) and analogous sultone (35) show[126] marked rate enhancements ($\sim 10^6$) for *hydrolysis* of the ring-oxygen bond to sulfur (relative to the acyclic analogs), while the six-membered-ring sultone (36) is some four orders of magnitude less reactive[127] than its five-membered-ring counterpart and not much more reactive than the acyclic analog.

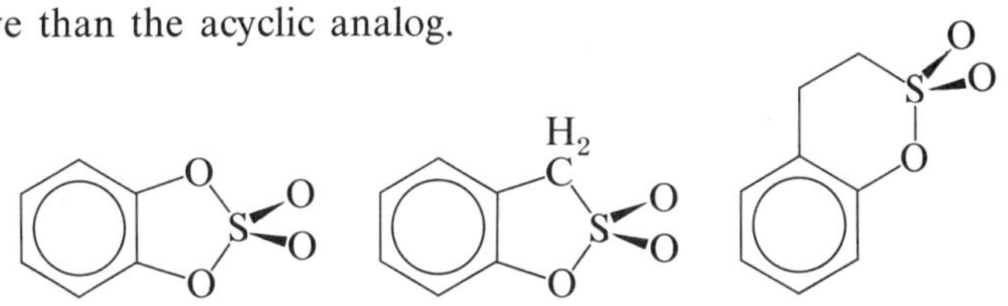

[Structures 34, 35, 36]

This behavior is reminiscent of the phosphates and phostonates. A more perplexing situation arises with the corresponding cyclic *sulfites*. Ethylene sulfite and catechol sulfite experience acid hydrolysis at rates not much different from those of their acyclic counterparts, while their base hydrolyses occur with rate constants only two or three orders of magnitude greater.[128] The failure of the sulfites to experience a marked ring-strain effect on the hydrolysis rate constant is undoubtedly related to the fact that there is no ground state ring strain in these compounds. This follows from the observation[129] that the hydrolysis reaction enthalpies of the cyclic sulfites are little

---

[126] E. T. Kaiser, *et al., J. Amer. Chem. Soc.,* **87,** 3781 (1965); O. R. Zaborsky and E. T. Kaiser, *ibid.,* **88,** 3084 (1966).
[127] E. T. Kaiser, *et al., J. Amer. Chem. Soc.,* **89,** 1393 (1967).
[128] C. A. Bunton, *et al., J. Chem. Soc.,* 1766 (1959); 4754 (1958); R. E. Davis, *J. Amer. Chem. Soc.,* **84,** 599 (1962); P. B. D. de la Mare, *et al., J. Chem. Soc.,* 4888 (1962); J. G. Tillet, *J. Chem. Soc.,* 5138 (1960); C. A. Bunton and G. Schwerin, *J. Org. Chem.,* **31,** 842 (1966).
[129] N. Pagdin, *et al., J. Chem. Soc.,* 3835 (1962).

different from those of the acyclic compounds, while cyclic sulfates (and phosphates) show a greater exothermicity of 5 to 10 kcal/mole.[130] Clearly, the principal difference between the sulfites and sulfates is the conversion of a sulfur lone pair to an S—O bond. The significance of this change may be that the sulfur lone pair is better able to accept increased s character than is the sulfur hybrid to oxygen, as an aid to "pinching down" the two hybrids to the ring oxygens.

Isotopic oxygen *exchange* studies with these compounds reveal some interesting insights into the mechanistic features of reactions at sulfinyl and sulfonyl sulfur. For example, the exchange reaction[131] for sulfoxides

$$\text{ArRSO} + \text{R}'_2\text{SO}^* \rightarrow \text{ArRSO}^* + \text{R}'_2\text{SO}$$

appears to occur with complete retention of chirality. The intermediate or transition state structure **37**

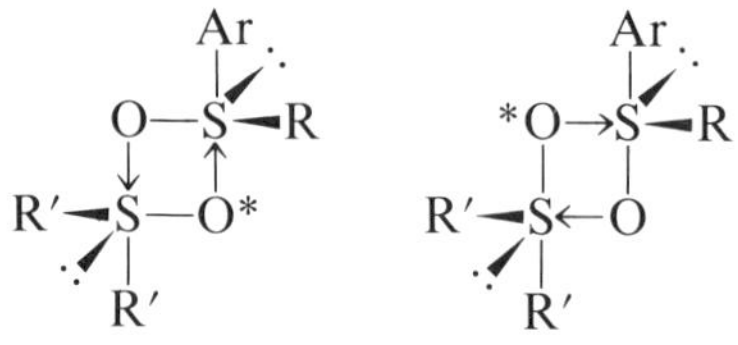

accounts for the retention feature and, if this mechanism is to be valid, the PMR imposes a dual-reaction-channels constraint. If the exchange involves only the species above as a transition state or intermediate structure, there must be (strictly, for Ar = R′ = R) a mirror image reaction coordinate involving the entering oxygen at an equatorial position (**37a**). In either event, points of entry and departure are different. To avoid different entry/departure points would require either a di-equatorial structure[132] **38** (about both sulfurs) or (double) pseudorotations[133] **39** about the sulfur lone pairs. The former event would lead to inversion and may be excluded. In the latter event, however, two mirror image reaction channels are still required by the chirality of ArRSO. Though this mechanism preserves retention of sulfur chirality, now is a good time to recall the earlier discussion (p. 432) of the relative pseudorotation rates of trigonal bipyramidal phosphorus and sulfur compounds containing fluorine. That is, pseudorotation about sulfur *may* be much too slow to allow this as a viable path.

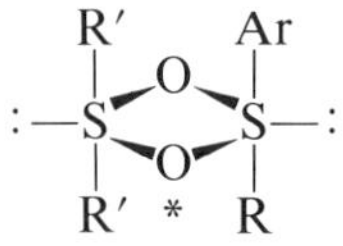

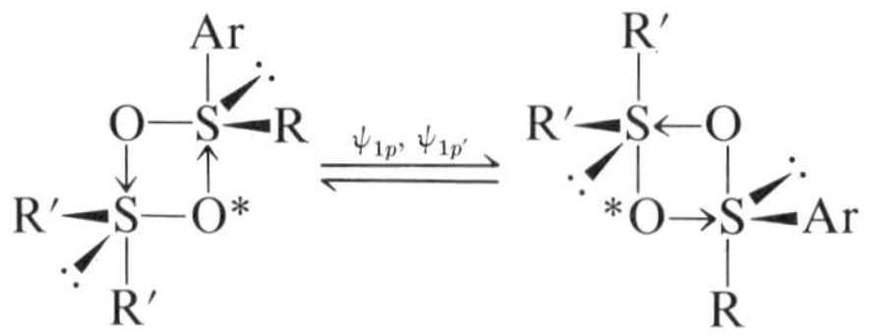

In contrast to this retention mechanism for exchange of terminal oxygens of the sulfinyl group, substitution reactions of other groups bound to S=O and SO_2 seem to proceed by an inversion mechanism. No clear evidence to distinguish the occurrence

---

[130] E. T. Kaiser, *et al.*, *J. Amer. Chem. Soc.*, **85**, 602 (1963).
[131] S. Oae, *et al.*, *Tetrahedron Lett.*, 4131 (1968).
[132] Such a structure could be accommodated by either a transition state model or an intermediate.
[133] This model requires the existence of an intermediate.

of the I or A mechanisms is available for these reactions, however. Analogy with phosphorus and silicon pathways would suggest that five-coordinate intermediates should be possible, but even the oxygen tracer studies have yielded ambiguous results. In marked contrast to the behavior of phosphates, the striking feature of hydrolyses of cyclic sulfites, sulfates, and sultones in $H_2{}^{18}O$

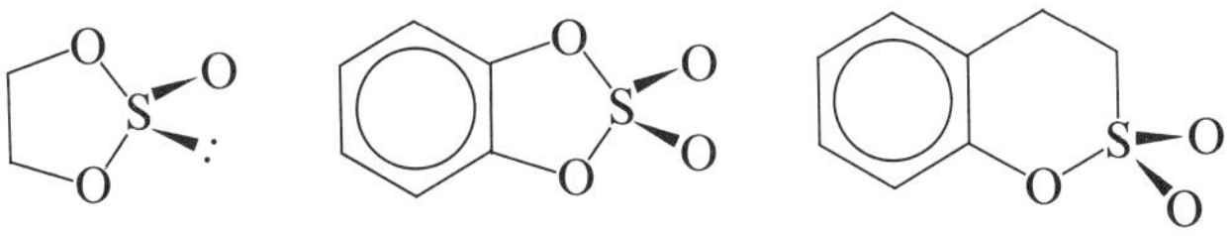

is that regardless of the pH conditions, there is *no incorporation* of ^{18}O into the unreacted ester[134] (*i.e.*, no oxygen exchange)! This is certainly consistent with a transition state structure and I mechanism for ring opening, but the existence of an intermediate such as

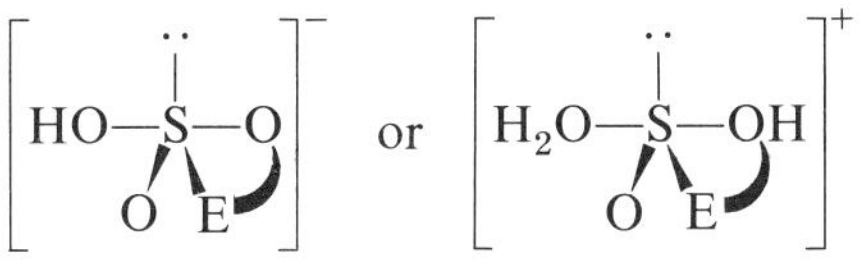

$$\left[\begin{array}{c} \ddot{}\\ HO-S-O \\ O\quad E \end{array} \right]^{-} \quad \text{or} \quad \left[\begin{array}{c} \ddot{}\\ H_2O-S-OH \\ O\quad E \end{array} \right]^{+}$$

cannot be ruled out at present because of the uncertainty surrounding the rate of pseudorotation about sulfur.

To explain this, let us begin with the failure[135] of an oxygen exchange to occur in the acid hydrolysis of phenyl benzenesulfonate; this does establish a link with phosphorus chemistry (*cf.* the scheme on p. 437). Formation of the five-coordinate intermediate **40** could be followed by protonation at either the phenoxide oxygen (**41**) or the terminal oxygen (**42**). A very slow loss of phenol from the (assumed) axial position in **41** without oxygen exchange is consistent with experimental observations.

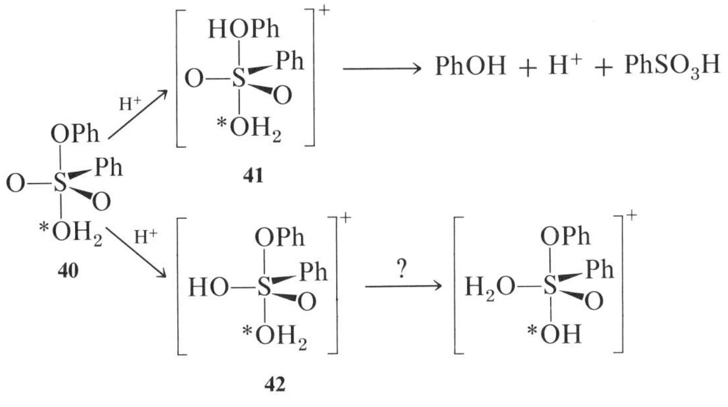

To eliminate H_2O and effect *exchange* via **42**, several things must happen. First there must be proton transfer between the axial and equatorial OH groups. This would produce a higher-energy intermediate, for axial OH^-/equatorial H_2O is not the preferred structure (for electronegativity and π bonding reasons), and the 1,3 proton transfer might be much slower than otherwise expected.[136] Secondly, and in obvious similarity to the behavior of phosphates, leaving group departure from the equatorial position could be very slow. If so, a pseudorotation about the phenyl group as pivot

---

[134] C. A. Bunton, *et al.*, *J. Chem. Soc.*, 4751 (1958); E. T. Kaiser and O. R. Zaborsky, *J. Amer. Chem. Soc.*, **90**, 4626 (1968).

[135] S. Oae, *et al.*, *Bull. Chem. Soc. Jap.*, **36**, 346 (1963).

[136] If not very slow, then it is reasonable that the equilibrium constant is unfavorable; coupled with a very slow (as with the phosphorus esters) loss of H_2O from an equatorial site, this leads us to expect no *O exchange.

(placing a terminal oxygen in an axial position!) would be required to allow more rapid loss of H_2O from an axial position.

In this way you see that an unfavorable 1,3 proton shift and/or an unfavorable pseudorotation within **42** would incur activation beyond that encountered with **41**, and therefore slower O exchange than OPh solvolysis.

Entirely analogous comments apply to the possibility for a five-coordinate intermediate in the hydrolysis of ethylene sulfite, which differs from the phenyl benzenesulfonate example primarily in the replacement of the phenyl group of the latter by a sulfur lone pair and the conversion of a terminal oxygen of the latter to a ring oxygen.

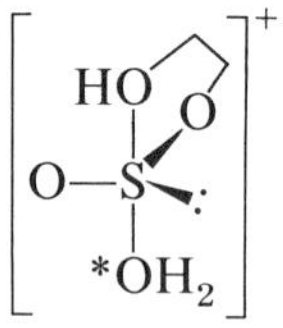

$$\begin{bmatrix} \text{HO} \diagup \text{O} \\ \text{O—S} \diagdown \\ *\text{OH}_2 \end{bmatrix}^+ \text{ and } \begin{bmatrix} \text{O} \diagup \text{O} \\ \text{HO—S} \diagdown \\ *\text{OH}_2 \end{bmatrix}^+$$

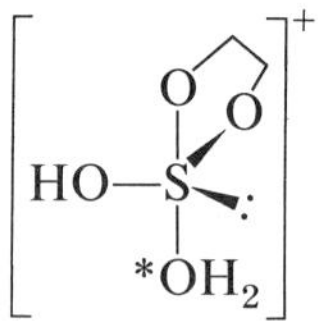

cf. **41**   cf. **42**

STUDY QUESTIONS

33. Why is it reasonable [see discussion of MePO(*i*-Pr)X] that $O_2N\text{—}C_6H_4\text{—}O^-$ would be a better leaving group than $C_6H_5\text{—}O^-$?

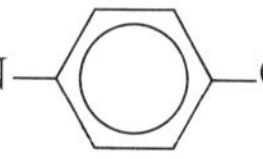

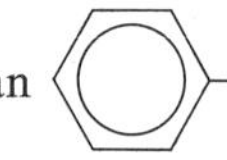

34. Suggest a series of steps for the reduction of chlorate by sulfite:

$$ClO_3^- + 3SO_3^{2-} \rightarrow Cl^- + 3SO_4^{2-}$$

35. Characterize the exchange reaction

$$*SO_3^{2-} + RSSO_3^- \rightleftharpoons RS*SO_3^- + SO_3^{2-}$$

according to the energy profiles of Figure 7-8.

36. Does it surprise you that the five-membered-ring phosphinate

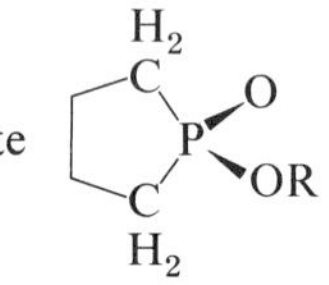

$$\begin{array}{c} H_2C\text{—}O \\ | \quad\quad \diagdown \\ | \quad\quad P \\ | \quad\quad \diagup \diagdown \\ H_2C\text{—}\quad OR \end{array}$$

does not hydrolyze any faster than its acyclic analog, in marked contrast to the hydrolyses of phosphate and phostonate esters?

EPILOG

In this chapter you have been confronted with two important *theoretical* concepts: the symmetry contraints of the *principle of microscopic reversibility* for isotope exchange reactions and their implications for "essentially symmetric" substitution reactions make possible the elimination of certain mechanistic possibilities; similarly, efficient *orbital following* of nuclear motion is a requisite condition for maintaining a low energy profile. You have seen that rate laws must be carefully interpreted in conjunction with other studies of stereochemical changes, isotopic labeling, solvent effects, rate laws for reverse reactions, and substituent effects. Any one such study alone is rarely sufficient to yield an "unambiguous" mechanistic model. Even when all of this information is available, there are often "holes" in a surmised pathway that are inaccessible to direct experimental study. Perhaps most important to your development of "chemical intuition" are the application of the structural and bonding concepts of earlier chapters and the sometimes unreliable method of analogy. As is true of all rational approaches to a subject, extensive study of a class of compounds eventually leads to formulation of generally applicable principles of their reactivities and to the definition of circumstances surrounding deviate behavior.

To summarize the most often encountered mechanistic concepts, we cite the following:
1) electrophilic/nucleophilic attack,
2) intermediates of higher or lower coordination number,
3) transition state species of higher coordination number achieved with more or less of a balance between bond-making and bond-breaking,
4) the CB step for deprotonation of a species by proton transfer prior to the activation step or by hydrogen bonding during activation (either by solvent molecules or by added donors),
5) the closely related atom transfer step for oxidation,
6) intramolecular rearrangement by pseudorotation or group migration (often a proton shift), and
7) elimination of two originally non-bonded groups as a bonded unit.

SPECIAL TECHNIQUES

    The Chemical Vacuum Line
    Plasmas
    Photochemical Apparatus
    Electrolysis

AN OVERVIEW OF STRATEGY

SYNTHESIS OF FLUORIDES AND CHLORIDES

    Boron and Aluminum
    Silicon
    Nitrogen and Phosphorus
    Oxygen and Sulfur
    Fluorine and Chlorine
    Xenon

FLUORINATING AGENTS

    SbF_3 and SbF_5
    ClF, ClF_3, BrF_3
    Sulfurtetrafluoride
    Dinitrogen Tetrafluoride
    Chloropentafluorosulfur

CHLORIDE SUBSTITUTION BY HYDROGEN AND ORGANICS

    Organometal Reagents
    Boron and Aluminum
    Silicon
    Nitrogen and Phosphorus
    Oxygen and Sulfur
    Halogens
    Hydrometalation and Others

CATENATION BY COUPLING

SOLVOLYSIS REACTIONS

    Boron and Aluminum
    Silicon
    Nitrogen and Phosphorus
    Oxygen and Sulfur
    Halogens and Xenon

EPILOG

8. SYNTHESIS OF IMPORTANT CLASSES OF NON-METAL COMPOUNDS

In this chapter we will present, insofar as is possible, a general view of the syntheses of the important classes of non-metal compounds. The preceding chapters have introduced you to the main characters of the cast in the "inorganic" story and have developed for you some appreciation of their personalities. While establishing the structural, bonding, and reactivity principles, these chapters also treated you to a few of the more important non-metal reactions. You are now well equipped to confront some of the important preparative uses of these compounds; accordingly, this chapter gathers together some systematics of inorganic syntheses.

SPECIAL TECHNIQUES

Your experiences in organic and physical chemistry laboratories have taught you many important techniques for synthesis and handling of compounds, and these find direct application in an "inorganic" laboratory. There are, however, several techniques commonly used in the manipulation of air- or moisture-sensitive compounds with which you may be unfamiliar. Furthermore, the inorganic chemist is sometimes called upon to deal with electrolytic techniques and high-energy electrical discharge methods for producing reactive species, and it is these methods we wish to introduce to you before you see their use in synthesis. Our overview will be limited in scope, for all we wish to achieve is your awareness of the equipment and fundamental techniques. For further details you should consult the original literature and the several excellent texts for use in inorganic laboratory courses.[1]

THE CHEMICAL VACUUM LINE

The vacuum line, a very basic version of which is shown in Figure 8–1, is one of the most useful apparatuses available to the synthetic chemist. Using such an

---

[1] R. J. Angelici, "Synthesis and Technique in Inorganic Chemistry," 2nd Ed., W. B. Saunders, Philadelphia, 1977; D. F. Shriver, "The Manipulation of Air-sensitive Compounds," McGraw-Hill, New York, 1969; W. L. Jolly, "The Synthesis and Characterization of Inorganic Compounds," Prentice-Hall, Englewood Cliffs, 1970.

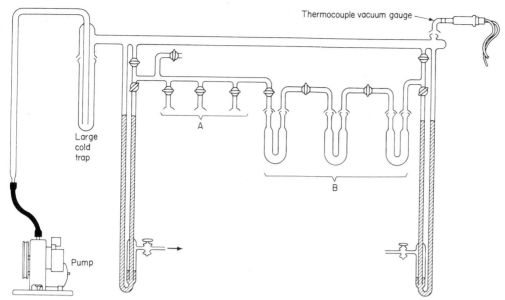

Figure 8-1. A basic chemical vacuum line. [From W. L. Jolly, "The Synthesis and Characterization of Inorganic Compounds," Prentice-Hall, Inc., Englewood Cliffs, N.J., 1970.]

apparatus, one can measure small quantities of volatile liquids and gases,[2] carry out reactions, and separate and measure products. Working on the millimolar scale, measurements of quantities can be made quite accurately and, since everything can be handled within the line, a complete material balance can often be achieved, even in the most complex reactions. Although these advantages are available to many synthesis apparatuses, the obvious advantage of the vacuum line is that one is able to handle air-sensitive liquids and gases such as the boron hydrides or halides and aluminum alkyls.

The pressure in a vacuum line is usually maintained at about 10^{-6} torr ($=10^{-6}$ mm Hg) by a combination of a mechanical pump (as shown in Figure 8-1) and an oil or mercury diffusion pump (though not shown in the figure, it is placed between the mechanical pump and the large trap[3]). Pressure is monitored with a thermocouple vacuum gauge[4] or mercury or oil manometers. In order to introduce material into the line, one can attach a gas cylinder or flask to one of the standard taper joints in section A. These are isolated from other sections of the line by "key" type stopcocks coated with special greases or by Teflon needle valves.

Section B of the line consists of a series of U-traps often called a "fractionation train." Together they compose a crude distilling column, with which one can separate compounds having sufficiently different volatilities. (This usually means that two compounds must have vapor pressures that differ by a factor of at least 100 at a given temperature.) For example, in order to purify BCl_3 (*i.e.,* separate other volatiles) one must remove from it the most likely impurity, HCl, and other possible compounds such as $SiCl_4$. In order to effect the purification, one passes the impure mixture through a first trap held at $-78°$ (by immersing the first U-trap in a Dewar flask

---

[2] This is achieved with a mercury or oil pump, driven by atmospheric pressure, to move a volatile sample into a bulb of pre-calibrated volume (not shown in Figure 8-1). There the gas pressure is measured and the ideal gas equation is used to compute the number of moles of gas at hand. The sample subsequently can be returned to the synthesis section of the line.

[3] This trap is normally held at $-196°C$ by immersion in a Dewar containing liquid nitrogen; its purpose is to protect the pumps from contamination.

[4] The temperature of a length of resistance wire depends on thermal conduction by gases to the line walls. The higher the pressure, the cooler the wire. A thermocouple measures the wire temperature.

containing a Dry Ice-acetone mixture) and then into a second held at $-196°$ (liquid nitrogen). The Dry Ice trap condenses $SiCl_4$ (m.p. = $-70°$) and allows the BCl_3 (v.p. = 2.4 torr at $-78°$) and HCl to pass through to be collected at $-196°$. If the mixture of HCl and BCl_3 is then passed through a $-112°$ trap,[5] the BCl_3 is retained at this low temperature, while the HCl is allowed to pass (v.p. of HCl is 124 torr at $-112°$) to be collected in a final trap at $-196°$.

PLASMAS

The preparation of materials of marginal or poor thermodynamic stability but reasonable kinetic stability usually requires that we have techniques available for producing "unstable" reactants. Two common techniques are those that utilize an electron flux or electric/magnetic fields to fragment stable molecules. Examples of such apparatuses are shown in Figure 8-2. An "electrode" discharge tube (A) is characterized by an alternating (60 Hz) voltage of several keV applied to the exposed electrodes, thereby maintaining a glowing discharge through a stream of one reactant flowing through the U-tube. The effluent stream of atoms and ions intercepts a flux of the other reactant a short distance downstream. To maintain the discharge and to minimize recombination of plasma particles, this experiment is conducted at plasma pressures of only a few mm Hg. A characteristic problem with these bare electrode techniques is that the electrode material often enters into reaction with the plasma. This can be turned to advantage in certain cases (see the later synthesis of B_2Cl_4) by using an electrode material that efficiently scavenges unwanted fragments.

---

[5] In order to achieve a temperature of $-112°$, one uses a CS_2 "slush bath." This is made by adding liquid nitrogen to CS_2 until the liquid just turns to a "slush." That is, the temperature of CS_2 is lowered to its normal freezing point. When contained in a Dewar flask, such a bath will maintain a trap at this temperature for 30 minutes or more. Obviously, other cold temperatures can be achieved in the same way, that is, by making a slush of an appropriate solvent. The most common are C_6H_{12}, $+7°$; CCl_4, $-23°$; $CHCl_3$, $-63°$; CH_2Cl_2, $-97°$; CS_2, $-112°$; and methylcyclohexane, $-126°$.

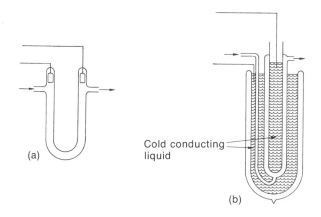

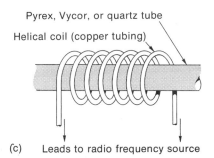

Figure 8-2. (a) Electrode discharge tube. (b) Ozonizer. (c) Inductive discharge tube. [From W. L. Jolly, in "Technique of Inorganic Chemistry," vol. 1, H. B. Jonassen and A. Weissberger (Eds.), Wiley-Interscience, New York, 1963.]

One way to avoid altogether the bare electrode problem is to utilize an "ozonizer" as shown in Figure 8–2(B). Here the electrodes are separated from the plasma by thin glass walls. This arrangement requires voltages on the order of several tens of keV and sometimes of higher frequency than 60 Hz. Notice that a larger electrode surface is designed into the apparatus than that used in Figure 8–2(A).

A technique that avoids the use of electrodes altogether is the microwave or radiofrequency discharge tube. In its simplest form, Figure 8–2(C), a reactant gas is converted to a plasma by passing it through a region of high-frequency electric/magnetic fields, generated by a high-power radiofrequency source.

All these techniques are carried out in conjunction with a chemical vacuum line, for the control of gas pressures, flow rates, and sample collections is easily accommodated within such a system. Very often these high-energy devices are very inefficient in terms of product yield and so place great demands on one's patience and vacuum line technique for collecting and isolating the desired product.

PHOTOCHEMICAL APPARATUS

Yet another way to "destabilize" stable molecules is to supply them with internal energy by photon absorption.[6] This method is somewhat more selective than the plasma methods, for it requires that a reactant molecule be capable of experiencing electronic excitation by photon capture. In this process a normally bonding or non-bonding electron is excited into an anti-bonding orbital, thereby weakening some bonds; fragmentation may occur directly, or an insurmountable activation barrier for a ground electronic state molecule may be more easily overcome. A difficulty with the method is that an excited state molecule often returns quickly to the ground state, before it encounters a reaction partner or experiences fragmentation. Therefore, in contrast to the plasma methods, a static system rather than a flow system is photolyzed with improvement in yield. Two photosynthesis arrangements are shown in Figure 8–3; (A) is typical for solution work, and (B) is characteristic of the set-up for gases.[7] The UV lamp generates a lot of energy not used in the reaction *per se,* which can cause severe heating problems. Thus, in (A) you see a water-cooled jacket serving the purpose of cooling both lamp and reaction mixture; a purging stream of some inert gas such as nitrogen stirs the mixture and also helps to remove any product gases. In (B) a rapid stream of air serves as a coolant. (A) is inherently more efficient than (B) because in the former configuration the reaction mixture completely surrounds the lamp. A final point: Pyrex strongly absorbs certain frequencies of radiation in the UV region, whereas quartz does not, so quartz containers are more generally useful.

ELECTROLYSIS

This is a "high-energy" technique you may remember from your freshman chemistry laboratory. As with the plasma techniques discussed earlier, energy is supplied to the reactants in the form of an electric current; but there the similarity stops. The electrolysis technique can be much more selective about which reactant is energized; it can be used in both static and flow configurations; it is limited to much smaller voltages, and therefore requires a conducting medium to support current flow. The arrangement shown in Figure 8–4 is used in the synthesis of peroxydisulfate

---

[6] The energy equivalents of 200, 300, and 400 nm photons are 143, 96, and 72 kcal/mole, respectively. Generally this energy is not completely isolated in one "bond," but you see that the energy content is sufficient to break "single" chemical bonds. Which bonds break is a dynamical problem determined only in part by the molecular orbital configuration and the location of the weakest bond.

[7] Total sample pressures much less than 1 atm are used, to allow for sample expansion by heating and/or a change in the number of moles of gases during reaction.

AN OVERVIEW OF STRATEGY 451

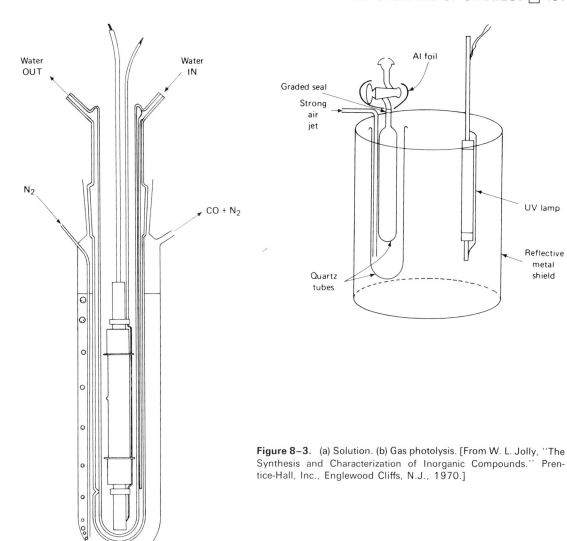

Figure 8–3. (a) Solution. (b) Gas photolysis. [From W. L. Jolly, "The Synthesis and Characterization of Inorganic Compounds." Prentice-Hall, Inc., Englewood Cliffs, N.J., 1970.]

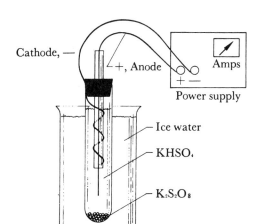

Figure 8–4. Electrolytic synthesis of peroxydisulfate. [From R. J. Angelici, "Synthesis and Technique in Inorganic Chemistry," 2nd Ed., W. B. Saunders Company, Philadelphia, 1977.]

by oxidation of bisulfate. Notice the use of electrical energy (2.1 eV) to generate the *radical* $HSO_4\cdot$ by the reactions

$$HSO_4^- \rightarrow HSO_4\cdot + e^-$$

$$2HSO_4\cdot \rightarrow S_2O_8^{2-} + 2H^+$$

An important part of the apparatus is the temperature bath, in this case used to cool the reaction mixture and facilitate the precipitation of $K_2S_2O_8$. Other applications to be encountered later in this chapter involve both high (molten salts) and low (liquid HF) temperature electrolysis.

AN OVERVIEW OF STRATEGY

In confronting descriptive "inorganic" chemistry for the first time, students are commonly awed (and not infrequently dismayed) at the diversity of compounds and reaction types. Reflecting upon the fact that you devoted the equivalent of a two-semester course to the study of the reactions of one class of compounds (those containing carbon bound to oxygen, hydrogen, and nitrogen), you can appreciate the difficulty of learning useful reactions of the other non-metals in a few months. Your study of organic chemistry has prepared you to some extent for a study of non-metal chemistry, but the greater variety of structure and bonding found in elements other than carbon has forced us to develop (in earlier chapters) greater sophistication and generality in concepts of how molecules do what is of interest to us. You are now well equipped to confront some of the important preparative uses of these compounds. The emphasis here is on both the words "some" and "important." The realities of the situation dictate the former, and the prejudices of your authors dictate the latter. This brings us to a final important point. Systematization of the chemistries of the other non-metal elements has not been carried to as high a level as has that of carbon. Should you be seeking challenge and opportunity in chemistry, the area of non-metal chemistry is a good place to look.

We will, of course, be thwarted at the outset to some extent in our attempts to present a few good synthetic methods for *all* non-metal halides, for example. Using the perspective already gained for the previous chapters, you know that the techniques used for the synthesis of nitrogen halides, for example, are different from those used to prepare the halides of boron. Similar comments apply to greater or lesser degree to all the important classes of non-metal compounds.

Probably the single most important class of non-metal compounds, at least in a synthetic sense, is that of the halides. The fluorides are useful for introducing fluorine as a substituent in other compounds, and the chlorides find great use through substitution of chlorine by other groups [for such reactions the chlorides tend to be more reactive than the fluorides (*cf.* Figure 6–1); in fact, you have already seen in Chapter 6 that *a general approach to synthesis of non-metal fluorides is to replace chlorine by fluorine, using certain metal fluoride salts*].

The great utility of the non-metal chlorides is summarized by the following reaction types, which forecast the organization of this chapter (E and M are general symbols for non-metal and metal elements, and R and A carry their usual representations of organo and electronegative groups).

$$E-Cl + MR \rightarrow E-R \qquad \text{alkyl or arylation}$$

$$+ MH \rightarrow E-H \qquad \text{hydrogenation}$$

$$+ M \rightarrow \frac{1}{x}[E-E]_x \qquad \text{coupling}$$

$$+ HA \rightarrow E-A + HCl \qquad \text{solvolysis}$$

These are the primary reaction types with which we will be concerned. Other, perhaps less general, synthetic strategies will be mentioned as appropriate. Reactions subsequent to those above also will be examined when appropriate.

As you read through the following discussion, frequently refer to the appropriate sections in Chapters 5, 6, and 7 to help systematize and contrast the non-metal compound properties that make possible the syntheses given here. Do not, however, expect to find the mechanism for each synthesis outlined in Chapter 7. In fact, the mechanisms have not been determined in all cases discussed here; we encourage you to use your imagination in such instances.

SYNTHESIS OF FLUORIDES AND CHLORIDES

In what follows, you will see examples of syntheses that take advantage of the need for easily separated products (gases + solids or liquids, or liquids + solids). In addition to the expediency of having mixed phases of products, entropy effects for gases and lattice energies for solids can help "drive" a reaction to completion. This, of course, is a statement of our general need to choose reactants with greater free energy content than products. A general summary of the techniques to be discussed is:

1. *Direct halogenation of the elements.* This is an obvious method based on the high oxidizing power (free energy content of F_2 and Cl_2 with respect to most fluoride and chloride compounds as products) of the halogens. In fact, F_2 often presents a problem in this regard because the low D_{F_2} ($\sim$38 kcal/mole) often results in reactions that are difficult to control. F_2 is a dangerous chemical in the presence of organic materials and is often not the reagent of choice for general synthesis work. Even when direct fluorination is non-violent, it is sometimes difficult to control the extent of incorporation of fluorine. Consequently, techniques have been developed for chlorine-fluorine exchange starting with the appropriate non-metal chloride and a fluoride salt (refer to the discussion of this general reaction in Chapter 6, p. 292). Consequently, it is important to know how to prepare non-metal chlorides. It is generally true that the non-metal chlorides are sufficiently stable for convenient synthesis, and direct chlorination reactions usually can be made to proceed smoothly so that the degree of elemental oxidation can be controlled by the stoichiometry of the reactants.

2. Two techniques for production of halo compounds from *non-metal oxides* are reaction *with HX* and *direct chlorination in the presence of carbon* to remove oxygen as carbon monoxide.

$$E_mO_n + 2nHX \rightarrow mEX_{2n/m} + nH_2O$$

$$E_mO_n + nCl_2 + nC \rightarrow mECl_{2n/m} + nCO$$

The first of these reactions is less applicable to chlorides than to fluorides because of the hydrolytic reactivity of many of the non-metal chlorides (the reverse reaction is favored).

3. Even more useful reagents than the elements themselves, in *direct halogenation,* are *anionic forms of the elements.* For example, you will find that silicides (Ca_2Si) and at least one oxide (HgO) are susceptible to halogenation.

4. As you expect from the high bond energies of the dinitrogen and dioxygen molecules, more energetic techniques than simple halogenation of the elements must be used. In addition, orbital following constraints can impose a significant activation barrier to the cleavage of the multiple bonds. General techniques are difficult to find, but *electrolysis* seems to be effective in some instances and *photochemical* methods are gaining in importance.

BORON AND ALUMINUM

Of these two elements, aluminum is generally the more reactive. This is nicely illustrated by the synthesis of AlF_3 and $AlCl_3$ from

$$Al + 3HX \rightarrow AlX_3 + 3H_2$$

(8-1)   and

$$Al_2O_3 + 6HX \rightarrow 2AlX_3 + 3H_2O$$

Boron is less reducing than Al, and BCl_3 is quite susceptible to hydrolysis, so that a convenient alternate synthesis of BCl_3 is achieved by heating elemental boron in flowing Cl_2. Yet another technique (suggesting metal/metalloid character for aluminum and boron) is the conversion of oxides to chlorides with Cl_2 in the presence of carbon.

(8-2)
$$M_2O_3 + 3Cl_2 + 3C \rightarrow 2MCl_3 \text{ (or } M_2Cl_6) + 3CO$$

This reaction has the advantage of removing oxygen via highly stable carbon monoxide; also, for Al, the room temperature products are a solid and a gas. For boron, of course, the gases CO and BF_3 or BCl_3 are easily separated on a fractionation train as in Figure 8-1.

In an interesting example of the role of lattice energy in assisting an anion exchange reaction, $AlCl_3$ may be reacted with BF_3 to form AlF_3 (a solid) and BCl_3 (a gas).

(8-3)
$$AlCl_{3(c)} + BF_{3(g)} \rightarrow AlF_{3(c)} + BCl_{3(g)}$$

A useful technique, which avoids the handling of HF as a reagent, for preparation of BF_3 is the fluorination of B_2O_3 with CaF_2 in the presence of H_2SO_4. This is actually an example of the $M_2O_3 + 3HX$ reaction; the H_2SO_4 serves the dual purpose of acting as a H^+ source and strongly solvating the product water to inhibit formation of $BF_3 \cdot OH_2$. The gaseous BF_3, of course, may be removed and collected, free of H_2O, under vacuum.

More unusual techniques are required to produce the marginally stable diboron tetrachloride (**1**).

$$\underset{\mathbf{1}}{Cl_2B-BCl_2} \quad \begin{bmatrix} \text{m.p. } -98° \\ \text{b.p. } 55° \end{bmatrix}$$

Though it is conceivable that this compound is produced in the synthesis of BCl_3 as given above, the high thermal conditions attendant to that reaction would cause B_2Cl_4 to decompose. The techniques used to synthesize diboron tetrachloride are somewhat difficult to control for efficient production of the desired product, since they likely involve generation of BCl_2 radicals. The usual method of preparation is to pass gaseous BCl_3 through a high-voltage discharge, and most studies[8] in which the compound has been prepared have used a set-up somewhat similar (except for the pumping arrangement) to that shown in Figure 8-5. The reaction that presumably occurs involves the initial formation of BCl_2 radicals, followed by radical combinations.

$$BCl_3 \xrightarrow[\text{discharge}]{\text{electric}} BCl_2 + Cl$$

(8-4)
$$2BCl_2 \rightarrow B_2Cl_4$$

$$2Cl + 2Hg \rightarrow Hg_2Cl_2$$

---

[8] J. P. Brennan, *Inorg. Chem.*, **13**, 491 (1974); A. G. Massey, *et al.*, *J. Inorg. Nucl. Chem.*, **28**, 365 (1966); J. Kotz and E. W. Post, *Inorg. Chem.*, **9**, 1661 (1970).

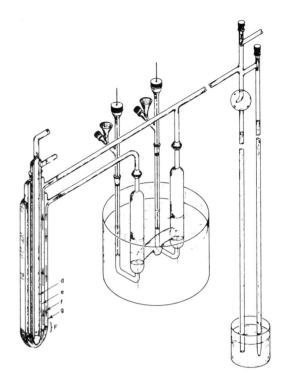

Figure 8–5. Apparatus for the continuous preparation of B_2Cl_4 in a high voltage discharge. The discharge is carried out in the "H" tube immersed in the temperature bath. The reservoir at the left is immersed in a dry ice bath ($-78°C$) to collect liquid BCl_3 and B_2Cl_4 in the region F after they enter through outer tube g. BCl_3 is much more volatile than B_2Cl_4 so liquid BCl_3 is vaporized to exit tube f by passing dry air down the innermost tube d and out tube e. Only tubes g and f are open in region F. [From J. P. Brennan, *Inorganic Chemistry*, **13**, 491 (1974).]

Since a chlorine atom is a highly reactive by-product of the initial step, a scavenger is usually included in the system; the apparatus shown in Figure 8–5 uses mercury pool electrodes for this reason. As expected for a reaction that relies on radical combination, yields are relatively poor, although adjustment of conditions has led to conversions of 100 to 200 mg per hour. Furthermore, the reaction is messy (because of its radical nature) in that an almost bewildering variety of by-products has been produced. Perhaps the best characterized[9] of these is the pale yellow tetrahedral boron cluster B_4Cl_4 (**2**), but there are others such as orange B_9Cl_9 and purple B_8Cl_8.[10]

$$\text{Cl}-\text{B}\cdots\overset{\overset{\overset{\text{Cl}}{|}}{\text{B}}}{\underset{\underset{\underset{\text{Cl}}{|}}{\text{B}}}{|}}\cdots\text{B}-\text{Cl} \qquad \left[\begin{array}{l}\text{pale yellow}\\ \text{m.p. }95°\end{array}\right]$$

2

Fortunately, these by-products are relatively involatile, and B_2Cl_4 may be readily separated, although the separation and subsequent handling of the compound is complicated by its ready disproportionation to BCl_3 and polymeric $(BCl)_n$.

$$B_2Cl_{4(l)} \xrightarrow{t > -78°} BCl_{3(l)} + (BCl)_{n(s)} \qquad (8\text{-}5)$$

Numerous attempts have been made to find chemical agents capable of reducing BX_3 to B_2X_4, and Timms has recently described such a system.[11] Using techniques

---

[9] G. Urry, et al., *Inorg. Chem.*, **2**, 396 (1963).
[10] G. F. Lanthier, et al., *J. Inorg. Nucl. Chem.*, **32**, 1807 (1970); *ibid.*, **33**, 1569 (1971).
[11] P. L. Timms, *J. Chem. Soc. Dalton*, 830 (1972); *Adv. Inorg. Chem. Radiochem.*, **14**, 121 (1972).

developed for studying reactions of atomic vapors, he found that B_2Cl_4 may be produced on a 10 gram scale by allowing BCl_3 and copper vapor to co-condense on a surface cooled with liquid nitrogen. The rather simple apparatus necessary for such preparations is shown in Figure 8-6.

Diboron tetrafluoride has been synthesized by halogen exchange.[12]

(8-6)
$$B_2Cl_{4(l)} + SbF_{3(s)} \rightarrow B_2F_{4(g)} + SbCl_{3(s)}$$

B_2F_4 appears to be considerably more stable to disproportionation than B_2Cl_4 and may be stored for some time in the gas phase (m.p. $-56°$) at room temperature without noticeable decomposition (whereas B_2Cl_4 must be stored at Dry Ice temperature). The reason for this stability contrast is not entirely clear from a thermodynamic viewpoint, for the B—B bond dissociation energy is quite large in both cases[13] (B_2Cl_4, 88 kcal/mole; B_2F_4, 103 kcal/mole). A clue to the reason might lie in the fact that both molecules have D_{2d} symmetry in fluid phases (D_{2h}, planar in the solid). As a possible explanation of this difference (here your authors are "practicing what they preach"—letting our imaginations work), a 1,2 halide shift within the D_{2d} geometry would produce a species

$$\begin{matrix} X \\ \diagdown \\ B-B \\ \diagup\diagdown \\ XX \end{matrix} \overset{X}{\diagup} \rightarrow X-B-B\overset{X}{\underset{X}{\diagdown}}X$$

from which BX_3 and BX may easily derive as products. The primary factor that favors such a rearrangement for X = Cl relative to X = F is B—X π bonding. Within the

---

[12] A. Finch and H. I. Schlesinger, *J. Amer. Chem. Soc.*, **80**, 3573 (1958); R. J. Botherton, *et al.*, *Inorg. Chem.*, **2**, 41 (1963); J. J. Ritler and T. D. Coyle, *J. Chem. Soc.*, (*A*), 1303 (1970).

[13] V. H. Dibler, *et al.*, *Inorg. Chem.*, **7**, 1742 (1968); *ibid.*, **8**, 50 (1969).

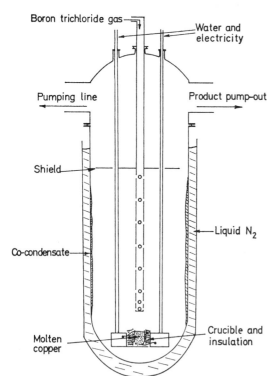

Figure 8-6. Apparatus for the preparation of B_2Cl_4. Copper is evolved from the electrically heated crucible in the form of a monatomic gas. The gas is condensed along with BCl_3 on the walls of the flask. [From P. L. Timms, *J. Chem. Soc. Dalton*, 830 (1972).]

BX$_2$ units of B$_2$X$_4$, B—X π bonding disfavors F transfer; within the BX$_3$ unit of the intermediate "XB—BX$_3$," B—X π bonding again disfavors X = F (review the discussion of the acidities of the BX$_3$ series in Chapter 5).

SILICON

Silicon is recognized as a metalloid because it shows a reactivity in many ways comparable to that of boron. For example,

$$Si_{(c)} + Cl_2 \xrightarrow{\Delta} SiCl_4$$

$$SiO_{2(c)} + 2Cl_2 + 2C \xrightarrow{\Delta} SiCl_4 + 2CO \tag{8-7}$$

$$SiO_2 + 4HF \rightarrow SiF_4 + 2H_2O$$

Just as with boron, the SiO$_2$/HCl reaction does not work for synthesis of the chlorosilane (b.p. $-57°$) because of its hydrolytic instability, and the fluorination of SiO$_2$ with HF must be carried out while avoiding an excess of HF, with which the SiF$_4$ gas (b.p. $\sim -70°$) rapidly forms the stable complex SiF$_6^{2-}$.

As a technique for preparation of chloropolysilanes, the reaction of calcium silicide with chlorine yields a mixture of products that must be subsequently separated by fractional distillation or chromatography.

$$Ca_2Si + xCl_2 \rightarrow CaCl_2 + SiCl_4, Si_2Cl_6, Si_3Cl_8, \ldots \tag{8-8}$$

NITROGEN AND PHOSPHORUS

Chemical oxidation of elemental nitrogen is, quite obviously, not a good method for laboratory synthesis of nitrogen halides. In fact, you have learned in previous chapters that the heavier NX$_3$ compounds are thermodynamically and kinetically unstable. The techniques for synthesizing the nitrogen halides raise some interesting mechanistic questions, since seemingly analogous techniques produce varying products. As noted in Chapter 7, p. 413, chlorination of NH$_3$ by Cl$_2$ under aqueous acid conditions (pH $<$ 4) produces NCl$_3$, which is isolated by extraction into CCl$_4$ or CHCl$_3$. For the large scale production required by the flour industry (for bleaching and sterilization), NCl$_3$ is generated by electrolysis of NH$_4$Cl solution (pH $<$ 4) and swept (b.p. 71°) from solution (and diluted) with an air current.[14]

The direct fluorination of NH$_3$ does not, however, yield NF$_3$; yet in the presence of a Cu metal catalyst[15] one finds a variety of NF compounds as products of this reaction:

$$2NH_3 + 3F_2 \rightarrow 6HF + N_2 \quad \text{(no catalyst)}$$

$$NH_3 + F_2 \xrightarrow{N_2, Cu} NF_3, N_2F_4, N_2F_2, NHF_2$$

Here N$_2$ acts as a carrier gas (diluent) to avoid a violent reaction. On the other hand, in closer analogy with the synthesis of NCl$_3$, electrolysis[16] of molten NH$_4$HF$_2$ or HF solutions of NH$_4$F yields NF$_3$ (and some N$_2$F$_2$) as the primary product

$$NH_4HF_2 \xrightarrow{elec.} NF_3 \text{ (mainly)} \tag{8-9}$$

$$\begin{bmatrix} \text{b.p.} -119° \\ \text{m.p.} -209° \end{bmatrix}$$

---

[14] J. D. Richards, in "Mellors Comprehensive Treatise on Inorganic and Theoretical Chemistry," Vol. 8, Suppl. II, *Nitrogen*, Part II, Longmans, London, 1967.
[15] E. W. Lawless and I. C. Smith, "Inorganic High Energy Oxidizers," Edward Arnold, London, 1968.
[16] O. Ruff, *et al.*, *Z. anorg. allgem. Chem.*, **172**, 417 (1928).

In view of this last reaction, it is somewhat perplexing that electrolysis of molten NH_4F favors production of N_2F_2. An interesting (unanswered) question centers on the role of HF (as solvent or in molten NH_4HF_2) in the electrolysis and the change in product distribution with its presence.

Trifluoramine is a useful precursor to a series of important haloamine compounds. In a later section of this chapter we will discuss some important reactions of tetrafluorohydrazine, which can be prepared[17] from NF_3 by fluorination of active metals:

(8-10)
$$2NF_3 + 2M \rightarrow 2MF + \underset{[b.p. -73°]}{N_2F_4}$$

(this is an example of the "coupling reaction" with which you are familiar from organic chemistry, and which is to be considered here in a later section). One of the important properties of N_2F_4 is its dissociation into radicals, meaning it may be combined with a hydrogen source[18] for production of NHF_2, a dangerously explosive gas (b.p. $\sim -70°$):

(8-11)
$$N_2F_4 + 2PhSH \rightarrow 2HNF_2 + (SPh)_2$$

This leads us to the synthesis[19] of chlorodifluoramine by chlorination of HNF_2 with OCl^-:

(8-12)
$$HNF_2 + OCl^- \xrightarrow{HgO} ClNF_2 + OH^-$$

The mechanism here is undoubtedly analogous to the chlorination of NH_3. The presence of HgO facilitates the formation of OCl^- from Cl_2/H_2O and simultaneously avoids the use of added OH^- for this purpose (refer to the discussion of this use of HgO in Chapter 6, p. 353).

Phosphorus is much more accommodating in the ease with which it forms halides. The reaction with chlorine can be controlled so as to produce either the trichloride or the pentachloride[20]

(8-13)
$$P_4 + 6Cl_2 \rightarrow 4PCl_3$$
$$P_4 + 10Cl_2 \rightarrow 4PCl_5$$

by the simple expedient of adjusting the ratio of P_4 to Cl_2. The trifluoride and pentafluoride are then synthesized by fluorination of the corresponding chlorides, for example[21]

(8-14)
$$PCl_3 \xrightarrow{AsF_3} PF_3$$
$$PCl_5 \xrightarrow[300°]{CaF_2} PF_5$$

---

[17] See ref. 15.
[18] J. P. Freeman, et al., J. Amer. Chem. Soc., **82**, 5304 (1960).
[19] See ref. 15.
[20] M. C. Forbes, et al., "Inorganic Syntheses," Vol. 2, W. C. Fernelius (Ed.), McGraw-Hill, New York, 1946; R. N. Maxson, ibid., Vol. 1, H. S. Booth (Ed.), 1939.
[21] A. A. Williams, "Inorganic Syntheses," Vol. 5, T. Moeller (Ed.), McGraw-Hill, New York, 1957; E. L. Muetterties, et al., J. Inorg. Nucl. Chem., **16**, 52 (1960).

An additional way to prepare the pentafluoride is to chlorinate[22] PF_3, forming the mixed chlorofluoride, and then to fluorinate dichlorotrifluorophosphorane with, say, NaF:

$$PF_3 \xrightarrow{Cl_2} PF_3Cl_2 \xrightarrow{NaF} PF_5 \qquad (8-15)$$

Finally, we note that the phosphoryl halides, highly useful as precursors to other reagents, may be synthesized in a variety of ways. Perhaps the two most convenient paths to $POCl_3$ are

$$PCl_3 + SO_2Cl_2 \rightarrow SOCl_2 + POCl_3$$
$$P_4O_{10} + 6PCl_3 \xrightarrow{Cl_2} 10POCl_3 \qquad (8-16)$$

The former takes advantage of the nucleophilicity of P(III) to abstract an oxygen atom, and the latter actually represents a P_4O_{10}/PCl_5 reaction to exchange O/Cl. As in (8-15), POF_3 is readily obtained from $POCl_3$ by use of a metal fluoride:

$$POCl_3 \xrightarrow[MeCN]{NaF} POF_3 \qquad (8-17)$$

OXYGEN AND SULFUR

The high free energy contents of the oxygen halides present difficulties similar to those encountered with the nitrogen halides. Again you find it necessary to start with reactants more stable than the products sought and to supply energy from external sources to achieve the products sought, taking care to avoid conditions favoring rapid reversion to the elements. For example, oxygen difluoride is synthesized by electrolysis of KHF_2 in 80% HF/H_2O

$$KHF_2 \xrightarrow[electrol.]{HF, H_2O} OF_{2(g)} \qquad (8-18)$$

$$\begin{bmatrix} \text{pale yellow} \\ \text{b.p.} -145° \end{bmatrix}$$

or, better, by direct fluorination of oxygen in anionic form[23]:

$$2F_2 + 2OH^- \rightarrow OF_2 + 2F^- + H_2O \qquad (8-19)$$

Direct oxidation of O_2 by F_2 in a glowing discharge apparatus[24] or by photolysis of O_2/F_2 mixtures are examples of techniques designed to produce nascent F, which readily attacks O_2 to produce dioxygen difluoride:

$$O_2 + F_2 \xrightarrow{discharge} O_2F_{2(g)} \qquad (8-20)$$

$$\begin{bmatrix} \text{yellow-orange} \\ \text{m.p.} -163° \end{bmatrix}$$

---

[22] E. L. Muetterties and J. E. Castle, *J. Inorg. Nucl. Chem.*, **18**, 148 (1961).
[23] G. H. Cady, *J. Amer. Chem. Soc.*, **57**, 246 (1935).
[24] O. Ruff and W. Menzel, *Z. anorg. allgem. Chem.*, **211**, 204 (1933); *ibid.*, **217**, 85 (1934); S. Aoyama and S. Sakuraba, *J. Chem. Soc. Japan*, **59**, 1321 (1938).

In contrasting the syntheses of OF_2 and O_2F_2, you should find it interesting that O_2F_2 has a short, strong O—O bond (the stretching force constant of ~ 10 md Å^{-1} and $D_{O-O} = 104$ kcal/mole are suggestive of O=O) and weak O—F bonds ($D_{OF} \approx 18$ kcal/mole).[25] Thus, it appears that initial formation of OOF will not lead to cleavage of the OO bond by ensuing attack of a second F nor by attack of F upon FOOF. To produce OF_2, it seems necessary to produce first a reactant not containing a strong dioxygen bond, *viz.* (8-19). In direct analogy with the fluorination of OH^-, oxygen dichloride has been synthesized from HgO (oxygen in the dinegative form) by direct chlorination with Cl_2.

(8-21)
$$2HgO + 2Cl_2 \rightarrow OCl_{2(g)} + HgCl_2 \cdot HgO$$

The use of HgO is another interesting instance of the choice [see (18-12)] of Hg^{2+} in a reaction producing Cl^- ion so as to form $HgCl_2$. (Review in Chapter 6 the desirable consequences of "covalent" salt lattices on a reaction free energy.)

Sulfur, like phosphorus, presents fewer difficulties for direct oxidation than does its lighter congener, oxygen. Direct fluorination of sulfur fully oxidizes S_8 to the hexafluoride

(8-22)
$$S_8 + 24F_2 \rightarrow 8SF_{6(g)}$$

and an alternate route is provided by fluorination of SF_4 (prepared by a method to be given in a moment), whereas chlorine is not nearly as effective and yields the golden yellow disulfur dichloride from molten sulfur.

(8-23)
$$S_8 + 4Cl_2 \rightarrow 4S_2Cl_{2(l)}$$

We noted in the discussion of sulfur halides in Chapter 6 (p. 346) that sulfur dichloride is less stable than S_2Cl_2 and reverts to the latter on standing. There it was not mentioned that Friedel-Crafts catalysts[26] ($FeCl_3$, $SnCl_4$) assist the oxidation of S_8 to the bright red dichloride.

(8-24)
$$S_2Cl_2 + Cl_2 \xrightarrow{MCl_n} 2SCl_{2(l)}$$

Sulfur dichloride is an important intermediate for the synthesis of sulfur tetrachloride and tetrafluoride. Oxidation of SCl_2 by chlorine produces the tetrachloride (which reverts to S_2Cl_2 above $-30°$).

(8-25)
$$SCl_2 + Cl_2 \xrightarrow{-80°} SCl_4$$
$$\begin{bmatrix} \text{yellow solid} \\ \text{m.p. } -31° \end{bmatrix}$$

The low temperature serves the dual purpose of retarding the reversion of both SCl_2 and SCl_4 to S_2Cl_2 and Cl_2. The other important use of SCl_2 is its easily controlled fluorination by NaF in acetonitrile solvent as an effective synthesis of sulfur tetrafluoride[27] [the expected product (SF_2) is (Chapter 6, p. 346) unstable with respect to

---

[25] T. J. Malone and H. A. McGee, Jr., *J. Phys. Chem.*, **69**, 4338 (1965); *ibid.*, **70**, 316 (1966).

[26] It is not known just how the catalyst works. Formation of chloronium ion (Cl^+) from Cl_2 for attack upon S_2Cl_2 is one possibility (contrast this with the comments on O_2F_2 above). Another possibility is formation of S_2Cl^+ from S_2Cl_2. SCl_2 is known to act as a halide donor toward $AlCl_3$, $FeCl_3$, $SbCl_5$; N. S. Nabi and M. A. Khaleque, *J. Chem. Soc.*, 3626 (1965).

[27] C. W. Tullock, *et al.*, *J. Amer. Chem. Soc.*, **82**, 539 (1960).

disproportionation into sulfur and a higher oxidation state product, although the appearance of S$_2$Cl$_2$ suggests a more complex pathway for this reaction].

$$3SCl_2 + 4NaF \xrightarrow{CH_3CN} SF_{4(g)} + S_2Cl_2 + 4NaCl \qquad (8\text{-}26)$$

In a seemingly unusual heterogeneous reaction, AgF directly fluorinates[28] S$_8$ to form a mixture of *two isomers* of S$_2$F$_2$:

$$S_8 \xrightarrow[\text{vacuum}]{\text{AgF, 125°}} FSSF_{(g)} + F_2SS_{(g)} \qquad (8\text{-}27)$$

Lack of complete characterization of this reaction lets us temporarily speculate that the relatively high oxidizing strength of Ag(I) in a fluoride lattice and/or the possibility of Ag$_2$S formation (layer lattice) may be the thermodynamic basis for the overall oxidative fluorination reaction. It would be well for you to review (in Chapter 6, p. 343) the discussion of the thermodynamic/structural aspects of the disulfurdifluoride → thiothionyl fluoride isomerization and the catalytic effects of ionic fluorides. In this vein, it is quite interesting that fluorination of ClSSCl with KSO$_2$F produces thiothionyl fluoride.

$$2KSO_2F + S_2Cl_2 \rightarrow SSF_2 + 2KCl + 2SO_2 \qquad (8\text{-}28)$$

The properties of SO$_2$F$^-$ (a weak adduct of SO$_2$ + F$^-$ having a measurable vapor pressure of SO$_2$ at room temperature) are consistent with greater catalytic activity than that of AgF.

As suggested in Chapter 7 (p. 431), ClF can be added to SF$_4$ in the presence of CsF catalyst (via formation of SF$_5^-$ as the kinetically active species) to form SF$_5$Cl. This reagent is useful for the preparation of compounds containing the SF$_5$ substituent through reactions directed at the S—Cl bond. Preparation of S$_2$F$_{10}$ is achieved through photolysis of SF$_5$Cl in the presence of H$_2$, whose role is likely to be that of a scavenger of Cl atoms produced by photolytic cleavage of the S—Cl bond.

$$SF_{4(g)} + ClF_{(g)} \xrightarrow{CsF} SF_5Cl_{(g)} \qquad (8\text{-}29)$$

$$2SF_5Cl + H_2 \xrightarrow{h\nu} S_2F_{10} + 2HCl \qquad (8\text{-}30)$$

$$\begin{bmatrix} \text{m.p.} & -55° \\ \text{b.p.} & 29° \end{bmatrix}$$

Before leaving this discussion of sulfur halides, we need to acknowledge the *sulfuryl* ($\supset$SO$_2$) and *thionyl* ($\supset$SO) halides as important classes of compounds. The chlorides are best synthesized by the following schemes[29]:

$$SO_2 + Cl_2 \rightarrow SO_2Cl_2 \qquad (8\text{-}31)$$

$$SO_2Cl_2 + PCl_3 \rightarrow SOCl_2 + POCl_3 \qquad (8\text{-}32)$$

---

[28] F. Seel and R. Budenz, *Chimia*, **17**, 355 (1963); *ibid.*, **22**, 79 (1968).
[29] F. Feher, in "Handbook of Preparative Inorganic Chemistry," Vol. I, G. H. Brauer (Ed.), Academic Press, New York, 1963.

where (8-32) is the same reaction as (8-16), this time cited for synthesis of thionyl chloride. Similar techniques based on oxygen atom transfer are

(8-33)
$$SCl_2 + SO_3 \rightarrow SOCl_2 + SO_2$$
$$3S_2Cl_2 + 3SO_3 \rightarrow SOCl_2 + \frac{1}{2}S_8$$

You can share the question with your authors about the possibility that this reaction actually proceeds through oxygen transfer from SO_3 to SCl_2, which might have arisen through the intermediacy of $SSCl_2$, the unknown analog of SSF_2. [Alternately, refer to the footnote for reaction (8-24) and note that SO_3 is a voracious Lewis acid.] The fluorides are simply produced from the chlorides by halogen exchange, as in

(8-34)
$$SOCl_2 \xrightarrow{SbF_3} SOF_2$$
$$SOCl_2 \xrightarrow[MeCN]{NaF} SOF_2$$

or by fluorination (limited to SO_2F_2)

(8-35)
$$SO_2 + F_2 \rightarrow SO_2F_2$$

FLUORINE AND CHLORINE[30]

Halogenation of the diatomic second row elements (O_2, N_2) requires, as you fully understand, rather energetic conditions—with resultant formation of several products, which must then be separated by distillation or some other means. The fluorination of the diatomic Cl_2 also suffers from orbital following difficulties and requires a temperature in the range between 200°C and 300°C to produce mixtures of ClF, ClF_3, and ClF_5. The best approach to synthesis of these compounds seems to be to form ClF_3 in good yield by controlling the reactant stoichiometry (and temperature at ~250°C). ClF_5 exists in equilibrium with ClF_3 at these temperatures, so that limiting the amount of F_2 available seems to reduce the amount of ClF_5 formed. The ClF also formed may be separated from the trifluoride by low temperature fractionation. Purification of ClF_3 is then achieved by reaction with KF to form the involatile tetrafluorochlorate salt

(8-36)
$$ClF_{3(g)} + KF \rightarrow KClF_{4(c)}$$

followed by reversal of (8-36) at about 140°C.

The ClF_3 prepared in this manner can then be *reduced* by Cl_2 to form ClF at ~250°C.

(8-37)
$$ClF_3 + Cl_2 \rightarrow 3ClF_{(g)}$$

The best synthesis yet developed for ClF_5 seems to be direct fluorination of chlorine in the *anionic form* as $KClF_4$ in a steel bomb at ~100°C, although the less severe thermal/pressure requirements of the photochemical synthesis (8-39) from ClF_3/F_2 mixtures are attractive.

(8-38)
$$KClF_4 + F_2 \xrightarrow{200°} ClF_{5(g)} + KF$$

---

[30] The many investigations into the syntheses of interhalogen compounds are reviewed and summarized most effectively by A. J. Downs and C. J. Adams in "Comprehensive Inorganic Chemistry," A. F. Trotman-Dickenson, *et al.* (Eds.), Vol. 2, Chapter 26, Pergamon Press, New York, 1973.

$$\text{ClF}_3 + \text{F}_2 \xrightarrow[\text{room temp.}]{h\nu} \text{ClF}_5 \tag{8-39}$$

XENON

The only rare gas with an extensive halogen chemistry is xenon. Direct bomb fluorination of Xe usually results in formation of XeF_2, XeF_4, and XeF_6, much as is the case for the fluorination of chlorine. Careful control of the reactant stoichiometries, the total pressure, and the temperature (>250° needed to produce useful reaction rates) is needed to maximize the yield of any one of these compounds.[31] Figure 8-7 bears out your expectation that cooler temperatures and higher F_2:Xe ratios favor XeF_6, while higher temperatures and lower F_2:Xe partial pressures gradually permit its dissociation into XeF_2. The optimum conditions for synthesis of any one fluoride can be worked out from knowledge of the respective equilibrium constants and their temperature dependence.

$$2\text{Xe} + \text{F}_2 \xrightarrow[400°]{} \text{XeF}_{2(c)} \tag{8-40}$$

---

[31] B. Weinstock, *et al., Inorg. Chem.,* **5,** 2189 (1966); N. Bartlett and F. O. Sladky, pp. 282ff, in "Comprehensive Inorganic Chemistry," A. F. Trotman-Dickenson, *et al.* (Eds.), Pergamon Press, New York, 1973; C. L. Chernick, *et al.,* in "Inorganic Syntheses," Vol. 8, H. F. Holzclaw, Jr. (Ed.), McGraw-Hill, New York, 1966.

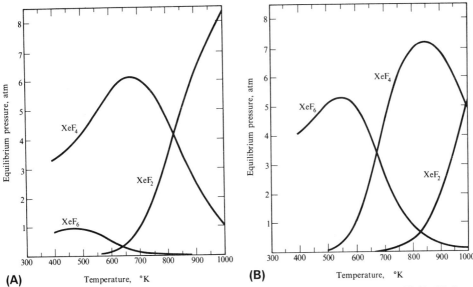

Figure 8-7. Pressure and temperature influence on XeF_2, XeF_4, and XeF_6 formation. (A) Equilibrium pressures of xenon fluorides as a function of temperature. Initial conditions: 125 mmoles Xe, 275 mmoles F_2 per 1000 ml. (B) Equilibrium pressures of xenon fluorides as function of temperature. Initial conditions: 125 mmoles Xe, 1225 mmoles F_2 per 1000 ml. [From N. Bartlett and F. O. Sladky, in "Comprehensive Inorganic Chemistry," A. F. Trotman-Dickenson, *et al.* (Eds.), Pergamon Press, New York, 1973.]

(8-41)[32]
$$Xe + \text{"}2\text{"}F_2 \xrightarrow[400°]{} XeF_{4(c)}$$

(8-42)
$$Xe + \text{"}20\text{"}F_2 \xrightarrow[\substack{300° \\ 50\ \text{atm}}]{} XeF_{6(c)}$$

The stoichiometries given in (8-40) to (8-42) suggest the ratios of reactants giving the best yield of each product.

In an interesting use of the reactive dioxygendifluoride, O_2F_2, its fluorination of xenon at *low temperature* ($-120°$) produces XeF_2:[33]

$$Xe + O_2F_2 \xrightarrow[-120°]{} XeF_2 + O_2$$

---

[32] The major impurity is XeF_6, which is removed by formation of the $NaXeF_7$ salt (see Chapters 5 and 6) or by treatment with AsF_5 in BrF_5 to form $[XeF_5][AsF_6]$ (this also removes any XeF_2 as $[Xe_2F_3][AsF_6]$).
[33] S. A. Morrow and A. R. Young, *Inorg. Chem.*, **4**, 759 (1965).

STUDY QUESTIONS

1. Give steps for syntheses of the following compounds from the elements. In (a), start from Me_4NF.

   a) $[Me_4N][SiF_5]$
   b) SF_5Cl
   c) $SiFCl_3$
   d) B_2F_4
   e) PF_2Cl
   f) POF_3
   g) NOF_3
   h) SOF_2
   i) SOF_4
   j) PF_3Cl_2
   k) PCl_3F_2
   l) PCl_4F

2. a) In the discharge synthesis of O_2F_2, which element, O_2 or F_2, is chosen to flow through the discharge tube?
   b) What compound does one smell when an electric discharge (from a faulty electric motor) occurs in air?
   c) How might the values of m and n in the O_mF_n product be altered if O_2 and F_2 were passed through separate discharge tubes, with the effluents to be mixed "down stream"? Does **17** in Chapter 4 fit into an explanation of the product?

3. a) What electronic transition within Cl_2 should be excited to *most* directly produce $2Cl$?
   b) What electronic transition in Cl_2 is responsible for its yellow-green hue (see Chapter 5, p. 210)?
   c) Convert the Cl—Cl bond dissociation energy into an equivalent photon wavelength ($h = 6.63 \times 10^{-27}$ erg·sec·molecule^{-1}; 1 erg $= 2.39 \times 10^{-8}$ cal). Is the electronic transition of Cl_2 that is responsible for the yellow-green color sufficiently energetic to effect dissociation of Cl_2; does the transition involve a net bond weakening or strengthening?

4. The following equilibrium constants apply for Xe/F_2 mixtures at 250°C.

   $$K_2 = 8.8 \times 10^4 \text{ atm}^{-1} \text{ for } Xe + F_2 \rightleftharpoons XeF_2$$
   $$K_4 = 1.1 \times 10^8 \text{ atm}^{-1} \text{ for } Xe + 2F_2 \rightleftharpoons XeF_4$$
   $$K_6 = 1.0 \times 10^8 \text{ atm}^{-1} \text{ for } Xe + 3F_2 \rightleftharpoons XeF_6$$

   a) In designing a synthesis of XeF_4 contaminated by very little XeF_2 and XeF_6, would use of stoichiometric amounts of Xe and F_2 achieve the desired result? (Hint: consider the extent of the disproportionation $2XeF_4 \rightleftharpoons XeF_2 + XeF_6$.) How much XeF_4 could be obtained?
   b) In a synthesis of XeF_2, suppose that you require 1 atm XeF_2 with less than 0.01 mole ratio XeF_4 contaminant (from $2XeF_2 \rightleftharpoons Xe + XeF_4$). Starting with 1 atm F_2, what *initial* pressure of Xe is required?

FLUORINATING AGENTS[34]

An important aspect of the chemistries of certain non-metal fluorides is their utility as fluorinating agents for both metal and other non-metal compounds. There is a great deal of current interest in their use to prepare fluorocarbon compounds for pharmaceutical and industrial application.

Rather arbitrarily, we will recognize two classes of fluorinating agents.

Harsh: Highly reactive, often oxidizing, and non-discriminating reagents that tend to degrade as well as fluorinate other compounds; in this category are F_2, ClF, ClF_3, BrF_3, and MF_n where the metal is in a high oxidation state (CoF_3, MnF_3, CeF_4, UF_6).

Moderate: Reagents that are more selective in their attack of functional groups and show less tendency to degrade molecules (in particular, leaving hydrogen bonds intact); important examples are SF_4, SbF_3, SbF_5, and MF_n where the metal is in a lower oxidation state (HgF_2, CaF_2, AgF, CsF).

The classes of reactions and reagents of most interest are:

Oxidation of metals: ClF_3, BrF_3
Substitution of H: ClF, ClF_3 (often with degradation)
Substitution of O: SF_4, ClF_3, BrF_3
Substitution of Cl: SF_4, SbF_3, SbF_5, AsF_3 (catalyst $SbCl_5$)
Addition to multiple bonds: ClF, ClF_3, BrF_3, SF_4 (F^- catalyzed addition to perfluoroolefins only)

Some general comments about the use of non-metal fluorides as fluorinating agents are in order before we begin discussing individual reagents. Many of these reagents are gases at room temperature and thus require the special techniques of manipulating vapor state reagents. The construction of special vacuum lines[35] from stainless steel, Kel-F, or Teflon is required because of the corrosive action of these materials; also for this reason, mercury manometers, float valves, diffusion pumps, and other parts of a conventional line must be replaced with substitutes fashioned from less reactive materials. All these reagents seem to be quite toxic in themselves, and can be particularly obnoxious when moisture is present because hydrolysis generally produces HF. In contact with skin, HF causes severe, painful burns that are slow to heal.

To avoid working with heterogeneous reaction mixtures, however, fluorination reactions *are* sometimes carried out in HF as the solvent. Very often the solvent HF and fluoride ion acceptors such as BF_3 (when they have been added) act in a fundamentally critical way as catalysts for a reaction. This role for the HF or BF_3 acids in the ensuing reactions is often not clear, because one can imagine them acting through formation of an adduct with a basic substrate to be fluorinated *or* they could act in the capacity of F^- acceptor from the fluorinating agent so as to enhance its electrophilicity:

$$HF + MF_n \rightarrow MF_{n-1}^+ + HF_2^-$$
$$BF_3 + MF_n \rightarrow MF_{n-1}^+ + BF_4^-$$

(8-43)

---

[34] E. W. Lawless and I. C. Smith, "Inorganic High Energy Oxidizers," Edward Arnold, London, 1968; "Advances in Fluorine Chemistry," M. Stacey, *et al.* (Eds.), Butterworths, London, an ongoing series (1960): W. A. Sheppard and C. M. Sharts, "Organic Fluorine Chemistry," Benjamin, New York, 1969; R. D. Chambers, "Fluorine in Organic Chemistry," Wiley, New York, 1973.

[35] D. F. Shriver, "The Manipulation of Air-sensitive Compounds," McGraw-Hill, New York, 1969.

or even, in the case of HF, act as a F^- donor.

(8-44)
$$2HF + MF_n \rightarrow H_2F^+ + MF_{n+1}^-$$

In fact, you will recall that many non-metal fluorides tend to undergo (in condensed phases) analogous *auto*-ionization reactions to form what are usually strongly oxidizing cations:

(8-45)
$$2MF_n \rightarrow MF_{n-1}^+ + MF_{n+1}^-$$

Given that the MF_n compounds are subject to hydrolysis and that the HF produced may act as a catalyst for the ensuing fluorination reaction, it is often not clear that a fluorination reaction involves just the substrate and molecular MF_n, when these are *intended* to be the only reagents. While the catalytic role of HF can often be used to advantage, its presence can have a detrimental effect by catalyzing an *undesired* reaction. In such instances NaF is sometimes added as a scavenger for HF through formation of the bifluoride salt, $NaHF_2$.

SbF_3 AND SbF_5

These reagents are "classical" fluorinating agents and have found their greatest application in the *conversion of chlorides into* the corresponding *fluorides* (refer to the bond energies in Chapter 6 and note that Sb—Cl > Sb—F). Of the two reagents, SbF_5 is the stronger; as a liquid (SbF_3 is a white solid) it is readily miscible with many organic solvents and often is more convenient to work with. The pentafluoride has long been used to replace chlorines stepwise in trichloromethyl groups, and this represents its most important use in organo-fluorine chemistry.[36]

(8-46)
$$R-CCl_3 \xrightarrow{SbF_5} RCCl_2F \rightarrow RCClF_2 \rightarrow RCF_3$$
$$RR'CCl_2 \longrightarrow RR'CClF \rightarrow RR'CF_2$$

One of the most important properties of SbF_5 is that it acts as a powerful F^- ion acceptor by forming salts of SbF_6^-. In combination with HF as a solvent, an equilibrium causing formation of HF_2^+ [as in (8-44)] renders the solutions conducting.

$$2HF + SbF_5 \rightleftharpoons HF_2^+ + SbF_6^-$$

In some instances the species HF_2^+ might play an important role in Cl/F exchange reactions. The question naturally arises (there seems to be no answer yet) as to the thermodynamic importance of the reaction

$$2SbF_5 \rightleftharpoons SbF_4^+ + SbF_6^-$$

and the catalytic role of even a minute amount of the cation SbF_4^+ in Cl/F exchange reactions such as (8-46).

As an interesting extension of this last idea, it is found that SbF_3 alone is a fairly weak fluorinating agent for Cl/F exchange with organic chlorides, yet addition of small amounts of $SbCl_5$ to reaction mixtures smoothly effects the Cl/F exchange. The

---

[36] The ease of these reactions is very sensitive to the nature of geminal and vicinal groups; the last step of the terminal $RCCl_3$ exchange is often difficult to effect, and exchange increases in difficulty in the series: —$CHCl_2$, —$CHFCl$, —CH_2Cl. It appears that sterically large and π electron releasing groups assist the exchange by supporting formation of a carbonium ion intermediate. See W. A. Sheppard and C. M. Sharts (p. 74) and R. D. Chambers (pp. 16ff) in ref. 34.

chemically active species, as far as is now known, could arise from any of the following:

$$SbF_3 + SbCl_5 \rightleftharpoons SbCl_4F + SbF_2Cl$$
$$\rightleftharpoons SbF_2^+ + SbCl_5F^-$$
$$\rightleftharpoons SbCl_4^+ + SbF_3Cl^-$$

or could simply be due to a Friedel-Crafts-like acid/base interaction between the substrate Cl and $SbCl_5$.

ClF, ClF$_3$, BrF$_3$[37]

As noted earlier, these reagents tend to exhibit harsh action and so are good for producing *high oxidation state metal fluorides* from metals and metal oxides. Generally speaking, the order of reactivity is $ClF_3 > ClF > BrF_3$ and the tendency for substrate degradation accompanying fluorination decreases in this order. An important handling consideration is that ClF is a gas at room temperature while BrF_3 is a liquid, and ClF_3 is a gas with a boiling point of only 11°C. Inadvertent (because of the "high" boiling point) condensation of ClF_3 with organic materials often results in violent explosions; its use must be accompanied by care that the reagents are not co-condensed as pure materials.

With organic compounds, the suitably diluted halogen fluorides have been found useful for *adding X—F across double bonds* and *replacement of other halogens* by fluorine, but a problem with their use can be the accompanying substitution of hydrogen by halogen. In this respect, however, the halogen fluorides are preferred to elemental fluorine. There are occasions, though, when this is not a "problem," but a useful tendency. Dehalogenation of the original addition product reinstates the olefin linkage with the net result of halogen replacing hydrogen, as in the non-destructive halogenation of benzene:

$$C_6H_6 + ClF_3 \xrightarrow{-6HX} C_6F_nCl_{(12-n)} \xrightarrow{Na} C_6F_nCl_{(6-n)} \tag{8-47}$$

An interesting example of BrF_3 fluorination is the reaction with keto groups. The initial reaction *exchanges* two fluorine atoms for an oxygen atom, and the reaction can be terminated at this stage if the initial product is sufficiently volatile to be swept from the reaction mixture by a stream of an inert gas.

$$Me_2C=O \xrightarrow[HF]{BrF_3} Me_2CF_2 \tag{8-48}$$

Degradation of the initial exchange product will follow, however, if it is not sufficiently volatile to be removed as it is formed:

$$\underset{H}{\overset{Me}{>}}C^*-\underset{Me}{\overset{O}{\underset{\|}{C}}}-Me \xrightarrow[HF]{BrF_3} \underset{Me}{\overset{Me}{>}}C^*\overset{F}{\underset{F}{<}} + CF_3CH_3 \tag{8-49}$$

---

[37] A good summary is to be found in W. A. Sheppard and C. M. Sharts (pp. 56ff, 178), in ref. 34. T. E. Stevens, *J. Org. Chem.*, **26**, 1627 (1961).

Other studies with BrF_3 have shown that it can be used to convert a nitrile into the corresponding trifluoromethyl compound (again volatility of the desired product is required).

(8-50)
$$RCN + BrF_3 \xrightarrow{HF} RCF_3$$

The halogen fluorides also exhibit the fluoride acceptor/donor property noted in the last section for antimony. Salts of Cl_2F^+, ClF_2^-, ClF_2^+, ClF_4^-, BrF_2^+, and BrF_4^- are all known and raise the possibility that in condensed phases, at least, the reactive species might be the cations listed; they could arise from the equilibria:

$$3ClF \rightleftharpoons Cl_2F^+ + ClF_2^-$$
$$2ClF_3 \rightleftharpoons ClF_2^+ + ClF_4^-$$
$$2BrF_3 \rightleftharpoons BrF_2^+ + BrF_4^-$$

Further complicating the characterization of fluorination pathways for $X'X_n$ is the specific role (if any) played by the solvent HF, when used.

SULFUR TETRAFLUORIDE[38]

Of all the fluorinating agents studied, SF_4 seems to be particularly useful, for among its variety of reactions are some of high specificity. For example, the conversions of phosphoryl groups to difluorophosphorane and of carbonyl groups to difluoromethylene groups are valuable reactions and illustrate the important reaction for SF_4 of *O/F exchange*.

(8-51)
$$\text{>}C=O + SF_4 \rightarrow \text{>}CF_2 + SOF_2$$

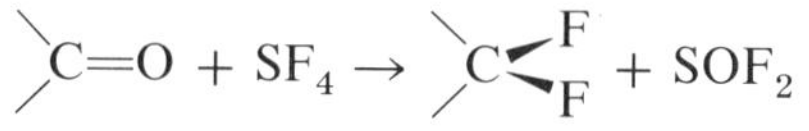

(8-52)
$$\text{>}P=O + SF_4 \rightarrow -PF_2\text{<} + SOF_2$$

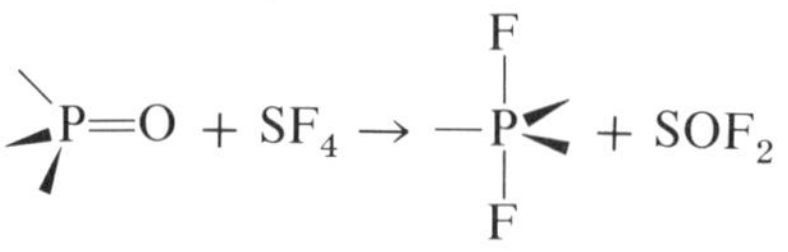

[Reaction (8-52) is important because it is one of the few ways available of chemically attacking the stable P=O group (see Chapter 6).] In organic chemistry, the selectivity of SF_4 fluorination of CO moieties follows the hierarchy:

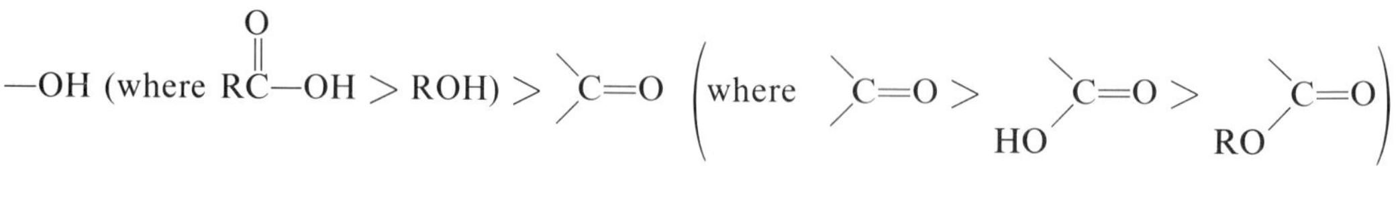

$$-OH \text{ (where } R\overset{O}{\underset{\parallel}{C}}-OH > ROH) > \text{>}C=O \left(\text{where } \text{>}C=O > \underset{HO}{\text{>}}C=O > \underset{RO}{\text{>}}C=O \right)$$

as illustrated by the following reactions:

(8-53)
$$EOH + SF_4 \rightarrow EF + HF + SOF_2$$

(8-54)
$$\text{>}C=O + SF_4 \rightarrow \text{>}CF_2 + SOF_2$$

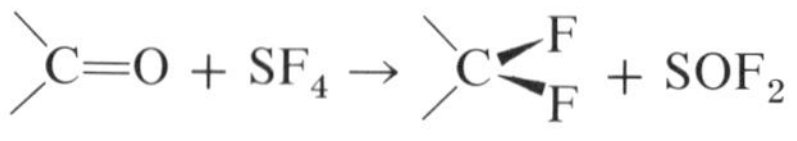

[38] See R. D. Chambers (pp. 48ff) and W. A. Sheppard and C. M. Sharts (pp. 164ff, 169ff) in ref. 34. Also W. R. Hasek, *et al., J. Amer. Chem. Soc.*, **82**, 543, (1960); W. C. Smith, *Angew. Chem. Intern. Ed. Engl.*, **1**, 467 (1962).

$$\begin{array}{c}\diagdown\\ \diagup\end{array}\!\!C\!\!=\!\!O + SF_4 \rightarrow \begin{array}{c}\diagdown\\ \diagup\end{array}\!\!C\!\!=\!\!O \xrightarrow{SF_4} -C\!\!\begin{array}{c}F\\ F\\ F\end{array} + SOF_2 \quad (8\text{-}55)$$
$$\text{OH} \qquad\qquad\qquad F$$
$$(+ SOF_2 + HF)$$

$$\begin{array}{c}\diagdown\\ \diagup\end{array}\!\!C\!\!=\!\!O + SF_4 \rightarrow \begin{array}{c}\diagdown\\ \diagup\end{array}\!\!C\!\!\begin{array}{c}F\\ F\end{array} \xrightarrow{SF_4} -C\!\!\begin{array}{c}F\\ F\\ F\end{array} + RF \quad (8\text{-}56)$$
$$\text{OR} \qquad\qquad \text{OR} \qquad\qquad\qquad + SOF_2$$
$$(+SOF_2)$$

In comparing (8-55) and (8-56) you see that esterification of the carboxyl group alters the order in which the oxygens are attacked. The alcohol reaction (8-53) is particularly effective for acidic alcohols, of both organic and inorganic types,[39] such as the following:

$$\begin{array}{c}\diagdown\\ \diagup\end{array}\!\!P\!\!-\!\!: + SF_4 \rightarrow \begin{array}{c}\diagdown\\ \diagup\end{array}\!\!P\!\!-\!\!: + HF + SOF_2 \quad (8\text{-}57)$$
$$\text{OH} \qquad\qquad\qquad F$$

The mechanism of (8-57) has not been established, although it is possible that the reaction proceeds through an intermediate like **3**:

$$\begin{array}{c}\diagdown\\ \diagup\end{array}\!\!E\!\!-\!\!O\!\!-\!\!S\!\!\begin{array}{c}F\\ |\\ F\\ |\\ F\end{array}\!\!\begin{array}{c}F\\ \\ :\end{array}$$
3

considering that RSF_3 compounds are known.[40] Whether the O/F exchange is completed by an intramolecular rearrangement (1,3 F shift)

$$\begin{array}{c}\diagdown\\ \diagup\end{array}\!\!E\!\!-\!\!O\!\!-\!\!S\!\!\begin{array}{c}F\\ |\\ F\\ |\\ F\end{array}\!\!\begin{array}{c}F\\ \\ :\end{array}$$

is not known for certain (an alternate scheme for F transfer might involve attack at E by the HF produced in the initial solvolysis of SF_4).

The exchange of terminal oxygen is thought to proceed through an $SF_4 \cdot O{=}E$ intermediate similar to that (**3**) for the OH group. Though it is not necessary, these reactions are frequently run in the presence of BF_3 or anhydrous HF as a catalyst. The role of the catalyst is not certain; however, since such catalysts are Lewis acids, they could polarize the terminal E=O bond prior to SF_4 attack upon E or could even act as a F^- acceptor from SF_4, whereupon SF_3^+ would form a cationic intermediate like **3**.

In contrast to the harsher halogen fluorides, SF_4 is known to *add to olefins* only in the presence of F^- as a catalyst, and then it will add only to perfluoroolefins.[41] Again, the mechanism of this reaction has not been determined, but it is reasonable to postulate that SF_5^- is more reactive than SF_4 (the presence of the fifth fluoride should labilize the S—F bonds as well as enhance the S lone pair nucleophilicity by decreasing its s character and by greater VSEPR).

$$\begin{array}{c}CF_3\\ \diagdown\\ F\end{array}\!\!C\overset{*}{=}C\!\!\begin{array}{c}F\\ \diagup\\ F\end{array} + SF_4 \xrightarrow{CsF} \begin{array}{c}F\\ \diagdown\\ F_3C\end{array}\!\!\overset{*}{C}\!\!-\!\!SF_3 \text{ and } \left(\begin{array}{c}F\\ \diagdown\\ F_3C\end{array}\!\!C\!\!-\!\!\right)_2\!\!SF_2 \quad (8\text{-}58)$$

---

[39] W. C. Smith, *J. Amer. Chem. Soc.*, **82**, 6176 (1960).
[40] W. A. Sheppard, *J. Amer. Chem. Soc.*, **84**, 3058 (1962); R. M. Rosenberg and E. L. Muetterties, *Inorg. Chem.*, **1**, 756 (1962).
[41] See ref. 40.

The general unreactivity of olefins toward SF_4 is not shared by the more polar π electrons of nitrile compounds, which react to produce imine sulfurdifluorides.[42]

(8-59) $$R-C\equiv N + SF_4 \rightarrow R-CF_2-N=SF_2$$

(8-60) $$O=C=N-R + SF_4 \rightarrow R-N=SF_2 + COF_2$$

It is supposed that attack upon SF_4 by the nitrogen leads to an intermediate like that of the alcohol and carbonyl reactions, and that this step is followed by intramolecular rearrangement (fluorine shifts). If so, it is somewhat surprising that the cyanate of (8-60) does not yield SOF_2 and $F_2C=NR$ in analogy with (8-54).

Before leaving the discussion of SF_4 reactions, we should note that a disadvantage to working with SF_4 (in addition to the dangers associated with production of HF from hydrolysis) is that it is a gas at room temperature. To avoid the attendant handling difficulty, it is often possible to substitute a solid aryl sulfurtrifluoride (like $C_6F_5SF_3$) for SF_4 in many reactions. Such aryl-SF_3 compounds are easily synthesized by fluorination of disulfides with AgF_2.[43]

(8-61) $$Ar_2S_2 \xrightarrow{AgF_2} ArSF_3$$

DINITROGEN TETRAFLUORIDE[44]

Perfluorohydrazine illustrates some useful radical reactions (recall from p. 458 that a sample of N_2F_4 contains NF_2 radicals), and an interesting photochemistry is beginning to evolve for this reagent. Thus, many of the high-temperature reactions of N_2F_4 are simply *radical* reactions of NF_2 rather than fluorination *per se*. Two characteristic reactions of most radicals are the "trapping" reaction, illustrated by (8-62) for NF_2:

(8-62) $$\underset{}{\overset{O}{\underset{}{\|}}}C-C\underset{}{\overset{O}{\underset{}{\|}}} \xrightarrow{h\nu} -\dot{C}=O \xrightarrow{N_2F_4} \underset{NF_2}{\overset{}{>}}C=O$$

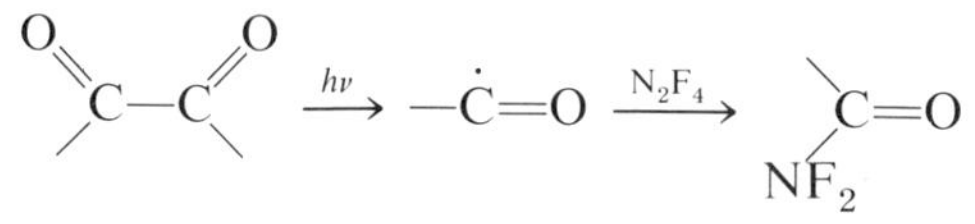

$$R-I \xrightarrow{h\nu} R\cdot \xrightarrow{N_2F_4} R-NF_2$$

and *hydrogen abstraction,* as in the series (8-63):

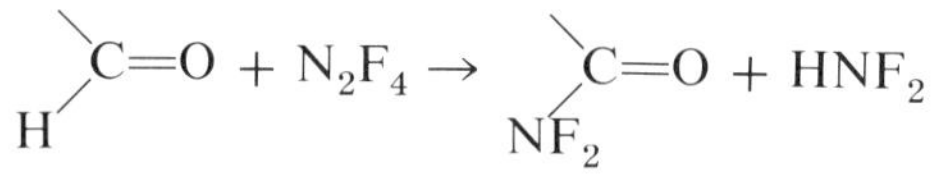

$$\underset{H}{\overset{}{>}}C=O + N_2F_4 \rightarrow \underset{NF_2}{\overset{}{>}}C=O + HNF_2$$

(8-63) $$2RSH + N_2F_4 \rightarrow R_2S_2 + 2HNF_2$$

$$R-H + N_2F_4 \rightarrow R-NF_2 + HNF_2$$

Addition of N_2F_4 to olefins[45] is another example of an NF_2 reaction channel:

(8-64) $$>C=C< + N_2F_4 \rightarrow \underset{NF_2}{\overset{F_2N}{>}}C-C<$$

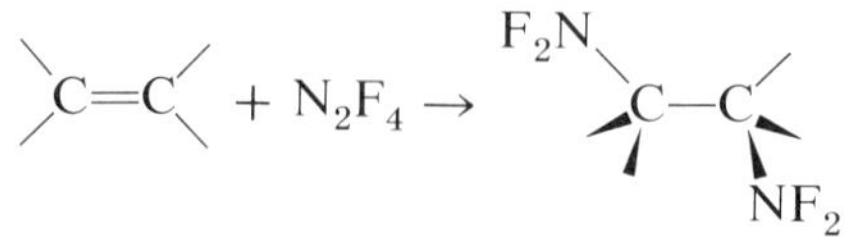

[42] W. C. Smith, *et al., J. Amer. Chem. Soc.,* **82,** 551 (1960).
[43] See ref. 40.
[44] The reaction types exhibited by N_2F_4 are thoroughly outlined with examples by E. W. Lawless and I. C. Smith in ref. 34.
[45] Such reactions are often accompanied by hydrogen abstraction at temperatures above 200°C.

Addition of N_2F_4 to alkynes is quickly followed by an interesting intramolecular rearrangement (a 1,3 shift)[46] to form the fluoroimine **4**:

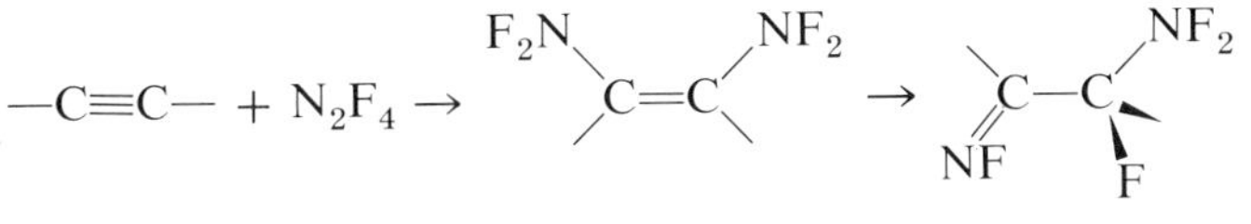

$$-C\equiv C- + N_2F_4 \rightarrow \underset{}{\overset{F_2NNF_2}{\underset{}{C=C}}} \rightarrow \underset{NFF}{\overset{NF_2}{C-C}} \quad \textbf{4}$$

(8-65)

made possible by the presence of the olefin linkage (orbital following is efficient, too) and the greater energy of the C—F than of the N—F bond (from Chapter 6, $E_{NF} \approx 67$ kcal/mole, $E_{CF} \approx 117$ kcal/mole). The only known class of acetylenes that are "stable"[47] to the 1,3 rearrangement are the perfluoroacetylenes, of which $CF_3-C\equiv C-CF_3$ is an example. Given the large difference between E_{NF} and E_{CF}, it is not surprising that the shift does occur on gentle heating but that there is some kinetic inhibition associated with the bulkier perfluoro groups.[48]

The *photochemistry* alluded to in the opening paragraph on N_2F_4 is believed to involve excitation of NF_2 to effect its dissociation into F and the *nitrene*, NF (a heteronuclear diatomic analog of the diradical O_2):

$$NF_2 \xrightarrow{h\nu} F + NF$$

(8-66)

In this way we conveniently explain the production, by "trapping" reactions, of NF_3 and N_2F_2 on photolysis of N_2F_4:

$$F + NF_2 \rightarrow NF_3$$

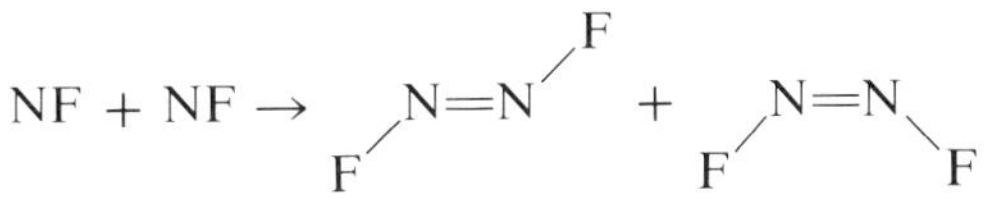

$$NF + NF \rightarrow \underset{anti}{\overset{}{\underset{F}{N}=\underset{}{N}\overset{F}{}}} + \underset{syn}{\overset{}{\underset{F}{N}=\underset{F}{N}}}$$

(8-67)

and the fact that photolysis of N_2F_4 in the presence of olefins effects the "addition" of NF_3 to the olefin (a two-step "trapping" reaction):

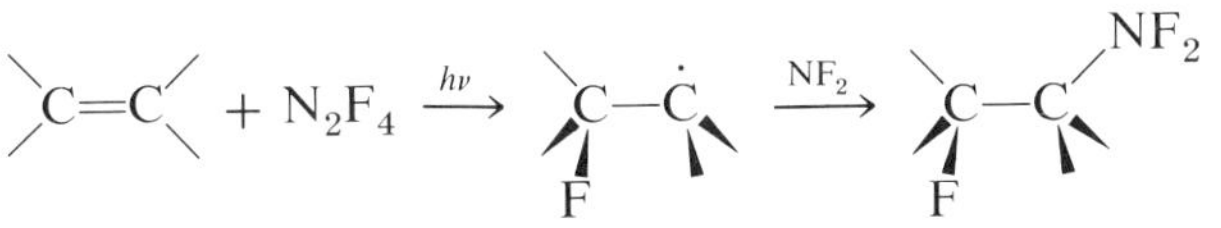

$$\overset{}{\underset{}{C=C}} + N_2F_4 \xrightarrow{h\nu} \underset{F}{\overset{}{C-\dot{C}}} \xrightarrow{NF_2} \underset{F}{\overset{NF_2}{C-C}} \quad \textbf{5}$$

(8-68)

to give **5**, among, of course, other products [contrast the slower thermal reaction (8-64)]. Additionally, the formation of RNF_2 compounds from photolyzed mixtures of alkanes and N_2F_4 is believed to result from initial abstraction of H from the alkane by the very reactive fluorine atom and trapping of R· by NF_2 as in (8-63).

$$R-H + N_2F_4 \xrightarrow{h\nu} R-H + \left\{\begin{array}{c} F \\ NF \\ NF_2 \end{array}\right\} \rightarrow R-NF_2 + HF$$

---

[46] G. N. Sausen and A. L. Logothetis, *J. Org. Chem.*, **32**, 2261 (1967); R. C. Petry, et al., *ibid.*, **32**, 1534 (1967).
[47] Gentle heating will effect the 1,3 shift.
[48] For some ideas on orbital following constraints on the stereochemistry of the 1,3 shift in general, see R. B. Woodward and R. Hoffman, "The Conservation of Orbital Symmetry," Academic Press, New York, 1970, p. 114.

The electronic structure of NF$_2$ and its photochemistry are interesting to consider in terms of the mo concepts of Chapter 4. Using the methods described there (and see study question 6-64), it is expected that the odd electron resides (see **6**) in the π^* mo

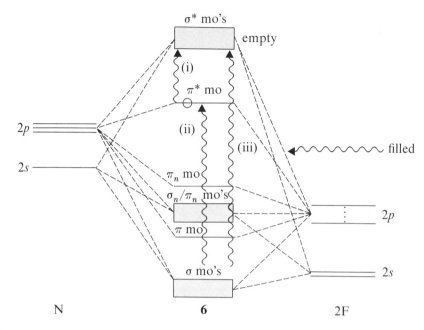

derived from the nitrogen and two fluorine p ao's perpendicular to the plane of the molecule. As this mo is expected to be polarized toward the less electronegative nitrogen atom, with lesser contributions from the fluorine ao's, the radical could be considered a "nitrogen-centered" radical. Abbreviating the ground electron configuration as $(occ)^{18}(\pi^*)^1(unocc)^0$, you may distinguish three classes of excited states:

(i) $(occ)^{18}(\pi^*)^0(unocc)^1$ spin multiplicity $(2S + 1) = 2$ only
(ii) $(occ)^{17}(\pi^*)^2(unocc)^0$ ″   ″   ″   ″ ″ ″
(iii) $(occ)^{17}(\pi^*)^1(unocc)^1$ ″   ″   ″   ″ 2 or 4

The first two possibilities are probably not important, because excited states of the same spin multiplicity as the ground state are thought to return to the ground state too quickly to permit dissociation of an NF bond. Any of the series of states in (iii) with $(2S + 1) = 4$ could serve as a precursor to loss of F. All such excited states, moreover, arise from configurations related to that of the ground state by an increase in anti-bonding density (and *sigma* anti-bonding at that) by one electron at the expense of one electron serving a bonding or non-bonding role in ground state NF$_2$. The N—F bonds should be considerably weaker in the precursor than in the ground state. There is an interesting tie-in with thermodynamics here, because we expect the ground state N—F bond energy to be in the range from 60 to 70 kcal/mole (Chapter 6) and 250 nm radiation (used to effect the photolysis of NF$_2$) provides a photon energy of ~115 kcal/mole [incidentally, such high-energy radiation is consistent with formation of excited states like (iii)].

CHLOROPENTAFLUOROSULFUR

Like perfluorohydrazine, SF$_5$Cl is not a fluorinating agent in the sense of introducing fluorine atoms at susceptible functional groups, but rather it introduces the SF$_5$ group as a substituent. Also like N$_2$F$_4$, many of its reactions are effected *photochemically*[49] and the reactions are of the *radical* type. The initial step is

$$SF_5Cl \xrightarrow{h\nu} SF_5 + Cl$$

followed by attack of the SF_5 radical on the substrate or another radical. Examples that are fairly straightforward are

$$SF_5Cl + H_2 \rightarrow S_2F_{10} + 2HCl$$

$$+ O_2 \rightarrow SF_5-OO-SF_5 \text{ and } SF_5-O-SF_5$$

$$+ N_2F_4 \rightarrow SF_5NF_2$$

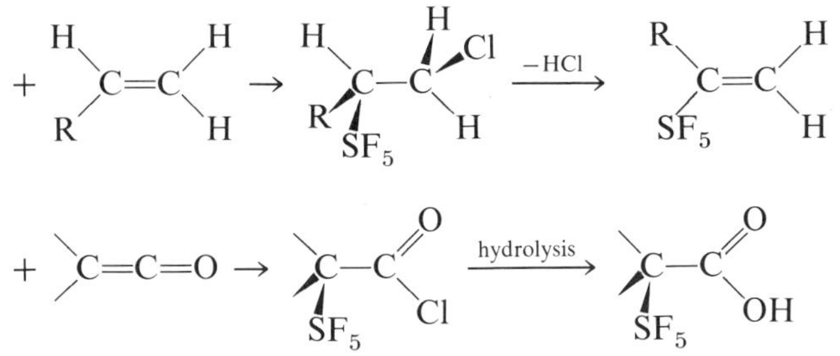

(8-69)

---

[49] Examples, with references, for the photochemical radical reactions of SF_5Cl are given by W. A. Sheppard and C. M. Sharts (pp. 234ff) in ref. 34.

STUDY QUESTIONS

5. Predict whether stepwise fluorination of C_2Cl_6 by SbF_5 proceeds in symmetric or unsymmetric steps; that is, whether the second product is CCl_2FCCl_2F or $CClF_2CCl_3$ (emphasize geminal steric and electronic effects in your considerations). Account for the fact that Cl_3CCHCl_2 is fluorinated unsymmetrically.

6. Contrast the products formed by reaction of the following with BrF_3 and with SF_4:

    a) Me_2CO
    b) $MeCN$
    c) $(C_5F_5)_2P(OH)$
    d) H_2NOH
    e) SO_2

7. Devise synthesis steps or predict reaction products for each of the following. For synthesis, start with elements and use any special reagent given in parentheses.

    a) $[NH_4][SO_3F]$ (NH_3)
    b) SF_5NF_2 (NH_3)
    c) SF_5CRCH_2 (any organic compound)
    d) N,N-difluoroacetamide (NH_3 and any oxyhydrocarbon)
    e) $(t\text{-}Bu_FSF_3)$ (any perfluorinated organic)
    f) $NF_3 + F_2 + SbF_5 \xrightarrow[85 \text{ atm}]{200°}$ (product a salt; remember that SbF_5 is a potent F^- acceptor)
    g) $N_2F_4 + Br_2 \xrightarrow{h\nu}$ (Br_2 is efficient at "trapping" F atoms)

8. Match reactants and products.

a) $OF_2 + SO_3 \xrightarrow{h\nu}$
b) $CO + F_2 \rightarrow ? \xrightarrow{CsF} ?$
c) $CO + F_2 \xrightarrow{AgF_2} ? \xrightarrow{F_2CO} ?$
d) $CF_3OF + SO_2 \rightarrow$
e) $OF_2 + 2OH^- \rightarrow$
f) $3SF_4 + PhAsO(OH)_2 \rightarrow$

1) F_2CO
2) $CF_3O_2CF_3$
3) CF_3OSO_2F
4) $Cs[CF_3O]$
5) $PhAsF_4$
6) $FSO_2(O_2F)$
7) CF_3OF
8) O_2

CHLORIDE SUBSTITUTION BY HYDROGEN AND ORGANICS

While it is true that some reagents are so active that they are unable to discriminate between E—F and E—Cl bonds, you have seen in Chapter 6 that fluorides tend to be less reactive than chlorides. The ease of preparation and ready availability of non-metal chlorides make many of them reagents of choice for preparation of a wide variety of non-metal compounds. You have already seen a few examples of chloride substitution in the fluorination reactions of certain non-metal chlorides. The other two most important substitution reactions are *hydrogenation* and *alkylation/arylation*.

Before we proceed to examine the application of chlorine substitutions, we need to note two important exceptions. First of all, the difficulty of handling and performing reactions with OCl_2 and NCl_3 means that these reagents are not useful precursors to H_2O, NH_3, ethers, and amines! The other important exception is that high oxidation state compounds of third row elements such as SR_4, SR_6, PR_5 (where R = H or alkyl group) have never been synthesized. The instability of SCl_4 and the non-existence of SCl_6 seem to preclude their use as precursors for SR_4 and SR_6. While PCl_5 is a stable compound of synthetic utility, it is likely that hydrogen and alkyl groups are generally too reducing to confer high stability on the higher oxidation states of phosphorus and sulfur [see the introductory comments in Chapter 6 (p. 271) on the beneficial effect of high terminal group electronegativity]. This is not to say that such compounds, particularly mixed forms containing hydrogen and halogens or organic groups and halogens, are not possible. In fact, you have encountered compounds such as $ArSF_3$ and PR_nF_{5-n} in previous chapters. Others will be discussed later in this chapter.

The replacement of chlorine by hydrogen or organic groups may be represented in simplest terms as an *"anion exchange"* reaction (compare the fluorination of chlorides) *using highly reactive forms of hydridic hydrogen or metal alkyls/aryls*.

$$ECl_n + \text{"MH"} \rightarrow EH_n + n\text{"MCl"}$$
$$+ \text{"MR"} \rightarrow ER_n + n\text{"MCl"}$$

The most convenient metal hydrides for this purpose are *lithium aluminum hydride* ($LiAlH_4$) and *sodium borohydride* ($NaBH_4$), but in certain cases binary hydrides like LiH and NaH are used. For the alkylation and arylation reactions, *Grignard reagents* (RMgX), *lithium alkyls* (LiR), and *organo mercury compounds* (HgR_2) are effective and commonly used. Many of these "anion exchange" reactions can be performed with pure reagents; but a more common technique, allowing better control of the reactions, is to use a weakly basic/acidic ethereal solvent [tetrahydrofuran

⟨tetrahydrofuran structure⟩; diethyl ether; monoglyme ($CH_3OC_2H_4OCH_3$); or diglyme ($CH_3OC_2H_4OC_2H_4OCH_3$)].

In the survey of methods used to produce the non-metalhydrides, you will encounter the technique of *acid hydrolysis of an anionic form of the non-metal* (*cf.* the use of non-metal anions in halogenation of non-metals); this method is included for its historical interest and also because it can be particularly convenient for the introduction of deuterium in place of normal hydrogen (as first mentioned in Chapter 7, p. 383).

An alternate method for synthesis of organo derivatives will also be mentioned for those hydrides of more hydridic nature (the hydrides of boron, aluminum, and silicon). This technique is one of *addition of a non-metal–hydrogen bond across an olefinic linkage* and is called *hydrometallation*:

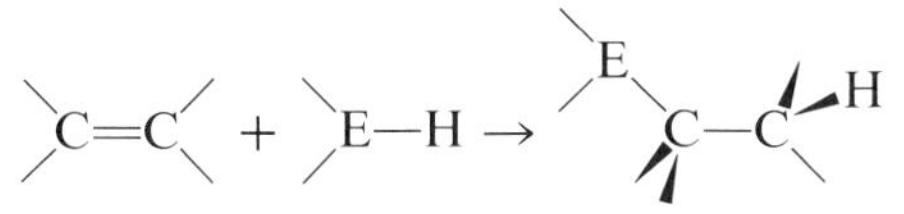

$$\text{C=C} + \text{E-H} \rightarrow \text{E-C-C-H}$$

To close this overview, be aware that the production of mixed organohydrides of the non-metals is easily achieved in most cases by tandem application of the methods used to introduce organic functionalities and hydrogen in place of chlorine. The reactions of non-metal chlorides with organometals can usually be controlled by stoichiometrically limiting the amount of organometal used, so as to replace chlorine only partially with the organic moiety. Reduction of the mixed organochloride compound with, say, $LiAlH_4$ produces the mixed organohydride.

$$ECl_n + \text{"MR"} \rightarrow ER_xCl_{n-x} \xrightarrow{\text{"MH"}} ER_xH_{n-x}$$

ORGANOMETAL REAGENTS‡

To add a measure of continuity to the ensuing descriptive chemistry, we should pause briefly to mention the source of the alkylating/arylating reagents used in some of the substitution reactions. As the preparation of these reagents and their properties can be more systematically approached in a general treatment of organometal synthesis, it is there (Chapter 16) that you will find a more complete treatment.

Grignard reagents are already familiar to you from organic chemistry. There you learned of their synthesis from Mg turnings and the appropriate organohalide

$$Mg + RX \xrightarrow{\text{"ether"}} RMgX$$

In an entirely analogous way, one obtains *organolithium* reagents

$$2Li + RX \xrightarrow{B} RLi + LiX\downarrow \quad (B = \text{benzene or hydrocarbon})$$

(As organolithium reagents tend to be reactive with ethers, such solvents are usually avoided.) An alternate route is afforded by use of a *diorganomercury* with Li

$$R_2Hg + 2Li \rightarrow 2RLi + Hg\downarrow$$

which is also useful for preparing *diorganomagnesium* compounds by a similar reaction

$$R_2Hg + \underset{\text{(excess)}}{Mg} \rightarrow R_2Mg + Hg$$

This reaction is usually conducted in the absence of a solvent, followed by extraction

---

‡A more comprehensive discussion of organometal synthesis is given in Chapters 16 and 17.

of R_2Mg with the desired solvent. To come full circle, a source of R_2Hg is the reaction

$$HgCl_2 + 2RMgCl \rightarrow R_2Hg + MgCl_2\downarrow$$

BORON AND ALUMINUM

Of all the metallic hydrides, the two MH_4^- anions from this group are the most important. For aluminum, the important compound is $LiAlH_4$ (dec. $\sim 150°$) prepared from LiH and $AlCl_3$:

(8-70)
$$8LiH + Al_2Cl_6 \xrightarrow{\text{"ether"}} 2LiAlH_4 + 6LiCl\downarrow$$

in an ether solvent (in which LiCl is insoluble). The tetrahydroaluminate anion is important both as a powerful reducing agent (this solid hydride becomes *incandescent* upon hydrolysis by a small amount of H_2O!) and as a precursor to the production of other aluminum hydride species. In Chapter 6 (p. 310) you learned of the unusual reaction (8-71):

(8-71)
$$2LiAlH_4 + H_2SO_4 \xrightarrow{\text{"ether"}} 2AlH_3 + 2H_2\uparrow + Li_2SO_4\downarrow$$

for the production of the simple binary hydride, alane. At the same time we noted the interesting reaction of lithium aluminum hydride with strong aprotic bases, like Me_3N, to form the hexahydroaluminate anion and the 2:1 adduct of AlH_3 with Me_3N, an intriguing reaction involving the conversion of a four-coordinate aluminum into five- and six-coordinate products.

(8-72)
$$3LiAlH_4 + 4Me_3N \rightarrow Li_3AlH_6 + 2AlH_3(NMe_3)_2$$

In addition to (8-72) as a route to the hexahydroaluminate anion, note that a more direct synthesis has been achieved by simple adduct formation between NaH and $NaAlH_4$:

(8-73)
$$2NaH + NaAlH_4 \rightarrow Na_3AlH_6$$

The tetrahydroborate anion is an important reducing agent, particularly since it is a less vigorous reducing agent than AlH_4^- and so presents the synthetic chemist with an alternate, easily handled reductant. Its production as a lithium salt from LiH and BF_3 is analogous to the formation of $LiAlH_4$:

(8-74)
$$4LiH + BF_3 \xrightarrow{\text{"ether"}} 3LiF\downarrow + LiBH_4$$

The use of the fluoride instead of boron trichloride is not required, for combination of almost any strongly hydridic reagent with either BF_3 or BCl_3 can be recommended. It is useful, but not really unexpected, to note that control of the reacting stoichiometries leads to production of the gas B_2H_6 (diborane).

(8-75)
$$6NaH + 8BF_3 \xrightarrow{\text{"ether"}} 6NaBF_4 + B_2H_6\uparrow$$

This binary hydride of boron is highly soluble in tetrahydrofuran, and such solutions are very convenient sources of borane for synthetic work. (As noted in Chapter 7, p. 400, ethers symmetrically cleave B_2H_6, so that such solutions contain the weak adduct $BH_3 \cdot THF$ rather than B_2H_6.)

Of some historical importance is the reaction of metal borides with acid to produce a variety of binary boron/hydrogen compounds. First discovered by Alfred

Stock[50] (the developer, for his work with the gaseous products of this reaction, of the modern vacuum line and associated techniques), the mixture of products can be thermally degraded to the simplest member of the boron hydride series, B_2H_6.‡

$$MgB_2 \xrightarrow{acid} B_nH_m \xrightarrow{\Delta} B_2H_6 \quad (8\text{-}76)$$

We will have much more to say about the components of Stock's mixture in Chapter 18 on polyhedral structures in inorganic chemistry.

SILICON

The known binary hydrides of silicon, Si_nH_{2n+2} ($n \leq 8$), are highly pyrophoric gases and liquids easily prepared from the corresponding chlorosilanes, Si_nCl_{2n+2}, by *reduction* with $LiAlH_4$:[51]

$$Si_nCl_{2n+2} + LiAlH_4 \rightarrow Si_nH_{2n+2} \quad (8\text{-}77)$$
$$\begin{bmatrix} n = 1: \text{b.p.} -112° \\ n = 4: \text{b.p.} 109° \end{bmatrix}$$

and by reduction, under more strenuous conditions, of SiO_2(!)

$$SiO_2 + LiAlH_4 \rightarrow Si_nH_{2n+2} \text{ mixture}$$

In analogy with his work with boron, Stock found it possible to synthesize many members of the silane series by *acid hydrolysis* of magnesium silicide

$$Mg_2Si \xrightarrow{acid} Si_nH_{2n+2} \text{ mixture} \quad (8\text{-}78)$$

The interesting feature of this reaction is that it points up rather emphatically that the silanes are not readily susceptible to acid hydrolysis; it suggests that the hydrogens of the silanes are not nearly as hydridic in nature as those of the boron analogs. (Based on the concepts of Chapter 7, however, be alert to the fact that silicon compounds are highly susceptible to attack by base, so the silanes are easily hydrolyzed in aqueous base.)

NITROGEN AND PHOSPHORUS

Trichloramine is, of course, not a very useful precursor for production of ammonia or amines. Probably the most convenient synthesis of ammonia or amines in the laboratory is from neutralization of NH_4^+ or $NH_xR_{4-x}^+$ salts with strong base.

$$NH_4X + OH^- \rightarrow NH_3 + X^- + H_2O \quad (8\text{-}79)$$

Use of ammonium salts does presuppose the prior formation of NH_3, and a more elementary reaction, the *acidification* of magnesium nitride, is convenient for this purpose. This reaction of the anionic form of a non-metallic element is useful[52] for the production of ND_3 in the laboratory (at this point you might wish to review study question 6-44 on the similar hydrolysis of Be_2C and Al_2C_3).

---

[50] A. Stock, "Hydrides of Boron and Silicon," Cornell University Press, Ithaca, 1933.
[51] A. E. Finholt, *et al., J. Amer. Chem. Soc.,* **69**, 2692 (1947).
[52] The alkali and alkaline earth metals all form nitrides from N_2 at elevated temperatures (Li does so at room temperature). For the preparation of ND_3 in this manner, see G. H. Payn, in "Mellors Comprehensive Treatise on Inorganic and Theoretical Chemistry," Vol. 8, Suppl. I, *Nitrogen,* Part I, Section V, Longmans, London, 1964.
‡See Chapter 18 for more on this subject.

(8-80)
$$Mg_3N_2 \xrightarrow{D_2O} ND_3 + Mg(OD)_2$$

The mechanistic details of the Raschig method for generation of hydrazine, the other important hydride of nitrogen, were discussed at some length in Chapter 7. Here we simply recall that the reaction of hypochlorite with ammonia proceeds through chloramine as an intermediate and that care must be taken to minimize the degradation of hydrazine by its counter-productive reaction with chloramine

(8-81)
$$2NH_3 + OCl^- \rightarrow N_2H_4 + Cl^- + H_2O \xrightarrow{2NH_2Cl} N_2 + 2NH_4Cl$$

Phosphorus trichloride is much more easily synthesized and handled than its nitrogen analog and can be used with a reducing agent such as LiAlH$_4$ to produce phosphine[53]

(8-82)
$$PCl_3 \xrightarrow{LiAlH_4} PH_3 \begin{bmatrix} \text{air stable} \\ \text{b.p. } -88° \end{bmatrix}$$

As mentioned in the introduction to this section on chloride replacements, PCl$_5$ reduction does not lead to PH$_5$ (an as yet unprepared compound) but rather produces PH$_3$. Mixed H and F compounds of P(V) are known, however; in fact, on p. 271 the P—F/Si—H exchange reaction was presented as an example of using bond energies to guide a synthesis. This strategy has been used to prepare (CF$_3$)PF$_3$H, (CF$_3$)$_2$PF$_2$H, and (CF$_3$)$_3$PH$_2$ from the corresponding (CF$_3$)$_n$PF$_{5-n}$. These materials, however, do tend to decompose, probably via HF elimination. Yet other examples, using SnH/PF "exchange" have been reported[54] for the syntheses of HPF$_4$ and H$_2$PF$_3$:

(8-83)
$$PF_5 + Me_3SnH \rightarrow HPF_4, H_2PF_3 + [Me_3Sn][PF_6]$$

These P(V) hydride species are also unstable and tend to eliminate HF slowly according to

(8-84)
$$HPF_4 \rightarrow HF + PF_3$$

Three pentaorgano derivatives of P(V) are known at present. The first prepared was Ph$_5$P (by Wittig[55] in 1948) according to the multi-step scheme:

(8-85)
$$Ph_3PO \xrightarrow[\text{2. HCl}]{\text{1. PhLi}} [Ph_4P]Cl \xrightarrow{HI} [Ph_4P]I \xrightarrow{PhLi} Ph_5P$$
$$[\text{dec. } 124°]$$

The starting material, Ph$_3$PO, can be obtained by Ph substitution of Cl in Cl$_3$PO. More recently, the mixed CF$_3$/CH$_3$ phosphoranes (CF$_3$)$_2$PMe$_3$ and (CF$_3$)$_3$PMe$_2$ were obtained[56] through reactions of Me$_4$Pb with (CF$_3$)$_2$PCl$_3$ and (CF$_3$)$_3$PCl$_2$. Both are low melting solids (44° and 64°, respectively) and have non-fluxional trigonal bipyramidal structures with Me groups equatorial. The importance of the electronegative (approximately the same as Cl) CF$_3$ groups in axial positions is readily apparent against a background of unsuccessful alkylation attempts. That is, R$_5$P species appear to be unstable with respect to reductive elimination of alkane and formation of an ylid:

---

[53] S. R. Gunn and L. G. Green, *J. Phys. Chem.*, **65**. 779 (1961).
[54] P. M. Treichel, *et al.*, *J. Amer. Chem. Soc.*, **89**, 2017 (1967).
[55] G. Wittig and M. Rieber, *Annalen*, **562**, 187 (1955); *Naturwiss.*, **35**, 345 (1948).
[56] K. I. The and R. G. Cavell, *J. Chem. Soc., Chem. Comm.*, 716 (1975).

$$\text{``}\underset{\underset{\text{Me}}{|}}{\overset{\overset{\text{Me}}{|}}{\text{Me}}}\text{P}-\underset{\underset{\text{H}}{|}}{\overset{\overset{\text{H}}{|}}{\text{C}}}\text{H}\text{''} \rightarrow \text{Me}_3\text{P}=\text{CH}_2 + \text{CH}_4$$

(a few pages later you will learn of the general route to ylids and their important use in organic synthesis). At this point you may wish to reread the section near the beginning of Chapter 6 on stabilities of PR_5 and SR_6 and to review the mechanistic comments in Chapter 7, p. 420, on the Arbusov reaction. The origin of the "stabilities" of $(CF_3)_2PMe_3$ and $(CF_3)_3PMe_2$ essentially boils down to a positive value for $(E_{PC\sigma} - E_{PC\pi})$. Apparently, $E_{PC\sigma}$ is greater with CF_3 axial than with CH_3 axial, by enough to render the reductive elimination of CF_3H thermodynamically, and probably kinetically, less easy. Finally, note that "reductive elimination" of C_6H_6 from Ph_5P would lead to a rather strained $Ph_3P\triangleleft\bigcirc$ structure and is not likely to occur.

Wittig did determine, however, that Ph_5P can catalyze the polymerization of styrene, and this strongly suggests Ph_5P acts as a source of Ph radicals.

It should come as no surprise that alkali and alkaline earth phosphides can be *hydrolyzed* to form phosphine.[57] This alternate route to PH_3 suffers from the simultaneous production of some contaminating diphosphine.

$$\text{Na}_3\text{P} + 3\text{H}_2\text{O} \rightarrow \text{PH}_3 + 3\text{NaOH} \; (+ \text{ some } \text{P}_2\text{H}_4, \text{H}_2) \tag{8-86}$$

This is important to remember since, unlike PH_3, P_2H_4 is a low boiling (52°) pyrophoric liquid and the spontaneous combustion of its vapor in air is sufficient to cause ignition of PH_3. The m.p. ($-99°$) of P_2H_4 does provide, however, for its ready separation from PH_3 in a vacuum fractionation train.

As you could anticipate by extrapolation from the weakly hydridic character of silanes, the phosphines exhibit even less hydridic character. PH_3 in *aqueous* solution is neither acidic nor basic but does show reducing character toward the heavier transition metal ions in aqueous systems, where formation of stable phosphorus oxyacids thermodynamically assists the oxidation of PH_3.

OXYGEN AND SULFUR

It is highly unlikely that you would be faced with preparing H_2O in the laboratory. At the risk of sounding ridiculous, reduction of OCl_2 would be a most undesirable way of making H_2O. As a reminder of the general scheme exemplified by (8-86), (8-80), (8-78), and (8-76), H_2O (and, more importantly, D_2O) can be synthesized by treatment of a simple metal oxide with (H,D)X (but even D_2O is readily available from commercial sources). In a similar vein, the strongly basic character of ionic metal oxides is of further use to the practicing chemist for removal of trace amounts of H_2O from organic solvents:

$$\text{BaO} \xrightarrow{\text{H}_2\text{O}} \text{Ba(OH)}_2$$

The synthesis of hydrogen peroxide can be approached in either of two ways. As sodium and barium readily combust in air to form appreciable amounts of the corresponding peroxide salts, their *hydrolysis* affords a useful synthesis of H_2O_2. The separation of H_2O_2 from the aqueous reaction mixtures requires, in both cases, distillation with utmost care to avoid a disastrous explosion.

---

[57] M. Baudler, *et al.*, *Z. anorg. allgem. Chem.*, **353**, 122 (1967).

480 SYNTHESIS OF IMPORTANT CLASSES OF NON-METAL COMPOUNDS

(8-87)
$$Na_2O_2 + 2H_2O \xrightarrow{H_2O} H_2O_2 + 2NaOH$$

$$BaO_2 + H_2SO_4 \xrightarrow{H_2O} H_2O_2 + BaSO_4\downarrow$$

The use of Na_2O_2 requires greater care and technique than use of BaO_2 because the production of OH^- in the Na_2O_2 hydrolysis means that the pH dependent decomposition of H_2O_2 (refer to Chapter 7) proceeds more rapidly than under the acid conditions of BaO_2 hydrolysis. The convenience of being able to separate the insoluble $BaSO_4$ by filtration prior to distillation should not go unnoticed.[58]

The second widely known technique for synthesizing hydrogen peroxide is related to those of (8-87), but with the distinction that a *non-metal* peroxide is hydrolyzed. Peroxydisulfate ion is electrolytically generated[59] from aqueous $KHSO_4$ at a voltage of -2.2 v at a platinum electrode (see Figure 8-4). (The O_2 overvoltage prevents oxidation of H_2O.)

$$K^+ + HSO_4^- \xrightarrow{H_2O} \tfrac{1}{2} K_2S_2O_{8(c)} + H^+ + e^-$$

The potassium salt is preferred because of its insolubility at the temperature (maintained at about 5°C) of the electrolysis. This defers the formation of $H_2S_2O_8$ and its hydrolysis to form Caro's acid (H_2SO_5), which would subsequently hydrolyze to form the desired hydrogen peroxide.

(8-88)
$$H_2S_2O_8 \xrightarrow{H_2O} H_2SO_5 + H_2SO_4$$
$$\xrightarrow{H_2O} H_2O_2 + H_2SO_4$$

Postponement of this hydrolysis is necessary to avoid leaving the hydrogen peroxide in contact with Caro's acid, for the two react to produce O_2 (the mechanism was discussed in Chapter 7, p. 424).

$$H_2O_2 + H_2SO_5 \to O_2 + H_2O + H_2SO_4$$

Thus, it is preferred that the isolated salt $K_2S_2O_8$ be hydrolyzed with dilute H_2SO_4 and the hydrogen peroxide quickly removed by *vacuum* distillation (the boiling point of H_2O_2 is 150°C at 1 atmosphere pressure).

The hydrosulfanes H_2S_n ($n = 2$ to 5) are probably best prepared, in analogy with hydrogen peroxide, from acid *hydrolysis* of alkali metal polysulfides followed by vacuum distillation of the "crude oil" to obtain individual components[60]

(8-89)
$$M_2S_n \xrightarrow[H_2O]{acid} H_2S_n$$

(the vacuum distillation technique permits fractionation at lower temperatures to minimize "cracking"). An alternate synthesis makes use of the fact that the chlorosulfanes readily react with H_2S and H_2S_2 to produce higher hydrosulfanes, with chain length increased by two and four, respectively:

(8-90)
$$mH_2S_x + nS_yCl_2 \to 2nHCl + (m-n)H_2S_z \qquad (m > n)$$

---

[58] This reaction is the one used by L. J. Thenard, *Ann. chim. phys.*, **8**, 306 (1818) when he reported the discovery of H_2O_2. A more general reference to hydrogen peroxide chemistry is W. L. Schumb, *et al.*, "Hydrogen Peroxide," Reinhold, New York, 1955.

[59] W. L. Jolly, "The Synthesis and Characterization of Inorganic Compounds," Prentice-Hall, Englewood Cliffs, 1970.

[60] K. W. C. Burton and P. Machmer, in "Inorganic Sulfur Chemistry," G. Nickless (Ed.), Elsevier, New York, 1968.

The "crude oil" product must then be fractionated to isolate individual products. The metal polysulfides used in the hydrolysis reactions are prepared by direct combination of the metal and sulfur in liquid ammonia solvent.[61] In this context, perhaps you are already aware of the reactivity of mercury and sulfur that simplifies the cleanup of mercury spills in the lab.

Finally, we should mention a despicable wartime use of S_2Cl_2. Ethylene, in a complex reaction, forms "mustard gas" when combined with disulfurdichloride by S—Cl addition to C=C.[62]

$$2C_2H_4 + S_2Cl_2 \rightarrow \frac{1}{8}S_8 + S(C_2H_4Cl)_2 \tag{8-91}$$

Not really a gas, di(2-chloroethyl)sulfide (m.p. 13°, b.p. 215°) is a dense liquid that could be sprayed into the atmosphere; staying close to the ground, the mist would be driven by gentle winds in a desired direction. This was the use to which mustard gas was put at the front lines in World War I. The toxic action of $(ClCH_2CH_2)_2S$ seems to derive from its oxidation, in the living cell, to $(ClCH_2CH_2)SO_2$, followed by HCl elimination to produce the divinylsulfide $(CH_2CH)_2S$. This latter molecule enters into disruptive reactions with the protein structures of the cell.

HALOGENS[63]

The hydrogen halides, in principle, could be synthesized by the technique of *hydrolysis* of the element in the anionic form. Thus, acidification of metal fluorides and chlorides or hydrolysis of non-metal fluorides and chlorides leads directly to the hydrogen halide.

$$MF_n \xrightarrow{acid} HF$$

$$SOCl_2 \xrightarrow{H_2O} HCl + SO_2 \tag{8-92}$$

To isolate the pure HX gas, you must avoid H_2O as a solvent; sulfuric acid would seem a good choice, for not only does it provide a source of hydrogen ion but it also acts as a desiccant. This technique works well for HF and HCl, but H_2SO_4 oxidizes HBr and HI to Br_2 and I_2, respectively. A better method for synthesis of HBr is careful hydrolysis of a non-metal bromide such as PBr_3, keeping the latter in excess.

$$PBr_3 + 3H_2O \rightarrow H_3PO_3 + 3HBr \tag{8-93}$$

The best methods for production of HI (interestingly enough, gaseous HI is thermodynamically unstable relative to the elements) involve reduction of elemental iodine

$$I_2 + H_2S \xrightarrow{H_2O} 2HI + \frac{1}{8}S_{8(c)} \quad \text{(for aqueous solutions of HI)} \tag{8-94}$$

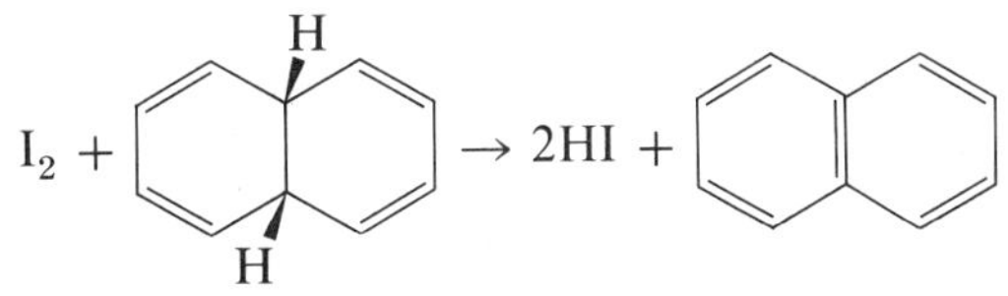

$$I_2 + \text{(dihydronaphthalene)} \rightarrow 2HI + \text{(naphthalene)}$$

---

[61] G. Schwarzenbach and A. Fischer, *Helv. Chim. Acta*, **43**, 1365 (1960); M. Schmidt, in "Inorganic Polymers," F. G. A. Stone and W. A. G. Graham (Eds.), Academic Press, New York, 1962.
[62] L. A. Wiles and Z. S. Ariyan, *Chem. Ind.*, **1962**, 2102.
[63] "Inorganic Syntheses," Vols. 1, 3, and 7, McGraw-Hill, New York, 1939-63.

It is instructive to pause at this point, return to the "orbital following" section of Chapter 7, and inquire into the possibility of direct combination of H_2 and X_2 as a means of synthesis of HX. For a smooth addition of H_2 to X_2 we are likely to suppose a symmetrical (C_{2v}) approach geometry like **7**.

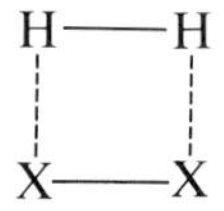

$$\begin{array}{c} H \text{———} H \\ | \quad\quad\quad | \\ | \quad\quad\quad | \\ X \text{———} X \end{array}$$

7

The HOMO's and LUMO's most suited to follow such an interaction are the σ and σ^* of H_2 and the π^* and σ^* of X_2:

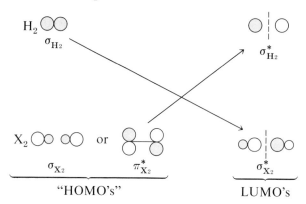

The X_2 LUMO is not of the proper symmetry for this approach geometry to receive electron density from σ_{H_2}. The only "sticky" interaction possible between H_2 and X_2 is $\pi^*_{X_2} \to \sigma^*_{H_2}$; while such charge flow weakens the H—H bond, it unfortunately also strengthens the X_2 bond, and you conclude that **7** is not a reasonable transition state structure. For the reaction to proceed, it requires electrical or photochemical initiation through X radicals as in (8-95),

(8-95)
$$\begin{array}{c} X_2 \to 2X\cdot \\ X\cdot + H_2 \to HX + H\cdot \\ H\cdot + X_2 \to HX + X\cdot \end{array}$$

or their production naturally as in (8-96):

(8-96)
$$H\text{—}H\text{------}X\text{—}X \to H\cdot + HX + X\cdot$$

Such a $C_{\infty v}$ (or C_s) pathway does provide for ($\sigma_{H_2} \to \sigma^*_{X_2}$) and ($\sigma_{X_2} \to \sigma^*_{H_2}$) charge flows to facilitate cleavage of H—H and X—X and formation of H—X. We will leave it to you to explore the similarity of orbital following in the $I_2 + H_2S$ and I_2 + 9,10-dihydronaphthalene reactions (8-94) to the double hydrogen transfer reaction of dihydrodiazene (Chapter 7, p. 377), where the X_2 LUMO and HOMO are as above and the H_2S HOMO‡ and LUMO, for example, are:

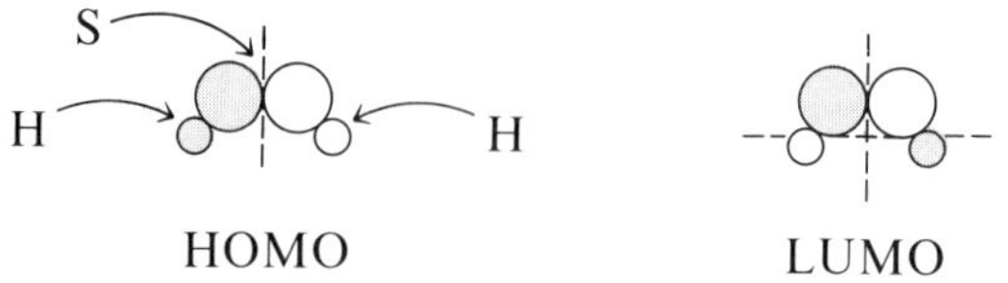

HOMO   LUMO

---

‡HOMO is used here to mean the highest energy occupied mo exhibiting significant S—H bonding character. As noted in Chapter 7, the HOMO's and LUMO's of an orbital following model must involve the bonds to be broken. The actual HOMO of H_2S is the out-of-plane p ao lone pair, which would not directly participate in dihydrogen transfer.

STUDY QUESTIONS

9. Give steps for syntheses of the following from elements and any special reagents identified in parentheses.

   a) $Al(BH_4)_3$
   b) $Al_2Me_3Cl_3$ (any alkyl halide)
   c) $AlH_3(NMe_3)_2$ (NMe_3)
   d) Me_2SiH_2
   e) $[PCl_3R][AlCl_4]$ (any alkyl chloride; the only phosphine cation to appear does so as the product)
   f) Me_2PH (any methyl halide)
   g) $LiNH_2$ (NH_3)
   h) $Et_2PO(OH)$
   i) D_2O
   j) $[R_3PNH_2]Cl$ (NH_3)

10. Match reactants with products.

    a) $[MeNBCl]_3 + NaBH_4 \rightarrow$ 1) Al_2Et_6
    b) $[MeNBCl]_3 + PhMgCl \rightarrow$ 2) n-BuLi
    c) $2Al + 3Et_2Hg \rightarrow$ 3) $Si(C_2H_5)_4$
    d) $SiCl_4 + 2ZnEt_2 \rightarrow$ 4) $[(CH_3)_3Si]_3N$
    e) $Me_3SiCl + Na[N(SiMe_3)_2] \rightarrow$ 5) $[MeNBH]_3$
    f) $ClNF_2 + R_2Hg \rightarrow$ 6) $EtCl + EtNF_2$
    g) $Li + n$-$BuCl \rightarrow$ 7) $[MeNBPh]_3$

11. What analogy can you find between the instabilities of compounds like "PMe_5" (p. 479), "$Me_3Al \cdot NH_3$" (p. 493), "$P(NH_2)_5$", and "$P(OH)_5$"? Review the discussion on p. 273.

HYDROMETALATION AND OTHERS

As anticipated in the introduction to this section, organoboranes, alanes, and silanes (under proper conditions) can be synthesized by the addition of E—H to olefins. These reactions can be utilized, in the cases of boron and aluminum, to introduce one, two, or three alkyl groups in place of hydrogen about the Group III atoms:

hydroboration:[64]

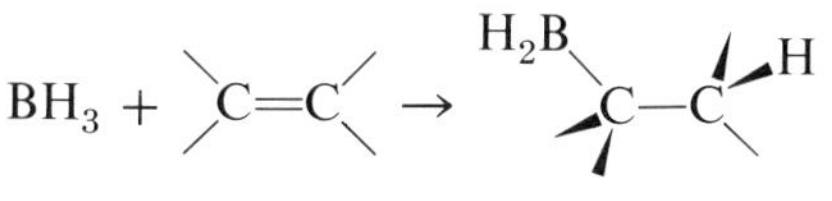

$$BH_3 + \;\;C=C\;\; \rightarrow \;\; H_2B-C-C-H \tag{8-97}$$

hydroalumination:

$$LiAlH_4 + 4RCH=CH_2 \rightarrow Li[Al(CH_2CH_2R)_4] \tag{8-98}$$

The important aspects of these reactions were discussed in Chapter 7, so at this point we simply recall the important conclusions:

(1) In hydroboration it is BH_3, not B_2H_6, that is reactive, a fact that dictates the use of weakly basic ethers as solvent.

---

[64] H. C. Brown's recounting ("Boranes in Organic Chemistry," Cornell University Press, Ithaca, 1972) of his and his students' experiences in this area is not only informative but highly entertaining reading. See also K. Wade, "Electron Deficient Compounds," Nelson, London, 1971.

484 □ SYNTHESIS OF IMPORTANT CLASSES OF NON-METAL COMPOUNDS

(2) The addition is *anti*-Markownikoff, as in

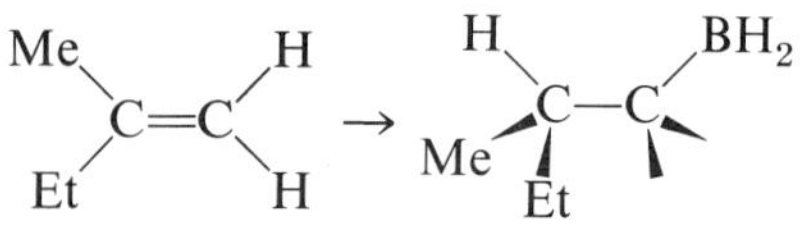

$$\underset{Et}{\overset{Me}{>}}C=C\underset{H}{\overset{H}{<}} \rightarrow \underset{\underset{Et}{Me}}{\overset{H}{>}}C-C\underset{}{\overset{BH_2}{<}}$$
8

(3) Hydrolysis of the addition product **8** with aqueous mineral acid introduces H in place of BH_2 (**9a**)

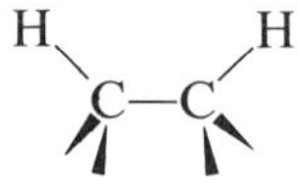

$$\overset{H}{>}C-C\overset{H}{<} \qquad \overset{H}{>}C-C\overset{OH}{<}$$
9a 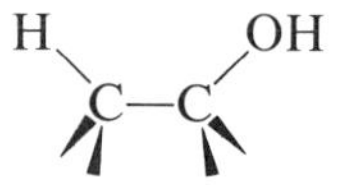 **9b**

whereas the more facile H_2O_2-assisted hydrolysis produces the alcohol **9b** (both techniques received comment in Chapter 7, p. 392.)

(4) Addition is *cis*-, as exemplified by

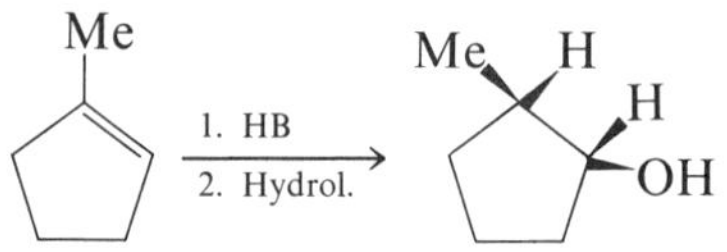

[cyclopentene with Me substituent] $\xrightarrow[\text{2. Hydrol.}]{\text{1. HB}}$ [cyclopentane with Me, H, H, OH substituents]

As we mentioned earlier in discussing the synthesis of silane, the Si—H bond appears to be considerably less hydridic than B—H and Al—H bonds. Accordingly, triorganosilanes can be caused to add to olefin functions, *if* a so-called "coordinatively unsaturated" transition metal complex is used as a catalyst. In Chapter 5, p. 206, we discussed the addition of O_2 to one such complex, Vaska's compound $Ir(CO)Cl(PPh_3)_2$. There the O—O multiple bond suffered a bond order reduction to one, whereas reduction of the Si—H bond order incurs its cleavage with formation of M—H and M—Si bonds:

(8-99)
$$ML_n \xrightarrow{H-SiR_3} L_nM\underset{SiR_3}{\overset{H}{<}} \xrightarrow{\text{[olefin]}} ML_n + \overset{H}{>}\!\!\!\!\!\!\overset{SiR_3}{<}$$

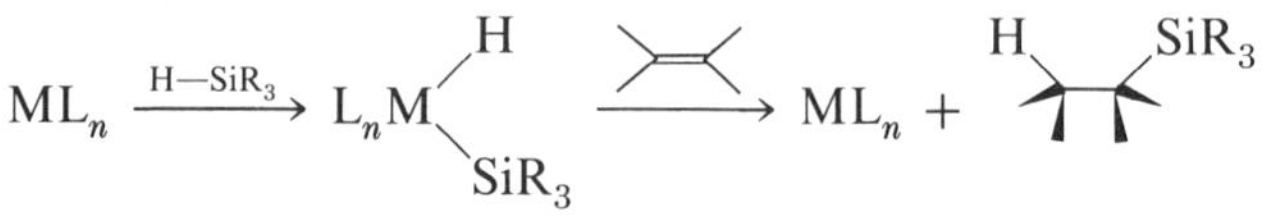

You will note that "ML_n" behaves in a general way like an unsaturated group, which easily adds and transfers the H and SiR_3 fragments. Clearly, the effect of ML_n is a kinetic one; therefore, you correctly surmise that orbital following is much more facile for the initial addition of H—Si to ML_n *and* for transfer of H and SiR_3 to the olefin than is that for direct addition of H—Si to doubly bound carbons. In a striking way, you should see an analogy with the H_2/X_2 reaction based on the facts that (i) the Si—H bond is not greatly polar, (ii) the symmetric transition state **7** (with Si—H replacing H—H) does not lead to the required bond-making and bond-breaking, and (iii) reactions (8-94) occur readily.

No similar techniques have been reported yet for aiding the addition of P—H[65] and S—H bonds to olefins; presumably the unaided addition is even less likely than for SiH because these bonds are not sufficiently polar or have the wrong polarity to effect the addition. As a unifying point for distinguishing Si—H, P—H, and S—H from BH_3, note that the former do not present the low-energy LUMO at E of EH that BH_3 does.

---

[65] See, however, study question 13.

While on the subject of organosilanes, we should mention the interesting reduction of organodisilanes by sodium to form triorganosilane anions[66]:

$$Si_2R_6 \xrightarrow{Na} NaSiR_3 \xrightarrow{R'-Cl} R_3SiR' + NaCl \qquad (8-100)$$

These silicon anions are good nucleophiles and readily displace chloride from organochlorides, thereby affording a useful means of producing unsymmetric organosilanes.

Entirely analogous reactions for the isoelectronic organosulfides and organophosphines can be cited[67]:

$$R_{2,3}E + R-I \rightarrow [R_{3,4}E]I \qquad (8-101)$$

The reaction of organophosphines to produce phosphonium salts is very important in the case of the triorganomethyl phosphonium cation because of its reaction with n-butyl lithium to form **10**, the phosphorus *ylid* (alkylidene phosphorane).[68]

$$R_3P-CH_3^+ \xrightarrow{BuLi} C_4H_{10} + \left[R_3P=CH_2 \leftrightarrow R_3\overset{+}{P}-\overset{-}{\ddot{C}}H_2\right] \qquad (8-102)$$
$$\phantom{R_3P-CH_3^+ \xrightarrow{BuLi} C_4H_{10} + \quad}\mathbf{10}$$

The polar nature of the phosphorus-methylene bond renders the ylid very reactive, and its use as an *in situ* reagent for replacing the carbonyl oxygen of a ketone with methylene is very important. We are reminded here of the affinity of phosphorus for oxygen in forming the phosphoryl linkage, P=O.

$$R_2C=O + Ph_3P=CH_2 \rightarrow Ph_3PO + R_2CCH_2 \qquad (8-103)$$

Recalling the discussion of phostonate (five-membered ring containing P) reactions in the last chapter (**32** on p. 436), it is quite possible that a synergic donor/acceptor interaction ($\pi_{CO} \rightarrow \pi_{PC}^*, \pi_{PC} \rightarrow \pi_{CO}^*$)

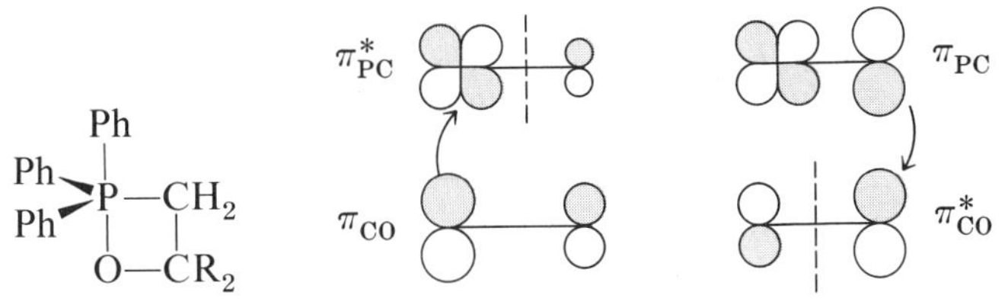

leads to a highly strained four-membered-ring intermediate, which then undergoes ring cleavage to form the products. You might be tempted to draw an analogy here with the addition of two olefins to form a cyclobutane. Such reactions have *high* activation energies because of orbital following restrictions (like the $H_2 + X_2$ reaction discussed in the preceding section). The extreme polarities of the $>C=O$ and $\geqq P=CH_2$ orbitals, however, considerably facilitate the ylid/carbonyl reaction, as

---

[66] A. G. MacDiarmid, *Adv. Inorg. Chem. Radiochem.*, **3**, 207 (1961).

[67] D. H. Chadwick and R. S. Watt, in "Phosphorus and its Compounds," Vol. 2, J. R. van Wazer (Ed.), Interscience, New York, 1961.

[68] H. Schmidbauer and W. Tronich, *Chem. Ber.*, **101**, 595, 604 (1968); A. Schmidt, *ibid.*, **101**, 4015 (1968); J. C. J. Bart, *J. Chem. Soc.* (*B*), 350 (1969). Note that n-Bu$^-$ should initially attack the electrophilic phosphorus, followed by reductive elimination of C_4H_{10} (see p. 479).

486 ☐ SYNTHESIS OF IMPORTANT CLASSES OF NON-METAL COMPOUNDS

discussed more generally in the section on "orbital following" in Chapter 7. The other interesting feature of the methylidene phosphorane is that it is a *formally* pentavalent phosphorus compound containing all organic substituents.

CATENATION BY COUPLING

The mixed organochlorides of the non-metals make useful reagents for preparing various organo compounds containing other substituents. Among the more interesting utilizations of these reagents are the coupling reactions, which have been extensively studied, particularly for silicon and phosphorus. Some examples will serve to illustrate the versatility of these reagents. Synthesis of asymmetric disilanes, presumably via a "Grignard-like" intermediate, can be achieved[69] by reacting different triorganosilylchlorides in the presence of Mg:

(8-104) $$Ph_3SiCl + Me_3SiCl \xrightarrow{Mg} Ph_3SiSiMe_3$$

Both silicon and phosphorus monochlorides undergo a Wurtz-type coupling on reduction with Na [compare (8-100)]:

(8-105) $$2R_{3,2}ECl \xrightarrow{Na} R_{6,4}E_2 + 2NaCl$$

An extension of this reaction, using the dichloro compounds, has made possible the synthesis of the penta (**11a**), hexa, and heptasilyl compounds[70]:

(8-106) $$Me_2SiCl_2 + 2Na \rightarrow \frac{1}{n}\{Me_2Si\}_n + 2NaCl$$

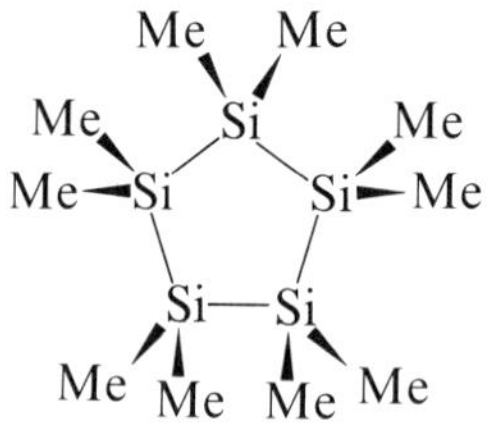

11a 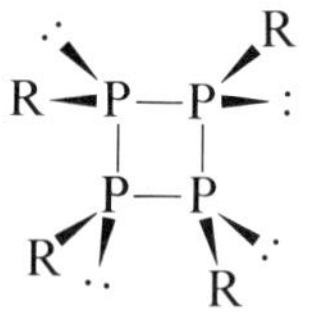   **11b**

Mercury, too, has been found to be an effective reductant for this type of reaction, particularly for the synthesis of cyclic polyphosphines[71] (**11b**).

(8-107) $$RPI_2 \xrightarrow{Hg} \frac{1}{n}\{RP\}_n + HgI_2$$

In yet another reaction type, familiar to you from your study of organic chemistry, mixtures of monochloro and monohydrido phosphines may be caused to eliminate HCl to form the corresponding asymmetric diphosphine[72]

(8-108) $$R_2PCl + R'_2PH \xrightarrow{\Delta} R_2PPR'_2 + HCl$$

---

[69] N. Duffant, et al., *Compt. Rend.*, **268C**, 967 (1969).
[70] F. S. Kipping and J. E. Sands, *J. Chem. Soc.*, 830 (1921); E. Carberry and R. W. West, *J. Amer. Chem. Soc.*, **91**, 5440 (1969).
[71] W. Mahler and A. B. Burg, *J. Amer. Chem. Soc.*, **80**, 6161 (1958).
[72] E. Fluck, *Compounds Containing P—P Bonds*, in "Preparative Inorganic Reactions," Vol. 5, W. L. Jolly (Ed.), Interscience, New York, 1968.

while use of dichloro and dihydrido phosphines leads to the expected unsymmetric polycyclic analogs of **11b**:

$$RPCl_2 + R'PH_2 \rightarrow \frac{1}{n}\{RPPR'\}_n + 2HCl \qquad (8\text{-}109)$$

An interesting extension of this type of reaction has led to formation of a chain triphosphine:

$$2R_2PCl + R'PH_2 \xrightarrow[\text{base}]{\text{organic}} R_2P\underset{\underset{R'}{|}}{-}P-PR_2 \qquad (8\text{-}110)$$

You should note the obvious resemblance of (8–108) through (8–110) to the NH_2Cl/NH_3 and NH_2Cl/N_2H_4 reactions discussed at length in Chapter 7, and also to the H_2S_n/SCl_2 reactions of the preceding section, all of which establishes a certain degree of generality for the coupling technique of generating catenated species.

STUDY QUESTIONS

12. Give steps for the syntheses of the following from the elements and any special reagents given in parentheses.

    a) $Na[B(OCH_2R)_4]$ (any carbonyl compound)

    b) $1,2\text{-}C_2H_4(BCl_2)_2$ and $1,2\text{-}C_2H_4(BMe_2)_2$ (any hydrocarbon and methyl halide)

    c) $Li[Al(C_2H_4R)_4]$ (any hydrocarbon)

    d) $Cl_3SiC_2H_4R$ (do *not* use an MR alkylating reagent, but any hydrocarbon)

    e) $Me_3Si_2(n\text{-}Pr)_3$

    f) 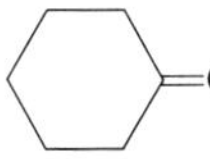 ⬡=CH_2 (use any ketone and organohalide)

    g) $[Me_3S]I$

13. Match reactants and products.

    a) $NaBH_4 + \frac{1}{2}I_2 \rightarrow$
    b) $\frac{1}{2}B_2H_6 + 1,5\text{-}C_8H_{12} \rightarrow$ not isolated $\xrightarrow{H_2O_2}$
    c) $B_2H_6 + 2C_4H_6 \rightarrow$ polymer $\xrightarrow{\text{distill}}$
    d) $BMe_3 + O_2 \xrightarrow[\text{flow system}]{N_2}$
    e) $\frac{1}{2}Al_2Me_6 + C_2H_4 \rightarrow$
    f) $KSiH_3 + MeCl \rightarrow$
    g) $2PF_2I + Hg \rightarrow$
    h) $PH_3 + HCl + 4R_2CO \rightarrow$
    i) $Me_3SiNR_2 + Et_3SiH \rightarrow$

    1) Me_2BO_2Me
    2) $Al_2Me_4(n\text{-}Pr)_2$
    3) P_2F_4
    4) $MeSiH_3$
    5) $\frac{1}{2}B_2H_6$
    6) $Me_3Si_2Et_3$
    7)
    8) $[P(CH_2OH)_4]Cl$
    9) 1,5- and $1,6\text{-}C_8H_{14}(OH)_2$

SOLVOLYSIS REACTIONS[73]

The so-called solvolysis reaction is a general reaction for non-metal halides and organohydrides (at least the hydridic ones). This reaction is analogous to the coupling reactions noted for phosphorus in the preceding section, but it leads to replacement of chloride or hydrogen about the non-metal atom by oxygen or nitrogen. *The reactions are simple in form; elimination of HCl or H_2 by reaction of E—Cl or E—H* (in those cases where the hydrogen is sufficiently hydridic) *with H—O or H—N leads to formation of the corresponding oxyacid, its ester, or the amide of the parent non-metal compound.* In those instances of ECl_2 functions reacting with H_2O or NH_3 or RNH_2, a common process is the elimination of two or more moles of HCl with formation of multiple bonds between E and O or N. The perspective from Chapter 6 alerts you to the possibility that such multiple bonds may be less stable than an equivalent number of single bonds and, in fact, one finds in the laboratory that *polymerization to form chains, rings, and polyhedra is a common feature of non-metal halide solvolysis reactions.* We have already discussed some of these in Chapters 6 and 7; at this time we will probe the subject more deeply.

The reaction types to be discussed are given by the following general equations:

$$R_xECl_n + nHA \rightarrow R_xEA_n + nHCl$$

$$+ \frac{n}{2}H_2A \rightarrow \underbrace{\frac{1}{3}(R_xEA)_3, \frac{1}{4}(R_xEA)_4,}_{\text{cyclic or polyhedral}} \underbrace{(R_xEA)_\infty}_{\text{1-, 2-, and 3-dimensional chains}} + nHCl$$

BORON AND ALUMINUM

The boron halides react with alcohols to form the simple esters of boric acid. Particularly interesting among this class of reactions are those of the monoorgano and diorgano chlorides and hydrides with water:

(8–111)
$$RBCl_2 \xrightarrow{H_2O} \underset{12}{RB(OH)_2} \xrightarrow{\Delta} \underset{14}{\frac{1}{3}(RBO)_3} + H_2O$$

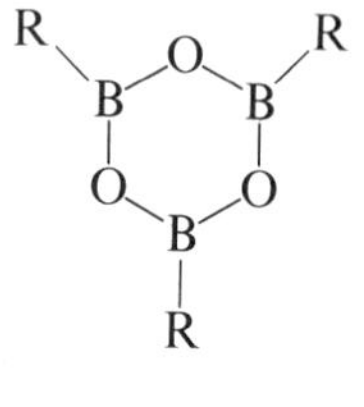

$\begin{bmatrix} R = Me \\ b.p. \quad 79° \\ m.p. \quad -68° \end{bmatrix}$

(8–112)
$$R_4B_2H_2 \xrightarrow{H_2O} \underset{13}{2R_2BOH} \xrightarrow{\Delta} \underset{15}{R_2BOBR_2} + H_2O$$

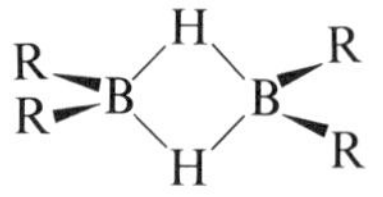

---

[73] As many of the non-metal solvolyses yield polymeric products (chains, rings, cages), this has been a generally active field. We recommend D. A. Armitage, "Inorganic Rings and Cages," Edward Arnold, London, 1972 for a more general survey of this field than we can give here. Armitage's survey includes not only more compound classes but also preparative schemes other than solvolysis.

The products, boronic **(12)** and borinic **(13)** acids, are acids in the Brönsted sense and have the interesting property (like the parent H_3BO_3) that they can be thermally dehydrated to form the anhydrides[74] **14** and **15**. The anhydrides of boronic acids are cyclic trimers called *boroxines,* with a distinct similarity in structure (but not chemistry) to benzene.

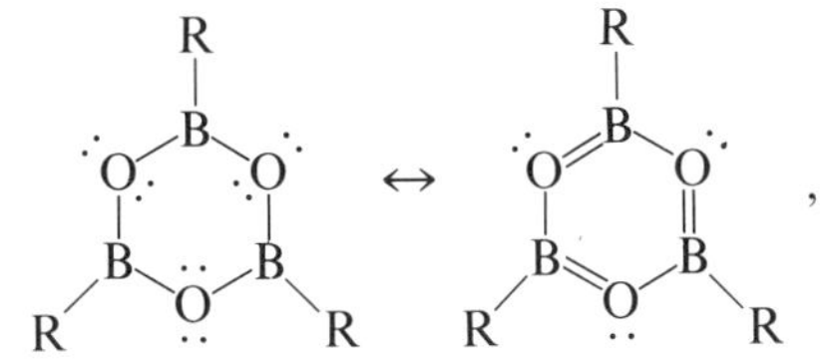

[As an aside, note the nomenclature system that designates $(RBO)_3$ as bor*o*xines and $(RBNR')$ as bor*a*zines.]

You know from the extensive discussion in Chapter 7 of their mechanism of formation that analogous reactions are known for amines, although it is often the case that more forcing conditions (presence of a tertiary amine base or elevated temperature) are required to effect HCl or H_2 elimination. The nature of the products (monomers, dimers, trimers) is highly dependent upon the steric bulk of groups bound to the boron and to the amine nitrogen. Both the rate of polymerization and the thermodynamic stability of the condensation reaction product depend on steric and electronic factors associated with the substituents. To review those discoveries:

(i) You can assume that tertiary amines lead to formation of 1:1 adducts called *amineboranes*.

$$BX_3 + NR_3 \rightarrow X_3B \cdot NR_3 \tag{8-113}$$

(ii) Secondary amines may lead to either monomeric or dimeric products, depending on the nature of the boron and nitrogen substituents through their influence on the stability of the $(B=N \leftrightarrow B-N)$ linkage in the monomer and on steric interactions between "eclipsed" X and R groups in the nearly planar dimer **(16)**.

$$BX_3 + NHR_2 \rightarrow BX_3 \cdot NHR_2 \tag{8-114}$$

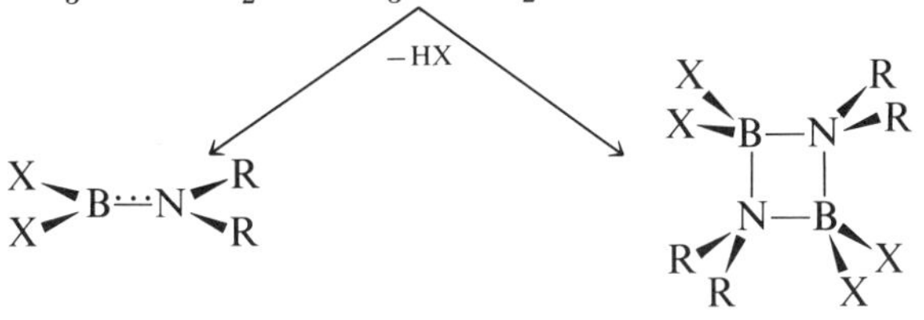

16

Trimerization of the *aminoboranes* (X_2BNR_2) appears uncommon, for the steric interaction between R and X may become too severe when $X \neq H$. When $X = H$, trimerization to the cyclohexane analog is apparently no problem.

(iii) Primary amines can lead to trimeric structures (*borazines*) analogous to the boroxines.

$$BX_3 + NH_2R \rightarrow X_3B \cdot NH_2R \xrightarrow{\Delta} (XBNR)_3 \tag{8-115}$$

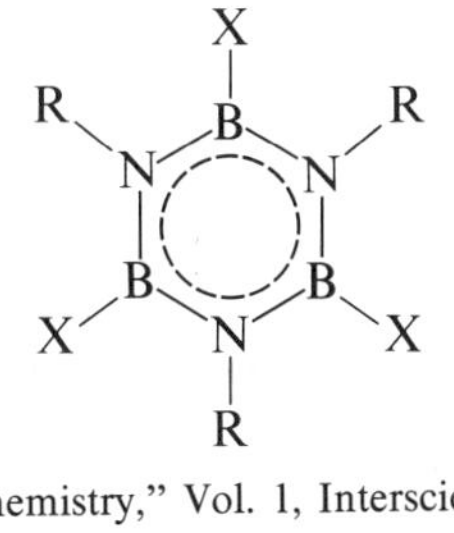

---

[74] H. Steinberg, "Organoboron Chemistry," Vol. 1, Interscience, New York, 1964.

490 ◻ SYNTHESIS OF IMPORTANT CLASSES OF NON-METAL COMPOUNDS

(iv) That stoichiometry can have a great deal to do with the products formed in unhindered cases is illustrated by the complete solvolysis of BCl_3 by liquid ammonia

(8-116)
$$BCl_3 + 3NH_3 \xrightarrow[NH_3]{} B(NH_2)_3 + 3[NH_4]Cl$$

and the reaction of BCl_3 with excess aniline

(8-117)
$$BCl_3 + \underset{\text{excess}}{PhNH_2} \rightarrow B(NHPh)_3 + 3[PhNH_3]Cl$$

leading not to dimeric or trimeric products, but to the fully substituted amides.

An interesting point of demarcation of the reaction pathways of BCl_3 and BH_3 (actually B_2H_6) arises in their reactions with liquid ammonia and derives from the difference (monomer vs. dimer) in structure of these acids. As noted above, BCl_3 leads to the fully substituted amide whereas, as discussed in Chapter 7, diborane leads to borazine via unsymmetric cleavage of diborane to produce initially the salt $[H_2B(NH_3)_2][BH_4]$.

As you might have anticipated by now, phosphine adducts such as

$$2Me_2PH + B_2H_6 \rightarrow 2Me_2PH \cdot BH_3$$

may be pyrolyzed to eliminate H_2 and thereby produce not a monomeric "boraphosphene," but rather the cyclic *phosphinoborane* trimer with a "cyclohexane" chair conformation (similar to the "saturated" aminoborane trimers).[75]

(8-118)
$$Me_2PH \cdot BH_3 \xrightarrow{\Delta}$$

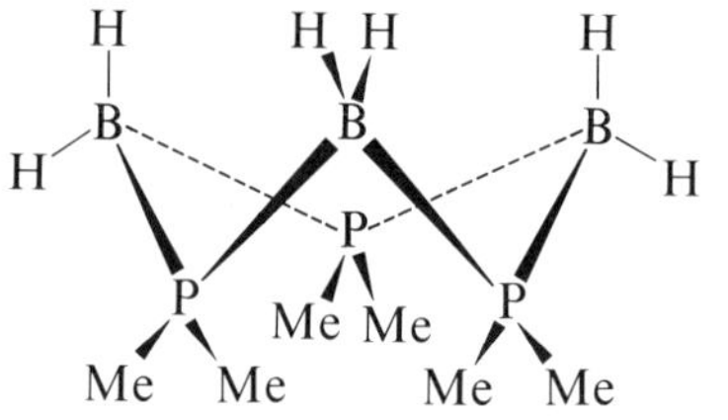

Note that the larger size of phosphorus than of nitrogen should facilitate the formation of trimers, as should weaker P═B character in the monomer. What is unexpected is the unusual thermal and chemical stability of this trimeric structure. The origin of the inertness is not completely understood, although arguments have been made that hyperconjugation of B—H bond density with the phosphorus d atomic orbitals (**17**) may be an important contributor to reduced B—H hydridic character.[76] (You will encounter this phosphorus acceptor concept again, in the section after next, where an NR group occurs in the place of BH_2.) It would be well to recall at this point the analogous stability of H_3PBH_3 to acid hydrolysis (p. 398).

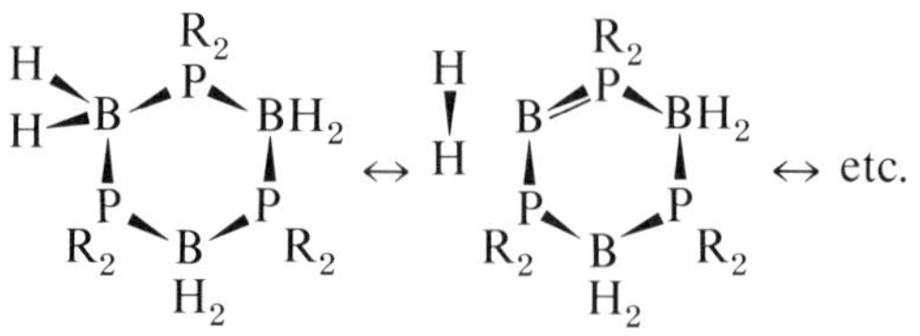

17

---

[75] A. B. Burg and G. Brendel, *J. Amer. Chem. Soc.*, **80**, 3198 (1958); R. I. Wagner and C. O. Wilson, Jr., *Inorg. Chem.*, **5**, 1009 (1966); W. Gee, et al., *J. Chem. Soc.*, 3171 (1965).

[76] See ref. 73; relevant structural data are given by W. C. Hamilton, *Acta Cryst.*, **8**, 199 (1955); G. J. Bullen and P. R. Mallinson, *Chem. Comm.*, 132 (1969). It is most interesting that P → B π bonding does not stabilize the monomer R_2PBH_2, relative to P → B σ bonding and olgimer formation, so that BH_2 → P π interaction becomes possible in the polymer.

Whatever the reason, the high chemical and thermal stability of the phosphinoborane trimer has spurred interest in parlaying this tendency into development of phosphinoborane polymers for various commercial uses. These efforts have been only partially successful, and the study of this class of compounds continues.

Many of the solvolysis products of boron halides and hydrides find counterparts when aluminum is the central atom. An important difference between boron and aluminum, as you have already seen, is the propensity of aluminum to increase its coordination number. Thus, the boroxine and borazine structures are not found in Al^{3+} chemistry. Rather, the *alkoxides* of aluminum, for example, are dimeric and even tetrameric. A sterically hindered alkoxide like aluminum *t*-butoxide achieves only the dimeric structure

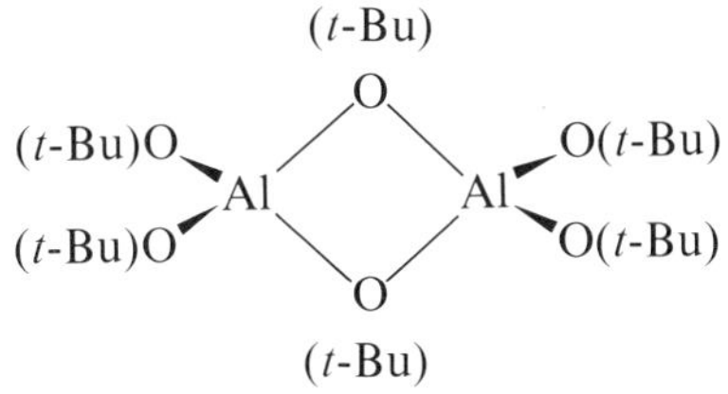

$$(t\text{-Bu})O\diagdown\!\!\!\!\underset{(t\text{-Bu})O}{}\text{Al}\diagup\!\!\!\!\underset{O}{\overset{O}{\diagdown}}\text{Al}\diagdown\!\!\!\!\overset{O(t\text{-Bu})}{\underset{O(t\text{-Bu})}{}}$$

whereas the important aluminum isopropoxide exhibits both trimeric and tetrameric (**18**) structures, the latter containing both four-coordinate and chiral six-coordinate aluminum centers.[77]

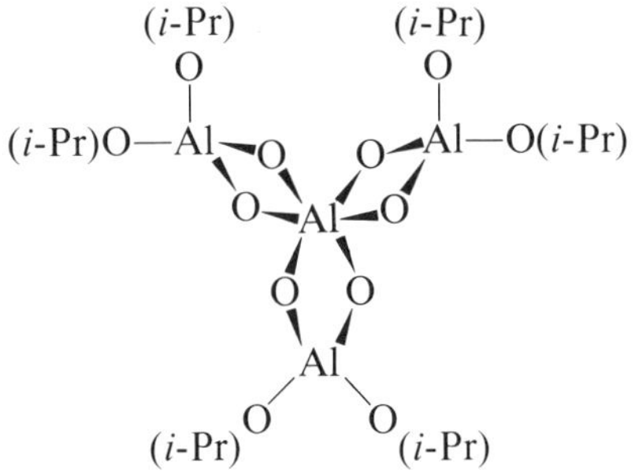

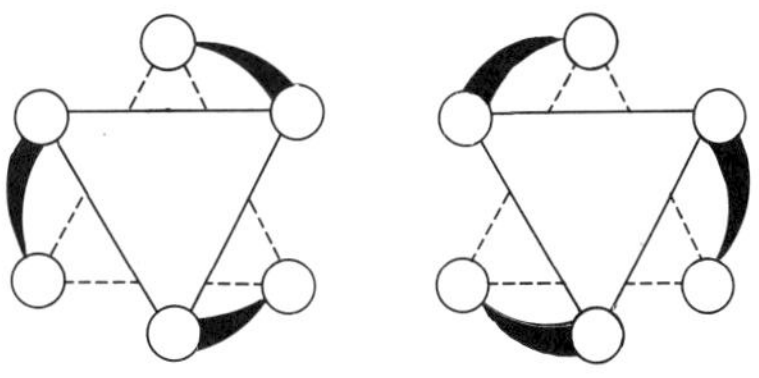

18 $\begin{bmatrix} \text{m.p. } 125° \\ \text{b.p. } 242°/10 \text{ torr} \end{bmatrix}$

The best methods for synthesizing the alkoxides appear to be the reaction of the chloride with the appropriate sodium alkoxide or alcohol (with NH_3 present) or even (aluminum being a very active metal) the reduction of the corresponding alcohol using $HgCl_2$ as a catalyst:

$$AlCl_3 + 3NaOR \rightarrow \frac{1}{n}\{Al(OR)_3\}_n + 3NaCl$$

$$Al + 3ROH \xrightarrow{HgCl_2} \frac{1}{n}\{Al(OR)_3\}_n + \frac{3}{2}H_2 \qquad (8\text{-}119)$$

As you have already seen from the earlier (Chapter 6) contrast of boron and aluminum halide hydrolyses, there is a marked difference between boron and aluminum halides in regard to ease of solvolysis in general. This is manifest in the fact that more forcing conditions (use of NaOR or ROH/NH_3) are needed for synthesis of $Al(OR)_3$.

---

[77] J. G. Oliver, et al., *J. Inorg. Nucl. Chem.*, **31**, 1609 (1969); *Chem. Comm.*, 918 (1968); *Inorg. Nucl. Chem. Letters*, **5**, 749 (1969).

These ideas presented for the alkoxides of aluminum pertain to amides also. Amines solvolyze aluminum halides even less readily than do alcohols, so that alkali metal amides are useful reagents, with stoichiometry control, for preparing the polymeric aluminum amides.[78] As you might expect from the earlier discussion concerning use of $LiAlH_4$ as a very active reducing agent, amines and phosphines tend to be reduced (with H_2 evolution) when brought in contact with lithium aluminum hydride,[79] but note that formation of polymeric materials is difficult to achieve in this way since four moles of NR_2^- (in the case of amines) are formed per mole of Al^{3+}.

(8-120)
$$LiAlH_4 \xrightarrow{R_2NH} LiAl(NR_2)_4 + 4H_2$$

$$\xrightarrow{PH_3} LiAl(PH_2)_4 + 4H_2$$

$$\xrightarrow{ROH} LiAlH(OR)_3 + 3H_2$$

One of the more intriguing examples of aminoalane or phosphinoalane condensations results from thermolyzing the *primary* amine and phosphine adducts of organoalanes.[80] The highly condensed tetrameric structure **19** sometimes encountered for these compounds is reminiscent of the hydrocarbon cubane (*cf.* the discussion in Chapter 6, p. 312).

(8-121)
$$R'NH_2 \cdot AlR_3 \rightarrow \tfrac{1}{4}(R'NAlR)_4 + 2RH$$

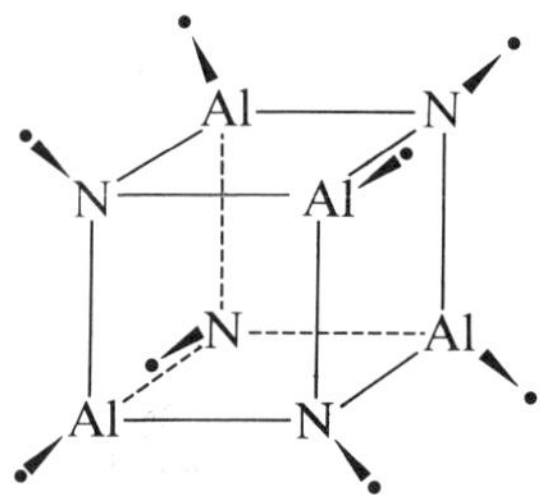

19

Quite generally, these elimination reactions are much more facile than the corresponding reactions of boron. This is dramatically illustrated by the fact that borazine is produced by pyrolysis of H_3BNH_3 or $[(H_3N)_2BH_2][BH_4]$ at 250°, while the following scheme[81] reveals the more facile and more extensive elimination of H_2 from H_3AlNH_3:

(8-122)
$$AlH_3 + NH_3 \xrightarrow[-80°]{\text{ether}} AlH_3 \cdot NH_3$$
$$H_2 + (H_2Al-NH_2)_3 \xleftarrow{-40°}$$
$$\xrightarrow{20°} H_2 + (HAlNH)_x$$

---

[78] These reactions are actually reductions by alkali metals in liquid ammonia: R. Brec and J. Rouxel, *Bull. Soc. Chim. France*, 2721 (1968); *Compt. Rend. Ser. C*, **270**, 491 (1970).

[79] These and other related reactions are cited and discussed by K. Wade and A. J. Banister in "Comprehensive Inorganic Chemistry," Vol. 1, pp. 1039ff, 1007ff, A. F. Trotman-Dickenson, *et al.* (Eds.), Pergamon Press, New York, 1973.

[80] J. K. Ruff, in "Developments in Inorganic Nitrogen Chemistry," Vol. 1, C. B. Colburn (Ed.), Elsevier, New York, 1968.

[81] E. Wiberg and A. May, *Z. Naturforsch.*, **10B**, 229 (1955).

The polymer obtained at room temperature is a non-volatile white solid that is insoluble in most solvents. Although the empirical formula is formally analogous to borazine, the unsaturation of $(HAlNH)_3$ is avoided by condensation (as in the tetramer **19**) into a polymeric solid, perhaps resembling **20**, for which $(CF)_n$ is a model[82] (see study question 6-43).

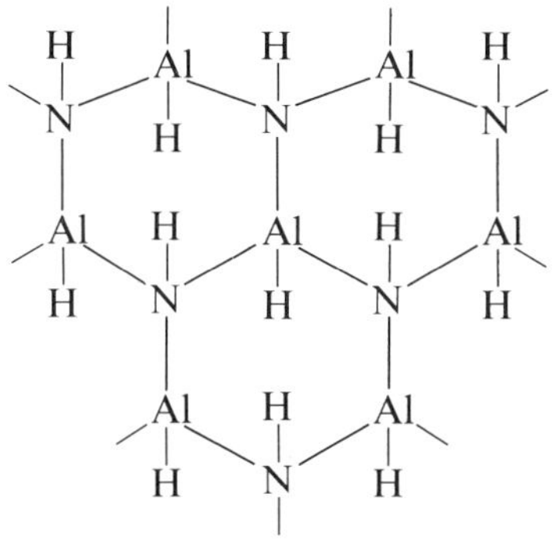

20

The reactions above illustrate a fundamental fact of Al—N adduct chemistry: *if both the aluminum and nitrogen have hydrogens attached, H_2 can be eliminated as above;* or, if either has an alkyl or aryl group and the other a hydrogen, then *elimination of hydrocarbon* can be observed. Some specific examples are as follows[83]:

$$Me_3Al \cdot NH_3 \xrightarrow{70°} CH_4 + (Me_2AlNH_2)_3 \xrightarrow{200°} CH_4 + (MeAlNH)_x$$
[mp 56.7°]   [mp 134.2°]   [glassy solid]

$$Et_2AlCl \cdot NH_2Me \xrightarrow{72-76°} C_2H_6 + (EtAlCl \cdot NHMe)_{2,3} \xrightarrow[54\ hr.]{210°} C_2H_6 + (ClAlNMe)_x$$
[mp −11.5°]   [mp 91°]   [nonvolatile white solid]

(8-123)

As implied in (8-122) and (8-123), even unsaturated Al—N compounds such as the aminoalanes (R_2Al-NR_2') are very prone to association to dimers and trimers, unlike the aminoboranes, many derivatives of which are monomeric or only weakly associated (Chapter 7, p. 389). As pointed out above, this is due, in part at least, to the fact that B—N π bonding more efficiently relieves the coordinative unsaturation of the boron atom (better p-$p\pi$ overlap) and also to the fact that there is less steric hindrance to trimerization of (R_2Al-NR_2').

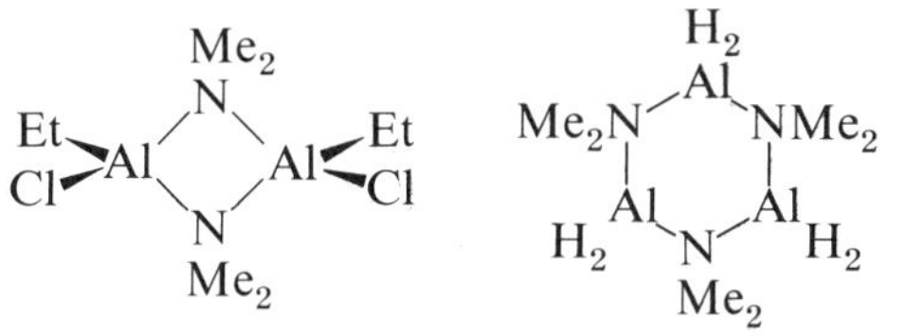

---

[82] L. B. Ebert, et al., *J. Amer. Chem. Soc.*, **96**, 7841 (1974).
[83] A. W. Laubengayer, et al., *J. Amer. Chem. Soc.*, **83**, 542 (1961).

SILICON

Generally speaking, the Si—C bond is reasonably stable (the order of bond energies B—C > Si—C > Al—C is inversely related to the order of polarities and susceptibility to attack); because of this, certain organosilanes and their derivatives have found commercial usage. Probably the most important class of organosilanes consists of the *organosiloxanes,* commonly known as silicones. These materials are polymers, with highly stable Si—O backbones, formed by hydrolyses of R_nSiX_{4-n}. For example:

(8-124)
$$R_3SiX + H_2O \rightarrow (R_3Si)_2O \quad \text{(disilyl ethers)}$$
$$R_2SiX_2 + H_2O \rightarrow (R_2SiO)_n \quad \text{(ring and chain polymers)}$$
$$RSiX_3 + H_2O \rightarrow (RSiO_2)_n \quad \text{(cross-linked chains)}$$

All these reactions proceed through formation of the corresponding silanol and, even though some diols and triols have moderate stability, they tend to eliminate H_2O and condense (as with the boronic and borinic acids, the E=O linkage affords less stability than oligomer formation). While somewhat of an art, it is possible to control the nature of the polymer formed by hydrolyzing mixtures of mono-, di-, and trihalosilanes; the dihalides tend to establish linear chains, the trihalides cross-link the chains, and the monohalides act as chain terminators.

In reactions analogous to the hydrolysis reactions, the organosilicon halides react with amines to form *silyl amines,* which bear a *formal* resemblance to alkyl amines. You will recall from Chapter 5, p. 228, that the compound trisilyl amine has the structurally interesting feature of a planar Si_3N frame.

(8-125)
$$3SiH_3X + NH_3 \rightarrow (SiH_3)_3N + 3HX$$

With respect to the VSEPR model of structures, you might initially think that trisilyl amine would have a pyramidal geometry about nitrogen. This conclusion is predicated on the assumption that the nitrogen atom is considered to have a *lone* pair. Many believe that the planarity of this compound is good evidence for the existence of $p\pi$-$d\pi$ interaction of the lone pair with the d orbitals of the silicon (recall the near planarity of the nitrogen in organic amides). Others have speculated that the electron-

$$H_3Si-\ddot{N}\begin{matrix}SiH_3\\SiH_3\end{matrix} \leftrightarrow H_3Si=N\begin{matrix}SiH_3\\SiH_3\end{matrix} \text{, etc.}$$

releasing nature of the silane facilitates the flattening of the nitrogen atom. Both the π overlap and σ factors act in the same direction, so an experimental answer as to which is more important is unlikely. In fact, the σ/π factors are synergically coupled (greater σ release to nitrogen facilitates retro-bonding from N to Si and *vice versa*). Given the larger size and lower electronegativity of phosphorus, the pyramidal structure of $P(SiH_3)_3$ can be rationalized by less σ bp/bp repulsion and correspondingly less retro-π interaction.

The compound *hexamethyldisilazane* or *bis*(trimethylsilyl)amine has attracted some attention as a "silylating" reagent. Reactions of a large variety of polar E—H groups (E = C, O, N, S) introduce the E—$SiMe_3$ linkage. For example:

(8-126)
$$(Me_3Si)_2NH + \begin{Bmatrix} ROH \\ RCO_2H \\ RSH \\ RCONH_2 \\ HCN \\ H_2SO_4 \end{Bmatrix} \rightarrow \begin{Bmatrix} ROSiMe_3 \\ RCO_2SiMe_3 \\ RSSiMe_3 \\ RCONH(SiMe_3) \\ Me_3SiCN \\ HSO_3(SiMe_3) \end{Bmatrix}$$
[b.p. 126°]

The facility of these reactions is due in large part to the Si—N π bonding in Me_3SiNH^- as a leaving group (from the other $SiMe_3$), for otherwise the amide group would not cleave very easily.

The reactions of dichlorosilanes with amines produce products generally in line with what you anticipate by now[84] [note that NH is isoelectronic with O, and refer to (8-124)].

$$RR'SiCl_2 + NH_3 \rightarrow [RR'SiNH]_x \quad \text{oligomers} \tag{8-127}$$

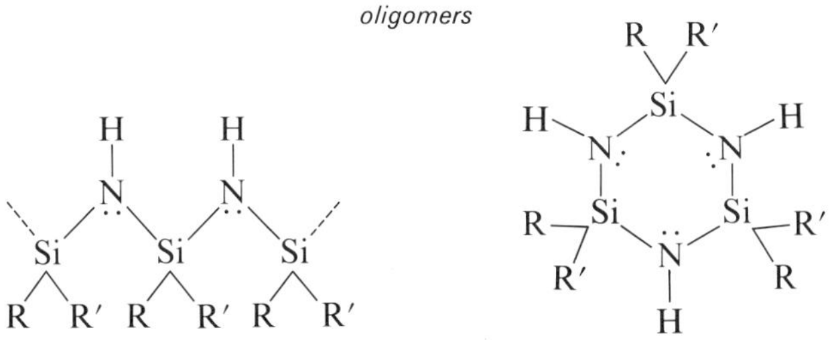

The progression of increased steric bulk represented by (R = R' = H; R' = H, R = Me, Et, Ph; R, R' = organic) results in less tendency for chain polymers and increasing yields of *tri-* and *tetrasilazanes*, $\{RR'SiNH\}_{3,4}$. When primary amines are used in place of NH_3, the steric congestion worsens and *diaminosilanes* are produced:

$$Me_2SiCl_2 + 2NH_2R \rightarrow Me_2Si(NHR)_2 \tag{8-128}$$

Deamination of these materials (compare the dehydration of silanols) by NH_4^+ catalysis is effective when R = Me, but the reaction becomes less facile as the bulk of R increases:

$$Me_2Si(NHMe)_2 \xrightarrow{NH_4^+} \{Me_2SiNMe\}_3 \tag{8-129}$$

Like the aluminum analogs, but unlike the borazines, the cyclotriazanes exhibit non-planar boat and chair structures; nevertheless, the Si—N bond distances and large Si—N—Si angles (up to 130°) suggest that some $N \rightarrow Si$ $p\text{-}d\pi$ bonding still takes place. We will defer a description of how this π interaction may take place in a non-planar ring to the following section on analogous PN rings.

Finally, we should recall that the silanes are marginally hydridic and do not rapidly hydrolyze in acidic media. This is dramatically illustrated by the fact that hydrogen halides will not react with silanes in the absence of aluminum halide (or other halide acceptor) as a catalyst[85] ($HAlCl_4$ should be a

$$SiH_4 + nHX \xrightarrow{Al_2X_6} SiX_nH_{4-n} + nH_2 \tag{8-130}$$

potent acid). Reaction of silane with alcohols occurs in the presence of alkoxide ion; when R = H the product $Si(OH)_4$ eliminates H_2O to form silicon dioxide.

$$SiH_4 + 4ROH \xrightarrow{OR^-} Si(OR)_4 + 4H_2 \tag{8-131}$$

This behavior is nothing more than base catalyzed hydrolysis of silanes, where you are again reminded of the characteristic ease of nucleophilic attack upon silicon.

---

[84] See ref. 73.
[85] See ref. 50, and F. G. A. Stone, "Hydrogen Compounds of the Group IV Elements," Prentice-Hall, New York, 1962.

NITROGEN AND PHOSPHORUS

The bonds to hydrogen of nitrogen and of phosphorus (though much less so) are weakly acidic and consequently do not experience solvolytic reactions as do the hydrides of the preceding elements in their periods. With regard to the reactions of the halides, recall that the only useful haloamine is chloramine ("useful" meaning that NH_2Cl is not an inordinately dangerous chemical to handle and thus might be used by chemists as a general synthetic material). As detailed in Chapter 7, chloramine experiences reactions with fairly strong nucleophiles through displacement of Cl^-. Thus, base hydrolysis and treatment with ammonia illustrate normal solvolysis behavior:

(8-132)
$$NH_2Cl + OH^- \rightarrow NH_2OH + Cl^-$$
$$NH_2Cl + 2NH_3 \rightarrow N_2H_4 + NH_4^+Cl^-$$

The halides of phosphorus solvolyze in the customary fashion, and several of these reactions are worthy of special comment. The trivalent halides undergo alcoholysis (vigorous) and aminolysis (less vigorous) to form the corresponding alkoxides and amides.

(8-133)
$$PCl_3 + 3H_2O \rightarrow H_3PO_3 + 3HCl$$
$$+ 3NH_3 \rightarrow P(NH_2)_3 + 3HCl$$

You should remember from the bond energy analysis in Chapter 6 that a characteristic of phosphorous acid and its trimethyl ester is a "keto-enol" type of rearrangement to produce a phosphoryl linkage:

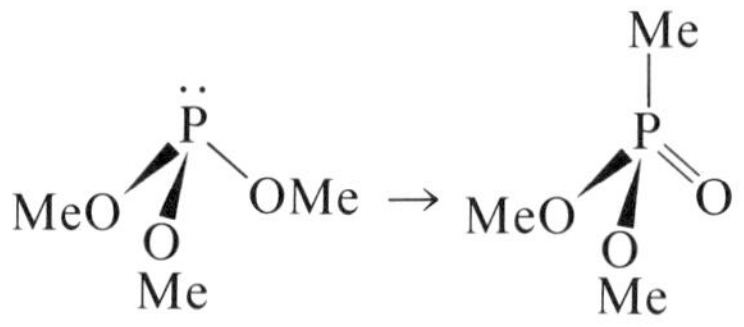

$$\text{MeO–P(OMe)(OMe)} \rightarrow \text{MeO–P(=O)(OMe)(Me)}$$

The influence of this stable structural arrangement is further evident in the analogous Arbusov reactions of phosphite esters with alkyl halides (see p. 420, Chapter 7):

(8-134)
$$\text{RO–P(OR)(OR)} + R'X \rightarrow \text{RO–P(=O)(OR)(R')} + RX$$

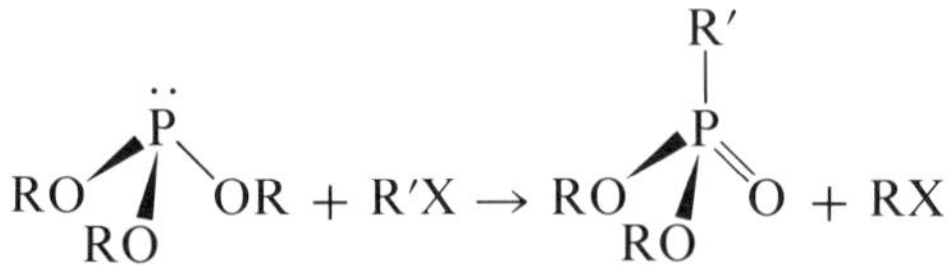

and in the fact that in the alcoholysis of PCl_3 failure to remove the HCl by the presence of some base (like an amine) leads to $(RO)_2PHO$ as a product.[86] In a most interesting example of substituent effects at phosphorus, the *bis*(pentafluorophenyl)phosphinous acid exhibits both "keto-enol" forms in solution[87]:

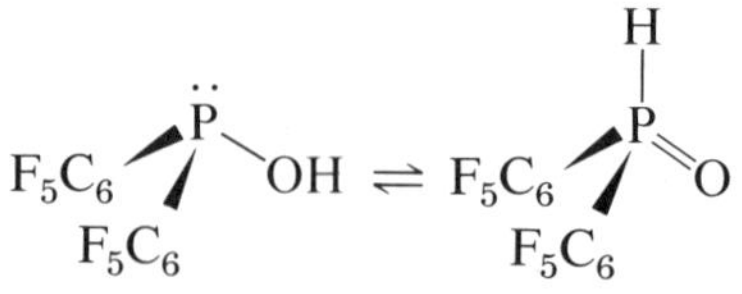

$$\text{F}_5\text{C}_6\text{–P(OH)(C}_6\text{F}_5\text{)} \rightleftharpoons \text{F}_5\text{C}_6\text{–P(=O)(H)(C}_6\text{F}_5\text{)}$$

---

[86] A. D. F. Toy, in "Comprehensive Inorganic Chemistry," Vol. 2, A. F. Trotman-Dickenson, *et al.* (Eds.), Pergamon Press, New York, 1973.

[87] D. D. Magnelli, *et al., Inorg. Chem.*, **5**, 457 (1966). This compound is rather unusual among phosphinous acids in that it does not disproportionate into $(F_5C_6)_2PH$ and $(F_5C_6)_2PO(OH)$.

These observations suggest two interesting features of the "keto-enol" rearrangement. The reaction is probably not an *intra*molecular 1,2 shift of an R group (R = H, organic) from O to P, but an intermolecular process like the reaction of phosphines with organohalides to produce phosphonium salts (*cf.* the discussion of phosphorus ylids on p. 485). Secondly, the energy of the P—(H,R) bond is subject to considerable substituent (on phosphorus) influence. That is, O—P multiple bond character may actually enhance [as in (8-134)] the phosphine basicity/nucleophilicity as noted for *tris*(dimethylamino)phosphine in Chapter 5 (p. 231).

The solvolysis products from the reactions of phosphorus halides with amines define an extensive class of polymeric materials. In a moment we will consider the reactions of PCl_5, but first consider the reactions of PCl_3 and $POCl_3$. The latter reacts with NH_3 and with primary and secondary amines to form stable *triamides*[88] wherein the nitrogen atoms adopt a planar or nearly planar geometry:

$$POCl_3 + 3NH_3 \rightarrow PO(NH_2)_3 \qquad (8\text{-}135)$$

$$P\text{—}\ddot{N}\text{\textless} \leftrightarrow P\text{=}N\text{\textless}$$

You should recall that this structural feature played an important role in the stereochemistry of $PF_4(NMe_2)$ and its Berry pseudorotation process (Chapter 7) and the basicity of $(Me_2N)_3P$ (Chapter 5, p. 231).

PCl_3 presents an aminolysis chemistry more in line with your expectations, as the following scheme shows[89]:

$$PCl_3 + 6NH_3 \xrightarrow[-78°]{CHCl_3} P(NH_2)_3 + 3NH_4Cl$$

$$\downarrow \text{deamination on warming}$$

$$PCl_3 + 5NH_3 \xrightarrow[-20°]{R_2O} \frac{1}{n}\{HN\text{=}PNH_2\}_n + 3NH_4Cl \qquad (8\text{-}136)$$
$$\text{[unstable at R.T.]}$$

$$PCl_3 + 4NH_3 \xrightarrow{\text{liq. } NH_3} \frac{1}{n}(PN)_n + 3NH_4Cl$$

$$4PCl_3 + 18MeNH_2 \rightarrow P_4(NMe)_6 + 12NMeH_3Cl$$

This last compound, a crystalline material, bears a striking resemblance to P_4O_6 in structural and chemical properties. To fit its formation into the scheme of NH_3 solvolysis, you may find it helpful to *imagine*[90] formation of $P(NHMe)_3$ with subsequent deamination (four times) to yield[91] $2\{MeN\text{=}P\text{—}\overset{\overset{\displaystyle Me}{|}}{N}\text{—}P\text{=}NMe\}$, where the four P=N links are less stable than four P—N links arising from condensation. This behavior leads us directly to the behavior of PCl_5 in contact with amines.

Following a pattern established by boron and aluminum, pentavalent phosphorus halides form a series of polymeric *phosphazenes*‡ with ammonia (the class of compounds commonly referred to as phosphonitrilic compounds).

$$PCl_5 + NH_4Cl \xrightarrow[C_2H_2Cl_4]{146°} \frac{1}{x}[Cl_2PN]_x + 4HCl \qquad (8\text{-}137)$$

---

[88] R. Klement and O. Koch, *Chem. Ber.*, **87**, 333 (1954); "Inorganic Syntheses," Vol. 6, E. G. Rochow (Ed.), McGraw-Hill, New York, 1960.

[89] E. Fluck, *Topics in Phosphorus Chemistry*, **4**, 293 (1967); ref. 86.

[90] The mechanism of the reaction has not been determined.

[91] Compare this monomer to a P_2O_3 monomer: O=P—O—P=O.

‡A recent short review of phosphazenes and their technological applications is H. R. Allcock, *Science*, **193**, 1214 (1976).

Perhaps best prepared by treating PCl_5 with NH_4Cl in a chlorinated hydrocarbon solvent, primarily trimeric **(22)** and tetrameric **(23)** dichlorophosphazene, $(NPCl_2)_n$, are formed.[92] These are useful for synthesis of various substituted phosphazenes, since the chloride substitution reactions common for non-metal chlorides work for this particular phosphonitrile.

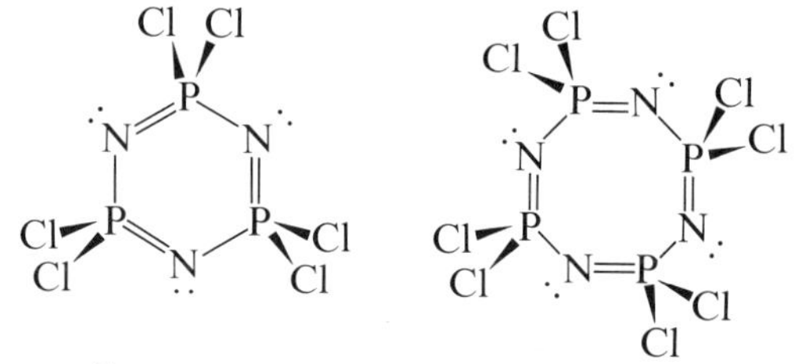

22 [m.p. 114°; b.p. 127°/13 torr]   **23** [m.p. 124°; b.p. 188°/13 torr]

The reason for the condensation reaction is most apparent if we inquire into the structure of the monomeric unit, Cl_2PN.

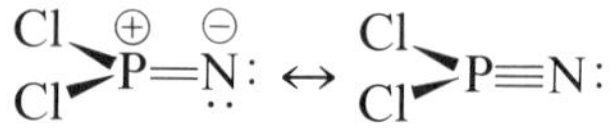

This structure is not so bizzare as it may first seem to you, for a good structural analog is to be found in carbon chemistry. Phosgene (Cl_2CO) is valence shell isoelectronic with this monomer, but, of course (and this is an important difference), the nuclear charges are distributed quite differently in CO and PN. In mo terms the $P\equiv N$ resonance structure[93] can be handled by delocalization of a nitrogen $p\pi$ lone pair into the anti-symmetric P—Cl σ^* orbitals; or, if you wish to include phosphorus d ao's in the scheme, the in-plane phosphorus d ao (say d_{yz}) can function to accept this pair. At any rate, the polarity of the PN bond will remain, the phosphorus atom will exhibit significant acid character, and the nitrogen will exhibit significant base character; polymerization seems inevitable. In **22** and **23** you still find the necessity of PN double bond character, and the bond distances are observed[94] to be shorter than "normal" (1.58 Å instead of 1.78 Å). Furthermore, the trimer structure appears to have D_{3h} symmetry, so all PN bonds are equivalent. The ring electron pairs are fully delocalized. You might note that, even in the ring structure, there *may* also be residual delocalization of the in-plane nitrogen lone pairs onto the phosphorus atoms.

You are probably already alert to the likeness of the trimeric phosphonitrile ring to that of the borazines, but there is a very important difference that must not be overlooked. The boron atom of the borazines has but a single out-of-plane ao (the $2p\pi$ ao) for involvement in π mo's. Phosphorus, if we admit the presence of d ao's, has two such ao's. Using the approach of Chapter 4, the nitrogen $2p\pi$ ao's are of the form

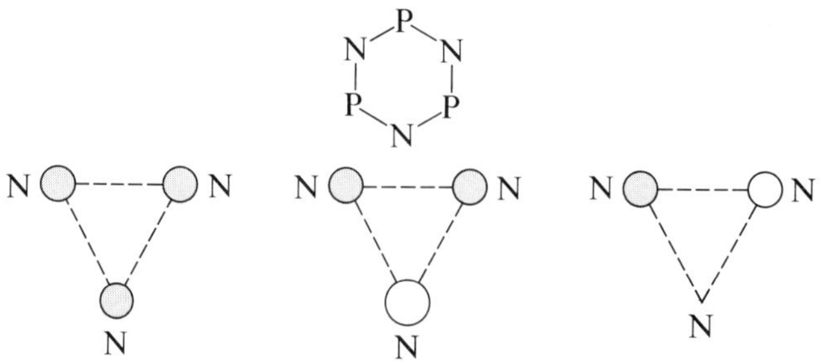

24a   **24b**   **24c**

---

[92] H. R. Allcock, "Phosphorus-Nitrogen Compounds," Academic Press, New York, 1972; *Scientific American*, **230**, 66 (1974). A good survey is also given in ref. 86. A more recent source, with photographs of products made from polyphosphazenes, is H. R. Allcock, *Science*, **193**, 1214 (1976).

[93] This structure is the basis for the name "phosphonitriles."

[94] G. J. Bullen and P. A. Tucker, *Chem. Comm.*, 1185 (1970).

while two phosphorus d ao's would appear as follows (the sketches represent a view down the z-axis of the d_{xz} and d_{yz} ao's; the highly analogous MX_3 example of Chapter 4 can be consulted for further details):

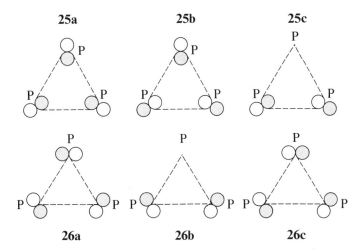

In this way, you see that five of the six phosphorus $d\pi$ TASO's may be taken in a bonding combination with the nitrogen lone pairs (**26a** is "non-bonding," since it fails to have non-zero overlap with any of **24a–c**). An energy level diagram would appear as follows.

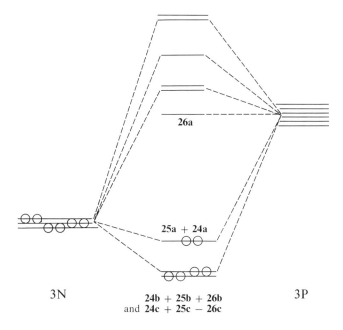

A further significant difference between the boron and phosphorus ao's used in this analysis is that boron has only the one $p\pi$ ao not heavily involved in the B—N σ electron system, while phosphorus has five d ao's. When the full set of P d ao's is available for π interaction with the nitrogen lone pairs, the spherical angular distribution of the set[95] incurs no constraint on the π electron system for a planar structure.

---

[95] Perhaps reminding you that a set of three p ao's, five d ao's, etc. constitute a spherically symmetric probability distribution about an atom will help you understand that the orientation of the nitrogen lone pair is not highly critical to implementation of the p-d π interaction.

When the phosphorus is symmetrically substituted (as in **22**) it seems that a planar structure is preferred, but unsymmetric phosphorus substitution could well introduce a tendency for the ring to pucker, and this can be accommodated by the π electron system. An example of puckering is found with $[NP(OMe)_2]_3$, which exhibits a planar structure as formed (as a solid) but which undergoes a thermal rearrangement (an extension of the "keto-enol" migration discussed on p. 496) to form the boat structure[96] **27**:

(8-138) $[NP(OMe)_2]_3 \xrightarrow{\Delta} [MeN\overset{O}{\underset{\|}{P}}(OMe)]_3$ like **22**

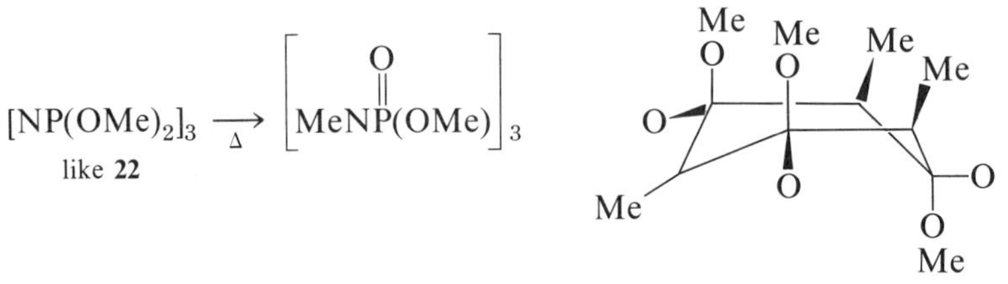

27

Apparently the larger tetrameric ring **23** introduces enough strain under D_{4h} symmetry that $(NPCl_2)_4$ and $[NP(NMe_2)_2]_4$ exhibit nonplanar structures,[97] without serious deterioration of the PN π bonding. In the context of these remarks on the N—P π interaction, you may wish to review the earlier comments about the possibility for B=P hyperconjugation in $(H_2BPH_2)_3$ and N=Si character in the cyclosilazanes.

As noted a few paragraphs back, the chloro-derivatives are useful for introducing other substituents at phosphorus atoms in the ring. Here we will mention one recently discovered, unusual reaction illustrating bicyclic ring formation.[98] Treatment of the *bis*(ethylamino) derivative of **23** with excess Me_2NH in $CHCl_3$ not only resulted in full amination of both (PCl_2) groups and one of the (PCl) groups but also caused one of the NHEt groups to bridge to the opposite (PCl) group, with HCl elimination.

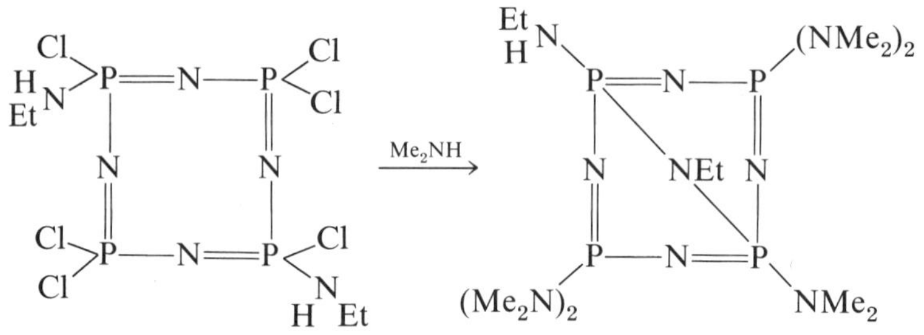

As implied by the "wedge" perspective, the bicyclo structure is reminiscent of adamantane.

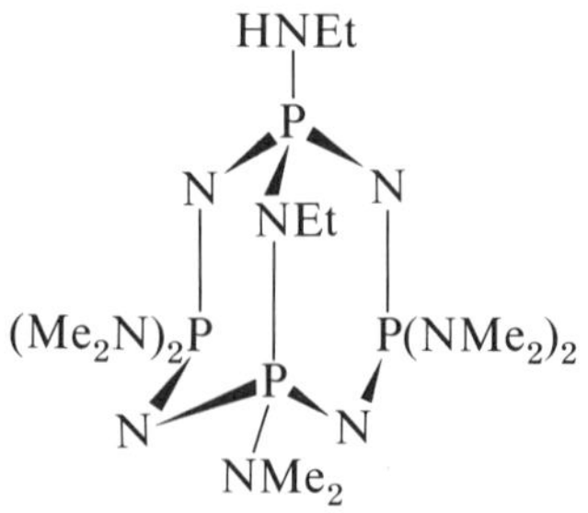

---

[96] G. B. Ansell and G. J. Bullen, *J. Chem. Soc. (A)*, 3026 (1968).
[97] See ref. 94.
[98] T. S. Cameron, *et al.*, *J. Chem. Soc., Chem. Comm.*, 975 (1975).

It is useful that one can stoichiometrically control the hydrolysis of the penta-halides (X = F, Cl, Br) of phosphorus to form the important *phosphoryl halides*

$$PCl_5 + H_2O \rightarrow Cl_3PO + 2HCl$$

From these reagents one can form organophosphine oxides (R_3PO), esters of phosphoric acid [$(RO)_3PO$], phosphoryl amides [$(R_2N)_3PO$], and so on. Complete hydrolysis, of course, leads to phosphoric acid, indicating that $P(OH)_5$ is not a stable compound.[99] Similarly, esters of $P(OH)_5$ are not known to arise from alcoholysis of PCl_5.

You may be puzzled at this point, wondering whether hydrolysis of phosphorus halides, in analogy with the aminolysis reactions, produces *polymeric* P—O compounds. The answer is emphatically "yes" in that a confusingly large number of linear and cyclic phosphorus-oxygen compounds is known.[100] Rather than embark on a lengthy discussion of these compounds, let us note that the simplest species of this type are *diphosphoric acid* (28) (obtained by dehydration of phosphoric acid at >200° or by its reaction with phosphoryl chloride)

$$2H_3PO_4 \rightarrow H_4P_2O_7 + H_2O$$
$$5H_3PO_4 + POCl_3 \rightarrow 3HCl + 3H_4P_2O_7$$

(8-139)

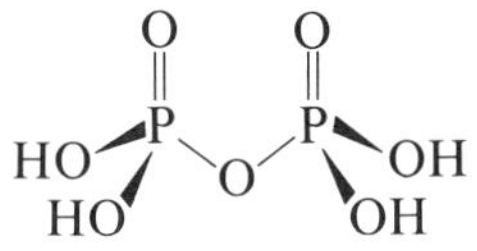

28
[m.p. 61°]

and the polymers of *metaphosphoric acid,* with empirical formula HPO_3. Best characterized are the salts of metaphosphoric acid, which contain a non-planar trimeric anion (**29**) [the acid itself is found either as an amorphous (glassy) solid or as a viscous liquid].

$$nH_3PO_4 \xrightarrow{\text{dehyd.}} \frac{1}{n}(HPO_3)_n + nH_2O$$

(8-140)

$$nNaH_2PO_4 \xrightarrow{\text{dehyd.}} \frac{1}{n}(NaPO_3)_n + nH_2O$$

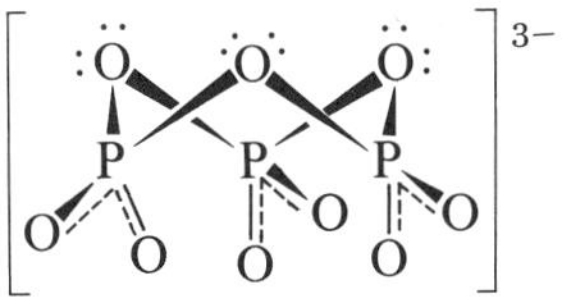

29

---

[99] Note that deaquation here, $P(OH)_5 \rightarrow PO(OH)_3 + H_2O$, bears a strong resemblance to the deamination reactions of P(III) amides as in equation (8-136) and to ylid formation (p. 479), and achieves the very stable P=O link.

[100] A good summary may be found in ref. 86 (pp. 469-534).

The chain polyphosphates, of which diphosphate is the simplest example, are of great significance in biosystems. In fact, the exothermic hydrolysis of adenosine triphosphate (ATP, **30**) to adenosine diphosphate (ADP, **31**) is of fundamental significance as a source of energy in living muscle cells. (See Chapter 19.)

(8-141)

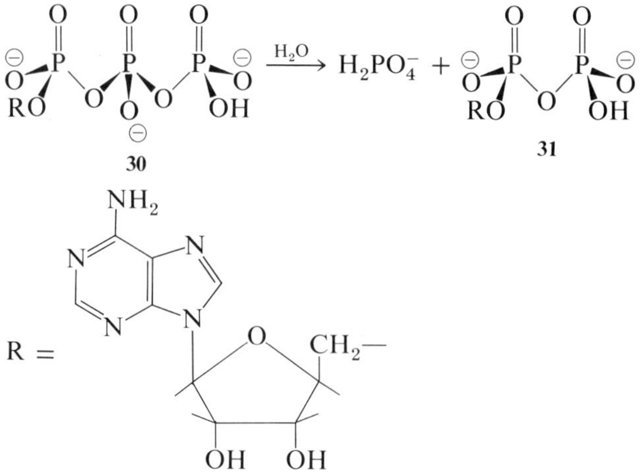

$$30 \xrightarrow{H_2O} H_2PO_4^- + 31$$

R = adenosine group (adenine + ribose –CH$_2$–)

Note that this reaction is only mildly exothermic ($\Delta G° \simeq -7$ kcal/mole) and is thermally neutral on an "additive bond energy" basis. Solvation energy and entropy changes provide considerable driving force for the reaction. The condensed phosphate best known to the general public is sodium *triphosphate* ($Na_5P_3O_{10}$)—used to control water hardness and pH in laundry detergents and one compound responsible for phosphate pollution of natural waters (the other main offender in this category is $NH_4H_2PO_4$, the field fertilizer).

The ultimate in condensed phosphate polyhedra is found in the anhydrides of phosphorus and phosphoric acids, commonly called phosphorus trioxide (**32**) (phosphorus oxide or tetraphosphorus hexoxide) and phosphorus pentoxide (**33**) (phosphoric oxide or tetraphosphorus decoxide). The molecular forms of these anhydrides (P_4O_{10} is also found in several polymeric forms) have the composition P_4O_6 and

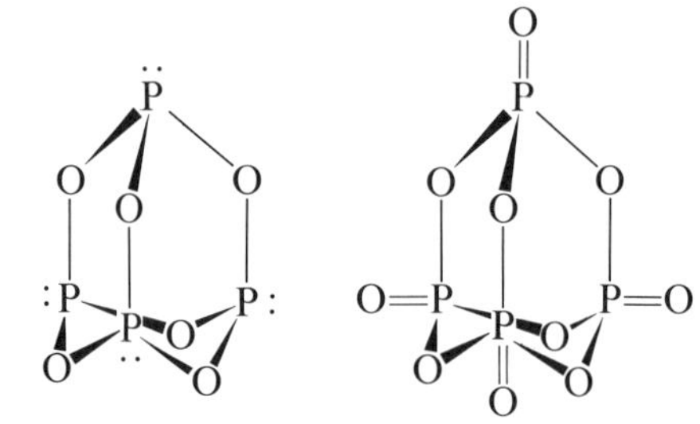

32 [m.p. 24°, b.p. 174°]   **33** [subl. 360°]

P_4O_{10}, as you will recall from the discussion of their structures in Chapter 6. Their synthesis, since they are *very* strong desiccants, is not best achieved by dehydration of H_3PO_3 and H_3PO_4 but by direct combination of elemental (white P_4) phosphorus and oxygen. Use of excess O_2 easily results in formation of P_4O_{10} but the yield of P_4O_6 from use of limited O_2 is only about half of the theoretical yield. (You learned about a better synthesis in the section on SO_2 as a solvent in Chapter 5, p. 258.) As P_4O_6 and P_4 are soluble in organic solvents but polymeric red phosphorus is not, conversion of unreacted white phosphorus to red by UV radiation[101] permits the isolation of the trioxide.

---

[101] Remember that UV radiation (200 to 300 nm wavelength) supplies 140 to 100 kcal/mole to a molecule that absorbs a photon in this range.

$$P_4 + 5O_2 \rightarrow P_4O_{10}$$
$$P_4 + 3O_2 \rightarrow P_4O_6 + (P_4O_7, P_4O_8, P_4O_9)^{102}$$
(8-142)

As mentioned above, a most important property of P_4O_{10} is its strong desiccating ability, making this oxide useful in producing the anhydrides of other oxyacids (for example, by dehydration of $HClO_4$, HNO_3, and even H_2SO_4). These reactions with P_4O_{10} are typical of compounds with acidic hydrogens. We can illustrate this general reaction by mentioning that HF reacts to form both difluorophosphoric and fluorophosphoric acids.[103]

$$P_4O_{10} + 6HF \rightarrow 2F_2PO(OH) + 2FPO(OH)_2$$
(8-143)

(You might recognize that a popular "fluoride" toothpaste uses the fluorophosphate anion as its anti-caries agent.)

A similar reaction for P_4O_6 and HCl

$$P_4O_6 + 6HCl \rightarrow 2PCl_3 + 2H_3PO_3$$
(8-144)

surprisingly shows no distribution of Cl between phosphorus atoms. Thus, the mechanism of this last reaction appears[104] to be markedly different from (8-143) for P_4O_{10} + HF and also from a related reaction, (8-145), of white phosphorus with water [be careful to note that (8-145) is quite slow in the absence of OH^-; in Chapter 6 we mentioned that white phosphorus can be protected from violent air oxidation by storing it under water at pH = 7]:

$$P_4 + 6H_2O \xrightarrow{OH^-} PH_3 + 3H_2PO(OH)$$
(8-145)

hypophosphorous acid

or

$$P_4 + 3OH^- + 3H_2O \rightarrow PH_3 + 3H_2PO_2^-$$

and[105]

$$3H_3PO_2 \xrightarrow{200°} 2H_3PO_3 + PH_3$$
(8-146)

That competing pathways operate in this disproportionation of P_4 in the presence of base is evidenced by the fact that diphosphine and phosphorous acid are also formed in the hydrolysis of white phosphorus.

$$P_4 + 4OH^- + 2H_2O \rightarrow P_2H_4 + 2H_2PO_3^-$$
(8-147)

The occurrence of (8-147) requires that PH_3 synthesized by (8-145) be purified *before* it is exposed to air!

OXYGEN AND SULFUR

As with nitrogen, the use of most oxygen halides as starting materials for synthesis of other materials is not to be recommended for other than the most

---

[102] These products are simply intermediates between P_4O_6 and P_4O_{10} where 1, 2, and 3 phosphorus atoms have terminal oxygens.

[103] W. Lange, in "Inorganic Syntheses," Vol. 2, C. W. Fernelius (Ed.), McGraw-Hill, New York, 1946, gives a more convenient synthesis based on substitution of F for OH by HF + H_3PO_4.

[104] We must be careful here; it is possible that $Cl_2PO(OH)$ and $ClPO(OH)_2$ are produced by the same steps as in the HF reaction, but that X/OH exchange occurs more readily for Cl than for F.

[105] S. D. Gokhale and W. L. Jolly, in "Inorganic Syntheses," Vol. 9, S. Y. Tyree (Ed.), McGraw-Hill, New York, 1967.

experienced and careful chemist. Even so, most reactions of O_mX_n are not typical solvolysis reactions, in the sense of preceding examples in this section. The appearance of oxygen in solvolytic products is therefore essentially restricted to the roles of H_2O and ROH as the solvolyzing agents. Even the reactions of the most important oxygen halide, OCl^- (again in analogy with nitrogen chemistry), are usually not considered solvolysis reactions. Here you should be thinking of the oxygen transfer reactions like

$$NO_2^- + OCl^- \rightarrow NO_3^- + Cl^-$$

$$SO_3^{2-} + OCl^- \rightarrow SO_4^{2-} + Cl^-$$

Consequently, our attention will focus primarily on sulfur compound solvolysis.

For sulfur we would be somewhat restricted in considering only binary halides for, as you have seen in earlier chapters, SF_6 is highly inert, SF_4 is useful primarily as a fluorinating agent, and SF_2 is not attainable in useful quantities. Of the chlorides, only S_2Cl_2 and SCl_2 are common (actually, SCl_4 is known, but it decomposes at $\sim -50°C$ into SCl_2 and S_2Cl_2). Of some interest is the product of the reaction[106] of S_2Cl_2 with NH_4Cl or NH_3:

(8-148)
$$6S_2Cl_2 + 4NH_4Cl \rightarrow (SN)_4 + S_8 + 16HCl$$

Notice the parallel between this reaction and that [equation (8-136)] of PCl_3 with liquid ammonia. Furthermore, by noting that NS is valence shell isoelectronic with NO you are confronted once again with the characteristic condensation of "unsaturated" species containing third period elements. That is, NO only *weakly* dimerizes in the solid state (Chapter 6, p. 333).

The compactness of the $NS)_3$ structure is the intriguing feature of this compound; if you were to assume a close analogy of S_4N_4 with the phosphonitrilic compounds, you might devise a structural/bonding model involving (as a single resonance structure) alternate S—N double and single bonds within an eight-membered ring. In doing so, however, you come up with an electronic structure with an unpaired electron on each sulfur (sulfur has one more valence electron than phosphorus).[107]

Your intuition should tell you that such a structure is probably not very stable with respect to further condensation; the experimental finding[108] that samples of the compound are *essentially* diamagnetic at room temperature confirms this suspicion. By analogy with the bicyclic structure of the S_8^{2+} cation (as discussed in Chapter 6), a reasonable choice for the structure would be the cage **34** (the structure actually found by X-ray analysis of the crystalline material[109]):

---

[106] The product tetrasulfur tetranitride is actually formed in $\sim 25\%$ yield. Consult M. Schmidt and W. Siebert in "Comprehensive Inorganic Chemistry," Vol. 2, A. F. Trotman-Dickenson, *et al.* (Eds.), Pergamon Press, New York, 1973, for entries to the literature of this fascinating reaction and its products.

[107] To be consistent with a D_{4h} molecular symmetry you must include both π resonance structures, as in the Kekule structures of benzene (the situation is not unlike that of the phosphonitriles). *Assuming* the ring to be planar, develop the cyclic out-of-plane π mo pattern (as in Figure 4-15) to show that such a structure would actually have only two unpaired electrons.

[108] R. C. Brasted, *J. Chem. Soc.*, 2297 (1965).

[109] D. Clark, *J. Chem. Soc.*, 1615 (1952).

[structures of S₄N₄ shown as resonance forms and cage structure]

34

[orange, shock sensitive m.p. 178°]

The S—S bond distances (2.58 Å) are long relative to the nominal values for S—S bonds (~2.1 Å) but considerably shorter than the non-bonded, van der Waals distance for two sulfur atoms (~3.6 Å). The analogy with S_8^{2+} is obvious. Consistent with the idea that the S—S bonding is weak is the thermochromic property of the solid: it is colorless at $-190°$C, orange at room temperature, and deep red at $100°$C. This color behavior is reminiscent of elemental sulfur which, when heated, undergoes ring rupture to produce colored (reddish) paramagnetic chains of sulfur atoms. Thus, heating of S_4N_4 is likely to lead to some opening of the cage, making possible the electronic transitions responsible for absorption of light in the visible region. To be consistent with the C_{2v} symmetry of **34** you must describe the four "π" electron pairs as fully delocalized over the "cage" framework (the situation is not unlike that of the phosphonitriles).

One of the most interesting reactions [110] of $(SN)_4$ is the controlled thermolysis to produce cage opening and eventually the chain oligimer $(SN)_x$ called "polythiazyl." Using the apparatus shown in Figure 8-8, the orange $(SN)_4$ crystals (at A) are gently warmed (85°) so that the vapors pass through Ag wool (heated to 220° at B) and are condensed onto the liquid nitrogen cold finger (C). The colorless solid obtained at this point is S_2N_2 (this material is shock sensitive). Careful sublimation of the impure $(SN)_2$ crystals onto an ice-water cold finger (D) produces well formed crystals that eventually become reflective and *bronze-colored* on annealing at room temperature. The chain oligimer $(SN)_x$ can be safely handled, for it is neither shock nor thermally sensitive.

$$(SN)_4 \xrightarrow[\Delta]{\text{Ag wool}} (SN)_2 \xrightarrow{\Delta} (SN)_x \tag{8-149}$$

---

[110] C. M. Mikulski, *et al., J. Amer. Chem. Soc.*, **97**, 6358 (1975); M. J. Cohen, *ibid.*, **98**, 3844 (1976); D. R. Salahub and R. P. Messmer, *J. Chem. Phys.*, **64**, 2039 (1976); R. H. Baughman, *et al., ibid.*, **64**, 1869 (1976).

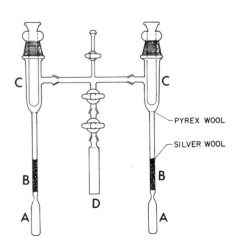

Figure 8-8. Apparatus for synthesis of $(SN)_x$ crystals. [Reproduced with permission from C. M. Mikulski, *et al., J. Amer. Chem. Soc.*, **97**, 6358 (1975). Copyright by the American Chemical Society.]

The exciting property of this polymer is its high electrical conductivity (approaching that of Hg). In color and conductivity, the non-metal polythiazyl has "metallic" character. To understand the origin of the metallic "band" of orbital levels (Figure 6–8) we first need to examine the bonding forces in the (NS)$_2$ dimer. As the NS monomers approach closely in a planar configuration, the π_i (i means in-plane) and π_i^* mo's describe a cyclic four-orbital situation, as in **35a–d.** An entirely analogous representation of the out-of-plane mo's is found in **36a–d.** As the total number of π/π^* electron pairs in two NS units is 5, five of the mo's from **35** and **36** are to be filled. The energy ordering of these mo's, based on the net number of σ and π type p/p overlaps, should be:

$$35a < 35b < 36a < 36b < 36c < 36d < 35c < 35d$$

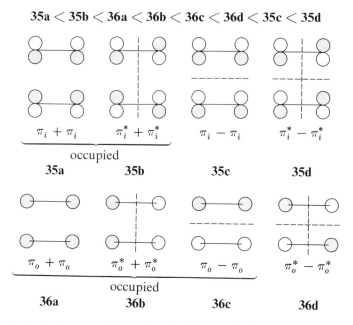

The appropriate electron configuration for these mo's is:

$$(35a)^2(35b)^2(36a)^2(36b)^2(36c)^2$$

There is, consequently, net in-plane and out-of-plane bonding between NS units. In fact, the (NS)$_2$ structure [111] has D_{2h} symmetry with all internal angles of 90° and could be represented by the following Lewis resonance forms (in each structure there

are three $p\pi$ electron pairs, equivalently described by **36a–c**). What must be considerable ring strain in this structure is relieved by ring opening on warming and subsequent condensation into a chain polymer:

---
[111] See ref. 110.

Notice that these resonance structures imply all equal N—S bond distances, as is, in fact, found (the out-of-plane π mo's of a four ao N_2S_2 "chain" when occupied by the three π electron pairs give this same result). The origin of the metallic conduction band can now be described in any of several equivalent ways:

i) $Np\pi$ ao's give an orbital energy band to be occupied by $\frac{3}{4}N$ electron pairs;
ii) N four-orbital chain mo's [for $(SNSN)_N$] give rise to four interpenetrating energy bands to be occupied by $3N$ electron pairs;
iii) N two-orbital chain mo's [π and π^* of $(NS)_N$] give rise to two interpenetrating energy bands to be occupied by $\frac{3}{2}N$ electron pairs.

This description provides for conductivity along the chain direction. In fact, the conductivity is not restricted to just this direction, so there must be "communication" between chains (in the form of π mo overlap) to permit electron motion between chains as well.

Tetrasulfur tetranitride is the starting reagent for preparation of a host of substituted, cyclic S—N compounds; we mention here only three interesting reactions. For example, oxidative fluorination of $(SN)_4$ with AgF_2 results in cage opening by fluorination of the sulfur atoms[112] (consistent with the ideas that the S—S bonds of **34** are the weak "links" and that S—F bonds are stronger than N—F bonds):

$$S_4N_4 \xrightarrow{AgF_2} \quad \text{(structure with 1.54 Å and 1.66 Å bonds)} \tag{8-150}$$

The analogy with $(F_2PN)_4$ seems obvious enough, yet the C_2 symmetry reported[113] for $(FSN)_4$ precludes the possibility of fully equivalent NS bonds. The tetramer of NSF does not have a planar structure [in analogy with an *unsymmetrically* substituted phosphonitrile tetramer $(NPXX')_4$] and is most simply described as SF groups lying along four edges of a "tetrahedron" of N atoms (**37**).

37

Reduction of tetrasulfur tetranitride with ethanolic stannous chloride disrupts the N—S multiple bonding altogether by creating NH bonds and giving a product $(SNH)_4$ isoelectronic with S_8 and having the analogous "tiara" structure[114] (**38**).

---

[112] It is a highly fascinating kinetic effect that monomeric FSN is known and that it trimerizes, rather than forms {FSN}$_4$!

[113] C. A. Wiegers and A. Vos, *Acta Cryst.*, **16**, 152 (1963). This symmetry lowering would be most perplexing were the sulfur atoms to possess only a single acceptor orbital like the borazines. In fact, the N—S "single bond" in {FSN}$_4$ is estimated to have a bond order of ~1.5. As with the phosphonitriles, ring distortions are not likely to greatly inhibit multiple bond formation so that intramolecular steric factors, sulfur substituent asymmetry, and even intermolecular steric effects in the solid state may cause ring distortions.

[114] D. Chapman and T. D. Waddingtion, *Trans. Faraday Soc.*, **58**, 1679 (1962).

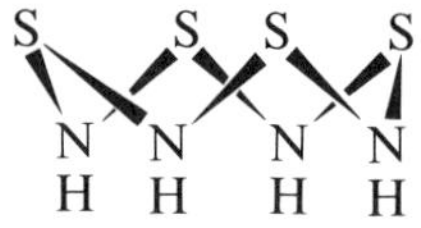

508 ☐ SYNTHESIS OF IMPORTANT CLASSES OF NON-METAL COMPOUNDS

38

Oxidative chlorination of S_4N_4 cleaves the eight-membered ring to produce the (planar, D_{3h}) trimeric form of NSCl (**39**).

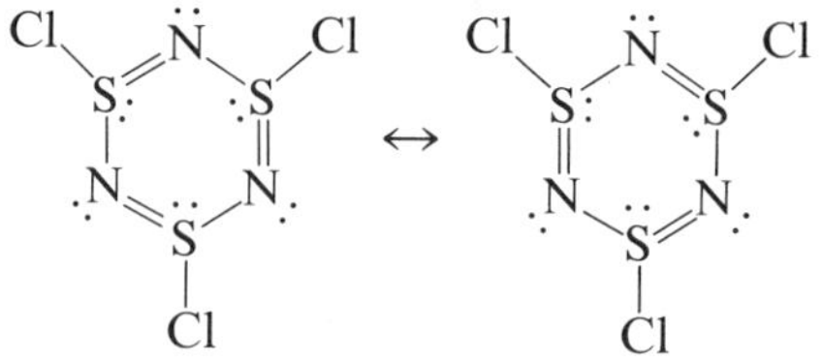

39

It is intriguing (and unexplained) that chlorination produces the ClSN trimer, while fluorination with AgF_2 retains the tetramer unit.[115]

Although S_2Cl_2 should hydrolyze to form $H_2S_2O_2$, this acid is apparently unstable with respect to disproportionation:

(8-151)
$$2\text{“}H_2S_2O_2\text{”} \rightarrow \frac{3}{8} S_8 + \text{“}H_2SO_3\text{”} + H_2O$$

The reaction is complex, and S_8 and $SO_2 \cdot H_2O$ are only the primary products.

Of greater interest and utility than the binary sulfur chlorides are the oxychlorides, since they provide an entry to a large number of important, naturally occurring sulfur compounds. While both thionyl and sulfuryl chlorides ($SOCl_2$ and SO_2Cl_2) are prepared [see (8-31) and (8-32), p. 461] from SO_2 by oxidative chlorination, the lower oxidation state compound can also be prepared by the exchange chlorination

(8-152)
$$SO_2 + PCl_5 \rightarrow SOCl_2 + POCl_3$$

The S—Cl bonds of thionyl and sulfuryl chlorides undergo most of the reactions we have identified as being typical of non-metal chloride bonds. The compounds are quite reactive in solvolytic reactions and so provide convenient reagents for synthesis of *sulfonamides* [$(R_2N)_2SO$] and *sulfamides* [$(R_2N)_2SO_2$], *esters* of sulfurous and sulfuric acids [$(RO)_2SO$ and $(RO)_2SO_2$], and *sulfoxides* [R_2SO] and *sulfones* [R_2SO_2].

Following the pattern of aminolysis behavior typical of the other non-metal chlorides, the sulfamide obtained from sulfuryl chloride and ammonia[116] can be thermally deaminated to yield the trimer[117] of NSO_2^- (**40**):

(8-153)
$$SO_2Cl_2 + 4NH_3 \rightarrow SO_2(NH_2)_2 + 2NH_4Cl$$
$$\xrightarrow{200°} (NH_4)_3(SO_2N)_3$$

---

[115] M. Becke-Goehring, "Ergebnisse und Probleme der Chemie der Schwefel-stickstoff-verbindungen," Academie-Verlag, Berlin, 1957. Here's an opportunity to use your imagination. Is it possible (based on your understanding of ring size effects for earlier elements) that steric interactions and/or weaker N → S bonds in the chloro substituted tetramer make its ring opening more facile? Perhaps you will find it important that {NSCl}$_3$ passes into the gas phase as monomers at 110°C.

[116] W. Traube, *Chem. Ber.*, **25**, 2472 (1892).

[117] See ref. 106.

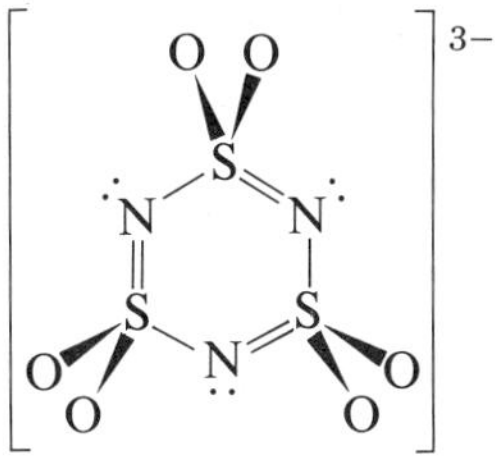

$$\begin{bmatrix} \text{structure 40: six-membered ring of alternating S and N atoms, each S bearing two O substituents} \end{bmatrix}^{3-}$$

40

By noting that N^- is isoelectronic with O, you see that the NSO_2^- trimer fits nicely into the structural pattern for one form of solid SO_3 (*cf.* Chapter 6). Various salts of this anion are known, perhaps the most useful being that with silver; from it, various nitrogen-substituted derivatives can be formed by the ingenious strategy of simple metathesis with a halide of the groups to be added to the nitrogen atoms[117]

$$Ag_3(NSO_2)_3 + 3RX \rightarrow 3AgX_{(c)} + \text{(trimeric ring with N-R substituents and SO}_2\text{ groups)} \qquad (8\text{-}154)$$

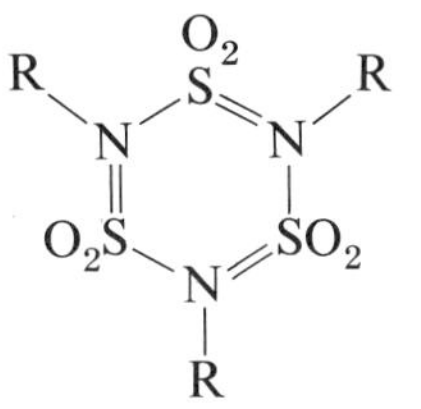

R = H, alkyl, etc.

These reactions expand the range of accessible trimeric S—N compounds beyond those of sulfur-substituted derivatives of $(ClSN)_3$, **39**.

The hydrolysis and alcoholysis reactions of thionyl and sulfuryl chlorides are extremely vigorous. The useful aspect of controlled thionyl hydrolysis is that SO_2 and HCl are products

$$SOCl_2 + 2H_2O \rightarrow SO_2 + 2HCl$$

and, being gases, are easily eliminated from a reaction mixture. For this reason, anhydrous metal chlorides are frequently prepared[118] from their hydrated forms by reaction with $SOCl_2$:

$$MCl_n \cdot xH_2O + xSOCl_2 \rightarrow MCl_n + 2xHCl + xSO_2 \qquad (8\text{-}155)$$

Recall also that thionyl chloride hydrolysis is a good reaction for synthesis of HCl (see the earlier section on HX synthesis).

Hydrolysis of $SOCl_2$ in excess water produces, of course, solutions of "sulfurous acid." You may recall from Chapter 6 that "H_2SO_3" is not known in the form $(HO)_2SO$ but rather the formulation $H_2O \cdot SO_2$ seems more appropriate, and, from Chapter 5, that SO_2 is generally a very weak Lewis acid. Thus, the anhydride, SO_2, of

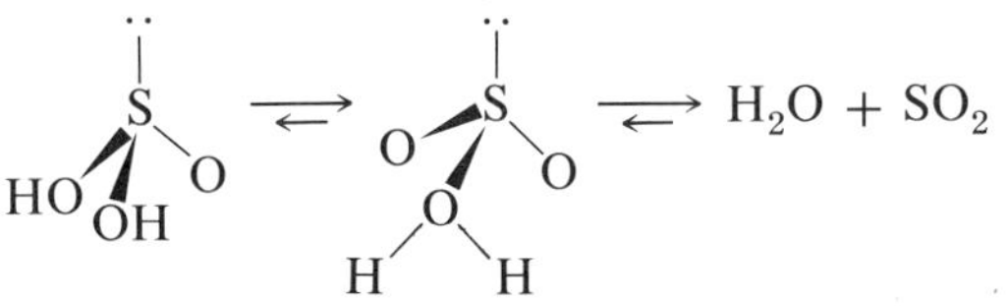

$$\text{(HO)}_2\text{S=O} \rightleftharpoons \text{H}_2\text{O} \cdot \text{SO}_2 \rightleftharpoons H_2O + SO_2$$

H_2SO_3 is so easily obtained that H_2SO_3 has never been isolated. Nevertheless, hydrolysis of $OSCl_2$ in an *excess* of the stronger (than H_2O) base OH^- forms bisulfite,

---

[118] A. R. Pray, "Inorganic Syntheses," Vol. 5, T. Moeller (Ed.), McGraw-Hill, New York, 1957.

which is well known in salt form. In solution HSO_3^- exhibits a 1,2 tautomerization in analogy with phosphorous acid.

$$\left[\begin{array}{c} \ddot{S}\text{---}O\text{---}H \\ O \quad O \end{array}\right]^- \rightleftharpoons \left[\begin{array}{c} H \\ | \\ S \\ O \quad O \\ O \end{array}\right]^-$$

As with some of the triesters of phosphorous acid, dialkylsulfites also exhibit tautomeric forms:

$$\begin{array}{c} \ddot{S} \\ RO \quad O \\ RO \end{array} \quad \text{and} \quad \begin{array}{c} R \\ | \\ S \\ RO \quad O \\ O \end{array}$$

While it is possible to alkylate thionyl chloride to generate sulfoxides, an often-used synthesis relies on oxygenation (oxygen atom transfer) of the appropriate dialkylsulfide. Hydrogen peroxide is a useful reagent for this kind of oxidation, and again you find an analogy with organophosphorus chemistry (organophosphines and sulfides possess heavy atom lone pairs and their nucleophilic character renders oxygen atom transfer an effective reaction mode, as discussed in the last chapter, pp. 412 and 423).

Given the "solvate" nature of "H_2SO_3," you should not be surprised to learn that deamination of $SO(NH_2)_2$ (actually unknown) must be facile [also not particularly surprising in view of the behavior of $P(NH_2)_3$ as noted by reactions (8–136):

(8–156) $SOCl_2 + 2NH_3 \rightarrow [NH_4][NSO]$[119] $+ 2HCl$

The conjugate acid, $HN{=}S{=}O$, may be isolated ($< -94°C$) from a vapor phase reaction; on warming to $-70°C$ the monomer polymerizes, and then depolymerizes on further heating[120]! An interesting change in the $SOCl_2/NH_3$ reaction channel arises when the reaction is performed in organic solvents ($CHCl_3$) with CaO present[121]:

(8–157) $SOCl_2 + NH_3 + CaO \rightarrow (HO)S{\equiv}N + CaCl_2 + H_2O$

This fascinating isomerization of HNSO has not been fully characterized; whether H_2O alone or in conjunction with unreacted CaO is responsible for the 1,3 proton shift can only be guessed. Of further interest is the effect of using excess $SOCl_2$ as solvent; apparently a Cl/OH exchange takes place, for $(ClSN)_3$ is formed (recall the previously mentioned preparation of this trimer, **39**, by chlorination of S_4N_4).

Sulfuryl chloride reacts with water in a fashion analogous to thionyl chloride. If an excess of water is used, the SO_3 liberated dissolves to form a sulfuric acid solution. Again, alkylation reactions are possible to produce sulfones, although these can better be obtained by oxygenation of sulfides or sulfoxides.

$$SO_2Cl_2 + H_2O \rightarrow SO_3 + 2HCl \quad \text{(but see } HSO_3Cl \text{ below)}$$

(8–158) $SO_2Cl_2 \xrightarrow{H_2O} H_2SO_4 + 2HCl$

$$R_2S \xrightarrow{O} R_2SO \xrightarrow{O} R_2SO_2$$

This last reaction was illustrated on p. 481 as an *in vivo* step in the toxic action of mustard gas.

Before leaving this discussion of sulfur chemistry, we need to note that there are some important, useful *halogenated sulfur acids*. Their syntheses are based on the Lewis acid characters of SO_2 (weak) and SO_3 (strong), as are those of $S_2O_5^{2-}$ and

---

[119] Note again the isostery of N^- and O.
[120] W. P. Schrenk and E. Krone, *Angew. Chem.*, **73**, 762 (1961).
[121] M. Becke-Goehring, *et al.*, *Z. anorg. Allgem. Chem.*, **293**, 294 (1957).

$S_2O_7^{2-}$. For SO_2, probably the most useful (for mild fluorination of other materials) of these reagents are actually the salts of *fluorosulfurous* acid. Prepared from alkali fluoride and SO_2, the salts are an interesting example of lattice stabilization of a molecular species.

$$SO_2 + MF \rightarrow MSO_2F \qquad (8\text{-}159)$$

The parent acid is not known and even the salts exhibit some instability by virtue of the fact that they have, even at room temperature, a detectable dissociation pressure of SO_2. This is a key concept for the reaction (8-28).

For the higher oxidation state of sulfur, both *fluorosulfuric* and *chlorosulfuric acid* are known. Both are prepared by the simple treatment of SO_3 with HX. You recognize HSO_3Cl as the reagent useful for sulfonation of aromatic compounds, and the fluoroacid you have encountered several times in previous chapters (particularly Chapter 5) of this text. HSO_3F is particularly useful as a solvent for high oxidation state species like the S_8^{2+} ions (see the discussion of sulfur and halogen cations in Chapter 6). Other important properties in its practical application as a solvent are that it can be removed from a reaction mixture by vacuum distillation (b.p. 163°C), and it does not strongly *auto*-ionize via

$$2HSO_3F \rightarrow H_2SO_3F^+ + SO_3F^-$$

As noted in Chapter 5, HSO_3F can be successfully used to exchange fluorine for oxygen in high oxidation state oxyanions. Impressive examples are

$$CrO_4^{2-} + 2HSO_3F \rightarrow CrO_2F_2 + 2HSO_4^-$$
$$ClO_4^- + HSO_3F \rightarrow ClO_3F + HSO_4^- \qquad (8\text{-}160)$$

HALOGENS AND XENON

The solvolysis reactions of the interhalogens and rare gas halides that are worth discussing at this point are their hydrolysis reactions. These reagents are strongly oxidizing, and it is not recommended that one casually combine them with oxidizable reagents (in particular, organo compounds).

It appears that one can be very general in viewing the hydrolysis reactions of the interhalogen compounds; *the lighter, terminal halogen atoms are displaced as simple anions while the heavier, central halogen becomes oxygenated to the corresponding* (same oxidation state) *oxyacid*. Examples are

$$ClF + H_2O \rightarrow HOCl + H^+ + F^-$$
$$BrF_5 + 3H_2O \rightarrow HBrO_3 + 5H^+ + 5F^- \qquad (8\text{-}161)$$

Of course, you might recall from Chapter 6 that halogen oxyacids of intermediate oxidation state rapidly disproportionate into higher and lower oxidation state forms under acid conditions. For example,

$$ClF \xrightarrow{H_2O} [HOCl] \xrightarrow{acid} Cl^- + HClO_3$$

$$BrF_3 \xrightarrow{H_2O} [HBrO_2] \xrightarrow{acid} Br^- + HBrO_3 \qquad (8\text{-}162)$$

Attempts to prepare esters and amides of halogen oxyacids by reaction of a higher oxidation state interhalogen with a normal alcohol or amine will be very inefficient (and dangerous) because of degradation of the alcohol or amine. This appears to be a problem with XF_n in particular, because of reactions catalyzed by HF. Simple solvolysis reactions may be possible with perfluoroalcohols and amines and certain inorganic acids like SF_5OOH and $HN(SO_2F)_2$, but generally one is limited to the preparation of substituted hypohalites and haloamines. Even so, these syntheses are often achieved by less energetic synthetic schemes than aminolysis/alcoholysis. You have studied the chlorination of ammonia; substituted chloramines are similarly prepared. Probably the most useful ester of hypochlorous acid is (t-Bu)OCl, where the absence of α-hydrogen atoms and the steric bulk of the (t-Bu) group confer a marginal stability on the ester:[122]

(8-163)
$$RCH_2OH + HOCl \rightarrow [RCH_2OCl] \rightarrow \underset{H}{\overset{R}{>}}C=O + HCl + H_2O$$

but

$$(t\text{-Bu})OH + HOCl \rightarrow (t\text{-Bu})OCl + H_2O$$

Of the rare gas halides, only those of xenon have a moderately developed chemistry. The three fluorides of xenon do hydrolyze, but only the *hexafluoride* (surprisingly enough, from a thermodynamic view!) does so without immediate accompanying oxidation of the water (even in this case, though, the kinetically "stable" intermediate product of complete hydrolysis, xenon trioxide, can be dangerously explosive).[123]

$$XeF_6 + H_2O \rightarrow XeOF_4 + 2HF$$
$$XeF_6 + 3H_2O \rightarrow XeO_3 + 6HF$$

The simple product $Xe(OH)_6$ appears to dehydrate easily (*cf.* "H_2SO_3" and "$P(OH)_5$") to form XeO_3, just as $Xe(OH)F_5$ apparently eliminates HF. Both xenon difluoride and xenon tetrafluoride oxidize water.[124] However, the difluoride undergoes only slow hydrolysis in acid media, while the reaction is much faster when base is present.

$$XeF_2 + 2OH^- \rightarrow Xe + \frac{1}{2}O_2 + 2F^- + H_2O$$

$$XeF_4 + 2H_2O \rightarrow \frac{1}{3}XeO_3 + \frac{2}{3}Xe + \frac{1}{2}O_2 + 4HF$$

The tetrafluoride does *not* produce the *unknown* dioxide (just as the difluoride does not produce the unknown monoxide[125]) on hydrolysis. The mechanisms of these reactions are not known, so one may only surmise that the monoxide and dioxide, if formed, rapidly disproportionate under the hydrolysis conditions to effect an oxidation reaction. The products from XeF_4 are suggestive of this view, but the stoichiometric relationship of XeO_3 and Xe suggests that a fairly complex mechanism prevails.

Finally, we might note that a compound with a Xe—N bond has been reported[126] from the "solvolysis" reaction

$$XeF_2 + HN(SO_2F)_2 \rightarrow FXe(NSO_2F)_2 + HF$$

XeF_2 is a good fluorinating agent, particularly in the presence of HF; it is considered that the fluorocarbon solvent used in this reaction plays the vital role of separating the HF produced as an immiscible layer so as to inhibit the degradation of $HN(SO_3F)_2$ [a side reaction that seems to dominate the reaction between XeF_2 and $HN(SO_2F)_2$ when the pure reagents are used].

---

[122] "Gmelins Handbuch der Anorganischen Chemie," 8 Auflage: "Chlor," Systemnummer 6, Teil B-Lieferung 2, Verlag Chemie, Berlin, 1969, p. 394.

[123] B. Jaselskis, *et al., J. Amer. Chem. Soc.,* **88,** 2149 (1966); D. F. Smith, *Science,* **140,** 899 (1963); J. Shamir, *et al., J. Amer. Chem. Soc.,* **87,** 2359 (1965).

[124] S. M. Williamson and C. W. Koch, *Science,* **139,** 1046 (1963); E. H. Appelman and J. G. Malm, *J. Amer. Chem. Soc.,* **86,** 2297 (1964).

[125] C. D. Cooper, *et al., J. Mol. Spec.,* **7,** 223 (1961) detected the diatomic molecule in the gas phase, and the data suggest a bond energy <9 kcal/mole.

[126] R. D. LeBlond and D. D. DesMarteau, *Chem. Comm.,* **1974,** 555.

STUDY QUESTIONS

14. Give steps for syntheses of the following from the elements, any normal organic compound, and any special reagents given in parentheses.

    a) $B(O_2CR)_3$
    b) $H[B(OSO_3H)_4]$ (H_2O)
    c) $Na[HB(OR)_3]$ (NaH)
    d) $[MeNBCl]_3$
    e) $[ROBO]_3$ (H_2O)
    f) $AlH(OR)_2$
    g) $[R_2AlPR'_2]$ (what are likely R, R'?)
    h) $[Me_2AlNMe_2]_2$, $[Me_2AlNHMe]_3$ (why is one a dimer, the other a trimer?)
    i) $[PhAlNMe]_4$
    j) $[H_2SiO]_n$ (H_2O)
    k) SiH_3NH_2 (NH_3)
    l) $MeOSiEt_3$
    m) $[NPF_2]_3$ (NH_3)
    n) $[ClBNH]_3$, and from it, $[Cl_2BNH_2]_3$ (NH_3)
    o) $MeCO(O_2H)$ (use an anhydride)

15. Match reactants and products.

    a) $\frac{1}{2} B_2H_6 + 3ROH \rightarrow$ 1) $Al_2Et_4H_2$
    b) $\frac{1}{2} Al_2Ph_6 + H_2NNMe_2 \rightarrow$ 2) $Si(NH)_2$
    c) $Al_2Et_6 + HCl \rightarrow$ 3) $Li[Al(PMe_2)_4]$
    d) $Al_2Et_6 \xrightarrow{\Delta}$ 4) $(Cl_3Si)_2O$
    e) $LiAlH_4 + 4HPMe_2 \rightarrow$ 5) Me_3SiOEt
    f) $2H_3SiNH_2 \xrightarrow{\Delta}$ 6) $P(OR)_3$
    g) $2SiCl_4 + H_2O \xrightarrow[-80°]{\text{"ether"}}$ 7) $ClNF_2$
    h) $Si(NH_2)_4 \xrightarrow{\Delta}$ 8) $B(OR)_3$
    i) $PCl_3 + 3ROH \xrightarrow{NR_3}$ 9) $LiNMe_2$
    j) $PCl_3 + 3ROH \rightarrow$ 10) $HOCl$
    k) $n\text{-}BuLi + Me_2NH \rightarrow$ 11) Al_2Et_5Cl
    l) $Me_3SiNR_2 + EtOH \rightarrow$ 12) $Na[N_2H_3]$
    m) $EtNa + N_2H_4 \rightarrow$ 13) $(H_3Si)_2NH$
    n) $HNF_2 + t\text{-}BuOCl \rightarrow$ 14) $HPO(OR)_2$
    o) $SF_5Cl + 4H_2O \rightarrow$ 15) $[PhAlN_2Me_2]_4$

EPILOG

The direction in this chapter has been to expand upon the descriptive chemistry of the important structural types of Chapter 6 and their reaction pathways of Chapter 7. The reactions given in this chapter have been broadly organized as substitution reactions, for most often we as chemists are faced with a need for converting one reagent into another by simple exchange or substitution of one group for another. In the course of these discussions you have learned to replace a chlorine atom about a central atom by H, C, N, O, and F in nucleophilic forms. This obviously does not cover all possibilities (!), but then our goal has been to gain for you a maximum yield in organization and perspective of non-metal reactions, under the time constraints surrounding a non-specialist first course. Should you decide that non-metal chemistry offers you the excitement and challenge you seek as a professional chemist, be assured there is plenty of opportunity awaiting your creative talents.

A SAMPLER
  The Language of Coordination Chemistry
  Spectral/Magnetic Characteristics of Transition Ion Complexes
  Structure Puzzles

THE TRANSITION METALS AND PERIODIC PROPERTIES

THE MOLECULAR ORBITAL MODEL
  General View of MD_6 and MD_4 Structures
  $MD_6(O_h)$
  $MD_4(D_{4h})$
  $MD_4(T_d)$
  The Angular Overlap Model Tenets

  d^* Orbital Energies and Occupation Numbers
  Complex Structures and Preferences
  Experimental Evidence for a Structural Preference Energy and the Ligand Field Stabilization Energy
  Jahn-Teller Distortions from O_h Geometry

ELECTRONIC STATES AND SPECTRA
  Ground and Excited States
  The Energies of Electronic Transitions
  The Spectrochemical Series
  Paramagnetism of Complexes

EPILOG

APPENDIX

9. FUNDAMENTAL CONCEPTS FOR TRANSITION METAL COMPLEXES

This chapter is fundamental to the succeeding five chapters on the structures and reactivities of the transition group compounds. Many of the concepts developed earlier for main group compounds are directly applicable here: the concepts of effective nuclear charge and its relation to radial wave functions, electronic configuration and associated terms (Chapter 1), the VSEPR and LCAO-MO philosophies (Chapters 2 and 4), and the donor/acceptor concepts (Chapter 5). With a thorough understanding of the application of these working tools for atoms with s and p ao's, the extension to those elements with valence d ao's is all that is required. In particular, you will find these concepts helpful in understanding the properties of color, paramagnetism, and structure which form the basis for transition metal chemistry and the thrust of this chapter.

A SAMPLER

This last half of the text addresses the most significant aspects of the chemistry of the *d*-block or transition elements. Until recently, progress in this area has occurred across two broad fronts—classical coordination chemistry and organometallic chemistry—which often have been tended by two separate schools of chemists. It now seems clear that further significant progress in transition metal chemistry will require intimate knowledge of both areas.

By classical coordination chemistry is meant the chemistry of adducts formed by metals in their higher oxidation states (formally $\gtrsim +2$) bonded to inorganic or organic ions or molecules. A companion characteristic is that the ligands bound to the metal ion are predominately sigma donors with moderate to weak π-acceptor or π-donor tendencies. Therefore, ions or molecules such as $[NiCl_4]^{-2}$, $[Co(H_2N{-}C_2H_4{-}NH_2)_2Cl_2]^+$, or $[(C_6H_5)_3P]_2NiCl_2$ can properly be called classical

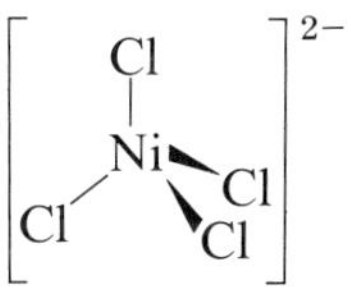

tetrachloronickelate (2−)	dichlorobis(ethylenediamine) cobalt (+)	dichlorobis(triphenyl phosphine)nickel
1	**2**	**3**

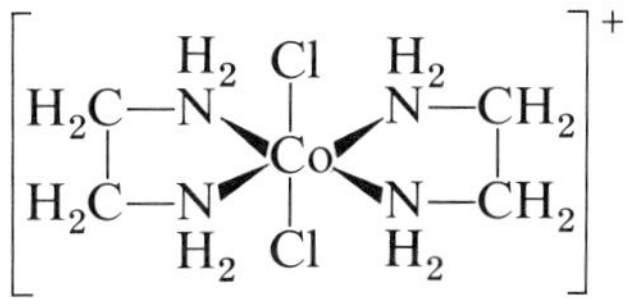

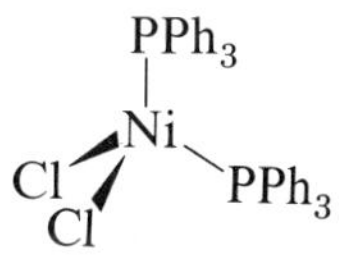

coordination compounds. Interest in both basic and technological research with these materials continues at a rapid pace. Synthetic and structural determination studies are certainly important and have been spurred by progress on the theory of electronic structure. The latter has provided insight into the spectroscopic, magnetic, structural, thermodynamic, and kinetic properties of complexes.

Paralleling the progress in classical coordination chemistry have been the rapid advances made in the coordination chemistry of the transition metals in their lower valence states, particularly in the area of organometallic chemistry. Impetus for these developments was provided by the synthesis of ferrocene (4) in 1951,[1] and homogeneous catalysts such as Vaska's compound, IrCl(CO) [P(C_6H_5)$_3$]$_2$ (5), in the 1960's.[2] Progress in the field has been so rapid and complete that generalizations concerning reactivity and stereochemistry similar to those found in organic chemistry are beginning to emerge.[3] This fascinating subject looms so large on the horizon of inorganic chemistry that Chapters 15–17 are devoted to it.

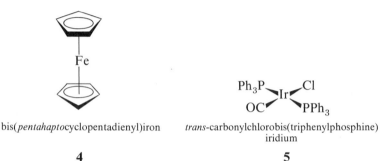

bis(*pentahapto*cyclopentadienyl)iron

4

trans-carbonylchlorobis(triphenylphosphine)iridium

5

The further development of inorganic chemistry requires that new areas be explored and new directions taken. One such area of transition metal chemistry under development is the synthesis and study of compounds containing metal-metal bonds, particularly metal cluster compounds; these are exemplified by structures 6 and 7.[4] Such compounds will be covered in more detail in Chapter 18 on the subject of molecular polyhedra.

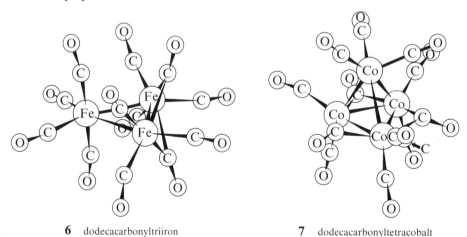

6 dodecacarbonyltriiron

7 dodecacarbonyltetracobalt

---

[1] T. J. Kealy and P. L. Pauson, *Nature,* **168**, 1039 (1951); S. A. Miller, J. A. Tebboth, and J. F. Tremaine, *J. Chem. Soc.,* 632 (1952).

[2] L. Vaska and J. W. DiLuzio, *J. Amer. Chem. Soc.,* **83**, 2784 (1961).

[3] C. A. Tolman, *Science,* **181**, 501 (1973); *Chem. Soc. Revs.,* **1**, 337 (1972); R. Heck, "Organotransition Metal Chemistry: A Mechanistic Approach," Academic Press, New York, 1974.

[4] M. C. Baird, *Prog. Inorg. Chem.,* **9**, 1–159 (1968); F. A. Cotton, *Acc. Chem. Res.,* **2**, 240 (1969).

Another very important field for new development is the study of inorganic compounds of biological interest (see Chapter 19). For example, the coenzyme of vitamin B_{12} **(8)** and a model compound of this vitamin, methylbis(dimethylglyoximato)(pyridine)cobalt **(9)**, have been the subject of extensive research.[5]

8

methylbis(dimethylglyoximato) (pyridine) cobalt

9

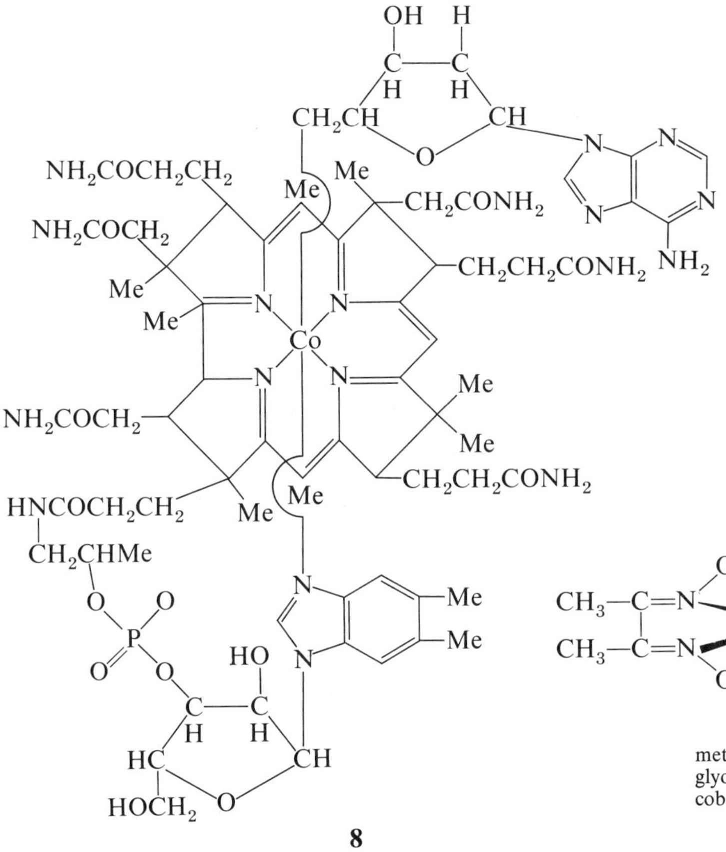

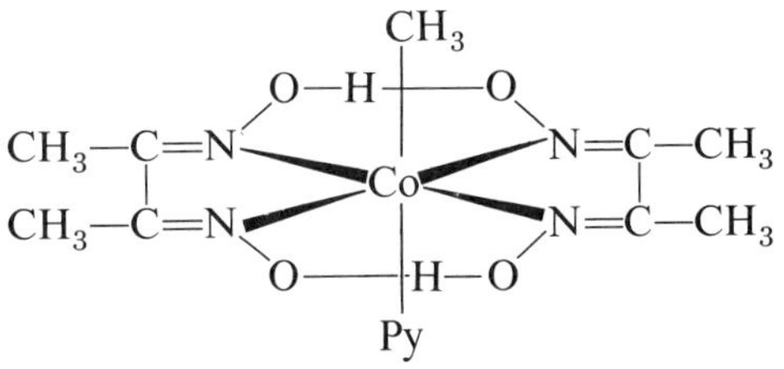

It is clear that, in order to understand the chemistry of vitamin B_{12} coenzyme, metal clusters, and other systems of future importance, chemists should be as knowledgeable as possible in both of the heretofore separate areas of classical coordination chemistry and organometallic chemistry.

THE LANGUAGE OF COORDINATION CHEMISTRY

Compounds **4** through **10** can all be referred to as organometallic compounds, as they have direct metal-carbon bonds. In most of these examples the metal is in a low formal oxidation state (1+, if the C_5H_5 group is considered to be anionic[6]). Compounds **1, 2, 3,** and **11** are classical coordination compounds, compounds wherein the

[5] J. M. Pratt, "Inorganic Chemistry of Vitamin B_{12}," Academic Press, New York, 1971; H. A. O. Hill, J. M. Pratt, and R. J. P. Williams, *Chemistry in Britain,* **5,** 156 (1969); G. N. Schrauzer, *Acc. Chem. Res.,* **1,** 97 (1968).

[6] As either a π radical (C_5H_5) or an anion ($C_5H_5^-$), this species presents another example (Chapter 5) of a π HOMO donor. There is always more or less ambiguity in assigning charges to metal atoms.

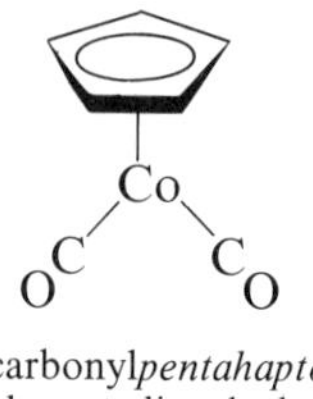

dicarbonyl*pentahapto*-
cyclopentadienylcobalt

10

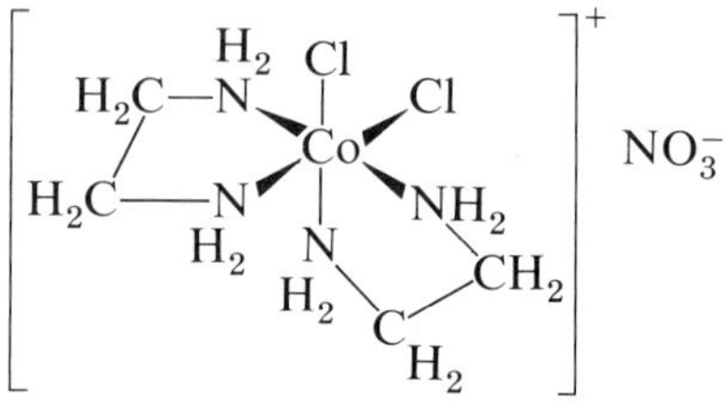

cis-dichlorobis(ethylenediamine)
cobalt (+) nitrate

11

metal is in a "normal" oxidation state (2+ or 3+) and there are no metal bonds to carbon. In all cases, the metal atom is referred to as the *central metal* or *coordinated metal* atom or ion. All of the groups attached directly to the central metal—whether ions or molecules—are the coordinating groups or *ligands*. A ligand attached directly through only one coordinating atom (or using only one coordination site on the metal) is called a *monodentate* ("one-toothed") ligand; carbon monoxide and chloride ion in the complexes above are monodentate ligands. A ligand that may be attached through more than one atom is *multidentate*, the number of actual coordinating sites being indicated by the terms bidentate, tridentate, tetradentate, and so forth. Multidentate ligands attached to a central metal by more than one coordinating atom are called *chelating ligands*. Their complexes are called *chelates*. The ethylenediamine molecule in compound **11** is a bidentate chelating ligand. Ligands such as the *pentahapto*cyclopentadienyl group,[7] C_5H_5, in compound **10** are not usually referred to as multidentate, although, strictly speaking, they are. A multidentate ligand that is attached to more than one central metal is called *bridging*. Therefore, the acetate ion in compound **12** is a bidentate chelating ligand, whereas it is a bidentate bridging group in **13**.[8,9] The whole assembly of one or more central metal atoms with their attached ligands is referred to as a *coordination complex* or, more simply, a complex.

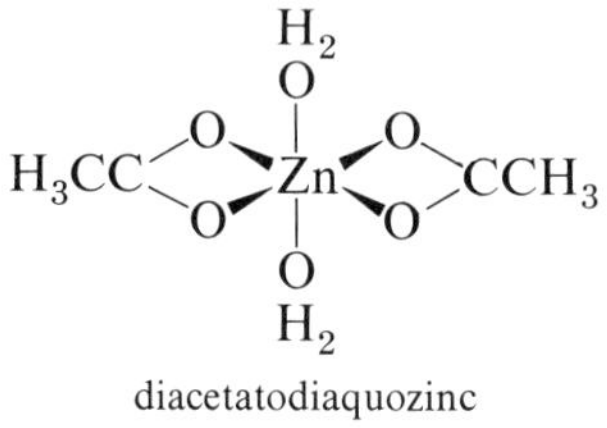

diacetatodiaquozinc

12

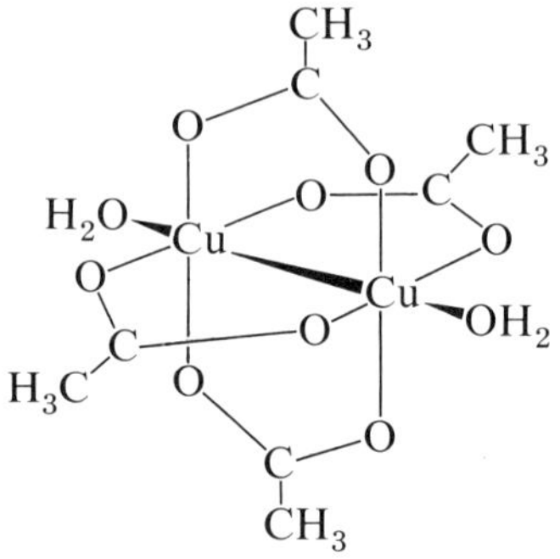

diaquo-μ-tetraacetatodicopper

13

The ligands attached to the central metal may be anionic, neutral, or (less frequently) cationic. In classical coordination chemistry, the number of donor atoms (with one Lewis base electron pair per donor atom) bound directly to the central metal defines the *coordination number*. The number is clearly six for **11** and **12**, but in organometallic compounds containing π Lewis base ligands the coordination number of the central metal is less clearly defined in terms of the number of nearest atoms, for

---

[7] The Greek prefix "hapto" means fasten, with an obvious connotation in the context of **10**.
[8] J. N. Van Niekirk, F. R. L. Schoening, and J. F. de Wett, *Acta Cryst.*, **6**, 720 (1953).
[9] F. A. Cotton, B. G. DeBoer, M. D. LaPrade, J. R. Pipal, and D. A. Ucko, *Acta Cryst.*, Sect. B, **27**, 1664 (1971).

not every such atom possesses a Lewis base electron pair. In such cases (cf. Chapter 15) we fall back to counting the number of donor electron pairs (rather than donor atoms from each ligand). The ligands directly bound to the metal are said to be in the *inner coordination sphere,* and the counter-ions that balance out the charge remaining on the complex after the coordination number of the central metal has been "satisfied" are said to be *outer sphere ions* (*e.g.,* NO_3^- in **11**).

An important aspect of the language of coordination chemistry is the application of a systematic nomenclature system. Given the huge variety of compounds known, it is inevitable that a complete systematization of nomenclature involves idiosyncracies that are difficult to remember. A few simple rules do handle most cases; familiarity with them will make it possible for you to read comfortably this text and the inorganic literature (check their application to the compounds already given):

1. Cations are named first, then anions.
2. Within the coordination complex, the ligands are named before the metal.
3. The charge on the complex, if an ion, is given in parentheses following the name of the metal; if the charge is negative, the suffix "ate" is added to the name of the metal (this is the Ewing-Basset convention).
4. Ligands are given in alphabetical order, for the most part.
5. Anionic ligands are given an "o" suffix; organic ligands (carbon the donor atom) and other neutral ligands are given their customary names as neutral species.
6. To indicate the number of times a simple ligand appears, use a prefix such as di, tri, tetra, penta, or hexa; for a complicated ligand, the ligand name is set off in parentheses with a prefix such as bis, tris, or tetrakis.
7. Bridging ligands are given the prefix μ-.

Probably the most difficult part of transition metal complex nomenclature arises with the organic ligands. Rarely are the IUPAC names used; rather, the practice is to use common names, which seem best learned through example and usage. Some fairly common ligands can be singled out here:

Name	Charge (abbr.)	Formula
ethylenediaminetetraacetato	4− (edta)	$[(O_2CCH_2)_2N(C_2H_4)N(CH_2CO_2)_2]^{4-}$
ethylenediamine	0 (en)	$H_2N(C_2H_4)NH_2$
acetylacetonato	1− (acac)	$[CH_3C(O)CHC(O)CH_3]^-$
glyoximato	1− (glyox)	$[HON(C_2H_2)NO]^-$
ammine	0	NH_3
amine	0	RNH_2, R_2NH, R_3N
aquo	0	H_2O
carbonyl	0	CO
nitrosyl	1+	NO^+
2,2′-bipyridine	0 (bipy)	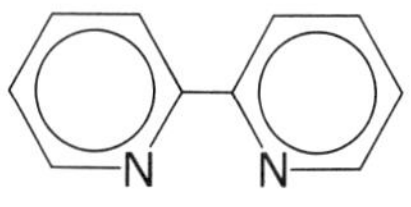
1,10-phenanthroline	0 (phen)	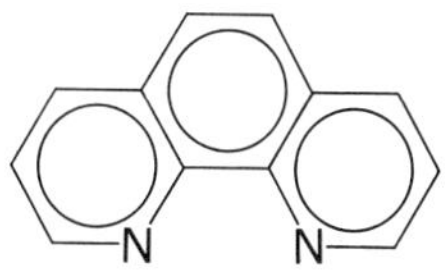

With reference to point 3 above, there is a long-standing alternate convention, called the Stock system, for indicating charge. The Stock system specifies the presumed oxidation state of the metal ion by a Roman numeral in parentheses following the name of the metal. For example, complex **11** appears as ···cobalt(III) nitrate in the Stock system. Generally, there is no ambiguity in applying the charge count procedure to determine the oxidation number of the metal:

$$\text{metal charge} = \text{complex ion charge} - \text{ligand charges}$$

There are times, however, when considerable ambiguity is associated with this method (notably when an easily oxidized or reduced ligand is coordinated to the metal). On the other hand, the Ewing-Basset method of specifying the complex ion charge is experimentally rigorous, free of arbitrary electron counting conventions, and, therefore, to be preferred.

STUDY QUESTIONS

1. Match the formulas with the names.

   a) $[Cr(NH_3)_6]Cl_3$

   b) $K[Co(edta)]$

   c) $K_2[FeO_4]$

   d) $[Cr(NH_3)_2(H_2O)_3(OH)](NO_3)_2$

   e) $[Pt(py)_4][PtCl_4]$

   f) 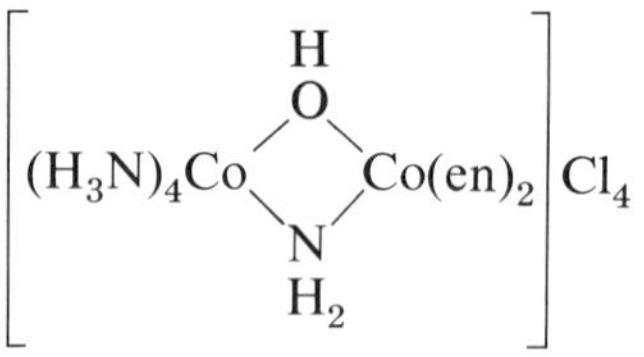 $\left[(H_3N)_4Co \begin{smallmatrix} H \\ O \\ \\ N \\ H_2 \end{smallmatrix} Co(en)_2 \right] Cl_4$

   g) $[Ni(en)_2Cl_2]$

   h) $K[PtCl_3(C_2H_4)]$

   1) Dichlorobis(ethylenediamine)nickel
   2) Potassium trichloro(ethylene)platinate(1−)
   3) Potassium ethylenediaminetetraacetatocobaltate(1−)
   4) Tetrakis(pyridine)platinum(2+) tetrachloroplatinate(2−)
   5) Diamminetriaquohydroxochromium(2+) nitrate
   6) Tetraamminebis(ethylenediamine)-μ-hydroxo-μ-amidodicobalt(4+) chloride
   7) Potassium tetraoxoferrate(2−)
   8) Hexaamminechromium(3+) chloride

2. Give the Stock convention names for the complexes in question 1.

3. How are the names of H_2O and NH_3, in the above examples, exceptions to the rules? What is the difference in spelling for NH_3 and RNH_2?

SPECTRAL/MAGNETIC CHARACTERISTICS OF TRANSITION ION COMPLEXES

Both the colors and the magnetic properties (unpaired electrons) of transition metal compounds were noticed very early and are striking in contrast to those of non-metal compounds. Observations of color and reaction-related color changes of coordination complexes have been important throughout the history of the field. Indeed, Fremy introduced a system of nomenclature based on color in 1852;[10] a few examples are collected in Table 9-1. Inspection of these complexes and their colors shows that only small changes in the coordination sphere apparently can lead to large changes in color. None of the early theories could account for these changes, much less for the fact that the complexes were colored to begin with.

---

[10] E. Fremy, *Ann. Chim. Phys.*, **35**, 257 (1852).

TABLE 9-1

COLORS AND CLASSICAL NAMES OF COORDINATION COMPLEXES

Color	Name	Formula
yellow	luteocobalt chloride	$[Co(NH_3)_6]Cl_3$
yellow-brown	xanthocobalt chloride	$[Co(NH_3)_5NO_2]Cl_2$
yellow	croceocobalt chloride	trans-$[Co(NH_3)_4(NO_2)_2]Cl$
brown	flavocobalt chloride	cis-$[Co(NH_3)_4(NO_2)_2]Cl$
purple-red	chloropurpureocobalt chloride	$[Co(NH_3)_5Cl]Cl_2$
green	chloropraseocobalt chloride	trans-$[Co(NH_3)_4Cl_2]Cl$
violet	chlorovioleocobalt chloride	cis-$[Co(NH_3)_4Cl_2]Cl$
rose	roseocobalt chloride	$[Co(NH_3)_5H_2O]Cl_3$

Any chemical compound is colored if light of a wavelength conjugate to that color is absorbed; the color of the complex derives, of course, from the wavelength of light *transmitted*. The absorption of light occurs because there are accessible higher energy electronic states of the system.[11] In the case of colored coordination complexes, these states must lie between 40 kcal (167 kJ) and 72 kcal (300 kJ) above the ground state (Table 9-2). Therefore, in order to understand the spectroscopic properties of coordination complexes, a theory is required that takes into account the excited electronic states of such systems.

Inspection of the absorption spectra of some coordination complexes shows further qualitative aspects of the absorption process for which an adequate theory must account. The spectra of the oxalate and tris(oxalato)aluminate(3−) ions (Figure 9-1) show that both ions undergo absorption only in the ultraviolet region of the spectrum (both are colorless), and that the oxalate bands undergo only small shifts to the red upon complexation by the Al(3+) ion. Additional features, however, are observed in the spectrum of tris(oxalato)chromate(3−) (Figure 9-2); two bands occur at approximately 17 and 24 kK (the orange and blue regions), so the complex

---

[11] You might wish to review at this point the discussion in Chapter 5 of the colors of I_2 adducts.

TABLE 9-2

WAVELENGTHS AND ENERGIES APPROPRIATE TO COLORS*

Color	Wavelength (nm)	Reciprocal Wavelength (kK)	Energy
Red	714–625	14–16	40–46 kcal 167–191 kJ
Yellow	556	18	51.5 kcal 215 kJ
Green	500	20	57 kcal 238 kJ
Blue	476–400	21–25	60–71.5 kcal 251–298 kJ
Violet	400	25	71.5 kcal 298 kJ

*Units and conversion: 1 nm = 10^{-9} m; 1 kK = 10^3 cm^{-1}; 0.350 kK = 1 kcal = 4.184 kJ.

522 ☐ **FUNDAMENTAL CONCEPTS FOR TRANSITION METAL COMPLEXES**

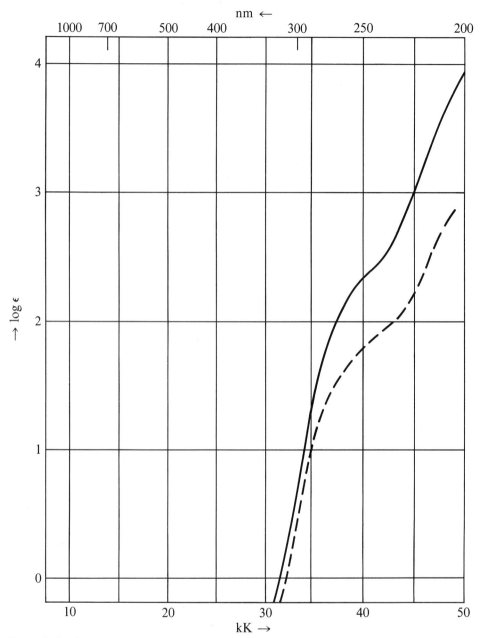

Figure 9-1. Absorption spectra of the oxalate ion (dotted line) and $[Al(C_2O_4)_3]^{3-}$ (solid line). [From H. L. Schlafer and G. Glieman, "Basic Principles of Ligand Field Theory," John Wiley and Sons, Inc., New York, 1969.]

appears red. The principal difference between the Al(3+) and Cr(3+) ions is that the latter has *three d electrons* in its valence shell (beyond an inert gas core). Therefore, it seems safe to conclude that it is the presence of *d* electrons that leads to the absorption of light in the visible region, and, consequently, to a colored complex. Note that the Cr(3+) complex has *two* excited electronic states in the visible range.

Further interest in the problem is stimulated by examination of the spectra of $Ti(H_2O)_6^{3+}$ and $V(H_2O)_6^{3+}$ and of several other Cr(3+) complexes (Figures 9-3, 9-4, and 9-5). The hexaaquotitanium(3+) ion is pale purple-red, since the absorption band is so placed that red and purple are the only two portions of the visible spectrum not absorbed. The hexaaquovanadium(3+) ion, however, is green, because wavelengths to either side of green are absorbed. These spectra show that the number of

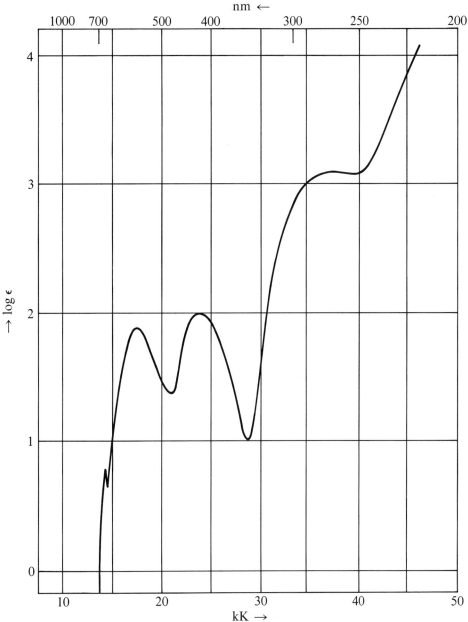

Figure 9-2. Absorption spectrum of $[Cr(C_2O_4)_3]^{3-}$. [From H. L. Schlafer and G. Glieman, "Basic Principles of Ligand Field Theory," John Wiley and Sons, Inc., New York, 1969.]

bands observed and their positions depend on the metal ion in question.[12] Finally, the spectra of complexes of the type $Cr(NH_3)_5X^{2+}$ illustrate the effect of changing the ligand (Figure 9-5).

On the basis of the absorption spectra discussed above, you should draw the following very tentative conclusions: colors of transition metal complexes depend on the metal *and* ligand, and ultimately on the fact that the central metal ion has d electrons in the valence shell. Therefore, a model of electronic structure is needed that will account for these spectral characteristics as well as for certain magnetic characteristics to be mentioned next.

---

[12] Do not be misled into thinking that, since Ti^{3+} is a d^1 ion and V^{3+} is a d^2 species, the number of bands is equal to the number of d electrons. As you shall see, the relationship between the number of bands and number of d electrons is not that simple.

524 ☐ FUNDAMENTAL CONCEPTS FOR TRANSITION METAL COMPLEXES

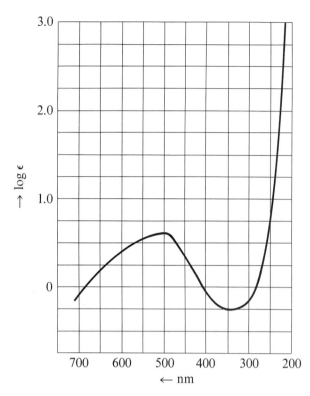

Figure 9-3. Absorption spectrum of $[Ti(H_2O)_6]^{3+}$. [From H. L. Schlafer and G. Glieman, "Basic Principles of Ligand Field Theory," John Wiley and Sons, Inc., New York, 1969.]

A characteristic of almost all the common oxidation states of the transition metal ions is that they contain several d electrons. The effects of spin correlation (Hund's first rule) require that the lowest energy electronic configuration for each ion has a *maximum* number of unpaired electrons. For example, Ti^{3+} should have one unpaired electron, V^{3+} should have two unpaired electrons, Fe^{3+} and Mn^{2+} should have five unpaired electrons, and so forth. While these expectations are realized for the bare ions, such is not always found to be the case in complexes. For example, magnetic studies show that $Mn(H_2O)_6^{2+}$ has a ground electronic state with five unpaired electrons but $Mn(CN)_6^{4-}$ has a configuration with only one unpaired elec-

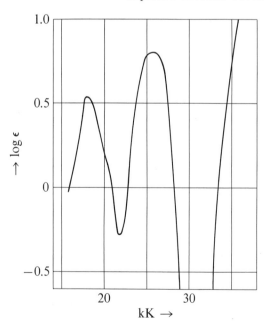

Figure 9-4. Crystalline spectrum of $[V(H_2O)_6]^{3+}$. [From H. L. Schlafer and G. Glieman, "Basic Principles of Ligand Field Theory," John Wiley and Sons, Inc., New York, 1969.]

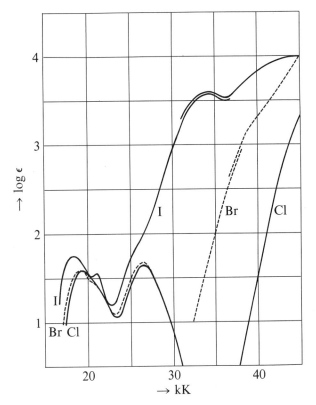

Figure 9-5. Absorption spectra of [Cr(NH$_3$)$_5$X]$^{2+}$ complexes where X = Cl, Br, I. [From H. L. Schlafer and G. Glieman, "Basic Principles of Ligand Field Theory," John Wiley and Sons, Inc., New York, 1969.]

tron! Obviously, there is some major difference in the Mn—OH$_2$ and Mn—CN bonds that accounts for this observation. Upon completing the section on the molecular orbital model for the electronic structures of transition metal complexes, you will see that this model does account for the dependence of metal ion magnetic properties on the ligands.

STRUCTURE PUZZLES

One of the more perplexing characteristics of transition metal complexes is structure. In Chapter 10 you will learn of geometries derived from stoichiometries MD$_n$, where n = 2 to 8 and higher, but for the moment you need to reflect only on the common stereochemistries of MD$_6$ and MD$_4$. For the latter, two limiting structures of T_d and D_{4h} symmetry are found, as well as the in-between D_{2d} form (see Table 9-3). According to the VSEPR model, which worked so well for the non-metal compounds, all four-coordinate complexes should be tetrahedral *if one considers only the metal-ligand bond pairs.* To do so means to ignore the presence of the d valence electrons of the metal and their destabilizing (anti-bonding) role.

This situation is not really so different from that of the non-metal compounds. For example, you will recall that the structure of SF$_4$ is *not* that of a tetrahedral arrangement of fluorines about sulfur; the double occupation of a σ^* mo destabilizes the tetrahedral structure (see Chapter 4, p. 153). The SF$_4$ molecule in fact adopts a C_{2v} structure in which the σ^* pair of electrons of the tetrahedral geometry is converted into a stereochemically active "non-bonding" lone pair at sulfur. A similar analysis (Chapter 4) led to the preferred D_{4h} structure of XeF$_4$ with two Xe pairs in excess of the octet.

Entirely analogous considerations apply to transition metal compounds, with the important distinction that there frequently are too many d electrons present for the complex entirely to avoid population of anti-bonding orbitals. Not all of the central atom electron pairs can become stereochemically active lone pairs. This leads to an

TABLE 9-3

CORRELATION BETWEEN d^n AND SYMMETRY FOR FOUR-COORDINATE STRUCTURES

Complex	Structure	d^n	
$Ni(CO)_4$	T_d	10	
$Cu(NH_3)_4^{2+}$	D_{4h}	9	
$CuCl_4^{2-}$	D_{4h}	9	(Counter-ion = NH_4^+)
	D_{2d}	9	(Counter-ion = Cs^+)
$Ni(CN)_4^{2-}$	D_{4h}	8	
$NiCl_4^{2-}$	D_{2d}	8	
$CoCl_4^{2-}$	T_d	7	
$FeCl_4^-$	T_d	5	
$TiCl_4$	T_d	0	

T_d D_{2d} D_{4h}

important generalization that metal-ligand bonds are often weaker than those associated with non-metal compounds. To illustrate this idea you might find it helpful to think of SF_4 as arising from the combination of S^{4+} and $4F^-$. The central "ion" has one pair of electrons in excess of the four pairs (one from each fluoride) to be used in forming S—F bonds. With $NiCl_4^{2-}$, which may be thought of as arising from combination of Ni^{2+} with $4Cl^-$, the central ion has eight valence electrons in excess of the four pairs to be used in the Ni—Cl adduct bonds. Thus you must recognize that while the bond pair–bond pair repulsions favor a tetrahedral geometry for both SF_4 and $NiCl_4^{2-}$, the excess (over the number of bond pairs) central atom electrons may be given a less destabilizing role to play in a structure other than tetrahedral.

To illustrate for you the idea that there is a definite connection between number of d electrons and structure, a few interesting examples are cited in Table 9-3. In examining the contents of this table (and ignoring the d electrons) note that the VSEPR model works well when the d ao's are empty ($TiCl_4$), half full ($FeCl_4^-$), and full ($Ni(CO)_4$). Appearing somewhat strange is the d^7 case ($CoCl_4^{2-}$), which also follows the VSEPR model, if you ignore the presence of the seven d electrons. The other cases, d^9 and d^8, exhibit either a square planar geometry or a structure intermediate between square planar and tetrahedral (the D_{2d} geometry is referred to as "squashed tetrahedral" or "buckled square planar"). The dependence of the structure of the tetrachlorocuprate anion on the lattice counter-ion suggests that in this case, at least, the energy difference between D_{2d} and D_{4h} structures is small, because the counter-ion size appears to determine the anion geometry through ion packing effects.

As much as possible, we shall treat the chemistry of both high and low valence states of the transition metals as a unified whole. After some preliminary discussion of the general properties of the elements, we shall begin a systematic development of bonding theory as applied to transition metal chemistry. With the bonding model fully developed, you shall be able to understand the factors that determine the structures of transition metal compounds, the many forms of isomeric behavior, and the idiosyncracies of transition metal reactivities.

THE TRANSITION METALS AND PERIODIC PROPERTIES

The transition elements display all of the properties common to metals: they have relatively low ionization potentials, they are good conductors of heat and electricity, and they are lustrous, malleable, and ductile. All of these properties are consistent with the band theory of bonding in elemental solids (Chapter 6). However, many of these properties are quite different in magnitude from those of the non-transition metals; also, we have not discussed their periodicity. It is well worth your time to investigate some of these physical properties for the insight they provide into the unique chemistry of the transition elements.

The most revealing properties of the elements are their binding energies, radii, and ionization potentials. The binding energy of an element can be expressed in terms of the enthalpy change for converting the element in its standard state at 298°K to an ideal gas at the same temperature. That is, the binding energy is equivalent to the heat of sublimation at 298°K. Binding energies[13] for the elements of the three transition metal periods are plotted in Figure 9-6. In all three periods, the binding energy at first increases, reaching a maximum value several times greater than those of the alkali metals, and then begins to fall off somewhere near the middle of the period. This behavior is not so strange as might first appear and may be explained reasonably well on the basis of the band theory. You encountered behavior similar to

---

[13] L. Brewer, *Science,* **161,** 115 (1968).

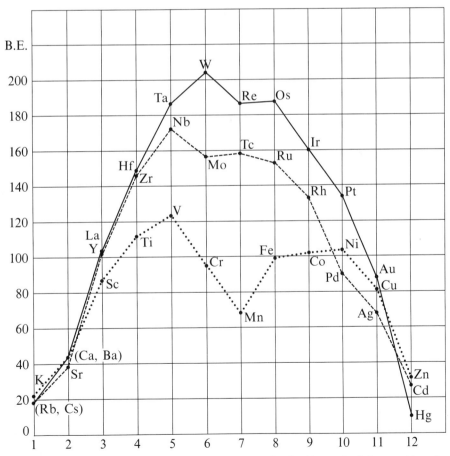

Figure 9-6. Binding energies (kcal/g-atom) of the transition elements as a function of the number of valence electrons.

this in the discussion of homonuclear diatomics in Chapter 4. In proceeding across the second row of elements (Li$_2$, ..., Ne$_2$), a maximum in atomization energy arises near the center of the series (N$_2$). Clearly, this behavior is characteristic of filling a set of bonding, non-bonding, and anti-bonding orbitals, with anti-bonding electron density introducing atom repulsion (in the directed orbital model this corresponds to lone pair repulsions). A similar situation is at hand here; important anomalies, however, prevent one from deriving more than a qualitative understanding of the trends.

First of all, the metals are macromolecules with a near *continuum* of orbital levels. Secondly, it is unclear to what extent the $(n + 1)s$ and $(n + 1)p$ ao's are important. Their separation from the nd ao's becomes smaller in the succeeding transition series ($n = 3, 4, 5$) so as to create additional bonding levels among the less strongly bonding and non-bonding d levels. This rationalizes the next important qualitative feature of Figure 9-6: the binding energies of the heavier elements tend to be greater than those of the corresponding lighter elements.

The radii[14] of the metals are plotted in Figure 9-7. The outstanding feature of this plot is obvious—there is a general decrease in radius for the metals of all three periods as the atomic number increases across a row, a minimum being reached with the Group VIII elements. This change is clearly related to d orbital penetration of core electrons (see Chapter 1); as the atomic number of an element in a period

---

[14] A. F. Wells, "Structural Inorganic Chemistry," Third Edition, Oxford University Press, New York, 1962, p. 983. Remember that problems were encountered in Chapter 6 when we discussed the definition of ionic radii. Much the same difficulties arise here.

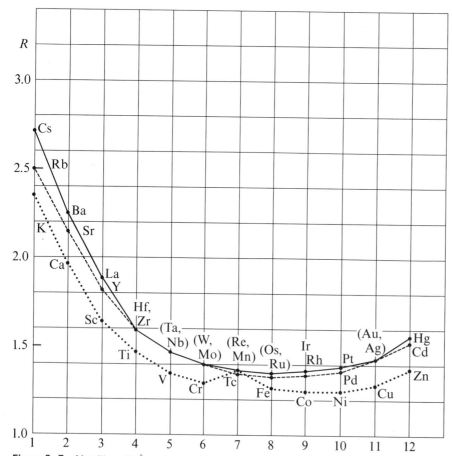

Figure 9-7. Metallic radii (Å) of the transition elements as a function of the number of valence electrons.

increases, there is a general contraction of the core and *d* orbitals, and a size decrease results. Anti-bonding orbital occupation in the later elements in each series tends to increase the M—M distance. Another observation that may be made is that there is a normal increase in size for the heavier elements in any given group; however, the increase is smaller between the fifth and sixth periods than between the fourth and fifth. This difference is again ascribed to the so-called "lanthanide contraction" (refer

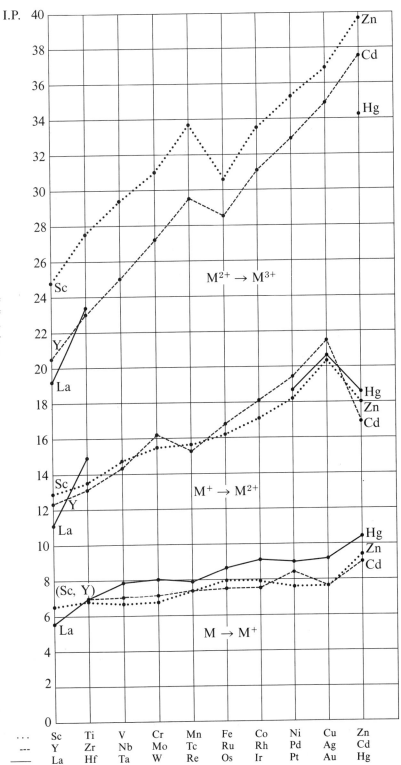

Figure 9-8. Ionization potentials (eV) of the transition elements as a function of the number of valence electrons. The abcissa values range from 3 to 12 for $M \rightarrow M^+$, from 2 to 11 for $M^+ \rightarrow M^{2+}$, from 1 to 10 for $M^{2+} \rightarrow M^{3+}$.

to Chapter 1) and as a consequence of the latter may also be ascribed to greater involvement of $(n + 1)s$ and p ao's in bond formation.

The ionization potential of a metal is an important characteristic of its chemistry; therefore, the variation in ionization potentials in a related series of elements can give you a feeling for the variation in chemical properties in that series. Plotted in Figure 9-8 are the first three ionization potentials[15] for each element in the transition metal series. (a) In each period, there is a general increase owing to the increase in effective nuclear charge across a period. (b) For the M^{2+} ions, all of which have d^n configurations (refer to Chapter 1), a break in this trend occurs at the d^6 case, just as is found for the p^4 cases of the p block elements. (c) Another significant trend is that for elements of a given group, the normal tendency for decreasing I.P. down the group is observed for the process $M^{2+} \to M^{3+}$ and for all ionization steps of Group III elements. The first and second ionization energies of the $5d$ elements following the lanthanide elements, however, are greater than those of the $3d$ congeners. Clearly, the effects of $5d$ penetration of the $4f^{14}$ shell are most pronounced early in the $5d$ period and for low ion charge. There are two rough generalizations to be made from Figure 9-8: (1) The elements early in the transition metal series have lower ionization potentials and should, therefore, reveal a more extensive chemistry of compounds in

[15] C. E. Moore, "Ionization Potentials and Ionization Limits Derived From the Analyses of Optical Spectra," NSRDS-NBS 34, National Bureau of Standards, Washington, D.C. (1970).

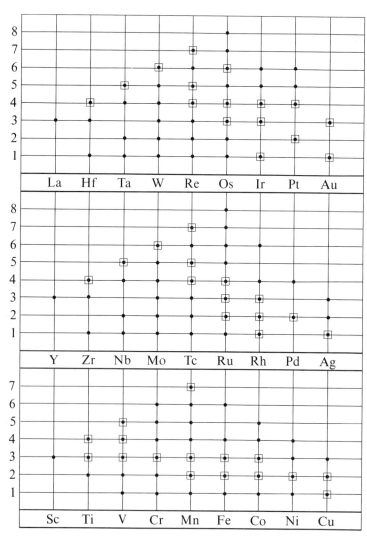

Figure 9-9. Known formal oxidation states of the transition elements; □ designates the most common states.

which the formal oxidation state of the metal is equal to the group number. The later elements have higher ionization potentials, however, and should reveal chemistries dominated by species containing the 2+ ions. (2) The second ionization is so easy that M^+ chemistries are likely to be less common than those of M^{2+}.

Examination of trends in actual oxidation states of the transition metals verifies that statements (1) and (2) are indeed true (Figure 9-9), and the chemistry of the iron sub-group elements (the Fe, Co, and Ni groups) further exemplifies these statements. The most common oxidation state of iron is 3+, although, by oxidizing suspensions of $Fe_2O_3 \cdot xH_2O$ in concentrated base with chlorine, an iron(VI) species, FeO_4^{2-}, can be isolated. The ion is, of course, an extremely powerful oxidizing agent, being stronger than the permanganate ion; FeO_4^{2-} can oxidize NH_3 to N_2 and primary amines and aldehydes to alcohols. No iron compounds with the metal in higher oxidation states have apparently been isolated. Osmium, on the other hand, can be readily converted to an osmium(VIII) species; the normal product of combustion of osmium metal in air is OsO_4, the most common commercial form of osmium. Tetraoxoosmium is a highly selective oxidizing agent, being used in organic chemistry to oxidize olefinic double bonds to give a *cis*-ester that can be converted to the *cis*-dihydroxo compound by treatment with Na_2SO_3.[16]

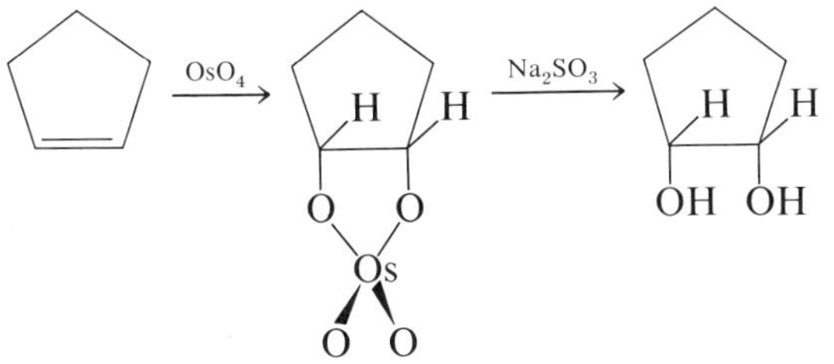

The properties of the elements (Figures 9-6 and 9-7) can be rationalized to a considerable extent in terms of relatively familiar atomic and band theory (mo) arguments. Rationalization of the properties of the ionized elements in their compounds derives from the same viewpoint. This will be our next major topic for discussion.

STUDY QUESTIONS

4. Contrast the variation in E—E bond energy down a main group with that down a transition group. How might the comparison be interpreted in terms of (a) a difference in the number of valence ao types important to E—E bonding, or, equivalently, (b) the number of anti-bonding electrons in balance with the number of bonding electrons?

5. If the *d*-type bonding and anti-bonding mo's are completely filled for a crystal of Zn, how do you explain the fact that Zn is not a gas?

6. Why is there a "curve dislocation" at d^6 in Figure 9-8 for the $M^{2+} \rightarrow M^{3+}$ processes?

THE MOLECULAR ORBITAL MODEL

Several decades ago, theoretical physicists began a search for a model with which to interpret the spectral and magnetic properties of metal ions in solids, *e.g.*, $CoCl_2$. Since the molecular orbital model was just being developed at that time, its applica-

---

[16] L. F. Fieser and M. Fieser, "Reagents for Organic Synthesis," Vols. 1 and 2, Wiley, 1967, 1969.

tion to systems with many electrons was beyond reach. A simplified model that emphasized the nature of the metal was called for, and Hans Bethe, in the 1930's, introduced the crystal field theory (CFT). This is the model you likely learned as a freshman chemistry student. The model is a paradoxical blend of quantum mechanical and classical concepts (metal ion electrons are treated according to wave mechanical principles while ligand electrons and nuclei are represented as point charges). Placing emphasis on the metal ion and the symmetry of the ligand cage, the model was, nevertheless, highly successful in promoting a qualitative understanding of the spectral and magnetic properties of metal ions in ligand cages. This is true in spite of the fact that the model is woefully inadequate when subjected to rigorous examination. As inorganic chemists enhanced their theoretical sophistication and their experimental results began to surpass the ability of the CFT to account for those results, there began a natural evolution toward the molecular orbital theory as a theoretical model.

In the spirit of genuine scientific progress, C. K. Jørgensen has taken a refreshing "tongue-in-cheek" view of this evolution (scientists have a sense of humor, too!). To quote from his essay, "Time is Going—Fixists and Mobilists," in his book "Modern Aspects of Ligand Field Theory" (American Elsevier Publishing Co., N.Y., 1971): "The electrostatic theory of the 'ligand field' combined the rather unusual properties of giving an excellent phenomenological classification of the energy levels of partly filled d and f shells of transition group complexes and having an absolutely unreasonable physical basis!" You should find the following excerpt from his essay on the evolution of scientific and social systems entertaining.

> It is well-known that ideas crystallize (or fossilize) as we grow older, and one of the reasons of recent unrest in young students is that the world changes so rapidly that the experience of old people is of negligible, if not negative, value, and it has become a justified working hypothesis always to do the opposite of what people did 50 years earlier. A clear-cut case was Svante Arrhenius having great troubles as a young chemist of getting the idea of ionic dissociation of salts in aqueous solution admitted; and older, he believed so strongly in Kohlrausch's law for conductivity that he did not realize that a strong sodium chloride solution is sufficiently different from water as bulk solvent that, already for this reason, it cannot be ascribed physical significance to determine a degree of dissociation 0.83 from measurements of the electric conductivity. In a recent paper on Teilhard de Chardin, Hourdin (1968) cites the previous author for the opinion that people are either *fixists* (wanting the rigid *status quo*) or *mobilists* (enjoying change) and that this dichotomy runs its way through all organizations and political parties. This opinion can readily be construed to mean that John XXIII and Mao Tse-Tung belong to the same class, what is true anyhow for no other reason than the proof given on p. 181 of O.N.O.S. that two things cannot be entirely different in all properties. Returning to science, "ligand field" theory has produced a most instructive debate between fixists and mobilists. When the electrostatic model of the "ligand field" collapsed in 1956, there were immediately fixists (and very intelligent ones) to propose that the old model had something to be said for it; it was simple (because familiar) and closely connected with the principles of physics. At the next level of defence, it was a hypothetico-deductive working model to be tried to see how long it would work. The fixists continued to write books and review articles about the electrostatic model and established a reasonably large shelf of a library of which at most a-tenth was written before the break-down. Band-wagons are sometimes funeral carriages. To be honest, one has a certain sympathy with pedagogues. It is not easy not to over-simplify, and one of the dark-coloured angels is always there to whisper that the young people would understand a clear-cut simple model more rapidly, and that, after all, they have all the rest of the life to be told about the subsequent developments. Hence, the electrostatic model finished its career as a starting point for text-books; it has been digested almost as many times as fertile soils by earthworms, and students tell us today how the success of "ligand field" theory clearly shows that NiF_6^{-2} is almost ionic, and how interesting it is that iron (III) acetylacetonate is covalent (because it can be sublimed) but nevertheless has $S = \frac{5}{2}$. It is a typical example of its kind (and Bernard Shaw said that it is better to be a bad example than to be of no use at all) but among its precursors were N. Bohr's hydrogen atom haunting the secondary schools and used by managers of nuclear energy commissions for their letterheads, including Sommerfeld's elliptic orbits. Another relic is the hybridization

theory applied to d-group complexes; text-books state that $Co(NH_3)_6^{+3}$ and NiF_6^{-2} obviously have d^2sp^3 hybridization since they have $S = 0$. One of the sad things about ignorance of formal logics is that many scientists no longer realize that (A implies B) does not imply that (B implies A). There is no doubt that many mobilists would have been more popular, though less efficient, if they had been more diplomatic and patient toward the fixists. The prototype of a fixist is the decimal classifyer of books always having the (perhaps subconscious) dream of having all these discoveries finished with, so a nice orderly system could finally be constructed.

With an eye on the present and toward the future, it is fortunate that significant strides have been made in the qualitative application of molecular orbital concepts. It would help the general advance of science little if the molecular orbital model were to remain so complex that only the specialist could use it. In the first part of this section you will find an extension of the general mo approach of Chapter 4, to be followed in the next part by a powerful, semi-quantitative simplification (the angular overlap model, AOM) of the general approach.

NOTE TO INSTRUCTORS

The material in the immediately following section is an application of the TASO/MO concepts of Chapter 4 to central elements with d, s, and p ao's; it should be treated in a manner consistent with your choice of level in Chapter 4. Time and interest may dictate passing over this section to the following one on the Angular Overlap Model.

GENERAL VIEW OF MD_6 AND MD_4 STRUCTURES

As will be discussed in more detail in the next chapter, transition metal compounds exhibit a wide range of structural types. Concern here will be with the most widely encountered geometries—the octahedral structure for six-coordinate and the square planar and tetrahedral structures for four-coordinate complexes. The TASO (terminal atom symmetry orbital) patterns for the ligands were developed for these structural types back in Chapter 4, so at this point the symmetry properties of the valence orbitals for the transition metal atoms need only be examined in relation to the symmetries of the ligand orbitals (see study question 4–30). *Like-symmetry metal ao's and TASO's will overlap and give rise to metal-ligand bonding.*

$MD_6(O_h)$

For the "first row" ($3d$) transition elements, the valence orbitals are the filled $3s$ and $3p$ orbitals, the partially occupied $3d$ and, of variable importance, the empty $4s$ and $4p$ orbitals. As you saw in Chapter 1, the $3s$ and $3p$ orbitals are quite stable (low in energy) and radially contracted so that they are rather ineffective in their overlap with the TASO's. It is generally a good approximation to treat the $3s$ and $3p$ electron pairs as core electrons. The $4s$ and $4p$ orbitals lie at somewhat higher energy than the $3d$ and are fairly diffuse so that their overlaps with the TASO's are reduced from that shown by the non-metal cases, where the s and p ao's of the central atom are the primary valence orbitals (the s/p-TASO interaction is discussed in Chapter 4, Figure 4–19). Thus, you may adopt the implications of Figure 4–19 for the $4s$ and $4p$ ao interactions with the TASO's, with the modification that the a_{1g} and t_{1u} bonding and anti-bonding mo energies are not so strongly displaced from the M and D ao energies. At this point you need to consider how the $3d$-TASO interactions modify the energy diagram and the extent of the M—D interaction.

Figure 9–10 shows sketches of the metal d orbitals which, by comparison with the TASO sketches in Chapter 4 for the EX_6 case, are seen to match symmetries with the

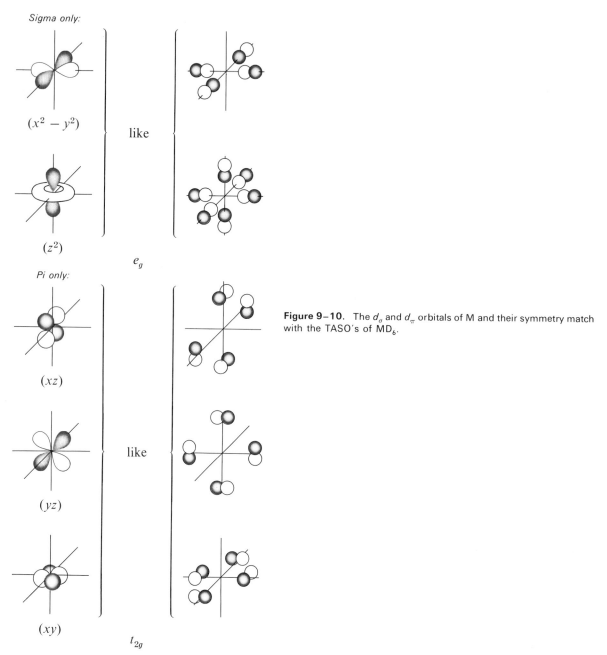

Figure 9-10. The d_σ and d_π orbitals of M and their symmetry match with the TASO's of MD_6.

e_g TASO pair ($d_{x^2-y^2}$ and d_{z^2}) and the t_{2g} TASO set (d_{xz}, d_{yz}, and d_{xy}). Thus, bonding and anti-bonding mo's between metal and ligands are to be formed for both the e_g and t_{2g} sets. Note that use is made of the concept that π interaction is weaker than σ. Figure 9-11 is a modification of Figure 4-19 showing the interactions of the metal d orbitals with the (non-bonding in Figure 4-19) e_g and t_{2g} TASO's. When the terminal atoms satisfy the octet rule, so that the TASO's are completely filled, the metal $3d$ electrons remain to be distributed among the t_{2g}^* and e_g^* mo's.

Owing to the weakness of the interaction of the metal $4s$ and $4p$ ao's with the appropriate TASO's, the principal M—D bonding arises from the two σ and three π mo's labelled e_g and t_{2g}. Thus, five electron pairs (supplemented to some extent by the four pairs of a_{1g} and t_{1u} symmetry) are responsible for the binding of six ligands to the metal. *Depending on the electron populations of the anti-bonding t_{2g}^* and e_g^* mo's, the M—D bonds will exhibit energies ranging from "normal" to weak.*

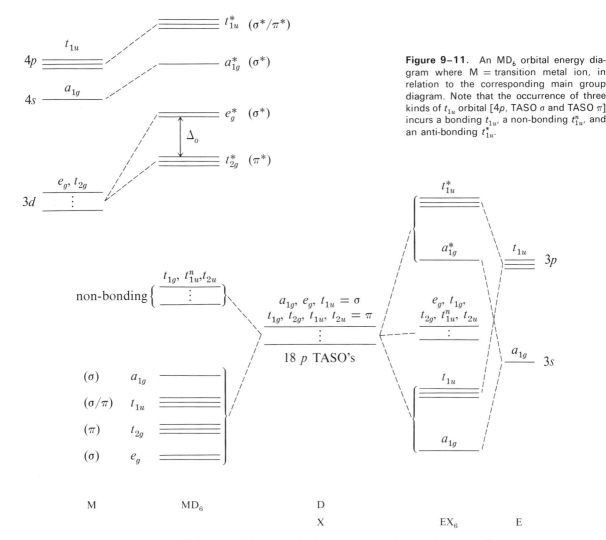

Figure 9-11. An MD_6 orbital energy diagram where M = transition metal ion, in relation to the corresponding main group diagram. Note that the occurrence of three kinds of t_{1u} orbital [$4p$, TASO σ and TASO π] incurs a bonding t_{1u}, a non-bonding t_{1u}^n, and an anti-bonding t_{1u}^*.

An important feature of Figure 9-11 is that the bonding e_g and t_{2g} mo's are well described as primarily the e_g and t_{2g} TASO's with some mixing in of metal e_g and t_{2g} ao's; *these bonding molecular orbitals have considerably more ligand character than metal character.* The electrons in these orbitals have greater probability at the ligands than at the metal and thus reflect the polarity of the metal-ligand bonds. The converse applies to the anti-bonding e_g^* and t_{2g}^* mo's, which are *more metal-like in character.* In cases of better metal-ligand energy match and/or overlap, the covalency is higher and the e_g, t_{2g}, e_g^*, and t_{2g}^* electrons become more evenly distributed between M and D. A second important feature of Figure 9-11 is the energy gap between the t_{2g}^* orbitals and the e_g^* orbitals. *This energy gap is crucial to the theory of the color and magnetism of transition metal complexes because it determines which is the HOMO set.* By commonly accepted convention, this energy splitting of t_{2g}^* and e_g^* orbitals is called Δ_o; it has a magnitude of 25 to 100 kcal/mole. According to the LCAO theory, this splitting arises from the difference in degree of M/D σ and M/D π bonding.

Now you are in a position to give a preliminary interpretation to the spectrum of $Ti(OH_2)_6^{3+}$ shown in Figure 9-3. The principal feature of this spectrum is the occurrence of a single absorption band at approximately 500 nm, which is not present in the spectrum of pure H_2O or gaseous Ti^{3+}. The energies of the filled molecular orbitals of the coordinated water molecules all lie below that of the t_{2g}^* metal orbitals, while the O—H σ^* molecular orbitals of the coordinated water molecules all lie above the e_g^* anti-bonding orbitals. As the Ti^{3+} ion has a $3d^1$ configuration, the complex ion molecular orbital occupations are ... $(t_{2g}^*)^1(e_g^*)^0$; that is, the highest-energy occupied molecular orbitals (HOMO) are the triply degenerate set of Ti—O π^* mo's. In addition, the lowest-energy unoccupied molecular orbitals (LUMO) of the complex are the Ti—O σ^* anti-bonding pair of molecular orbitals. Even less-than-sophisticated molecular orbital treatments indicate that the difference between the t_{2g}^* and e_g^* orbitals corresponds to the energy of a 500 nm photon. A logical interpretation of the occurrence of the band in the middle of the visible region is that the absorption process corresponds to an electronic transition from the configuration $(t_{2g}^*)^1(e_g^*)^0$ to the configuration $(t_{2g}^*)^0(e_g^*)^1$.

You may wonder at this point why we have not dealt with energy *terms* of the complex ion, rather than configuration energies, since the distinction between orbital configurations and ion states is of great importance (Chapter 1) in dealing with the absorption spectra of free atoms and ions. In fact, you must also think in terms of states instead of configurations for complex ions. However, because the ions are not free, with degenerate d ao's (a situation to which the vector model is applicable), but are perturbed by the surrounding ligands, the vector model must be modified. To deduce *all* the terms arising from a given complex ion configuration, resort must be made to some advanced theoretical principles, beyond the reach of a first inorganic course. Later you will see that you do not need to know *all* of these terms; just the terms of the same spin multiplicity as the ground term are needed, and this less formidable derivation is pursued in a later section of this chapter. At that time you will also learn how to estimate the relative energies of the various complex ion terms. For the moment, you need only appreciate that the energies of transitions involving a change in the numbers of t_{2g}^* and e_g^* electrons will depend on the magnitude of the splitting, Δ_o, as well as on a change in interelectron repulsions. Furthermore, as this splitting is directly related to the difference in anti-bonding characters of the t_{2g}^* and e_g^* mo's, Δ_o responds oppositely to M—D σ and π bonding; in short, the positions of complex ion absorption bands provide information about the metal-ligand interaction. It suffices for the present to say that the configuration $(t_{2g}^*)^1(e_g^*)^0$ gives rise to only a single ground term (six microstates) called $^2T_{2g}$ while the configuration $(t_{2g}^*)^0(e_g^*)^1$ gives rise to only a single excited term (four microstates) called ^2E_g. Consequently, only one so-called "d-d" transition is expected for an octahedral complex of Ti^{3+}. Transitions from the ground state of an O_h complex of a d^2 ion, such as V^{3+} (Figure 9-4), are expected to lead to two or three such d-d bands in the visible region.

More needs to be said about the role of the t_{2g}^* orbitals in metal ion complexes, and some generalizations are possible. When the ligands possess *occupied* π TASO's (either $p\pi$ lone pairs or π bonding ligand mo's), the ligand TASO's are generally found at lower energy than the metal orbitals. (This is the situation depicted in Figure 9-11.) It is important also to recognize that some ligands, such as CO, not only have stable occupied π mo's but also have corresponding π^* mo's that are empty and form TASO's of the same symmetry patterns as the occupied π mo's. When these π^* TASO's are low enough in energy, their t_{2g} component can significantly interact with the metal t_{2g} ao's. If this happens, the π anti-bonding nature of the metal t_{2g}^* mo's is reduced by virtue of the stabilizing (π bonding) effect of mixing a higher-energy orbital into a lower-energy one. *Ligands that have only occupied π TASO's (halides, water, etc.) are called π donor ligands because their π electrons, by virtue of covalency, are delocalized onto the metal.* The ligands with empty π^* TASO's can experience back-donation of electron density from the metal into the π^* TASO's, by virtue of the mixing of metal t_{2g}^* and the π^* TASO's. In this sense, the π^* TASO's play the role of acceptor orbitals. *Ligands for which t_{2g}^*/π^* TASO interaction is appreciable are called π acceptor ligands.* The π donor and π acceptor cases are compared in Figure 9-12 [this concept is entirely equivalent to the retrobonding phenomenon in BH$_3$CO and BH$_3$PF$_3$, as discussed in Chapter 5, and you may wish to refer to Chapter 5 for preliminary discussion of the M—C bonding in M(CO)$_x$]. Be sure to note that the π acceptor interaction tends to increase Δ_o and, if strong enough, could shift the visible region electronic transitions into the ultraviolet region and render the compounds colorless, as with Cr(CO)$_6$. Additionally, such interactions tend to strengthen the

Figure 9-12. Comparison of π donor and π acceptor ligands.

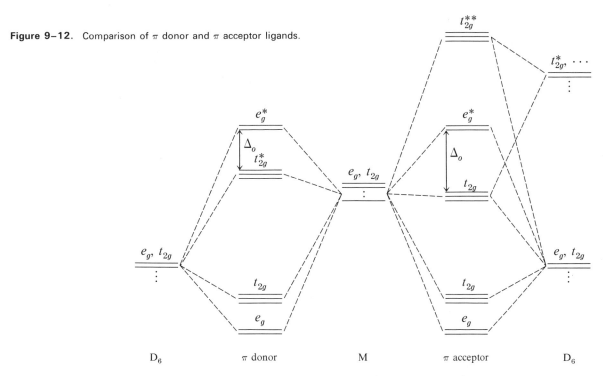

metal-ligand bonds by superimposing metal-ligand π bonding character upon the normal π anti-bonding nature of the t_{2g}^* mo's.

$MD_4(D_{4h})$

Another important structure in transition metal chemistry is that of the square planar arrangement of ligands about a central metal. To describe the orbital energy diagram for such a case, you may begin with the results of Chapter 4 for the non-metal EX_4 case. Realizing, as just discussed for the MD_6 structure, that the metal $4s/4p$ ao's weakly play the role of the non-metal valence s and p, you may now modify Figure 4-18 for the inclusion of central atom d ao's as the primary valence orbitals. Sketches of the central d ao's are given in Figure 9-13, where they are matched with the corresponding TASO's. In the non-metal case of Chapter 4, the TASO's labeled b_{1g}, b_{2g}, and e_g were all non-bonding orbitals and only a_{1g} was involved in bonding the central atom (through its valence s ao). Here you find the primary interaction of the a_{1g} TASO to be with the d_{z^2} ao, and the other, formerly non-bonding TASO's now enter into the bonding picture.

Before attempting to draw the energy level diagram, first compare the two σ (a_{1g} and b_{1g}) and three π interactions. It should be obvious that the $d_{x^2-y^2}$ interaction with the b_{1g} TASO will be greater than that of the d_{z^2} with the a_{1g} TASO. The energy diagram and concept of the metal-ligand bonding should reflect this expectation. Similarly, the in-plane π interaction (b_{2g}) should be greater than the out-of-plane π interaction (e_g). The diagram should reflect this difference, too. However, there is an ambiguity regarding the placement of the relatively weak a_{1g} σ interaction with respect to the b_{2g} and e_g π interactions. In practice, the relative placements of these mo's varies with the metal/ligand combination, and you can go no further at this point with regard to attempts to place the final mo's. (What a_{1g}, b_{2g} order is favored with a strong π donor?)

Figure 9-14 is the result of the construction of the orbital energy diagram, and from it you should draw several important conclusions. First of all, in relation to the $MD_6(O_h)$ case, the lower symmetry of $MD_4(D_{4h})$ results in less degeneracy of the anti-bonding mo's. No single orbital splitting factor, as Δ_o in the MD_6 case, will suffice. Owing to the fact that the metal-ligand sigma bonding is confined to the plane of the

538 □ FUNDAMENTAL CONCEPTS FOR TRANSITION METAL COMPLEXES

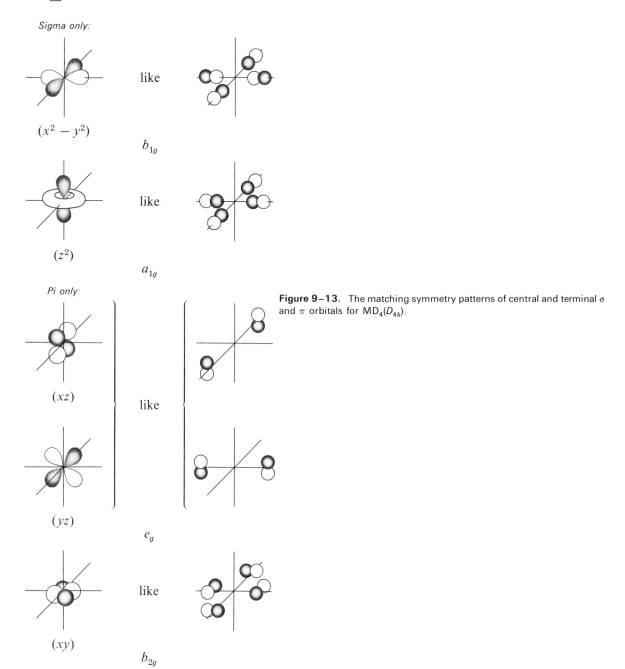

Figure 9-13. The matching symmetry patterns of central and terminal σ and π orbitals for $MD_4(D_{4h})$.

molecule, the $d_{x^2-y^2}$ and d_{z^2} orbitals no longer do an equally good job of binding the ligands; so the degenerate e_g^* pair for MD_6 correlates with non-degenerate σ^* mo's in MD_4. The loss of ligands from the apical positions also results in a partial loss of degeneracy for the t_{2g}^* mo's of the MD_6 case. Secondly, you find two σ bonding TASO pairs of electrons and three TASO π bonding pairs. The bonding that is due to these five pairs and to the three pairs[17] involving the s and p ao's is offset by metal electrons occupying the anti-bonding mo's. The most commonly occurring $MD_4(D_{4h})$ complexes are those with metal ions with eight d electrons ($Ni^{2+}, Pt^{2+}, Pd^{2+}$); in these cases,

---

[17] There are only three bonding mo's from the $4s/4p$ set because the $4s$ contribution is superimposed upon the a_{1g} mo already in existence because of the $3d_{z^2}$ overlap with the a_{1g} TASO. The additional mo incurred by inclusion of the $4s$ ao in the mo diagram is nominally "non-bonding" in character.

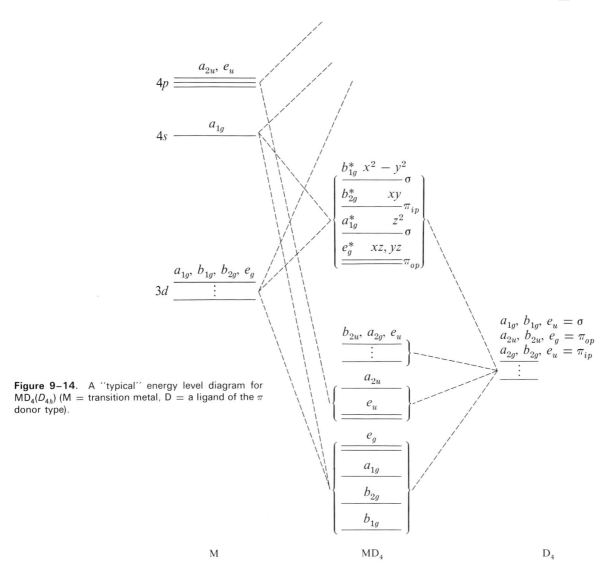

Figure 9-14. A "typical" energy level diagram for $MD_4 (D_{4h})$ (M = transition metal, D = a ligand of the π donor type).

then, you find a net of four bonding pairs. Finally, the MD_4 electronic spectrum will be richer than that of MD_6 because electron excitation from the ground configuration $\ldots (e_g^*)^4 (a_{1g}^*)^2 (b_{2g}^*)^2 (b_{1g}^*)^0$ should lead to successively higher energy configurations in which an electron is promoted into (b_{1g}^*) from (b_{2g}^*), from (a_{1g}^*), and from (e_g^*).

A particularly interesting and important example of metal-ligand bonding occurs with the ligand ethylene (C_2H_4). While the chemical consequences of the bonding will be explored in the chapters on organometallic complexes, the stage is now set for an examination of how this ligand is bound to metal ions. The best known example of ethylene as a ligand in a transition metal compound is Zeise's salt, $PtCl_3(C_2H_4)^-$, where C_2H_4 has replaced Cl^- in the also well-known anion, $PtCl_4^-$. The overall structure of this anion is that of the square plane as shown below.

540 ☐ **FUNDAMENTAL CONCEPTS FOR TRANSITION METAL COMPLEXES**

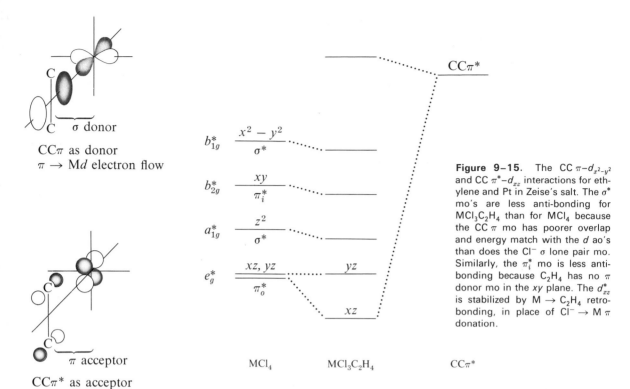

Figure 9–15. The CC π–$d_{x^2-y^2}$ and CC π^*–d_{xz} interactions for ethylene and Pt in Zeise's salt. The σ^* mo's are less anti-bonding for $MCl_3C_2H_4$ than for MCl_4 because the CC π mo has poorer overlap and energy match with the d ao's than does the Cl^- σ lone pair mo. Similarly, the π_i^* mo is less anti-bonding because C_2H_4 has no π donor mo in the xy plane. The d_{xz}^* is stabilized by M $\rightarrow$ C_2H_4 retro-bonding, in place of $Cl^- \rightarrow$ M π donation.

Note that the C—C axis is perpendicular to the square plane. This most interesting structural feature needs to be explained, and the background of the square planar orbital system provides a good beginning. Ask yourself which orbitals on the ethylene molecule might be available for interaction with the metal d orbitals. As the CC and CH sigma orbitals are either of very low energy (the occupied bonding orbitals) or of very high energy (the unoccupied, anti-bonding orbitals), only the CC π and π^* orbitals (HOMO and LUMO) have energies similar enough to those of the metal d orbitals to be effective in the binding of ethylene to the metal. The orbital sketches in Figure 9–15 show that the CC π bonding orbital, occupied by a pair of electrons, can play the role normally ascribed to a ligand σ orbital (σ, that is, with respect to the metal-ligand interaction). The ethylene π^* orbital has the symmetry normally associated with π-type interactions between metal and ligand. Consequently, the ethylene molecule interacts with the metal both as a σ *donor* (through the CC π orbital) and as a π *acceptor* (through the CC π^* orbital). While there could be interaction between the CC π^* orbital and the d_{xy} orbital if the ethylene were rotated 90° (placing the CC axis in the square plane), presumably steric interactions with the adjacent ligands make this structure less favorable than the one with the CC axis perpendicular to the square plane. An important consequence of the π interaction between the metal d_{xz} orbital, say, and the CC π^* orbital is to stabilize the xz component of e_g^* (cf. Figures 9–12 and 9–15) and remove the degeneracy of the d_{xz}, d_{yz} pair. This stabilization is due to the lower energy of the d_{xz} than of the CC π^* orbital and may be contrasted with the π donor character of the Cl^- ligands.[18]

---

[18] Notice that were the CC π^* to lie in the xy plane, its role would be to stabilize the b_{2g}^* mo ($\sim d_{xy}$). This would have a greater stabilizing effect on the molecule as a whole because b_{2g}^* is anti-bonding between M and all three Cl^-, while the d_{xz} component of e_g^* is anti-bonding between M and only one Cl^-. Apparently, steric congestion is energetically more important than the additional stabilization of having CC π^* in-plane. A more complete analysis of this is to be found in Chapter 16.

MD$_4$(T$_d$)

In addition to the square planar structure for complexes of MD$_4$ stoichiometry, the tetrahedral geometry is sufficiently prevalent to require special mention. As before, the previous discussion in Chapter 4 provides a starting point for assessment of the role of metal d orbitals in this structure. Initially limiting the bonding interactions to those of the sigma type, Figure 9-16 shows sketches of the metal d ao's and their symmetry likenesses to the ligand $p\sigma$ TASO's. The originally degenerate d ao's are separated, in the presence of the ligands, into two sets; three ao's (xz, yz, and xy) form bonding/anti-bonding combinations with the $p\sigma$ TASO's of t_2 symmetry and a pair (z^2, $x^2 - y^2$) become non-bonding and are labeled e. Allowing for the fact that the metal possesses higher-lying s and p ao's (of a and t_2 symmetry; consult Figure

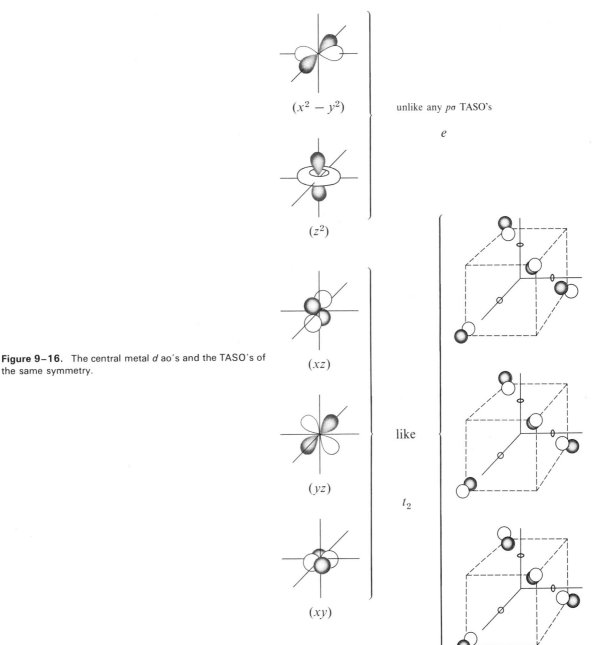

Figure 9-16. The central metal d ao's and the TASO's of the same symmetry.

4-16), you recognize two additional features of the M—D bonding: first, the totally symmetric $p\sigma$ TASO (of a symmetry) is somewhat bonding owing to the weak interaction with the metal s ao, and, second, that the t_2 and t_2^* mo's experience some contribution from the metal p ao's. The fact that the metal p ao's are higher-lying than the d ao's means that inclusion of the p ao's in a refinement of the theory will result in some stabilization of the t_2 and, more importantly, the t_2^* mo's. The ideas to this point are summarized in Figure 9-17.

For those ligands with all TASO's filled with electron pairs, a net of four bonding pairs of electrons is predicted (one s type and three d/p type), the stabilizing effect of which is offset by any metal electrons occupying the t_2^* mo's. Also observe that a separation is found for the e and t_2^* mo's; this energy gap is labeled Δ_t, in analogy with the MD_6 case. It is important to note that, for a given metal and ligands, Δ_t *should be considerably smaller than* Δ_o. This follows from the fact that the overlaps between the metal ao's (xz, yz, xy) and the t_2 TASO's cannot be nearly as good as those incurred by the head-on relation between the e_g metal ao's and TASO's of the MD_6 case. This has important consequences when one compares, as you shall shortly, the spectroscopic, magnetic, and thermodynamic properties of $MD_4(T_d)$ and $MD_6(O_h)$ complexes.

Having commented on the role of M—D π interactions in the square planar and octahedral structures, a few observations are in order for the tetrahedral case as well. In Chapter 4 we commented on the fact that the $p\pi$ TASO's are difficult to draw effectively, and simply stated that the eight $p\pi$ ligand ao's give rise to $p\pi$ TASO's of e, t_1, and t_2 symmetry. With respect to the problem at hand, this means that the metal ao's (z^2, $x^2 - y^2$) of e symmetry are not strictly non-bonding but are weakly π^* anti-bonding mo's (weakly because the metal ao/TASO alignment is not optimum, as just brought out for the σ orbitals). Furthermore, the t_2^* mo's, described up to this point as σ anti-bonding mo's, take on pi anti-bonding character by virtue of the fact that the $t_2 \pi$ TASO's lie lower in energy. Accordingly, Figure 9-17 is modified, with the result shown in Figure 9-18. Finally, recalling the concept of π donor and π acceptor ligands as used in the MD_6 case, when the ligands D possess low-lying, empty π^* mo's (giving ligand π^* TASO's of e, t_1, and t_2 symmetry), involvement of these TASO's with the e^* and t_2^* mo's will stabilize the latter by endowing them with M—D π bonding character. As found in Chapter 2 for tetrahedral non-metal compounds, it is not strictly possible to identify the t_2 and t_2^* mo's as purely σ or purely π; relative to the MD_6 case, it is much more difficult to interpret changes in Δ_t from one complex to another in terms of metal-ligand sigma and pi interactions. The e mo's are purely π in character, but the t_2^* mo's are a mix of M—D σ and π interactions.

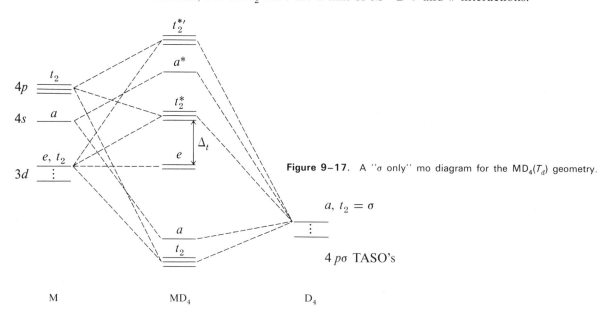

Figure 9-17. A "σ only" mo diagram for the $MD_4(T_d)$ geometry.

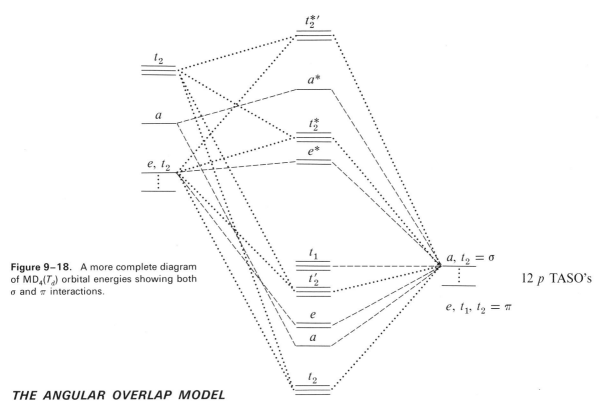

Figure 9-18. A more complete diagram of $MD_4(T_d)$ orbital energies showing both σ and π interactions.

THE ANGULAR OVERLAP MODEL

Tenets

The approach we choose to follow from this point on is that called the *Angular Overlap Model*[19] (abbreviated AOM), for it is a simple first approximation to the full mo model and embodies all the characteristics of metal-ligand interactions important to an understanding of the principles of complex structure, magnetism, and color. Its primary value lies in its aid to estimating the TASO stabilization and central atom ao destabilization in MD_n structures. Therefore, it guides us to the key orbital interactions as a basis for molecular structure and reactivity.

To introduce the application of the model to d valence ao's, we will consider a ligand hybrid ao located along the $+z$ axis and directed toward the metal ion centered at the origin:

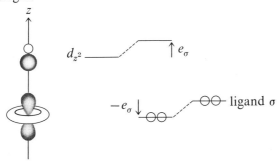

The overlap of the ligand σ ao and the metal d_{z^2} ao results in the creation of bonding and anti-bonding mo's, as you are well aware by now. The extent of metal-ligand

---

[19] The references which can be cited for the AOM all presuppose at least some familiarity on the part of the reader with the mathematical machinery of LCAO-MO theory. Probably the least demanding in this regard is E. Larsen and G. N. LaMar, *J. Chem. Educ.*, **51**, 633 (1974), which may be consulted for references to C. K. Jørgensen's and C. E. Schäffer's writings in the research literature. Unfortunately, the *J. Chem. Educ.* reference has typographical errors on page 635. *Caveat emptor!*

orbital interaction (the downward shift in energy of the ligand ao and the upward shift in energy of the metal ao)[20] will be represented[21] by the quantity e_σ (here defined as + signed). The magnitude of e_σ depends on the magnitude of the metal-ligand overlap integral (the detailed characteristics of the metal and ligand ao radial and angular functions) and on the energy match between the ao's. *Therefore, e_σ is a property of both the metal and the ligand bound to it.*

It is of great interest to understand why some complexes adopt a square planar and others a tetrahedral geometry; it is important to know, then, how the metal-ligand interaction depends on the orientation of the ligand about the metal d_{z^2} and other d ao's. For example, moving the ligand down into the xy plane should reduce

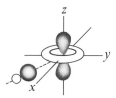

the d_{z^2}–ligand interaction because it is now the less highly directed collar of the metal ao that overlaps the ligand ao. Accordingly, this *angular* change in the position of the ligand, without changing the metal-ligand distance, scales down the stabilization of the ligand ao, and the destabilization of the metal ao, to a fraction of e_σ. Even more dramatic is the effect of moving the ligand off the z axis by half the tetrahedral angle,

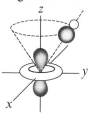

for this is exactly the angle that the d_{z^2} nodal cone makes with the z axis. In this event, the metal-ligand interaction goes to zero because the *angular* part of the overlap integral goes to zero.

The mathematical techniques actually used to compute the variation of the interaction between a metal ao and the ligand ao as a function of the angles θ and ϕ (Figure 1-2) about the metal ion are clearly beyond the purpose of the text. This in no way interferes with your use of the results (Table 9-4), which are evaluated for you for commonly encountered positions of ligands about a metal ion. Each ligand is considered to have one sigma ao (an hao in the spirit of item 5.c, p. 164) and two pi type ao's (or mo's in some cases), and the table entries are the scaling factors for e_σ, the unit of σ interaction, and for e_π, the unit of π interaction, both defined by the standard orientations:

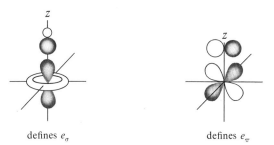

defines e_σ defines e_π

Recall the generalization from Chapter 4 that π interactions are normally less than the corresponding σ interactions; here $e_\pi < e_\sigma$. Additionally, e_σ is positive, whereas e_π may be positive (*for π donors*) or even negative (*for good π acceptor ligands*).

---

[20] In actuality, the downward shift is slightly smaller than the upward shift; if you have forgotten this, consult your physical chemistry text.

[21] Our use of e_σ corresponds to the use of l_σ in reference 19.

THE MOLECULAR ORBITAL MODEL 545

TABLE 9-4

ANGULAR SCALING FACTORS FOR e_σ AND e_π WHEN EACH LIGAND HAS 1 σ AND 2 π ORBITALS.

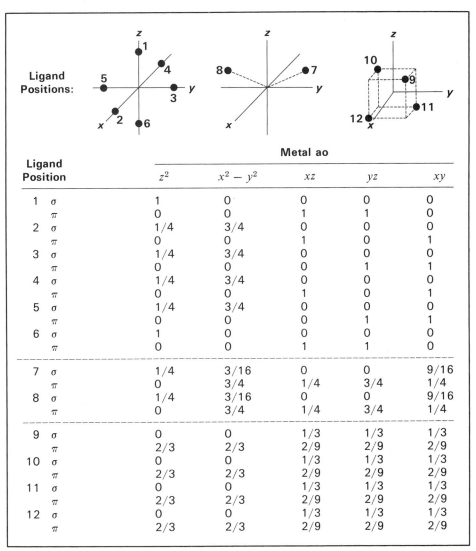

Ligand Position		Metal ao				
		z^2	$x^2 - y^2$	xz	yz	xy
1	σ	1	0	0	0	0
	π	0	0	1	1	0
2	σ	1/4	3/4	0	0	0
	π	0	0	1	0	1
3	σ	1/4	3/4	0	0	0
	π	0	0	0	1	1
4	σ	1/4	3/4	0	0	0
	π	0	0	1	0	1
5	σ	1/4	3/4	0	0	0
	π	0	0	0	1	1
6	σ	1	0	0	0	0
	π	0	0	1	1	0
7	σ	1/4	3/16	0	0	9/16
	π	0	3/4	1/4	3/4	1/4
8	σ	1/4	3/16	0	0	9/16
	π	0	3/4	1/4	3/4	1/4
9	σ	0	0	1/3	1/3	1/3
	π	2/3	2/3	2/9	2/9	2/9
10	σ	0	0	1/3	1/3	1/3
	π	2/3	2/3	2/9	2/9	2/9
11	σ	0	0	1/3	1/3	1/3
	π	2/3	2/3	2/9	2/9	2/9
12	σ	0	0	1/3	1/3	1/3
	π	2/3	2/3	2/9	2/9	2/9

For quick reference, the following list summarizes the common structures for two through six ligands and the location of ligands with reference to Table 9-4.

Structure	Atoms
Linear	1 and 6
Trigonal planar	2, 7, and 8
Square planar	2-5
Tetrahedron	9-12
Trigonal bipyramid	1, 2, 7, 8, and 6
Square pyramid	1-5
Octahedron	1-6

d^* Orbital Energies and Occupation Numbers

To determine the first approximation values of the d ao energy shifts arising from overlap with the ligand orbitals, you simply sum over all ligands the coefficients of the appropriate type (σ,π) in each column of Table 9-4. For the octahedron, the upward shift in the $(z^2)^*$ mo energy is simply $1 + \frac{1}{4} + \frac{1}{4} + \frac{1}{4} + \frac{1}{4} + 1 = 3$ (the unit is e_σ). The same result is obtained for $(x^2 - y^2)^*$. Similarly, the d_{xz}, d_{yz}, and d_{xy} ao's shift as a degenerate set of three by an amount $4e_\pi$. The energy gap between the σ^* and π^* mo's is $3e_\sigma - 4e_\pi$, and this is the quantity defined earlier as Δ_o.

The d ao shifts for the D_{4h} and T_d structures are obtained similarly, and those shifts are summarized below in a form useful for future reference. You should find it instructive to compare these diagrams with those of the preceding section and note the σ^*/π^* nature of the various mo's.

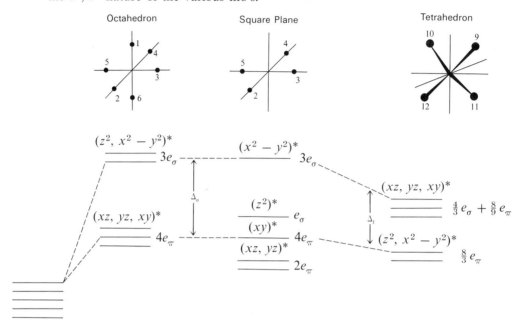

Several interesting features of these diagrams are immediately obvious. First of all, the ligand environment of the $(x^2 - y^2)^*$ and $(xy)^*$ mo's of the square planar geometry are the same as in the octahedral geometry, so the energy gap between them is simply Δ_o in both cases. For the square planar geometry, the position of the $(z^2)^*$ mo relative to the $(xy)^*$ orbital is dependent upon the relative magnitudes of e_σ and e_π, for if e_σ is smaller than $4e_\pi$ then the $(z^2)^*$ mo will lie below the $(xy)^*$ mo. Examination of the tetrahedral case shows the orbital splitting to be much smaller than in the octahedral case, and in fact this simple AOM approximation yields the important result (assuming no difference in M—L distance) that Δ_t is only about $\frac{1}{2}\Delta_o$:

$$\Delta_t = \frac{4}{3}e_\sigma + \frac{8}{9}e_\pi - \frac{8}{3}e_\pi = \frac{4}{3}e_\sigma - \frac{16}{9}e_\pi = \frac{4}{9}(3e_\sigma - 4e_\pi)$$

$$\Delta_t = \frac{4}{9}\Delta_o$$

So far, the issue of assigning valence electrons to the t_{2g}^* and e_g^* mo's of the MD_6 case and to the t_2^* and e^* mo's of the $MD_4(T_d)$ case has been skirted. The critical question surrounding the assignment of a ground state configuration is whether or not Hund's rule is still applicable to the d^* mo's. In other words, if Δ_o and Δ_t are fairly small, the spin correlation forces between electrons may determine that Hund's rule is to be followed. On the other hand, if the splitting of anti-bonding mo's is large enough, the lowest-energy configuration will result by half-filling, and then filling, the t_{2g}^* (e^* in the T_d case) mo's before assigning electrons to the e_g^* (t_2^*) mo's. The Hund's rule case, where electron repulsions (not the energy splitting) dominate, is called the *high spin* case; the half-filling/filling situation is called the *low spin* case.

For octahedral complexes, the concept of high and low spin cases is without meaning when there are fewer than four and greater than seven d electrons to be assigned to the anti-bonding mo's. Furthermore, in practice, it appears that the Δ_t *is usually too small for anything but the high spin case to result*[22] for MD_4 (although the distinction would be meaningless for metal ions with fewer than three and greater than six valence electrons).

Pursuing these ideas in a slightly more quantitative way, you want to determine whether a high spin configuration $\cdots (t_{2g}^*)^m(e_g^*)^n$ is more or less stable than the low spin configuration $\cdots (t_{2g}^*)^{m+n}(e_g^*)^0$. The energy difference between the two configurations may be approximately thought of as arising from two contributions; energy may be released on permitting the ne_g^* electrons to occupy the t_{2g}^* mo's of lower energy, but at the expense of greater repulsions between the mt_{2g}^* electrons and the n electrons in question (because the spin correlation energy for like-spin electrons is lost and because the charge correlation energy is increased when two electrons possess the same orbital function). Representing the (average) increase in repulsion energy for each of the n electrons as P (the pairing energy) and the difference in t_{2g}^* and e_g^* orbital energies as Δ_o, the energy change on passing from high to low spin is $n(P - \Delta_o)$. Thus, if $P > \Delta_o$, the energy change will be positive and the high spin configuration is favored; whereas if $\Delta_o > P$, the energy change is negative and the low spin configuration is favored.

$$\Delta E = E_{HS} - E_{LS} = (P - \Delta_o)$$

$$\Delta E = 2(P - \Delta_o)$$

The question now is, "Can one predict for a given complex whether a high or low spin configuration is appropriate?" To do so requires that you be able to estimate the relative magnitudes of P and Δ_o for any complex. Since molecular orbitals are delocalized over metal *and* ligand atoms, P quite obviously depends not only on the metal but also on the ligands in question. A similar dependency on the metal/ligand combination is expected for Δ_o. It is fortunate that in most cases, where the metal-ligand bonds are highly polar, the anti-bonding mo's are mainly metal d ao's with some moderate to small contamination by the ligand orbitals, and the electron repulsions for the anti-bonding mo's are highly characteristic of the metal and the ion charge. For the estimation of Δ_o you must recognize that some metal/ligand combinations produce strong bonding and large Δ_o's while others result in small Δ_o's. *The terms "large" and "small" are used in relation to the size of P. At least one important generalization is possible: high metal ion charge leads to "large" Δ_o, and ligand properties of strong σ donor (e_σ large) and weak π donor (or strong π acceptor) character ($e_\pi \lesssim 0$) lead to "large" Δ_o.*

To help clarify the competition between P and Δ_o in determining the low or high spin character of O_h complexes, the following tables present some values of P (these are nominal values calculated from spectra of *free* ions and so are probably up to 20 per cent higher than in complexes, where the repulsion energies are reduced by delocalization of the electrons onto the ligands) and some Δ_o values for H_2O as ligand (extracted from the electronic spectra of the complexes by methods to be detailed later). With the exception of Mn^{3+} and Co^{3+}, the Δ_o values fall in the range from 30

---
[22] Only when extremely bulky, strong ligands are involved is this generalization excepted. $Cr(N(SiMe_3)_2)_3NO$ is a rare example: D. C. Bradley, *Chem. Britain*, **11**, 393 (1975).

to 70 per cent of the *free ion P* value, and it is not surprising that the $M(OH_2)_6^{n+}$ are high spin cases. This is consistent with the expectation that H_2O does not form very strong adduct σ bonds and is a π donor (albeit with but a single $p\pi$ electron pair). The Mn^{3+} and Co^{3+} Δ_o's are 80 and 90 per cent of the respective *free ion P* values. Depending on how extensive the covalencies of the M—D bonds are, the *complex ion P* values could be reduced below the Δ_o's. In fact, $Mn(OH_2)_6^{3+}$ is a high spin case while $Co(OH_2)_6^{3+}$ is a low spin example. Thus, Mn^{3+} behaves as does Fe^{3+}, and it is not until Co^{3+} that the energy match between metal and H_2O orbitals is sufficiently good to produce a low spin complex.

P VALUES[23] (IN UNITS OF kK[24])			
d^4	d^5	d^6	d^7
Cr^{2+} 20.4	Mn^{2+} 23.8	Fe^{2+} 19.2	Co^{2+} 20.8
Mn^{3+} 25.2	Fe^{3+} 29.9	Co^{3+} 23.6	

Δ_o VALUES[25] WITH H_2O (UNITS OF kK[24])				
d^3	d^4	d^5	d^6	d^7
V^{2+} 11.8	Cr^{2+} 14.0	Mn^{2+} 7.5	Fe^{2+} 10.0	Co^{2+} 10.0
Cr^{3+} 17.6	Mn^{3+} 21.0	Fe^{3+} 14.0	Co^{3+} ~21 (oxidizes water)	

At the other extreme of ligand behavior you find CN^-, a strong σ donor (e_σ large) also capable of acting as a π^* acceptor ($e_\pi < 0$). For example, Δ_o values with Fe^{2+} and Co^{3+} are large (33.0 and 33.5 kK) and CN^- complexes are generally low spin. Now you can appreciate the facts revealed on p. 524 that $Mn(H_2O)_6^{2+}$ has a ground electronic configuration with five unpaired electrons [Δ_o small $\Rightarrow$ a $(t_{2g}^*)^3(e_g^*)^2$ configuration to follow Hund's rule] while $Mn(CN)_6^{4-}$ has only one unpaired electron [Δ_o large $\Rightarrow$ a $(t_{2g}^*)^5(e_g^*)^0$ configuration]. NH_3 is an example of an intermediate ligand that gives low spin $Co(NH_3)_6^{3+}$ but high spin $Co(NH_3)_6^{2+}$ complexes. Here again [cf. $Co(OH_2)_6^{2+}$ and $Co(OH_2)_6^{3+}$] is the effect of increasing the metal ion charge. NH_3 is a fairly strong σ donor but a poor π donor ($e_\pi \approx 0$, since the NH_3 π TASO's are actually N—H bonding orbitals).

Two interesting trends, which can be anticipated, are observed in P and Δ_o. For pairs of isoelectronic ions (Cr^{2+},Mn^{3+}/Mn^{2+},Fe^{3+}/Fe^{2+},Co^{3+}), the ion of greater charge has greater effective nuclear charge (refer to Chapter 1) for the valence shell electrons. As a result, there is an increase in both P and Δ_o for a given ligand.[26] Similarly, for a series of ions of given charge (Cr^{2+},Mn^{2+},Fe^{2+},Co^{2+};Mn^{3+},Fe^{3+},Co^{3+}), the value of P is maximum for those ions with five $3d$ electrons. In other words, the pairing energy reaches a maximum for the half-filled shell, a trend you should recognize as having seen before, when you analyzed (in Chapter 1) the discontinuity in ionization potentials for the elements (B,C,N,O,F), (Al,Si,P,S,Cl), and the I.P. trend for $M^{2+} \to M^{3+}$ (p. 530). Both the ionization potential trends for these elements and the trend in P for transition metal ions have a common origin in the spin and charge correlation energies of electrons in the same valence shell.

---

[23] L. E. Orgel, *J. Chem. Phys.*, **23**, 1819 (1955); J. S. Griffith, *J. Inorg. Nucl. Chem.*, **2**, 1, 229 (1956).
[24] The symbol K stands for the unit 1 Kaiser = 1 cm^{-1}. From the relationship $\lambda \nu = c$, we have $1/\lambda = \nu/c \equiv \tilde{\nu}$. $\tilde{\nu}$ is the wave number of a photon of frequency ν and is generally expressed in Kaisers.
[25] T. M. Dunn, *et al.*, "Some Aspects of Crystal Field Theory," Harper and Row, New York, 1965.
[26] P and Δ_o increase because of radial contraction of the d ao's, and Δ_o increases because of better energy match between the metal ($3d,4s,4p$) and ligand orbitals. In fact, because of better overlap, the $d\sigma^*$ mo's should be more sensitive to energy match improvement than the $d\pi^*$ mo's.

STUDY QUESTIONS

(These questions will take on added meaning in Chapter 12, Electron Transfer Reactions.)

7. Focusing your attention on the HOMO energies of $Co(NH_3)_6^{2+}$ and $Co(OH_2)_6^{2+}$, which half-cell potential should be more positive or less negative?

$$Co(NH_3)_6^{2+} \to Co(NH_3)_6^{3+} + e^-$$

or
$$Co(OH_2)_6^{2+} \rightarrow Co(OH_2)_6^{3+} + e^-$$

8. Which reaction is accompanied by the smaller change in metal-ligand distance?

$$Fe(CN)_6^{3-} \rightarrow Fe(CN)_6^{4-}$$

or

$$Co(NH_3)_6^{3+} \rightarrow Co(NH_3)_6^{2+}$$

(Answer on the basis of the change in numbers of t_{2g}^* and e_g^* electrons.)

9. Describe by words or sketch the metal and ligand orbitals that contribute to the e^* and t_2^* mo's of Figure 9-18. Does the inclusion of ligand π TASO's tend to increase Δ_t relative to its value in the "σ only" scheme (Figure 9-17)?

10. Using 34 kK as a nominal value of Δ_o for CN^- with a transition metal ion, identify any metals on p. 548 for which $\Delta_t > P$.

Complex Structures and Preferences

Now the tools are at hand for a discussion of the important consequences of d electron occupation of the σ^* and π^* mo's. To begin with, examine the d^0 case with an octahedral structure. Since the $(z^2)^*$ mo is destabilized by $3e_\sigma$, *the corresponding TASO of the same symmetry[27] must be stabilized by the same amount* (see Figure 9-10), and this TASO is doubly occupied with a pair of electrons; therefore, *the TASO contributes $-6e_\sigma$ to the electronic energy* of the complex. Similarly, the TASO with a symmetry matching that of the $d_{x^2-y^2}$ ao is stabilized by $3e_\sigma$ and the associated electron pair also contributes $-6e_\sigma$ units to the molecular energy. The total energy of the bonding pairs is therefore 12 units of sigma interaction—a value equal to twice the number of ligands. This is a general result you can depend on. *The sum total σ bonding electron energies is equal to twice the number of ligands* (12 units for the octahedron, 8 for both tetrahedral and square planar geometries). By the same token, if each ligand possesses two π-type TASO electron pairs, the stabilization of these electrons is *four times the number of ligands* (in units of e_π).

It is important to realize that the TASO stabilization of $8e_\sigma$ is *the same for both the square planar and tetrahedral geometries. That a given metal ion may favor one or the other structure derives from the way in which the d electrons are distributed among the anti-bonding (d^*) mo's*. While it is *not necessary* to do so, the analysis may be simplified by assuming that the e_π factor is no more important than other minor corrections that a more thorough treatment makes to the e_σ parameter; π effects shall, therefore, be ignored. In this case the square planar and tetrahedral orbital diagrams reduce to

```
   (x² − y²)*
   ─────────── 3eσ
                              (xz, yz, xy)*  4/3 eσ
     (z²)*                    ───────────
   ─────────── eσ
                              ───────────

   ───────────  zero          ───────────
   ───────────

        D₄ₕ                        Tᵈ
```

Now for each d^n case, put in the appropriate number of electrons and add the cumulative energies of these anti-bonding electrons to the TASO electron energies of

---

[27] The "z^2" TASO has the quantitative form $(1/\sqrt{12})[2\phi_1 - \phi_2 - \phi_3 - \phi_4 - \phi_5 + 2\phi_6]$ and is *lowered* in energy by $3e_\sigma$. Similarly, the "$x^2 - y^2$" TASO is $(\frac{1}{2})[\phi_2 - \phi_3 + \phi_4 - \phi_5]$ and is *lowered* by $3e_\sigma$.

$-8e_\sigma$. Further, recognize the two extremes of electron configuration—high spin (complexes with weakly basic ligands) and low spin (complexes containing strongly basic ligands).[28] As noted earlier, the orbital splitting of $\frac{4}{3}e_\sigma$ for the tetrahedral geometry is not likely to be greater than the d electron pairing energy, so you need only consider the high spin cases for this geometry—that is, each d^n configuration is achieved by first placing one electron in each orbital, starting from the bottom and working up, before assigning any orbital double occupancy. With the D_{4h} geometry you need to note that the 1.0 e_σ splitting of the $(z^2)^*$ mo from the lower triad should be less than the pairing energy, but that the splitting of the $(x^2 - y^2)^*$ mo from the $(z^2)^*$ mo by 2.0 e_σ may be greater than P.[29] Thus, low and high spin configurations must be distinguished for the D_{4h} geometry in the following way. In the low spin case, the lowest four orbitals are half-filled and then filled before occupancy is assigned to the $(x^2 - y^2)^*$ mo. For the high spin case, even $3e_\sigma$ is less than P and the filling of orbitals with electrons proceeds as in the tetrahedral case. A plot of the results in the form of a "square planar preference energy" $[E(D_{4h}) - E(T_d)]$ is given in Figure 9-19(A).

$$E(D_{4h}) = (n_{z^2} + 3n_{x^2-y^2} - 8)e_\sigma$$

$$E(T_d) = \left(\frac{4}{3}n_\sigma - 8\right)e_\sigma$$

$$E(D_{4h}) - E(T_d) = \left(n_{z^2} + 3n_{x^2-y^2} - \frac{4}{3}n_\sigma\right)e_\sigma$$

---

[28] More correctly, complexes of large or small $(e_\sigma - \frac{4}{3}e_\pi)$, respectively.

[29] Assuming that $e_\pi \approx 0$, the generalization that $P > \frac{4}{9}\Delta_o = 1.3e_\sigma$ for all known cases establishes the guideline that orbitals split by less that $1.3e_\sigma$ are to be treated as "degenerate" by electron correlation forces.

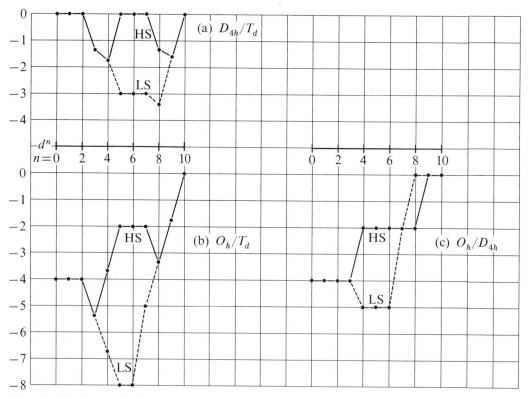

Figure 9-19. Structural preference energies. (A) $E(D_{4h}) - E(T_d)$; (B) $E(O_h) - E(T_d)$; (C) $E(O_h) - E(D_{4h})$; units = e_σ. A negative value means that the first named geometry is preferred.

A negative value for the difference means that the D_{4h} structure better minimizes the anti-bonding effects of the d electrons and is favored.

It is apparent from Figure 9–19(A) that the VSEPR/steric forces that favor the T_d geometry are unopposed for the d^n ions with $n = 0, 1, 2$, and 10 only. For all others, the magnitude of e_σ is highly important in determining the preference for D_{4h}. For sufficiently weak ligands (e_σ small, high spin case) it is clear that VSEPR forces are unopposed for the $n = 5, 6$, and 7 ions; for the $n = 3, 4, 8$, and 9 ions the observed structures are likely to be intermediate (D_{2d}) between square planar and tetrahedral when e_σ is small. The geometries in Table 9–3 for Cl^- as a ligand are completely explained in this way. For the stronger ligands NH_3 and CN^- you find examples of a preference for a low spin, square planar geometry—as predicted by the low spin graph of Figure 9–19(A).

By way of conclusion, *the formation of a T_d or D_{2d} complex is likely for the $n > 2$ cases only by ligands that weakly interact with the metal ion because of intrinsically weak M—D bonds and/or because of unusual ligand steric requirements. More basic or less hindered ligands will prefer to bind the metal with a planar geometry.*

Some useful conclusions can also be drawn from a comparison of the six- and four-coordinate structures. This has been done graphically in Figure 9–19(B) and (C), where it is immediately seen that both high and low spin octahedral complexes are preferred over the high spin tetrahedral structures [Figure 9–19(B)]. This suggests that in addition to the basicity and steric factors listed above for the preparation of tetrahedral complexes, you must realize that stoichiometry control at a 4:1 ligand-to-metal-ion ratio is an important consideration for formation of tetrahedral complexes.

A summary of the effects of d^* occupation (from Figure 9–19) is as follows:

I. $MD_6 \geq D_{4h} \geq T_d$ as a general preference order.

II. Large e_σ; low spin
   a) O_h and $D_{4h} > T_d$
   b) $O_h > D_{4h}$, except for d^8 and d^9; $[Ni(CN)_4]^{2-}$ and $[Cu(NH_3)_4]^{2+}$ are examples.
   c) VSEPR and/or steric congestion favors D_{4h} over O_h in d^8 and d^9 cases.
   The formation of two additional M—D bonds favors O_h in the other cases. Extreme VSEPR and steric congestion favor T_d over D_{4h}.

   Small e_σ; high spin
   a) $O_h \geq D_{4h} \geq T_d$
   b) Among the MD_4 structures, $d^{3,4,8,9}$ are likely to be D_{2d} (competition between d^* occupation and VSEPR/steric bulk); $d^{5,6,7}$ should favor T_d (no d^* occupation preference).
   d^5: $FeCl_4^-$ (T_d)
   d^7: $CoCl_4^{2-}$ (T_d)
   d^8: $NiCl_4^{2-}$ (D_{2d})
   d^9: $CuCl_4^{2-}$ (D_{2d} or D_{4h}, depending on cation present)

EXPERIMENTAL EVIDENCE FOR A STRUCTURAL PREFERENCE ENERGY AND THE LIGAND FIELD STABILIZATION ENERGY

To further illustrate applicability of the structural preference energy, consider the semi-quantitative estimates[30] of the energies of the reactions

$$6H_2O_{(g)} + MCl_4^{2-}{}_{(g)} \rightarrow 4Cl^-_{(g)} + M(OH_2)_6^{2+}{}_{(g)}$$
$$\quad\quad\quad\quad\quad\quad T_d \quad\quad\quad\quad\quad\quad\quad\quad O_h$$

for M = Mn through Zn. It is very difficult to determine the appropriate gas phase energies for such reactions because the raw data are collected for the reactions in

---

[30] A. B. Blake and F. A. Cotton, *Inorg. Chem.*, **3**, 9 (1964).

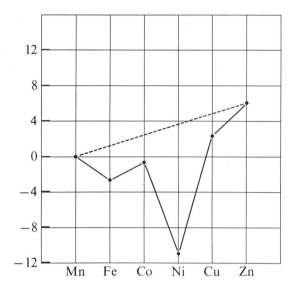

Figure 9-20. Enthalpies (kcal/mole) for the reaction $6H_2O + MCl_4^{2-} \rightarrow 4Cl^- + M(OH_2)_6^{2+}$, relative to that for $M = Mn^{2+}$.

aqueous media. This means that allowance must be made for the differential solvation energy terms (Chapter 6) to obtain the energies of the reactions in the gas phase. In addition, one must be able to estimate accurately the lattice energies of the MCl_4^{2-} salts. Nevertheless, by not being too demanding in your expectations of such calculations, you may safely presume that the errors will vary smoothly from one reaction to the next. A plot of the reaction energies, taking that for M = Mn as the reference, is given in Figure 9-20. If you compare the deviations of the points in this graph from the smoothed data (dotted line), you see a variation with the number of d electrons that is remarkably similar to that on the right side of the high spin curve in Figure 9-19(B). Of particular importance is that both plots exhibit strong minima for Ni^{2+}, the d^8 case.

Further illustration of the utility of the AOM model arises in application to lattice and hydration energies and radii of the divalent transition metal ions.[31] Graphs of these quantities as a function of atomic number are shown in Figure 9-21, where you will notice (i) an underlying smooth trend in values, as evidenced by the d^0, d^5, and d^{10} cases, and (ii) rather characteristic deviations from this trend by the intervening cases. The directions of the smooth trends are exactly what you should expect from the increase in ion effective nuclear charge across the series. The effect of increasing effective nuclear charge, in terms of the AOM model, is to increase progressively the magnitudes of e_σ and e_π by virtue of the fact that increasing effective nuclear charge lowers the energies of the d ao's so that they more effectively interact with the ligand TASO's. Implicit in such an expectation, however, is an assumption that *each d electron has the same average anti-bonding role* to play in opposing the coordination of six ligands to the metal ion. A more specific statement of this background role for the d electrons is that they are evenly distributed among the five d orbitals so that the d^n charge cloud has a spherical shape. However, it is clear from preceding discussions that such is not, in fact, the case, for only the d^0, d^5 (high spin), and d^{10} configurations can have such a spherical charge distribution. In terms of the AOM, the idea that each d electron is spherically distributed in the d shell amounts to assigning each d electron an average anti-bonding energy

$$\frac{1}{5}[2 \cdot 3e_\sigma + 3 \cdot 4e_\pi] = \frac{6}{5}e_\sigma + \frac{12}{5}e_\pi$$

---

[31] Lattice and hydration energies from P. George and D. S. McClure, *Prog. Inorg. Chem.*, (F. A. Cotton, Ed.), **1**, 381 (1959); radii from R. D. Shannon and C. T. Prewitt, *Acta Cryst.*, **B25**, 925 (1969), **B26**, 1076 (1970).

THE MOLECULAR ORBITAL MODEL 553

Regardless of the value of n in d^n for the metal ion, each MD_6^{2+} complex has an energy of $-12e_\sigma - 24e_\pi$ from TASO electron stabilization. To this is added $n(\frac{6}{5}e_\sigma + \frac{12}{5}e_\pi)$ for the *reference* state, in which each of the n electrons plays an average anti-bonding role. For an *actual* high spin or low spin configuration, the correct energy is

$$-12e_\sigma - 24e_\pi + n_\pi[4e_\pi] + n_\sigma[3e_\sigma]$$

The difference between the energy of a given $t_{2g}^m e_g^n$ configuration and the energy of the reference state is simply

$$E - E_r = n_\sigma \left[3e_\sigma - \frac{6}{5}e_\sigma - \frac{12}{5}e_\pi\right] + n_\pi \left[4e_\pi - \frac{6}{5}e_\sigma - \frac{12}{5}e_\pi\right]$$
$$= [3n_\sigma - 2n_\pi]\Delta_o/5$$

The following table reflects this relative electronic configuration energy brought on by the different destabilizing effects of the $d\sigma^*$ and $d\pi^*$ electrons. In the language developed along with the crystal field theory, these energies were called *ligand field stabilization energies*, LFSE.

In units of $(\Delta_o/5)$:

	d^0	d^1	d^2	d^3	d^4	d^5	d^6	d^7	d^8	d^9	d^{10}
Low spin	0	−2	−4	−6	−8	−10	−12	−9	−6	−3	0
High spin	0	−2	−4	−6	−3	0	−2	−4	−6	−3	0

A plot of LFSE against n of d^n has the appearance of Figure 9-22, where for simplicity Δ_o is presumed constant for all M^{2+} of a given charge (this is not a poor approximation when all M come from the same period). Quite remarkably, the shape of the LFSE curve closely resembles those in Figure 9-21. While the LFSE diagram is an energy diagram, there is also a simple direct relationship between orbital occupation and metal-ligand bond distance (or metal ion radius). Each electron to occupy a $d\pi^*$ mo should lengthen the distance by an amount less than average. The deviations are cumulative as the $d\pi^*$ mo's are occupied, so that minima occur at the high spin d^3 and d^8 cases and at the low spin d^6 case [Figure 9-21(C) and (D)].

JAHN-TELLER DISTORTIONS FROM O_h GEOMETRY

Even though idealized MD_6 and MD_4 structures have been assumed for the comparisons of the preceding sections, comment is needed on the long established phenomenon of distorted MD_6 structures of d^4 (high spin), d^7 (low spin), and d^9 complexes. These complexes are all characterized by the configurational property of half-filled or filled t_{2g}^* mo's but an odd number of electrons in the e_g^* mo's of $O_h MD_6$.

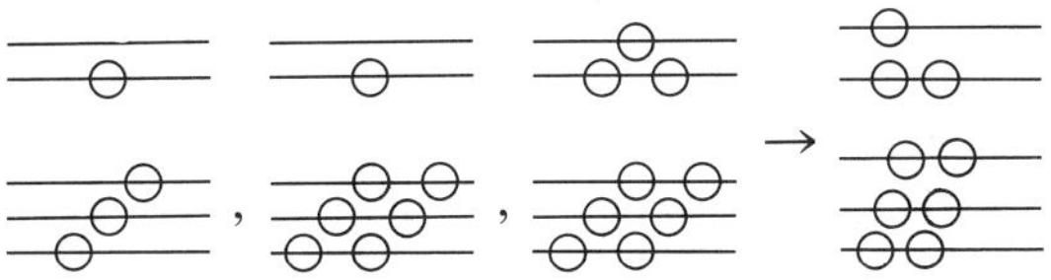

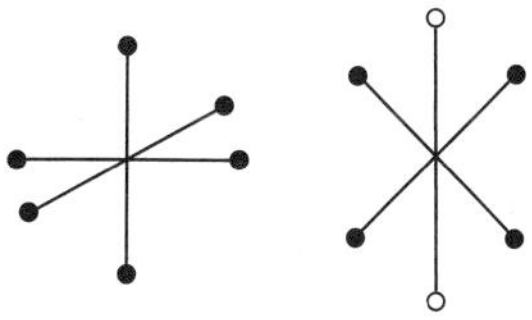

554 □ FUNDAMENTAL CONCEPTS FOR TRANSITION METAL COMPLEXES

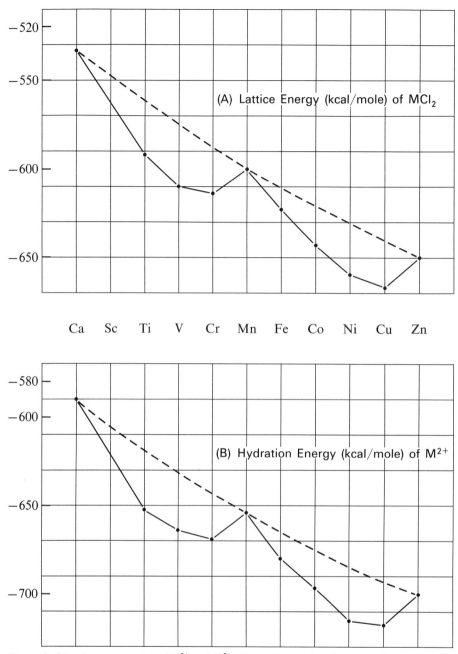

Figure 9–21. Some properties of M^{2+} and M^{3+}. *Illustration continued on opposite page.*

The lowest energy configurations are, therefore, orbitally doubly degenerate (the symbol is E_g). Generally, the distortions from O_h symmetry take the form of elongation of a pair of bonds *trans* to one another, with adoption of a D_{4h} symmetry.

The distortions inherent to the d^4, d^7, and d^9 O_h cases are examples of a general theorem recognized and proved by Jahn and Teller,[32] to the effect that any structure producing an orbitally degenerate electronic configuration of a non-linear molecule is intrinsically unstable; in short, there exists another structure for the molecule that eliminates the orbital degeneracy and leads to lower molecular energy. Accordingly, the distortions are often referred to as *Jahn-Teller distortions*. The occurrence of an

---

[32] H. A. Jahn and E. Teller, *Proc. Roy. Soc.,* **A161,** 220 (1937); H. A. Jahn, *ibid.,* **A164,** 117 (1938).

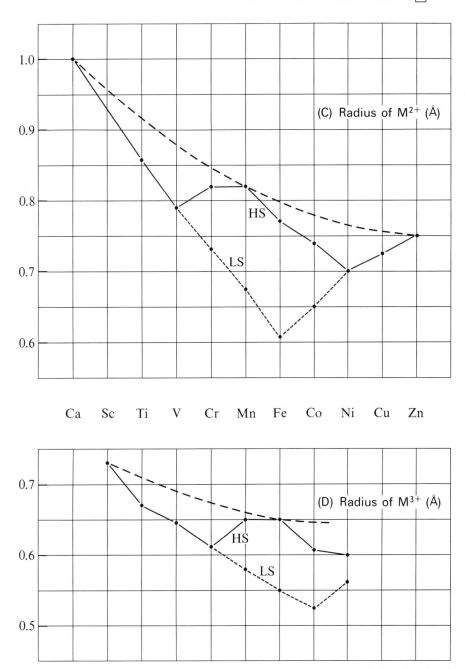

Figure 9-21. *Continued.*

electronic degeneracy for a structure is identified by the possibility of a non-unique assignment of electrons to degenerate (equi-energetic) orbitals. In this vein, degeneracies also occur for t_{2g}^* mo's, as in the d^2 case, and a perfect O_h structure is not possible. With t_{2g}^* degeneracies, however, the driving force for the distortion arises

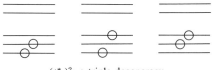

$(t_{2g}^*)^2$: a triple degeneracy

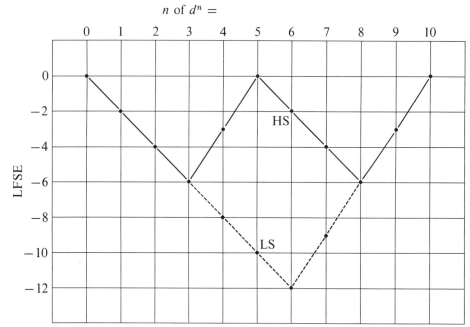

Figure 9-22. The relative electronic configuration energy of MD_6 complexes as a function of the number of d electrons.

from M—Dπ interactions, which are much weaker than the σ interactions; these Jahn-Teller distortions in ground electronic states have so far escaped conclusive experimental detection.

Returning to the E_g ground state complexes, the energy of the distorted D_{4h} MD_6 geometry must be lower than that of the O_h structure, and an interpretation of this preference is needed.[33] In the d^4 and d^7 cases, under rigorous O_h geometry, the single e_g^* electron must be evenly distributed between the $(x^2 - y^2)^*$ and $(z^2)^*$ mo's. This is indicated below by showing each e_g^* level occupied by a half electron (◖). With all six M—D distances the same, the e_σ parameter for each ligand is the same and the complex realizes an electronic energy of $-12e_\sigma + 3e_\sigma = -9e_\sigma$. The question now is, "Will partial withdrawal of the axial ligands lead to a lower energy for the complex?"

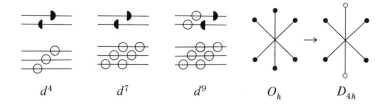

The key to understanding how such a stabilization can arise comes from recognizing that the electron distribution within the e_g^* pair of orbitals affects the equatorial and axial M—D bond distances and the associated e_σ parameters. To be specific, an electron in the $(x^2 - y^2)^*$ mo serves to lengthen *only* the equatorial bonds. In fact, the anti-bonding effect of an electron in this mo is dispersed evenly among the four equatorial ligands and has a net effect of 1/4 electron on each ligand. On the other hand, an electron in the $(z^2)^*$ mo affects *both* axial and equatorial ligands, but not equally. The interaction of each axial ligand with the d_{z^2} ao is four times as large (Table 9-4) as that of an equatorial ligand. The anti-bonding effect of a $(z^2)^*$ electron is, therefore, distributed 1/3 for each axial ligand and 1/12 for each equatorial ligand. Returning to

---

[33] This section is an elaboration of ideas suggested by J. K. Burdett, *Inorg. Chem.,* **14,** 931 (1975).

the d^4 and d^7 O_h cases, the distribution of one electron *evenly* between the $(x^2 - y^2)^*$ and $(z^2)^*$ mo's means that each M—D bond is destabilized by the effect of 1/6 of an anti-bonding electron.

A D_{4h} distortion achieved by lengthening the axial bonds reduces the e_σ parameter for those ligands so that the $(z^2)^*$ mo moves to lower energy. The consequence of this is that the electron becomes localized in the lower energy orbital, $(z^2)^*$. The associated change in anti-bonding electron distribution amounts to partial removal of such density from the equatorial bond region and its greater concentration in the axial bond region. Each equatorial bond is now weakened by only 1/12 of an anti-bonding electron while each axial bond is weakened by 1/3 of an anti-bonding electron. The equatorial bonds should shorten and cause their e_σ parameter to *increase to* e'_σ, say. Meanwhile, the axial bonds lengthen and their e_σ *decreases* to e''_σ. Now, total the electronic energy of this D_{4h} geometry for comparison to that of $-9e_\sigma$ for the O_h structure.

The $x^2 - y^2$ TASO *electron pair* endows the complex with $-6e'_\sigma$ units of stabilization, and the z^2 TASO *pair* adds $-2e'_\sigma - 4e''_\sigma$ units. Against this are the $e'_\sigma + 2e''_\sigma$ units of destabilization from the $(z^2)^*$ electron. The net energy is $-7e'_\sigma - 2e''_\sigma$. Therefore, the change in energy for distorting the O_h structure is

$$\Delta E(O_h \to D_{4h}) = 7\underbrace{(e_\sigma - e'_\sigma)}_{<0} + 2\underbrace{(e_\sigma - e''_\sigma)}_{>0}$$

The first term in this expression is negative (favoring the distortion) while the second is positive (opposing the axial elongation). Which is larger, $(e_\sigma - e''_\sigma)$ or $(e_\sigma - e'_\sigma)$? The first term is associated with the loss for each equatorial ligand of $(1/6 \to 1/12 =) 1/12$ of an anti-bonding electron. Each axial ligand experiences an increase of $(1/6 \to 1/3 =) 1/6$ of an anti-bonding electron. It is logical to expect the quantity $(e_\sigma - e''_\sigma)$ to be larger in magnitude than $(e_\sigma - e'_\sigma)$. Nevertheless, as long as the former is less than 3.5 times the latter, stabilization will accompany axial elongation. An important point is that *the elongation cannot be too extreme,* for when $(e_\sigma - e''_\sigma)$ becomes greater than $3.5(e_\sigma - e'_\sigma)$ the distortion is opposed by an unfavorable change in the total energy.

An entirely analogous analysis of the d^9 case results in the distortion energy expression (Can you show this?):

$$\Delta E(O_h \to D_{4h}) = 3\underbrace{(e_\sigma - e'_\sigma)}_{<0}$$

where e_σ is the parameter appropriate to all ligands in the O_h structure and e'_σ is the new equatorial parameter for the D_{4h} geometry. In contrast to the preceding analysis of the d^4 and d^7 cases, there is no σ bond energy opposition to the complete loss of the axial ligands (to the accuracy of the model as applied here). This means that the axial metal-ligand distance in d^9 complexes is likely to be influenced by ion packing effects in solids and solvation forces in solution. This is generally what is found—in the solid state CuD_6^{2+} complexes exhibit a variety of tetragonal distortions (the symmetry is reduced to D_{4h}); axial elongations apparently are more frequently reported, but axial compressions are also known. For example, $CuCl_2$ exhibits four Cu—Cl distances of 2.3 Å and two Cu—Cl distances of 2.95 Å, while K_2CuF_4 exhibits four Cu—F distances of 2.08 Å and two Cu—F distances of 1.95 Å. (In these examples the octahedron of halide ions about Cu^{2+} is achieved by the sharing of halides between copper ions; see Chapter 6.)

Another significant implication is that the formation of octahedral complexes in solution, as represented by the equilibrium

$$MD_4^{n+} + 2D \rightleftharpoons MD_6^{n+}$$

may be expected to have a small equilibrium constant for the d^9 case. You have seen (Table 9-3) that Cl^- forms the complex anion, $CuCl_4^{2-}$; only with a large cation like $Cr(NH_3)_6^{3+}$ can one obtain $CuCl_5^{3-}$ (another example of solid state stabilization of an ion through choice of the proper counter-ion, Chapter 6), and no $CuCl_6^{4-}$ is known. Of course, when the solvent is the ligand, the inherently high ligand concentration tends to drive the equilibrium in the direction of the $MD_6(D_{4h})$ complex. Examples are $Cu(OH_2)_6^{2+}$ in water, $Cu(NH_3)_4(OH_2)_2^{2+}$ in aqueous ammonia, and $Cu(NH_3)_6^{2+}$ in liquid ammonia.

558 ☐ FUNDAMENTAL CONCEPTS FOR TRANSITION METAL COMPLEXES

Though possessing an even number of electrons in the e_g^* mo's, Ni^{2+} has an interesting behavior somewhat related to the d^9 case. Application of the preceding analysis of variable equatorial and axial e_σ parameters to the d^8 case suggests correctly that there should be an energy activation barrier associated with the loss of two axial ligands. For small tetragonal distortions, the $d_{x^2-y^2}^*$ and $d_{z^2}^*$ orbital occupations remain at one electron each. Decreasing the axial e_σ to e_σ'' does not appreciably change the equatorial e_σ. Therefore, no D_{4h} distortion from O_h symmetry is expected, since the distortion energy increases according to

$$\Delta E(O_h \to D_{4h}) = 4(e_\sigma - e_\sigma') + 2(e_\sigma - e_\sigma'') \sim 2(e_\sigma - e_\sigma'') > 0$$

However, when the axial ligands are completely removed and if $2e_\sigma > P$, the $(x^2 - y^2)^*$ and $(z^2)^*$ mo's become so different in energy that the $(x^2 - y^2)^*$ electron moves into $(z^2)^*$; in this event the complex becomes diamagnetic and square planar, the equatorial e_σ becomes larger (e_σ'), and ΔE becomes negative:

$$\Delta E(O_h \to D_{4h}) = 6(e_\sigma - e_\sigma') < 0, \text{ for } e_\sigma' > e_\sigma$$

In fact, complexes that are rigorously O_h are commonly found; but attempts to form mixed complexes of general formula $NiD_4L_2^{2+}$, where D is a much better ligand than L, are only moderately successful. This is to be expected according to the energy change associated with conversion of a diamagnetic, square planar NiD_4^{2+} complex to a paramagnetic "octahedral" complex $NiD_4L_2^{2+}$:

$$E_{sp} = -8e_\sigma + 2e_\sigma = -6e_\sigma$$

$$E_o = -8e_\sigma' - 4e_\sigma'' + 4e_\sigma' + 2e_\sigma'' = -4e_\sigma' - 2e_\sigma''$$

$$\Delta E = 4(e_\sigma - e_\sigma') + 2(e_\sigma - e_\sigma'') > 0$$

The weakening of equatorial bonds as $(x^2 - y^2)^*$ becomes occupied renders $e_\sigma' \lesssim e_\sigma$ and, by definition, $e_\sigma'' < e_\sigma$. (What is the sign of ΔE if the two axial ligands, L, are stronger donors than the equatorial ligands, D?)

Specific examples of this special behavior are the Lifschitz salts with ethylenediamines (en) in a square planar orientation about the Ni^{2+} ion. The cations are characteristically diamagnetic ($2e_\sigma > P$) and have yellow or red colors.

When Lifschitz salts are dissolved in reasonably good donor solvents, the solutions are blue or green in color and contain the paramagnetic "octahedral" $Ni(en)_2S_2^{2+}$ cation. Consistent with the expected weak binding of the solvent ligands, the diamagnetic, planar complex can be regenerated by simply heating the isolated octahedral complex.

Although the comments in this section have aimed specifically at ground state degeneracies and their structural effects, Jahn-Teller distortions arise for excited states as well. For example, both the ground and excited configurations, $(t_{2g}^*)^3(e_g^*)^1$ and

$(t_{2g}^*)^2(e_g^*)^2$, of a high spin d^4 ion are subject to distortion; the distortion in the excited state is directly observable through its effect on the electronic absorption spectrum, the subject to be broached next.

STUDY QUESTIONS

11. Confirm the energies of the d^* mo's on page 546.

12. Determine which of the following might be expected to exhibit distorted structures (relative to O_h for MD_6 and T_d or D_{4h} for MD_4): $Cr(OH_2)_6^{3+}$, $Ti(OH_2)_6^{3+}$, $Fe(CN)_6^{4-}$, $CoCl_4^{2-}$, $Pt(CN)_4^{2-}$, $ZnCl_4^{2-}$, $Ni(en)_3^{2+}$, $FeCl_4^-$, $Mn(OH_2)_6^{2+}$.

13. Determine which of the following ligands is most likely to yield a tetrahedral complex MD_4 (rather than MD_6 or square planar MD_4) with Mn^{2+}: CN^-, NO_2^-, 2,6-dimethylpyridine, or NH_3. (Use the concepts of donor atom effective nuclear charge and hybridization to estimate an order based on e_σ. Then check to see that the relative e_π values do not alter the order.)

14. Consider axial compression of the octahedron for MD_6^{n+} when M^{n+} is a d^4 or d^7 ion. Follow an analogous procedure to that given in the text for axial elongation. Is it, or is it not, possible to say that elongation is favored over compression?

15. Characterize the structural/magnetic properties of $Ni(en)_3^{2+}$, $NiCl_4^{2-}$, and $Ni(CN)_4^{2-}$.

16. Using the information in Table 9-4, construct a plot of "structural preference energy" vs. n of d^n for the square pyramidal and trigonal bipyramidal structures. For what d^n and high-or-low spin condition is the "C_{4v}" geometry preferred most strongly? What does the fact that $Fe(CO)_5$ has D_{3h} geometry tell you about the importance of VSEPR vs. e_σ?

ELECTRONIC STATES AND SPECTRA

With the foregoing concepts of complex structure and orbital sequencing well in hand, it is time to explore the link between those ideas and the striking color and magnetic properties of complexes. Specific examples of the challenge are found in Figures 9-2 through 9-5. How are the numbers and relative positions of the absorption bands to be explained? How does changing a ligand cause shifts in these bands? As anticipated on page 536, once you have written down the ground and excited *configurations* for a complex, the next step is to decipher the molecular *terms* that arise from those configurations. For a d^2 complex like $V(OH_2)_6^{3+}$, the situation is described as follows:

$(t_{2g}^*)^2(e_g^*)^0$, $(t_{2g}^*)^1(e_g^*)^1$ or $(t_{2g}^*)^0(e_g^*)^2$

Terms $\xrightarrow{h\nu}$ Terms   Terms

As you shall see, only transitions from the ground term to certain excited terms are observed. For the d^2 octahedral case, three excited terms are accessible by photon absorption.

Perhaps it is best to begin by reviewing the principles of Chapter 1 for finding free ion term symbols. Using a d^2 configuration ion like V^{3+} (there are two electrons to be assigned to five degenerate ao's), the ground term is quickly found by following

Hund's rules: maximum spin multiplicity is assured by half-filling ao's with like-spin electrons before pairing electrons; electrons are assigned so as to give the largest L quantum number. For d^2:

1	1				$\Rightarrow {}^3F$
2	1	0	−1	−2	

Though only one microstate (that with largest S and L) need be written for the ground term derivation, in the case at hand a total of 21 microstates $[= (2S + 1)(2L + 1)]$ are required to completely define the ${}^3F$ term. Of the 45 total microstates $(= 10 \cdot 9/1 \cdot 2)$, 24 remain to define *excited terms* arising from this *ground configuration* (an *excited configuration* would be $3d^14s^1$, for example). Application of the technique of Chapter 1 reveals the excited terms to be ${}^1D$ (five microstates), ${}^3P$ (nine microstates), ${}^1G$ (nine microstates), and ${}^1S$ (one microstate). This sorting of microstates into terms of different energies is a consequence of electron/electron repulsions. Furthermore, note that only one excited term (${}^3P$) has the same spin multiplicity as the ground term (${}^3F$). For the excited configuration d^1s^1, two terms are possible from the twenty possible microstates $[= (10/1)(2/1)]$. These are simply ${}^3D$ and ${}^1D$. The ground term (${}^3D$) of the excited configuration has the same value of $2S + 1$ as does the ground term of the ground configuration.

With respect to a d^2 octahedral complex, this process must be repeated for the ground configuration $(t_{2g}^*)^2(e_g^*)^0$ and the excited configurations $(t_{2g}^*)^1(e_g^*)^1$ and $(t_{2g}^*)^0(e_g^*)^2$. Note that the d^* mo's of the complex are not all degenerate, so some of the 45 microstates of the free ion ground configuration represent excited configurations in a complex! To begin with, the number of accessible excited terms must be discovered. That is the subject of the next few pages. After that you will learn the techniques for extracting chemical information from the absorption band positions.

GROUND AND EXCITED STATES

To go through the configuration → term analysis for each d^n case and each structural type means that the term analysis must be performed 18 times to cover all d^1 through d^9 cases under octahedral and tetrahedral geometries. Fortunately, nature kindly provides considerable redundance in these electronic terms—consequently, you need to learn the term symbols for only a few cases and then remember some inversion rules relating d^n to d^{10-n} cases[34] and relating the octahedral to the tetrahedral geometry.

Table 9-5 summarizes the results of the analysis of metal ion ground configurations and terms. Notice that the spin multiplicity values ($= 2S + 1$) are appended to the term symmetry labels in the familiar fashion. As it is the lowest energy terms that are of interest, the spin multiplicities are consistent with Hund's rule about states of maximum spin multiplicity lying lowest. The other part of the state label denotes the symmetry of the orbital part of the wavefunction. For the *free ion* the L quantum number serves the purpose of identifying the orbital part of the electron motions; in fact, the L quantum number and the symbols $S, P, D, \ldots$ are also symmetry labels [for the point group (symbol $= K$) representing spherical symmetry]. Since the ion no longer has a spherically symmetric environment, the orbital symmetry labels appropriate to the O_h point group must be used.

The meanings to be associated with the symmetry labels A, E, and T are simply that the orbital wavefunctions for the states are, respectively, non-degenerate [$A \Rightarrow$ only one way (one wavefunction) of occupying the metal orbitals with like-spin electrons], two-fold degenerate [$E \Rightarrow$ two ways (two equivalent wavefunctions) of

---

[34] Recall from the discussion of *free* atom terms in Chapter 1 that p^n configurations and p^{6-n} configurations have the same terms. Similarly, d^n and d^{10-n} configurations give identical terms.

TABLE 9-5
GROUND CONFIGURATIONS AND TERMS FOR $MD_6(O_h)$ AND $MD_4(T_d)$ COMPLEXES

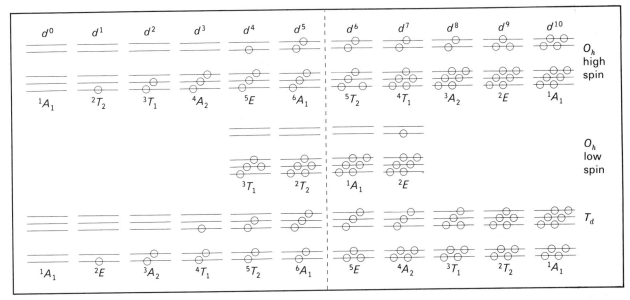

assigning the like-spin electrons to the orbitals], and three-fold degenerate [$T \Rightarrow$ three ways (three equivalent wavefunctions) of assigning the like-spin electrons to the orbitals].[35] For example, considering the d^1 octahedral case, it is obvious that there are three ways of assigning the one electron among the triply degenerate orbitals of the metal (to xy^*, xz^*, or yz^*)—hence the $^2T_2$ symbolism. For the d^2 octahedral case there are also just three ways of assigning *like-spin* electrons among the three low-energy orbitals ($^3T_1$). For the d^4 case the ground term symbol depends on whether the configuration is high or low spin. For high spin there is only one way to fill in the first three electrons (like the preceding d^3 case) but two ways for assigning the fourth electron—hence the label is ^5E. For the low spin case there are three ways to assign the fourth electron, and the ground state term symbol is accordingly $^3T_1$. For the d^5 case, the high and low spin situations must be distinguished. The high spin case is orbitally non-degenerate and is called $^6A_1$, while the low spin ground term is $^2T_2$. As an exercise we suggest that you verify the A, E, and T symmetry labels for the remaining cases.

Now you come to the first of the redundancies nature has so thoughtfully provided. The orbital labels for the d^6 through d^{10} sequence in the high spin part of the table are repetitions of the d^1 through d^5 section. The second part of the redundancy concerns the ground state labels for the tetrahedral geometry—the sequence is exactly *inverted* from that for the octahedral case. This is quite obviously due to the inversion relation between the orbital splitting of the O_h and T_d geometries. Thus, not only is it easy to just write down from inspection the ground term symbols for any d^n case, but you have as a check the fact that the $d^n(O_h)$ and $d^{10-n}(T_d)$ labels are identical.

Having established the ground term symbols, the next step is to inquire into the number of electronic excited states to which the molecule may go upon absorption of a photon. Were you to take a course in the theory of electronic absorption processes,

---

[35] According to the O_h and T_d character tables, there are two kinds of A and T wavefunction symmetries possible: A_1, A_2 and T_1, T_2. (As the terms derive from electron occupation of *gerade d* orbitals, all O_h terms from d^n configurations also carry the g subscript.) To determine which subscript is appropriate in any situation requires greater development of the uses of character tables than is appropriate here. The most important aspect for now is to recognize and distinguish A from E from T. For accuracy, we have added the correct subscript to each state symmetry label; do not let this concern you, for understanding how to arrive at the correct subscript is not very important for the purposes of this text.

TABLE 9-6
THE EXCITED CONFIGURATIONS AND TERMS OF THE SAME SPIN MULTIPLICITY AS THE GROUND TERM OF $MD_6(O_h)$ COMPLEXES

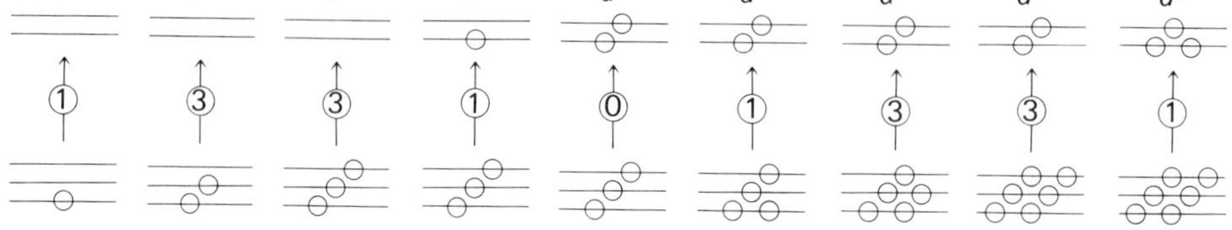

	d^1	d^2	d^3	d^4	d^5	d^6	d^7	d^8	d^9
Number of Transitions	①	③	③	①	⓪	①	③	③	①
Second Excited Configuration		$^3A_2$	$^4T_1$				$^4A_2$	$^3T_1$	
First Excited Configuration		$^3T_1$	$^4T_1$				$^4T_1$	$^3T_1$	
First Excited Configuration	^2E	$^3T_2$	$^4T_2$	$^5T_2$		^5E	$^4T_2$	$^3T_2$	$^2T_2$
Ground Configuration	$^2T_2$	$^3T_1$	$^4A_2$	^5E	$^6A_1$	$^5T_2$	$^4T_1$	$^3A_2$	^2E

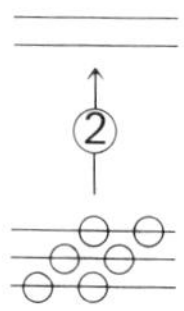

Low spin d^6:

Number of Transitions: ②

First Excited Configuration	$^1T_2$
	$^1T_1$
Ground Configuration	$^1A_1$

one of the most important concepts you would learn is that transitions between terms (ground and excited) of the same spin multiplicity are by far the most probable (will possess the greatest extinction coefficient). Therefore, let us consider for each of the high spin O_h d^n cases what excited terms of the *same multiplicity as the ground term* are possible for each excited configuration. The possibilities are given in Table 9-6. The d^1 and d^9 cases are very straightforward. For both there is only one excited configuration possible and that arises from excitation of an electron from a t_2 orbital to an e orbital. For the d^1 case there are two ways that the single electron can be assigned to the e orbitals, and the excited state is ^2E. For the d^9 case the excitation produces a configuration with filled e orbitals and one-less-than-filled t_2 orbitals. This "hole" in the t orbitals may be placed in any of three ways, so the configuration is triply degenerate ($^2T_2$). The customary notation for the single electronic transition in the d^1 case is $^2E \leftarrow {}^2T_2$ and that for the d^9 case is $^2T_2 \leftarrow {}^2E$. Note that these two cases of d^1 and $d^{10-1} = d^9$ *are inversely related!* As you will see, nature has been kind to us again, for *all high spin d^n and d^{10-n} cases are inversely related*.

For d^2 and d^8 you encounter for the first time the possibility of two excited configurations achieved by excitation of one and then two electrons from the t_2 to the e orbitals. Here also is encountered for the first time the need to consider microstate resolution by electron/electron repulsion. The first excited configuration of d^2 (i.e., $t_2^1 e^1$) incurs a six-fold orbital degeneracy. For each of three ways of assigning the t_2 electron, there are two ways of assigning the e electron *of the same spin*. Electron/electron repulsions separate these six microstates into two groups—each of which is orbitally triply degenerate. Both are therefore ^3T states ($^3T_1 + {}^3T_2$).[36] For the doubly excited configuration (e^2) there is only one way to assign the two electrons *and* maintain the triplet spin multiplicity of the ground state. The term symbol is, therefore, $^3A_2$.

For the d^8 case, the first excitation produces a configuration $t_2^5 e^3$. The t orbitals possess a "hole" that may be assigned in three ways, and for each of these assignments the "hole" in the e orbitals may be assigned in two ways. Again, the six-fold degeneracy of this configuration is resolved by electron/electron repulsions into two triply degenerate sets—both ^3T ($^3T_1 + 3T_2$). The double excitation (the configuration is $t_2^4 e^4$) produces a triple degeneracy in the t_2 orbitals (the two "holes" may be assigned in only three ways) and the term for this configuration is also $^3T_1$. To summarize the possible transitions, we write

d^2	d^8
$^3T_2 \leftarrow {}^3T_1$	$^3T_2 \leftarrow {}^3A_2$
$^3T_1 \leftarrow$	$^3T_1 \leftarrow$
$^3A_2 \leftarrow$	$^3T_1 \leftarrow$

Be sure to note the inversion relation between the lowest and highest terms of these two cases: for d^2, in order of increasing energy,

$$^3T_1 < ({}^3T_2 < {}^3T_1) < {}^3A_2$$

and for d^8 the order is

$$^3A_2 < ({}^3T_2 < {}^3T_1) < {}^3T_1$$

For d^3 and d^7 a situation entirely analogous to that for d^2 and d^8 arises. The ground term of d^3 is $^4A_2$. The first excited configuration ($t_2^2 e^1$) gives two terms—both ^4T ($^4T_2 + {}^4T_1$). The second excited configuration ($t_2^1 e^2$) gives only one quartet state, and that is orbitally triply degenerate so the term symbol is $^4T_1$. The inverse arrangement of high and low quartet states is encountered for d^7 case: $^4T_1 < ({}^4T_2 < {}^4T_1) < {}^4A_2$.

The d^4 and d^6 cases are as easy as the d^1 and d^9 cases. Only one *quintet* excited configuration is possible ($t_2^2 e^2$ for d^4 and $t_2^3 e^3$ for d^6). The excited term symbols are, logically enough, $^5T_2$ for d^4 and ^5E for d^6. Again the ground and excited terms are inversely ordered.

For the d^5 high spin case, no excited configurations of the same spin multiplicity (6) as the ground state are possible! This is a critical prediction of electronic state theory, which is well supported by experiment—*high spin d^5 complexes exhibit no strong transitions in the visible region* and are "colorless." [The spectrum of $Mn(OH_2)_6^{2+}$ in Figure 9-23 exhibits only *weak* transitions to terms of different $(2S + 1)$ than the $^6A_{1g}$ ground term.]

All this information about the ground and excited electronic states of O_h complexes can now be conveniently summarized with the following diagrams:

for d^1, d^6 for d^4, d^9

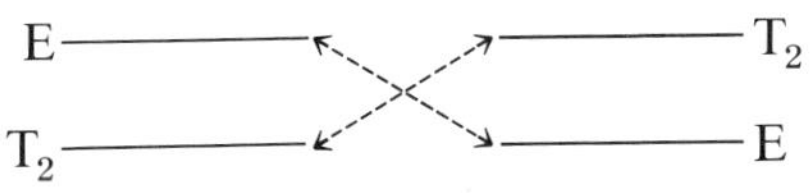

E ———————↙   ↗——————— T_2

T_2 ———————↖   ↘——————— E

---

[36] Note that an orbital degeneracy greater than 3 is not permitted under cubic symmetry. While the six-fold microstate degeneracy could have been resolved by electron correlation into $A + E + T$ terms, in fact the resolution is into $T_1 + T_2$. A graduate course in transition metal chemistry would show you how this is predicted from group theoretical principles.

564 ☐ **FUNDAMENTAL CONCEPTS FOR TRANSITION METAL COMPLEXES**

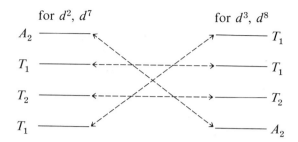

From these diagrams it is apparent that the $d^{1,6,4,9}$ octahedral complexes are characterized by the occurrence of only one absorption band in the visible region. The $d^{2,7,3,8}$ octahedral complexes should exhibit three bands in this region of the spectrum. Actual spectra of $M(OH_2)_6^{n+}$ complexes bear out these predictions; see Figure 9-23.

Even a cursory examination of these spectra should suggest to you, however, that the foregoing derivation of accessible excited terms is rudimentary. All the bands show evidence of asymmetry or even resolved splitting. This means, of course, that

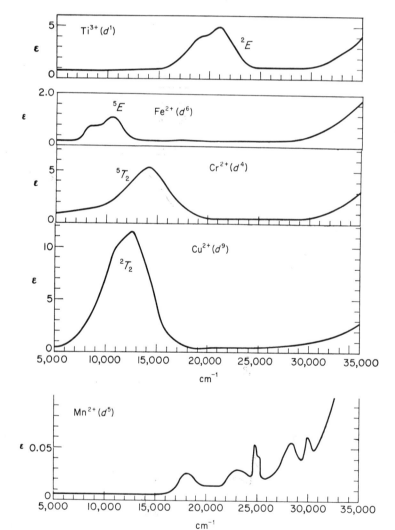

Figure 9-23. Electronic spectra of $M(OH_2)_6^{n+}$ in the region from 5 to 35 kK. Note that the ϵ scale is two orders of magnitude smaller for $Mn(OH_2)_6^{n+}$ than for the others. [From B. N. Figgis, "Introduction to Ligand Fields," Interscience, New York (1966).]

Illustration continued on opposite page.

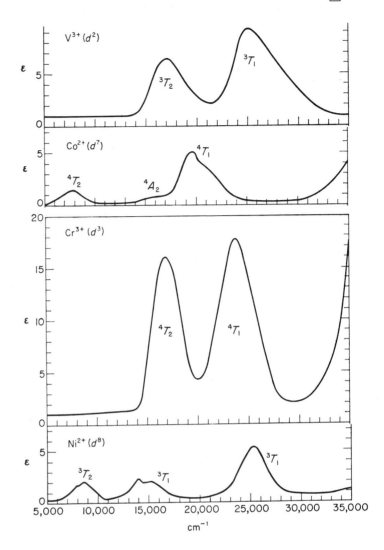

Figure 9-23. Continued.

where a single transition has been predicted there are actually several transitions in close proximity. One source of such splitting is spin-orbit coupling, like that discussed in Chapter 1 for the free ions. There you found that coupling of the spin and orbital angular momenta (or magnetic moments) could further resolve the atomic term microstates into spin-orbit terms. As the spin-orbit coupling energies are generally small for first row transition metals, the small splittings observed in Figure 9-23 could be due to such coupling.[37] Another source of term splitting derives from the Jahn-Teller distortion of molecular structure. Just as each ground term has an associated lowest energy structure, each excited term has a distinct molecular architecture, which may be different from those of the ground term and other excited terms. (Why, in terms of the change in $t_{2g}^* e_g^*$ occupancies, should this be the case?) In the event of an orbitally degenerate term, Jahn-Teller distortion will cause there to be at least two excited terms, one of which will be slightly lower in energy than the other. Transitions to these molecular terms could account for the splitting of a single band predicted to arise only by ignoring the occurrence of spin-orbit and Jahn-Teller effects (cf. $Ti(OH_2)_6^{3+}$ in Figure 9-23).

---

[37] Note that only terms with orbital angular momentum (orbital degeneracy) can exhibit such spin-orbit degeneracy resolution.

Additionally, you notice that only two, not three, bands are found in the $n = 2$ and 3 spectra. The third transition occurs at such a high energy that it appears in the ultraviolet region, where it is lost among more intense transitions (above 35 kK) of the charge transfer (CT) type. These CT transitions are very much like those discussed in Chapter 5 for I_2 adducts; in both cases the CT excitation is associated[38] with the promotion of an electron in a ligand (donor) TASO (σ or π) to a metal (acceptor) orbital (σ or π d^* mo). In the next section you will consider the quantitative aspects of all three electronic transitions and verify at that time that the third band of the $d^{2,3}$ ions often is lost among the more intense CT bands.

To close this preliminary examination of high spin complex spectra, we will simply state that the inversion relation already noted for the ground electronic terms of octahedral and tetrahedral structures carries over into the excited states as well (you will have an opportunity to verify the statement by working study question 18). Thus, the energy level diagram for d^1 and d^6 octahedral geometry given above is appropriate for the $d^{4,9}$ tetrahedral cases as well. Similarly, the $d^{2,7}$ octahedral diagram is the same as the $d^{3,8}$ tetrahedral diagram.

> Thus far, the energy terms for low spin complexes have been ignored. In fact, the determination of the excited states for the d^4, d^5, and d^7 cases is quite difficult, requiring techniques far beyond the level of this introduction. To illustrate for you the complexity of the situation for the d^4 case, consider deriving the terms for the first excited configuration ($t_2^3 e^1$). You have already encountered this configuration for the *high spin* case, where only the lowest-energy term (with $2S + 1 = 5$) was of interest. From Hund's spin multiplicity rule you were able to stipulate that all four electrons have the same spin, and this considerably simplified the problem (only one quintet term, ^5E, was possible). For the low spin case, the lowest-energy ^5E state from the $t_2^3 e^1$ configuration is *not* of interest, because the ground state of t_2^4 is a spin triplet. Thus, for the low spin case you need to know the symmetry labels for all *triplet* terms of $t_2^3 e^1$ lying higher than ^5E, and there are no rules to help with the task. The total number of microstates for a $t_2^3 e^1$ configuration is 80 ($6 \cdot 5 \cdot 4/1 \cdot 2 \cdot 3 = 20$ for the three t_2 electrons, and for each of these there are four microstates for the e electron)! Of these, 10 are required to define the lowest-energy ^5E term. From the remaining 70 microstates you are to discover the triplet terms. This is a problem for the serious student of electronic spectra.
>
> Perhaps fortunately, the d^6 case is simple enough to discuss here, and it is just this case which is more commonly encountered—the examples from the first transition series being Fe^{2+} and Co^{3+}. The ground configuration of the low spin d^6 octahedral structure is simply $^1A_1$, since the t_2 orbitals are completely filled with electrons. The first excited configuration is $t_2^5 e^1$, a configuration conveniently viewed as one electron in the e orbitals and one hole in the t_2 orbitals (see Table 9-6). For each of the three ways of assigning the t_2 hole there are two ways of assigning the e electron; the total of six ways means that there are two T states ($^1T_1 + {}^1T_2$). Thus, there are two excited terms of the same spin multiplicity as the ground term, and two absorption bands arise from the configuration change $t_2^6 \to t_2^5 e^1$. It is unlikely that you would have to go beyond these two excited terms, for you should remember that the orbital energy gap (Δ_o) is large in low spin complexes and the electronic transitions will be of high energy. Terms arising from two-electron (and more) promotions are likely to be lost among the charge transfer bands in the ultraviolet portion of the spectrum. Two cases in point are ferrocyanide [$Fe(CN)_6^{4-}$] and $Co(en)_3^{3+}$ [tris(ethylenediamine)cobalt(3+)], the spectra for which are shown in Figure 9-24. You will notice that $Fe(CN)_6^{4-}$ shows one readily apparent low-energy band as a weak shoulder on the tail of a more intense CT band. $Co(en)_3^{3+}$ shows two visible bands with no strong CT absorptions in the blue. These spectra bear out the expectation that no more than the two states arising from the $t_2^5 e^1$ configuration need concern you.

---

[38] With ligands of the π acceptor type, the ligand π^* mo's lie low enough in energy to also afford metal $\to$ ligand CT bands of the $d^* \to \pi^*$ type.

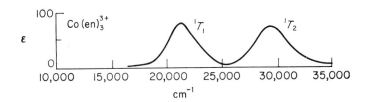

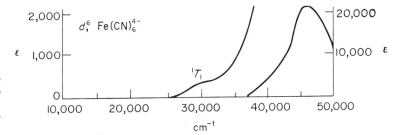

Figure 9-24. Electronic spectra of low spin d^6 cases from 10 to 50 kK. Note the difference in ϵ scales. [From B. N. Figgis, "Introduction to Ligand Fields," Interscience, New York (1966).]

THE ENERGIES OF ELECTRONIC TRANSITIONS

Now that you have seen how to account for the number of d-d bands as a function of structure and d^{*n} configuration, the extraction of chemical information from the positions of those bands is the next important step. In the next few pages you will find the rudiments of spectral analysis to extract the value of Δ_t or Δ_o, which reflect the metal-ligand σ and π bonding.

In spite of the fact that actually doing such an analysis is of great value, the more important goal is chemical information. To emphasize this, we will begin at the end, so to speak, by examining the results to be obtained later.

(i) For d^1, d^4, d^6, and d^9 cases (O_h or T_d), only a single $d^* \leftarrow d^*$ band is observed, and it has an energy Δ_o or Δ_t.

(ii) For d^2, d^3, d^7, and d^8 cases, three bands are expected. For the d^3, d^8 (O_h) and d^2, d^7 (T_d) cases, the lowest-energy transition gives Δ_o and Δ_t. For d^3, d^8 (T_d) and d^2, d^7 (O_h), it is the separation between the first and third bands that gives Δ_o and Δ_t. (Unfortunately, the third band cannot always be located, as noted in the last section; in such events more complete analysis is required.)

To illustrate the chemical interpretation process, consider Δ_o and B for $V(OH_2)_6^{3+}$ and $Ni(OH_2)_6^{2+}$ (d^2 and d^8 cases). The quantity Δ_o has been defined earlier and B, as you will see on page 568, is a term that reflects the magnitude of the electron/electron repulsion energy. Both affect the energy separation of molecular terms.

	Δ_o	B
$V(OH_2)_6^{3+}$	18.5 kK	0.65 kK
$Ni(OH_2)_6^{2+}$	8.7 kK	0.97 kK

The most striking difference is in Δ_o, which is less than half as large for Ni^{2+} as for V^{3+}. The ligand is the same in these complexes, so the difference must lie with the metal ions. Clearly the vanadium complex has a greater charge, and this should lead to stronger metal-ligand bonds (greater e_σ, e_π, and Δ_o), but this is not the whole story. The V^{3+} complex has only two d electrons in t_{2g}^* orbitals, so they play a π^* role in the complex; Ni^{2+}, however, has eight d electrons, and these are distributed so that six play the π^* role (located in t_{2g}^*) and two play a σ^* role in e_g^* orbitals. Again you conclude that the Ni^{2+} complex should exhibit longer, weaker bonds and thus a smaller Δ_o.

The B values of the complexes give further useful information about the complexes. That the B value for Ni^{2+} is larger than that for V^{3+} is consistent with the greater number of electrons in the former, and this is also reflected by the free ion values of 1.04 and 0.86 kK, respectively. Of greater significance is the fact that complex formation reduces the interelectronic repulsions to 75% of the free ion value in V^{3+} but only to 93% in Ni^{2+}. That there is a reduction at all is diagnostic of covalency in the metal-ligand interactions and reminds you that the d^* mo's are just that, molecular orbitals by which the d electrons radially expand onto the ligand nuclei. That the reduction is so much greater for V^{3+} is certainly consistent with the conclusion (based on Δ_o values) about the relative extents of metal-ligand interaction.

The methods that theoreticians use to determine the energies of the ground and excited terms of a complex are far beyond the requirements of a text such as this. Rather, our purpose is to use their results in the more pragmatic study of complexes. For the interpretation of electronic spectra, algebraic forms for the *transition* energies are required (one can determine experimentally only the energies of excited terms *relative* to the ground term). For the d^1 case the situation is particularly simple, since

(9-1)
$$^2E \leftarrow {}^2T_2: \quad \tilde{\nu} = \Delta_o$$

there is only one transition and its energy is a function only of the orbital splitting—that is, there is no contribution to the energy change from a change in interelectronic repulsions. This is a trivial statement for the one-electron case (!) but also holds for the high spin d^4, d^6, and d^9 cases. Thus, the position of the single absorption band for such complexes gives directly the magnitude of Δ_o.

For the d^2, d^3, d^7 (high spin), and d^8 cases you can generally expect to find three absorption bands, and these are a function of both Δ_o and a change in interelectronic repulsion energy (the reason one configuration can give several terms). The algebraic expressions for the d^2 cases are given below.

(9-2)
$$^3A_2 \leftarrow {}^3T_1: \quad \nu_3' = \frac{3}{2}\Delta_o' - \frac{15}{2} + \frac{1}{2}[(15)^2 + 18\Delta_o' + \Delta_o'^2]^{1/2}$$

$$^3T_1 \leftarrow {}^3T_1: \quad \nu_2' = [(15)^2 + 18\Delta_o' + \Delta_o'^2]^{1/2}$$

$$^3T_2 \leftarrow {}^3T_1: \quad \nu_1' = \frac{1}{2}\Delta_o' - \frac{15}{2} + \frac{1}{2}[(15)^2 + 18\Delta_o' + \Delta_o'^2]^{1/2}$$

You will notice that these expressions contain only the single unknown Δ_o'. This is because it is particularly convenient to express the energy of a transition and Δ_o in relation to a theoretically defined unit of electron/electron repulsion called B. This simply means that ν' and Δ_o' are defined as $\nu' = \tilde{\nu}/B$ and $\Delta_o' = \Delta_o/B$. To help you visualize the dependence of the transition energies on Δ_o', Figure 9-25 is a plot[39] of these functions. The right side of this figure represents the d^2 O_h case; the left is the "inverted" d^8 O_h case to be discussed shortly.

In a moment we will illustrate how these equations can be utilized to determine Δ_o and the unit of electron repulsion B from the spectrum of a d^2 complex, but we will digress for a moment to assure you that these equations behave properly in the limit of

---

[39] Such diagrams have been called Orgel diagrams, after L. E. Orgel, who introduced them.

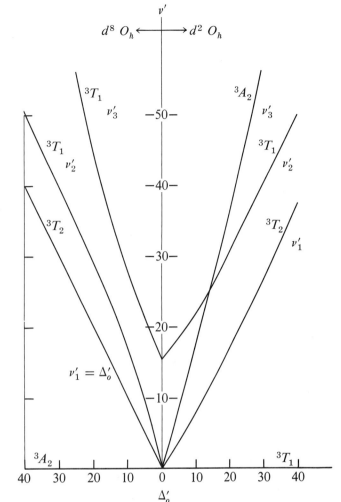

Figure 9-25. The transition energies of MD_6 complexes of d^2, d^8 ions as a function of Δ'_o.

large Δ'_o and small Δ'_o. When Δ'_o is sufficiently large that terms not containing Δ'_o may be ignored, the transition energies are of the form

$$\nu'_3 = \frac{3}{2}\Delta'_o + \frac{1}{2}[18\,\Delta'_o + \Delta'^2_o]^{1/2}$$

$$\nu'_2 = [18\,\Delta'_o + \Delta'^2_o]^{1/2}$$

$$\nu'_1 = \frac{1}{2}\Delta'_o + \frac{1}{2}[18\,\Delta'_o + \Delta'^2_o]^{1/2}$$

The square root term approaches $\frac{1}{2}\Delta'_o$ so that the transition energies become, in the limit of large Δ'_o,

$$\nu'_3 = 2\,\Delta'_o \quad \text{(from the } e^2 \text{ configuration)}$$

$$\left.\begin{array}{l} \nu'_2 = \Delta'_o \\ \nu'_1 = \Delta'_o \end{array}\right\} \quad \text{(from the } t^1_2 e^1 \text{ configuration)}[40]$$

corresponding correctly to the energies of the e^2 and $t^1_2 e^1$ configurations. A more interesting result arises for the limit of $\Delta'_o = 0$. In this limit the metal ion is free of its

---

[40] When Δ_o is so large that B can be ignored, one in effect ignores the basis for resolution of the $t^1_2 e^1$ microstates into $^3T_1 + {}^3T_2$ states!

ligand cage and is in actuality a free ion, as discussed in Chapter 1. The molecular terms should then collapse into the free ion terms of the same spin multiplicity. With $\Delta'_o = 0$, the transition energies become

$$\nu'_3 = 0$$
$$\nu'_2 = 15$$
$$\nu'_1 = 0$$

where you see that the $^3A_2$ and the $^3T_2$ terms are degenerate with the ground $^3T_1$ term. These degenerate terms should now agree in spin and orbital multiplicity with the d^2 free ion ground term, which is ^3F. The total degeneracy for the molecular terms $^3A_2$, $^3T_2$, and $^3T_1$ is 21, agreeing with that for the free ion ^3F term (number of microstates = orbital degeneracy · spin degeneracy). The remaining $^3T_1$ term of the complex transforms into an excited term of the free ion at an energy of $+15$ (remember that the unit of energy is B). There should be a triplet excited term for the free ion of total degeneracy 9. The introductory paragraphs of the last section pointed out the presence of a ^3P term above the ^3F term. The theoretical details also confirm its energy as $15B$ above ^3F. You should find it reassuring that the theoretician's equations for the electronic transition energies behave properly at the limits of metal-ligand interaction.

Now, how do you go about using the equations to extract Δ_o and B? It is sometimes helpful to note [equation (9–2)] that the separation between ν'_3 and ν'_1 is just Δ'_o or, equivalently, the difference $\tilde{\nu}_3 - \tilde{\nu}_1 = \Delta_o$. Unfortunately, $\tilde{\nu}_3$ often occurs at such large energy that it is lost among the more intense charge transfer bands, so you cannot depend on using this difference to extract Δ_o from a spectrum. In such cases, after Δ_o and B have been determined from the positions of $\tilde{\nu}_1$ and $\tilde{\nu}_2$ you may easily predict the position of $\tilde{\nu}_3$ as a check.

The transition energy equations are given for ν', while it is $\tilde{\nu}$ that is measured. A convenient method for solving the problem employs the fact that $\tilde{\nu}_1/\tilde{\nu}_2 = \nu'_1/\nu'_2$, for it makes no difference to the ratio what unit of energy is used. The theoretical expression for the ratio ν'_1/ν'_2 is, from equation (9–2),

$$\nu'_1/\nu'_2 = \frac{1}{2} + \frac{\Delta'_o - 15}{2[(15)^2 + 18\Delta'_o + \Delta'^2_o]^{1/2}} = \tilde{\nu}_1/\tilde{\nu}_2$$

and a plot of this function of Δ'_o is given in Figure 9–26. For any octahedral d^2 complex the experimental value of $\tilde{\nu}_1/\tilde{\nu}_2$ is used to interpolate from the graph the appropriate value of Δ'_o.

We will illustrate this technique with the spectrum of $V(OH_2)_6^{3+}$ in Figure 9–23. The centroids of the two bands are located at 17.2 kK and 25.6 kK, and we assign these to $\tilde{\nu}_1$ and $\tilde{\nu}_2$, respectively.[41] The ratio $\tilde{\nu}_1/\tilde{\nu}_2$ is 0.672, and interpolating from Figure 9–26 you find a value for Δ'_o of 28.6. To find the numerical values of Δ_o and B, one further step is needed. Returning to equation (9–2) for ν'_1 or ν'_2, the unknown B is introduced by simply multiplying both sides of either equation by B. Either choice produces one equation in one unknown; using the ν'_2 equation:

$$\nu'_2 B = \tilde{\nu}_2 = [(15)^2 + 18\Delta'_o + \Delta'^2_o]^{1/2} B$$
$$25.6 = 39.5B$$

where we have substituted $\tilde{\nu}_2 = 25.6$ kK on the left and $\Delta'_o = 28.6$ on the right.

---

[41] There is some risk in this assignment, for you can see in Figure 9–25 that the $^3A_2$ and $^3T_1$ terms cross at $\Delta'_o = 13.5$. The assignment used here is predicated on the assumptions (i) that there is no band below 5 kK (refer to Figure 9–23) which could be assigned to $\tilde{\nu}_1$ and (ii) that $\tilde{\nu}_3 > \tilde{\nu}_2$ (i.e., $\Delta'_o > 13.5$). Generally, you may be confident that an erroneous assignment will lead to unrealistic values of Δ_o, B, and/or the predicted position of the third band.

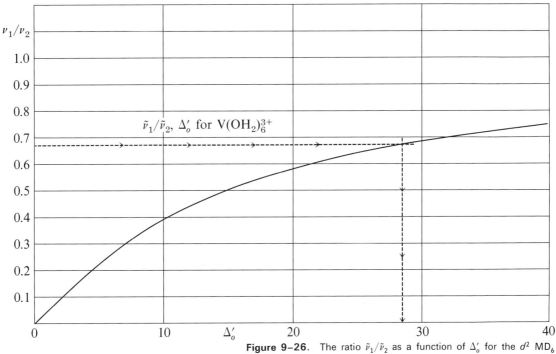

Figure 9-26. The ratio $\tilde{\nu}_1/\tilde{\nu}_2$ as a function of Δ'_o for the d^2 MD_6 complex.

Solution gives $B = 0.649$, and since $\Delta_o = \Delta'_o B$ we obtain

$$\Delta_o = 18.5 \text{ kK}$$
$$B = 0.65 \text{ kK}$$

As a check on the reasonableness of these values, you may use them to calculate the position of $\tilde{\nu}_3$, for which

$$\tilde{\nu}_3 = \tilde{\nu}_1 + \Delta_o = 35.7 \text{ kK}$$

Thus, the values of Δ_o and B are at least consistent with the fact that $\tilde{\nu}_3$ is lost among the CT bands in the ultraviolet region (not shown in Figure 9-23).

A straightforward modification of the transition energy equations for the d^2 case leads to the appropriate equations for the d^8 case. Remember the earlier conclusion that the d^8 term sequence is an inverted d^2 sequence. Two modifications of the d^2 equations must be made to account for this. First of all, $^3T_1$ is no longer the ground term; $^3A_2$ is. This means that the transition energies must be written *as though* the transitions take place from $^3A_2$ to the ^3T terms. Accordingly,

$^3T_1 \leftarrow {}^3A_2$: $\quad -\nu'_3 = -\dfrac{3}{2}\Delta'_o + \dfrac{15}{2} - \dfrac{1}{2}[(15)^2 + 18\Delta'_o + \Delta'^2_o]^{1/2}$

$^3T_2 \leftarrow {}^3A_2$: $\quad -\nu'_3 + \nu'_1 = -\Delta'_o$

$^3T_1 \leftarrow {}^3A_2$: $\quad -\nu'_3 + \nu'_2 = -\dfrac{3}{2}\Delta'_o + \dfrac{15}{2} + \dfrac{1}{2}[(15)^2 + 18\Delta'_o + \Delta'^2_o]^{1/2}$

572 FUNDAMENTAL CONCEPTS FOR TRANSITION METAL COMPLEXES

To effect the inversion of terms, simply replace Δ'_o by $-\Delta'_o$. The final results are

(9–3)
$$^3T_1 \leftarrow {}^3A_2: \quad \nu'_3 = \frac{3}{2}\Delta'_o + \frac{15}{2} + \frac{1}{2}[(15)^2 - 18\,\Delta'_o + \Delta'^2_o]^{1/2}$$

$$^3T_1 \leftarrow {}^3A_2: \quad \nu'_2 = \frac{3}{2}\Delta'_o + \frac{15}{2} - \frac{1}{2}[(15)^2 - 18\,\Delta'_o + \Delta'^2_o]^{1/2}$$

$$^3T_2 \leftarrow {}^3A_2: \quad \nu'_1 = \Delta'_o$$

In this final set of equations we have arranged the transitions in decreasing order of energy and *relabeled them with ν'_3 the highest*. These equations are plotted on the left side of Figure 9–25.

Perhaps the most striking thing about these d^8 equations is that the lowest-energy transition gives Δ_o directly! That is, the lowest-energy $^3T_2$ term from the $t^5_{2}e^3$ excited configuration has the same interelectronic repulsion energy as the ground $^3A_2$ term. Since Δ_o is directly given in this way, we simply rewrite the equation for ν'_2 as

$$\nu'_2 B = \tilde{\nu}_2 = \frac{3}{2}\Delta_o + \frac{15}{2}B - \frac{1}{2}[(15B)^2 - 18\,\Delta_o B + \Delta^2_o]^{1/2}$$

so that, after inserting the numerical value for Δ_o, you have one equation in the unknown B and there is no need for the simultaneous solution of two equations in two unknowns (the reason behind the graphical solution in the d^2 case). Similarly, converting the expression for ν'_3 gives an expression for the energy of the highest transition as a check on values of Δ_o and B.

$$\nu'_3 B = \tilde{\nu}_3 = \frac{3}{2}\Delta_o + \frac{15}{2}B + \frac{1}{2}[(15B)^2 - 18\,\Delta_o B + \Delta^2_o]^{1/2}$$

The tools are now at hand to analyze the $Ni(OH_2)_6^{2+}$ spectrum of Figure 9–23. You will notice that three bands are observed for this complex but that only two of them are in the visible region. The lowest-energy band occurs in what spectroscopists call the near-infrared region. The centroids of the three bands are as follows:

$$\tilde{\nu}_3 = 25.3 \text{ kK}$$
$$\tilde{\nu}_2 = 14.5 \text{ kK}$$
$$\tilde{\nu}_1 = 8.7 \text{ kK}$$

The lowest band gives the value of $\Delta_o = 8.7$ kK. The equation to be solved for B is

$$\tilde{\nu}_2 = 14.5 = \frac{3}{2}(8.7) + \frac{15}{2}B - \frac{1}{2}[(15B)^2 - 18(8.7)B + (8.7)^2]^{1/2}$$

from which $B = 0.97$ kK. Using these values of Δ_o and B, the third band is predicted to lie at 26.1 kK. The error in the calculated position of the third band is on the order of 3 per cent, a value which is certainly acceptable for most purposes.

Before leaving this section on the interpretation of the electronic spectra of high spin octahedral complexes, remember the inversion relation between O_h and T_d term energies. The equations (9–2) given for the transition energies for d^2 and d^7 O_h cases can be used directly for d^3 and d^8 T_d geometries, where Δ'_o is to be replaced by Δ'_t. Similarly, the d^3 and d^8 O_h transition energy equations (9–3) are directly applicable to the d^2 and d^7 T_d spectra.

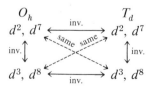

Recall also that the magnitude of Δ_t should be approximately half that of Δ_o for any given metal-ligand combination. This relationship is often of great help in analyzing the spectrum of a T_d complex for which only two of three bands may be observed. For example, if the lowest-energy visible band is assigned as the lowest-energy transition, when in fact this band is actually the second transition, then too large a Δ_t will result. The ion $FeCl_4^{2-}$, for example, has such a low Δ_t that its electronic transition occurs in the *infrared* at 4 kK! (What is the value of Δ_t in this case?)

To conclude this introduction to the molecular terms of complexes, a qualification is necessary to the earlier claim that the transition energy diagrams for the d^1 and d^2 cases are applicable to the d^4, d^6, and d^7 cases. The claim is true up to a certain value of Δ_o' equal to the pairing energy (P' in units of B) for the d electrons. For Δ_o' beyond this magnitude, there will be a change in the ground term corresponding to the *change from a high to a low spin configuration.* For each d^{4-7} case this change of ground term does not arise from a discontinuous change in term energies. Rather, when Δ_o' is small there is present a term of lower spin multiplicity at higher energy than the high spin ground term. As Δ_o' increases, the energy gap between this excited term and the ground term decreases as $(P' - \Delta_o')$ decreases, until the two terms cross one another. This is illustrated as follows:

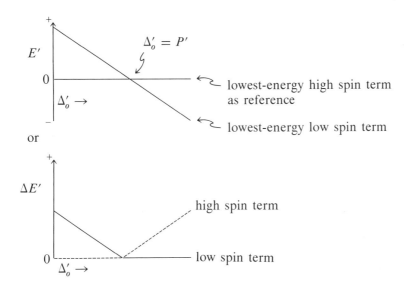

In the appendix to this chapter, you will find diagrams similar to Figure 9-25 for the $d^2 \rightarrow d^8$ O_h cases. The state energies as a function of Δ_o' are given there only for states of the same spin multiplicity as the ground state, whether high or low spin.

The pairing energy is actually the difference between the average repulsion energies for the high and low spin configurations. For the $d^{4,5,6,7}$ configurations, the theoretical difference in repulsion energies is:

$$(d^4) \quad 6B + 5C = P$$
$$(d^5) \quad 15B + 10C = 2P; \quad P = 7.5B + 5C$$
$$(d^6) \quad 5B + 8C = 2P; \quad P = 2.5B + 4C$$
$$(d^7) \quad 4B + 4C = P$$

where B is the unit of repulsion energy used earlier, and C is an additional repulsion factor that did not appear in the previous equations. The crossover of high and low spin terms occurs when Δ_o is equal to the pairing energy. While there is no theoretical relationship between the magnitudes of B and C, an empirical relationship of usefully approximate validity is $C = 4.5B$. With such an approximation, the pairing energies become $28.5B$, $30B$, $20.5B$, and $22B$, and the crossover values of Δ_o' are approximately

574 ☐ **FUNDAMENTAL CONCEPTS FOR TRANSITION METAL COMPLEXES**

$$(d^6)\ 20.5$$
$$(d^7)\ 22$$
$$(d^4)\ 28.5$$
$$(d^5)\ 30$$

Obviously, not too much can be made out of these values, but there is a clear distinction to be made between the $d^{6,7}$ and $d^{4,5}$ cases in that *the crossover requires a stronger metal-ligand interaction in the $d^{4,5}$ ions.* This expectation is realized in several comparisons. For example, among the aquo complexes of the trivalent metals of the first transition series [$M(OH_2)_6^{3+}$], only Co^{3+} (d^6) exhibits a low spin complex.[42] That none of the divalent complexes $M(OH_2)_6^{2+}$ exhibit low spin configurations (including the d^6 Fe^{2+} and d^7 Co^{2+} complexes) indicates the importance of the metal ion charge in increasing Δ_o. Furthermore, the only known high spin complexes of Co^{3+} are CoF_6^{3-} and $CoF_3(OH_2)_3$, both characterized by the presence of the F^- ligand (complexes of which have characteristically low Δ_o's). Similarly, Mn^{2+} (d^5) exhibits a marked preference to be high spin, and to a greater extent than the more highly charged Fe^{3+} ion (also d^5). The striking contrast between the tendencies for low spin configurations of Fe^{3+} and Co^{3+} is but another example of the role that pairing energy has to play in the ground electron configuration.

Analysis of the spectra of low spin octahedral complexes presents some difficulties not inherent in the high spin cases. In particular, the occurrence of the additional electron repulsion parameter C in the equations for the transition energies requires that more than two absorption bands be observed. With reference to the spectrum of $Co(en)_3^{3+}$ given in Figure 9-24, two prominent bands arise in the visible region.

(9-4)
$$\begin{aligned}\nu_2' &= \Delta_o' - C' + 16 \quad (^1T_2 \leftarrow {}^1A_1) \\ \nu_1' &= \Delta_o' - C' \quad\quad\quad\;\; (^1T_1 \leftarrow {}^1A_1)\end{aligned}$$

The energies of these transitions are characterized by three parameters (Δ_o, B, and C) from theory. To completely solve the problem, at least one additional absorption band must be *located and correctly assigned.* This is a formidable task, since the additional bands are necessarily much weaker than the others surrounding them. For example, transitions to terms of different spin multiplicity than the ground term occur in the same region as ν_1 and ν_2 (see the appendix to this chapter) but are difficult to locate with accuracy. Similarly, transitions to higher-lying states of the same spin multiplicity as the ground state are well into the ultraviolet region, where they are masked by the intense charge transfer bands.

From equations (9-4), the separation of the ν_1 and ν_2 bands is just $16B$. For the complex $Co(en)_3^{3+}$, for which $\tilde{\nu}_2 = 29.4$ kK and $\tilde{\nu}_1 = 21.4$ kK, one arrives at $B = 0.5$ kK. From the ratio of $\tilde{\nu}_2/\tilde{\nu}_1$,

$$\frac{\nu_2'}{\nu_1'} = 1.38 = 1 + 16(\Delta_o' - C')^{-1}$$

or

$$\Delta_o' - C' = 42 \text{ kK}$$

To determine Δ_o', additional information is needed. The assumption (p. 573) that C' lies in the range between 4 and 5 yields

$$46 < \Delta_o' < 47$$
$$\Delta_o \approx 23 \text{ kK}$$

---

[42] In fact, $Co(OH_2)_6^{3+}$ has Δ_o just barely greater than P, for it is believed that there is present a small amount of high-spin complex in room temperature samples (prior to the generation of high spin $Co(OH_2)_6^{2+}$ by the Co^{3+} oxidation of H_2O).

Since $P < 23$ kK, the spectrum of $Co(en)_3^{3+}$ fits the theory for a low spin Co^{3+}. This is more directly confirmed from the d^6 term diagram in the appendix; the high spin case permits only *one* fairly low energy transition [see the $Fe(OH_2)_6^{2+}$ spectrum in Figure 9-23]. Nevertheless, a word of caution about making the correlation of one band $\Rightarrow$ high spin, two bands $\Rightarrow$ low spin is in order. Since the separation of ν_1 and ν_2 is given by $16B$, it is possible that a complex with a very low value of B will exhibit overlapping of ν_1 and ν_2, as may be the case for $Fe(CN)_6^{4-}$ (Figure 9-24).

STUDY QUESTIONS

17. If CoF_6^{3-} is a high spin complex, what do you predict for $CoBr_6^{3-}$? (Base your estimate on the relative σ bond strengths of HF and HBr.) In fact, CoF_6^{3-} is the only known binary Co^{3+}/halogen complex. Refer to footnote 42 and comment on the "non-existence" of CoX_6^{3-}, X = Cl, Br, I.

18. Show by arguments like those in the text for the high spin octahedral cases that for $d^{4,9}$ tetrahedral complexes the ground and excited terms are simply T and E, while those for $d^{3,8}$ tetrahedral complexes are the same as the terms for $d^{2,7}$ octahedral ones (exclusive of spin multiplicity).

19. Refer to Figure 9-23 and notice that the band for $Fe(OH_2)_6^{2+}$ falls at lower energy than that of $Ti(OH_2)_6^{3+}$. Estimate from the spectra the values of Δ_o for these complexes and explain why one is greater than the other (consider both cation charge and d^* mo occupation numbers).

20. Referring to Figure 9-23, compare estimated Δ_o's for $Cr(OH_2)_6^{2+}$ and $Cu(OH_2)_6^{2+}$ and account for the fact that one is greater than the other.

21. Using the spectra in Figure 9-23, determine Δ_o and B for $Cr(OH_2)_6^{3+}$ for comparison with the text values for $Ni(OH_2)_6^{2+}$. Rationalize the relative Δ_o and relative B values.

22. Explain why FeF_6^{3-} is colorless while CoF_6^{3-} exhibits only a single absorption band in the visible region.

23. Give the number of t_2^* and e^* electrons and spin multiplicity for the ground configuration of each of the following; then predict the number of electronic absorption bands deriving from $t_2^* \rightarrow e^*$ electron promotion.

$$Fe^{3+} \ (O_h, HS)$$
$$Cu^{2+} \ (O_h)$$
$$Zn^{2+} \ (T_d)$$
$$Cr^{3+} \ (O_h)$$
$$Fe^{2+} \ (T_d)$$
$$Co^{2+} \ (T_d)$$

24. Which ion(s) in the preceding question shows the greatest preference for six-coordination?

THE SPECTROCHEMICAL SERIES

Through analysis of complex spectra by the techniques just illustrated, transition metal chemists have compiled considerable data on the parameters Δ_o and B. Inspection of the results has made possible several generalizations of use to the practicing chemist. For example, it is observed that if the ligands of MD_6 complexes are arranged in order of increasing Δ_o for each metal ion, the order is generally independent of the metal ion. It would appear[43] that a rough single-scale ordering of metal-ligand interactions is possible (recall your previous encounter with single-scale concepts in Chapter 5 on donor/acceptor interactions, of which transition metal complexes are special cases). A partial series, including members of commonly occurring ligand types, is (in order of increasing Δ_o):

$$I^- < Br^- < Cl^- < {}^*SCN^- < F^- \sim \text{urea} < OH^- < \text{acetate} < \text{oxalate}$$
$$< H_2O < {}^*NCS^- < \text{glycine} < \text{pyridine} \sim NH_3 < \text{en} < SO_3^{*2-}$$
$$< \text{dipyridine} \sim o\text{-phen} < {}^*NO_2^- < CN^-$$

(* labels the donor atom)

From analysis of the magnitude of Δ_o in terms of e_σ and e_π, you expect to find that combinations of weak sigma and strong pi donor properties produce small Δ_o, while strong sigma donor and weak pi donor or strong pi acceptor character of a ligand should produce a large Δ_o. It comes as no surprise, then, that the halides appear early in the series and NO_2^- and CN^- appear late. Other comparisons, which might not make any sense at all if only relative sigma donor properties needed to be considered, make good sense upon including metal-ligand pi interactions. For example, the higher position of H_2O than of OH^- is probably due to the fact that OH^- has two pi donor electron pairs while H_2O has only one.

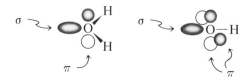

Similarly, the higher positions of dipyridine and o-phenanthroline than of pyridine are believed to be due to the weaker pi donor character and stronger pi acceptor nature of the more delocalized amines (delocalization over an extended carbon frame tends to lower the energies of both π HOMO and π LUMO).

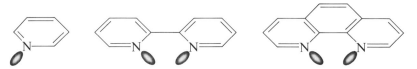

Quite generally, you see a relationship among nitrogen donors between Δ_o and p character of the donor lone pairs, although the certainty of this correlation is confused by the variation in π interactions (for example, dipyridine and o-phenanthroline appear higher than NH_3 and en). In addition to the sp character of the nitrogen lone pair, a factor in the low position of *NCS^- relative to the other nitrogen donors may be its pi donor nature. Conversely, NO_2^- appears higher than the other "sp^2" hybridized nitrogen donors because of an appreciable π^* acceptor nature. It is reassuring that the overall order

$$(\text{halogen}) < (\text{oxygen}) < (\text{nitrogen}) < (\text{carbon})$$

is in keeping with our expectations of sigma donor strength.

Of course, the metal ion influences the magnitude of Δ_o through the overlap and

---

[43] Don't forget that even in the simple AOM model, Δ_o is a function of two parameters, e_σ and e_π.

energy match criteria; the following metal ion spectrochemical series generally follows expectations:

$$(Mn^{2+} < Ni^{2+} < Co^{2+} < Fe^{2+} < V^{2+}) < (Fe^{3+} < Cr^{3+} < V^{3+} < Co^{3+})$$
$$< Mn^{4+} < (Mo^{3+} < Rh^{3+} < Ru^{3+}) < Pd^{4+} < Ir^{3+} < Re^{4+} < Pt^{4+}$$

Here you find increasing Δ_o with increasing metal charge; within a series of like charged ions, the effect of improved metal-ligand orbital energy match (improves as the d orbitals are stabilized by increasing *nuclear* charge) is in competition with the weakening effect of increasing occupation of the π^* and σ^* mo's; the heavier elements of a group generate a larger Δ_o than the lightest element of the same charge for two reasons. *Generally, the pairing energies of the heavier elements are smaller, meaning that low spin complexes are the rule.* This enhances Δ_o because low spin complexes minimize the number of electrons in the σ^* mo's. Additionally, the "transition metal contraction," symptomatic of a disproportionately high effective nuclear charge acting on the ligand electrons and greater involvement of the $(n+1)\,s$ and p ao's facilitates a stronger metal-ligand interaction, with e_σ being enhanced to a greater extent than e_π.

STUDY QUESTIONS

25. The Δ_o values for $Ru(OH_2)_6^{2+}$ and $RuCl_6^{3-}$ are nearly the same. Is this consistent with the positions of H_2O and Cl^- in the spectrochemical series? If not, how do you explain the similarity in Δ_o values?

26. Following are spectral data for some NiD_6^{2+} complexes. From the given band centroid frequencies (in kK) for each complex, obtain Δ_o and B. Then arrange a ligand spectrochemical series and interpret trends in both Δ_o and B. (DMSO = dimethylsulfoxide; DMA = N,N-dimethylacetamide.)

$Ni(DMSO)_6^{2+}$	7.7	13.0	24.0
$Ni(DMA)_6^{2+}$	7.6	12.7	23.8
$Ni(OH_2)_6^{2+}$	8.7	14.5	25.3
$Ni(NH_3)_6^{2+}$	10.8	17.5	28.2

PARAMAGNETISM OF COMPLEXES

In addition to the visible region spectral properties of transition metal complexes, there is another unique, valuable property of transition metal compounds. Their paramagnetism arises from the presence of unpaired electrons, and you have seen how this property is derived from n of d^{*n} and structure and, in the d^{4-7} cases, depends on the strength of metal-ligand bonding.

There is, of course, a characteristic magnetic field associated with the motion of a charged particle. The operation of solenoids is an example from everyday experience. Current passing through a coil of wire induces a magnetic field along the axis of the coil, and this field may be thought of as arising from a bar magnet at the coil center. For the electron there are two types of motion that may give rise to an associated magnetic field. The first of these is a quantum relativistic phenomenon for which no classical analogy is strictly possible. The intrinsic *spin angular momentum* of the electron is loosely thought of as arising from motion of the electron about its own axis and has associated with it a magnetic field represented by a magnetic dipole or moment, μ_s. It is not surprising that there is a relation between the magnitude of the angular momentum and the magnitude of the magnetic moment,

$$\mu_s = g_s \beta_e \sqrt{s(s+1)}$$

where g_s has the magnitude 2.0023 (the difference from 2.00 is of little practical significance) and $\sqrt{s(s+1)}$ gives the magnitude of the spin angular momentum. β_e, the *Bohr magneton,* is the only dimensioned quantity on the right side; its value is 9.27×10^{-21} erg/gauss (see Table 1-1). Therefore, μ_s has these units. The proper name given to the g factor is *magnetogyric ratio,** a name to remind us that g_s is defined as the ratio of the magnetic moment to the angular momentum.

The experimental determination of the magnetic moment of an electron or electrons *in a molecule* yields a result consisting of contributions from both the spin and orbital moments, the *orbital motion* being the second way an electron acquires magnetism. For purposes of generality and simplicity in reporting results, the magnetogyric ratio in such cases can still be defined in terms of the ratio of total, observed moment to the spin (*not* total) angular momentum.

$$g = \frac{\mu_{obs}}{\beta_e \sqrt{s(s+1)}}$$

In this way, the orbital contribution to the magnetic moment appears through an alteration of the spin-only factor, g_s. Thus, g becomes an important property of the molecule, in contrast to g_s which is one of the universal constants and, therefore, unrelated to molecular structure and the orbital motion of electrons.

Table 9-7 presents some typical g value ranges for the first row transition ion complexes so that you may see how significant are the orbital contributions. To completely account for these deviations, one must consider the following:

a) Does the geometry deviate from O_h and T_d? If so, how does this affect the orbital contribution?
b) Does the *coupling* of spin and orbital motions act to increase or decrease the magnetic moment?
c) How is the spin-orbital coupling affected by covalency in the metal-ligand bonds?

---

*Also called the Landé g factor.

TABLE 9-7

RANGES OF g VALUES FOR O_h AND T_d COMPLEXES

Ion	Configuration	g	μ_s/β_e (spin only)
O_h			
Ti^{3+}, V^{4+}	d^1	1.85–2.08	1.7
Cu^{2+}	d^9	1.96–2.54	
V^{3+}, Cr^{4+}	d^2	1.91–2.05	2.8
Ni^{2+}	d^8	2.05–2.33	
Cr^{3+}, Mn^{4+}	d^3	1.91–2.06	3.9
Co^{2+}	d^7	2.22–2.68	
Co^{2+}, Ni^{3+}	d^7 (LS)	2.08–2.31	1.7
Cr^{2+}, Mn^{3+}	d^4	1.92–2.04	4.9
Fe^{2+}	d^6	2.08–2.33	
Cr^{2+}, Mn^{3+}	d^4 (LS)	2.26–2.33	2.8
Mn^{2+}, Fe^{3+}	d^5	1.89–2.06	5.9
Mn^{2+}, Fe^{3+}	d^5 (LS)	2.08–2.89	1.7
T_d			
Cr^{5+}, Mn^{6+}	d^1	1.96–2.08	1.7
Cr^{4+}, Mn^{5+}	d^2	1.84–1.98	2.8
Ni^{2+}	d^8	2.61–2.83	
Fe^{5+}	d^3	1.86–1.91	3.9
Co^{2+}	d^7	2.16–2.47	
Fe^{2+}	d^6	2.16–2.24	4.9
Mn^{2+}	d^5	1.99–2.09	5.9

d) Does the presence of a magnetic field alter the magnetic properties of the electronic terms of a complex by causing term mixing?
e) In the presence of a magnetic field, the microstates of the lowest-energy term are of different energy, as the magnetic moments of these microstates are in various degrees of alignment with the magnetic field (the microstates represent the quantization of the angular momentum or magnetic moment in the direction of the field). It is important to know whether or not all these levels are within thermal energy of each other, because a macroscopic sample will exhibit a Boltzmann distribution of molecules among these energy levels (microstates).

The theory for explaining these deviations has been satisfactorily developed but is so involved that only the specialist interested in the operation of the above list of influences will be sufficiently moved to pursue the details. Fortunately, the deviations of g from the spin-only value are usually sufficiently small that the measurement of the magnetic moment of a complex reveals the number of unpaired electrons present. Magnetic measurements can, for example, help one determine whether a given complex is to be considered high- or low-spin octahedral. In at least one case (Ni^{2+}) the range of g values helps to distinguish O_h from T_d complexes, and to some extent this is true of Co^{2+} complexes.

Let us now sharpen somewhat our understanding of the measurement of magnetic moments, and in the process achieve some understanding of the origin of $g \neq g_s$. The concept of a molecular magnetic moment implicitly assumes that each molecule possesses a tiny bar magnet whose magnetic field strength is derived from the orbital and spin motions of the electrons in the molecule. Only unpaired electrons can contribute spin magnetism because paired electrons contribute offsetting spin magnets. Similarly, completely filled orbital shells contribute no orbital magnetism because for every electron revolving about the nucleus in one direction ($m_l = +n$) there is a partner revolving in the opposite sense ($m_l = -n$). (Recall from Chapter 1 the interpretation given to the m_l quantum number.) For transition metal complexes it is the unfilled d^* mo's that are responsible for the presence of the molecular magnet.

For a sample of the complex in free space, thermal agitation causes complete disordering of these magnets; but in a magnetic field this chaos is partially resolved as the molecular magnets prefer to align their lines of force with those of the laboratory

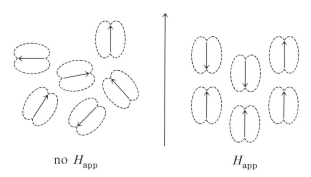

no H_{app} H_{app}

magnet. The extent to which this alignment occurs depends on the energy of the coupling between the laboratory and molecular magnets, relative to the thermal energy available [see item (e) above]. Consequently, the sample achieves a net magnetic polarization, in the field of the laboratory magnet, in proportion to the strength of the latter and to the strength of the molecular magnet, and inversely related to the temperature. This tendency of a sample to experience net magnetic polarization in proportion to the strength of an applied field is called its *susceptibility,* κ (kappa). The greater this susceptibility, the greater the interaction between the molecular and laboratory magnets. This provides a convenient experimental method

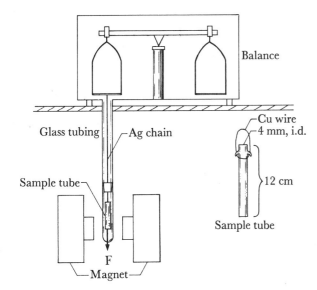

Figure 9-27. The Gouy apparatus and the force acting on a net paramagnetic sample. [From R. J. Angelici, "Synthesis and Technique in Inorganic Chemistry," 2nd ed., W. B. Saunders Company, Philadelphia, 1977.]

for determining the susceptibility. If the sample were uniformly within a region of constant field strength, there would be no net force acting on the sample. On the other hand, if one end of the sample were in a region of high field strength and the other end in a region of low field strength, there would be a net force acting so as to draw the sample into (paramagnetism) or out of (diamagnetism) the region of greater strength. An experimental arrangement such as that shown in Figure 9-27 illustrates the so-called Gouy technique.

Quite clearly, the susceptibility is concentration dependent; a low concentration of molecular magnets in the sample will result in a low susceptibility, all other factors being equal. Thus, the susceptibility κ is identified as a *volume susceptibility* with units of erg gauss^{-2} vol^{-1}. A *gram susceptibility*, χ (chi), is obtained from κ by dividing the latter by the density of the sample. Finally a *molar susceptibility*,[44] χ_M, is defined by multiplying χ by the molecular weight. χ_M is a macroscopic property of a bulk sample, and some connection between it and the molecular magnetic moment, μ, is needed.

$$\chi_M = \chi \times (\text{MW}) = \kappa \times \left(\frac{\text{MW}}{d}\right) \text{erg gauss}^{-2} \text{mole}^{-1}$$

Before looking into this relationship between χ_M and μ, you must realize that every molecule contains a great number of paired electrons that do not, by arguments given earlier, contribute to μ but do affect the magnitude of χ_M from which μ is to be determined. This contribution from paired electrons *opposes* that from the unpaired electrons and so distinguishes the *diamagnetic behavior of the paired electrons in the molecule from the paramagnetic* property of unpaired electrons. The application of a magnetic field induces a circulation of all the electrons about the field direction, setting up, in loose parlance, a weak intramolecular electric current (note the analogy with an electric current generator). This induced current itself generates a magnetic field in opposition to the applied field (recall the right- and left-hand rules from your physics course). Thus, the field applied to a sample *induces* a molecular diamagnetism in opposition to the paramagnetism of the unpaired electrons. This, in effect, reduces the susceptibility of the sample below that due to the paramagnetism alone. Fortunately, the molecular diamagnetism is usually very much smaller than the paramag-

---

[44] The cgs susceptibility unit erg G^{-2} mole^{-1} is equivalent to the unit cm^3 mole^{-1}. Custom has been, however, to report χ_M as a number followed by the acronym "cgs".

netism, but pesky errors accrue in large molecules if its presence is neglected. Generally, it is satisfactory to assume the additivity of nominal atomic diamagnetic susceptibilities, and tables of these have been published for convenience. After correcting the measured χ_M for the molecular diamagnetic susceptibility, the stage is set for inquiring how the g value may differ from 2.00.

Of the full list (a) through (e) on p. 578, only points (b) and (e) will be considered here. With regard to (b), it would be inappropriate to explain in full in this text "how" spin-orbital coupling (that is, interaction of the spin and orbital magnetic moments) affects the energy of a molecular state; the effects can be illustrated for the d^1 case by using plausible arguments based on the concepts of Chapter 1. For the d^1 octahedral case you have seen that the ground term is $^2T_2$, a symbol that has the physical significance that this term consists of six microstates (for each of three orbital possibilities there are two spin possibilities). Spin-orbital coupling renders these six microstates no longer degenerate by resolving the six-fold degenerate $^2T_2$ term into a doubly degenerate term at an energy of $+\lambda$ and a four-fold degenerate set at $-\frac{1}{2}\lambda$, where λ is a unit of spin-orbit coupling energy. The resolution is diagrammed in Figure 9-28. (This behavior is entirely analogous to the resolution of the ^3P free ion term microstates as explored in Chapter 1.) An important point here is that the orbital magnetism of each of these new terms is different from the other and from the $^2T_2$ term. In fact, the g values of the spin-orbit levels of $^2T_2$ work out to be 2 and 0. It is an interesting example of the vagaries of spin-orbit coupling that the orbital and spin moments *cancel* in the four lowest microstates.

Now point (e) can be considered. In a macroscopic sample of d^1 O_h complexes, there will be a thermal distribution (Boltzmann distribution) of molecules among these spin-orbital terms because the separation of $\frac{3}{2}\lambda$ between their energies is on the order of only a few tenths of 1 kK, while at room temperature the thermal energy available (kT) is ~0.2 kK. The magnetic susceptibility one measures in the Gouy apparatus is therefore a Boltzmann population average over the different magnetic susceptibilities of the spin-orbit terms. The g value obtained from the Gouy measurement is also, therefore, a weighted average value between 0 and 2.[45]

A complete, general, theoretical treatment[46] of the Boltzmann distribution of

---

[45] The values 2 and 0 result from assuming (i) perfect O_h symmetry, (ii) no magnetic field alteration of the spin-orbit resolved states, and (iii) that the ^2E excited state lies at very high energy (Δ_o large). In a real sample these "ideal" conditions are not well met and typical g values are larger than the simple Boltzmann average of 0.66.

[46] See, for example, S. F. A. Kettle, "Coordination Compounds," Nelson, London, 1969; B. N. Figgis, "Introduction to Ligand Fields," Interscience, New York, 1966.

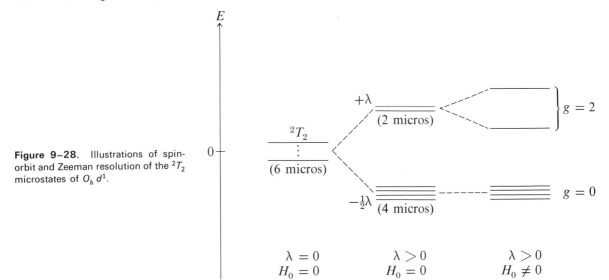

Figure 9-28. Illustrations of spin-orbit and Zeeman resolution of the $^2T_2$ microstates of O_h d^1.

molecules among the spin-orbit and magnetic field states (e.g., those at the right side of Figure 9-28) so as to determine the weighted average μ for a sample and, therefore, the susceptibility yields the following result (χ_M^P is χ_M corrected for diamagnetic contributions):

$$\chi_M^P = \frac{N}{3kT}\mu^2$$

The form of this equation ($\chi = C/T$) is that determined empirically by Curie and is known as the *Curie law*. For most paramagnetic materials, a plot of χ vs. T^{-1} yields a straight line, in agreement with the theory. When one does not go to the trouble of experimentally verifying that the sample follows the Curie law, but only determines its susceptibility at one temperature, the accepted procedure is to simply report μ as calculated from

$$\mu = \sqrt{\frac{3k}{N}} \cdot \sqrt{\chi_M^P \cdot T}$$
$$= 2.84\sqrt{\chi_M^P \cdot T}$$

In most magnetic susceptibility studies by chemists, the magnetic moment is not reported in units of erg gauss^{-1} but as a dimensionless multiple of β_e (as in Table 9-7). Accordingly, one usually finds in the literature reports of μ_{eff}, a pure number, defined by

$$\mu_{\text{eff}} = \frac{\mu}{\beta_e} = \frac{2.84}{\beta_e}\sqrt{\chi_M^P \cdot T} \qquad \text{(dimensionless)}$$

STUDY QUESTIONS

27. The magnetic susceptibility of a sample of Ni(PPh$_3$)$_2$Cl$_2$ was found to be 6.02×10^{-6} cgs. Is the complex square planar or "tetrahedral"? The diamagnetic correction factors ($\times 10^6$ cgs) for the atoms are Ni(12.8), P(26.3), Cl$^-$(23.4), H(2.93), and C(6.0). What is the value of χ_M^P in SI units (see inside front cover)?

28. The complex (solid) Hg[Co(SCN)$_4$] is frequently used to calibrate a Gouy susceptibility apparatus because its magnetic moment is 4.5 Bohr magnetons. Can you eliminate any of the following as possibilities for the anion: T_d(high spin), T_d(low spin), D_{4h}(high spin), D_{4h}(low spin)?

29. Characterize the following complexes as high or low spin:

	μ (B.M.)		μ (B.M.)
Co(NH$_3$)$_6^{2+}$	5.0	Pt(NH$_3$)$_2$Cl$_2$	0.0
Co(NO$_2$)$_6^{4-}$	1.9	Fe(CN)$_6^{3-}$	2.3
CoF$_6^{3-}$	5.3	Fe(OH$_2$)$_6^{2+}$	5.3

Grouping these complexes into two categories based on whether their moments deviate from the spin-only value by more or less than 0.4 B.M., examine the t_{2g}^* occupations and draw a conclusion.

30. Give spin-only magnetic moments for the following:

RuF$_6^{3-}$ (O_h)	Cu(NH$_3$)$_4$(OH$_2$)$_2^{2+}$ (D_{4h})	Fe(NO)(CN)$_5^{2-}$ (C_{4v})
Pd(CN)$_4^{2-}$ (D_{4h})	Ru(CO)$_5$ (D_{3h})	AuCl$_4^{2-}$ (D_{4h})
NiCl$_4^{2-}$ (D_{2d})	Co(NH$_3$)$_3$(OH$_2$)$_3^{3+}$ (C_{3v})	AlF$_6^{3-}$ (O_h)

31. Because of their close similarity to the naturally occurring vitamin B_{12} series of compounds, the "cobaloximes" have received considerable attention in recent years. Cyanobis(dimethylgloximato)(pyridine)cobalt is an example of such model complexes.

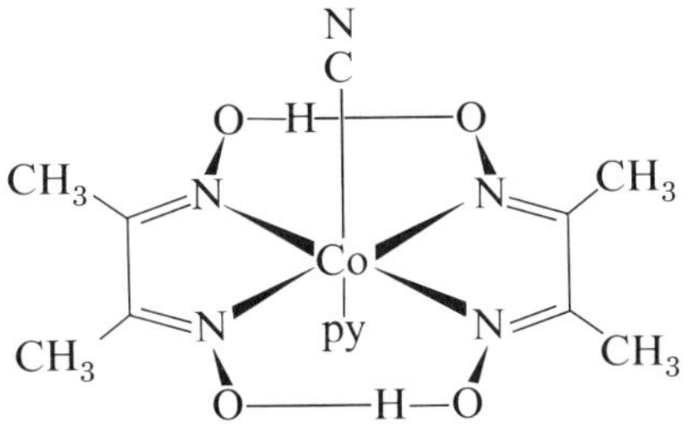

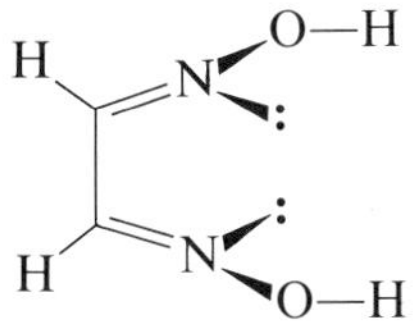

a) Predict a magnetic moment for this "cobaloxime" (the cobalt oxidation state is determined from the fact that the complex bears no net charge).
b) Of greater similarity to the natural cobalamins are the cobaloximes with CH_3^- in place of CN^-. Related to the role of the cobalamins as biological electron transfer reagents, methylcobaloximes experience irreversible chemical and polarographic reduction with loss of the CH_3 group. It is believed that the precursor to the electron transfer is the square pyramidal $Co(DMG)_2CH_3$. Identify the LUMO for this complex and, from its metal-ligand characteristics, explain how the loss of CH_3 is facilitated by the reduction of Co(III).

32. The methyl cobaloximes are among the most stable organo-transition metal compounds known, but they are photochemically reactive. Exposure to light leads to reactions via formation of CH_3 radicals. Suggest an excited electronic configuration that could explain this property of the methyl(pyridine)cobaloximes (though formally octahedral, these complexes do present different ligand environments along the x, y, and z axes).

EPILOG

Considerable spade-work has been performed in this chapter with the goal of acquainting you with the fundamental properties and concepts of transition metal complex structures and stabilities. Not only have the d^* occupation and VSEPR concepts proved valuable to understanding the structural preferences for the MD_6, MD_5, and MD_4 stoichiometries, but the mo model is the best we have at the present time for interpretation of the unique spectral and magnetic properties of such compounds. The time spent developing the ideas of atomic terms and states in Chapter 1 was not an end in itself but was necessary as a foundation for the development of the spectral and magnetic properties of complexes. Similarly, the discussion in Chapter 5 on the general donor/acceptor interaction has yielded further returns in this chapter. With this solid background of descriptive and theoretical material at hand, you are well prepared to investigate the fascinating world of transition complex structures and reactivities awaiting you in the following chapters.

Appendix to Chapter 9
State Energy Diagrams for MD_6 (O_h)

Where appropriate, low spin states are distinguished by dotted lines.

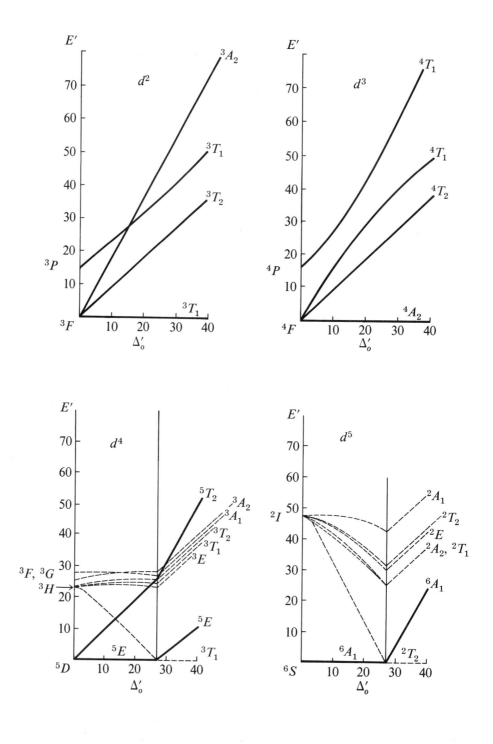

APPENDIX TO CHAPTER 9 □ 585

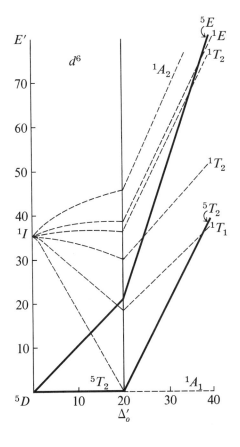

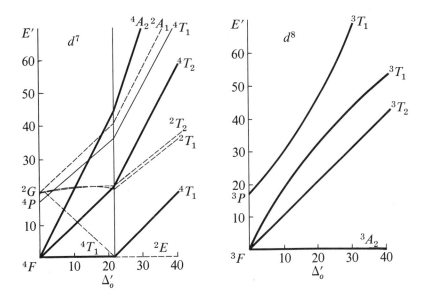

GENERAL CONSIDERATIONS

LOW COORDINATION NUMBERS
   Two-Coordinate Complexes
   Three-Coordinate Complexes
   Four-Coordinate Complexes
     Tetrahedral Complexes
     Square Planar Complexes
   Five-Coordinate Complexes

Six-Coordinate Complexes

POLYHEDRA OF HIGH COORDINATION NUMBER
   Seven-Coordinate Complexes
   Eight-Coordinate Complexes
   Complexes Having Coordination Numbers of Nine or Higher

EPILOG

10. COORDINATION CHEMISTRY: STRUCTURAL ASPECTS

> *The transition metals may form complexes with coordination numbers from two through nine, and even higher coordination numbers are accessible if the metal ion involved is a lanthanide or actinide. In the preceding chapter you explored the conditions under which a metal complex may be six-coordinate and octahedral or four-coordinate and either square planar or tetrahedral. In this chapter we examine in a more general way the several stereochemical arrangements found for coordination numbers from two through six and a few of the topologies found for higher coordination numbers. As you saw in Chapter 9, the application of molecular orbital methods, and especially the angular overlap model, should allow you to systematize to a great extent the structural chemistry of the transition metals in their higher valence states.*

GENERAL CONSIDERATIONS

In earlier sections of this book you found that the structures of non-metal compounds can usually be accounted for by the relatively simple VSEPR theory. However, very early in the development of this theory, it became apparent that it unfortunately did not apply to transition metal complexes. Since these metal compounds are characterized by a rich variety of stereochemical arrangements, the lack of a simple, consistent theory for the rationalization and prediction of structures has heretofore been a serious handicap. However, in Chapter 9 we showed that the stereochemistry of transition metal compounds is mainly determined by two factors: (i) the number of $d\sigma^*$ electrons and (ii) ligand-ligand repulsion forces.[1] That is, if several structures are possible for a given complex, the one adopted is that which minimizes the role of the $d\sigma^*$ anti-bonding electrons; if this does not lead to a clear-cut structural preference, the structure adopted is the one that minimizes repulsions between ligands.

A further structural constraint sometimes observed is that of ligand rigidity, wherein the intrinsic structure of a ligand may be such that the donor atoms have definite positions and so define a "pocket" into which the metal atom or ion must fit. Such characteristics are peculiar to each ligand and will be explored as their complexes come into our discussions. For the present we shall examine only the constraints that metal ions place on the stereochemistry of complex ions.

Figure 10-1 is a series of plots of structural preference energies for the limiting two-, three-, four-, five-, and six-coordinate geometries to be discussed later. As

---

[1] For the sake of convenience, we shall refer to ligand-ligand and bond pair-bond pair repulsion forces collectively as VSEPR forces. You should be aware that they are not strictly the same, however. As used in earlier sections of the book, VSEPR forces arise from repulsions between stereochemically active electron pairs, not atoms or groups of atoms.

588 ☐ COORDINATION CHEMISTRY: STRUCTURAL ASPECTS

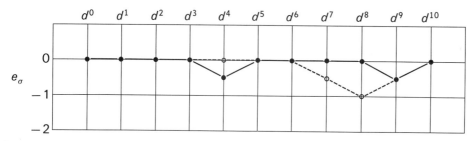

(A) Two-coordination. E(linear) − E(bent). Linear geometry favored over bent.

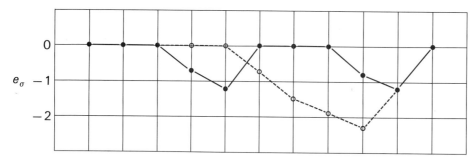

(B) Three-coordination. $E(\top) - E(\curlywedge)$. $\top(C_{2v})$ favored over $\curlywedge(D_{3h})$.

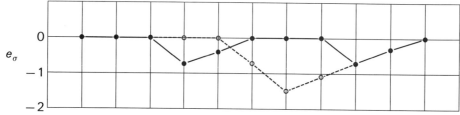

(C) Three-coordination. $E(\perp) - E(\curlywedge)$. $\perp(C_{3v})$ favored over $\curlywedge(D_{3h})$.

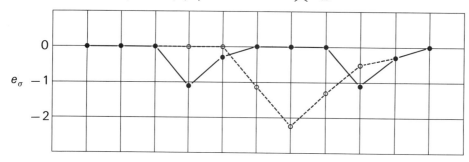

(D) Five-coordination. $E(\star) - E(\star)$. $\star(C_{4v})$ favored over $\star(D_{3h})$.

Figure 10–1. Structure preference energy plots for coordination numbers two through six. See Figure 9–19 for the corresponding plots for four-coordinate complexes. Plot (F) is the energy of the geometry favored by the angular overlap model relative to the octahedron. All are plotted in units of e_σ. In all cases, high spin is shown by the solid line and low spin by the dashed line.

Illustration continued on opposite page.

demonstrated in Chapter 9, these plots are constructed on the basis of the angular overlap model, and they consider metal-ligand σ interactions only, since these generally dominate the selection of one structure in preference to another. On the basis of the plots in Figure 10–1, one can make some general observations:

(a) For two-coordinate complexes the limiting geometries are linear and 90° bent. VSEPR and steric forces favor the linear geometry in all cases, but the $d\sigma^*$ occupation factors significantly favor this geometry only for low spin d^8 metal ions. Therefore, MD$_2$ compounds are generally found to have linear geometries, while

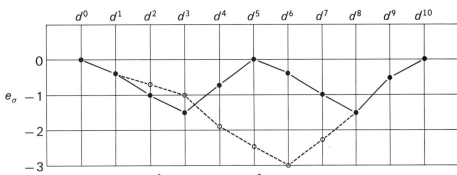

(E) Six-coordination. $E(O_h) - E(\text{pentagonal})$. ✳ favored over ▢.

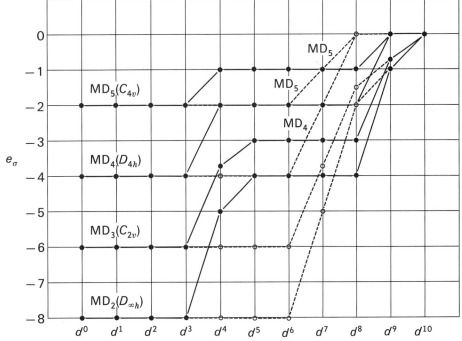

(F) $E(MD_6) - E(MD_x)$, where MD_x has the symmetry preferred on the basis of the angular overlap model.

Figure 10–1. *Continued*

any ligand structural features requiring a bent geometry will be largely unopposed by the metal.

(b) Three geometries have been considered for three-coordinate complexes: trigonal planar (D_{3h}), T-shaped (C_{2v}), and 90°-pyramidal (C_{3v}). The last may be considered a simple variant of the D_{3h} symmetry wherein the metal is out of the plane of the donor atoms. Generally speaking, the D_{3h} geometry is *not* preferred by the $d\sigma^*$ occupation factors, and competition between these and the VSEPR/steric forces will be important. The most pronounced discrimination is found for the low spin cases of six to nine electrons. The C_{3v} geometry is most likely to be found for d^6 and d^7 complexes, while d^6 to d^9 metals show greatest preference for the T-shape. Among high spin complexes, the structure preference is not large, but it reaches a maximum for the T-shape for d^4 and d^9 metals.

(c) For four-coordinate cases, square planar geometry is clearly favored (in opposition to VSEPR/steric forces) for strongly complexing ligands. Less strongly bonding ligands still show a preference for the square planar structure for d^3, d^4, d^8, and d^9. This behavior was discussed at length and examples were given in Chapter 9 (see Figure 9–19).

(d) Limiting MD_5 structures are the trigonal bipyramid (D_{3h}) and the square pyramid (C_{4v}; all bond angles 90°). The important conclusion in this case is that VSEPR/steric factors will produce a trigonal bipyramidal structure without much opposition from $d\sigma^*$ occupation in all cases except high spin d^3 and d^8 and low spin d^5, d^6, and d^7. In fact, the preference for C_{4v} geometry is quite large for the low spin case d^6.

(e) For the MD_6 geometries we have considered the usual O_h structure and the trigonal prism of D_{3h} symmetry. For all d^n cases, except d^{10} and high spin d^5, the octahedron is preferred by both VSEPR/steric factors and $d\sigma^*$ occupation factors, a preference that reaches a maximum with low spin d^6 complexes.

The final series of plots in Figure 10–1 depicts the structural preference energies for each of the lower coordination numbers relative to the MD_6 O_h geometry. Clearly, the greater number of bonds formed in the MD_6 case favors its formation. However, this is of great importance only early in the transition metal series; it diminishes considerably for low spin configurations greater than d^6 and high spin configurations greater than d^3. As a generalization, one can say that the largest number of low coordination number geometries should be found for the later members of the transition metal series. Other factors that assist the formation of less than six-coordinate complexes are stoichiometry control in synthesis, crystal packing forces in the solid, the entropy change accompanying formation of the complex, special ligand characteristics, VSEPR/steric forces, and weak metal–ligand interactions (*i.e.*, low e_σ).

With this general background, we shall now examine each coordination number to sample known complexes in each class and to identify their geometries.

STUDY QUESTIONS

1. Calculate structure preference energies for the three possible geometries for three-coordinate complexes, and verify Figures 10–1(B) and 10–1(C).

2. Calculate structure preference energies for the D_{3h} and C_{4v} geometries of five-coordinate complexes, and verify Figure 10–5 (p. 600). Prepare a similar plot of $E(O_h) - E(MD_5$, square pyramid). Comment on any differences between this new plot and Figure 10–5.

3. Based on the angular overlap model and other considerations such as those outlined in this chapter and in Chapter 9, predict the structure of each of the following molecules or molecular fragments (all are low spin):

    a) $Fe(CO)_5$
    b) $Re(CO)_5$ (See footnote 21 in J. K. Burdett, *Inorg. Chem.*, **14**, 375 (1975).)
    c) $Ni(CO)_3$ (J. K. Burdett, *J. C. S. Chem. Commun.*, 763 (1973).
    d) $Cr(CO)_3$ (See footnote 13 in J. K. Burdett, *Inorg. Chem.*, **14**, 375 (1975).)
    e) $Fe(CO)_3$ (M. Poliakoff, *J. C. S. Dalton*, 210 (1974).)
    f) $Cr(CO)_4$ (same reference as $Cr(CO)_3$ above)
    g) $Fe(CO)_4$ (M. Poliakoff and J. J. Turner, *J. C. S. Dalton*, 1351 (1973); *ibid.*, 2276 (1974).)

LOW COORDINATION NUMBERS

TWO-COORDINATE COMPLEXES

Complexes having a metal ion or atom coordinated to just two ligands are quite unusual, even though one of the best known complexes in inorganic chemistry is the two-coordinate species $[Ag(NH_3)_2]^+$. As expected on the basis of both structure preference energies and VSEPR factors, this ion is linear. Indeed, Figure 10–1(F) allows you to predict that, if such two-coordinate complexes are to form, they will be most often observed for higher $d\sigma^*$ populations, especially d^{10}. To date this has been borne out experimentally, since, with one exception, all of the two-coordinate complexes known are based on d^{10} species such as Au(I) [e.g., $Au(PPh_3)_2^+ I^-$][2] and Pd(0) [e.g., $Pd[PPh(t-Bu)_2]_2$].[3] The one exception is a d^7 complex, $Co[N(SiMe_3)_2]_2$.[4] Although it is not yet known whether this complex (with a magnetic moment of 4.83 B.M.) is linear or bent, structure preference energies and steric factors lead to the prediction of a linear geometry.

THREE-COORDINATE COMPLEXES

Complexes in which a transition metal is coordinated to three groups have only recently been examined in detail. For example, although the well-characterized two-coordinate Au(I) complex noted above was described some time ago, only recently were $[Au(PPh_3)_3]^+$ and $Au(PPh_3)_2Cl$ discovered and shown clearly to contain trigonally coordinated Au(I).[5,6] Again, these are d^{10} ions, a configuration most favorable to lower coordination numbers. However, a d^{10} species should show no electronic preference for any of the three possible three-coordinate geometries, so the ultimate atomic arrangement should depend solely on VSEPR and/or steric factors; this is, of course, clearly operative in the case of these gold complexes.

Other d^{10} species also form three-coordinate complexes having a trigonal, planar geometry. A particularly rich source of such materials has been Cu(I), which forms numerous complexes with sulfur-containing ligands in a variety of geometries (often four-coordinate with tetrahedral or distorted tetrahedral geometry).[7,8] $Cu(SPMe_3)_3$ **(1)** is one example of a planar, trigonally coordinated Cu(I).

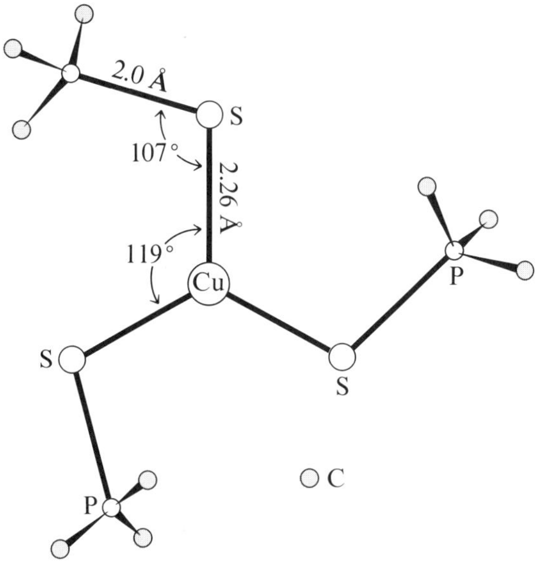

1 $Cu(SPMe_3)_3$

Reaction of $LiN(SiMe_3)_2$ with transition metal halides, usually in the form of their coordination complexes with ethers [$MCl_3(THF)_3$; M = Cr, Fe] or phosphines

---

[2] J. W. Collier, A. R. Fox, I. G. Hinton, and F. G. Mann, *J. Chem. Soc.*, 1819 (1964).
[3] A. Immirzi and A. Musco, *J. C. S. Chem. Commun.*, 400 (1974); M. Matsumoto, H. Yoshioka, K. Nakatsu, T. Yoshida, and S. Otsuka, *J. Amer. Chem. Soc.*, **96**, 3322 (1974). See also M. C. Mazza and C. G. Pierpont, *Inorg. Chem.*, **12**, 2955 (1973).
[4] D. C. Bradley and K. J. Fisher, *J. Amer. Chem. Soc.*, **93**, 2058 (1971).
[5] F. Klanberg, E. L. Muetterties, and L. G. Guggenberger, *Inorg. Chem.*, **7**, 2273 (1968).
[6] N. C. Baenziger, K. M. Dittemore, and J. R. Doyle, *Inorg. Chem.*, **13**, 805 (1974).
[7] J. A. Tiethof, A. T. Hetey, and D. W. Meek, *Inorg. Chem.*, **13**, 2505 (1974).
[8] P. G. Eller and P. W. R. Corfield, *Chem. Commun.*, 105 (1971).

592 ☐ COORDINATION CHEMISTRY: STRUCTURAL ASPECTS

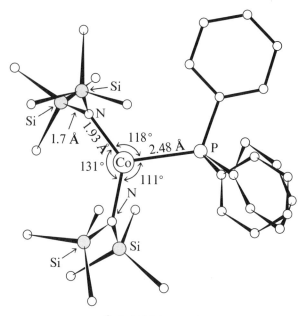

2 Co[N(SiMe$_3$)$_2$]$_2$PPh$_3$

[MCl$_2$(PR$_3$)$_2$; M = Co, Ni], gives compounds such as Fe[N(SiMe$_3$)$_2$]$_3$ and Co[N(SiMe$_3$)$_2$]$_2$PPh$_3$ **(2)**.[9,10] Indeed, similar compounds have been prepared for all of the first row transition metals except Mn(III), a fact which should strike you as unusual in view of the SPE plot in Figure 10-1(F). That is, the octahedron is seen to be strongly favored over three-coordination for d^n occupations of 8 or less. This prediction, of course, presumes that the steric bulk of the ligand does not interfere, a presumption apparently not valid with the —N(SiMe$_3$)$_2$ ligand.

All of the compounds of the type M[N(SiMe$_3$)$_2$]$_3$ (M = Ti, V, Cr, and Fe) or Co[N(SiMe$_3$)$_2$]$_2$PPh$_3$ or Ni[N(SiMe$_3$)$_2$](PPh$_3$)$_2$ are high spin. Therefore, a few—Cr(III) and Ni(I)—are predicted to favor the T-shaped or pyramidal structures on the basis of structure preference energies. However, this preference is not great (less than one unit of e_σ), and experience indicates that this is easily overcome by ligand-ligand repulsions or steric effects. Therefore, the trigonal planar geometry observed for these complexes is not surprising.

The three-coordinate complexes containing the disilylamido ligand are also all colored.[11] For example, Ti[N(SiMe$_3$)$_2$]$_3$ is bright blue and Cr[N(SiMe$_3$)$_2$]$_3$ is bright green. Their absorption spectra may be interpreted using the angular overlap model; this is left as a study question.

---

[9] E. C. Alyea, D. C. Bradley, and R. G. Copperthwaite, *J. C. S. Dalton*, 1580 (1972); see also D. C. Bradley, *Chem. in Britain*, **11**, 393 (1975).

[10] D. C. Bradley, M. B. Hursthouse, R. J. Smallwood, and A. J. Welch, *J. C. S. Chem. Commun.*, 872 (1972).

[11] E. C. Alyea, D. C. Bradley, R. G. Copperthwaite, and K. D. Sales, *J. C. S. Dalton*, 185 (1973).

STUDY QUESTIONS

4. It was noted above that complexes of the type M[N(SiMe$_3$)$_2$]$_3$ (M = Ti, V, Cr, and Fe) are all colored and that their spectra could be interpreted using the angular overlap method.

    a) Calculate the energies of the $d\sigma^*$ and $d\pi^*$ orbitals using the methods of Chapter 9.
    b) Give the symbols for the ground and first excited states for the complexes where M = Ti, V, Cr.
    c) In the case of the titanium complex, how many absorption bands are expected altogether? Give the symbol for any excited state beyond the first.

d) If the bands observed for the titanium case are assigned as follows: 4.8 kK ($\varepsilon = 10$), $^2E'' \leftarrow {}^2A_1'$, and 17.4 kK ($\varepsilon = 122$), $^2E' \leftarrow {}^2A_1'$, what does this imply about the energy level ordering given by the angular overlap model? Can you account for any differences between the AOM and experiment?

e) The iron complex has a magnetic moment of 5.94 B.M. Show the electron configuration that can account for this.

FOUR-COORDINATE COMPLEXES

On moving from three- to four-coordinate complexes, the phenomenon of "polytopism" or "polytopal isomerism" is first observed.[12] *Polytopal isomers are compounds of the same constitution, but in which the ligands are arranged about the metal ion or atom to describe different coordination polyhedra.* Therefore, tetrahedral and square planar compounds are polytopal isomers. In Chapter 9 (page 551) we concluded that the formation of tetrahedral (T_d) or distorted tetrahedral (D_{2d}) complexes was likely for metal ions or atoms with more than two d electrons only when the ligands interact weakly with the metal; this weak interaction may be due to intrinsically weak M-D bonds and/or unusual ligand steric requirements. More basic or less hindered ligands will prefer to bind the metal with a planar geometry. Since the latter are the types of ligands perhaps most often used, square planar complexes are more commonly observed than tetrahedral or distorted tetrahedral ones.

Tetrahedral Complexes

Generally speaking, only neutral or anionic complexes have tetrahedral or distorted tetrahedral (D_{2d}) geometries. Most of those which are known are formed with weakly basic ligands interacting with elements of the first row of the transition metal series, particularly those of Group VIII. [Indeed, more tetrahedral complexes are known for Co(II) than for any other ion.] This combination of factors, of course, produces low e_σ values, a situation favorable to the tetrahedron. As noted on page 551, halide complexes of Fe(III) and Cu(II) (*e.g.*, [FeCl$_4$]$^-$ and [CuCl$_4$]$^{2-}$) are good examples of such behavior.

Cu(II) in particular gives rise to an astounding variety of complexes.[13,14] For example, a solution of CuBr$_2$ and CsBr produces both Cs$_2$CuBr$_4$ and CsCuBr$_3$. In the former, [CuBr$_4$]$^{2-}$ is clearly a distorted tetrahedron (**3**) as is the [CuCl$_4$]$^{2-}$ analog.[15] (On the other hand, the [CuBr$_3$]$^-$ anion is a three-dimensional network of [Cu$_2$Br$_9$]$^{5-}$ dimeric units, each dimer consisting of two face-shared octahedra.)

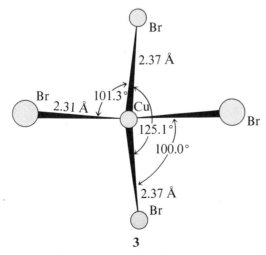

3

---

[12] E. L. Muetterties, *J. Amer. Chem. Soc.*, **91**, 1636 (1969); see correction, **91**, 4943 (1969).
[13] B. J. Hathaway and A. A. G. Tomlinson, *Coord. Chem. Rev.*, **5**, 1 (1970).
[14] B. J. Hathaway and D. E. Billig, *Coord. Chem. Rev.*, **5**, 143 (1970).
[15] T. I. Li and G. D. Stucky, *Inorg. Chem.*, **12**, 441 (1973).

The D_{2d} geometry [CuX$_4$]$^{2-}$ is expected on the basis of the angular overlap model (Chapter 9, page 550). In fact, more complete molecular orbital calculations for [CuCl$_4$]$^{2-}$ show that the flattened tetrahedron (D_{2d}) is about 2 kcal more stable than the T_d geometry (which is in turn about 18 kcal more stable than the planar D_{4h} configuration).[16] As a consequence, the distorted tetrahedron is often observed for four-coordinate Cu(II).

Just as you saw with three-coordinate complexes, ligand steric effects play a very important role in determining the geometry of four-coordinate complexes. For example, salicylaldimine complexes of Ni(II) and Cu(II) of the type shown as structure **4** are planar when the substituent R is an *n*-propyl group.[17]

4

However, when R is a isopropyl group, there is apparently an equilibrium in solution between the *trans*-planar form and tetrahedral geometry. The magnetic moments of the Ni(II) complexes vary from about 1.8 to 2.3 B.M. depending on the ring substituent, corresponding to approximately 30 to 50 per cent tetrahedral form (if it is assumed that the tetrahedral molecule would have a magnetic moment of 3.3 B.M.). When the *t*-butyl group is used, however, steric hindrance forces the ligands considerably more out of plane and results in magnetic moments of at least 3.2 B.M., indicating ~95 per cent tetrahedral form.

Ligand-metal orbital interactions, as well as ligand steric effects, are important in deciding whether the complexes of another series, L$_2$NiX$_2$ (L = phosphine, X = halide) are square planar or tetrahedral. We shall discuss this interesting series of compounds in more detail in Chapter 14.

Square Planar Complexes

Planar complexes have been very important to the development of coordination chemistry. Alfred Werner[18] first postulated their existence in 1893 to explain the existence of isomers of Pt(NH$_3$)$_2$Cl$_2$, and the deep red dimethylglyoxime complex of Ni(II) (**5**) was discovered by Chugaev in 1905; the latter compound is still commonly used in the gravimetric analysis of nickel.

5

The overwhelming majority of square planar complexes are those having a d^8 metal ion, that is, Rh(I), Ir(I), Ni(II), Pd(II), Pt(II), and Au(III). Examples of d^8 square planar complexes include **5** and **6** as well as two complexes of great importance in organometallic chemistry, **7** (Wilkinson's catalyst)[19] and **8** (Vaska's com-

---

[16] J. Demuynck, A. Veillard, and U. Wahlgren, *J. Amer. Chem. Soc.*, **95**, 5563 (1973).
[17] R. H. Holm, G. W. Everett, and A. Chakravorty, *Prog. Inorg. Chem.*, **7**, 83 (1966).
[18] A. Werner, *Z. anorg. Chem.*, **3**, 267–330 (1893); this paper and others of Werner's papers of great importance to coordination chemistry have been published in translated and annotated form: G. B. Kauffman, *Classics in Coordination Chemistry*, Dover Publications, Inc., New York, 1968.
[19] D. Evans, G. Yagupsky, and G. Wilkinson, *J. Chem. Soc.* (A), 2660, 2665 (1968).

pound).[20] The reason for the importance of the square planar geometry in d^8 complexes, at least relative to the tetrahedron, is presumably the stabilization achieved by reducing the HOMO anti-bonding energy for the square planar geometry (see Chapter 9, page 550).

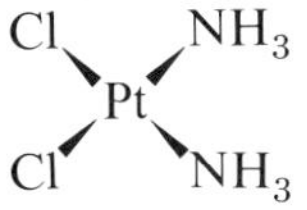

6
cis-diamminedichloroplatinum

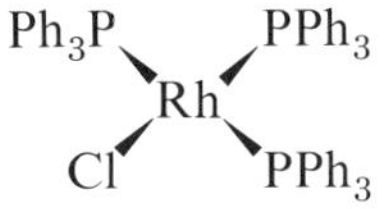

7
chlorotris(triphenylphosphine)-rhodium

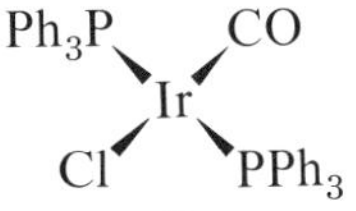

8
trans-carbonylchlorobis(triphenylphosphine) iridium

Other metals can also form square planar complexes with the proper ligand. For example, Gray reasoned that if the metal p_z orbital of a square planar complex could be fully utilized in π bonding with the ligand, the square planar configuration could be stabilized further with respect to the tetrahedron or octahedron.[21] Appropriate ligands were found to be 1,2-dithiolates such as maleonitriledithiolate (**9**) and toluene-3,4-dithiolate (**10**). Using these ligands and other similar ones, square planar complexes of Fe(II) (**11**), Co(II), Cu(II), and Re(II), among others, were readily isolated.[22]

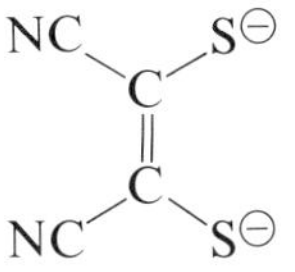

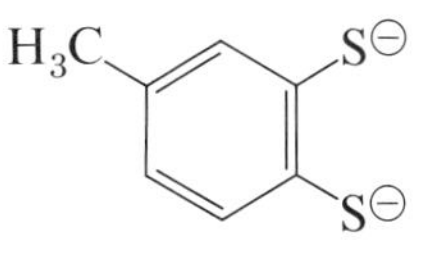

---

[20] L. Vaska, *Acc. Chem. Res.*, **1**, 335 (1968).
[21] H. B. Gray in *Transition Metal Chemistry*, Vol. 1, ed. by R. L. Carlin, Marcel Dekker, New York, 1965.
[22] J. A. McCleverty, *Prog. Inorg. Chem.*, **10**, 49–221 (1969).

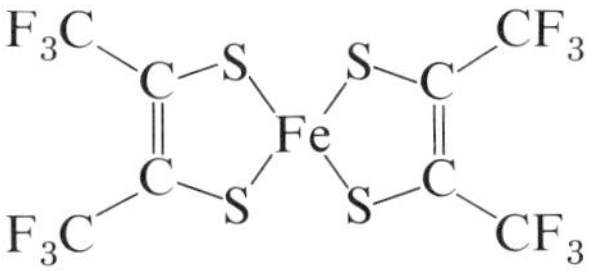

STUDY QUESTIONS

5. The addition of PEtPh$_2$ to NiBr$_2$ at $-78°C$ in CS$_2$ gives a red complex with the formula (PEtPh$_2$)$_2$NiBr$_2$, which is converted to a green complex of the same formula on standing at room temperature. The red complex is diamagnetic, but the green complex has a magnetic moment of 3.2 B.M.

   a) Which of these complexes is square planar and which is tetrahedral? State reasons for your choice.
   b) Rationalize the colors of these complexes in view of your structural choice.

6. NiCl$_2$ and NiI$_2$ also form complexes with the ligand in problem 5. These complexes have the following characteristics: (i) (PEtPh$_2$P)$_2$NiCl$_2$ is red and is diamagnetic at all temperatures. (ii) (PEtPh$_2$)NiI$_2$ is red-brown and is paramagnetic at all temperatures.

   a) Which of these complexes is square planar and which is tetrahedral?
   b) Attempt to rationalize why the (PEtPh$_2$)$_2$NiX$_2$ complexes assume different structures depending on the halide ion, X.

7. On page 591 the magnetic moment of Co[N(SiMe$_3$)$_2$]$_2$ was given as 4.83 B.M. Why doesn't this moment tell us whether the geometry of the molecule is linear or bent? How could you experimentally determine the geometry of the molecule (aside from x-ray techniques)? (Hint: Compare predicted electronic spectra.)

8. The [Cu$_2$Br$_9$]$^{5-}$ ion mentioned on page 593 is said to consist of two face-shared octahedra. Sketch such a structure.

FIVE-COORDINATE COMPLEXES

One of the most active areas of research in structural chemistry in recent years has concerned five-coordination.[23-25] We have already discussed examples of this in non-metal chemistry (*e.g.*, Chapters 6 and 7)—such compounds being most prevalent for Groups VA and VIIA—and now turn to those in transition metal chemistry.

Five-coordinate complexes are found for all of the first row transition metals from titanium to zinc, and occur about as frequently as square planar or tetrahedral four-coordinate complexes. Figure 10–1(F) illustrates the fact that six-coordinate complexes are almost always favored relative to lower coordinate ones (if only σ effects are considered), simply on the basis of thermodynamics. However, five-coordinate complexes are only about one to two e_σ units less favored than six-coordinate octahedral complexes, so there is often a very real competition between five- and six-coordination. We hasten to add, though, that this statement really applies only to the first row metals; five-coordination for second and third row metals is much less common. In part, this may be due to the fact that the heavier metals are characterized by larger e_σ values for ligand-metal interaction, and there is, therefore, a greater difference in energy between five- and six-coordination. Yet another reason for the preference of heavier metals for higher coordination may be simply their size relative to the first row metals; the heavier metals are larger, and six ligands can be accommodated with lower ligand-ligand repulsion. On the other hand, as in all areas of chemistry, these rationalizations may be upset as more research is done on ways of building five-coordinate complexes of the heavier metals.

As noted at the beginning of this chapter, there are two limiting structures for five-coordinate complexes: the trigonal bipyramid (D_{3h}) and the square-based pyramid (C_{4v}). The latter is rarely found in non-transition metal compounds, whereas

---

[23] J. S. Wood, *Prog. Inorg. Chem.*, **16**, 227 (1972).
[24] B. F. Hoskins and F. D. Williams, *Coord. Chem. Rev.*, **9**, 365 (1972–1973).
[25] C. Furlani, *Coord. Chem. Rev.*, **3**, 141 (1968).

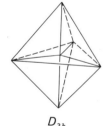
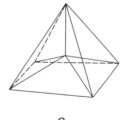

(A)

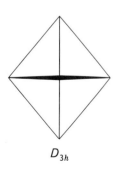

(B)

Figure 10–2. Views of the two idealized five-coordination polyhedra; (A) conventional perspective framework diagrams; (B) comparison of the square pyramid viewed along the four-fold axis and the trigonal bipyramid viewed along one of its twofold axes.

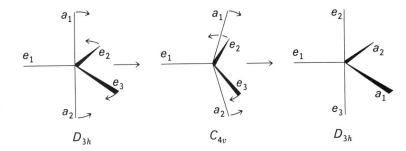

Figure 10–3. Schematic representation of the Berry intramolecular rearrangement for the trigonal bipyramid (also know as the pseudorotation process) involving a square pyramidal intermediate.

both geometries exist with about equal frequency for transition metal compounds. Indeed, the angular overlap model predicts that the square-based pyramid is significantly favored over the trigonal bipyramid for d^3 and d^8 high spin complexes or d^5 to d^7 low spin complexes. Otherwise, VSEPR and/or steric effects will generally favor the trigonal bipyramid.

In practice, the geometry actually observed for five-coordinate complexes is often intermediate between the two limiting structures. The reason for this is readily seen on viewing the polyhedra down their C_2 and C_4 axes as shown in Figure 10-2; a small pivoting of the basal bonds converts the square pyramid into a trigonal bipyramid. Indeed, the square pyramid is simply an intermediate state in the interchange of axial and equatorial groups in the trigonal bipyramid. This rearrangement—the Berry pseudorotation (Figure 10–3)—gives rise to the stereochemical non-rigidity of five-coordinate complexes, a phenomenon found important in the chemistry of silicon and phosphorus compounds (Chapter 7).

An interesting illustration of the fact that the trigonal bipyramid and square pyramid are energetically close to one another is found in the $[Ni(CN)_5]^{3-}$ anion. In fact, this complex is also an example of the importance of the nature of the counter-ion in the synthesis of complexes (see Chapter 6). It has been observed that $[Ni(CN)_5]^{3-}$ cannot be isolated from aqueous solution at room temperature with K^+ as a counter-ion, but addition of $[Cr(en)_3]^{3+}$ does lead to precipitation of the anion as a stable salt, $[Cr(en)_3][Ni(CN)_5]\cdot 1.5H_2O$.[26,27]

$$[Cr(en)_3][Ni(CN)_5] \text{ stable solid}$$
$$\uparrow [Cr(en)_3]^{3+}$$
$$[Ni(CN)_4]^{2-} + CN^- \rightleftharpoons [Ni(CN)_5]^{3-} \xrightarrow{K^+, \text{conc.}} K_2Ni(CN)_4 + KCN$$
$$-15° \downarrow K^+$$
$$K_3Ni(CN)_5 \xrightarrow{\text{room temp.}}$$

The interesting feature of this salt is that there are two independent $[Ni(CN)_5]^{3-}$ ions per unit cell, one of which is a distorted trigonal bipyramid **(12)** and the other a distorted square pyramid **(13)**. This should not be too surprising, however, as the angular overlap model predicts that the square-based pyramid is favored over the trigonal bipyramid by only about $1/2 e_\sigma$ unit. Experience shows that $1/2 e_\sigma$ is not decisive when in competition with VSEPR forces, and the structure may in fact be determined by crystal lattice forces.

Ligand geometry not only can frequently influence the structure of a complex or series of complexes, but it also can lead to complexes of differing stoichiometry.[28,29] For example, the ligand tris(2-dimethylaminoethyl)amine, $N(CH_2CH_2NMe_2)_3$, forms complexes of the type [M(ligand)Br]Br with all of the metals from manganese to zinc,

---

[26] K. N. Raymond, P. W. R. Corfield, and J. A. Ibers, *Inorg. Chem.*, **7**, 1362 (1968).
[27] F. Basolo, *Coord. Chem. Rev.*, **3**, 213–223 (1968).
[28] J. L. Shafer and K. N. Raymond, *Inorg. Chem.*, **10**, 1799 (1971).
[29] L. Sacconi, C. A. Ghilardi, C. Mealli, and F. Zanobini. *Inorg. Chem.*, **14**, 1380 (1975).

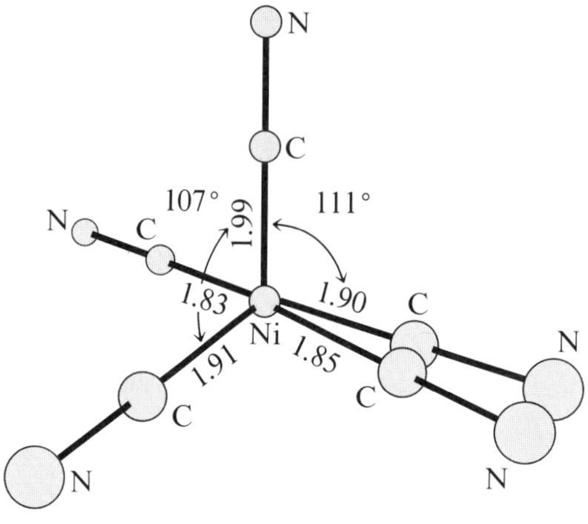

12 [Ni(CN)$_5$]$^{3-}$, trigonal bipyramid

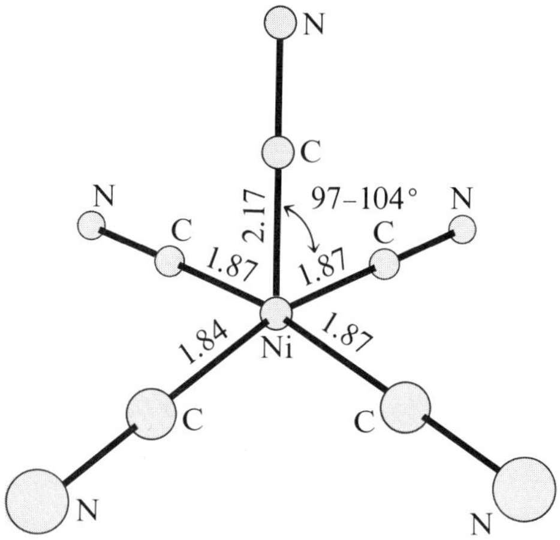

13 [Ni(CN)$_5$]$^{3-}$, square pyramid

and all of these complexes have trigonal bipyramidal cations of the general structure shown in **14**.

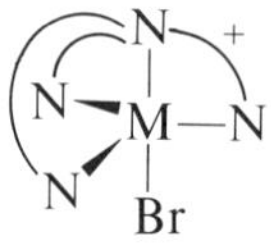

14

Adding one more methylene group to the "arm" of the ligand, thereby forming a ligand of the type N(CH$_2$CH$_2$CH$_2$NR$_2$)$_3$, gives a more "facultative" ligand, that is, a ligand that does not restrict the complex to a given geometry; for example, this N—C$_3$—N ligand is capable of spanning four positions of either a trigonal bipyramid *or* a tetrahedron. Indeed, the ligand tris(3-aminopropyl)amine, N(CH$_2$CH$_2$CH$_2$NH$_2$)$_3$, apparently forms both tetrahedral and trigonal bipyramidal complexes with Co(II) (**15**).[30]

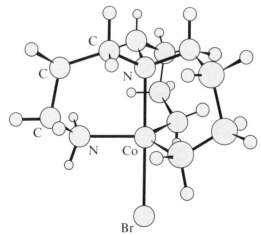

15 [Co{N(CH$_2$CH$_2$CH$_2$NH$_2$)$_3$} Br]$^+$

---

[30] See ref. 28.

Finally, the thermodynamics of the formation of five-coordinate complexes have been examined, with results that bear importantly on the predictions of the angular overlap model.[31] In Figure 10-4 values of ΔH plotted are for the following reactions:

Curve A: $M[HN(CH_2CH_2NH_2)_2]_3^{2+} + MeN(CH_2CH_2NMe_2)_2 \rightarrow M[MeN(CH_2CH_2NMe_2)_2]Br_2 + 2HN(CH_2CH_2NH_2)_2 + 2Br^-$

Curve B: $M[HN(CH_2CH_2NH_2)_2]_3^{2+} + N(CH_2CH_2NMe_2)_3 \rightarrow [M\{N(CH_2CH_2NMe_2)_3\}Br]^+ + 2HN(CH_2CH_2NH_2)_2 + Br^-$

That is, curves A and B essentially represent structure preference energies for the two types of high spin trigonal bipyramidal complexes relative to the octahedron. The most striking observation is the very large preference for octahedral geometry for the high spin d^8 Ni(II) ion; also, five-coordination is even slightly favored for Cu(II) and Zn(II). Both of these observations would in fact have been predicted by the structure preference energy plots in Figures 10-1F and 10-5.

---

[31] P. Paoletti and M. Ciampolini, *Inorg. Chem.*, **6**, 64 (1967).

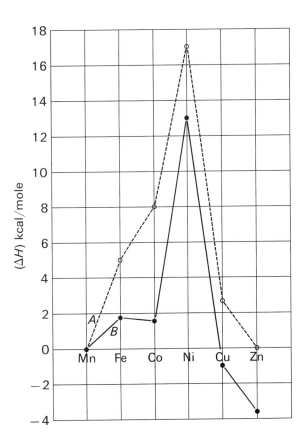

Figure 10-4. Enthalpy changes for the transformation of the octahedral complex $M[HN(CH_2CH_2NH_2)_2]_3^{2+}$ to the trigonal bipyramidal complex $M[MeN(CH_2CH_2NMe_2)_2]Br_2$ (curve A) or to the trigonal bipyramidal cation $[M\{N(CH_2CH_2NMe_2)_3\}Br]^+$ (curve B). Adapted from P. Paoletti and M. Ciampolini, Inorg. Chem., **6**, 64 (1967).

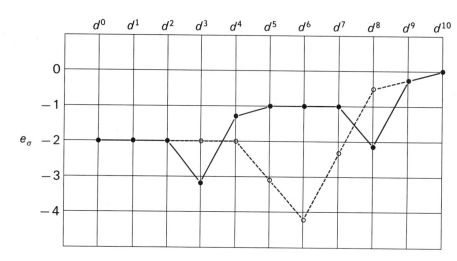

Figure 10-5. $E(O_h) - E(MD_5,$ trigonal bipyramid). Structure preference energy for the five-coordinate trigonal bipyramid relative to the octahedron. High spin is shown by the solid line and low spin by the dashed line.

SIX-COORDINATE COMPLEXES

There are several possible geometries for six-coordinate complexes: planar hexagonal, trigonal prismatic, and octahedral, among others. One of Alfred Werner's greatest contributions to inorganic stereochemistry was his finding that octahedral geometry was adopted by the six-coordinate complexes known at that time. Proof was provided by the observation that, when three bidentate ligands are coordinated to a metal center, two optically active enantiomers may be isolated **(16)** (see the discussion of stereoisomerism in Chapter 11). Had the geometry been planar hexagonal, no isomers would have been found **(17)** (there is an improper rotation-reflection axis, S_3, the existence of which precludes generation of enantiomers), while only two diastereomers would be observed for the trigonal prism **(18)**.

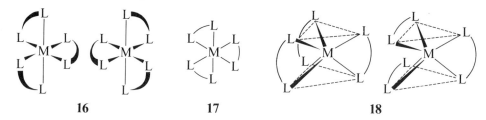

16   **17**   **18**

Therefore, six-coordinate complexes have always been assumed to be octahedral since Werner's work. However, in 1965 this long-held truism was invalidated: a trigonal prismatic rhenium compound, $Re[S_2C_2Ph_2]_3$ **(19)**, was reported.[32] The

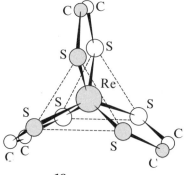

19 $Re[S_2C_2Ph_2]_3$

---

[32] R. Eisenberg and J. A. Ibers, *Inorg. Chem.*, **5**, 411 (1966).

Figure 10-6. The twist mechanism for the racemization of tris-bidentate octahedral complexes, and illustration of the twist angle θ.

quadrilateral sides of the trigonal prism are almost perfectly square, there being little difference between intra- and interligand S—S distances (3.03 and 3.05 Å, respectively); further, the top and bottom of the prism are nearly eclipsed.

A trigonal prism may, of course, be formed from an octahedron by twisting one face relative to the other (Figure 10-6). Indeed, Bailar has suggested this as a mechanism for the racemization of tris-bidentate complexes (this phenomenon is discussed in detail in Chapter 14).[33] Therefore, it is not surprising that, with continuing work on trigonal prismatic geometry, many additional examples have been found with the twist angle, θ, varying from the pure trigonal prism (TP, $\theta = 0°$) to the trigonal antiprism or octahedron (TAP or O_h, $\theta = 60°$).[34-37]

The geometry of the ligand is apparently one of the most critical factors in the formation of trigonal prismatic complexes. According to Wentworth, "a ligand is required which will maximize all bonding interactions and minimize all non-bonded contacts when the donor atoms are in TP array."[35] This has been achieved by rigid hexadentate ligands such as that illustrated in compound **20**. Using this ligand

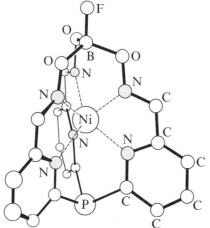

20 trigonal prismatic coordination in [fluoroborotris(2-aldoximo-6-pyridyl)phosphine]nickel(1+)

---

[33] J. C. Bailar, *J. Inorg. Nucl. Chem.,* **8,** 165 (1958).
[34] E. Larsen, G. N. LaMar, B. E. Wagner, J. E. Parks, and R. H. Holm, *Inorg. Chem.,* **11,** 2652 (1972).
[35] R. A. D. Wentworth, *Coord. Chem. Rev.,* **9,** 171 (1972-73); W. O. Gillum, R. A. D. Wentworth, and R. F. Childers, *Inorg. Chem.,* **9,** 1825 (1970).
[36] E. B. Fleischer, A. E. Gebala, D. R. Swift, and P. A. Tasker, *Inorg. Chem.,* **11,** 2775 (1972).
[37] M. R. Churchill and A. H. Reis, Jr., *Inorg. Chem.,* **11,** 1811 (1972).

(abbreviated PccBF) with Co(II), Ni(II), Zn(II), and Fe(II), the twist angle, θ, is only 0° to 2° for the first three, while it is 21° to 22° for the Fe(II) complex. Other ligands giving trigonal prismatic complexes are illustrated by compounds **21** through **23**.

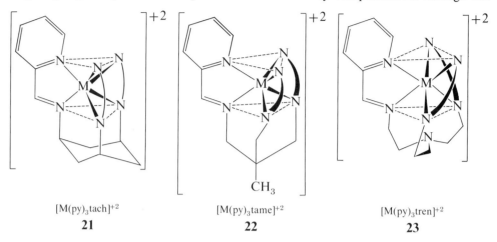

[M(py)$_3$tach]$^{+2}$
21

[M(py)$_3$tame]$^{+2}$
22

[M(py)$_3$tren]$^{+2}$
23

With these ligands, however, there is less rigidity, and the tendency to approach trigonal prismatic coordination varies as

$$\text{PccBF} > (\text{py})_3\text{tach (}\mathbf{21}\text{)} > (\text{py})_3\text{tame (}\mathbf{22}\text{)} > (\text{py})_3\text{tren (}\mathbf{23}\text{)}$$

As with other coordination numbers and their polytopes, the nature of the metal ion is also clearly important in determining whether a complex will be octahedral or trigonal prismatic, and the angular overlap model can be used here to analyze structure preference. If the trigonal prism is placed in the coordinate system shown in

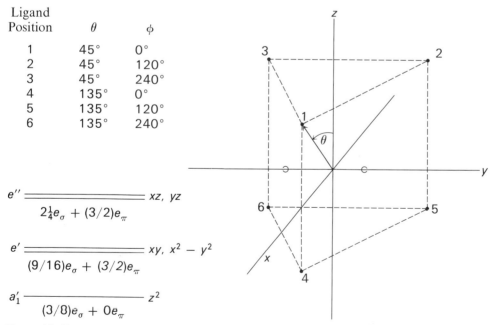

Ligand Position	θ	ϕ
1	45°	0°
2	45°	120°
3	45°	240°
4	135°	0°
5	135°	120°
6	135°	240°

e'' ——————— xz, yz
$2\frac{1}{4}e_\sigma + (3/2)e_\pi$

e' ——————— $xy, x^2 - y^2$
$(9/16)e_\sigma + (3/2)e_\pi$

a'_1 ——————— z^2
$(3/8)e_\sigma + 0e_\pi$

Figure 10-7. (A) Coordinate system used in defining the ligand positions in the trigonal prism. (B) d^* orbital energies calculated from the angular overlap model.

Figure 10-7, the d-like anti-bonding orbitals are split into three sets; $d_{z^2} < d_{xy}$, $d_{x^2-y^2} < d_{xz}$, d_{yz}. The energy shifts for these sets of orbitals, taking both σ and π bonding into account, are also illustrated in Figure 10-7. Using only σ bonding for the moment, the structure preference energy for the octahedron relative to the trigonal prism may be calculated; the results are plotted in Figure 10-1.[38] Note again that the octahedron is the preferred geometry for all configurations except d^0, high spin d^5, and d^{10}; moreover, the preference is most decided for low spin d^6 complexes. This is an especially interesting result in terms of complexes of the type M(PccBF)$^+$ (*e.g.*, **20**) which were mentioned above. The Fe(II) complex is of interest, because Fe(II) is d^6 and low spin, a configuration predicted to show maximum favoritism toward the octahedron relative to the trigonal prism. Be sure to note that the Fe(II) complex is the only one involving the PccBF ligand that has a twist angle ($= 22°$) significantly greater than zero; that is, the arrangement of ligating atoms about the metal ion is about $\frac{1}{3}$ of the way toward the octahedron. The other complexes of Co(II) (d^7, high spin), Ni(II) (d^8), and Zn(II) (d^{10}) all have twist angles of about 0°, and it is complexes of these ions that the angular overlap model predicts will show much less preference for the octahedron (see Figure 10-1).

---

[38] Holm has published a very similar plot based on more extensive calculations (see ref. 34). He also considered the more realistic case in which the so-called "bite angle"—that is, the angle L_1-M-L_4 in Figure 10-7—is less than 90°, the situation most often observed.

---

9. The magnetic moment of the trigonal prismatic molecule Re[S$_2$C$_2$Ph$_2$]$_3$ is 1.79 B.M. Is it more reasonable to consider this as a complex between Re(0) and three neutral dithiolene ligands (*A*) or as a complex between Re(VI) and three anionic dithiolates (*B*)?

(*A*) Re(0) +   (*B*) Re(VI) +

STUDY QUESTIONS

POLYHEDRA OF HIGH COORDINATION NUMBER[39]

Examination of the stereochemistry of coordination complexes has certainly been one of the most active areas of investigation in inorganic chemistry in recent years. With the advent of automated x-ray methods for the determination of crystal and molecular structure, complexes of higher coordination number—seven, eight, nine, and more—have been increasingly studied. Such complexes are extremely interesting because numerous geometries are possible, and, as the energies of the various polytopes are often quite similar, interconversion is feasible and frequently observed.[39a] Before taking up coordination numbers seven and eight in particular, it

---

[39] Much of the pioneering work in this area has been done by J. L. Hoard, and his definitive paper on eight-coordination should be consulted [J. L. Hoard and J. V. Silverton, *Inorg. Chem.*, **2**, 235 (1963)]. In addition, there are several excellent reviews of the subject: R. V. Parish, *Coord. Chem. Rev.*, **1**, 439 (1966); S. J. Lippard, *Prog. Inorg. Chem.*, **8**, 109 (1967); E. L. Muetterties and C. W. Wright, *Quart. Revs.*, **21**, 109 (1967).

[39a] You are reminded here of the extreme non-rigidity of XeF$_6$, which was discussed in Chapter 2.

is worthwhile to discuss briefly the conditions leading to the formation of high coordinate complexes.

With a given ligand, the ability of a metal ion to achieve a high coordination number appears to depend in part on a subtle balance of charge and ion size. That is, in order to attain optimum bonding conditions, the charge-to-radius ratio must be one that maximizes metal-ligand attraction and minimizes ligand-ligand repulsions. Eight-coordinate complexes, for example, are found predominantly for fifth and sixth period transition metal ions in Groups III to VI in oxidation states $+3$ to $+5$; furthermore, eight-coordinate complexes are very common for lanthanide ions in $+3$ oxidation states. It is, of course, these ions that are large enough to accommodate seven or more ions or molecules in a coordination sphere without severe ligand-ligand repulsions. Furthermore, these relatively highly charged ions exercise an attraction on the ligands that is sufficient to overcome the force of ligand-ligand repulsions. (Of course, the metal oxidation state can be too high to give high-coordinate complexes; in a very high formal oxidation state the metal ion will be highly polarizing, too much electron density may be transferred to the metal, and fewer ligands will be necessary to achieve electroneutrality.[40])

The $3+$, $4+$, and $5+$ ions of Groups III to VIB have electron configurations of d^0, d^1, or d^2 (a $3+$ ion of a Group VI element would of course be d^3, but high-coordinate complexes of such ions are quite uncommon). This implies that d orbital occupation has some influence on the formation of high-coordinate complexes, and that arguments similar to those used with low-coordinate complexes also apply here. That is, low d orbital populations will minimize the σ anti-bonding role of the d electrons. In any complex of coordination number greater than six, there will be at least one d-like molecular orbital that is not strongly anti-bonding; therefore, population by one or two electrons will not seriously affect the σ bonding in the complex.

The nature of the ligand is another important factor to be considered. If the polarizability of an ionic ligand is high, a large amount of charge density is easily transferred to the metal ion, and fewer ligands will be necessary to arrive at electroneutrality. As a consequence, complexes of higher coordination number are formed readily by ligands that are small and/or of low polarizability (*e.g.*, H^-, F^-, OH^-). However, most high-coordinate complexes are formed with chelating ligands having low steric requirements and donor atoms such as nitrogen and oxygen. The β-diketones, oxalate ion, EDTA, tropolonate (**24**), and nitrilotriacetate (**25**, NTA) are particularly good examples. In addition, the diarsine **26** has been found to be an important ligand in forming seven- and eight-coordinate complexes.

SEVEN-COORDINATE COMPLEXES

Seven-coordinate complexes present a very complicated situation, and the literature concerning such compounds is often confusing. The three idealized polyhedra usually considered for seven-coordination are the C_{3v} capped octahedron (**27**), the D_{5h} pentagonal bipyramid (**28**), and the C_{2v} monocapped trigonal prism (**29**).[41-45]

---

[40] Electroneutrality: in forming bonds to build a molecule, the bonding electrons are spread throughout the molecule in such a way that the constituent atoms are as close to electrical neutrality as possible. Therefore, a highly charged ion can so strongly polarize attached ligands (with a resulting transfer of electron density from ligands to the metal ion) that only six ligands, for example, would be needed to bring the ion near neutrality instead of seven or more.

[41] E. L. Muetterties and L. J. Guggenberger, *J. Amer. Chem. Soc.*, **96**, 1748 (1974).
[42] H. B. Thompson and L. S. Bartell, *Inorg. Chem.*, **7**, 488 (1968).
[43] T. A. Claxton and G. C. Benson, *Canad. J. Chem.*, **44**, 157 (1966).
[44] D. L. Kepert, *J. Chem. Soc., Dalton*, 617 (1974).
[45] D. Wester and G. J. Palenik, *J. Amer. Chem. Soc.*, **96**, 7565 (1974).

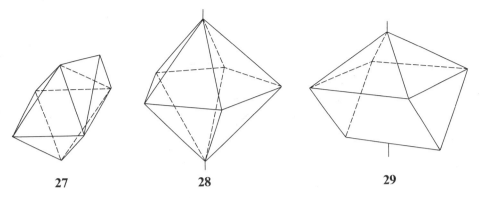

27 **28** **29**

In addition, there is a geometry intermediate between the pentagonal bipyramid and the capped trigonal prism—the C_s polyhedron (**30**).

30 [From E. L. Muetterties and C. M. Wright, Quart. Rev., **21**, 109 (1967).]

Although seemingly quite different, all of these polyhedra are actually very similar when viewed in a different perspective as in Figure 10-8, and the difference in energy

Figure 10-8. Perspectives illustrating structural similarities in the idealized heptacoordinate models. From left to right are the C_{2v} capped trigonal prism, C_s tetragonal base–trigonal base, D_{5h} pentagonal bipyramid, and C_{3v} capped octahedron. [From E. L. Muetterties and C. M. Wright, Quart. Rev., **21**, 109 (1967).]

between them is small. Therefore, which polyhedron a complex may adopt will depend on a subtle interplay of metal ion size, metal ion electron configuration, and ligand character and geometry. In most instances, however, complexes do not adopt precisely idealized geometry, and their description in terms of one geometry or another is arbitrary; hence the confusion in the literature.

Complexes that closely resemble each of the idealized polyhedra are known. The heptacyanovanadate(4−) ion, for example, is a slightly distorted pentagonal bipyramid,[46] a geometry that also appears to be the preferred configuration for complexes

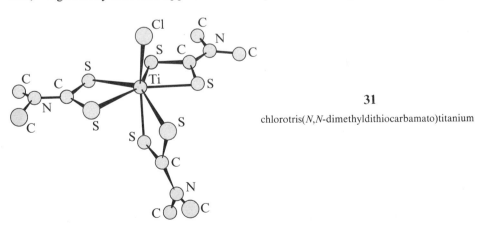

31
chlorotris(*N,N*-dimethyldithiocarbamato)titanium

---

[46] R. L. R. Towns and R. A. Levenson, *J. Amer. Chem. Soc.*, **94**, 4345 (1972); R. A. Levenson, R. J. G. Dominguez, M. A. Willis, and F. R. Young III, *Inorg. Chem.*, **13**, 2761 (1974).

of the type M(chelate)$_3$X when X is a ligand which forms a relatively strong, covalent M—X bond. Chlorotris(*N,N*-dimethyldithiocarbamato)titanium (**31**) is an example of this type of complex.[47] In contrast, when X is a dipolar ligand (such as water), the few structures available suggest that there is a preference for the monocapped octahedron or monocapped trigonal prism. An especially good example of the former is tris-(diphenylpropanedionato)aquoholmium, Ho(C$_6$H$_5$COCHCOC$_6$H$_5$)$_3$·H$_2$O,[48] while *trans*-1,2-diaminocyclohexane-*N,N'*-tetraacetatoferrate(1−) can best be described as a distorted capped trigonal prism (**32**).[49] The water molecule is the cap of the polyhedron, and the oxygen atoms form the vertices of the capped quadrilateral face (this face is outlined by a dotted line); the nitrogen atoms describe the remaining vertices.

32

EIGHT-COORDINATE COMPLEXES

Before beginning the discussion of high-coordinate complexes, we mentioned that such complexes are generally found to involve the heavier metals of the earlier groups of the transition metal series. Eight-coordinate complexes have been observed rather commonly for heavier metal ions of Groups IV, V, and VI in oxidation states of +4 and +5. That is, ions such as Zr(IV), Mo(IV,V) and Re(V,VI) having d^0, d^1, and d^2 electron configurations are often involved. An explanation for this can, of course, be found in the angular overlap model of metal-ligand binding: the heavier metals are involved because they generally give rise to larger e_σ values owing to stronger M—L bonding, and the metals are from the earlier groups because such metal ions have low $d\sigma^*$ occupancy.

Several idealized geometries are possible for eight-coordinate complexes, the cube being the simplest and most symmetrical of these. However, although cubic geometry is observed in ionic crystal lattices, it has never been observed for molecular polyhedra. One reason for this, at least in the case of transition metals, is that only seven (of the nine) *d*, *s*, and *p* valence orbitals of the metal possess the proper symmetry for bonding. The reason for this can be seen by noting that the cube is equivalent to two interlocked tetrahedra. From Table 9-4 (p. 545) you see that the d_{z^2} and $d_{x^2-y^2}$ orbitals (which would be located at the cube faces) would have zero overlap with ligands at the cube corners. Therefore, only seven metal atomic orbitals [one *ns*, three *np*, and three (*n* − 1)*d* orbitals: d_{xy}, d_{xz}, and d_{yz}] are able to stabilize the eight electron pairs in the ligand σ TASO's. This, of course, does not preclude the formation of cubic complexes by the lanthanides or actinides, as these elements have *f* orbitals available. In any event, even if the proper orbitals were available, the cube

---

[47] D. F. Lewis and R. C. Fay, *J. Amer. Chem. Soc.*, **96**, 3843 (1974); see also A. N. Bhat, R. C. Fay, D. F. Lewis, A. F. Lindmark, and S. H. Strauss, *Inorg. Chem.*, **13**, 886 (1974).
[48] A. Zalkin, D. H. Templeton, and D. C. Karraker, *Inorg. Chem.*, **8**, 2680 (1969).
[49] G. H. Cohen and J. L. Hoard, *J. Amer. Chem. Soc.*, **88**, 3228 (1966).

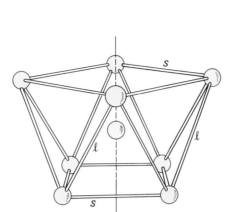
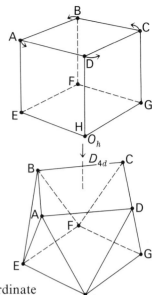

Figure 10-9. The square antiprismatic polyhedron and its mode of formation from the cube. [From F. A. Cotton, *Prog. Inorg. Chem.*, **8**, 109 (1967).]

maximizes interligand repulsions relative to the other possible eight-coordinate polyhedra.

The only two polyhedra actually observed for eight-coordinate complexes of the transition metals are simple distortions of the cube, both of which lead to a lessening of interligand repulsions in comparison to those found in the cubic geometry. The D_{4d} square antiprism may be obtained by simply rotating one face of a cube by 45° relative to the opposite face (Figure 10-9), while the D_{2d} dodecahedron may be formed by closing the cube along opposite faces (Figure 10-10); the net result of the latter is that the D_{2d} dodecahedron may be visualized as two interlocking trapezoids.[49a]

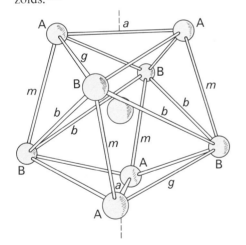
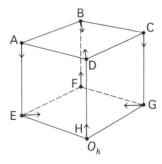
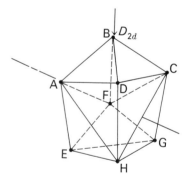

Figure 10-10. The dodecahedron and its mode of formation from the cube. [From F. A. Cotton, *Prog. Inorg. Chem.*, **8**, 109 (1967).]

The polyhedra in Figures 10-9 and 10-10 are labeled according to the notation of Hoard and Silverton.[50] In the D_{4d} polyhedron, the sixteen edges are divided equally between two symmetry types, l and s ($l = s$ for the idealized polyhedron). In the dodecahedron, on the other hand, there are eighteen edges divided into four classes: two a edges, four b edges, four m edges, and eight g edges ($g = m = a$ and $b = 1.250a$ for the idealized polyhedron). In both cases, in contrast to the cube, there are sufficient ns, np, and $(n-1)d$ orbitals available for σ bonding.[51]

As in the case of the seven-coordinate complexes, there is little difference in interligand repulsions for the two polyhedra, and slight bending motions may convert

---

[49a] Alternatively, the D_{2d} dodecahedron may be viewed as a tetra-capped tetrahedron.
[50] See Hoard and Silverton, ref. 39.
[51] See Lippard, ref. 39.

one into the other as shown in Figure 10-11. It is easily seen that the square antiprism may be obtained from the dodecahedron by simply opening up opposing b edges (B_1-B_4 and B_2-B_3). Yet another possible eight-coordinate polyhedron, the C_{2v} 4,4-bicapped trigonal prism,[52] may also be obtained by opening the B_2-B_3 b edge of the dodecahedron or by closing the square antiprism along a B_1-B_4 line. The accessibility of the C_{2v} structure implies that it may be adopted by eight-coordinate complexes, and there is some indication that at least a few complexes do approach this geometry.[53]

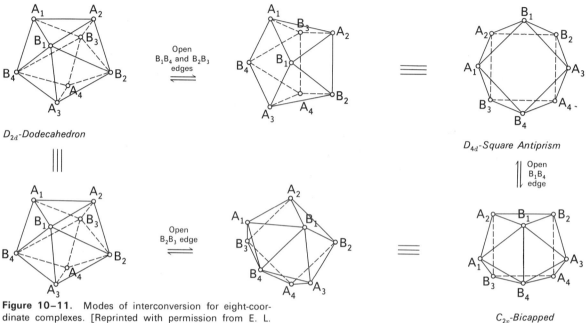

Figure 10-11. Modes of interconversion for eight-coordinate complexes. [Reprinted with permission from E. L. Muetterties and L. J. Guggenberger, *J. Amer. Chem. Soc.*, **96**, 1748 (1974). Copyright by the American Chemical Society.]

Hoard has been largely responsible for opening up the field of high coordination number stereochemistry,[54] and it was he who first established D_{2d} dodecahedral geometry for eight-coordinate complexes in his study of $[Mo(CN)_8]^{4-}$.[55] Numerous other complexes have since been found to be dodecahedral [*e.g.*, the tetrakis-(oxalato)zirconate(4−) ion (**33**)],[56] and it has begun to appear that the D_{2d} dodecahedron will be more common than the square antiprism. Among the best confirmed examples of the latter geometry are $[TaF_8]^{3-}$ and $[ReF_8]^{3-}$. Indeed, these octafluoro-

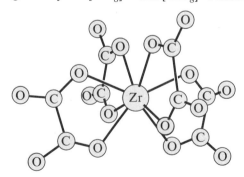

33 tetrakis(oxalato)zirconate(4−)

---

[52] This polyhedron is described as a 4,4-bicapped trigonal prism because there are ligands "capping" each quadrilateral face of a trigonal prism.
[53] See ref. 41.
[54] See Hoard and Silverton, ref. 39.
[55] J. L. Hoard and H. H. Nordsieck, *J. Amer. Chem. Soc.*, **61**, 2853 (1939).
[56] G. L. Glenn, J. V. Silverton, and J. L. Hoard, *Inorg. Chem.*, **2**, 250 (1963).

metallates, and others such as [MoF$_8$]$^{2-}$, are the only square antiprismatic complexes with monodentate ligands.

Before leaving this brief discussion of eight-coordinate complexes, we can point out one other general principle concerning their stereochemistry. Bidentate ligands often give rise to complexes of higher coordination number, and numerous complexes of the type [M(bidentate)$_4$]^x are known. Both experimental and theoretical work indicate that as the ligand bite—the distance between the two donor atoms when the ligand is complexed—increases, there is a smooth and continuous shift from the dodecahedron to the square antiprism.[57]

COMPLEXES HAVING COORDINATION NUMBERS OF NINE OR HIGHER

Transition metal ions surrounded by nine ligands or donor atoms are rare; perhaps the best examples of the anions are [ReH$_9$]$^{2-}$ and [TcH$_9$]$^{2-}$,[58,59] the structure of the former having been established by x-ray and neutron diffraction techniques. As shown in **34**, the hydride ligands clearly define a symmetrically tricapped trigonal prism. Although there are a number of possible idealized geometries for nine-coor-

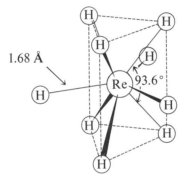

34 [ReH$_9$]$^{2-}$

dinate complexes, it is interesting that only the symmetrically tricapped trigonal prism has been reported for such complexes. Further, it is important to note that, if a transition metal is to be nine-coordinate and all ligand σ TASO's are to be stabilized, then full utilization of the d, s, and p orbitals of the metal valence shell is required; Re(VII) and Tc(VII), with no d electrons, meet this criterion and avoid populating the $d\sigma^*$ molecular orbitals.

There are a few complexes known with coordination numbers of ten or higher, but they are all based on lanthanide or actinide elements. This is not surprising, since, from the point of view of ligand σ TASO stabilization, it is only these elements that have a sufficient number of valence shell orbitals.

EPILOG

In this chapter we have applied the methods of molecular orbital theory, and especially the idea of d^* orbital occupation, to rationalize in large part the structural chemistry of coordination compounds. That is, using these methods and the effects of ligand-ligand repulsion and steric effects, one can often specify which of several geometries (or polytopal isomers) a complex of given coordination number may assume. In the next chapter we turn to other types of isomerism, especially stereoisomerism.

---

[57] D. G. Blight and D. L. Kepert, *Inorg. Chem.,* **11,** 1556 (1972).
[58] K. Knox and A. P. Ginsberg, *Inorg. Chem.,* **1,** 945 (1962) **3,** 555 (1964).
[59] S. C. Abrahams, A. P. Ginsberg, and K. Knox, *Inorg. Chem.,* **3,** 558 (1964).

CONSTITUTIONAL ISOMERISM

Hydrate Isomerism
Ionization Isomerism
Coordination Isomerism
Polymerization Isomerism
Linkage Isomerism

STEREOISOMERISM

General Aspects of Stereochemical Notation in Coordination Chemistry
Four-Coordinate Complexes
Six-Coordinate Complexes
Isomerism From Ligand Distribution and Unsymmetrical Ligands; Isomer Enumeration
Isomerism From Ligand Chirality and Conformation
Chirality and the Special Nomenclature of Chiral Coordination Complexes
Optical Activity, ORD, CD
Absolute Configuration of Chiral Coordination Complexes
Ligand Conformation

EPILOG

11. COORDINATION CHEMISTRY: ISOMERISM

> In the previous chapter we discussed the various coordination numbers assumed by transition metal complexes, and you saw that for a given coordination number there can often be polytopal isomers; that is, for compounds of the same formula with the same ligating atoms, two or more different polyhedra can be described by the ligands. The trigonal prism and the octahedron (or trigonal antiprism) for coordination number six would be examples of polytopal isomerism.
>
> In this chapter we turn to other types of isomerism: constitutional isomerism and stereoisomerism. In the former, the empirical formulas of the isomers are identical, but the atom-to-atom bonding sequences differ. For example, the thiocyanate ion, NCS⁻, may be bound either through N or through S in a complex ion such as $[Co(CN)_5(SCN)]^{3-}$. Stereoisomers, on the other hand, have the same empirical formulas and the same atom-to-atom bonding sequences; the difference between stereoisomers comes in the spatial arrangement of the atoms.

Chemical isomerism is a pervasive phenomenon. Since most of you were exposed to a rigorous study of organic chemistry before examining other fields of chemistry in any detail, your ideas of isomerism were developed in that context. You know, therefore, that there are two basic forms of isomerism: *constitutional isomerism* and *stereoisomerism*.[1-3]

Constitutional isomerism, which has often been called structural or position isomerism, results when two or more molecules have the same empirical formula but the constituents of the molecules are arranged differently; that is, there is a difference in the atom-to-atom bonding sequence. For example, 1-butene (**1**) and 2-butene (**2**) differ in the placement of the double bond, while acetone (**3**) and propionaldehyde (**4**) differ in the position of the carbonyl group.[4]

$$H_3C-CH_2-CH=CH_2 \quad\quad H_3C-CH=CH-CH_3$$
$$\mathbf{1} \quad\quad\quad\quad\quad\quad\quad \mathbf{2}$$

$$H_3C-\overset{\overset{O}{\|}}{C}-CH_3 \quad\quad H_3C-CH_2-\overset{\overset{O}{\|}}{C}-H$$
$$\mathbf{3} \quad\quad\quad\quad\quad\quad \mathbf{4}$$

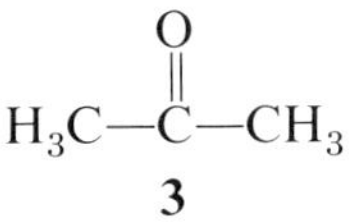

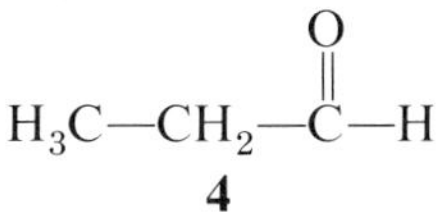

---

[1] K. Mislow, "Introduction to Stereochemistry," W. A. Benjamin, Inc., New York, 1965.

[2] R. T. Morrison and R. N. Boyd, "Organic Chemistry," 3rd edition, Allyn and Bacon, Inc., Boston, 1975.

[3] A. L. Ternay, Jr., "Contemporary Organic Chemistry," W. B. Saunders Company, Philadelphia, 1976.

[4] The type of isomerism exhibited by acetone and propionaldehyde has sometimes been called "functional isomerism," since movement of the carbonyl group changes the chemical function of the molecule.

In contrast, *stereoisomerism* arises when two or more compounds have the same empirical formula and the same atom-to-atom bonding sequence but the atoms differ in their arrangement in space. Examples from organic chemistry would include the pair of enantiomers **R**- and **S**-3-bromo-2-methylpentane (**5** and **6**, respectively) or the pair of diastereomers *cis*- and *trans*-2-butene (**7** and **8**).[5] In addition, there are conformational isomers such as diequatorial (**9**) and diaxial (**10**) *trans*-1,2-dimethylcyclohexane.[6]

Considerable effort was spent on understanding stereochemistry when studying organic chemistry, because it was clear that some organic reactions could produce different structural isomers, either or both enantiomers of a pair, or one or more diastereoisomers. To pursue the study of organic chemistry to the fullest, you had to learn ways to detect these isomers, to separate them, and to understand the mechanistic implications of the isolation of a particular isomeric form. The same rationale applies to inorganic chemistry, since there are numerous examples of isomerism in this field as well. For instance, there are many forms of constitutional isomerism, such as the so-called linkage isomers **11a** and **11b**, and a great many stereoisomers have been isolated and examined.

[5] The nomenclature of stereochemistry is reviewed later in this chapter.
[6] Stereoisomerism is important in literature as well. Lewis Carroll's story "Through the Looking Glass and What Alice Found There" is a journey into a mirror image world. The characters that best exemplify this world, in fact, are Tweedledum and Tweedledee, two inhabitants of the Looking Glass world that Carroll meant to be enantiomers. See "The Annotated Alice," Martin Gardner, ed., Clarkson N. Potter Publisher, New York, 1960.

One recent, unique example of the isolation of an enantiomer of a coordination compound is the report that the bacterium *Pseudomonas stutzeri* selectively removes one enantiomer, **12a,** from a racemic mixture of $[Co(en)_2(phen)]^{3+}$; after 60 hours under aerobic conditions the remaining enantiomer, **12b,** has an optical purity of 82%.[7]

12a   **12b**

Stereoisomerism in particular has been studied very extensively in coordination chemistry, in part because it can be much more complex than in organic chemistry and, therefore, provides considerable challenge. For example, only one pair of enantiomers may be generated by a given chiral carbon center, whereas as many as 15 pairs may be generated by a chiral octahedral complex.

The importance of isomerism in inorganic chemistry, and in coordination chemistry in particular, means that you must be prepared to deal with it in the same general way as in organic chemistry. Therefore, in this chapter we shall examine systematically the various forms of isomerism displayed by coordination compounds and the special nomenclature by which they can be described. In subsequent chapters you shall have cause to refer back to this material; for example, linkage isomers can be produced in some electron transfer reactions (Chapter 12), stereochemical change can be effected by substitution reactions (Chapter 13), and enantiomeric octahedral complexes can interconvert spontaneously (Chapter 14).

Before beginning our discussion, we might point out that the organization of this chapter according to isomer type follows, in large part, a classification scheme laid down by Alfred Werner in his work early in this century.[8-10]

CONSTITUTIONAL ISOMERISM

Constitutional isomers have identical empirical formulas, but the atom-to-atom connections differ. There are numerous forms of such isomerism, but only five of the most common are described below.

HYDRATE ISOMERISM

The classical example of hydrate isomerism is the series of three hydrated chromium chloride complexes having the empirical formula $CrCl_3 \cdot 6H_2O$. The green commercial form of this compound (called Recoura's Green Chloride) is obtained from concentrated hydrochloric acid solutions and is assigned the molecular formula $[Cr(H_2O)_4Cl_2]Cl \cdot 2H_2O$. On dissolving this complex in water, however, the metal-bound Cl^- ions are successively replaced by water molecules, giving first blue-green $[Cr(H_2O)_5Cl]Cl_2 \cdot H_2O$ and finally violet $[Cr(H_2O)_6]Cl_3$.

---

[7] L. S. Dollimore, R. D. Gillard, and I. H. Mather, *J. C. S. Dalton,* 518 (1974).

[8] Alfred Werner's papers of greatest importance to coordination chemistry have been translated and annotated by G. B. Kauffman, "Classics in Coordination Chemistry," Dover Publications, Inc., New York, 1968.

[9] G. B. Kauffman, *Coord. Chem. Rev.,* **11,** 161 (1973).

[10] G. B. Kauffman, *Coord. Chem. Rev.,* **12,** 105 (1974).

IONIZATION ISOMERISM

This type of isomerism is somewhat similar to hydrate isomerism. However, instead of the exchange of coordinated ions for water, ionization isomers arise from the interchange of two ions between inner and outer coordination spheres. For instance, the complex formed from a cobalt(III) ion, five ammonia molecules, a bromide ion, and a sulfate ion exists in two forms, one dark violet (**A**) and the other violet-red (**B**). The dark violet form (**A**) gives a precipitate with barium chloride but none with silver nitrate. Form (**B**) behaves in an opposite manner. Therefore, the two complexes may be formulated as $\mathbf{A} = [Co(NH_3)_5Br]SO_4$, and $\mathbf{B} = [Co(NH_3)_5SO_4]Br$. Another example of ionization isomerism is the pair of complexes *trans*-$[Co(en)_2Cl_2]NO_2$ (green) and *trans*-$[Co(en)_2(NO_2)Cl]Cl$ (red).

COORDINATION ISOMERISM

This form of isomerism is possible only for *salts* wherein both the anion and cation contain a metal ion that can function as a coordination center. Isomerism arises from a different distribution of ligands between the two metal ions. For example, if two different metal ions (M and M′) and two different ligands (*a* and *b*) are involved, two of the possible coordination isomers would be $[Ma_x][M'b_x]$ and $[M'a_x][Mb_x]$.

One real example of coordination isomerism is found in two of the salts that have the empirical formula $CoCr(NH_3)_6(CN)_6$. They may be prepared, and their constitutions determined, by the following reactions:

$$[Co(NH_3)_6]Cl_3 + K_3[Cr(CN)_6] \rightarrow \underset{13}{[Co(NH_3)_6][Cr(CN)_6]} + 3\ KCl$$

Compound **13** + $AgNO_3 \rightarrow$ insoluble salt identified as $Ag_3[Cr(CN)_6]$

$$[Cr(NH_3)_6]Cl_3 + K_3[Co(CN)_6] \rightarrow \underset{14}{[Cr(NH_3)_6][Co(CN)_6]} + 3\ KCl$$

Compound **14** + $AgNO_3 \rightarrow$ insoluble salt identified as $Ag_3[Co(CN)_6]$

Additional examples of coordination isomerism include the following:
(a) different metal ions and different bidentate ligands:

$$[Co(en)_3][Cr(C_2O_4)_3] \text{ and } [Co(en)_2C_2O_4][Cr(en)(C_2O_4)_2]$$

(b) identical metal ions but different ligands:

$$[Co(NH_3)_6][Co(NO_2)_6] \text{ and } [Co(NH_3)_4(NO_2)_2][Co(NH_3)_2(NO_2)_4]$$

(c) identical metal ions in different oxidation states; different ligands:

$$[Pt^{II}(NH_3)_4][Pt^{IV}Cl_6] \text{ and } [Pt^{IV}(NH_3)_4Cl_2][Pt^{II}Cl_4]$$

POLYMERIZATION ISOMERISM

In organic chemistry the term polymer implies the linking of small radicals or molecules into chains, layers, or lattices. In coordination chemistry, however, the term polymerization isomerism has a slightly different meaning. Each member of a series of polymerization isomers has the same empirical formula, and the molecular formula of each is some multiple of the simplest formula. However, unlike organic polymers, inorganic polymerization isomers are not formed by the direct chemical

TABLE 11-1

POLYMERIZATION ISOMERS OF A PLATINUM(II) COMPLEX.

Formula	Multiple of Simplest Formula Weight	Comments
[Pt(NH$_3$)$_2$Cl$_2$]	1	Yellow; *cis* isomer known as Peyrone's chloride, and the *trans* isomer often called Reiset's chloride.
[Pt(NH$_3$)$_4$][PtCl$_4$]	2	Known as Magnus's Green Salt. First Pt(II)-ammine discovered.
[Pt(NH$_3$)$_3$Cl][Pt(NH$_3$)Cl$_3$]	2	—
[Pt(NH$_3$)$_4$][Pt(NH$_3$)Cl$_3$]$_2$	3	Orange-yellow.
[Pt(NH$_3$)$_3$Cl]$_2$[PtCl$_4$]	3	—

bonding of small units to give larger units. To illustrate this, a series of polymerization isomers is listed in Table 11-1. (Notice also that the second and third and the fourth and fifth compounds are pairs of coordination isomers.)

LINKAGE ISOMERISM

Although Sophus Mads Jørgensen prepared two forms of the coordination compound [Co(NH$_3$)$_5$(NO$_2$)]Cl$_2$ in 1894, it has only been within the past fifteen years that they have been clearly established as linkage isomers.[11] Of all of the forms of constitutional isomerism, linkage isomers have recently attracted the greatest interest.[12,13] As you shall see, the factors that control the isolation of a particular isomer are subtle, and results both confusing and interesting are often obtained; considerable work must be done in this area in the coming years.

In organic chemistry it is well established that substituent groups such as NO$_2$ can form nitro $\left(R-N\begin{matrix}O\\O\end{matrix} \right)$ compounds or nitrites (R—O—N—O), CN can form nitriles (R—C≡N) or isonitriles (R—N≡C), SCN can form thiocyanates (R—S—C≡N) or isothiocyanates (R—N=C=S), and so on. In coordination chemistry these same groups can behave as ambidentate ligands, using either end of the ion in binding to a metal ion. So, in the case of the NO$_2^-$ ion, it can form nitrito **(15)** or nitro **(16)** complexes.

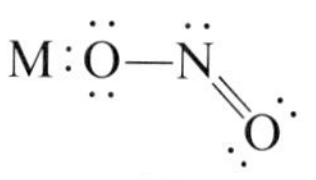

$$M:\ddot{O}-\ddot{N}\diagdown\underset{\cdot\cdot}{\overset{\cdot\cdot}{O}}\cdot \qquad M:\ddot{N}\diagup\overset{\ddot{O}:}{\underset{\ddot{O}:}{}}$$

       15 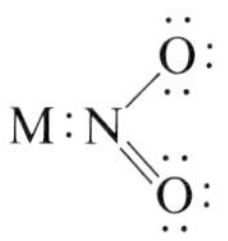   16

---

[11] Alfred Werner recognized this form of isomerism in his 1905 paper "Neuere Anschauungen auf dem Gebiete der anorganischen Chemie," but he originally called it "salzisomerie" or salt isomerism. See ref. 9.

[12] J. L. Burmeister, *Coord. Chem. Rev.*, **3**, 225 (1968); **1**, 205 (1966).

[13] A. H. Norbury and A. I. P. Sinha, *Quart. Revs.*, **24**, 69 (1970).

The two cobalt(III)-NO_2 linkage isomers mentioned above may be prepared according to the reactions:[14]

$$[Co(NH_3)_5Cl]^{2+} + H_2O \rightarrow [Co(NH_3)_5H_2O]^{3+} + Cl^-$$
$$[Co(NH_3)_5H_2O]^{3+} + NO_2^- \rightarrow [Co(NH_3)_5ONO]^{2+} + H_2O$$

light red
pentaamminenitritocobalt(2+) ion

$$[Co(NH_3)_5ONO]^{3+} \xrightarrow{H^+} [Co(NH_3)_5NO_2]^{2+}$$

yellow
pentaamminenitrocobalt(2+) ion

The yellow N-bonded or nitro isomer is stable for months if stored out of direct sunlight, but irradiation with ultraviolet light leads to isomerism to the light red O-bonded or nitrito form. Upon standing, the nitrito complex reverts completely to the nitro form again. The mechanism of interconversion, even in the solid state, is believed to be intramolecular, proceeding through the three-center intermediate **17**.[15]

$$[Co(NH_3)_5NO_2]^{2+} \xrightleftharpoons{h\nu} \left[(NH_3)_5Co \begin{smallmatrix} O \\ \diagdown \\ N \\ \diagup \\ O \end{smallmatrix} \right]^{2+} \rightleftharpoons [Co(NH_3)_5OONO]^{2+}$$

17

As mentioned above, considerable effort has been expended recently to uncover the factors that determine which linkage isomer is formed in a particular situation.[16] Although there is still some confusion, several things are clearly important, even though their interplay is often subtle and predictions are difficult to make. The important factors seem to be (1) the nature of the metal ion, (2) the electronic and steric effects of the ambidentate ligand or other ligands in the complex, and (3) the solvent system used in the isomer synthesis or in recrystallization.[17-20]

In 1958 Ahrland, Chatt, and Davies pointed out that metal ions in their common oxidation states could be divided roughly into two classes, depending on their abilities to coordinate with certain donor atoms.[21] Those metals in class *a* (see Figure 11-1)

---

[14] W. L. Jolly, "The Synthesis and Characterization of Inorganic Compounds," Prentice-Hall Inc., Englewood Cliffs, N.J., 1970, pp. 462–463.
[15] V. Balzani, R. Ballardini, N. Sabbatini, and L. Moggi, *Inorg. Chem.*, **7**, 1398 (1968).
[16] See ref. 12.
[17] R. T. M. Fraser, *Adv. Chem. Ser.*, **62**, 295 (1967).
[18] N. J. Stefano and J. L. Burmeister, *Inorg. Chem.*, **10**, 998 (1971).
[19] J. L. Burmeister, R. L. Hassel, and R. J. Phelan, *Inorg. Chem.*, **10**, 2032 (1971).
[20] J. L. Burmeister, E. A. Deardorff, A. Jensen, and V. H. Christiansen, *Inorg. Chem.*, **9**, 58 (1970).
[21] S. Ahrland, J. Chatt, and N. R. Davies, *Quart. Revs.*, **11**, 265–276 (1958).

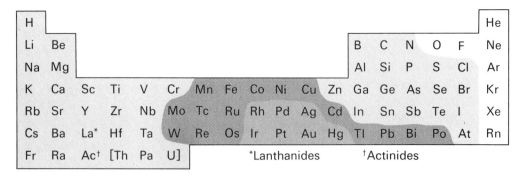

Figure 11-1. Classification of acceptor ions in their "normal" valence states into *a*, *b*, and *borderline* types.

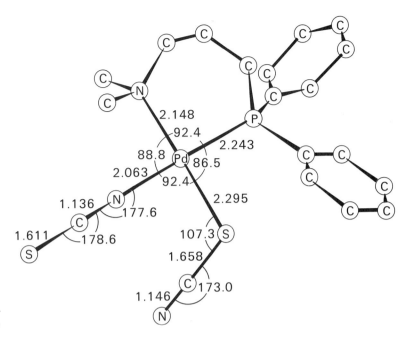

Figure 11-2. Mixed thiocyanate bonding in [Pd(Ph$_2$P-C$_3$H$_6$-NMe$_2$)(SCN)(NCS)].

generally bind most strongly with ligands whose donor atoms are second row elements (N, O, or F), and those in class *b* form their most stable complexes with elements of subsequent rows of the periodic table (P, S, Cl, etc.).[22] Since, with the exception of NO$_2^-$ and CN$^-$,[23] all of the ligands exhibiting ambidentate behavior possess one donor atom from the second row of the Periodic Table and one from the third or higher row, the nature of the metal ion acceptor is clearly important. For example, the most stable complex formed between SCN$^-$ and Pd(II) is the S-bonded linkage isomer [Pd(SCN)$_4$]$^{2-}$, while the analogous complex formed by Cd(II), an ion classified as being on the border between *a* and *b*, has both N- and S-bonded SCN$^-$ ligands. On the other hand, the class *a* metal ion Co(III) forms the N-bonded complex ion [Co(NH$_3$)$_5$NCS]$^{2+}$.[24]

The steric and electronic effects of the other ligands attached to the acceptor metal ion have a very great effect on the coordination behavior of ambidentate ligands. For example, in contrast with the N-bonded complex [Co(NH$_3$)$_5$NCS]$^{2+}$ **(18)**, replacing the NH$_3$ ligands with CN$^-$ affords the S-bonded complex [Co(CN)$_5$SCN]$^{3-}$ **(19)**, and both modes of bonding are observed in the compound in Figure 11-2. There are two possible explanations, one electronic and one steric, with the truth probably lying somewhere between.

---

[22] We shall explore in more detail the question of the thermodynamic stability of coordination complexes in Chapter 13.

[23] Another potentially ambidentate ligand is NCO$^-$. Although there was a report several years ago that this ion could bond through either end, the complex (η^5-C$_5$H$_5$)$_2$Ti(NCO)$_2$ was recently shown to be N-bonded. However, there is still some question regarding the possibility that it is both N- and O-bonded in solution. S. J. Anderson, D. S. Brown, and A. H. Norbury, *J. C. S. Chem. Commun.*, 996 (1974).

[24] You should recognize these phenomena as examples of reversals in ligand basicity, as discussed in Chapter 5. Be aware that the *a/b* classification system must be used with caution, for it tends to oversimplify in the sense of a single-scale concept of acidity/basicity. The compound in Figure 11-2 is a good example.

The electronic explanation for the difference in bonding of SCN⁻ in **18** and **19** is clear on examining the π mo's of the ion **(20)**.

$$\underset{\textbf{20}}{\begin{array}{c} \text{S—C—N} \quad \pi^* \\ \text{S—C—N} \quad \pi_n \quad \bullet\bullet \\ \text{S—C—N} \quad \pi \quad \bullet\bullet \end{array}}$$

The most important observation to make is that the π-bonding mo is polarized toward the nitrogen atom (N is the most electronegative atom), so the π^* mo is polarized toward sulfur. All three mo's (π, π_n, and π^*) can overlap with a metal orbital of π symmetry using either end of the SCN⁻ ion. However, if M—N overlap occurs, the ligand would function as a stronger π-donor than π-acceptor. Conversely, in M—SCN complexes the ligand would be a strong π-acceptor but a weak π-donor. Therefore, it is to be expected that SCN⁻ will preferentially use the sulfur end in binding to a metal ion having filled (or partially filled) π-symmetry orbitals. On the other hand, if the metal ion has no such orbitals or if another ligand competes successfully for them, then SCN⁻ may prefer to use the π-donor, or nitrogen, end of the ligand. In the case of the Co(III) complexes above, the negative charge placed on the metal by the cyanide ligand in **19** can be diffused better if the SCN⁻ ion is π-accepting or S-bonded; therefore, [Co(CN)₅SCN]³⁻ is the more stable form. Arguments similar to these can also be used to explain the fact that SCN⁻ is N-bonded when it is *trans* to the potentially π-bonded phosphorus atom, while it is S-bonded when it is *trans* to the purely σ-bonding amino group in Pd(Ph₂—C₃H₆—NMe₂)(SCN)(NCS) (Figure 11-2).[25] However, further research has indicated that electronic arguments are only part of the explanation for the bonding mode in SCN⁻ complexes; steric effects are also important.

As seen in the palladium compound in Figure 11-2, the SCN⁻ ligand is linear when N-bonded and bent when S-bonded.[26] Generally, the steric requirement of the bent SCN⁻ ion should be greater than that of the linear ion. Thus, in **21** for example, both

21

[25] G. R. Clark and G. J. Palenik, *Inorg. Chem.*, **9**, 2754 (1970).
[26] G. L. Palenik, M. Mathew, W. L. Steffen, and G. Beran, *J. Amer. Chem. Soc.*, **97**, 1059 (1975).
[27] The bond distances in the SCN⁻ ligands in Figure 11-2 suggest that the metal-ligand system can be represented by the following electron dot formulas:

$$\overset{\ominus}{M} \leftarrow :\overset{..}{\underset{..}{S}}-C \equiv N: \qquad :\overset{\ominus}{\underset{..}{S}}-C \equiv \overset{\oplus}{N}: \rightarrow \overset{\ominus}{M}$$

The S-bonded form should be bent, since there are two bond pairs and two lone pairs about the sulfur. Conversely, the N-bonded form should be linear. These electron dot pictures also suggest that the C—S stretching frequency should be less in the S-bonded isomer than in the N-bonded form. Indeed, it has been found that the C—S stretch in M—NCS is usually in the range from 860 to 780 cm⁻¹, while that for M—SCN is usually between 720 and 690 cm⁻¹. Further, the N-bonded isomer usually gives rise to a more intense C—N stretching band than does the S-bonded isomer. See ref. 12.

SCN⁻ ligands are bent, apparently because the P—Pd—P angle (73.33°) is smaller than the P—Pd—N angle in the compound in Figure 11-2, and the SCN⁻ ligands can assume the sterically more demanding bent form.[27] As the number of carbon atoms between the two phosphorus atoms in **21** is increased (on going from $Ph_2P-CH_2-PPh_2$ to $Ph_2P-C_2H_4-PPh_2$ to $Ph_2P-C_3H_6-PPh_2$) the P—Pd—P angle is enlarged, and one and then finally both SCN⁻ ligands assume the linear, N-bonded form. A similar explanation may apply to **18** and **19**. The NH_3 must require more space about Co(III) than does a CN⁻ ligand. Therefore, a more stable complex is formed if SCN⁻ is N-bonded when in the presence of five ammonia ligands.

STUDY QUESTIONS

1. How many coordination isomers could be formed starting with $[Cu(NH_3)_4][PtCl_4]$? Write the formula for each isomer.

2. Alfred Werner first cited the nine known compounds of empirical formula $Co(NH_3)_3(NO_2)_3$ as examples of polymerization isomerism.

   a) Write the formulas for at least four polymerization isomers based on $Co(NH_3)_3(NO_2)_3$.
   b) Give the systematic name for each of these four isomers.

3. *Linkage isomerism.* Hexafluorophosphoric acid and KSCN were added to $[(C_5H_5)Fe(CO)_2]_2$ in absolute ethanol, and air was bubbled through the solution for one hour at

room temperature. After hydrolysis, the components of the organic layer were separated on a silicic acid column into a yellow band and a red band; the eluting solvents were CH_2Cl_2 followed by acetone. The yellow band yielded golden yellow crystals (mp 119°), which had a C—S stretching band ascribed to the thiocyanate ligand at 830 cm⁻¹ (medium relative intensity). The red band yielded dark red crystals with a much lower melting point (36°), and the infrared absorption band appropriate for the thiocyanate group was found at 698 cm⁻¹ (weak relative intensity). On warming the red isomer to a temperature just below the melting point, isomerization to the yellow form occurs. On the basis of this information, decide in which of the compounds, the red or the yellow form, the SCN ligand is N-bonded. (See ref. 27.)

STEREOISOMERISM[28]

Stereoisomerism has played a decisive role in coordination chemistry, since it was largely on the basis of the isolation of stereoisomers of four- and six-coordinate complexes that Werner was able to prove his revolutionizing concepts of inorganic stereochemistry.[29,30] For this reason, and because stereoisomerism continues to be studied intensively,[28c] we will describe this form of isomerism in some detail.

---

[28] It is difficult to recommend any one or several books on the stereochemistry of coordination complexes, since so many have been written on this topic. However, the three listed below provide a good introduction.
   a. This book and the next provide a rather dated but excellent account of the early work in the field: "The Chemistry of Coordination Compounds," J. C. Bailar, Jr., ed., Reinhold Publishing Corp., New York, 1956.
   b. J. Lewis and R. G. Wilkins, "Modern Coordination Chemistry," Interscience, New York, 1960.
   c. An excellent survey of more recent work is given by C. J. Hawkins, "Absolute Configuration of Metal Complexes," Interscience, New York, 1971.

[29] A. Werner, *Zeitschrift für anorganische Chemie*, **3**, 267-330 (1893); this paper is included in ref. 8.
[30] See ref. 10.

A GLOSSARY OF STEREOCHEMICAL TERMS

Stereoisomers: Two or more molecules that have the same empirical formula and the same atom-to-atom bonding sequence, but in which the atoms differ in their arrangement in space.

Enantiomer: A stereoisomer that has a non-superimposable mirror image.

Diastereoisomers: Stereoisomers that are not enantiomers; diastereoisomerism includes *cis-trans* isomerism as a sub-class.

Asymmetric: Applied to a molecule totally lacking in symmetry.

Dissymmetric: Applied to a molecule that lacks an S_n axis, that is, a rotation-reflection axis. Because it lacks an S_n axis, a dissymmetric molecule has neither a plane of symmetry (S_1) nor a center of symmetry (S_2). In order for a molecule to have an enantiomer, it must *not* have an S_n axis. However, a dissymmetric molecule may have a proper axis of rotation C_n (with $n > 1$).

Chiral compound: A molecule that is asymmetric or dissymmetric is chiral, and it is therefore not superimposable on its mirror image. Chirality denotes "handedness," and molecules of opposite chirality (enantiomers) are related to one another as the left hand is to the right.

Optical activity: The ability of chiral molecules to rotate the plane of plane polarized light.

The general nomenclature of stereoisomerism has undergone a profound change in the past decade. In an attempt to relate your probably firmer grasp of organic stereochemistry and its terminology to inorganic stereochemistry, we shall use as much as possible the terms developed so carefully by organic chemists.[31-34]

Stereoisomers are compounds having the same molecular formula and the same atomic sequence; however, their atoms differ in their spatial arrangement. They generally fall into two classes: *enantiomers* and *diastereomers*.

Enantiomers are molecules that are non-superimposable mirror images of one another; that is, they have opposite chirality. The term chirality (allegedly coined by Lord Kelvin in 1893)[35] denotes "handedness," and molecules of opposite chirality are related to one another as the left hand is to the right. Chirality can arise from the presence of an "asymmetric" or chiral center such as a carbon atom (**5, 6,** and **22**) or a metal atom (**23**). To be a center of chirality, a carbon atom must be surrounded by four different groups, but a chiral metal atom in an octahedral complex may be produced by the appropriate arrangement of at least three different types of monodentate ligands; for example, the Pt(II) compound $[Pt(NH_3)_2(NO_2)_2Cl_2]$ has been resolved into its enantiomers (**24a** and **24b**).[36] However, just as there are other ways for carbon compounds to be chiral [for example, an appropriately tetra-substituted biphenyl (**25**) does not have a chiral carbon, yet such compounds are chiral], there are six ways for chirality to be introduced into an octahedral complex; some of these are discussed in more detail in a later section of this chapter.

A necessary and sufficient condition for chirality in a molecule is that it lacks an improper axis of rotation—that is, a rotation-reflection axis, S_n. However, a less

---

[31] See refs. 1, 2, and 3.
[32] E. L. Eliel, *J. Chem. Educ.*, **48,** 163 (1971).
[33] J. A. Hirsch, "Concepts in Theoretical Organic Chemistry," Allyn and Bacon, Boston, 1974.
[34] N. L. Allinger and E. L. Eliel, eds., "Topics in Stereochemistry," Vol. 1, Interscience, New York, 1967.
[35] See ref. 2, p. 124.
[36] See ref. 28 and I. I. Chernyaev, T. N. Fedotova, and O. N. Adrianova, *Russ. J. Inorg. Chem.*, **10,** 567 (1965).

rigorous criterion often applied to decide whether or not a molecule is chiral is the lack of a center or plane of symmetry. The two enantiomers **24a** and **24b** certainly possess neither of these. However, a center of symmetry is equivalent to an S_2 axis and a plane to an S_1 axis, so the main criterion is fulfilled. The less rigorous criterion should be applied with caution, however, since it is a well-recognized fact that a molecule may lack both a center and a plane of symmetry but still have a superimposable mirror image; compound **26** is an example of such a case.

The fact that chiral molecules lack the all-important S_n axis does not mean that they must also be *asymmetric,* that is, totally lacking in symmetry. Rather, in coordination chemistry, many of the compounds that have been resolved into their enantiomers have proper axes of rotation as in the tris(bidentate)metal complex **27**. In summary, the most important criterion for the existence of an enantiomer of a molecule is that it be *dissymmetric;* that is, that, while it may possess other elements of symmetry, it must lack an S_n axis.

All stereoisomers that are not enantiomers are *diastereoisomers* (sometimes referred to as *diastereomers*). Diastereoisomerism includes as a sub-class the well-known phenomenon of *cis-trans* or geometrical isomerism. For example, compounds **24c-f** are diastereoisomers of **24a** and **24b**. Diastereoisomers may be achiral **(24c-f)** or they may be chiral **(24a,b)**.

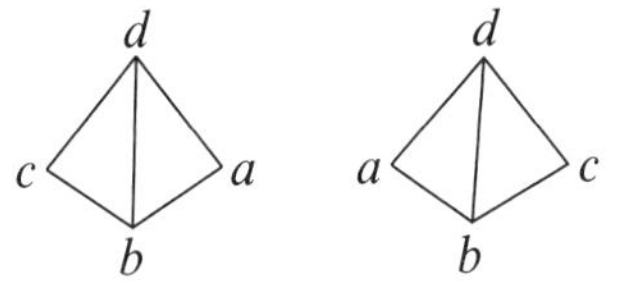

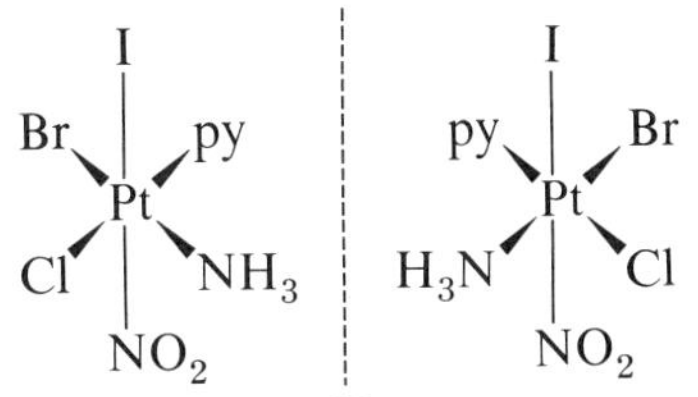

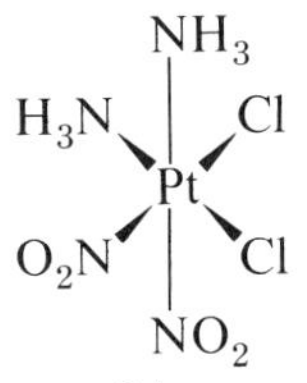

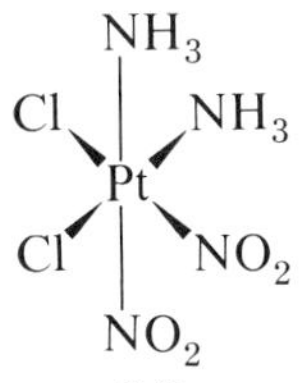

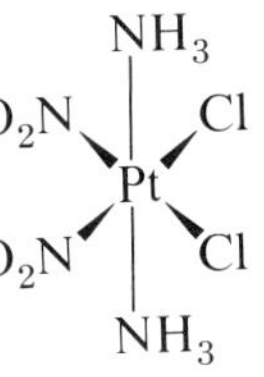

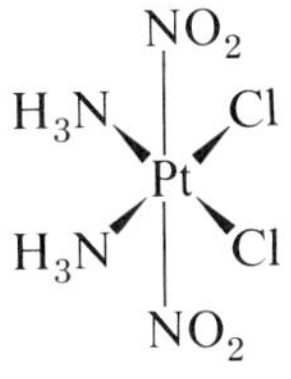

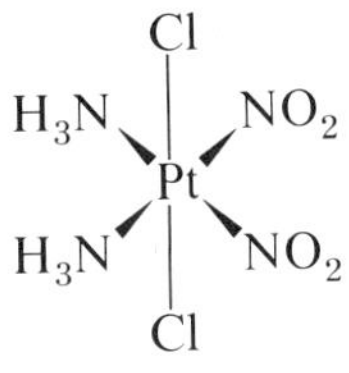

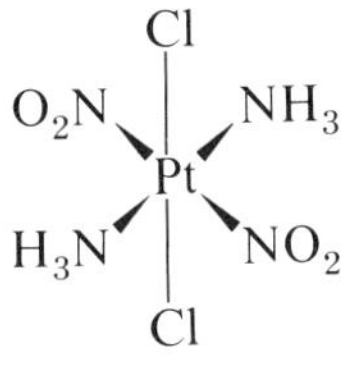

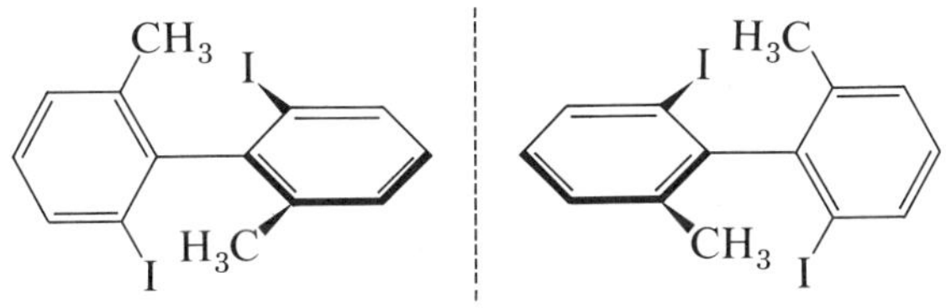

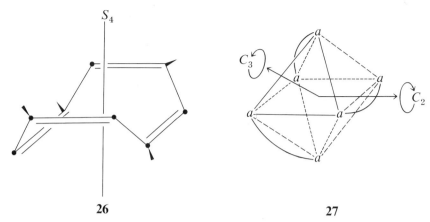

26   **27**

A very important difference between enantiomers and diastereomers is that the latter usually differ appreciably in their chemical and physical properties (in fact, in any property depending on molecular shape). However, two enantiomers differ only in their abilities to rotate plane polarized light.[37] Therefore, in order to separate two enantiomers (recall that with certain exceptions[38] a racemic mixture will arise in any normal chemical reaction producing chiral products), they must be first converted to diastereomers. For example, the enantiomers of cis-$[Co(en)_2(NO_2)_2]^+$ **(28)** can

28

be separated into optically pure components by adding the chiral salt potassium antimonyl-d-tartrate to an aqueous solution of the racemic mixture.[39] The chiral tartrate forms salts with both the left-handed and right-handed isomers of the cobalt complex, these salts being diastereomeric because only one enantiomer of the anion is used. As diastereomers, they have different solubilities, and the salt cis-$(-)_D$-$[Co(en)_2(NO_2)_2][(+)_D$-$SbOC_4H_4O_6]$ precipitates as yellow needles before the other diastereomer, cis-$(+)_D$-$[Co(en)_2(NO_2)_2][(+)_D$-$SbOC_4H_4O_6]$.[40]

In the sections that follow, we shall examine briefly the types of stereoisomerism observed for the most important coordination numbers—four and six—and the consequences of that isomerism.

GENERAL ASPECTS OF STEREOCHEMICAL NOTATION IN COORDINATION CHEMISTRY

From the discussions in this and the previous two chapters, it should be evident that coordination chemists have produced new compounds in many stereochemical variations, but it is only recently that nomenclature systems have been proposed that can cope with the tremendous variety of compounds. Outlined below is, in part, the notation system espoused by *Chemical Abstracts* to specify the stereochemistry of both diastereomers and enantiomers;[41] in a later section of this chapter we shall discuss further the special nomenclature of complexes that are chiral because of multidentate

---

[37] When such rotation is observed experimentally, the system is said to be optically active. The origin and measurement of optical activity are explored in more detail in the course of the discussion of the absolute configuration of metal complexes, page 644 and following.

[38] G. R. Brubaker, *J. Chem. Educ.*, **51**, 608 (1974).

[39] S. Kirschner, "Optically Active Coordination Compounds," in "Preparative Inorganic Reactions," W. L. Jolly, ed., Vol. 1, Interscience, New York, 1964, p. 29.

[40] G. B. Kauffman and E. V. Lindley, Jr., *J. Chem. Educ.*, **51**, 424 (1974). This paper describes the resolution of $[Co(en)_2(NH_3)Br]^{2+}$, an experiment suitable for the undergraduate laboratory.

[41] M. F. Brown, B. R. Cook, and T. E. Sloan, *Inorg. Chem.*, **14**, 1273 (1975).

ligands. In both instances, we shall illustrate only the application of these systems to six-coordinate complexes.

Chemical Abstracts uses the basic approach of the Cahn-Ingold-Prelog (CIP) system, which has been so successfully applied to chiral organic molecules.[42,43] The CIP approach uses a set of rules to assign a sequence of priority to the four atoms arranged about a chiral carbon center. In its simplest form, the sequence rule is that atoms of higher atomic number have higher priority, the highest priority atom being 1, the atom of next highest priority being 2, and so on. This is illustrated in the example below. In this case, the Br has a priority number of 1 and H has a priority number of 4. This means that one of the remaining C atoms must be 2 and the other 3. To determine which number is to be assigned to which carbon, move to the next "shell" of atoms beyond the so-called branch points, that is, beyond the carbons directly attached to the chiral center. There you see that the isopropyl carbon has two C's and one H attached, whereas the ethyl carbon is bound to two H's and one C. Therefore, the isopropyl group is assigned priority number 2 because it has more atoms of higher atomic number attached to the branch point carbon. The priority numbers having been assigned, the stereochemistry of the chiral center is then specified by noting whether the priority

descends ($1 \rightarrow 2 \rightarrow 3$) in a clockwise ($= \mathbf{R}$) or counterclockwise ($= \mathbf{S}$) direction when one views the chiral center from the face of the molecule opposite the lowest priority atom (the atom of largest priority number).

Chemical Abstracts has applied a system similar to the CIP system to define the configuration of diastereomers and enantiomers formed by complexes of any geometry.[44] Space permits us to present only the briefest outline of this system, and we shall use only relatively simple examples. To understand properly more complicated cases, you should consult reference 41.

Chemical Abstracts gives a coordination compound a set of stereochemical descriptors composed of four parts: (i) a symmetry site term; (ii) a configuration number; (iii) a chirality symbol; and (iv) a ligand segment.

The *symmetry site term* refers to the gross symmetry of the metal site; that is, any difference in ligating atoms or slight distortions are ignored temporarily. The symmetry site terms used are:

Symmetry Site Term	Gross Geometry
T-4	tetrahedral
SP-4	square planar
TB-5	trigonal bipyramid
SP-5	square pyramid
OC-6	octahedral
TP-6	trigonal prism

The *configuration number* is a two-digit number that is assigned as follows:[41] "*The first digit of the two-digit number (for the octahedral case) is the CIP priority number of the ligating atom trans to the . . . ligating atom of priority number 1. These two atoms define the principal axis of the octahedron. The second digit of the configuration number is the CIP priority number of the ligating atom trans to the . . . ligating atom of smallest CIP priority number in the plane perpendicular to the principal axis.*"

A complex of the type Ma_2bcde has a total of 15 stereoisomers: there are nine diastereomers, six of which have an enantiomer. (See page 629 for a method for the enumeration of possible isomers.) Two isomers of this hypothetical complex, **29a** and **29b**, are shown with their configuration numbers.

---

[42] R. S. Cahn, C. Ingold, and V. Prelog, *Angew. Chem., Int. Ed. Engl.,* **5,** 385 (1966); *J. Org. Chem.,* **35,** 2849 (1970), Appendix.

[43] See refs. 2, 3, and 32.

[44] See ref. 41.

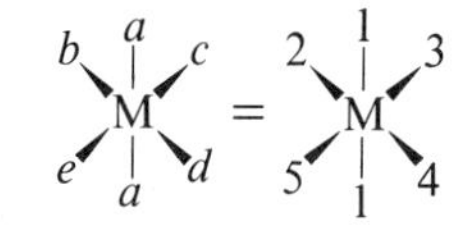

29a
configuration number = 14

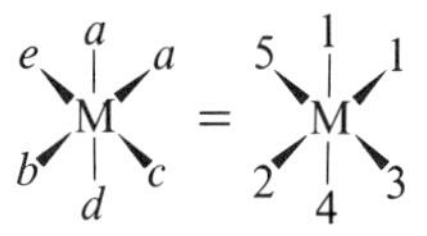

29b
configuration number = 42

In assigning priority numbers, we have assumed that the priority descends in the order $a = 1, \ldots, e = 5$. The principal axis is defined by the ligand of highest priority and the ligand *trans* to it. Therefore, the aMa axis is the principal axis in **29a**. Since a ligand of priority number 1 is *trans* to a ligand of priority 1, the first digit of the configuration number is 1. The highest priority ligand in the plane perpendicular to the principal axis is $b = 2$, and the ligand *trans* to it is $d = 4$. Therefore, the second digit of the configuration number is 4.

Compound **29b** presents a problem because there are two ligands of highest priority *cis* to one another and, therefore, two possible principal axes, aMd and aMb. When there is an ambiguity as to the principal axis, preference is arbitrarily given to the axis with the largest CIP priority number opposite 1. Thus, the principal axis is aMd, and the first digit of the configuration number is 4. The ligand of highest priority in the plane perpendicular to the principal axis is 1, and the ligand *trans* to it is 2. This means that the second digit of the configuration number is 2.

Two isomers of **24** are illustrated below with their systematic names. The ligand priority numbers can be assigned just as they are to the atoms or groups attached to the chiral center of a carbon compound. That is, Cl is the element of highest atomic number and so has the highest priority ($= 1$). Since the remaining ligands are all attached using the same atom, we must examine the atoms bound to the ligating atoms. Oxygen has a higher atomic number than hydrogen, so the NO_2^- ligand is assigned a higher priority ($= 2$) than NH_3 ($= 3$). The assignment of the configuration numbers then follows as in the hypothetical example above.

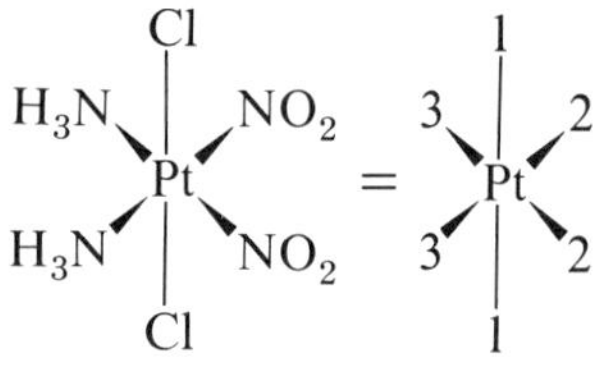

24e′
configuration number = 13
(OC-6-13)-dichlorodinitro-
diammineplatinum

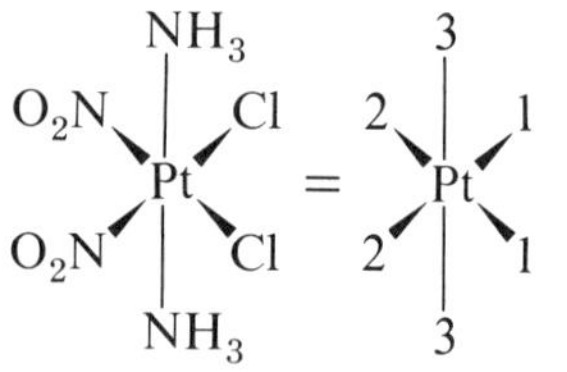

24c′
configuration number = 22
(OC-6-22)-dichlorodinitro-
diammineplatinum

Although the system just described may seem cumbersome at first, it does eliminate the ambiguities (for six-coordinate complexes) of the more commonly used *cis/trans* and *mer/fac* systems of nomenclature. For example, it is not clear at all whether the two isomers of compound **24** should be labeled *cis* or *trans*, but the CA system leads to an unambiguous label for both compounds. As another example, consider compound **30**. **30a** might have been given the stereochemical description *fac* in the older nomenclature system to designate the fact that the three Cl's are located on a *face* of the octahedron. On the other hand, compound **30b** might have had the label *mer* to indicate that the

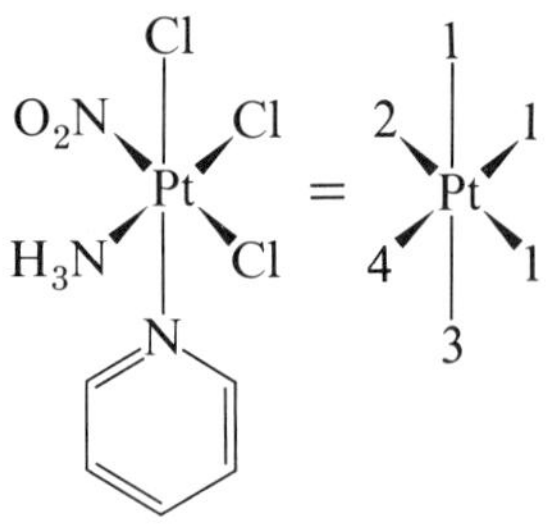

30a
configuration number = 43

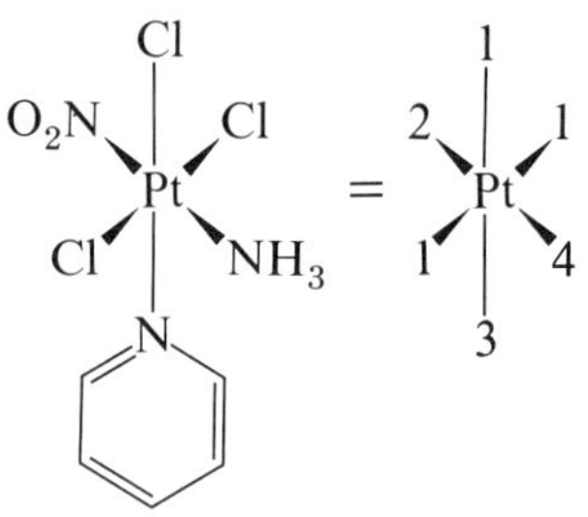

30b
configuration number = 31

three Cl's are located on a *meri*dian of the octahedron. As you can readily appreciate, there will be cases where this simple system can lead to ambiguous nomenclature, whereas the CA system gives an accurate and unambiguous description.

Because of their ligand distributions, complexes can be chiral, a condition discussed in much greater detail beginning on page 638. For example, the two hypothetical complexes below are enantiomers. To specify their chirality, *Chemical Abstracts* assigns the chirality symbol *C* to that configuration in which the CIP priority numbers of the ligating atoms in the plane perpendicular to the principal axis increase proceeding in a clockwise direction when viewing the plane from the most preferred atom of CIP priority number 1 on the principal axis. The opposite configuration is assigned the chirality symbol *A* (for anticlockwise).

configuration number = 32
chirality = *A*

configuration number = 32
chirality = *C*

Octahedral complexes that are chiral because two or three bidentate ligands are bound to the central metal ion are an exception to the nomenclature system described above. It is usually simpler to describe their chirality by using the Δ and Λ nomenclature suggested by the IUPAC; the latter system is discussed later in this chapter (page 638) when you can better use it.

STUDY QUESTIONS

4. There are five diastereoisomers of compound **24**, one of them having an enantiomer. For each of the achiral isomers of **24**, locate the S_n axis to prove that it is not chiral.

5. Refer back to compound **30** for the following questions:

    a) Give the chirality symbol for **30a**.
    b) Draw the enantiomer of **30a**. Give its configuration number and chirality symbol.
    c) The stereochemistry of **30a** and **30b** was indicated by configuration numbers and the *mer-fac* nomenclature system. At times such isomers would be referred to as *cis* and *trans*. Apply these latter terms to **30a** and **30b**.

6. Use the *Chemical Abstracts* system of configuration notation to assign configuration numbers and chirality symbols to:

    a) All of the isomers of compound **24**.
    b) **40a** and **40b** (page 628).
    c) All of the isomers of **44** (page 629).

FOUR-COORDINATE COMPLEXES

An understanding of the stereochemistry of four-coordinate complexes was extremely important to the development of coordination chemistry.[45] Alfred Werner's revolutionary, original paper in coordination chemistry (published in 1893)[46] was a reply to Blomstrand (1826–1897) and Jørgensen (1837–1914), whose formulation of four-coordinate complexes was based on ideas of organic structure prevalent in the middle of the nineteenth century. For example, they recognized that there were two complexes having the empirical formula $Pt(NH_3)_2Cl_2$, to which they assigned struc-

---

[45] See D. P. Mellor, "Stereochemistry of Square Complexes," *Chem. Rev.*, **33**, 137 (1943) for a summary of the early work in this area.

[46] See Ref. 8.

tures **31** and **32**. However, Werner rejected these structures for many reasons, but largely because they could not be used to rationalize the chemistry of the compounds.

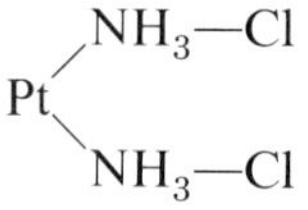

$$\text{Pt}\begin{array}{c}\diagup\text{NH}_3-\text{Cl}\\ \diagdown\text{NH}_3-\text{Cl}\end{array} \qquad \text{Pt}\begin{array}{c}\diagup\text{NH}_3-\text{NH}_3-\text{Cl}\\ \diagdown\text{Cl}\end{array}$$

31   **32**

platosammine chloride   platosemidiammine

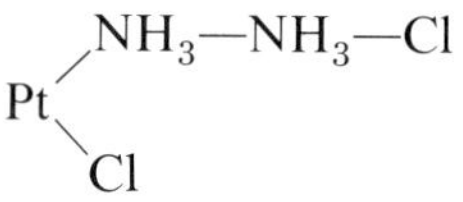

Instead, he proposed that the platosammine chloride was in fact the square planar complex *trans*-diamminedichloroplatinum, and that the platosemidiammine was the

$$\mathbf{31} = \begin{array}{c}\text{H}_3\text{N}\diagdown\quad\diagup\text{Cl}\\ \text{Pt}\\ \text{Cl}\diagup\quad\diagdown\text{NH}_3\end{array} \qquad \mathbf{32} = \begin{array}{c}\text{H}_3\text{N}\diagdown\quad\diagup\text{Cl}\\ \text{Pt}\\ \text{H}_3\text{N}\diagup\quad\diagdown\text{Cl}\end{array}$$

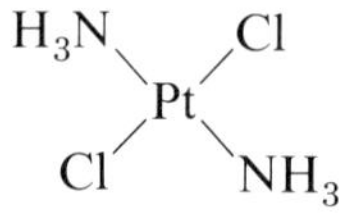

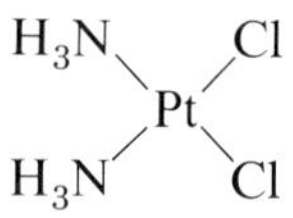

square planar *cis* isomer. Proof of the correctness of Werner's proposal came in work by Grinberg in 1931.[47] Grinberg made the very reasonable assumption that when a chelating agent such as the oxalate ion coordinates to two positions of a square planar complex, then these two positions must be *cis*. Therefore, *cis*- and *trans*-diamminedichloroplatinum are predicted to react with oxalic acid as shown below, a prediction borne out by experiment.

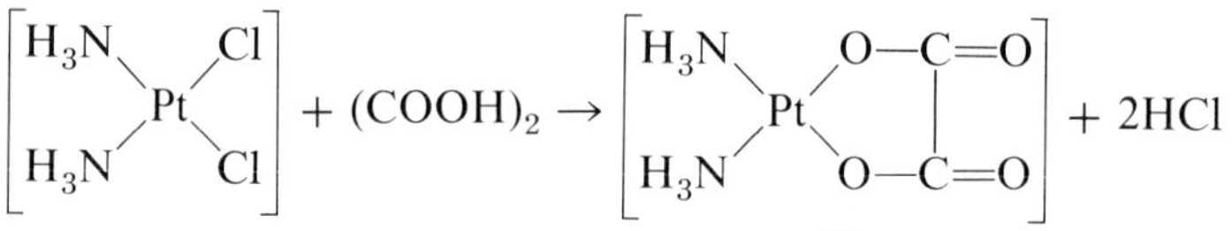

$$\left[\begin{array}{c}\text{H}_3\text{N}\diagdown\quad\diagup\text{Cl}\\ \text{Pt}\\ \text{H}_3\text{N}\diagup\quad\diagdown\text{Cl}\end{array}\right] + (\text{COOH})_2 \rightarrow \left[\begin{array}{c}\text{H}_3\text{N}\diagdown\quad\diagup\text{O}-\text{C}=\text{O}\\ \text{Pt}\quad\quad\quad\quad\;|\\ \text{H}_3\text{N}\diagup\quad\diagdown\text{O}-\text{C}=\text{O}\end{array}\right] + 2\text{HCl}$$

33

$$\left[\begin{array}{c}\text{H}_3\text{N}\diagdown\quad\diagup\text{Cl}\\ \text{Pt}\\ \text{Cl}\diagup\quad\diagdown\text{NH}_3\end{array}\right] + 2(\text{COOH})_2 \rightarrow \left[\begin{array}{c}\text{H}_3\text{N}\diagdown\quad\diagup\text{OOC}-\text{COOH}\\ \text{Pt}\\ \text{HOOC}-\text{COO}\diagup\quad\diagdown\text{NH}_3\end{array}\right] + 2\text{HCl}$$

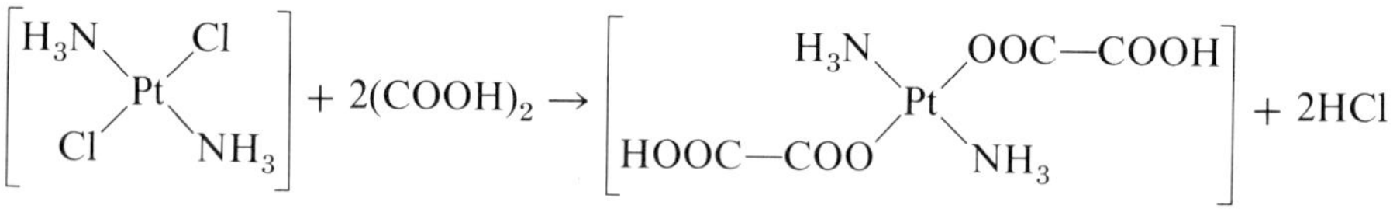

34

Because they have a plane of symmetry (the molecular plane), square planar complexes cannot have enantiomeric forms unless the ligand itself is chiral, as discussed below. Contrariwise, a tetrahedral molecule cannot have *cis* and *trans* isomers, again with the exception of complexes with unsymmetrical ligands (see study question 7). Therefore, an elegant proof of the fact that Pt(II) is normally found in square planar complexes was Chernyaev's synthesis[48,49] in 1926 of three diastereomeric forms of the ion [Pt(NH$_3$)(NH$_2$OH)(py)(NO$_2$)]$^+$ (**35a–c**). (In carrying out the synthesis of these isomers, Chernyaev formulated the concept now called the *trans* effect, a kinetic effect outlined in Chapter 13.)

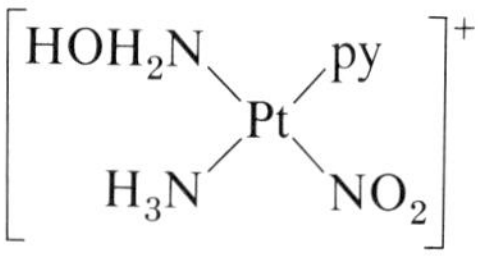

$$\left[\begin{array}{c}\text{HOH}_2\text{N}\diagdown\quad\diagup\text{py}\\ \text{Pt}\\ \text{H}_3\text{N}\diagup\quad\diagdown\text{NO}_2\end{array}\right]^+ \quad \left[\begin{array}{c}\text{py}\diagdown\quad\diagup\text{NH}_3\\ \text{Pt}\\ \text{HOH}_2\text{N}\diagup\quad\diagdown\text{NO}_2\end{array}\right]^+ \quad \left[\begin{array}{c}\text{H}_3\text{N}\diagdown\quad\diagup\text{py}\\ \text{Pt}\\ \text{HOH}_2\text{N}\diagup\quad\diagdown\text{NO}_2\end{array}\right]^+$$

35a 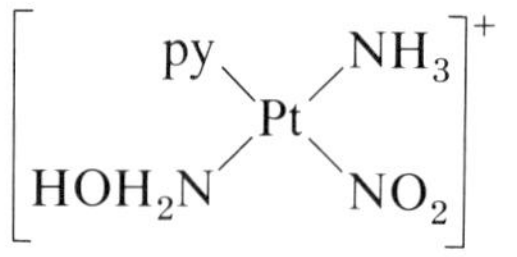   **35b** 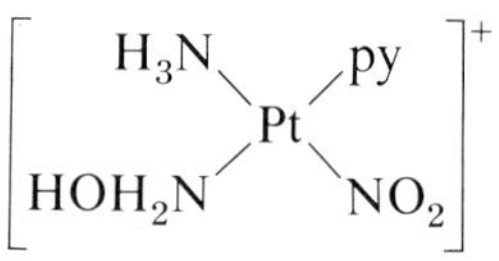   **35c**

---

[47] A. A. Grinberg, *Helv. Chim. Acta*, **14**, 455 (1931); see M. M. Jones, "Elementary Coordination Chemistry," Prentice-Hall, Inc., Englewood Cliffs, N.J., 1964, p. 57.

[48] M. M. Jones, "Elementary Coordination Chemistry," Prentice-Hall, Inc., Englewood Cliffs, N.J., 1964, p. 59.

[49] F. Basolo and R. G. Pearson, "The *trans* Effect in Metal Complexes," *Prog. Inorg. Chem.*, **4**, 381 (1962).

The only methods for introducing chirality into square planar complexes are to use two different, specially chosen bidentate ligands or to use a chiral ligand. In a classic paper, Mills and Quibell reported the synthesis of a platinum(II) complex containing one molecule of *meso*-stilbenediamine and one molecule of isobutylenediamine.[50] When it is square planar, the molecule (**36a**) has no S_n axis and is therefore chiral. If the complex is tetrahedral (**36b**), however, a plane of symmetry is clearly present (the plane of the ring on the right side of the structure as drawn). In this particular case, the Pt(II) complex was successfully resolved into its enantiomers, thereby conclusively showing the square planarity of the complex.

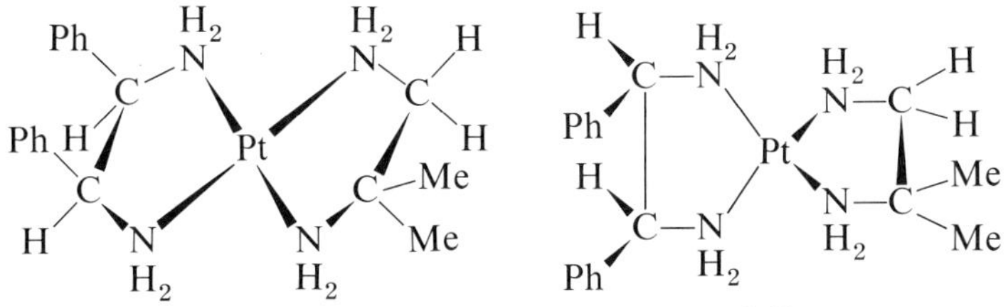

36a   **36b**

Ligand chirality is the only other method by which square planar complexes may become chiral. Compound **37**, which has been resolved into its enantiomers, is an example.[51]

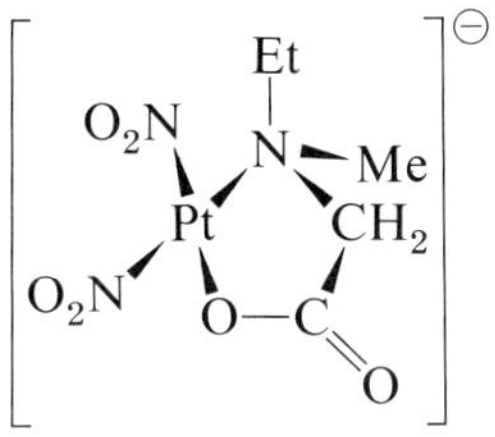

37

In contrast to carbon-based compounds, there are no examples of tetrahedral transition metal complexes with four different ligands. Therefore, the only way to observe chirality for tetrahedral complexes is to use unsymmetrical, bidentate ligands, and for this reason the syntheses and stereochemistries of salicylaldimine, unsymmetrical β-diketone, and β-ketoamine (*e.g.*, **38**) complexes have been investigated extensively.[52,53]

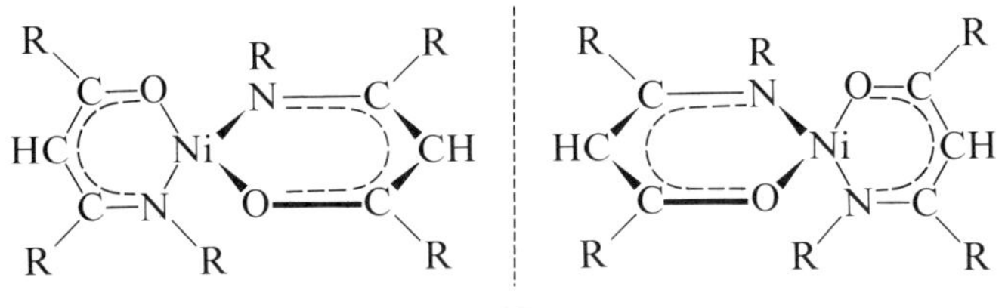

38

Only a very few tetrahedral bis-chelate complexes have been even partially resolved into their enantiomers, and most of them involve non-transition elements such as beryllium, boron, and zinc.[54] The few transition metal complexes that have been partially resolved racemize quite rapidly.[55,56]

---

[50] W. H. Mills and T. H. H. Quibell, *J. Chem. Soc.*, 839 (1935).
[51] J. R. Kuebler, Jr., and J. C. Bailar, Jr., *J. Amer. Chem. Soc.*, **74**, 3535 (1952).
[52] S. S. Eaton and R. H. Holm, *Inorg. Chem.*, **10**, 1446 (1971).
[53] R. H. Holm, G. W. Everett, Jr., and A. Chakravorty, *Prog. Inorg. Chem.*, **7**, 83 (1966).
[54] F. Basolo and R. G. Pearson, "Mechanisms of Inorganic Reactions," 1st edition, Wiley, New York, 1958, p. 284.
[55] T.-M. Hseu, D. F. Martin, and T. Moeller, *Inorg. Chem.*, **2**, 587 (1963).
[56] T. M. Lowry and R. C. Traill, *Proc. Roy. Soc., Ser.A*, **132**, 398 (1931).

SIX-COORDINATE COMPLEXES

The most elegant proof that many complexes are octahedral came from Werner's stereochemical studies, primarily on cobalt compounds. At the time Werner began work in this area, several isomeric series of salts of the type Coa_4b_2 were known. The Blomstrand-Jørgensen explanation for the existence of isomers of $[Co(NH_3)_4Cl_2]Cl$, for example, was that cobalt possessed three types of valence—α, β, and γ— and that the two isomers of the compound could be formulated as **39a** and **39b**. However,

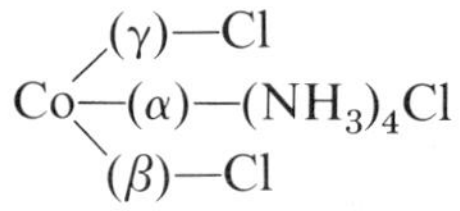

$$\text{Co} \begin{matrix} (\gamma)\text{—Cl} \\ (\alpha)\text{—}(NH_3)_4Cl \\ (\beta)\text{—Cl} \end{matrix} \qquad \text{Co} \begin{matrix} (\gamma)\text{—Cl} \\ (\alpha)\text{—Cl} \\ (\beta)\text{—}(NH_3)_4Cl \end{matrix}$$

39a 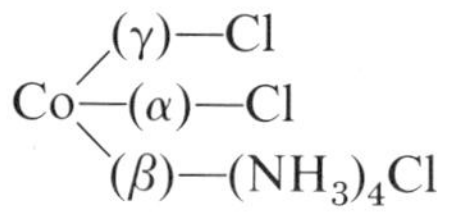   **39b**

Werner proved that such compounds were actually the *cis* and *trans* isomers **40a** and **40b**. Further, he found that only two isomers could be isolated for complexes of this type, a fact that the Blomstrand-Jørgensen approach could not rationalize.[57]

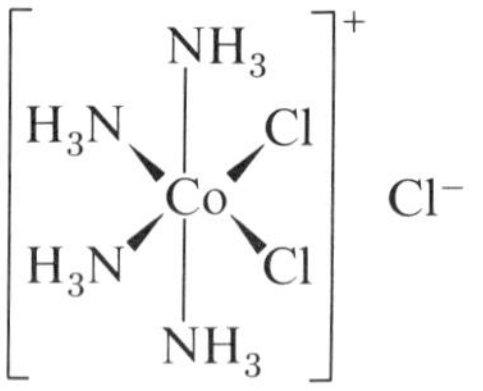

40a 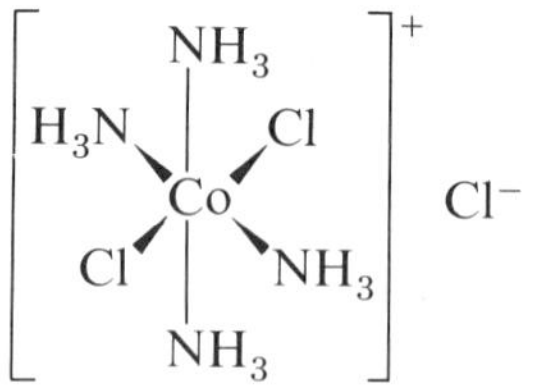   **40b**

Perhaps the best proof of the overwhelming importance of the octahedron came with the resolution of *cis*-amminechlorobis(ethylenediamine)cobalt(2+) salts into their enantiomers (**41a** and **41b**), a feat accomplished in 1911 by an American, King, working in Werner's laboratory.[58]

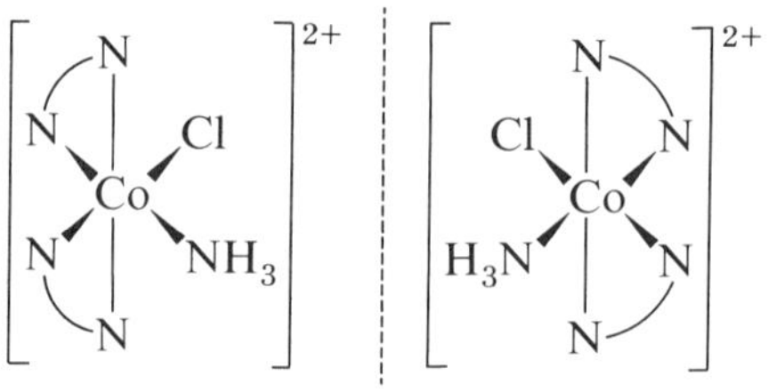

41a   **41b**

Some detractors of Werner's work, however, argued that the optical activity observed for these enantiomers arose in some mysterious way from the organic ligand. To answer these critics, therefore, Werner resolved into its enantiomers the completely inorganic complex ion tris[tetraammine-μ-dihydroxocobalt]cobalt(6+) (**42**); the inorganic ligand in this case clearly cannot be the center of chirality.

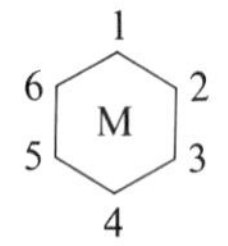

As you have seen from the brief discussion thus far, both "geometrical" and "optical" isomerism are possible for octahedral complexes, and there are several different ways of introducing chirality into six-coordinate complexes just as there are for four-coordinate ones. Indeed, there are six ways for octahedral complexes to be dissymmetric:[59]

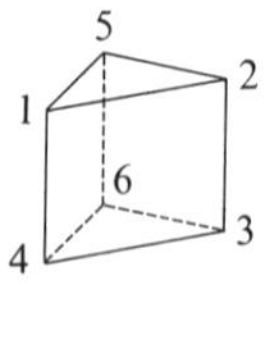

---

[57] Another consequence of this is that planar and trigonal prismatic geometries are virtually ruled out, as three isomers should be possible for Ma_4b_2 in each of these geometries. For a hexagonal planar geometry the isomers would have b in the 1,2, and 1,3, or the 1,4 positions, and the same would be true for a trigonal prismatic complex (see structures at left).

[58] A. Werner, *Berichte der Deutschen Chemischen Gesellschaft*, **44**, 1887 (1911); see page 159 of ref. 8.

[59] See ref. 28c.

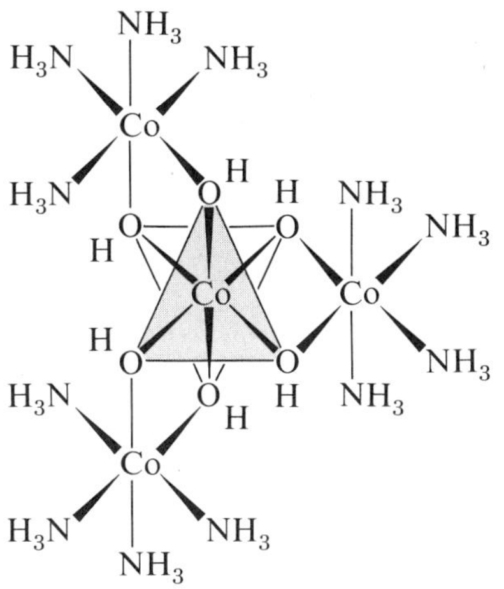

42

a. The distribution of unidentate ligands about the central metal.
b. The distribution of chelate rings about the central metal.
c. The coordination of unsymmetrical multidentate ligands.
d. The conformation of chelate rings.
e. The coordination of a chiral ligand.
f. The coordination of a donor atom that is asymmetric.

Our plan is to discuss first types a, b, and c along with ways of systematically enumerating the isomers that may arise from a given situation. Finally, we shall take up the more complicated forms of chirality along with the special nomenclature of chiral coordination complexes and the determination of their absolute configuration.

Isomerism from Ligand Distribution and Unsymmetrical Ligands; Isomer Enumeration

As you shall see in Chapter 17, alkyl halides undergo so-called "oxidative addition" reactions with square planar metal complexes such as $IrCl(CO)(PPh_3)_2$ to give octahedral products.[60] Although a single isomer is isolated from the particular reaction below,[61] eight stereoisomers are possible: six diastereoisomers, two with

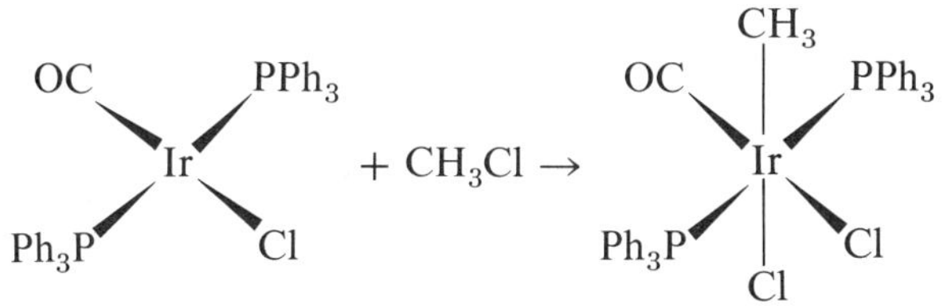

43 + CH_3Cl → **44**

enantiomers. Although it is possible to define these eight isomers by systematically drawing their structures, many people have attempted to describe more "foolproof" methods of isomer enumeration. Perhaps the most successful of these was developed by Bailar.[62,63] To illustrate this method, we shall derive the possible geometries for

---

[60] Oxidative addition reactions involving non-metals were discussed in Chapter 7.
[61] J. P. Collman and C. T. Sears, Jr., *Inorg. Chem.*, **7**, 27 (1968).
[62] J. C. Bailar, Jr., *J. Chem. Educ.*, **34**, 334 (1957); see also S. A. Meyper, *ibid.*, **34**, 623 (1957).
[63] Professor John C. Bailar, Jr., was born in 1904 in Golden, Colorado; he received his B.A. (1924) and M.A. (1925) degrees from the University of Colorado and his Ph.D. (in organic chemistry) from the University of Michigan (1928). He has served on the faculty at Illinois since that time, and he has contributed immeasurably to coordination chemistry, both through his own work and through the work of students trained in his laboratory. A symposium, consisting of papers given by former students of Professor Bailar, was held in 1969. The papers have been collected in a single volume: S. Kirschner, ed., "Coordination Chemistry," Plenum Press, New York, 1969.

630 ☐ COORDINATION CHEMISTRY: ISOMERISM

	L	M	N
1	ab cd ef	ab ce df	ab cf de
2	ac bd ef	ac be df	ac bf de
3	ad bc ef	ad be cf	ad bf ce
4	ae bc df	ae bd cf	ae bf cd
5	af bc de	af bd ce	af be cd

TABLE 11-2
ENUMERATION OF THE ISOMERS POSSIBLE FOR A COMPLEX OF THE TYPE M$abcdef$

the diastereoisomers of molecules having only monodentate ligands and symmetrical and unsymmetrical, bidentate ligands.

The basis of the Bailar method is Table 11-2, a chart that lists systematically all of the diastereoisomers possible for a complex having six different monodentate ligands, M$abcdef$, of which [Pt(py)(NH$_3$)(NO$_2$)(Cl)(Br)(I)] (**23**) is the only known example. Such a complex will have the maximum number of isomers possible for an octahedral complex: fifteen diastereomers, each with an enantiomer. Since replacement of some of these ligands by duplicates or by multidentate ligands will result in more "symmetrical" molecules with fewer numbers of isomers, Table 11-2 can later be adapted to all other situations.

Table 11-2 is built up in the following manner:
(i) Each different ligand is assigned a different letter, $a \ldots f$.

$$[\text{Pt}(\text{py})(\text{NH}_3)(\text{NO}_2)(\text{Cl})(\text{Br})(\text{I})]$$
$$\text{M} \quad a \quad\; b \quad\;\; c \quad d \;\; e \;\; f$$

(ii) Taking ligand a as a fixed point of reference, we place the remaining five ligands in the other possible coordination positions to give, for example, **45**. To denote the configuration of **45**, the letters for the ligands *trans* to one another are grouped alphabetically in pairs (ab, cd, and ef), and these pairs of letters are listed one under the other in alphabetical order. The resulting symbol ($ab/cd/ef$) is found in the first row (= 1) and first column (= L) of Table 11-2 and is therefore identified as "1L".

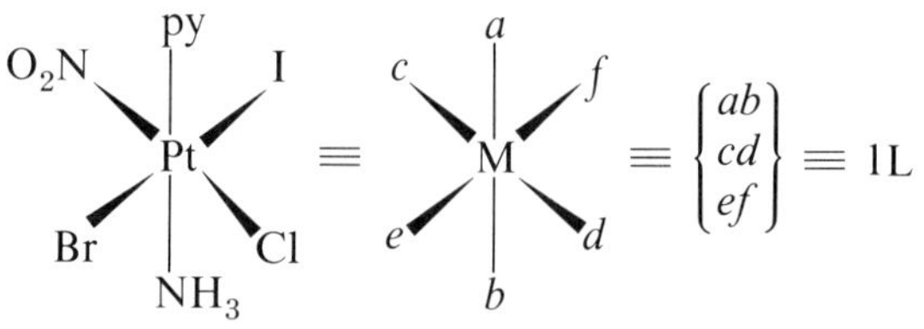

$$\underset{\mathbf{45}}{\begin{array}{c}\text{py}\\ \text{O}_2\text{N}\diagdown \;\; | \;\; \diagup \text{I}\\ \text{Pt}\\ \text{Br}\diagup \;\; | \;\; \diagdown \text{Cl}\\ \text{NH}_3\end{array}} \equiv \begin{array}{c}a\\ c\diagdown \;\; | \;\; \diagup f\\ \text{M}\\ e\diagup \;\; | \;\; \diagdown d\\ b\end{array} \equiv \left\{\begin{array}{c}ab\\ cd\\ ef\end{array}\right\} \equiv 1\text{L}$$

(iii) Another isomer of M$abcdef$ can be generated by keeping the ab pair of ligands fixed and interchanging d and e. This gives structure **46**. The pairs of *trans* ligands are now ab, ce, and df. The symbol for this isomer is identified as 1M in Table 11-2.

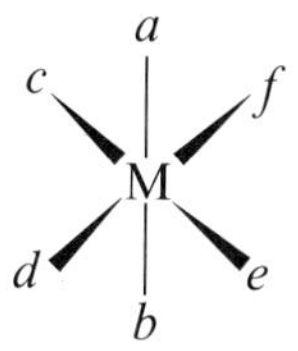

$$\begin{Bmatrix} ab \\ ce \\ df \end{Bmatrix} \equiv 1\text{M}$$

46

(iv) Again keeping the *ab* pair of ligands fixed in position, only one more isomer can be generated. That is, starting with **45** and interchanging *d* and *f* leads to **47**. This isomer is identified as 1N in Table 11-2.

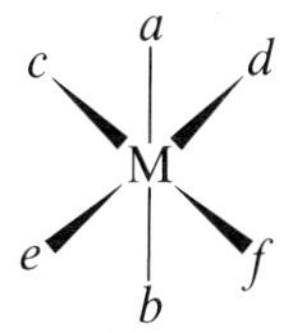

$$\begin{Bmatrix} ab \\ cf \\ de \end{Bmatrix} \equiv 1\text{N}$$

47

(v) In the steps above, we generated the three isomers that are possible when *a* and *b* are left fixed and the remaining ligands are permuted among the other coordination positions. Now, if *b* is replaced by *c*, for example, and the other four ligands (*b, d, e,* and *f*) are permuted among the remaining positions, three isomers will again be generated; they are those of the second row in this case (*e.g.*, **48**). Continuing this process of replacing the ligand *trans* to *a* and permuting the remaining four ligands will lead to the isomers of the third, fourth, and fifth rows of the Table. Thus, for each of the five possible *ax* combinations of *trans* ligands (which are presented by the five rows of the table), there are three possible isomers (represented by the three columns of the table).

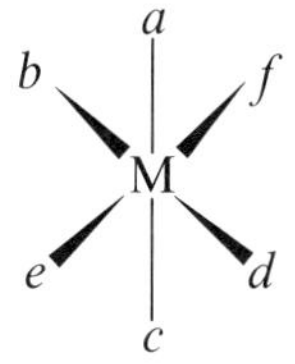

$$\begin{Bmatrix} ac \\ bd \\ ef \end{Bmatrix} \equiv 2\text{L}$$

48

Table 11-2 lists all of the 15 diastereoisomers possible for M*abcdef*. However, it gives no indication that each of these has an enantiomeric form (*e.g.*, **45a′** and **45b′**).

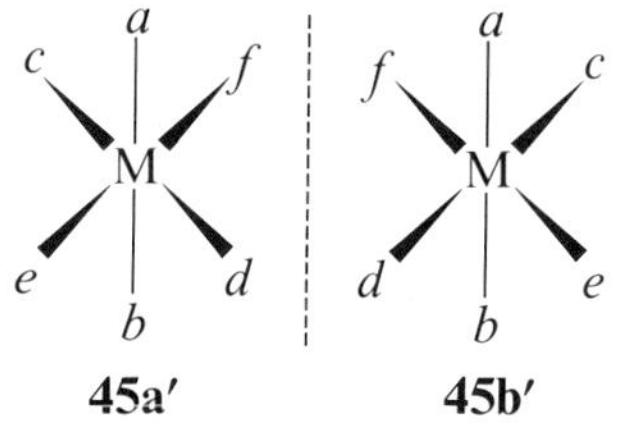

45a′   **45b′**

In using Table 11-2 and in adapting it to other complexes, you should be aware of the fact that only the possible diastereoisomers for a given complex will be generated. Each diastereomer will have to be inspected to see if it is dissymmetric.

You are now in a position to apply Table 11-2 to other complexes, **44** for example. Since there are two sets of identical ligands, we shall identify both ligands (Cl) of one set by *a* (*i.e.*, *b* = *a* in M*abcdef*) and both ligands of the other set (PPh$_3$)

632 ☐ COORDINATION CHEMISTRY: ISOMERISM

	L	M	N
1	aa cc ef	aa ce cf	[aa cf ce]
2	ac ac ef	ac ae cf	ac af ce
3	[ac ac ef]	[ac ae cf]	[ac af ce]
4	[ae ac cf]	[ae ac cf]	ae af cc
5	[af ac ce]	[af ac ce]	[af ae cc]

TABLE 11-3
ENUMERATION OF THE ISOMERS POSSIBLE FOR A COMPLEX OF THE TYPE Ma_2c_2ef. Symbols in [ ] are duplicates of other configurations in the table.

by c (i.e., $d = c$). Only ligands e and f (i.e., CO and CH_3) are unique. Therefore, we substitute a for b and c for d in Table 11-2 and generate Table 11-3, the latter an

$Mabcdef = Ir(Cl)_2(PPh_3)_2(CO)(CH_3) = M(ab)(cd)ef$
but since $a = b$ and $c = d$. $Ir(Cl)_2(PPh_3)_2(CO)(CH_3) = Ma_2c_2ef$

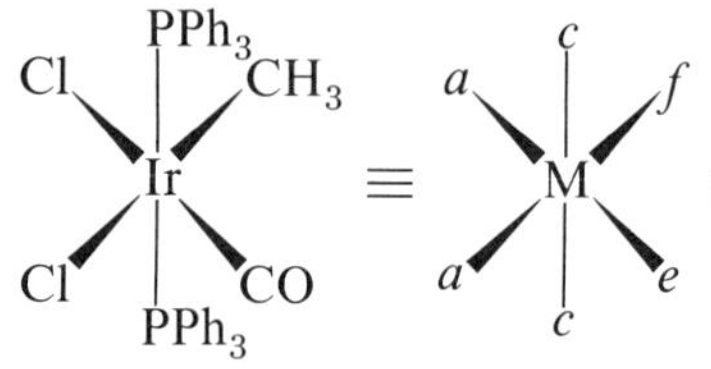

$\equiv \begin{Bmatrix} ae \\ cc \\ af \end{Bmatrix} = 4N$ in Table 11-3

enumeration of the diastereoisomers possible for Ma_2c_2ef. It is clear that duplicate configurations will be listed in Table 11-3, because there must be fewer isomers of Ma_2c_2ef than of $Mabcdef$. The duplicates are indicated in Table 11-3 by square brackets and are as follows:

$$1M = 1N \qquad 2M = 3M, 4L, \text{ and } 4M \qquad 4N = 5N$$
$$2L = 3L \qquad 2N = 3N, 5L, \text{ and } 5M$$

This leaves only six distinguishable configurations or diastereoisomers. On drawing these, you see that two of them are clearly dissymmetric and have enantiomeric forms.

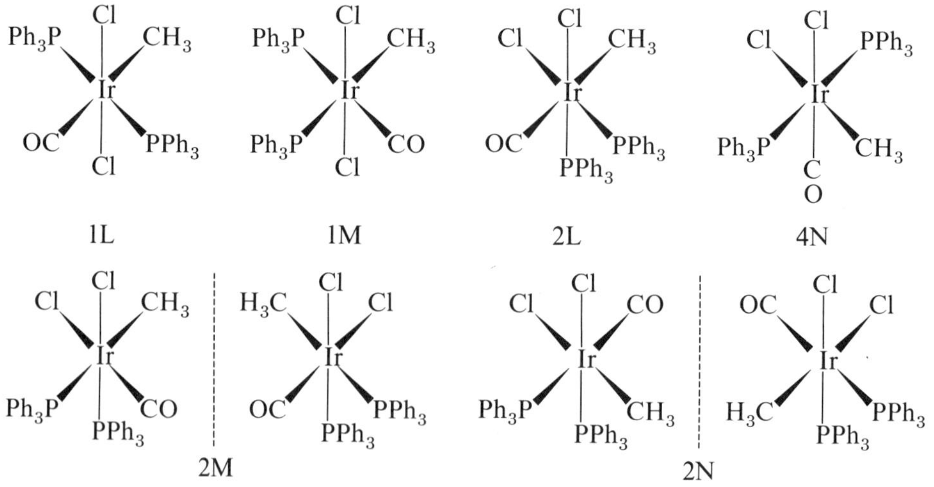

1L   1M   2L   4N

2M   2N

TABLE 11-4
ENUMERATION OF THE ISOMERS POSSIBLE FOR A COMPLEX OF THE TYPE M(AA)c_2ef (WHERE AA IS A SYMMETRICAL BIDENTATE LIGAND). Symbols in [ ] are duplicates of other configurations in the table. Configurations in ( ) are eliminated because they imply that the bidentate ligand spans *trans* positions.

	L	M	N
1	$\begin{pmatrix} AA \\ cc \\ ef \end{pmatrix}$	$\begin{pmatrix} AA \\ ce \\ cf \end{pmatrix}$	$\begin{pmatrix} AA \\ cf \\ ce \end{pmatrix}$
2	Ac Ac ef	Ac Ae cf	Ac Af ce
3	$\begin{bmatrix} Ac \\ Ac \\ ef \end{bmatrix}$	$\begin{bmatrix} Ac \\ Ae \\ cf \end{bmatrix}$	$\begin{bmatrix} Ac \\ Af \\ ce \end{bmatrix}$
4	$\begin{bmatrix} Ae \\ Ac \\ cf \end{bmatrix}$	$\begin{bmatrix} Ae \\ Ac \\ cf \end{bmatrix}$	Ae Af cc
5	$\begin{bmatrix} Af \\ Ac \\ ce \end{bmatrix}$	$\begin{bmatrix} Af \\ Ac \\ ce \end{bmatrix}$	$\begin{bmatrix} Af \\ Ae \\ cc \end{bmatrix}$

To extend the method to chelating ligands is not difficult. Consider, for example, the hypothetical ion [Co(en)(NH$_3$)$_2$(Cl)(H$_2$O)]$^{2+}$ **(49)**. In this case the chelating ligand

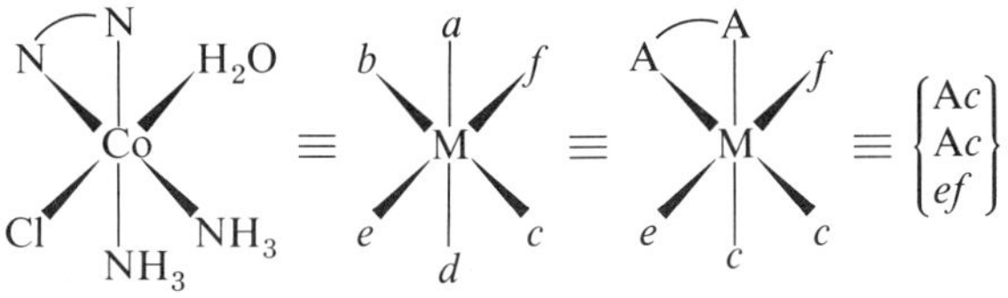

$$\text{49'} \equiv \text{M}(b,f,e,d,a,c) \equiv \text{M}(A,f,e,A,c) \equiv \begin{Bmatrix} Ac \\ Ac \\ ef \end{Bmatrix} \equiv \text{2L and 3L in Table 11-4}$$

[Co(en)(NH$_3$)$_2$(Cl)(H$_2$O)]$^{2+}$ = M(AA)c_2ef

will be represented by upper case letters and the monodentate ligands by lower case letters. The complex may be related to M*abcdef* by equating *a* and *b* to AA and *c* and *d* to *c*. Making the appropriate substitutions in Table 11-2, Table 11-4 is generated. The configurations in the first row can be eliminated immediately, since they imply that the bidentate ligand spans *trans* positions. Of the remaining configurations, the following equivalencies are noted:

2L = 3L   2M = 3M, 4L, and 4M   2N = 3N, 5L, and 5M   5N = 4N

This leaves a total of four diastereoisomers, two of which have enantiomers.

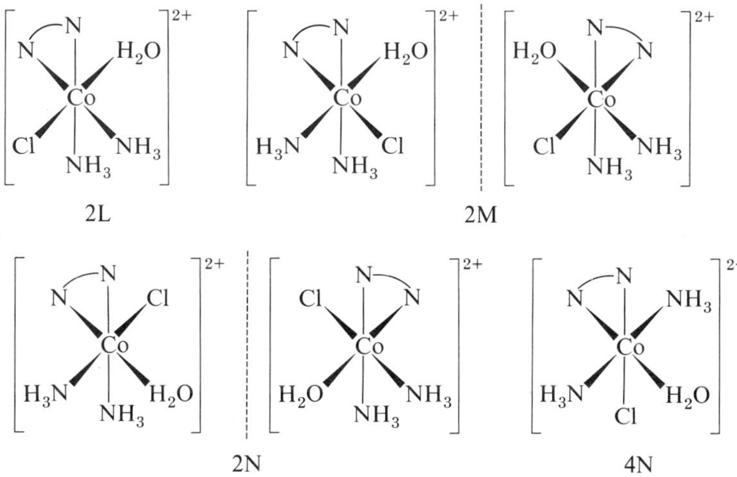

[structures 2L, 2M, 2N, 4N shown]

634 ☐ COORDINATION CHEMISTRY: ISOMERISM

	L	M	N
1	(AB / A'B' / ee)	(AB / A'e / B'e)	(AB / A'e / B'e)
2	AA' / BB' / ee	AA' / Be / B'e	[AA' / Be / B'e]
3	AB' / BA' / ee	AB' / Be / A'e	[AB' / Be / A'e]
4	[Ae / BA' / B'e]	Ae / BB' / A'e	(Ae / Be / A'B')
5	[Ae / BA' / B'e]	[Ae / BB' / A'e]	(Ae / Be / A'B')

TABLE 11-5
ENUMERATION OF THE ISOMERS POSSIBLE FOR A COMPLEX OF THE TYPE M(AB) (A'B')e_2 WHERE AB AND A'B' ARE TWO IDENTICAL UNSYMMETRICAL BIDENTATE LIGANDS. Symbols in [ ] are duplicates of other configurations in the table. Configurations in ( ) are eliminated because they imply that the bidentate ligand spans *trans* positions.

The third way to obtain dissymmetric complexes is to use unsymmetrical multidentate ligands, and the enumeration of the isomers of complexes containing such ligands can also be treated by Bailar's method. For example, in a complex of the glycinate ion, $Cu(H_2NCH_2COO)_2(H_2O)_2$, you must take into account the fact that, while the nitrogen and oxygen of one chelate must be *cis* to one another, they can be *trans* to the oxygen and nitrogen, respectively, of the other ligand. Therefore, in substituting symbols into Table 11–2, you must differentiate the two bidentate ligands. To accomplish this, represent the copper complex by M(AB)(A'B')e_2 where AB = a and b, A'B' = c and d, and $e = e$ and f. In the resulting table (Table 11–5), 1L, 1M, 1N, and

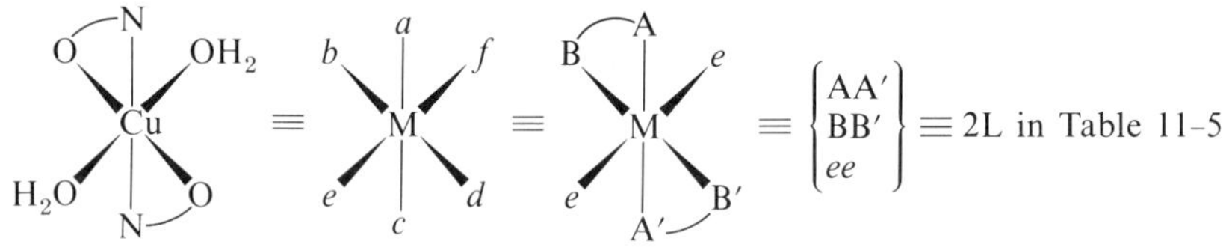

$$Cu(H_2NCH_2COO)_2(H_2O)_2 \equiv M(AB)(A'B')e_2 \equiv \begin{Bmatrix} AA' \\ BB' \\ ee \end{Bmatrix} \equiv 2L \text{ in Table 11-5}$$

50

4N (= 5N) can be eliminated immediately, as they involve a ligand spanning *trans* positions. Further, the following equivalencies are noted:

    2M = 2N   3M = 4L and 5L (because $A = A'$, $B = B'$‡)
    3M = 3N   5M = 4M

This leaves only five diastereoisomers, three of which have enantiomers, as seen below.

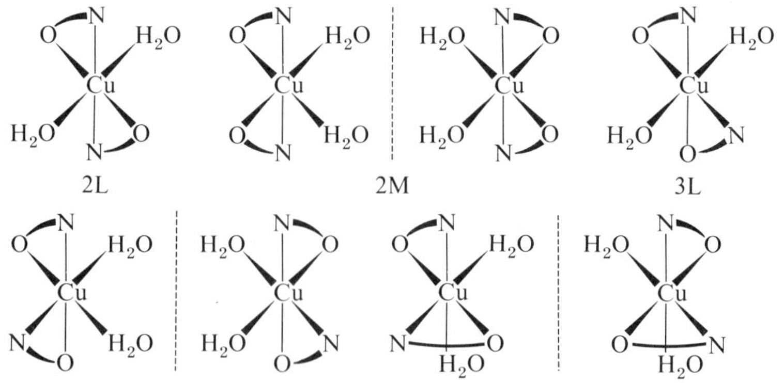

    2L   2M   3L

    3M   **50**   4M

---

‡Notice that 4L and 5L are enantiomers of 3M.

TABLE 11-6
ISOMERS OF OCTAHEDRAL COMPLEXES[a,b]

General Formula	Total Number of Stereoisomers	Pairs of Enantiomers
Ma_6	1	0
Ma_5f	1	0
Ma_4e_2	2	0
Ma_3d_3	2	0
Ma_4ef	2	0
Ma_3def	5	1
Ma_2cdef	15	6
$Mabcdef$	30	15
$Ma_2c_2e_2$	6	1
Ma_2c_2ef	8	2
Ma_3d_2f	3	0
M(AA)(BC)ef	10	5
M(AB)(AB)ef	11	5
M(AB)(CD)ef	20	10
M(AB)(AB)(AB)	4	2
M(ABA)def	9	3
M(ABC)(ABC)	11	5
M(ABBA)ef	7	3
M(ABCBA)f	7	3

[a] Lower case letters indicate monodentate ligands and upper case letters represent the donor atoms of chelating ligands.
[b] Table compiled from the following sources: W.E. Bennett, *Inorg. Chem.*, **8**, 1325 (1969); B.A. Kennedy, D.A. McQuarrie, and C.H. Brubaker, *Inorg. Chem.*, **3**, 265 (1964).

Bailar's method can clearly be extended to more complex cases, and a computer program has been devised to simplify the task.[64] Results for other cases of interest have been obtained by using this program and are included in Table 11-6.

---

[64] W. E. Bennett, *Inorg. Chem.*, **8**, 1325 (1969).

---

7. *Cis* and *trans* isomerism is possible for a four-coordinate, tetrahedral complex if an unsymmetrical ligand such as **R**-1,2-diaminopropane is used. Make a sketch or build a model of a complex of the type $(H_2N-CH_2CH(NH_2)CH_3)_2M$ to show how this is possible.

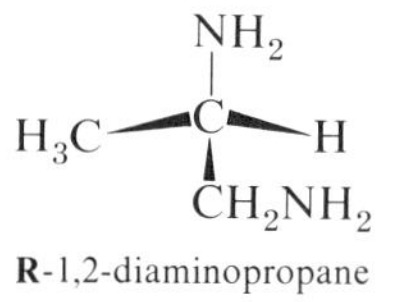

$$H_3C - \underset{CH_2NH_2}{\overset{NH_2}{\underset{|}{\overset{|}{C}}}} - H$$

R-1,2-diaminopropane

8. Table 11-6 lists the number of stereoisomers possible for a number of different types of octahedral complexes. Verify the data for each of the following and draw the structure of each isomer generated.

   a) Ma_3def
   b) Ma_2cdef
   c) M(AB)(AB)ef

9. For each of the complexes listed below, derive the number of stereoisomers possible and indicate which are chiral. Draw the structure for each isomer.

STUDY QUESTIONS

a) M(ABA)(CDC)   (two non-identical, tridentate ligands are attached to the metal)
b) M(ABCA)e_2 (a tetradentate ligand and two monodentate ligands are attached to the metal ion)

10. The ligand 1,8-diamino-3,6-dithiaoctane, $H_2N-C_2H_4-S-C_2H_4-S-C_2H_4-NH_2$, forms numerous complexes with the Co(III) ion. Draw the isomers possible for a complex such as $[Co(NSSN)Cl_2]^+$, carefully showing each enantiomer and diastereomer.

11. The square antiprism is often formed by eight-coordinate complexes. As discussed in Chapter 10, and as illustrated below, the square antiprism has two types of edges, l and s. How many stereoisomers will a square antiprismatic complex of the type $M(AA)_4$ have (AA is a symmetrical bidentate ligand)? Indicate which are enantiomeric. For a recent paper on such complexes, see T. J. Pinnavaia, B. L. Barnett, G. Podolsky, and A. Tulinsky, *J. Amer. Chem. Soc.*, **97**, 2712 (1975).

12. In their very important paper on ligand conformation, Corey and Bailar [*J. Amer. Chem. Soc.*, **81**, 2620 (1959)] stated that "there are 20 possible isomers of the ion $[Co(stien)_3]^{3+}$ (stien = stilbenediamine)." This statement has recently been challenged by R. E. Tapscott, *Inorg. Chem.*, **14**, 216 (1975). He noted that stilbenediamine

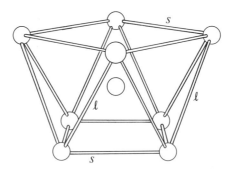

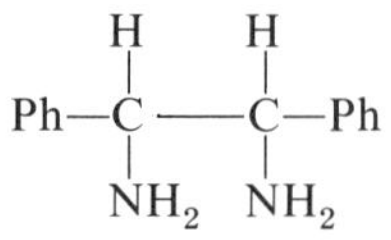

has two chiral carbon atoms. Therefore, the ligand may bind to the cobalt(III) in the **R,R**; **R,S**; or **S,S** form, and because of this many more than 20 stereoisomers are possible. How many are there?

Isomerism from Ligand Conformation and Chirality[65]

We turn now to the last three ways described on page 629 for a coordination complex to be dissymmetric.[66] One of these is to attach a multidentate ligand so that the chelate rings thereby formed can exist in more than one conformation.[67] For example, ethylenediamine bound to a metal ion creates a five-membered, puckered ring with a two-fold axis of symmetry (through M and the mid-point of the C—C bond) as shown in Figure 11-3. Two conformations are clearly possible for the

---

[65] In order to make sense out of this section and those that follow, you are *strongly* urged to use the best available set of molecular models. "Framework Molecular Models" by Prentice-Hall are adequate, but Dreiding models, distributed in the U.S.A. by Brinkmann Instruments, are highly recommended.
[66] See ref. 28c.
[67] A. M. Sargeson, "Conformations of Coordinated Ligands," in R. L. Carlin, ed., "Transition Metal Chemistry," Vol. 3, Marcel Dekker, New York, 1966, pp. 303–343.

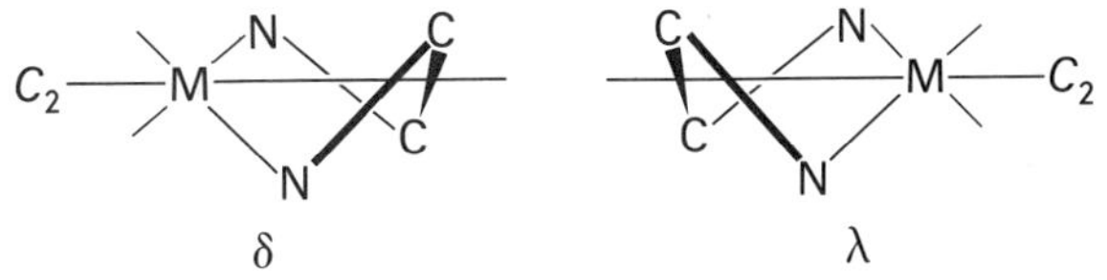

Figure 11–3. The enantiomeric conformations of a diamine chelate ring.

ring—one labeled λ and the other δ.[68,69] It is also clear from Figure 11–3 that these conformations are non-superimposable mirror images. However, they have little conformational stability (*i.e.*, a low barrier to ring inversion), and it is not possible to resolve a complex such as $[Co(en)(NH_3)_4]^{3+}$. The situation is somewhat different, though, for complexes with two or three ethylenediamine rings. For example, for a complex with two ethylenediamine rings *trans* to one another, there are three stereochemical possibilities: the enantiomeric and energetically equivalent δδ and λλ forms [only the latter is shown in Figure 11–4(A)] and the diastereomeric λδ form [Figure 11–4(B)]. On the basis of an estimation of the interaction between the

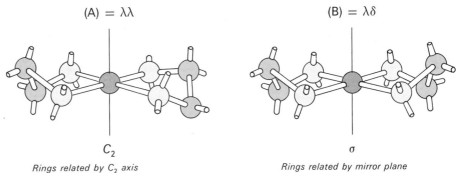

(A) = λλ
C_2
Rings related by C_2 axis
H's on adjacent N's are staggered

(B) = λδ
σ
Rings related by mirror plane
H's on adjacent N's are eclipsed

Figure 11–4. Conformation of chelate rings in a planar bidentate ethylenediamine complex. [Reprinted with permission from E. J. Corey and J. C. Bailar, Jr., *J. Amer. Chem. Soc.*, **81**, 2620 (1959). Copyright by the American Chemical Society.]

hydrogens of adjacent —NH_2 groups, Corey and Bailar suggested that the λλ and δδ forms should be more stable than the λδ conformation by about 1 kcal; in the λδ conformation the adjacent —NH_2 hydrogens are eclipsed, whereas they are staggered in the other two conformations. In reality, however, the vast majority of the *trans*-bis(ethylenediamine) complexes are found in the λδ conformation in the solid state, a fact that reflects the small energy difference between the conformations and, perhaps, the importance of crystal packing factors. Several other examples of ring conformation will be discussed in the next section.

Yet another way to introduce chirality into a coordination complex is to use a chiral ligand.[70] For example, optical activity is observed for complexes of the type $[Co(NH_3)_5(S\text{—}amH)]^{3+}$ **(51)**, where **S—amH** is the zwitterionic form of an amino acid such as S-alanine **(52)**.

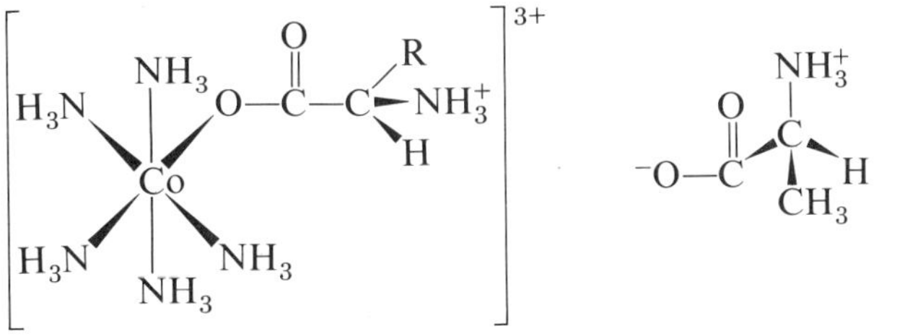

51
52

---

[68] E. J. Corey and J. C. Bailar, Jr., *J. Amer. Chem. Soc.*, **81**, 2620 (1959).

[69] The symbols δ and λ are explained in the section on "Chirality and the Special Nomenclature of Coordination Compounds" that follows. Note, however, that in the nomenclature of Corey and Bailar (see ref. 68) λ = k and δ = k'.

[70] See ref. 28c, pp. 176–195.

CHIRALITY AND THE SPECIAL NOMENCLATURE OF CHIRAL COORDINATION COMPLEXES

The situation regarding the nomenclature of chiral molecules is both confused and confusing. Numerous systems have been proposed, some of which have symbolism similar to others but use it in an opposite sense. The systems most carefully worked out are those proposed by Hawkins,[71] by the IUPAC Commission,[72] and by *Chemical Abstracts*.[73] The latter two are essentially complementary, and we shall adopt their approach. In summary, the nomenclature system used in this book is as follows:

a. Δ and Λ will be used to describe the absolute configuration of a dissymmetric complex when the dissymmetry is due to chelate ring distribution. As described below, Δ is used when the rings describe a right-handed helix and Λ when they describe a left-handed helix.[74] For octahedral complexes not having bidentate ligands, the C and A nomenclature system introduced on page 625 is used.

b. δ and λ are used to describe the conformations of individual chelate rings. δ denotes a ring of right-handed helicity and λ denotes a ring of left-handed helicity.[75]

c. **R** and **S** are used to describe the absolute configurations of dissymmetric ligands according to the Cahn-Ingold-Prelog (CIP) system now commonly used in organic chemistry.[76]

d. $(+)_\lambda$ and $(-)_\lambda$ are used to designate the sign of rotation of plane polarized light at the specified wavelength, λ. When the sign appears without the wavelength noted, the wavelength is assumed to be that of a sodium D line (5890 and 5896 Å).

Many things natural and unnatural are chiral. The former is illustrated by the words of a song from the delightful Broadway show, "At the Drop of a Hat."

"Misalliance" from "At the Drop of a Hat"
by Michael Flanders and Donald Swan

The fragrant honeysuckle spirals clockwise to the sun,
And many other creepers do the same.
But some climb anti-clockwise,
The bindweed does for one (or convolvulus to give her proper name).

Rooted on either side a door, one of each species grew,
And raced toward the window ledge above.
Each corkscrewed to the lintel in the only way it knew,
Where they stopped, touched tendrils, smiled, and fell in love.

Said the right-handed honeysuckle to the left-handed bindweed,
"Oh, let us get married if our parents don't mind.
We'd be loving and inseparable, inextricably entwined.
We'd live happily ever after," said the honeysuckle to the bindweed.

To the honeysuckle's parents it came as a great shock.
"The bindweeds," they cried, "are inferior stock.
They are uncultivated, of breeding bereft.
We twine to the right and they twine to the left."

Said the anti-clockwise bindweed to the clockwise honeysuckle.
"We'd better start saving, many a mickle mak's a muckle.
Then run away for a honeymoon and hope that our luck'll
Take a turn for the better," said the bindweed to the honeysuckle.

---

[71] See ref. 28c, pp. 20–31.
[72] *Inorg. Chem.*, **9**, 1 (1970).
[73] See ref. 41.
[74] Hawkins (see ref. 28c) uses D and L for Λ and Δ, respectively.
[75] The symbols originally suggested by Corey and Bailar (see ref. 68), k and k', are equivalent to λ and δ, respectively.
[76] See refs. 1, 2, 3, and 42.

A bee who was passing remarked to them then,
"I said it before and I'll say it again,
Consider your offshoots if offshoots there be,
They'll never receive any blessing from me."

Poor little sucker, how will it learn,
When it is climbing which way to turn,
Right - left - what a disgrace.
Or it may go straight up and fall flat on its face.

Said the right-hand thread honeysuckle,
To the left-hand thread bindweed,
"It seems that against us all fate has combined,
Oh my darling, oh my darling, oh my darling columbine,
Thou art lost and gone for ever,
We shall never intertwine."

Together they found them the very next day,
They had pulled up their roots and just shriveled away,
Deprived of that freedom for which we must fight,
To veer to the left or to veer to the right!

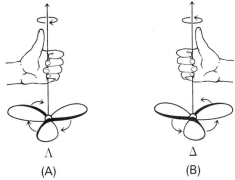

(A)   (B)

Illustration of the left- (Λ) and right-handedness (Δ) of enantiomeric three-bladed propellors.

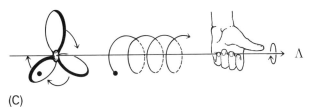

(C)

A left-handed propellor traces out a left-handed helix on traveling through space.

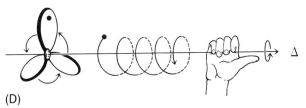

(D)

A right-handed propellor traces out a right-handed helix on traveling through space.

Figure 11–5. Illustration of the handedness of enantiomeric three-bladed propellors.

More germane to the topic of inorganic stereochemistry is the three-bladed propellor in Figure 11–5. These propellors are enantiomers of each other and as such can describe a left-handed (A) or right-handed (B) helix on rotating and moving through space, a fact that requires some explanation for those not familiar with airplane and boat propellors. Imagine that the propellor in Figure 11–5 is a dimestore pinwheel and that the "blades" of the pinwheel curl toward you; that is, the heavy edge in the figure is toward you when you look down the major axis of rotation. If you move the pinwheel quickly through space, the pinwheel will turn in the direction indicated. If you fix your eye on one point on one blade, that point will describe a helix as the propellor rotates while moving through space. However, the two propellors in Figure 11–5 are different in that they have different "handedness." One propellor (A) describes a left-handed helix on moving through space. That is, if you use the thumb of your left hand to indicate the direction of motion, the fingers of your left hand curl in the same direction as the motion of the propellor blades. By the same test, the propellor at the bottom of Figure 11–5 is right-handed.[77]

In exactly the same way that a propellor or pinwheel can have "handedness" and describe left- or right-handed helices, a molecule such as a tris(bidentate)metal complex can trace out a left- or right-handed helix on rotating and moving through space (Figure 11–6). It is on the basis of the opposite helicities of enantiomeric molecules with multidentate ligands that the IUPAC system of nomenclature is based. As noted above, molecules having right-handed helicity are labeled Δ and those with left-handed helicity are labeled Λ.

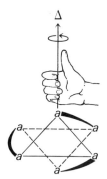 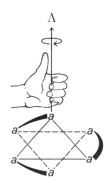

Figure 11–6. Illustration of the right- (Δ) and left-handed (Λ) helicity of a tris(bidentate) complex. The pseudo-threefold axis of the octahedron is used as the defining axis.

The approach described above works very well in describing the helicity of tris(bidentate)metal complexes. However, to decide on the helicity of a complex that has, for example, two bidentate ligands *cis* to one another, the more general approach described in the IUPAC report is more useful.[78] Figure 11–7(A) illustrates two helices, one left-handed (Λ) and one right-handed (Δ). The axis of propagation (no direction of propagation is implied) of either helix is AA, and BB is a tangent to the cylinder described by the helix. In a Δ or right-handed helix, the tangent BB is skew to AA in a left-to-right direction from bottom to top; the opposite is true of a Λ or left-handed helix. The way to apply this criterion to a chiral molecule is quite straightforward (Figure 11–8). The first step is to replace the bidentate ligands by lines, ignoring any asymmetry the ligands themselves may possess.[78] At least one pair

---

[77] If you still have difficulty telling the chirality or handedness of a propellor or a propellor-like molecule, try this. Imagine that you cup the propellor in your hand so that your hand is around the perimeter of the object, and your thumb is then parallel with the axis of rotation. Now imagine that you "walk" the fingers of this hand up the edge of one blade (from the back of the blade to the front), starting with your little finger and moving toward your first finger. If you can accomplish this operation with your left hand, the propellor will describe a left-handed helix or will have a left-handed screw axis; its chirality is given the symbol Λ. If, on the other hand, your right hand is required to carry out the "cupping and walking" exercise, the propellor is right-handed, and the object is given the chirality symbol Δ.

[78] A bidentate amino acid, for example, is asymmetric in that the donor atoms are nitrogen and oxygen, and not two nitrogen atoms as in a symmetric ligand such as ethylenediamine. Whether or not the ligand is asymmetric has nothing to do with determining the basic chirality of the coordination complex.

STEREOISOMERISM □ 641

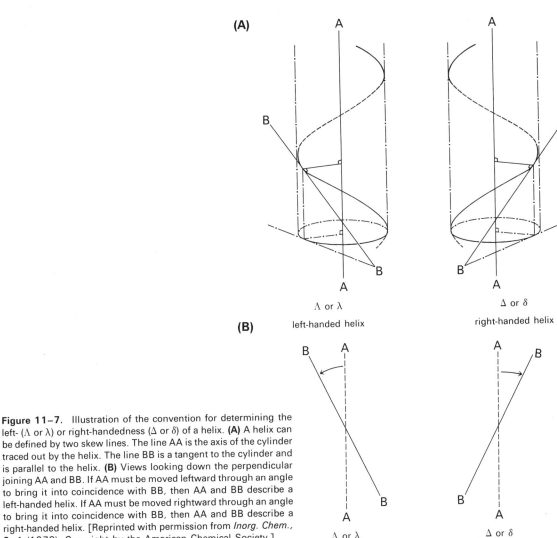

Figure 11-7. Illustration of the convention for determining the left- (Λ or λ) or right-handedness (Δ or δ) of a helix. **(A)** A helix can be defined by two skew lines. The line AA is the axis of the cylinder traced out by the helix. The line BB is a tangent to the cylinder and is parallel to the helix. **(B)** Views looking down the perpendicular joining AA and BB. If AA must be moved leftward through an angle to bring it into coincidence with BB, then AA and BB describe a left-handed helix. If AA must be moved rightward through an angle to bring it into coincidence with BB, then AA and BB describe a right-handed helix. [Reprinted with permission from *Inorg. Chem.*, **9**, 1 (1970). Copyright by the American Chemical Society.]

of lines will *always* be skew to one another in a chiral molecule and will describe a helix; the direction of skewness defines the direction of helicity.‡

It is also straightforward to extend the system described in Figure 11-7 to denote the conformation of chelate rings. If one looks down the two-fold rotational axis of an ethylenediamine-metal ring system, for example (Figure 11-3), the C—C bond and the imaginary line joining the two nitrogen donor atoms are skew and describe a helix, the handedness of the helix being defined by δ or λ (Figure 11-9).‡

In the preceding section we discussed the introduction of dissymmetry through chelate ring conformation. In addition to the case of *trans*-bis(bidentate) complexes that we previously discussed, Corey and Bailar described the interrelationship between *configuration (molecular helicity)* and *conformation (ring helicity)* for tris (bidentate) complexes.[79] If the ligands are thought of as planar, only two enantiomeric forms—Δ and Λ—are possible for a complex of the type $[M(en)_3]^{3+}$. However, owing to the fact that the ligands can have δ or λ conformations, there are now four pairs of enantiomers possible. That is, the configuration of the molecule may be Δ or Λ, and, for a given configuration, the ring conformations may be δδδ, δδλ, δλλ, or λλλ.

---

‡You need to notice that it makes no difference which line is called *AA* and which is called *BB*. To define helicity, it is necessary, however, that the *AA* line be placed *behind* the *BB* line.

[79]See ref. 68.

642 ☐ COORDINATION CHEMISTRY: ISOMERISM

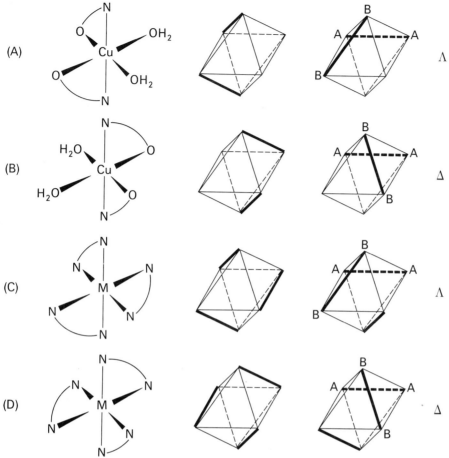

Figure 11-8. Illustration of the method for determining the helicity of chiral molecules. The first step is to replace the bidentate ligands by lines. At least one pair of lines will always be skew to one another in a chiral molecule. Therefore, as in Figure 11-7, one line will define the axis of a cylinder traced out by a helix (AA), and the other line will be the tangent to that helix (BB). AA and BB are designated on each molecule above. The helicity of the molecule is then defined by the criterion of Figure 11-7. Note that in the amino acid complexes (a,b) the absolute configuration would be the same whether the ligands are attached as shown or with the oxygens *trans*.

Since it is obvious that a δ ring conformation is the enantiomer of a λ conformation (Figure 11-3), the four isomer pairs are

$\Lambda\delta\delta\delta$	$\Delta\lambda\lambda\lambda$
$\Lambda\delta\delta\lambda$	$\Delta\lambda\lambda\delta$
$\Lambda\delta\lambda\lambda$	$\Delta\lambda\delta\delta$
$\Lambda\lambda\lambda\lambda$	$\Delta\delta\delta\delta$

Corey and Bailar discussed the $\Delta\lambda\lambda\lambda$ and $\Delta\delta\delta\delta$ forms in particular.[80] The former was called the *lel* form, because the C—C bonds are paral*lel* with the pseudo-threefold axis that passes through the octahedral face (Figure 11-10). The other form, $\Delta\delta\delta\delta$, is called the *ob* form, because the C—C bonds are *ob*lique to the three-fold axis. Corey and Bailar calculated that, for the Δ configuration, the $\lambda\lambda\lambda$ conformation was about 1.8 kcal/mole more stable than the $\delta\delta\delta$ form owing to nonbonded N—H···N—H interactions[81] (see Figure 11-4). X-ray studies have generally confirmed this view for

---

[80] Note that there is an error in the original paper. The upper drawing in Figure 3 in the Corey-Bailar paper is actually $\Lambda\delta\delta\delta$. The figure appears correctly in B. E. Douglas and D. H. McDaniel, "Concepts and Models of Inorganic Chemistry," Blaisdell Publishing Co., New York, 1965, page 373.

[81] For the enantiomeric Λ form the situation is reversed; that is, the $\delta\delta\delta$ conformation is more stable than the $\lambda\lambda\lambda$ conformation.

STEREOISOMERISM 643

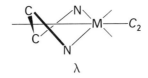
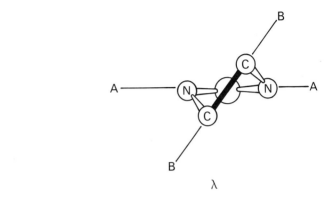
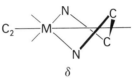
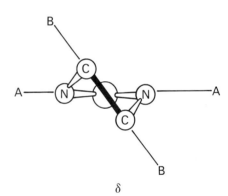

Figure 11-9. Illustration of the convention for designating the helicity of chelate ring conformations. λ = left-handed and δ = right-handed.

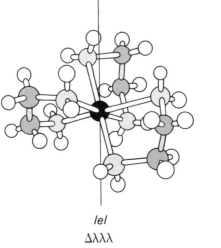
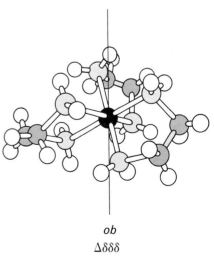

Figure 11-10. The $\lambda\lambda\lambda$ (= *lel*) and $\delta\delta\delta$ (= *ob*) conformations of the Δ configuration of a complex of the type $[M(en)_3]^{n+}$. [Reprinted with permission from E. J. Corey and J. C. Bailar, Jr., *J. Amer. Chem. Soc.*, **81**, 2620 (1959). Copyright by the American Chemical Society.]

lel
$\Delta\lambda\lambda\lambda$

ob
$\Delta\delta\delta\delta$

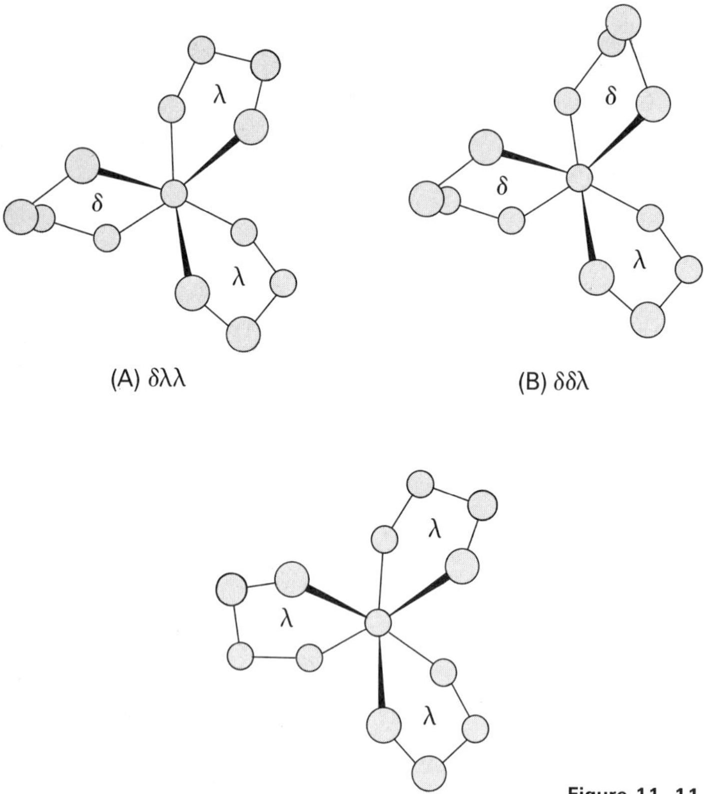

(A) δλλ

(B) δδλ

(C) λλλ

Figure 11-11. Perspective drawings of three conformations of Λ-[Cr(en)$_3$]$^{3+}$.

complexes in the solid state, but there are examples of other conformations. For instance, in [Cr(en)$_3$][Ni(CN)$_5$] · 1.5H$_2$O there are two crystallographically independent [Cr(en)$_3$]$^{3+}$ ions and a third such ion in [Cr(en)$_3$][Co(CN)$_6$] · 6H$_2$O.[82] In no case was the most stable Λδδδ conformation observed; instead, all three of the other possible conformations were observed in the Λ configuration (Figure 11-11).

In solution, however, the conformational situation is somewhat different.[83] Although the δ conformation is more stable than the λ conformer in the Λ configuration, there is rapid inversion between the two conformations. The result is that, because of a statistical entropy effect, the most abundant configuration in solution is Λδδλ rather than Λδδδ.

OPTICAL ACTIVITY, ORD, AND CD

A knowledge of the absolute configuration of a metal complex and the conformation of any chelate rings is of great interest to the chemist concerned with fundamental concepts of structure and bonding. Furthermore, such knowledge is of use to the reaction chemist, as it has been recognized for some years that many organic reactions are stereospecific, and the reactions of coordination compounds are no exception.[84,85] Therefore, methods for the determination of absolute configuration are of interest, and we turn to a very brief discussion of them now.

---

[82] K. N. Raymond, P. W. R. Corfield, and J. A. Ibers., *Inorg. Chem.*, **7**, 842 (1968).
[83] J. K. Beattie, *Acc. Chem. Res.*, **4**, 253 (1971).
[84] See ref. 38.
[85] One example of a stereospecific reaction is the use of stereospecific metabolism by a microorganism to resolve a racemic mixture into enantiomeric forms. This technique was first used by Pasteur when he found that *Penicillium glaucum* metabolized the **R**-(+) half of racemic ammonium tartrate. Another example is the isolation of one enantiomer of [Co(en)$_2$(phen)]$^{3+}$, which was cited on page 613.

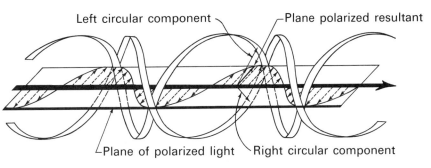

Figure 11-12. Plane polarized light is composed of left and right circularly polarized light as illustrated. Addition of the electric field vectors of each of the components at any instant gives a resultant field in a plane. [From E. J. Bair, "Introduction to Chemical Instrumentation," McGraw-Hill Book Co., New York (1962).]

The method of determining absolute configuration rests on three techniques: x-ray diffraction, optical rotatory dispersion (ORD), and circular dichroism (CD). Although you may well have had some introduction to these latter techniques in courses in instrumental analysis[86] or organic chemistry,[87] it is well worth the time and space here to at least outline their physical basis and the information available from them.[88]

Light consists of electric and magnetic fields oscillating at right angles to one another and at right angles to the direction of propagation. In ORD and CD one is concerned with three forms of light: plane, circularly, and elliptically polarized light. In plane polarized light the magnetic and electric field vectors are constrained to oscillate in perpendicular planes, the plane of polarization usually being taken as the plane of the electric field. Plane polarized light may actually be thought of as composed of two interfering, circularly polarized waves of equal amplitude and opposite rotation. In Figure 11-12 the right circularly polarized component traces out a right-handed helix, and the opposite is true for the left circularly polarized component. However, addition of the electric field vectors at any given instant gives an oscillating electric field in a plane. A chiral molecule is said to be optically active when it rotates the plane of polarized light. This occurs by a mechanism originally described by Fresnel in 1825. That is, while traversing a chiral medium—a quartz crystal or a test tube filled with one enantiomer of a substance—the velocity of propagation of one of the circularly polarized components of the plane polarized light is not equal to the velocity of the other.

---

[86] H. H. Willard, L. L. Merritt, Jr., and J. A. Dean, "Instrumental Methods of Analysis," 5th edition, D. Van Nostrand Co., New York, 1974.

[87] See refs. 1 and 3.

[88] If additional information on ORD and CD is desired, the following reviews are recommended in addition to the book by Hawkins (see ref. 28c).

a. F. Woldbye, "Technique of Optical Rotatory Dispersion and Circular Dichroism," in "Technique of Inorganic Chemistry," H. B. Jonassen and W. Weissberger, eds., Vol. 4, Interscience Publishers, New York, 1965.

b. R. D. Gillard, *Prog. Inorg. Chem.*, **7,** 215 (1966); R. D. Gillard, "Optical Rotatory Dispersion and Circular Dichroism," in "Physical Methods in Inorganic Chemistry," H. A. O. Hill and P. Day, eds., Interscience Publishers, New York, 1968.

c. J. Fujita and Y. Shimura, "Optical Rotatory Dispersion and Circular Dichroism," in "Spectroscopy and Structure of Metal Chelate Compounds," K. Nakamoto and P. J. McCarthy, eds., John Wiley and Sons, New York, 1968. The stereochemical notation used in this article is apparently a mix of several systems. Read with care!

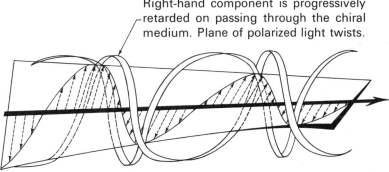

Figure 11-13. The right circularly polarized component of plane polarized light is retarded relative to the left circularly polarized component. The net result is a rotation of the plane of polarization to the right or clockwise as seen on looking back along the direction from which the light has come. [From E. J. Bair, "Introduction to Chemical Instrumentation," McGraw-Hill Book Co., New York (1962).]

Since the index of refraction of any medium is given by the ratio of the velocity of light in a vacuum to that in the medium

$$n = c_{vacuum}/c_{medium}$$

this means that the index of refraction of right circularly polarized light in a chiral medium will not be equal to that of left circularly polarized light. The difference between the indices of refraction, $n_{right} - n_{left}$, is known as *circular birefringence,* and the net result of this difference is a rotation of the plane of polarization of the light (Figure 11-13). The relationship between the angle of rotation, α (in degrees per decimeter), and the index of refraction is given by

$$\alpha = \frac{1800(n_{left} - n_{right})}{\lambda_{vac}}$$

The angle of rotation can be used to characterize a chiral object if it is measured at a specified wavelength or as a function of wavelength. The latter technique is useful, as α often changes in an especially dramatic way in the region of a molecular absorption band, particularly if the chromophore is also the center of chirality. The general effect of the change in α as a function of wavelength is called *optical rotatory dispersion,* and the effect on α in the region of an absorption band is *part* of what is known as the Cotton effect (after Aimé Cotton, who first observed the phenomenon in 1895) (see Figure 11-14). As an absorption band is approached from the long wavelength side, the rotation may increase, pass through a maximum, and become zero at the maximum of the absorption band. Passing on to shorter wavelengths, the opposite effect is observed.

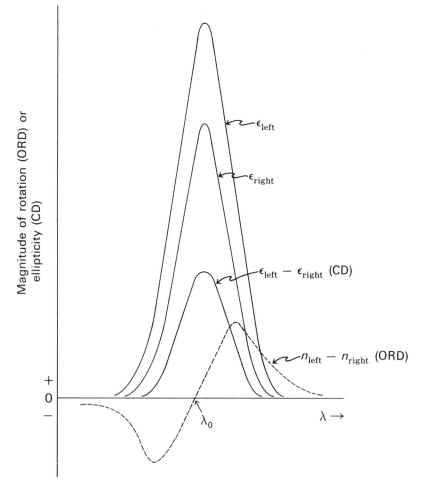

Figure 11-14. Idealized curves illustrating a *positive* Cotton effect in optical rotatory dispersion (ORD, -----) and circular dichroism (CD, ———). The ORD curve reflects the difference in the refractive indices of an enantiomer toward the left and right circularly polarized components of plane polarized light in the region of a molecular absorption band. The CD curve results from a difference in molar extinction coefficients of the enantiomer toward the two circularly polarized components. The center of the absorption band is given by λ_o.

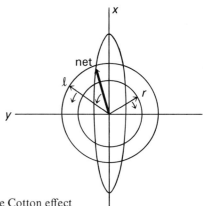

Figure 11–15. The cause of the circular dichroism effect. Left and right circularly polarized components of plane polarized light are traveling away from you in a chiral medium. The extinction coefficient of the enantiomer for the right-hand component (ε_r) is assumed greater than the left-hand component, and the right circularly polarized light is therefore absorbed to a greater extent. The net effect is that addition of the transmitted electric vectors l and r will give a resultant vector which traces out an ellipse. For the sake of simplicity here, it is assumed that the l and r components are in phase. In reality, however, they are not in phase in a chiral medium due to circular birefringence; although not illustrated in this Figure, the result is that the major axis of the ellipse is rotated by an angle α.

The particular case illustrated is called a positive Cotton effect; a negative Cotton effect would be observed if α were to become negative on moving from longer to shorter wavelengths. Since enantiomers have opposite chirality or helicity, they affect polarized light in opposite ways, and, therefore, have mirror image Cotton effects.

The molar extinction coefficient for a particular band is a measure of the ability of the molecule to absorb light *via* the transition giving rise to the band. Just as enantiomeric molecules have different indices of refraction toward right or left circularly polarized light, they have different molar extinction coefficients toward light rays of opposite chirality. This difference in extinction coefficients (which may be positive or negative) is called *circular dichroism*. The result of this difference is that, in the region of an absorption band, one circularly polarized component will be absorbed more than the other. The experimental observation is that the plane polarized light will be converted into elliptically polarized light (Figure 11-15). Just as in ORD, CD can be positive (as illustrated in Figure 11-14) or negative (a mirror image of the positive effect).

In summary, if plane polarized light is passed through a chiral medium *in the region of an absorption band,* and if the chromophore giving rise to the band is the center of chirality or is affected by a "vicinal" (or nearby) center of chirality, three effects will be observed: an inverse curve of rotatory dispersion, circular dichroism, and a change to elliptically polarized light; the sum of these effects is called the *Cotton effect*.

ABSOLUTE CONFIGURATIONS OF CHIRAL COORDINATION COMPLEXES

Optical rotatory dispersion and circular dichroism can be used to assist in the determination of the absolute configuration of a metal complex. The essence of the method is to recognize that molecules of identical configuration should have the same Cotton effect for electronic transitions of the same origin, whereas enantiomers will have mirror image Cotton effects. The problem is to determine the absolute configuration of one or more coordination compounds, which can then be used as standards. This was solved by Bijvoet, who used the method of anomalous diffraction of x-rays to determine the absolute configuration of sodium rubidium *d*-tartrate.[89,90] The method was then applied to $\{(+)\text{-}[Co(en)_3]Cl_3\}_2 \cdot NaCl \cdot 6H_2O$, and it was found that the $(+)\text{-}[Co(en)_3]^{3+}$ ion had the $\Lambda\delta\delta\delta$ configuration (*i.e.*, the enantiomer of the $\Delta\lambda\lambda\lambda$ configuration in Figure 11-10).[91] Therefore, it is now possible for this complex to be used to assist in determining the absolute configuration of other cobalt compounds by ORD and CD.[92]

Tris(aminoacido)cobalt complexes illustrate the application of CD spectra to the determination of configuration, and show as well a number of interesting features of coordination chemistry and inorganic stereochemistry in general.[93] If three glycinate

---

[89] A. F. Peerdeman, A. J. van Bommel, and J. M. Bijvoet, *Proc. Roy. Soc., Amsterdam,* **54,** 16 (1951).
[90] See ref. 28c.
[91] K. Kakatsu, M. Shiro, Y. Saito, and H. Kuroya, *Bull. Chem. Soc. Jap.,* **30,** 158 (1957).
[92] A. J. McCaffery and S. F. Mason, *Mol. Phys.,* **6,** 359 (1963).
[93] R. G. Denning and T. S. Piper, *Inorg. Chem.,* **5,** 1056 (1966).

ions are bound to a metal ion, two diastereoisomers (**53** and **54**) are possible, and each of these has an enantiomer.

fac isomer
configuration number = 22
53

mer isomer
configuration number = 21
54

If a substituted amino acid in a given **R** or **S** configuration is used, *fac* and *mer* isomerism is still possible,[94] and the arrangement of ligands about the metal ion can still result in Λ or Δ chirality. However, the right-handed *fac* isomer can no longer be enantiomeric with the left-handed *fac* isomer, for example. That is, the use of **S**-alanine can give the *fac* Λ isomer shown in Figure 11-16. A mirror image of this compound cannot be generated with **S**-alanine because the enantiomer necessarily contains **R**-alanine. Therefore the *fac* Δ isomer is a diastereoisomer of the *fac* Λ isomer. The net result is that, while *fac* and *mer* isomers have an enantiomeric partner if glycine is used, an optically active amino acid will give a diastereoisomer for each chirality of each *fac* or *mer* isomer. Because each of the isomers in Figure 11-16 is a diastereoisomer, each of the forms should be separable from the others under achiral conditions, and each should be different from the others in physical properties.

---

[94] The term *fac* designates a compound such as **53**, in which three identical donor atoms are on a *fac*e of the octahedron. The term *mer* designates the other isomer, **54**, in which the three identical groups are on a *mer*idian of the octahedron. See page 624.

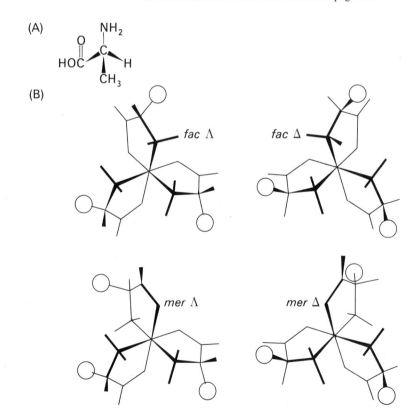

Figure 11-16. Isomerism in tris(S-alaninato)cobalt. (A) The configuration of S-alanine. (B) The four diastereomeric forms of tris(S-alaninato)cobalt.

STEREOISOMERISM 649

The four diastereomeric forms of tris(S-alaninato)cobalt (Figure 11-16) are prepared by boiling a solution of **S**-alanine with freshly prepared cobalt(III) hydroxide or with hexamminecobalt(3+)chloride. The isomers can be separated by their solubility differences. In general, *fac* isomers are less soluble than *mer* isomers, and (+) isomers are much less soluble than (−) isomers. For example, after boiling an aqueous suspension of $Co(OH)_3$ in a solution of $CH_3CH(NH_2)COOH$ for eight hours, the *mer*-(+) isomer separates as violet needles; the *fac*-(−) and *mer*-(−) isomers, which are left in solution, can be separated by extraction and chromatography. The *fac*-(+) isomer, obtained from the $[Co(NH_3)_6]^{3+}$ reaction, will dissolve only in 50% sulfuric acid or stronger acids.

The *fac* and *mer* isomers show distinct spectroscopic differences from one another. On the one hand, the *fac* isomers give rise to similar UV-visible spectra (Figure 11-17) (they are both red) and similar nmr spectra. The methyl groups of these isomers appear as a simple doublet (due to spin-spin coupling to the geminal

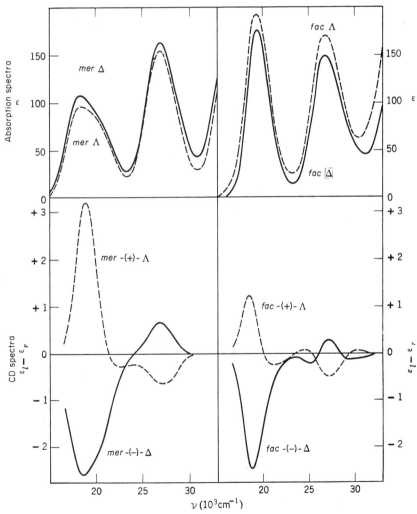

Figure 11-17. The absorption and CD spectra for the four diastereoisomers of [Co(**S**-alaninate)$_3$]. See Figure 11-16 for configurations of the isomers. Figure adapted from J. Fujita and Y. Shimura, "Optical Rotatory Dispersion and Circular Dichroism" in "Spectroscopy and Structure of Metal Chelate Compounds," K. Nakamoto and P. J. McCarthy, John Wiley and Sons, New York, 1968, page 193.

proton) in their ^{1}H nmr spectra, thereby indicating that all of the methyl groups are equivalent. In contrast, the *mer* isomers, which are both violet and give UV-visible spectra similar to one another, have ^{1}H nmr spectra that indicate that the methyl groups are different from each other; their spectra show three doublets for these groups.

The absolute configurations of these isomers can be assigned on the basis of chemical evidence and by a comparison of their CD spectra with each other and with that of [Co(en)$_3$]$^{3+}$. First, the sign of the Cotton effect for the lowest energy band of the *fac*-(+) and *mer*-(+) isomers is positive just as it is for Λ-(+)-[Co(en)$_3$]$^{3+}$. This result strongly suggests (but does not conclusively prove) that these complexes have the same absolute configuration, Λ.[95]

The fact that the *fac*-(−) and *mer*-(−) isomers have a negative Cotton effect for the lowest energy CD band suggests that they have an absolute configuration opposite to that for the (+) isomers. Be careful to note, however, that the CD curves for the (+) and (−) isomers are not exactly mirror images of one another. This must certainly be due to the fact that, while the complexes have opposite absolute configurations based on the positioning of chelate rings, they are in fact diastereoisomers; as mentioned above, this arises because only **S**-alaninate was used as a ligand (Figure 11-16).

LIGAND CONFORMATION

Complexes of the chiral ligand 1,2-diaminopropane (= pn) have been studied extensively with regard to their vicinal effects and because they provide interesting examples of stereospecificity and conformational analysis.[96-98] The configuration of the (+) enantiomer of the ligand has been shown to be **S (55)**.

$$\text{S-(+)-pn} \qquad \textbf{(55)} \qquad \text{R-(−)-pn}$$

---

[95] An x-ray analysis of *mer*-(+)-[Co(**S**-alaninate)$_3$] proved that it did indeed have the Λ configuration. See M. G. B. Drew, J. H. Dunlop, R. D. Gillard, and D. Rogers, *Chem. Commun.*, 42 (1966); B. E. Douglas and S. Yamada, *Inorg. Chem.*, **4**, 1561 (1965).
[96] See ref. 28c.
[97] See ref. 67.
[98] J. H. Dunlop and R. D. Gillard, *Adv. Inorg. Chem. Radiochem.*, **9**, 185 (1966).

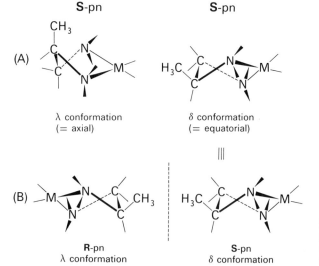

Figure 11-18. (A) The λ and δ conformations of **S**-(+)-1,2-diaminopropane when bound to a metal ion. (B) The mirror image conformations of 1,2-diaminopropane with the methyl group in the more stable equatorial position.

TABLE 11-7
THE ISOMERS OF TRIS(1,2-DIAMINOPROPANE)COBALT(III) ION.[a,b]

Compound	Molecular Configuration	Ligand Conformation	Specific Rotation $[\alpha]_D$
[Co(**S**-pn)$_3$]I$_3 \cdot$ H$_2$O	Λ	δδδ	+24°
[Co(**R**-pn)$_3$]I$_3 \cdot$ H$_2$O	Δ	λλλ	−24°
[Co(**S**-pn)$_3$]I$_3 \cdot$ H$_2$O	Δ	δδδ	−214°
[Co(**R**-pn)$_3$]I$_3 \cdot$ H$_2$O	Λ	λλλ	+214

[a] F.P. Dwyer, F.L. Garvan, and A. Shulman, *J. Amer. Chem. Soc.*, **81**, 290 (1959).
[b] C.J. Hawkins, "Absolute Configurations of Metal Complexes," Interscience, New York, 1971, p. 130.

If the molecule in this configuration is attached to a metal, two conformations are possible (Figure 11-18); the conformation with the CH$_3$ group in an equatorial position (*i.e.*, δ) is clearly more likely. Note that the **S**-pn δ conformation is the mirror image of the **R**-pn λ conformation [Figure 11-18(B)]; therefore, if the diamine of **R** configuration is used, the conformation of greater stability should be λ.

If **R**-1,2-diaminopropane (**R**-pn) is combined with cobalt(III) sulfate and oxygen, two isomers of empirical formula [Co(**R**-pn)$_3$]$^{3+}$ are isolated, and they are clearly not enantiomers of one another.[99] Instead of having equal but opposite rotations as would a pair of enantiomers, they have specific rotations of −24° and +214°. **S**-pn also gives two diastereoisomers, but these have rotations of +24° and −214°. It is clear that **R**-pn and **S**-pn have given two pairs of enantiomers, one pair having **R**- or **S**-ligands with specific rotations of ± 24° and another pair, again with **R**- or **S**-ligands, with specific rotations of ± 214°. An x-ray analysis of the (−)-[Co(**S**-pn)$_3$]$^{3+}$ isomer indicates that (1) all three chelate methyl groups are *cis* to one another, (2) the absolute configuration is Δ, and (3) the pn chelate rings are in the expected δ configuration.[100] Therefore, assuming that the methyl groups are *cis* to one another in all of the complexes, the enantiomeric (+)-[Co(**R**-pn)]$^{3+}$ must have the Λ configuration with rings arranged in the λ conformation. Likewise, the configurations and ring conformations of the re-

---

[99] F. P. Dwyer, F. L. Garvan, and A. Shulman, *J. Amer. Chem. Soc.*, **81**, 290 (1959).
[100] See ref. 28c, p. 130.

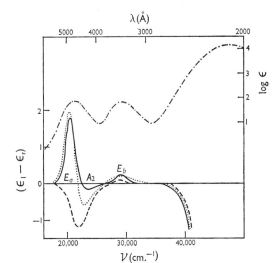

Figure 11-19. The electronic absorption spectra of (+)- and (−)-[Co(**S**-pn)$_3$]$^{3+}$ (−·−·−), and the CD spectra of Λ-(+)-[Co(en)$_3$]$^{3+}$ (———), (−)-[Co(**S**-pn)$_3$]$^{3+}$ (− − − − −), and (+)-[Co(**S**-pn)$_3$]$^{3+}$ (·········). A. J. McCaffery, S. F. Mason, and B. J. Norman, *Chem. Commun.*, 49 (1965).

maining isomers are determined as summarized in Table 11-7. The correctness of this analysis is suggested by the CD spectrum of $(+)$-$[Co(S\text{-}pn)_3]^{3+}$, which is similar to that of $(+)$-Λ-$[Co(en)_3]^{3+}$ (Figure 11-19); in contrast, the CD spectrum of $(-)$-$[Co(S\text{-}pn)_3]^{3+}$ is approximately opposite to that of the tris(ethylenediamine) complex.[101,102]

---

[101] A. J. McCaffery, S. F. Mason, and B. J. Norman, *Chem. Commun.*, 49 (1965).
[102] For recent work on ligand conformations see, for example, G. W. Everett, Jr., and R. R. Horn, *J. Amer. Chem. Soc.*, **96**, 2087 (1974).

STUDY QUESTIONS

13. Give the correct chirality symbol—Λ or Δ—to each enantiomer generated by compound **50**.

14. Give the correct chirality symbol for each enantiomer found for the compound in study question 10.

15. In the discussion of tris(aminoacido)cobalt complexes and their configurations, it was pointed out that, because a dissymmetric amino acid was used as a ligand, the *fac* Λ isomer is not the mirror image of the *fac* Δ isomer.

    a) Draw the enantiomer of the *fac* Λ isomer.
    b) Since the *fac* (or *mer*) Λ and Δ isomers are not enantiomers because of the ligand, how must the ligand be changed so that you could synthesize the enantiomer of *fac* Λ sketched in part a?

16. What are the chirality symbols for the two-bladed propellors below?

    (A)   (B)

17. Pictured below and on page 653 are the dissymmetric conformers of *trans*-dichlorotriethylenetetraminecobalt$(+)$. Give the chirality symbol for each five-membered chelate ring.

    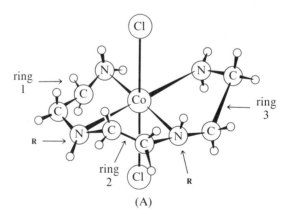

    (A)

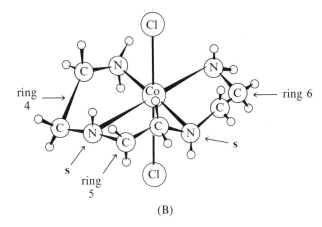

(B)

18. The following question concerns a molecule such as [Co(diamine)(NH$_3$)$_2$Cl$_2$]$^+$. Assume that the bidentate diamine ligand is **R**-propylenediamine, structure **55**.

    a) Neglecting for the moment the question of the diamine ring conformation, how many diastereomers are possible?
    b) What is the most likely conformation for the **R**-propylenediamine ring?
    c) Now, considering the diamine to exist in the more favored conformation, how many diastereoisomers are possible for [Co(diamine)(NH$_3$)$_2$Cl$_2$]$^+$?

EPILOG

With this chapter we have completed the discussion of the important aspects of structure and bonding in coordination compounds. We have been concerned chiefly with the molecular orbital approach to bonding in these compounds, with the use of the MO-AOM approach to predict and rationalize structures and spectra, and with the various polytopal and constitutional isomers and stereoisomers possible for this class of compounds. All of these are areas with which you should be familiar in order to understand the chemistry of coordination compounds. Therefore, we can now turn, in the next three chapters, to an examination of the mechanisms of the reactions of coordination compounds and to the use of these reactions in synthesis.

MECHANISMS OF ELECTRON TRANSFER REACTIONS

   Key Ideas Concerning Electron Transfer Between Transition Metals
   Outer Sphere Electron Transfer Reactions
      Chemical Activation and Electron Transfer
      Cross Reactions and Thermodynamics
   Inner Sphere Electron Transfer Reactions
      Formation of Precursor Complexes
      Rearrangement of the Precursor Complex and Electron Transfer
      Electronic Structure of the Oxidant and Reductant
      The Nature of the Bridge Ligand
      Fission of the Successor Complex
   Two-Electron Transfers
   Non-Complementary Reactions

SYNTHESIS OF COORDINATION COMPOUNDS USING ELECTRON TRANSFER REACTIONS

EPILOG

12. COORDINATION CHEMISTRY: REACTION MECHANISMS AND METHODS OF SYNTHESIS; ELECTRON TRANSFER REACTIONS

This is the first of three chapters that examine the main forms of reactivity displayed by coordination compounds and the usefulness of this reactivity in the synthesis of other coordination compounds.[1-4] *The approach adopted here is to begin with changes occurring within the metal ion itself; that is, changes in oxidation state brought about by electron transfer from the metal ion of one compound to the metal ion of another.*[5-11] *Subsequent chapters deal with reactions involving changes in the inner coordination sphere of a complex; that is, the substitution of one ligand for another, rearrangement of ligands, and chemical reactions of the ligands themselves.*

In this chapter you shall see that electron transfer reactions may occur by either or both of two mechanisms: (1) **Outer sphere mechanism**—*the transfer of an electron between two complexes whose inner or first coordination spheres remain intact.* (2) **Inner sphere mechanism**—*the transfer of an electron through a ligand that is simultaneously a member of the inner or first coordination spheres of both oxidant and reductant. Atom or group transfer may or may not accompany this electron transfer. However, it is often the case that, as an electron is transferred from reductant to oxidant through a bridging ligand, that ligand is transferred in the opposite direction.*

The following books and review article represent a selection of the literature available on mechanistic inorganic chemistry in general.

[1] F. Basolo and R. G. Pearson, "Mechanisms of Inorganic Reactions," 2nd Edition, John Wiley and Sons, New York (1967).

[2] M. L. Tobe, "Inorganic Reactions Mechanisms," Nelson, London (1972).

[3] R. G. Wilkins, "The Study of the Kinetics and Mechanism of Reactions of Transition Metal Complexes," Allyn and Bacon, Inc. Boston (1974).

[4] R. G. Pearson and P. C. Ellgen, "Mechanisms of Inorganic Reactions in Solution," in "Physical Chemistry, An Advanced Treatise," Vol. VII, H. Eyring, ed., Academic Press, New York (1975). An excellent, up-to-date, but brief review.

The next group of books and review articles are representative of the material available on electron transfer reactions.

[5] H. Taube, *J. Chem. Educ.*, **45**, 452 (1968).

[6] H. Taube, "Electron Transfer Reactions of Complex Ions in Solution," Academic Press, New York (1970).

[7] W. L. Reynolds and R. W. Lumry, "Mechanisms of Electron Transfer," Ronald Press, New York (1966).

[8] N. Sutin, "Free Energies, Barriers, and Reactivity Patterns in Oxidation-Reduction Reactions," *Acc. Chem. Res.*, **1**, 225 (1968).

[9] J. H. Espenson, "Oxidation of Transition Metal Complexes by Cr(VI)," *Acc. Chem. Res.*, **3**, 347 (1970).

[10] H. Taube and E. S. Gould, "Organic Molecules as Bridging Ligands in Electron Transfer Reactions," *Acc. Chem. Res.*, **2**. 321 (1969).

[11] A. Haim, "Role of the Bridging Ligand in Inner Sphere Electron Transfer Reactions," *Acc. Chem. Res.*, **8**, 265 (1975).

One of the most important topics in any introductory chemistry course is a discussion of oxidation-reduction reactions, reactions wherein there is a change in the formal oxidation states of the substances involved. However, it is often not pointed out that there are formally two types of reactions falling within the scope of redox processes: (i) reactions involving simple *electron transfer,* and (ii) reactions that can be considered as *atom transfer* reactions, both with and without electron transfer.

An example of a reaction that truly occurs by *electron transfer* is that between two different oxidation states of osmium.

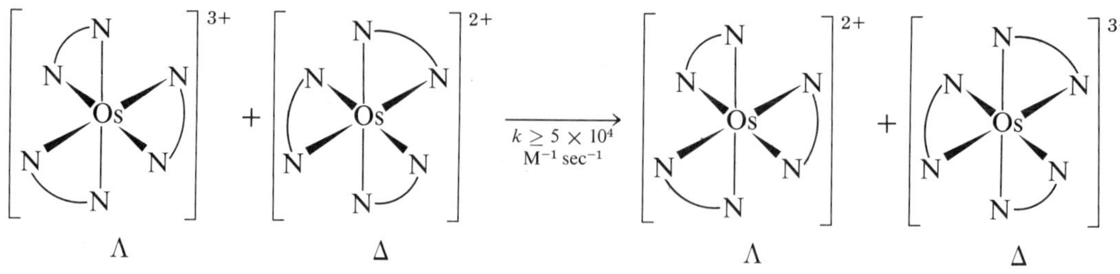

(12-1)

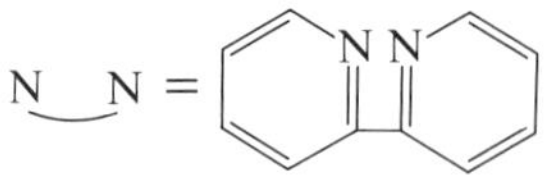

[Os(N−N)₃]³⁺ (Λ) + [Os(N−N)₃]²⁺ (Δ) $\xrightarrow{k \geq 5 \times 10^4 \text{ M}^{-1} \text{sec}^{-1}}$ [Os(N−N)₃]²⁺ (Λ) + [Os(N−N)₃]³⁺ (Δ)

N−N = (2,2'-bipyridine)

The two chiral reactants do not themselves racemize, but, when they are mixed, electron transfer leads to a rapid loss of optical activity since the Δ and Λ forms are equally distributed between the two oxidation states at equilibrium.[12,13] A great many other reactions also occur by direct transfer of electrons, and we shall turn to them momentarily.

In an *atom transfer reaction,* the fact that oxidation-reduction has occurred is a matter of the way in which one formally counts electrons. For example, in the oxidation of NO_2^- by OCl^-, the nitrogen oxidation state changes from +3 to +5, while chlorine changes from +1 to −1.

(12-2)
$$NO_2^- + OCl^- \rightarrow NO_3^- + Cl^-$$

By labeling the oxygen in OCl^-, the mechanism of this reaction has been shown to involve oxygen atom transfer (many such reactions were discussed in Chapter 7).[14]

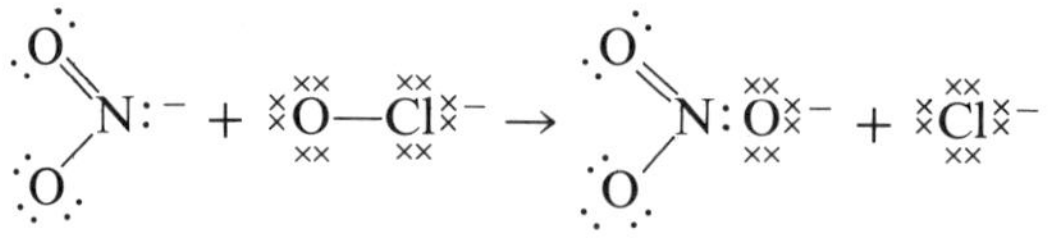

$$O_2N{:}^- + {:}\ddot{O}-\ddot{Cl}{:}^- \rightarrow O_2N{:}\ddot{O}{:}^- + {:}\ddot{Cl}{:}^-$$

The change in formal oxidation states comes about because electrons that were formally non-bonding (the N lone pair) now become bonding (the N—O σ bond pair), and those that were previously counted as bonding (the O—Cl bond pair) are now non-bonding (the Cl^- lone pair).

Although this description of the NO_2^-/OCl^- reaction is simple enough, it may be useful to see the orbital-following model for the reaction in order to set the stage for our later discussions. In Figure 12-1 we have depicted the mo's of the reactants, the

---

[12] F. P. Dwyer and E. C. Gyarfas, *Nature,* **166,** 481 (1950).
[13] M. W. Dietrich and A. C. Wahl, *J. Chem. Phys.,* **38,** 1591 (1963).
[14] H. Taube, *Record Chem. Progr. Kresge-Hooker Sci. Lib.,* **17,** 25 (1956).

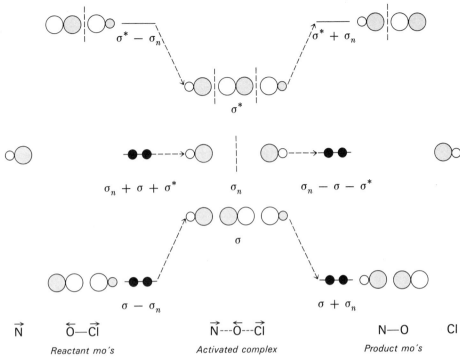

Figure 12-1. Orbital following in the $NO_2^- + OCl^-$ oxidation-reduction reaction. The mo's of the activated complex are labeled σ, σ_n, and σ^*, and the reactant and product mo's are linear combinations of those of the activated complex. (See Chapter 4 for the principles of mo construction.)

activated complex, and the product. The mo's of the activated complex are labeled σ, σ_n, and σ^*. The figure shows that linear combination (*i.e.,* mixing) of these mo's, as the central nucleus shifts to the left or right, leads to either the product or the reactant mo's, respectively. In the NO_2^-/OCl^- example, the lowest-energy mo of the product, the N—O σ bonding mo, evolves as an in-phase combination of σ and σ_n, whereas the highest-energy empty mo of the product, the σ^* N—O orbital, arises from $\sigma^* + \sigma_n$. Furthermore, the occupied σ_n mo smoothly evolves into the lone pair of the other product, Cl^-.[15] An equally important point to notice is that the reactant bonding and non-bonding mo's, each of which contains a pair of electrons, directly correlate with the product bonding and non-bonding orbitals, each of which also contains a pair of electrons. Therefore, there can be a smooth change in the function of the two pairs of electrons as described in the preceding paragraph.

Oxidation-reduction reactions in transition metal chemistry can also occur by atom transfer. For example, in acid solution

$$[Co^{III}(NH_3)_5Cl]^{2+} + [Cr^{II}(H_2O)_6]^{2+} \xrightarrow[-H_2O]{} [(NH_3)_5Co^{III}-Cl: \rightarrow Cr^{II}(H_2O)_5]$$

$$\downarrow$$

$$[Co(H_2O)_6]^{2+} + 5NH_4^+ + [Cr(H_2O)_5Cl]^{2+} \xleftarrow[+H_2O]{H^+} [(NH_3)_5Co^{II} \leftarrow :Cl-Cr^{III}(H_2O)_5]$$

(12-3)

This reaction is similar to the NO_2^-/OCl^- reaction in that anion transfer has occurred. However, there has also been electron transfer in a direction opposite to the direction of ion transfer. In reaction (12-3), water on the Cr(II) center is substituted by a chloride [already attached to Co(III)] utilizing a lone pair of electrons; as the chloride

---

[15] The phasings of σ, σ_n, and σ^* along the reaction coordinate are dictated by the principle that the lowest mo stays lowest and the highest stays highest. The orbital following is successful because, under this constraint, the occupied reactant mo's evolve correctly into occupied product mo's.

moves from being cobalt-bound to being chromium-bound, there is a net flow of one electron in the opposite direction.[16]

$$Co^{+++}-Cl: + Cr^{++} \rightarrow [Co^{+++}-Cl: \xrightarrow{e^-} Cr^{++}] \rightarrow Co^{++} + :Cl-Cr^{+++}$$

Reactions such as this occur by what is commonly known as an "inner sphere mechanism," because electron flow occurs by virtue of an atom or group held in common in the inner or first coordination spheres of the two metal centers; we shall discuss such reactions as a separate topic later.

Another type of atom transfer redox reaction is illustrated by the addition of molecular chlorine [equation (12-4)] or oxygen to an iridium(I) compound. In such an addition reaction there is no necessity to postulate the formal transfer of an electron or electrons, but, because of the way in which one formally counts electrons, the iridium is said to be oxidized to iridium(III) and chlorine is reduced to Cl^-. Reactions of this type have come to be known as "oxidative additions," and they are now recognized as an important class of reactions. They were first introduced in Chapter 5 (formation of O_2 adducts) and discussed again in Chapter 7 (phosphorus chemistry); one further example will be described in the section on Synthesis of Coordination Compounds later in this chapter, and we shall describe them in detail at a point where they are most pertinent, that is, in Chapter 17.

Study of redox reactions, especially those with electron transfer, has generally accelerated over the past decade; though far from complete, the scientific community's understanding has increased to the point where a reasonable interpretive framework is now in place. Indeed, it is now possible to begin an attack on redox reactions of ultimate importance—i.e., biological electron transfer.[17] The functions of hemoglobin (O_2 transport in plasma), myoglobin (oxygen storage in muscle), transferrin (iron storage in plasma), ferritin (iron storage in cells), and cytochrome c (electron transport) all involve in some manner reversible changes in the oxidation states of one or more iron atoms. For example, the transfer of iron from the gut to plasma is thought to involve the series of steps shown in Figure 12-2. Because of the obvious importance of these biological redox reactions, the case for a thorough study of electron transfer reactions in general hardly needs further justification.

(12-4)

$$\begin{array}{c} Ph_3P \diagdown \diagup Cl \\ Ir^I \\ OC \diagup \diagdown PPh_3 \end{array} + Cl_2 \rightarrow \begin{array}{c} Cl \\ Ph_3P \diagdown | \diagup Cl \\ Ir^{III} \\ OC \diagup | \diagdown PPh_3 \\ Cl \end{array}$$

1   2

---

[16] This reaction may be considered as either Cl atom transfer or Cl^- ion transfer + transfer of an electron. The two points of view are equivalent, but the latter is generally more consistent with the donor role of Cl^- ion complexes.

[17] L. E. Bennett, *Prog. Inorg. Chem.*, **18**, 1 (1973).

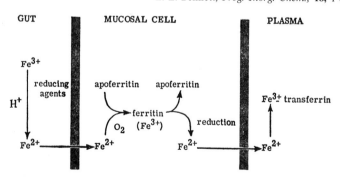

Figure 12-2. A possible mechanism for the absorption of iron from the gut into plasma. Following reduction from the +3 state, iron is taken up by the mucosal cell and is converted, with oxidation, into ferritin. Ferritin is an iron storage protein that consists of a shell of protein subunits surrounding a core of ferric hydroxyphosphate. Iron is then released and passed through the cell wall in the reduced form into the plasma, where it is stored in the +3 form in transferrin. [From R. R. Crichton, *Structure and Bonding*, **17**, 67 (1973).]

MECHANISMS OF ELECTRON TRANSFER REACTIONS

Electron transfer reactions may occur by either or both of two mechanisms: outer sphere and inner sphere. In principle, an **outer sphere mechanism** *involves electron transfer from reductant to oxidant, with the coordination shells or spheres of each staying intact.* That is, one reactant becomes involved in the outer or second coordination sphere of the other reactant, and an electron flows from reductant to oxidant. For example,

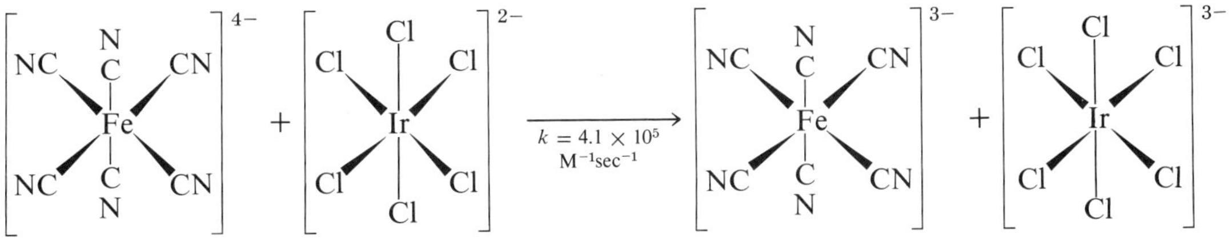

$$[Fe(CN)_6]^{4-} + [IrCl_6]^{2-} \xrightarrow{k = 4.1 \times 10^5 \text{ M}^{-1}\text{sec}^{-1}} [Fe(CN)_6]^{3-} + [IrCl_6]^{3-} \quad (12\text{-}5)$$

Such a mechanism is established when rapid electron transfer occurs between two substitution-inert complexes.[18] *An **inner sphere mechanism**, on the other hand, is one in which the reductant and oxidant share a ligand in their inner or primary coordination spheres, the electron being transferred across a bridging group* as in reaction (12-6).[19]

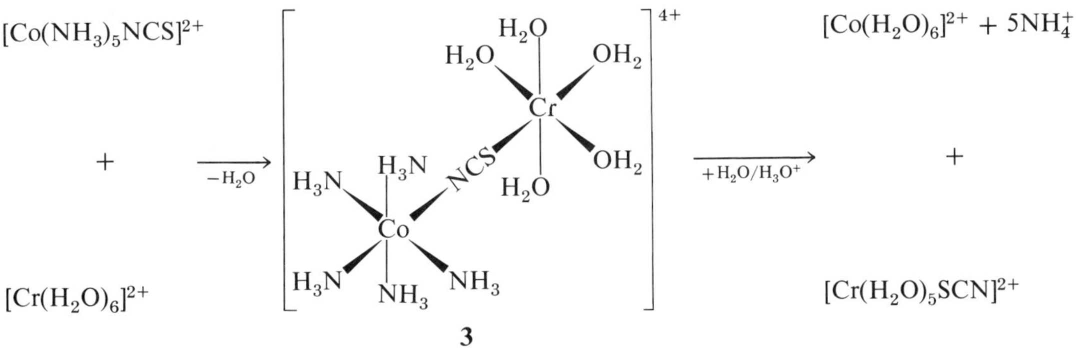

$$[Co(NH_3)_5NCS]^{2+} + [Cr(H_2O)_6]^{2+} \xrightarrow{-H_2O} \left[(H_3N)_5Co\text{-}NCS\text{-}Cr(OH_2)_5 \right]^{4+} \xrightarrow{+H_2O/H_3O^+} [Co(H_2O)_6]^{2+} + 5NH_4^+ + [Cr(H_2O)_5SCN]^{2+} \quad (12\text{-}6)$$

3

There are clearly two prerequisites for the operation of an *inner sphere mechanism*. The *first requirement* is that *one reactant* (usually the oxidant) *possess at least one ligand capable of binding simultaneously to two metal ions,* however transiently. (As you shall see later, this requirement suggests a way to synthesize linkage isomers.) Although this bridging ligand is frequently transferred from oxidant to reductant in the course of electron transfer, this need not be the case: ligand transfer is not a

---

[18] Coordination complexes may also undergo substitution reactions; that is, one or more of the ligands on the original complex may be substituted by other ligands. Generally, a complex is classified as inert to substitution if the rate of substitution is low, that is, a half-life greater than 1 minute. The opposite of inertness is kinetic lability; a complex is labile if its substitution reactions have half-lives less than 1 minute. In this chapter on electron transfer reactions, a substitution-inert complex is one that undergoes substitution at a rate substantially less than the rate of electron transfer.

[19] Reaction (12-6) is an example of "remote attack." That is, the reductant, Cr^{2+}, is attached to an atom "remote" from the oxidant instead of the atom directly bound to, or adjacent to, the oxidant, the nitrogen. You shall see other examples of both "adjacent" and "remote" attack (see page 676, for example).

requirement of an inner sphere mechanism. The *second requirement* for the inner sphere mechanism is that *one ligand of one reactant* (usually the reductant) *be substitutionally labile; that is, one ligand must be capable of being replaced by a bridging ligand in a facile substitution process.*[20]

There are numerous examples of both mechanistic types. It is useful to keep in mind the fact that the outer sphere mechanism is always possible, but, in certain cases, the inner sphere mechanism becomes competitive or even dominant.

KEY IDEAS CONCERNING ELECTRON TRANSFER BETWEEN TRANSITION METALS

Before beginning a discussion of the two main mechanistic types of electron transfer reactions, you should recognize that there are several key ideas that apply to both inner sphere and outer sphere reactions. These ideas will recur in our discussions, and you should be alerted at the outset to their importance and generality.

1. In most chemical reactions there is a chemical activation process that gets either or both of the reactants into the proper configuration for reaction. In order to effect electron transfer, a requirement seems to be that the electron donor mo of the reductant and the electron receptor mo of the oxidant be of the same type. It will soon become apparent that efficient outer sphere electron transfer requires both donor and receptor mo's to both be π^* orbitals, whereas inner sphere reactions require donor and receptor mo's to both be π^* or σ^*. (Ordinarily one would expect the HOMO of the reductant to be the donor orbital and the LUMO of the oxidant to be the receptor orbital.) However, if either of these does not meet the symmetry requirements for effective transfer, greater chemical activation—encompassing both structural deformation and electron configuration change—will be required. (We might mention that there is not much we can do with this concept in a quantitative sense at this time. Rather, we can only use it as a qualitative guide to decide when reactions are relatively fast or slow.)

2. With regard to chemical activation, one has the tendency to think of electron transfer reactions as involving an intermediate. For example, consider a Co(III) complex [where the metal ion has a $(\pi^*)^6$ configuration in the ground state] as an oxidant in an outer sphere reaction. In order to effect electron transfer from a π^* orbital of a reductant to a π^* orbital of Co(III), the latter must be chemically activated to a configuration such as high spin $(\pi^*)^4(\sigma^*)^2$ in order to change the receptor orbital from σ^* to π^*. It is not at all clear whether this activation occurs as a discrete step prior to electron transfer or whether the activation takes place as a blurred part of the overall electron transfer process. Unfortunately, there is no good experimental evidence regarding this question. *Therefore, although our language will imply that a configurationally activated precursor exists prior to electron transfer, we really have no evidence for this. You should recognize this gap in our knowledge and be careful not to jump to conclusions.*

3. Finally, there is a thermodynamic influence on the activation energy barrier. As you shall see, reactions that may require a HOMO and/or LUMO change (as in paragraph 2 above) can have lower barriers than might be predicted, because the reaction has a negative free energy change.

OUTER SPHERE ELECTRON TRANSFER REACTIONS

The elementary steps involved in the outer sphere mechanism are:

(12-7)

Formation of a precursor (cage) complex	$Ox + Red \rightleftharpoons Ox\|Red$
Chemical activation of the precursor, electron transfer, and relaxation to the successor complex	$Ox\|Red \rightleftharpoons {}^-Ox\|Red^+$
Dissociation to the separated products	${}^-Ox\|Red^+ \rightleftharpoons Ox^- + Red^+$

---

[20] See ref. 18.

The first step of the mechanism is the formation of the so-called precursor complex, wherein the distance between the reactant centers (the metal ions) is approximately that required for electron transfer, but their relative orientations and internal structures do not yet permit transfer.

The second step then involves both solvent cage and precursor structural changes to accommodate electron transfer. Within the precursor there must be both a reorientation of the oxidant and reductant complexes and, within those complexes, structural changes that define the chemical activation process for electron transfer. As the transition state is passed, there follows the completion of the electron transfer and final relaxation of the oxidant and reductant structures.

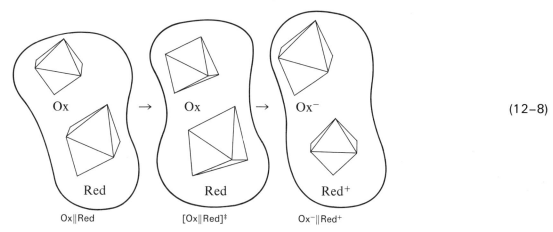

(12-8)

In the simplified schematic above, the oxidant is shown to enlarge while the reductant contracts in size. You know that these general changes are required because metal d electrons usually play an anti-bonding role.[21] To accommodate the incoming electron, the oxidant must increase its metal-ligand distances, while the metal-ligand distances of the reductant (which loses a d^* electron) shorten. (In specific cases, there may be other, angular distortions of the ligands about the metal ions as well.) In short, partial bond-breaking occurs in the oxidant, while partial bond-making follows in the reductant. The schematic is also drawn to remind you that solvent reorganization to accommodate these precursor structural changes can be an important contributor to the activation free energy, $\Delta G^\ddagger$.

The final step in the outer sphere process, as in any reaction, is the separation of the product ions or molecules.

> In all of the discussion that follows, you should keep in mind the fact that measurement of a rate constant for an outer sphere electron transfer reaction generally gives a measure only of the overall rate of the reaction; it is difficult if not impossible to ascribe it to any one elementary step. This is because the first and last steps are assumed to be rapid equilibria compared with the second step. For example, in the reaction below, the formation of the precursor complex is fast:

step a: formation of precursor

$$[\text{Co(NH}_3)_5\text{OH}_2]^{3+} + [\text{Fe(CN)}_6)]^{4-} \xrightarrow[K=1500\,\text{M}^{-1}]{\text{rapid}} [\text{Co(NH}_3)_5\text{OH}_2^{3+}\|\text{Fe(CN)}_6^{4-}]$$

(12-9)

step b: chemical activation with electron transfer and fission of successor complex

$$[\text{Co(NH}_3)_5\text{OH}_2^{3+}\|\text{Fe(CN)}_6^{4-}] \rightarrow [\text{Co(NH}_3)_5\text{OH}_2^{2+}\|\text{Fe(CN)}_6^{3-}] \rightarrow \text{products}$$

$$k = 1.9 \times 10^{-1}\,\text{sec}^{-1} \qquad t_{1/2} = 4\,\text{sec}$$

(12-10)

---

[21] An exception to this generalization arises when the $d\pi^*$ ao's are stabilized by the presence of several π acceptor ligands, such as CN^-, CO, NO_2^-, or phen.

whereas electron transfer within the precursor complex is slow.[22] It is of course the chemical activation-electron transfer step which has attracted the greatest attention, and we turn now to a fairly detailed discussion of this portion of the mechanism.

Chemical Activation and Electron Transfer

Probably the most important determinant of the rate of electron transfer processes between complexes, whether outer- or inner sphere, is the π^* or σ^* nature of the electron donor mo of the reductant (the orbital from which the electron is transferred) and the receptor mo of the oxidant (the orbital into which the electron moves). Quite generally, you could expect more facile electron transfer when both donor and receptor mo are of the π^* type. One reason for this is that reductant/oxidant activation (changing M—L distance) is usually less for a change in $d\pi^*$ than for a change in $d\sigma^*$ electron density.

Another general principle is that the better the donor-receptor mo overlap and mixing are, the easier is the electron transfer. Since $d\pi^*$ electrons in O_h complexes are exposed everywhere but at the vertices of the octahedron (e.g., the d_{xy} orbital, **4**), while $d\sigma^*$ electrons at the vertices (e.g., the $d_{x^2-y^2}$ orbital, **5**) are more shielded from the surroundings, $\pi^* \to \pi^*$ electron transfer should be faster than $\sigma^* \to \sigma^*$ transfer.

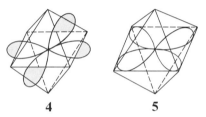

4   **5**

The representative data in Table 12-1 for outer sphere electron transfer reactions largely bear out such expectations. The first four examples require little chemical activation for M(II) → M(III) electron exchange because there is no change in the σ^* orbital configuration in either complex. For example, the change in metal-ligand distance in the Ru(III)-Ru(II) exchange illustrated below is only 0.04 Å.[23] For this reason, reactions such as these will generally be relatively rapid.

(12-11)

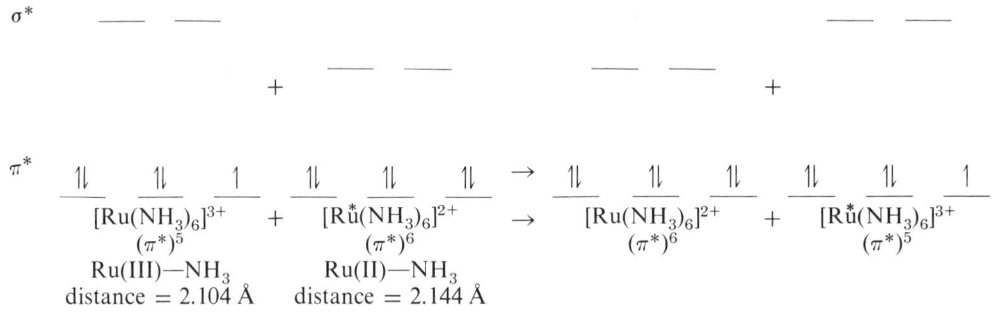

$$\sigma^* \quad \underline{}\ \underline{} \qquad\qquad\qquad\qquad\qquad\qquad \underline{}\ \underline{}$$

$$+ \qquad\qquad\qquad\qquad\qquad\qquad\qquad +$$

$$\pi^* \quad \underline{\uparrow\downarrow}\ \underline{\uparrow\downarrow}\ \underline{\uparrow}\quad \underline{\uparrow\downarrow}\ \underline{\uparrow\downarrow}\ \underline{\uparrow\downarrow} \;\to\; \underline{\uparrow\downarrow}\ \underline{\uparrow\downarrow}\ \underline{\uparrow\downarrow}\quad \underline{\uparrow\downarrow}\ \underline{\uparrow\downarrow}\ \underline{\uparrow}$$

$$[\text{Ru(NH}_3)_6]^{3+} \;+\; [\overset{*}{\text{Ru}}(\text{NH}_3)_6]^{2+} \;\to\; [\text{Ru(NH}_3)_6]^{2+} \;+\; [\overset{*}{\text{Ru}}(\text{NH}_3)_6]^{3+}$$

$$(\pi^*)^5 \qquad\qquad (\pi^*)^6 \qquad\qquad (\pi^*)^6 \qquad\qquad (\pi^*)^5$$

Ru(III)—NH$_3$ Ru(II)—NH$_3$
distance = 2.104 Å   distance = 2.144 Å

The apparent exception to the statement at the end of the previous paragraph is the $[\text{Fe(H}_2\text{O})_6]^{2+}/[\text{Fe(H}_2\text{O})_6]^{3+}$ reaction, which is about 10^2 times slower than the Ru(II)-Ru(III) exchange. However, you should recognize that it is dangerous to inter-

---

[22] This is a particularly favorable case in which the precursor complex can be observed and K for its formation can be measured (probably because of the opposite ion charges). Otherwise, when K is small, the observed rate constant is Kk and K is not measured. When comparing k_{obs} for different electron transfer reactions, it is well to remember that differences may be, at least partially, due to differences in K's.

[23] H. C. Stynes and J. A. Ibers, *Inorg. Chem.*, **10**, 2304 (1971).

TABLE 12-1

SECOND ORDER RATE CONSTANTS FOR SOME OUTER SPHERE ELECTRON TRANSFER REACTIONS

Reaction			Rate Constant k ($M^{-1}sec^{-1}$)	Reference
NET $\pi^* \to \pi^*$				
$Fe(H_2O)_6^{2+}$ $(\pi^*)^4(\sigma^*)^2$	+	$Fe(H_2O)_6^{3+}$ $(\pi^*)^3(\sigma^*)^2$	4.0	a
$Fe(phen)_3^{2+}$ $(\pi^*)^6$	+	$Fe(phen)_3^{3+}$ $(\pi^*)^5$	$\geq 3 \times 10^7$	b
$Ru(NH_3)_6^{2+}$ $(\pi^*)^6$	+	$Ru(ND_3)_6^{3+}$ $(\pi^*)^5$	8.2×10^2	c
$Ru(phen)_3^{2+}$ $(\pi^*)^6$	+	$Ru(phen)_3^{3+}$ $(\pi^*)^5$	$\geq 10^7$	c
NET $\sigma^* \to \sigma^*$				
$Co(H_2O)_6^{2+}$ $(\pi^*)^5(\sigma^*)^2$	+	$Co(H_2O)_6^{3+}$ $(\pi^*)^6$	~ 5	d
$Co(NH_3)_6^{2+}$ $(\pi^*)^5(\sigma^*)^2$	+	$Co(NH_3)_6^{3+}$ $(\pi^*)^6$	$\leq 10^{-9}$	e
$Co(en)_3^{2+}$ $(\pi^*)^5(\sigma^*)^2$	+	$Co(en)_3^{3+}$ $(\pi^*)^6$	1.4×10^{-4}	f
$Co(phen)_3^{2+}$ $(\pi^*)^5(\sigma^*)^2$	+	$Co(phen)_3^{3+}$ $(\pi^*)^6$	1.1	g

[a] J. Silverman and R. W. Dobson, *J. Phys. Chem.*, **56**, 846 (1952).
[b] D. W. Larsen and A. C. Wahl, *J. Chem. Phys.*, **43**, 3765 (1965).
[c] T. J. Meyer and H. Taube, *Inorg. Chem.*, **7**, 2369 (1968).
[d] N. A. Bonner and J. P. Hunt, *J. Amer. Chem. Soc.*, **82**, 3826 (1960).
[e] D. R. Stranks, *Discuss. Faraday Soc.*, **29**, 73 (1960).
[f] F. P. Dwyer and A. G. Sargeson, *J. Phys. Chem.*, **65**, 1892 (1961).
[g] B. R. Baker, F. Basolo, and H. M. Neuman, *J. Phys. Chem.*, **63**, 371 (1959).

pret a difference in rate constants of only 10^2 in terms of a fundamental difference in the chemical activation step. A 100-fold change in k_{obs} implies (see Chapter 7) only about a 3 kcal/mole change in $\Delta G^{\ddagger}$; the latter is determined not only by complex activation energy but also by complex activation entropy, the free energy of solvent cage reorganization, and the rate of precursor formation. The rate enhancement by five to seven orders of magnitude for *o*-phenanthroline (**6**) as a ligand, however, has a physicochemical interpretation. Phenanthroline is a potential π acceptor ligand,[24] which means

6

[24] W. L. Reynolds and R. W. Lumry, "Mechanisms of Electron Transfer," Ronald Press, New York (1966). On page 45 it is reported that calculations by Mataga and Reynolds support the transfer of an electron from the $[Fe(phen)_3]^{2+}$ to $[Fe(phen)_3]^{3+}$ by overlap of π molecular orbitals.

that the complex π^* mo's can be highly delocalized onto the phenanthroline skeleton. Thus, donor-receptor mixing can be achieved, without stringent inter-complex orientation requirements, by simple overlap of oxidant/reductant π^* mo's. In short, the radial diffuseness of the electron donor and receptor orbitals affects the rate of electron transfer.

One of the most striking features in Table 12-1 is the slow rate of the $[Co(NH_3)_6]^{2+}/[Co(NH_3)_6]^{3+}$ exchange in relation to the Fe(II)/Fe(III) and Ru(II)/Ru(III) systems. (Notice, however, the rate enhancement when phen is the ligand rather than NH_3 and ethylenediamine.) The most obvious distinction between the Co(II)/Co(III) systems and those of Fe(II)/Fe(III) and Ru(II)/Ru(III) in Table 12-1 is that the former requires change in the high-spin/low-spin conditions of the reductant and oxidant. This is just another way of saying that a change in the number of σ^* electrons in both oxidant and reductant must accompany electron transfer. And we should point out that, because of this, the change in Co—NH_3 distance is 0.178 Å, a change more than four times that in the analogous Ru(II)/Ru(III) exchange.[25] Presumably, major changes in Co—L bond distances must accompany electron

$$[Co(NH_3)_6]^{3+} \quad + \quad [Co^*(NH_3)_6]^{2+} \quad \rightarrow \quad [Co(NH_3)_6]^{2+} \quad + \quad [Co^*(NH_3)_6]^{3+}$$
$$(\pi^*)^6 \qquad\qquad (\pi^*)^5(\sigma^*)^2 \qquad\qquad (\pi^*)^5(\sigma^*)^2 \qquad\qquad (\pi^*)^6$$
Co(III)—NH_3 Co(II)—NH_3
distance = 1.936 Å   distance = 2.114 Å

transfer. As the associated energy changes (to some degree accompanying or preceding electron transfer) are also large, we find a logical explanation for the rather large difference in velocities of high spin/high spin and low spin/low spin electron exchange on the one hand, and high spin/low spin exchange on the other. Furthermore, the remarkable effect of delocalized π acceptor ligands (*e.g.,* phen) on electron transfer rates is basically understood and confirms the idea that the outer sphere mechanism involves $\pi^* \rightarrow \pi^*$ electron transfer.

The details of the chemical activation and electron transfer step are difficult to conceptualize for the Co(II)/Co(III) exchange reactions. One pathway, chosen to emphasize the chemical activation by Co(II)—L bond distance shortening and Co(III)—L bond distance lengthening, postulates the occurrence of Co(III) and Co(II) electronically excited *intermediates* prior to and following electron transfer. Two such possibilities exist:[26]

---

[25] See ref. 23.

[26] If, in an electron transfer reaction, an ion is produced in an excited state, it must relax to the ground state. For example, in curve A, Co^{3+} is produced in the $(\pi^*)^5(\sigma^*)^1$ configuration and must change to the final state $(\pi^*)^6$. This has been shown as requiring an activation energy; indeed, in curves A through D all configuration relaxations have been depicted as requiring an activation energy. However, it is not at all clear that this need be the case. In the gas phase, an ion in an excited state does not require energy to relax to the ground state. However, in the case of a complex ion in solution, ligand and solvent changes must occur, so an activation energy may in fact be required.

MECHANISMS OF ELECTRON TRANSFER REACTIONS 665

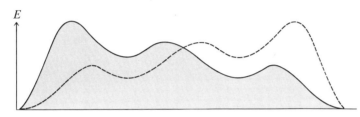

A Co^{2+} $(\pi^*)^5(\sigma^*)^2$ $\rightleftharpoons$ $(\pi^*)^6(\sigma^*)^1$ $\xrightarrow[\text{e.t.}]{\pi^* \to \pi^*}$ $(\pi^*)^5(\sigma^*)^1$ $\rightleftharpoons$ $(\pi^*)^6$ Co^{3+}
   Co^{3+} $(\pi^*)^6$ $(\pi^*)^5(\sigma^*)^1$ $(\pi^*)^6(\sigma^*)^1$ $(\pi^*)^5(\sigma^*)^2$ Co^{2+}

B Co^{2+} $(\pi^*)^5(\sigma^*)^2$ $\rightleftharpoons$ $(\pi^*)^5(\sigma^*)^2$ $\xrightarrow[\text{e.t.}]{\pi^* \to \pi^*}$ $(\pi^*)^4(\sigma^*)^2$ $\rightleftharpoons$ $(\pi^*)^6$ Co^{3+}
   Co^{3+} $(\pi^*)^6$ $(\pi^*)^4(\sigma^*)^2$ $(\pi^*)^5(\sigma^*)^2$ $(\pi^*)^5(\sigma^*)^2$ Co^{2+}

Path A is a single-path mechanism involving *both* Co(II) and Co(III) activation, while B entails *only* Co(III) activation. [It is necessary by the dictates of the "principle of microscopic reversibility" (Chapter 7) that Co(II) and Co(III) both lose their identity in a single-path mechanism. The indistinguishability of Co(II)/Co(III) occurs in both A and B during the electron transfer step.]

A second possibility for separate activation and electron transfer steps, in which two paths are necessary because Co(II) and Co(III) retain their identities along an asymmetric path, is the following:

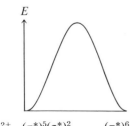

Solid curve

C Co^{2+} $(\pi^*)^5(\sigma^*)^2$ $\rightleftharpoons$ $(\pi^*)^6(\sigma^*)^1$ $\xrightarrow[\text{e.t.}]{\pi^* \to \pi^*}$ $(\pi^*)^5(\sigma^*)^1$ $\rightleftharpoons$ $(\pi^*)^6$ Co^{3+}
   Co^{3+} $(\pi^*)^6$ $(\pi^*)^4(\sigma^*)^2$ $(\pi^*)^5(\sigma^*)^2$ $(\pi^*)^5(\sigma^*)^2$ Co^{2+}

Dashed curve

D Co^{2+} $(\pi^*)^5(\sigma^*)^2$ $\rightleftharpoons$ $(\pi^*)^5(\sigma^*)^2$ $\xrightarrow[\text{e.t.}]{\pi^* \to \pi^*}$ $(\pi^*)^4(\sigma^*)^2$ $\rightleftharpoons$ $(\pi^*)^6$ Co^{3+}
   Co^{3+} $(\pi^*)^6$ $(\pi^*)^5(\sigma^*)^1$ $(\pi^*)^6(\sigma^*)^1$ $(\pi^*)^5(\sigma^*)^2$ Co^{2+}

In pathway C, both Co(II) and Co(III) must be activated, whereas only one product ion [Co(III), $(\pi^*)^5(\sigma^*)^1$] must relax to the ground state following electron transfer. The dashed curve D is the mirror image of the solid curve C, and so the activation/relaxation steps in D are the mirror image of those represented by C. That is, following curve D from left to right involves the activation of one reactant ion prior to electron transfer and the relaxation of two product ions following transfer. (The number of reactant ions activated and product ions relaxed is the difference between curves A/B and C/D. In the former pair, two ions must be activated prior to electron transfer, and two product ions must relax following transfer.)

While the above two stepwise mechanisms offer certain conceptual advantages, there is no evidence at present for their correctness. In fact, single-step paths are also possible, wherein activation and electron transfer occur synchronously (either single or dual paths are possible):

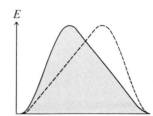

Co^{2+} $(\pi^*)^5(\sigma^*)^2$ $(\pi^*)^6$ Co^{3+} Co^{2+} $(\pi^*)^5(\sigma^*)^2$ $(\pi^*)^6$ Co^{3+}
Co^{3+} $(\pi^*)^6$ $\rightleftharpoons$ $(\pi^*)^5(\sigma^*)^2$ Co^{2+} Co^{3+} $(\pi^*)^6$ $\rightleftharpoons$ $(\pi^*)^5(\sigma^*)^2$ Co^{2+}

TABLE 12-2
KINETIC PARAMETERS FOR SOME OUTER SPHERE REACTIONS[a-c]

Reducing Agent	$\Delta E^{\ddagger}$ Required for Red. Ag.?	Oxidizing Agent	$\Delta E^{\ddagger}$ Required for Oxid. Ag.?	Reduction Potential of Oxid. Ag. (Volts vs. S.H.E.[d])	$\Delta E°$ (volts)	Exchange Rate for Oxid. Ag.	k_{obs} ($M^{-1}sec^{-1}$)	k_{calc}
Net $\sigma^* \to \sigma^*$								
Cr^{2+}	yes	$[Co(NH_3)_6]^{3+}$	yes	+0.1	0.51	$\leq 10^{-9}$	1.0×10^{-3}	$\leq 1.6 \times 10^{-3}$
Cr^{2+}	yes	$[Co(en)_3]^{3+}$	yes	-0.24	0.17	2×10^{-5}	3.4×10^{-4}	$\leq 5.1 \times 10^{-4}$
Cr^{2+}	yes	$[Co(phen)_3]^{3+}$	yes	+0.42	0.83	5.0	3.0×10^1	$\leq 1.1 \times 10^4$
Net $\sigma^* \to \pi^*$								
Cr^{2+}	yes	$[Ru(NH_3)_6]^{3+}$	no	+0.1	0.51	4×10^3	2×10^2	$\leq 1.5 \times 10^3$
Net $\pi^* \to \sigma^*$								
V^{2+}	no	$[Co(NH_3)_6]^{3+}$	yes	+0.1	0.355	$\leq 10^{-9}$	1.0×10^{-2}	$\leq 2.3 \times 10^{-3}$
V^{2+}	no	$[Co(en)_3]^{3+}$	yes	-0.24	0.015	2×10^{-5}	7.2×10^{-4}	5.8×10^{-4}
V^{2+}	no	$[Co(phen)_3]^{3+}$	yes	+0.42	0.675	5.0	3.8×10^3	2.3×10^4
Net $\pi^* \to \pi^*$								
V^{2+}	no	$[Ru(NH_3)_6]^{3+}$	no	+0.1	0.355	4×10^3	80	4.2×10^3

[a] Oxidation potentials: $Cr^{2+} = Cr^{3+} + e^-$; $E° = +0.41$ v. $V^{2+} = V^{3+} + e^-$; $E° = +0.255$ v.
[b] Exchange rates for $Cr^{2+}/Cr^{3+} \leq 2 \times 10^{-5}$; for $V^{2+}/V^{3+} = 1 \times 10^{-2}$
[c] Data taken from: T. Przystas and N. Sutin, J. Amer. Chem. Soc., **95**, 5545 (1973); J. F. Endicott and H. Taube, ibid., **86**, 1686 (1964); T. J. Meyer, and H. Taube, Inorg. Chem., **7**, 2369 (1968).
[d] S.H.E. = standard hydrogen electrode.

Given that low spin Co(III) complexes have energies not far removed from the low spin/high spin crossover (review the analysis of $[Co(en)_3]^{3+}$ in Chapter 9, page 574, where $\Delta_0 \geq P$), it would seem that the greater part of the total chemical activation energy accompanying electron transfer in any one Co(II)/Co(III) system must derive from the Co(II) complex. For $[CoL_6]^{2+}$ complexes, a rough upper estimate of activation energy may be derived from the nominal values of $\Delta_0 \approx 10$ kK and $P \approx 20$ kK; this means that $|\Delta_0 - P| \approx 10$ kK ≈ 30 kcal/mole. This is, of course, a gross overestimate of $\Delta E^{\ddagger}$[Co(II)] because it is the high spin/low spin energy difference for a Co—L distance appropriate to the high spin structure alone.†

With respect to the variation in k in Table 12-1 as L varies from H_2O to NH_3 to en, we may make the following observations: The Co(III) activation energy should increase while the Co(II) activation energy should decrease. In this situation a minimum in k at NH_3 is not too surprising, although one is hard-pressed to estimate the relative importance of the *changes* in $\Delta E^{\ddagger}$[Co(III)] and $\Delta E^{\ddagger}$[Co(II)] to the total $\Delta E^{\ddagger}$ as the ligands are varied. It is also impossible to predict at present the changes in $\Delta S^{\ddagger}$ from the complexes, $\Delta G^{\ddagger}$ from the solvent cage reorganization, and differences in K for precursor formation.

Cross Reactions and Thermodynamics

The reactions we have discussed thus far are *exchange reactions* occurring between two different oxidation states of the same compound. But what about *cross reactions,* reactions between two completely different ions or molecules? It is clear that, in order to achieve the maximum rate in an exchange reaction, one should have systems that effect $\pi^* \rightarrow \pi^*$ electron transfer. Cross reactions, however, are often rapid in spite of an otherwise large chemical activation (transfer involving at least one σ^* electron) because the large $\Delta E^{\ddagger}$ is partly "neutralized" by an attendant negative free energy change. Some examples of cross reactions are given in Table 12-2 along with standard electrochemical cell potentials; you will notice that there is a rough correlation between the rate of reaction and its free energy change (as given by $\Delta G° = -nFE°$).

The exchange reactions discussed in the preceding section arise when reactants and products are identical; $\Delta G° = 0$ for such reactions. However, in cross reactions (Table 12-2) the products are more stable than the reactants ($\Delta G° < 0$), and, all other factors being equal, there is a distortion of the energy profile diagram so as to lower $\Delta E^{\ddagger}$ (Figure 12-3). In the simplest possible terms, the thermodynamic spontaneity

---

†The rates of low spin/high spin interconversion for Co(II) complexes in solution have recently been measured. For ^2E(low spin) $\underset{k_{-1}}{\overset{k_1}{\rightleftarrows}}$ ^4T(high spin), $k_1 = 3.2 \times 10^6$ sec^{-1} and $k_{-1} = 9.1 \times 10^6$ sec^{-1}. M. G. Simmons and L. J. Wilson, *Inorg. Chem.*, **16**, 126 (1977).

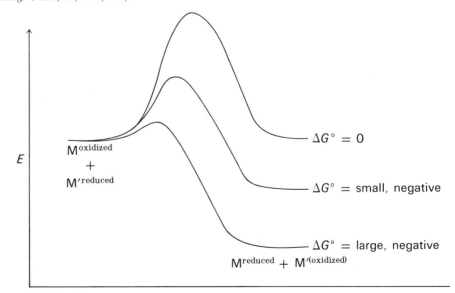

Figure 12-3. Activation energy profile for an electron transfer reaction in which the free energies of the products may be less than those of the reactants. Notice that the transfer of electrons now occurs "earlier" along the reaction coordinate.

alters the activation energy by supporting electron transfer "earlier" in the chemical activation process (less oxidant/reductant structure deformation).

The outer sphere reductions of $[Ru(NH_3)_6]^{3+}$ and $[Co(NH_3)_6]^{3+}$ by Cr(II) and V(II) (see Table 12-2) provide some insight into the thermodynamics of the activation process. The Cr(II) outer sphere reductions should require greater $\Delta E^{\ddagger}$ than the V(II) reductions because the high spin ion $[Cr(H_2O)_6]^{2+}$ with a $(\pi^*)^3(\sigma^*)^1$ configuration requires activation toward the low spin $(\pi^*)^4$ configuration. On the other hand, little activation of $[V(H_2O)_6]^{2+}$ $[(\pi^*)^3]$ should be required, as there is no change in σ^* occupation on electron transfer. This expectation is realized in Cr(II)/Cr(III) exchange ($k \leq 10^{-5}$) compared with V(II)/V(III) exchange ($k \approx 10^{-2}$), and is generally seen in outer sphere Cr(II) and V(II) cross reactions even though $\Delta G°$ for Cr(II) reductions is more negative than for V(II) reductions.[‡]

The rate constants of cross reactions can be predicted reasonably well by using the approach of Marcus[27] and Hush.[28,29] That is, the rate constant, k_{12}, of a cross reaction is given by the equation

(12-12a)
$$k_{12} = (k_1 k_2 K_{12} f)^{1/2}$$

where K_{12} is the equilibrium constant for a cross reaction such as

(12-13)
$$[Fe(CN)_6]^{4-} + [Mo(CN)_8]^{3-} \rightarrow [Fe(CN)_6]^{3-} + [Mo(CN)_8]^{4-}$$

and k_1 and k_2 are the rate constants for the exchange reactions

(12-14)
$$[Fe(CN)_6]^{4-} + [\overset{*}{Fe}(CN)_6]^{3-} \xrightarrow{k_1} [Fe(CN)_6]^{3-} + \overset{*}{Fe}(CN)_6]^{4-}$$

$$[Mo(CN)_8]^{4-} + [\overset{*}{Mo}(CN)_8]^{3-} \xrightarrow{k_2} [Mo(CN)_8]^{3+} + [\overset{*}{Mo}(CN)_8]^{4-}$$

The parameter f, which is usually close to 1, is given by

(12-12b)
$$\log f = \frac{(\log K_{12})^2}{4 \log (k_1 k_2/z^2)}$$

where z is a measure of collision frequency.[30] As seen in Table 12-2, reasonably good agreement between theory (k_{calc}) and experiment (k_{obs}) is found.

An important conclusion can be drawn from the discussion of the Cr(II) and V(II) reductions in Table 12-2 and from the Marcus-Hush equation for k_{12}: the barrier to electron transfer consists of two parts, a so-called intrinsic contribution [the difference in chemical activation for transferring a σ^* electron in the case of Cr(II) and a π^* electron in the case of V(II)] and a thermodynamic part ($\Delta G°$ for the reaction). That is, the free energy of activation for an electron transfer reaction, $\Delta G^{\ddagger}_{12}$, is given by

(12-12c)
$$\Delta G^{\ddagger}_{12} = \underbrace{\frac{(\Delta G^{\ddagger}_1 + \Delta G^{\ddagger}_2)}{2}}_{\text{intrinsic contribution}} + \underbrace{\frac{\Delta G°_{12}}{2}}_{\text{thermodynamic contribution}}$$

This equation, where $\Delta G^{\ddagger}_1$ and $\Delta G^{\ddagger}_2$ are the activation free energies appropriate to the exchange reactions and $\Delta G°$ is the free energy for the cross reaction, follows directly

---

[‡] If V(II) reductions are faster than Cr(II) reductions of a common oxidant, this is considered indicative of an outer sphere mechanism. See, for example, M. Hery and K. Wieghardt, *Inorg. Chem.*, **15**, 2315 (1976).
[27] R. A. Marcus, *J. Phys. Chem.*, **72**, 891 (1968).
[28] N. S. Hush, *Trans. Faraday Soc.*, **57**, 557 (1961).
[29] See refs. 7 and 8.
[30] The diffusion controlled rate of collision between uncharged particles in solution is 10^{11} M^{-1}sec^{-1}.

from equation (12–12a) (assuming that $f = 1$). If equation (12–12c) is dominated by the free energy change in the cross reaction, then there may be a linear relationship between log k and $\Delta G°$. A rough correlation of this type was noted above (Table 12–2) and an excellent one was reported recently.[31,32]

STUDY QUESTIONS

1. The outer sphere electron exchange between $[IrCl_6]^{2-}$ and $[IrCl_6]^{3-}$ has been examined. Would you expect the reaction to be relatively fast or relatively slow? That is, in comparison with the rate constants in Table 12–1, will the iridium system have a rate constant close to that for the ruthenium compounds in Table 12–1 or closer to the cobalt compounds in that table?

2. In the table below are given the rate constants for the exchange of an electron between the hexaaquo metal ions listed. All occur by outer sphere transfer. Rationalize these rate constants in terms of the general rules for relative rates of outer sphere transfers.

Metal Ion Pair	Rate Constant ($M^{-1}sec^{-1}$)
Fe^{2+}/Fe^{3+}	4
Cr^{2+}/Fe^{3+}	2.3×10^3
V^{2+}/Fe^{3+}	1.8×10^4
Cr^{2+}/Cr^{3+}	2×10^{-5}
V^{2+}/V^{3+}	1×10^{-2}

(You will need some electrochemical information in your answer. The oxidation potentials for V^{2+} and Cr^{2+} are in Table 12–2; the oxidation potential for Fe^{2+} is -0.771 v vs. S.H.E.)

INNER SPHERE ELECTRON TRANSFER REACTIONS

The reduction of hexaamminecobalt(3+) by hexaaquochromium(2+) occurs rather slowly ($k = 10^{-3}$ $M^{-1}sec^{-1}$) by an outer sphere mechanism.

$$[Co(NH_3)_6]^{3+} + [Cr(H_2O)_6]^{2+} \xrightarrow{H^+} [Co(H_2O)_6]^{2+} + 6NH_4^+ + [Cr(H_2O)_6]^{3+} \quad (12\text{–}15)$$
$$(\pi^*)^6 \qquad (\pi^*)^3(\sigma^*)^1 \qquad (\pi^*)^5(\sigma^*)^2 \qquad\qquad\qquad (\pi^*)^3$$

However, if one ammonia ligand on Co(III) is substituted by Cl^-, reaction (12–16) now occurs,

$$[Co(NH_3)_5Cl]^{2+} + [Cr(H_2O)_6]^{2+} \xrightarrow{H^+} [Co(H_2O)_6]^{2+} + [Cr(H_2O)_5Cl]^{2+} + 5NH_4^+ \quad (12\text{–}16)$$
$$(\pi^*)^6 \qquad (\pi^*)^3(\sigma^*)^1 \qquad (\pi^*)^5(\sigma^*)^2 \qquad (\pi^*)^3$$

and with a substantially greater rate ($k = 6 \times 10^5$ $M^{-1}sec^{-1}$). It is clear that different mechanistic pathways are probably utilized by these two reactions, a fact that Taube

---

[31] D. P. Rillema, J. F. Endicott, and R. C. Paul, *J. Amer. Chem. Soc.*, **94**, 394 (1972).

[32] Before leaving this discussion of cross reactions and thermodynamics, there are several additional points to be made concerning the data in Table 12–2. You should be alert to the facts that, as in Co^{2+} reductions, the Co^{3+} contribution to the activation should be small, that $[Ru(NH_3)_6]^{3+}$, with a $(\pi^*)^5$ configuration, should require little activation, and that phen, as before, enhances the rate constant and emphasizes the $\pi^* \to \pi^*$ nature of the electron transfer process.

has elegantly demonstrated.[33-35] Indeed, it is likely that an inner sphere mechanism is followed in the reduction of $[Co(NH_3)_5Cl]^{2+}$; that is, the Cl^- ligand, while still attached to the Co(III), replaces an H_2O at Cr(II) to give an intermediate such as **7**, and electron transfer follows.

$$\left[\begin{array}{c} (H_3N)_4(Cl)Co\text{—}Cl\text{—}Cr(OH_2)_5 \end{array} \right]^{4+}$$

7

The proof that this is the correct pathway is as follows: The reductant, $[Cr(H_2O)_6]^{2+}$, is substitution labile,[36] as is the product $[Co(H_2O)_6]^{2+}$ (the specific rate constant for the exchange of chromium(II)-coordinated water with solvent water is greater than 10^9 sec^{-1}, while that for exchange with $[Co(H_2O)_6]^{2+}$ is $2 \times 10^5 \text{ sec}^{-1}$). On the other hand, the two chloro complexes are substitution inert.[36] Therefore, the only way in which chloride could be transferred from Co(III) to Cr(II) in a reaction where $k \approx 10^5 \text{ M}^{-1}\text{sec}^{-1}$ would be through some complex resulting from attack of $CoCl^{2+}$ on Cr^{2+}. The Cl^- must have been passed *directly* from one ion to the other. An alternative mechanism might have been outer sphere electron transfer followed by release of Cl^- by Co(II) to the solution and substitution of the free chloride onto Cr(III). However, this is quite unlikely since the specific rate constant for the replacement of water in $[Cr(H_2O)_6]^{3+}$ by Cl^- is $3 \times 10^{-8} \text{ M}^{-1}\text{sec}^{-1}$. Just as convincing, though, was another experiment; upon adding free, radioactive Cl^- to the $[Co(NH_3)_5Cl]^{2+}/[Cr(H_2O)_6]^{2+}$ solution, no radio-chloride was found in the product, $[Cr(H_2O)_5Cl]^{2+}$.

The general conclusion that can be drawn from Taube's experiment is that *an inner sphere mechanism can be unequivocally assigned when both the oxidant and the oxidized reducing agent are substitution inert and when ligand transfer from oxidant to reductant has accompanied electron transfer.* This criterion has been successfully applied to numerous reactions of $[Cr(H_2O)_6]^{2+}$ with $[Co(NH_3)_5X]^{n+}$ and $[Cr(H_2O)_5X]^{n+}$. However, we must emphasize at the outset that *ligand transfer is not a requirement of the inner sphere mechanism.* The reduction of the hexachloroiridate(2−) ion by Cr(II)

(12-17)

$$[IrCl_6]^{2-} + [Cr(H_2O)_6]^{2+} \rightarrow [IrCl_6]^{3-} + [Cr(H_2O)_6]^{3+}$$

$$(\pi^*)^5 \qquad (\pi^*)^3(\sigma^*)^1 \qquad (\pi^*)^6 \qquad (\pi^*)^3$$

is apparently such a case. Although the two reacting metal ions may well be bridged, the bond between the reduced metal ion and the bridge [Ir(III)—Cl] may be stronger than the bond between the oxidized metal ion and the bridge [Cr(III)—Cl], and the latter will be broken before the former.

The elementary steps in a generalized inner sphere mechanism (in aqueous solution) can be represented by:

---

[33] See refs. 5 and 6.
[34] H. Taube, H. Myers, and R. L. Rich, *J. Amer. Chem. Soc.*, **75**, 4118 (1953).
[35] H. Taube and H. Myers, *J. Amer. Chem. Soc.*, **78**, 2103 (1954).
[36] See ref. 18.

Formation of precursor complex $\quad$ Ox—X + Red(H$_2$O) $\rightleftharpoons$ Ox—X$\cdots$Red + H$_2$O $\quad$ (12–18)

Activation of precursor complex and electron transfer $\quad$ Ox—X$\cdots$Red $\rightleftharpoons$ $\overset{-}{\text{Ox}}\cdots$X—$\overset{+}{\text{Red}}$

Dissociation to separated products $\quad$ $\overset{-}{\text{Ox}}\cdots$X—$\overset{+}{\text{Red}}$ + H$_2$O $\rightleftharpoons$ Ox(H$_2$O)$^-$ + RedX$^+$

Before beginning to discuss the mechanistic details of inner sphere processes, we note that such reactions usually show second order kinetic behavior.[37] Therefore, for the net process

$$\text{Ox—X + Red} \underset{k_2}{\overset{k_1}{\rightleftharpoons}} [\text{Ox—X—Red}] \underset{k_4}{\overset{k_3}{\rightleftharpoons}} \text{Ox}^- + \text{RedX}^+ \quad (12\text{–}19)$$

the rate law is given by

$$\text{Rate}_f = \frac{k_1 k_3}{k_2 + k_3}[\text{Ox—X}][\text{Red}] \quad (12\text{–}12\text{d})$$

where k_3 is an overall rate constant for the second and third steps. In some cases $k_3 \gg k_2$, and the rate determining step is formation of the precursor complex (that is, substitution of the bridge ligand X for H$_2$O on Red). Therefore, the rate law would be Rate$_f = k_1[\text{Ox—X}][\text{Red}]$. However, there are many instances in which the rate determining step is rearrangement and electron transfer within the intermediate, or fission of the successor complex. This means that $k_3 < k_2$ and the rate law becomes Rate$_f = Kk_3[\text{Ox—X}][\text{Red}]$.

Formation of Precursor Complexes

As you shall see in the next chapter on substitution reactions, octahedral complexes with a d^3 metal ion are relatively inert to substitution, while high spin d^4 and d^5 complexes are labile to substitution.[38] For example, compare the following reactions for V^{2+} [a $(\pi^*)^3$ ion], Cr^{2+} [a $(\pi^*)^3(\sigma^*)^1$ ion], and Fe^{2+} [a $(\pi^*)^4(\sigma^*)^2$ ion].

[V(H$_2$O)$_6$]$^{2+}$ + H$_2$O* $\rightarrow$ [V(H$_2$O)$_5$(H$_2$O*)]$^{2+}$ + H$_2$O $\quad k = 1 \times 10^2 \text{ sec}^{-1}$ $\quad$ (12–20)

[V(H$_2$O)$_6$]$^{2+}$ + NCS$^-$ $\rightarrow$ [V(H$_2$O)$_5$NCS]$^+$ + H$_2$O $\quad k = 28 \text{ M}^{-1}\text{sec}^{-1}$ $\quad$ (12–21)

[Cr(H$_2$O)$_6$]$^{2+}$ + H$_2$O* $\rightarrow$ [Cr(H$_2$O)$_5$(H$_2$O*)]$^{2+}$ + H$_2$O $\quad k = 10^9 \text{ sec}^{-1}$ $\quad$ (12–22)

[Fe(H$_2$O)$_6$]$^{2+}$ + H$_2$O* $\rightarrow$ [Fe(H$_2$O)$_5$(H$_2$O*)]$^{2+}$ + H$_2$O $\quad k = 10^6 \text{ sec}^{-1}$ $\quad$ (12–23)

Such differences in substitution rates will clearly be important in determining relative inner sphere electron transfer rates for V(II) and Cr(II). In fact, the data for V(II) reductions in Table 12-3 suggest that when this ion functions as an inner sphere reductant, the reaction rate is controlled by the rate at which water is lost from the coordination sphere of [V(H$_2$O)$_6$]$^{2+}$. This conclusion is reasonable because (i) rates of inner sphere reductions by V(II) are generally quite comparable to the rate of water substitution on the ion, and (ii) just as importantly, there is little variation in the rate of the electron transfer reaction with the nature of the bridging ligand supplied by the oxidant.

The conclusions regarding V(II) reductions in Table 12-3 are in direct contrast with those concerning Cr(II) and Fe(II) reductions. The rates of their electron transfer reactions are considerably less than their rates of water substitution.

---

[37] See Chapter 7 for a complete discussion of kinetics; recall that the second order behavior of the reaction does not imply a particular mechanism.

[38] J. M. Malin and J. H. Swinehart, *Inorg. Chem.*, **7**, 250 (1968).

TABLE 12-3
SECOND ORDER RATE CONSTANTS (k, $M^{-1}sec^{-1}$) FOR THE INNER SPHERE REDUCTION OF VARIOUS Co(III) COMPLEXES AND SOME Ru(III) COMPLEXES AT 25°. NOTE THAT R = $Co(NH_3)_5$

Oxidant	Reductant [HOMO]		
	$V^{2+}[(\pi^*)^3(\sigma^*)^0]$ net π^* to σ^*	$Cr^{2+}[(\pi^*)^3(\sigma^*)^1]$ net σ^* to σ^*	$Fe^{2+}[(\pi^*)^4(\sigma^*)^2]$ net σ^* to σ^*
RF^{2+}	—	2.5×10^5	6.6×10^{-3}
RCl^{2+}	7.6	6×10^5	1.4×10^{-3}
RBr^{2+}	25	1.4×10^6	7.3×10^{-4}
RI^{2+}	13	3.4×10^6	—
RN_3^{2+}	13	3.0×10^5	8.7×10^{-3}
$RNCS^{2+}$	0.3	1.9×10^1	3.0×10^{-3}
$RSCN^{2+}$	—	see text	1.2×10^{-1}
$RC_2O_4H^{2+}$	12.5	4.0×10^2	4.3×10^{-1}
$Ru(NH_3)_5Cl^{2+}$	—	3.5×10^4	—
$Ru(NH_3)_5Br^{2+}$	—	2.2×10^3	—
$Ru(NH_3)_5I^{2+}$	—	$<5 \times 10^2$	—

Data taken from: H. Taube, "Electron Transfer Reactions of Complex Ions in Solution," Academic Press, New York (1970), page 51; J. A. Stritar and H. Taube, *Inorg. Chem.*, **8**, 2284 (1969); J. Espenson, *Inorg. Chem.*, **4**, 121 (1965).

In a few instances there is information on inner sphere complexes and their stability. For example, the nitrilotriacetic acid (= NTA) complex of penta-amminecobalt(III) reacts with Fe(II) according to the equation[39]

(12-24) [structure 8] ⇌ [structure 9] $\xrightarrow{k_3}$ Co(II) + Fe^{III} NTA
    $K = k_1/k_2$

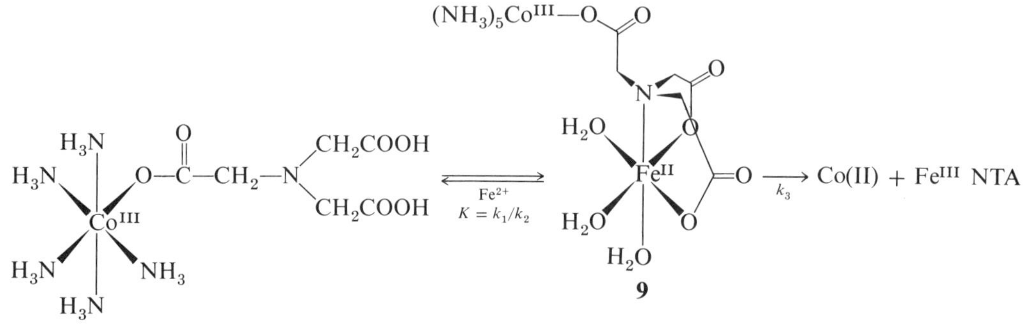

The overall rate constant, $k_f = Kk_3$, was found to be 1×10^5 $M^{-1}sec^{-1}$ at 25°. The nitrilotriacetic acid ligand apparently greatly enhances the reaction, since Fe(II) reductions of $[Co(NH_3)_5Cl]^{2+}$ and $[Co(NH_3)_5C_2O_4H]^{2+}$ are considerably slower ($k_f = 1.35 \times 10^{-3}$ $M^{-1}sec^{-1}$ and 0.43 $M^{-1}sec^{-1}$, respectively). It was suggested that a complex such as **9** is formed when the nitrilotriacetic acid ligand is used and that this complex is extremely stable. Its formation constant (= K) was found to be 1.1×10^6 M^{-1}, meaning that k_3 is 9.4×10^{-2} sec^{-1}; thus, the lifetime of the complex in which electron transfer occurs is about 10 seconds.[40]

---

[39] R. D. Cannon and J. Gardiner, *J. Amer. Chem. Soc.*, **92**, 3800 (1970).
[40] Another example of a relatively long-lived precursor complex is given by D. Gaswick and A. Haim, *J. Amer. Chem. Soc.*, **96**, 7845 (1974).

Rearrangement of the Precursor Complex and Electron Transfer

More often than not, the formation of the reactant pair is very fast and reversible and the rate of an inner sphere reaction is controlled by the rearrangement and electron transfer steps, as in reaction 24 above. Indeed, it is these portions of the mechanism which have received the greatest amount of study in the past few years.

Clearly the rate of an inner sphere electron transfer can depend on the nature of the oxidant, the reductant, and the bridging species. We turn first to the connection between rate and the electronic structure of the metal ions involved.

Electronic Structure of the Oxidant and Reductant

The outer sphere reduction of $[Co(NH_3)_6]^{3+}$ by $[Cr(H_2O)_6]^{2+}$ is quite slow ($k = 1.6 \times 10^{-3}$ M^{-1}sec^{-1}). However, when one ammonia ligand is replaced by a halide, the following reaction, cited earlier, occurs:

$$[Co(NH_3)_5Cl]^{2+} + [Cr(H_2O)_6]^{2+} \xrightarrow{H^+} [Co(H_2O)_6]^{2+} + [Cr(H_2O)_5Cl]^{2+} + 5NH_4^+$$
$$(k = 6 \times 10^5 \text{ M}^{-1} \text{ sec}^{-1}) \quad (12\text{-}16')$$

As you have already seen, it is presumed that the mechanism is inner sphere since the halide ion has been transferred to the oxidized reducing agent. However, what is more important now is that the reaction has been tremendously accelerated relative to the outer sphere mechanism for the similar reaction. Since a similar effect is observed for numerous other acidopentaamminecobalt(2+) complexes, some general explanation is called for.

The answer is found in a consideration of the symmetries of the reductant orbital from which the electron is lost and the oxidant orbital into which the electron moves. Examination of the data in Tables 12-2 and 12-3 for outer sphere and inner sphere reactions, respectively, gives the following general observations:

HOMO	LUMO	Example	Approximate acceleration on going from outer sphere to inner sphere mechanism for similar reaction
σ^*	σ^*	Cr^{2+}/Co^{3+}	10^{10}
σ^*	π^*	Cr^{2+}/Ru^{3+}	10^2
π^*	σ^*	V^{2+}/Co^{3+}	10^4
π^*	π^*	V^{2+}/Ru^{3+}	no data—all reactions occur by outer sphere mechanism

The obvious conclusion is that, while some rate acceleration has generally occurred on going from an outer sphere to an inner sphere mechanism in comparable reactions, the effect is most marked when both HOMO and LUMO are σ^*. This is somewhat surprising, since, as you saw in the discussion of outer sphere mechanisms, there is a requirement of considerable chemical activation because of bond distance changes in transfers to or from σ^* orbitals. Furthermore, effective mixing of σ^* donor and receptor mo's should not be as readily achieved as if both were of the more "exposed" π^* type. These impediments to rapid reaction are apparently largely overcome when the bridging ligand is intimately involved with donor and receptor mo's and when bridging group transfer accompanies electron transfer, as it does in reactions such as (12-16). Orbital following arguments (Chapter 7 and Figure 12-1) are particularly effective here, and the essential features of the inner sphere mechanism can be clearly illustrated using a simple M^{ox}—X$\cdots$M'^{red} fragment. In the transition state, a three-center/three-orbital model is appropriate (and bears a strong resemblence to the orbital pattern for a compound such as XeF_2; see Chapter 4). The sketches in Figure 12-4 illustrate qualitatively the σ mo system for such a model, where the important ao's are the d_{z^2} of M^{ox} and M'^{red} and the $p\sigma$ of the bridging Cl^-.

674 ☐ REACTION MECHANISMS AND METHODS OF SYNTHESIS

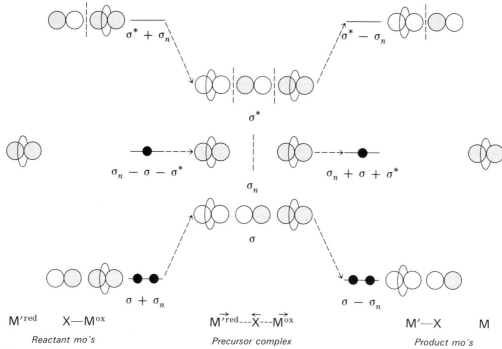

Figure 12-4. Orbital following for an inner sphere electron transfer reaction $M'^{red} + M^{ox}$—X to give M'—X (where the oxidation state of M' has increased by 1) and M (where the oxidation state of M has decreased by 1) The σ mo's of the precursor or activated complex are labeled σ, $σ_n$, and σ*. The mo's of the reactants and the products are linear combinations of σ, $σ_n$, and σ* brought about by movement of X toward M'^{red}. (See Chapter 4 for the principles of mo construction.)

On the left are the orbitals appropriate for the reactants, in the middle those for the precursor, and on the right those for the products. The critical feature of this model is that as Cl^- transfers from M to M', the odd electron orbital becomes concentrated on M. The orbital following argument is as follows:

As in Figure 12-1, the mo's of the precursor complex are labeled σ, $σ_n$, and σ*. Figure 12-4 shows that linear combinations of these mo's can lead to either the product or reactant mo's. [For example, the odd electron orbital on the reactant (i.e., on M'^{red}) arises from the orbital following combination of $σ_n - σ - σ^*$.] The most important feature of this model is that the reactant and product mo's directly correlate with one another. That is, the odd electron orbital on the reductant correlates with the odd electron orbital on the product, and the M^{ox}—Cl σ bonding mo correlates with the M'—Cl σ bonding mo. Transfer of the odd electron from M'^{red} to M^{ox} and transfer of the chloride from M^{ox} to M'^{red} is accomplished smoothly through $σ_n$.

The smoothness of the orbital following, and the consequent observed rate enhancement, is attributable to two important features: the donor mo of the reductant and the receptor mo of the oxidant are of the same type (i.e., σ*), and ligand transfer accompanies electron transfer. Precursor formation itself weakens the original M^{ox}—Cl bond and partially transfers the odd electron to M^{ox} because $σ_n$ is delocalized. Further progress along the reaction coordinate sees the odd electron move smoothly out of the Cl—M'^{red} region (where it would otherwise play an anti-bonding role) into the M^{ox}—Cl region, where its anti-bonding role becomes necessarily weaker until it is non-bonding.

One final comment concerning the relative slowness of the V(II) and Fe(II) reductions in Table 12-3 is called for. This slowness is believed to be due to several factors. For V(II), which is substitution inert, formation of the precursor complex will be slower than with Cr(II). Furthermore, to take advantage of σ orbital following, V(II) must activate to a $(π^*)^2(σ^*)^1$ configuration. Finally, the thermodynamic lowering of $\Delta E^{\ddagger}$ is not as great for V(II) as for Cr(II). You will also notice that Fe(II) $[(π^*)^4(σ^*)^2]$ does not suffer from either slow precursor complex formation or the need for electronic activation to take advantage of the σ orbital following; rather, Fe(II) reductions are slow because of a marked thermodynamic disadvantage ($Fe^{2+} = Fe^{3+} + e^-$, $E° = -0.77$ v, compared with $Cr^{2+} = Cr^{3+} + e^-$, $E° = +0.41$ v; $\Delta G° > 0$).

The Nature of the Bridge Ligand

A glance at Table 12-3 shows that the nature of the bridge ligand is important, and extensive research has been done to understand its importance.[41] As Haim recently pointed out, "The role of the bridging ligand [in an inner sphere mechanism] is . . . dual. It brings the metal ions together (thermodynamic contribution) and mediates the transfer of the electron (kinetic contribution)."[41] The thermodynamic contribution arises from factors important to the stability of the intermediate complex, and the kinetic contribution arises from factors such as oxidant-reductant reorganization and matching of donor and receptor mo types. Both inorganic and organic bridges have been studied, and we shall discuss them in that order.

In studying electron transfer reactions, a major difficulty is often encountered in deciding whether a particular reaction is inner or outer sphere. However, one relatively simple experimental test has been devised and was found to apply to a limited number of reactions of the type[42]

$$[Co(NH_3)_5X]^{2+} + [Cr(H_2O)_6]^{2+} \xrightarrow{H^+} [Cr(H_2O)_5X]^{2+} + [Co(H_2O)_6]^{2+} + 5NH_4^+ \qquad (12\text{-}25)$$

where the bridge ligand X is N_3^- or NCS^-; $k_{N_3^-}^{o.s.}/k_{NCS^-}^{o.s.} \approx 1$, while $k_{N_3^-}^{i.s.}/k_{NCS^-}^{i.s.} \gg 1$ (see Tables 12-3 and 12-4). The reason for this may be the preferential formation of the intermediate in the inner sphere mechanism by one bridge ligand or the other. Considering the resonance structures that should in principle contribute to the N_3^- (**10**) and NCS^- (**11**) precursors,

$$\overset{\ominus}{M^{ox}}-\overset{..}{\underset{}{N}}=\overset{\oplus}{\underset{}{N}}=\overset{\ominus}{\underset{}{\ddot{N}}}: \qquad \overset{\ominus}{M^{ox}}-\overset{..}{\underset{}{N}}=C=\overset{..}{\underset{}{\ddot{S}}}:$$
$$\qquad\qquad \textbf{10} \qquad\qquad\qquad \textbf{11}$$

an azide bridge can better form an intermediate like **12** than can thiocyanate.

12

Further work using NCS^- as a bridging ligand has uncovered the fact that the situation is more complicated and even more interesting than just indicated. When

---

[41] See refs. 10 and 11.
[42] J. H. Espenson, *Inorg. Chem.*, **4**, 121 (1965).

TABLE 12-4
RELATIVE RATES OF REDUCTION OF AZIDO AND THIOCYANATO COMPLEXES AT 25°C[a]

Oxidant	Reductant	$k_{N_3^-}/k_{NCS^-}$	Reaction Type
$[Co(NH_3)_5X]^{2+}$	Cr^{2+}	10^4	inner sphere
$[Co(NH_3)_5X]^{2+}$	V^{2+}	27	not determined[b]
$[Co(NH_3)_5X]^{2+}$	Fe^{2+}	$\geq 3 \times 10^3$	inner sphere
$[Co(NH_3)_5X]^{2+}$	$Cr(bipy)_3^{2+}$	4	outer sphere
$[Cr(H_2O)_5X]^{2+}$	Cr^{2+}	4×10^4	inner sphere

[a] Data taken from J. Espenson, *Inorg. Chem.*, **4**, 121 (1965).
[b] This is probably an inner sphere reaction that is controlled by the rate of water replacement on the V^{2+}.

NCS⁻ is S-bound to the oxidant, $[Co(NH_3)_5SCN]^{3+}$, the nitrogen end of the ligand can be attacked by the reductant, say $[Cr(H_2O)_6]^{2+}$, and the rate of this reaction is comparable to the case in which azide is the bridge (1.9×10^5 M⁻¹sec⁻¹ vs. 3.0×10^5 M⁻¹sec⁻¹ for the N_3^- bridge). However, S-bound thiocyanate allows another possibility for attack by the reducing agent. When S-bound to the oxidant, the ligand is bent[43] and there are two lone pairs of electrons on the sulfur atom; the reductant can attack at the sulfur atom directly bound to the oxidant metal ion, and the

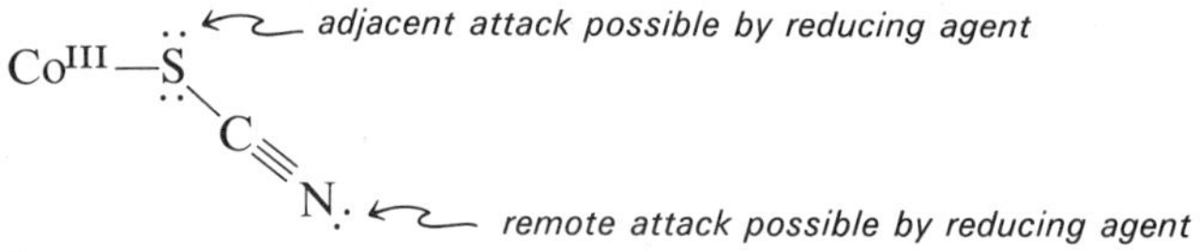

$Co^{III}—\ddot{S}$ ← adjacent attack possible by reducing agent
    \
     $C≡N:$ ← remote attack possible by reducing agent

observed rate is nearly as rapid (8×10^4 M⁻¹sec⁻¹) as when attack occurs at the more remote nitrogen end. That is, both *remote attack* at the nitrogen and *adjacent attack* at the sulfur are possible and have been observed to occur at nearly comparable rates, at least when $[Cr(H_2O)_6]^{2+}$ is the reducing agent.[44]

Studies of inner sphere reactions with *organic bridging agents* are extremely interesting and informative, in that they show that reaction rates can be controlled by bridge steric effects, the point of attack by the reductant on the bridge, and the electronic structure of the bridge, including its reducibility.[45,46] Most such studies have centered on carboxylate complexes (such as **13**) and amine complexes (such as **14**). Second order kinetics are still observed for these reactions, and the proposed mechanisms all involve attack at a remote site rather than at the atom attached directly to the oxidant metal center.

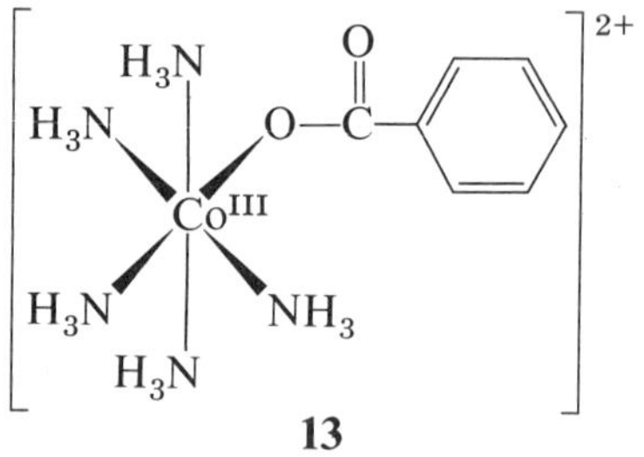

[pentaamminecobalt(III) benzoate complex] **13** (charge 2+)

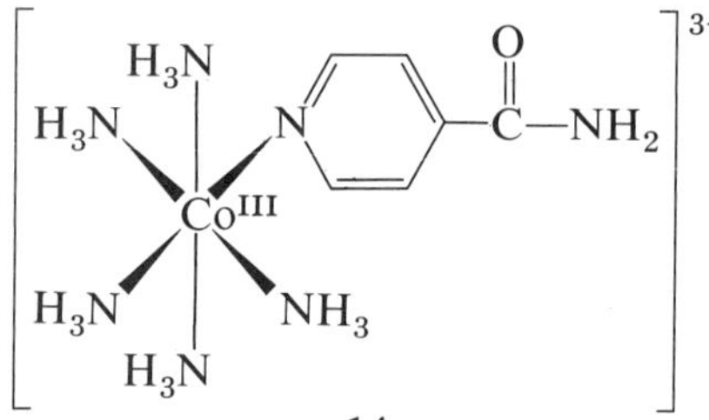

[pentaamminecobalt(III) isonicotinamide complex] **14** (charge 3+)

Pictured below are a few of the many acidopentaamminecobalt(III) complexes that have been observed to react with Cr(II). Attack by the Cr(II) ion is at the carbonyl oxygen of the carboxylate group, and it is clear that the rate constants for reduction by $[Cr(H_2O)_6]^{2+}$ fall with increasing steric bulk of the organic group (compounds **15** to **17**). Further, if another basic group is added, chelation of the reducing agent [Cr(II) in this case] is possible, and the rate constants increase on going from **16** to **19**, for example.[47]

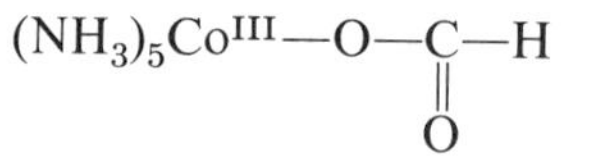

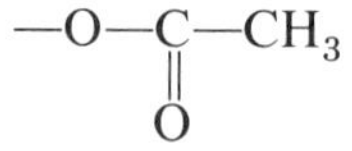

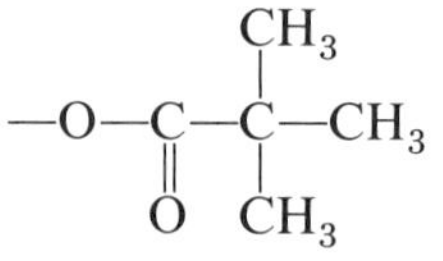

$(NH_3)_5Co^{III}—O—\underset{\underset{O}{\|}}{C}—H$ $—O—\underset{\underset{O}{\|}}{C}—CH_3$ $—O—\underset{\underset{O}{\|}}{C}—\underset{\underset{CH_3}{|}}{\overset{\overset{CH_3}{|}}{C}}—CH_3$

$k_f = 7.2$ M⁻¹sec⁻¹ for Cr(II) $k_f = 0.35$ $k_f = 9.6 \times 10^{-3}$

15   **16**   **17**

---

[43] When thiocyanate is S-bound to a metal ion, the two major resonance structures are $M—\overset{\ominus}{\underset{..}{\ddot{S}}}—C≡N:$ and $M—\overset{\oplus}{\underset{..}{\ddot{S}}}=C=\overset{\ominus}{\ddot{N}}$, both of which lead to bending (by VSEPR forces) at the sulfur. See Figure 11-2, where the M—S—C angle is 107.3°, thereby implying that the first of the two resonance structures above is the more important (the M—S—C angle should be closer to the tetrahedral angle in the first structure).

[44] C. Shea and A. Haim, *J. Amer. Chem. Soc.*, **93**, 3056 (1971); *Inorg. Chem.*, **12**, 3013 (1973).

[45] See ref. 10.

[46] E. S. Gould, *J. Amer. Chem. Soc.*, **96**, 2373 (1974). This paper is one of the latest in a series dealing with electron transfer through organic structural units.

[47] You might note that these carboxylate bridging ligands do not lead to the same degree of rate enhancement as the halides and N_3^-. The carboxylates have an effect on the rate much like SCN⁻. This feature could arise from both precursor formation rates and orbital following difficulties (*e.g.*, bond breaking not synchronous with bond making and electron transfer).

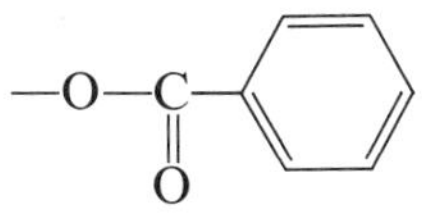

$$\begin{array}{cc} \text{—O—C—}\bigcirc & \text{—O—C—C—H} \\ \| & \|| \\ \text{O} & \text{O}\text{OH} \\ k_f = 0.15 & k_f = 3.1 \\ \textbf{18} & \textbf{19} \end{array}$$

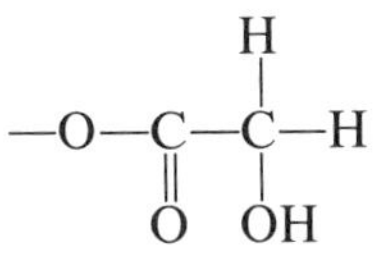

Electron transfer *via* attack at a site even more remote than the carboxyl carbonyl oxygen is possible and has clearly been observed in the reactions of various Co(III) and Ru(III) complexes of substituted pyridines.[48,49] For example, penta-amminepyridinecobalt(3+) (**20**) reacts relatively slowly with Cr(II)

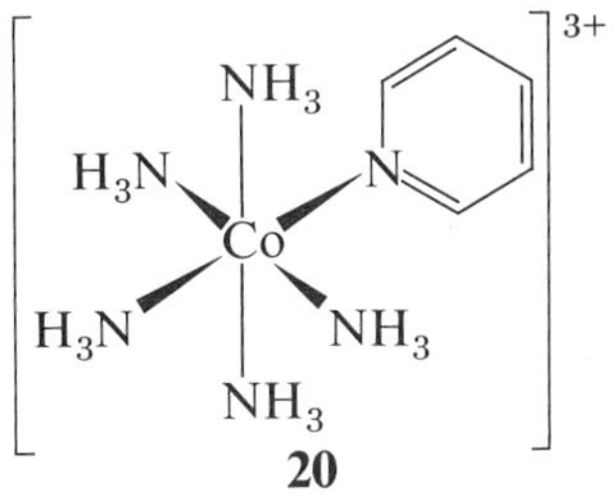

20

($k_f = 4 \times 10^{-3}$ M^{-1}sec^{-1}) to give [Cr(H$_2$O)$_6$]$^{3+}$ *by an outer sphere pathway;* however, the isonicotinamide complex (**21**) reacts much more rapidly ($k_f = 17.4$ M^{-1}sec^{-1}), and the first Cr(III) product observed is **22**. These latter observations suggest that the attack of Cr(II) is at the remote carbonyl group in **21**. This attack site is expected from the Lewis base properties of amides (as discussed in Chapter 5), where the carbonyl oxygen is usually the preferred donor site when compared with the —NR$_2$ group.

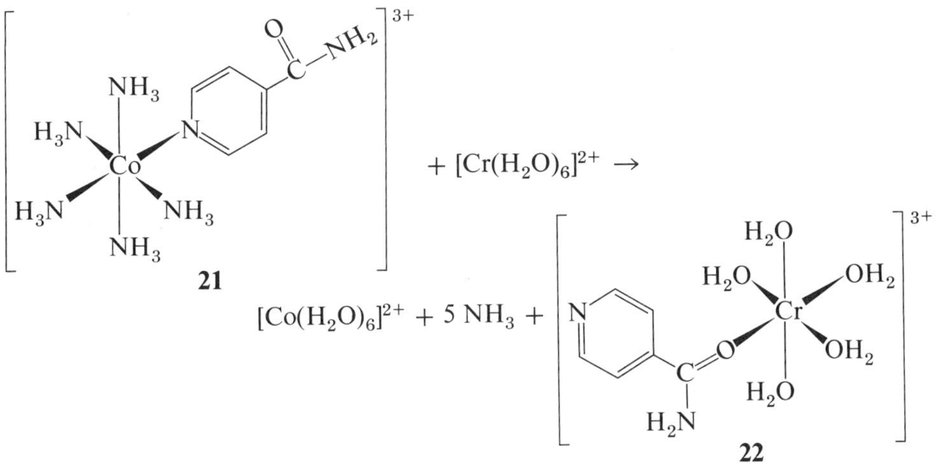

21 + [Cr(H$_2$O)$_6$]$^{2+}$ → (12-26)

[Co(H$_2$O)$_6$]$^{2+}$ + 5 NH$_3$ + **22**

The mechanism of electron transfer across bridging ligands is of great interest, and two extreme mechanisms can be proposed. In one—often called the chemical, radical, or stepwise mechanism—an electron is transferred from M^{red} to the bridge ligand to produce a radical anion; then, in a subsequent step, the electron is transferred from the radical ion to the metal ion of the oxidizing agent.[50,51] In the other mechanism—the

---

[48] R. G. Gaunder and H. Taube, *Inorg. Chem.*, **9**, 2627 (1970).
[49] F. Nordmeyer and H. Taube, *J. Amer. Chem. Soc.*, **90**, 1162 (1968).
[50] Another possibility, generally considered less likely, is that the bridging ligand could transfer an electron to the oxidant to leave the bridge ligand a radical cation (M^{ox}—L—M$'^{red}$ → $\overset{\ominus}{\text{M}}$—$\overset{\oplus}{\text{L}}\cdot$—M$'^{red}$). This would then be followed by reduction of the bridge back to its original state by the reducing agent ($\overset{\ominus}{\text{M}}$—$\overset{\oplus}{\text{L}}\cdot$—M$'^{red}$ → $\overset{\ominus}{\text{M}}$—$\overset{\oplus}{\text{L}}$—M$'$).
[51] See Ref. 10.

TABLE 12-5
RATES OF REDUCTION OF VARIOUS Co(III), Cr(III), AND Ru(III) COMPLEXES WITH Cr^{2+}(aq).[a,b]
SECOND ORDER RATE CONSTANTS IN UNITS OF $M^{-1}sec^{-1}$

Bridging Ligand X	Oxidizing Agent		
	$[Co^{III}(NH_3)_5X]^{n+}$	$[Cr^{III}(H_2O)_5X]^{n+}$	$[Ru^{III}(NH_3)_5X]^{n+}$
F^-	2.5×10^5	2.64×10^{-2}	—
Cl^-	6×10^5	90	3.5×10^4
$-O-\underset{\underset{O}{\|}}{C}-CH_3$	3×10^{-1}	10^{-4}	2.6×10^4
isonicotinamide ($-N\langle\text{py}\rangle C(O)NH_2$, 4-position)	1.74×10^1	1.8	3.9×10^5
nicotinamide ($-N\langle\text{py}\rangle C(O)NH_2$, 3-position)	1.4×10^{-2} (outer sphere); 3.3×10^{-2} (inner sphere)	—	0.84×10^4 (outer sphere); 1.35×10^4 (inner sphere)
pyridine ($-N\langle\text{py}\rangle$)	4.0×10^{-3} (outer sphere)	—	3.4×10^3 (outer sphere)

[a] Unless otherwise noted, all rate constants are for inner sphere reduction.
[b] Data taken from: F. Nordmeyer and H. Taube, *J. Amer. Chem. Soc.*, **90**, 1162 (1968); R. G. Gaunder and H. Taube, *Inorg. Chem.*, **9**, 2627 (1970); J. A. Stritar and H. Taube, *Inorg. Chem.*, **8**, 2281 (1969).

resonance or exchange mechanism—the bridge acts simply as a mediator of electron flow; this type of mechanism was illustrated in the discussion of σ "orbital following" in inner sphere electron transfer reactions with bridge transfer (p. 674). Unfortunately, it is not at all simple to distinguish the two mechanisms in practice, but the data in Table 12-5 illustrate one approach to answering this important mechanistic question.

From the data in Table 12-5 you see that, while the relative reduction rates of Co(III) and Cr(III) complexes with a simple bridge (F^-, Cl^-, CH_3COO^-) are widely different, the isonicotinamide complexes of these two metal ions are reduced at very similar rates. It is obvious that the factors (such as overall thermodynamic favorability) favoring Co(III) reductions over analogous Cr(III) reductions have been circumvented or are no longer of importance when isonicotinamide is the bridging ligand. One explanation for this is that bridges such as isonicotinamide are reducible, and reductions proceeding through such bridges occur by a chemical or radical mechanism. The idea is that the first step after precursor formation is Cr(II) activation $[(\pi^*)^3(\sigma^*)^1 \rightarrow (\pi^*)^4]$ and transfer of an electron into a π^* orbital of isonicotinamide, thereby forming Cr(III) and the radical anion $H_2NC(O)C_5H_4N^{\overline{\cdot}}$; this step is thought to be rate-determining, and convincing experimental evidence has been obtained to support the free radical nature of the bridge.[52,53] On the other hand, reductions of Co(III) and Cr(III) through simpler bridges such as Cl^- occur by the concerted σ orbital following path, because these species are not reducible. Thus, reductions through simple inorganic bridges are generally controlled by oxidant activation and the thermodynamics of the reaction.

---

[52] J. E. French and H. Taube, *J. Amer. Chem. Soc.*, **91**, 6951 (1969).
[53] Two recent pieces of evidence for radical intermediates are: a. E. S. Gould, *J. Amer. Chem. Soc.*, **94**, 4360 (1972). b. M. Z. Hoffman and M. Simic, *J. Amer. Chem. Soc.*, **94**, 1757 (1972).

To further support the contention that reductions of Co(III) and Cr(III) occurring through isonicotinamide bridges proceed by a radical mechanism, it is instructive to compare the family of complexes of the type $[Co(NH_3)_5(NC_5H_4R)]^{3+}$ with those based on Ru(III) (Table 12-5). While the rates of reduction of the Co(III) and Ru(III) complexes are similar for bridging Cl^-, the differences in their reduction rates with organic bridging ligands are dramatic. Furthermore, there is little change in this situation when X = pyridine (when the reaction occurs by an outer sphere mechanism) and when X = nicotinamide (and the reaction proceeds by an inner sphere mechanism). Apparently, it is the electronic structure of the oxidant metal ion [Ru(III) compared with Co(III)] that is again coming into play, just as you saw in the comparison of outer sphere electron transfers involving Co(II)/Co(III) and Ru(II)/Ru(III) (see p. 662). The LUMO or acceptor orbital on Ru(III) $[(\pi^*)^5]$ is a π^* orbital, whereas it is a σ^* orbital on Co(III). Therefore, the activation step for Ru(III) reduction is determined by the reductant activation $[(\pi^*)^3(\sigma^*)^1 \rightarrow (\pi^*)^4$ for Cr(II)], since transfer of a π^* electron to Ru(III) can occur with little activation of the oxidant. The nicotinamide bridge facilitates the electron transfer because its π^* mo's couple with the $d\pi^*$ orbitals of M^{ox} and M^{red} in a manner analogous to the σ orbital mixing described on page 674. In short, you have encountered a π orbital following pathway.

Fission of the Successor Complex[54,55]

Having discussed two of the factors controlling inner sphere reduction rates—rate of formation of the precursor and electron transfer within the binuclear intermediate—we come finally to the fact that the overall rates of electron transfer reactions can be controlled by the rate of fission of the successor complex. [From equation (12-12d), page 671, if $k_3 \ll k_2$, then $k_{obs} = Kk_3$.] A particularly interesting example is provided by the Ru(III)-nicotinamide complexes in Table 12-5.

When pentaammineisonicotinamideruthenium(3+) reacts with Cr(II), precursor formation and electron transfer to give Ru(II) and Cr(III) are complete within the time of mixing.[56] However, the yellow-orange binuclear successor complex (**23**) is quite stable and decomposes only slowly to give, ultimately, $[Ru(NH_3)_5(isoamide)]^{2+}$ and Cr(III) by three different routes.

$$\left[(H_3N)_5Ru^{III}N\underset{}{\diagup\!\!\!\diagdown}C\underset{NH_2}{\overset{O}{\diagup}}\right]^{3+} + [Cr(H_2O)_6]^{2+} \quad (12\text{-}27)$$

$$\downarrow k = 3.92 \times 10^5 \text{ M}^{-1}\text{sec}^{-1}$$

$$\left[(H_3N)_5Ru^{II}N\underset{}{\diagup\!\!\!\diagdown}C\underset{NH_2}{\overset{O-Cr^{III}(H_2O)_5}{\diagup}}\right]^{5+}$$

23

| $k_0 = 3.07 \times 10^{-6}$ sec^{-1} | $k_1 = 2.77 \times 10^{-7}$ M^{-1}sec^{-1} (OH$^-$ catalyzed) | k_2 (OH$^-$, Cr^{II}-catalyzed) $= 4.74 \times 10^{-4}$ sec^{-1} |

$$\left[(H_3N)_5Ru^{II}N\underset{}{\diagup\!\!\!\diagdown}C\underset{NH_2}{\overset{O}{\diagup}}\right]^{2+} + [Cr(H_2O)_6]^{3+}$$

---

[54] R. W. Craft and R. G. Gaunder, *Inorg. Chem.*, **14**, 1283 (1975); *ibid.*, **13**, 1005 (1974).
[55] R. G. Gaunder and H. Taube, *Inorg. Chem.*, **9**, 2627 (1970).
[56] The isonicotinamide-Ru(III) complex discussed here is apparently reduced by a process that is inner sphere only. In contrast, the nicotinamide complex (nicotinamide has the pyridine N and the amide group *meta* to one another) is reduced by both inner sphere and outer sphere processes with quite similar rate constants (see Table 12-5).

680 □ REACTION MECHANISMS AND METHODS OF SYNTHESIS

The reason for the stability of **23** is that Cr(III) and Ru(II) are quite inert to substitution. Since, as you shall see in the next chapter, substitution reactions of octahedral transition metal complexes usually occur by dissociative activation, the dissociation of the binuclear complex **23** and formation of the final products are slow.

In view of the substitutional inertness of Cr(III) in particular, the three routes for the decomposition of **23** are quite interesting. The k_0 path represents the spontaneous dissociation of the complex to give products. The k_1 path has an inverse dependence on [H$^+$]. This is thought to indicate that **23** loses a proton from Cr(III)-bound water to give **24**, which then dissociates to products.[56a]

$$\left[(H_3N)_5Ru^{II}N \!\!\!\bigcirc\!\!\!\! \begin{array}{c} O-Cr^{III}(H_2O)_4(OH) \\ C \\ NH_2 \end{array} \right]^{4+}$$

24

The k_2 path also has an inverse dependence on [H$^+$] but it is, just as importantly, directly dependent on excess [Cr(II)]. The explanation for this is that the OH$^-$ ligand in **24** functions as a bridge ligand between **24** and unreacted Cr(II). That is, a bridge of the type —C=O—Cr^{III}—O—Cr^{II} is formed, and inner sphere electron transfer
           | H
of an electron from Cr(II) to Cr(III) ensues. This means that the species —C=O—Cr^{II}—O—Cr^{III} results, and, since Cr(II) is more susceptible to substitution
     | H
reactions than Cr(III), complex **24** [which now contains Cr(II) instead of Cr(III)] can decompose to products. Finally, because of this catalysis by excess Cr(II), the k_2 path is the most rapid of the three decomposition paths.

TWO-ELECTRON TRANSFERS

To this point we have limited the discussion to reactions in which the oxidant and reductant change by only one unit in their formal oxidation state. However, two-electron changes have certainly been observed for the transition metals, and it may be shown that such reactions fit within the general framework presented thus far.

One of the best understood two-electron transfer reactions in transition metal chemistry is the Pt(II)-catalyzed exchange of free Cl$^-$ for chloride bound to a Pt(IV) compound.[57-59] That is,

(12–28) $$\textit{trans-}[Pt(en)_2Cl_2]^{2+} + {}^*Cl^- \xrightarrow[{[Pt(en)_2]^{2+}}]{} \textit{trans-}[Pt(en)_2^*ClCl]^{2+} + Cl^-$$

The rate law for this reaction is

$$\text{Rate}_f = k[Pt(II)][Pt(IV)][Cl^-]$$

and the mechanism to account for this involves rapid addition of free chloride to the Pt(II) compound to form a five-coordinate, monopositive ion that then forms a six-coordinate, inner sphere complex with the Pt(IV) reactant. The two platinum atoms are now in very similar environments, and the transfer of two σ^* electrons, with accompanying anion transfer in the opposite direction, can readily occur. Notice that the orbital following model illustrated in Figure 12-4 can also apply here with the minor change that the mo labeled σ_n contains two electrons.

[56a] Compare with the CB mechanism on page 410, Chapter 7, and in Chapter 13.
[57] F. Basolo, P. H. Wilks, R. G. Pearson, and R. G. Wilkins, *J. Inorg. Nucl. Chem.*, **6**, 161 (1958).
[58] R. C. Johnson and F. Basolo, *J. Inorg. Nucl. Chem.*, **13**, 36 (1960).
[59] F. Basolo, M. L. Morris, and R. G. Pearson, *Disc. Faraday Soc.*, **29**, 80 (1960).

$$[Pt(en)_2]^{2+} + {}^*Cl^- + trans\text{-}[Pt(en)_2Cl_2]^{2+}$$

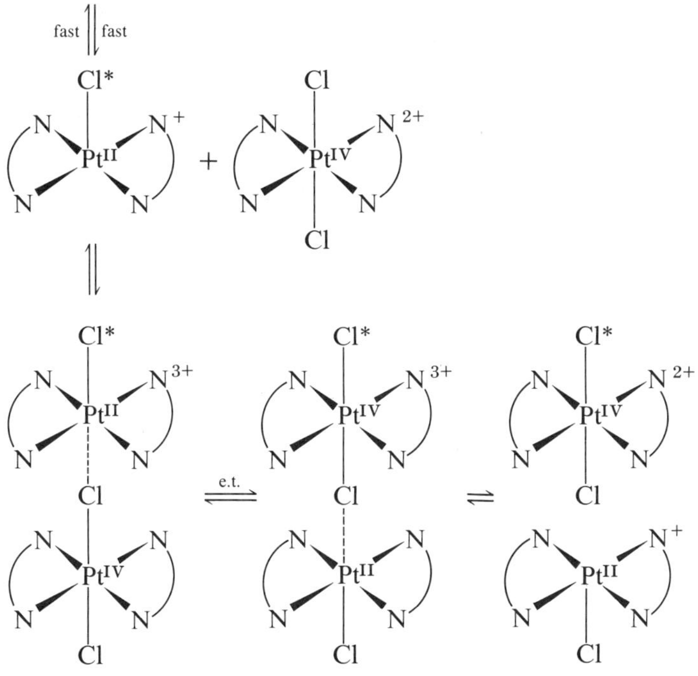

fast ⇅ fast

[structure showing [Pt(en)_2]^+ with Cl* axial label and trans-[Pt(en)_2Cl_2]^{2+} with Cl axial labels]

⇅

[structures 25a and 25b showing the symmetric bridged intermediates with e.t. (electron transfer) equilibrium between them]

25a   **25b**

⇅

$$Cl^- + trans\text{-}[Pt(en)_2Cl^*Cl]^{2+} + [Pt(en)_2]^{2+}$$

(12-29)

This reaction has a great many interesting features. First, the mechanism was in large part established through the use of a radioactive tracer and a chiral Pt(IV) complex. That is, the exchange of radiochloride was found to occur at the same rate as the change in *optical rotation* in the reaction of trans-[Pt(ℓ-pn)$_2$Cl$_2$]$^{2+}$ with [Pt(en)$_2$]$^{2+}$.[60] Clearly, the rate determining step in the reaction

$$[Pt(en)_2]^{2+} + trans\text{-}[Pt(\ell\text{-pn})_2Cl_2]^{2+} \rightarrow [Pt(\ell\text{-pn})_2]^{2+} + trans\text{-}[Pt(en)_2Cl_2]^{2+}$$

(12-30)

involves both the initially free chloride and the chiral, six-coordinate Pt(IV) compound.

Further evidence that the reaction passes through a symmetric transition state resembling structure **25** is the fact that the exchange is prevented when the ligand on Pt(IV) is tetramethylethylenediamine, Me$_2$N—C$_2$H$_4$—NMe$_2$. Apparently, the very bulky ligand prevents approach of the Pt(II) species.

NON-COMPLEMENTARY REACTIONS

The reactions that we have discussed thus far in this section are called *complementary reactions.* That is, the formal oxidation states of the oxidant and reductant

---

[60] ℓ-pn is the ℓ-form of H$_2$N—CHMe—CH$_2$—NH$_2$. The idea behind the experiment described in the text is that the optical rotations of the reactant, trans-[Pt(ℓ-pn)$_2$Cl$_2$]$^{2+}$, and of the product, [Pt(ℓ-pn)$_2$]$^{2+}$, are different. Therefore, the rate of change of optical rotation from that characteristic of reactant to that appropriate for the product should give the rate of the reaction.

both change by the same number of units. The counterparts of these reactions are non-complementary processes such as

$$\text{Mn(VII)} + 5\text{Fe(II)} \rightarrow \text{Mn(II)} + 5\text{Fe(III)}$$
$$\text{Cr(VI)} + 3\text{Fe(II)} \rightarrow \text{Cr(III)} + 3\text{Fe(III)}$$
$$2\text{Cr(II)} + \text{Pt(IV)} \rightarrow 2\text{Cr(III)} + \text{Pt(II)}$$

(12-31)

*In a **non-complementary reaction**, the oxidation states of the reactants change by unequal amounts and the stoichiometries are not 1:1.*

Some of the best examples of non-complementary processes in transition metal chemistry are chromate ion oxidations.[61] Many, if not all, of these oxidations proceed in a series of three elementary steps, each involving a single electron transfer.

$$\text{Cr(VI)} [= \text{HCrO}_4^-] + \text{Red} \underset{k_{56}}{\overset{k_{65}}{\rightleftharpoons}} \text{Cr(V)} + \text{Ox}$$

$$\text{Cr(V)} + \text{Red} \xrightarrow{k_{54}} \text{Cr(IV)} + \text{Ox}$$

$$\text{Cr(IV)} + \text{Red} \xrightarrow{\text{fast}} \text{Cr(III)} + \text{Ox}$$

Such a mechanism will lead to rate law (12-12e) if a steady state concentration is assumed for Cr(V) (see Chapter 7, p. 370).

(12-12e)
$$-d[\text{HCrO}_4^-]/dt = \frac{k_{65}k_{54}[\text{Cr(VI)}][\text{Red}]^2}{k_{54}[\text{Red}] + k_{56}[\text{Ox}]}$$

In general, some modification of this limiting rate law is often observed. In cases where an inner sphere process is possible (with $[\text{Fe(H}_2\text{O)}_6]^{2+}$ or VO^{2+}, for example), the rate determining step is the second of the three; *i.e.*, $k_{54}[\text{Red}] \ll k_{56}[\text{Ox}]$. This is generally attributed to the slowness of the separation of the inner sphere complex after electron transfer in the second step; the difficulty ultimately has its origin in the necessity for the Cr(IV) ion to change from a four-coordinate species in the inner sphere complex to a free, six-coordinate Cr(IV) ion.

For chromate oxidations that involve substitution inert reducing agents, the rate law becomes $-d[\text{HCrO}_4^-]/dt = k[\text{HCrO}_4^-][\text{Red}]$, since $k_{54}[\text{Red}] \gg k_{56}[\text{Ox}]$. Now the first step is rate determining because the difficulty mentioned above is no longer present and because the first step apparently has the least thermodynamic driving force of the three steps. This has been observed for such reducing agents as $[\text{Fe(CN)}_6]^{4-}$ and $[\text{Fe(bipy)}_3]^{2+}$.[62] However, the reaction does not proceed to the products predicted by the simple stoichiometry

(12-32)
$$\text{HCrO}_4^- + 3[\text{Fe(CN)}_6]^{4-} + 7\text{H}^+ = \text{Cr}^{3+} + 3[\text{Fe(CN)}_6]^{3-} + 4\text{H}_2\text{O}$$

Rather, the final product in the $[\text{Fe(CN)}_6]^{4-}$ reaction is apparently the species $(\text{H}_2\text{O})_5\text{Cr}-\text{N}\equiv\text{C}-\text{Fe(CN)}_5$! It is likely that the final step in the three-step sequence involves inner sphere reduction of Cr(IV), and as such represents one of the few cases, if not the only case, in which the bridging ligand is supplied by the reducing agent in an inner sphere process.

---

[61] See ref. 9.
[62] J. P. Birk, *J. Amer. Chem. Soc.*, **91**, 3189 (1969).

STUDY QUESTIONS

3. Review the discussions of σ orbital following models for the $\text{NO}_2^-/\text{OCl}^-$ reaction (Figure 12-1) and for the inner sphere mechanism (Figure 12-4). On the basis of these models, consider the other possibility—π orbital following for inner sphere electron transfer.

   a) Using a Cl π ao and M^{ox} and M'^{red} $d\pi$ ao's derive the three-center mo pattern for the reactant $\text{M}^{\text{ox}}-\text{X} + \text{M}'^{\text{red}}$ and trace the orbital evolution as Cl moves toward M'^{red}.

   b) Apply this orbital picture to $\text{Co}^{3+} [(\pi^*)^6]/\text{V}^{2+} [(\pi^*)^3]$ to show that no transfer from $\text{V}^{2+} (\pi)$ to $\text{Co}^{3+} (\pi)$ is possible. What if $\text{M}^{\text{ox}} = \text{Ru}^{3+} [(\pi^*)^5]$?

4. In the spirit of question 3b, should inner sphere $\pi^* \to \pi^*$ electron transfer be fast? Why are no such transfers known (all known $\pi^* \to \pi^*$ transfers occur by the outer sphere mechanism)?

5. Given the reaction:

$$[Fe(H_2O)_6]^{2+} + [Fe(H_2O)_5X]^{n+} \to \text{electron exchange}$$

consider the rate constants given below.

X in $[Fe(H_2O)_5X]^{n+}$	$k(M^{-1}sec^{-1})$
H_2O	4
F	40
Cl	38
Br	17
N_3^-	1×10^4

As indicated in Table 12-1, electron exchange between the two hexaaquo ions is outer sphere. Can you draw any conclusions as to the mechanism when X is a halide? Why is there rate acceleration when $X = N_3^-$? (Note the rate constant for $[Fe(phen)_3]^{2+/3+}$ in Table 12-1.)

6. Three complexes of dicarboxylic acids are illustrated below. Each reacts with Cr(II) with reduction of the bound Co(III). Complexes **A** and **B** react by an inner sphere mechanism, and complex **C** reacts by an outer sphere mechanism. What does this imply about the point of attack of Cr(II) on **A** and **B**? Why does **A** react more slowly than **B**?

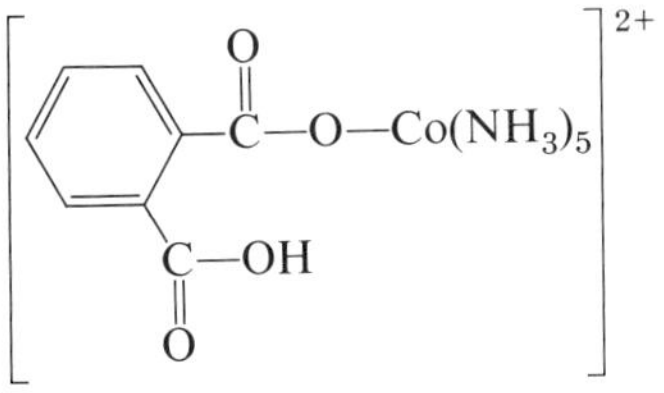

A $k_f = 0.075\ M^{-1}sec^{-1}$ 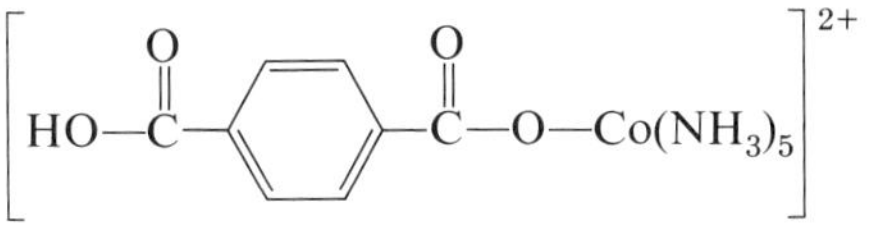   **B** $k_f = 0.2\ M^{-1}sec^{-1}$ 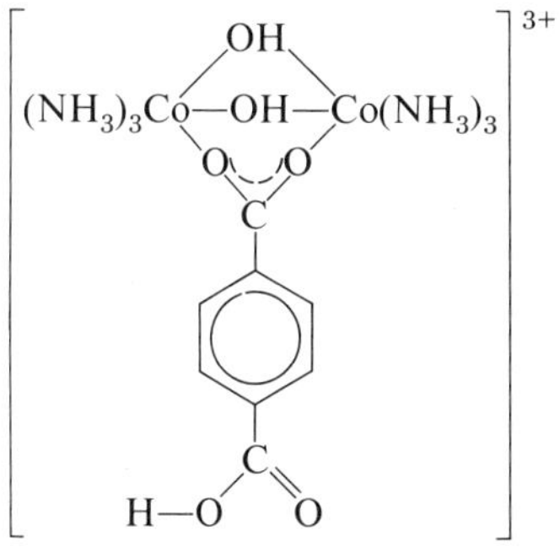   **C** $k_f = 1.9 \times 10^{-3}\ M^{-1}sec^{-1}$

7. The possibility that water may function as a bridging ligand in an inner sphere electron transfer has been much debated. The following questions probe this problem.

   a) Consider a reaction such as $[Co(NH_3)_5OH_2]^{3+} + Cr(II)$. The intermediate in an inner sphere process would resemble

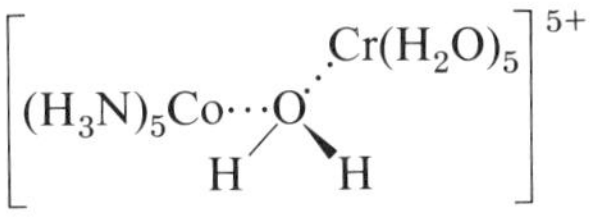

   $$\left[(H_3N)_5Co\cdots O\underset{H}{\overset{Cr(H_2O)_5}{\diagup}}H \right]^{5+}$$

   Is it reasonable to expect water to function as a bridge in such an intermediate?
   b) There is a strong inverse dependence on $[H^+]$ in this reaction. That is, the rate is slowed with increasing hydrogen ion concentration. Suggest a mechanism of the inner sphere type that will account for this.
   c) Having answered part (b), consider and explain the following observations. The water of $[Co(NH_3)_5OH_2]^{3+}$ may be labeled with oxygen-18, and the rate of reaction of the labeled compound with various reductants may be compared with the rates for the unlabeled compound. It was found that the labeled compound reacted more slowly than the unlabeled compound with Cr(II), whereas $[Ru(NH_3)_6]^{2+}$ reacted at approximately the same rate with both labeled and unlabeled compounds.

SYNTHESIS OF COORDINATION COMPOUNDS USING ELECTRON TRANSFER REACTIONS[63]

Electron transfer reactions have been studied largely for the insight they give into the actual process of transfer. However, there are examples of their use in chemical synthesis, and we shall discuss a few of the situations in which they have been used or observed.

As described in the previous chapter, the resolution of chiral compounds into their optically active enantiomers has been enormously important to the development of coordination chemistry. The resolution of $[Co(en)_3]^{3+}$, for example, was first described by Alfred Werner in 1912.[64] This was achieved by adding one enantiomer of a chiral anion [say, the (+) form of the chiral tartrate ion] to the solution of the racemic complex. The pair of enantiomeric complex ions will then form a pair of salts, (+)(+) and (−)(+). These salt pairs are diastereomeric, and differ in their solubility as a result. In this particular case, the (+)(+) salt precipitates from solution, leaving the (−)(+) form in the mother liquor. A similar process was described in Chapter 11 (page 622).

(12-33)

$$[Co(en)_3]^{3+} \quad (+)-\Lambda \qquad [Co(en)_3]^{3+} \quad (-)-\Delta$$

$\downarrow$ (+)-tartrate

Λ-(+)-[Co(en)$_3$][(+)-tartrate] Cl $\qquad$ Δ-(−)-[Co(en)$_3$]$^{3+}$
precipitate

Assuming that the (+) and (−) enantiomers were originally formed in equal amounts, half of the original material can be isolated in one form or the other. However, by clever use of an electron transfer reaction one can convert *all* of the original material—both (+) and (−) enantiomers—to his choice of one *or* the other of the enantiomers. That is, a 100% yield can be obtained of one or the other of the enantiomers. The tris-ethylenediamine complexes of Co(II) and Co(III) undergo electron transfer by an outer sphere mechanism (see Table 12-1). More important, though, for our present purposes is the fact that $[Co(en)_3]^{2+}$ is configurationally labile; that is, if it could be obtained optically pure in one enantiomer or the other, it would racemize rapidly. Therefore, combination of the lability of the 2+ oxidation state and electron transfer between the oxidation states can be used in the isolation of (+)- or (−)-$[Co(en)_3]^{3+}$ in yields approaching 100%.[65,66]

After precipitation of [(+)-Co(en)$_3$]Cl[(+)-tartrate] by the procedure outlined in reaction sequence (12-33), the mother liquor contains a preponderance of the more soluble diastereomer [(−)-Co(en)$_3$]Cl[(+)-tartrate]. At this point a small amount of ethylenediamine and $CoCl_2 \cdot 6H_2O$, which presumably form Co(en)$_3$Cl$_2$, are added to the mother liquor. There then ensues a reaction between $[Co(en)_3]^{2+}$, formed with equal amounts of the (+) and (−) enantiomers, and (−)-$[Co(en)_3]^{3+}$.

---

[63] J. L. Burmeister and F. Basolo, "The Application of Reaction Mechanisms to Synthesis of Coordination Compounds," in "Preparative Inorganic Reactions," Vol. 5, W. L. Jolly, ed., Interscience Publishers, New York (1968).
[64] A. Werner, *Chemische Berichte*, **45**, 121, 3061 (1912).
[65] D. H. Busch, *J. Amer. Chem. Soc.*, **77**, 2747 (1955).
[66] C. S. Lee, E. M. Gorton, H. M. Neumann, and H. R. Hunt, Jr., *Inorg. Chem.*, **5**, 1397 (1966).

$$\frac{1}{2}(+)\text{-}[Co(en)_2]^{2+} + \frac{1}{2}(-)\text{-}[Co(en)_3]^{2+} + (-)\text{-}[Co(en)_3]^{3+} \rightarrow$$

$$\frac{1}{2}(+)\text{-}[Co(en)_3]^{3+} + \frac{1}{2}(-)\text{-}[Co(en)_3]^{3+} + (-)\text{-}[Co(en)_3]^{2+}$$

(12-34)

The latter forms $(-)$-$[Co(en)_3]^{2+}$, which then rapidly racemizes to $(+)$- and $(-)$-$[Co(en)_3]^{2+}$, in effect replenishing the stock of $[Co(en)_3]^{2+}$ introduced by adding ethylenediamine and $CoCl_2 \cdot 6H_2O$. The other product of the electron transfer reaction is a racemic mixture of the enantiomers of the Co(III) complex. The $(+)$ isomer of the newly formed 3+ complex is precipitated from solution to leave the $(-)$ enantiomer, which can again undergo electron transfer with the Co(II) complex to again produce a racemic mixture of $[Co(en)_3]^{3+}$, the $(+)$ enantiomer of which again precipitates from solution. Continuation of this cycle can obviously be used to convert virtually all of the $(-)$ enantiomer originally present to the $(+)$ enantiomer; alternatively, by starting with a tartrate anion of opposite chirality, all of the $(+)$ enantiomer can be converted to $(-)$ enantiomer. That is, the cycle of reactions can in principle lead ultimately to complete conversion of one enantiomer to another. In practice, however, the yield is only somewhat in excess of 75%.

The resolution of $[Co(en)_3]^{3+}$ as described above is a process catalyzed by a lower valent state of the same complex. Although this may at first glance seem to be an isolated case, reactions catalyzed in this matter are not at all uncommon, and some have synthetic implications.[67] For example, electrochemical reduction of the *ortho*-phenanthroline or bipyridyl complexes of Cr(III) [reaction (12-35)] leads ultimately to the production of diaquobis(diamine)chromium(3+) ion in a reaction sequence (12-36) involving electron transfer catalysis.[68]

$$\left[\begin{array}{c}\text{Cr(N)}_6\end{array}\right]^{3+} + e^- \Longleftrightarrow \left[\begin{array}{c}\text{Cr(N)}_6\end{array}\right]^{2+} \qquad N\frown N = \text{bipy} \qquad E_{1/2} = -0.49 \text{ v}$$

(12-35)

That is, the Cr(II) species initially produced is labile to substitution, and one diamine is replaced by two water molecules. Presuming that the bulk of the tris(diamine)chromium(3+) ion is still present, this ion then undergoes electron transfer with the diaquo Cr(II) species to produce the final product, diaquobis(diamine)chromium(3+).

$$[Cr(L)_3]^{2+} + 2 H_2O \underset{k_2}{\overset{k_1}{\rightleftharpoons}} L + [Cr(L)_2(H_2O)_2]^{2+}$$

$$[Cr(L)_2(H_2O)_2]^{2+} + [Cr(L)_3]^{3+} \rightarrow [Cr(L)_3]^{2+} + [Cr(L)_2(H_2O)_2]^{3+}$$

(12-36)

Yet another example of catalysis by lower valent complexes is the Pt(II)-Pt(IV) two-electron transfer reaction described previously (p. 680). This has obvious synthetic implications as well, but the subject is left for you to explore in a Study Question at the close of this section. Instead, we move on to examine a somewhat similar catalytic process that has opened up a new aspect of rhodium chemistry.

A not uncommon observation in the laboratory is that some substance thought to be relatively inert will enter into a reaction in some way, either to divert the reaction to different products or to catalyze some previously unimportant reaction. Ethyl alcohol has been observed to serve just those functions on occasion. For example, you

---

[67] M. C. Hughes and D. J. Macero, *Inorg. Chem.*, **13**, 2739 (1974); D. M. Soignet and L. G. Hargis, *ibid.*, **11**, 2921 (1972); B. R. Baker and B. Dev Mehta, *ibid.*, **4**, 849 (1965).

[68] The symbol ⇔ indicates that the process is electrochemically reversible at the potential indicated.

shall see in Chapter 17 (p. 944) that the presence of alcohol allows a reaction to produce a molecular nitrogen complex instead of a "less interesting" product. Another example is relevant here.

Some years ago Delepine set out to synthesize rhodium(III) complexes of amine ligands.[69] Although pyridine reacts with Na_3RhCl_6 in aqueous solution to give such complexes, the reaction stops with $Rh(py)_3Cl_3$. The latter compound is insoluble, and the substitution of a chloride by a fourth molecule of pyridine to give $[Rh(py)_4Cl_2]^+$ requires lengthy reflux. To circumvent this difficulty, Delepine did what many of us are likely to do: he added another solvent, ethyl alcohol in this case, with the expectation that the solubility of the tris-pyridine intermediate would be increased with the rate of formation of $[Rh(py)_4Cl_2]^+$ enhanced as a result. Indeed, the reaction of Na_3RhCl_6 and pyridine in water-ethanol solution proceeded immediately and quantitatively to the desired compound at room temperature; none of the intermediate species were observed.

(12-37)

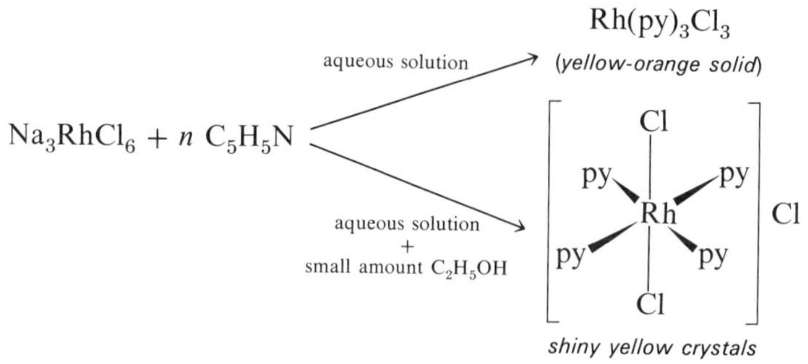

$$Na_3RhCl_6 + n\ C_5H_5N \xrightarrow{\text{aqueous solution}} Rh(py)_3Cl_3 \text{ (yellow-orange solid)}$$

$$\xrightarrow[\text{small amount } C_2H_5OH]{\text{aqueous solution}} \begin{bmatrix} & Cl & \\ py & | & py \\ & Rh & \\ py & | & py \\ & Cl & \end{bmatrix} Cl$$

shiny yellow crystals

More recently, Delepine's synthesis was studied with regard to its kinetics and mechanism, and the result was a proposed mechanism (12-38) quite similar to that for the Pt(II)-Pt(IV) reaction discussed earlier.[70] The first step is apparently the reduction of the rhodium(III) chloride to a rhodium(I) species by the alcohol. Although the initial form in which the Rh(I) species is produced is not known, it is likely that $[Rh(py)_4]^+$ forms rapidly since the reaction is run in excess pyridine. The stage is now set for the formation of $[Rh(py)_4Cl_2]^+$ by a mechanism identical to that postulated earlier for the isoelectronic Pt(II)-Pt(IV) two-electron transfer reaction. (Notice that $Rh(py)_3Cl_3$ is not involved in this mechanism. Thus, the alcohol enhances the formation of $[Rh(py)_4Cl_2]^+$ not by solubilizing $Rh(py)_3Cl_3$ but rather by functioning as a reducing agent that promotes a very different reaction.)

(12-38)

$$[Rh(H_2O)Cl_5]^{2-} + C_2H_5OH \rightarrow Rh(I) \text{ complex} + CH_3CHO$$

$$Rh(I) + 4\ py \xrightarrow{\text{fast}} [Rh(py)_4]^+$$

$$[Rh(py)_4]^+ + [Rh(H_2O)Cl_5]^{2-} + H_2O \xrightarrow{\text{slow}} \begin{bmatrix} & H_2O & \\ py & | & py \\ & Rh^I & \\ py & | & py \\ & Cl & \\ Cl & | & Cl \\ & Rh^{III} & \\ Cl & | & Cl \\ & H_2O & \end{bmatrix}$$

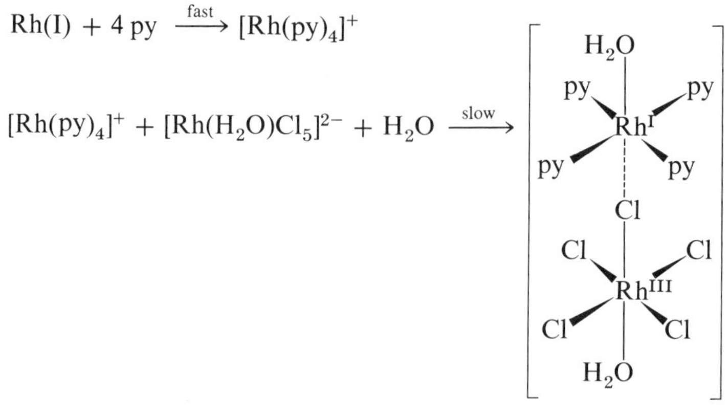

[69] M. Delepine, *Bull. Soc., Chim. France*, **45**, 235 (1929); *Compt. rend.*, **236**, 559 (1953).
[70] J. V. Rund, F. Basolo, and R. G. Pearson, *Inorg. Chem.*, **3**, 658 (1964).

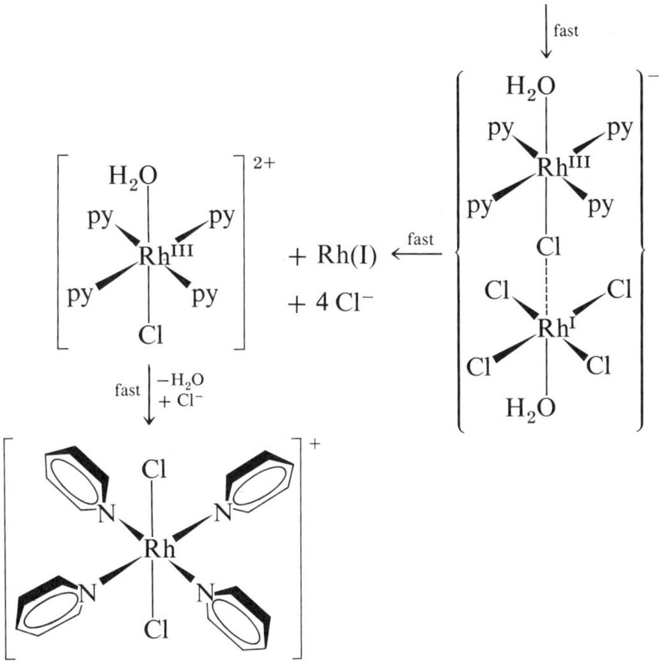

$$\text{(12-38)}$$
(Continued)

The mechanism outlined above clearly demands that Rh(I) be present only in trace amounts. This was experimentally verified by the fact that vapor phase chromatography of the volatile material present upon completion of the reaction showed that no perceptible amount of alcohol had disappeared; more importantly, though, it was found that traces of acetaldehyde, the oxidation product of alcohol, were present. Finally, to show that it is a rhodium(I) compound that serves as a catalyst, authentic Rh(I) compounds were added to solutions not containing alcohol; $[Rh(CO)_2Cl]_2$ was observed to be a very effective catalyst.

Other rhodium(III) compounds can be prepared by the same technique, but the exact nature of the product often depends critically on the ligands involved and on the reaction conditions.[71,72] For example, even in the presence of alcohol, $[Rh(SCN)_6]^{3-}$ reacts with pyridine to give only $Rh(py)_3(SCN)_3$ (wherein the thiocyanate groups are thought to be S-bonded).[71] The fact that the reaction does not form the *trans*-$[Rh(py)_4(SCN)_2]^+$ ion is thought to be due to the potential for serious steric hindrance between pyridine and SCN^- in this ion. Four pyridine molecules cannot lie flat in a plane about Rh(III); rather, the pyridine ligands presumably describe a four-bladed propellor about the metal ion [see reaction (12-38)]. Since the M—SCN linkage is bent (see Figure 11-2), apparently serious repulsions arise between the non-planar pyridines and the bent SCN ligand.

One of the best ways to prove that an inner sphere electron transfer reaction has occurred is to be able to show that atom or group transfer has accompanied oxidation state change. As you have previously seen, the group that is transferred (which must be at least potentially ambidentate) is usually attached to the oxidizing agent, and in the course of the reaction it also becomes attached to the reducing agent. This means that one may be able to begin the reaction with an unsymmetrical ligand such as NCS^-, attached to the oxidizing agent through either N or S, and finish the reaction with the ligand attached to the now-oxidized reducing agent through either S or N, respectively [reaction (12-39)].

---

[71] R. D. Gillard, J. A. Osborn, and G. Wilkinson, *J. Chem. Soc.*, 1951 (1965).
[72] D. J. Baker and R. D. Gillard, *Chem. Commun.*, 520 (1967).

688 ☐ REACTION MECHANISMS AND METHODS OF SYNTHESIS

(12-39)
$$\left.\begin{array}{l}M^{III}-S\diagdown\\ \phantom{M^{III}-}C\!\!\equiv\!\!N\\ M^{III}-N\!\!\equiv\!\!C-S\end{array}\right\} + M'(II) \rightarrow \left\{\begin{array}{l}S-C\!\!\equiv\!\!N-M'^{III}\\ \diagdown C\diagup S-M'^{III}\\ N\!\!\equiv\!\!\end{array}\right\} + M(II)$$

In principle, it is possible to run the reaction both ways, and, given that the product ions or molecules are inert to substitution, linkage isomers could be produced. No one has yet discovered a pair of linkage isomers that will produce another pair of linkage isomers by electron/group transfer. However, reactions approximating this have been studied with interesting results.

When discussing the nature of bridge influences on inner sphere reactions, we pointed out that an S-bonded thiocyanate ligand could be attacked at either the sulfur or the nitrogen end (p. 676), while an N-bonded NCS$^-$ should be attacked only at the remote end of the ligand, the sulfur end. The latter situation is observed in the following inner sphere electron transfer reaction (12-40), where the product is the S-bonded linkage isomer.[73,74]

(12-40)
$$[Co(NH_3)_5NCS]^{2+} + [Co(CN)_5]^{3-} \xrightarrow[k = 1.1 \times 10^6 \, M^{-1}sec^{-1}]{}$$
$$[Co(CN)_5SCN]^{3-} + Co^{2+}(aq) + 5NH_3$$

The fact that it is indeed the S-bonded isomer that comes out of the electron transfer reaction is substantiated by the observation that it is quite different from the N-bonded isomer previously isolated from a substitution reaction (12-41).[75]

(12-41)
$$[Co(CN)_5(OH_2)]^{2-} + NCS^- \rightarrow [Co(CN)_5NCS]^{3-} + H_2O$$
$$\downarrow \text{heat at } 150°/6 \text{ hr}$$
$$[Co(CN)_5SCN]^{3-}$$

This N-bonded isomer is stable in solution for several hours (but if the solid isomer stands for several months at room temperature, or if it is heated to 150° for 6 hours, it isomerizes to the S-bonded form). This stability precludes the possibility that the N-bonded form was initially obtained in some manner, and then isomerized, in the electron transfer/group transfer reaction. In addition, it indicates that the product from reaction (12-41) is kinetically and not thermodynamically controlled. Apparently, the nitrogen end of free thiocyanate is a better nucleophile *in this reaction* than the sulfur end.

We have briefly described the use of $[Co(CN)_5]^{3-}$ as an electron transfer agent, but it has a much broader chemical behavior.[76] For example, it has been widely studied as a reagent in reactions of the general type

(12-42)
$$2[Co(CN)_5]^{3-} + X-Y \rightarrow [Co(CN)_5X]^{3-} + [Co(CN)_5Y]^{3-}$$

where X—Y can be H_2, H_2O, Br_2, I_2, H_2O_2, CH_3I, or RX.[77-81] Although oxidation-

---

[73] I. Stotz, W. K. Wilmarth, and A. Haim, *Inorg. Chem.*, **7**, 1250 (1968).
[74] C. J. Shea and A. Haim, *Inorg. Chem.* **12**, 3013 (1973).
[75] J. L. Burmeister, *Inorg. Chem.*, **3**, 919 (1964).
[76] B. M. Chadwick and A. G. Sharpe, "Transition Metal Cyanides and Their Complexes," *Adv. Inorg. Chem. Radiochem.*, **8**, 84 (1966).
[77] J. Halpern and M. Pribanic, *Inorg. Chem.*, **9**, 2616 (1970).
[78] P. B. Chock, R. B. K. Dewar, J. Halpern, and L-Y. Wong, *J. Amer. Chem. Soc.*, **91**, 82 (1969).
[79] J. Halpern and J. P. Maher, *J. Amer. Chem. Soc.*, **86**, 2311 (1964).
[80] J. Halpern and J. P. Maher, *J. Amer. Chem. Soc.*, **87**, 5361 (1965).
[81] P. B. Chock and J. Halpern, *J. Amer. Chem. Soc.*, **91**, 582 (1969).

reduction has occurred and the cobalt in both complexes is best described as Co(III), this reaction is not a pure electron transfer reaction such as we have been discussing to this point. Rather, it is better described as an atom transfer redox reaction analogous with reaction (12-2). Alternatively, the reaction has been depicted as an "oxidative addition" since the Co(II) has been oxidized to Co(III) and a group has added to the metal center. Although it is apropos to discuss the chemistry of $[Co(CN)_5]^{3-}$ at this point, the *general* discussion of "oxidative addition" reactions is delayed to Chapter 17.

Pentacyanocobaltate(3−) is generally prepared and used *in situ* by dissolving a Co(II) salt, a soluble source of cyanide, and potassium or sodium hydroxide in water. Although the formula is usually written as $[Co(CN)_5]^{3-}$, there is experimental evidence that the species existing in aqueous solution is $[Co(CN)_5(H_2O)]^{3-}$;[82] however, since the water molecule is apparently lost readily prior to reaction (see Chapter 13), the five-coordinate representation is not entirely inaccurate. This is especially so because the five-coordinate complex can be isolated from non-aqueous solvents under the following conditions:[83]

$$CoCl_2 + 5Et_4N^+CN^- \xrightarrow{\text{dry, deoxygenated DMF}} [Et_4N]_3[Co(CN)_5] \qquad (12\text{-}43)$$

yellow, translucent needles
$\mu_{eff} = 1.77$ *B.M.*

The structure of the pentacyanocobaltate(3−) ion in the salt $[Et_4N]_3[Co(CN)_5]$ is square pyramidal,[84] as predicted by the structure preference energy plots in Figure 10-1.

The salt is air-sensitive, as is sometimes the case with Co(II) complexes, and a red-brown crystalline solid with the formula $[Et_4N]_3[Co(CN)_5O_2] \cdot 1.5H_2O$ precipitates if a solution of the salt in reaction (12-43) is allowed to stand in the air for several days. This dioxygen complex, which has the O_2 bonded endwise to the Co(II) (*i.e.*, Co—O⟍O

with a Co—O—O angle of 153.4°) was mentioned in Chapter 5 (p. 205).[85] Yet another, somewhat similar dioxygen complex of cobalt has also been studied in some detail, and its formation is found to depend on an electron transfer reaction.[86] Dicyanotris-(dimethylphenylphosphine)cobalt reacts with oxygen to form a complex having structure **26**.

26

The point of interest here is the mechanism of formation of the complex. That is, it is thought that O_2 reacts with the original compound to displace a phosphine ligand. This intermediate is then attacked by unreacted $Co(CN)_2(PMe_2Ph)_3$

---

[82] J. M. Pratt and R. J. P. Williams, *J. Chem. Soc.*, A, 1291 (1967).
[83] D. A. White, A. J. Solodar, and M. M. Baizer, *Inorg. Chem.*, **11**, 2160 (1972).
[84] L. D. Brown and K. N. Raymond, *Inorg. Chem.*, **14**, 2590 (1975).
[85] L. D. Brown and K. N. Raymond, *Inorg. Chem.*, **14**, 2595 (1975).
[86] J. Halpern, B. L. Goodall, G. P. Khare, H. S. Lim, and J. J. Pluth, *J. Amer. Chem. Soc.*, **97**, 2301 (1975).

(12-44)
$$Co^{II}(CN)_2(PMe_2Ph)_3 + O_2 \rightarrow Co^{II}(CN)_2(PMe_2Ph)_2(O_2) + PMe_2Ph$$
$$Co^{II}(CN)_2(PMe_2Ph)_2(O_2) + Co^{II}(CN)_2(PMe_2Ph)_3 \rightarrow$$
$$[(Me_2PhP)_3(NC)_2Co^{II}-NC-Co^{II}(CN)(PMe_2Ph)_2(O_2)] \rightarrow \mathbf{26}$$
$$\mathbf{27}$$

to give **27**, through which electron transfer may occur in the usual inner sphere manner. The fact that two electrons have been transferred to the dioxygen ligand is apparent from the bond length of the coordinated O_2. That is, this ligand has an O—O bond length of 1.441 Å, a value quite close to the 1.49 Å observed for peroxide, O_2^{2-} (see Chapter 5, p. 206). For this reason, the formal oxidation states in compound **26** correspond to $[Co^{III}-NC-Co^{III}(O_2^{2-})]$.

The reactions of $[Co(CN)_5]^{3-}$ that are of greatest interest in terms of synthesis are those with organic compounds.[87] For example, if benzyl bromide is added to a water-methanol solution of $[Co(CN)_5]^{3-}$ prepared *in situ*, the observed reaction is

(12-45)
$$2[Co(CN)_5]^{3-} + PhCH_2Br \xrightarrow{\text{inert atmosphere}} [PhCH_2Co(CN)_5]^{3-} + [Co(CN)_5Br]^{3-}$$

The yellow crystalline salt $Na_3[PhCH_2Co(CN)_5] \cdot 2H_2O$ is stable to alkali in the absence of air, but it reacts with acid to give an unidentified intermediate which, upon addition of base, releases phenylacetonitrile.[88] The following scheme was devised to account for this reaction, which is also observed when R is *n*-alkyl, benzyl, or phenyl.

(12-46)
$$\overset{R}{\underset{|}{Co-C \equiv N}} \xrightarrow{H^+} \overset{R}{\underset{|}{Co-\overset{+}{C}=NH}} \rightarrow \overset{+}{Co}-CR=NH \xrightarrow{OH^-} RC \equiv N$$

When R is an olefinic residue, however, reaction with acid leads directly to the olefin; for example,

(12-47)
$$2[Co(CN)_5]^{3-} + CH_2=CHCH_2I \rightarrow [Co(CN)_5(CH_2CH=CH_2)]^{3-} + [Co(CN)_5I]^{3-}$$
$$\downarrow H^+$$
$$H_3C-CH=CH_2$$

Examination of the kinetics of reactions such as (12-45) indicates that they are first order in both the cobalt compound and the organic halide. Further, the rate is dependent on the strength of the carbon-halogen bond; that is, the rate constants increase by roughly a factor of 10^3 to 10^4 on going from RCl to RBr to RI. The mechanism proposed to account for these facts is an atom transfer redox reaction involving free radical intermediates.

(12-48)
$$[Co^{II}(CN)_5]^{3-} + RX \rightarrow [Co^{III}(CN)_5X]^{3-} + R\cdot \quad \textit{(rate determining)}$$
$$[Co^{II}(CN)_5]^{3-} + R\cdot \rightarrow [RCo^{III}(CN)_5]^{3-} \quad \textit{(fast)}$$

The free radical nature of the reaction was later established in the course of studying the $H_2O_2 + [Co(CN)_5]^{3-}$ reaction.[89] Based on the mechanism proposed for the alkyl halide reactions, the mechanism for the peroxide reaction should be

(12-49)
$$[Co^{II}(CN)_5]^{3-} + H_2O_2 \rightarrow [Co^{III}(CN)_5OH]^{3-} + HO\cdot \quad \textit{(rate determining)}$$
$$[Co^{II}(CN)_5]^{3-} + HO\cdot \rightarrow [Co^{III}(CN)_5OH]^{3-}$$

---

[87] See refs. 79, 80, and 81.
[88] J. Kwiatek and J. K. Seyler, *J. Organometal. Chem.*, **3**, 433 (1965).
[89] See ref. 78.

In support of this mechanism it was found that the addition of I^- caused the reaction stoichiometry to change to

$$2[Co(CN)_5]^{3-} + H_2O_2 + I^- \rightarrow [Co(CN)_5OH]^{3-} + [Co(CN)_5I]^{3-} + OH^- \qquad (12\text{-}50)$$

This observation can be explained readily by the capture of the $HO\cdot$ radical by I^- to give $I\cdot$, the latter then being consumed by the Co(II) reagent.

The pentacyanocobaltate(3−) ion also reacts with H_2, but in a concerted process and not a free radical one such as those discussed above.[90] The reaction proceeds according to equation (12-51) to give hydridopentacyanocobaltate(3−),

$$2[Co^{II}(CN)_5]^{3-} + H_2 \rightarrow 2[Co^{III}(CN)_5H]^{3-} \qquad (12\text{-}51)$$

and obeys the rate law

$$-d[Co(CN)_5^{3-}]/dt = 2k_1[Co(CN)_5^{3-}][H_2] - 2k_{-1}[Co(CN)_5H^{3-}]^2$$

The equilibrium constant is approximately $10^5 \, M^{-1}$ over a wide range of H_2 partial pressures. This has important implications, because $[Co(CN)_5]^{3-}$ has been used as a catalyst for homogeneous hydrogenations such as reaction (12-52).[91-95]

$$H_2C=CH-CH=CH_2 + H_2 \xrightarrow[{[Co(CN)_5]^{3-}}]{} H_2C=CH-CH_2-CH_3 + H_3C-CH=CH-CH_3 \qquad (12\text{-}52)$$

Indeed, Pratt and Craig state that "there is probably no other catalyst which can hydrogenate such a wide variety of substrates under such mild conditions."[96] The following conversions, among others, may be effected:

    conjugated olefins → olefins

    $PhCH=CH_2$ → $PhCH_2CH_3$

    PhCHO   → $PhCH_2OH$

    $PhNO_2$ → $PhN=NPh + PhNH-NHPh + PhNH_2$

    $C_6H_4O_2$ (quinone) → $C_6H_6O_2$ (hydroquinone)

Two mechanisms have been proposed to account for such hydrogenations [reactions (12-53) and (12-54)], and both involve the preliminary formation of $[Co^{III}(CN)_5H]^{3-}$.

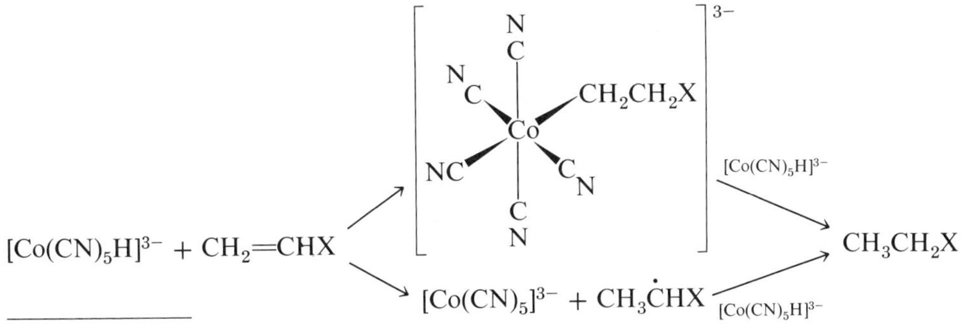

$$[Co(CN)_5H]^{3-} + CH_2=CHX \longrightarrow \begin{bmatrix} \text{Co with 5 CN ligands and } CH_2CH_2X \end{bmatrix}^{3-} \xrightarrow{[Co(CN)_5H]^{3-}} CH_3CH_2X \qquad (12\text{-}53)$$

$$\searrow [Co(CN)_5]^{3-} + CH_3\dot{C}HX \xrightarrow{[Co(CN)_5H]^{3-}} \qquad (12\text{-}54)$$

---

[90] See ref. 77.

[91] J. Kwiatek, I. L. Mador, and J. K. Seyler, *J. Amer. Chem. Soc.*, **84**, 304 (1962).

[92] J. Kwiatek and J. K. Seyler, *J. Organometal. Chem.*, **3**, 421 (1965).

[93] J. M. Pratt and P. J. Craig, "Preparation and Reactions of Organocobalt(III) Complexes," *Adv. Organometal. Chem.*, **11**, 331 (1973).

[94] M. G. Burnett, P. J. Connolly, and C. Kemball, *J. Chem. Soc.*, A, 991 (1968).

[95] By "homogeneous hydrogenation" we mean to imply that the reaction is carried out with all reactants in the same phase, that is, all in solution.

[96] See page 434 of ref. 93.

In one mechanism the elements of Co—H add across the C=C double bond to give an organocobalt complex, and the now-reduced olefin is released by further reaction with the cobalt hydride (12–53). The other mechanism is a free radical process involving two successive hydrogen transfers from $[Co^{III}(CN)_5H]^{3-}$ (12–54).

STUDY QUESTIONS

8. Refer back to the discussion of two-electron transfer, and suggest a method for the synthesis of *trans*-$[Pt(en)_2Br_2]^{2+}$ from *trans*-$[Pt(en)_2Cl_2]^{2+}$. [See R. Johnson and F. Basolo, *J. Inorg. Nucl. Chem.*, **13**, 36 (1960).]

9. On page 686 the preparation of $[Rh(py)_4Cl_2]Cl$ was discussed. Attempts to apply this synthetic procedure to other halide or pseudohalide ligands were often not successful, however. For example, both I^- and SCN^- led to $Rh(py)_3X_3$. Suggest one or more reasons for this observation.

10. Consider further the Rh(III) thiocyanate complex mentioned in problem 9. Two compounds having the empirical formula $Rh(py)_3X_3$ (where X is SCN^-) were isolated, one yellow and the other orange.

    a) How might the thiocyanate ligand be bound in these complexes?
    b) Suggest structures for the yellow and orange complexes.

11. In the text we discussed the atom transfer reactions of $[Co(CN)_5]^{3-}$. In a very interesting variation on that mechanism, the reactions of complexes of the type below with organic halides were studied. The over-all reaction is

    Co(saloph)B =

    saloph = N,N'-bis(salicylidene)-*o*-phenylenediimino

    B = pyridine, imidazole, triphenylphosphine

    $2\ Co(saloph)B + RX + B \rightarrow Co(saloph)BR + Co(saloph)B_2^+ + X^-$

    and the rate law is $-d[Co(saloph)B]/dt = 2k[Co(saloph)B][RX]$, where k (for RX = p-$CNC_6H_4CH_2Br$) ranges from $1.3 \times 10^{-3} M^{-1} sec^{-1}$ for B = 3-chloropyridine to $0.9\ M^{-1}sec^{-1}$ for B = triphenylphosphine. Suggest a mechanism for this reaction and comment on the dependence on the basicity of B ($pK_a = 2.8$ for 3-chloropyridine and $pK_a = 8.7$ for trimethylphosphine). [See L. G. Marzilli, P. A. Marzilli, and J. Halpern, *J. Amer. Chem. Soc.*, **93**, 1374 (1971).]

12. Consult the literature for a method of preparation of the linkage isomers of $[Co(NH_3)_5(SCN)]^{2+}$. [D. A. Buckingham, I. I. Creaser, and A. M. Sargeson, *Inorg. Chem.*, **9**, 655 (1970).] If $[Co(NH_3)_5OH_2]^{3+}$ is allowed to react with SCN^-, which linkage isomer should predominate? (Hint: consult p. 688.)

13. The anion $[Co(CN)_5(SCN)]^{3-}$ was prepared by an electron/atom transfer reaction (p. 688). When the preparation of the other linkage isomer, $[Co(CN)_5(NCS)]^{3-}$, was attempted by the inverse of reaction (12–41) (that is, $[Co(NH_3)_5(SCN)]^{2+} + [Co(CN)_5]^{3-}$), only the S-bonded isomer was obtained [C. J. Shea and A. Haim, *Inorg. Chem.*, **12**, 3013 (1973)]. Does this imply adjacent or remote attack by Co(II) on the SCN^- ligand? Explain fully.

EPILOG

We have just examined in some detail the fact that transfer of electrons between metal ions may occur by either or both of two mechanisms: (1) the outer sphere mechanism, a mechanism wherein the coordination spheres of two complexes remain intact, and electron transfer occurs by an interaction of these coordination spheres, and (2) the inner sphere mechanism, a mechanism in which the transfer of electrons is mediated by a ligand held in common by the oxidant and reductant. The ligand bridging the two reacting ions is usually provided by the oxidant, and this ligand, with the oxidant ion attached, substitutes for another ligand attached to the reductant. As one or more electrons flow from reductant to oxidant, the bridge ligand often transfers in the opposite direction, and the result is both electron and ligand transfer.

A main feature of the inner sphere mechanism is the substitution of the ligand of one compound for a ligand of another. Although we have glossed over this step in this chapter, a fundamental understanding of the substitution step is clearly important. Indeed, an understanding of substitution reactions in general is of paramount importance, since many reactions of coordination complexes involve ligand substitution. Therefore, we turn in the next chapter to a survey of substitution reactions and to their use in coordination compound synthesis.

REPLACEMENT REACTIONS AT FOUR-COORDINATE REACTION CENTERS

The General Mechanism of Square Planar Substitution Reactions
Factors Affecting the Reactivity of Square Planar Complexes of Pt(II) and Other d^8 Metal Ions
Influence of the Entering Group
Influence of Other Groups in the Complex—Ligands *trans* to the Entering Group
Influence of Other Groups in the Complex—Ligands *cis* to the Entering Group
Nature of the Leaving Group
Effect of the Central Metal
The Intimate Mechanism for Replacement at Four-Coordinate Planar Reaction Centers

SUBSTITUTION REACTIONS OF OCTAHEDRAL COMPLEXES

Replacement of Coordinated Water
The Mechanism of Water Replacement
Rates of Water Replacement
Orbital Occupation Effects on Substitution Reactions of Octahedral Complexes
Solvolysis or Hydrolysis
Hydrolysis under Acidic Conditions
Base-Catalyzed Hydrolysis: The Conjugate Base or CB Mechanism

SYNTHESIS OF COORDINATION COMPOUNDS BY SUBSTITUTION REACTIONS

Thermodynamic Stability of Coordination Compounds
The Synthesis and Chemistry of Some Cobalt Compounds
The Synthesis and Chemistry of Some Platinum Compounds

EPILOG

13. COORDINATION CHEMISTRY: REACTION MECHANISMS AND METHODS OF SYNTHESIS; SUBSTITUTION REACTIONS

In the first of the three chapters devoted to the mechanisms of the reactions of coordination compounds, we focused on reactions that occur by transfer of electrons to or from coordinated metal ions. In electron transfer reactions that proceed by the inner sphere mechanism, one or more ligands of one reactant (usually the reductant) are substituted by a ligand from the other reactant (usually the oxidant), and the substituting ligand forms a bridge between the two metal ions. The details of this substitution step were not stressed in the previous chapter, although it was clearly important to the successful functioning of the mechanism. Moreover, it should be evident that substitution reactions in general are of utmost importance, since many of the reactions of transition metal complexes involve ligand substitutions. It is for this reason that we turn now to a discussion of the principles of substitution reactions and their use in the synthesis of coordination compounds.

The focus in this chapter is on substitution reactions—reactions wherein ligand-metal bonds are broken and new ones formed in their stead.[1-4] We can reach back more than three-quarters of a century, to Alfred Werner's second paper on coordination compounds, to illustrate substitution reactions.[5]

If dichrocobalt chloride is dissolved in water at room temperature, a solution is obtained which is colored green for a moment, but *only for a moment;* the color changes very quickly to blue and then to violet, a reliable indication that the dichro salt does not remain unchanged in solution. According to the observations which we have made with praseo salts, it seems unquestionable to us that the change in color is caused by the transformation of the triaminepraseo salt into the triaminepurpureo salt and the triamineroseo salt.

---

[1] M. L. Tobe, "Inorganic Reaction Mechanisms," Thomas Nelson and Sons, London, 1972.
[2] F. Basolo and R. G. Pearson, "Mechanisms of Inorganic Reactions," 2nd edition, John Wiley and Sons, New York, 1967.
[3] C. H. Langford and H. B. Gray, "Ligand Substitution Processes," W. A. Benjamin Inc., 1965.
[4] R. G. Pearson and P. C. Ellgen, "Mechanisms of Inorganic Reactions in Solution" in "Physical Chemistry, An Advanced Treatise," volume VII, H. Eyring, ed., Academic Press, New York, 1975.
[4a] R. G. Wilkins, "The Study of Kinetics and Mechanism of Reactions of Transition Metal Complexes," Allyn and Bacon, Inc., Boston, 1974.
[5] A. Werner and A. Miolati, *Z. phys. Chem.*, **14**, 506-21 (1894). This paper is available in translation in G. B. Kauffman, "Classics in Coordination Chemistry," Dover Publications, Inc., New York, 1968.

That is, the following substitution reactions occur:

(13-1) $$[Co(NH_3)_3(H_2O)Cl_2]Cl + H_2O \rightarrow [Co(NH_3)_3(H_2O)_2Cl]Cl_2$$
dichrocobalt chloride
or
triaminepraseocobalt chloride
triaminepurpureocobalt chloride

(13-2) $$[Co(NH_3)_3(H_2O)_2Cl]Cl_2 + H_2O \rightarrow [Co(NH_3)_3(H_2O)_3]Cl_3$$
triamineroseocobalt chloride

To introduce the principles of substitution reactions as they apply to coordination compounds, we shall turn first to those typical of square planar complexes (reaction 13-3) and then to those of octahedral complexes.

(13-3)

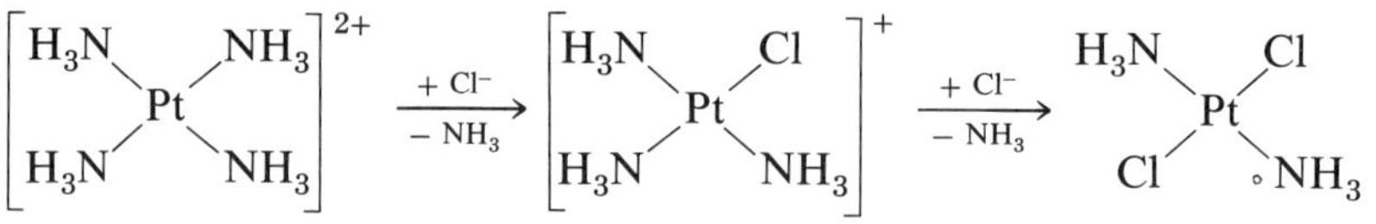

$$\begin{bmatrix} H_3N & & NH_3 \\ & Pt & \\ H_3N & & NH_3 \end{bmatrix}^{2+} \xrightarrow[-NH_3]{+Cl^-} \begin{bmatrix} H_3N & & Cl \\ & Pt & \\ H_3N & & NH_3 \end{bmatrix}^{+} \xrightarrow[-NH_3]{+Cl^-} \begin{array}{c} H_3N \quad\; Cl \\ Pt \\ Cl \quad\; NH_3 \end{array}$$

          1   **2**   **3**

Reiset's salt

Far less is known about the substitution reactions of coordination compounds of other geometries or of other coordination numbers, so we shall introduce material on such compounds only peripherally.

REPLACEMENT REACTIONS AT FOUR-COORDINATE PLANAR REACTION CENTERS[6,7]

As was pointed out in Chapter 9, by far the greatest number of square planar complexes known are based on transition metal ions with d^8 electron configurations, especially those metal ions listed below.

	Group VIII		Group IB
—	—	Ni(II)	
—	Rh(I)	Pd(II)	
—	Ir(I)	Pt(II)	Au(III)

Of these, Pt(II) offers the most useful combination of properties for the study of substitution reactions: (i) Pt(II) is more stable to oxidation than Rh(I) or Ir(I). (ii) Pt(II) complexes are always square planar, unlike Ni(II) complexes which can often be tetrahedral. (iii) Pt(II) chemistry has been thoroughly studied, and its substitution reactions proceed at rates convenient for laboratory study. This latter property is in contrast with Ni(II) complexes, for example, which can undergo substitution 10^6 more rapidly than Pt(II) complexes. For these reasons and others, the remainder of this discussion is largely concentrated on complexes of Pt(II).[8,9]

---

[6] L. Cattalini, *Prog. Inorg. Chem.*, **13**, 263 (1970).
[7] A. Peloso, *Coord. Chem. Rev.*, **10**, 123–181 (1973).
[8] U. Belluco, "Organometallic and Coordination Chemistry of Platinum," Academic Press, New York, 1974.
[9] F. R. Hartley, "The Chemistry of Platinum and Palladium," John Wiley and Sons, New York, 1973.

THE GENERAL MECHANISM OF SQUARE PLANAR SUBSTITUTION REACTIONS

In almost all of the substitution reactions studied with square planar complexes, the observed rate law has the form

$$-\frac{d[ML_3X]}{dt} = (k_s + k_Y[Y])[ML_3X]$$

for the general reaction

$$ML_3X + Y \rightarrow ML_3Y + X$$

This rate law has been rationalized in terms of two parallel pathways, both involving an associative or A mechanism (Figure 13-1). In the k_Y pathway the nucleophile Y attacks the metal complex and the reaction passes through a five-coordinate transition state and intermediate, this complex presumably having a regular trigonal bipyramidal structure. Since *the geometry of the original complex is invariably maintained*—that is, the ligands *cis* and *trans* to the substituted ligand remain in that arrangement—the entering and leaving groups as well as the ligand originally *trans* to the leaving group are in the trigonal plane. The k_s pathway also involves the formation of a trigonal bipyramidal transition state, except that the solvent is the entering group. Therefore, the latter pathway is also associative, and k_s is actually $k[\text{solvent}]$. A great deal of evidence has accumulated in favor of this mechanism, and it is most instructive to outline some of it.

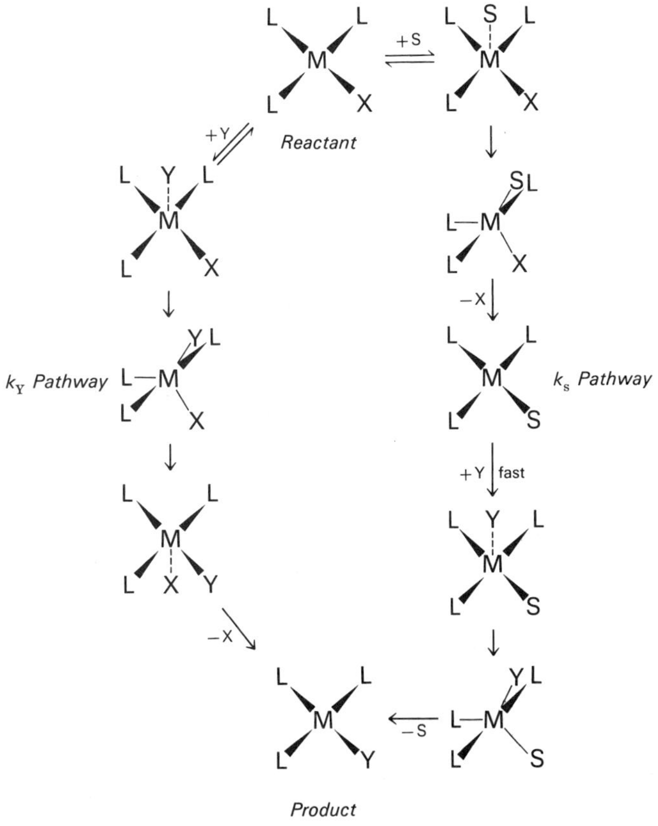

Figure 13-1. The associative pathways utilized in the substitution of one ligand for another at a square planar reaction center.

698 ☐ REACTION MECHANISMS AND METHODS OF SYNTHESIS; SUBSTITUTION REACTIONS

a. Likelihood of Five-Coordinate Transition State. A molecular orbital diagram for a square planar complex is given in Figure 9-14. The feature of greatest interest for our present purpose is that the metal p_z and s orbitals are not heavily used in metal-ligand σ bonding; they are therefore available for the initiation of the addition of a fifth ligand to the square planar substrate, and an associative mechanism is easily envisioned.

In this connection, recall that numerous d^8 compounds are in fact five-coordinate, especially compounds of Fe(0), Ru(0), Os(0), Co(I), Rh(I), Ir(I), and Ni(II); as examples, consider $Fe(CO)_5$, $[CoL_2(CO)_3]^+$, and $[Ni(CN)_5]^{3-}$ (see Chapter 10). Furthermore, in several cases, five-coordinate intermediates have actually been detected.[10,11]

b. Verification of Rate Law. The usual way to follow substitution reactions at planar reaction centers is to run the reaction under pseudo-first order conditions. That is, for a reaction such as

(13-4)

Pt(bipy)Cl$_2$ + py $\xrightarrow{\text{MeOH}}$ [Pt(bipy)(py)Cl]$^+$

the concentration of the platinum complex was held at about 10^{-5} to 10^{-4} M and the pyridine concentration was varied from 0.122 to 0.030 M.[12] The observed pseudo-first order rate constant, k_{obs}, is given by

$$k_{obs} = k_s + k_Y[\text{pyridine}]$$

---

[10] J. P. Fackler, Jr., W. C. Seidel, and J. A. Fetchin, *J. Amer. Chem. Soc.*, **90**, 2708 (1968).
[11] R. G. Pearson and D. A. Sweigert, *Inorg. Chem.*, **9**, 1167 (1970).
[12] L. Cattalini, A. Orio, and A. Doni, *Inorg. Chem.*, **5**, 1517 (1966).

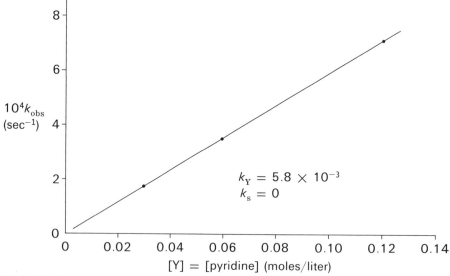

Figure 13-2. Verification of the rate law $k_{obs} = k_s + k_Y[\text{pyridine}]$ for the reaction of Pt(bipy)Cl$_2$ with pyridine to give Pt(bipy)(py)Cl$^+$. Data taken from L. Cattalini, A. Orio, and A. Doni, *Inorg. Chem.*, **5**, 1517 (1966).

The correctness of this expression is indicated by the fact that a plot of k_{obs} vs. [pyridine] is a straight line (Figure 13-2). The slope of the line gives k_Y and the intercept is equal to k_s, the latter being nearly zero in this case.

c. The Nature of the Two Reaction Pathways.
The next reaction involves unchanged reactants and products and so could be studied in a wide variety of solvents ranging from polar to non-polar.[13-15] If the reaction follows the usual parallel path mechanism, the solvent-dependent path should become less important in solvents of lower coordinating ability, and k_s should approach zero. As seen in Figure 13-3, this is the observation made. In hexane the intercept of the plot is effectively zero. In methanol, however, the reaction proceeds completely by the solvent-dependent path, as there is no amine-dependent contribution to the observed rate.[16]

$$\text{Cl}\diagdown\!\!\!\!\!\underset{\text{Pr}_3\text{P}\diagup\!\!\!\!\!}{\text{Pt}}\!\!\!\!\!\diagup\!\!\!\!\!\diagdown\!\!\text{Cl}^{\text{NHEt}_2} + {}^*\text{NHEt}_2 \xrightarrow{\text{MeOH}} \text{Cl}\diagdown\!\!\!\!\!\underset{\text{Pr}_3\text{P}\diagup\!\!\!\!\!}{\text{Pt}}\!\!\!\!\!\diagup\!\!\!\!\!\diagdown\!\!\text{Cl}^{{}^*\text{NHEt}_2} + \text{NHEt}_2 \qquad (13\text{-}5)$$

A comparison of reactions 13-6 and 13-7 also provides information concerning the nature of the two paths.[17] On going from reaction 13-6 to 13-7, *both* k_s and k_Y decrease by a factor of approximately 1000. Since both rate constants are equally affected, this suggests that both paths occur by the same mechanism.

---

[13] A. L. Odell and H. A. Raethel, *Chem. Commun.*, 1323 (1968).

[14] T. P. Cheeseman, A. L. Odell, and H. A. Raethel, *Chem. Commun.*, 1497 (1968); note that the figure captions are reversed in this paper.

[15] A. L. Odell and H. A. Raethel, *Chem. Commun.*, 87 (1969).

[16] Apparently steric hindrance of the entering and leaving groups prevents attack of the Pt(II) by diethylamine when there is an alternative path of considerably lower steric requirement, that is, the attack of methanol on the metal center.

[17] See ref. 13.

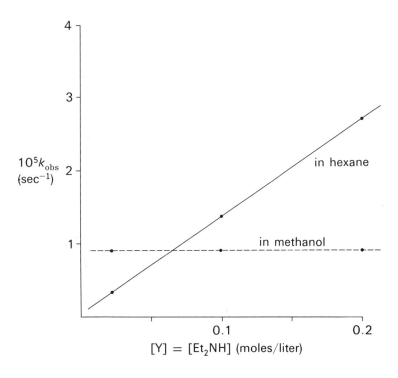

Figure 13-3. An illustration of the importance of solvent on the substitution pathways for square planar reaction centers. These data are for the reaction of *trans*-Pt(Pr$_3$P)(N*HEt$_2$)Cl$_2$ with NHEt$_2$. Data from T.P. Cheeseman, A.L. Odell, and H.A. Raethel, *Chem. Commun.*, 1496 (1968).

700 REACTION MECHANISMS AND METHODS OF SYNTHESIS; SUBSTITUTION REACTIONS

(13-6)

$$\text{trans-}[Pt(Pr_3P)_2Cl_2] + \text{pyridine} \xrightarrow[\substack{k_s = 0.83 \times 10^{-2}\ \text{sec}^{-1} \\ \text{in EtOH}}]{k_Y = 1.66\ M^{-1}\ \text{sec}^{-1}} [Pt(Pr_3P)_2(py)Cl]^+$$

(13-7)

$$[Pt(Pr_3P)(NHEt_2)Cl_2] + \text{pyridine} \xrightarrow[\substack{k_s = 0.86 \times 10^{-5}\ \text{sec}^{-1} \\ \text{in MeOH}}]{k_Y = 0.63 \times 10^{-3}\ M^{-1}\ \text{sec}^{-1}} [Pt(Pr_3P)(py)Cl_2]$$

The experiments cited above, and many others, all point to the correctness of the postulated mechanism, and it is now recognized as the "normal" reactivity mode for substitution reactions of square planar complexes of low spin d^8 metal ions.

FACTORS AFFECTING THE REACTIVITY OF SQUARE PLANAR COMPLEXES OF Pt(II) AND OTHER d^8 METAL IONS

Now that the general mechanism for substitution reactions at square planar reaction centers has been established, the effects of various factors on the reactivity for such complexes must be considered. These factors are:
  a. the nature of the group entering the complex
  b. the effect of other groups in the complex
     i. ligands *trans* to the leaving group
     ii. ligands *cis* to the leaving group
  c. the nature of the leaving group
  d. the nature of the central metal ion

At least for Pt(II) complexes, the first three factors have been arranged in order of decreasing dominance, although this is not always the case for other d^8 ions. Nonetheless, these factors will be discussed in the order listed above. Since examination of all of the influences on reactivity can provide information on the nature of the transition state, the discussion of substitutions at square planar complexes will conclude with a brief inquiry into the intimate mechanism of these reactions.

Influence of the Entering Group

Rates of reactions proceeding by an associative mechanism must depend in some manner on the nature of the entering group. The usual parameter to which one refers in substitution reactions of this type is the *nucleophilicity* of the reactant. You will recall that this is different from the *basicity* of a compound, the latter term being thermodynamic in origin and being defined, for example, by the pK_a of the conjugate acid of a Lewis base. On the other hand, the nucleophilicity is a parallel, but not necessarily an equivalent, concept which refers to the ability of the Lewis base to act as the entering group and influence the reaction rate in a nucleophilic substitution; that is, nucleophilicity is a kinetic term.

Attempts to measure and predict entering group nucleophilicity are widespread, especially in organic chemistry.[18-20] One of the earliest successful equations relating reactivity and nucleophilicity is the Swain-Scott equation:[21,22]

---

[18] K. M. Ibne-Rasa, *J. Chem. Educ.*, **44**, 89 (1967).
[19] P. R. Wells, "Linear Free Energy Relationships," Academic Press, New York, 1968.
[20] R. D. Gilliom, "Introduction to Physical Organic Chemistry," Addison-Wesley Publishing Company, Reading, Massachusetts, 1970.
[21] C. G. Swain and C. B. Scott, *J. Amer. Chem. Soc.*, **75**, 141 (1953).
[22] Notice that the Swain-Scott equation is a two-parameter equation and is of a form that prohibits unique solution. (Recall our discussion of E and C parameters and four-parameter equations in Chapter 5.) The fact that $n = 0$ for water is dictated by using $\log k/k_0$, but $s = 1$ for MeBr is an arbitrary reference.

$$\log (k/k_0) = sn \qquad (13\text{-}8)$$

This equation compares the ability of a given nucleophile to replace Br^- from CH_3Br (in an associative process) with the ability of water to act as a nucleophile in this reaction. That is, the log of the ratio of the rate constant for CH_3Br + nucleophile ($= k$) to the rate constant for $CH_3Br + H_2O$ ($= k_0$) is equal to the product of two constants, s and n. The constant n is the nucleophilic constant, a number characteristic of a given nucleophile; $n = 0.0$ for H_2O by definition. The quantity s is a discrimination constant, a number that "measures" the sensitivity of the substrate (CH_3Br or some other electrophile; $s = 1.0$ for CH_3Br) to various nucleophiles. Data resulting from the Swain-Scott equation are of interest for two reasons. (i) Knowledge of s and n parameters for various electrophiles and nucleophiles, respectively, can be used to predict reaction rates. (ii) Values of n for various nucleophiles toward a common electrophile can be examined to understand the factors determining relative nucleophilicities. It is for the second of these two reasons that the Swain-Scott approach has been carried over to substitutions at square planar d^8 metal ion centers. In this case, the nucleophilicity parameter for ligands attacking platinum(II) complexes is defined by[23]

$$\log (k_Y/k_s) = sn_{Pt} \qquad (13\text{-}9)$$

---

[23] Equation 13-9 follows from the Swain-Scott equation if $k = k_Y$ and $k_0 = k_s$.

TABLE 13-1
NUCLEOPHILIC CONSTANTS AND pK_a VALUES FOR VARIOUS NUCLEOPHILES ATTACKING trans-$[Pt(py)_2Cl_2]$ AND CH_3I[a]

Nucleophile	n°_{Pt} [b]	n_{CH_3I}	pK_a
CH_3OH	0.0	0.0	-1.7
F^-	<2.2	~ 2.7	3.45
Cl^-	3.04	4.37	-5.7
NH_3	3.07	5.50	9.25
piperidine	3.13	7.30	11.21
pyridine	3.19	5.23	5.23
NO_2^-	3.22	5.35	3.37
N_3^-	3.58	5.78	4.74
Br^-	4.18	5.79	-7.7
$(CH_2)_4S$	5.14	5.66	-4.8
I^-	5.46	7.42	-10.7
Ph_3Sb	6.79	<2.0	—
Ph_3As	6.89	4.77	—
CN^-	7.14	6.70	9.3
Ph_3P	8.93	7.00	2.73

[a] Data taken from R. G. Pearson, H. Sobel, and J. Songstad, *J. Amer. Chem. Soc.*, **90**, 319 (1968). All reactions done in methanol.

[b] Since k_Y in equation 13-9 is a second order constant (and has units of M^{-1} sec^{-1}) and k_s is a pseudo-first order constant (and has units of sec^{-1}), the nucleophilicity parameter would have units of M^{-1}. In order to produce a unitless parameter, it is customary to divide k_s by the solvent concentration. (This same dimensional change is made in the Swain-Scott approach.) Therefore,

$$sn^\circ_{Pt} = \log\left[\frac{k_Y}{k_s/[\text{MeOH}]}\right] = \log(k_Y/k_s^\circ)$$

($s = 1$ for trans-$Pt(py)_2Cl_2$)

where $[\text{MeOH}] = 24.9$ M at $30°$

As noted previously, k_Y is the rate constant for the reaction of a given Pt(II) complex with a given nucleophile, while k_s is the rate constant for reaction of the same Pt(II) complex with the solvent, say methyl alcohol.

Nucleophilicity parameters toward both *trans*-Pt(pyridine)$_2$Cl$_2$ and CH$_3$I are given in Table 13-1 along with other data pertaining to properties of the nucleophiles.[24] There are several points that may be made, even after a cursory inspection of the table.

    a. The nucleophilicity of the halide ions decreases in the order I$^-$ > Br$^-$ > Cl$^-$ ≫ F$^-$.

    b. The Group VA bases, except amines, are all excellent nucleophiles toward Pt(II), but their nucleophilicity declines in the order phosphines > arsines > stibines ≫ amines. (Sulfur donors are also better nucleophiles than oxygen donors.)

Such trends as these can obviously be quite important when considering the synthesis of a particular complex.

    The possibility of a correlation between nucleophilicity parameters and other physical constants is also of interest. However, plotting the nucleophilicity parameters for Pt(II) and CH$_3$I against each other gives little or no correlation. This is not altogether surprising, since the reactivity of an sp^3 carbon should be different from that of the more polarizable Pt(II) center.[25]

    There is also no correlation netween n°_{Pt} and nucleophile basicity, as measured by the pK_a of the conjugate acid of the nucleophile. Indeed, numerous chemists have pointed out that the nucleophilicity of amines (and other reagents as well) is not particularly dependent upon their basicity.[26] Instead, there seems to be only a correlation between n°_{Pt} and the presumed polarizability of the entering ligand.[27]

    Finally, there is no correlation between n°_{Pt} and other parameters such as the E and C values (Chapter 5). This is not surprising, though, since these parameters, like pK values, are thermodynamic parameters, and the kinetic and thermodynamic behaviors of bases need not necessarily be related.

Influence of Other Groups in the Complex—Ligands *trans* to the Entering Group[28-30]

An understanding of the stereochemistry of coordination compounds played an important role in the development of chemical theory. For example, one important stereochemical question, which was not settled to everyone's satisfaction until the 1920's, was whether the complexes of platinum(II) were square planar or tetrahedral. Since some understanding of kinetics—the effect of ligands on the ease of substitution of groups to which they are *trans*—played an important role in the answer to the basic stereochemical question, it is worthwhile to discuss briefly the experiments involved and the conclusions that may be drawn.

One approach to solving the problem of the correct structure of platinum(II) complexes was to synthesize deliberately the isomers of a complex of the general formula Pt*abcd*.[31] If the complex were square planar, three diastereoisomers (**4**) should be isolated, while only two enantiomers (**5**) should arise if the complex were tetrahedral. In the 1920's the Russian chemist Chernyaev believed, as did most chemists, that Pt(II) complexes were square planar, and he recognized that the stereospecific synthesis of the three diastereoisomers of Pt*abcd* should be possible

---

[24] R. G. Pearson, H. Sobel, and J. Songstad, *J. Amer. Chem. Soc.*, **90**, 319 (1968).
[25] This difference in reactivity presumably arises from (i) the presence of non-bonding or π^* electrons on Pt(II) and (ii) the great difference between the LUMO's of Pt(II) and CH$_3$I.
[26] R. Romeo and M. L. Tobe, *Inorg. Chem.*, **13**, 1991 (1974).
[27] See ref. 6 (page 268).
[28] See refs. 2 and 4.
[29] T. G. Appleton, H. C. Clark, and L. E. Manzer, *Coord. Chem. Rev.*, **10**, 335–422 (1973).
[30] F. Basolo and R. G. Pearson, *Prog. Inorg. Chem.*, **4**, 381 (1962).
[31] See also our previous discussion of this problem on p. 626 of Chapter 11.

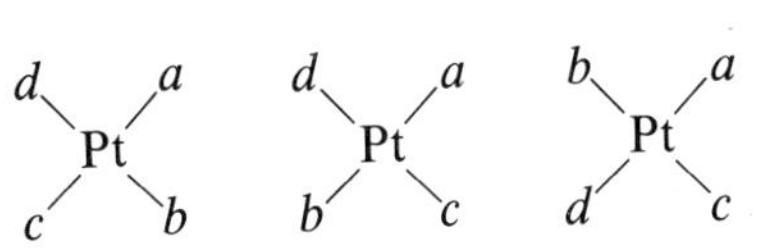

4 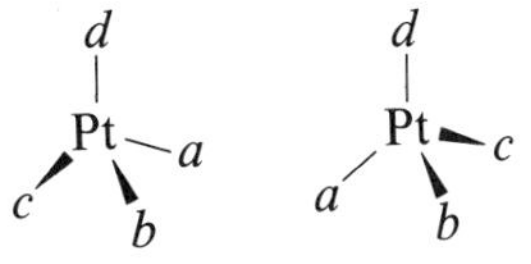 **5**

using an effect first mentioned by Alfred Werner—"*trans* elimination."[32] That is, assuming Pt(II) complexes are square planar, certain groups can more readily than others cause the elimination and substitution of groups *trans* to themselves. If it is assumed that the *trans* labilizing effect of the following groups decreases in the order given—$NO_2^- > Cl^- > NH_3 \sim$ pyridine $\sim NH_2OH$—one can design synthetic paths to each of the three isomers, **6a-c**.

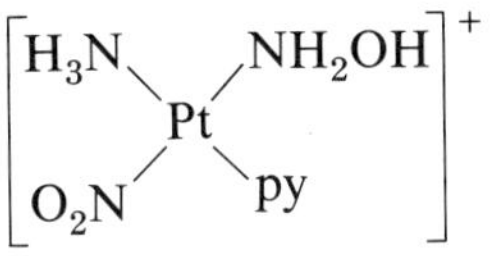

6a 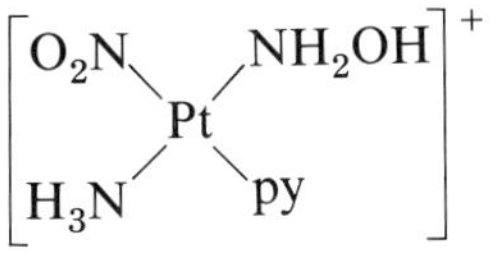 **6b** 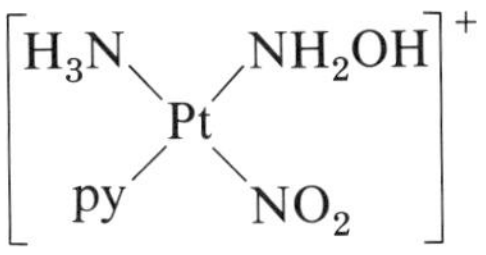 **6c**

For example,

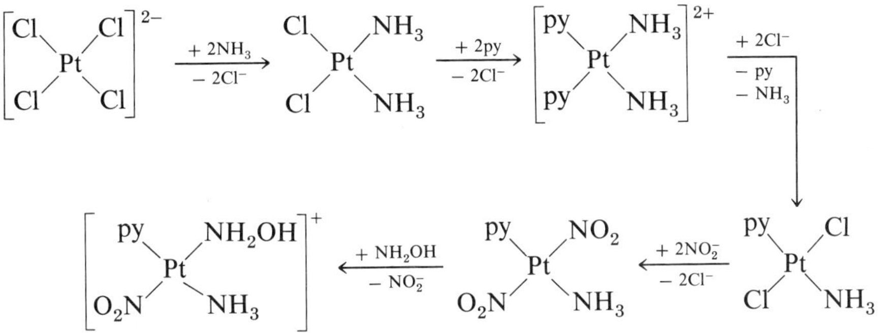

(13-10)

6a

All of the steps in this synthesis follow the assumed order of *trans* labilizing effect—except step 2. However, if you accept the order of *trans* effects for what it is—an empirical order—you can accept another empirical guide: other things being equal, a metal-halogen bond is generally more labile than a metal-nitrogen bond.[33]

Since Werner's and Chernyaev's original work, considerable effort has gone into probing the effects wrought by *trans* ligands in square planar (and octahedral) complexes and into understanding the causes of such effects. It has come to the point where the inclusion of the *"trans* effect" in an inorganic chemistry course is virtually mandatory, even if it is, as Tobe has said,[34] "among the legendary beasts of modern inorganic chemistry." We shall therefore proceed to discuss the more "theoretical" ideas surrounding the beast and return later in the chapter to its use in synthesis.

The *trans* effect is perhaps best defined as *the effect of a coordinated ligand upon the rate of substitution of ligands opposite to it*.[35] For the particular case of Pt(II) complexes, the labilizing effect of typical ligands is generally in the order

$$H_2O \sim OH^- \sim NH_3 \sim \text{amines} < Cl^- \sim Br^- < SCN^- \sim I^- \sim NO_2^- \sim C_6H_5^-$$
$$< CH_3^- \sim S{=}C(NH_2)_2 < \text{phosphines} \sim \text{arsines} \sim H^- < \text{olefins} \sim CO \sim CN^-$$

---

[32] A. Werner, *Z. anorg. Chem.*, **3**, 267–330 (1893). This important paper in inorganic chemistry has been translated and annotated by G. B. Kauffman, "Classics in Coordination Chemistry, Part I: The Selected Papers of Alfred Werner," Dover Publications, Inc., New York, 1968.

[33] See the section on "The Thermodynamic Stability of Coordination Complexes" beginning on page 733.

[34] See ref. 1 (page 54).

[35] See ref. 30.

Notice that the term "labilizing effect" is used above to emphasize the fact that we are discussing a kinetic phenomenon. For its explanation we shall have to examine the effect of these groups on the activation energy barrier. That is, the labilization may arise because of the destabilization (a thermodynamic term!) of the ground state and/or by stabilization of the transition state.

The language used above was chosen carefully to differentiate the *trans effect*—a kinetic phenomenon—from the *trans influence*—a thermodynamic phenomenon.[36] That is, ligands can *influence* the ground state properties of groups to which they are *trans* such as the *trans* metal-to-ligand bond distance, the vibrational frequency or force constant, the nmr coupling constant between the metal and the *trans* ligand donor atom (say ^{195}Pt and ^{31}P), or a host of other parameters. The relationship of these observable ground state properties to the *trans effect* is not entirely clear as yet, although, as is described below, recent theoretical work indicates a definite correlation.

As an example of the kinetic *trans* effect, consider the following reaction.[37]

(13-11)
$$\begin{array}{c}\text{Et}_3\text{P}\diagdown\hspace{-6pt}\diagup\text{Cl}\\\text{Pt}\\\text{T}\diagup\hspace{-6pt}\diagdown\text{PEt}_3\end{array} + \text{Y} \xrightarrow{\text{MeOH}} \begin{array}{c}\text{Et}_3\text{P}\diagdown\hspace{-6pt}\diagup\text{Y}\\\text{Pt}\\\text{T}\diagup\hspace{-6pt}\diagdown\text{PEt}_3\end{array} + \text{Cl}^-$$

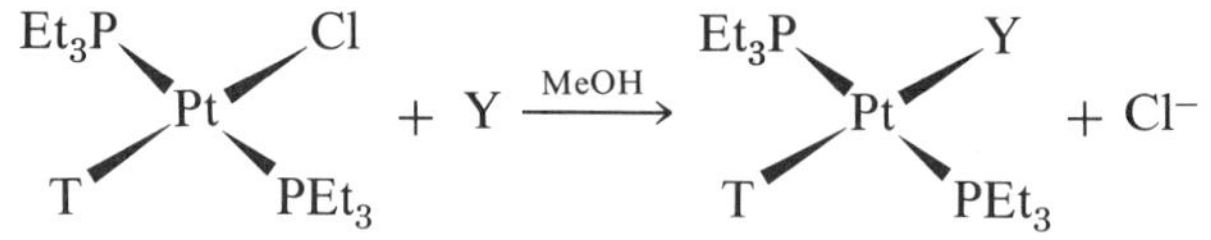

$$\text{T} = \text{Cl}^-, \text{CH}_3^-, \text{ and } \text{C}_6\text{H}_5^-$$

When Y is Br^-, N_3^-, NO_2^-, or pyridine (all relatively weak nucleophiles according to Table 13-1), the *trans* effect is in the order given above: $\text{CH}_3^- > \text{C}_6\text{H}_5^- > \text{Cl}^-$ (see Table 13-2). However, when the better nucleophile I^- is used, the *trans* effect order now becomes $\text{CH}_3^- > \text{Cl}^- > \text{C}_6\text{H}_5^-$. Although the reason for this inversion in *trans* effects is not at all clear, two other observations can be made with assurance: (i) the *trans* effect depends in part on the nature of the entering group, and (ii) the effects of *trans* ligands and entering groups cannot be neatly compartmentalized.‡

The *trans* influence of a wide variety of ligands has been "measured" with techniques such as x-ray crystallography and infrared, nmr, nuclear quadrupole resonance, photoelectron, and Mössbauer spectroscopy.[38,39] For instance, more precise techniques of x-ray crystallography have led to more accurate information on

---

[36] See ref. 29.
[37] U. Belluco, M. Graziana, and P. Rigo, *Inorg. Chem.*, **5**, 1123 (1966).
‡ Notice that $\Delta S^\ddagger$ has a significant influence on the relative k_Y's.
[38] G. M. Bancroft and K. D. Butler, *J. Amer. Chem. Soc.*, **96**, 7203 (1974).
[39] See ref. 3.

TABLE 13-2
KINETIC DATA FOR THE REACTION OF
trans-Pt(PEt$_3$)$_2$ClT WITH VARIOUS
NUCLEOPHILES IN METHANOL[a,b]

Nucleophile	T = Cl$^-$			T = C$_6$H$_5^-$			T = CH$_3^-$		
	$10^2 k_Y$	$\Delta H^\ddagger$	$\Delta S^\ddagger$	$10^2 k_Y$	$\Delta H^\ddagger$	$\Delta S^\ddagger$	$10^2 k_Y$	$\Delta H^\ddagger$	$\Delta S^\ddagger$
CH$_3$OH	—	—	—	0.85×10^{-2}	12	−36	4×10^{-2}	12	−31
N$_3^-$	0.02	15.5	−24	0.8	17	−11	7	15.5	−13
Br$^-$	0.093	—	—	1.8	16.5	−12	11.6	14	−16
I$^-$	23.6	—	—	6.0	12	−24	40	12	−21

[a] Data taken from U. Belluco, M. Graziani, and R. Rigo, *Inorg. Chem.*, **5**, 1123 (1966).
[b] Rate constants are in units of M^{-1} sec^{-1}, $\Delta H^\ddagger$ in units of kcal/mol, and $\Delta S^\ddagger$ in units of eu/mol.

the variation in bond lengths of Pt—Cl bonds *trans* to a variety of ligands. In the series below, for example, the *trans* Pt—Cl bond length decreases on going from **7** to **9**.[40] If the usual assumption can be made concerning the relation between bond

$$\begin{bmatrix} \text{Cl} & \overset{2.382\,\text{Å}}{\diagup} \text{Cl} \\ & \text{Pt} \\ \text{Et}_3\text{P} & \diagdown \text{Cl} \end{bmatrix}$$
7

$$\begin{bmatrix} \text{Cl} & \overset{2.327\,\text{Å}}{\diagup} \text{Cl} \\ \text{H}_2\text{C} & \text{Pt} \\ \| & \diagdown \text{Cl} \\ \text{CH}_2 & \end{bmatrix}^{-}$$
8

$$\begin{bmatrix} \text{Cl} & \overset{2.317\,\text{Å}}{\diagup} \text{Cl} \\ & \text{Pt} \\ \text{Cl} & \diagdown \text{Cl} \end{bmatrix}^{2-}$$
9

length and bond strength, the Pt—Cl bond increases in strength in the order: *trans* ligand = phosphine < olefin < chloride. Based on more extensive data, the structural *trans* influence order has been given as

$$\sigma\text{—R} \sim \text{H}^- \geq \text{carbenes} \sim \text{PR}_3 \geq \text{AsR}_3 > \text{CO} \sim \text{RNC} \sim \text{C}=\text{C} \sim \text{Cl}^- \sim \text{NH}_3$$

The kinetic *trans* effect and the thermodynamic *trans* influence require theoretical explanation, and hopefully the explanation of both can rest on the same foundation. You should notice again, however, that the former is presumably related in some way to the nature of the transition state, while the latter is a ground state phenomenon (*cf.* the difference in nucleophilicity and basicity, page 700). Any explanation must take this difference into account. Furthermore, any explanation must also account for the fact that some ligands with high *trans* influence also have a high *trans* effect, while others—especially olefins—have a high *trans* effect but a low *trans* influence.

Perhaps the explanation of the *trans* effect that has best stood the test of time is that of Langford and Gray,[41] an idea examined more recently by others through molecular orbital calculations.[42,43] For σ bonding ligands, these calculations show that the strength of the Pt—N bond in *trans*-PtCl$_2$(NH$_3$)T decreases as T is changed from H$_2$O to CH$_3^-$ in the order H$_2$O ≥ H$_2$S ≥ Cl$^-$ ≫ PH$_3$ ≥ H$^-$ ≥ CH$_3^-$.[42]

$$\begin{array}{c} \text{Cl} \quad \text{NH}_3 \\ \diagdown \quad \diagup \\ \text{Pt} \\ \diagup \quad \diagdown \\ \text{T} \quad \text{Cl} \end{array}$$
10

In general, the *trans*-Pt—N bond is weakened as T becomes a better labilizer of the *trans*-ammonia. Therefore, at least for σ bonding ligands, there must be a connection between the ability of T to act as a labilizer of the *trans* position and its effectiveness in the sense of the *trans* influence. However, Drago argues that the weakening on going from the poor *trans* labilizer H$_2$O to the good *trans* labilizer H$^-$ is only a relatively small fraction of the total bond strength, so bond weakening cannot be the total argument.[42] Rather, stabilization of the transition state must also be an important contribution to the *trans* effect.

As depicted in Figure 13-1, square planar substitution reactions are thought to proceed through a trigonal bipyramidal transition state. As the entering ligand approaches the complex, the symmetry of its donor orbital is appropriate for overlap with the unoccupied metal p_z orbital. We have designated this as phase A in the reaction illustrated in the margin. At this point along the reaction coordinate, the entering ligand (= E) can move in any of four possible directions: toward either of the *cis* ligands (= C), toward the group assumed to eventually be leaving (= L), or toward the group *trans* to L (= T). However, based on thermodynamic reasoning, it is a safe assumption that E will move toward the *least* strongly bound ligand (assumed to be L) and away from the most strongly bound ligand (assumed to be T) (phase B). This act of discrimi-

Phase A

Phase B

Phase C

Phase D

---

[40] G. Bushnell, A. Pidcock, and M. A. R. Smith, *J. C. S. Dalton*, 572 (1975).
[41] See ref. 3.
[42] S. S. Zumdahl and R. S. Drago, *J. Amer. Chem. Soc.*, **90**, 6669 (1968).
[43] D. R. Armstrong, R. Fortune, and P. G. Perkins, *Inorganica Chimica Acta*, **9**, 9 (1974).

nation explains how there can be a connection between the *trans* effect and the *trans* influence. That is, certain ligands cause the preferential loss of ligands *trans* to themselves by weakening the bond to the *trans* ligand, a weakening that shows up in the ground state properties of the molecule. Yet another aspect of phase B is crucial to an understanding of the *trans* effect. If it is assumed that the entering and leaving ligands will be separated by about 120° in the transition state (i.e., the transition state structure is a trigonal bipyramid), then L must move away from the y axis faster than E approaches the y axis. That is, L moves through 60° in about the same time that E moves through 30°. Therefore, the e_σ parameter for L interacting with the p_y orbital decreases faster than the increase in e_σ for E + p_y. The net effect is that the overlap of E and L together with p_y is less than the overlap of L with p_y in the parent square planar complex. Most importantly, because of this, the preoccupation of T with the p_y orbital can increase, especially if T is a particularly effective σ donor such as H^- or CH_3^-. The result will be greater transition state stabilization for the more strongly bonding T groups, and the reaction rate will be enhanced.

These same points may be illustrated further by the simple mo diagram below. If the *trans* ligand T is a more effective σ donor than L, then the energy of the T donor

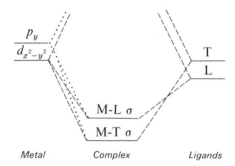

orbital lies higher than that of L. This means that there will be a better energy match between T and M-p_y than between L and M-p_y. Therefore, greater stabilization will be afforded by T-p_y interaction than by L-p_y. According to the argument above, T and p_y can interact even more effectively in the transition state, thereby lowering the energy of the resulting mo even further, and giving rise to transition state stabilization. Similar arguments using the metal $d_{x^2-y^2}$ orbital lead to similar results. However, the metal p_y orbital is probably more effective in communicating the effect of T to L.

The effectiveness of ligands such as olefins in labilizing *trans* ligands can also be explained by transition state stabilization. As you saw in Chapter 9 (Figure 9-15), an olefin binds to a metal in two ways: donation of π electron density from the olefin into a metal orbital of σ symmetry (structure **11**) and back-donation of metal electron density from a metal orbital of π symmetry (say d_{yz}) into an olefin orbital of the same symmetry, the olefin π^* orbital (**12**). As the entering group in a substitution reaction approaches, the electron density on the metal is increased, and any metal-T atomic orbital interaction that can remove this excess density should stabilize the transition state and lead to rate enhancement. This is, of course, precisely the effect of metal-to-ligand $d\pi$-π^* bonding, and ligands capable of this—CO, CN^-, and olefins—will have large *trans* effects but not necessarily a large *trans* influence.

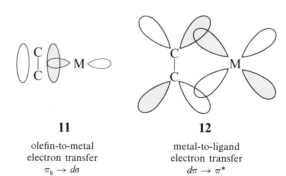

11
olefin-to-metal
electron transfer
$\pi_b \to d\sigma$

12
metal-to-ligand
electron transfer
$d\pi \to \pi^*$

Influence of Other Groups in the Complex—Ligands *cis* to the Entering Group

In cases where a relatively poor nucleophile acts as the entering group, the relative ability of ligands to act as *trans* labilizers is the same as their ability to act as *cis* labilizers, but the effect is less marked. For example, in the reaction[44]

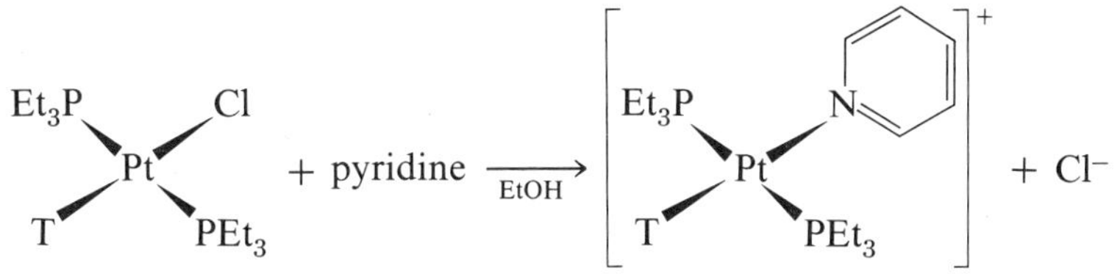

$$\text{[Et}_3\text{P, T, Cl, PEt}_3\text{]Pt} + \text{pyridine} \xrightarrow{\text{EtOH}} \text{[Et}_3\text{P, T, N-pyridine, PEt}_3\text{]Pt}^+ + \text{Cl}^- \quad (13\text{-}12)$$

the relative ordering of *trans* labilizing effects, measured against T = Cl as the standard case, is given by

$$T = H^- > CH_3^- > C_6H_5^- > Cl^-$$

$$\frac{k_Y(T)}{k_Y(Cl)} = >10^4 \quad 1700 \quad 400 \quad 1.0$$

In contrast, the *cis* effect of these same ligands is in the same order but is considerably less.[44]

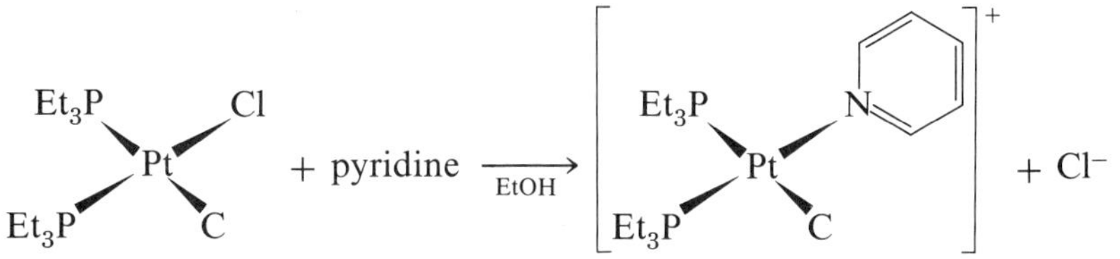

$$\text{[Et}_3\text{P, Et}_3\text{P, Cl, C]Pt} + \text{pyridine} \xrightarrow{\text{EtOH}} \text{[Et}_3\text{P, Et}_3\text{P, N-pyridine, C]Pt}^+ + \text{Cl}^- \quad (13\text{-}13)$$

$$C = CH_3^- > C_6H_5^- > Cl^-$$

$$\frac{k_Y(C)}{k_Y(Cl)} = 3.6 \quad 2.3 \quad 1.0$$

The most likely explanation for this observation is that the *cis* effect is similar in origin to the *trans* effect. Indeed, molecular orbital calculations show that ligands that weaken bonds *trans* to themselves also weaken *cis* M—L bonds, but not by as large an amount.[45] (This is presumably due to the fact that there is direct communication between *trans* ligands *via* the metal p_y and $d_{x^2-y^2}$ orbitals, whereas *cis* ligand communication depends only on one orbital, $d_{x^2-y^2}$.) Therefore, bond breaking in the transition state becomes somewhat easier for good *cis* directors, and the reaction rate increases, although not quite as much as if the *cis* ligand were in the *trans* position.

The Nature of the Leaving Group

The nature of the leaving group can also affect the rate of the substitution reaction. Among the many examples observed is the following reaction:[46]

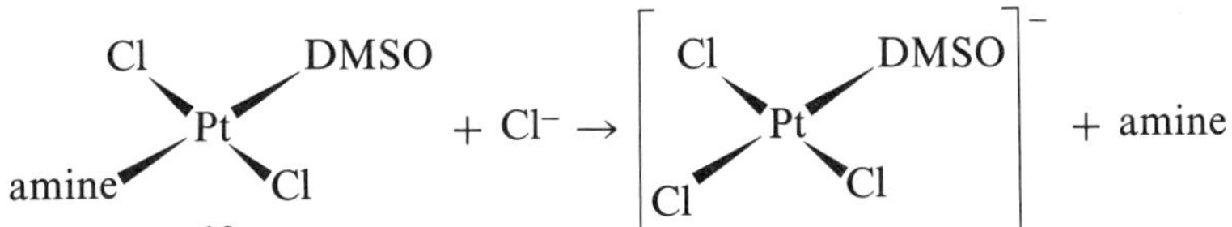

$$\text{[Cl, amine, Pt, DMSO, Cl]} + \text{Cl}^- \rightarrow \text{[Cl, Cl, Pt, DMSO, Cl]}^- + \text{amine} \quad (13\text{-}14)$$

13

---

[44] F. Basolo, J. Chatt, H. B. Gray, R. G. Pearson, and B. L. Shaw, *J. Chem. Soc.*, 2207 (1961).
[45] See ref. 42.
[46] See ref. 26.

The displacement of a wide variety of amines was found to proceed by both the k_Y and k_s paths. More interesting, however, is the fact that the rate constant k_s is very sensitive to the basicity of the leaving group (so is k_Y, but less so). In fact, there was an excellent correlation between the pK_a of the leaving group and log k_s; as the basicity of the amine increases, its replacement becomes more difficult. This could imply that bond breaking is important in the rate determining step.[47]

Effect of the Central Metal

Reaction 13-15 illustrates the effect of the central metal ion on the rate of square planar substitutions.[48] In this case the tendency to undergo substitution is in the order Ni(II) > Pd(II) ≫ Pt(II). This order of reactivity is in the same order as the tendency to form five-coordinate complexes. More ready formation of a five-coordinate intermediate complex leads to stabilization of the transition state and to rate enhancement.

(13-15)

for M = Ni $k_Y = 33$ M^{-1} sec^{-1}
for M = Pd $k_Y = 0.58$ M^{-1} sec^{-1}
for M = Pt $k_Y = 6.7 \times 10^{-6}$ M^{-1} sec^{-1}

THE INTIMATE MECHANISM FOR REPLACEMENT AT FOUR-COORDINATE PLANAR REACTION CENTERS

Substitution reactions at square planar reaction centers largely proceed by the k_Y path. If the reaction is simply an exchange of a ligand—as in reaction 13-5—then the principle of microscopic reversibility for an A mechanism dictates a reaction energy profile qualitatively like that in Figure 13-4A. This profile, of course, applies only to a case in which the energies of the reactant and product are the same and in which the energies of the transition states are identical. With different entering and leaving groups, and therefore with different energies for product and reactant, the reaction profiles will change. Profile B is appropriate to a reaction for which bond breaking is rate determining (the mechanism is D in the intermediate), whereas profile C depicts the opposite situation. Langford and Gray postulate that profile C is appropriate to a situation in which the entering group (Y) lies higher in the *trans* effect series than the leaving group; profile B is then characteristic of the opposite situation.[49]

In the cases discussed above, the intermediate is of higher energy than either the reactant or product. In such instances it is often difficult to determine which profile is most appropriate. However, when the intermediate is reasonably stable (its decom-

---

[47] The k_s/K_a correlation is not conclusive evidence for Pt-amine bond rupture in the transition state because rate retardation with increasing base strength could arise from decreasing platinum nucleophilicity. More conclusive, however, is the fact that ln k_Y = ln K + constant for reaction 13-14. This can happen only if k^r, the overall rate constant for the reverse of 13-14, is constant and therefore independent of amine; that is, since $K = k_Y/k^r$, and so ln k_Y = ln K + ln k^r, then ln k^r equals the constant noted above.

An alternative approach to this same conclusion begins with the well-known relations ln K = $-\Delta G°/RT$ and ln K_Y = $-(\Delta G^{\ddagger}/RT)$ + ln (kT/h). Writing the change in ln k_Y and ln K with substituents as δ_s ln k_Y and δ_s ln K, the linear relationship of ln $k_Y = \alpha$ ln K + β implies that $\delta_s\Delta G^{\ddagger} = \alpha\delta_s\Delta G°$. Since $\delta_s\Delta G^{\ddagger}$ is defined as $\delta_s G^{\ddagger} - \delta_s G_R°$ and $\delta_s\Delta G°$ is defined by $\delta_s G_P° - \delta_s G_R°$, then $\delta_s G^{\ddagger} = \alpha\delta_s G_P° + (1 - \alpha)\delta_s G_R°$. In the present case, $\delta_s G_P° = 0$ and $\alpha = 1$ for reaction 13-14, and $\delta_s G^{\ddagger} = 0$. That is to say, the transition state free energy is independent of the amine! The transition state is characterized by essentially no Pt-amine bonding. Whether treated as an I mechanism or as a mechanism with an intermediate (see pp. 366-368), dissociative activation is called for in both the forward and reverse directions of 13-14.

[48] See ref. 44.
[49] See ref. 3.

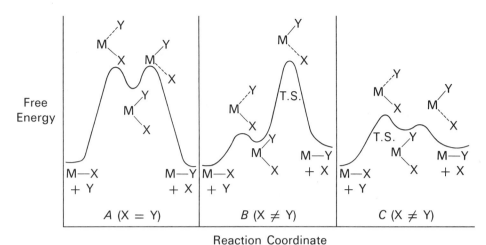

Figure 13-4. Reaction profiles for the substitution reaction of square planar complexes. Profile A represents a simple exchange of ligands; that is, Y = X. Profile B is appropriate to a reaction wherein bond breaking is rate determining; the transition state (=T.S.) follows the intermediate. Profile C is for the case where bond making is rate determining; the transition state precedes the intermediate.

position rate is slow), it is possible to decide more confidently the shape of the reaction profile. On mixing the reactants in reaction 13-16, there is an immediate

$$\text{[Rh(Cl)(SbR}_3\text{)]} + \text{pyridine} \rightarrow \text{[Rh(Cl)(py)]} + \text{SbR}_3 \quad (13\text{-}16)$$

14

change in the electronic spectrum of the solution, followed by a slow change to give the spectrum of the final product.[50] If the reaction is run in a variety of solvents, and an excess of amine is used, the slow step is first order with respect to the original Rh(I) complex. Further, the rate constant depends on the nature of the amine (if other than pyridine is used) but not on the nature of the solvent. All of these observations suggest that the usual parallel path mechanism is not followed. That is, the solvent-dependent path is apparently not utilized. Instead, a relatively stable, five-coordinate intermediate is formed rapidly between the original Rh(I) complex and pyridine; this intermediate then expels SbR$_3$ in a slower, rate determining step. The reaction would be characterized, therefore, by the profile shown in Figure 13-5. This mechanism is in agreement with the observed proclivity of Rh(I) to form complexes of coordination number higher than four.

---

[50] L. Cattalini, R. Ugo, and A. Orio, *J. Amer. Chem. Soc.,* **90,** 4800 (1968).

Figure 13-5. A reaction profile for a reaction where a quasi-stable intermediate is formed before the transition state; bond breaking in the five-coordinate intermediate is more important than bond making to form the intermediate.

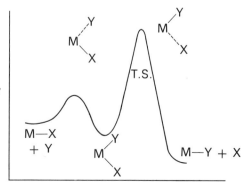

710 REACTION MECHANISMS AND METHODS OF SYNTHESIS; SUBSTITUTION REACTIONS

STUDY QUESTIONS

1. Outline the preparation of the *cis* and *trans* isomers of [Pt(NH$_3$)(NO$_2$)Cl$_2$]$^-$, given that the substituent *trans* effects are in the order NO$_2^-$ > Cl$^-$ > NH$_3$.

2. a. On page 707 (reaction 13-14) we discussed the effects of leaving groups on substitutions at square planar centers. The compound needed in these studies was *trans*-[Pt(amine)(DMSO)Cl$_2$]. Given that the *trans* effects of the ligands are in the order DMSO (S-bonded) > Cl$^-$ > amine, describe a preparation of the *cis* and *trans* isomers from [PtCl$_4$]$^{2-}$.

   b. If *cis*-[Pt(amine)$_2$(DMSO)Cl]$^+$ is heated with one equivalent of HCl, one ligand is substituted with chloride. What is the composition of the final product and what is its structure?

3. The basicities (as measured by pK_a values) of trimethyl-, triethyl-, and tri-*n*-propyl-phosphine are nearly identical. However, in reaction 13-5, if tri-*n*-propylphosphine is replaced by Et$_3$P the rate is enhanced; even greater enhancement is gained by replacement with Me$_3$P. Why should this be the case?

4. Of what mechanistic significance are the $\Delta S^{\ddagger}$ values in Table 13-2?

5. Why should the rate constants in reaction 13-6 be so much greater than those in 13-7?

6. Why do phosphines have a high *trans* effect? (Hint: Consider the way in which they are bonded to a transition metal ion.)

7. Explain more fully why a ligand such as an olefin or CN$^-$ can have a small *trans* influence but a large *trans* effect.

SUBSTITUTION REACTIONS OF OCTAHEDRAL COMPLEXES[51-53]

The vast majority of the complexes formed by the transition metals are octahedral, so an examination of their substitution reactions is expecially important. Before beginning our discussion, however, we might point out several general features of octahedral substitutions and some similarities and differences with the reactions of square planar complexes.

(1) Most of the research on the rates and mechanisms of substitution reactions of square planar complexes has been done with Pt(II) complexes. Compounds of this ion combine several factors favorable to kinetic studies: they may be readily prepared (as you shall see later in the chapter), and their reactions are slow enough to be followed by conventional means. Similarly, in kinetic studies of octahedral complexes, the majority of the work has been done with Co(III) complexes, that is, until the advent of methods for the examination of fast reactions (half-lives less than one minute).[54,55] Further, based on Alfred Werner's extensive work, numerous complexes of Co(III) are known,[56] and many undergo reaction at convenient rates.

---

[51] See refs. 1 to 4.
[52] N. Sutin, *Ann. Rev. Phys. Chem.*, **17**, 119 (1966).
[53] A. McAuley and J. Hill, *Quart. Rev.*, **23**, 18 (1969).
[54] K. Kustin and J. Swinehart, *Prog. Inorg. Chem.*, **13**, 107-158 (1970).
[55] C. M. Frey and J. Stuehr, "Kinetics of Metal Ion Interactions with Nucleotides and Base Free Phosphates," in "Metal Ions in Biological Systems," Vol. 1, pp. 51-116, ed. H. Sigel, Marcel Dekker, Inc., New York, 1974.
[56] F. Morral, *Adv. Chem. Ser.*, **62**, 70-77 (1967).

(2) Reactions at square planar centers occur by associative activation; only a few examples of dissociative activation are known. In contrast, substitution reactions of the complexes of divalent, first row transition metals and of Co(III) occur by dissociative activation. This change from a to d activation with change in coordination number is not surprising because additional coordination positions are not as available in octahedral complexes as they are in square planar complexes. However, evidence has been accumulating in recent years that substitution reactions of complexes of trivalent ions—except for Co(III)—may occur by associative activation (in particular, I_a).[56a] This apparent change in mechanism with metal ion charge is not unreasonable, since bond-making could occur more readily than bond-breaking with a trivalent ion than with a divalent ion.[56b]

(3) Studies on octahedral complexes have been largely limited to two types of reactions:

(i) *Replacement of coordinated solvent (water)*. Perhaps the most thoroughly studied replacement reaction of this type is the formation of a complex ion from the hydrated metal ion in solution (reactions 13-17 and 13-18). When the entering group is an anion, as in reaction 13-17,

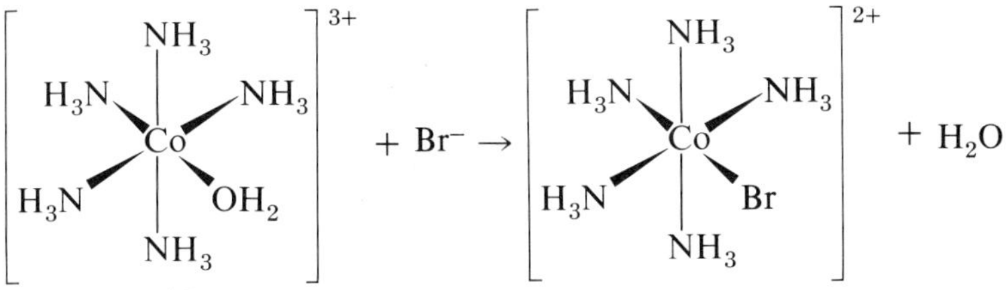

$$[\text{Co(NH}_3)_5\text{OH}_2]^{3+} + \text{Br}^- \rightarrow [\text{Co(NH}_3)_5\text{Br}]^{2+} + \text{H}_2\text{O} \quad (13\text{-}17)$$
15

the reaction is often called *anation*.[57]

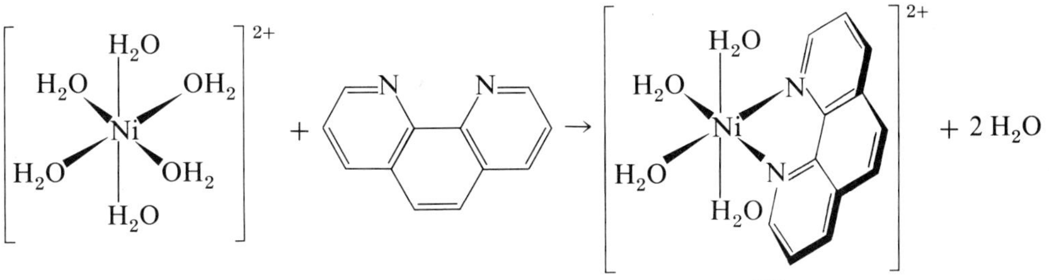

$$[\text{Ni(H}_2\text{O})_6]^{2+} + \text{phen} \rightarrow [\text{Ni(H}_2\text{O})_4\text{(phen)}]^{2+} + 2\,\text{H}_2\text{O} \quad (13\text{-}18)$$
16

(ii) *Solvolysis*. Since the majority of such reactions have been carried out in aqueous solution, *hydrolysis* is a more appropriate term. Hydrolysis reactions have been done under both acidic (13-19) and basic (13-20) conditions.[58,59]

[56a] T. W. Swaddle, *Coord. Chem. Rev.*, **14**, 217 (1974); see also M. Maestri, F. Bolletta, N. Serpone, L. Moggi, and V. Balzani, *Inorg. Chem.*, **15**, 2048 (1976). Although both of these papers conclude that a activation is consistent with experiment in many cases, especially for Cr(III) complexes, the matter of associative activation substitution reactions of octahedral complexes in general is still a matter of conjecture and discussion.

[56b] Reactions of Co(III) have long been considered as typical of other transition metal ions. However, it would now appear that complexes of this ion may in fact be anomalous in their behavior; reasons for this are uncertain. See ref. 56a.

[57] W. L. Reynolds, I. Murati, and S. Ašperger, *J. C. S. Dalton*, 719 (1974).

[58] W. L. Reynolds, M. Biruš, and S. Ašperger, *J. C. S. Dalton*, 716 (1974).

[59] D. A. Buckingham, I. I. Olsen, and A. M. Sargeson, *J. Amer. Chem. Soc.*, **90**, 6654 (1968).

(13-19) [Co(NH₃)₅(OS(Me)₂)]³⁺ + H₂O → [Co(NH₃)₅(OH₂)]³⁺ + Me₂SO

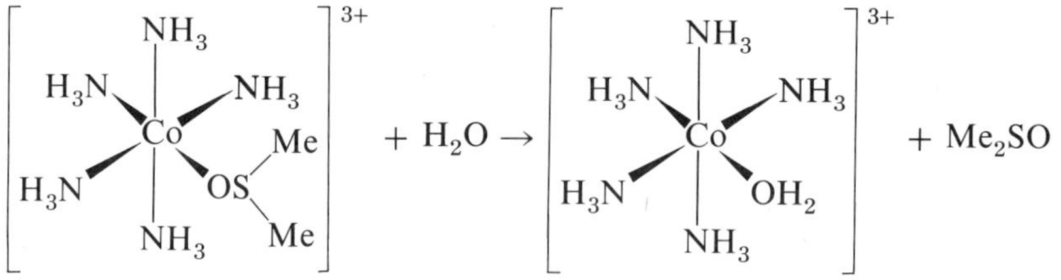

(structure **17** → structure **15′**)

There is a very interesting and fundamental difference in the mechanisms under these two pH ranges (13-19 and 13-20), a fact that we shall explore later in the chapter.

(13-20) [Co(en)₂(NH₃)Cl]²⁺ $\xrightarrow[-H_2O]{+OH^-}$ [Co(en)(en-H)(NH₃)Cl]⁺ $\xrightarrow[-Cl^-]{+H_2O}$ [Co(en)₂(NH₃)(OH)]²⁺

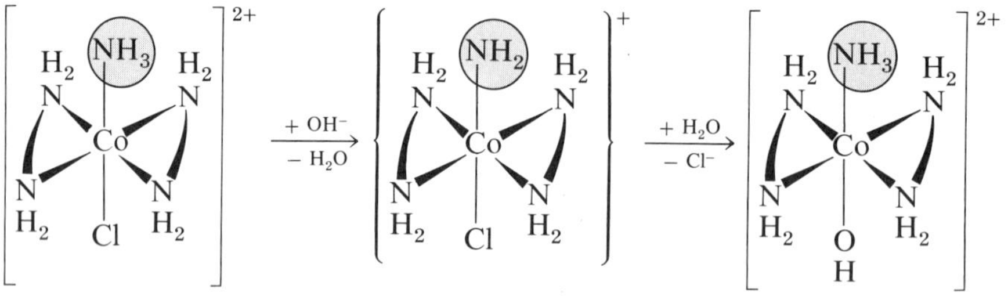

(structures **18** → → **19**)

(the **Conjugate Base** or **CB** mechanism)

You will notice that no mention is made of reactions involving the *direct* interchange of two anions, a reaction type commonly observed with four-coordinate Pt(II) complexes. Instead, an octahedral complex must first lose a coordinated anion by hydrolysis and then replace the newly coordinated solvent by the other anion.[60] (This sequence is, of course, akin to the k_s path for square planar Pt(II) complexes.) This

(13-21) [Co(NH₃)₅Br]²⁺ $\underset{Br^-}{\overset{H_2O}{\rightleftharpoons}}$ [Co(NH₃)₅(OH₂)]³⁺ $\underset{H_2O}{\overset{NCS^-}{\rightleftharpoons}}$ [Co(NH₃)₅NCS]²⁺

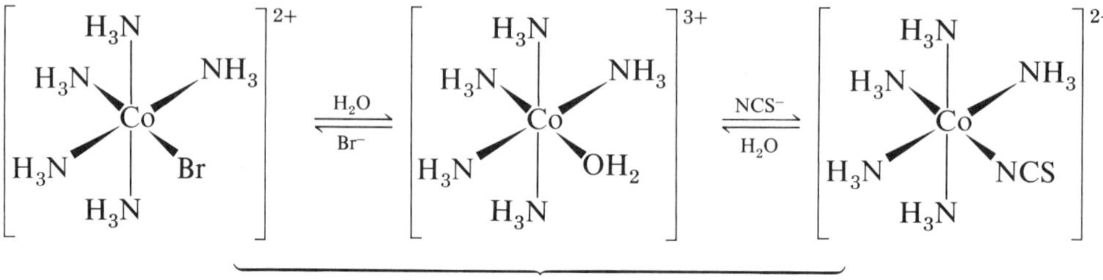

net reaction

$[Co(NH_3)_5Br]^{2+} + NCS^- \rightleftharpoons [Co(NH_3)_5NCS]^{2-} + Br^-$

observation is an important one for you to ponder. Firstly, hydrolysis and anation are the reverse of one another, so the two reactions follow the same reaction coordinate in opposite directions.[61] Secondly, both reactions apparently occur by dissociative activation. That is, since one anion does not directly replace another, but is instead replaced by water, this means either that water is always the superior nucleophile *or* that bond formation plays an insignificant role in the activation process. The superior nucleophilicity of water may generally be safely discounted.

---

[60] R. G. Pearson and J. W. Moore, *Inorg. Chem.*, **3**, 1334 (1964).
[61] The principle of microscopic reversibility, applied to ligand exchange, requires that forward and reverse directions be mirror images. When reactant and product are different this is no longer true; however, the loss of symmetry may or may not appreciably destroy the mirror image feature.

REPLACEMENT OF COORDINATED WATER

A great deal of information regarding substitution reactions of octahedral complexes has been obtained by a thorough examination of the simple process of the exchange of bound water for bulk water[62] or for other ligands when the metal ion is placed in aqueous solution.[63] We shall first discuss evidence concerning the mechanism of water replacement on divalent ions and then turn to an analysis of the dependence of rates on the nature of the metal ion.

The Mechanism of Water Replacement

The alternative mechanisms for the replacement of coordinated water by another ligand are illustrated in Figure 13-6. In one of these—the dissociative or D mechanism—the original metal complex dissociates a water molecule to produce a five-coordinate, solvent caged intermediate in the rate determining step.

The other two mechanisms both involve, as a first step, the formation of an ion- or reactant-pair. Manfred Eigen, who shared the Nobel Prize in 1967 for his work on fast reactions, found that the following stoichiometric mechanism generally best describes the replacement of water by another ligand:

$$[M(H_2O)_6]^{n+} + X \rightleftharpoons [M(H_2O)_6]^{n+} \cdots X \qquad (13\text{-}22a)$$

$$[M(H_2O)_6]^{n+} \cdots X \to [M(H_2O)_5X]^{m+} + H_2O \qquad (13\text{-}22b)$$

The first step in this process is the formation of the reactant-pair. Since the incoming ligand is held in the solvent layer immediately surrounding the metal ion, this is often called an outer sphere complex to convey the fact that the new ligand is contained in the outer coordination sphere of the metal ion. [Recall that we referred to this as the cage complex in Chapter 7 (pp. 359–371) and that the equilibrium constant for its formation was labeled K_{dif}.] Following the formation of the reactant-pair, a seven-coordinate intermediate may be formed in the rate determining step. This pathway is associative and is labeled A.

If the experimental data clearly indicate the presence of an intermediate, the reaction can be labeled D or A as appropriate. However, if there is no apparent intermediate, the reaction is assigned an interchange mechanism ($= I$) in which there is a smooth interchange of entering and leaving groups (the pathway in the middle of Figure 13-6). As you have already seen in this chapter and in Chapter 7, there is the additional possibility that the interchange activation may be predominantly associative or dissociative. That is, the mechanism is labeled I_a if bond making is apparently more important than bond breaking in the transition state; the opposite situation, when bond breaking is more important than bond making, is labeled I_d. Experimental tests for such distinctions were discussed in Chapter 7, as were the derivations of the rate laws for D, A, and I mechanisms.

The rate law observed for the substitution of an aquo complex is first order in both metal ion and entering ligand; that is, Rate $= k_{obs}$[complex][X]. Unfortunately, all of the rate laws given in Figure 13-6 for the various possible mechanisms can reduce to the simple, observed law under easily attained conditions. (The circumstances under which this can occur are left to you to explore as a Study Question later in this chapter.) Nonetheless, the accumulated evidence strongly favors I_d as the mechanism of substitution of water on an aquated, divalent transition metal ion in most cases. The evidence for this conclusion is well worth discussion.

---

[62] The number of coordinated waters for almost all metal ions has been determined. Although almost all of these ions—whether from the main groups, the transition metals, or the lanthanides—have coordination numbers of six, some alkaline earth ions (*e.g.*, Be^{2+} and Ca^{2+}) have coordination numbers of four, and some lanthanide ions have higher coordination numbers. See ref. 54.

[63] See ref. 53.

714 ☐ REACTION MECHANISMS AND METHODS OF SYNTHESIS; SUBSTITUTION REACTIONS

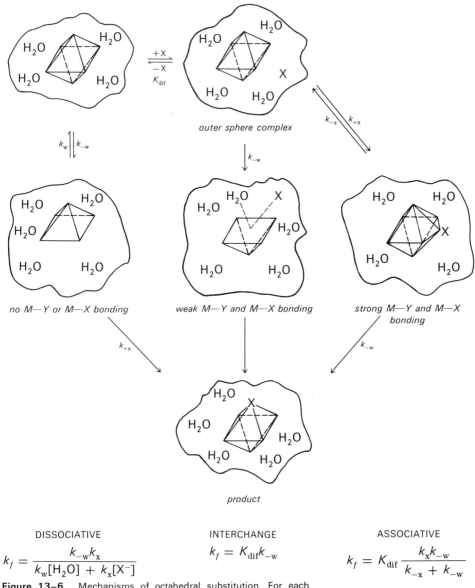

Figure 13-6. Mechanisms of octahedral substitution. For each mechanism the rate law Rate = k_f[complex] [X] applies, where k_f is given for each mechanistic type. The group Y in the figure is the leaving group, H_2O.

DISSOCIATIVE
$$k_f = \frac{k_{-w}k_x}{k_w[H_2O] + k_x[X^-]}$$

INTERCHANGE
$$k_f = K_{dif}k_{-w}$$

ASSOCIATIVE
$$k_f = K_{dif}\frac{k_x k_{-w}}{k_{-x} + k_{-w}}$$

1. The formation of Ni(II) complexes from $[Ni(H_2O)_6]^{2+}$ has been studied with a great many ligands, and some of these data are included in Table 13-3. It would appear that the observed rate constants show considerable variation with incoming ligand, but closer inspection shows that this is largely due to the effect of charge. That is, k tends to increase with increasing negative change on the ligand. If the first kinetically important step is the formation of the outer sphere complex (reaction 13-22a), this is the expected result for the reaction of M^{n+} and X^{m-} (that is, k_{forward} for outer sphere complex formation increases and k_{backward} for outer sphere complex dissociation decreases as m increases). On the other hand, it is generally observed that, within a group of ligands of the same charge, there is no relationship between observed rate constants and ligand basicity or the nucleophilicity of the ligands toward other substrates. This latter observation effectively rules out the possibility of mechanisms depending on bond making in the rate determining step; that is, A and I_a paths are apparently quite unlikely.

TABLE 13-3
RATE CONSTANTS FOR THE REACTIONS OF $[M(H_2O)_6]^{2+}$ WITH VARIOUS LIGANDS IN AQUEOUS SOLUTION[a]

Substituting Ligand	log k_{obs} (M^{-1} sec^{-1})				log k_{-w} (sec^{-1})[b]			
	Co^{2+}	Ni^{2+}	Cu^{2+}	Zn^{2+}	Co^{2+}	Ni^{2+}	Cu^{2+}	Zn^{2+}
H_2O	6.4	4.5	>9.9	7.5	—	—	—	—
bipyridyl	4.9	3.2	7	6.0	—	—	—	—
o-phenanthroline	5.5	3.6	7.3	6.3	—	—	—	—
glycinate	—	4.3	9.6	—	—	4.0	9.3	—
SCN^-	>4.0	3.7	—	—	>3.7	3.4	—	—
$C_2O_4^{2-}$	—	4.8	—	—	—	3.7	—	—
dithiooxalate	—	4.8	—	—	—	3.7	—	—
maleonitrile dithiolate	—	4.8	—	—	—	3.7	—	—
H_2EDTA^{2-} [c]	—	3.3	5.6	—	—	2.2	4.5	—
$HEDTA^{3-}$	7	5.2	9	9	5.0	3.2	7	7

[a] All data taken from R. G. Pearson and P. C. Ellgen in "Physical Chemistry, An Advanced Treatise," Vol. 7, ed. H. Eyring, Academic Press, New York, 1975, p. 232.
[b] Calculated from log k_{obs} according to the procedure given in the text.
[c] $EDTA^{4-}$ = ethylenediaminetetraacetate.

The data in Table 13-3 show that a more important influence on the observed rate constant is the nature of the metal ion. That is, k_{obs} for the reaction of any ligand with a particular metal ion is within one or two orders of magnitude of the rate of water exchange for that ion, while a change in metal ion causes three or more orders of magnitude change in k_{obs} and k_{-w}. (We shall discuss the influence of the metal ion on rates in the next section.) Therefore, since there is no dependence on the entering group but there is a dependence on the central metal ion, it is likely that the substitution of water by another ligand on an aquometal ion proceeds by a D or I_d mechanism; in other words, the rate determining step exclusively or predominantly involves loss of water.

Eigen and others have found it possible to measure or calculate the equilibrium constants for the formation of outer sphere complexes, K_{dif}. If an I mechanism is assumed, then the rate constants in Table 13-3 are in fact equal to $K_{dif}k_{-w}$, and k_{-w} can be calculated when K_{dif} is known. A glance at Table 13-3 shows that this leads to rate constants that are quite similar to one another and to the rate constant for water exchange. Therefore, the variation in observed rate constants with ligand can be accounted for almost solely by pre-association equilibria.

Although an I mechanism was assumed just above, still unanswered is the interesting question of whether the mechanism is in fact I_d or D. (The dilemma exists because the rate law for the D mechanism can reduce to $Kk[\text{complex}][X]$ at small values for $k_x[X]$ and large values of $k_w[H_2O]$ (a very fast pre-equilibrium loss of water; see equation 7-4b and Figure 13-6).) However, there is no good evidence for the D mechanism in the vast majority of cases; therefore, the I_d mechanism is usually invoked to rationalize the kinetic observations.

2. Cobalt(III) complexes provide another interesting test of the possibility of the I_d mechanism. The larger positive charge on a Co(III) complex can lead to larger values of K_{dif}. Therefore, given that equation 13-23 is the rate law for an I_d mechanism,

$$\text{Rate} = \underbrace{\frac{K_{dif}k_{-w}}{1 + K_{dif}[X]}[X]}_{\text{pseudo-first order constant}}[\text{complex}]_t \tag{13-23}$$

where $[\text{complex}]_t = [CoL_6^{3+}] + [CoL_6^{3+} \cdots X^{m-}]$

the pseudo-first order rate constant observed in the presence of a large amount of X^-

should depend in a non-linear way on the concentration of X.[64] This is indeed the case for the anation reaction[65]

(13-24)
$$[Co(NH_3)_5H_2O]^{3+} + X^- \rightleftharpoons [Co(NH_3)_5X]^{2+} + H_2O$$

There are two observations regarding this reaction that are of considerable importance. Firstly, the values of K_{dif} are substantial;[66] for example, $K_{dif} = 1.9 \times 10^3 \, M^{-1}$ for $X^- = SO_4^{2-}$ or 3.1 for $X^- = Cl^-$. Secondly, the derived values of k_{-w} are nearly identical for a variety of X^- ligands (SO_4^{2-}, Cl^-, NCS^-), and, more importantly, k_{-w} is a constant fraction of the rate constant for water exchange on $[Co(NH_3)_5H_2O]^{3+}$. That is, $k_{-w}/k_{water\,exchange} \sim \frac{1}{5}$. This value is perhaps the strongest argument yet for the I_d mechanism. If the anation reaction above occurs by an I_d mechanism, this means that there is considerable weakening of the Co—OH$_2$ bond in the transition state. Stabilization is achieved, not by forming some five-coordinate intermediate, but rather by adding the first available ligand. It would appear that the outer sphere complex contains one X^- ligand and five water molecules arrayed about the substitution site of the substrate complex, so there is only about one chance in five or six of the X^- ligand entering the substrate in place of the leaving group.

Further evidence for an I_d mechanism comes from the study of the reverse of reaction 13-24, that is, the hydrolysis of $[Co(NH_3)_5X]^{2+}$. This is discussed later in this chapter (page 721).

3. There is no experimental evidence for intermediates of lower coordination number except in a few instances of unusual circumstance. For example, the anation of $[Co(CN)_5H_2O]^{2-}$ proceeds in the following kinetically important steps:

(13-25)
$$[Co(CN)_5H_2O]^{2-} \underset{k_w}{\overset{k_{-w}}{\rightleftharpoons}} [Co(CN)_5]^{2-} + H_2O$$

$$[Co(CN)_5]^{2-} + X^- \underset{k_{-x}}{\overset{k_x}{\rightleftharpoons}} [Co(CN)_5X]^{3-}$$

In this case, the build-up of an outer sphere complex is not kinetically important (as it would be in the case of an I_d mechanism), largely because an anionic substrate reacts with an anionic ligand. Instead, the rate determining step is loss of water from the substrate to produce an intermediate, $[Co(CN)_5]^{2-}$, of finite lifetime; that is, a D mechanism is proposed.[67,68]

Rates of Water Replacement[69]

First order rate constants for the substitution of inner sphere water molecules by other ligands on both transition metal and non-transition metal ions fall over a very wide range of values. Figure 13-7 is a tabulation of these constants by periodic group for a large number of ions, and, in Figure 13-8, the constants have been plotted against the number of d electrons for the di- and trivalent ions of the first row transition metals. Several observations can be made regarding these data.

1. The rate constants for the substitution reactions of a given ion are approximately the same no matter what the nature of the entering group. (In the discussion

---

[64] See the footnote to the discussion of equation 7-3b. Notice that, in that footnote, $K = K_{dif}K_{eq}$, but in the present context K_{eq} is redundant.

[65] C. H. Langford and W. R. Muir, *J. Amer. Chem. Soc.*, **89**, 3141 (1967).

[66] We should point out that reaction 13-24 is slow enough so that $[Co(NH_3)_5H_2O]^{3+} \cdots X^-$ can be observed and K_{dif} measured.

[67] A. Haim, R. J. Grassi, and W. K. Wilmarth, *Adv. Chem. Ser.*, **49**, 31 (1965).

[68] Unfortunately, it is impossible to discriminate between the D mechanism for reaction 13-25 and an I_d mechanism simply on the basis of the form of their rate laws. Therefore, other criteria must be used. One hallmark of a D mechanism is that the intermediate is sufficiently long-lived to discriminate between several nucleophiles. In the anation of $[Co(CN)_5H_2O]^{2-}$ the nucleophiles are water and X^-, so the ratio of rate constants k_x/k_w is expected to differ depending on the nature of X^-. This is indeed the case, as values of relative reactivities show: that is, if k_x/k_w for $H_2O = 1.0$, then k_x/k_w is 6.0 for Cl^-, 11.7 for I^-, 20.4 for NCS^-, 422 for I_3^-, and 23.9 for pyridine.

[69] See refs. 4, 53, and 54.

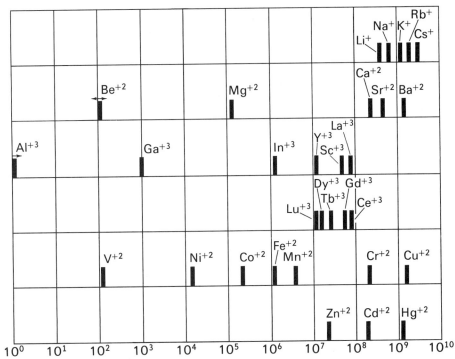

Figure 13-7. Characteristic rate constants (sec^{-1}) for the substitution of inner sphere water molecules in a number of metal ions. [From C.M. Frey and J. Stuehr, "Kinetics of Metal Ion Interactions with Nucleotides and Base Free Phosphates," in "Metal Ions in Biological Systems," ed., H. Sigel, Marcel Dekker, Inc., New York, vol. 1, 1974, p. 69.]

of substitution mechanism above, this was taken as evidence for a mechanism involving dissociative activation.)

2. The metal ions of Figures 13-7 and 13-8 can be placed in four reasonably distinct classes depending on their substitution rates.

Class I. Exchange of water is very fast and is essentially diffusion controlled ($k \gtrsim 10^8$ sec^{-1}). The class encompasses ions of Periodic Groups IA, IIA (except Be^{2+} and Mg^{2+}), and IIB (except Zn^{2+}), plus Cr^{2+} and Cu^{2+}.

Class II. Rate constants are in the range from 10^4 to 10^8 sec^{-1}. This class includes most of the first row transition metal divalent ions (the exceptions are V^{2+}, Cr^{2+}, and Cu^{2+}) and Mg^{2+} and the lanthanide 3+ ions.

Class III. Rate constants are roughly in the range from 1 to 10^4 sec^{-1}. This class includes Be^{2+}, Al^{3+}, V^{2+}, and some of the 3+ ions of the first row transition metals.

Class IV. Rate constants are roughly in the range from 10^{-3} to 10^{-6} sec^{-1}. The following ions fall within the class: Cr^{3+}, Co^{3+}, Rh^{3+}, Ir^{3+}, and Pt^{2+}.

Like so many other thermodynamic and kinetic properties, there are numerous factors, originating with the metal ion, that affect the rate of exchange of bound water. In the case of Class I ions, for instance, ionic size is obviously of considerable importance. As the ions become smaller, the substitution rate slows, suggesting that the rates are controlled by the effective charge on the ion or some other related property (e.g., orbital overlap between the metal ion and the departing ligand). The same general trend is observed in Group IIA and Group IIB, although the rate for Be^{2+} is considerably slower than might be expected. This latter effect is apparently due to hydrolysis reactions, which complicate the kinetics of simple substitution. Notice the large difference in Mg^{2+} and Ca^{2+} rates, a difference that is thought to play a role in their different behavior as enzymatic activators.

718 ☐ REACTION MECHANISMS AND METHODS OF SYNTHESIS; SUBSTITUTION REACTIONS

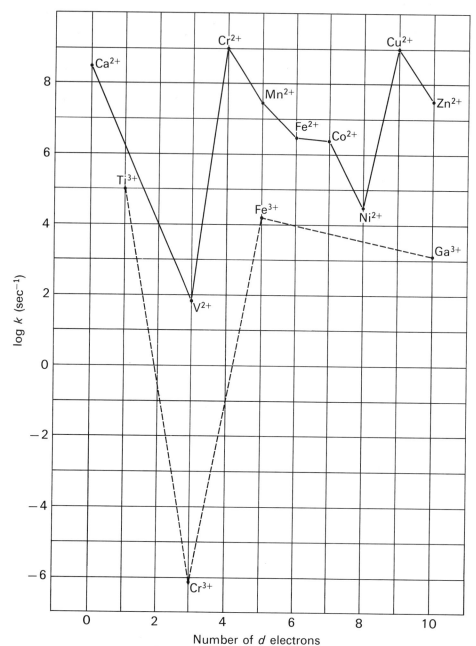

Figure 13-8. Rate constants for the exchange of water molecules in the first coordination sphere of $[M(H_2O)_6]^{n+}$ in aqueous solution. Data largely taken from R.G. Pearson and P.C. Ellgen in "Physical Chemistry, An Advanced Treatise," Volume VII, H. Eyring, ed., Academic Press, New York, 1975, Chapter 5.

There is obviously less correlation between rate and ion size for ions outside of Groups I and II. For example, Cr^{2+}, Ni^{2+}, and Cu^{2+} have nearly identical radii. However, Cr^{2+} and Cu^{2+} are in Class I whereas Ni^{2+} as in Class II. Fortunately, there is a rather simple explanation for this observation. That is, Cu^{2+} (d^9) and Cr^{2+} (d^4) complexes are often structurally distorted, with bonds to axial groups longer and weaker than bonds to equatorial groups (see the discussion of the Jahn-Teller effect in Chapter 9). Therefore, the ground state structures for Cr^{2+} and Cu^{2+} are not far removed from the transition state structures, and the axial water molecules, which are held less tightly, can exchange more rapidly.

To explain the behavior of the other 2+ ions of the first row transition metals, we shall pursue shortly a more careful analysis of the correlation between rate and d electron configuration. That such a correlation may exist is seen in Figure 13–8; these plots resemble some of the plots of physico-chemical data *vs.* electron configuration that were discussed in Chapter 9.

Trivalent metal ions react somewhat more slowly than divalent ions, so the trivalent ions largely constitute Classes III and IV. In the case of transition metal ions, however, there must again be an orbital occupation effect, as the pattern of rate constants *vs.* d electron configuration is approximately the same as that for the divalent ions of the same configuration.

Orbital Occupation Effects on Substitution Reactions of Octahedral Complexes

Basolo and Pearson introduced the idea that there is a connection between d orbital energies and the relative kinetic inertness or lability of transition metal complexes.[70] The basic notion behind this approach is that a significant contribution to the activation energy in a substitution reaction is the change in d orbital energy on going from the ground state of the complex ion to the transition state. Any loss in

[70] See ref. 2.

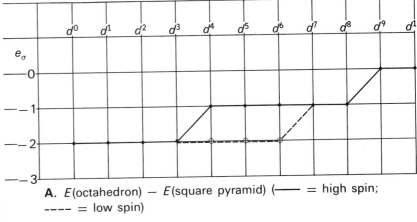

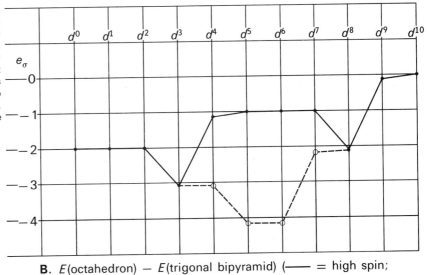

Figure 13–9. Structure preference energy plots for the octahedron compared with the two possible transition state structures in a substitution reaction occurring by the D mechanism. Notice that electron distribution in the d^* orbitals was left unchanged on going from the octahedron to the transition state structure. This has no effect on the calculations *except* for the d^8 case. A d^8 ion may have TBP or SP structures that are either low spin or high spin, in contrast with the octahedron where there is no choice. Therefore, in order to maintain electron spin unchanged, only the high spin transition state structures were considered.

A. E(octahedron) − E(square pyramid) (——— = high spin; - - - - = low spin)

B. E(octahedron) − E(trigonal bipyramid) (——— = high spin; - - - - = low spin)

energy is presumed to be proportional to the activation energy. Using this approach one can, in principle, answer the following questions: (i) Will the complex be kinetically labile or inert? If the structure preference energy, SPE, of the octahedron (i.e., the difference in d electron energies between the octahedron and a transition state structure) is less than some arbitrary value, the complex is predicted to be labile; if the SPE is more than that value, the complex will be inert. (ii) What is the most probable transition state structure? For a given d orbital occupation, the difference in energy between the octahedron and one of several transition state structures may be less than that for the others; this will be the favored pathway for reaction.

In Figure 13-9 we have plotted, in terms of the AOM e_σ parameter, the difference in energy between the octahedron and the possible transition state structures.[‡] Since complexes of divalent metal ions of the first transition series react by dissociative activation in octahedral substitution reactions, assume, for the sake of simplifying the calculations, a D mechanism and trigonal bipyramidal or square pyramidal structure for the transition state. Further, assume that a difference of about one e_σ unit or less between the octahedron and the five-coordinate transition state means that the activation energy barrier is low and that the complex can be kinetically labile; conversely, a difference of more than about two e_σ units means that the complex is kinetically inert. The results of such calculations are summarized in Table 13-4, and they are of considerable interest when compared with experimental data.

According to Figure 13-8, the high spin d^3 through d^{10} hexaaquo complexes show relative labilities in the order: V^{2+}(inert) $\ll$ $Cr^{2+} > Mn^{2+} > Fe^{2+} > Co^{2+} > Ni^{2+} \ll Cu^{2+} > Zn^{3+}$. This order can now be rationalized on the basis of the predictions of the angular overlap model and other considerations of transition metal chemistry.

1. The SPE plots both indicate that a d^3 ion should be substitutionally inert. This agrees with the observation of inertness for $[V(H_2O)_6]^{2+}$.

2. Except for Ni^{2+} (d^8), the high spin d^4-d^{10} M^{2+} ions are predicted to be labile

---

[‡]A similar analysis recently appeared in the literature: J.K. Burdett, *J.C.S. Dalton*, 1725 (1976).

TABLE 13-4

PREDICTIONS OF INERTNESS AND LABILITY FOR OCTAHEDRAL TRANSITION METAL COMPLEXES

d Electron Configuration	Reaction by Dissociative Activation	
	SP Intermediate C_{4v}	TBP Intermediate D_{3h}
d^0	inert	inert
d^1	inert	inert
d^2	inert	inert
d^3	inert	inert
d^4 low spin	inert	inert
high spin	labile	labile
d^5 low spin	inert	inert
high spin	labile	labile
d^6 low spin	inert	inert
high spin	labile	labile
d^7 low spin	labile	inert
high spin	labile	labile
d^8	labile	inert
d^9	labile	labile
d^{10}	labile	labile

by either pathway (C_{4v} or D_{3h} transition state structures). The Ni^{2+} ion, however, is predicted to be labile if the transition state structure is a square pyramid but inert if it is a trigonal bipyramid. The latter is in keeping with experiment, since Ni^{2+} complexes are the least labile of the d^4-d^{10} M^{2+} ions.

3. Considering the comment about Ni^{2+} above, the SPE plots predict the relative labilities for the high spin d^4-d^{10} M^{2+} ions to be $Cr^{2+} \leq Mn^{2+} = Fe^{2+} = Co^{2+} > Ni^{2+} \ll Cu^{2+} \leq Zn^{2+}$. Considering the crudeness of the predictive model, the agreement between theory and experiment is satisfying. However, there are some deviations that must be explained. First, Cr^{2+} (d^4) and Cu^{2+} (d^9) complexes are observed to be especially labile. However, as mentioned previously, these two ions show pronounced Jahn-Teller distortions, which lead to more rapid reaction. Second, Mn^{2+}, Fe^{2+}, and Co^{2+} are predicted to have equal labilities, when in fact substitution rates decrease across the series. The reason for this decline is clearly the increase in $Z_{effective}$ on going from Mn^{2+} to Fe^{2+}, an increase leading to greater values of e_σ and greater $\Delta E^{\ddagger}$.

4. Low spin complexes having d^4, d^5, and d^6 configurations are predicted to be relatively inert to substitution by either the D_{3h} or the C_{4v} pathway. Most of the complexes discussed in this chapter are low spin d^6 Co(III) complexes, and all undergo substitution quite slowly. In addition, ions such as $[Cr(CN)_6]^{4-}$ (low spin d^4) and $[Mn(CN)_6]^{4-}$ (low spin d^5) are substitutionally inert.

5. Complexes having d^0, d^1, and d^2 configurations are predicted to be inert, but this is not generally found to be the case. Recall, though, that an arbitrary decision was made to limit labile complexes to those having differences in d orbital energies of about one e_σ unit or less. Furthermore, the value of e_σ for complexes early in the transition metal series is liable to be low, owing primarily to low values of $Z_{effective}$ for the early metals. Therefore, one to two e_σ units may in fact represent a relatively small difference in energy between the octahedral complex and one of the possible five-coordinate transition states for the metals early in the transition series.

SOLVOLYSIS OR HYDROLYSIS

The hydrolysis of metal complexes under acidic conditions has been extensively studied, as has base- and metal-ion-catalyzed hydrolysis. The first two types will be discussed in the following sections, while metal-ion-catalyzed reactions are mentioned in the next chapter.

Hydrolysis under Acidic Conditions

As mentioned earlier, hydrolysis reactions are the reverse of anation reactions and, as suggested by the principle of microscopic reversibility, they traverse the same reaction coordinate in opposite directions. The hydrolysis of acidopentaammine-

$$[Co(NH_3)_5X]^{2+} + H_2O \underset{k_{anation}}{\overset{k_{hydrolysis}}{\rightleftharpoons}} [Co(NH_3)_5H_2O]^{3+} + X^- \qquad (13\text{-}26)$$

cobalt(3+) salts has been studied extensively, with results that bear further on the finding that substitution reactions of octahedral metal complexes generally occur by an I_d mechanism.[71]

Figure 13-10 is a plot of $\log k_{hydrolysis}$ against $\log K$, where K is the equilibrium constant for the hydrolysis reaction and is equal to $k_{hydrolysis}/k_{anation}$. An important conclusion can be drawn from this plot. Earlier in this chapter (p. 716) we pointed out that the rate of the anation reaction does not depend on the nature of X^-. This and other information were used to support an I_d mechanism for anation. On the other hand, Figure 13-10 shows that the rate of hydrolysis is dependent on the nature of X^-. This dependence on the nucleophilicity of the leaving group is taken as evidence

---

[71] C. H. Langford, *Inorg. Chem.*, **4**, 265 (1965).

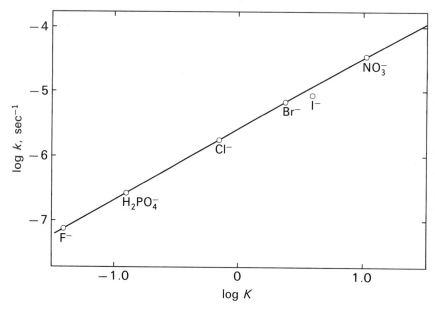

Figure 13-10. Correlation of the acid hydrolysis rates with the equilibrium constants in the reaction $Co(NH_3)_5X^{2+} + H_2O \rightarrow Co(NH_3)_5H_2O^{3+} + X^-$. [Reproduced from "Ligand Substitution Processes," by Cooper H. Langford and Harry B. Gray, with permission of publishers, Addison-Wesley/W. A. Benjamin Inc., Advanced Book Program, Reading, Mass.]

for dissociative activation. As there is no experimentally apparent intermediate, the mechanism is assigned an I_d label.[72]

To this point we have given no indication of the steric course of the octahedral substitutions. However, acid hydrolysis reactions of molecules such as **20** give a

$$\left(\underset{\underset{A}{|}}{\overset{\underset{X}{|}}{\underset{N}{N}\diagdown\overset{N}{\underset{Co}{}}\diagup\underset{N}{N}}}\right)^{n+} \qquad N\frown N = H_2NC_2H_4NH_2$$

20

detailed picture of the molecule in the transition state. In Table 13-5 are listed a variety of *cis* and *trans* molecules of the type $[Co(N_4)AX]^{n+}$, where N_4 represents a pair of ethylenediamine molecules or a quadridentate ligand such as **21** or **22**. The

cyclam
1,4,8,11-tetraazacyclotetradecane
21

trans-[14]diene
5,7,7,12,14,14-hexamethyl-1,4,8,11-tetraazacyclo-tetradeca-4,11-diene
22

---

[72] The fact that the slope of the log k vs. log K plot is about 1 also supports an I_d mechanism. Recall the discussion of a similar plot for square planar substitution (ref. 47).

TABLE 13-5

COMPARISON OF RATES AND STEREOCHEMISTRY FOR THE AQUATION OF *cis* AND *trans* ISOMERS OF COBALT(III) COMPLEXES OF THE TYPE $Co(N)_4ACl^{n+}$. [$(N)_4$ = A PAIR OF ETHYLENEDIAMINE MOLECULES OR A QUADRIDENTATE LIGAND; SEE TEXT.]

Compound	A group	k (sec^{-1})	$\Delta H^\ddagger$ kcal/mol	$\Delta S^\ddagger$ eu	% *cis* isomer in product
trans Isomers					
$Co(en)_2ACl^{n+}$	OH^-	1.6×10^{-3}	25.9	+20	75
	N_3^-	2.2×10^{-4}	22.5	0	20
	Cl^-	3.5×10^{-5}	26.2	+14	35
	Br^-	4.5×10^{-5}	24.9	+3	50
	NH_3	3.4×10^{-7}	23.2	−11	0
	NO_2^-	1.0×10^{-3}	20.9	−2	0
$Co(cyclam)ACl^{n+}$	Cl^-	1.1×10^{-6}	24.6	−3	0
$Co(trans[14]diene)ACl^{n+}$	Cl^-	3.62×10^{-2}	23.2	+12.3	0
cis Isomers					
$Co(en)_2ACl^{n+}$	OH^-	1.2×10^{-2}	23.0	+10	100
	N_3^-	2.0×10^{-4}	21.3	−4	100
	Cl^-	2.4×10^{-4}	21.5	−5	100
	Br^-	1.4×10^{-4}	23.5	+5	100
	NH_3	5×10^{-7}	22.9	−11	100
	NO_2^-	1.1×10^{-4}	21.8	−3	100
$Co(cyclam)ACl^{n+}$	Cl^-	1.6×10^{-2}	18.3	−6	100

Data taken from: N. Sutin, *Ann. Rev. Phys. Chem.*, **17**, 133 (1966); M. L. Tobe, *Inorg. Chem.*, **7**, 1260 (1968); J. A. Kernohan and J. F. Endicott, *Inorg. Chem.*, **9**, 1504 (1970); C. K. Poon and M. L. Tobe, *J. Chem. Soc. (A)*, 1549 (1968).

point of these data is that they show that some of these molecules undergo stereochemical change upon hydrolysis, an observation needing explanation.

Several years ago, Tobe noticed that only those compounds with a positive $\Delta S^\ddagger$ underwent stereochemical change.[73] He concluded that positive entropies of activation seem to indicate an incipient trigonal bipyramidal intermediate, while negative entropies of activation indicate an incipient tetragonal pyramid. That is, no matter what the stereochemical result, the reactions proceed by a D or I_d mechanism, with only the incipient intermediate differing. Although this connection between $\Delta S^\ddagger$ and stereochemical change has since been shown to have limited validity,[74] the suggestion of differing intermediates is apparently correct.

In Figure 13-11 we have outlined the stereochemical course that the hydrolysis of *trans*-$[Co(en)_2AX]^{n+}$ may take; a D mechanism is presumed. If the reaction proceeds through a *square pyramidal intermediate, retention of configuration* will be observed. Proceeding through a *trigonal bipyramid*, however, can lead to *stereochemical change*. As X is lost, the molecular structure can relax in either of two ways: either the 3,5 N's or the 2,4 N's can move toward one another. There are now three edges of the trigonal bipyramid at which the ligating atoms are separated by 120°,

---

[73] M. L. Tobe, *Inorg. Chem.*, **7**, 1260 (1968).
[74] J. A. Kernohan and J. F. Endicott, *Inorg. Chem.*, **9**, 1504 (1970); M. C. Couldwell and D. A. House, *Inorg. Chem.*, **13**, 2949 (1974).

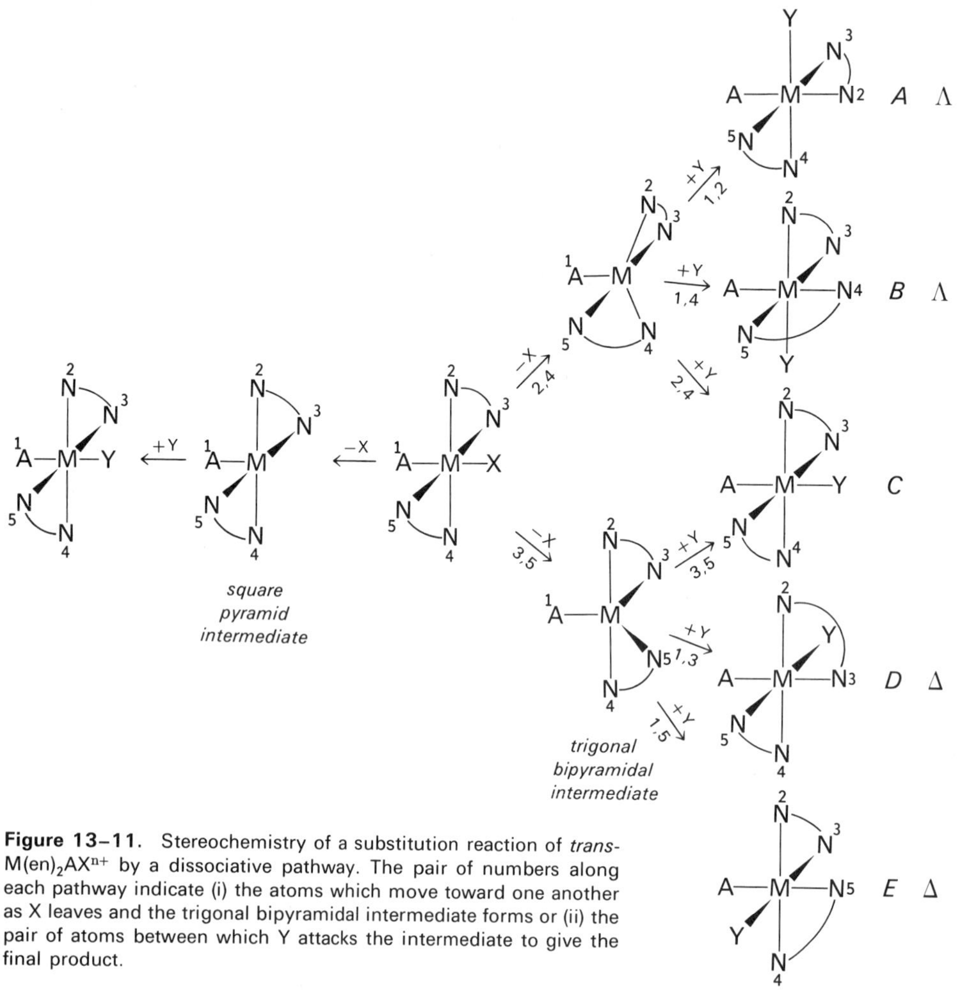

Figure 13-11. Stereochemistry of a substitution reaction of trans-$M(en)_2AX^{n+}$ by a dissociative pathway. The pair of numbers along each pathway indicate (i) the atoms which move toward one another as X leaves and the trigonal bipyramidal intermediate forms or (ii) the pair of atoms between which Y attacks the intermediate to give the final product.

and it is these edges that offer opportunity for attack by the entering group. (Attack along the other edges, where the ligands are separated by 90°, is less likely because of steric crowding.) Since there are three edges, there are three possibilities for attack in each of the two possible trigonal bipyramidal intermediates. Both "branches" can lead to a complex wherein stereochemistry is retained; that is, either intermediate can form product C, and the principle of microscopic reversibility (see Chapter 7) is simply satisfied by both paths. However, one branch leads to two cis products (A and B) that are the same and would have the Λ absolute configuration. The other branch also leads to two cis products (D and E), but these have the Δ absolute configuration. Therefore, rather severe stereochemical change may be produced easily. If the product is cis, it is clear that the reaction proceeds through an incipient trigonal bipyramid (see Table 13-5). However, if the product is trans, the intermediate is not altogether certain, although the trigonal bipyramid is not as likely as the alternative square pyramid because there is only a $\frac{1}{3}$ chance of producing a product with retention of configuration along the former route.

Statistical analysis of the relative amounts of various products of substitution for cis-$[Co(en)_2AX]^{n+}$ is more complex and is left as an exercise for you. As a check, the expected results for both cis and trans complexes are listed in Table 13-6.

In general, trans complexes of the type $[Co(en)_2AX]^{n+}$ often undergo stereochemical change upon hydrolysis, whereas their cis counterparts react with retention of the original chirality, implying that cis hydrolysis may well proceed through a

TABLE 13-6
AMOUNTS OF PRODUCTS (AS PERCENTAGES) PREDICTED TO ARISE BY VARIOUS MECHANISMS FOR OCTAHEDRAL SUBSTITUTION

Compound	D or I_d Mechanism			
	Tetragonal or Square Pyramid Intermediate		Trigonal Bipyramid Intermediate	
	cis product	trans product	cis product	trans product
trans-ML_4AX^{n+}	0	100	66.6	33.3
cis-ML_4AX^{n+}	100	0	83.3	16.6
trans-$M(L-L)_2AX^{n+}$	0	100	66.6	33.3
Λ cis-$M(L-L)_2AX^{n+}$*	100	0	$\begin{pmatrix}\Lambda = 58.3\\ \Delta = 25.0\end{pmatrix}$**	16.7

*Starting material is arbitrarily assumed to have a Λ configuration.
**These results arise from the fact that one half of the starting material may give an intermediate that can in turn give either of two products having opposite chiralities; the other half of the starting material can give three products, two having the same configuration as the starting material and the third having trans geometry.

square pyramidal intermediate. These stereochemical results, and the fact that the cis isomers generally react more rapidly than the trans isomers, have been explained in terms of transition state stabilization. The ligands A that lead to stereochemical changes in the trans series are those with lone pairs of electrons in $p\pi$ orbitals. It has been argued that such electrons can interact through ligand-to-metal π bonding with an empty metal orbital—but only if the complex assumes trigonal bipyramidal geometry (Figure 13-12). Therefore, π donor ligands such as A = Cl^- and N-bonded NCS^- favor the trigonal bipyramid, an intermediate clearly leading to the possibility for stereochemical change. On the other hand, in the cis isomer the $p\pi$ lone pairs of electrons on the A ligands can strongly interact with the metal p orbital previously preoccupied with σ bonding to the leaving group, without rearrangement to the trigonal bipyramid. Therefore, the tetragonal pyramid can be stabilized, and the reaction can be more rapid since there is no need for rearrangement.

Base-Catalyzed Hydrolysis: The Conjugate Base or CB Mechanism.[75,75a]

It is fair to say that octahedral substitution reactions are not sensitive to the nature of the entering group—with one exception. In basic media, Co(III) complexes having ligands of the type NH_3, RNH_2, or R_2NH are sensitive to the nature of the entering group. In this case OH^- is the apparent entering ligand, and *base catalyzed reactions are generally much more rapid than anations or hydrolyses in acid solution.* Furthermore, the rate law for the general reaction

$$R_5MX + OH^- \to R_5MOH + X^- \tag{13-27}$$

is

$$\frac{-d[\text{complex}]}{dt} = k_{\text{obs}}[\text{complex}][OH^-] \tag{13-28}$$

---

[75] M. L. Tobe, *Acc. Chem. Res.*, **3**, 377 (1970).
[75a] The CB mechanism has already been discussed in connection with nitrogen chemistry; see page 410.

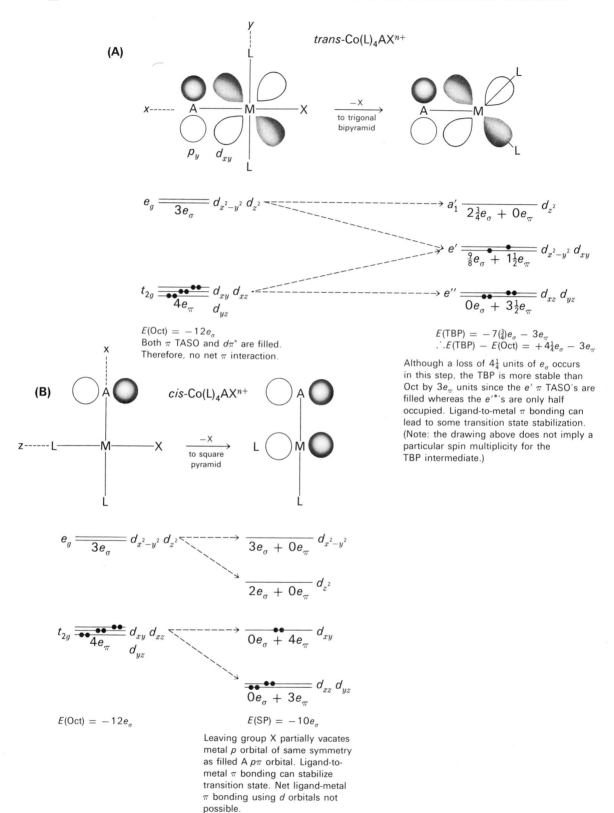

Figure 13-12. (A) Ligand-metal interactions in the octahedral reactant and trigonal bipyramidal transition state for substitution reactions of *trans*-$[Co(en)_2AX]^{n+}$. (B) Ligand-metal interactions in the octahedral reactant and square pyramidal transition state for substitution reactions of *cis*-$[Co(en)_2AX]^{n+}$.

The clear-cut second order kinetics of the reaction early prompted many people to suggest that hydrolyses in basic media occurred by an A or I_a mechanism, a suggestion clearly at odds with the previously observed generality of D or I_d mechanisms. Such seeming contradictions need explanation.

The mechanism of base-catalyzed hydrolyses is controversial. However, it is now generally agreed that the correct stoichiometric mechanism, illustrated for the case of pentaamminechlorocobalt(2+), is

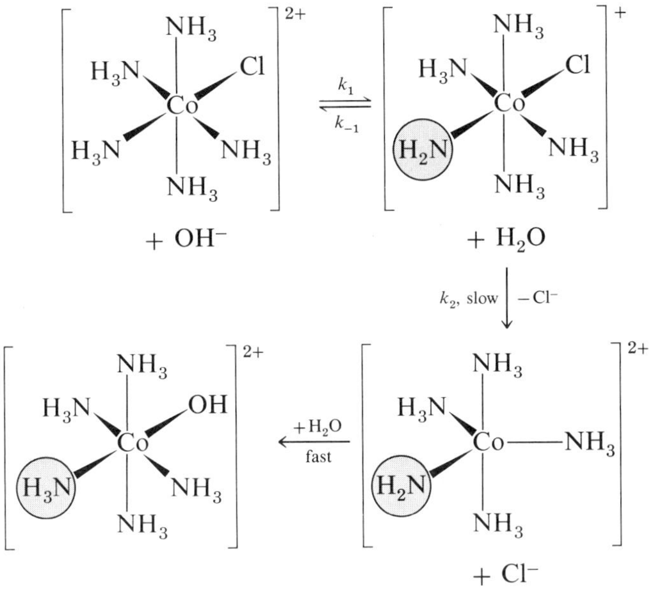

$$\begin{array}{c}\left[\text{Co(NH}_3)_5\text{Cl}\right]^{2+} + \text{OH}^- \underset{k_{-1}}{\overset{k_1}{\rightleftharpoons}} \left[\text{Co(NH}_3)_4(\text{NH}_2)\text{Cl}\right]^+ + \text{H}_2\text{O} \\ \downarrow k_2,\ \text{slow}\ -\text{Cl}^- \\ \left[\text{Co(NH}_3)_4(\text{NH}_3)\text{OH}\right]^{2+} \underset{\text{fast}}{\overset{+\text{H}_2\text{O}}{\longleftarrow}} \left[\text{Co(NH}_3)_4(\text{NH}_2)\right]^{2+} + \text{Cl}^- \end{array}$$

(13-29)

You will notice that the chief feature of this mechanism is the reversible removal of a proton from an ammonia ligand in the first step. This same mechanism was discussed for substitution at amine nitrogen in Chapter 7 (page 410). Here, as there, this creates the conjugate base of the original complex, hence the name "conjugate base" or CB mechanism. This step is generally very rapid, perhaps 10^5 times faster than the release of Cl^- in the second step. Therefore, presuming that the first step represents a very rapid pre-equilibrium to the rate determining step, and using the usual steady state assumption, the rate expression that has been applied is the same in form as that found for the anation of $[\text{Co(NH}_3)_5\text{H}_2\text{O}]^{3+}$:

$$\frac{-d[\text{complex}]}{dt} = \frac{k_2 K[\text{complex}][\text{OH}^-]}{1 + K[\text{OH}^-]} \qquad (13\text{-}30)$$

If, however, K is quite small [it is apparently in the range from 0.01 to 0.20 for Pt(IV) complexes],[76] so that $K[\text{OH}^-] \ll 1$, the rate law would reduce to that experimentally observed (see the footnote to equation 7-3b).

Given that the stoichiometric mechanism is generally correct, we need to explain three features of the reaction: (i) What is the exact site from which the proton is removed in the first step? (ii) What is the shape of the intermediate? (iii) If K is small, the rate is large only if $k_2 K$ is very large. What is it that leads to such large values of k_2? These questions have not been answered unequivocally, so only the currently fashionable hypotheses are briefly presented.

---

[76] L. E. Erickson, *J. Amer. Chem. Soc.*, **91**, 6284 (1969).

728 ☐ **REACTION MECHANISMS AND METHODS OF SYNTHESIS; SUBSTITUTION REACTIONS**

Perhaps the first question is open to experimental test, and the answer seems to be that it is the proton of the amino group *trans* to the leaving group that is affected. For example, the red isomer of [Co(tren)(NH$_3$)Cl]$^{2+}$ **(23)** is hydrolyzed much more rapidly (about 10^4 times more reactive) than the purple isomer **(24)**.[77] Only the red isomer has a removable proton on the nitrogen atom *trans* to the leaving group. Extrapolation of this result to other compounds, however, may be dangerous, as there are excellent arguments for deprotonation of the *cis* amino group in [Co(NH$_3$)$_5$Cl]$^{2+}$ and in *cis*- and *trans*-[Co(en)$_2$AX]$^{n+}$.[78]

(13-31)

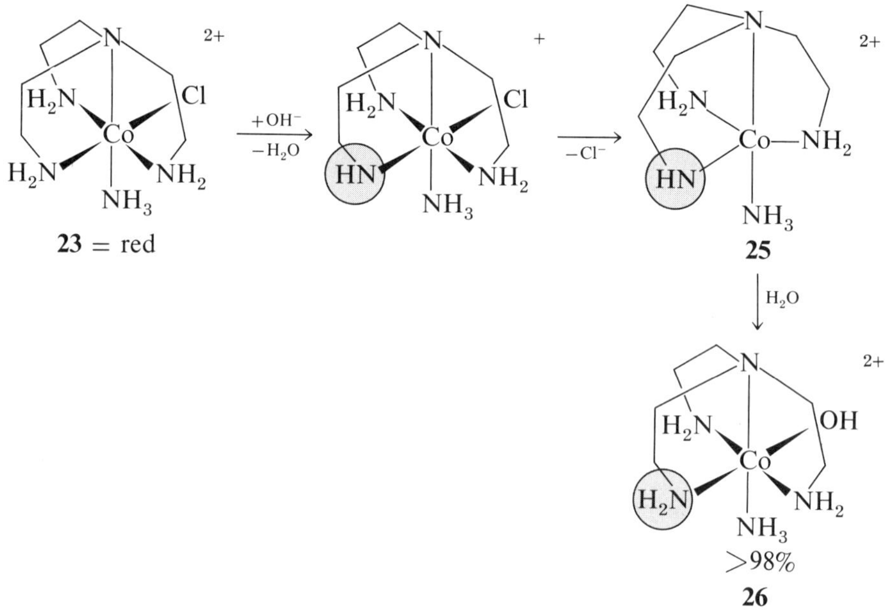

(13-32)

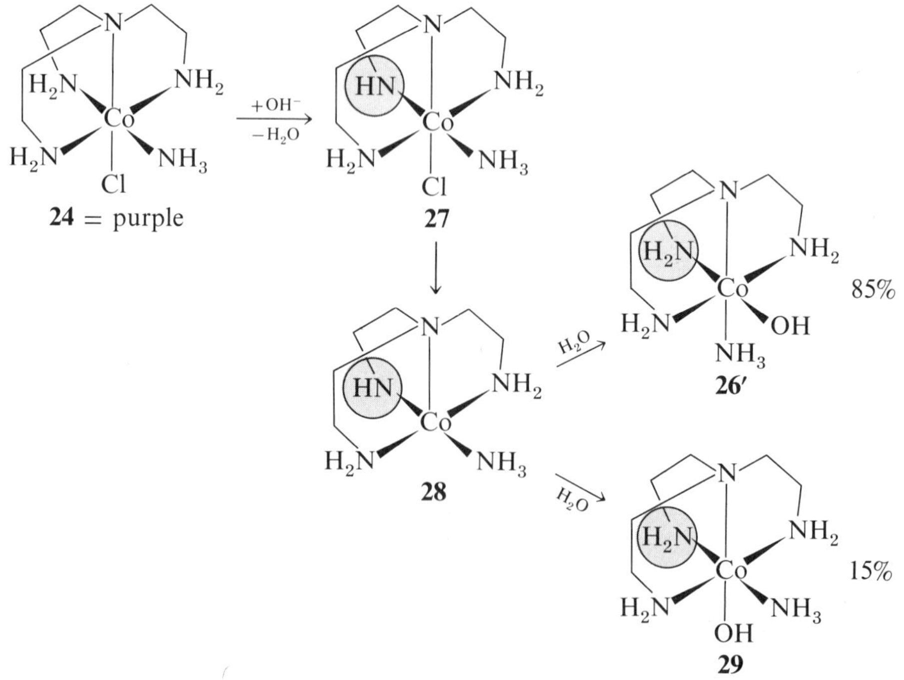

[77] D. A. Buckingham, P. J. Cressell, and A. M. Sargeson, *Inorg. Chem.*, **14**, 1485 (1975).
[78] F. R. Nordmeyer, *Inorg. Chem.*, **8**, 2780 (1969).

No matter what the site of deprotonation of the amino group, the intermediate in a D or I_d mechanism is apparently a trigonal bipyramid. Hydrolysis of **23** leads to *complete retention of configuration.* Since tren is a ligand well known to lead to stabilization of trigonal bipyramidal complexes (Chapter 10, page 598), an intermediate such as **25** is easily possible. Attack by hydroxide along any of the three equatorial edges then leads to the observed product **(26).** On the other hand, purple $[Co(tren)(NH_3)Cl]^{2+}$ **(24)** leads to two products, the major product being the same as that seen in the hydrolysis of the red compound. The most reasonable rationalization of this is that the initial square pyramidal intermediate from **24** (*i.e.,* **28**) is unstable; some of it is trapped and gives **29,** while the majority rearranges to give ultimately **26.**

Finally, why does base catalysis lead to rate enhancement? The usual explanation, which resembles that used above to account for stereochemical changes in *trans*-$[Co(en)_2AX]^{n+}$, is tentative at best. That is, it has been suggested that the amido group can stabilize the transition state by a ligand-to-metal π type interaction between the $-\ddot{N}R_2$ group lone pair and an appropriate metal orbital (see Figure 13–12). Such an explanation would clearly be most effective if the $M-N\begin{smallmatrix}R\\R\end{smallmatrix}$ grouping is planar. However, there is experimental evidence that planarity is not necessarily achieved in the course of the reaction.[79] Therefore, an alternative explanation is that the amido group, a very strong σ base, can stabilize the trigonal bipyramidal transition state in a σ interaction in the same way that strong *trans* directors stabilize the trigonal bipyramidal transition state in substitution reactions of square planar complexes.

STUDY QUESTIONS

8. Table 13-6 presents the predicted stereochemical results for the substitution of the group X in the complex *cis*-$[Co(en)_2AX]^{n+}$. Verify these results.

9. On page 713 it was noted that the rate laws given in Figure 13-6 can all reduce to the simple form Rate = k_{obs}[Complex][X] under certain circumstances. Show how this may be possible. Consult Chapter 7, Table 7-1.

10. In the text you learned of the rate laws for the substitution of a monodentate ligand on an octahedral complex by another monodentate ligand. However, another very important process is the substitution of two monodentate ligands by a chelating ligand. Wilkins [*Acc. Chem. Res.,* **3,** 408 (1970)] shows that the steps currently considered important when the substitution ligand is bidentate are

$$(H_2O)_5M(H_2O) + L\text{-}L \underset{k_{-x}}{\overset{k_x}{\rightleftharpoons}} (H_2O)_5M(H_2O)\cdots L\text{-}L$$

$$(H_2O)_5M(H_2O)\cdots L\text{-}L \underset{k_{-w}}{\overset{k_w}{\rightleftharpoons}} (H_2O)_5M(L\text{-}L) + H_2O$$

$$(H_2O)_5M(L\text{-}L) \underset{k_{-b}}{\overset{k_b}{\rightleftharpoons}} (H_2O)_4M\begin{pmatrix}L\\L\end{pmatrix} + H_2O$$

That is, after formation of an outer sphere complex, one end of the bidentate ligand substitutes for one water molecule, followed by substitution of the other end for yet another water. Assuming that the first reaction is faster than the other two (and that $k_x/k_{-x} = K_{dif}$), and that a steady state exists for the intermediate $(H_2O)_5M(L-L)$, derive the rate constant expression. Show that when $k_b \gg k_w$, the rate constant expression reduces to that commonly observed for the substitution of a monodentate ligand by another monodentate ligand.

11. Perhaps the chief conclusion to be reached from the discussion of reaction mechanisms in this chapter is that four-coordinate planar complexes react predominantly by an associative mechanism, whereas octahedral complexes react predominantly by dissocia-

---

[79] D. A. Buckingham, P. A. Marzilli, and A. M. Sargeson, *Inorg. Chem.,* **8,** 1595 (1969).

tive pathways. The question that is now of interest is the mechanism of substitution in five-coordinate complexes. Clearly, both associative and dissociative mechanisms are possible, since a five-coordinate complex can potentially either add or lose a ligand. Some research has recently been published on such reactions, and this question concerns one of those publications.

The replacement of a phosphine ligand on an iron- or cobalt-dithiolene complex proceeds according to the stoichiometry:

[Structural diagram of reaction: Co-dithiolene complex with X ligand + L → Co-dithiolene complex with L ligand + X]

X and L = phosphines

a. For a dissociative mechanism, the reactions involved would be

$$M(S_2C_2Ph_2)_2X \underset{k_2}{\overset{k_1}{\rightleftharpoons}} M(S_2C_2Ph_2)_2 + X$$

$$M(S_2C_2Ph_2)_2 + L \overset{k_3}{\longrightarrow} M(S_2C_2Ph_2)_2L$$

Write the rate law for this mechanism, assuming that the four-coordinate intermediate is in a steady state. What is the expression for k_{obs} when excess L is used (that is, you operate under pseudo-first order conditions)? To what does this expression reduce when $k_3 \gg k_2$?

b. For an associative mechanism, the reactions involved would be

$$M(S_2C_2Ph_2)_2X + L \underset{k_2}{\overset{k_1}{\rightleftharpoons}} M(S_2C_2Ph_2)_2XL \overset{k_3}{\longrightarrow} M(S_2C_2Ph_2)_2L + X$$

Assuming steady state conditions for the six-coordinate intermediate, write a rate law for the associative mechanism. Again assume pseudo-first order conditions in L and write the expression for k_{obs}.

c. Given below is a plot of k_{obs} vs. [L] for the reaction of $Co(S_2C_2Ph_2)_2PPh_3 + L$. When $L = P(OEt)_3$, $k_{obs} = 0.05 + 42.5[L]$.

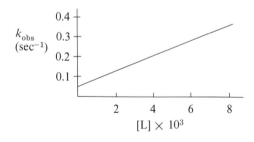

From this plot, and the fact that added X (the leaving group) has a small effect on the reactions of the cobalt complex, what can you conclude about the mechanism of the reaction? That is, is the reaction strictly dissociative or strictly associative, or are both pathways used? If both pathways are used, which predominates? [If you wish further information on reactions of five-coordinate complexes, consult the following, more recent references: D. A. Sweigart and P. Heidtmann, *J. C. S. Chem. Commun.*, 556 (1973); *J. C. S. Dalton*, 1686 (1975).]

12. Cd^{2+} and Hg^{2+} exchange coordinated water very rapidly and are in Class I. Why is Zn^{2+} in Class II and not Class I?

13. Why does Ga^{3+} exhibit a greater rate of water exchange than Al^{3+}? What would the rate of exchange be for Tl^{3+} with respect to Al^{3+} and Ga^{3+}? What about Tl^+?

14. The values of $\Delta S^{\ddagger}$ in Table 13-5 are generally positive or are only slightly negative. Why should this be the case? In contrast, the values of $\Delta S^{\ddagger}$ in Table 13-2 are quite negative. Why?

15. The values of $\Delta H^{\ddagger}$ in Table 13-5 are, in general, greater than the corresponding values in Table 13-2. Why?

SYNTHESIS OF COORDINATION COMPOUNDS BY SUBSTITUTION REACTIONS[80-86]

In this chapter and the previous ones on coordination chemistry, we have discussed a great variety of square planar and octahedral coordination complexes as well as some of the other geometries and coordination numbers. However, you must realize that you cannot normally purchase these compounds from Fisher Scientific Company, Alfa Products, or some other supplier of inorganic chemicals. Rather, most compounds must be synthesized from starting materials such $CrCl_3$, $CoCl_2 \cdot 6H_2O$, $NiCl_2 \cdot 6H_2O$, K_2PtCl_4, and so on, and most of these syntheses involve the substitution of water and/or halide ion in the primary coordination sphere of the starting material. As examples, consider the following preparations:

I. Preparation of tetraamminecarbonatocobalt(1+) nitrate and pentaamminechlorocobalt(2+) chloride.[83] The starting material for these syntheses is cobalt(II) nitrate. The formula of the compound is $Co(NO_3)_2 \cdot 6H_2O$, and it is therefore more properly formulated as $[Co(H_2O)_6](NO_3)_2$. The d^7 high spin Co(II) ion is predicted to be substitutionally labile (Table 13-4), so in the presence of NH_3 and CO_3^{2-} the formation of $Co(NH_3)_4CO_3$ should be quite possible. If an oxidizing agent is also present, the intermediate Co(II) carbonato complex can then be oxidized to the desired final product. The overall reaction (unbalanced) is

$Co(NO_3)_2 + NH_{3(aq)} + (NH_4)_2CO_3 + H_2O_2 \rightarrow$

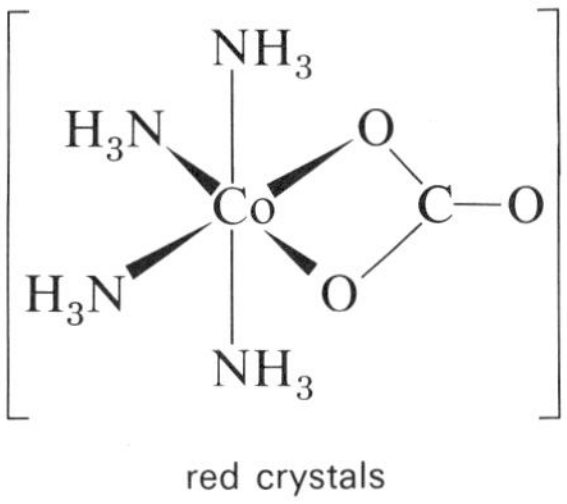

$$\left[\begin{array}{c} H_3N \diagdown \overset{NH_3}{\underset{\underset{NH_3}{|}}{|}} \diagup O \diagdown \\ H_3N \diagup \overset{Co}{\underset{|}{|}} \diagdown \underset{O}{\diagup} C-O \end{array}\right] NO_3 + NH_4NO_3 + H_2O \qquad (13\text{-}33)$$

red crystals
30

---

[80] The best source of information on the synthesis of coordination compounds is *Inorganic Syntheses*, published by the McGraw-Hill Book Company.

[81] Another excellent source of information on synthesis is the "Handbook of Preparative Inorganic Chemistry," G. Brauer, ed., Vols. 1 and 2, Academic Press, New York, 1965.

[82] The following is a review of coordination compound synthesis in general: J. L. Burmeister and F. Basolo, "Synthesis of Coordination Compounds," in *Preparative Inorganic Chemistry*, **5**, 1 (1968).

[83] This reference and the next three are primarily textbooks of preparative inorganic chemistry and contain much useful information on coordination compound synthesis. R. J. Angelici, "Synthesis and Technique in Inorganic Chemistry," 2nd Ed., W. B. Saunders Company, Philadelphia, 1977.

[84] W. L. Jolly, "The Synthesis and Characterization of Inorganic Compounds," Prentice-Hall, Inc., Englewood Cliffs, N.J., 1970.

[85] G. Pass and H. Sutcliffe, "Practical Inorganic Chemistry," 2nd Ed. Chapman and Hall Ltd., London, 1974.

[86] G. C. Schlessinger, "Inorganic Laboratory Preparations," Chemical Publishing Co., Inc., New York, 1962. This is a particularly good source of the syntheses of many "common" coordination compounds.

Although Co(III) complexes are relatively inert to substitution, a variety of complexes can be prepared from the carbonato complex by replacement of the CO_3^{2-} ligand.[83] For example, the pentaamminechlorocobalt(2+) ion is formed in the next series of reactions:

(13-34)

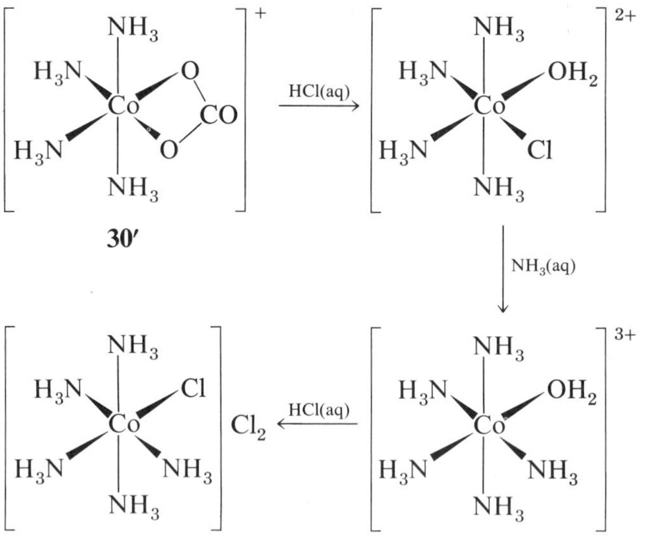

$[Co(NH_3)_4(CO_3)]^+$ **30'** $\xrightarrow{HCl(aq)}$ $[Co(NH_3)_4(OH_2)(Cl)]^{2+}$ $\xrightarrow{NH_3(aq)}$ $[Co(NH_3)_5(OH_2)]^{3+}$ $\xrightarrow{HCl(aq)}$ $[Co(NH_3)_5Cl]Cl_2$

31 purple-red crystals

II. Preparation of hexaamminechromium(3+) nitrate.[87] This preparation is interesting because it illustrates the application of the CB mechanism to chemical synthesis. If anhydrous $CrCl_3$ is allowed to react with liquid ammonia, $[Cr(NH_3)_5Cl]Cl_2$ will form as the major product. The remaining chromium-bound chloride ion is displaced very slowly. However, you might recall that base catalyzed hydrolysis reactions occur very rapidly compared with those in neutral solution. Therefore, it is possible to replace the remaining Cl^- by a base catalyzed solvolysis in liquid ammonia where the amide ion, NH_2^-, is the base and the entering ligand is NH_3. Mechanistically, the reaction probably follows a path similar to that shown on page 727.

(13-35)

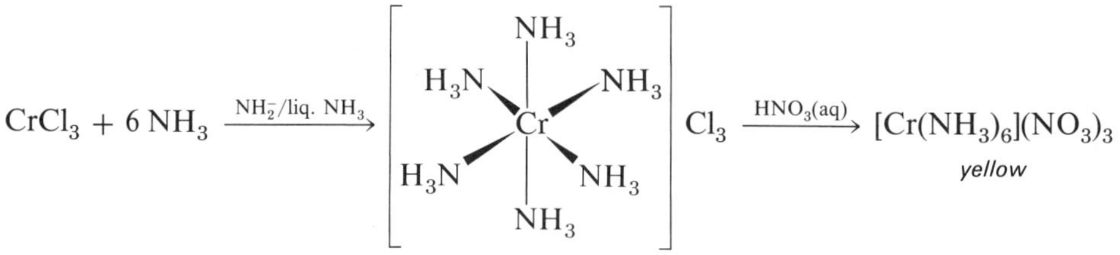

$CrCl_3 + 6NH_3 \xrightarrow{NH_2^-/\text{liq. }NH_3} [Cr(NH_3)_6]Cl_3 \xrightarrow{HNO_3(aq)} [Cr(NH_3)_6](NO_3)_3$

32 brown — yellow

(13-36)

$[Cr(NH_3)_5Cl]^{2+} + NH_2^- \xrightleftharpoons{\text{fast}} [Cr(NH_3)_4(NH_2)Cl]^+ + NH_3$

$[Cr(NH_3)_4(NH_2)Cl]^+ \xrightarrow{\text{slow}} [Cr(NH_3)_4NH_2]^{2+} + Cl^-$

$[Cr(NH_3)_4NH_2]^{2+} + 2NH_3 \xrightarrow{\text{fast}} [Cr(NH_3)_6]^{3+} + NH_2^-$

---

[87] See ref. 83.

Before we proceed to discuss the syntheses of coordination compounds of two representative transition elements—cobalt and platinum—there is one additional topic that we should take up, a discussion of the thermodynamic stability of coordination compounds. We have already seen that complexes of some transition metal ions, Co(III) for example, are notably kinetically inert with respect to substitution. However, this does not necessarily mean that the complex will persist in solution under ordinary conditions for an indefinite length of time. Such would be the case only for a complex of high thermodynamic stability. For example, $[Co(NH_3)_6]^{3+}$ may be inert to substitution, but it is completely unstable thermodynamically in acid solution with respect to formation of NH_4^+ and $[Co(H_2O)_6]^{3+}$. On the other hand, although the d^8 ion $[Ni(CN)_4]^{2-}$ has a dissociation constant for replacement of CN^- by H_2O of 10^{-30} (indicating great thermodynamic stability), bound CN^- is exchanged very rapidly for free CN^- in solution (indicating substitutional lability).

THERMODYNAMIC STABILITY OF COORDINATION COMPOUNDS[88-90]

The thermodynamic stability of coordination compounds is usually defined in terms of the equilibrium constant for the reaction

$$M^{m+}_{(aq)} + L^{n-} \rightleftharpoons ML^{(m-n)+}_{(aq)} \tag{13-37}$$

$$K = \frac{[ML^{(m-n)+}]}{[M^{m+}][L^{n-}]} \tag{13-38}$$

For example, for the reaction of aqueous Cu(II) with NH_3:

$$[Cu(H_2O)_4]^{2+} + NH_3 \rightleftharpoons [Cu(H_2O)_3NH_3]^{2+} + H_2O \tag{13-39}$$
$$K = 1.66 \times 10^4$$

The constant K is usually referred to as the stoichiometric stability constant. You will notice that we have written it in terms of concentrations. It would, of course, be more correct to give the expression in terms of activities or molal concentrations and activity coefficients (since $a_i = m_i \gamma_i$). However, the measurement of activity coefficients is experimentally difficult, so it is generally assumed that stability constants of sufficient accuracy may be obtained by measuring K in solutions in which the ionic strength is maintained at a high value through addition of an inert salt.

More often than not, more than one water molecule on the original metal ion may be substituted by another ligand, and it is assumed that this may occur in a sequence of steps, each step characterized by its own stoichiometric equilibrium constant, K_n. For example, the reaction of tetraaquocopper(2+) with ammonia (13-39) is just the first of four possible steps, and the stoichiometric equilibrium constant for the first step would be K_1. The subsequent steps would then be as follows:

Step 2 $[Cu(H_2O)_3NH_3]^{2+} + NH_3 \rightleftharpoons [Cu(H_2O)_2(NH_3)_2]^{2+} + H_2O$ (13-40)

$$K_2 = \frac{[Cu(H_2O)_2(NH_3)_2^{2+}]}{[Cu(H_2O)_3NH_3^{2+}][NH_3]} = 3.16 \times 10^3 \tag{13-41}$$

---

[88] S. J. Ashcroft and C. T. Mortimer, "Thermochemistry of Transition Metal Complexes," Academic Press, New York, 1970. All of the thermodynamic data quoted in this section, and all of the figures and tables in this section, are derived from data in this source.
[89] R. J. Angelici in G. L. Eichhorn, "Inorganic Biochemistry," Elsevier Scientific Publishing Company, New York, 1973, p. 63.
[90] F. J. C. Rossotti in J. Lewis and R. G. Wilkins, "Modern Coordination Chemistry," Interscience Publishers, New York, 1960, p. 1.

734 ☐ REACTION MECHANISMS AND METHODS OF SYNTHESIS; SUBSTITUTION REACTIONS

(13-42)   *Step 3* $[Cu(H_2O)_2(NH_3)_2]^{2+} + NH_3 \rightleftharpoons [Cu(H_2O)(NH_3)_3]^{2+} + H_2O$

(13-43) $$K_3 = \frac{[Cu(H_2O)(NH_3)_3^{2+}]}{[Cu(H_2O)_2(NH_3)_2^{2+}][NH_3]} = 8.31 \times 10^2$$

(13-44)   *Step 4* $[Cu(H_2O)(NH_3)_3]^{2+} + NH_3 \rightleftharpoons [Cu(NH_3)_4]^{2+} + H_2O$

(13-45) $$K_4 = \frac{[Cu(NH_3)_4^{2+}]}{[Cu(H_2O)(NH_3)_3^{2+}][NH_3]} = 1.51 \times 10^2$$

Summing up these four steps, the overall reaction that occurs in this case is:

(13-46) $$[Cu(H_2O)_4]^{2+} + 4NH_3 \rightleftharpoons [Cu(NH_3)_4]^{2+} + 4H_2O$$

and an equilibrium constant for the overall process is:

(13-47) $$K_{total} = \frac{[Cu(NH_3)_4^{2+}]}{[Cu(H_2O)_4^{2+}][NH_3]^4} = 6.58 \times 10^{12}$$

The overall equilibrium constant or K_{total} is, of course, just the product of the stepwise constants; that is, $K_{total} = K_1 \cdot K_2 \cdot K_3 \cdot K_4$. Usually, K_{total} is given the symbol β and is equal to $K_1 \cdot K_2 \cdot K_3 \cdot \cdots K_n$.

In general, it is to be expected that the addition of each successive ligand to a metal ion in place of water will be less favored than the previous addition. This is

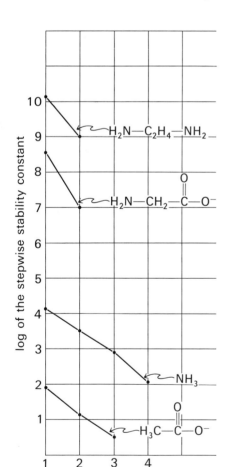

Figure 13-13. Log K values for the reactions of $Cu(H_2O)_4^{2+}$ with various ligands to illustrate the effect of the addition of successive ligands on complex stability. Ethylenediamine and glycine are bidentate and ammonia and acetate are monodentate.

SYNTHESIS OF COORDINATION COMPOUNDS BY SUBSTITUTION REACTIONS 735

TABLE 13-7
THE THREE CLASSES OF METAL IONS

Class a	Borderline	Class b
H^+, Li^+, Na^+, K^+	Fe^{2+}, Co^{2+}, Ni^{2+}	Cu^+, Ag^+, Au^+, Tl^+
Be^{2+}, Mg^{2+}, Ca^{2+}, Sr^{2+}	Cu^{2+}, Zn^{2+}, Pb^{2+}	Hg_2^{2+}, Hg^{2+}, Pd^{2+}, Pt^{2+}
Mn^{2+}, Al^{3+}, Sc^{3+}, Ga^{3+}		Pt^{4+}, Tl^{3+}
In^{3+}, La^{3+}		
Cr^{3+}, Co^{3+}, Fe^{3+}		
Ti^{4+}, Sn^{4+}		

illustrated by the constants for $Cu(H_2O)_4^{2+}$ and by Figure 13-13. There are clearly many reasons—statistical, steric, and electrostatic—for this observed decrease.

The main factors affecting the thermodynamic stability of transition metal complexes are outlined below, and each will be discussed in turn.[91]

Factors affecting thermodynamic stability
- Nature of the metal ion
  - Class a or class b
  - Number of d^* electrons
- Nature of the ligand
  - Basicity
  - Chelate effect

In Chapter 11 we discussed the phenomenon of linkage isomerism. In order to rationalize the preference of a metal ion for one end or another of a ligand that is potentially ambidentate, we used the Ahrland-Chatt-Davies *empirical* classification of metal ions into three groups: class a, class b, and borderline (see Figure 11-1 and Table 13-7).[92,93] As mentioned previously, this classification was made on the basis of the relative stabilities of complexes with ligands having donor atoms from Groups VA, VIA, or VIIA. If the stability of the complexes is greatest with the lightest element of each of these Groups as the donor atom (see Table 13-8), the ions are placed in class a. Conversely, class b ions form the least stable complexes with the lightest element of each Group as donor atom. Furthermore, class b ions form stable complexes with CO and olefins, while those of class a do not form such complexes.

Some ions frequently form complexes whose stabilities cannot be predicted on the basis of the order of generally observed stabilities in Table 13-8, and these ions are placed in the borderline class. The 2+ transition metal ions in this class (Table

---

[91] The measurement and interpretation of stability constants is an immense subject and an extremely interesting one. We cannot hope to do it justice in the short space available here. There are times when our explanations are perhaps shorter than we would like and when we have not followed up on apparently interesting correlations or anomalies. We hope, however, that this will only whet your appetite for more information and that you are led to discover on your own the almost endless fascination of inorganic thermochemistry.

[92] S. Ahrland, J. Chatt, and N. R. Davies, *Quart. Rev.*, **11**, 265–276 (1958).
[93] S. Ahrland, *Structure and Bonding*, **5**, 118 (1968).

TABLE 13-8
METAL ION CLASSIFICATION AND LIGAND DONOR ATOM TRENDS

The trend in complex stability with ligand donor atom type for class a metal ions.	The trend in complex stability with ligand donor atom type for class b metal ions.
$F > Cl > Br > I$	$F < Cl < Br < I$
$O \gg S > Se > Te$	$O \ll S \sim Se \sim Te$
$N \gg P > As > Sb$	$N \ll P > As > Sb$

13-8) are, of course, just the ones in which we are often most interested. Fortunately, it has been known for some time that the stability of complexes of the borderline ions with a given ligand is almost invariably in the order

$$Mn^{2+} < Fe^{2+} < Co^{2+} < Ni^{2+} < Cu^{2+} > Zn^{2+}$$

This order is known as the *Irving-Williams series,* and it is illustrated in Figure 13–14 for several different ligands. Although the figure shows only the trend for K_1, the Irving-Williams series usually holds for K_2 and K_3 (at least) as well.

The stability constant for a metal ion and its ligands is related to $\Delta G°$ for the reaction ($\Delta G° = -2.303 \, RT \log K$). Since $\Delta G° = \Delta H° - T\Delta S°$, the next obvious question is which of the two thermodynamic functions, $\Delta H°$ or $\Delta S°$, is the controlling factor in the Irving-Williams series. The plots in Figure 13–15 provide an answer. For the two ligands having an amino donor group, ethylenediamine and glycine, the plot of ΔH against metal ion shows the same trend as the Irving-Williams series. Therefore, at least for ligands having the amino donor group, the reaction is enthalpically controlled. On the other hand, the malonate ion reactions are all endothermic, so the stability of its complexes must arise from a large, positive ΔS of reaction. (The more detailed data in Table 13–9 back up this conclusion.) The positive ΔS apparently arises from the extensive desolvation of the carboxylate group upon complexation of that group with the metal ion; an anion in aqueous solution is a good "organizer" of solvent, and this organization is lost when the carboxylate binds to the metal ion. The positive ΔH of the carboxylate ion complexation reaction also has its origin in ligand desolvation. That is, water is bound more strongly to the anionic carboxylate than to the dipolar amine group. Therefore, more energy is necessary to desolvate the carboxylate than the amine, and the net reaction is endothermic.[94]

---

[94] This endothermicity is a bit surprising in view of the fact that, in the reaction products, the ion-ion M^{2+}-carboxylate interaction should be more exothermic than the ion-dipole M^{2+}-amine interaction.

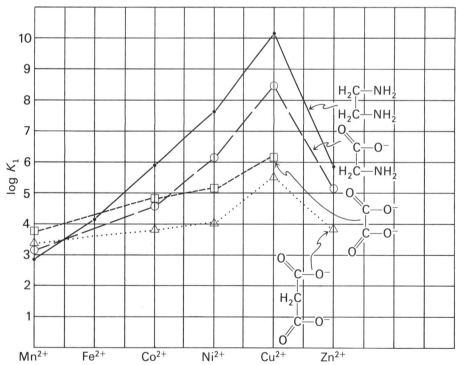

Figure 13–14. Illustration of the Irving-Williams series, the trend in log K_1 values as M^{2+} is changed.

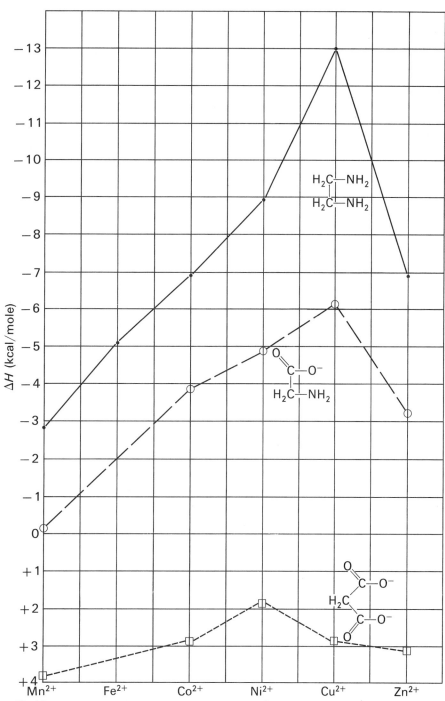

Figure 13-15. The trend in ΔH values (for the reaction $M^{2+} + L \rightleftharpoons ML^{n+}$) as a function of M^{2+}.

The plots of log K or ΔH against metal ion that were discussed above should look familiar to you. They have the same general form as other plots of thermodynamic functions against number of d^* electrons that were discussed in Chapter 9 [see Figure 9–21(A), for example]. Therefore, it is evident that the Irving-Williams series arises from many of the same factors (*e.g.*, d orbital occupation, $Z_{effective}$ of the metal ion, d/TSAO overlap and energy match) that control the previously discussed trends.

TABLE 13-9
ΔH AND ΔS FOR THE REACTION $M^{2+} + L \rightleftharpoons ML^{n+}$ WHERE L IS ETHYLENEDIAMINE, GLYCINATE, OR MALONATE[a]
(ΔH = kcal/mol and ΔS = e.u.)

		Metal Ion, M^{2+}					
		Mn^{2+}	Fe^{2+}	Co^{2+}	Ni^{2+}	Cu^{2+}	Zn^{2+}
H_2C-NH_2 | H_2C-NH_2	ΔH	−2.8	−5.1	−6.9	−8.9	−13.0	−6.7
	ΔS	3.0	3.0	4.0	5.5	5.4	4.0
O=C−O⁻ | H_2C-NH_2	ΔH	−0.3	—	−2.8	−4.9	−6.2	−3.3
	ΔS	13.5	—	13.7	11.9	18.4	12.7
O=C−O⁻ | H_2C | O=C−O⁻	ΔH	3.68	—	2.9	1.88	2.85	3.13
	ΔS	27.4	—	27.0	25.0	35.4	28.0

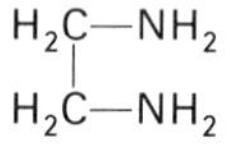

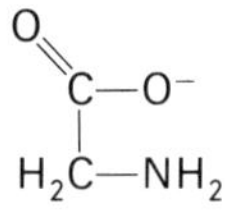

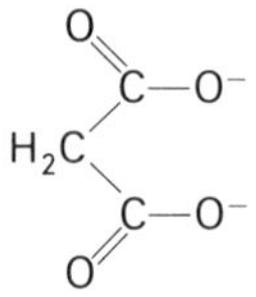

[a] All data taken from S. J. Ashcroft and C. T. Mortimer, "Thermochemistry of Transition Metal Complexes," Academic Press, New York, 1970.

The basicity of a ligand will most certainly be related to complex stability, but this effect is not always as predictable as one might wish.[95] At first glance it would seem reasonable to presume that complex stability should, in general, be directly related to ligand basicity, more stable complexes being formed from more basic ligands. Indeed, there are reasonable correlations between ligand basicity (as measured by the pK_a of the ligand conjugate acid) and log K—but good correlations exist only for series of closely related ligands.[96] Little or no correlation of any predictive value exists even for a series of ligands such as NH_3, pyridine, and imidazole, where the ligating atom is only nitrogen;[97] and the situation is even worse when considering ligands having different donor atoms (e.g., RNH_2, RCO_2^-, and RS^-).

When a series of complexes do show a correlation between log K and pK_a, several very interesting observations may be made.[98] By saying that a correlation exists, we mean that a plot of log K vs. pK_a will obey the equation for a straight line: log K = apK_a + **b**. The constant **a** is the slope of the line and will be a positive number because stronger complexes are generally formed from more basic ligands. However, a fact that must not be overlooked is that the metal ion is competing for the ligand with an H^+ ion

---

[95] By "basicity" here we mean the free energy change or equilibrium constant for $H^+ + L = HL^+$.
[96] You should be reminded of the concept of donor reversals as pursued in Chapter 5.
[97] As an example, consider the following data for Cu^{2+} complexes of the type $[CuL_4]^{2+}$:

	Pyridine	Imidazole	Ammonia
pK_a	5.25	6.95	9.3
$\log \beta_4$	6.63	12.5	12.8

[98] J. G. Jones, J. B. Poole, J. C. Tomkinson, and R. J. P. Williams, *J. Chem. Soc.*, 2001 (1958).

that may be in the solution (*e.g.*, from water dissociation). That is, there are actually two equilibria to be considered in aqueous solution:

$$H^+ + L^- \rightleftharpoons HL \quad \text{equilibrium constant} = 1/K_a$$
$$M^{m+} + L^- \rightleftharpoons ML^{(m-1)+} \quad \text{equilibrium constant} = K$$

and the overall reaction to be considered is

$$HL + M^{m+} \rightleftharpoons ML^{(m-1)+} + H^+ \quad \text{equilibrium constant} = K \cdot K_a$$

This means that if one starts with equal concentrations of ligand and metal ion, the more basic ligand (smaller K_a) can lead to a lower concentration of complex in solution.[98a] For example, consider the overall equilibrium constants for Cu(II) reacting with formate and acetate:

$$Cu^{2+} + HCO_2H \quad K \cdot K_a = (10^{2.8})(10^{-4.75}) = 10^{-1.95}$$
$$Cu^{2+} + CH_3CO_2H \quad K \cdot K_a = (10^{3.36})(10^{-6.01}) = 10^{-2.65}$$

In summary, the amount of complex actually present in aqueous solution will depend on both the complex stability constant and the ligand basicity.

In many of the complexes we have discussed in this and previous chapters, there are one or more ligands that span two or more coordination positions; that is, they are chelating ligands. It is fair to ask, finally, why such complexes are so numerous and have been studied so thoroughly. The answer lies in thermodynamics—*complexes of chelating ligands are in general more stable thermodynamically than those with an equivalent number of monodentate ligands*. This special effect, which is strikingly illustrated in Figure 13–13, is called the *chelate effect*. It deserves more detailed comment.

In order to analyze the origins of the chelate effect, consider first the reactions of Cu(II) with two ammonia molecules or one ethylenediamine molecule:

$$[Cu(H_2O)_4]^{2+} + 2NH_3 \rightleftharpoons [Cu(NH_3)_2(H_2O)_2]^{2+} \tag{13-48}$$
$\Delta G = -10.5 \text{ kcal/mol} \quad \Delta H = -11.1 \text{ kcal/mol} \quad \Delta S = -2 \text{ e.u.}$
$\log \beta_2 = 7.65$

$$[Cu(H_2O)_4]^{2+} + en \rightleftharpoons [Cu(en)(H_2O)_2]^{2+} \tag{13-49}$$
$\Delta G = -14.6 \text{ kcal/mol} \quad \Delta H = -13.0 \text{ kcal/mol} \quad \Delta S = +5.4 \text{ e.u.}$
$\log \beta_1 = 10.64$

The reaction of Cu(II)$_{(aq)}$ with ethylenediamine is clearly more favorable than reaction with two ammonia molecules. This means that if ethylenediamine is allowed to react with diamminediaquocopper(2+), the reaction is thermodynamically favorable and the ammonia molecules can be displaced in favor of the chelating ligand.

$$[Cu(NH_3)_2(H_2O)_2]^{2+} + en \rightleftharpoons [Cu(en)(H_2O)_2]^{2+} + 2NH_3 \tag{13-50}$$
$\Delta G = -4.1 \text{ kcal/mol} \quad \Delta H = -1.9 \text{ kcal/mol} \quad \Delta S = +7.4 \text{ e.u.}$
$\log K = 3.0$

As the enthalpy and entropy changes for the NH_3/en exchange show, these factors both favor the reaction and provide some insight into the chelate effect. ΔS is evidently positive because two particles (the copper complex and en) have been converted into three particles (the new copper complex and two NH_3 molecules).

[98a] Only if the ratio of K's is in the sense opposite to that of the ratio of K_a's, but larger, will more complex arise for the more basic ligand.

This, of course, should generally be the case when a chelating ligand displaces monodentate ligands, and this provides a driving force for the displacement. The enthalpy change also favors the NH_3/en exchange. However, ΔH is small because the donor groups of the two ligands are somewhat similar (pK_a for NH_4^+ is 9.3 and pK_a for $H_2NC_2H_4NH_3^+$ is 9.6), and the Cu—L bonds are almost isoenergetic. The entropy change doubles the thermodynamic driving force in this ligand exchange at room temperature.

The *size of the chelating ring* is also important. Generally, five-membered chelate rings are more stable than six-membered rings, which are in turn more stable than seven-membered rings. The importance of ring size is illustrated in Figure 13-16 for a series of dicarboxylate ligands.

Ring size effects are also evident in Figure 13-14, which we introduced earlier to illustrate the Irving-Williams series. That is, for most metal ions, five-membered ring chelates formed by ethylenediamine, oxalate, and glycinate are more stable than the six-membered ring chelates of the malonate ion. However, Figures 13-14 and 13-15 and Table 13-9 illustrate another important point. That is, for chelate rings of the same size with the same metal ion, the log K values will usually be in the order $(O-O)^{2-} < (O-N)^- < N-N$. *In general,* diamine complexes will be more stable than amino acid complexes, which will in turn be more stable than dicarboxylate complexes, at least for the first row 2+ transition metal ions from Co^{2+} to Zn^{2+}. The explanation for this is found in the data in Table 13-9. As mentioned earlier (page 736), carboxylate complexes derive stability from large changes in entropy, an effect attributable, in part, to desolvation of the anion upon complexation. On the other hand, diamines form quite stable complexes because of a very favorable ΔH.

Finally, the number of chelate rings also affects complex stability, as illustrated in Figure 13-17. On going from ethylenediamine to diethylenetriamine

$$H_2N-C_2H_4-NH-C_2H_4-NH_2$$

to triethylenetetramine

$$H_2N-C_2H_4-NH-C_2H_4-NH-C_2H_4-NH_2$$

complex stability increases both because the entropy of reaction increases (more water

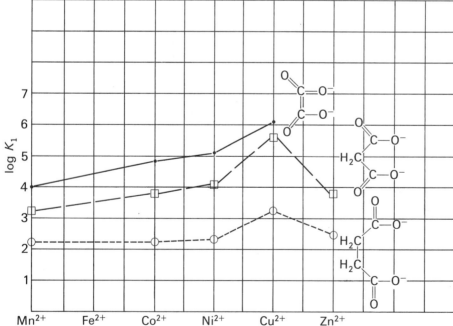

Figure 13-16. Dependence of complex stability (log K_1) on the size of the chelate ring.

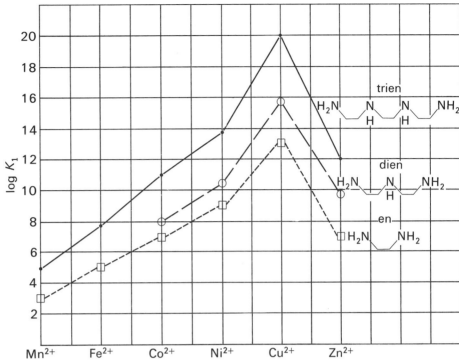

Figure 13-17. Influence of the number of chelate rings formed on complex stability. Triethylenetetramine (= trien) can form three chelate rings, diethylenetriamine (= dien) can form two rings, and ethylenediamine (= en) one ring.

molecules are released per metal ion upon complexation) and because the enthalpy change becomes increasingly negative.

STUDY QUESTIONS

16. Given below are thermodynamic data for the reactions of halides and pseudo-halides with Fe(III) and Hg(II) in water.

Metal Ion	Halide	$\Delta H°$(kcal/mol)	$\Delta G°$(kcal/mol)	$\Delta S°$(e.u.)
Hg^{2+}	F^-	0.85	−1.4	8
	Cl^-	−5.9	−9.19	11.1
	Br^-	−9.6	−12.8	10.9
	I^-	−18.0	−17.5	−2
	CN^-	−23.0	−23.2	0.7
Fe^{3+}	F^-	7.5	−6.68	48
	Cl^-	8.5	−2.02	35
	Br^-	6.1	−0.82	23
	SCN^-	−1.5	−4.1	8.7
	N_3^-	−1.6	−6.8	17.3

Plot ΔG, ΔH, and $T\Delta S$ against the halide in the order given, and notice the great difference in the plots for the two different metal ions. Comment on these differences. For example, which factor—ΔH or ΔS—controls the reactions of Fe(III) and Hg(II)? Does complex stability reflect the factors that went into the classification of metal ions into a, b, and borderline groups? (See Tables 13-7 and 13-8).

17. Figure 13-15 shows the trend in ΔH for diamine, amino acid, and dicarboxylate ligands. Supportive data are presented in Table 13-9. Speculate on the reason that ΔH for dicarboxylate ligands is greater than zero, whereas that for diamine ligands is less than zero.

18. One of the most important ligands in inorganic chemistry is the acetylacetonate ion, $[H_3C-CO-CH-CO-CH_3]^-$, a bidentate ligand that forms complexes with virtually all metal ions. The thermodynamic data for the formation of 1:1 complexes (reaction below) are given in tabular form below. Assuming room temperature, calculate $\Delta G°$ or log K. Plot $\Delta H°$ and $\Delta G°$ or log K vs. metal ion as illustrated in Figures 13-14 and 13-15,

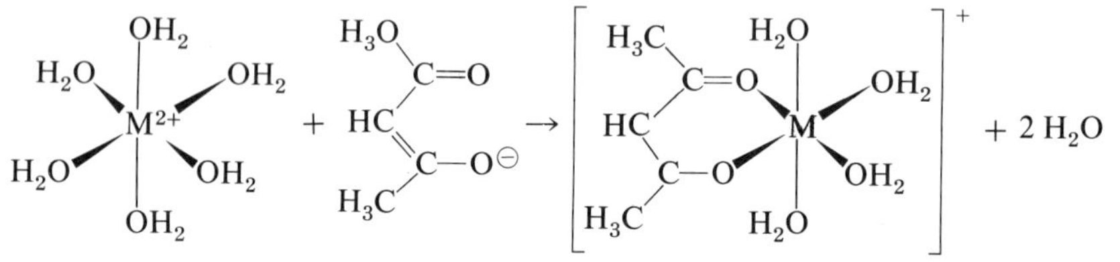

and compare the resulting plots with those for diamine and dicarboxylate ligands in Figures 13-14 and 13-15. Also compare the data below with those in Table 13-9. Comment on similarities and differences.

	Mn^{2+}	Fe^{2+}	Co^{2+}	Ni^{2+}	Cu^{2+}	Zn^{2+}	
$\Delta H°$	−2.5	—	−1.2	−6.7	−4.7	−1.9	(kcal/mol)
$\Delta S°$	11	—	21	12	22	17	(e.u.)

THE SYNTHESIS AND CHEMISTRY OF SOME COBALT COMPOUNDS[99]

To discuss the chemistry of cobalt is to discuss the history of chemistry in modern times. The element has, of course, been known for centuries in the form of "cobalt glass," the beautiful blue color coming from the presence of Co(II) in a tetrahedral oxide environment. A systematic development of cobalt chemistry, however, has occurred within the past two centuries. In 1798, B. M. Tassaert observed that on placing a cobalt(II) salt in aqueous ammonia, the solution gradually turned brown in the presence of air and finally wine red on boiling. From this solution at least two compounds, **36** and **37**, can be isolated, and others such as **38** can be obtained if excess chloride ion is added.

$CoCl_3 \cdot 6NH_3$
luteocobaltic chloride
yellow

$CoCl_3 \cdot 5NH_3 \cdot H_2O$
roseocobaltic chloride
red

$CoCl_3 \cdot 5NH_3$
purpureocobaltic chloride
purple

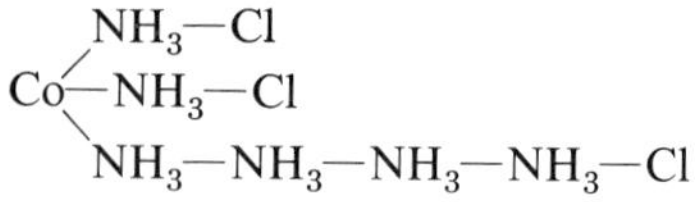

33

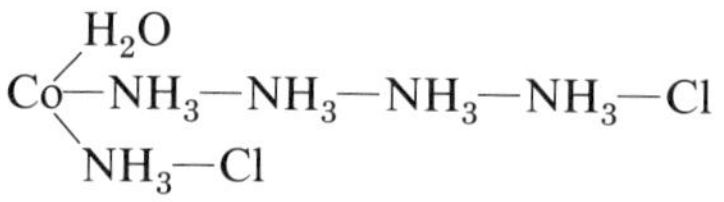

34

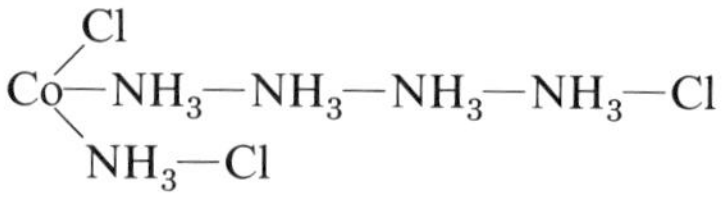

35

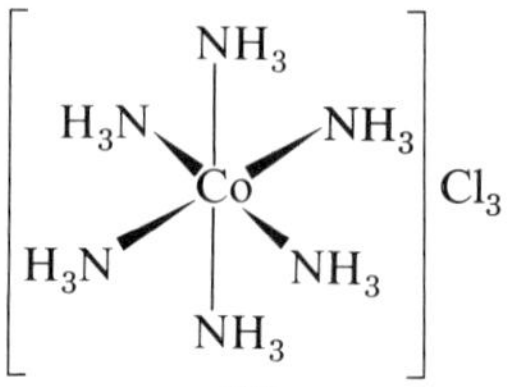

36

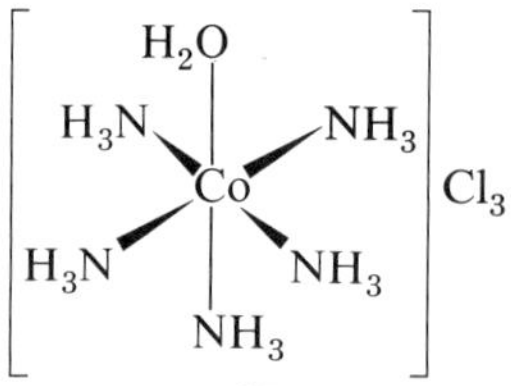

37

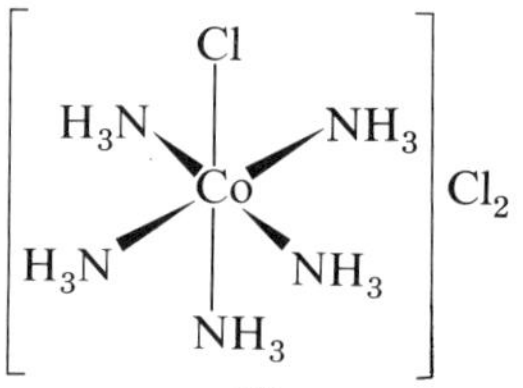

38

---

[99] D. Nicholls in "Comprehensive Inorganic Chemistry," J. C. Bailar, Jr., H. J. Emeleus, R. Nyholm, and A. F. Trotman-Dickenson, eds., Pergamon Press, New York, 1973, vol. 3, p. 1053.

Although chemists in the earlier part of the 19th century formulated them as **33, 34,** and **35,** Alfred Werner recognized in 1893 that their structures were probably based on the octahedron. From that point on, Co(III) compounds played an important role in the development of Werner's ideas on coordination chemistry, and he and his students synthesized about 700 cobalt compounds and published approximately 127 papers on these compounds before his career was prematurely ended with his death in 1919 at the age of 52.[100,101]

There are probably two reasons for the usefulness of cobalt(III) compounds in the development of chemical theory: (i) As you have already seen in this chapter, Co(II) compounds are labile to substitution, whereas those of Co(III) are inert. (ii) Cobalt complexes of great variety are readily prepared. We now turn to that aspect of transition metal chemistry.

As is often the case in synthetic chemistry in general (see Chapter 8), much of the preparative work in cobalt chemistry can begin with the halides. Reaction of cobalt(II) oxide or carbonate with aqueous HCl gives the pink compound $CoCl_2 \cdot 6H_2O$, a salt that presumably contains the aquated, octahedral Co^{2+} ion. Dehydration at

$$CoO_{(s)} + 2HCl + 5H_2O \rightarrow \left[\begin{array}{c} H_2O \\ H_2O\diagdown \;\; \big|\;\; \diagup OH_2 \\ Co \\ H_2O \diagup \;\;\big|\;\; \diagdown OH_2 \\ H_2O \end{array}\right] Cl_2 \qquad (13\text{-}51)$$

$$\downarrow {\substack{\text{heat } in\ vacuo \\ -H_2O \;\; \text{or react with } SOCl_2 \\ (H_2O + SOCl_2 \rightarrow 2HCl + SO_2)}}$$

$$Co_{(s)} + Cl_2 \rightarrow CoCl_2$$
blue

150 °C *in vacuo* or by reaction with $SOCl_2$ gives anhydrous $CoCl_2$, the solid state structure of which is of the $CdCl_2$ type (see page 288). Cobalt(II) chloride, which is blue, can also be prepared by the direct reaction of cobalt metal with chlorine. It is this anhydrous salt that is the indicator in silica gel desiccants; as the desiccant absorbs water and loses its effectiveness, the cobalt(II) compound hydrates to the blue-violet monohydrate, the violet dihydrate, and finally the pink hexahydrate.

Aside from the 0 and +1 oxidation states of cobalt, which are more typical of organometallic compounds of the element, the common oxidation states of cobalt in coordination compounds are 2+ and 3+. Of the two possible hexaaquo complexes of cobalt in these latter oxidation states, the 2+ ion is thermodynamically the more stable form. Hexaaquocobalt(3+) is a powerful oxidant capable of oxidizing water

$$4Co^{3+}_{(aq)} + 2H_2O \rightarrow 4Co^{2+}_{(aq)} + 4H^+ + O_2 \qquad (13\text{-}52)$$

to O_2, and this capability is presumably the reason that simple salts of Co(III) can be isolated only with anions that are not readily oxidized (see pp. 291, 465). For example, Co(III) fluoride, which is stable in the absence of water, can be prepared from CoF_2 and F_2 at 250°C (the reaction is reversible above 350°C). On the other hand, the chloride, bromide, and iodide of Co(III) are not known. The most common form of cobalt (III) in a simple salt is the sulfate, $Co_2(SO_4)_3 \cdot 18H_2O$; the compound can be prepared by oxidation of the corresponding Co(II) compound with O_3 or F_2 in

---

[100] See ref. 32.
[101] F. R. Morral, *Adv. Chem. Ser.*, **62**, 70 (1967).

8 N H_2SO_4. The salt is stable when dry, but it decomposes in water with evolution of O_2.

In the presence of coordinating ligands other than water, Co(III) is stabilized relative to Co(II); that is, the higher valent species is no longer a strong oxidizing agent.

(13-53) $\quad$ $[Co(H_2O)_6]^{3+} + e^- \rightleftharpoons [Co(H_2O)_6]^{2+}$ $\qquad$ $E° = +1.84$ volts vs. S.H.E.

(13-54) $\quad$ $[Co(NH_3)_6]^{3+} + e^- \rightleftharpoons [Co(NH_3)_6]^{2+}$ $\qquad$ $E° = +0.108$ volts

(13-55) $\quad$ $[Co(CN)_6]^{3-} + e^- \rightleftharpoons [Co(CN)_5]^{3-} + CN^-$ $\qquad$ $E° = -0.8$ volts

This change in reduction potential, a common phenomenon in coordination chemistry, has been related to complex stability.[102] The hexaammine complexes of Co(II) and Co(III) are both more stable than the corresponding hexaaquo complexes; $\Delta H°$ for formation of $[Co(NH_3)_6]^{2+}$ from the hexaaquo ion is -13 kcal/mol, whereas the corresponding value for $[Co(NH_3)_6]^{3+}$ is -56.8 kcal/mol. This indicates that the ammonia molecules are more tightly bound to Co(III) or Co(II) than are the water molecules in the corresponding aquo complexes. As a result, the effective positive charge on the metal ion in the oxidizing agent $[Co(NH_3)_6]^{3+}$ is lower than that on Co(III) in $[Co(H_2O)_6]^{3+}$. In terms of the electrochemical properties of these complexes, this means that the oxidizing ability of the Co(III)-ammine complex is less than that of the Co(III)-aquo complex. Of course, using the same argument, the reducing ability of the Co(II)-ammine complex should be greater than that of the Co(II)-aquo complex. However, it may be argued that the greater relative charge neutralization occurs in the Co(III) complexes, and the decrease in oxidizing ability of the Co(III) complexes is greater than the increase in reducing ability of the Co(II)-ammine complexes (both changes being relative to their corresponding aquo ions). The net result is that $\Delta G°$ for $[Co(H_2O)_6]^{3+} + e^- \rightleftharpoons [Co(H_2O)_6]^{2+}$ will be more negative than that for similar reactions where the ligands have been changed to, for example, ammonia. This point is illustrated further in Figure 13-18. (See also Study Questions 9-7 and 9-8.)

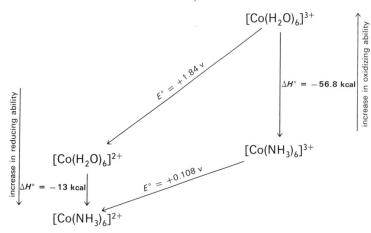

Figure 13-18. The change in electrochemical properties of Co(II) and Co(III) with change in coordinating ligand.

Whatever the explanation for changes in electrochemical properties, it is clear that coordination of a transition metal ion by ligands other than water can usually change the oxidizing and reducing ability of the metal ion, a property important to consider in synthetic chemistry. For example, if this idea is combined with your knowledge of kinetic behavior, the synthesis of an ion such as $[Co(NH_3)_6]^{3+}$ is rationalized. That is, you now realize that you cannot simply place a Co(III) salt in aqueous ammonia and expect the desired hexaamminecobalt(III) ion; much of the Co(III) would be converted to Co(II). For this reason one must begin with a Co(II) salt

---
[102] V. Gutmann, *Structure and Bonding*, **15**, 141 (1973).

in water. Because high spin d^7 ions are labile to substitution, water coordinated to Co(II) can be displaced in favor of ammonia in an ammoniacal solution to give $[Co(NH_3)_6]^{2+}$, an ion thermodynamically more stable than the starting material, $[Co(H_2O)_6]^{2+}$. Oxidation of hexaamminecobalt(II) ion is then easily accomplished by any of a host of common laboratory oxidants. The usual laboratory preparation of $[Co(NH_3)_6]^{3+}$, of course, combines all of these steps into a single operation. This compound is commonly prepared by air oxidizing $CoCl_2 \cdot 6H_2O$ in a strongly ammoniacal solution in the presence of activated charcoal as a catalyst.[103-105] After several hours the initially red solution has turned yellow-brown and the orange product, $[Co(NH_3)_6]Cl_3$, can be isolated by filtration. Separation of the charcoal catalyst from the product is achieved by re-dissolving the complex in hot, dilute HCl and filtering to remove the charcoal; the product precipitates from solution on cooling. Other syntheses of Co(III) complexes can be done in very similar ways.

Although $[Co(NH_3)_6]Cl_3$ is the ultimate product from the oxidation of ammoniacal solutions of $CoCl_2 \cdot 6H_2O$, it has been known for more than 100 years that a brown intermediate product can be isolated.[106,107] This brown material has two cobalt ions in the same molecule and has the formula $[(NH_3)_5Co \cdot O_2 \cdot Co(NH_3)_5]X_4$. Although Alfred Werner and his students did extensive research on this compound and others like it,[108] the true nature of these compounds was not understood until recently.

We mentioned above that water molecules coordinated to Co(II) can be displaced by ammonia. One of the possible products of this displacement is $[Co(NH_3)_5H_2O]^{2+}$, and it is now known that the last remaining water can be displaced by molecular oxygen according to the equation:[109,110]

$$[Co(NH_3)_5H_2O]^{2+} + O_2 \underset{k_{-1}}{\overset{k_1}{\rightleftharpoons}} [Co(NH_3)_5O_2]^{2+} \tag{13-56}$$
$$k_1 = 2.5 \times 10^4 \text{ M}^{-1} \text{ sec}^{-1}$$

The mononuclear oxygen complex (which is presumably similar to $[Co(CN)_5(O_2)]^{3-}$ discussed on pages 205 and 689[111]) then reacts with the remaining pentaammineaquo complex to give the final binuclear cobalt compound.[109,110]

$$[Co(NH_3)_5O_2]^{2+} + [Co(NH_3)_5H_2O]^{2+} \underset{k_{-2}}{\overset{k_2}{\rightleftharpoons}} \underset{\text{brown}}{[(NH_3)_5Co \cdot O_2 \cdot Co(NH_3)_5]^{4+}} \tag{13-57}$$
$$k_{-2} = 56 \text{ sec}^{-1} \qquad\qquad\qquad\qquad \mathbf{39}$$

The overall process, with an equilibrium constant of 6.3×10^5 M^{-2} and a $\Delta H°$ of about 30 kcal/mol, is reversible over short periods of time, say ten minutes. After a longer period of time, however, decomposition to give mononuclear Co(III) complexes such as $[Co(NH_3)_6]^{3+}$ is observed. The binuclear complex, **39**, is reasonably

---

[103] J. Bjerrum and J. P. McReynolds, *Inorg. Syn.*, **2**, 216 (1946).
[104] See ref. 81.
[105] The catalytic activity of charcoal may be a general phenomenon for reactions involving Co—N bonds. For example, the racemization of $[Co(en)_3]^{3+}$, which we noted in Chapter 12 as being catalyzed by electron transfer, can also be catalyzed by charcoal. See D. Sen and W. C. Fernelius, *J. Inorg. Nucl. Chem.*, **10**, 269 (1959); F. P. Dwyer and A. M. Sargeson, *Nature*, **187**, 1022 (1960); J. A. Broomhead, F. P. Dwyer, and J. W. Hogarth, *Inorg. Syn.*, **6**, 186 (1960).
[106] A. G. Sykes and J. A. Weil, *Prog. Inorg. Chem.*, **13**, 1 (1970).
[107] The syntheses of a wide range of binuclear cobalt compounds are given by R. Davies, M. Mori, A. G. Sykes, and J. A. Weil, *Inorg. Syn.*, **12**, 197 (1970).
[108] A. W. Chester, *Adv. Chem. Ser.*, **62**, 78 (1967).
[109] M. Mori, J. A. Weil, and M. Ishiguro, *J. Amer. Chem. Soc.*, **90**, 615 (1968).
[110] J. Simplicio and R. G. Wilkins, *J. Amer. Chem. Soc.*, **91**, 1325 (1969).
[111] L. D. Brown and K. N. Raymond, *Inorg. Chem.*, **14**, 2595 (1975).

stable as a solid and in strongly ammoniacal solutions, but it is decomposed in acid solution according to the reaction

(13-58) $$[(NH_3)_5Co \cdot O_2 \cdot Co(NH_3)_5]^{4+} + 10H^+ \rightarrow 2Co^{2+} + 10\,NH_4^+ + O_2$$

This decomposition to return molecular oxygen, and the reversibility of the process over short periods of time, are of obvious interest, and such complexes have been thoroughly studied as synthetic carriers of molecular oxygen.

The structure of the binuclear complex has only recently been determined (**40**),[112] and it is now clear that the cobalt ions are bridged by an O—O linkage. Of greatest interest is the length, 1.47 Å, of the O—O bond. On comparing this with O—O distances in some diatomic oxygen species in Table 13-10, it is obvious that the bridging O—O is very similar to a peroxide ion, O_2^{2-}. (Therefore, although the bonding in the Co—O—O—Co bridge is polar covalent, one way to count electrons is to assume that two Co^{3+} ions are bound to a peroxide ion, O_2^{2-}.) Indeed, the

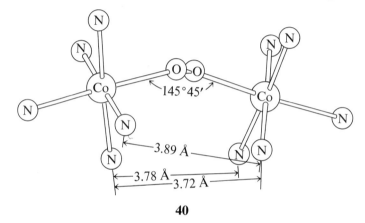

40

structure of the Co—O—O—Co unit somewhat resembles that of H_2O_2. Because of this clear similarity to peroxides, the systematic name of the binuclear complex is decaammine-μ-peroxo-dicobalt(4+) ion.

The peroxo complex can be oxidized with, for example, peroxydisulfate ($S_2O_8^{2-}$) to a green paramagnetic compound, $[(NH_3)_5Co \cdot O_2 \cdot Co(NH_3)_5]^{5+}$, where the O—O bridge bond length is now 1.31 Å.

---

[112] W. P. Schaefer, *Inorg. Chem.*, **7**, 725 (1968).

TABLE 13-10

BOND DISTANCES FOR SOME DIATOMIC OXYGEN SPECIES[a]

Species	O—O Bond Distance, Å
O_2^+	1.12
O_2	1.20
O_2^- (superoxide)	1.32
O_2^{2-} (peroxide)	1.49

[a] A. G. Sykes and J. A. Weil, *Prog. Inorg. Chem.*, **13**, 1 (1970).

$$2[(NH_3)_5Co \cdot O_2 \cdot Co(NH_3)_5]^{4+} + S_2O_8^{2-} \rightarrow 2[(NH_3)_5Co \cdot O_2 \cdot Co(NH_3)_5]^{5+} + 2SO_4^{2-} \qquad (13\text{-}59)$$

brown, diamagnetic   *green, paramagnetic* $(S = \tfrac{1}{2})$
41

Comparison with the data in Table 13-10 indicates that the bridge now resembles the paramagnetic superoxide ion, O_2^-, and this green complex may be thought of as having a $Co^{3+} \cdot O_2^- \cdot Co^{3+}$ unit where the paramagnetism formally arises from the superoxide bridge; however, recent electron spin resonance studies show that the odd electron is delocalized onto two equivalent Co^{3+} ions.

There are numerous other bridged, binuclear cobalt complexes in which there are two or three bridges, and the bridging groups can be NH_2^-, O_2H^-, OH^-, SO_4^{2-}, NO_2^-, and Cl^- among others. Perhaps the greatest number of complexes are known with μ-amido bridges, NH_2^-, and of these the best known are the components of "Vortmann's sulfate." If air is passed through an ammoniacal solution of $Co(NO_3)_2$ at 35°C and then the mixture is neutralized with dilute H_2SO_4, Vortmann's sulfate—a mixture of $[(NH_3)_4Co \cdot NH_2, O_2 \cdot Co(NH_3)_4](SO_4)_2$ [= octaammine-μ-amido-μ-superoxo-dicobalt(4+) disulfate] and $[(NH_3)_4Co \cdot OH, NH_2 \cdot Co(NH_3)_4](SO_4)_2$ [= octaammine-μ-amido-μ-hydroxo-dicobalt(4+) disulfate]—can be isolated. The first of these two salts is paramagnetic $(S = \tfrac{1}{2})$ owing to the presence of the superoxide bridge, O_2^-. It can be obtained free of the other component by a two-step process that further illustrates the chemistry of these binuclear complexes. That is, KOH is added to a solution of the μ-peroxo compound **41** in 15 M NH_3 to give the μ-peroxo-μ-amido(3+) salt (**42**) (13-60a); **42** can be readily oxidized to the paramagnetic μ-superoxo-μ-amido(4+) salt (13-60b).

$$[(NH_3)_5Co \cdot O_2 \cdot Co(NH_3)_5]^{4+} \xrightarrow{KOH} [(NH_3)_4Co \cdot NH_2, O_2 \cdot Co(NH_3)_4]^{3+} + NH_4^+ \qquad (13\text{-}60a)$$

41'   *brown, diamagnetic*
                                        42

$$[(NH_3)_4Co \cdot NH_2, O_2 \cdot Co(NH_3)_4]^{3+} \xrightarrow{Ce^{4+}} [(NH_3)_4Co \cdot NH_2, O_2 \cdot Co(NH_3)_4]^{4+} \qquad (13\text{-}60b)$$

42'   *green, paramagnetic*

The brown, diamagnetic intermediate in this process, **42**, is interesting in that the $\underline{Co-N-Co-O-O}$ ring is non-planar and the $O-O$ bond distance is 1.41 Å, the latter indicating that the complex is best considered as having a peroxo bridge.[113] The μ-peroxo-μ-amido compound (**42**) undergoes an interesting series of reactions (13-61) in acid solution to give, among other things, a complex having a hydroperoxo bridge, O_2H^-.[114]

$$\begin{bmatrix} \text{structure of compound 42 with two Co centers bridged by } NH_2 \text{ and } O_2, \text{ each Co coordinated to four } NH_3 \end{bmatrix}^{3+}$$

42

---

[113] G. G. Christoph, R. E. Marsh, and W. P. Schaefer, *Inorg. Chem.*, **8**, 291 (1969).

[114] This series of reactions is known to occur when L_4 is two ethylenediamine molecules, and it is presumed to occur when L_4 is four NH_3 molecules.

(13-61)

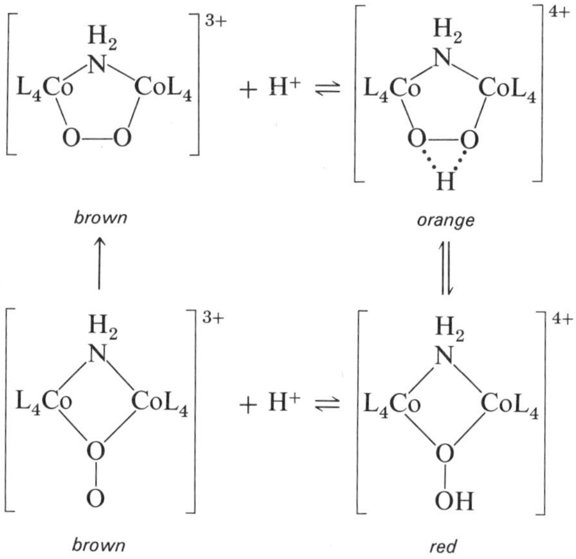

$$\left[L_4\text{Co}\underset{O\!-\!O}{\overset{\underset{|}{\overset{H_2}{N}}}{}}\text{CoL}_4\right]^{3+} + H^+ \rightleftharpoons \left[L_4\text{Co}\underset{\underset{H}{\overset{\cdot\cdot}{O\!-\!O}}}{\overset{\underset{|}{\overset{H_2}{N}}}{}}\text{CoL}_4\right]^{4+}$$

brown　　　　　　　　　orange

$$\left[L_4\text{Co}\underset{\underset{O}{O}}{\overset{\underset{|}{\overset{H_2}{N}}}{}}\text{CoL}_4\right]^{3+} + H^+ \rightleftharpoons \left[L_4\text{Co}\underset{\underset{OH}{O}}{\overset{\underset{|}{\overset{H_2}{N}}}{}}\text{CoL}_4\right]^{4+}$$

brown　　　　　　　　　red

In addition to the extensive series of cobalt-ammine complexes, cobalt also forms an important group of complexes with the bidentate ligand ethylenediamine. The discussion of chiral compounds in Chapter 11 used [Co(en)$_3$]$^{3+}$ and cis-[Co(en)$_2$(NH$_3$)Br]$^{2+}$ as examples, because these compounds played such an important part in Alfred Werner's proof of the octahedron as the appropriate coordination geometry for many complex compounds. Ethylenediamine complexes can often be prepared by starting with a Co(III)-NH$_3$ complex and using the thermodynamic driving force of the "chelate effect" to displace NH$_3$ in favor of the bidentate, chelating diamine. Generally, however, the tris(ethylenediamine)cobalt(3+) ion is prepared by bubbling air through a solution of CoCl$_2 \cdot$ 6H$_2$O and ethylenediamine in dilute HCl. By adjusting the amount of HCl and ethylenediamine used, trans-dichlorobis(ethylenediamine)cobalt(1+) is obtained instead.[115,116]

(13-62)

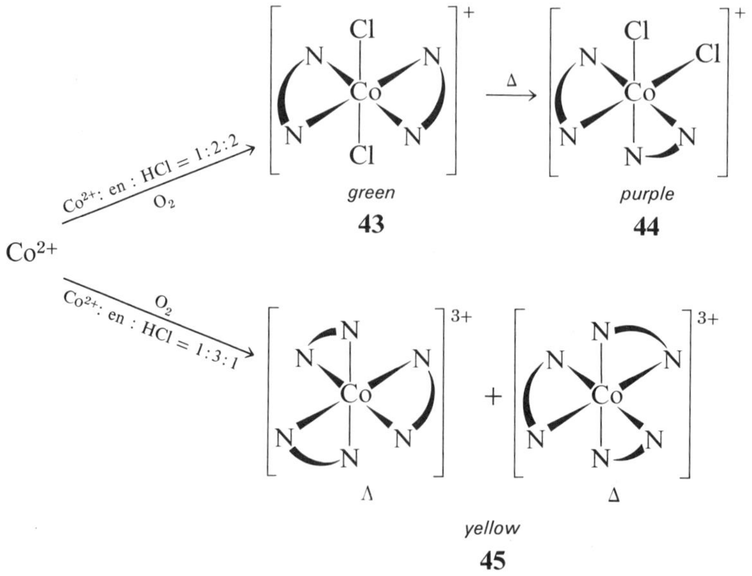

$$\text{Co}^{2+} \xrightarrow[O_2]{\text{Co}^{2+}:\text{en}:\text{HCl}=1:2:2} \underset{\substack{\text{green}\\ \mathbf{43}}}{[\text{Co(en)}_2\text{Cl}_2]^+} \xrightarrow{\Delta} \underset{\substack{\text{purple}\\ \mathbf{44}}}{[\text{Co(en)}_2\text{Cl}_2]^+}$$

$$\text{Co}^{2+} \xrightarrow[\text{Co}^{2+}:\text{en}:\text{HCl}=1:3:1]{O_2} [\text{Co(en)}_3]^{3+}_\Lambda + [\text{Co(en)}_3]^{3+}_\Delta$$

yellow
45

---

[115] Cis- and trans-dichlorobis(ethylenediamine)cobalt chloride: J. C. Bailar, Jr., *Inorg. Syn.*, **2**, 222 (1946).

[116] Tris(ethylenediamine)cobalt(3+) ion: J. A. Broomhead, F. P. Dwyer, and J. W. Hogarth, *Inorg. Syn.*, **6**, 183 (1960).

SYNTHESIS OF COORDINATION COMPOUNDS BY SUBSTITUTION REACTIONS □ 749

In the discussion of mechanisms of hydrolysis of Co(III) complexes, we pointed out that hydrolysis in basic media is considerably faster (by perhaps 10^6) than hydrolysis in acidic or neutral solution. For example, the half-life of the hydrolysis of *trans*-$[Co(en)_2Cl_2]^+$ in neutral or acidic solution is a matter of many minutes to hours at room temperature or slightly above, whereas the hydrolysis of the same ion is virtually instantaneous in base.

$$\textit{trans-}[Co(en)_2Cl_2]^+ \text{ (green)} \begin{array}{c} \xrightarrow[t_{1/2} = \text{minutes to hours}]{H_2O,\, pH\,=\,7} [Co(en)_2(H_2O)Cl]^{2+} \text{ (pink)} \\ \\ \xrightarrow[\text{instantaneous}]{H_2O,\, pH\,>\,7} [Co(en)_2(OH)(Cl)]^+ \text{ (purple)} \end{array} \quad (13\text{-}63)$$

The latter reaction is presumed to occur by the CB mechanism which may be outlined as follows:

$$[Co(en)_2Cl_2]^+ + OH^- \xrightarrow{\text{fast}} [Co(en)(en\text{—}H)Cl_2] + H_2O \qquad (13\text{-}64a)^\ddagger$$

$$[Co(en)(en\text{—}H)Cl_2] \xrightarrow{\text{slow}} [Co(en)(en\text{—}H)Cl]^+ + Cl^- \qquad (13\text{-}64b)$$

$$[Co(en)(en\text{—}H)Cl]^+ + H_2O \xrightarrow{\text{fast}} [Co(en)_2(OH)Cl]^+ \qquad (13\text{-}64c)$$

This mechanism suggests that the base catalyzed reaction may be turned to preparative uses *if* the reaction is run in a non-aqueous solvent. That is, if the five-coordinate intermediate can be generated as in reaction 13-64b by base catalysis, an added anion can compete effectively for the open site in a non-aqueous solvent. Such a reaction has actually been carried out. If *trans*-$[Co(en)_2Cl(NO_2)]^+$ is dissolved in DMSO (dimethylsulfoxide) and a small amount of OH^- is added, the Cl^- ligand is rapidly replaced by added anions such as NO_2^-.[117] The first two steps of this reaction (13-65) are presumed to be similar to those above (13-64a and b).

$$[Co(en)_2(NO_2)Cl]^+ + \text{Base} \xrightarrow{\text{fast}} [Co(en)(en\text{—}H)(NO_2)Cl] + \text{BaseH}^+ \qquad (13\text{-}65)^\ddagger$$

$$[Co(en)(en\text{—}H)(NO_2)Cl] \xrightarrow{\text{slow}} [Co(en)(en\text{—}H)(NO_2)]^+ + Cl^-$$

$$[Co(en)(en\text{—}H)(NO_2)]^+ + NO_2^- \xrightarrow{\text{fast}} [Co(en)(en\text{—}H)(NO_2)_2]$$

$$[Co(en)(en\text{—}H)(NO_2)_2] + \text{BaseH}^+ \xrightarrow{\text{fast}} [Co(en)_2(NO_2)_2]^+ + \text{Base}$$

---

‡The symbol en—H stands for the $H_2NC_2H_4NH^+$ ion.
[117] R. G. Pearson, H. H. Schmidtke, and F. Basolo, *J. Amer. Chem. Soc.*, **82**, 4434 (1960).

THE SYNTHESIS AND CHEMISTRY OF SOME PLATINUM COMPOUNDS[118-120]

Although there is evidence that platinum was known to the ancients, it was apparently confused with silver, so there is no historical record of a separate metal called platinum until the early part of the 18th century. Originally discovered in South America, platinum was found in relative abundance in Russia in 1819. As a result, much of the early work on platinum chemistry was done by the Russian school. Currently, however, platinum is produced chiefly in South Africa and Canada as well as the Soviet Union.

The chemistry of platinum encompasses three oxidation states—0, II, and IV—and some representative compounds in these oxidation states are **46, 47,** and **48.**

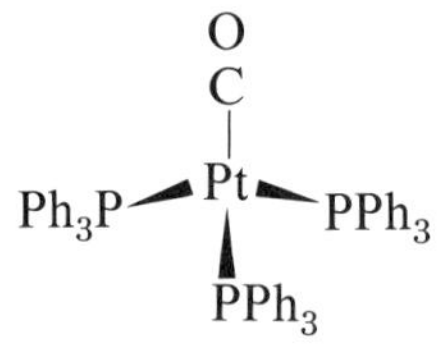

46

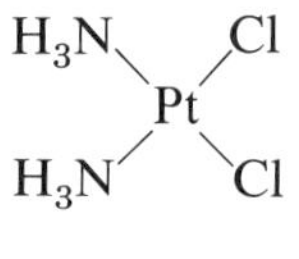

47
Pt—Cl = 2.33 Å
Pt—N = 2.01 Å

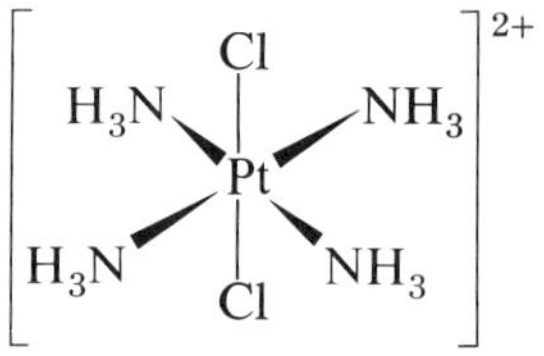

48
Pt—Cl = 2.25 Å
Pt—N = 2.03 Å

Platinum(0) compounds are a very recent development, having been first synthesized about 1957.[121] They are representative of the tremendous development in the chemistry of low valent compounds in general. Many platinum(0) compounds have catalytic activity, and some are clusters of the metal. Both of these topics are among the more important recent developments in inorganic chemistry and their discussion is best postponed to the chapters on organometallic chemistry, where the chemistry of low-valent metals in general is discussed.

Platinum(II) compounds are especially abundant, and, as seen from previous discussions of their importance in the development of chemical theory (see Chapter 11, p. 626, and this chapter, p. 702), they have an interesting and rather systematic chemistry. Entrance into platinum(II) chemistry is often gained through the halides as starting materials. The following reactions are typical of metallic platinum and lead to just the starting materials needed.

(13-66)

$$\text{Pt sponge} \xrightarrow[\text{2. add KCl}]{\text{1. dissolve in aqua regia}} K_2PtCl_6 \xrightarrow{(N_2H_6)(SO_4)_2} K_2PtCl_4$$

yellow potassium hexachloroplatinate(2−)

red (86% yield) potassium tetrachloroplatinate(2−)

49

---

[118] S. E. Livingston in "Comprehensive Inorganic Chemistry," J. C. Bailar, Jr., H. J. Emeleus, R. Nyholm, and A. F. Trotman-Dickenson, eds., Pergamon Press, New York, 1973, vol. 3, p. 1330.

[119] See refs. 8 and 9.

[120] So many of the fundamental ideas of chemistry were developed in Europe in the 19th and early 20th centuries that we tend to forget that American chemists made significant contributions as well. One example is James Lewis Howe (1859-1955), who did especially important laboratory and bibliographic research in the chemistry of the Group VIII metals, particularly on ruthenium. In view of Howe's contributions, it is significant that he spent much of his career at a small college in the South, Washington and Lee University, where for a number of years he was the chemistry department. One of the authors had the privilege of attending that college as an undergraduate shortly after Howe's death. He has fond memories of seeing Howe's laboratory equipment and papers in a dusty corner of the attic of the chemistry building, Howe Hall. Another memory is of Howe's dauther, Guendolen, who carried on the family's association with the University after her father's death; although not a chemist herself, she frequently attended weekly seminars, bringing cookies and cakes in a wicker basket. For a short biography of Howe and a discussion of his work see G. B. Kauffman, *J. Chem. Educ.*, **45**, 804 (1968).

[121] L. Malatesta and M. Angoletta, *J. Chem. Soc.*, 1186 (1957); R. Ugo, *Coord. Chem. Rev.*, **3**, 319 (1968).

$$Pt + xCl_2 \xrightarrow{\begin{array}{c}250\text{-}300°C\\ \\500°C\end{array}} \begin{array}{l} PtCl_4 \text{ (red-brown; sol. } H_2O\text{, acetone)} \\ \quad \downarrow \Delta \\ PtCl_2 \text{ (olive green; insol. } H_2O\text{)} \end{array} \quad (13\text{-}67)$$

With M_2PtX_4 and PtX_2 as starting materials, a wealth of other Pt(II) compounds may then be synthesized; just a few of these are outlined below.

a. Preparation of $[PtL_4]^{2+}$. There are several routes to ions of this type; two of them are

$$PtL_2X_2 + 2L + 2AgNO_3 \rightarrow [PtL_4](NO_3)_2 + 2AgX \quad (13\text{-}68)$$
$$2[PtCl_4]^{2-} + 4L \rightarrow [PtL_4][PtCl_4] + 4Cl^- \quad (13\text{-}69)$$

The second reaction gives $[PtL_4][PtCl_4]$, a class of compounds generally known as Magnus' salts, and it should be apparent that these can also be prepared by direct reaction of the cation with the anion. For example, the product of reaction 13-70 is "Magnus' green salt."

$$[Pt(NH_3)_4]Cl_2 + K_2PtCl_4 \rightarrow 2KCl + [Pt(NH_3)_4][PtCl_4] \quad (13\text{-}70)$$

One of the starting materials for this reaction, $[Pt(NH_3)_4]Cl_2$, was synthesized by Magnus in 1828 and is presumably the first metal-ammine complex ever prepared.

b. Preparation of $PtLL'X_2$. The following reaction is useful when $L = L'$.

$$M_2PtX_4 + 2L \rightarrow 2MCl + PtL_2X_2 \quad (13\text{-}71)$$

A specific example is the preparation of the bright yellow complex $Pt(en)Cl_2$.[122]

$$K_2PtCl_4 + 2H_2N\text{—}C_2H_4\text{—}NH_2 \rightarrow 2KCl + \underset{\mathbf{50}}{\begin{array}{c}H_2C\text{—}NH_2\\ \quad\quad\quad\quad\diagdown \\ H_2C\text{—}NH_2 \end{array}}\!\!\!Pt\!\!\begin{array}{c}Cl\\ \\Cl\end{array} \quad (13\text{-}72)$$

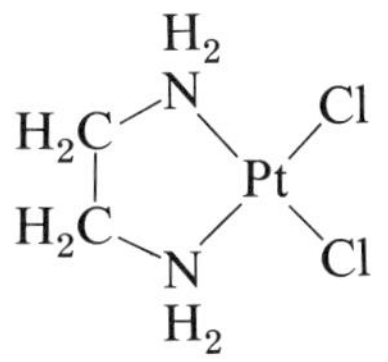

As might be anticipated from reaction 13-69, however, a by-product is the violet complex $[Pt(en)_2][PtCl_4]$. Fortunately, the by-product and the desired compound have different solubilities in liquid ammonia (the salt is not soluble), and this fact allows their separation.

Two of the most important complexes in platinum chemistry from an historical point of view are *cis-* and *trans-*$Pt(NH_3)_2Cl_2$.[123] The *cis* isomer is prepared by the direct reaction of K_2PtCl_4 with aqueous ammonia. Notice that this reaction is apparently controlled by the *trans* effect. Once one NH_3 has been attached to the Pt(II) ion, the ligand most readily replaced is one of the Cl^- ions *cis* to the NH_3.

$$K_2PtCl_4 + NH_3 \rightarrow \left\{\begin{bmatrix}Cl & NH_3\\ \diagdown Pt \diagup \\ Cl & Cl\end{bmatrix}^-\right\} \rightarrow \underset{\mathbf{47'}}{\begin{array}{c}Cl\\ \diagdown \\ Cl\end{array}}\!Pt\!\begin{array}{c}NH_3\\ \diagup \\ NH_3\end{array} + 2KCl \quad (13\text{-}73)$$

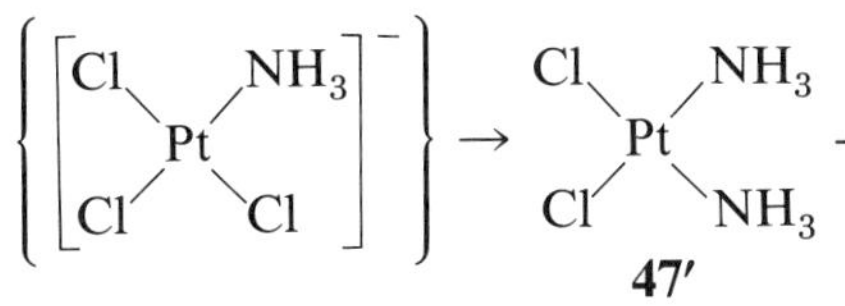

---

[122] G. L. Johnson, *Inorg. Syn.*, **8**, 242 (1966).
[123] G. B. Kauffman and D. O. Cowan, *Inorg. Syn.*, **7**, 239 (1963).

752 ▢ REACTION MECHANISMS AND METHODS OF SYNTHESIS; SUBSTITUTION REACTIONS

In the reaction above, one must be careful to avoid the formation of Magnus' green salt as in reaction 13-69. The synthesis of the *trans* isomer, on the other hand, takes advantage of the fact that $[Pt(NH_3)_4]^{2+}$ is formed in high concentrations of ammonia. This tetraammine complex is first prepared from $[PtCl_4]^{2-}$,

(13-74)
$$K_2PtCl_4 + 4NH_3 \rightarrow [Pt(NH_3)_4]Cl_2 + 2KCl$$

and it is then heated in the presence of excess Cl^- (13-75).

(13-75)

$$\begin{bmatrix} H_3N & NH_3 \\ & Pt & \\ H_3N & NH_3 \end{bmatrix} Cl_2 \xrightarrow{-NH_3} \begin{bmatrix} H_3N & NH_3 \\ & Pt & \\ Cl & NH_3 \end{bmatrix} Cl \xrightarrow{-NH_3} \begin{array}{c} H_3N \quad Cl \\ Pt \\ Cl \quad NH_3 \end{array}$$
$$\mathbf{51}$$

This is, of course, yet another illustration of the *trans* effect. Once the first NH_3 has been replaced by Cl^-, the *trans* effect leads to the desired *trans* isomer.

The preparation of $PtLL'X_2$ where $L \neq L'$ is more difficult and generally follows the sequence of reactions

(13-76)
$$PtL_2X_2 + HX \rightarrow [PtLX_3]^- + HL^+$$
$$\xrightarrow{L'} PtLL'X_2 + X^-$$

Compounds of the type PtL_2X_2 are clearly important, both historically and as starting materials for other platinum(II) compounds. However, they have recently taken on an added importance, because Rosenberg found that *cis*-$Pt(NH_3)_2Cl_2$ and *cis*-$Pt(en)Cl_2$ are effective anti-tumor agents.[124] This has led to a veritable avalanche of research on platinum chemistry in the past few years,[125-127] and new facets of its chemistry are being uncovered. In particular, it is now recognized that only the *cis* isomer has any biological activity, apparently because the complex PtL_2X_2 must lose both of the X ligands and form a complex with some biochemically important polymer that can function as a bidentate ligand. Although the "bidentate ligand" is currently thought to be a portion of the DNA molecule, the exact binding site is not known, nor is the reason for the effectiveness of *cis*-PtL_2X_2 compounds in general known. Nonetheless, testing of various platinum(II) complexes continues in order to find the most effective anti-cancer drug.[128,129] Since biochemical functioning can often be controlled by seemingly minor changes in a drug, it is important to be able to tailor-make a variety of new platinum compounds in order to more carefully probe their functioning. It is on this synthetic problem that we can bring to bear much of what has been discussed thus far in this chapter on substitution reactions.

If we presume that the platinum compound functions by forming a complex with some site in a protein that can act as a bidentate ligand, structure **52**, then we can discuss some requirements for the type of platinum-containing drug that will be most effective.

---

[124] B. Rosenberg, L. Van Camp, J. E. Trosko, and V. H. Mansour, *Nature*, **222**, 385 (1969).
[125] B. Rosenberg, *Naturwissenschaften*, **60**, 399-406 (1973).
[126] B. Rosenberg, *Platinum Metals Review*, **15**, 42-51 (1971).
[127] A. J. Thomson, R. J. P. Williams, and S. Reslova, *Structure and Bonding*, **11**, 1 (1972).
[128] M. J. Cleare and J. D. Hoeschele, *Platinum Metals Review*, **17**, 2-13 (1973).
[129] M. J. Cleare, *Coord. Chem. Rev.*, **12**, 349-405 (1974).

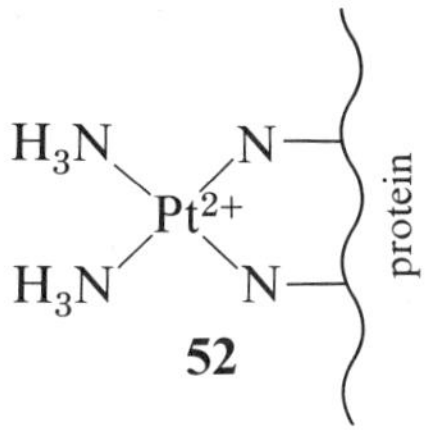

$$\begin{array}{c} H_3N \diagdown \quad \diagup N \\ \quad Pt^{2+} \\ H_3N \diagup \quad \diagdown N \\ \mathbf{52} \end{array} \Bigg\} \text{protein}$$

(i) A *cis* compound of the type PtL_2X_2 is required with two L ligands that are good *trans* labilizing groups.

(ii) The two ligands L that remain after loss of X should be reasonably strongly bound.

(iii) If it is assumed that the final Pt-protein complex must remain undissociated once formed, the Pt-protein bonds must be thermodynamically stable. Further, the remaining L ligands must not be such good *trans* labilizers that the Pt-protein bonds are easily broken; that is, the final complex should be kinetically stable.

Against the requirements for a Pt-containing drug laid out above, consider the following facts:

(i) Pt(II) is a class *b* metal ion, and the following order of metal-ligand bond strength is generally observed in the laboratory:

and
$$CN^- > OH^- > NH_3 > SCN^- > I^- > Br^- > Cl^- > F^- \geq H_2O$$
$$RS^- > R_2S \geq NH_3$$

A ligand in this series cannot be replaced by one to the right of it in a thermodynamically favorable reaction, provided that the two ligand concentrations are approximately equal. Therefore, a good X ligand in *cis*-PtL_2X_2 drug would be Cl^-, since it can be replaced in a thermodynamically favorable reaction by biologically important groups such as RS^-, R_2S, and RNH_2.

(ii) The *trans* effect generally decreases down the series

$$CN^-, \text{olefin} > SR_2, PR_3 > NO_2^- > I^- > Br^- > Cl^- > NH_3, \text{py}, RNH_2 > OH^- > H_2O$$

At first glance one would say that L should be a group well to the left of X in *trans* effect series. In fact, however, this is not desirable, since an L ligand with a large *trans* effect will also labilize the Pt-protein bond to dissociation. Because of this factor, a compound such as *cis*-$Pt(R_3P)_2Cl_2$ would be a poor choice as an anti-tumor drug.

Two of the compounds found most effective thus far as anti-tumor agents are two of the most common, **47** and **50**. Both generally meet the criteria spelled out above.

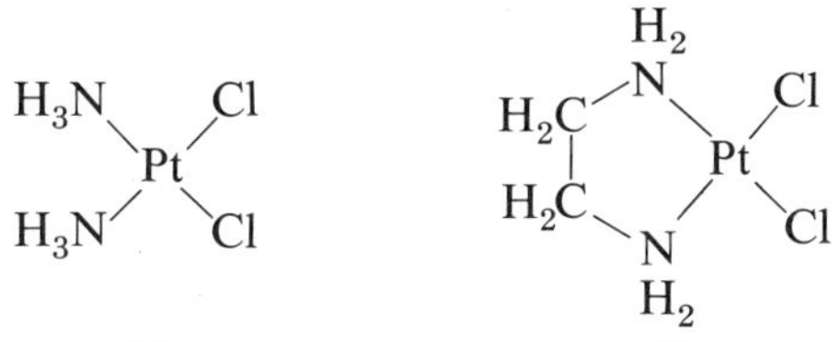

$$\begin{array}{cc} H_3N \diagdown \quad \diagup Cl & \quad H_2C\!-\!\overset{H_2}{N}\diagdown \quad \diagup Cl \\ \quad Pt & \quad \quad \quad \quad Pt \\ H_3N \diagup \quad \diagdown Cl & \quad H_2C\!-\!\underset{H_2}{N} \diagup \quad \diagdown Cl \\ \mathbf{47'} & \mathbf{50'} \end{array}$$

The Cl^- ligand is a good leaving group because the thermodynamic and kinetic stability of the Pt—Cl bond is low. Both ethylenediamine and NH_3 have a relatively low *trans* effect and so do not labilize the Pt-protein bond, once formed. Finally, ethylenediamine is a chelating ligand and effectively blocks the two remaining positions on the drug to further substitution by monodentate ligands; the "chelate effect" helps render its displacement thermodynamically unfavorable.

754 ☐ REACTION MECHANISMS AND METHODS OF SYNTHESIS; SUBSTITUTION REACTIONS

The effectiveness of compounds 47 and 50 as cancer chemotherapy agents suggests that a fairly wide range of compounds may in fact be effective. We can expect considerably greater research efforts in this area in the coming years.

Platinum(IV) is a d^6 ion that forms neutral, cationic, and anionic octahedral complexes. The complexes are always diamagnetic because heavier transition metals, especially those in higher formal oxidation states, have large e_σ values. The preparation of Pt(IV) complexes is usually done by so-called oxidative addition reactions

(13-77)
$$[Pt^{II}(NH_3)_2Cl_2] + Cl_2 \rightarrow [Pt^{IV}(NH_3)_2Cl_4]$$

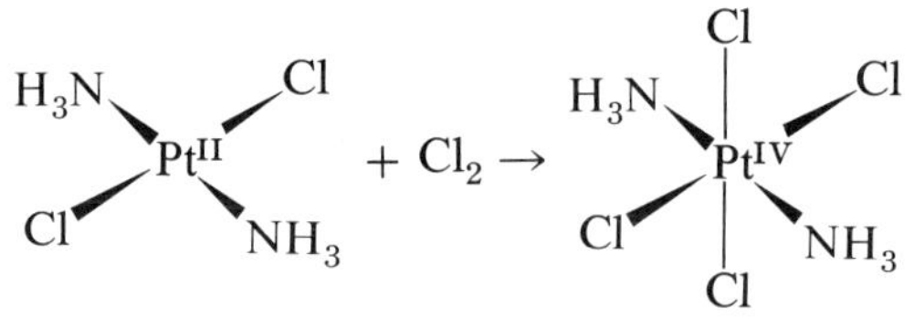

(such as 13-77) or by ligand exchange. The most notable feature of Pt(IV) complexes is their kinetic stability. For example, the ion $[Pt(en)_3]^{4+}$ was resolved into its enantiomers by Alfred Werner in 1917. Once separated, the enantiomers are extremely stable to racemization, a feature that you will appreciate more after the discussion of mechanisms of rearrangement in Chapter 14. Yet another example of the kinetic stability of Pt(IV) complexes is illustrated by the isolation of two of the fifteen possible (see Table 11-2) diastereomers of the compound $Pt(NH_3)(NO_2)pyClBrI$.[130]

STUDY QUESTIONS

19. In the course of experiments on the biological functioning of platinum compounds, *cis*- and *trans*-[Pt(NH$_3$)$_2$Cl$_2$] were allowed to react with methionine, with the results outlined below. What can these reactions tell about the ability of Cl$^-$ to act as a leaving group and about the *trans* effect of the thioether sulfur?

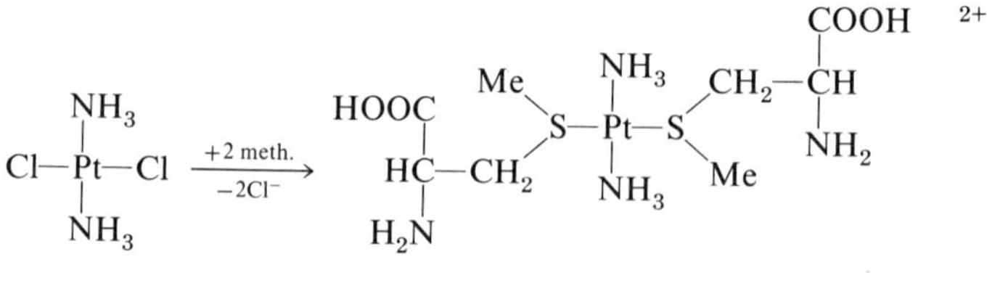

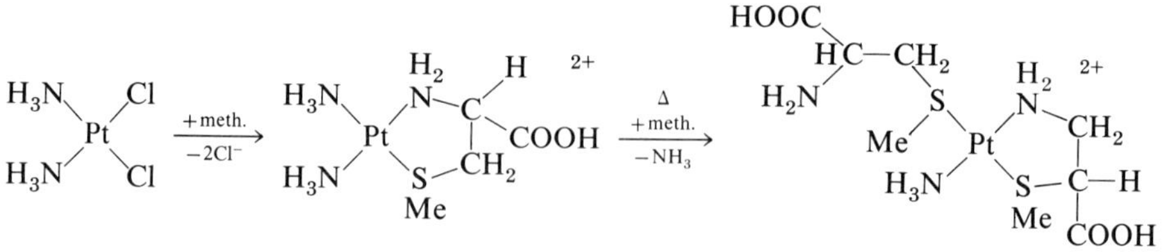

20. What spectral transition(s) gives rise to the blue color of cobalt glass? Why are octahedral Co(III) complexes usually not blue, in contrast with tetrahedral Co(II) complexes?

---

[130] L. N. Essen and A. D. Gel'man, *Zh. Neorg. Khim.*, **1**, 2475 (1956); see also page 1367 of ref. 118.

21. What is the chirality of compound **44**?

22. Outline the synthesis of *trans*-Pt(py)(PEt$_3$)Cl$_2$ starting with [PtCl$_4$]$^{2-}$.

23. The compound below has been found to be effective against certain types of cancer. Synthesize the compound starting with [PtCl$_4$]$^{2-}$.

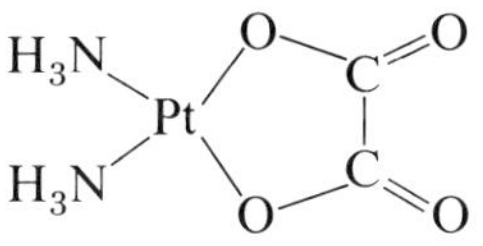

EPILOG

The main conclusion that can be drawn from this discussion of substitution mechanisms is that all four of the mechanistic pathways possible for such reactions—D, I_d, I_a, and A—are utilized by transition metal complexes. Since coordination positions are available on square planar complexes, and a fifth ligand can be added to form an intermediate or activated complex, associative mechanisms are generally observed for ligand substitution on square planar complexes. In contrast, six-coordinate complexes form seven-coordinate intermediates or activated complexes only with difficulty. Therefore, it is reasonable that six-coordinate complexes of divalent metal ions and of Co(III) are generally observed to react by dissociative mechanisms, that is, I_d or, less commonly, D. I_a pathways, however, may be typical of octahedrally coordinated trivalent ions [except for Co(III)], although this is still a matter of some controversy. Finally, the CB mechanism is observed for transition metal complexes just as it is for non-metal compounds (*cf*. Chapter 6, page 410).

MOLECULAR REARRANGEMENTS

    Four-Coordinate Complexes
    Six-Coordinate Octahedral Complexes

REACTIONS AT COORDINATED LIGANDS

    Reactions Due to Metal Ion Polarization of Coordinated Ligands
    Hydrolysis of Amino Acid Esters and Amides and of Peptides
    Aldol Condensation
    Imine Formation, Hydrolysis, and Substituent Exchange
    The Template Effect and Macrocyclic Ligands

EPILOG

14. COORDINATION CHEMISTRY: REACTION MECHANISMS AND METHODS OF SYNTHESIS; MOLECULAR REARRANGEMENTS AND REACTIONS OF COORDINATED LIGANDS

In the previous two chapters, we first examined reactions that intimately involved the metal ion—electron transfer reactions—and then reactions wherein ligands of the first coordination sphere were replaced by others—substitution reactions. In this, the final chapter on reactions of coordination complexes, we turn to reactions in which the ligands remain attached to the central metal ion but in which these ligands rearrange within the coordination sphere or undergo addition, substitution, or oxidation-reduction.

MOLECULAR REARRANGEMENTS

The first part of this chapter is devoted to reactions in which the ligands of the first coordination sphere remain attached to the central metal ion but rearrangement of these ligands occurs. As you shall see, this may lead either to *cis-trans* isomerization or to a change in chirality, or both. The possibility of rearrangement obviously has a bearing on the synthesis of coordination compounds. Although you may set out to synthesize one isomer of a particular complex, another may in fact result.

The study of molecular rearrangements is an area of very current research interest, and great advances have been made in the 1970's.[1-5] However, the field may also be complex to the uninitiated. It is our intention to provide you with only a cursory look at the major types of rearrangements of four- and six-coordinate complexes and, briefly, how these rearrangements may be analyzed in terms of mechanistic pathway.

FOUR-COORDINATE COMPLEXES

When ethyldiphenylphosphine is added to nickel(II) bromide (14-1), a green complex, **1a,** is precipitated.[6] On dissolution of this product in chloroform or nitromethane, a green solution with a hint of red is obtained, while dissolution in benzene or carbon disulfide gives a red solution with no green color. On cooling to $-78°C$, the red solution will give dark red crystals that slowly turn green again on standing at room temperature.

---

[1] N. Serpone and D. G. Bickley, *Prog. Inorg. Chem.,* **17,** 391 (1972).
[2] J. J. Fortman and R. E. Sievers, *Coord. Chem., Rev.,* **6,** 331–375 (1971).
[3] R. H. Holm and M. J. O'Connor, *Prog. Inorg. Chem.,* **14,** 241 (1971).
[4] E. L. Muetterties, *Acc. Chem. Res.,* **3,** 266 (1970).
[5] J. P. Jesson, "Stereochemistry and Stereochemical Nonrigidity in Transition Metal Hydrides," in "Transition Metal Hydrides," E. L. Muetterties, ed., Marcel Dekker, New York (1971).
[6] R. G. Hayter and F. S. Humiec, *Inorg. Chem.,* **4,** 1701 (1965).

758 ☐ MOLECULAR REARRANGEMENTS AND REACTIONS OF COORDINATED LIGANDS

(14-1)

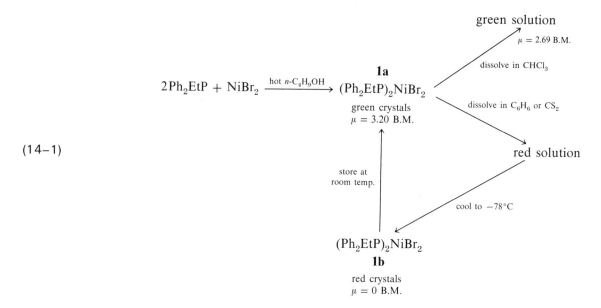

The color changes observed in the reaction above are due to a molecular rearrangement, one that we shall explore as an illustration of the types of rearrangements possible for four-coordinate complexes and of the experimental methods used to attack the structural and mechanistic problem.

The thermochromic behavior of compound **1** is understood upon examination of the effective magnetic moments and UV-visible spectra of the two solids and the magnetic moment of the green compound in solution. The green solid (**1a**) has a magnetic moment of 3.20 B.M., and, as shown in Figure 14-1, clearly absorbs at longer wavelengths than does the red solid (**1b**). On the other hand, the magnetic moment of the red solid is 0.0 B.M. The obvious conclusion is that the green solid is the tetrahedral isomer, whereas the red solid is the square planar isomer. The lowest energy d-d transition for tetrahedral complexes always occurs at lower energies than the corresponding band for square planar complexes, since $\Delta_t < \Delta_{sp}$ (page 546). Further, a tetrahedral d^8 complex (such as **1a**) would be expected to have two unpaired electrons, whereas a square planar d^8 complex (such as **1b**) should be diamagnetic.

In solution in CH_2Cl_2, however, both isomers exist together, as indicated by the fact that the effective magnetic moment of the compound under these conditions is

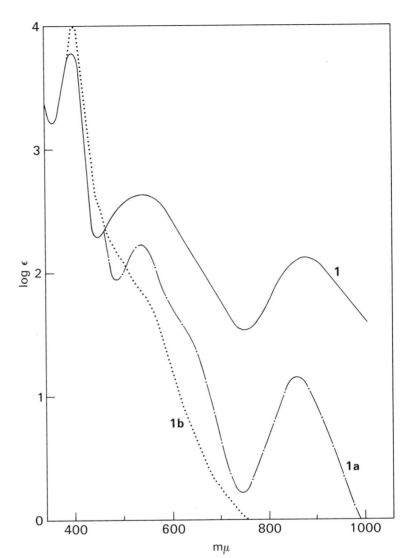

Figure 14-1. Absorption spectra of NiBr$_2$(PPhEt$_2$)$_2$: **1**, benzene solution of either isomer; **1a**, solid tetrahedral isomer—green; **1b**, solid square planar isomer—red. Units of absorption for spectra **1a** and **1b** are arbitrary. [Reprinted with permission from R. G. Hayter and F. S. Humiec, *Inorg. Chem.*, **4**, 1701 (1965). Copyright by the American Chemical Society.]

only 2.69 B.M., less than that expected for two unpaired electrons. Additional information comes from the proton nmr spectrum of the solution; the spectrum for the comparable compound, (Ph$_2$MeP)$_2$NiBr$_2$, is illustrated in Figure 14-2.[7] At +40°C only an averaged signal is seen for the *ortho, meta,* and *para* protons of the phenyl groups. At −65°C, however, interconversion of the polytopes slows and two sets of phenyl protons are seen—one set having very great chemical shifts that arise from the paramagnetic tetrahedral isomer (labeled with a subscript T) and another set from the diamagnetic square planar isomer (labeled with a subscript S). The chemical shifts of the resonance lines assigned to the tetrahedral isomer are very large owing to an *isotropic shift*.[8] This can arise, in part, from delocalization of unpaired

---

[7] G. N. LaMar and E. O. Sherman, *J. Amer. Chem. Soc.*, **92**, 2691 (1970).

[8] A consideration of isotropically shifted nmr spectra is well beyond the scope of this text. For more information, the following books or articles can be consulted:
  a. G. N. LaMar, W. DeW. Horrocks, Jr., and R. H. Holm, "NMR of Paramagnetic Molecules," Academic Press, New York (1973).
  b. R. S. Drago, "Physical Methods in Inorganic Chemistry," W. B. Saunders Company, Philadelphia (1977).
  c. R. S. Drago, J. I. Zink, R. M. Richman, and W. D. Perry, *J. Chem. Educ.*, **51**, 371, 464 (1974).

760 ☐ MOLECULAR REARRANGEMENTS AND REACTIONS OF COORDINATED LIGANDS

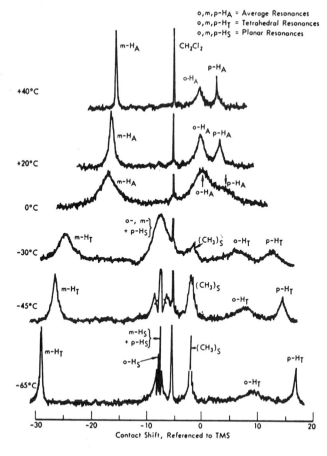

Figure 14-2. 100 MHz proton nmr spectra of $(Ph_2MeP)_2NiBr_2$ in CD_2Cl_2 as a function of temperature. Resonances for the paramagnetic tetrahedral isomer (≡T) and for the diamagnetic square planar isomer (≡S) are frozen out at low temperature, while only an average of these resonances (≡A) is observed at room temperature. [Reprinted with permission from G. N. LaMar and E. O. Sherman, *J. Amer. Chem. Soc.*, **92**, 2691 (1970). Copyright by the American Chemical Society.]

electrons from the metal ion onto the coordinated ligands. A great deal of information can be obtained from isotropically shifted nmr spectra, and in this case the free energy for the equilibrium

$$\text{planar complex } (S = 0) \rightleftharpoons \text{tetrahedral complex } (S = 1)$$

could be measured as well as the rate constant and activation parameters for the isomerization.[9] Furthermore, the mechanism of the rearrangement can be examined. Apparently, two mechanistic pathways are followed by this system: intra- and intermolecular. As only a small flattening motion along the pseudo-S_4 axis is necessary to convert one isomer into the other, this intramolecular route is the one

---

[9] The fraction of tetrahedral isomer in the solution can be calculated from the equation

$$N_t = (\mu_{obs}/\mu_t)^2$$

where μ_{obs} is the observed magnetic moment for a compound in solution at a particular temperature, and μ_t is the magnetic moment for pure tetrahedral isomer. Alternatively, N_t may be obtained from the ratio of the observed averaged isotropic shift for a given proton in the limit of rapid isomerization to the shift for the tetrahedral isomer at that temperature. The value of N_t thus obtained may then be used to calculate the ΔG for the equilibrium from the equation

$$N_t = [\exp(\Delta G/RT) + 1]^{-1}$$

ΔH and ΔS are found by plotting ΔG vs. $1/T$. Kinetic parameters are obtained by line width analysis methods widely used in nmr spectroscopy (see ref. 8a).

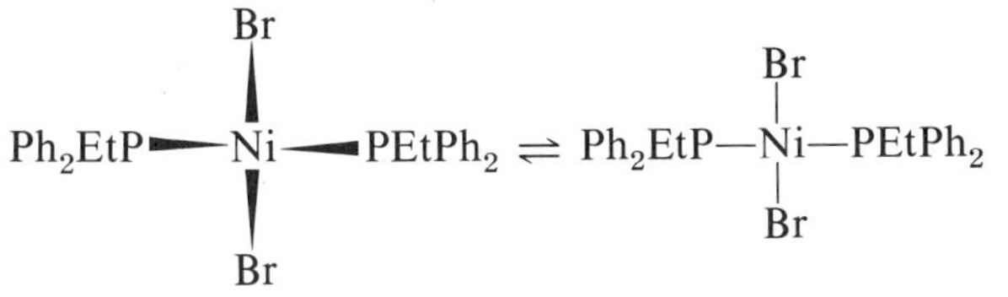

$$\text{Ph}_2\text{EtP}-\underset{\underset{\text{Br}}{|}}{\overset{\overset{\text{Br}}{|}}{\text{Ni}}}-\text{PEtPh}_2 \rightleftharpoons \text{Ph}_2\text{EtP}-\underset{\underset{\text{Br}}{|}}{\overset{\overset{\text{Br}}{|}}{\text{Ni}}}-\text{PEtPh}_2$$

ca. 70% tetrahedral at 40°C

$k^{298} = 1.6 \times 10^5 \text{ sec}^{-1}$
$\Delta H^{\ddagger} = 9.0 \text{ kcal/mol}$
$\Delta S^{\ddagger} = -4.7 \text{ e.u.}$

1a ⇌ **1b**

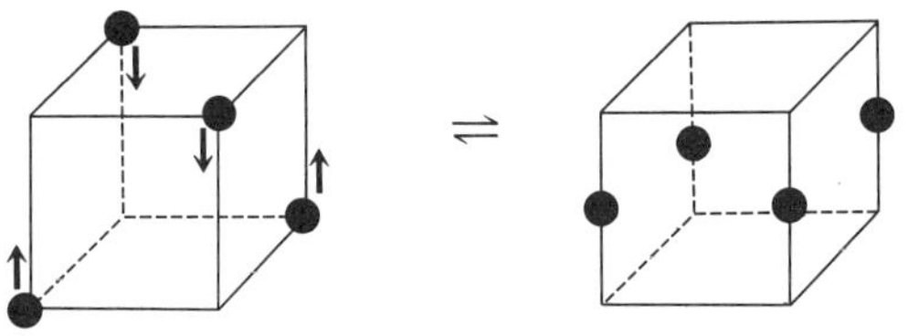

(14-2)

normally followed. However, in the presence of excess ligand, a bimolecular path, presumably similar to that used in substitution reactions of square planar complexes (Chapter 13), becomes more important than the intramolecular path.

Several general conclusions may be drawn regarding the thermodynamics of the tetrahedral ⇌ square planar equilibrium for complexes of the type $L_2\text{NiX}_2$. First, the fraction of tetrahedral isomer increases in the order $X = \text{Cl} < \text{Br} < \text{I}$ and $L = R_3P < \text{ArR}_2P < \text{Ar}_2\text{RP} < \text{Ar}_3P$ where Ar is an aryl group such as phenyl and R is an alkyl group such as methyl or ethyl). Both of these trends can be understood readily in terms of steric effects. As you saw in Chapters 9 and 10 (see Figure 9-19), the square planar geometry is generally favored by d^8 complexes, whether high spin or low spin (if only σ bonding is considered). However, the fact that a tetrahedral ⇌ square planar interconversion exists at all means that these isomers are close to one another in energy, and the steric bulk of the ligands and their mutual repulsion can strongly influence the geometry of the complex. This is apparently the case for the phosphine complexes in question. As the steric bulk of the ligand increases, the fraction of the tetrahedral isomer increases. Because the tetrahedron has greater bond angles, it is better able to accommodate bulky ligands.

Keeping the halide and the steric bulk of the ligand constant, the phosphine electronic properties are varied by changing the groups Z and Z' on $(p\text{-ZC}_6\text{H}_4)(p\text{-Z}'\text{C}_6\text{H}_4)\text{MeP}$.[10] The result is that the fraction of tetrahedral isomer of $L_2\text{NiX}_2$ changes dramatically with changing Z and Z'. For example, when both Z and Z' are changed from OMe to CF_3, the fraction of tetrahedral isomer decreases seven-fold. In general, as the substituent on the phosphine ligand becomes more electron-withdrawing, the fraction of tetrahedral isomer decreases. One explanation for this is based on the possibility of metal-to-ligand π bonding. As the groups attached to phosphorus become more electronegative (by changing Z and/or Z'), the phosphorus d orbitals become lower in energy and contract radially so as to overlap better with a metal orbital of π symmetry such as d_{xy} (as illustrated below) in the square planar complex. Since metal ion d orbitals in a tetrahedral complex overlap much less effectively with ligand π orbitals, in contrast to those in square planar complexes (see the values of the angular scaling factors in Table 9-4), metal-to-ligand π bonding can better stabilize the square planar isomer, and the latter will predominate.

P $d\pi \leftarrow$ M $d\pi$ (d_{xy})

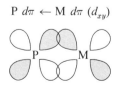

---

[10] L. H. Pignolet, W. D. W. Horrocks, Jr., and R. H. Holm, *J. Amer. Chem. Soc.*, **92**, 1855 (1970).

Other studies of configurational change have been based on the very clever use of stereochemical concepts.[11,12] One example is introduced here with the following objectives in mind: (a) to exemplify one experimental approach; (b) to show in an example simpler than a six-coordinate complex the possible paths for optical inversion in coordination complexes; and (c) to draw some general conclusions regarding stereochemical change in four-coordinate complexes.

If a tetrahedral complex is formed from unsymmetrical ligands [*e.g.,* (A—B)$_2$M] it will be chiral, and a pair of enantiomers with Δ and Λ configurations will be generated.

As mentioned in Chapter 10, many such complexes cannot be isolated in optically pure form, as they racemize quickly. Therefore, in order to study the mechanism of conversion of one enantiomer to the other, one cannot take one enantiomer and follow its rate of change to the racemic mixture by an ordinary method such as polarimetry [although this can be used for some tris complexes of Cr(III) and Co(III), as seen later in the discussion]. Instead, other methods must be used. Such a procedure is outlined for the case of complexes such as **2** and **3**.

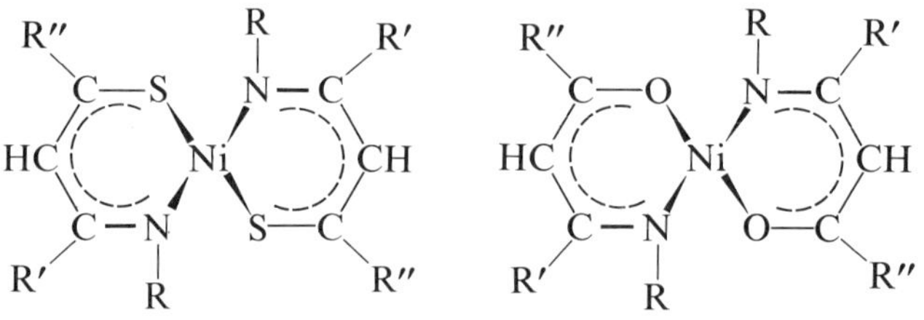

In order to follow the change in chirality of a compound such as **2**, Holm synthesized the complex with an isopropyl group on the coordinated imine nitrogen.[13]

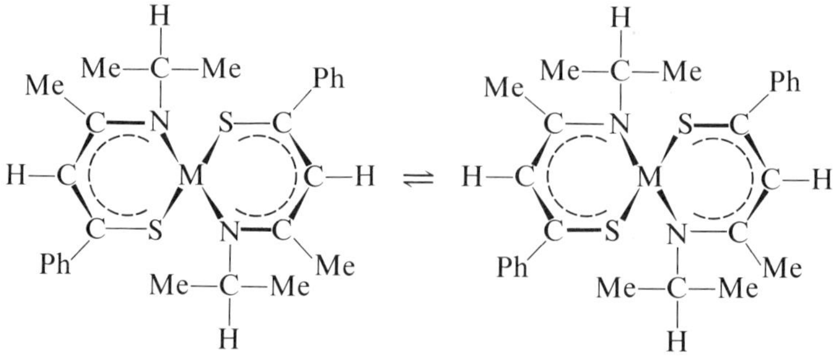

This allows the kinetics of the interconversion between **2a** and **2b** to be followed by proton nmr spectroscopy because the methyl groups of the isopropyl substituents are

---
[11] R. H. Holm, *Acc. Chem. Res.*, **2**, 307 (1969).
[12] S. S. Eaton and R. H. Holm, *Inorg. Chem.*, **10**, 1446 (1971).
[13] See ref. 8.

Figure 14-3. The isopropyl group configuration in the enantiomers of compound **2**. The isopropyl group is viewed down the $C_{isoprop}$—N_{imine} bond. Note that when the Λ enantiomer is converted to the Δ enantiomer, the methyl group environments are exchanged.

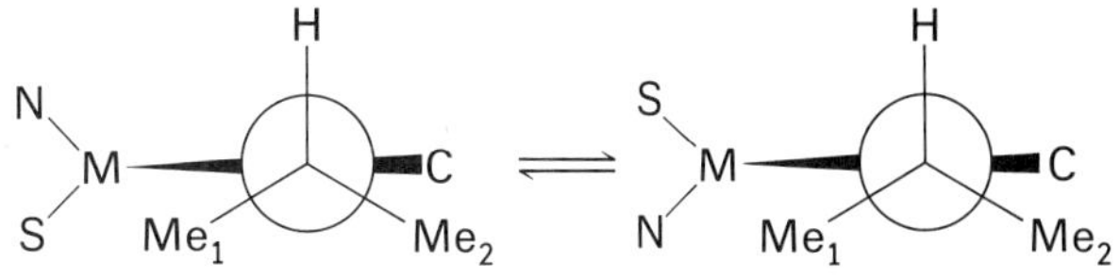

Λ enantiomer
Compound **2a**

Δ enantiomer
Compound **2b**

diastereotopic. Diastereotopic groups or atoms cannot be interchanged by any symmetry operation.[14] Thus, you see in Figure 14-3 that the isopropyl methyl groups in a particular enantiomer of **2** are diastereotopic. Because of this they cannot be magnetically equivalent (note their positions relative to S, for example), and so one methyl gives rise to a resonance line with a chemical shift different from the other methyl. However, when the Λ enantiomer is converted to the Δ enantiomer (and the isopropyl groups rotate), the methyl group environments are exchanged. Most importantly, though, if a twisting mechanism is assumed for the racemization, and the isopropyl groups are free to rotate, the isopropyl methyls will be enantiotopic[14,‡] in a planar transition state and will, therefore, have the same chemical shift. Thus, when the rate of Λ ⇌ Δ interconversion is rapid, the isopropyl methyl groups will have averaged environments, and the separate resonance lines for the different methyls in the two enantiomers will collapse to a single line. The rate of interconversion of methyl group environments, and therefore of Λ ⇌ Δ interconversion, may be obtained by the usual methods of nmr spectroscopy.

Three pathways for the inversion of configuration of compounds such as **2** and **3** are thought possible (Figure 14-4), and the first two of these are applicable to the octahedral case to be discussed in the next section. (i) The first pathway involves a twisting or flattening of the tetrahedral complex to a square planar complex and then

---

[14] A. L. Ternay, Jr., "Contemporary Organic Chemistry," W. B. Saunders Company, Philadelphia, 1976; K. Mislow and M. Raban, *Topics in Stereochemistry*, **1**, 1 (1967).

‡In rotamers where the isopropyl H is in the molecular plane of the planar transition state, the methyls above and below the plane can be exchanged by a plane of symmetry. They are now said to be enantiotopic, and are in equivalent environments where they will have the same chemical shift.

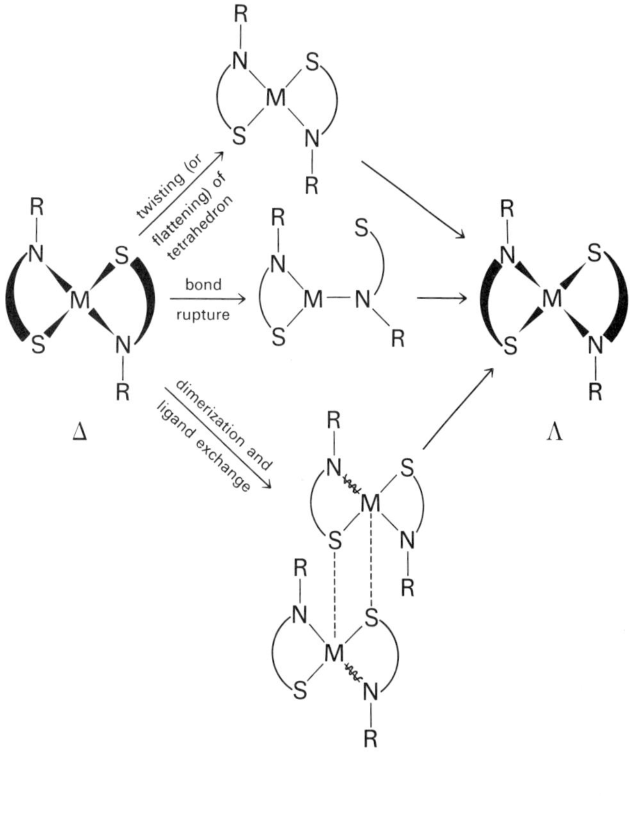

Figure 14-4. Three pathways leading to configuration inversion for a bis-chelate complex of the type (A—B)₂M.

further twisting to re-form the tetrahedron. (ii) The second path involves breaking a metal-to-ligand bond to give a three-coordinate intermediate; re-formation of the bond may occur to give a complex having either the original or the opposite chirality. (iii) Finally, the third path involves the formation of a complex between two square planar units *via* intermolecular metal-sulfur interactions. Breaking of a metal-nitrogen bond and re-formation of a metal-nitrogen bond with the other metal, followed by dissociation of the dimer, could lead to inversion with concomitant ligand exchange.

As in almost all studies of rearrangements, it is difficult if not impossible to determine the mechanism explicitly. The mechanism of rearrangement of compound **2** is no exception. In the case of complexes such as **2,** those of Zn(II) have rates of inversion that are considerably more rapid than the rate of ligand exchange between two complexes. This means that inversion is probably an intramolecular process, a conclusion that rules out significant contribution from the third path in Figure 14-4. On the other hand, Cd(II) complexes of type **2** have comparable kinetic parameters for ligand exchange between complexes and for optical inversion. Although this does not rule out any of these pathways, the fact that ligand exchange (and presumably optical inversion) in the Cd(II) complexes is concentration-dependent suggests the occurrence of the third path, perhaps in addition to either or both of the others.

The general conclusions that may be drawn from this example of stereochemical change in complexes such as **2** and **3** relate to the rate of rearrangement. First, it has been observed that the rate increases with increasing radius of the coordinated metal ion. That is, Zn(II) complexes ($r = 0.74$ Å) rearrange less rapidly than Cd(II) complexes ($r = 0.97$ Å). Indeed, this has also been observed for many d^0 and d^{10} octahedral complexes. Second, on comparing inversion rates of Zn—N_2S_2 complexes with Zn—N_2O_2, the latter invert more slowly than the former. This suggests a twisting mechanism because N—S ligands generally stabilize the square planar geometry to a greater extent than do N—O ligands. Third, if Ni(II) complexes such as **3** are involved in the tetrahedral–square planar equilibrium, it has not yet proved possible to slow the rate of inversion, even at low temperatures, to a point at which the nmr spectrum of each isomer may be observed.[15] This is, of course, in decided contrast with the L_2NiX_2 complexes discussed above (see Figure 14-2). Fourth, ligands such as the β-ketoamines (the ligand in **3**) which stabilize a measurable amount of Ni(II) in the tetrahedral form will produce Co(II) complexes that are 100% tetrahedral; conversely, these same ligands will produce Ni(II) complexes that are 100% planar if they give measurable amounts of planar Co(II) complexes.[15] This is in agreement with the prediction of the angular overlap model. That is, Co(II) is a d^7 ion and, as such, there should be no distinction between square planar and tetrahedral geometries based on d_σ orbital energies (see Figure 9-19). You recall that when this is the case, ligand-ligand repulsions determine geometry, and the tetrahedron is preferred for four-coordinate complexes. It is for this reason that Co(II) forms more tetrahedral complexes than any other 2+ transition metal ion. Ni(II) complexes, on the other hand, are predicted to show some preference for square planarity. However, this preference is not overwhelming (ca. 1.3 e_σ units), so planar and tetrahedral isomers are observed.

SIX-COORDINATE OCTAHEDRAL COMPLEXES

Intramolecular rearrangements in six-coordinate complexes are more complicated than those of four-coordinate ones, and great experimental ingenuity is required to obtain useful information.[16-21] Much of the work that has given some

---

[15] G. W. Everett, Jr., and R. H. Holm, *Inorg. Chem.*, **7,** 776 (1968); the planar Co(II) complexes are low spin.

[16] See refs. 1 and 2.

[17] J. G. Gordon, II, and R. H. Holm, *J. Amer. Chem. Soc.*, **92,** 5319 (1970).

[18] R. C. Fay and T. S. Piper, *J. Amer. Chem. Soc.*, **85,** 500 (1963).

[19] A. Y. Girgis and R. C. Fay, *J. Amer. Chem. Soc.*, **92,** 7061 (1970).

[20] M. C. Palazzotto, D. J. Duffy, B. L. Edgar, L. Que, Jr., and L. H. Pignolet, *J. Amer. Chem. Soc.*, **95,** 4537 (1973).

[21] J. I. Musher, *Inorg. Chem.*, **11,** 2335 (1972); S. S. Eaton and G. R. Eaton, *J. Amer. Chem. Soc.*, **95,** 1825 (1973). These papers and the references therein discuss molecular rearrangements in a more theoretical and mathematical way.

definition to the problem of intramolecular rearrangements in six-coordinate octahedral complexes has been done with metal hydrides such as **4** or with complexes having unsymmetrical, chelating ligands attached to a trivalent transition metal ion such as Co(III) or Cr(III) or to a main group metal such as Al(III). Compound **5** is an example of this latter type,[17] the type that we shall discuss as an example of molecular rearrangements of octahedral complexes.

For a complex such as **5**, two diastereoisomers, each with an enantiomer, are possible: *cis* Δ and *cis* Λ and *trans* Δ and *trans* Λ. In principle, each can be converted by an intramolecular process into any or all of the others.

The possible interconversions are mapped out in Figure 14–5. These interconversions can occur by twisting and/or bond-breaking/bond-making processes (cf. the mechanisms for rearrangement in four-coordinate complexes in Figure 14–4), and the question with which we are now confronted concerns the mechanism of interconversion. Later, as an aside, we shall discuss an actual case study.

Before looking at the intramolecular twist, the first of several possible rearrangement mechanisms, you must be aware of the symmetry differences between the

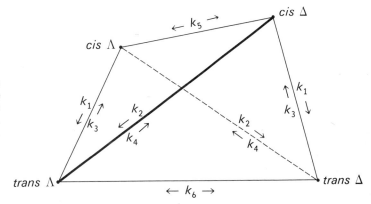

Figure 14–5. Interconversions between the possible stereoisomers of a complex of the type (unsym-chel)₃M. Rate constant designations given according to J. G. Gordon, II, and R. H. Holm, *J. Amer. Chem. Soc.*, **92**, 5319 (1970).

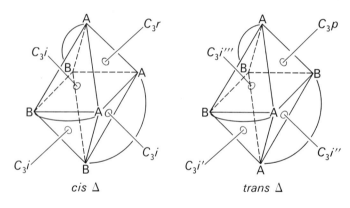

Figure 14-6. The *cis* Δ and *trans* Δ enantiomers of a complex of the type (unsym-chel)$_3$M = (A-B)$_3$M showing the real (C_3r) and imaginary (C_3i) axes of the *cis* complex and the pseudo (C_3p) and imaginary (C_3i') axes of the *trans* complex.

cis and *trans* diastereoisomers of M(AB)$_3$ (Figure 14–6). *Cis* complexes have a real C_3 axis (= C_3r) passing through the octahedral faces at which the three A's and the three B's are located. There are also three symmetry-equivalent, "imaginary" C_3 axes (= C_3i) passing through the other three sets of faces. On the other hand, the *trans* isomer has only a pseudo-C_3 axis (= C_3p) and three non-equivalent, imaginary C_3 axes (= C_3i', for example).

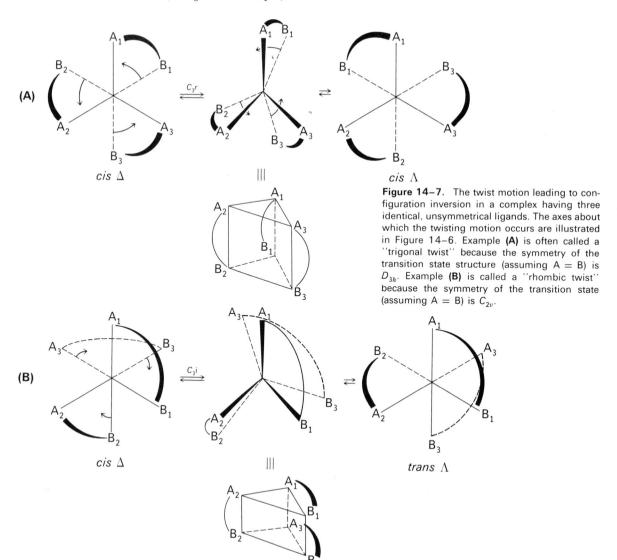

Figure 14–7. The twist motion leading to configuration inversion in a complex having three identical, unsymmetrical ligands. The axes about which the twisting motion occurs are illustrated in Figure 14–6. Example **(A)** is often called a "trigonal twist" because the symmetry of the transition state structure (assuming A = B) is D_{3h}. Example **(B)** is called a "rhombic twist" because the symmetry of the transition state (assuming A = B) is C_{2v}.

In the twist mechanism for rearrangement, one trigonal face of the octahedron is rotated with respect to the opposite face about one of the C_3 axes.[22] The transition state structure is the other six-coordinate polytope, the trigonal prism, and further rotation about the same axis leads to a complex having the opposite chirality. Figure 14-7 depicts this process for rotation about the real and imaginary axes of the *cis* Δ enantiomer. Rotation about C_3r results in inversion of chirality, and rotation about C_3i gives rise to both chirality inversion and *cis-trans* isomerization. These results are summarized on a topological map in Figure 14-8, along with the results for twisting motions about the four different C_3 axes of a *trans* isomer. A very important conclusion may be drawn from Figure 14-8: *The intramolecular twist always leads to an inversion of chirality but not necessarily to cis-trans isomerization.* Isomerization without racemization is not possible by the twist mechanism. To accomplish this, another process, bond rupture, must be postulated.

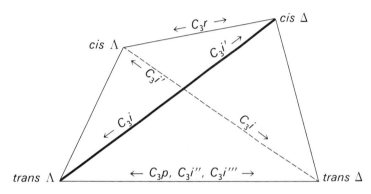

Figure 14–8. The possible intramolecular rearrangements for the *cis* Δ and *trans* Δ enantiomers of a complex $(A-B)_3M$ according to the twist mechanism.

Bond rupture can lead to two different intermediate structures: the trigonal bipyramid (TBP) and the square pyramid (SP). In the case of bond rupture leading to a TBP intermediate, the five-coordinate complex may have either a "dangling axial ligand" or a "dangling equatorial ligand." These choices are illustrated in Figure 14-9 for the *cis* Λ enantiomer. Because of the real C_3 symmetry of a *cis*-M(AB)$_3$ complex, there are two types of bonds that may be broken (*a* and *b*), and, as a result, there are four possible intermediates; only two (formed by breaking bond *b*) are illustrated. (In the case of the *trans* isomer, there are six distinguishable bonds, and 12 intermediates may form; see Study Question 2.) Complete analysis (an admittedly lengthy and laborious process) of the path proceeding through a TBP intermediate leads to several important conclusions. Among these are (see Figure 14-10 for a summary):

(i) Each of the enantiomers may directly convert to any of the others by a TBP rupture, except *cis* Λ to *cis* Δ.

(ii) "Equatorial" intermediates give only isomerization; no inversion can occur.

Intramolecular rearrangement by the path involving square pyramidal intermediates is illustrated in Figure 14-11, also for the *cis* Λ enantiomer. In this mechanism there are so-called "primary" and "secondary" routes, these being distinguished by which end of the ligand migrates to fill the coordination position opened by bond rupture. The results of a complete analysis of this route are also included in Figure 14-10. Most importantly, notice that the SP routes can lead to interconversion of all isomers. This is in direct contrast with the other paths.

Now we can turn to an actual example of the determination of a rearrangement mechanism. The main questions are (i) how is this to be done and (ii) does rearrangement proceed by all paths or by one or two in particular? Experimental details are considered in the small type section that follows.

The first requirement for the determination of a rearrangement mechanism in a compound of the type $M(A-B)_3$ is to isolate, in as pure a form as possible, the four possible isomers: *cis* and *trans* Δ and *cis* and *trans* Λ. Complexes of Co(III) and Cr(III)

---

[22] J. C. Bailar, Jr., *J. Inorg. Nucl. Chem.*, **8**, 165 (1958).

768 ▢ MOLECULAR REARRANGEMENTS AND REACTIONS OF COORDINATED LIGANDS

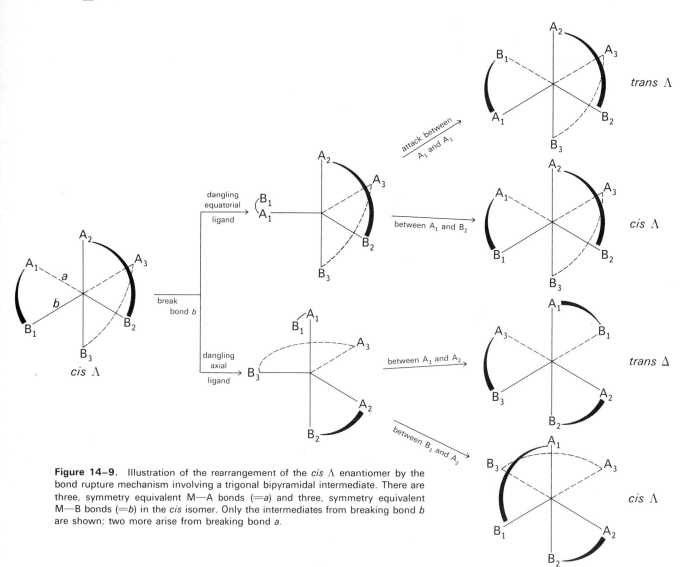

Figure 14-9. Illustration of the rearrangement of the *cis* Λ enantiomer by the bond rupture mechanism involving a trigonal bipyramidal intermediate. There are three, symmetry equivalent M—A bonds (=a) and three, symmetry equivalent M—B bonds (=b) in the *cis* isomer. Only the intermediates from breaking bond b are shown; two more arise from breaking bond a.

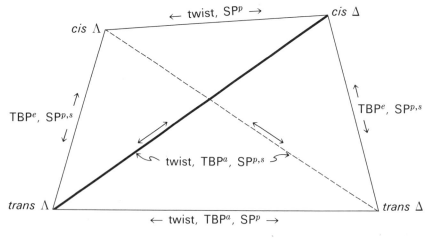

Figure 14-10. A summary of the interconversions possible for a complex of the type (A—B)$_3$M by three possible mechanisms.

twist = twist mechanism (Figures 14-7 and 8)

TBP = rearrangement through bond rupture with trigonal bipyramidal intermediates (Figure 14-9).

  a = dangling axial ligand and e = dangling equatorial ligand

SP = rearrangement through bond rupture with square pyramidal intermediates (Figure 14-11).

  p = primary process and s = secondary process

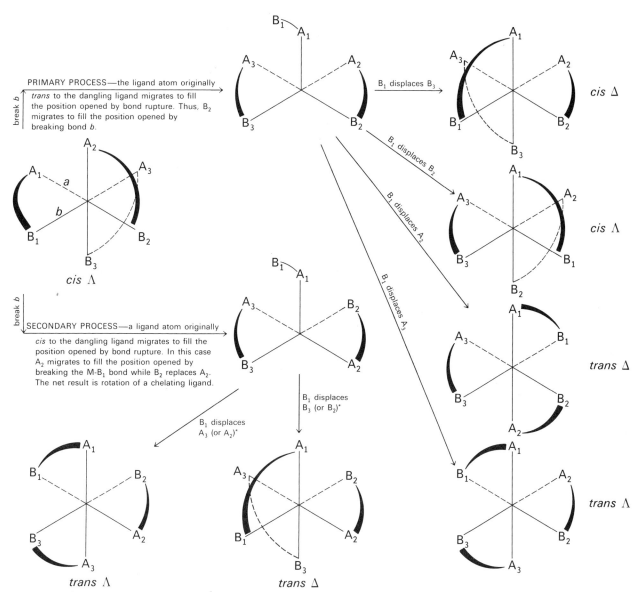

Figure 14-11. Illustration of the rearrangement of the *cis* Δ enantiomer by the bond rupture mechanism involving a square pyramidal intermediate. As in Figure 14-9, only intermediates from breaking bond *b* are shown. (*If B_1 displaces A_2 or B_2, *trans* Λ and *trans* Δ isomers, respectively, are generated. However, A_1 is now *trans* to A_2 or B_2, respectively.)

are often relatively stable to isomerization and racemization [unlike those of main group metals such as Al(III)] and so, in some cases, separation of diastereoisomers, and partial separation of enantiomers, has been achieved. As an example, consider the cobalt(III) complex of 5-methylhexane-2,4-dione (**6**).[23] The ligand, a yellow oil, is allowed to react with $Na_3Co(CO_3)_3 \cdot 3H_2O$ in refluxing acetone/water; the Co(III) ion is released from the salt by careful addition of HNO_3 to give $Co(mhd)_3$ (**7**) as a green oil. The *cis* and *trans* isomers of the complex are separated by chromatography on acid-washed alumina with a hexane-benzene mixture. As in all previously observed cases, the *trans* isomer

---

[23] See ref. 17.

(smaller dipole moment) moves down the column more rapidly than the *cis* isomer. The enantiomers of each diastereomer can then be separated to a small extent (1 to 10% optical purity) by column chromatography with hexane on a column packed with D-(+)-lactose.

$$3 \quad \begin{array}{c} H_3C \\ HC-C-C=C-CH_3 \\ H_3C \end{array} + Co(III) \xrightarrow[\text{ligand}]{-H^+ \text{ from}}$$

6

7 = Co(mhd)$_3$

The rate of *cis-trans* isomerization can be measured by nmr methods. The reason for the efficacy of the method is related to the symmetry of the two diastereomers. As noted in Figure 14-6, *cis* isomers have a real C_3 axis, so the CH$_3$ groups in **7** or **8a**, for example, are magnetically equivalent and should give rise to a single resonance line in the nmr spectrum of the complex.

8a **8b**

cis- and *trans*-[Ru(F$_3$C—CO—CH=CO—CH$_3$)$_3$]

On the other hand, the *trans* isomer has only a pseudo-C_3 axis, and each of the CH$_3$ groups is magnetically different; three resonance lines should therefore be observed. As seen in Figure 14-12, this is indeed the case for the *cis* and *trans* isomers of **8**.[24] At higher

---

[24] J. G. Gordon, II, M. J. O'Connor, and R. H. Holm, *Inorg. Chem. Acta*, **5**, 381 (1971).

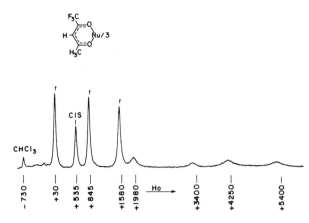

Figure 14-12. 100 MHz proton nmr spectrum of an equilibrium mixture of *cis*- and *trans*-[Ru(CF$_3$—CO—CH=CO—CH$_3$)$_3$] in CDCl$_3$ at 30°C (compounds **8a** and **8b**). The three CH$_3$ resonance lines for the *trans* isomer are marked with a *t*. The chemical shifts (in Hz) are very large because the ligand is coordinated to the paramagnetic center, Ru(III) (cf. Figure 14-2). [Adapted from J. G. Gordon, II, M. J. O'Connor, and R. H. Holm, *Inorg. Chim. Acta*, **5**, 381 (1971).]

temperatures the rate of *cis-trans* isomerization becomes more rapid, and the methyl groups swap environments so rapidly that they are no longer differentiated by the nmr spectrometer; the previously different methyl resonance lines collapse to a single line. Well-defined methods of nmr spectroscopy can be used to derive the rates of isomerization at various temperatures between the limits of very slow and very rapid isomerization.

The rate of racemization can be measured only by methods that detect chiral molecules or changes in chirality. Thus, such rates are generally measured by polarimetry.

Nmr and polarimetric methods were used to obtain the rates of racemization and isomerization by various paths for compound **7**. Taking the rate constant for the *cis* Λ → *trans* Δ (or *cis* Δ → *trans* Λ) conversion as equal to 100, relative values of rate constants were calculated and are summarized in Figure 14-13. Matching these values with the summary of mechanistic pathways in Figure 14-10 leads to some conclusions as to the probable mechanism of rearrangement.

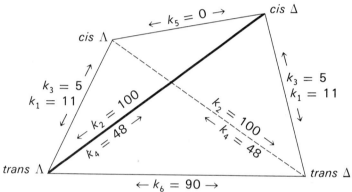

Figure 14-13. Relative rate constants for the stereoisomerization of Co(mhd)$_3$, compound **7**. Compare with the summary of possible mechanistic pathways in Figure 14-10.

(i) k_5 is zero. Thus, the twist mechanism is not used. If this mechanism were operative, the *cis* Δ to *cis* Λ conversion would be possible, and a finite value of k_5 would have been observed.

(ii) The bond rupture mechanism with SP intermediates is relatively unimportant. Certain branches of this path can lead to the unobserved ($k_5 = 0$) *cis* Δ to *cis* Λ conversion. Furthermore, other branches (k_1 and k_3) of the path lead to isomerization without racemization, but the small relative values of k_1 and k_3 imply that these branches are little used.

(iii) The bond rupture mechanism with trigonal bipyramidal–equatorial intermediates (TBP^e in Figure 14-10) is also relatively unimportant. Isomerization without racemization can occur only through this mechanism and those already deemed unimportant in paragraph (ii).

(iv) Having completely or largely eliminated the twist and SP pathways, it can be concluded that isomerization and racemization of **7** occurs primarily through bond rupture with a TBP-axial intermediate.

STUDY QUESTIONS

1. The twisting mechanism for rearrangements of octahedral complexes was shown only for a *cis* isomer. Taking the *trans* Δ isomer below, show the result of the twisting mechanism for the C_3p and C_3i' axes.

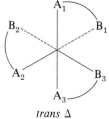

trans Δ

2. Figure 14-9 illustrates the rearrangement of the *cis* Λ enantiomer by the bond rupture mechanism with TBP intermediates. What products arise if bond *a* is broken in this enantiomer and a TBP-equatorial intermediate forms?

3. Taking the *trans* Λ enantiomer below, and considering the bond rupture mechanism with TBP intermediates, show that only axial intermediates will lead to inversion of chirality and that equatorial intermediates will result in isomerization.

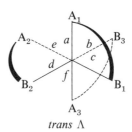

trans Λ

4. The rates of *cis-trans* isomerization of metal complexes of trifluoroacetylacetonate (*e.g.*, compound **5**) have been studied extensively, and these rates decrease in the order (Fe,In) > Mn > Ga > Al > V > Co > Ru > Rh. Assuming for the moment that the isomerization occurs through the twist mechanism (*i.e.*, it passes through a trigonal prismatic transition state), relate the relative rates of isomerization to the structure preference energies for the two possible six-coordinate polytopes (see Chapters 9 and 10).

5. A considerable amount of work has also been done on complexes of the type (A—A)(B—B)$_2$M, of which the two below are examples.

a) If one examines the fluorine nmr spectrum of **A** or the proton nmr spectrum of **B**, two different types of CF$_3$ or CH$_3$ groups, respectively, are seen at lower temperatures. Why should this be the case?

b) At higher temperatures, the separate lines for the two different types of methyl groups coalesce into a single line, indicating that the methyl groups are rapidly exchanging their environments. Originally this was assumed to be due to optical inversion of the complex [J. J. Fortman and R. E. Sievers, *Inorg. Chem.*, **6**, 2022 (1967)]. More recently, however, it has been found that methyl group scrambling can occur both with and without inversion, and the most likely mechanism is thought to be the SP bond rupture path using primary processes only. The objective of this study question is to discover how this may occur. To help you along the way, first recognize that there are four possible ways of depicting a Δ enantiomer of an (A—A)(B—B)$_2$M complex. One way is shown on p. 773 and is called the Δ$_{23}$ enantiomer because the B groups labeled 2 and 3 are *trans* to the ligand A—A.

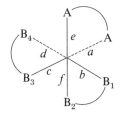

As a first step, write down the three other possible "isomers" of this enantiomer, and notice that each leads to a different arrangement of the B groups with respect to Δ_{23}. Next, break one of the bonds in Δ_{23} to form an SP intermediate by allowing the end of the ligand *trans* to the dangling group to migrate to the coordination site vacated by breaking the bond. Allow products to form by the so-called primary process (Figure 14-11). Notice that breaking bonds b, c, d, or f generates all of the other "isomers" that you drew, some in the Δ form and others in the Λ form. Breaking any of the six bonds, however, will give an "isomer" labeled Λ_{14}, and the reaction $\Delta_{23} \rightleftharpoons \Lambda_{14}$ is thought to be the major process in B group scrambling in complexes of the type (A—A)(B—B)$_2$M.

REACTIONS AT COORDINATED LIGANDS[25-29]

We come finally to the periphery of the coordination complex and to reactions that occur at the coordinated ligands themselves. It is obvious that a metal ion should be able to influence strongly the reactions at coordinated ligands, as compared with the uncoordinated ligand, by causing changes in intra-ligand charge distribution (both by σ withdrawal and by π donation or withdrawal). This may lead to enhancement of a particular reaction of the coordinated ligand, force a given reaction to occur by a different mechanism, or give a product not attainable without prior metal coordination. In addition, metal coordination may have a "template effect"; that is, the coordination of ligands may place them in the correct geometry for their later union, thereby overcoming, by means of a favorable ΔH for reaction, an unfavorable ΔS for reaction in the absence of coordination. Alternatively, metal coordination may stabilize a particular tautomer, a form not stable or of little consequence without metal coordination.

In the discussion that follows, we intend to outline only a few examples of the ways in which reaction can occur at coordinated ligands. These have been grouped into two broad classifications: (i) reactions occurring because of metal ion polarization of the coordinated ligand and (ii) those in which a "template effect" is clearly operative. The one class of reactions not really covered below consists of those involving true catalysis, that is, reactions involving the activation and transformation of small molecules as in the Wacker process (14-3), for example.

$$[PdCl_4]^{2-} + C_2H_4 + H_2O \rightarrow CH_3CHO + Pd + 4Cl^- + 2H^+ \tag{14-3}$$

Instead, these are covered in considerably more detail in Chapter 16 on reactions of organometallic compounds.

---

[25] J. P. Collman, "Reaction of Ligands Coordinated with Transition Metals," in "Transition Metal Chemistry," Vol. 2, R. L. Carlin, ed., Marcel Dekker, New York (1966).
[26] Q. Fernando, *Adv. Inorg. Chem. Radiochem.*, **7**, 185–261 (1965).
[27] M. M. Jones, "Ligand Reactivity and Catalysis," Academic Press, New York (1968).
[28] L. F. Lindoy and D. H. Busch, "Complexes of Macrocyclic Ligands," in "Preparative Inorganic Reactions," Vol. 6, W. L. Jolly, ed., Wiley-Interscience, New York (1971).
[29] L. F. Lindoy, *Chem. Soc. Rev.*, **4**, 421 (1975).

REACTIONS DUE TO METAL ION POLARIZATION OF COORDINATED LIGANDS

Hydrolysis of Amino Acid Esters and Amides and of Peptides

Carboxypeptidase A is a zinc-containing digestive enzyme that hydrolyzes the carboxyl-terminal peptide bond in polypeptide chains.[30]

(14-4)

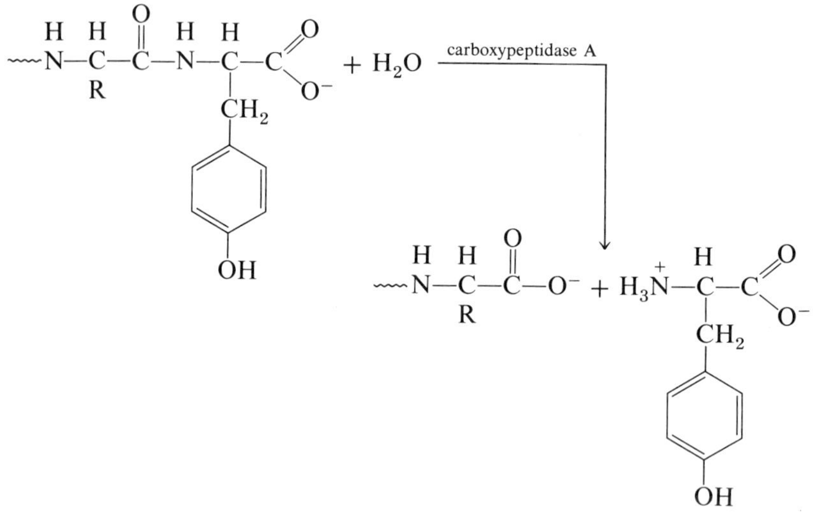

Its structure has recently been determined and its functioning has been probed.[31] On the basis of this work, one suggested mechanism involves metal ion polarization of a peptide carbonyl group, labilizing the carbon to attack by a neighboring carboxyl oxygen (from glutamine, residue number 270).

(14-5)

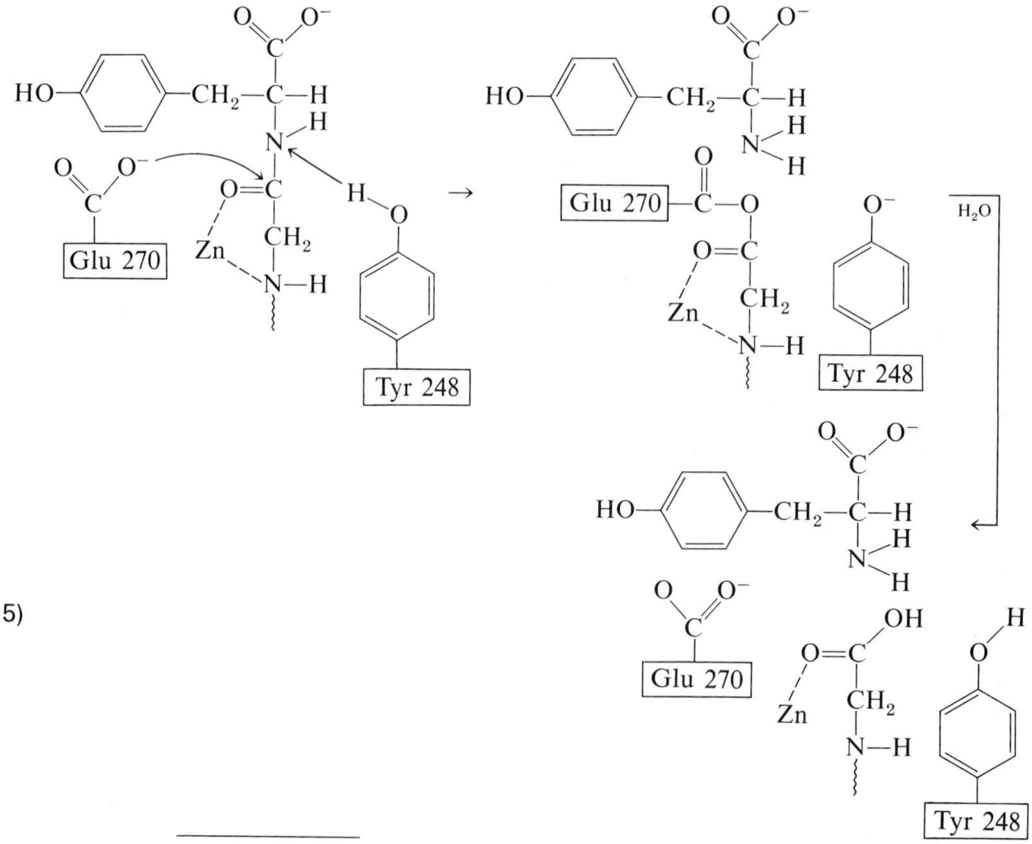

---

[30] L. Stryer, "Biochemistry," W. H. Freeman and Company, San Francisco (1975).
[31] W. N. Lipscomb, *Acc. Chem. Res.*, **3**, 81 (1970).

This is by no means yet the proved mechanism by which the enzyme operates, so a great deal of work is continuing in enzyme studies; the discussion that follows relates to inorganic model systems.

Esters of organic acids hydrolyze under acidic and basic conditions according to the overall reaction 14-6:

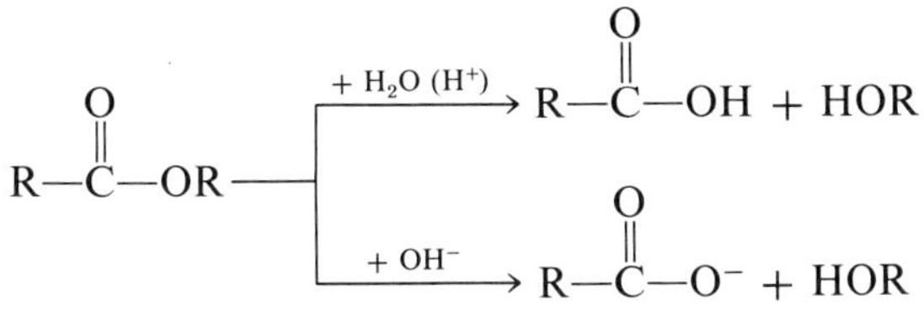

$$R-C(=O)-OR \xrightarrow{+H_2O\,(H^+)} R-C(=O)-OH + HOR \quad\quad R-C(=O)-OR \xrightarrow{+OH^-} R-C(=O)-O^- + HOR \tag{14-6}$$

The detailed mechanism for this process involves, as a first step, the attack of water on the carbonyl carbon (14-7) to produce an unstable intermediate. Although the next step (14-8) then appears to involve intramolecular proton transfer, in reality the coordinated water probably transfers a proton to the solvation shell, and the carbonyl group obtains a proton from the solvent. The formation of the intermediate by reactions 14-7 and 14-8 usually represents the rate determining step in the overall hydrolysis reaction. From this point on, the formation of products is rapid (reaction 14-9).

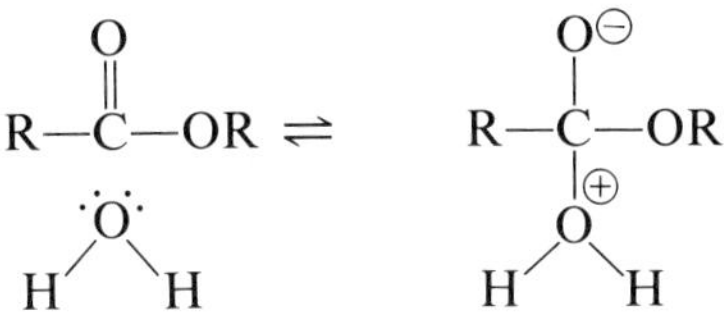

$$\underset{\underset{H\,\,\,\,H}{\overset{\cdot\cdot}{O}}}{R-\overset{O}{\underset{\|}{C}}-OR} \rightleftharpoons \underset{\underset{H\,\,\,\,H}{\overset{\oplus}{O}}}{R-\overset{O^{\ominus}}{\underset{|}{C}}-OR} \tag{14-7}$$

very unstable intermediate

$$\underset{\underset{H\,\,\,\,H}{\overset{\oplus}{O}}}{R-\overset{O^{\ominus}}{\underset{|}{C}}-OR} \rightleftharpoons \underset{OH}{R-\overset{OH}{\underset{|}{C}}-OR} \tag{14-8}$$

unstable intermediate

$$\underset{OH}{R-\overset{OH}{\underset{|}{C}}-OR} \xrightleftharpoons{1,3\text{-shift}} \underset{\underset{OH}{\overset{\oplus}{}}}{R-\overset{O^{\ominus}}{\underset{|}{C}}-\overset{H}{\underset{}{O}}R} \rightarrow R-\overset{O}{\underset{\|}{C}}-OH + HOR \tag{14-9}$$

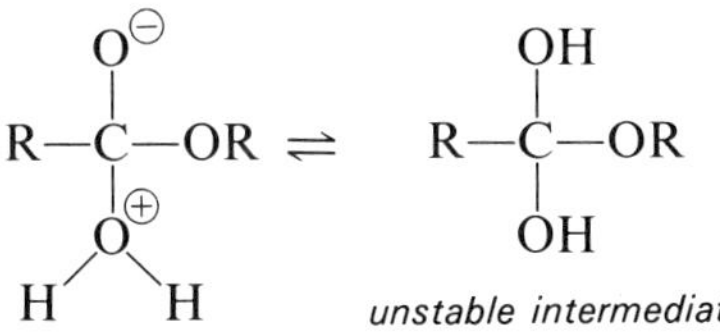

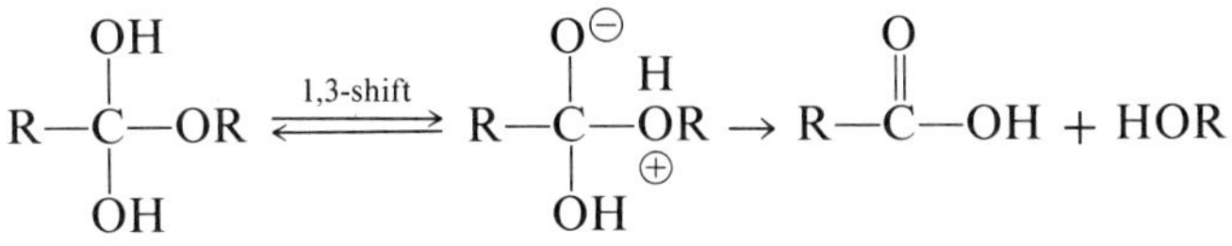

Acids and bases can catalyze the ester hydrolysis reaction in several ways. An acid, say the proton, may attack the carbonyl oxygen (reaction 14-10) to give a

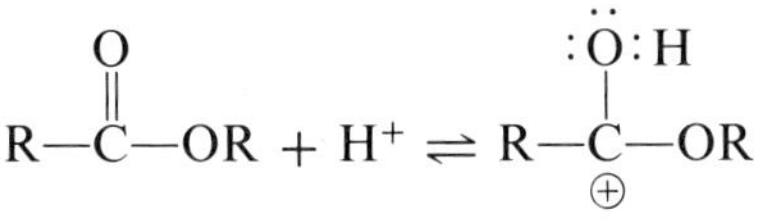

$$R-\overset{O}{\underset{\|}{C}}-OR + H^+ \rightleftharpoons R-\overset{\overset{\cdot\cdot}{O}:H}{\underset{\oplus}{\underset{|}{C}}}-OR \tag{14-10}$$

$$R-\underset{+\delta}{\overset{\overset{-\delta}{\overset{\cdot\cdot}{:O:}}}{\underset{\|}{C}}}-OR \qquad\qquad R-\underset{\underset{H}{\overset{\oplus}{O}-H\cdots OH^-}}{\overset{\overset{\cdot\cdot}{:O:^{\ominus}}}{\underset{|}{C}}}-OR$$

$$\,\searrow OH^-$$

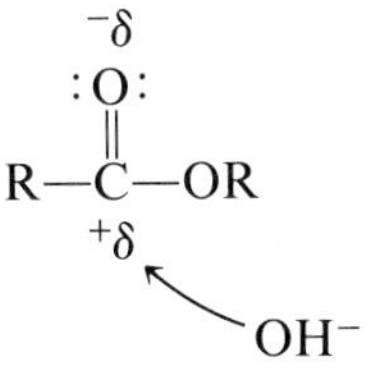

9 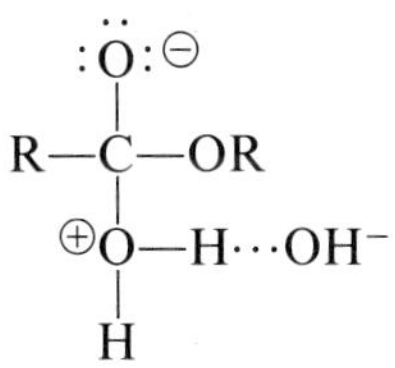 10

complex in which the electrophilicity of the carbonyl carbon is enhanced. This facilitates the formation of the intermediate in reaction 14-7. On the other hand, a base such as OH^- can be the nucleophile in reaction 14-7 (see structure **9**), or it can assist in removal of the proton in reaction 14-8 (see structure **10**).

A metal ion such as Co^{2+} or Cu^{2+} is a Lewis acid, and it might also be expected to catalyze the hydrolysis of esters. In fact, since many common ions have a $+2$ charge and are therefore more highly charged than a proton, they might be expected to enhance ester hydrolysis rates relative to H^+. However, this is not the case with simple organic esters. To explain this, two effects should be considered: the σ and the π polarization of $\rangle C=O$ by M^{2+} as compared with H^+. The σ polarization of the carbonyl group will depend on the extent of HOMO (oxygen lone pairs) and LUMO (H^+ or M^{2+} σ acceptor orbital) interaction. This interaction is clearly greater with H^+ than with M^{2+}, as there will be better HOMO/LUMO energy match between the O lp and H^+ as well as better overlap (the H $1s$ orbital is less diffuse than a metal σ acceptor orbital). And then there is the π effect. The diagram below illustrates the effect on the π system of M^{2+} coordination at the carbonyl oxygen. The most important feature to notice is the π^* LUMO, because it is this orbital that is under attack by water or hydroxide ion in the hydrolysis reaction. When attached to a transition metal with occupied π orbitals, the LUMO is raised in energy and delocalized onto $d\pi$. Both factors tend to inhibit "bond formation" between the nucleophile HOMO and the π^* LUMO, thereby offsetting the otherwise rate-enhancing effect of π polarization: $\overset{\oplus}{\rangle}C-\overset{..}{\underset{\ominus}{O}}$

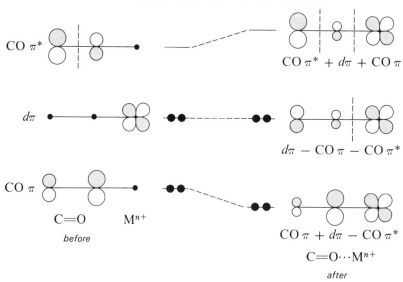

Metal $d\pi$ + carbonyl orbitals before and after interaction

In contrast with simple organic esters, amino acid esters are hydrolyzed very rapidly in the presence of ions such as Cu^{2+}, Co^{2+}, and Mg^{2+}, even at approximately neutral pH's where the amino acid esters are normally quite stable to hydrolysis.[32-38] As an example, consider the results in reaction 14-11 for the hydrolysis of phenylalanine ethyl ester.

---

[32] H. Kroll, *J. Amer. Chem. Soc.*, **74**, 2036 (1952).

[33] M. L. Bender and B. W. Turnquest, *J. Amer. Chem. Soc.*, **79**, 1889 (1957).

[34] H. L. Conley, Jr., and R. B. Martin, *J. Phys. Chem.*, **69**, 2914, 2923 (1965).

[35] J. E. Hix and M. M. Jones, *Inorg. Chem.*, **5**, 1863 (1966).

[36] M. L. Bender, "Mechanisms of Homogeneous Catalysis from Protons to Proteins," Wiley-Interscience, New York (1971).

[37] See ref. 27.

[38] In contrast with the catalysis of amino acid ester hydrolysis by metal ions, the hydrolysis of simple esters is not affected by metal ions. For example, Cu(II) neither interacts with ethyl acetate nor effects its hydrolysis. Apparently a simple ester cannot displace H_2O from the M^{2+} coordination sphere as effectively as a chelating amino acid ester (*cf.* Chapter 13). M. L. Bender and B. W. Turnquest, *J. Amer. Chem. Soc.*, **79**, 1656 (1957).

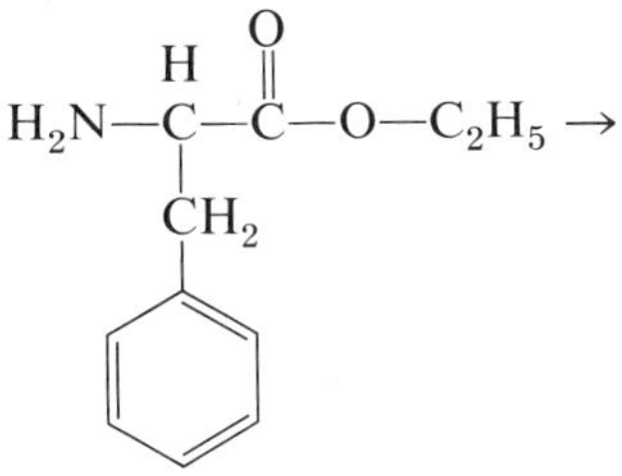

$$H_2N-\underset{CH_2(C_6H_5)}{\overset{H}{C}}-\overset{O}{\underset{}{C}}-O-C_2H_5 \rightarrow \begin{Bmatrix} \text{Catalyst} & k\ (\text{sec}^{-1})\ (\text{when [ester]} = 0.0138\ M) \\ H^+\ (pH = 2.3) & 1.46 \times 10^{-11} \\ OH^-\ (pH = 7.3) & 5.8 \times 10^{-9} \\ Cu^{2+}\ (0.0775\ M) & 2.7 \times 10^{-3} \end{Bmatrix} \quad (14\text{-}11)$$

11

$$H_2N-\underset{CH_2(C_6H_5)}{\overset{H}{C}}-\overset{O}{\underset{}{C}}-OH$$

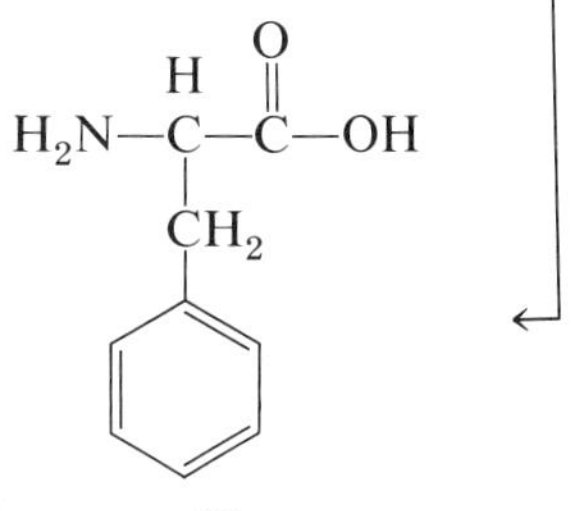

12

Such hydrolysis reactions have been studied extensively, and the evidence is overwhelming that a metal ion–amino acid complex such as **13** is the kinetically important species and that the rate determining step in the hydrolysis is the attack of water or hydroxide on the 1:1 complex.

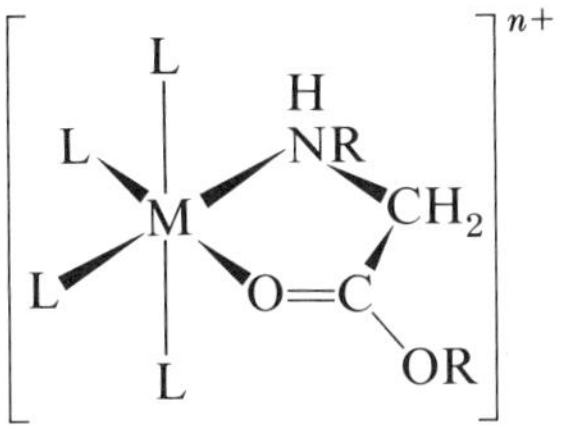

13

For a variety of metal ion catalyzed hydrolyses of amino acid esters in the pH range from 7 to 9, the overall rate is given by

$$\text{Rate} = k_{\text{OH}}[\text{ME}^{2+}][\text{OH}^-] + k_{\text{H}_2\text{O}}[\text{ME}^{2+}][\text{H}_2\text{O}]$$

where ME^{2+} is the 1:1 metal-ester complex. The specific rate constants, k_{OH} and $k_{\text{H}_2\text{O}}$, have been determined for ethyl glycinate hydrolysis with the metal ions Co^{2+}, Ni^{2+}, Cu^{2+}, and Zn^{2+}. These are listed, along with the stability constants for the 1:1 metal-ester complexes, in Table 14–1. There are two main points of interest concerning these data: (i) the values of k_{OH} are about 10^9 greater than $k_{\text{H}_2\text{O}}$ (because OH^- is a better nucleophile than H_2O), and (ii) the values of k_{OH} more or less increase with

TABLE 14–1
SPECIFIC RATE CONSTANTS FOR THE Co^{2+}, Ni^{2+}, Cu^{2+}, AND Zn^{2+} CATALYZED HYDROLYSIS OF ETHYL GLYCINATE AND THE STABILITY CONSTANTS FOR THE M^{2+}:ETHYL GLYCINATE COMPLEXES[a]

Metal Ion	$10^{-4}\ k_{\text{OH}}$ $M^{-1}\ \text{sec}^{-1}$	$10^6\ k_{\text{H}_2\text{O}}$ $M^{-1}\ \text{sec}^{-1}$	$K_{\text{stability}}$ (for 1:1 complex)
Co^{2+}	0.99	8.3	26.8
Ni^{2+}	0.398	3.5	198
Cu^{2+}	7.64	0.77	7400
Zn^{2+}	2.33	7.07	62

[a] J. E. Hix, Jr., and M. M. Jones, *Inorg. Chem.*, **5**, 1863 (1966).

complex stability. It is the latter point that may enable us to propose an explanation for the fact that great rate acceleration for hydrolysis is observed only when an amino acid ester chelates with a metal ion. When chelation with a metal ion occurs, resonance structures **14** and **15** are important. The C—O—M angle in **14** is in theory (*i.e.,* VSEPR) close to 120°, an angle leading to considerable ring strain. On the other hand, the C—O—M angle in **15** is about 109°, an angle much closer to the theoretical angle (108°) for a symmetric five-membered chelate ring. Thus, ring strain is reduced in **15** relative to **14**, and we presume that **15** becomes the more important resonance structure as complex stability increases.[38a] It seems logical to conclude that, as **15** becomes more important, the electrophilicity of the carbonyl carbon increases, and the rate of OH⁻ attack is increased.

More recent work on the hydrolysis of amino acid esters and amides and of peptides has been done with more kinetically inert Co(III) complexes. For example,[39] complexes such as **16** are known to be stable to hydrolysis for several hours in aqueous solution, even at low pH.

$X = Cl^-, Br^-$

$N\ N = H_2NCH_2CH_2NH_2$

However, when Hg^{2+} is added (at pH = 1), rapid hydrolysis of the ester function ensues. Hg(II) is an ion known to catalyze aquation of halometal complexes,[40] so the first step in the ester hydrolysis (Figure 14–14) is apparently the loss of halide to form a five-coordinate intermediate **(17)**. Since retention of configuration is observed, this five-coordinate intermediate has structure **17** rather than **18**, since the latter would lead to racemization (see Study Question 12).

The next steps in the reaction are the rapid formation of the intermediate **19** and the rate determining attack of water or OH⁻ at the carbonyl carbon. The intermediate **19** gives itself away by exhibiting a band in the infrared spectrum at 1610 cm⁻¹ assigned to the stretching of a coordinated carbonyl group (v_{CO} for **16** is found at 1740 cm⁻¹). The rate of the subsequent hydrolysis of **19** can then be followed by measuring the rate of disappearance of the 1610 cm⁻¹ band. The reaction is relatively slow in the pH range from 1 to 4, where the attacking species is H_2O, but the rate accelerates by at least 10^2 at higher pH's, where the attacking nucleophile is now OH⁻.

The hydrolysis of amino acid amides is more closely related to biological problems, so the study of the hydrolysis of the amide C—N bond in a compound such as **21** (see Figure 14–15) is important.[41,42] Under basic conditions (pH = 9), coordi-

---

[38a] That **15** is relatively more important than **14** is also expected from the σ inductive effect of M^{2+}.
[39] M. D. Alexander and D. H. Busch, *J. Amer. Chem. Soc.,* **88**, 1130 (1966).
[40] Hg^{2+} promotes the hydrolysis of compounds such as **16** by assisting in halide removal via the reaction $[M—Cl]^{n+} + Hg^{2+} \rightleftharpoons M^{(n+1)+} + HgCl^+$. For a discussion of such metal ion-assisted processes, see R. G. Wilkins, "The Study of Kinetics and Mechanism of Reactions of Transition Metal Complexes," Allyn and Bacon, Inc., Boston (1974), page 213.
[41] D. A. Buckingham, D. M. Foster, and A. M. Sargeson, *J. Amer. Chem. Soc.,* **92**, 6151 (1970).
[42] D. A. Buckingham, C. E. Davis, D. M. Foster, and A. M. Sargeson, *J. Amer. Chem. Soc.,* **92**, 5571 (1970).

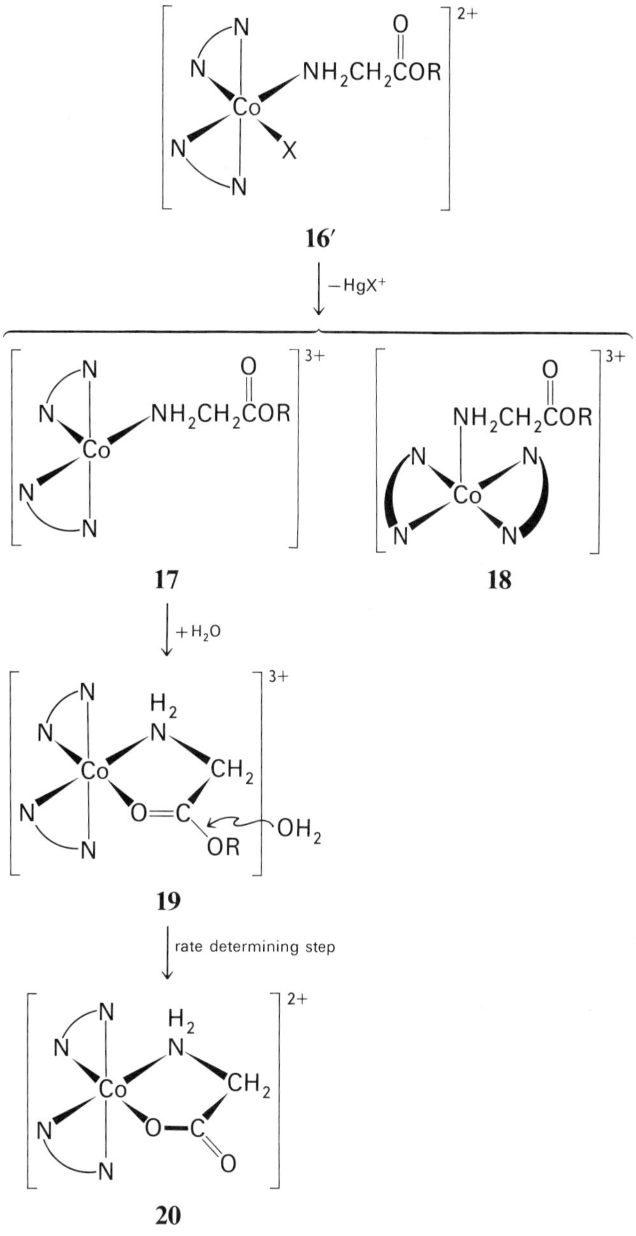

Figure 14–14. Possible mechanism for hydrolysis of a Co(III) complex of an amino acid ester.

nated halide is first removed in a base-assisted or CB mechanism (see Chapter 13). This produces a five-coordinate intermediate (compound **22**) just as in the hydrolysis of amino acid esters (Figure 14–14). However, unlike amino acid ester hydrolysis, the five-coordinate intermediate is relatively long-lived (as compared with **17**) and two nucleophiles—water and the amide carbonyl group—compete for the sixth coordination position.[43] This competition means that amide hydrolysis may occur by two paths, A and B in Figure 14–15.[44-46] Indeed, ^{18}O tracer studies indicate (i) that

---

[43] If anions are added in quantity, they can compete for the open coordination position in **22**. Further reaction is then blocked. This observation, of course, indicates the finite existence of a five-coordinate intermediate.

[44] Both paths A and B are observed under the experimental conditions described in the text, but only A is observed under acid conditions with Hg(II)-assisted removal of the bromide ion (see refs. 45 and 46).

[45] D. A. Buckingham, D. M. Foster, L. G. Marzilli, and A. M. Sargeson, *Inorg. Chem.,* **9,** 11 (1970).

[46] D. A. Buckingham, D. M. Foster, and A. M. Sargeson, *J. Amer. Chem. Soc.,* **91,** 5102 (1969).

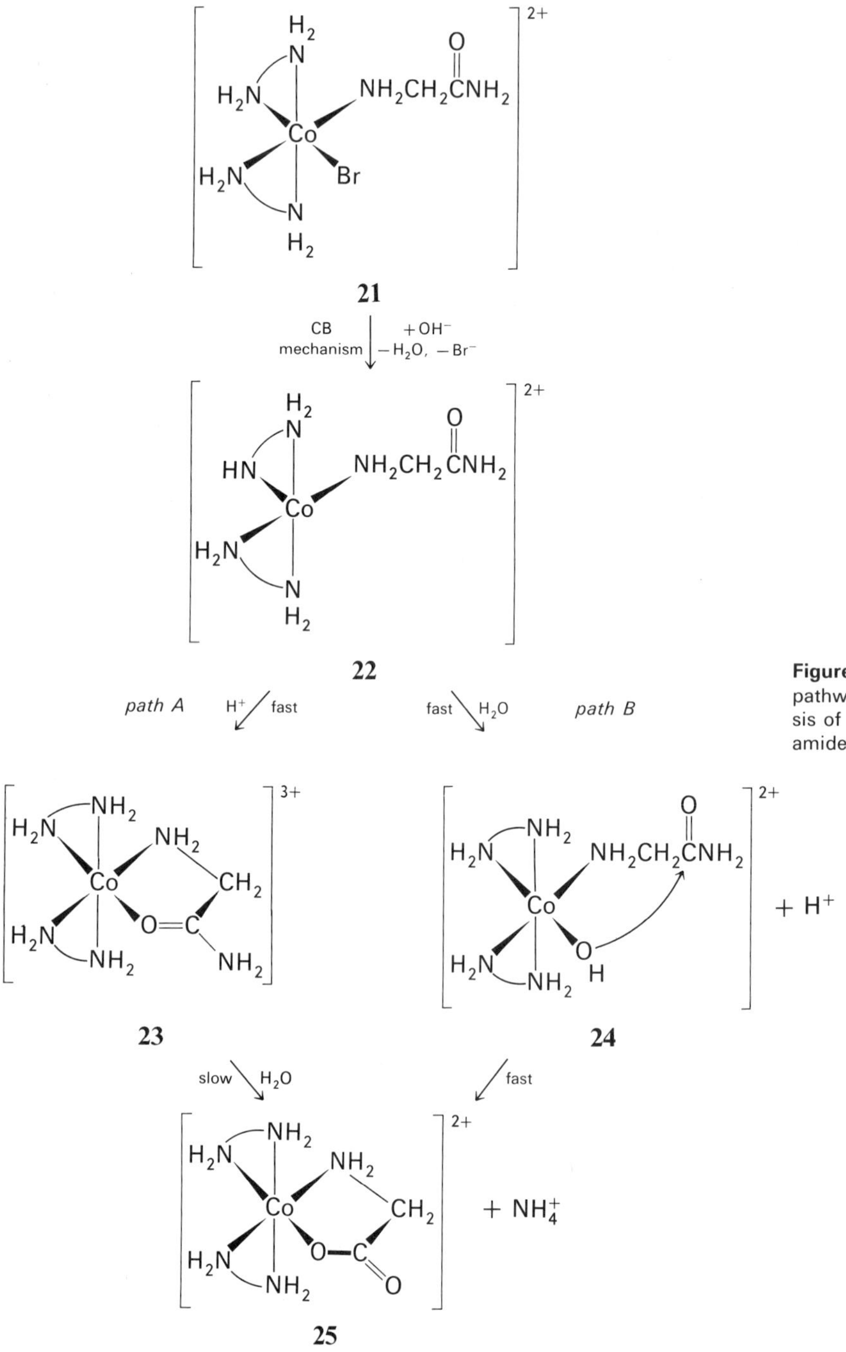

Figure 14-15. The two mechanistic pathways for the base-catalyzed hydrolysis of a Co(III) complex of an amino acid amide.

path A accounts for 46% of the product and path B for 54%, and (ii) that the product distribution is pH independent. Perhaps of greater importance, however, is the observation that paths A and B have decidedly different efficiencies. *Inter*molecular hydrolysis by path A is approximately 10^4 times more rapid that the rate of hydrolysis of the uncoordinated amino acid amide; in contrast, *intra*molecular hydrolysis by path B is *at least* 10^7 times more rapid than hydrolysis of the uncoordinated molecule.

Related to the hydrolysis of amino acid amide is the observation that β-[Co(trien)(OH)(H$_2$O)]$^{2+}$ (**26**) (trien = triethylenetetramine) (the α form is compound **27**) will effect the hydrolysis of peptides.[47]

---

[47] D. A. Buckingham, J. P. Collman, D. A. R. Happer, and L. G. Marzilli, *J. Amer. Chem. Soc.*, **89**, 1082 (1967).

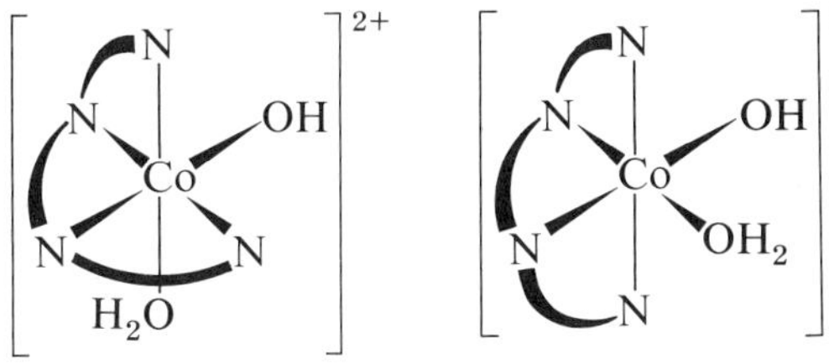

[Structures **26** and **27**: Co(III) complexes with tetradentate N-donor ligand, with OH, H₂O (26) and OH, OH₂ (27) coordinated]

26 **27**

That is, the Co(III) compound will act as a promoter of the hydrolysis of N-terminal amino acids from peptides according to the following scheme:

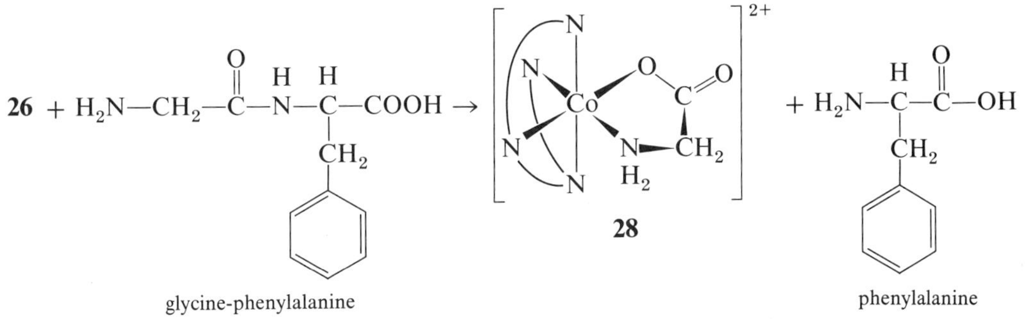

26 + H₂N—CH₂—C(O)—N(H)—C(H)(CH₂C₆H₅)—COOH → [Co complex **28**]²⁺ + H₂N—CH(CH₂C₆H₅)—C(O)—OH   (14–12)

glycine-phenylalanine   phenylalanine

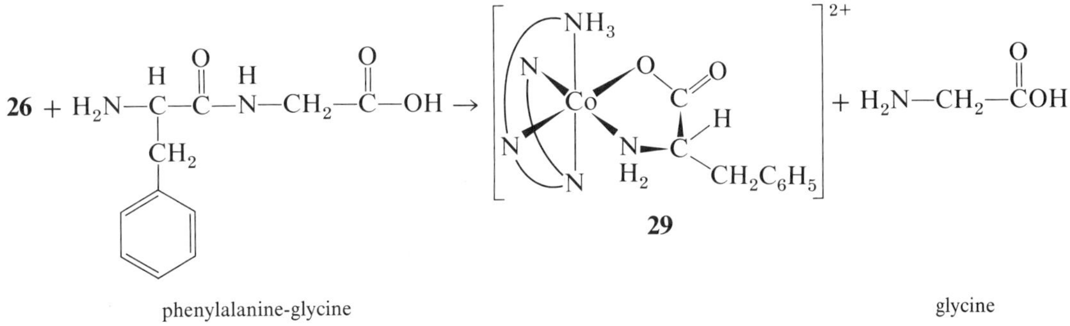

26 + H₂N—C(H)(CH₂C₆H₅)—C(O)—N(H)—CH₂—C(O)—OH → [Co complex **29**]²⁺ + H₂N—CH₂—COOH   (14–13)

phenylalanine-glycine   glycine

You will notice that the Co(III) compound (**26**) is not a true catalyst in this case (nor was it in the preceding cases), since the original reactant is not regenerated. Rather, **26** is a stoichiometric reagent, because the N-terminal amino acid ends up as a group chelated to the metal ion. The mechanisms of reactions 14–12 and 14–13 are presumably similar to that described for amino acid amide hydrolysis in Figure 14–15. That is, the terminal amino group can replace the water in **26**, and the amide linkage is then in position to undergo intramolecular hydrolysis.

There are a large number of other nucleophilic reactions that occur because the ligand is polarized by a coordinated metal ion. Two important ones are mentioned below.

Aldol Condensation

When an amino acid is coordinated to a metal ion, the α-C—H bond is polarized and therefore susceptible to attack by a nucleophile. For example, OH⁻ can apparently lead to formation of an enolate ion, because reaction of bis(glycinato) copper (**30**) with acetaldehyde in base ultimately gives a complex of threonine (**31**). The reaction can be generalized to other amino acids.[48]

---

[48] J. C. Drabrowiak and D. W. Cooke, *Inorg. Chem.*, **14**, 1305 (1975), and references therein.

782 ◻ COORDINATION CHEMISTRY: REACTION MECHANISMS AND METHODS OF SYNTHESIS; MOLECULAR REARRANGEMENTS AND REACTIONS OF COORDINATED LIGANDS

(14-14)

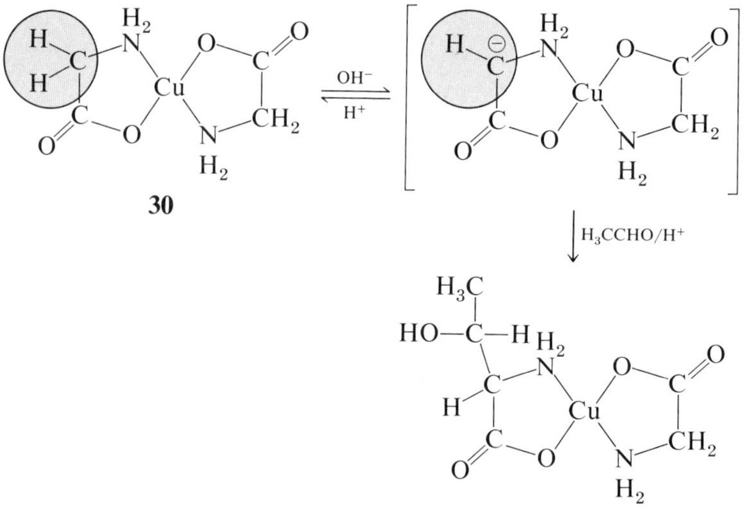

[structure **30** ⇌ (OH⁻/H⁺) carbanion intermediate ↓ H₃CCHO/H⁺ → structure **31**]

Imine Formation, Hydrolysis, and Substituent Exchange[49]

You should recall from your courses in organic chemistry that aldehydes and ketones may be converted to Schiff bases (imines) by an acid-catalyzed reaction with a primary amine:

(14-15)

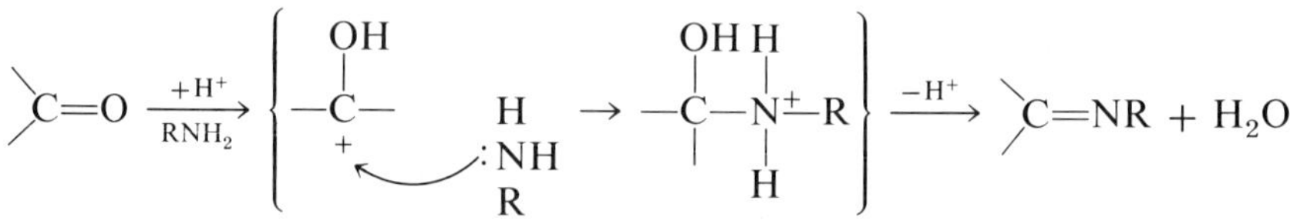

$$\text{C=O} \xrightarrow[\text{RNH}_2]{+\text{H}^+} \left\{ -\overset{\text{OH}}{\underset{+}{\text{C}}}- \quad \overset{\text{H}}{\underset{\text{R}}{:\text{NH}}} \rightarrow -\overset{\text{OH}}{\underset{|}{\text{C}}}-\overset{\text{H}}{\underset{\text{H}}{\text{N}^+}}-\text{R} \right\} \xrightarrow{-\text{H}^+} \text{C=NR} + \text{H}_2\text{O}$$

Since a metal ion can sometimes function as a catalyst in place of a proton, it is possible that an imino complex (**33**) may be formed by reaction of an amine with an aldehyde complex, **32**.

(14-16)

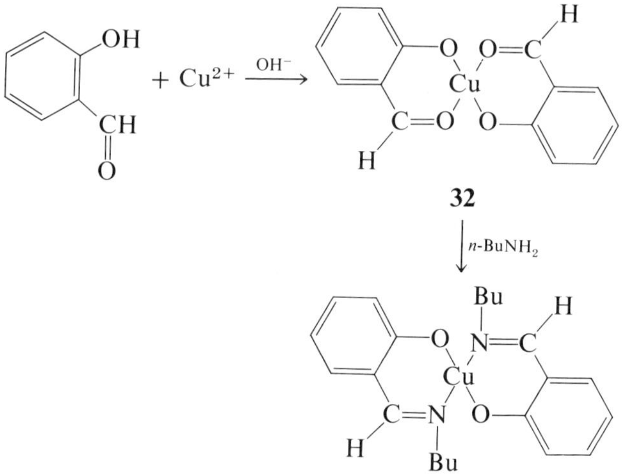

[salicylaldehyde + Cu²⁺ →(OH⁻) complex **32** ↓ n-BuNH₂ → complex **33**]

[dark green
m.p. 79–81°
planar]

---

[49] D. F. Martin, "Metal Complexes of Ketimine and Aldimine Compounds," in "Preparative Inorganic Reactions," Vol. 1, W. L. Jolly, ed., Wiley-Interscience, New York (1964).

Aldimines (the ligand in **33**; it would be called a ketimine if formed from a ketone) constitute an important class of ligands, particularly as it is possible to interchange nitrogen substituents by a transamination reaction, the process connecting **33** to **34**.

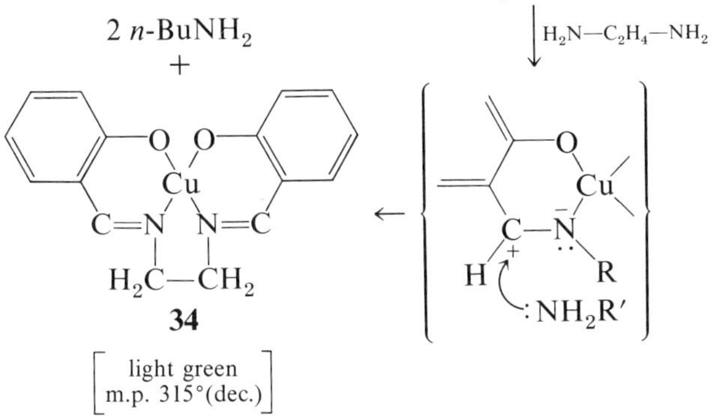

Yet another path to the synthesis of imine complexes was described recently—the intramolecular condensation of coordinated ammonia with a carbonyl group of a ligand coordinated *cis* to the ammonia.[50] Such reactions may be useful in the preparation of many new coordination compounds in the future.

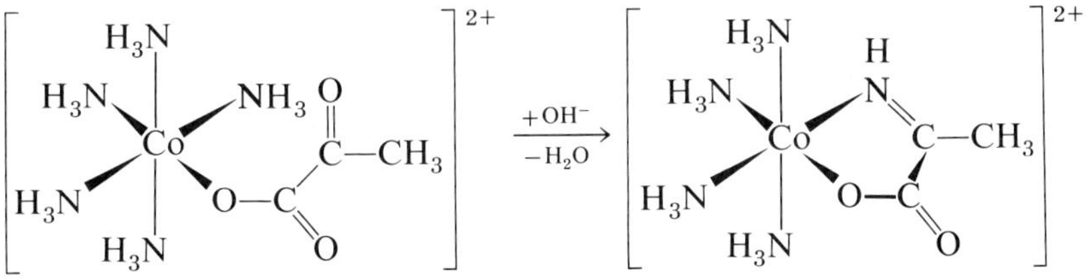

(14-17)

THE TEMPLATE EFFECT AND MACROCYCLIC LIGANDS

Imine-forming reactions are extremely important in the synthesis of multidentate ligands such as the one in compound **34** and in the formation of macrocyclic ligands, that is, ligands that are at least quadridentate and that can completely encircle the metal ion.[51,52] In fact, the formation of macrocycles, as in reaction 14-18 on p. 784 (or in the synthesis of the ligand in compound **20**, Chapter 10), serves to illustrate another important feature of the reactions of coordinated ligands—the coordination template effect.

---

[50] J. MacB. Harrowfield and A. M. Sargeson, *J. Amer. Chem. Soc.*, **96**, 2634 (1974).
[51] See ref. 28.
[52] M. Green and P. A. Tasker, *Chem. Commun.*, 518 (1968).

784 ☐ COORDINATION CHEMISTRY: REACTION MECHANISMS AND METHODS OF SYNTHESIS; MOLECULAR REARRANGEMENTS AND REACTIONS OF COORDINATED LIGANDS

(14-18)

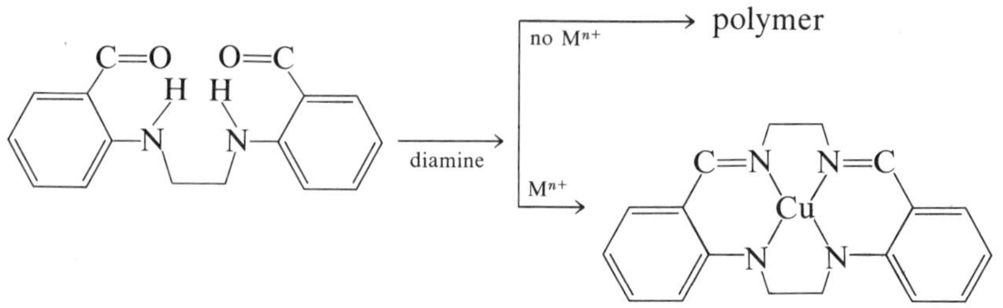

One of the most systematically studied syntheses[53] of a macrocyclic ligand illustrates the template effect as well as the points we have discussed thus far in this chapter. Busch wished to synthesize the potentially quadridentate ligand **37**, but direct reaction of an α-diketone with 2-aminoethanethiol gave a thiazoline (**36**) instead.[54]

(14-19)

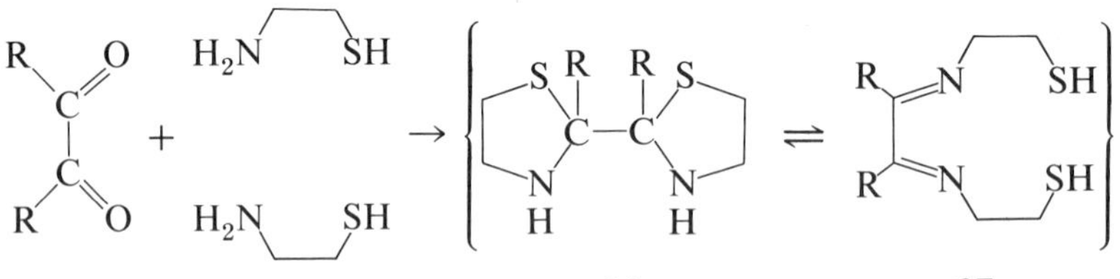

However, chemical tests for mercaptan indicated that a small amount of the desired compound, **37**, existed in equilibrium with **36**. Therefore, the diketone and the aminothiol were allowed to react in the presence of nickel(II) acetate, and the Ni(II) complex of the desired ligand (**38**) was produced in 70% yield. It was theorized that **37** was actually an intermediate in the formation of the thermodynamically more stable thiazoline, **36**. However, Ni(II) apparently stabilizes this intermediate by chelation, an effect sometimes referred to as a *thermodynamic template effect*.[55]

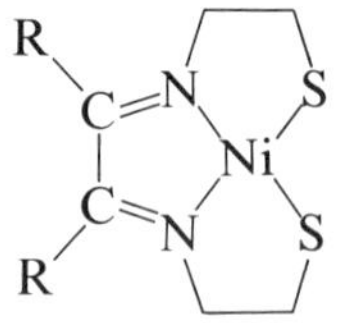

In earlier work Busch had found that coordinated RS⁻ could function as a nucleophile and would readily undergo alkylation. Therefore, in order to utilize **37** as the core of a macrocyclic ligand, **38** was allowed to react with an organic dihalide (**39**); complexes of the type illustrated by **40** were produced in good yield.[53] Busch has pointed out that such reactions illustrate the *kinetic template effect;* that is, a metal ion has been utilized to hold reactive groups in the proper geometry so that "a stereochemically selective multistep reaction" may occur.

---

[53] M. C. Thompson and D. H. Busch, *J. Amer. Chem. Soc.*, **86**, 3651 (1964).
[54] M. C. Thompson and D. H. Busch, *J. Amer. Chem. Soc.*, **86**, 213 (1964).
[55] For a recent example of the "thermodynamic template effect," see L. F. Lindoy and D. H. Busch, *Inorg. Chem.*, **13**, 2494 (1974).

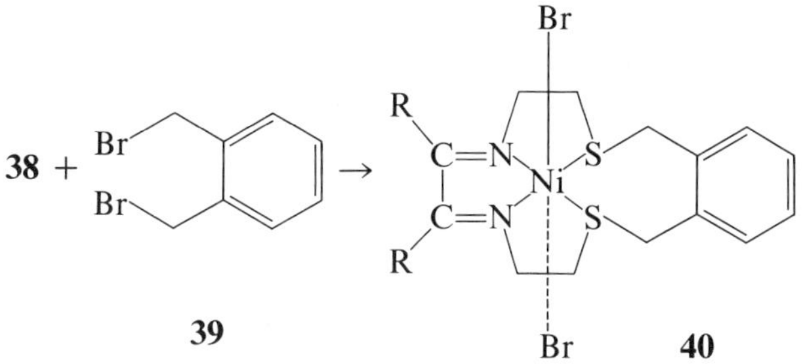

(14-20)

As just demonstrated, template reactions can lead to macrocyclic ligands, a class of molecules or ions that has been increasingly studied for its possible relevance to biochemically important ligands such as the porphyrins (41).[56]

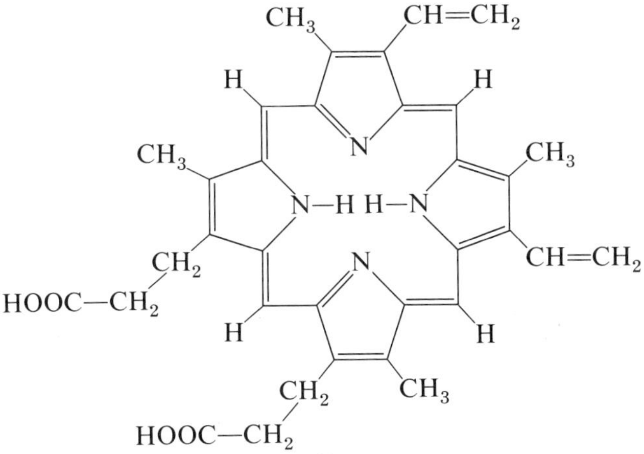

41

protoporphyrin IX

Aside from that, macrocyclic ligands are of interest because they often give complexes having unusual metal-donor atom bond angles and distances, and because they are capable of stabilizing unusual oxidation states of metals such as Cu(III). Further, metal complexes of macrocyclic ligands often possess considerably greater thermodynamic and kinetic stability (with respect to ligand dissociation) than their open chain analogs (these effects have been referred to collectively as the "macrocyclic effect"; see Study Question 7).[57] In fact, macrocyclic ligands can impart kinetic stability even when the central metal ion is thought to lead normally to labile complexes. This property makes macrocyclic ligands ideal for the study of ligand reactions as influenced by metal ion, since the possibility of metal-ligand dissociation during reaction is much reduced. The reactions most studied are those involving chemical and electrochemical oxidation and reduction; both are discussed below.

Because of their relationship to the porphyrins, perhaps the most thoroughly studied macrocyclic ligands have been those containing four nitrogen atoms capable of coordination. Of the ligands of this type, those having 14- and 16-membered rings are most common (compounds **42** to **48**).[58-60]

---

[56] See ref. 29.

[57] F. P. Hinz and D. W. Margerum, *J. Amer. Chem. Soc.*, **96**, 4993 (1974); *Inorg. Chem.*, **13**, 2941 (1974).

[58] Each compound is designated by the abbreviation commonly used in the literature. Taking compound **43** as an example, Me$_6$ indicates the number of substituent methyl groups, [14] is the ring size, and 4,11 is the location of the C=N double bonds. The full, systematic name of this ligand would be 5,7,7,12,14,14-hexamethyl-1,4,8,11-tetraazacyclotetradeca-4,11-diene. See V. L. Goedken, P. H. Merrell, and D. H. Busch, *J. Amer. Chem. Soc.*, **94**, 3397 (1972), for the details of the naming system used.

[59] A. B. P. Lever, *Adv. Inorg. Chem. Radiochem.*, **7**, 28 (1965).

[60] N. F. Curtis, *Coord. Chem. Rev.*, **3**, 3–47 (1968).

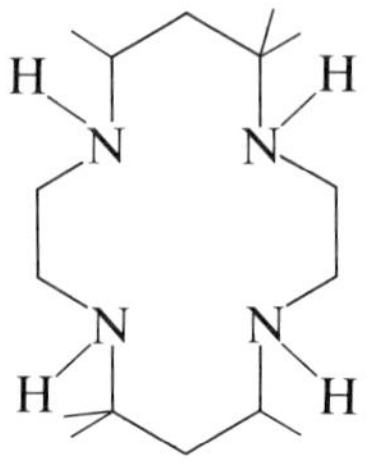

786 ☐ COORDINATION CHEMISTRY: REACTION MECHANISMS AND METHODS OF SYNTHESIS; MOLECULAR REARRANGEMENTS AND REACTIONS OF COORDINATED LIGANDS

Me$_6$[14]ane N$_4$
42

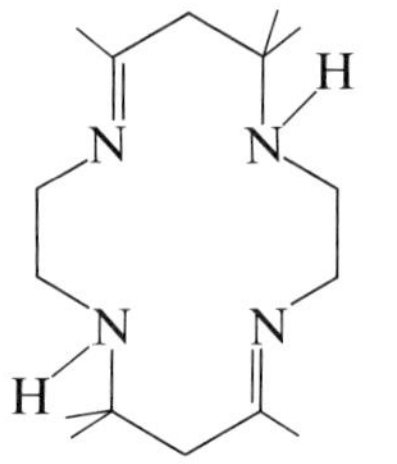

Me$_6$[14]4,11-diene N$_4$
43

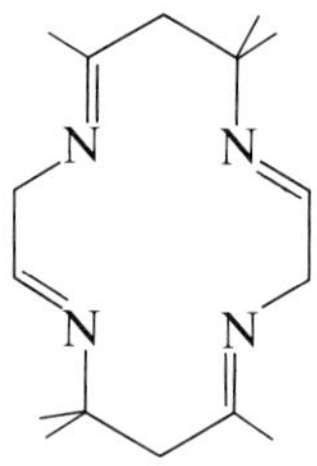

Me$_6$[14]1,4,8,11-tetraene N$_4$
44

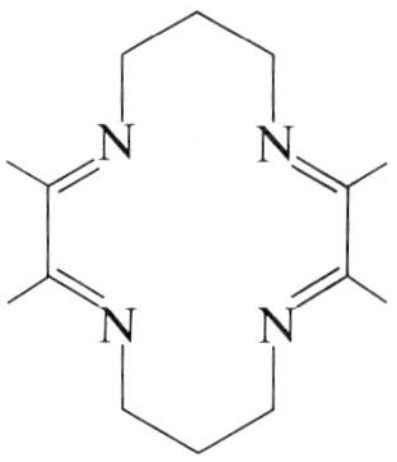

Me$_4$[14]1,3,8,10-tetraene N$_4$
45

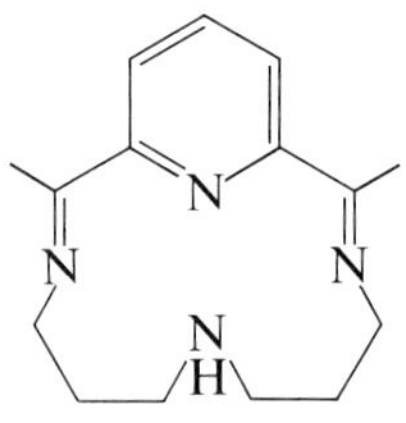

46

14-membered ring macrocyclic ligands

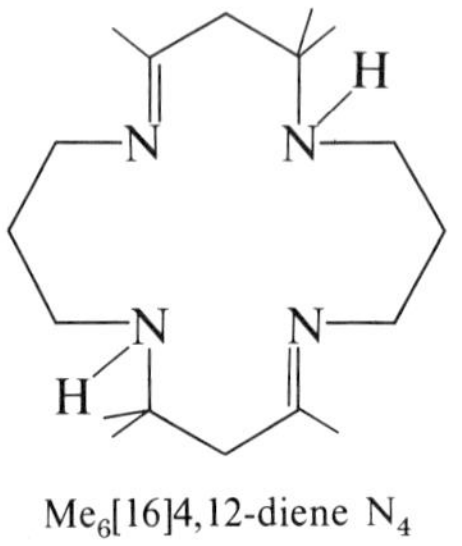

Me$_6$[16]4,12-diene N$_4$
47

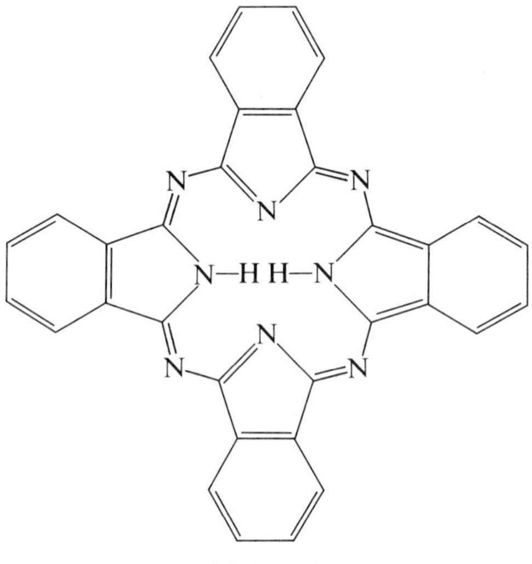

phthalocyanine
48

16-membered ring macrocyclcic ligands

Curtis first reported a Ni(II) complex of **43** as a reaction product of tris(ethylenediamine)nickel(2+) perchlorate and acetone.[61] If the reaction is carried out at room temperature, the violet color of the [Ni(en)$_3$]$^{2+}$ ion disappears rapidly and is replaced by a brown color; the yellow crystalline product is isolated from this brown solution. Approximately 70% of the final product is isolated in the *trans* form **(49)** and about 30% in the *cis* form **(50).** A similar procedure using bis(ethylenediamine)copper(2+) perchlorate leads to analogous Cu(II) complexes, which are orange.

---

[61] N. F. Curtis, *J. Chem. Soc.*, 4409 (1960).

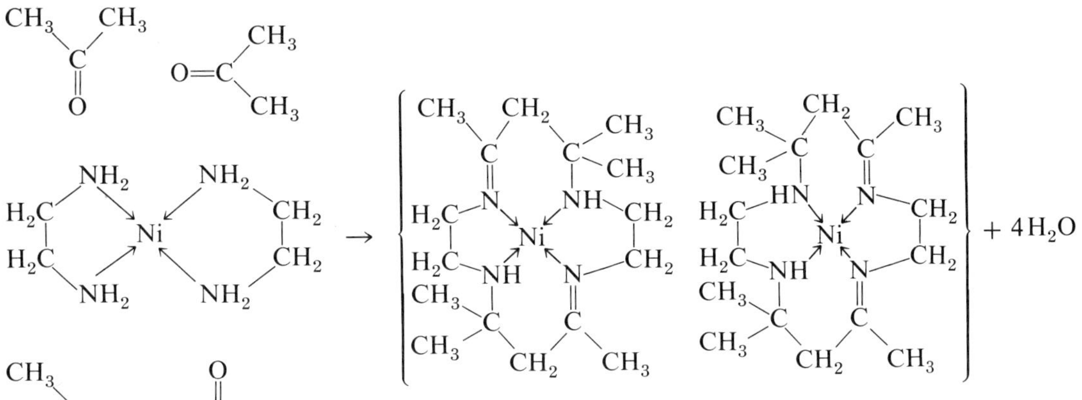

REACTIONS AT COORDINATED LIGANDS 787

Only Ni(II) and Cu(II) diamine complexes are known to react with acetone to give macrocycles **49** and **50**. Similar reactions do not occur with metal-diamine complexes that are stable with respect to ligand dissociation, *e.g.,* Co(III) complexes (see the discussion of complex stability in Chapter 13). This suggests that the acetone reacts with an —NH$_2$ group temporarily dissociated from the metal ion to give an imino linkage according to the scheme below. From this point on, a very plausible, but lengthy, mechanism may be proposed.

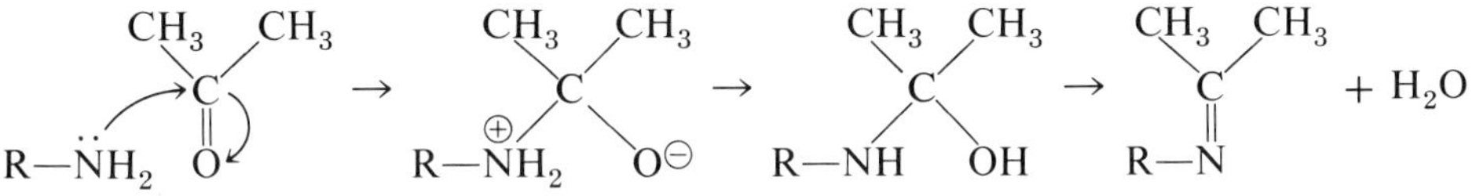

Regarding the mechanism, it is important to point out that the *trans* ligand (**49**) may be formed in the *absence* of a metal ion, so the synthesis of this macrocycle is *not* a template reaction.[62] On the other hand, the *cis* macrocycle, **50**, forms only in the presence of a metal ion, and its synthesis is therefore properly classified as a kinetic template reaction.

The nickel(II) complex, **49**, may be hydrogenated (with NaBH$_4$, H$_2$/Pt, or cathodic reduction) to give macrocycles with no double bonds, while oxidation with nitric acid or sulfurous acid, for example, leads to compounds with three or four double bonds.

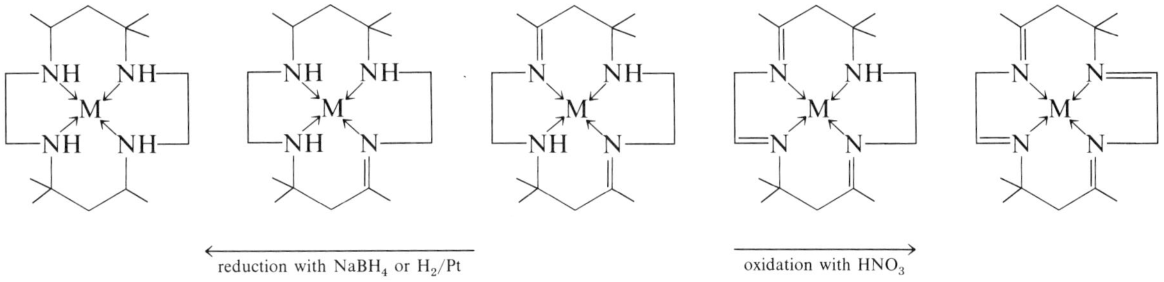

$\longleftarrow$ reduction with NaBH$_4$ or H$_2$/Pt   oxidation with HNO$_3$ $\longrightarrow$

Oxidative dehydrogenation has been more thoroughly studied with the Fe(II) complex of **43**.[63] Apparently the first step is the oxidation of the bound iron(II) to iron(III)

---

[62] For this reason, complexes of **49** with metal ions other than Ni(II) and Cu(II) may be synthesized and studied. Compound **51** is an example.

[63] V. L. Goedken and D. H. Busch, *J. Amer. Chem. Soc.,* **94**, 7355 (1972). Note again that a compound such as **51** cannot be synthesized directly from an iron-ethylenediamine complex and acetone. Rather, **51** is made by reacting a salt of **43**, formed in a non-template synthesis (see ref. 60), with an iron(II) salt.

(to give **52**). If a trace of water is present, however, the metal ion is reduced by intramolecular charge transfer of an electron from an amino nitrogen to Fe(III) with concomitant loss of H⁺; any oxidizing agent present may then further oxidize this radical intermediate with loss of a C—H proton to give **53**.

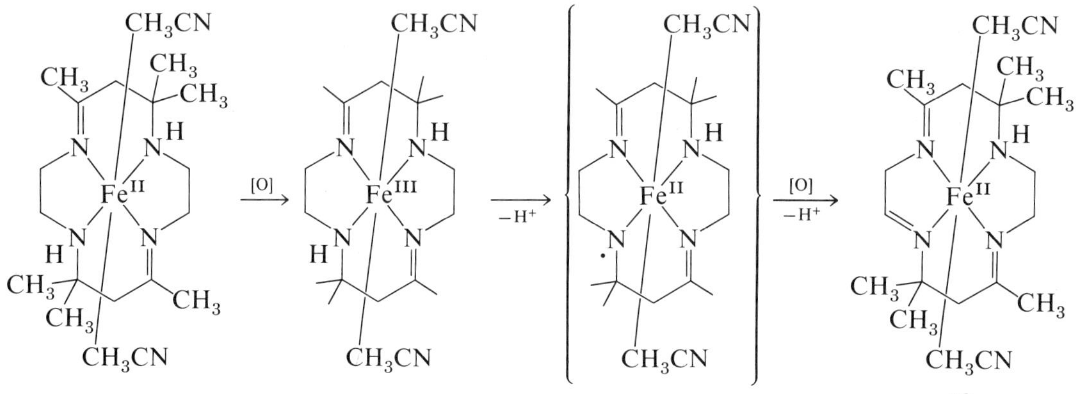

51   **52**   **53**

Very extensive electrochemical work has been done on complexes of these N_4 ligands, with results that bear importantly on biochemical systems.[64,65] It was suggested above that oxidation of the Fe(II) macrocycle complex resulted from removal of an electron from the central metal ion. However, it is not always the case that oxidation and reduction occur by simple removal or addition of an electron from or to the coordinated metal ion, respectively. Instead, it is possible that it is actually the ligand that is oxidized or reduced. That is, if we consider only localized changes in oxidation state, an electron may be added to either the ligand *or* the metal upon reduction, or it may be removed from either of these sites on oxidation.[66]

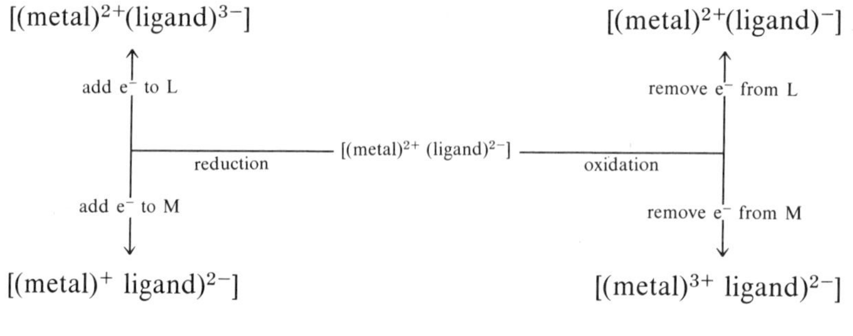

$$[(metal)^{2+}(ligand)^{3-}] \quad\quad\quad [(metal)^{2+}(ligand)^{-}]$$
$$\uparrow \text{add } e^- \text{ to L} \quad\quad\quad \uparrow \text{remove } e^- \text{ from L}$$
$$\xleftarrow{\text{reduction}} [(metal)^{2+}(ligand)^{2-}] \xrightarrow{\text{oxidation}}$$
$$\downarrow \text{add } e^- \text{ to M} \quad\quad\quad \downarrow \text{remove } e^- \text{ from M}$$
$$[(metal)^{+}(ligand)^{2-}] \quad\quad\quad [(metal)^{3+}(ligand)^{2-}]$$

The actual behavior observed has, in fact, been found to be critically dependent on the extent of ligand unsaturation, among other factors. For example, the electroactive site in the Ni(II) and Cu(II) complexes of **43** has been found to be the metal; in each

---

[64] F. V. Lovecchio, E. S. Gore, and D. H. Busch, *J. Amer. Chem. Soc.*, **96**, 3109 (1974).

[65] J. C. Drabrowiak, F. V. Lovecchio, V. L. Goedken, and D. H. Busch, *J. Amer. Chem. Soc.*, **94**, 5502 (1972).

[66] It is somewhat less restrictive to consider oxidation and reduction from a molecular orbital point of view. That is, oxidation is the loss of an electron from the HOMO and reduction is the gain of an electron by the LUMO. These mo's may, of course, have varying degrees of delocalization. That is, the HOMO may be completely delocalized over the entire complex or, more likely, polarized toward either the metal ion or the ligand. Therefore, oxidation may in effect be the loss of an electron from the metal ion or the ligand, respectively. Similarly, the LUMO may be polarized toward the metal ion or the ligand, so reduction may be the gain of an electron by the metal ion or the ligand, respectively.

of these both M(I) and M(III) have been observed.[67] On the other hand, the more highly unsaturated ligand in complex **54** is the site of oxidation and reduction.[68-70]

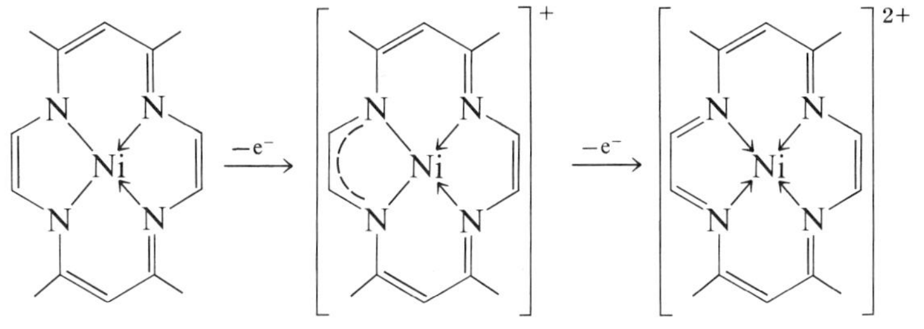

54 **55** **56**

This behavior more closely parallels that of the metalloporphyrins, *e.g.*, **57**,[71] as the primary mode of oxidation of these complexes is electron removal from the ligand.

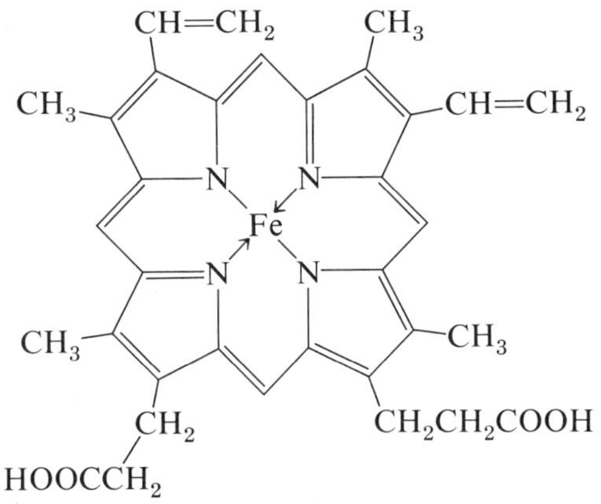

57

---

[67] D. C. Olson and J. Vasilevskis, *Inorg. Chem.*, **8**, 1611 (1969); *Inorg. Chem.*, **10**, 463 (1971).

[68] T. J. Truex and R. H. Holm, *J. Amer. Chem. Soc.*, **94**, 4529 (1972); M. Millar and R. H. Holm, *J. Amer. Chem. Soc.*, **97**, 6052 (1975).

[69] Notice that the number of π electrons on the ligands in these complexes are as follows: **54** = 16 π electrons; **55** = 15 π electrons; and **56** = 14 π electrons. The terminal complex, **56**, should be especially stable, as a 14 π electron system is a $(4n + 2)$ π electron aromatic species.

[70] E. B. Fleischer, *Acc. Chem. Res.*, **3**, 105 (1970).

[71] G. M. Brown, F. R. Hopf, J. A. Ferguson, T. J. Meyer, and D. G. Whitten, *J. Amer. Chem. Soc.*, **95**, 5939 (1973), and references therein.

STUDY QUESTIONS

6. Suggest a possible mechanism for the formation of **35**. (Hint: Consult the section on hydrolysis under basic conditions in Chapter 13.)

7. On page 785 the special thermodynamic and kinetic stability of complexes of macrocyclic ligands is noted. Considering the following reaction for the formation or dissociation of

790 ☐ MOLECULAR REARRANGEMENTS AND REACTIONS OF COORDINATED LIGANDS

such a complex, and considering the thermodynamic and kinetic factors involved in such a reaction, develop an explanation for the "macrocyclic effect."

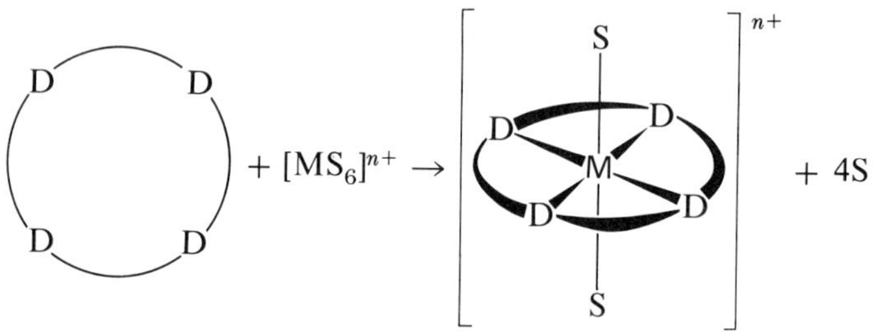

$$\text{(cyclic-D}_4\text{)} + [MS_6]^{n+} \rightarrow [\text{M(D}_4\text{)S}_2]^{n+} + 4S$$

8. A possible mechanism for the hydrolysis of a peptide by carboxypeptidase A is given on page 774. Why does the interaction of tyrosine, residue 248, play an important role in this reaction?

9. In compound **23,** Figure 14-15, the amino acid amide uses the carbonyl oxygen and not the amide nitrogen to bind to the cobalt ion. On the basis of the discussion in Chapter 5, is it reasonable to expect this bonding mode?

10. It was noted in the text that water and the amide carbonyl group are the competitors for the open coordination site in the five-coordinate intermediate **22** in Figure 14-15. What evidence is there that OH^- is not a competitor?

11. Suggest a method for the synthesis of the macrocyclic ligand shown below.

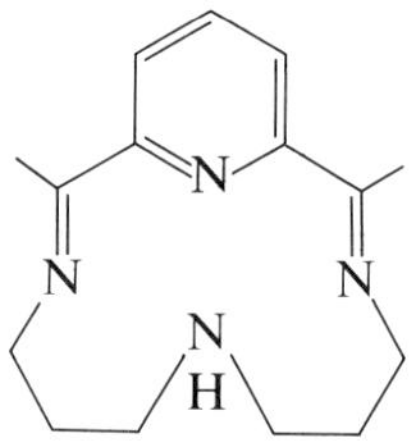

If you run out of patience, consult the papers by Busch and his coworkers: *Inorg. Chem.*, **11,** 283 (1972), and *Inorg. Chem.*, **8,** 1149 (1969).

12. In the hydrolysis reaction outlined in Figure 14-14, retention of configuration is observed. It was noted that this suggests that **18** is not the intermediate, since formation of **18** could lead to inversion of configuration.

    a) What is the chirality of **16,** Δ or Λ?
    b) How can configuration inversion occur if **18** is the intermediate?

13. In reaction 14-13, the terminal amino group of the peptide displaces water from the complex, and the amide linkage is then in position to undergo hydrolysis. Why does the amide nitrogen not coordinate to the metal ion (cf. Study Question 9)?

EPILOG

With this chapter we complete the series of chapters on the compounds of the higher valence states of the transition metals—on their modes of bonding, their structures, their syntheses, and the mechanisms of their reactions. Clearly, we have not covered all of the nooks and corners of this field, which is so immense both in scope and interest. Instead, we hope that you now have an appreciation for, and an excitement over, the fact that although much is now known, we are only just beginning to understand in a truly fundamental way the structural dynamics of these molecules and the courses of their reactions. There are many new areas concerning the higher valence states of the transition metals yet open to research by the serious chemist. We shall now turn our attention to the lower valence states of these metals, that is, to a discussion of the organometallic chemistry of the transition metals and of the main group metals.

THE SIXTEEN AND EIGHTEEN ELECTRON RULE	THEORETICAL ASPECTS OF THE 16 AND 18 ELECTRON RULE

15. FROM CLASSICAL TO ORGANOMETALLIC TRANSITION METAL COMPLEXES AND THE SIXTEEN AND EIGHTEEN ELECTRON RULE

> *This chapter is a bridge between the previous six chapters on the chemistry of transition metal complexes involving the metals in their higher oxidation states—with ligands such as amines, halides, and oxygen donors—and chapters treating those complexes wherein the metal is in a lower formal oxidation state and is bound through carbon to ligands such as CO, CH_3, C_6H_6, and C_2H_4. That is, our next topic is organometallic chemistry. Although many of the compounds we shall discuss are based on main group metals, a large part of organometallic chemistry is concentrated on transition metal complexes. We shall, therefore, first present an empirical rule, the "sixteen and eighteen electron rule" or the "effective atomic number rule" (= EAN), a device that will be an invaluable aid in systematizing the structural and reaction chemistry of transition metal complexes. Secondly, we shall discuss why the 16 and 18 electron rule should be valid for organometallic compounds but not necessarily for classical compounds.*

THE SIXTEEN AND EIGHTEEN ELECTRON RULE[1]

Before beginning the chapters on organometallic chemistry, there are several matters to be attended to, and the first of these is the "sixteen and eighteen electron rule" or "effective atomic number rule" (EAN). This rule allows the systematization of much of the structural chemistry of transition metal organometallic compounds (Chapter 16) and is a useful guide to their reactions (Chapter 17).

The transition metal complexes dealt with thus far in this book are often referred to as "classical" because their discovery and characterization reach back into the history of modern chemistry. Generally speaking, such complexes contain the metal in an oxidation state of 2+ or higher and involve ligands such as amines and other nitrogen donors, oxygen donors, and halides. However, in the subsequent two chapters, we deal with metal compounds having ligands that are capable of, and even require, synergic ligand → metal σ/metal → ligand π electron flow.

First introduced in Chapter 5, the synergic interaction arises when the σ donor has a low energy π acceptor orbital (LUMO) while the σ acceptor in the adduct configuration has a π donor orbital (HOMO). An important consequence of this two-way donor-acceptor bonding (**1** and **2**) is that a bonding role can be given to

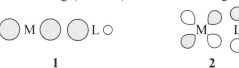

1   **2**

d_{z^2} metal ← ligand σ bonding   metal → ligand π bonding

---

[1] C. A. Tolman, *Chem. Soc. Rev.*, **1**, 337 (1972).

THE SIXTEEN AND EIGHTEEN ELECTRON RULE

otherwise non-bonding or even σ^* electrons on the metal. It is an interesting, empirical fact that, as compared with classical complexes, M-L adducts of this type more closely follow rules relating structure to number of valence electrons. Consider, as an example, $[Cr(NH_3)_6]^{3+}$ and $Cr(CO)_6$. The total number of valence electrons about Cr^{3+} in the hexaamminechromium(3+) ion is 15 (three from Cr^{3+} and 12 from the six NH_3's), whereas the total for hexacarbonylchromium is 18 (six from Cr(0) and 12 from the six CO's). Since there are a total of nine d, s, and p valence orbitals available to a transition metal, all orbitals can be utilized fully in the carbonyl complex. In molecular orbital terms, this means that, for $Cr(CO)_6$, the six bonding mo's that arise from the two d orbitals of σ symmetry, the s orbital, and the three p orbitals are all full. In addition, and most importantly, the three metal $d\pi$ orbitals are full, and it is these orbitals that participate in metal → ligand π electronic flow and stabilize the complex. Otherwise, $Cr(CO)_6$, a complex formed from a zero-valent metal, would suffer too great a build-up of negative charge on the metal.

In general, organic complexes of the transition metals are formed with those metals in low oxidation states (-1, 0, and $+1$), since it is only in that condition that the metal will have populated orbitals of π symmetry and a low Z^* to allow metal → ligand π electronic flow. We shall account for such complexes in more complete molecular orbital terms in the next chapter. However, for the moment it is sufficient to state that, with certain well-defined exceptions, *stable organometallic compounds of the transition metals will have a total of 18 valence electrons about the metal*; in other words, *they will have the "effective atomic number" (EAN) of the next higher inert gas*. The *exception* mentioned is the fact that *molecules having only 16 valence electrons can often be just as stable* as, or even more stable than, 18-electron molecules of the same metal. This is especially true of metals at the bottom right corner of the transition metal series, that is, Rh, Pd, Ir, and Pt. Recall from Chapters 9 and 10 (*cf.* Figure 10-1) that it is just these elements that have the least tendency to be six-coordinate. Thus, in oxidation states such as Ir(I) and Pt(II), such metals can form stable 16-electron molecules.

The application of the EAN rule is quite simple. If a metal is to accommodate six adduct bond pairs and follow the EAN rule, the metal must have three electron pairs of its own. That is, the metal atom or ion involved will have a d^6 configuration **(3)**, and the structures of such complexes will be expected to be roughly octahedral. Metals with four valence electron pairs of their own require the presence of five adduct bond pairs and should exhibit structures based on the trigonal bipyramid or square pyramid **(4)**. Metals with five valence electron pairs should adopt tetrahedral or square planar structures by acquisition of four ligand electron pairs **(5)**.

3
d^6 + 6 ligand bond pairs = MD_6
(V^-, Cr^0, Mn^+, Fe^{2+})

4
d^8 + 5 ligand bond pairs = MD_5
(Mn^-, Fe^0, Co^+)

5
d^{10} + 4 ligand bond pairs = MD_4
(Co^-, Ni^0, Cu^+)

Some compounds that illustrate the 18 electron rule are given on p. 795. In counting electrons, ligands such as CO, PPh_3, halides, H^-, and alkyl and aryl groups (e.g., CH_3^- and $C_6H_5^-$) are considered two electron donors. NO and the π bonded organic ligands deserve special comment, and we return to that momentarily.

MD_6: $M = d^6 = V^-, Cr^0, Mn^+, Fe^{2+}$

$V(CO)_6^-$ $Cr(CO)_6$ $Mn(CO)_6^+$ $Fe(CO)_4H_2$

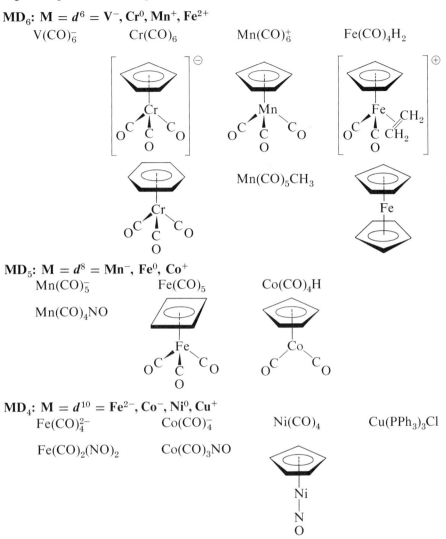

$Mn(CO)_5CH_3$

MD_5: $M = d^8 = Mn^-, Fe^0, Co^+$

$Mn(CO)_5^-$ $Fe(CO)_5$ $Co(CO)_4H$

$Mn(CO)_4NO$

MD_4: $M = d^{10} = Fe^{2-}, Co^-, Ni^0, Cu^+$

$Fe(CO)_4^{2-}$ $Co(CO)_4^-$ $Ni(CO)_4$ $Cu(PPh_3)_3Cl$

$Fe(CO)_2(NO)_2$ $Co(CO)_3NO$

As mentioned before, there are numerous stable organometallic compounds that have only 16 valence electrons about the metal. Such complexes are invariably based on d^8 metal ions and are square planar (four ligand bond pairs), or they are based on d^{10} metal atoms with three ligands (three ligand bond pairs) and are presumably trigonal. Examples include those shown below.

MD_4: $d^8 = Rh^+$ or Ir^+; Pd^{2+} or Pt^{2+}

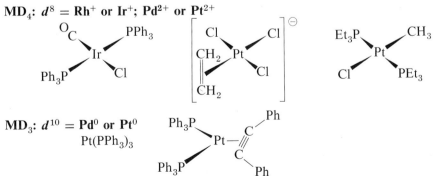

MD_3: $d^{10} = Pd^0$ or Pt^0

$Pt(PPh_3)_3$

We turn now to the convention used to determine the number of electrons contributed by a ligand to the valence shell or the metal. As mentioned previously, this is no problem for ligands of the classical type (amines, phosphines, halides, etc.), but some of the π donor ligands encountered in organometallic chemistry are less familiar to you at this point and deserve special comment. Our point of view for electron counting—and it is but one of several that we could take—is simply to retain the convention of Chapter 5; that is, view the complexes as donor/acceptor adducts wherein each ligand is thought of as an electron pair donor and the metal as an electron pair acceptor. Little needs to be added at this point about the donors of one electron pair: CO, H^-, PF_3, PPh_3, and C_2H_4. (Recall that C_2H_4 donates a pair of π electrons in binding to Pt^{2+} in $[Pt(C_2H_4)Cl_3]^-$; see Chapter 9, page 539). However, ligands such as NO^+, $C_5H_5^-$, C_6H_6, or $H_2C{=}CH{-}CH{=}CH_2$ require comment.

As described in Chapter 4, NO is an odd-electron molecule possessing not only a nitrogen donor lone pair but also one electron in the π^* mo's. Our convention is to treat NO as coordinated NO^+ with the extra electron counted among the metal or other ligand electrons. Other conventions may treat NO as a three-electron donor or as NO^-, the latter formed by taking an electron from the metal and assigning it to the ligand. However, the convention used here is based on the analogy of NO^+ to CO, CN^-, and N_2, which constitute an isoelectronic series. The assignment of the π^* electron to the metal for electron counting purposes facilitates this comparison but generally has little or no physical significance (but neither do the other two conventions). The NO^+ ligand *in the complex* will involve its π^* mo's with the metal $d\pi$ ao's to a considerable extent, and there will be, in the end, some net electron density distributed between the metal and the NO^+ π^* mo's that is independent of electron counting devices. Electron counting in NO complexes is illustrated by the following examples.

$$Mn(CO)_4NO \qquad\qquad Fe(CO)_2(NO)_2$$

tetracarbonylnitrosylmanganese   dicarbonyldinitrosyliron

$$
\begin{array}{ll}
NO^+ = 1\ \sigma\ \text{pair} & 2NO^+ = 2\ \sigma\ \text{pairs} \\
4CO = 4\ \sigma\ \text{pairs} & 2CO = 2\ \sigma\ \text{pairs} \\
\underline{Mn^- = 4\ \text{pairs}} & \underline{Fe^{2-} = 5\ \text{pairs}} \\
\quad 9\ \text{pairs or} & \quad 9\ \text{pairs or} \\
\quad 18\ \text{electrons} & \quad 18\ \text{electrons}
\end{array}
$$

Before proceeding further with this discussion of electron counting conventions, we must insert a word concerning *nomenclature*. The systematic name of compound **6** is tricarbonyl(η^5-cyclopentadienyl)manganese and that of **7** is (η^6-benzene)tricar-

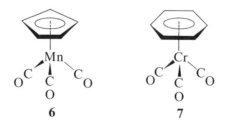

6   7

bonylchromium. The notation η^n is a shorthand designating the number of ligand atoms ($=n$) formally bound to the metal. The letter η (Greek *eta*) stands for *hapto*, from the Greek word *haptein* meaning "to fasten." Thus, the $C_5H_5^-$ ligand in **6** is said to be a *pentahapto*cyclopentadienyl group, whereas C_6H_6 in **7** is *hexahapto*benzene.[2]

---

[2] F. A. Cotton, *J. Organometal. Chem.*, **100**, 29 (1975).

The utility of this system will become more obvious as you study the examples in this chapter and subsequent ones. Finally, we call your attention to the fact that the nomenclature of transition organometallic compounds is otherwise no different from that for classical compounds as laid out in Chapter 9 (p. 519).

Now to return to electron counting conventions. The ligand designated η^1-C_5H_5 is the cyclopentadiene molecule with a metal in place of a hydrogen. The η^1 structure of the cyclopentadienyl places it in the category of σ bonded alkyl and aryl groups,

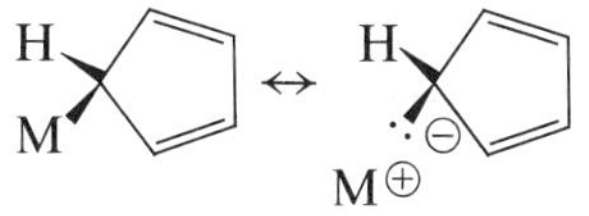

which must be thought of as carbanions (R^-) in order to conform to the electron pair scheme.

*Pentahapto*cyclopentadienyl is one of the most important ligands in organometallic chemistry. In this configuration the cyclopentadienyl group is positioned so as to have all five carbons equidistant from the metal.

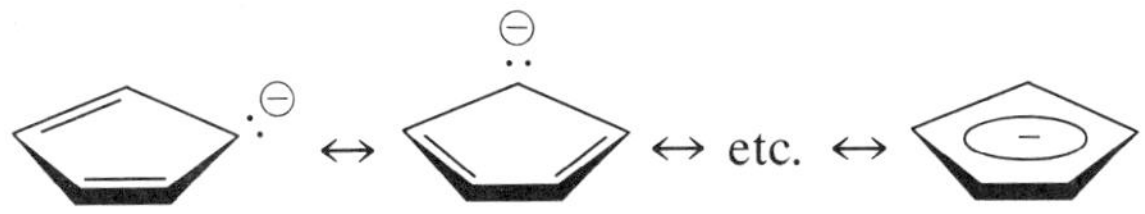

In this structure you are to think of the ligand as a carbanion and a three-pair π donor. $Fe(CO)_2(\eta^5$-$C_5H_5)(\eta^1$-$C_5H_5)$ has two kinds of $C_5H_5^-$ ligands, and the electrons are counted as follows.

$$Fe(CO)_2(\eta^5\text{-}C_5H_5)(\eta^1\text{-}C_5H_5)$$

dicarbonyl(*monohapto*cyclopentadienyl)(*pentahapto*cyclopentadienyl)iron

$$\begin{array}{rl} 2CO & = 2\ \sigma \text{ pairs} \\ \eta^5\text{-}C_5H_5^- & = 3\ \pi \text{ pairs} \\ \eta^1\text{-}C_5H_5^- & = 1\ \sigma \text{ pair} \\ Fe^{2+} & = 3 \text{ pairs} \\ \hline & 9 \text{ pairs or} \\ & 18 \text{ electrons} \end{array}$$

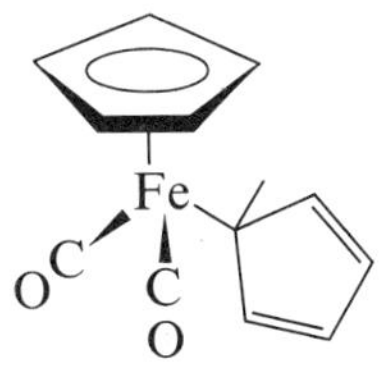

8

Another three-pair π donor is benzene, as illustrated by bis(η^6-benzene)chromium.

$$\begin{array}{rl} 2C_6H_6 & = 6\ \pi \text{ pairs} \\ Cr & = 3 \text{ pairs} \\ \hline & 9 \text{ pairs} \end{array}$$

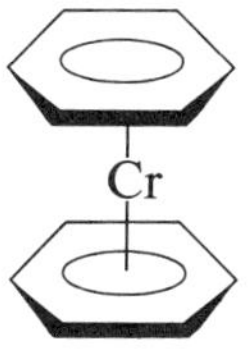

9

The simplest two-pair π donor is derived from allyl, H_2C=CH—CH_2. Like cyclopentadienyl, C_3H_5 contains an odd number of electrons when considered as a neutral hydrocarbon. However, our convention is to treat it as an anion and therefore as a two-pair π donor. In this configuration all three carbons are bound to the metal, and it is designated as η^3-$C_3H_5^-$.

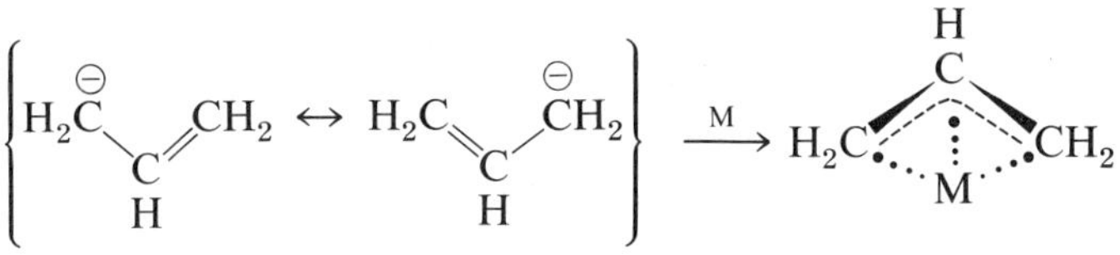

For example, valence electrons in Mn(CO)$_4$(η^3-C$_3$H$_5$) and Co(CO)$_3$(η^3-C$_3$H$_5$) are counted as follows.

Mn(CO)$_4$(η^3-C$_3$H$_5$)   Co(CO)$_3$(η^3-C$_3$H$_5$)

(*trihapto*allyl)tetracarbonylmanganese   (*trihapto*allyl)tricarbonylcobalt

$$\begin{array}{ll} \eta^3\text{-C}_3\text{H}_5^- = 2\ \pi \text{ pairs} \\ 4\text{CO} = 4\ \sigma \text{ pairs} \\ \underline{\text{Mn}^+ = 3 \text{ pairs}} \\ \quad\quad\quad 9 \text{ pairs} \end{array} \quad \begin{array}{ll} \eta^3\text{-C}_3\text{H}_5^- = 2\ \pi \text{ pairs} \\ 3\text{CO} = 3\ \sigma \text{ pairs} \\ \underline{\text{Co}^+ = 4 \text{ pairs}} \\ \quad\quad\quad 9 \text{ pairs} \end{array}$$

Among the neutral two-pair donors are

η^4-C$_4$H$_4$ η^4-C$_4$H$_6$ η^4-C$_7$H$_8$
*tetrahapto*cyclobutadiene   *tetrahapto*butadiene   *tetrahapto*norbornadiene

An example of a diene-metal complex is Fe(CO)$_3$(η^4-C$_4$H$_4$) on page 795.

The interesting hydrocarbon cycloheptatriene, C$_7$H$_8$, is an isomer of norbornadiene, and has an intriguing flexibility in the number of donor pairs it can provide.

C$_7$H$_8$ = 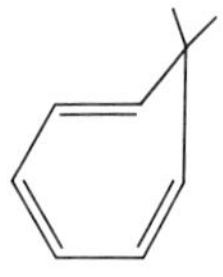 = cycloheptatriene

Most obviously it could act as a neutral, three-pair π donor as in (η^6-C$_7$H$_8$)Cr(CO)$_3$ **(10)**.

10

tricarbonyl(*hexahapto*cycloheptatriene)chromium

In Fe(CO)$_3$(η^4—C$_7$H$_8$), however, it appears as a neutral, two-pair π donor like the dienes mentioned above.

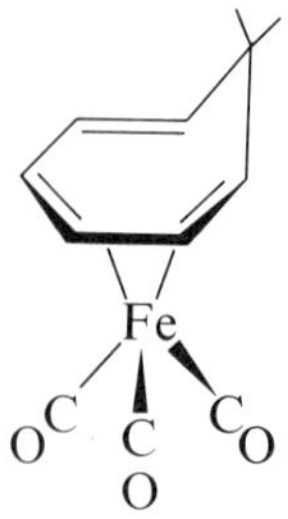

11

Loss of H$^+$ or H$^-$ from C$_7$H$_8$ leads to ions that can also function as ligands in a most interesting manner. For example, H$^-$ abstraction (using Ph$_3$C$^+$, for example) from (η^6-C$_7$H$_8$)Mo(CO)$_3$ gives a complex of the tropylium ion, C$_7$H$_7^+$, **12**. The *heptahapto*-C$_7$H$_7^+$ ion is considered to be a three-pair or six-electron donor, so the central molybdenum atom is surrounded by 18 valence electrons. On the other hand, proton abstraction from C$_7$H$_8$ can lead to a complex such as **13** wherein the C$_7$H$_7^-$ ligand is analogous with the η^3-allyl ligand; that is, C$_7$H$_7^-$ is *trihapto* and is considered to donate four electrons.

The final example of a hydrocarbon π donor is cyclooctatetraene, C$_8$H$_8$. Generally this ligand functions in a two-pair donor capacity as in Fe(CO)$_3$(η^4-C$_8$H$_8$) and Co(η^5-C$_5$H$_5$)(η^4-C$_8$H$_8$).[3]

14
tricarbonyl(1,2,5,6-*tetrahapto*cyclooctatetraene)iron

15
(1,2,5,6-*tetrahapto*cyclooctatetraene)(*pentahapto*-cyclopentadienyl)cobalt

Finally, there is one example of C$_8$H$_8$ as a three-pair donor, Cr(CO)$_3$(η^6-C$_8$H$_8$), **16**.

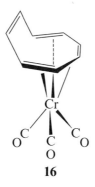

16

FROM CLASSICAL TO ORGANOMETALLIC TRANSITION METAL COMPLEXES AND THE SIXTEEN AND EIGHTEEN ELECTRON RULE

TABLE 15-1
THE DONOR/COORDINATION CHARACTERISTICS OF CERTAIN HYDROCARBON π DONORS

Ligand	Donor Pairs	Coordination
η^2-C_2H_4	1	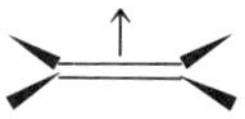
η^3-C_3H_5 (allyl)	2	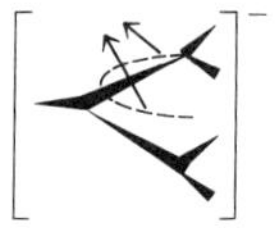
η^3-C_7H_7 (cycloheptatrienyl)		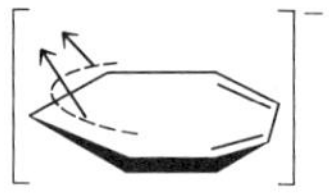
η^4-C_4H_6 (butadiene)		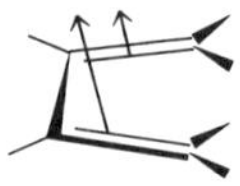
η^4-C_7H_8 (norbornadiene)		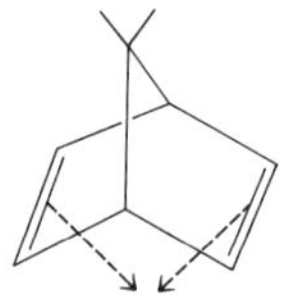
η^4-C_7H_8 (cycloheptatriene)		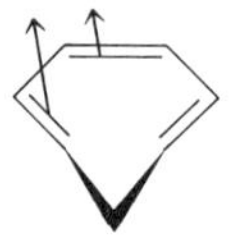
η^4-C_8H_8 (cyclooctatetraene)		
η^5-C_5H_5 (cyclopentadienyl)	3	
η^6-C_6H_6 (benzene)		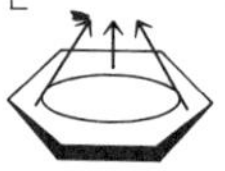
η^6-C_7H_8 (cycloheptatriene)		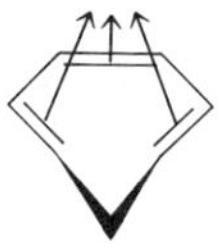
η^6-C_8H_8 (cyclooctatetraene)		
η^7-$C_7H_7^+$ (tropylium)		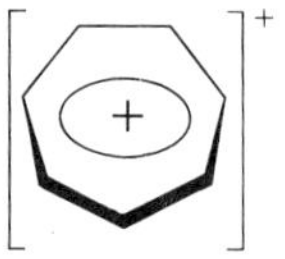

For your reference when the chemistries of these complexes are discussed in Chapters 16 and 17, Table 15-1 summarizes the donor and coordination characteristics of hydrocarbon ligands.

Among the π ligands discussed above, you have surely noticed that cycloheptatriene and cyclooctatetraene are unique in that they act as polyene rather than aromatic π systems. A structural characteristic of aromatic systems is that they are planar with all C—C distances equal; conversely, polyene molecules are generally not planar and have alternately long and short C—C distances. $C_7H_7^-$ and C_8H_8 are no exception. To see why these rings are not planar, refer to the techniques of Chapter 4 for deriving the π mo's of cyclic systems to determine that D_{7h} and D_{8h} (aromatic) geometries for these two species would have the following π orbital energy diagrams and electron configurations:

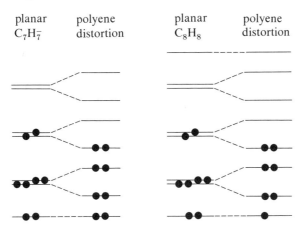

A structural distortion of the planar systems will have two important consequences. First, the symmetry-imposed degeneracy of the π HOMO's will be lifted, with the consequent loss of diradical character and stabilization of the π electron configuration. Such a distortion must be limited, of course, by the demands of the σ electrons. It is these distortions that carry the π mo's from a fully delocalized, aromatic situation into the more localized π orbitals of a polyene. Properly viewed as polyenes, these ligands have the flexibility of using less than all the π electrons as donor pairs and are not, therefore, observed to form η^7-$C_7H_7^-$ and η^8-C_8H_8 structures. On the other hand, the tropylium ion $C_7H_7^+$ (in a planar, D_{7h} geometry) does not realize a π stabilization by a degeneracy-removing distortion and is expected to act as an aromatic, η^7 ligand, which it does.

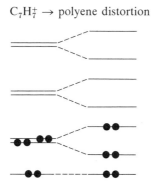

$C_7H_7^+ \rightarrow$ polyene distortion

You may have noticed that we have made no mention of MD_x cases in which an odd number of electrons surround the metal, a situation of course not "permitted" by the 16 and 18 electron rule. However, there are cases wherein the empirical formula of the molecule would lead one to believe that there is an odd number of valence electrons. For example, there are formally 17 electrons about cobalt in $Co(CO)_4$. Such a species does not exist as a stable entity under normal conditions, though, as it "dimerizes" to form $(OC)_4Co-Co(CO)_4$ **(17)**, an even-electron molecule with a metal-metal bond. Assigning one of the two electrons in the Co—Co bond to each metal, each now has a share in 18 valence electrons. Another example of metal-metal bond formation to satisfy the 18 electron rule is **18**.

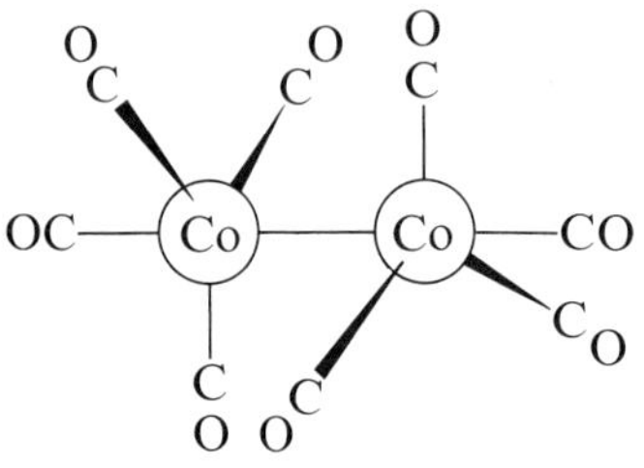

17
octacarbonyldicobalt

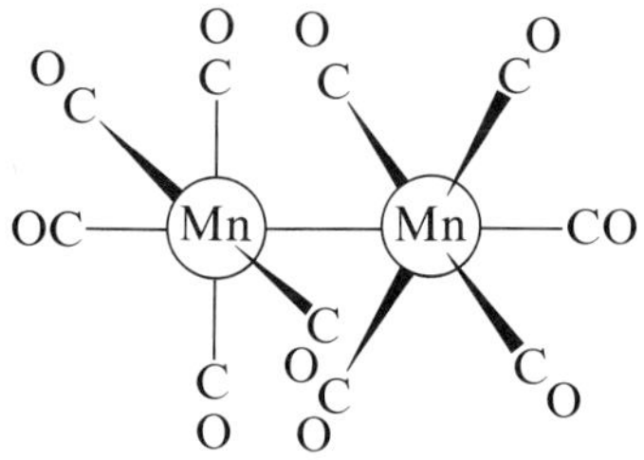

18
decacarbonyldimanganese

The subject of metal-metal bonds is a relatively new and fascinating area of transition metal chemistry, and it will be discussed in more detail in Chapter 18. For now, simply note that dimerization of formally odd electron fragments is not the only way in which metal-metal bonded species can arise. Mononuclear metal carbonyls can form complexes of higher nuclearity by losing CO and forming M—M bonds, both with and without CO bridges between the metals. The best known examples are the iron carbonyls.

$$2Fe(CO)_5 \rightarrow Fe_2(CO)_9 + CO$$
$$3Fe(CO)_5 \rightarrow Fe_3(CO)_{12} + 3CO$$

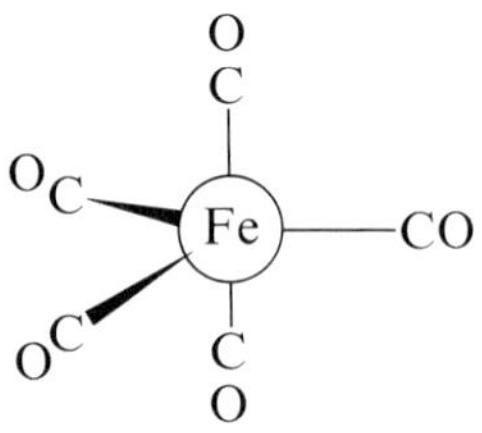

$Fe(CO)_5$
pentacarbonyliron
(D_{3h})

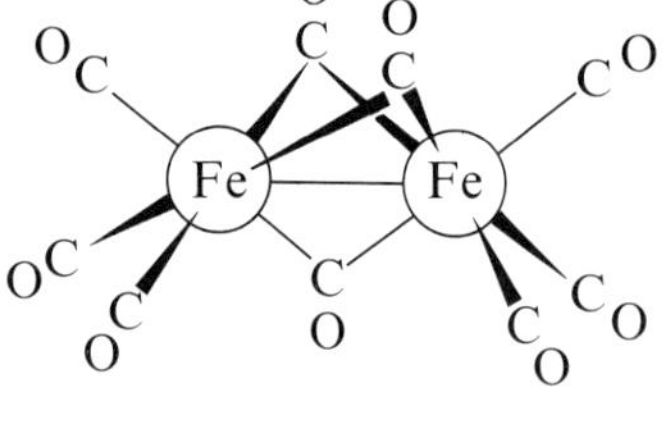

$Fe_2(CO)_9$
nonacarbonyldiiron
(D_{3h})

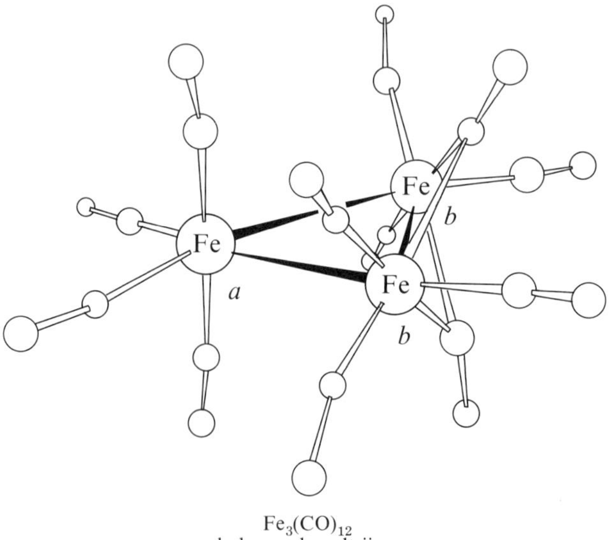

$Fe_3(CO)_{12}$
dodecacarbonyltriiron
(C_{2v})

The electron count for the bridging CO group derives from the description of the Fe—(CO)—Fe bridge bond as a three-center, two-electron case. The carbon lone pair of electrons is delocalized over both iron atoms and, therefore, makes a contribution of $\frac{1}{2}$ pair to each iron valence shell. For $Fe_2(CO)_9$ the electron count for each iron is the same by symmetry and arises as follows.

$3CO_t = 3\ \sigma$ pairs   (the subscript t means terminal)
$3CO_b = 1\frac{1}{2}\ \sigma$ pairs   (the subscript b means bridging)
$1Fe-Fe = \frac{1}{2}$ pair
$Fe = 4$ pairs
─────────
9 pairs

For $Fe_3(CO)_{12}$ there are two kinds of Fe atoms:

apex Fe ($= a$)

$4CO_t = 4\ \sigma$ pairs
$2Fe-Fe = 1$ pair
$Fe = 4$ pairs
─────────
9 pairs

base Fe ($= b$)

$3CO_t = 3\ \sigma$ pairs
$2CO_b = 1$ pair
$2Fe-Fe = 1$ pair
$Fe = 4$ pairs
─────────
9 pairs

As a final note, it is sometimes found that structures with bridging rather than terminal CO's are favored where none are "required" by the EAN. The conversion takes the general forms

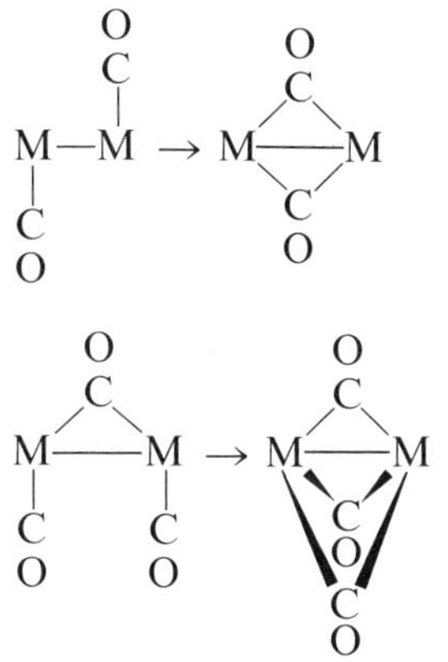

and it occurs by pairs, in accordance with the "rule of 18." Cases in point are $Fe_2(CO)_9$ and $Fe_3(CO)_{12}$. A more striking example is octacarbonyldicobalt. In hydrocarbon solvents this complex has all terminal CO groups, but in the solid state the structure exhibits two bridging CO groups. This example makes the important point that often the energy distinction between two bridging and two terminal CO groups is subtle and easily influenced by environmental factors such as solvation and lattice forces.

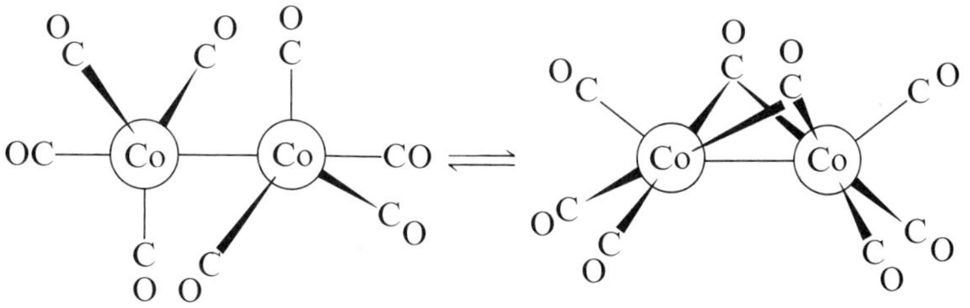

$Co_2(CO)_8$ in solution $Co_2(CO)_8$ in solid state

Occasionally, 14-electron molecules have been postulated as intermediates in reactions of organometallic compounds,[1] but such species were never adequately confirmed. Recently, however, a 14-electron *bis*(acetylene)platinum(0) complex, $Pt(PhC\equiv CPh)_2$, was reported.[3] Although more such molecules are certain to be isolated in the future, for the moment 14-electron molecules are the exception rather than the rule. Nonetheless, their existence as reaction intermediates may be more probable than was once thought.

───────────────
[3] M. Green, D. M. Grove, J. A. K. Howard, J. L. Spencer, and F. G. A. Stone, *J.C.S. Chem. Commun.*, 759 (1976).

FROM CLASSICAL TO ORGANOMETALLIC TRANSITION METAL COMPLEXES AND THE SIXTEEN AND EIGHTEEN ELECTRON RULE

THEORETICAL ASPECTS OF THE 16 AND 18 ELECTRON RULE

Having treated the 16 and 18 electron rule as an empirical fact, it is useful to turn to a more theoretical treatment of the structural foundations of the rule and thereby answer some of the following questions:

(1) What is the chief distinction between classical and organometallic complexes, and why do the latter tend to obey the 16 and 18 electron rule more frequently than the former?

(2) Organometallic complexes of the transition metals generally are based on metal atoms or ions with d^6, d^8, and d^{10} configurations. Why do d^6 metals adopt structures best described as six-coordinated (octahedral), while d^8 and d^{10} complexes are four- or five-coordinate (square planar or trigonal bipyramidal)?

(3) Why are certain 16-electron complexes stable?

(4) Why are d^{10} complexes stable in view of the fact that, with all of the d orbitals full, there is a filled, anti-bonding "d" mo for every bonding "d" mo?

The key to the difference between classical and organometallic complexes lies in the nature of the ligands. Organometallic complexes are characterized by the use of π acceptor ligands and metals of low oxidation state (and thus of low Z^*). For this reason, metal-to-ligand $d\pi \rightarrow$ ligand π^* electronic flow is possible, an effect that lowers the energy of the previously non-bonding "d" mo's, and contributes to the stabilization of the complex as illustrated in the scheme below.

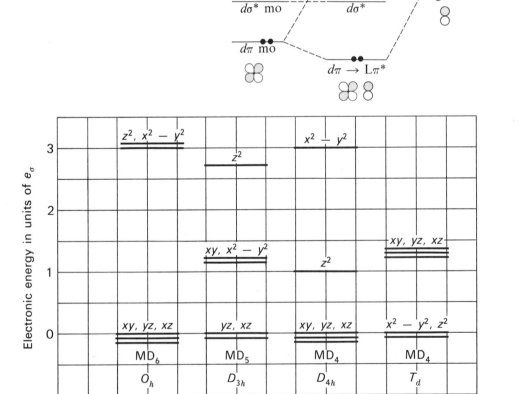

Figure 15-1. Orbital energies for MD_6 (octahedral), MD_5 (trigonal bipyramidal), MD_4 (T_d, tetrahedral), and MD_4 (D_{4h}, square planar). Units of e_σ.

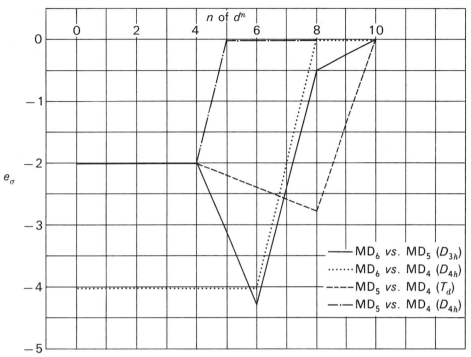

Figure 15-2. Structure preference energies for geometries typical of transition metal organometallic compounds. Units of $e\sigma$.

Furthermore, this lowering ensures electron pairing in the lowest energy "d" mo's. As seen in Figure 15-1, this means that the six d electrons (in an O_h d^6 case) occupy the three non-bonding or slightly bonding "d" mo's, leading to maximum stabilization of the complex, even when only the σ mo electrons are used to estimate the electronic energy of the complex. As explained previously (and in more detail below), d^8 complexes are generally trigonal bipyramidal or square planar. Thus, stabilization of the lower "d" mo's by ligand-to-metal π bonding would ensure that the eight d electrons pair up in the lowest energy "d" mo's, thereby again ensuring maximum complex stabilization.

Finally, as mentioned previously, the reason that organometallic compounds are formed with metals in low oxidation states is two-fold: (i) the Z^* is low enough that metal-to-ligand π electronic flow, a factor supportive of compound formation, is possible, and (ii) the necessary metal $d\pi$ mo's are sufficiently populated.

We can now turn to the structural preferences of d^6 and d^8 complexes. Figure 15-2 is a structure preference energy plot (*cf.* Chapter 9, p. 549) comparing the various structures possible for MD_6, MD_5, and MD_4. (Only the σ bonding factors from Table 9-4 are considered, so the results are independent of ligand π characteristics.) It is immediately apparent from this plot that MD_6 octahedral complexes are favored for d^6 atoms or ions relative to the other possible geometries. This is true not only because the greatest number of adduct bonds have been formed but because, at least with respect to the trigonal bipyramid and tetrahedron, none of the six electrons enters a $d\sigma^*$ mo. However, if the metal possesses an additional one or two pairs of electrons—the d^8 and d^{10} cases—you see that σ bonding effects alone markedly decrease the favored status of the MD_6 complex. This is because for *all* structures the fourth and fifth pairs must enter σ^* mo's. On the basis of σ bonding alone, the orbital characteristics create a less discriminatory bias toward the MD_6 structure. In all cases the bias is further reduced by smaller e_σ as the σ^* mo's are populated and ligand-ligand repulsions and entropy factors, which favor lower coordination numbers, gain in stature as influences on the number of coordinated ligands. (You should note that these conclusions are general and apply to all complexes without regard to the π donor or π^* acceptor character of the ligand.)

Another important conclusion that can be drawn from Figure 15-2 is that, for the d^8 case, the square planar complex is not discriminated against relative to the trigonal bipyramid. Therefore, a 16-electron MD_4 complex (eight d electrons and eight ligand electrons) is neither more nor less favored than the 18-electron MD_5 complex (eight d electrons and ten ligand electrons). (Notice, however, that the MD_5 trigonal bipyramid is considerably more stable than tetrahedral MD_4 for a d^8 metal or metal ion.) This is apparently the origin of the observation that d^8 metals or metal ions can form stable 16-electron square planar complexes. However, such complexes are known to undergo a formal oxidation as in the reaction below to give the very stable d^6 18-electron MD_6 configuration.

$$\underset{\text{16 electrons}}{\begin{array}{c}L\diagdown\;\;\;\diagup Cl\\ Ir^I\\ OC\diagup\;\;\;\diagdown L\end{array}} + HCl \rightarrow \underset{\text{18 electrons}}{\begin{array}{c}L\;\;\;H\;\;\;Cl\\ \diagdown |\diagup\\ Ir^{III}\\ \diagup|\diagdown\\ OC\;\;\;Cl\;\;\;L\end{array}}$$

Next we turn to the question of the stability of d^{10} complexes. In Chapters 9 and 10 you saw that there should be no distinction between square planar and tetrahedral geometry on the basis of the angular overlap model. However, ligand-ligand repulsions undoubtedly lead to the general observation of tetrahedral geometry for such complexes. The more important question at this time, though, is why such complexes should be stable. For every bonding electron pair there is a counterpart anti-bonding pair. If you were to conclude that a complex such as $Ni(CO)_4$ or $Pt(PEt_3)_4$ is not very stable with respect to dissociation, you would be correct; $Ni(CO)_4$ readily decomposes to metallic nickel and CO gas, and $Pt(PEt_3)_4$ loses one PEt_3 to give the more stable 16-electron molecule $Pt(PEt_3)_3$. The reason for the fact that tetrahedral d^{10} MD_4 complexes are even marginally stable lies in something we have thus far assumed to be unimportant: the active role of the metal $4s$ and $4p$ orbitals. Although these orbitals are at a considerably higher energy that the $3d$ levels for metals early in the transition series, the $4s$, $4p$, and $3d$ energies become comparable for later metals, e.g., the Ni and Cu subgroups. Thus, for the later metals, mixing the $4s$ and/or $4p$ orbitals into the d functions can lead to d mo stabilization—if orbital symmetry allows such mixing. For the MD_4 tetrahedral symmetry, the $4p$ ao's do indeed have the same symmetry properties as the metal d_{xz}, d_{yz}, and d_{xy} orbitals (cf. character tables in Chapter 2) and will serve to lower the energies of the d^* mo's that arise from these d ao's. However, no stabilization is achieved by s-d mixing in a tetrahedral molecule, since these orbitals do not have the same symmetry. In summary, d^{10} tetrahedral MD_4 molecules can exist, although their stability may be slight, because of $nd/(n+1)p$ mixing.

Finally, we wish to account for odd-electron species such as $Mn(CO)_5$ and $Co(CO)_4$ and the fact that they dimerize. For $Mn(CO)_5$ the seven metal electrons will occupy the first four orbitals of the MD_5 diagram, with the odd electron in the fourth highest ($d\sigma^*$) orbital. With a *square pyramidal* structure for MD_5 this HOMO corresponds to the $(d_{z^2})^*$ orbital, which is geometrically disposed along the C_4 axis. This is important, in that the electron is exposed to overlap with a corresponding electron of another Mn(CO)$_4$ fragment to form the dimer (OC)$_4$Mn—Mn(CO)$_4$. In the process, the anti-bonding character of the $(d_{z^2})^*$ orbital, which is due to the axial CO, is offset by virtue of the fact that the two $(d_{z^2})^*$ mo's are combined into a bonding (the metal-metal bond) and an anti-bonding pair of mo's, with the odd electrons pairing to populate the bonding member of the newly created pair of mo's.

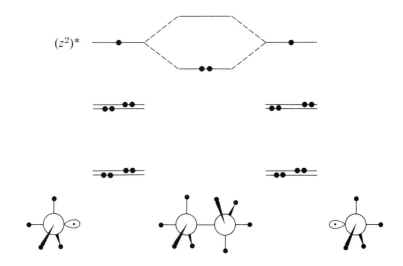

Entirely analogous arguments account for the dimerization of Co(CO)$_4$, a d^9 case in which the orbital containing the odd electron is again $(d_{z^2})^*$ and the appropriate MD$_4$ fragment for dimerization is the C_{3v} structure.[4]

---

[4] It has been assumed, for the sake of this analysis, that the structures of Mn(CO)$_5$ and Co(CO)$_4$ are C_{4v} and C_{3v}, respectively. The actual structures may well be different (trigonal bipyramidal and tetrahedral, respectively), but they are not known.

EPILOG

Stable organometallic compounds of the transition metals will have 16 or 18 valence electrons about the metal. This fact is of considerable assistance in organizing the structural and reaction chemistry of these interesting and important compounds. The purpose of this chapter was to familiarize you with the conventions for counting valence electrons and to give you some insight into the reasons for the 16 and 18 electron rule.

STUDY QUESTIONS

1. Apply the 18 electron rule to the following combinations of metals and ligands to predict possible complex stoichiometries (consider cationic, anionic, and dimeric species).

   a) Fe, CO, NO
   b) Ni, PF_3, CO
   c) Co, $C_5H_5^-$, NO
   d) Fe, norbornadiene, CO
   e) Pd, cyclooctatetraene, Cl
   f) Mn, cycloheptatriene, PF_3
   g) Fe, PF_3, H

2. Show how the valence electrons are counted in each transition metal species in the following reactions:

   a) $Ni(CO)_4 \xrightarrow{-CO} [Ni(CO)_3] \xrightarrow{PPh_3} Ni(PPh_3)(CO)_3$

   b) 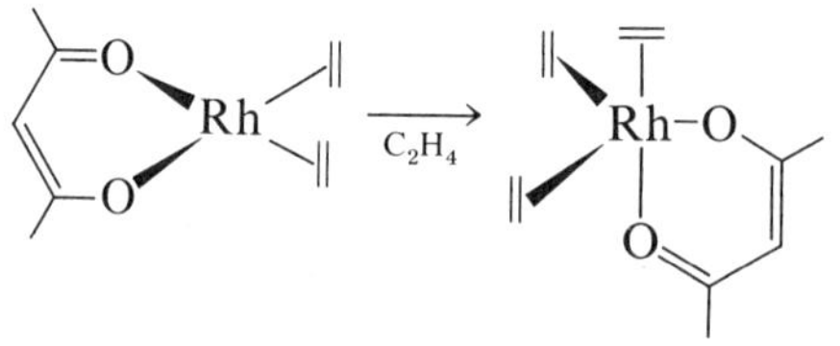 [Rh(acac)(ethylene)$_2$ type complex] $\xrightarrow{C_2H_4}$ [Rh–O complex]

   c) [Fe(CO)$_3$ diene complex] $+ Ph_3C^+ \rightarrow$ [cationic Fe(CO)$_3$ complex] $+ Ph_3CH$

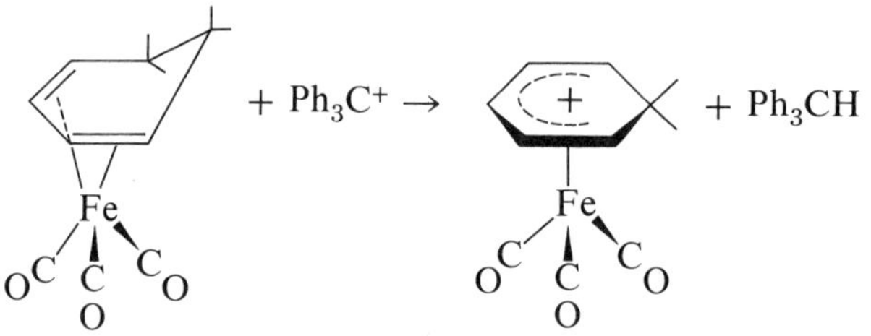

   d) 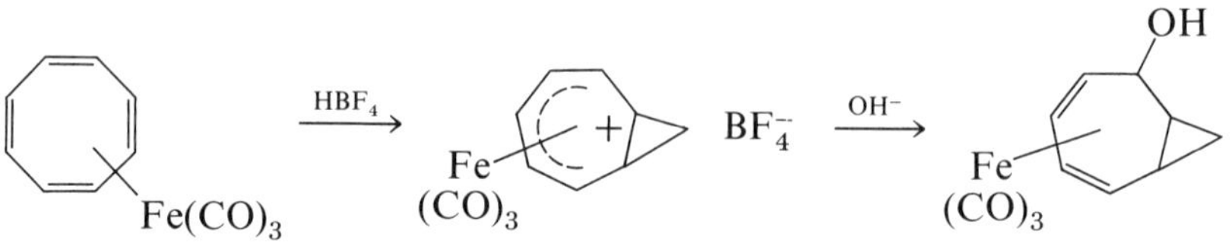 cyclooctatetraene·Fe(CO)$_3$ $\xrightarrow{HBF_4}$ [cationic Fe(CO)$_3$ complex] $\xrightarrow{BF_4^-,\ OH^-}$ [OH-substituted Fe(CO)$_3$ complex]

   e) $Cr(CO)_6 + PhLi \rightarrow (OC)_5Cr-C(OLi)(Ph) \xrightarrow[CH_2N_2]{H^+} (OC)_5Cr-C(OMe)(Ph)$

3. Metals of the first transition series form an interesting series of compounds containing $C_5H_5^-$ and CO. For the metals Cr through Ni, write the formulas for the neutral compounds containing both $C_5H_5^-$ and CO, and draw the structure of each.

4. It is believed that the catalytic role of organometallic complexes in effecting organic reactions may be understood in terms of successive steps generating 16- and 18-electron complexes from 18- and 16-electron molecules, respectively. Thus, the "hydroformylation" of olefins

$$CH_2=CHR + H_2 + CO \rightarrow RCH_2CH_2CHO$$

is catalyzed by $HCo(CO)_4$ according to the following scheme. Identify each cobalt-containing species as a 16- or 18-electron molecule.

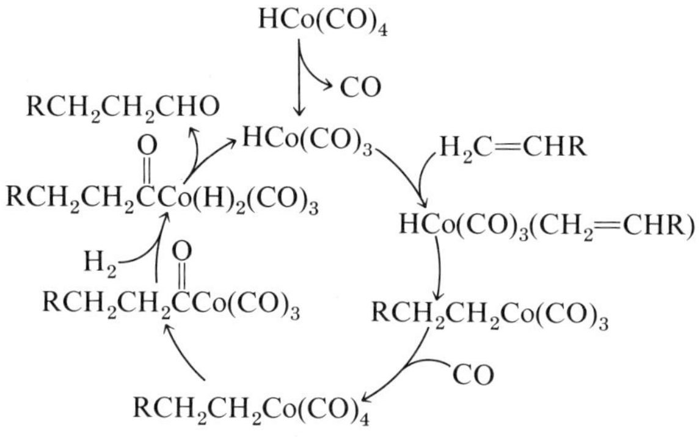

$$HCo(CO)_4 \xrightarrow{-CO} HCo(CO)_3 \xrightarrow{H_2C=CHR} HCo(CO)_3(CH_2=CHR) \rightarrow RCH_2CH_2Co(CO)_3 \xrightarrow{CO} RCH_2CH_2Co(CO)_4 \rightarrow RCH_2CH_2\overset{O}{\overset{\|}{C}}Co(CO)_3 \xrightarrow{H_2} RCH_2CH_2\overset{O}{\overset{\|}{C}}Co(H)_2(CO)_3 \rightarrow RCH_2CH_2CHO + HCo(CO)_3$$

16. ORGANOMETALLIC CHEMISTRY: SYNTHESIS, STRUCTURE, AND BONDING

INTRODUCTION

A NOTE ON THE ORGANIZATION OF ORGANOMETALLIC CHEMISTRY

THE LITERATURE OF ORGANOMETALLIC CHEMISTRY

CARBON σ DONORS

  The Synthesis of Metal Alkyls and Aryls
    Direct Reaction of a Metal with an Organic Halide
      Thermodynamic Considerations
      Experimental Considerations
    Reactions of Anionic Alkylating Agents with Metal Halides or Oxides
    Reaction of a Metal with a Mercury Alkyl or Aryl
    Metalation Reactions: Metal-Hydrogen Exchange
    Oxidative Addition Reactions
    Reactions of Metal-Containing Anions with Organic Halides
    1,2-Addition of Metal Complexes to Unsaturated Substrates
      Hydrometalations
      Oxymetalations
      Halometalations
      Organometalations
  Structure and Bonding in Metal Alkyls and Aryls
  Metal Carbonyls
    The Synthesis of Metal Carbonyls
    Metal Carbonyls: Properties and Structures
    Bonding in Metal Carbonyls
  Metal-Carbene and -Carbyne Complexes

CARBON π DONORS

  Chain π Donor Ligands (Olefins, Acetylenes, and π-Allyl)
    Synthesis of Olefin, Acetylene, and π-Allyl Complexes
    Olefin Complexes
    Acetylene Complexes
    π-Allyl Complexes
    Structure and Bonding in Olefin, Acetylene, and π-Allyl Complexes
  Complexes with Cyclic π Donors
    Synthesis and Properties
    Structure and Bonding

EPILOG

APPENDIX: METAL CARBONYLS AND INFRARED SPECTROSCOPY

*In the Prologue to this book, we surveyed the various areas of inorganic chemistry to arrive at some feeling for their current relative importance. The sub-area of organometallic chemistry was clearly prominent. To see why this should be the case, we turn now to a systematic look into this fascinating field. In this chapter we shall survey the various types of compounds with metal-carbon bonds, with particular emphasis being given to the methods by which they are synthesized, the peculiarities of the metal-carbon bond, and their structural chemistry. In organizing the field, we break the topic down into an examination of those compounds formed by σ donors—both anionic (**A**, ligand L = CH_3^-) and neutral donors (**B**, L = CO)—and compounds formed by chain (**C**, L = $H_2C{=}CH{-}CH{=}CH_2$) or cyclic (**D**, L = C_6H_6) π donors. In Chapter 17 we shall give more complete attention to the reactivity of the various types of compounds.*

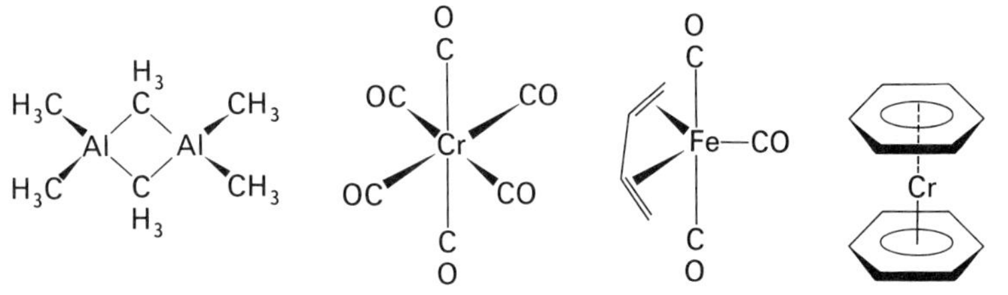

We turn now to one of the most interesting and important areas in which modern inorganic chemists are involved: organometallic chemistry, a broad, interdisciplinary field whose sphere of interest includes all compounds wherein a metal, usually in a low valence state, is bonded through carbon to an organic molecule, radical, or ion.

The beginnings of the field can be traced to the discovery of Zeise's salt in 1831,[1] the synthesis of zinc alkyls by Frankland in 1848,[2] and Victor Grignard's Nobel Prize winning synthesis and exploitation of organomagnesium halides in 1900.[3]

$$C_2H_4 + K_2PtCl_{4(aq)} \rightarrow K^+[(C_2H_4)PtCl_3]^- + KCl \qquad (16\text{-}1)$$

---

[1] W. C. Zeise, *Pogg. Ann.*, **21**, 497 (1831).
[2] E. Frankland, *J. Chem. Soc.*, **2**, 263 (1848).
[3] V. Grignard, *Compt. rend.*, **130**, 1322 (1900). Grignard received the Nobel Prize in 1912 for his work. See also E. C. Ashby, *Quart. Rev.*, **21**, 259 (1967).

(16-2) $$Zn + C_2H_5I \rightarrow C_2H_5ZnI \rightarrow \tfrac{1}{2}(C_2H_5)_2Zn + \tfrac{1}{2}ZnI_2$$

(16-3) $$Mg + C_2H_5Br \xrightarrow{ether} C_2H_5MgBr$$

Of greater importance to the more recent development of the field, however, has been the synthesis of such compounds as ferrocene[4,5] and the extensive exploration of their chemistry that followed.[6,7,8]

(16-4) $$2 \; \text{(cyclopentadienyl-MgBr)} + FeCl_2 \rightarrow \text{ferrocene (Fe(C}_5\text{H}_5)_2\text{)}$$

1

The fascination of organometallic chemistry arises not only because of the tremendously varied chemistry and structural and bonding forms displayed by these compounds, but also because of the actual and potential practical uses of such materials. Alkylaluminum compounds form the basis of the Ziegler-Natta catalyst system, which is widely used in industry for the homogeneous polymerization of ethylene and propylene. Trialkyltin oxides and acetates have found wide application as fungicides, and dialkyltin compounds are used as stabilizers—anti-oxidants and ultraviolet radiation filters—for polyvinyl chloride or rubber.[9] Finally, silicon compounds are extensively used in the form of silicones, or organosilicon-oxygen polymers, which have, among others, unique lubricating properties.[10]

In the context of a discussion of the 16 and 18 electron rule in Chapter 15, we examined some of the ligands frequently seen in organometallic chemistry and some aspects of transition metal chemistry. However, before beginning a detailed discussion of organometallic chemistry, we feel it might be useful to discuss very briefly some aspects of tin and iron chemistry so that you can place the remainder of this chapter in the proper perspective.

As illustrated in Figure 16-1, tin forms a variety of organometallic derivatives that may be somewhat arbitrarily divided into three main types: (i) The tetraalkyls and -aryls; (ii) compounds of the type $R_{4-x}SnX_x$ where X = halide, NR_2, H, OH, OR, etc.; and (iii) tin(II) compounds. Like many main group or transition metal alkyls, compounds of the type R_4Sn can be synthesized by the reaction of a metal halide, in this case $SnCl_4$, with a Grignard reagent, alkyllithium, or organoaluminum compound. Such compounds share many physical properties with their hydrocarbon analogs; for example, both Me_4Sn and 2,2-dimethylpropane are volatile (their boiling points are 76.8° and 9.5°, respectively), both are inert to air and water, and both are stable to about 400°. Chemically, however, they differ in that Me_4Sn reacts readily with halogens and $SnCl_4$; this difference may be due to the greater polarity of the

---

[4] T. J. Kealy and P. L. Pauson, *Nature*, **168**, 1039 (1951).

[5] S. A. Miller, J. A. Tebboth, and J. F. Tremaine, *J. Chem. Soc.*, 632 (1952).

[6] For a very personal discussion of the activity in the chemical community immediately following the announcement of the discovery of ferrocene, see G. Wilkinson, *J. Organometal. Chem.*, **100**, 273 (1975). As mentioned in the Prologue to this book, Wilkinson (now at Imperial College, London) and Fischer at the University of Munich shared the Nobel Prize in 1973 for their work on the chemistry of metallocenes.

[7] M. Rosenblum, "Chemistry of the Iron Group Metallocenes," Interscience Publishers, New York (1965).

[8] A very nice accounting of the history of organometallic chemistry is given by J. S. Thayer, "Organometallic Chemistry: A Historical Perspective," *Adv. Organometal. Chem.*, **13**, 1–49 (1975).

[9] P. Smith and L. Smith, *Chem. in Brit.*, **11**, 208 (1975).

[10] J. A. C. Watt, *Chem. in Brit.*, **6**, 519 (1970).

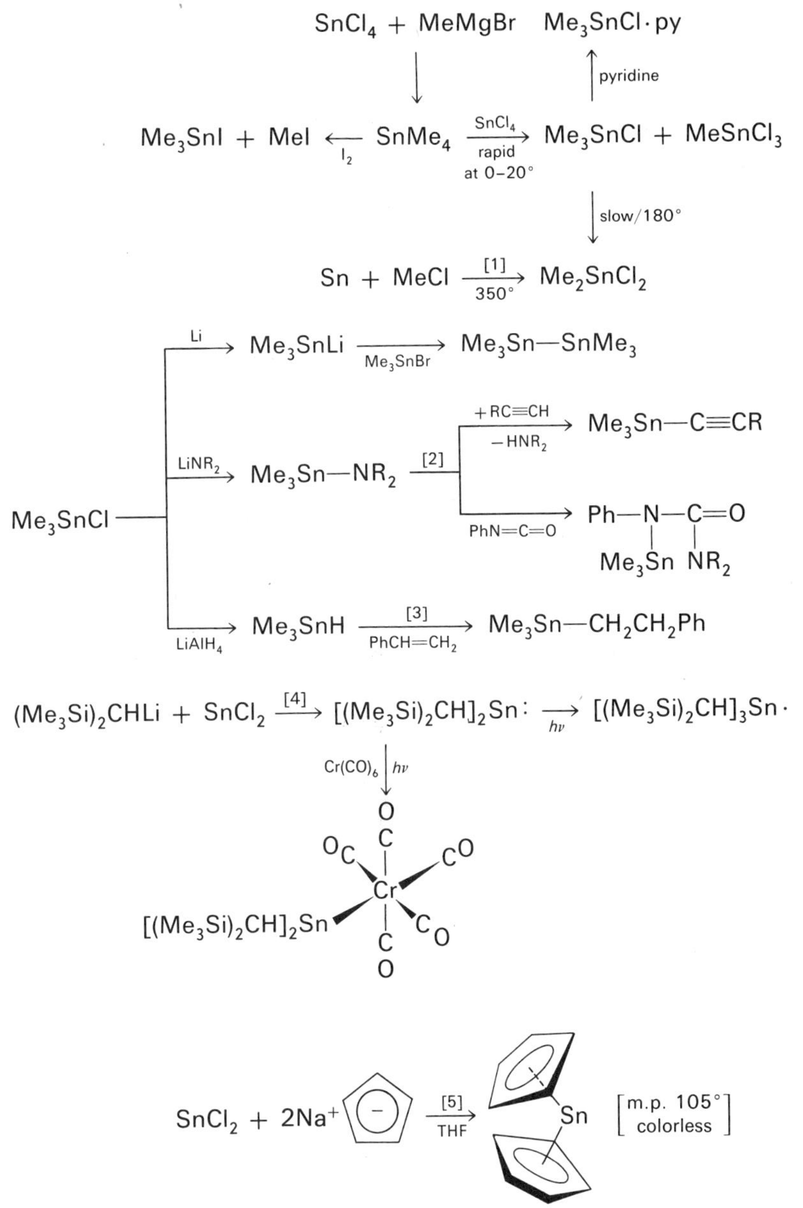

Figure 16-1. Representative organometallic chemistry of tin. Numbers in brackets indicate the following references:
1. E. G. Rochow and A. C. Smith, Jr., *J. Amer. Chem. Soc.*, **75**, 4103 (1953); J. J. Zuckerman, *Adv. Inorg. Chem. Radiochem.*, **6**, 383–432 (1964).
2. K. Jones and M. F. Lappert, *Organometal. Chem. Rev.*, **1**, 76–92 (1966).
3. H. G. Kuivila, *Acc. Chem. Res.*, **1**, 299 (1968).
4. P. J. Davidson, M. F. Lappert, and R. Pearce, *Acc. Chem. Res.*, **7**, 209 (1974).
5. P. G. Harrison and J. J. Zuckerman, *J. Amer. Chem. Soc.*, **92**, 2577 (1970).

Sn—C bond and/or the ease with which tin experiences valence shell expansion. The organotin halides, which share some chemical properties with organic halides, provide access to a great variety of other tin derivatives. The aminostannanes, R_3Sn-NR_2, and tin hydrides, R_3SnH, are of greatest synthetic importance. For example, aminostannanes undergo metal-hydrogen exchange with weakly acidic compounds and add to a variety of unsaturated 1,2-dipolar compounds. In a reaction somewhat similar to hydroboration (except in mechanism, see page 391), organotin hydrides also add to olefins. The reaction of the $(Me_3Si)_2CH^-$ ion with tin(II) chloride leads to the very interesting compound $[(Me_3Si)_2CH]_2Sn$.[11] Organometallic Sn(II) compounds usually polymerize to form cyclic stannanes with the general formula $[R_2Sn]_n$, a behavior reminiscent of carbenes. In the case of $[(Me_3Si)_2CH]_2Sn$, though, the steric bulk of the substituent apparently prevents polymerization, and the red

---

[11] P. J. Davidson, M. F. Lappert, and R. Pearce, *Acc. Chem. Res.*, **7**, 209 (1974).

crystalline compound is stable up to its melting point of 135°C. Photolysis of $[(Me_3Si)_2CH]_2Sn$ leads to the novel disproportionation reaction $2\ M^{II} \rightleftharpoons M^{I} + M^{III}$, where the Sn(III) species is the very stable free radical $[(Me_3Si)_2CH]_3Sn\cdot$. However, if a mixture of $[(Me_3Si)_2CH]_2Sn$ and $Cr(CO)_6$ is photolyzed, CO is lost by the metal carbonyl and the tin(II) compound behaves as a Lewis base in forming a Sn→Cr bond.[12] Finally, reaction of $Na^+C_5H_5^-$ with $SnCl_2$ gives bis(η^5-cyclopenta-

---

[12] J. D. Cotton, P. J. Davidson, D. E. Goldberg, M. F. Lappert, and K. M. Thomas, *Chem. Commun.*, 893 (1974).

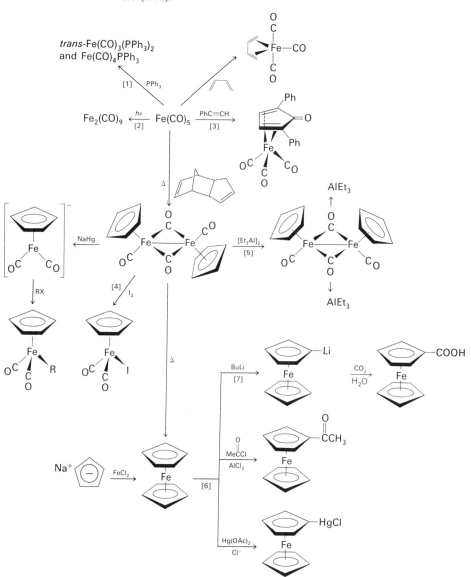

Figure 16-2. Representative organometallic chemistry of iron. Numbers in brackets indicate the following references:

1. A. F. Clifford and A. K. Mukherjee, *Inorg. Syn.*, **8**, 185 (1966). F. A. Cotton, *Acc. Chem. Res.*, **1**, 257 (1968).
2. J. J. Eisch and R. B. King, eds., "Organometallic Syntheses" (Volume 1—Transition Metal Compounds), Academic Press, New York (1965).
3. F. L. Bowden and A. B. P. Lever, *Organometal. Chem. Rev.*, **3**, 227 (1968).
4. E. C. Johnson, T. J. Meyer, and N. Winterton, *Inorg. Chem.*, **10**, 1673 (1971).
5. A. Alich, N. J. Nelson, D. Strope, and D. F. Shriver, *Inorg. Chem.*, **11**, 2976 (1972).
6. M. Rosenblum, "Chemistry of the Iron Group Metallocenes," John Wiley and Sons, New York (1965).
7. D. W. Slocum, T. R. Engelmann, C. Ernst, C. A. Jennings, W. Jones, B. Koonsvitsky, J. Lewis, and P. Shenkin, *J. Chem. Educ.*, **46**, 144 (1969).

dienyl)tin, one of the very few complexes of a main group metal with a π donor ligand.

The organometallic chemistry of iron is dominated by the reactions of pentacarbonyliron and bis(*pentahapto*cyclopentadienyl)iron or ferrocene. At the top of Figure 16-2, you see that iron carbonyl may react with phosphines or olefins to replace carbon monoxide. Generally, a maximum of two CO's are replaced. The reason for this is that, as the strongly π-acceptor CO ligand (*i.e.*, the CO ligand forms metal-to-carbon $d\pi$-$p\pi$ bonds) is replaced by ligands less capable of back-bonding, the bonds between the metal and the remaining CO's become stronger, and CO replacement becomes more difficult. The reaction with dicyclopentadiene (a source of the *pentahapto*cyclopentadienyl group), however, does result in the loss of more than two CO's from $Fe(CO)_5$, since the reaction is apparently driven by the formation of the very stable $[(\eta^5\text{-}C_5H_5)Fe(CO)_2]_2$. This dimer may be both oxidized and reduced to give derivatives from which a great number of other organometallic iron compounds may be synthesized. A reaction of great interest to the continued development of metal carbonyl chemistry in general is the addition of triethylaluminum to the oxygen atom of the bridging carbonyl groups. Pyrolysis of $[(\eta^5\text{-}C_5H_5)Fe(CO)_2]_2$ leads to the formation of ferrocene in small yields. However, commercially useful amounts of the latter are better obtained by reaction of the cyclopentadienide ion ($C_5H_5^-$, generated in any of a number of ways) and an iron(II) salt. The most outstanding property of ferrocene is the great ease with which it undergoes electrophilic substitution. For example, it may be acetylated 10^6 times faster than benzene, it reacts with $Hg(OAc)_2$ at room temperature, and it may be lithiated readily with *n*-butyllithium. These latter derivatives may in turn be used as intermediates in the syntheses of a very large number of other ferrocene derivatives.

A NOTE ON THE ORGANIZATION OF ORGANOMETALLIC CHEMISTRY

The field of organometallic chemistry is relatively new; systematic study in the area by large numbers of people did not really begin until the 1950's. When this systematic work did get underway, chemists naturally worked on the chemistry of a particular element or group of elements, and, furthermore, divided themselves rather naturally into main group or transition metal chemists. Therefore, the literature of the field and texts in the area have often sub-divided organometallic chemistry by the part of the periodic table being studied. (Indeed, we, too, have been guilty of doing this in Figures 16-1 and 16-2.) At least initially, this made some sense, because there is some periodicity in the types of metal-carbon bonds formed (see Figure 16-3). In the past five years or so, however, it has become increasingly obvious that such divisions are artificial. Heretofore it was thought that so-called electron deficient compounds such as dimeric trimethylaluminum, Me_6Al_2, would be limited to a few elements in Groups IA, IIA, and IIIA; however, recent work has uncovered examples in the *d*-block elements.[13] Until recently it was accepted as a truism that simple alkyl derivatives of the transition metals of the type MR_x (called homoleptic metal alkyls by Lappert)[14] were intrinsically unstable in comparison with analogous main group derivatives. However, recent work, especially by Lappert and Wilkinson,[13] has shown that transition metal alkyls of great stability may in fact be obtained rather easily. The only ligands that seem to favor strongly one type of metal over another are cyclic π donors such as benzene and σ acids/π bases such as CO. Complexes of such ligands among the main group elements are unstable.

Because there no longer seems to be any reason to organize organometallic chemistry by element type, we have chosen to organize these chapters by ligand and

---

[13] See ref. 11.
[14] A homoleptic compound is one in which all groups bound to the metal are the same (*e.g.*, $SnMe_4$); in contrast, a heteroleptic compound contains two or more different groups (*e.g.*, Me_3SnCl).

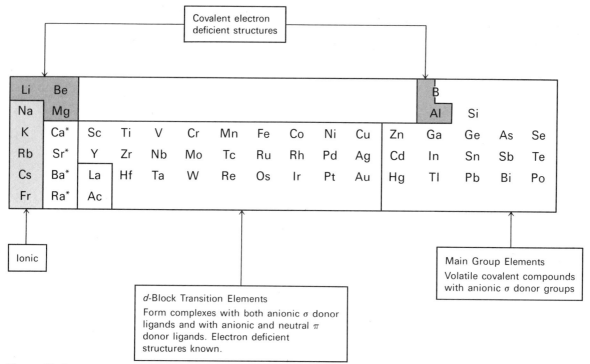

Figure 16-3. Types of bonding in organometallic compounds. (*The nature of metal-carbon bonding in alkaline earth compounds has not been clearly established.)

reaction type. As we mentioned in Chapter 5, there are in general two types of carbon donor ligands: σ and π. Therefore, in this chapter we shall first discuss molecules formed by σ donors—metal alkyls and aryls, metal carbonyls, and metal carbenes and carbynes—and then turn to complexes formed between metals and π donors (that is, molecules such as ferrocene, **1**; see also Figure 16-2).

Within each class of compounds, our concern will be with the structures of typical compounds, with current ideas of bonding, and with synthetic pathways and methods. We have given synthesis some emphasis, just as we did in our previous discussions of non-metal chemistry and coordination compounds. The "how to do it" aspect is all too often missing in discussions of descriptive chemistry, and we hope this deficiency is rectified here. Therefore, before discussing the details of structure and bonding, we have first outlined methods for the syntheses of various types of compounds and have attempted to convey some impression of their general physical and chemical properties. A more detailed discussion of the chemical reactions of organometallic compounds is given in Chapter 17.

One last note before beginning your study of organometallic chemistry. The 16 and 18 electron rule (or "effective atomic number," EAN, rule) will be used extensively in the discussions of the organometallic chemistry of transition metal compounds. This topic was covered in Chapter 15, and we urge you to review the appropriate sections. Further, a facility with nomenclature is important, and you might re-read Chapter 15 (and the nomenclature section of Chapter 9) in this regard.

THE LITERATURE OF ORGANOMETALLIC CHEMISTRY

The Journal of Organometallic Chemistry (Elsevier Sequoia, SA) is a journal solely dedicated to compounds having metal-carbon bonds. This journal publishes original research papers, area reviews, book reviews, and, very importantly, an annual survey of organometallic chemistry. The *Journal of the American Chemical Society,* the *Journal of the Chemical Society* (*Dalton Transactions* and *Chemical Communications*), *Inorganic*

Chemistry, and *Chemische Berichte* also publish research papers of importance in organometallic chemistry. Another invaluable source of information is the annual review series, *Advances in Organometallic Chemistry,* edited by R. West and F. G. A. Stone (Academic Press). Finally, the review series *Advances in Inorganic Chemistry and Radiochemistry,* edited by H. J. Emeleus and A. G. Sharpe (Academic Press) and *Progress in Inorganic Chemistry,* edited by S. J. Lippard (John Wiley and Sons) publish occasional articles on special aspects of organometallic chemistry.

In addition to journals and review series, a general source of information is the third edition of *Organometallic Compounds* by G. E. Coates, M. L. H. Green, and K. Wade (Methuen) (Volume I, The Main Group Elements, G. E. Coates and K. Wade, 1967; Volume II, The Transition Elements, M. L. H. Green, 1968). These two volumes have been condensed into a shorter book useful as a text: *Principles of Organometallic Chemistry* by G. E. Coates, M. L. H. Green, P. Powell, and K. Wade, Methuen, 1968.

Finally, the synthesis of common organometallic compounds has been very well detailed in *Organometallic Syntheses* (Volume I—Transition-Metal Compounds, R. B. King, 1965) edited by J. J. Eisch and R. B. King (Academic Press).

CARBON σ DONORS

Organometallic compounds formed from carbon σ donor ligands may be subdivided into two classes: (a) compounds in which the organic group can be considered an anionic σ donor, and (b) compounds in which the organic group is a neutral σ donor and π acceptor. Compounds of the first type include metal alkyls and aryls such as phenyl sodium (an ionic compound according to Figure 16-3); dimeric trimethylaluminum $[(CH_3)_6Al_2]$ (2) (a volatile "electron deficient" compound according to Figure 16-3);[15] a volatile, covalent compound such as $(CH_3)_4Sn$ (see Figure 16-1); or an involatile transition metal compound such as $(\eta^5\text{-}C_5H_5)Fe(CO)_2(CH_3)$ (see Figure 16-2). Type (b) compounds are almost exclusively formed by the transition metals in their low valence states, and they are exemplified by complexes with carbon monoxide and carbene (3)[16] and with isonitriles (4).[17]

Because of the natural division of carbon σ donor ligands into two sub-types, this first section will follow this general organizational scheme; that is, we shall consider first the synthesis, structure, and bonding in compounds of type (a) and then those same aspects for complexes with carbon monoxide.

THE SYNTHESIS OF METAL ALKYLS AND ARYLS[18]

In the sections that follow, we attempt to introduce to you, very briefly, the variety of ways currently available for the synthesis of metal-carbon σ bonds. As you

---

[15] For the structure of trimethylaluminum, see R. G. Vranka and E. L. Amma, *J. Amer. Chem. Soc.,* **89**, 3121 (1967).

[16] M. Green, J. R. Moss, I. W. Nowell, and F. G. A. Stone, *Chem. Commun.,* 1339 (1972).

[17] J. K. Stalick and J. A. Ibers, *J. Amer. Chem. Soc.,* **92**, 5333 (1970).

[18] The best general source of information on metal alkyl and aryl synthesis is the set of books by Coates, Green, and Wade.
  a. "Organometallic Compounds" (Volume I, G. E. Coates and K. Wade, "The Main Group Elements"), Methuen, London (1967).
  b. "Organometallic Compounds" (Volume II, M. L. H. Green, "The Transition Elements"), Methuen, London (1968).

read through these sections, note the relative advantages and disadvantages of the different methods. Notice also the similarities and differences between main group metals and transition metals. Their syntheses and chemistry show many similarities, but they are often different in some important ways. It is these differences that make the comparative study of main group and transition metal compounds interesting.

Direct Reaction of a Metal with an Organic Halide

The direct reaction of metals with alkyl and aryl halides may be used to produce numerous organometallic derivatives. Because of the simplicity of the reaction, the products are frequently of industrial importance. Furthermore, such reactions often produce many of the more reactive representatives of each class of compounds. For example, direct reaction of alkyl halides with Li, Mg, Al, or Zn (*e.g.*, reactions 16-2 and 16-3) leads to some of the most reactive metal alkyls known; these compounds may then be used to produce alkyl derivatives of metals for which the direct reaction does not work.

In 1966, Rochow published a short article in the *Journal of Chemical Education* in which he described the synthesis of organometallic compounds by the direct reaction of organic halides with metals.[19] At that time, few if any transition metals could be included in his survey. However, very recently the technique of "metal atom" synthesis has come into use (see page 824),[20,21] and a number of transition metals are now known to react directly with organic halides. Therefore, a more current listing of metals that can react directly with organic halides is given in Figure 16-4. (This figure also includes, for future reference, those metals that can react directly with CO or with hydrocarbons such as olefins and arenes. Such reactions are discussed later in the chapter.)

The synthesis of metal alkyls and aryls by the direct reaction of metals with organic halides is often affected in a simple and direct way by the thermodynamics of the reaction. Therefore, before discussing the experimental conditions used in the direct reaction, it is useful to examine briefly the thermodynamic parameters of the M + RX reaction to see of what value they may be in a predictive and evaluative sense.

Thermodynamic Considerations. Considerable thermodynamic data have been accumulated for reactions of the metals of Groups IIB, IIIA, and IVA with organic halides, and some are plotted in Figure 16-5.[22]

For the following general reaction:

(16-5)
$$2 M_{(s)} + x\, CH_3Cl_{(g)} \rightarrow M(CH_3)_{x(g)} + MCl_{x(g)}$$

we have plotted in Figure 16-5 the heats of formation of the metal alkyl and the metal halide, as well as the heat of reaction. In general, the reaction becomes less exothermic as the atomic weight of the metal increases. Figure 16-5 shows that there are several reasons for this: both $\Delta H_f^\circ[MCl_x]$ and $\Delta H_f^\circ[M(CH_3)_x]$ become less exo-

---

[19] E. G. Rochow, *J. Chem. Educ.*, **43**, 58 (1966).
[20] P. L. Timms, *Adv. Inorg. Chem. Radiochem.*, **14**, 121 (1972).
[21] K. J. Klabunde, *Acc. Chem. Res.*, **8**, 393 (1975).
[22] The thermodynamic data plotted in Figure 16-5 are taken from the following sources:
   a. H. A. Skinner, *Adv. Organometal. Chem.*, **2**, 49–114 (1964).
   b. D. D. Wagner, W. H. Evans, V. B. Parker, I. Halow, S. M. Bailey, and R. A. Schumm, "Selected Values of Chemical Thermodynamic Properties," National Bureau of Standards, Technical Notes 270-3 and 270-4.
   c. M. F. Lappert, J. B. Pedley, J. Simpson, and T. R. Spalding, *J. Organometal. Chem.*, **29**, 195 (1971).
   d. J. C. Baldwin, M. F. Lappert, J. B. Pedley, and J. S. Poland, *J. C. S. Dalton*, 1943 (1972) and references therein.
   e. J. D. Cox and G. Pilcher, "Thermochemistry of Organic and Organometallic Compounds," Academic Press, New York (1970).

1 H																		2 He
3 Li ◆	4 Be ◇											5 B	6 C	7 N	8 O	9 F	10 Ne
11 Na ◆□	12 Mg ◇□											13 Al ◆□	14 Si ◆	15 P ◆	16 S	17 Cl	18 A
19 K ◆□	20 Ca ◇	21 Sc	22 Ti □	23 V □	24 Cr □	25 Mn	26 Fe ■	27 Co ■□	28 Ni ■□	29 Cu ◆■	30 Zn ◆	31 Ga	32 Ge ◆	33 As ◆	34 Se ◇	35 Br	36 Kr
37 Rb ◆□	38 Sr ◇	39 Y	40 Zr □	41 Nb	42 Mo ■□	43 Tc	44 Ru ■	45 Rh ■	46 Pd ◆□	47 Ag ◆	48 Cd ◆	49 In	50 Sn ◆	51 Sb ◆	52 Te ◆	53 I ◆	54 Xe
55 Cs ◆□	56 Ba ◇	57 La *	72 Hf	73 Ta	74 W ■□	75 Re	76 Os	77 Ir	78 Pt ◆□	79 Au ◆	80 Hg ◆	81 Tl ◆	82 Pb ◆	83 Bi ◆	84 Po	85 At	86 Rn
87 Fr	88 Ra	89 Ac *															

◆ react with RX with or without solvent
◇ react with RX in a solvent
■ react with CO
□ react with unsaturated hydrocarbon

Figure 16–4. The chemical elements that react directly with organic halides, carbon monoxide, or unsaturated hydrocarbons. [Adapted from E. G. Rochow, *J. Chem. Educ.*, **43**, 58 (1966).]

820 ORGANOMETALLIC CHEMISTRY: SYNTHESIS, STRUCTURE, AND BONDING

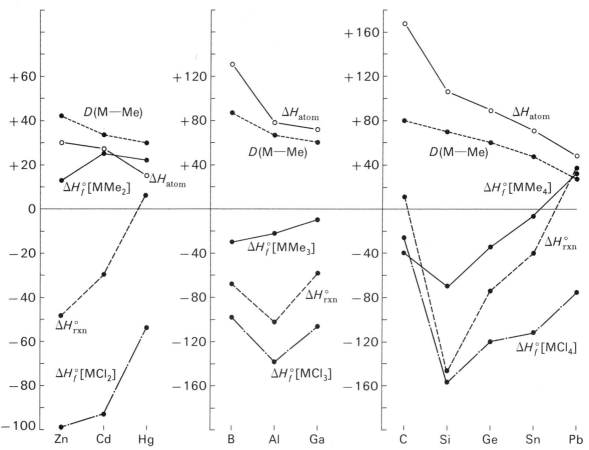

Figure 16-5. Periodic plots of thermodynamic quantities involved in $2 M_{(s)} + x\, CH_3Cl_{(g)} \rightarrow M(CH_3)_{x(g)} + MCl_{x(g)}$. Note that $\Delta H_f^\circ[CH_3Cl_{(g)}] = -19.32$ kcal/mol and that $\Delta H_f^\circ[MCl_2]$ is referred to the solid phase for Group IIB metals.

thermic with increasing metal atomic weight. In fact, it would appear that the heat of formation of the metal halide is more important in determining the heat of reaction than is the heat of metal alkyl formation. That is, in general there is a closer parallel between ΔH_{rxn}° and $\Delta H_f^\circ[MCl_x]$ than between ΔH_{rxn}° and $\Delta H_f^\circ[M(CH_3)_x]$. However, even a reasonably exothermic heat of formation of the metal halide cannot offset the effects of a weak metal-carbon bond. You will observe that metal-carbon σ bond strengths generally decline with increasing atomic weight and forecast a less exothermic heat of formation of the metal alkyl.

The fact that the methyl derivatives of the lighter elements have more exothermic heats of formation and greater bond energies—and are therefore thermodynamically more stable than the derivatives of the heavier elements—does not necessarily imply greater ease of preparation; quite the contrary. Apparently, another quantity plotted in Figure 16-5—the heat of metal atomization, ΔH_{atom}—plays a role in the kinetics of metal alkyl formation. For example, lithium and magnesium have low heats of atomization (38.4 and 35.1 kcal/g.at., respectively), and their alkyls may be prepared by reaction of the metal and organic halide in a solvent at room temperature. In contrast, direct reaction of Al with CH_3Cl requires heating the mixture to about 120°C, the Group IVA elements Si, Ge, and Sn require temperatures of about 300°C, and boron reacts with methyl chloride (to produce BCl_3 as the only isolable product) at about 500°C.

Experimental Considerations. **Alkyl lithium compounds** are usually prepared by the reaction of a lithium metal dispersion with an alkyl chloride or bromide (reaction 16-6) in a solvent such as petroleum ether, cyclohexane, benzene, or ether; the reaction must be carried out in an atmosphere of dry nitrogen because of the air-sensitivity of the organometallic product. Generally speaking, alkyl iodides cannot be used, because unreacted RI can quickly react with LiR to produce the coupling product as shown in reaction 16–7.

$$2\ Li + n\text{-}C_4H_9Cl \rightarrow n\text{-}C_4H_9Li + LiCl \tag{16-6}$$

$$2\ Li + C_2H_5I \rightarrow C_2H_5Li + LiI$$
$$\xrightarrow{C_2H_5I} C_4H_{10} + LiI \tag{16-7}$$

The choice of solvent for the synthesis of alkyl and aryl lithium compounds is important. The most useful commercially available compound, n-butyllithium, is sold as a 1 molar solution in heptane or some other hydrocarbon solvent; if properly protected from air and moisture, these solutions are stable for extended periods of time. On the other hand, a solution of n-butyllithium in diethyl ether is halved in concentration in one week at 25°, because alkyl and aryl lithium compounds attack ethers, although to different degrees. Nucleophilic attack by the alkyl or aryl carbanion at the hydrogen α to the ether oxygen gives rise to butane and an intermediate lithium complex, which then decomposes by olefin elimination (16–8).[23]

$$n\text{-}C_4H_9Li + (C_2H_5)_2O \text{---} CH_3CH(Li)OC_2H_5 + C_4H_{10}$$
$$\downarrow$$
$$LiOC_2H_5 + C_2H_4 \tag{16-8}$$

Sodium alkyls and aryls are not usually made by direct reaction of sodium with organic halides for the simple reason that the very reactive organosodium compound can react rapidly with any excess organic halide to produce a coupled product in a reaction sequence similar to that illustrated by reaction 16–7. This is, of course, the familiar Wurtz-Fittig coupling reaction. In order to prepare sodium alkyls or aryls on a laboratory scale, the method of choice is the reaction of sodium with a mercury alkyl or aryl, a reaction used to produce organic derivatives of many different metals; this reaction is discussed in more detail on page 825.

The class of organometallic compounds perhaps most familiar to chemists in general is that of the Grignard reagents; *i.e.,* **alkyl- or arylmagnesium halides.** These compounds are also made directly from the metal and organic halides; again, the chloride and bromide are used in preference to the iodide in order to reduce coupling reactions. In contrast with organolithium reagents, however, Grignard reagents do not attack ethers to any appreciable extent; indeed, the solvent of choice in the preparation and use of organomagnesium halides is an ether.

Organic derivatives of mercury are extremely important as synthetic reagents. Alkylmercury halides and dialkylmercury compounds may be prepared by reaction of mercury with the organic halide, but only if the metal is in the form of a sodium amalgam. That this reaction requires the use of amalgamated mercury is further evidence for the importance of thermodynamics in determining the course of some organometallic reactions.[23a] From the following equations, you see that there is little driving force for the direct reaction of mercury with methyl chloride. However, if the chloride is taken up in the form of its sodium salt, the very high heat of formation of NaCl (-98.23 kcal/mole) provides the driving force for the reaction.

$$2\ Hg_{(liq)} + 2\ CH_3Cl_{(g)} \rightarrow Hg(CH_3)_{2(g)} + HgCl_{2(s)} \quad \Delta H°_{rxn} = +7.3\ \text{kcal}$$
$$\underline{HgCl_{2(s)} + 2\ Na_{(s)} \rightarrow Hg_{(liq)} + 2\ NaCl_{(s)} \quad \Delta H°_{rxn} = -142.9\ \text{kcal}} \tag{16-9}$$

$$Hg_{(liq)} + 2\ Na_{(s)} + 2\ CH_3Cl_{(g)} \rightarrow$$
$$Hg(CH_3)_{2(g)} + 2\ NaCl_{(s)} \quad \Delta H°_{rxn} = -135.6\ \text{kcal} \tag{16-10}$$

---

[23] R. L. Burwell, *Chem. Rev.*, **54**, 615 (1954); S. C. Honeycutt, *J. Organometal. Chem.*, **29**, 1 (1971).
[23a] Here again you find application of the ideas discussed on p. 290 and following.

Since the preparation of an organometallic compound by the direct reaction of a metal with an organic halide is formally an oxidation-reduction reaction, it is evident that the reaction will work only when the metal is sufficiently electropositive. It is partly for this reason that organoboron compounds cannot be prepared by the direct method. However, the reaction is possible for the other members of group IIIA, especially aluminum. Aluminum metal—in the form of foil or turnings—reacts smoothly with alkyl halides in a hydrocarbon solvent to give alkylaluminum sesquihalides. Indeed, this reaction has been suggested as a useful, if somewhat spectacular, lecture demonstration.[24]

(16-11)
$$2\ Al + 3\ C_2H_5I \rightarrow (C_2H_5)_3Al_2I_3$$

The sesquihalides are, like all organoaluminum compounds, extremely air-sensitive, reacting readily with oxygen and water; however, they do not have sharp boiling points [for example, $(CH_3)_3Al_2Cl_3$ has a boiling range of 127° to 148°], as they are in reality an equilibrium mixture of organoaluminum halides.

(16-12)
$$2\ (CH_3)_3Al_2Cl_3 \rightleftharpoons (CH_3)_2Al_2Cl_4 + (CH_3)_4Al_2Cl_2$$

Aluminum trialkyls may be produced from the sesquihalides by reduction or disproportionation. In the reduction method, metallic sodium reacts with the sesquihalide to give the trialkylaluminum compound and metallic aluminum.

(16-13)
$$2\ (CH_3)_3Al_2Cl_3 + 6\ Na \rightarrow [(CH_3)_3Al]_2 + 2\ Al + 6\ NaCl$$

The disproportionation method involves the addition of some reagent, e.g., $[(CH_3)_3Al]_2$, that causes the exclusive formation of the dimeric dialkylaluminum halide as a first step (16-14).

(16-14)
$$2\ (CH_3)_3Al_2Cl_3 + [(CH_3)_3Al]_2 \rightleftharpoons 3[(CH_3)_2AlCl]_2$$

The addition of sodium fluoride can then produce trimethylaluminum in a continuous process involving three steps (16-15). This represents a highly efficient method of obtaining trimethylaluminum, since very little metal is lost in the process.

(16-15)
$$(CH_3)_4Al_2Cl_2 + 2\ NaF \rightarrow (CH_3)_4Al_2F_2 + 2\ NaCl$$
$$(CH_3)_4Al_2F_2 + 2\ NaF \rightarrow 2\ Na^+[(CH_3)_2AlF_2]^-$$
$$3\ Na^+[(CH_3)_2AlF_2]^- \rightarrow Na_3AlF_6 + [(CH_3)_3Al]_2$$

Group IVA elements also react directly with alkyl halides, but only when catalyzed by copper or some other suitable metal. The most important and widely studied system is the reaction of elemental **silicon** with methyl chloride in the presence of a copper catalyst at 285°C (16-16). This industrially important reaction was first described in the literature by Rochow after World War II, although the problem of the efficient large-scale production of organosilanes was solved in 1940 by Rochow's group at General Electric.[25]

The overall reaction of silicon and methyl chloride is simple,

(16-16)
$$2\ CH_3Cl + Si \xrightarrow[235°C]{Cu} (CH_3)_2SiCl_2$$

but many by-products such as $SiCl_4$, CH_3SiCl_3, and $(CH_3)_3SiCl$ are obtained. Under appropriate conditions, the product consists of approximately 65% of $(CH_3)_2SiCl_2$,

---

[24] E. Krahe and E. G. Rochow, *J. Chem. Educ.*, **43**, 63 (1966).
[25] E. G. Rochow and A. C. Smith, Jr., *J. Amer. Chem. Soc.*, **75**, 4103 (1953); J. J. Zuckerman, *Adv. Inorg. Chem. Radiochem.*, **6**, 383-432 (1964).

25% of CH_3SiCl_3, and 5% of $(CH_3)_3SiCl$; the remainder is a complex mixture of products. After careful separation of dimethyldichlorosilane from the other products, the former is hydrolyzed under controlled conditions to yield commercially important polyorganosiloxanes or silicones. The function of the copper catalyst is not entirely clear, but the most recent view is that it brings about the formation of methyl radicals and transforms methyl chloride into a more reactive form by a reaction such as 16–17.

$$2\ Cu + CH_3Cl \rightarrow CuCl + CuCH_3 \tag{16-17}$$

Methylcopper, like many other simple transition metal alkyls, is quite unstable toward decomposition to the free metal and methyl radicals.

Germanium and **tin** react with methyl chloride to produce methylchlorogermanes and methylchlorostannanes, even in the absence of a copper catalyst.[26] However, as expected, the catalyst lowers the temperature required for the germanium reaction from over 400° to about 350°. No catalyst is required for the tin reaction; 50% yields of $(CH_3)_2SnCl_2$ may be obtained by passing CH_3Cl through molten tin (300–350°). This is in agreement with your expectations based on the thermodynamic arguments discussed previously; the data in Figure 16–5 show that the heats of atomization of the Group IVA elements decrease in the order Si (+108 kcal/g.at.) > Ge(+90 kcal/g.at.) > Sn(+72 kcal/g.at.) > Pb(+47 kcal/g.at.).

The direct reaction of **lead** with ethyl chloride is used to produce the currently important and controversial gasoline additive, tetraethyllead. This reaction, which occurs only if the lead is in the form of a sodium alloy, is one of the best examples of the importance of thermodynamics in a commercial process.

$$\begin{array}{ll} 4\ C_2H_5Cl_{(liq)} + 2\ Pb_{(s)} \rightarrow Pb(C_2H_5)_{4(liq)} + PbCl_{4(liq)} & \Delta H°_{rxn} = +68.3\ kcal \\ \underline{4\ Na_{(s)} + PbCl_{4(liq)} \rightarrow 4\ NaCl_{(s)} + Pb_{(s)}} & \underline{\Delta H°_{rxn} = -317.9\ kcal} \\ 4\ C_2H_5Cl_{(liq)} + Pb_{(s)} + 4\ Na_{(s)} \rightarrow & \\ \quad 4\ NaCl_{(s)} + Pb(C_2H_5)_{4(liq)} & \Delta H°_{rxn} = -249.6\ kcal \end{array} \tag{16-18}$$

[26] See ref. 25.

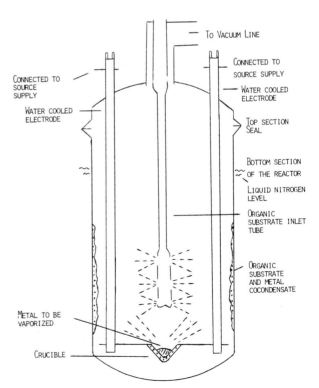

Figure 16–6. Apparatus used for the vaporization of metals and the co-deposition of the metal vapor with an organic substrate. [Reprinted with permission from K. J. Klabunde, *Acc. Chem. Res.*, **8**, 394 (1975). Copyright by the American Chemical Society.]

824 ORGANOMETALLIC CHEMISTRY: SYNTHESIS, STRUCTURE, AND BONDING

In the past decade, techniques have been developed for the direct reaction of metal atoms with inorganic and organic substrates.[27,28] The apparatus that is currently available commercially is sketched in Figure 16-6. The entire apparatus is evacuated to $<10^{-5}$ torr and the bottom portion is immersed in liquid nitrogen ($t = -196°C$). A sample of metal held in the crucible is then vaporized by heating it resistively, and the resulting metal atom vapor is condensed on the cold outer walls of the apparatus. At the same time, a volatile substrate is codeposited on the walls. When metal deposition is complete, the walls are allowed to warm slowly. Reaction generally occurs at liquid nitrogen temperatures or slightly above.

As indicated in Figure 16-4, a number of transition metals have been vaporized and may then react directly with organic halides and hydrocarbons. For example, palladium (which has a heat of vaporization of 88 kcal/mol) can be readily vaporized. If palladium atoms are cocondensed with C_6F_5Br, an orange-brown powder is isolated (reaction 16-19).[29]

(16-19) $C_6F_5Br + Pd(\text{condensed vapor}) \rightarrow [C_6F_5\text{-PdBr}]_x$

$\downarrow 2PR_3$

$C_6F_5\text{-Pd(PR}_3)_2\text{-Br}$

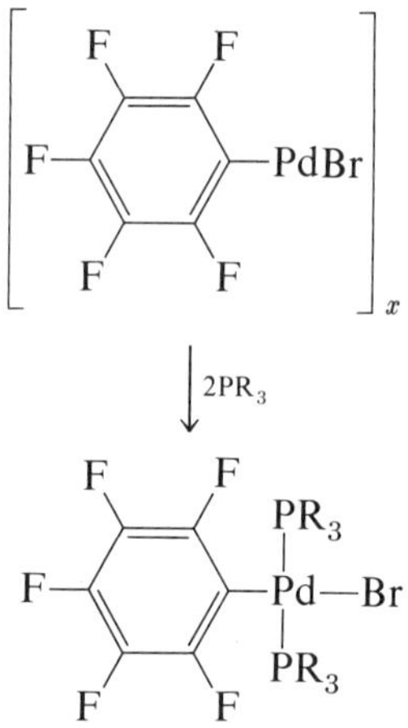

This solid proves to be C_6F_5-Pd-Br, a compound that is reasonably stable to air and moisture, soluble in organic solvents (where it is found to be dimeric and trimeric), and very reactive with Lewis bases. Later in this chapter we shall discuss briefly the cocondensation of metal vapors and arenes to give metal-arene complexes. This is an immense area of chemistry, and the next decade should see significant advances both in technique and in the new areas of chemistry available.

Reactions of Anionic Alkylating Agents with Metal Halides or Oxides

Perhaps the most useful and generally applied method of synthesis of metal-carbon σ bonds is the reaction of a metal halide or oxide with an alkyl or aryl derivative of lithium, sodium, magnesium, or aluminum.

(16-20) (norbornadiene)PtI$_2$ + 2PhMgI → (norbornadiene)PtPh$_2$ + 2MgI$_2$

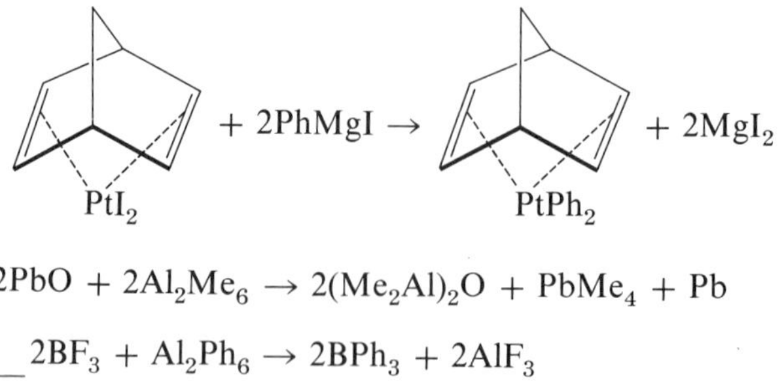

(16-21) $2PbO + 2Al_2Me_6 \rightarrow 2(Me_2Al)_2O + PbMe_4 + Pb$

(16-22) $2BF_3 + Al_2Ph_6 \rightarrow 2BPh_3 + 2AlF_3$

---

[27] a. M. P. Silvon, E. M. Van Dam, and P. S. Skell, *J. Amer. Chem. Soc.*, **96**, 1945 (1974).
   b. E. M. Van Dam, W. N. Brent, M. P. Silvon, and P. S. Skell, *J. Amer. Chem. Soc.*, **97**, 465 (1975).
[28] See refs. 20 and 21.
[29] K. J. Klabunde and J. Y. F. Low, *J. Amer. Chem. Soc.*, **96**, 7674 (1974).
[30] C. R. Kistner, J. H. Hutchinson, J. R. Doyle, and J. C. Storlie, *Inorg. Chem.*, **2**, 1255 (1963).
[31] M. Boleslawski and S. Pasynkiewicz, *J. Organometal. Chem.*, **43**, 81 (1972).

Of these methods, the most useful one for preparing transition metal–carbon σ bonds is the Grignard reaction, although the products are not always predictable. For example, reaction of diiodo(η^4-norbornadiene)platinum(II) with methylmagnesium iodide instead of phenylmagnesium iodide (as in 16-20) leads only to the monomethylated product.[30]

Alkyl and aryl aluminum compounds are also excellent reagents for transferring organic groups to other metals (reactions 16-21 and 16-22).[31] Reaction 16-21, for example, is thought to proceed by the electrophilic attack of the aluminum-containing Lewis acid on the oxygen. The ultimate product of this first step is dimethyllead, a compound that is unstable to disproportionation to tetramethyllead and lead metal.

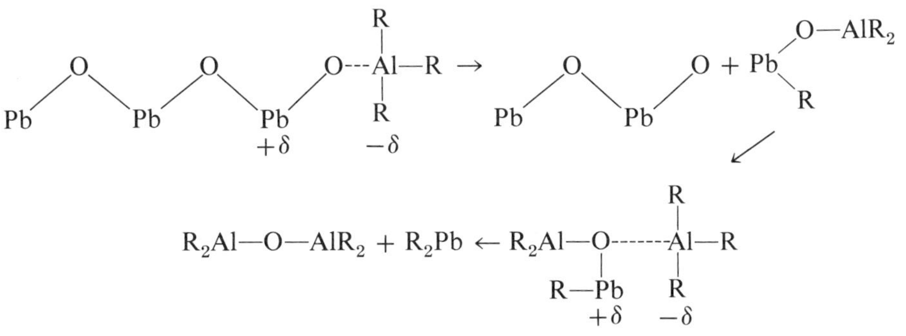

$$2R_2Pb \rightarrow Pb + PbR_4$$

(16-21′)

Reaction of a Metal with a Mercury Alkyl or Aryl

Just as with the first two methods of metal–carbon bond synthesis, the efficacy of this method is due in large part to thermodynamics. For example, ΔH°_{rxn} for the interaction of metallic aluminum with dimethylmercury is -114 kcal.

$$3\ HgMe_{2(liq)} + 2\ Al_{(s)} \rightarrow 3\ Hg_{(liq)} + Al_2Me_{6(liq)} \qquad (16\text{-}23)$$

There are two ways to think of the M + R_2Hg reaction; that is, two ways to predict whether it will apply to a particular metal. Firstly, recognize that it is formally an oxidation–reduction reaction. Therefore, it should apply best to the more electropositive metals; indeed, only the elements of Groups IA, IIA, and IIIA (except boron) have been observed to participate. Alternatively, one would predict that the reaction will form any metal alkyl whose M—C bond energy is greater than the Hg—C bond energy; since mercury–carbon bonds are notoriously weak (Figure 16-5), it should be possible to transfer the mercury alkyl or aryl group to a great many other metals. In practice, however, this is not always the case; experimentally it is found that only those metals that react directly with organic halides under relatively mild conditions will also react with organomercury compounds. Again, as with M + RX reactions, this may simply be because the Group IVA metals and boron, for example, have relatively high heats of atomization, a factor that apparently affects the kinetics of the reaction.

One of the great advantages of the M + R_2Hg reaction is that it can be run in non-ethereal solvents. Therefore, it can be used to prepare ether-free compounds of aluminum, gallium, or beryllium, compounds that are exceptionally good Lewis acids (see Chapters 5 and 17) and would bind ether in the product. In addition, it may be used for the preparation of organosodium compounds [recall that the direct reaction (M + RI) could not be used in this case because of radical coupling reactions] and organolithium compounds.

Metalation Reactions: Metal-Hydrogen Exchange

Reactions of this type have been studied for a number of years and are more generally used to prepare compounds with metal-carbon σ bonds than would appear at first glance.[32] However, the method deserves continuing study, as the full scope of such reactions is surely not yet known, nor are their mechanisms completely understood.

Lithiation reactions (reactions 16–24 to 16–26) are thought to involve nucleophilic attack of the hydrocarbon portion of the Li-containing reagent (*e.g.*, the butyl group) on a hydrogen atom of the compound undergoing metalation.[33]

(16–24) [phenoxathiin] + $n\text{-}C_4H_9Li$ → [lithiated phenoxathiin] + $n\text{-}C_4H_{10}$

(16–25) [dibenzofuran] $\xrightarrow{n\text{-}C_4H_9Li}$ [lithiated dibenzofuran] $\xrightarrow[H_2O]{CO_2}$ [dibenzofuran-COOH]

(16–26)[34] [ferrocene] + $n\text{-}C_4H_9Li$ → [lithioferrocene] + $n\text{-}C_4H_{10}$

Thus, in order for metalation to occur, the H atom must be relatively acidic. Several lines of evidence point in this direction. First, the lithiation of phenoxathiin with *n*-butyllithium (reaction 16–24) occurs *ortho* to the oxygen rather than to the sulfur owing to the greater $-I$ effect (inductive electron withdrawal) of the oxygen atom.[32]

---

[32] H. Gilman, *Adv. Organometal. Chem.*, **7**, 1–152 (1968).
[33] H. Gilman and J. W. Morton, *Organic Reactions*, **8**, 258 (1954).
[34] D. W. Slocum, T. R. Engelmann, C. Ernst, C. A. Jennings, W. Jones, B. Koonvitsky, J. Lewis, and P. Shenkin, *J. Chem. Educ.*, **46**, 144 (1969).

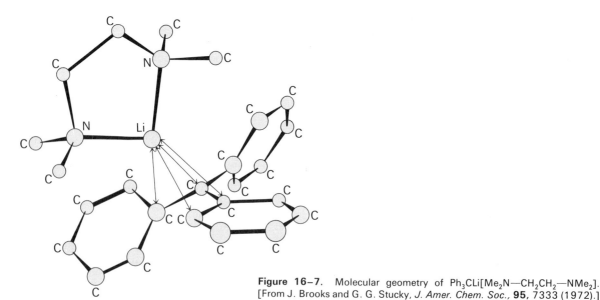

Figure 16–7. Molecular geometry of $Ph_3CLi[Me_2N\text{—}CH_2CH_2\text{—}NMe_2]$. [From J. Brooks and G. G. Stucky, *J. Amer. Chem. Soc.*, **95**, 7333 (1972).]

Second, considerably higher yields of lithiated products are obtained in polar, coordinating solvents. Third, remarkable metalations can be performed if tetramethylethylenediamine (TMED) is added to the system. Such solutions will monolithiate benzene and will dilithiate ferrocene (one Li on each ring) in high yield in a few hours at room temperature.[35,36] The reason for this is that the lithium ion is strongly complexed by the diamine, thereby rendering the organic group even more carbanionic.‡ Several such complexes have been isolated and their structures determined crystallographically; the structure of the TMED-LiC(C_6H_5)$_3$ complex is shown in Figure 16-7.[37] This complex is best thought of as a contact ion pair between a triphenylmethyl carbanion and a lithium ion coordinated to a TMED group.

Additional examples of metal-hydrogen exchange are those in reactions 16-27[38] and 16-28.[39]

$$Ph_3Al + PhC\equiv CH \xrightarrow[\text{benzene}]{20-50°} Ph_2Al-C\equiv C-Ph + PhH \qquad (16-27)$$

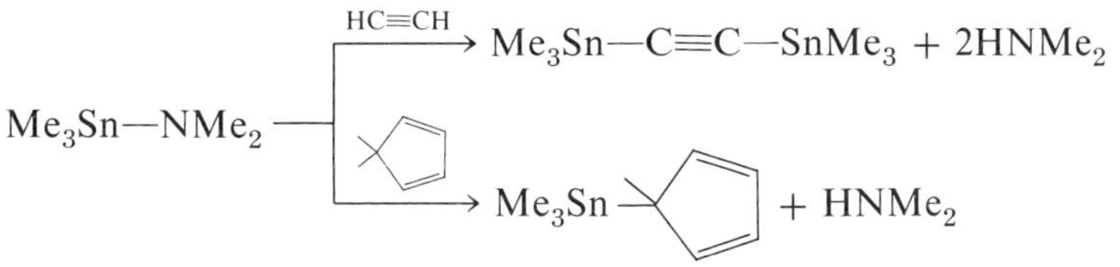

$$Me_3Sn-NMe_2 \begin{array}{c} \xrightarrow{HC\equiv CH} Me_3Sn-C\equiv C-SnMe_3 + 2HNMe_2 \\ \\ \longrightarrow Me_3Sn\text{–}C_5H_5 + HNMe_2 \end{array} \qquad (16-28)$$

Their mechanisms, however, may be somewhat different from that proposed for the lithiation reaction. For example, in contrast to the facile metalation of phenylacetylene with (C_6H_5)$_3$Al, indene (5), a hydrocarbon having an acidity comparable to that of phenylacetylene ($pk_a = 21$), reacts with triphenylaluminum only to a very small extent.[38] Therefore, a four-center transition state (6), which is apparently formed by electrophilic attack of the metal on the C≡C bond, has been suggested for the acetylene reaction. Such an intermediate is not possible with indene, since the latter has no donor π-electrons at the benzylic carbon (i.e., the carbon bearing the normally acidic hydrogens). The reactions of aminostannanes with compounds bearing "acidic" hydrogens also presumably proceed through a four-center transition state.[39]

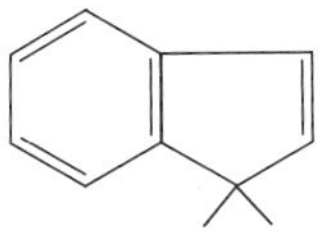

5 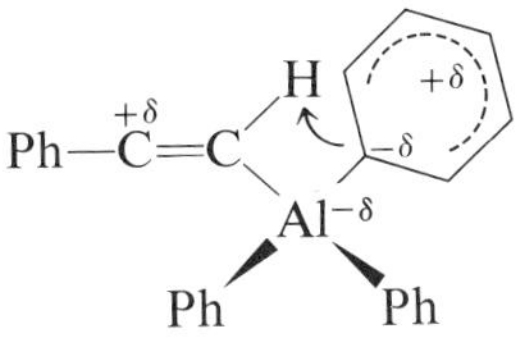   **6**

Intramolecular metalation reactions (e.g., 16-29 and 16-30) involving transition metals have only recently been discovered and hold great promise for providing numerous new, stable molecules containing metal-carbon σ bonds.[40]

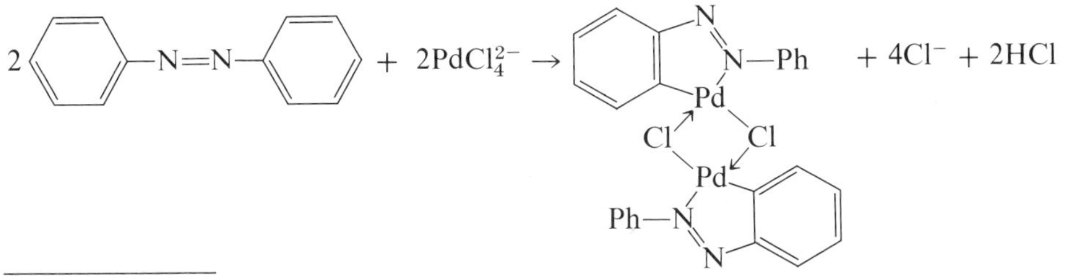

$$2\,Ph-N=N-Ph + 2PdCl_4^{2-} \rightarrow [\text{dimeric Pd-azobenzene chloride bridged complex}] + 4Cl^- + 2HCl \qquad (16-29)$$

---

[35] M. D. Rausch and D. J. Ciappenelli, *J. Organometal. Chem.*, **10**, 127 (1967).
[36] See ref. 34.
‡ Furthermore, TMED breaks up the less reactive RLi tetramer.
[37] J. Brooks and G. D. Stucky, *J. Amer. Chem. Soc.*, **94**, 7333 (1972).
[38] J. J. Eisch and W. Kaska, *J. Organometal. Chem.*, **2**, 184 (1964).
[39] K. Jones and M. F. Lappert, *Organometal. Chem. Rev.*, **1**, 76-92 (1966).
[40] G. W. Parshall, *Acc. Chem. Res.*, **3**, 139 (1970).

(16-30) $[(PhO)_3P]_4RuHCl \rightarrow$ [(PhO)₃P]₃Ru(Cl)(κ²-P,C-OC₆H₄-P(OPh)₂) $+ H_2$

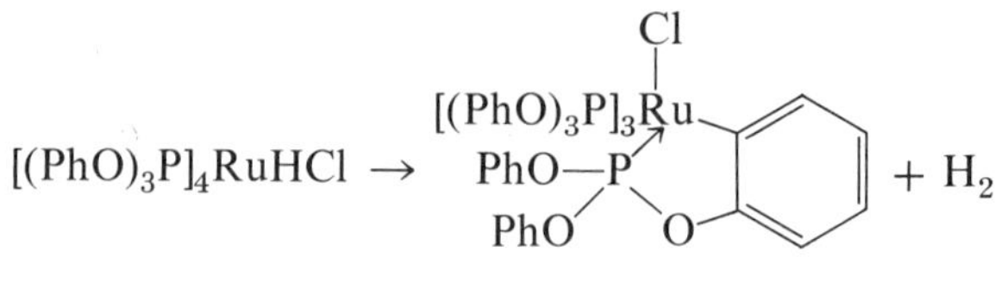

They are formally analogous to the reactions discussed above, but there is evidence that they differ mechanistically. Using azobenzene with a substituent in the *para* position on one ring (reaction 16-29), preferential substitution was observed on the ring bearing the electron-donating substituent.[41] Therefore, intramolecular metalations are probably electrophilic substitutions, and a mechanism reflecting that fact has been proposed.[40] In this mechanism, the formation of a normal coordination complex is followed by the elimination of H^+, the latter step being facilitated by the formation of a π-arene complex.

(16-29′)

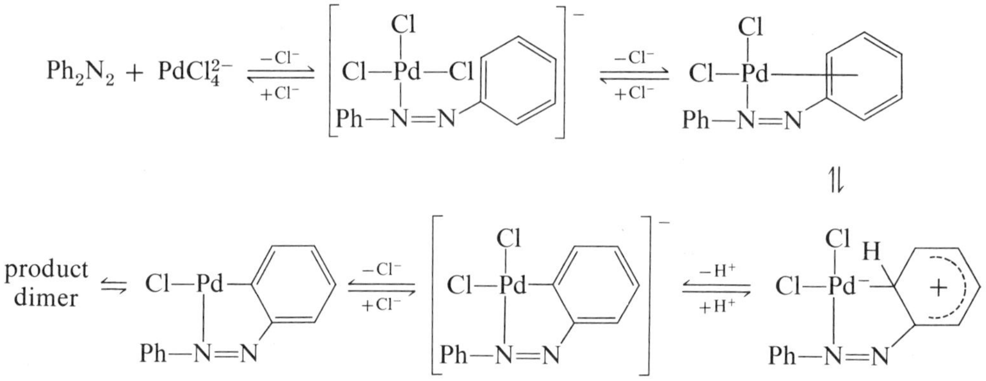

Mercuration reactions (*e.g.*, 16-31) are important examples of metal-hydrogen exchange,[42] and they represent a particularly good way of introducing a metal into an aromatic molecule. The reaction of benzene with mercuric acetate in acetic acid at elevated temperatures gives phenylmercuric acetate in good yield; ferrocene, however, gives ferrocenylmercuric acetate under much milder conditions. As ferrocene is known to undergo electrophilic substitutions with considerably more ease than benzene, these results, and others, lead to the conclusion that mercurations are electrophilic substitutions that follow a classical mechanistic pattern (16-31′). The product of a mercuration reaction can, of course, be used as an intermediate in the synthesis of other organometallics.

(16-31) $C_6H_6 + Hg(OAc)_2 \rightarrow C_6H_5-Hg-OAc + HOAc$

(16-31′)

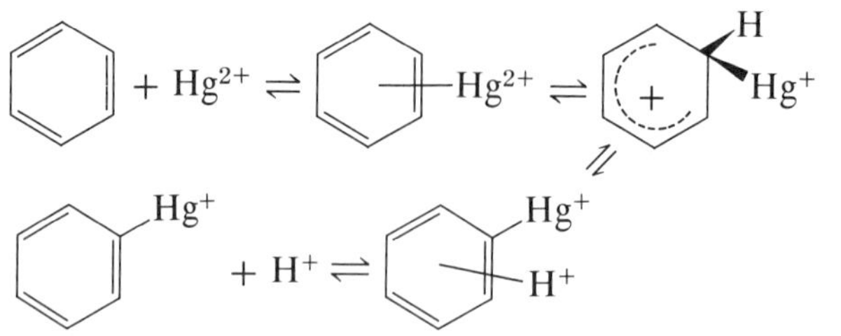

---

[41] H. Takahashi and J. Tsuji, *J. Organometal. Chem.*, **10**, 511 (1967).
[42] W. Kitching, *Organometal. Chem. Rev.*, **3**, 35 (1968).

Oxidative Addition Reactions

Oxidative addition reactions have generally come to mean a class of reactions of transition metal complexes in which oxidation, *i.e.*, an increase in *formal* oxidation number of the central metal, is accompanied by an increase in the coordination number.[43-45] Such reactions are commonly observed for compounds having metals in low spin d^7, d^8, or d^{10} configurations. Examples of oxidative addition reactions that lead to the formation of metal-carbon σ bonds are outlined below. In each case, the number of d electrons of the central metal is given, as well as the number (in parentheses) of valence electrons surrounding that metal.

d^7 (17 valence shell electrons) → d^6 (18 valence shell electrons)

$$ML_5 \xrightarrow[+L']{-1e^- \text{ from M}} ML_5L'$$

$$2\,[Co(CN)_5]^{3-} + MeI \rightarrow [Co(CN)_5Me]^{3-} + [Co(CN)_5I]^{3-} \qquad (16\text{-}32)^{46}$$

d^8 (16 valence shell electrons) → d^6 (18 valence shell electrons)

$$ML_4 \xrightarrow[+2L']{-2e^- \text{ from M}} ML_4(L')_2$$

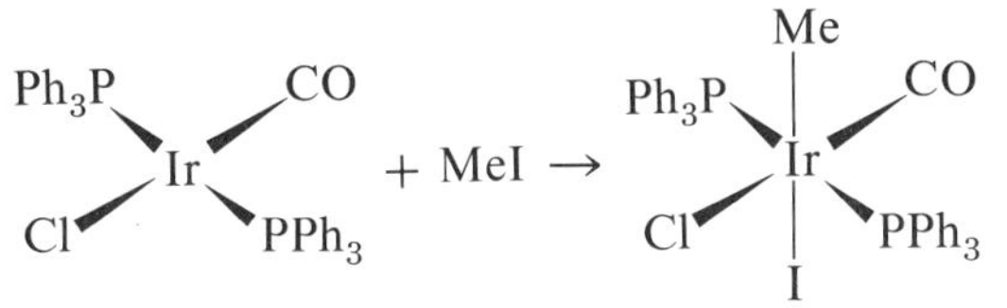

$$(16\text{-}33)^{47}$$

$$ML_5 \xrightarrow[-L + 2L']{-2e^- \text{ from M}} ML_4(L')_2$$

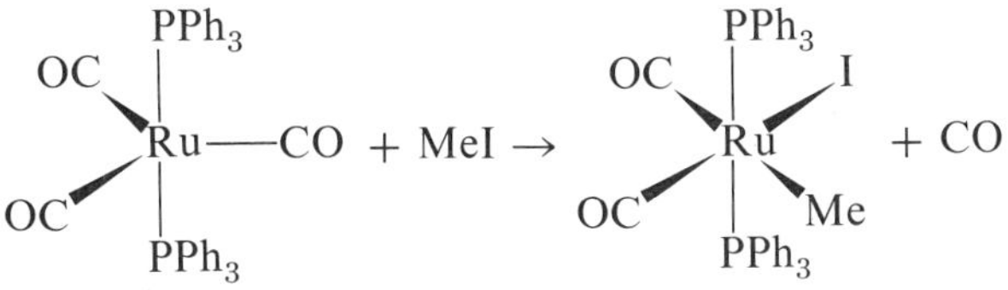

$$(16\text{-}34)^{43}$$

d^{10} (16 valence shell electrons) → d^8 (16 valence shell electrons)

$$ML_3 \xrightarrow[-L + 2L']{-2e^- \text{ from M}} ML_2(L')_2$$

$$M(PEt_3)_4 \rightleftharpoons PEt_3 + M(PEt_3)_3$$

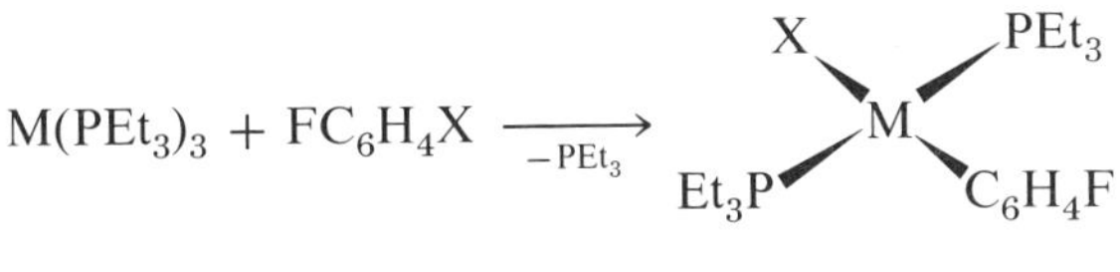

(M = Ni, Pd, Pt)
$$(16\text{-}35)^{48}$$

---

[43] J. Halpern, *Acc. Chem. Res.*, **3**, 386 (1970).
[44] J. P. Collman and W. R. Roper, *Adv. Organometal. Chem.*, **7**, 53 (1968).
[45] Oxidative addition reactions were first encountered in Chapters 5 and 7.
[46] See page 688ff. in Chapter 12.
[47] J. P. Collman and C. T. Sears, Jr., *Inorg. Chem.*, **7**, 27 (1968).
[48] G. W. Parshall, *J. Amer. Chem. Soc.*, **96**, 2360 (1974).

Although oxidative addition reactions will be discussed in much greater detail in Chapter 17, we can make the point here that an obvious driving force for such reactions is the attainment of the highly favored 18-electron configuration (see Chapter 15). In addition, it might be noted that the reactions of the d^7 complex $[Co(CN)_5]^{3-}$ are thought to follow a free radical mechanism (see Chapter 12, p. 690), while those of the d^8 and d^{10} complexes involve either a free radical mechanism or an A or I_a type of attack by the metal complex on the electrophilic alkyl or aryl halide.[49,50]

Reactions of this type are now recognized as being far more general than previously realized. For example, reaction 16-30, an intramolecular metalation discussed earlier, can be thought of as proceeding *via* an oxidative addition followed by a reductive elimination step.

(16-30')

$$[(PhO)_3P]_4RuHCl \xrightarrow{-(PhO)_3P} [(PhO)_3P]_3RuHCl$$

$Ru^{2+} = d^6$ $Ru^{2+} = d^6$
18 valence e^- total   16 valence e^- total

[Reaction scheme showing intermediate structures with Ru complexes, oxidative addition to form $Ru^{4+} = d^4$, 18 valence e^- total species, then $-H_2$ (reductive elimination) to give $Ru^{2+} = d^6$, 16 valence electrons species, then $+(PhO)_3P$ to give $Ru^{2+} = d^6$, 18 valence electrons product.]

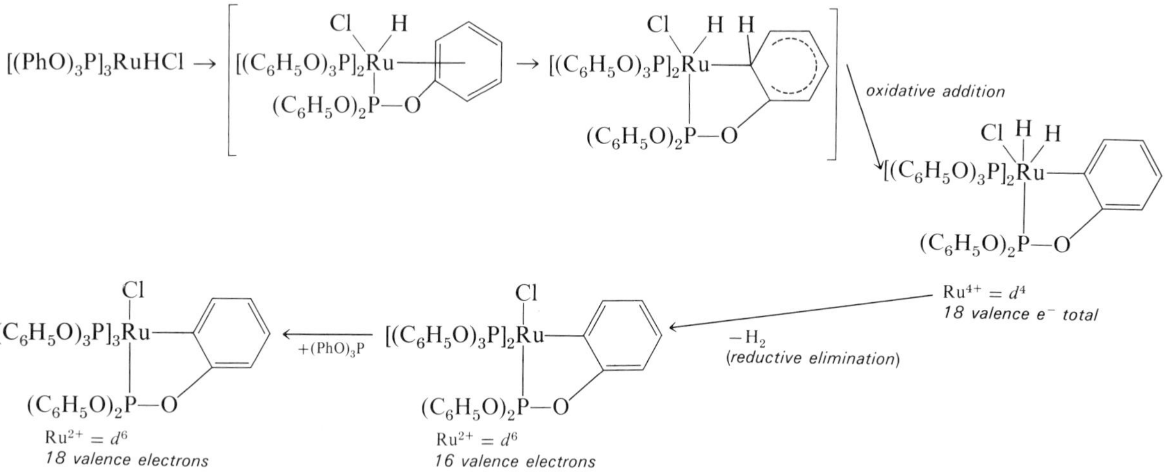

Furthermore, the reactions of the next section that involve transition metal-containing anions can also be thought of as oxidative additions. In any event, because of their obvious generality and importance to the whole of organometallic chemistry and homogeneous catalysis, we shall examine oxidative additions in detail in Chapter 17.

Reactions of Metal-Containing Anions with Organic Halides

Reactions such as these are most commonly observed for anions containing Group IVA metals and those of the transition metal series. However, complex ions containing transition metals are of greater importance, with metal carbonylates, or anions derived from metal carbonyls, being especially important.[51,52] Some reactions of the molybdenum-containing anion $[\eta^5\text{-}C_5H_5Mo(CO)_3]^-$ are outlined in the next three equations. The first step in each of them is the straightforward metathetical exchange of anions. In 16-38, however, CO is lost to produce a π-benzyl derivative (*cf.* allyl).

---

[49] See ref. 43, and M. F. Lappert and P. W. Lednor, *Adv. Organometal. Chem.*, **14**, 345 (1976).
[50] See Chapter 7 for kinetic nomenclature.
[51] R. B. King, *Adv. Organometal. Chem.*, **2**, 157–256 (1964).
[52] R. B. King, *Acc. Chem. Res.*, **3**, 417–427 (1970).

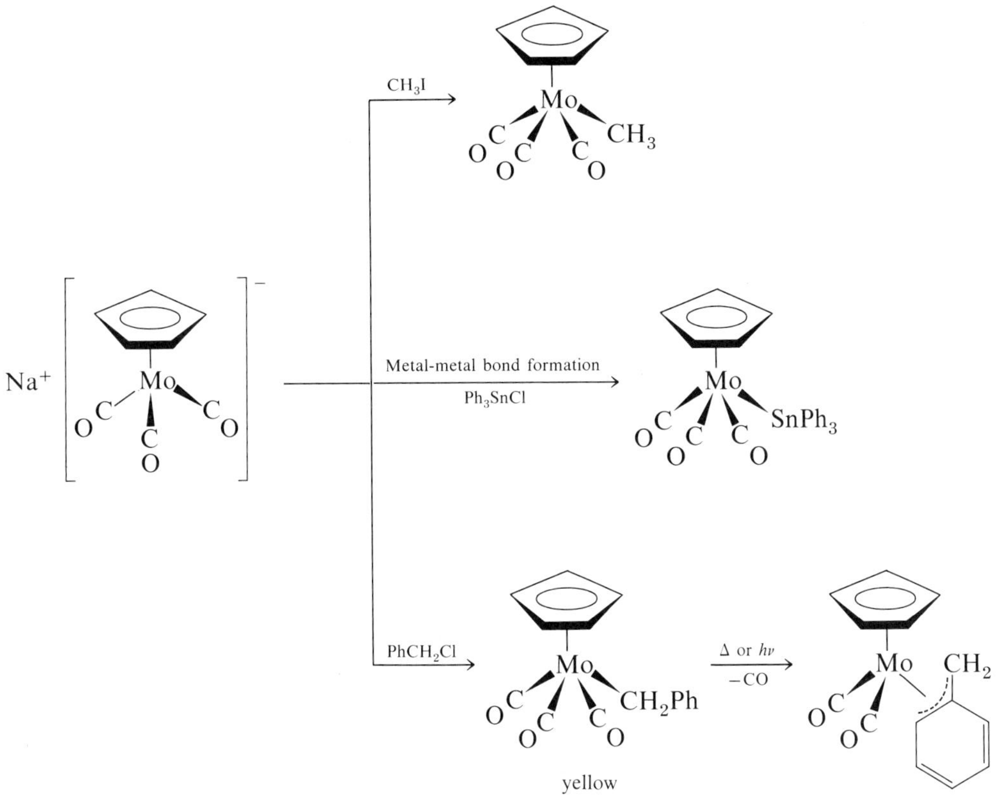

CARBON σ DONORS   831

(Scheme: Na⁺[CpMo(CO)₃]⁻ reacts with:
- CH₃I → CpMo(CO)₃CH₃ (16-36)[53]
- Ph₃SnCl (metal–metal bond formation) → CpMo(CO)₃SnPh₃ (16-37)[54]
- PhCH₂Cl → CpMo(CO)₃CH₂Ph (yellow) →(Δ or hν, −CO) CpMo(CO)₂(η³-CH₂Ph) (16-38)[55])

The Group IVA elements—Si, Ge, Sn, and Pb—also form anions of the type R_3M^-. The organic groups R are generally aromatic rather than aliphatic, as the latter do not allow the formation of particularly stable anions, presumably because alkyl groups do not stabilize the negative charge as well as do aryl groups. These anions are synthesized by the cleavage of metal-metal bonds (16-39), or by the reaction of an alkali metal with a triorganometal halide (reactions 16-40 and 16-41). The alkali metal must be in a very reactive form. Therefore, the preparations are sometimes carried out in liquid ammonia (see Chapter 5 for a discussion of metal-liquid ammonia solutions), or sodium and potassium may be in the form of the highly reactive liquid alloy NaK.

$$Ph_3Ge\text{—}GePh_3 + NaK \xrightarrow{Et_2O + THF} Ph_3GeK + Ph_3GeNa \qquad (16\text{-}39)[56]$$

$$Me_3SnBr + 2Na \xrightarrow{liq.\ NH_3} Me_3Sn\text{—}SnMe_3 + 2NaBr \qquad (16\text{-}40)[57]$$
$$\xrightarrow{2Na} 2Me_3SnNa$$

$$Me_3SnCl + 2C_{10}H_8Na \rightarrow Me_3SnNa + NaCl + 2C_{10}H_8 \qquad (16\text{-}41)[58]$$

---

[53] T. S. Piper and G. Wilkinson, *J. Inorg. Nucl. Chem.*, **3**, 104 (1956).
[54] W. Jetz, P. B. Simons, J. A. J. Thompson, and W. A. G. Graham, *Inorg. Chem.*, **5**, 2217 (1966).
[55] R. B. King and A. Fronzaglia, *J. Amer. Chem. Soc.*, **88**, 709 (1966).
[56] A. G. Brook and H. Gilman, *J. Amer. Chem. Soc.*, **76**, 77 (1954).
[57] See page 456 of ref. 18a.
[58] H. G. Kuivila, J. L. Considine, and J. D. Kennedy, *J. Amer. Chem. Soc.*, **94**, 7206 (1972).

832 ☐ ORGANOMETALLIC CHEMISTRY: SYNTHESIS, STRUCTURE, AND BONDING

A very convenient alternative is the use of the sodium-naphthalene ion pair, $Na^+C_{10}H_8^-$, or some other metal-arene ion pair.[59,60]

1,2-Addition of Metal Complexes to Unsaturated Substrates[61,62]

In this section we shall consider the general process

(16-42)
$$M-G + \!\!\!{>}\!C\!\!=\!\!C{<}\!\!\! \rightarrow M-\underset{|}{\overset{|}{C}}-\underset{|}{\overset{|}{C}}-G$$

where G is hydrogen, oxygen, a halogen, or an organic group.

Hydrometalations. Reactions of this type are among the most widely studied in chemistry, and they have been discussed briefly in Chapter 8 (p. 483). Most important of these is the *hydroboration* reaction:[63] the anti-Markownikoff, *cis* addition of the elements of diborane to olefins and acetylenes.

(16-43)
$$3\ \text{HC(Me)}\!=\!\text{CH(Me)} + \tfrac{1}{2}B_2H_6 \rightarrow \left[\text{HC(Me)(H)}\!-\!\text{C(Me)(H)}\!-\!B\right]_3$$

(16-44)
$$2\ \text{MeC(Me)}\!=\!\text{CH(Me)} + \tfrac{1}{2}B_2H_6 \rightarrow \left[\text{MeC(Me)(H)}\!-\!\text{C(Me)(H)}\!-\!BH\right]_2$$

disiamylborane

---

[59] Sodium dissolves readily in THF or 1,2-dimethoxyethane solutions of aromatic molecules containing two or more joined (biphenyl), fused (naphthalene or anthracene), or conjugated (1,4-diphenylbutadiene) rings. The sodium reduces the arene to the highly reactive radical anion and is itself taken into

$$Na + C_{10}H_8 \xrightarrow{\text{1,2-dimethoxyethane}} [\text{Na(dimethoxyethane)}_2]^+ [C_{10}H_8]^{\bar{\cdot}}$$

solution as the solvated cation. The efficacy of the reaction depends on (i) the ability of the solvent to solvate the metal cation (therefore, diethyl ether is a poor solvent, whereas the chelating ether 1,2-dimethoxyethane is especially good) and (ii) the energy of the LUMO of the arene. Such solutions are invariably colored, because electronic transitions are possible between the narrowly spaced π^* levels of the arene. See E. deBoer, *Adv. Organometal. Chem.*, **2**, 115–156 (1964).

[60] The reactions of metal-containing anions with inorganic and organic halides depend on the ability of the anion to act as a nucleophile. The rates of reactions of metal-containing anions with organic halides have been studied in detail, and the factors influencing the nucleophilicity of the anions discussed. We shall treat this topic in the context of the discussion of substitution reactions in Chapter 17.

[61] There are three general reviews of reactions of this type:
   a. M. F. Lappert and B. Prokai, *Adv. Organometal. Chem.*, **5**, 225 (1967).
   b. R. F. Heck, *Adv. Chem. Series*, **49**, 181 (1965).
   c. J. P. Candlin, K. A. Taylor, and D. T. Thompson, "Reactions of Transition Metal Complexes," Elsevier Publishing Company, New York (1968), pp. 119–151.

[62] The 1,2-additions of metal-containing molecules to olefins and acetylenes are frequently called "insertion reactions" by organometallic chemists, since such reactions can be considered as the insertion of the unsaturated molecule into a metal-element bond. The use of this terminology emphasizes the role of the metal-containing compound and allows the inclusion of reactions such as (16-42) in a much larger class of reactions. In addition to the reactions discussed in this section, insertion reactions include those wherein the molecule being inserted may be CO_2, SO_2, CS_2, or organic compounds such as ketones, aldehydes, or isocyanates, among many others.

[63] Unless otherwise noted, the information on hydroborations was taken from the following sources.
   a. H. C. Brown, "Hydroboration," W. A. Benjamin, Inc., New York (1962).
   b. G. Zweifel and H. C. Brown, in "Organic Reactions," Volume 13, A. C. Cope, ed., John Wiley and Sons, New York (1963), Chap. 1.
   c. H. C. Brown, *Acc. Chem. Res.*, **2**, 65 (1969).
   d. H. C. Brown, "Organic Syntheses via Boranes," Wiley-Interscience, New York (1975).
   e. H. C. Brown, "Boranes in Organic Chemistry," Cornell University Press, Ithaca, New York (1972).

$$\text{(2,4-dimethyl-1,3-pentadiene)} + \tfrac{1}{2}B_2H_6 \rightarrow \text{3,5-dimethylborinane} \tag{16-45}$$

3,5-dimethylborinane

The hydroboration reaction has been developed primarily by H. C. Brown and his students at Purdue University.[63] During the early 1940's Brown was a member of H. I. Schlesinger's group at the University of Chicago; it was this group that so greatly expanded Alfred Stock's pioneering studies[64] of the boron hydrides. One of their more important discoveries was a method for the synthesis of alkali metal borohydrides.[65]

$$4 \text{ NaH} + \text{B(OMe)}_3 \xrightarrow{250°} \text{NaBH}_4 + 3 \text{ NaOMe} \tag{16-46}$$

Sodium borohydride, for example, is soluble in water (at pH > 7; see p. 385) and alcohols; only in methanol, however, does it undergo rapid solvolysis to produce hydrogen.

$$\text{NaBH}_4 + 4 \text{ MeOH} \rightarrow \text{Na[B(OMe)}_4] + 2 \text{ H}_2 \tag{16-47}$$

In any of these solvents the borohydride reacts rapidly with aldehydes and ketones to produce alcohols; but no reaction is observed with other functional groups under these conditions. More important to the further development of borohydride chemistry, however, was the finding that, while $NaBH_4$ is insoluble in diethyl ether, it is slightly soluble in THF, and very soluble (3 M at 40°) in diglyme (dimethyl ether of diethylene glycol, $CH_3OCH_2CH_2OCH_2CH_2OCH_3$) and triglyme (dimethyl ether of triethylene glycol).[66] In ethereal solvents, controlled reaction of $NaBH_4$ with a Brønsted acid (HCl) or a Lewis acid (BF_3 or BCl_3) produces diborane, B_2H_6, the *essential* ingredient of hydroborations, in high yield.[67] The most convenient conditions for carrying out hydroborations have been found to be the addition of a diglyme solution of $NaBH_4$ to a diglyme solution of $F_3B:OEt$.[68]

$$3 \text{ NaBH}_4 + 4 \text{ F}_3\text{B:OEt}_2 \rightarrow 2 \text{ B}_2\text{H}_6 + 3 \text{ NaBF}_4 + 4 \text{ OEt}_2 \tag{16-48}$$

The diborane thereby generated is then carried in a nitrogen stream into an ether solution of the compound undergoing hydroboration.

Although it is possible to hydroborate an olefin when both the substrate and the diborane are in the gas phase, the reaction is exceedingly slow.[69] This is presumably because only very small amounts of BH_3, the reactive species in hydroborations, are formed by the gas phase dissociation of diborane (see Chapter 18). Therefore, hydroborations are carried out in weakly basic solvents such as diethyl ether or THF, which promote the dissociation of B_2H_6 to BH_3 (page 399).

The boron-containing products of hydroborations are not usually isolated; rather, they are converted *in situ* into a variety of organic derivatives. For example, oxidation of the intermediate alkylboranes with alkaline peroxide replaces the boron with an OH group with retention of configuration. Analysis of the oxidation products

---

[64] A. Stock, "The Hydrides of Boron and Silicon," Cornell University Press, Ithaca, New York (1933).
[65] H. I. Schlesinger, H. C. Brown, and A. E. Finholt, *J. Amer. Chem. Soc.*, **75**, 205 (1953).
[66] This increased solubility in glymes is undoubtedly due to the complexation of the sodium ion by the solvent. See ref. 59.
[67] There are many possible variations on this reaction. Diborane will be produced by any hydride ($LiBH_4$, $NaBH_4$, KBH_4, $LiAlH_4$, LiH, NaH) reacting with an acid (BF_3, BCl_3, $AlCl_3$, $TiCl_4$, HCl, etc.). The only requirement is that one of the two reactants should contain boron.
[68] Adding the Lewis acid to the borohydride is not as efficient. The generated diborane reacts with the borohydride, which is in excess, and forms NaB_2H_7.
[69] D. T. Hurd, *J. Amer. Chem. Soc.*, **70**, 2053 (1948).

from reactions such as 16-49 through 16-51 indicates that the addition of the elements of diborane to olefins is *cis* and largely anti-Markownikoff; furthermore,

(16-49)[70]
$$CH_3(CH_2)_3CH=CH_2 \xrightarrow[\text{2. [O]}]{\text{1. HB}} CH_3(CH_2)_3\underset{\underset{OH}{|}}{CH}-CH_3 + CH_3(CH_2)_3CH_2CH_2OH$$
$$\phantom{CH_3(CH_2)_3CH=CH_2 \xrightarrow[\text{2. [O]}]{\text{1. HB}} }6\% \phantom{CH_3(CH_2)_3\underset{\underset{OH}{|}}{CH}-CH_3 + } 94\%$$

(16-50)

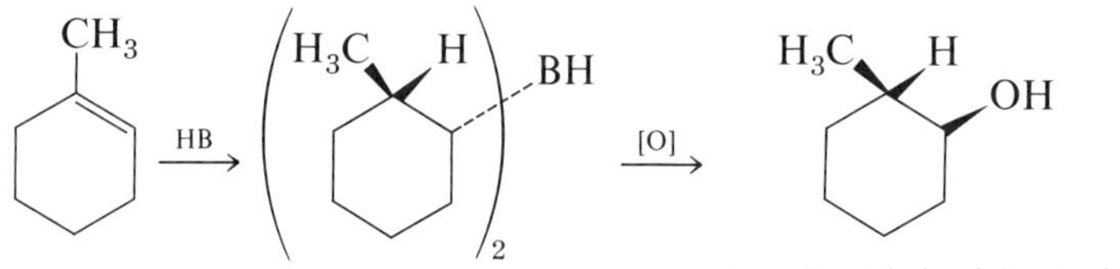

trans-2-methylcyclohexanol

(16-51)

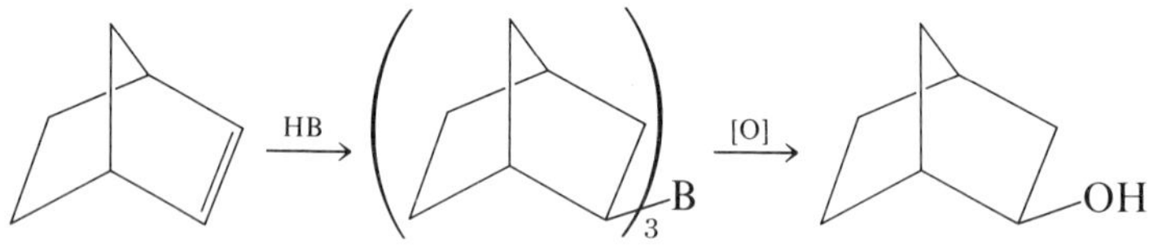

exo-2-norborneol

attack is always at the less hindered face of a double bond (see reaction 16-51). Based on these results, and many others, the mechanism must involve a concerted 1,2-addition, the end of the double bond to which the boron adds being dictated by the carbon substituents.

(16-52)
$$H_3C-CH=CH_2 \xrightarrow{B-H} H_3C-\overset{+\delta}{\underset{\underset{-\delta}{H}}{C}}\cdots\cdots\overset{-\delta}{\underset{\underset{+\delta}{B}}{C}}H_2$$

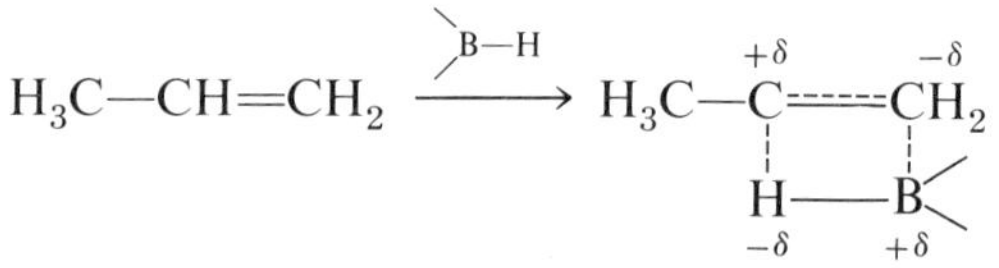

A major thrust of Brown's work has been to devise methods whereby the intermediate alkylboranes can be converted easily and directly into organic compounds of all types. Such reactions were discussed briefly in Chapter 7 (p. 392), but a complete discussion is not within the province of this text. If you are interested in additional information, Brown's books on the subject are both entertaining and informative.[71]

Hydroaluminations (the addition of Al—H to unsaturated systems) are generally similar to hydroborations in that both are anti-Markownikoff and *cis*.[72-74] However, there are some important differences.

(16-53)
$$(i\text{-Bu})_2\text{AlH} + \text{MeC}\equiv\text{CPh} \xrightleftharpoons[100°]{50°} \underset{H}{\overset{Me}{}}C=C\underset{Al(i\text{-Bu})_2}{\overset{Ph}{}}$$

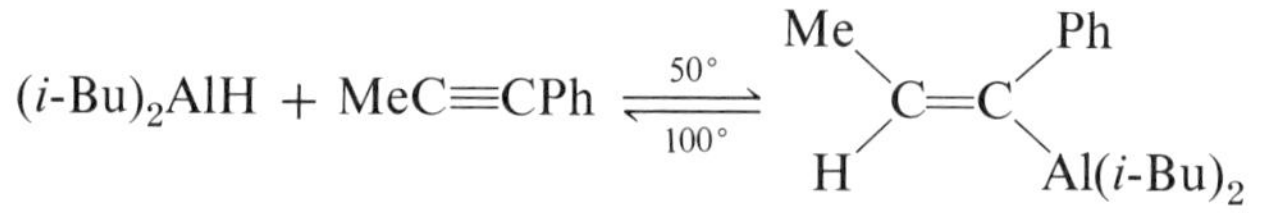

Boron is the only element of Group IIIA that forms a discrete, nonpolymeric hydride of the type M_2H_6 under normal conditions. The reduction of $AlCl_3$ with

---

[70] HB is a symbol for the addition of the elements B and H (from B_2H_6) to an olefin in an ethereal solution at temperatures of 0° to 25°. The symbol [O] designates *in situ* oxidation with H_2O_2 under basic conditions.
[71] See ref. 63.
[72] J. J. Eisch and W. C. Kaska, *J. Amer. Chem. Soc.*, **88**, 2213 (1966).
[73] R. Koster and P. Binger, *Adv. Inorg. Chem. Radiochem.*, **7**, 263 (1965).
[74] H. Reinheckel, K. Haage, and D. Jahnke, *Organometal. Chem. Rev.*, **4**, 1 (1969).

LiAlH$_4$ results in the formation of (AlH$_3$)$_x$, an insoluble substance that is thought to be a hydrogen-bridged polymer.[75] In the presence of base, however, the hydride can be trapped as a monomeric complex such as Me$_3$N:AlH$_3$;[76] the usefulness of this material in hydroaluminations has not been demonstrated.

$$3x \text{ LiAlH}_4 + x \text{ AlCl}_3 \rightarrow 4 \text{ (AlH}_3)_x + 3x \text{ LiCl} \tag{16-54}$$

As is the case in reaction 16–53, hydroaluminations are most commonly carried out with diorganoaluminum hydrides such as (i-Bu)$_2$AlH or Et$_2$AlH. Such compounds, in fact, are an integral part of the industrial synthesis of trialkylaluminum compounds by the *direct synthesis* method developed by Zeigler.[77] In the first step of this process (16-55), activated aluminum and hydrogen react with preformed R$_3$Al to form a dialkylaluminum hydride. This then reacts with an olefin in a hydroalumination reaction to give additional aluminum trialkyl.

$$\text{Al} + \frac{3}{2} \text{H}_2 + 2 \text{ Et}_3\text{Al} \rightleftharpoons 3 \text{ Et}_2\text{AlH} \tag{16-55}$$

$$3 \text{ Et}_2\text{AlH} + 3 \text{ CH}_2\text{=CH}_2 \rightleftharpoons 3 \text{ Et}_3\text{Al} \tag{16-56}$$

Olefin hydroalumination is an extremely rapid reaction requiring only a very small activation energy (4 to 6 kcal/mole of monomeric R$_2$AlH).[78-80] [By comparison, the reaction of R$_2$BH with olefins has an activation energy of about 12 kcal/mole of monomeric R$_2$BH, and E_a for HBr + 1-butene is 26.9 kcal.] The heats of hydroalumination reactions are in the range from -20 to -25 kcal/mole of monomeric R$_2$AlH, strongly suggesting that trialkylaluminum compounds are thermodynamically more stable than dialkylaluminum hydrides. Nonetheless, an important feature of the hydroalumination reaction is its ready reversibility. For example, heating (i-Bu)$_3$Al at 160-180° in a nitrogen atmosphere results in the smooth evolution of isobutene and formation of (i-Bu)$_2$AlH.[78,79,81]

Reversibility is an important aspect of hydrometalations in general. Although dialkylboron hydrides cannot be isolated by the thermal disproportionation of trialkylboranes, the reaction must be reversible to some extent, as the olefin isomerization process (reaction 16-57) proves.[82]

$$\tag{16-57}$$

---

[75] Refer to the sketch of (AlH$_3$)$_x$ on p. 310. See also S. Cucinella, A. Mazzei, and W. Marconi, *Inorg. Chim. Acta. Rev.*, **4**, 51 (1970).
[76] J. K. Ruff and M. F. Hawthorne, *J. Amer. Chem. Soc.*, **82**, 2141 (1960).
[77] K. Ziegler, in "Organometallic Chemistry," ACS Monograph No. 147, H. Zeiss, ed., Reinhold, New York (1960), pp. 194-269.
[78] K. W. Egger, *J. Amer. Chem. Soc.*, **91**, 2867 (1969).
[79] K. W. Egger and A. T. Cocks, *Trans. Faraday Soc.*, **67**, 2629 (1971).
[80] K. W. Egger and A. T. Cocks, *J. Amer. Chem. Soc.*, **94**, 1810 (1972).
[81] See ref. 72.
[82] See ref. 63.

On the other hand, as you shall see in Chapter 17, it is frequently difficult to isolate transition metal alkyls because of their very rapid decomposition to metal hydride and olefin.

Terminal olefins and acetylenes very readily add the elements of tin and hydrogen in the so-called *hydrostannation* reaction.[83]

(16-58)[84,85]
$$2Me_3SnCl + 2NaBH_4 \xrightarrow{glyme} 2Me_3SnH + 2NaCl + B_2H_6$$

$$Me_3SnH + \underset{\text{(4-vinylcyclohexene)}}{\overset{HC=CH_2}{\bigcirc}} \rightarrow \underset{\text{}}{\overset{CH_2-CH_2-SnMe_3}{\bigcirc}}$$

Since a range of functional groups—keto, cyano, hydroxyl, and pyridyl, among others—do not interfere, this is a good method for the synthesis of functionally substituted tin compounds. Unlike hydroboration and hydroalumination, however, hydrostannations are usually free radical reactions, presumably occurring according to the mechanism of reaction 16-59. Hydrostannations are initiated by heat, ultraviolet light, or radical sources such as azo*bis*isobutyronitrile.

$$\text{initiator} \rightarrow 2R'\cdot$$

$$R'\cdot + R_3SnH \rightarrow R_3Sn\cdot + R'H$$

(16-59)
$$R_3Sn\cdot + \overset{}{\underset{}{>}}C=C\overset{}{\underset{}{<}} \rightarrow R_3Sn-\overset{|}{\underset{|}{C}}-\overset{|}{\underset{|}{C}}\cdot$$

$$R_3Sn-\overset{|}{\underset{|}{C}}-\overset{|}{\underset{|}{C}}\cdot + R_3SnH \rightarrow R_3Sn-\overset{|}{\underset{|}{C}}-\overset{|}{\underset{|}{C}}-H + R_3Sn\cdot$$

Hydrometalations involving transition metal–hydrogen bonds are also important examples of this reaction type,[86-90] because the insertion of olefins into transition metal–hydrogen bonds is implicated in numerous industrially important (both actual and potential) processes that are catalyzed by transition metals. Examples of catalytic processes wherein M—H/olefin interaction is presumed to occur include hydroformylation of olefins by $HCo(CO)_4$ (basically the addition of H_2 and CO to an tion, and hydroformylation; presumably, olefin insertion into a Pt—H bond plays genation by $RhCl(PPh_3)_3$.[93] A very recent example is the molten salt system $R_4N^+SnCl_3^- + 7$ wt% $PtCl_2$, which can be used for olefin hydrogenation, isomerization, and hydroformylation; presumably, olefin insertion into a Pt—H bond plays an important part in these processes.[94] In general, important catalytic systems involve Co(I), Pt(II) (both d^8 systems), or Rh(III) (d^6) complexes. However, many other metal

---

[83] H. G. Kuivila, *Acc. Chem. Res.*, **1**, 299 (1968).
[84] R. H. Fish, H. G. Kuivila, and I. J. Tyminski, *J. Amer. Chem. Soc.*, **89**, 5861 (1967).
[85] Note that, just as in hydroboration and hydroalumination reactions, only the *exo* double bond in 4-vinylcyclohexene is hydrometalated. See K. Zeigler, F. Krupp, and K. Zesel, *Ann. Chem. Liebigs*, **629**, 241, 248, 249 (1960).
[86] R. F. Heck, *Acc. Chem. Res.*, **2**, 10 (1969).
[87] R. F. Heck, *Adv. Chem. Ser.*, **49**, 181 (1965).
[88] C. A. Tolman, *Chem. Soc. Rev.*, **1**, 337 (1972).
[89] R. Cramer, *Acc. Chem. Res.*, **1**, 186 (1968).
[90] R. F. Heck, "Organotransition Metal Chemistry," Academic Press, New York (1974).
[91] A. J. Chalk and J. F. Harrod, *Adv. Organometal. Chem.*, **6**, 119 (1968).
[92] C. A. Tolman, *J. Amer. Chem. Soc.*, **94**, 2994 (1972).
[93] J. A. Osborne, F. H. Jardine, J. F. Young, and G. Wilkinson, *J. Chem. Soc.* (A), 1711 (1966).
[94] G. W. Parshall, *J. Amer. Chem. Soc.*, **94**, 8716 (1972).

complexes [*e.g.*, the Ni(II) complex above] are being investigated. The subject of homogeneous catalysis by transition metal compounds is of such importance that a separate section is devoted to this topic in Chapter 17. However, it is instructive to introduce at this point one example of a hydrometalation involving a transition metal.

At 95° and 80 atm of ethylene, the olefin reacts with *trans*-PtHCl(PEt$_3$)$_2$ to give a 25% conversion to a product containing a σ-bonded ethyl group (reaction 16–60).[95]

$$\begin{array}{c}\text{PEt}_3\\|\\\text{Cl—Pt—H}\\|\\\text{PEt}_3\end{array} + \text{C}_2\text{H}_4 \underset{\text{heat}/\textit{in vacuo}}{\overset{\text{heat/pressure}}{\rightleftharpoons}} \begin{array}{c}\text{PEt}_3\\|\\\text{Cl—Pt—C}_2\text{H}_5\\|\\\text{PEt}_3\end{array} \qquad (16\text{–}60)$$

The reverse reaction occurs if the ethyl derivative is heated to about 180° *in vacuo*. One of the main reasons for studying this reaction was to determine which of the ethyl group H's was left on the metal when the reaction was reversed. This was solved in large part by heating the deuterated complex *trans*-Pt(CD$_2$CH$_3$)Cl(PEt$_3$)$_2$ *in vacuo* and observing that the product contained both Pt—H and Pt—D bonds. It was therefore concluded that the addition or elimination of olefin proceeded through the symmetrical transition state illustrated below.

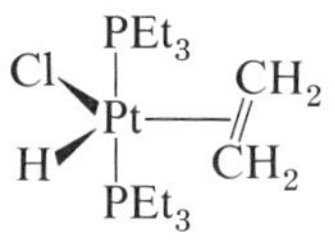

$$\begin{array}{c}\text{Cl}\diagdown\;\;\text{PEt}_3\\\;\;\;\;\;\;\;\;\;|\;\;\;\;\;\text{CH}_2\\\;\;\;\;\;\;\;\;\text{Pt}\text{—}||\\\;\;\;\;\;\;\;\;\;|\;\;\;\;\;\text{CH}_2\\\text{H}\diagup\;\;\text{PEt}_3\end{array}$$

Another important conclusion of this study was that olefin size determines the equilibrium position in reaction 16–60. When propene is used instead of ethylene, only a very small yield of *trans*-Pt(*n*-C$_3$H$_7$)Cl(PEt$_3$)$_2$ is obtained. With octene-1, no Pt(*n*-octyl) complex is formed. However, when *trans*-PtDCl(PEt$_3$)$_2$ was heated with octene-1 at 130° for 48 hours, all of the platinum deuteride was converted into *trans*-PtHCl(PEt$_3$)$_2$. It must be concluded, therefore, that octene addition occurs, but that the equilibrium lies far to the left.

Oxymetalations. Oxymercuration—the addition of the elements Hg and O to olefins—is one of the most widely studied forms of this reaction type.[96,97] It is a synthetically useful reaction that may be used to introduce the hydroxy, acetoxy, and alkoxy groups, among others, into unsaturated systems.[98]

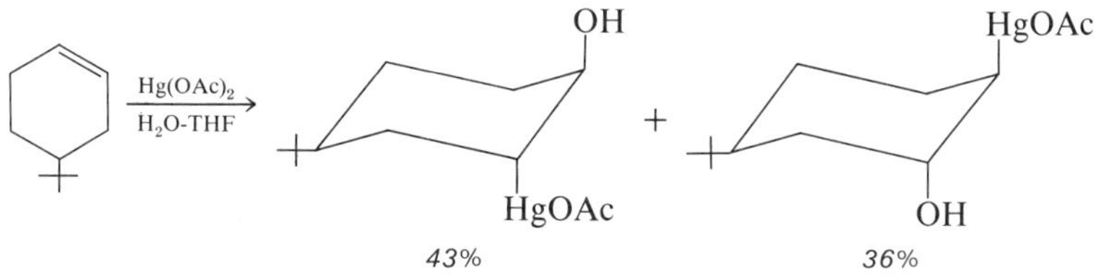

<chemical reaction 16-61: cyclohexene + Hg(OAc)$_2$ / H$_2$O-THF → trans-2-hydroxycyclohexyl mercuric acetate (OH axial/HgOAc equatorial) 43% + trans isomer (HgOAc/OH) 36%>

(16–61)

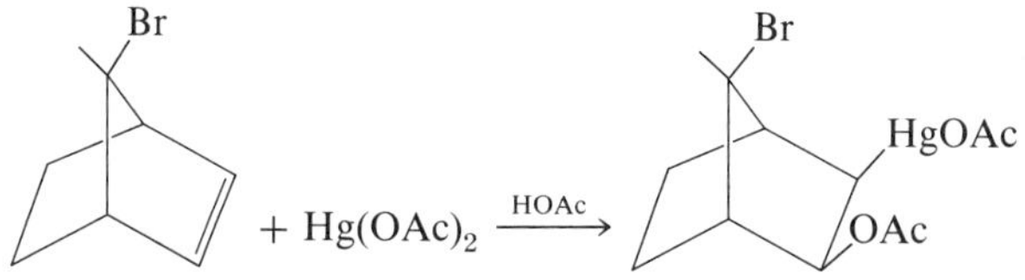

<chemical reaction 16-62: 7-bromonorbornene + Hg(OAc)$_2$ / HOAc → bromo-norbornyl acetate mercuric acetate adduct>

(16–62)[99]

---

[95] J. Chatt, R. S. Coffey, A. Gough, and D. T. Thompson, *J. Chem. Soc.* (A), 190 (1968).
[96] D. J. Pasto and J. A. Gontarz, *J. Amer. Chem. Soc.*, **93**, 6902 (1971).
[97] W. Kitching, *Organometal. Chem. Rev.*, **3**, 61 (1968).
[98] H. C. Brown and M-H. Rei, *J. Amer. Chem. Soc.*, **91**, 5646 (1969).
[99] T. T. Tidwell and T. G. Traylor, *J. Org. Chem.*, **33**, 2614 (1968).

Oxymercurations are electrophilic addition reactions that seem to follow different mechanistic pathways in different situations. A great number of studies have been done to elucidate the mechanism of these reactions, but controversy remains.[100,101] The oxymercuration of cyclohexenes, for example, produces exclusively *trans*-addition products, as noted in reaction 16-61.[100] On the other hand (reaction 16-62), oxymercuration of a variety of norbornenes results in highly stereospecific *cis, exo* addition.[101] In the case of acyclic olefins and non-strained cyclic olefins (such as cyclohexene), it seems generally agreed that the so-called "mercurinium ion" mechanism is the most viable explanation of the observed stereochemistry and other experimental observations.[100,102]

$$Hg(OAc)_2 \rightleftharpoons HgOAc^+ + OAc^-$$

(16-61′)

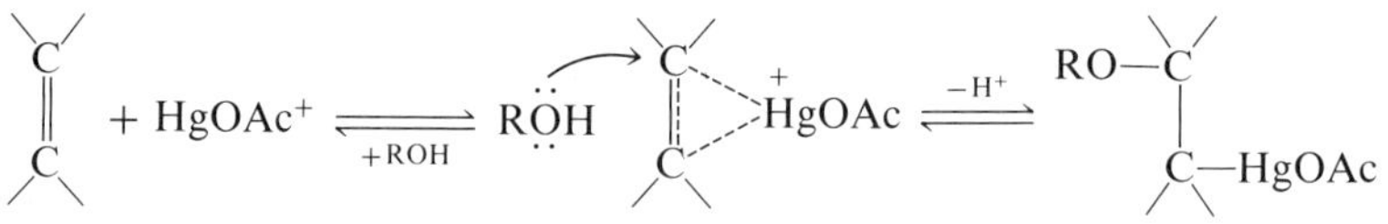

It would appear, however, that oxymercuration of strained olefins such as bicyclic norbornenes proceeds not by the "mercurinium ion" pathway but by a "non-classical" mechanism, 16-62′.

(16-62′)

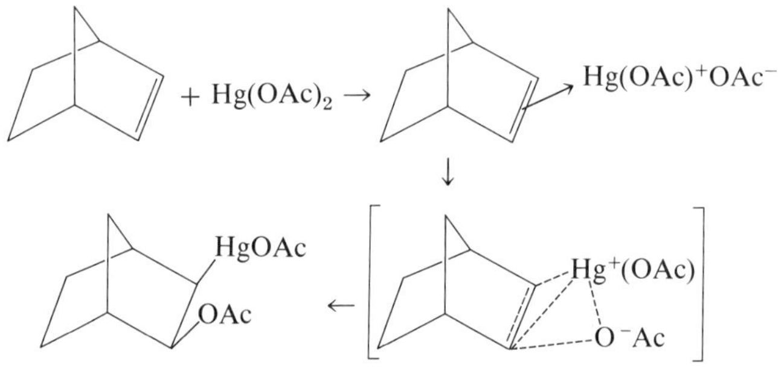

Of greater interest to the organometallic chemist are oxypalladation and oxyplatination reactions.[102,104-106] The former is of special importance, because oxypalladation presumably occurs in the Wacker process, the industrial-scale palladium-catalyzed oxidation of ethylene to acetaldehyde (16-63).[106]

(16-63)
$$PdCl_4^{-2} + C_2H_4 + H_2O \rightarrow Pd^0 + CH_3CHO + 2\ HCl + 2\ Cl^-$$

The mechanism of the reaction has been extensively investigated and will be discussed in detail in the section on catalysis in Chapter 17.[107]

Halometalations. The insertion of olefins and acetylenes into metal-halogen bonds has not been examined nearly as extensively as hydro-, oxy-, or organometalations. Only haloboration and halopalladation have been studied in any detail.

In general, haloborations are more complex than hydroborations and differ from them considerably in several respects.[108-110] (i) Except in the case of acetylene (16-64),

---

[100] See ref. 96.
[101] T. G. Traylor, *Acc. Chem. Res.*, **2**, 152 (1969).
[102] See ref. 97.
[103] P. M. Henry and G. A. Ward, *J. Amer. Chem. Soc.*, **93**, 1494 (1971), and references therein.
[104] S. Wolfe and P. G. C. Campbell, *J. Amer. Chem. Soc.*, **93**, 1497, 1499 (1971).
[105] J. K. Stille and R. A. Morgan, *J. Amer. Chem. Soc.*, **88**, 5135 (1966).
[106] A. Aguilo, *Adv. Organometal. Chem.*, **5**, 321 (1967).
[107] P. M. Henry, *J. Amer. Chem. Soc.*, **86**, 3246 (1964); *ibid.*, **94**, 4437 (1972); *Acc. Chem. Res.*, **6**, 16 (1973).
[108] M. F. Lappert and B. Prokai, *J. Organometal. Chem.*, **1**, 384 (1964).
[109] F. Joy, M. F. Lappert, and B. Prokai, *J. Organometal. Chem.*, **6**, 506 (1966).
[110] J. J. Eisch and L. J. Gonsior, *J. Organometal. Chem.*, **8**, 53 (1967).

the haloboration reaction apparently gives a *cis* addition product (16-65 and 16-66).

$$HC\equiv CH + BBr_3 \underset{pyridine}{\rightleftarrows} \underset{H}{\overset{Br}{>}}C=C\underset{BBr_2}{\overset{H}{<}} \qquad (16\text{-}64)$$

$$PhC\equiv CH + BCl_3 \rightarrow \underset{Cl}{\overset{Ph}{>}}C=C\underset{BCl_2}{\overset{H}{<}} \xrightarrow{PhC\equiv CH} [Ph(Cl)C=CH]_2BCl \qquad (16\text{-}65)$$

$$\text{no reaction} \xleftarrow{BCl_3} PhC\equiv CPh \underset{pyridine}{\overset{BBr_3}{\rightleftarrows}} \underset{Br}{\overset{Ph}{>}}C=C\underset{BBr_2}{\overset{H}{<}} \qquad (16\text{-}66)$$

This implies that the reactions usually proceed through the same type of four-center transition state postulated for hydroborations. The differing stereochemistries for acetylene and substituted acetylenes, however, imply that haloborations are both kinetically and thermodynamically controlled. Hydroborations, in contrast, are apparently kinetically controlled and give only *cis* addition products. An explanation for this difference may be found in the fact that haloborations are considerably more reversible than hydroborations. Therefore, after formation of the kinetically favored *cis* product, rearrangement could result in the thermodynamically favored *trans* adduct. (You shall see the same effect in oxidative addition reactions in Chapter 17). (ii) It is well known that BBr_3 is a stronger Lewis acid than BCl_3. Since haloborations such as 16-66 are electrophilic addition reactions, it is not surprising that they depend to some extent on the Lewis acidity of the borane.[111] (iii) Whereas hydroboration of an internal acetylene (*e.g.*, 3-hexyne) utilizes all three B—H bonds of BH_3, at most two B—X bonds are used in haloboration (16-65).

Several examples of chloropalladation have been studied.[112,113] Kinetic analysis of allylic trifluoroacetate exchange with chloride in the presence of $PdCl_2$, for example,

$$CH_2=CHCH_2OOCCF_3 + LiCl \xrightarrow{Pd(II)} CH_2=CHCH_2Cl + LiOOCCF_3 \qquad (16\text{-}67)$$

suggests that the rate determining step is chloropalladation and that the products are formed in a subsequent detrifluoroacetoxypalladation step.

$$\underset{\overset{|}{OOCCF_3}}{RCH=CHCHR'} + -PdCl \rightarrow \underset{\overset{|}{Cl}\;\overset{|}{Pd}}{RCH-\overset{\overset{|}{OOCCF_3}}{C}HCHR'} \rightarrow \underset{\overset{|}{Cl}}{RCHCH=CHR'} + -PdOOCCF_3 \qquad (16\text{-}68)$$

It must be emphasized here that such addition-elimination reactions are very common in transition metal chemistry, and will be discussed further in Chapter 17.

Organometalations. Reactions such as these involve the insertion of alkenes and alkynes into metal-carbon σ bonds. The most important of these are **carbalumination** reactions—the addition of the elements Al and C to unsaturated systems. The discovery that olefins insert into transition metal-carbon σ bonds is more recent, largely as a result of the fact that transition metal-carbon σ bonds have only recently been studied in any detail.

If triethylaluminum is heated to 90 to 120° with ethylene at a pressure of 100 atmospheres, ethylene is inserted into the Al—C bond at the rate of about one mole of ethylene added per mole of R_3Al in one hour.[114,115] This so-called *growth reaction*

---

[111] See Chapter 5, page 232.
[112] J. Tsuji, M. Morikawa, and J. Kiji, *J. Amer. Chem. Soc.*, **86**, 8451 (1964); J. Tsuji, *Acc. Chem. Res.*, **2**, 144 (1969).
[113] P. M. Henry, *Inorg. Chem.*, **11**, 1876 (1972); *Acc. Chem. Res.*, **6**, 16 (1973).
[114] K. W. Egger, *Trans. Faraday Soc.*, **67**, 2638 (1971).
[115] See refs. 77 and 79.

(16-69a), discovered by Karl Ziegler, is in competition with dehydroalumination, the β-elimination of $>$Al—H to form a 1-alkene (16-69b).

(16-69a)
$$Et_3Al + 3m\, C_2H_4 \rightarrow Al\begin{array}{c}(C_2H_4)_nC_2H_5\\-(C_2H_4)_pC_2H_5\\(C_2H_4)_qC_2H_5\end{array} \quad (n+p+q)/3 = m$$

(16-69b)
$$R_2Al(C_2H_5) \rightarrow R_2AlH + C_2H_4$$

Even at relatively low temperatures, the number of addition reactions does not exceed eliminations by more than 100; i.e., the maximum chain length that can be formed is only about C_{200}. In practice, the growth reaction is more suitable for the synthesis of straight chain aliphatic compounds between C_4 and about C_{30}.

As we have already discussed in the hydrometalation section above, one of the most important features of organoaluminum chemistry is the essential reversibility of the dehydroalumination reaction. In combination with carbalumination, several reactions of practical importance can be performed. For example, the net result of the reaction of propene with tri-n-propylaluminum is the catalytic dimerization of propene. The carbalumination step leads to a β-branched alkylaluminum compound that is unstable with respect to dehydroalumination; this, of course, is similar to tri-isobutylaluminum, mentioned in the hydrometalation section (page 835). The product of this process, 2-methyl-1-pentene, is important, because it can be converted to the industrially useful material isoprene, $CH_2=CHC(CH_3)=CH_2$.

(16-70)

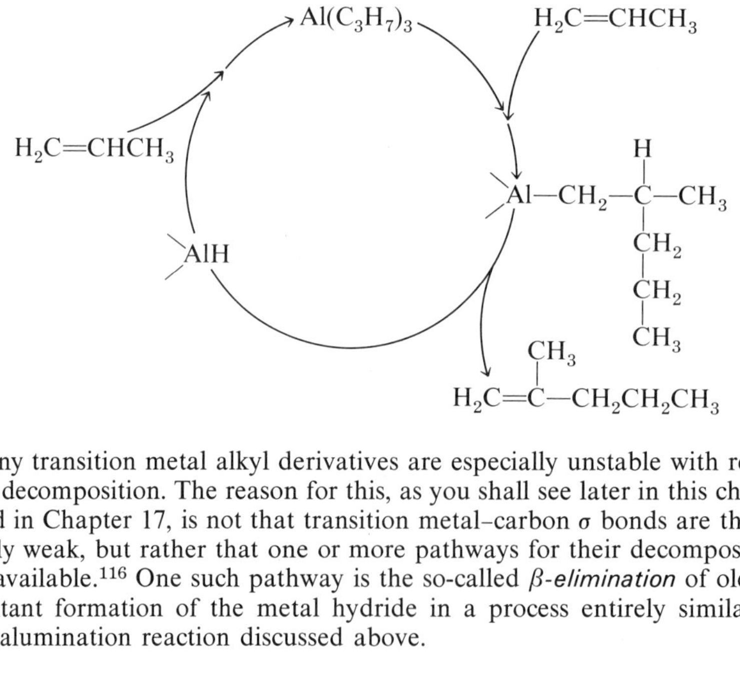

Many transition metal alkyl derivatives are especially unstable with respect to thermal decomposition. The reason for this, as you shall see later in this chapter (p. 855) and in Chapter 17, is not that transition metal–carbon σ bonds are thermodynamically weak, but rather that one or more pathways for their decomposition are readily available.[116] One such pathway is the so-called *β-elimination* of olefin with concomitant formation of the metal hydride in a process entirely similar to the dehydroalumination reaction discussed above.

(16-71)
$$L_nM-CH_2CH_2R \rightleftharpoons \left[L_nM\cdots\begin{array}{c}H\quad CHR\\ \|\\ CH_2\end{array}\right] \rightleftharpoons L_nMH + H_2C-CHR$$

---

[116] P. S. Braterman and R. J. Cross, *J. C. S. Dalton*, 657 (1972); *Chem. Soc. Revs.*, **2**, 271 (1973).

N-butyl(tri-n-butylphosphine)copper(I), for example, decomposes completely in 4 hours at 0° to give 1-butene, n-butane, copper metal, and free P(n-Bu)$_3$ (16–72);[117] the 1-butene undoubtedly arises from β-elimination (16–72a), while n-butane comes from binuclear reductive elimination (16–72b) or some other similar pathway, which we shall discuss in Chapter 17.

β-elimination

$$CH_3CH_2CH_2CH_2CuP(n\text{-Bu})_3 \rightarrow CH_3CH_2CH=CH_2 + HCuP(n\text{-Bu})_3 \quad (16\text{–}72a)$$

reductive elimination

$$CH_3CH_2CH_2CH_2CuP(n\text{-Bu})_3 + HCuP(n\text{-Bu})_3 \rightarrow \\ CH_3CH_2CH_2CH_3 + 2\ Cu + 2\ P(n\text{-Bu})_3 \quad (16\text{–}72b)$$

In any event, the behavior of transition metal alkyls is very similar to that of aluminum alkyls, but with the important difference that olefin elimination frequently occurs so much more readily with transition metal alkyls.

The ready β-elimination of olefins has been used to produce many new organic compounds in a cyclic process catalyzed by palladium salts.[118]

Overall process:

$$R'HgCl + RCH=CR_2 \xrightarrow[CuCl_2/HCl/O_2]{PdCl_2} RR'C=CR_2$$

(a) $PdCl_2 + RHgCl \rightarrow \text{``RPdCl''} + HgCl_2$

(b) 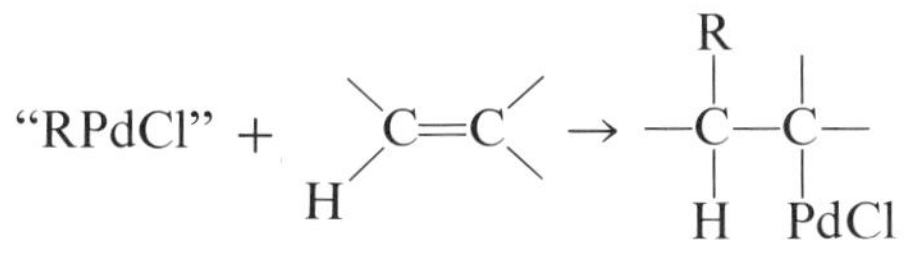 "RPdCl" + $\begin{array}{c}H\\ \end{array}$C=C $\rightarrow$ —C(R)(H)—C(PdCl)—

(c) —C(R)(H)—C(PdCl)— $\xrightarrow{\beta\text{-elimination}}$ $\begin{array}{c}R\\ \end{array}$C=C + "HPdCl"

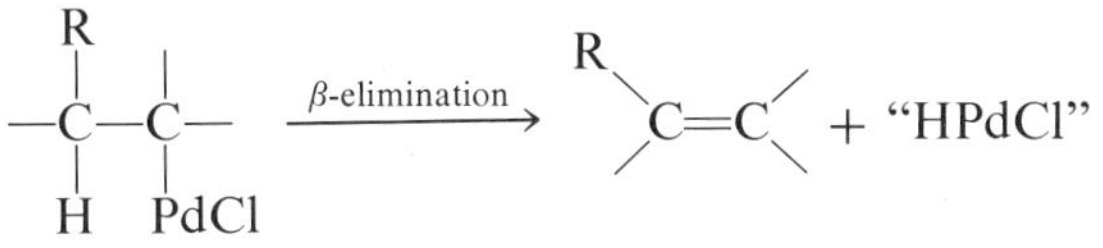

(d) "HPdCl" $\xrightarrow{\text{reductive elimination}}$ HCl + Pd^0

(e) catalyst regeneration $Pd^0 + 2CuCl_2 \rightarrow PdCl_2 + 2CuCl$

(f) $2CuCl + 2HCl + \tfrac{1}{2}O_2 \rightarrow 2CuCl_2 + H_2O$

In the first step (a), PdCl$_2$ (in the form of a salt such as Li$_2$PdCl$_4$) is arylated with another organometallic compound; the latter is usually an arylmercuric halide, but it may also be a diarylmercury or tetraaryltin compound. The intermediate arylpalladium compound, the nature of which is not really known (but see reaction 16–19), then adds to the olefin (b). The β-elimination mechanism provides a pathway for the immediate decomposition of the new organopalladium compound (c), leaving a Pd-H species. This hydride disproportionates to HX and metallic palladium (d). The reaction can be made catalytic in palladium, however, if a good oxidizing agent such as CuCl$_2$ is added to the system. After regenerating the catalyst (e), the reduced copper salt is reoxidized with atmospheric oxygen (f) in a manner similar to the

---

[117] G. M. Whitesides, E. R. Stedronsky, C. P. Casey, and J. San Filippo, Jr., *J. Amer. Chem. Soc.*, **92**, 1426 (1970).

[118] R. F. Heck, *J. Amer. Chem. Soc.*, **90**, 5518, 5526, 5535, 5538, 5542, 5546 (1968); *ibid.*, **91**, 6707 (1969); *ibid.*, **93**, 6896 (1971); *ibid.*, **94**, 2712 (1972).

Wacker process of olefin oxidation (see reaction 16–63 and Chapter 17). This basic reaction scheme, and variations of it, have been used in the arylation of olefins, of allylic alcohols and halides, and of enol esters, ethers, and halides; and for aromatic haloethylation and the decarboalkoxylation of olefins and acetylenes.

(16–73)

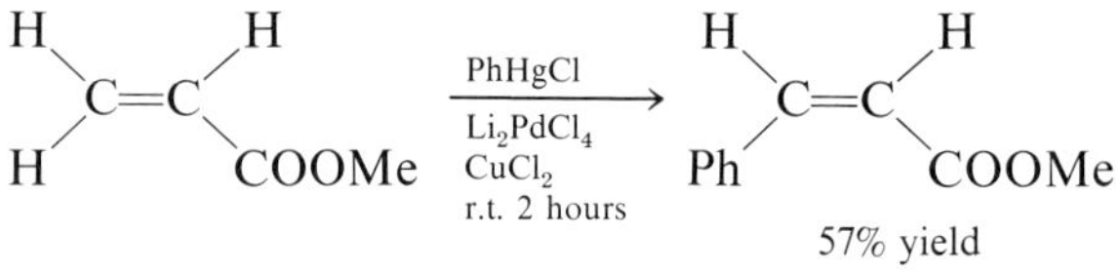

$$\underset{H}{\overset{H}{>}}C=C\underset{COOMe}{\overset{H}{<}} \quad \xrightarrow[\substack{Li_2PdCl_4 \\ CuCl_2 \\ \text{r.t. 2 hours}}]{PhHgCl} \quad \underset{Ph}{\overset{H}{>}}C=C\underset{COOMe}{\overset{H}{<}}$$
57% yield

STUDY QUESTIONS

1. Synthesis of metal alkyls:

   a) In order to carry out an infrared study of the structure of trimethylaluminum, suppose you wish to have the fully deuterated compound $Al_2(CD_3)_6$. Outline a synthesis of the ether-free material from appropriate starting materials. (The compound CD_3I can be purchased.)
   b) Synthesize Ph_4Sn from appropriate starting materials.
   c) Outline a synthesis of a catenated Group IVA compound such as Ph_6Ge_2 from appropriate materials.
   d) Synthesize Me_2Cd and Me_3Ga from appropriate starting materials.

2. A useful synthetic method for many organometallic compounds is to let a metal react directly with a dialkyl- or diarylmercury compound (see reaction 16–23). Can this method be applied to B? to Si?

3. Lithioferrocene is prepared as in reaction 16–26. How may this compound be used to prepare the two compounds illustrated below?

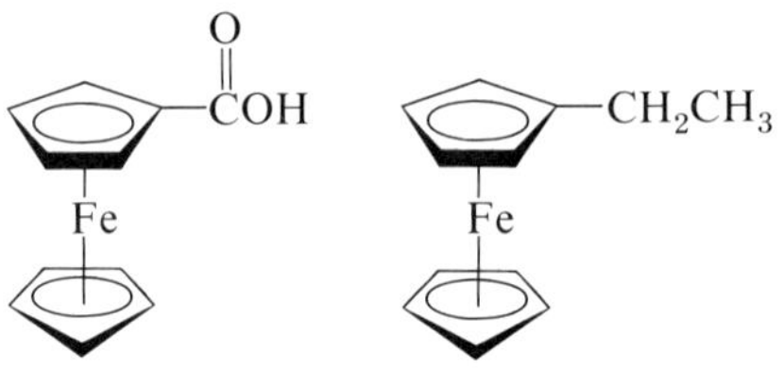

4. Group IIIA alkyls and aryls are considerably more air-sensitive than Group IVA alkyls and aryls. Suggest why this may be the case.

5. Use a hydrometalation reaction to prepare $HO-CH_2CH_2CH_2CH(CH_3)_2$ from $H_3C-CH=CH-CH(CH_3)_2$.

6. Karl Ziegler observed that the absorption rate of ethylene by dissolved triethylaluminum to form diethylbutylaluminum increases when more solvent is added. The rate of absorption of olefin is proportional to the square root of the total solution volume for a given amount of starting material. What does this observation suggest about the nature of the reacting species?

7. Refer to the oxidative addition reactions on page 829. Suggest a method for the synthesis of the two compounds below.

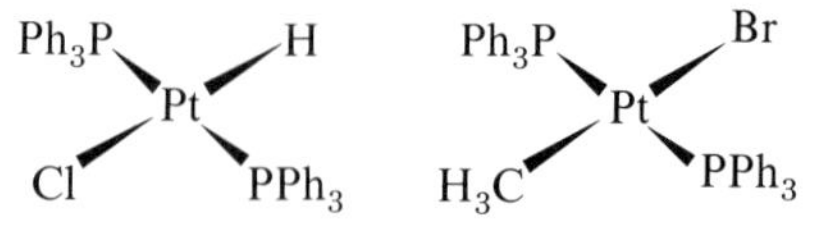

8. Osmium(0) and iridium(I) compounds most readily undergo oxidative addition reactions. However, reactivity falls off rapidly as one goes to the right in the Periodic Table; that is, Pt(II) and Au(III) compounds are much less susceptible to this type of reaction. Suggest why this may be the case.

STRUCTURE AND BONDING IN METAL ALKYLS AND ARYLS[118a]

The majority of metal alkyls and aryls are constructed and bonded in entirely predictable ways, except for the so-called "electron deficient" compounds of Groups IA, IIA, and IIIA and some transition metal derivatives.

Many of the simpler alkyls and aryls of Group IA, IIA, and IIIA metals are self-associated—they form dimers, trimers, and higher molecular weight species. Self-association of compounds through bridging atoms is a phenomenon that pervades inorganic chemistry, and we shall discuss here those aspects that apply specifically to organometallic chemistry.

Diborane, dimeric trimethylaluminum, and the dimeric aluminum halides can be considered the prototypes of self-associated species. We have already discussed the structure and bonding in the "electron deficient" molecule diborane in Chapter 5 (pp. 193–194). This compound was said to be "electron deficient," but not because there are insufficient electrons available to account for bonding; rather, the term is used in the topological sense to indicate that there are not enough valence electron pairs to account for all of the nearest neighbor atom-atom connections. The same approach used to account for bonding in diborane can be applied to "electron deficient" organometallic molecules.

The interaction of trialkylboranes with diborane leads to a variety of alkylboranes, all of which are associated through B—H—B bridges; alkyl bridging does not occur in boron chemistry for reasons discussed below (and in Chapter 6, p. 305).

$$Me_3B + B_2H_6 \rightarrow \{\ldots\} \quad (16\text{-}74)$$

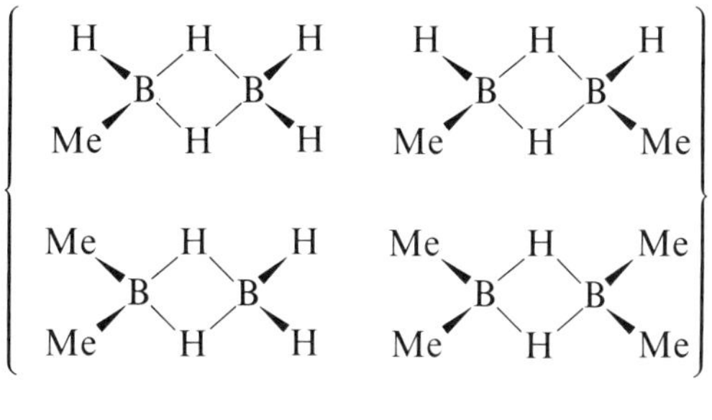

In contrast with diborane, aluminum hydride is a polymeric substance, the aluminum being surrounded octahedrally by hydrogen bridges to other aluminum atoms (Chapter 6, p. 310). Bridging also occurs with aluminum halides, but, unlike the hydride, $AlBr_3$ (7) and AlI_3 are dimeric (as is $AlCl_3$ in the gas phase) and are not electron deficient.[119,120] In this case, the octet of electrons surrounding each aluminum atom is completed by donation of a lone pair of electrons from a halogen of the other AlX_3 unit in the dimer pair (8).[120a]

[118a] P. J. Davidson, M. F. Lappert, and R. Pearce, *Chem. Rev.*, **76**, 219 (1976).
[119] M. J. Bigelow, *J. Chem. Educ.*, **46**, 495 (1969).
[120] Aluminum fluoride and aluminum chloride in the solid state are similar to aluminum hydride.
[120a] Refer to the orbital sketches on p. 311.

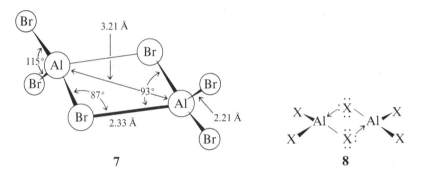

Organoaluminum chlorides, bromides, and iodides are also dimeric, the halogen atoms continuing to be the bridging elements (9).[121]

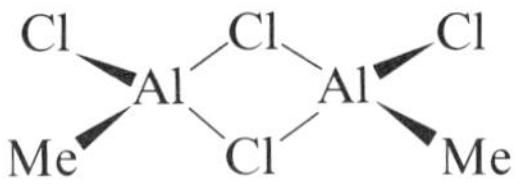

$$\text{Me} \diagdown \underset{\underset{\text{Cl}}{\diagup}}{\overset{\overset{\text{Cl}}{\diagdown}}{\text{Al}}} \underset{\underset{\text{Cl}}{\diagup}}{\overset{\overset{\text{Cl}}{\diagdown}}{\text{Al}}} \diagdown \text{Me}$$
9

Organoaluminum hydrides and fluorides, however, are unusual. Because of the non-directional character of the hydrogen $1s$ orbital that is utilized in bridging, an Al—H—Al bridge can assume almost any angle and still result in good overlap. Therefore, it should not be surprising that diorganoaluminum hydrides (10) are found to be trimeric, with very obtuse Al—H—Al angles.[122]

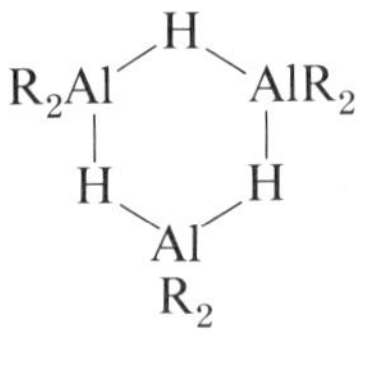

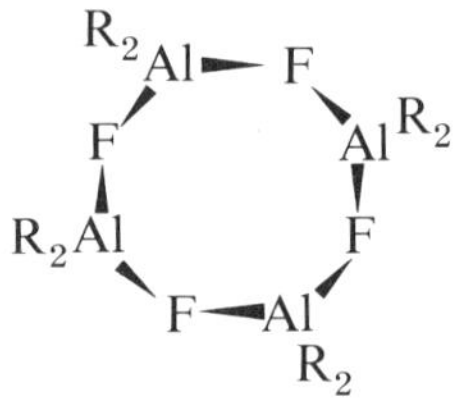

(R = Me)
Al—F = 1.808 Å
Al—C = 1.942 Å
Al—F—Al = 148°
F—Al—F = 94°

In the case of the fluorides, the halogen is also the bridging element, but the Al—F—Al angle is probably greater than 140° (11).[123] It would appear that bridging fluoride characteristically adopts very obtuse angles.[124]

Trialkyl- or triarylaluminum compounds are associated as well, but bridging is accomplished through an organic group. The majority of such compounds are dimeric, except in a few cases where steric hindrance prevents bridging. Trimethylaluminum is perhaps the best characterized organoaluminum compound. This compound, which has a melting point of 15.0° and a boiling point of 132°, is a dimer as a pure liquid, in the vapor phase, or in solution in non-basic solvents. Like all other organoaluminum compounds, it is extremely sensitive to both water and oxygen, and reacts readily with a variety of Lewis bases (see Chapters 5 and 17).

Thermodynamic values for trimethylaluminum and other organoaluminum compounds are listed in Table 16-1. All of the compounds listed, with the possible exception of tri-isobutylaluminum, are dimers at normal temperatures in non-basic

---

[121] G. Allegra, G. Perego, and A. Immirzi, *Makromol. Chem.,* **61,** 69 (1963).
[122] E. G. Hoffmann, *Annalen,* **629,** 104 (1960).
[123] G. Gundersen, T. Haugen, and A. Haaland, *Chem. Commun.,* 708 (1972).
[124] For a general discussion of bridge angles in inorganic, four-membered, heterocyclic rings, see V. R. Magnuson and G. D. Stucky. *J. Amer. Chem. Soc.,* **90,** 3269 (1968); *ibid.,* **91,** 2544 (1969).

TABLE 16-1
THERMOCHEMICAL DATA FOR
ALUMINUM ALKYLS[a]

Compound	$-\Delta H_f^\circ$ (liq)	ΔH_{vap}°	ΔH_{diss}° (liq)	ΔH_{diss}° (g)	$-\Delta H_f^\circ$ (g)	K_{diss}[h]
$[(CH_3)_3Al]_2$	72.0	9.88[b]	19.4[c]	20.4[b,d,e]	20.8	$2.5 \times 10^{-6c,f}$
$[(C_2H_5)_3Al]_2$	111.8	16.0	16.9[c]	18.17[c]	40.9	3.5×10^{-4g}
$[(C_3H_7)_3Al]_2$	154.0	21.9	20.9	—	55.8	—
$[(C_4H_9)_3Al]_2$	178.0	27.8	9.0	—	70.6	—
$[(i\text{—}C_4H_9)_3Al]_2$	185.6	27.8	8.0	—	76.8	—

[a] Unless otherwise specified, all data are from K. W. Egger and A. T. Cocks, *J. Amer. Chem. Soc.*, **94**, 1810 (1972). All heats in kcal.
[b] C. H. Henrickson and D. P. Eyman, *Inorg. Chem.*, **6**, 1461 (1967).
[c] M. B. Smith, *J. Organometal. Chem.*, **46**, 31 (1972).
[d] A. W. Laubengayer and W. F. Gilliam, *J. Amer. Chem. Soc.*, **63**, 477 (1941).
[e] K_{diss} (gas) = 5.1×10^{-9}; ΔG° = 11.3 kcal; ΔS° = 30.6 eu. H. M. Neumann, J. Laemmle, and E. C. Ashby, *J. Amer. Chem. Soc.*, **95**, 2597 (1973).
[f] In hexadecane.
[g] In hexadecane. M. B. Smith, *J. Organometal. Chem.*, **46**, 211 (1972).
[h] K_{diss} for $(Ph_3Al)_2$ in benzene is 5.7×10^{-4}. J. J. Eisch and C. K. Hordis, *J. Amer. Chem. Soc.*, **93**, 4496 (1971).

solvents. However, as indicated by the heats of dissociation of the dimers, trimethylaluminum is more highly associated than the others.

All but one of the triorganoaluminum compounds whose molecular structures have been determined resemble Me_6Al_2 (structure **2**).[125] Features of the Me_6Al_2 structure that are important to understanding the bonding in dimeric organoaluminum compounds and their uniqueness among other Group IIIA organometallics are: the Al-C_{bridge}-Al angle, the $C_{terminal}$-Al-$C_{terminal}$ angle, and the Al—Al distance.[126]

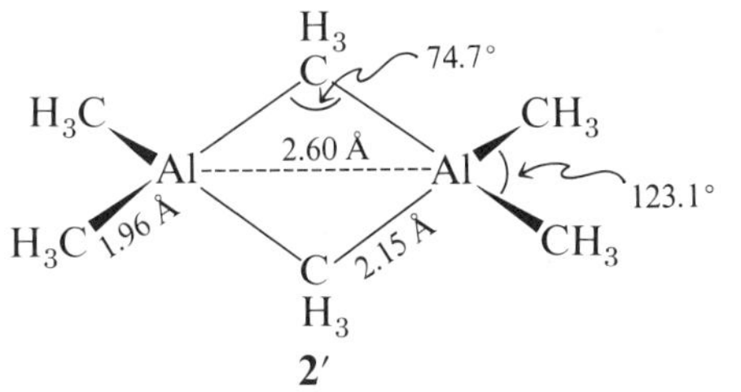

2'

Perhaps the most useful theoretical approach to the bonding in these dimers is to use diborane as a model (see Chapter 5, pp. 193-194). First, we will assume that the Al atoms are sp^2 hybridized. The bridge framework, then, is constructed from two Al sp^2 orbitals, two Al p ao's, and two carbon sp^3 orbitals (**12**).

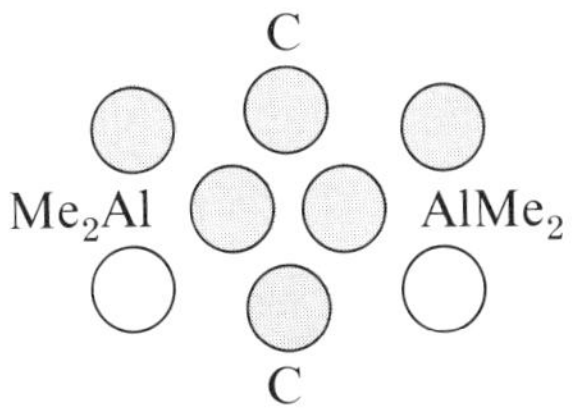

12

---

[125] The exception is the dimer of diphenyl(phenylethynyl)aluminum. G. D. Stucky, A. M. McPherson, W. E. Rhine, J. J. Eisch, and J. L. Considine, *J. Amer. Chem. Soc.*, **96**, 1941 (1974).
[126] R. G. Vranka and E. L. Amma, *J. Amer. Chem. Soc.*, **89**, 3121 (1967).

This will produce, as it did in the case of diborane, two bonding orbitals (which can accommodate two electrons each), two non-bonding orbitals, and two anti-bonding orbitals.

One of the features of the diborane-like model for organoaluminum dimerization is that one of the occupied molecular orbitals (13) involves Al—Al bonding.

$$\text{Al} \underset{\underset{\text{C}}{\bigcirc}}{\overset{\overset{\text{C}}{\bigcirc}}{\bigcirc \ \bigcirc}} \text{Al}$$

13

In fact, more detailed molecular orbital calculations indicate that the σ overlap population between the aluminum atoms in Me_6Al_2 is about 0.6 electron.[127] Furthermore, the geometry of the molecule (2′) suggests that such bonding may occur. Not only are the C_t-Al-C_t bond angles very close to those expected for sp^2 hybridization, but the Al—Al distance is only slightly larger than twice the covalent radius of Al (1.26 Å). [Notice that the Al—Al distance is considerably longer in dimeric aluminum bromide (7), a compound for which it is not necessary to postulate a diborane-like bonding scheme.] Finally, the very acute Al-C_{bridge}-Al angle may also imply the existence of an Al—Al bond; the bond distance requirements for both Al—Al and Al—C—Al bonding could force such a geometry on the bridge framework. The acute bridging angle is, incidentally, a characteristic of organo-bridged electron deficient compounds of Al, Be, and Mg. Another characteristic of such compounds is that M—C bridge bonds are longer than M—C terminal bonds. Although this could be ascribed to the electron deficiency of the bridging framework, it must be noted that such a difference in bridge and terminal bond lengths occurs in the aluminum halides, which are not electron deficient.

Molecular orbital calculations also show that boron alkyls may dimerize in the same manner if only the extent of B—B overlap is considered.[127] However, if the boron atoms move close enough to form a B—B bond (the B—B bond length in B_2Cl_4 is 1.7 to 1.8 Å),[128] serious steric interactions between bridge and terminal alkyl groups would destabilize the dimer.[129]

None of the heavier metals of Group IIIA form associated organometallics. In order to achieve the appropriate bridge geometry, the metal atoms may have to move too close together and strong metal-metal inner shell repulsions would result.

Because they have approximately the same charge-to-radius ratio, Be^{2+} and Al^{3+} resemble each other chemically, and their compounds are frequently structurally similar. Beryllium has only two valence electrons, and so forms compounds of the type Me_2Be. The beryllium atom can satisfy its electron deficiency and coordinative unsaturation by self-association or by forming complexes with Lewis bases such as NMe_3. Thus, Me_2Be is a self-associated, polymeric substance in the solid phase, while other alkyl- and arylberyllium compounds are predominantly dimeric. The structure of $(Me_2Be)_x$ strongly suggests that bonding follows the Me_6Al_2 pattern.[130] The Be-C_{bridge}-Be angle is very acute (66°) and the Be—C bond length is slightly longer (1.93 Å) than the sum of the covalent radii (Be = 0.91 Å; C = 0.77 Å), thereby implying that the Be—C bridge bond is less than a full two-electron bond. Both of these features are shared with Me_6Al_2. However, the angles about the Be atoms are more nearly tetrahedral than was the case in Me_6Al_2.

---

[127] A. H. Cowley and W. D. White, *J. Amer. Chem. Soc.*, **91**, 34 (1969).
[128] L. Trefonas and W. N. Lipscomb, *J. Chem. Phys.*, **28**, 54 (1958).
[129] In this connection, recall the discussion of F-strain in Chapter 5.
[130] A. I. Snow and R. E. Rundle, *Acta Cryst.*, **4**, 348 (1951).

Me   Me   Me   Me   Me   Me
 \\ / \\ / \\ / \\ /
  Be   Be   Be   Be   Be
 / \\ / \\ / \\ / \\
Me   Me   Me   Me   Me   Me

14

A very interesting difference between Me_6Al_2 and $(Me_2Be)_x$ is that the latter is apparently not associated in the gas phase at about 180°, whereas trimethyl- and triethylaluminum are still largely dimeric at temperatures in this same range.[131] However, the asymmetric Be—C stretching frequency of monomeric Me_2Be gas (1081 cm^{-1}) is quite high in comparison with other beryllium alkyls,[131] implying a very strong Be—C bond in the methyl derivative. It has been suggested that this may be due to hyperconjugation, an effect that would explain the complete formation of monomer in the gas phase.

$$H_3C—\overset{\ominus}{Be}=\overset{H^{\oplus}}{CH_2}$$

15

Owing to the difference in $2p$ vs. $3p$ ao's, Be—C π overlap is more favorable than Al—C π overlap. Therefore, hyperconjugation competes successfully with polymerization as a stabilization process for beryllium compounds, but not for aluminum compounds. Carrying this argument one step further, hyperconjugation is also one more factor favoring the stability of trialkylborane monomers over dimers.[132]

Dialkylmagnesium compounds can be prepared by the reaction of alkyllithium compounds with the corresponding alkylmagnesium halides (i.e., Grignard reagents) in ether. Although Grignard reagents are highly solvated (page 850), Me_2Mg precipitates from ether without attached solvent. The compound has been found to assume a polymeric structure very similar to that of Me_2Be.[133] Unlike dimethylberyllium, however, dimethylmagnesium polymerization is favored over complexation with a weak Lewis base such as Et_2O. Whereas dimethylmagnesium can be sublimed only with difficulty, a 1:2 molar mixture of Me_2Mg and Me_3Al yields a complex that sublimes at room temperature to give colorless, rod-shaped crystals.[134] An x-ray diffraction study of these crystals provides further insight into electron deficient compounds (**16**). Notice

[$Me_2Al(\mu$-$Me)_2Mg(\mu$-$Me)_2AlMe_2$]

16

again that the M—C$_{bridge}$ bond lengths are greater than the M—C$_{terminal}$ bond lengths and that the Mg-Me-Al angle is very acute. Further, the M—M bond length is only slightly longer than the sum of the covalent (Mg = 1.38 Å; Al = 1.26 Å).

The other Group II metal alkyls such as Me_2Zn, Me_2Cd, and Me_2Hg are all monomeric. This lack of self-association can be explained by noting that, outside of

---

[131] R. A. Kovar and G. L. Morgan, *Inorg. Chem.*, **8**, 1099 (1969).
[132] In Chapter 5 (page 230), hyperconjugation was used to explain the low acidity of BMe_3 relative to the aluminum analog.
[133] G. Stucky and R. E. Rundle, *J. Amer. Chem. Soc.*, **86**, 4825 (1964).
[134] J. L. Atwood and G. Stucky, *J. Amer. Chem. Soc.*, **91**, 2538 (1969).

the transition metals of Groups VI to VIII, metal-metal bond formation is a rare occurrence with heavier metals; even when it does occur (as it does for metals of Group IVA), the bond stability drops drastically on descending the Periodic Table. In part, the reason for this is that, as the heavier metals move within bonding distance, metal-metal inner shell repulsions become relatively more destabilizing as the ao overlaps decrease from increasing radial diffuseness.

As you might have anticipated from the behavior of the organometallic derivatives of the lighter, main group metals, alkyl- and aryllithium compounds are also associated, although they form only tetramers and hexamers in the solid or vapor phase or in non-basic solvents.[135] Methyllithium, for example, is tetrameric in the solid and in ether. A powder x-ray diffraction study suggests that the structure (17) is composed of a tetrahedron of lithium atoms with methyl groups located symmetrically in the faces of the tetrahedron.[136] As in the other electron deficient molecules examined thus far, the Li-C-Li angle is very acute (68.30°), and the Li—C distances (2.31 Å) are only slightly larger than the sum of their respective covalent radii (Li = 1.34 Å; C = 0.77 Å).

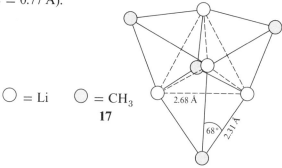

○ = Li   ○ = CH_3

17

Although several bonding models for organolithium tetrahedra have been proposed, the one that perhaps best agrees with experiment involves the formation of localized four-center bonds in the faces of the tetrahedron. In part (A) of Figure 16-8, the tetrahedron of lithium atoms is viewed down a three-fold axis, and an sp^3 hybridized lithium is placed at this apex and at each of the other apices. Therefore, extending into the region above each triangular face of the tetrahedron are three hybrid orbitals, one from each of three lithium atoms [Figure 16-8(B)]. These three hao's can form three TASO's that may interact with a carbon sp^3 hao extending into the tetrahedral face [part (C)]. The net result is the formation of molecular orbitals involving the four atoms common to each tetrahedral face [part (D)]. The bonding mo can accommodate two electrons. Therefore, each face of the Li_4C_4 tetrahedron involves a two-electron/four-center bond, and the eight valence electrons available may be accounted for.

If the picture in Figure 16-8 is adopted, one implication is that the tetrahedral cluster may be acidic. That is, sp^3 hybridized Li atoms were used in the bonding scheme, leaving one empty, radially-directed hao on each metal atom. An apparent result of this is that methyllithium tetramers are associated in the solid state. The unit cell of Li_4Me_4 (18) is body-centered cubic, with Li_4 tetrahedra at the eight corners and the center.

△ = Li_4 ○ = CH_3

18 Li_4Me_4 *unit cell*

[From E. Weiss and E. A. C. Lucken, *J. Organometal. Chem.*, **2**, 197 (1964).]

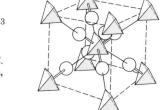

---

[135] The best source of general information on organolithium compounds is B. J. Wakefield, "The Chemistry of Organolithium Compounds," Pergamon Press, New York (1974).
[136] E. Weiss and E. A. C. Lucken, *J. Organometal. Chem.*, **2**, 197 (1964).

(A) Li sp^3 hybrid

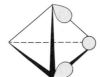

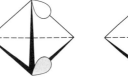

(B) TASO's generated by Li sp^3 orbitals in one face of Li_4 tetrahedron.

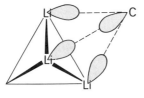

(C) Bonding mo formed by overlap of C sp^3 hybrid with the symmetric TASO. Same type of mo formed in other three faces.

(D)

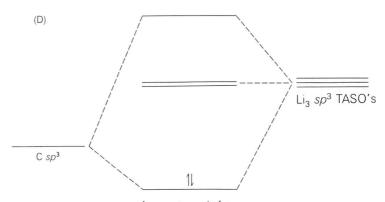

four-center mo's for one face of Li_4C_4 tetrahedron

Figure 16-8. A scheme for bonding between one carbon and three lithiums in each of the four faces of the Li_4C_4 tetrahedron.

Each of the methyl groups of the center tetrahedron interacts with a Li atom of a corner Li_4 cluster *via* C—H ⋯ Li three-center bonds; similarly, methyl groups from the four corner tetrahedra interact with lithium atoms of the center tetrahedron. The effect of this association is that methyllithium is non-volatile and insoluble in hydrocarbon solvents. In contrast, the individual clusters of $(t\text{-BuLi})_x$ and $(EtLi)_4$ are not as strongly associated in the solid state as $(MeLi)_4$ (the former are volatile and soluble in hydrocarbons),[137] presumably owing to intra-cluster Li-alkyl hydrogen interactions or to the steric bulk of the alkyl groups.[138] Finally, the effect of the orbital unused in tetramer bonding is seen in the fact that $(MeLi)_4$ and $(n\text{-BuLi})_4$ can both be dissolved in basic solvents, and the tetrameric unit is not disrupted. This is in decided contrast to other electron deficient organometallic compounds such as $(R_3Al)_2$, where basic solvents cause dissociation of the dimeric unit.

Several organolithium compounds are hexameric in the solid (*e.g.*, cyclohexyllithium), in solution (*e.g.*, ethyllithium), and even in the vapor phase. The Li atoms in the hexameric units are arranged in a distorted octahedron, with organic groups occupying six of the eight faces.[138] The bonding in such clusters may be accounted for in a manner similar to that used for the tetramer.[138a]

---

[137] The degree of association of *t*-BuLi in the solid state has not been determined. However, the compound is tetrameric in benzene.

[138] R. Zerger, W. Rhine, and G. Stucky, *J. Amer. Chem. Soc.*, **96**, 6048 (1974).

[138a] See Chapter 18 for further discussion of bonding in metal clusters.

Grignard reagents are one of the most important classes of organometallic compounds having M—C σ bonds.[139] Although they are commonly formulated as RMgX (*cf.* page 475), the nature of Grignard reagents in solution is considerably more complicated, and there are many unanswered questions surrounding their constitution.

All of the accumulated experimental evidence suggests that RMgX is the first species formed upon reaction of RX with Mg in a basic solvent.

(16-75)
$$yRX + yMg + xyS \rightleftharpoons (RMgXS_x)_y$$

However, the extent to which RMgX is solvated, its degree of association (= observed molecular weight ÷ formula weight), and the possibility of disproportionation to R_2Mg and MgX_2 all depend on the nature of the solvent, the organic group, and X. For example, ethylmagnesium bromide and chloride are highly associated in diethyl ether (Figure 16-9), whereas EtMgBr is a monomer in tetrahy-

---

[139] The best source of information on the composition of Grignard reagents is the review by E. C. Ashby, *Quart. Revs.*, **21**, 259 (1967).

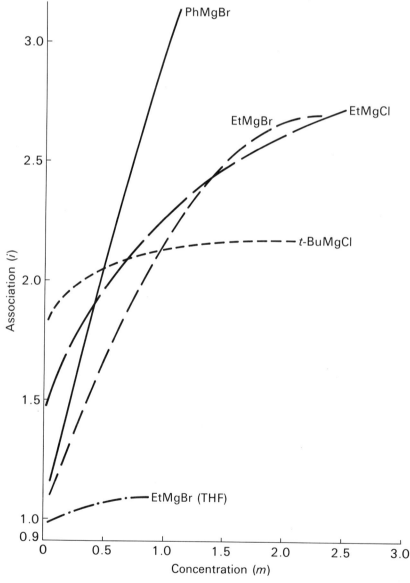

Figure 16-9. Degree of association of various organomagnesium halides in diethyl ether. [From E. C. Ashby, *Quart. Revs.*, **21**, 259 (1967).]

drofuran (THF). Nonetheless, concentration of a diethyl ether solution of EtMgBr precipitates the disolvated monomer, **19**.[140] In contrast, if triethylamine is added to a di-*n*-butyl ether solution of the same Grignard reagent, the solvated dimer **20** is isolated.[141] Because of the steric demands of triethylamine, halide bridging competes successfully with disolvation (which would lead to an analog of **19**).

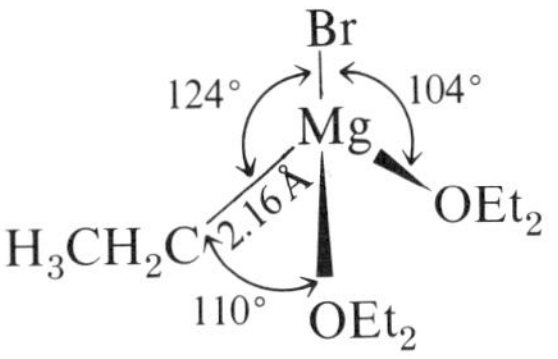

The behavior of EtMgCl in THF is unique; concentration of the solution leads to essentially quantitative yields of Et_2Mg and the very interesting self-associated species $[EtMg_2Cl_3 \cdot 3THF]_2$ (**21**).[142]

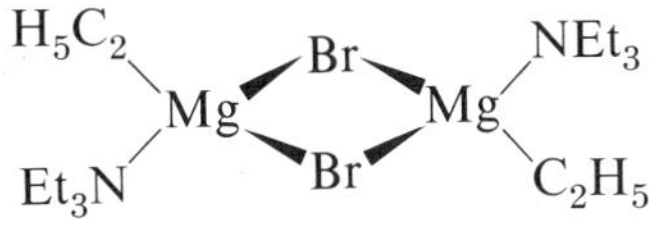

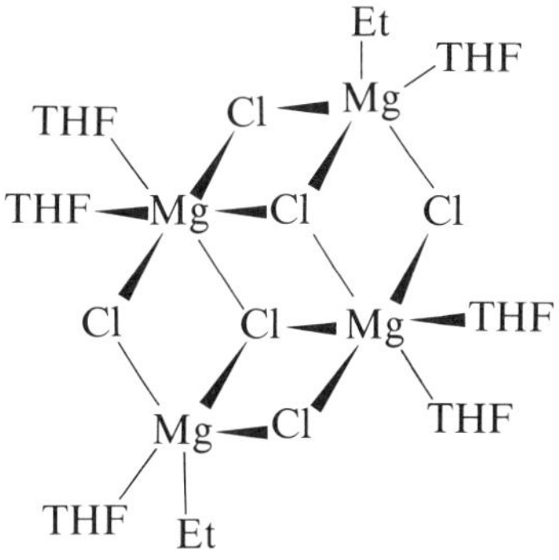

The existence of **21** implies that equilibria such as 16-76 and 16-77 must exist; reaction 16-76 is further substantiated by the isolation of MgI_2 when Grignard reagents are prepared from Mg and RI in THF.

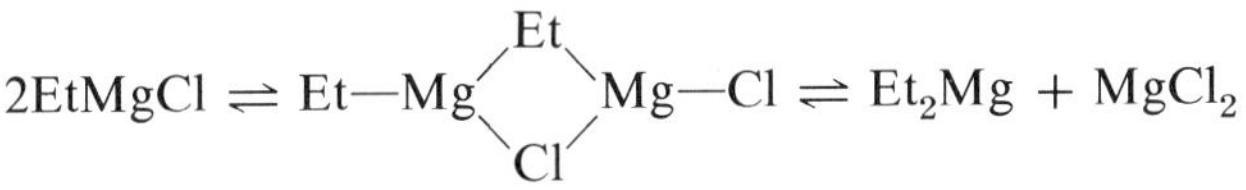

$$2EtMgCl \rightleftharpoons Et-Mg\underset{Cl}{\overset{Et}{\diamond}}Mg-Cl \rightleftharpoons Et_2Mg + MgCl_2 \qquad (16\text{-}76)$$

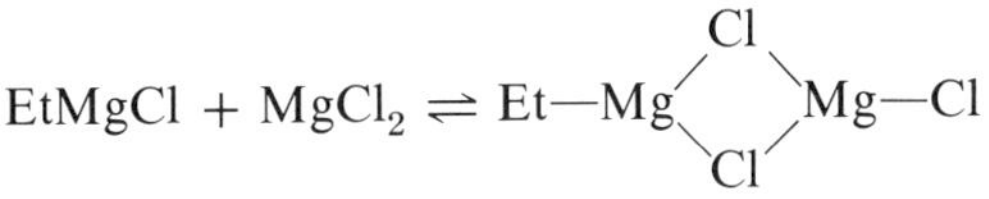

$$EtMgCl + MgCl_2 \rightleftharpoons Et-Mg\underset{Cl}{\overset{Cl}{\diamond}}Mg-Cl \qquad (16\text{-}77)$$

As noted above, the precise nature of Grignard reagents is determined to some extent by the steric effect of the organic group. As an example, compare *t*-BuMgCl and EtMgCl in Figure 16-9 and also note that concentration of a THF solution of MeMgBr affords the trisolvated complex $MeMgBr \cdot 3THF$, whereas PhMgBr gives only a disolvated species, $PhMgBr \cdot 2THF$.

Based on numerous solution studies and examination of solids precipitated from Grignard solutions, it has been proposed that a Grignard reagent may be involved in some or all of the equilibria outlined in Figure 16-10. However, as we mentioned

---

[140] G. D. Stucky and R. E. Rundle, *J. Amer. Chem. Soc.*, **86**, 4825 (1964).
[141] J. Toney and G. D. Stucky, *Chem. Commun.*, 1168 (1967).
[142] J. Toney and G. D. Stucky, *J. Organometal. Chem.*, **28**, 5 (1971).

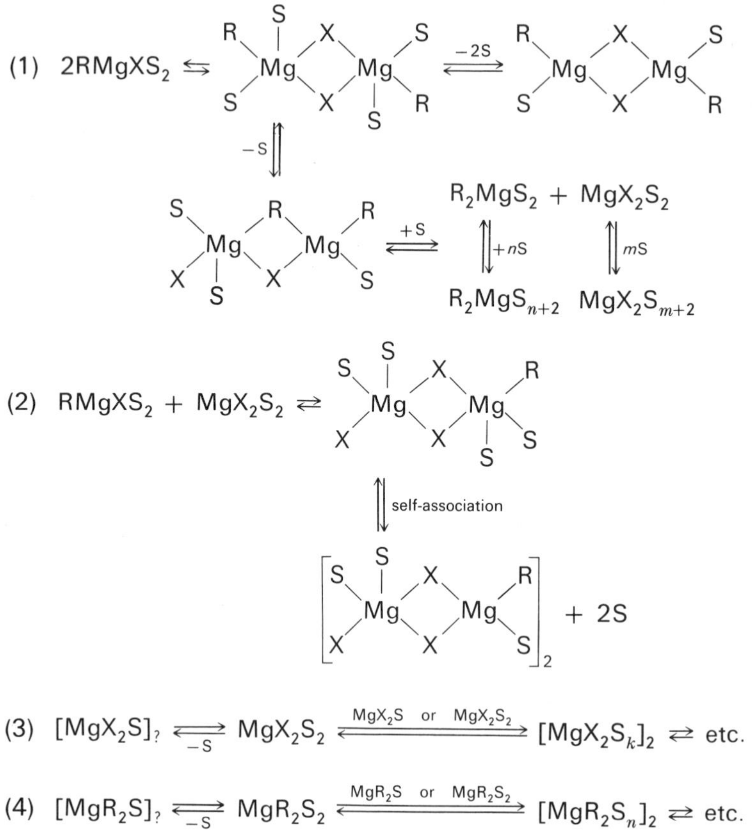

(1) $2RMgXS_2 \rightleftharpoons$ [dimer structure] $\xrightleftharpoons{-2S}$ [dimer structure]

$\Updownarrow -S$

[dimer structure] $\xrightleftharpoons{+S}$ $R_2MgS_2 + MgX_2S_2$

$\Updownarrow +nS \quad \Updownarrow mS$

$R_2MgS_{n+2} \quad MgX_2S_{m+2}$

(2) $RMgXS_2 + MgX_2S_2 \rightleftharpoons$ [dimer structure]

$\Updownarrow$ self-association

$[\text{dimer structure}]_2 + 2S$

(3) $[MgX_2S]_? \xrightleftharpoons{-S} MgX_2S_2 \xrightleftharpoons{MgX_2S \text{ or } MgX_2S_2} [MgX_2S_k]_2 \rightleftharpoons$ etc.

(4) $[MgR_2S]_? \xrightleftharpoons{-S} MgR_2S_2 \xrightleftharpoons{MgR_2S \text{ or } MgR_2S_2} [MgR_2S_n]_2 \rightleftharpoons$ etc.

Figure 16-10. An outline of the equilibria possible for a Grignard reagent in solution in a basic solvent. [From J. Toney and G. D. Stucky, *J. Organometal. Chem.*, **28**, 5 (1971).]

previously, the equilibria appropriate to a particular reagent will depend on the solvent and the nature of the organic group and the halide.

One of the most important transition metal organometallic compounds is the coenzyme of vitamin B_{12} (**22**).[143] For our present purposes, the most interesting property of this molecule, and that of vitamin B_{12} model complexes such as bis-(dimethylglyoximato)methyl(pyridine)cobalt (often abbreviated methylpyridine-cobaloxime) (**23**),[144] is the striking stability of the Co—C σ bond. In neutral, aqueous solution, or as solids at room temperature, the coenzyme, other alkylcobalamins, and alkylcobaloximes are thermally and oxidatively stable.[144] However, heating methyl-cobalamin (similar to **22** except that the 5′-deoxyadenosyl group is replaced by methyl) above 200°C gives methane and ethane in equal amounts. Ethylcobalamin produces ethylene and traces of ethane and butane on pyrolysis. Photolysis at room temperature gives the same hydrocarbon products from the alkylcobalamins. In acidic or basic solution, Co—C cleavage occurs even in the absence of light.

The very great stability of the Co—C bond in vitamin B_{12} coenzyme and model systems is in decided contrast to numerous other transition metal–carbon σ-bonded alkyl compounds, as mentioned on page 840. Originally, it was thought that this instability was an intrinsic thermodynamic property, that the isolation of homoleptic[145] alkyls of the early transition metals would be especially difficult, and that ligands such as (η^5-C_5H_5), CO, and R_3P are required to achieve stability. However, Lappert and Wilkinson have recently isolated compounds such as $Cr(CH_2CPhMe_2)_4$

---

[143] J. M. Pratt, "Inorganic Chemistry of Vitamin B_{12}," Academic Press, New York (1972).
[144] G. N. Schrauzer, *Acc. Chem. Res.*, **1**, 97 (1968).
[145] See ref. 14.

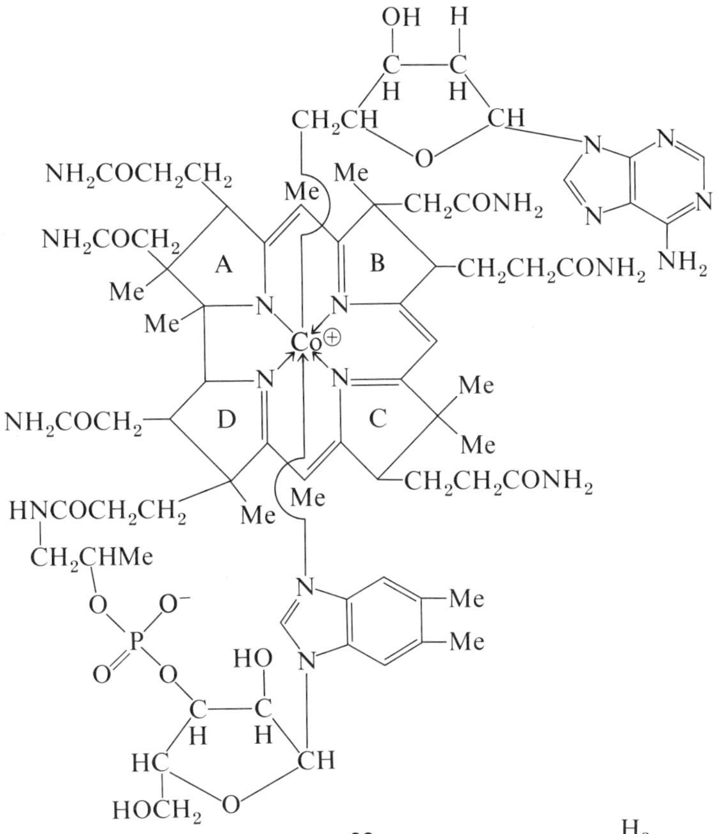

22

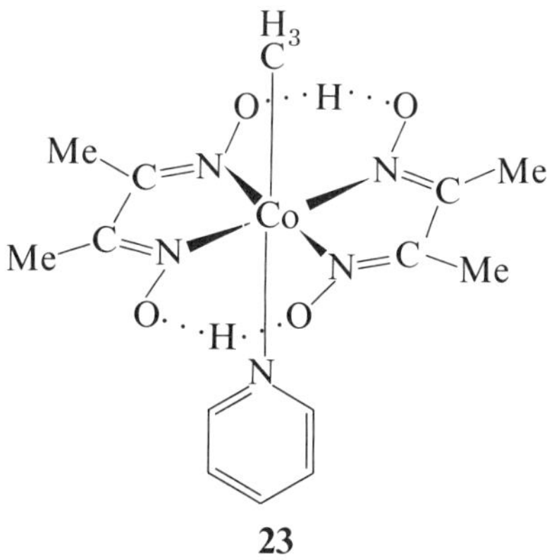

23

(24), Ti(CH$_2$SiMe$_3$)$_4$, and Mo$_2$(CH$_2$SiMe$_3$)$_4$ **(25)**, which are quite stable to decomposition.[146,147,147a] Most importantly, the availability of such compounds has permitted the determination of average metal–carbon bond strengths, some of which are plotted in Figure 16-11 for Ti and Zr compounds of the type MX$_4$.[148] When X is a σ-bonded alkyl group, the metal–carbon bond energies are about 40 to 80 kcal, a range similar to that observed for main group metal–carbon bonds (see Figure 16-5). In addition, it should be noted that while the M—O, M—N, and M—Cl bond

---

[146] See ref. 11.
[147] W. Mowat, A. Shortland, G. Yagupsky, N. J. Hill, M. Yagupsky, and G. Wilkinson, *J. Chem. Soc.* (A), 533 (1972).
[147a] R. R. Schrock and G. W. Parshall, *Chem. Rev.*, **76**, 243 (1976); see also ref. 118a.
[148] M. F. Lappert, D. A. Patil, and J. B. Pedley, *Chem. Commun.*, 830 (1975).

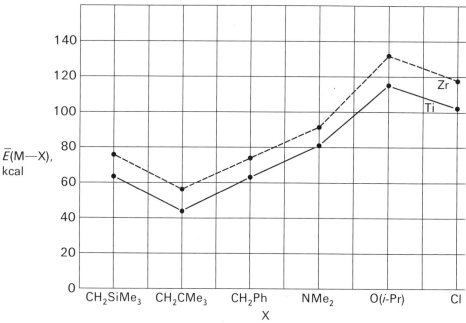

Figure 16-11. Average M—X bond energies in molecules of the type TiX$_4$ and ZrX$_4$. [After M. F. Lappert, D. A. Patil, and J. B. Pedley, *Chem. Commun.*, 830 (1975).]

energies are greater than that of M—C, this is the same general trend (M—O > M—Cl > M—N > M—C) observed for main group metals such as Si, Ge, and Sn.[149] However, in contrast with the main group metals, the bond between carbon and the heavier metal (Zr—C)[150] is stronger than the bond between carbon and the lighter metal (Ti—C).

24[151]

Cr(CH$_2$CPhMe$_2$)$_4$

25[152,153]

Mo$_2$(CH$_2$SiMe$_3$)$_6$

---

[149] J. C. Baldwin, M. F. Lappert, and J. B. Pedley, *J. Chem. Soc.* (A), 1943 (1972).

[150] You should also be aware that the order of Hf-X bond strengths is Hf-O > Hf-Cl > Hf-N and Hf-C. Furthermore, the Hf-X bonds are generally stronger than Zr-X bonds, in keeping with the general trend observed in transition metal chemistry.

[151] V. Gramlich and K. Pfefferkorn, *J. Organometal. Chem.*, **61**, 247 (1973).

[152] F. Huq, W. Mowat, A. Shortland, A. C. Skapski, and G. Wilkinson, *Chem. Commun.*, 1079 (1971).

[153] A most interesting feature of **25** is that there is a Mo-Mo triple bond. We shall discuss multiple bonding between metals in Chapter 18.

From the discussion above, it would appear that the instability associated with some transition metal alkyls is probably kinetic in origin, since several low-energy pathways are available for their decomposition.[154] Although homolytic M—C bond cleavage can occur (*i.e.*, free radical decomposition, as in the case of methylcobalamin pyrolysis above), a more important decomposition mechanism is β-elimination of an olefin and formation of a metal hydride. As illustrated in reaction 16–71, elimination of an olefin from a metal alkyl (*e.g.*, in the pyrolysis of ethylcobalamin above) proceeds *via* an interaction between the metal and the hydrogen on the carbon β to the metal. If such an interaction can occur easily, decomposition of the metal alkyl by olefin elimination can occur. However, if the interaction is prevented by steric effects, by placing elements other than H on the β carbon (as in **24** and **25**), or by replacing the β carbon by an element unlikely to form π bonds in the eliminated, unsaturated species (as in **25**), decomposition can occur only by homolytic cleavage; as you have seen, such cleavages may be prevented by a reasonably strong M—C bond.

STUDY QUESTIONS

9. Structure **12** illustrates the atomic orbitals involved in formation of the bridge mo structure in a dimeric aluminum alkyl or aryl. Structure **13** is one of the mo's. Draw the remainder of the mo's responsible for bridge bonding in Al_2R_6.

10. Knowing that you can purchase Me_3SiCH_2Cl, suggest a preparation for $Ti(CH_2SiMe_3)_4$ from chloromethyltrimethylsilane and other appropriate starting materials.

11. The structure of $Cr(CH_2CPhMe_2)_4$ is given on page 854. Does the observed structure agree with the prediction of the angular overlap model? What structure do you predict for $Ti(CH_2SiMe_3)_4$?

METAL CARBONYLS[155-159]

Transition metal carbonyls are frequent starting materials for building other compounds.[160] Not only can the carbon monoxide be replaced by a wide variety of ligands, but any carbonyl groups that remain frequently stabilize the molecule with respect to oxidation and/or thermal decomposition.[161] Furthermore, the CO ligands can be used as a probe of the electronic and molecular structure of a compound; such information can be obtained by examining the frequency and intensity of the C—O stretching modes in the infrared region (see Appendix to this chapter). For these reasons and others, the study of the chemical and physical properties of metal carbonyls and their derivatives has been extensively pursued.

In this section we shall discuss the preparation and physical properties of simple binary carbonyls and their anions and cations. Subsequent sections of this chapter will deal with derivatives of metal carbonyls.

---

[154] See ref. 116.
[155] J. C. Hileman, "Metal Carbonyls," in "Preparative Inorganic Reactions," Vol. 1, W. L. Jolly, ed., Interscience, New York (1964).
[156] E. W. Abel and F. G. A. Stone, *Quart. Rev.*, **23**, 325 (1969).
[157] E. W. Abel and F. G. A. Stone, *Quart. Rev.*, **24**, 498 (1970).
[158] P. S. Braterman, *Structure and Bonding*, **10**, 57 (1972).
[159] F. A. Cotton, *Prog. Inorg. Chem.*, **21**, 1 (1976).
[160] See, for example, Figure 16-2.
[161] For example, dibenzenechromium is very susceptible to oxidation, whereas $(\eta^6\text{-}C_6H_6)Cr(CO)_3$ can be handled in the air.

The Synthesis of Metal Carbonyls[155,157]

The known neutral, binary metal carbonyls are listed in Table 16-2. The shaded elements have not been reported to form simple, isolable binary carbonyls;[162] these metals form carbonyl complexes only when the metal is complexed with other ligands as well, or they form only anionic complexes such as $[Pt_6(CO)_{12}]^{2-}$. Almost all of the carbonyls listed in Table 16-2 are commercially available.[163]

Two methods of preparation are used for the simplest metal carbonyls:[155] (1) direct reaction of the metal with CO; and (2) reductive carbonylation. The only metals that can directly react with CO under mild conditions are iron and nickel (see Figure 16-4).[164] The remainder require the use of the second method, that is, reduction of a salt of the metal (using an active metal or H_2) in the presence of CO. For example,

(16-78)
$$VCl_3 + 4\,Na + 6\,CO(200\text{ atm}) \xrightarrow[160°C]{\text{diglyme}} [Na(\text{diglyme})_2][V(CO)_6] + 3\,NaCl$$
$$\downarrow \text{HCl-Et}_2\text{O}$$
$$V(CO)_6$$

or

(16-79)
$$2\,Co(H_2O)_4(OAc)_2 + 8\,(Ac)_2O + 8\,CO(160\text{ atm}) + 2\,H_2(40\text{ atm}) \longrightarrow Co_2(CO)_8 + 20\,HOAc$$

Methods for the preparation of polynuclear carbonyls depend on the particular metal. For example, photolysis of $Fe(CO)_5$ in glacial acetic acid gives an excellent yield of $Fe_2(CO)_9$.[165,166] One way in which $Fe_3(CO)_{12}$ may be obtained is oxidation of

---

[162] This statement should be qualified. Matrix isolation techniques have led to the observation of such otherwise unstable species as $Re(CO)_5$ and $Pt(CO)_2$: G. A. Ozin and A. Vander Voet, *Acc. Chem. Res.,* **6,** 313 (1973); H. Huber, E. P. Kundig, and G. A. Ozin, *J. Amer. Chem. Soc.,* **96,** 5585 (1974).

[163] $Fe(CO)_5$ costs about \$25/kg, $Co_2(CO)_8$ about \$62/100 grams, and $Ru_3(CO)_{12}$ about \$140/5 grams.

[164] Other metals also react directly with CO (Figure 16-4), although they require considerably more severe conditions than do Fe and Ni.

[165] W. L. Jolly, "The Synthesis and Characterization of Inorganic Compounds," Prentice-Hall, Englewood Cliffs, N. J. (1970), p. 472.

[166] See Chapter 8, page 450 for a description of the apparatus used in this experiment.

TABLE 16-2

THE KNOWN NEUTRAL BINARY METAL CARBONYLS. SHADED ELEMENTS FORM ONLY ANIONIC CARBONYL COMPLEXES OR ONLY AFFORD M—CO BONDS WHEN THE METAL IS COORDINATED TO OTHER LIGANDS.

III	IV	V	VI	VII	VIII			I	II
	Ti	$V(CO)_6$	$Cr(CO)_6$	$Mn_2(CO)_{10}$	$Fe(CO)_5$ $Fe_2(CO)_9$ $Fe_3(CO)_{12}$	$Co_2(CO)_8$ $Co_4(CO)_{12}$	$Ni(CO)_4$	Cu	
	Zr	Nb	$Mo(CO)_6$	$Tc_2(CO)_{10}$ $Tc_3(CO)_{12}$	$Ru(CO)_5$ $Ru_3(CO)_{12}$	$Rh_2(CO)_8$ $Rh_4(CO)_{12}$ $Rh_6(CO)_{16}$	Pd	Ag	
	Hf	Ta	$W(CO)_6$	$Re_2(CO)_{10}$	$Os(CO)_5$ $Os_3(CO)_{12}$	$Ir_2(CO)_8$ $Ir_4(CO)_{12}$	Pt	Au	

a solution of the anion $[HFe(CO)_4]^-$, which is prepared in turn from the pentacarbonyl:[167]

$$Fe(CO)_5 + 2\ OH^- \rightarrow [HFe(CO)_4]^- + HCO_3^-$$
$$3\ HFe(CO)_4^- + 3\ MnO_2 \rightarrow Fe_3(CO)_{12} + 3\ OH^- + 3\ MnO$$

(16-80)[168]

Metal carbonyl anions and their derivatives are especially important as intermediates in organometallic chemistry, as you have already seen on page 831 and in Figure 16-2. However, they are also interesting in themselves. Not only are the simple metal carbonyl anions much more numerous than neutral carbonyls, but for some metals only anionic carbonyls (*e.g.*, $[Nb(CO)_6]^-$ and $[Ta(CO)_6]^-$) have been isolated. Methods for the preparation of metal carbonyl anions include:

1. Reaction of a metal carbonyl with a base such as an amine or OH^-.

$$Fe(CO)_5 \begin{cases} \xrightarrow{Na} Na_2Fe(CO)_4{}^{169} \\ \xrightarrow[Et_4NI]{NH_3/H_2O} (Et_4N)(HFe(CO)_4) \\ \xrightarrow{Et_3N/H_2O} (Et_3NH)(HFe_3(CO)_{11}) \end{cases}$$

(16-81)

2. Reaction with an alkali metal.

$$2\ Na + Co_2(CO)_8 \xrightarrow{THF} 2\ Na[Co(CO)_4]$$

(16-82)

$$[(\eta^5\text{-}C_5H_5)Mo(CO)_3]_2 + 2\ Na \rightarrow Na[(\eta^5\text{-}C_5H_5)Mo(CO)_3]$$

(16-83)

3. Displacement of CO from a metal carbonyl with an anion.

$$Mo(CO)_6 + NaC_5H_5 \rightarrow Na[(\eta^5\text{-}C_5H_5)Mo(CO)_3] + 3\ CO$$

(16-84)

Reactions 16-83 and 16-84 represent two methods of synthesizing the same compound. However, the latter is the more advantageous, as it circumvents the preparation of the intermediate species $[(\eta^5\text{-}C_5H_5)Mo(CO)_3]_2$, a synthesis that has been described as "somewhat of an art."[170]

One of the very important observations to be made concerning these reactions is that the nature of the product often depends critically on reaction conditions. Furthermore, although it may often appear that a widely disparate group of complexes is isolated [*e.g.*, the products of the $Fe(CO)_5$ reactions], they are just as often interrelated. For example, if a CO group is removed from a given metal carbonyl, the

---

[167] J. J. Eisch and R. B. King, eds., "Organometallic Syntheses" (Volume 1—Transition Metal Compounds by R. B. King), Academic Press, New York (1965).

[168] Although the mechanism of this reaction is not known, an interesting possibility is as follows:

$$\left[\text{Fe-C}\begin{matrix}O\\OH\end{matrix}\right]^- \xrightarrow{\beta\text{-elim.}} \left[\text{Fe-H}\right]^- + CO_2 \xrightarrow{OH^-} HCO_3^-$$

The first step involves nucleophilic attack at a carbonyl carbon; as you shall see in the section on carbenes that follows, such a reaction can lead to other interesting compounds. The second reaction step involves a β-elimination to give the final products.

[169] The $Fe(CO)_4^{2-}$ ion has found extensive use as a reagent in organic chemistry: J. P. Collman, *Acc. Chem. Res.*, **8**, 342 (1975); see also Chapter 17.

[170] See page 110 of ref. 167.

CO may be replaced by (i) two electrons, (ii) a hydrogen atom and one electron, or (iii) two hydrogen atoms. This relationship is well illustrated by the following isoelectronic series of iron carbonyls and their anions.

$Fe(CO)_5$	$Fe_2(CO)_9$	$Fe_3(CO)_{12}$	$Fe_4(CO)_{14}$ (unknown)
$[Fe(CO)_4]^{2-}$	$[Fe_2(CO)_8]^{2-}$	$[Fe_3(CO)_{11}]^{2-}$	$[Fe_4(CO)_{13}]^{2-}$
$[Fe(CO)_4H]^-$	$[Fe_2(CO)_8H]^-$	$[Fe_3(CO)_{11}H]^-$	$[Fe_4(CO)_{13}H]^-$
$Fe(CO)_4H_2$	$Fe_2(CO)_8H_2$	$Fe_3(CO)_{11}H_2$	$Fe_4(CO)_{13}H_2$

Metal carbonyl cations are not nearly so numerous as anions. The probable reason for this is that anions can disperse the excess negative charge over the entire molecule by back-bonding electron density from the metal to the CO ligand (see below; also Chapter 5, p. 199, and Chapter 9), an effect that strengthens the M—CO bond. On the other hand, removal of an electron from a metal carbonyl to produce a cation leads to a decrease in M—CO $d\pi$-$p\pi$* bonding, a bond-weakening effect. Carbonyl cations can be produced in several ways, one of them being the disproportionation of the parent carbonyl:

(16–85) $$Co_2(CO)_8 + 2\,PPh_3 \rightarrow [(Ph_3P)_2Co(CO)_3^+][Co(CO)_4^-] + CO$$

Metal Carbonyls: Properties and Structures

Some properties of the simplest carbonyl compounds, primarily of the first row elements, are collected in Table 16–3. You should note that, with the exception of the mononuclear carbonyls of the iron sub-group and $Ni(CO)_4$, all of the known metal carbonyls are solids; and that all of the mononuclear carbonyls are colorless, while the polynuclear compounds are colored, the color becoming darker as the number of metal atoms increases. For example, while $Fe(CO)_5$ is a colorless liquid, $Fe_2(CO)_9$ forms golden-yellow plates, and $Fe_3(CO)_{12}$ is a very dark green-black.

Investigation of the structures of metal carbonyls has occupied chemists for many years. The structure of $Ni(CO)_4$ was first established in 1935, but many others have only recently been defined unambiguously. The mononuclear carbonyls all have structures expected on the basis of their formulas; that is, O_h for $M(CO)_6$ where M = V, Cr, Mo, and W; D_{3h} for $M(CO)_5$ where M = Fe, Ru, and Os; and T_d for $Ni(CO)_4$.

It is the polynuclear carbonyls that are of the greatest structural interest. You found in Chapter 15 that, in order to satisfy the EAN rule for a neutral metal carbonyl, metals with odd atomic numbers must form either paramagnetic compounds [e.g., $V(CO)_6$] or polynuclear compounds with metal-metal bonds [e.g., $Mn_2(CO)_{10}$], supplemented in some cases with bridging carbonyl groups [e.g., $Co_2(CO)_8$]. In the case of polynuclear carbonyls, the 18 electron rule suggests that two arrangements are equally likely for the hypothetical carbonyl $M_2(CO)_2$; the two metals, bonded to each other, may each have a terminal CO ligand, or the two metals may share two CO ligands as bridging groups. $Co_2(CO)_8$ adopts both of these configurations, the former (26a) in solution and the latter (26b) in the solid state.

26a
$Co_2(CO)_8$ in solution

26b
$Co_2(CO)_8$ in the solid state

TABLE 16-3
PROPERTIES OF METAL CARBONYLS

Compound	Color	Melting Point (°C)	Structure	IR ν(CO)(cm^{-1})	Comments
V(CO)$_6$	black-green	70(d)	O_h	1976[a]	Paramagnetic by 1e^-
Cr(CO)$_6$	white	130(d)	O_h	2000[b]	Sublimes; Cr—C = 1.92 Å[c] Δ_0 = 32,200 cm^{-1}[c]
Mo(CO)$_6$	white	—	O_h	2004[b]	Sublimes; Mo—C = 2.06 Å[c] Δ_0 = 32,150 cm^{-1}[c]
W(CO)$_6$	white	—	O_h	1998[b]	Sublimes; W—C = 2.07 Å[c] Δ_0 = 32,200 cm^{-1}[c]
Mn$_2$(CO)$_{10}$	golden yellow	154	D_{4d}	2044(m)[d] 2013(s) 1983(m)	Mn—Mn = 2.93 Å ΔG_f° = −400.9 kcal
Tc$_2$(CO)$_{10}$	white	160	D_{4d}	2065(m)[d] 2017(s) 1984(m)	
Re$_2$(CO)$_{10}$	white	177	D_{4d}	2070(m)[d] 2014(s) 1976(m)	
Fe(CO)$_5$	colorless	−20	D_{3h}	2034(s)[e] 2013(vs)	b.p. 103°; Fe—C$_{axial}$ = 1.810 Å Fe—C$_{eq}$ = 1.833 Å ΔG_f° = −182.6 kcal
Ru(CO)$_5$	colorless	−22	D_{3h}	2035(s)[f] 1999(vs)	Unstable with respect to light-catalyzed decomposition to Ru$_3$(CO)$_{12}$
Os(CO)$_5$	colorless	−15	D_{3h}	2034(s)[f] 1991(vs)	Very unstable with respect to Os$_3$(CO)$_{12}$
Fe$_2$(CO)$_9$	golden yellow	d	D_{3h}	2082(m) 2019(s) 1829(s) (bridging CO)	Fe—Fe = 2.46 Å
Co$_2$(CO)$_8$	orange-red	51(d)	C_{2v} (solid) D_{3d} (solution)	C_{2v}[g] D_{3d} 2112 2107 2071 2069 2059 2042 2044 term. 2031 2031 2023 2001 1991 1886 bridge 1857	Co—Co = 2.54 Å
Ni(CO)$_4$	colorless	−25	T_d	2057	b.p. 43; Ni—C = 1.84 Å Very toxic due to ready decomp. to Ni + 4 CO

[a] In cyclohexane: W. Hicker, J. Peterhaus, and E. Winter, *Chem. Ber.*, **94**, 2572 (1961).
[b] Gas phase: L. H. Jones, *Spectrochim. Acta*, **19**, 329 (1963).
[c] N. A. Beach and H. B. Gray, *J. Amer. Chem. Soc.*, **90**, 5713 (1968).
[d] N. Flitcroft, D. K. Huggins, and H. D. Kaesz, *Inorg. Chem.*, **3**, 1123 (1964).
[e] Gas phase: L. H. Jones and R. S. McDowell, *Spectrochim. Acta*, **20**, 248 (1964).
[f] F. Calderazzo and F. L'Eplattenier, *Inorg. Chem.*, **6**, 1220 (1967).
[g] K. Noack, *Spectrochim. Acta*, **19**, 1925 (1963); *Helv. Chim. Acta*, **47**, 1555 (1964). This reference should be consulted for relative band intensities.

The reasons for the preference for one form or another are not entirely clear. However, it would appear that the larger the metal radius, the less likely it is that CO bridging will occur. Therefore, owing to the decrease in metal radius on moving across the transition metal series, $Mn_2(CO)_{10}$ (**27**) does not have CO bridges, whereas $Co_2(CO)_8$ does, at least in the solid state. $Fe(CO)_9$ (**28**) can satisfy the 18 electron rule with two structures, one with three CO bridges and an Fe—Fe bond (the observed structure), or another with only one CO bridge and an Fe—Fe bond. Although the latter is not observed for iron, there is evidence that the osmium analogue, $Os_2(CO)_9$, has only one bridging CO.[171]

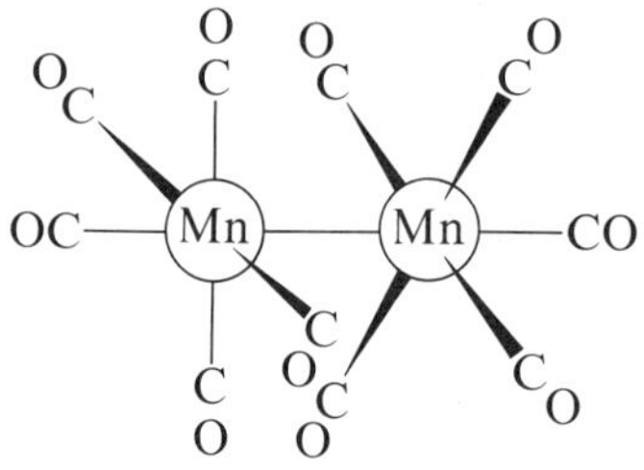

27
$Mn_2(CO)_{10}$

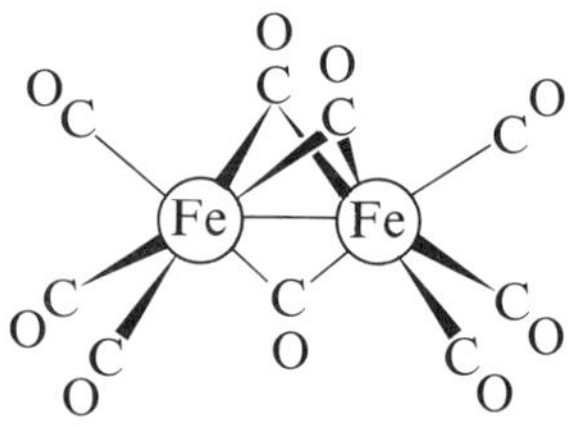

28
$Fe_2(CO)_9$

The more complex polynuclear carbonyls are metal clusters (**29–31**), a structural form that we shall discuss in detail in Chapter 18. A careful analysis of these structures shows that all of them except $Rh_6(CO)_{16}$ (**31**) obey the 18 electron rule.[172] A comparison of the iron- and cobalt-carbonyl clusters with their heavier congeners shows that the generalization that metals with larger radii do not allow CO bridging is true for the polynuclear clusters as well.

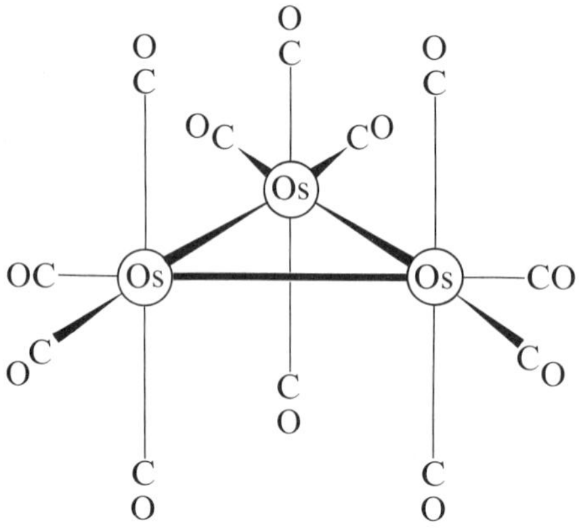

29
$Os_3(CO)_{12}$

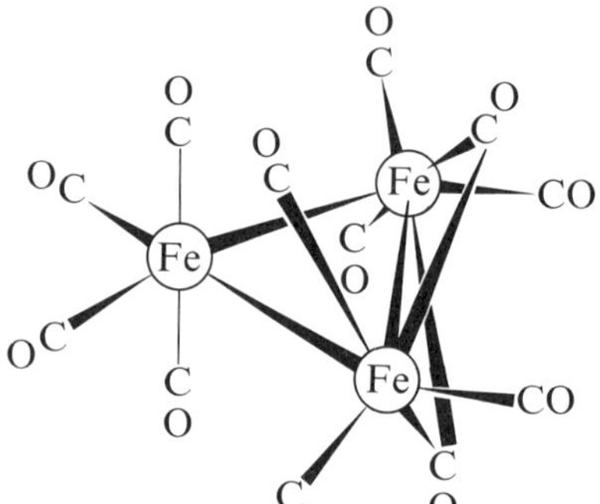

30
$Fe_3(CO)_{12}$

---

[171] J. R. Moss and W. A. G. Graham, *Chem. Commun.*, 835 (1970).
[172] This exception is discussed in the context of metal cluster chemistry in Chapter 18.

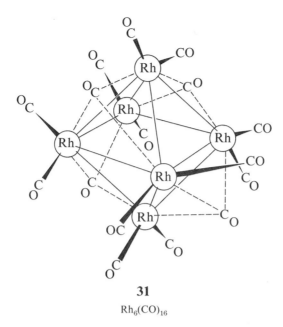

31

$Rh_6(CO)_{16}$

Bonding in Metal Carbonyls

Carbon monoxide is a notoriously poor σ donor. Nonetheless, you have seen that it will react with transition metals in low oxidation states (usually −1, 0, or +1) to form complexes that are often quite stable with respect to oxidation, dissociation, and substitution. The feature of metal-CO bonding that is presumed to be responsible for this stability is the interaction of filled metal d orbitals of π symmetry (t_{2g} in an O_h molecule) with the empty π^* or anti-bonding orbitals of the CO (Chapter 5, p. 199) **(32)**.

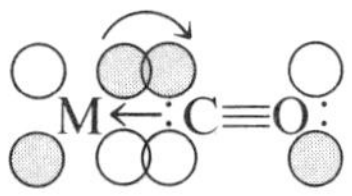

32

You have already seen in Chapter 9 that the consequence of such an interaction is an increase in Δ owing to a lowering of the energy of the t_{2g} orbitals (Figure 9-12). In other terms, delocalization of electron density into ligand π^* orbitals—a process called backbonding—causes an increase in the metal effective nuclear charge, an increase that should eventually limit the extent of metal $t_{2g} \to \pi^*$ electron flow. However, the increase in Z^* has the further effect of producing, by the simple expedient of increased attraction by the metal, a stronger ligand → metal σ bond than might otherwise be possible. As the amount of electron density delocalized from the CO σ orbital onto the metal increases, though, this diminishes Z^* and further metal $t_{2g} \to \pi^*$ electron flow is possible. The result of this two-way electron flow is that the metal-ligand bond is stronger than the sum of isolated ligand-to-metal σ bonding and metal-to-ligand π bonding effects. This mutual strengthening of two effects to produce a result greater than the sum of the two acting individually is called *synergism*.

The plausibility of the M—CO bonding model and its consequences have been examined in a great many ways, both theoretically and experimentally. One such study was an examination of the electronic spectra of the isoelectronic series $[V(CO)_6]^-$, $Cr(CO)_6$, and $[Mn(CO)_6]^+$, as well as $Mo(CO)_6$ and $W(CO)_6$.[173] The most important

---

[173] N. A. Beach and H. B. Gray, *J. Amer. Chem. Soc.*, **90**, 5713 (1968).

result of this was confirmation of the basic mo scheme, illustrated qualitatively below,

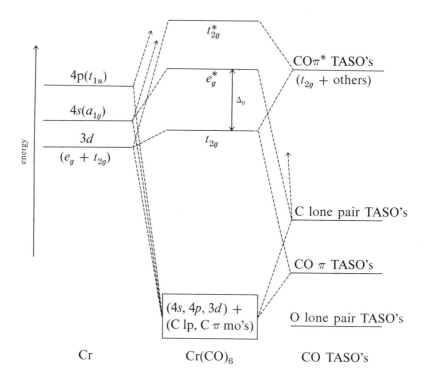

that has been previously assumed for M—CO bonding. In agreement with such a scheme, the values of Δ_0 for the Group VIB metal carbonyls are quite high ($\sim$32,000 cm^{-1}) and relatively constant down the series. However, in the isoelectronic series [V(CO)$_6$]$^-$, Cr(CO)$_6$, and [Mn(CO)$_6$]$^+$ a decrease in Δ_0 might be anticipated. In an attempt to delocalize the negative charge in [V(CO)$_6$]$^-$, extensive back-bonding and concomitant stabilization of t_{2g} with an increase in the t_{2g}-e_g separation would be predicted from π considerations alone. However, the reverse is actually found—the Δ_0 values increase in the series [V(CO)$_6$]$^-$ (25,500 cm^{-1}), Cr(CO)$_6$ (32,200 cm^{-1}) and [Mn(CO$_6$)]$^+$ (41,050 cm^{-1}). Apparently, any effect of decreasing metal-ligand back-bonding is overridden by an increase in the effective nuclear charge of the metal with increasing nuclear charge, an effect that results in lowering the energy of the metal d orbitals and less t_{2g} stabilization but greater e_g^* destabilization.

Further evidence for the back-bonding model of metal-CO bonding is the reduction in the stretching frequency (and presumably the force constant; see Appendix on infrared spectroscopy as the end of this chapter) of the C—O bond upon coordination with a metal as well as the variation in stretching frequency with metal oxidation state. The stretching frequency of free CO is 2140 cm^{-1} (force constant = 18.5 mdyne/Å), whereas it is only 2000 cm^{-1} (force constant = 17.87 mdyne/Å) in Cr(CO)$_6$. The explanation of this is that back-bonding populates the CO π^* orbital, thereby resulting in C—O bond weakening and a lowering of the force constant.[174] [Alternatively, you can consider it in terms of the canonical forms **(33)**. Increased metal-CO double bonding leads to decreased C-O multiple bonding.]

$$\overset{-\delta}{\text{M}}-\overset{+\delta}{\text{C}}\equiv\text{O}: \leftrightarrow \text{M}=\text{C}=\overset{..}{\underset{..}{\text{O}}}$$

33

---

[174] K. G. Caulton and R. F. Fenske, *Inorg. Chem.*, **7**, 1273 (1968).

CARBON σ DONORS

TABLE 16-4
INFRARED RESULTS FOR THE
ISOELECTRONIC SERIES [V(CO)$_6$]$^-$, Cr(CO)$_6$, [Mn(CO)$_6$]$^+$

	Stretching Frequencies (cm^{-1})		
	[V(CO)$_6$]$^-$	Cr(CO)$_6$	[Mn(CO)$_6$]$^+$
M—C	460	441	416
C—O	1859	1981	2101

If the M—CO bonding model is correct, the C—O stretching frequency (and presumably the force constant) should increase in the isoelectronic series [V(CO)$_6$]$^-$ < Cr(CO)$_6$ < [Mn(CO)$_6$]$^+$, while the M—C stretching frequency should decrease. The experimental results in Table 16-4 support the model extremely well.

METAL-CARBENE AND -CARBYNE COMPLEXES[175-178]

Two other types of metal-carbon σ-bonded complexes are important: metal-carbenes (*e.g.*, **34**) and metal-carbynes (*e.g.*, **35**). The synthesis, structure, and chemistry of these types of complexes are of great current interest, although they have not yet achieved the importance of metal-alkyls/aryls and metal carbonyls.

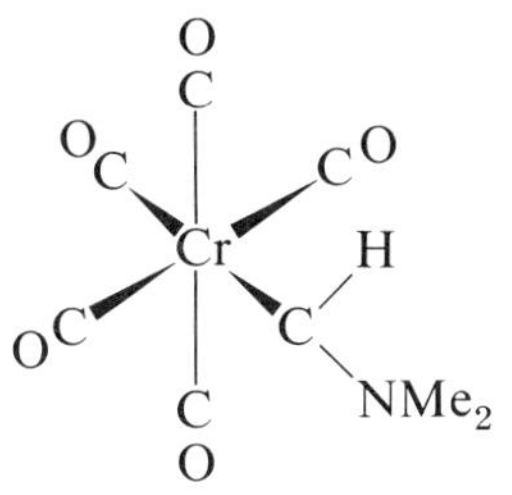

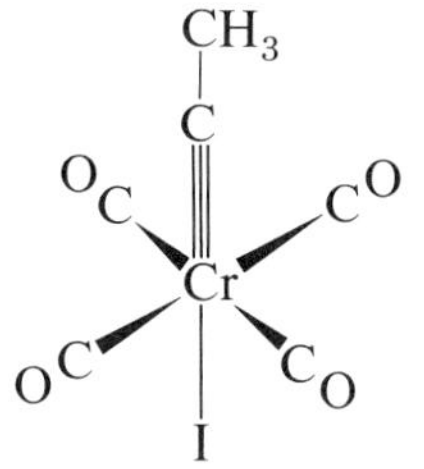

Molecular orbital calculations indicate that the carbon atom of CO becomes more positive upon formation of a metal carbonyl,[179] a feature that should make the carbonyl carbon susceptible to nucleophilic attack. Such attack has indeed been observed, especially with carbanions from lithium alkyls (16–86).

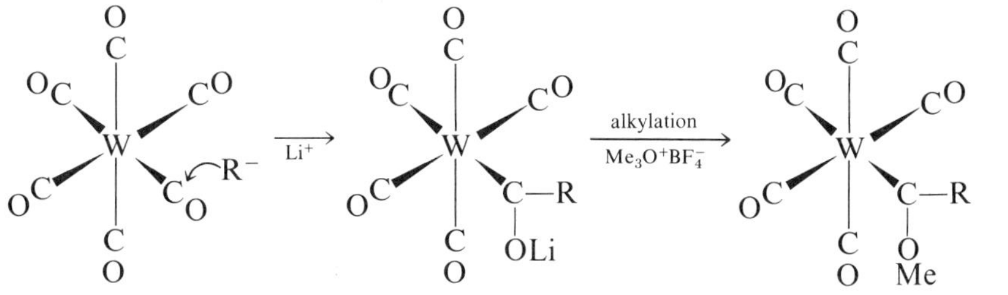

(16-86)

Reaction 16–86, which represents one of the earliest routes to metal-carbenes, has been extended to a number of metals, and other synthetic routes have since been devised (16–87).

---

[175] D. J. Cardin, B. Cetinkaya, and M. F. Lappert, *Chem. Rev.*, **72**, 545 (1972).
[176] D. J. Cardin, B. Cetinkaya, M. J. Doyle, and M. F. Lappert, *Chem. Soc. Rev.*, **2**, 99 (1973).
[177] F. A. Cotton and C. M. Lukehart, *Prog. Inorg. Chem.*, **16**, 487 (1972).
[178] E. O. Fischer, *Adv. Organometal. Chem.*, **14**, 1 (1976).
[179] See ref. 174.

(16-87)

$$\underset{Et_3P}{\overset{Cl}{\underset{|}{Pt}}}\underset{Cl}{\overset{Cl}{\underset{|}{\diagdown}}}\underset{PEt_3}{\overset{Cl}{\underset{|}{Pt}}} + \underset{\underset{Ph}{N}}{\overset{\underset{Ph}{N}}{\diagdown}}C=C\underset{\underset{Ph}{N}}{\overset{\underset{Ph}{N}}{\diagup}} \rightarrow 2\,Et_3P-\underset{Cl}{\overset{Cl}{\underset{|}{Pt}}}-C\underset{\underset{Ph}{N}}{\overset{\underset{Ph}{N}}{\diagup}}$$

There are now well in excess of 300 metal-carbene complexes known, and most of the later transition metals are involved. Metal electron configurations range from d^3 to d^{10}, with d^5 and d^9 as yet unrepresented. Metal oxidation states range from 4+ to 0 and coordination numbers from 2 to 7.

The carbene ligand may be considered as being based on an sp^2 carbon atom with an empty orbital capable of interacting with a π base. This being the case, it is interesting to realize that all of the authenticated examples of metal-carbene complexes isolated thus far[180] involve carbenes in which one or both of the carbon substituents is (are) capable of π bonding with the carbene carbon. Therefore, resonance structure **36a** is assumed to be important, but **36b** and **36c** are certainly plausible.

$$L\bar{M}-C\underset{Y}{\overset{\overset{+}{X}}{\diagup}} \leftrightarrow L\bar{M}-\overset{+}{C}\underset{Y}{\overset{\ddot{X}}{\diagup}} \leftrightarrow LM=C\underset{Y}{\overset{\ddot{X}}{\diagup}}$$

     36a   **36b**   **36c**

The relative importance of these representations is difficult to assess. However, x-ray structural data have indicated that the metal-carbene carbon bond is not especially short (0.2 Å longer than a metal—CO bond), and nmr experiments involving compounds of the type M—C(R)NR$_2$ show that there is restricted rotation about the C—N bond. The former observation suggests resonance structure **36c** is not of great importance, and the latter observation argues for **36a**. Recently reported x-ray photoelectron spectra have somewhat confused the situation, however, by indicating that **36c** does indeed have considerable importance.[181]

Before passing on to carbyne complexes, we call your attention to a complex (**37**) discussed in connection with Figure 16-1. The tin-containing ligand in this complex (**38**) can, of course, be considered an inorganic carbene.[182]

37: $[(Me_3Si)_2CH]_2Sn$—$Cr(CO)_5$

38: $(Me_3Si)_2CH$—$\ddot{S}n$—$CH(SiMe_3)_2$

Among the reactions of carbene complexes is that with boron trihalide, a reaction that has led to a most startling development: the synthesis of metal-carbyne complexes such as **35**.[183] In this case, the metal-ligand bond is quite short (as opposed to the situation with metal-carbenes), and the M-C-R angle is usually about 180°.

---

[180] There is one exception to this statement: $Ta(\eta^5\text{-}C_5H_5)_2(CH_3)(CH_2)$. See R. R. Schrock, *J. Amer. Chem. Soc.*, **97**, 6577 (1975).
[181] W. B. Perry, T. F. Schaaf, W. L. Jolly, L. J. Todd, and D. L. Cronin, *Inorg. Chem.*, **13**, 2038 (1974).
[182] See ref. 11.
[183] See ref. 178.

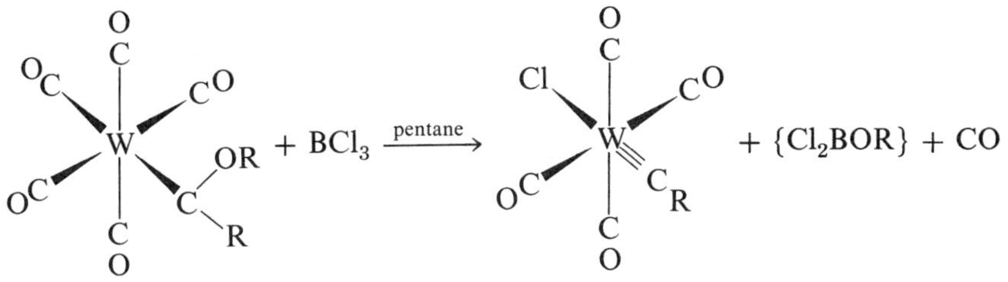

$$\text{(OC)}_4\text{W}(\text{OR})(\text{CR}) + BCl_3 \xrightarrow{\text{pentane}} Cl(OC)_4W\equiv CR + \{Cl_2BOR\} + CO \qquad (16\text{-}88)$$

Both features suggest electronic structure **39**, a structure that may be derived by assuming that an sp hybridized carbyne carbon can donate a pair of electrons to the metal. The metal may in turn back-donate electrons from two orthogonal π-type orbitals (*e.g.*, d_{xz} and d_{yz}).

$$XL_4M\equiv C-R$$

39

STUDY QUESTIONS

12. Given below are three metal carbonyl derivatives, and C—O stretching frequencies for each:

Compound	C—O Stretching Frequency (cm^{-1})
$F_3Si\text{-}Co(CO)_4$	2128, 2073, 2049
$Cl_3Si\text{-}Co(CO)_4$	2125, 2071, 2049
$Me_3Si\text{-}Co(CO)_4$	2100, 2041, 2009

    (the relative band intensities are always weak, medium, and strong on going from high to low energy)

    a) There are two possible structures for the compounds above, both based on the trigonal bipyramid. Draw these structures and, assuming that the substituent group R_3Si is just a sphere, assign each structure to a point group.
    b) Using Table 16–8 in the Appendix, decide which of the structures in part (a) is correct.
    c) Suggest a method for the synthesis of $Me_3Si\text{-}Co(CO)_4$.
    d) Rationalize the changes in the C—O stretching frequencies as the substituent group is changed.

13. What structures are predicted by the angular overlap model for $Fe(CO)_5$, $Co(CO)_4^-$, and $Ni(CO)_4$? Do these predictions agree with experiment?

14. Verify that the metal atoms in $Os_3(CO)_{12}$, $Fe_3(CO)_{12}$, and the two forms of $Co_2(CO)_8$ satisfy the "effective atomic number" rule.

15. Propose a method for the synthesis of compound **38**, $Sn[CH(SiMe_3)_2]_2$.

16. Make drawings of the metal and ligand orbitals involved in metal-carbene and -carbyne complexes.

17. The data given below pertain to the hexacarbonyls of Group VIB metals.

	Ionization Potential of the Gaseous Metal	Ionization Potential of Gaseous $M(CO)_6$	Bond Energy per M—CO Bond	C—O Stretching Frequency
Cr	6.76 eV	8.15 eV	27 kcal	2000 cm^{-1}
Mo	7.38	8.23	36	1984
W	7.98	8.56	42	1960

a) Why is the IP of the metal carbonyl so close to that of the free metal?

b) Although the IP of $M_{(g)}$ and $M(CO)_{6(g)}$ are close, that for the metal carbonyl is always slightly higher than that for the free metal. Suggest a reason for this.

c) Is there a correlation between the $M(CO)_{6(g)}$ IP, the M—CO bond energy, and the C—O stretching frequency? If so, rationalize your correlation.

CARBON π DONORS

In this, the second main portion of the chapter, we shall consider compounds whose existence clearly depends on the formation of complexes between transition metals in low valence states and ligands that can act as π donors. Many complexes of this type are implicated in metal-catalyzed processes such as hydrogenation, polymerization, hydroformylation, and cyclization. Thus, metal π donor complexes constitute an important class of organometallic compounds that deserve intensive study.

Hydrocarbon π donor ligands were discussed briefly in Chapter 15, and examples were listed in Table 15-1. However, before proceeding to a more detailed discussion of the synthesis and properties of their metal complexes, we might list again examples of the types of compounds observed.

1. Olefins are thought of as donating an even number of electrons to a metal or metals; that is, 2 **(40)**, 4 **(41)**, 6 **(42)**, or 8 **(43)**.[184-186]

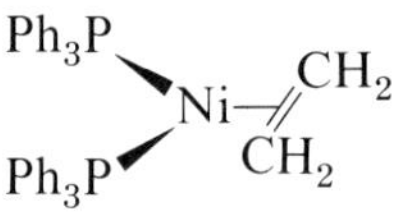

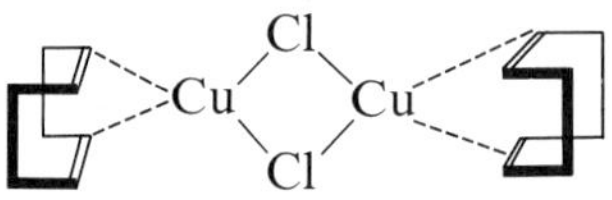

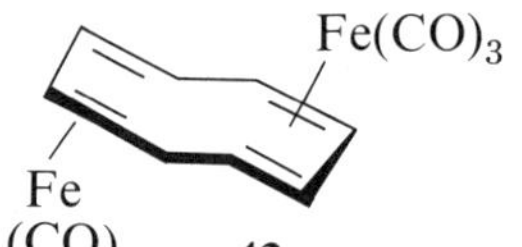

2. In addition to being σ-bonded through a single carbon, an allyl group may be π-bonded, the allyl anion effectively donating four electrons.[187]

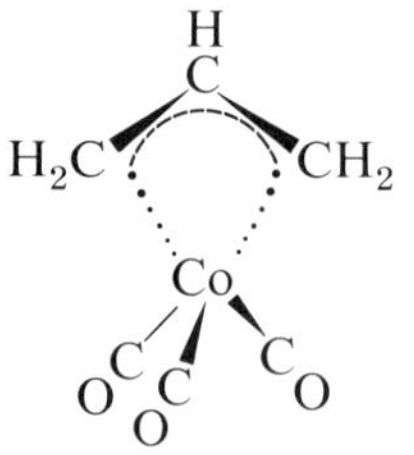

3. Acetylenes may donate one pair of π electrons **(45)** or both pairs **(46)**.[185,186,188]

---
[184] H. W. Quinn and J. H. Tsai, *Adv. Inorg. Chem. Radiochem.*, **12**, 217 (1969).
[185] J. H. Nelson and H. B. Jonassen, *Coord. Chem. Rev.*, **6**, 27 (1971).
[186] F. R. Hartley, *Chem. Rev.*, **69**, 799 (1969).
[187] M. L. H. Green and P. L. I. Nagy, *Adv. Organometal. Chem.*, **2**, 325 (1964).
[188] F. L. Bowden and A. B. P. Lever, *Organometal. Chem. Rev. A*, **3**, 227 (1968).

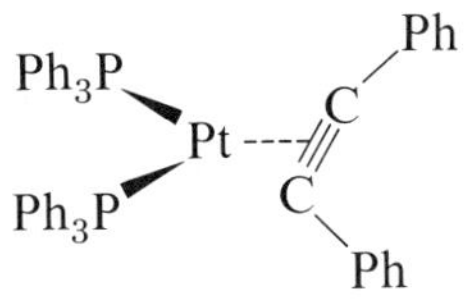

4. For the purpose of counting electrons, a *pentahapto*cyclopentadienyl group (that is, a C_5H_5 ligand in which all five carbon atoms are involved in bonding to the metal) is considered as an anion donating six π electrons.

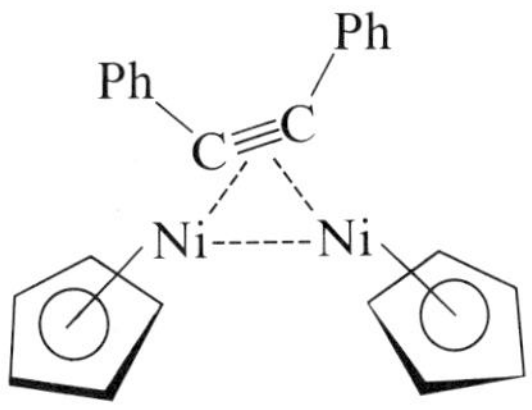

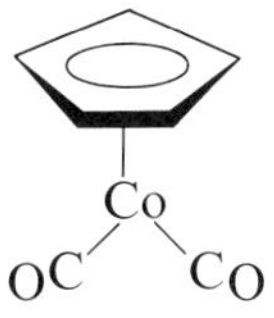

5. Benzene and its derivatives can clearly donate six π electrons to a bound metal.[189,190]

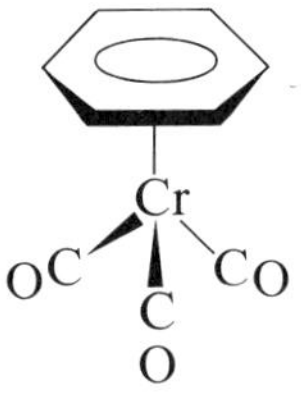

Other potential donors of three π pairs include the η^7-$C_7H_7^+$ ion (tropylium ion or cycloheptatrienyl ion) and its precursor, cycloheptatriene.[191] The existence of the tropylium ion complex **49** illustrates the fascination of organometallic chemistry: the stabilization of organic species that are not stable enough to be isolated in the absence of a complexing metal.

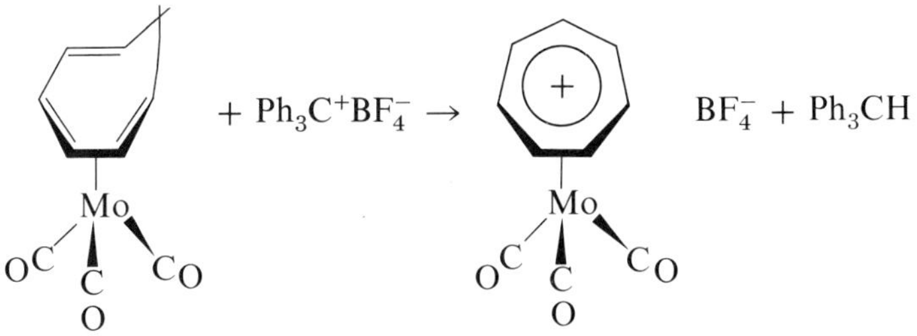

$$\text{42'} + Ph_3C^+BF_4^- \rightarrow \text{49} \; BF_4^- + Ph_3CH \tag{16-89}$$

CHAIN π DONOR LIGANDS (OLEFINS, ACETYLENES, AND π-ALLYL)

Except for a few scattered reports,[192,193] the vast majority of metal complexes with π donor ligands involve low valent transition metals, especially those of Groups

---

[189] H. Zeiss, P. J. Wheatley, and H. J. S. Winkler, "Benzenoid-Metal Complexes," Ronald Press, New York (1966).
[190] W. E. Silverthorn, *Adv. Organometal. Chem.*, **13**, 47 (1975).
[191] M. A. Bennet, *Adv. Organometal. Chem.*, **4**, 353 (1966).
[192] J. St. Denis, T. Dolzine, and J. P. Oliver, *J. Amer. Chem. Soc.*, **94**, 8260 (1972).
[193] P. H. Kasai and D. McLeod, Jr., *J. Amer. Chem. Soc.*, **97**, 5609 (1975).

VIB through VIII. The reason for the necessity of a low valent metal (*i.e.,* oxidation states −1 to +2) becomes clear on examining the currently accepted model for bonding between metals and olefins. This model, which was discussed briefly in Chapter 9 (p. 539), is the Dewar-Chatt-Duncanson (or DCD) model.[194] Its essence is that the ligand donates π electron density to a metal orbital of σ symmetry directed to the center of the ligand π system **(50a),** and the metal in turn back-bonds electron density into a ligand π* orbital **(50b).** The result is a synergism that, as in the case of the metal carbonyls, leads to relatively strong bonding. Refinements to this simple model are described below. However, even in this simple form it is clear that such bonding can occur effectively only with a low valent (low Z^*) metal with populated π symmetry orbitals (*i.e.,* a metal late in the transition series).

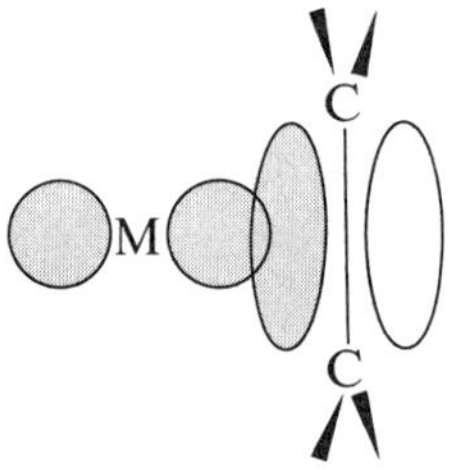

$$M \leftarrow L\pi \qquad M \rightarrow L\pi^*$$

50a   **50b**

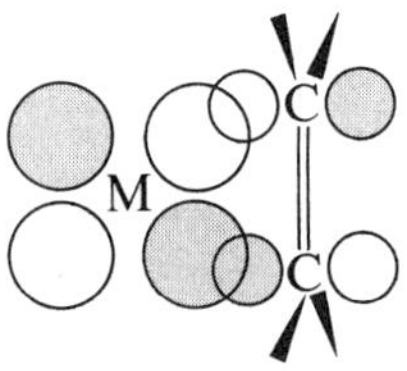

Synthesis of Olefin, Acetylene, and π-Allyl Complexes

Although a few metal complexes of olefins and acetylenes may be prepared by direct addition of the ligand to the metal-containing substrate, most complexes are prepared by displacement of some other, more weakly bound ligand. π-Allyl complexes, however, are prepared by quite different methods, as you shall see.

Olefin Complexes. There are three methods that may be used for the preparation of olefin complexes.

i. DIRECT REACTION. Some coordinatively unsaturated metal complexes can react directly with olefins. For example, Vaska's compound forms complexes with ethylene, tetrafluoroethylene, or tetracyanoethylene (reaction 16-90), albeit with varying degrees of effectiveness, as indicated by the equilibrium constants.[195]

(16-90) $IrCl(CO)(PPh_3)_2 \ + \ R_2C{=}CR_2 \ \rightleftharpoons$ **51**

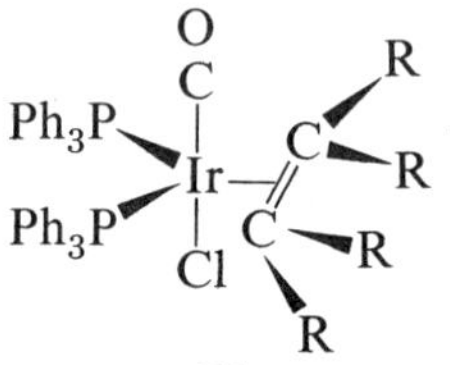

Olefin	Relative Equilibrium Constant
$H_2C{=}CH_2$	1
$(NC)HC{=}CH(CN)$	1,500
$(NC)_2C{=}C(CN)_2$	140,000

---

[194] M. J. S. Dewar, *Bull. Soc. Chim. Fr.,* **18,** C79 (1951); J. Chatt and L. A. Duncanson, *J. Chem. Soc.,* 2939 (1953).
[195] L. Vaska, *Acc. Chem. Res.,* **1,** 335 (1968).

The fact that tetracyanoethylene forms the most stable complexes is especially interesting in view of the discussion of its properties in Chapter 5 (p. 197). It was noted in that discussion that this olefin could function as a very good π acid. This, coupled with the fact that molecular orbital calculations indicate that tetracyanoethylene has a C=C π bond order of only about 0.35,[196] suggests that iridium-to-olefin electron flow is quite important in determining the stability of metal-olefin complexes.

Earlier in this chapter (p. 824), we mentioned that metal atoms may react directly with organic halides to give metal alkyls and aryls. Not surprisingly, it has been found recently that metal atoms may also react directly with olefins, especially dienes, to give metal π donor complexes. Especially interesting examples are tris(1,3-butadiene)molybdenum and -tungsten, where the C=C double bonds are situated in a trigonal prismatic arrangement in such a way that all twelve carbon atoms are equidistant from the metal atom.[197]

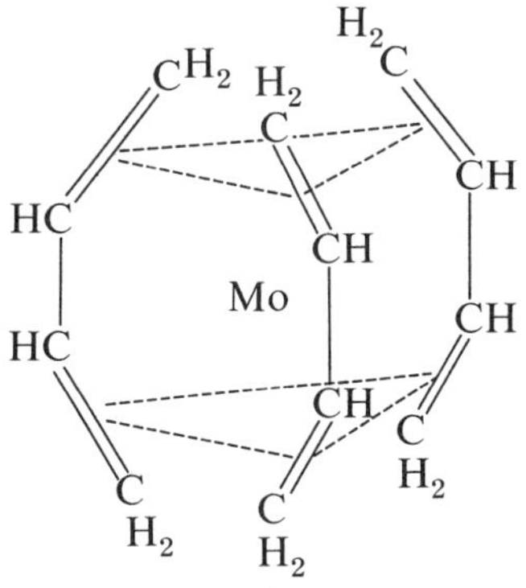

52

ii. DISPLACEMENT OF A WEAKLY BOUND LIGAND. The ligands displaced may include CO (reaction 16-91),[198] non-conjugated or non-chelating olefins (reaction 16-92),[199] benzene,[200] and others. A particularly efficacious method for preparing Pd(II) or Pt(II) olefin complexes is by displacement of the weakly coordinating benzonitrile (reaction 16-93).[201]‡

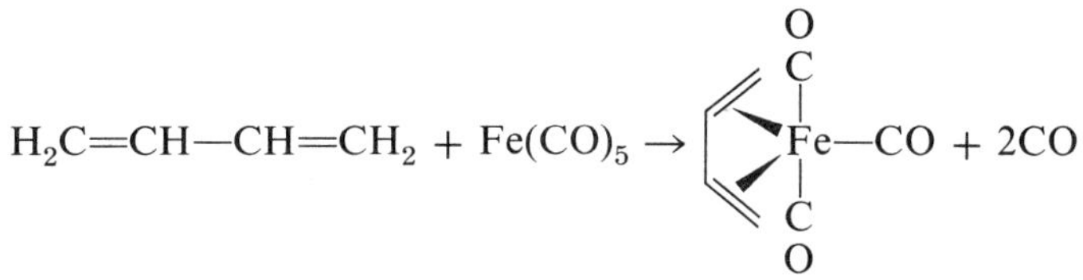

$$H_2C=CH-CH=CH_2 + Fe(CO)_5 \rightarrow \text{[complex]}Fe-CO + 2CO \qquad (16\text{-}91)$$

$$\text{[Rh-Cl-Rh complex with } CR_2\text{]} + 2\text{[COD]} \rightarrow \text{[Rh-Cl-Rh complex]} \qquad (16\text{-}92)$$

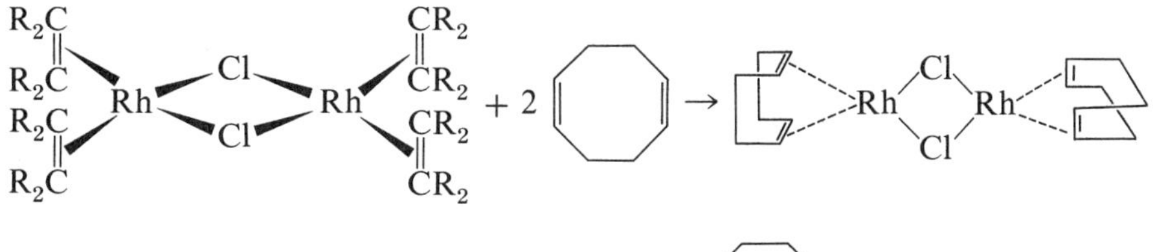

$\Delta H_{rxn} = -10.3$ kcal/mole when $R_2C=CR_2 =$ [cyclooctene]

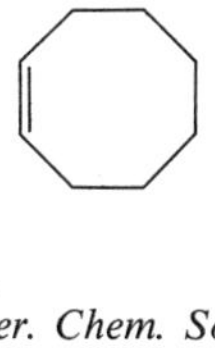

[196] W. H. Baddley, *J. Amer. Chem. Soc.*, **90**, 3705 (1968).
[197] P. S. Skell, E. M. Van Dam, and M. P. Silvon, *J. Amer. Chem. Soc.*, **96**, 627 (1974).
[198] See ref. 167.
[199] W. Partenheimer and E. F. Hoy, *J. Amer. Chem. Soc.*, **95**, 2840 (1973).
[200] R. G. Solomon and J. K. Kochi, *J. Amer. Chem. Soc.*, **95**, 1889 (1973).
[201] W. Partenheimer, *Inorg. Chem.*, **11**, 743 (1972).

‡You might notice that in each of the illustrated reactions, a monodentate ligand is replaced by a chelating ligand; *cf.* the chelate effect in Chapter 13.

(16-93) $[(PhC\equiv N)_2PdCl_2]$ + [norbornadiene] → [norbornadiene-PdCl$_2$ complex] + $2PhC\equiv N$

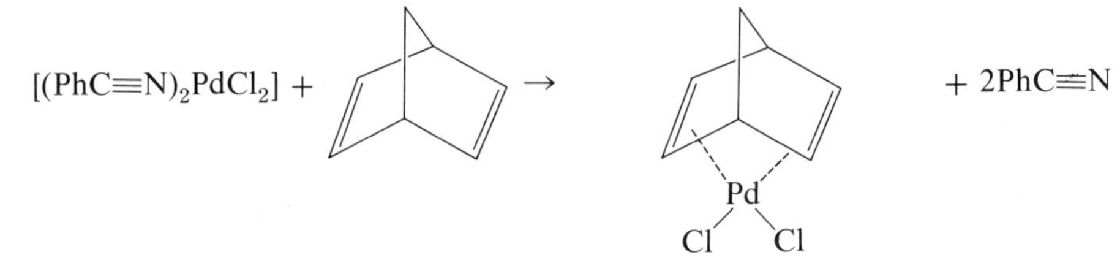

$\Delta H_{rxn} = -13.3$ kcal/mole

iii. REDUCTIVE ADDITION. This method is used especially with platinum group metals, where one may start with a higher-valent salt of the metal and produce an olefin complex of the zero-valent metal.[202]

(16-94) $(Ph_3P)_2PtCl_2 + N_2H_4 \cdot H_2O + EtOH + C_2H_4 \rightarrow (Ph_3P)_2Pt(C_2H_4)$

Acetylene Complexes. In general, acetylene complexes may be prepared in ways entirely analogous to those used for olefins. In addition, there are two other modes of reactivity frequently exhibited in the synthesis of acetylene complexes.

i. As noted previously (compound **46**), mono-acetylenes can function as four-electron donors by utilizing both π bonds.

(16-95) $Co_2(CO)_8 + PhC\equiv CPh \xrightarrow[benzene]{\Delta} (CO)_3Co\text{-}\overset{PhC\equiv CPh}{\text{-}\text{-}\text{-}\text{-}}Co(CO)_3 + 2\,CO$

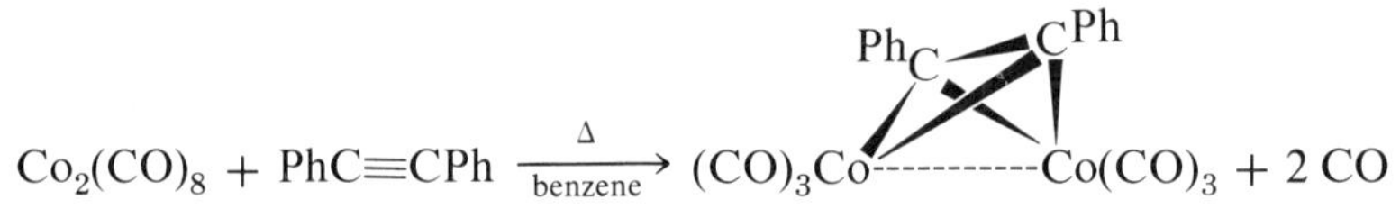

ii. Unless the acetylene has bulky substituents or substituents that can interact with the metal, acetylene complexes with divalent platinum group metals are frequently unstable, giving rise to a variety of other products. For example,[203]

(16-96)

(16-97) $trans\text{-}PtClMeQ_2 + RC\equiv CR' + AgPF_6$

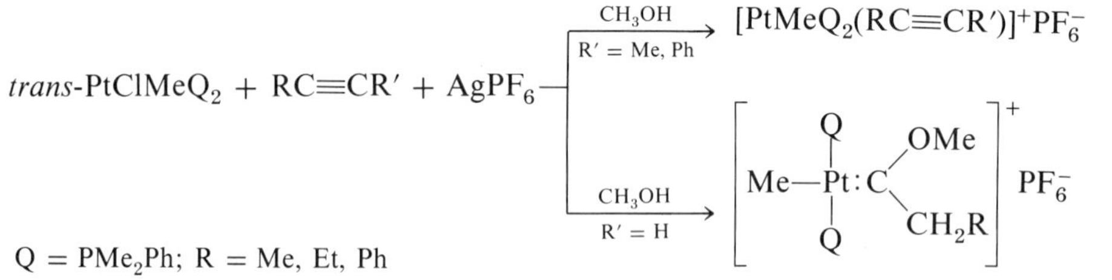

$\xrightarrow[R' = Me, Ph]{CH_3OH} [PtMeQ_2(RC\equiv CR')]^+PF_6^-$

$\xrightarrow[R' = H]{CH_3OH} \left[Me\text{-}Pt\overset{Q}{\underset{Q}{:}}C\overset{OMe}{\underset{CH_2R}{}} \right]^+ PF_6^-$

$Q = PMe_2Ph; R = Me, Et, Ph$

In many other cases, acetylene polymerization is observed. At least eight compounds have been isolated, for example, from the reaction of diphenylacetylene with $Fe(CO)_5$.[204]

(16-98) $Fe(CO)_5 + PhC\equiv CPh$ → [tetraphenylcyclobutadiene Fe(CO)$_3$] ; [tetraphenylcyclopentadienyl-Fe(CO)$_3$ with Fe(CO)$_3$] ; [tetraphenylcyclopentadienone Fe(CO)$_3$]

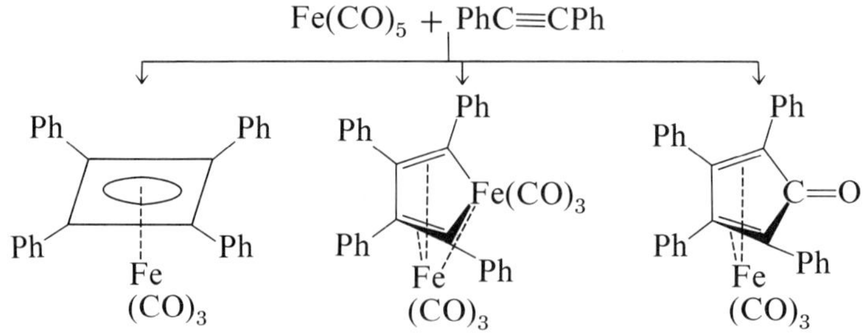

[202] F. R. Hartley, *Organometal. Chem. Rev. A*, **6**, 119 (1970).
[203] M. H. Chisholm and H. C. Clark, *Inorg. Chem.*, **10**, 2557 (1971); M. H. Chisholm, H. C. Clark, and D. H. Hunter, *Chem. Commun.*, 809 (1971).
[204] See ref. 188.

π-Allyl Complexes. Metal carbonyl anions react with allyl chloride to give a σ-bonded allyl derivative (53). Heat or light then causes expulsion of a CO ligand and the formation of a π-allyl or η^3-allyl complex. Although a number of other π-allyl complexes are now known, their preparation invariably proceeds through a σ-bonded compound, even though the latter may be only a transient intermediate.

$$[\text{CpMo(CO)}_3]^- + CH_2=CH-CH_2Cl \rightarrow \text{CpMo(CO)}_3-CH_2-CH=CH_2 \xrightarrow{h\nu} \text{CpMo(CO)}(\eta^3-C_3H_5) \quad (16\text{-}99)$$

[yellow oil, mp, −5°]   **53**

[yellow solid, mp, 134° (dec)]   **54**

Structure and Bonding in Olefin, Acetylene, and π-Allyl Complexes[205]

The structure of Zeise's salt, $K[PtCl_3(C_2H_4)]$, the original metal-olefin complex, was described in Chapter 9, as was the classical Dewar-Chatt-Duncanson (DCD) model for bonding in metal-olefin and metal-acetylene complexes (see structure **50**).[206] This model, which was developed to account for the stability of olefin complexes of Ag(I) and Pt(II), has been generally accepted as an excellent starting place to rationalize the properties of olefin and acetylene complexes of other metals. In the past several years there has been further, very extensive discussion of this topic. This has come about in large part because of the availability of crystallographic facilities, which has resulted, in turn, in an increase in the number of accurately known structures of metal π donor complexes. This renewed discussion has led to the major conclusion that the DCD model is substantially correct. However, there have been some refinements, and, after a discussion of the structures of metal complexes of chain π donor molecules and a description of some of their physical properties, we shall return to discuss briefly the modifications and extensions of the DCD model.

The molecular structures of numerous mono-olefin and -acetylene complexes have been determined by x-ray techniques.[205] Representative of these are Zeise's salt, $K[PtCl_3(C_2H_4)]$ (Figure 16-12), and $(PPh_3)_2Pt(PhC\equiv CPh)$ (Figure 16-13). The important features to notice in these structures are: (i) The length of the coordinated

---

[205] S. D. Ittel and J. A. Ibers, *Adv. Organometal. Chem.*, **14**, 33 (1976).
[206] See ref. 194; see also N. Rosch, R. P. Messmer, and K. H. Johnson, *J. Amer. Chem. Soc.*, **96**, 3855 (1974).

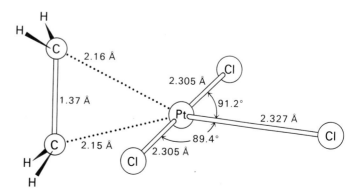

Figure 16-12. Molecular configuration of Zeise's salt, $K[PtCl_3(C_2H_4)]H_2O$. Adapted from J. A. J. Jarvis, B. T. Kilbourn, and P. G. Owston, *Acta Cryst.*, **B27**, 366 (1971); M. Black, R. H. B. Mais, and P. G. Owston, *ibid.*, **B25**, 1753 (1969). See also R. A. Love, *et al.*, *Inorg. Chem.*, **14**, 2653 (1975) for a neutron diffraction structure.

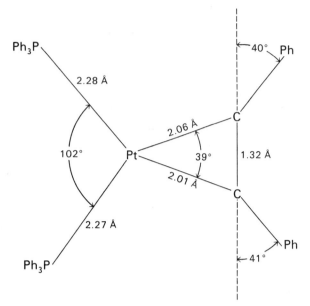

Figure 16-13. The molecular structure of bis(triphenylphosphine)diphenylacetyleneplatinum. [From J. O. Glanville, J. M. Stewart, and S. O. Grim, *J. Organometal Chem.*, **7**, P9 (1967).]

olefinic or acetylenic C—C bond is greater than the C—C bond in the free hydrocarbon. The C=C bond in ethylene has gone from 1.337 Å in free C_2H_4 to 1.37 Å in Zeise's salt, whereas the C≡C bond in Ph_2C_2 has increased from about 1.20 Å to 1.32 Å in the platinum complex. (ii) The C=C bond (or C≡C bond) is perpendicular to the molecular plane in 16-electron molecules of the type L_3M(olefin) (*e.g.*, Zeise's salt; **55a**), whereas it is in the molecular plane in 16-electron molecules such as L_2M(olefin) (*e.g.*, the Pt-acetylene complex in Figure 16-13; **55b**) and in 18-electron molecules of the type L_4M(olefin) (*e.g.*, $Ir(PPh_3)_2Br(CO)[(NC)_2C=C(CN)_2]$; **55c**). (iii) Upon complexation, the originally planar olefin becomes non-planar, with the substituents bending away from the metal and other ligands; acetylenes become non-linear upon complexation. (iv) Complexation is symmetrical. That is, the olefinic or acetylenic carbons are nearly equidistant from the metal. Observations (i) and (ii) are discussed further below.

The DCD model assumes a synergic relation between ligand-to-metal $\pi \rightarrow \sigma$ bonding (**50a**) and metal-to-ligand $d\pi \leftarrow \pi^*$ electron flow (**50b**). Both donation of π electron density from the olefin to the metal and accumulation of electron density in the olefin π^* orbital would be expected to lower the π bond order of the coordinated olefin. As a consequence, the coordinated olefinic or acetylenic bond would be lengthened, an effect that has been observed in most cases. The lowered bond order is also manifested by a decrease in the C=C or C≡C stretching frequency; for example, $\nu_{C=C}$ drops from 1623 cm^{-1} for the free olefin to 1526 cm^{-1} in Zeise's salt,[207,208] and $\nu_{C≡C}$ drops from 1657 cm^{-1} in the uncomplexed acetylene to

---

[207] M. J. Grogan and K. Nakamoto, *J. Amer. Chem. Soc.*, **88**, 5454 (1966).

[208] The C=C stretching mode in ethylene couples very strongly with the CH$_2$ scissoring mode. Therefore, the frequency noted above is not a "pure" C=C stretching frequency, and it should not be used as a quantative measure of bond strength; see M. H. Chisholm and H. C. Clark, *Inorg. Chem.*, **12**, 991 (1973).

1230 cm^{-1} in (Ph$_3$P)$_2$Pt(PhC≡CH).[209] In general, $\Delta\nu_{C=C}$ and $\Delta\nu_{C\equiv C}$ are in the range from -50 to -250 cm^{-1} for divalent platinum and palladium complexes, whereas decreases of 400 to 500 cm^{-1} accompany complexation of acetylene to zero-valent metals; the latter has been taken as evidence of a much stronger metal-ligand interaction in the zero-valent complexes.

One of the more interesting properties of metal-olefin complexes, and one that gives us some further insight into metal-olefin bonding, is the fact that the olefin has in some cases been found to rotate about the metal-olefin bond axis (56).[210,211] (Some

$$M \leftarrow \updownarrow \quad \underset{C}{\overset{C}{\|}}$$

56

experimental evidence is outlined below.) Reflection on the DCD model of metal-olefin bonding shows how this may happen. The ligand-to-metal σ bond, which accounts for about 75% of the bonding interaction in Zeise's salt,[212] can occur whether the C=C bond is in the molecular plane or perpendicular to it (see **50a**). Similarly, since out-of-plane $d\pi \to \pi^*$ back-bonding can take place through overlap of a d_{xz} orbital (**57**), the same interaction can, in principle, take place between the metal d_{yz} orbital and the C=C π^* orbital when the olefin is in the molecular plane (**58**).

$d_{xz} \to \pi^*$ $d_{yz} \to \pi^*$
57   **58**

---

[209] E. O. Graves, C. J. Lock, and P. M. Maitlis, *Can. J. Chem.*, **46**, 3879 (1968).
[210] R. Cramer, J. B. Kline, and J. D. Roberts, *J. Amer. Chem. Soc.*, **91**, 2519 (1969); R. Cramer, *ibid.*, **86**, 217 (1964).
[211] J. Ashley-Smith, I. Douek, B. F. G. Johnson, and J. Lewis, *J. C. S. Dalton*, 1776 (1972), and references therein.
[212] See ref. 206.

Rotation of an Olefin About a Metal-Olefin Bond Axis: Some Experimental Evidence. The first thoroughly studied example of olefin rotation about a metal-olefin bond axis was in (η^5-C$_5$H$_5$)Rh(C$_2$H$_4$)$_2$.[213] A portion of its temperature-dependent ^{1}H nmr spectrum is given in Figure 16-14. At about $-20°$C the structure of the molecule is static by nmr criteria (*i.e.*, the olefins are rotating so slowly that the spectrometer is able to distinguish the various proton environments), and the spectrum can be described by the nmr notation AA'XX'. [This notation reflects the fact that there are two nearly equivalent protons of one type (A and A') and two of another type (X and X').] The A and A' protons, for example, may correspond to the protons labeled *o* for "outer" in Figure 16-14, and it is these protons that give rise to the downfield multiplet at $-19°$C. The X and X' protons would then correspond to the protons labeled *i* for "inner," and they give rise to the upfield multiplet. As the temperature increases, ethylene rotation about the metal-olefin bond axis causes all of the ethylene protons to become magnetically equivalent, and the separate lines collapse to a single line, the

[213] See ref. 210.

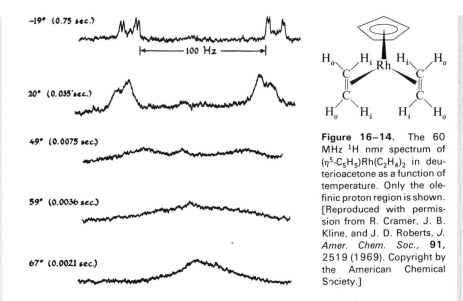

Figure 16-14. The 60 MHz ^{1}H nmr spectrum of $(\eta^5\text{-}C_5H_5)Rh(C_2H_4)_2$ in deuterioacetone as a function of temperature. Only the olefinic proton region is shown. [Reproduced with permission from R. Cramer, J. B. Kline, and J. D. Roberts, *J. Amer. Chem. Soc.*, **91**, 2519 (1969). Copyright by the American Chemical Society.]

chemical shift of which is the average of the separated lines. Line shape analysis gives an activation energy for the rotation process of about 15 kcal.

Besides rotation of the ethylene about the metal-ligand axis, at least two other mechanisms may lead to the observed magnetic equivalence of the protons: (i) exchange of C_2H_4 between two different complex molecules, and (ii) rotation of the olefin about the olefin axis (**56′**).

$$M - \overset{C}{\underset{C}{\rightleftharpoons}}$$

56′

The former is ruled out by the high metal-olefin bond energies (the Rh-C_2H_4 dissociation energy is 31 kcal[214]). Both mechanisms are more conclusively eliminated by a ^{13}C nmr study of the compound below.[215]

At +20°C only one ^{13}C resonance line is observed for the olefin (at 49.3 ppm). However, two lines of equal intensity (at 54.4 and 43.6 ppm) are seen at lower temperatures. Furthermore, at −90°C the ^{1}H nmr spectrum shows two separate resonance lines of equal intensity (one for the two olefinic protons nearest the CO and one for the two protons nearest the NO), whereas warming to −65°C caused their coalescence to a single line (whose shift is the average of the shifts of the previously separate lines). Most importantly, this single ^{1}H nmr line resolves at higher temperatures to a 1:2:1 triplet owing to spin-spin coupling between the four now-equivalent olefinic protons and two equivalent ^{31}P nuclei ($I = \frac{1}{2}$). Both the ^{13}C and ^{1}H results conclusively identify the metal-olefin bond as the rotational axis and, because the olefin protons are spin coupled to the phosphorus atoms at higher temperatures, eliminate an intermolecular exchange mechanism.

---

[214] R. Cramer, *J. Amer. Chem. Soc.*, **94**, 5681 (1972).
[215] B. F. G. Johnson and J. A. Segal, *Chem. Commun.*, 1312 (1972).

Using the DCD model, we have been able to explain in a rather straightforward way why an olefin in a molecule such as $PtCl_2L(R_2C=CR_2)$ is able to rotate about the metal-ligand axis: $(C=C){\rightarrow}M\sigma$ bonding with $M{\rightarrow}(C=C)$ $d\pi{\rightarrow}\pi^*$ bonding are satisfactory whether the C=C axis is in the molecular plane or perpendicular to it. However, we must confront another, related observation. That is, in all known cases of 16-electron complexes of the type $L_2M(C=C)$ (where M is a zero-valent metal), the C=C bond axis is in the molecular plane; whereas the C=C axis is perpendicular to the molecular plane in 16-electron complexes of the type $L_3M(C=C)$ (where M is a higher-valent metal) (see **55a** and **b**). The fact that $L_3M(C=C)$ complexes prefer a perpendicular arrangement can be explained on steric grounds: ligand-ligand repulsion is clearly lessened when the C=C axis is perpendicular to the molecular plane. But why, then, is the planar geometry observed for $L_2M(C=C)$ complexes?

To find an answer to the dilemma posed above, consider first the $L_2M(C=C)$ type of complex. The critical feature to consider is the role of the olefin π^* orbital and its relation to the metal $d\pi$ orbitals, that is, d_{xz} (**57**) and d_{yz} (**58**). A simple molecular orbital approach is the most efficacious. That is, there will be nine molecular orbitals generated by the five metal d orbitals, the two L hybrid σ orbitals, and the C=C π and π^* orbitals. Sixteen electrons fill eight of these mo's; one orbital, an anti-bonding orbital that most closely resembles the metal d_{z^2} orbital, is left unfilled. The critical mo's "d_{xz}" and "d_{yz}" are occupied, however, and we now have to explore more precisely their function. A more complete treatment of the σ bonding in a trigonal planar ML_3 complex shows that, while "d_{xz}" is a non-bonding mo, "d_{yz}" is actually an anti-bonding σ^* mo with respect to the two L ligands (**59**).[216] However, stabilization of the electron pair in the "d_{yz}" orbital may be achieved by a bonding interaction between the "d_{yz}" orbital and the C=C π^* orbital *if* the C=C bond is in the plane of the $L_2M(C=C)$ complex.

59

Consider next 16-electron complexes of the type $L_3M(C=C)$. Again there is no reason to expect that the olefin π to metal d_{z^2} donor action should be affected by rotation of the olefin about the metal-olefin bond axis. However, in this case the function of the d_{xz} and d_{yz} orbitals has changed somewhat from the $L_2M(C=C)$ case. That is, both orbitals are non-bonding if we consider only M-ligand σ interactions. Therefore, there is no reason to prefer a d_{xz}-π^* interaction (**57**) over a d_{yz}-π^* interaction (**58**), and the complex may have the same stability whether the olefin is in the L_3M plane or perpendicular to it. For this reason, it is left to steric interactions to cause the olefin to favor a perpendicular conformation in 16-electron complexes of the type $L_3M(C=C)$, a prediction amply confirmed by experiment. At the same time, this bonding model predicts that, if the steric requirements for the ligands L and the olefin are small, there should be only a small energy barrier to displacement of the olefin from its perpendicular conformation, and olefin rotation can occur. This has also been amply confirmed by experiment, as has the fact that rotation becomes increasingly unlikely as the steric requirements of the ligands or olefin increase.

The η^3-allyl ligand is bound to a metal, as in bis(η^3-2-methylallyl)nickel,[217] in such a way that the three allylic carbon atoms are equidistant from the metal. Because of this structure, it should not be surprising that the bonding model applied to π-allyl complexes is similar to the metal-olefin DCD model.[218]

---

[216] N. Rösch and R. Hoffman, *Inorg. Chem.*, **13**, 2656 (1974).
[217] See ref. 187; see also R. Uttech and H. Dietrich, *Z. Krist.*, **122**, 60 (1965).
[218] See ref. 216.

876 □ ORGANOMETALLIC CHEMISTRY: SYNTHESIS, STRUCTURE, AND BONDING

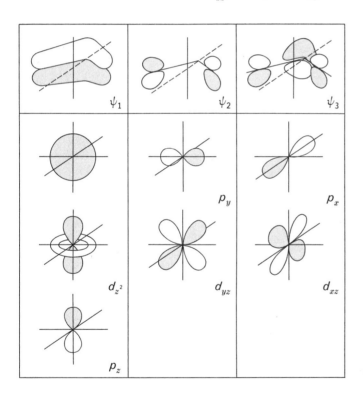

The three $p\pi$ orbitals of the allyl group combine to form three mo's in the pattern familiar from Chapter 4 (see Figure 16–15 and Table 4–1). Ligand-to-metal electron donation may then occur by overlap of ψ_1 with the metal s, d_{z^2}, and p_z orbitals and by overlap of ψ_2 with the metal p_y and d_{yz} orbitals. The metal-to-ligand back-bond then derives from d_{xz} overlap with ψ_3.

Figure 16–15. π-Allyl molecular orbitals and the metal orbitals with which they can most effectively overlap to produce substantial metal-ligand bonding. The metal lies below the C_3 plane, along the vertical axis. [Adapted from M. L. H. Green and P. L. I. Nagy, *Adv. Organometal. Chem.*, **2**, 325 (1964).]

COMPLEXES WITH CYCLIC π DONORS

The discussion of complexes formed with cyclic π donors will center largely on those of the six π electron donors, the *pentahapto*cyclopentadienyl ion and the *hexahapto*benzene molecule. There are basically three types of complexes formed by η^5-$C_5H_5^-$, η^6-C_6H_6, and related molecules:

(i) $(\pi\text{-R})_2M$. Symmetrical "sandwich" type complexes represented by ferrocene **(61)**, dibenzenechromium **(62)**, and (η^5-cyclopentadienyl)(η^6-benzene)manganese **(63)**.

(ii) $(\eta^5\text{-}C_5H_5)_2ML_x$. Bent metallocenes where L represents some other ligand such as H^-, R^-, halide, olefin, or NO; examples include compounds **64** and **65**.

(iii) $(\pi\text{-R})ML_x$. Complexes in this class contain only one cyclic π donor ligand in addition to other ligands L; compounds **66** and **67** are examples.

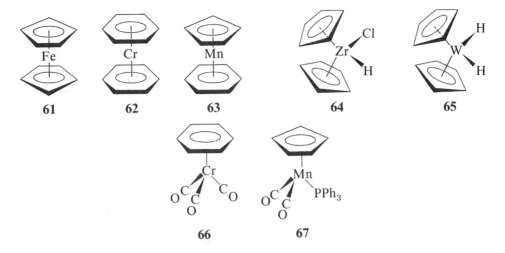

Synthesis and Properties

The preparation of symmetrical bis(η^5-cyclopentadienyl)metal compounds[219]—often called metallocenes[220]—begins with the preparation of the parent molecule C_5H_6, a diene that is, in turn, obtained by thermally cracking dicyclopentadiene (**68**).

$$\text{68} \rightarrow 2 \quad \text{(cyclopentadiene)} \tag{16-100}$$

The basic problem, then, in any synthetic procedure is to remove H^+ from the diene, and several useful methods are available (16-101).

$$\begin{array}{c} \xrightarrow{Na/THF} Na^+ \, C_5H_5^- \\ \xrightarrow{Tl_2SO_4/KOH} Tl^+ \, C_5H_5^- \end{array} \tag{16-101}$$

The hydrolytically unstable $C_5H_5^-$ salts are not usually isolated but are used directly in a reaction with an anhydrous salt of the metal to produce the desired metallocene. As an alternative, it is possible to use a hydrated metal salt if the cyclopentadienide ion can be generated *in situ* as in reaction 16-103. (See Table 16-5 for a summary of properties of the common metallocenes.)

$$2NaC_5H_5 + CoCl_2 \rightarrow (\eta^5\text{-}C_5H_5)_2Co + 2NaCl \tag{16-102}$$

$$8KOH + 2\,C_5H_6 + FeCl_2\cdot 4H_2O \rightarrow (\eta^5\text{-}C_5H_5)_2Fe + 2KCl + 6KOH\cdot H_2O \tag{16-103}[221]$$

---

[219] R. L. Pruett, "Cyclopentadienyl and Arene Metal Carbonyls," in "Preparative Inorganic Reactions," Volume 2, W. L. Jolly, ed., Interscience, New York (1965).

[220] The name "metallocene" was applied to bis(η^5-cyclopentadienyl)metal compounds soon after the discovery of ferrocene. Hence, the congeners of ferrocene are called ruthenocene and osmocene, for example.

[221] W. L. Jolly, *Inorg. Syn.*, **11**, 120 (1968).

TABLE 16-5
PROPERTIES OF SOME METALLOCENES

Compound	Color	Melting Point (°C)	Remarks
"$(C_5H_5)_2Ti$"	dark green	>200(d)	Actually exists as $[(C_5H_5)(C_5H_4)TiH]_2$.[a]
$(C_5H_5)_2V$	purple	167	Very air-sensitive, $\mu = 3.84$ B.M.
$(C_5H_5)_2Cr$	dark red	173	Very air-sensitive, $\mu = 3.20$ B.M.
$[(C_5H_5)_2Mo]_x$[b]	red-brown solid		Diamagnetic.
$[(C_5H_5)_2W]_x$[b]	red solid		
$(C_5H_5)_2Mn$	dark brown	173	Air-sensitive. Hydrolyzed with characteristic crackling sound. Heating under N_2 to 158° turns the brown solid to pale pink solid, which melts at 173°. $\mu = 5.86$ B.M.
$(C_5H_5)_2Fe$	orange	173	Air-stable. Can be chemically oxidized to blue-green solid.
$(C_5H_5)_2Co$	purple-black	174	Air-sensitive. Oxidized to stable yellow $[(C_5H_5)_2Co]^+$ salts. $\mu = 1.73$ B.M.
$(C_5H_5)_2Ni$	dark green	173	Oxidizes slowly in air to give rather unstable yellow-orange $[(C_5H_5)_2Ni]^+$. $\mu = 2.86$ B.M.

[a] H. H. Brintzinger and J. E. Bercaw, *J. Amer. Chem. Soc.*, **92**, 6182 (1970); A. Davison and S. S. Wreford, *ibid.*, **96**, 3017 (1974).

[b] J. L. Thomas, *J. Amer. Chem. Soc.*, **95**, 1838 (1973).

(16-104)[222]

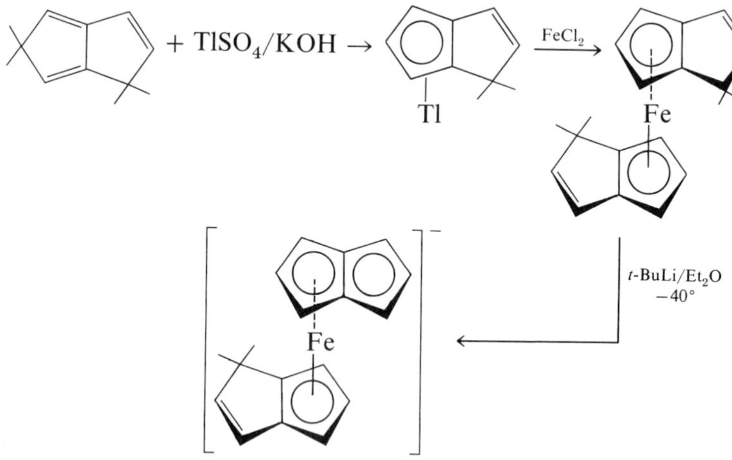

[222] T. J. Katz and J. J. Mrowca, *J. Amer. Chem. Soc.*, **89**, 1105 (1967).

The preparation of dibenzenechromium and similar molecules (Table 16-6) presents a somewhat different problem.[223,224] Since these are clearly complexes between neutral ligands and a metal in a zero (or perhaps +1) oxidation state, the problem is to bring a reduced metal in a reactive form in contact with the ligand. Therefore, the bis-arenes have usually been prepared by the aluminum reduction method developed by Fischer (*e.g.*, reaction 16-105).[223]

$$3\ CrCl_3 + 2\ Al + AlCl_3 + 6\ C_6H_6 \rightarrow 3\ [(\eta^6\text{-}C_6H_6)_2Cr][AlCl_4]$$

$$2\ [(\eta^6\text{-}C_6H_6)_2Cr]^+ + S_2O_4^{2-} + 4\ OH^- \rightarrow (\eta^6\text{-}C_6H_6)_2Cr + 2\ SO_3^{2-} + 2\ H_2O$$

(16-105)

More recently, bis-arenes have been synthesized by an obvious method—the co-condensation of metal atom vapor with benzene or benzene derivatives. (The metal vapor method was mentioned earlier as a preparation for metal alkyls; see page 824). Fischer's aluminum reduction method suffers from the serious disadvantage that arenes having halogen or basic substituents cannot be used. The metal

---

[223] E. O. Fischer, *Inorg. Syn.*, **6**, 132 (1960).
[224] H. Zeiss, P. J. Wheatley, and H. J. S. Winkler, "Benzenoid-Metal Complexes," Ronald Press, New York (1966).

TABLE 16-6
PROPERTIES OF SOME BIS-π-ARENE METAL COMPLEXES[a]

Compound	Color	Melting Point (°C)	Remarks
$(C_6H_6)_2V$	black	227	Oxidized rapidly in air to the red-brown cation $[(C_6H_6)_2V]^+$.
$(\eta^6\text{-}C_6H_5F)_2V$	red	—	Sublimes, air-sensitive.
$(C_6H_6)_2Cr$	brown-black	284	Oxidized readily to the yellow cation $[(C_6H_6)_2Cr]^+$.
$(\eta^6\text{-}C_6H_5Cl)_2Cr$	olive green	89–90	Sublimes, air-stable.
$(C_6H_5F)_2Cr$	yellow	96–98	Sublimes, air-stable.
$[1,4\text{-}C_6H_4(CF_3)_2]_2Cr$	amber	150–152	Air-stable to 266°.
$(C_6H_6)_2Mo$	green	115	Very air-sensitive.
$(C_6H_6)_2W$	yellow-green	160(d)	Less air-sensitive than the Mo analog.
$[\eta^6\text{-}C_6(CH_3)_6]_2Mn^+$	pink-white	—	Diamagnetic.
$[\eta^6\text{-}C_6(CH_3)_6]_2Fe^{2+}$	orange	—	Can be reduced with dithionite to the deep violet Fe(I) complex and to the extremely air-sensitive black, paramagnetic (2e^-) Fe(0) complex.
$[\eta^6\text{-}C_6(CH_3)_6]_2Co^+$	yellow	—	Paramagnetic (2e^-).

[a] H. Zeiss, P. J. Wheatley, and H. J. S. Winkler, "Benzenoid-Metal Complexes," Ronald Press Company, New York, 1966; M. L. H. Green, "Organometallic Compounds, Volume Two: The Transition Elements," G. E. Coates, M. L. H. Green, and K. Wade, eds., Methuen, London, 3rd edition, 1968; K. J. Klabunde and H. F. Efner, *Inorg. Chem.*, **14**, 789 (1975); P. S. Skell, D. L. Williams-Smith, M. J. McGlinchey, *J. Amer. Chem. Soc.*, **95**, 3337 (1973).

vapor method, on the other hand, has no such disadvantage. Therefore, Cr, Mo, and W complexes (and V complexes in some cases) have been synthesized with C_6H_4XY where X is F, Cl, CF_3, NMe_2, C(O)OMe, and so forth, and Y is H or one of the aforementioned groups (see Table 16-6).[225-227]

(16-106)

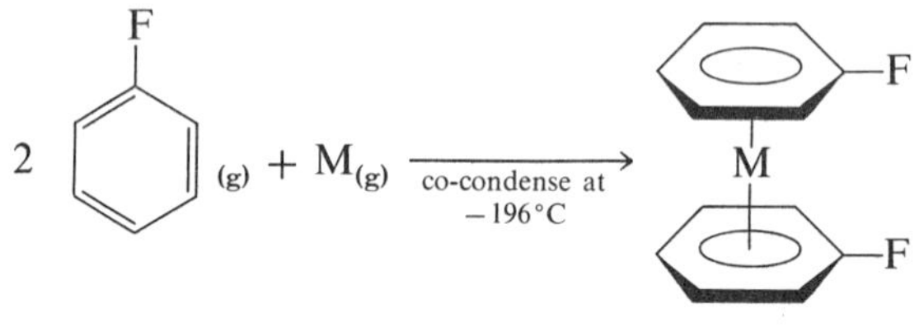

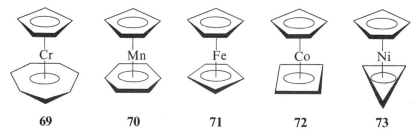

A number of mixed sandwich compounds are also now known. For example, all of the members of the isoelectronic series **69-73** have been prepared,[228] and their syntheses illustrate the imaginative methods used in organometallic chemistry.

For example, (η^5-cyclopentadienyl)(η^6-benzene)manganese is prepared from (η^5-C_5H_5)MnCl and PhMgBr in THF. Reaction 16-107 presumably proceeds through formation of a σ-phenyl derivative;[229] however, upon hydrolysis this intermediate is converted to the π-benzene complex, and the metal is reduced. Similar reactions leading to the formation of dibenzenechromium have been studied by Zeiss.[230]

(16-107)

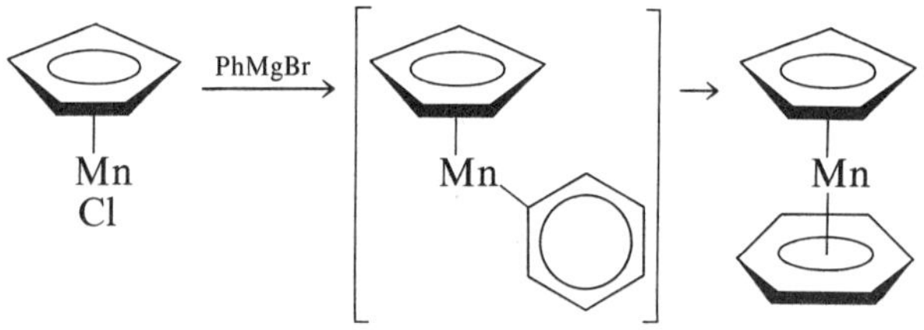

The η^4-cyclobutadiene complex of cobalt **(72)** is prepared in a photolytic reaction.[231] However, a similar tetrasubstituted cyclobutadiene complex may be synthesized by acetylene condensation;[228] indeed, the condensation of disubstituted acetylenes to give tetrasubstituted cyclobutadienes and hexasubstituted benzene in the presence of metal is known to occur in many other instances.

---

[225] P. L. Timms, *J. Chem. Educ.*, **49**, 782 (1972).
[226] P. S. Skell, D. L. Williams-Smith, and M. J. McGlinchey, *J. Amer. Chem. Soc.*, **95**, 3337 (1973).
[227] K. J. Klabunde and H. F. Efner, *Inorg. Chem.*, **14**, 789 (1975); see also ref. 21.
[228] M. D. Rausch, *Pure and Appl. Chem.*, **30**, 523 (1972).
[229] T. H. Coffield, V. Sandel, and R. D. Closson, *J. Amer. Chem. Soc.*, **79**, 5826 (1957).
[230] H. Zeiss, in "Organometallic Compounds," H. Zeiss, ed., A. C. S. Monograph 147, Reinhold, New York (1960), page 380.
[231] M. Rosenblum, B. North, D. Wells, and W. P. Giering, *J. Amer. Chem. Soc.*, **94**, 1239 (1972).

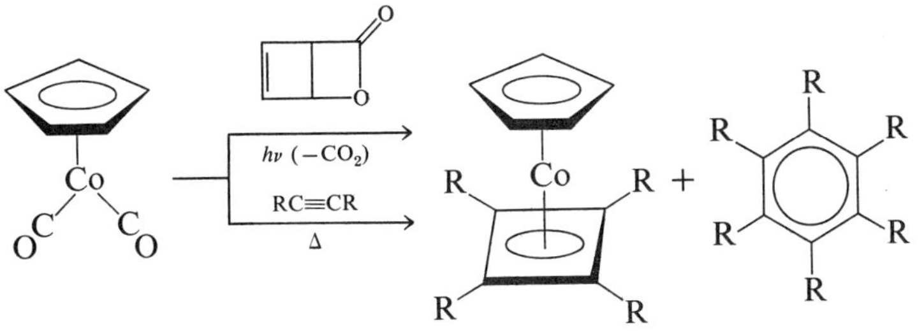

$$\text{CpCo(CO)}_2 \xrightarrow[\Delta]{\substack{h\nu\ (-CO_2)\\ RC\equiv CR}} \text{CpCo(C}_4R_4) + C_6R_6 \tag{16-108}$$

The preparation of bent metallocenes of the type $(\eta^5\text{-}C_5H_5)_2ML_x$ has not yet been systematized. However, considerable work has been done with the bis(η^5-cyclopentadienyl)metal complexes of Group VIB (chromocene, molybdenocene, and tungstenocene), and some of their reactions lead to compounds of the desired type.[232,233]

Preparation of Molybdenocene.

$$MoCl_5 + 3NaC_5H_5 + 2NaBH_4 \rightarrow Cp_2MoH_2 + 5NaCl + \tfrac{1}{2}C_{10}H_{10} + B_2H_6$$

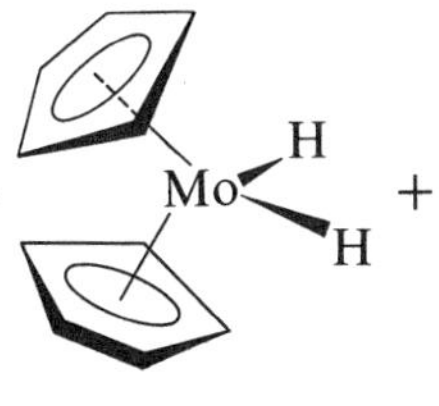

$$(C_5H_5)_2MoH_2 \xrightarrow{\Delta,\ CHCl_3} (C_5H_5)_2MoCl_2 \xrightarrow{Na/Hg,\ THF} [(C_5H_5)_2Mo]_x \tag{16-109}$$
red-brown solid;
sparingly sol. in THF

Reactions of Molybdenocene.

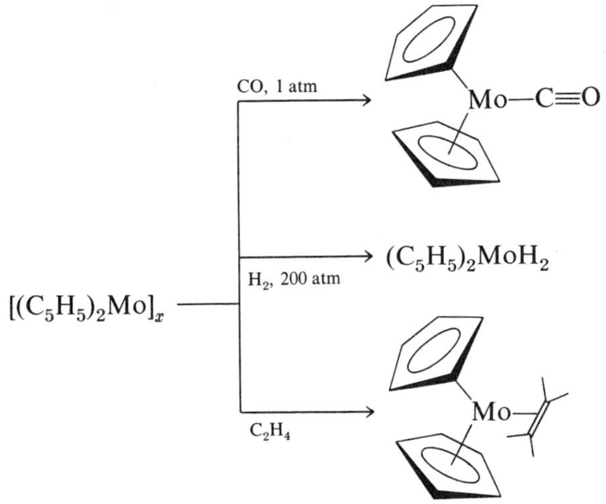

$$[(C_5H_5)_2Mo]_x \begin{cases} \xrightarrow{CO,\ 1\ atm} Cp_2Mo-C\equiv O \\ \xrightarrow{H_2,\ 200\ atm} (C_5H_5)_2MoH_2 \\ \xrightarrow{C_2H_4} Cp_2Mo(C_2H_4) \end{cases} \tag{16-110}$$

With the exception of ferrocene and bis-arenes with very electronegative substituents, most metallocenes and bis-arenes are oxidatively unstable. Therefore, considerable research has been done with the oxidatively more stable η^5-cyclopenta-

---

[232] K. L. Tang Wong and H. H. Brintzinger, *J. Amer. Chem. Soc.*, **97**, 5143 (1975).
[233] J. L. Thomas, *J. Amer. Chem. Soc.*, **95**, 1838 (1973).

882 ☐ ORGANOMETALLIC CHEMISTRY: SYNTHESIS, STRUCTURE, AND BONDING

dienylmetal and η^6-benzenemetal carbonyls. In general, π-bonded cyclopentadienylmetal carbonyls are prepared from metallocenes and CO or from the metal carbonyl and $C_5H_5^-$ (reaction 16-111) or dicyclopentadiene.[233] Reaction 16-112 is a particularly good example of the latter. Although the intermediates in 16-112 have been isolated in separate reactions, the pathway depicted was not confirmed until recently, when the important $(\eta^5\text{-}C_5H_5)M(CO)_2H$ was isolated in the reaction of $Ru_3(CO)_{12}$ with cyclopentadiene.[235]

(16-111)[234]

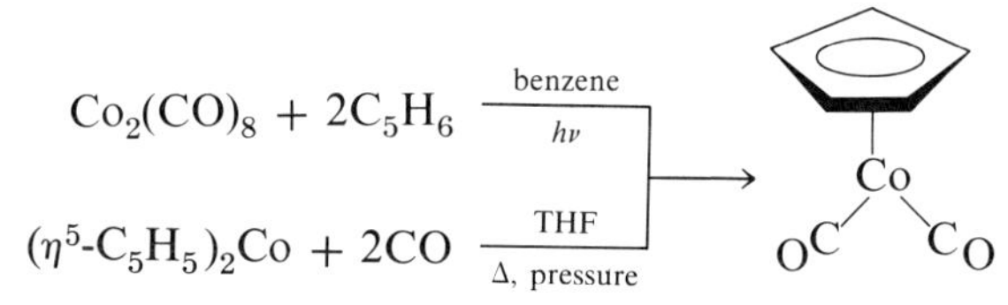

$$Co_2(CO)_8 + 2C_5H_6 \xrightarrow[h\nu]{\text{benzene}}$$

$$(\eta^5\text{-}C_5H_5)_2Co + 2CO \xrightarrow[\Delta,\text{ pressure}]{\text{THF}}$$

(16-112)

$$Fe(CO)_5 \text{ or } Ru_3(CO)_{12} + \text{(cyclopentadiene)} \rightarrow \text{intermediate} \xrightarrow{-CO} \text{intermediate} \rightarrow \text{dimer}$$

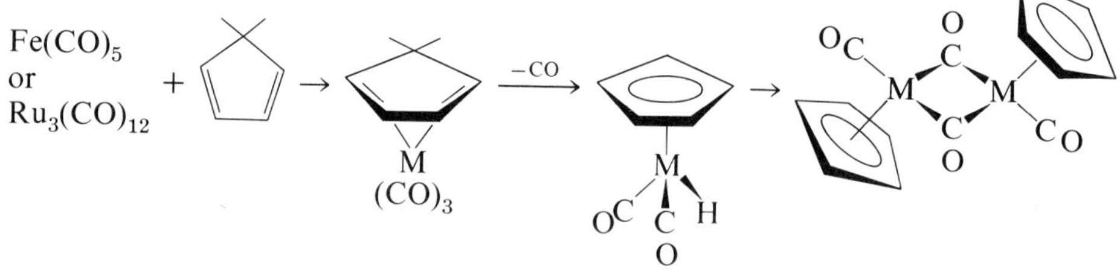

Arenemetal carbonyls, especially those of Group VIB, are readily synthesized from the appropriate metal carbonyl and the arene. If the potential ligand is a high boiling liquid (*e.g.*, mesitylene), the carbonyl may be refluxed under N_2 in the liquid arene; heat brings about the dissociation of the carbonyl and its replacement by the arene.[236]

(16-113)

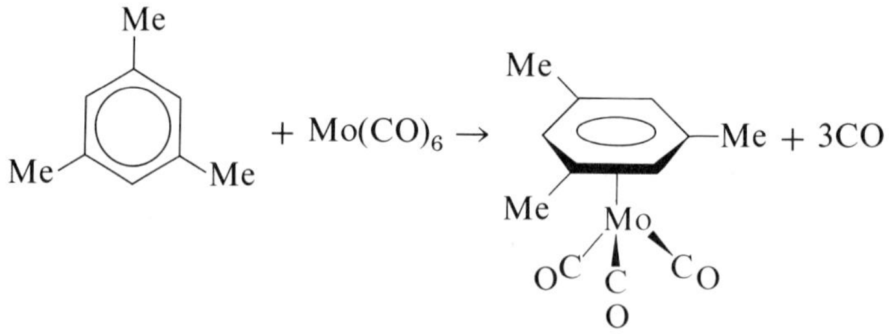

$$\text{mesitylene} + Mo(CO)_6 \rightarrow \text{(mesitylene)Mo(CO)}_3 + 3CO$$

74

A more useful method, however, is the prior formation of a labile complex of the carbonyl with ligands such as diglyme or acetonitrile.[237] Neither of these ligands is capable of forming strong metal-to-ligand $d\pi\text{-}p\pi$ bonds; therefore, the metal-ligand bond is weak, and the ligand is easily replaced. The intermediate may be isolated and then treated with arene, or the metal carbonyl may simply be refluxed in diglyme, for example, already containing the arene.[238]

---

[234] See ref. 167.
[235] A. P. Humphries and S. A. R. Knox, *Chem. Commun.*, 326 (1973).
[236] R. J. Angelici, "Synthesis and Technique in Inorganic Chemistry," 2nd ed., W. B. Saunders Co., Philadelphia (1977).
[237] R. B. King and A. Fronzaglia, *Inorg. Chem.*, **5**, 1837 (1966).
[238] B. Nicholls and M. C. Whiting, *J. Chem. Soc.*, 351 (1959).

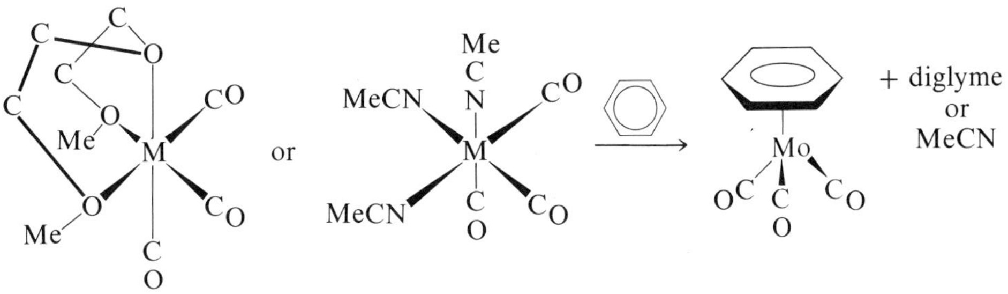

One of the most sought-after molecules in organic chemistry has been cyclobutadiene. Although it was recently isolated at low temperatures by photolytic methods, it was first prepared in 1965 in the form of its iron carbonyl complex.[239] The reaction is important for two reasons: (i) It illustrates the fact that *gem* and *vicinal* dihalides may be dehalogenated with metal carbonyls.[240] (ii) It represents an increasingly important direction being taken by organic and organometallic chemists—the isolation and study of highly reactive organic molecules in the form of their metal complexes.

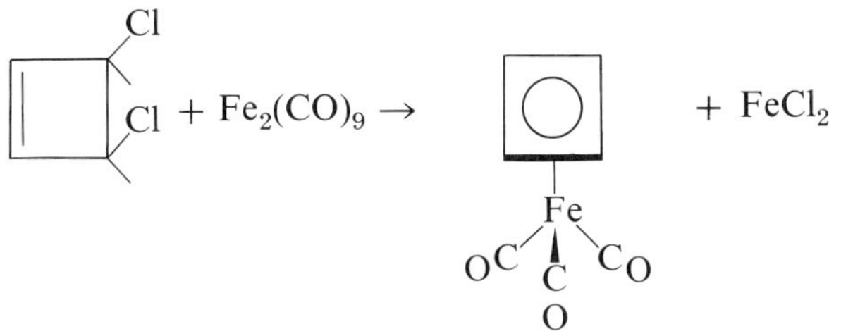

[pale yellow prisms]

Structure and Bonding

The fundamental questions of structure and bonding can be approached by examining a few representative compounds: ferrocene, bis(η^5-cyclopentadienyl)-rhenium hydride, bis(η^6-benzene)chromium, (η^6-benzene)tricarbonylchromium, and bis(η^8-cyclooctatetraene)uranium.

In the solid state the cyclopentadienyl rings of ferrocene are staggered (D_{5d} symmetry), whereas those of ruthenocene and osmocene are eclipsed (D_{5h} symmetry).[241]

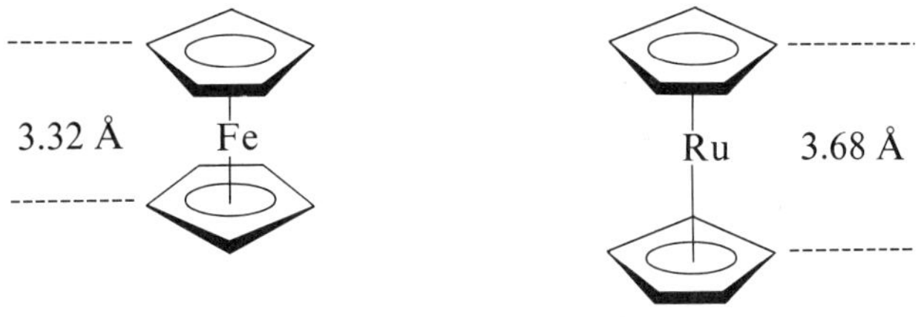

These arrangements are apparently the result of crystal packing forces, because the barrier to rotation of the ferrocene rings, measured in the gas phase, is quite small

---

[239] R. Pettit and J. Henery, *Organic Syntheses,* **50,** 21 (1970); see also R. Pettit *et al., J. Amer. Chem. Soc.,* **87,** 131, 3253, 3254 (1965).
[240] M. Ryang, *Organometal. Chem. Rev., A,* **5,** 67 (1970).
[241] See ref. 7.

(~1 kcal).²⁴² Furthermore, there is abundant evidence that the rings rotate freely when ferrocene and its derivatives are placed in solution.

The rotational freedom of the cyclopentadienyl rings in ferrocene was proved not long after the discovery of the compound in some very elegant experiments by Woodward.²⁴³ Ferrocene undergoes Friedel-Crafts catalyzed acetylation (see Chapter 17) very readily on one ring and only slightly less readily on both rings. If the two rings are free to rotate, only one 1,1'-disubstituted compound would be isolated, whereas three 1,1'-disubstituted isomers could be formed in the absence of rotation. As anticipated, only one compound was indeed isolated.²⁴⁴

(16-116)

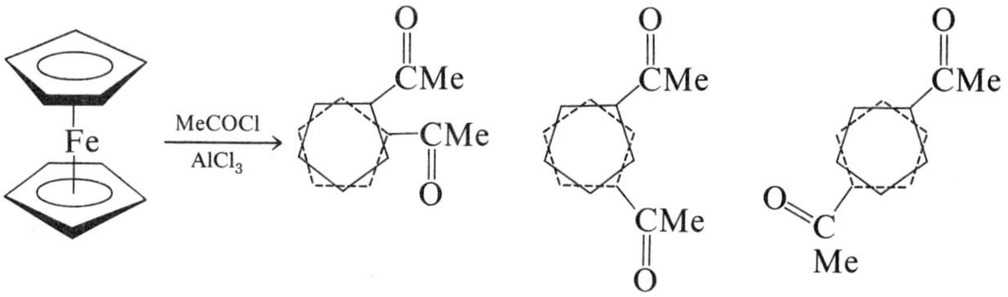

The electrochemical properties of the metallocenes will ultimately give some insight into the bonding in these molecules. Ferrocene, for example, can lose an electron to give the deep blue-green, paramagnetic ferricenium ion.

(16-117)

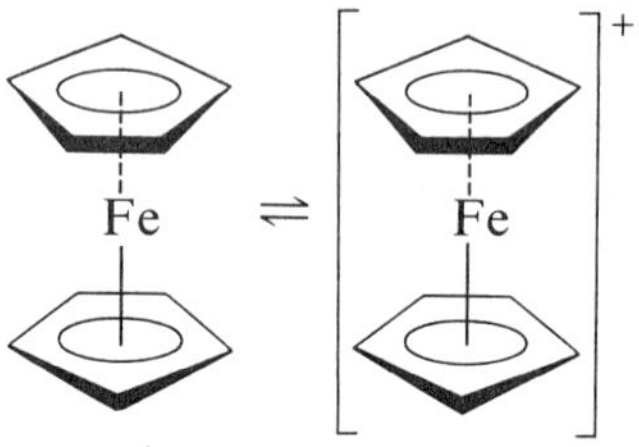

$$E° = +0.56 \text{ v. } vs. \text{ S.H.E.}$$

orange   blue-green

The oxidation is chemically and electrochemically reversible, and the potential shows a sensitivity to the nature of substituents that may be correlated with well-known substituent parameters such as the Hammett and Taft functions.²⁴⁵ The other metallocenes generally oxidize more readily than ferrocene. For example, cobaltocene, $(C_5H_5)_2Co$, in which cobalt exceeds the "effective atomic number" of the next higher inert gas by one electron, readily loses that electron to give the cobalticenium ion, $(\eta^5\text{-}C_5H_5)_2Co^+$. At low temperatures, however, oxygen is absorbed by the purple metallocene and a very unusual orange solid dioxygen adduct **(76)** is isolated.²⁴⁶

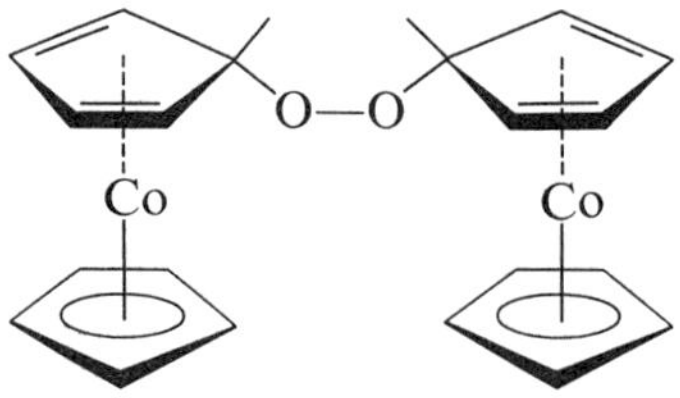

76

Ferrocene serves as a prototype for metal binding to π donor ligands, and we can develop a suitable molecular orbital picture using the methods of Chapter 4.

---

²⁴²R. K. Bohn and A. Haaland, *J. Organometal. Chem.*, **5**, 470 (1966).

²⁴³M. Rosenblum and R. B. Woodward, *J. Amer. Chem. Soc.*, **80**, 5443 (1958).

²⁴⁴You should notice that this is really negative evidence. That is, the non-isolation of potential compounds is very poor evidence to use in any chemical argument. However, in this case the experiment is corroborated by many others. It must also be mentioned that a compound having two acetyl groups on one ring is also isolated, but it is formed in a ratio of only 1:60 to the 1,1'-disubstituted compound.

²⁴⁵S. P. Gubin, *Pure and Appl. Chem.*, **23**, 463 (1970).

²⁴⁶H. Kojima, S. Takahashi, and N. Hagihara, *Chem. Commun.*, 230 (1973).

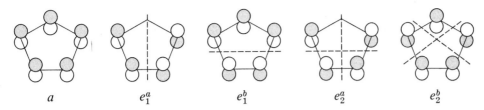

Orbitals a through e_2^b are the π mo's of $C_5H_5^-$. To construct the ten TASO's of ferrocene, the π mo's of $C_5H_5^-$ are put together in $(+)(+)$ ($= u$ for ungerade or anti-symmetric to inversion) and $(-)(+)$ ($= g$ for gerade or symmetric to inversion) combinations as illustrated below. This gives rise to three sets of TASO's: a filled pair of a_{1g} and a_{2u} symmetry; a higher energy, filled set of e_{1g} and e_{1u} symmetry; and an even higher energy, unfilled set of e_{2g} and e_{2u} symmetry. Also shown below are the iron orbitals with which the TASO's may effectively overlap.

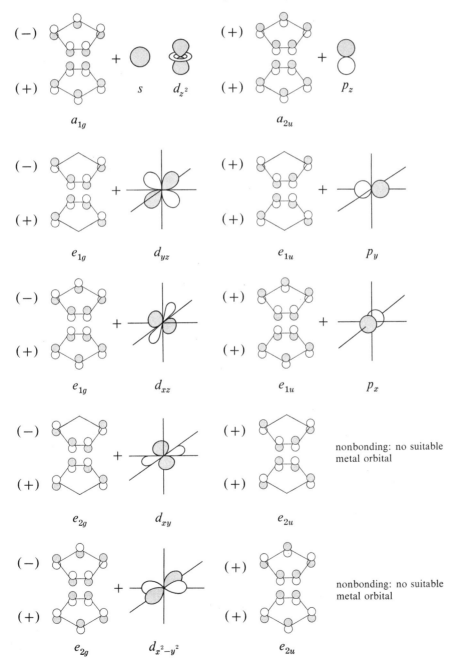

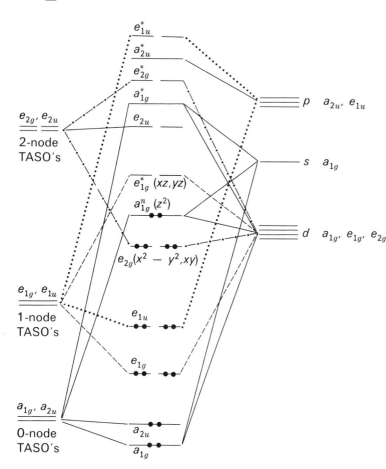

Figure 16-16. A qualitative molecular orbital diagram for ferrocene.

The most recent molecular orbital calculations for ferrocene indicate that the ordering of mo's is that illustrated in Figure 16-16.[247] Notice that the metal d_{xz} and d_{yz} orbitals (e_{1g} symmetry) have been found to interact more strongly with the cyclopentadienyl TASO's than the metal s (a_{1g}), p_x and p_y (e_{1u}), and p_z (a_{2u}) orbitals. The a_{1g}^n and e_{2g} sets of orbitals are primarily d_{z^2} and ($d_{x^2-y^2}, d_{xy}$) in character.[248]

Based on this simple molecular orbital picture, it is not surprising that ferrocene is the most stable metallocene yet discovered. The molecule has 18 valence electrons (six from each $C_5H_5^-$ and six from Fe^{2+}), and these nine pairs of electrons are just accommodated in the bonding and non-bonding mo's. However, cobaltocene (a 19-electron molecule based on d^7 Co^{2+}) and nickelocene (a 20-electron molecule based on d^8 Ni^{2+}) have electrons in the degenerate, anti-bonding e_{1g}^* orbitals. This fact is reflected in their electrochemical properties as noted above: both the cobalt and nickel compounds are readily oxidized. Vanadocene (15 valence electrons, d^3 V^{2+}) and chromocene (16 valence electrons, d^4 Cr^{2+}), on the other hand, are electron

---

[247] J. W. Lauher and R. Hoffmann, *J. Amer. Chem. Soc.*, **98**, 1729 (1976). This paper summarizes much of what is currently known about "bent metallocenes." It is also a very lucid and readable paper and is especially recommended as a starting place for study of the area.

[248] The e_{2g} pair is weakly bonding and the orbitals are nearly localized on Fe. The d_{z^2} ao constitutes the middle mo of a three-orbital combination and so is not absolutely non-bonding.

deficient and, as you shall see momentarily, they will add additional ligands that can contribute more electrons.

To see how metallocenes with only partially filled e_{2g} and a_{1g} mo's may relieve their electron deficiency by adding extra ligands, we must turn back to an early attempt by Wilkinson to prepare a cyclopentadienyl sandwich of rhenium by reaction of $ReCl_5$ and NaC_5H_5 in THF.[249] By analogy with the previously isolated manganocene [= $(C_5H_5)_2Mn$; see Table 16-6 and the discussion below], the expected product was $(C_5H_5)_2Re$. This molecule, if it is to be isostructural with ferrocene, would be paramagnetic because the a_{1g}^n level would be occupied by only one electron. However, it was quickly apparent that the molecule was diamagnetic and that it could not be formulated as $(C_5H_5)_2Re$. The ^1H nmr spectrum of the actual product, a yellow, air-stable solid, consisted of a doublet for the cyclopentadienyl rings at $\tau 6.36$ and another broad resonance line at very high field, $\tau 23.5$. As the ratio of intensities of the low and high field resonance lines was $10:1$, the compound was formulated as the hydride $(\eta^5\text{-}C_5H_5)_2ReH$, the proton of the Re—H group being responsible for the high field resonance line.[250] This formulation, of course, means that the molecule has the favored 18-electron configuration: six electrons from each $C_5H_5^-$, two electrons from H^-, and four electrons from Re^{3+}. The structure of the compound, the first such metal hydride, was thought to involve a tilted arrangement of cyclopentadienyl rings (**77**), an arrangement later confirmed in isoelectronic molecules such as $(\eta^5\text{-}C_5H_5)_2MoH_2$ (**78**).

77 **78**

The bonding in compounds **77** and **78** and similar molecules has been the subject of considerable discussion in the literature. Recently, however, a model has been proposed that can rationalize the bonding in molecules of the general formulation $(\eta^5\text{-}C_5H_5)_2ML_x$ (which are now known in sufficient number to constitute a general class of organometallic compounds).[251] The essence of the bonding model is that, as the cyclopentadienyl rings are tilted back from their parallel positions in ferrocene, there is a mixing of the three previously non-bonding or weakly bonding a_{1g}^n and e_{2g} orbitals (as well as the metal s and p orbitals in that plane) to form three new, more highly directed orbitals, the shapes of which are shown in Figure 16-17. It is these three new orbitals that may be used to bind additional ligands to the metal to raise the electron count to 18. For example, in $(\eta^5\text{-}C_5H_5)_2ReH$ there are six electrons beyond the 12 already accommodated in the six lowest-energy orbitals in Figure 16-16. The three new orbitals described in Figure 16-17 combine with the s orbital of the H^- ion as shown in Figure 16-18, and the six electrons not already in the mo's that are primarily C_5H_5 in character can now be accommodated in the new molecular orbitals.

The bonding picture just discussed accounts very nicely for the fact that ferrocene is basic in the Lewis sense.[252] That is, on placing ferrocene in strong acid media, the proton nmr spectrum of the solution shows the presence of a metal hydride, $(\eta^5\text{-}C_5H_5)_2FeH^+$, an ion isoelectronic with $(\eta^5\text{-}C_5H_5)_2ReH$ (reaction 16-118, p. 889).

---

[249] G. Wilkinson and J. M. Birmingham, *J. Amer. Chem. Soc.*, **77**, 3421 (1955); M. L. H. Green, L. Pratt, and G. Wilkinson, *J. Chem. Soc.*, 3916 (1958).

[250] Note that ^1H nmr resonance lines at very high fields (τ 12-40) are characteristic of transition metal hydrides and often offer the best proof of their existence.

[251] See ref. 247.

[252] Molecules such as **79**, **80**, and **82** are discussed in a general review of metal basicity: J. Kotz and D. Pedrotty, *Organometal. Chem. Rev. A*, **4**, 479 (1969).

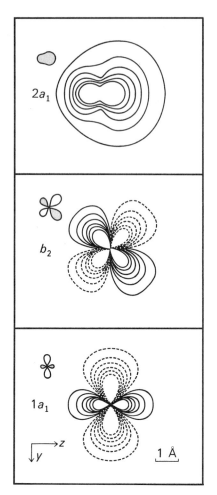

Figure 16-17. Contour diagrams for the three new bonding orbitals obtained as the η^5-C_5H_5 rings are tilted back from their parallel positions in ferrocene. The b_2 orbital is mainly d_{yz} in character, while the two a_1 orbitals are formed from the metal s, p_z, d_{z^2}, and $d_{x^2-y^2}$ ao's. [Reprinted with permission from J. W. Lauher and R. Hoffmann, *J. Amer. Chem. Soc.*, **98**, 1729 (1976). Copyright by the American Chemical Society.]

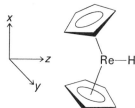

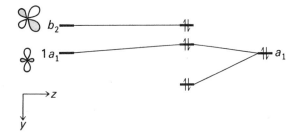

Figure 16-18. A qualitative mo diagram involving the three new mo's for a bent metallocene. The molecule is one of the type $(\eta^5$-$C_5H_5)_2$ReH. The six mo's that are chiefly responsible for metal-C_5 ring bonding are not shown. [Reprinted with permission from J. W. Lauher and R. Hoffman, *J. Amer. Chem. Soc.*, **98**, 1729 (1976). Copyright by the American Chemical Society.]

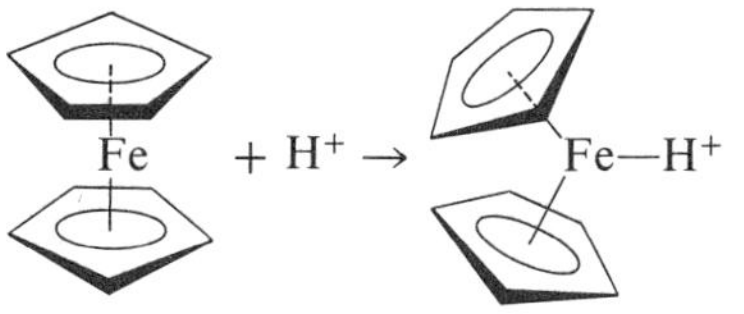

$$\text{Fe} + \text{H}^+ \rightarrow \text{Fe—H}^+$$

79 (16-118)

Similarly, the infrared spectrum of β-ferrocenylethanol shows that the alcoholic proton is hydrogen-bonded to the metal (**80**). In both instances, if the cyclopentadienyl rings are tilted back, the orbitals described in Figure 16-17 become available. All are filled with electrons, so a proton may bind to the metal, and the mo scheme is just that described for the isoelectronic $(\eta^5\text{-}C_5H_5)_2\text{ReH}$.

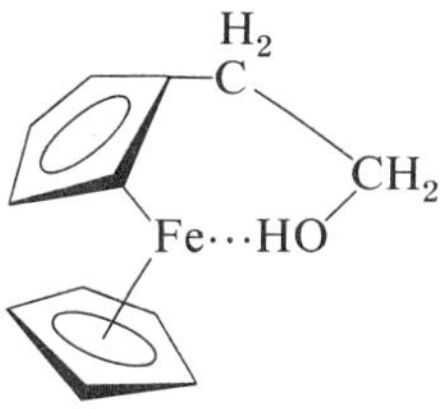

80

As we mentioned above, metallocenes with only partially filled e_{2g} and a_{1g}^n orbitals can remove this electron deficiency by adding extra ligands, as in the reactions of molybdenocene (reaction 16-110). This has been increasingly recognized as a general phenomenon in the chemistry of the earlier transition metals. For example, compounds **64, 81,** and **82** have been described, and their bonding may be rationalized by an extension of the ideas discussed above.

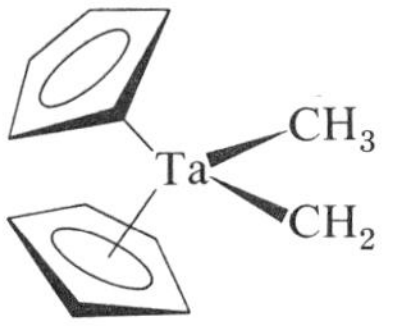

81[253] 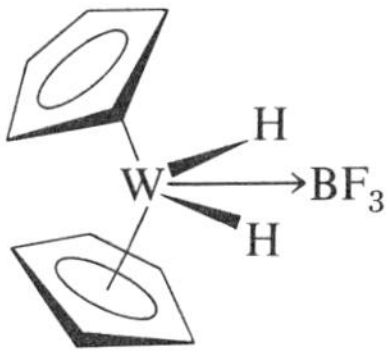   **82**

Besides ferrocene, the metallocenes that have provoked perhaps the greatest amount of discussion are titanocene and manganocene. Reduction of $(\eta^5\text{-}C_5H_5)_2\text{TiCl}_2$ gives a green compound with the empirical formula $(C_5H_5)_2\text{Ti}$, and it was included as such in the earliest tabulations of metallocenes as titanocene. Later work, however, showed the molecule to be dimeric, and numerous proposals were advanced to account for this. Very recently the structure of the dimer was found to be **83,** a structure that can be accounted for by using the mo picture of a bent metallocene.[254]

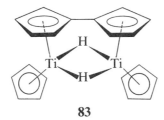

83

[titanocene = $\mu\text{-}(\eta^5\text{:}\eta^5\text{-fulvalene})\text{-di-}\mu\text{-hydrido-}$
bis(cyclopentadienyltitanium)]

---

[253] See ref. 180.
[254] A. Davison and S. S. Wreford, *J. Amer. Chem. Soc.*, **96,** 3018 (1974).

Manganocene is an amber solid that can be prepared from $Na^+C_5H_5^-$ and anhydrous $MnCl_2$ in THF. However, early investigations of the chemistry of $(C_5H_5)_2Mn$ led chemists to believe that it may be in a class apart from the covalent metallocenes such as ferrocene. That is, manganocene reacts rapidly with ferrous salts to give ferrocene, and its cyclopentadienyl groups exchange rapidly with $C_5H_5^-$ added to its solutions. These properties, coupled with the fact that the compound has a magnetic moment of 5.86 ± 0.05 B.M. (very close to the spin-only moment for five electrons) in dilute solid solution, indicated to early workers that the compound was ionic, *i.e.*, $(C_5H_5^-)_2Mn^{2+}$. More recently, however, this idea was put to rest as further, extensive studies of the magnetic behavior of manganocene strongly suggest that the overall degree of Mn—C_5H_5 orbital mixing is similar to that in ferrocene. The magnetic moment simply arises from the fact that the energies of the five metal-like mo's (e_{1g}^*, e_{2g}, and a_{1g}^n) are very similar, and Hund's rule is obeyed in filling electrons into the orbitals.[255]

Dibenzenechromium **(62)** is the best example of the other type of symmetrical metal sandwiches.[256] One of the major questions that arose after the discovery of the compound was whether it could best be described as a triene complex **(62′)** or a complex of benzene **(62″)**.

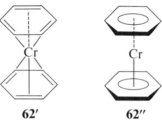

62′ **62″**

Some early x-ray studies, in fact, suggested the former; but more careful work recently has established reasonably well that dibenzenechromium is indeed a benzene complex of true D_{6h} symmetry.[257] Because of this, we can build a bonding model for dibenzenechromium in much the same way as we did for ferrocene. The π mo's of benzene are shown below. As in the case of ferrocene, one can place one ring on top of the other in such a way as to generate gerade and ungerade combinations.

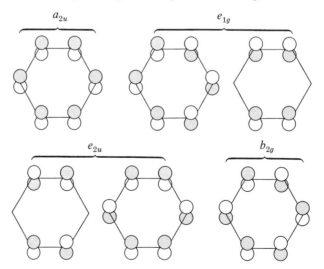

The u combinations may combine with metal p orbitals (*e.g.*, **84**) and the g combinations with metal s and d orbitals (*e.g.*, **85**). The result is the mo diagram given in Figure 16-19.

---

[255] M. E. Switzer, R. Wang, M. F. Rettig, and A. H. Maki, *J. Amer. Chem. Soc.*, **96**, 7669 (1974); recall (page 574) that the pairing energy is greatest for a d^5 case.

[256] See ref. 190.

[257] F. A. Cotton, W. A. Dollase, and J. S. Wood, *J. Amer. Chem. Soc.*, **85**, 1543 (1963).

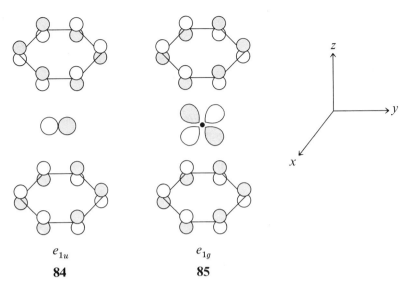

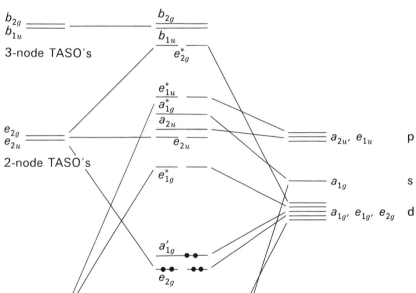

Figure 16-19. Molecular orbital diagram for dibenzenechromium.

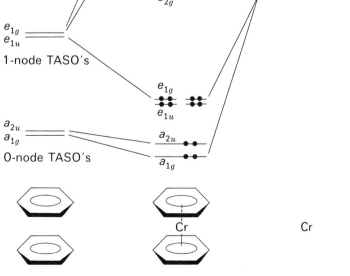

One of the chief differences between ferrocene and dibenzenechromium concerns their stabilities with respect to oxidation. As discussed previously, the iron compound (with Fe formally in the 2+ oxidation state) may be oxidized chemically or electrochemically to $[(\eta^5\text{-}C_5H_5)_2Fe^{III}]^+$, but it is nonetheless stable in air in the solid state or in solution under normal conditions. However, dibenzenechromium and many other bis-arenes of Cr, Mo, or W (wherein the metal is formally in the 0 oxidation state) oxidize very readily in air to give the paramagnetic species $[(\text{arene})_2M]^+$.[258] The reason for this is that Z^* for Fe^{II} is greater than for Cr^0.

You can extrapolate from the conclusion above to the concept that any effect that enhances the charge on the central metal should render it less likely to lose electrons. Indeed, replacing one arene ligand in a bis-arene with three carbonyl groups leads to a great increase in oxidative stability. For example, the Group VIB arenemetal tricarbonyls (86) are usually oxidatively stable as solids, although they often oxidize slowly in solution.

86

There is considerable evidence that carbonyl groups are more efficient π acceptors of electron density than are cyclopentadienyl or benzene (viz., the e_{2g} orbital system in Figure 16-19) ligands. Therefore, inclusion of CO ligands in place of C_6H_6 may indirectly lower the energy of all of the metal based mo's and lessen the reducing ability of the metal.

Lanthanide and actinide organometallic chemistry is a rapidly growing area of interest.[259] Tricyclopentadienides of all of the lanthanides and many of the actinides have been prepared by ligand exchange reactions such as 16-119.

(16-119) $$2 \text{ PuCl}_3 + 3 \text{ (C}_5\text{H}_5)_2\text{Be} \xrightarrow[\text{melt}]{65°} 2 \text{ (C}_5\text{H}_5)_3\text{Pu} + 3 \text{ BeCl}_2$$

All of the compounds thus far isolated are air- and moisture-sensitive, but they are thermally stable. In general, they display properties characteristic of ionic materials. Still open to question is the idea of utilization of f orbitals in these complexes, as the f orbitals lie too low in energy to be used in the case of a lanthanide or actinide 3+ ion.

In order to utilize f orbitals in metal-ligand bonding, it would appear that an actinide ion should be combined with a ligand whose π TASO's have symmetries suitable for overlap with the full range of f orbitals (Figure 16-20; see also Appendix, Chapter 1). Cyclooctatetraene dianion (COT^{2-}), which is prepared by reaction of the parent tetraene with K in THF, is a planar, $4n + 2$ aromatic ion having 10 π electrons. These fill the no-node a_{2u} mo, the degenerate one-node e_{1g} mo's, and the degenerate two-node e_{2u} mo's.

---

[258] Note that there are exceptions to this statement. However, all of the air stable bis-arene compounds have very electronegative groups attached to the benzene ring; see Table 16-6.

[259] E. C. Baker, G. W. Halstead, and K. N. Raymond, *Structure and Bonding*, **25**, 23 (1976).

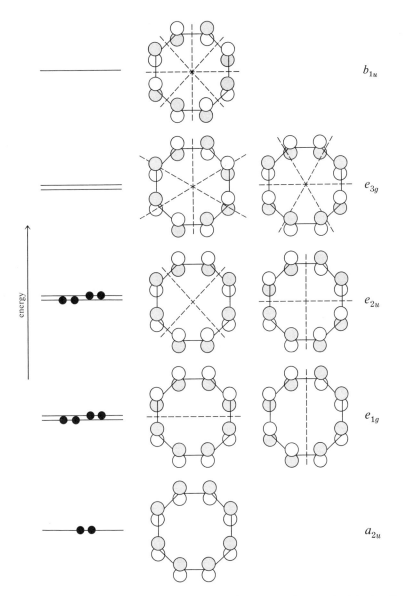

If two of the COT^{2-} ions are placed parallel with one another and with the C atoms eclipsed, 16 TASO's will be generated. Of these 16, half will be *ungerade*, and the seven *ungerade* f orbitals match all but the b_{1u} TASO in symmetry. For example, the

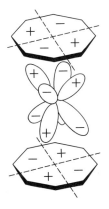

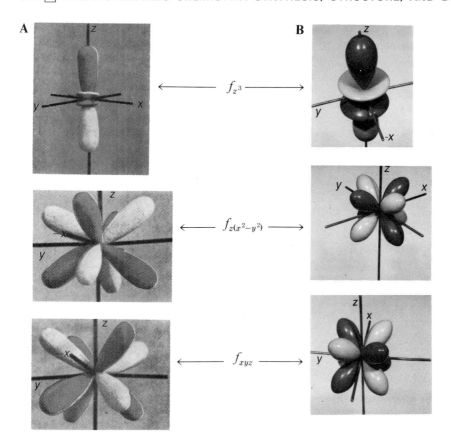

Figure 16-20. Set **A** is the "general" set of f ao's given in Appendix A to Chapter 1 and appropriate for the uranocene discussion in the text. Set **B** is the "cubic" set of f ao's (see the O_h character table in the Appendix to Chapter 2), which are most useful for structures based on cubic symmetry (O_h and T_d). [Set **A** from C. Becker, *J. Chem. Educ.*, **41**, 358 (1964).]

[U(η^8-C$_8$H$_8$)$_2$]

87

f_{xyz} [or $f_{z(x^2-y^2)}$] orbital can effectively interact with the e_{2u} TASO as illustrated below.

The important conclusion to be drawn from the mo analysis above is that an element with f orbitals at least partially filled may form a symmetrical sandwich with parallel cyclooctatetraene dianions. Therefore, it should come as no surprise that reaction of COT^{2-} with UCl$_4$ at 0°C in THF, or direct reaction of metallic uranium with cyclooctatetraene itself, gives "uranocene," U(COT)$_2$ **(87)**.[260]

[260] A. Streitwieser, Jr., U. Muller-Westerhoff, G. Sonnichsen, F. Mares, D. G. Morrell, K. O. Hodgson, and C. A. Harmon, *J. Amer. Chem. Soc.*, **95**, 8644 (1973). This paper includes a character table for D_{8h}.

CARBON π DONORS 895

Figure 16-20. (continued).

The green crystalline compound is pyrophoric (*i.e.*, it burns) in air, but it is stable in water. X-ray crystallography has shown that the molecule is definitely a sandwich compound with the metal located equidistant from the two planar *octahapto*cyclooctatetraene dianions. The planarity of the COT^{2-} ligands, the equality of their C—C bond lengths (average 1.392 Å), and the symmetrical positioning of the U^{4+} ion are consistent with the possibility that the metal is covalently bonded to an aromatic system. Based on this supposition, a very qualitative molecular orbital scheme has been suggested (Figure 16-21). As predicted by this scheme, the molecule is indeed paramagnetic (2.43 B.M.).

896 ORGANOMETALLIC CHEMISTRY: SYNTHESIS, STRUCTURE, AND BONDING

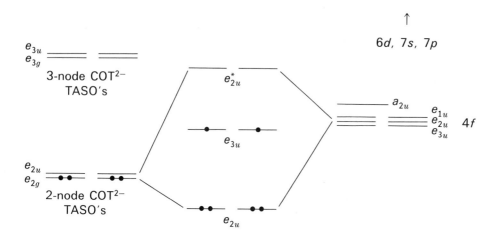

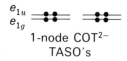

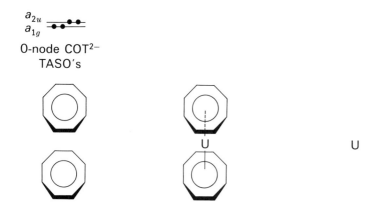

Figure 16–21. Qualitative molecular orbital diagram for uranocene. Only the main 4f-TASO interaction is shown. [Adapted from A. Streitwieser, Jr., et al., J. Amer. Chem. Soc., **95**, 8644 (1973).]

STUDY QUESTIONS

18. The first row transition metals from vanadium through nickel can form complexes of the type $[(\eta^5-C_5H_5)M(CO)_x]_n$. Using the EAN rule, predict the molecular formula for the compound formed by each first row transition metal. Draw the structure for each of the compounds.

19. Suggest syntheses for each of the following:

a) 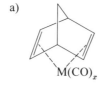   where M = Fe or Mo

b)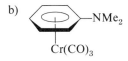

c) compound **42**

d)

20. Verify that the EAN rule is satisfied for compounds **69–73** (p. 880).

21. Predict the products for the reaction of molybdenocene, $[(C_5H_5)_2Mo]_x$, with N_2 and RX (*e.g.*, CH_3I).

22. Olefins can rotate about the metal-olefin bond axis as described on page 873. Would you expect acetylene complexes to behave similarly?

23. The bis(*pentahapto*cyclopentadienyl) compounds of V, Cr, Co, and Ni are all paramagnetic; the magnetic moment for each compound is listed in Table 16–5. Account for each of these compounds using the mo diagram suggested for ferrocene (Figure 16–16).

24. What is the function of KOH in reaction 16–103 (p. 877)?

25. Dibenzenechromium, $(C_6H_6)_2Cr$, is quite susceptible to air-oxidation. However, some other bis-arenes listed in Table 16–6 are apparently stable to air-oxidation. Why is a compound such as $(C_6H_5F)_2Cr$ more stable to oxidation than $(C_6H_6)_2Cr$?

26. Using Figure 16–18 as a guide, derive a similar mo diagram for $(\eta^5\text{-}C_5H_5)_2MoH_2$, structure **78**.

27. The bent metallocene $(\eta^5\text{-}C_5H_5)_2ReH$ has a Lewis basicity comparable to that of ammonia and can add a proton to give the molecule illustrated below.

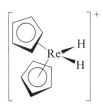

Similarly, $(\eta^5\text{-}C_5H_5)_2MH_2$ can add a proton or BX_3 at the metal (see structure **82**, p. 889). Account for this behavior in terms of the molecular orbital picture of such compounds developed in Figure 16–18 and in the previous question.

28. Draw the a_{1g} and a_{2u} TASO's for dibenzenechromium and decide with which metal orbitals these TASO's may overlap.

29. Draw the e_{1u} and e_{1g} TASO's for uranocene. With which metal f orbitals of the general set can they overlap, if any? Are any other metal orbital overlaps allowed with these TASO's?

EPILOG

Organometallic chemistry is clearly an immensely rich area and one that is very much alive. Although organometallic compounds have been known for more than 100 years, systematic study of the field is only about 20 years old, and many totally new types of compounds have been isolated in that brief time, some only recently. (In fact, metal-carbyne complexes were discovered while the writing of this book was in progress.) More new types of complexes are surely yet to come, and synthetic chemists will continue their work along that line. However, perhaps a more important direction in which the field should move is toward the systematization of what is currently known. For example, "bent metallocenes" have been known for fifteen or more years, but only recently were the structural and reaction chemistry of that class of molecules systematized to the extent that we can predict new advances and more fully understand current results.

One of the richest areas for systematization is the reaction chemistry of organometallic compounds in general. Beginning perhaps with the discovery of the general reaction type called "oxidative addition," there has been increasing recognition of common reactivity forms for organometallic compounds. The hope is that one can erect an organizational framework similar to that in organic chemistry, since only then will you be able to predict the course of new reactions and begin to search more systematically for commercial applications. With this goal in mind, we turn to the next chapter, Organometallic Compounds: Reaction Pathways.

Appendix to Chapter 16

Metal Carbonyls and Infrared Spectroscopy[261-263]

It should be obvious from the discussion in this chapter that infrared spectroscopy is useful in defining the bonding of metal carbonyls. Indeed, it is such a powerful tool that it deserves more detailed comment for those wishing to do more extensive work in organometallic chemistry.

C—O STRETCHING FREQUENCIES AND MOLECULAR STRUCTURE

In this discussion we shall be referring only to the C—O stretching modes of the carbonyl ligands, since it is these that provide the greatest amount of information about molecular structure and bonding. The stretching frequencies of terminal CO's generally lie in the region from 2140 to 1800 cm^{-1}, while bridging CO ligands have lower frequencies. For a CO group bridging two metals, the stretching frequency will be found in the region from 1700 to 1850 cm^{-1} (**88**),[264] whereas a CO bridging three metal atoms may have a frequency as low as about 1620 cm^{-1} (**89**).[265]

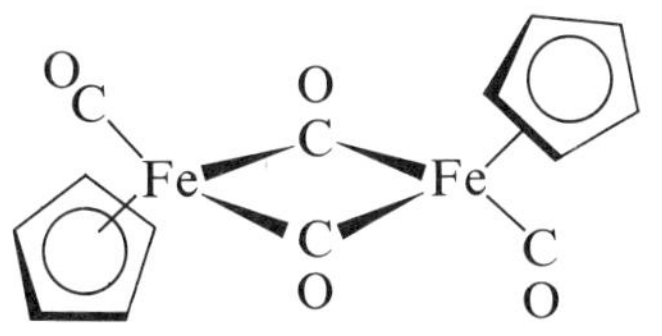

ν_{CO}(bridge) = 1781 cm^{-1}
88

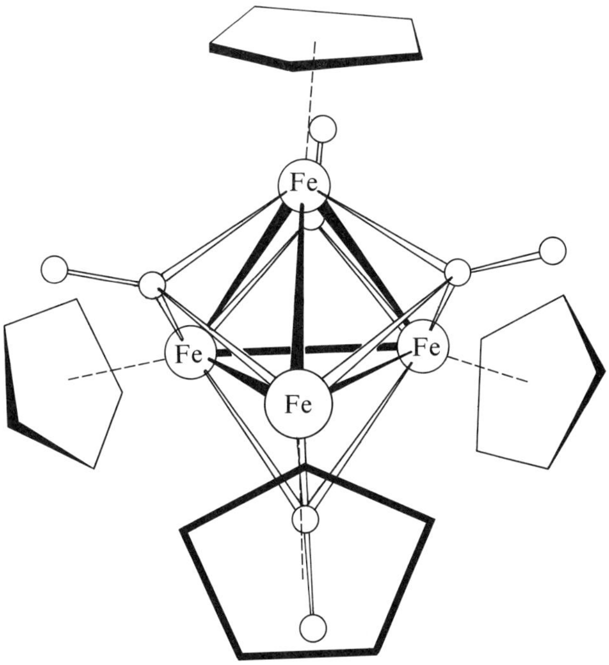

ν_{CO} = 1620 cm^{-1}
89

---

[261] D. M. Adams, "Metal-Ligand and Related Vibrations," St. Martin's Press, New York (1968).
[262] L. M. Haines and M. H. B. Stiddard, *Adv. Inorg. Chem. Radiochem.*, **12**, 53 (1969).
[263] P. S. Braterman, *Structure and Bonding*, **26**, 1 (1976).
[264] J. G. Bullitt, F. A. Cotton, and T. J. Marks, *Inorg. Chem.*, **11**, 671 (1972).
[265] R. B. King, *Inorg. Chem.*, **5**, 2227 (1966).

In examining a spectrum, however, you must be aware that other types of ligands (albeit very few) may have fundamental stretching frequencies in this region; for example, ν_{N-N} in M—N_2, ν_{M-H} in metal hydrides, and ν_{N-O} in metal nitrosyls. However, it is frequently obvious from the preparation or chemistry of the compound in question whether or not other ligands are present.

Once the formula of the complex has been established, its gross molecular structure may often be determined by examining the spectrum of the ν_{C-O} region.[266] The number and intensity of the bands depend largely on the local symmetry about the metal to which the carbonyls are attached. The symmetry of the other ligands—and their influence on the true molecular symmetry—is usually not important. For example, group theory leads us to predict that a complex such as cis-$L_2M(CO)_4$ (which has C_{2v} symmetry if the ligands are treated as "points") will have four infrared-active CO stretching vibrations: 2 a_1, b_1, and b_2. The vibrations that correspond to these labels may be illustrated if the CO stretching motion is depicted as a vector.

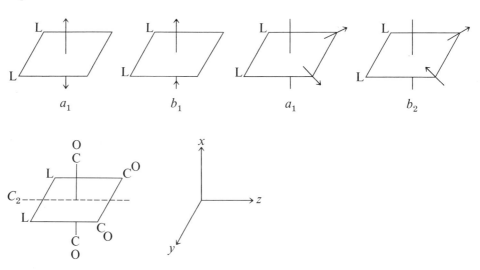

The vibrations labeled a_1 are totally symmetric with respect to all of the operations of the C_{2v} point group (see Appendix to Chapter 2 for character tables) and therefore belong to (or transform according to) the a_1 irreducible representation. The b_1 vibration is asymmetric with respect to rotation around the C_2 axis and to reflection in the yz plane; the b_2 vibration is also asymmetric with respect to C_2 rotation and to reflection in the xz plane. The next step is to decide whether or not these vibrations are infrared-active; that is, whether they can interact with, and absorb energy from, light of appropriate wavelengths. The simple "rule of thumb" by which this decision can be made is as follows: If a vibration transforms according to (i.e., is in the same symmetry class as) a translational degree of freedom (indicated by x, y, and z in the first column to the right of a character table), that vibration will be active. In the case of the C_{2v} metal carbonyl derivative, all of the CO stretching modes are active.

For your future reference, the number, symmetry, and activity of the carbonyl stretching bands expected for other common geometries are summarized in Table 16–7.

Another consequence of group theory is that the more symmetric the molecule, the fewer the number of distinct bands expected. Therefore, trans-$L_2M(CO)_4$, which has D_{4h} symmetry (more symmetry elements are present in this point group than in C_{2v}), has only one infrared-active band compared to four for the cis compound (Table 16–7 and Figure 16–22). You will notice in Figure 16–22, however, that additional bands can be observed; the asymmetry of the ligand lowers the symmetry of the molecule as a whole to a considerably lower symmetry than that around the metal. Previously inactive bands may become weakly active in this case.

The intensity of bands is not always easy to predict. However, a general rule of thumb is that the more symmetric vibrations have smaller extinction coefficients. Therefore, the a_1 bands are less intense than the b_1 and b_2 bands in the example above. The reason for this is that the dipole moment of the molecule changes less during a symmetric vibration than during an asymmetric vibration. Since intensity depends roughly on the magnitude of the dipole

---

[266] M. Y. Darensbourg and D. J. Darensbourg, J. Chem. Educ., **47**, 33 (1970); ibid., **51**, 787 (1974).

TABLE 16-7
NUMBER AND TYPE OF INFRARED STRETCHING FREQUENCIES EXPECTED FOR COMMON METAL CARBONYL COMPLEXES

Molecule	Point Group	Number of ν_{CO} Expected	Symmetry of ν_{CO} Bands
1.  M(CO)$_5$L	C_{4v}	3	$2a_1 + e$
2. cis-[M(CO)$_4$L$_2$] 	C_{2v}	4	$2a_1 + b_1 + b_2$
3.  trans-[M(CO)$_4$L$_2$]	D_{4h}	1	e_u
4. cis-[M(CO)$_3$L$_3$] 	C_{3v}	2	$a_1 + e$
5.  trans-[M(CO)$_3$L$_3$]	C_{2v}	3	$2a_1 + b_2$
6. cis-[M(CO)$_2$L$_2$X$_2$] 	C_{2v}	2	$a_1 + b_2$
7. 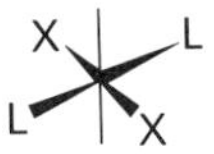 trans-[M(CO)$_2$L$_2$X$_2$]	D_{2h}	1	b_{1u}
8. M(CO)$_4$L 	C_{3v}	3	$2a_1 + e$
9.  M(CO)$_4$L	C_{2v}	4	$2a_1 + b_1 + b_2$
10. M(CO)$_3$L$_2$ 	D_{3h}	1	e'
11.  M(CO)$_3$L$_2$	C_s	3	$2a' + a''$

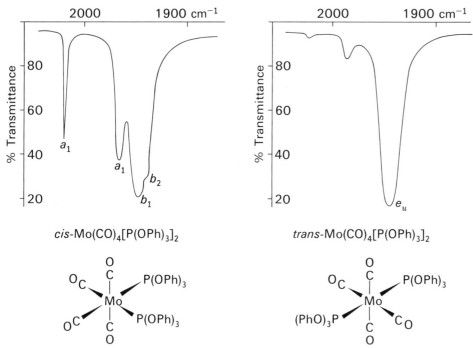

Figure 16-22. Infrared spectra in the ν_{CO} region for two metal carbonyl complexes. [Adapted from M. Y. Darensbourg and D. J. Darensbourg, *J. Chem. Educ.*, **47**, 33 (1970).]

change, it is evident that asymmetric modes will give rise to more intense absorptions. Further, it has been claimed that infrared intensities of the CO modes in transition metal carbonyl compounds (and in adsorbed CO, which has a frequency less than 2140 cm^{-1}) are determined solely by π bonding effects.[267] In agreement with this, the infrared intensities of the allowed CO stretching modes decrease dramatically in the isoelectronic series $[V(CO)_6]^-$, $Cr(CO)_6$, and $[Mn(CO)_6]^+$.

You should also be aware of the fact that the number of bands observed and their width can depend on the medium in which the molecule finds itself. In polar solvents such as $CHCl_3$, bands are frequently broadened and shifted slightly in position. In the solid phase (*e.g.*, a KBr pellet), there can be intermolecular vibrational coupling or pressure-induced molecular distortion, and band splitting can occur. Therefore, the best way to obtain a spectrum of a metal carbonyl is in the vapor phase or in a non-interacting solvent such as hexane.

In addition to changes in spectra due to interactions with the medium, it is also quite possible to observe spectral changes due to changes in the molecular structure from solid to solution. We have already mentioned the example of $Co_2(CO)_8$ (**26a** and **b**). Another well-studied case is $[(\eta^5-C_5H_5)Fe(CO)_2]_2$ (**88**).[268] In the solid phase, the latter has C_{2h} symmetry (*trans* CO-bridged structure; **88b**), and exhibits only one band for terminal and one band for bridging CO's (although both are split owing to intermolecular coupling). In solution, however, the terminal CO band is split into two bands, a result that has been interpreted as indicating the presence of four interconverting isomers in solution: *cis* bridged (**88a**), *trans* bridged (**88b**), *trans* non-bridged (**88c**), and a polar non-bridged structure (probably **88d**).[268]

88a   **88b**

---

[267] T. L. Brown and D. J. Darensbourg, *Inorg. Chem.*, **6**, 971 (1967).
[268] See ref. 264 and Chapter 17.

88c **88d**

C—O STRETCHING FREQUENCIES AND FORCE CONSTANTS

In the discussion thus far, we have implied that the C—O stretching frequency is related to bond order. However, this is not always the case. In a molecule, the distortion of one bond must be countered by other motions in the molecule so that the center of gravity of the molecule is maintained. This means that the stretching frequency of the CO group in F_2CO, for example, reflects not only the C—O stretching motion but F—C motion as well. Therefore, to be rigorously correct, the bond order should be related only to the stretching force constant. Unfortunately, the derivation of force constants is not always an easy task. However, C—O vibrations are reasonably well isolated in metal carbonyls, since only small motions of the heavy metal center are necessary to preserve the center of gravity, and there is little coupling of CO and non-CO motions.‡

It is possible, then, that one can relate C—O stretching frequency and bond order with some degree of confidence, and therefore use changes in vibrational frequencies as a probe of the electronic structure of a series of compounds. Consider the following examples:

i. For molecules of the type $RMn(CO)_5$, the C—O stretching frequencies should generally increase as R becomes more electronegative. For $(CF_3)Mn(CO)_5$ the observed frequencies (in cm^{-1}) are 2134 (a_1), 2019 (a_1), and 2043 (e), whereas $(CH_3)Mn(CO)_5$ has frequencies of 2111 (a_1), 1990 (a_1), and 2012 (e).[269]

ii. When BF_3 is added to $(Ph_3P\text{-}C_5H_4)Mo(CO)_3$ to form a 1:1 complex, there is a question as to the site of BF_3 attack, i.e., the molybdenum or an oxygen of the CO.[270] The spectrum of the product shows an increase in the number of C—O absorption bands and a general shift of the pattern to higher frequencies. Both observations are consistent with attachment to the metal, because the local symmetry of the metal is lowered, and the Lewis acid withdraws electron density from the metal, thereby lowering the amount available for back-donation to the carbonyl groups and strengthening C—O σ bonding.

90

For the most meaningful comparisons, however, it is best to compare force constants, and the derivation of such constants in metal carbonyls has been greatly simplified by the Cotton-Kraihanzel model.[271] This model is based on the premise of the overriding importance of variations in π bonding between the metal and CO in determining the variations in force constants, and on the assumption that there is negligible mixing of the CO stretching

---

‡For more discussion of these concepts, see page 110ff.
[269] F. A. Cotton, A. Musco, and G. Yagupsky, *Inorg. Chem.*, **6**, 1357 (1967).
[270] J. Kotz and D. Pedrotty, *J. Organometal. Chem.*, **22**, 425 (1970).
[271] F. A. Cotton and C. S. Kraihanzel, *J. Amer. Chem. Soc.*, **84**, 4432 (1962).

coordinates into the other normal modes of the molecule. Simple equations that allow the calculation of force constants have been derived on the basis of these assumptions. Therefore, for the important and much studied case of $LM(CO)_5$ (**91**; D_{4h} symmetry), for example, one can calculate k_c, the force constant for the CO groups *cis* to L, and k_t, the force constant for the CO group *trans* to L.

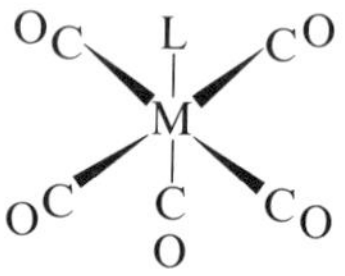

91

If L is an organic group as in $RMn(CO)_5$ mentioned above, it is found that k_t drops from 16.58 to 16.13 mdynes/Å and that k_c drops from 17.34 to 16.83 mdynes/Å when $R = CF_3$ is replaced by $R = CH_3$. This drop in force constants, of course, parallels the drop in absorption frequencies. Both results are expected because a highly electronegative group has been replaced with one that is less electronegative, thereby allowing more electron density to flow into CO π^* orbitals.

One especially important use of force constant analyses has been the evaluation of ligand-metal σ and π effects in molecules of the type $LM(CO)_5$; *e.g.*, $(CF_3)Mn(CO)_5$ or $(Ph_3P)Mo(CO)_5$. Graham[272] has argued that the σ donor (or acceptor) ability of a ligand can manifest itself equally well in the stretching frequencies of the carbonyls both *cis* and *trans* to it, while any π bonding between ligand and metal will influence the CO *trans* to the ligand twice as much as the *cis* CO's. The reason for the difference in σ and π influences is that *trans* ligands may interact through two metal $d\pi$ symmetry orbitals, while *cis* ligands can interact through only one metal π orbital.

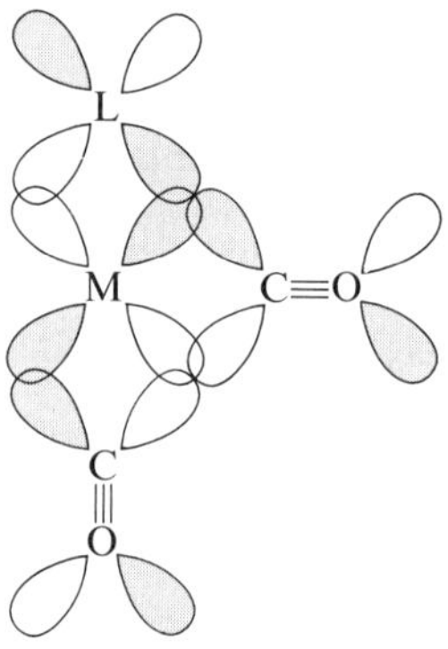

On this basis, Graham concluded that, on changing from one ligand to another:

$$\Delta k_t = \text{change in force constant for CO } trans \text{ to } L = \Delta\sigma_L + 2\Delta\pi_L$$

$$\Delta k_c = \text{change in force constant for CO } cis \text{ to } L = \Delta\sigma_L + \Delta\pi_L$$

where σ_L is a measure of the σ inductive effect of a ligand and π_L is a measure of the ability of the ligand to act as a π acid. Choosing the methyl group as a reference point, Graham has calculated σ and π parameters for a great variety of ligands and plotted these values (Figure 16-23). The important feature of this plot is the obvious fact that good donor ligands are also good π acids, a conclusion in agreement with your notion of the σ-π synergism model of σ donor–π acceptor ligands.

---

[272] W. A. G. Graham, *Inorg. Chem.*, **7**, 315 (1968).

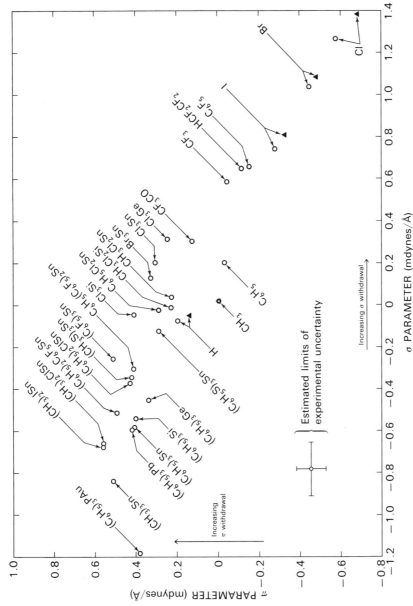

Figure 16-23. Relative σ and π parameters for the ligand L in LMn(CO)$_5$. The σ parameter is a measure of the σ donor effect of L, and the π parameter is a measure of the ability of L to act as a π acceptor. [Reprinted with permission from W. A. G. Graham, *Inorg. Chem.*, **7**, 315 (1968). Copyright by the American Chemical Society.]

THE 16 AND 18 ELECTRON RULE
AND REACTIONS OF TRANSITION
METAL ORGANOMETALLIC
COMPOUNDS

ASSOCIATION REACTIONS

   The Lewis Acidity and Basicity of
      Organometallic Compounds
   Ligand Protonation

SUBSTITUTION REACTIONS

   Nucleophilic Ligand Substitutions
   Electrophilic and Nucleophilic Attack
      on Coordinated Ligands

ADDITION AND ELIMINATION
REACTIONS

   1,2-Additions to Double Bonds
   1,1-Addition to CO: Carbonylation
      and Decarbonylation
   Oxidative Addition Reactions
     General Considerations
     Stereochemistry

     Influence of Central Metal, Ligands,
       and Addend
     Mechanism
   Elimination Reactions and the
      Stability of Metal-Carbon
      σ Bonds

REARRANGEMENT REACTIONS

   Redistribution Reactions
   Fluxional Isomerism or
      Stereochemical Non-rigidity

CATALYSIS INVOLVING
ORGANOMETALLIC COMPOUNDS

   Olefin Hydrogenation
   The Oxo Reaction
   The Wacker Process
   Polymerization
   Cyclooligomerization, Olefin
      Isomerization and Metathesis,
      and Polymer-Bound Catalysts

EPILOG

17. ORGANOMETALLIC COMPOUNDS: REACTION PATHWAYS

> In this chapter the types of reactions usually encountered in organometallic chemistry are surveyed. Briefly, these are
> i. *Association/dissociation reactions*: primarily reactions involving the formation of Lewis acid-base type complexes or ligand protonation reactions.
> ii. *Electrophilic and nucleophilic substitution reactions*: the substitution of one ligand, atom, or group for another.
> iii. *Addition/elimination reactions*: including 1,2-additions of M—R to double bonds, 1,1-additions of M—R to the carbon of CO (carbonylation), and the 1,1-addition of R—X to an unsaturated metal center (oxidative additions). Also discussed are the reverses of these reaction types—β-elimination of M—H from alkyls, decarbonylation, and reductive elimination.
> iv. *Rearrangements*: including intermolecular ligand redistribution and the intramolecular phenomenon of stereochemical non-rigidity.
> v. *Catalysis*: the homogeneous catalysis by organometallic compounds of addition and polymerization reactions.

The previous chapter on organometallic chemistry opened with a discussion of some representative reactions of iron- and tin-containing compounds. In Figures 16–1 and 16–2 were illustrated, for example:

an acid-base combination reaction

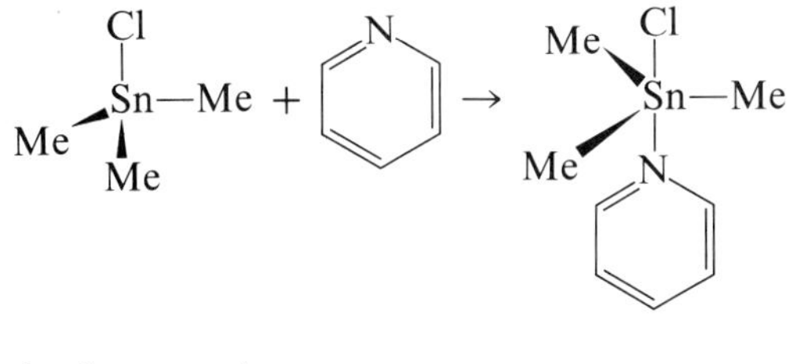

$$\text{Me}_3\text{SnCl} + \text{py} \rightarrow \text{Me}_3\text{SnCl}(\text{py}) \tag{17-1}$$

a ligand substitution reaction

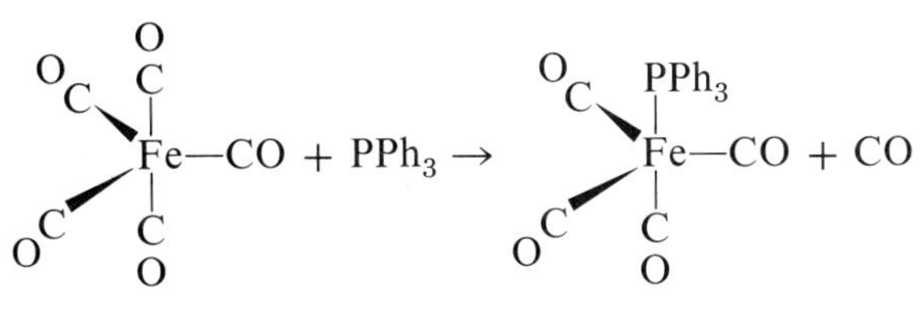

$$\text{Fe(CO)}_5 + \text{PPh}_3 \rightarrow \text{Fe(CO)}_4(\text{PPh}_3) + \text{CO} \tag{17-2}$$

an electrophilic attack on a coordinated ligand

(17-3)
$$\text{Cp}_2\text{Fe} + \text{Cl}-\overset{\text{O}}{\underset{\|}{\text{C}}}-\text{CH}_3 \xrightarrow{\text{AlCl}_3} \text{CpFe}(\text{C}_5\text{H}_4\text{COCH}_3) + \text{HCl}$$

a 1,2-addition of an organometallic compound across a double bond (alternatively, insertion of an olefin into a metal—X bond)

(17-4)
$$\text{Me}_3\text{SnH} + \text{H}_2\text{C}=\text{CHPh} \rightarrow \text{Me}_3\text{Sn}-\text{CH}_2\text{CH}_2\text{Ph}$$

an olefin or acetylene coupling with CO insertion

(17-5)
$$\text{Fe(CO)}_5 + 2\text{PhC}\equiv\text{CH} \rightarrow \text{(Ph}_2\text{C}_4\text{H}_2\text{CO)Fe(CO)}_3$$

a redistribution reaction

(17-6)
$$\text{SnMe}_4 + \text{SnCl}_4 \rightleftharpoons \text{Me}_3\text{SnCl} + \text{MeSnCl}_3$$

It is these types of reactions, and others, that are described in more detail in this chapter. Our focus is generally on their mechanisms and, where possible, on their usefulness in organic synthesis or in actual or potential industrial processes.

Before beginning this discussion, you should realize that no attempt is made to be exhaustive in this coverage of organometallic reactivity. Rather, we plan to examine in detail only a few examples of each major reaction type. Broader and more complete coverage is found in review articles[1] and monographs.[2-15] Finally, you

---

[1] *Advances in Organometallic Chemistry,* F. G. A. Stone and R. West, eds., Academic Press, New York. Issued approximately annually.

[2] G. E. Coates, M. L. H. Green, and K. Wade, "Organometallic Compounds," Methuen, London. Vol. 1, "The Main Group Elements," G. E. Coates and K. Wade (1967); Vol. 2, "The Transition Elements," M. L. H. Green (1968).

[3] B. J. Wakefield, "The Chemistry of Organolithium Compounds," Pergamon Press, New York (1974).

[4] T. Mole and E. A. Jeffery, "Organoaluminum Compounds," Elsevier, New York (1972).

[5] U. Belluco, "Organometallic and Coordination Chemistry of Platinum," Academic Press, New York (1974).

[6] F. R. Hartley, "The Chemistry of Platinum and Palladium," John Wiley and Sons, New York (1973).

[7] J. P. Candlin, K. A. Taylor, and D. T. Thompson, "Reactions of Transition Metal Complexes," Elsevier, New York (1968).

[8] R. F. Heck, "Organotransition Metal Chemistry," Academic Press, New York (1974).

[9] R. C. Poller, "The Chemistry of Organotin Compounds," Academic Press, New York (1970).

[10] P. M. Maitlis, "The Organic Chemistry of Palladium," Vols. I and II, Academic Press, New York (1971).

[11] D. S. Matteson, "Organometallic Reaction Mechanisms of the Nontransition Elements," Academic Press, New York (1974).

[12] P. W. Jolly and G. Wilke, "The Organic Chemistry of Nickel," Academic Press, New York (1974).

[13] P. C. Wailes, R. S. P. Coutts, and H. Weigold, "Organometallic Chemistry of Titanium, Zirconium, and Hafnium," Academic Press, New York (1974).

[14] B. R. James, "Homogeneous Hydrogenation," John Wiley and Sons, New York (1973).

[15] M. Rosenblum, "Chemistry of the Iron Group Metallocenes," John Wiley and Sons, New York (1965).

should also recognize that the type of reaction may not always be clearly defined, and its assignment to a particular category within this chapter may be somewhat arbitrary.

THE 16 AND 18 ELECTRON RULE AND REACTIONS OF TRANSITION METAL ORGANOMETALLIC COMPOUNDS

First, we must discuss again one of the more important organizing devices in inorganic chemistry: the 16 and 18 electron rule. You will recall from Chapter 15 that this rule states that thermodynamically or kinetically stable, diamagnetic organometallic compounds of the transition metals have a total of 16 or 18 valence electrons. In this chapter we add a corollary to this rule: *Reactions of transition metal organometallic compounds proceed by elementary steps that involve only 16 or 18 valence electron species as intermediates.* As an example, consider reaction 13-11 (p. 704). The 16 electron platinum(II) compound passes through an intermediate having 18 electrons, which in turn leads to a 16 electron product.

$$\begin{array}{c}\text{Et}_3\text{P}\diagdown\;\;\diagup\text{Cl}\\ \text{Pt}\\ \text{R}\diagup\;\;\diagdown\text{PEt}_3\end{array} + \begin{array}{c}\diagup\!\!\!\diagdown\\ \diagdown\!\!\!\!\diagup\\ \text{N}\end{array} \rightarrow \left\{\begin{array}{c}\text{PEt}_3\\ \text{py}\\ \text{R}-\text{Pt}\\ \text{Et}_3\text{P}\;\;\;\text{Cl}\end{array}\right\} \rightarrow \left[\begin{array}{c}\text{Et}_3\text{P}\diagdown\;\;\diagup\text{py}\\ \text{Pt}\\ \text{R}\diagup\;\;\diagdown\text{PEt}_3\end{array}\right]^+ + \text{Cl}^- \qquad (17\text{--}7)$$

Tolman has expanded this basic notion into a more general scheme for organizing the reactions of organic compounds of the transition metals.[16] He has demonstrated that the reactions of such compounds can be broken down into five elementary reaction types, each with its microscopic reverse. Examples are given in Table 17-1. The first type is an acid association/dissociation reaction, and it can occur without change in the total number of valence electrons surrounding the metal. Furthermore, since a Lewis acid does not bring or carry away a pair of electrons in a reaction, acid association or dissociation can occur with either 16 or 18 electron molecules. However, Lewis base dissociation, the second reaction type, is restricted to 18 electron molecules, because the base carries away two electrons to give a 16 electron species. Conversely, Lewis base association can occur only with a 16 electron molecule. Reductive elimination, insertion, and oxidative coupling are restricted to 18 electron molecules because all three reactions lead to a loss of two valence electrons by the metal.[17] The reverse of each of these processes—oxidative addition, deinsertion, and reductive coupling—can occur, therefore, only with 16 electron molecules, because all lead to an increase in the number of valence electrons.

Because this chapter is meant to cover the reactions of both main group and transition metal organometallic compounds, Tolman's scheme is not the most convenient for organizing the chapter. Rather, we have put the chemistry of organometallic compounds as a whole into a somewhat more general framework, which only broadly resembles Tolman's scheme. Thus, the first two sections, on association reactions and on substitution, contain examples of the first two reaction types in Table 17-1. The third section, on addition/elimination reactions, encompasses both inser-

---

[16] C. A. Tolman, *Chem. Soc. Rev.*, **1**, 337 (1972).
[17] To say that Lewis base dissociation and reductive elimination, insertion, and oxidative coupling are restricted to 18-electron molecules may be too dogmatic. There may be exceptions. See, for example, S. Komiya, P. A. Albright, R. Hoffmann, and J. K. Kochi, *J. Amer. Chem. Soc.*, **98**, 7255 (1976).

TABLE 17-1
CLASSIFICATION OF REACTIONS OF TRANSITION METAL ORGANOMETALLIC COMPOUNDS[a]

Reaction	ΔNVE	ΔOS	ΔN	Example	Reverse Reaction	ΔNVE	ΔOS	ΔN
Lewis acid dissociation	0	−2	−1	$HCo(CO)_4 \rightleftarrows H^+ + Co(CO)_4^-$	Lewis acid association	0	+2	+1
ligand dissociation	0	0	−1	$Cp_2WH_2 \cdot BF_3 \rightleftarrows BF_3 + Cp_2WH_2$		0	0	+1
Lewis base dissociation	−2	0	−1	$Pt(PPh_3)_4 \rightleftarrows PPh_3 + Pt(PPh_3)_3$	Lewis base association	+2	0	+1
Reductive elimination	−2	−2	−2	$H_2IrCl(CO)L_2 \rightleftarrows H_2 + IrCl(CO)L_2$	Oxidative addition	+2	+2	+2
Insertion	−2	0	−1	$MeMn(CO)_5 \rightleftarrows Me\text{—}\overset{O}{\overset{\|}{C}}\text{—}Mn(CO)_4$	Deinsertion	+2	0	+1
Oxidative coupling	−2	+2	0	$\overset{\diagup}{\underset{\diagup}{\diagdown}}Fe(CO)_3 \rightleftarrows \bigcirc Fe(CO)_3$	Reductive decoupling	+2	−2	0

[a] NVE = number of valence electrons, OS = oxidation state, and N = coordination number.

tion/deinsertion and oxidative addition/reductive elimination reactions. Following a section on rearrangements, the final section deals with catalysis, a reaction type especially characteristic of transition metals and a type wherein the 16 and 18 electron rule will find its greatest usefulness.

ASSOCIATION REACTIONS

THE LEWIS ACIDITY AND BASICITY OF ORGANOMETALLIC COMPOUNDS

The interaction of Lewis acids and bases pervades chemistry, so much so that the principles of such reactions were discussed separately in Chapter 5. There you were introduced to those types of interactions typical of organometallic compounds. In this chapter the consequences of the acidity and basicity of such compounds are explored more fully.

One important aspect of the chemistry of organometallic compounds of Groups IA, IIA, and IIIA to some extent (aluminum in particular) is the tendency to self-association in order to reduce the metal electron deficiency (Chapter 16, p. 844). This objective also may be achieved by accepting electrons from a normal Lewis base, a reaction that most frequently leads to dissociation of the organometallic compound as in reaction 17-8 (see also Grignard reagents in Chapter 16, p. 850).

(17-8) $$2Me_3N + Me_6Al_2 \rightarrow 2Me_3N \cdot AlMe_3$$

Alkyllithium compounds, however, may associate with ethers, for example, without dissociation of the tetrameric unit; for instance, $(n\text{-butyllithium})_4$ accepts up to four molecules of Me_2O at $-25°C$. If it is assumed that each lithium atom uses one sp^3 hybrid orbital in each face of the RLi tetramer to form a four-center, two-electron bond with an alkyl group (see Figure 16-8), then one empty, radially directed orbital remains on each lithium, and base complexation is possible.[18]

---

[18] P. D. Bartlett, C. V. Goebel, and W. P. Weber, *J. Amer. Chem. Soc.*, **91**, 7425 (1969).

Complexes of the organometallic derivatives of Group IIIA metals have been extensively studied because all of the metals form isolable, characterizable complexes with a great variety of bases.[19-21] Reaction 17-9 is especially important, because the structures of the reactants (Me$_3$N and monomeric Me$_3$Al) and the product have all been determined in the gas phase.[22] As you might expect, the planar Me$_3$Al monomer (D_{3h}) rearranges to a pyramidal structure, and the amine groups fold back slightly; in addition, the N—C and Al—C bonds are all lengthened. The overall molecular geometry of the complex is ethane-like with staggered methyl groups (C_{3v} skeletal symmetry). It is interesting to note that the Al—N bond is longer in this complex than in Me$_3$N · AlCl$_3$ (1.96 Å); the electronegative chlorines obviously lead to bond strengthening as compared with the methyl derivative.

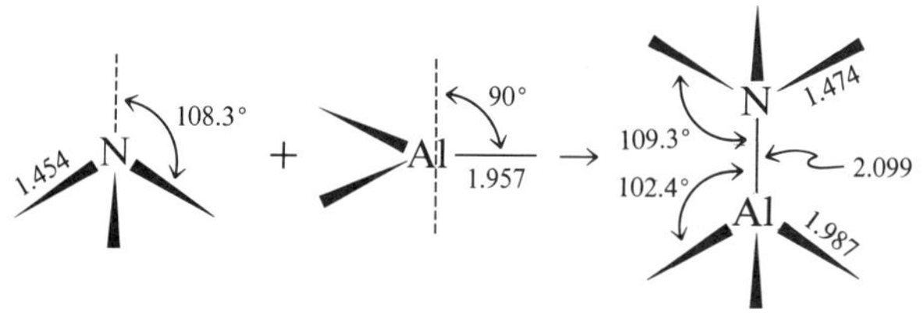

$$\text{(structural diagram)} \tag{17-9}$$

Trimethylamine-trimethylaluminum is thermodynamically the most stable of the Me$_3$N complexes of Group IIIA trimethyl derivatives. The heats of dissociation of the Me$_3$N · MMe$_3$ series of complexes (Table 17-2) show that the order of thermodynamic stability—and the order of Lewis acid strength—is B < Al > Ga > In > Tl (cf. Chapter 5, p. 229ff.). Many complexes analogous to these are isolated when other tertiary amines or phosphines, ethers, or sulfides are used. Either direct addition or the exchange of a stronger base for a weaker one may be used for their synthesis. Indeed, base exchange experiments involving trimethylaluminum, for example

$$\text{Me}_3\text{P} \cdot \text{AlMe}_3 + \text{Me}_3\text{N} \rightarrow \text{Me}_3\text{N} \cdot \text{AlMe}_3 + \text{Me}_3\text{P} \tag{17-10}$$

---

[19] *Boron:* T. D. Coyle and F. G. A. Stone, in "Progress in Boron Chemistry," Vol. 1, H. Steinberg and A. L. McCloskey, eds., Macmillan, New York (1964).
[20] *Aluminum:* R. Köster and P. Binger, *Adv. Inorg. Chem. Radiochem.,* **7,** 263 (1965).
[21] *Gallium, Indium:* K. Yasuda and R. Okawara, *Organometal. Chem. Rev.,* **2,** 255 (1967).
[22] G. A. Anderson, F. R. Forgaard, and A. Haaland, *Chem. Commun.,* 480 (1971); M. Hargittai, I. Hargittai, and V. P. Spiridonov, *ibid.,* 750 (1973).

TABLE 17-2
GAS PHASE DISSOCIATION DATA FOR COMPOUNDS OF THE TYPE Me$_3$N · MMe$_3$[a]

Compound	ΔH_{diss} (kcal/mol)	K_{diss} (atm, 100°C)
Me$_3$N · BMe$_3$[b]	18	0.472
Me$_3$N · AlMe$_3$	30	—
Me$_3$N · GaMe$_3$	21	—
Me$_3$N · InMe$_3$	20	—
Me$_3$N · TlMe$_3$	—	appreciably dissociated at 0°

[a] Data from M. F. Lappert in "The Chemistry of Boron and its Compounds," E. L. Muetterties, ed., John Wiley, New York, 1967, p. 523.
[b] $\Delta S = 45.7$ cal/deg mol and $\Delta G = 0.56$ kcal/mol at 100°C.

indicate that the order of donor strength toward this acid is

$$Me_3N > Me_3P > Me_2O > Me_2S > Me_2Se > Me_2Te$$

That is, the basicity decreases on descending a group and on moving to the right in the Periodic Table, a characteristic of main group Lewis bases. As a point of reference, a very careful determination of the heat of formation of $Me_2S \cdot AlMe_3$ from monomeric constituents in the gas phase gave -18.05 ± 0.5 kcal/mol.[23]

If the Lewis base contains an acidic hydrogen atom—primary and secondary amines and phosphines or alcohols—a *condensation reaction*[24] may occur after the initial formation of the donor-acceptor complex, especially in the case of beryllium,[25] aluminum, and gallium compounds, but *not* with boron alkyls. The initial complex may or may not have sufficient kinetic stability to permit its isolation. The products of such condensation reactions are usually associated to dimers, trimers, tetramers, or higher polymers, the degree of association frequently depending on the steric requirements of the organic substituents. For example, 17-11 describes the tetrameric Al—N cubane-like molecule produced when the aniline has no *ortho* substituent;[26]

(17-11)

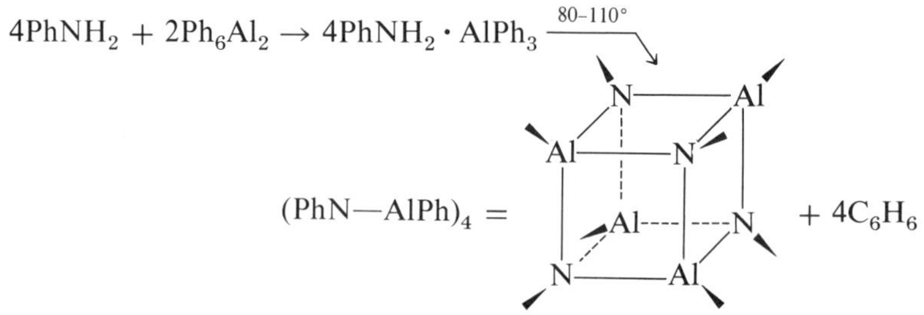

$$4PhNH_2 + 2Ph_6Al_2 \rightarrow 4PhNH_2 \cdot AlPh_3 \xrightarrow{80-110°}$$

$$(PhN-AlPh)_4 = \text{[cubane structure]} + 4C_6H_6$$

otherwise, only dimeric molecules of the type shown in reaction 17-12 are isolated. μ-Diphenylamino-μ-methyl-tetramethyldialuminum (**1**) is especially interesting because it is the first confirmed example of a dibridged molecule having only one electron-deficient type bridge.[27] More importantly for present purposes, **1** illustrates the difference between a four-electron bridge of the type Al—N→Al and a two-electron M—C—M bridge. Bridges of the former type generally have bridge angles considerably greater than those for M—C—M.

(17-12)[28]

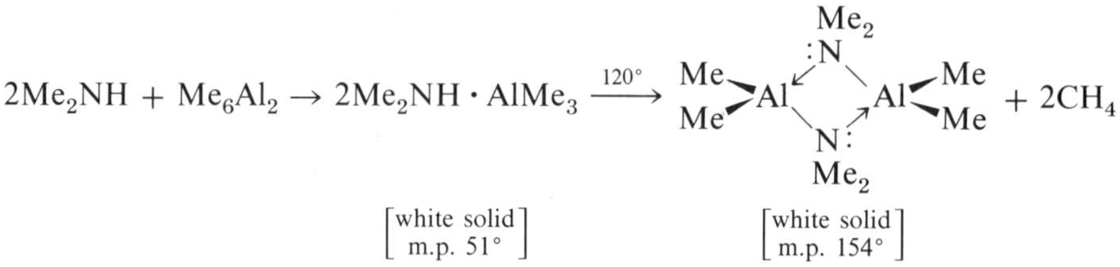

$$2Me_2NH + Me_6Al_2 \rightarrow 2Me_2NH \cdot AlMe_3 \xrightarrow{120°} \text{[dimer structure]} + 2CH_4$$

$$\begin{bmatrix} \text{white solid} \\ \text{m.p. } 51° \end{bmatrix} \qquad \begin{bmatrix} \text{white solid} \\ \text{m.p. } 154° \end{bmatrix}$$

(17-13)

$$Ph_2NH + Me_6Al_2 \xrightarrow[-CH_4]{100°} \text{[dimer structure with angles 85.8° and 77.4°]}$$

$$\begin{bmatrix} \text{white solid} \\ \text{sublimes } in \ vacuo \end{bmatrix}$$

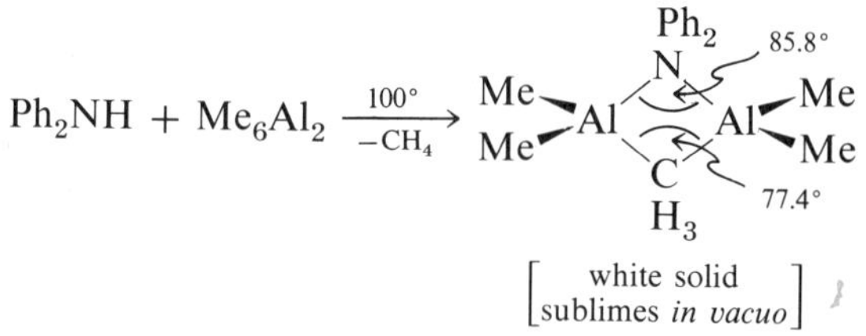

1

---

[23] C. H. Henrickson and D. P. Eyman, *Inorg. Chem.*, **6**, 1461 (1967).
[24] *Cf.* discussion of solvolysis reactions in Chapter 8.
[25] N. R. Fetter, *Organometal. Chem. Rev.*, **3**, 1 (1968).
[26] J. I. Jones and W. S. McDonald, *Proc. Chem. Soc.*, 366 (1966); J. I. Jones, *Chem. Ind.*, 159 (1964).
[27] V. R. Magnuson and G. D. Stucky, *J. Amer. Chem. Soc.*, **90**, 3269 (1968); **91**, 2544 (1969).
[28] N. Davidson and H. C. Brown, *J. Amer. Chem. Soc.*, **64**, 316 (1942).

Finally, the reaction of trimethylaluminum with methanol (17–14) illustrates the reason that almost all alkyl and aryl derivatives of Group I–IIIA metals must be handled in a dry atmosphere.

$$3\text{MeOH} + \tfrac{3}{2}\text{Me}_6\text{Al}_2 \rightarrow \text{Me}_2\text{Al}\begin{array}{c}\text{Me}\quad\text{Me}_2\\ \text{O}\!-\!\text{Al}\\ \diagup\quad\diagdown\\ \diagdown\quad\diagup\\ \text{O}\!\rightarrow\!\text{Al}\\ \text{Me}\quad\text{Me}_2\end{array}\text{OMe} + 3\text{CH}_4 \qquad (17\text{–}14)$$

Tetraorgano derivatives of Group IVA metals, on the other hand, are stable to water and do not form ordinary coordination complexes. However, replacement of one organic group by a halogen on Me_4Sn, for example, increases the metal Z^*, and complexes such as $\text{Me}_3\text{SnCl}\cdot\text{pyridine}$ may be isolated (cf. reaction 17–1). Indeed, the Lewis acid properties of Me_3SnCl have been thoroughly examined;[29] it is found to react more strongly with oxygen and nitrogen donors than with third row donors, the usual behavior for main group Lewis acids.

That main group metal compounds function as acids is easily accepted, and you are accustomed to thinking of transition metals as acids. But then, one is used to considering only transition metal compounds in which the metal is in a formal oxidation state of 2+ or higher. In the low oxidation states characteristic of organometallic compounds, however, the metal may function as a site of Lewis basicity in a reaction that is the first of Tolman's five elementary reaction types. For example, in reaction 17–15 the tungsten has a share in 18 valence electrons and donates a pair of those to a Lewis acid such as H^+, BF_3, or AlMe_3.

$$\text{Cp}_2\text{WH}_2 + \text{A} \rightarrow \text{Cp}_2\text{WH}_2\!\rightarrow\!\text{A} \qquad (17\text{–}15)$$

NVE = 18   A = H⁺, BF₃, AlMe₃   NVE = 18
OS = 4+   OS = 4+ for BF₃, AlMe₃
2   OS = 6+ for H⁺

(Recall that the bonding in such compounds as **2** was discussed in Chapter 16, p. 887, and metal basicity was mentioned in Chapter 5, p. 213). There are currently many other examples of metal basicity,[30,31] and such compounds are of great interest because of possible importance in metal-catalyzed reactions. In particular, Vaska's compound—carbonylchlorobis(triphenylphosphine)iridium—and its rhodium analog react with boron trihalides to form 1:1 complexes with a M → B donor-acceptor linkage.[32,33]

$$\underset{\substack{\text{NVE}=16\\ \text{OS}=1+}}{\text{Ir(CO)(PPh}_3)_2\text{Cl}} + \text{BF}_3 \rightarrow \underset{\substack{\text{NVE}=16\\ \text{OS}=1+}}{\text{Ir(CO)(PPh}_3)_2\text{Cl}\cdot\text{BF}_3} \qquad (17\text{–}16)$$

---

[29] T. F. Bolles and R. S. Drago, J. Amer. Chem. Soc., **88**, 3921, 5730 (1966).
[30] J. Kotz and D. G. Pedrotty, Organometal. Chem. Rev. A, **4**, 479 (1969).
[31] D. F. Shriver, Acc. Chem. Res., **3**, 321 (1970).
[32] P. Powell and H. Nöth, Chem. Commun., 637 (1966).
[33] R. N. Scott, D. F. Shriver, and L. Vaska, J. Amer. Chem. Soc., **90**, 1079 (1968).

Most importantly, however, the rhodium compound reacts only with BCl_3 and BBr_3 but not with BF_3, whereas the iridium compound reacts with BF_3; since the order of acid strength is clearly $BF_3 < BCl_3 < BBr_3$ (see Chapter 5), this implies that the iridium compound is more basic than the rhodium derivative. Although there are exceptions, this same trend—an increase in basicity on descending a group—is observed for other metal sub-groups; the reason for the inversion in behavior when compared with main group bases is not yet clear.

Metal anions, in which the metal is formally in a -1 oxidation state, may also be considered bases (e.g., $[Co(CO)_4]^-$ in Table 17-1).[34,35] They react with the proton to form conjugate Brönsted acids such as $HMn(CO)_5$, $H_2Fe(CO)_4$, and $HCo(CO)_4$. The acid dissociation constants increase on proceeding across the Periodic Table; thus, $HCo(CO)_4$ is a strong acid, whereas the first dissociation constant for $H_2Fe(CO)_4$ is approximately equal to that of acetic acid.

Metal-containing anions also function as nucleophiles in displacement reactions of several types. We discussed in Chapter 16 the use of such reactions as a method of synthesis of metal-carbon bonds (p. 830).

(17-17)
$$CH_3\overset{O}{\overset{\|}{C}}-Cl + [Fe(CO)_2(\eta^5-C_5H_5)]^- \rightarrow Cl^- + CH_3\overset{O}{\overset{\|}{C}}-Fe(CO)_2(\eta^5-C_5H_5)$$

Indeed, the relative nucleophilicities of a number of metal-containing anions have been measured by determining their rates of reaction with an alkyl halide (Table 17-3). On the basis of these data and experiments with other acids, some general statements regarding metal basicity or nucleophilicity can be made:[36,37]

1. With the exception of some compounds of the iron sub-group, metal basicity and nucleophilicity generally increase on descending a given group. This is clearly opposite the usual trend noted in main group chemistry.

2. CO is one of the strongest known π acceptor ligands. Therefore, replacement of CO by less capable π acids (such as PPh_3 and η^5-C_5H_5) increases the relative nucleophilicity of the metal.

---

[34] See ref. 30.
[35] R. B. King, *Adv. Organometal. Chem.*, **2**, 157 (1964)
[36] See ref. 31.
[37] R. B. King, *Acc. Chem. Res.*, **3**, 417 (1970).

TABLE 17-3
RELATIVE NUCLEOPHILICITIES OF SOME METAL CARBONYL ANIONS[a,b]

Anion	Relative Nucleophilicity
$(C_5H_5)Fe(CO)_2^-$	70,000,000
$(C_5H_5)Ru(CO)_2^-$	7,500,000
$(C_5H_5)Ni(CO)^-$	5,500,000
$Re(CO)_5^-$	25,000
$(C_5H_5)W(CO)_3^-$	500
$Mn(CO)_5^-$	77
$(C_5H_5)Mo(CO)_3^-$	67
$(C_5H_5)Cr(CO)_3^-$	4
$Co(CO)_4^-$	1

[a] The data are taken from R. E. Dessy, R. L. Pohl, and R. B. King, *J. Amer. Chem. Soc.*, **88**, 5121 (1966).
[b] The nucleophilicity of $Co(CO)_4^-$ has arbitrarily been set at 1.

3. In contrast to main group bases (see reaction 17–9), metal bases frequently undergo large geometrical changes upon reaction with an acid.[38]

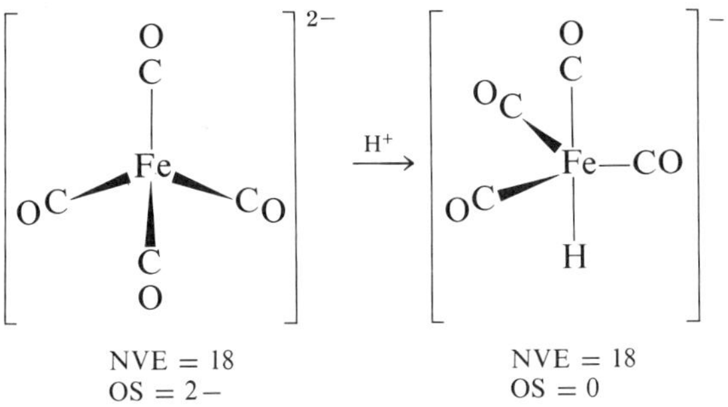

$$\left[\begin{array}{c} \text{OC} \\ \text{OC}-\text{Fe}-\text{CO} \\ \text{CO} \end{array}\right]^{2-} \xrightarrow{H^+} \left[\begin{array}{c} \text{OC} \\ \text{OC}-\text{Fe}-\text{CO} \\ \text{H} \end{array}\right]^{-} \qquad (17\text{–}18)$$

NVE = 18 NVE = 18
OS = 2− OS = 0

4. Main group anions of the type Ph_3M^- (M = Sn, Ge) are *more* nucleophilic than those containing transition metals. This is not unexpected, since the negative charge of the Group IVA anions is not stabilized as well as in a transition metal anion with π acceptor ligands.

In addition to reacting with BX_3 and H^+, $(\eta^5\text{-}C_5H_5)_2Fe_2(CO)_4$ and other organometal carbonyls form complexes with triorganoaluminum compounds.[39,40] However, instead of attacking the metal, aluminum alkyls bind to the oxygen of the CO ligand as in $(\eta^5\text{-}C_5H_5)_2Fe_2(CO)_4 \cdot 2AlEt_3$ **(3)**.

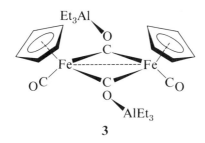

3

Although CO functions as a Lewis base toward BH_3 and transition metals through the carbon atom, the iron-containing Et_3Al adduct **(3)** represents one of the first observations that the carbon monoxide molecule could similarly utilize the oxygen lone pair, albeit only when the carbon is bound to a metal.

The study of transition metal basicity and nucleophilicity is very important to an understanding of certain metal-catalyzed reactions. The catalytic behavior of Vaska's compound may be understood conceptually in terms of acid-base properties,[41] and there is evidence that metal or CO basicity may be important in reactions co-catalyzed by BF_3 or $AlCl_3$.[42]

---

[38] This does not necessarily imply a large reorganization energy. Since a metal CO π bonding pair is converted to a metal-H^+ σ bonding pair, the reorganization energy is in fact small. (Contrast this with a main group base, where a lone pair is converted to a bond pair.) See Chapter 5, page 213.
[39] J. Kotz and C. Turnipseed, *Chem. Commun.*, 41 (1970).
[40] D. F. Shriver and A. Alich, *Coord. Chem. Rev.*, **8**, 15 (1972).
[41] L. Vaska, *Acc. Chem. Res.*, **1**, 335 (1968).
[42] G. N. Schrauzer, B. N. Bastian, and G. A. Franklin, *J. Amer. Chem. Soc.*, **88**, 4890 (1966).

LIGAND PROTONATION

Treatment of a transition metal organometallic compound with strong protonic acids can result in ligand protonation as well as metal protonation, but the former can give rise to subsequent ligand rearrangement as in 17-19.[43,44]

(17-19)

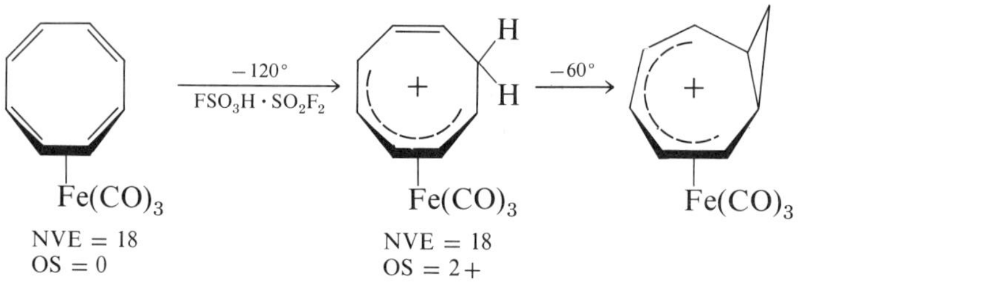

The mechanism and stereochemistry of ligand protonation, an understanding of which is important to the later discussion of electrophilic substitutions on ligands, have been the subjects of several studies.[45]

The mechanism of ligand protonation that seems best to explain the observed results comes from a study of the behavior of bicyclo[6.1.0]nonatrienetricarbonyl-molybdenum (17-20) in $HSO_3F\text{-}SO_2F_2$ at low temperatures.[46] The ligand changes from a 3π pair donor to a 2π pair/1σ pair donor upon protonation of the metal.

(17-20)

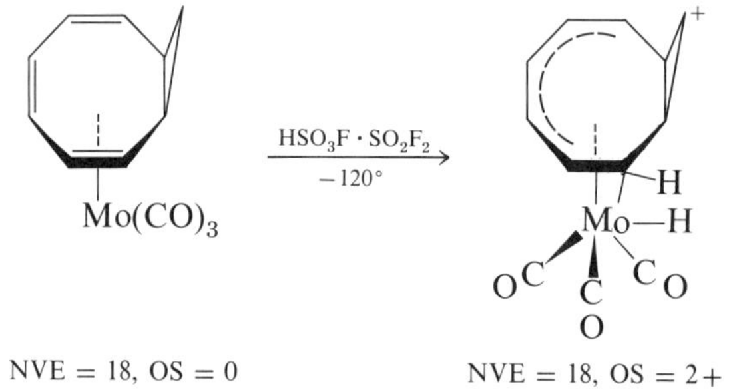

A structure similar to **4** can be postulated as an intermediate to explain the fact that butadienetricarbonyliron (**5**) eventually gives **7b** upon reaction with anhydrous HCl.

(17-21)

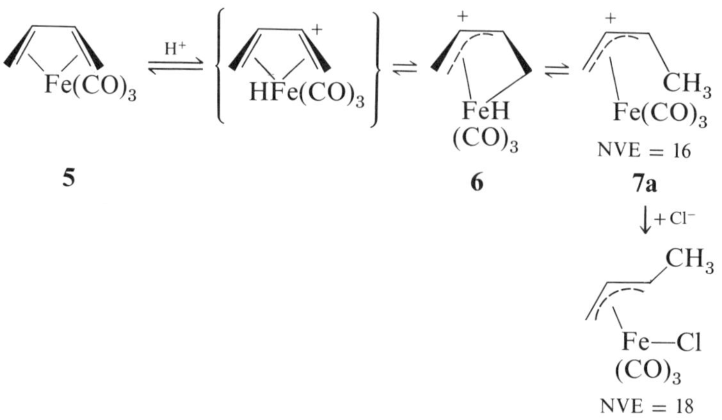

---

[43] M. A. Haas, *Organometal. Chem. Rev.*, **4**, 307 (1969).
[44] M. Brookhart, E. R. Davis, and D. L. Harris, *J. Amer. Chem. Soc.*, **94**, 7853 (1972).
[45] M. Brookhart, T. H. Whitesides, and J. M. Crockett, *Inorg. Chem.*, **15**, 1550 (1976), and references therein.
[46] M. Brookhart and D. L. Harris, *Inorg. Chem.*, **13**, 1541 (1974).

Furthermore, the equilibrium between **6** and **7a** rationalizes the fact that the added proton and the ligand terminal CH$_3$ protons in **7** scramble with solvent protons when butadienetricarbonyliron is dissolved in HSO$_3$F/SO$_2$.

The reversible deuteration of tricarbonyl-η^4-1,3-cyclohexadieneiron (17-22) suggests that ligand protonation is stereospecific. That is, reaction with deuterated trifluoroacetic acid shows that only the *endo* protons (those on the same side of the C$_6$ ring as the metal) exchange for deuterium; the *endo,endo*-dideuterodiene is obtained as the final product of complete exchange. A mechanism similar to that in reaction 17-21 can readily account for this observation.

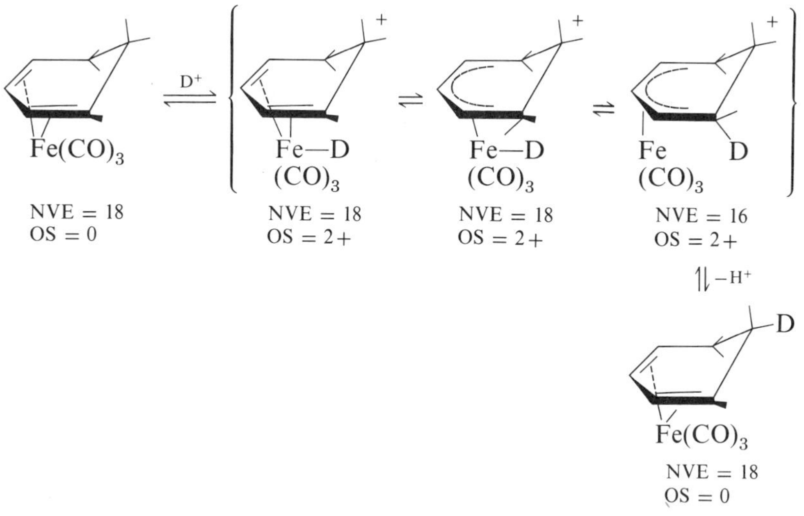

(17-22)

---

1. On page 911 it is noted that the Al—N bond in Me$_3$N · AlCl$_3$ is shorter than in the AlMe$_3$ analog. Account for this in terms of orbital hybridization arguments as in Chapter 3.

2. Di-*iso*-propylberyllium reacts in a 1:2 ratio with dimethylamine to give [Be(NMe$_2$)$_2$]$_3$. One mole of amine per mole of R$_2$Be is taken up in the first step, carried out at 25°, and a second mole of amine is taken up in a second step at 40°.

    a) Write balanced equations to depict the course of these two steps.
    b) A possible structure of the final product is a cyclic trimer. Draw the structure of such a molecule.
    c) The trimer drawn in part (b) is not the actual structure, however. Rather, the observed structure involves one four-coordinate Be and two three-coordinate Be's. Draw a possible structure based on three- and four-coordinate Be atoms. [See J. L. Atwood and C. D. Stucky, *Chem. Commun.*, 1169 (1967).]

3. In reaction 17-16, would the stretching frequency of the CO ligand be expected to increase or decrease upon adduct formation? Does this reaction fit Tolman's 16/18 electron rule? Calculate ΔNVE, ΔOS, and ΔN for this reaction.

4. The iron atom in Fe(CO)$_5$ is a weak Lewis base. However, replacement of a CO ligand with a phosphine to give, for example, Fe(CO)$_4$PPh$_3$ causes the metal basicity to be enhanced. Why should this be the case? (If you need help, refer back to Chapter 5, page 214.)

5. In reaction 17-22, show that only two, non-olefinic hydrogens can be exchanged for deuterium.

STUDY QUESTIONS

6. Propose a mechanism for

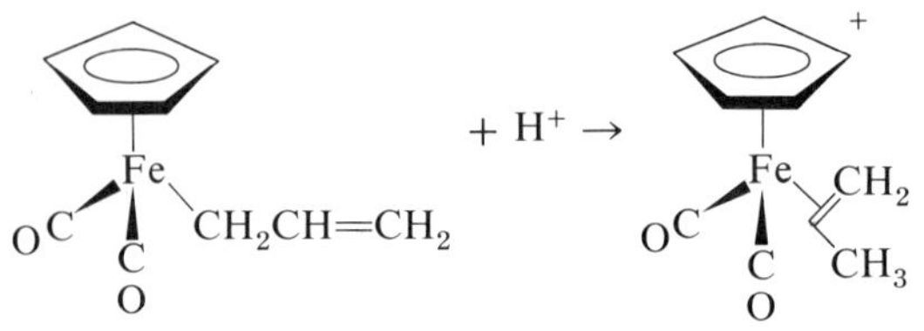

$$[\text{CpFe(CO)}_2\text{CH}_2\text{CH}=\text{CH}_2] + \text{H}^+ \rightarrow [\text{CpFe(CO)}_2(\text{CH}_2=\text{CHCH}_3)]^+$$

Show ΔNVE and ΔOS for each step.

7. Turn back to Figures 16-1 and 16-2 and identify other reactions in the following catagories: ligand substitution, attack on a coordinated ligand, 1,2-addition, acid-base association.

8. Place the steps of each of the following reactions into one of the categories in Table 17-1 and calculate ΔNVE, ΔOS, and ΔN for each step.

a) 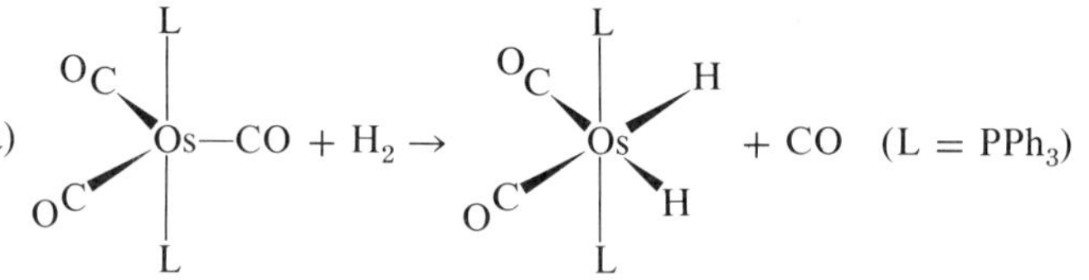 $\text{(OC)}_3\text{L}_2\text{Os-CO} + \text{H}_2 \rightarrow \text{(OC)}_2\text{L}_2\text{Os(H)}_2 + \text{CO}$ (L = PPh$_3$)

b) 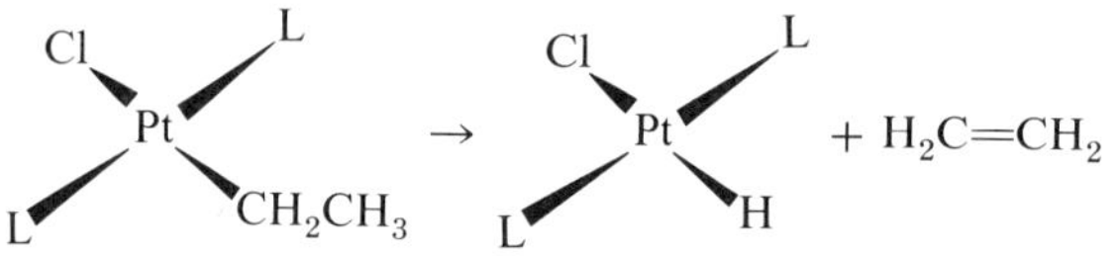 $trans\text{-PtCl(CH}_2\text{CH}_3)\text{L}_2 \rightarrow trans\text{-PtClHL}_2 + \text{H}_2\text{C}=\text{CH}_2$

c) $\text{CpRh(CO)L} \xrightarrow{\text{MeI}} [\text{CpRh(CO)(L)(Me)}]^+ \text{I}^-$

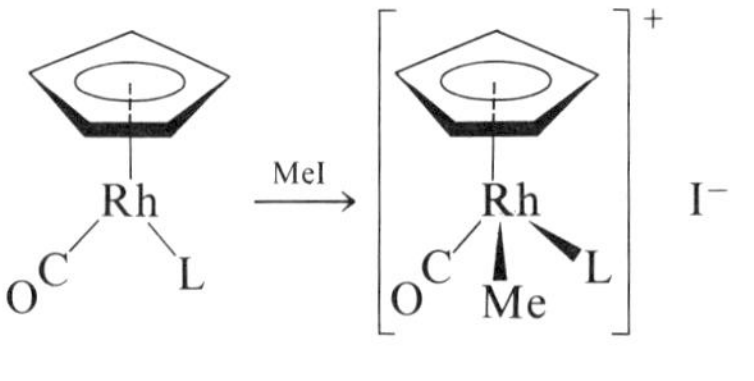

d) 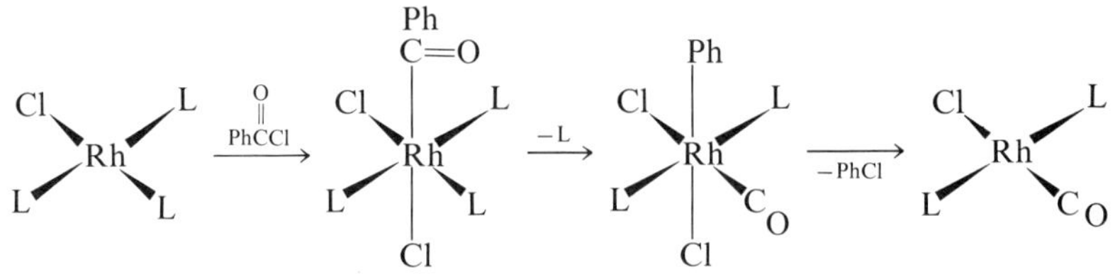 $\text{RhClL}_3 \xrightarrow{\text{PhCOCl}} \text{RhCl}_2\text{L}_2(\text{COPh}) \xrightarrow{-L} \text{RhCl}_2\text{L}_2(\text{COPh}) \xrightarrow{-\text{PhCl}} \text{RhClL}_2(\text{CO})$

SUBSTITUTION REACTIONS[47,48]

NUCLEOPHILIC LIGAND SUBSTITUTIONS[49,50]

In discussing metal carbonyls in Chapter 16, it was pointed out that their importance lies in the fact that they are the starting place for the syntheses of so many

---

[47] See ref. 7.
[48] F. Basolo and R. G. Pearson, "Mechanisms of Inorganic Reactions," 2nd ed., John Wiley, New York (1967).
[49] R. J. Angelici, *Organometal. Chem. Rev. A*, **3**, 173 (1968).
[50] A. Z. Rubezhov and S. P. Gubin, *Adv. Organometal. Chem.*, **10**, 347 (1972).

other organometallic compounds. Therefore, the mechanism and stereochemistry of the nucleophilic substitution of CO by another ligand or ligands have been studied extensively, and some general principles have evolved that help in understanding these reactions and in predicting methods for the syntheses of new compounds.[50]

Many substitution reactions of metal carbonyls proceed according to a rate law that is first order in metal carbonyl and zero order in the substituting ligand; fewer substitutions are first order in both. This generalization was partly established by a very careful examination of the substitution of CO by isotopically labeled CO or phosphine in $Ni(CO)_4$ (17–23).[51]

$$2Ni(CO)_4 + 3L\ (= CO \text{ or } PPh_3) \rightarrow Ni(CO)_3L + Ni(CO)_2L_2 + 3CO \tag{17-23}$$

The rate of exchange is followed by observing the decrease in the intensity of ν_{CO} for $Ni(CO)_4$ (2046 cm^{-1}) and the appearance of bands for the two products. The products are formed in a stepwise manner, as indicated by the disappearance of the 2046 cm^{-1} band and the concomitant appearance of 2003 and 2071 cm^{-1} bands;

$$Ni(CO)_4 + L\ (= PPh_3) \rightarrow Ni(CO)_3L\ (\nu_{CO} = 2003, 2071 \text{ cm}^{-1}) + CO$$
$$Ni(CO)_3L + L\ (= PPh_3) \rightarrow Ni(CO)_2L_2\ (\nu_{CO} = 2005, 1950 \text{ cm}^{-1}) + CO \tag{17-24}$$

these latter bands then slowly decay in intensity and those of the disubstituted product increase. Most importantly, the absorbance of $Ni(CO)_4$ follows first order decay for more than 95 per cent of the reaction. Therefore,

$$\frac{-d[Ni(CO)_4]}{dt} = k[Ni(CO)_4]$$

This same technique was applied to the exchange of $C^{18}O$ with $Ni(C^{16}O)_4$. The rate of formation and subsequent exchange of $Ni(C^{18}O)(C^{16}O)_3$ are also found to follow a first order rate law. Finally, in an experiment very important to the synthetic chemist, PPh_3 (in a mixture of PPh_3 and $C^{18}O$) is observed to react five times more rapidly than CO with the presumed intermediate $Ni(CO)_3$, although the sum of the rates of formation of $Ni(C^{18}O)(C^{16}O)_3$ and $Ni(CO)_3(PPh_3)$ is equal to the rate of disappearance of $Ni(CO)_4$.

Based on the results outlined above (see Table 17-4 for rate constants and activation parameters) the following dissociative ($= D$) mechanism was proposed:

$$\begin{aligned} Ni(CO)_4 &\xrightarrow{k_{sub}} Ni(CO)_3 + CO \quad \text{(slow)} \\ Ni(CO)_3 + L &\rightarrow Ni(CO)_3L \quad \text{(fast)} \end{aligned} \tag{17-25}$$

The hypothesis that Ni—CO dissociation is rate determining is consistent with a $\Delta H^\ddagger$ of about 25 kcal/mole (the *mean* Ni—CO bond dissociation energy is 35 kcal/mol) and a positive $\Delta S^\ddagger$.

The kinetic behavior of ligand substitution reactions such as

$$Cr(CO)_6 + (n\text{-Bu})_3P \rightarrow [(n\text{-Bu})_3P]Cr(CO)_5 + CO \tag{17-26}$$

$$\text{(piperidine)(CO)}_4\text{Mo} + PPh_3 \rightarrow (PPh_3)(CO)_4\text{Mo} + \text{piperidine} \tag{17-27}$$

---

[51] J. P. Day, F. Basolo, and R. G. Pearson, *J. Amer. Chem. Soc.*, **90**, 6927 (1968).

TABLE 17-4
RATE CONSTANTS AND ACTIVATION PARAMETERS FOR REACTIONS OF SOME METAL CARBONYLS AND THEIR COMPLEXES[a]

Compound	Temp., °C	k_1, sec^{-1}	$\Delta H_1^{\ddagger}$, kcal/mole	$\Delta S_1^{\ddagger}$, eu	k_2, M^{-1}sec^{-1}	$\Delta H_2^{\ddagger}$, kcal/mole	$\Delta S_2^{\ddagger}$, eu
		$M(CO)_x + {}^*CO \rightarrow M(CO)_{x-1}({}^*CO) + CO$					
$Cr(CO)_6$	117	0.2×10^{-4}	38.7	18.5	—	—	—
$Mo(CO)_6$	116	0.75×10^{-4}	30.2	−0.4	—	—	—
$W(CO)_6$	142	0.026×10^{-4}	39.8	11	—	—	—
$Ni(CO)_4$	35	183×10^{-4}	22.1	7.7	—	—	—
		$M(CO)_x + R_3P \rightarrow M(CO)_{x-1}(R_3P) + CO$					
$Cr(CO)_6$[b]	131	1.38×10^{-4}	40.2	22.6	0.8×10^{-4}	25.5	−14.6
$Mo(CO)_6$[b]	112	2.1×10^{-4}	31.7	6.7	20.5×10^{-4}	21.7	−14.7
$W(CO)_6$[b]	165	1.1×10^{-4}	39.9	13.7	7.1×10^{-4}	29.2	−6.9
$Ni(CO)_4$[c]	35	203×10^{-4}	22.3	8.4	—	—	—
		$Mo(CO)_5(amine) + PPh_3 \rightarrow Mo(CO)_5(PPh_3) + amine$					
Amine =							
cyclohexylamine	35	9.9×10^{-5}	24.5	2.2	3.9×10^{-3}	15.9	−18.6
piperidine	35	1.8×10^{-5}	25.8	2.8	1.24×10^{-3}	16.5	−20.1
quinuclidine	35	0.9×10^{-5}	26.9	5.5	0.03×10^{-3}	13.2	−38

[a] Data taken from R. J. Angelici, *Organometal. Chem. Rev. A*, **3**, 173 (1968); J. P. Day, F. Basolo, and R. G. Pearson, *J. Amer. Chem. Soc.*, **90**, 6927 (1968); W. D. Covey and T. L. Brown, *Inorg. Chem.*, **12**, 2820 (1973).
[b] $R_3P = Bu_3P$
[c] $R_3P = PPh_3$

lies between that for $Ni(CO)_4/L$ and that usually observed for substitution reactions of octahedral complexes of higher valent metal ions (cf. Chapter 13).[52] That is, the rate law for 17-26 and 17-27 is given by the general expression

(17-28) $$\text{Rate of disappearance of starting material} = k_1[\text{substrate}] + k_2[\text{nucleophile}][\text{substrate}]$$

This rate law is reminiscent of those for substitution reactions of square planar complexes, but its interpretation is very different.[53-55] A very careful study of reaction 17-27 by Brown clearly shows that the rate law is consistent with the following overall mechanism.[53]

$$L + A-M \underset{k_{-1}}{\overset{k_1}{\rightleftharpoons}} [M + A] \overset{+L, k_3}{\rightarrow} M-L + A$$
$$L + A-M \overset{+L, k_2}{\rightarrow} [L-M-A] \rightarrow M-L + A$$

$L = R_3P$; A = amine; M = $Mo(CO)_5$ or $W(CO)_5$

That is, the upper pathway is dissociative [just as in the $Ni(CO)_4$ case] and leads to an $M(CO)_5$ intermediate. The steady state rate expression for this path is

---

[52] G. R. Dobson, *Acc. Chem. Res.*, **9**, 300 (1976).
[53] W. D. Covey and T. L. Brown, *Inorg. Chem.*, **12**, 2820 (1973).
[54] G. R. Dobson, *Inorg. Chem.*, **13**, 1790 (1974).
[55] For the most recent work in this area, see: J. Ewen and D. J. Darensbourg, *J. Amer. Chem. Soc.*, **97**, 6874 (1975); B. J. McKerley, G. C. Faber, and G. R. Dobson, *Inorg. Chem.*, **14**, 2275 (1975).

$-d[\text{A—M}]/dt = k_1 k_3 [\text{A—M}][\text{L}]/(k_{-1}[\text{A}] + k_3[\text{L}])$, which, when L is present in excess and [A] is about zero, reduces to the first term in the observed rate law.[56]

The lower pathway in the overall mechanism can be either an associative ($= A$) or an interchange ($= I$) process.[56] However, as there is no evidence for a seven-coordinate intermediate, the mechanism is classified as I. You will recall that this is exactly the conclusion to which you were led on the basis of a second order rate law for substitution reactions of octahedral complexes in Chapter 13. The only decision left, then, is whether the activation process in the interchange involves primarily bond making ($= I_a$) or bond breaking ($= I_d$).

There are two pieces of evidence that strongly suggest that the second order path in reaction 17–27 is an I_d process. (i) The second order rate constants are sensitive to the amine substituent, the reaction being most rapid with the least basic amine. (ii) These same rate constants are insensitive to the nature of the entering ligand.

You should notice that the conclusion that both pathways are dissociatively activated (D and I_d) is in agreement with Tolman's rules (Table 17–1) for reactions of transition metal organometallic compounds. That is, a dissociatively activated pathway will see the starting material change from an 18 electron molecule to a 16 electron species and then back to an 18 electron molecule. An associatively activated path would necessarily involve a 20 electron species, a most unlikely event based on the current state of understanding of organometallic reactions.

The activation parameters for reactions 17–26 and 17–27 give you further insight into these processes. Firstly, the values of $\Delta H_1^\ddagger$ for reaction 17–26 are in the order Mo < Cr ≃ W. Since this is the activation enthalpy for the dissociative path, it should reflect the strength of the M—C bond being broken. Indeed, the M—C stretching force constants, another measure of metal-carbon bond strength, are in the same order as the activation enthalpies. This trend in activation enthalpies also reflects an important experimental fact: *it is commonly observed in Group VIB metal carbonyl chemistry that molybdenum derivatives are most readily prepared.*

Secondly, $\Delta H_2^\ddagger$ for reactions 17–26 and 17–27 (see Table 17–4) is considerably less than $\Delta H_1^\ddagger$. This apparently implies that, even though the second order path is a dissociatively activated interchange, there must be sufficient involvement of the entering ligand in the transition state so that the activation parameters are altered. In other words, although bond breaking is more important than bond making, the latter is still significant.

On the basis of the study of reaction 17–27, Brown concluded that all instances of substitution at metal carbonyl centers in which there is a distribution of 18 electrons probably occur by dissociative pathways, either D or I_d.[57] However, he noted that this rule may be circumvented when a ligand rearrangement can occur to give a 16 electron species; the substitution process can then be associatively activated. In fact, reaction 17–29 may be an example of such an exception.[58]

(17-29)[59]

NVE = 18   NVE = 16   NVE = 18

---

[56] Refer to Chapter 7 for a complete discussion of rate laws and the various mechanisms possible for a chemical reaction.

[57] See ref. 53.

[58] H. G. Schuster-Woldan and F. Basolo, *J. Amer. Chem. Soc.*, **88**, 1657 (1966).

[59] If the Co or Ir analogs of the Rh compound are used, the reaction rates are in the order Rh ≫ Ir > Co. Notice that, as in reaction 17–26, complexes of metals in the second transition series are most readily substituted.

This reaction is observed to be first order in both the rhodium complex and the entering ligand, the entropy of activation is negative, and, most importantly, the rate depends on the nature of the entering phosphine. Thus, one might conclude that an A or I_a mechanism is plausible for reaction 17-29. (However, an A mechanism cannot be proved, as no intermediate was observed.) To rationalize an associatively activated process, it is proposed that the approach of the nucleophile causes the metal atom to move to one side of the C_5 ring and a pair of electrons to concentrate on a carbon of the cyclopentadienyl ring, thereby increasing the metal Z^* and opening a metal coordination site to the entering ligand.

Although numerous examples of the exchange of π bonded ligands themselves have been observed (e.g., reaction 17-30),

(17-30)[60]

their mechanisms have rarely been examined.[61] One example that seems to provide some insight into the general principles of ligand substitution is reaction 17-31.[62]

(17-31)

The rate of this reaction is first order in both the arene complex and the entering phosphine, and the reaction rates decrease in the order arene = toluene $\simeq$ p-xylene $\gg$ mesitylene and L = $(n\text{-Bu})_3\text{P} \gg \text{PCl}_2\text{Ph} > \text{PCl}_3$. The fact that the rate is greatest with the most nucleophilic phosphine [i.e., $(n\text{-Bu})_3\text{P}$] and with the least sterically demanding arene (i.e., toluene) implies an associatively activated mechanism very similar to that for reaction 17-29. That is, as the phosphine approaches, the arene becomes only a 2π pair donor, and then a 1π pair donor with the approach of the second phosphine. Such a mechanism preserves the integrity of molybdenum as an 18 electron metal center. Similar mechanisms have been proposed for the replacement of one olefin by another.[63]

Associative mechanisms are also the preferred pathway for olefin exchange in four-coordinate, 16 electron molecules. Expansion of the coordination sphere is readily accomplished and rapid substitution results. For example, whereas Zeise's salt, $[\text{PtCl}_3(\text{C}_2\text{H}_4)]^-$, completely exchanges olefin in 1 minute at 0°C, no olefin exchange is observed for the 18 electron molecule $(\eta^5\text{-C}_5\text{H}_5)\text{Rh}(\text{C}_2\text{H}_4)_2$ after six days at room temperature.

Many other examples of interesting ligand substitution have been studied,[64] and a few examples were mentioned as methods for the synthesis of olefin complexes in Chapter 16.

---

[60] C. C. Lee, R. G. Sutherland, and B. J. Thomson, *Chem. Commun.*, 907 (1972).
[61] For a recent study of such a mechanism, see M. E. Switzer and M. F. Rettig, *Inorg. Chem.*, **13**, 1975 (1974).
[62] F. Zingales, A. Chiesa, and F. Basolo, *J. Amer. Chem. Soc.*, **88**, 2707 (1966).
[63] See ref. 50.
[64] See refs. 49 and 50.

ELECTROPHILIC AND NUCLEOPHILIC ATTACK ON COORDINATED LIGANDS[65,66]

In the previous section, the substitution of one ligand for another was discussed. In this part, we examine briefly substitutions on the ligands themselves, reactions that most frequently occur by electrophilic mechanisms.

The chemistry of ferrocene and its derivatives has been studied more extensively than that of perhaps any other organometallic compound.[67] The cyclopentadienyl rings are clearly aromatic, and much of the chemistry of ferrocene or its derivatives may be predicted on this basis. However, there are certainly exceptions to this generalization, because the metal has a clear influence on substitutions occurring on the rings. The major reaction types observed with ferrocene (and to some extent ruthenocene, osmocene, and tricarbonyl-η^5-cyclopentadienylmanganese, among others) are as follows:

1. **Acetylation.** This reaction is of some historical importance, because it established the aromatic character of ferrocene.

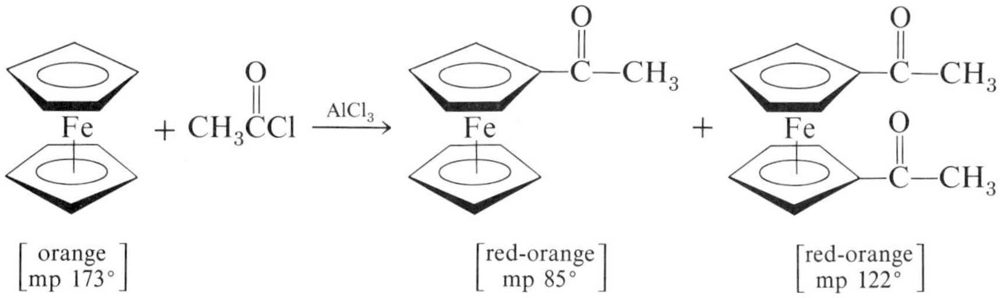

(17-32)

The reaction may be catalyzed by any Lewis acid (most often $AlCl_3$, but H_3PO_4 is very effective because it limits the amount of disubstituted product) and proceeds to give both mono- and 1,1'-disubstituted products, with the latter being formed in very small amounts [see experiments described in Chapter 16 (p. 884) that proved the rotation of the C_5 rings in ferrocene].[68] If disubstitution on one ring does occur, it is most frequently *ortho*.

2. **Aminomethylation.** Dimethylamine and formaldehyde undergo a Mannich reaction with ferrocene to give dimethylaminomethylferrocene, a compound useful in the preparation of many other derivatives.

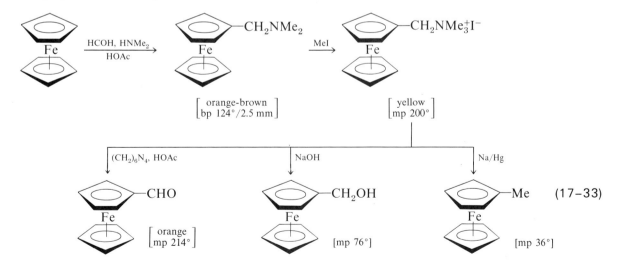

(17-33)

---

[65] D. A. White, *Organometal. Chem. Rev. A*, **3**, 497 (1968).
[66] J. P. Collman, *Transition Metal Chemistry*, **2**, 1 (1966).
[67] D. E. Bublitz and K. L. Rinehart, Jr., *Organic Reactions*, **17**, 1 (1969).
[68] R. J. Angelici, "Synthesis and Technique in Inorganic Chemistry," 2nd ed., W. B. Saunders Company, Philadelphia (1977).

3. **Metalation.** Lithioferrocene (Chapter 16, p. 826) and chloromercuriferrocene (Chapter 16, p. 828), two readily prepared organometallic compounds, are widely used as intermediates in the preparation of other ferrocene derivatives.

(17-34)

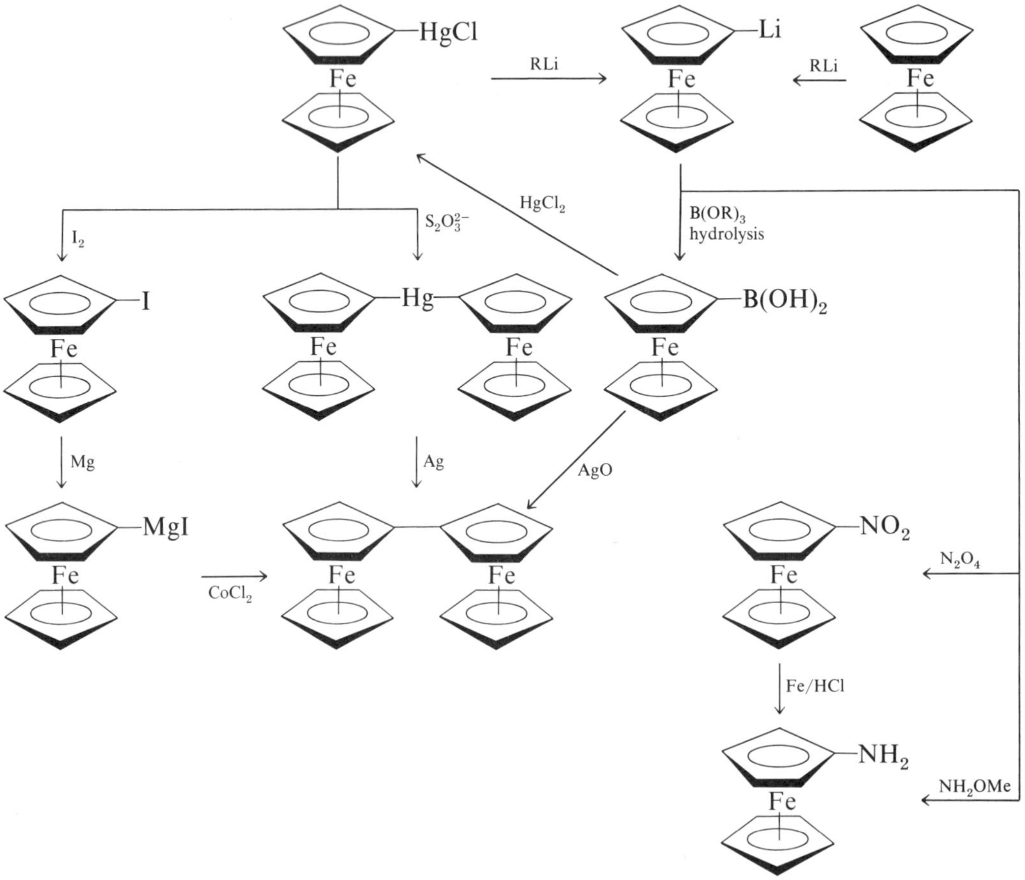

4. **Nitration and Halogenation.** The direct introduction of —NO_2 or —X into ferrocene is not possible, as the metallocene is oxidized to ferricenium ion or other degradation products under the reaction conditions. Instead, both nitro and halo derivatives must be prepared from the metalated intermediates above.

The most important chemical characteristic of ferrocene is its ability to act as an electron releasing substituent. For example, comparison of the Taft σ_I ($= -0.05$) and Hammett σ_m ($= -0.15$) values for the ferrocenyl group with those of the butyl group (-0.07 and -0.13, respectively) indicate that the metallocenyl group is among the strongest inductive electron releasing agents known.[69,70] In agreement with this, aminoferrocene is found to be a stronger base than aniline by a factor of about 20, and ferrocene carboxylic acid is a weaker acid than benzoic acid. Finally, α-ferrocenylcarbinyl acetates undergo solvolysis (reaction 17-35)

(17-35)

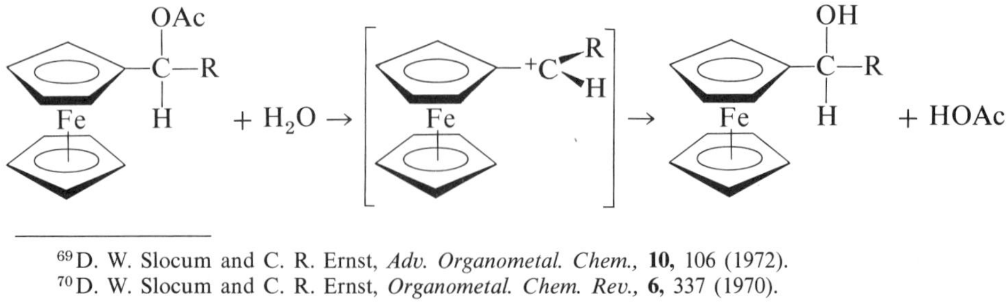

---

[69] D. W. Slocum and C. R. Ernst, *Adv. Organometal. Chem.*, **10**, 106 (1972).
[70] D. W. Slocum and C. R. Ernst, *Organometal. Chem. Rev.*, **6**, 337 (1970).

much more rapidly than their phenyl analogs, and, more importantly, the intermediate α-ferrocenylcarbonium ions are stable enough to be isolated.[71,72]

The electron donor ability of ferrocene is further exemplified by its great propensity to undergo electrophilic substitution. It has been estimated that ferrocene may be acetylated 10^6 times more rapidly, and mercurated 10^9 more rapidly, than benzene. To explain the high rate of substitution, and the stability of the α-ferrocenylcarbonium ions, direct metal participation has been suggested.[71,72] Given the fact that ferrocene may be protonated at the metal in strong acid media (Chapter 16, p. 887), it has been argued that an electrophile could similarly attack the weakly bonding electrons of the iron (reaction 17–36);

$$\text{Fe} \underset{}{\overset{+E^+}{\rightleftharpoons}} \text{Fe}\rightarrow E^+ \rightarrow \text{Fe} \xrightarrow{-H^+} \text{Fe}-E \tag{17-36}$$

this would be followed by transfer of the group to the C_5 ring in an *endo* manner and by proton elimination. Similarly, it has been proposed that the carbonium ion center in the α-ferrocenylcarbonium ions (in reaction 17–35) could interact with the weakly bonding e_{2g} iron orbitals (*cf.* Figure 16–16).[72,73]

The chemistry of $(\eta^5\text{-}C_5H_5)\text{Mn(CO)}_3$ is similar to that of ferrocene, although the manganese compound is less susceptible to electrophilic substitution. Relative to the $(\eta^5\text{-}C_5H_5)\text{Fe}$ and C_6H_5 groups, the Mn(CO)_3 group is more electron withdrawing, as indicated by the fact that the relative base strengths of a series of related aromatic amines (RNH_2) are in the order $(\eta^5\text{-}C_5H_4NH_2)\text{Mn(CO)}_3 < C_6H_5NH_2 < (\eta^5\text{-}C_5H_5)\text{Fe}(\eta^5\text{-}C_5H_4NH_2)$.

A Group VIB metal tricarbonyl group also lowers the electron releasing ability of C_6H_5X in $(\eta^6\text{-}C_6H_5X)M(CO)_3$ (M = Cr, Mo, W). [For example, aniline is a stronger base ($pK_b = 11.70$) than $(\eta^6\text{-}C_6H_5NH_2)\text{Cr(CO)}_3$ ($pK_b = 13.31$), and benzoic acid ($pK_a = 5.68$) is a weaker acid than $(\eta^6\text{-}C_6H_5\text{COOH})\text{Cr(CO)}_3$ ($pK_a = 4.77$).[74]] Most importantly, this suggests a new synthetic route to benzene derivatives. Most reactions occurring at an aromatic ring (*e.g.*, Friedel-Crafts acylation and alkylation) are electrophilic substitutions. To carry out a nucleophilic substitution requires the attachment of a strongly electron-withdrawing group to the ring prior to the desired substitution; the electron-withdrawing group must then be removed later, a process not often easily done. However, an electron-withdrawing group such as the π-bonded —$M(CO)_3$ group can be easily attached to benzene and then removed simply by mild oxidizing agents. For example, π-bonded benzene or chlorobenzene is converted in high yield to phenylisobutyronitrile after a few hours at room temperature (17–37).[75,76]

$$\underset{\text{Cr(CO)}_3}{\text{C}_6\text{H}_5\text{Cl}} + {}^{\ominus}\text{C(CH}_3)_2\text{CN} \rightarrow \underset{\text{Cr(CO)}_3}{\text{C}_6\text{H}_5\text{C(CH}_3)_2\text{CN}} \xrightarrow{I_2/H_2O} \text{C}_6\text{H}_5\text{C(CH}_3)_2\text{CN} + \text{Cr(III)} \tag{17-37}$$

---

[71] M. Cais, *Organometal. Chem. Rev. A*, **1**, 435 (1966).

[72] R. L. Sime and R. J. Sime, *J. Amer. Chem. Soc.*, **96**, 892 (1974).

[73] α-Ferrocenylcarbonium ion stability has also been ascribed to a more general delocalization of charge. G. H. Williams, D. F. Traficante, and D. Seyferth, *J. Organometal. Chem.*, **60**, C53 (1973).

[74] H. Zeiss, P. J. Wheatley, and H. J. S. Winkler, "Benzenoid-Metal Complexes," Ronald Press, New York (1966).

[75] M. F. Semmelhack and H. T. Hall, *J. Amer. Chem. Soc.*, **96**, 7091, 7092 (1974).

[76] M. F. Semmelhack, H. T. Hall, M. Yoshifuji, and G. Clark, *J. Amer. Chem. Soc.*, **97**, 1248 (1975).

926 ORGANOMETALLIC COMPOUNDS: REACTION PATHWAYS

No reaction is observed when just chlorobenzene, for example, is used under the same conditions. The proposed intermediates (reaction 17-38) resemble those thus far discussed for ligand protonation or ligand substitution. Notice that the mechanism again follows the 16 and 18 electron rules laid down at the beginning of this chapter.

(17-38)

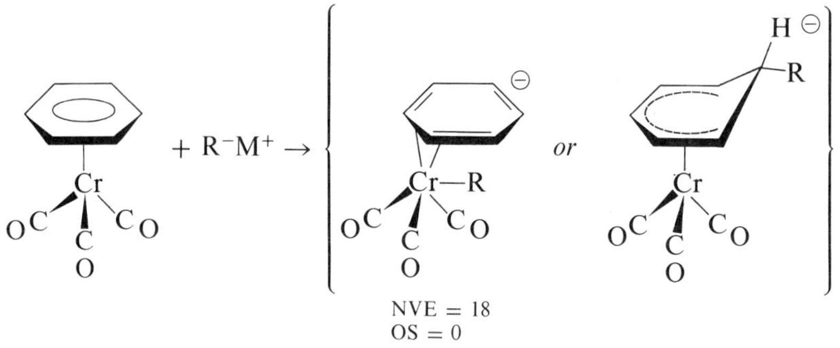

$$(\eta^6\text{-}C_6H_6)Cr(CO)_3 + R^-M^+ \rightarrow \left\{ [(\eta^6\text{-}C_6H_6)Cr(R)(CO)_3]^- \text{ or } [(\eta^5\text{-}C_6H_6R)Cr(CO)_3]^- H^- \right\}$$

NVE = 18
OS = 0

STUDY QUESTIONS

9. Two pieces of evidence were cited to suggest that the second order path in reaction 17-27 is an I_d path (see page 921). Explain more fully why this evidence indicates a dissociatively activated path.

10. If reaction 17-26 is carried out using $Mo(CO)_6$ or $W(CO)_6$ instead of the chromium compound, comparison of k_1 and k_2 values (obtained at the same temperature) is quite informative. For example, while the k_2/k_1 ratio is 0.7 for $Cr(CO)_6$, it is 9.6 for $Mo(CO)_6$ and 34.8 for $W(CO)_6$ (all data having been obtained at 112°C). [See J. R. Graham and R. J. Angelici, *Inorg. Chem.*, **6**, 2082 (1967).] Considering the structure of the transition state in an interchange process, suggest a reason for the relatively greater importance of the interchange mechanism as the metal atomic weight increases.

11. Explain in your own words how the reactivity orders observed for reaction 17-31 suggest an associatively activated mechanism.

12. Tricarbonyl-η^6-cycloheptatrienechromium exchanges its triene ligand for free cycloheptatriene. The rate law for the reaction is

$$\text{Rate} = k_1[(\eta^6\text{-}C_7H_8)Cr(CO)_3] + k_2[(\eta^6\text{-}C_7H_8)Cr(CO)_3][C_7H_8^*]$$

where k_1 is greater than k_2. Suggest a mechanism consistent with this rate law.

13. Refer to the discussion of ligand protonation and reactions 17-37 and 17-38 (p. 925) and then outline a mechanism that will produce the following diene from the illustrated reaction. (Hint: the diene is obtained only after quenching the initial reaction with H_2O followed by I_2.)

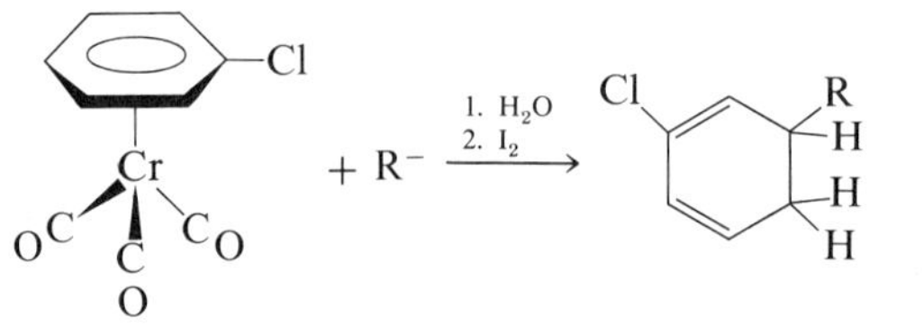

14. Refer to reactions 17-32 and 17-34, and outline:
    a) three ways to prepare ferrocenyldiphenylphosphine, $(C_5H_5)Fe(C_5H_4PPh_2)$.
    b) a preparation of ferrocenyl carboxylic acid.

ADDITION AND ELIMINATION REACTIONS[77,78]

The 1,1- and 1,2-addition of metal-containing compounds to unsaturated substrates, or the 1,1-oxidative addition of organic halides and other molecules to low-valent metals in compounds of low coordination number, constitutes one of the most important forms of organometallic reactivity. Such reactions have been implicated in the homogeneous hydrogenation or polymerization of olefins, in the formation of metal-carbene complexes, in rearrangement reactions of metal carbonyls, and in many others. In Chapter 16 we discussed the usefulness of such reactions (*i.e.*, hydro-, oxy-, halo-, and organometalations) in the synthesis of organometallic compounds. In this section we shall discuss briefly a few examples of addition reactions of organometallic compounds themselves, with a special emphasis on mechanism and stereochemistry.

1,2-ADDITIONS TO DOUBLE BONDS

In this portion we are interested in reactions of the type

$$X'_n M-X + Z=Y \rightarrow X'_n M-Z-Y-X \qquad (17\text{-}39)$$

and such reactions are exemplified by additions to the following types of compounds.

a) Alkenes:

$$Et_6Al_2 + R-CH=CH_2 \rightarrow [R-CH(Et)-CH_2-AlEt_2]_2 \qquad (17\text{-}40)$$

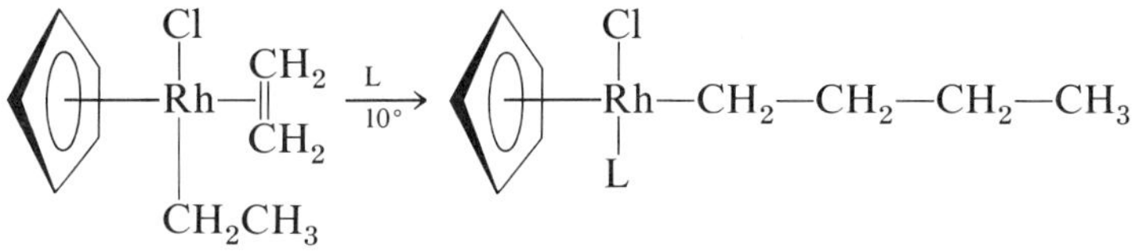

$$(17\text{-}41)^{79}$$

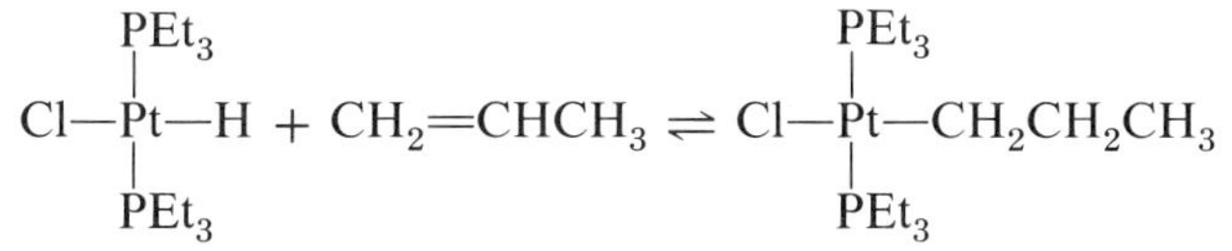

$$(17\text{-}42)^{80}$$

b) Alkynes:

$$(17\text{-}43)^{81}$$

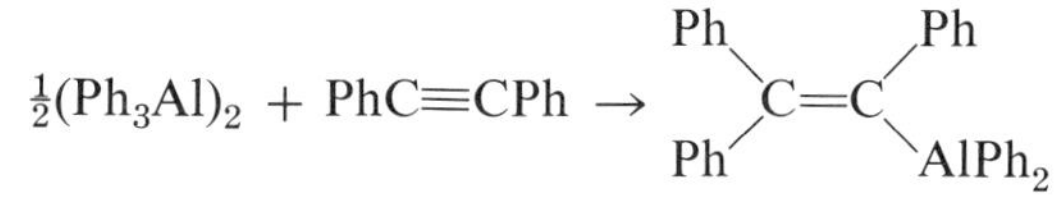

$$\tfrac{1}{2}(Ph_3Al)_2 + PhC{\equiv}CPh \rightarrow \underset{Ph}{\overset{Ph}{>}}C=C\underset{AlPh_2}{\overset{Ph}{<}}$$

c) Carbon dioxide:

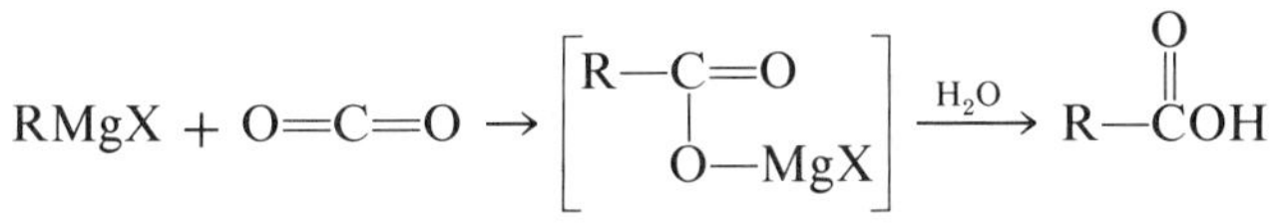

$$RMgX + O=C=O \rightarrow \left[\begin{array}{c} R-C=O \\ | \\ O-MgX \end{array}\right] \xrightarrow{H_2O} R-\overset{O}{\overset{\|}{C}}OH \qquad (17\text{-}44)$$

---

[77] R. F. Heck, *Adv. Chem. Series*, **49**, 181 (1965).
[78] M. F. Lappert and B. Prokai, *Adv. Organometal. Chem.*, **5**, 225 (1967).
[79] R. Cramer, *J. Amer. Chem. Soc.*, **87**, 4717 (1965); *Acc. Chem. Res.*, **1**, 186 (1968).
[80] J. Chatt, R. S. Coffey, A. Gough, and D. T. Thompson, *J. Chem. Soc.* (A), 190 (1968).
[81] J. J. Eisch and C. K. Hordis, *J. Amer. Chem. Soc.*, **93**, 2974, 4496 (1971).

d) Ketones:

(17-45)[82]

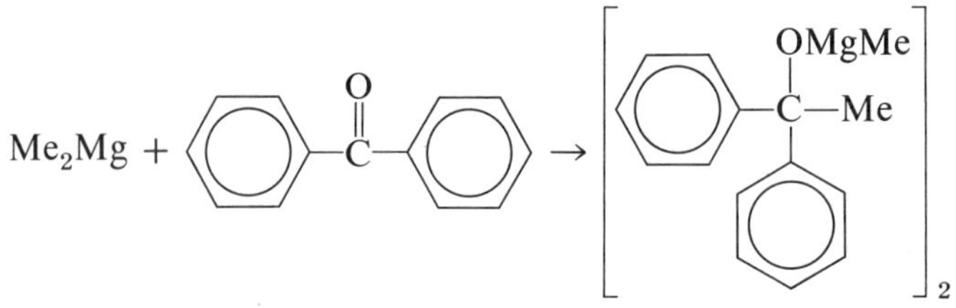

Me$_2$Mg + Ph-C(=O)-Ph → [Ph$_2$C(OMgMe)Me]$_2$

e) Cyanides:

(17-46)[83]

Me$_6$Al$_2$ + 2PhCN → cyclic dimer with Al$_2$N$_2$ ring, each N double-bonded to C(Me)(Ph), each Al bearing two Me groups

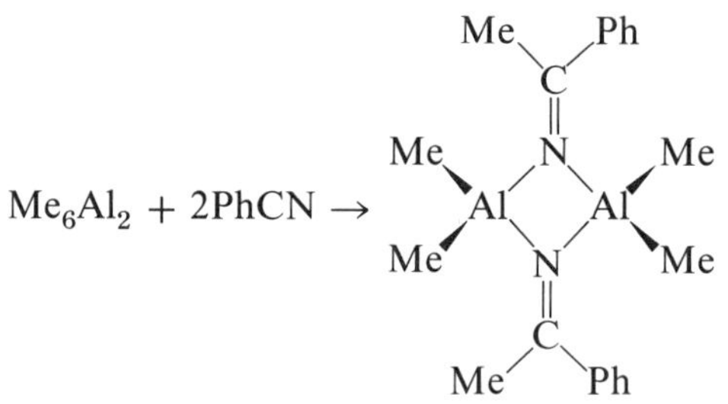

f) Isocyanates:

(17-47)[84]

$$Me_3Sn-NMe_2 + PhN=C=O \rightarrow \begin{array}{c} PhN-C=O \\ | \quad | \\ Me_3Sn \;\; NMe_2 \end{array}$$

All of these reactions apparently occur by the same general mechanism: initial coordination of the organometallic $X'_n M-X$ to the substrate ZY followed by migration of X to the substrate.

(17-48)

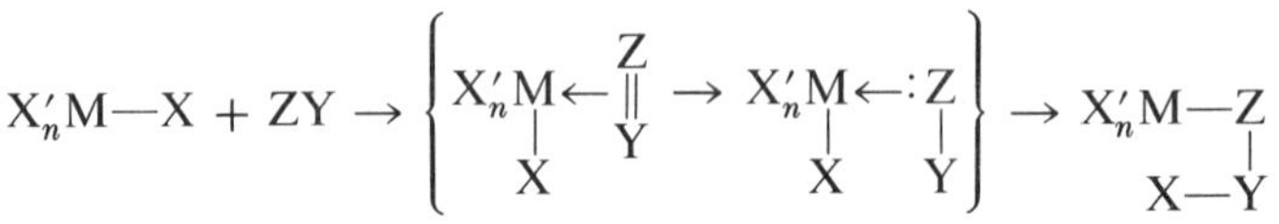

$$X'_n M-X + ZY \rightarrow \left\{ \begin{array}{c} X'_n M \leftarrow \|{{Z}\atop{Y}} \\ | \\ X \end{array} \rightarrow \begin{array}{c} X'_n M \leftarrow :Z \\ | \quad | \\ X \quad Y \end{array} \right\} \rightarrow \begin{array}{c} X'_n M-Z \\ | \\ X-Y \end{array}$$

As illustrations of this general scheme, we shall examine a few mechanisms in detail. However, one last point is pertinent: as noted in Chapter 16, addition reactions are often referred to by organometallic chemists as insertion reactions, because the substrate ZY is inserted into the M—X bond.

Because of the importance of the alkylaluminum-catalyzed "growth reaction," that is, the polymerization of olefins in the presence of aluminum alkyls (Chapter 16, p. 839), the kinetics and mechanism of the addition of organoaluminum compounds to C—C unsaturated systems has been extensively explored.[85-87] One important piece of mechanistic evidence, for example, is that the rate of olefin or acetylene absorption

---

[82] E. C. Ashby, J. Laemmle, and H. M. Neumann, *Acc. Chem. Res.,* **7,** 272 (1974).
[83] J. R. Jennings, J. E. Lloyd, and K. Wade, *J. Chem. Soc.,* 5083 (1965).
[84] See ref. 78.
[85] K. Ziegler in "Organometallic Chemistry," H. H. Zeiss, ed., Reinhold Publishing Company, New York (1960), pp. 194 ff.
[86] K. W. Egger and A. T. Cocks, *J. Amer. Chem. Soc.,* **94,** 1810 (1972).
[87] J. N. Hay, P. G. Hooper, and J. C. Robb, *J. Organometal. Chem.,* **28,** 193 (1971).

by triethylaluminum is proportional to the square root of the Et_6Al_2 concentration. Similarly, the triphenylaluminum reaction with diphenylacetylene (17–43) is one-half order in the arylaluminum compound.[88] On the other hand, tri-*iso*-butylaluminum absorbs ethylene at a rate proportional to the concentration of $(i\text{-Bu})_3Al$. These observations make sense when you recall that almost all organoaluminum compounds, except for tri-*iso*-butylaluminum,[89] are largely dimeric in solution in non-basic solvents and in the gas phase at normal temperatures (see Chapter 16, p. 845). Therefore, the reactions are really first order in organoaluminum monomer. This, of course, has a parallel in hydroborations (*cf.* Chapter 7, p. 391), where the reactive species is BH_3 and not B_2H_6.

Because the addition of organoaluminum compounds to double bonds must involve monomer, the rate determining reaction of the monomer with the substrate must be preceded by a dissociative pre-equilibrium between organoaluminum dimer (= D) and monomer (= M).
The basic rate law then is

$$D \underset{k_{-1}}{\overset{k_1}{\rightleftharpoons}} 2M$$

$$M + \text{substrate}\ (= B) \xrightarrow{k_2} \text{product}\ (= P)$$

$$\frac{d[P]}{dt} = -\frac{d[B]}{dt} = k_2[M][B]$$

Substituting the monomer concentration from the equilibrium expression

$$K = \frac{k_1}{k_{-1}} = \frac{[M]^2}{[D]_0 - [M]/2}$$

you see that the reaction will obey $\frac{3}{2}$ order kinetics if $16[D]_0/K \gg 1$, as is usually observed (*cf.* equation 7–4b):

$$-\frac{d[B]}{dt} = \frac{k_2 K}{4}\left(-1 + \sqrt{1 + (16[D]_0/K)}\right)[B] = k[D]_0^{1/2}[B]$$

Triphenylaluminum reacts with diphenylacetylene to give, ultimately, an aluminole (17–49).

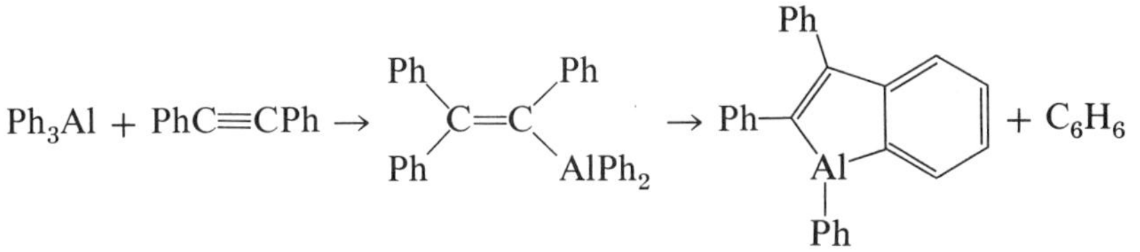

$$Ph_3Al + PhC{\equiv}CPh \rightarrow \underset{Ph}{\overset{Ph}{}}C{=}C\underset{AlPh_2}{\overset{Ph}{}} \rightarrow Ph\text{-indole-Al(Ph)} + C_6H_6 \qquad (17\text{–}49)$$

However, the first identifiable product is the addition compound $Ph_2Al{-}C(Ph){=}CPh_2$.[90] If an unsymmetrical acetylene is used, though, three different products are possible. Compound **10** can arise from either *cis* or *trans* addition, but **8** and **9** can only be the result of *cis* or *trans* addition, respectively.

---

[88] See ref. 81.
[89] There is evidence that $(i\text{-Bu})_3Al$ dimerizes to a very small extent when dissolved in non-basic solvents: M. B. Smith, *J. Organometal. Chem.*, **22**, 273 (1970).
[90] See ref. 81.

(17-50)

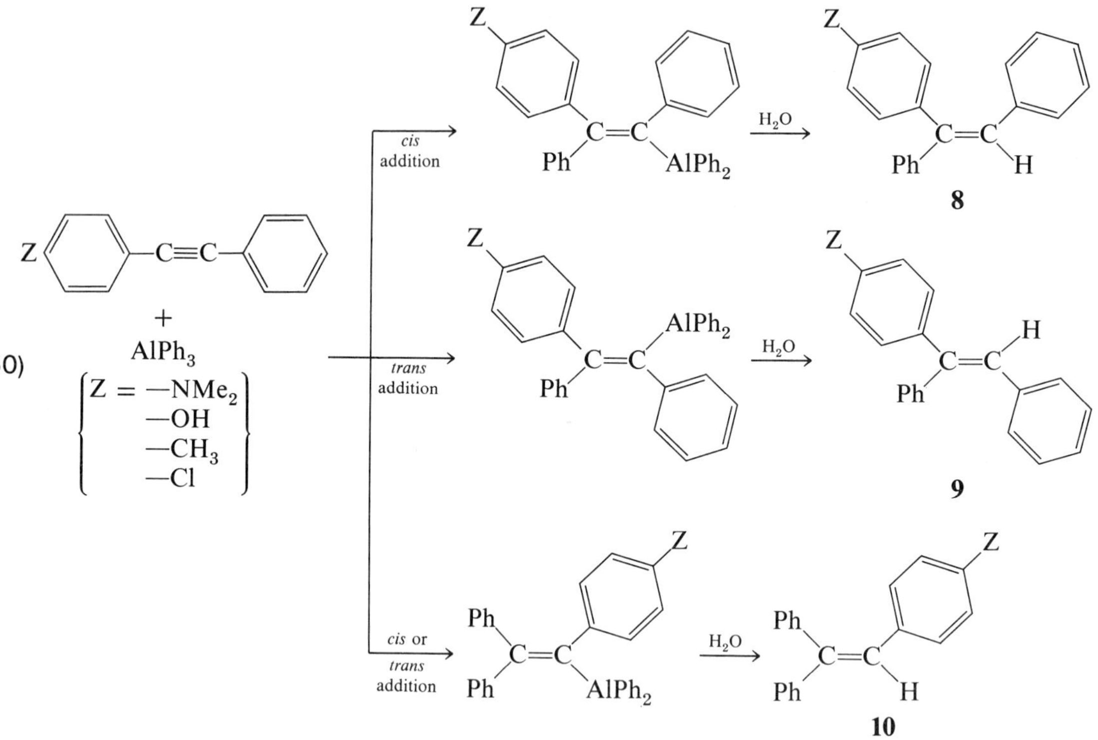

After hydrolysis of the addition product (which is known to proceed with retention of configuration), mixtures of **8** and **10** are isolated, thereby proving that addition proceeds only in a *cis* manner as in hydroborations (*cf.* Chapters 7 and 16). Furthermore, the observation that the relative amount of compound **8** increases as Z becomes more electron-releasing supports the idea of electrophilic attack at an acetylenic carbon and suggests a transition state such as **11**. In this case, attack of the more positively charged Al occurs preferentially at the carbon β to the ring, and the incipient carbonium center α to the ring is then stabilized by electron-releasing substituents, Z, in the transition state.

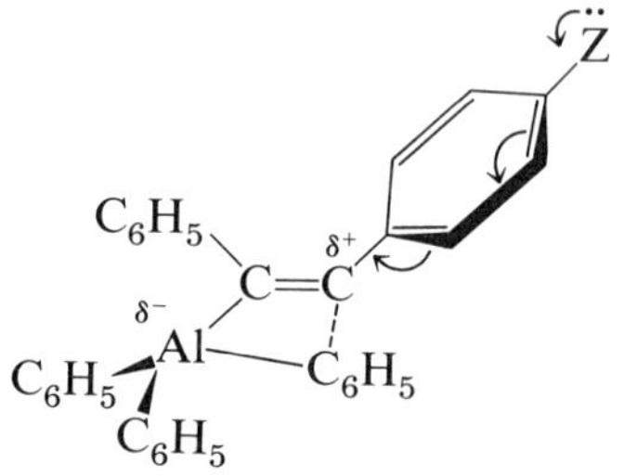

The nature of the transition state and possible intermediates in carbalumination and hydroalumination reactions has been discussed extensively.[91] Although the situation is not yet resolved, one thing is clear: that a π complex of some type between the olefin or acetylene and the aluminum compound probably exists as a transient intermediate. One piece of evidence for such a complex is the molecular structure of diphenyl(phenylethynyl)aluminum dimer, **12**.[92]

---

[91] J. J. Eisch and N. E. Burlinson, *J. Amer. Chem. Soc.*, **98**, 753 (1976); J. J. Eisch and S-G. Rhee, *ibid.*, **96**, 7276 (1974).

[92] G. D. Stucky, A. M. McPherson, W. E. Rhine, J. J. Eisch, and J. L. Considine, *J. Amer. Chem. Soc.*, **96**, 1941 (1974).

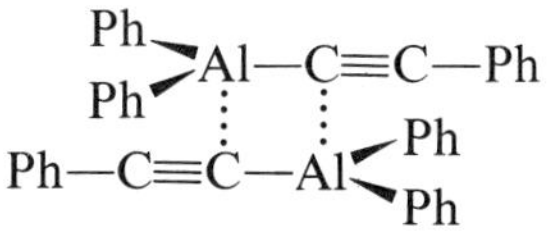

$$\begin{array}{c} \text{Ph} \diagdown \text{Al} - \text{C} \equiv \text{C} - \text{Ph} \\ \text{Ph} \diagup \phantom{\text{Al}} \vdots \phantom{-\text{C}\equiv\text{C}-} \diagup \text{Ph} \\ \text{Ph} - \text{C} \equiv \text{C} - \text{Al} \diagdown \text{Ph} \end{array}$$

12

The Al—C≡C— grouping is linear, with an Al—C—Al bridge angle of 91.73°. This suggests that the bridge is formed by overlap of the aluminum $3p_z$ orbital and a $p\pi$ orbital of the acetylenic linkage.

Olefin arylation has been accomplished under mild conditions by the reaction of an olefin with an "aryl palladium salt." The detailed composition of the arylating agent, which is made from phenylmercuric acetate and palladium acetate (17-51), is not known, but it is represented as "Ph—Pd—OAc."[93]

$$\text{Ph-Hg-OAc} + \text{Pd(OAc)}_2 \rightarrow \text{"Ph-Pd-OAc"} + \text{Hg(OAc)}_2 \tag{17-51}$$

The olefin arylation reaction itself undoubtedly proceeds by addition of this "aryl palladium salt" to the olefin (17-52), followed by elimination of a palladium hydride.

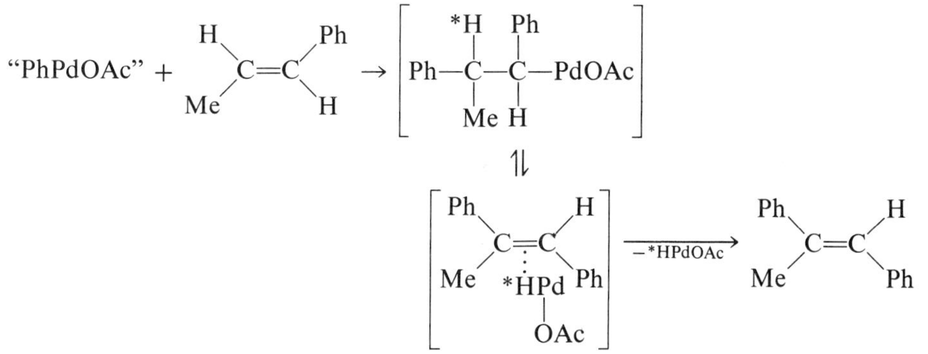

(17-52)

Notice that the palladium generally becomes bound to a hydrogen atom from a β carbon in the course of the elimination reaction; hence, the process is often called a β-elimination, a reaction mentioned in Chapter 16 (p. 840) and described later in this chapter. Investigation of the mechanism and stereochemistry of olefin arylation shows that the reaction is apparently controlled more by steric effects than by electronic factors. That is, the organic group predominantly adds to the less substituted carbon, regardless of the electronic nature of the substituents. Furthermore, the stereochemistry of addition is most likely *cis*, as is elimination of the "hydridopalladium salt."

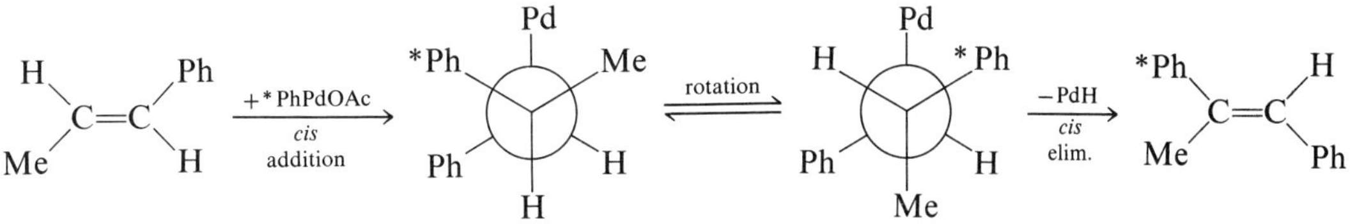

(17-53)

Trans addition followed by *trans* elimination would also account for the observed stereochemistry, but this would require an ionic or radical mechanism. An ionic mechanism is ruled out by competitive rate studies, which show that ethylene reacts more rapidly than propylene; opposite behavior would be expected for an ionic

---

[93] R. F. Heck, *J. Amer. Chem. Soc.*, **91**, 6707 (1969); *ibid.*, **93**, 6896 (1971); see also ref. 8.

mechanism. Therefore, the mechanism of organopalladation presumably resembles in large part the mechanism proposed for the addition of Ph_3Al to acetylenes.

Grignard reagents,[94] diorganomagnesium compounds,[95] triorganoaluminum compounds,[96] and similar materials add to aldehydes and ketones to produce secondary and tertiary alcohols, respectively, after hydrolysis. The mechanism of these reactions is of great interest, and a recent study of the trimethylaluminum/benzophenone reaction

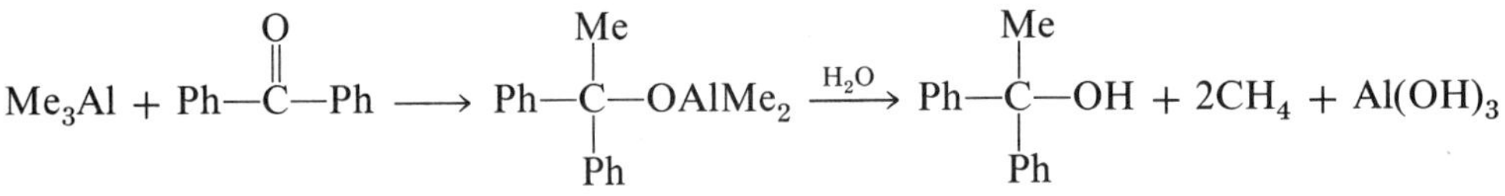

$$Me_3Al + Ph-\overset{O}{\underset{\|}{C}}-Ph \longrightarrow Ph-\underset{Ph}{\overset{Me}{\underset{|}{C}}}-OAlMe_2 \xrightarrow{H_2O} Ph-\underset{Ph}{\overset{Me}{\underset{|}{C}}}-OH + 2CH_4 + Al(OH)_3$$

is especially important in this regard. In benzene (reaction ratio = 1:1) or diethyl ether, the reactions are first order in alkylaluminum monomer and benzophenone; however, the reaction is slower in ether than in benzene. The mechanism proposed for the reaction (17-55) again involves a pre-equilibrium between monomer and dimer as in the "growth reaction" discussed above.

(17-55)

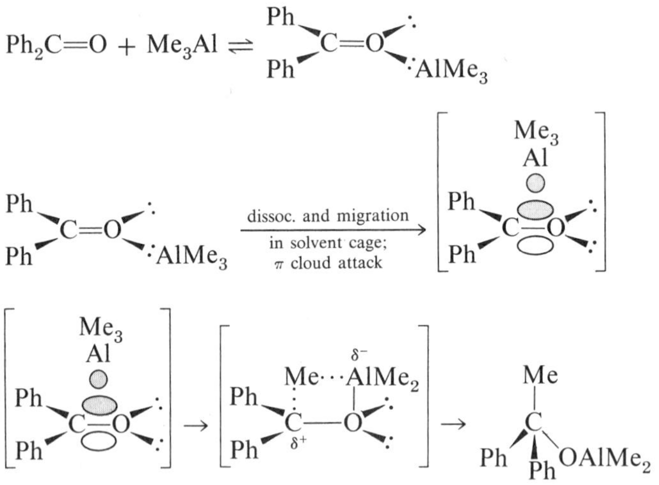

This step is followed by formation of a benzophenone-$AlMe_3$ complex, evidence for the existence of which has been obtained. (The relative slowness of the addition reaction in ether is explained, then, by realizing that formation of $Et_2O \cdot AlMe_3$ is in competition with $Ph_2CO \cdot AlMe_3$ formation.) From this point on, there are two possible paths open to the reactions. The original proposal was that the initial complex dissociated within the solvent cage with formation of a π complex. This places the alkyl group in a more favorable position for attack on the ketonic carbon. Alternatively, the reaction could simply proceed by a 1,3-sigmatropic shift after formation of the initial $Ph_2CO \cdot AlMe_3$ adduct.[97]

Attack of nucleophiles on coordinated olefins can also lead to 1,2-addition products. For example, addition of various nucleophiles to cationic complexes of the

---

[94] See ref. 82.
[95] L. Laemmle, E. C. Ashby, and H. M. Neumann, *J. Amer. Chem. Soc.,* **93,** 5120 (1971).
[96] H. M. Neumann, J. Laemmle, and E. C. Ashby, *J. Amer. Chem. Soc.,* **95,** 2597 (1973).
[97] Notice that configuration inversion should be observed for the migrating carbon in a 1,3 shift but not in a 1,2 shift. Apparently, no experiment has yet been done to clarify this mechanistic point.

type $[(\eta^5\text{-}C_5H_5)Fe(CO)_2(\text{olefin})]^+$ gives products wherein the nucleophile and iron have added to the olefinic bond.[98]

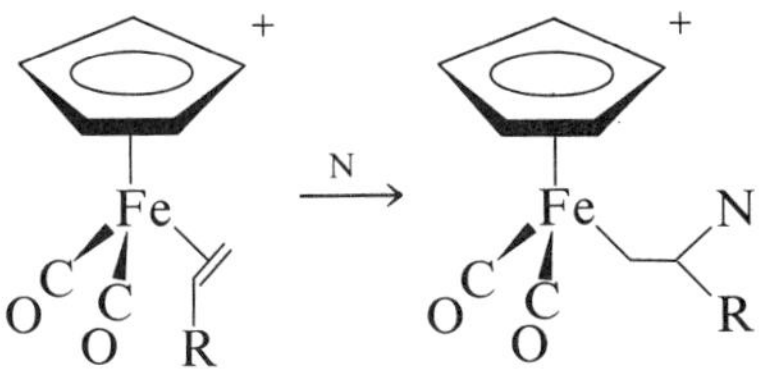

$$N = \overset{\ominus}{C}H(COOEt)_2, PPh_3 \qquad (17\text{-}56)$$

Such reactions have definite applications to organic synthesis, since reaction of the addition product with halogens or other oxidizing agents leads to cleavage of the Fe—C bond and introduction of a halide or carboxylate group, respectively.

Yet another very important example of a 1,2-addition or insertion reaction was also mentioned in Chapter 16, that is, reaction 17-57. In this instance, the elements of Pt and H add across the double bond.

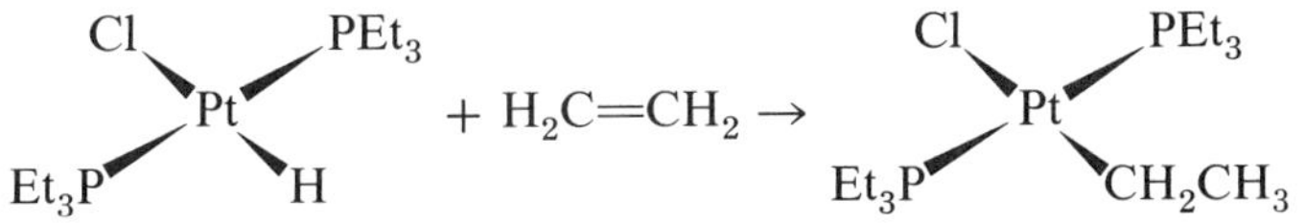

$$(17\text{-}57)$$

Such reactions are undoubtedly involved in metal-catalyzed olefin hydrogenations, and their mechanism is pursued in a section on that subject later in the chapter. However, before dismissing the topic, we should mention *hydrozirconation*,[99] the addition of the elements of zirconium and hydrogen across double bonds, which has potential in organic synthesis.

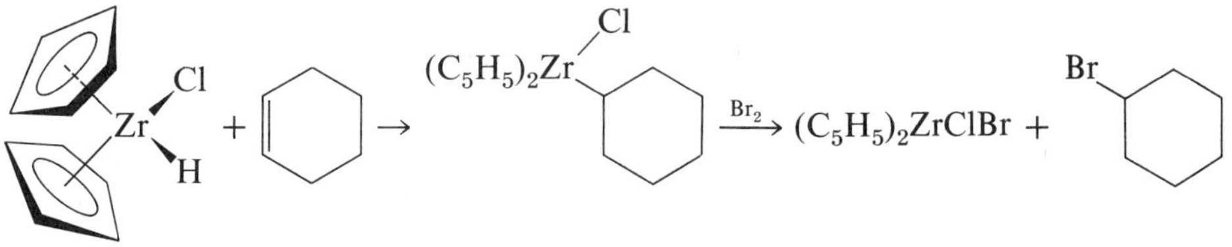

$$(17\text{-}58)$$

There are two distinct advantages of the hydrozirconation method when compared with hydroboration or hydroalumination: (i) unlike the boron and aluminum alkyls, dry air does not decompose the zirconium alkyl (but the latter can be hydrolyzed to give the appropriate alkane), and (ii) cleavage of the Zr—C bond with halogen or acyl halide to provide the organic derivative leaves the compound $(\eta^5\text{-}C_5H_5)_2ZrClX$, which can be recycled back to the starting metal hydride.[100]

1,1-ADDITION TO CO: CARBONYLATION AND DECARBONYLATION

The *oxo process* is one of the most important examples of homogeneous catalysis. In a system composed of a cobalt salt, CO, H_2, and an olefin, the main

---

[98] M. Rosenblum, *Acc. Chem. Res.*, **7**, 122 (1974).
[99] D. W. Hart and J. Schwartz, *J. Amer. Chem. Soc.*, **96**, 8115 (1974).
[100] Cp_2ZrClH can be produced from Cp_2ZrCl_2 (commercially available) by reduction of the latter with Vitride in THF. [Vitride = $NaAlH_2(OCH_2CH_2OCH_3)_2$, Eastman Organic Chemicals.] See Chapter 8 for a discussion of the replacement of X by H^-.

reaction that occurs is hydroformylation, the conversion of the olefin to an aldehyde.[101]

(17-59)
$$\diagdown C = C \diagup + H_2 + CO \xrightarrow{\text{Co catalyst}} H-\underset{|}{\overset{|}{C}}-\underset{|}{\overset{|}{C}}-\overset{O}{\overset{\|}{C}}-H$$

Under the reaction conditions (temperatures of 90° to 200°C and pressures of 100 to 400 atm), the cobalt is presumed to be in the form of $Co_2(CO)_8$ and $HCo(CO)_4$, these compounds being related by an equilibrium reaction involving hydrogen.

(17-60)
$$Co_2(CO)_8 + H_2 \rightleftharpoons 2HCo(CO)_4$$

The aldehyde is thought to be formed in three steps, of which the first involves 1,2-addition of the hydridocarbonyl to the olefin (reaction 17-61) the second is an insertion of CO into the cobalt-alkyl bond (carbonylation; reaction 17-62), and the third is a reaction of the acyl derivative with H_2 to give the aldehyde and regenerate catalyst (reaction 17-63).[102] It is the second of these steps in which we are now interested.

(17-61)
$$HCo(CO)_4 + H_2C{=}CH_2 \rightarrow H_3C-CH_2-Co(CO)_4$$

(17-62)
$$H_3C-CH_2-Co(CO)_4 + CO \rightarrow H_3C-CH_2-\overset{O}{\overset{\|}{C}}-Co(CO)_4$$

(17-63)
$$H_3C-CH_2-\overset{O}{\overset{\|}{C}}-Co(CO)_4 + H_2 \rightarrow H_3C-CH_2\overset{O}{\overset{\|}{C}}H + HCo(CO)_4$$

The insertion of CO into metal-carbon bonds has been extensively studied,[103,104] and one example is reaction 17-64 involving methylpentacarbonylmanganese (reversible when L = CO).

(17-64)
$$H_3C-Mn(CO)_5 + L \rightleftharpoons H_3C-\overset{O}{\overset{\|}{C}}-Mn(CO)_4L$$

(L = CO, amine, phosphine)

In general, a two-stage mechanism seems to operate (17-65); the reaction passes through an intermediate in which the acetyl group is formed by migration of the R group to one of the *cis* CO ligands.

(17-65)

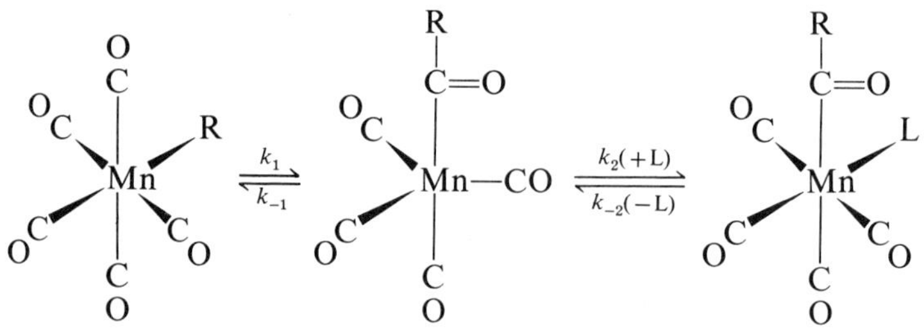

[Structural diagram showing three Mn complexes in equilibrium: first with R and five CO ligands (NVE = 18), converting via k_1/k_{-1} to an acetyl intermediate Mn–CO with C=O–R group (NVE = 16), then via $k_2(+L)/k_{-2}(-L)$ to the final complex with L ligand (NVE = 18)]

---

[101] R. F. Heck, *Adv. Organometal. Chem.*, **4**, 243 (1966).
[102] Reactions 17-61 to 17-63 are not meant to convey mechanistic details. This aspect of the overall process is considered on page 966.
[103] A. Wojcicki, *Adv. Organometal. Chem.*, **11**, 87 (1973).
[104] See ref. 8.

(In spite of the fact that such reactions apparently occur quite generally by R migration, formation of M—C(=O)—R from R—M—CO is often referred to as a CO insertion reaction,[103] a terminology used in this chapter.) Experiments supporting this mechanism are outlined below.

If the steady state approximation is applied to the concentration of the intermediate in reaction 17-65, the following expression is obtained for the rate of appearance of the insertion product [M = $Mn(CO)_4$] (cf. page 366):

$$\frac{d[\text{RCOML}]}{dt} = k_2[L]\frac{k_1[\text{RM(CO)}] + k_{-2}[\text{RCOML}]}{k_{-1} + k_2[L]} - k_{-2}[\text{RCOML}] \tag{17-66}$$

This expression can be expanded, and then rearranged, to give

$$\frac{d[\text{RCOML}]}{dt} = k_f[\text{RM(CO)}][L] - k_r[\text{RCOML}] \tag{17-67}$$

where

$$k_f = \frac{k_1 k_2}{k_{-1} + k_2[L]} \quad \text{and} \quad k_r = \frac{k_{-1}k_{-2}}{k_{-1} + k_2[L]}$$

as in equation 7-4 on page 370. By assuming that reaction 17-64 goes to completion and that k_r is small, equation 17-67 reduces to

$$\frac{d[\text{RCOML}]}{dt} = \frac{k_1 k_2}{k_{-1} + k_2[L]}[\text{RM(CO)}][L] \tag{17-68}$$

That 17-65 may be a reasonable assessment of the stoichiometric mechanism of the process is indicated by the fact that the reaction goes from a first order dependence on both RM(CO) and L at low concentrations of L ($k_f \approx k_1 k_2/k_{-1} = K_{eq}k_2$) to pseudo-first order kinetics at high concentrations of L ($k_f = k_1/[L]$). Further support for the RCOM intermediate comes from studies of a related pair of reactions. Rate constants were found to be identical for the substitution of CO in CH_3—CO—$Mn(CO)_5$ and for the decarbonylation of the acetyl compound.

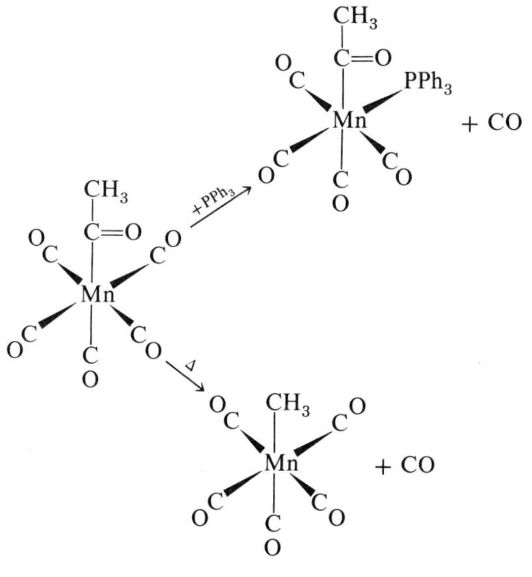

(17-69)

This implies that the rate determining step in both reactions is loss of CO to give the RCOM intermediate, which can either be trapped by the nucleophile PPh_3 or rearrange to the alkylmanganese compound. That is, the rate controlling step in both processes is the k_{-2} step in reaction 17-65 (when L = CO).

One of the most elusive aspects of the CO insertion reaction has been its stereochemistry. That is, does the CO insert into the R—Mn bond, or does the methyl group migrate to the carbonyl ligand? The latter has been shown to be correct for $H_3CMn(CO)_5$ in a very clever series of experiments.[105] Among these experiments are the following:

On adding ^{13}CO to $CH_3Mn(CO)_5$, only cis-$CH_3COMn(CO)_4(^{13}CO)$ is formed (reaction 17-70a). Furthermore, on heating the labeled complex $CH_3(^{13}CO)Mn(CO)_5$, the product is exclusively cis-$CH_3Mn(CO)_4(^{13}CO)$ (reaction 17-70b). Among other things, these experiments show (i) that CO insertion or deinsertion (no decision regarding CO or R migration yet implied) is intramolecular, and (ii) that methyl or CO migration is regiospecific in the forward (carbonylation) and reverse (decarbonylation)

(17-70a)

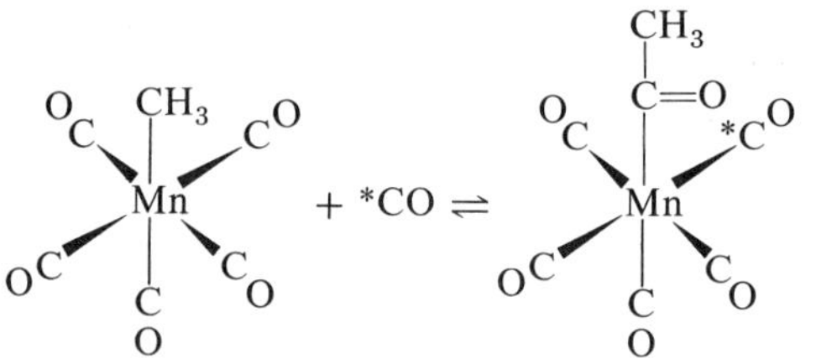

(17-70b)

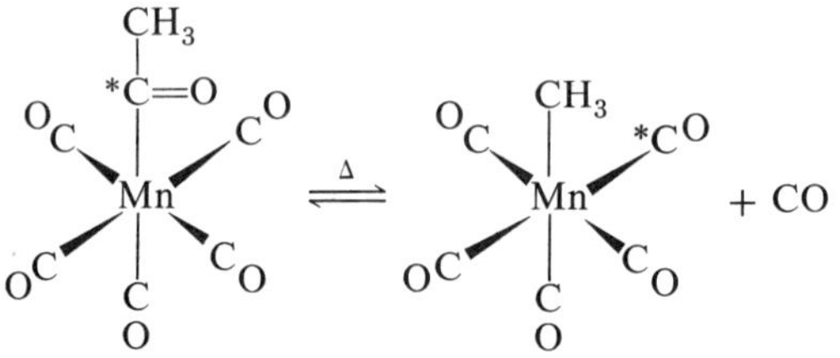

directions. Furthermore, these carbonylation and decarbonylation experiments define an overall CO exchange process proceeding through an acetyltetracarbonylmanganese intermediate (reaction 17-65).

As a result of the experiments just discussed, the definitive stereochemical experiment can be done. That is, an analysis of the decarbonylation products of $CH_3COMn(CO)_4(^{13}CO)$ allows a decision to be made between carbonyl and methyl migration. According to the scheme in reaction 17-71, the two pathways would give significantly different product distributions. In fact, cis- and trans-$CH_3Mn(CO)_4(^{13}CO)$ are obtained in a 2:1 ratio, thereby proving that methyl migration is the correct mechanism.

Carbonyl insertion

(17-71a)

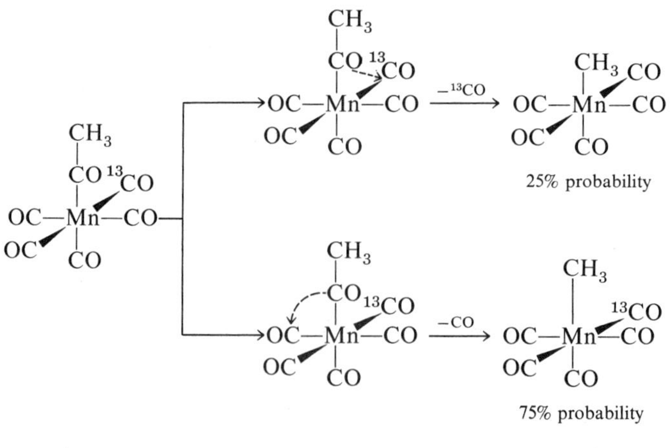

25% probability

75% probability

---

[105] K. Noack and F. Calderazzo, *J. Organometal. Chem.*, **10**, 101 (1967).

Methyl migration

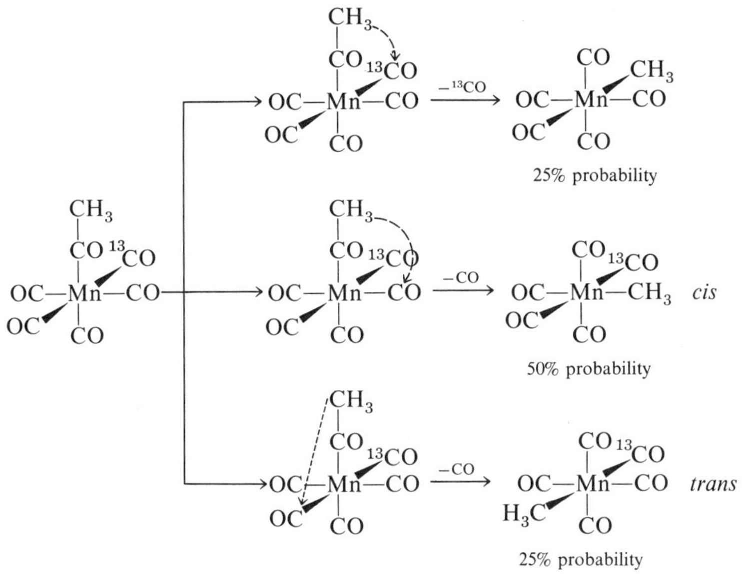

[Scheme showing three pathways from $CH_3Mn(CO)_4(^{13}CO)$ precursor:]

— Top pathway: $-{}^{13}CO$, gives product, **25% probability**
— Middle pathway: $-CO$, gives *cis* product, **50% probability**
— Bottom pathway: $-CO$, gives *trans* product, **25% probability**

(17-71b)

The experiments just outlined clearly show that the decarbonylation of a metal acyl occurs by migration of the alkyl group to a metal coordination site that has been vacated by loss of a ligand, usually CO. (A further implication of these experiments is that carbonylation, the formation of a metal acyl, occurs by the reverse of this process: an alkyl group migrates to a metal-bound CO, and a new ligand such as CO or phosphine takes up the coordination site vacated by the migrating alkyl group.) The decarbonylation of $(\eta^5\text{-}C_5H_5)Fe(CO)(PPh_3)(COEt)$ (reaction 17-72) provides another, elegantly simple proof of the decarbonylation mechanism.[106]

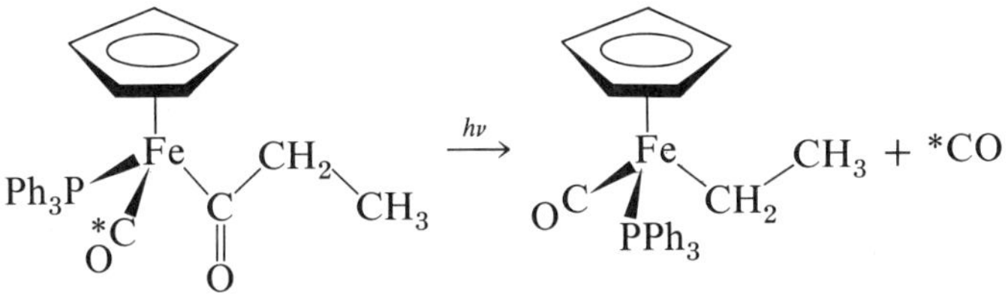

[Structure: Cp-Fe complex with PPh₃, *C(=O)CH₂CH₃ ligand, and CO] $\xrightarrow{h\nu}$ [Cp-Fe complex with OC, PPh₃, and CH₂CH₃] + *CO   (17-72)

The important aspect of the metal acyl complex in reaction 17-72 is that it is chiral, the center of chirality being the iron atom. When CO is lost upon photolyzing the metal acyl, the chirality of the complex is inverted.[107] The only way to account for configuration inversion is to assume that the CO marked with an asterisk has been lost and that the alkyl group has migrated to the newly opened coordination site.

As mentioned at the beginning of this section, the migration of organic groups to produce acyl-metal complexes is an important component of certain catalyzed processes. In addition, it has quite recently been applied to organic synthesis. Collman found that reaction of pentacarbonyliron with sodium in dioxane (to which benzophenone is added to produce $Ph_2CO^-Na^+$) gives an excellent yield of the white dioxanate complex of $Na_2Fe(CO)_4$.[108]

$$Fe(CO)_5 + Na \xrightarrow{\text{dioxane, }100°} Na_2Fe(CO)_4 \cdot 1.5 \text{ dioxane} \qquad (17\text{-}73)$$

---

[106] A. Davison and N. Martinez, *J. Organometal. Chem.*, **74**, C17 (1974) and references therein.

[107] We should also note that other experiments have shown that the configuration of the α carbon atom in a metal alkyl does not change upon carbonylation: P. L. Bock, D. J. Boschetto, J. R. Rasmussen, J. P. Demers, and G. M. Whitesides, *J. Amer. Chem. Soc.*, **96**, 2814 (1974).

[108] J. P. Collman, *Acc. Chem. Res.*, **8**, 342 (1975).

Because of its chemistry, this compound has been called a transition metal Grignard reagent. For example, it reacts with alkyl halides to give an alkyltetracarbonyliron complex, which may then be decomposed with acid or O_2/H_2O to give, respectively, the alkane or carboxylic acid. More important to the current discussion of addition reactions is the reaction with a Lewis base such as CO or a phosphine to give **13** *via* an alkyl migration. This acyl-metal complex can then be used to produce a variety of organic compounds.

(17-74)

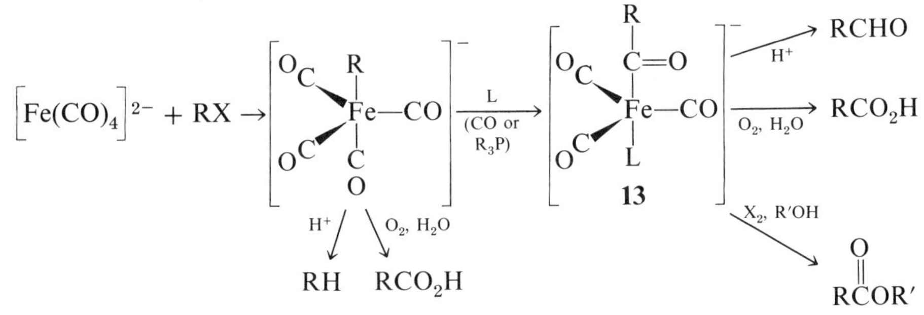

Finally, 1,1-addition to CO can be combined with nucleophilic attack at a coordinated olefin to give a cyclic metallo-ketone.[109] This reaction may also have applications in organic synthesis, but they are as yet unexplored.

(17-75)

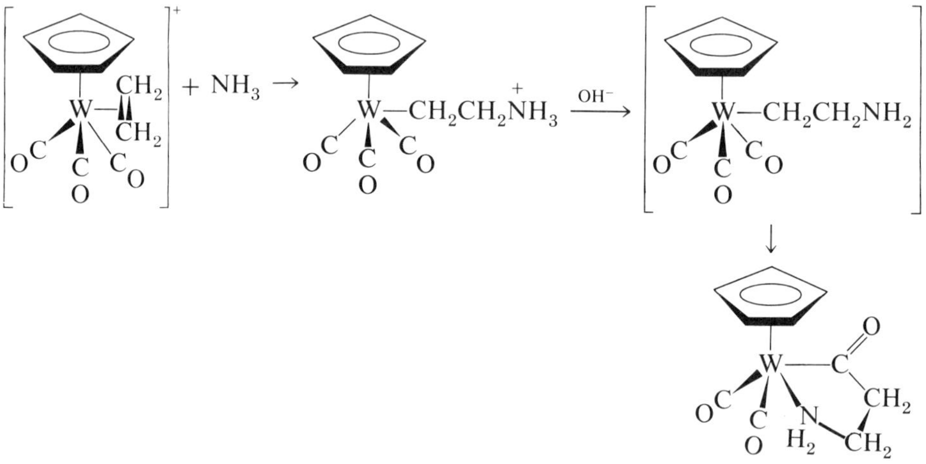

There are many other examples of the insertion of carbon monoxide into metal-carbon bonds. We shall discuss a few additional cases in the remainder of this chapter; for a particularly complete outline of these reactions, though, you should consult the book by Heck.[110]

OXIDATIVE ADDITION REACTIONS[111-115]

Generalizations concerning the reactivity of organometallic compounds have begun to emerge over the past few years, just as they have over the past forty or fifty years in organic chemistry. One of the earliest recognized in organometallic chemistry was the idea of an oxidative addition reaction. As seen in Table 17-1, such reactions involve an increase in formal oxidation state and in coordination number; hence the

---

[109] W. H. Knoth, *Inorg. Chem.*, **14**, 1566 (1975).
[110] See ref. 8.
[111] J. P. Collman and W. R. Roper, *Adv. Organometal. Chem.*, **7**, 53 (1968).
[112] L. Vaska, *Acc. Chem. Res.*, **1**, 335 (1968).
[113] J. Halpern, *Acc. Chem. Res.*, **3**, 386 (1970).
[114] J. P. Collman, *Acc. Chem. Res.*, **1**, 136 (1968).
[115] See ref. 16.

term oxidative addition. Additional examples are given in reactions 17-76 through 17-78, in which all steps are labeled with their reaction types.

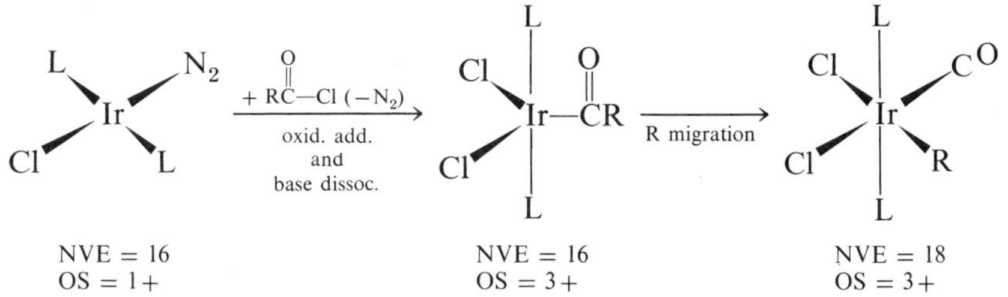

(17-76)[116]

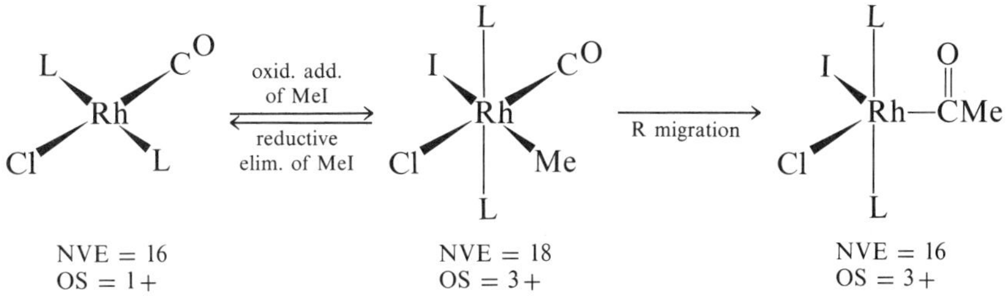

(17-77)[117]

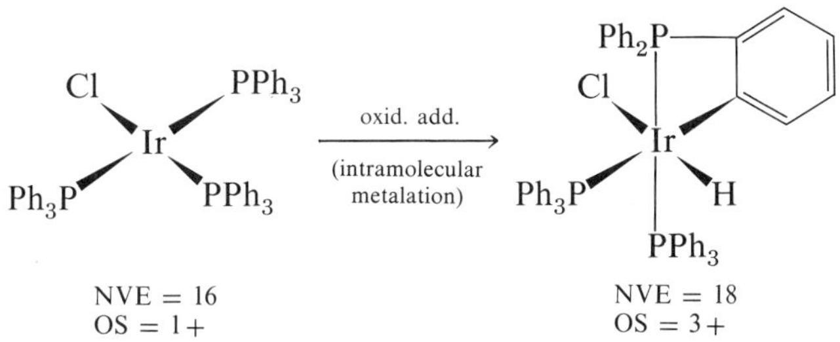

(17-78)[118]

Oxidative addition reactions of $[Co(CN)_5]^{3-}$ were mentioned in Chapter 12 (p. 690) and were discussed as a preparative method for organometallic compounds in Chapter 16 (p. 829). From this and the reactions above, however, you should not get the notion that oxidative additions are restricted to metal compounds; recall that such reactions are characteristic of three-coordinate phosphorus (Chapter 7, p. 419).

Before examining in greater detail oxidative additions of d^8 and d^{10} transition metal organometallic compounds, we should make one additional point. That is, the addition reactions discussed in the previous sections involve addition of M—X to some unsaturated substrate (alternatively, the substrate inserts into the M—X bond); no metal oxidation occurs. In contrast, in oxidative addition reactions it is the metal complex that is unsaturated and can add various substrates (alternatively, the metal complex inserts into an E—E′ bond).

General Considerations

Usually, d^8 complexes of Fe(0), Co(I), and Ni(II) and their congeners are four-coordinate, square planar, 16 electron complexes or five-coordinate, trigonal bipyramidal, 18 electron complexes (cf. Chapter 15). Both polar and non-polar species such as H_2, halogens, hydrogen halides, alkyl and acyl halides, organotin halides, mercuric halides, organosilicon halides and hydrides, and other similar compounds can add to these transition metal compounds, sometimes reversibly. In general, addition to square planar, 16 electron, d^8 complexes gives six-coordinate, 18 electron, d^6 complexes (17-77), whereas one of the original ligands must be lost in the oxidative addition of two new ligands to five-coordinate, 18 electron, d^8 complexes to

---

[116] M. Kubota, D. M. Blake, and S. A. Smith, *Inorg. Chem.*, **10**, 1430 (1971); M. Kubota and D. M. Blake, *J. Amer. Chem. Soc.*, **93**, 1368 (1971).
[117] I. C. Douek and G. Wilkinson, *J. Chem. Soc.* (A), 2604 (1969).
[118] See Chapter 16, page 827 for a discussion of intramolecular metalation.

give a six-coordinate, 18 electron, d^6 complex as the final product (17-79); in some cases the intermediate in a reaction such as 17-79 may be isolated.

(17-79)[119]

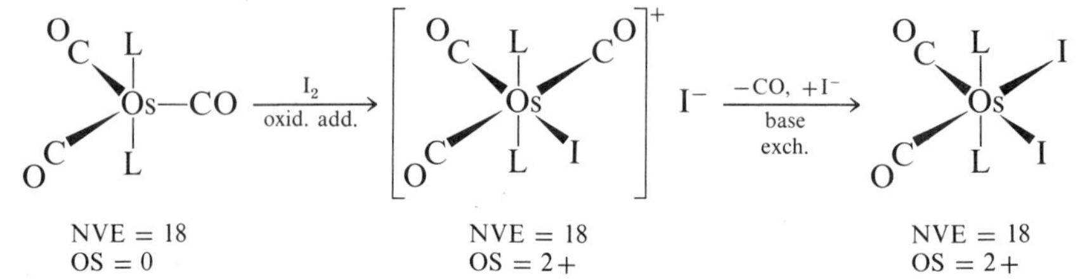

$$\text{OC-Os(L)(CO)(L)(CO)-CO} \xrightarrow[\text{oxid. add.}]{I_2} \left[\text{OC-Os(L)(CO)(L)(CO)-I}\right]^+ I^- \xrightarrow[\text{base exch.}]{-CO,\ +I^-} \text{OC-Os(L)(I)(L)(CO)-I}$$

NVE = 18   NVE = 18   NVE = 18
OS = 0   OS = 2+   OS = 2+

Frequently, d^{10} complexes are coordinatively unsaturated, being only three- or four-coordinate, so addition occurs readily.[120,121]

(17-80)

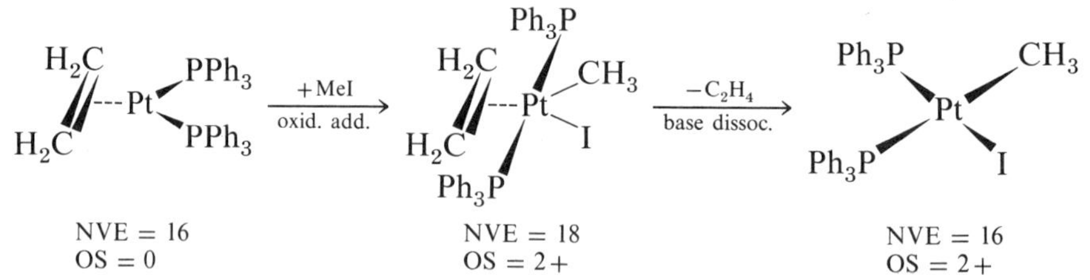

$$(\eta^2\text{-}C_2H_4)_2\text{Pt}(PPh_3)_2 \xrightarrow[\text{oxid. add.}]{+MeI} (\eta^2\text{-}C_2H_4)\text{Pt}(PPh_3)_2(CH_3)(I) \xrightarrow[\text{base dissoc.}]{-C_2H_4} \text{Pt}(PPh_3)_2(CH_3)(I)$$

NVE = 16   NVE = 18   NVE = 16
OS = 0   OS = 2+   OS = 2+

Additions to d^8 complexes have been most thoroughly studied using IrCl(CO)(PPh$_3$)$_2$, frequently called Vaska's compound after Lauri Vaska, who first synthesized the material in 1961 and recognized its importance as a model compound

---

[119] J. P. Collman and W. R. Roper, *J. Amer. Chem. Soc.*, **88**, 3504 (1966).
[120] J. P. Birk, J. Halpern, and A. L. Pickard, *J. Amer. Chem. Soc.*, **90**, 4491 (1968).
[121] R. Ugo, *Coord. Chem. Rev.*, **3**, 319 (1968).

TABLE 17-5
KINETIC DATA FOR OXIDATIVE ADDITION REACTIONS OF MX(CO)L$_2$[a]

Metal	Halogen	Ligand	Addend	Rate Constant ($M^{-1}sec^{-1}$)	$\Delta H^{\ddagger}$ (kcal/mole)	$\Delta S^{\ddagger}$ (eu)
Ir	Cl	PPh$_3$	H$_2$	0.67	10.8	−23
	Br			10.5	12.0	−14
	I			>10^2		
Ir	Cl	PPh$_3$	O$_2$	3.36 × 10^{-2}	13.1	−21
	Br			0.74 × 10^{-1}	11.8	−24
	I			0.30	10.9	−24
Ir	Cl	PPh$_3$	CH$_3$I	3.5 × 10^{-3}	5.6	−51
	Br			1.6 × 10^{-3}	7.6	−46
	I			0.91 × 10^{-3}	8.8	−43
Ir	Cl	P(p-CH$_3$OC$_6$H$_4$)$_3$	CH$_3$I	3.5 × 10^{-2}	8.8	−35
		P(p-ClC$_6$H$_4$)$_3$		3.7 × 10^{-5}	14.9	−28
Rh	Cl	PPh$_3$	CH$_3$I	12.7 × 10^{-4}	9.1	−44
		P(p-CH$_3$OC$_6$H$_4$)$_3$		51.5 × 10^{-4}	10.2	−43

[a] P. B. Chock and J. Halpern, *J. Amer. Chem. Soc.*, **88**, 3511 (1966); R. Ugo, A. Pasini, A. Fusi, and S. Cenini, *ibid.*, **94**, 7364 (1972); I. C. Douek and G. Wilkinson, *J. Chem. Soc.* (A), 2604 (1964). With a few exceptions, the reactions were run in benzene at 25°C.

for studies in homogeneous catalysis.[122,123] The reactions of this compound and its analogs (the chloride may be changed to another halogen or pseudo-halide, and PPh_3 to a variety of other phosphines) with H_2, O_2, and CH_3I are first order in both reactants.[124-127] Depending on the halide, the phosphine, and the solvent, the values of $\Delta H^\ddagger$ can vary from 5 to 17 kcal/mol and $\Delta S^\ddagger$ from -10 to -60 e.u. (Table 17-5). Although these activation parameters and the rate law suggest an A or I_a mechanism wherein the metal complex functions as a nucleophile, there is still considerable uncertainty concerning the mechanism of oxidative additions, a topic taken up briefly later in this section.

Stereochemistry of Oxidative Additions

The stereochemistry of oxidative additions has been closely scrutinized for clues to the reaction mechanism and the structure of the transition state. In general, non-polar substances such as H_2 add in a *cis* fashion, whereas alkyl halide addition, for example, has been reported to be both *cis* and *trans*, but most often *trans*. The following example serves to illustrate the experimental approach to determination of the stereochemistry about the metal.[128]

If methyl iodide is added to $IrCl(CO)L_2$, nine isomeric products are possible:

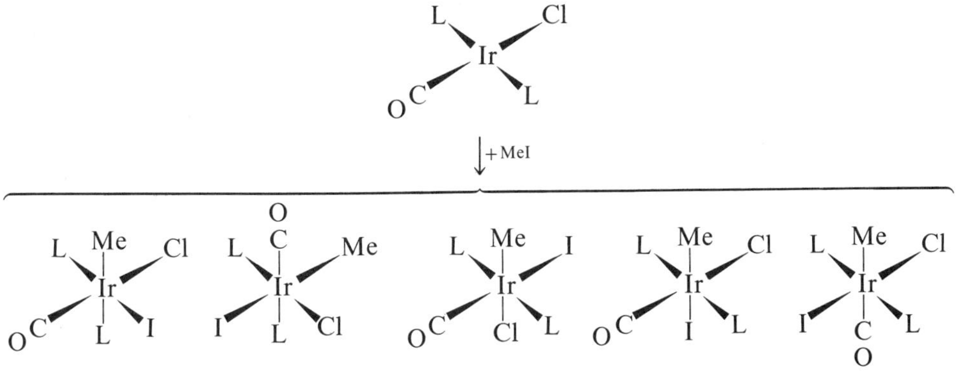

(17-81)

14 **15** — + 4 other diastereomers with L ligands *cis*

cis add. **16**; *trans* add. **17**; *cis* add. **18** — L ligands *trans*

there are six isomers in which the L ligands are *cis* (only two, **14** and **15,** are shown) and three isomers (**16, 17,** and **18**) in which the L ligands remain *trans* (see Chapter 11, p. 635). To decide which isomer or isomers arise in a given reaction, first the phosphine (= L) positions and then the halide positions are determined.

The positions of the phosphine ligands in $IrCl(Me)(I)(CO)L_2$ can be determined by a simple nmr method, and, fortunately, chemists have consistently found that the phosphine ligands remain *trans* in the octahedral addition products. Thus, six isomer possibilities (*e.g.,* **14** and **15**) are eliminated. The three remaining isomer possibilities **(16-18)** may be differentiated by their Ir—Cl stretching frequencies in the following manner. Addition of Cl_2 to Vaska's compound shows that the stretching frequency of the Ir—Cl bond when *trans* to CO is in the range from 300 to 310 cm^{-1} **(19)**, and the

---

[122] L. Vaska and J. W. DiLuzio, *J. Amer. Chem. Soc.,* **83,** 2784 (1961); *ibid.,* **84,** 679 (1962).

[123] Although better preparations are now available [J. Chatt, N. P. Johnson, and B. L. Shaw, *J. Chem. Soc.* (A), 604 (1967)], Vaska's compound may be synthesized simply by refluxing $IrCl_3$ with PPh_3 in an alcohol or ether (see ref. 122); the CO apparently arises from the solvent.

[124] P. B. Chock and J. Halpern, *J. Amer. Chem. Soc.,* **88,** 3511 (1966).

[125] See ref. 117.

[126] R. Ugo, A. Pasini, A. Fusi, and S. Cenini, *J. Amer. Chem. Soc.,* **94,** 7364 (1972).

[127] J. F. Harrod, C. A. Smith, and K. A. Than, *J. Amer. Chem. Soc.,* **94,** 8321 (1972).

[128] J. P. Collman and C. T. Sears, Jr., *Inorg. Chem.,* **7,** 27 (1968).

942 ORGANOMETALLIC COMPOUNDS: REACTION PATHWAYS

methyl chloride addition product shows two bands at 305 cm^{-1} and 257 cm^{-1}, the latter attributed to ν_{Ir-Cl} when *trans* to a methyl group **(20)**.

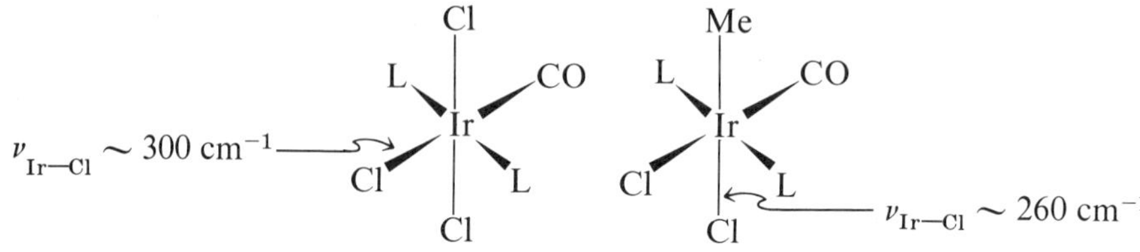

$\nu_{Ir-Cl} \sim 300$ cm^{-1} ⟶ [structure **19**]   [structure **20**] ← $\nu_{Ir-Cl} \sim 260$ cm^{-1}

Addition of MeBr or MeI to a benzene solution of IrCl(CO)L$_2$ (when L = PPh$_2$Me or PPhMe$_2$) gives only one isomer in each case.[129,130] That isomer has the L ligands *trans* and has only one band for ν_{Ir-Cl} at about 300 cm^{-1}. Thus, addition is stereospecific and *trans* (compound **17**) in these particular cases.

Unfortunately, the stereochemical question is not simple, since there is evidence that the stereochemistry of addition is, at the very least, solvent dependent. For example, addition of MeBr or MeI to IrCl(CO)(PPhMe$_2$)$_2$ gives both *cis* and *trans* products (**16** and **17** but not **18**) when the iridium compound is dissolved in methanol.[130] Furthermore, addition of MeCl to IrBr(CO)(PPh$_2$Me)$_2$ gives a *trans* addition product **(21)** if the reaction is carried out in benzene, but, on refluxing in MeOH—C$_6$H$_6$, **21** is converted to the complex **(22)** that would have resulted if the original oxidative addition had been *cis*.[129] This result suggests that the stereochemistry of oxidative addition is the result of kinetic and not thermodynamic control.

(17-82)

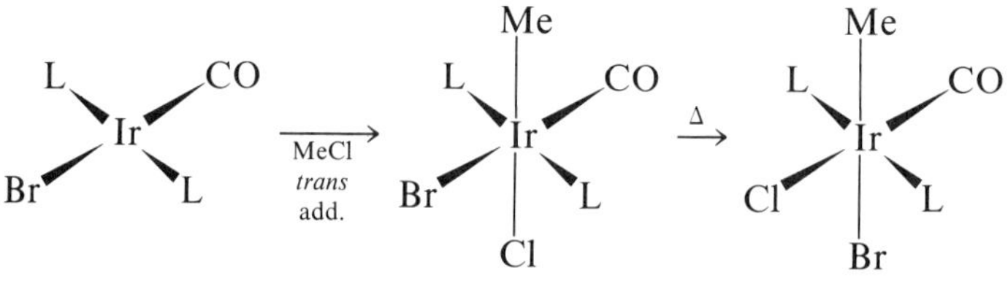

[structures showing MeCl trans add., then Δ, giving compounds **21** and **22**]

Another aspect of the stereochemistry of oxidative additions is the regiochemistry of the carbon atom newly bound to the metal in alkyl halide reactions. That is, by noting the configuration about this carbon atom before and after addition, some insight may be gained into the nature of the transition state. One very thorough study of this aspect of oxidative additions involved the series of reactions 17-83 and 17-84.[131] S-(−)-α-phenethylbromide oxidatively adds to carbonyltris(triphenylphosphine)-palladium to give, after a subsequent alkyl migration (with retention of configuration), compound **23** with inversion of configuration. Notice that compound **23** can also be produced by reaction of the acyl halide of known configuration, **24**, with Pd(PPh$_3$)$_4$. This proves that configuration inversion occurred in the addition of Ph(Me)(Br)CH, since addition of **24** can occur only with retention.

(17-83)

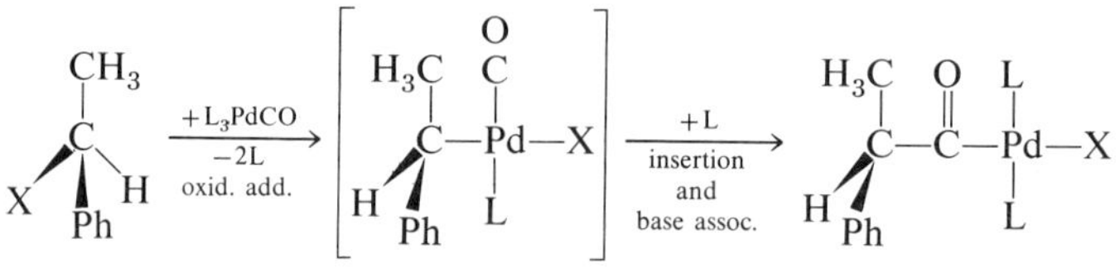

[reaction scheme: S-(−) starting material + L$_3$PdCO, −2L, oxid. add. → R-(+) intermediate → +L insertion and base assoc. → R-(+) **23**]

---

[129] See ref. 128.
[130] A. J. Deeming and B. L. Shaw, *J. Chem. Soc.* (A), 1128 (1969).
[131] K. S. Y. Lau, R. W. Fries, and J. K. Stille, *J. Amer. Chem. Soc.*, **96**, 4983 (1974); *ibid.*, **96**, 5956 (1974).

$$\underset{\underset{23}{R\text{-}(+)}}{\overset{H_3C}{\underset{H}{\overset{|}{\underset{Ph}{C}}}}\text{—}\overset{O}{\overset{\|}{C}}\text{—}\overset{L}{\underset{L}{\overset{|}{Pd}}}\text{—}X} \xrightarrow[\substack{\text{base dissoc.} \\ \text{and} \\ \text{oxid. add.}}]{-2L} \underset{\underset{24}{R\text{-}(-)}}{\overset{CH_3}{\underset{H}{\overset{|}{\underset{Ph}{C}}}}\text{—}COX} + PdL_4 \quad (17\text{-}84)$$

In addition to this and other examples of configuration inversion, several instances of racemization have been reported.[132,133] The mechanistic implications of these results are discussed below.

Influence of Central Metal, Ligands, and Addend on Oxidative Addition

The one aspect of oxidative addition reactions that has been emphasized from the beginning is that the metal complex can be thought of as functioning as a nucleophile, and anything that affects the nucleophilicity (or basicity) of the metal therefore influences the course of the reactions.[134,135] For this reason, the effect of the metal itself, the electronic and steric properties of the ligands, and the addend (the electrophile) have been extensively examined, and a reasonably consistent reactivity picture has begun to emerge.

The Lewis basicity of low valent metals in organometallic compounds was mentioned in Chapter 16 (p. 887) and earlier in this chapter (p. 913).[136] Especially noteworthy was the fact that, whereas Vaska's compound reacts with BF_3 to form a 1:1 adduct with an Ir—B bond (reaction 17-16), the rhodium analog will form adducts only with the stronger Lewis acids BCl_3 and BBr_3. In addition to this example, there are many other indications that the lighter metals in a given subgroup are less basic or nucleophilic than the heavier elements (see p. 913). For example, while Vaska's compound reacts completely and irreversibly with CH_3I, the reaction is quite reversible for the rhodium analog (reaction 17-77). Similarly, the relative rates of reaction of CH_3I with $(\eta^5\text{-}C_5H_5)M(CO)PPh_3$ are 1.0 for M = Co, 1.4 for M = Rh, and about 8 for M = Ir.[137]

$$\text{(Structure with Cp, M, Ph}_3\text{P, CO)} + CH_3I \xrightarrow{\text{oxid. add.}} \left[\text{(Cp-M complex with Ph}_3\text{P, CH}_3\text{, CO)} \right]^+ \xrightarrow[\substack{\text{R migration} \\ \text{and} \\ \text{base assoc.}}]{I^-} \text{(Cp-M complex with Ph}_3\text{P, I, COCH}_3\text{)} \quad (17\text{-}85)$$

M = Co, Rh, Ir
NVE = 18
OS = 1+

NVE = 18
OS = 3+

NVE = 18
OS = 3+

Although there are exceptions, the tendency to undergo oxidative additions parallels the trends normally observed in Group VIII. That is, the largest metals (the third row metals) in the lowest oxidation states show the greatest tendency to undergo oxidative additions.[138]

---

[132] For leading references see ref. 131.
[133] M. F. Lappert and P. W. Lednor, *Adv. Organometal. Chem.*, **14**, 345 (1976).
[134] C. K. Brown, D. Georgian, and G. Wilkinson, *J. C. S. Dalton*, 929 (1973) and references therein.
[135] A. J. Deeming and B. L. Shaw, *J. Chem. Soc. (A)*, 1802 (1969).
[136] See refs. 30 and 31.
[137] A. J. Hart-Davis and W. A. G. Graham, *Inorg. Chem.*, **9**, 2658 (1970).
[138] R. S. Nyholm and K. Vrieze, *J. Chem. Soc.*, 5337 (1965).

944 ORGANOMETALLIC COMPOUNDS: REACTION PATHWAYS

Fe(0)	Co(I)	Ni(II)	↓ tendency to undergo oxidative addition
Ru(0)	Rh(I)	Pd(II)	
Os(0)	Ir(I)	Pt(II)	

← tendency to undergo oxidative addition

A very complete study of the various effects on oxidative addition reactions arose in a most unexpected way. Collman began a study of the reaction of Vaska's compound with acyl azides, expecting to observe an adduct, **25**. Instead, a molecular nitrogen adduct, **26**, was isolated![139] As this work was done shortly after the first molecular nitrogen complexes were discovered, the isolation of $IrCl(N_2)(PPh_3)_2$ was an especially important result and warranted further study. Therefore, the original reaction was repeated with carefully purified reagents and solvent (chloroform)—to the detriment of the reaction; **27** was formed instead. By observing the infrared spectrum of the reaction mixture during the course of the reaction, it was found that the ethyl alcohol that is present in commercial chloroform as an oxidation inhibitor traps the acyl isocyanate that is the product of the first stage of the reaction. If this is not trapped, the acyl isocyanate reacts further with the N_2 complex to produce **27**.[140]

(17-86)

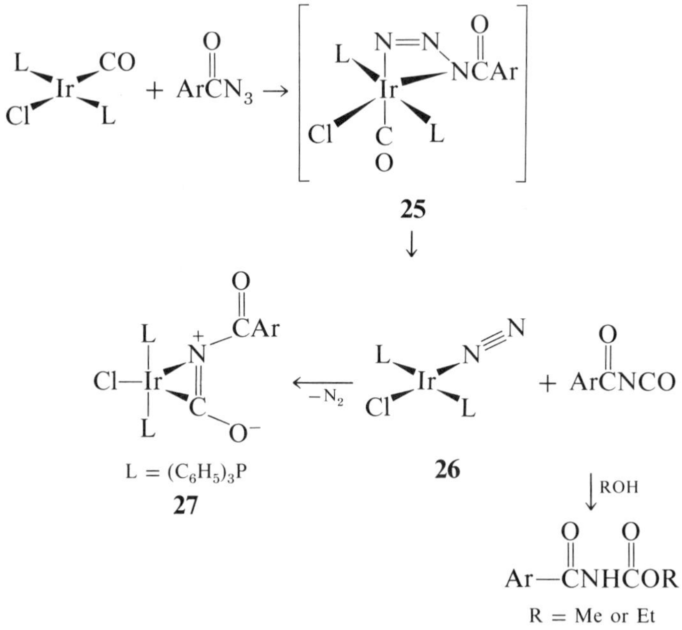

The importance of reaction 17-86 to the current discussion is the change in the rate of formation of **25** with changes in metal, halogen, and azide. In all cases, the rate law indicates a first order dependence on both reactants; this and the activation parameters ($\Delta H^\ddagger = 6.7$ to 8.8 kcal/mol; $\Delta S^\ddagger = -37$ to -42 e.u.) suggest that the rate determining step is a concerted, bimolecular formation of **25**. More importantly, the rate data show that the rate decreases as

---

[139] J. P. Collman, M. Kubota, F. D. Vastine, J. Y. Sun, and J. W. Wang, *J. Amer. Chem. Soc.*, **90**, 5430 (1968).

[140] Students take notice! There must be a moral somewhere in this tale.

i. the metal is changed from iridium to rhodium
ii. the halogen is changed in the order $I > Br > Cl > N_3$
iii. the phosphine is changed in the order $PhEt_2P > Ph_2MeP > Ph_3P$
iv. the azide is changed in the order $PhCON_3 > PhN_3 > p\text{-}MeC_6H_4N_3$

It is readily apparent from these observed reactivity trends that the central metal functions as a base and the azide as an acid; any change in the metal complex that leads to an increase in electron density at the central metal leads to an increase in reactivity. These same general trends have been found, with a few exceptions,[141] in numerous other studies of the kinetics of oxidative addition reactions as a function of ligand.

The influence of other ligands on oxidative additions is also evident from the reaction of O_2 with Vaska's compound and its analogs. If O_2 is bubbled into a benzene solution of Vaska's compound, the yellow color changes to orange owing to the formation of the O_2 adduct, **28**.[142]

$$IrCl(CO)(PPh_3)_2 + O_2 \rightleftharpoons \underset{\mathbf{28}}{\begin{array}{c}\text{Cl}\quad\text{PPh}_3\quad\text{O}\\ \diagdown\;|\;\diagup\\ \text{Ir}\\ \diagup\;|\;\diagdown\\ \text{OC}\quad\text{PPh}_3\quad\text{O}\end{array}}$$

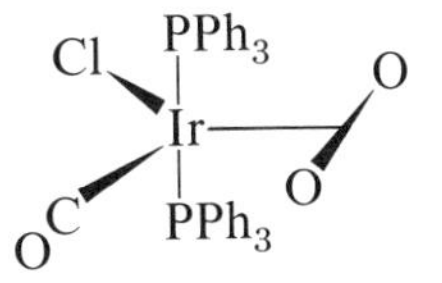

(17-87)

Most importantly, the reaction is reversible (as in the case of the hemoglobin-O_2 interaction); the O_2 may be removed by sweeping the solution with N_2. However, the fine control exerted by ligands in reactions of this type is apparent in the irreversibility of O_2 adduct formation when the Cl in Vaska's compound is changed to I. The reason for this is readily explained in terms of a bonding model described below.

On page 206, the $IrCl(CO)L_2 \cdot O_2$ complex was treated briefly as an O_2^{2-} adduct of Ir(III). That approach is amplified here to explain the increasing irreversibility of the complexation reaction as the halide is changed from Cl to I and to give further insight into the structure of the O_2 complex (**28**).

Applying the angular overlap model to a unit of the type IrL_4O_2, and considering only σ bonding, the d mo energies and filling are illustrated below. However, the O_2 HOMO's (**29**) must be added to the picture, and it is crucial to the argument to

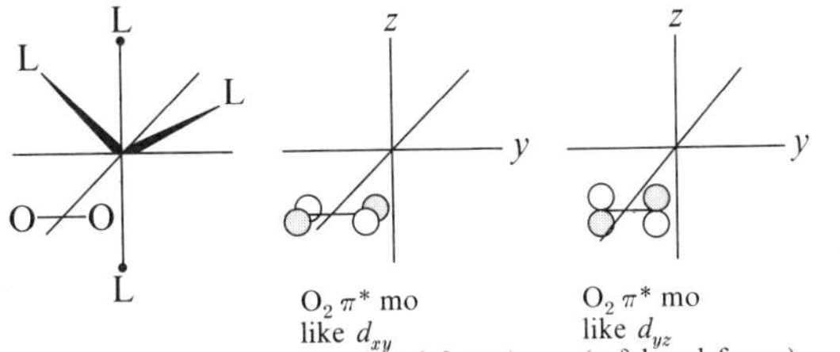

$O_2 \pi^*$ mo like d_{xy} (a π bond forms)

$O_2 \pi^*$ mo like d_{yz} (a δ bond forms)

29

recognize that these HOMO's are the $O_2 \pi^*$ mo's and that they are of a much lower energy than the π^* mo's of an olefin. As illustrated by **29**, the $O_2 \pi^*$ mo's can interact with metal d_{xy} and d_{yz} orbitals, and the result is illustrated in the energy level diagram on page 946. The most important result of this mo picture is that a pair of electrons has in essence been transferred from the metal to the O_2 ligand. Thus, the O_2 complex **28** is best considered as a complex of O_2^{2-} and d^6 Ir(III), and the term "oxidative addition" is meaningful.

---

[141] L. Vaska, L. S. Chen., W. V. Miller, *J. Amer. Chem. Soc.*, **93**, 6671 (1971).
[142] L. Vaska, *Acc. Chem. Res.*, **9**, 175 (1976).

mo's for IrL_4O_2
σ bonding only

mo's for IrL_4O_2
after considering
Ir—O_2 π* interaction

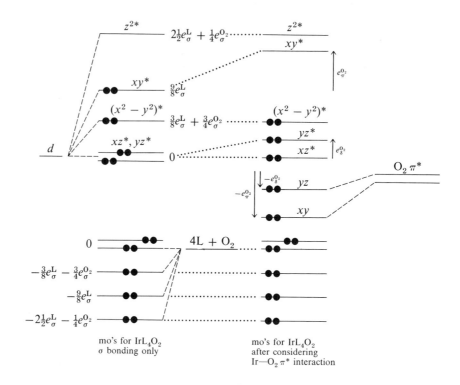

The effect of changing the halide ion can now be rationalized. As the ligand ao's are changed from those of Cl to the more diffuse ones of I, the Ir—X overlap is decreased, and the metal d_{xy} orbital takes on more Ir(d_{xy})/O_2(π*) character. This has two effects: (i) The metal—O_2 bond is strengthened. (ii) As the iridium d_{xy} pair becomes more localized on O_2, it plays an antibonding role and leads to O—O bond lengthening.

The angular overlap approach can also be used to explain the fact that the O—O bond axis is not parallel with the z axis. In this case the O_2 π* mo's would interact with metal d_{xz} and d_{yz} orbitals **(30)**, and this would result in the mo picture below and the energy level diagram on page 947.

O_2 π* mo
like d_{xz}
(π bond forms)

O_2 π* mo
like d_{yz}
(δ bond forms)

30

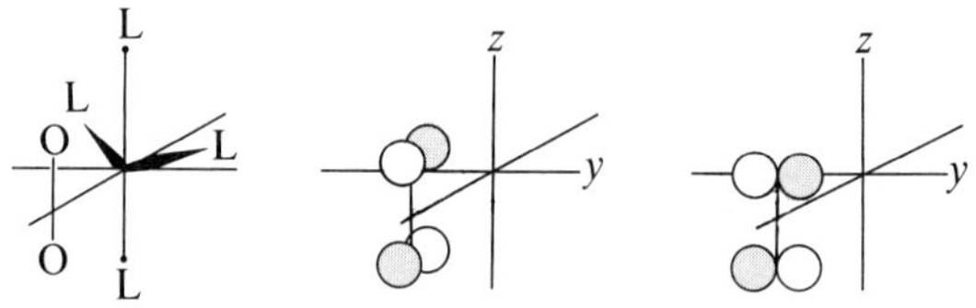

The most important aspect of this scheme is that metal—O_2 π bonding leads to destabilization of the previously non-bonding d_{xz} electron pair. Thus, as opposed to the mo scheme for **29**, where the net M-$\pi^*_{O_2}$ interaction is stabilizing, that interaction is lost when the O_2 axis is parallel with the z axis **(30)**. Whereas the stabilization of $\pi^*_{O_2}(xy)$ in **29** is not offset by occupation of the d_{xy} mo, in **30** the stabilization of $\pi^*_{O_2}(xz)$ is offset by the occupation of d^*_{xz}.

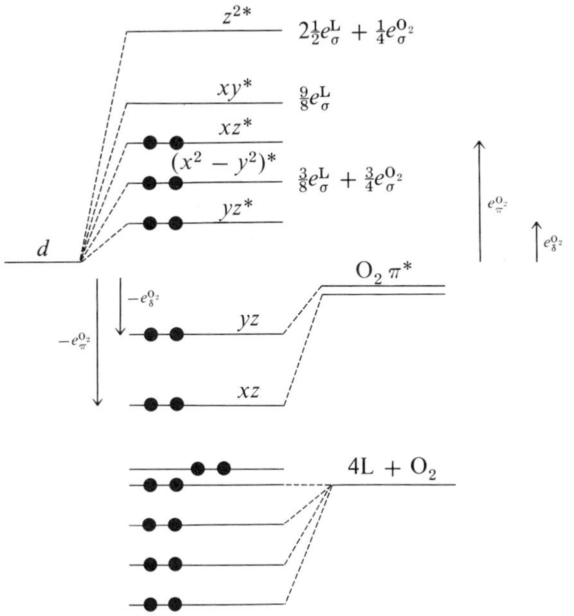

[Energy level diagram with labels: z^{2*} $2\frac{1}{2}e_\sigma^L + \frac{1}{4}e_\sigma^{O_2}$; xy^* $\frac{9}{8}e_\sigma^L$; xz^*; $(x^2-y^2)^*$ $\frac{3}{8}e_\sigma^L + \frac{3}{4}e_\sigma^{O_2}$; yz^*; d; $O_2\pi^*$; yz; xz; $4L + O_2$; with arrows $e_\pi^{O_2}$, $e_\delta^{O_2}$, $-e_\delta^{O_2}$, $-e_\pi^{O_2}$]

Mechanism of Oxidative Addition

The intimate mechanism of oxidative additions is a matter of considerable speculation, and we can only reflect the current uncertainty in the literature.[143] Basically, two mechanistic pathways—not necessarily mutually exclusive—have been proposed: (i) the nucleophilic attack of the metal complex on RX, resulting in a three-center transition state with varying degrees of asymmetry and polarity,[144] and (ii) free radical paths.[143,145,145a]

Prior to about 1972 only the first of these two paths was considered. In general, oxidative additions involving alkyl halides are first order in both reactants, the entropy of activation is usually quite negative, the activation parameters are dependent on the solvent polarity, and the rate is dependent on the apparent nucleophilicity of the metal complex. Such observations suggest that oxidative additions involve nucleophilic attack on the alkyl halide and pass through a transition state of considerable polarity (reaction 17-88).

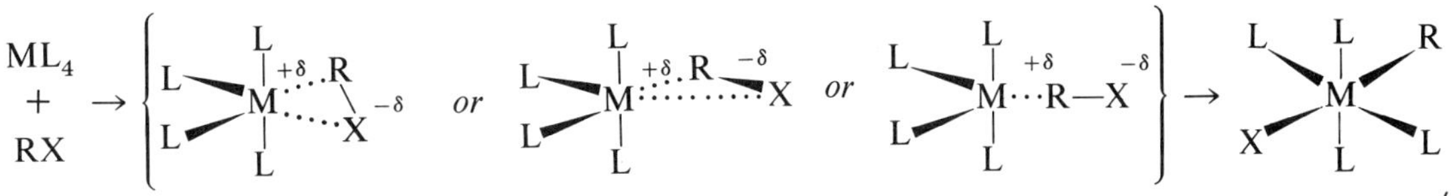

$$ML_4 + RX \rightarrow \left\{ \text{three transition state structures} \right\} \rightarrow \text{product} \tag{17-88}$$

In fact, in its extreme, the transition state resembles that usually proposed for the attack of a nucleophile on an alkyl halide, a suggestion that rationalizes the observations of configuration inversion cited above and is consistent with *trans* addition.

More recently, considerable experimental evidence has accumulated which suggests that radical mechanisms may compete with nucleophilic attack. For example, the reaction of $IrCl(CO)(PMe_3)_2$ with $PhCHFCH_2Br$ is initiated by radical

---

[143] See ref. 133.
[144] R. Ugo, A. Pasini, A. Fusi, and S. Cenini, *J. Amer. Chem. Soc.*, **94**, 7364 (1972).
[145] A. V. Kramer and J. A. Osborne, *J. Amer. Chem. Soc.*, **96**, 7832 (1974).
[145a] J. A. Connor and P. I. Riley, *Chem. Commun.*, 634 (1976). This is the first unambiguous evidence for the formation of R· and a metal center of intermediate valence.

sources such as O_2 or benzoyl peroxide, and it is inhibited by radical scavengers (*e.g.*, hydroquinone).[146] Furthermore, racemization is observed upon addition of chiral $MeCHBrCO_2Et$ to this same Ir(I) compound.[147] Most importantly, however, CIDNP effects are observed in the proton nmr spectrum of a reacting mixture of $Pt(PEt_3)_3$ and isopropyl iodide.[148] (CIDNP stands for "chemically induced dynamic nuclear polarization." It is an effect observed only if free radicals are present.)[149,150] To explain these clear-cut evidences for the involvement of radicals in oxidative addition, it is suggested that radical processes may be in competition with the first pathway discussed.[148] That is, the following scheme has been proposed for the reaction of PtL_3 with an alkyl halide:

(17-89a)

(17-89b)

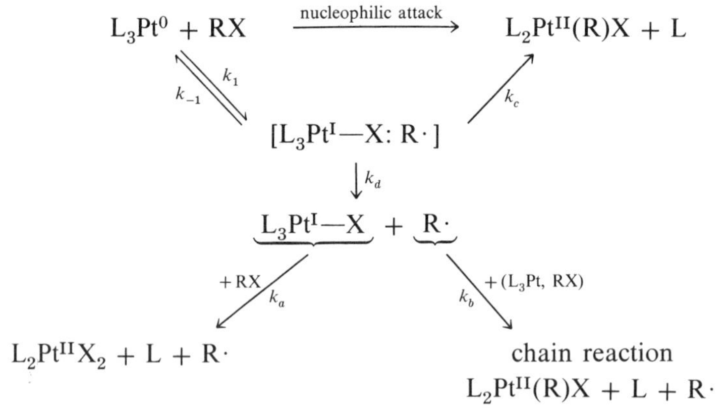

$$L_3Pt^0 + RX \xrightarrow{\text{nucleophilic attack}} L_2Pt^{II}(R)X + L$$

$$\underset{k_{-1}}{\overset{k_1}{\rightleftarrows}} [L_3Pt^I-X: R\cdot] \xrightarrow{k_c}$$

$$\downarrow k_d$$

$$\underbrace{L_3Pt^I-X} + \underbrace{R\cdot}$$

$$\underset{+RX}{\overset{k_a}{\swarrow}} \quad \underset{+(L_3Pt, RX)}{\overset{k_b}{\searrow}}$$

$$L_2Pt^{II}X_2 + L + R\cdot \qquad \qquad \text{chain reaction}$$

$$L_2Pt^{II}(R)X + L + R\cdot$$

In addition to nucleophilic attack of the metal complex on the alkyl halide (17-89a), another path (17-89b) involves halide abstraction by the metal complex to form the radical pair $[L_3Pt^I-X: R\cdot]$. This cage complex can collapse to the product expected in the nucleophilic displacement, or it can diffusively separate to form L_3Pt^I-X and $R\cdot$. The course of the reaction now apparently depends on the reactivity of the alkyl halide. For example, because a C—Br bond is stronger than a C—I bond, the chain mechanism predominates for isopropyl bromide, whereas k_a and k_b are competitive for isopropyl iodide.

ELIMINATION REACTIONS AND THE STABILITY OF METAL–CARBON σ BONDS[150a,b]

Until recently, one of the least understood phenomena in organometallic chemistry was the wide range of stabilities observed for metal–carbon σ bonds.[151,152] For example, whereas tetramethyltitanium decomposes at $-65°C$ in ether to give methane, quite stable metal-carbon σ bonds are observed in $Pt(PPh_3)_2Cl(CH_3)$ and other Pt(II) complexes, vitamin B_{12} (a Co—C bond), $Cr(CH_2CPhMe_2)_4$, and $Mo_2(CH_2SiMe_3)_6$, among others.[153] Many had contended that transition metal-carbon σ bonds in particular were simply intrinsically unstable in a thermodynamic

---

[146] J. S. Bradley, D. E. Connor, D. Dolphin, J. A. Labinger, and J. A. Osborne, *J. Amer. Chem. Soc.*, **94**, 4043 (1972).
[147] A. V. Kramer, J. A. Labinger, J. S. Bradley, and J. A. Osborne, *J. Amer. Chem. Soc.*, **96**, 7145 (1974).
[148] See ref. 145.
[149] S. H. Pine, *J. Chem. Educ.*, **49**, 664 (1972).
[150] T. H. Lowry and K. S. Richardson, "Mechanism and Theory in Organic Chemistry," Harper and Row, New York (1976).
[150a] P. J. Davidson, M. F. Lappert, and R. Pearce, *Chem. Rev.*, **76**, 219 (1976).
[150b] R. R. Schrock and G. W. Parshall, *Chem. Rev.*, **76**, 243 (1976).
[151] P. S. Braterman and R. J. Cross, *J. C. S. Dalton*, 657 (1972); *Chem. Soc. Rev.*, **2**, 271 (1973).
[152] G. Wilkinson, *Pure and Appl. Chem.*, **30**, 627 (1972).
[153] See Chapter 16, pp. 852–855.

sense, with kinetic stability sometimes being achieved by the presence of π bonding or bulky ligands.[154] However, recent thermodynamic measurements have shown that transition metal–carbon σ bonds are often more stable thermodynamically than their main group counterparts.[153] The wide range of stabilities of transition metal compounds having M—C σ bonds must therefore be explicable, at least in part, in terms of relative kinetic stability. That is, unstable compounds must have low energy pathways available for their decomposition.

Two possible pathways for transition metal–carbon bond scission have been proposed: multi-step, radical mechanisms and single-step, non-radical mechanisms.[151] These are discussed briefly at this point, not only because this question is important to metal chemistry in general, but also because the latter mechanism can involve a metal-hydrogen deinsertion reaction and/or reductive elimination, reactions that are apparently important in some metal-catalyzed processes.[155]

The non-concerted pathway is apparently preferred by metals early in the transition metal series and especially by the later metal Hg(II). For example, pyrolysis of diphenylmercury in alcohols leads to formation of benzene and aldehydes, a reaction presumed to follow a radical pathway such as[156]

$$Ph_2Hg \rightarrow 2Ph\cdot + Hg$$
$$Ph\cdot + RCH_2OH \rightarrow PhH + RCH_2O\cdot$$
$$Ph\cdot + RCH_2O\cdot \rightarrow PhH + RCHO$$

The more general decomposition pathways for metals later in the transition metal series[157] are concerted, non-radical mechanisms such as β-elimination (17-90)[158] (called 1,2- and vicinal elimination in Chapter 7;) binuclear elimination (17-91),

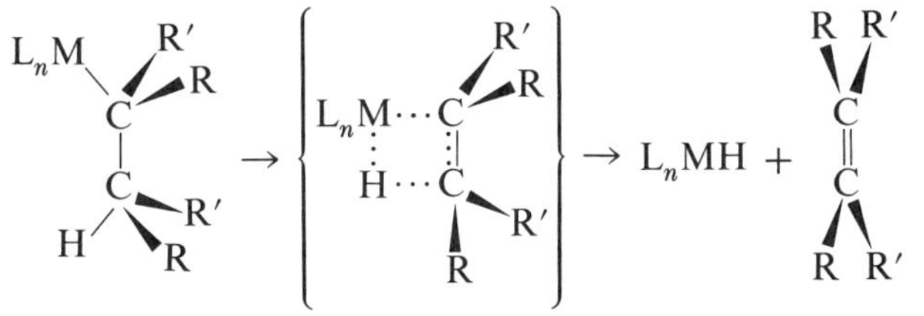

$$\text{(structural diagram)} \rightarrow \{\text{transition state}\} \rightarrow L_nMH + \text{alkene} \qquad (17\text{-}90)$$

or

$$L_nM\text{—}H + R\text{—}ML_n \rightarrow R\text{—}H + 2M + 2nL$$
$$L_nM\text{—}R + R'\text{—}ML_n \rightarrow R\text{—}R' + 2M + 2nL$$
$$\qquad (17\text{-}91)$$

and reductive elimination (17-92)[159] (which can also be called 1,1- or geminal elimination on comparison with 17-90).

---

[154] G. W. Parshall and J. J. Mrowca, *Adv. Organometal. Chem.*, **7**, 157 (1968).
[155] Refer back to Chapter 7, pp. 375 ff., for an orbital-following view of H_2 elimination.
[156] K. C. Bass, *Organometal. Chem. Rev.*, **1**, 391 (1966).
[157] See ref. 151.
[158] For a discussion of the stereochemistry of such reactions, see N. A. Dunham and M. C. Baird, *J. C. S. Dalton*, 774 (1975), and references therein. Elimination is generally thought to be *cis*, but experimental tests thus far have not been unambiguous.
[159] For a recent, leading reference, see E. L. Muetterties and P. L. Watson, *J. Amer. Chem. Soc.*, **98**, 4665 (1976).

(17-92)
$$L_nM\begin{matrix}R\\R'\end{matrix} \rightarrow L_nM + R-R' \quad \text{where R and R'} = \text{organic group or H}$$

$$\text{cis-}H_2CoL_4^+ \rightarrow H_2 + CoL_4^+$$
$$\begin{matrix}\text{NVE} = 18 & & \text{NVE} = 16\\ \text{OS} = 3+ & & \text{OS} = 1+\end{matrix}$$

An excellent example of the first two of these paths is the thermal decomposition of *n*-butyl (tri-*n*-butylphosphine)copper (17-93).[160]

(17-93)
$$2(H_3CCH_2CH_2CH_2)Cu(PBu_3) \xrightarrow[\text{ether}]{0°C}$$
$$H_3C-CH_2-CH=CH_2 + H_3CCH_2CH_2CH_3 + 2Cu + 2PBu_3$$

The proposed reaction sequence involves as a first step β-elimination of the olefin and a copper hydride (17-94a), followed by binuclear elimination in the second step (17-94b). This mechanism demands the formation of equal amounts of *n*-butene and *n*-butane, as is observed.

(17-94a) $\quad CH_3CH_2CH_2CH_2Cu(PBu_3) \rightarrow CH_3CH_2CH=CH_2 + HCu(PBu_3)$

(17-94b) $\quad CH_3CH_2CH_2CH_2Cu(PBu_3) + HCu(PBu_3) \rightarrow CH_3CH_2CH_2CH_3 + 2Cu + 2PBu_3$

The position from which the hydrogen is removed in the first step was established by decomposition of compounds having deuterium atoms on the carbons α and β to the copper (17-95),

(17-95)
$$CH_3-CH_2-CH_2-CD_2-CuL \rightarrow CH_3-CH_2-CH=CD_2 + HCuL$$
$$CH_3-CH_2-CD_2-CH_2-CuL \rightarrow CH_3-CH_2-CD=CH_2 + DCuL$$

and by the presence of HD in the hydrolysis products of the thermolysis residue (*i.e.*, $H_2O + DCuL \rightarrow HD + \cdots$). Finally, by observing the reaction of DCuL with $C_4H_9Cu(PBu_3)$, it was established that *n*-butane is a product of binuclear elimination (17-94b).

Concerted binuclear elimination may also account for the results of classical coupling reactions such as the Ullmann reaction (the coupling of aryl groups of aryl halides by metallic copper).[161]

(17-96)
$$2ArI \xrightarrow{Cu^0} Ar-Ar$$

Evidence for this point of view comes from a study of the decomposition of *m*-(trifluoromethyl)phenylcopper octamer, $(RCu)_8$,[162] and vinylic copper(I) and silver(I) compounds.[163] In the former case, only 3,3'-bis(trifluoromethyl)biphenyl was observed; as there was no evidence for $F_3C-C_6H_4$ radicals, apparently the *m*-(trifluoromethyl)phenyl groups are evolved pairwise in a unimolecular reaction. Also observed is a green species thought to be a Cu(I)/Cu(0) complex; since the original $(RCu)_8$ complex apparently loses two R groups, this green complex may be $(RCu)_6Cu_2$. This latter complex also loses R_2 slowly. The overall process is accounted for by the following equation.

(17-97)
$$(RCu)_8 \xrightarrow{\text{fast}} R-R + (RCu)_6Cu_2 \xrightarrow{\text{slow}} (RCu)_4Cu_4 + R-R$$
$$R = m\text{-}CF_3C_6H_4$$

---

[160] G. M. Whitesides, E. R. Stedronsky, C. P. Casey, and J. San Filippo, Jr., *J. Amer. Chem. Soc.*, **92**, 1426 (1970).

[161] J. March, "Advanced Organic Chemistry: Reactions, Mechanisms, and Structure," McGraw-Hill, New York (1968).

[162] A. Cairncross and W. A. Sheppard, *J. Amer. Chem. Soc.*, **93**, 247 (1971).

[163] G. M. Whitesides, C. P. Casey, and J. K. Krieger, *J. Amer. Chem. Soc.*, **93**, 1379 (1971).

In the case of the vinylic copper(I) and silver(I) compounds, the important observation is that the configuration about the olefinic bonds is retained when the final diene product is formed. Radical inversion (17-98) is known to be extremely rapid ($k = 10^7$ to 10^9 sec^{-1}), so the requirement of configuration retention demands that coupling of vinylic radicals be even more rapid. This is not a reasonable assumption, so it is presumed that the vinylic groups are not released as radicals but couple after formation of a metal cluster of some type.

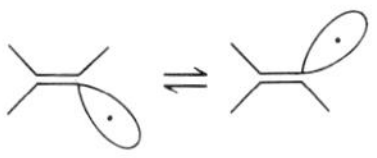

$$\text{（vinyl radical）} \rightleftharpoons \text{（vinyl radical）} \tag{17-98}$$

$$\text{（vinyl-Ag）} \xrightarrow[\text{4-6 hrs.}]{\text{AgPBu}_3 \text{ room temp.}} \text{（diene）} + \text{Ag}(0) + \text{PBu}_3 \tag{17-99}$$

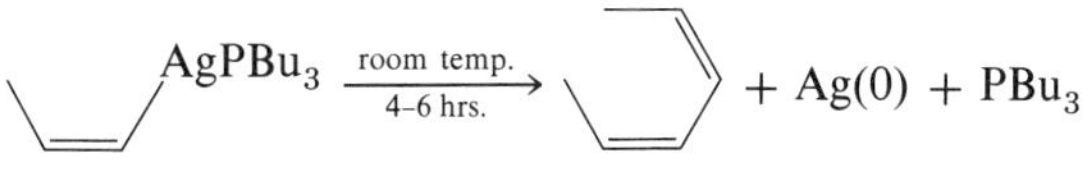

Yet another way for metal–carbon bonds to be broken is the 1,1-reductive elimination process, the opposite of 1,1-oxidative addition. The reaction of chloromethylbis(triphenylphosphine)platinum with HCl is a well-understood example of such a reaction.[164]

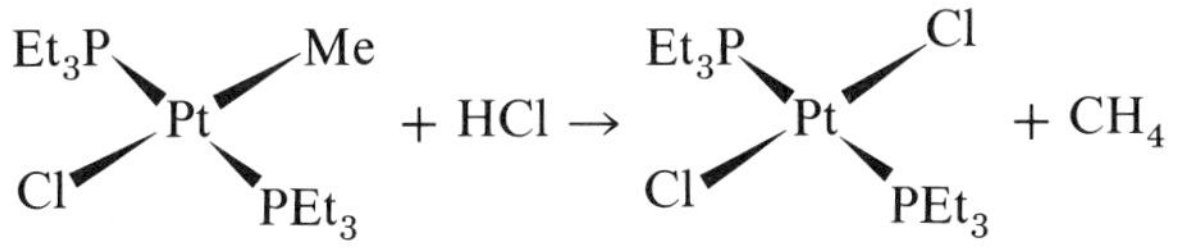

$$\underset{\substack{\text{Et}_3\text{P} \\ \text{Cl}}}{\overset{\text{Me}}{\text{Pt}}}\underset{\text{PEt}_3}{} + \text{HCl} \rightarrow \underset{\substack{\text{Et}_3\text{P} \\ \text{Cl}}}{\overset{\text{Cl}}{\text{Pt}}}\underset{\text{PEt}_3}{} + \text{CH}_4 \tag{17-100}$$

The rate law for the process is especially interesting in that it involves both second and third order terms.

$$-\frac{d[\text{PtL}_2\text{ClMe}]}{dt} = \{k_2[\text{H}^+] + k_3[\text{H}^+][\text{Cl}^-]\}[\text{PtL}_2\text{ClMe}]$$

The second order term is thought to reflect the formation of an intermediate from the addition of H$^+$ and solvent to the starting material, while the third order term indicates that both H$^+$ and Cl$^-$ are involved in the formation of an intermediate. In both cases, the addition of H$^+$ is an oxidative addition; that is, the H$^+$ becomes a hydride ligand, H$^-$, while platinum is oxidized to 4+. The total number of electrons is raised to 18 by addition of Cl$^-$ or a basic solvent molecule (MeOH). No matter which intermediate forms, the rate determining step is concerted, 1,1-reductive elimination of CH$_4$ to give the final, Pt(II) complex.

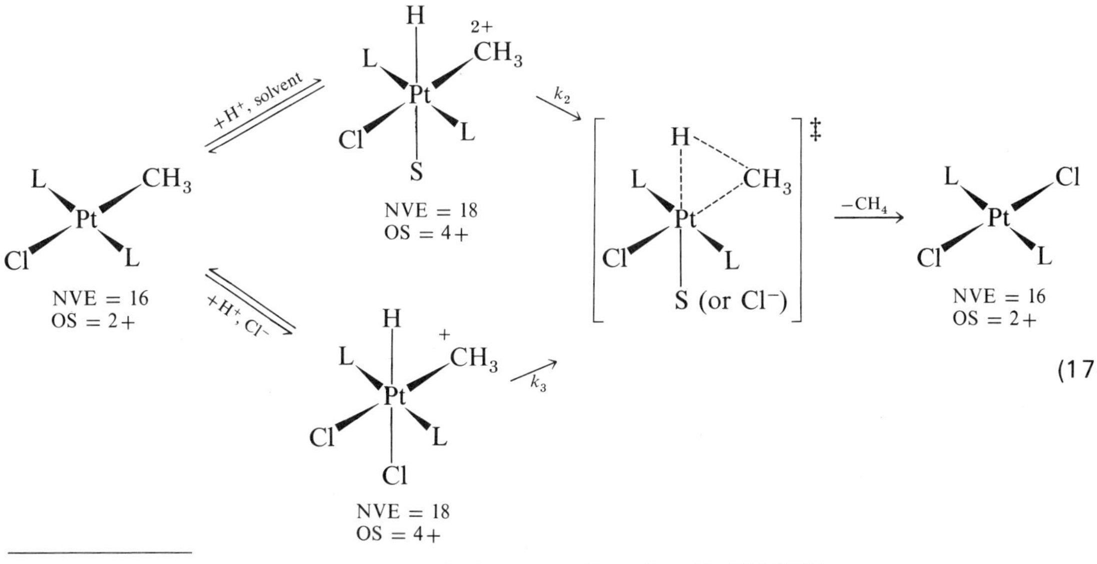

(17-101)

[164] U. Belluco, M. Giustiniani, and M. Graziani, *J. Amer. Chem. Soc.*, **89**, 6494 (1967).

STUDY QUESTIONS

15. Propose an intermediate for the reaction

    $$PtL_2(R_2C=CR_2) + L \rightarrow PtL_3 + R_2C=CR_2 \quad (L = \text{a phosphine})$$

16. Consider the data in Table 17-5 and answer the following questions:

    a) Why should oxidative addition of H_2 generally be accelerated as X is changed from Cl to I?

    b) Methyl iodide reacts faster with $IrCl(CO)L_2$ when the L ligand is changed from $P(p\text{-}ClC_6H_4)_3$ to $P(p\text{-}CH_3OC_6H_4)_3$. Why?

17. Suggest a mechanism for the reaction

    [Structure: CpMn(CO)(PPh$_3$)(CO) + HSiPh$_3$ ⇌ CpMn(CO)(CO)(H)(SiPh$_3$) + PPh$_3$]

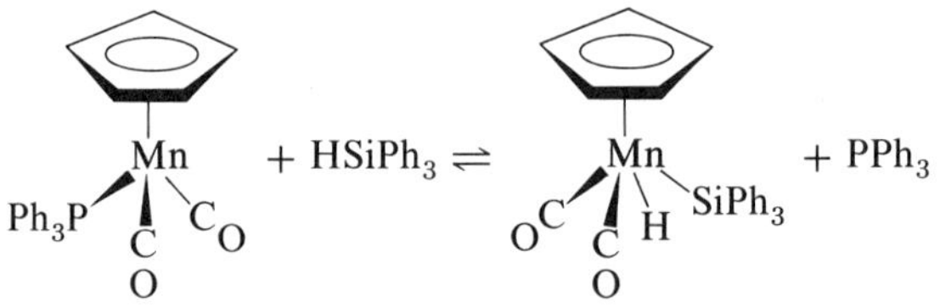

    Calculate ΔNVE, ΔOS, and ΔN. Into which category in Table 17-1 does this process best fit? [If you need help, see A. J. Hart-Davies and W. A. G. Graham, *J. Amer. Chem. Soc.*, **93**, 4388 (1971).]

18. Based on your knowledge of substitution reactions in square planar complexes from Chapter 13 and the discussion of insertion reactions in this chapter, propose two mechanistic pathways for the following reaction that are first order in both starting materials.

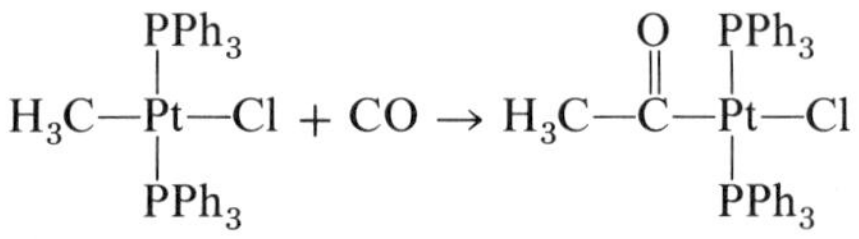

    $$H_3C\text{-}Pt(PPh_3)_2\text{-}Cl + CO \rightarrow H_3C\text{-}C(O)\text{-}Pt(PPh_3)_2\text{-}Cl$$

    [If you need some help, see P. E. Garrou and R. F. Heck, *J. Amer. Chem. Soc.*, **98**, 4115 (1976).]

19. Cl^- reacts with *trans*-$[Pt(PEt_3)_2(C_6H_5)_2]$ in the presence of acid to give *trans*-$[Pt(PEt_3)_2(C_6H_5)Cl]$ and benzene. No reaction occurs with LiCl in the absence of acid, and the rate decreases drastically on changing the solvent from methanol to anhydrous diethyl ether. The overall rate expression is

    $$-\frac{d[PtL_2Ph_2]}{dt} = k[H^+][PtL_2Ph_2]$$

    Suggest a mechanism for the reaction consistent with the information above. [If you need help, see U. Belluco, et. al., *Inorg. Chem.*, **6**, 718 (1967).]

20. Using addition reactions, suggest syntheses for the following compounds:

    a) 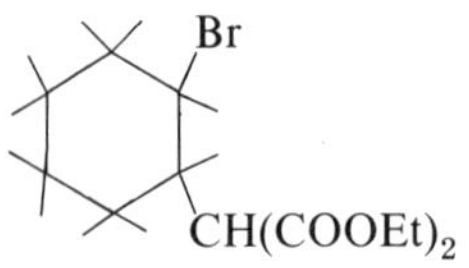 [cyclohexane with Br and CH(COOEt)$_2$ substituents]

    b) 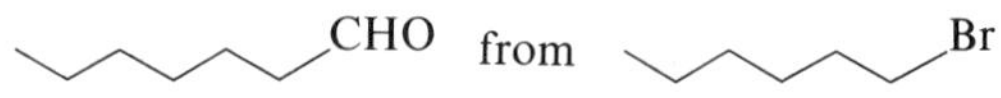 [CH$_3$(CH$_2$)$_n$CHO] from [CH$_3$(CH$_2$)$_n$Br]

c) 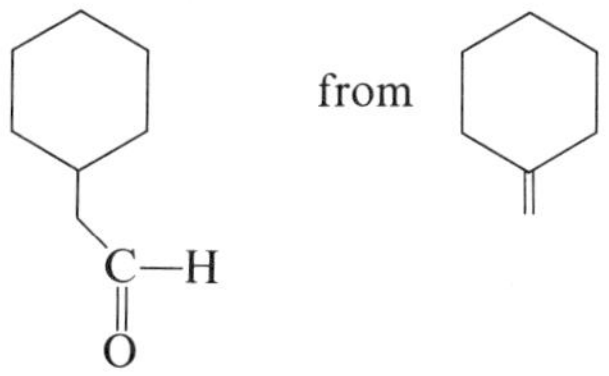

21. If MeCl is added to $IrCl(CO)L_2$, six isomers are possible. Draw the structures of these isomers.

22. Nine isomers arise in reaction 17–81, but only five are illustrated. Draw the remaining four.

23. Account for the course of the following reaction.

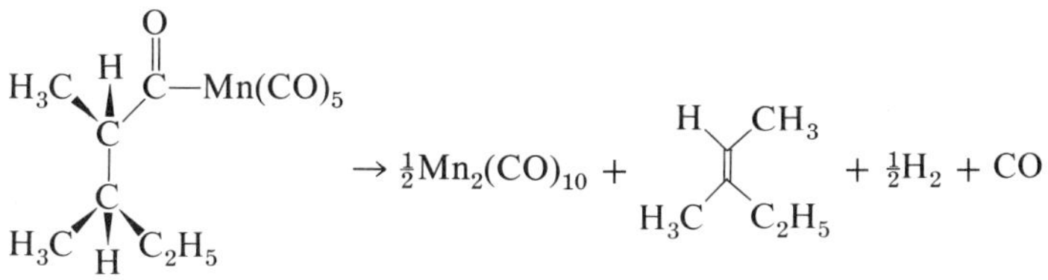

24. In addition to the Ullmann coupling reaction mentioned on page 950, one can couple organic groups by reacting an alkyl or aryl halide with Mg to give the Grignard reagent and then adding a metal halide such as $CoCl_2$ to the Grignard. From your knowledge of elimination reactions, suggest a stoichiometric mechanism for such a reaction.

25. Write equations to show how the "chain reaction" mechanism would operate after formation of L_xPt^I—X in scheme 17–89b.

26. With regard to the mechanism of oxidative addition, why is racemization of an alkyl halide taken as evidence for free radical mechanisms?

REARRANGEMENT REACTIONS

In this section we take up another reaction type that is prevalent in organometallic chemistry as well as in classical coordination chemistry and main group chemistry. That is, there are redistribution or exchange reactions, which are intermolecular processes, and fluxional isomerism or stereochemical non-rigidity, the intramolecular time-averaging of several possible molecular configurations. You have previously encountered exchange reactions in boron and silicon chemistry (Chapter 7, pp. 390 and 401), and you learned of the importance of non-rigidity to silicon and phosphorus compounds in Chapter 7 and to classical coordination compounds in Chapter 14.

REDISTRIBUTION REACTIONS[165–168]

There are at least three distinct types of organometallic redistribution reactions:
a) Exchange of a ligand between two molecules of the same type.

$$Me_3B \cdot NMe_3 + BMe_3^* \rightleftharpoons Me_3B + Me_3N \cdot BMe_3^* \qquad (17\text{–}102)^{169}$$

[165] J. C. Lockhart, "Redistribution Reactions," Academic Press, New York, (1970).
[166] J. P. Oliver, *Adv. Organometal. Chem.,* **8,** 167 (1970).
[167] T. L. Brown, *Acc. Chem. Res.,* **1,** 23 (1968).
[168] K. Moedritzer, *Organometal. Chem. Rev.,* **1,** 179 (1966); *Adv. Organometal. Chem.,* **6,** 171 (1968).
[169] A. H. Cowley and J. L. Mills, *J. Amer. Chem. Soc.,* **91,** 2915 (1969).

b) Scrambling of a group within a molecule.

(17-103)[170]

$$\text{Me}_2\text{Al}(\mu\text{-Me})(\mu\text{-Me})\text{Al}(\text{Me})(\text{Me*}) \rightleftharpoons \text{Me}_2\text{Al}(\mu\text{-Me})(\mu\text{-Me*})\text{Al}(\text{Me})_2$$

c) Interchange of groups or ligands between two molecules of the same or different types.

(17-104)[171] $a\text{Me}_4\text{Sn} + b\text{SnCl}_4 \rightleftharpoons x\text{Me}_3\text{SnCl} + y\text{Me}_2\text{SnCl}_2 + z\text{MeSnCl}_3$

(17-105)[172] $a\text{Ni(CO)}_4 + b\text{Ni(PF}_3)_4 \rightleftharpoons x\text{Ni(CO)}_3(\text{PF}_3) + y\text{Ni(CO)}_2(\text{PF}_3)_2 + z\text{Ni(CO)}(\text{PF}_3)_3$

(17-106)[173] $\frac{1}{2}(\text{LiMe})_4 + \text{Me}_2\text{Mg} \rightleftharpoons \text{Li}_2\text{MgMe}_4 \overset{+\text{LiMe}}{\rightleftharpoons} \text{Li}_3\text{MgMe}_5$

(17-107) $\text{Me}_6\text{Al}_2 + \text{Me}_2\text{Al}_2\text{Cl}_4 \rightleftharpoons 2\text{Me}_4\text{Al}_2\text{Cl}_2$

Since reactions such as 17-102 have already been discussed in detail in Chapter 7 (p. 410ff.), we shall look only briefly at the second and third types of redistribution reactions.

---

[170] K. C. Williams and T. L. Brown, *J. Amer. Chem. Soc.*, **88**, 5460 (1966).
[171] D. Grant and J. R. Van Wazer, *J. Organometal. Chem.*, **4**, 229 (1965).
[172] R. J. Clark and E. O. Brimm, *Inorg. Chem.*, **4**, 651 (1965).
[173] L. M. Seitz and T. L. Brown, *J. Amer. Chem. Soc.*, **89**, 1607 (1967).

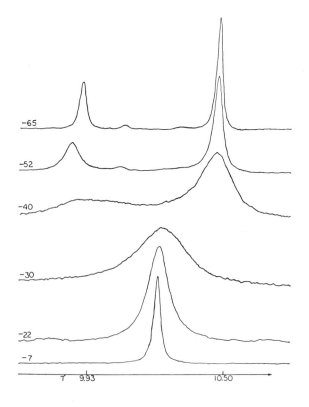

Figure 17-1. Proton nmr spectrum of Me_6Al_2 in toluene. [Reprinted with permission from K. C. Williams and T. L. Brown, *J. Amer. Chem. Soc.*, **88**, 5460 (1966). Copyright by the American Chemical Society.]

One of the most thoroughly studied exchange reactions is the scrambling of the bridge and terminal methyl groups in dimeric trimethylaluminum and analogous organoaluminum compounds.[174-176] At room temperature the proton nmr spectrum of Me_6Al_2 in toluene consists of a single resonance line for all methyl groups, but cooling to $-40°$ and finally to $-65°$ causes the line to split into two resonance lines, one at higher field for the terminal methyl groups and one at lower field for the bridging groups (Figure 17-1). There are two possible mechanisms for the bridge-terminal exchange that occurs at room temperature.

i) Rate determining dissociation of the dimer to monomers, followed by monomer recombination in a new orientation or combination of a new pair of monomers.

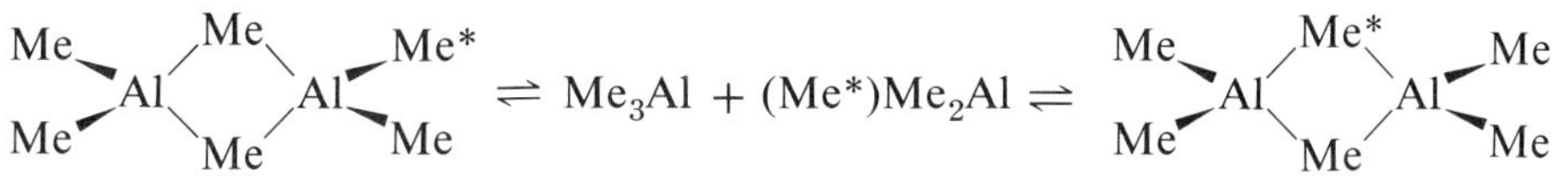

$$\text{Me}_2\text{Al}(\mu\text{-Me})_2\text{AlMe}\text{Me}^* \rightleftharpoons \text{Me}_3\text{Al} + (\text{Me}^*)\text{Me}_2\text{Al} \rightleftharpoons \text{Me}_2\text{Al}(\mu\text{-Me},\mu\text{-Me}^*)\text{AlMe}_2 \tag{17-108}$$

ii) Intramolecular processes involving either the breaking of only one Al—Me—Al bridge followed by rotation about the surviving Al—C_{bridge} bond and reformation of the doubly bridged dimer,

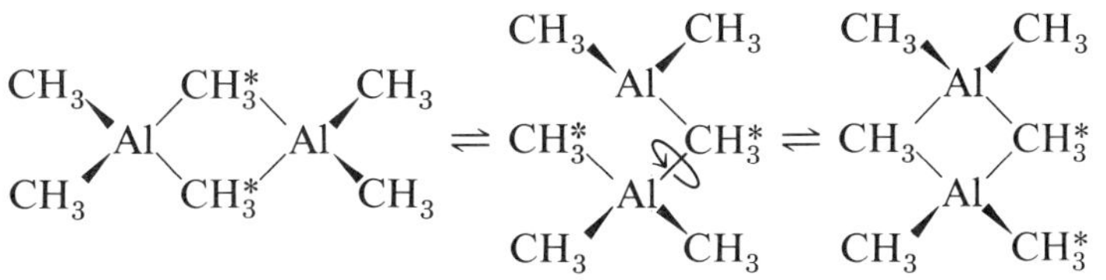

$$\tag{17-109}$$

or formation and subsequent collapse of a species with four bridging groups.

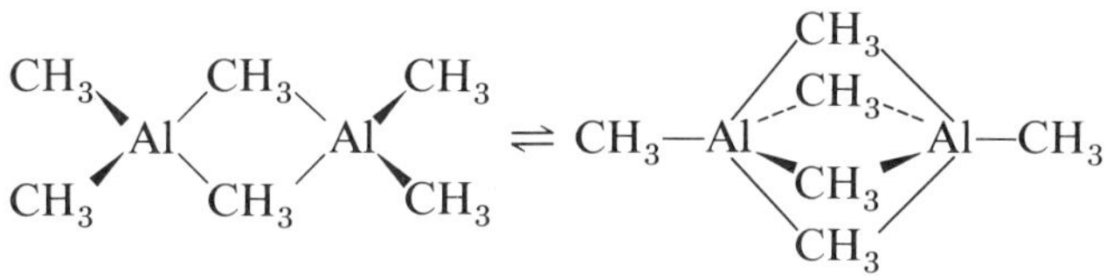

$$\tag{17-110}$$

All of the available evidence points to mechanism(i). The activation energy for the concentration-independent exchange is 15.4 ± 2 kcal/mole in toluene or cyclopentane, a value somewhat less than the heat of dissociation of the trimethylaluminum dimer (20.4 kcal/mole; Table 16-1). While the discrepancy between $\Delta E^{\ddagger}$ and ΔH_{diss} may suggest mechanisms 17-109 or 17-110, these possibilities are largely ruled out by the observation that the exchange of methyl groups between Al_2Me_6 and monomeric Me_3Ga or Me_3In

$$Al_2Me_6 + EMe_2(Me^*) \rightleftharpoons Al_2Me_5(Me^*) + EMe_3 \quad (E = Ga, In) \tag{17-111}$$

also has an activation energy of about 15 kcal. This implies that bridge-terminal exchange in Al_2Me_6 and reaction 17-111 involve the same rate determining step: dissociation of dimeric Al_2Me_6. Thus, the following sequence of reactions has been proposed to account for the type of exchange observed in reaction 17-111:

$$Al_2Me_6 \underset{k_{-1}}{\overset{k_1}{\rightleftharpoons}} 2AlMe_3 \tag{17-112}$$

$$AlMe_3 + EMe_2(Me^*) \underset{k_{-2}}{\overset{k_2}{\rightleftharpoons}} AlMe_2(Me^*) + EMe_3 \tag{17-113}$$

---

[174] See ref. 170.
[175] E. A. Jeffery and T. Mole, *Aust. J. Chem.*, **22**. 1129 (1969).
[176] D. S. Matteson, *Inorg. Chem.*, **10**, 1555 (1971).

As mentioned in the introduction to Chapter 16, tin compounds, often having several different substituents, are important as polymerization catalysts, agricultural fungicides and growth stimulators, and so on. Redistribution reactions such as 17-104 play a major role in their manufacture.[177] As an example of such reactions, consider that between $SnMe_4$ and $SnCl_4$.[177a] In principle, redistribution of Me and Cl between two Sn atoms can lead to three mixed species: $SnMe_3Cl$, $SnMe_2Cl_2$, and $SnMeCl_3$. In order to describe the equilibria that can involve these mixed compounds and the starting materials, three basic equilibrium expressions are necessary; that is

(17-114a) $\qquad SnCl_4 + SnMe_2Cl_2 \rightleftharpoons 2SnMeCl_3 \qquad K_{ideal} = 2.67$

(17-114b) $\qquad SnMeCl_3 + SnMe_3Cl \rightleftharpoons 2SnMe_2Cl_2 \qquad K_{ideal} = 2.25$

(17-114c) $\qquad SnMe_4 + SnMe_2Cl_2 \rightleftharpoons 2SnMe_3Cl \qquad K_{ideal} = 2.67$

The equilibrium constants for these reactions, K_{ideal}, can be obtained by statistical methods. In addition, it is important for you to note that, since the number and type of bonds do not change in redistribution reactions, ΔH should ideally be zero, and the reaction is driven by the entropy of mixing.[177b]

All possible equilibrium processes involving the five $SnMe_iCl_{4-i}$ ($i = 0 \ldots 4$) compounds can be expressed by the equations above or combinations of them. For example, addition of the three equations gives

(17-115) $\qquad SnCl_4 + SnMe_4 \rightleftharpoons SnMeCl_3 + SnMe_3Cl$

Indeed, this reaction is that first observed in mixtures of $SnCl_4$ and $SnMe_4$ of any composition in the temperature range from 0° to 50°C. This redistribution goes to completion before any other reactions are seen.

At higher temperatures, a 1:1 mixture of $SnCl_4$ and $SnMe_4$ ultimately gives $SnMe_2Cl_2$ by 17-115 followed by 17-114b. However, the equilibrium constant observed for 17-114b ($K_{obs} = 10^4$) is much larger than expected for an "ideal random distribution" in a reaction where it is presumed that $\Delta H = 0$. As a result, after 17 hours at 175°C, the product of a 1:1 mixture of $SnCl_4$ and $SnMe_4$ is almost completely $SnMe_2Cl_2$.

When an excess of either $SnCl_4$ or $SnMe_4$ is used, the equilibria represented by equations 17-114a and 17-114c can be observed. With $SnMe_4$ and an excess of $SnCl_4$, for example, 17-115 still occurs rapidly and completely. However, this is followed by 17-116 by which the excess $SnCl_4$ begins to be consumed. As the concentration of $SnMe_2Cl_2$ builds up, 17-114a comes into play, further consuming $SnCl_4$.

(17-116) $\qquad SnCl_4 + SnMe_3Cl \rightleftharpoons SnMe_2Cl_2 + SnMeCl_3$

(17-114a) $\qquad SnCl_4 + SnMe_2Cl_2 \underset{K_{obs} = 14.3}{\rightleftharpoons} 2SnMeCl_3$

The opposite situation, $SnCl_4$ and an excess of $SnMe_4$, leads to the following reactions after the rapid production of mixed species by 17-115.

(17-117) $\qquad SnMe_4 + SnMeCl_3 \rightleftharpoons SnMe_2Cl_2 + SnMe_3Cl$

(17-114c) $\qquad SnMe_4 + SnMe_2Cl_2 \underset{K_{obs} = 330}{\rightleftharpoons} 2SnMe_3Cl$

---

[177] K. Moedritzer, *Organometal. Chem. Rev.*, **1**, 179 (1966); *Adv. Organometal. Chem.*, **6**, 171 (1968).
[177a] D. Grant and J. R. Van Wazer, *J. Organometal. Chem.*, **4**, 229 (1965).
[177b] The notion that $\Delta H = 0$ for redistribution reactions is based on the premise that the Sn—Cl and Sn—C bond energies, for example, are ideally the same in all of the species $SnMe_iCl_{4-i}$. However, it is safe to say that, in fact, this will not be true. Compare, for example, the C—Cl bond dissociation energies in CCl_4 ($= 70 \pm 5$ kcal/mol) and in CH_3Cl ($= 81 \pm 5$ kcal/mol).

All of the redistribution reactions involving the species $SnMe_iCl_{4-i}$ are not enthalpically zero, as indicated by their substantial equilibrium constants.[177b] Apparently, the Sn—Cl or Sn—C bond energies are greater in the mixed products than in the reactants, perhaps owing to an increase in bond ionicity or s character. In any event, the thermodynamic favorability of these and other redistribution reactions has important commercial consequences.

FLUXIONAL ISOMERISM OR STEREOCHEMICAL NON-RIGIDITY

Another form of molecular rearrangement is the intramolecular time-averaging of two or more possible molecular configurations. This phenomenon has been called fluxional isomerism by Cotton[178] and stereochemical non-rigidity by Muetterties.[179] We have already discussed this form of isomerism in the case of classical coordination complexes (Chapter 14) and for XeF_6, Si and P compounds (Chapters 2 and 7), and we now turn to this phenomenon as it manifests itself in organometallic chemistry.

There are at least four types of stereochemical non-rigidity observable in organometallic chemistry:

a) Rearrangement of σ-bonded polyenes.[180,181]

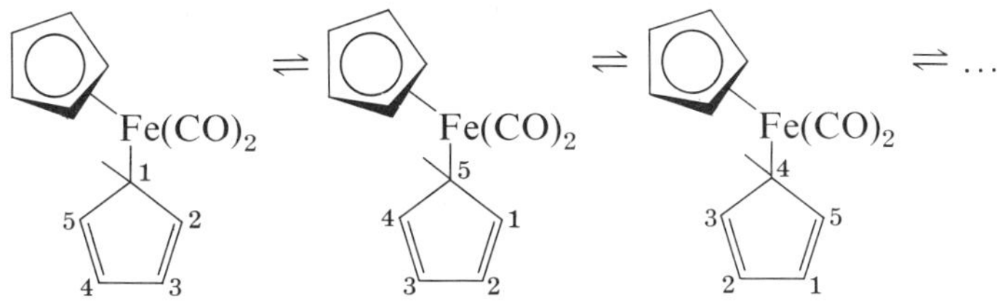

$$\text{(17-118)}$$

b) Acyclic π-bonded polyenes.

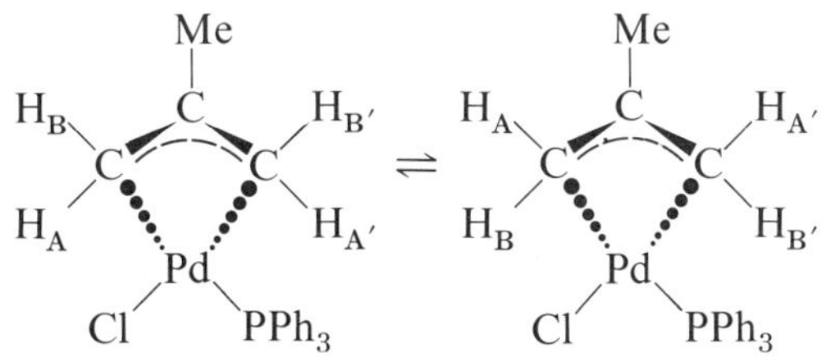

$$\text{(17-119)}[182]$$

c) Cyclic π-bonded polyenes.

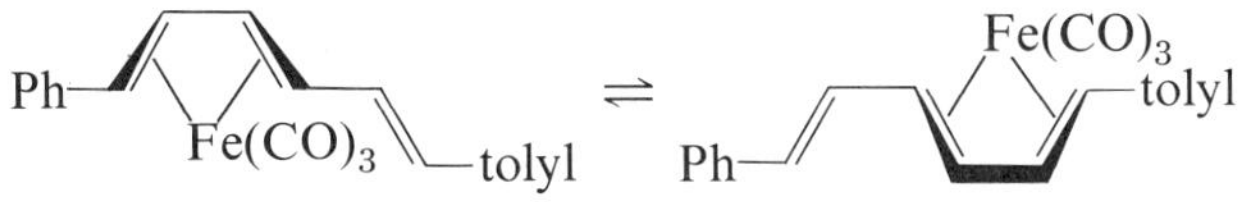

$$\text{(17-120)}[183]$$

---

[178] F. A. Cotton, *Acc. Chem. Res.,* **1,** 257 (1968).
[179] E. L. Muetterties, *Acc. Chem. Res.,* **3,** 266 (1970).
[180] This and the next reference provide entry into the earlier literature on polyenes: F. A. Cotton and T. J. Marks, *J. Amer. Chem. Soc.,* **91,** 7524 (1969).
[181] A. J. Campbell, C. A. Fyfe, R. G. Goel, E. Maslowsky, Jr., and C. V. Senoff, *J. Amer. Chem. Soc.,* **94,** 8387 (1972).
[182] F. A. Cotton, J. W. Faller, and A. Musco, *Inorg. Chem.,* **6,** 182 (1967).
[183] H. W. Whitlock and Y. N. Chuah, *J. Amer. Chem. Soc.,* **87,** 3605 (1965).

958 ☐ ORGANOMETALLIC COMPOUNDS: REACTION PATHWAYS

d) Metal carbonyls.

(17-122)

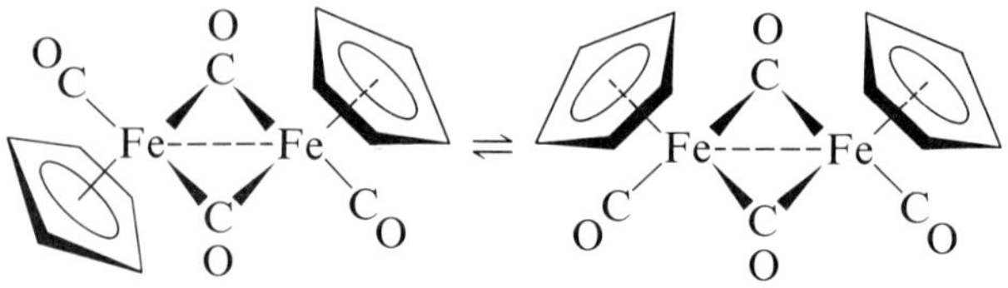

[Structure showing equilibrium between two isomers of bridged Fe-Fe cyclopentadienyl carbonyl complexes]

One of the first examples of fluxional isomerism to be recognized was $(\eta^1\text{-}C_5H_5)(\eta^5\text{-}C_5H_5)Fe(CO)_2$ (reaction 17–118), a compound first synthesized by Piper and Wilkinson in 1956.[184] It was recognized immediately as unusual in that only two sharp lines were observed in the ^1H nmr spectrum at room temperature; one would, of course, expect the *pentahapto*cyclopentadienyl group to give a single line, and the *monohapto* group to give rise to a pair of doublets or triplets. That the *monohapto* C_5H_5 ligand is seen only as a single sharp line at room temperature in this compound, and in many other similar systems isolated since, demands that there be available some low energy pathway for rapid rearrangement that renders all of the protons of $\eta^1\text{-}C_5H_5$ magnetically equivalent. Careful examination of the nmr spectra of compounds such as $(\eta^1\text{-}C_5H_5)(\eta^5\text{-}C_5H_5)Fe(CO)_2$ and $(H_3C)_3Ge(\eta^1\text{-}C_5H_5)$ has ruled out mechanisms whereby all protons simultaneously (through a dissociative mechanism, for example) or randomly become equivalent. The two most reasonable pathways for proton averaging that remain are either a 1,2- or a 1,3-metal shift; that is, movement of a metal atom around the C_5 ring from one carbon to the carbons α or β to it. Experiment more strongly supports the 1,2-shift as the correct mechanism (reaction 17–118).[185,186] (See Study Question 28.)

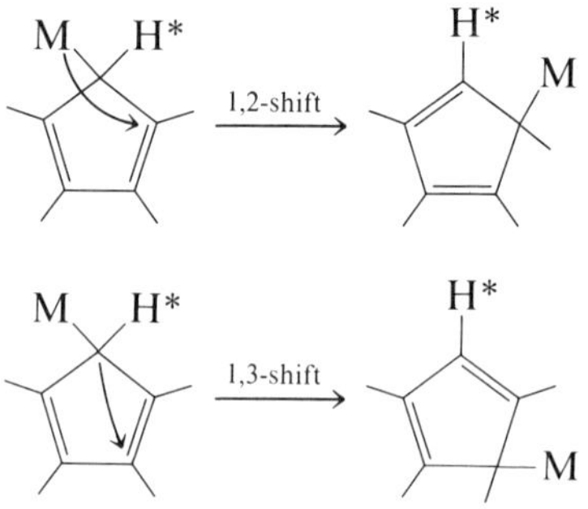

[Diagram showing 1,2-shift and 1,3-shift mechanisms on cyclopentadienyl ring]

One of the more complicated forms of stereochemical non-rigidity is found in metal-η^3-allyl complexes (see Chapter 16, p. 875). A stable complex of this type can exhibit a "static" proton nmr spectrum such as that illustrated in Figure 17–2. (A molecule corresponding to such a spectrum is often labeled AA'BB'X where, in the case of the η^3-allyl complex, X is the proton on carbon 2, A and A' are protons quite different from X and are those located *anti* to X, and B and B' are protons again quite different from X but similar to A/A' and located *syn* to X.) At higher temperatures or in the presence of a base, however, this spectrum collapses to a "dynamic" A_4X spectrum, the chief feature of which is the change of the doublets for the A and B protons to a single doublet, presumably owing to a rapid (on the nmr time scale) intramolecular rearrangement that causes the A/A' (*anti*) and B/B' (*syn*) protons to have the same chemical shift.

The mechanism of η^3-allyl rearrangement has been extensively examined, and a

---

[184] T. S. Piper and G. Wilkinson, *J. Inorg. Nucl. Chem.*, **3**, 104 (1956).
[185] See ref. 180.
[186] A. Davison and P. E. Rakita, *Inorg. Chem.*, **9**, 289 (1970).

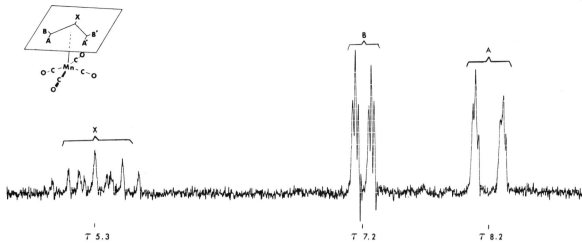

Figure 17-2. Proton nmr spectrum of (π^3-allyl)tetracarbonylmanganese. [Reprinted with permission from F. A. Cotton, J. W. Faller, and A. Musco, *Inorg. Chem.*, **6**, 181 (1967). Copyright by the American Chemical Society.]

number of pathways for the scrambling of *syn* and *anti* protons have been proposed. For example, three of the transition states or intermediates that can lead to *syn,anti* proton equilibration in a complex such as **31** are illustrated below.[187]

In the first of these **(32a)**, the two terminal carbon atoms are momentarily bonded to the palladium by normal σ bonds, and the middle carbon flips from a position above

---

[187] J. W. Faller, M. E. Thomsen, and M. J. Mattina, *J. Amer. Chem. Soc.*, **93**, 2642 (1971) and references therein. See also J. W. Faller and M. E. Thomsen, *ibid.*, **91**, 6871 (1969).

the plane to one below; in the process, the distinction between *syn* and *anti* protons is lost. The same result is achieved if the allyl group is converted to one of its localized forms, $H_2C=C(Me)-CHR$, and rotation of the group occurs about a metal-olefin bond **(32b)** (*cf.* olefin rotation in Chapter 16). Finally, a third way is for the allyl to be σ-bonded to the metal through one of its terminal carbons, *i.e.*, $RCH=C(Me)-CH_2-Pd$ **(32c)**; rotation and re-formation of the η^3 form of bonding can lead to the observed terminal proton equilibration. This latter mechanism is in fact the one that best accounts for proton equilibration in **31**, and in most other cases where a static proton nmr spectrum for a η^3-allyl is observed to change to a dynamic spectrum.[188-190]

Among the most thoroughly studied examples of stereochemical non-rigidity is compound **33** and its derivatives.[191] The *trans* form of compound **33**, for example, is the only form isolated after allowing a warm solution of the material to cool under a nitrogen atmosphere. The *cis* form is obtained by crystallization under reduced pressure or by careful fractional sublimation.[192]

33 *trans*   **33** *cis*

This implies that both *cis* and *trans* isomers exist in solution, a supposition proved by nmr and infrared spectroscopy. Indeed, these techniques show that two processes occur in solution: (i) *cis-trans* isomerization and (ii) scrambling of the bridge and terminal CO ligands. The point of greatest interest is the mechanism of these processes, and a study of isomerization and scrambling in **34,** an unsymmetrical derivative of **33,** provides an unambiguous answer.[193]

34

The key observation is that the rates of isomerization and bridge-terminal CO scrambling are identical. Thus, *cis-trans* isomerization involves bridge opening, rotation, and bridge re-formation to give the other isomer (reaction 17–23, p. 961). Notice that both bridges open and then close simultaneously. In this way, the *cis* isomer of **34** can isomerize to the *trans* isomer, and the process of bridge-terminal CO scrambling is begun. Continuation of the process leads ultimately to complete bridge-terminal CO scrambling. Finally, you should contrast the CO scrambling in **34** with bridge-terminal R group scrambling in aluminum alkyls (p. 955). In the latter case, dissociation to monomers is required, but the Fe—Fe bond in **34** allows retention of molecular integrity in the course of the bridge-terminal CO scrambling.

---

[188] See ref. 182.
[189] D. L. Tibbetts and T. L. Brown, *J. Amer. Chem. Soc.*, **92**, 3031 (1970).
[190] For an apparent exception, see J. W. Faller and M. J. Incorvia, *Inorg. Chem.*, **7**, 840 (1968).
[191] J. G. Bullitt, F. A. Cotton, and T. J. Marks, *Inorg. Chem.*, **11**, 671 (1972); R. D. Adams and F. A. Cotton, *Inorg. Chim. Acta*, **7**, 153 (1973).
[192] R. F. Bryan and P. T. Green, *J. Chem. Soc.* (A), 3064 (1970). See also ref. 193.
[193] F. A. Cotton, L. Kruczynski, and A. J. White, *Inorg. Chem.*, **13**, 1402 (1974).

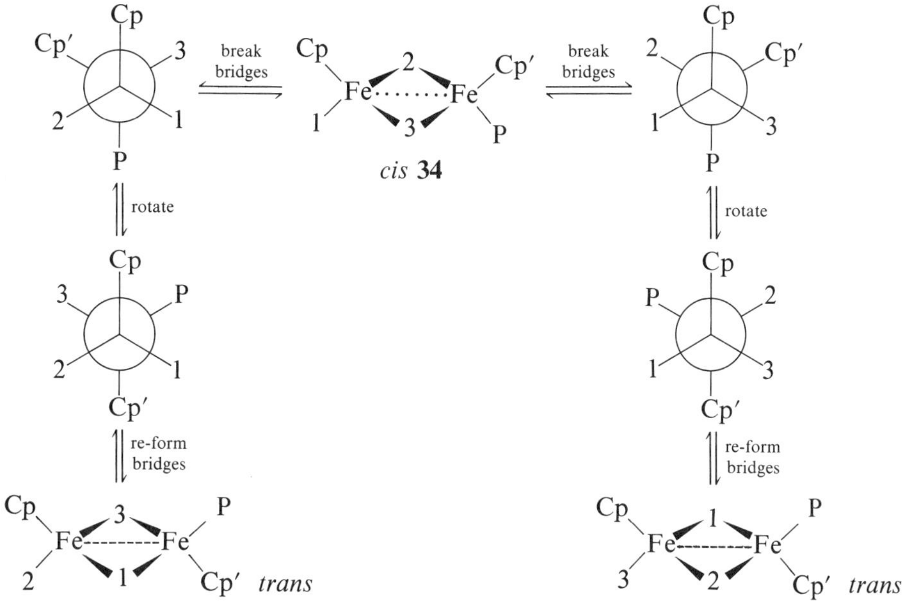

(17-123)

Stereochemical non-rigidity is observed in five-coordinate organometallic complexes just as it is in trigonal bipyramidal main-group compounds (*cf.* Chapter 7).[194] For example, compound **35** displays three different carbonyl stretching bands in its infrared spectrum, as would be expected for the geometry indicated.[195]

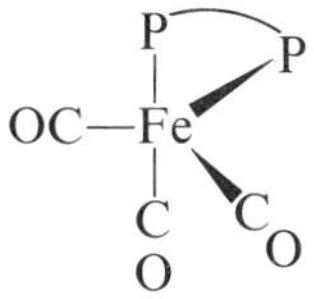

$P\frown P = Me_2P-C_2H_4-PMe_2$

35

However, the ^{13}C nmr spectrum of the compound in the ^{13}CO region consists of a 1:2:1 triplet.[196] It was suggested that this arises from spin-spin coupling between two equivalent ^{31}P nuclei (nuclear spin = $\frac{1}{2}$) and three equivalent CO groups, and that the P and CO equivalence arises because of a very rapid intramolecular averaging process. This process may in fact be the Berry pseudo-rotation process described in Chapter 7 and illustrated below for compound **35**.

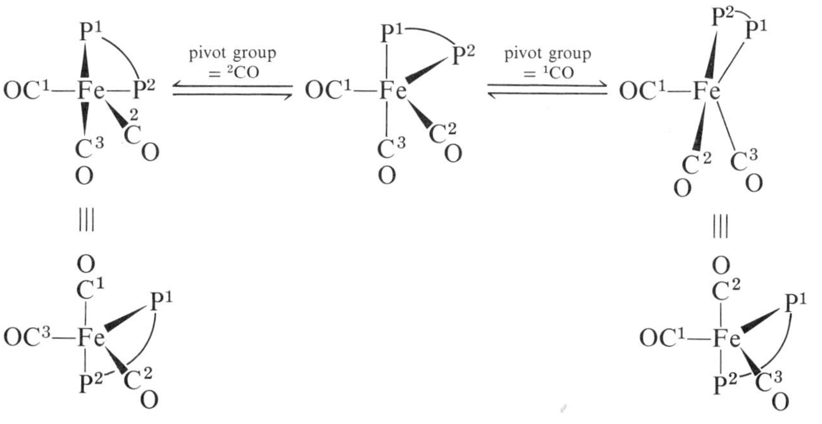

(17-124)

---

[194] J. R. Shapley and J. A. Osborn, *Acc. Chem. Res.*, **6**, 305 (1973).

[195] M. Akhtar, P. D. Ellis, A. G. MacDiarmid, and J. D. Odom, *Inorg. Chem.*, **11**, 2917 (1972).

[196] Recall that the infrared technique can record events occurring over a much shorter time span than nmr. For example, infrared can show that there are three different carbonyl groups in **35**, whereas nmr cannot distinguish them because the molecule rearranges in a shorter time span than that needed for the nmr observation; in contrast, the time span needed for an infrared observation is much shorter than the time of rearrangement. See E. L. Muetterties, *Inorg. Chem.*, **4**, 769 (1965).

962 ORGANOMETALLIC COMPOUNDS: REACTION PATHWAYS

STUDY QUESTIONS

27. Refer back to the mechanism for isomerization and bridge-terminal CO scrambling in scheme 17-123. If the three CO groups are differentiated as shown there, the other two possible *cis* isomers are drawn as follows:

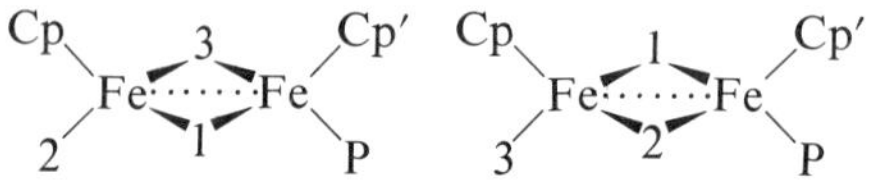

$$\begin{array}{cc} Cp\diagdown3\diagup Cp' & Cp\diagdown1\diagup Cp' \\ Fe\cdots Fe & Fe\cdots Fe \\ 2\diagup1\diagdown P & 3\diagup2\diagdown P \end{array}$$

   a) Show that each of these *cis* isomers gives one of the isomers generated in scheme 17-123 and a new *trans* isomer, and that both of the new *cis* isomers give the same new *trans* isomer.
   b) Show that the isomerization-scrambling process constitutes a cycle. That is, a *cis* isomer gives a *trans* isomer, which gives a *cis* isomer, and so on until you return to the original *cis* isomer.

28. Consider again the case of proton equilibration by 1,2- and 1,3-metal shifts in a *monohapto*cyclopentadienyl group. According to nmr convention, the protons of the ligand can be labeled as illustrated below.

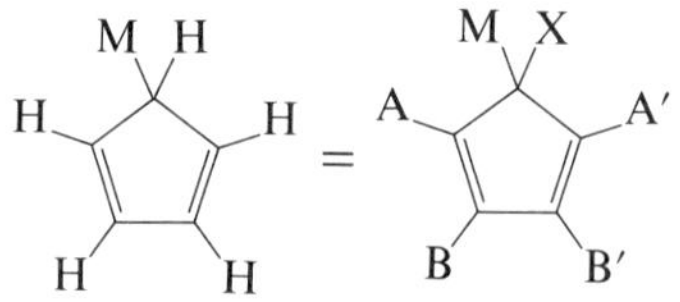

At low temperatures where the rate of metal shifting is slow, the three types of protons (A and A', B and B', and X) can be differentiated by an nmr spectrometer, and a spectrum consisting of basically three lines (one for each proton type) is seen. However, as the temperature is raised, the metal shifting becomes more rapid, the line for the X proton, and that for the A and A' protons, begins to collapse before that for the B and B' protons. That is, the X, A, and A' protons are more affected by metal shifting than are the B and B' protons. To account for this observation, and to decide between 1,2- and 1,3-shifts, show to what type of proton the X, A, A', B, and B' protons are transformed in the first 1,2- or 1,3-shift.

CATALYSIS INVOLVING ORGANOMETALLIC COMPOUNDS

Ethylene, acetylene, 1,4-butadiene, and other olefins are common products of commercial processes and are building blocks for the synthesis of many other compounds of commerce after hydrogenation, oxidation, rearrangement, or polymerization. However, severe conditions are often required to effect such reactions. For example, ethylene must be subjected to a series of vigorous addition-elimination reactions to produce commercially useful acetaldehyde. Such conditions may be circumvented by using a catalyst such as palladium chloride, as you shall see below.

(17-125)
$$\text{ethylene} + H_2O \xrightarrow[\text{or gas phase over phosphoric acid absorbed on diatomaceous earth}]{\text{liquid phase in the presence of } H_2SO_4} C_2H_5OH$$

$$C_2H_5OH \xrightarrow{260-290°C;\ \text{copper catalyst}} CH_3CHO + H_2$$

The function of a catalyst—whether homogeneous or heterogeneous—is to increase the rate of a thermodynamically allowed reaction by lowering the activation energy barrier for the process. In addition, if several reaction paths are possible, a catalyst may increase product specificity by lowering the barrier for one path or by raising it for another. In the case of the hydrocarbons mentioned above, all form well-known complexes with a variety of low-valent transition metals, especially those late in the transition series. Coordination of the organic substrate can result in a lower activation energy in either of two ways: coordination of an olefin may make it more susceptible to nucleophilic attack (as in the $PdCl_2$-catalyzed oxidation of ethylene to acetaldehyde), or both reactants may be brought into proximity by being coordinated to adjacent sites on the catalytic metal (as in hydrogenations, polymerizations and cyclooligomerizations, and nitrogen fixation).

The importance of the products of catalyzed reactions has brought about a great increase in research on catalysts, and understanding has increased immensely in the last few years. We have already seen some examples of catalyzed reactions in Chapter 16: the mercury-catalyzed formation of Et_4Pb (reaction 16-18) and the aluminum-catalyzed growth reaction (reaction 16-69). In addition, some of the best known catalytic processes involving organometallic compounds are:

a) hydrogenation of olefins in the presence of compounds of low-valent metals such as rhodium [*e.g.*, $RhCl(PPh_3)_3$, Wilkinson's catalyst].

b) hydroformylation of olefins using a cobalt or rhodium catalyst (oxo process).

c) oxidation of olefins to aldehydes and ketones (Wacker process).

d) polymerization of propylene using an organoaluminum-titanium catalyst (Ziegler-Natta catalyst) to give stereoregular polymers.

d) cyclooligomerization of acetylenes using nickel catalysts (Reppe's or Wilke's catalysts).

f) olefin isomerization using nickel catalysts.

The chemistry of some of these systems is discussed below. Unfortunately, space does not permit us to explore this field as thoroughly as we would like, so you are referred to various general references.[197-204]

OLEFIN HYDROGENATION

A great many metals having d^5 to d^{10} configurations [Cu(II), Cu(I), Ag(I), Hg(II), Ru(II), Ru(III), Co(I), Co(II), Rh(I), Rh(III), Ir(I), Pd(II), and Pt(II)] are known to activate hydrogen, some of the best known of these being ruthenium(III) chloride (d^5), $[Co(CN)_5]^{3-}$ (d^7), and $RhCl(PPh_3)_3$ (d^8). These three compounds illustrate the three ways in which molecular hydrogen may be activated for hydrogenation:[197]

a) oxidative addition

$$RhCl(PPh_3)_3 + H_2 \rightarrow Rh(H)_2Cl(PPh_3)_3 \qquad (17-126)$$

b) heterolytic splitting

$$[Ru^{III}Cl_6]^{3-} + H_2 \rightarrow [Ru^{III}HCl_5]^{3-} + H^+ + Cl^- \qquad (17-127)$$

---

[197] See refs. 8, 14, and 16.

[198] M. L. Bender, "Mechanisms of Homogeneous Catalysis from Protons to Proteins," Wiley-Interscience, New York (1971).

[199] A lower level version of ref. 198, but also an excellent introduction to catalysis, is: M. L. Bender and L. J. Brubacher, "Catalysis and Enzyme Action," McGraw-Hill, New York (1973).

[200] A. J. Chalk and J. F. Harrod, *Adv. Organometal. Chem.*, **6**, 119 (1968).

[201] R. Cramer, *Acc. Chem. Res.*, **1**, 186 (1968).

[202] R. F. Heck, *Acc. Chem. Res.*, **2**, 10 (1969).

[203] P. M. Henry, *Acc. Chem. Res.*, **6**, 16 (1973).

[204] J. Tsuji, *Acc. Chem. Res.*, **2**, 144 (1969).

c) homolytic splitting

(17-128)
$$2[Co^{II}(CN)_5]^{3-} + H_2 \rightarrow 2[Co^{III}H(CN)_5]^{3-}$$

The first of these reactions offers some insight into catalytic processes in general. While studying the oxidative addition reactions of a rhodium(I) compound, Wilkinson found that $RhCl(PPh_3)_3$ is an effective homogeneous hydrogenation catalyst.[205] [Figure 17-3 compares the rates of hydrogenation of 1-heptene using 10^{-3} M $RhCl(PPh_3)_3$ and a 10^{-3} M suspension of Adam's catalyst, a heterogeneous catalyst based on platinum oxide.] Although there is still considerable speculation, the pathway for the Rh(I)-catalyzed hydrogenation is thought to resemble that illustrated by the scheme below.[206] The first step of importance is formation of the oxidative addition product **36**. This is followed by formation of an olefin complex and addition to Rh and H across the C=C bond. The process is completed by reductive elimination of the now-hydrogenated olefin and regeneration of the catalytic species.

(17-129a)

(17-129b)

---

[205] J. A. Osborne, F. H. Jardine, J. F. Young, and G. Wilkinson, *J. Chem. Soc.* (A), 1711 (1966).
[206] See refs. 8 and 16.

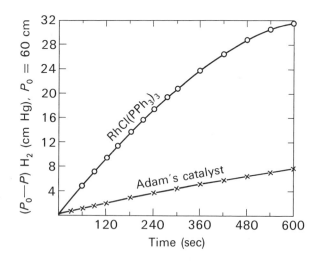

Figure 17-3. Comparison of the rates of homogeneous and heterogeneous hydrogenation. ○ = $RhCl(PPh_3)_3$ and × = Adam's catalyst. [From G. Wilkinson, *Proc. R. A. Welch Found. Conf. Chem. Res.*, **9**, 139 (1966).]

$$\text{RhL}_2\text{HCl(R)} \rightleftharpoons \underset{\substack{\text{NVE}=18\\ \text{OS}=3+}}{\begin{array}{c}\text{Cl}\quad\;\;\text{L}\\ \diagdown\;\;\Big|\\ \text{Rh}-\text{C}-\text{CH}\\ \diagup\;\;\Big|\\ \text{H}\quad\;\;\text{L}\end{array}} \xrightarrow{+S} \underset{\substack{\text{NVE}=18\\ \text{OS}=3+}}{\begin{array}{c}\text{Cl}\quad\;\;\text{L}\quad\;\;\text{S}\\ \diagdown\;\;\Big|\;\diagup\\ \text{Rh}\\ \diagup\;\;\Big|\;\diagdown\\ \text{H}\quad\;\;\text{L}\quad\;\;\text{C}-\text{CH}\end{array}} \xrightarrow{\text{red. elim.}} \underset{\substack{\text{NVE}=16\\ \text{OS}=1+}}{\begin{array}{c}\text{Cl}\quad\;\;\text{L}\\ \diagdown\;\diagup\\ \text{Rh}\\ \diagup\;\diagdown\\ \text{L}\quad\;\;\text{S}\end{array}} + \text{HC}-\text{CH} \qquad (17\text{-}129\text{c})$$

The details of the hydrogen-transfer step in catalytic hydrogenation (*e.g.*, reaction 17-129b) are extremely important, and have been elucidated for the case of a ruthenium(II) chloride complex.[207] The addition of an olefin such as ethylene, maleic acid, or fumaric acid to a solution of the metal chloride results in the formation of a 1:1 metal-olefin π complex. Hydrogenation (at H_2 pressures up to 1 atm) then occurs in a slow step at 60° to 90°C in the case of those olefins in which the double bond is activated (*e.g.*, by the presence of a carboxylic acid group). The overall result is

$$H_2 + \underset{\text{fumaric acid}}{\begin{array}{c}\text{H}\quad\quad\text{COOH}\\ \diagdown\quad\diagup\\ \text{C}=\text{C}\\ \diagup\quad\diagdown\\ \text{HOOC}\quad\text{H}\end{array}} \xrightarrow{[\text{RuCl}_6]^{4-}} \begin{array}{c}\text{HOOC}-\text{CH(H)}\\ \;\\ \text{(H)HC}-\text{COOH}\end{array} \qquad (17\text{-}130)$$

Rate = k[Ru(II)-olefin complex][H_2]

The proposed mechanism involves heterolytic H_2 splitting to form a metal hydride–olefin complex in the rate determining step.[208] This suggestion is confirmed and the stereochemistry of the hydrogenation step elucidated in deuterium tracer experiments as follows.

When the reaction is run in H_2O and in the presence of D_2, no deuterium appears in the product. Moreover, only 2,3-dideuterosuccinic acid is produced from fumaric or maleic acid when the solvent is D_2O, regardless of whether H_2 or D_2 is used. Clearly, *the source of the hydrogen in the reduction process is the solvent.* It is presumed, however, that the H_2 (or D_2) forms a metal-hydride (or deuteride). Relatively rapid H—H, H—D, D—H, or D—D exchange with the solvent would lead to the observed result. In fact, the rate of exchange, measured by using an olefin that is not hydrogenated (5-norbornene-2,3-dicarboxylic anhydride), is approximately equal to the rate of hydrogenation when maleic or fumaric acid is used.

The deuterium studies also show unambiguously that the hydrogenation is stereospecifically *cis. Cis* addition will produce D- or L-2,3-dideuterosuccinic acid, whereas *trans* addition would give the *meso* form.

---

[207] J. Halpern, J. F. Harrod, and B. R. James, *J. Amer. Chem. Soc.*, **88**, 5150 (1966).
[208] The metal hydride has not been detected in this case, but is assumed to exist by analogy with the known Ru(II) hydrides HRuCl(CO)(PPh$_3$)$_3$ and HRuCl(Et$_2$PC$_2$H$_4$PEt$_2$)$_2$.

(17-131)

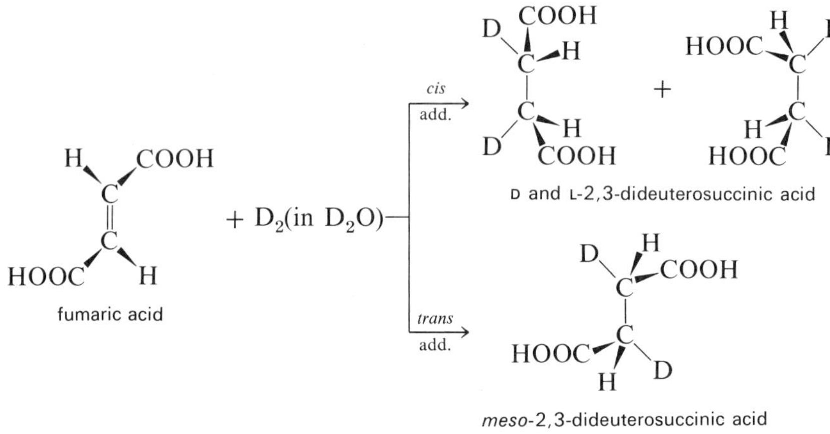

[Fumaric acid + D$_2$ (in D$_2$O) → via cis addition gives D and L-2,3-dideuterosuccinic acid; via trans addition gives meso-2,3-dideuterosuccinic acid]

Only optically active product was in fact isolated. This result implies that the hydrogenation in reaction 17-131 (and presumably in 17-129b) proceeds through a four-center intermediate such as that illustrated in reaction 17-132. The initial formation of the ruthenium alkyl in 17-132 must then be followed by electrophilic attack (with retention of configuration) of H$^+$ of D$^+$ at the Ru—C σ bond.

(17-132)

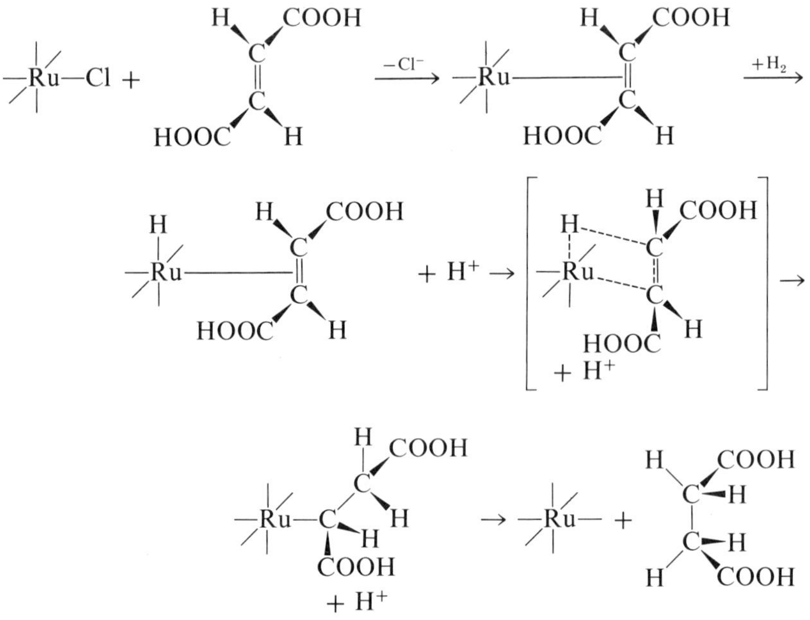

THE OXO REACTION[209,210]

In a system composed of cobalt salts, CO, H$_2$, and an olefin, the most important reaction that occurs is hydroformylation—the addition of the elements of formaldehyde to an olefin.

$$RHC=CH_2 + CO + H_2 \xrightarrow{Co_2(CO)_8} RH_2C-CH_2-\overset{O}{\underset{\|}{C}}-H$$

---

[209] See refs. 200 and 202.

[210] A new rhodium-based oxo process was recently put into operation [Chemical and Engineering News, April 26, 1976]. If propene is used as a feedstock in the oxo process, for example, both normal and iso-butyraldehyde are produced. The advantage of the Rh-based process is that it produces a higher normal-to-iso ratio than the Co-based reaction.

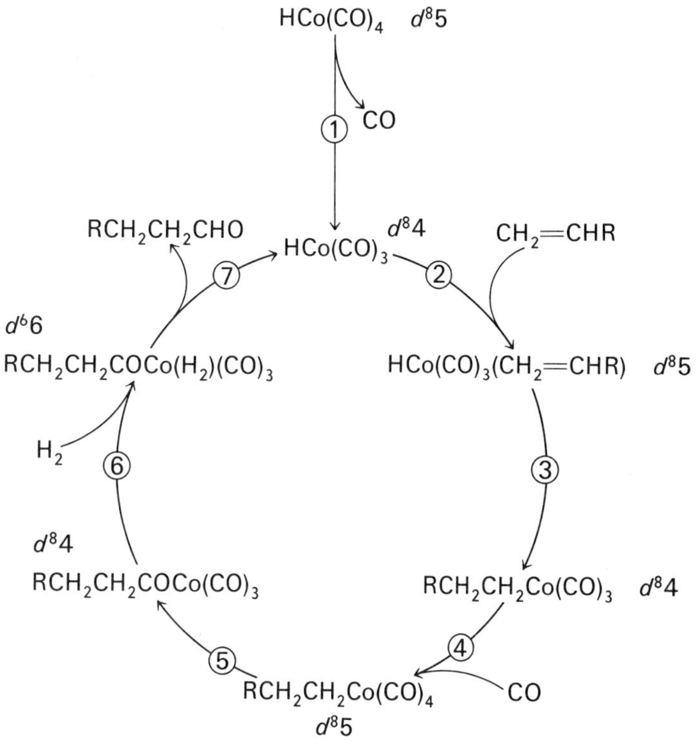

Figure 17-4. Suggested pathway for the hydroformylation of a terminal olefin ["OXO" reaction] by $HCo(CO)_4$. [From C. A. Tolman, *Chem. Soc. Revs.*, **1**, 337 (1972).] (The number of d electrons and the coordination number are given for each transition metal species.)

Cycle (clockwise from top):
$HCo(CO)_4$ $d^8 5$ → (1) $-CO$ → $HCo(CO)_3$ $d^8 4$ → (2) $CH_2=CHR$ → $HCo(CO)_3(CH_2=CHR)$ $d^8 5$ → (3) → $RCH_2CH_2Co(CO)_3$ $d^8 4$ → (4) CO → $RCH_2CH_2Co(CO)_4$ $d^8 5$ → (5) → $RCH_2CH_2COCo(CO)_3$ $d^8 4$ → (6) H_2 → $RCH_2CH_2COCo(H_2)(CO)_3$ $d^6 6$ → (7) → RCH_2CH_2CHO + $HCo(CO)_3$

Clues to the mechanism of the reaction are provided by the observations that the reaction is first order in olefin and approximately first order in the amount of cobalt present; the rate is faster for terminal olefins than for internal olefins; and excess CO inhibits the reaction. One possible scheme explaining these observations is shown in Figure 17-4. As we have previously mentioned (p. 934), the primary cobalt-containing species in the system is hydridotetracarbonylcobalt, $HCo(CO)_4$, a five-coordinate, 18 electron molecule. If one is to build a pathway involving only 16 and 18 electron species, the most rational first step is loss of CO to give a coordinatively unsaturated 16 electron species, $HCo(CO)_3$. Indeed, this initial loss of CO rationalizes the observation that the reaction is inhibited by high CO pressures. Addition of olefin then returns the molecule to an 18 electron complex. Insertion of olefin into the Co—H bond [step (3)] then presumably occurs *via* a four-center transition state such as that postulated for the H—Ru—C_2R_4 case. Step (5) is the now familiar alkyl migration reaction discussed for H_3C—$Mn(CO)_5$ in the section on insertion and addition. From this point on, the pathway is less clear. Oxidative addition of H_2 in step (6) may produce the six-coordinate (d^6), 18 electron species, which can then undergo reductive elimination to give the aldehyde and return the cycle to the catalytically active $HCo(CO)_3$.

THE WACKER PROCESS (SMIDT REACTION)[211-213]

Acetaldehyde, vinyl acetate, and vinyl chloride are important industrial intermediates. Although they may be produced in a variety of ways, many of these methods suffer major disadvantages. For example, acetylene is considerably more expensive than ethylene, but addition of water to acetylene to give acetaldehyde was found earlier to be somewhat easier than production of acetaldehyde from ethylene.

---

[211] See ref. 203.
[212] A. Aguilo, *Adv. Organometal. Chem.*, **5**, 321 (1967).
[213] F. R. Hartley, *J. Chem. Educ.*, **50**, 263 (1973).

However, it was recently discovered that acetaldehyde may be produced by bubbling ethylene into an aqueous solution of a soluble palladium(II) salt in the presence of O_2 (17–133). Moreover, if the reaction is run in the presence of acetate, vinyl acetate is obtained instead (17–134), whereas vinyl chloride is among the products formed when ethylene is passed into a suspension of $PdCl_2$ in a non-aqueous solvent (17–135).

(17–133) $$H_2C=CH_2 + \frac{1}{2}O_2 \xrightarrow{PdCl_4^{2-}/H_2O} H_3CCHO$$

(17–134) $$H_2C=CH_2 + PdCl_2 + OAc^- \rightarrow H_2C=CHOAc + Pd + H^+ + 2Cl^-$$

(17–135) $$H_2C=CH_2 + PdCl_2 + 2L \rightarrow H_2C=CHCl + [HPdClL_2]$$

The oxidation of ethylene to acetaldehyde with platinum group metals has been known since 1894. When ethylene is bubbled into a solution of K_2PtCl_4, the first product is, of course, Zeise's salt, $K[PtCl_3(C_2H_4)]$, but heating this compound in water results ultimately in the formation of acetaldehyde and metallic Pt.

(17–136) $$[PtCl_3(C_2H_4)]^- + H_2O \xrightarrow{heat} H_3C-\overset{\overset{O}{\|}}{C}-H + Pt(0) + 3Cl^- + 2H^+$$

In contrast, palladium chloride does not give an isolable intermediate analogous to Zeise's salt;[214] rather, the process moves rapidly to final products.

The palladium chloride-catalyzed production of acetaldehyde was first exploited commercially by Smidt at Wacker Chemie in Germany.[215] Basically, three reactions are involved in the overall process:

product formation

(17–137) $$C_2H_4 + PdCl_4^{2-} + H_2O \rightarrow CH_3CHO + Pd(0) + 2HCl + 2Cl^-$$

catalyst regeneration

(17–138) $$Pd(0) + 2CuCl_2 + 2Cl^- \rightarrow PdCl_4^{2-} + 2CuCl$$

co-catalyst regeneration

(17–139) $$2CuCl + 2HCl + \frac{1}{2}O_2 \rightarrow 2CuCl_2 + H_2O$$

The mechanism of the first step in the Wacker process has attracted a great deal of attention. Kinetic studies in aqueous solution suggest the following rate law:

$$\frac{d[H_3CCHO]}{dt} = \frac{k[C_2H_4][PdCl_4^{2-}]}{[H^+][Cl^-]^2}$$

wherein there is a first order dependence on ethylene and the palladium halide, an inverse dependence on (an inhibition by) H^+, and an inverse square dependence on Cl^-. When D_2O is used as a solvent, furthermore, the rate is slowed. However, no deuterium appears in the product, thereby indicating that all of the hydrogen atoms in the product must have originated in the olefin. Only a very small isotope effect is

---

[214] Note that the dimeric complex $[PdCl_2(C_2H_4)]_2$ may be isolated in non-aqueous solvents.
[215] J. Smidt, R. Jira, J. Sedlmeier, R. Sieber, R. Turringer, and H. Kojer, *Angew. Chemie Int. Ed. Engl.*, **1**, 80 (1962).

seen, though, when C_2D_4 is substituted for C_2H_4, suggesting that no C—H bonds are broken during the rate determining step. On the basis of this information, and other experiments, the following sequence of reactions is proposed:

a) Formation of a metal-olefin π complex.

$$\begin{bmatrix} Cl & Cl \\ \diagdown\!Pd\!\diagup \\ Cl^{\nearrow} & {}^{\searrow}Cl \end{bmatrix}^{2-} + C_2H_4 \rightleftharpoons \begin{bmatrix} Cl & Cl \\ \diagdown\!Pd\!\diagup \\ H_2C^{\nearrow} & {}^{\searrow}Cl \\ \| \\ H_2C \end{bmatrix}^{-} + Cl^- \qquad (17\text{-}137\text{a})$$

b) Substitution of another Cl^- by H_2O. Since Cl^- is eliminated in this step and the preceding one, this accounts for the inverse square dependence on chloride concentration.

$$\begin{bmatrix} Cl & Cl \\ \diagdown\!Pd\!\diagup \\ H_2C^{\nearrow} & {}^{\searrow}Cl \\ \| \\ H_2C \end{bmatrix}^{-} + H_2O \rightleftharpoons \begin{bmatrix} Cl & Cl \\ \diagdown\!Pd\!\diagup \\ H_2C^{\nearrow} & {}^{\searrow}OH_2 \\ \| \\ H_2C \end{bmatrix} + Cl^- \qquad (17\text{-}137\text{b})$$

c) Loss of H^+ to form a hydroxo-metal complex. It is this step that accounts for rate retardation in D_2O and for the inverse dependence on H^+ concentration.

$$\begin{bmatrix} Cl & Cl \\ \diagdown\!Pd\!\diagup \\ H_2C^{\nearrow} & {}^{\searrow}OH_2 \\ \| \\ H_2C \end{bmatrix} \rightleftharpoons \begin{bmatrix} Cl & Cl \\ \diagdown\!Pd\!\diagup \\ H_2C^{\nearrow} & {}^{\searrow}OH \\ \| \\ H_2C \end{bmatrix}^{-} + H^+ \qquad (17\text{-}137\text{c})$$

d) The rate determining step is migration of the hydroxyl ligand to the coordinated olefin.

$$\begin{bmatrix} Cl & Cl \\ \diagdown\!Pd\!\diagup \\ H_2C^{\nearrow} & {}^{\searrow}OH \\ \| \\ H_2C \end{bmatrix}^{-} \xrightarrow{+ \text{ solvent}} \begin{bmatrix} Cl & Cl \\ \diagdown\!Pd\!\diagup \\ HOCH_2CH_2^{\nearrow} & {}^{\searrow}S \end{bmatrix}^{-} \qquad (17\text{-}137\text{d})$$

e) An important feature of palladium chemistry is the affinity of the metal for hydrogen.[216] (Refer to Chapter 16, p. 841, where you saw that an important step in Pd-catalyzed arylation of olefins is β-elimination of an olefin to form "HPdCl.") Therefore, the next step in the Wacker process is thought to be β-elimination to produce a π-vinyl alcohol, which then reinserts into the Pd—H bond in the opposite sense.

$$\begin{bmatrix} Cl & Cl \\ \diagdown\!Pd\!\diagup \\ CH_2^{\nearrow} & {}^{\searrow}S \\ | \\ CH_2 \\ | \\ OH \end{bmatrix}^{-} \xrightleftharpoons{- S} \begin{bmatrix} Cl & Cl \\ \diagdown\!Pd\!\diagup \\ H_2C^{\nearrow} & {}^{\searrow}H \\ \| \\ HC \\ | \\ OH \end{bmatrix}^{-} \xrightleftharpoons{+ S} \begin{bmatrix} Cl & Cl \\ \diagdown\!Pd\!\diagup \\ H_3C^{\nearrow} & {}^{\searrow}S \\ | \\ HC \\ | \\ OH \end{bmatrix}^{-} \qquad (17\text{-}137\text{e})$$

f) Finally, acetaldehyde is produced in another β-elimination, followed by a base dissociation reaction.

---

[216] See ref. 204.

970 ☐ ORGANOMETALLIC COMPOUNDS: REACTION PATHWAYS

(17-137f)

$$\left[\begin{array}{c} Cl \quad\quad Cl \\ H_3C \diagdown Pd \diagup \\ HC \diagup \quad \diagdown S \\ \| \\ OH \end{array} \right]^{-} \xrightarrow{-S} \left[\begin{array}{c} Cl \quad\quad Cl \\ H_3C \diagdown Pd \diagup \\ HC \diagup \quad \diagdown H \\ \| \\ O \end{array} \right]^{-} \rightarrow H_3C\overset{O}{\overset{\|}{C}}H + [HPdCl_2]^{-}$$

or

$$\left[\begin{array}{c} Cl \quad\quad Cl \\ \diagdown Pd \diagup \\ H_3C \diagdown \quad\quad \diagdown H \\ \quad\; C=O \\ \quad\; | \\ \quad\; H \end{array} \right]^{-}$$

POLYMERIZATION[217,218]

The polymerization of olefins to give commercially useful fibers, resins, and plastics is a thermodynamically allowed process. However, the olefin must first be activated, or each step of the polymerization reaction must be activated.[219] Compounds having metal-carbon bonds are frequently essential to this activation or catalytic process, and the reactions that occur are those we have been discussing in this chapter. Therefore, the discussion that follows outlines the important metal-olefin interactions leading to polymerization and the function of such interactions in producing poly-olefins with unique properties.

There are basically two types of polymers—addition and condensation. The familiar material polystyrene is an addition polymer formed by the stepwise addition of each monomer unit to the growing polymer.

(17-140)

$$n \; \text{PhCH=CH}_2 \rightarrow \text{—CH(Ph)—CH}_2\text{—CH(Ph)—CH}_2\text{—CH(Ph)—CH}_2\text{—}$$

Nylon, however, is a condensation polymer; water is eliminated in the reaction of an amine with a carboxylic acid.[220] Addition polymers are the only type to be pursued in this chapter.

(17-141)

$$n H_2N(CH_2)_6NH_2 + n HO\overset{O}{\overset{\|}{-C}}(CH_2)_4\overset{O}{\overset{\|}{C}}-OH \rightarrow \{N(CH_2)_6\overset{H}{\overset{|}{N}}-\overset{O}{\overset{\|}{C}}(CH_2)_4\overset{O}{\overset{\|}{C}}\} + 2n H_2O$$

Addition polymerizations are chain reactions for which three stages may be defined: initiation, propagation, and termination. The all-important initiation step may occur in any of several ways, some of which are associated intimately with the propagation step.

a) *Free radical initiation.* A free radical source is added to the system; the radical adds to one monomer, which forms the first unit of the growing polymer.

---

[217] See refs. 198 and 199.
[218] M. L. Tobe, "Inorganic Reaction Mechanisms," Thomas Nelson and Sons Ltd., London (1972).
[219] Such a reaction, with strict olefin-substituent requirements, was noted in the H_2SO_4 solvent system in Chapter 5, p. 255. The key principle there was carbonium ion stabilization.
[220] Note the analogy here with the general "solvolysis" polymerization of non-metal compounds (discussed in the last half of Chapter 8).

$$\underset{\text{(benzoyl peroxide)}}{\text{PhC(O)O-OC(O)Ph}} \rightarrow 2\ \text{PhC(O)O}\cdot$$

$$\text{PhC(O)O}\cdot + \text{PhCH=CH}_2 \rightarrow \text{PhC(O)O-CH}_2\text{-CH(Ph)}\cdot \tag{17-142}$$

b) *Acid catalysis.* A very strong Lewis acid is added to a scrupulously dry non-aqueous solvent; the strong protonic acid that is formed protonates a monomer unit, thereby forming a carbonium ion that is the first unit of a polymer.[221]

$$\text{TiCl}_4 + \text{RH}\ (e.g.,\ \text{ethylene}) \rightarrow \text{H}^+\text{TiCl}_4\text{R}^-$$

$$\text{H}^+\text{TiCl}_4\text{R}^- + \underset{H}{\overset{H}{}}\text{C=C}\underset{CH_3}{\overset{CH_3}{}} \rightarrow \text{TiCl}_4\text{R}^- + \text{H}_3\text{C-}\overset{+}{\text{C}}\underset{CH_3}{\overset{CH_3}{}} \tag{17-143}$$

c) *Base catalysis.* The base first attacks a monomer unit, thereby forming a carbanion, which is again the first unit of the polymer.

$$\text{Na}^+ + \text{NH}_2^- + \underset{H}{\overset{H}{}}\text{C=C}\underset{CN}{\overset{H}{}} \rightarrow \text{Na}^+ + \text{H}_2\text{N-CH}_2\text{-}\overset{H}{\underset{CN}{\text{C}}}{:}^- \tag{17-144}$$

d) *Organometallic initiators and templates for chain propagation.* Examples of compounds that function as initiators and/or propagation templates are nickel salts, alkyllithium compounds, and alkylaluminum compounds with transition metal salts as co-catalysts, the latter being called Ziegler-Natta catalysts. It is these types of compounds we wish to turn to now.

One of the features of organometallic polymerization catalysts is that they produce stereoregular polymers.[222] Since this regularity must be related to the nature of the interaction between metal catalyst, growing polymer, and monomer, we must say something about polymer stereochemistry first. Stereoregular polymerization is not new. Nature has accomplished it beautifully for millions of years. Quartz, the polysaccharides starch and cellulose, and the polynucleotides DNA (deoxyribonucleic acid) and RNA (ribonucleic acid) are all examples of stereoregular polymerization. Poly-olefins may also be polymerized with stereoregularity. Propylene, for example, may form two stereoregular polymers—either isotactic or syndiotactic—or a non-regular or atactic polymer (Figure 17-5). In atactic polypropylene the methyl groups have no regularity along the chain, whereas those of an isotactic polymer always lie on the same side of the chain. (Note that an isotactic polymer cannot lie "flat"; rather, it is forced into a helical arrangement so that the methyl groups do not interfere with one another.) Syndiotactic polymers are formed when the methyl groups alternate along the chain.

Stereoregularity often introduces some very desirable properties into a polymer. For example, stereoregular polypropylene is harder and tougher, and has a higher melting point than polyethylene, and it is stronger than nylon.

---

[221] See ref. 219.
[222] G. Natta, *Scientific American*, **205** (Aug.), 33 (1961).

972 ORGANOMETALLIC COMPOUNDS: REACTION PATHWAYS

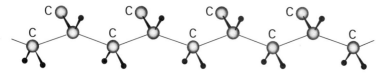

(A) Isotactic—all chain carbons have the same configuration

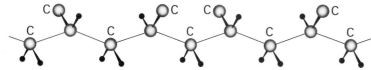

Figure 17-5. Possible configurations for linear polymers of propene.

(B) Syndiotactic—regular alternation of configuration

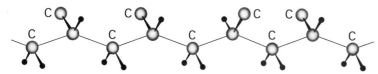

(C) Atactic—no regularity of configuration

Alkyllithium compounds are commonly used as initiators in olefin polymerization, and the reaction can be viewed as the now familiar 1,2-addition of R—M to a C=C bond. Most importantly, the stereochemistry of addition can be controlled and stereoregular polymerization achieved.[223-225] After initial formation of an addition product between R—Li and styrene, for example (Figure 17-6), the most rational way for the addition product and additional styrene to interact is for the organolithium compound to approach the monomer as illustrated. This arrangement minimizes steric interactions over all other modes of attack. The resulting polymer should be syndiotactic, as indeed it is if the reaction is carried out below $-40°$C. However, if the temperature is raised to $0°$, an atactic polymer results, since steric control of the transition state is less effective at higher temperatures.

One of the products of the "growth reaction" in alkylaluminum chemistry is isoprene (Chapter 16, p. 840). This compound is of great commercial value, as it may

---

[223] H. Gilman and J. J. Eisch, *Scientific American,* **208** (Jan.), 88 (1963).
[224] P. D. Bartlett, S. J. Tauber, and W. P. Weber, *J. Amer. Chem. Soc.,* **91,** 6362 (1969); P. D. Bartlett, C. V. Goebel, and W. P. Weber, *ibid.,* **91,** 7425 (1969).
[225] See ref. 3.

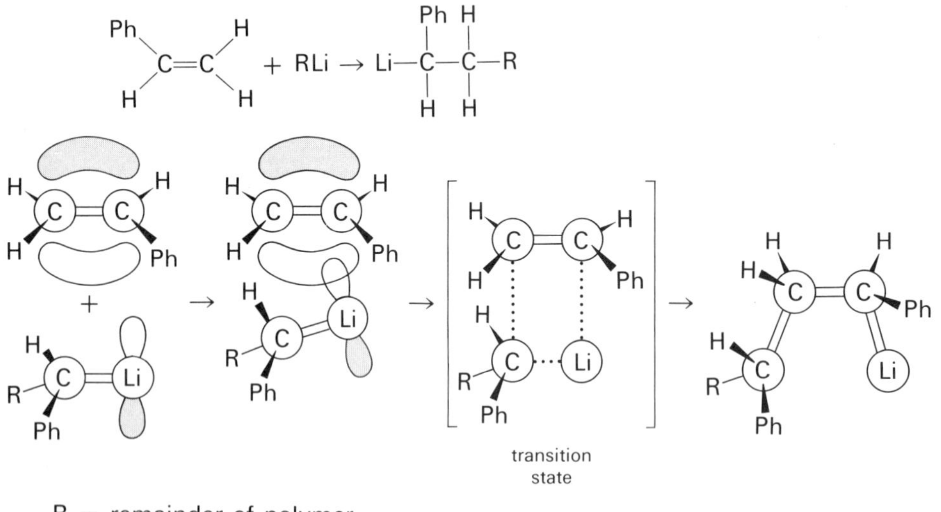

Figure 17-6. A suggested mechanism for the alkyllithium-promoted polymerization of styrene.

R = remainder of polymer

be polymerized using organolithium initiators—or Ziegler-Natta catalysts of the type to be discussed momentarily—to produce *cis*-1,4-polyisoprene, a polymer identical to natural rubber.

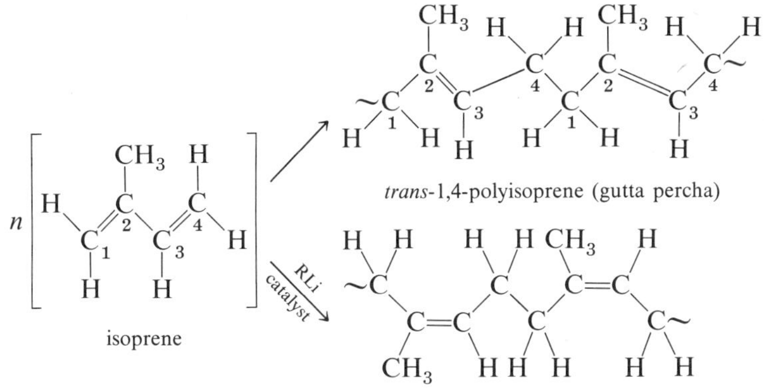

In 1954, Ziegler discovered that, on attempting to reproduce experiments on "growth reactions," ethylene gave only 1-butene, not the long chain hydrocarbon that was the usual product of such reactions.[226] Careful investigation revealed that a nickel salt had contaminated the autoclave used for the reaction, the nickel salt apparently catalyzing chain termination after formation of the C_4 chain. Ziegler guessed that if nickel catalyzed chain termination before substantial chain growth, perhaps another metal would catalyze chain lengthening to form linear polyethylene with a usable molecular weight (ca. 30,000 to 50,000 instead of weights considerably less than 28,000 that are normally produced in the growth reaction). Indeed, Ziegler soon found that a system composed of ethylene, triethylaluminum, and *tris*(acetylacetonato)zirconium(III) gave linear polyethylene, and substitution of titanium chlorides for the zirconium compound gave stereoregular polypropylene from propylene.[227] Continuing work over the years since 1954 has shown that stereoregular polymerization of olefins may be accomplished readily by catalysts (both homo- and heterogeneous) formed from a Group I-III metal alkyl (usually Et_3Al or Et_2AlCl) and some transition metal compound (usually a halide of a metal early in the series; *e.g.*, $TiCl_3$ or VCl_4). In fact, Ziegler cites the following rule: "Take any organometallic, combine it with a compound of a transition metal, and you will have a good chance of finding a suitable catalyst for your special polymerization problem." For their work on metal-catalyzed, stereoregular polymerization, Karl Ziegler and Giulio Natta shared the Nobel Prize in Chemistry in 1963.

The functioning of a Ziegler-Natta catalyst is outlined in the reactions that follow.[228,229] The first step is clearly the alkylation of the transition metal compound with the soluble Group I-III metal alkyl, thereby producing the catalytically active transition metal alkyl.

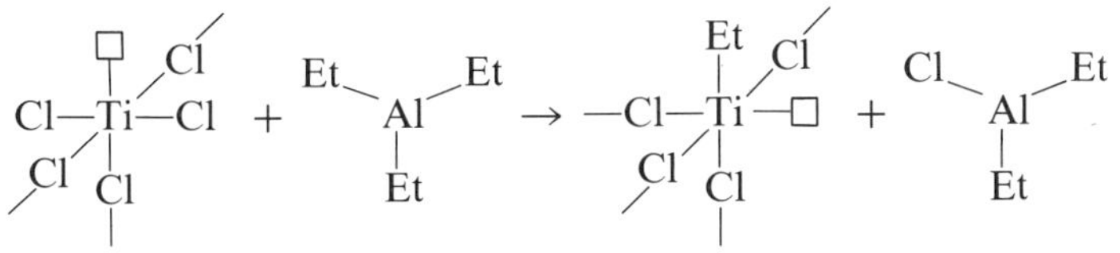

(17-145)

□ = ligand vacancy

---

[226] K. Ziegler, *Adv. Organometal. Chem.*, **6**, 1 (1968).

[227] Serendipitous results such as these are quite common in chemistry. It is often the lucky accident that, in the hands of a scientist of insight, imagination, and intuition, turns into an important and useful discovery.

[228] P. Cossee, *Trans. Faraday Soc.*, **58**, 1226 (1962); P. Cossee, *J. Catalysis*, **3**, 80 (1964); E. J. Arlman and P. Cossee, *J. Catalysis*, **3**, 99 (1964).

[229] G. Henrici-Olivé and S. Olivé, *Angew. Chem. Int. Ed. Engl.*, **6**, 790 (1967).

974 ☐ ORGANOMETALLIC COMPOUNDS: REACTION PATHWAYS

(When TiCl$_3$, VCl$_4$, or similar salts are used, a heterogeneous catalyst is generated, since these compounds are not soluble in the reaction medium; therefore, polymerization occurs at a solid surface. However, homogeneous or soluble catalysts can be formed from Cp$_2$TiCl$_2$ or Cp$_2$TiEtCl.) If the alkylated metal halide has an empty coordination site available (or one occupied only by solvent), olefin can bind to the metal in the usual manner. As you have seen so often before, the alkyl group migrates to the olefin, thereby forming a new metal alkyl, that is, a growing polymer chain. A coordination site is therefore again vacated on the metal, and olefin binding may again occur. The polymerization process then continues by successive alkyl (= polymer) migration to the olefin and olefin occupation of the vacated site.

(17–146)

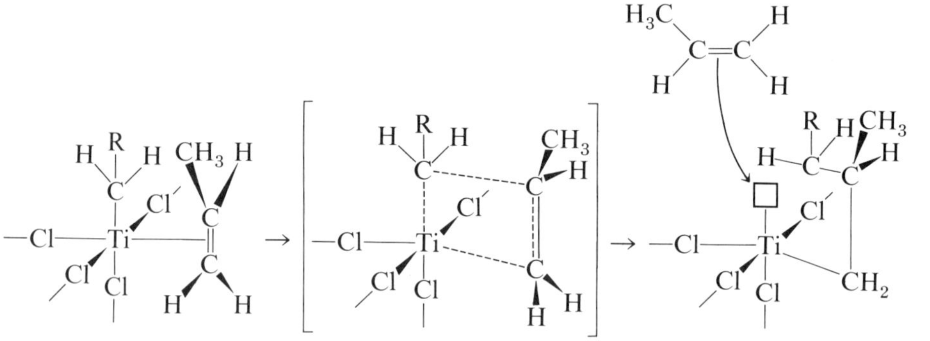

The process may be interrupted by β-elimination of a metal hydride or by hydrolysis.

(17–147)

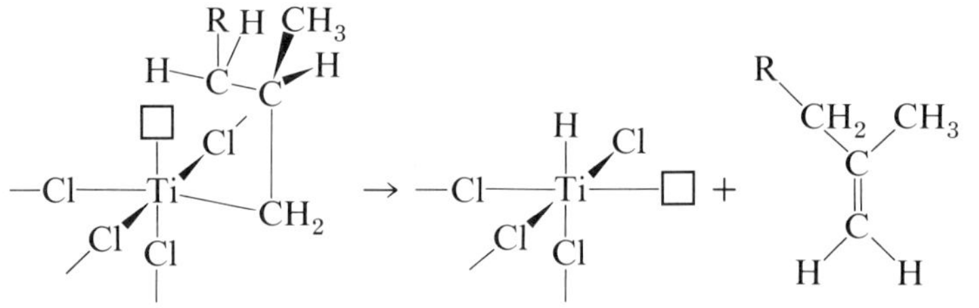

Highly charged metal ions early in the transition series are of special significance in the Ziegler-Natta system for two reasons. α-TiCl$_3$, which produces an isotactic polymer, has a layer lattice structure (see Chapter 6, p. 286ff.) and consists of a hexagonal close-packed array of Cl$^-$ ions, with Ti^{3+} ions occupying two-thirds of the octahedral holes in each alternate pair of layers [Figure 17–7(A)]. However, while the internal titanium ions are octahedrally coordinated, the surface ions are not. Therefore, the process of alkylation of a surface ion and coordination of an olefin to that ion [Figure 17–7(B)] brings the reactants into proximity for the ligand transfer reaction. More importantly, when an asymmetric olefin such as propene is used, the R group and five chloride ions of the hexagon allow the propene to approach the titanium ion in only one way, that is, with the methyl group protruding from the cavity [Figure 17–7(B)]. However, the propene position is even more restricted. The orientation in Figure 17–7(B) is considered more likely than any other orientation; for example, if the olefin C=C bond is rotated by 90° about the Ti$\cdots\overset{C}{\underset{C}{\|}}$ axis, the CH$_2$ group of the olefin would protrude directly into the neighboring titanium. Therefore, addition of the alkyl group or growing polymer chain to the coordinated olefin will always occur with the same stereochemistry. In the next alkyl transfer step [Figure 17–7(C)], the olefin position is again restricted to the one shown, and alkyl transfer occurs with the same stereochemistry as in the previous step. The end result is the formation of an isotactic polymer.

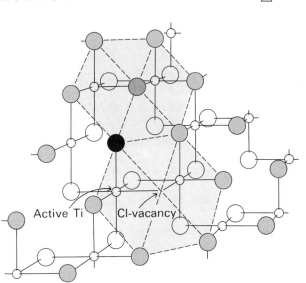

○ Ti
🔘 and ○ Cl
● Alkyl

Figure 17-7. (A) One layer of the crystal structure of α-TiCl₃ showing the Ti-alkyl bond and the chlorine vacancy that forms the "active" center in the surface. The dotted line shows the hexagonal packing of the Cl⁻ ions. Note that the R group was introduced by reaction with AlR₃ in the initiation step.

(B) A propene molecule has been inserted into the chlorine vacancy, acting as a π donor toward a titanium. An alkyl group or polymer chain can now migrate to the olefin. After migration, the sequence of groups is R → CH₃ → H clockwise as seen from the carbon attached to the metal.

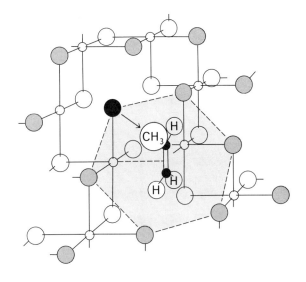

(C) Following step (B), another propene is inserted into the newly vacated metal coordination position. Migration of the growing polymer chain can occur, the sequence of groups again being R → CH₃ → H clockwise as seen from the carbon attached to the metal; thus, polymerization is stereoregular. Notice the alternation of olefin coordination between the two possible coordination sites on Ti.

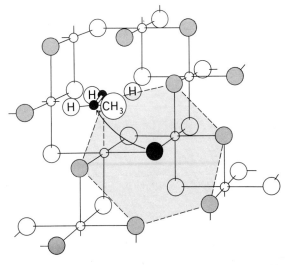

976 ORGANOMETALLIC COMPOUNDS: REACTION PATHWAYS

Finally, another reason for the importance of higher valent metal ions early in the transition series is that such ions bind olefins only weakly. Thus, while the olefin is bound sufficiently to ensure proper stereochemical alignment, it remains susceptible to addition.

CYCLOOLIGOMERIZATION, OLEFIN ISOMERIZATION AND METATHESIS, AND POLYMER-BOUND CATALYSTS

Unfortunately, space does not permit us to discuss fully examples of homogeneous and heterogeneous catalysis such as:

a) Metal-Catalyzed Cyclooligomerization of Olefins and Acetylenes.[230] The classical examples, discovered by Reppe in 1940,[231] involve nickel(II) salts with ligands such as acetylacetonate and salicylaldehyde. Such ligands can be partially displaced by the acetylene, the number of vacatable sites determining the type of product formed as shown below (where L—L is a bidentate ligand such as those mentioned, S is solvent, and P is a tertiary phosphine).

(i) catalyst with four vacatable sites → cyclooctatetraene

(17-148)

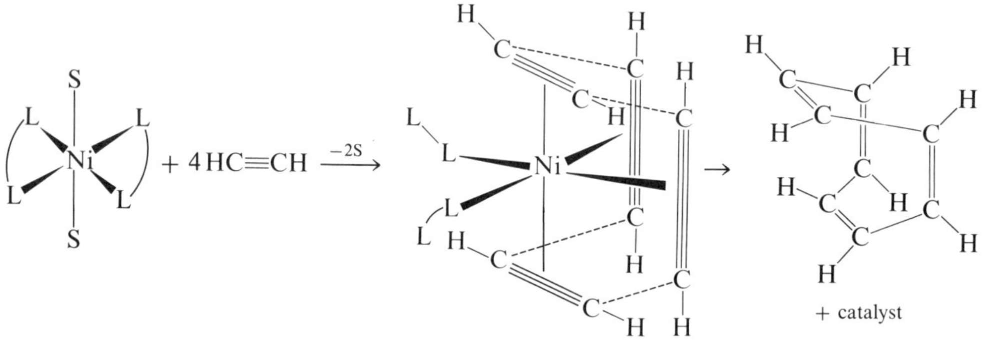

(ii) three vacatable sites → benzene

(17-149)

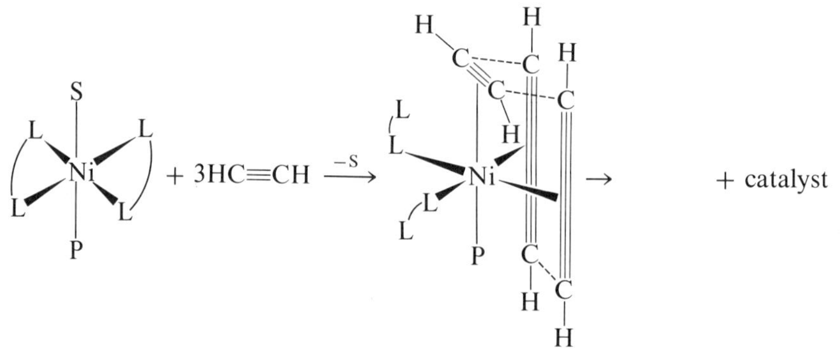

(iii) two *trans* vacatable sites → no reaction

(17-150)

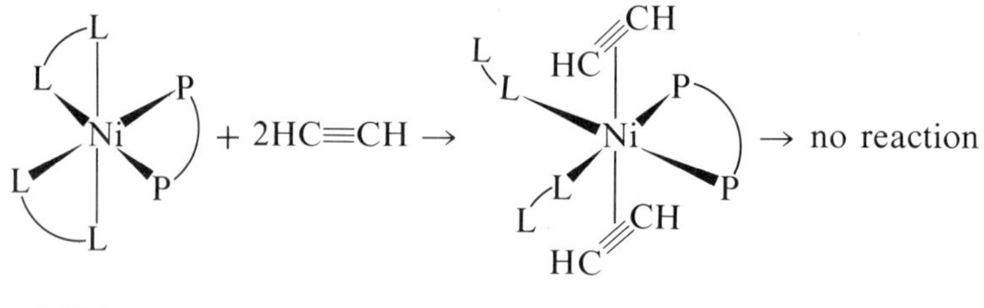

---

[230] P. M. Maitlis, *Acc. Chem. Res.*, **9**, 93 (1976).
[231] W. Reppe, O. Schlichting, K. Klager, and T. Topel, *Justus Liebigs Ann. Chem.*, **560**, 1 (1948).

b) Olefin Isomerization with Ni[P(OEt)$_3$]$_4$ and H$_2$SO$_4$.[232] In the example below, 1-butene (1-B) is isomerized to *cis*-2-butene (*c*-2-B) and *trans*-2-butene (*t*-2-B). (The number of d electrons and Ni coordination number are given for each species.)

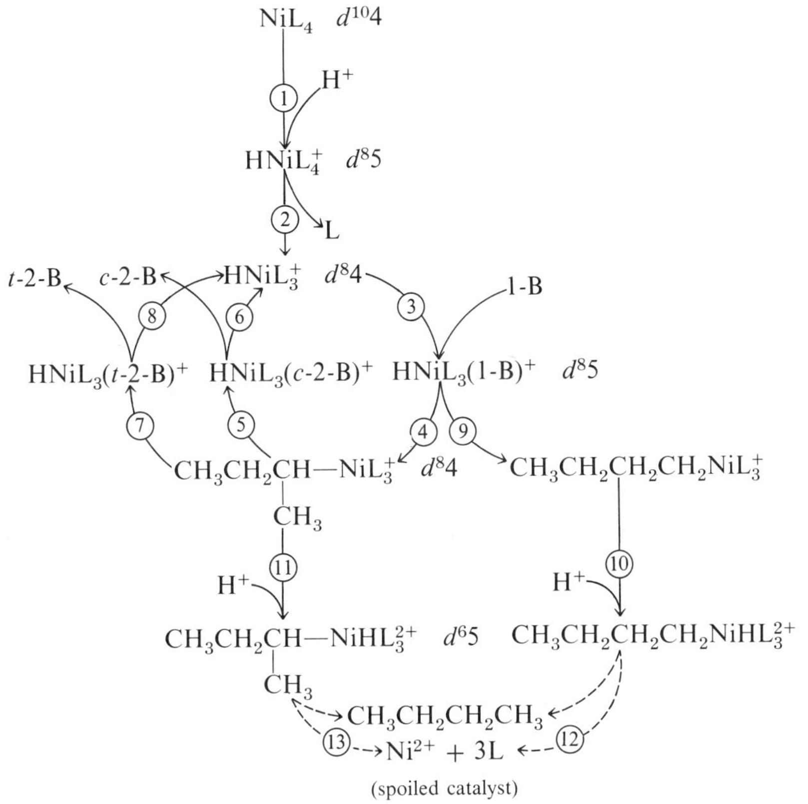

$$\begin{array}{c}
\text{NiL}_4 \quad d^{10}4 \\
\downarrow \text{(1)} \; H^+ \\
\text{HNiL}_4^+ \quad d^8 5 \\
\downarrow \text{(2)} \; L \\
t\text{-2-B} \leftarrow c\text{-2-B} \leftarrow \text{HNiL}_3^+ \quad d^8 4 \quad \xleftarrow{\text{(3)}} \quad 1\text{-B} \\
\text{HNiL}_3(t\text{-2-B})^+ \quad \text{HNiL}_3(c\text{-2-B})^+ \quad \text{HNiL}_3(1\text{-B})^+ \quad d^8 5 \\
\text{(7)} \quad \text{(5)} \quad \text{(4)} \quad \text{(9)} \\
\text{CH}_3\text{CH}_2\text{CH}-\text{NiL}_3^+ \quad d^8 4 \quad \text{CH}_3\text{CH}_2\text{CH}_2\text{CH}_2\text{NiL}_3^+ \\
\quad | \\
\text{CH}_3 \\
H^+ \downarrow \text{(11)} \quad H^+ \downarrow \text{(10)} \\
\text{CH}_3\text{CH}_2\text{CH}-\text{NiHL}_3^{2+} \quad d^6 5 \quad \text{CH}_3\text{CH}_2\text{CH}_2\text{CH}_2\text{NiHL}_3^{2+} \\
\quad | \\
\text{CH}_3 \\
\dashrightarrow \text{CH}_3\text{CH}_2\text{CH}_2\text{CH}_3 \dashleftarrow \\
\text{(13)} \dashrightarrow \text{Ni}^{2+} + 3L \dashleftarrow \text{(12)}
\end{array}$$

(spoiled catalyst)

(17–151)

[Reprinted with permission from C. A. Tolman, *J. Amer. Chem. Soc.*, **94**, 2994 (1972). Copyright by the American Chemical Society.]

c) Olefin Methathesis.[233,234] In 1931 it was observed that propylene can be converted into ethylene and butene at 725°C.

$$2H_3C-CH=CH_2 \rightarrow H_2C=CH_2 + H_3C-CH=CH-CH_3 \qquad (17\text{–}152)$$

However, some years later it was reported that introduction of a catalyst (activated molybdenum hexacarbonyl supported on alumina) brings about the same conversion at considerably lower temperatures and pressures. Indeed, catalyzed olefin methathesis is of considerable industrial importance, and it is currently used to convert propylene into polymerization-grade ethylene and high-purity butene. More recently, homogeneous catalysts (*e.g.*, mixtures of WCl$_6$ + EtOH + 2EtAlCl$_2$) have been formed that allow one to carry out olefin metatheses at room temperature. The mechanism of such reactions is the subject of considerable debate, but it would appear that agreement is near on some key points.[235]

---

[232] C. A. Tolman, *J. Amer. Chem. Soc.*, **94**, 2994 (1972).
[233] R. J. Haines and G. J. Leigh, *Chem. Soc. Rev.*, **4**, 155 (1975).
[234] N. Calderon, *Acc. Chem. Res.*, **5**, 127 (1972).
[235] M. T. Mocella, M. A. Busch, and E. L. Muetterties, *J. Amer. Chem. Soc.*, **98**, 1283 (1976).

d) Polymer-Bound Catalysts.[236-239] One of the problems of homogeneous catalysis has been the separation of the catalyst from the products. However, when the "homogeneous" catalyst is bound to a polymer, it is possible to simply decant the products and free the catalyst for further use.

(17-153)

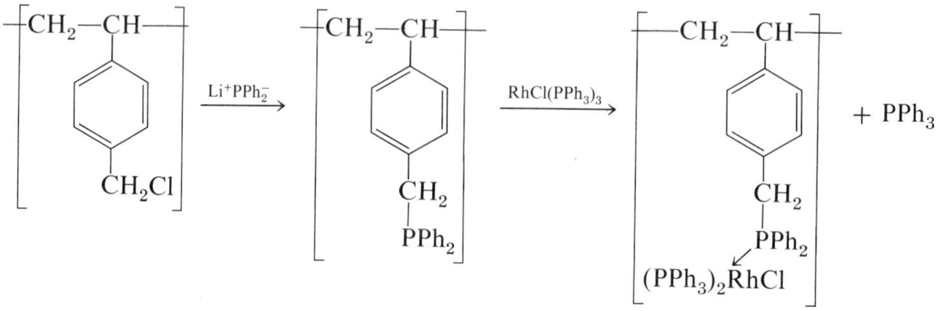

$$\left[\begin{array}{c} CH_2-CH- \\ | \\ \phi \\ | \\ CH_2Cl \end{array} \right] \xrightarrow{Li^+PPh_2^-} \left[\begin{array}{c} CH_2-CH- \\ | \\ \phi \\ | \\ CH_2 \\ | \\ PPh_2 \end{array} \right] \xrightarrow{RhCl(PPh_3)_3} \left[\begin{array}{c} CH_2-CH- \\ | \\ \phi \\ | \\ CH_2 \\ | \\ PPh_2 \\ \downarrow \\ (PPh_3)_2RhCl \end{array} \right] + PPh_3$$

The binding of catalysts to polymers is a new area that shows great promise for further development, particularly because metal-containing biological catalysts, such as DNAase, are essentially polymer-bound catalysts.

Obviously, full elucidation of heterogeneous and homogeneous catalysis is an enormous area that calls upon the talents of chemists of every background and inclination. Many important new developments can be expected in the area in the future.

---

[236] C. U. Pittman, Jr., and G. O. Evans, "Polymer-Bound Catalysts and Reagents," *Chem. Tech.*, page 560, September, 1973.

[237] R. H. Grubbs, C. Gibbons, L. C. Kroll, W. D. Bonds, Jr., and C. H. Brubaker, Jr., *J. Amer. Chem. Soc.*, **95**, 2373 (1973).

[238] T. J. Pinnavaia and P. K. Welty, *J. Amer. Chem. Soc.*, **97**, 3819 (1975); see Chapter 6, p. 326.

[239] C. U. Pittman, Jr., L. R. Smith, and R. M. Hanes, *J. Amer. Chem. Soc.*, **97**, 1742 (1975).

STUDY QUESTIONS

29. At about 160°C and with about 100 atm of CO, $Co_2(CO)_8$ in aqueous acetone will convert cyclohexene to cyclohexanecarboxylic acid in high yield. Assuming that $HCo(CO)_4$, formed from $Co_2(CO)_8$ and water, is the actual catalytic species, propose a mechanism for the olefin-to-acid conversion.

30. Show how two molecules of norbornadiene can dimerize in the presence of $Fe(CO)_5$. The steps involved in the reaction are ligand substitution, oxidative coupling, and reductive elimination.

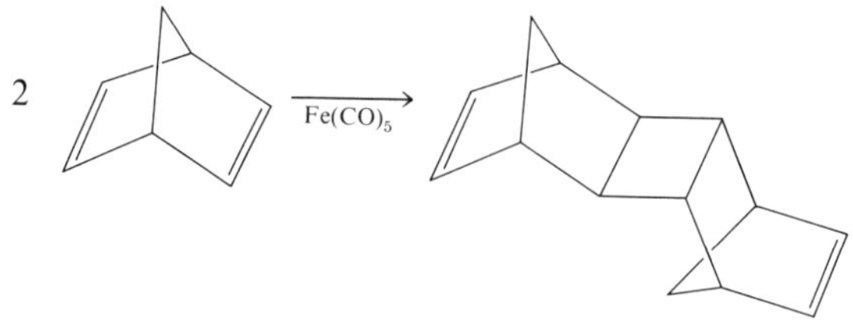

31. Show how 1-butene may be isomerized to 2-butene in the presence of catalytic amounts of $HCo(CO)_4$.

32. $Pd[P(OPh)_3]_4$ catalyzes the addition of HCN to olefins:

$$H_2C=CH_2 + HCN \rightarrow H_3C-CH_2-CN$$

Write out the various steps involved if they are base dissociation, oxidative addition, base association followed by insertion, and reductive elimination of product.

33. Cyclohexene is dehydrogenated to benzene in the presence of two molecules of Pd(OAc)₂ in acetic acid. The first stage of the process gives cyclohexadiene and is outlined below. Propose a mechanism for the conversion of the diene to benzene.

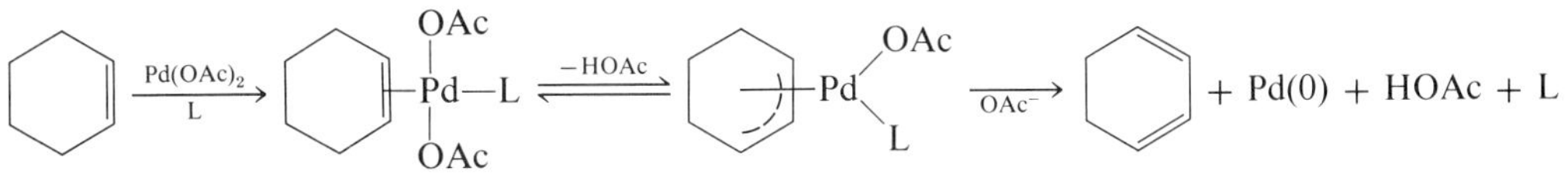

$$\text{cyclohexene} \xrightarrow[L]{Pd(OAc)_2} \text{[Cy-Pd(OAc)_2L]} \xrightleftharpoons{-HOAc} \text{[Cy-Pd(OAc)L]} \xrightarrow{OAc^-} \text{benzene} + Pd(0) + HOAc + L$$

34. Write a mechanism for the homogeneous hydrogenation of olefins, using the Rh(I) compound below as the catalyst. As in reaction 17–129, the first step is base dissociation. However, the second step in this case is association of the rhodium complex with the olefin. With this beginning, complete the mechanism of hydrogenation.

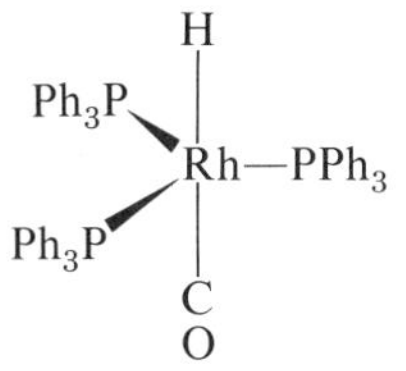

35. Ethylene and CO give the lactone illustrated below in the presence of dichlorobis(triphenylphosphine)palladium. Assuming that the catalytically active species is Pd(PPh₃)₂HCl, write a mechanism for the production of the lactone.

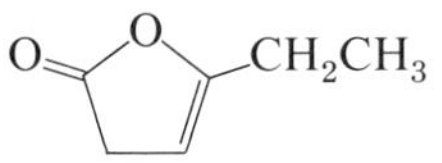

EPILOG

These last three chapters have only scratched the surface of organometallic chemistry. All of the field has not been surveyed by any means. Rather, only some representative compounds or very interesting forms of reactivity have been discussed. Moreover, the field is advancing so rapidly that many new discoveries will have been made between the time of writing and the time of appearance of this text (ca. 9 months). We do hope, however, that we have conveyed the importance of the field and its dynamic character. Further, we wish to emphasize the many new directions in which the field may grow. The "energy crisis" will force the chemical community to search for alternatives to petroleum products and may restrict development in some areas of organometallic chemistry. On the other hand, revitalized research on coal byproducts may give rise to new discoveries in areas not yet foreseen. Similarly, a major thrust for coordination and organometallic chemistry may be in the field of hydride chemistry and metal clusters. Therefore, in the chapter that follows, we apply the concepts of coordination and organometallic chemistry to a discussion of the chemistry of boron hydrides and metal clusters.

THE BORON HYDRIDES

The Neutral Boron Hydrides, $(BH)_pH_q$
  Structure and Bonding
    The Topological Approach to Boron Hydride Structure: the *styx* Numbers
    Molecular Orbital Concepts
  Synthesis and Reactivity of the Neutral Boron Hydrides
A General Organizational Scheme for the Neutral Boron Hydrides, the *Closo* Polyhedral Hydroborate Ions, and the Carboranes
A Molecular Orbital View of *Closo*-Hydroborate Anions and Carboranes
  Closo-Hydroborate Anions
  The Carboranes
  Metallocarboranes

METAL-METAL BONDS AND METAL CLUSTERS

  Binuclear Compounds
  Three-Atom Clusters
  Four-Atom, Tetrahedral Clusters
  Five- and Six-Atom Clusters

EPILOG

APPENDIX: A BONDING MODEL FOR M_6 CLUSTERS

18. MOLECULAR POLYHEDRA: BORON HYDRIDES AND METAL CLUSTERS

The study of the boron hydrides is one of the oldest areas of modern inorganic chemistry; Alfred Stock carried out his classical work from about 1910 to 1930. At the present time, the examination of neutral boron hydrides (e.g., B_5H_9), hydroborate anions (e.g., $[B_{10}H_{10}]^{2-}$) and carboranes (e.g., $B_{10}C_2H_{12}$) remains one of the most interesting areas, not only because of the unique chemistry of these compounds, but also because of the structural and bonding problems that they offer. For our present purposes, the most important structural feature that you should recognize is that the B—H compounds are molecular polyhedra or fragments thereof.

More recently, the field of boron chemistry has taken a new direction with the synthesis of metalloboranes, in which one or more BH vertices of a boron hydride polyhedron are replaced by a metal. The ion $[(B_9C_2H_{11})_2Fe^{II}]^{2-}$, a compound similar to ferrocene in that both are molecular sandwiches, is an example. Metal-containing boron hydrides are also molecular polyhedra and, as such, are part of another new and potentially large area of inorganic chemistry: metal clusters. Although progress in the study of metal clusters is slow because of experimental difficulties, such unusual materials as $Co_4(CO)_{12}$, $[Pt_9(CO)_{18}]^{2-}$, and $[Mo_6Cl_8]^{4+}$ have been characterized.

From the standpoint of a broad spectrum of chemists—structural and theoretical chemists, synthetic chemists, and materials scientists—one of the most interesting and important new areas in inorganic chemistry is the study of molecular polyhedra. By a *molecular polyhedron* we mean, for the purposes of this chapter, any compound that can be considered *a network of atoms wherein each atom is bound to at least two other atoms of the same kind*,[1,2] the net result of which is a polyhedron or a fragment of a polyhedron.[3] The most extensive series of inorganic molecular polyhedra are formed by boron—neutral boron hydrides such as pentaborane(9) (**1**), anions such as decahydrodecaborate(2—) (**2**), and the carboranes (*e.g.*, **3**)—and by the transition metals (*e.g.*, **4**).[4] Although there are some examples of molecular polyhedra in non-metal chemistry (*e.g.*, P_4 and S_8 discussed in earlier chapters), only boron

---

[1] This is a somewhat more general definition than that given by R. B. King, *J. Amer. Chem. Soc.*, **94**, 95 (1972).

[2] Note that the word "bound" has a very general meaning here. Bonding in molecular polyhedra is not restricted to two-center/two-electron bonds.

[3] The requirement that each atom be bound to two others of the same kind eliminates such interesting polyhedra as P_4O_{10}, the Al-N cube mentioned in Chapter 17 (reaction 17-11), silicates, and other heteroatomic cage compounds. Also not discussed in this chapter are homoatomic chains.

[4] The lines joining boron atoms or cobalt atoms in structures **1-4** (and in many subsequent structures) do not necessarily represent bonds. Rather, these lines simply outline the polyhedron.

hydrides and metal clusters are of concern in this chapter. However, as a prelude to the section on metal clusters, the equally important topic of bonds formed between just two metals is discussed.

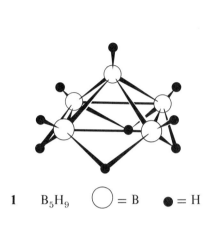

1   B_5H_9 ◯ = B   ● = H

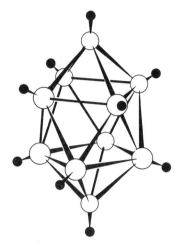

2   $B_{10}H_{10}^{2-}$ ◯ = B   ● = H

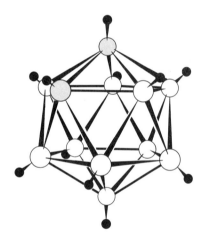

3   $B_{10}C_2H_{12}$ ◯ = B   ⬤ = C   ● = H

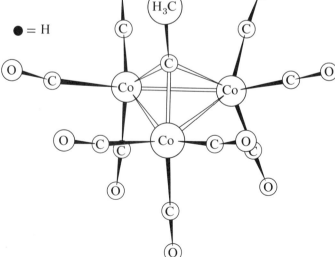

4   $CH_3-CCo_3(CO)_9$

Before proceeding, it might be mentioned that there are two reasons for placing this chapter near the end of the text. (i) Various possible coordination polyhedra have already been examined in Chapter 10, and some ideas can be carried over to the present material, of course with the recognition that polyhedra found in more "classical" coordination compounds and in some non-metal compounds are defined by the positions assumed by ligands bound to a central atom, while in a true molecular polyhedron there is no central atom and the vertices are defined by atoms bound to one another. (ii) Many metal clusters are formed by metals in low formal oxidation states and are stabilized by bonds to unsaturated organic ligands and to carbon monoxide. These are, of course, just the same types of ligands important in "normal" organometallic compounds (Chapters 15–17).

THE BORON HYDRIDES[5-9]

Boron hydride chemistry underwent a tremendous expansion in the 1950's, with a great deal of the impetus provided by government funding of a high energy fuels program.[10] Although a number of the neutral hydrides had been known since Alfred Stock's work in the 1920's, the more recent research resulted in extension of the neutral series, and the synthesis and characterization of anionic hydrides of the type $B_pH_p^{2-}$, the carboranes, and, even more recently, metallocarboranes. Progress in boron hydride chemistry continues unabated, and there have been important advances in the organization of the field.

Alfred Stock (1876–1946), a German chemist, first considered taking up the study of boron chemistry about 1900. However, he was dissuaded from doing so by Emil Fischer, the director of the University of Berlin laboratories in which Stock was then working. Fischer said that correspondence with William Ramsay indicated that Ramsay's laboratory had already thoroughly investigated the subject, the implication being that further work in the area would not lead to any especially interesting results. So the plan was abandoned until 1912, when Stock began work on the chemistry of boron and silicon hydrides in his own laboratory in the Technische Hochschule of Breslau. This work was extremely fruitful, and it is summarized in a book important to inorganic chemistry.[11,12]

One reads Stock's book in amazement and awe. It was Stock and his co-workers who developed designs for vacuum lines (p. 448) and vacuum techniques still in use today. Moreover, he discovered and fully characterized innumerable hydrides of boron and silicon, all without the aid of the infrared and nmr spectrometers that we consider so indispensable today. Unfortunately, Stock contracted mercury poisoning, an affliction that prevented him from working with complete effectiveness. To quote his book (page 18):

"Among the unpleasant accidents must be reckoned the fact that, through years of work with mercury apparatus, my collaborators and I contracted chronic mercurial poisoning. Its cause went unrecognized for a long time and its psychic-nervous effects impaired our working ability. This experience led to a detailed study of insidious mercurial poisoning, a phenomenon that had been almost forgotten in the course of time. The writer has made numerous reports on the latter subject in the last few years."

---

[5] E. L. Muetterties and W. H. Knoth, "Polyhedral Boranes," Marcel Dekker, New York (1968).
[6] K. Wade, "Electron Deficient Compounds," Thomas Nelson and Sons Ltd., London (1971). For the newcomer to the field, this is the best source of information.
[7] W. N. Lipscomb, "Boron Hydrides," W. A. Benjamin, Inc., New York (1963).
[8] E. L. Muetterties, "Boron Hydrides," Academic Press, New York (1975).
[9] N. N. Greenwood in "Comprehensive Inorganic Chemistry," J. C. Bailar, Jr., H. J. Emeleus, R. Nyholm, and A. F. Trotman-Dickenson, eds., Pergamon Press, New York (1973), Chapter 11.
[10] R. F. Gould, ed., "Borax to Boranes," Advances in Chemistry Series, No. 32, American Chemical Society, Washington, D.C. (1961); R. T. Holzmann, ed., "Production of the Boranes and Related Research," Academic Press, New York (1967).
[11] A. Stock, "Hydrides of Boron and Silicon," Cornell University Press, Ithaca (1933).
[12] For further details of the career of this important early inorganic chemist, see V. Bartow, *Adv. Chem. Ser.*, **32**, 5 (1961).

And on page 203, Stock goes on to say:

> "The vacuum apparatus requires that its manipulators constantly handle considerable amounts of mercury. Mercury is a strong poison, particularly dangerous because of its liquid form and noticeable volatility even at room temperature. Its poisonous character has been rather lost sight of during the present generation. My co-workers and myself found from personal experience . . . that protracted stay in an atmosphere charged with only 1/100 of the amount of mercury required for its saturation, sufficed to induce chronic mercury poisoning. This first reveals itself only as an affection of the nerves, causing headaches, numbness, mental lassitude, depression, and the loss of memory; such are very disturbing to one engaged in an intellectual occupation."

In attempting to present an overview of the important aspects of polyhedral borane chemistry in a short chapter, organization is of great importance. An almost bewildering array of compounds have been synthesized; without systematic study, the first impression of the newcomer is that compounds are seemingly formed with random structures and that there is no relation of one to another. However, Lipscomb

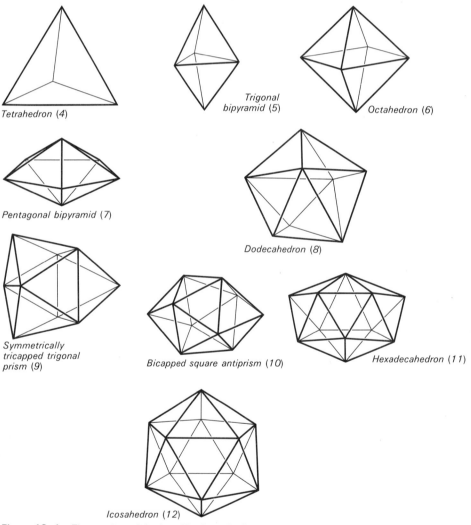

Figure 18–1. The regular polyhedra with triangular faces on which the neutral boranes, hydroborate anions, and carboranes are based. Boron atoms (or carbon atoms in the carboranes) are located at the polyhedral vertices. *Note that the lines connecting vertices **do not** represent bonds.*

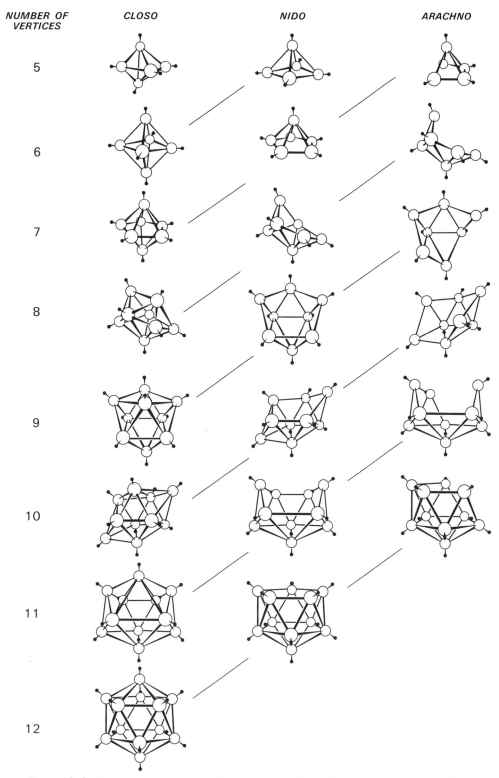

Figure 18-2. The vertical columns give the basic *closo, nido,* and *arachno* frameworks used in boron hydride chemistry. Bridging hydrogens and BH_2 groups are not shown where they exist in *nido-* and *arachno-*boranes. However, they are placed around the open face of the polyhedron in as symmetrical a manner as possible. The diagonal lines define series of related *closo, nido,* and *arachno* structures. Removing the most highly connected vertex of a *closo* framework generates the related *nido* framework, and the *arachno* framework is generated by removing the most highly connected vertex in the open face of the related *nido* framework. [Reprinted with permission from R. W. Rudolph and W. R. Pretzer, *Inorg. Chem.,* **11**, 1974 (1972). Copyright by the American Chemical Society.]

very early suggested an "organizational" scheme based on a simple topological theory,[7] and more recent papers by others have presented even more general schemes.[13-16]

Although early workers in boron hydride chemistry thought that all of the compounds were fragments of the icosahedron, it is now clear that all neutral compounds, anions, and carboranes are based on the full range of regular polyhedra with triangular faces (sometimes called deltahedra) shown in Figure 18–1. Those hydrides that have a boron atom at each vertex of one of these polyhedra (or with one or more vertices substituted by some other element such as carbon, phosphorus, or a metal) are called *closo* compounds. Such compounds are formed only by hydroborate anions, $(BH)_p^{2-}$, and the carboranes, $(CH)_a(BH)_p$. Note that there is only a single terminal hydrogen attached at each vertex; there are no extra terminal or bridging hydrogens.

If the atom at the most highly connected vertex of a *closo* polyhedron is removed, a *nido* compound is formed (*nido* coming from the Greek for "nest"). The relationship between a *nido* compound and the *closo* polyhedron from which it arises is clearly seen in Figure 18–2.[17] Ascending one of the diagonal lines in Figure 18–2 carries one from a *closo* polyhedron to the corresponding *nido*-borane and finally to an *arachno*-borane (*arachno* coming from the Greek for "spider web"), the latter being formed by removing the most highly connected vertex on the "open" face of a *nido* compound.

Nido and *arachno* compounds are formed primarily by neutral boranes of general formula $(BH)_pH_q$.[18] Since, in comparison with *closo*-hydroborate anions and carboranes, the bonding in these compounds can be more easily represented by localized two- and three-center bonds, and their structures represented by a simple topological theory, we shall begin by describing them in some detail. Before reading on, however, be sure to note the nomenclature rules summarized in Figure 18–3.

---

[13] I. R. Epstein and W. N. Lipscomb, *Inorg. Chem.*, **10**, 1921 (1971). Two equations in this paper are incorrect; see I. R. Epstein, *ibid.*, **12**, 709 (1973) for the correction.

[14] R. W. Rudolph and W. R. Pretzer, *Inorg. Chem.*, **11**, 1974 (1972); R. W. Rudolph, *Accts. Chem. Res.*, **9**, 446 (1976).

[15] K. Wade, *Chem. Commun.*, **11**, 1974 (1972); *Chem. Britain*, **11**, 177 (1975); *Adv. Inorg. Chem. Radiochem.*, **18**, 1 (1976).

[16] R. E. Williams, *Adv. Inorg. Chem. Radiochem.*, **18**, 67 (1976).

[17] See refs. 14 and 16.

[18] Although formulas for neutral boron hydrides are usually written B_pH_{p+q}, we shall use the equivalent formula $(BH)_pH_q$ to be consistent with the "electron counting rules" to be developed later in the chapter.

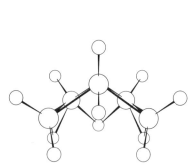

Neutral Boron Hydrides

pentaborane(11), B_5H_{11}
penta-: indicates the number of boron atoms
borane: indicates a neutral compound containing only B and H
(11): indicates number of H atoms

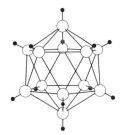

Anionic Boron Hydrides

dodecahydrododecaborate(2−), $[B_{12}H_{12}]^{2-}$ or dodecahydro-*closo*dodecaborate(2−)
dodeca-: indicates 12 hydrogens or borons
hydro and borate: indicates an anion
closo: indicates a closed polyhedron
(2−): indicates the anion charge

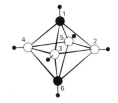

Carboranes

1,6-dicarba-*closo*-hexaborane(6), $B_4C_2H_6$
1,6: indicates the location of the C atoms
hexa-: indicates the number of vertices
closo: indicates a closed polyhedron
borane: indicates a neutral species
(6): indicates the number of H atoms

Figure 18–3. Examples of the nomenclature used in boron hydride chemistry. See *Inorg. Chem.*, **7**, 1945 (1968), and R. M. Adams, *Pure Appl. Chem.*, **30**, 683 (1972).

TABLE 18-1
PHYSICAL PROPERTIES OF SOME BORON HYDRIDES*

Compound	Name	Melting Point (°C)	Boiling Point (°C)	$\Delta H_f^{\circ a}$ (kcal/mol)	ΔG_f° (kcal/mol)	$S°$ (25°C) (cal/deg mol)	Structural Type
B_2H_6	diborane(6)	−164.9	−92.6	8.5	20.7	55.45	—
B_4H_{10}	tetraborane(10)	−120	18	15.8	—	—	arachno
B_5H_9	pentaborane(9)	−46.6	48	17.5	41.8	65.92	nido
B_5H_{11}	pentaborane(11)	−123	63	24.7	—	—	arachno
B_6H_{10}	hexaborane(10)	−62.3	108	22.6	—	—	nido
B_6H_{12}	hexaborane(12)[b]	−82.3	80–90	—	—	—	arachno (?)
B_8H_{12}	octaborane(12)[c,d]	−20	(very unstable at −20°)	—	—	—	nido
B_8H_{14}	octaborane(14)[d]	very unstable		—	—	—	arachno (?)
B_8H_{18}	octaborane(18)[e]	very unstable		—	—	—	2 tetraboranes linked through BB bond (?)
$n\text{-}B_9H_{15}$	normal-nonaborane(15)[f]	2.6	—	—	—	—	arachno[g]
$i\text{-}B_9H_{15}$	iso-nonaborane(15)[c]	—	—	—	—	—	?
$B_{10}H_{14}$	decaborane(14)	99.7	213	7.54	51.66	84.4	nido

*The compounds listed in this table are the better characterized representatives of this class of compounds. However, a number of others have been isolated. See, for example, J. Rathke and R. Schaeffer, Inorg. Chem., **13**, 3008 (1974) ($B_{15}H_{23}$ and $B_{14}H_{22}$) and J. Rathke, D. C. Moody, and R. Schaeffer, Inorg. Chem., **13**, 3040 (1974) ($B_{13}H_{19}$).
[a] All thermodynamic values taken from the National Bureau of Standards Tables.
[b] D. F. Gaines and R. Schaeffer, Inorg. Chem., **3**, 438 (1964).
[c] J. Dobson, P. C. Keller, and R. Schaeffer, Inorg. Chem., **7**, 399 (1968).
[d] J. Dobson and R. Schaeffer, Inorg. Chem., **7**, 402 (1968).
[e] J. Dobson, D. F. Gaines, and R. Schaeffer, J. Amer. Chem. Soc., **87**, 4072 (1965).
[f] A. B. Burg and R. Kratzer, Inorg. Chem., **1**, 725 (1962); R. Schaeffer and L. G. Sneddon, Inorg. Chem., **11**, 3102 (1972).
[g] P. G. Simpson and W. N. Lipscomb, J. Chem. Phys., **35**, 1340 (1961).

988 ☐ MOLECULAR POLYHEDRA: BORON HYDRIDES AND METAL CLUSTERS

THE NEUTRAL BORON HYDRIDES, $(BH)_pH_q$

Structure and Bonding

Many of the known boron hydrides are listed in Table 18–1, and the structures of the more important ones are shown in Figure 18–4. Notice that all of these compounds are either *nido*- or *arachno*-boranes. As an example of their formation according to the scheme outlined in Figure 18–2, consider the series beginning with *closo*-$B_7H_7^{2-}$ [$= (BH)_7^{2-}$]. Removal of a BH^{2+} vertex from this compound would give the hypothetical ion *nido*-$B_6H_6^{4-}$ [$= (BH)_6^{4-}$]. Such a highly charged ion is unlikely to exist and so, in a formal sense, H^+ ions are added to neutralize the charge; these additional protons stitch up the open pentagonal face of the $(BH)_6^{4-}$ ion with BHB bridges (the extra bridging H's are always placed about the open face as symmetrically as possible), and the neutral *nido* compound $(BH)_6H_4$ or B_6H_{10} results. Removal

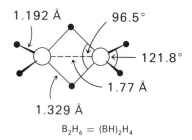

$B_2H_6 = (BH)_2H_4$

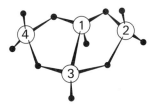

$B_4H_{10} = (BH)_4H_6$
$B_1–B_2 = 1.85$ Å
$B_1–B_3 = 1.72$ Å

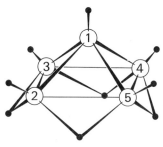

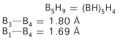

$B_5H_9 = (BH)_5H_4$
$B_3–B_4 = 1.80$ Å
$B_1–B_4 = 1.69$ Å

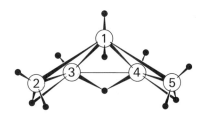

$B_5H_{11} = (BH)_5H_6$
$B_3–B_4 = 1.77$ Å $B_1–B_5 = 1.87$ Å
$B_1–B_4 = 1.72$ Å

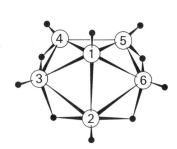

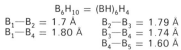

$B_6H_{10} = (BH)_6H_4$
$B_1–B_2 = 1.7$ Å $B_2–B_3 = 1.79$ Å
$B_1–B_4 = 1.80$ Å $B_3–B_4 = 1.74$ Å
                     $B_4–B_5 = 1.60$ Å

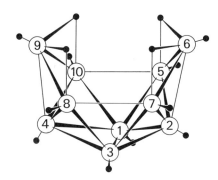

$B_{10}H_{14} = (BH)_{10}H_4$
$B_1–B_3 = 1.71$ Å
$B_2–B_6 (= B_4–B_9) = 1.72$ Å
$B_6–B_7 (= B_9–B_8, \ldots) = 1.77$ Å

Figure 18–4. Structures of the more common neutral boron hydrides with their numbering conventions and some interatomic distances. Note that the lines connecting boron atoms are not meant to depict bonds; rather, their purpose is to define the polyhedral framework.

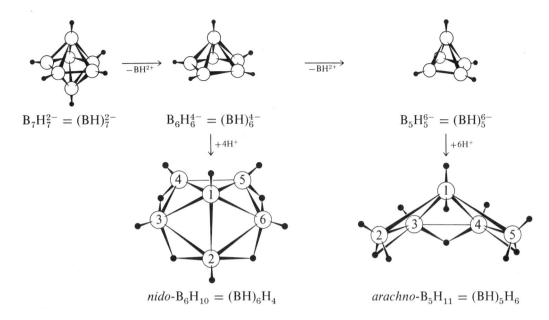

of yet another BH^{2+} vertex from the edge of the open face of $(BH)_6^{4-}$ leads to the hypothetical *arachno* ion $(BH)_5^{6-}$, but formal addition of six protons (to form three BHB bridges and three additional BH terminal bonds) gives the neutral *arachno* species $(BH)_5H_6$ or B_5H_{11}.

The series of transformations discussed above illustrates a most important fact: *Nido-boranes have the general formula* $(BH)_pH_4 [= B_pH_{p+4}]$ *and arachno-boranes have the general formula* $(BH)_pH_6 [= B_pH_{p+6}]$. For the moment this can be accepted as an empirical fact, but its generality and origins will be explored more carefully later in the chapter.

All of the structures in Figure 18-4 have been determined by x-ray methods, and, in the case of decaborane(14), the hydrogen positions were determined by neutron diffraction. For other, more unstable boranes, spectroscopic methods—chiefly ^{11}B nmr spectroscopy—must be relied upon. As an example of this technique, consider the spectrum of tetraborane(10), B_4H_{10} (Figure 18-5).[19] There are clearly two sets of boron atoms in this structure: B_1 and B_3 are symmetry equivalent, as are the pair B_2 and B_4. For this reason the spectrum consists of two multiplets; that for B_1 and B_3 is a doublet owing to spin coupling between each ^{11}B and its terminal hydrogen, while B_2 and B_4 each give rise to a triplet of triplets owing to $^{11}B-^1H_{bridge}$ and $^{11}B-^1H_{terminal}$ spin coupling. Other examples of the use of ^{11}B nmr spectroscopy will be introduced as a means of inferring probable structures of products of reactions with the neutral boranes and polyhedral anions (see Study Question 4).

A description of the bonding in the so-called "electron-deficient" molecule diborane has been a classical problem in chemistry. As described in Chapter 5, the unique feature of this molecule is B—H—B bridging, and a useful view of the bonding in this bridge is that of two three-center B—H—B mo's. In the early 1960's Lipscomb applied the concept of the three-center bond to the higher boron hydrides and developed a simple method for describing and predicting the topology of such compounds.[20] That is, he developed a scheme that describes a boron hydride in terms

---

[19] R. C. Hopkins, J. D. Baldeschwieler, R. Schaeffer, F. N. Tebbe, and A. Norman, *J. Chem. Phys.*, **43**, 975 (1965).

[20] See refs. 6 and 7.

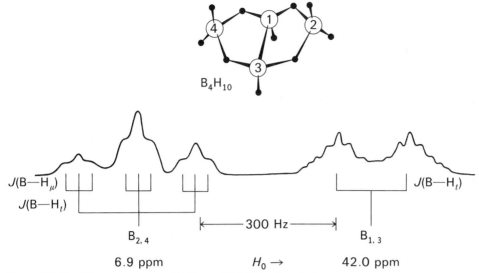

Figure 18-5. ^{11}B nmr spectrum of B_4H_{10} at 15.1 MHz. Chemical shifts relative to $BF_3 \cdot O(C_2H_5)_2$. The fine structure in the high-field doublet arises from extensive ^{11}B—H spin coupling, which is not explained by a simple first-order scheme. [After R. C. Hopkins, J. D. Baldeschwieler, R. Schaeffer, F. N. Tebbe, and A. D. Norman, *J. Chem. Phys.*, **43**, 975 (1965).

of the interconnections between atoms; from it you can gain some idea of the gross geometry of the molecule.

The Topological Approach to Boron Hydride Structure: the *styx* Numbers.[20] In all neutral boron hydrides, hydroborate anions, or carboranes, each boron has at least one H (or other substituent) attached by a normal two-electron σ bond. Lipscomb proposed that, in addition to these, there could be three-center B—H—B bonds (labeled *s* in Figure 18–6; *cf.* B_2H_6 in Chapter 5), closed and/or open three-center B—B—B bonds (both labeled *t*), additional B—$H_{terminal}$ bonds (labeled *x*) and B—B two-center bonds (labeled *y*). Subsequent theoretical studies have shown that open B—B—B bonds need not be considered.[21] However, it is apparent that three-center bonds of some type are necessary to account for boron hydride bonding because each B—H unit at a polyhedral vertex can supply three orbitals but only two electrons to the framework. Consequently, relative to the "octet" concept, each boron is responsible for a "bonding electron deficiency" of one electron, and the molecule would be "deficient" by one electron per boron if only two-center bonds were formed. How-

[21] E. Switkes, W. N. Lipscomb, and M. D. Newton, *J. Amer. Chem. Soc.*, **92**, 3847 (1970).

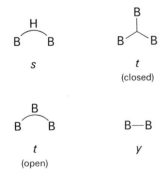

Figure 18-6. Topological representations of the localized interactions commonly found in the boranes.

ever, *if one three-center bond* (holding two electrons) *is formed per boron, the "bonding electron deficiency" in the boron hydride is eliminated.* For example, in B_2H_6 there are 12 valence electrons and 14 valence orbitals; therefore, two three-center BHB bridges are formed in addition to two extra two-center B—H bonds. In an even more extreme case, B_6H_{10} has 28 valence electrons and 34 valence orbitals. Its 14 pairs can be accommodated in eight two-center bonds (BH and BB) and six three-center bonds (BHB and BBB).

Utilizing the concept of localized, three-center BBB and BHB bonds, in addition to two-center BB and additional BH bonds, it is possible to describe the topology of the boron hydrides. The relationship between the formula of the boron hydride $(BH)_pH_q$ and the number and kinds of bonds in the molecule is described by simple *equations of balance:*[22]

(i) *Three-center orbital balance.*

$$\boxed{p = s + t} \qquad (18\text{-}1)$$

As stated above, the "electron deficiency" in a boron hydride is eliminated if one three-center bond is formed per boron atom. Therefore, the sum of the number of three-center BHB bonds (s) and the number of three-center BBB bonds (t) must equal the number of BH units (p).

(ii) *Hydrogen balance.*

$$\boxed{q = s + x} \qquad (18\text{-}2)$$

It is assumed that each boron has *at least one* "terminal" hydrogen attached. Therefore, the remaining hydrogen atoms, q, must be divided between bridges (s) and *additional* B—H terminal bonds (x).

(iii) *Electron balance.*
no. of electron pairs = no. of bonds

$$\boxed{p + (q/2) = s + t + y + x} \qquad (18\text{-}3)$$

or more simply [by substitution of (i) and (ii)]

$$\boxed{y = \frac{1}{2}(s - x)} \qquad (18\text{-}4)$$

This rule arises from the fact that each of the p (BH) groups can contribute two electrons or one pair to the framework, and each of the additional q hydrogens can contribute one electron or half of a pair. All of these electron pairs must then be used in bonding, the total number of bond pairs being $s + t + y + x$.

With this set of equations relating the formula of a boron hydride to the types of bonds formed, the topology of a boron hydride may be described in terms of the so-called *styx* number. For example, consider the simplest possible case, diborane, $B_2H_6 = (BH)_2H_4$.

---

[22] You are reminded of the equations developed in Chapter 2 that allowed the allocation of the valence electrons in a molecule (= V) to σ bonds, π bonds, and lone pairs. Here we present equations that relate the number of valence electrons to the different types of σ bonds that can form.

Step a. For diborane, $p = 2$ and $q = 4$.

Step b. In general, s, the number of BHB bridges, must always be >0 and $\geq x$ but $\leq q$.[23] Furthermore, s must be $\geq q/2$.[24] That is, s must have values in a range from $q/2$ to q. For diborane, $q = 4$ and $2 \leq s \leq 4$, so $s = 2, 3,$ or 4.

Step c. As a result of the three possible values for s, there are three sets of solutions for the values of s, t, y, and x (called the *styx* numbers).

s	t	y	x
2	0	0	2
3	−1	1	1
4	−2	2	0

Only the first of these three solutions is physically reasonable, since the latter two involve negative parameters. From your knowledge of the diborane structure, the *styx* number 2002 is, of course, seen to be correct; that is, there are two BHB bridges ($= s$), no three-center BBB bonds ($= t$) or two-center BB bonds ($= y$), and two B—H terminal bonds in addition to those already assumed.

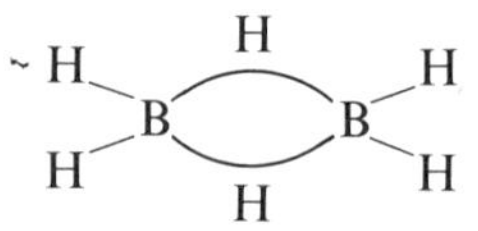

B_2H_6 *styx* = 2002

5

Other boron hydrides with their *styx* numbers are pictured below. You might note at this point that there are four variables (s, t, y, and x) but only three independent equations relating them, so you can arrive only at sets of possible *styx* numbers (defined by the range $q/2 \leq s \leq q$). However, by applying the empirical rules given in Study Question 3 (where they are applied to tetraborane), reasonable topologies may be predicted, and it is these that are shown.

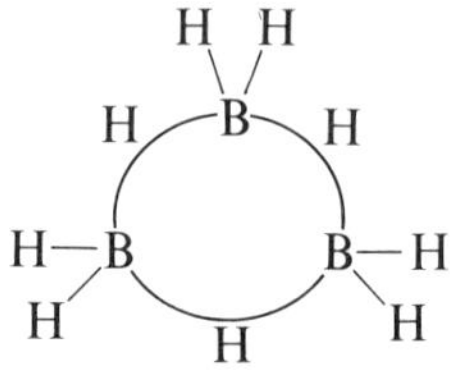

B_3H_9 (hypothetical)
styx = 3003

6

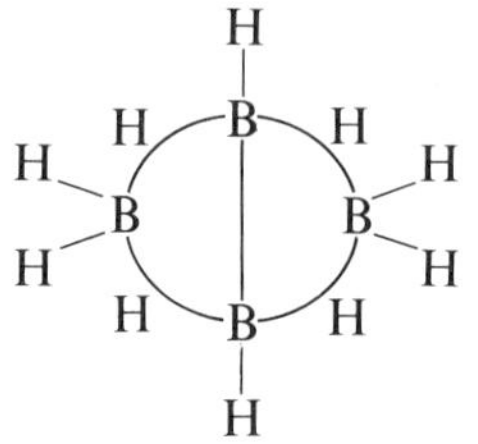

B_4H_{10} *styx* = 4012

7

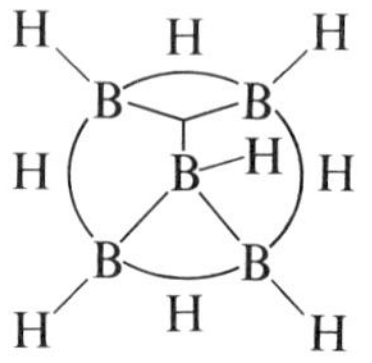

B_5H_9 *styx* = 4120

8

---

[23] The statement that $s \leq q$ is true because the additional q H's form BHB bridges ($= s$) and/or additional terminal BH bonds ($= x$). From equation 18–2, you see that the number of BHB bridges cannot exceed the number of additional H's. The statement that $s \geq x$ follows from equation 18–4, since y cannot be negative (*i.e.*, $y \geq 0$).

[24] This statement is derived as follows: Since $p + (q/2) = (q - x) + t + y + x$, then $p = t + y + (q/2)$. From equation 18–1, $(s + t) = t + y + (q/2)$, so $y = s - (q/2)$, and $s = (q/2) + y$.

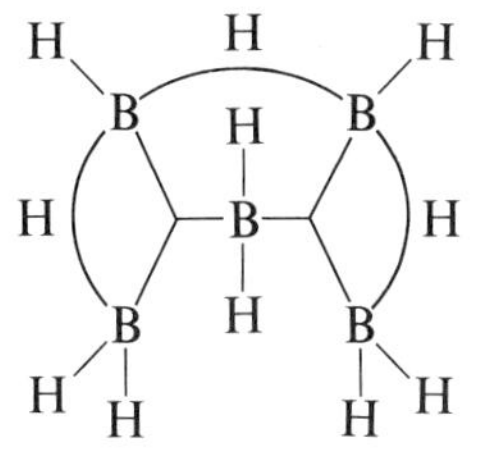

B_5H_{11} styx = 3203
9

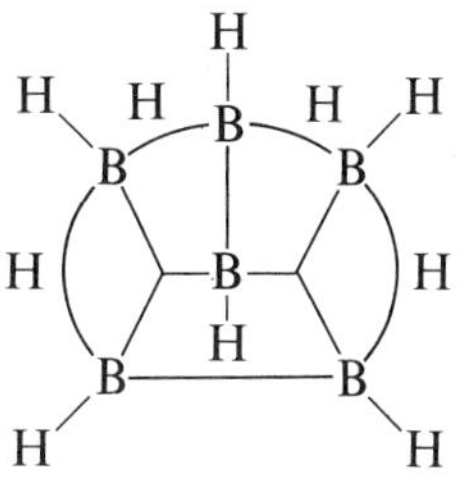

B_6H_{10} styx = 4220
10

STUDY QUESTIONS

1. Derive the *styx* number for B_3H_9 (**6**) and show that it is the only one possible. On a sketch of **6** show to which parts of the molecule each of the *styx* letters applies.

2. B_3H_9 (**6**) and B_3H_7 are thought to be kinetically unstable intermediates in the pyrolysis of diborane. They may arise according to the reaction

$$B_2H_6 + BH_3 \rightarrow B_3H_9 \rightarrow B_3H_7 + H_2$$

   a) Derive the possible sets of styx numbers for B_3H_7.
   b) Draw the structure corresponding to each *styx* number.
   c) If you have drawn two or more structures, which is the most reasonable and why?

3. There are usually several *styx* numbers for a given boron hydride, and therefore several possible topologies. In order to choose among the several possibilities, empirical rules have been developed and are given as follows:

   1. All known boron hydrides have at least a two-fold element of symmetry, so it is assumed that any new hydride probably would have at least a plane, center, or two-fold axis of symmetry. Low symmetry appears to provide centers of reactivity.
   2. Only one terminal hydrogen, and no bridging hydrogen, may be attached to a boron that is bound to five neighboring borons. This restricts BHB bridges and BH_2 groups to the open edges of boron frameworks.
   3. If a boron is bound to four other boron atoms, it will probably not make use of more than one BHB bridge.
   4. A boron atom that is bound to only two other boron atoms will be involved in at least one BHB bridge.

   a) Derive the different *styx* numbers possible for B_4H_{10}.
   b) Draw the structure corresponding to each *styx* number.
   c) On the basis of the rules above, choose the most likely structure.

4. Although B_6H_{10} was known from Alfred Stock's work (see page 983), B_6H_{12} was only recently isolated, the compound having been synthesized by the reaction of $B_3H_8^-$ with polyphosphoric acid.

   a) Derive the possible sets of *styx* numbers.
   b) To which *styx* numbers do the two structures below correspond?

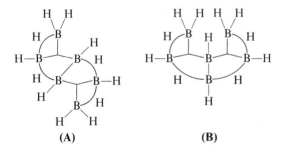

(A)   (B)

c) B_6H_{12} is thermally unstable, so structural information comes from the low temperature ^{11}B nmr spectrum that is sketched below. Note that there is no spin coupling between ^{11}B and bridging H; only $B-H_{terminal}$ spin coupling is seen.

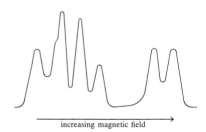

increasing magnetic field

The spectrum consists of two doublets of equal intensity and a triplet. Identify these parts of the spectrum. How many different types of boron atoms are there in B_6H_{12}? How many borons are bound to just one terminal H and how many to two terminal H's? Of the two topologies possible for B_6H_{12} drawn in part (b), which is the more likely? For a recent spectrum, see J. B. Leach, et al., Inorg. Chem., **9**, 2170 (1970).]

5. It is also possible to derive structures for anionic boron hydrides such as $B_2H_7^-$ by modifying the *styx* equations to take the charge into account. For a molecule of the general formula $[(BH)_pH_{q+c}]^{c+}$, the new equations are:

$$x = q + c - s$$
$$t = p + c - s$$
$$y = \frac{1}{2}(s - 3c - x)$$

For example, for $B_2H_7^-$, the structure of which is given below, $p = 2$, $q = 6$ (note that $q + c = 5$, and so $q = 6$ since $c = -1$), $c = -1$, $x = 4$, $s = 1$, $t = 0$, and $y = 0$.

$$\begin{bmatrix} H & & H \\ H-B & \overset{H}{-} B-H \\ H & & H \end{bmatrix}^-$$

a) Derive the sets of *styx* numbers for $B_3H_8^-$, and sketch a structure based on each *styx* number.
b) Derive *styx* numbers and structures for $B_5H_{11}^{2-}$ (see also Study Question 6).
c) The nmr spectra of some simple BH anions were recently reported [see S. G. Shore, et al., J. Amer. Chem. Soc., **97**, 5395 (1975)], among them $B_4H_9^-$. Derive a reasonable structure for this anion.

Molecular Orbital Concepts. Pentaborane(9), B_5H_9 or $(BH)_5H_4$, is a *nido*-borane. As such, it illustrates the variety of problems with which you are confronted when considering boranes wherein at least one boron atom is formally "connected" to more than three other boron atoms.

The most common *nido*-boranes are B_5H_9, B_6H_{10} [= $(BH)_6H_4$], and $B_{10}H_{14}$ [= $(BH)_{10}H_4$]. As illustrated on page 989 and in Figure 18–4, *nido*-boranes all have the general formula $(BH)_pH_4$ (*i.e.*, $q = 4$) and all of the q hydrogens are used to stitch up the open, non-triangular face of the borane with BHB bridges. Further, the compounds all have one or more "apical" (or non-facial) boron atoms that are formally connected to four or five other boron atoms, in addition to forming a terminal B—H bond. Since each apical boron has only three atomic or hybrid orbitals remaining to be used in framework bonding (one orbital having been used to form a terminal BH bond), each apical boron must be involved in at least one three-center BBB bond. In fact, it can be shown that, for neutral *nido*-boranes of general formula $(BH)_pH_4$, the number of three-center BBB bonds ($= t$) must be $p - 4$.[25] Therefore, for the *nido*-boranes being considered, $t = 1$ for B_5H_9, $t = 2$ for B_6H_{10}, and $t = 6$ for $B_{10}H_{14}$.

Based on the analysis above, pentaborane(9) must have the topology depicted by structure **8'**. However, this is not the final answer, since it is clear from the x-ray structural studies on this molecule that all of the B_{apical}—B_{edge} distances are equal (Figure 18-4). The only way to rationalize this is to recognize that **8** is only one of four possible canonical resonance structures (**8a–8d**) for the 4120 topology of B_5H_9. Since electron delocalization often confers special kinetic and/or thermodynamic stability (*cf.* benzene), B_5H_9 should be a "stable" molecule, as indeed it is. (We shall explore the question of reactivity very shortly.)

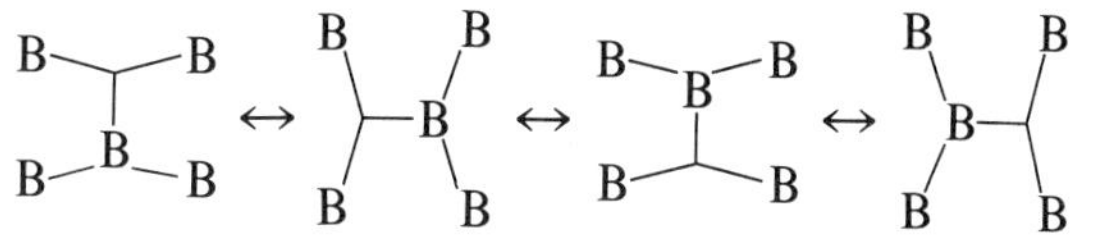

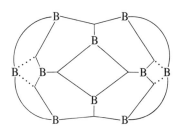

The fact that $B_{10}H_{14}$ must have more three-center bonds ($p = s + t = 10$) than pentaborane(9) means that it may have numerous canonical structures. In fact, Lipscomb has calculated that there will be 24 topologically allowed valence structures if open three-center BBB bonds are not considered.[26] Four of the 24 possible resonance structures are shown below as **a** through **d**, while the last structure best conveys the picture of decaborane bonding that has been derived from localized molecular orbital calculations (see discussion of such calculations at the end of Chapter 4).[27,28]

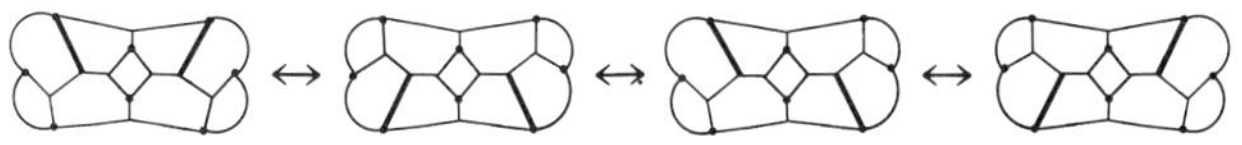

To avoid the problem of writing out numerous resonance structures to describe the bonding in the more symmetrical boranes (*closo* and *nido* compounds), it is clear that molecular orbital theory presents a distinct advantage. Therefore, we turn now to a molecular orbital treatment of pentaborane(9).

---

[25] This statement is derived as follows: Since, in general, $q/2 \leq s \leq q$, then s must be 2, 3, or 4 when $q = 4$. However, in *nido*-boranes, all extra q H's are used in BHB bridges, so $x = 0$; since $q = s + x$, then q must equal $s = 4$. Since $t = p - s$ (equation 18-1), t must equal $p - 4$.

[26] See ref. 13.

[27] E. A. Laws, R. M. Stevens, and W. N. Lipscomb, *J. Amer. Chem. Soc.*, **94**, 4467 (1972).

[28] W. N. Lipscomb, *Pure and Appl. Chem.*, **29**, 493 (1972). This paper presents a summary of the theoretical work done by Lipscomb and his students.

996 ☐ MOLECULAR POLYHEDRA: BORON HYDRIDES AND METAL CLUSTERS

Applying the techniques learned in Chapter 4, you first seek sets of symmetry-related atoms. In the case of B_5H_9 (C_{4v} symmetry), there are two symmetry-related sets of boron atoms: the four basal borons and the apical boron. The basal boron atoms define a set of "σ" TASO's as shown in Figure 18–7(A)[29] (cf. Figure 4–15 and Table 4–1); the apical boron atom will utilize the orbitals in Figure 18–7(B). The next step is to decide which of the central atom ao's (the apical boron in this case) will have

---

[29] These are shown as "sp" hao's following Chapter 4, p. 163, item 2.

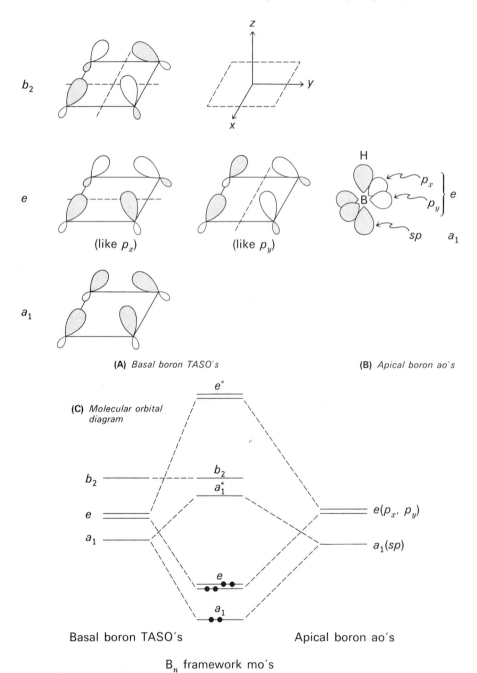

Figure 18–7. Formation of B_5H_9 framework molecular orbitals from basal boron TASO's and apical boron ao's.

non-zero overlaps with the TASO's, and to form bonding and anti-bonding mo's accordingly. The molecular orbital diagram that results is shown in Figure 18-7(C). The six electrons remaining after allowing for five B—H$_{terminal}$ pairs and four BHB pairs just fill the cage or framework bonding molecular orbitals and give the required diamagnetic molecule.

In addition to pentaborane(9),[30] detailed molecular orbital calculations have been carried out for the majority of the neutral boron hydrides,[31,32] and the following general conclusions can be drawn:

a) There is little or no evidence for open, three-center BBB bonding.

b) Apical boron atoms carry more negative charge than others. (As seen below, such borons are therefore more susceptible to electrophilic attack.)

c) A characteristic of BHB bridge units is that there is very little electron density along a line connecting the two boron atoms.

Synthesis and Reactivity of the Neutral Boron Hydrides[33,34]

Since the neutral boranes are polyhedral fragments, it might be supposed that they can be synthesized from the appropriate polyhedral species. However, nothing could be further from the truth; with a few exceptions, the neutral boranes are in fact synthesized from smaller boranes. Indeed, the best way to synthesize B_4H_{10}, B_5H_9, B_5H_{11}, and $B_{10}H_{14}$ is to pyrolyze diborane under carefully controlled conditions, and the fact that such an approach is feasible becomes clearer if we organize the borane family somewhat differently than we have done thus far.

Up to this point the general borane formula has been used in the form B_pH_{p+q}.

---

[30] See ref. 21.
[31] See ref. 28.
[32] W. N. Lipscomb, *Acc. Chem. Res.*, **6**, 257 (1973).
[33] R. W. Parry and M. K. Walter, "The Boron Hydrides," in "Preparative Inorganic Reactions," Vol. 5, W. L. Jolly, ed., Interscience Publishers, New York (1968).
[34] A potentially powerful topological approach to the systematization of boron hydride chemistry is given by R. W. Rudolph and D. A. Thompson, *Inorg. Chem.*, **13**, 2779 (1974).

TABLE 18-2

THE NEUTRAL BORANES WRITTEN ACCORDING TO THE GENERAL FORMULATIONS $(BH)_pH_q$ AND $(BH)_n(BH_3)_x$. Unknown boranes are indicated by *.

n	$q = 2, x = 1$	Nido $q = 4, x = 2$	Arachno $q = 6, x = 3$
0	BH_3	B_2H_6 $(BH_3)_2$	$B_3H_9^*$ $(BH_3)_3$
1		$B_3H_7^*$ $(BH)(BH_3)_2$	B_4H_{10} $(BH)(BH_3)_3$
2		$B_4H_8^*$ $(BH)_2(BH_3)_2$	B_5H_{11} $(BH)_2(BH_3)_3$
3		B_5H_9 $(BH)_3(BH_3)_2$	B_6H_{12} $(BH)_3(BH_3)_3$
4		B_6H_{10} $(BH)_4(BH_3)_2$	$B_7H_{13}^*$ $(BH)_4(BH_3)_3$
5		$B_7H_{11}^*$ $(BH)_5(BH_3)_2$	B_8H_{14} $(BH)_5(BH_3)_3$
6		B_8H_{12} $(BH)_6(BH_3)_2$	B_9H_{15} $(BH)_6(BH_3)_3$
7		$B_9H_{13}^*$ $(BH)_7(BH_3)_2$	$B_{10}H_{16}$ $(BH)_7(BH_3)_3$
8		$B_{10}H_{14}$ $(BH)_8(BH_3)_2$	$B_{11}H_{17}$ $(BH)_8(BH_3)_3$

This was done to clarify the prediction of numbers of different σ bonds and topologies. However, the chemical relationships between boranes may be seen better if we combine the B_pH_{p+q} formulation with another general formula $(BH)_n(BH_3)_x$, where n can assume values from 0 through 10, and $x = 1, 2$, or 3.[35] In Table 18-2 the known (and some unknown) boranes through B_{11} are listed according to their n, q, and x numbers. The usefulness of this tabular form becomes apparent when it is recognized that it shows that two boranes may be interconverted, or higher boranes made from simpler ones, by the application of one or more of the following reactions in the proper sequence:[36]

(A) Gain or loss of BH_3. Used to convert a borane of a given n and x to a borane of the same n but different x.

(B) Gain or loss of H_2. Used to convert a borane of a given n and x to one of the next higher or lower n and x. Can be thought of in terms of the reaction $BH_3 \rightleftharpoons BH + H_2$. Thus,

$$(BH)_n(BH_3)_x \rightleftharpoons (BH)_{n+1}(BH_3)_{x-1} + H_2$$

or

$$(BH)_pH_q \rightleftharpoons (BH)_pH_{q-2} + H_2$$

(C) Gain or loss of a BH unit. Used to convert a borane of a given n and x to a borane of higher n but the same x. Can also be thought of as occurring by the hypothetical reaction $BH_3 \rightarrow BH + H_2$.

These reactions are illustrated by a portion of Table 18-2 reproduced below.

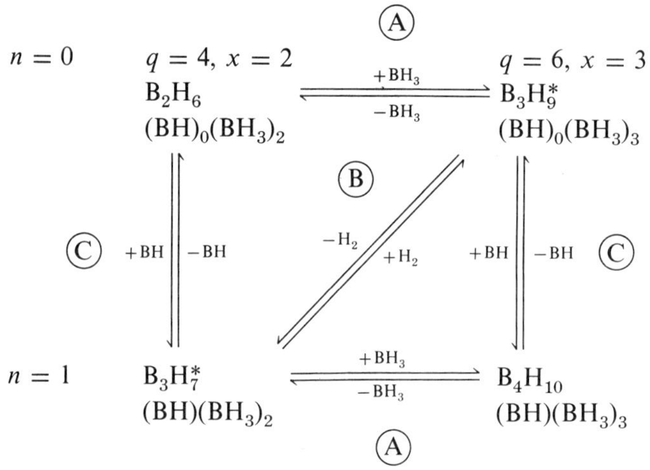

```
                              (A)
  n = 0      q = 4, x = 2    +BH₃      q = 6, x = 3
             B₂H₆           ⇌          B₃H₉*
             (BH)₀(BH₃)₂    -BH₃       (BH)₀(BH₃)₃
                ↕                         ↕
                          (B)
      (C) +BH ‖ -BH   -H₂ ╲  +H₂    +BH ‖ -BH  (C)
                ↕                         ↕
                             +BH₃
  n = 1      B₃H₇*           ⇌          B₄H₁₀
             (BH)(BH₃)₂     -BH₃        (BH)(BH₃)₃
                              (A)
```

The interrelations between boranes have been recognized for some time, and a great deal of effort has gone into writing reaction routes for the production of higher boranes from the simpler ones. However, as Parry and Walter have commented, "Up to the present time [i.e., 1968] such studies have frequently generated more heat than light."[37] One of the latest schemes linking diborane to B_3, B_4, and B_5 boranes incorporates the ideas noted above in addition to several important experimental facts.[38]

$$2B_2H_6 \rightleftharpoons BH_3 + B_3H_9$$
$$BH_3 + B_2H_6 \rightleftharpoons B_3H_9$$

---

[35] See ref. 6.
[36] See ref. 33.
[37] See ref. 33.
[38] L. H. Long, *J. Inorg. Nucl. Chem.*, **32**, 1097 (1970); *Prog. Inorg. Chem.*, **15**, 1 (1972).

$$B_3H_9 \rightleftharpoons B_3H_7 + H_2$$
$$2B_3H_9 \rightarrow 3B_2H_6$$
$$BH_3 + B_3H_7 \rightleftharpoons B_4H_{10}$$
$$B_2H_6 + B_3H_7 \rightarrow BH_3 + B_4H_{10} \rightleftharpoons B_5H_{11} + H_2$$
$$B_3H_9 + B_3H_7 \rightarrow B_2H_6 + B_4H_{10}$$
$$2B_3H_7 \rightarrow B_2H_6 + B_4H_8$$
$$B_3H_9 + B_4H_{10} \rightarrow B_2H_6 + B_5H_{11} + H_2$$
$$B_3H_7 + B_4H_{10} \rightarrow B_2H_6 + B_5H_{11}$$
$$H_2 + B_4H_8 \rightarrow B_4H_{10}$$
$$BH_3 + B_4H_8 \rightleftharpoons B_5H_{11}$$
$$B_3H_9 + B_4H_8 \rightarrow B_2H_6 + B_5H_{11}$$

The enthalpy change for the dissociation of B_2H_6 into two BH_3 moieties is now thought to be about 35.5 kcal.[39] Since the activation energy for the recombination of $2BH_3$'s to give B_2H_6 is approximately zero,[40] the activation energy for the dissociation of diborane must also be about 35 kcal. As this seems an excessively high activation energy for the first step in the decomposition process, the alternative bimolecular reaction of two diborane molecules to give BH_3 and B_3H_9 has been suggested. Triborane(9) is also formed in the second reaction, so that the sum of the first two processes is $3B_2H_6 \rightarrow 2B_3H_9$. This agrees with the experimental observation that the rate of disappearance of diborane is 3/2 order in diborane. The third reaction generates still another reactive intermediate, B_3H_7, and the scheme continues on to the formation of B_4H_{10}, B_5H_9, and B_5H_{11} using basically reaction types (A), (B), and (C). The intermediates BH_3, B_3H_7, and B_3H_9 are important to all of the steps in the scheme, and their great reactivity is indicated by the fact that they have never been isolated or characterized as the simple boranes (although all are known in the form of adducts with Lewis bases).

Tetraborane(10) may be prepared in several ways, but perhaps the most effective on a larger scale is the pyrolysis of diborane.[41,42] This is usually done in a "hot-cold" reactor such as that illustrated in Figure 18-8. The diborane is first admitted to the evacuated reactor and is frozen onto the outer walls by immersing the reactor in liquid nitrogen ($-196°C$). The liquid nitrogen is then removed and replaced by a

---

[39] D. A. Dixon, I. M. Pepperberg, and W. N. Lipscomb, *J. Amer. Chem. Soc.*, **96**, 1325 (1974).
[40] G. W. Mappes, S. A. Fridmann, and T. P. Fehlner, *J. Phys. Chem.*, **74**, 3307 (1970).
[41] See ref. 33.
[42] M. J. Klein, B. C. Harrison, and I. J. Solomon, *J. Amer. Chem. Soc.*, **80**, 4149 (1958).

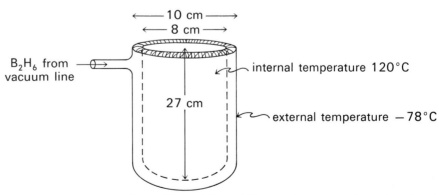

Figure 18-8. A "hot-cold" reactor for the pyrolytic condensation of B_2H_6 to B_4H_{10}.

bath at $-78°C$ (crushed Dry Ice in a suitable medium such as acetone or trichloroethylene). The diborane boils off from the now "warmer" outer wall and comes into contact with the hot inner well (which contains ethylene glycol heated to $120°C$ by an electric immersion heater), where pyrolysis occurs to give B_4H_{10}. Approximately 80% yields of B_4H_{10} can be obtained in 1 to 3 hours.

Alfred Stock first isolated tetraborane—and smaller amounts of the other hydrides B_5H_9, B_5H_{11}, B_6H_{10}, and $B_{10}H_{14}$—from the reaction of magnesium boride with aqueous acid.[43]

$$Mg_3B_2 + HCl + H_2O \rightarrow B_4H_{10} + \cdots$$

Most importantly, however, no diborane can be isolated from this reaction. The reason for this apparently is that tetraborane, unlike diborane, is much less susceptible to hydrolysis by cold water. With hot water, however, the complete degradation of tetraborane proceeds readily.

$$B_4H_{10} + 12H_2O \xrightarrow{\Delta} 4B(OH)_3 + 11H_2$$

Like all of the liquid boranes, tetraborane(10) is quite susceptible to oxidation (this excludes $B_{10}H_{14}$, which is a solid and reasonably stable to air and water) so, like the others, tetraborane must be handled in a chemical vacuum line.[44]

> We might add parenthetically that another reason for working with the hydrides in a vacuum line is that they have a rather disagreeable odor. After smelling decaborane, for example, one can "taste and smell" the compound for some time afterward. Further, the boranes have been classed as "very toxic." As Levinskas says in a review of borane toxicology, "Animals dosed with decaborane showed listlessness, incoordination, tremors, convulsions, and coma. . . . Small, repeated doses of decaborane were not well tolerated. Rats and rabbits given small doses daily, by various routes, died from cumulative doses less than twice the LD_{50} values for the respective routes of administration."[45,46]

One of the most thoroughly studied reactions of tetraborane is that with Lewis bases. Schaeffer has found that, no matter what the base, the first product observable by ^{11}B nmr has the spectral properties of $B_3H_8^-$ (11). This being the case, the other product must then be $[H_2B(base)_2]^+$ (12).[47,48]

---

[43] See ref. 11.

[44] D. F. Shriver, "The Manipulation of Air-Sensitive Compounds," McGraw-Hill, New York (1969). See also Chapter 8.

[45] G. J. Levinskas, "Toxicology of Boron Compounds," in "Boron, Metallo-Boron Compounds and Boranes," R. M. Adams, ed., Interscience, New York (1964), page 713.

[46] The LD_{50} number is defined as the dose of any agent, given under specified conditions, that causes death in 50% of the organisms being observed.

[47] R. Schaeffer, F. Tebbe, and C. Phillips, *Inorg. Chem.*, **3**, 1475 (1964).

[48] This mechanism is in accord with mo calculations that indicate that nucleophilic attack should occur at B_2 (or B_4) (charge $= +0.08$) rather than at B_1 (or B_3) (charge $= -0.02$); E. Switkes, I. R. Epstein, J. A. Tossell, R. M. Stevens, and W. N. Lipscomb, *J. Amer. Chem. Soc.*, **92**, 3837 (1970).

With ammonia, **12** is stable and the reaction products (**11** and **12**) can be isolated at this stage. However, with more hindered bases, **12** is unstable and accepts a hydride ion from $B_3H_8^-$ to form base $\cdot BH_3$ and B_3H_7. The latter is extremely unstable, having a coordinatively unsaturated boron atom, and rapidly forms base $\cdot B_3H_7$ (**13**).[48a]

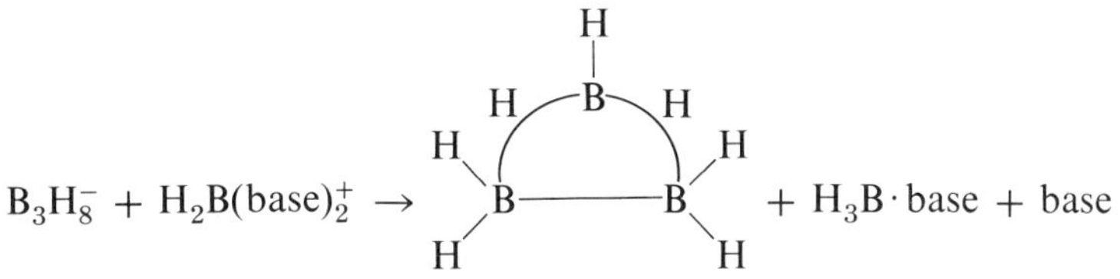

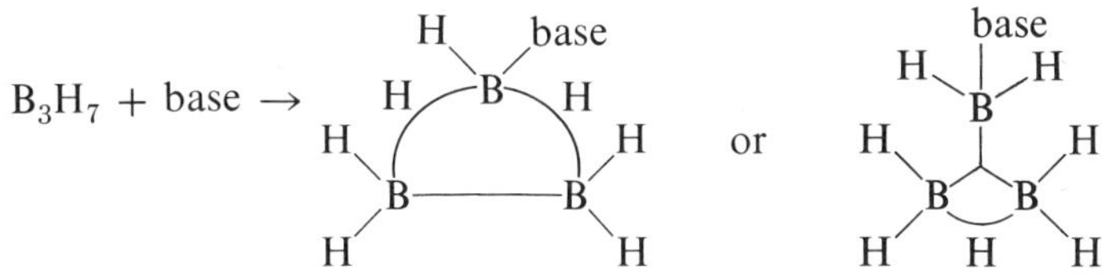

$$B_3H_8^- + H_2B(\text{base})_2^+ \rightarrow \quad \text{[structure]} \quad + H_3B \cdot \text{base} + \text{base}$$

$$B_3H_7 + \text{base} \rightarrow \quad \textbf{13a} \ (styx = 2013) \quad \text{or} \quad \textbf{13b} \ (styx = 1104)$$

Pentaborane(9) is the most stable of the neutral boranes to heat, and it can be prepared in excellent yield by diborane pyrolysis. This combination of factors has made it one of the most widely studied boranes (perhaps third after diborane and decaborane).[49] One of the characteristic reactions of pentaborane(9) is the Friedel-Crafts alkylation of the apical boron to give 1-alkylpentaborane (**14**).

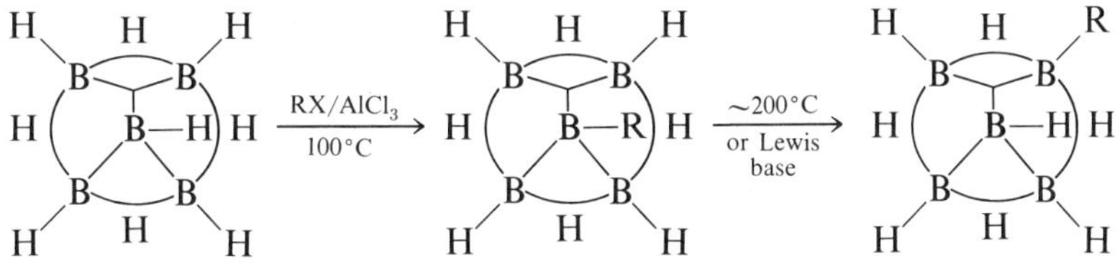

$$\textbf{8} \xrightarrow[100\,°C]{RX/AlCl_3} \textbf{14} \xrightarrow[\text{or Lewis base}]{\sim 200\,°C} \textbf{15}$$

The fact that an electrophilic substitution should occur at the apical boron is predicted from molecular orbital calculations, which indicate that the apical boron is negatively charged relative to the basal boron.[50] Furthermore, it is apparent that apical substitution is more a kinetic than a thermodynamic preference for substitution site. That is, heating the 1-alkylpentaboranes or adding a Lewis base (such as 2,6-dimethylpyridine) causes the alkyl group to migrate to the 2-position (structure **15**). Such rearrangements are not uncommon in boron hydride chemistry; you shall see extensive framework rearrangements in the carborane series to be discussed below.

One of the most exciting developments in boron chemistry has been the discovery of metalloboranes.[51] Such compounds are far more prevalent in the carborane series, but it is with pentaborane that examples of metalloboranes are first encountered. For example, salts of $B_5H_8^-$ react with Group IV trialkyl halides to give compounds such as **16**.[49]

[48a] See L. D. Brown and W. N. Lipscomb, *Inorg. Chem.*, **16**, 1 (1977) for a theoretical study of base $\cdot B_3H_7$ adducts.
[49] D. F. Gaines, *Acc. Chem. Res.*, **6**, 416 (1973).
[50] See ref. 48.
[51] N. N. Greenwood and I. M. Ward, *Chem. Soc. Revs.*, **3**, 231 (1974).

$$B_5H_9 \xrightarrow[-HR]{+MR (= LiR, NaH)/R_2O} M^+B_5H_8^- \xrightarrow[\substack{-MX \\ (M^{IV} = Si, Ge, Sn)}]{+XM^{IV}R_3}$$

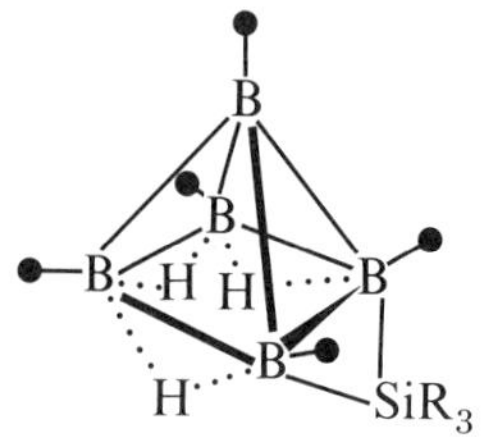

16

The Group IV element, which is equidistant from the two nearest borons, has clearly taken the place of a bridging hydrogen, and the bonding is undoubtedly not unlike that of Al_2R_6 dimers (Chapter 16, p. 845).

Yet another type of metalloborane (**17**) is formed upon reaction of $Fe(CO)_5$ with B_5H_9 in a hot-cold reactor.[52] This compound is quite analogous to $(\eta^4\text{-}C_4H_4)Fe(CO)_3$ (Chapter 16, p. 883), where $\eta^4\text{-}C_4H_4$ is a two-pair donor. The basal boron atoms of **17** have a set of TASO's (see Figure 18-7) that are entirely similar to the $p\pi$ set of cyclobutadiene; thus, the B_4H_8 fragment is also a two-pair donor.

$$Fe(CO)_5 + B_5H_9 \xrightarrow{200°C}$$

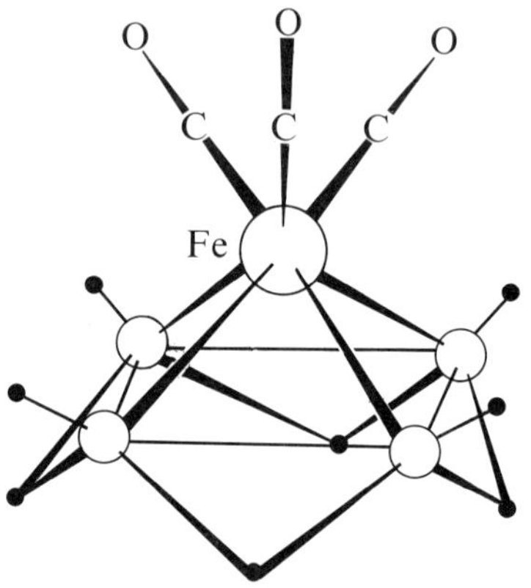

17

Of the commonly known boranes, decaborane(14), $B_{10}H_{14}$, is the only one stable enough toward water and air to be handled in the open. Although it has, unfortunately, a disagreeable odor, its stability has led to an extensive examination of its chemistry.[53,54] The molecule can form a wide variety of derivatives of the type $B_{10}H_{13}X$, $B_{10}H_{12}X_2$, ... (where X is a one-electron donor like a hydrogen or halogen atom) or $B_{10}H_{12}L_2$ (where L is a Lewis base or a hydride ion, H^-), or an extensive series of ions. Although at first glance it may seem that there is a bewildering variety of molecules and ions, they are often interrelated by loss or gain of H^+ or H_2. A few of these interrelationships are illustrated on p. 1003 (where B = BH). Note that the molecule $B_{10}H_{12}L_2$ is isoelectronic and isostructural with $B_{10}H_{14}^{2-}$, and other such relationships exist.

---

[52] N. N. Greenwood, C. G. Savory, R. N. Grimes, L. G. Sneddon, A. Davison, and S. S. Wreford, *Chem. Commun.*, 718 (1974).
[53] See ref. 8.
[54] M. F. Hawthorne, *Adv. Inorg. Chem. Radiochem.*, **5**, 307 (1963).

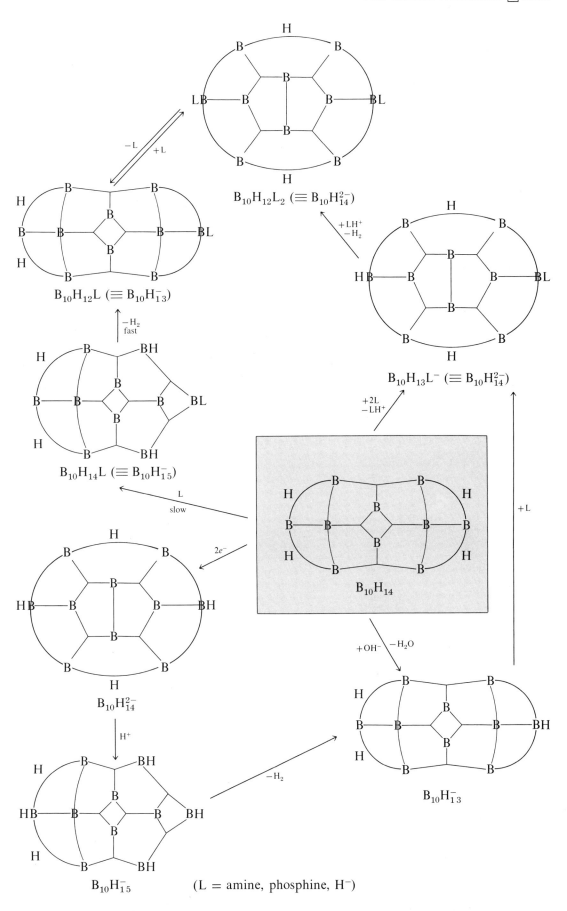

TABLE 18-3
SOME REACTIONS OF DECABORANE(14)

Reaction	Reference
A. *Electrophilic Substitutions* (in the 1 ... 4 positions)	a,b
$RBr + B_{10}H_{14} \xrightarrow{AlCl_3/CS_2} 2\text{-}RB_{10}H_{13} (+ \ldots R_4B_{10}H_{10})$	
$I_2 + B_{10}H_{14} \rightarrow 2,4\text{-}B_{10}H_{12}I_2$	c
B. *Proton Abstraction*	
$B_{10}H_{14} \underset{+H^+}{\overset{OH^- \text{ or } L}{\rightleftharpoons}} B_{10}H_{13}^- + H_2O \text{ or } LH^+$	d,e
$B_{10}H_{14} \underset{-H_2}{\overset{+NaH/Et_2O}{\longrightarrow}} NaB_{10}H_{13} \underset{-H_2}{\overset{xs\ NaH}{\longrightarrow}} Na_2B_{10}H_{12}$	f
$B_{10}H_{14} + RMgX \rightarrow RH + B_{10}H_{13}MgX \xrightarrow{RX} 6\text{-}RB_{10}H_{13}$	g
C. *Electron Transfer*	
$B_{10}H_{14} + 2Na \rightarrow 2Na^+ + B_{10}H_{14}^{2-}$	h
$4H_2O + B_{10}H_{14} + KBH_4 \xrightarrow{H_2O} K^+ + B_{10}H_{14}^{2-} + 3H_2 + H_3O^+ + B(OH)_3$	i
D. *Degradation*	
$B_{10}H_{14} + OH^- + 2H_2O \rightarrow B_9H_{14}^- + B(OH)_3 + H_2$	j
E. *Reactions Involving Lewis Bases* (at positions 6 and 9)	
$B_{10}H_{14} + CH_3CN \rightarrow B_{10}H_{12}(CH_3CN)_2 + H_2$	k
$B_{10}H_{12}L_2 + 2L' \rightarrow B_{10}H_{12}L'_2 + 2L$	l
(displacement tendency: $R_2S = RCN < R_2NCN < R_2NCOR < R_3N < R_3P =$ pyridine)	
$B_{10}H_{14} \xrightarrow{NaH} B_{10}H_{13}^- \xrightarrow{L} B_{10}H_{13}L^-$	m
(L = $EtNH_2$, Et_2NH, Me_3N, py, PPh_3)	
$B_{10}H_{13}L^- + 2H_2O + H_3O^+ \rightarrow 2H_2 + B(OH)_3 + B_9H_{13}L$	n
(L = RCN, R_2NH, PPh_3) $\quad +L \uparrow -SEt_2$	
$B_{10}H_{12}(SEt_2)_2 + 3MeOH \rightarrow H_2 + B(OMe)_3 + B_9H_{13}(SEt_2) + SEt_2$	n
$+OH^- \downarrow$	
$B_9H_{12}^-$	n
$B_{10}H_{12}L_2 + 2R_3N \rightarrow 2R_3NH^+ + B_{10}H_{10}^{2-} + 2L$	o,p
$B_{10}H_{12}(SEt_2)_2 + HC\equiv CH \xrightarrow{n\text{-propyl ether}} B_{10}C_2H_{12} + H_2 + SEt_2$	q,r

[a] N. J. Blay, I. Dunstan, and R. L. Williams, *J. Chem. Soc.*, 430 (1960).
[b] R. L. Williams, I. Dunstan, and N. J. Blay, *J. Chem. Soc.*, 5006 (1960).
[c] R. Schaeffer, *J. Amer. Chem. Soc.*, **79**, 2726 (1957).
[d] G. A. Guter and G. W. Schaeffer, *J. Amer. Chem. Soc.*, **78**, 3546 (1956).
[e] M. F. Hawthorne, *J. Amer. Chem. Soc.*, **80**, 3480 (1958).
[f] J. J. Miller, and M. F. Hawthorne, *J. Amer. Chem. Soc.*, **81**, 4501 (1959).
[g] B. Siegel, J. L. Mack, J. V. Lowe, and J. Gallaghan, *J. Amer. Chem. Soc.*, **80**, 4523 (1958); I. Dunstan, N. J. Blay, and R. L. Williams, *J. Chem. Soc.*, 5016 (1960); E. L. Muetterties and V. D. Aftandilian, *Inorg. Chem.*, **1**, 731 (1962).
[h] E. B. Rupp, D. E. Smith, and D. F. Shriver, *J. Amer. Chem. Soc.*, **89**, 5562 (1967); D. E. Smith, E. B. Rupp, and D. F. Shriver, *J. Amer. Chem. Soc.*, **89**, 5568 (1967).
[i] E. L. Muetterties, *Inorg. Chem.*, **2**, 647 (1963).
[j] L. E. Benjamin, S. F. Stafiej, and E. A. Takacs, *J. Amer. Chem. Soc.*, **85**, 2674 (1963).
[k] R. Schaeffer, *J. Amer. Chem. Soc.*, **79**, 1006 (1957).
[l] R. J. Pace, J. Williams, and R. L. Williams, *J. Chem. Soc.*, 2196 (1961).
[m] B. M. Graybill, A. R. Pitochelli, and M. F. Hawthorne, *Inorg. Chem.*, **1**, 622 (1962).
[n] B. M. Graybill, A. R. Pitochelli, and M. F. Hawthorne, *Inorg. Chem.*, **1**, 626 (1962).
[o] E. L. Muetterties, J. H. Balthis, Y. T. Chia, W. H. Knoth, and H. C. Miller, *Inorg. Chem.*, **3**, 444 (1964).
[p] M. F. Hawthorne and R. L. Pilling, *Inorg. Syn.*, **9**, 16 (1967).
[q] T. L. Heying and co-workers, *Inorg. Chem.*, **2**, 1089 (1963). This article is followed by numerous others on carborane chemistry.
[r] M. F. Hawthorne and co-workers, *Inorg. Syn.*, **10**, 91 (1967).

Perhaps the most important discovery in decaborane(14) chemistry, in terms of its overall development, is the finding that weak bases—nitriles, phosphines, dialkylsulfides—react to give H_2 and the disubstituted borane $B_{10}H_{12}L_2$ (**18** depicts the structure of the acetonitrile adduct, the first such compound discovered).[55]

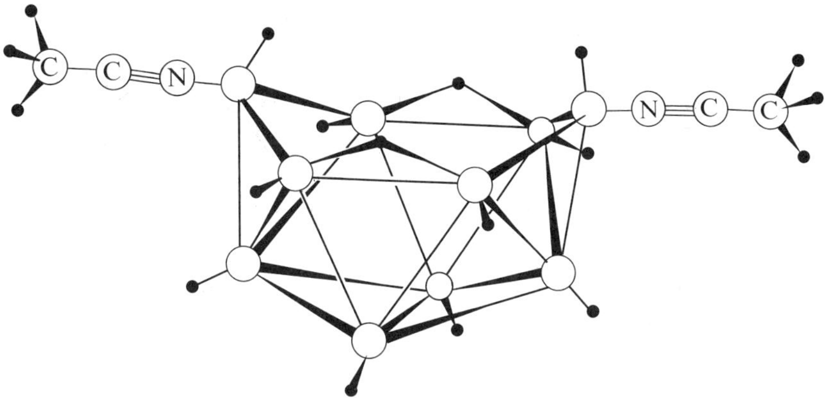

18 $B_{10}H_{12}(NCCH_3)_2$

As noted above, this compound is isoelectronic and isostructural with $B_{10}H_{14}^{2-}$, an ion formed by reacting decaborane(14) with sodium amalgam. The chemistry of these two species—$B_{10}H_{12}L_2$ and $B_{10}H_{14}^{2-}$—is extensive, and some reactions are included in the outline of decaborane reactions in Table 18-3. One of the most important reactions is that producing the *closo* ion $B_{10}H_{10}^{2-}$, an extremely stable ion that has an extensive chemistry of its own; another is the reaction of $B_{10}H_{12}L_2$ with an acetylene to give a 1,2-dicarba*closo*dodecaborane, commonly called *ortho*-carborane. The chemistry of the *closo* ions and of carboranes in general is discussed in a succeeding section.

Finally, one of the predictions of a molecular orbital treatment of $B_{10}H_{14}$ is that the 2,4 boron atoms bear more negative charge than the 1,3 borons (see Figure 18-4 for the numbering convention). The usual interpretation of this is that electrophilic substitutions will occur first at the 2,4 positions of $B_{10}H_{14}$, with the 1,3 positions next in preference. This prediction is completely in accord with experiment. For example, reaction of $B_{10}H_{14}$ with alkyl halides in the presence of the usual Friedel-Crafts catalysts gives some 1-$RB_{10}H_{13}$ but predominantly 2-$RB_{10}H_{13}$.

A GENERAL ORGANIZATIONAL SCHEME FOR THE NEUTRAL BORON HYDRIDES, THE CLOSO POLYHEDRAL HYDROBORATE IONS, AND THE CARBORANES

Having completed a discussion of the neutral boranes, we wish to turn to the polyhedral hydroborates such as $B_{10}H_{10}^{2-}$ (**2**), carboranes such as $B_{10}C_2H_{12}$ (**3**), and the metalloboranes. Since there are many more examples of compounds in these classes than of neutral boron hydrides, it is useful to introduce at this point a scheme for the organization of the whole of polyhedral borane chemistry.

Given the molecular formula, a method is needed for predicting the probable structural classification of a boron compound—*closo, nido,* or *arachno*. As described earlier, the $(BH)_pH_q$ symbolism is best for structural purposes, and the scheme has been extended by Wade[56] and by Rudolph[57] (see also Williams[58]).

---

[55] R. Schaeffer, *J. Amer. Chem. Soc.,* **79**, 1006 (1957).
[56] See ref. 15.
[57] See ref. 14.
[58] See ref. 16.

> The formula of any neutral borane, hydroborate, or carborane can be written as
>
> $$[(CH)_a(BH)_pH_q]^{c-}$$
>
> where the number of vertices of the polyhedron or polyhedral fragment is $a + p = n$, and q H's are involved in BHB or extra B—$H_{terminal}$ bonds.
>
> The number of framework electrons is given by
>
> $$3a + 2p + q + c = 2n + a + q + c$$
>
> and so the number of electron pairs is $\frac{1}{2}(2n + a + q + c)$. It is assumed that a structure is adopted in which all of these electrons completely occupy only the available bonding molecular orbitals. (There are $3n + q$ valence ao's at hand, so at most $\frac{1}{2}(3n + q)$ bonding mo's can arise.) Thus,
>
> if $a + q + c = 2$, the compound assumes a *closo* structure.
>
> if $a + q + c = 4$, the compound assumes a *nido* structure.
>
> if $a + q + c = 6$, the compound assumes an *arachno* structure.

Consider hexahydrohexaborate(2−), $(BH)_6^{2-}$ or $B_6H_6^{2-}$, as an example of these rules. There are 26 electrons available for bonding in this ion. After 12 are used in terminal B—H bonding, 14 are left for bonding in the B_6 cage. Since the structure must contain six vertices ($2n = 12$), then $2n + a + q + c = 14$ and $a + q + c = 2$, and $B_6H_6^{2-}$ assumes the *closo* geometry.

We have already considered the molecular orbital scheme of pentaborane(9) or $(BH)_5H_4$ (p. 996). Twenty-four valence electrons are available for bonding in this molecule, but after allocation of 10 of them to terminal BH bonding, only 14 remain for bonding within the framework or for forming BHB bridges or additional B—$H_{terminal}$ bonds. This same electron total comes from $2n + a + q + c$ where $2n = 10$, and so $a + q + c = 4$. Therefore, you are led to the *nido* structure for pentaborane(9).

> To more completely describe the structure of pentaborane(9), you would of course like to know where the "extra" four ($= q$) hydrogens are placed. A partial answer to this comes from the topological theory previously outlined (pp. 990–993). From equations 18-2 and 18-3 on page 991, $q = s + x$ and so s must be $\geq x$ [from $y = \frac{1}{2}(s - x)$]. Therefore, there can never be more extra terminal hydrogens ($= x$) than bridging groups ($= s$) (for B_5H_9, $x = 0$ and $s = 4$ as it always does for a *nido* borane), and a reasonable structure can be worked out on this basis.

There are 26 electrons available for bonding in pentaborane(11), $(BH)_5H_6$, but after 10 of them are allotted to terminal BH bonding, 16 remain for framework bonding and for the formulation of additional terminal BH bonds and BHB bonds. Therefore, with a five-vertex structure required, $2n = 10$ and $a + q + c = 6$; the *arachno* structure is predicted and observed.[59]

Very early in this chapter (p. 989), you learned of the correlation between *closo*-$B_7H_7^{2-}$, *nido*-B_6H_{10}, and *arachno*-B_5H_{11}. There it was concluded that a *nido* bo-

---

[59] As suggested in the paragraph above, the number of BHB bridges ($= s$) formed by the "extra" six hydrogens must be less than or equal to the number of extra terminal BH groups ($= x$); in fact, $s = 3$ and $x = 3$ for B_5H_{11}.

rane has the general formula $(BH)_pH_4$ and an *arachno* borane has the general formula $(BH)_pH_6$. That is, *nido* boranes are fully protonated $(BH)_p^{4-}$ ions, and *arachno* boranes are fully protonated $(BH)_p^{6-}$ ions, the extra protons forming BHB bridges or extra BH terminal groups. With reference to the new scheme, you see now that a $(BH)_p^{4-}$ ion will have $2p + 4$ electrons ($a + q + c = 4$) and a $(BH)_p^{6-}$ ion will have $2p + 6$ electrons ($a + q + c = 6$), so *nido* and *arachno* geometries are predicted for B_6H_{10} and B_5H_{11} by the new, more general scheme. This observation presents yet another way of making structural predictions. That is,[60]

a) compounds isoelectronic with $(BH)_p^{2-}$ will have a *closo* structure.

b) compounds isoelectronic with $(BH)_p^{4-}$ will have a *nido* structure.

c) compounds isoelectronic with $(BH)_p^{6-}$ will have an *arachno* structure.

This is, of course, exactly equivalent to the method previously outlined. However, in certain cases you may find this second way of thinking to be of more general applicability or somewhat easier to use.

The $(BH)_p^{c-}$ formulation suggests that a great range of compounds containing the BH unit may be prepared by simply replacing a BH unit with some other group also capable of donating two electrons to the polyhedral framework. That is, a BH may be replaced by CH^+, P^+, S^{2+}, and so on. For example, *closo*-$B_4C_2H_6$ may be generated from *closo*-$B_6H_6^{2-}$ by removing two BH units and adding two CH^+ groups.

$$(BH)_6^{2-} \xrightarrow[-2BH]{} [(BH)_4^{2-}] \xrightarrow[+2CH^+]{} [(BH)_4^{2-} + (CH^+)_2] \ (= B_4C_2H_6)$$

As a more extensive example of the use of the structural prediction rules, consider the chain of reactions illustrated on p. 1008.[61] Compound **3**, the well known icosahedral carborane 1,2-$B_{10}C_2H_{12}$, reacts with bases such as ethanolic ethoxide ion to give a degradation product, the *nido* ion 7,8-$B_9C_2H_{12}^-$ (**19**). The formula of this ion may be written alternatively as $[(CH)_2(BH)_9H]^-$, considering it as a $[(BH)_{11}H]^{3-}$ ion with two CH^+ units replacing two BH units and with a proton stitching up part of the open face. Note that $a + q + c = 4$ for this ion, indicating that it is a *nido* species. Deprotonation with hydride gives *nido*-7,8-$B_9C_2H_{11}^{2-}$ (**20**), a carborane isoelectronic with *nido*-$(BH)_{11}^{4-}$. On the other hand, protonation of **19** gives 7,8-$B_9C_2H_{13}$ (**21**), a neutral *nido*-carborane. Pyrolysis of this latter molecule in solution in turn gives the eleven-particle *closo* polyhedron, 2,3-$B_9C_2H_{11}$ (**22**). As indicated by the sum $a + q + c \ (= 2 + 0 + 0 = 2)$, **22** must be a *closo*-carborane isoelectronic with $(BH)_{11}^{2-}$. Finally, oxidative degradation of either **19** or **22** produces the neutral species 1,3-$B_7C_2H_{13}$ (**23**), a molecule predicted to be an *arachno*-carborane based on the sum of $a + q + c \ (= 2 + 4 + 0 = 6)$.

---

[60] Two examples of a new class of boranes, the *hypho* class (*hypho* = net), have recently been isolated [A. V. Fratini, G. W. Sullivan, M. L. Denniston, R. K. Hertz, and S. G. Shore, *J. Amer. Chem. Soc.*, **96**, 3013 (1974); M. Mangion, R. K. Hertz, M. L. Denniston, J. R. Long, W. R. Clayton, and S. G. Shore, *J. Amer. Chem. Soc.*, **98**, 449 (1976)]. One of these is $B_5H_9(PMe_3)_2$, a molecule that is isoelectronic with the hypothetical ion $B_5H_{11}^{2-}$, for which the structure below is postulated.

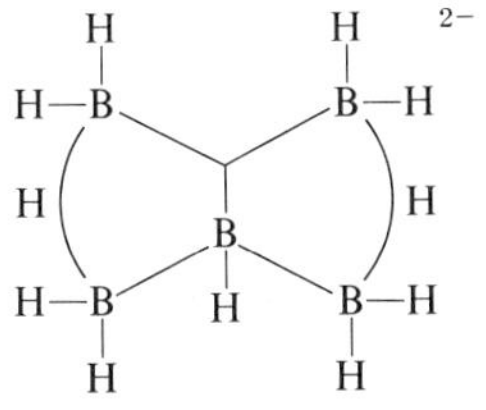

This species clearly does not fit into any of the series yet discussed, as it is equivalent to $(BH)_5^{8-}(H^+)_6$. That is, it is a $2n + 8$ molecule ($a + q + c = 8$). It is considered the forerunner of the new *hypho* series.

[61] G. B. Dunks and M. F. Hawthorne, *Acc. Chem. Res.*, **6**, 124 (1973).

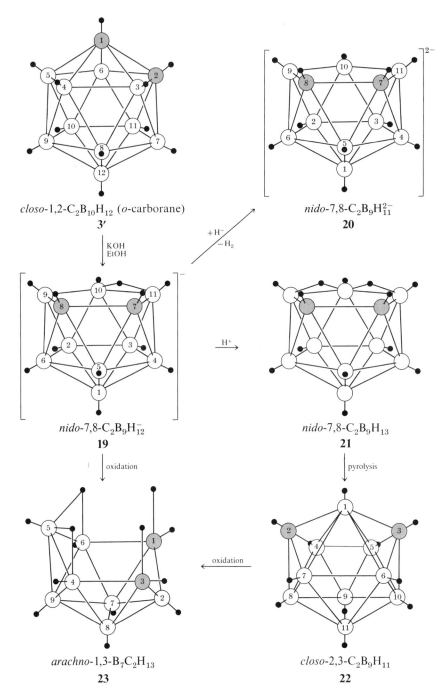

closo-1,2-C$_2$B$_{10}$H$_{12}$ (*o*-carborane)
3′

nido-7,8-C$_2$B$_9$H$_{11}^{2-}$
20

nido-7,8-C$_2$B$_9$H$_{12}^{-}$
19

nido-7,8-C$_2$B$_9$H$_{13}$
21

arachno-1,3-B$_7$C$_2$H$_{13}$
23

closo-2,3-C$_2$B$_9$H$_{11}$
22

Before leaving this topic, you need to note one last important consequence of the structural rules: the polyhedra in a given *row* in Figure 18-2 are related by oxidation-reduction. That is, as electrons are added to a *closo* polyhedron with n vertices and $2n + 2$ framework electrons, the polyhedron opens to form the *nido* compound of the same number of vertices and $2n + 4$ electrons; addition of yet another two electrons obviously then leads to the *arachno* compound with $2n + 6$ electrons. As an example, *closo*-B$_6$H$_6^{2-}$ forms *nido*-B$_6$H$_6^{4-}$ upon addition of two electrons. (The latter ion is not observed, but is converted to the known B$_6$H$_{10}$ by protonation.) At least one complete series of compounds related by gain or loss of electrons and protons is known.[62-64]

---

[62] See ref. 14.
[63] W. R. Pretzer and R. W. Rudolph, *J. Amer. Chem. Soc.*, **95**, 931 (1973).
[64] Note that B$_9$H$_9$S is equivalent to B$_{10}$H$_{10}^{2-}$, since S^{2+} has replaced a BH unit.

$$closo\text{-}B_9H_9S \xrightarrow{+2e + 2H^+} nido\text{-}B_9H_{11}S \xrightarrow{+2e + H^+} arachno\text{-}B_9H_{12}S^-$$

$$2n + 2 = 22 \qquad\qquad 2n + 4 = 24 \qquad\qquad 2n + 6 = 26$$

and many other pairs have been observed.

STUDY QUESTIONS

6. Shore recently reported [*J. Amer. Chem. Soc.,* **97,** 5395 (1975)] the nmr spectra of a number of simple BH anions.

    a) Into which structural class does each of the following ions fit? $B_4H_9^-$, $B_5H_{12}^-$, $B_6H_{11}^-$, and $B_7H_{12}^-$.
    b) Derive a possible structure for $B_4H_9^-$ (*cf.* Study Question 5).

7. Place each of the following molecules or ions in the correct structural class:

    a) $B_5CH_6^-$
    b) $B_4C_2H_6$
    c) B_5H_9
    d) $B_3C_2H_7$
    e) $B_6H_{10}(PMe_3)_2$. (In thinking about this compound, realize that PMe_3 is a substituent on the cage; an H^- ion could be put in its place.)

8. In the preceding question, what are the relationships between (a) and (b), and between (c) and (d)?

9. Show that $B_9H_{11}S$ and $B_9H_{12}S^-$ fall into the *nido* and *arachno* classes, respectively.

10. Give the correct structural classifications for:

    a) $B_{10}C_2H_{12}$ and $B_{10}CH_{11}P$
    b) NB_8H_{13}
    c) $NC_2B_8H_{11}$ [*cf.* Schaeffer and co-workers, *Chem. Commun.,* 935 (1975).]

A MOLECULAR ORBITAL VIEW OF CLOSO-HYDROBORATE ANIONS AND CARBORANES

The reason for the efficacy of the organizational schemes just discussed is found in molecular orbital theory. As an example, consider the $B_6H_6^{2-}$ ion, and the approach taken to constructing the molecular orbital picture of this ion is similar to that outlined in Chapter 4.

After accounting for B—H bonding, each BH moiety presents three ao's for cage bonding: one $sp\sigma$ hao directed to the center of the cage and two (p_x and p_y) ao's tangent to the cage (**24**). Thus, the $(BH)_6$ moiety is highly analogous to the F_6 "group" in SF_6, for example.

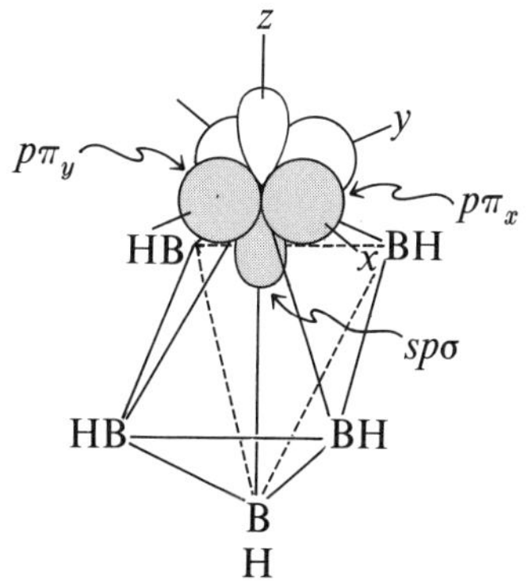

24

1010 □ MOLECULAR POLYHEDRA: BORON HYDRIDES AND METAL CLUSTERS

In spite of the facts that $(BH)_6$ contains no central atom counterpart of the S in SF_6 and that there are only seven electron pairs for cage bonding (F_6 has 15), the orbital pattern for the $(BH)_6$ aggregate is no different from that for the F_6 TASO's of Chapter 4 (Appendix). Here we need simply apply the σ and π TASO's from the earlier discussion of SF_6.

The inter-boron distances in these cages are small enough so that significant boron-boron interactions can occur. The σ mo's that can arise are shown below. They increase in energy in the order $a_{1g} < t_{1u} < e_g$, with a_{1g} fully bonding, t_{1u} weakly anti-bonding, and e_g strongly anti-bonding.

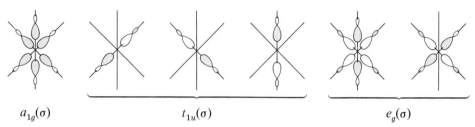

There are 12 boron ao's tangent to the B_6 cage (6 borons × 2/boron). Hence, there are 12 possible bonding and anti-bonding combinations. As in the SF_6 case, there are four sets of triply degenerate mo's: the t_{2g} bonding set and its anti-bonding partner t_{1g}^*, and the t_{1u} set and its anti-bonding counterpart t_{2u}^*. One member of each of these sets is illustrated below. The energy order is $(t_{2g} < t_{1u}) < (t_{2u}^* < t_{1g}^*)$.

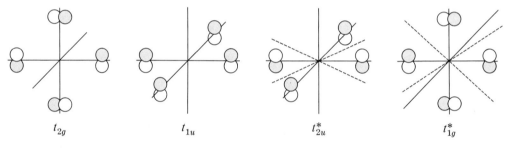

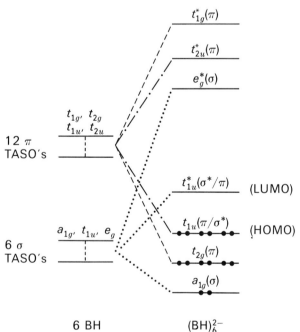

Figure 18-9. Framework molecular orbitals for $B_6H_6^{2-}$.

The cage mo diagram for $B_6H_6^{2-}$ appears in Figure 18-9. With seven valence electron pairs for cage bonding, $(BH)_6^{2-}$ has a configuration of $(a_{1g})^2(t_{2g})^6(t_{1u})^6$, and the anion is diamagnetic. [Notice that the $t_{1u}^*(\sigma)$ and $t_{1u}(\pi)$ mo's have identical symmetry; their overlap causes them to mix and leads to stabilization of the $t_{1u}(\pi)$ electron pairs. Edge bonding using a $t_{1u}(\pi/\sigma^*)$ mo is improved over that accomplished by a $t_{1u}(\pi)$ mo alone, but this improvement comes at the expense of diagonal bonding.]

Finally, it is important to notice that the total number of bonding mo's for $B_6H_6^{2-}$ is 7 or $n + 1$, and the number of electrons required to just fill those orbitals is $2n + 2$. This will always be the case for a *closo* compound and is the basis for the empirical rules developed in the previous section.

CLOSO-*HYDROBORATE IONS*[65,66]

The term "aromaticity" has a special meaning in chemistry. It is usually given to those compounds that, because of extensive electron delocalization, have a special kinetic and/or thermodynamic stability and a unique reaction chemistry. Both the *closo*-hydroborate anions and the carboranes fit into this category. The higher members of both classes of compounds have an extraordinary *kinetic* stability (in general the boron hydrides are thermodynamically unstable) to thermal and oxidative degradation, and they have an extensive chemistry based on the replacement of exopolyhedral B—H or C—H bonds.

Although there are boron-containing polyhedra based on every regular deltahedron from the tetrahedron through the icosahedron (see Figure 18-1), the known *closo*-hydroborate ions of general formula $(BH)_p^{2-}$ run only from $B_6H_6^{2-}$ through $B_{12}H_{12}^{2-}$. The fact that these ions are polyhedra does not mean that they are particularly large; the long dimension in the $B_{10}H_{10}^{2-}$ ion (2) is comparable to the cross-ring distance in benzene, and the volume of the $B_{12}H_{12}^{2-}$ ion (Figure 18-3) is comparable to that swept out by a benzene molecule spinning about an in-plane C_2 axis.

Although the structure of a particular ion has been determined ultimately by x-ray crystallography in most cases, ^{11}B nmr spectroscopy has been used extensively for obtaining symmetry information and for determining the probable shapes of reaction products. As an example, the ^{11}B nmr spectrum of $B_9H_9^{2-}$ is sketched in Figure 18-10. This spectrum allows a choice to be made between the two possible structures for the $B_9H_9^{2-}$ ion:[67] a square antiprism with a cap on one square face, or

---

[65] See ref. 5.
[66] F. R. Scholer and L. J. Todd, "Polyhedral Boranes and Heteroatom Boranes," in "Preparative Inorganic Chemistry," Vol. 7, W. L. Jolly, ed., Wiley-Interscience, New York (1971).
[67] F. Klanberg and E. L. Muetterties, *Inorg. Chem.*, **5**, 1955 (1966).

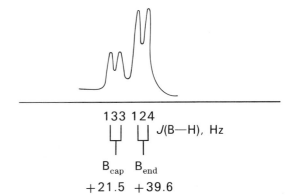

Figure 18-10. ^{11}B nmr spectrum of $B_9H_9^{2-}$. Chemical shifts relative to external $B(OMe)_3$.

(A)

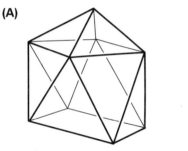
Capped square antiprism
C_{4v}

Symmetrically tricapped trigonal prism
D_{3h}

Figure 18–11. (A) Two possible geometries for the boron framework of $B_9H_9^{2-}$. (B) Closure of the square face opposite the capped face of the capped square antiprism leads to the symmetrically tricapped trigonal prism. As only small motions are required, the structures can be readily interconverted.

(B)

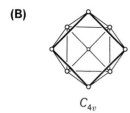 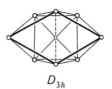

C_{4v} $\qquad\qquad\qquad D_{3h}$

a trigonal prism with a cap on each square face [Figure 18–11(A)]. Furthermore, as illustrated in Figure 18–11(B), these two polyhedra are quite similar and a very small motion converts one to the other.[68] Nonetheless, the ^{11}B nmr spectrum of the B_9 ion shows two types of boron atoms in a 2:1 ratio, clearly indicating that the tricapped trigonal prism is the structure adopted. This is in agreement with all other structural work on the $B_pH_p^{2-}$ ions, which indicates that all are deltahedra; that is, they all adopt structures with triangular faces only.

Syntheses of all of the polyhedral hydroborate anions are outlined in *Inorganic Syntheses* or *Preparative Inorganic Reactions*.[69–72] With the exception of $B_7H_7^{2-}$ and $B_8H_8^{2-}$, the *closo* ions are formed from the lower hydrides in one of two ways: by the so-called "B—H condensation" process or by pyrolysis.[73] The "B—H condensation" route involves the following general reaction.

$$2BH_4^- + B_xH_y \xrightarrow{\text{heat}} [(2+x)/p]B_pH_p^{2-} + [(y-x+6)/2]H_2$$

where B_xH_y represents some number of neutral boron hydride molecules such as B_2H_6 or $B_{10}H_{14}$. As an example, $B_6H_6^{2-}$ may be prepared directly by the reaction of $BH_4^- + B_2H_6$:

$$2BH_4^- + 2B_2H_6 \xrightarrow[162°/16\ hr]{\text{diglyme}} B_6H_6^{2-} + 7H_2$$

$$\begin{cases} BH_4^- + B_2H_6 \rightarrow B_3H_8^- + H_2 \\ 20CsB_3H_8 \xrightarrow[230°/30\ min]{} 2Cs_2B_9H_9 + 2Cs_2B_{10}H_{10} + Cs_2B_{12}H_{12} + 10CsBH_4 + 34H_2 \end{cases}$$

---

[68] For a general paper on intramolecular rearrangements in boron clusters, see: E. L. Hoel, C. G. Salentine, and M. F. Hawthorne, *Inorg. Chem.*, **14**, 950 (1975).
[69] M. F. Hawthorne and R. L. Pilling, *Inorg. Syn.*, **9**, 16 (1967).
[70] H. C. Miller and E. L. Muetterties, *Inorg. Syn.*, **10**, 81 (1967).
[71] F. Klanberg and E. L. Muetterties, *Inorg. Syn.*, **11**, 24 (1968).
[72] See ref. 66.
[73] H. C. Miller, N. E. Miller, and E. L. Muetterties, *Inorg. Chem.*, **3**, 1456 (1964).

whereas the B_9, B_{10}, and B_{12} ions come from the pyrolysis of a $B_3H_8^-$ salt, the latter having been isolated from a condensation reaction.[74,75]

The fact that the *closo*-hydroborate ions form a continuous series from $B_6H_6^{2-}$ to $B_{12}H_{12}^{2-}$ presents a unique situation wherein it is possible to compare trends in reactivity within a closely related series of molecules. As a result of extensive research,[76] one important generalization can be made: the most symmetrical of the ions, $B_{12}H_{12}^{2-}$ (I_h point group), is the most stable and least reactive of the series. For example, $Cs_2B_{12}H_{12}$ can be heated to about 800°C *in vacuo* in a quartz tube without change, whereas $Cs_2B_7H_7$ begins to decompose at about 400°C. The other ions form salts of intermediate stability, and this same order carries over to hydrolytic stability. The $B_{12}H_{12}^{2-}$ ion is stable to strong base and can be heated without change in 3N HCl at 95°C. Although the B_{10} ion is slowly degraded (to boric acid and H_2) under acid conditions, both this ion and $B_{12}H_{12}^{2-}$ are converted, by passage through an ion exchange column, to $[H_3O^+]_2[B_pH_p^{2-}]$, compounds that have acidities comparable to that of sulfuric acid. The other $B_pH_p^{2-}$ ions are all somewhat less stable than the B_{10} and B_{12} ions, partly owing to the fact that, because there are fewer boron atoms with the same 2− charge, there is greater charge density on each boron, and the possibility of electrophilic attack is increased.

All of the *closo* ions are comparable to aromatic hydrocarbons in that the delocalized bonding mo's are completely full and the LUMO's lie considerably higher in energy. Therefore, none of the ions may be reduced, although some such as $B_8H_8^{2-}$ and $B_{10}H_{10}^{2-}$ may be oxidized. In fact, the oxidative behavior of the latter is especially interesting because it leads to new ions having coupled B_{10} cages, as illustrated below.[77–79]

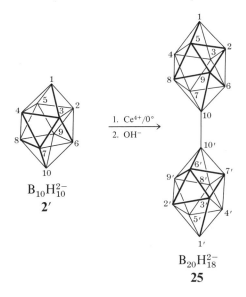

[74] J. L. Boone, *J. Amer. Chem. Soc.*, **86**, 5036 (1964).
[75] See ref. 71.
[76] See ref. 5.
[77] M. F. Hawthorne, R. L. Pilling, and P. F. Stokely, *J. Amer. Chem. Soc.*, **87**, 1893 (1965).
[78] A. P. Schmitt and R. L. Middaugh, *Inorg. Chem.*, **13**, 163 (1974).
[79] The following potentials (*vs.* a saturated calomel reference electrode) have been reported for hydroborate oxidation in acetonitrile: $B_{10}H_{10}^{2-}$, +0.4 v; $B_{11}H_{11}^{2-}$, +0.05 v; and $B_{12}H_{12}^{2-}$, +1.4 v. Such potentials mean that the B_{11} may be oxidized by many common oxidizing agents, B_{10} should require somewhat stronger oxidizers, and B_{12} can be oxidized only by electrochemical methods. R. L. Middaugh and R. J. Wiersma, *Inorg. Chem.*, **10**, 423 (1971).

Substitution of hydrogen on these "aromatic," polyhedral $B_pH_p^{2-}$ ions leads to many other new and interesting compounds. With regard to the chemistry of these ions, the following generalizations may be made:

a) Because the polyhedra are negatively charged, the observed substitutions are electrophilic.

b) As mentioned previously, the B_{12} ion is more stable than the B_{10} ion.

c) The apical or 1,10 borons (see numbering convention above) are generally substituted before the equatorial borons in $B_{10}H_{10}^{2-}$.

d) Reactions of $B_{10}H_{10}^{2-}$ and $B_{12}H_{12}^{2-}$ often take place with a change in the net charge. For example, $B_{10}H_{10}^{2-}$ reacts with excess nitrous acid to give 1,10-$B_{10}H_8(N_2)_2$.[80] This can best be thought of as replacement of a hydride ion, H^-, by the substituent.[81]

$$B_{10}H_{10}^{2-} \rightarrow [B_{10} + 2H^-] \xrightarrow[{[+2N_2]}]{HONO} N\equiv N-B\cdots B-N\equiv N$$

Diazonium salts and carbonyl compounds are important classes of organic compounds, ones that allow the synthetic chemist entry into other classes of organic compounds. These two classes are also of interest in the chemistry of the *closo*-hydroborate anions. As illustrated above, excess nitrous acid gives the colorless solid 1,10-$B_{10}H_8(N_2)_2$, a diazonium analog. This particular compound is important because the two N_2 substituents may be replaced by other nucleophilic groups.[82]

$$1,10\text{-}B_{10}H_8(N_2)_2 \xrightarrow{NH_3} 1,10\text{-}B_{10}H_8(NH_3)_2$$
$$\xrightarrow{pyridine} 1,10\text{-}B_{10}H_8(C_5H_5N)_2$$
$$\xrightarrow{CO} 1,10\text{-}B_{10}H_8(CO)_2$$

The last reaction provides the carbonyl derivative of $B_{10}H_{10}^{2-}$, another valuable intermediate in the preparation of still other derivatives. For example, the carbon of the CO group is apparently susceptible to nucleophilic attack just as is CO attached to a transition metal (Chapter 16, p. 863); for example,[83,84]

$$1,10\text{-}B_{10}H_8(CO)_2 \xrightarrow{NH_3} (NH_4^+)_2[1,10\text{-}B_{10}H_8(CONH_2)_2^{2-}]$$

The parent carbonyl, 1,10-$B_{10}H_8(CO)_2$ [and the related 1,12-$B_{12}H_{10}(CO)_2$], is a colorless, sublimable solid. Most interestingly, the vibrational frequency of the CO group stretching motion is quite high (2147 cm^{-1}), indicating that the CO retains much triple bond character. Therefore, the compound is best thought of as a neutral $B_{10}H_8$ cage with CO groups bound through a simple acid–base interaction [$B\leftarrow C\equiv O:$]; apparently, little or no back-donation from the cage into the CO group occurs, as it seems to do in transition metal–carbon monoxide complexes (where the CO stretching frequencies are often less than 2000 cm^{-1}).

---

[80] W. H. Knoth, *J. Amer. Chem. Soc.*, **88**, 935 (1966).

[81] The species in brackets in this reaction are not meant to convey mechanistic information, but only a formal way of thinking.

[82] See ref. 80.

[83] W. H. Knoth, J. C. Sauer, J. H. Balthis, H. C. Miller, and E. L. Muetterties, *J. Amer. Chem. Soc.*, **89**, 4842 (1967).

[84] W. H. Knoth, J. C. Sauer, H. C. Miller, and E. L. Muetterties, *J. Amer. Chem. Soc.*, **86**, 115 (1964).

THE CARBORANES[85-88]

Perhaps the first evidence for the class of compounds now called carboranes was obtained in 1953 when mixtures of diborane and acetylene were ignited with a hot wire. Since that time, many examples of carbon-boron-hydrogen polyhedral molecules have been isolated, some in large enough amounts that their chemistry may be fully studied. Indeed, there is sufficient information to recognize the study of such molecules as a sub-area in its own right, and it is important enough that a recent book summarized activity in the field.[85]

According to the organization scheme on page 1006, carboranes have the general formula $[(CH)_a(BH)_pH_q]^{c-}$, where $a + q + c = 2$ for a *closo* structure, 4 for a *nido* structure, and 6 for an *arachno* structure. As seen in Table 18-4 and Figure 18-12, all of the known *neutral closo*-carboranes, with one exception (CB_5H_7), have two carbon atoms in their framework and therefore have the general formula

---

[85] R. N. Grimes, "Carboranes," Academic Press, New York (1970).
[86] R. E. Williams, "Carboranes," in "Progress in Boron Chemistry," Vol. 3, R. J. Brotherton and H. Steinberg, eds., Pergamon Press, New York (1970).
[87] This and the next paper are the first publications in carborane chemistry: M. M. Fein, J. Bobinski, N. Mayes, N. N. Schwartz, and M. S. Cohen, *Inorg. Chem.*, **2**, 111 (1963).
[88] T. L. Heying, J. W. Ager, S. L. Clark, D. J. Mangold, H. L. Goldstein, M. Hillman, R. J. Polak, and J. W. Szymanski, *Inorg. Chem.*, **2**, 1089 (1963).

TABLE 18-4

THE *CLOSO*- AND *NIDO*-CARBORANES, $[(CH)_a(BH)_pH_q]^{c-}$

A. The *Closo* Series: $2n + a + q + c = 2n + 2$ (where $n = a + p$)

The $[(CH)_2(BH)_p]$ Series $a = 2, p = n - 2,$ $q = 0, c = 0$	The $[(CH)_1(BH)_p]^{1-}$ Series $a = 1, p = n - 1,$ $q = 0, c = 1$	Isoelectronic and Isostructural $(BH)_p^{2-}$ Ion
$B_3C_2H_5$	$[B_4CH_5^-]^a$	$[B_5H_5^{2-}]$
$B_4C_2H_6$	$B_5CH_6^-$	$B_6H_6^{2-}$
$B_5C_2H_7$	$[B_6CH_7^-]$	$B_7H_7^{2-}$
$B_6C_2H_8$	$[B_7CH_8^-]$	$B_8H_8^{2-}$
$B_7C_2H_9$	$[B_8CH_9^-]$	$B_9H_9^{2-}$
$B_8C_2H_{10}$	$B_9CH_{10}^-$	$B_{10}H_{10}^{2-}$
$B_9C_2H_{11}$	$B_{10}CH_{11}^-$	$B_{11}H_{11}^{2-}$
$B_{10}C_2H_{12}$	$B_{11}CH_{12}^-$	$B_{12}H_{12}^{2-}$

B. The *Nido* Series: $2n + a + q + c = 2n + 4$

Carborane	Isoelectronic and Isostructural Neutral *Nido*-Borane	Number of BHB Bridges in Carborane[b]
$B_3C_2H_7$	B_5H_9	2
B_5CH_9	B_6H_{10}	3
$B_4C_2H_8$		2
$B_3C_3H_7$		1
$B_2C_4H_6$		0
$B_9CH_{12}^-$	$B_{10}H_{14}$	2
$B_{10}CH_{13}^-$	$B_{11}H_{15}$	2
$B_9C_2H_{13}$		2

[a] Compounds not yet isolated are given in brackets.
[b] The number of BHB bridges (= s in the topological theory equations) is always 4 for a neutral *nido*-borane.

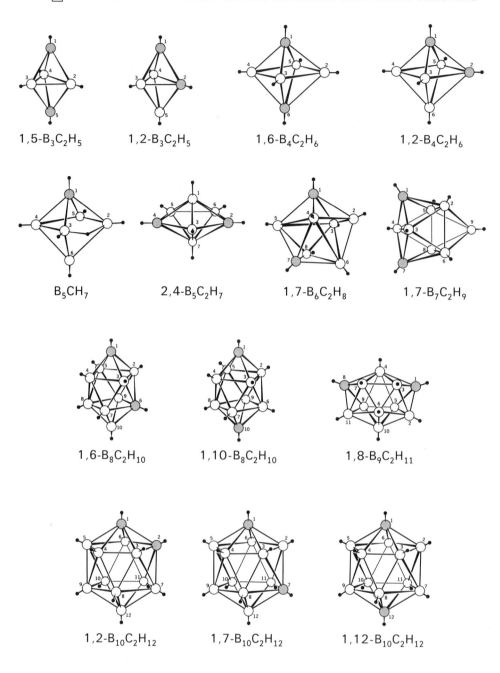

Figure 18-12. Structures of the known neutral *closo*-carboranes.

$[(CH)_a(BH)_p]$, where p varies from 3 through 10. In several cases, at least two isomers are known. As also seen in Table 18-4, the *closo* anions ($c = 1$) have only one carbon-hydrogen unit, as they must if $a + q + c$ is to equal 2.

The neutral *nido*-carboranes all have the same general formula $[(CH)_a(BH)_pH_q]$, where $a + q = 4$. Several are known with a ranging from 1 to 4 and q from 3 to 0; compounds **26** and **27** exemplify this series. Apparently, only one neutral *arachno*-carborane is known, $1,3\text{-}B_7C_2H_{13}$ (**23**).

2-B$_5$CH$_9$
26

2,3,4,5-B$_2$C$_4$H$_6$
27

The icosahedral 1,2-dicarba-*closo*-dodecaborane, B$_{10}$C$_2$H$_{12}$, is often called simply *ortho-carborane* (3). As this compound is by far the best known, we shall dwell almost exclusively on its preparation and properties.

The usual method of synthesis for σ-carborane on a large scale is a two-step process starting with decaborane.[89]

$$B_{10}H_{14} + 2Et_2S \xrightarrow{n\text{-propyl ether}} B_{10}H_{12}(Et_2S)_2 + H_2$$

$$B_{10}H_{12}(Et_2S)_2 + C_2H_2 \xrightarrow{n\text{-propyl ether}} C_2B_{10}H_{12} + H_2 + 2Et_2S$$

Careful control of conditions and exclusion of air and moisture are imperative. In the first stage of the reaction, diethyl sulfide is added to purified decaborane in *n*-propyl ether, and the resulting solution is heated to 65–67° for several hours. Acetylene is then passed through the solution now containing the diadduct for 35 hours at 85 ± 2°. Workup of the reaction is messy, and fires have been reported.[90]

Decaborane is a rather unpleasant smelling, volatile solid. However, upon putting a "C—C handle" on the decaborane "basket," the odor is changed completely—1,2-C$_2$B$_{10}$H$_{12}$ smells rather like camphor. The compound is very soluble in aromatic solvents but only sparingly so in aliphatic solvents. It melts at 295°C and, in an inert atmosphere, is not affected until the temperature reaches 400–500°C. Over a period of 24 to 48 hours at 400–500°, however, 1,2-dicarba-*closo*-dodecaborane (*i.e.*, *o*-carborane) is converted to the 1,7-isomer or *meta*-carborane in nearly quantitative yields (see Figure 18–12). Finally, at about 620° decomposition sets in and further rearrangement occurs to give the most thermodynamically stable isomer, the 1,12- or *para*-carborane.

The mechanism of the rearrangement process has been the subject of much discussion and experiment. A very attractive possible route from the *ortho* to the *meta* isomer is the so-called "diamond-square-diamond" or "dsd" mechanism illustrated in Figure 18–13.[91] That is, the icosahedron changes smoothly to a cubo-octahedron, which then converts to an icosahedron by reclosing the square faces along diagonals other than those along which they formed. Unfortunately, this cannot account for the *meta* to *para*

---

[89] M. F. Hawthorne, *et al.*, *Inorg. Syn.*, **10**, 91 (1967).

[90] An apparently safer procedure is given by C. R. Kutal, D. A. Owen, and L. J. Todd, *Inorg. Syn.*, **11**, 19 (1968).

[91] W. N. Lipscomb, *Science*, **153**, 373 (1966); see Study Question 6-32.

Figure 18–13. The "dsd" or "diamond - square - diamond" mechanism for the rearrangement of a 1,2-carborane to a 1,7-isomer.

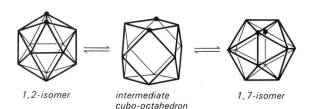

1,2-isomer   *intermediate cubo-octahedron*   *1,7-isomer*

1018 ☐ MOLECULAR POLYHEDRA: BORON HYDRIDES AND METAL CLUSTERS

conversion, so other mechanisms alone or in combination with the icosahedron–cubo-octahedron conversion have been suggested. Among these is the possibility that a B_2C face undergoes rotation. However, Muetterties and Knoth have commented that they "see no experiment that will realistically distinguish between these possibilities."[92]

The chemistry of *ortho*-carborane has been thoroughly explored, and some generalizations can be drawn and some comparisons can be made with the *closo*-hydroborate anions:

a) In view of the fact that *o*-, *m*-, and *p*-carboranes are neutral, whereas $B_{10}H_{10}^{2-}$ is negatively charged, the carboranes are much less susceptible to electrophilic attack. When it does occur, electrophilic reagents attack as far from the carbon atoms as possible, that is, at positions 8, 9, 10, or 12 (see Figure 18-12 for the numbering convention).

b) Nucleophilic attack on $B_{10}C_2H_{12}$ is much more possible than on $B_{10}H_{10}^{2-}$ and generally occurs as close to the carbon atoms as possible, that is, at borons 3 or 6.

c) The $B_{10}C_2$ cage is strongly electron-withdrawing with respect to organic substituents.

One consequence of the electron-withdrawing effect of the carborane cage is that 1-carboxy-1,2-dicarba-*closo*-carborane [HC($B_{10}H_{10}$)C—COOH] is a somewhat stronger acid ($pK_a = 2.48$) than benzoic acid ($pK_a = 4.2$). Further, it is interesting

---

[92] E. L. Muetterties and W. H. Knoth, "Polyhedral Boranes," Marcel Dekker, New York (1968), page 70.

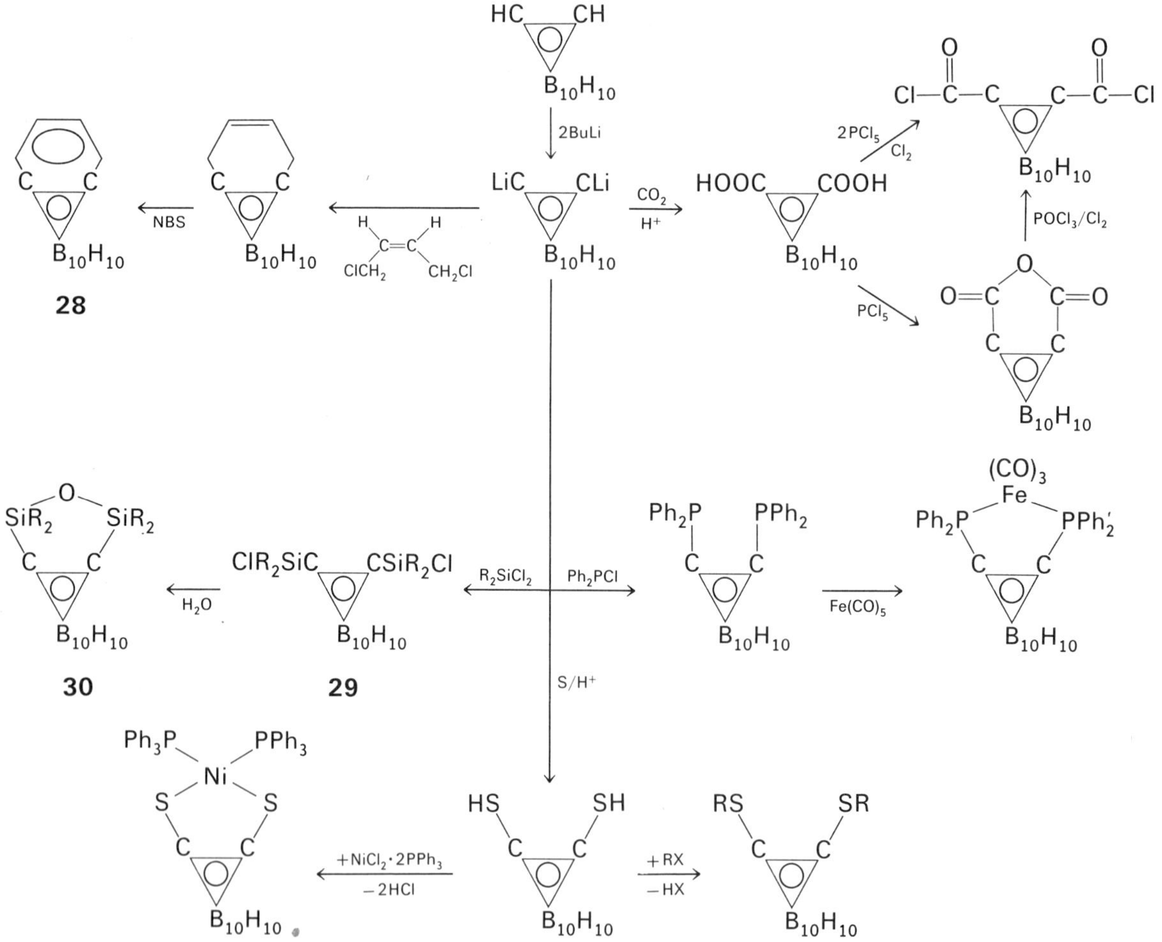

Figure 18-14. Some representative reactions of *ortho*-carborane, 1,2-$C_2B_{10}H_{12}$.

that the acidity of the mono-carboxy-carborane changes if the carboxy group is attached to one of the carbons in *meta*-carborane. The pK_a of 1-carboxy-1,3-dicarba-*closo*-carborane increases, relative the 1,2-isomer, to 3.20, the drop in acid strength presumably occurring because a cooperative withdrawing effect of two *ortho* carbons is diminished in the *meta* derivative.

Still another consequence of the electron-withdrawing effect of the carborane cage is that the carbon hydrogens are weakly acidic and may be removed by carbanionic reagents such as alkyllithiums or Grignard reagents. In fact, this represents one of the most useful ways of gaining access to carborane derivatives. Numerous reactions are outlined in Figure 18–14, and some deserve comment.

Compound **28** in Figure 18–14 was synthesized to test the possibility that electron delocalization could occur by a π mechanism from the carborane cage to the fused diene ring, and thereby render the fused ring aromatic.[93] Although the fused ring was found to be extremely stable (very slow reaction with Br_2/CCl_4 or $KMnO_4$, no appreciable reaction with concentrated H_2SO_4 at 100°), there is little concrete evidence for aromaticity. Instead, the lack of reactivity of the fused ring was ascribed to the large steric bulk of the cage and to its inductive effect.

The great thermal and oxidative stability of the carboranes has attracted the interest of polymer chemists for some time, and very recently the Olin Corporation has begun to produce, under the trade name Dexsil, siloxane polymers incorporating carborane into the backbone.[94-96] If a silicon-substituted compound such as **29** is hydrolyzed, the product is the cyclized material **30**; such internal cyclizations are common for 1,2-disubstituted *ortho*-carboranes. Therefore, to avoid this complication, *meta*-carborane is usually used as the building block if carborane is to be incorporated into a polymer backbone. Reaction of $Cl—Si(Me)_2—O—Si(Me)_2—Cl$ with the dilithium salt of *meta*-carborane gives the monomer **31**. Simple hydrolytic condensation at ice-bath temperatures gives polymers such as **32** with molecular weights in the range from 16,000 to 30,000. Although the elastomeric properties of such polymers leave something to be desired, they are oxidatively and thermally stable to 300–500°C.

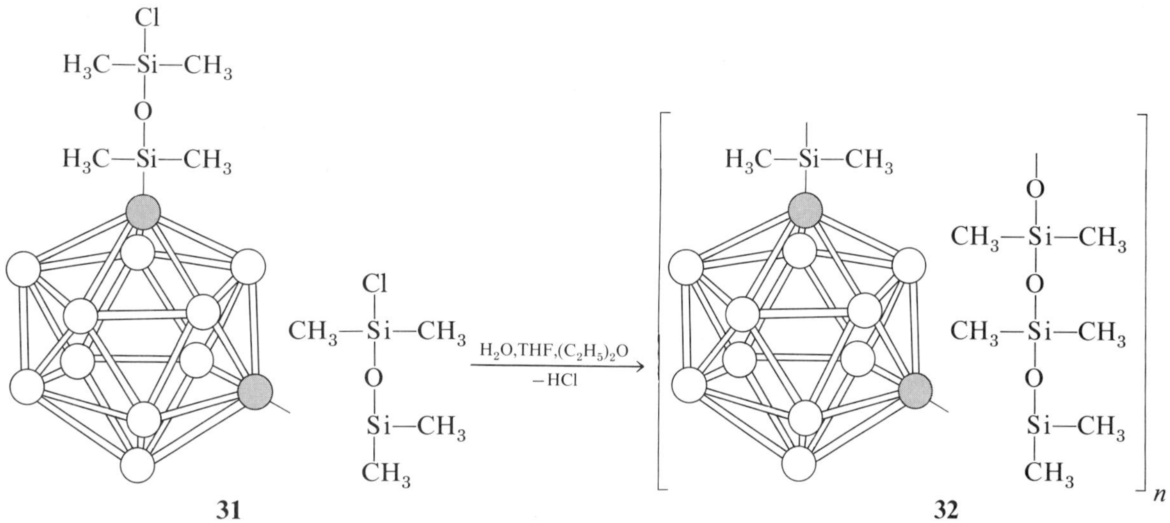

---

[93] D. S. Matteson and N. K. Hota, *J. Amer. Chem. Soc.*, **93**, 2893 (1971); D. S. Matteson and R. A. Davis, *Chem. Commun.*, 669 (1970).
[94] *Chemical and Engineering News*, March 22, 1971, page 46.
[95] R. E. Williams, *Pure and Appl. Chem.*, **29**, 569 (1972).
[96] Union Carbide is also involved in the investigation of carborane-siloxane polymers. E. F. Peters, *et al.*, Abstracts, 172nd National Meeting of the American Chemical Society, San Francisco, Sept. 1976, No. INORG-25.

In terms of the next topic—metallocarboranes—one of the most important reactions of the carboranes is cage degradation, a reaction already illustrated on page 1008. The essence of the reaction is the attack of a base at the most electropositive site, the boron atoms closest to the cage carbon atoms (*i.e.*, the 3 or 6 borons in *o*-carborane and the 2 or 3 borons in *m*-carborane). This reaction will be discussed further in the next section.

The smaller *closo*-carboranes have been studied to a much smaller extent, although more can be expected as reasonably efficient syntheses are now available. In general their syntheses fall into two groups: The B_6—B_9 carboranes can be synthesized by degradation of $B_{10}C_2H_{12}$,[97,98] while the smaller carboranes (B_3—B_5) arise from high energy borane-alkyne interactions.[99]

METALLOCARBORANES[100-103]

Earlier in this chapter we hinted at the possibility of metal-containing boron hydrides. For example, it was noted that B_5H_9 reacts with $Fe(CO)_5$ to give **17**. One point of view regarding the formation of this complex is that the B_5H_9 molecule loses the apical boron in the form of a BH moiety to give B_4H_8, which then binds to an $Fe(CO)_3$ group. Formulating the reaction in this way means that four of the six electrons originally contained within the boron framework remain in three of the four basal boron TASO's illustrated in Figure 18-7. Overlap of these basal TASO's with metal *d*, *s*, and *p* ao's of appropriate symmetry can lead to complexation.

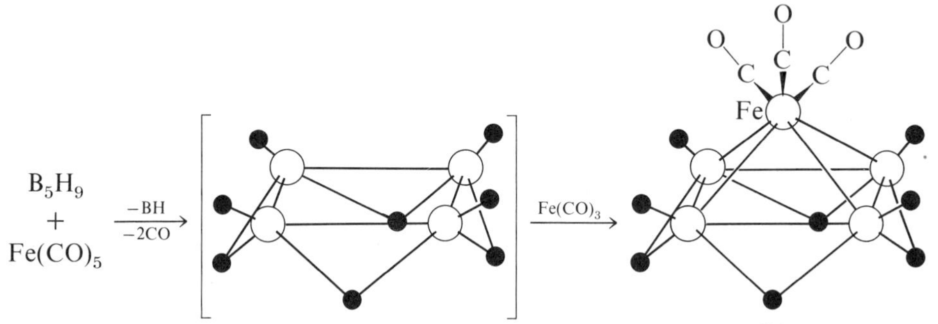

$$B_5H_9 + Fe(CO)_5 \xrightarrow[-2CO]{-BH} [\ldots] \xrightarrow{Fe(CO)_3} \mathbf{17'}$$

From the example above, it should be evident that a BH vertex, which is formally attached to four or five other cage atoms in a boron hydride or carborane, can be removed and replaced by some other element or group, the only constraint being that the new group must contribute sufficient electrons and orbitals to the cage. For example, mention was also made earlier of *closo*-B_9H_9S (p. 1009). This compound can be considered as having been formed by loss of a BH group (which contributed three orbitals and two electrons) from $B_{10}H_{10}^{2-}$ and replacement by the isoelectronic S^{2+} ion (which has a lone pair in place of the B—H bond and can contribute three hao's and two electrons to the cage). There are other examples of BH (or CH^+ in a

---

[97] See ref. 61.
[98] P. M. Garrett, J. C. Smart, G. S. Ditta, and M. F. Hawthorne, *Inorg. Chem.*, **8**, 1907 (1969).
[99] T. Onak, R. P. Drake, and G. B. Dunks, *Inorg. Chem.*, **3**, 1686 (1964).
[100] K. P. Callahan and M. F. Hawthorne, *Adv. Organometal. Chem.*, **14**, 145 (1976).
[101] M. F. Hawthorne, *Acc. Chem. Res.*, **1**, 281 (1968).
[102] L. J. Todd, *Adv. Organometal. Chem.*, **8**, 87 (1970).
[103] See refs. 61, 66, and 85.

carborane) replacement by main group elements (*e.g.*, $B_{10}CH_{11}P$ from $B_{10}C_2H_{12}$) to give interesting new compounds. However, of greater interest and much wider scope is replacement by a transition metal.

In the course of the discussion of a general organizational scheme for boron hydrides and carboranes, you learned that the carborane $B_{10}C_2H_{12}$ can be degraded by strong base to give, ultimately, the *nido* ion $7,8\text{-}B_9C_2H_{11}^{2-}$ (**20**; p. 1008).

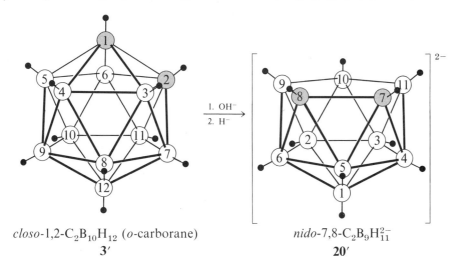

closo-1,2-$C_2B_{10}H_{12}$ (*o*-carborane)
3′

nido-7,8-$C_2B_9H_{11}^{2-}$
20′

That is, a BH^{2+} group has been removed from either the 3 or the 6 position to give an ion that has six electrons in the five TASO's formed by the boron and carbon hao's that point into the open face (**33**).

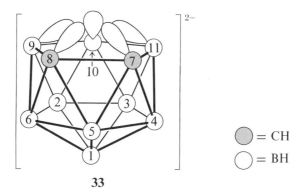

● = CH
○ = BH

33

Because of this, the *nido*-carborane anion is formally analogous to the $C_5H_5^-$ ion, a three π pair donor and a ligand of such great importance in organometallic chemistry. Indeed, the analogy between $C_5H_5^-$ and $7,8\text{-}B_9C_2H_{11}^{2-}$ led Hawthorne to attempt the reaction of the *nido*-carborane anion with Fe^{2+}, a reaction that led to the preparation of the pink, diamagnetic ferrocene analog **34**.[104,105]

---

[104] See refs. 100 and 101.
[105] M. F. Hawthorne, *et al.*, *J. Amer. Chem. Soc.*, **90**, 879 (1968); this is the first paper to describe completely the initial work on metallocarboranes.

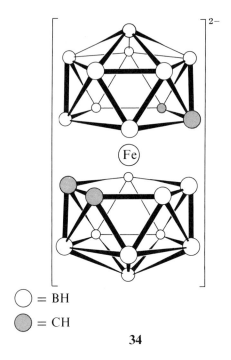

34

Just as ferrocene can be oxidized to the ferricenium ion [$(\eta^5\text{-}C_5H_5)_2Fe^+$], **34** is very easily oxidized to a maroon Fe(III) complex. Nmr studies of this paramagnetic Fe(III) compound have revealed that the unpaired electron is not extensively delocalized into the carborane cage.[106]

Complexes of **20** have now been synthesized with all of the transition metals of the first row between Cr and Cu, and complexes of Mo, W, Pd, and Au are also known. The more recently discovered ligand $1,2\text{-}B_{10}C_2H_{12}^{2-}$ (described in detail below) affords complexes with Ti and Zr as well.[107] In general, these complexes are quite stable to common acids and bases and to oxidizing agents. In fact, their stability generally exceeds that of the corresponding cyclopentadienyl analogs. Particularly impressive is the fact that sandwich-type complexes of Pd(II–IV), Cu(II–III), and Au(II–III) can be formed, whereas cyclopentadienyl analogs of these metals are not known.

A further important feature of carboranyl complexes in general is that such ligands stabilize high oxidation states of the later transition metals [Co(IV), Ni(IV), and Cu(III)] and lower oxidation states of the earlier metals [Ti(II), Zr(II), and V(II)]; such states are rare for organometallic complexes. The ability to stabilize high oxidation states, and to form sandwich complexes with the later metals, is perhaps best ascribed to the great electronegativity of the carboranyl ligands.

Structurally, the sandwich complexes of **20** fall basically into two classes: the "symmetrical sandwich" typified by **34** and **35** and formed by metal ions having d^3[Cr(III)], d^5[Fe(III)], d^6[Fe(II),Co(III),Ni(IV), Pd(IV)], and d^7[Co(II),Ni(III),Pd(III)] configurations; and the "slipped sandwich" type typified by **36** and formed by d^8[Cu(III),Ni(II),Pd(II),Au(III)] and d^9[Cu(II),Au(II)] ions.

---

[106] R. J. Wiersma and M. F. Hawthorne, *J. Amer. Chem. Soc.*, **96**, 761 (1974).
[107] C. G. Salentine and M. F. Hawthorne, *J. Amer. Chem. Soc.*, **97**, 426 (1975).

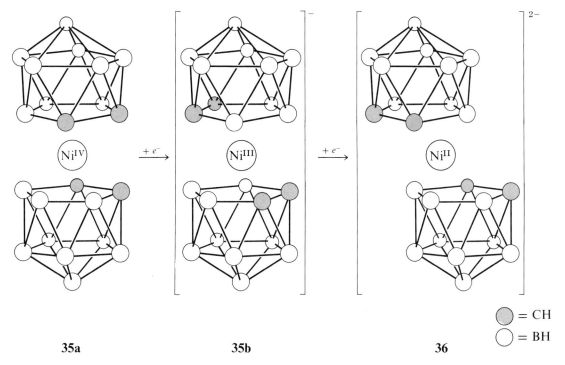

○ = CH
○ = BH

35a **35b** **36**

In addition to the synthesis of metallocarboranes by direct reaction of a carboranyl anion and metal cation, there are at least three other synthetic methods: cage degradation, polyhedral expansion, and direct insertion.[108] The first of these is illustrated by the $[Co(1,2-B_9C_2H_{11})_2]^-$ ion. Treatment with 30% NaOH and excess $CoCl_2$ for several hours at 100° gives **37**, wherein one ligand cage has had yet another boron atom removed and now functions as a "bidentate" ligand.[109]

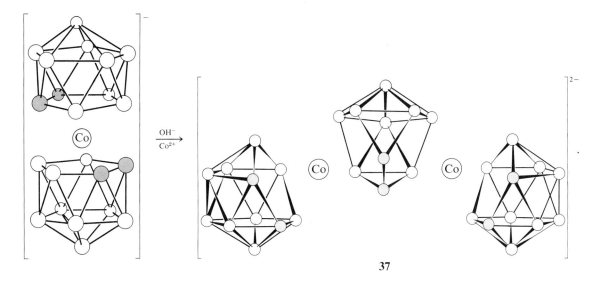

37

---

[108] See refs. 61 and 100.
[109] J. N. Francis and M. F. Hawthorne, *J. Amer. Chem. Soc.*, **90**, 1663 (1968).

Perhaps the most promising methods for the generation of a whole host of new complexes are polyhedral expansion and direct metal insertion. The general procedure used in polyhedral expansion involves reduction of the carborane with sodium in some active form.[110] This is followed by addition of metal ion or metal ion and cyclopentadienide, and the resulting product has one more vertex than the original polyhedron. In the reactions illustrated, the $B_{10}C_2$ icosahedron has been expanded to a *13-vertex* polyhedron (**38**),[110] and the 1,7-carborane has been expanded twice to include two metal atoms in a 10-vertex polyhedron (**39**).[111] Extensive studies of such compounds by Hawthorne have turned up the fact that their chemistry closely resembles that of the carboranes themselves with regard to polyhedral expansion, rearrangement, and contraction.

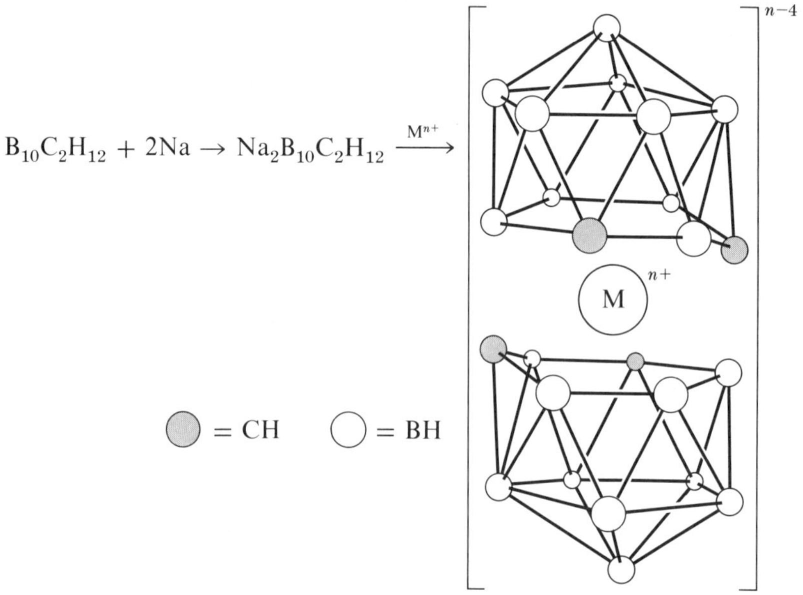

$$B_{10}C_2H_{12} + 2Na \rightarrow Na_2B_{10}C_2H_{12} \xrightarrow{M^{n+}}$$

● = CH   ○ = BH

38

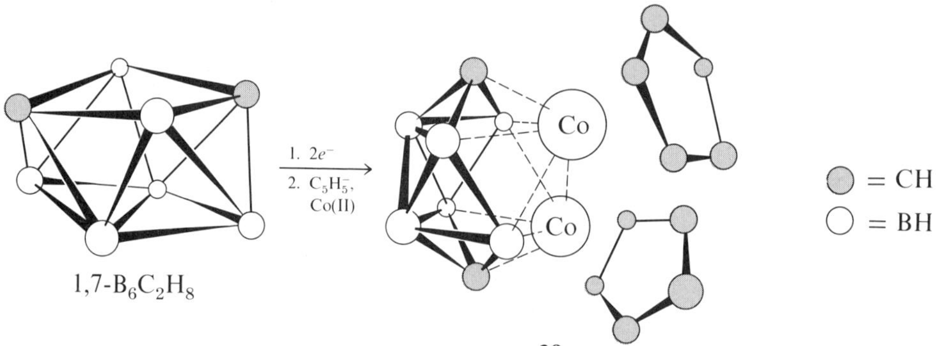

1,7-$B_6C_2H_8$ $\xrightarrow[\text{2. } C_5H_5^-, \text{Co(II)}]{\text{1. } 2e^-}$

● = CH
○ = BH

39

---

[110] D. F. Dustin, G. B. Dunks, and M. F. Hawthorne, *J. Amer. Chem. Soc.*, **95**, 1109 (1973).
[111] W. J. Evans, G. B. Dunks, and M. F. Hawthorne, *J. Amer. Chem. Soc.*, **95**, 4565 (1973).

Direct insertion of metals into carboranes promises to be of broad applicability, especially for the smaller carboranes. For example, the reaction of 1,5-$C_2B_3H_5$ with $Fe(CO)_5$ or (η^5-C_5H_5)$Co(CO)_2$ at temperatures somewhat over 200°C produced the mono- and di-metallocarboranes illustrated below (**40–42**).[112]

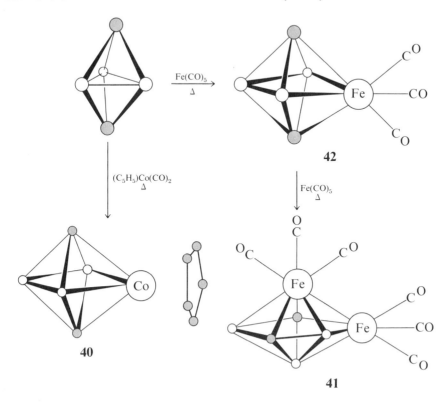

The scope of metallocarborane chemistry is clearly almost without limit. In the near future one can expect further advances in our understanding of their bonding modes, rearrangement mechanisms, and chemistry. Of particular interest will be those polyhedra containing several metals of the same or different kinds (*e.g.*, compound **41**). Such complexes form a link between the pure boron hydrides and carboranes and metal clusters, the next topic.

---

[112] V. R. Miller, L. G. Sneddon, D. C. Beer, and R. N. Grimes, *J. Amer. Chem. Soc.*, **96**, 3090 (1974).

STUDY QUESTIONS

11. Sketch the structures of the carboranes $B_4C_2H_8$ and $B_3C_3H_7$.

12. Sketch the ^{11}B nmr spectrum that might be expected for the $B_9H_9^{2-}$ ion if it were a capped square antiprism.

13. Polyhedral hydroborates can be made by the "B—H condensation" route. Show how the general equation for this process applies to the synthesis of $B_6H_6^{2-}$ on page 1012. $NaB_{11}H_{14}$ can be prepared by the "B—H condensation" method starting with either B_2H_6 or $B_{10}H_{14}$ and other appropriate materials. Write a balanced equation to depict this synthesis.

14. Table 18-4 lists two carborane anions belonging to the *nido* series. Using the structural classification formulas developed earlier in the chapter, show that their *nido* classification is correct.

15. It was said that the "dsd" mechanism in Figure 18-13 cannot account for the conversion of *meta*-carborane to *para*-carborane. Prove that this is indeed true.

16. Tell what reagents would be used in the following reactions:

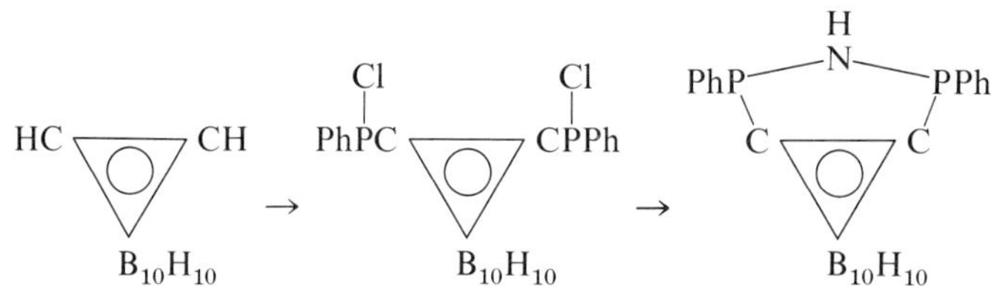

17. Write out a method for the synthesis of the following compound.

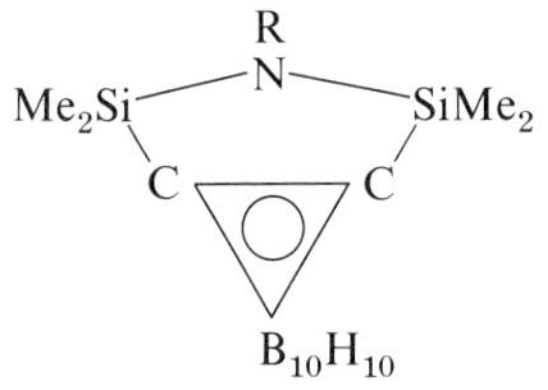

18. If the compound $(\eta^5\text{-}C_5H_5)CoB_4C_2H_6$ is to be prepared by the direct insertion method, with which carborane would one begin and what are the possible structures of the product?

19. Compounds **40** and **42** are synthesized by direct insertion into $B_3C_2H_5$. Discuss the bonding in these compounds. (Hint: refer back to the discussion of bonding in **17**.)

METAL-METAL BONDS AND METAL CLUSTERS[113-118]

Until about fifteen years ago very few compounds having metal-metal bonds were known or confirmed, save a few bimetallic compounds such as the mercury(I) halides. Since then, however, numerous compounds having a bond between at least

---

[113] D. L. Kepert and K. Vrieze in "Comprehensive Inorganic Chemistry," J. C. Bailer, Jr., H. J. Emeleus, R. Nyholm, and A. F. Trotman-Dickenson, eds., Pergamon Press, New York (1973), Chapter 47 in Volume 4.

[114] F. A. Cotton, *Chem. Soc. Revs.*, **4**, 27 (1975); *Acc. Chem. Res.*, **2**, 240 (1969).

[115] M. C. Baird, *Prog. Inorg. Chem.*, **9**, 1 (1968).

[116] R. B. King, *Prog. Inorg. Chem.*, **15**, 287 (1972).

[117] B. R. Penfold, "Stereochemistry of Metal Cluster Compounds," in "Perspectives in Structural Chemistry," J. D. Dunitz and J. A. Ibers, eds., John Wiley and Sons, New York (1968), Volume 2.

[118] P. Chini, G. Longoni, and V. G. Albano, *Adv. Organometal. Chem.*, **14**, 285 (1976).

METAL-METAL BONDS AND METAL CLUSTERS 1027

two metals have been reported, and representative of this now important class are octachlorodirhenate(2−) $[Re_2Cl_8^{2-}]$[113,119,120] with a quadruple Re—Re bond, the cluster R—CCo$_3$(CO)$_9$ (**4**),[121,122] and the organometallic chromium compound **43** with a triple Cr—Cr bond.[123]

Cr≡Cr structure with Cp and CO ligands

43

Just like the boron hydrides, metal clusters can be organized according to structural types. As seen in Table 18–5, there are bridged and unbridged compounds with simple metal-metal bonds—with orders ranging from 1 to 4—as well as triangular, tetrahedral, and octahedral clusters. A few other types, including an example of a cubic cluster, are mentioned in Table 18–5.

Understanding of the structures of compounds with metal-metal bonds has recently increased to the point where some generalizations can be made:

a) Unlike the boron hydrides, smaller metal clusters are usually not electron deficient, and the edges of metal cluster polyhedra can frequently be considered as normal two-electron bonds. There are, of course, exceptions, chiefly among the six-atom clusters to be discussed below.

b) The position of a metal in the Periodic Table, its oxidation state, and the attached ligands are important in determining the number and type of metal-metal bonds. Organometallic compounds with metal-metal bonds are common to almost all of the transition metals. On the other hand, metal halides with M—M bonds are based only on elements earlier in the second and third transition metal series.

The second of these observations can be explained by arguments based on valence orbital size. In general, as the oxidation state of a metal increases, or as Z^* increases, the valence orbitals contract. Such contraction should improve the ability of these orbitals to overlap. However, overlap between contracted orbitals on two metals can be effective only if the internuclear distance is shortened, and there are two factors opposing a decrease in M—M distance: (i) increasing atomic core repulsions, and (ii) increasing ligand-ligand repulsions. Core repulsions are disproportionately higher for $3d$ than for $4d$ or $5d$ metals, so metals in the first transition series do not form metal-metal bonds in higher oxidation states. However, in the low oxidation states characteristic of organometallic compounds, the first transition series metals are able to form metal-metal bonds.

BINUCLEAR COMPOUNDS

Some of the most thoroughly studied compounds of Class I-A1 (see Table 18–5) are $[Re_2Cl_8]^{2-}$ and $[Mo_2Cl_8]^{2-}$ and their derivatives (Figure 18–15). The preparation

---

[119] The three main papers concerning the $[Re_2Cl_8]^{2-}$ ion are:
a. *Preparation:* F. A. Cotton, N. F. Curtis, B. F. G. Johnson, and W. R. Robinson, *Inorg. Chem.,* **4**, 326 (1965).
b. *Structure:* F. A. Cotton and G. B. Harris, *Inorg. Chem.,* **4**, 330 (1965).
c. *Bonding:* F. A. Cotton, *Inorg. Chem.,* **4**, 334 (1965).
[120] For the latest work on $[Re_2Cl_8]^{2-}$ and similar compounds, the following is a leading reference: F. A. Cotton and E. Pedersen, *Inorg. Chem.,* **14**, 383 (1975).
[121] B. R. Penfold and B. H. Robinson, *Acc. Chem. Res.,* **6**, 73 (1973).
[122] D. Seyferth, *Adv. Organometal. Chem.,* **14**, 97 (1976).
[123] J. Potenza, P. Giordano, D. Mastropaolo, and A. Efraty, *Inorg. Chem.,* **13**, 2540 (1974).

TABLE 18-5
METAL CLUSTERS

Cluster Type	Compound	Structure/Properties	Reference
I. Binuclear			
A. Homoatomic			
1. Inorganic unbridged	$[Re_2Cl_8]^{2-}$	quadruple bond; Figure 18-16	a
	$Mo_2(NMe_2)_6$	triple Mo—Mo bond	b
2. Inorganic bridged	$Cr_2(O_2CCH_3)_4 \cdot 2H_2O$	quadruple Cr—Cr bond	c
	$[Mo_2(SO_4)_4]^{3-}$	quadruple Mo—Mo bond	d
3. π-Organometallic bridged	$(\eta^5\text{-}C_5H_5)_2V_2(CO)_5$		e
4. π-Organometallic unbridged	$(\eta^5\text{-}C_5Me_5)_2Cr_2(CO)_2$	triple Cr—Cr bond	f
	$(\eta^5\text{-}C_5H_5)_2Cr_2(CO)_6$	single Cr—Cr bond	g
5. σ-Organometallic unbridged	$Mo_2(CH_2SiMe_3)_6$	triple Mo—Mo bond	h
	$Mn_2(CO)_{10}$	single Mn—Mn, D_{2d} symmetry	i
6. σ-Organometallic bridged	$Fe_2(CO)_9$	single Fe—Fe bond, three CO bridges	j
	$(C_{10}H_{18})_2Fe_2(CO)_4$	Fe—Fe double bond with two $Me_3C\text{—}C{\equiv}C\text{—}CMe_3$ bridges	k
B. Heteroatomic	$(\eta^5\text{-}C_5H_5)Co(CO)_2 \cdot HgCl_2$	Co functions as a Lewis base	l
	$(\eta^5\text{-}C_5H_5)W(CO)_3SnMe_3$		m
	$Me_3Ge\text{—}Mn(CO)_5$		n
II. Trinuclear	$[Re_3Cl_{12}]^{3-}$	dark red, structure **48**	a
	$Fe_3(CO)_{12}$	green-black, structure **50**	a
	$H_2Os_3(CO)_{10}$	46 valence electrons; unsaturated with apparent Os—Os double bond; OsH_2Os bridging unit.	o
	$(\eta^5\text{-}C_5H_5)_3Ni_3(CO)_2$	green; triply bridging CO's on each side of metal triangle; paramagnetic	p
III. Tetranuclear			
A. Four metal atoms	$[(\eta^5\text{-}C_5H_5)Fe(CO)]_4$	dark green, structure **52**	a
	$H_2Ru_4(CO)_{13}$	red	q
	$H_2Re_4(CO)_{14}$	deep red; 58 valence electrons; unsaturated with two Re—Re double bonds	r
	$Co_4(CO)_{12}$	black — note color change with increasing metal weight; 9 terminal CO + 3 bridging CO for Co and Rh, all terminal for Ir; structures **53** and **54**	s, t
	$Rh_4(CO)_{12}$	red	
	$Ir_4(CO)_{12}$	yellow	
B. Three metal atoms + one heteroatom	$RC\text{—}Co_3(CO)_9$	deep purple; all CO terminal; structure **55**	u
	$SCo_3(CO)_9$	black	v
IV. Five-atom cluster	$[M_2Ni_3(CO)_{16}]^{2-}$	(M = Cr, Mo, W) one of the very few 5-metal atom clusters; trigonal bipyramid with 3 Ni in equatorial plane	w

TABLE 18-5 (Continued)

Cluster Type	Compound	Structure/Properties	Reference
V. Six-atom cluster	$CFe_5(CO)_{15}$	square pyramid of $Fe(CO)_3$ units with carbide ion in square base; structure **59**	x
	$Rh_6(CO)_{16}$	black; 6 $Rh(CO)_2$ units in octahedral arrangement with 4 facial CO's; structure **56**	y
	$Os_6(CO)_{18}$	bicapped trigonal prism of Os	z
	$[(Nb_6Cl_{12})Cl_6]^{3-}$		aa
VI. Seven-atom cluster	$CRh_6(CO)_{17}$	deep red; C atom in center of Rh_6 octahedron	bb
VII. Eight-atom cluster	$Ni_8(CO)_8(\mu_4\text{-PPh})_6$	a cube of Ni atoms with 2-electron Ni—Ni bonds along each edge; a PPh ligand symmetrically caps each face; structure **65**	cc
VIII. Nine-atom cluster	$[Pt_9(CO)_9(\mu_2\text{-CO})_9]^{2-}$	structure **58**	dd

[a] See text.
[b] M. Chisholm, F. A. Cotton, B. A. Frenz, and L. Shive, *Chem. Commun.*, 480 (1974).
[c] F. A. Cotton, B. G. DeBoer, M. D. LaPrade, J. R. Pipal, and D. A. Ucko, *J. Amer. Chem. Soc.*, **92**, 2926 (1970).
[d] F. A. Cotton, B. A. Frenz, and T. R. Webb, *J. Amer. Chem. Soc.*, **95**, 4431 (1973).
[e] F. A. Cotton, B. A. Frenz, and L. Kruczynski, *J. Amer. Chem. Soc.*, **95**, 951 (1973).
[f] J. Potenza, P. Giordano, D. Mastropaolo, and A. Efraty, *Inorg. Chem.*, **13**, 2540 (1974).
[g] R. D. Adams, D. E. Collins, and F. A. Cotton, *J. Amer. Chem. Soc.*, **96**, 749 (1974).
[h] F. Huq, W. Mowat, A. Shortland, A. C. Skapski, and G. Wilkinson, *Chem. Commun.*, 1079 (1971).
[i] L. F. Dahl and R. E. Rundle, *Acta Cryst.*, **16**, 419 (1963).
[j] H. M. Powell and R. V. G. Evans, *J. Chem. Soc.*, 286 (1939).
[k] K. Nicholas, L. S. Bray, R. E. Davis, and R. Pettit, *Chem. Commun.*, 608 (1971).
[l] I. M. Nowell and D. R. Russell, *J. Chem. Soc., A*, 817 (1967).
[m] T. A. George and C. D. Turnipseed, *Inorg. Chem.*, **12**, 394 (1973).
[n] A. Terzis, T. C. Strekas and T. G. Spiro, *Inorg. Chem.*, **13**, 1346 (1974).
[o] H. Kaesz, *Chemistry in Britain*, **9**, 344 (1972).
[p] M. R. Churchill and R. Bau, *Inorg. Chem.*, **7**, 2606 (1968).
[q] D. B. W. Yawney and R. J. Doedens, *Inorg. Chem.*, **11**, 838 (1972).
[r] R. Saillant, G. Barcelo, and H. Kaesz, *J. Amer. Chem. Soc.*, **92**, 5740 (1970).
[s] C. H. Wei, *Inorg. Chem.*, **8**, 2384 (1969).
[t] S. H. H. Chaston and F. G. A. Stone, *J. Chem. Soc., A*, 500 (1969).
[u] B. R. Penfold and B. H. Robinson, *Acc. Chem. Res.*, **6**, 73 (1973).
[v] C. H. Wei and L. F. Dahl, *Inorg. Chem.*, **6**, 1229 (1967); C. E. Strouse and L. F. Dahl, *J. Amer. Chem. Soc.*, **93**, 6032 (1971).
[w] J. K. Ruff, R. P. White, Jr., and L. F. Dahl, *J. Amer. Chem. Soc.*, **93**, 2159 (1971).
[x] E. H. Braye, W. Hübel, L. F. Dahl, and D. L. Wampler, *J. Amer. Chem. Soc.*, **84**, 4633 (1962).
[y] E. R. Corey, L. F. Dahl, and W. Beck, *J. Amer. Chem. Soc.*, **85**, 1202 (1963).
[z] R. Mason, K. M. Thomas, and D. M. P. Mingos, *J. Amer. Chem. Soc.*, **95**, 3802 (1973).
[aa] F. W. Koknat and R. E. McCarley, *Inorg. Chem.*, **13**, 295 (1974).
[bb] A. Sirigu, M. Bianchi, and E. Benedetti, *Chem. Commun.*, 596 (1969).
[cc] L. D. Lower and L. F. Dahl, *J. Amer. Chem. Soc.*, **98**, 5046 (1976).
[dd] P. Chini, G. Longoni, and V. G. Albano, *Adv. Organometal. Chem.*, **14**, 285 (1976).

of the two octahalodimetallates is illustrative of methods used to produce metal-metal bonds in general.[124,125] That is, a higher valent compound is reduced to a lower valent species wherein metal-metal bond formation is more favorable.

The chief feature of interest with regard to these ions is their molecular geometry and what this implies about the bonding between the metals (Figure 18-16).[125,126] Two aspects of their structures are especially important: (i) in both cases the Cl atoms

---

[124] See ref. 119a.
[125] J. V. Brencic and F. A. Cotton, *Inorg. Chem.*, **9**, 346, 351 (1970).
[126] See ref. 119b.

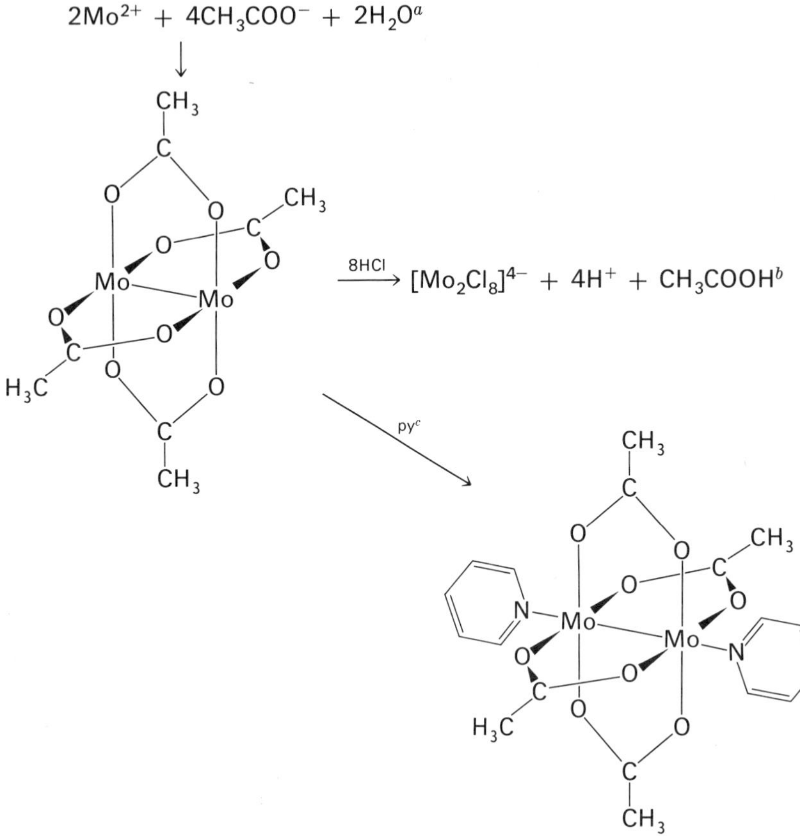

1030 ☐ MOLECULAR POLYHEDRA: BORON HYDRIDES AND METAL CLUSTERS

$2Mo^{2+} + 4CH_3COO^- + 2H_2O^a$

[Mo₂(OAc)₄ structure]

$\xrightarrow{8HCl}$ $[Mo_2Cl_8]^{4-} + 4H^+ + CH_3COOH^b$

$\xrightarrow{py^c}$

[Mo₂(OAc)₄(py)₂ structure]

Figure 18–15. The preparation and chemical properties of $[Re_2Cl_8]^{2-}$, $[Mo_2Cl_8]^{2-}$, and related compounds. References: [a] T. A. Stephenson, E. Bannister, and G. Wilkinson, *J. Chem. Soc.*, 2538 (1964). [b] J. V. Brencic and F. A. Cotton, *Inorg. Chem.*, **9**, 346 (1970). [c] F. A. Cotton and J. G. Norman, Jr., *J. Amer. Chem. Soc.*, **94**, 5697 (1972). [d] F. A. Cotton, N. F. Curtis, B. F. G. Johnson, and W. R. Robinson, *Inorg. Chem.*, **4**, 326 (1965). [e] F. A. Cotton, C. Oldham, and R. A. Walton, *Inorg. Chem.*, **6**, 216 (1967). [f] F. A. Cotton, W. R. Robinson, R. A. Walton, and R. Whyman, *Inorg. Chem.*, **6**, 929 (1967). [g] F. A. Cotton and E. Pedersen, *Inorg. Chem.*, **14**, 383 (1975).

are eclipsed, and (ii) both have a considerably shorter M—M distance [Re—Re = 2.24 Å and Mo—Mo = 2.14 Å] than in the metals themselves [Re = 2.714 Å and Mo = 2.725 Å]. Both of these features were recognized very early as being due to quadruple M—M bonds! To account for this, the following formalism may be adopted: The $d_{x^2-y^2}$ orbitals on the two metals are utilized for bonding to Cl⁻ ions. This leaves four orbitals—d_{z^2}, d_{xz}, d_{yz}, and d_{xy}—and the four electrons on each Re^{3+} (d^4). The σ bond between the metals derives from overlap of the d_{z^2} ao's, and the d_{xz} and d_{yz} orbitals form two π bonds. Finally, overlap of the d_{xy} orbitals on the two metals forms a δ bond (Figure 18–16). The new result is a quadruple bond with a bond strength estimated at 300 to 400 kcal/mole! Although the δ component is thought to be the weakest portion of the bond, it is of course just this component that dictates the eclipsed configuration.[127,128]

The bonding scheme outlined above means that there will be a d_{z^2} LUMO (**44**)

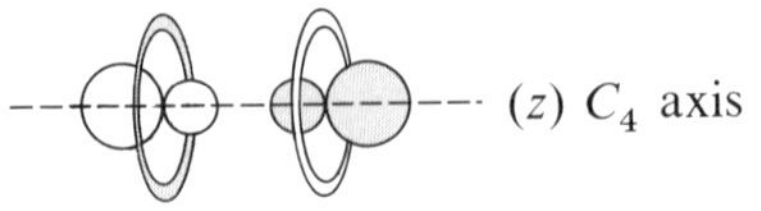

[orbital diagram] --- (z) C_4 axis

44

---

[127] The bonding in the rhenium salt has been described in two papers:
a. See ref. 119c.
b. F. A. Cotton and G. B. Harris, *Inorg. Chem.*, **6**, 924 (1967).
[128] J. G. Norman, Jr., and H. J. Kolari, *J. Amer. Chem. Soc.*, **97**, 33 (1975). This is a complete mo calculation that confirms the results of the earlier, more qualitative calculations.

[Figure 18-15 reaction scheme showing synthesis and reactions of $[Re_2Cl_8]^{2-}$]

Figure 18-15. (Continued).

available for adduct formation. Therefore, it is not surprising that a molecule such as $Mo_2(O_2CR)_4(py)_2$ can be prepared (see Figure 18-15).[129] The further significance of this diadduct is that the Mo—Mo bond is lengthened by only 0.039 Å [from 2.080 Å in $Mo_2(O_2CCF_3)_4$ to 2.129 Å in the diadduct] and that the Mo—N bond (2.548 Å) is much longer than the Mo—O bond (2.116 Å).[130] These facts suggest that the metals

---

[129] Recall the Lewis acid behavior of I_2 (Chapter 5, p. 210), where a base can interact with an analogous p_z LUMO.

[130] F. A. Cotton and J. G. Norman, *J. Amer. Chem. Soc.*, **94**, 5697 (1972).

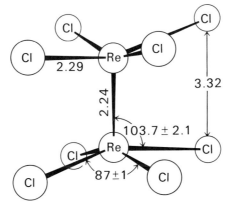 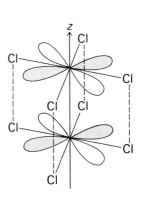

Figure 18-16. (A) The structure of the $[Re_2Cl_8]^{2-}$ ion, and (B) a sketch showing the formation of a δ bond in $[Re_2Cl_8]^{2-}$ by overlap of the metal d_{xy} orbitals. [Reprinted with permission from F. A. Cotton and G. B. Harris, *Inorg. Chem.*, **6**, 924 (1967). Copyright by the American Chemical Society.]

(A)   (B)

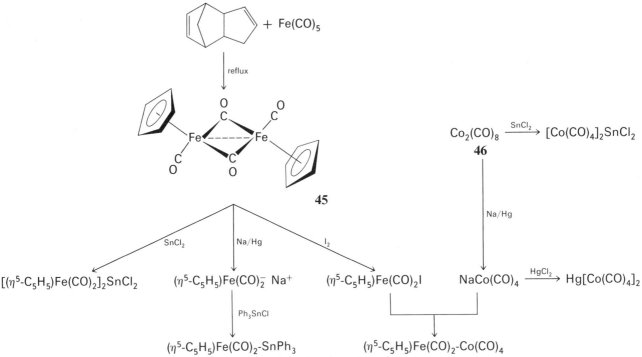

Figure 18-17. Some reactions leading to the formation of metal-metal bonds.

prefer to maintain their strong interaction rather than trade off any bond strength for a more "normal" Mo—N bond. This, of course, follows directly from the fact that it is a $d\sigma^*$ mo that is occupied by the pyridine lone pairs.

In general, there are probably more organometallic compounds having metal-metal bonds than there are of the type just discussed. More than likely this comes as a result of the fact that organometallic compounds are formed only by metals in low oxidation states, a factor that leads to greater net valence orbital availability as discussed earlier (p. 1027). Furthermore, at the moment there appear to be more straightforward synthetic methods available for organometallic compounds than for compounds such as the M—M bonded metal halides. For example, the organometallic compounds **45** and **46** in Figure 18-17 may be cleaved with sodium amalgam to give their respective mono-metal anions, and these anions may then react with a metal halide to give a host of unique compounds.

In addition to organometallic compounds with simple σ bonds between the same or different metals, there are those with multiple bonds. For example, reaction of 5-acetyl-1,2,3,4,5-pentamethylcyclopentadiene with $Cr(CO)_6$ gives **43**,[131] a green, air-stable compound with a Cr—Cr triple bond.[132]

As might reasonably be expected, the Cr—Cr distance of 2.280 Å in **43** is significantly shorter than the Cr—Cr bond distance in **47** (3.281 Å), the latter clearly having only a single Cr—Cr bond.[133]

---

[131] R. B. King and A. Efraty, *J. Amer. Chem. Soc.*, **94**, 3773 (1972).
[132] See ref. 123.
[133] R. D. Adams, D. E. Collins, and F. A. Cotton, *J. Amer. Chem. Soc.*, **96**, 749 (1974).

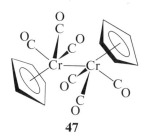

47

However, the Cr—Cr distance in **43** is also shorter than the quadruple metal-metal bond in $Cr_2(O_2CCH_3)_4 \cdot 2H_2O$ (2.362 Å) (see Figure 18-15 for the Mo analog). The apparent conclusion that can be drawn from these comparisons is that metal-metal bond lengths cannot always be relied upon to reflect accurately the bond order; as discussed earlier, bond lengths are a compromise between attractive M—M forces and ligand-ligand repulsions.

THREE-ATOM CLUSTERS

Three metal atoms may form a chain or a triangular arrangement. However, in contrast to the *sp* block elements, the triangle appears to be the more favorable for transition metals; even in clusters with more than three metal atoms, the polyhedral faces are triangular just as in the boron hydrides. Numerous clusters containing three metals in a triangular arrangement are known, and among the most systematically studied are $[Re_3Cl_{12}]^{3-}$, its derivatives, and $Fe_3(CO)_{12}$ and its Ru and Os analogs.

The dark red salt of empirical formula $CsReCl_4$ is prepared by the following sequence of reactions:[134,135]

$$2\,Re + 5\,Cl_2 \rightarrow 2\,ReCl_5$$

$$ReCl_5 \xrightarrow{\Delta} ReCl_3 + Cl_2$$

$$ReCl_3 + CsCl \xrightarrow{HCl} CsReCl_4$$

X-ray crystallography showed, however, that it is not the simple tetrahedral ion $ReCl_4^-$, but rather the trimeric ion $[Re_3Cl_{12}]^{3-}$ **(48)**.[136] The chief structural feature of interest is the Re—Re bond distance of 2.47 Å in the triangulated compound. This is somewhat longer than the quadruple bond in $[Re_2Cl_8]^{2-}$ (2.24 Å), but it is considerably shorter than the simple single bond in $(OC)_5Re—Re(CO)_5$. The Re—Re bond length in **48** may be explained by postulating a double bond between the rhenium atoms.

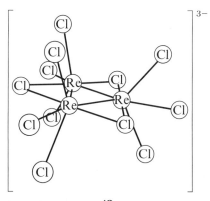

48

---

[134] L. C. Hurd and E. Brimm, *Inorg. Syn.*, **1**, 180 (1939).
[135] W. Geilman and F. W. Wrigge, *Z. Anorg. Allgem. Chem.*, **223**, 144 (1935).
[136] J. A. Bertrand, F. A. Cotton, and W. A. Dollase, *Inorg. Chem.*, **2**, 1166 (1963).

Photolysis of $Fe(CO)_5$ in glacial acetic acid gives excellent yields of golden-yellow $Fe_2(CO)_9$, a compound having an Fe—Fe bond in addition to three bridging CO groups **(49)**;[137,138]

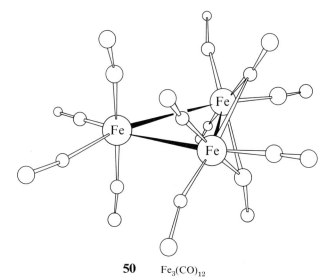

$$2Fe(CO)_5 \rightarrow CO + \mathbf{49}$$

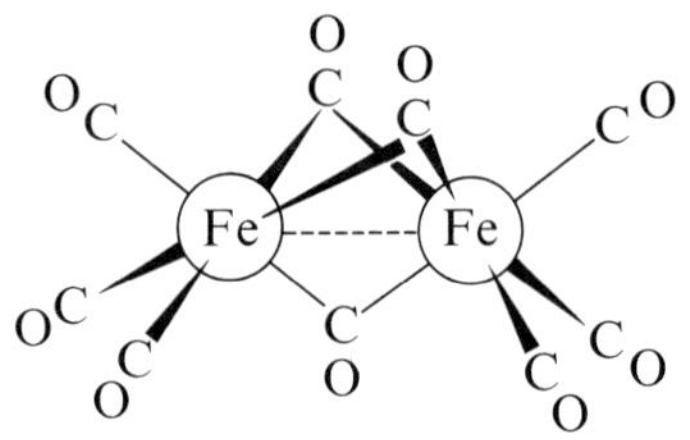

and mild pyrolysis of this di-iron compound, or use of the following sequence of reactions, gives the greenish-black compound $Fe_3(CO)_{12}$.[139]

$$3Fe(CO)_5 + NR_3 + 2H_2O \rightarrow [R_3NH][HFe_3(CO)_{11}] + 2CO_2 + 2CO + H_2$$
$$12[R_3NH][HFe_3(CO)_{11}] + 18HCl \rightarrow 11Fe_3(CO)_{12} + 15H_2 + 3FeCl_2 + 12R_3NHCl$$

Numerous speculations on the structure of this molecule fill the literature. However, the correct structure was finally obtained in 1966, and, not surprisingly, it displays the triangular arrangement that is now known to be most favored by three metal atoms **(50)**.[140]

50 $Fe_3(CO)_{12}$

Reductive carbonylation of $RuCl_3$ gives $Ru_3(CO)_{12}$,[141]

$$6RuCl_3 + 9Zn + 24CO \xrightarrow{methanol} 2Ru_3(CO)_{12} + 9ZnCl_2$$

the structure of which is similar to that of its iron analog *except* that there are no bridging CO groups **(51)**.

---

[137] W. L. Jolly, "The Synthesis and Characterization of Inorganic Compounds," Prentice-Hall, Englewood Cliffs, N.J. (1970), p. 472.

[138] R. B. King, "Organometallic Syntheses, Volume 1, Transition Metal Compounds," Academic Press, New York (1965), pp. 93–98.

[139] W. McFarlane and G. Wilkinson, *Inorg. Syn.*, **8**, 181 (1966).

[140] F. A. Cotton, *Prog. Inorg. Chem.*, **21**, 1 (1976); see also C. H. Wei and L. F. Dahl, *J. Amer. Chem. Soc.*, **88**, 1821 (1966).

[141] M. I. Bruce and F. G. A. Stone, *J. Chem. Soc. (A)*, 1238 (1967).

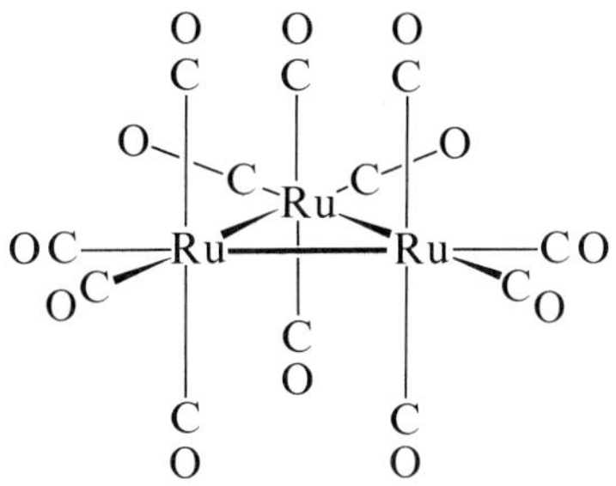

[Structure **51**: Ru₃(CO)₁₂ with bridging carbonyls depicted]

51

This change is illustrative of a central feature in metal carbonyl cluster chemistry—in general, bridging becomes less prevalent down a group. Further, the colors of the $M_3(CO)_{12}$ compounds—Fe, greenish-black; Ru, orange; and Os, yellow—illustrate the trend to less deeply colored compounds with heavier metals. Finally, owing to weaker core and ligand-ligand repulsions, the Ru and Os clusters are more strongly bonded than the iron cluster and, therefore, less easily broken down in chemical reactions. For example,[142]

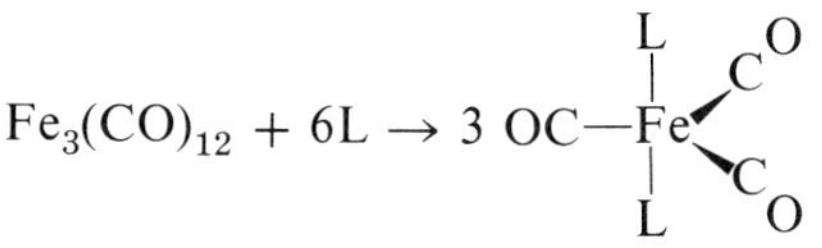

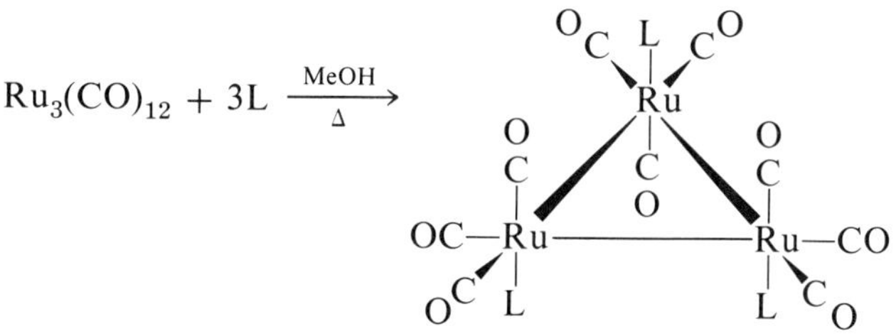

$$Fe_3(CO)_{12} + 6L \rightarrow 3\ OC-Fe(L)(L)(CO)(CO)$$

$$Ru_3(CO)_{12} + 3L \xrightarrow[\Delta]{MeOH} [Ru_3(CO)_9L_3]$$

L = Ph₃P, Et₃P

From the discussion above, and the literature, another generalization can be made: for low valent metals in particular, three-metal-atom clusters are stable if they have a total of 48 valence electrons ($= 3 \times 16$) supplied by the three metals and their ligands. That is, if we count up the total number of electrons supplied by the ligands, divide by the number of metal atoms, and then add to the quotient the number of electrons supplied by the metal, the total should be 16 for a metal in a triangular cluster. The favorable "effective atomic number" of 18 may be gained by requiring each metal to form two metal-metal bonds. This analysis is outlined in Table 18-6 along with those for tetrahedral and octahedral clusters. In general, the electron counting rule works well, except for six-atom clusters; the latter will require molecular orbital methods, as seen below.

FOUR-ATOM, TETRAHEDRAL CLUSTERS

Clusters having four, five, or six vertices represent some of the most interesting cluster species, and research in this area is quite active, partly because it is thought

---

[142] F. Piacenti, M. Bianchi, E. Benedetti, and G. Braca, *Inorg. Chem.*, **7**, 1815 (1968).

TABLE 18-6
ELECTRON COUNTING IN METAL CLUSTERS

Compound	Structure	Total Number of Valence Electrons (CO + M)	Electrons per Metal Atom	Number of M—M Bonds per M Atom Required to give EAN = 18
$Cr(CO)_6$	octahedron	12 + 6 = 18	18	0
$Mn_2(CO)_{10}$	two octahedra connected by Mn—Mn bond	20 + 14 = 34	17	1
$Fe_3(CO)_{12}$	structure **50**	24 + 24 = 48	16	2
$Co_4(CO)_{12}$	tetrahedron (structure **53**)	24 + 36 = 60	15	3
$[Co_6(CO)_{14}]^{4-}$	octahedron (structure **60**)	28 + 54 + 4 = 86	14.33	(see text)

that such clusters may resemble metals themselves and thereby offer an opportunity to study, after a fashion, reactions at metal surfaces.[143]

If $Fe(CO)_5$ is heated with dicyclopentadiene for several hours, an excellent yield of the purple metal-metal bonded compound $(\eta^5-C_5H_5)_2Fe_2(CO)_4$ is obtained (Figure 18-17). However, if this compound is refluxed for an additional two weeks(!), the green-black tetrahedral cluster $[(\eta^5-C_5H_5)Fe(CO)]_4$ **(52)** is obtained,[144] and this compound has several interesting features that can be used to illustrate general aspects of metal clusters.

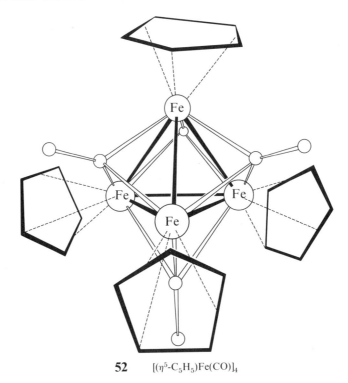

52 $[(\eta^5-C_5H_5)Fe(CO)]_4$

---

[143] M. Primet, J. M. Basset, E. Garbowski, and M. V. Mathieu, *J. Amer. Chem. Soc.*, **97**, 3655 (1975); H. D. Kaesz, *Chemistry in Britain*, **9**, 344 (1972).
[144] R. B. King, *Inorg. Chem.*, **5**, 2227 (1966).

Figure 18-18. Cyclic voltammogram of $[(\eta^5\text{-}C_5H_5)Fe(CO)]_4$ in acetonitrile; 0.1 M tetrabutylammonium hexafluorophosphate used as supporting electrolyte. Potentials measured at a Pt electrode *vs.* a saturated sodium chloride calomel electrode. [Reprinted with permission from J. A. Ferguson and T. J. Meyer, *J. Amer. Chem. Soc.*, **94**, 3409 (1972). Copyright by the American Chemical Society.]

The electronic properties of compounds can often be probed by examining their electrochemical properties. One electrochemical technique that is particularly useful in this regard is cyclic voltammetry.[145] The cyclic voltammogram of **52** is illustrated in Figure 18-18, and it clearly shows that four molecular oxidation states (2+, 1+, 0, and 1−) are reversibly accessible for the cluster.[146] The accessibility of these states, and their apparent stability, suggests that the bonding in this cluster is quite delocalized, in spite of the fact that the 18 electron rule can be satisfied by assuming localized Fe—Fe single bonds along each tetrahedral edge. Delocalization of bonding electrons is further suggested by the fact that *all* of the Fe—Fe distances shorten slightly on going to the 1+ cluster.[147,148] This shortening suggests that the electron is removed

---

[145] In cyclic voltammetry, the potential of an electrode in an unstirred solution is scanned from an initial value to a second value and back to the first to constitute a cycle. If an electroactive species is present in the solution, current will flow as the potential for reduction (or oxidation) is approached and will increase until the concentration of electroactive species in the immediate vicinity of the electrode is depleted. At this point the current drops to a constant value that is maintained by diffusion of the electroactive species from the bulk solution. A plot of current *vs.* potential shows a peak for this process. If the electron transfer process is reversible, the product from the initial electron transfer will also be electroactive and will undergo oxidation (or reduction) on the return sweep. The cyclic voltammogram will resemble one of the coupled peaks of Figure 18-18. For a thermodynamically reversible process, the current flowing on oxidation is equal to that flowing on reduction, and the average of the two peak potentials is the electrode potential for the process. If the product of the electron transfer is not stable, it may decompose to an electroinactive species or to another electroactive species; the shape of the cyclic voltammogram will reflect this.

Although cyclic voltammetry is frequently employed to determine reversibility of an electron transfer process, the technique has also been used in studies of the rate and mechanism of electron transfer and subsequent reactions of oxidized and reduced species.

For further information on cyclic voltammetry and other electrochemical techniques of value to the inorganic chemist, see:

    a. D. T. Sawyer and J. L. Roberts, Jr., "Experimental Electrochemistry for Chemists," Wiley-Interscience, New York (1974).

    b. J. B. Headridge, "Electrochemical Techniques for Inorganic Chemists," Academic Press, New York (1969).

[146] J. A. Ferguson and T. J. Meyer, *J. Amer. Chem. Soc.*, **94**, 3409 (1972); T. J. Meyer, *Prog. Inorg. Chem.*, **19**, 1 (1975).

[147] M. A. Neuman, Trinh-Toan, and L. F. Dahl, *J. Amer. Chem. Soc.*, **94**, 3383 (1972).

[148] Trinh-Toan, W. P. Fehlhammer, and L. F. Dahl, *J. Amer. Chem. Soc.*, **94**, 3389 (1972).

from an mo that is essentially non-bonding or weakly antibonding, a suggestion confirmed by a simple mo approach.[147-149]

One other feature of $[(\eta^5\text{-}C_5H_5)Fe(CO)]_4$ is the triply bridging CO group, a form of bonding often found in metal cluster chemistry. As is usual for bridging CO ligands (Chapter 16), the more metals with which a CO group interacts, the lower will be the range within which the CO stretching frequency is found. In this particular case of **52**, ν_{CO} is 1620 cm^{-1}, one of the lowest ever observed.

Cobalt, rhodium, and iridium form a structural well-defined series of compounds with the general formula $M_4(CO)_{12}$.[150] This series also illustrates some general features of cluster chemistry. The black cobalt cluster **53** can be prepared by warming (slightly above room temperature) the commercially available, red, metal-metal bonded dimer $Co_2(CO)_8$. You will recall from Chapter 16 that the color change from red to black is typical of metal carbonyls: the greater the number of metals in the molecule, the deeper the color.

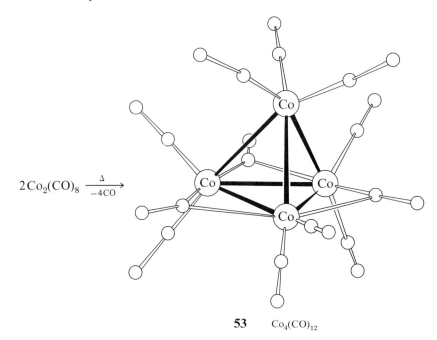

$$2Co_2(CO)_8 \xrightarrow[-4CO]{\Delta}$$

53 $Co_4(CO)_{12}$

The kinetics of the thermolysis of $Co_2(CO)_8$ have been studied, and the rate law suggests that the coordinatively unsaturated species $Co_2(CO)_6$ is formed rapidly by loss of CO from $Co_2(CO)_8$, and then, in a slow step, $Co_2(CO)_6$ dimerizes to the observed cluster.

Red $Rh_4(CO)_{12}$ or yellow $Ir_4(CO)_{12}$ can be made by reductive carbonylation reactions such as

$$RhCl_3 \cdot xH_2O \xrightarrow{CO} [Rh(CO)_2Cl]_2 \xrightarrow[Cu]{CO,\ 200\ atm} Rh_4(CO)_{12}$$

As was also pointed out above, it is another characteristic of metal clusters that those of the first transition series are most deeply colored. In changing from black (Co) to red (Rh) or yellow (Ir), the $M_4(CO)_{12}$ series again illustrates this trend.

---

[149] This mo treatment is essentially confirmed by a recent Mössbauer study of the neutral and 1+ $[(\eta^5\text{-}C_5H_5)Fe(CO)]_4$ clusters. R. B. Frankel, W. M. Reiff, T. J. Meyer, and J. L. Cramer, *Inorg. Chem.*, **13**, 2515 (1974).

[150] See ref. 116.

The three $M_4(CO)_{12}$ clusters are based on the tetrahedron, and, assuming localized electron-pair M—M bonds along each edge, each metal atom obeys the 18 electron rule. However, the CO ligands are arranged differently in the iridium cluster than in the cobalt or rhodium clusters. While there are nine terminal and three bridging CO's in the cobalt and rhodium clusters, all CO's are terminal in the iridium compound (**54**). This is another example of a general feature of metal carbonyls: within a related series, compounds of the heavier metals have fewer bridging CO's than those of the lighter metals.

54 $Ir_4(CO)_{12}$

Just as in borane chemistry, one can remove a vertex from a metal cluster and replace it with another group as long as the replacement contributes a sufficient number of electrons and orbitals of appropriate symmetry. Therefore, it is not surprising that a $Co(CO)_3$ vertex in $Co_4(CO)_{12}$ can be replaced by a $\begin{array}{c} Y \\ | \\ C \end{array}$ to give $Y\text{—}CCo_3(CO)_9$ (**55**).

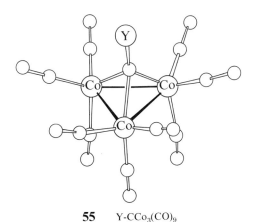

55 $Y\text{-}CCo_3(CO)_9$
Y = H, halide, or organic group

Although the CO ligands play a slightly different role in **55** than in $Co_4(CO)_{12}$, each vertex again obeys the inert gas rule. These alkylidynetricobalt nonacarbonyl clusters represent a large and thoroughly studied class of organometallic compounds.[151] Many can be prepared in high yield by reaction of $Co_2(CO)_8$ with geminal trihalides such as $Cl-CCl_3$ and $Cl_3C-COOR$.[152]

$$Co_2(CO)_8 + Cl_3C-R \rightarrow R-CCo_3(CO)_9$$

The resulting CCo_3 cage complexes are usually quite stable to atmospheric oxidation and the compounds are highly colored, most often purple or deep brown-purple. The cage is generally considered electron-withdrawing with respect to the apical Y group, and, as in **52**, there is experimental evidence for electron delocalization within the cage.[152a]

The reactions of the apical carbon in **55**, and of groups attached to it, are certainly not ordinary, as they are strongly influenced by the severe stereochemical constraints arising from the equatorial CO groups that make flank-side attack difficult; backside attack at the apical carbon is, of course, impossible. In fact, it is apparently because of these stereochemical constraints that the simple acid-catalyzed hydrolysis of $(OC)_9Co_3C-COOR$ esters does not occur. Instead, hydrolysis to give $(OC)_9Co_3C-COOH$ may be accomplished only by dissolving the purple ester in concentrated sulfuric acid.[153] By analogy with the known organic chemistry of highly hindered carboxylic acids, it is presumed that this reaction proceeds through an intermediate acylium ion, $(OC)_9Co_3C-C\equiv O^+$. This fact was substantiated by the finding that the acylium ion (after having been isolated in the form of its PF_6^- salt) could react with a host of different nucleophiles (among them water to give the carboxylic acid), as shown below.[154]

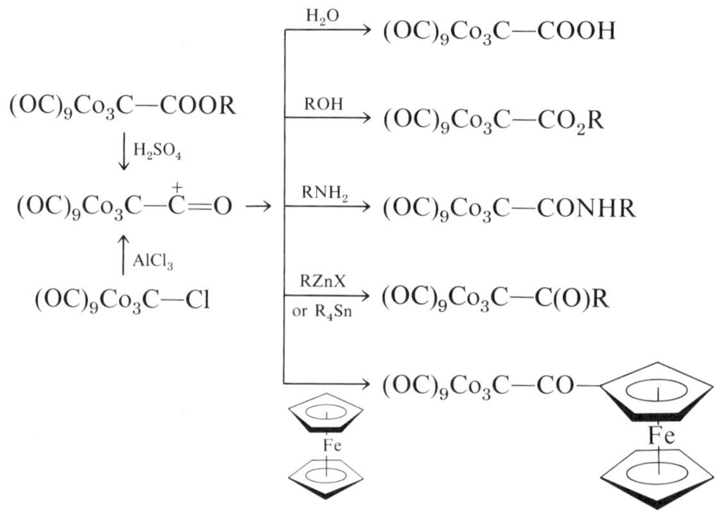

FIVE- AND SIX-ATOM CLUSTERS

Numerous clusters containing five or more metal atoms in the cage are known. For example, there are more than 50 clusters based on metal carbonyls: neutral

---

[151] See ref. 122.
[152] D. Seyferth, J. E. Hallgren, and P. L. K. Hung, *J. Organometal. Chem.*, **50**, 265 (1973). It is unfortunate that the mechanism of this interesting synthesis is not yet known.
[152a] J. Kotz, J. V. Petersen, and R. C. Reed, *J. Organometal. Chem.*, **120**, 433 (1976).
[153] D. Seyferth, J. E. Hallgren, and C. S. Eschbach, *J. Amer. Chem. Soc.*, **96**, 1730 (1974).
[154] The acylium ion may also be generated by reaction of the readily available $(OC)_9Co_3C-Cl$ with $AlCl_3$. D. Seyferth and G. A. Williams, *J. Organometal. Chem.*, **38**, C11 (1972).

compounds (56), anions (57 and 58), carbides (59), and hydrides [*e.g.,* $H_2Rh_6(CO)_{18}$].
All of the clusters of this type contain Group VIII metals.

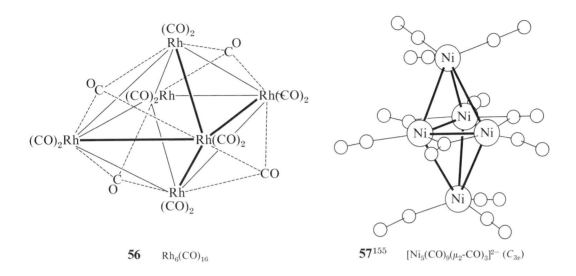

56 $Rh_6(CO)_{16}$

57[155] $[Ni_5(CO)_9(\mu_2\text{-}CO)_3]^{2-}$ (C_{3v})

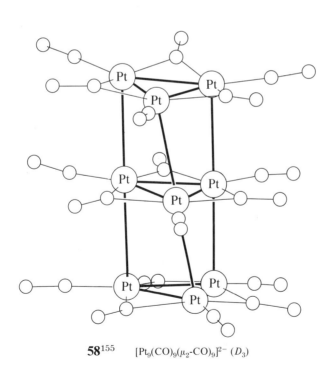

58[155] $[Pt_9(CO)_9(\mu_2\text{-}CO)_9]^{2-}$ (D_3)

---

[155] The symbol μ designates a bridging ligand, and the subscript indicates the number of metal atoms bridged.

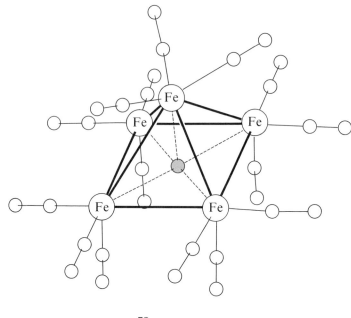

59 $Fe_5(CO)_{15}C$

Cobalt carbonyl clusters are among the more thoroughly studied of the larger clusters. As seen in the following scheme, a key species is the yellow-green anion $[Co_6(CO)_{15}]^{2-}$. This anion can be obtained by reduction of the previously discussed $Co_4(CO)_{12}$ cluster (**53**) or better by heating an ethanolic solution of $[Co(EtOH)_x][Co(CO)_4]_2$ *in vacuo*. Reduction of the $[Co_6(CO)_{15}]^{2-}$ anion or direct reduction of $Co_4(CO)_{12}$ gives the deep red species $[Co_6(CO)_{14}]^{4-}$ (**60**).[156]

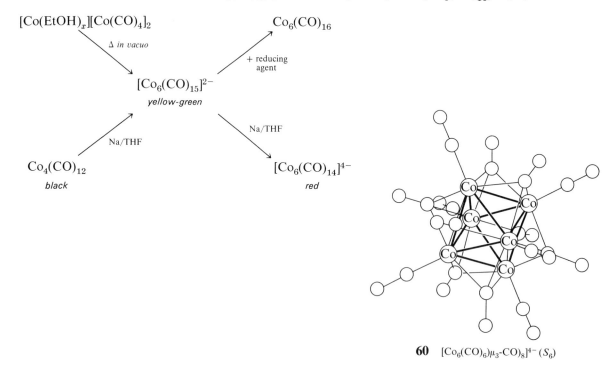

60 $[Co_6(CO)_6)\mu_3\text{-}CO)_8]^{4-}$ (S_6)

---

[156] See ref. 118.

The mode of bonding in clusters such as $[Co_6(CO)_{14}]^{4-}$ has attracted considerable interest.[157] As indicated in Table 18-6, M_3 clusters are characterized by 48 total valence electrons, and bonding can be adequately represented by localized electron pair bonds between metals. The same is true of M_4 clusters, where there are 60 valence electrons. In $[Co_6(CO)_{14}]^{4-}$, however, a total of 86 electrons is contributed by the metal atoms, the CO's, and the charge. Were the "edge-bond" description possible for this compound as for $[(\eta^5\text{-}C_5H_5)Fe(CO)]_4$ or $Co_4(CO)_{12}$, only 84 electrons would be necessary.[158] To account for this discrepancy, Wade has suggested that bonding within the cage is quite delocalized and is formally analogous to that of the stoichiometrically equivalent borane compounds.[159] In the present case, $[Co_6(CO)_{14}]^{4-}$ (or written alternatively as $[(CoCO)_6(CO)_8]^{4-}$) is equivalent to $(BH)_6^{2-}$, the molecular orbital diagram for which was presented in Figure 18-9. A very similar mo scheme can be built for the cobalt cluster by recognizing that the Co $d\sigma$ ao is analogous to the *endo* h ao of each boron, and the Co $d\pi$ ao's replace the $p\pi$'s of boron.

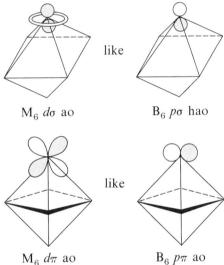

M_6 $d\sigma$ ao B_6 $p\sigma$ hao

M_6 $d\pi$ ao B_6 $p\pi$ ao

In both $(BH)_6^{2-}$ and $[Co_6(CO)_{14}]^{4-}$, M_6 cage bonding is determined by seven occupied, strongly bonding mo's $[a_{1g}(\sigma) + t_{2g}(\pi) + t_{1u}(\pi)]$ and not by 12, as required by the "edge-bond" description. The bonding of the 14 CO ligands implies, as in Figure 18-9, the stabilization of the 14 electron pairs in the CO σ TASO's. Thus, M_6 and CO bonding account for 21 of the 43 valence pairs associated with the cage. The remaining 22 pairs ($3\frac{2}{3}$ per cobalt) apparently fill cage non-bonding mo's stabilized by cage-CO π retrobonding. Accordingly, the "EAN" of each cobalt is just $14\frac{1}{3}$ ($= 86/6$). This same analysis is true for $[Ru_6(CO)_{18}]^{2-}$, which is isoelectronic with $[Co_6(CO)_{14}]^{4-}$. In the case of the ruthenium compound there are again seven cage bond pairs. 18 CO σ bond pairs, and 18 "non-bonding" pairs (three per Ru) for an "EAN" of $14\frac{1}{3}$ again. In both the Co and Ru clusters, the failure of the usual two-center electron pair bond concept to account for M—M bonding causes the breakdown of the 18 electron rule. Finally, note that the discrepancy between the prediction of the 18 electron rule and the observed electron count amounts to a deficiency of $3\frac{2}{3}$ electrons per metal—not an excess of 1/3 as implied by the "edge-bond" counting procedure.

---

[157] D. M. P. Mingos, *J. C. S. Dalton*, 133 (1974).

[158] The need for 84 electrons is arrived at as follows: If each Co is to have an EAN of 18 electrons, and edge-bonding is to exist, the nine electrons of each Co must be supplemented by four electrons from M—M bonds to nearest neighbors, by two electrons from a bond to a terminal CO, by two and two-thirds electrons from the four triply bridging CO's that each cobalt shares with its four nearest neighbors, and by one-third of an extra electron per Co for a cluster charge of 2—. That is, the formula of the cluster should be $[Co_6(CO)_{14}]^{2-}$, an anion having a total of 84 electrons.

[159] K. Wade, *Chem. Britain*, **11**, 177 (1975); *Adv. Inorg. Chem. Radiochem.*, **18**, 1 (1976).

Although it is accelerating because of interest in the area, the publication of research on metal clusters is slow. This is largely due to the difficulty of the synthesis, separation, and structural characterization of clusters. Thus, relatively little has yet been done in a systematic way regarding their chemistry, aside from examination of such basic processes as[160]

oxidation: $[Ir_6(CO)_{15}]^{2-} + 2H^+ + CO \rightarrow Ir_6(CO)_{16} + H_2$

reduction: $2[Pt_9(CO)_{18}]^{2-} + 2Li \rightarrow 3[Pt_6(CO)_{12}]^{2-} + 2Li^+$

ligand substitution (which frequently occurs with cage degradation):

$$Rh_6(CO)_{16} + 12PPh_3 \xrightarrow[C_6H_6]{25°C} 3[Rh(CO)_2(PPh_3)_2]_2 + 4CO$$

Just as the heavier metals at the left of the transition metal series form binuclear halide complexes with metal-metal bonds, these same metals form larger clusters. Two of the best known are those based on the octahedron: $[Mo_6Cl_8]^{4+}$ (**61**)[161] and $[M_6X_{12}]^{n+}$ (M = Nb, Ta; X = F, Cl, I; n = 2,3,4) (**62**).[162,163]

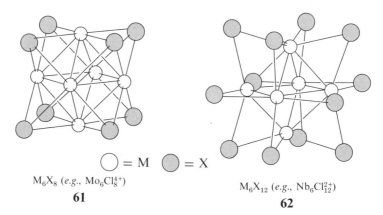

M_6X_8 (*e.g.*, $Mo_6Cl_8^{4+}$)
61

$\bigcirc$ = M $\bullet$ = X

M_6X_{12} (*e.g.*, $Nb_6Cl_{12}^{2+}$)
62

These also present a problem regarding a description of their bonding. If the halide ions are removed from the clusters, a core of the type Mo_6^{12+} or Nb_6^{14+} remains. In the case of Mo_6^{12+}, the cluster is composed of Mo^{2+} ions just as in $[Mo_2Cl_8]^{4-}$, which was discussed earlier. This means that there are $6 \times 4 = 24$ electrons available for cluster bonding in Mo_6^{12+}; if all 12 pairs occupy bonding mo's, there is a net of one electron pair for each Mo—Mo pair, and each Mo may be considered to be bonded to its four nearest neighbors by localized two-electron bonds. In the Nb_6^{14+} core, however, there are $6 \times 5 - 14 = 16$ electrons or only eight pairs for cage bonding. Thus, as in $[Co_6(CO)_{14}]^{4-}$, there is a deficiency of electrons and an "edge-bond" description is not appropriate. It is best to resort again to mo methods.

A more detailed mo approach to bonding in M_6 clusters such as $[Nb_6Cl_{12}]^{2+}$ is outlined in an Appendix to this chapter. The most important result of this approach is that $d\delta$ ao's are present at each vertex (**63** and **64**). That is, metal $d_{x^2-y^2}$

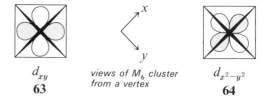

d_{xy}
63

views of M_6 cluster from a vertex

$d_{x^2-y^2}$
64

---

[160] See ref. 118.
[161] J. C. Sheldon, *J. Chem. Soc.*, 1007, 3106 (1960).
[162] F. W. Koknat and R. E. McCarley, *Inorg. Chem.*, **13**, 295 (1974).
[163] B. G. Hughes, J. L. Meyer, P. B. Fleming, and R. E. McCarley, *Inorg. Chem.*, **9**, 1343 (1970).

ao's lie so that their lobes extend over octahedral edges, while metal d_{xy} lobes extend over octahedral faces. In the $[Mo_6Cl_8]^{4+}$ case, the last six electrons of the 12 cage bonding electron pairs are accommodated in a t_{2u}^6 mo composed of $d_{x^2-y^2}$ orbitals, and the mo's arising from the d_{xy} orbitals are unused. It is for this reason that eight Cl^- ions are associated with the Mo_6^{12+} core and that these Cl^- ions are located at the faces where the Cl^- electron pairs can overlap the empty d_{xy} mo's! On the other hand, the $[Nb_6Cl_{12}]^{2+}$ cluster has eight cage electron pairs, and the last pair enters an a_{2u}^8 mo composed of d_{xy} ao's; the $d_{x^2-y^2}$ mo's, lying along octahedral edges, are unused. Therefore, there are $12Cl^-$ ions bridging the 12 octahedral edges where they can effectively interact with the empty $d_{x^2-y^2}$ mo's.

STUDY QUESTIONS

20. Unlike clusters with six metal atoms, compound **65** (p. 1046) with a cube of metal atoms obeys the 18 electron rule. Verify that this is indeed the case and that localized, two-electron bonds can be postulated between adjacent nickel atoms.

21. By reference to the 18 electron rule, verify the fact that it is necessary to postulate a Cr—Cr triple bond in **43** if the compound is to be diamagnetic.

22. Suggest a way to form $W-SnR_3$, $W-GeR_3$, or $W-PbR_3$ bonds if the starting materials are the tungsten compound below and a suitable germanium, tin, or lead compound.

[If you need a hint, see T. A. George and C. D. Sterner, *Inorg. Chem.*, **15**, 165 (1976).]

23. What is the most likely order for the Mo—Mo bond in the ion $[Mo_2(CH_3)_8]^{4-}$? Sketch a possible structure of the ion.

24. Two of the best characterized metal-metal bonded compounds are $Mo_2(CH_2SiMe_3)_6$ and $Mo_2(NMe_2)_6$. What is the Mo—Mo bond order in these two compounds? Sketch possible structures for the molecules.

EPILOG

This discussion of the structure, bonding, and reactivity of boron hydrides and metal clusters was intended to be an overview; as such, it is far from a complete treatment of an area that promises to be one of increasing importance in the next few years. The discovery of metalloboranes has led to a resurgence of interest in boron hydride chemistry, and, as metalloboranes show promise as catalysts,[164] research in this area is certain to continue.

As the structural chemistry of metal clusters becomes better understood one can expect more attempts at their deliberate synthesis, with increasing probability of success.[164a] The recent synthesis of $Ni_8(CO)_8(\mu_4\text{-}PPH)_6$ (**65**) is a case in point.[165] Finally,

---

[164] T. E. Paxson and M. F. Hawthorne, *J. Amer. Chem. Soc.*, **96**, 4676 (1974); E. H. S. Wong and M. F. Hawthorne, *Chem. Commun.*, 257 (1976).

[164a] For example, see G. L. Geoffroy and W. L. Gladfelter, *J. Amer. Chem. Soc.*, **99**, 304 (1977).

[165] L. D. Lower and L. F. Dahl, *J. Amer. Chem. Soc.*, **98**, 5046 (1976).

we would emphasize that we have by no means covered the topic of molecular polyhedra completely. Perhaps arbitrarily, we have chosen to restrict ourselves to those compounds in which metal-metal bonding is a virtual certainty. However, there are many other molecular polyhedra that do not involve metal-metal bonding and yet have interesting and important properties. For example, compound **66** has been described as a model for iron-sulfur proteins, and it will be discussed in considerably more detail in the next chapter on biochemical aspects of inorganic chemistry.

In general, it is certain that molecular polyhedra will come more and more to the attention of the chemical community as practical uses for them continue to be found.

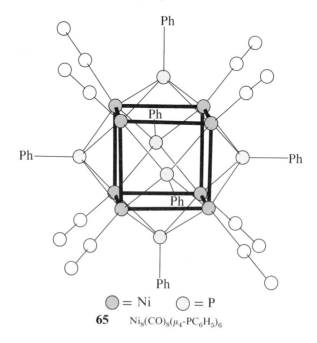

65 $Ni_8(CO)_8(\mu_4\text{-}PC_6H_5)_6$

66 $[Fe_4S_4(SC_6H_5)_4]^{2-}$

Appendix to Chapter 18

Appendix A: Bonding Model for M_6 Clusters

It is not difficult to build a molecular orbital diagram for an M_6 cluster, only tedious. The mo's are directly attainable from those for the TASO's of the EX_6 case of Chapter 4. To review quickly, each metal atom has a $d\sigma$ ao (d_{z^2}) directed to the center of the octahedron. These six $d\sigma$ ao's give rise to six TASO's of a_{1g}, t_{1u}, and e_g symmetry (the relative phasings are given in Chapter 4, page 181).

a_{1g} $(d\sigma_1 + d\sigma_2 + d\sigma_3 + d\sigma_4 + d\sigma_5 + d\sigma_6)$

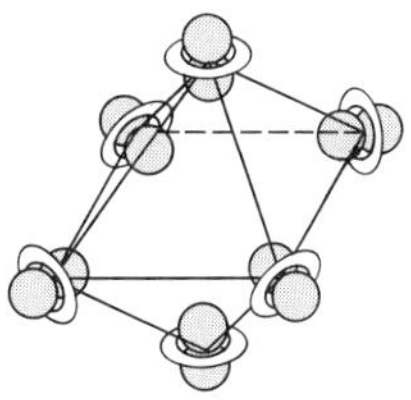

t_{1u} $(d\sigma_1 - d\sigma_3)$
       $(d\sigma_2 - d\sigma_4)$ etc.
       $(d\sigma_5 - d\sigma_6)$

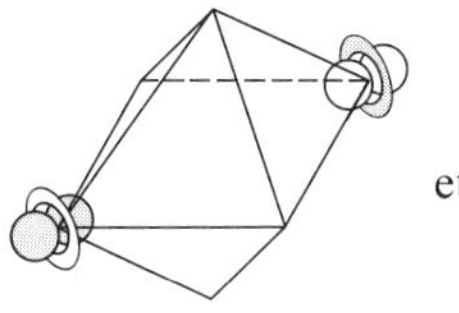

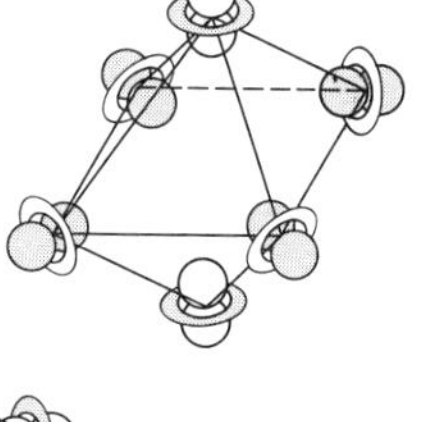

e_g $(d\sigma_1 + d\sigma_2 + d\sigma_3 + d\sigma_4 - d\sigma_5 - d\sigma_6)$

      $(d\sigma_1 - d\sigma_2 + d\sigma_3 - d\sigma_4)$

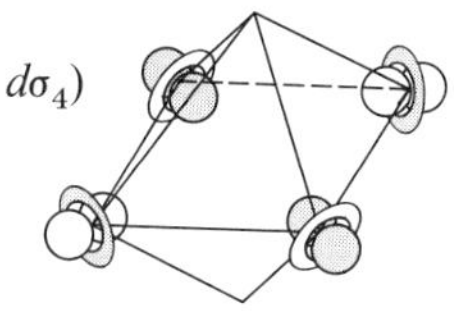

Counting the number of nearest- and cross-neighbor interactions, you quickly determine that a_{1g} is the only strongly bonding mo of the set.

In an entirely analogous way, the metal $d\pi$ ao's (d_{xz}, d_{yz}) behave as do the $p\pi$ ao's of the EX_6 case. Again you find t_{1g}, t_{2g}, t_{1u}, and t_{2u} TASO's constructed in analogy with the $p\pi$ TASO's shown on page 180 of Chapter 4. The bonding TASO's are again t_{1u} and t_{2g}, with the t_{2u} and t_{1g} playing anti-bonding roles. Up to this point the bonding mo's are a_{1g}^σ, t_{2g}^π, and t_{1u}^π while the anti-bonding mo's are e_g^σ, t_{1u}^σ, t_{1g}^π, and t_{2u}^π, exactly as in the $(BH)_6^{2-}$ case of Figure 18-9!

An important new twist for the M_6 cluster is the presence of the $d\delta$ ao's ($d_{x^2-y^2}, d_{xy}$) on each metal vertex. Starting with the "standard" ao orientations shown below, the six TASO's in each case result from the identical linear combinations given above for the $d\sigma$ TASO's.

$d_{x^2-y^2}$ d_{xy}

The symmetries of these TASO's are not the same as those of the $d\sigma$ set, of course, but the symmetries are quickly determined by inspection once you have drawn the TASO's according to the phasings of the $d\sigma$ equations above. The results (which you can verify if so inclined) are as follows:

	$d_{x^2-y^2}$	d_{xy}
$(1 + 2 + 3 + 4 + 5 + 6)$ $(1 - 2 + 3 - 4)$	e_g (4 bonding interactions)	e_u (4 anti-bonding interactions)
$(1 + 2 + 3 + 4 - 5 - 6)$	a_{2g} (12 anti-bonding interactions)	a_{2u} (12 bonding interactions)
$(1 - 3)$ $(2 - 4)$ $(5 - 6)$	t_{2u} (non-bonding interactions)	t_{2g} (non-bonding interactions)

If you construct the TASO sketches for the $d\delta$ mo's and then count the bonding/anti-bonding interactions between nearest neighbors, you will find the results given in parentheses alongside the representation symbols above. The net result of all of this is that there are 10 TASO's that are cluster bonding [three more than for $(BH)_6^{2-}$] and six that are non-bonding; the remainder (14) are antibonding. A summary is as follows:

	$d\sigma$	$d\pi$	$d\delta$	
			$d_{x^2-y^2}$	d_{xy}
anti-bonding	t_{1u} e_g	t_{2u} t_{1g}	a_{2g}	e_u
non-bonding	—	—	t_{2u}	t_{2g}
bonding	a_{1g}	t_{1u} t_{2g}	e_g	a_{2u}

For a molecule like $Mo_6Cl_8^{4+}$, where the Mo_6^{12+} cluster has 12 pairs for cage bonding (each Mo^{2+} has two valence electron pairs), the lowest 12 mo's should be occupied; specifically, this suggests a configuration of $(a_{1g}^\sigma)^2(t_{1u}^\pi)^6(t_{2g}^\pi)^6(e_g^\delta)^4(t_{2u}^\delta)^6$. It is important to notice here that the $d_{x^2-y^2}$ cage mo's, not the d_{xy} set, are required to accommodate all 12 pairs. Since these mo's are concentrated about the edges of the octahedron, while the d_{xy} set is concentrated at the faces, it is not surprising that only eight Cl^- ions are associated with the Mo_6^{12+} cluster, and that these ions are located at the faces where the Cl^- electron pairs can overlap the empty d_{xy} mo's.

For a cluster derived from Nb_6^{14+} with only eight cage electron pairs, the eight pairs are exactly accommodated in the configuration $(a_{1g}^\sigma)^2(t_{1u}^\pi)^6(t_{2g}^\pi)^6(a_{2u}^\delta)^2$, where it is the d_{xy} set (concentrated in the eight faces of the octahedron) that is used for cluster bonding; the $d_{x^2-y^2}$ mo's (concentrated along the 12 edges) are empty and available for bonding of Cl^- ions. Accordingly, the Nb_6^{14+} cluster is found associated with 12 chloride ions in an $Nb_6Cl_{12}^{2+}$ cluster, and the Cl^- ions are attached at the edges!

The cage bonding in $[Co_6(CO)_{14}]^{4-}$ (represented as $[(CoCO)_6(CO)_8]^{4-}$) is analogous to that of $Mo_6Cl_8^{4+}$ in that eight CO ligands coordinate at the eight octahedral faces. Accordingly, the $d_{x^2-y^2}$ mo's are involved in Co_6 cluster bonding, and there are 12 pairs of the bonding and/or non-bonding type $(a_{1g}^\sigma, t_{1u}^\pi, t_{2g}^\pi, e_g^\delta, t_{2u}^\delta)$. Since the Co_6^{4-} cluster must accommodate 29 pairs of electrons ($4\frac{1}{2}$ from each Co and 2 for the charge), there must be 17 "anti-bonding" pairs; this, if unaltered, would be disastrous to the cluster stability. Two features of $Co_6(CO)_{14}^{4-}$, not operative in the chloride complexes discussed above, save the day. First of all, the 4s and 4p

TASO's of the Co_6^{4-} cluster lie much closer in energy to the $3d$ TASO's than is the case for the Mo_6^{12+} and Nb_6^{14+} and mix with them more strongly. This mixing confers bonding character on the $3d$ TASO's to offset their anti-bonding nature. Secondly, the presence of the strongly π-accepting CO ligands serves to stabilize further the cage anti-bonding pairs by endowing them with Co—CO bond character and delocalizing them from the cage. Entirely similar comments apply to the Ru_6^{2-} cage with 25 cluster pairs in the compound $[Ru_6(CO)_{18}]^{2-}$ (also written as $[Ru(CO)_3]_6^{2-}$).

THE CELL

PROCESSES COUPLED TO PHOSPHATE HYDROLYSIS

   Nucleotide Transfer—DNA Polymerase
   Phosphate Transfer
     General Comments
     Pyruvate Kinase
     Glucose Storage—Phosphoglucomutase
     Phosphate Storage in Muscle—Creatine Kinase
   Na^+/K^+ Ion Pump—ATPase

OXYGEN CARRIERS—HEMOGLOBIN AND MYOGLOBIN

COBALAMINS; VITAMIN B_{12} COENZYME

ELECTRON TRANSFER AGENTS

   Cytochromes
   Iron-Sulfur Proteins
     1-Fe—S and 2-Fe—S
     2-Fe—S*
     8-Fe—S*
     N_2 Fixation
   Chlorophyll

EPILOG

19. BIOCHEMICAL APPLICATIONS

One of the more rapidly developing areas of inorganic chemistry, indeed in all of chemistry, is that of its applications to living systems. The name "inorganic" applied to metal ions and polyphosphate anions, in particular, is certainly a misnomer. Such reagents are vital to the existence of living systems as sources of chemical potential to drive unfavorable organic reactions and as catalysts to ensure that the required reactions proceed with sufficient velocity. Even organometallic reactions have found a place in living systems, as have more traditional concepts of coordination chemistry in the areas of transport of O_2, electrons, and metal ions. In what follows you will find general overviews of selected in vivo processes, presented from the organizational viewpoint of the biochemist, followed by discussions of vital "inorganic" links in the processes. We have chosen this format to emphasize the truly interdisciplinary nature of life chemistry, for the inorganic aspects take on significance only in relation to the general biochemical processes of which they are a part. A key role for the inorganic chemist is to model as closely as possible the active prosthetic groups of metal-requiring in vivo enzymes. In this context, many of the key concepts of earlier chapters (mechanisms of substitution reactions at non-metal centers, electron transfer, ligand effects on metal ion coordination geometry and electron structure, organometallic reactions, and photochemical concepts) are intimately involved in attempts to understand the working pieces of the living system. The subjects chosen for this chapter reflect varying degrees of incomplete understanding by the interdisciplinary chemical community, and so constitute a fitting conclusion for a text about a subject with an exciting future.

THE CELL[1]

The reaction "flask" of living organisms is the cell (Figure 19-1). Each cell is characterized by an *outer membrane* whose function is to contain the highly organized chemical system and to monitor the influx of needed reagents. In a later section of this chapter you will be introduced to the Na^+/K^+ ion pump function of the membrane, a function that raises the concentration of K^+ within the cell while diminishing the concentration of Na^+. Mg^{2+} is another ion that is necessary for enzymatic action within the cell. Ca^{2+}, on the other hand, is excluded from the cell (but it is found in bones, teeth, and as an activator of extracellular enzymes).

Within the cell membrane are several organelles immersed in cellular fluid or

---

[1] An excellent, comprehensive treatment of this subject, and one to which we will frequently refer, is "Inorganic Biochemistry," G. L. Eichhorn, ed., Vols. 1 and 2, American Elsevier, New York (1973).

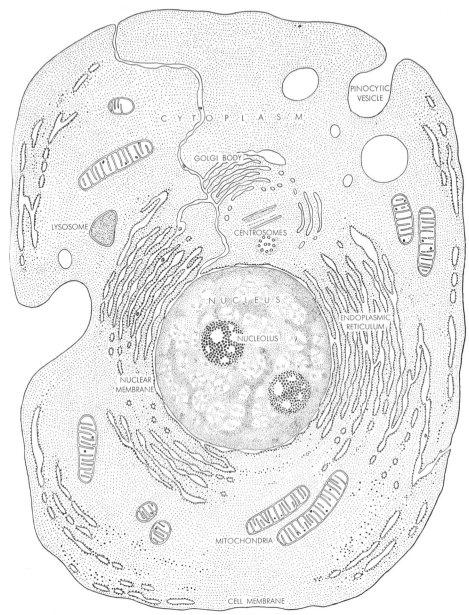

Figure 19-1. A typical cell. [From J. Brachet, *Scientific American*, p. 55, September, 1961.]

cytoplasm. Two of these organelles—the nucleus and the mitochondria—are the focus of most of the chemistry discussed in this chapter. The *nucleus*[2] is surrounded by a nuclear membrane, within which occur the processes concerned with DNA: DNA replication, RNA synthesis, and membrane synthesis. The other important focus in this chapter is on the *mitochondria,* organelles within which occur the redox/electron-transfer processes so important in the combustion of glucose and the synthesis of adenosine triphosphate (ATP). Each mitochondrion consists of an outer

---

[2] Bacterial cells are different from animal and plant cells. For example, the former do *not* have a nucleus.

membrane and a folded, inner membrane to which a complex (but necessarily highly organized) system of enzymes, their cofactors, and electron transfer agents is attached. In animals the mitochondria are the sole centers of energy generation in the cell; the cells of green plants contain, in addition, *chloroplasts*. These organelles contain chlorophyll, which makes possible the light-sensitized phosphorylation reactions of ATP regeneration. (In the dark the mitochondria of plant cells maintain this regeneration, though at a lower rate than in animals because the mitochondria are sparser in green plant cells.)

The lysosome and Golgi bodies shown in Figure 19-1 are involved in cell digestion and excretion, respectively, and will not be further mentioned. The endoplasmic reticulum defines an intracellular network of channels for transport of proteins synthesized by the *ribosomes* that stud the surfaces of the channels.

PROCESSES COUPLED TO PHOSPHATE HYDROLYSIS

A key requirement of *in vivo* reactions is that they be spontaneous in a thermodynamic sense—ΔG must be negative. While this statement seems trivial, many reactions that are essential to living systems are in fact not spontaneous ($\Delta G > 0$) when isolated in a reaction flask at standard state concentrations. Nature has solved this problem by coupling such *endergonic* reactions with others that are strongly *exergonic*.[3] Since two such "half-reactions" are individually complete, they are "coupled" by common involvement with one or more other reagents (or intermediates). For example, it is an empirical observation that endergonic reactions are frequently coupled to hydrolysis of a triphosphate ester ($R-OPO_2OPO_2OPO_3^-$). *Adenosine triphosphate* (ATP, **1**) is an important example.

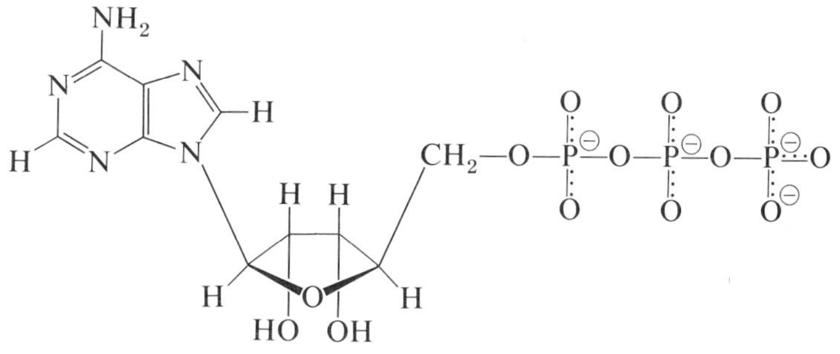

ATP
adenosine triphosphate
(a nucleotide diphosphate)

1

The nomenclature derives from that of the ester group, *adenosine* (**2**). Adenosine is the *nucleoside*[4] composed of the base adenine (**3**) and the five-carbon sugar, ribose, in its cyclic form (**4**).[5]

---

[3]Endergonic (= *end* + ergonic) and exergonic (= *ex* + ergonic) refer to reactions with $\Delta G > 0$ and < 0, respectively.

[4]A nucleo*s*ide is a nitrogenous base/sugar combination. A nucleo*t*ide is a nucleoside phosphate ester. ATP is both a nucleoside triphosphate and a nucleotide diphosphate. Nucleotides are the structural/chemical units of DNA (a nucleotide polymer, in which the sugar component is 2'-deoxyribose bound to separate phosphates at the 3'- and 5'-carbons in **4**).

[5]2'-deoxyribose is ribose with H replacing OH at the 2'-carbon.

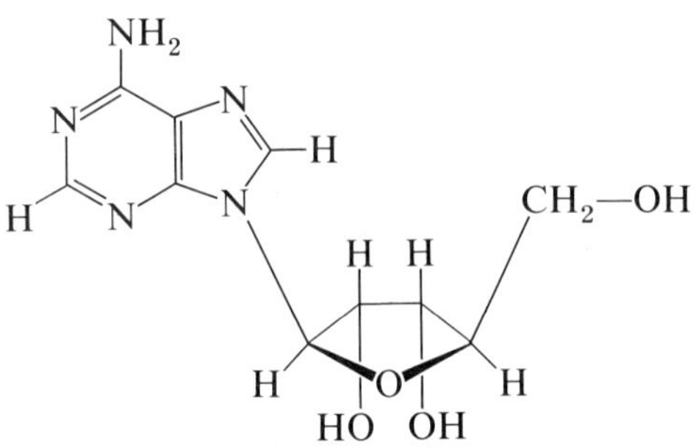

1054 ☐ BIOCHEMICAL APPLICATIONS

adenosine
(a nucleoside)
2

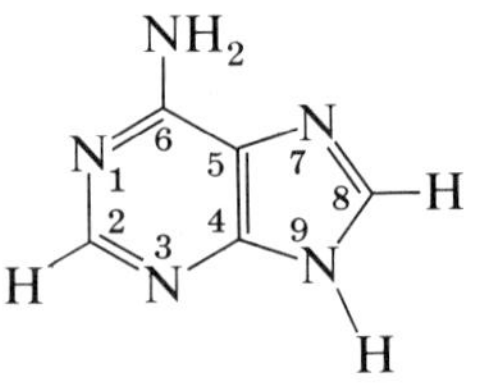

adenine
3

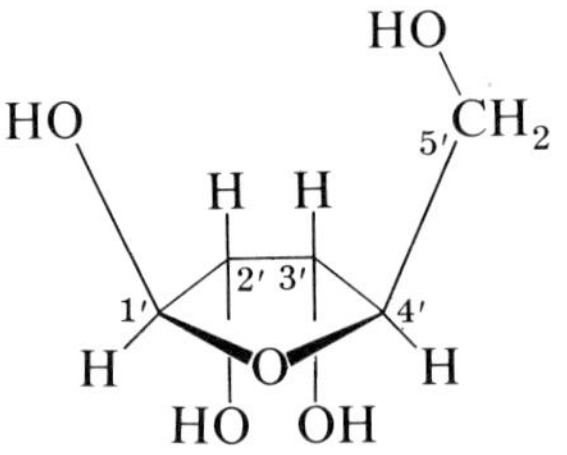

β-D-ribose
4

There seems to be a very good thermodynamic reason for the association of triphosphate ester hydrolysis with important reactions in living systems; that is, $\Delta G°$ for such hydrolyses are quite large:[6]

(19-1)[6] $\quad ATP^{4-} + H_2O \rightarrow ADP^{3-} + HPO_4^{2-} + H^+ \qquad \begin{array}{l}\Delta G°' = -9.7 \text{ kcal/mole} \\ K' = 1.3 \times 10^7\end{array}$

ADP = adenosine *di*phosphate

(19-2) $\quad ATP^{4-} + H_2O \rightarrow AMP^{2-} + HP_2O_7^{3-} + H^+ \qquad \begin{array}{l}\Delta G°' = -10.4 \text{ kcal/mole} \\ K' = 4.2 \times 10^7\end{array}$

AMP = adenosine *mono*phosphate

(19-3) $\quad AMP^{2-} + H_2O \rightarrow AH + HPO_4^{2-} \qquad \begin{array}{l}\Delta G°' = -3.0 \text{ kcal/mole} \\ K' = 1.6 \times 10^2\end{array}$

It is of some significance that the ATP hydrolysis reactions 19-1 and 19-2 are nearly the same thermochemically, and both are much favored with respect to solvolysis of the ester group. The coupling of the ATP hydrolysis reactions with other reactions proceeds through one or more steps of phosphate or *nucleotide*[7] transfer to the other reagent. In addition, it is always the case that an *enzyme* (catalyst) is required for the intermediate reaction(s) to proceed at a sufficiently rapid rate or to decide the regiochemical course of the hydrolysis (*i.e.*, to select 19-1 or 19-2). Accordingly, *in vivo* phosphate hydrolysis/transfer reactions have two "inorganic" features, for not only does the reaction occur at a phosphorus center but often a metal ion is required by the enzyme for its catalytic function. In Chapter 14 the role of metal ion catalysis

---

[6] Notice that $\Delta G°'$, rather than $\Delta G°$, is given in equations 19-1 through 19-3. This signifies standard state conditions defined by biochemists, which are quite different from those of chemists. In this case, pH = 7.4, $T = 25°C$, and 10^{-4} M Mg^{2+} ion is present; all of these more nearly approximate *in vivo* conditions than do the conditions (pH = 0.0 and no Mg^{2+} present) of the chemist. $\Delta G°$ is very sensitive both to pH and to the presence of Mg^{2+}, which is complexed by oxygen atoms of the second and third phosphate groups of ATP. For further information, see R. W. McGilvery, "Biochemical Concepts," W. B. Saunders, Philadelphia (1975); A. L. Lehninger, "Biochemistry," 2nd Ed., Worth Publishers, New York (1974).

[7] See ref. 4.

of reactions at organic molecules was discussed, and amino acid and amide hydrolyses were given as examples. In this and the following section you will examine two analogous metal-catalyzed hydrolyses of the phosphate group.

NUCLEOTIDE TRANSFER—DNA POLYMERASE

The polymerization of nucleoside phosphates produces the nucleic acids, DNA (deoxyribonucleic acid) and RNA (ribonucleic acid). DNA consists of two strands, each of which is a condensation polymer of nucleotides containing 2′-deoxyribose as the sugar.[8]

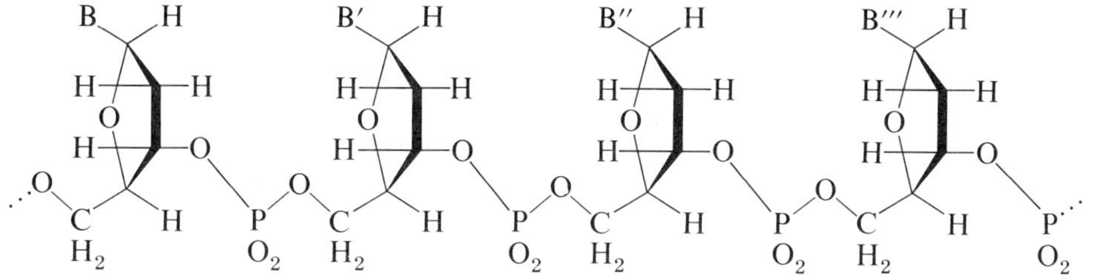

a generalized deoxyribonucleic acid
B, B′, B″, B‴ = bases (adenine, cytosine, etc.)

RNA is a single-strand polymer of nucleotides containing ribose sugars (**4**) rather than deoxyribose. The mechanism of polymerization involves the DNA molecule, a nucleoside triphosphate, and an enzyme (either DNA polymerase or RNA polymerase, depending upon whether it is DNA or RNA that is to be synthesized):

$$(NA)_n^{(n+1)-} + \text{nucleoside}-P_3O_{10}^{4-} \xrightarrow[\text{DNA}]{\text{enzyme}} (NA)_{n+1}^{(n+2)-} + HP_2O_7^{3-}$$

NA = nucleic acid

The enzyme plays a regioselective role by effecting cleavage of pyrophosphate from the nucleoside triphosphate (see equation 19–2) with addition of the nucleotide to the

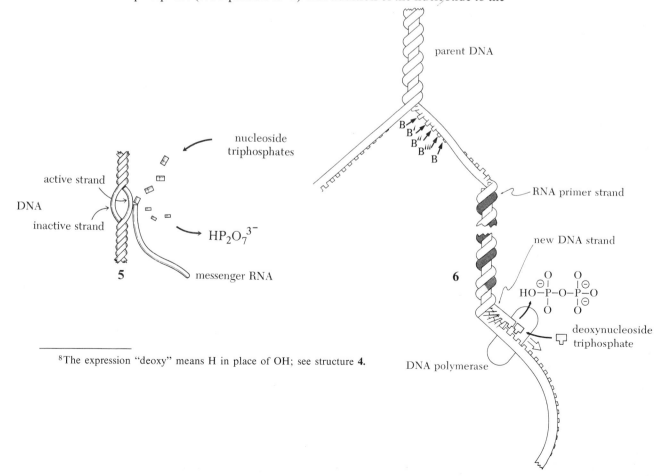

---

[8] The expression "deoxy" means H in place of OH; see structure **4**.

growing polymer. DNA is the template that assures that the proper nucleotide is next added to the growing RNA or DNA strand. That is, the sequencing B, B', ... in the parent DNA is copied by the developing nucleic acid strand; the nucleotide is bound to the strand by "key-lock" (hydrogen bonding/electrostatic) interactions between the base of the parent DNA and that of the incoming nucleotide.

The role of mRNA (messenger RNA) in **5** is to diffuse from the nucleus to the ribosomes to direct the assembly of amino acids into proteins. The DNA reproduction in **6** is necessary prior to cell division, in order to provide the daughter cell with the proper DNA sequences in its nucleus.

A simplified schematic of the replication appears as follows:

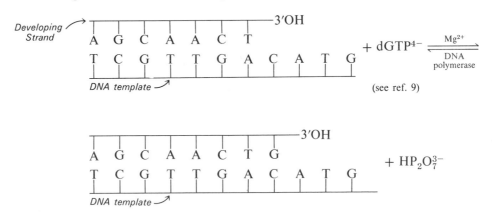

Both Mg^{2+} and DNA polymerase are required as catalysts for the reaction. Furthermore, it has been established[10] that Zn^{2+} is coordinated both by DNA polymerase and by the developing DNA strand. Zn^{2+} does not appear to simply hold the two pieces together; a mechanistic role for it has been established by the fact that o-phenanthroline coordination of Zn^{2+} in the Zn^{2+}(oligomer)(enzyme) complex inhibits the polymerization. That is, not all the coordination sites about Zn^{2+} are occupied by DNA and enzyme donor atoms. Others, probably occupied by H_2O, become blocked by coordination of the strong ligand o-phen and thereby prevent completion of the polymerization reaction. Furthermore, it is known that binding of the entering deoxynucleoside triphosphate to the oligomer/enzyme complex occurs not at Zn^{2+} but at Mg^{2+}. A proposed mechanism incorporating these features is shown in Figure 19-2. The important concepts to notice are:

a) the enzyme contains both Zn^{2+} and Mg^{2+};

b) The Zn-enzyme molecule is coordinated to the oligomer/DNA molecule at the sugar 3'-OH group of the oligomer (as in **I**);

c) in step ①, the incoming nucleoside triphosphate is coordinated to Mg^{2+} at the α-phosphate unit (as in **II**);

d) the 3'-OH group acts (step ②) as a nucleophile upon the α-phosphate unit to displace $P_2O_7^{4-}$ (it may also be that a nearby base in the enzyme assists the attack by deprotonation of the 3'-OH group);

e) migration (step ③) of the Zn enzyme complex to the 3'-OH group of the added nucleoside sets the stage for the next nucleoside addition.

---

[9] A, G, C, and T refer to the nucleoside units containing adenine, guanine, cytosine, and thymine, respectively, as the nitrogenous base components. The lower case d, in dGTP, designates the "deoxy" ribose sugar unit.

[10] A. S. Mildvan and C. M. Grisham, *Structure and Bonding,* **20**, 1 (1974).

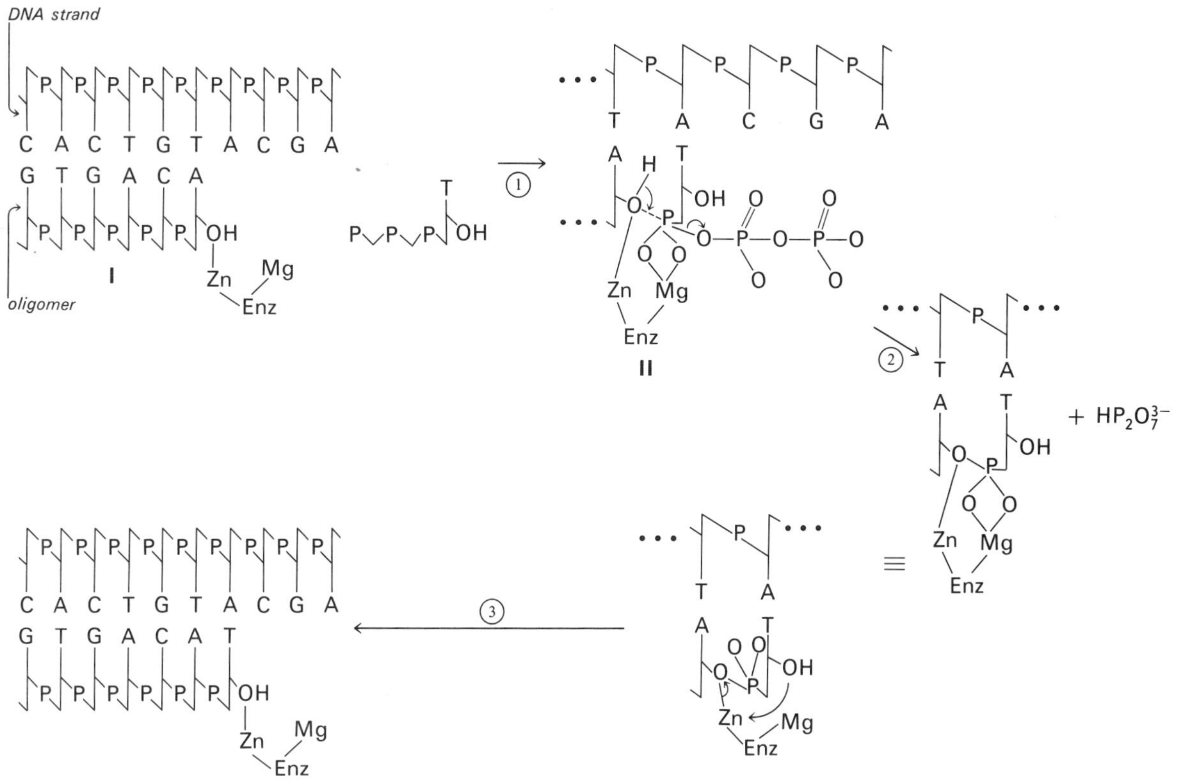

Figure 19-2. Proposed mechanism of *E. coli* DNA polymerase reaction.

PHOSPHATE TRANSFER

Aerobic organisms derive their "energy" from sugar oxidation by O_2 ($\rightarrow CO_2 + H_2O$). In man, it is glucose (**7**) that is oxidized to CO_2 in several steps (see Scheme I, p. 1058), many of which involve phosphate transfer (phosphorylation) *from* nucleoside triphosphates and *to* nucleoside diphosphates. Some of these steps simply involve substitution reactions at phosphorus, while in others, electron transfer is a necessary co-reaction.[11] There are numerous examples of metal ion requirements in both kinds of "phosphorylations"; in this section we will consider just the non-redox metal ion processes, and postpone the redox systems.

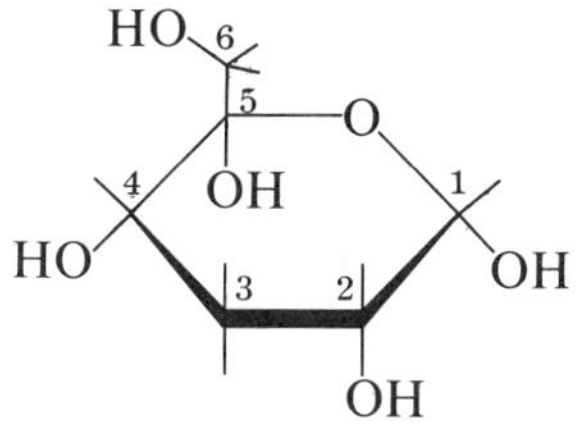

General Comments

A general description of the glucose oxidation might be represented as in Scheme II (p. 1059). In this sequence, glucose is converted to pyruvate in the cell cytoplasm; within the cell mitochondria, pyruvate is oxidized to CO_2.

---

[11] The formation of ATP^{4-} from AMP^{2-} or ADP^{3-} is, of course, strongly *endergonic*. Thus, ATP^{4-} synthesis is "coupled" to the glucose oxidation by various intermediate reactions, most of which, in fact, require electron transfer (redox) steps involving transition metal ion complexes.

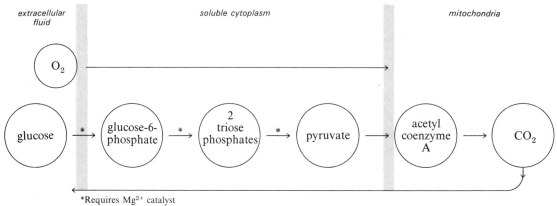

*Requires Mg^{2+} catalyst

Scheme I

Step 1 involves phosphorylation of the 6-OH group of glucose by ATP through an enzyme requiring Mg^{2+}. Step 2 involves isomerization of glucose-6-phosphate to fructose-6-phosphate (using an enzyme with no known metal ion requirement) followed by Mg^{2+}-catalyzed phosphorylation at the 1-OH group. The six-carbon fructose diphosphate is cleaved by another enzyme (no metal necessary) to the triose phosphates—dihydroxyacetone phosphate and glyceraldehyde 3-phosphate; the latter is the precursor to pyruvate (the triose phosphates are equilibrated by yet another enzyme). Step 4 represents the oxidative phosphorylation of glyceraldehyde-3-phosphate by $NAD^{\oplus}$ using HPO_4^{2-}.[12] In step 5, another enzyme requiring Mg^{2+} transfers the 1-phosphate group of 1,3-diphosphoglycerate to ADP to form ATP and 3-phosphoglycerate. Step 6 actually involves two steps: an isomerization of 3-phosphoglycerate to 2-phosphoglycerate (no metal ion), followed by Mg^{2+}-catalyzed dehydration of the latter to *phosphoenolpyruvate* (PEP). Mg^{2+} is again encountered in the enzyme *pyruvate kinase*, which dephosphorylates PEP by phosphate transfer to ADP with formation of pyruvate.

This rather elaborate process requires Mg^{2+} in five different enzymes, all of which involve phosphate transfer *from* ATP or *to* ADP. The net overall reaction (called the Embden-Meyerhof pathway) is

$$\text{glucose} + 2NAD^{\oplus} + 2ADP^{3-} + 2HPO_4^{2-} \rightarrow$$
$$2 \text{ pyruvate}^- + 2NADH + 2ATP^{4-} + 2H^+ + 2H_2O$$

and partially accounts for the replenishment of the ATP consumed by other cell reactions.

Pyruvate Kinase

The last step in the Embden-Meyerhof pathway for the breakdown to pyruvate is catalyzed by the enzyme pyruvate kinase, which requires Mg^{2+} (or Mn^{2+}) and is activated by K^+. A proposed mechanism[13] for the transfer of the 3-phosphate of PEP to ADP is shown in Figure 19-3. The principal features of the reaction are:

a) coordination of Mg^{2+} by the enzyme;
b) coordination of the terminal phosphate of ADP and the phosphate of PEP to Mg^{2+};
c) "initiation" of the transfer of phosphate from PEP to ADP by K^+ coordination to the PEP carboxylate group (K^+ is enzyme coordinated, too);
d) nucleophilic attack by an ADP oxygen upon the PEP phosphate;
e) concomitant or subsequent H^+ transfer (from an adjacent enzyme site) to the vinyl carbon of PEP to generate the keto form of pyruvate.

---

[12]$NAD^{\oplus}$ is an organic oxidizing agent (nicotinamide adenine dinucleotide), which acts as a H^- acceptor at the *para*-position of a pyridyl ring. Nicotinamide (*m*-$CONH_2$pyridine) and adenine are the base units of ribose nucleosides bound to a $P_2O_7^{2-}$ unit. See p. 1079.

[13]See ref. 10.

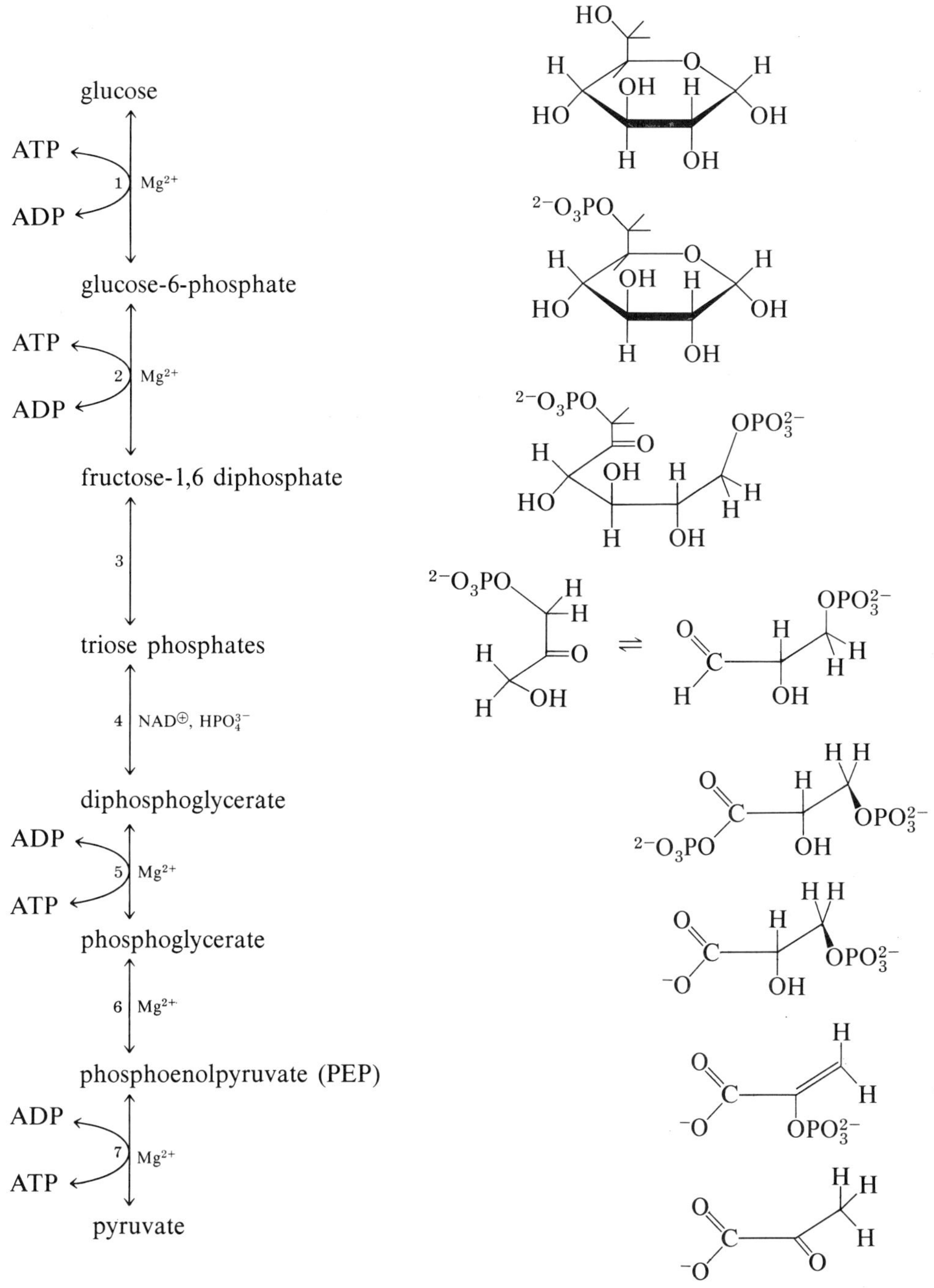

PROCESSES COUPLED TO PHOSPHATE HYDROLYSIS   1059

Scheme II

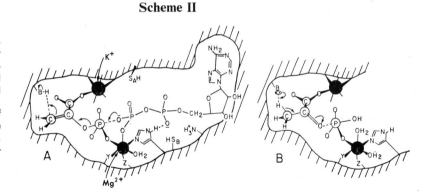

Figure 19–3. A proposed mechanism for phosphate transfer from PEP to ADP by pyruvate kinase. The enzyme is represented by the shaded boundary, and some of its attached functional groups are shown. (A) shows the PEP → ADP transfer of phosphate; (B) shows a regeneration of PEP from pyruvate + phosphate. [From A. S. Mildvan and C. M. Grisham, Structure and Bonding, **20**, 1 (1974).]

Glucose Storage—Phosphoglucomutase

When it is not required for immediate combustion, the glucose-6-phosphate from step 1 of the Embden-Meyerhof pathway may be converted to *glycogen* in muscle tissue for later use. Glycogen is an amylose polymer (**8**) made by linking glucose carbons 1 and 4 through oxo-bridges.

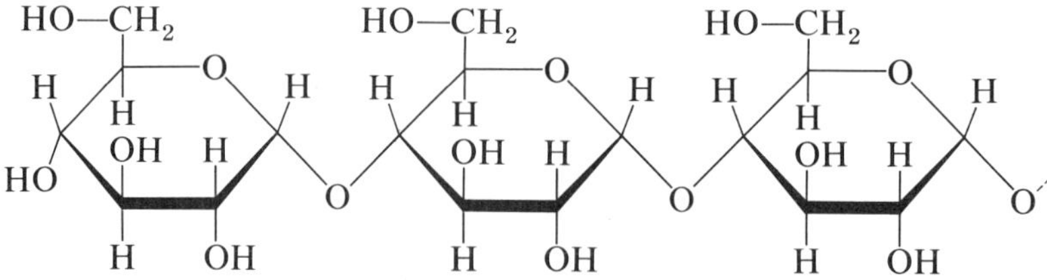

8

There are three steps in this process, each of which requires Mg^{2+}. The first of these, to be outlined in the next paragraph, is an interesting net rearrangement of glucose-6-phosphate to glucose-1-phosphate by the Mg^{2+}-requiring enzyme *phosphoglucomutase*.

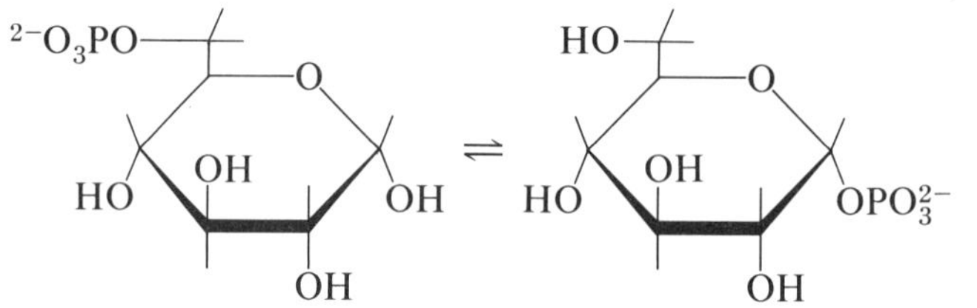

The second step involves displacement by glucose-1-phosphate of pyrophosphate ($HP_2O_7^{3-}$) from uridine triphosphate [UTP is a nucleoside triphosphate like ATP, where uridine (**9**) occurs in place of adenosine].

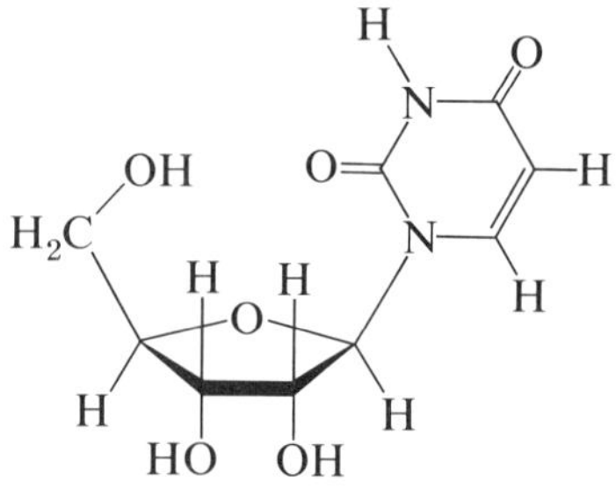

9

The product of this second step, uridine diphosphate glucose (**10**), is cleaved by another Mg^{2+}-requiring enzyme to form UDP, with addition of another sugar unit to the amylose polymer (glycogen).[14]

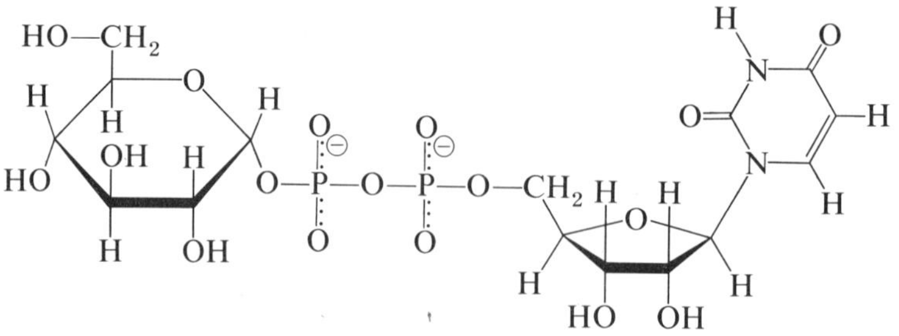

10

---

[14] Note the regiospecificity of this reaction; the glucose ester bond, not the ribose ester bond, is cleaved. Note also that, in spite of equations 19-1 through 19-3, the $P_2O_7^{2-}$ unit is not cleaved.

The first step (the 1- → 6-OPO$_3^{2-}$ "migration") in this process actually proceeds through small amounts of glucose-1,6-diphosphate.[15] The formation of the diphosphate and the net "migration" require Mg^{2+} and a serine unit (**11**) in the enzyme.

$$\left(\text{HO—CH}_2\text{—}\overset{\overset{\displaystyle H}{|}}{\underset{\underset{\displaystyle NH—}{|}}{C}}\text{—}\overset{\overset{\displaystyle O}{\|}}{C}\text{—} \right)$$

11

As shown in Figure 19-4, the serine OH group is likely coordinated to Mg^{2+}, as is the 3-OH group of the glucose diphosphate. Conveniently, the enzyme contains two sites at which the 1,6-phosphate groups may also be "bound"—the one at ~5.5 Å from the metal is where the transfer to serine occurs, and the more remote position (10 Å) serves to orient the glucose for the transfer to serine. A proposed sequence of events is as follows:

a) phosphorylation of the enzyme serine unit by, for example, ATP (this step is not shown in Figure 19-4);

b) coordination of glucose-6-phosphate of Mg^{2+} at the 3-OH group and "binding" of the 6-phosphate to the enzyme site 10 Å from Mg^{2+} (this step is not shown);

c) phosphorylation of the glucose 1-OH by serine phosphate (Figure 19-4 depicts the situation after the phosphorylation);

d) rotation of the glucose-1,6-diphosphate about the Mg—OH bond;

e) displacement of the 6-phosphate by the serine OH;

f) release from the enzyme of glucose-1-phosphate for the reaction with UTP, leaving behind a serine phosphate ready for the next glucose-6-phosphate.

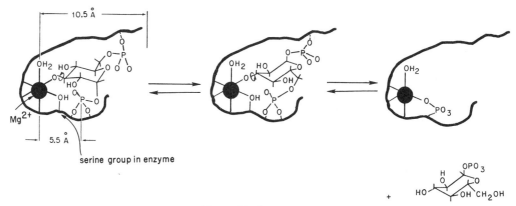

Figure 19-4. A schematic of the proposed action of phosphoglucomutase and Mg^{2+} on glucose-1,6-diphosphate. [From A. S. Mildvan and C. M. Grisham, Structure and Bonding, **20**, 1 (1974).]

Phosphate Storage in Muscle—Creatine Kinase

It is known that hydrolysis of ATP to ADP in muscle provides a "half-reaction" with sufficient free energy loss to drive the "half-reaction(s)" for muscle contraction. However, the concentration of ATP normally found in cells (a few millimoles per kilogram of tissue) is rapidly depleted by muscle contraction, and the body needs a quickly accessible store of "high energy" phosphate; this store is provided by the *phosphagens*. *Phosphocreatine* (**12**) is the phosphagen found in most vertebrates.

---

[15] See ref. 10.

12

Phosphocreatine, in conjunction with the Mg^{2+}-requiring enzyme creatine kinase, effects the interconversion of ADP and ATP (see **12'**). Figure 19-5 depicts the mechanism[16] of this reaction (top to bottom indicates the formation of phosphocreatine from ATP, while the reverse sequence is the pathway used when the ADP concentration builds up during muscle contraction).

---

[16] A. S. Mildvan and M. Cohn, in "Advances of Enzymology," F. F. Nord, ed., Vol. 33, Interscience Publishers, New York (1970).

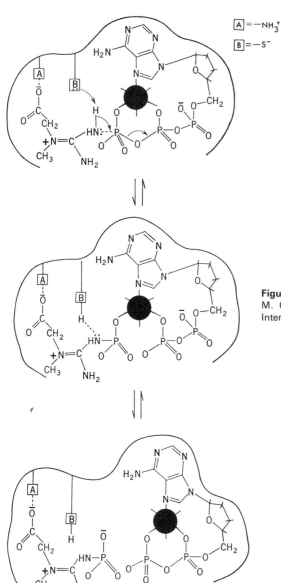

Figure 19-5. The creatine kinase mechanism. [From A. S. Mildvan and M. Cohn, in "Advances in Enzymology," F. F. Nord, Ed., Vol. 33, Interscience Publishers, New York (1970).]

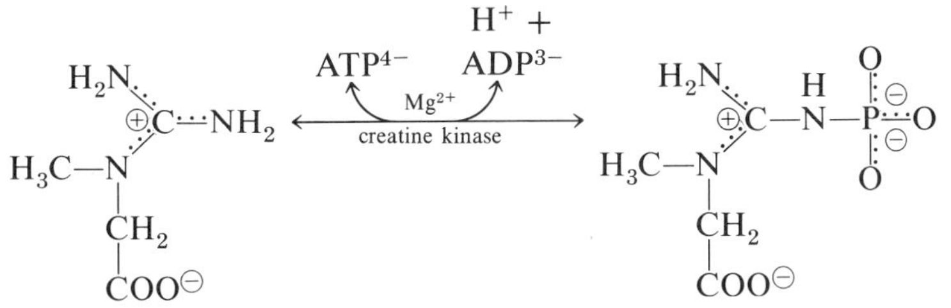

$$\text{[creatine]} \underset{\text{creatine kinase}}{\overset{\text{ATP}^{4-} \quad \text{H}^+ + \text{ADP}^{3-}}{\rightleftharpoons}} \text{[phosphocreatine]}$$

12′

The enzyme features of note in Figure 19-5 are a cationic group (NH_3^+), A, to bind the carboxylate of creatine, and a sulfhydryl group, B (the SH group is essential to the enzyme action), to assist attack by NH_2 upon the terminal phosphate of ATP. Note the coordination of the ATP *terminal* and *central* phosphates to Mg^{2+}. This should also assist the NH_2 attack.

Na^+/K^+ Ion Pump—ATPase

One of the more interesting applications of ATP hydrolysis is brought into play by the cell to "pump" Na^+ out of the cytoplasm and K^+ into the cytoplasm.[17] The K^+ concentration within a cell is maintained at ~ 0.1 M so that K^+ is available for glucose combustion (recall the discussion of pyruvate kinase) and for protein synthesis by ribosomes. The Na^+/K^+ gradients are particularly important to the operation of nerve cells, too. This pumping must be done against unfavorable Na^+ and K^+ concentration gradients and so constitutes a good example of coupling (by ATPase) of a highly exergonic half-reaction (ATP → ADP) to an otherwise endergonic process. The rather complex overall process (Figure 19-6) has several key elements.

Species A has been identified[18] as an intermediate and is characterized by the

---

[17]See ref. 10.
[18]R. L. Post, *et al., J. Gen. Physiol.*, **54,** 306s (1969).

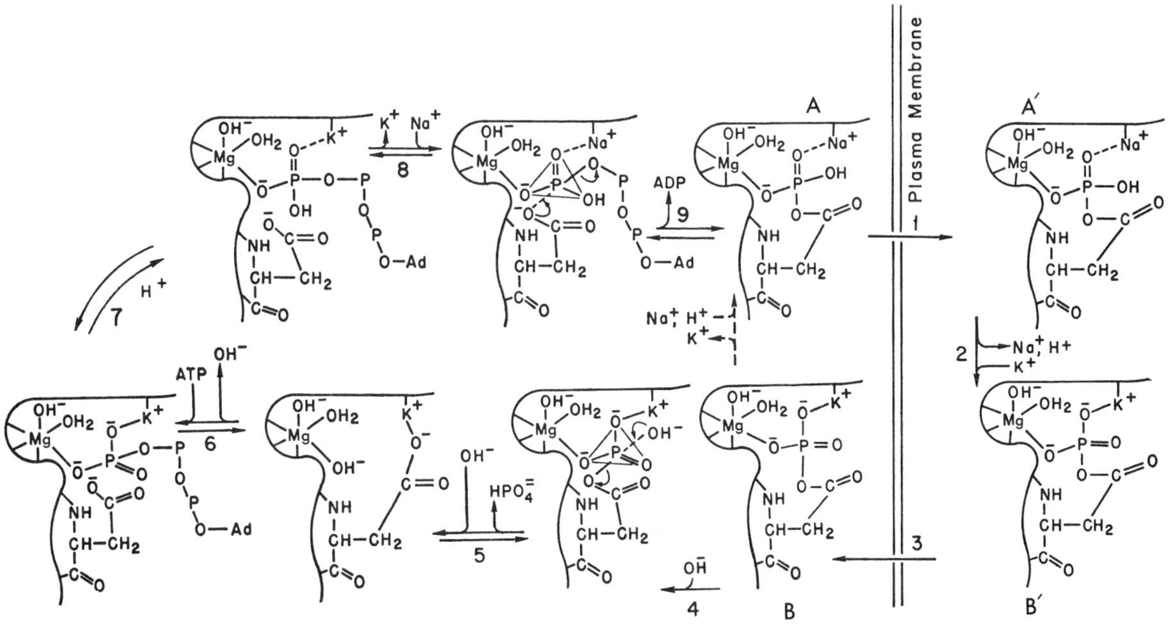

Figure 19-6. A mechanism for Na^+/K^+ transport by ATPase. [From A. S. Mildvan and C. M. Grisham, Structure and Bonding, **20,** 1 (1974).]

mixed phosphoanhydride with aspartic acid from the enzyme. This intermediate is believed to be responsible for Na^+ transport to the outside of the cell. The second key feature is the deprotonation of the phosphate group in the extracellular plasma. As implied by step 2, this favors the replacement of Na^+ by K^+. The K^+ complex returns to the cytoplasm, whereupon phosphate is returned to the cytoplasm (step 5) and ATP enters (step 6). Protonation of the terminal phosphate of ATP (now bridging Mg^{2+} and K^+) leads to a reversal of the Na^+/K^+ exchange (step 8) experienced outside the cell (step 2). Displacement of adenosine diphosphate from the terminal phosphate by aspartate returns the sequence to the intermediate species A, and the cycle may then be repeated. The role of steps 4 through 9 is simply to reverse (in the cytoplasm) the exergonic Na^+, H^+/K^+ exchange that occurs outside the plasma membrane. Currently puzzling are the triggering of Na^+/K^+ exchange by H^+, and the phosphoanhydride hydrolysis in B but not in A. There is some interesting "inorganic" chemistry at work here!

STUDY QUESTIONS

1. At what point in the reaction sequence of Figure 19-2 does o-phen coordination of Zn^{2+} block the oligomerization phenomenon?

2. Consult a biochemistry book in your library, if necessary, and depict the structures of dATP, dGTP, dCTP, and dTTP.

3. Refer to Chapter 7 for ideas on cleavage of non-metal oxyanion E—O bonds. Do you suppose that at pH > 7.4 the ATP hydrolysis of equation 19-1 is favored only in a thermodynamic sense?

4. Draw structures for $NAD^{\oplus}$ and NADH (see footnote 12).

5. The following data for $\Delta G°'$ of hydrolysis at pH = 7.0 and T = 35°C are taken from Lehninger (see footnote 6):

PEP	−14.8 kcal/mole
1,3-diphosphoglycerate	−11.8 kcal/mole
ATP^{4-}	−7.30 kcal/mole

What are the $\Delta G°'$ values for phosphate transfer from PEP and 1,3-diphosphoglycerate to ADP^{3-} (steps 5 and 7 of the Embden-Meyerhof pathway)?

6. Regarding the mechanism for phosphoglucomutase, it was suggested in the text that the enzyme serine unit is prepared for the entry of glucose-6-phosphate by prior phosphorylation ($H_2PO_4^-$ or ATP^{4-}). This is not necessarily the case. Describe the enzyme action upon glucose-1,6-diphosphate, presuming the latter to have arisen elsewhere in the biosystem.

7. Refer to Figure 19-6 for the action of ATPase transport of Na^+/K^+. Do structures A and B carry the same charge? Write a balanced equation for the net chemical reaction *inside* the cell membrane. Factor this equation into two simpler equations. One of the latter is an equation for the simple, direct conversion of species B to species A. Why, then, is it necessary to invoke steps 4 through 9?

OXYGEN CARRIERS—HEMOGLOBIN AND MYOGLOBIN[19]

Probably the most familiar examples of bioinorganic chemistry are the O_2 carriers so necessary for getting oxygen to the cell (hemoglobin) and into the cell

---

[19] J. M. Rifkind in Vol. 2, p. 832 of ref. 1; a recent review of model systems is G. McLendon and A. E. Martell, *Coord. Chem. Revs.*, **19**, 1 (1976).

(myoglobin) for glucose oxidation. Before beginning a discussion of their modes of action, we need to characterize their composition. The iron complexes defining the heme group are those of *protoporphyrin IX* (13) (abbreviated PIX).[20] The porphyrins are a class of nitrogen macrocycles derived from *porphin* (14), an unsubstituted tetra-pyrrole connected at the α-carbons by methylidyne (CH) bridges.[21]

Coordination of Fe^{2+} by the four nitrogens of PIX produces uncharged *heme* (heme carries a dinegative charge at pH 7 because the carboxyl groups are ionized). Since heme is irreversibly air oxidized in H_2O to the Fe^{3+} complex (called *hemin*), heme by

---

[20] The dimethylester (PIXDME), with the carboxylic acid groups esterified, is a commonly encountered derivative.

[21] These methylidyne carbon positions are variously labeled the α, β, γ, δ and 5, 10, 15, 20 positions of porphin and porphyrins. A commonly used synthetic porphyrin is the 5,10,15,20-tetraphenyl derivative (TPP), desirable for *in vitro* studies because of its ease of synthesis and purification.

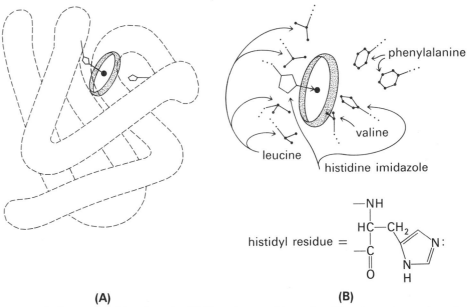

Figure 19-7. Hemoglobin schematic. (A) Heme in a peptide crevice; (B) a closer view to show the immediate hydrocarbon environment of Fe.

itself cannot act as an O_2 carrier *in vivo*. Nature avoids this catastrophe by embedding heme in a protein crevice where it is surrounded by hydrophobic groups from amino acids. A schematic of heme buried in peptide is shown in Figure 19-7, where you see that the Fe^{2+} in hemoglobin finds itself close to a nitrogen of one imidazole group (the proximal histidine) and further from another (the distal histidine). The proximal histidine is coordinated to the Fe^{2+}, which actually lies ~ 0.8 Å above (toward the proximal histidine) the mean porphyrin plane (the five-coordinate Fe^{2+} finds itself in a square pyramidal environment of idealized C_{4v} symmetry).

The role of the peptide portion of hemoglobin in preventing the irreversible O_2 oxidation of Fe^{2+} is thought to be twofold: (i) the hydrocarbon environment of the heme pocket has a low dielectric constant and so acts as a nonpolar "solvent" that cannot support the ionic charge separation developed at the transition state of Fe^{2+} oxidation,[22] and (ii) the sterically protected environment of the heme prevents formation of an intermediate μ-oxo-heme dimer, which appears to be a mechanistically necessary intermediate in the irreversible oxidation.[23] (The μ-oxo dimer may well be preceded by a μ-peroxo dimer; the formation of $[(H_3N)_5CoO_2Co(NH_3)_5]^{4+}$ (Chapter 13, p. 746) is an analog from conventional coordination chemistry.)

Various studies support both concepts. The 1-(2-phenylethyl) imidazole heme diethylester embedded in an amorphous mixture of polystyrene shows reversible O_2 uptake.[24] Similarly, Fe(TPP) bound to an imidazole, itself attached to a silica gel support (as in **15**), also exhibits reversible O_2 uptake.[25] Recent work with the "picket fence" porphyrin **16** shows that it, too, provides a proper environment for reversible O_2 coordination of Fe^{2+} without oxidation.[26]

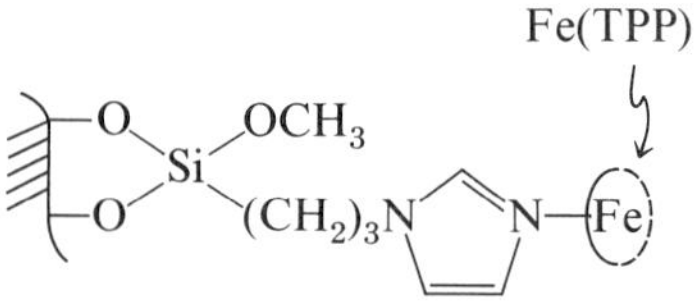

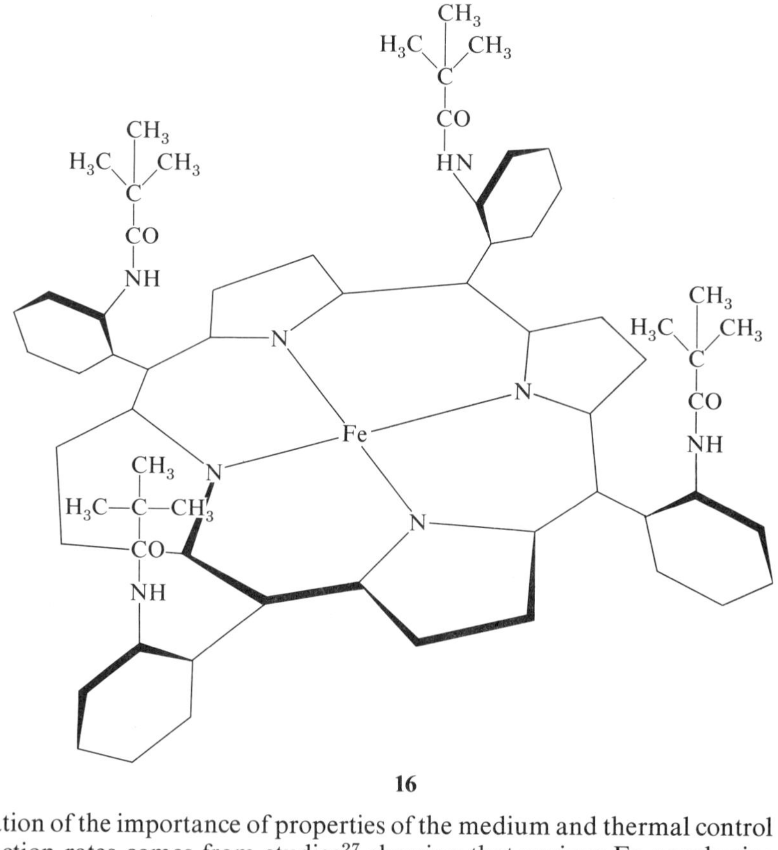

Further indication of the importance of properties of the medium and thermal control of relative reaction rates comes from studies[27] showing that various Fe porphyrins

[22] J. H. Wang, et al., *J. Amer. Chem. Soc.*, **80**, 1109 (1958).
[23] J. O. Alben, et al., *Biochemistry*, **7**, 624 (1968); I. A. Cohen and W. S. Caughey, *ibid.*, **7**, 636 (1968).
[24] J. H. Wang, *J. Amer. Chem. Soc.*, **80**, 3168 (1958).
[25] O. Leal, et al., *J. Amer. Chem. Soc.*, **97**, 5125 (1975).
[26] J. P. Collman, et al., *J. Amer. Chem. Soc.*, **96**, 6522 (1974).
[27] C. K. Chang and T. G. Traylar, *J. Amer. Chem. Soc.*, **95**, 5810 (1973); see also ref. 32.

may be caused to reversibly bind O_2, without formation of the μ-oxo Fe^{3+} dimer, at low temperatures in non-aqueous solvents.

The protein structures in hemoglobin consist of a peptide backbone with various side chains. These side chains define a "surface" for the polymer consisting of a variety of nonpolar (hydrocarbon), cationic (*e.g.*, $-NH_3^+$), and anionic (*e.g.*, $-CO_2^-$) groups. Hydrogen bonding between N—H and O=C groups of the peptide backbone units and interactions between nonpolar regions of the polymer surface determine the polymer structure (generally, of two kinds—a helix or a pleated sheet). Myoglobin (Mb) has a peptide surface not conducive to self-association, but hemoglobin (Hb) does. Consequently, Mb is monomeric whereas Hb is a tetrameric unit (of C_2 symmetry) consisting of two α- and two β-peptide chains (actually folded helices; Figure 19-7 depicts one of these units). Figure 19-8 depicts the tetramer segments as protein masses at the corners of a distorted tetrahedron.

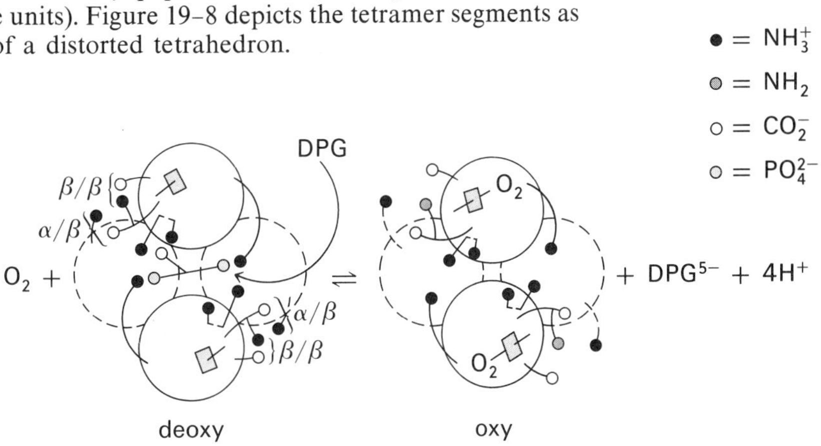

Figure 19-8. An idealization of the deoxyHb $\rightleftharpoons$ oxyHb structure change, including the role of DPG^{4-} and H^+ ionization. Two of the four protons released on forming oxyHb arise from disruption of α/α salt bridges not shown. Each protein mass contains a heme unit.

● = NH_3^+
◉ = NH_2
○ = CO_2^-
○ = PO_4^{2-}

$O_2 +$ (deoxy) $\rightleftharpoons$ (oxy) $+ DPG^{5-} + 4H^+$

X-ray studies of deoxyHb and oxyHb have revealed what appear to be highly significant changes in $-NH_3^+ \cdots -O_2C-$ interactions within and between peptide chains.[28] Specifically, there are eight of these salt bridges in deoxyHb, which are disrupted upon coordination of O_2 to Fe. Two disruptions occur *within* the β-chains, four *between* the α-units, and one *between* each α,β pair. Only the β/β and α/β bridges are depicted in Figure 19-8. Also shown is diphosphoglycerate (DPG, see p. 1059), bridging the β helices in deoxyHb. To condense an otherwise complex description, the disruption of the DPG and $-NH_3^+ \cdots -O_2C-$ hydrogen bond interactions accompanies oxygenation of Hb, and the β chains move closer together along their common edge of the tetrahedron.

It is presently believed that the peptide chains are constrained by these salt bridges (all eight of them) in deoxyHb. In this "tense" form the sixth coordination position about Fe^{2+} is directed "inward" and sterically blocked by a valine[29] group [Fig. 19-7(B)]. The bridge disruption incurred on coordination of one or two heme units by O_2 is associated with a conformational change in which this sixth position is exposed and the inter- and intra-protein salt bridges are broken. The entire "tense" tetramer then relaxes so as to expose the sixth positions about the other heme iron atoms. This phenomenon is referred to as *cooperativity* between heme units. For example, myoglobin, a single-heme protein found within cells, takes up O_2 in a 1:1 ratio with Fe^{2+} according to a normal relationship between the degree of complexation and O_2 pressure. This is shown in Figure 19-9 as a plot of percentage oxymyoglobin as a function of O_2 pressure, p_{O_2}. By way of contrast, Figure 19-9 also depicts the same sort of plot for hemoglobin. Comparison of the curves should

---

[28] R. E. Dickerson, *Ann. Rev. Biochem.*, **41**, 807 (1972).
[29] Valine is a terminal group here, with a structure

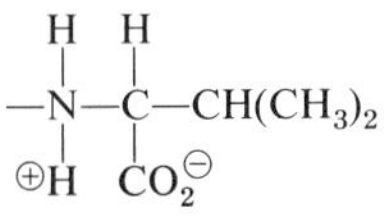

$$-\underset{\overset{|}{\oplus H}}{\overset{\overset{H}{|}}{N}}-\underset{\overset{|}{CO_2^{\ominus}}}{\overset{\overset{H}{|}}{C}}-CH(CH_3)_2$$

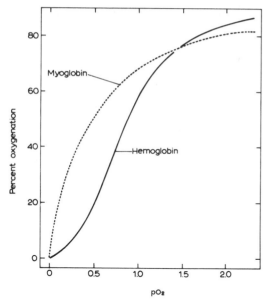

Figure 19–9. Oxygen saturation curves of myoglobin and hemoglobin. [From J. M. Rifkind, in "Inorganic Biochemistry," G. L. Eichhorn (Ed.), Vol. 2, Elsevier, New York (1973) p. 832.

convince you that greater p_{O_2} is required to oxygenate Hb than Mb, or that hemoglobin is less efficient at O_2 uptake under low O_2 pressures, where myoglobin is very efficient. In muscle tissue, for example, where p_{O_2} is small, there is a thermodynamically favorable O_2 transfer from oxyhemoglobin to myoglobin to pass O_2 into the cell.

The oxygenation equilibrium for myoglobin is represented as

$$\text{Fe} + \text{O}_2 \rightleftharpoons \text{FeO}_2 \quad \text{with } K = \frac{[\text{FeO}_2]}{[\text{Fe}]p_{O_2}}$$

Defining the fraction of O_2-bearing iron as $f = [\text{FeO}_2]/\{[\text{Fe}] + [\text{FeO}_2]\}$, then

$$K = f/\{(1-f)p_{O_2}\}$$
$$f = Kp_{O_2}/(1 + Kp_{O_2})$$
$$\frac{1}{f} = 1 + \left(\frac{1}{K}\right)\frac{1}{p_{O_2}}$$

The myoglobin curve in Figure 19–9 corresponds to this equation. The hemoglobin curve does not follow such an equation, but it does follow an empirically modified form with p_{O_2} replaced by $p_{O_2}^n$, where the exponent n is called the Hill constant. For hemoglobin, $n = 2.8$; this reflects the degree of cooperativity between heme units.[30]

The salt bridges between and within the peptide chains provide a means for control of O_2 transfer from oxyHb to Mb. First of all, the cooperativity effect is pH-dependent, as shown in Figure 19–10(A). This is called the *Bohr effect,* and it is known not to involve acid dissociation of ligands bound to Fe^{2+} in hemoglobin. Rather, $-\text{NH}_3^+$ sites involved in α/α and β/β bridges (see Figure 19–8 for the β/β sites) are more strongly acidic when they no longer form hydrogen bonds to the carboxylate groups. Consequently, basic conditions remove these protons and shift the deoxyHb $\rightleftharpoons$ oxyHb equilibrium in favor of the oxy-form. The acid condition (lactic acid) in working muscle tissues therefore facilitates the release of O_2 from oxyHb. Secondly, deoxyHb has a greater affinity for phosphate ester anions (such as 2,3-diphosphoglycerate, DPG) and EDTA than does oxyHb. Both anions bear several negatively charged oxygen atoms and are large enough to span cationic sites (*e.g.*, $-\text{NH}_3^+$) between the β-protein chains. The effect of the hydrogen bonding of these

---

[30] See ref. 19.

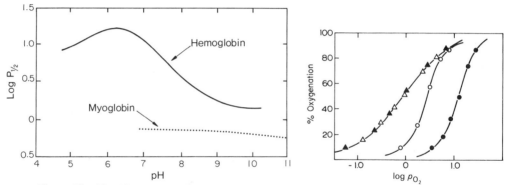

Figure 19—10. (A) The Bohr effect; (B) the effect of 2,3-diphosphoglyceric acid on Hb-O_2 uptake [the two leftmost curves are for α and β chains alone, both with and without 2,3-DPG; the two rightmost curves are for hemoglobin without (○) and with (●) 2,3-DPG]. [From J. M. Rifkind, in "Inorganic Biochemistry," G. L. Eichhorn (ED.), Elsevier, New York (1973), p. 832.]

anions to the cationic sites is to hold the β-chains apart, in their deoxy configuration, and facilitate O_2 loss. Again, in working tissues the DPG concentration is fairly high (refer to the discussion on page 1058 of glucose combustion) and favors the transfer of O_2 to myoglobin. Figure 19-10(B) confirms that these effects are peculiar to the tetrameric peptide chains of Hb, for in neither Mb nor the individual Hb peptide units is O_2 uptake affected by pH or DPG.

The link between O_2 coordination and salt bridge disruption is currently being studied. One hypothesis is centered on the characteristic feature that formation of oxyhemoglobin or oxymyoglobin causes the conversion of five-coordinate, high spin Fe^{2+} to six-coordinate, low spin Fe^{2+}, with Fe^{2+} being pulled into the porphyrin plane.

It was originally thought that the proximal histidine, in following this Fe^{2+} displacement of ~0.8 Å, also shifted ~0.8 Å and so caused a conformational change throughout the peptide; this in turn was postulated to alter the interpeptide salt bridges (as noted above) and result in crevice opening.[31] More recent work[32] with Co^{2+} in place of Fe^{2+} (coboglobin, CoHb) reveals, however, the same degree of cooperativity ($n \approx 2.5$) as found in hemoglobin, with less than one-half the histidine displacement. Consequently, the problem is not solved and the search continues.

Of more direct interest to the coordination chemist is the nature of the O_2-Fe^{2+} interaction and its stereochemistry. It would be desirable to study the Fe-O_2 interaction in oxyHb and various models, but definitive studies have proven difficult. As a result, studies of Co^{2+} analogs have been pursued, with interesting results. In both the Fe and Co systems the M—O—O group is angular[33] and a simple mo

---

[31] M. F. Perutz and C. F. Ten Eyck, *Cold Spring Harbor Symp. Quant. Biol.*, **36**, 315 (1971).
[32] F. Basolo, B. M. Hoffman, and J. A. Ibers, *Accts. Chem. Res.*, **8**, 384 (1975) and references therein.
[33] L. D. Brown and K. N. Raymond, *Inorg. Chem.*, **14**, 2595 (1975); see also ref. 32.

model[34] depicts a σ interaction between the metal d_{z^2} ao and one of the O_2 π^* mo's. A π interaction between d_{xz}, say, and the orthogonal O_2 π^* mo is also possible. (Other discussions of M—O_2 bonding can be found in Chapter 5, p. 206, and Chapter 17, p. 945.)

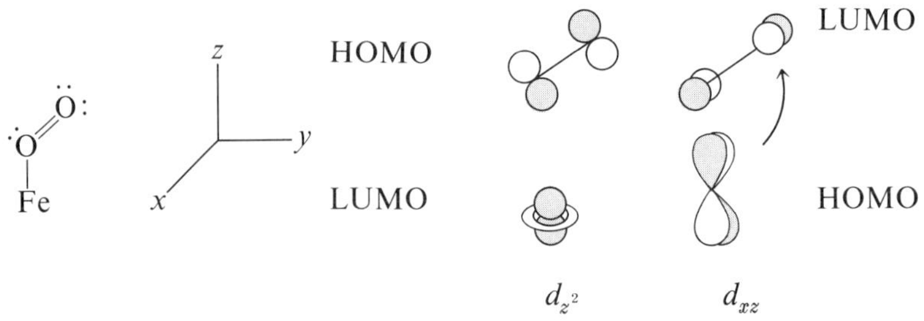

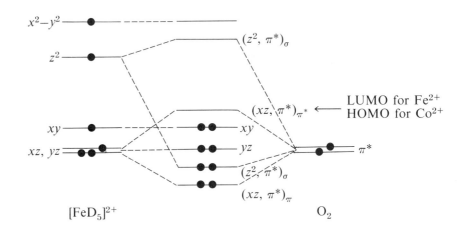

Viewing the oxy-complexes as adducts of O_2 with a square pyramidal complex, these M—O_2 orbital interactions result in the following orbital scheme:

With Fe^{2+} as the metal, there are eight electrons to populate these mo's, giving a configuration ... $(xy)^2$. Notice that both the $(z^2/\pi^*)_\sigma$ and $(xz/\pi^*)_\pi$ mo's are occupied. Since the σ and π bonding electron pairs are polarized in the sense $Fe^{\delta+}/O_2^{\delta-}$, there is a partial oxidation of Fe^{2+}, and the π interaction would seem to be an important feature of the M—O_2 bonding. An equivalent view of the Fe—O_2 interaction is that of coordination of low spin, d^6 Fe^{2+} by a spin-singlet O_2 molecule with a configuration ... $(\pi^*)^2(\pi^*)^0$. The π^* pair then serves a normal Lewis base function to form a σ bond to Fe^{2+}, while the empty π^* mo acts as a π acceptor orbital.

On the other hand, with Co^{2+} you find nine electrons to occupy the mo's of the scheme, and this means that the $(xz/\pi^*)_{\pi^*}$ mo is half-occupied. The higher nuclear charge of Co^{2+}, however, seems to cause a subtle, but important, change in the nature of the orbital interactions. Because of the higher Z^* for cobalt, the metal d ao's are contracted and lowered in energy, relative to those of iron. This translates into weaker (xz/π^*) interaction (weaker π bonding) from both overlap and energy match criteria. The (z^2/π^*) overlap is similarly diminished, but the energy match improves (it's hard to tell whether the σ bonding changes much).

---

[34] A. Dedieu, et al., J. Amer. Chem. Soc., **98**, 5789 (1976).

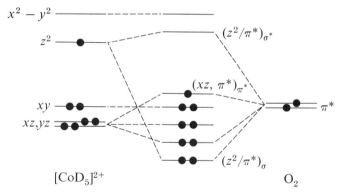

With this background, it is not terribly surprising that electron spin resonance studies[35] of Co^{2+}—O_2 reveal the unpaired electron to reside in a molecular orbital nearly localized on O_2. That is, the xy, yz, and $(xz,\pi^*)_\pi$ electron pairs are completely or highly localized on Co, while the $(z^2/\pi^*)_\sigma$ and $(xz,\pi^*)_{\pi^*}$ electrons are more highly concentrated on O_2. This makes it appear as though Co^{2+} has been oxidized to Co^{3+} (low spin, d^6), so that an alternate view of the Co—O_2 interaction is that of Co^{3+} with O_2^-.

The role of the other ligands bound to the metal has received some consideration. Again with cobalt, studies[36] with model five- and six-coordinate systems support the above concept, for they reveal an over-all correlation between O_2 uptake, ease of

$$Co(L)B^{2+} + O_2 \rightleftharpoons Co(L)B(O_2)^{2+}$$

Co^{2+} oxidation, and (roughly) the pK_a of the group axially coordinated to Co^{2+} (Figure 19-11).

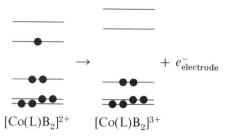

[see Figure 19-11(A)]   [see Figure 19-11(B)]

These correlations can be interpreted as follows. $E_{1/2}$ measures the ease of removal of the z^2 electron from *six*-coordinate $Co(L)B_2^{2+}$ and roughly corresponds to the electron transfer from z^2 to $O_2\pi^*$ noted above

(the oxidation amounts to intermolecular electron transfer to an orbital in the electrode, rather than intramolecular transfer to the $O_2\pi^*$ mo). Thus, the correlation in Figure 19-11 is not surprising and reflects the destabilization of the z^2 electron as the donor ability of B increases.

---

[35] B. M. Hoffman and D. H. Petering, *Proc. Nat. Acad. Sci. U.S.A.*, **67**, 637 (1970); J. C. W. Chien and L. C. Dickinson, *ibid.*, **69**, 2787 (1973); L. C. Dickinson and J. C. W. Chien, *J. Biol. Chem.*, **284**, 5005 (1973); B. S. Tovrog et al., *J. Amer. Chem. Soc.*, **98**, 5144 (1976).

[36] See ref. 32.

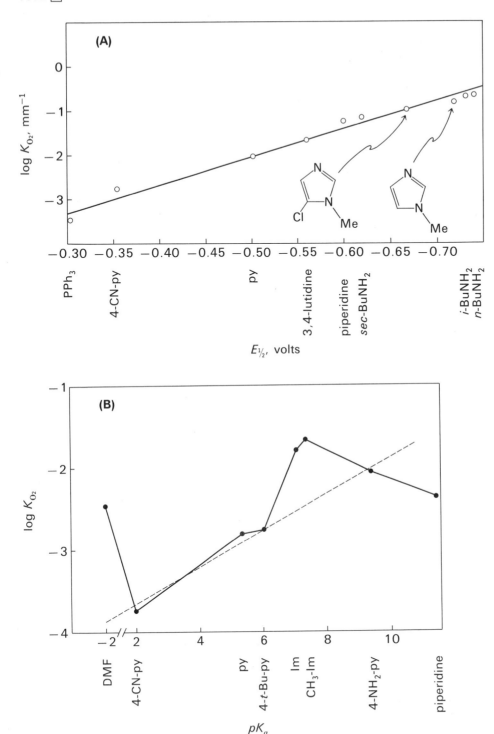

Figure 19-11. Correlations between K_{O_2} for Co(L)B (ordinate of both graphs) and (A) $E_{\frac{1}{2}}$ (polarographic half-wave oxidation potentials) for Co(L)B$_2$, and (B) pK_a of the fifth ligand in Co^{2+}(L)B.

The order of decreasing K_{O_2} and $E_{1/2}$ of n-BuNH$_2$ > 1-methylimidazole > piperidine does not, however, follow the order of amine bascities toward the proton (pK_a's: 10.6 > 7.3 < 11.3). Thus, 1-methylimidazole promotes O$_2$ coordination and Co(L)B$_2$ oxidation to an extent greater than anticipated from its proton basicity, and this is attributed to its π-donor nature. The ease of z^2 electron removal derives not only from direct σ interaction with the axial base but also from an indirect destabilization of that electron by axial-base π-donor effects. The correlation of K_{O_2} with pK_a is generally less satisfactory, but is excellent for a series of related *para*-substituted pyridine ligands. Here again, increased axial ligand donor ability does tend to enhance O_2 uptake.

STUDY QUESTIONS

8. CO and CN^- bind to heme with similar affinities. However, CO has greater affinity for hemoglobin and myoglobin than for heme, while CN^- does *not* bind well to the heme group in the proteins [as an aside, CN^- does bind to oxidized Fe(III)Hb, called methemoglobin]. How might solvation energies and the nature of the "heme pocket" in the proteins explain this marked reversal in CN^- and CO ligation? In the context of your answer, does the heme pocket environment support CN^- coordination of metHb?

9. Further regarding solvent effects on O_2 uptake, the K_{O_2} values for Co(L)B in toluene increase on changing the solvent to the more polar *N,N*-dimethylformamide. How does this support the idea that there is net Co→O_2 electron drift in the oxygenated complex?

10. With reference to the low spin, *five*-coordinate Co^{2+}(L)B complexes of Figure 19-11, is the metal-centered HOMO a σ or π type mo (axial bonding by B is weaker than equatorial by L, so that $e_\sigma^{ax} < e_\sigma^{eq}$)? How does the energy of this mo vary as the σ donor strength of B increases? If B is a good π donor, will the HOMO electron be easier or harder to remove? Should the HOMO energy also depend on the donor strengths of the equatorial ligands? If so, to a greater or lesser extent than on the axial ligands?

11. The fact that organically grown plants may contain high levels of nitrate/nitrite was recently realized in the case of an infant fed carrot juice from carrots grown in the Florida Everglades. The nitrite stored in the carrots oxidized most of the infant's hemoglobin to methemoglobin (Fe^{3+}Hb). This oxidation is dangerous because O_2 has little affinity for methemoglobin. What is there about the Fe—O_2 interaction that explains this reduced affinity?

COBALAMINS; VITAMIN B_{12} COENZYME[37]

Cobalamin has a structure that is in some ways similar to that of myoglobin, but there are significant differences. The macrocycle (**17**) binding Co^{3+} in this case is like a porphyrin but is less regular. Specifically, one of the methylidyne carbons of porphyrin is missing and there is only one NH group at the center. This basic tetrapyrrole macrocycle is called a *corrin*.

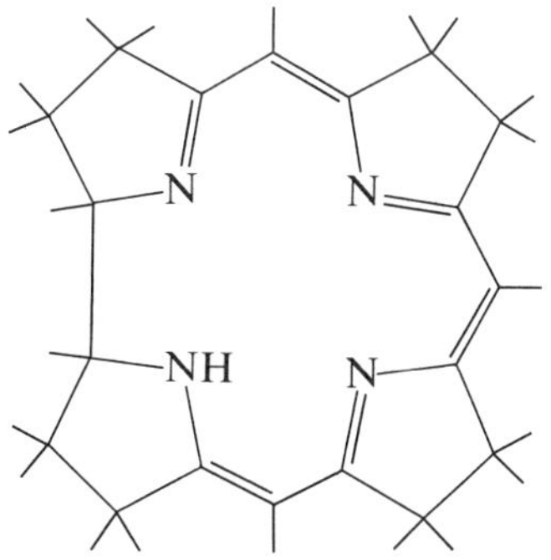

17

---

[37] H. A. O. Hill in Vol. 2, p. 1067 of ref. 1; J. M. Pratt, "Inorganic Chemistry of Vitamin B_{12}," Academic Press, New York (1972).

Furthermore (see **18**), the carbon atoms at the periphery of the corrin ring are saturated and bear seven amide groups as substituents; three of these (*a,b,d*) are acetamide and three (*c,e,f*) are propionamide. The last amide (*g*) is an N-substituted propionamide. The amide substituent is a propyl group connected through the 2-carbon to an unusual ribonucleotide. The unusual feature of this nucleotide[38] is that the base unit is 5,6-dimethylbenzimidazole. The cobalt corrin ring and its substituents down to and including the ribose phosphate are collectively called *cobamide*. Including the benzimidazole, the structure is a *cobalamin*. As in heme, the Co^{3+} ion finds itself coordinated to four "pyrrole" and one imidazole nitrogen atoms. In hemes, the sixth position normally is vacant or occupied by H_2O or O_2; in natural cobalamins, the sixth position is normally occupied by a carbon donor atom!

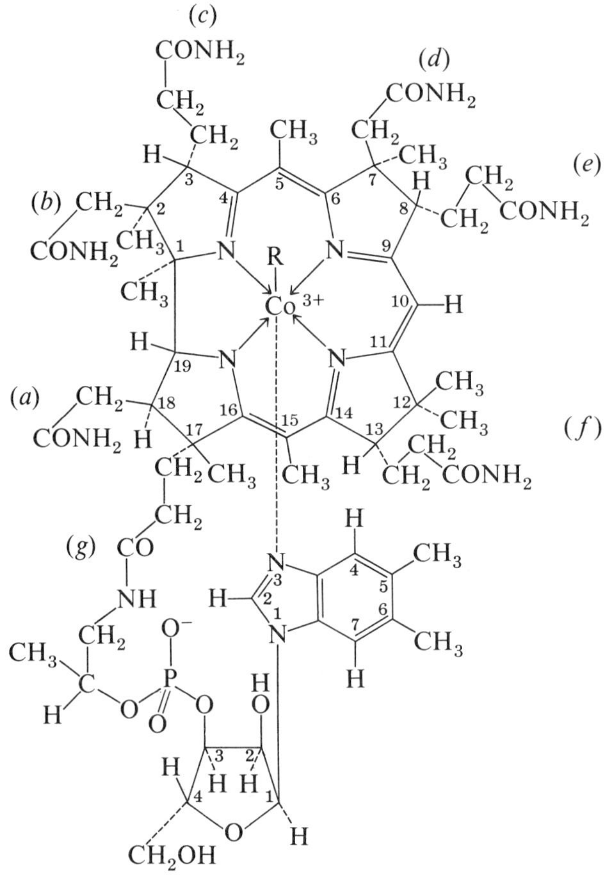

18

It was initially thought[39] that pernicious anemia arose from a deficiency of the red, diamagnetic cyanocobalamin (vitamin B_{12}), where R in **18** is CN^-. Many years later it was learned[40] that the enzyme cofactor actually encountered *in vivo* is orange-yellow and diamagnetic, and has R = 5'-deoxyadenosine (**19**); the cyanocobalamin arose during the earlier isolation as an artifact of the isolation technique.

---

[38] See ref. 4.
[39] E. L. Rickes, *et al.*, *Science*, **107**, 396 (1948); E. L. Smith and L. F. J. Paker, *Biochem. J.*, **43**, viii (1948); G. R. Minot and W. P. Murphy, *J. Amer. Med. Assoc.*, **87**, 470 (1962).

This coenzyme (5'-deoxyadenosine cobalamin, otherwise known as vitamin B_{12} coenzyme) is the only known naturally occurring organometallic compound.

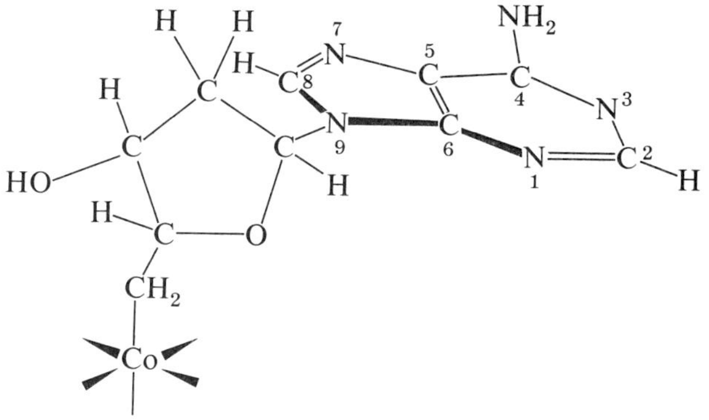

19

At present, the only known mammalian requirement for B_{12} coenzyme is as a cofactor for methylmalonyl coenzyme A mutase in the conversion of methylmalonyl coenzyme A into succinyl coenzyme A.

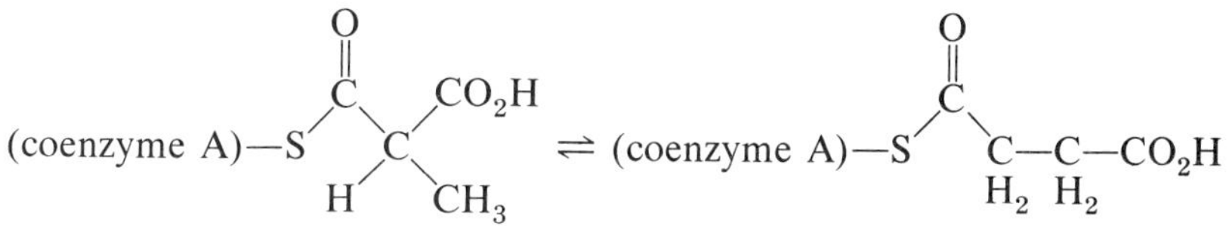

$$(\text{coenzyme A})-S-\overset{O}{\underset{}{C}}-\underset{H}{\overset{CO_2H}{C}}-CH_3 \;\rightleftharpoons\; (\text{coenzyme A})-S-\overset{O}{\underset{}{C}}-\underset{H_2}{C}-\underset{H_2}{C}-CO_2H$$

This is one of several reactions catalyzed by the B_{12} coenzyme in which there is, in effect, a swapping of H and X substituents on adjacent carbons:

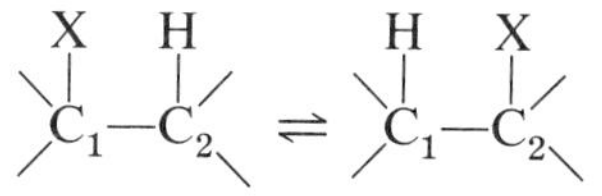

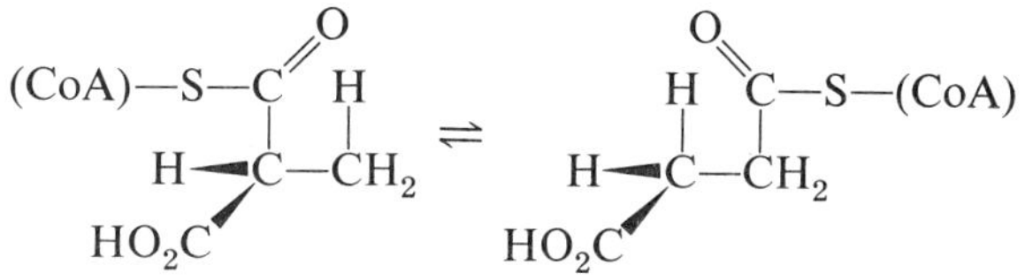

$$\overset{X}{\underset{}{C_1}}-\overset{H}{\underset{}{C_2}} \;\rightleftharpoons\; \overset{H}{\underset{}{C_1}}-\overset{X}{\underset{}{C_2}}$$

$$(\text{CoA})-S-\overset{O}{C}-\overset{H}{\underset{HO_2C}{C}}-CH_2 \;\rightleftharpoons\; \overset{O}{C}-S-(\text{CoA})-\overset{H}{\underset{HO_2C}{C}}-CH_2$$

The mechanism of this unusual reaction is being vigorously pursued at present. A current proposal,[41] which draws on experience from metal catalysis of organic reactions (Chapters 16 and 17), is illustrated on p. 1076 for the methylmalonyl mutase reaction.

---

[40] H. A. Barker, et. al., *Proc. Nat. Acad. Sci. U.S.A.*, **44**, 1093 (1958).
[41] R. H. Abeles and D. Dolphin, *Accts. Chem. Res.*, **9**, 114 (1976); R. B. Silverman and D. Dolphin, *J. Amer. Chem. Soc.*, **98**, 4626 (1976).

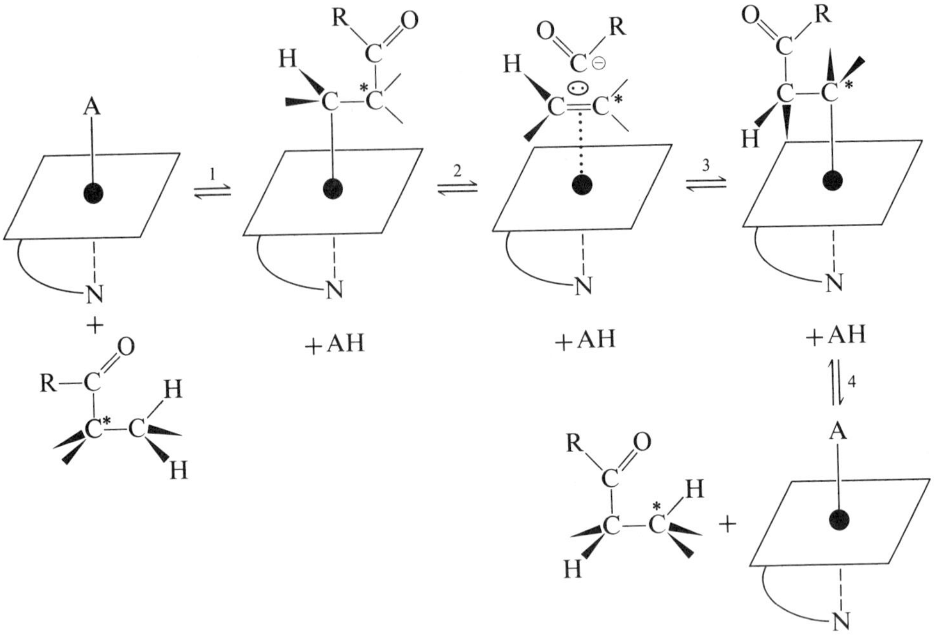

Step 1 in this scheme entails dissociation of the 5′-deoxyadenosyl group (as a radical or carbanion, it is not known for certain; recall, however, the known radical reactions of Co(CN)$_5^{3-}$ in Chapter 12) and consequent hydrogen or proton abstraction from the substrate, yielding transalkylated Co(III) and adenosine. Drawing from the well-known β-activation pathways in organometallic chemistry and the stability of olefin complexes, step 2 may involve dissociation of the β acyl group. Step 3 accomplishes the migration of this group by its attack on the olefin complex at what was initially the α-carbon to yield a different alkylated cobalamin. Finally in step 4, in a reaction like step 1, the rearranged substrate-cobalamin dissociates, followed by hydrogen or proton abstraction from 5′-deoxyadenosine and regeneration of the B$_{12}$ coenzyme. One role for the enzyme associated with the B$_{12}$ cofactor, then, is to "bind" the 5′-deoxyadenosine so it may not diffuse away from the vicinity of the cobalamin after step 1.

In addition to applying the concepts from organometallic reactions, the last step of this mechanism is consistent with the fact that labeling of the 5′-deoxyadenosine carbon of the B$_{12}$ cofactor with tritium (^{3}H) causes tritium to appear in the rearranged substrate.

This is not to say, however, that radical or carbanion pathways (*i.e.,* cleavage of the Co—C bond) are not possible for step 2, and it is not necessarily true that all substrates react by the same mechanism. In radical and carbanion pathways, β-group migration would occur in *uncoordinated* substrate radicals or carbanions after step 1 and in place of steps 2 and 3:

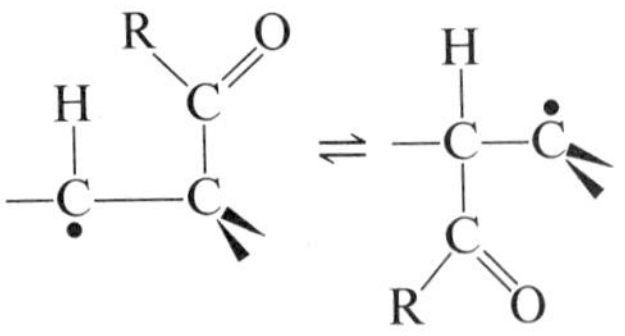

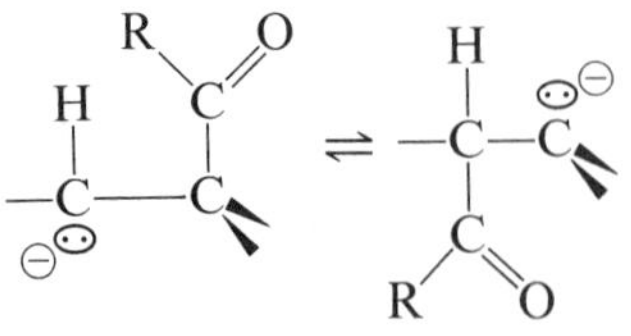

A variety of B_{12} derivatives are now known; the hydroxocobalamin (called B_{12b}) and aquocobalamin (B_{12a}) can be catalytically reduced[42] to hydroxo-Co(II)cobalamin (B_{12r}). Further reduction[43] of B_{12r} yields a green Co(I)cobalamin (d^8) complex (B_{12s}), which is a strong nucleophile and will reduce O_2 and (below pH 8) H_2O. Its nucleophilic character is useful in the synthesis of various cobalamins (including the B_{12} coenzyme). Some examples follow:

$$\begin{array}{c} & & Co-CH=CH_2 & & Co-\overset{R}{\underset{|}{C}}=O \\ & \underset{C_2H_2,}{\nearrow} & & \underset{RCOX}{\nearrow} & \\ CH_3-Co & \underset{H^+}{\overset{CH_2N_2}{\longleftarrow}} & B_{12s} & \overset{RX}{\longrightarrow} & Co-R \\ & & \downarrow \overset{O}{\triangle}, H^+ & & \\ & & Co-C_2H_4OH & & \end{array}$$

Since H^+ must be shown as an additional reagent in some of these reactions for material balance, it may well be that H^+ forms an adduct with B_{12s}, presenting another example (Chapters 5 and 17) of metal Lewis basicity,[44]

$$[Co^{III}-H]^+ \rightleftharpoons [Co^I] + H^+$$

Either species [five-coordinate Co(I) or six-coordinate Co(III)-H] could be effective in the reaction scheme above; the precedent for hydrometalation reactions was presented in Chapters 16 and 17, and the nucleophilicity of Co(II) in $Co(CN)_5^{3-}$ was established in Chapter 12. Knowing that B_{12r} and B_{12s} exist raises interesting questions, to which there are no firm answers as yet, concerning their appearance in *in vivo* reactions.

---

[42] This reduction is mild and readily achieved by $SnCl_2$, Cr^{2+} (pH ~ 5), and H_2/PtO_2, for example.
[43] Any one of Na, BH_4^-, Cr^{2+} (pH ~ 10), and electrochemical reduction (-1.4 v *vs.* S.C.E.) is sufficient for $B_{12r} \to B_{12s}$.
[44] D. Dolphin, *et al., Ann. N.Y. Acad. Sci.,* **112,** 590 (1964); G. N. Schrauzer and R. J. Holland, *J. Amer. Chem. Soc.,* **93,** 4060 (1971); see also ref. 41.

STUDY QUESTIONS

12. Devise a laboratory synthesis of methylcobalamin from aquocobalamin. A few years ago a great stir was caused by the realization that many bacteria contain Me-cobalamin; in natural waters these bacteria can convert otherwise relatively harmless (insoluble) Hg^{2+} salts into highly toxic (soluble) organomercury compounds, RHg^+. Describe a possible reaction mechanism for Me-cobalamin + Hg^{2+} that is consistent with the experimental rate law:

    $$v = (3.7 \times 10^2 \text{ M}^{-1}\text{sec}^{-1})[\text{Me-Co}][\text{Hg}^{2+}]$$

13. Refer to study question 9-28 and explain why exposure of B_{12} coenzyme to light produces 5′-8-cycloadenosine (the ribose 5′-carbon is bound to the 8-carbon of adenine). What cobalamin product is produced by this reaction?

14. Describe the steps for propane-1,2-diol $\to$ propionaldehyde and ethanolamine $\to$ acetaldehyde conversions by enzymes requiring B_{12} coenzyme.

15. What is the implication regarding a *trans* effect in cobalamins from the fact that the benzimidazole pK_a in Me-cobalamin and aquo-Co(II)cobalamin are the same (~2.5)? The pK_a of benzimidazole in B_{12a} (aquocobalamin) is -2.4.

16. Discuss the thesis that in aqueous solution B_{12s} is only four-coordinate (you may wish to refer to Chapter 9 and the discussion there of Lifschitz complexes).

ELECTRON TRANSFER AGENTS

CYTOCHROMES[45]

The compounds called *cytochromes* contain heme-like prosthetic groups[46] and are utilized in the membranes of cell mitochondria and chloroplasts to assist electron transfer reactions. Each cytochrome has an associated protein structure (the apoenzyme) by which the protein is bound to the prosthetic group as a substituent on a carbon of the porphyrin, by protein coordination of Fe as in myoglobin, or by both methods.[47] The *cytochromes b* have the protoporphyrin IX structure (13) and their proteins are not covalently bonded to the porphyrin. The *cytochromes a* are hemes with a formyl group as a ring substituent. The *cytochromes c* have the same prosthetic group as the *b* cytochromes, but the protoporphyrin IX vinyl groups have condensed with cysteinyl SH groups of the apoenzyme, as in 20. Again, as in hemoglobin, the prosthetic group resides in a protein pocket, where the fifth coordination position about the Fe is occupied by an imidazole nitrogen of a protein histidyl unit; the sixth coordination site about Fe is occupied by a tightly bound sulfur of a methionyl side chain of the protein. For this reason, the Fe does not take up O_2 or CO. The iron ion is low spin Fe^{2+}/Fe^{3+} in the reduced/oxidized forms, respectively.

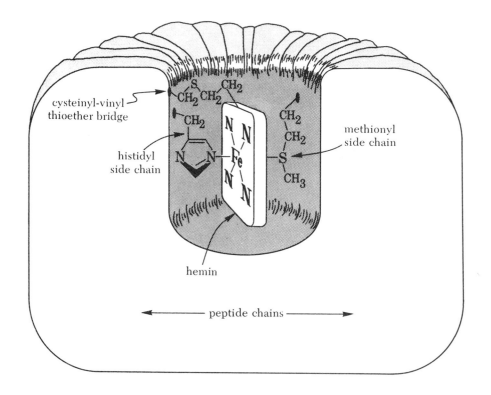

20

---

[45] H. A. Harbury and R. H. L. Marks in Vol. 2, p. 902, and D. C. Wharton in Vol. 2, p. 955, of ref. 1.

[46] A prosthetic group is a compound required by an enzyme to facilitate a particular reaction; such compounds are also called coenzymes. The peptide portion of an enzyme that requires a prosthetic group is called an apoenzyme.

[47] These prosthetic group differences, as well as variations in prosthetic group/apoenzyme binding, cause important changes in the electronic spectral and redox properties of the cytochromes. In fact, the nomenclature *a*, *b*, and *c* is based on the early recognized spectral differences as a means for distinguishing cytochromes.

An important chemical role for the cytochromes arises in *oxidative phosphorylation* of ADP to form ATP. Earlier you learned of the glucose → pyruvate conversion (Embden-Meyerhof pathway) in which ATP was regenerated by *non-oxidative phosphorylation* of ADP. Oxidative phosphorylation is an even more efficient mechanism for regeneration of ATP, according to the following overall reaction:

$$\text{glucose} + 6\,O_2 + 36\,ADP^{3-} + 36\,HPO_4^{2-} + 36\,H^+ \rightarrow$$
$$6\,CO_2 + 36\,ATP^{4-} + 42\,H_2O$$

Quite obviously this is a complicated reaction consisting of many steps, and in fact it involves a variety of *electron and/or hydrogen transfer agents*. Some of these agents are strictly organic in nature, some involve iron in a tetrahedral environment of sulfur, and others the cytochromes. Detailed mechanistic descriptions of these steps cannot, and would not, be presented in a chapter such as this. However, Figure 19-12 is a schematic of the complete electron transport chain, starting (at the left of Figure 19-12) with substrate dehydrogenation to form an olefin by the organic transfer agent FAD or starting with an alcohol that is dehydrogenated by NAD$^\oplus$ to form a ketone. The sequence of electron transfers culminates in reduction of O_2 to H_2O (extreme right of Figure 19-12) and, depending on the "high" (alcohol) or "low" (alkyl) redox potential of the initial substrate, results in formation of three or two moles, respectively, of ATP formed per pair of electrons transferred.

NAD$^\oplus$ (nicotinamide adenine dinucleotide) is a diester of pyrophosphate, ROPO$_2$OPO$_2$OR′, where R and R′ are ribose nucleosides. R is adenosine (2), and the base of R′ is nicotinamide:

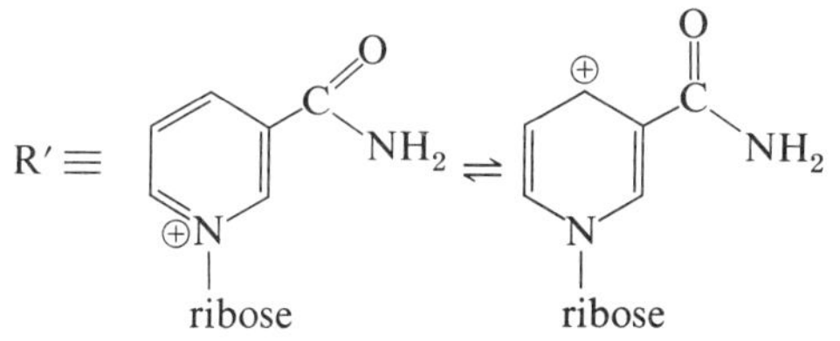

NAD$^\oplus$ acts as a $2e^-$ oxidant through hydride transfer reactions. The *para*-position of the pyridine ring is the site of H$^-$ addition. FAD (flavin adenine dinucleotide) is also a pyrophosphate diester, where R′ in this case is a flavin polyalcohol:

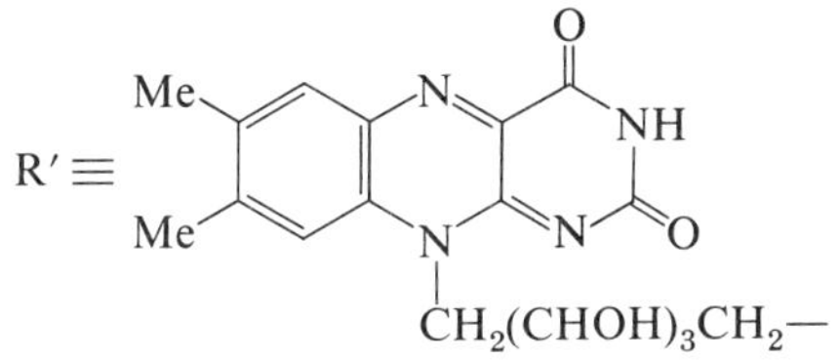

It is a stronger $2e^-$ oxidant than NAD$^\oplus$ and acts as a dihydrogen transfer agent. Thus, FAD appears in the "low potential substrate" branch of Figure 19-12. The name *dinucleotide* is actually a misnomer in this case (R′OPO$_2$O— is not a nucleotide) because the polyhydroxyl substituent is not a sugar. FAD acts as a 2H· transfer agent by converting the two ring-imine nitrogens to N—H.

In Figure 19-12 you see that the first appearance of a cytochrome is in the oxidation of ubihydroquinone by cytochrome *b*.

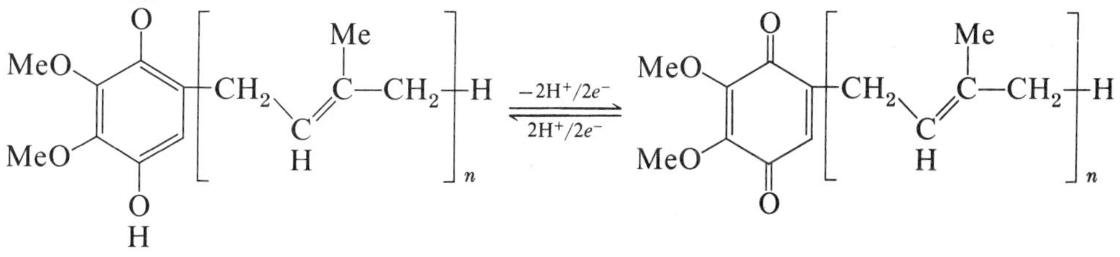

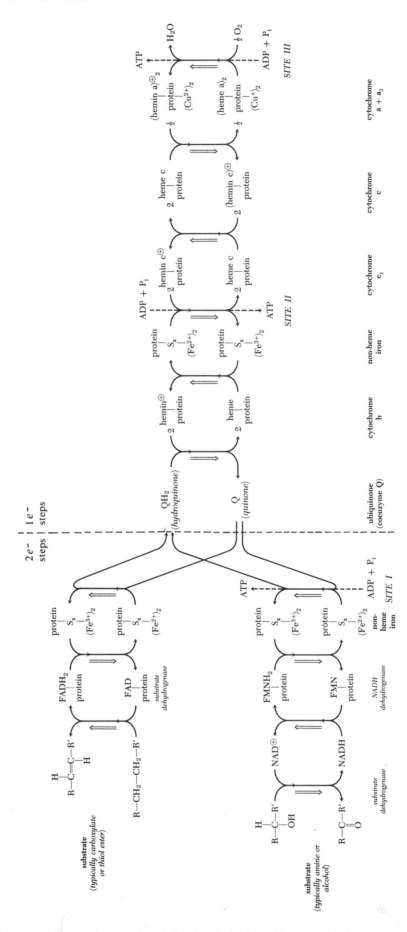

Figure 19-12. The oxidative phosphorylation electron transport chain for O_2 oxidation of substrates of "low chemical potential" (top) and "high chemical potential" (bottom). [From R. W. McGilvery, "Biochemical Concepts," W. B. Saunders Co., Philadelphia (1975).]

From this branch-point onward in Figure 19-12, the electron transfer involves only one-electron inorganic transfer agents, and it is in this chain of reactions that either all or two-thirds of the oxidative phosphorylation occurs, depending upon the initial substrate.

Among these steps are the three oxidative phosphorylations; unfortunately, little is known about them at present. In Figure 19-12 you see that the first phosphorylation (site I of the lower branch) occurs during reduction of ubiquinone by a *non-heme* iron sulfide complex; the second phosphorylation arises during electron transfer between a *non-heme* iron catalyst and a cytochrome c (site II); and the final phosphorylation arises upon reduction of O_2 by a complex structure (called cytochrome oxidase) involving two a cytochromes and Cu^+/Cu^{2+} in a 1:1 ratio of Fe:Cu (site III).

An attractive hypothesis regarding the oxidative phosphorylation process is called *chemiosmosis*.[48] This model bears some resemblance to the Na^+/K^+ pump discussed earlier, with the important distinction that both a concentration gradient *and* an electric potential gradient are established across a membrane. The electron transfer agents are known to be distributed about the *inner* surface of the inner mitochondrial membrane. In the course of electron transfer from glucose to O_2, protons are considered to be transported by enzymes from the fluid inside the inner membrane to the fluid between the inner and outer membranes. In this way the free energy of the glucose oxidation is "stored" as chemical potential in the form of a proton concentration gradient and electric potential across the inner membrane. The extent of these gradients is determined by the concentrations of ATP and ADP within the inner membrane. Should the ATP concentration there drop, the "stored" potential is released by driving the formation of ATP from ADP and PO_4^{3-}. To maintain the gradients it must be postulated that the membrane is impermeable to H_3O^+; how, then, is the gradient relieved in the process of converting ADP to ATP? One symbolic scheme is the following:

$$I_o^- + XO_o^- + 2H_o^+ \rightleftharpoons IH_o + XOH_o$$

$$IH_o + XOH_o \rightleftharpoons I\text{—}X_o + H_2O_o$$

$$IX_o + PO_{4i}^{3-} + ADP_i^{3-} \rightleftharpoons ATP_i^{4-} + I_i^- + XO_i^-$$

where the subscripts o and i designate species outside and inside the inner membrane. The net reaction is

$$I_o^- + XO_o^- + 2H_o^+ + PO_{4i}^{3-} + ADP_i^{3-} \rightleftharpoons H_2O_o + ATP_i^{4-} + I_i^- + XO_i^-$$

or

$$2H_o^+ + PO_{4i}^{3-} + ADP_i^{3-} \rightleftharpoons ATP_i^{4-} + H_2O_o$$

In this scheme, I^-, XO^-, and I-X are intermediates at sites I, II, III (Figure 19-12) that traverse the membrane and carry the burden of coupling the chemical potential of the proton gradient to the formation of ATP. Similarly, enzymes are required in the initial transport of H^+ to the outer surface of the inner membrane.

Very little is actually known about the mechanisms of the electron transfer at the membrane inner surface. From the coordination chemists' point of view, some or all of the reactions could be of the outer-sphere variety. According to such a view, no ligand substitution about Fe^{n+} is required. At least for the c *cytochromes* (which do not bind O_2) this is consistent. With reference to site III of Figure 19-12, it is known that CN^- does not complex cytochrome a, but O_2 coordination to the Fe atom of cytochrome a_3 or to the Cu atoms remains distinctly possible. In fact, the toxicity of CN^- is due to inhibition of the respiratory chain at site III, presumably by coordination of Fe or Cu.

---

[48] A good discussion of this and other hypotheses is to be found in A. L. Lehninger, "Biochemistry," 2nd Ed., Worth Publishers, New York (1974).

IRON-SULFUR PROTEINS[49]

The non-heme iron proteins, as they are sometimes called, are known to participate in redox reactions in all forms of life tested to date; in particular, they play important roles in photosynthesis, nitrogen fixation, and mitochrondial respiration (as in Figure 19-12). More importantly, the Fe—S proteins exhibit reversible redox couples within a few hundred millivolts of the H^+/H_2 couple, in contrast to Fe^{3+}/Fe^{2+} redox potential of 0.78 v.

The sulfur in iron-sulfur proteins is from cisteinyl units in the protein chain and from coordinated S^{2-} ions. Thus, iron-sulfur proteins are often referred to by the

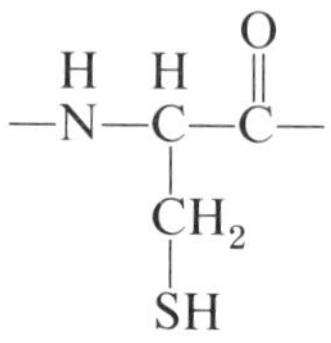

$$\begin{array}{c} \text{H} \quad \text{H} \quad \text{O} \\ | \quad\;\; | \quad\;\; \| \\ \text{—N—C—C—} \\ | \\ \text{CH}_2 \\ | \\ \text{SH} \end{array} \quad \textit{cysteinyl residue in proteins}$$

abbreviation n-Fe—S*, where n represents the number of Fe cations per protein and S* designates the presence of labile sulfur, usually also n in number. The labile sulfur is generally identified as the S^{2-} ions that are coordinated to Fe^{n+}; they are labile in the sense that (i) they generate H_2S on acidification of the protein and (ii) they are *readily* air-oxidized to elemental sulfur. Also, as far as is known, these proteins feature each Fe^{n+} in an approximately tetrahedral environment of sulfur atoms, at least one of which is a cysteinyl sulfur from the protein. Another potentially important structural feature of the iron-sulfur proteins is that n-Fe—S* proteins of the same n, but from different sources, have the same number of cysteinyl groups, in spite of the fact that the protein sequences are different. For example, the 2-Fe—S* proteins all have five cysteinyl residues, while the 8-Fe—S* proteins have eight such residues.

1-Fe—S and 2-Fe—S

The so-called *rubredoxins* (1-Fe—S and 2-Fe—S proteins) lack labile sulfur, an example being the 1-Fe—S protein **21**. Here the protein chain is folded to create a pocket of four sulfur donor atoms to coordinate the Fe^{n+} (the chain termini are marked *A* and *B*).

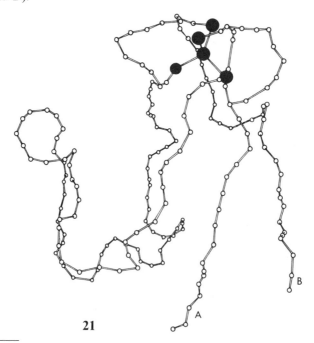

21

---

[49] W. H. Orme-Johnson, in Vol. 2, p. 710 of ref. 1.

The rubredoxin **21** is a one-electron transfer agent, with both Fe^{2+} and Fe^{3+} having high spin d^{*n} configurations. One of the interesting structural features of this complex is the C_1 symmetry of the S atoms about Fe^{n+}. As indicated by **22**, S_1 and S_4 appear radially equivalent while S_2 and S_3 are distinctly unique.[50] A "Jahn-Teller" distorted structure (Chapter 9, p. 553) is, of course, expected for T_d Fe^{2+} but not for T_d Fe^{3+}; the distortion in **21** is ligand-imposed, and some believe it to be critical for the low redox potential and fast electron transfer properties of rubredoxin (Chapter 12, p. 662). That is, the ligand-imposed coordination geometry imposes little "chemical activation" on the process $Fe^{2+} \rightleftharpoons Fe^{3+}$. Similarly, the distortion structure may render the HOMO neither strongly bonding nor anti-bonding, thereby facilitating (both kinetically and thermodynamically) the electron transfer.

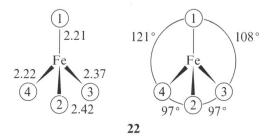

22

2-Fe—S*

These proteins are also one-electron transfer agents (in spite of having $2Fe^{n+}$/molecule) and one (or two) of them (see Figure 19-12) is known to be involved in mitochondrial respiration. The oxidized protein bears two high spin Fe^{3+} ions in the form of a S^{2-} bridged dimer (the bridge sulfurs are labile); each iron is also coordinated by two terminal cysteinyl sulfurs **(23)**.

$$\begin{array}{c}RS \\\diagdown \\ \end{array} \begin{array}{c} S \\ \diagup \diagdown \\ Fe^{III} \quad Fe^{III} \\ \diagup \diagdown \\ S \end{array} \begin{array}{c} SR \\ \diagup \\ \end{array}$$

23

The oxidized proteins are *diamagnetic* at very low temperatures, while the reduced form is paramagnetic to the extent of a single electron. How can two high spin Fe^{3+} ions yield a diamagnetic dimer? The diamagnetism indicates that the electrons of one ion have spin orientations opposite to those of the other, and so the electrons appear spin-paired. Normally this arises in covalent electron pair bond formation, but here the Fe^{3+} ions are not likely to be *directly* bonded by *five* electron pair bonds. Rather, they are *indirectly* coupled through the bridging sulfide ligands.[51] A consequence of spin correlation forces, this phenomenon is well known and is called "superexchange" or "anti-ferromagnetic coupling" of electrons. It is entirely analogous to the coupling of nuclear spins (Chapter 3, p. 117). Upon one-electron reduction, the total number of d^* electrons in the protein is odd ($= 11$) and so the *reduced protein* behaves as a spin doublet ($2S + 1 = 2$). It is intriguing that the Fe^{n+} ions of the reduced protein are non-equivalent,[52] so the Fe_2S_6 unit cannot have an idealized D_{2h} symmetry; other-

---

[50] J. R. Herriott, *et al., J. Mol. Biol.,* **50**, 391 (1970).

[51] This is not to say, however, that there is *no* direct overlap of Fe *d* electrons. In fact, the Fe ions in a structure like **23** are close enough (~2.78 Å) for such interaction to occur, superimposed upon reinforcing spin polarization of the sulfide-iron bond pairs.

[52] R. Dunham, *et al., Biochim. Biophys. Acta,* **253**, 134 (1971); M. Poe, *et al., Proc. Nat. Acad. Sci. U.S.A.,* **68**, 68 (1971).

wise, the added electron would be found with equal probability at the two iron ions. The fact that the Fe^{3+} ions of the oxidized form are equivalent indicates that some structural non-equivalence is induced upon reduction and that the added electron is "localized" about one iron ion.[53]

Synthetic models of the 2-Fe—S* proteins are beginning to appear[54] (*e.g.,* **24**). Their study over the next few years will hopefully augment and clarify the findings of biochemists regarding the natural proteins.

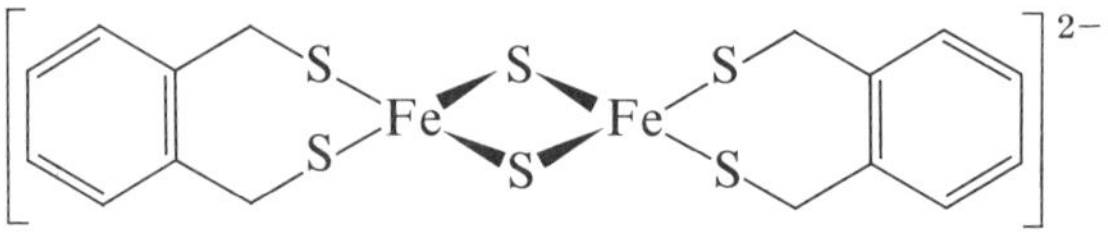

24

8-Fe—S*

Escalating in complexity, you encounter the important class of electron transfer agents called *ferredoxins*.[55] These proteins are found only in lower organisms and are most important constituents in the nitrogen fixation mechanism of plants. In such cases the ferredoxin is usually associated in some way with a molybdenum-containing protein (the Mo is also essential to the nitrogen fixation action). The ferredoxin protein contains two $Fe_4(cysteine)_4S_4^*$ clusters, each of which acts as a *one-electron* transfer site; the *complete protein,* therefore, can act as a *two-electron* agent. In what may represent a finding of great significance to the mechanistic action of ferredoxin, the non-cysteinyl sulfur atoms undergo exchange with radioactive S^{2-} and are labile in the usual sense of H_2S evolution on acidification. Recently, a good model of the Fe—S cluster has been synthesized.[56] The synthetic **(25)** and natural clusters (compare the 2-Fe—S* proteins) feature terminal cysteine and bridging sulfide groups with four Fe^{n+} and four S^{2-} at the eight corners of a D_{2d} "cube."

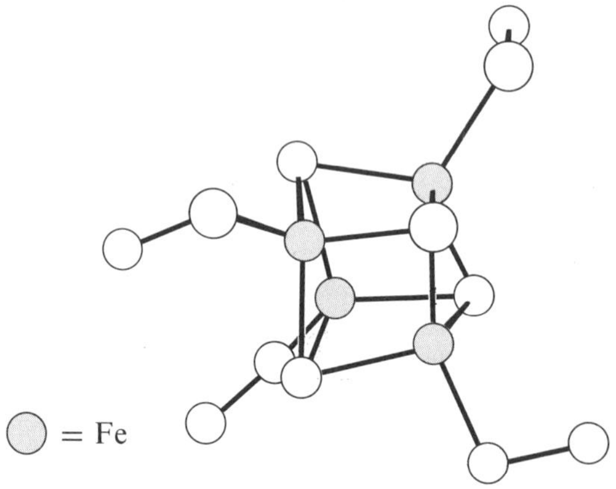

◯ = Fe

25

One way to view the Fe_4S_8 unit (so as to emphasize its relationship to the 2-Fe—S* dimer) is to look at any two opposite faces of the tetramer as 2-Fe—S* dimers now

---

[53] At this stage of the community's understanding, there remains the possibility of a "slow" dynamical structure change to permit e^- transfer back and forth between Fe ions.

[54] J. J. Mayerle, *et al., J. Amer. Chem. Soc.,* **97,** 1032 (1975); T. Herskovitz, *et al., Inorg. Chem.,* **14,** 1426 (1975).

[55] Unfortunately for communication clarity, all Fe—S proteins are sometimes referred to as "ferredoxins," where "fer" = iron and "redoxin" = redox protein.

[56] B. V. DePamphilis, *et al., J. Amer. Chem. Soc.,* **96,** 4159 (1974); L. Que, *et al., ibid.,* **96,** 4168 (1974).

associated by substitution of one terminal cysteine at each iron by a sulfide of the other dimer. This view has the advantage of relating the "anti-ferromagnetic coupling" behaviors of 2-Fe—S* and 4-Fe—S* clusters. A view that two such facial sub-units are like two 2-Fe—S* units cannot be taken literally, however, because such a model would imply that each cube could act as a $2e^-$ redox unit—in clear contradiction with fact. Furthermore, the average Fe^{n+} oxidation states in the 2-Fe—S* protein are $Fe^{3+} \rightleftharpoons Fe^{2.5+}$, while the following changes have been found for $[Fe_4S_4(SR)_4]^z$:

$$[Fe_4S_4(SR)_4]^- \rightleftharpoons [Fe_4S_4(SR)_4]^{2-} \rightleftharpoons [Fe_4S_4(SR)_4]^{3-}$$

$$Fe^{2.75+} \underset{\text{like HIPIP}}{\rightleftharpoons} Fe^{2.5+} \underset{\text{like ferredoxin}}{\rightleftharpoons} Fe^{2.25+}$$

The second redox couple mimics the situation in ferredoxin; the first couple corresponds to the redox couple for another type of non-heme iron protein, the so-called "high potential iron protein" (HIPIP).[57] HIPIP is a 4-Fe—S* protein, of unknown function, containing a *single* $[Fe_4S_4(SR)_4]^z$ cluster. The fact that HIPIP and ferredoxin exhibit different redox couples reveals an important effect on the prosthetic group by the protein surrounding it. Whether this is a specific structural phenomenon or a general "medium" effect as in hemoglobin is unknown.

Another point of demarcation for 2-Fe—S* and the tetrameric cluster concerns the equivalence of Fe ions in the intermediate state. The D_{2d} symmetry of the cage (which is actually a smaller Fe_4 tetrahedron within a larger S_4 tetrahedron) insures the equivalency of the Fe atoms and S atoms. Though too complex to pursue here, the mo scheme for such interpenetrating tetrahedra reveals the absence of an electronic degeneracy, which would incur a Jahn-Teller distortion.

---

[57] The reduction $E^{\circ\prime}$ values (E° at pH 7) are $+220$ mV (COQ-cytochrome reductase), $+350$ mV (HIPIP), and ~ -330 to -490 mV (ferredoxins). These values are relative to $E^{\circ\prime}_{H^+/H_2} = -420$ mV.

STUDY QUESTIONS

17. Review the text comments on the molecular and electronic structures of *c* cytochromes and argue whether their electron transfer reactions should be inner or outer sphere. How might the porphyrin ring facilitate one of these mechanisms (Chapter 12); would the advantage be realized if the heme group appeared as in **20**?

18. The enzyme cytochrome oxidase (Figure 19-12) contains two heme prosthetic units. It has been suggested that the electron transfer process in the oxidase involves the sequencing cytochrome $a \rightarrow$ "copper site" $\rightarrow$ cytochrome a_3, where a and a_3 designate different heme types. Cytochrome a does not react with CO while a_3 does, and the latter also reacts with CN^-. Draw what conclusions you can about the coordination geometries of Fe in these two hemes. Which heme is likely to be involved in the reduction of O_2? What analogy do you find here with the difference in the reactions of heme (not hemoglobin) and myoglobin with O_2?

19. Reasonably facile RS^- exchange with $[Fe_4S_4(SR')_4]^{2-}$ has made possible the synthesis of various ferredoxin analogs. Is such exchange unexpected, given the oxidation state and spin condition of Fe^{n+} in such complexes? Does the concept of an $\{Fe_4S_4\}$ cage cluster make reasonable the possibility that its electron transfer reactions are of the "outer sphere" variety? Does the occurrence of RS^- exchange complicate this view?

N₂ Fixation[58]

As noted in the preceding section on ferredoxin, an 8-Fe—S* protein is essential to nitrogenase activity. Also essential to the activity of this enzyme is a Mo/Fe protein, which can be separated from the ferredoxin protein. In *Azotobacter* nitrogenase, the $Mo:Fe:S^{2-}:$ cysteine ratio is $\sim 1:20:20:>20$ and the *separate* Mo/Fe and ferredoxin proteins are ineffective at N₂ fixation.

These two protein types appear to be completely general in the sense that they are common to all nitrogen fixing organisms. In fact, the chemical action of the nitrogen fixing enzymes in these widely variant organisms is so completely similar, the following definition of *nitrogenase* (N_2ase) has recently been made[58]:

"... N₂ases are complexes of Mo—Fe and Fe proteins whose syntheses are repressed by fixed nitrogen[59] and whose activity couples ATP hydrolysis to electron transfer for the reduction of N_2 to $2NH_3$, N_3^- to $N_2 + NH_3$, N_2O to $N_2 + H_2O$, RCCH to $RCHCH_2$, RCN to $RCH_3 + NH_3$, RNC to RNH_2 + alkanes and alkenes, and/or $2H_3O^+$ to $2H_2O + H_2$, with H_2 a competitive inhibitor of N_2 reduction and CO an inhibitor of all reductions except that of H_3O^+."

This rather comprehensive definition encompasses the known crucial chemical features of N₂ases, features that have led (as you will see in a moment) to a basic model for the workings of N₂ase. These features are summarized in the following diagram:

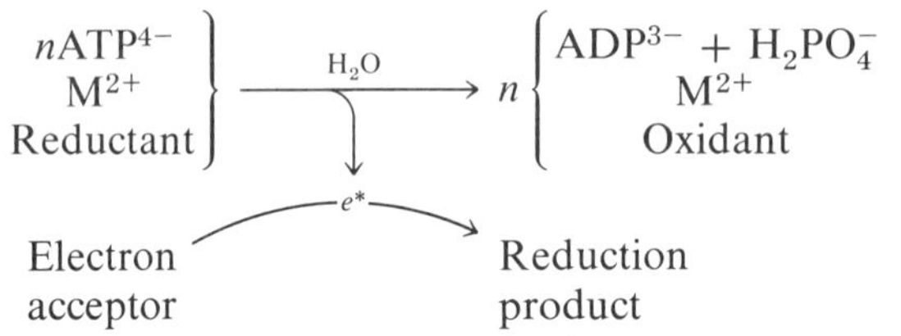

$$\left.\begin{array}{c} nATP^{4-} \\ M^{2+} \\ \text{Reductant} \end{array}\right\} \xrightarrow{H_2O} n \left\{\begin{array}{c} ADP^{3-} + H_2PO_4^- \\ M^{2+} \\ \text{Oxidant} \end{array}\right.$$

Electron acceptor $\xrightarrow{e^*}$ Reduction product

	Electron acceptor → Reduction product	Inhibitors
Endogenous	$H_3O^+ \longrightarrow H_2$	
Exogenous	$N_2 \longrightarrow NH_3$	H_2, CO
	$N_3^- \longrightarrow N_2, NH_3$	CO
	$N_2O \longrightarrow N_2, H_2O$	CO
	$RCN \longrightarrow RCH_3, NH_3$	CO
	$RNC \longrightarrow RNH_2, CH_4, C_2H_4, C_2H_6$, etc.	CO
	$RCCH \longrightarrow RCHCH_2$	CO

The Mo/Fe protein of *Azotobacter* (molecular weight $\sim 10^5$) appears to consist of two Mo-containing units, each with a $Mo:Fe:S^{2-}:$ cysteine ratio of $1:16:\sim 13:20$ (note the nearly equivalent amounts of Fe, S^{2-}, and cysteine). The electronic spectral properties of this protein bear a marked resemblance to those of the Fe—S proteins of the last section, although electron spin resonance has shown the occurrence of at least two unique Fe sites. Thus, it may well be that the Fe sites of the Mo/Fe protein are specific to that protein and to nitrogen fixation enzymes. In the oxidized state the Fe^{n+} ions appear to have $n = 3$ and to have high spin d^{*n} configurations. Reduction with HSO_3^- converts about half the iron to high spin Fe^{2+} (or Fe^+). The reduced protein is paramagnetic to the extent of an average of two unpaired electrons per Fe. At best this is only sketchy information about the Fe^{n+} sites and their reduction; spectroscopic characterization of the Mo^{n+} sites has been even more difficult. It is established, however, that Mo participates in electron transfer processes, just as the Fe sites do, and it seems that the processes are $2e^-$ in nature at Mo ($Mo^{6+} \rightleftharpoons Mo^{4+}$,

---

[58] R. W. F. Hardy, R. C. Burns, and G. W. Parshall, in Vol. 2, p. 745 of ref. 1.
[59] That is, NH_3.

$Mo^{5+} \rightleftharpoons Mo^{3+}$). Furthermore, recent evidence suggests that the presence of sulfur in the coordination sphere of Mo.

The Fe—S protein component of *Azotobacter* (molecular weight 5×10^4) consists of two identical sub-units, each containing $2Fe^{n+}$ and $2S^{2-}$. In both the oxidized and reduced forms the Fe^{3+} and Fe^{2+} ions are high spin.

The chemist's understanding of these proteins and their necessary interdependence is especially incomplete at present. The Mo/Fe protein appears to contain ferredoxin or a ferredoxin-like protein, while the Fe protein appears to contain 2-Fe—S* residues.

As noted in the definition of N_2ase, an interesting, crucial feature of N_2ases is their strict requirement for ATP^{4-} and a divalent metal ion (Mg^{2+} works best). ADP^{3-} and $H_2PO_4^-$ are products of the N_2 fixation process, so it appears that ATP^{4-} hydrolysis is intimately involved in driving the electron transfer to Fe and Mo. Such an ATP/redox-agent relationship is inverted with respect to those in Figure 19-12 for glucose oxidation; furthermore, nitrogenase is ATP-specific—the other commonly occurring nucleotide diphosphates are ineffective! The electron source varies from organism to organism; some rely on oxidation of pyruvate, and others depend on photocatalyzed transfers, while the free-living *Azotobacter* and *Rhizobia* bacteria utilize electrons from the ordinary electron transport chain (*cf.* Figure 19-12).

There are several current hypotheses concerning the N_2ase mechanism, only one of which will we outline here. The N_2 fixation is proposed[60] to occur at a Mo—S—Fe site after ATP^{4-}-induced reduction of Mo^{n+} to $Mo^{(n-2)+}$ or Mo—H. A possibility is:

$$Fe\diagdown_S\diagup Mo-OH + ATP^{4-} \rightarrow ADP^{3-} + MoOPO_3H^- \xrightarrow{2e^-\ PO_4^{3-}} Fe\diagdown_S\diagup Mo-H$$

(The reduction of Mo^{n+} here is reminiscent of the $B_{12a} \rightarrow B_{12s}$ reduction, for reduced N_2ase is also capable of reducing H_3O^+ to H_2; *cf.* the definition of N_2ase.) This scheme accounts for the ATP-required reductive dephosphorylation step. The Fe at this site (presumably Fe^{3+}) is prepared for the entry of N_2 by reduction to Fe^+. Upon coordination of N_2 to Fe^+, the active site is suggested to appear as in **26a**. Formation of the dinuclear Mo—N=N—Fe complex of bridging diazene (**26b**) then arises by addition of Mo—H to the π system of N≡N—Fe. Subsequent one- or two-electron reductions (and accompanying protonations) of this bridged diazene-like intermediate results in bridging hydrazine, **26c**. Further reduction and protonation of hydrazine cleaves the N—N bond, and two NH_3 molecules are released.

26a   26b   26c

One of the intriguing features about N_2ase chemistry is that H_2 inhibits the reduction of dinitrogen. This is a startling finding, in view of the fact that N_2 fixation amounts to the addition of H_2 to N_2! In the presence of D_2, this inhibition incurs the formation of HD; a reasonable mechanism for this in terms of the Mo—NN—Fe bridged species along the path of **26** is as follows:[61]

---

[60] R. W. F. Hardy, *et al.*, *Adv. Chem.*, **100**, 219 (1971); *Biochem. Biophys. Res. Commun.*, **20**, 539 (1965); G. E. Hoch, *et al.*, *Biochem. Biophys. Acta*, **37**, 273 (1960); also, ref. 58.
[61] *Chem. Eng. News*, **54**, No. 18, 17 (1976).

$$\underset{\substack{H \\ N=N \\ \text{(Mo)} \quad \text{(Fe)} \\ S}}{\overset{H}{}} + D_2 \rightarrow \left[\underset{\substack{H \\ N=N \\ \text{(Mo)} \quad \text{(Fe)} \\ S}}{\overset{D-D}{\overset{H}{}}}\right] \rightarrow \underset{\substack{N\equiv N \\ \text{(Mo)} \quad \text{(Fe)} \\ S}}{} + 2HD$$

You might wish to review in this context the orbital following arguments of Chapter 7, where the analogous diazene reaction $2N_2H_2 \rightarrow N_2 + N_2H_4$ was used to illustrate the electron flow and nuclear motion behind the hydrogen transfer. This phenomenon reminds you again of the need for transition metal chemists to be conversant with principles of non-metal chemistry.

Particularly critical to the viability of this model of N_2ase action are the formation of Fe—N_2 and Fe—N_2—Mo complexes and their reactions. For example, the addition of Mo—H to bound N_2 in this model is analogous to the known reaction of chlorohydridobis(triethylphosphine)platinum to the benzenediazonium cation and subsequent reduction steps to produce phenylhydrazinium cation.

$$(PEt_3)_2ClPt-H + N_2-Ph^+ \rightarrow \left[(PEt_3)_2ClPt-\overset{H}{N}=N-Ph\right]^+$$

$$\downarrow H_2/Pt$$

$$(Et_3P)_2ClPt-H + [H_3NNHPh]^+ \underset{H_2/Pt}{\longleftarrow} \left[(PEt_3)_2ClPt-\overset{HH}{\underset{H}{N-N}}-Ph\right]^+$$

For many years it was generally thought that N_2 was ineffective as a ligand toward metal ions in general, in spite of the fact that N_2 was found bound to acids such as N^- (N_3^-), O (N_2O), and organic cations (as in N_2Ph^+ above). The fact that NO, CO, CN^-, and isonitriles (CNR) have lone pair HOMO's and π acceptor LUMO's much like those of N_2 should have forecast the existence of metal-N_2 complexes, since the former ligands are known to form complexes with a variety of metals. In 1965, the Canadian workers Allen and Senoff reported[62] the first N_2 complex, pentaaminedinitrogenruthenium(2+), in which N_2 is bound "end-on"[63] as with NO, CO, and similar ligands.

$$\left[\begin{array}{c} H_3N\overset{NH_3}{}NH_3 \\ H_3N-Ru-N\equiv N \\ H_3NNH_3 \end{array}\right]^{2+}$$

This ruthenium complex was synthesized by reduction of $RuCl_3$ with hydrazine in aqueous solution, a reaction representing the antithesis of nitrogen fixation. The first example of direct ligation by N_2 came from Russian scientists[64] who reduced $RuCl_3$ with Zn in aqueous tetrahydrofuran under an atmosphere of N_2. Their complex proved to have the composition $RuCl_2(H_2O)_2(THF)(N_2)$. In this country, Taube

---

[62] A. D. Allen and C. Senoff, *Chem. Comm.*, 621 (1965); be sure to read A. D. Allen and F. Bottomley, *Accts. Chem. Res.*, **1**, 360 (1968) for comments on the discovery.

[63] F. Bottomley and S. C. Nyburg, *Chem. Comm.*, 897 (1966).

[64] A. E. Shilov, *et al.*, *Kinet. Catalysis*, **7**, 768 (1966); A. K. Shilova and A. E. Shilov, *ibid.*, **10**, 267 (1969).

demonstrated[65] even more clearly the ligating ability of N_2 by the reduction of $[(H_3N)_5RuCl]^{2+}$ with Zn under N_2:

$$[(H_3N)_5RuCl]^{2+} \xrightarrow{Zn/H_2O} [(H_3N)_5RuOH_2]^{2+} \xrightarrow{N_2} [(H_3N)_5RuN_2]^{2+} + H_2O$$

wherein N_2 is found to displace H_2O from Ru^{2+}. An interesting development in this work, particularly in view of the bridged species in **26**, is the appearance of an N_2-bridged dinuclear complex:

$$[(H_3N)_5RuN_2]^{2+} + [(H_3N)_5RuOH_2]^{2+} \rightleftharpoons [(H_3N)_5Ru-NN-Ru(NH_3)_5]^{4+}$$

The formation of the linear[66] Ru—NN—Ru bridge from $2[(H_3N)_5Ru(OH_2)]^{2+}$ and N_2 is *exo*thermic by 22 kcal/mole and is characterized by an $N\equiv N$ stretching frequency (2100 cm^{-1}) that is depressed from those of N_2 itself (2331 cm^{-1}) and $[(H_3N)_5RuN_2]^{2+}$ (2140 cm^{-1}). As discussed in Chapter 3, $N_2 \rightarrow$ metal σ donation and metal $\rightarrow N_2$ π retrobonding have competing effects on the diatomic force constant, which complicate interpretation of the coordination-induced frequency shift.[67]

Concurrent with these developments, many N_2 complexes containing phosphines as ligands have been prepared. Of particular bearing on the action of N_2ase are the complexes of Mo and Fe. Structures **27** and **28** illustrate two examples.

bis(dinitrogen)bis(diphosphine)molybdenum

27

dihydridodinitrogentris(phosphine)iron

28

Thus, there is ample precedent for the proposed formation of N_2 complexes in N_2ase; at such time as the identities of the ligands in the first coordination spheres of Mo and Fe in N_2ase become known, it should be possible to develop realistic *in vitro* model complexes to test the hypotheses regarding N_2 fixation. These developments can be expected in the next few years and hopefully will lead to development of commercial processes for N_2 fixation.

---

[65] D. E. Harrison and H. Taube, *J. Amer. Chem. Soc.*, **89**, 5706 (1967); D. E. Harrison, *et al.*, *Science*, **159**, 320 (1968).

[66] I. M. Treitel, *et al.*, *J. Amer. Chem. Soc.*, **91**, 6512 (1969).

[67] K. F. Purcell, *Inorg. Chim. Acta*, **3**, 540 (1969); K. G. Caulton, *et al.*, *J. Amer. Chem. Soc.*, **92**, 515 (1970).

STUDY QUESTIONS

20. $E°'$ for the H_3O^+/H_2 couple is quite negative (-0.42 v). Does this suggest to you a need for coupling of ATP hydrolysis to the reduction of N_2ase (the Mo reduction step)? (If you have forgotten the meaning of $E°'$, see ref. 6, p. 1054.)

21. Why is it that CO inhibition of all N_2ase reductions, except H_3O^+/H_2, requires two active metal sites? (Help can be found in ref. 58.)

22. Refer to Chapter 6 (study question 6-58) and comment on the features of the N_2ase model that are likely to lower the ΔG barrier to reduction of N_2.

23. Draw a reasonable conclusion about the nature of the Mo—Fe "pocket" from the fact that CH_3CN and CH_3CCH, but not *i*-BuCN or CH_3CCCH_3, are reduced by N_2ase.

24. Regarding the definition of N_2ases, it was noted there that NH_3 inhibits the synthesis by the organism of its N_2ase. Is Nature working against itself here, or has the evolutionary process achieved perfection?

25. Given the different coordination characteristics of low valent and high valent metal ions, how does reduction of N_2ase appear to facilitate the coordination of N_2 and the release of NH_3?

26. Estimate, from the polarizations of the σ HOMO's and LUMO's of N_2 and CO, which diatomic should most tightly bind to a low spin d^6 metal ion (estimate the relative metal-ligand orbital overlaps). Does the relative Z^* of N *vs.* C reinforce or oppose this predicted ordering of ligating ability (consider the orbital energy matches)?

CHLOROPHYLL[68]

Plants *and* animals effect glucose oxidation by O_2. Plants, however, are unique in synthesizing glucose from CO_2 and H_2O by a photo-initiated reaction. In this system, creation of ATP from ADP and $H_2PO_4^-$ is also accomplished in the process. Both cytochromes and ferredoxins are involved in the electron transfer that is initiated by oxidation of photo-excited chlorophyll.

Structure **29** shows that chlorophyll (Chl) is a tetrapyrrole belonging to the porphyrin family, but with some significant modifications to the porphyrin ring

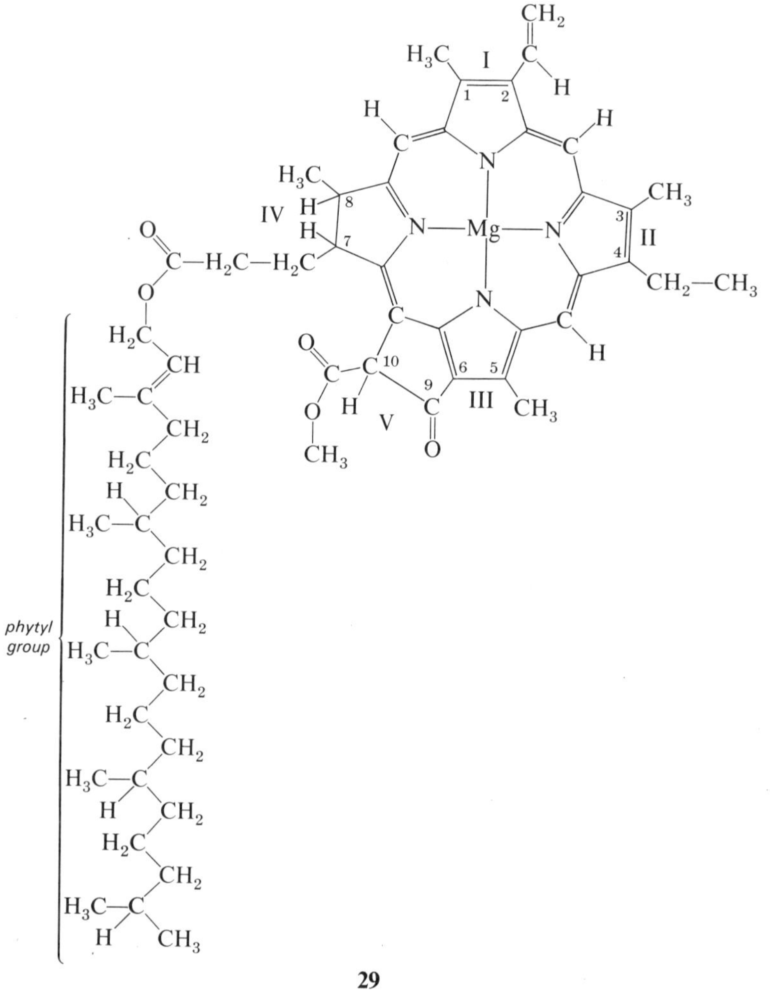

29

---

[68] For a global overview of photosynthesis in relation to energy and material resources, the article by M. Calvin, *American Scientist,* **64,** 270 (1976) is recommended. For a technical analysis of data and model evolution, see J. J. Katz, in Vol. 2, p. 1022 of ref. 1.

complex (the various chlorophylls differ by the ring substituents at the 2, 3, and 10 positions).

First of all, Chl is not a heme because the metal at the tetrapyrrole coordination site is Mg^{2+} (which, incidently, also is 0.3 to 0.5 Å above the non-planar, "ruffled" macrocycle plane), and Mg^{2+} is mysteriously unique among metal ions in conferring just the right physico-chemical properties necessary for the photocatalysis of glucose synthesis. The porphyrin modification that distinguishes Chl from other porphyrins is the connection of the methylidyne carbon between pyrrole rings III and IV with the 6-carbon in ring III by the $H\overset{|}{C}(CO_2Me)\overset{|}{C}{=}O$ group. This group exhibits *keto-enol* equilibration strongly favoring the keto form, as shown in **29**. The long phytyl group serves to bind Chl to membranes in the cell.

Before delving into what little is known about the photo-induced electron transport mechanism, let us pause to view the overall process.[69] Basically, light quanta are utilized to photo-excite a chlorophyll *a* molecule to a strongly reducing excited state, from which electron transfer proceeds through an unknown species X (perhaps an Fe—S protein), to a ferredoxin (green plants appear to utilize a 2Fe—S ferredoxin of the type **23** or **24,** while bacteria use a 4Fe—S of the type **25**), to $NADP^{\oplus}$. $NADP^{\oplus}$ is an oxidizing agent that differs from $NAD^{\oplus}$ (p. 1079) by the presence of a phosphate group at the 3′-carbon of adenosine.

NADP is but one more intermediate along the path to reduction of CO_2. The important consequence of this process (called *photosystem I* in Figure 19-13) is that Chl *a* has been oxidized to Chl *a*$^+$. To sustain the action of Chl *a*, Chl *a*$^+$ must be reduced. Water, unassisted, has too high a redox potential to reduce Chl a_1^+; thus, a second photo-induced electron transfer process (*photosystem II*) enters the picture. The Chl *a* site of this system absorbs quanta of higher energy than that of system I, so the two Chl sites are distinguished as Chl a_2 and Chl a_1 or, alternatively, P680 and P700, where 680 and 700 designate excitation photon wavelengths. Analogously with photosystem I, Chl a_2^* reduces some yet-to-be-identified reagent (Q) and the electron is transported through a system of a quinone (plastoquinone), two cytochromes, and a Cu^{2+} protein known as plastocyanin. The journey culminates at Chl a_1^+, which is reduced to Chl a_1. In short, there are two photo-redox processes coupled by a series of "dark" electron transfer steps. At this point, oxidized Chl a_2^+ remains, and it must be reduced to Chl a_2 for the process to continue. Chl a_2^+ is strongly oxidizing and, with the assistance of some Mn^{2+} protein,[70] is reduced by H_2O. This entire sequence of "dark" and "light" reactions is illustrated in Figure 19-13, where the reduction potentials of the various intermediates are roughly indicated by the E_0' scale to the left of the figure.

---

[69] A very readable account of the critical experiments and their interpretation is given in Chapters 7 and 8 of D. W. Krogmann, "The Biochemistry of Green Plants," Prentice-Hall, Englewood Cliffs, N.J. (1973).

[70] About all that is known about this prosthetic group is that the Mn^{2+} complex is labile, for attempts to isolate the protein with Mn^{2+} intact have failed. The principles of Chapter 13 regarding the lability of metal ion complexes seem to be followed by this bio-complex.

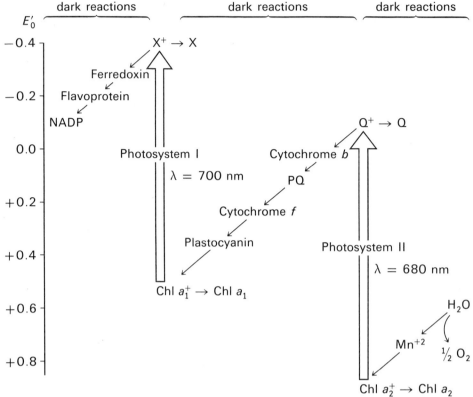

Figure 19–13. The "Z diagram" of the photosynthetic electron transport sequence. The photo processes are endergonic, while the dark reactions are exergonic. E_0' is the reduction potential of the oxidized form of each couple at pH = 7.

The highly conjugated prophyrin ring of chlorophyll is ultimately linked to the absorption of red light (transmission of green light) that initiates the electron transfer processes. It is known, however, that the concentration of Chl at the cell "active site" is actually too small (because of the required presence of the other proteins at that site) to capture photons efficiently. This explains the presence of carotenes in the cell; carotenes are highly conjugated hydrocarbon derivatives that undergo photo-excitation at higher frequencies (mid-visible region) than Chl. The photo-excitation energy of the carotenes is then passed on to the Chl oligomers (see later) about the "active site" and then on to the "active site" Chl itself.

One of the more interesting chemical/structural features of Chl is that it dimerizes in solvents that poorly coordinate Mg^{2+} (poor Lewis basicity) and poorly solvate the porphyrin moiety. In such circumstances, the special keto group at carbon 9 of ring V coordinates the Mg^{2+} of a second Chl through the carbonyl oxygen. Structure **30** is a schematic of a possible dimer structure.

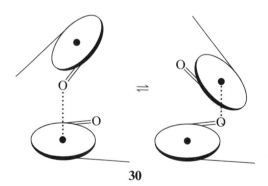

30

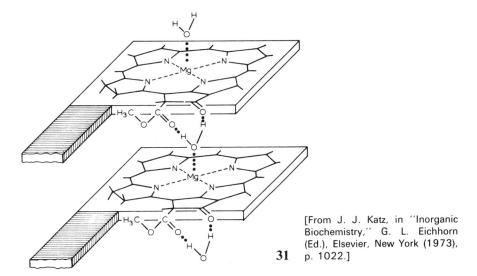

In aqueous solution these dimers dissociate to form $(Chl \cdot H_2O)_n$ oligimers. Because H_2O can act as both a Lewis base and acid, it may bridge Chl monomers as shown in **31**.

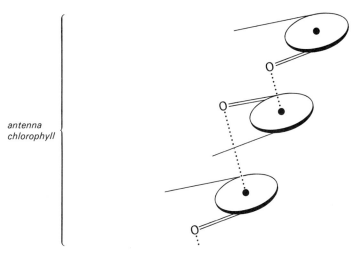

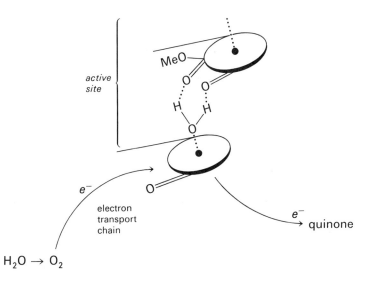

Figure 19-14. An idealization of photoelectron transfer in the cell.

The critical feature of such H_2O-bridged Chl molecules is that they (and not monomers or dimers) are the only Chl species known to become paramagnetic on photolysis. The origin of this paramagnetism seems to arise from photo-induced H atom transfer to $>C=O$ of Chl from a "sandwiched" H_2O. Electron transfer to the enzyme electron transport chain leaves an unpaired electron in the porphyrin ring system (actually, the odd electron appears to "hop" rapidly from one ring to the next).

Very simply, the current hypothesis for the photo-oxidation of Chl is as follows (see Figure 19-14). The "antenna" Chl consist of $(Chl)_n$ chains terminating in a $(Chl \cdot H_2O \cdot Chl)$ unit. The antenna Chl either harvest light quanta by direct photoexcitation or are sensitized by energy transfer from a carotene that was previously photo-excited. This excitation energy is passed on to the $(Chl \cdot H_2O \cdot Chl)$ unit at the "active site" containing the ferredoxins, cytochromes, and so forth. There, in photosystem I, the transfer noted above initiates the oxidation of $(Chl \cdot H_2O \cdot Chl)$ by quinones, which diffuse away with the electron to begin the conversion of CO_2 to glucose.

This concept of chlorophyll has been successfully applied to the formulation[71] of a "synthetic leaf." The construction of the leaf (Figure 19-15) is as follows: $(Chl \cdot H_2O)_n$ adduct is impregnated in a polymer membrane or deposited on a metal foil; this membrane or foil separates the two compartments of a photoelectric cell, of which one compartment contains an oxidizing agent (such as tetramethylphenylenediamine) and the other contains a reducing agent (sodium ascorbate). Each compartment is connected to an external circuit by electrodes. The cell responds to red light by generating a potential difference of 422 mV (the optimum obtained to date) and a current of 24 μamp.

---

[71] J. J. Katz, et al., Chem. Eng. News., **54**, No. 7, 32 (1976).

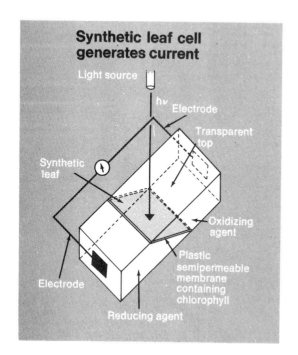

Figure 19-15. The "synthetic leaf" photovoltaic cell under development at Argonne National Laboratory. [From J. J. Katz, et al., Chem. Eng. News, **54**, No. 7, 32 (1976).]

STUDY QUESTIONS

27. The addition of H_2O to hydrocarbon solutions of $(Chl)_2$ causes important changes in the infrared spectrum. The two keto stretching vibrations at 1655 cm^{-1} and 1700 cm^{-1} are depressed to a common band at 1638 cm^{-1}. The C$\cdots$O ester vibrations at $\sim$1735 cm^{-1} are clearly resolved into bands at higher (1745 cm^{-1}) and lower (1727 cm^{-1}) energies. Identify the features of the $(Chl \cdot H_2O)_n$ oligomer that account for the directions of these i.r. band shifts. Similarly account for the appearance of O—H stretching vibrations at 3460 cm^{-1} and 3240 cm^{-1}.

28. The absorption of a photon with $\lambda = 740$ nm by the "synthetic leaf" generated a potential difference of 422 mV. Compare the energy of the 740 nm photon to that of the potential drop. What is the percentage energy conversion?

EPILOG

It should be clear to you at this point that a great deal remains to be discovered and unraveled about the role of "inorganic" processes in organic systems. There are a great many more areas of interest to the bioinorganic chemist than could be presented here, such as the storage and release of iron *in vivo,* the role of metals in amino acid/peptide synthesis, metal ion requirements in nerve tissues, and so on. Extensions of the understanding of how living systems effect energy conversion and the fixation of nitrogen are certainly two of the more pressing scientific issues for mankind today. A truly interdisciplinary effort will be required to push forward; and it is with an eye to the future, and the promise it holds, that we conclude this chapter and text.

COMPOUND INDEX

If the point of interest in a compound is a metal atom, the compound is listed under the metal's heading. Compounds containing one atom of the metal per formula are listed first, followed by compounds with 2, 3, ..., n metal atoms. Within these groups, compounds are arranged by coordination number of the metal and, within coordination groups, alphabetically.

Compounds based on non-metals are grouped by number of substituents and, within those groups, alphabetically.

Ag

AgF, as fluorinating agent, 293
   stability, 292, 293
AgO, stability, 292

Al

Al^{3+}, aquated halides, 313
   hydroxo complexes, 315
$Al(BH_4)_3$, 192
$[Al(C_2O_4)_3]^{3-}$, spectrum of, 522
$Al(C_6H_5)_3$, addition to triple bonds, 929
   phenyl bridge bonding in, 311
Na_3AlF_6, structure, 282
AlH_3, amine adducts of, 311
   structure, 310, 835, 843
   synthesis, 310, 476
$AlHO_2$, structure, 314
$Al(OH)_3$, structure, 314
$Al(OR)_3$, synthesis, 491
$RAlNR'$, 312, 492
$R_2AlNR'_2$, 312
AlX_3, self-association, 195
   structures, 843
AlH_4^-, 311
   hydrolytic stability, 311
   synthesis and mechanism, 390, 476
$AlMe_3 \cdot NMe_3$, structure, 911
$Al(OH)_4^-$, 315
AlH_6^{3-}, 311
   synthesis, 476
$Al_2(CH_3)_6$, bonding, mo's, 845
   bridge-terminal methyl scrambling, 955
$Al_2(CH_3)_2R_4$, methyl bridging in, 195
$Al_2Cl_7^-$, 313
$Al_2Mg(CH_3)_8$, structure, 847
Al_2O_3, structure, 313
$(Al_2Si_4O_{12}^{4-})_n$, 325
$(Al_2Si_5O_{14}^{2-})_n$, 325
$(Ph_2Al-C\equiv C-Ph)_2$, structure, 931
$(Al_{12}Si_{12}O_{48}^{12-})_n$, 326

Au

AuX, stability, 292
$Au(PPh_3)_2Cl$, 591
$Au(PPh_3)_2^+I^-$, 591
$[Au(PPh_3)_3]^+$, 591

B

$B(CH_3)_3N(CH_3)_3$, amine exchange, 393
   electrophilic substitution at nitrogen, 410
BCl_4^-, isolation, 293
$BF_3 \cdot OH_2$, ionization, 305, 387
$BF_3 \cdot 2H_2O$, 305
BH_4^-, as ligand, 192
   hydrolytic stability, 304
   mechanism of H exchange, 385
   mechanism of hydrolysis, 385
   synthesis, 476
$BH(SO_4H)_4$, formation in H_2SO_4, 252
$B(C_6H_5)_3$, as Lewis acid, 231
BCl_3, formation of metal carbynes, 865
BCl_3/BBr_3, addition to acetylenes, 839
BF_3, mo's, 151
$BF_x(CH_3)_{3-x}$ adducts, 232
BH_3 adducts, hyperconjugation in, 232
$B(NH_2)_3$, synthesis, 490
$B(OH)_3$, ionization, 309
$B(OR)_3$, alkoxy exchange, 390
   formation, 309
$B(O_2CR)_3$, synthesis, 309
BX_3, Lewis acidities of, 232
   π mo's and Lewis acidity, 195
HBO_2, structure, 308
$RB(OH)_2$, synthesis, 309, 488
R_2BOH, synthesis and bonding, 309, 488
R_2NBX_2, synthesis, 489
BH_3CO, 201
B_2, mo's, 146

B_2Cl_4, structure and stability, 307
   synthesis, 454
B_2F_4, structure and stability, 307
   synthesis, 456
$B_2H_2R_4$, bridge bonds in, 195
B_2H_6, 191
   dissociation energy, 302
   localized mo's, 194
   mo's for, 193
   synthesis, 304, 476
      mechanism, 390
B_2O_3, structure and bonding, 307
$B_2O_4^{2-}$, 308
$B_2(OH)_4$, formation of, 307
$B_2O_5^{4-}$, 307
$B_2O(OH)_6^{2-}$, 307
$B_2H_7^-$, 192
$B_2H_6 \cdot NH_3$, 399
$B_2H_6 \cdot 2NH_3$, 399
$B_3H_6N_3$, π mo's for, 306
   synthesis, 306, 489
$B_3N_3H_9X_3$, 388
$B_3O_6^{3-}$, 308
$(RBO)_3$, as anhydride of borinic acids, 309
   from oxygen transfer in SO_2, 258
   synthesis, 488
B_4Cl_4, synthesis, 455
B_4H_{10}, ^{11}B nmr spectrum, *990*
   chemistry, 999
   structure, 303
   synthesis, 304
$B_4O_7^{2-}$, 308
B_5H_9, bonding, 303, 992
   chemistry, 1001
   structure, 303, 982
B_6 clusters, in borides, 302
arachno-1,3-$B_7C_2H_{13}$, preparation, 1008
B_8Cl_8, synthesis, 455
closo-2,3-$B_9C_2H_{11}$, preparation, 1008
nido-7,8-$B_9C_2H_{11}^{2-}$, preparation, 1008, 1021
   reaction with Fe^{2+}, 1021
B_9Cl_9, synthesis, 455
$B_9H_9^{2-}$, structure and ^{11}B nmr spectrum, 1011
closo-B_9H_9S, reaction, 1009
$B_{10}H_{10}^{2-}$, structure, 982
$B_{10}H_{14}$, chemistry, 1002 ff.
   structure and bonding, 304
$B_{10}C_2H_{12}$, *o*-carborane, base degradation, 1008
   structure, 982
B_{12} clusters, in borides, 301
   in boron, 300

Ba

BaF_2, fluorinating agent, 293

Be

$Be(CH_3)_2$, structure and bonding, 846
$BeCl_2$, structure, 289
BeH_2, mo's, 135
$Be(OH_2)_4SO_4$, structure, 281

Br

Br_2, hydrolysis mechanism, 429
Br_n^+, 350
BrF_2^+, 204
BrF_3, fluorinating agent, 467
   as F^- donor, 212, 352

BrF_3 (*Continued*)
   hydrolysis, 511
   self-association, 212
HBr, synthesis, 482
BrNO, 335

C

CN^-, as a Lewis base, 199
   mo's, 147
   ν_{CN} in adducts, 113
CsCN, structure, 281
KCN, structure, 281
CO, as retro-bonding ligand, 201
   dipole moment, 116
   mo's, 147
   photoelectron spectrum, *149*
   proton affinity of, 232
CO_2, bond energy, 120
   π mo's, 156
   σ mo's, 158
CO_3^{2-}, mo's of, *167*
$CaCO_3$, structure, 283
C_2, mo's, 146
CaC_2, structure, 283
C_2H_4, localized mo's, *185*
$(CN)_2$, 328
CH_3CO^+, 196
C_5H_5, π mo's, 159
C_6H_6, as a π donor, 197
   π mo's, 159
$(CF)_n$, 327
C_nO_2, 328

Ca

CaF_2, structure, 282

Cd

$CdCl_2$, structure, 286
Cd_2^{2+}, isolation of, 293

Cl

ClF, as fluorinating agent, 467
   bond energy, 269
   solvolysis, 511
   synthesis, 462
ClF_3, as fluorinating agent, 467
   bond energy, 269
   mo's, 151
   synthesis, 462
   VSEPR and, 77
ClF_4^-, synthesis, 462
ClF_5, synthesis, 462
HCl, mo's, 137
   photoelectron spectrum, *140*
   synthesis, 481
ClNO, 335
$ClNO_2$, 335
ROCl, 512
ClO_3F, 511
Cl_2, as Lewis acid, 210
   as Lewis base, 209
   hydrolysis mechanism, 429
Cl_3^+, 209, 350
Cl_3^-, 351
Cl_nO_m, 352

Co

Co(II)/Co(III), electrochemical properties, 744
$[Co(B_9C_2H_{11})_2]^-$, cage degradation, 1023
$Co[N(SiMe_3)_2]_2$, 591
$Co[N(SiMe_3)_2]_2 PPh_3$, structure, 592
$Co(CO)_4^-$, as Lewis base, 214
$HCo(CO)_4$, catalyst in oxo reaction, 967
$(\eta^5\text{-}C_5H_5)Co(CO)_2$, preparation, 882
$Co(H_2O)_6^{2+}$, spectrum, 564
$CoCl_4^{2-}$, structure, 526
$[Co(CN)_5]^{3-}$, electron transfer agent, 688 ff.
$[Co(CN)_5O_2]^{3-}$, structure, 689
$[Co(CN)_5SCN]^{3-}$, linkage isomers, 617
  synthesis by electron transfer, 688
$[Co(NH_3)_4CO_3]NO_3$, synthesis, 731
$[Co(NH_3)_4Cl_2]^+$, isomers, 628
$[Co(NH_3)_6]^{2+/3+}$, electron transfer, 664
$[Co(NH_3)_5Cl]^{2+}$, inner sphere reaction with $[Cr(H_2O)_6]^{2+}$, 657, 669 ff.
$[Co(NH_3)_5Cl]Cl_2$, synthesis, 732
$[Co(NH_3)_5NCS]^{2+}$, linkage isomers, 617
  electron transfer, 688
$[Co(NH_3)_5(NO_2)]Cl_2$, linkage isomerism, 615
$[Co(NH_3)_5O_2]^{2+}$, 745
$[Co(NH_3)_6][CuCl_5]$, 294
$Co(NH_3)_6I_3$, structure, 282
$[Co(NH_3)_6][TiCl_6]$, structure, 281
$[Co(en)_2(NH_3)Br]^{2+}$, resolution of enantiomers, 622
trans-$[Co(en)_2Cl_2]^+$, synthesis and reactions, 748
$[Co(en)_2(NH_3)Cl]^{2+}$, resolution of enantiomers, 628
$[Co(en)_2(NO_2)_2]^+$, resolution of enantiomers, 622
$[Co(en)_2(phen)]^{3+}$, bacterial resolution, 613
$[Co(en)_3]^{3+}$, absolute configuration, 647
  resolution by electron transfer reaction, 684
  spectrum, 567
$[Co(1,2\text{-diaminopropane})_3]^{3+}$, conformational isomerism, 651
$[Co\{N(CH_2CH_2CH_2NH_2)_3\}Br]^+$, structure, 598
$Co_2(CO)_8$, electron counting, 802
  in hydroformylation, 934
  in oxo reaction, 966 ff.
  structure, 858
  synthesis, 856
$[(NH_3)_4Co \cdot NH_2, O_2 \cdot Co(NH_3)_4]^{3+/4+}$, 747
$[(NH_3)_5Co \cdot O_2 \cdot Co(NH_3)_5]^{4+/5+}$, 745–747
$Co_3(CO)_9C-Y$, structure and chemistry, 1039
$Co_3(CO)_9CCH_3$, structure, 982
$Co_4(CO)_{12}$, structure, 1038
$[Co_6(CO)_6(\mu_3\text{-}CO)_8]^{4-}$, structure, 1042

Cr

$Cr[N(SiMe_3)_2]_3$, 592
$Cr(CH_2CPhMe_2)_4$, structure, 854
$Cr_2(\eta^5\text{-}C_5Me_5)_2(CO)_4$, preparation and structure, 1027, 1032
$Cr(C_5H_5)_2$, properties, 878 (t)
$Cr(C_6H_6)_2$, mo's for, 890 ff.
$(\eta^6\text{-}C_6H_5Cl)Cr(CO)_3$, nucleophilic substitution, 925
$Cr(CO)_6$, bonding in, 199
  CO substitution mechanism, 919
$[Cr(C_2O_4)_3]^{3-}$, spectrum of, 523
$CrCl_3$, structure, 288
$CrCl_3 \cdot 6H_2O$, hydrate isomerism, 613
$CrCl_6^{3-}$, isolation, 294
$[Cr(NH_3)_6]Cl_3$, synthesis, 732, 745
$(Cr(NH_3)_5X)^{2+}$, spectra of, 525
$[Cr(en)_3]^{3+}$, conformations, 644
$Cr(H_2O)_6^{2+}$, spectrum, 564
$Cr(H_2O)_6^{3+}$, spectrum, 564

Cs

CsCl lattice, 280
CsF, as fluorinating agent, 292
CsO_2, 293

Cu

$[CuBr_4]^{2-}$, structure, 593
$CuCl_2$, structure, 289
$CuCl_4^{2-}$, structure, 526
$CuCl_5^{3-}$, isolation of, 294
CuH, stability, 268
$Cu(NH_3)_4^{2+}$, structure, 526
$[Cu(H_2O)_4]^{2+}$, stability constants for reactions with various ligands, 734
$Cu(H_2O)_6^{2+}$, spectrum, 564
$Cu(SPMe_3)_3$, structure, 591
CuX, stability, 292

F

HF, as catalyst in fluorination reactions, 465
  as F^- acceptor, 255
  as F^- donor, 255
  as solvent, 249–255
  synthesis, 481
H_2F^+, 255
F_2, hydrolysis mechanism, 429
  mo's, 145
HF_2^-, 255

Fe

$Fe[N(SiMe_3)_2]_3$, 592
$[Fe(CO)_4]^-$, Collman's reagent, 937
$FeCl_4^-$, structure, 526
FeO_4^{2-}, 531
$Fe(CO)_5$, as Lewis base, 214
  bonding in, 199
$Fe(Me_2PC_2H_4PMe_2)(CO)_3$, stereochemical non-rigidity, 961
$[Fe(B_9C_2H_{11})_2]^{2-}$, structure, 1022
$(\eta^4\text{-}B_4H_8)Fe(CO)_3$, structure, 1002, 1020
$(\eta^4\text{-}C_4H_4)Fe(CO)_3$, preparation, 883
$Fe(C_5H_5)_2$, see Ferrocene in Subject Index.
$(\eta^1\text{-}C_5H_5)(\eta^5\text{-}C_5H_5)Fe(CO)_3$, fluxional isomerism, 958
$[(\eta^5\text{-}C_5H_5)Fe(CO)_2(\text{olefin})]^+$, addition of nucleophiles, 933
$(\eta^5\text{-}C_5H_5)Fe(CO)(PPh_3)(COC_2H_5)$, decarbonylation, 937
$Fe(CN)_6^{4-}$, spectrum, 567
$FeCl_3$, behavior in solvents, 237
$FeCl_6^{3-}$, isolation of, 294
$Fe(H_2O)_6^{2+}$, spectrum, 564
$Fe_2(CO)_9$, electron counting, 802
  preparation, structure, 1034
  structure, 860
  synthesis, 856
$[(\eta^5\text{-}C_5H_5)Fe(CO)_2]_2$, chemistry, 815, 1032
  complex with $AlEt_3$, 915
  fluxional isomerism, 960

[$(\eta^5$-$C_5H_5)Fe(CO)_2]_2$ (*Continued*)
  infrared spectrum, 902
  linkage isomerism, 619
$(\eta^5$-$C_5H_5)_2Fe_2(CO)_3P(OPh)_3$, fluxional
  isomerism, 960
$Fe_3(CO)_{12}$, electron counting, 802
  preparation, structure, 1034
  structure, 860
  synthesis, 856
[η^5-$C_5H_5FeCO]_4$, infrared and structure, 899
  structure, electrochemistry, 1036–1037
[$Fe_4S_4(SR)_4]^{z+}$, structure, 1046, 1084
$Fe_5(CO)_{15}C$, structure, 1042

Ga

$Ga(CH_3)_3 \cdot$ amine, amine exchange, 393
  $Ga(CH_3)_3$ exchange, 393, 410
$Ga(GaCl)_4$, 292

Ge

$GeCl_4$, 238
GeH_3^-, formation in liquid ammonia, 248

H

NaH, stability, 268
H_2, mechanisms of reduction by, 384
  mo's, 134

Hg

$HgBr_2$, structure, 288
$HgCl_2$, bond energy, 120
  structure, 288
HgF_2, as fluorinating agent, 293
  stability, 292
  structure, 288
HgI_2, structure, 288
HgO, stability, 292
HgR_2, synthesis, 476

I

ICl_2^-, 211
ICl_3, as Cl^- acceptor, 211
ICl_4^-, 211
  isolation of, 293
  mo's for, 175
  VSEPR and, 76
IF_n, 351
HI, synthesis of, 481
NH_4I, structure, 281
IO_2^+, formation in HF, 252
IO_2F, formation in HF, 252
$IO_2(OSO_3H)$, formation in H_2SO_4, 252
HIO_3, reaction with HF, 252
H_5IO_6, 353
$I(SO_3F)_3$, 261
I_2, as a Lewis acid, 210
  hydrolysis mechanism, 429
  mo's, 146
I_2^+, 261

I_3^+, 261
I_3^-, 210
  synthesis in SO_2, 259
I_4^{2+}, 261
I_n^+, 350
I_n^-, 351

In

$In(CH_3)_3 \cdot$ amine, amine exchange, 393
  $In(CH_3)_3$ exchange, 393

Ir

$IrCl(CO)(PPh_3)_2$, Vaska's compound, 594, 629
  addition of MeI, isomers, 941
  addition of O_2, mo's for, 945
  basicity, 913
  reaction with olefins, 868
$Ir(Cl)_2(PPh_3)_2(CO)(CH_3)$, isomer enumeration, 632
[$Ir(NH_3)_5(NH_2OH)]^{3+}$, 414
$Ir_4(CO)_{12}$, structure, 1039

K

KF, as fluorinating agent, 293

Kr

KrF^+, 356
KrF_2, 354
$Kr_2F_3^+$, 356

Li

$Ph_3CLi(Me_2N$—C_2H_4—$NMe_2)$, *826*
LiF, as fluorinating agent, 293
$LiN(SiMe_3)_2$, 591
LiR, synthesis of, 475
Li_2O, 293
$Li_4(CH_3)_4$, structure and bonding, 848

Mg

Mg^{2+}, in chlorophyll, 1090
  in phosphate hydrolysis, 1053
MgF_2, as fluorinating agent, 293
[$EtMg_2Cl_3 \cdot 3THF]_2$, structure, 851

Mn

$Mn(CO)_5^-$, as Lewis base, 213
  bonding in, 213
$Mn(C_5H_5)_2$, properties, 878(t), 890
$MnH(CO)_5$, as a Bronsted acid, 213
$MnCl_6^{3-}$, isolation of, 294
$Mn(H_2O)_6^{2+}$, spectrum, *564*
$Mn_2(CO)_{10}$, electron counting, 802

Mo

$Mo(CO)_6$, catalyst in olefin metathesis, 977
cis- and *trans*-$Mo(CO)_4[P(OPh)_3]_2$, infrared
  spectra, *902*

COMPOUND INDEX □ 1101

$[Mo(C_5H_5)_2]_x$, properties, 878(t), 881
$(\eta^5\text{-}C_5H_5)_2MoH_2$, structure and bonding, 887
$(\eta^6\text{-}C_6H_3Me_3)Mo(CO)_3$, preparation, 882
$[Mo(CN)_8]^{4-}$, 607
$Mo_2Cl_8^{2-}$, preparation, reactions, 1030
$Mo_2(CH_2SiMe_3)_6$, structure, 854
$Mo_6Cl_8^{4+}$, structure and bonding, 1044

N

$N(CH_3)_3$, adducts with Group III acids, 229
$N(CH_3)_3B(CH_3)_3$, thermodynamic data for, 216
$N(C_2H_5)_3$, steric hindrance of basicity, 229
NH_2Cl, disproportionation, 415
  solvolysis of, 496
  synthesis and mechanisms of reactions, 413–415
$NHCl_2$, mechanisms of reactions, 416
  synthesis, 413
NCl_3, mechanisms of reactions, 416
  synthesis, 413, 457
NF_2, mo's for, 472
  photochemistry of, 471
  radical reactions of, 470
NHF_2, reaction mechanisms of, 417
  synthesis, 457–458
$ClNF_2$, synthesis, 458
NF_3, dipole moment, 116
  kinetic stability, 417
  Lewis basicity, 203
  mo's, 151
  proton affinity of, 232
  synthesis, 457
NH_3, dipole moment, 116
  laboratory synthesis of, 477
NH_3, liquid, as solvent, 241 ff.
NH_2X, nucleophilic attack upon, 410
NO, 333
  as a ligand, EAN rule, 796
  mo's, 147
NO^+, from N_2O_4, 335
$NOCl$, hydrolysis mechanism, 417
FNO, 335
NF_3O, 203, 336
NH_2NO_2, hydrolysis of, 413
NH_2OH, 336
  oxygen transfer oxidation of phosphines, 412
  mechanism of synthesis, 418
XNO, hydrolytic stability, 335
NO_2, 333
NO_2^+, formation in H_2SO_4, 252
  from N_2O_5, 335
  π mo's, 156
$NO_2^- + OCl^-$, orbital following model, 657
NO_2^z ($z = -1, 0, +1$), VSEPR and, 78
FNO_2, 335
HNO_2, 337
  reaction mechanisms, 417
NO_3^-, mo's of, 167
HNO_3, 337
HNO_3/H_2SO_4 mixed solvent, 252
$NaNO_3$, structure, 283
XNO_3, 336
$HN(SiMe_3)_2$, reactions of, 494
$N(SiH_3)_3$, basicity of, 228
N_2, fixation, 1086–1089
  mo's, 144
N_2F_2, synthesis, 457

N_2F_4, as fluorinating agent, 470–473
  photochemistry of, 471
  synthesis, 457–458
N_2H_2, disproportionation, orbital control of, 377
  in nitrogen fixation, 1087
N_2H_4, in nitrogen fixation, 1087
  synthesis of, 414, 478
N_2O, 334
$N_2O_2^{2-}$, 338
N_2O_3, and nitrogen atom exchange, 334
N_2O_4, structure and bonding, 333
  isomeric forms, 335
N_2O_5, isomeric forms, 335
N_3^-, bridging ligand in electron transfer, 675
$N_3H_6^+$, as an intermediate, 414
N_nH_{n+2}, stability of, 269

Na

NaCl lattice, 278
NaF, as fluorinating agent, 293

Nb

$Nb_6Cl_{12}^{2+}$, structure and bonding, 1044

Ni

$Ni[N(SiMe_3)_2](PPh_3)_2$, 592
$Ni(CO)_4$, bonding in, 199
  ligand substitution mechanism, 919
$Ni(CN)_4^{2-}$, formation in liquid ammonia, 247
  structure, 526
$[Ni(CN)_5]^{3-}$, structure, 598
$NiCl_4^{2-}$, structure, 526
$Ni(en)_2^{2+}$, 558
$Ni(PF_3)_4$, 204
$Ni[P(OEt)_3]_4$, olefin isomerization catalyst, 977
$Ni[P(OR)_3]_4$, as Lewis base, 214
NiX_2L_2 (L = phosphine), rearrangement, 757 ff.
$Ni(B_9C_2H_{11})_2$, structure, 1023
$Ni(C_5H_5)_2$, properties, 878 (t)
$Ni(H_2O)_6^{2+}$, spectrum, 564
  analysis of, 572
$[Ni(OH_2)_6][SnCl_6]$, structure, 281
$[Ni_5(CO)_9(\mu_2\text{-}CO)_3]^{2-}$, structure, 1041
$Ni_8(CO)_8(\mu_4\text{-}PPh)_6$, structure, 1046

O

HOBr, mechanism of disproportionation, 429
OCl_2, stability, 340
  synthesis, 460
HOF, mechanism of decomposition, 425, 429
OF_2, stability, 340
  synthesis, 459
H_2O, localized mo's, 184
HOX, mechanisms of reactions, 425
KOH, structure, 281
$O_2^{0/+/-/2-}$, bond distances, 746(t)
O_2, adducts, 205
  mo's, 144
  mo's for complexes with CoD_5^{2+}, 1071
  mo's for Fe binding in hemoglobin, 1070
O_2^-, as a ligand, 206
  formation in liquid ammonia, 247
O_2^{2-}, as a ligand, 206
O_2F_2, synthesis, 459
H_2O_2, synthesis, 480

O_2R_2, mechanism of reactions, 423
O_2X_2, 341
O_3, π mo's, 156
O_3X_2, 341

Os

OsO_4, 531
$[Os(bipy)_3]^{2+/3+}$, electron transfer, 656
$Os_3(CO)_{12}$, structure, 860

P

PCl_3, aminolysis of, 497
  solvolysis of, 496
  synthesis, 458
PF_3, as Lewis base, 203
  synthesis, 458
PH_3, synthesis, 478
PH_4^+ salts, 238
PX_3, hydrolysis mechanisms, 430
$P(CH_3)_3B(CH_3)_3$, thermodynamic data for, 216
PF_3BH_3, 204
PCl_3F_2, structure, 110
PCl_5, adducts, 204
  structure of solid, 204
  synthesis, 458
$PF_2(CH_3)_3$, intermolecular fluorine exchange, 432
$PF_3(CH_3)_2$, intermolecular fluorine exchange, 432
PF_3H_2, intermolecular fluorine exchange, 432
  synthesis, 478
HPF_4, stability, 478
  synthesis, 478
$PF_4N(CH_3)_2$, pseudorotation of, 431
PF_5, synthesis, 458
PF_nR_{5-n}, mechanism of pseudorotation, 433
PH_5, stability, 274
$POCl_3$, as Lewis acid, 205
  solvolysis of, 497
  synthesis, 459
POF_3, synthesis, 459
R_2POH, 496
R_3PO, synthesis, 501
$PF_2O(OH)$, synthesis, 503
$PFO(OH)_2$, synthesis, 503
$H_2PO(OH)$, 332
  synthesis, 503
$HPO(OH)_2$, 332
$P(OR)_3$, tautomerism, 332
  synthesis, 496
H_3PO_4, reaction with HF, 253
  reaction with H_2SO_4, 253
$PO(OCH_3)_3$, hydrolysis and oxygen exchange, 435
$PO(OC_6H_5)_3$, hydrolysis mechanism, 435
$PO(OR)_3$, 332, 501
PPh_5, synthesis of, 478
PR_5, reductive elimination reaction, 479
PX_5, hydrolysis mechanisms, 430
P_2H_4, pyrophoric nature, 413
  synthesis, 503
$H_4P_2O_7$, 501
$Na_3P_3O_{10}$, 502
$[(MeO)_2PN]_3$, 500
$(R_2PBH_2)_3$, synthesis, 490
P_4, 330

P_4O_6, from PCl_3 and SO_3^{2-} in SO_2, 258
  structure, 331
  synthesis, 502
P_4O_{10}, structure, 331
  synthesis, 502
  oxygen-chlorine exchange, 459
$(Cl_2PN)_n$, synthesis, 497
$(HPO_3)_n$, 501
P_nH_{2n}, cyclic phosphines, 331

Pd

$PdCl_2$, α-form, structure, 289
  β-form, structure, 290
$Pd(Ph_2PC_3H_6PPh_2)(SCN)(NCS)$, structure, 617

Pt

$PtCl_{2/4}$, synthesis, 751
$[PtCl_3(C_2H_4)]^-$, mo's for, 539
  olefin exchange, 922
  structure and bonding, 871
$PtHCl(PEt_3)_2$, reaction with PhN_2^+, 1088
$trans$-$PtHCl(PEt_3)_2$, reaction with olefins, 837
$PtClMe(PEt_3)_2$, reductive elimination, 951
PtF_4, stability, 292
$[PtL_4]^{2+}$, synthesis, 751
$PtLL'X_2$, synthesis, 751
$Pt(NH_3)_4$, formation in liquid ammonia, 247
$Pt(NH_3)_2Cl_2$, isomers, 625
cis-$Pt(NH_3)_2Cl_2$, cancer chemotherapy, 753
$trans$-$Pt(NH_3)_2Cl_2$, Reiset's salt, synthesis, 696, 703
cis- and $trans$-$Pt(NH_3)_2Cl_2$, synthesis, 751
$[Pt(NH_3)(NH_2OH)(py)(NO_2)]^+$, 626
$[Pt(NH_3)_4][PtCl_4]$, Magnus' green salt, 751
$Pt(en)Cl_2$, synthesis, 751
  cancer chemotherapy, 753
$trans$-$[Pt(en)_2Cl_2]^{2+}$, electron transfer, 680
$PtO_2 \cdot H_2O$, stability, 292
$Pt(PhC{\equiv}CPh)(PPh_3)_2$, structure, 872
K_2PtCl_4, synthesis, 750
$Pt(py)(NH_3)(NO_2)(Cl)(Br)(I)$, isomer enumeration, 630
K_2PtCl_6, synthesis, 750
$[Pt_9(CO)_9(\mu_2\text{-}CO)_9]^{2-}$, structure, 1041

Pu

$Pu(C_5H_5)_3$, 892

Re

$(\eta^5\text{-}C_5H_5)_2ReH$, structure and bonding, 887
$Re[S_2C_2Ph_2]_3$, 600, 603
$[ReH_9]^{2-}$, structure, 609
$Re_2Cl_8^{2-}$, preparation, reactions, structure, 1031
$Re_3Cl_{12}^{3-}$, structure, 1033

Rh

Rh_4^{2+}, heterogeneous catalysis with silicates, 327
$RhCl(PPh_3)_3$, Wilkinson's catalyst, 595
  mechanism in homogeneous hydrogenation, 964
$(\eta^5\text{-}C_5H_5)Rh(C_2H_4)_2$, nmr spectrum, 874
$(\eta^5\text{-}C_5H_5)Rh(CO)_2$, CO substitution, 921

[Rh(py)$_4$Cl$_2$]$^+$, synthesis by electron transfer reaction, 686
Rh(py)$_3$X$_3$ (X = Cl$^-$, SCN$^-$), 686, 687
Rh$_4$(CO)$_{12}$, structure, cf. Ir$_4$(CO)$_{12}$, 1039
Rh$_6$(CO)$_{16}$, structure, 861, 1041

Ru

RuCl$_2$(H$_2$O)$_2$(THF)(N$_2$), 1088
[Ru(NH$_3$)$_6$]$^{2+/3+}$, electron transfer, 662
Ru(NH$_3$)$_5$(N$_2$)$^{2+}$, 1088
[(H$_3$N)$_5$Ru-NN-Ru(NH$_3$)$_5$]$^{4+}$, 1089
Ru$_3$(CO)$_{12}$, preparation, structure, 1034

S

SCN$^-$, π mo's, 618
   bridging ligand in electron transfer, 657
SO, 345
SCl$_2$, 346
   synthesis, 460
SF$_2$, 346
CsSH, structure, 281
KSH, structure, 281
SO$_2$, as Lewis acid and base, 207
   as solvent, 255–259
   auto-ionization controversy, 257
   oxygen coordination, 208, 256
   oxygen exchange, 257
   reactions in, as solvent, 258
   sulfur coordination, 208, 256
SO$_2^-$, 344
SCl$_3^+$, 204
RNSF$_2$ compounds, synthesis, 470
SOCl$_2$, aminolysis, 510
   hydrolysis, 430, 509
   synthesis, 461, 508
SOF$_2$, synthesis, 462
SO(NR$_2$)$_2$, 508
SO(OR)$_2$, tautomerism, 510
SOX$_2$, 347
SO$_2$BF$_3$, 208, 256
SO$_2$F$^-$, 347, 511
SO$_2$SbF$_5$, 256
SO$_3$, 345
   adducts as sulfonating agents, 208
   as Lewis acid and base, 207
SCl$_4$, 346
   as Cl$^-$ donor, 204
   synthesis, 460
SF$_4$, 346
   addition to double bonds, 469
   as F$^-$ acceptor, 209
   as fluorinating agent, 468–470
   hydrolysis of, 430
   mo's, 153
   pseudorotation, 432
   reaction with Cl$_2$, 431
   synthesis, 461
SF$_2$R$_2$, pseudorotation, 432
SF$_3$R, pseudorotation, 432
ArSF$_3$, synthesis, 470
SH$_4$, stability, 274
SO$_2$Cl$_2$, aminolysis, 508
   hydrolysis, 430
   oxygen transfer reaction, 459
   synthesis, 461
SO$_2$F$_2$, synthesis, 462

SO$_2$(NR$_2$)$_2$, 508
SO$_2$X$_2$, 347
S(CH$_3$)$_2$BH$_3$, 207
SF$_5$ radicals, 473
SF$_5^-$, reaction with Cl$_2$, 431
SOF$_4$, 347
H$_2$SO$_3$, 345
HSO$_3$Cl, 347, 511
SO$_3$F$^-$, 347
HSO$_3$F, 511
   as solvent, 259–262
HSO$_3$SiMe$_3$, 494
SF$_6$, 346
   hydrolysis, 430
   mo's for, 182
   synthesis, 460
SF$_5$Cl, photochemistry, 473
   reactions, 473
   synthesis, 461
SH$_6$, stability, 274
H$_2$SO$_4$, as solvent, 249–255
S$_2$Cl$_2$, 346
   solvolysis, 508
   and synthesis of mustard gas, 481
   synthesis, 460
S$_2$F$_2$, 343
   synthesis, 461
S$_2$F$_{10}$, 343
   synthesis, 461
S$_2$O, 345
S$_2$O$_4^{2-}$, 344
S$_2$O$_5^{2-}$, 345
S$_2$O$_6$F$_2$, 261, 341
H$_2$S$_2$O$_7$, formation in H$_2$SO$_4$, 251
S$_2$O$_8^{2-}$, 341
(NSCl)$_3$, synthesis, 508
(NSO)$_3^{3-}$, 508
(SO$_2$NR)$_3$, 509
S$_4^{2+}$, 262
S$_4$N$_4$, 504
(NSF)$_4$, synthesis, 507
(HNS)$_4$, synthesis, 507
S$_8^+$, 262
S$_8^{2+}$, 262
S$_n^{2+}$, 342
S$_n^{2-}$, 342
   formation in liquid ammonia, 248
S$_n$Cl$_2$, 346
H$_2$S$_n$, 342
   synthesis, 480
HS$_n$SO$_3$H, 344
S$_{n+2}$O$_6^{2-}$, 344
(SN)$_n$, mo's for, 506
   synthesis, 505

Sb

SbF$_3$, as fluorinating agent, 467
SbCl$_5$, as Cl$^-$ acceptor, 204
SbF$_5$, as F$^-$ acceptor, 204
   as fluorinating agent, 467
SbF$_x$(OSO$_2$F)$_{5-x}$, 260
SbCl$_6^{3-}$, formation in SO$_2$, 259
SbF$_2$(OSO$_2$F)$_6^-$, 260

Si

SiCl$_4$, synthesis, 457
SiF$_4$, as Lewis acid, 202
   hydrolysis of, 320

SiF$_4$ (*Continued*)
  mo's, 153
  synthesis, 457
SiO$_2$, chirality of, 321
"SiO$_3^{2-}$", 321
SiO$_4^{4-}$, 321
R$_3$SiOH, 319
SiF$_5^-$, 202
  structure, 320
SiF$_6^{2-}$, 202
Si$_2$, 318
Si$_2$O$_7^{6-}$, 321
(R$_2$SiO)$_n$, 319
(Si$_2$O$_5^{2-}$)$_n$, 325
(Si$_4$Al$_2$O$_{12}^{4-}$)$_n$, 325
(Si$_4$O$_{11}^{6-}$)$_n$, 322
(Si$_5$Al$_2$O$_{14}^{2-}$)$_n$, 325
(Si$_{12}$Al$_{12}$O$_{48}^{12-}$)$_n$, 326
Si$_n$Cl$_{2n+2}$, 317
Si$_n$F$_{2n+2}$, 317
Si$_n$H$_{2n+2}$, synthesis, 317, 477

Sn

SnCl$_2$, as Lewis acid and base, 203
SnCl$_3^-$, as Lewis base, 202
SnR$_3$, radicals in liquid ammonia, 248
Sn(CH$_3$)$_3$Cl adducts, 202
Sn$_n$R$_{2n+2}$, formation in liquid ammonia, 248

Sr

SrF$_2$, as fluorinating agent, 293

Ti

TiCl$_3$, Ziegler-Natta co-catalyst, 974
Ti[N(SiMe$_3$)$_2$]$_3$, 592
TiCl$_4$, structure, 526
TiO$_2$, structure, 282
[Ti(H$_2$O)$_6$]$^{3+}$, spectrum of, *524, 564*
[(C$_5$H$_5$)(C$_5$H$_4$)TiH]$_2$, titanocene, 878(t), 889

Tl

TlC$_5$H$_5$, in metallocene preparation, 877

U

U(C$_8$H$_8$)$_2$, preparation and properties, 893 ff.
  mo's for, *896*

V

V(C$_5$H$_5$)$_2$, properties, 878(t)
V(CO)$_6$, synthesis, 856
[V(H$_2$O)$_6$]$^{3+}$, spectrum of, *524, 564*
  analysis of, 570
[V(CN)$_7$]$^{4-}$, 605

Xe

Xe, as Lewis acid and base, 212–213
XeF$_2$, aminolysis, 512
  as Lewis acid and base, 212
  hydrolysis, 512
  σ mo's, 158
  synthesis, 463
XeF$_2^+$, 212
XeFN(SO$_2$F)$_2$, 354
XeF$_3^+$, 212
XeF$_4$, as Lewis acid and base, 212
  hydrolysis, 512
  mo's, 153, 175
  synthesis, 464
XeF$_6$, as Lewis acid and base, 212
  hydrolysis, 512
  synthesis, 464
  VSEPR and, 80
XeOF$_4$, as Lewis base, 212
XeO$_2$F$_2$, as Lewis base, 212
XeO$_3$, as Lewis acid, 213
  synthesis, 512
XeO$_3$F$^-$, 213
XeF$_8^{2-}$, 213
XeO$_4^{2-}$, 355
XeO$_6^-$, 355
XeF$_n$, 354
XeO$_n$, 354
XeO$_n$F$_m$, 355
Xe$_2$F$_3^+$, 212
Xe$_2$F$_{11}^+$, 213

Zn

ZnS, lattice, 281

Zr

[Zr(C$_2$O$_4$)$_4$]$^{4-}$, structure, 608

SUBJECT INDEX

Items are listed alphabetically letter by letter, disregarding spaces and prefixes. *Italic* page numbers indicate figures; (t) indicates tables.

Absolute configurations, 647 ff.
Acetonitrile, complexes of, in organometallic synthesis, 882
Acetylacetonate complexes, thermodynamic data for formation of, 742
Acetylation, organometallic, 923
Acetylcholine, inhibition by phosphorus toxins, 433
Acetylene-metal complexes, 867–876
　bonding and structure, 871 ff.
　synthesis, 870
Acetylenes, as π donors, 198
Acid-base strengths, thermodynamic definition, 216–217
Acid catalysis, polymerization, 971
Acidity, as a factor in solvation, 236
　and basicity, organometallic compounds, 910 ff.
Actinide organometallic chemistry, 892
Activated complex, 362
Activation parameters, 371
Addition polymers, 970
Addition reactions, organometallic, 927–953
　1,1-addition to CO, 933 ff.
　1,2-addition to double bonds, 832, 927 ff.
Adduct stabilities, prediction of, 218
Adenine, structure, 1054
Adenosine triphosphate, structure, 1053
Adjacent attack, inner sphere electron transfer, 676
Alane adducts, reductive elimination, 492
Aldehydes, production from olefins, Wacker process, 967
Aldimine complexes, 782
Aldol condensation, metal ion assisted, 782
Alkali fluorides, analysis of stabilities, 275
Alkylating agents, reaction with metal halides or oxides, 824
Allen, A. D., 1088
Allyl, π mo's, 156
Allyl-metal complexes, 867–876
　bonding and structure, 871 ff.
　mo's for, 876
　synthesis, 871
η^3-Allyl-metal complexes, fluxional isomerism, 958 ff.
α-peptide chains, hemoglobin, 1067
Alumina, 313
Aluminum, organo compounds, synthesis, 476
　tri-*iso*-butyl-, 835, 840
　triphenyl-, addition to triple bonds, 929
Aluminum acids and bases, 194–196

Aluminum alkoxides, synthesis, 491
Aluminum alkyls, acidity, 910
　addition to ketones, mechanism, 932
　polymerization of olefins, 839
　synthesis, 822
　Ziegler-Natta catalysts, 973 ff.
Aluminum compounds, structures and stabilities, 310–315
Aluminum halides, synthesis, 454–457
Aluminum hydride, 310, 476, 835, 843
Aluminum organometallic compounds, addition to double and triple bonds, 928 ff.
Aluminum trimethyl, synthesis, 822
Ambidentate ligands, 615
Amine basicities, solvation and, 239
Amino acid amides, Co(III) complexes, hydrolysis of, 779
Amino acid ester/amide hydrolysis, metal ion catalyzed, 774 ff.
Amino acid esters, Co(III) complexes, hydrolysis of, 778
Aminoalanes, 312
　polymerization, 493
　synthesis, 492
Aminoboranes, oligomerization, 305
　synthesis, 489
Aminomethylation, ferrocene, 923
Ammonia, solutions of metals in, 234
　synthesis in, 247
Ammoniated electron, 246
Amphiboles, 322
Anation, 711
Angular functions, 17
　and orbital shapes, 26
Angular Overlap Model, 543–545
Anions, metal carbonyl, synthesis, 857
　metal-containing, reactions with organic halides, 830
Anodic oxidations, in HF, 255
Antenna chlorophyll, *1093*
Anti-bonding electrons, and structure, 525
Anti-bonding mo's, 133
Anti-cryolite structure, 282
Anti-ferromagnetic coupling, in Fe-S proteins, 1083
Anti-fluorite lattice, 281, 282
Antimony fluorides, as fluorinating agents, 467
Aquocobalamin, B_{12a}, 1077
Arachno, definition, 986
Arbusov reaction, 420
Arenes, bis-π-, 879(t)
　metal complexes, 879(t)

1105

Arrhenius acids and bases, 187
Association reactions, organometallic, 910
Associative activation mode, 362
Associative mechanism, 360
Asymmetric, definition, 620
Atactic polymers, *972*
Atomic orbitals, directed, 99–107
Atomic screening, 34
Atomic terms, 47–54
Atom transfer reaction, 656
Atom valence, "typical", 68, 69(t)
ATP, structure, 1053
ATPase, 1063
Aufbau principle, 34
Average potential energy, 23
Azotobacter, 1086

B, definition, 567
Back-bonding, 201
Back-strain, 226
Bailar, John C., Jr., 629
Band theory of metals, 296
Base catalysis, polymerization, 971
Basicity, as factor in solvation, 236
   thermodynamic effect, 700
Basolo, F., 719
Bent metallocenes, bonding, 887 ff.
   preparation, 881
Berry pseudorotation, 403, 597
β, overall stability constant, 734
β-elimination, 840, 931, 949
   in homogeneous hydrogenation, 964
   vitamin B_{12} coenzyme, 1076
   Wacker process, 969
   Ziegler-Natta catalysis, 974
β-peptide chains, hemoglobin, 1067
Binuclear elimination, 949
Body-centered lattice, 278, 283
Bohr effect, hemoglobin, 1068
Bohr magneton, 14(t), 578
Bond dissociation energies, 120, 268
Bond distances, and hao, 108
Bond energies, 120, 270(t), *272*
   and hao, 119–123
   definition, 268
   metal-carbon, 820, 854
   Ti-X and Zr-X, *854*
Bonding, metal alkyls and aryls, 843
   metal carbonyls, 861 ff.
Bonding mo's, 133
Bond order, 73
   olefinic and acetylenic, in metal complexes, 872
Bond pair dipole moments, 114
Borane, aminolysis of, 389
Borane adducts, of phosphines, stability, 398
   solvolysis mechanisms, 398
   substitution by phosphines, 393
Boranes, structures and bonding, 302
Borate esters, alkoxy exchange, 390
   formation of, 309
Borax, 308
Borazine, addition of HX to, 388
   mechanism of formation, 387
   synthesis, 306, 489
Borides, 301
Borinic acids, synthesis and bonding, 309, 488
Born equation, 277
Boron, elemental, 300
   mechanisms of reaction at, 386–392, 393–401
   organo compounds, synthesis, 476

Boron acids and bases, 194–196
Boron compounds, structures and stabilities, 300–310
Boron halides, aminolysis mechanisms, 387
   steric effects on, 388
   solvolysis mechanisms, 386
   solvolysis of adducts, 395
   synthesis, 454–457
Boron hydrides, *see also* Carboranes *and* Hydroborate ions.
   hydroborate ions, chemistry, 1011 ff.
   neutral, 988–1008
      molecular orbital concepts, 994
      organizational scheme, 1004
      structure and bonding, 988 ff.
      synthesis and reactivity, 997
      topology and *styx* numbers, 990
      toxicity, 1000
Boronic acids, 309
   phenyl, bromination of, 389
   synthesis, 488
Boroxines, synthesis, 488
Bridging halide, 195, 205
Bridging ligands, definition, 518
   in inner sphere electron transfer, 675
Bromine trifluoride, as fluorinating agent, 467
Brönstedt acids and bases, 187
Brown, H.C., 833
Brown, T.L., 920
Busch, D.H., 784

C—O stretching frequencies, metal carbonyls, 899 ff.
   number of bands expected, 901(t)
Cage complex, 360
Cage degradation, metallocarboranes, 1023
Cancer, and Pt chemistry, 752
Capped octahedron, 605
Carbalumination, 839
Carbene complexes, CH_2 complex, 889
   transition metals, 863 ff.
Carbon acids and bases, 196–203
Carbon fluoride, poly, 327
Carbonium ions, 196
   in H_2SO_4, 253
   in SO_2, 259
   in superacid media, 261
Carbon monoxide, metal complexes, 855–863
Carbon π donors, metal complexes, 866–898
Carbon σ donors, 817
Carbonylation, mechanism, 933 ff.
Carbonyl insertion (alkyl migration), 934 ff.
Carboranes, metallo-, 1020–1025
   mo's, 1009 ff.
   organizational scheme, 1004
   polyhedral rearrangements, 1017
   polymers, 1019
   preparation and chemistry, 1017 ff., *1018*
   structures, *1016*
   table of, 1015
Carboxypeptidase A, 774
Carbyne complexes, transition metals, 863 ff.
Caro's acid, 341
   in synthesis of H_2O_2, 480
   oxygen exchange, 424
Carotenes, 1092
Carroll, Lewis, 612
Catalysis, organometallic compounds, 962–979
Catenation, boron, 300
   halogens, 350

Catenation (*Continued*)
  in non-metals, synthesis, 486
  nitrogen, 329, 331
  oxygen, 341
  phosphorus, 330, 331
  silicon, 317
  sulfur, 342
Cations, metal carbonyl, 858
CB mechanism, 712, 725-729
CD (circular dichroism), 644 ff.
Cell, *1052*
Central metal, effect on square planar substitution reactions, 708
C—H coupling constants, 119(t)
Chain phosphines, 331
Characters, 90
Character tables, 89-97, 96-97(t)
Charge correlation of electron motions, 36
Charge transfer absorption bands, 198
  in I_2 adducts, 211
Charge transfer complexes, 198
Chazabite, 325
Chelate effect, 739
  in organometallic synthesis, 869
Chelates, definition, 518
Chemical activation, electron transfer, 662
Chemiosmosis, 1081
Chernyaev, 702
Chiral compound, definition, 620
  resolution, by electron transfer reactions, 684
Chiral coordination complexes, special nomenclature, 638 ff.
Chirality, changes in, for octahedral complexes, 765 ff.
    for tetrahedral complexes, 762 ff.
  in silane substitution, 403-409
Chloramines, solvolysis of, 496
  synthesis and reaction mechanisms, 413-417
Chlorides, of non-metals, synthesis, 453-464
Chlorine fluorides, as fluorinating agents, 467
  synthesis, 462-463
Chlorine nitrate, 336
Chlorine oxides, 352
Chloropentafluorosulfur, reactions of, 473
Chlorophyll, 1090
Chloroplasts, 1053
Chlorosilanes, 317
Chlorosulfuric acid, 511
Chromate ion oxidations, 682
Chromocene, electronic structure, 886
Circular birefringence, definition, 646
Circular dichroism (CD), 644
Cis effect, square planar replacement reactions, 707
Class, of symmetry elements, 85
Class *a* and *b*, metal classification, 616, 735
Closo, definition, 986
Closo-boranes, 304
Cobalamins, 1073
Cobalt compounds, bis(dimethylglyoximato)-methyl(pyridine)cobalt, 852
  dicyanotris(dimethylphenylphosphine)cobalt, reaction with oxygen, 689
  pentacyanocobaltate(3−), electron transfer agent, 688 ff.
  synthesis and chemistry of, 742-749
  tris(s-alaninato)cobalt, absorption and CD spectra, *649*
    stereoisomers, 648
  tris[tetraammine-μ-dihydroxocobalt]cobalt-(6+), 628
Cobaltocene, electronic structure, 886
  properties, 878(t)

Cobaltocene (*Continued*)
  reaction with oxygen, 884
Cobamide, 1074
Coboglobin, 1069
Collman, J.P., 937, 944
Collman's reagent, 937
Common ligands, 519(t), inside cover
Complexes, electronic states and spectra, 559-575
  g factor for, 578(t)
  high spin, definition, 546
  low spin, definition, 546
  magnetic susceptibility, 580
  no model for, 531-543
  octahedral, mo's for, *535*
  predicting structures of, 549
  spin-orbit coupling, 581
  square planar, mo's for, *539*
  state energy diagrams, *584*
  structure preference energies, *550*
    experimental evidence, 551
  Tanabe-Sugano diagrams for, *584*
  tetrahedral, mo's for, *542*
Compton effect, 14
Condensation energies, and adduct stabilities, 217
  in thermodynamics, 274
Condensation polymers, 970
Condensation reactions, organometallic, 912
Conductor, definition of, 297
Configuration number, 623
Conjugate base mechanism, 712, 725-729
  for haloamines, 410
  for NH_2NO_2, 413
Constitutional isomerism, 613 ff.
Contour maps, 30
Cooperativity, hemoglobin, 1067
Coordinated ligands, organometallic, electrophilic and nucleophilic attack, 923-926
  reactions of, 773-789
Coordinated metal, definition, 518
Coordination chemistry, classical, definition, 515
Coordination complex, definition, 518
Coordination isomerism, 614
Coordination number, definition, 518
Copper, *n*-butyl(tri-*n*-butylphosphine)-, thermal decomposition, 950
Corrin, 1073
Corundum, 314
Cotton, F.A., 957
Cotton effect, definition, 646
Cotton-Kraihanzel model, C-O force constants in metal carbonyls, 903
Counter ions, use in stabilizing complex ions, 294
Creatine kinase, 1061
Cross reactions, electron transfer, 667
Cryolite structure, 282
Crystal field theory, 532
Cubic close packing, 287
Cubic packing, 287
Curie law, 582
Curtis, N.F., 786
Cyanocobalamin, 1074
Cyanogen, 328
Cyclic mo's, ao phasing, 162
Cyclic phosphate esters, hydrolysis mechanisms, 436
Cyclic phosphines, 331
Cyclic π donors, metal complexes, 876-898
Cyclic voltammetry, description, 1037

Cycloheptatriene, complexes of, electron counting, 798
Cyclooctatetraene dianion, metal complexes, 893 ff.
mo's for, 893
Cyclooligomerization, 976
Bis(η^5-cyclopentadienyl)metal compounds, 877 ff.
Cysteine, in *azotobacter* nitrogenase, 1086
Cytochromes, 1078–1081
Cytoplasm, 1052

$d^{6/8/10}$ complexes, EAN rule, 794
d^{10}, stability of organometallic complexes, 806
DABCO, adducts with alanes, 312
Dangling axial ligand, 767
Dangling equatorial ligand, 767
Dative bonding, 68, 70(t)
Davisson and Germer experiment, 15
DCD model, bonding in metal-olefin complexes, 868, 871
de Broglie, L., 15
Decaborane, structure and bonding, 304
Decaborane(14), chemistry, 1002 ff.
Decarbonylation, mechanism, 933 ff.
Δ (delta), stereochemical notation, 638 ff.
Δ_0, definition, 535, 545
values for H_2O, 548(t)
values for metal carbonyls, 862
Δ_t, definition, 542, 546
5'-Deoxyadenosine, structure, 1075
DeoxyHb, deoxyhemoglobin, 1067
Deoxyribonucleic acid, composition, 1055
Desargues-Levi topological maps, for silane substitution, 404
Dewar-Chatt-Duncanson model, metal-olefin complexes, 868, 871
Dialkylsulfites, tautomerism, 510
Diamond-square-diamond, carborane rearrangement, 1017
Diastereoisomer, definition, 620
Diastereotopic groups, 763
Diatomic molecular orbitals, 141–149
Diazene, in nitrogen fixation, 1087
Dibenzenechromium, mo's for, 890 ff.
preparation, 879
Diborane, dissociation energy, 302
symmetrical cleavage, 399, 400(t)
unsymmetrical cleavage of, 399
use in hydroboration, 391, 483, 832
Dicyclopentadiene, 877
Dielectric constant, as factor in solvation, 236
Diffusion controlled reactions, 365
Difluoramine, reaction mechanisms of, 417
Diglyme, complexes of, in organometallic synthesis, 882
Dinitrogen tetrafluoride, as fluorinating agent, 470–473
Diorganoaluminum hydrides and fluorides, structure, 844
Dioxygen, mo's for complexes with CoD_5^{2+}, 1071
mo's for Fe binding in hemoglobin, 1070
Dioxygen bridges, in cobalt complexes, 745–748
Diphosphoglycerate, DPG, 1058, 1067
Diphosphoric acid, 501
Dipole moments, and hao, 114
bond pair, 114
lone pair, 114
Direct metal insertion, metallocarboranes, 1024

Direct reaction, metal and CO, 856
metal and organic halide, 818 ff., *819*
metal complex with olefin, 868
Direct synthesis, trialkylaluminum compounds, 835
Discharge tube, electrode, 449
microwave, 450
Disiamylborane, 832
Dissociative activation mode, 362
Dissociative mechanism, 360
Dissymmetric, definition, 620
Disulfides, exchange reaction mechanisms, 426
Disulfite ion, 345
Dithiolates, 595
Dithionite, 344
DNA, composition, 1055
reproduction, 1056
DNA polymerase, 1055 ff.
E. coli, 1057
Dodecahedron, *608*
DPG, diphosphoglycerate, 1058, 1067
Drago, R.S., 705
Dualistic nature of light, 15

E and C parameters, 219, 220(t), 221(t)
Effective atomic number (EAN) rule, 792–809
definition, 793
organometallic reactions, 909 ff.
Effective nuclear charge, 35
Eight-coordinate complexes, 606 ff.
modes of interconversion, *608*
Eighteen Electron Rule, 792–809
organometallic reactions, 909 ff.
Electrochemistry, of $[(\eta^5-C_5H_5)Fe(CO)]_4$, *1037*
Electrolysis, in synthesis, 450
Electron affinity, 40, 63(t)
Electron configurations, periodic table and, 33–44
Electron counting, common ligands, 800(t)
organometallic complexes, 792–809
Electronegativity, 54, 59(t)
atom, 54
Mulliken, 55, 57(t)
orbital, 54
Pauling, 58, 348
Electroneutrality, 604
Electronic states, and spectra of complexes, 559–575
Electron matter waves, 13–17
Electron motions, charge correlation, 36
spin correlation, 36
Electron pair repulsion model, 75–81
Electron point probability density, 17
Electron transfer, key ideas, 660
orbital following model, *657, 674*
Electron transfer agents, biochemical, 1078–1094
Electrophilic attack on coordinated ligands, organometallic, 923–926
Electrophilic substitution, at nitrogen, 410
Elimination, β, 840
reductive, 479, 492, 841, 950
Elimination reactions, organometallic, 927–953
Embden-Meyerhof pathway, glucose oxidation, 1059
Enantiomer, definition, 620
Enantiotopic groups, 763
Encounter complex, 360
Endergonic, definition, 1053
Energy and nodes, 29
Energy profile diagrams, 361

Entering group, influence in square planar replacement reactions, 700
Enzyme, definition, 1054
Equilibrium, review of, 265
Ester carbonyl/metal ion interaction, mo's for, 776
Ester hydrolysis, mechanism, 775
Esters, ν_{CO} in adducts, 112
Tris-ethylenediamine complexes, *ob* and *lel* conformations, *643*
Ethyl glycinate, metal ion catalyzed hydrolysis, 777(t)
Ewing-Basset convention, 519
Exchange energy, 40
Exchange reactions, alkoxy groups between borate esters, 390
    amines with amine boranes, 393
    between $Ga(CH_3)_3$ and its amine adducts, 393
    between $In(CH_3)_3$ and its amine adducts, 393
Exergonic, definition, 1053

Face-centered lattice, 278, 279
fac isomers, 624, 648
Facultative ligand, 598
FAD, flavin adenine dinucleotide, 1079
Faujasite, 325
Fermi contact coupling, 118
Ferrate(1−), *trans*-1,2-diaminocyclohexane-N,N'-tetraacetato-, structure, 606
Ferredoxins, 1084
Ferrocene, barrier to ring rotation, 883
    chemistry, *814, 923–924*
    electrochemistry, 884
    metal basicity, 889
    mo's for, 885 ff.
    preparation, 877
    properties, 878
    substitution of C_5H_5, 922
α-Ferrocenylcarbonium ions, 925
Fieser's reagent, 344
First order rate law, derivation and interpretation, 369
Fischer, E.O., 812, 879
Five-atom metal clusters, 1040–1045
Five-coordinate complexes, 596
    in square planar substitution reactions, 698
    substitution reaction, 730
Flavin adenine dinucleotide, FAD, 1079
Fluorides, of non-metals, synthesis, 453–464
Fluorinating agents, 465–474
    ionic, lattice energies and, 292
Fluorination, of carbonyl group, 467, 468
    of carboxylate groups, 468
    of phosphoryl group, 468
    reactions of HSO_3F and superacids, 260
Fluorine, bonds to, stability of, 349
Fluorine nitrate, 336
Fluorite structure, 282
Fluorosulfuric acid, 511
Fluorosulfurous acid, 511
Fluxional isomerism, organometallic, 957
Fock equation, 129
f orbitals, *894–895*
Force constants, and hao, 110
    C—O stretching, 903
Formal charge, atom, 68, 72(t)
Forward rate constant, general expression, 364
Four-atom metal clusters, 1035–1040
Four-coordinate complexes, 593
    stereoisomerism, 625 ff.
Frankland, E., 811

Free energy, review of, 265
Free radical initiation, polymerization, 970
Free radicals, in oxidative addition reactions, 948
Friedel-Crafts acylation, of ferrocene, 884
Friedel-Crafts catalysts, 313
Friedel-Crafts reactions, in SO_2, 259
Front-strain, 226

g factor, for complexes, 578(t)
Glucose storage, 1060 ff.
Glycogen, 1060
Graham, W.A.G., 904
Gray, H.B., 705
Grignard, V., 811
Grignard reagents, degree of association in ether, *850*
    equilibria possible in solution, *852*
    structures and composition, 821, 850 ff.
Group theory, 82–95
    infrared spectroscopy of metal carbonyls, 900 ff.
Growth reaction, organoaluminum compounds, 839, 928

Halates, 353
Halide transfer reactions, 204
Haloamines, nucleophilic attack upon, 410
Halogenation, ferrocene, 924
Halogen cations, formation in superacids, 261
Halogen compounds, structure and stability, 349
Halogen fluorides, 351
Halogens, addition to double bonds, 467
    as Lewis acids and bases, 209–213
    mechanisms of reactions at, 429–430
Halometalation, 838
Hapto, definition, 796
Hawthorne, M.F., 1024
Heat of atomization, 820
Heat of formation, metal alkyls, 820
Hectorite, 327
Heme, 1065
Hemin, 1065
Hemoglobin, 1064–1072
    O_2 binding, mo's for, 1070
Heterogeneous catalysis, silicates and, 326
    with Rh_4^{2+} and silicates, 327
Hexamethyldisilazane, 494
High coordination numbers, 603
Highest occupied molecular orbital (HOMO), 137
High potential iron protein, HIPIP, 1085
High spin complexes, definition, 546
High spin–low spin crossover, 573
Hill constant, for hemoglobin, 1068
HIPIP, high potential iron protein, 1085
HOMO, 137
Homoleptic, organometallic compounds, definition, 815
Howe, James Lewis, 750
Hund's rule, 51
Hybrid atomic orbitals, 99, 104(t)
    electronegativities, 57
Hydrate isomerism, 613
Hydration energies, of divalent transition metals, *554*
Hydrazine, in nitrogen fixation, 1087
Hydrides, Lewis basicity, 191

1110 □ SUBJECT INDEX

Hydrides (*Continued*)
  mechanisms of reactions, 383, 385
  of non-metals, synthesis, 474–488
Hydroalumination, in synthesis, 483, 834
Hydroborate ions, chemistry, 1011 ff.
  mo's, 1009
  polyhedral, organizational scheme, 1004
Hydroboration reaction, 391, 832
  in synthesis, 483
Hydroformylation, stoichiometric mechanism, 934
Hydrozirconation, 933
Hydrogen, as Lewis acid and base, 188–194
  exchange reaction, 383
  mechanisms of reactions at, 383–386
  methods of catalytic activation, 963
Hydrogen affinity, 226
Hydrogenation of olefins, catalyzed, 963 ff.
Hydrogen atom wavefunctions, 61(t)
Hydrogen bonding, 188
  mo's for, 189
Hydrogen halides, mechanism of synthesis, 482
  synthesis, 481
Hydrometalation reactions, 483–486, 832
  use in catalytic processes, 836
Hydrostannation, 836
Hydroxocobalamin, B_{12b}, 1077
Hydroxo-Co(II)cobalamin, 1077
Hydroxylamine, complex with Ir, 414
  mechanism of synthesis, 418
    oxygen atom transfer, oxidation of phosphines, 412
      oxidation of sulfites, 419
Hydroxylamine-O-sulfonate, hydrolysis of, 410
  nucleophilic attack upon, 410
Hyperconjugation, in $B(CH_3)_3$, 230
Hypohalous acids, 352
  mechanisms of reactions, 425
Hyponitrous acid, 338
Hypophosphorus acid, 332, 503

Imidoalanes, 312, 492
Imine complexes, 782
Imine sulfurdifluorides, 470
Inert, to substitution, 659
Infrared spectroscopy, metal carbonyls, 899 ff.
Inner coordination sphere, definition, 519
Inner sphere mechanism, definition, 659
  discussion, 669 ff.
  oxidant/reductant electronic structure, 673
  table of rate constants, 672, 678
Insertion reaction, of O_2, 412
  of $SnCl_2$, 203
  of SO_2, 208
  organometallic, 832 ff.
Insulator, definition of, 297
Interchange mechanism, 360
Interhalogens, 349
  as Lewis acids and bases, 211
  solvolysis, 511
Intermediate, 362
Intimate mechanism, square planar replacement reactions, 708 ff.
Intramolecular acid/base interaction, 195
Intramolecular metalations, 827
Ionization isomerism, 614
Ionization potential, 40, 62(t)
Ion radii, 283–286, 284(t)
Ion solvation energies, 294, 295(t)
Iron, absorption from gut into plasma, electron transfer in, 658

Iron compounds, organometallic, *814*
Iron-sulfur proteins, 1082
Irreducible representations, 90
Irving-Williams series, 736
Isomer enumeration, 629 ff.
Isomerism, constitutional, 613 ff.
  coordination, 614
  hydrate, 613
  ionization, 614
  linkage, 615
  polymerization, 614
  polytopal, 593
Isomerization, olefin, 977
Isoprene, 840
  polymerization, 972
Isotactic polymers, *972*
Isotropic shift, 759
  definition, *760*
Isovalent hybridization, 109

Jahn-Teller distortions, 553
Jørgensen, C.K., 532
Jørgensen, Sophus Mads, 615, 625

Kapustinskii equation, 290
Ketimine complexes, 782
Ketones, ν_{CO} in adducts, 112
Kinetic data, tables of, *see* Rate constants.
Kinetic template effect, 784

Labile, to substitution, 659
Λ (lambda), stereochemical notation, 638 ff.
Lande *g* factor, 578
Langford, C.H., 705
Lanthanide contraction, 60, 529
Lanthanide organometallic chemistry, 892
Lappert, M.F., 815, 852
Lattice energies, 276–296
  MCl_2, *554*
  synthesis principles and, 290–296
Lattice structure and stoichiometry, 279–283
Layer lattices, 286–290
LCAO, 128
Leaving group, in square planar replacement reactions, 707
lel form, octahedral complexes, *642*
Lewis acidity and basicity, organometallic compounds, 910 ff.
Lewis acids and bases, 187
Lewis structures, 65
Lifschitz salts, 558
Ligand, definition, 518
Ligand conformation, 636–637, *643*, 650–652
Ligand field stabilization energies, 553
Ligand protonation, 916
Ligand substitutions, organometallic, 918–922
Linear combination of atomic orbitals (LCAO), 128
Linkage isomerism, 615
Linkage isomers, synthesis by electron transfer, 688
Lipscomb, W.N., 986, 990
Liquid ammonia solvent, 241–248
Literature, organometallic chemistry, 816
Lithiation reactions, 826
Lithium alkyls, acidity, 910
  polymerization initiators, 972

Lithium alkyls (*Continued*)
  synthesis, 820
Localized mo's, from delocalized mo's, 183
Lone pair dipoles, 114
Low coordination numbers, 590
Lowest unoccupied molecular orbital (LUMO), 135
Low spin complexes, definition, 546
LUMO, 135

Macrocyclic effect, 785
Macrocyclic ligands, 783–789
Madelung constant, 278, 279(t)
Magnesium organohalides (Grignard reagents), structures and composition, 850 ff.
Magnetic dipole model, 577
Magnetic moment, 577
Magnetic susceptibility, 580
Magnetogyric ratio, 578
Magnus' green salt, 751
Manxine, proton affinity of, 227
Marcus-Hush equation, 668
Mechanisms, basic concepts, 359–364
  dual channel, 378–382
  single channel, 378–382
  survey of, 382–444
Mercuration reactions, 828
Mercury, diphenyl-, decomposition, 949
  organic derivatives, 821
Mercury alkyls and aryls, reactions with metals, 825
mer isomers, 624, 648
Messenger RNA, 1056
Metaboric acid, 308
Metal alkyls and aryls, structure and bonding, 843
  synthesis, 817
Metalation, ferrocene, 924
Metalation reactions, 826
Metal atom reactions, 823, 869, 880
Metal basicity, 213, 312 ff., 913
  ferrocene, 889
  in cobalamins, 1077
  use of infrared spectroscopy in diagnosis, 903
Metal-carbon bond energies, 820, 854
Metal-carbon σ bonds, stability, 948
Metal carbonyl anions, nucleophilicity, 914
Metal carbonyls, 855–863
  CO substitution, rate constants, 920(t)
  infrared spectroscopy, 899 ff.
  properties, 859(t)
Metal clusters, 1026–1049
  electron counting, 1036
  three-atom, 1033–1035
  four-atom, 1035–1040
  five-atom, 1040–1045
  six-atom, 1040–1045, 1047
    bonding, 1043, 1047
  table of, 1028–1029
Metal-containing anions, reactions with organic halides, 830
Metal-containing solids, 276–300
Metal halides and oxides, reactions with alkylating agents, 824
Metal-hydrogen exchange, 826
Metallocarboranes, 1020–1025
Metallocenes, 877 ff.
  bent, structure and bonding, 887 ff.
  properties, 878(t)
  structure and bonding, 883
Metalloporphyrins, redox, 789

Metal-metal bonds, 1026–1049
  Cr-Mn, 213
  In-Co, 214
  Si-Mn, 213
  Sn-Mn, 202
Metals, 296–300
  band theory of, 296
  oxidative stabilities, 299
  solutions in ammonia, 243
  structures of, 296
Metaphosphoric acid, 501
Metasilicates, 321
Metathesis, olefin, 977
Methyllithium, structure and bonding, 848
Methylmalonyl mutase, 1075
Micas, 322
Microstates, 49
Mitochondria, 1052
Mitochondrial respiration, 1082
Molecular orbital model for complexes, 531–543
Molecular orbitals, 127–185
  B_5H_9, 996
  $B_6H_6^{2-}$, 1010
  construction and interpretation, 134–155
Molecular terms, 139
Molecular topography, guidelines, 70
Molybdenocene, preparation and reactions, 881
Molybdenum, in *Azotobacter* nitrogenase, 1086 ff.
Monodentate ligand, definition, 518
Muetterties, E.L., 957
Multiple bonds, AlN, 312
  BN, 306
  BO, 307
  NN, 331
  NO, 336
  OO, 340
  PF, 333
  PN, 333
  PO, 332
  PP, 331
  SS, 340
  SiO, 319
  SiSi, 318
Mustard gas, synthesis, 481
Myoglobin, 1064–1072

Na^+/K^+ ion pump, 1063
NAD^+, composition, 1058, 1079
$NADP^+$, structure, 1091
Natta, G., 973
N-H coupling constants, 119(t)
Nickel, bis(η^3-2-methylallyl)-, structure, 876
Nickelocene, electronic structure, 886
Nicotinamide adenine dinucleotide, NAD^+, 1058, 1079
Nido, definition, 986
Nine-coordinate complexes, 609
Nitramide, hydrolysis of, 413
Nitration, ferrocene, 924
Nitric acid, 337
Nitrogen, as Lewis acid and base, 203
  electrophilic substitution at, 410
  mechanisms of reactions at, 410–422
Nitrogenase, definition, 1086
Nitrogen compounds, structures and stabilities, 329–339
Nitrogen fixation, 1086–1089
Nitrogen halides, synthesis, 457–459
Nitrosation reactions, mechanisms of, 417
Nitrosyl chloride, mechanism of hydrolysis, 417

Nitrosyl halides, 335
Nitrous acid, 337
   reaction mechanisms, 417
Nitryl chloride, 335
Nitryl fluoride, 335
Noble gas compounds, structure and stability, 354
Nodes, and energy, 29
   matter wave, 21
Nomenclature, of complexes, 519
   organometallic, 796
Non-bonding mo's, 136
Non-complementary reactions, electron transfer, 681
Non-metal acids and bases, 194–213
Non-specific solute/solvent interactions, 235
Normalization condition, 16
NO single bond, weakness of, 336
Nuclear spin coupling, 117–119
Nucleophilic attack on coordinated ligands, organometallic, 923–926
Nucleophilic constants, for trans-[Pt(py)$_2$Cl$_2$], 701(t)
Nucleophilicity, kinetic effect, 700
   metal carbonyl anions, 914
Nucleophilic ligand substitutions, 918–922
Nucleoside, definition, 1053
Nucleotide, definition, 1053
Nucleotide transfer, 1055 ff.
Nucleus, cell, 1052

ob form, octahedral complexes, 642
Occupation number, 17
Octahedral complexes, isomers of, 635(t)
   mo's for, 535
   substitution reactions, 710–731
      rate laws for, 714
Octahedral structures, mo's for, 177, 182
Olefin, as π donor, 198, 254
Olefin hydrogenation, catalyzed, 963 ff.
Olefin isomerization, 835, 977
Olefin-metal complexes, 867–876
   bonding and structure, 871 ff.
   synthesis, 868 ff.
Olefin metathesis, 977
Olefin polymerization in H$_2$SO$_4$, 255
One-dimensional molecules, mo's, 155–158, 157
Optical activity, 644 ff.
   definition, 620
Optical rotatory dispersion, (ORD), 644
Orbital contraction, 43
Orbital energies and radial density functions, 18
Orbital following, in chemical reactions, 373–378
   in disproportionation of N$_2$H$_2$, 377
   in electron transfer, 657, 674
   in nucleophilic attack at terminal atom, 375
   in 1,2-elimination reactions, 375
ORD (optical rotatory dispersion), 644 ff.
Organoboranes, solvolysis of, 389
   structures and bonding, 305
Organohypochlorites, 512
Organometal reagents, in non-metal compound synthesis, 475
Organometalations, 839 ff.
Organometallic compounds, definition, 517
Organonon-metal compounds, synthesis, 474–488
Organophosphates, hydrolysis mechanisms, 434
Organophosphine oxides, synthesis, 501
Organophosphorus cations, synthesis, 485

Organosilane anions, in synthesis, 485
Organosiloxanes, synthesis, 494
Organosulfur cations, synthesis, 485
Organotin compounds, 813
   commercial uses, 812
Orthoperiodic acid, 353
Outer coordination sphere, definition, 519
Outer sphere mechanism, definition, 659
   discussion, 660 ff.
   rate constants, 663(t), 666(t)
Overlap and symmetry, 131
Overlap integral, 107, 130
Overlap probability, 131
Oxidation number, of metals, convention, 520
Oxidation states, of cations, lattice energies and, 291
   stabilization of, 271
Oxidative addition, 206
   in homogeneous hydrogenation, 964
   to phosphorous esters, 420
   to phosphorus trihalides, 419
   transition metal compounds, addition of acyl azides, 944
      electron counting, 806
      general discussion, 938 ff.
      influence of metal, ligands, and addend, 943 ff.
      kinetic data, 940(t)
      mechanism, 947
      stereochemistry, 941
      to carbonyltris(triphenylphosphine)palladium, 942
      use in synthesis, 829
Oxidative phosphorylation, 1079
Oxo reaction, 966 ff.
Oxygen, as Lewis acid and base, 205–209
   mechanisms of reactions at, 422–426
Oxygen atom transfer, formation of POCl$_3$ from PCl$_5$ and SO$_2$, 258
   to SO$_2$, 258
   with HOBr, 430
   with HOCl, 425
   with hydroxylamine, 412, 419
   with hypohalous acids, 425
   with peroxides, 423
Oxygen carriers, 1064–1072
Oxygen compounds, structures and stabilities, 340–342
Oxygen halides, synthesis, 459–462
OxyHb, oxyhemoglobin, 1067
Oxymercuration, 837
Oxymetalations, 837
Oxypalladation, 838
Ozonizer, in synthesis, 450

Pairing energy, definition, 547, 548(t)
Palladium, addition of organopalladium to double bond, 931
   carbonyltris(triphenylphosphine)-, oxidative addition to, 942
Palladium η^3-allyl complexes, fluxional isomerism, 959
Palladium salts, addition to olefins, 841
   catalyst in Wacker process, 968
Paramagnetism of complexes, 577–582
Parry, R.W., 998
Pauli exclusion principle, 36
Pearson, R.G., 719
Penetration, orbital, 22
Pentaborane, structure and bonding, 303
Pentaborane(9), chemistry, 1001

Pentagonal bipyramid, 605
Pentavalent phosphorus, esters, hydrolysis mechanism, 433–441
  pseudorotation, 431
PEP, phosphoenolpyruvate, 1058
Peptide hydrolysis, promoted by Co(III) complex, 781
Perhalates, 353
Pernicious anemia, 1074
Peroxides, 341
  disproportionation, 424
  mechanisms of reactions, 423–425
P-H coupling constants, 119(t)
Phosphagens, 1061
Phosphate esters, 332
  cyclic, hydrolysis mechanisms, 436
  pseudorotation in hydrolysis, 437
Phosphate hydrolysis, metal-ion catalyzed, 1053 ff.
Phosphate storage in muscle, 1061
Phosphate transfer, 1057 ff.
Phosphazenes, synthesis, 497
Phosphine(s), catenated, synthesis, 486
  oxidation, by hydroxylamine, 412
  substitution of amines at boron, 393
Phosphinoalanes, synthesis, 492
Phosphinoboranes, synthesis, 490
Phosphinous acids, 496
Phosphocreatine, 1061
Phosphoenolpyruvate, PEP, 1058
Phosphoglucomutase, 1060
Phosphonitriles, 497
  mo's for, 498
Phosphorous acid, 332
  esters of, synthesis, 332, 496
Phosphorus, black, 330
  mechanisms of reactions at, 410–422, 430–441
  organo compounds, synthesis, 477
  red, 330
  white, 330
Phosphorus compounds, as Lewis acids and bases, 203
  structures and stabilities, 329–339
Phosphorus halides, hydrolysis mechanisms, 430
  hydrolytic stability, 332
  synthesis, 457–459
Phosphorylation, 1057
Phosphoryl halides, solvolysis of, 497
Photochemical apparatus, in synthesis, 450
Photoelectron spectra, 139
Photosystems I and II, in photosynthesis, 1091
Phthalocyanine, structure, 786
Phytyl group, in chlorophyll, 1090
π acceptor ligands, definition, 536, *537*
π acids and bases, 197
π donor ligands, definition, 536, *537*
π donors, cyclic, metal complexes, 876–898
  synthesis and properties, 877 ff.
π parameter, metal carbonyl bonding, 904
π resonance effects, on Lewis basicity and acidity, 230
"Picket fence" porphyrin, structure, 1066
Picolines, basicities of, 229
Plasmas, in chemical synthesis, 449–450
Plastocyanin, in photosynthesis, 1091
Plastoquinone, in photosynthesis, 1091
Platinum compounds, synthesis and chemistry, 750–755
Point group character tables, 96–97
Point groups, common, *87*
Polarized light, *645*
Polyhedra, deltahedra, *984*
Polyhedra (*Continued*)
  molecular, definition, 981
Polyhedral expansion, metallocarboranes, 1024
Polyiodides, 351
Polymer-bound catalysts, 978
Polymerization, metal catalyzed, 970 ff.
Polymerization isomerism, 614
  platinum complexes, 616(t)
Polymers, stereochemistry, *972*
Polythiazyl, 505
Polythionate ions, 344
Polytopal isomerism, 593
Porphin, structure, 1065
Porphyrins, as macrocyclic ligands, 785
Precursor complexes, inner sphere electron transfer, 671
Principle of microscopic reversibility, 378–382
  in phosphate ester hydrolysis, 439
  in silane substitution, 408
Probability density functions, 13–33
Prosthetic group, 1078
Proton affinities, 226–227
Protonation, coordinated organic ligands, 916
Proton transfer reactions, 188, 365
Protoporphyrin IX, structure, 1065
Proximal histine, 1069
Pseudo-first order rate law, derivation and interpretation, 368
Pseudorotation, Berry, 403, 597
  in pentavalent phosphorus compounds, 431–433
Pyrophoric reaction, of phosphines with O_2, 412
Pyrosilicate, 321
Pyroxenes, 321
Pyruvate kinase, 1058

Quartz, 320
  chirality of, 321
Quinuclidine, as Lewis base, 229
  proton affinity of, 227

Radial density functions, and orbital energies, 18
Radial function, 17, 18(t)
Radii, divalent transition metals, *555*
  ionic, 283–286, 284(t)
  nonpolar covalent, inside back cover
  transition metals, *528*
  trivalent transition metals, *555*
Raschig synthesis of hydrazine, 414, 478
Rate constants and kinetic data, tables of
  aquation of *cis*- and *trans*-Co(N$_4$)ACl^{n+}, 723
  for substitution reactions of $[M(H_2O)_6]^{2+}$, 715
  inner sphere reduction of Co(III) complexes, 672, 678
  M^{2+}-catalyzed hydrolysis of ethyl glycinate, 777
  nucleophilic displacement at boron, 394
  outer sphere electron transfer, 663, 666
  oxidative additions of MX(CO)L$_2$, 940
  trans-Pt(PEt$_3$)$_2$Cl(Ligand), 704
  reactions of metal carbonyls, 920
Rate law, information from, 360
  square planar replacement reactions, 698
  substitution in octahedral complexes, *714*
  Wacker process, 968

Rate law expressions, derivations and interpretations, 364–371
  general, 364
  summary of, 362, 370(t)
Reaction coordinate, 361
Reaction potential, review of, 265
Reaction profiles, square planar replacement reactions, 709
Rearrangements, four-coordinate complexes, 757–764
  organometallic, 953 ff.
  six-coordinate complexes, 764–773
Redistribution reactions, organometallic, 953–957
Reductive addition, 870
Reductive carbonylation, 856
Reductive elimination, of alane adducts, 492
  of PR_5, 479
  organometallic, 479, 492, 841, 950
Remote attack, inner sphere electron transfer, 676
Reppe, W., 976
Resonance structures, guidelines, 73
Retro-bonding, 201, 232–234
Reversals in acidity and basicity, 219
  prediction of, 222
Reverse rate constant, general expression, 364
Rhizobia bacteria, 1087
Rhombic twist, racemization of octahedral complexes, 766
Ribonucleic acid, composition, 1055
Ribose, structure, 1054
Ribosomes, 1053
RNA, composition, 1055
Rochow, E. G., 818, 822
Rotation, olefins about metal-olefin bond, 873
Rotation axis, improper, 82
  proper, 82
Rubredoxins, 1082
Ruby, 314
Rudolph, R.W., 1004
Russell–Saunders coupling, 48
Rutile structure, 282

Sapphire, 313
Sarin, mechanisms of hydrolysis, 433
Scaling factors, for the AOM, 545(t)
Schaeffer, R., 1000
Schlesinger, H.I., 833
Schoenflies symbols, 84(t)
Second order rate law, derivation and interpretation, 366
Semiconductor, extrinsic, 298
  intrinsic, 298
  n-type, 298
  p-type, 298
Senoff, C., 1088
Serine, 1061
Seven-coordinate complexes, 604
σ parameter, metal carbonyl bonding, 904
Si-H coupling constants, 119(t)
Silanes, 317
  cyclic, hydrolysis of, 401
  pseudorotation in substitution, 403–409
  racemization in substitution, 402–409
  retention of chirality in substitution, 402–409
  solvolysis, 495
  synthesis, 477
Silanols, 319
Silazane(s), 495
  hexamethyl-, 494

Silicates, 321
Silicon, acids and bases, 196–203
  catenated, synthesis, 486
  halides, synthesis, 457
  mechanisms of reactions at, 401–409
  organic derivatives, 822
Silicon compounds, structures and stabilities, 316–327
Silicones, 319
  synthesis, 494
Silicon-silicon double bond, 318
Siloxanes, 319
Silylamines, synthesis, 494
Simple cubic lattice, 279
Single scale acid/base concept, 218
Six-atom metal clusters, 1040–1045, 1047
  bonding, 1043–1047
Six-coordinate complexes, 600 ff.
  stereoisomerism, 628 ff.
Sixteen electron rule, 792–809
  organometallic reactions, 909 ff.
Slater orbitals, 44–47
Smidt reaction, Wacker process, 967
Sodium, reaction with naphthalene, 832
Sodium alkyls and aryls, 821
Sodium borohydride, synthesis and properties, 833
Solvated electron, in ammonia, 246
Solvation phenomena, 235–240
Solvent effects, on mechanisms and rates of reactions, 372
Solvent levelling, 239
Solvolysis, acidic conditions, 721
  conjugate base (CB) mechanism, 725 ff.
  hydrolysis, octahedral complexes, 711, 721 ff.
  rates of aquation for Co(III) complexes, 723(t)
Solvolysis reactions of non-metals, 488–512
Specific solute/solvent interactions, 235
Spectra, and electronic states of complexes, 559–575
Spectrochemical series, ligands, 576
  metals, 577
Spherical harmonics, 17, 26(t)
Spin correlation of electron motions, 36
Spin-orbit coupling, 50
  in complexes, 581
Spin polarization of electron pairs, 118
Spontaneity, reaction, 266
Square antiprism, 607
Square planar complexes, mo's for, 539
  replacement reactions, 696–710
  structures, 594
Square planar structures, mo's for, 174, 176
Square pyramidal complexes, 596 ff.
  intermediate in racemization and isomerization of octahedral complexes, 767 ff.
"Stability", meaning, 267
  metal-carbon σ bonds, 948
Stability constants, dependence on chelate ring size, 740
  dependence on number of chelate rings, 741
  trend in, with different M^{2+}, 736
Steady-rate approximation, 364
Stereochemical non-rigidity, organometallic, 957 ff.
Stereochemical notation, Chemical Abstracts, 622 ff.
Stereochemical terms, glossary, 620
Stereochemistry, substitution reaction of *trans*-$[M(en)_2AX]^{n+}$, 724
Stereoisomerism, 619 ff.
  definition, 620
Stock, Alfred, 833

Stock (*Continued*)
   biographical information, 983
Stock convention, 520
Stoichiometric mechanism, 360
Stoichiometric stability constant, 733
Structure, and anti-bonding electrons, 525
Structure(s), metal alkyls and aryls, 843
Structure preference energies, 549 ff.
   geometries typical of organometallic compounds, *805*
   octahedron *vs.* trigonal bipyramid, *600*
   substitution reactions of octahedral complexes, *719*
   two- through six-coordinate complexes, *588–589*
styx numbers, 990 ff.
Substituent effects on mechanisms, 363
Substitution, of amines by phosphines at boron, 393
Substitution reactions, five-coordinate complexes, 730
   octahedral complexes, 710–731
      predictions of lability and inertness, 720(t)
   organometallic, 918–926
   square planar complexes, 696–710
Successor complexes, electron transfer, 679
Sulfamides, 508
Sulfanemonosulfuric acids, 344
Sulfanes, 342
   synthesis, 480
Sulfate esters, hydrolysis mechanisms, 441
Sulfenyl sulfur, reaction mechanisms, 426
Sulfides, mechanism of reactions, 426
Sulfinyl sulfur, 348
   reaction mechanisms, 441
Sulfite esters, hydrolysis mechanisms, 441
Sulfonamides, 508
Sulfonating agents, SO_3 adducts as, 208
Sulfonation of secondary alcohols, stereochemistry of, 253
Sulfone, 348
Sulfonyl sulfur, 348
   reaction mechanisms, 441
Sulfoxides, 348
   as ambidextrous ligands, 113
   oxygen exchange mechanisms, 442
   ν_{SO} in adducts, 113
Sulfur, as Lewis acid and base, 205–209
   cations, 342
      formation in superacids, 261
   halides, synthesis, 459–462
   mechanisms of reactions at, 422, 430–444
Sulfur compounds, structures and stabilities, 342–348
Sulfur fluorides, 347
Sulfur trioxide, 345
Sulfuric acid, esters, 348
Sulfurnitride, 504
Sulfurous acid, 345
   esters, 348
Sulfurtetrafluoride, as fluorinating agent, 468–470
Sulfuryl, 348
Superacid solvents, 259–262
Superexchange, in Fe-S proteins, 1083
Surface density function, 18
Swain-Scott equation, 700
Symmetry, and infrared spectroscopy of metal carbonyls, 900 ff.
   and overlap, 131
Symmetry concepts, 82–97
Symmetry control of mechanisms, 373–378
Symmetry site term, 623

Syndiotactic polymers, *972*
Synergic interaction, metal + ligand, 793
Synergism, in metal carbonyls, 861
   in metal-olefin bonding, 868, 872
Synthesis, of coordination complexes, electron transfer reactions, 684
      substitution reactions, 731 ff.
   of metal alkyls and aryls, 817
   of metal carbonyls, 856 ff.
Synthetic leaf, *1094*

Talc, 327
Tanabe-Sugano diagrams, for complexes, *584*
TASO, 135
TCNE, as a π acid, 198
Template effect, 784
Terminal atom symmetry orbitals, 135
Terms, for complexes, *561, 562*
Tetraborane, structure of, 303
Tetraborane(10), chemistry, 999
Tetracyanoethylene, metal complex, 869
Tetraethyllead, synthesis, 823
Tetrahedral complexes, mo's for, *542*
   racemization, 762
   structures, 593
Tetrahedral structures, mo's for, *169*
Tetramethylphenylenediamine, in synthetic leaf, 1094
Tetrasulfurtetranitride, 504
Thermodynamics, formation of $M(CH_3)_x$, 818 ff.
   formation of mercury alkyls, 821
   formation of $PbEt_4$, 823
Thermodynamic stability, coordination complexes, 733 ff.
Thermodynamic template effect, 784
Thermodynamic values, absolute enthalpies of hydration, 295
   bond energies, 270
      boron and aluminum, 300
      carbon and silicon, 316
      halogens, 349
      nitrogen and phosphorus, 329
      oxygen and sulfur, 340
      Ti-X and Zr-X, *854*
   boron hydrides, 987(t)
   ΔH_f° and ΔG_f° for some common compounds, 266
   ΔH° and ΔG° for some reactions, 267
   ΔH for $H_2O + MCl_4^-$, *552*
   ΔH and ΔS for $M^{2+} + L$, *737*, 738
   enthalpy terms affecting metal E°'s, 299
   free energies of solvation for common ions, 245
   heats of dissociation for $Me_3N \cdot MMe_3$ (M = B, Al, Ga, In, Tl), 229, 911
   heats of solution and solubilities in liquid NH_3, 246
   heats of solution in H_2O and NH_3, 243
   heats of solution of crystalline alkali halide, 294
   Hg^{2+} and Fe^{3+} + various halides, 741
   hydration energy for M^{2+}, *554*
   lattice energies, 287
      for MCl_2, *554*
   M^{2+} + acetylacetonate, 742
   Me_3Ga and I_2 complexes, heats of dissociation, 218
   "nominal bond energies," graphical presentation, 272
   periodic plots for M + CH_3Cl, *820*

Thermodynamic values (*Continued*)
  stability constants, as function of chelate ring size, *740*
    as function of number of chelate rings, *741*
    Cu^{2+} complexes, *734*
    M^{2+} complexes, *736*
  triorganoaluminum compounds, 845
Thionyl, 348
Three-atom metal clusters, 1033–1035
Three-center bridge bond, mo's for, 192
Three-coordinate complexes, 591
Three-dimensional molecules, mo's, 163–183
Three-dimensional mo's, stratagem for, 163
Tin compounds, organometallic, *813*
  redistribution reactions, 956
Titanium, chlorotris(*N,N*-dimethyldithiocarbamato)-, structure, 605
Titanium halides, Ziegler-Natta co-catalysts, 973 ff.
Titanocene, structure, 889
Tobe, M.L., 703, 723
Tolman, C.A., 909
Topological maps, for silane substitution, 404
Total density functions, 30
trans-effect, Pt compound synthesis, 751
  square planar replacement reactions, 626, 703 ff.
trans influence, square planar complexes, 704
Transition metals, binding energies, *527*
  ionization potentials, 529
  oxidation states, *530*
  radii, *528*
Transition state, complex, 362
  theory, 371
Triazane, as an intermediate, 414
Triethylaluminum, addition to olefins, 839
Trimethylamine, complexes with Group IIIA MMe_3, heats of dissociation, 911(t)
Trigonal bipyramid, axial bonding, mo description, 172
  equatorial bonding, mo description, 172
  equatorial *vs.* axial π bonding in, 173
  intermediate in racemization and isomerization of octahedral complexes, 767 ff.
  mo's for, 171, *173*
  substituent electronegativity and position, 110
  transition metal complexes, 596 ff.
Trigonal planar structures, mo's for, 164, *167*
Trigonal prismatic complexes, 600 ff.
Trigonal twist mechanism, racemization of octahedral complexes, 601, *766*
Trimethylaluminum, synthesis, 822
Trioxides, 341
Triphosphate, 502
Tweedledum and Tweedledee, 612
Twisting mechanism, racemization, tetrahedral complexes, 762 ff.
  octahedral complexes, 766
Two-coordinate complexes, 590
Two-dimensional molecules, mo's, 158–162, *160*
Two-electron transfer, 680

Ubihydroquinone, 1079
Ullmann reaction, 950
Unit cell, definition, 279

Units/conversions, inside front cover
Uridine triphosphate, UTP, 1060

Vacuum line, chemical, in synthesis, 447
Valence electron pair repulsion model, 75–81
Valence state, atomic, 56
  promotion energy, 56, 121
Vanadocene, electronic structure, 886
Vaporization of metals, reactions with organic substrate, 823
Vaska, Lauri, 940
Vaska's compound, $IrCl(CO)(PPh_3)_2$, 594
  addition of O_2, mo's for, 945
  reactions with olefins, 868
Vicinal elimination, 949
Vitamin B_{12} coenzyme, 1073
  structure, 853
Vortmann's sulfate, 747
VSEPR, 75–81
  and LCAO-MO, 149–155

Wacker process, 838, 967
Wade, K., 1004
Water replacement, octahedral complexes, 713 ff.
Wave form properties, 15
Wave interference, constructive, 102
  destructive, 102
Wavelength-energy conversion, 521(t)
Werner, Alfred, 594, 600, 613, 619, 625, 628, 684, 695, 703, 743, 745, 754
Wilkinson, G., 812, 852
Wilkinson's catalyst, $RhCl(PPh_3)_3$, 595
  mechanism in homogeneous hydrogenation, 964
Williams, R.E., 1004
Woodward, R.B., 884
Wurtz coupling, of phosphorus compounds, 486
  of silanes, 486

Xenon fluoride cations, 355
Xenon fluorides, synthesis, 463–464
Xenon oxyfluorides, 355

Ylid reactions, orbital following in, 485
Ylids, synthesis of, 485

Zeise, W.C., 811
Zeise's salt, mo's for, 539
  structure and bonding, *871*
Zeiss, H., 880
Zeolite A, 326
Ziegler, K., 835, 840, 973
Ziegler-Natta catalysts, 973 ff.
Zinc blend structure, 281
Zirconium hydrides, addition to olefins, 933

Periodic Table of the Elements

IA																	VIIIA
1 $1s^1$ **H** 1.0080	IIA											IIIA	IVA	VA	VIA	VIIA	2 $1s^2$ **He** 4.0026
3 (He)$2s^1$ **Li** 6.941	4 (He)$2s^2$ **Be** 9.012											5 (He)$2s^2 2p^1$ **B** 10.81	6 (He)$2s^2 2p^2$ **C** 12.011	7 (He)$2s^2 2p^3$ **N** 14.007	8 (He)$2s^2 2p^4$ **O** 15.999	9 (He)$2s^2 2p^5$ **F** 18.998	10 (He)$2s^2 2p^6$ **Ne** 20.179
11 (Ne)$3s^1$ **Na** 22.99	12 (Ne)$3s^2$ **Mg** 24.31	IIIB	IVB	VB	VIB	VIIB		VIIIB		IB	IIB	13 (Ne)$3s^2 3p^1$ **Al** 26.98	14 (Ne)$3s^2 3p^2$ **Si** 28.09	15 (Ne)$3s^2 3p^3$ **P** 30.974	16 (Ne)$3s^2 3p^4$ **S** 32.06	17 (Ne)$3s^2 3p^5$ **Cl** 35.453	18 (Ne)$3s^2 3p^6$ **Ar** 39.948
19 (Ar)$4s^1$ **K** 39.102	20 (Ar)$4s^2$ **Ca** 40.08	21 (Ar)$3d^1 4s^2$ **Sc** 44.96	22 (Ar)$3d^2 4s^2$ **Ti** 47.90	23 (Ar)$3d^3 4s^2$ **V** 50.94	24 (Ar)$3d^5 4s^1$ **Cr** 52.00	25 (Ar)$3d^5 4s^2$ **Mn** 54.94	26 (Ar)$3d^6 4s^2$ **Fe** 55.85	27 (Ar)$3d^7 4s^2$ **Co** 58.93	28 (Ar)$3d^8 4s^2$ **Ni** 58.71	29 (Ar)$3d^{10} 4s^1$ **Cu** 63.55	30 (Ar)$3d^{10} 4s^2$ **Zn** 65.37	31 (Ar)$3d^{10} 4s^2 4p^1$ **Ga** 69.72	32 (Ar)$3d^{10} 4s^2 4p^2$ **Ge** 72.59	33 (Ar)$3d^{10} 4s^2 4p^3$ **As** 74.92	34 (Ar)$3d^{10} 4s^2 4p^4$ **Se** 78.96	35 (Ar)$3d^{10} 4s^2 4p^5$ **Br** 79.904	36 (Ar)$3d^{10} 4s^2 4p^6$ **Kr** 83.80
37 (Kr)$5s^1$ **Rb** 85.47	38 (Kr)$5s^2$ **Sr** 87.62	39 (Kr)$4d^1 5s^2$ **Y** 88.91	40 (Kr)$4d^2 5s^2$ **Zr** 91.22	41 (Kr)$4d^4 5s^1$ **Nb** 92.91	42 (Kr)$4d^5 5s^1$ **Mo** 95.94	43 (Kr)$4d^5 5s^2$ **Tc** (99)^a	44 (Kr)$4d^7 5s^1$ **Ru** 101.07	45 (Kr)$4d^8 5s^1$ **Rh** 102.91	46 (Kr)$4d^{10}$ **Pd** 106.4	47 (Kr)$4d^{10} 5s^1$ **Ag** 107.87	48 (Kr)$4d^{10} 5s^2$ **Cd** 112.40	49 (Kr)$4d^{10} 5s^2 5p^1$ **In** 114.82	50 (Kr)$4d^{10} 5s^2 5p^2$ **Sn** 118.69	51 (Kr)$4d^{10} 5s^2 5p^3$ **Sb** 121.75	52 (Kr)$4d^{10} 5s^2 5p^4$ **Te** 127.60	53 (Kr)$4d^{10} 5s^2 5p^5$ **I** 126.90	54 (Kr)$4d^{10} 5s^2 5p^6$ **Xe** 131.30
55 (Xe)$6s^1$ **Cs** 132.91	56 (Xe)$6s^2$ **Ba** 137.34	57 (Xe)$5d^1 6s^2$ **La*** 138.91	72 (Xe)$4f^{14} 5d^2 6s^2$ **Hf** 178.49	73 (Xe)$4f^{14} 5d^3 6s^2$ **Ta** 180.95	74 (Xe)$4f^{14} 5d^4 6s^2$ **W** 183.85	75 (Xe)$4f^{14} 5d^5 6s^2$ **Re** 186.2	76 (Xe)$4f^{14} 5d^6 6s^2$ **Os** 190.2	77 (Xe)$4f^{14} 5d^7 6s^2$ **Ir** 192.2	78 (Xe)$4f^{14} 5d^9 6s^1$ **Pt** 195.09	79 (Xe)$4f^{14} 5d^{10} 6s^1$ **Au** 196.97	80 (Xe)$4f^{14} 5d^{10} 6s^2$ **Hg** 200.59	81 (Xe)$4f^{14} 5d^{10} 6s^2 6p^1$ **Tl** 204.37	82 (Xe)$4f^{14} 5d^{10} 6s^2 6p^2$ **Pb** 207.2	83 (Xe)$4f^{14} 5d^{10} 6s^2 6p^3$ **Bi** 208.98	84 (Xe)$4f^{14} 5d^{10} 6s^2 6p^4$ **Po** (210)	85 (Xe)$4f^{14} 5d^{10} 6s^2 6p^5$ **At** (210)	86 (Xe)$4f^{14} 5d^{10} 6s^2 6p^6$ **Rn** (222)
87 (Rn)$7s^1$ **Fr** (223)	88 (Rn)$7s^2$ **Ra** (226)	89 (Rn)$6d^1 7s^2$ **Ac†** (227)	104 (Rn)$5f^{14} 6d^2 7s^2$ **?**^{b} (260)	105 (Rn)$5f^{14} 6d^3 7s^2$ **Ha**^{b}													

^{a} Value in parentheses denotes mass number of most stable known isotope.
^{b} Name and symbol are not officially accepted. Kurchatovium, Ku, has been proposed by Russian investigators and rutherfordium, Rf, by American investigators for element 104.

***Lanthanide Series**

58 (Xe)$4f^2 6s^2$ **Ce** 140.12	59 (Xe)$4f^3 6s^2$ **Pr** 140.91	60 (Xe)$4f^4 6s^2$ **Nd** 144.24	61 (Xe)$4f^5 6s^2$ **Pm** (147)	62 (Xe)$4f^6 6s^2$ **Sm** 150.4	63 (Xe)$4f^7 6s^2$ **Eu** 151.96	64 (Xe)$4f^7 5d^1 6s^2$ **Gd** 157.25	65 (Xe)$4f^9 6s^2$ **Tb** 158.93	66 (Xe)$4f^{10} 6s^2$ **Dy** 162.50	67 (Xe)$4f^{11} 6s^2$ **Ho** 164.93	68 (Xe)$4f^{12} 6s^2$ **Er** 167.26	69 (Xe)$4f^{13} 6s^2$ **Tm** 168.93	70 (Xe)$4f^{14} 6s^2$ **Yb** 173.04	71 (Xe)$4f^{14} 5d^1 6s^2$ **Lu** 174.97

†Actinide Series

90 (Rn)$6d^2 7s^2$ **Th** 232.04	91 (Rn)$5f^2 6d^1 7s^2$ **Pa** (231)	92 (Rn)$5f^3 6d^1 7s^2$ **U** 238.03	93 (Rn)$5f^4 6d^1 7s^2$ **Np** (237)	94 (Rn)$5f^6 7s^2$ **Pu** (244)	95 (Rn)$5f^7 7s^2$ **Am** (243)	96 (Rn)$5f^7 6d^1 7s^2$ **Cm** (245)	97 (Rn)$5f^8 6d^1 7s^2$ **Bk** (247)	98 (Rn)$5f^{10} 7s^2$ **Cf** (249)	99 (Rn)$5f^{11} 7s^2$ **Es** (254)	100 (Rn)$5f^{12} 7s^2$ **Fm** (255)	101 (Rn)$5f^{13} 7s^2$ **Md** (256)	102 (Rn)$5f^{14} 7s^2$ **No** (254)	103 (Rn)$5f^{14} 6d^1 7s^2$ **Lr** (257)